BIG IDEAS

MATH®

Pre-Algebra

A Common Core Curriculum

FLORIDA TEACHING EDITION

Ron Larson
Laurie Boswell

Erie, Pennsylvania
BigIdeasLearning.com

Big Ideas Learning, LLC
1762 Norcross Road
Erie, PA 16510-3838
USA

For product information and customer support, contact Big Ideas Learning at **1-877-552-7766** or visit us at ***BigIdeasLearning.com***.

About the Cover
The cover images on the *Big Ideas Math* series illustrate the advancements in aviation from the hot-air balloon to spacecraft. This progression symbolizes the launch of a student's successful journey in mathematics. The sunrise in the background is representative of the dawn of the Common Core era in math education, while the cradle signifies the balanced instruction that is a pillar of the *Big Ideas Math* series.

Printed in the U.S.A.

ISBN 13: 978-1-60840-591-6
ISBN 10: 1-60840-591-5

2 3 4 5 6 7 8 9 10 WEB 17 16 15 14 13

AUTHORS

Ron Larson is a professor of mathematics at Penn State Erie, The Behrend College, where he has taught since receiving his Ph.D. in mathematics from the University of Colorado. Dr. Larson is well known as the lead author of a comprehensive program for mathematics that spans middle school, high school, and college courses. His high school and Advanced Placement books are published by Holt McDougal. Ron's numerous professional activities keep him in constant touch with the needs of students, teachers, and supervisors. Ron and Laurie Boswell began writing together in 1992. Since that time, they have authored over two dozen textbooks. In their collaboration, Ron is primarily responsible for the pupil edition and Laurie is primarily responsible for the teaching edition of the text.

Laurie Boswell is the Head of School and a mathematics teacher at the Riverside School in Lyndonville, Vermont. Dr. Boswell received her Ed.D. from the University of Vermont in 2010. She is a recipient of the Presidential Award for Excellence in Mathematics Teaching. Laurie has taught math to students at all levels, elementary through college. In addition, Laurie was a Tandy Technology Scholar, and served on the NCTM Board of Directors from 2002 to 2005. She currently serves on the board of NCSM, and is a popular national speaker. Along with Ron, Laurie has co-authored numerous math programs.

ABOUT THE BOOK

The *Florida Big Ideas Math* series uses the same research-based strategy of a balanced approach to instruction that made the first *Florida Big Ideas Math* series so successful. This approach opens doors to abstract thought, reasoning, and inquiry as students persevere to answer the Essential Questions that introduce each section. The foundation of the program is the Florida Common Core Standards for Mathematical Content and Standards for Mathematical Practice. Students are subtly introduced to "Habits of Mind" that help them internalize concepts for a greater depth of understanding. These habits serve students well not only in mathematics, but across all curricula throughout their academic careers.

The *Florida Big Ideas Math* series exposes students to highly motivating and relevant problems. Woven throughout the series are the depth and rigor students need to prepare for career-readiness and other college-level courses. In addition, the *Florida Big Ideas Math* series prepares students to meet the challenge of PARCC testing.

We are excited to bring the *Florida Big Ideas Math* series to the students in Florida as they prepare for college and career readiness.

Ron Larson   Laurie Boswell

TEACHER REVIEWERS

- Lisa Amspacher
  Milton Hershey School
  Hershey, PA
- Mary Ballerina
  Orange County Public Schools
  Orlando, FL
- Lisa Bubello
  School District of Palm Beach County
  Lake Worth, FL
- Sam Coffman
  North East School District
  North East, PA
- Kristen Karbon
  Troy School District
  Rochester Hills, MI
- Laurie Mallis
  Westglades Middle School
  Coral Springs, FL
- Dave Morris
  Union City Area School District
  Union City, PA
- Bonnie Pendergast
  Tolleson Union High School District
  Tolleson, AZ
- Valerie Sullivan
  Lamoille South Supervisory Union
  Morrisville, VT
- Becky Walker
  Appleton Area School District
  Appleton, WI
- Zena Wiltshire
  Dade County Public Schools
  Miami, FL

STUDENT REVIEWERS

- Mike Carter
- Matthew Cauley
- Amelia Davis
- Wisdom Dowds
- John Flatley
- Nick Ganger
- Hannah Iadeluca
- Paige Lavine
- Emma Louie
- David Nichols
- Mikala Parnell
- Jordan Pashupathi
- Stephen Piglowski
- Robby Quinn
- Michael Rawlings
- Garrett Sample
- Andrew Samuels
- Addie Sedelmyer
- Tyler Steffy
- Erin Taylor
- Reid Wilson

CONSULTANTS

- Patsy Davis
  Educational Consultant
  Knoxville, Tennessee

- Bob Fulenwider
  Mathematics Consultant
  Bakersfield, California

- Linda Hall
  Mathematics Assessment Consultant
  Norman, Oklahoma

- Ryan Keating
  Special Education Advisor
  Gilbert, Arizona

- Michael McDowell
  Project-Based Instruction Specialist
  Fairfax, California

- Sean McKeighan
  Interdisciplinary Advisor
  Norman, Oklahoma

- Bonnie Spence
  Differentiated Instruction Consultant
  Missoula, Montana

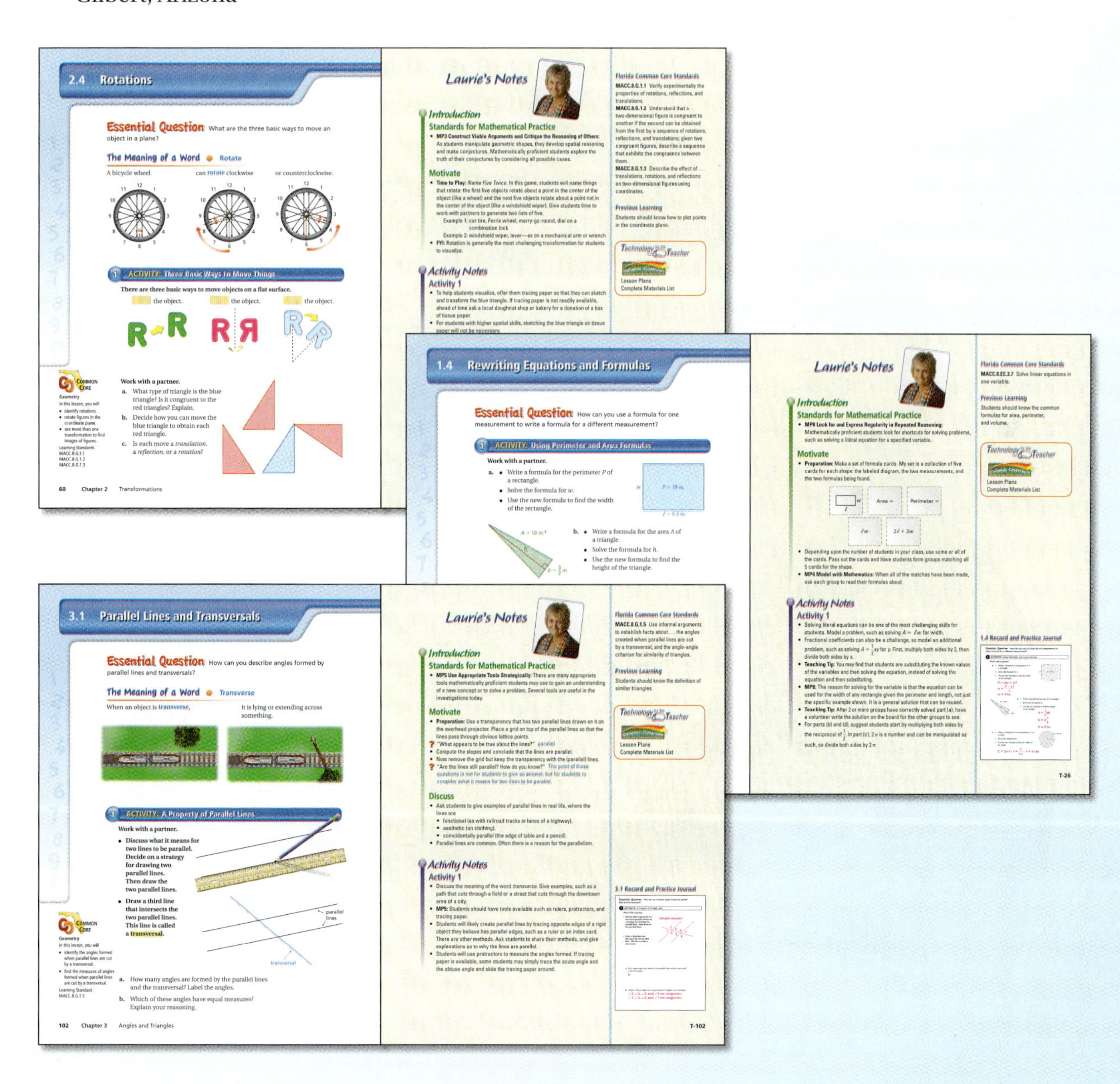
2.4 Rotations

Essential Question What are the three basic ways to move an object in a plane?

Laurie's Notes

1.4 Rewriting Equations and Formulas

Essential Question How can you use a formula for one measurement to write a formula for a different measurement?

Laurie's Notes

3.1 Parallel Lines and Transversals

Essential Question How can you describe angles formed by parallel lines and transversals?

Laurie's Notes

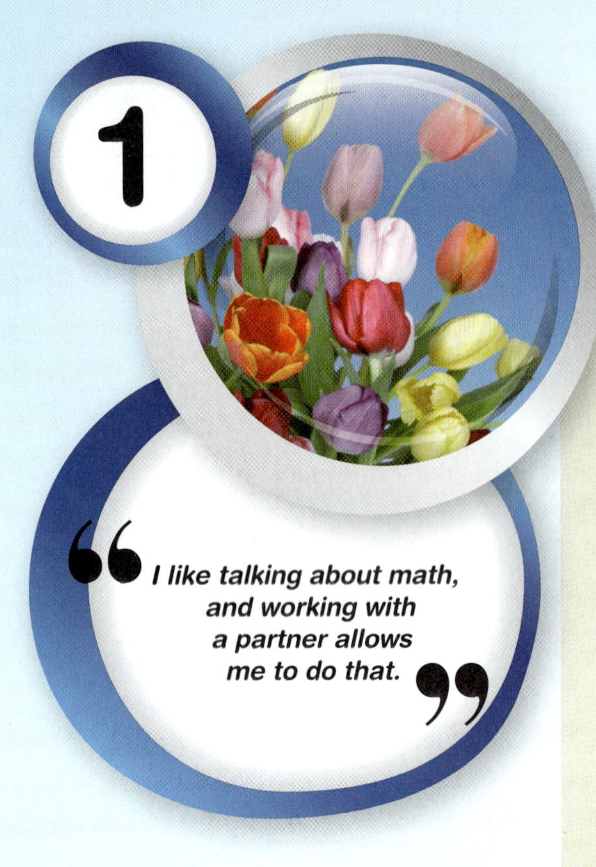
1
I like talking about math, and working with a partner allows me to do that.

Equations

Transformations

"*With my eBook, I get to decide when I use technology and when I use print.*"

Angles and Triangles

Graphing and Writing Linear Equations

"I really enjoy the projects at the end of the book because they help connect the math to other subjects, like science or art."

Systems of Linear Equations

Functions

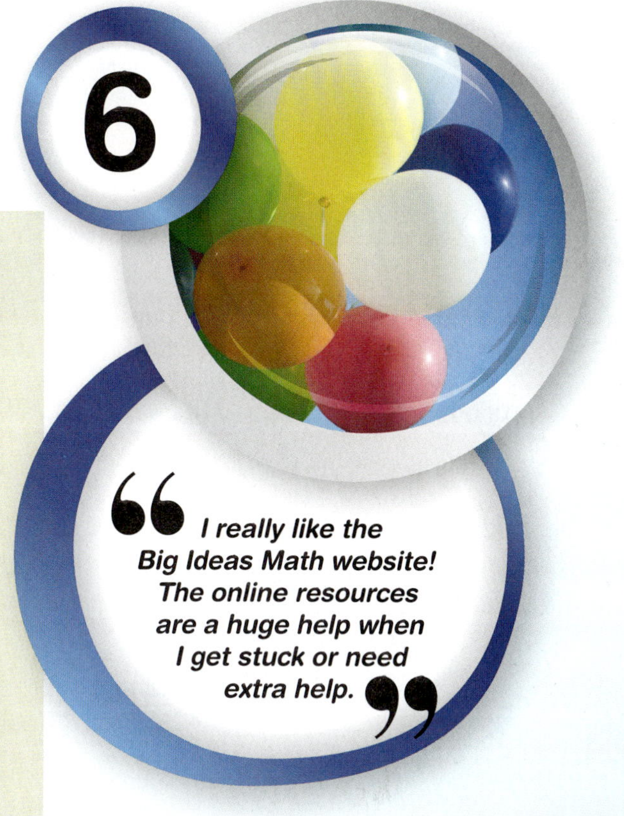

"I really like the Big Ideas Math website! The online resources are a huge help when I get stuck or need extra help."

7 Real Numbers and the Pythagorean Theorem

“I like the real-life application exercises because they show me how I can use the math in my own life.”

Volume and Similar Solids

"I like playing the games in the Game Closet! They are a fun way to practice concepts we are learning in class."

Data Analysis and Displays

Exponents and Scientific Notation

"I like that I can print the Glossary Flashcards to review before a test."

Appendix A: My Big Ideas Projects

PROGRAM OVERVIEW

Print

Also available online and in digital format

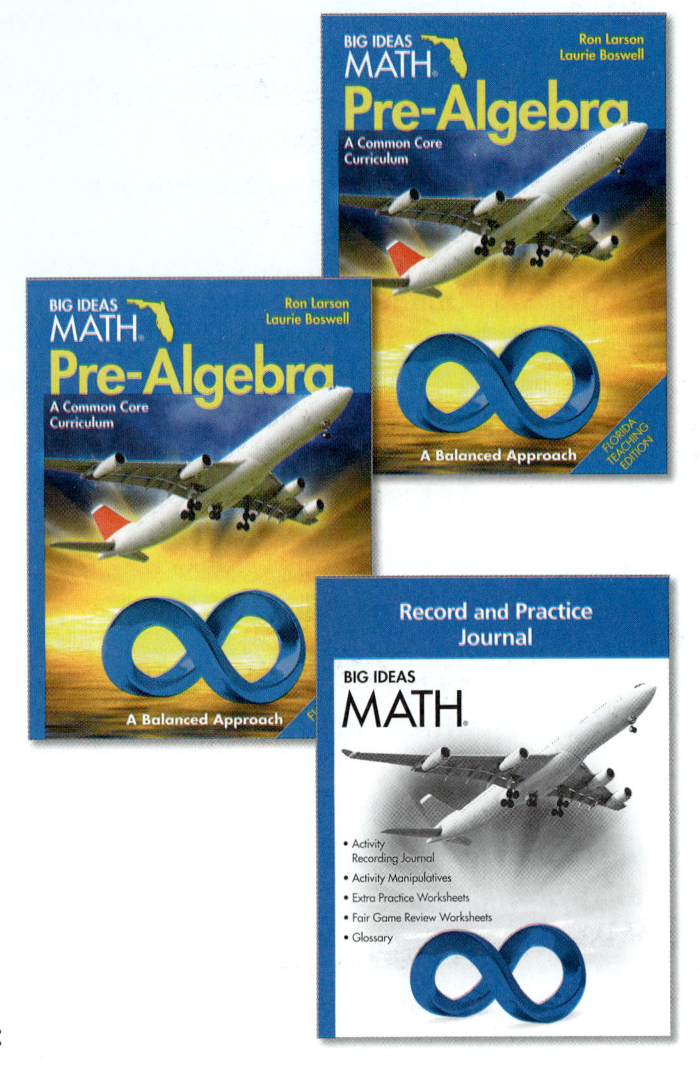

- **Pupil Edition**
  Also available in eReader format
- **Teaching Edition**
- **Record and Practice Journal: English and Spanish**
- **Assessment Book**
  - Pre-Course Test
  - Quizzes
  - Chapter Tests
  - Standards Assessment
  - Alternative Assessment
  - End-of-Course Tests
- **Resources by Chapter**
  - Family and Community Involvement: English and Spanish
  - Start Thinking! and Warm Up
  - Extra Practice
  - Enrichment and Extension
  - Puzzle Time
  - Technology Connection
  - Projects with Rubrics

Technology

Online Resources at *BigIdeasMath.com*

Teach Your Lesson
- Dynamic Classroom
  - Whiteboard Classroom Presentations
  - Interactive Manipulatives
  - Support for Mathematical Practices
  - Answer Presentation Tool
- Multi-Language Glossary
- Teaching Edition
- Vocabulary Flash Cards
- Worked-Out Solutions

Plan Your Lesson
- Editable Resources
  - Lesson Plans
  - Assessment Book
  - Resources by Chapter
- Math Tool Paper
- Pacing Guides
- Project Rubrics

Response to Intervention
- Differentiating the Lesson
- Game Closet
- Lesson Tutorials
- Skills Review Handbook
- Basic Skills Handbook

Additional Support for Florida Common Core Standards
- Florida Common Core Standards
- Performance Tasks by Standard

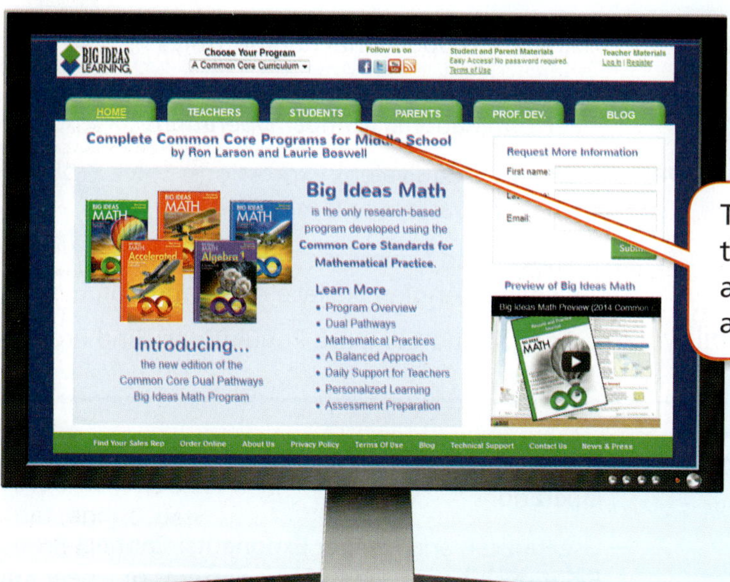

The Student tab includes the online textbook, tutorials, and help with homework and remediation.

DVDs
- Dynamic Assessment Resources
  - ExamView® Assessment Suite
- Dynamic Teaching Resources
- Dynamic Student Edition

eInstruction® EXAMVIEW® ASSESSMENT SUITE

SCOPE AND

Regular Pathway

Grade 6		
	Ratios and Proportional Relationships	– Understand Ratio Concepts; Use Ratio Reasoning
	The Number System	– Perform Fraction and Decimal Operations; Understand Rational Numbers
	Expressions and Equations	– Write, Interpret, and Use Expressions, Equations, and Inequalities
	Geometry	– Solve Problems Involving Area, Surface Area, and Volume
	Statistics and Probability	– Summarize and Describe Distributions; Understand Variability

Grade 7		
	Ratios and Proportional Relationships	– Analyze Proportional Relationships
	The Number System	– Perform Rational Number Operations
	Expressions and Equations	– Generate Equivalent Expressions; Solve Problems Using Linear Equations and Inequalities
	Geometry	– Understand Geometric Relationships; Solve Problems Involving Angles, Surface Area, and Volume
	Statistics and Probability	– Analyze and Compare Populations; Find Probabilities of Events

Grade 8		
	The Number System	– Approximate Real Numbers; Perform Real Number Operations
	Expressions and Equations	– Use Radicals and Integer Exponents; Connect Proportional Relationships and Lines; Solve Systems of Linear Equations
	Functions	– Define, Evaluate, and Compare Functions; Model Relationships
	Geometry	– Understand Congruence and Similarity; Apply the Pythagorean Theorem; Apply Volume Formulas
	Statistics and Probability	– Analyze Bivariate Data

SEQUENCE

Advanced Pathway

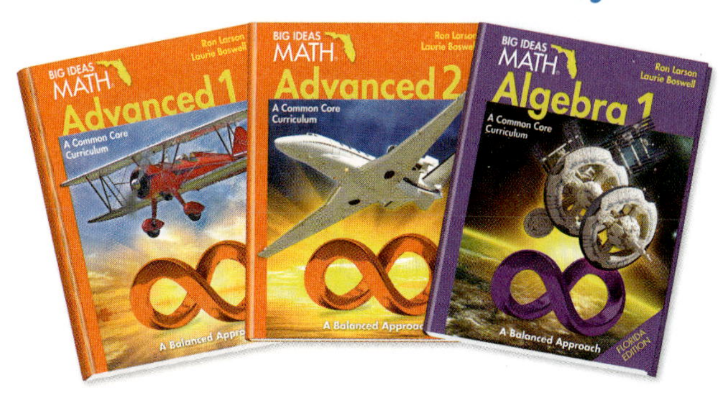

Grade 6 Advanced

Ratios and Proportional Relationships	– Understand Ratio Concepts; Use Ratio Reasoning; Analyze Proportional Relationships
The Number System	– Perform Fraction and Decimal Operations; Understand Rational Numbers; Perform Rational Number Operations
Expressions and Equations	– Write, Interpret, and Use Expressions, Equations, and Inequalities; Generate Equivalent Expressions; Solve Problems Using Linear Equations
Geometry	– Solve Problems Involving Area, Surface Area, and Volume
Statistics and Probability	– Summarize and Describe Distributions; Understand Variability

Grade 7 Advanced

The Number System	– Approximate Real Numbers; Perform Real Number Operations
Expressions and Equations	– Solve Problems Using Linear Inequalities; Use Radicals and Integer Exponents; Connect Proportional Relationships and Lines; Solve Systems of Linear Equations
Functions	– Define, Evaluate, and Compare Functions; Model Relationships
Geometry	– Understand Geometric Relationships; Solve Problems Involving Angles, Surface Area, and Volume; Understand Congruence and Similarity; Apply the Pythagorean Theorem
Statistics and Probability	– Analyze and Compare Populations; Find Probabilities of Events; Analyze Bivariate Data

Algebra 1

Number and Quantity	– Use Rational Exponents; Perform Real Number Operations
Algebra	– Solve Linear and Quadratic Equations; Solve Inequalities and Systems of Equations
Functions	– Define, Evaluate, and Compare Functions; Write Sequences; Model Relationships
Geometry	– Apply the Pythagorean Theorem
Statistics and Probability	– Represent and Interpret Data; Analyze Bivariate Data

FLORIDA COMMON CORE STANDARDS TO BOOK CORRELATION

After a standard is introduced, it is revisited many times in subsequent activities, lessons, and exercises.

Domain: The Number System

Standards (MACC)

8.NS.1.1 Know that numbers that are not rational are called irrational. Understand informally that every number has a decimal expansion; for rational numbers show that the decimal expansion repeats eventually, and convert a decimal expansion which repeats eventually into a rational number.

- **Section 7.4** Approximating Square Roots
- **Extension 7.4** Repeating Decimals

8.NS.1.2 Use rational approximations of irrational numbers to compare the size of irrational numbers, locate them approximately on a number line diagram, and estimate the value of expressions.

- **Section 7.4** Approximating Square Roots

Domain: Expressions and Equations

Standards (MACC)

8.EE.1.1 Know and apply the properties of integer exponents to generate equivalent numerical expressions.

- **Section 10.1** Exponents
- **Section 10.2** Product of Powers Property
- **Section 10.3** Quotient of Powers Property
- **Section 10.4** Zero and Negative Exponents

8.EE.1.2 Use square root and cube root symbols to represent solutions to equations of the form $x^2 = p$ and $x^3 = p$, where p is a positive rational number. Evaluate square roots of small perfect squares and cube roots of small perfect cubes. Know that $\sqrt{2}$ is irrational.

- **Section 7.1** Finding Square Roots
- **Section 7.2** Finding Cube Roots
- **Section 7.3** The Pythagorean Theorem
- **Section 7.4** Approximating Square Roots
- **Section 7.5** Using the Pythagorean Theorem

8.EE.1.3 Use numbers expressed in the form of a single digit times an integer power of 10 to estimate very large or very small quantities, and to express how many times as much one is than the other.

- **Section 10.5** Reading Scientific Notation
- **Section 10.6** Writing Scientific Notation
- **Section 10.7** Operations in Scientific Notation

8.EE.1.4 Perform operations with numbers expressed in scientific notation, including problems where both decimal and scientific notation are used. Use scientific notation and choose units of appropriate size for measurements of very large or very small quantities. Interpret scientific notation that has been generated by technology.

- **Section 10.5** Reading Scientific Notation
- **Section 10.6** Writing Scientific Notation
- **Section 10.7** Operations in Scientific Notation

8.EE.2.5 Graph proportional relationships, interpreting the unit rate as the slope of the graph. Compare two different proportional relationships represented in different ways.

- **Section 4.1** Graphing Linear Equations
- **Section 4.3** Graphing Proportional Relationships

8.EE.2.6 Use similar triangles to explain why the slope m is the same between any two distinct points on a non-vertical line in the coordinate plane; derive the equation $y = mx$ for a line through the origin and the equation $y = mx + b$ for a line intercepting the vertical axis at b.

- **Section 4.2** Slope of a Line
- **Extension 4.2** Slopes of Parallel and Perpendicular Lines
- **Section 4.3** Graphing Proportional Relationships
- **Section 4.4** Graphing Linear Equations in Slope-Intercept Form
- **Section 4.5** Graphing Linear Equations in Standard Form

8.EE.3.7 Solve linear equations in one variable.

a. Give examples of linear equations in one variable with one solution, infinitely many solutions, or no solutions. Show which of these possibilities is the case by successively transforming the given equation into simpler forms, until an equivalent equation of the form $x = a$, $a = a$, or $a = b$ results (where a and b are different numbers).

- **Section 1.1** Solving Simple Equations
- **Section 1.2** Solving Multi-Step Equations
- **Section 1.3** Solving Equations with Variables on Both Sides
- **Section 1.4** Rewriting Equations and Formulas
- **Extension 5.4** Solving Linear Equations by Graphing

b. Solve linear equations with rational number coefficients, including equations whose solutions require expanding expressions using the distributive property and collecting like terms.

- **Section 1.1** Solving Simple Equations
- **Section 1.2** Solving Multi-Step Equations
- **Section 1.3** Solving Equations with Variables on Both Sides
- **Section 1.4** Rewriting Equations and Formulas
- **Extension 5.4** Solving Linear Equations by Graphing

8.EE.3.8 Analyze and solve pairs of simultaneous linear equations.

a. Understand that solutions to a system of two linear equations in two variables correspond to points of intersection of their graphs, because points of intersection satisfy both equations simultaneously.

- **Section 5.1** Solving Systems of Linear Equations by Graphing
- **Section 5.4** Solving Special Systems of Linear Equations
- **Extension 5.4** Solving Linear Equations by Graphing

b. Solve systems of two linear equations in two variables algebraically, and estimate solutions by graphing the equations. Solve simple cases by inspection.

- **Section 5.1** Solving Systems of Linear Equations by Graphing
- **Section 5.2** Solving Systems of Linear Equations by Substitution
- **Section 5.3** Solving Systems of Linear Equations by Elimination
- **Section 5.4** Solving Special Systems of Linear Equations
- **Extension 5.4** Solving Linear Equations by Graphing

c. Solve real-world and mathematical problems leading to two linear equations in two variables.

- **Section 5.1** Solving Systems of Linear Equations by Graphing
- **Section 5.2** Solving Systems of Linear Equations by Substitution
- **Section 5.3** Solving Systems of Linear Equations by Elimination
- **Section 5.4** Solving Special Systems of Linear Equations
- **Extension 5.4** Solving Linear Equations by Graphing

Domain: Functions

Standards (MACC)

8.F.1.1 Understand that a function is a rule that assigns to each input exactly one output. The graph of a function is the set of ordered pairs consisting of an input and the corresponding output.

- **Section 6.1** Relations and Functions
- **Section 6.2** Representations of Functions

8.F.1.2 Compare properties of two functions each represented in a different way (algebraically, graphically, numerically in tables, or by verbal descriptions).

- **Section 6.3** Linear Functions

8.F.1.3 Interpret the equation $y = mx + b$ as defining a linear function, whose graph is a straight line; give examples of functions that are not linear.

- **Section 6.3** Linear Functions
- **Section 6.4** Comparing Linear and Nonlinear Functions

8.F.2.4 Construct a function to model a linear relationship between two quantities. Determine the rate of change and initial value of the function from a description of a relationship or from two (x, y) values, including reading these from a table or from a graph. Interpret the rate of change and initial value of a linear function in terms of the situation it models, and in terms of its graph or a table of values.

- **Section 4.6** Writing Equations in Slope-Intercept Form
- **Section 4.7** Writing Equations in Point-Slope Form
- **Section 6.3** Linear Functions

8.F.2.5 Describe qualitatively the functional relationship between two quantities by analyzing a graph. Sketch a graph that exhibits the qualitative features of a function that has been described verbally.

- **Section 6.5** Analyzing and Sketching Graphs

Domain: Geometry

Standards (MACC)

8.G.1.1 Verify experimentally the properties of rotations, reflections, and translations:

a. Lines are taken to lines and line segments to line segments of the same length.
- **Section 2.2** Translations
- **Section 2.3** Reflections
- **Section 2.4** Rotations

b. Angles are taken to angles of the same measure.
- **Section 2.2** Translations
- **Section 2.3** Reflections
- **Section 2.4** Rotations

c. Parallel lines are taken to parallel lines.
- **Section 2.2** Translations
- **Section 2.3** Reflections
- **Section 2.4** Rotations

8.G.1.2 Understand that a two-dimensional figure is congruent to another if the second can be obtained from the first by a sequence of rotations, reflections, and translations; given two congruent figures, describe a sequence that exhibits the congruence between them.
- **Section 2.1** Congruent Figures
- **Section 2.2** Translations
- **Section 2.3** Reflections
- **Section 2.4** Rotations

8.G.1.3 Describe the effect of dilations, translations, rotations, and reflections on two-dimensional figures using coordinates.
- **Section 2.2** Translations
- **Section 2.3** Reflections
- **Section 2.4** Rotations
- **Section 2.7** Dilations

8.G.1.4 Understand that a two-dimensional figure is similar to another if the second can be obtained from the first by a sequence of rotations, reflections, translations, and dilations; given two similar two-dimensional figures, describe a sequence that exhibits the similarity between them.
- **Section 2.5** Similar Figures
- **Section 2.6** Perimeters and Areas of Similar Figures
- **Section 2.7** Dilations

8.G.1.5 Use informal arguments to establish facts about the angle sum and exterior angle of triangles, about the angles created when parallel lines are cut by a transversal, and the angle-angle criterion for similarity of triangles.
- **Section 3.1** Parallel Lines and Transversals
- **Section 3.2** Angles of Triangles
- **Section 3.3** Angles of Polygons
- **Section 3.4** Using Similar Triangles

8.G.2.6 Explain a proof of the Pythagorean Theorem and its converse.

- **Section 7.3** The Pythagorean Theorem
- **Section 7.5** Using the Pythagorean Theorem

8.G.2.7 Apply the Pythagorean Theorem to determine unknown side lengths in right triangles in real-world and mathematical problems in two and three dimensions.

- **Section 7.3** The Pythagorean Theorem
- **Section 7.5** Using the Pythagorean Theorem

8.G.2.8 Apply the Pythagorean Theorem to find the distance between two points in a coordinate system.

- **Section 7.3** The Pythagorean Theorem
- **Section 7.5** Using the Pythagorean Theorem

8.G.3.9 Know the formulas for the volumes of cones, cylinders, and spheres and use them to solve real-world and mathematical problems.

- **Section 8.1** Volumes of Cylinders
- **Section 8.2** Volumes of Cones
- **Section 8.3** Volumes of Spheres
- **Section 8.4** Surface Areas and Volumes of Similar Solids

Domain: Statistics and Probability

Standards (MACC)

8.SP.1.1 Construct and interpret scatter plots for bivariate measurement data to investigate patterns of association between two quantities. Describe patterns such as clustering, outliers, positive or negative association, linear association, and nonlinear association.

- **Section 9.1** Scatter Plots
- **Section 9.2** Lines of Fit
- **Section 9.4** Choosing a Data Display

8.SP.1.2 Know that straight lines are widely used to model relationships between two quantitative variables. For scatter plots that suggest a linear association, informally fit a straight line, and informally assess the model fit by judging the closeness of the data points to the line.

- **Section 9.2** Lines of Fit

8.SP.1.3 Use the equation of a linear model to solve problems in the context of bivariate measurement data, interpreting the slope and intercept.

- **Section 9.2** Lines of Fit

8.SP.1.4 Understand that patterns of association can also be seen in bivariate categorical data by displaying frequencies and relative frequencies in a two-way table. Construct and interpret a two-way table summarizing data on two categorical variables collected from the same subjects. Use relative frequencies calculated for rows or columns to describe possible association between the two variables.

- **Section 9.3** Two-Way Tables

BOOK TO FLORIDA COMMON CORE STANDARDS CORRELATION

Chapter 1

Equations

Expressions and Equations

8.EE.3.7a–b

Chapter 2

Transformations

Geometry

8.G.1.1a–c
8.G.1.2
8.G.1.3
8.G.1.4

Chapter 3

Angles and Triangles

Geometry

8.G.1.5

Chapter 4

Graphing and Writing Linear Equations

Expressions and Equations

8.EE.2.5
8.EE.2.6

Functions

8.F.2.4

Chapter 5

Systems of Linear Equations

Expressions and Equations

8.EE.3.7a–b
8.EE.3.8a–c

Chapter 6

Functions

Functions

8.F.1.1
8.F.1.2
8.F.1.3
8.F.2.4
8.F.2.5

Chapter 7

Real Numbers and the Pythagorean Theorem

The Number System

8.NS.1.1
8.NS.1.2

Expressions and Equations

8.EE.1.2

Geometry

8.G.2.6
8.G.2.7
8.G.2.8

Chapter 8

Volume and Similar Solids

Geometry

8.G.3.9

Chapter 9

Data Analysis and Displays

Statistics and Probability

8.SP.1.1
8.SP.1.2
8.SP.1.3
8.SP.1.4

Chapter 10

Exponents and Scientific Notation

Expressions and Equations

8.EE.1.1
8.EE.1.3
8.EE.1.4

EMBEDDED MATHEMATICAL PRACTICES THROUGH A BALANCED APPROACH TO INSTRUCTION

Essential Question How can you identify congruent triangles?

The Florida Common Core Standards require students to do more than memorize how to solve problems. They define skills and knowledge that young people need to succeed academically in credit-bearing, college entry courses and in workforce training programs. Mastering the skills reflected in the Florida Common Core Standards prepares students for college and career.

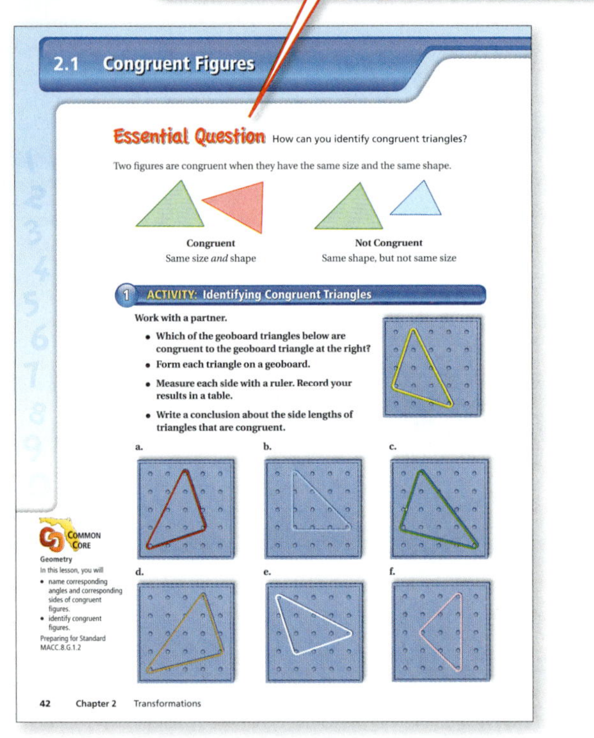

2.1 Congruent Figures

Essential Question How can you identify congruent triangles?

Two figures are congruent when they have the same size and the same shape.

Congruent
Same size *and* shape

Not Congruent
Same shape, but not same size

1 ACTIVITY: Identifying Congruent Triangles

Work with a partner.

- Which of the geoboard triangles below are congruent to the geoboard triangle at the right?
- Form each triangle on a geoboard.
- Measure each side with a ruler. Record your results in a table.
- Write a conclusion about the side lengths of triangles that are congruent.

a. b. c. d. e. f.

Common Core
Geometry
In this lesson, you will
- name corresponding angles and corresponding sides of congruent figures.
- identify congruent figures.

Preparing for Standard MACC.8.G.1.2

42 Chapter 2 Transformations

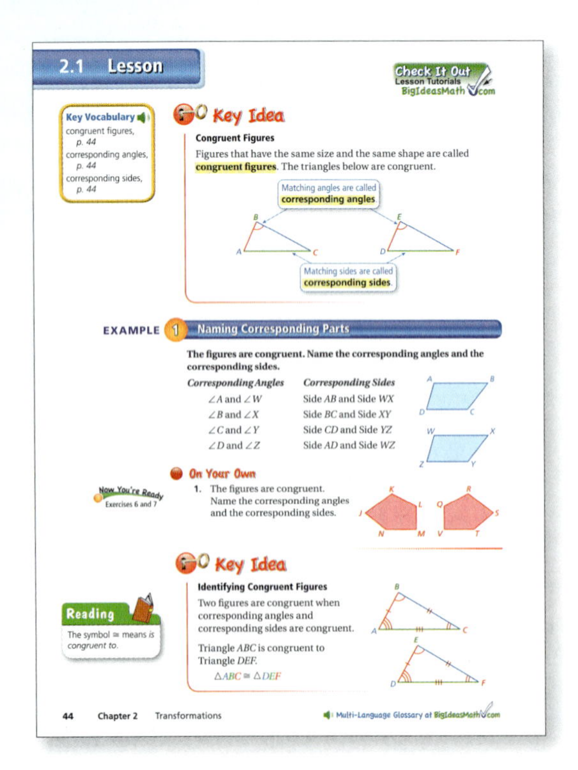

2.1 Lesson

Check It Out
Lesson Tutorials
BigIdeasMath.com

Key Vocabulary
congruent figures, p. 44
corresponding angles, p. 44
corresponding sides, p. 44

Key Idea

Congruent Figures

Figures that have the same size and the same shape are called **congruent figures**. The triangles below are congruent.

Matching angles are called **corresponding angles**

Matching sides are called **corresponding sides**

EXAMPLE 1 Naming Corresponding Parts

The figures are congruent. Name the corresponding angles and the corresponding sides.

Corresponding Angles	*Corresponding Sides*
$\angle A$ and $\angle W$	Side AB and Side WX
$\angle B$ and $\angle X$	Side BC and Side XY
$\angle C$ and $\angle Y$	Side CD and Side YZ
$\angle D$ and $\angle Z$	Side AD and Side WZ

On Your Own

Now You're Ready
Exercises 6 and 7

1. The figures are congruent. Name the corresponding angles and the corresponding sides.

Key Idea

Identifying Congruent Figures

Two figures are congruent when corresponding angles and corresponding sides are congruent.

Triangle ABC is congruent to Triangle DEF.

$\triangle ABC \cong \triangle DEF$

Reading
The symbol $\cong$ means *is congruent to*.

44 Chapter 2 Transformations

Multi-Language Glossary at BigIdeasMath.com

Research shows that students benefit from a program that includes equal exposure to discovery and direct instruction. By beginning each lesson with an inquiry-based activity, *Big Ideas Math* allows students to explore, question, explain, and persevere as they seek to answer Essential Questions that encourage abstract thought.

These rich activities are followed by direct instruction lessons, allowing for procedural fluency, modeling, and the opportunity to use clear, precise mathematical language.

DAILY SUPPORT FOR TEACHERS IN MANAGING THE FLORIDA COMMON CORE STANDARDS

Standards for Mathematical Practice

- **MP5 Use Appropriate Tools Strategically:** Geoboards allow students to change the shape or orientation of a figure with little effort. There is no erasing to do! When you change the location of a vertex, the two sides meeting at that vertex change automatically to meet at the new vertex.

The *Big Ideas Math* Teaching Edition is unique in its organization. It provides teachers with complete support as they teach the program. Using side-by-side pages, teachers have access to the full student page as they teach. Throughout the book Laurie Boswell shares insights into the Florida Common Core Standards and the Standards for Mathematical Practice.

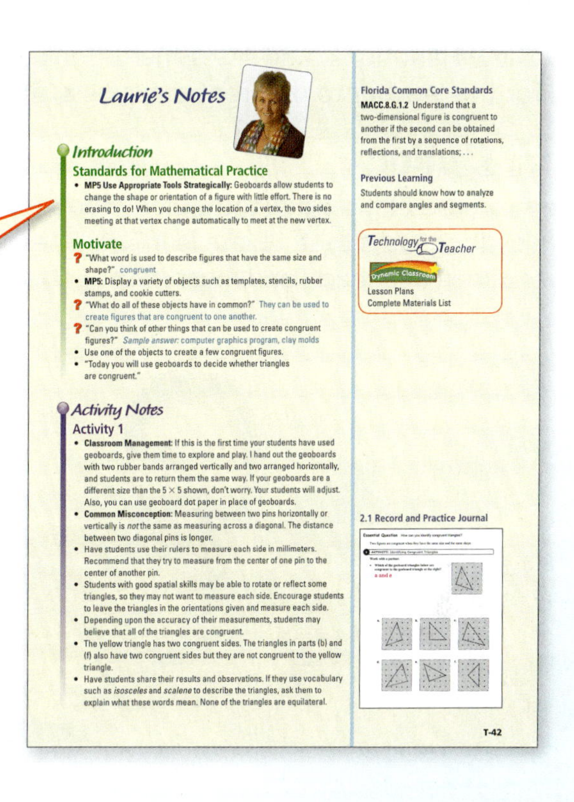
Laurie's Notes

Introduction

Standards for Mathematical Practice
- MP5 Use Appropriate Tools Strategically: Geoboards allow students to change the shape or orientation of a figure with little effort. There is no erasing to do! When you change the location of a vertex, the two sides meeting at that vertex change automatically to meet at the new vertex.

Motivate
- ? "What word is used to describe figures that have the same size and shape?" congruent
- MP5: Display a variety of objects such as templates, stencils, rubber stamps, and cookie cutters.
- ? "What do all of these objects have in common?" They can be used to create figures that are congruent to one another.
- ? "Can you think of other things that can be used to create congruent figures?" *Sample answer:* computer graphics program, clay molds
- Use one of the objects to create a few congruent figures.
- "Today you will use geoboards to decide whether triangles are congruent."

Activity Notes

Activity 1
- Classroom Management: If this is the first time your students have used geoboards, give them time to explore and play. I hand out the geoboards with two rubber bands arranged vertically and two arranged horizontally, and students are to return them the same way. If your geoboards are a different size than the 5 × 5 shown, don't worry. Your students will adjust. Also, you can use geoboard dot paper in place of geoboards.
- Common Misconception: Measuring between two pins horizontally or vertically is *not* the same as measuring across a diagonal. The distance between two diagonal pins is longer.
- Have students use their rulers to measure each side in millimeters. Recommend that they try to measure from the center of one pin to the center of another pin.
- Students with good spatial skills may be able to rotate or reflect some triangles, so they may not want to measure each side. Encourage students to leave the triangles in the orientations given and measure each side.
- Depending upon the accuracy of their measurements, students may believe that all of the triangles are congruent.
- The yellow triangle has two congruent sides. The triangles in parts (b) and (f) also have two congruent sides but they are not congruent to the yellow triangle.
- Have students share their results and observations. If they use vocabulary such as *isosceles* and *scalene* to describe the triangles, ask them to explain what these words mean. None of the triangles are equilateral.

Florida Common Core Standards

MACC.8.G.1.2 Understand that a two-dimensional figure is congruent to another if the second can be obtained from the first by a sequence of rotations, reflections, and translations; . . .

Previous Learning

Students should know how to analyze and compare angles and segments.

Technology for the Teacher

Dynamic Classroom

Lesson Plans
Complete Materials List

2.1 Record and Practice Journal

T-42

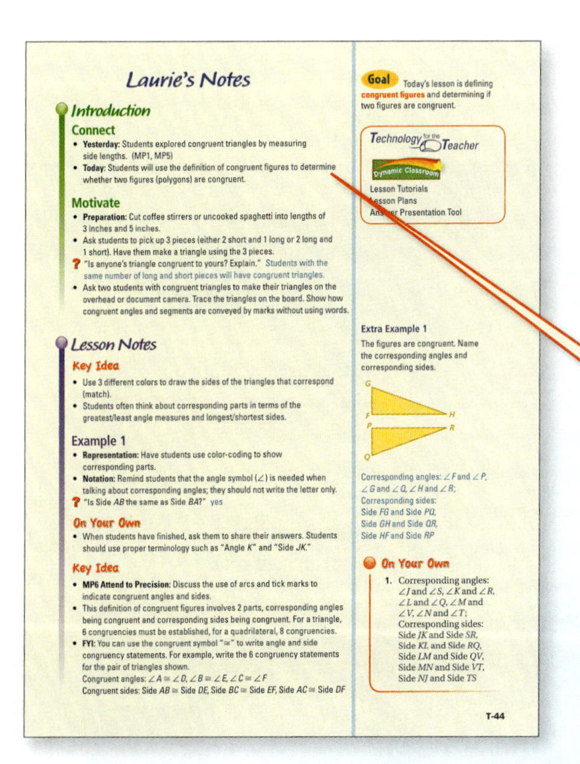
Laurie's Notes

Introduction

Connect
- Yesterday: Students explored congruent triangles by measuring side lengths. (MP1, MP5)
- Today: Students will use the definition of congruent figures to determine whether two figures (polygons) are congruent.

Motivate
- Preparation: Cut coffee stirrers or uncooked spaghetti into lengths of 3 inches and 5 inches.
- Ask students to pick up 3 pieces (either 2 short and 1 long or 2 long and 1 short). Have them make a triangle using the 3 pieces.
- ? "Is anyone's triangle congruent to yours? Explain." Students with the same number of long and short pieces will have congruent triangles.
- Ask two students with congruent triangles to make their triangles on the overhead or document camera. Trace the triangles on the board. Show how congruent angles and segments are conveyed by marks without using words.

Lesson Notes

Key Idea
- Use 3 different colors to draw the sides of the triangles that correspond (match).
- Students often think about corresponding parts in terms of the greatest/least angle measures and longest/shortest sides.

Example 1
- Representation: Have students use color-coding to show corresponding parts.
- Notation: Remind students that the angle symbol (∠) is needed when talking about corresponding angles; they should not write the letter only.
- ? "Is Side *AB* the same as Side *BA*?" yes

On Your Own
- When students have finished, ask them to share their answers. Students should use proper terminology such as "Angle *K*" and "Side *JK*."

Key Idea
- MP6 Attend to Precision: Discuss the use of arcs and tick marks to indicate congruent angles and sides.
- This definition of congruent figures involves 2 parts, corresponding angles being congruent and corresponding sides being congruent. For a triangle, 6 congruencies must be established, for a quadrilateral, 8 congruencies.
- FYI: You can use the congruent symbol "≅" to write angle and side congruency statements. For example, write the 6 congruency statements for the pair of triangles shown.

Congruent angles: $\angle A \cong \angle D$, $\angle B \cong \angle E$, $\angle C \cong \angle F$
Congruent sides: Side $AB \cong$ Side DE, Side $BC \cong$ Side EF, Side $AC \cong$ Side DF

Goal Today's lesson is defining **congruent figures** and determining if two figures are congruent.

Technology for the Teacher

Dynamic Classroom

Lesson Tutorials
Lesson Plans
Answer Presentation Tool

Extra Example 1

The figures are congruent. Name the corresponding angles and corresponding sides.

Corresponding angles: ∠*F* and ∠*P*, ∠*G* and ∠*Q*, ∠*H* and ∠*R*;
Corresponding sides: Side *FG* and Side *PQ*, Side *GH* and Side *QR*, Side *HF* and Side *RP*

On Your Own

1. Corresponding angles: ∠*J* and ∠*S*, ∠*K* and ∠*R*, ∠*L* and ∠*Q*, ∠*M* and ∠*V*, ∠*N* and ∠*T*; Corresponding sides: Side *JK* and Side *SR*, Side *KL* and Side *RQ*, Side *LM* and Side *QV*, Side *MN* and Side *VT*, Side *NJ* and Side *TS*

T-44

Each student page is accompanied by a support page. Laurie includes connections to previous learning, support for the Mathematical Practices, and closure opportunities.

Connect

- **Yesterday:** Students explored congruent triangles by measuring side lengths. (MP1, MP5)
- **Today:** Students will use the definition of congruent figures to determine whether two figures (polygons) are congruent.

The Teaching Edition also provides Differentiated Instruction, Response to Intervention, and English Language Learner support.

CUSTOMIZED INSTRUCTION

Print Option

The print Teaching Edition provides teachers with help through Laurie's Notes and other features that help manage the classroom. The Chapter Resource Book, Skills Review Handbook, and Assessment Book complete a teaching array that makes it easy for the teacher to differentiate, assess, and teach.

Digital Option

Teachers can use 21st century technology tools found throughout the program to provide exciting ways to stimulate learning. These tools provide innovative electronic activities, timely feedback, and measures for accurate assessment.

Blended Option

Teachers will find that using the blended option provides them a multitude of ways to teach, differentiate, and assess. Teachers and students can customize their teaching and learning by blending the power of creative technology tools with the accessibility of print resources. Rich content and the combination of creative print and online resources allows for an engaging and challenging approach to teaching the Florida Common Core Standards.

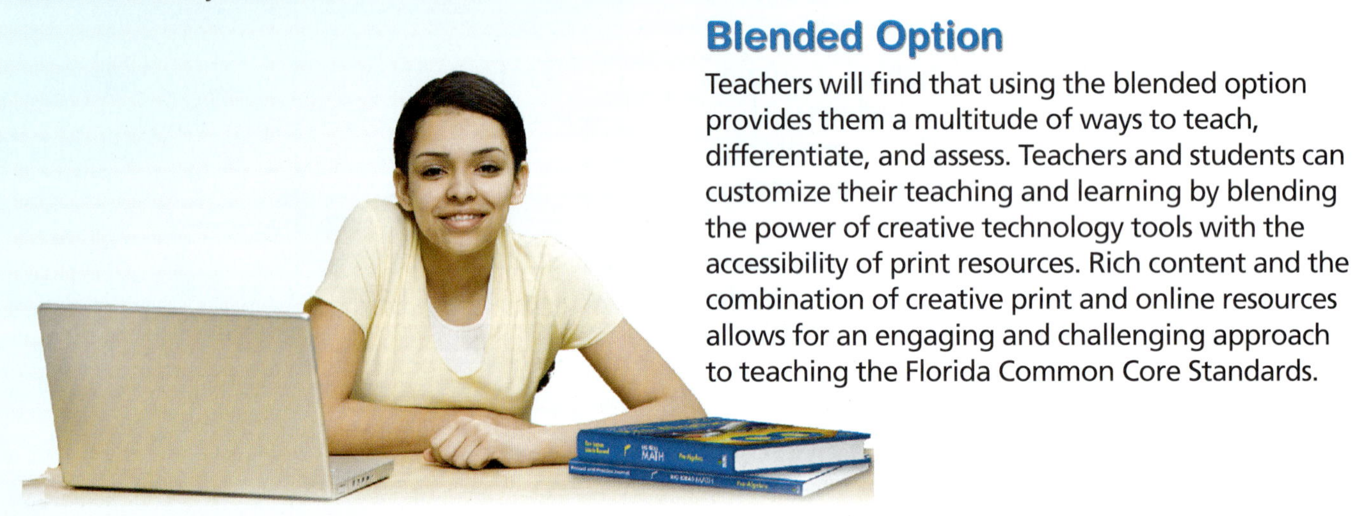

The Dynamic Classroom

Regardless of the option you choose for your students, *The Dynamic Classroom* will be one of your most valued tools in the *Big Ideas Math* program. This powerful tool can be used with interactive whiteboards and includes the following:

- Chapter Openers
- Start Thinking!
- Warm Ups
- Record and Practice Journal pages
- Virtual Manipulatives
- *On Your Own* Exercises
- Extra Examples
- Mini Assessments
- Closure Activities
- Graphic Organizers
- Mathematical Practices
- Answer Presentation Tool

PERSONALIZED LEARNING

The *Big Ideas Math* program offers teachers and students many ways to personalize and enrich the learning experience of all levels of learners.

Lesson Tutorials Online

Two- to three-minute lesson tutorials provide colorful visuals and audio support for every example in the textbook. The Lesson Tutorials are valuable for students who miss a class, need a second explanation, or just need some help with a homework assignment. Parents can also use the tutorials to stay connected or to provide additional help at home.

The Dynamic Student Edition

This unique tool provides students with 21st century learning tools making it easier to directly interact with the underlying mathematics. From the dynamic student textbook to engaging tutorials, students use electronic manipulatives, flashcards, and games to enhance their learning and understanding of math.

Differentiated Instruction

Through print and digital resources, the *Big Ideas Math* program completely supports the 3-Tier Response to Intervention model. Using research-based instructional strategies, teachers can reach, challenge and motivate each student with germane, high quality instruction targeted to individual needs.

Tier 3: Customized Learning Intervention

- Intensive Intervention Lessons
- Activities

Tier 2: Strategic Intervention

- Lesson Tutorials
- Basic Skills Handbook
- Skills Review Handbook
- Differentiated Instruction
- Game Closet

Tier 1: Daily Intervention

- Record and Practice Journal
- Fair Game Review
- Graphic Organizers
- Vocabulary Support
- Mini Assessments
- Game Closet
- Lesson Tutorials
- On Your Own

CONTINUOUS PREPARATION FOR

In the Textbook

Activities

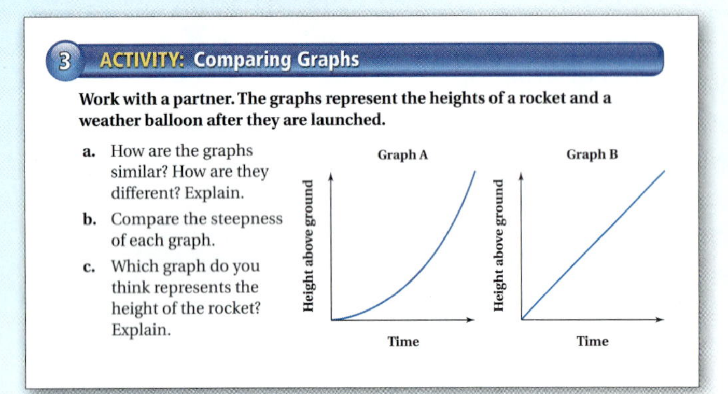

3 ACTIVITY: Comparing Graphs

Work with a partner. The graphs represent the heights of a rocket and a weather balloon after they are launched.

a. How are the graphs similar? How are they different? Explain.

b. Compare the steepness of each graph.

c. Which graph do you think represents the height of the rocket? Explain.

The new Common Core assessments will require a higher level of thinking. In the Activities, students are asked to explain their reasoning.

Exercises

16. MODELING The table shows the cost y (in dollars) of x pounds of sunflower seeds.

Pounds, x	Cost, y
2	2.80
3	?
4	5.60

a. What is the missing y-value that makes the table represent a linear function?

b. Write a linear function that represents the cost y of x pounds of seeds. Interpret the slope.

c. Does the function have a maximum value? Explain your reasoning.

The new Common Core assessments will ask for multiple representations. The *Big Ideas Math* program provides students opportunities to use multiple approaches to solve problems.

Quizzes and Tests

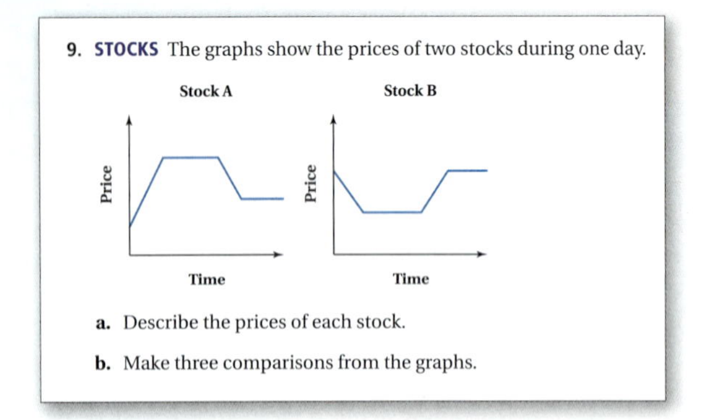

9. STOCKS The graphs show the prices of two stocks during one day.

a. Describe the prices of each stock.

b. Make three comparisons from the graphs.

The assessments in the textbook include problems that extend concepts.

Standards Assessment

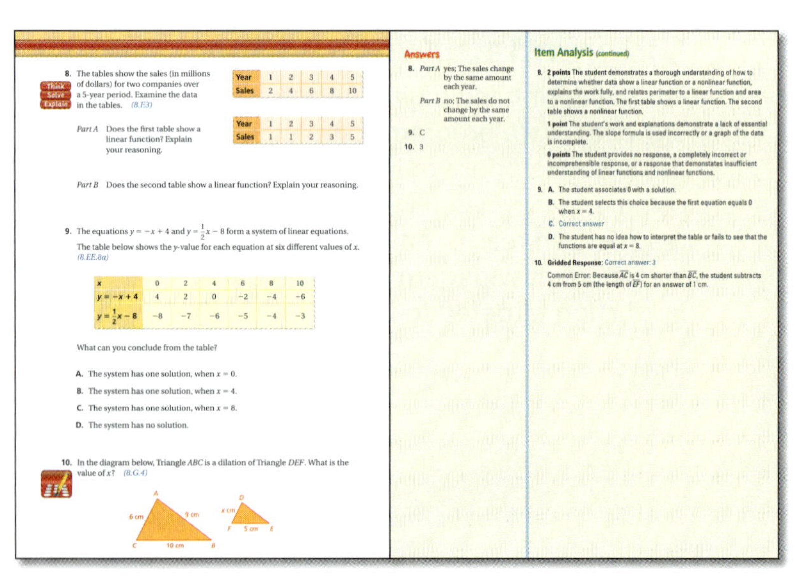

8. The tables show the sales (in millions of dollars) for two companies over a 5-year period. Examine the data in the tables. (8.F.3)

Year	1	2	3	4	5
Sales	2	4	6	8	10

Year	1	2	3	4	5
Sales	1	1	2	3	5

Part A Does the first table show a linear function? Explain your reasoning.

Part B Does the second table show a linear function? Explain your reasoning.

9. The equations $y = -x + 4$ and $y = \frac{1}{2}x - 8$ form a system of linear equations. The table below shows the y-value for each equation at six different values of x. (8.EE.8a)

x	0	2	4	6	8	10
$y = -x + 4$	4	2	0	−2	−4	−6
$y = \frac{1}{2}x - 8$	−8	−7	−6	−5	−4	−3

What can you conclude from the table?

A. The system has one solution, when $x = 0$.

B. The system has one solution, when $x = 4$.

C. The system has one solution, when $x = 8$.

D. The system has no solution.

10. In the diagram below, Triangle ABC is a dilation of Triangle DEF. What is the value of x? (8.G.4)

Answers

8. *Part A* yes; The sales change by the same amount each year.

Part B no; The sales do not change by the same amount each year.

9. C

10. 3

Item Analysis (continued)

8. 2 points The student demonstrates a thorough understanding of how to determine whether data show a linear function or a nonlinear function, explains the work fully, and relates perimeter to a linear function and area to a nonlinear function. The first table shows a linear function. The second table shows a nonlinear function.

1 point The student's work and explanations demonstrate a lack of essential understanding. The slope formula is used incorrectly or a graph of the data is incomplete.

0 points The student provides no response, a completely incorrect or incomprehensible response, or a response that demonstates insufficient understanding of linear functions and nonlinear functions.

9. A. The student associates 0 with a solution.

B. The student selects this choice because the first equation equals 0 when $x = 4$.

C. Correct answer

D. The student has no idea how to interpret the table or fails to see that the functions are equal at $x = 8$.

10. Gridded Response: Correct answer: 3

Common Error: Because $\overline{AC}$ is 4 cm shorter than $\overline{BC}$, the student subtracts 4 cm from 5 cm (the length of $\overline{EF}$) for an answer of 1 cm.

The Standards Assessments include questions in multiple formats to help prepare students for the Common Core assessments. A detailed Item Analysis for each question in the Pupil Edition is available in the Teaching Edition.

COMMON CORE ASSESSMENT

In the Technology

Online Test Practice

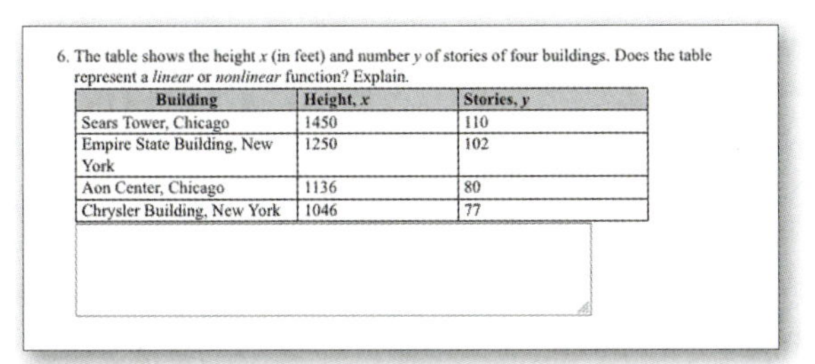

6. The table shows the height x (in feet) and number y of stories of four buildings. Does the table represent a *linear* or *nonlinear* function? Explain.

Building	Height, x	Stories, y
Sears Tower, Chicago	1450	110
Empire State Building, New York	1250	102
Aon Center, Chicago	1136	80
Chrysler Building, New York	1046	77

Students will find self-grading quizzes and tests on the student side of the website.

Performance Tasks

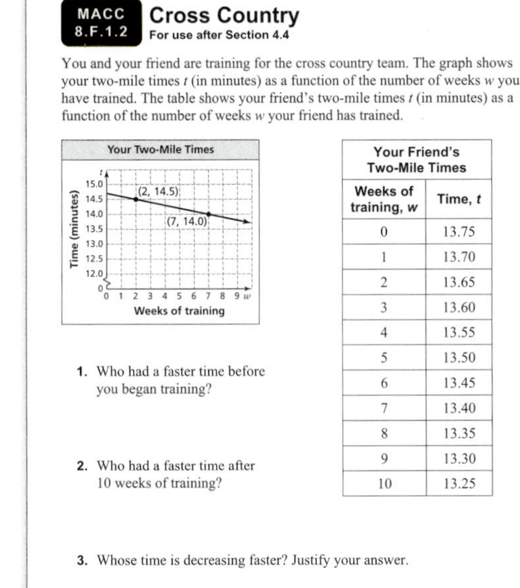

MACC 8.F.1.2 **Cross Country**
For use after Section 4.4

You and your friend are training for the cross country team. The graph shows your two-mile times t (in minutes) as a function of the number of weeks w you have trained. The table shows your friend's two-mile times t (in minutes) as a function of the number of weeks w your friend has trained.

Your Friend's Two-Mile Times

Weeks of training, w	Time, t
0	13.75
1	13.70
2	13.65
3	13.60
4	13.55
5	13.50
6	13.45
7	13.40
8	13.35
9	13.30
10	13.25

1. Who had a faster time before you began training?
2. Who had a faster time after 10 weeks of training?
3. Whose time is decreasing faster? Justify your answer.
4. Is either function a linear function? If so, write the linear function(s).
5. If the patterns continue, after what week will you and your friend have the same two-mile time?

Performance Tasks for every Florida Common Core Standard are available at *BigIdeasMath.com*.

Resources

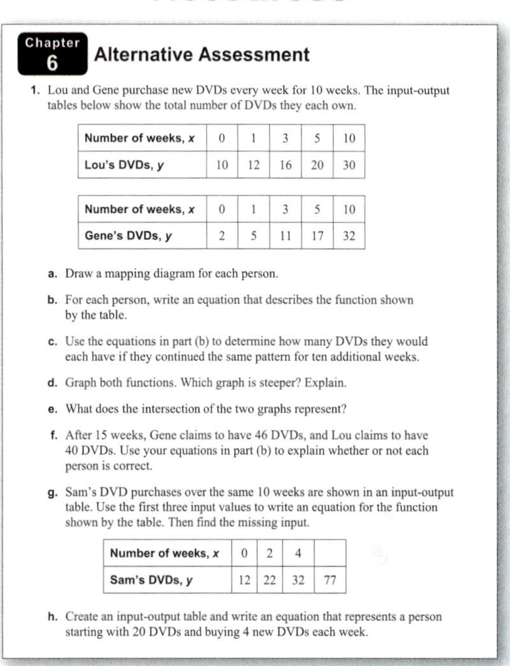

Chapter 6 **Alternative Assessment**

1. Lou and Gene purchase new DVDs every week for 10 weeks. The input-output tables below show the total number of DVDs they each own.

Number of weeks, x	0	1	3	5	10
Lou's DVDs, y	10	12	16	20	30

Number of weeks, x	0	1	3	5	10
Gene's DVDs, y	2	5	11	17	32

a. Draw a mapping diagram for each person.

b. For each person, write an equation that describes the function shown by the table.

c. Use the equations in part (b) to determine how many DVDs they would each have if they continued the same pattern for ten additional weeks.

d. Graph both functions. Which graph is steeper? Explain.

e. What does the intersection of the two graphs represent?

f. After 15 weeks, Gene claims to have 46 DVDs, and Lou claims to have 40 DVDs. Use your equations in part (b) to explain whether or not each person is correct.

g. Sam's DVD purchases over the same 10 weeks are shown in an input-output table. Use the first three input values to write an equation for the function shown by the table. Then find the missing input.

Number of weeks, x	0	2	4	
Sam's DVDs, y	12	22	32	77

h. Create an input-output table and write an equation that represents a person starting with 20 DVDs and buying 4 new DVDs each week.

Other resources include the Assessment Book, Resources by Chapter, and Additional Support at *BigIdeasMath.com*.

ExamView® Test Generator

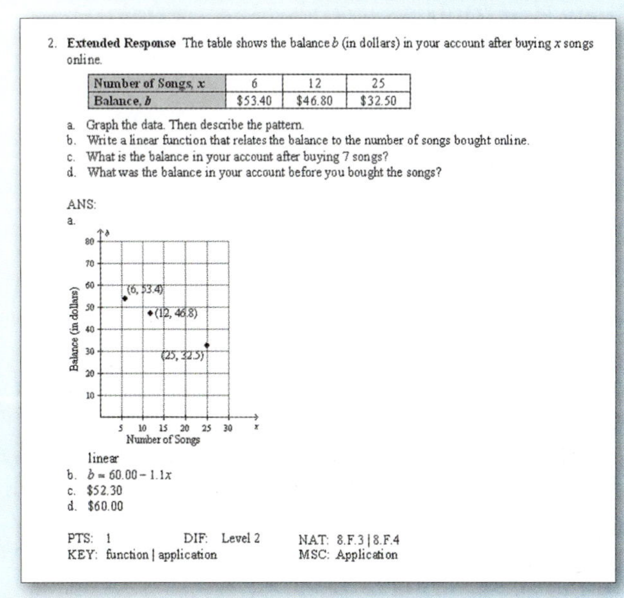

2. **Extended Response** The table shows the balance b (in dollars) in your account after buying x songs online.

Number of Songs, x	6	12	25
Balance, b	\$53.40	\$46.80	\$32.50

a. Graph the data. Then describe the pattern.
b. Write a linear function that relates the balance to the number of songs bought online.
c. What is the balance in your account after buying 7 songs?
d. What was the balance in your account before you bought the songs?

ANS:
a.

linear
b. $b = 60.00 - 1.1x$
c. \$52.30
d. \$60.00

PTS: 1 DIF: Level 2 NAT: 8.F.3 | 8.F.4
KEY: function | application MSC: Application

The *Big Ideas Math* assessment package includes enhanced ExamView® testing.

PACING GUIDE

Chapters 1–10 **149 Days**

Scavenger Hunt (1 Day)

Chapter 1 (12 Days)

Chapter Opener	1 Day
Activity 1.1	1 Day
Lesson 1.1	1 Day
Activity 1.2	1 Day
Lesson 1.2	1 Day
Study Help/Quiz	1 Day
Activity 1.3	1 Day
Lesson 1.3	1 Day
Activity 1.4	1 Day
Lesson 1.4	1 Day
Chapter Review/Chapter Tests	2 Days

Chapter 2 (19 Days)

Chapter Opener	1 Day
Activity 2.1	1 Day
Lesson 2.1	1 Day
Activity 2.2	1 Day
Lesson 2.2	1 Day
Activity 2.3	1 Day
Lesson 2.3	1 Day
Activity 2.4	1 Day
Lesson 2.4	1 Day
Study Help/Quiz	1 Day
Activity 2.5	1 Day
Lesson 2.5	1 Day
Activity 2.6	1 Day
Lesson 2.6	1 Day
Activity 2.7	1 Day
Lesson 2.7	2 Days
Chapter Review/Chapter Tests	2 Days

Chapter 3 (13 Days)

Chapter Opener	1 Day
Activity 3.1	1 Day
Lesson 3.1	1 Day
Activity 3.2	1 Day
Lesson 3.2	1 Day
Study Help/Quiz	1 Day
Activity 3.3	1 Day
Lesson 3.3	1 Day
Activity 3.4	1 Day
Lesson 3.4	2 Days
Chapter Review/Chapter Tests	2 Days

Chapter 4 (19 Days)

Chapter Opener	1 Day
Activity 4.1	1 Day
Lesson 4.1	1 Day
Activity 4.2	1 Day
Lesson 4.2	1 Day
Extension 4.2	1 Day
Activity 4.3	1 Day
Lesson 4.3	1 Day
Study Help/Quiz	1 Day
Activity 4.4	1 Day
Lesson 4.4	1 Day
Activity 4.5	1 Day
Lesson 4.5	1 Day
Activity 4.6	1 Day
Lesson 4.6	1 Day
Activity 4.7	1 Day
Lesson 4.7	1 Day
Chapter Review/Chapter Tests	2 Days

Chapter 5 (13 Days)

Chapter Opener	1 Day
Activity 5.1	1 Day
Lesson 5.1	1 Day
Activity 5.2	1 Day
Lesson 5.2	1 Day
Study Help/Quiz	1 Day
Activity 5.3	1 Day
Lesson 5.3	1 Day
Activity 5.4	1 Day
Lesson 5.4	1 Day
Extension 5.4	1 Day
Chapter Review/Chapter Tests	2 Days

Chapter 6 (14 Days)

Chapter Opener	1 Day
Activity 6.1	1 Day
Lesson 6.1	1 Day
Activity 6.2	1 Day
Lesson 6.2	1 Day
Activity 6.3	1 Day
Lesson 6.3	1 Day
Study Help/Quiz	1 Day
Activity 6.4	1 Day
Lesson 6.4	1 Day
Activity 6.5	1 Day
Lesson 6.5	1 Day
Chapter Review/Chapter Tests	2 Days

Chapter 7 (15 Days)

Chapter Opener	1 Day
Activity 7.1	1 Day
Lesson 7.1	1 Day
Activity 7.2	1 Day
Lesson 7.2	1 Day
Activity 7.3	1 Day
Lesson 7.3	1 Day
Study Help/Quiz	1 Day
Activity 7.4	1 Day
Lesson 7.4	1 Day
Extension 7.4	1 Day
Activity 7.5	1 Day
Lesson 7.5	1 Day
Chapter Review/Chapter Tests	2 Days

Chapter 8 (13 Days)

Chapter Opener	1 Day
Activity 8.1	1 Day
Lesson 8.1	1 Day
Activity 8.2	1 Day
Lesson 8.2	1 Day
Study Help/Quiz	1 Day
Activity 8.3	1 Day
Lesson 8.3	1 Day
Activity 8.4	1 Day
Lesson 8.4	2 Days
Chapter Review/Chapter Tests	2 Days

Chapter 9 (12 Days)

Chapter Opener	1 Day
Activity 9.1	1 Day
Lesson 9.1	1 Day
Activity 9.2	1 Day
Lesson 9.2	1 Day
Study Help/Quiz	1 Day
Activity 9.3	1 Day
Lesson 9.3	1 Day
Activity 9.4	1 Day
Lesson 9.4	1 Day
Chapter Review/Chapter Tests	2 Days

Chapter 10 (18 Days)

Chapter Opener	1 Day
Activity 10.1	1 Day
Lesson 10.1	1 Day
Activity 10.2	1 Day
Lesson 10.2	1 Day
Activity 10.3	1 Day
Lesson 10.3	1 Day
Activity 10.4	1 Day
Lesson 10.4	1 Day
Study Help/Quiz	1 Day
Activity 10.5	1 Day
Lesson 10.5	1 Day
Activity 10.6	1 Day
Lesson 10.6	1 Day
Activity 10.7	1 Day
Lesson 10.7	1 Day
Chapter Review/Chapter Tests	2 Days

Florida Common Core Standards for Mathematical Practice

Make sense of problems and persevere in solving them.

- Multiple representations are presented to help students move from concrete to representative and into abstract thinking
- *Essential Questions* help students focus and analyze
- *In Your Own Words* provide opportunities for students to look for meaning and entry points to a problem

Reason abstractly and quantitatively.

- Visual problem solving models help students create a coherent representation of the problem
- Opportunities for students to decontextualize and contextualize problems are presented in every lesson

Construct viable arguments and critique the reasoning of others.

- *Error Analysis*; *Different Words, Same Question*; and *Which One Doesn't Belong* features provide students the opportunity to construct arguments and critique the reasoning of others
- *Inductive Reasoning* activities help students make conjectures and build a logical progression of statements to explore their conjecture

Model with mathematics.

- Real-life situations are translated into diagrams, tables, equations, and graphs to help students analyze relations and to draw conclusions
- Real-life problems are provided to help students learn to apply the mathematics that they are learning to everyday life

Use appropriate tools strategically.

- *Graphic Organizers* support the thought process of what, when, and how to solve problems
- A variety of tool papers, such as graph paper, number lines, and manipulatives, are available as students consider how to approach a problem
- Opportunities to use the web, graphing calculators, and spreadsheets support student learning

Attend to precision.

- *On Your Own* questions encourage students to formulate consistent and appropriate reasoning
- Cooperative learning opportunities support precise communication

Look for and make use of structure.

- *Inductive Reasoning* activities provide students the opportunity to see patterns and structure in mathematics
- Real-world problems help students use the structure of mathematics to break down and solve more difficult problems

Look for and express regularity in repeated reasoning.

- Opportunities are provided to help students make generalizations
- Students are continually encouraged to check for reasonableness in their solutions

Go to *BigIdeasMath.com* for more information on the Florida Common Core Standards for Mathematical Practice.

Florida Common Core Standards for Mathematical Content for Grade 8

Chapter Coverage for Standards

Domain The Number System

- Know that there are numbers that are not rational, and approximate them by rational numbers.

Domain Expressions and Equations

- Work with radicals and integer exponents.
- Understand the connections between proportional relationships, lines, and linear equations.
- Analyze and solve linear equations and pairs of simultaneous equations.

Domain Functions

- Define, evaluate, and compare functions.
- Use functions to model relationships between quantities.

Domain Geometry

- Understand congruence and similarity using physical models, transparencies, or geometry software.
- Understand and apply the Pythagorean Theorem.
- Solve real-world and mathematical problems involving volume of cylinders, cones, and spheres.

Domain Statistics and Probability

- Investigate patterns of association in bivariate data.

Go to *BigIdeasMath.com* for more information on the Florida Common Core Standards for Mathematical Content.

How to Use Your Math Book

- Read the **Essential Question** in the activity.

  Discuss the **Math Practice** question with your partner.

  Work with a partner to decide **What Is Your Answer?**

  Now you are ready to do the **Practice** problems.

- Find the **Key Vocabulary** words, **highlighted in yellow**.

  Read their definitions. Study the concepts in each **Key Idea**.

  If you forget a definition, you can look it up online in the

  Multi-Language Glossary at BigIdeasMath.com.

- After you study each **EXAMPLE**, do the exercises in the **On Your Own**.

  Now You're Ready to do the exercises that correspond to the example.

  As you study, look for a 

  or a **Common Error** .

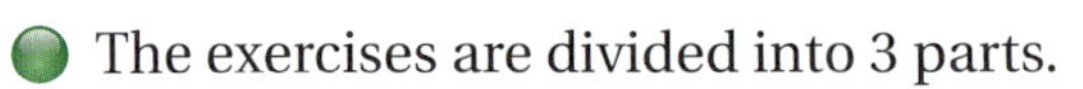

- The exercises are divided into 3 parts.

   **Vocabulary and Concept Check**

   **Practice and Problem Solving**

  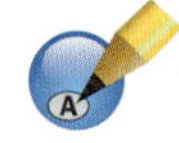 **Fair Game Review**

  If an exercise has a 1 next to it, look back at Example 1 for help with that exercise.

  More help is available at 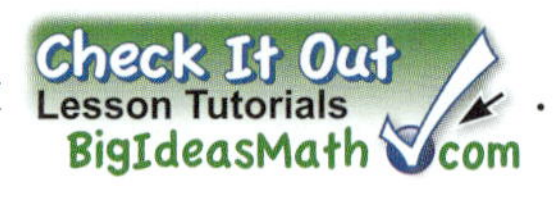

.

- To help study for your test, use the following.

  Chapter Review **Chapter Test**

SCAVENGER HUNT

Use this *Scavenger Hunt* to find where things are in **Chapter 1**.

1 Equations

1.1 Solving Simple Equations

1.2 Solving Multi-Step Equations

1.3 Solving Equations with Variables on Both Sides

1.4 Rewriting Equations and Formulas

"Dear Sir: Here is my suggestion for a good math problem."

"A box contains a total of 30 dog and cat treats. There are 5 times more dog treats than cat treats."

"How many of each type of treat are there?"

"Push faster, Descartes! According to the formula $R = D \div T$, the time needs to be 10 minutes or less to break our all-time speed record!"

Florida Common Core Progression

6th Grade

- Apply the properties of operations to generate equivalent expressions.
- Write and solve simple equations in which the coefficients and solutions are nonnegative rational numbers.
- Write an equation to represent one quantity in terms of another quantity.

7th Grade

- Add, subtract, factor, and expand linear expressions with rational coefficients.
- Solve multi-step problems posed with positive and negative rational numbers.
- Solve word problems leading to equations of the form $px + q = r$ and $p(x + q) = r$, where p, q, and r are rational numbers.

8th Grade

- Solve linear equations with rational number coefficients, including equations whose solutions require expanding expressions using the distributive property and collecting like terms.
- Show that a linear equation in one variable has one solution, infinitely many solutions, or no solution by transforming the equation into simpler forms.

Chapter Summary

Section	Florida Common Core Standard (MACC)	
1.1	Learning	8.EE.3.7a, 8.EE.3.7b
1.2	Learning	8.EE.3.7a, 8.EE.3.7b
1.3	Learning	8.EE.3.7a, 8.EE.3.7b
1.4	Applying	8.EE.3.7
★ Teaching is complete. Standard can be assessed.		

Pacing Guide for Chapter 1

Chapter Opener	1 Day
Section 1	
Activity	1 Day
Lesson	1 Day
Section 2	
Activity	1 Day
Lesson	1 Day
Study Help / Quiz	1 Day
Section 3	
Activity	1 Day
Lesson	1 Day
Section 4	
Activity	1 Day
Lesson	1 Day
Chapter Review/ Chapter Tests	2 Days
Total Chapter 1	12 Days
Year-to-Date	13 Days

Technology for the Teacher

BigIdeasMath.com
Chapter at a Glance
Complete Materials List
Parent Letters: English, Spanish, and Haitian Creole

Florida Common Core Standards

MACC.6.EE.1.3 Apply the properties of operations to generate equivalent expressions.

MACC.7.NS.1.1d Apply properties of operations as strategies to add and subtract rational numbers.

Additional Topics for Review

- Equation
- Equivalent Equations
- Inverse Operations
- Solution
- Checking a Solution
- Order of Operations
- Reciprocal
- Two-Step Equations
- Formulas

Try It Yourself

1. $4m$ **2.** $14g - 30$

3. $18 - 6y$ **4.** $12a - 48$

5. $7n - 1.3$ **6.** $7k + 88$

7. -7 **8.** -13

9. 8 **10.** 32

Record and Practice Journal Fair Game Review

1. $14x$ **2.** $-11b - 4$

3. $90 - 15g$ **4.** $2y - 42$

5. $5m + 48$ **6.** $13a - 2$

7. $1.5p + 3p + 2.5p; 7p$

8. -4 **9.** -12

10. -8 **11.** -8

12. 7 **13.** -7

14. 6 **15.** -12

16. 58°F

Math Background Notes

Vocabulary Review

- Expression
- Variable
- Like Terms
- Constant
- Coefficient
- Simplest Form
- Distributive Property
- Commutative Property of Addition
- Integer
- Opposite
- Absolute Value

Simplifying Algebraic Expressions

- Students should know how to simplify algebraic expressions.
- Remind students that an algebraic expression contains numbers, operations, and variables.
- Remind students that like terms are terms that have the same variables raised to the same exponents. A term without a variable, such as 5, is a constant. Constant terms are also like terms.
- **Common Error:** When identifying terms, make sure students include the sign of the term.
- An algebraic expression is in simplest form if it has no like terms and no parentheses. To combine like terms that have variables, use the Distributive Property to add or subtract the coefficients.

Adding and Subtracting Integers

- Students should know how to add and subtract integers.
- Remind students how to add integers with the same sign. They should add the absolute values of the integers and then use the common sign.
- Remind students how to add integers with different signs. They should subtract the lesser absolute value from the greater absolute value and then use the sign of the integer with the greater absolute value.
- **Common Error:** Students may ignore the signs and just add the integers. Remind them of the meaning of absolute value. Make sure they understand that they should use the sign of the number that is farther from zero.
- Subtraction problems can be rewritten as addition problems. When subtracting an integer, add its opposite.
- **Common Error:** Students may change the sign of the first number, or forget to change the problem from subtraction to addition when changing the sign of the second number. Remind them that the first number is a starting point and will never change. Also remind students that the sign of the second number and the operation may change.

Reteaching and Enrichment Strategies

If students need help...	If students got it...
Record and Practice Journal • Fair Game Review Skills Review Handbook Lesson Tutorials	Game Closet at *BigIdeasMath.com* Start the next section

What You Learned Before

Simplifying Algebraic Expressions (MACC.6.EE.1.3)

Example 1 **Simplify $10b + 13 - 6b + 4$.**

$10b + 13 - 6b + 4 = 10b - 6b + 13 + 4$	Commutative Property of Addition
$= (10 - 6)b + 13 + 4$	Distributive Property
$= 4b + 17$	Simplify.

Example 2 **Simplify $5(x + 4) + 2x$.**

$5(x + 4) + 2x = 5(x) + 5(4) + 2x$	Distributive Property
$= 5x + 20 + 2x$	Multiply.
$= 5x + 2x + 20$	Commutative Property of Addition
$= 7x + 20$	Combine like terms.

Try It Yourself

Simplify the expression.

1. $9m - 7m + 2m$
2. $3g - 9 + 11g - 21$
3. $6(3 - y)$
4. $12(a - 4)$
5. $22.5 + 7(n - 3.4)$
6. $15k + 8(11 - k)$

Adding and Subtracting Integers (MACC.7.NS.1.1d)

Example 3 **Find $4 + (-12)$.**

$$4 + (-12) = -8$$

$|-12| > |4|$. So, subtract $|4|$ from $|-12|$.

Use the sign of -12.

Example 4 **Find $-7 - (-16)$.**

$-7 - (-16) = -7 + 16$	Add the opposite of -16.
$= 9$	Add.

Try It Yourself

Add or subtract.

7. $-5 + (-2)$
8. $0 + (-13)$
9. $-6 + 14$
10. $19 - (-13)$

1.1 Solving Simple Equations

Essential Question How can you use inductive reasoning to discover rules in mathematics? How can you test a rule?

1 ACTIVITY: Sum of the Angles of a Triangle

Work with a partner. Use a protractor to measure the angles of each triangle. Copy and complete the table to organize your results.

a.

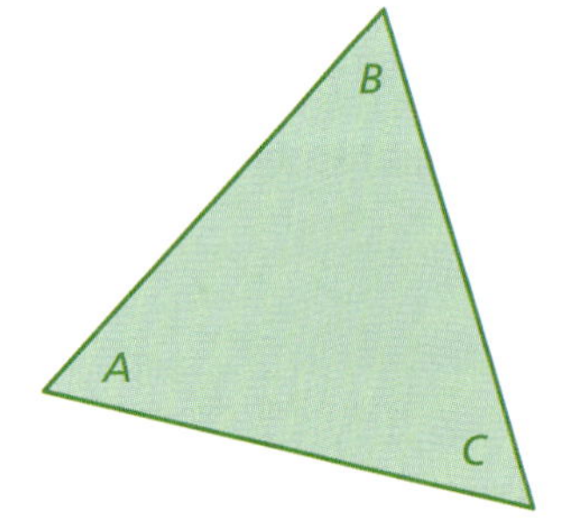

b.

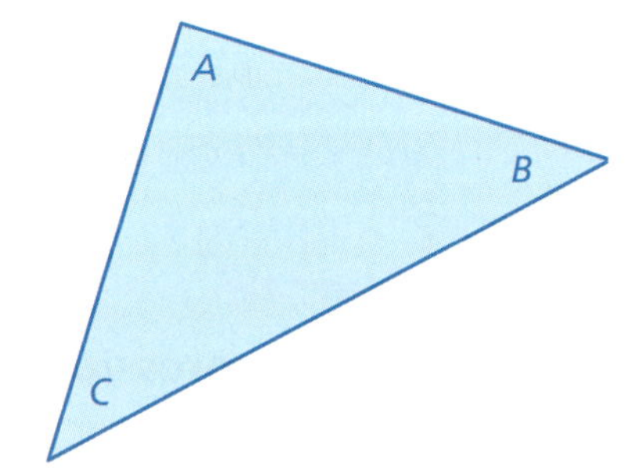

c.

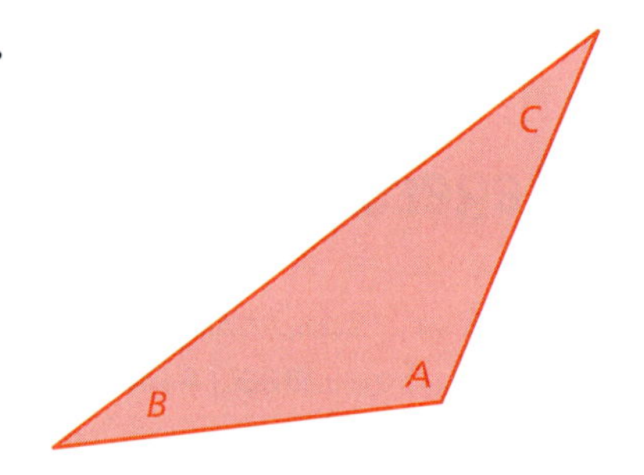

d.

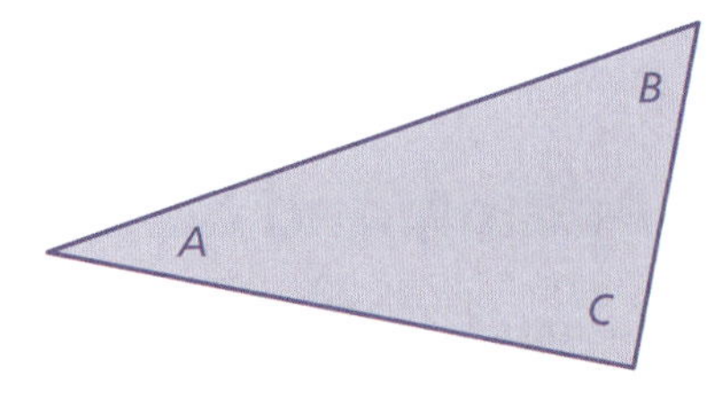

Solving Equations
In this lesson, you will
- solve simple equations using addition, subtraction, multiplication, or division.

Learning Standards
MACC.8.EE.3.7a
MACC.8.EE.3.7b

Triangle	Angle A (degrees)	Angle B (degrees)	Angle C (degrees)	$A + B + C$
a.				
b.				
c.				
d.				

Laurie's Notes

Introduction

Standards for Mathematical Practice

- **MP7 Look for and Make Use of Structure:** Mathematically proficient students look closely to discern a pattern, which may lead to making a conjecture.

Motivate

? "What do Tony Hawk, Shaun White, and Rodney Mullen have in common?" All are famous skateboarders. Shaun White is also a snowboarder.

- Today's activity is about angle measures. Boarders know a lot about angle measure, in particular the multiples of 180°, because of the different tricks they perform.

Activity Notes

Words of Wisdom

? "What does it mean to measure an angle?" Listen for an understanding of the rotation from one ray to a second ray. Both angles shown have the same measure, although some students would say the angle on the left is greater.

- Review with students how to place the protractor on the angle and how to read the protractor.
- **Common Error:** Notice that 0° does not always align with the bottom edge of some protractors, nor does the vertex of the angle always align with the bottom edge. It is common for students to align the bottom edge of the protractor with one ray of the angle, producing an error of more than 5°.

Activity 1

? **MP3 Construct Viable Arguments and Critique the Reasoning of Others:** "Do you see any pattern(s) in the table? Describe the pattern(s)." The sum of the angle measures is 180°, or close to 180°. Students might also mention that in part (a), all of the angles are congruent (same measure) and in parts (b) and (c), two of the angles are congruent.

- **FYI:** If a sum is significantly different from 180°, the student may have read the protractor incorrectly (i.e., they recorded 150° instead of 30°).

Florida Common Core Standards

MACC.8.EE.3.7a Give examples of linear equations in one variable with one solution, . . . successively transforming the given equation into simpler forms, until an equivalent equation of the form $x = a$ . . . results . . ..

MACC.8.EE.3.7b Solve linear equations with rational number coefficients, including equations whose solutions require . . . collecting like terms.

Previous Learning

Students should know the vocabulary of angles, such as ray, vertex, acute, obtuse, right, and straight.

1.1 Record and Practice Journal

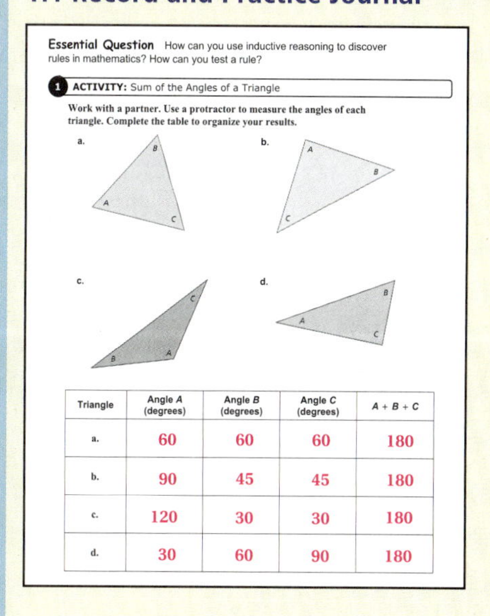

Essential Question How can you use inductive reasoning to discover rules in mathematics? How can you test a rule?

1 ACTIVITY: Sum of the Angles of a Triangle

Work with a partner. Use a protractor to measure the angles of each triangle. Complete the table to organize your results.

a. b. c. d.

Triangle	Angle *A* (degrees)	Angle *B* (degrees)	Angle *C* (degrees)	*A* + *B* + *C*
a.	60	60	60	180
b.	90	45	45	180
c.	120	30	30	180
d.	30	60	90	180

Differentiated Instruction

Kinesthetic

Ask two students to assist you at the board or overhead when solving equations. Assign one student to the left side of the equation and the other student to the right side. Each student is responsible for performing the operations on his or her side. Emphasize that to keep the equality, both students must perform the same operation to solve the equation.

1.1 Record and Practice Journal

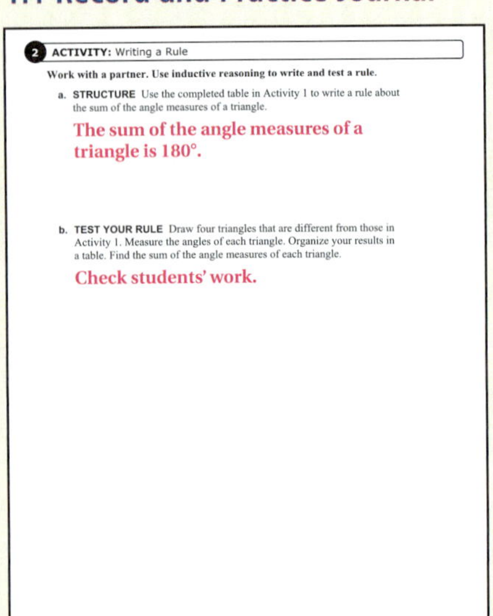

2 ACTIVITY: Writing a Rule

Work with a partner. Use inductive reasoning to write and test a rule.

a. **STRUCTURE** Use the completed table in Activity 1 to write a rule about the sum of the angle measures of a triangle.

The sum of the angle measures of a triangle is 180°.

b. **TEST YOUR RULE** Draw four triangles that are different from those in Activity 1. Measure the angles of each triangle. Organize your results in a table. Find the sum of the angle measures of each triangle.

Check students' work.

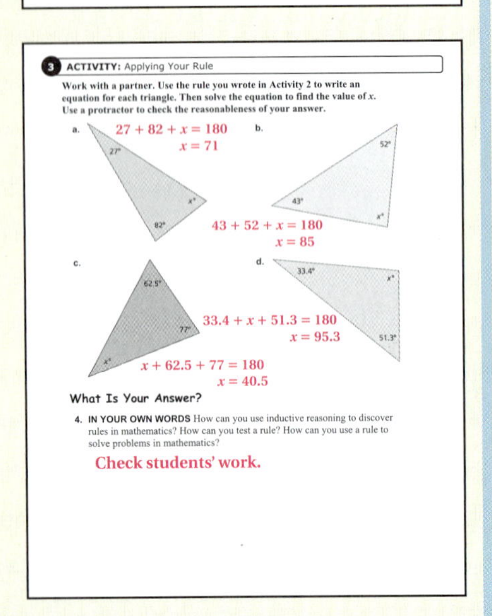

3 ACTIVITY: Applying Your Rule

Work with a partner. Use the rule you wrote in Activity 2 to write an equation for each triangle. Then solve the equation to find the value of x. Use a protractor to check the reasonableness of your answer.

a. $27 + 82 + x = 180$, $x = 71$

b. $43 + 52 + x = 180$, $x = 85$

c. $x + 62.5 + 77 = 180$, $x = 40.5$

d. $33.4 + x + 51.3 = 180$, $x = 95.3$

What Is Your Answer?

4. **IN YOUR OWN WORDS** How can you use inductive reasoning to discover rules in mathematics? How can you test a rule? How can you use a rule to solve problems in mathematics?

Check students' work.

Laurie's Notes

Activity 2

- **MP3:** "What rule did you write for the sum of the angle measures of a triangle?" Students should write that the sum of the angle measures of a triangle equals 180°.
- Suggest to students that they make their triangles larger so that it is easier to measure the angles. They should also use a straight edge to make straight lines.
- "Did you measure all three angles for each triangle?" Some students will not measure all three angles! They will only measure two and do a quick computation to find the third.

Activity 3

- You should model the first problem and write the equation. Otherwise, students will do a computation to find the missing angle.
- **Write:** $27 + 82 + x = 180$. Students have solved equations previously and may simply write the answer as the second step: $x = 71$. Focus on the representation of equation solving instead of the intuitive sense of how to solve this addition equation. Model the second step by showing 109 subtracted from each side of the equation.
- Note that parts (c) and (d) integrate decimal review. Their answers should be exact.
- Have students share their answers.

What Is Your Answer?

- "What is inductive reasoning and how was it used in the activities today?" *Sample answer:* Inductive reasoning is writing a general rule based on examples. Today I found that the sum of the angle measures of several triangles equals 180°, so I wrote a rule for triangles in general.
- Have students discuss their ideas to the questions posed in Question 4.

Closure

- **Exit Ticket:** Two angles of a triangle measure 48.2° and 63.8°. Make a reasonable sketch of the triangle. Write and solve an equation to find the measure of the third angle. $48.2 + 63.8 + x = 180$; $x = 68$

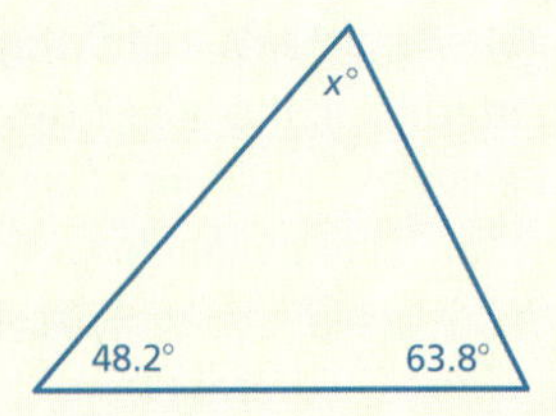

2 ACTIVITY: Writing a Rule

Work with a partner. Use inductive reasoning to write and test a rule.

a. **STRUCTURE** Use the completed table in Activity 1 to write a rule about the sum of the angle measures of a triangle.

b. **TEST YOUR RULE** Draw four triangles that are different from those in Activity 1. Measure the angles of each triangle. Organize your results in a table. Find the sum of the angle measures of each triangle.

3 ACTIVITY: Applying Your Rule

Math Practice 3

Analyze Conjectures

Do your results support the rule you wrote in Activity 2? Explain.

Work with a partner. Use the rule you wrote in Activity 2 to write an equation for each triangle. Then solve the equation to find the value of *x*. Use a protractor to check the reasonableness of your answer.

a.

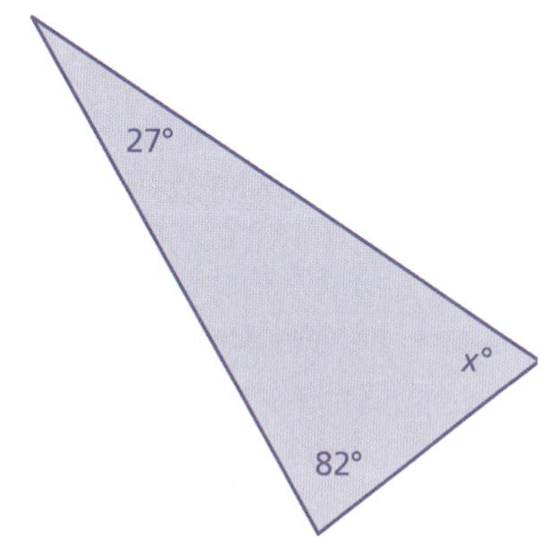

b.

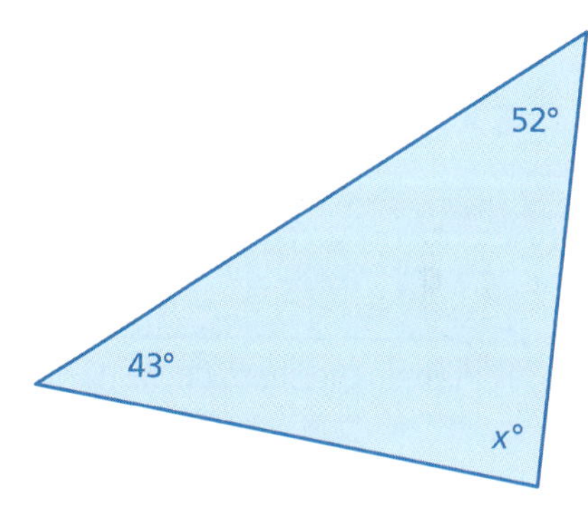

c.

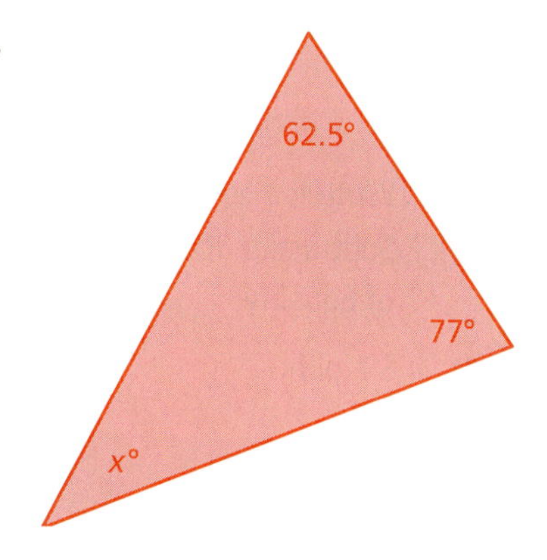

d.

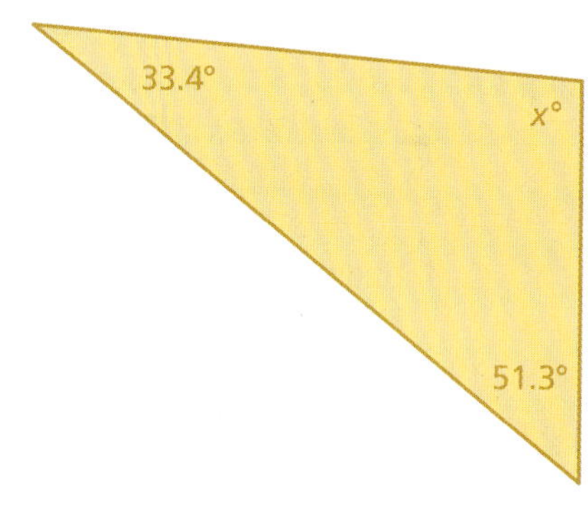

What Is Your Answer?

4. **IN YOUR OWN WORDS** How can you use inductive reasoning to discover rules in mathematics? How can you test a rule? How can you use a rule to solve problems in mathematics?

Use what you learned about solving simple equations to complete Exercises 4–6 on page 7.

1.1 Lesson

Remember

Addition and subtraction are inverse operations.

Key Ideas

Addition Property of Equality

Words Adding the same number to each side of an equation produces an equivalent equation.

Algebra If $a = b$, then $a + c = b + c$.

Subtraction Property of Equality

Words Subtracting the same number from each side of an equation produces an equivalent equation.

Algebra If $a = b$, then $a - c = b - c$.

EXAMPLE 1 Solving Equations Using Addition or Subtraction

a. Solve $x - 7 = -6$.

$x - 7 = -6$ — Write the equation.

Undo the subtraction. → $+7 \quad +7$ — Addition Property of Equality

$x = 1$ — Simplify.

The solution is $x = 1$.

Check

$x - 7 = -6$

$1 - 7 \stackrel{?}{=} -6$

$-6 = -6$ ✓

b. Solve $y + 3.4 = 0.5$.

$y + 3.4 = 0.5$ — Write the equation.

Undo the addition. → $-3.4 \quad -3.4$ — Subtraction Property of Equality

$y = -2.9$ — Simplify.

The solution is $y = -2.9$.

Check

$y + 3.4 = 0.5$

$-2.9 + 3.4 \stackrel{?}{=} 0.5$

$0.5 = 0.5$ ✓

c. Solve $h + 2\pi = 3\pi$.

$h + 2\pi = 3\pi$ — Write the equation.

Undo the addition. → $-2\pi \quad -2\pi$ — Subtraction Property of Equality

$h = \pi$ — Simplify.

The solution is $h = \pi$.

Laurie's Notes

Introduction

Connect

- **Yesterday:** Students used the sum of the angle measures of a triangle to explore simple equation solving. (MP3, MP7)
- **Today:** Students will use Properties of Equality to solve one-step equations.

Motivate

- Tell students that you are going to play a quick game of *REVERSO*. The directions are simple: give a command to a student and the student must give the reverse (inverse) command to undo the original command. For example, say, "Take 3 steps forward, " and the student would say, "Take 3 steps backward."
- Sample commands: turn lights on; step up on a chair; turn to your right; fold 2 sheets of paper; draw a square; open the door
- After you play a few rounds with the students, have students play with partners.
- The goal is for students to think about inverse operations.

Lesson Notes

Key Ideas

- Write the Key Ideas.
- Remind students that two equations that have the same solution are *equivalent equations.*
- **Teaching Tip:** Use an alternate color to show adding (subtracting) c to (from) each side of the equation.
- Remind students of the big idea. Whatever you do to one side of the equation, you must do to the other side of the equation.

Example 1

- Work through each part. Note that the vertical format of equation solving is used. The number being added to or subtracted from each side of the equation is written vertically below the number with which it will be combined.

? "What is the approximate value of π?" 3.14 "of 2π?" 6.28

- Remind students that 2π and 3π are (irrational) numbers, so you can treat these numbers as you would integers. It is common for students to think of π as a variable and they will say that there are two variables in $h + 2\pi$.

Goal Today's lesson is solving one-step equations.

Lesson Tutorials
Lesson Plans
Answer Presentation Tool

Extra Example 1

a. Solve $d - \frac{1}{4} = -\frac{1}{2}$. $-\frac{1}{4}$

b. Solve $m + 4.8 = 9.2$ 4.4

c. Solve $r - 6\pi = 2\pi$. 8π

On Your Own

1. $b = -7$ 2. $g = 0.8$
3. $k = -6$ 4. $r = 2\pi$
5. $t = -\frac{1}{2}$ 6. $z = -13.6$

Extra Example 2

a. Solve $\frac{2}{5}m = -4$. -10

b. Solve $3p = -\frac{2}{3}$. $-\frac{2}{9}$

On Your Own

7. $y = -28$ 8. $x = 6$
9. $w = 20$

English Language Learners

Vocabulary

In this section, students will learn to use inverse (or opposite) operations to solve equations. Students will use addition to solve a subtraction equation and use subtraction to solve an addition equation. Review these pairs of words that are essential to understanding mathematics. Give students one word of a pair and ask them to provide the opposite.

Examples:

odd, even	positive, negative
add, subtract	sum, difference
multiply, divide	product, quotient
plus, minus	

Laurie's Notes

On Your Own

- Circulate as students work on these six questions. Remind them that it is the practice of *representing* their work that is important in these questions.
- **MP7 Look for and Make Use of Structure:** Note that in Question 3, the variable is on the right side of the equation. Some students may be unclear about how to find the solution, because they often view the equal sign as a right-pointing arrow. By subtracting 3 from both sides of the equation, you isolate the variable on the right side. Check to make sure students are finding the solution correctly.
- **Common Error:** In Questions 1, 3, and 6, students can subtract two negative numbers to find the solution. Subtraction errors are common with problems such as these. For instance, in Question 6, students may come up with the solution $z = 2.4$. Encourage students to check their solutions as directed.

Key Ideas

- Write the Key Ideas.
- **Representation:** Review different ways in which multiplication is represented.

  $a(c) = b(c)$ $\quad ac = bc$ $\quad a \times c = b \times c$ $\quad a \cdot c = b \cdot c$
- Generally, when there are variables in equations, you do not want to use $\times$ to represent multiplication because it can be mistaken for a variable.
- **Representation**: Review different ways in which division is represented.

  $a \div c = b \div c$ $\quad \frac{a}{c} = \frac{b}{c}$ $\quad a/c = b/c$

Example 2

- Remind students that the goal is to solve for the variable so that it has a coefficient of 1.

? "What is the inverse operation of multiplying by $-\frac{3}{4}$?" dividing by $-\frac{3}{4}$

"What operation is equivalent to dividing by $-\frac{3}{4}$? Explain." multiplying by $-\frac{4}{3}$; When dividing by a fraction, multiply by the reciprocal of the fraction.

- Remind students that the Multiplicative Inverse Property states that the product of a number and its reciprocal is 1. In this case, $-\frac{4}{3} \cdot -\frac{3}{4} = 1$.
- The purpose of part (b) is to practice working with π in an algebraic expression.

On Your Own

- Circulate as students work on these three questions. Stress representation with students. Do not let them short-cut the process and simply record the answer. They should show what operation is being performed on each side of the equation.
- **Common Error:** In Question 8, students may try to subtract 6π from πx or subtract πx from 6π. Remind them that the variable they are solving for is x.

Solve the equation. Check your solution.

1. $b + 2 = -5$ **2.** $g - 1.7 = -0.9$ **3.** $-3 = k + 3$

4. $r - \pi = \pi$ **5.** $t - \frac{1}{4} = -\frac{3}{4}$ **6.** $5.6 + z = -8$

Key Ideas

Remember

Multiplication and division are inverse operations.

Multiplication Property of Equality

Words Multiplying each side of an equation by the same number produces an equivalent equation.

Algebra If $a = b$, then $a \cdot c = b \cdot c$.

Division Property of Equality

Words Dividing each side of an equation by the same number produces an equivalent equation.

Algebra If $a = b$, then $a \div c = b \div c$, $c \neq 0$.

EXAMPLE 2 Solving Equations Using Multiplication or Division

a. Solve $-\frac{3}{4}n = -2$.

$-\frac{3}{4}n = -2$ Write the equation.

Use the reciprocal. → $-\frac{4}{3} \cdot \left(-\frac{3}{4}n\right) = -\frac{4}{3} \cdot (-2)$ Multiplication Property of Equality

$n = \frac{8}{3}$ Simplify.

The solution is $n = \frac{8}{3}$.

b. Solve $\pi x = 3\pi$.

$\pi x = 3\pi$ Write the equation.

Undo the multiplication. → $\frac{\pi x}{\pi} = \frac{3\pi}{\pi}$ Division Property of Equality

$x = 3$ Simplify.

The solution is $x = 3$.

Check

$\pi x = 3\pi$

$\pi(3) \stackrel{?}{=} 3\pi$

$3\pi = 3\pi$ ✓

On Your Own

Now You're Ready
Exercises 18–26

Solve the equation. Check your solution.

7. $\frac{y}{4} = -7$ **8.** $6\pi = \pi x$ **9.** $0.09w = 1.8$

EXAMPLE 3 Identifying the Solution of an Equation

What value of k makes the equation $k + 4 \div 0.2 = 5$ true?

Ⓐ -15 Ⓑ -5 Ⓒ -3 Ⓓ 1.5

$k + 4 \div 0.2 = 5$	Write the equation.
$k + 20 = 5$	Divide 4 by 0.2.
$-20 \quad -20$	Subtraction Property of Equality
$k = -15$	Simplify.

The correct answer is Ⓐ.

EXAMPLE 4 Real-Life Application

The melting point of bromine is -7°C.

The *melting point* of a solid is the temperature at which the solid becomes a liquid. The melting point of bromine is $\frac{1}{30}$ of the melting point of nitrogen. Write and solve an equation to find the melting point of nitrogen.

Words The melting point of bromine is $\frac{1}{30}$ of the melting point of nitrogen.

Variable Let n be the melting point of nitrogen.

Equation $-7 = \frac{1}{30} \cdot n$

$-7 = \frac{1}{30}n$	Write the equation.
$30 \cdot (-7) = 30 \cdot \left(\frac{1}{30}n\right)$	Multiplication Property of Equality
$-210 = n$	Simplify.

So, the melting point of nitrogen is -210°C.

On Your Own

10. Solve $p - 8 \div \frac{1}{2} = -3$.

11. Solve $q + |-10| = 2$.

12. The melting point of mercury is about $\frac{1}{4}$ of the melting point of krypton. The melting point of mercury is -39°C. Write and solve an equation to find the melting point of krypton.

Laurie's Notes

Example 3

? "What is $10 + 4 \div 2$?" Listen for order of operations; Answer is 12, *not* 7.

- Students could use *Guess, Check, and Revise*. However, it is more efficient to use order of operations and then solve the equation.

Example 4

- Note the color-coding of the words and symbols. Discuss this feature with students. Students find it difficult to read a word problem and translate it into symbols. This skill is practiced throughout the text.
- **MP2 Reason Abstractly and Quantitatively:** The term $\frac{1}{30}n$ could also have been written as $\frac{n}{30}$. Make sure students understand why. It is how a fraction and number are multiplied.
- **FYI:** The final answer $-210 = n$ can also be written as $n = -210$.

On Your Own

- Remind students to follow the order of operations.
- **Common Error:** $8 \div \frac{1}{2} \neq 4$; $8 \div \frac{1}{2} = 16$

? "For Question 11, what does $|-10|$ mean?" absolute value of -10, which equals 10

Closure

- Describe in words how to solve a one-step equation.
- Write and solve a one-step equation.

Extra Example 3

Solve $w - 4 \div \frac{1}{2} = 5$. $w = 13$

Extra Example 4

The melting point of ice is $\frac{2}{9}$ of the melting point of candle wax. The melting point of ice is 32°F. Write and solve an equation to find the melting point of candle wax. $\frac{2}{9}x = 32$; 144°F

On Your Own

10. $p = 1$

11. $q = -8$

12. $-39 = \frac{1}{4}k$; -156°C

Vocabulary and Concept Check

1. $+$ and $-$ are inverses. $\times$ and $\div$ are inverses.
2. yes; The solution of each equation is $x = -3$.
3. $x - 3 = 6$; It is the only equation that does not have $x = 6$ as a solution.

Practice and Problem Solving

4. $x = 32$
5. $x = 57$
6. $x = 111$
7. $x = -5$
8. $g = 24$
9. $p = 21$
10. $y = -2.04$
11. $x = 9\pi$
12. $w = 10\pi$
13. $d = \frac{1}{2}$
14. $r = -\frac{7}{24}$
15. $n = -4.9$
16. $p - 14.50 = 53$; $67.50
17. a. $105 = x + 14$; $x = 91$

    b. no; Because $82 + 9 = 91$, you did not knock down the last pin with the second ball of the frame.

Assignment Guide and Homework Check

Level	Day 1 Activity Assignment	Day 2 Lesson Assignment	Homework Check
Basic	4–6, 45–48	1–3, 7–11 odd, 17–29 odd, 33, 35	7, 11, 21, 29, 33
Average	4–6, 45–48	1–3, 12–16 even, 23–43 odd	14, 16, 25, 31, 39
Advanced	1–6, 12, 14, 20, 26, 27, 28–44 even, 45–48		14, 20, 28, 38, 42

For Your Information

- **Exercise 17** Students may not be familiar with the bowling terms *frame* and *spare*. Each set of ten pins is called a frame. An entire game has ten frames. A spare means that all the pins were knocked down after the second ball of a frame. To calculate the score after a spare, you add the number of pins knocked down on your next ball to 10. For example, if you got a spare in the first frame and then knocked down 6 pins on your next ball, your score for the first frame would be 16.
- **Exercise 31** Simple interest will be covered in a later chapter.

Common Errors

- **Exercises 4–6** Students may struggle using the protractor to find the missing angle. Encourage them to trace the triangle and extend the sides so they can get a more accurate reading.
- **Exercises 7–15** Students may perform the same operation that they are trying to undo instead of the opposite operation. Remind them that to solve for the variable, they must use the opposite (or inverse) operation on both sides of the equation.

1.1 Record and Practice Journal

Solve the equation. Check your solution.

1. $x + 5 = 16$ — $x = 11$
2. $11 = w - 12$ — $w = 23$
3. $\frac{3}{4} + z = \frac{5}{6}$ — $z = \frac{1}{12}$
4. $3y = 18$ — $y = 6$
5. $\frac{k}{7} = 10$ — $k = 70$
6. $\frac{4}{5}n = \frac{9}{10}$ — $n = \frac{9}{8}$
7. $x - 12 + 6 = 9$ — $x = 11$
8. $h + |-8| = 15$ — $h = 7$
9. $1.3(2) + p = 7.9$ — $p = 5.3$
10. A coupon subtracts $5.16 from the price p of a shirt. You pay $15.48 for the shirt after using the coupon. Write and solve an equation to find the original price of the shirt.

    $p - 5.16 = 15.48$

    $p = \$20.64$

1.1 Exercises

Vocabulary and Concept Check

1. **VOCABULARY** Which of the operations $+$, $-$, $\times$, and $\div$ are inverses of each other?
2. **VOCABULARY** Are the equations $3x = -9$ and $4x = -12$ equivalent? Explain.
3. **WHICH ONE DOESN'T BELONG?** Which equation does *not* belong with the other three? Explain your reasoning.

$x - 2 = 4$ $\qquad$ $x - 3 = 6$ $\qquad$ $x - 5 = 1$ $\qquad$ $x - 6 = 0$

Practice and Problem Solving

CHOOSE TOOLS Find the value of x. Check the reasonableness of your answer.

4. 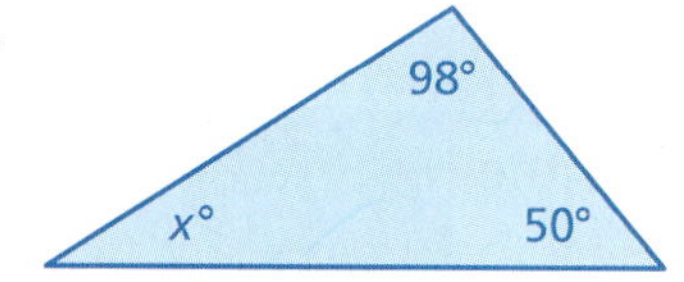

5. 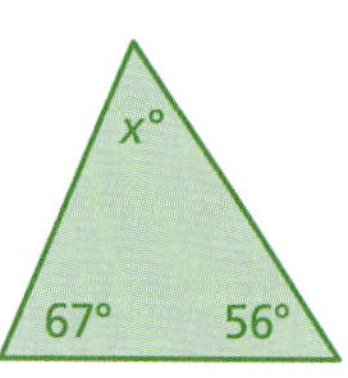

6. 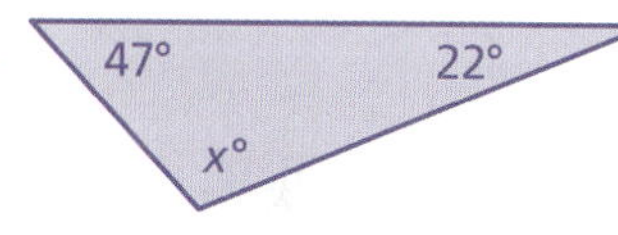

Solve the equation. Check your solution.

1

7. $x + 12 = 7$
8. $g - 16 = 8$
9. $-9 + p = 12$
10. $0.7 + y = -1.34$
11. $x - 8\pi = \pi$
12. $4\pi = w - 6\pi$
13. $\frac{5}{6} = \frac{1}{3} + d$
14. $\frac{3}{8} = r + \frac{2}{3}$
15. $n - 1.4 = -6.3$

16. **CONCERT** A discounted concert ticket costs \$14.50 less than the original price p. You pay \$53 for a discounted ticket. Write and solve an equation to find the original price.

17. **BOWLING** Your friend's final bowling score is 105. Your final bowling score is 14 pins less than your friend's final score.

    a. Write and solve an equation to find your final score.

    b. Your friend made a spare in the 10th frame. Did you? Explain.

Solve the equation. Check your solution.

2 **18.** $7x = 35$

19. $4 = -0.8n$

20. $6 = -\frac{w}{8}$

21. $\frac{m}{\pi} = 7.3$

22. $-4.3g = 25.8$

23. $\frac{3}{2} = \frac{9}{10}k$

24. $-7.8x = -1.56$

25. $-2 = \frac{6}{7}p$

26. $3\pi d = 12\pi$

27. ERROR ANALYSIS Describe and correct the error in solving the equation.

✗
$$-1.5 + k = 8.2$$
$$k = 8.2 + (-1.5)$$
$$k = 6.7$$

28. TENNIS A gym teacher orders 42 tennis balls. Each package contains 3 tennis balls. Which of the following equations represents the number x of packages?

$x + 3 = 42$ $3x = 42$ $\frac{x}{3} = 42$ $x = \frac{3}{42}$

MODELING In Exercises 29–32, write and solve an equation to answer the question.

29. PARK You clean a community park for 6.5 hours. You earn $42.25. How much do you earn per hour?

30. ROCKET LAUNCH A rocket is scheduled to launch from a command center in 3.75 hours. What time is it now?

Launch Time
11:20 A.M.

31. BANKING After earning interest, the balance of an account is $420. The new balance is $\frac{7}{6}$ of the original balance. How much interest did it earn?

Roller Coasters at Cedar Point	
Coaster	**Height (feet)**
Top Thrill Dragster	420
Millennium Force	310
Magnum XL-200	205
Mantis	?

32. ROLLER COASTER Cedar Point amusement park has some of the tallest roller coasters in the United States. The Mantis is 165 feet shorter than the Millennium Force. What is the height of the Mantis?

Common Errors

- **Exercises 18–26** Students may use the same operation that they are trying to undo instead of the opposite operation to get the variable by itself. Remind them that they must use the opposite (or inverse) operation. Demonstrate that using the same operation will not work. For example:

Incorrect	Correct
$7x = 35$	$7x = 35$
$7 \cdot 7x = 35 \cdot 7$	$\frac{7x}{7} = \frac{35}{7}$
$49x = 245$	$x = 5$

- **Exercise 32** Students may skip the step of writing the equation and just subtract the difference in height from the height of the Millennium Force. Encourage them to develop the problem solving technique of writing the equation before solving. This skill will be useful when solving more difficult problems.
- **Exercises 33–38** Students may forget to use the order of operations when solving for the variable. Remind them to use the order of operations before solving.

Practice and Problem Solving

18. $x = 5$ **19.** $n = -5$

20. $w = -48$ **21.** $m = 7.3\pi$

22. $g = -6$ **23.** $k = 1\frac{2}{3}$

24. $x = 0.2$ **25.** $p = -2\frac{1}{3}$

26. $d = 4$

27. They should have added 1.5 to each side.
$-1.5 + k = 8.2$
$k = 8.2 + 1.5$
$k = 9.7$

28. $3x = 42$

29. $6.5x = 42.25$; \$6.50 per hour

30. $x + 3.75 = 11\frac{1}{3}$; 7:35 A.M.

31. $420 = \frac{7}{6}b$, $b = 360$; \$60

32. $x + 165 = 310$; 145 ft

English Language Learners

Vocabulary

Have students create a table in their notebooks of the common words used to indicate addition, subtraction, multiplication, and division.
For instance,

Addition	Subtraction
added to	subtracted from
plus	minus
sum of	difference of
more than	less than
increased by	decreased by
total of	fewer than
and	take away
Multiplication	**Division**
multiplied by	divided by
times	quotient of
product of	
twice	
of	

Practice and Problem Solving

33. $h = -7$
34. $w = 19$
35. $q = 3.2$
36. $d = 0$
37. $x = -1\frac{4}{9}$
38. $p = -\frac{1}{12}$
39. greater than; Because a negative number divided by a negative number is a positive number.
40. *Sample answer:* $x - 2 = -4$, $\frac{x}{2} = -1$
41. 3 mg
42. See *Taking Math Deeper*.
43. 12 in.
44. **a.** \$18, \$27, \$45

    b. *Sample answer:* Everyone did not do an equal amount of painting.

Fair Game Review

45. $7x - 4$
46. $1.6b - 3.2$
47. $\frac{25}{4}g - \frac{2}{3}$
48. A

Mini-Assessment

Solve the equation. Check your solution.

1. $t + 17 = 3$ $t = -14$
2. $-2\pi + d = -3\pi$ $d = -\pi$
3. $-13.5 = 2.7s$ $s = -5$
4. $\frac{2}{3}j = 8$ $j = 12$
5. You earn \$9.65 per hour. This week, you earned \$308.80 before taxes. Write and solve an equation to find the number of hours you worked this week. $9.65x = 308.8$; You worked 32 hours this week.

Taking Math Deeper

Exercise 42

A nice way to organize the given information is to put it into a table.

Use a table to organize the information.

	Total	Retake
Girls	x	$\frac{1}{4}x = 16$
Boys	y	$\frac{1}{8}y = 7$

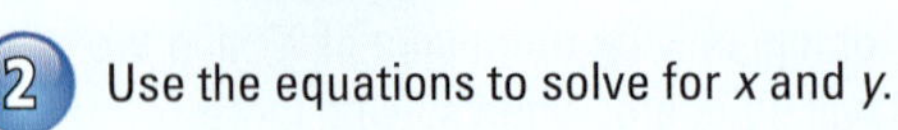

Use the equations to solve for x and y.

Girls: $\frac{1}{4}x = 16$

$x = 64$

Boys: $\frac{1}{8}y = 7$

$y = 56$

Answer the question.

There are $64 + 56 = 120$ students in the class.

Project

Find out how many retakes were done at your school last year. Do the given ratios work for your school? What do you think are some of the reasons students have retakes?

Reteaching and Enrichment Strategies

If students need help. . .	If students got it. . .
Resources by Chapter • Practice A and Practice B • Puzzle Time Record and Practice Journal Practice Differentiating the Lesson Lesson Tutorials Skills Review Handbook	Resources by Chapter • Enrichment and Extension • Technology Connection Start the next section

Solve the equation. Check your solution.

3 **33.** $-3 = h + 8 \div 2$

34. $12 = w - |-7|$

35. $q + |6.4| = 9.6$

36. $d - 2.8 \div 0.2 = -14$

37. $\frac{8}{9} = x + \frac{1}{3}(7)$

38. $p - \frac{1}{4} \cdot 3 = -\frac{5}{6}$

39. LOGIC Without solving, determine whether the solution of $-2x = -15$ is *greater than* or *less than* -15. Explain.

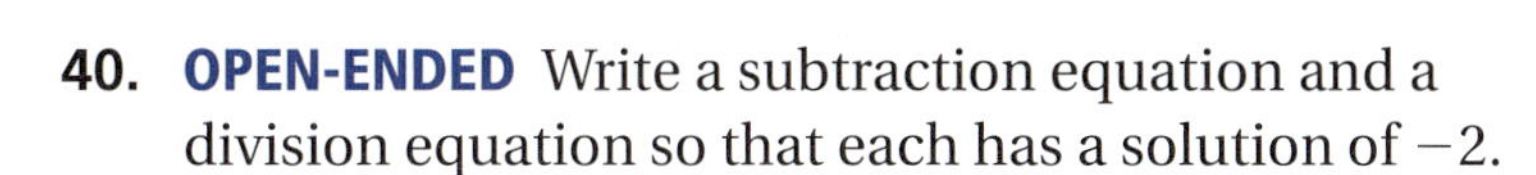

40. OPEN-ENDED Write a subtraction equation and a division equation so that each has a solution of -2.

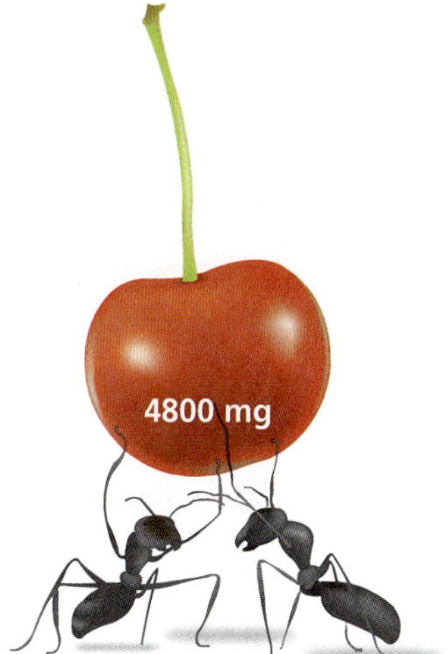

41. ANTS Some ant species can carry 50 times their body weight. It takes 32 ants to carry the cherry. About how much does each ant weigh?

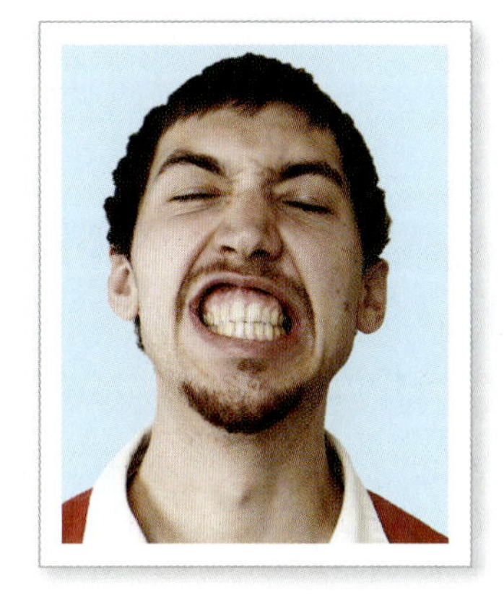

42. REASONING One-fourth of the girls and one-eighth of the boys in a class retake their school pictures. The photographer retakes pictures for 16 girls and 7 boys. How many students are in the class?

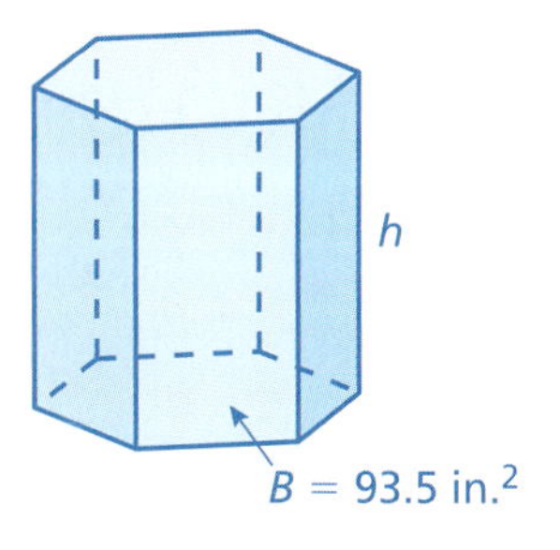

43. VOLUME The volume V of the prism is 1122 cubic inches. Use the formula $V = Bh$ to find the height h of the prism.

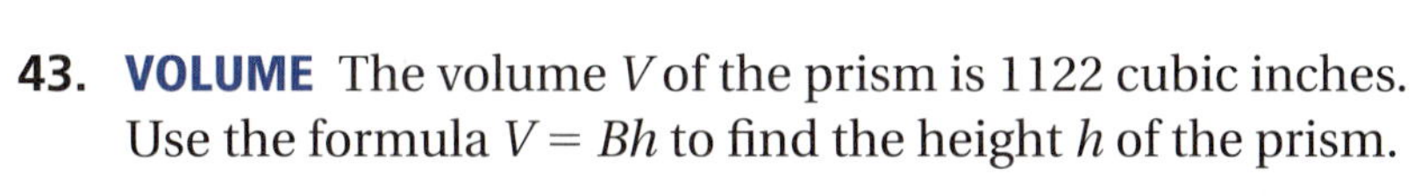

44. Critical Thinking A neighbor pays you and two friends \$90 to paint her garage. You divide the money three ways in the ratio 2 : 3 : 5.

a. How much does each person receive?

b. What is one possible reason the money is not divided evenly?

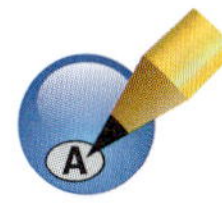

Fair Game Review What you learned in previous grades & lessons

Simplify the expression. *(Skills Review Handbook)*

45. $2(x - 2) + 5x$

46. $0.4b - 3.2 + 1.2b$

47. $\frac{1}{4}g + 6g - \frac{2}{3}$

48. MULTIPLE CHOICE The temperature at 4:00 P.M. was $-12\,^\circ\text{C}$. By 11:00 P.M., the temperature had dropped $14\,^\circ\text{C}$. What was the temperature at 11:00 P.M.? *(Skills Review Handbook)*

Ⓐ $-26\,^\circ\text{C}$ Ⓑ $-2\,^\circ\text{C}$ Ⓒ $2\,^\circ\text{C}$ Ⓓ $26\,^\circ\text{C}$

1.2 Solving Multi-Step Equations

Essential Question How can you solve a multi-step equation? How can you check the reasonableness of your solution?

1 ACTIVITY: Solving for the Angles of a Triangle

Work with a partner. Write an equation for each triangle. Solve the equation to find the value of the variable. Then find the angle measures of each triangle. Use a protractor to check the reasonableness of your answer.

a.

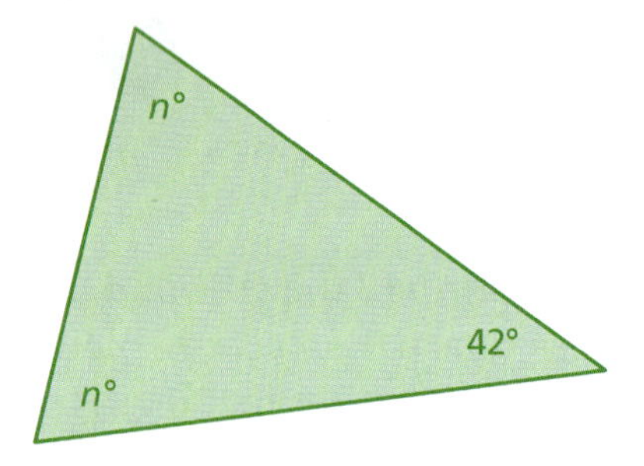

b.

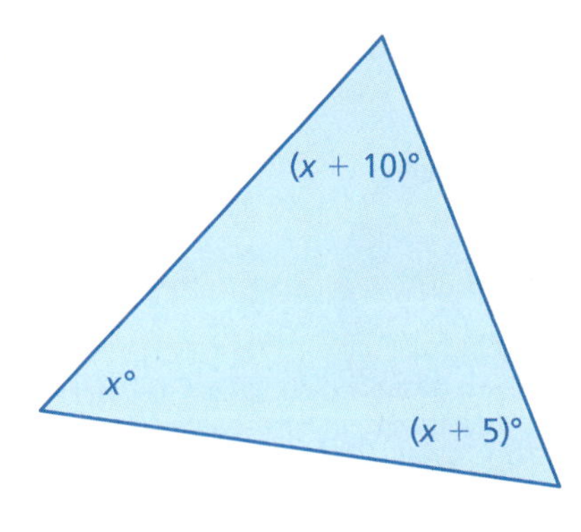

c.

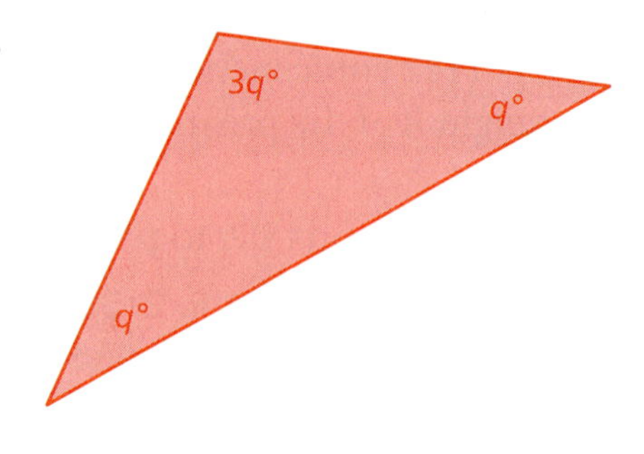

d.

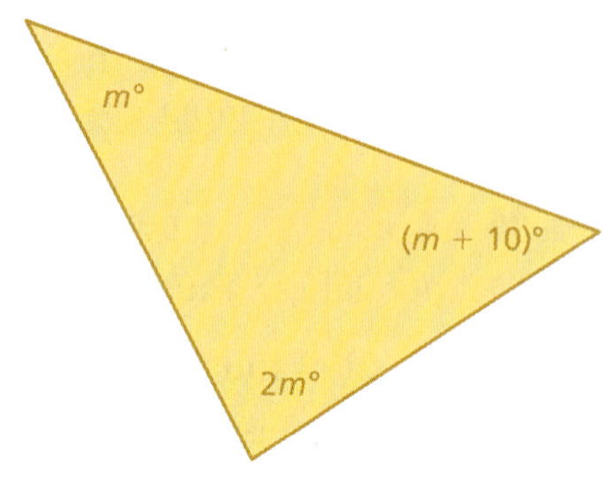

Solving Equations

In this lesson, you will

- use inverse operations to solve multi-step equations.
- use the Distributive Property to solve multi-step equations.

Learning Standards
MACC.8.EE.3.7a
MACC.8.EE.3.7b

e.

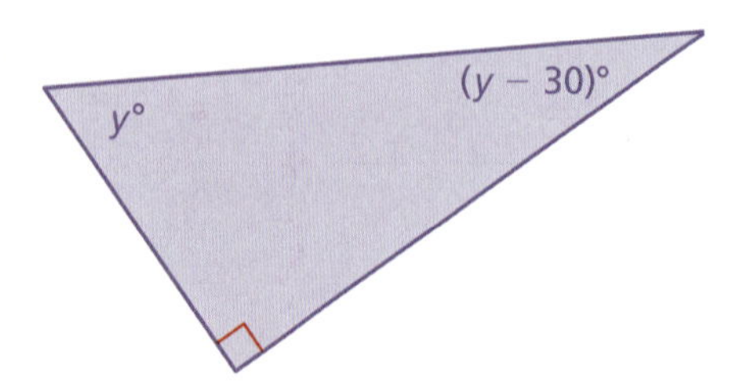

f.

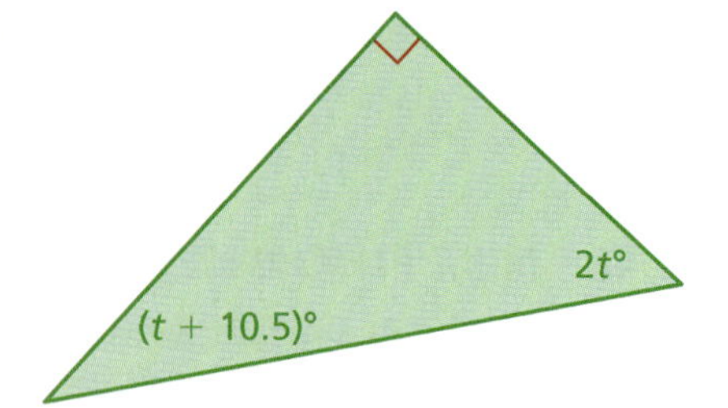

Laurie's Notes

Introduction

Standards for Mathematical Practice

- **MP1 Make Sense of Problems and Persevere in Solving Them:** Mathematically proficient students look for entry points in a problem. In solving a multi-step equation, they analyze the given information and determine how they can begin the solution.

Motivate

- Make a card for each student in your class. Write a variable term on each card. Students will walk around to find others with a card containing a *like term* to the one they are holding.

  Samples: $5x$, $-13x$, $5y$, $6xy$, x, $3.8x$, $\frac{1}{2}y$, $-3.8y$

- Ask students to explain what it means for terms to be *like* terms.

Activity Notes

Activity 1

? Ask a few questions to prepare students for the activity.
  - "In the previous lesson, what did you conclude about the sum of the angle measures of a triangle?" sum = 180°
  - "So if two angles measure 65° and 75°, what does the third angle measure?" 40°
  - "If the angles of a triangle measure $x°$, $2x°$, and $3x°$, could you determine the measure of each angle?" Students should say yes.
- Model how to write and solve the equation $x + 2x + 3x = 180$. Be sure to mention like terms when solving. Ask about the coefficient of x.
- **Common Error:** After solving the equation, you still need to substitute the value into each angle expression to solve for each angle measure. Students sometimes forget this step.
- **FYI:** The triangles are drawn to scale, so the angle measures can be checked using a protractor.
- Ask for volunteers to show a few of the solutions at the board.

? "Why are there only two angles with variable expressions written in parts (e) and (f)?" The third angle in each is a right angle.

Florida Common Core Standards

MACC.8.EE.3.7a Give examples of linear equations in one variable with one solution, . . . successively transforming the given equation into simpler forms, until an equivalent equation of the form $x = a$ . . . results . . ..

MACC.8.EE.3.7b Solve linear equations with rational number coefficients, including equations whose solutions require expanding expressions using the distributive property and collecting like terms.

Previous Learning

Students should know how to use inverse operations to solve one-step equations.

Lesson Plans
Complete Materials List

1.2 Record and Practice Journal

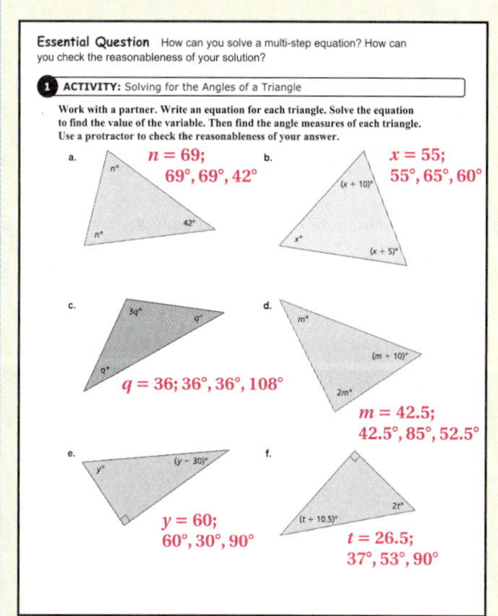

Essential Question How can you solve a multi-step equation? How can you check the reasonableness of your solution?

1 ACTIVITY: Solving for the Angles of a Triangle

Work with a partner. Write an equation for each triangle. Solve the equation to find the value of the variable. Then find the angle measures of each triangle. Use a protractor to check the reasonableness of your answer.

a. $n°$, $n°$, $42°$ — $n = 69$; 69°, 69°, 42°

b. $(x + 10)°$, $x°$, $(x + 5)°$ — $x = 55$; 55°, 65°, 60°

c. $3q°$, $q°$, $q°$ — $q = 36$; 36°, 36°, 108°

d. $m°$, $(m + 10)°$, $2m°$ — $m = 42.5$; 42.5°, 85°, 52.5°

e. $y°$, $(y - 30)°$ — $y = 60$; 60°, 30°, 90°

f. $(t + 10.5)°$, $2t°$ — $t = 26.5$; 37°, 53°, 90°

Differentiated Instruction

Auditory

Remind students that in order to solve an equation, the variable must be isolated on one side of the equation. The operations on the same side as the variable are those that need to be undone.

1.2 Record and Practice Journal

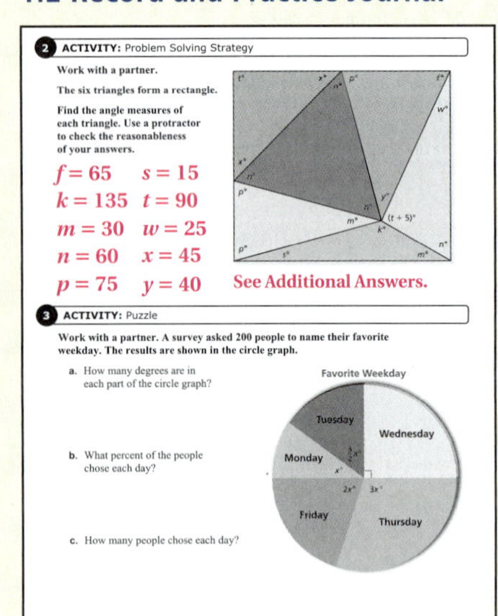

2 ACTIVITY: Problem Solving Strategy

Work with a partner.

The six triangles form a rectangle.

Find the angle measures of each triangle. Use a protractor to check the reasonableness of your answers.

$f = 65$ $s = 15$
$k = 135$ $t = 90$
$m = 30$ $w = 25$
$n = 60$ $x = 45$
$p = 75$ $y = 40$

See Additional Answers.

3 ACTIVITY: Puzzle

Work with a partner. A survey asked 200 people to name their favorite weekday. The results are shown in the circle graph.

a. How many degrees are in each part of the circle graph?

b. What percent of the people chose each day?

c. How many people chose each day?

d. Organize your results in a table.

a–d.

	Mon.	Tues.	Wed.	Thurs.	Fri.
Degrees	36°	54°	90°	108°	72°
Percent	10%	15%	25%	30%	20%
People	20	30	50	60	40

What Is Your Answer?

4. IN YOUR OWN WORDS How can you solve a multi-step equation? How can you check the reasonableness of your solution?

Use inverse operations.

Check by substituting the solution back into the original equation.

Laurie's Notes

Activity 2

- **MP1:** Students will have different strategies for solving this puzzle.

? "Define a straight angle." An angle that measures 180°.

- Remind students to look for a variety of ways to check their answers.
- Discuss results and strategies for finding the angle measures. Listen for: right angles at the vertices of the rectangle; sum of angle measures forming a straight angle equals 180°; sum of angle measures about a point equals 360°; sum of the angle measures of a triangle equals 180°.
- Most students will not write a formal equation, but the thinking involved is an equation. For example, $k + m + s = 180$. If you know k and m, you can use mental math to solve for s.
- **Extension:** Ask students to classify each triangle. For example, the purple triangle in the upper left is a right isosceles triangle.

Activity 3

- This example reviews fraction addition, mixed numbers, fraction division, and percents.

? Ask a few questions to help students begin the activity.

- "How many people were surveyed?" 200
- "What is the sum of the five central angle measures?" 360°
- "What is the angle measure of the sector labeled Wednesday?" 90°

- **MP1:** Some students may use all five angles and set the expression equal to 360, while other students may only consider the four angles represented by a variable expression and set it equal to 270.

? "How do you find the percent each angle measure represents?"

Convert $\frac{\text{angle measure}}{360}$ to a percent.

Words of Wisdom

- There are many steps in Activity 3, but it is possible to solve. This problem takes time and students will feel a sense of accomplishment when they finish.

Closure

- Find the angle measures in the right triangle. $x = 60$, $y = 30$, $z = 120$

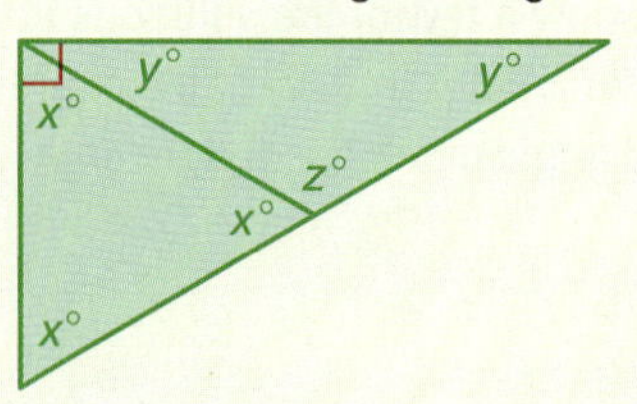

2 ACTIVITY: Problem-Solving Strategy

Math Practice 1

Find Entry Points

How do you decide which triangle to solve first? Explain.

Work with a partner.

The six triangles form a rectangle.

Find the angle measures of each triangle. Use a protractor to check the reasonableness of your answers.

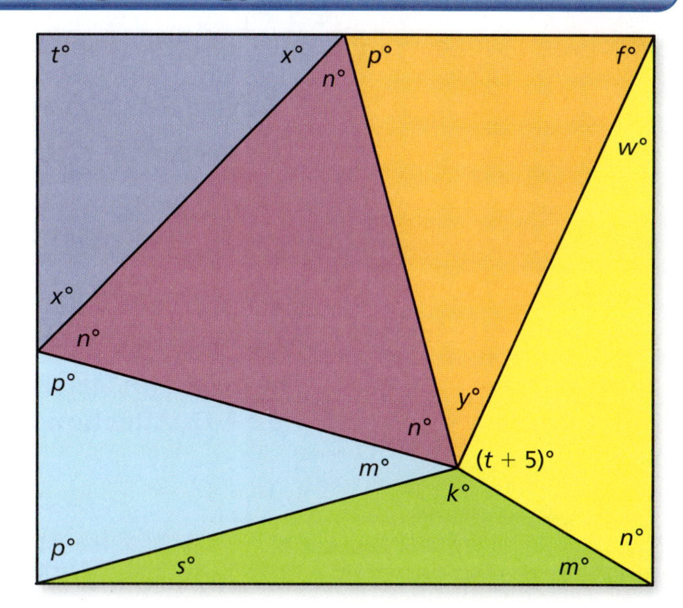

3 ACTIVITY: Puzzle

Work with a partner. A survey asked 200 people to name their favorite weekday. The results are shown in the circle graph.

a. How many degrees are in each part of the circle graph?

b. What percent of the people chose each day?

c. How many people chose each day?

d. Organize your results in a table.

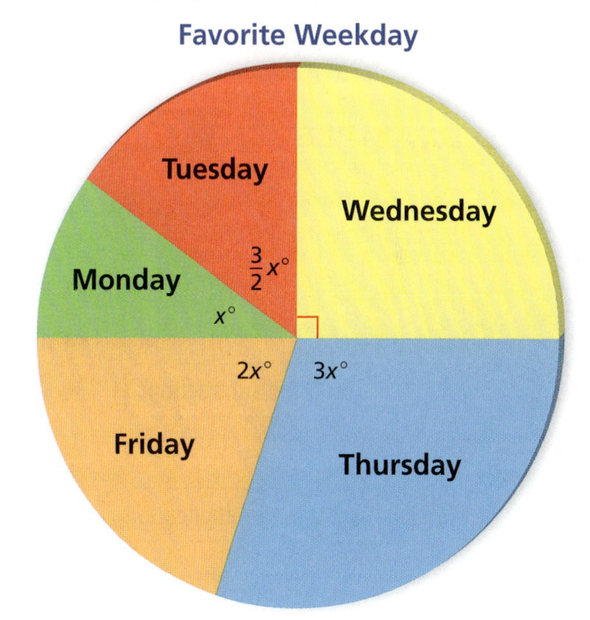

What Is Your Answer?

4. **IN YOUR OWN WORDS** How can you solve a multi-step equation? How can you check the reasonableness of your solution?

Practice Use what you learned about solving multi-step equations to complete Exercises 3–5 on page 14.

1.2 Lesson

Solving Multi-Step Equations

To solve multi-step equations, use inverse operations to isolate the variable.

EXAMPLE 1 Solving a Two-Step Equation

The height (in feet) of a tree after x years is $1.5x + 15$. After how many years is the tree 24 feet tall?

$1.5x + 15 = 24$ — Write an equation.

Undo the addition. → $-15 \quad -15$ — Subtraction Property of Equality

$1.5x = 9$ — Simplify.

Undo the multiplication. → $\dfrac{1.5x}{1.5} = \dfrac{9}{1.5}$ — Division Property of Equality

$x = 6$ — Simplify.

So, the tree is 24 feet tall after 6 years.

EXAMPLE 2 Combining Like Terms to Solve an Equation

Solve $8x - 6x - 25 = -35$.

$8x - 6x - 25 = -35$ — Write the equation.

$2x - 25 = -35$ — Combine like terms.

Undo the subtraction. → $+25 \quad +25$ — Addition Property of Equality

$2x = -10$ — Simplify.

Undo the multiplication. → $\dfrac{2x}{2} = \dfrac{-10}{2}$ — Division Property of Equality

$x = -5$ — Simplify.

The solution is $x = -5$.

On Your Own

Solve the equation. Check your solution.

1. $-3z + 1 = 7$
2. $\frac{1}{2}x - 9 = -25$
3. $-4n - 8n + 17 = 23$

Laurie's Notes

Introduction

Connect

- **Yesterday:** Students developed an intuitive understanding about solving multi-step equations. (MP1)
- **Today:** Students will solve multi-step equations by using inverse operations to isolate the variable.

Motivate

- Share information about the three man-made, palm-shaped islands built in Dubai, the self-proclaimed "Eighth Wonder of the World!"
  - Each of the islands is being built in the shape of a palm tree consisting of a trunk and a crown with fronds, and will have residential, leisure, and entertainment centers on them.
  - Each of the palm-shaped islands is surrounded by a crescent island that acts as a breakwater.

- Example 1 involves finding how long it takes a tree to reach a given height. **FYI:** Some palm trees are fast growing and are desirable for instant landscape impact.

Lesson Notes

Key Idea

- **Connection:** When you evaluate an expression, you follow the order of operations. Solving an equation undoes the evaluating, in reverse order. The goal is to isolate the variable term and then solve for the variable.

Example 1

- One way to explain the equation is to think of the tree as being 15 feet tall when being planted. It then grows at a rate of 1.5 feet each year.
- **MP4 Model with Mathematics:** Make a table to show the height of the tree from the first year to the sixth year.

Example 2

? "Why is $8x - 6x = 2x$?" Use the Distributive Property to subtract the terms; $8x - 6x = (8 - 6)x = 2x$.

On Your Own

- In Question 2, students may divide both sides by $\frac{1}{2}$ and get $x = -8$. Remind students that dividing by $\frac{1}{2}$ is the same as multiplying by 2.

Goal Today's lesson is solving multi-step equations.

Technology for the Teacher

Lesson Tutorials
Lesson Plans
Answer Presentation Tool

Extra Example 1

The height (in inches) of a plant after t days is $\frac{1}{2}t + 6$. After how many days is the plant 21 inches tall? 30 days

Extra Example 2

Solve $-2m + 4m + 5 = -3$. $m = -4$

1. $z = -2$ **2.** $x = -32$
3. $n = -0.5$

Extra Example 3

Solve $-4(3g - 5) + 10g = 19$. 0.5

Extra Example 4

You have scored 7, 10, 8, and 9 on four quizzes. Write and solve an equation to find the score you need on the fifth quiz so that your mean score is 8.
$\frac{x + 7 + 10 + 8 + 9}{5} = 8$; 6

On Your Own

4. $x = -1.5$ **5.** $d = -1$

6. $\frac{88 + 92 + 87 + x}{4} = 90$;

$x = 93$

English Language Learners

Vocabulary

English learners will benefit from understanding that a *term* is a number, a variable, or the product of a number and variable. *Like terms* are terms that have identical variable parts.
3 and 16 are like terms because they contain no variable.
$4x$ and $7x$ are like terms because they have the same variable x.
$5a$ and $5b$ are *not* like terms because they have different variables.

Laurie's Notes

Example 3

- Ask students to identify the operations involved in this equation. from left to right: multiplication (by 2), subtraction, multiplication ($5x$), addition
- **Note:** Combining like terms in the third step is not obvious to students. When the like terms are not adjacent, students are unsure of how to combine them. Rewrite the left side of the equation as $2 + (-10)x + 4$.

Words of Wisdom

- Take time to work through the Study Tip and discuss the steps. Instead of using the Distributive Property, both sides of the equation are divided by 2 in the third step. This will not be obvious to students, nor will they know why it is okay to do this.
- Explain to students that the left side of the equation is 2 times an expression. When the expression $2(1 - 5x)$ is divided by 2, it leaves the expression $1 - 5x$. In the next step, students want to add 1 to each side because of the subtraction operation shown. Again, it is helpful to write $1 - 5x$ as $1 + (-5)x$ so that it makes sense to students why 1 is subtracted from each side.

Example 4

- You may need to review *mean* with the students.
- Discuss the information displayed in the table and write the equation.

? "Is it equivalent to write $\frac{x + 3.5}{5} = 1.5$ instead of $\frac{3.5 + x}{5} = 1.5$? Explain." yes; Commutative Property of Addition

- **FYI:** It may be helpful to write the third step with parentheses: $5\left(\frac{3.5 + x}{5}\right)$.
- **MP2 Reason Abstractly and Quantitatively:** Ask students to explain the impact of trying to achieve a mean of 1.5 miles run per day when you ran 0 miles on two of the days.
- **Note:** This is a classic question. When all of the data are known except for one, what is needed in order to achieve a particular average? Students often ask this in the context of wanting to know what they have to score on a test in order to achieve a certain average.

On Your Own

- Encourage students to work in pairs. Students need to be careful with multi-step equations, and it is helpful to have a partner check each step.

Closure

- **Exit Ticket:** Solve $8x + 9 - 4x = 25$. Check your solution. $x = 4$

EXAMPLE 3 Using the Distributive Property to Solve an Equation

Solve $2(1 - 5x) + 4 = -8$.

$2(1 - 5x) + 4 = -8$	Write the equation.
$2(1) - 2(5x) + 4 = -8$	Distributive Property
$2 - 10x + 4 = -8$	Multiply.
$-10x + 6 = -8$	Combine like terms.
$\underline{-6} \quad \underline{-6}$	Subtraction Property of Equality
$-10x = -14$	Simplify.
$\dfrac{-10x}{-10} = \dfrac{-14}{-10}$	Division Property of Equality
$x = 1.4$	Simplify.

Study Tip

Here is another way to solve the equation in Example 3.

$2(1 - 5x) + 4 = -8$

$2(1 - 5x) = -12$

$1 - 5x = -6$

$-5x = -7$

$x = 1.4$

EXAMPLE 4 Real-Life Application

Use the table to find the number of miles x you need to run on Friday so that the mean number of miles run per day is 1.5.

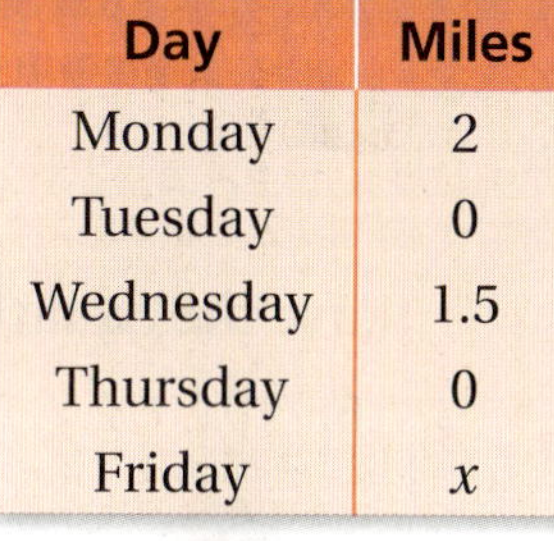

Day	Miles
Monday	2
Tuesday	0
Wednesday	1.5
Thursday	0
Friday	x

Write an equation using the definition of *mean*.

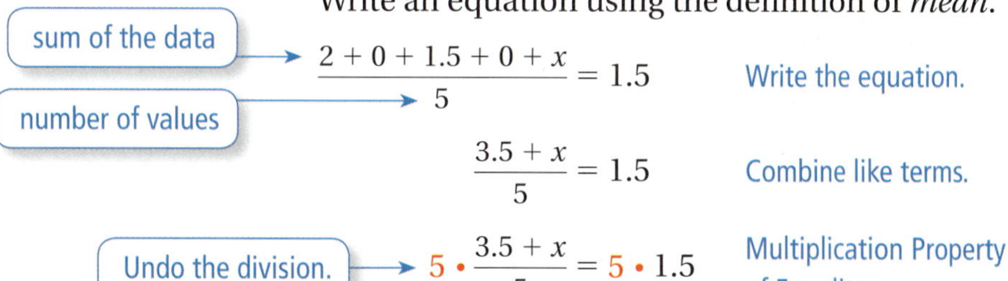

$\dfrac{2 + 0 + 1.5 + 0 + x}{5} = 1.5$	Write the equation.
$\dfrac{3.5 + x}{5} = 1.5$	Combine like terms.
$5 \cdot \dfrac{3.5 + x}{5} = 5 \cdot 1.5$	Multiplication Property of Equality
$3.5 + x = 7.5$	Simplify.
$\underline{-3.5} \quad \underline{-3.5}$	Subtraction Property of Equality
$x = 4$	Simplify.

Undo the addition.

So, you need to run 4 miles on Friday.

On Your Own

Exercises 10 and 11

Solve the equation. Check your solution.

4. $-3(x + 2) + 5x = -9$
5. $5 + 1.5(2d - 1) = 0.5$
6. You scored 88, 92, and 87 on three tests. Write and solve an equation to find the score you need on the fourth test so that your mean test score is 90.

1.2 Exercises

Vocabulary and Concept Check

1. **WRITING** Write the verbal statement as an equation. Then solve.

   2 more than 3 times a number is 17.

2. **OPEN-ENDED** Explain how to solve the equation $2(4x - 11) + 9 = 19$.

Practice and Problem Solving

CHOOSE TOOLS **Find the value of the variable. Then find the angle measures of the polygon. Use a protractor to check the reasonableness of your answer.**

3. 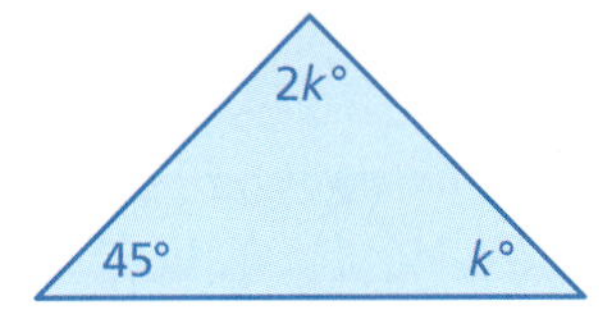

   Sum of angle measures: 180°

4. 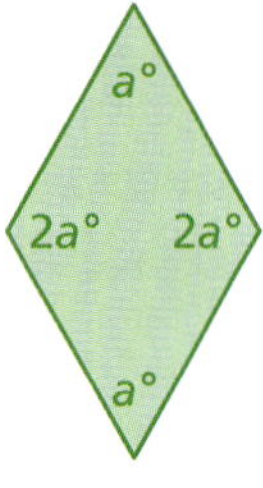

   Sum of angle measures: 360°

5. 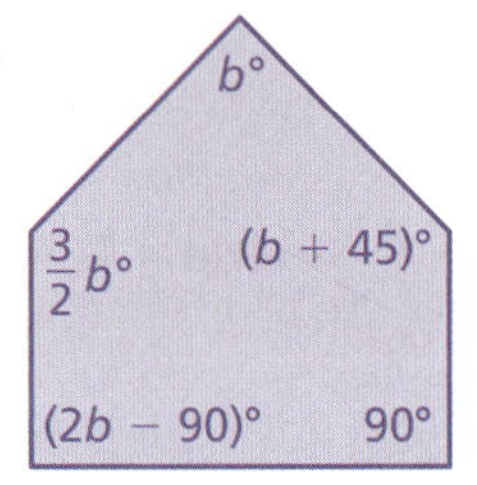

   Sum of angle measures: 540°

Solve the equation. Check your solution.

① ② 6. $10x + 2 = 32$

7. $19 - 4c = 17$

8. $1.1x + 1.2x - 5.4 = -10$

9. $\frac{2}{3}h - \frac{1}{3}h + 11 = 8$

③ 10. $6(5 - 8v) + 12 = -54$

11. $21(2 - x) + 12x = 44$

12. **ERROR ANALYSIS** Describe and correct the error in solving the equation.

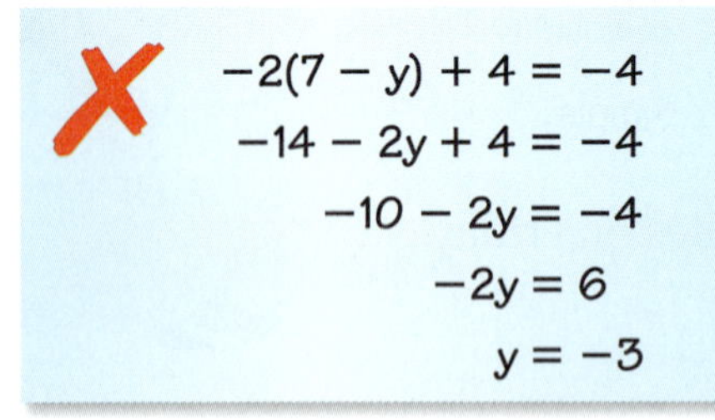

13. **WATCHES** The cost C (in dollars) of making n watches is represented by $C = 15n + 85$. How many watches are made when the cost is \$385?

14. **HOUSE** The height of the house is 26 feet. What is the height x of each story?

Assignment Guide and Homework Check

Level	Day 1 Activity Assignment	Day 2 Lesson Assignment	Homework Check
Basic	3–5, 19–22	1, 2, 6–14	6, 8, 10, 12, 14
Average	3–5, 19–22	1, 2, 7, 9, 11, 12–17	7, 9, 11, 14, 16
Advanced	1–5, 6–14 even, 15–22		10, 12, 14, 16

Common Errors

- **Exercises 8 and 9** When combining like terms, students may square the variable. Remind them that $x^2 = x \cdot x$, and in these exercises they are not multiplying the variables. Remind them that when adding and subtracting variables, they perform the addition or subtraction on the coefficient of the variable.
- **Exercises 10 and 11** When using the Distributive Property, students may forget to distribute to all the values within the parentheses. Remind them that they need to distribute to all the values and encourage them to draw arrows showing the distribution, if needed.
- **Exercise 16** Students may struggle with writing the equation for this problem because of the tip that is added to the total. Encourage them to write an expression for the cost of the food and then add on the tip.

1.2 Record and Practice Journal

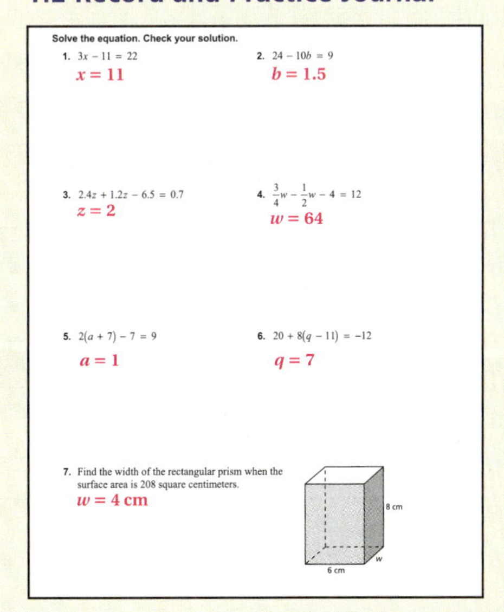
Solve the equation. Check your solution.

1. $3x - 11 = 22$
   $x = 11$
2. $24 - 10b = 9$
   $b = 1.5$
3. $2.4z + 1.2z - 6.5 = 0.7$
   $z = 2$
4. $\frac{3}{4}w - \frac{1}{2}w - 4 = 12$
   $w = 64$
5. $2(a + 7) - 7 = 9$
   $a = 1$
6. $20 + 8(q - 11) = -12$
   $q = 7$
7. Find the width of the rectangular prism when the surface area is 208 square centimeters.
   $w = 4$ cm

Vocabulary and Concept Check

1. $2 + 3x = 17; x = 5$
2. *Sample answer:* Subtract 9 from each side. Divide each side by 2. Add 11 to each side. Divide each side by 4.

Practice and Problem Solving

3. $k = 45$; 45°, 45°, 90°
4. $a = 60$; 60°, 120°, 60°, 120°
5. $b = 90$; 90°, 135°, 90°, 90°, 135°
6. $x = 3$
7. $c = 0.5$
8. $x = -2$
9. $h = -9$
10. $v = 2$
11. $x = -\frac{2}{9}$
12. They did not distribute the -2 properly.

$$\begin{aligned} -2(7 - y) + 4 &= -4 \\ -14 + 2y + 4 &= -4 \\ 2y - 10 &= -4 \\ 2y &= 6 \\ y &= 3 \end{aligned}$$

13. 20 watches
14. 10 ft

Practice and Problem Solving

15. $4(b + 3) = 24$; 3 in.

16. $1.15(2p + 1.5) = 11.5$; $4.25

17. $\frac{2580 + 2920 + x}{3} = 3000$; 3500 people

18. See *Taking Math Deeper.*

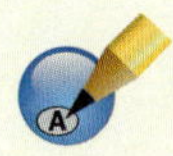

Fair Game Review

19. <	**20.** =
21. >	**22.** D

Mini-Assessment

Solve the equation. Check your solution.

1. $18 = 5a - 2a + 3$ $a = 5$
2. $2(4 - 2w) - 8 = -4$ $w = 1$
3. $2.3y + 4.4y - 3.7 = 16.4$ $y = 3$
4. $\frac{3}{4}z + \frac{1}{4}z - 6 = -5$ $z = 1$
5. The perimeter of the picture is 36 inches. What is the height of the picture? 10 in.

Taking Math Deeper

Exercise 18

This problem is an example of how algebra can be used in fields outside of mathematics.

Begin by translating the scoring system into a mathematical formula.

Final score = 0.6(degree of difficulty)(sum of countries' scores)

(sum of countries' scores: not including the highest and lowest)

Substitute the given information.

Let x = the degree of difficulty.

$$77.7 = 0.6(x)(7.5 + 8.0 + 7.0 + 7.5 + 7.0)$$
$$77.7 = 0.6x(37)$$
$$77.7 = 22.2x$$
$$3.5 = x$$

a. The degree of difficulty is 3.5.

This question has many answers.

Let x = sum of the five countries' scores.

$$97.2 = 0.6(4)(x)$$
$$97.2 = 2.4x$$
$$40.5 = x$$

One possibility is the following:

b. 8.0 + 8.0 + 8.0 + 8.0 + 8.5 with a low score of 7.5 and a high score of 9.0

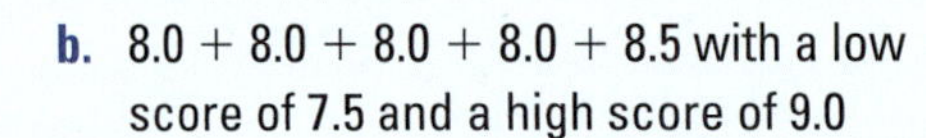

Project

Use the Internet or school library to find all the different dives that are scored in a diving competition. Find the degree of difficulty that goes with each dive.

Reteaching and Enrichment Strategies

If students need help. . .	If students got it. . .
Resources by Chapter • Practice A and Practice B • Puzzle Time Record and Practice Journal Practice Differentiating the Lesson Lesson Tutorials Skills Review Handbook	Resources by Chapter • Enrichment and Extension • Technology Connection Start the next section

In Exercises 15–17, write and solve an equation to answer the question.

15. POSTCARD The area of the postcard is 24 square inches. What is the width b of the message (in inches)?

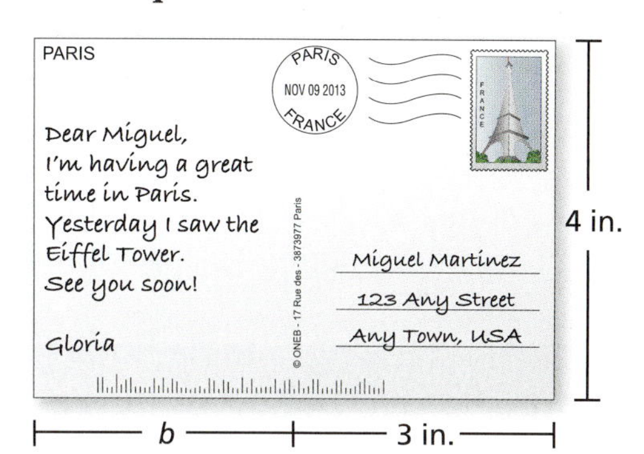

16. BREAKFAST You order two servings of pancakes and a fruit cup. The cost of the fruit cup is $1.50. You leave a 15% tip. Your total bill is $11.50. How much does one serving of pancakes cost?

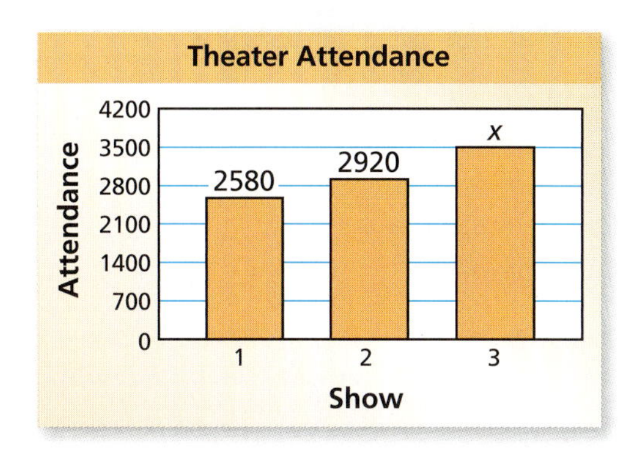

17. THEATER How many people must attend the third show so that the average attendance per show is 3000?

18. DIVING Divers in a competition are scored by an international panel of judges. The highest and the lowest scores are dropped. The total of the remaining scores is multiplied by the degree of difficulty of the dive. This product is multiplied by 0.6 to determine the final score.

a. A diver's final score is 77.7. What is the degree of difficulty of the dive?

Judge	Russia	China	Mexico	Germany	Italy	Japan	Brazil
Score	7.5	8.0	6.5	8.5	7.0	7.5	7.0

b. Critical Thinking The degree of difficulty of a dive is 4.0. The diver's final score is 97.2. Judges award half or whole points from 0 to 10. What scores could the judges have given the diver?

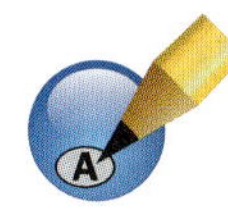

Fair Game Review What you learned in previous grades & lessons

Let $a = 3$ and $b = -2$. Copy and complete the statement using <, >, or =.
(Skills Review Handbook)

19. $-5a$ ☐ 4 **20.** 5 ☐ $b + 7$ **21.** $a - 4$ ☐ $10b + 8$

22. MULTIPLE CHOICE What value of x makes the equation $x + 5 = 2x$ true?
(Skills Review Handbook)

Ⓐ -1 Ⓑ 0 Ⓒ 3 Ⓓ 5

1 Study Help

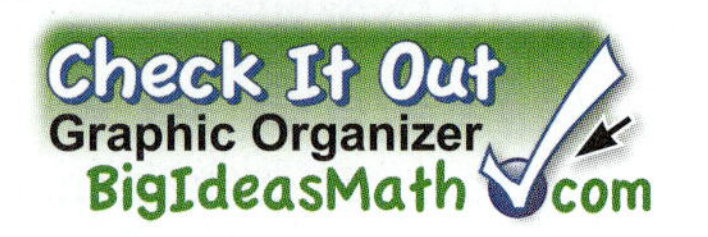

You can use a **Y chart** to compare two topics. List differences in the branches and similarities in the base of the Y. Here is an example of a Y chart that compares solving simple equations using addition to solving simple equations using subtraction.

Solving Simple Equations Using Addition	Solving Simple Equations Using Subtraction
• Add the same number to each side of the equation.	• Subtract the same number from each side of the equation.

• You can solve the equation in one step.
• You produce an equivalent equation.
• The variable can be on either side of the equation.
• It is always a good idea to check your solution.

On Your Own

Make Y charts to help you study and compare these topics.

1. solving simple equations using multiplication and solving simple equations using division
2. solving simple equations and solving multi-step equations

After you complete this chapter, make Y charts for the following topics.

3. solving equations with the variable on one side and solving equations with variables on both sides
4. solving multi-step equations and solving equations with variables on both sides
5. solving multi-step equations and rewriting literal equations

"I made a Y chart to compare and contrast Fluffy's characteristics with yours."

Sample Answers

1.

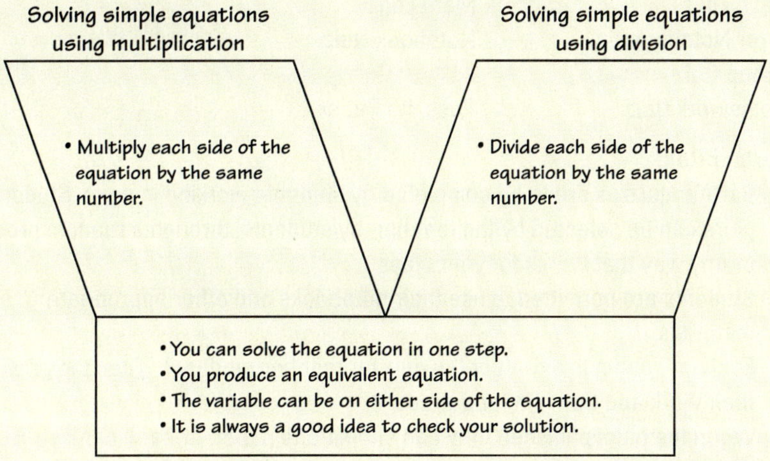

Solving simple equations using multiplication | Solving simple equations using division

- Multiply each side of the equation by the same number.

- Divide each side of the equation by the same number.

- You can solve the equation in one step.
- You produce an equivalent equation.
- The variable can be on either side of the equation.
- It is always a good idea to check your solution.

2.

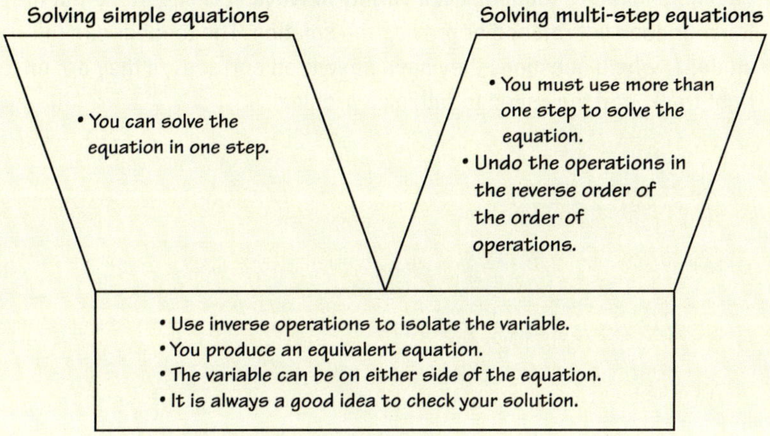

Solving simple equations | Solving multi-step equations

- You can solve the equation in one step.

- You must use more than one step to solve the equation.
- Undo the operations in the reverse order of the order of operations.

- Use inverse operations to isolate the variable.
- You produce an equivalent equation.
- The variable can be on either side of the equation.
- It is always a good idea to check your solution.

List of Organizers

Available at *BigIdeasMath.com*

Comparison Chart
Concept Circle
Definition (Idea) and Example Chart
Example and Non-Example Chart
Formula Triangle
Four Square
Information Frame
Information Wheel
Notetaking Organizer
Process Diagram
Summary Triangle
Word Magnet
Y Chart

About this Organizer

A **Y Chart** can be used to compare two topics. Students list differences between the two topics in the branches of the Y and similarities in the base of the Y. A Y chart serves as a good tool for assessing students' knowledge of a pair of topics that have subtle but important differences. You can include blank Y charts on tests or quizzes for this purpose.

Technology for the Teacher

Editable Graphic Organizer

Answers

1. $y = \frac{1}{2}$
2. $w = 5\pi$
3. $m = 0.5$
4. $q = -3.6$
5. $k = 4$
6. $z = 16$
7. $n = \frac{3}{2}$
8. $t = 8$
9. $x = 60$; 55°, 60°, 65°
10. $x = 126$; 63°, 80°, 126°, 91°
11. $32
12. 50 ft, 150 ft, 75 ft, 180 ft
13. $230x = 1265$; 5.5 hours
14. $\frac{25 + 15 + 18 + p}{4} = 20$;
    $p = 22$ points

Technology for the Teacher

Online Assessment
Assessment Book
ExamView® Assessment Suite

Alternative Quiz Ideas

100% Quiz
Error Notebook
Group Quiz
Homework Quiz
Math Log
Notebook Quiz
Partner Quiz
Pass the Paper

Partner Quiz

- Partner quizzes are to be completed by students working in pairs. Student pairs can be selected by the teacher, by students, through a random process, or any way that works for your class.
- Students are permitted to use their notebooks and other appropriate materials.
- Each pair submits a draft of the quiz for teacher feedback. Then they revise their work and turn it in for a grade.
- When the pair is finished they can submit one paper, or each can submit their own.
- Teachers can give feedback in a variety of ways. It is important that the teacher does not reteach or provide the solution. The teacher can tell students which questions they have answered correctly, if they are on the right track, or if they need to rethink a problem.

Reteaching and Enrichment Strategies

If students need help. . .	If students got it. . .
Resources by Chapter • Practice A and Practice B • Puzzle Time Lesson Tutorials *BigIdeasMath.com*	Resources by Chapter • Enrichment and Extension • Technology Connection Game Closet at *BigIdeasMath.com* Start the next section

1.1–1.2 Quiz

Solve the equation. Check your solution. *(Section 1.1)*

1. $-\frac{1}{2} = y - 1$

2. $-3\pi + w = 2\pi$

3. $1.2m = 0.6$

4. $q + 2.7 = -0.9$

Solve the equation. Check your solution. *(Section 1.2)*

5. $-4k + 17 = 1$

6. $\frac{1}{4}z + 8 = 12$

7. $-3(2n + 1) + 7 = -5$

8. $2.5(t - 2) - 6 = 9$

Find the value of *x*. Then find the angle measures of the polygon. *(Section 1.2)*

9.

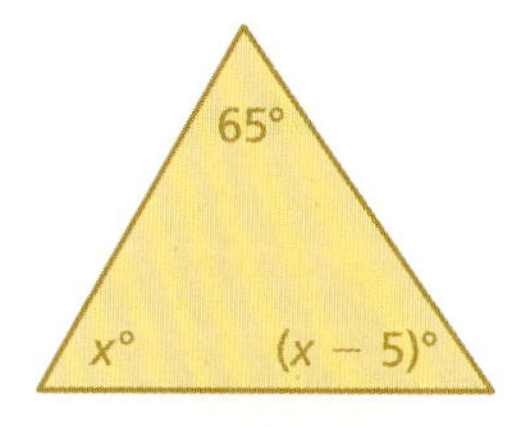

Sum of angle measures: 180°

10.

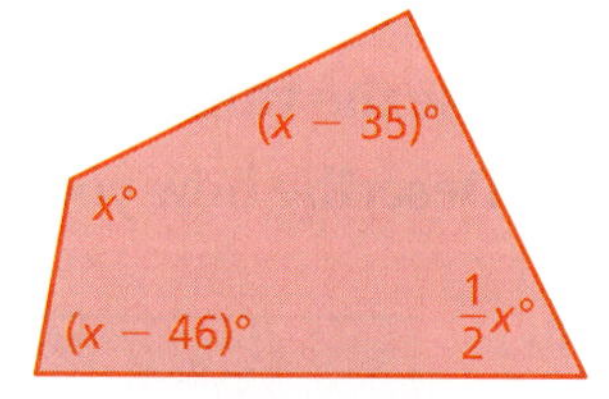

Sum of angle measures: 360°

11. **JEWELER** The equation $P = 2.5m + 35$ represents the price P (in dollars) of a bracelet, where m is the cost of the materials (in dollars). The price of a bracelet is \$115. What is the cost of the materials? *(Section 1.2)*

12. **PASTURE** A 455-foot fence encloses a pasture. What is the length of each side of the pasture? *(Section 1.2)*

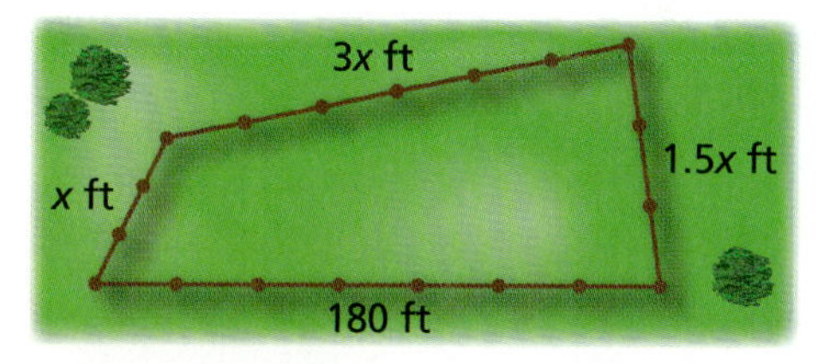

13. **POSTERS** A machine prints 230 movie posters each hour. Write and solve an equation to find the number of hours it takes the machine to print 1265 posters. *(Section 1.1)*

14. **BASKETBALL** Use the table to write and solve an equation to find the number of points p you need to score in the fourth game so that the mean number of points is 20. *(Section 1.2)*

Game	Points
1	25
2	15
3	18
4	p

1.3 Solving Equations with Variables on Both Sides

Essential Question How can you solve an equation that has variables on both sides?

1 ACTIVITY: Perimeter and Area

Work with a partner.

- **Each figure has the unusual property that the value of its perimeter (in feet) is equal to the value of its area (in square feet). Write an equation for each figure.**
- **Solve each equation for x.**
- **Use the value of x to find the perimeter and the area of each figure.**
- **Describe how you can check your solution.**

a.

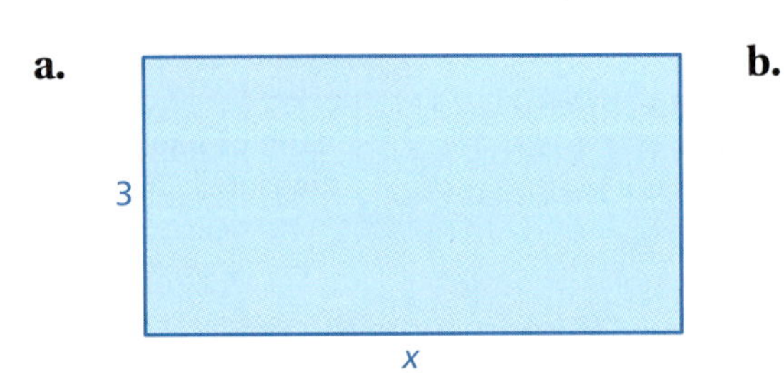

b.

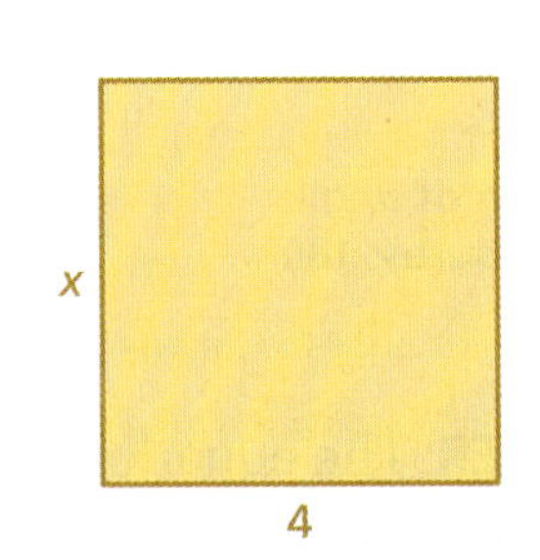

c.

d.

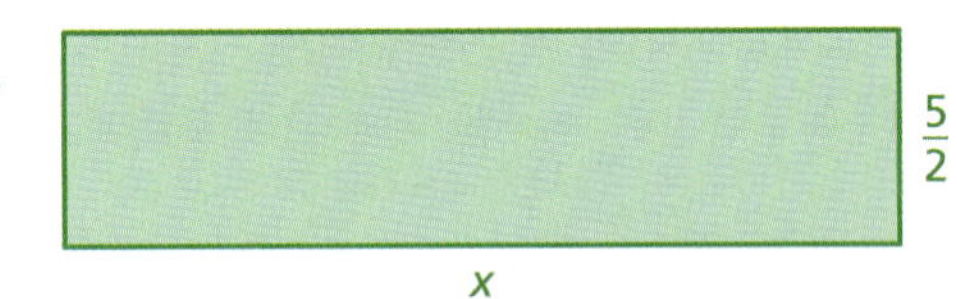

e.

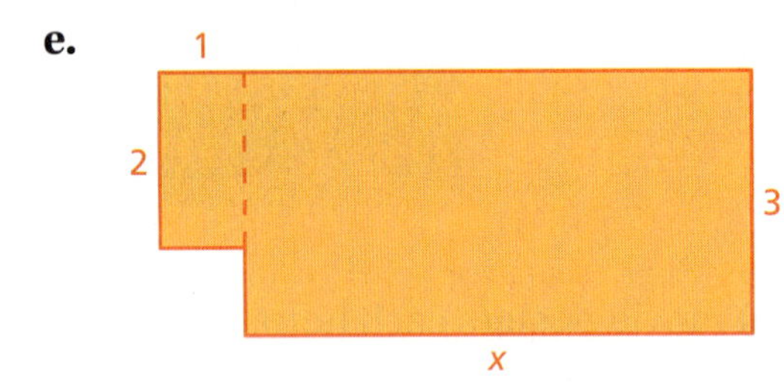

f.

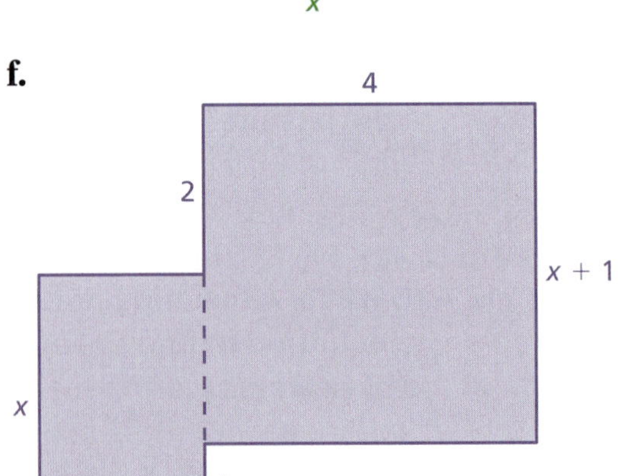

g.

3x

3

1

x x

COMMON CORE

Solving Equations

In this lesson, you will

- solve equations with variables on both sides.
- determine whether equations have no solution or infinitely many solutions.

Learning Standards
MACC.8.EE.3.7a
MACC.8.EE.3.7b

Laurie's Notes

Introduction

Standards for Mathematical Practice

- **MP6 Attend to Precision:** Mathematically proficient students try to communicate precisely with others. In today's activities, students will work with vocabulary related to measurement.

Motivate

? "What balances with the cylinder? Explain."

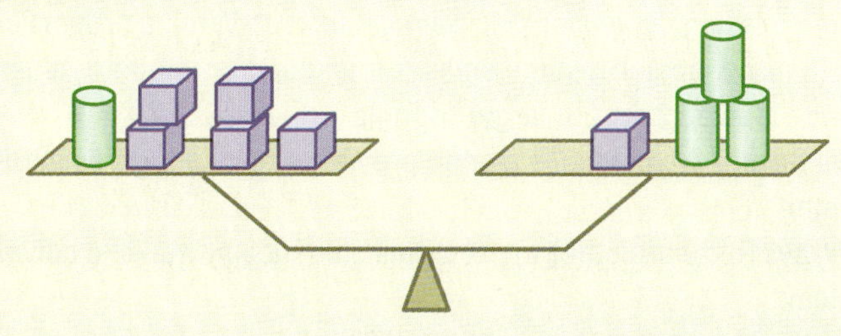

2 cubes; Remove one cube and one cylinder from each side. 2 cylinders balance with 4 cubes, so 1 cylinder would balance with 2 cubes.

- The balance problem is equivalent to $x + 5 = 3x + 1$, where x is a cylinder and the whole numbers represent cubes. This is an example of an equation with variables on both sides, the type students will solve today. Return to this equation at the end of class.
- **MP4 Model with Mathematics:** If students are familiar with algebra tiles, you can model the problem using the tiles. The cylinder is replaced with an x-tile, and the cubes are replaced with unit tiles.

Activity Notes

Activity 1

- Discuss with students the general concept of what it means to measure the attributes of a two-dimensional figure. In other words, what is the difference between a rectangle's perimeter and a rectangle's area? What type of units are used to measure each? linear units for perimeter and square units for area
- **MP6:** Be sure to make it clear that the directions are saying that perimeter and area are not the same, but their values are equal. For example, a square that measures 4 centimeters on each edge has a perimeter of 16 centimeters and an area of 16 square centimeters. The value (16) is the same, but the units of measure are not.

? Before students begin, ask a few review questions.

- "How do you find the perimeter P and the area A of a rectangle with length ℓ and width w?" $P = 2\ell + 2w$ and $A = \ell w$
- "How do you find the perimeter and the area of a composite figure?" Listen for students' understanding that perimeter is the sum of all of the sides. The area is found in parts and then added together.

- Have a few groups share their work at the board, particularly for part (d), fractions, and part (g), algebraic expressions.

Florida Common Core Standards

MACC.8.EE.3.7a Give examples of linear equations in one variable with one solution, infinitely many solutions, or no solutions. Show which of these possibilities is the case by successively transforming the given equation into simpler forms, until an equivalent equation of the form $x = a$, $a = a$, or $a = b$ results (where a and b are different numbers).

MACC.8.EE.3.7b Solve linear equations with rational number coefficients, including equations whose solutions require expanding expressions using the distributive property and collecting like terms.

Previous Learning

Students should know common formulas for perimeter, area, surface area, and volume.

1.3 Record and Practice Journal

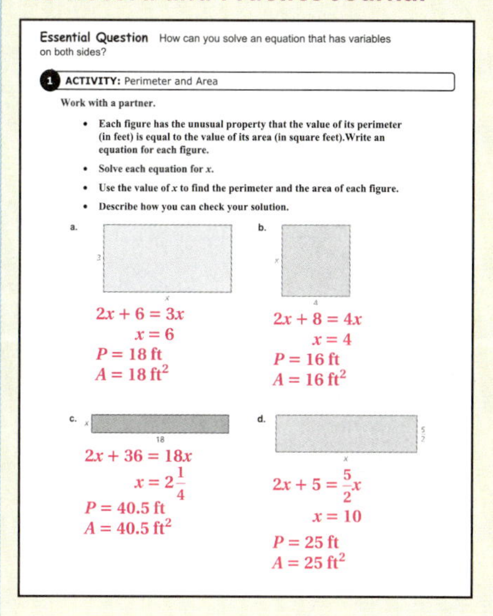

Essential Question How can you solve an equation that has variables on both sides?

1 ACTIVITY: Perimeter and Area

Work with a partner.

- Each figure has the unusual property that the value of its perimeter (in feet) is equal to the value of its area (in square feet). Write an equation for each figure.
- Solve each equation for x.
- Use the value of x to find the perimeter and the area of each figure.
- Describe how you can check your solution.

a. $2x + 6 = 3x$, $x = 6$, $P = 18$ ft, $A = 18$ ft^2

b. $2x + 8 = 4x$, $x = 4$, $P = 16$ ft, $A = 16$ ft^2

c. $2x + 36 = 18x$, $x = 2\frac{1}{4}$, $P = 40.5$ ft, $A = 40.5$ ft^2

d. $2x + 5 = \frac{5}{2}x$, $x = 10$, $P = 25$ ft, $A = 25$ ft^2

Differentiated Instruction

Auditory

Point out to students that skills used to solve equations in this lesson are the same skills they have used before. The goal is to isolate the variable on one side of the equation. Just as they used the Addition Property of Equality to remove a constant term from one side of the equation, they will use the same property to remove the variable term from one side of the equation.

1.3 Record and Practice Journal

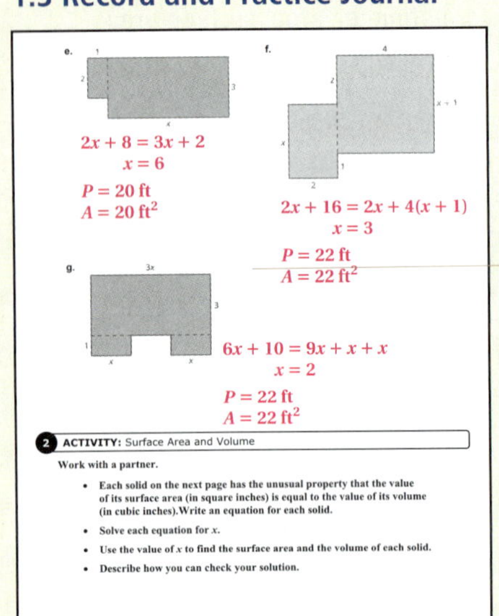

e. $2x + 8 = 3x + 2$
$x = 6$
$P = 20$ ft
$A = 20$ ft^2

f. $2x + 16 = 2x + 4(x + 1)$
$x = 3$
$P = 22$ ft
$A = 22$ ft^2

g. $6x + 10 = 9x + x + x$
$x = 2$
$P = 22$ ft
$A = 22$ ft^2

2 ACTIVITY: Surface Area and Volume

Work with a partner.

- Each solid on the next page has the unusual property that the value of its surface area (in square inches) is equal to the value of its volume (in cubic inches). Write an equation for each solid.
- Solve each equation for x.
- Use the value of x to find the surface area and the volume of each solid.
- Describe how you can check your solution.

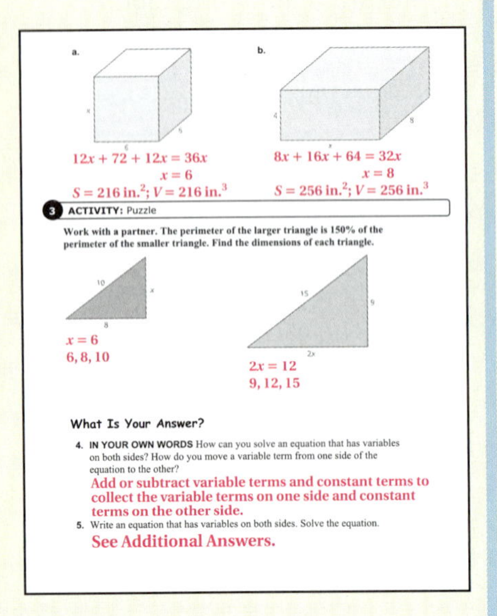

a. $12x + 72 + 12x = 36x$
$x = 6$
$S = 216$ in.^{2}; $V = 216$ in.^{3}

b. $8x + 16x + 64 = 32x$
$x = 8$
$S = 256$ in.^{2}; $V = 256$ in.^{3}

3 ACTIVITY: Puzzle

Work with a partner. The perimeter of the larger triangle is 150% of the perimeter of the smaller triangle. Find the dimensions of each triangle.

$x = 6$
6, 8, 10

$2x = 12$
9, 12, 15

What Is Your Answer?

4. IN YOUR OWN WORDS How can you solve an equation that has variables on both sides? How do you move a variable term from one side of the equation to the other?
Add or subtract variable terms and constant terms to collect the variable terms on one side and constant terms on the other side.
5. Write an equation that has variables on both sides. Solve the equation.
See Additional Answers.

Laurie's Notes

Activity 2

- This activity is similar to Activity 1.
- ? "How do you find the surface area S and volume V of a rectangular prism with length ℓ, width w, and height h?" $S = 2\ell w + 2\ell h + 2wh$ and $V = \ell wh$
- Students may guess that part (a) is a cube, suggesting $x = 6$. Ask students to verify their guesses.

Activity 3

- ? "The larger triangle is a scale drawing of the smaller triangle. How can you find the missing dimensions?" *Sample answer:* Write a proportion to solve for the side labeled x in the smaller triangle, then use this value to find the side labeled $2x$ in the larger triangle.
- The above described method is a good preview of similar figures in the next chapter.
- **Another Way:** There are other approaches students may take in solving this problem.
  - One method is to solve the equation: 150% of the smaller triangle's perimeter is equal to the perimeter of the larger triangle.

    $$150\% \text{ of } (18 + x) = 24 + 2x$$
    $$1.5(18 + x) = 24 + 2x$$

    This method reviews decimal multiplication.
  - Another method is to find the scale factor, and then use it to determine the missing dimensions.

What Is Your Answer?

- **Neighbor Check:** Have students work independently and then have their neighbors check their work. Have students discuss any discrepancies.

Closure

- Describe how to solve $x + 5 = 3x + 1$. *Sample answer:* Subtract x from both sides, subtract 1 from both sides, and then divide both sides by 2.

2 ACTIVITY: Surface Area and Volume

Math Practice 2

Use Operations

What properties of operations do you need to use in order to find the value of x?

Work with a partner.

- **Each solid has the unusual property that the value of its surface area (in square inches) is equal to the value of its volume (in cubic inches). Write an equation for each solid.**
- **Solve each equation for x.**
- **Use the value of x to find the surface area and the volume of each solid.**
- **Describe how you can check your solution.**

a.

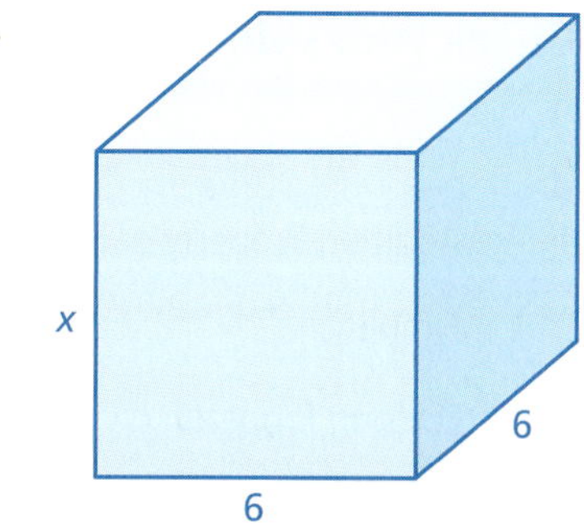

b.

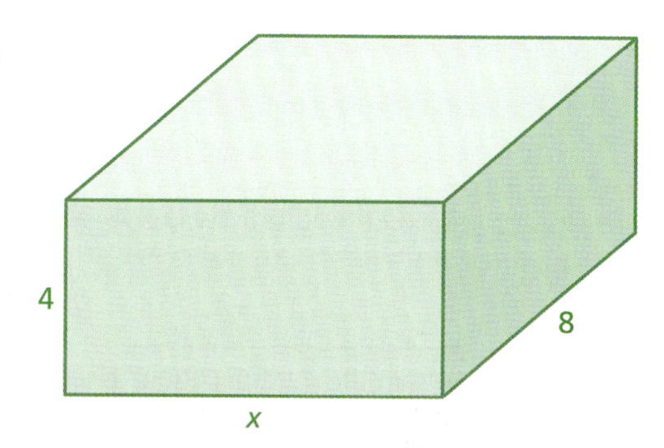

3 ACTIVITY: Puzzle

Work with a partner. The perimeter of the larger triangle is 150% of the perimeter of the smaller triangle. Find the dimensions of each triangle.

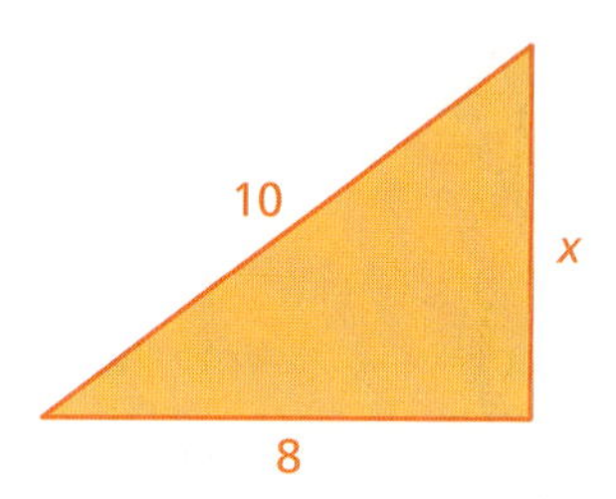

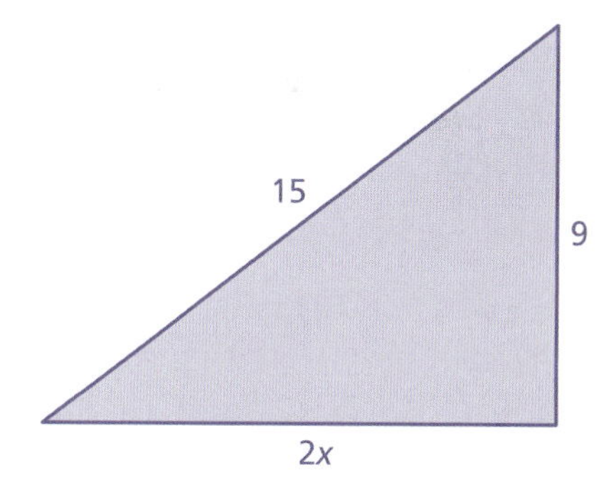

What Is Your Answer?

4. **IN YOUR OWN WORDS** How can you solve an equation that has variables on both sides? How do you move a variable term from one side of the equation to the other?

5. Write an equation that has variables on both sides. Solve the equation.

Use what you learned about solving equations with variables on both sides to complete Exercises 3–5 on page 23.

1.3 Lesson

Solving Equations with Variables on Both Sides

To solve equations with variables on both sides, collect the variable terms on one side and the constant terms on the other side.

EXAMPLE 1 Solving an Equation with Variables on Both Sides

Solve $15 - 2x = -7x$. Check your solution.

	$15 - 2x = -7x$	Write the equation.
Undo the subtraction.	$+2x \quad +2x$	Addition Property of Equality
	$15 = -5x$	Simplify.
Undo the multiplication.	$\frac{15}{-5} = \frac{-5x}{-5}$	Division Property of Equality
	$-3 = x$	Simplify.

The solution is $x = -3$.

Check

$15 - 2x = -7x$

$15 - 2(-3) \stackrel{?}{=} -7(-3)$

$21 = 21$ ✓

EXAMPLE 2 Using the Distributive Property to Solve an Equation

Solve $-2(x - 5) = 6\left(2 - \frac{1}{2}x\right)$.

	$-2(x - 5) = 6\left(2 - \frac{1}{2}x\right)$	Write the equation.
	$-2x + 10 = 12 - 3x$	Distributive Property
Undo the subtraction.	$+3x \qquad +3x$	Addition Property of Equality
	$x + 10 = 12$	Simplify.
Undo the addition.	$-10 \quad -10$	Subtraction Property of Equality
	$x = 2$	Simplify.

The solution is $x = 2$.

On Your Own

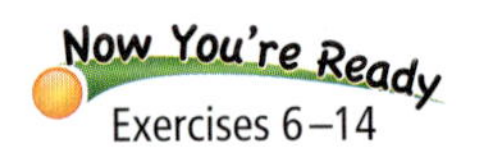

Solve the equation. Check your solution.

1. $-3x = 2x + 19$
2. $2.5y + 6 = 4.5y - 1$
3. $6(4 - z) = 2z$

Laurie's Notes

Introduction

Connect

- **Yesterday:** Students developed an intuitive understanding of solving equations with variables on both sides. (MP4, MP6)
- **Today:** Students will solve equations with variables on both sides by using Properties of Equality and collecting variable terms on one side. The equations could have one solution, no solution, or infinitely many solutions.
- **FYI:** This is a long lesson. Take your time and present the concepts well.

Motivate

- Share some information about the Mississippi River. It is the third longest river in North America, flowing 2350 miles from Lake Itasca in Minnesota to the Gulf of Mexico. The width of the river ranges from 20–30 feet at its narrowest to more than 11 miles at its widest.
- Explain that one of the examples in today's lesson involves a boat traveling upstream on the Mississippi River.

Lesson Notes

Key Idea

? "In the expression $5x - 2 - 9x + y + 4$, what are the variable terms? Constant terms?" $5x$, $-9x$, and y; -2 and 4

- **Common Error:** Students may forget to include the sign of the variable term.

Example 1

- Notice that a constant is on the left side only. For this reason, it makes sense to solve for the variable term on the right side of the equation. It is possible to solve for the variable term on the left side of the equation, but finding the solution involves an extra step. Show students this approach as well, and point out that it gives the same solution.

Example 2

- **Teaching Tip:** Before distributing on the left side, rewrite the inside expression as $x + (-5)$ (add the opposite). Students are likely to recognize that they are multiplying $(-2)(-5)$ to get a product of 10.

? "How do you multiply a whole number and a fraction like $6\left(\frac{1}{2}\right)$?" Listen for language such as, write the whole number over a denominator of 1 and multiply straight across.

- **FYI:** Solve for the variable on the side of the equation where the coefficient is the greatest. The coefficient of the variable term will be positive, a condition that generally renders fewer mistakes.

On Your Own

- **Neighbor Check:** Have students work independently and then have their neighbors check their work. Have students discuss any discrepancies.

Goal Today's lesson is solving equations with variables on both sides.

Technology for the Teacher

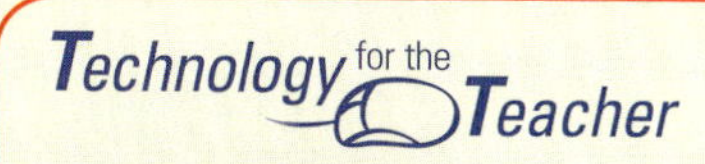

Lesson Tutorials
Lesson Plans
Answer Presentation Tool

Extra Example 1

Solve $r = -5r + 18$. Check your solution. 3

Extra Example 2

Solve $6\left(1 + \frac{1}{2}x\right) = 2(x + 1)$. -4

On Your Own

1. $x = -3.8$
2. $y = 3.5$
3. $z = 3$

Laurie's Notes

Discuss

? Ask a few questions about equation solving:
- "Does every equation have a solution?" no
- "Does every equation have just one solution?" no
- "Is it possible for an equation to have two solutions?" yes Students may say no, but using the third example below will convince them otherwise. They will encounter these types of equations in Chapter 7.

- Share some common equations and discuss the number of solutions:

  $x + 2 = 7$ one solution, 5
  $x + 2 = x + 7$ no solution
  $x^2 = 4$ two solutions, 2 and -2
  $x + 2 = x + 2$ infinitely many solutions
- Explain that in Examples 3 and 4, students will investigate equations that have no solution or infinitely many solutions. Assure students that they will use the same techniques for solving equations as before.

Extra Example 3

Solve $3x - 5 = 7 + 3x$. no solution

Example 3

- **Teaching Tip:** Instead of telling students when an equation has no solution, work through the example and ask students about the "solution" $3 = -7$.
- Work through the problem in two ways, as shown and by collecting the constant terms on one side of the equation as the first step. Show that the solution is the same both ways.
- For the final step, write $3 \neq -7$ to emphasize that there is no solution.
- **MP2 Reason Abstractly and Quantitatively** and **MP7 Look for and Make Use of Structure:** Students should notice that the same quantity, $4x$, is being subtracted from different numbers, 3 and -7. They should reason that the two sides of the equation can never be equal, so there is no solution.

? "How do you know when an equation has no solution?" Solve the equation normally and if you end up with a false statement, the equation has no solution.

Extra Example 4

Solve $\frac{1}{4}(8x - 12) = 2x - 3$ infinitely many solutions

Example 4

- **Teaching Tip:** Instead of telling students when an equation has infinitely many solutions, work through the example and ask students about the "solution" $4 = 4$.
- To check the solution, ask volunteers to choose several values for x. Substitute these values into the original equation and show that all result in true equations.
- **MP2** and **MP7:** Students should notice that in the second step, the expressions on both sides of the equal sign are the same. They should reason that both sides will be equal for any value of x, so there are infinitely many solutions.

On Your Own

4. no solution
5. infinitely many solutions
6. infinitely many solutions
7. no solution

On Your Own

- **Common Error:** In Exercises 5–7, students may forget to distribute the factor to the second term within the parentheses.
- **Neighbor Check:** Have students work independently and then have their neighbors check their work. Have students discuss any discrepancies.

English Language Learners

Vocabulary

Remind English learners that *like terms* are terms with the same variables raised to the same exponents. As the number of terms in an equation increases, an important skill is to identify and combine like terms.

Some equations do not have one solution. Equations can also have no solution or infinitely many solutions.

When solving an equation that has no solution, you will obtain an equivalent equation that is not true for any value of the variable, such as $0 = 2$.

EXAMPLE 3 Solving Equations with No Solution

Solve $3 - 4x = -7 - 4x$.

$$3 - 4x = -7 - 4x$$ Write the equation.

Undo the subtraction. → $\underline{+ 4x} \quad \underline{+ 4x}$ Addition Property of Equality

$$3 = -7 \;\; ✗$$ Simplify.

∴ The equation $3 = -7$ is never true. So, the equation has no solution.

When solving an equation that has infinitely many solutions, you will obtain an equivalent equation that is true for all values of the variable, such as $-5 = -5$.

EXAMPLE 4 Solving Equations with Infinitely Many Solutions

Solve $6x + 4 = 4\left(\frac{3}{2}x + 1\right)$.

$$6x + 4 = 4\left(\frac{3}{2}x + 1\right)$$ Write the equation.

$$6x + 4 = 6x + 4$$ Distributive Property

Undo the addition. → $\underline{- 6x} \quad \underline{- 6x}$ Subtraction Property of Equality

$$4 = 4$$ Simplify.

∴ The equation $4 = 4$ is always true. So, the equation has infinitely many solutions.

On Your Own

Now You're Ready
Exercises 18–29

Solve the equation.

4. $2x + 1 = 2x - 1$

5. $\frac{1}{2}(6t - 4) = 3t - 2$

6. $\frac{1}{3}(2b + 9) = \frac{2}{3}\left(b + \frac{9}{2}\right)$

7. $6(5 - 2v) = -4(3v + 1)$

EXAMPLE 5 Writing and Solving an Equation

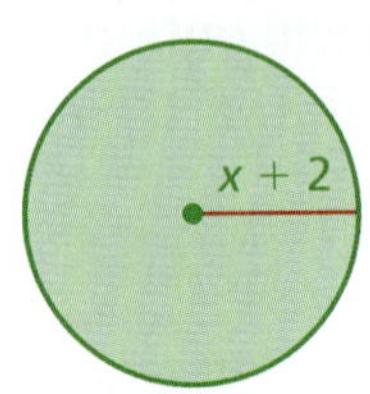

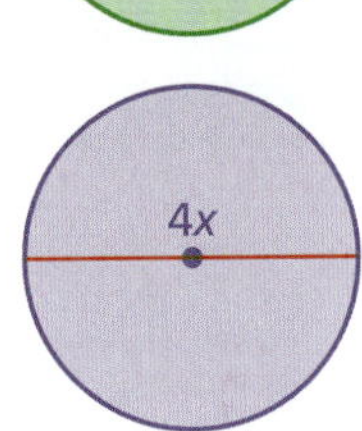

The circles are identical. What is the area of each circle?

Ⓐ 2 Ⓑ 4 Ⓒ 16π Ⓓ 64π

The circles are identical, so the radius of each circle is the same.

$x + 2 = 2x$	Write an equation. The radius of the purple circle is $\frac{4x}{2} = 2x$.
$-x \quad -x$	Subtraction Property of Equality
$2 = x$	Simplify.

Because the radius of each circle is 4, the area of each circle is $\pi r^2 = \pi(4)^2 = 16\pi$.

So, the correct answer is Ⓒ.

EXAMPLE 6 Real-Life Application

A boat travels x miles per hour upstream on the Mississippi River. On the return trip, the boat travels 2 miles per hour faster. How far does the boat travel upstream?

The speed of the boat on the return trip is $(x + 2)$ miles per hour.

Distance upstream = Distance of return trip

$3x = 2.5(x + 2)$	Write an equation.
$3x = 2.5x + 5$	Distributive Property
$-2.5x \quad -2.5x$	Subtraction Property of Equality
$0.5x = 5$	Simplify.
$\frac{0.5x}{0.5} = \frac{5}{0.5}$	Division Property of Equality
$x = 10$	Simplify.

The boat travels 10 miles per hour for 3 hours upstream. So, it travels 30 miles upstream.

On Your Own

8. **WHAT IF?** In Example 5, the diameter of the purple circle is $3x$. What is the area of each circle?

9. A boat travels x miles per hour from one island to another island in 2.5 hours. The boat travels 5 miles per hour faster on the return trip of 2 hours. What is the distance between the islands?

Laurie's Notes

Example 5

- ? "Define diameter and radius of a circle." Diameter is the distance across a circle through its center. Radius is half the diameter.
- Solve the equation as shown.
- This example reviews the formula for the area of a circle. You might also ask about the formula for the circumference of a circle. $C = 2\pi r$

Example 6

- ? "How do you find the distance traveled d when you know the rate r and time t?" multiply; $d = rt$
- Discuss with students that the distance both ways is the same, a simple but not obvious fact.
- ? "If you travel 40 miles per hour for 2 hours, how far will you go? How about 40 miles per hour for a half hour?" 80 mi; 20 mi
- ? "How far do you travel at x miles per hour for 3 hours?" $3x$ mi
- Students need to read the time from the illustration. The rates for each direction are x and $(x + 2)$.
- Write and solve the equation as shown.

On Your Own

- **Think-Pair-Share:** Students should read each question independently and then work in pairs to answer the questions. When they have answered the questions, the pair should compare their answers with another group and discuss any discrepancies.

Closure

- Give an example of an equation that has
  a. no solution. *Sample answer:* $x + 5 = x + 2$
  b. one solution. *Sample answer:* $x + 5 = 2x + 3$
  c. infinitely many solutions. *Sample answer:* $4x + 12 = 2(2x + 6)$

Extra Example 5

The legs of the right triangle have the same length. What is the area of the triangle? $1\frac{1}{8}$ square units

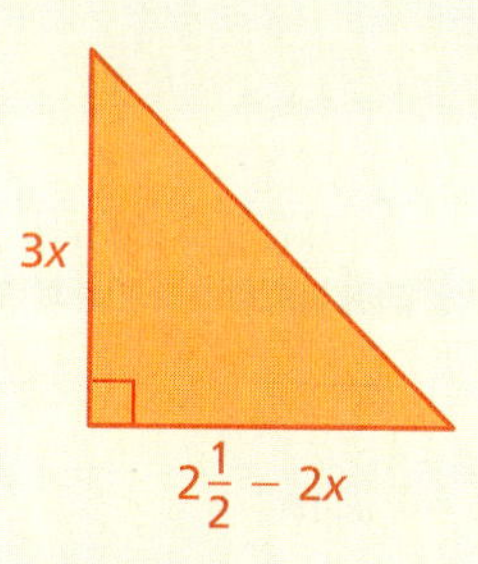

Extra Example 6

A boat travels 3 hours downstream at r miles per hour. On the return trip, the boat travels 5 miles per hour slower and takes 4 hours. What is the distance the boat travels each way? 60 mi

8. 36π

9. 50 mi

Vocabulary and Concept Check

1. no; When 3 is substituted for x, the left side simplifies to 4 and the right side simplifies to 3.
2. *Sample answer:* $4x + 1 = 3x - 2$

Practice and Problem Solving

3. $x = 13.2$ in.
4. $x = 7.2$ in.
5. $x = 7.5$ in.
6. $m = -4$
7. $k = -0.75$
8. $x = 8.2$
9. $p = -48$
10. $w = 2$
11. $n = -3.5$
12. $z = 1.6$
13. $x = -4$
14. $d = 14$
15. The 4 should have been added to the right side.
$$3x - 4 = 2x + 1$$
$$3x - 2x - 4 = 2x + 1 - 2x$$
$$x - 4 = 1$$
$$x - 4 + 4 = 1 + 4$$
$$x = 5$$
16. 2 lb
17. $15 + 0.5m = 25 + 0.25m$; 40 mi

Assignment Guide and Homework Check

Level	Day 1 Activity Assignment	Day 2 Lesson Assignment	Homework Check
Basic	3–5, 44–47	1, 2, 7–31 odd	7, 11, 17, 19, 25
Average	3–5, 44–47	1, 2, 13–29 odd, 30–42 even	13, 17, 23, 27, 34
Advanced	1–5, 15, 16–42 even, 43–47		20, 26, 34, 36, 40

For Your Information

- **Exercise 16** The equation represents a mixture problem in which peanuts are added to other ingredients, making trail mix. The equation shows that p pounds of peanuts that cost \$4.05 per pound are added to other ingredients that cost a total of \$14.40. This mixture creates $(p + 3)$ pounds of trail mix that costs \$4.50 per pound.

Common Errors

- **Exercises 6–14** Students may perform the same operation that they are trying to undo instead of the opposite operation when trying to get the variable or constant terms on the same side. Remind them that whenever a variable or constant term is moved from one side of the equal sign to the other, the opposite operation is used.
- **Exercises 10–14** When using the Distributive Property, students may forget to distribute to all the values within the parentheses. Remind them that they need to distribute to all the values and encourage them to draw arrows showing the distribution, if needed.

1.3 Record and Practice Journal

Solve the equation. Check your solution.

1. $x + 16 = 9x$
   $x = 2$
2. $4y - 70 = 12y + 2$
   $y = -9$
3. $5(p + 6) = 8p$
   $p = 10$
4. $3(g - 7) = 2(10 + g)$
   $g = 41$
5. $1.8 + 7n = 9.5 - 4n$
   $n = 0.7$
6. $\frac{3}{7}w - 11 = -\frac{4}{7}w$
   $w = 11$
7. One movie club charges a \$100 membership fee and \$10 for each movie. Another club charges no membership fee but movies cost \$15 each. Write and solve an equation to find the number of movies you need to buy for the cost of each movie club to be the same.
   $100 + 10x = 15x$; $x = 20$

1.3 Exercises

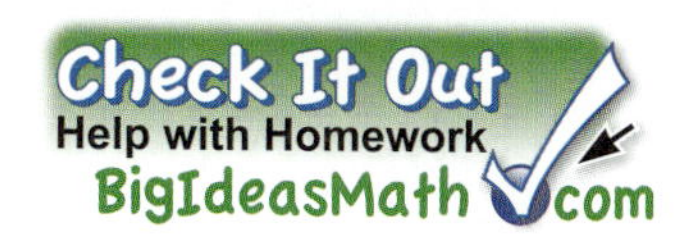

Vocabulary and Concept Check

1. **WRITING** Is $x = 3$ a solution of the equation $3x - 5 = 4x - 9$? Explain.
2. **OPEN-ENDED** Write an equation that has variables on both sides and has a solution of -3.

Practice and Problem Solving

The value of the solid's surface area is equal to the value of the solid's volume. Find the value of x.

3. 

4. 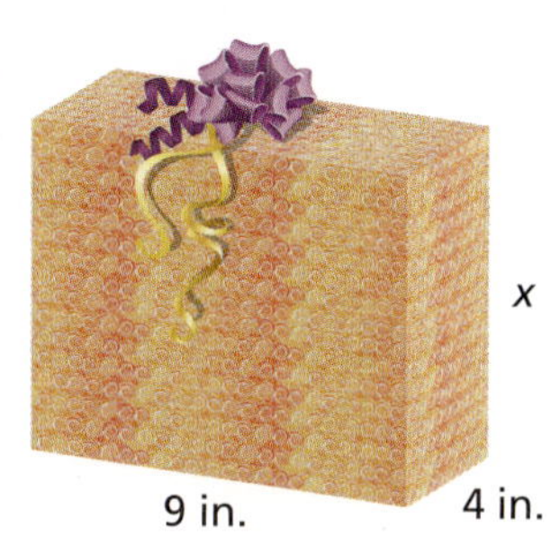

5. 

Solve the equation. Check your solution.

6. $m - 4 = 2m$
7. $3k - 1 = 7k + 2$
8. $6.7x = 5.2x + 12.3$
9. $-24 - \frac{1}{8}p = \frac{3}{8}p$
10. $12(2w - 3) = 6w$
11. $2(n - 3) = 4n + 1$
12. $2(4z - 1) = 3(z + 2)$
13. $0.1x = 0.2(x + 2)$
14. $\frac{1}{6}d + \frac{2}{3} = \frac{1}{4}(d - 2)$

15. **ERROR ANALYSIS** Describe and correct the error in solving the equation.

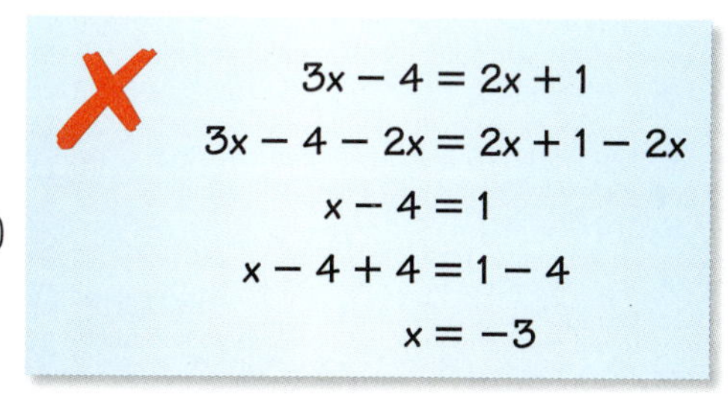

16. **TRAIL MIX** The equation $4.05p + 14.40 = 4.50(p + 3)$ represents the number p of pounds of peanuts you need to make trail mix. How many pounds of peanuts do you need for the trail mix?

17. **CARS** Write and solve an equation to find the number of miles you must drive to have the same cost for each of the car rentals.

$15 plus $0.50 per mile

$25 plus $0.25 per mile

Solve the equation. Check your solution, if possible.

③ ④ **18.** $x + 6 = x$ **19.** $3x - 1 = 1 - 3x$ **20.** $4x - 9 = 3.5x - 9$

21. $\frac{1}{2}x + \frac{1}{2}x = x + 1$ **22.** $3x + 15 = 3(x + 5)$ **23.** $\frac{1}{3}(9x + 3) = 3x + 1$

24. $5x - 7 = 4x - 1$ **25.** $2x + 4 = -(-7x + 6)$ **26.** $5.5 - x = -4.5 - x$

27. $10x - \frac{8}{3} - 4x = 6x$ **28.** $-3(2x - 3) = -6x + 9$ **29.** $6(7x + 7) = 7(6x + 6)$

30. ERROR ANALYSIS Describe and correct the error in solving the equation.

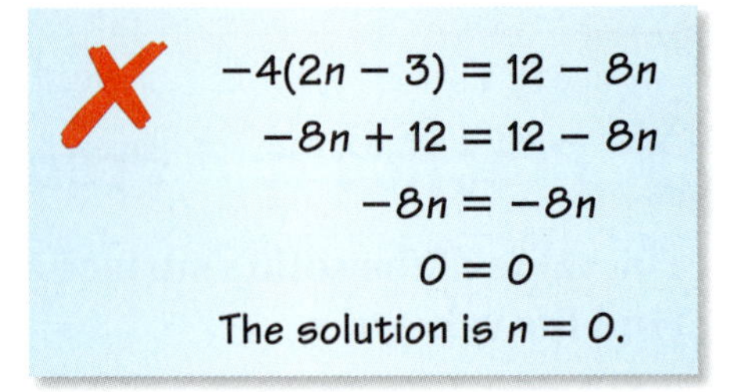

31. OPEN-ENDED Write an equation with variables on both sides that has no solution. Explain why it has no solution.

32. GEOMETRY Are there any values of x for which the areas of the figures are the same? Explain.

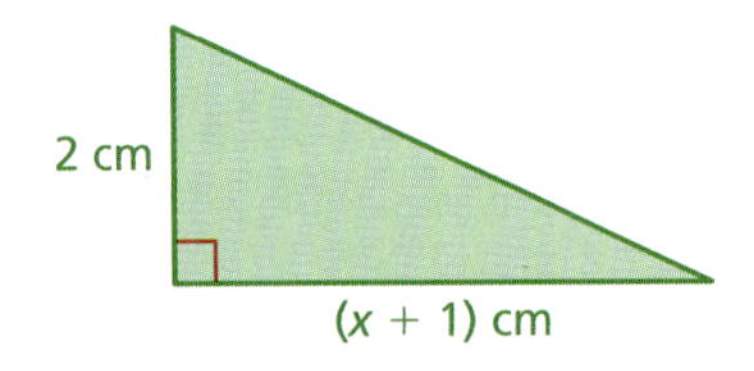

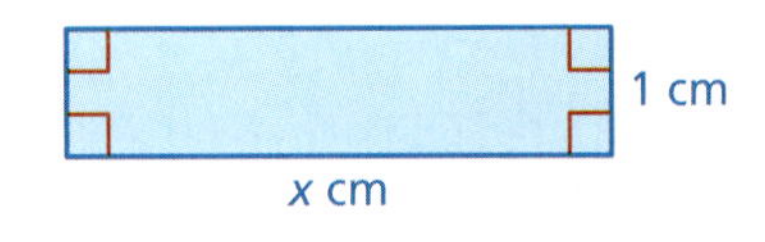

33. SATELLITE TV Provider A charges \$75 for installation and charges \$39.95 per month for the basic package. Provider B offers free installation and charges \$39.95 per month for the basic package. Your neighbor subscribes to Provider A the same month you subscribe to Provider B. After how many months is your neighbor's total cost the same as your total cost for satellite TV?

34. PIZZA CRUST Pepe's Pizza makes 52 pizza crusts the first week and 180 pizza crusts each subsequent week. Dianne's Delicatessen makes 26 pizza crusts the first week and 90 pizza crusts each subsequent week. In how many weeks will the total number of pizza crusts made by Pepe's Pizza equal twice the total number of pizza crusts made by Dianne's Delicatessen?

35. PRECISION Is the triangle an equilateral triangle? Explain.

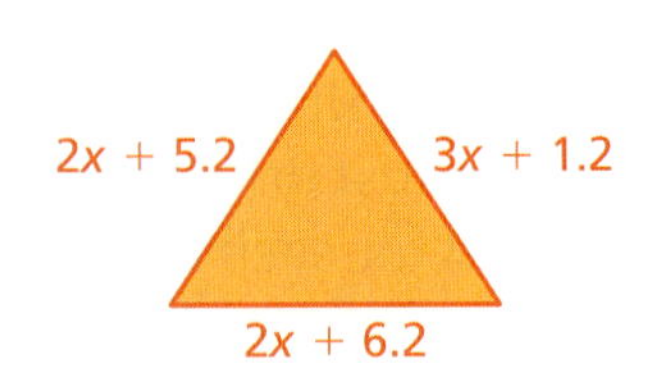

Common Errors

- **Exercises 18–29** Students may end up with an equivalent equation such as $6 = 0$ or $3x + 15 = 3x + 15$ and get confused about how to state the final answer. Remind students that an equation can have no solution or infinitely many solutions.
- **Exercises 18–29** When using the Distributive Property, students may forget to distribute to all the values within the parentheses. Remind them that they need to distribute to all the values and encourage them to draw arrows showing the distribution, if needed.
- **Exercise 20** Students may end up with an equivalent equation such as $0.5x = 0$ and not know how to proceed to a final answer. Remind them that dividing 0 by a nonzero number is permissible and gives a result of 0. Encourage students to check their solutions. The solution $x = 0$ checks in this exercise.
- **Exercise 25** Students may be confused about the meaning of the minus sign outside the parentheses. Remind students that it means -1 times the quantity in parentheses, and they can make this substitution in a solution step if it helps them see how to arrive at a solution.
- **Exercise 27** Students may combine like terms on the left side of the equation incorrectly, ending up with $14x$ instead of $6x$. Remind them to consider the signs of the variable terms.

Practice and Problem Solving

18. no solution

19. $x = \frac{1}{3}$

20. $x = 0$

21. no solution

22. infinitely many solutions

23. infinitely many solutions

24. $x = 6$ **25.** $x = 2$

26. no solution

27. no solution

28. infinitely many solutions

29. infinitely many solutions

30. When the equation is $0 = 0$, it means it is true for all values of n, not just 0; The equation has infinitely many solutions.

31. *Sample answer:* $8x + 2 = 8x$; The number $8x$ cannot be equal to 2 more than itself.

32. no; There is no solution to the equation stating the areas are equal, $x + 1 = x$.

33. It's never the same. Your neighbor's total cost will always be $75 more than your total cost.

34. The total number of crusts made by Pepe's Pizza is always twice the total number of crusts made by Dianne's Delicatessen.

35. no; $2x + 5.2$ can never equal $2x + 6.2$.

Differentiated Instruction

Kinesthetic

Some students may benefit from using algebra tiles to model and solve the equations. Check that the students have correctly modeled the equation before attempting to solve.

Practice and Problem Solving

36. 3 units **37.** 7.5 units

38. 232 units

39. See *Taking Math Deeper*.

40. fractions; Because $\frac{1}{3}$ written as a decimal is repeating.

41. 10 mL **42.** 25 grams

43. See Additional Answers.

Fair Game Review

44. 27 cm^3

45. 15.75 cm^3

46. 24 $in.^3$

47. C

Mini-Assessment

Solve the equation.

1. $n - 4 = 3n + 6$ $n = -5$
2. $0.3(w + 10) = 1.8w$ $w = 2$
3. $-3x + 15 = 3(5 - x)$ infinitely many solutions
4. $\frac{1}{2}(4x + 14) = 2(x - 7)$ no solution
5. The perimeter of the rectangle is equal to the perimeter of the square. What are the side lengths of each figure? rectangle: 4 units by 10 units; square: 7 units by 7 units

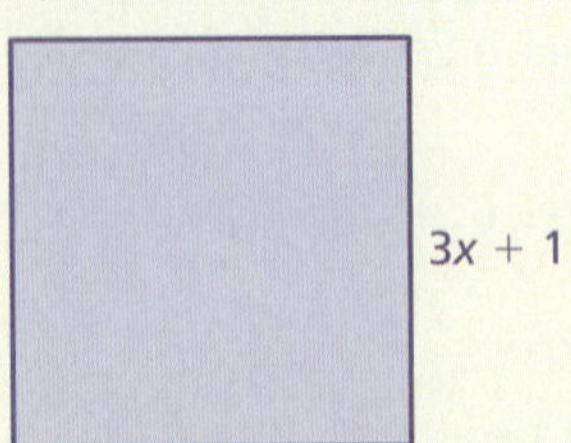

Taking Math Deeper

Exercise 39

This problem seems like it is easy, but it can actually be quite challenging.

1. Identify the key information in the table.

	Packing Material	Priority	Express
Box	\$2.25	\$2.50/lb	\$8.50/lb
Envelope	\$1.10	\$2.50/lb	\$8.50/lb

2. Write and solve an equation.

Let x = the weight of the DVD and packing material.

Cost of Mailing Box: $2.25 + 2.5x$

Cost of Mailing Envelope: $1.10 + 8.5x$

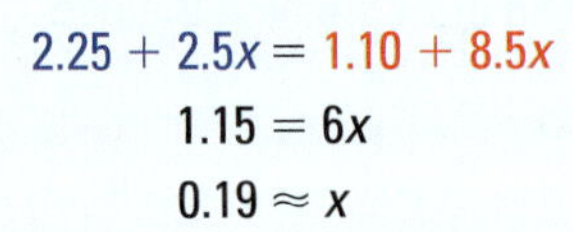

$$2.25 + 2.5x = 1.10 + 8.5x$$
$$1.15 = 6x$$
$$0.19 \approx x$$

3. Answer the question.

The weight of the DVD and packing material is about 0.19 pound, or about 3 ounces.

Project

Postage for special types of mail, such as Priority Mail®, is determined by the weight of the package and the distance it needs to travel. Find the cost of sending a 15-ounce package from your house to Los Angeles, Washington, D.C., and Albuquerque.

Reteaching and Enrichment Strategies

If students need help. . .	If students got it. . .
Resources by Chapter • Practice A and Practice B • Puzzle Time Record and Practice Journal Practice Differentiating the Lesson Lesson Tutorials Skills Review Handbook	Resources by Chapter • Enrichment and Extension • Technology Connection Start the next section

A polygon is *regular* if each of its sides has the same length. Find the perimeter of the regular polygon.

36.

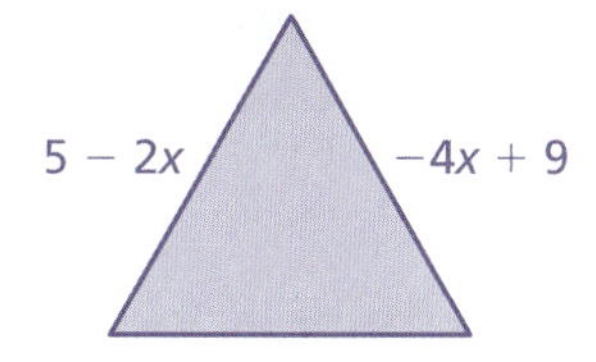

37.

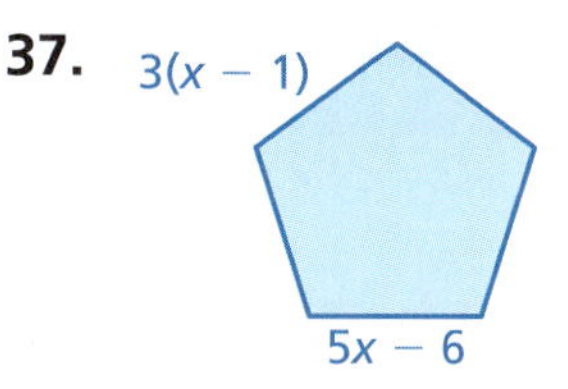

38.

$\frac{4}{3}x - \frac{1}{3}$ $x + 7$

39. PRECISION The cost of mailing a DVD in an envelope by Express Mail® is equal to the cost of mailing a DVD in a box by Priority Mail®. What is the weight of the DVD with its packing material? Round your answer to the nearest hundredth.

	Packing Material	Priority Mail®	Express Mail®
Box	\$2.25	\$2.50 per lb	\$8.50 per lb
Envelope	\$1.10	\$2.50 per lb	\$8.50 per lb

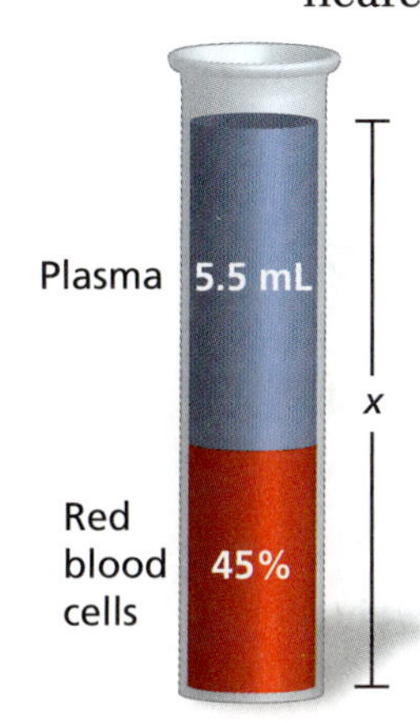

40. PROBLEM SOLVING Would you solve the equation $0.25x + 7 = \frac{1}{3}x - 8$ using fractions or decimals? Explain.

41. BLOOD SAMPLE The amount of red blood cells in a blood sample is equal to the total amount in the sample minus the amount of plasma. What is the total amount x of blood drawn?

42. NUTRITION One serving of oatmeal provides 16% of the fiber you need daily. You must get the remaining 21 grams of fiber from other sources. How many grams of fiber should you consume daily?

43. Geometry A 6-foot-wide hallway is painted as shown, using equal amounts of white and black paint.

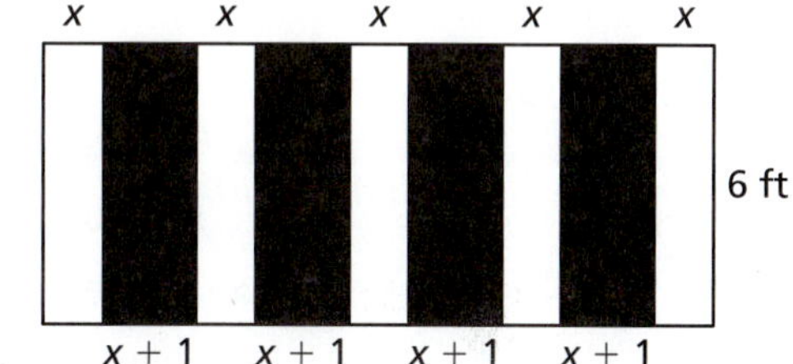

a. How long is the hallway?

b. Can this same hallway be painted with the same pattern, but using twice as much black paint as white paint? Explain.

Fair Game Review What you learned in previous grades & lessons

Find the volume of the solid. *(Skills Review Handbook)*

44.

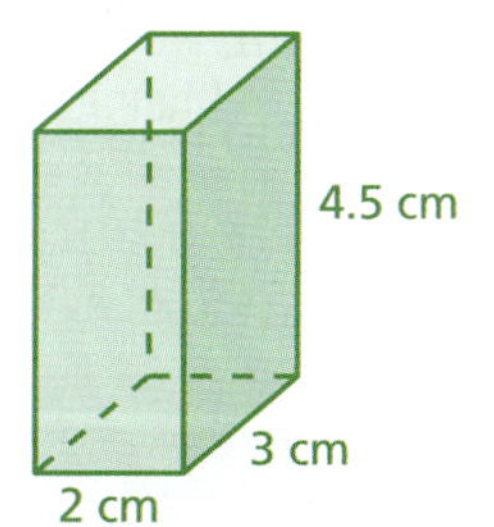

45.

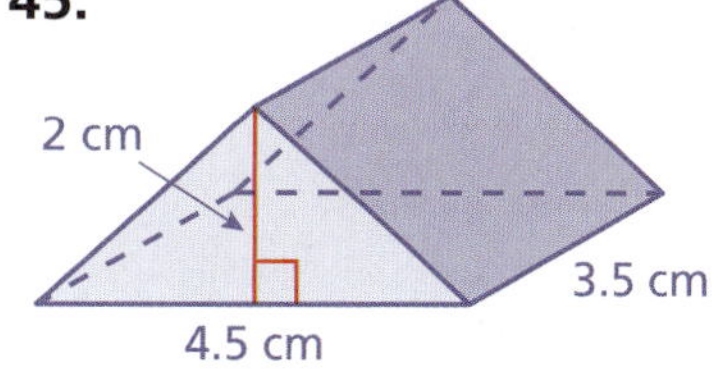

46.

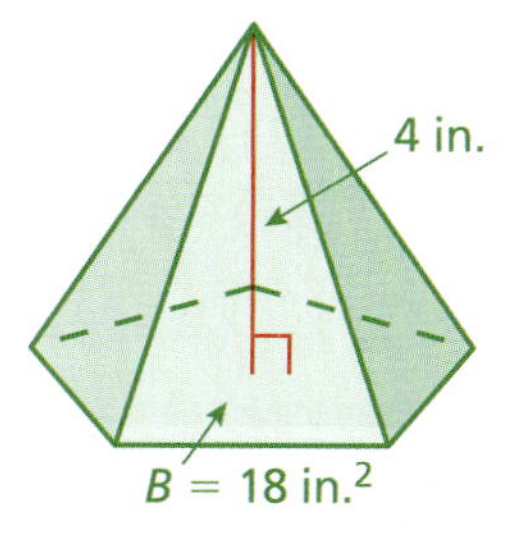

47. MULTIPLE CHOICE A car travels 480 miles on 15 gallons of gasoline. How many miles does the car travel per gallon? *(Skills Review Handbook)*

Ⓐ 28 mi/gal Ⓑ 30 mi/gal Ⓒ 32 mi/gal Ⓓ 35 mi/gal

1.4 Rewriting Equations and Formulas

Essential Question How can you use a formula for one measurement to write a formula for a different measurement?

1 ACTIVITY: Using Perimeter and Area Formulas

Work with a partner.

a. • Write a formula for the perimeter P of a rectangle.

• Solve the formula for w.

• Use the new formula to find the width of the rectangle.

w $P = 19$ in.

$\ell = 5.5$ in.

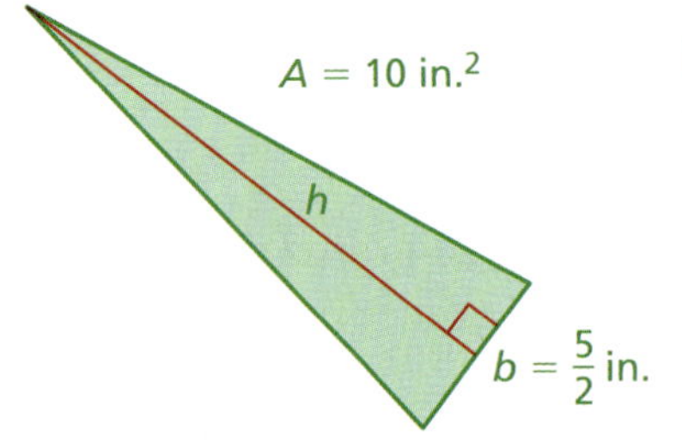

b. • Write a formula for the area A of a triangle.

• Solve the formula for h.

• Use the new formula to find the height of the triangle.

c. • Write a formula for the circumference C of a circle.

• Solve the formula for r.

• Use the new formula to find the radius of the circle.

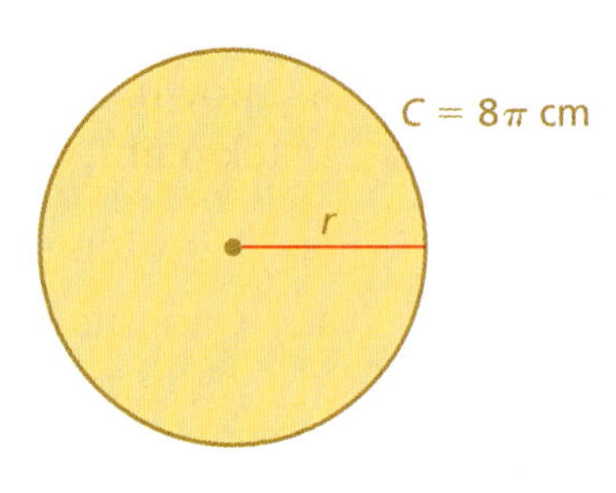

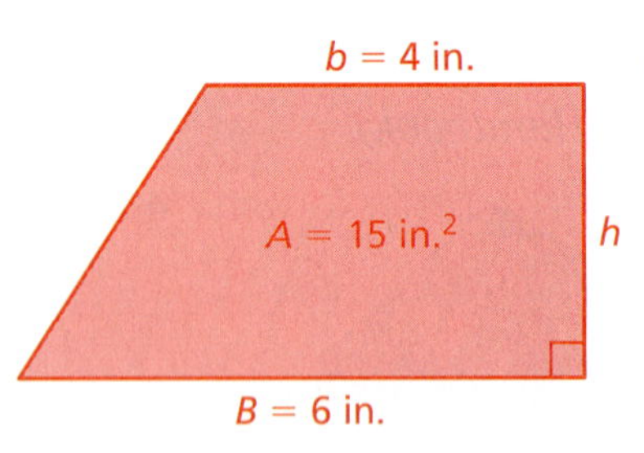

d. • Write a formula for the area A of a trapezoid.

• Solve the formula for h.

• Use the new formula to find the height of the trapezoid.

Solving Equations
In this lesson, you will
• rewrite equations to solve for one variable in terms of the other variable(s).
Applying Standard MACC.8.EE.3.7

e. • Write a formula for the area A of a parallelogram.

• Solve the formula for h.

• Use the new formula to find the height of the parallelogram.

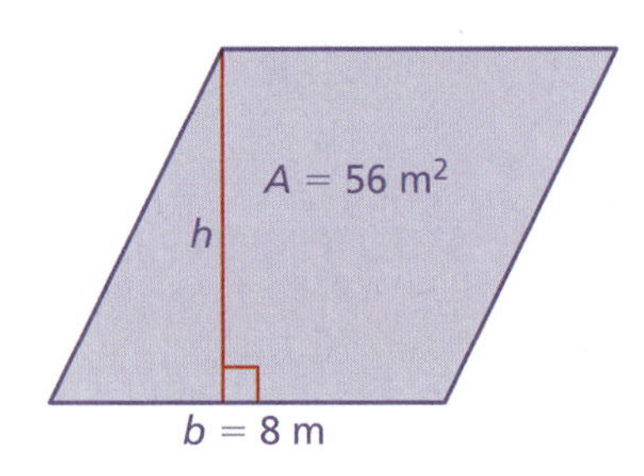

Laurie's Notes

Introduction

Standards for Mathematical Practice

- **MP8 Look for and Express Regularity in Repeated Reasoning:** Mathematically proficient students look for shortcuts for solving problems, such as solving a literal equation for a specified variable.

Motivate

- **Preparation:** Make a set of formula cards. My set is a collection of five cards for each shape: the labeled diagram, the two measurements, and the two formulas being found.

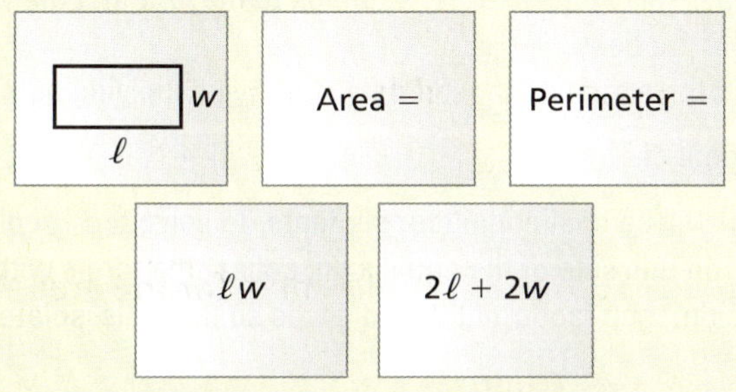

- Depending upon the number of students in your class, use some or all of the cards. Pass out the cards and have students form groups matching all 5 cards for the shape.
- **MP4 Model with Mathematics:** When all of the matches have been made, ask each group to read their formulas aloud.

Activity Notes

Activity 1

- Solving literal equations can be one of the most challenging skills for students. Model a problem, such as solving $A = \ell w$ for width.
- Fractional coefficients can also be a challenge, so model an additional problem, such as solving $A = \frac{1}{2}xy$ for y. First, multiply both sides by 2, then divide both sides by x.
- **Teaching Tip:** You may find that students are substituting the known values of the variables and then solving the equation, instead of solving the equation and then substituting.
- **MP8:** The reason for solving for the variable is that the equation can be used for the width of any rectangle given the perimeter and length, not just the specific example shown. It is a general solution that can be reused.
- **Teaching Tip:** After 2 or more groups have correctly solved part (a), have a volunteer write the solution on the board for the other groups to see.
- For parts (b) and (d), suggest students start by multiplying both sides by the reciprocal of $\frac{1}{2}$. In part (c), 2π is a number and can be manipulated as such, so divide both sides by 2π.

Florida Common Core Standards

MACC.8.EE.3.7 Solve linear equations in one variable.

Previous Learning

Students should know the common formulas for area, perimeter, and volume.

1.4 Record and Practice Journal

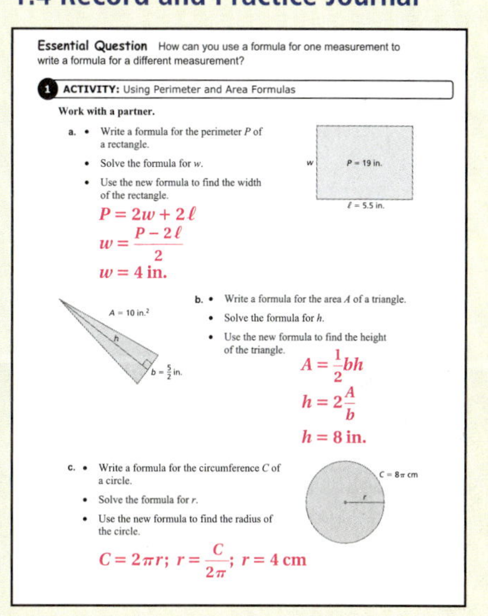

Essential Question How can you use a formula for one measurement to write a formula for a different measurement?

1 ACTIVITY: Using Perimeter and Area Formulas

Work with a partner.

a. • Write a formula for the perimeter P of a rectangle.
• Solve the formula for w.
• Use the new formula to find the width of the rectangle.

$P = 19$ in., $\ell = 5.5$ in.

$P = 2w + 2\ell$

$w = \frac{P - 2\ell}{2}$

$w = 4$ in.

b. • Write a formula for the area A of a triangle.
• Solve the formula for h.
• Use the new formula to find the height of the triangle.

$A = 10$ in.^{2}, $b = \frac{5}{2}$ in.

$A = \frac{1}{2}bh$

$h = 2\frac{A}{b}$

$h = 8$ in.

c. • Write a formula for the circumference C of a circle.
• Solve the formula for r.
• Use the new formula to find the radius of the circle.

$C = 8\pi$ cm

$C = 2\pi r$; $r = \frac{C}{2\pi}$; $r = 4$ cm

Differentiated Instruction

Kinesthetic

Have kinesthetic learners model the areas of polygons on grid paper. Then have them compare their answers with the answers found using area formulas.

1.4 Record and Practice Journal

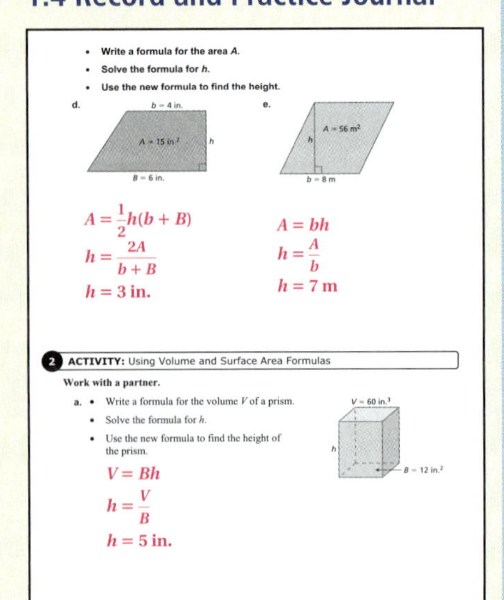

- Write a formula for the area A.
- Solve the formula for h.
- Use the new formula to find the height.

d. $b = 4$ in., $A = 15$ in.^{2}, h, $B = 6$ in.

$A = \frac{1}{2}h(b + B)$

$h = \frac{2A}{b + B}$

$h = 3$ in.

e. $A = 56$ m^2, h, $b = 8$ m

$A = bh$

$h = \frac{A}{b}$

$h = 7$ m

2 ACTIVITY: Using Volume and Surface Area Formulas

Work with a partner.

a. • Write a formula for the volume V of a prism.
• Solve the formula for h.
• Use the new formula to find the height of the prism.

$V = 60$ in.^{3}, h, $B = 12$ in.^{2}

$V = Bh$

$h = \frac{V}{B}$

$h = 5$ in.

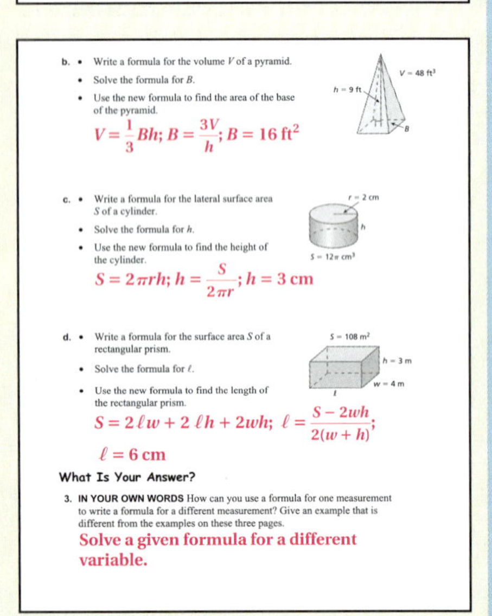

b. • Write a formula for the volume V of a pyramid.
• Solve the formula for B.
• Use the new formula to find the area of the base of the pyramid.

$V = 48$ ft^3, $h = 9$ ft, B

$V = \frac{1}{3}Bh$; $B = \frac{3V}{h}$; $B = 16$ ft^2

c. • Write a formula for the lateral surface area S of a cylinder.
• Solve the formula for h.
• Use the new formula to find the height of the cylinder.

$r = 2$ cm, h, $S = 12\pi$ cm^2

$S = 2\pi rh$; $h = \frac{S}{2\pi r}$; $h = 3$ cm

d. • Write a formula for the surface area S of a rectangular prism.
• Solve the formula for ℓ.
• Use the new formula to find the length of the rectangular prism.

$S = 108$ m^2, $h = 3$ m, $w = 4$ m, ℓ

$S = 2\ell w + 2\ell h + 2wh$; $\ell = \frac{S - 2wh}{2(w + h)}$;

$\ell = 6$ cm

What Is Your Answer?

3. **IN YOUR OWN WORDS** How can you use a formula for one measurement to write a formula for a different measurement? Give an example that is different from the examples on these three pages.

Solve a given formula for a different variable.

Laurie's Notes

Activity 2

- This activity is similar to Activity 1, where students worked with perimeter and area formulas. In Activity 2, students will work with volume and surface area formulas.
- Students should be familiar with each of the formulas they are being asked to write. More volume formulas, including those for cylinders, cones, and spheres, will be covered in Chapter 8.
- Use the *Teaching Tips* from Activity 1. Have students work in groups of 3 or 4 and post a correct solution on the board after 2 or more groups have been successful.
- **MP7 Look for and Make Use of Structure:** Note that parts (a) and (b) use B for the area of the base instead of having students use specific area formulas. This helps to draw attention to the fact that the volume is a factor of the base.
- For part (b), suggest that students start by multiplying both sides by the reciprocal of $\frac{1}{3}$.
- Part (d) will be challenging for students. To solve for ℓ, collect the terms with ℓ on one side of the equal sign, collect the terms without ℓ on the other side, then factor out ℓ and divide so that ℓ is isolated.

What Is Your Answer?

- **Neighbor Check:** Have students work independently and then have their neighbors check their work. Have students discuss any discrepancies.

Closure

- Describe how to solve $d = rt$ for t. *Sample answer:* Divide both sides of the equation by r.

2 ACTIVITY: Using Volume and Surface Area Formulas

Math Practice 8

Find General Methods

What do you have to do each time to solve for the given variable? Why does this process result in a new formula?

Work with a partner.

a. • Write a formula for the volume V of a prism.
• Solve the formula for h.
• Use the new formula to find the height of the prism.

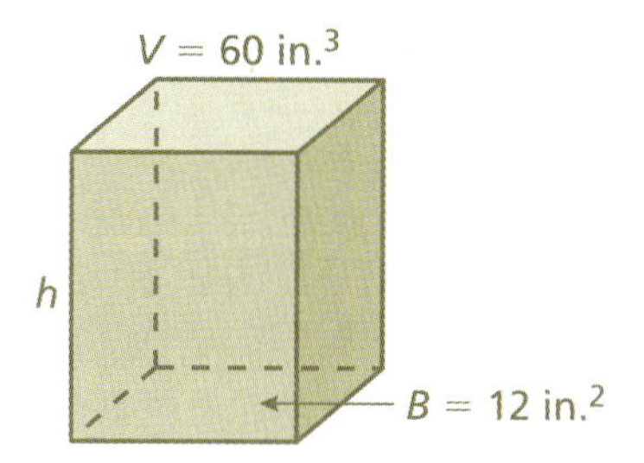

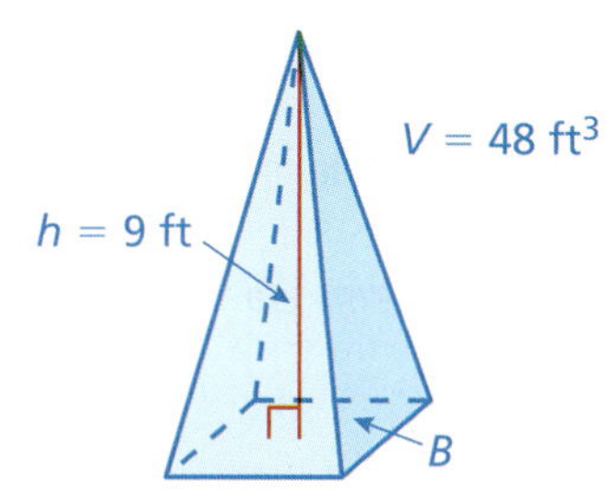

b. • Write a formula for the volume V of a pyramid.
• Solve the formula for B.
• Use the new formula to find the area of the base of the pyramid.

c. • Write a formula for the lateral surface area S of a cylinder.
• Solve the formula for h.
• Use the new formula to find the height of the cylinder.

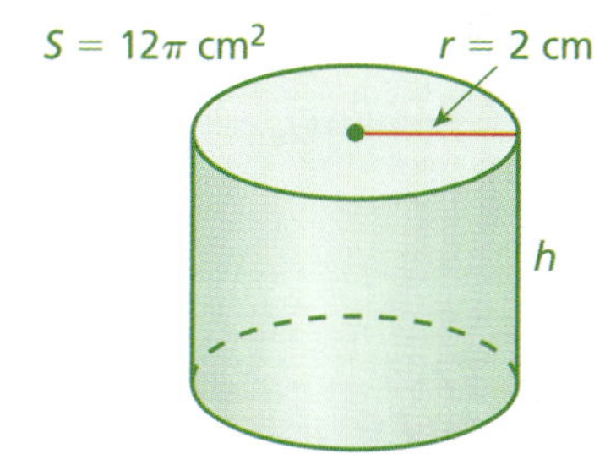

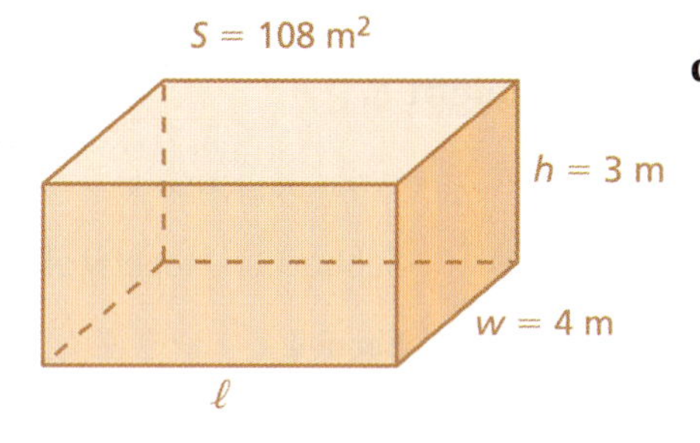

d. • Write a formula for the surface area S of a rectangular prism.
• Solve the formula for ℓ.
• Use the new formula to find the length of the rectangular prism.

What Is Your Answer?

3. **IN YOUR OWN WORDS** How can you use a formula for one measurement to write a formula for a different measurement? Give an example that is different from the examples on these two pages.

Use what you learned about rewriting equations and formulas to complete Exercises 3 and 4 on page 30.

1.4 Lesson

Key Vocabulary
literal equation, *p. 28*

An equation that has two or more variables is called a **literal equation**. To rewrite a literal equation, solve for one variable in terms of the other variable(s).

EXAMPLE 1 Rewriting an Equation

Solve the equation $2y + 5x = 6$ for y.

	$2y + 5x = 6$	Write the equation.
Undo the addition.	$2y + 5x - 5x = 6 - 5x$	Subtraction Property of Equality
	$2y = 6 - 5x$	Simplify.
Undo the multiplication.	$\frac{2y}{2} = \frac{6 - 5x}{2}$	Division Property of Equality
	$y = 3 - \frac{5}{2}x$	Simplify.

On Your Own

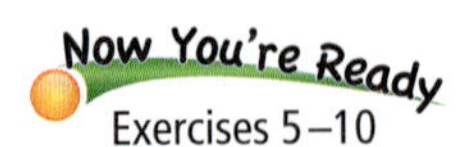

Solve the equation for y.

1. $5y - x = 10$
2. $4x - 4y = 1$
3. $12 = 6x + 3y$

EXAMPLE 2 Rewriting a Formula

Remember

A *formula* shows how one variable is related to one or more other variables. A formula is a type of literal equation.

The formula for the surface area S of a cone is $S = \pi r^2 + \pi r\ell$. Solve the formula for the slant height ℓ.

$S = \pi r^2 + \pi r\ell$	Write the formula.
$S - \pi r^2 = \pi r^2 - \pi r^2 + \pi r\ell$	Subtraction Property of Equality
$S - \pi r^2 = \pi r\ell$	Simplify.
$\frac{S - \pi r^2}{\pi r} = \frac{\pi r\ell}{\pi r}$	Division Property of Equality
$\frac{S - \pi r^2}{\pi r} = \ell$	Simplify.

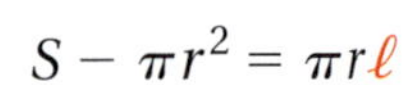

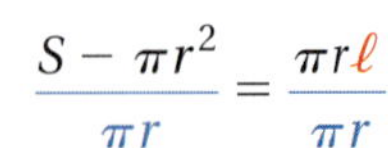

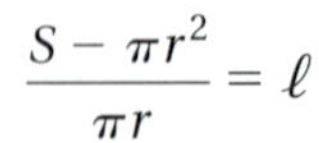

On Your Own

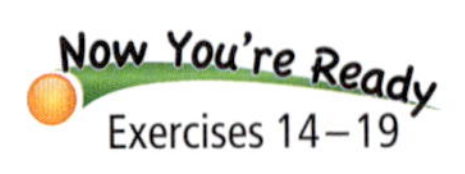

Solve the formula for the red variable.

4. Area of rectangle: $A = bh$
5. Simple interest: $I = Prt$
6. Surface area of cylinder: $S = 2\pi r^2 + 2\pi rh$

Laurie's Notes

Introduction

Connect

- **Yesterday:** Students rewrote geometric formulas. (MP4, MP7, MP8)
- **Today:** Students will rewrite literal equations.

Motivate

- To pique interest, share with students the highest and lowest recorded temperatures, in degrees Fahrenheit and degrees Celsius, in New Mexico.

Highest Recorded Temperature		Lowest Recorded Temperature	
122°F	50°C	−50°F	−46°C

Lesson Notes

Example 1

? "Can 6 and $5x$ be combined? Explain." no; They are not like terms.

- **MP7 Look for and Make Use of Structure:** Simplifying the last step is not obvious to all students. The expression $\frac{(6-5x)}{2}$ is a fraction with two terms in the numerator. You subtract the numerators and keep the same denominator. For example:

$$\frac{5-3}{7}=\frac{5}{7}-\frac{3}{7} \quad \text{and} \quad \frac{6-5x}{2}=\frac{6}{2}-\frac{5x}{2}=3-\frac{5}{2}x.$$

On Your Own

- Spend sufficient time on these problems. Students will encounter many equations in which they need to rewrite y in terms of x.
- Notice in Question 2 that the coefficient of y is -4. Suggest students rewrite the equation as $4x+(-4)y=1$.

Example 2

- **Teaching Tip:** Highlight the variable ℓ in red as shown in the textbook. Discuss the idea that everything except the variable ℓ must be moved to the left side of the equation using Properties of Equality.
- **MP7:** Structurally, the formula has the form $A = B + Cx$. It can be solved by subtracting B from both sides and then dividing both sides by C.

? "The term πr^2 is added to the term $\pi r\ell$. How do you move it to the left side of the equation?" Subtract πr^2 from each side of the equation.

- Discuss the technique of dividing by πr in one step, instead of dividing by π and then dividing by r.

On Your Own

- **Think-Pair-Share:** Students should read each question independently and then work in pairs to answer the questions. When they have answered the questions, the pair should compare their answers with another group and discuss any discrepancies.

Goal Today's lesson is solving **literal equations**.

Technology for the Teacher

Lesson Tutorials
Lesson Plans
Answer Presentation Tool

Extra Example 1

Solve the equation $-2x-3y=6$ for y.

$y=-\frac{2}{3}x-2$

On Your Own

1. $y=2+\frac{1}{5}x$
2. $y=x-\frac{1}{4}$
3. $y=4-2x$

Extra Example 2

The formula for the surface area of a square pyramid is $S=x^2+2x\ell$. Solve the formula for the slant height ℓ.

$\ell=\frac{S-x^2}{2x}$

On Your Own

4. $b=\frac{A}{h}$
5. $P=\frac{I}{rt}$
6. $h=\frac{S-2\pi r^2}{2\pi r}$

English Language Learners

Vocabulary

Have students start a *Formula* page in their notebooks with the formulas used in this section. Each formula should be accompanied by a description of what each of the variables represents and an example. In the case of area formulas, units of measure should be included with the description (e.g., units and square units). As students progress throughout the year, additional formulas can be added to the *Formula* notebook page.

Extra Example 3

Solve the temperature formula $F = \frac{9}{5}C + 32$ for C. $C = \frac{5}{9}(F - 32)$

Extra Example 4

Which temperature is greater, 400°F or 200°C? 400°F

7. greater than

Laurie's Notes

Key Idea

- Write the formula for converting from degrees Fahrenheit to degrees Celsius.
- Use this formula if you know the temperature in degrees Fahrenheit and you want to find the temperature in degrees Celsius.

? "You are traveling abroad, and the temperature is always stated in degrees Celsius. How can you figure out the temperature in degrees Fahrenheit, with which you are more familiar?" Students may recognize that you will want to have a different conversion formula that allows you to substitute for C and calculate F.

Example 3

? "What is the reciprocal of $\frac{5}{9}$?" $\frac{9}{5}$

- Remind students that multiplying by the reciprocal $\frac{9}{5}$ is more efficient than dividing by the fraction $\frac{5}{9}$.

Example 4

- **FYI:** The graphic on the left provides information about the temperature of a lightning bolt and the temperature of the surface of the sun. The two temperatures use different scales.

? "How can you compare two temperatures that are in different scales?" Listen for understanding that one of the temperatures must be converted.

? "How do you multiply $\frac{9}{5}$ times 30,000?" Students may recall that you can simplify before multiplying. Five divides into 30,000 six thousand times, so $6000 \times 9 = 54,000$.

? "Approximately how many times hotter is a lightning bolt than the surface of the sun?" 5 times This is a *cool* fact for students to know!

On Your Own

- **Neighbor Check:** Have students work independently and then have their neighbors check their work. Have students discuss any discrepancies.

Closure

- **Exit Ticket:** Solve $2x + 4y = 11$ for y. $y = -\frac{1}{2}x + \frac{11}{4}$

Temperature Conversion

A formula for converting from degrees Fahrenheit F to degrees Celsius C is

$$C = \frac{5}{9}(F - 32).$$

EXAMPLE 3 Rewriting the Temperature Formula

Solve the temperature formula for F.

$C = \frac{5}{9}(F - 32)$ — Write the temperature formula.

Use the reciprocal. → $\frac{9}{5} \cdot C = \frac{9}{5} \cdot \frac{5}{9}(F - 32)$ — Multiplication Property of Equality

$\frac{9}{5}C = F - 32$ — Simplify.

Undo the subtraction. → $\frac{9}{5}C + 32 = F - 32 + 32$ — Addition Property of Equality

$\frac{9}{5}C + 32 = F$ — Simplify.

The rewritten formula is $F = \frac{9}{5}C + 32$.

EXAMPLE 4 Real-Life Application

Which has the greater temperature?

Convert the Celsius temperature of lightning to Fahrenheit.

$F = \frac{9}{5}C + 32$ — Write the rewritten formula from Example 3.

$= \frac{9}{5}(30,000) + 32$ — Substitute 30,000 for C.

$= 54,032$ — Simplify.

Because 54,032 °F is greater than 11,000 °F, lightning has the greater temperature.

7. Room temperature is considered to be 70 °F. Suppose the temperature is 23 °C. Is this greater than or less than room temperature?

1.4 Exercises

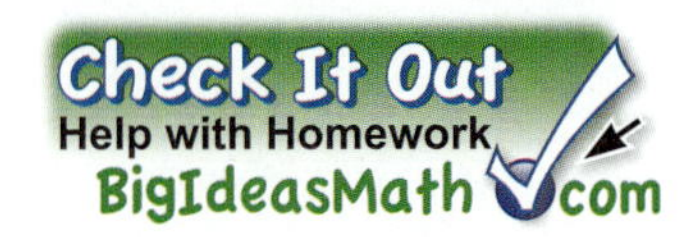

Vocabulary and Concept Check

1. **VOCABULARY** Is $-2x = \frac{3}{8}$ a literal equation? Explain.

2. **DIFFERENT WORDS, SAME QUESTION** Which is different? Find "both" answers.

Solve $4x - 2y = 6$ for y.	Solve $6 = 4x - 2y$ for y.
Solve $4x - 2y = 6$ for y in terms of x.	Solve $4x - 2y = 6$ for x in terms of y.

Practice and Problem Solving

3. **a.** Write a formula for the area A of a triangle.
   b. Solve the formula for b.
   c. Use the new formula to find the base of the triangle.

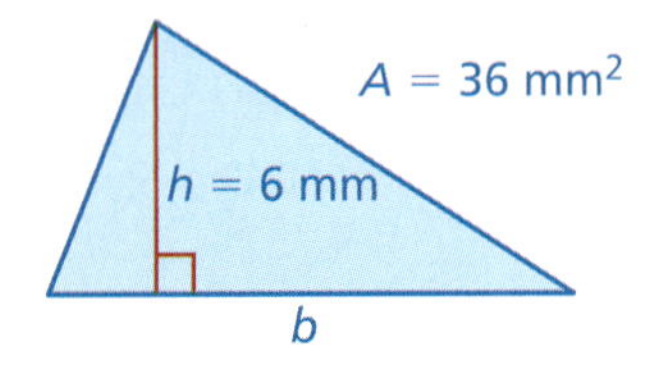

4. **a.** Write a formula for the volume V of a prism.
   b. Solve the formula for B.
   c. Use the new formula to find the area of the base of the prism.

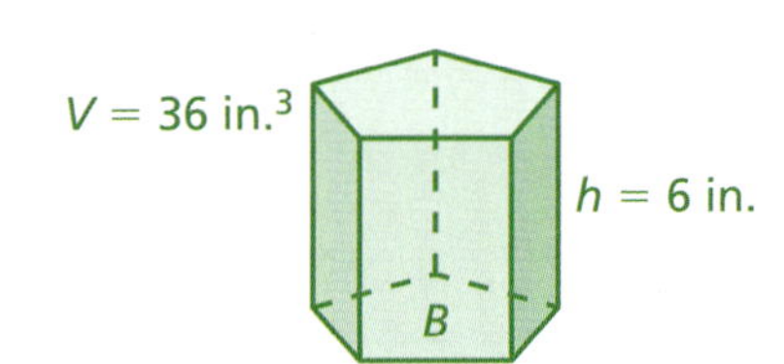

Solve the equation for *y*.

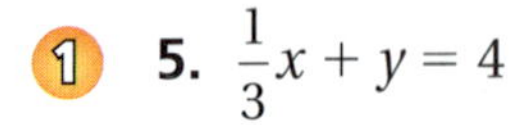

5. $\frac{1}{3}x + y = 4$

6. $3x + \frac{1}{5}y = 7$

7. $6 = 4x + 9y$

8. $\pi = 7x - 2y$

9. $4.2x - 1.4y = 2.1$

10. $6y - 1.5x = 8$

11. **ERROR ANALYSIS** Describe and correct the error in rewriting the equation.

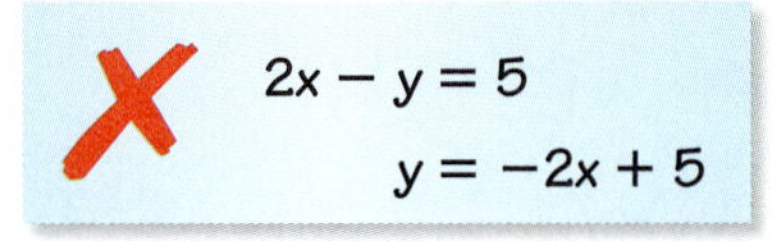

12. **TEMPERATURE** The formula $K = C + 273.15$ converts temperatures from Celsius C to Kelvin K.
    a. Solve the formula for C.
    b. Convert 300 Kelvin to Celsius.

13. **INTEREST** The formula for simple interest is $I = Prt$.
    a. Solve the formula for t.
    b. Use the new formula to find the value of t in the table.

I	\$75
P	\$500
r	5%
t	

Assignment Guide and Homework Check

Level	Day 1 Activity Assignment	Day 2 Lesson Assignment	Homework Check
Basic	3, 4, 24–28	1, 2, 5–21 odd	7, 9, 13, 17, 19
Average	3, 4, 24–28	1, 2, 5–13 odd, 14–22 even	7, 9, 13, 18, 22
Advanced	1–4, 6–10 even, 11, 12–22 even, 23–28		6, 10, 18, 20, 22

For Your Information

- **Exercise 2** In the *Different Words, Same Question* exercise, three of the four choices pose the same question using different words. The remaining choice poses a different question. So there are two answers.

Common Errors

- **Exercises 5–10** Students may solve the equation for the wrong variable. Remind them that they are solving the equation for *y*. Encourage them to make *y* a different color when solving so that it is easy to remember that they are solving for *y*.
- **Exercises 14–19** In each exercise, different steps are required to solve for the red variable, and some of these steps could confuse students. Remind them to take their time and think about the process for solving for a variable.

1.4 Record and Practice Journal

Solve the equation for *y*.

1. $2x + y = -9$
$y = -2x - 9$

2. $4x - 10y = 12$
$y = \frac{2}{5}x - \frac{6}{5}$

3. $13 = \frac{1}{6}y + 2x$
$y = -12x + 78$

Solve the formula for the bold variable.

4. $V = \ell wh$
$w = \frac{V}{\ell h}$

5. $f = \frac{1}{2}(r + 6.5)$
$r = 2f - 6.5$

6. $S = 2\pi r^2 + 2\pi rh$
$h = \frac{S - 2\pi r^2}{2\pi r}$

7. The formula for the area of a triangle is $A = \frac{1}{2}bh$.

   a. Solve the formula for *h*.
   $h = \frac{2A}{b}$

   b. Use the new formula to find the value of *h*.
   $h = 9$ in.

A = 54 in.²
h
12 in.

Vocabulary and Concept Check

1. no; The equation only contains one variable.
2. Solve $4x - 2y = 6$ for x in terms of y.; $x = \frac{3}{2} + \frac{1}{2}y$; $y = -3 + 2x$

Practice and Problem Solving

3. a. $A = \frac{1}{2}bh$
   b. $b = \frac{2A}{h}$
   c. $b = 12$ mm
4. a. $V = Bh$
   b. $B = \frac{V}{h}$
   c. $B = 6$ in.^{2}
5. $y = 4 - \frac{1}{3}x$
6. $y = 35 - 15x$
7. $y = \frac{2}{3} - \frac{4}{9}x$
8. $y = \frac{7}{2}x - \frac{\pi}{2}$
9. $y = 3x - 1.5$
10. $y = \frac{4}{3} + \frac{1}{4}x$
11. The y should have a negative sign in front of it.
$$2x - y = 5$$
$$-y = -2x + 5$$
$$y = 2x - 5$$
12. a. $C = K - 273.15$
    b. 26.85°C
13. a. $t = \frac{I}{Pr}$
    b. $t = 3$ yr

Practice and Problem Solving

14. $t = \dfrac{d}{r}$

15. $m = \dfrac{e}{c^2}$

16. $C = R - P$

17. $\ell = \dfrac{A - \frac{1}{2}\pi w^2}{2w}$

18. $V = \dfrac{Bh}{3}$

19. $w = 6g - 40$

20. The rewritten formula is a general solution that can be reused.

21. **a.** $F = 32 + \frac{9}{5}(K - 273.15)$

b. 32°F

c. liquid nitrogen

22. See *Taking Math Deeper*.

23. $r^3 = \dfrac{3V}{4\pi}$; $r = 4.5$ in.

Fair Game Review

24. $3\frac{3}{4}$ **25.** $-5\frac{1}{3}$

26. $\frac{1}{3}$ **27.** $1\frac{1}{4}$

28. D

Mini-Assessment

Solve the formula for the red variable.

1. Distance Formula: $d = rt$ $r = \dfrac{d}{t}$
2. Area of a triangle: $A = \frac{1}{2}bh$ $h = \dfrac{2A}{b}$
3. Circumference of a circle: $C = 2\pi r$ $r = \dfrac{C}{2\pi}$
4. The temperature in Portland, Oregon, is 37°F. The temperature in Mobile, Alabama, is 22°C. In which city is the temperature higher? Mobile, Alabama

Taking Math Deeper

Exercise 22

This problem reviews circles and percents, as well as distance, rate, and time. It also has a bit of history related to George Ferris, who designed the first Ferris wheel for the 1893 World's Fair in Chicago.

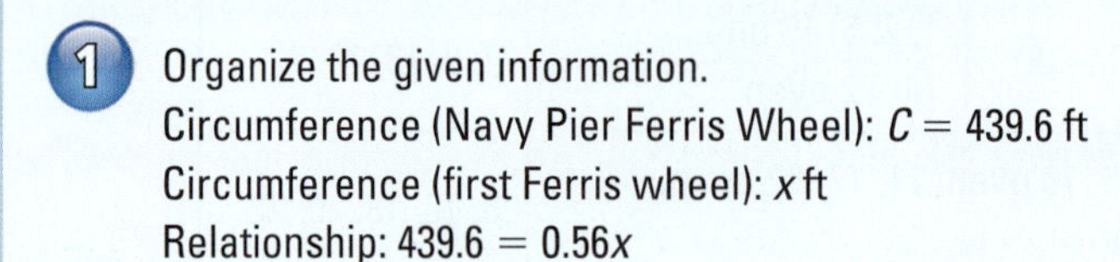

1 Organize the given information.
Circumference (Navy Pier Ferris Wheel): $C = 439.6$ ft
Circumference (first Ferris wheel): x ft
Relationship: $439.6 = 0.56x$

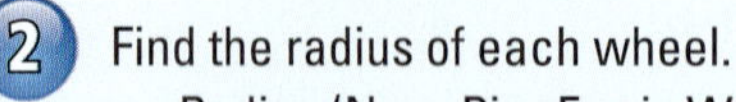

2 Find the radius of each wheel.

a. Radius (Navy Pier Ferris Wheel):

$$C = 2\pi r$$
$$439.6 \approx 2(3.14)r$$
$$70 = r$$

Circumference (first Ferris wheel):

$$439.6 = 0.56x$$
$$785 = x$$

b. Radius (first Ferris wheel):

$$785 \approx 2(3.14)R$$
$$125 = R$$

3 **c.** The first Ferris wheel made 1 revolution in 9 minutes. How fast was the wheel moving?

$$\text{rate} = \frac{785 \text{ ft}}{9 \text{ min}} \approx 87.2 \text{ ft per min}$$

It might be interesting for students to know that the first Ferris wheel had 36 cars, each of which held 60 people!

Project

Use your school's library or the Internet to find how long one revolution takes for the Ferris wheel on the Navy Pier in Chicago and the one in London, England. Which one has the greater circumference? Which one travels faster? How do you know?

Reteaching and Enrichment Strategies

If students need help. . .	If students got it. . .
Resources by Chapter • Practice A and Practice B • Puzzle Time Record and Practice Journal Practice Differentiating the Lesson Lesson Tutorials Skills Review Handbook	Resources by Chapter • Enrichment and Extension • Technology Connection Start the next section

Solve the equation for the red variable.

2 **14.** $d = rt$ **15.** $e = mc^2$ **16.** $R - C = P$

17. $A = \frac{1}{2}\pi w^2 + 2\ell w$ **18.** $B = 3\frac{V}{h}$ **19.** $g = \frac{1}{6}(w + 40)$

20. LOGIC Why is it useful to rewrite a formula in terms of another variable?

21. REASONING The formula $K = \frac{5}{9}(F - 32) + 273.15$ converts temperatures from Fahrenheit F to Kelvin K.

a. Solve the formula for F.

b. The freezing point of water is 273.15 Kelvin. What is this temperature in Fahrenheit?

c. The temperature of dry ice is −78.5 °C. Which is colder, dry ice or liquid nitrogen?

Navy Pier Ferris Wheel

22. FERRIS WHEEL The Navy Pier Ferris Wheel in Chicago has a circumference that is 56% of the circumference of the first Ferris wheel built in 1893.

a. What is the radius of the Navy Pier Ferris Wheel?

b. What was the radius of the first Ferris wheel?

c. The first Ferris wheel took 9 minutes to make a complete revolution. How fast was the wheel moving?

23. Repeated Reasoning The formula for the volume of a sphere is $V = \frac{4}{3}\pi r^3$. Solve the formula for r^3. Use Guess, Check, and Revise to find the radius of the sphere.

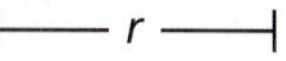

$V = 381.51$ in.^{3} r

Fair Game Review What you learned in previous grades & lessons

Multiply. *(Skills Review Handbook)*

24. $5 \times \frac{3}{4}$ **25.** $-2 \times \frac{8}{3}$ **26.** $\frac{1}{4} \times \frac{3}{2} \times \frac{8}{9}$ **27.** $25 \times \frac{3}{5} \times \frac{1}{12}$

28. MULTIPLE CHOICE Which of the following is not equivalent to $\frac{3}{4}$? *(Skills Review Handbook)*

(A) 0.75 (B) 3 : 4 (C) 75% (D) 4 : 3

1.3–1.4 Quiz

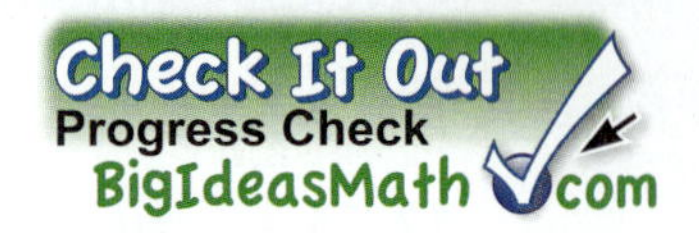

Solve the equation. Check your solution, if possible. *(Section 1.3)*

1. $2(x + 4) = -5x + 1$

2. $\frac{1}{2}s = 4s - 21$

3. $8.3z = 4.1z + 10.5$

4. $3(b + 5) = 4(2b - 5)$

5. $n + 7 - n = 4$

6. $\frac{1}{4}(4r - 8) = r - 2$

Solve the equation for *y*. *(Section 1.4)*

7. $6x - 3y = 9$

8. $8 = 2y - 10x$

Solve the formula for the red variable. *(Section 1.4)*

9. Volume of a cylinder: $V = \pi r^2 h$

10. Area of a trapezoid: $A = \frac{1}{2}h(b + B)$

11. TEMPERATURE In which city is the water temperature higher? *(Section 1.4)*

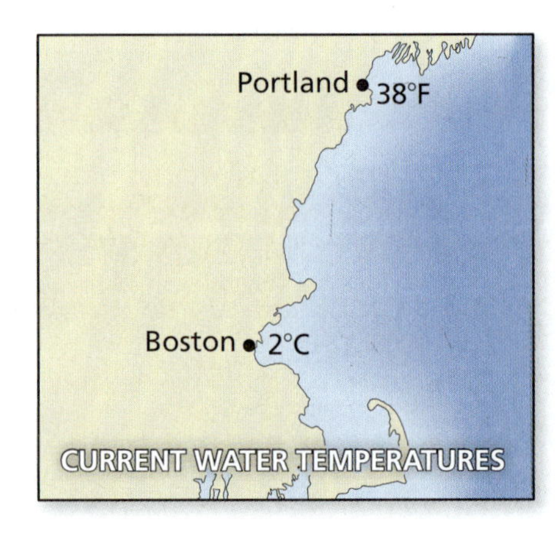

12. SAVINGS ACCOUNT You begin with \$25 in a savings account and \$50 in a checking account. Each week you deposit \$5 into savings and \$10 into checking. After how many weeks is the amount in checking twice the amount in savings? *(Section 1.3)*

13. INTEREST The formula for simple interest I is $I = Prt$. Solve the formula for the interest rate r. What is the interest rate r if the principal P is \$1500, the time t is 2 years, and the interest earned I is \$90? *(Section 1.4)*

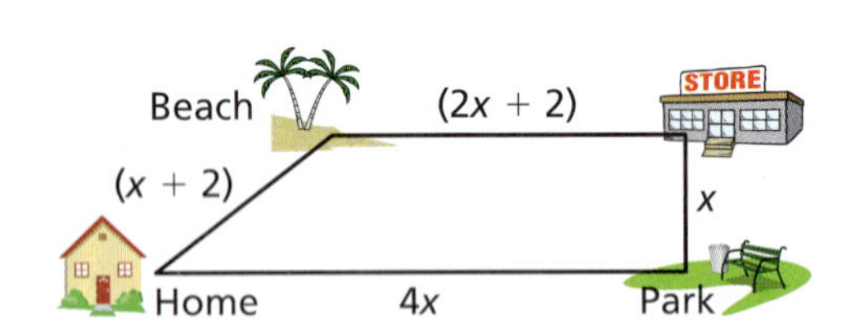

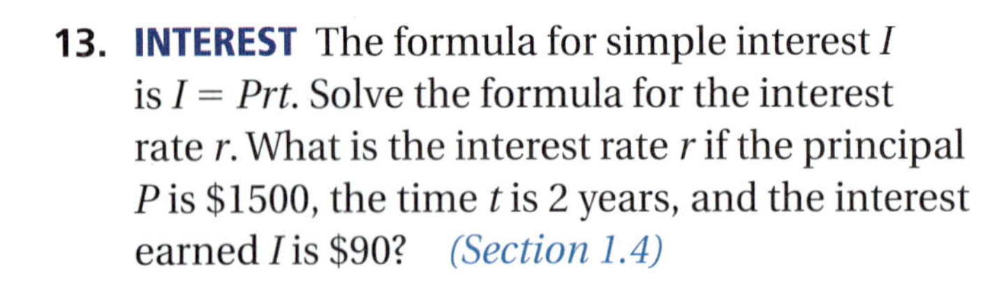

14. ROUTES From your home, the route to the store that passes the beach is 2 miles shorter than the route to the store that passes the park. What is the length of each route? *(Section 1.3)*

15. PERIMETER Use the triangle shown. *(Section 1.4)*

a. Write a formula for the perimeter P of the triangle.

b. Solve the formula for b.

c. Use the new formula to find b when a is 10 feet and c is 17 feet.

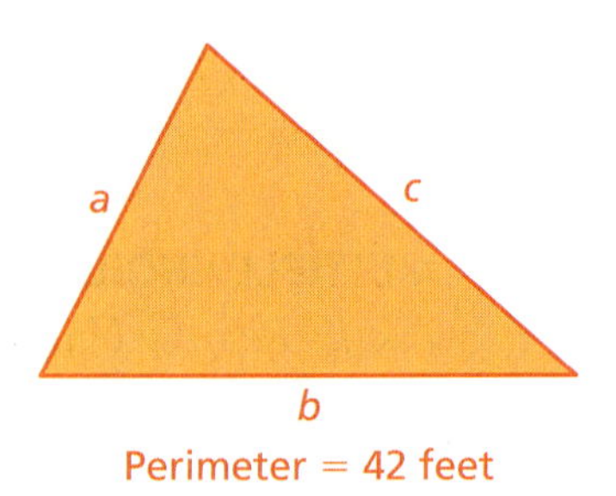

Alternative Assessment Options

Math Chat	Student Reflective Focus Question
Structured Interview	Writing Prompt

Math Chat

- Have students work in pairs. One student describes how to rewrite equations and formulas, giving examples. The other student probes for more information.
- The teacher should walk around the classroom listening to the pairs and asking questions to ensure understanding.

Study Help Sample Answers

Remind students to complete Graphic Organizers for the rest of the chapter.

3.

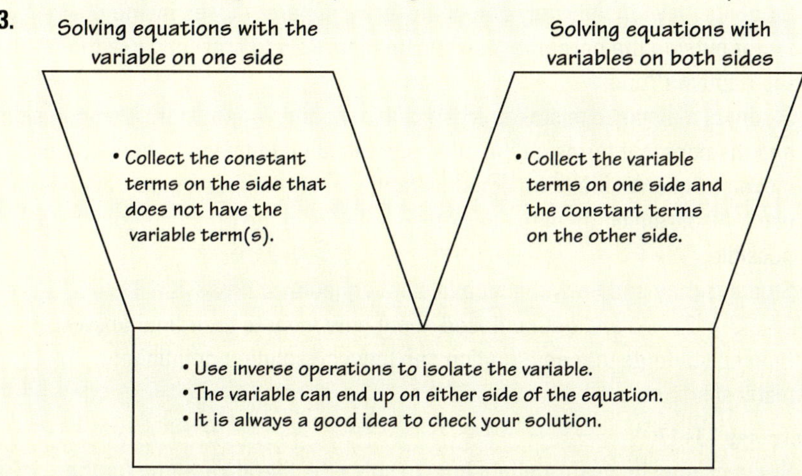

4.

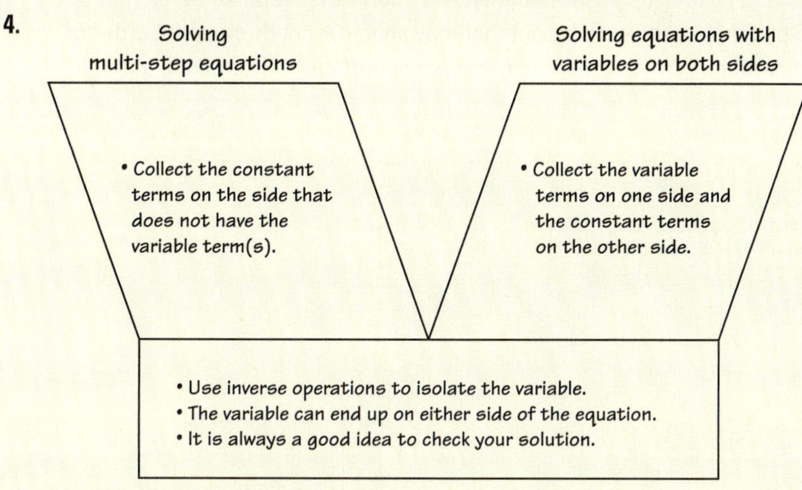

5. Available at *BigIdeasMath.com.*

Reteaching and Enrichment Strategies

If students need help...	If students got it...
Resources by Chapter • Practice A and Practice B • Puzzle Time Lesson Tutorials *BigIdeasMath.com*	Resources by Chapter • Enrichment and Extension • Technology Connection Game Closet at *BigIdeasMath.com* Start the Chapter Review

Answers

1. $x = -1$
2. $s = 6$
3. $z = 2.5$
4. $b = 7$
5. no solution
6. infinitely many solutions
7. $y = 2x - 3$
8. $y = 5x + 4$
9. $h = \frac{V}{\pi r^2}$
10. $b = \frac{2A}{h} - B$
11. Portland
12. The amount in checking is always twice the amount in savings.
13. 3%
14. passing beach: 13 miles
    passing park: 15 miles
15. **a.** $P = a + b + c$

    b. $b = P - a - c$

    c. $b = 15$ feet

Technology for the Teacher

Online Assessment
Assessment Book
ExamView® Assessment Suite

For the Teacher
Additional Review Options

- *BigIdeasMath.com*
- Online Assessment
- Game Closet at *BigIdeasMath.com*
- Vocabulary Help
- Resources by Chapter

Answers

1. $y = -19$
2. $n = -8$
3. $t = 12\pi$

Review of Common Errors

Exercises 1–3

- Students may perform the same operation that they are trying to undo instead of the inverse operation. Remind them that they must use an inverse operation to undo an operation. Also, remind them to check their solution in the original equation.

Exercises 4–6

- Students may change the exponent of the variable when combining like terms. For example, they may write the sum $x + x + \frac{1}{2}x + \frac{1}{2}x$ as $3x^3$. Remind them how to correctly combine like terms that have variables.

Exercises 7–12

- Students may multiply only one of the terms in parentheses by the factor outside the parentheses. Remind them how to correctly use the Distributive Property.
- Students may make mistakes when collecting the variable terms on one side and the constant terms on the other side. Remind them that when a term is moved from one side of an equation to the other, the inverse operation is used. Also, remind them to check their solution in the original equation, if possible.
- Students may end up with an equivalent equation such as $3 = 8$ or $2z - 3 = 2z - 3$ and get confused about how to state their final answer. Remind students that an equation can have no solution or infinitely many solutions.

Exercises 13–17

- Students may be unsure about how to solve the literal equation for the specified variable. Point out that they should work through the order of operations *backwards,* using inverse operations to isolate the variable.

1 Chapter Review

Check It Out
Vocabulary Help
BigIdeasMath.com

Review Key Vocabulary

literal equation, *p. 28*

Review Examples and Exercises

1.1 Solving Simple Equations *(pp. 2–9)*

The *boiling point* of a liquid is the temperature at which the liquid becomes a gas. The boiling point of mercury is about $\frac{41}{200}$ of the boiling point of lead. Write and solve an equation to find the boiling point of lead.

Let x be the boiling point of lead.

$$\frac{41}{200}x = 357 \quad \text{Write the equation.}$$

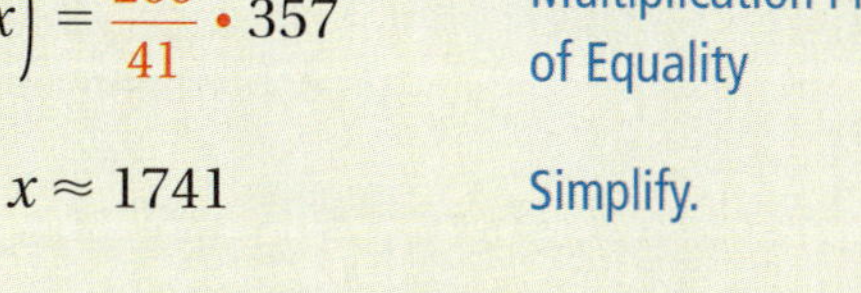

$$\frac{200}{41} \cdot \left(\frac{41}{200}x\right) = \frac{200}{41} \cdot 357 \quad \text{Multiplication Property of Equality}$$

$$x \approx 1741 \quad \text{Simplify.}$$

∴ The boiling point of lead is about 1741°C.

Mercury 357°C

Exercises

Solve the equation. Check your solution.

1. $y + 8 = -11$
2. $3.2 = -0.4n$
3. $-\frac{t}{4} = -3\pi$

1.2 Solving Multi-Step Equations *(pp. 10–15)*

Solve $-14x + 28 + 6x = -44$.

$$-14x + 28 + 6x = -44 \quad \text{Write the equation.}$$

$$-8x + 28 = -44 \quad \text{Combine like terms.}$$

$$\underline{-28} \qquad \underline{-28} \quad \text{Subtraction Property of Equality}$$

$$-8x = -72 \quad \text{Simplify.}$$

$$\frac{-8x}{-8} = \frac{-72}{-8} \quad \text{Division Property of Equality}$$

$$x = 9 \quad \text{Simplify.}$$

∴ The solution is $x = 9$.

Exercises

Find the value of x. Then find the angle measures of the polygon.

4.

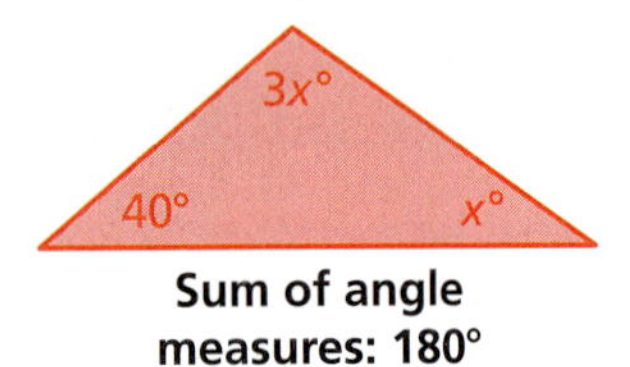

Sum of angle measures: 180°

5.

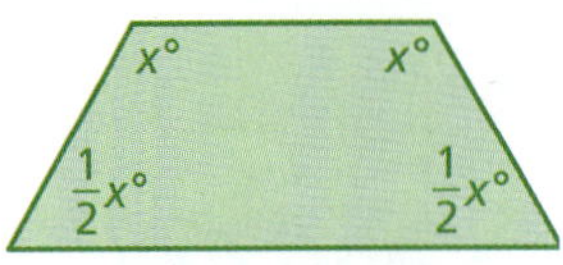

Sum of angle measures: 360°

6.

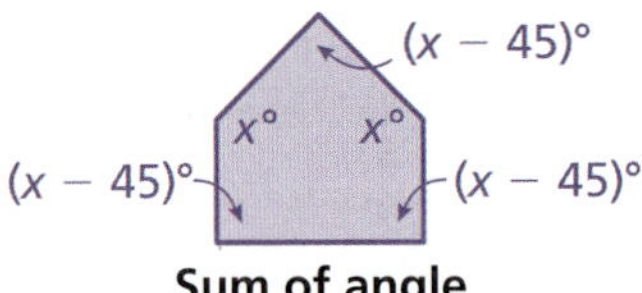

Sum of angle measures: 540°

1.3 Solving Equations with Variables on Both Sides *(pp. 18–25)*

a. Solve $3(x - 4) = -2(4 - x)$.

$3(x-4) = -2(4-x)$	Write the equation.
$3x - 12 = -8 + 2x$	Distributive Property
$\underline{-2x} \qquad \underline{-2x}$	Subtraction Property of Equality
$x - 12 = -8$	Simplify.
$\underline{+12} \qquad \underline{+12}$	Addition Property of Equality
$x = 4$	Simplify.

∴ The solution is $x = 4$.

b. Solve $4 - 5k = -8 - 5k$.

$4 - 5k = -8 - 5k$	Write the equation.
$\underline{+5k} \qquad \underline{+5k}$	Addition Property of Equality
$4 = -8$ ✗	Simplify.

∴ The equation $4 = -8$ is never true. So, the equation has no solution.

c. Solve $2\left(7g + \frac{2}{3}\right) = 14g + \frac{4}{3}$.

$2\left(7g + \frac{2}{3}\right) = 14g + \frac{4}{3}$	Write the equation.
$14g + \frac{4}{3} = 14g + \frac{4}{3}$	Distributive Property
$\underline{-14g} \qquad \underline{-14g}$	Subtraction Property of Equality
$\frac{4}{3} = \frac{4}{3}$	Simplify.

∴ The equation $\frac{4}{3} = \frac{4}{3}$ is always true. So, the equation has infinitely many solutions.

Review Game

Musical Toss

Materials

- soft object that can be tossed around
- a device to play music
- old homework, quiz, and test questions

Directions

Divide the class into pairs (groups of two).

Designate one pair of students to play the music and write the problems on the board. Pairs of students should be switched periodically.

The remaining members of the class will stand in a circle with each pair clearly identifiable.

When the music starts, the soft object is tossed to a pair of students and the problem is written on the board. That pair has to solve the problem and toss the object to another pair before the music stops.

Who wins?

The group holding the object when the music stops is eliminated. This will continue until there is one group remaining, the winner.

For the Student
Additional Practice

- Lesson Tutorials
- Multi-Language Glossary
- Self-Grading Progress Check
- *BigIdeasMath.com*
  Dynamic Student Edition
  Student Resources

Answers

4. $x = 35$; $40°, 105°, 35°$

5. $x = 120$; $60°, 120°, 120°, 60°$

6. $x = 135$; $90°, 135°, 90°, 135°, 90°$

7. $m = 6$

8. $p = 0.4$

9. $n = -19$

10. no solution

11. no solution

12. infinitely many solutions

13. $y = -\frac{1}{6}x + \frac{4}{3}$

14. $y = 2x - 3$

15. $y = -\frac{1}{2}x + 2$

16. **a.** $K = \frac{5}{9}(F - 32) + 273.15$

b. about $388.71\ K$

17. **a.** $A = \frac{1}{2}h(b + B)$

b. $h = \frac{2A}{b + B}$

c. $h = 6$ cm

My Thoughts on the Chapter

What worked. . .

Teacher Tip

Not allowed to write in your teaching edition? Use sticky notes to record your thoughts.

What did not work. . .

What I would do differently. . .

Exercises

Solve the equation. Check your solution, if possible.

7. $5m - 1 = 4m + 5$

8. $3(5p - 3) = 5(p - 1)$

9. $\frac{2}{5}n + \frac{1}{10} = \frac{1}{2}(n + 4)$

10. $7t + 3 = 8 + 7t$

11. $\frac{1}{5}(15b - 7) = 3b - 9$

12. $\frac{1}{6}(12z - 18) = 2z - 3$

1.4 Rewriting Equations and Formulas *(pp. 26–31)*

a. Solve $7y + 6x = 4$ for y.

$7y + 6x = 4$ — Write the equation.

$7y + 6x - 6x = 4 - 6x$ — Subtraction Property of Equality

$7y = 4 - 6x$ — Simplify.

$\frac{7y}{7} = \frac{4 - 6x}{7}$ — Division Property of Equality

$y = \frac{4}{7} - \frac{6}{7}x$ — Simplify.

b. The equation for a line in slope-intercept form is $y = mx + b$. Solve the equation for x.

$y = mx + b$ — Write the equation.

$y - b = mx + b - b$ — Subtraction Property of Equality

$y - b = mx$ — Simplify.

$\frac{y - b}{m} = \frac{mx}{m}$ — Division Property of Equality

$\frac{y - b}{m} = x$ — Simplify.

Exercises

Solve the equation for y.

13. $6y + x = 8$

14. $10x - 5y = 15$

15. $20 = 5x + 10y$

16. a. The formula $F = \frac{9}{5}(K - 273.15) + 32$ converts a temperature from Kelvin K to Fahrenheit F. Solve the formula for K.

b. Convert 240 °F to Kelvin K. Round your answer to the nearest hundredth.

17. a. Write the formula for the area A of a trapezoid.

b. Solve the formula for h.

c. Use the new formula to find the height h of the trapezoid.

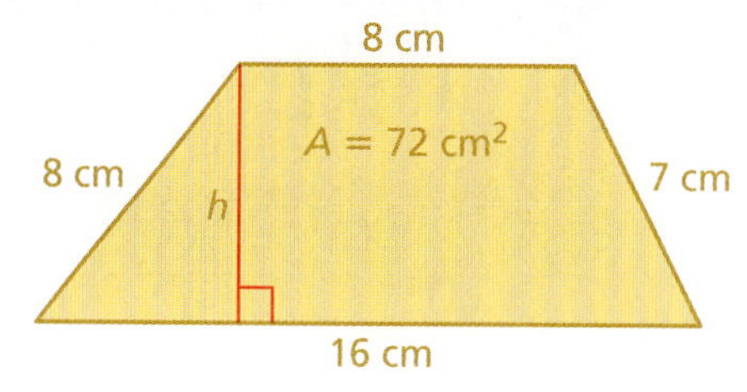

1 Chapter Test

Solve the equation. Check your solution, if possible.

1. $4 + y = 9.5$

2. $-\frac{x}{9} = -8$

3. $z - \frac{2}{3} = \frac{1}{8}$

4. $3.8n - 13 = 1.4n + 5$

5. $9(8d - 5) + 13 = 12d - 2$

6. $9j - 8 = 8 + 9j$

7. $2.5(2p + 5) = 5p + 12.5$

8. $\frac{3}{4}t + \frac{1}{8} = \frac{3}{4}(t + 8)$

9. $\frac{1}{7}(14r + 28) = 2(r + 2)$

Find the value of *x*. Then find the angle measures of the polygon.

10.

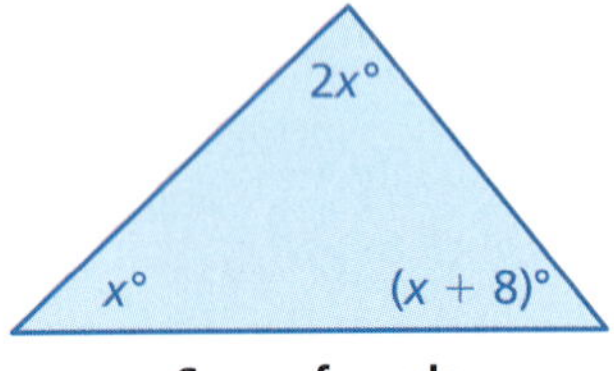

Sum of angle measures: 180°

11.

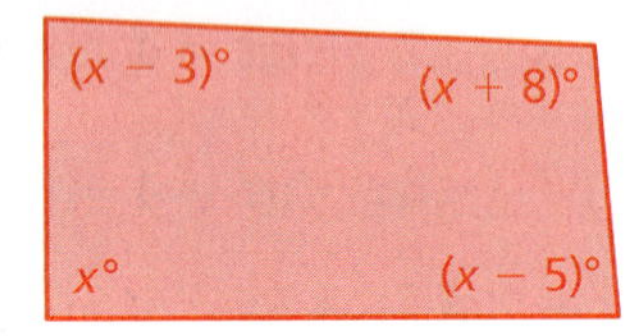

Sum of angle measures: 360°

Solve the equation for *y*.

12. $1.2x - 4y = 28$

13. $0.5 = 0.4y - 0.25x$

Solve the formula for the red variable.

14. Perimeter of a rectangle: $P = 2\ell + 2w$

15. Distance formula: $d = rt$

16. BASKETBALL Your basketball team wins a game by 13 points. The opposing team scores 72 points. Explain how to find your team's score.

17. CYCLING You are biking at a speed of 18 miles per hour. You are 3 miles behind your friend, who is biking at a speed of 12 miles per hour. Write and solve an equation to find the amount of time it takes for you to catch up to your friend.

18. VOLCANOES Two scientists are measuring lava temperatures. One scientist records a temperature of 1725°F. The other scientist records a temperature of 950°C. Which is the greater temperature? $\left(\text{Use } C = \frac{5}{9}(F - 32).\right)$

19. JOBS Your profit for mowing lawns this week is \$24. You are paid \$8 per hour and you paid \$40 for gas for the lawn mower. How many hours did you work this week?

Test Item References

Chapter Test Questions	Section to Review	Florida Common Core Standards (MACC)
1–3, 16	1.1	8.EE.3.7a, 8.EE.3.7b
10, 11, 19	1.2	8.EE.3.7a, 8.EE.3.7b
4–9, 17	1.3	8.EE.3.7a, 8.EE.3.7b
12–15, 18	1.4	8.EE.3.7

Test-Taking Strategies

Remind students to quickly look over the entire test before they start so that they can budget their time. When working with equations, students need to write all numbers and variables clearly, line up terms in each step, and not crowd their work. Have students use the **Stop** and **Think** strategy.

Common Errors

- **Exercises 1–9** Students may perform the same operation that they are trying to undo instead of the inverse operation. Remind them that they must use an inverse operation to undo an operation. Also, remind students to check their answers in the original equation, if possible.
- **Exercises 5 and 7–9** When using the Distributive Property, students may forget to distribute to all the values within the parentheses. Remind them that they need to distribute to all the values and encourage them to draw arrows showing the distribution, if needed.
- **Exercises 6–9** Students may end up with an equivalent equation such as $5p + 12.5 = 5p + 12.5$ or $\frac{1}{8} = 6$ and get confused about how to state their final answers. Remind students that an equation can have no solution or infinitely many solutions.
- **Exercises 10 and 11** Students may change the exponent of the variable when combining like terms that have variables. For example, they may write the sum $2x + x + (x + 8)$ as $4x^3 + 8$. Remind them how to correctly combine like terms that have variables.
- **Exercises 12–15 and 18** Students may be unsure about how to solve for the specified variable. Point out that they should work through the order of operations *backwards,* using inverse operations to isolate the variable.

Reteaching and Enrichment Strategies

If students need help. . .	If students got it. . .
Resources by Chapter • Practice A and Practice B • Puzzle Time Record and Practice Journal Practice Differentiating the Lesson Lesson Tutorials *BigIdeasMath.com* Skills Review Handbook	Resources by Chapter • Enrichment and Extension • Technology Connection Game Closet at *BigIdeasMath.com* Start Standards Assessment

Answers

1. $y = 5.5$
2. $x = 72$
3. $z = \frac{19}{24}$
4. $n = 7.5$
5. $d = 0.5$
6. no solution
7. infinitely many solutions
8. no solution
9. infinitely many solutions
10. $x = 43$; $43°, 86°, 51°$
11. $x = 90$; $90°, 87°, 98°, 85°$
12. $y = 0.3x - 7$
13. $y = 0.625x + 1.25$
14. $w = \frac{P}{2} - \ell$
15. $r = \frac{d}{t}$
16. *Sample answer:* Write and solve the equation $x - 13 = 72$; $x = 85$ points
17. $18x = 12x + 3$; $\frac{1}{2}$ hour
18. 950° C
19. 8 hours

Technology for the Teacher

Online Assessment
Assessment Book
ExamView® Assessment Suite

Test-Taking Strategies

Available at *BigIdeasMath.com*

After Answering Easy Questions, Relax
Answer Easy Questions First
Estimate the Answer
Read All Choices before Answering
Read Question before Answering
Solve Directly or Eliminate Choices
Solve Problem before Looking at Choices
Use Intelligent Guessing
Work Backwards

About this Strategy

When taking a multiple choice test, be sure to read each question carefully and thoroughly. Before answering a question, determine exactly what is being asked, then eliminate the wrong answers and select the best choice.

Answers

1. A
2. I
3. B
4. 12
5. G

Technology for the Teacher

Florida Common Core Standards Support
Performance Tasks
Online Assessment
Assessment Book
ExamView® Assessment Suite

Item Analysis

1. **A.** Correct answer
   B. The student subtracts 4 from 32 instead of dividing 32 by 4.
   C. The student adds 4 to 32 instead of dividing 32 by 4.
   D. The student multiplies 4 and 32 instead of dividing 32 by 4.

2. **F.** The student correctly subtracts 3 from 39, but then multiplies instead of dividing.
   G. The student correctly subtracts 3 from 39, but then subtracts 2 instead of dividing.
   H. The student incorrectly adds 3 and 39 instead of subtracting, then performs division correctly.
   I. Correct answer

3. **A.** The student makes an operation error when solving for the variable.
   B. Correct answer
   C. The student makes an operation error when solving for the variable.
   D. The student makes an operation error when solving for the variable.

4. **Gridded Response:** Correct answer: 12 units

   Common Error: The student correctly determines that $x = 3$, but states this as the final answer without finding the side lengths of the square.

5. **F.** The student moves r over in the equation instead of dividing both sides by r.
   G. Correct answer
   H. The student subtracts r instead of dividing by r.
   I. The student divides r by d and moves t to the other side of the equal sign instead of dividing d by r.

Standards Assessment Icons

Gridded Response

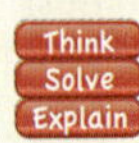

Short Response (2-point rubric)

Extended Response (4-point rubric)

1 Standards Assessment

1. Which value of x makes the equation true? *(MACC.8.EE.3.7b)*

$$4x = 32$$

A. 8 **C.** 36

B. 28 **D.** 128

2. A taxi ride costs \$3 plus \$2 for each mile driven. When you rode in a taxi, the total cost was \$39. This can be modeled by the equation below, where m represents the number of miles driven.

$$2m + 3 = 39$$

How long was your taxi ride? *(MACC.8.EE.3.7b)*

F. 72 mi **H.** 21 mi

G. 34 mi **I.** 18 mi

Test-Taking Strategy

Solve Directly or Eliminate Choices

When a cat wakes up, it's grumpy for x hours, where $2x-5x=x-4$. What's x?

(A) 0 (B) 1 (C) 2 (D) -3

Don't talk to me until I've had my morning milk.

"You can eliminate A and D. Then, solve directly to determine that the correct answer is B."

3. Which of the following equations has exactly one solution? *(MACC.8.EE.3.7a)*

A. $\frac{2}{3}(x + 6) = \frac{2}{3}x + 4$

B. $\frac{3}{7}y + 13 = 13 - \frac{3}{7}y$

C. $\frac{4}{5}\left(n + \frac{1}{3}\right) = \frac{4}{5}n + \frac{1}{3}$

D. $\frac{7}{8}\left(2t + \frac{1}{8}\right) = \frac{7}{4}t$

4. The perimeter of the square is equal to the perimeter of the triangle. What are the side lengths of the square? *(MACC.8.EE.3.7a)*

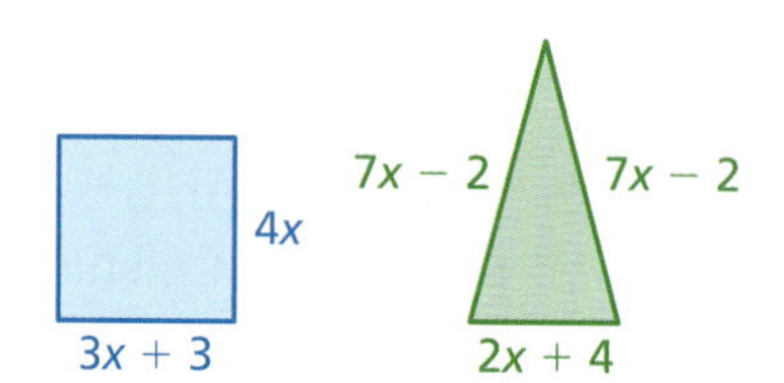

5. The formula below relates distance, rate, and time.

$$d = rt$$

Solve this formula for t. *(MACC.8.EE.3.7b)*

F. $t = dr$

G. $t = \frac{d}{r}$

H. $t = d - r$

I. $t = \frac{r}{d}$

6. What could be the first step to solve the equation shown below? *(MACC.8.EE.3.7b)*

$$3x + 5 = 2(x + 7)$$

A. Combine $3x$ and 5.

B. Multiply x by 2 and 7 by 2.

C. Subtract x from $3x$.

D. Subtract 5 from 7.

7. You work as a sales representative. You earn $400 per week plus 5% of your total sales for the week. *(MACC.8.EE.3.7b)*

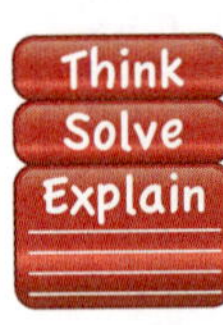

Part A Last week, you had total sales of $5000. Find your total earnings. Show your work.

Part B One week, you earned $1350. Let s represent your total sales that week. Write an equation that you could use to find s.

Part C Using your equation from Part B, find s. Show all steps clearly.

8. In 10 years, Maria will be 39 years old. Let m represent Maria's age today. Which equation can you use to find m? *(MACC.8.EE.3.7b)*

F. $m = 39 + 10$

G. $m - 10 = 39$

H. $m + 10 = 39$

I. $10m = 39$

9. Which value of y makes the equation below true? *(MACC.8.EE.3.7b)*

$$3y + 8 = 7y + 11$$

A. -4.75

B. -0.75

C. 0.75

D. 4.75

10. The equation below is used to convert a Fahrenheit temperature F to its equivalent Celsius temperature C.

$$C = \frac{5}{9}(F - 32)$$

Which formula can be used to convert a Celsius temperature to its equivalent Fahrenheit temperature? *(MACC.8.EE.3.7b)*

F. $F = \frac{5}{9}(C - 32)$

G. $F = \frac{9}{5}(C + 32)$

H. $F = \frac{9}{5}C + \frac{32}{5}$

I. $F = \frac{9}{5}C + 32$

Item Analysis (continued)

6. **A.** The student misunderstands that $3x$ and 5 are not like terms.

   B. Correct answer

   C. The student does not realize that the Distributive Property must first be used to multiply 2 by $x + 7$.

   D. The student does not realize that the Distributive Property must first be used to multiply 2 by $x + 7$.

7. **4 points** The student demonstrates a thorough understanding of evaluating expressions, writing equations, and solving equations, and presents his or her steps clearly. The following answers should be obtained: Part A: \$650; Part B: $0.05s + 400 = 1350$; Part C: \$19,000.

   3 points The student demonstrates an essential but less than thorough understanding. In particular, the correct equation or its equivalent should be given in Part B, but an arithmetic error may have been performed in Part C.

   2 points The student demonstrates a partial understanding of the processes of writing and solving equations. Part A should be correctly completed, but the equation in Part B may be written incorrectly. Alternatively, the correct equation could be written in Part B, but Part C might display misunderstanding of how to proceed.

   1 point The student demonstrates a limited understanding of equation writing and solving, as well as working with percents. The student's response is incomplete and exhibits many flaws.

   0 points The student provided no response, a completely incorrect or incomprehensible response, or a response that demonstrates insufficient understanding of percents and equations.

8. **F.** The student misunderstands the problem and decides to add the two numbers together.

   G. The student gets the idea that m and 39 are 10 apart, but chose subtraction instead of addition to relate them.

   H. Correct answer

   I. The student mistakes $10m$ for $10 + m$.

9. **A.** The student adds 8 and 11 instead of subtracting, and either misplaces a negative sign when subtracting $3x$ and $7x$ or performs a sign error when dividing.

   B. Correct answer

   C. The student subtracts 8 and 11 correctly, but then misplaces a negative sign when subtracting $3x$ and $7x$ or performs a sign error when dividing.

   D. The student adds 8 and 11 instead of subtracting.

10. **F.** The student simply interchanges the variables.

    G. The student "inverts" the variables, the fraction, and the subtraction.

    H. The student correctly multiplies both sides of the equation by $\frac{9}{5}$, but also incorrectly applies operations to 32.

    I. Correct answer

Answers

6. B
7. *Part A* \$650

   Part B $0.05s + 400 = 1350$

   Part C \$19,000
8. H
9. B
10. I

Answers

11. 14 weeks

12. B

13. I

14. A

Item Analysis (continued)

11. Gridded Response: Correct answer: 14 weeks

Common Error: The student adds 35 and 175 to get 210, then divides by 10 to get 21 as an answer.

12. A. The student finds x.

B. Correct answer

C. The student solves $5x = 180$ and uses $3x$ as the greatest angle measure.

D. The student solves $5x = 180 + 50$ and uses $3x$ as the greatest angle measure.

13. F. The student distributes correctly but then makes a mistake combining the constant terms, yielding $2x = -11$.

G. The student does not distribute the left side of the equation correctly, yielding $6x - 3$.

H. The student combines $6x$ and $4x$ incorrectly, yielding $10x$ instead of $2x$.

I. Correct answer

14. A. Correct answer

B. The student incorrectly uses the fact that there are 4 items on one side and 2 on the other to get the ratio $\frac{2}{4} = \frac{1}{2}$.

C. The student incorrectly uses the fact that there are 4 items on one side and 2 on the other to get the ratio $\frac{2}{4} = \frac{1}{2}$, and then misuses the order of the ratio.

D. The student gets the correct ratio of $\frac{1}{3}$, but misuses it.

11. You have already saved \$35 for a new cell phone. You need \$175 in all. You think you can save \$10 per week. At this rate, how many more weeks will you need to save money before you can buy the new cell phone? *(MACC.8.EE.3.7b)*

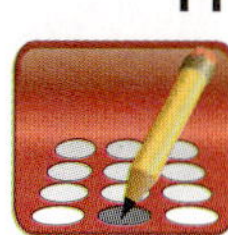

12. What is the greatest angle measure in the triangle below? *(MACC.8.EE.3.7b)*

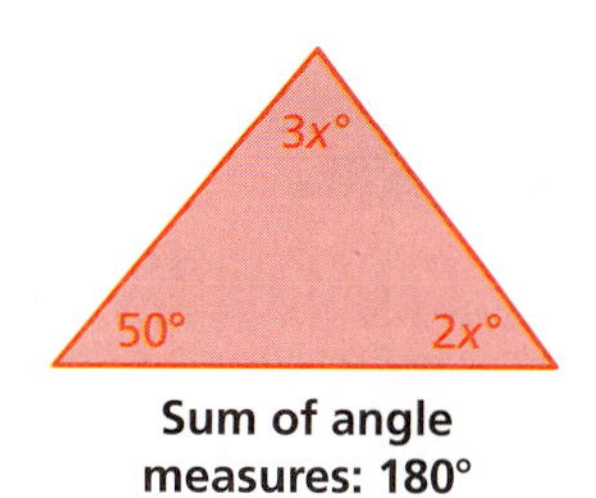

A. 26°

B. 78°

C. 108°

D. 138°

13. Which value of x makes the equation below true? *(MACC.8.EE.3.7b)*

$$6(x - 3) = 4x - 7$$

F. -5.5

G. -2

H. 1.1

I. 5.5

14. The drawing below shows equal weights on two sides of a balance scale.

What can you conclude from the drawing? *(MACC.8.EE.3.7b)*

A. A mug weighs one-third as much as a trophy.

B. A mug weighs one-half as much as a trophy.

C. A mug weighs twice as much as a trophy.

D. A mug weighs three times as much as a trophy.

2 Transformations

"Just 2 more minutes. I'm almost done with my 'cat tessellation' painting."

"If you hold perfectly still..."

"...each frame becomes a horizontal..."

"...translation of the previous frame..."

Florida Common Core Progression

6th Grade

- Draw polygons in the coordinate plane given vertices and find lengths of sides whose endpoints have the same x- or y-coordinate.
- Find areas of triangles and special quadrilaterals.
- Understand ratios and describe ratio relationships.

7th Grade

- Draw geometric shapes with given conditions.
- Represent proportional relationships with equations.
- Find unit rates associated with ratios of perimeters and areas.
- Use proportionality to solve ratio problems.
- Reproduce a scale drawing at a different scale.

8th Grade

- Verify the properties of translations, reflections, and rotations.
- Describe translations, reflections, rotations, and dilations using coordinates.
- Understand that figures are congruent (or similar) when they can be related by a sequence of translations, reflections, and rotations (and dilations).
- Describe a sequence that exhibits congruence or similarity between two figures.

Chapter Summary

Section	Florida Common Core Standard (MACC)	
2.1	Preparing for	8.G.1.2
2.2	Learning	8.G.1.1, 8.G.1.2, 8.G.1.3
2.3	Learning	8.G.1.1, 8.G.1.2, 8.G.1.3
2.4	Learning	8.G.1.1 ★, 8.G.1.2 ★, 8.G.1.3
2.5	Preparing for	8.G.1.4
2.6	Preparing for	8.G.1.4
2.7	Learning	8.G.1.3 ★, 8.G.1.4 ★

★ Teaching is complete. Standard can be assessed.

Pacing Guide for Chapter 2

Chapter Opener	1 Day
Section 1	
Activity	1 Day
Lesson	1 Day
Section 2	
Activity	1 Day
Lesson	1 Day
Section 3	
Activity	1 Day
Lesson	1 Day
Section 4	
Activity	1 Day
Lesson	1 Day
Study Help / Quiz	1 Day
Section 5	
Activity	1 Day
Lesson	1 Day
Section 6	
Activity	1 Day
Lesson	1 Day
Section 7	
Activity	1 Day
Lesson	2 Days
Chapter Review/ Chapter Tests	2 Days
Total Chapter 2	19 Days
Year-to-Date	32 Days

Technology for the Teacher

BigIdeasMath.com
Chapter at a Glance
Complete Materials List
Parent Letters: English, Spanish, and Haitian Creole

Florida Common Core Standards

MACC.6.NS.3.6b Understand signs of numbers in ordered pairs as indicating locations in quadrants of the coordinate plane; recognize that when two ordered pairs differ only by signs, the locations of the points are related by reflections across one or both axes.

MACC.6.G.1.3 Draw polygons in the coordinate plane given coordinates for the vertices; use coordinates to find the length of a side joining points with the same first coordinate or the same second coordinate. Apply these techniques in the context of solving real-world and mathematical problems.

Additional Topics for Review

- Geoboard
- Congruent Angles
- Congruent Sides
- Right Angle
- Perimeter
- Area
- Vertical
- Horizontal
- Quadrants
- Clockwise
- Counterclockwise
- Ratio
- Proportion

Try It Yourself

1. **a.** $(7, -3)$ **b.** $(-7, 3)$
2. **a.** $(-4, -6)$ **b.** $(4, 6)$
3. **a.** $(5, 5)$ **b.** $(-5, -5)$
4. **a.** $(-8, 3)$ **b.** $(8, -3)$

5–10. See Additional Answers.

Record and Practice Journal Fair Game Review

1. $(1, -1)$; $(-1, 1)$
2. $(-2, 4)$; $(2, -4)$
3. $(-3, -3)$; $(3, 3)$
4. $(4, 3)$; $(-4, -3)$

5–12. See Additional Answers.

Math Background Notes

Vocabulary Review

- Reflection
- Coordinate Plane
- Ordered Pair
- *x*-Axis
- *y*-Axis
- *x*-Coordinate
- *y*-Coordinate
- Opposite
- Polygon
- Quadrilateral
- Triangle
- Vertices

Reflecting Points

- Students should have studied reflecting points in Grade 6. If you sense that a review is necessary, then go over Example 1 in detail.
- You may also want to expand upon Example 1 and reflect the point in the *y*-axis as well as in both axes. When reflecting a second time, however, be sure that students use the reflected point and not the original point.
- If you choose to review reflection in both axes, then an excellent follow-up question would be whether reflection in the *x*-axis followed by reflection in the *y*-axis produces a different result than reflection in the *y*-axis followed by reflection in the *x*-axis. (The answer is no; the end result is the same.)
- **Common Error:** Students may confuse reflection in the *x*-axis with reflection in the *y*-axis (and vice versa). This is typically because algebraically, the coordinate substitutions necessary to come up with a reflected point do not seem natural to students. To reflect in the *x*-axis, you replace the *y*-coordinate with its opposite (and vice versa).
- To overcome confusion, have students actually plot the point as well as its reflection. Discourage rote memorization of algebraic procedures.

Drawing a Polygon in a Coordinate Plane

- Students have been studying polygons since elementary school. Also, students should be familiar with drawing a polygon in a coordinate plane from Grade 6.
- You may want to expand upon Example 2 and the exercises that follow and ask students to classify the polygons. As recently as Grade 7, they should have classified triangles and reviewed classifying quadrilaterals.
- Students may mistakenly classify the quadrilateral in Example 2 as a kite, because it may appear to them to have two pairs of congruent adjacent sides. In fact, there is only one pair of congruent adjacent sides, so the polygon cannot be classified any further than "quadrilateral." Students do not learn the distance formula until Chapter 7, so the best that students can do at this point is to use a ruler to measure the side lengths. The take-away lesson here is that students should not "eyeball" a polygon to come up with a classification.
- **FYI:** The triangle in Exercise 9 is obtuse scalene. The quadrilateral in Exercise 10 is a square.

Reteaching and Enrichment Strategies

If students need help...	If students got it...
Record and Practice Journal • Fair Game Review Skills Review Handbook Lesson Tutorials	Game Closet at *BigIdeasMath.com* Start the next section

What You Learned Before

"Did you know that when you look at yourself in the mirror, your left and right get switched?"

Reflecting Points (MACC.6.NS.3.6b)

Example 1 **Reflect (3, −4) in the *x*-axis.**

Plot (3, −4).

To reflect (3, −4) in the *x*-axis, use the same *x*-coordinate, 3, and take the opposite of the *y*-coordinate. The opposite of −4 is 4.

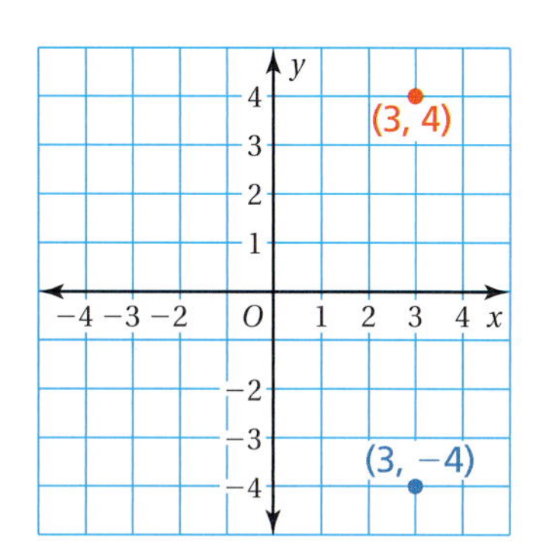

∴ So, the reflection of (3, −4) in the *x*-axis is (3, 4).

Try It Yourself

Reflect the point in (a) the *x*-axis and (b) the *y*-axis.

1. (7, 3) **2.** (−4, 6) **3.** (5, −5) **4.** (−8, −3)

5. (0, 1) **6.** (−5, 0) **7.** (4, −6.5) **8.** $\left(-3\frac{1}{2}, -4\right)$

Drawing a Polygon in a Coordinate Plane (MACC.6.G.1.3)

Example 2 **The vertices of a quadrilateral are *A*(1, 5), *B*(2, 9), *C*(6, 8), and *D*(8, 1). Draw the quadrilateral in a coordinate plane.**

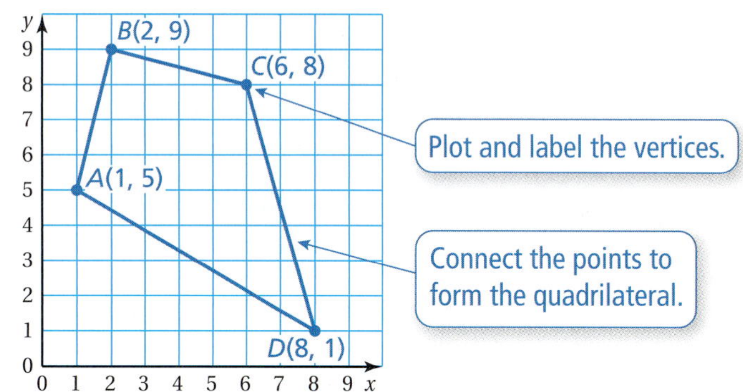

Try It Yourself

Draw the polygon with the given vertices in a coordinate plane.

9. *J*(1, 1), *K*(5, 6), *M*(9, 3) **10.** *Q*(2, 3), *R*(2, 8), *S*(7, 8), *T*(7, 3)

2.1 Congruent Figures

Essential Question How can you identify congruent triangles?

Two figures are congruent when they have the same size and the same shape.

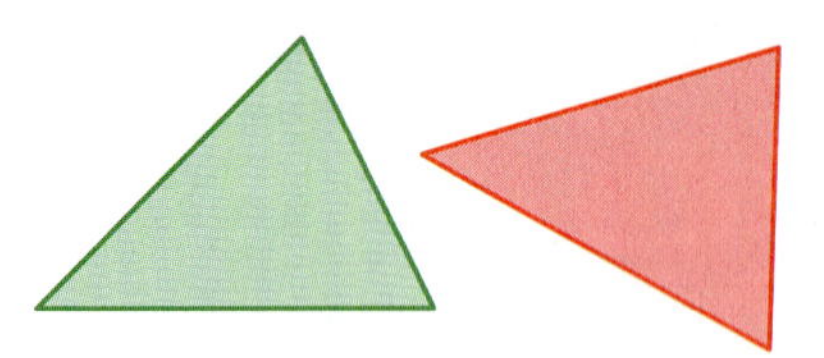

Congruent
Same size *and* shape

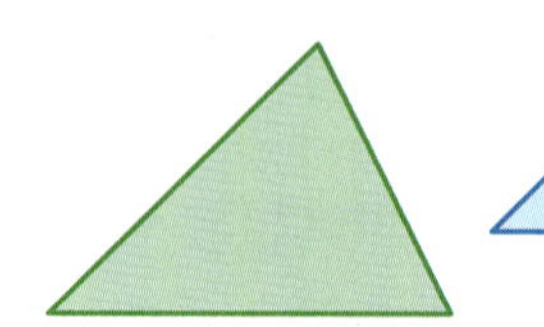

Not Congruent
Same shape, but not same size

ACTIVITY: Identifying Congruent Triangles

Work with a partner.

- **Which of the geoboard triangles below are congruent to the geoboard triangle at the right?**
- **Form each triangle on a geoboard.**
- **Measure each side with a ruler. Record your results in a table.**
- **Write a conclusion about the side lengths of triangles that are congruent.**

a.

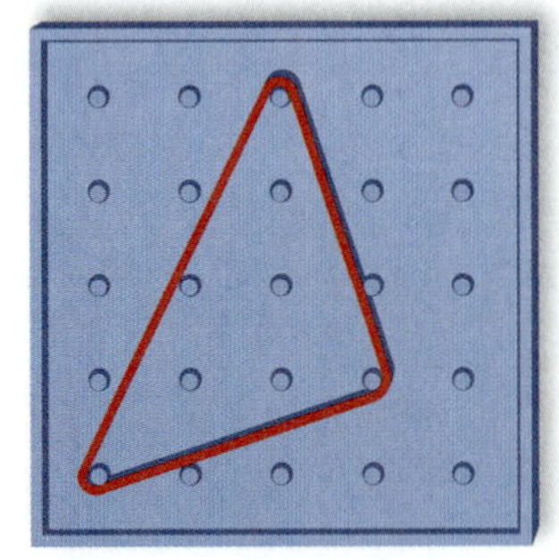

b.

c.

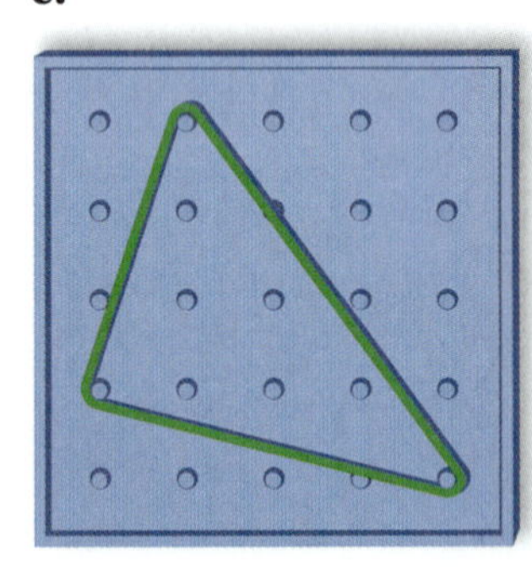

d.

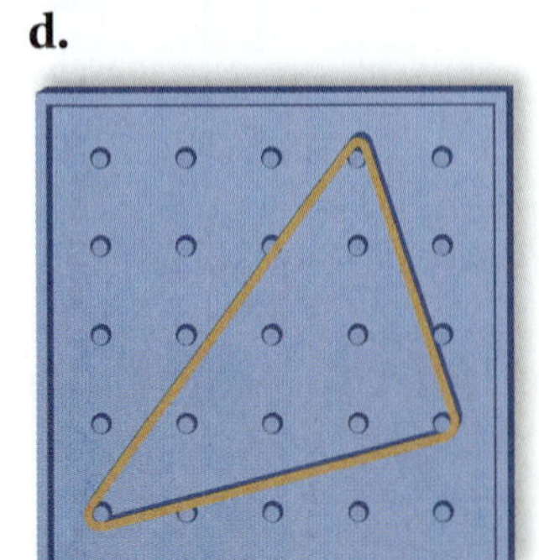

e.

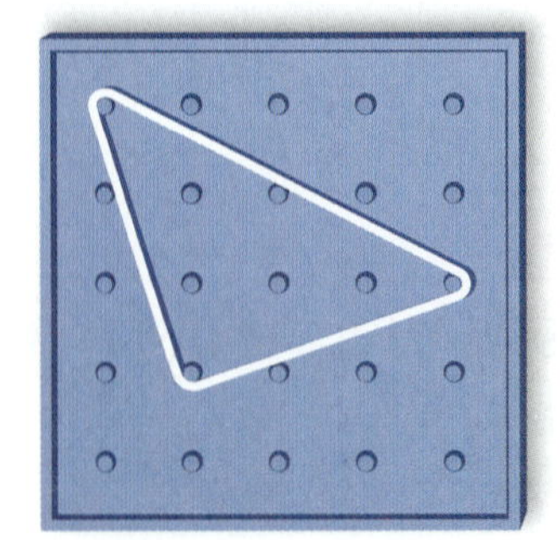

f.

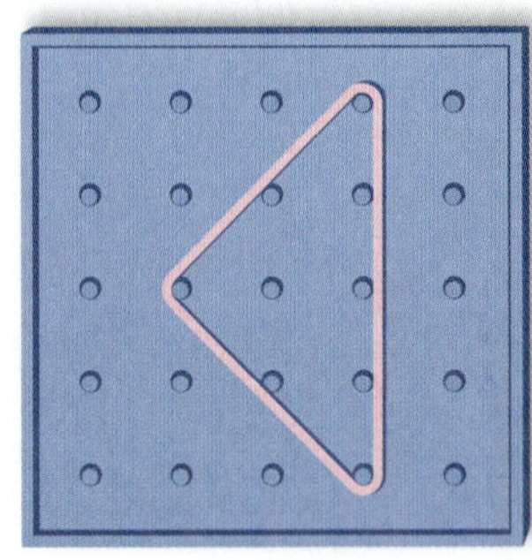

COMMON CORE

Geometry

In this lesson, you will

- name corresponding angles and corresponding sides of congruent figures.
- identify congruent figures.

Preparing for Standard MACC.8.G.1.2

Laurie's Notes

Introduction

Standards for Mathematical Practice

- **MP5 Use Appropriate Tools Strategically:** Geoboards allow students to change the shape or orientation of a figure with little effort. There is no erasing to do! When you change the location of a vertex, the two sides meeting at that vertex change automatically to meet at the new vertex.

Motivate

- ? "What word is used to describe figures that have the same size and shape?" congruent
- **MP5:** Display a variety of objects such as templates, stencils, rubber stamps, and cookie cutters.
- ? "What do all of these objects have in common?" They can be used to create figures that are congruent to one another.
- ? "Can you think of other things that can be used to create congruent figures?" *Sample answer:* computer graphics program, clay molds
- Use one of the objects to create a few congruent figures.
- "Today you will use geoboards to decide whether triangles are congruent."

Activity Notes

Activity 1

- **Classroom Management:** If this is the first time your students have used geoboards, give them time to explore and play. I hand out the geoboards with two rubber bands arranged vertically and two arranged horizontally, and students are to return them the same way. If your geoboards are a different size than the 5×5 shown, don't worry. Your students will adjust. Also, you can use geoboard dot paper in place of geoboards.
- **Common Misconception:** Measuring between two pins horizontally or vertically is *not* the same as measuring across a diagonal. The distance between two diagonal pins is longer.
- Have students use their rulers to measure each side in millimeters. Recommend that they try to measure from the center of one pin to the center of another pin.
- Students with good spatial skills may be able to rotate or reflect some triangles, so they may not want to measure each side. Encourage students to leave the triangles in the orientations given and measure each side.
- Depending upon the accuracy of their measurements, students may believe that all of the triangles are congruent.
- The yellow triangle has two congruent sides. The triangles in parts (b) and (f) also have two congruent sides but they are not congruent to the yellow triangle.
- Have students share their results and observations. If they use vocabulary such as *isosceles* and *scalene* to describe the triangles, ask them to explain what these words mean. None of the triangles are equilateral.

Florida Common Core Standards

MACC.8.G.1.2 Understand that a two-dimensional figure is congruent to another if the second can be obtained from the first by a sequence of rotations, reflections, and translations; . . .

Previous Learning

Students should know how to analyze and compare angles and segments.

Technology for the Teacher

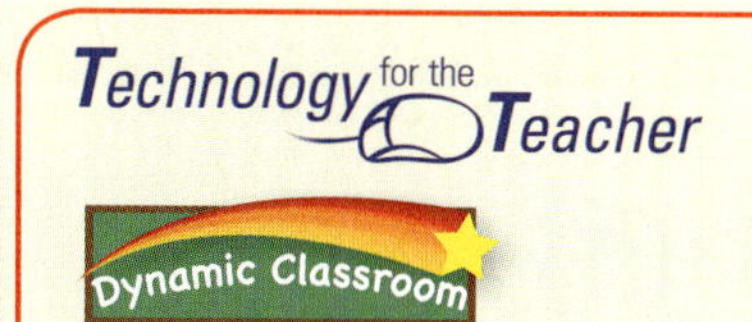

Lesson Plans
Complete Materials List

2.1 Record and Practice Journal

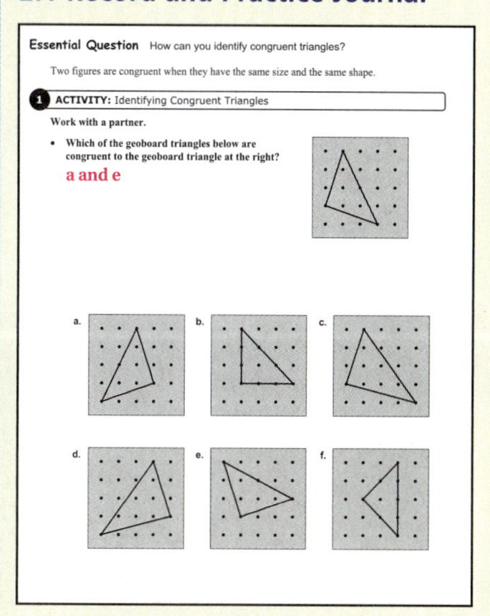

Essential Question How can you identify congruent triangles?

Two figures are congruent when they have the same size and the same shape.

1 ACTIVITY: Identifying Congruent Triangles

Work with a partner.

- Which of the geoboard triangles below are congruent to the geoboard triangle at the right?

a and e

a. b. c.

d. e. f.

Differentiated Instruction

Kinesthetic

Provide students with cutouts of congruent figures. Let them turn and flip the cutouts to discover that figures can be congruent even though they may have different orientations.

2.1 Record and Practice Journal

- Form each triangle on a geoboard.

  Check students' work.

- Measure each side with a ruler. Record your results in the table.

Sample answer:

	Side 1	Side 2	Side 3
Given Triangle	126 mm	126 mm	180 mm
a.	126 mm	126 mm	180 mm
b.	120 mm	120 mm	170 mm
c.	165 mm	126 mm	200 mm
d.	165 mm	126 mm	200 mm
e.	126 mm	126 mm	180 mm
f.	115 mm	115 mm	160 mm

- Write a conclusion about the side lengths of triangles that are congruent.

  Sample answer: The corresponding sides of triangles that are congruent have the same length.

2 ACTIVITY: Forming Congruent Triangles

Work with a partner.

a. Form the given triangle in Activity 1 on your geoboard. Record the triangle on geoboard dot paper. See Additional Answers.

b. Move each vertex of the triangle one peg to the right. Is the new triangle congruent to the original triangle? How can you tell?

   yes; The new triangle is the same shape and size as the original triangle.

c. On a 5-by-5 geoboard, make as many different triangles as possible, each of which is congruent to the given triangle in Activity 1. Record each triangle on geoboard dot paper. See Additional Answers.

What Is Your Answer?

3. **IN YOUR OWN WORDS** How can you identify congruent triangles? Use the conclusion you wrote in Activity 1 as part of your answer.

   Congruent triangles are the same size and shape. The corresponding sides of congruent triangles have the same length.

4. Can you form a triangle on your geoboard whose side lengths are 3, 4, and 5 units? If so, draw such a triangle on geoboard dot paper.

   yes;

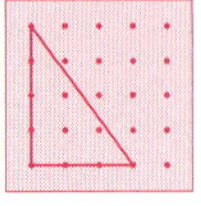

Laurie's Notes

Activity 2

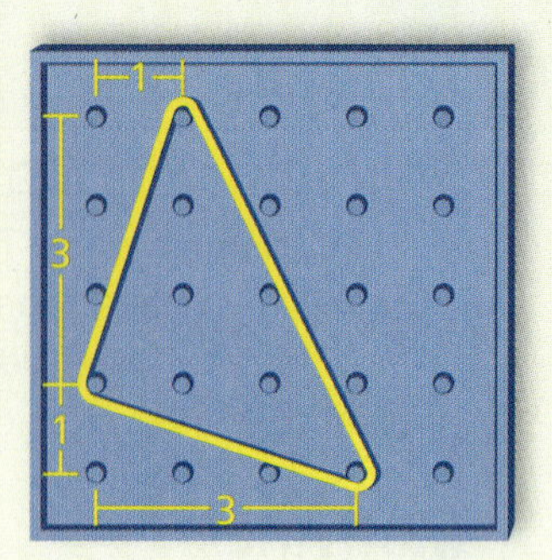

- In Activity 1, students concluded that the triangles in parts (a) and (e) are congruent to the yellow triangle and that each triangle has two congruent sides (the three triangles are isosceles).
- Point out that for each of these triangles, you can move three units (vertically or horizontally) and then one unit (horizontally or vertically) to get from one end of the congruent side to the other, as shown in the figure at the right.
- **MP1 Make Sense of Problems and Persevere in Solving Them:** To sketch additional triangles congruent to the yellow triangle, encourage students to consider different orientations.

What Is Your Answer?

- In answering Question 4, the units are not inches or centimeters, but *geoboard units*. This means that the vertical or horizontal distance between two pins is one geoboard unit. Determining the 3- and 4-unit side lengths is straightforward. Convincing yourself that the third side has a length of 5 units is less obvious. The triangle should be a 3-4-5 right triangle.
- Ask students to describe how they determined that the side lengths are 3, 4, and 5 units.

Closure

- Name 3 different examples of congruent figures or objects in the classroom. *Sample answer:* eraser, marker, chalkboard, bulletin board

Math Practice 5

Recognize Usefulness of Tools

What are some advantages and disadvantages of using a geoboard to construct congruent triangles?

The geoboard at the right shows three congruent triangles.

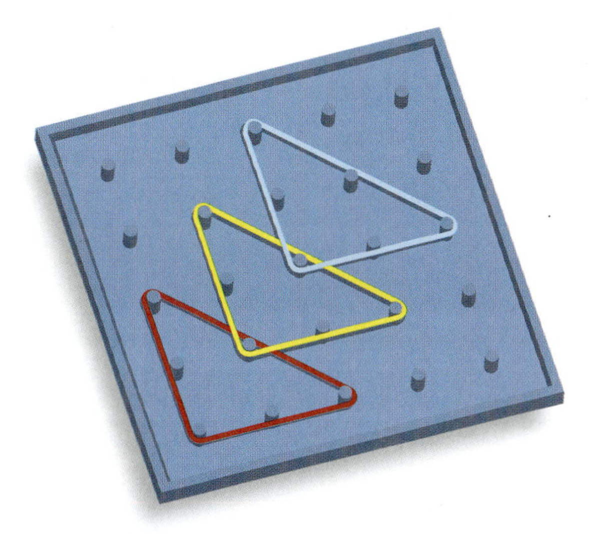

2 ACTIVITY: Forming Congruent Triangles

Work with a partner.

a. Form the yellow triangle in Activity 1 on your geoboard. Record the triangle on geoboard dot paper.

b. Move each vertex of the triangle one peg to the right. Is the new triangle congruent to the original triangle? How can you tell?

c. On a 5-by-5 geoboard, make as many different triangles as possible, each of which is congruent to the yellow triangle in Activity 1. Record each triangle on geoboard dot paper.

What Is Your Answer?

3. **IN YOUR OWN WORDS** How can you identify congruent triangles? Use the conclusion you wrote in Activity 1 as part of your answer.

4. Can you form a triangle on your geoboard whose side lengths are 3, 4, and 5 units? If so, draw such a triangle on geoboard dot paper.

Use what you learned about congruent triangles to complete Exercises 4 and 5 on page 46.

Key Vocabulary

congruent figures, p. 44

corresponding angles, p. 44

corresponding sides, p. 44

Key Idea

Congruent Figures

Figures that have the same size and the same shape are called **congruent figures**. The triangles below are congruent.

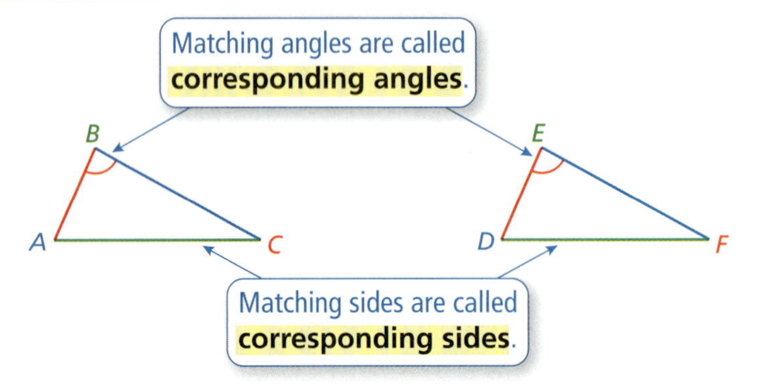

EXAMPLE 1 Naming Corresponding Parts

The figures are congruent. Name the corresponding angles and the corresponding sides.

Corresponding Angles	*Corresponding Sides*
$\angle A$ and $\angle W$	Side AB and Side WX
$\angle B$ and $\angle X$	Side BC and Side XY
$\angle C$ and $\angle Y$	Side CD and Side YZ
$\angle D$ and $\angle Z$	Side AD and Side WZ

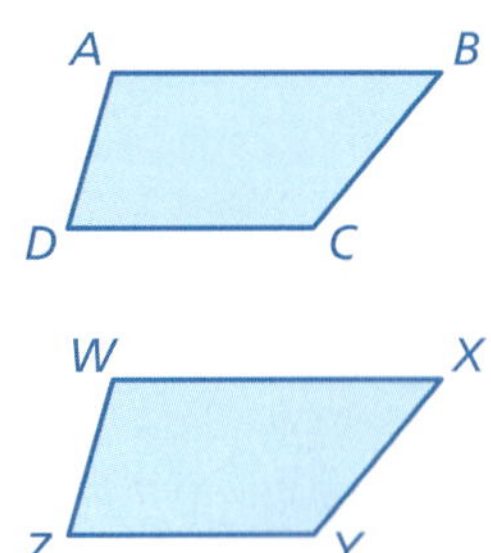

On Your Own

1. The figures are congruent. Name the corresponding angles and the corresponding sides.

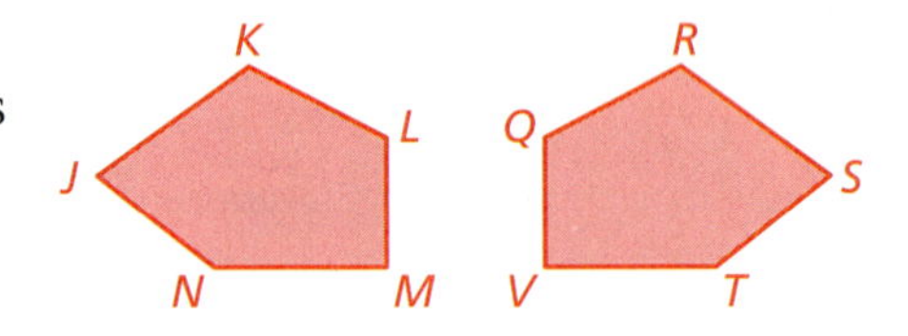

Key Idea

Identifying Congruent Figures

Two figures are congruent when corresponding angles and corresponding sides are congruent.

Triangle ABC is congruent to Triangle DEF.

$$\triangle ABC \cong \triangle DEF$$

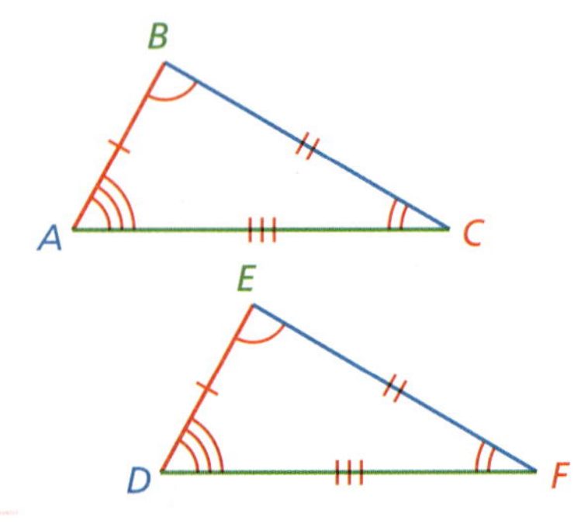

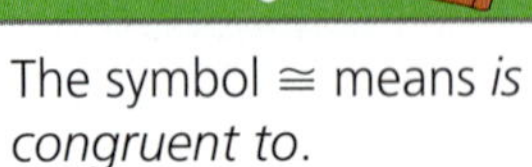

The symbol $\cong$ means *is congruent to*.

Laurie's Notes

Introduction

Connect

- **Yesterday:** Students explored congruent triangles by measuring side lengths. (MP1, MP5)
- **Today:** Students will use the definition of congruent figures to determine whether two figures (polygons) are congruent.

Motivate

- **Preparation:** Cut coffee stirrers or uncooked spaghetti into lengths of 3 inches and 5 inches.
- Ask students to pick up 3 pieces (either 2 short and 1 long or 2 long and 1 short). Have them make a triangle using the 3 pieces.
- ? "Is anyone's triangle congruent to yours? Explain." Students with the same number of long and short pieces will have congruent triangles.
- Ask two students with congruent triangles to make their triangles on the overhead or document camera. Trace the triangles on the board. Show how congruent angles and segments are conveyed by marks without using words.

Lesson Notes

Key Idea

- Use 3 different colors to draw the sides of the triangles that correspond (match).
- Students often think about corresponding parts in terms of the greatest/least angle measures and longest/shortest sides.

Example 1

- **Representation:** Have students use color-coding to show corresponding parts.
- **Notation:** Remind students that the angle symbol ($\angle$) is needed when talking about corresponding angles; they should not write the letter only.
- ? "Is Side *AB* the same as Side *BA*?" yes

On Your Own

- When students have finished, ask them to share their answers. Students should use proper terminology such as "Angle *K*" and "Side *JK*."

Key Idea

- **MP6 Attend to Precision:** Discuss the use of arcs and tick marks to indicate congruent angles and sides.
- This definition of congruent figures involves 2 parts, corresponding angles being congruent and corresponding sides being congruent. For a triangle, 6 congruencies must be established, for a quadrilateral, 8 congruencies.
- **FYI:** You can use the congruent symbol "$\cong$" to write angle and side congruency statements. For example, write the 6 congruency statements for the pair of triangles shown.
  Congruent angles: $\angle A \cong \angle D$, $\angle B \cong \angle E$, $\angle C \cong \angle F$
  Congruent sides: Side $AB \cong$ Side DE, Side $BC \cong$ Side EF, Side $AC \cong$ Side DF

Goal Today's lesson is defining **congruent figures** and determining if two figures are congruent.

Technology for the Teacher

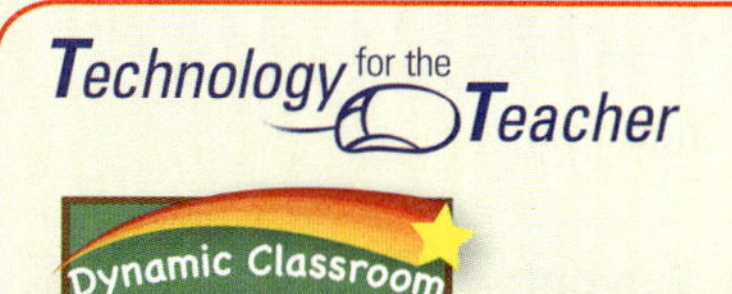

Lesson Tutorials
Lesson Plans
Answer Presentation Tool

Extra Example 1

The figures are congruent. Name the corresponding angles and corresponding sides.

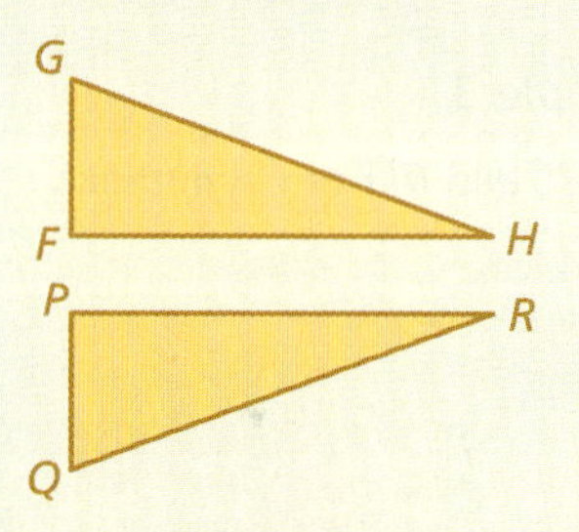

Corresponding angles: $\angle F$ and $\angle P$, $\angle G$ and $\angle Q$, $\angle H$ and $\angle R$; Corresponding sides: Side *FG* and Side *PQ*, Side *GH* and Side *QR*, Side *HF* and Side *RP*

On Your Own

1. Corresponding angles: $\angle J$ and $\angle S$, $\angle K$ and $\angle R$, $\angle L$ and $\angle Q$, $\angle M$ and $\angle V$, $\angle N$ and $\angle T$; Corresponding sides: Side *JK* and Side *SR*, Side *KL* and Side *RQ*, Side *LM* and Side *QV*, Side *MN* and Side *VT*, Side *NJ* and Side *TS*

English Language Learners

Symbols

Make sure that students understand the markings on congruent angles and sides. Use the figures in the *Key Ideas* to show which angles and sides are congruent.

Extra Example 2

Tell whether the two figures are congruent. Explain your reasoning.

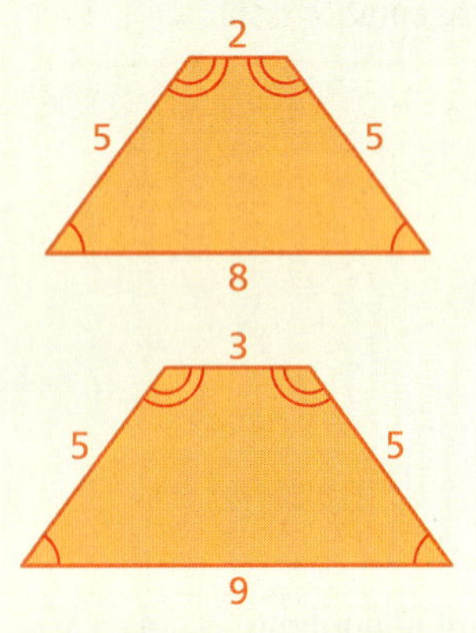

not congruent; The side lengths of the corresponding bases are not congruent.

Extra Example 3

Triangles *ABC* and *DEF* are congruent.

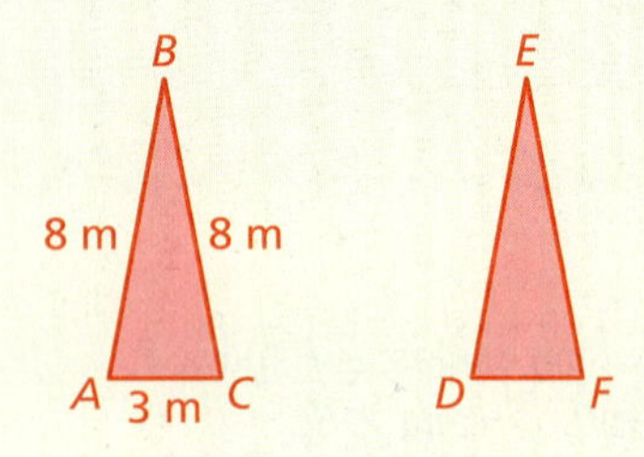

a. What is the length of Side *EF*? 8 m

b. What is the perimeter of *DEF*? 19 m

On Your Own

2. Square B
3. $\angle L$; 8 ft

Laurie's Notes

Example 2

- Work through the example. Point out that the orientation of the figure does not influence whether the two figures are congruent.
- **Common Error:** Students often refer to Square C as a diamond versus square.

Example 3

- Draw two congruent scalene triangles with different orientations. Label the triangles $\triangle PQR$ and $\triangle STU$.

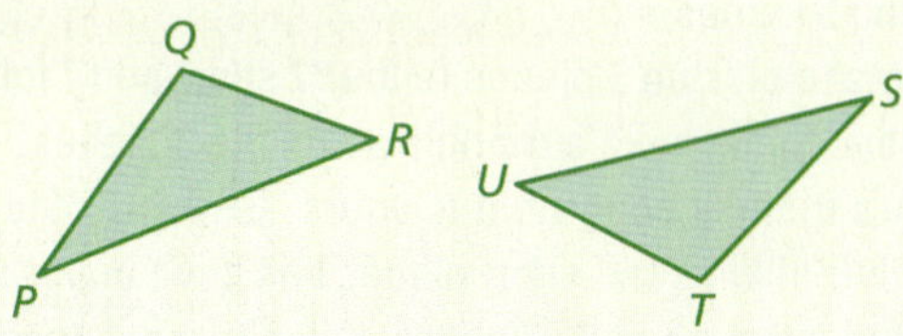

? "Which pairs of sides appear to be the same length?" Side *PQ* and Side *ST*, Side *QR* and Side *TU*, Side *RP* and Side *US* "Which pairs of angles appear to be the same measure?" $\angle P$ and $\angle S$, $\angle Q$ and $\angle T$, $\angle R$ and $\angle U$

- Students should always look at the figures and first determine if they are in the same orientation.

? "Are the two trapezoids in the same orientation?" yes

- Work through the example, as shown.

Discuss

- To preview transformations, ask students these questions.

? "How could you move Quadrilateral *ABCD* to get Quadrilateral *WXYZ* in Example 1?" Slide (move) the quadrilateral down.

? "What can you do to Pentagon *JKLMN* to get Pentagon *SRQVT* in On Your Own Question 1?" Reflect (flip) the pentagon in (or across) a line at the right of the pentagon to form a mirror image of the pentagon.

? "What can you do to Square A to get Square C in Example 2?" Rotate (turn) the square and slide it to the right.

On Your Own

- **Think-Pair-Share:** Students should read each question independently and then work in pairs to answer the questions. When they have answered the questions, the pair should compare their answers with another group and discuss any discrepancies.

Closure

- Draw a triangle and a quadrilateral. Exchange papers with your partner. Draw a figure congruent to your partner's figure. Then mark the corresponding sides and the corresponding angles of the congruent figures.

EXAMPLE 2 Identifying Congruent Figures

Which square is congruent to Square A?

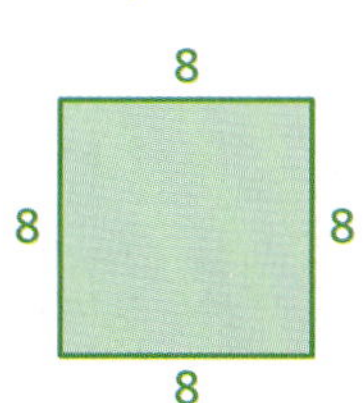

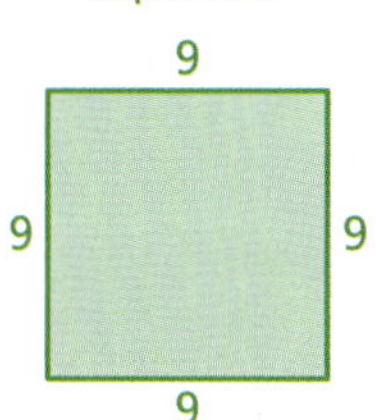

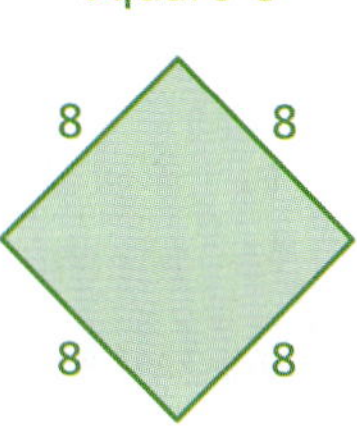

Each square has four right angles. So, corresponding angles are congruent. Check to see if corresponding sides are congruent.

Square A and Square B

Each side length of Square A is 8, and each side length of Square B is 9. So, corresponding sides are not congruent.

Square A and Square C

Each side length of Square A and Square C is 8. So, corresponding sides are congruent.

So, Square C is congruent to Square A.

EXAMPLE 3 Using Congruent Figures

Trapezoids *ABCD* and *JKLM* are congruent.

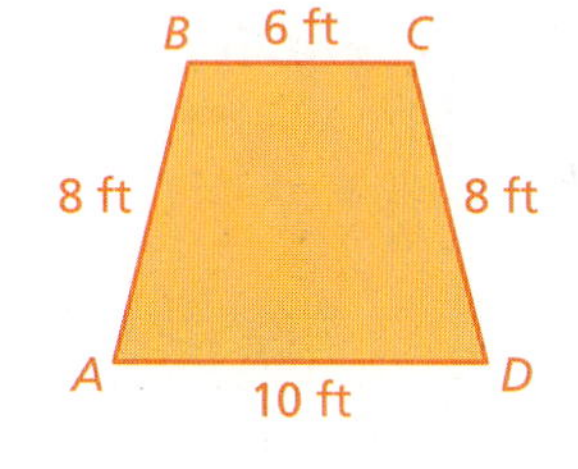

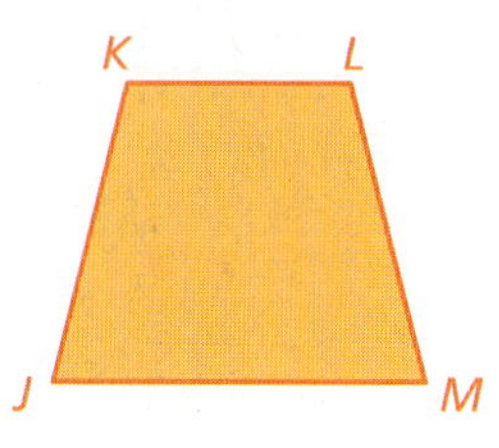

a. What is the length of side *JM*?

Side *JM* corresponds to side *AD*.

So, the length of side *JM* is 10 feet.

b. What is the perimeter of *JKLM*?

The perimeter of *ABCD* is $10 + 8 + 6 + 8 = 32$ feet. Because the trapezoids are congruent, their corresponding sides are congruent.

So, the perimeter of *JKLM* is also 32 feet.

On Your Own

Exercises 8, 9, and 12

2. Which square in Example 2 is congruent to Square D?

3. In Example 3, which angle of *JKLM* corresponds to $\angle C$? What is the length of side *KJ*?

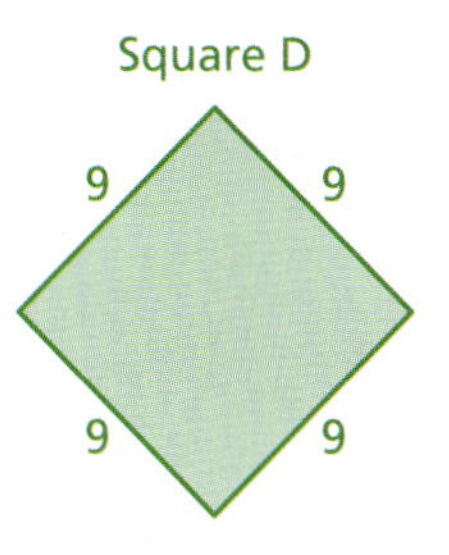

2.1 Exercises

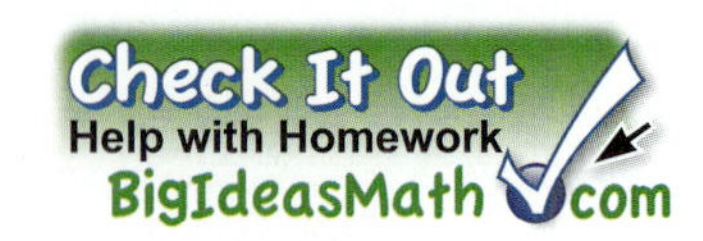

Vocabulary and Concept Check

1. **VOCABULARY** $\triangle ABC$ is congruent to $\triangle DEF$.

   a. Identify the corresponding angles.

   b. Identify the corresponding sides.

2. **VOCABULARY** Explain how you can tell that two figures are congruent.

3. **WHICH ONE DOESN'T BELONG?** Which one does *not* belong with the other three? Explain your reasoning.

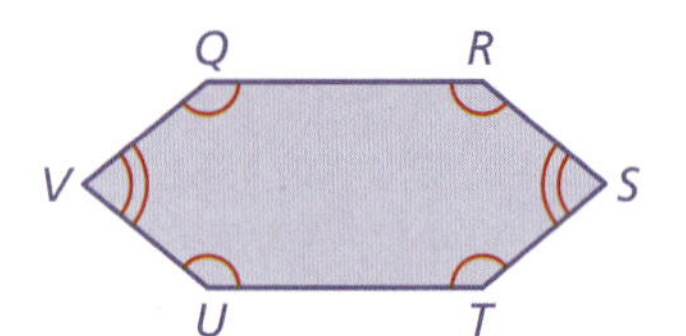

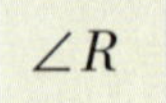

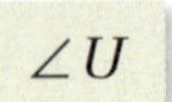

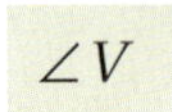

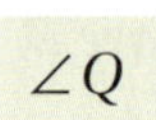

Practice and Problem Solving

Tell whether the triangles are *congruent* or *not congruent*.

4. 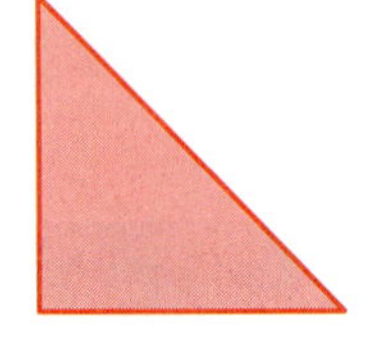 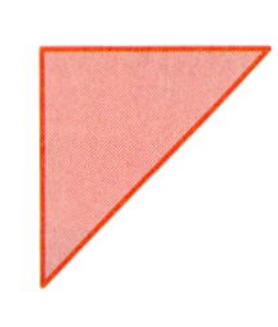

5. 

The figures are congruent. Name the corresponding angles and the corresponding sides.

6. 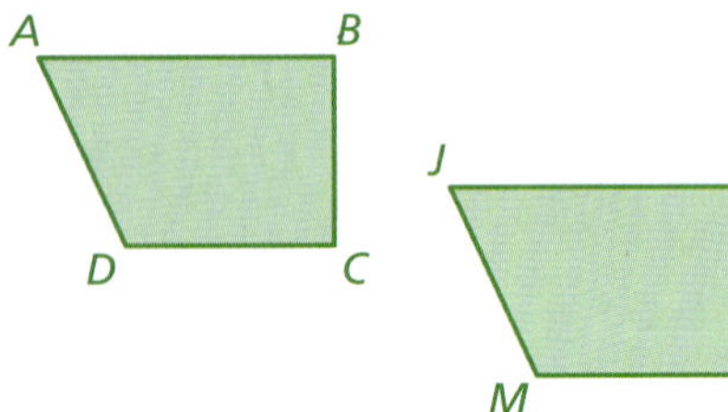

7. 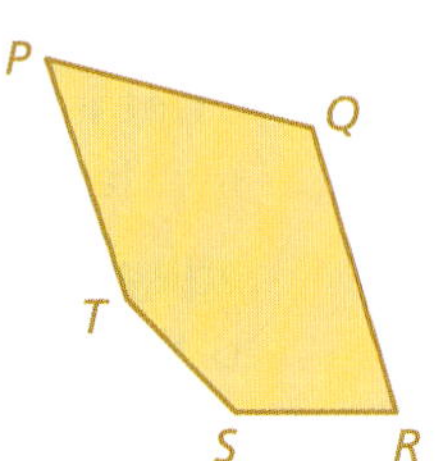

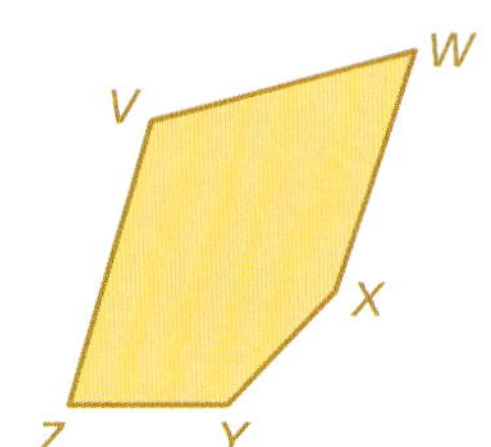

Tell whether the two figures are congruent. Explain your reasoning.

8. 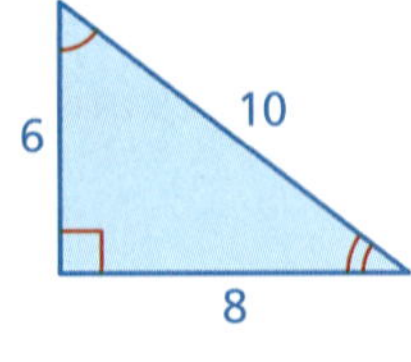

9. 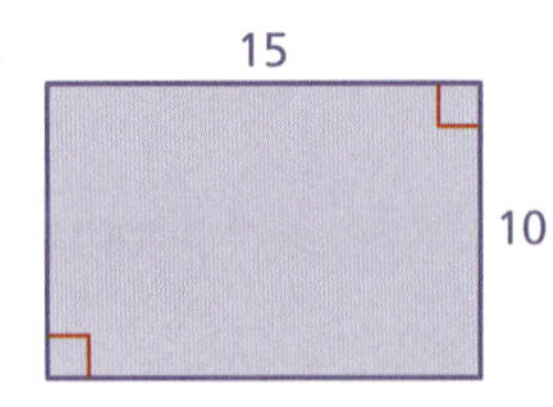

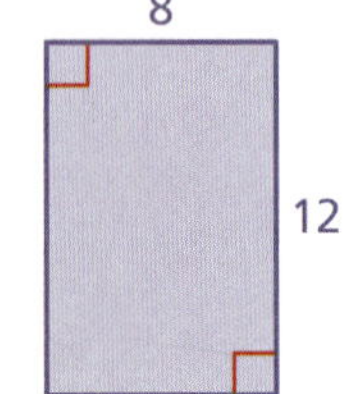

10. **PUZZLE** Describe the relationship between the unfinished puzzle and the missing piece.

Assignment Guide and Homework Check

Level	Day 1 Activity Assignment	Day 2 Lesson Assignment	Homework Check
Basic	4, 5, 16–20	1–3, 6–13	7, 9, 10, 11
Average	4, 5, 16–20	1–3, 7, 9, 10–14	7, 9, 11, 14
Advanced	1–6, 8, 11–20		6, 8, 11, 14

For Your Information

- **Exercise 1** Students may not be familiar with the symbol "△" used in this exercise. Mention that this symbol is used to represent a triangle.

Common Errors

- **Exercise 5** Students may think that the figures are not congruent because they do not have the same orientation. Remind them that as long as the figures have the same size and the same shape, they are congruent.
- **Exercises 6 and 7** Students may forget to write the angle symbol with the name. Remind them that A is a point and $\angle A$ is the angle.
- **Exercises 8, 9, and 11** Students may think that the figures are congruent because corresponding angles are congruent *or* corresponding sides are congruent. Remind them that there are two parts for determining that two figures are congruent, corresponding angles *and* corresponding sides.
- **Exercise 12** In part (b), students may have difficulty determining the corresponding angle. Point out that the fronts of the houses form congruent figures and remind them that corresponding angles in congruent figures have matching positions.

2.1 Record and Practice Journal

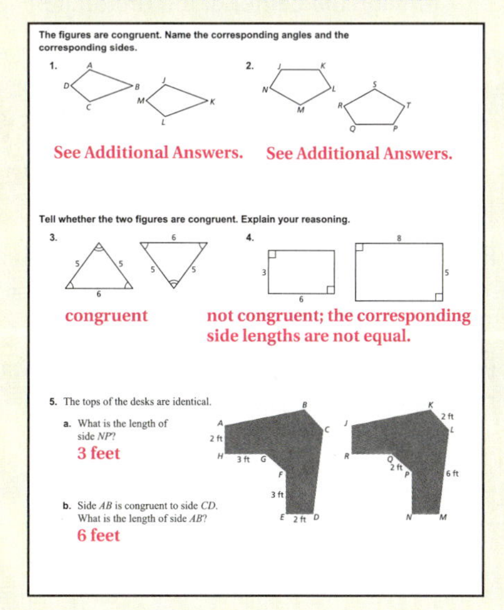
The figures are congruent. Name the corresponding angles and the corresponding sides.

1. See Additional Answers.

2. See Additional Answers.

Tell whether the two figures are congruent. Explain your reasoning.

3. congruent

4. not congruent; the corresponding side lengths are not equal.

5. The tops of the desks are identical.

   a. What is the length of side NP?
   3 feet

   b. Side AB is congruent to side CD. What is the length of side AB?
   6 feet

Vocabulary and Concept Check

1. a. $\angle A$ and $\angle D$, $\angle B$ and $\angle E$, $\angle C$ and $\angle F$

   b. Side AB and Side DE, Side BC and Side EF, Side AC and Side DF

2. Two figures are congruent if they have the same size and the same shape.

3. $\angle V$ does not belong. The other three angles are congruent to each other, but not to $\angle V$.

Practice and Problem Solving

4. not congruent

5. congruent

6. $\angle A$ and $\angle J$, $\angle B$ and $\angle K$, $\angle C$ and $\angle L$, $\angle D$ and $\angle M$; Side AB and Side JK, Side BC and Side KL, Side CD and Side LM, Side DA and Side MJ

7. $\angle P$ and $\angle W$, $\angle Q$ and $\angle V$, $\angle R$ and $\angle Z$, $\angle S$ and $\angle Y$, $\angle T$ and $\angle X$; Side PQ and Side WV, Side QR and Side VZ, Side RS and Side ZY, Side ST and Side YX, Side TP and Side XW

8. congruent; Corresponding side lengths and corresponding angles are congruent.

9. not congruent; Corresponding side lengths are not congruent.

10. The unfinished portion of the puzzle and the missing piece are congruent.

11. The corresponding angles are not congruent, so the two figures are not congruent.

Practice and Problem Solving

12. See Additional Answers.

13. See *Taking Math Deeper*.

14–15. See Additional Answers.

Fair Game Review

16–20. See Additional Answers.

Mini-Assessment

1. The figures are congruent. Name the corresponding angles and the corresponding sides.

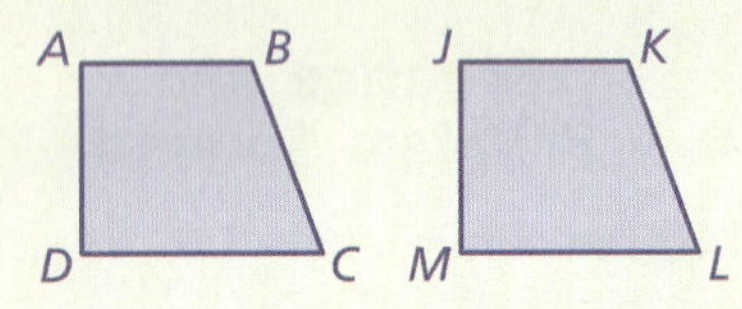

$\angle A$ and $\angle J$, $\angle B$ and $\angle K$, $\angle C$ and $\angle L$, $\angle D$ and $\angle M$; Side AB and Side JK, Side BC and Side KL, Side CD and Side LM, Side AD and Side JM

Tell whether the two figures are congruent. Explain your reasoning.

2.

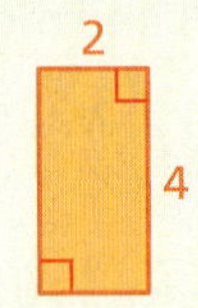

not congruent; Corresponding sides are not congruent.

3.

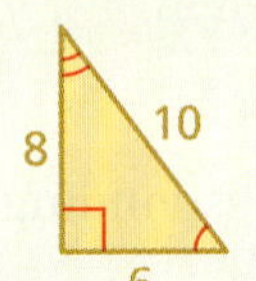

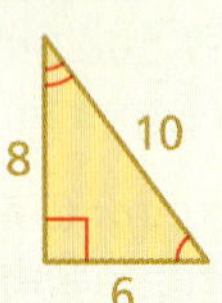

congruent; Corresponding angles and corresponding sides are congruent.

4. Trapezoids *ABCD* and *EFGH* are congruent.

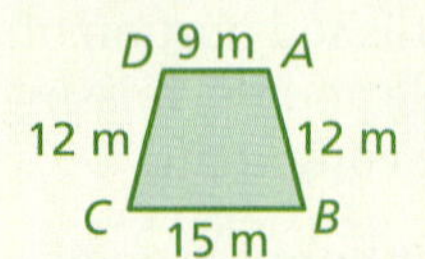

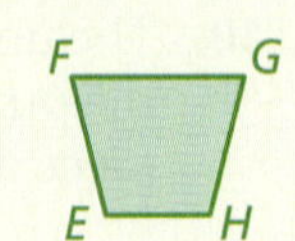

a. What is the length of Side *FG*? 15 m

b. What is the perimeter of Trapezoid *EFGH*? 48 m

Taking Math Deeper

Exercise 13

This problem requires students to visualize congruent shapes and requires "thinking outside the box."

1. The two basic ways to use a single line to divide a rectangle into two congruent figures are given in the problem.

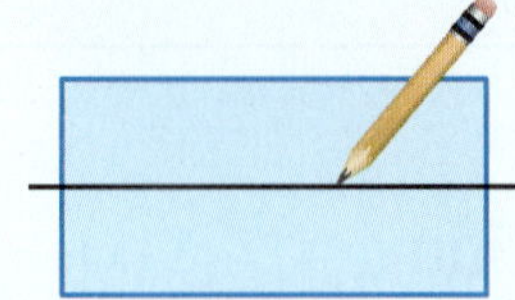

2. Two other ways are to draw the diagonal lines between opposite corners.

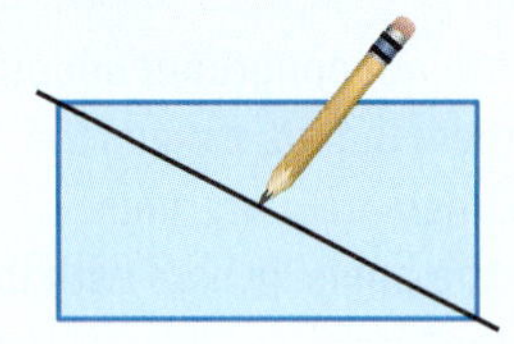

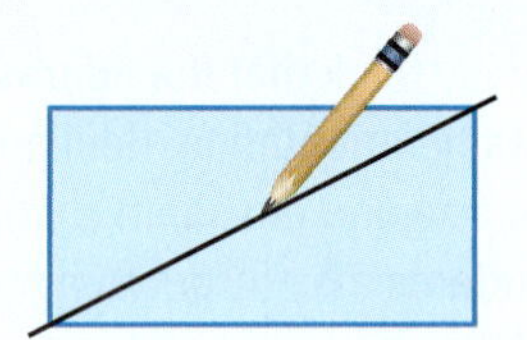

3. Some students may have trouble thinking of a third way to create two congruent figures. As it turns out, it is fun to realize that there are infinitely many ways to do this. Here are two of them.

For the trapezoids to be congruent, the shorter bases must be congruent and the longer bases must be congruent.

Note that in each case the figures are congruent when the line goes through the center of the rectangle.

Reteaching and Enrichment Strategies

If students need help. . .	If students got it. . .
Resources by Chapter • Practice A and Practice B • Puzzle Time Record and Practice Journal Practice Differentiating the Lesson Lesson Tutorials Skills Review Handbook	Resources by Chapter • Enrichment and Extension • Technology Connection Start the next section

11. ERROR ANALYSIS Describe and correct the error in telling whether the two figures are congruent.

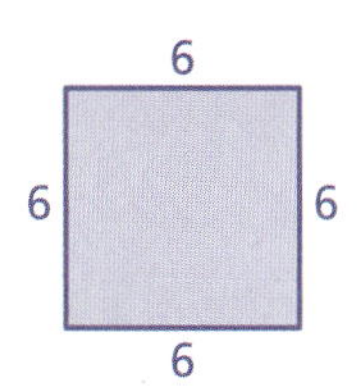

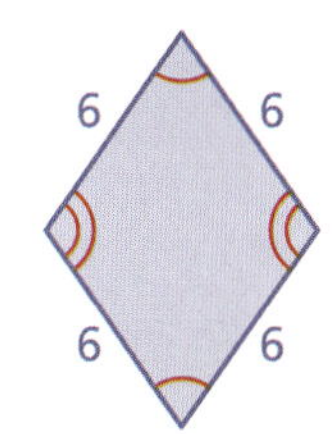

Both figures have four sides, and the corresponding side lengths are equal. So, they are congruent.

③ **12. HOUSES** The fronts of the houses are identical.

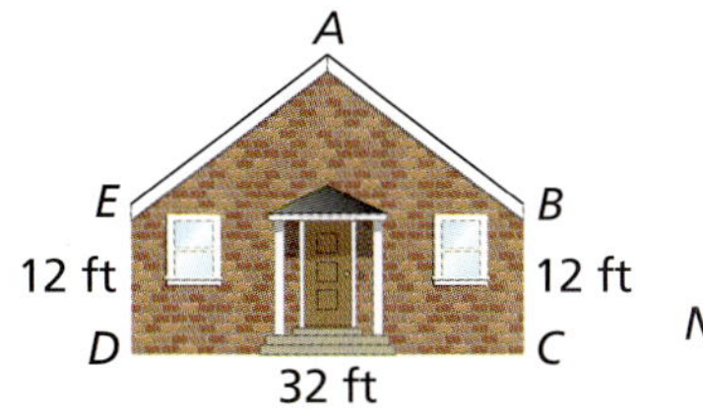

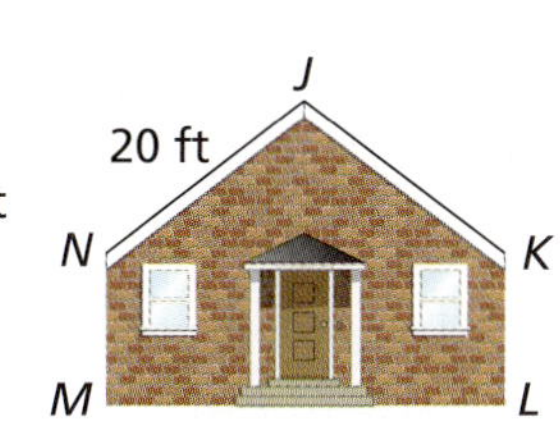

a. What is the length of side LM?

b. Which angle of $JKLMN$ corresponds to $\angle D$?

c. Side AB is congruent to side AE. What is the length of side AB?

d. What is the perimeter of $ABCDE$?

13. REASONING Here are two ways to draw *one* line to divide a rectangle into two congruent figures. Draw three other ways.

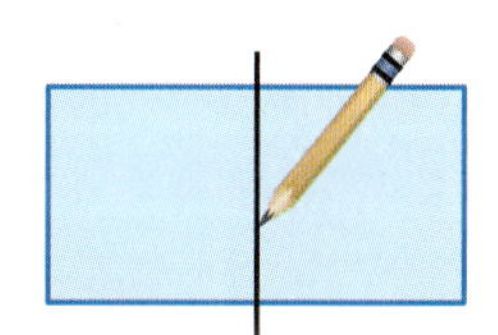

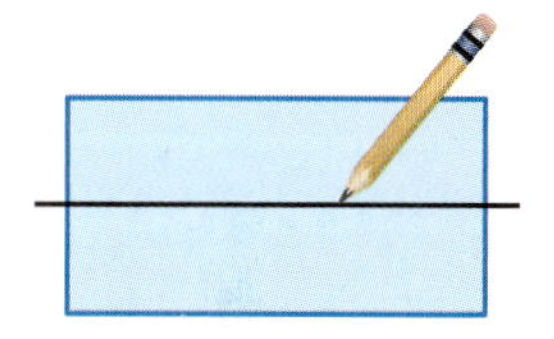

14. CRITICAL THINKING Are the areas of two congruent figures equal? Explain. Draw a diagram to support your answer.

15. True or False? The trapezoids are congruent. Determine whether the statement is *true* or *false*. Explain your reasoning.

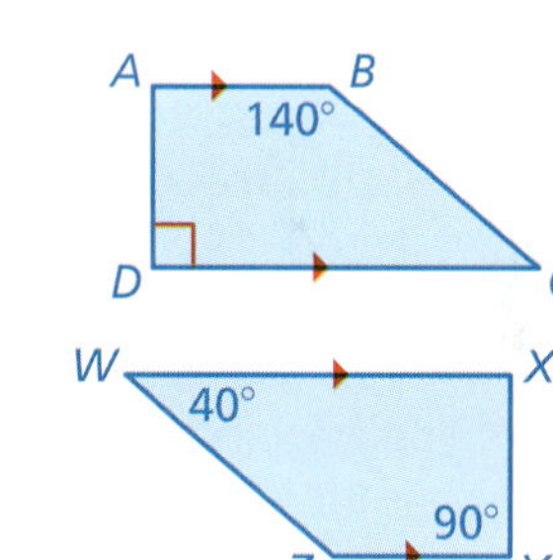

a. Side AB is congruent to side YZ.

b. $\angle A$ is congruent to $\angle X$.

c. $\angle A$ corresponds to $\angle X$.

d. The sum of the angle measures of $ABCD$ is 360°.

Fair Game Review What you learned in previous grades & lessons

Plot and label the ordered pair in a coordinate plane. *(Skills Review Handbook)*

16. $A(5, 3)$ **17.** $B(4, -1)$ **18.** $C(-2, 6)$ **19.** $D(-4, -2)$

20. MULTIPLE CHOICE You have 2 quarters and 5 dimes in your pocket. Write the ratio of quarters to the total number of coins. *(Skills Review Handbook)*

Ⓐ $\frac{2}{5}$ Ⓑ 2 : 7 Ⓒ 5 to 7 Ⓓ $\frac{7}{2}$

2.2 Translations

Essential Question How can you arrange tiles to make a tessellation?

The Meaning of a Word ● Translate

When you **translate** a tile, you slide it from one place to another.

When tiles cover a floor with no empty spaces, the collection of tiles is called a *tessellation*.

1 ACTIVITY: Describing Tessellations

Work with a partner. Can you make the tessellation by translating single tiles that are all of the same shape and design? If so, show how.

a. Sample:

Tile Pattern

Single Tiles

b.

c.

COMMON CORE

Geometry

In this lesson, you will

- identify translations.
- translate figures in the coordinate plane.

Learning Standards
MACC.8.G.1.1
MACC.8.G.1.2
MACC.8.G.1.3

2 ACTIVITY: Tessellations and Basic Shapes

Work with a partner.

a. Which pattern blocks can you use to make a tessellation? For each one that works, draw the tessellation.

b. Can you make the tessellation by translating? Or do you have to rotate or flip the pattern blocks?

Laurie's Notes

Introduction

Standards for Mathematical Practice

- **MP3 Construct Viable Arguments and Critique the Reasoning of Others:** As students manipulate geometric shapes, they develop spatial reasoning and make conjectures. Mathematically proficient students explore the truth of their conjectures by considering all cases.

Motivate

- **Whole Class Activity:** Model translations by having all students stand in an open area facing the same direction. Give directions such as: two steps right; three steps backward, etc. They should step backward or sideways as needed without turning their torso to keep their orientation.
- **The Meaning of a Word:** To **translate** a figure, you slide it to a new location in the plane. All points in the figure move the same distance and direction. The size, shape, and orientation of the figure do not change.
- Model a translation on the overhead (or document camera) by sliding a transparency with a shape or design on it.
- To help students develop spatial skills, have them explore transformations by moving and manipulating pattern blocks and designs traced on tracing paper or transparencies.

Activity Notes

Discuss

- Point out examples of tessellations. Tessellations may be found in the tile patterns on a floor, wall, or ceiling. They are also in wallpaper books.
- **FYI:** The word *tessellation* comes from the Latin word *tessellae*, which is what ancient Romans called small tiles used for pavements and walls.

Activity 1

- Have students trace on tracing paper the single tile they feel they can translate to form the tessellation.
- Remind students they are looking for a single tile that can be traced repeatedly to cover the surface without gaps or overlaps. Compare this to buying a box of tiles of the same size and shape to cover a floor.
- Make transparencies of each tessellation. When students share their solutions, have them trace the single tile on a clear transparency and then slide it over the tessellation transparency to show that the tile works.

Activity 2

- **Management Tip:** Place at least 10 triangles and 6 of the remaining pattern block shapes into a plastic bag for each pair of students.
- **Review:** "What is the name of each pattern block shape?" square, triangle, rhombus or parallelogram (tan and blue), hexagon, trapezoid
- **Discuss:** Four of the six shapes can be used to make a tessellation using translations only. The other two shapes (triangles and trapezoids) require both translations and rotations (or flips) to make a tesselation.

Florida Common Core Standards

MACC.8.G.1.1 Verify experimentally the properties of . . . translations.
MACC.8.G.1.2 Understand that a two-dimensional figure is congruent to another if the second can be obtained from the first by a sequence of . . . translations; given two congruent figures, describe a sequence that exhibits the congruence between them.
MACC.8.G.1.3 Describe the effect of . . . translations . . . on two-dimensional figures using coordinates.

Previous Learning

Students should know how to plot points in the coordinate plane.

Technology for the Teacher

Lesson Plans
Complete Materials List

2.2 Record and Practice Journal

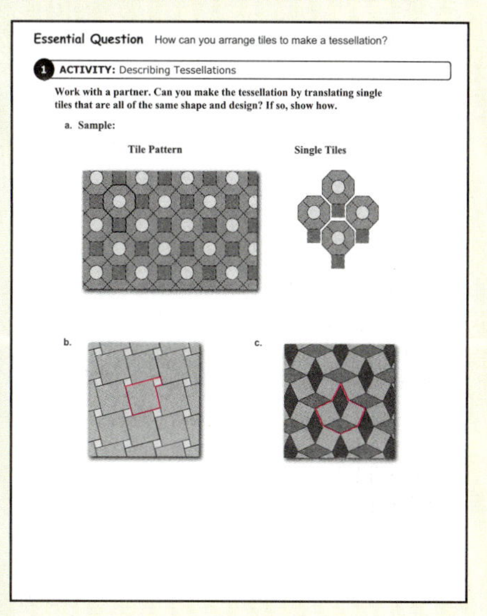
Essential Question How can you arrange tiles to make a tessellation?

1 ACTIVITY: Describing Tessellations

Work with a partner. Can you make the tessellation by translating single tiles that are all of the same shape and design? If so, show how.

a. Sample:

Tile Pattern Single Tiles

b.

c.

Differentiated Instruction

Auditory

To help students remember what each type of transformation is, use the words *slide* for translation, *flip* for reflection, and *turn* for rotation.

2.2 Record and Practice Journal

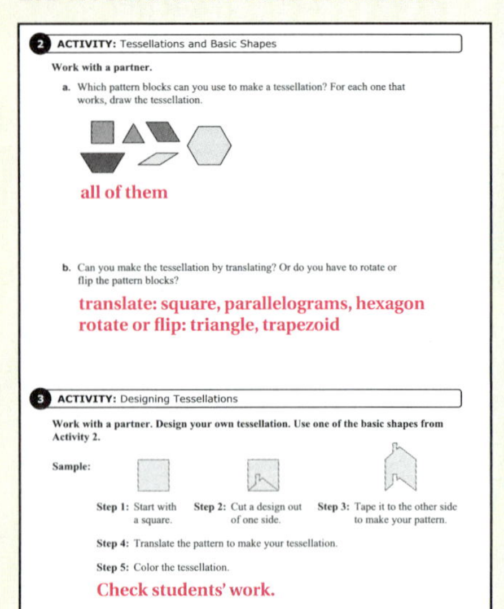

2 ACTIVITY: Tessellations and Basic Shapes

Work with a partner.

a. Which pattern blocks can you use to make a tessellation? For each one that works, draw the tessellation.

all of them

b. Can you make the tessellation by translating? Or do you have to rotate or flip the pattern blocks?

translate: square, parallelograms, hexagon
rotate or flip: triangle, trapezoid

3 ACTIVITY: Designing Tessellations

Work with a partner. Design your own tessellation. Use one of the basic shapes from Activity 2.

Sample:

Step 1: Start with a square. **Step 2:** Cut a design out of one side. **Step 3:** Tape it to the other side to make your pattern.

Step 4: Translate the pattern to make your tessellation.

Step 5: Color the tessellation.

Check students' work.

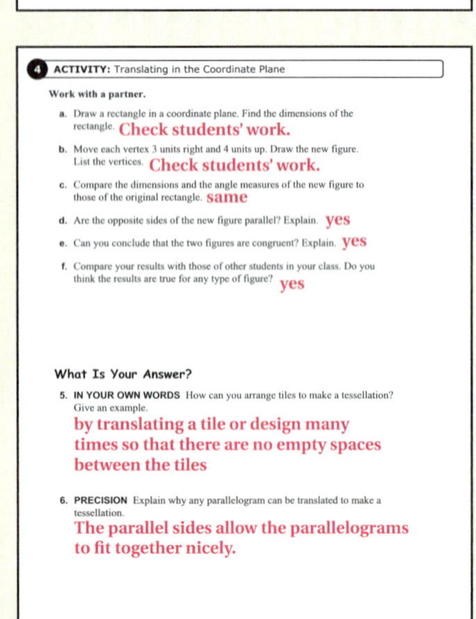

4 ACTIVITY: Translating in the Coordinate Plane

Work with a partner.

a. Draw a rectangle in a coordinate plane. Find the dimensions of the rectangle. Check students' work.

b. Move each vertex 3 units right and 4 units up. Draw the new figure. List the vertices. Check students' work.

c. Compare the dimensions and the angle measures of the new figure to those of the original rectangle. same

d. Are the opposite sides of the new figure parallel? Explain. yes

e. Can you conclude that the two figures are congruent? Explain. yes

f. Compare your results with those of other students in your class. Do you think the results are true for any type of figure? yes

What Is Your Answer?

5. **IN YOUR OWN WORDS** How can you arrange tiles to make a tessellation? Give an example.
by translating a tile or design many times so that there are no empty spaces between the tiles

6. **PRECISION** Explain why any parallelogram can be translated to make a tessellation.
The parallel sides allow the parallelograms to fit together nicely.

Laurie's Notes

Activity 3

- This activity could be started in class, and students could finish their designs for homework.
- Explain to students that they need to start with one of the basic shapes from Activity 2.
- Use the *cut and bump* method to create the tessellation. Whatever shape is *cut* from one edge of the shape must *bump out* on the edge to where the shape slides. After the initial shape is created, additional artwork (such as windows and roof coloring) can be added to provide additional details.
- **Management Tip:** Some students find it helpful to draw their designs on grid paper. The grid helps the students to be more accurate with their designs.
- This activity could be used as a project that students work on throughout the chapter. They may find other transformations in later sections that they want to incorporate into their tessellations.

Activity 4

- The *size* of the original rectangle does not matter, but encourage students to draw a rectangle that is not a square. A square is a special case of a rectangle, and their observations may only be true for the square.

? When students finish, ask a few questions.

  - "Did the second rectangle overlap with the original rectangle?" Answers will vary.
  - "Was the second rectangle in the same quadrant as the original rectangle?" Answers will vary.
  - "When you were locating the vertices of the second rectangle did you count 3 units right and 4 units up for each vertex?" Some may not have; they may have guessed at the location of the fourth vertex after locating the first three vertices.

? **MP3:** "What conjectures can you make about the two rectangles?" Students should share many conjectures, such as the rectangles are congruent; they have the same shape, size, perimeter, and area; and their opposite sides are parallel. They should also give evidence for their conjectures. Students may also make conjectures about the vertices when a rectangle is translated.

- **Discuss:** A translation of a line is a line. A translation of a line segment is a line segment of the same length as the original.

What Is Your Answer?

- **Think-Pair-Share:** Students should read each question independently and then work in pairs to answer the questions. When they have answered the questions, the pair should compare their answers with another group and discuss any discrepancies.

Closure

- Would *any* rectangle cover or tessellate a flat surface? Explain. Yes, because the sides are parallel.

3 ACTIVITY: Designing Tessellations

Work with a partner. Design your own tessellation. Use one of the basic shapes from Activity 2.

Sample:

Step 1: Start with a square.

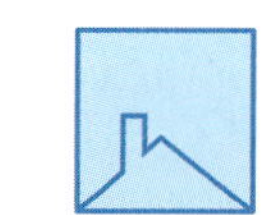

Step 2: Cut a design out of one side.

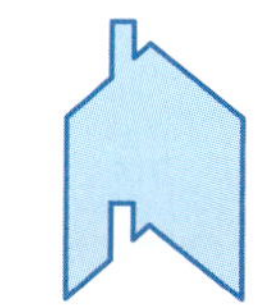

Step 3: Tape it to the other side to make your pattern.

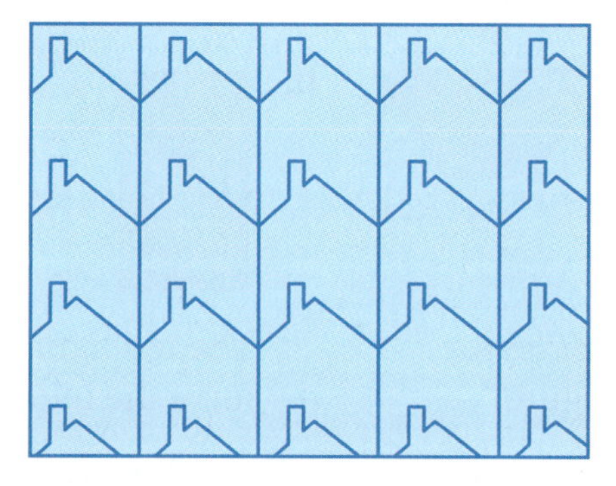

Step 4: Translate the pattern to make your tessellation.

Step 5: Color the tessellation.

4 ACTIVITY: Translating in the Coordinate Plane

Work with a partner.

a. Draw a rectangle in a coordinate plane. Find the dimensions of the rectangle.

b. Move each vertex 3 units right and 4 units up. Draw the new figure. List the vertices.

c. Compare the dimensions and the angle measures of the new figure to those of the original rectangle.

d. Are the opposite sides of the new figure still parallel? Explain.

e. Can you conclude that the two figures are congruent? Explain.

f. Compare your results with those of other students in your class. Do you think the results are true for any type of figure?

Math Practice 3

Justify Conclusions

What information do you need to conclude that two figures are congruent?

What Is Your Answer?

5. **IN YOUR OWN WORDS** How can you arrange tiles to make a tessellation? Give an example.

6. **PRECISION** Explain why any parallelogram can be translated to make a tessellation.

Practice Use what you learned about translations to complete Exercises 4–6 on page 52.

2.2 Lesson

Key Vocabulary
transformation, *p. 50*
image, *p. 50*
translation, *p. 50*

A **transformation** changes a figure into another figure. The new figure is called the **image**.

A **translation** is a transformation in which a figure *slides* but does not turn. Every point of the figure moves the same distance and in the same direction.

EXAMPLE 1 Identifying a Translation

Tell whether the blue figure is a translation of the red figure.

a.

The red figure *slides* to form the blue figure.

So, the blue figure is a translation of the red figure.

b.

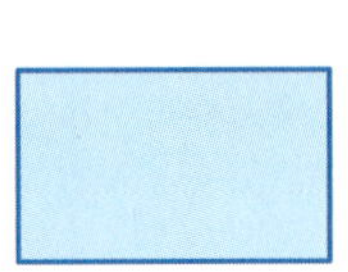

The red figure *turns* to form the blue figure.

So, the blue figure is *not* a translation of the red figure.

On Your Own

Now You're Ready
Exercises 4–9

Tell whether the blue figure is a translation of the red figure. Explain.

1.

2.

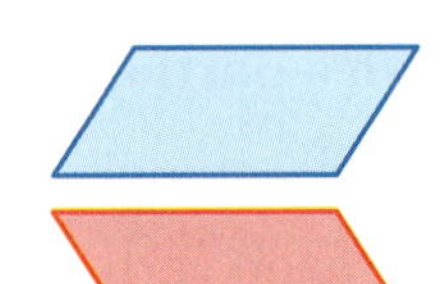

3.

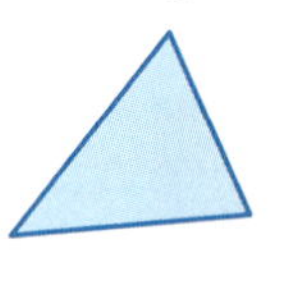

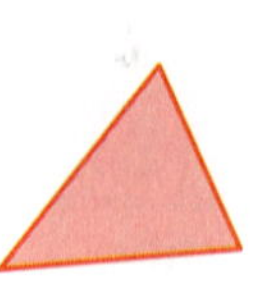

Key Idea

Translations in the Coordinate Plane

Words To translate a figure a units horizontally and b units vertically in a coordinate plane, add a to the x-coordinates and b to the y-coordinates of the vertices.

Positive values of a and b represent translations up and right. Negative values of a and b represent translations down and left.

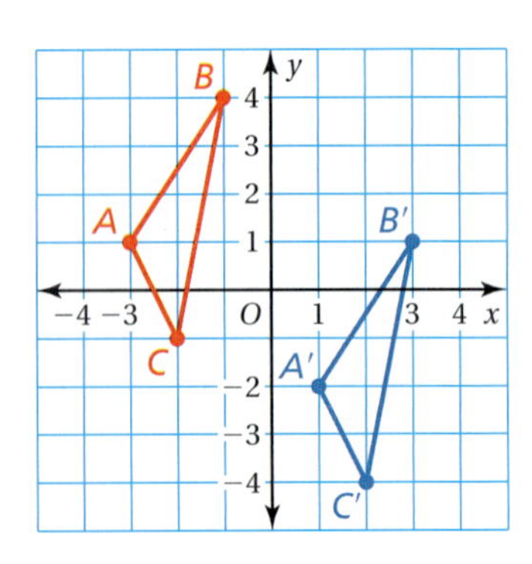

Algebra $(x, y) \rightarrow (x + a, y + b)$

In a translation, the original figure and its image are congruent.

Reading

A' is read "A prime." Use *prime* symbols when naming an image.

$A \rightarrow A'$
$B \rightarrow B'$
$C \rightarrow C'$

Multi-Language Glossary at BigIdeasMath.com

Laurie's Notes

Introduction

Connect

- **Yesterday:** Students explored translations by manipulating pattern blocks and sketching translations. (MP3)
- **Today:** Students will use their visual skills to draw translations in the coordinate plane.

Motivate

- Share a quick story about movie animation. Perhaps some of your students have made flipbooks where images are drawn at a slightly different location on each card, so that as you flip through the cards the image appears to move. If the image were a baseball, it would be translated to a new location on each card and appear to be moving as you flip the cards.
- Make a flipbook in advance to share with the class.

Lesson Notes

Discuss

- Discuss the introductory vocabulary: transformation, image, and translation.
- Relate a "translation" to the Motivate activity yesterday when each student would move (slide) the same distance in the same direction. Images in the flipbook are translations.

Example 1

- **Common Misconception:** The translation does not need to be in a horizontal or vertical direction. It can also be in a diagonal direction.
- Students generally have little difficulty identifying translations.

On Your Own

- **Think-Pair-Share:** Students should read each question independently and then work in pairs to answer the questions. When they have answered the questions, the pair should compare their answers with another group and discuss any discrepancies.
- For any questions that are translations, describe the direction of the translation.

Key Idea

- Write the Key Idea, using the language of A and A' (A prime) as you identify the coordinates of the original figure (A, B, C) and its image (A', B', C').
- In this example, red triangle ABC is the original figure and blue triangle $A'B'C'$ is the translated image.
- Use a third color to draw the translation arrow from A to A'.

? "How was vertex A translated to vertex A'?" It moved 4 units right and 3 units down.

Goal Today's lesson is identifying **translations**.

Technology for the Teacher

Lesson Tutorials
Lesson Plans
Answer Presentation Tool

Extra Example 1

Tell whether the blue figure is a translation of the red figure.

a.

a translation

b.

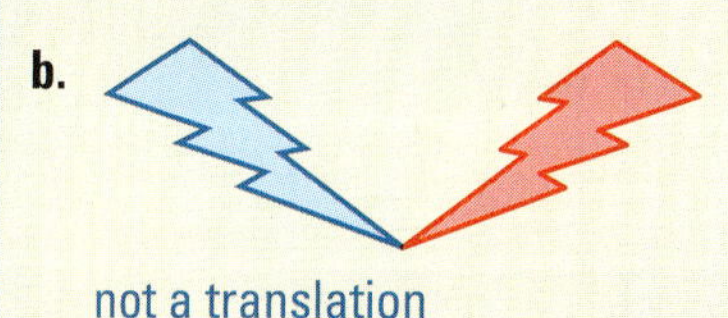

not a translation

On Your Own

1. no; The blue figure is larger than the red figure.
2. no; The red figure flips to form the blue figure.
3. yes; The red figure slides up and to the left to form the blue figure.

Extra Example 2

Translate the red triangle 2 units right and 5 units up. What are the coordinates of the image?

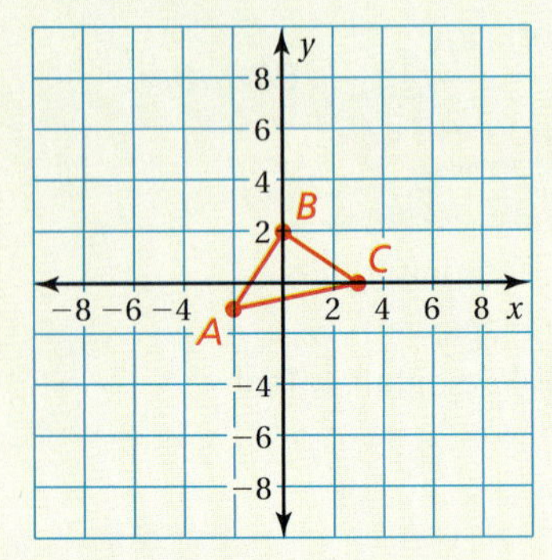

$A'(0, 4)$, $B'(2, 7)$, $C'(5, 5)$

On Your Own

4. $A'(-6, 3)$, $B'(-2, 7)$, $C'(-3, 4)$

Extra Example 3

The vertices of a rectangle are $A(1, 4)$, $B(3, 4)$, $C(3, 1)$ and $D(1, 1)$. Draw the figure and its image after a translation 3 units left and 4 units down.

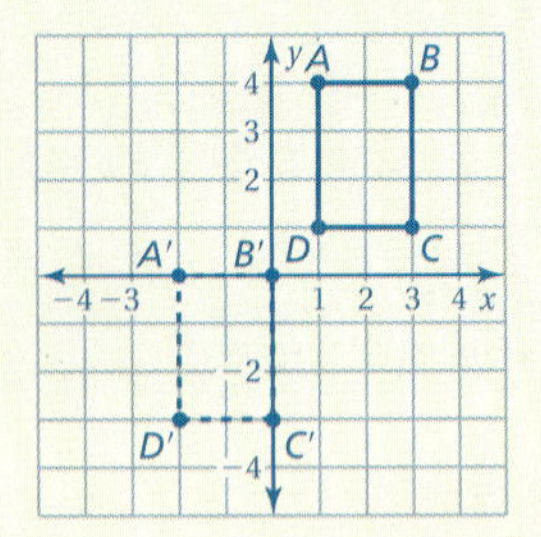

On Your Own

5. See Additional Answers.

English Language Learners

Visual

Make a poster in the classroom to illustrate the movement of a point in a coordinate plane based on the coordinate notation.

$x + h$	→
$x - h$	←
$y + k$	↑
$y - k$	↓

Laurie's Notes

Example 2

- Draw $\triangle ABC$ and label the vertices on a transparency. Slide the transparency 3 units to the right and 3 units down.
- Alternatively, you can model the translation on an interactive white board. Draw $\triangle ABC$. Copy $\triangle ABC$ and slide the copy to the new position.
- **Representation:** The **image** of a transformation (translation, reflection, rotation, or dilation) is written with the prime symbol. This helps to distinguish the image from the original figure, often referred to as the pre-image.
- After the result of the translation has been drawn, you can draw arrows from A to A', B to B', and C to C'. The resulting figure appears to be a 3-D diagram of a triangular prism.
- Explain that translating the triangle on a diagonal is equivalent to translating the triangle horizontally and then vertically. The two steps focus on what happens to each of the coordinates in an ordered pair.

? "Is the blue triangle the same size and shape as the red triangle?" yes

- Reinforce the concept of same size and shape by talking about the lengths of corresponding sides, the measures of the corresponding angles, and the perimeters and areas of the two triangles.

On Your Own

- Students should work in pairs.

Example 3

- Plot the four ordered pairs.
- **Common Error**: Students may interchange x- and y-directions in plotting the ordered pairs.

? Ask questions about the translation.

  - "In what quadrant is the original square?" IV
  - "If a figure is translated in the coordinate plane 4 units left, what will change, the x-coordinate or the y-coordinate?" x-coordinate
  - "If a figure is translated in the coordinate plane 6 units up, what will change, the x-coordinate or the y-coordinate?" y-coordinate

- **MP8 Look for and Express Regularity in Repeated Reasoning:** Explain the notation in the table. Use an alternate color to draw attention to the repeated pattern (subtracting 4 and adding 6) that occurs with each ordered pair.
- Draw the new image.

? "In what quadrant is the image?" II

? "Is the blue square the same size and shape as the red square?" yes

On Your Own

- **Neighbor Check:** Have students work independently and then have their neighbors check their work. Have students discuss any discrepancies.

Closure

- Draw a right triangle in Quadrant II. Translate the triangle so that the image is in Quadrant IV. Describe the translation.

EXAMPLE 2 Translating a Figure in the Coordinate Plane

Translate the red triangle 3 units right and 3 units down. What are the coordinates of the image?

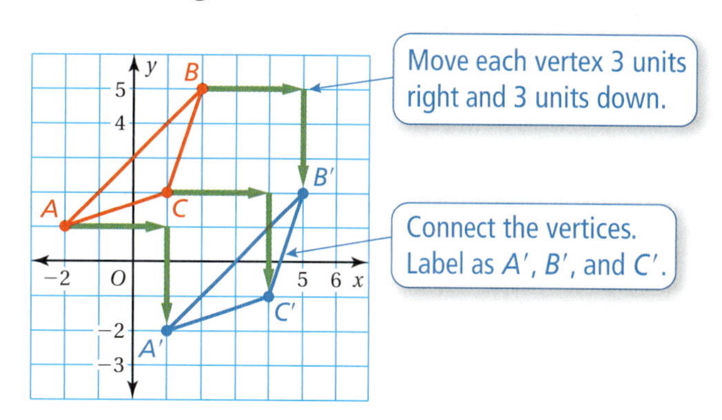

The coordinates of the image are $A'(1, -2)$, $B'(5, 2)$, and $C'(4, -1)$.

On Your Own

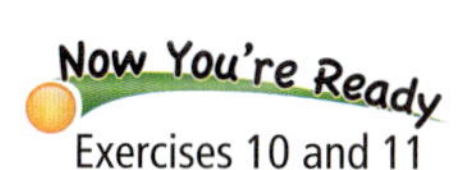

4. **WHAT IF?** The red triangle is translated 4 units left and 2 units up. What are the coordinates of the image?

EXAMPLE 3 Translating a Figure Using Coordinates

The vertices of a square are $A(1, -2)$, $B(3, -2)$, $C(3, -4)$, and $D(1, -4)$. Draw the figure and its image after a translation 4 units left and 6 units up.

Add −4 to each x-coordinate. So, subtract 4 from each x-coordinate.

Add 6 to each y-coordinate.

Vertices of $ABCD$	$(x - 4, y + 6)$	Vertices of $A'B'C'D'$
$A(1, -2)$	$(1 - 4, -2 + 6)$	$A'(-3, 4)$
$B(3, -2)$	$(3 - 4, -2 + 6)$	$B'(-1, 4)$
$C(3, -4)$	$(3 - 4, -4 + 6)$	$C'(-1, 2)$
$D(1, -4)$	$(1 - 4, -4 + 6)$	$D'(-3, 2)$

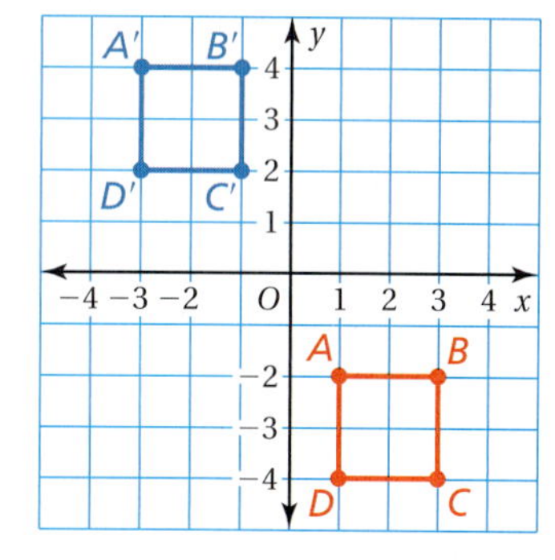

The figure and its image are shown at the above right.

On Your Own

5. The vertices of a triangle are $A(-2, -2)$, $B(0, 2)$, and $C(3, 0)$. Draw the figure and its image after a translation 1 unit left and 2 units up.

2.2 Exercises

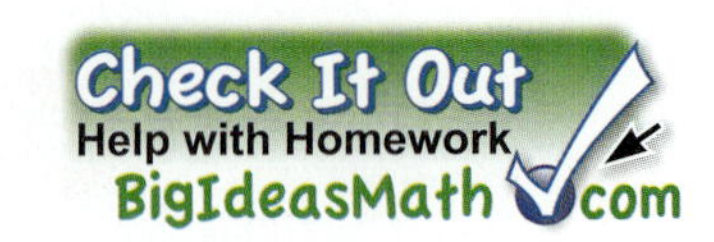

Vocabulary and Concept Check

1. **VOCABULARY** Which figure is the image?

2. **VOCABULARY** How do you translate a figure in a coordinate plane?
3. **WRITING** Can you translate the letters in the word TOKYO to form the word KYOTO? Explain.

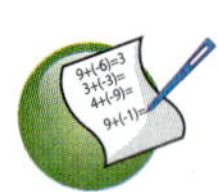

Practice and Problem Solving

Tell whether the blue figure is a translation of the red figure.

① **4.**

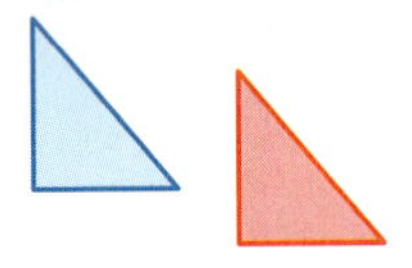

5.

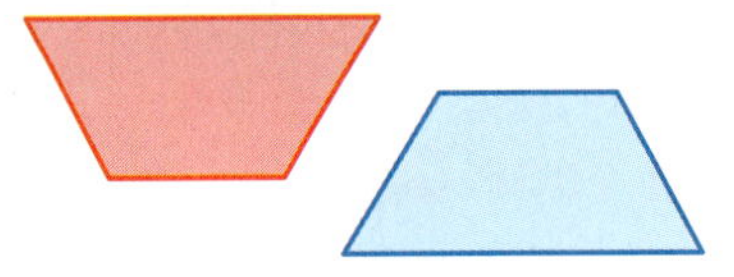

6.

7.

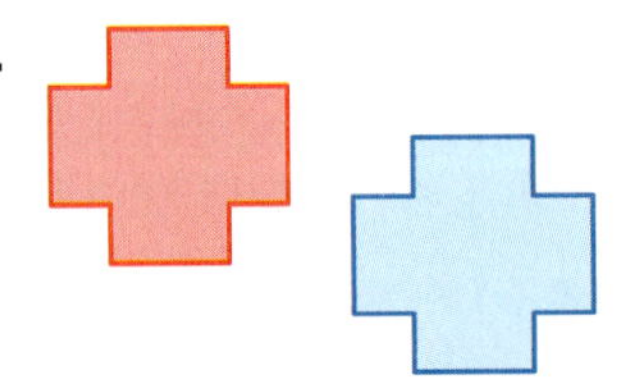

8.

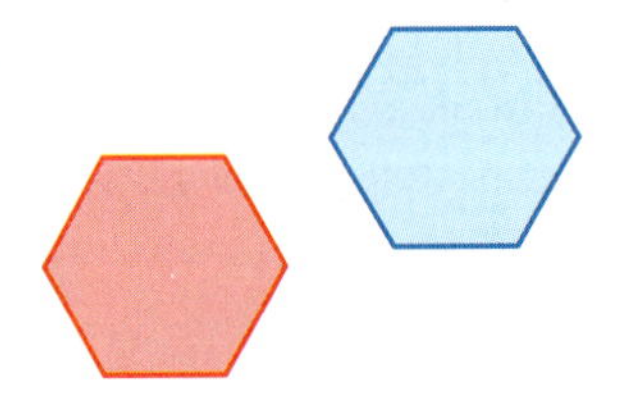

9.

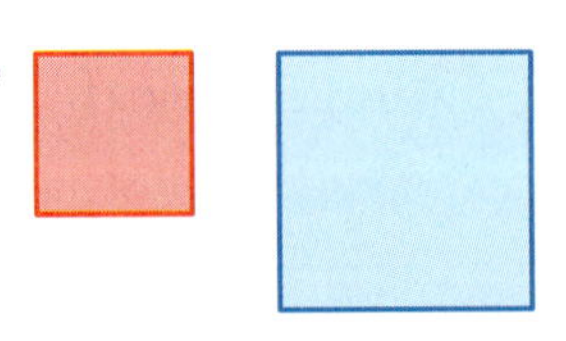

② **10.** Translate the triangle 4 units right and 3 units down. What are the coordinates of the image?

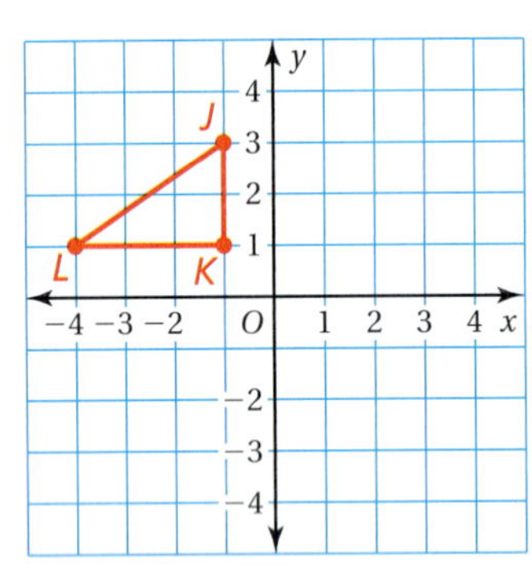

11. Translate the figure 2 units left and 4 units down. What are the coordinates of the image?

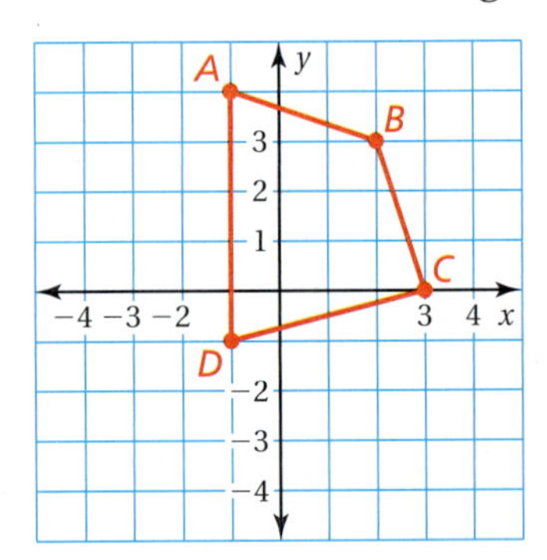

The vertices of a triangle are $L(0, 1)$, $M(1, -2)$, and $N(-2, 1)$. Draw the figure and its image after the translation.

③ **12.** 1 unit left and 6 units up

13. 5 units right

14. $(x + 2, y + 3)$

15. $(x - 3, y - 4)$

16. **ICONS** You can click and drag an icon on a computer screen. Is this an example of a translation? Explain.

Assignment Guide and Homework Check

Level	Day 1 Activity Assignment	Day 2 Lesson Assignment	Homework Check
Basic	4–6, 24–28	1–3, 7–15 odd, 16, 17, 19	7, 11, 13, 17
Average	4–6, 24–28	1–3, 8–16 even, 17–21 odd	8, 10, 12, 17
Advanced	1–6, 12–20 even, 21–28		12, 18, 20, 22

For Your Information

- **Exercise 3** The Japanese language is composed of symbols, not letters. KYO means *capitol* and TO means *new*. So, KYO TO was the ancient capitol of Japan, and TO KYO is the modern day capitol of Japan.

Common Errors

- **Exercises 4–9** Students may forget that the objects must be the same size to be a translation. Remind them that the size stays the same. Tell students that when the size is different, it is a scale drawing.
- **Exercises 10–15** Students may translate the shape the wrong direction or mix up the units for the translation. Tell them to redraw the original on graph paper. Also, tell students to write the direction of the translation using arrows to show the movement left, right, up, or down.
- **Exercises 17 and 18** Students may struggle finding the translation. Encourage students to plot the points in a coordinate plane and count the change left, right, up, or down.
- **Exercises 19 and 20** Students may count the translation to the wrong point. Ask them to label the red figure with points *A*, *B*, etc. and the corresponding points on the blue figure as *A′*, *B′*, etc.

2.2 Record and Practice Journal

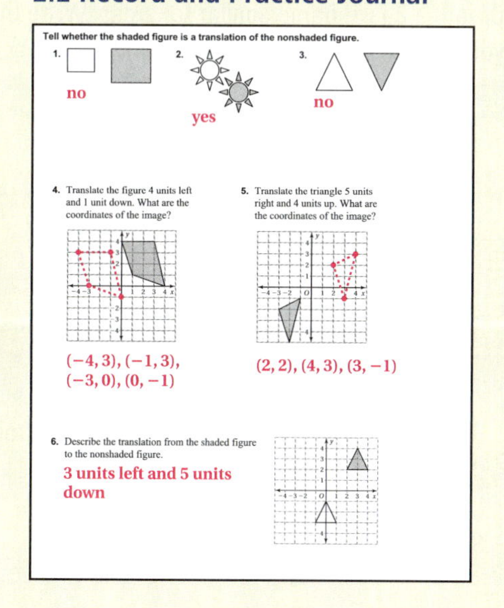
Tell whether the shaded figure is a translation of the nonshaded figure.

1. no
2. yes
3. no

4. Translate the figure 4 units left and 1 unit down. What are the coordinates of the image?

(−4, 3), (−1, 3), (−3, 0), (0, −1)

5. Translate the triangle 5 units right and 4 units up. What are the coordinates of the image?

(2, 2), (4, 3), (3, −1)

6. Describe the translation from the shaded figure to the nonshaded figure.

3 units left and 5 units down

Vocabulary and Concept Check

1. A
2. Move each vertex according to the translation.
3. yes; Translate the letters T and O to the end.

Practice and Problem Solving

4. yes
5. no
6. no
7. yes
8. yes
9. no
10. $J'(3, 0)$, $K'(3, -2)$, $L'(0, -2)$
11. $A'(-3, 0)$, $B'(0, -1)$, $C'(1, -4)$, $D'(-3, -5)$
12.

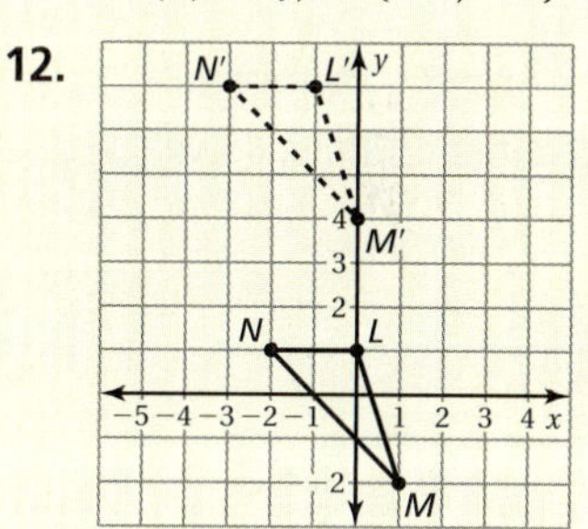

13.

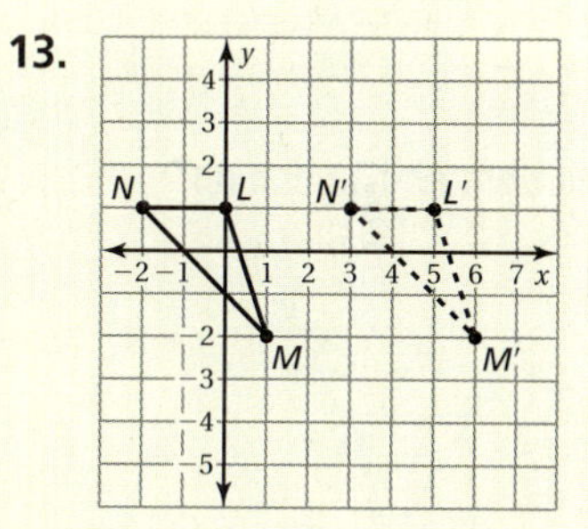

14.

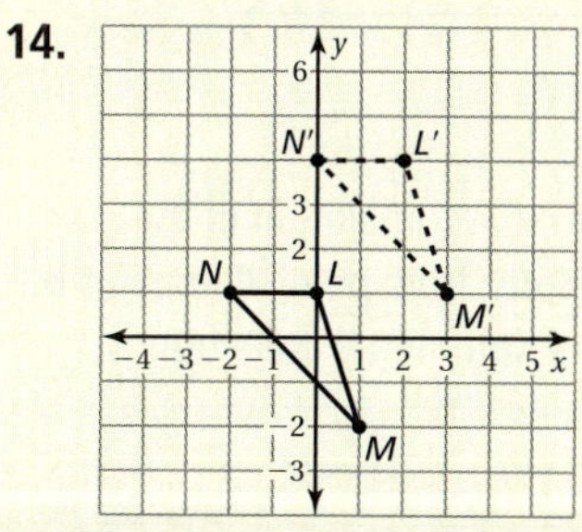

15. See Additional Answers.
16. Yes, because the figure slides.
17. 2 units left and 2 units up
18. 5 units right and 9 units up

19–22. See Additional Answers.

Practice and Problem Solving

23. See *Taking Math Deeper.*

Fair Game Review

24–28. See Additional Answers.

Mini-Assessment

The vertices of a triangle are $A(1, 3)$, $B(4, 3)$, and $C(3, 0)$. Draw the figure and its image after the translation.

1. 2 units left and 3 units down

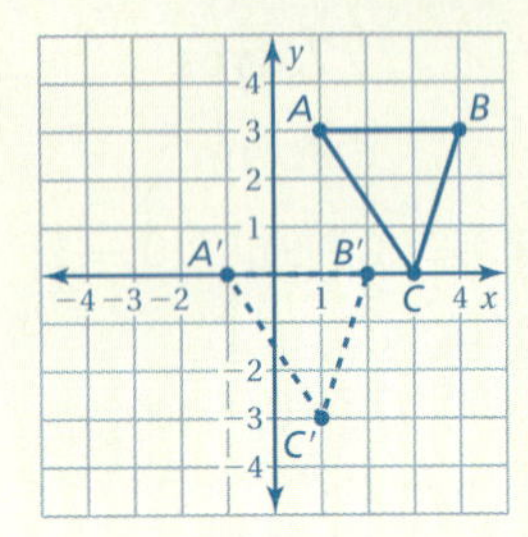

2. 1 unit left and 4 units down

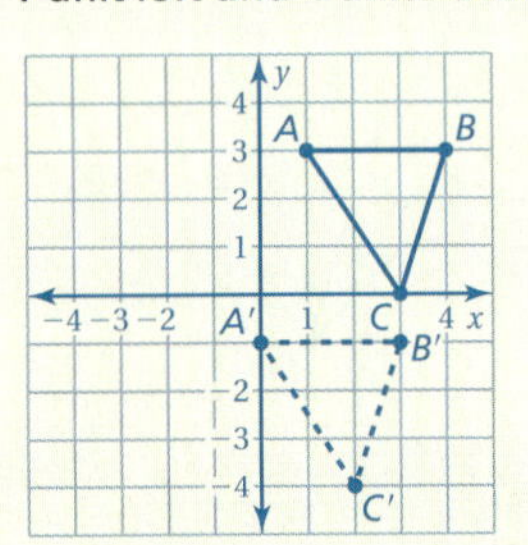

3. 1 unit right and 2 units up

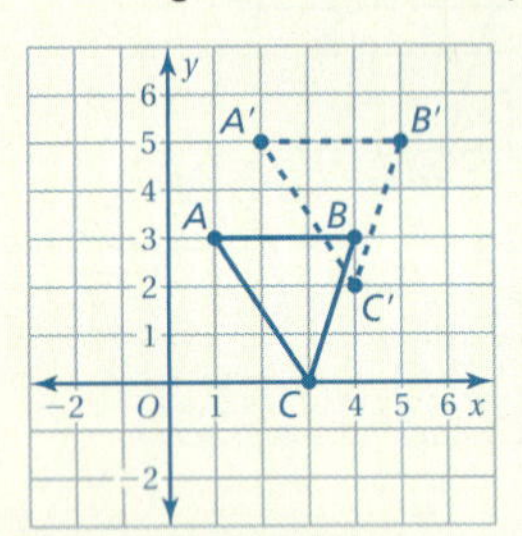

4. Describe a translation of the helicopter from point A to point B.
5 units right and 7 units up

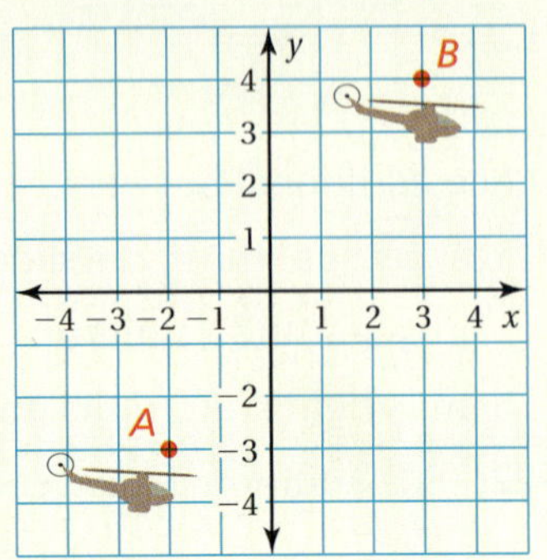

Taking Math Deeper

Exercise 23

There are thousands of correct answers to this question. This could be a nice discussion question for pairs or groups of students.

1. Here is one translation that takes 5 moves.

2. Here is one translation that takes only 3 moves.

3. It is not possible to move from g8 to g5 in less than 3 moves.

Project

Create a board game similar to chess. Write the rules for your game and play your game with another student.

Reteaching and Enrichment Strategies

If students need help. . .	If students got it. . .
Resources by Chapter • Practice A and Practice B • Puzzle Time Record and Practice Journal Practice Differentiating the Lesson Lesson Tutorials Skills Review Handbook	Resources by Chapter • Enrichment and Extension • Technology Connection Start the next section

Describe the translation of the point to its image.

17. $(3, -2) \rightarrow (1, 0)$

18. $(-8, -4) \rightarrow (-3, 5)$

Describe the translation from the red figure to the blue figure.

19.

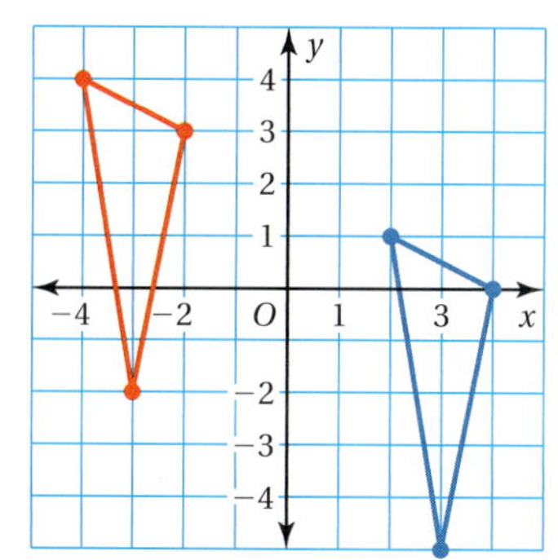

20.

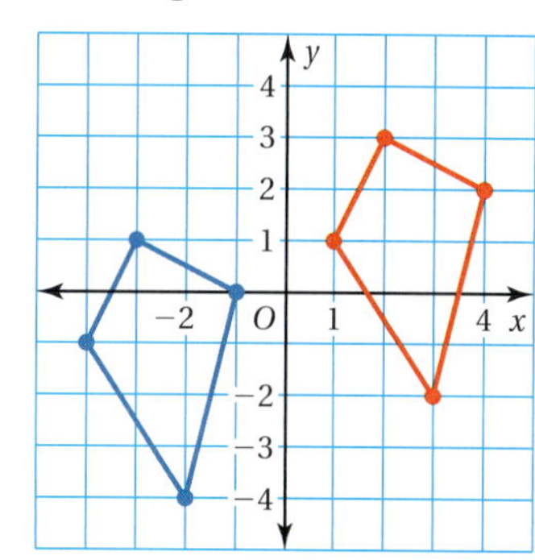

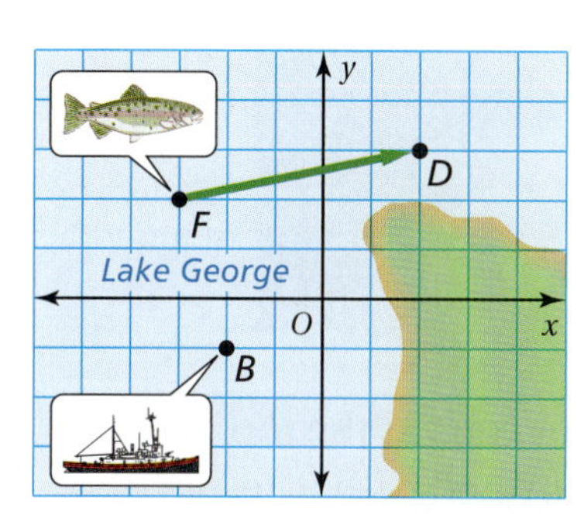

21. FISHING A school of fish translates from point F to point D.

a. Describe the translation of the school of fish.

b. Can the fishing boat make the same translation? Explain.

c. Describe a translation the fishing boat could make to get to point D.

22. REASONING The vertices of a triangle are $A(0, -3)$, $B(2, -1)$, and $C(3, -3)$. You translate the triangle 5 units right and 2 units down. Then you translate the image 3 units left and 8 units down. Is the original triangle congruent to the final image? If so, give two ways to show that they are congruent.

23. Problem Solving In chess, a knight can move only in an L-shaped pattern:

- *two* vertical squares, then *one* horizontal square;
- *two* horizontal squares, then *one* vertical square;
- *one* vertical square, then *two* horizontal squares; or
- *one* horizontal square, then *two* vertical squares.

Write a series of translations to move the knight from g8 to g5.

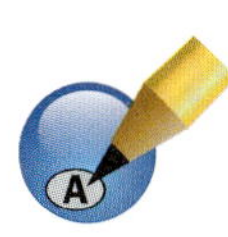

Fair Game Review What you learned in previous grades & lessons

Tell whether you can fold the figure in half so that one side matches the other. *(Skills Review Handbook)*

24. 

25. 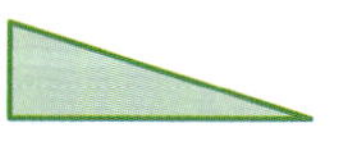

26. 

27. 

28. MULTIPLE CHOICE You put \$550 in an account that earns 4.4% simple interest per year. How much interest do you earn in 6 months? *(Skills Review Handbook)*

Ⓐ \$1.21 Ⓑ \$12.10 Ⓒ \$121.00 Ⓓ \$145.20

2.3 Reflections

Essential Question How can you use reflections to classify a frieze pattern?

The Meaning of a Word ● Reflection

When you look at a mountain by a lake, you can see the **reflection**, or mirror image, of the mountain in the lake.

If you fold the photo on its axis, the mountain and its reflection will align.

A *frieze* is a horizontal band that runs at the top of a building. A frieze is often decorated with a design that repeats.

- All frieze patterns are translations of themselves.
- Some frieze patterns are reflections of themselves.

1 ACTIVITY: Frieze Patterns and Reflections

Work with a partner. Consider the frieze pattern shown.

a. Is the frieze pattern a reflection of itself when folded horizontally? Explain.

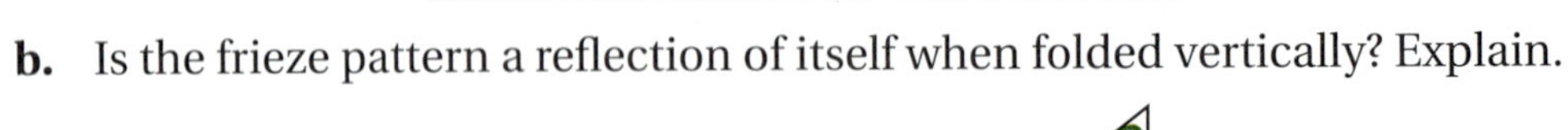

b. Is the frieze pattern a reflection of itself when folded vertically? Explain.

Geometry

In this lesson, you will
- identify reflections.
- reflect figures in the x-axis or the y-axis of the coordinate plane.

Learning Standards
MACC.8.G.1.1
MACC.8.G.1.2
MACC.8.G.1.3

Laurie's Notes

Introduction

Standards for Mathematical Practice

- **MP3 Construct Viable Arguments and Critique the Reasoning of Others:** As students manipulate geometric shapes they make conjectures and develop spatial reasoning. Mathematically proficient students explore the truth of their conjectures by considering all cases.

Motivate

- Before class, practice folding a long strip of scrap paper. Cut it to make a frieze pattern. A common design is the stick figure. Practice various folds so you can create reflections.

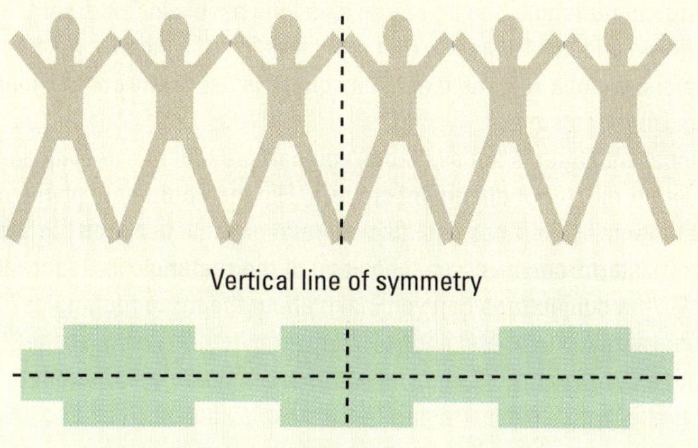

Vertical line of symmetry

Horizontal and vertical lines of symmetry

- **Teaching Tip:** Lay the cut-out, still folded, on the overhead projector. Ask students what it will look like when it is unfolded.

Activity Notes

Discuss

- Today's investigation involves working with a pattern known as a frieze. The text shows an example of an architectural frieze on a building. Friezes also occur on wallpaper borders, designs on pottery, ironwork railings, and the headbands and belts of the indigenous people of North America, to name a few.
- A frieze is a pattern which repeats in one direction and can always be translated onto itself. Friezes may also contain reflections, and that is the focus of this investigation.

Activity 1

- Provide students with tracing paper so that they can sketch the original design. Some students will not need the tracing paper, but it should be offered to all students.
- When students finish, listen to the evidence they give as they explain whether the pattern reflects onto itself horizontally or vertically.

Florida Common Core Standards

MACC.8.G.1.1 Verify experimentally the properties of . . . reflections and translations.

MACC.8.G.1.2 Understand that a two-dimensional figure is congruent to another if the second can be obtained from the first by a sequence of . . . reflections and translations; given two congruent figures, describe a sequence that exhibits the congruence between them.

MACC.8.G.1.3 Describe the effect of . . . reflections on two-dimensional figures using coordinates.

Previous Learning

Students should know how to plot points in the coordinate plane.

2.3 Record and Practice Journal

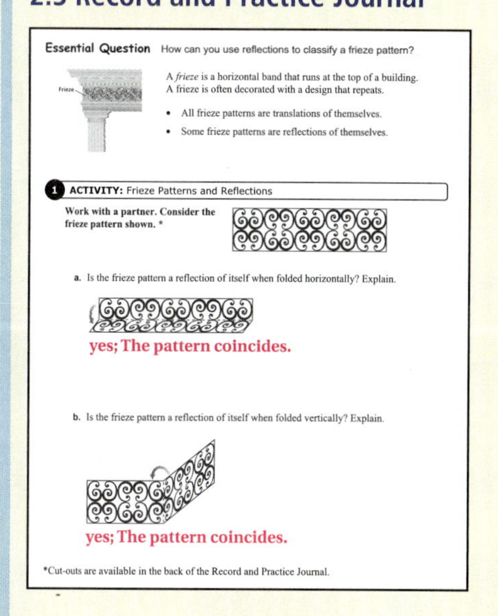

Essential Question How can you use reflections to classify a frieze pattern?

Frieze

A *frieze* is a horizontal band that runs at the top of a building. A frieze is often decorated with a design that repeats.

- All frieze patterns are translations of themselves.
- Some frieze patterns are reflections of themselves.

1 ACTIVITY: Frieze Patterns and Reflections

Work with a partner. Consider the frieze pattern shown. *

a. Is the frieze pattern a reflection of itself when folded horizontally? Explain.

yes; The pattern coincides.

b. Is the frieze pattern a reflection of itself when folded vertically? Explain.

yes; The pattern coincides.

*Cut-outs are available in the back of the Record and Practice Journal.

Differentiated Instruction

Kinesthetic

Have students fold paper in half or in quarters and use scissors to cut out various shapes. Open the paper and find the lines of symmetry. Depending on the cut out, there may be more than one line of symmetry.

2.3 Record and Practice Journal

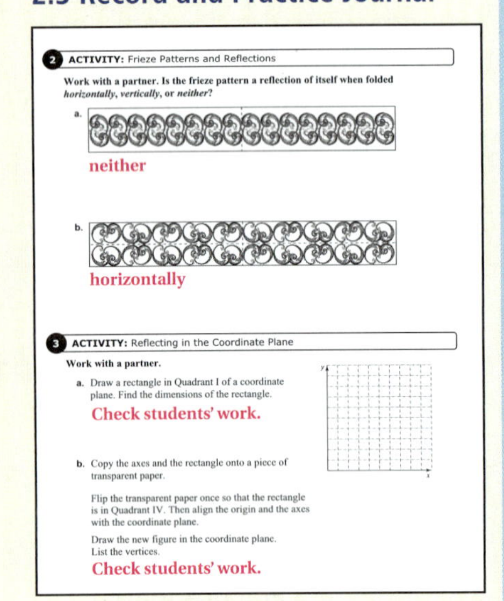

2 ACTIVITY: Frieze Patterns and Reflections

Work with a partner. Is the frieze pattern a reflection of itself when folded *horizontally*, *vertically*, or *neither*?

a. neither

b. horizontally

3 ACTIVITY: Reflecting in the Coordinate Plane

Work with a partner.

a. Draw a rectangle in Quadrant I of a coordinate plane. Find the dimensions of the rectangle.
Check students' work.

b. Copy the axes and the rectangle onto a piece of transparent paper.

Flip the transparent paper once so that the rectangle is in Quadrant IV. Then align the origin and the axes with the coordinate plane.

Draw the new figure in the coordinate plane. List the vertices.
Check students' work.

c. Compare the dimensions and the angle measures of the new figure to those of the original rectangle.
same

d. Are the opposite sides of the new figure still parallel? Explain.
yes

e. Can you conclude that the two figures are congruent? Explain.
yes

f. Flip the transparent paper so that the original rectangle is in Quadrant II. Draw the new figure in the coordinate plane. List the vertices. Then repeat parts (c)–(e).
Check students' work; same; yes; yes

g. Compare your results with those of other students in your class. Do you think the results are true for any type of figure?
yes

What Is Your Answer?

4. **IN YOUR OWN WORDS** How can you use reflections to classify a frieze pattern?
Fold a frieze pattern horizontally or vertically. If it coincides, then it is a reflection of itself.

Laurie's Notes

Activity 2

- Students should work with partners.
- Students may wish to have tracing paper to test their thinking about the patterns shown.
- Make an overhead transparency of the designs to help facilitate discussion. Have clear transparencies available for students to trace their answers.
- Remind students that a reflection that folds onto itself in a frieze must be horizontal or vertical.
- **Common Error:** Students will see the rotation in the pattern and identify it as a reflection.

Activity 3

- The original rectangle can be any size as long as it is in Quadrant I. Encourage students to draw a rectangle that is not a square. A square is a special case of a rectangle and if a square is used, any conclusions may only be true for a square.
- If transparent paper is not available, you can present this activity on the overhead projector or document camera. Flip the coordinate plane and have students list vertices and discuss relationships between the vertices.
- In part (f), start from the original position of the rectangle in Quadrant I.

? **MP3:** "What conjectures can you make about the three rectangles?" Students should share and support many conjectures, such as the rectangles are congruent; they have the same shape, size, perimeter, and area; and opposite sides are parallel. Students may also talk about the vertices when a rectangle is reflected about the *x*- or *y*-axis.

- **Extension:** Draw a frieze pattern for the following cases.
  - The pattern is a reflection of itself when folded horizontally.
  - The pattern is a reflection of itself when folded vertically.
  - The pattern is not a reflection of itself when folded horizontally or vertically.

What Is Your Answer?

- Students should consider their work in all three activities to answer the question.

Closure

- Imagine footprints in sand left by someone walking normally. Are the footprints a reflection? no
- Imagine footprints in mud left by a rabbit hopping normally. Are the footprints a reflection? yes

2 ACTIVITY: Frieze Patterns and Reflections

Work with a partner. Is the frieze pattern a reflection of itself when folded *horizontally*, *vertically*, or *neither*?

a.

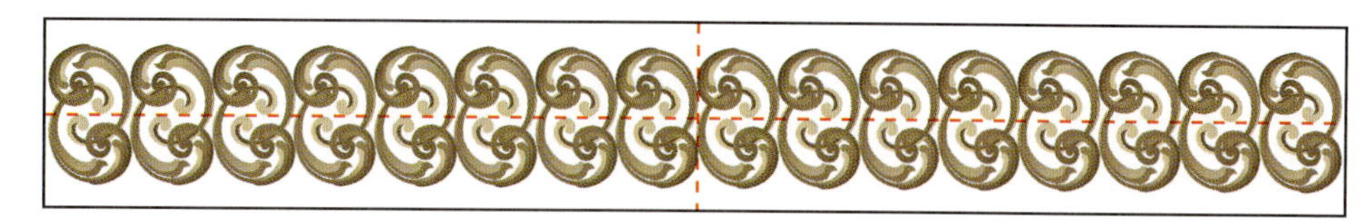

b.

3 ACTIVITY: Reflecting in the Coordinate Plane

Work with a partner.

Math Practice 7

Look for Patterns

What do you notice about the vertices of the original figure and the image? How does this help you determine whether the figures are congruent?

a. Draw a rectangle in Quadrant I of a coordinate plane. Find the dimensions of the rectangle.

b. Copy the axes and the rectangle onto a piece of transparent paper.

Flip the transparent paper once so that the rectangle is in Quadrant IV. Then align the origin and the axes with the coordinate plane.

Draw the new figure in the coordinate plane. List the vertices.

c. Compare the dimensions and the angle measures of the new figure to those of the original rectangle.

d. Are the opposite sides of the new figure still parallel? Explain.

e. Can you conclude that the two figures are congruent? Explain.

f. Flip the transparent paper so that the original rectangle is in Quadrant II. Draw the new figure in the coordinate plane. List the vertices. Then repeat parts (c)–(e).

g. Compare your results with those of other students in your class. Do you think the results are true for any type of figure?

What Is Your Answer?

4. **IN YOUR OWN WORDS** How can you use reflections to classify a frieze pattern?

Practice Use what you learned about reflections to complete Exercises 4–6 on page 58.

2.3 Lesson

Key Vocabulary
reflection, *p. 56*
line of reflection, *p. 56*

A **reflection**, or *flip*, is a transformation in which a figure is reflected in a line called the **line of reflection**. A reflection creates a mirror image of the original figure.

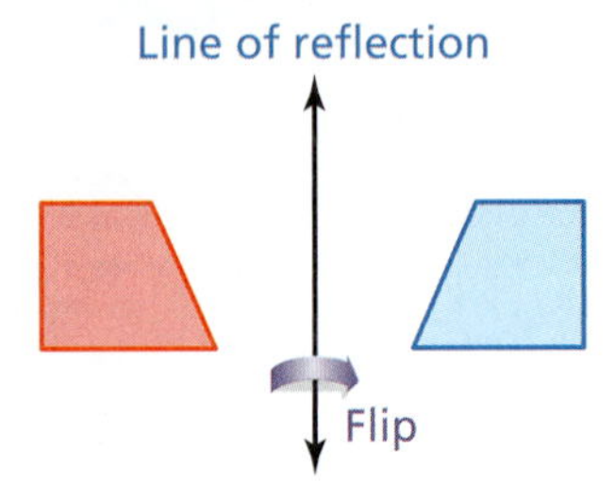

EXAMPLE 1 Identifying a Reflection

Tell whether the blue figure is a reflection of the red figure.

a.

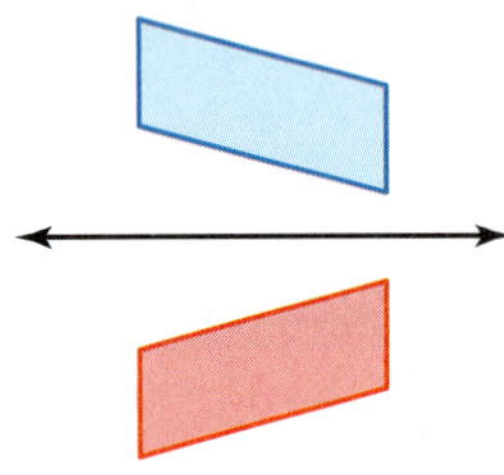

The red figure can be *flipped* to form the blue figure.

So, the blue figure is a reflection of the red figure.

b.

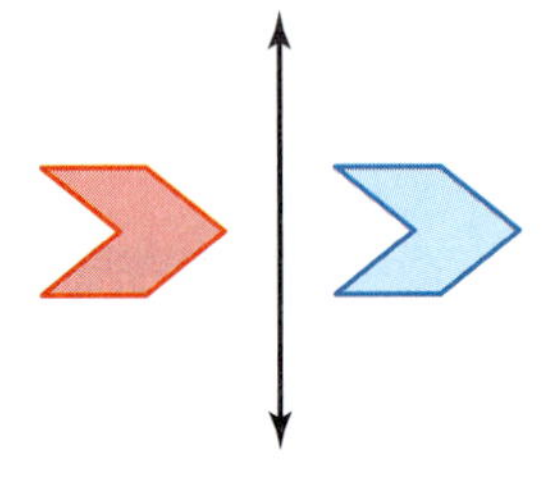

If the red figure were *flipped,* it would point to the left.

So, the blue figure is *not* a reflection of the red figure.

On Your Own

Now You're Ready
Exercises 4–9

Tell whether the blue figure is a reflection of the red figure. Explain.

1.

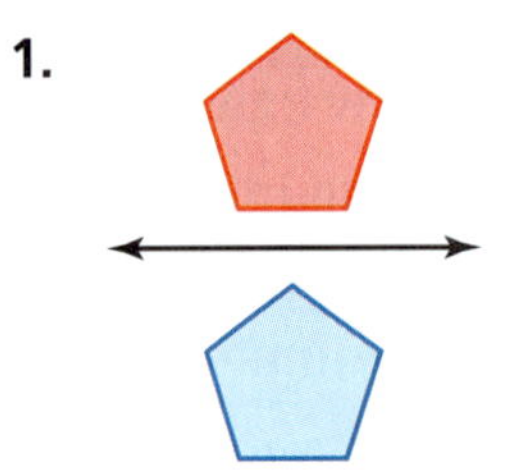

2.

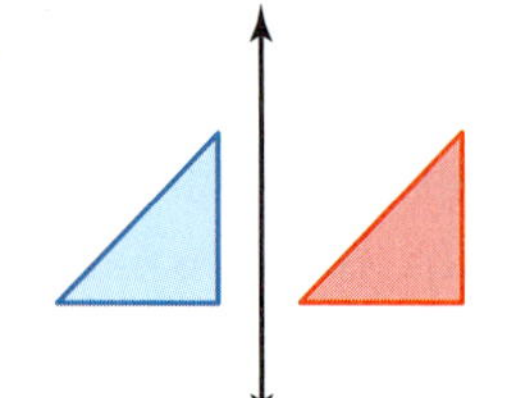

3.

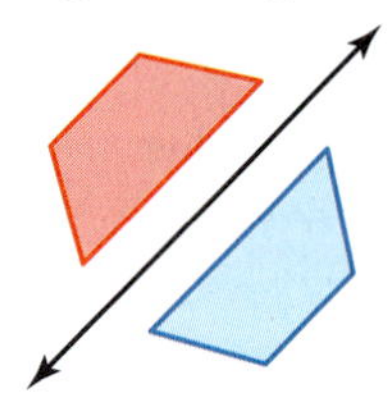

Key Idea

Reflections in the Coordinate Plane

Words To reflect a figure in the x-axis, take the opposite of the y-coordinate.

To reflect a figure in the y-axis, take the opposite of the x-coordinate.

Algebra Reflection in x-axis: $(x, y) \rightarrow (x, -y)$
Reflection in y-axis: $(x, y) \rightarrow (-x, y)$

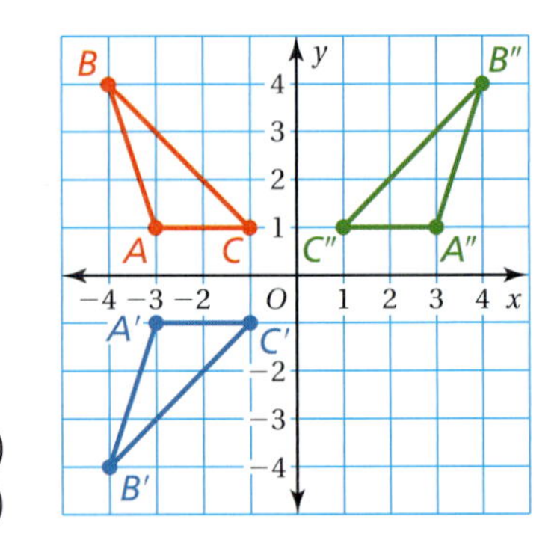

In a reflection, the original figure and its image are congruent.

Laurie's Notes

Introduction

Connect

- **Yesterday:** Students explored reflections in frieze patterns. (MP3)
- **Today:** Students will use their visual skills to draw reflections in the coordinate plane.

Motivate

? Write the word **MOM** on a transparency, and ask a few questions.
  - "What is special about this word?" Listen for ideas about reflection.
  - "Describe the result when the word is reflected in the red line." MOM
  - "Describe the result when the word is reflected in the green line." WOW

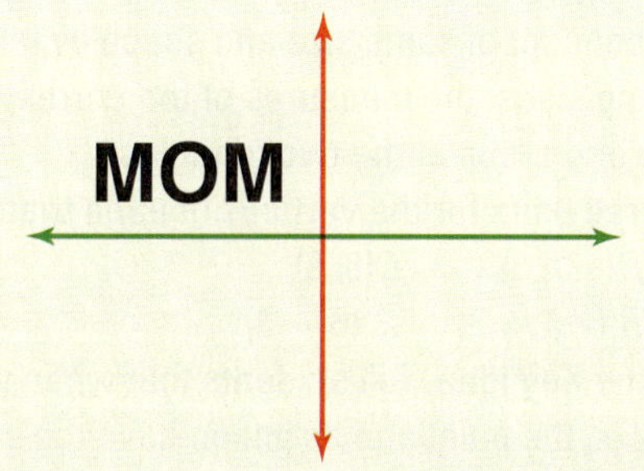

? "Can you think of other words that behave in a similar fashion?"

Lesson Notes

Discuss

- Discuss the introductory vocabulary: reflection (flip) and line of reflection.
- Relate lines of reflection to the red and green lines in the Motivate activity.

Example 1

- **Common Error:** Students may call part (b) a reflection because the shapes remain the same size and the orientation is the same. It is actually a translation.
- Offer tracing paper to students who struggle with spatial reasoning.

On Your Own

- **Neighbor Check:** Have students work independently and then have their neighbors check their work. Have students discuss any discrepancies.

Key Idea

- Write the Key Idea, using the language of A and A' as you identify the coordinates of the original figure (A, B, C) and its image (A', B', C') reflected in the x-axis. The reflection of (A, B, C) in the y-axis is the image (A'', B'', C''). Note that you read A'' as *A double prime*.
- Discuss how the coordinates change when you reflect a figure in each axis.
- **MP6 Attend to Precision:** Students may think $(x, -y)$ means there is a positive x-coordinate and a negative y-coordinate. Read the ordered pair as (x, the opposite of y) and explain that when y is negative, $-y$ is positive. Have students read the ordered pair this way also.

Goal Today's lesson is identifying and drawing **reflections**.

Lesson Tutorials
Lesson Plans
Answer Presentation Tool

Extra Example 1

Tell whether the blue figure is a reflection of the red figure.

a.

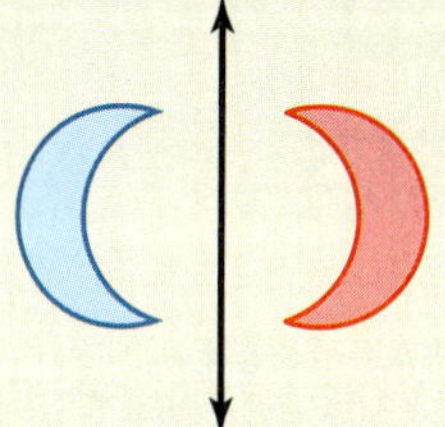

a reflection

b.

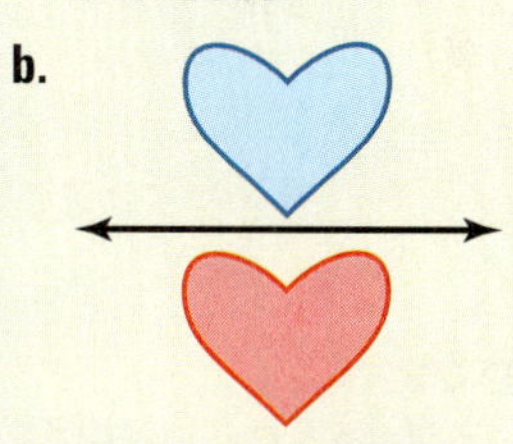

not a reflection

On Your Own

1. no; It is a translation.
2. no; It is a translation.
3. yes; The red figure can be flipped to form the blue figure.

Extra Example 2

The vertices of a parallelogram are $A(-1, -1)$, $B(2, -1)$, $C(4, -3)$, and $D(1, -3)$. Draw the figure and its reflection in the x-axis. What are the coordinates of the image?

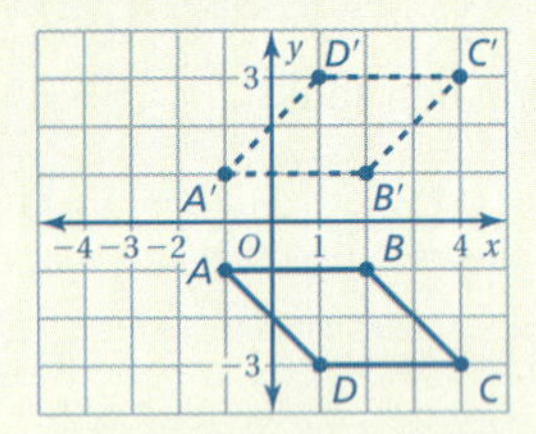

$A'(-1, 1)$, $B'(2, 1)$, $C'(4, 3)$, $D'(1, 3)$

Extra Example 3

The vertices of a triangle are $A(1, -2)$, $B(4, -2)$, and $C(1, 4)$. Draw the figure and its reflection in the y-axis.

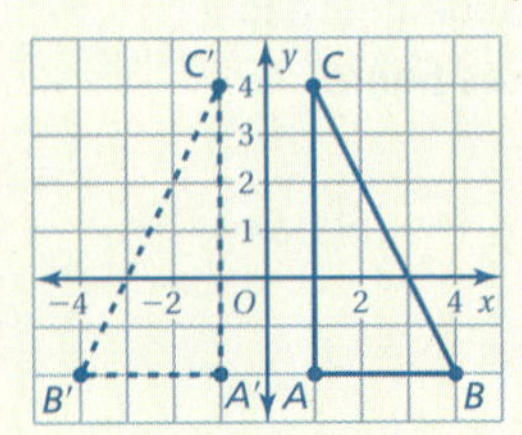

$A'(-1, -2)$, $B'(-4, -2)$, $C'(-1, 4)$

On Your Own

4. See Additional Answers.

English Language Learners

Vocabulary

Tell students that different words may be used to describe a reflection in a coordinate plane. For example, a figure is a reflection *in* the x-axis, *about* the x-axis, *across* the x-axis, or *over* the x-axis. The same words can be used to describe a reflection in the y-axis as well.

Laurie's Notes

Example 2

- Draw $\triangle ABC$ and label the vertices.
- ? "Which is the x-axis?" the horizontal axis
- We want to reflect the triangle from above the x-axis to below the x-axis.
- Note the suggestion boxes on the graph. Start with point A. Say, "Because A is 1 unit above the x-axis, it will be reflected to 1 unit below the x-axis." Repeat using similar language for points B and C. When students don't use this approach, they can easily translate the triangle instead of reflecting it.
- **Common Error:** The numbers written horizontally along the x-axis may cause students to be off by one number when they find the coordinates of each point in the blue triangle.
- ? "Is the blue triangle the same size and shape as the red triangle?" yes
- Reinforce the concept of same size and shape by talking about the lengths of corresponding sides, the measures of the corresponding angles, and the perimeters and areas of the two triangles.
- Write the ordered pairs for the vertices of each triangle.

  $A(-1, 1)$ $B(-1, 3)$ $C(6, 3)$

  $A'(-1, -1)$ $B'(-1, -3)$ $C'(6, -3)$

  Refer back to the Key Idea. Tell students that when you reflect a point across the x-axis, the point and its image have the same x-coordinates, and the y-coordinates are opposites.

Example 3

- This problem is similar to Example 2 except the original figure is a quadrilateral and it is reflected in the y-axis.
- **MP4 Model with Mathematics:** Set up a table to show the change in the x-coordinate when you reflect a point in the y-axis.
- ? "Is the blue quadrilateral the same size and shape as the red quadrilateral?" yes
- ? Ask questions like the following about each vertex and its image.
  - "How many units is P from the y-axis?" 2 units
  - "How many units is P' from the y-axis?" 2 units

On Your Own

- **Think-Pair-Share:** Students should read each question independently and then work in pairs to answer the questions. When they have answered the questions, the pair should compare their answers with another group and discuss any discrepancies.

Closure

- Draw a right triangle in Quadrant II. Reflect the triangle in the x-axis. Reflect the original triangle in the y-axis.

EXAMPLE 2 Reflecting a Figure in the x-axis

The vertices of a triangle are $A(-1, 1)$, $B(-1, 3)$, and $C(6, 3)$. Draw the figure and its reflection in the x-axis. What are the coordinates of the image?

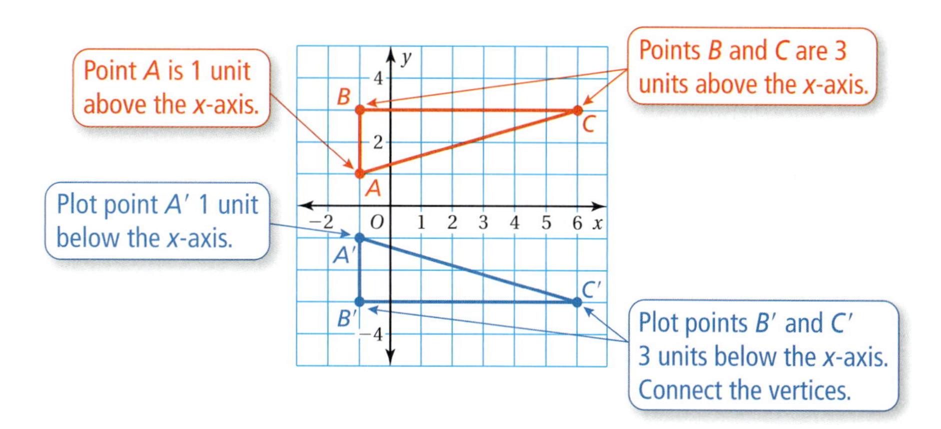

The coordinates of the image are $A'(-1, -1)$, $B'(-1, -3)$, and $C'(6, -3)$.

EXAMPLE 3 Reflecting a Figure in the y-axis

The vertices of a quadrilateral are $P(-2, 5)$, $Q(-1, -1)$, $R(-4, 2)$, and $S(-4, 4)$. Draw the figure and its reflection in the y-axis.

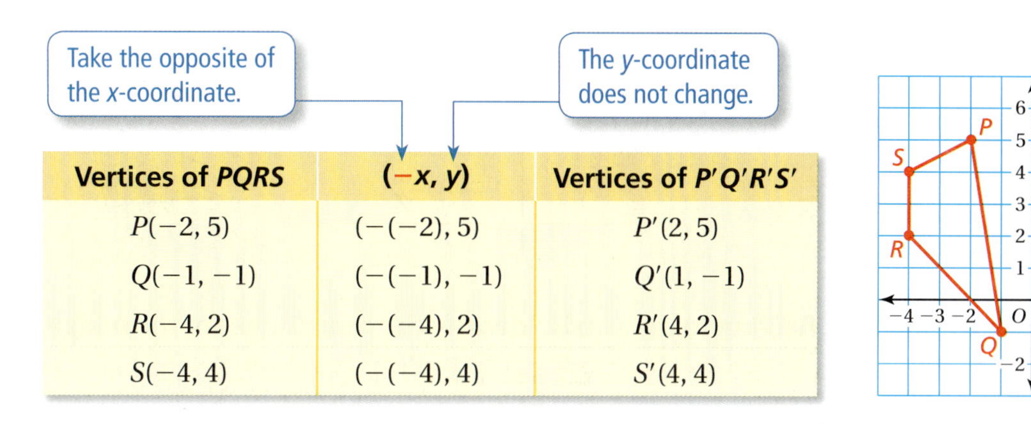

Vertices of $PQRS$	$(-x, y)$	Vertices of $P'Q'R'S'$
$P(-2, 5)$	$(-(-2), 5)$	$P'(2, 5)$
$Q(-1, -1)$	$(-(-1), -1)$	$Q'(1, -1)$
$R(-4, 2)$	$(-(-4), 2)$	$R'(4, 2)$
$S(-4, 4)$	$(-(-4), 4)$	$S'(4, 4)$

The figure and its image are shown at the above right.

On Your Own

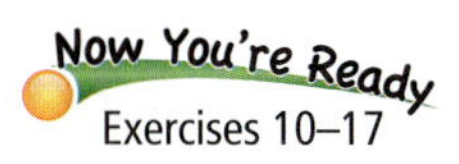

4. The vertices of a rectangle are $A(-4, -3)$, $B(-4, -1)$, $C(-1, -1)$, and $D(-1, -3)$.

   a. Draw the figure and its reflection in the x-axis.

   b. Draw the figure and its reflection in the y-axis.

   c. Are the images in parts (a) and (b) congruent? Explain.

2.3 Exercises

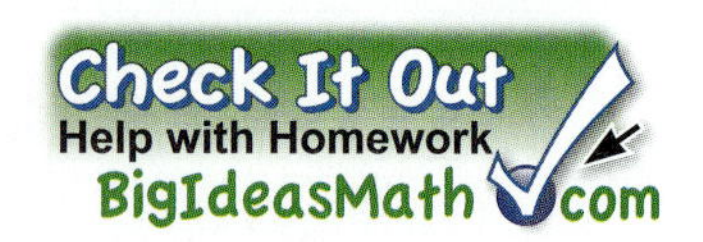

Vocabulary and Concept Check

1. **WHICH ONE DOESN'T BELONG?** Which transformation does *not* belong with the other three? Explain your reasoning.

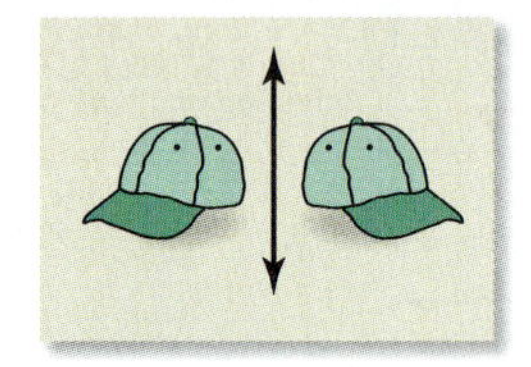 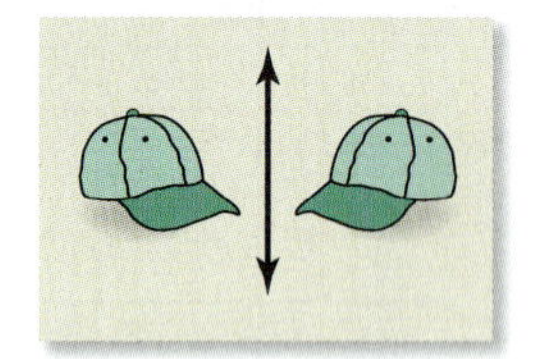 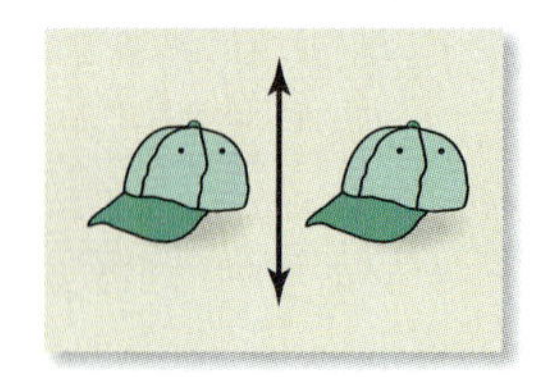 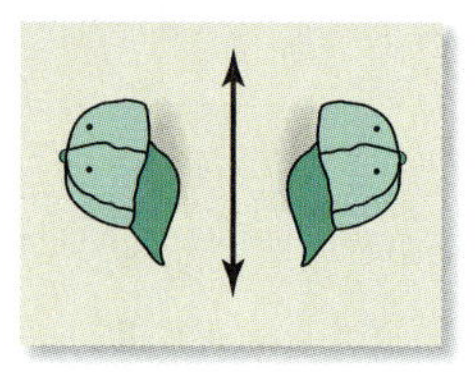

2. **WRITING** How can you tell when one figure is a reflection of another figure?

3. **REASONING** A figure lies entirely in Quadrant I. The figure is reflected in the x-axis. In which quadrant is the image?

Practice and Problem Solving

Tell whether the blue figure is a reflection of the red figure.

① 4. 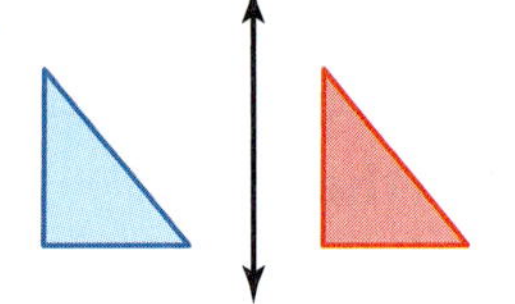

5. 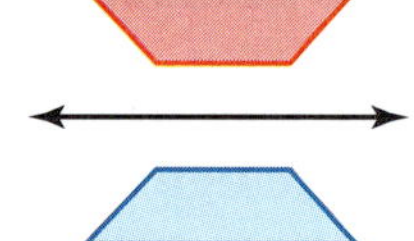

6. 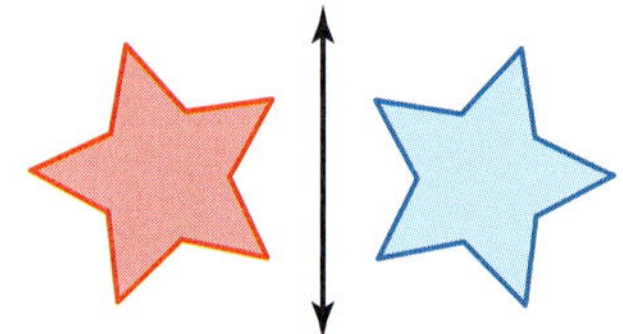

7. 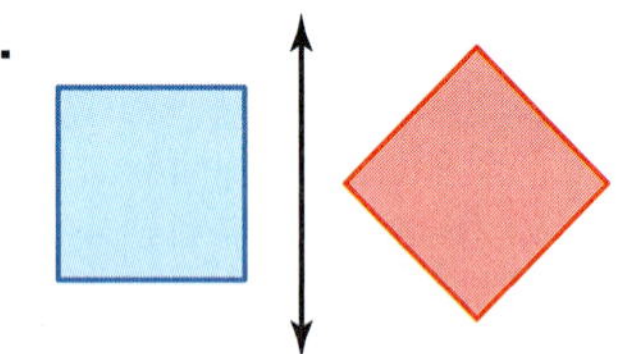

8. 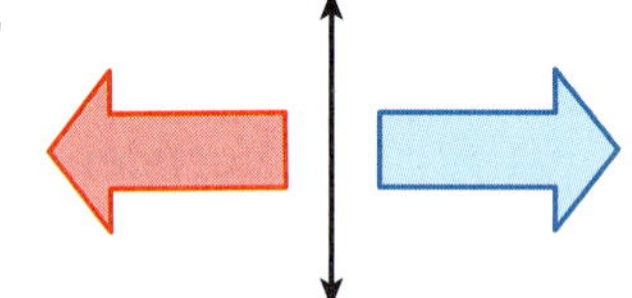

9. 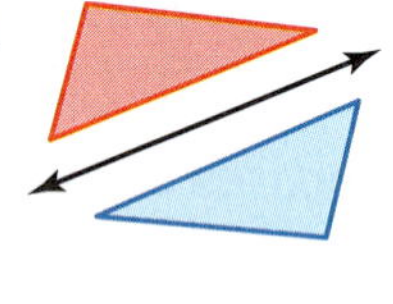

Draw the figure and its reflection in the x-axis. Identify the coordinates of the image.

② 10. $A(3, 2)$, $B(4, 4)$, $C(1, 3)$

11. $M(-2, 1)$, $N(0, 3)$, $P(2, 2)$

12. $H(2, -2)$, $J(4, -1)$, $K(6, -3)$, $L(5, -4)$

13. $D(-2, -1)$, $E(0, -1)$, $F(0, -5)$, $G(-2, -5)$

Draw the figure and its reflection in the y-axis. Identify the coordinates of the image.

③ 14. $Q(-4, 2)$, $R(-2, 4)$, $S(-1, 1)$

15. $T(4, -2)$, $U(4, 2)$, $V(6, -2)$

16. $W(2, -1)$, $X(5, -2)$, $Y(5, -5)$, $Z(2, -4)$

17. $J(2, 2)$, $K(7, 4)$, $L(9, -2)$, $M(3, -1)$

18. **ALPHABET** Which letters look the same when reflected in the line?

A B C D E F G H I J K L M N O P Q R S T U V W X Y Z

Assignment Guide and Homework Check

Level	Day 1 Activity Assignment	Day 2 Lesson Assignment	Homework Check
Basic	4–6, 29–33	1–3, 7–21 odd	7, 11, 15, 19
Average	4–6, 29–33	1–3, 8–18 even, 19–23 odd, 24, 25	8, 10, 14, 19
Advanced	1–6, 10–22 even, 24–33		10, 14, 20, 27

Common Errors

- **Exercises 4–9** Some students may struggle with the visual and think that a translation is actually a reflection. Give students tracing paper to trace the objects, and then fold the paper to see if the vertices line up.
- **Exercises 10–17** Students may reflect in the incorrect axis. Refer them back to Examples 2 and 3.
- **Exercise 18** Students may need to copy the alphabet and fold their paper on the line to see which letters look the same.

A B C D E F G H I J K L M N O P Q R S T U V W X Y Z

∀ B C D E Ⅎ ⅁ H I ſ K ⅂ W N O d Ò ꓤ S ⊥ ∩ Λ M X ⅄ Z

2.3 Record and Practice Journal

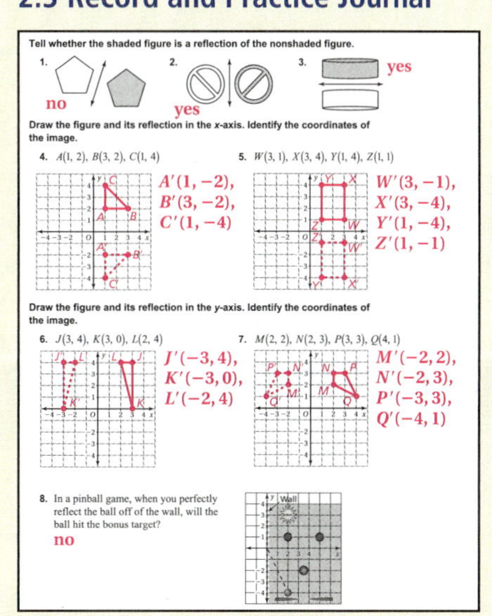
Tell whether the shaded figure is a reflection of the nonshaded figure.

1. no
2. yes
3. yes

Draw the figure and its reflection in the *x*-axis. Identify the coordinates of the image.

4. $A(1, 2)$, $B(3, 2)$, $C(1, 4)$ — $A'(1, -2)$, $B'(3, -2)$, $C'(1, -4)$
5. $W(3, 1)$, $X(3, 4)$, $Y(1, 4)$, $Z(1, 1)$ — $W'(3, -1)$, $X'(3, -4)$, $Y'(1, -4)$, $Z'(1, -1)$

Draw the figure and its reflection in the *y*-axis. Identify the coordinates of the image.

6. $J(3, 4)$, $K(3, 0)$, $L(2, 4)$ — $J'(-3, 4)$, $K'(-3, 0)$, $L'(-2, 4)$
7. $M(2, 2)$, $N(2, 3)$, $P(3, 3)$, $Q(4, 1)$ — $M'(-2, 2)$, $N'(-2, 3)$, $P'(-3, 3)$, $Q'(-4, 1)$

8. In a pinball game, when you perfectly reflect the ball off of the wall, will the ball hit the bonus target? no

Vocabulary and Concept Check

1. The third one because it is not a reflection.
2. A figure is a reflection of another figure if one is the mirror image of the other.
3. Quadrant IV

Practice and Problem Solving

4. no
5. yes
6. yes
7. no
8. yes
9. no
10. 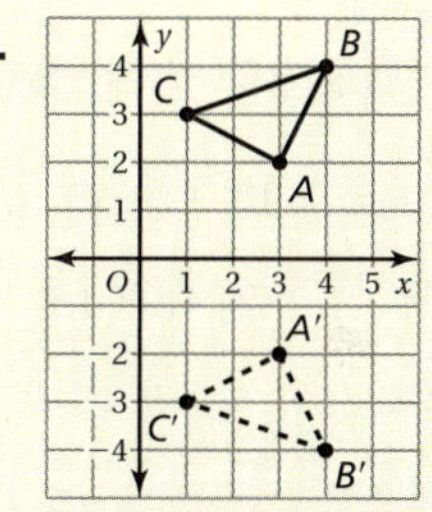

$A'(3, -2)$, $B'(4, -4)$, $C'(1, -3)$

11. 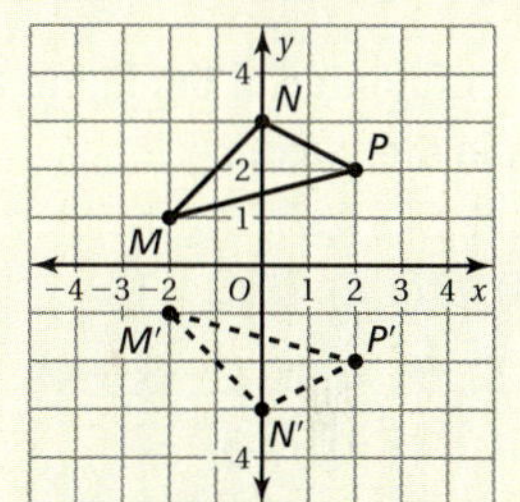

$M'(-2, -1)$, $N'(0, -3)$, $P'(2, -2)$

12. 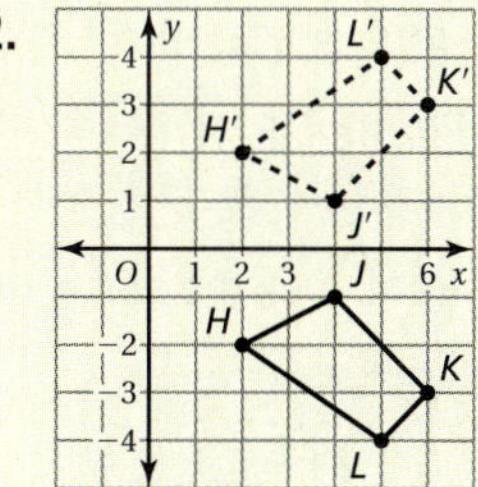

$H'(2, 2)$, $J'(4, 1)$, $K'(6, 3)$, $L'(5, 4)$

13–17. See Additional Answers.

18. B, C, D, E, H, I, K, O, X

Practice and Problem Solving

19. x-axis
20. y-axis
21. y-axis
22. x-axis
23. $R'(3, -4)$, $S'(3, -1)$, $T'(1, -4)$
24. $W'(-4, 5)$, $X'(-4, 2)$, $Y'(0, 2)$, $Z'(2, 5)$
25. yes; Translations and reflections produce images that are congruent to the original figure.
26. $(-x, -y)$
27. See *Taking Math Deeper*.
28. See Additional Answers.

Fair Game Review

29. obtuse
30. straight
31. right
32. acute
33. B

Mini-Assessment

Find the coordinates of the figure after reflecting in the y-axis.

1. $A(-2, 4)$, $B(-4, 2)$, $C(-1, -1)$
   $A'(2, 4)$, $B'(4, 2)$, $C'(1, -1)$
2. $A(-2, 5)$, $B(-5, 1)$, $C(-3, -4)$
   $A'(2, 5)$, $B'(5, 1)$, $C'(3, -4)$

Find the coordinates of the figure after reflecting in the x-axis.

3. $A(-4, -2)$, $B(4, -1)$, $C(1, -6)$
   $A'(-4, 2)$, $B'(4, 1)$, $C'(1, 6)$
4. $A(-2, 5)$, $B(4, 8)$, $C(5, 1)$
   $A'(-2, -5)$, $B'(4, -8)$, $C'(5, -1)$
5. Will the letter E look the same when reflected in the y-axis? no

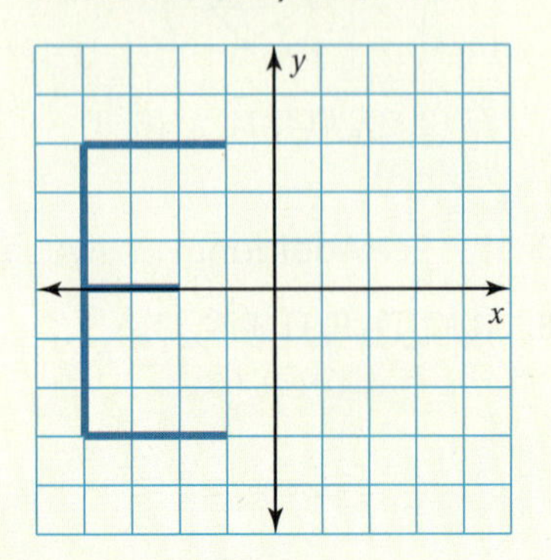

Taking Math Deeper

Exercise 27

Students need a mirror to see this one.

Looking straight on, this is the ambulance.

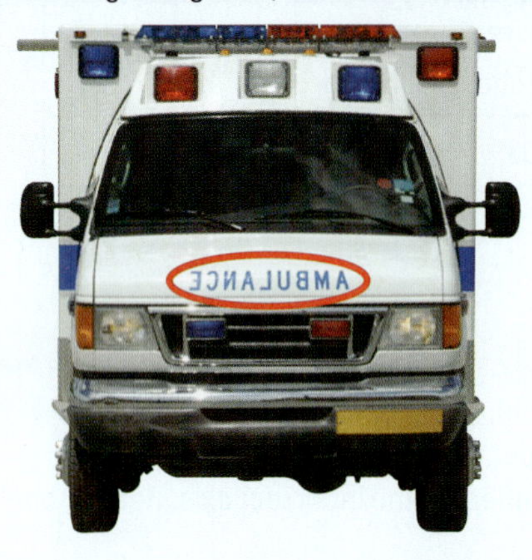

a. Looking in a mirror, this is what you see.

b. The word "AMBULANCE" is printed backwards so that when the ambulance comes up behind a car, the word will look correct in the rear-view mirror.

Reteaching and Enrichment Strategies

If students need help. . .	If students got it. . .
Resources by Chapter • Practice A and Practice B • Puzzle Time Record and Practice Journal Practice Differentiating the Lesson Lesson Tutorials Skills Review Handbook	Resources by Chapter • Enrichment and Extension • Technology Connection Start the next section

The coordinates of a point and its image are given. Is the reflection in the *x*-axis or *y*-axis?

19. $(2, -2) \rightarrow (2, 2)$

20. $(-4, 1) \rightarrow (4, 1)$

21. $(-2, -5) \rightarrow (2, -5)$

22. $(-3, -4) \rightarrow (-3, 4)$

Find the coordinates of the figure after the transformations.

23. Translate the triangle 1 unit right and 5 units down. Then reflect the image in the y-axis.

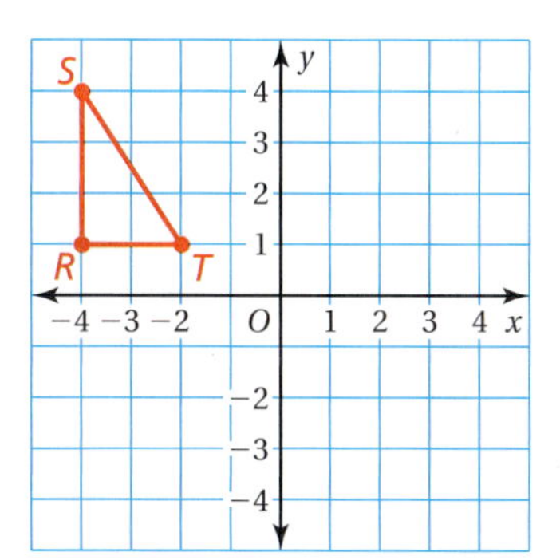

24. Reflect the trapezoid in the x-axis. Then translate the trapezoid 2 units left and 3 units up.

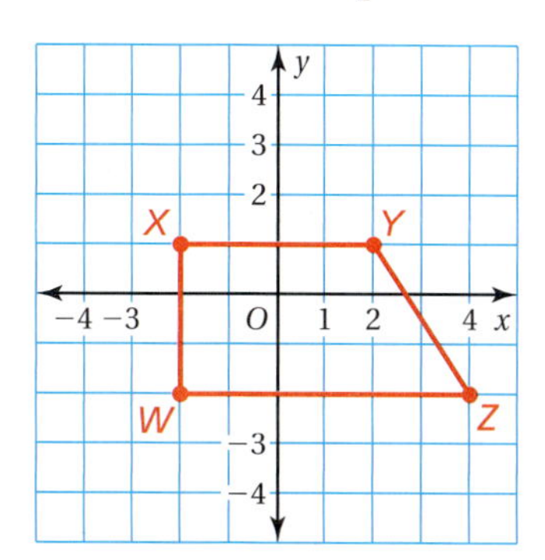

25. REASONING In Exercises 23 and 24, is the original figure congruent to the final image? Explain.

26. NUMBER SENSE You reflect a point (x, y) in the x-axis, and then in the y-axis. What are the coordinates of the final image?

27. EMERGENCY VEHICLE Hold a mirror to the left side of the photo of the vehicle.

a. What word do you see in the mirror?

b. Why do you think it is written that way on the front of the vehicle?

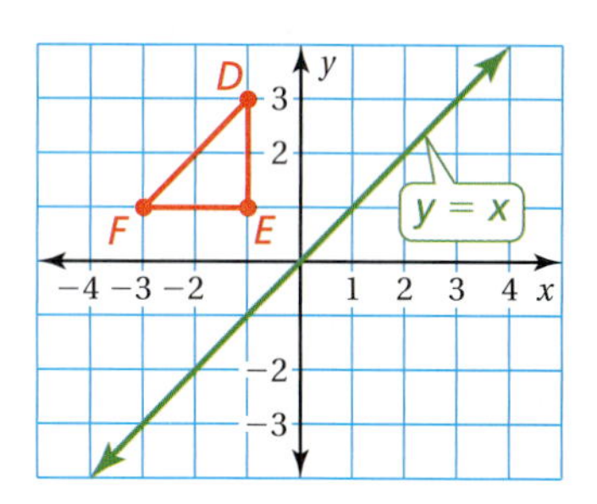

28. Critical Thinking Reflect the triangle in the line $y = x$. How are the x- and y-coordinates of the image related to the x- and y-coordinates of the original triangle?

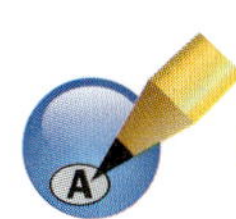

Fair Game Review What you learned in previous grades & lessons

Classify the angle as *acute, right, obtuse,* or *straight*. *(Skills Review Handbook)*

29.

30.

31.

32.

33. MULTIPLE CHOICE 36 is 75% of what number? *(Skills Review Handbook)*

Ⓐ 27 Ⓑ 48 Ⓒ 54 Ⓓ 63

2.4 Rotations

Essential Question What are the three basic ways to move an object in a plane?

The Meaning of a Word ● Rotate

A bicycle wheel can **rotate** clockwise or counterclockwise.

11 12 1 10 2 9 3 8 4 7 6 5

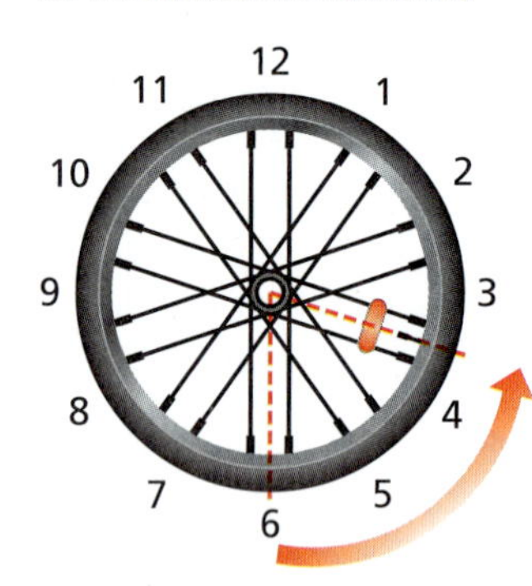

1 ACTIVITY: Three Basic Ways to Move Things

There are three basic ways to move objects on a flat surface.

_____ the object. _____ the object. _____ the object.

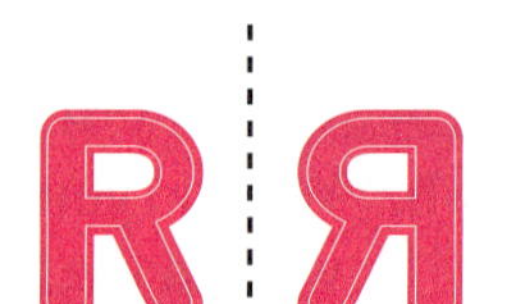

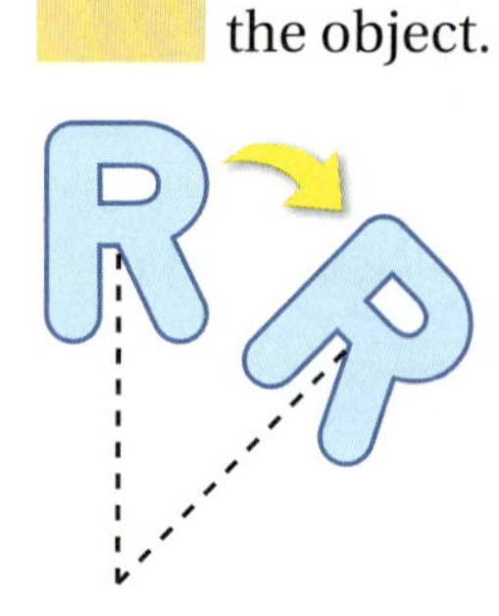

Geometry

In this lesson, you will

- identify rotations.
- rotate figures in the coordinate plane.
- use more than one transformation to find images of figures.

Learning Standards
MACC.8.G.1.1
MACC.8.G.1.2
MACC.8.G.1.3

Work with a partner.

a. What type of triangle is the blue triangle? Is it congruent to the red triangles? Explain.

b. Decide how you can move the blue triangle to obtain each red triangle.

c. Is each move a *translation*, a *reflection*, or a *rotation*?

Laurie's Notes

Introduction

Standards for Mathematical Practice

- **MP3 Construct Viable Arguments and Critique the Reasoning of Others:** As students manipulate geometric shapes, they develop spatial reasoning and make conjectures. Mathematically proficient students explore the truth of their conjectures by considering all possible cases.

Motivate

- **Time to Play:** *Name Five Twice.* In this game, students will name things that rotate: the first five objects rotate about a point in the center of the object (like a wheel) and the next five objects rotate about a point not in the center of the object (like a windshield wiper). Give students time to work with partners to generate two lists of five.
  Example 1: car tire, Ferris wheel, merry-go-round, dial on a combination lock
  Example 2: windshield wiper, lever—as on a mechanical arm or wrench
- **FYI:** Rotation is generally the most challenging transformation for students to visualize.

Activity Notes

Activity 1

- To help students visualize, offer them tracing paper so that they can sketch and transform the blue triangle. If tracing paper is not readily available, ahead of time ask a local doughnut shop or bakery for a donation of a box of tissue paper.
- For students with higher spatial skills, sketching the blue triangle on tissue paper will not be necessary.
- After students have finished the activity, ask them to share their results. If necessary, prompt them to explain their reasoning.

Florida Common Core Standards

MACC.8.G.1.1 Verify experimentally the properties of rotations, reflections, and translations.
MACC.8.G.1.2 Understand that a two-dimensional figure is congruent to another if the second can be obtained from the first by a sequence of rotations, reflections, and translations; given two congruent figures, describe a sequence that exhibits the congruence between them.
MACC.8.G.1.3 Describe the effect of . . . translations, rotations, and reflections on two-dimensional figures using coordinates.

Previous Learning

Students should know how to plot points in the coordinate plane.

2.4 Record and Practice Journal

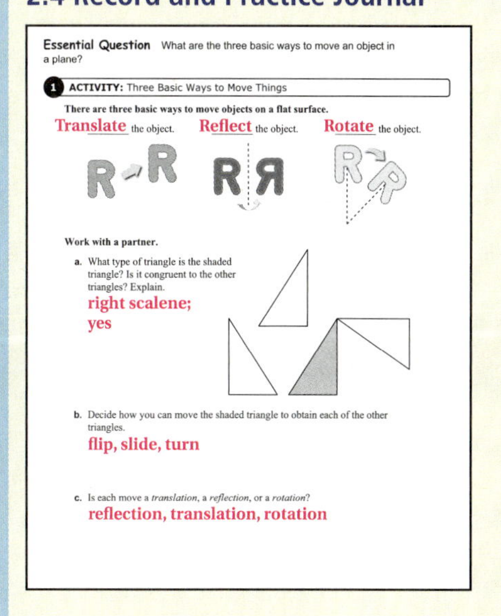

Essential Question What are the three basic ways to move an object in a plane?

1 ACTIVITY: Three Basic Ways to Move Things

There are three basic ways to move objects on a flat surface.

Translate the object. Reflect the object. Rotate the object.

Work with a partner.

a. What type of triangle is the shaded triangle? Is it congruent to the other triangles? Explain.
right scalene; yes

b. Decide how you can move the shaded triangle to obtain each of the other triangles.
flip, slide, turn

c. Is each move a *translation*, a *reflection*, or a *rotation*?
reflection, translation, rotation

Differentiated Instruction

Kinesthetic

Project a coordinate plane using an overhead projector. Give one student a geometric shape to place in one quadrant of the coordinate plane. Give another student a duplicate shape to place in another quadrant of the coordinate plane so that the shape is a rotation of the first shape. Discuss whether the duplicate shape could also be a translation or reflection of the first shape.

2.4 Record and Practice Journal

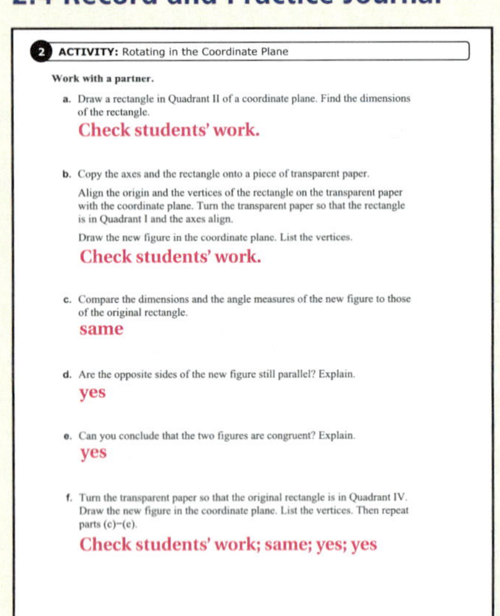

2 ACTIVITY: Rotating in the Coordinate Plane

Work with a partner.

a. Draw a rectangle in Quadrant II of a coordinate plane. Find the dimensions of the rectangle.
Check students' work.

b. Copy the axes and the rectangle onto a piece of transparent paper.
Align the origin and the vertices of the rectangle on the transparent paper with the coordinate plane. Turn the transparent paper so that the rectangle is in Quadrant I and the axes align.
Draw the new figure in the coordinate plane. List the vertices.
Check students' work.

c. Compare the dimensions and the angle measures of the new figure to those of the original rectangle.
same

d. Are the opposite sides of the new figure still parallel? Explain.
yes

e. Can you conclude that the two figures are congruent? Explain.
yes

f. Turn the transparent paper so that the original rectangle is in Quadrant IV. Draw the new figure in the coordinate plane. List the vertices. Then repeat parts (c)–(e).
Check students' work; same; yes; yes

g. Compare your results with those of other students in your class. Do you think the results are true for any type of figure?
yes

What Is Your Answer?

3. **IN YOUR OWN WORDS** What are the three basic ways to move an object in a plane? Draw an example of each.
translation, reflection, rotation

4. **PRECISION** Use the results of Activity 2(b).
   a. Draw four angles using the conditions below.
      - The origin is the vertex of each angle.
      - One side of each angle passes through a vertex of the original rectangle.
      - The other side of each angle passes through the corresponding vertex of the rotated rectangle.

      Check students' work.
   b. Measure each angle in part (a). For each angle, measure the distances between the origin and the vertices of the rectangles. What do you notice?
   All angle measures are 90°. The distances are equal.
   c. How can the results of part (b) help you rotate a figure?
   ***Sample answer:* You can use a ruler and a protractor to rotate each vertex of a figure.**

5. **PRECISION** Repeat the procedure in Question 4 using the results of Activity 2(f).
b. All angle measures are 180°. The distances are equal.

Laurie's Notes

Activity 2

- The size of the rectangle students draw is irrelevant, as long as they draw it in Quadrant II. Also, encourage students not to draw a square. Their observations with a square may not apply to a rectangle that is not a square.
- **Teaching Tip:** Before students turn the transparent paper, suggest that they hold it in place with the tip of a sharp pencil located at the origin. Then tell them to rotate the transparent paper a quarter turn clockwise. To make sure each student performs the rotation correctly, I tell them to rotate the paper so that the top edge (vertical) becomes the right edge (horizontal).
- If you are using an interactive white board and you have access to the Internet in the classroom, ahead of time find an applet online that demonstrates a 90° clockwise rotation. Share it with students during class.
- **MP3:** Ask students what conjectures they can make about the three rectangles. Make sure they provide evidence for their conjectures. Students may suggest, for instance, that the rectangles are congruent, have the same perimeter or area, all have opposite sides that are parallel, etc. They may also make conjectures about the coordinates of the vertices after rotation and their relation to those of the original rectangle.

What Is Your Answer?

- In Questions 4 and 5, make sure students know that the "original rectangle" is the rectangle they drew in part (a) of Activity 2.
- You may need to demonstrate part (a) of Question 4. After students have finished Question 4, ask a volunteer to share his or her results.

Closure

- Draw a right triangle in Quadrant I with the right angle at (0, 0). Rotate the triangle 90° clockwise about the origin.

2 ACTIVITY: Rotating in the Coordinate Plane

Work with a partner.

a. Draw a rectangle in Quadrant II of a coordinate plane. Find the dimensions of the rectangle.

b. Copy the axes and the rectangle onto a piece of transparent paper.

Align the origin and the vertices of the rectangle on the transparent paper with the coordinate plane. Turn the transparent paper so that the rectangle is in Quadrant I and the axes align.

Draw the new figure in the coordinate plane. List the vertices.

Math Practice 6

Calculate Accurately

What must you do to rotate the figure correctly?

c. Compare the dimensions and the angle measures of the new figure to those of the original rectangle.

d. Are the opposite sides of the new figure still parallel? Explain.

e. Can you conclude that the two figures are congruent? Explain.

f. Turn the transparent paper so that the original rectangle is in Quadrant IV. Draw the new figure in the coordinate plane. List the vertices. Then repeat parts (c)–(e).

g. Compare your results with those of other students in your class. Do you think the results are true for any type of figure?

What Is Your Answer?

3. **IN YOUR OWN WORDS** What are the three basic ways to move an object in a plane? Draw an example of each.

4. **PRECISION** Use the results of Activity 2(b).

   a. Draw four angles using the conditions below.
   - The origin is the vertex of each angle.
   - One side of each angle passes through a vertex of the original rectangle.
   - The other side of each angle passes through the corresponding vertex of the rotated rectangle.

   b. Measure each angle in part (a). For each angle, measure the distances between the origin and the vertices of the rectangles. What do you notice?

   c. How can the results of part (b) help you rotate a figure?

5. **PRECISION** Repeat the procedure in Question 4 using the results of Activity 2(f).

Use what you learned about transformations to complete Exercises 7–9 on page 65.

2.4 Lesson

Key Vocabulary

rotation, *p. 62*
center of rotation, *p. 62*
angle of rotation, *p. 62*

Rotations

A **rotation**, or *turn*, is a transformation in which a figure is rotated about a point called the **center of rotation**. The number of degrees a figure rotates is the **angle of rotation**.

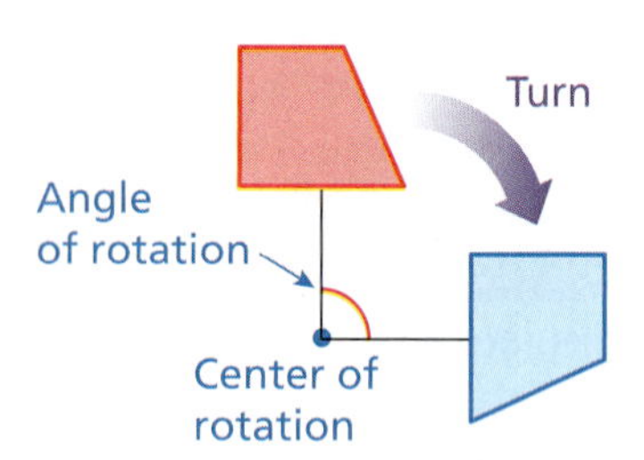

In a rotation, the original figure and its image are congruent.

EXAMPLE 1 Identifying a Rotation

You must rotate the puzzle piece 270° clockwise about point *P* to fit it into a puzzle. Which piece fits in the puzzle as shown?

Ⓐ Ⓑ Ⓒ Ⓓ

Rotate the puzzle piece 270° clockwise about point *P*.

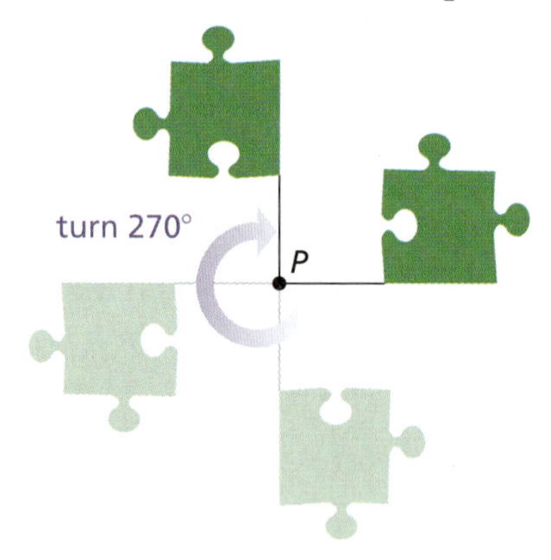

So, the correct answer is Ⓒ.

Study Tip

When rotating figures, it may help to sketch the rotation in several steps, as shown in Example 1.

On Your Own

1. Which piece is a 90° counterclockwise rotation about point *P*?
2. Is Choice D a rotation of the original puzzle piece? If not, what kind of transformation does the image show?

Laurie's Notes

Introduction

Connect

- **Yesterday:** Students explored and sketched rotations. (MP3)
- **Today:** Students will use their visual skills to draw rotations in the coordinate plane.

Motivate

- Use a marker to make two sizeable dots, one at the tip of your middle finger and one at the base of your palm. Anchor your elbow on a level surface. Wave at the class so that your elbow is the pivot.
- ? Do a "wave" of 90°, by starting in the horizontal position and "waving" to the vertical position. Ask the following questions.
  - "Through how many degrees did I wave my hand?" 90°
  - "Did my elbow move?" no
  - "Did the two points move the same distance?" no "If not, which point moved farther?" The point on the tip of the middle finger moved farther.
- Relate this motion to that of a windshield wiper. The farther a point on the wiper is from the point of rotation, the farther it travels.

Lesson Notes

Key Idea

- The rotation is hard to visualize because the center of rotation is generally not attached to the shape being rotated. My hand is connected to my forearm, which is connected to my elbow, so the "wave" is easier to see as a rotation. When a diagram only shows the original figure and the image, it is harder to see the angle of rotation.
- **MP3 Construct Viable Arguments and Critique the Reasoning of Others:** Note that the relationships between the coordinates of the vertices of a figure before and after rotation are not given. If you feel that your students are ready, explore this with them. Have students make conjectures about the relationships and explain their reasoning.

Example 1

- Model a rotation of 270° clockwise using a transparency with an arrow pointing to the right. Lightly place your finger on the middle to act as the center of rotation. Turn the transparency 90° clockwise, 3 times, stopping each time for students to see where the arrow is pointing. After 270° the arrow will be pointing up.

On Your Own

- If you have a puzzle piece that you can use to model these two questions, then it would help those students who have difficulty visualizing the movement of the pieces.

Goal Today's lesson is identifying and drawing **rotations**.

Lesson Tutorials
Lesson Plans
Answer Presentation Tool

Extra Example 1

Tell whether the blue figure is a 180° clockwise rotation of the red figure.

a.

Is a 180° clockwise rotation.

b.

Is not a 180° clockwise rotation.

On Your Own

1. C
2. no; reflection

Extra Example 2

The vertices of a triangle are $A(-4, 1)$, $B(-1, 6)$, and $C(-1, 1)$. Rotate the triangle 90° clockwise about the origin. What are the coordinates of the image?

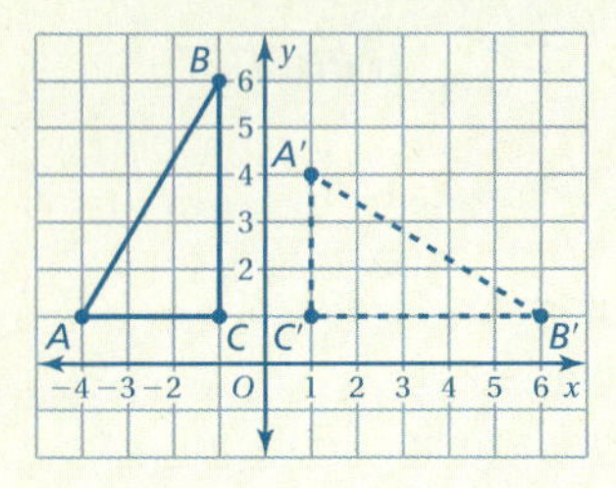

$A'(1, 4)$, $B'(6, 1)$, and $C'(1, 1)$

Extra Example 3

The vertices of a trapezoid are $A(1, 2)$, $B(4, 4)$, $C(4, 6)$, and $D(1, 6)$. Rotate the trapezoid 270° counterclockwise about vertex A. What are the coordinates of the image?

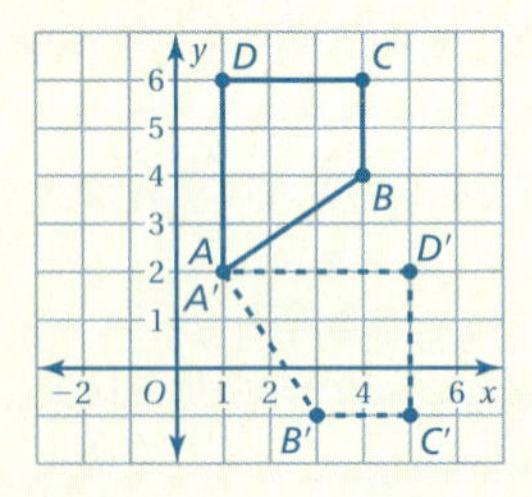

$A'(1, 2)$, $B'(3, -1)$, $C'(5, -1)$, $D'(5, 2)$

On Your Own

3. See Additional Answers.

English Language Learners

Vocabulary

Discuss the meanings of the words *translation*, *reflection*, and *rotation*. Students may think of translation as a process of writing text in another (parallel) language. Mathematically, a translation is when all points of a figure move along parallel lines. Have students visualize a sun setting on the horizon of the ocean. At the point where half of the sun has set, its reflection in the water gives the appearance of a full-circled sun. A rotation about a point in the plane is similar to a nail rotating around the wheel of a tire.

Laurie's Notes

Example 2

- Tracing paper or a transparency will be needed by many students for this example. Students need to *see* where the trapezoid rotates, before they can plot the ordered pairs.
- Draw the trapezoid and label the vertices.
- **Common Error:** Students will rotate the trapezoid about vertex Z instead of rotating about the origin.
- **Teaching Strategy:** Remind students that when a figure is rotated 180°, what was on the top will rotate to the bottom, and vice versa. Model this by holding a sheet of paper and rotating it 180°.
- **Extension:** List the coordinates of the original trapezoid and the image.

  $W(-4, 2) \longrightarrow W'(4, -2)$
  $X(-3, 4) \longrightarrow X'(3, -4)$
  $Y(-1, 4) \longrightarrow Y'(1, -4)$
  $Z(-1, 2) \longrightarrow Z'(1, -2)$

? **MP8 Look for and Express Regularity in Repeated Reasoning:** "What happens to the coordinates of the point (x, y) when you rotate the point 180° about the origin?" The coordinates become $(-x, -y)$.

Example 3

- In this example, a triangle is rotated 90° counterclockwise and the center of rotation is one of the vertices instead of the origin.
- Hold a sheet of paper facing the students.

? "When I rotate the sheet of paper 90° counterclockwise, which way will the top of the paper rotate?" to the left

- Work through the problem as shown.

? "Did all of the points on the triangle move? Explain." no; Vertex L did not move because it is the center of rotation.

- **Big Idea:** The lengths of the sides of a triangle do not change when you rotate the triangle.

On Your Own

- **Think-Pair-Share:** Students should read each question independently and then work in pairs to answer the questions. When they have answered the questions, the pair should compare their answers with another group and discuss any discrepancies.

EXAMPLE 2 Rotating a Figure

The vertices of a trapezoid are $W(-4, 2)$, $X(-3, 4)$, $Y(-1, 4)$, and $Z(-1, 2)$. Rotate the trapezoid 180° about the origin. What are the coordinates of the image?

Study Tip

A 180° clockwise rotation and a 180° counterclockwise rotation have the same image. So, you do not need to specify direction when rotating a figure 180°.

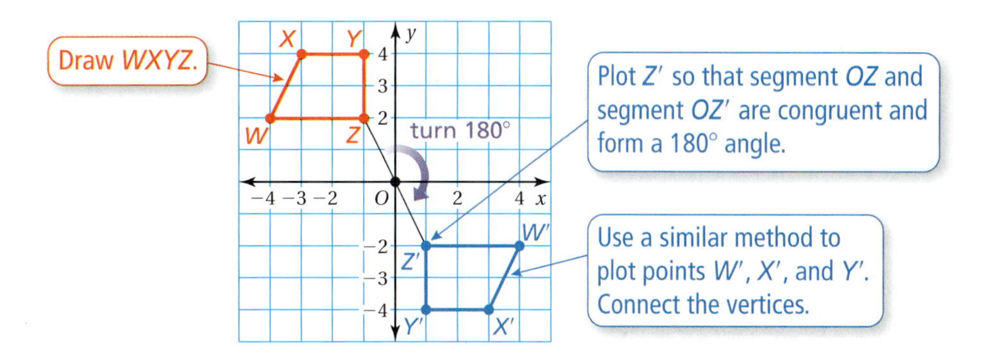

∴ The coordinates of the image are $W'(4, -2)$, $X'(3, -4)$, $Y'(1, -4)$, and $Z'(1, -2)$.

EXAMPLE 3 Rotating a Figure

The vertices of a triangle are $J(1, 2)$, $K(4, 2)$, and $L(1, -3)$. Rotate the triangle 90° counterclockwise about vertex L. What are the coordinates of the image?

Common Error

Be sure to pay attention to whether a rotation is clockwise or counterclockwise.

Plot K' so that segment KL and segment $K'L'$ are congruent and form a 90° angle.

Use a similar method to plot point J'. Connect the vertices.

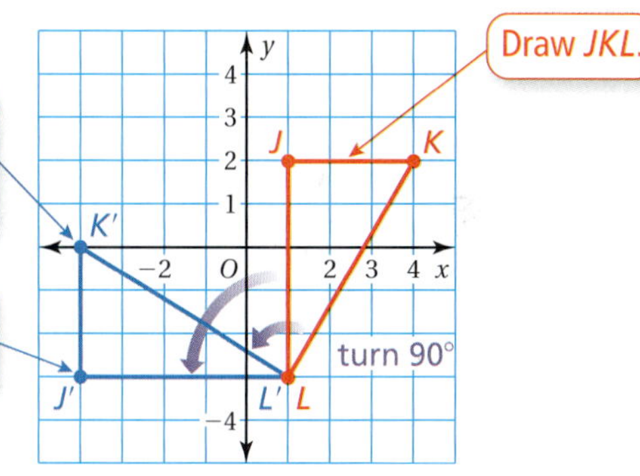

∴ The coordinates of the image are $J'(-4, -3)$, $K'(-4, 0)$, and $L'(1, -3)$.

On Your Own

3. A triangle has vertices $Q(4, 5)$, $R(4, 0)$, and $S(1, 0)$.

   a. Rotate the triangle 90° counterclockwise about the origin.

   b. Rotate the triangle 180° about vertex S.

   c. Are the images in parts (a) and (b) congruent? Explain.

EXAMPLE 4 Using More than One Transformation

The vertices of a rectangle are $A(-3, -3)$, $B(1, -3)$, $C(1, -5)$, and $D(-3, -5)$. Rotate the rectangle 90° clockwise about the origin, and then reflect it in the y-axis. What are the coordinates of the image?

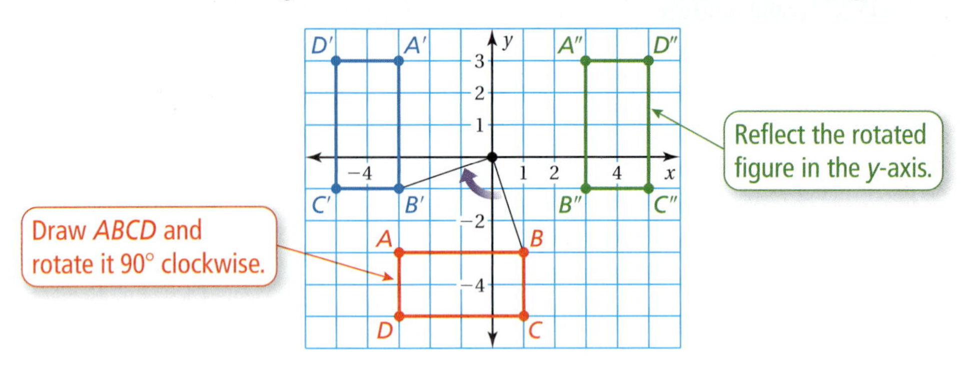

∴ The coordinates of the image are $A''(3, 3)$, $B''(3, -1)$, $C''(5, -1)$ and $D''(5, 3)$.

The image of a translation, reflection, or rotation is congruent to the original figure. So, two figures are congruent when one can be obtained from the other by a sequence of translations, reflections, and rotations.

EXAMPLE 5 Describing a Sequence of Transformations

The red figure is congruent to the blue figure. Describe a sequence of transformations in which the blue figure is the image of the red figure.

You can turn the red figure 90° so that it has the same orientation as the blue figure. So, begin with a rotation.

After rotating, you need to slide the figure up.

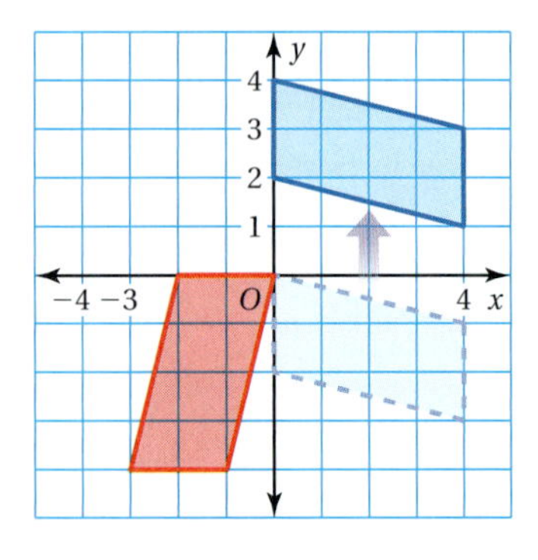

∴ So, one possible sequence of transformations is a 90° counterclockwise rotation about the origin followed by a translation 4 units up.

On Your Own

4. The vertices of a triangle are $P(-1, 2)$, $Q(-1, 0)$, and $R(2, 0)$. Rotate the triangle 180° about vertex R, and then reflect it in the x-axis. What are the coordinates of the image?
5. In Example 5, describe a different sequence of transformations in which the blue figure is the image of the red figure.

Laurie's Notes

Example 4

- This example involves two transformations. First the rectangle is rotated and then it is reflected. Work slowly and carefully through this example.
- ? "Can you visualize where the image will be after the two transformations?" Answers will vary.
- Graph the original rectangle $ABCD$.
- ? "Does it matter whether you rotate the rectangle 90° clockwise or 90° counterclockwise?" yes
- To construct the image after the rotation, draw the segment OB, where O represents a point at the origin. Next locate vertex B' by drawing OB' so that $\angle BOB'$ is 90° and OB and OB' are the same length. After you locate vertex B', the remaining vertices should be relatively easy to locate.
- Reflect rectangle $A'B'C'D'$ in the y-axis to obtain rectangle $A''B''C''D''$.
- **Teaching Tip:** Use transparent paper to help visualize the transformations.
- **Common Error:** Because the reflection resembles a translation, students may label the vertices of rectangle $A''B''C''D''$ incorrectly. Watch for this.
- ? **MP3 Construct Viable Arguments and Critique the Reasoning of Others:** "Are rectangles $ABCD$ and $A''B''C''D''$ congruent? Explain." yes; Rectangles $ABCD$ and $A'B'C'D'$ are congruent because in a rotation, the original figure and its image are congruent. Rectangles $A'B'C'D'$ and $A''B''C''D''$ are congruent because in a reflection, the original figure and its image are congruent. Because rectangles $ABCD$ and $A''B''C''D''$ are both congruent to rectangle $A'B'C'D'$, they are congruent to each other.
- **MP3 Construct Viable Arguments and Critique the Reasoning of Others** and **MP6 Attend to Precision:** If time allows, ask students whether the order in which you perform the transformations matters. (The answer is yes.) Have them thoroughly explain their reasoning.

Example 5

- Ask a volunteer to read the problem. Give students time to come up with an answer. Ask volunteers to share their answers with the class.
- Use transparent paper to help visualize the transformations.
- ? **Extension:** "Is there is a single transformation to get from the red figure to the blue figure?" 90° counterclockwise rotation about the point $(-2, 2)$. Students may need a hint to consider moving the center of rotation.

On Your Own

- **Think-Pair-Share:** Students should read each question independently and then work in pairs to answer the questions. When they have answered the questions, the pair should compare their answers with another group and discuss any discrepancies.

Closure

- Draw a right triangle in Quadrant II. Reflect the triangle in the x-axis. Rotate the original triangle about the origin 90° clockwise.

Extra Example 4

The vertices of a rectangle are $A(-1, 1)$, $B(-4, 1)$, $C(-4, 5)$, and $D(-1, 5)$. Rotate the rectangle 90° clockwise about the origin, and then reflect it in the y-axis. What are the coordinates of the image?

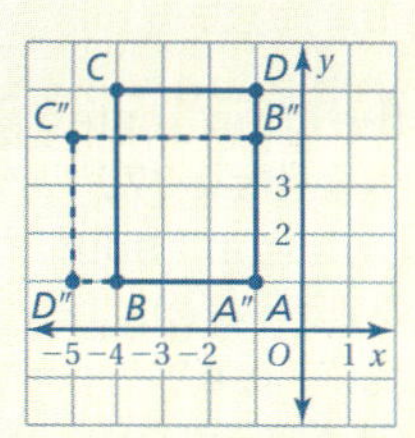

$A''(-1, 1)$, $B''(-1, 4)$, $C''(-5, 4)$, $D''(-5, 1)$

Extra Example 5

The red figure is congruent to the blue figure. Describe a sequence of transformations in which the blue figure is the image of the red figure.

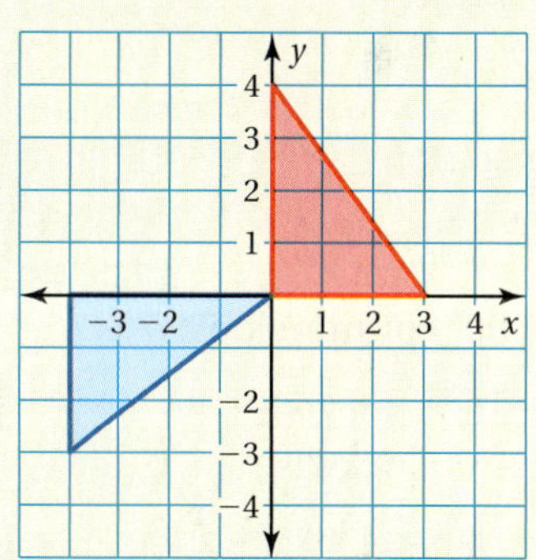

Sample answer: 90° clockwise rotation about the origin followed by a translation 4 units left

On Your Own

4. 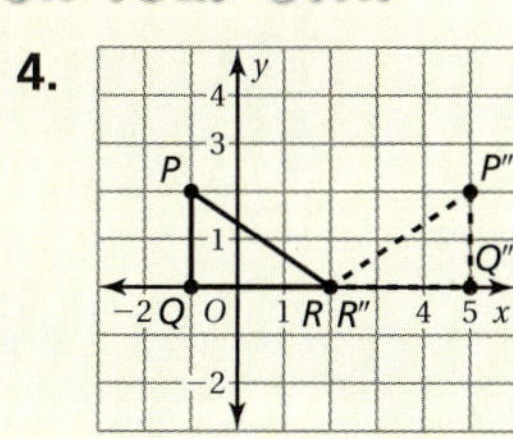

$P''(5, 2)$, $Q''(5, 0)$, $R''(2, 0)$

5. *Sample answer:* 90° clockwise rotation about the origin followed by a translation 4 units right and 1 unit up

Vocabulary and Concept Check

1. $(0, 0)$; $(1, -3)$
2. Quadrant I
3. Quadrant IV
4. Quadrant III
5. Quadrant II
6. What are the coordinates of the figure after a 270° clockwise rotation about the origin?; $A'(-4, 2)$, $B'(-4, 4)$, $C'(-1, 4)$, $D'(-1, 2)$; $A'(4, -2)$, $B'(4, -4)$, $C'(1, -4)$, $D'(1, -2)$

Practice and Problem Solving

7. reflection
8. rotation
9. translation
10. no
11. yes; 90° counterclockwise
12. yes; 180° clockwise or counterclockwise
13. $A'(2, 2)$, $B'(1, 4)$, $C'(3, 4)$, $D'(4, 2)$
14. $F'(-1, -2)$, $G'(-3, -5)$, $H'(-3, -2)$
15. $J'(0, -3)$, $K'(0, -5)$, $L'(-4, -3)$

Assignment Guide and Homework Check

Level	Day 1 Activity Assignment	Day 2 Lesson Assignment	Homework Check
Basic	7–9, 30–34	1–6, 11–27 odd	11, 15, 17, 19, 27
Average	7–9, 30–34	1–6, 11–21 odd, 22–28 even	11, 15, 17, 21, 26
Advanced	1–9, 16–26 even, 27–34		18, 20, 22, 26, 28

Common Errors

- **Exercises 7–12** Students with minimal spatial skills may not be able to tell whether a figure is rotated. Give them tracing paper and have them copy the red figure and rotate it.

2.4 Record and Practice Journal

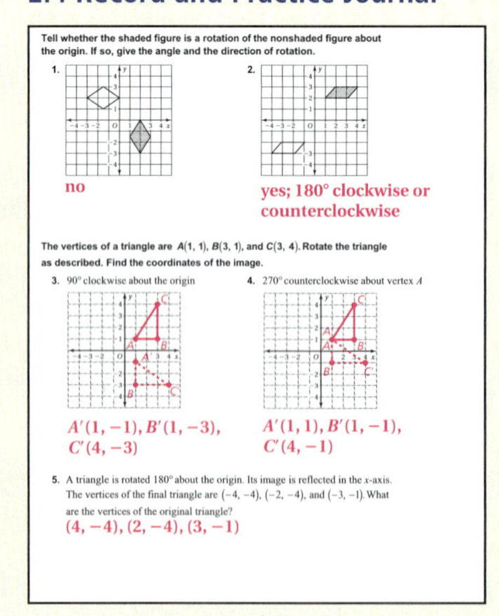

Tell whether the shaded figure is a rotation of the nonshaded figure about the origin. If so, give the angle and the direction of rotation.

1. no
2. yes; 180° clockwise or counterclockwise

The vertices of a triangle are $A(1, 1)$, $B(3, 1)$, and $C(3, 4)$. Rotate the triangle as described. Find the coordinates of the image.

3. 90° clockwise about the origin

   $A'(1, -1)$, $B'(1, -3)$, $C'(4, -3)$

4. 270° counterclockwise about vertex A

   $A'(1, 1)$, $B'(1, -1)$, $C'(4, -1)$

5. A triangle is rotated 180° about the origin. Its image is reflected in the x-axis. The vertices of the final triangle are $(-4, -4)$, $(-2, -4)$, and $(-3, -1)$. What are the vertices of the original triangle?

   $(4, -4)$, $(2, -4)$, $(3, -1)$

2.4 Exercises

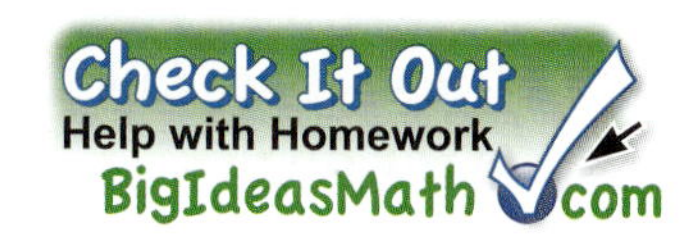

Vocabulary and Concept Check

1. **VOCABULARY** What are the coordinates of the center of rotation in Example 2? Example 3?

MENTAL MATH A figure lies entirely in Quadrant II. In which quadrant will the figure lie after the given clockwise rotation about the origin?

2. 90°
3. 180°
4. 270°
5. 360°

6. **DIFFERENT WORDS, SAME QUESTION** Which is different? Find "both" answers.

What are the coordinates of the figure after a 90° clockwise rotation about the origin?

What are the coordinates of the figure after a 270° clockwise rotation about the origin?

What are the coordinates of the figure after turning the figure 90° to the right about the origin?

What are the coordinates of the figure after a 270° counterclockwise rotation about the origin?

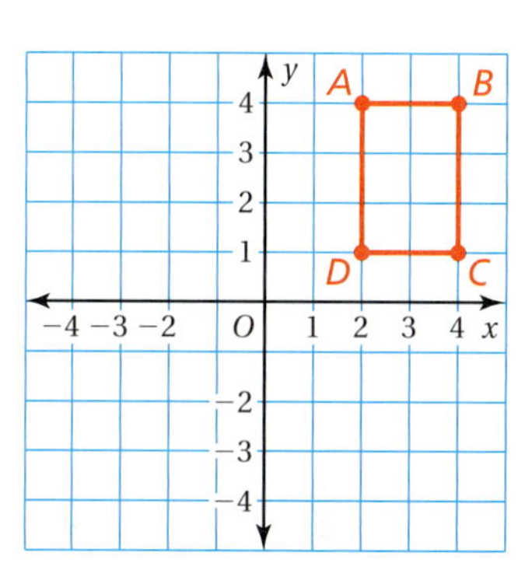

Practice and Problem Solving

Identify the transformation.

7.
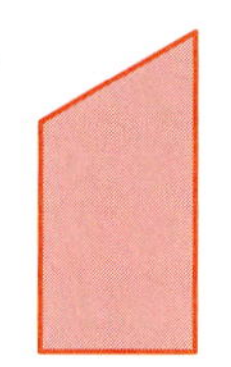
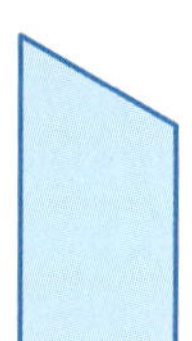

8.
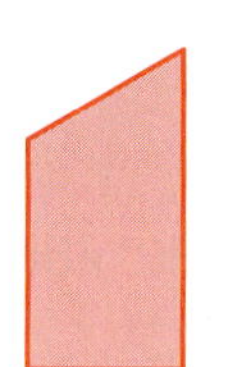
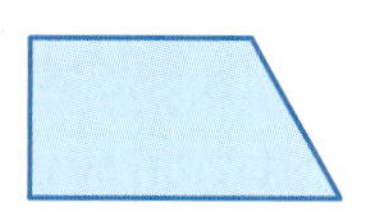

9.

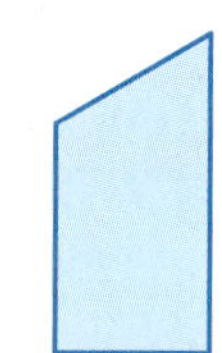

Tell whether the blue figure is a rotation of the red figure about the origin. If so, give the angle and direction of rotation.

1 10.
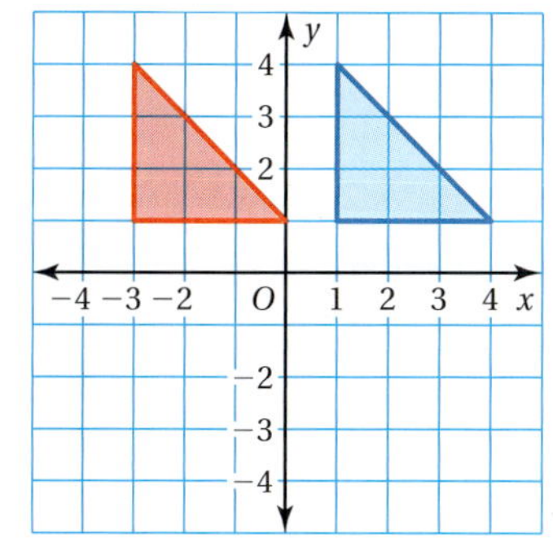

11.
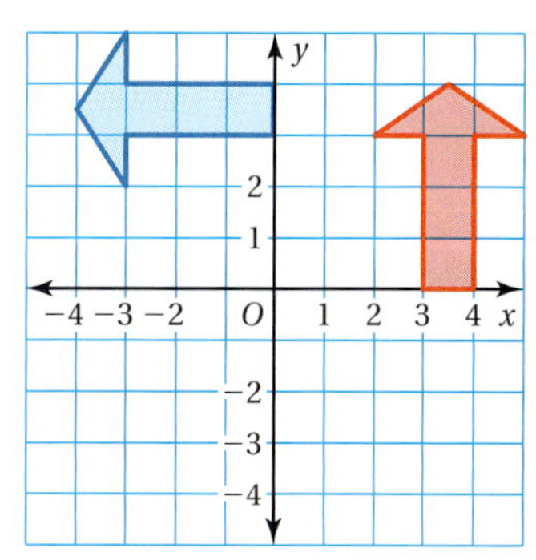

12.
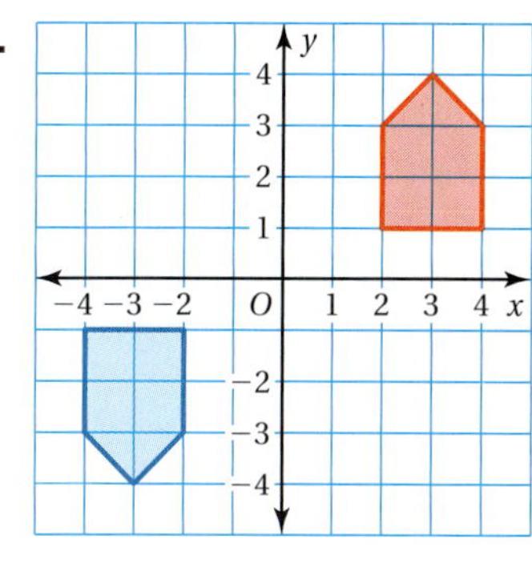

The vertices of a figure are given. Rotate the figure as described. Find the coordinates of the image.

13. $A(2, -2)$, $B(4, -1)$, $C(4, -3)$, $D(2, -4)$
90° counterclockwise about the origin

14. $F(1, 2)$, $G(3, 5)$, $H(3, 2)$
180° about the origin

15. $J(-4, 1)$, $K(-2, 1)$, $L(-4, -3)$
90° clockwise about vertex L

16. $P(-3, 4)$, $Q(-1, 4)$, $R(-2, 1)$, $S(-4, 1)$
180° about vertex R

17. $W(-6, -2)$, $X(-2, -2)$, $Y(-2, -6)$, $Z(-5, -6)$
270° counterclockwise about the origin

18. $A(1, -1)$, $B(5, -6)$, $C(1, -6)$
90° counterclockwise about vertex A

A figure has *rotational symmetry* if a rotation of 180° or less produces an image that fits exactly on the original figure. Explain why the figure has rotational symmetry.

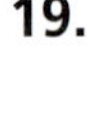

19.

20.

21.

The vertices of a figure are given. Find the coordinates of the figure after the transformations given.

22. $R(-7, -5)$, $S(-1, -2)$, $T(-1, -5)$

Rotate 90° counterclockwise about the origin. Then translate 3 units left and 8 units up.

23. $J(-4, 4)$, $K(-3, 4)$, $L(-1, 1)$, $M(-4, 1)$

Reflect in the x-axis, and then rotate 180° about the origin.

The red figure is congruent to the blue figure. Describe two different sequences of transformations in which the blue figure is the image of the red figure.

24.

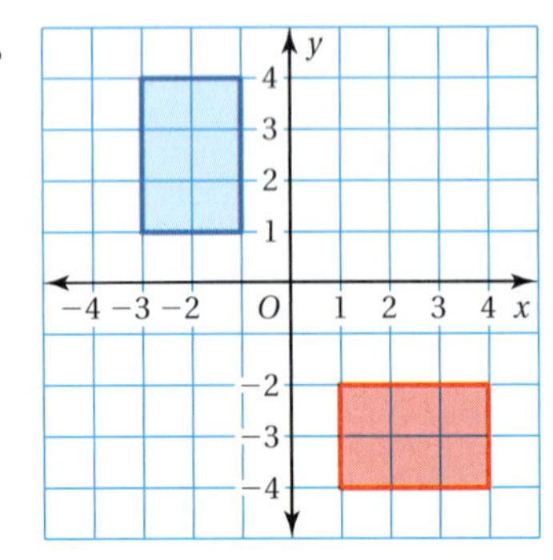

25.

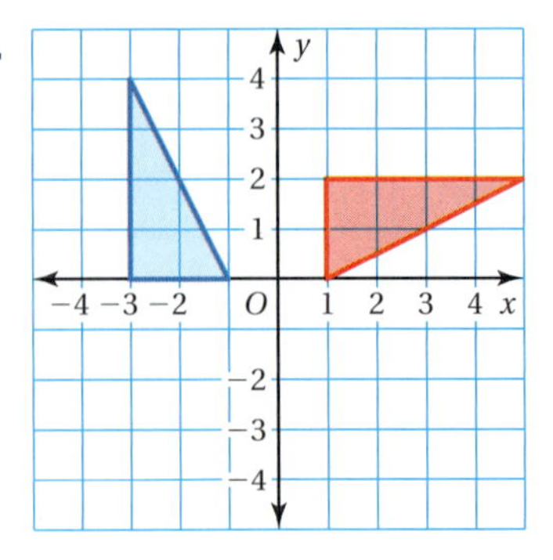

Common Errors

- **Exercises 13–18** Students may rotate the figure in the wrong direction. Remind them what clockwise and counterclockwise mean. It may be helpful for students to draw an arrow on a graph of the figure in the direction of rotation.

Practice and Problem Solving

16. $P'(-1, -2)$, $Q'(-3, -2)$, $R'(-2, 1)$, $S'(0, 1)$

17. $W'(-2, 6)$, $X'(-2, 2)$, $Y'(-6, 2)$, $Z'(-6, 5)$

18. $A'(1, -1)$, $B'(6, 3)$, $C'(6, -1)$

19. It only needs to rotate 120° to produce an identical image.

20. It only needs to rotate 90° to produce an identical image.

21. It only needs to rotate 180° to produce an identical image.

22. $R''(2, 1)$, $S''(-1, 7)$, $T''(2, 7)$

23. $J''(4, 4)$, $K''(3, 4)$, $L''(1, 1)$, $M''(4, 1)$

24. *Sample answer:* Rotate 90° counterclockwise about the origin and then translate 5 units left; Rotate 90° clockwise about the origin and then translate 1 unit right and 5 units up.

25. *Sample answer:* Rotate 180° about the origin and then rotate 90° clockwise about vertex $(-1, 0)$; Rotate 90° counterclockwise about the origin and then translate 1 unit left and 1 unit down.

26. **a.** $A'(6, 2)$, $B'(3, 2)$, $C'(1, 4)$, $D'(6, 4)$

b. Reflect the trapezoid in the x-axis and then in the y-axis, or reflect the trapezoid in the y-axis and then in the x-axis.

English Language Learners

Kinesthetic

Give students examples of translations, reflections, and rotations in the classroom. You can move a chair to show a translation, use a mirror to show a reflection, and spin the chair around to show a rotation.

Practice and Problem Solving

27. See *Taking Math Deeper*.

28. See Additional Answers.

29. (2, 4), (4, 1), (1, 1)

Fair Game Review

30. no **31.** yes

32. yes **33.** no

34. B

Mini-Assessment

Tell whether the blue figure is a rotation of the red figure about the origin. If so, give the angle and direction of rotation.

1. yes; 90° clockwise rotation

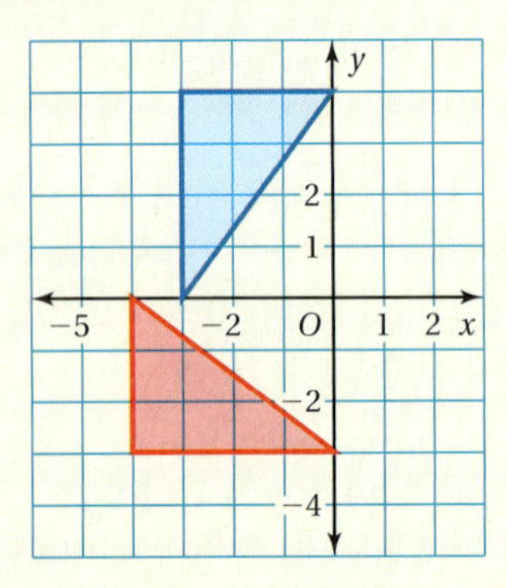

2. yes; 90° counterclockwise rotation

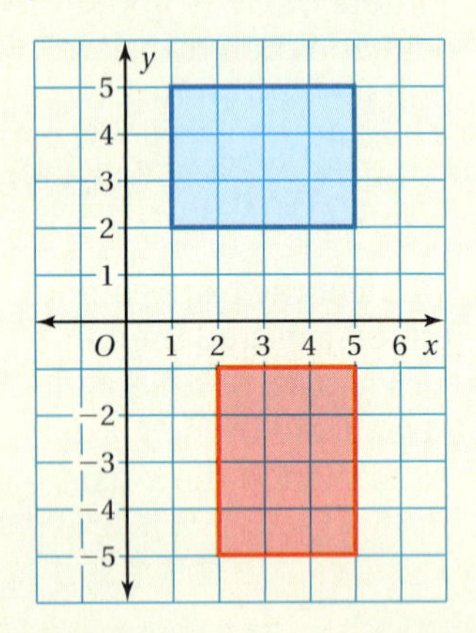

The vertices of a figure are given. Rotate the figure as described. Find the coordinates of the image.

3. $L(3, 2)$, $M(1, 1)$, $N(1, 5)$; 90° counterclockwise about the origin $L'(-2, 3)$, $M'(-1, 1)$, $N'(-5, 1)$

4. $T(2, 5)$, $U(5, 4)$, $V(6, 1)$, $W(2, 1)$; 270° clockwise about the origin $T'(-5, 2)$, $U'(-4, 5)$, $V'(-1, 6)$, $W'(-1, 2)$

Taking Math Deeper

Exercise 27

Students can use *Guess, Check, and Revise* to find a correct sequence of transformations. This is a good question for students to discuss in pairs or groups.

Do the rotations.

Original position

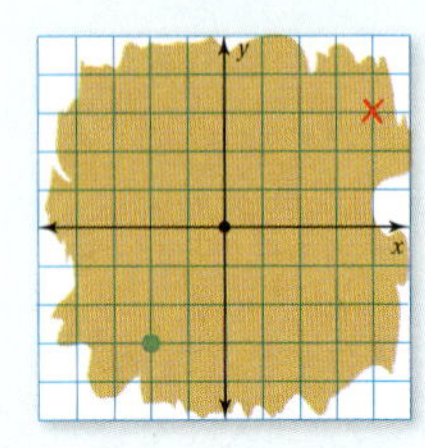

Rotate 180° about the origin.

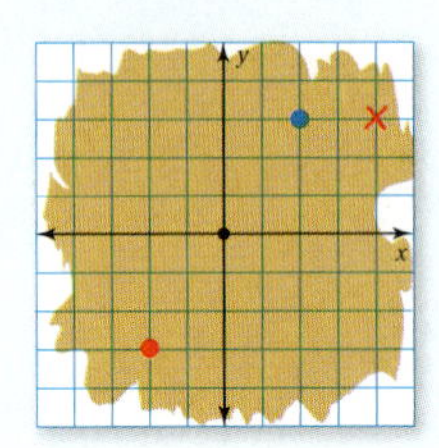

Rotate 90° counterclockwise about the origin.

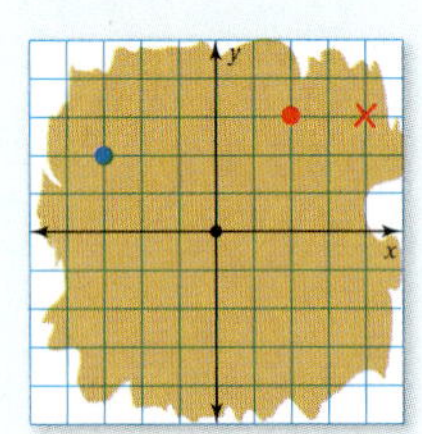

2 Do the reflection.

Position after rotations.

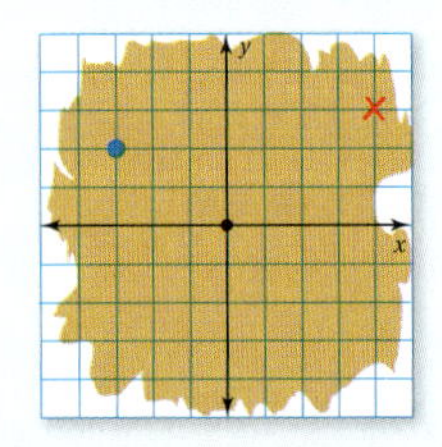

Reflect in y-axis.

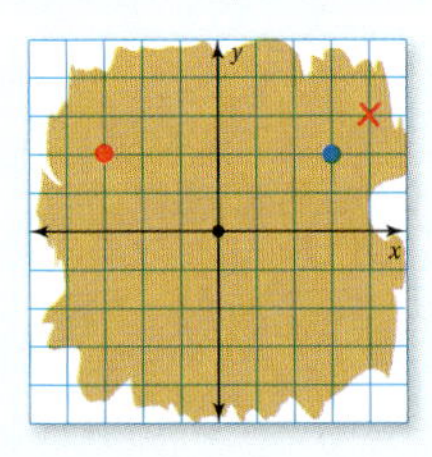

3 Do the translation.

Position after rotations and reflection.

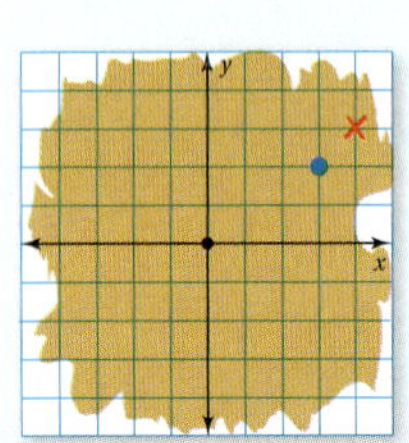

Translate 1 unit right and 1 unit up.

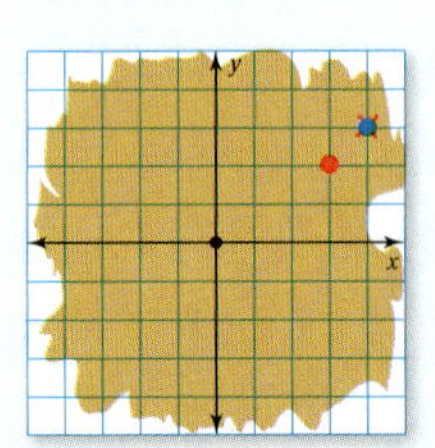

Reteaching and Enrichment Strategies

If students need help. . .	If students got it. . .
Resources by Chapter • Practice A and Practice B • Puzzle Time Record and Practice Journal Practice Differentiating the Lesson Lesson Tutorials Skills Review Handbook	Resources by Chapter • Enrichment and Extension • Technology Connection Start the next section

26. REASONING A trapezoid has vertices $A(-6, -2)$, $B(-3, -2)$, $C(-1, -4)$, and $D(-6, -4)$.

a. Rotate the trapezoid 180° about the origin. What are the coordinates of the image?

b. Describe a way to obtain the same image without using rotations.

27. TREASURE MAP You want to find the treasure located on the map at ×. You are located at ●. The following transformations will lead you to the treasure, but they are not in the correct order. Find the correct order. Use each transformation exactly once.

- Rotate 180° about the origin.
- Reflect in the y-axis.
- Rotate 90° counterclockwise about the origin.
- Translate 1 unit right and 1 unit up.

28. CRITICAL THINKING Consider $\triangle JKL$.

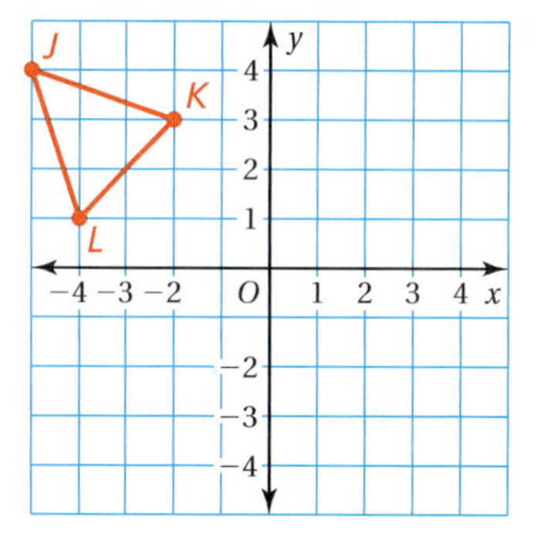

a. Rotate $\triangle JKL$ 90° clockwise about the origin. How are the x- and y-coordinates of $\triangle J'K'L'$ related to the x- and y-coordinates of $\triangle JKL$?

b. Rotate $\triangle JKL$ 180° about the origin. How are the x- and y-coordinates of $\triangle J'K'L'$ related to the x- and y-coordinates of $\triangle JKL$?

c. Do you think your answers to parts (a) and (b) hold true for any figure? Explain.

29. Reasoning You rotate a triangle 90° counterclockwise about the origin. Then you translate its image 1 unit left and 2 units down. The vertices of the final image are $(-5, 0)$, $(-2, 2)$, and $(-2, -1)$. What are the vertices of the original triangle?

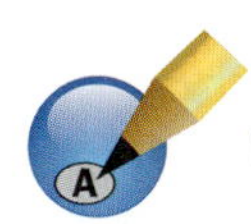

Fair Game Review What you learned in previous grades & lessons

Tell whether the ratios form a proportion. *(Skills Review Handbook)*

30. $\frac{3}{5}, \frac{15}{20}$ **31.** $\frac{2}{3}, \frac{12}{18}$ **32.** $\frac{7}{28}, \frac{12}{48}$ **33.** $\frac{54}{72}, \frac{36}{45}$

34. MULTIPLE CHOICE What is the solution of the equation $x + 6 \div 2 = 5$? *(Section 1.1)*

Ⓐ $x = -16$ Ⓑ $x = 2$ Ⓒ $x = 4$ Ⓓ $x = 16$

2 Study Help

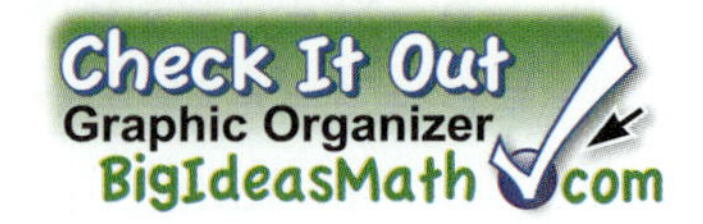

You can use a **summary triangle** to explain a concept. Here is an example of a summary triangle for translating a figure.

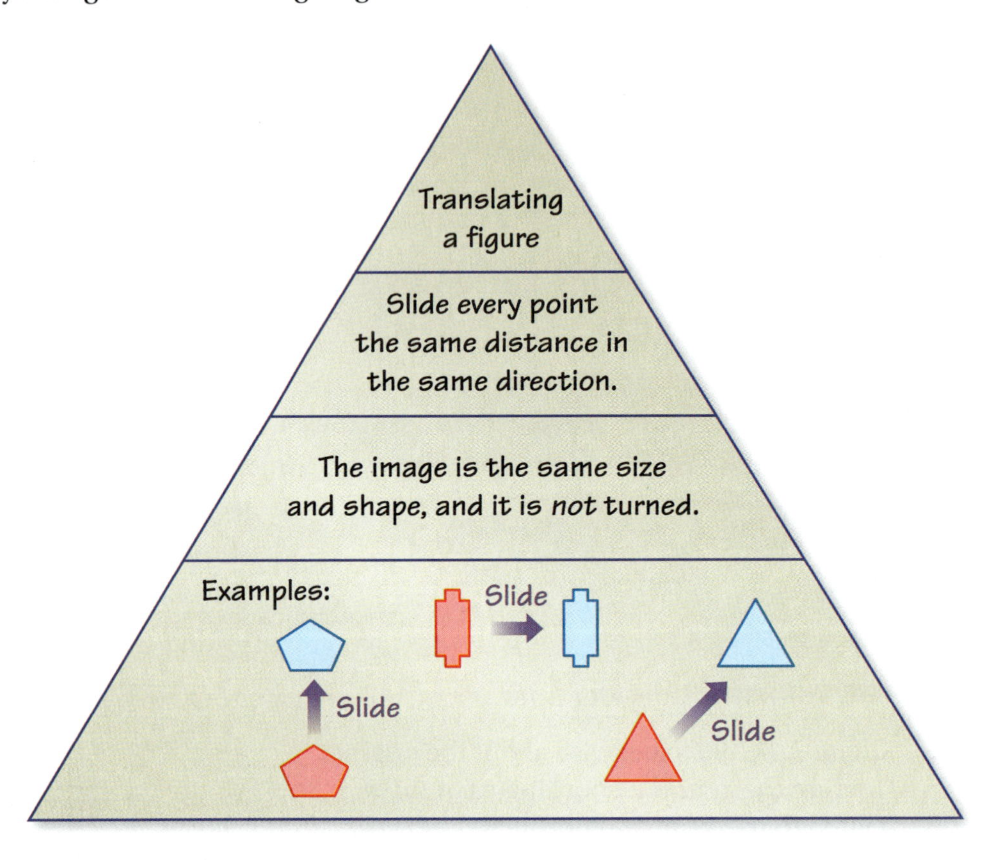

On Your Own

Make summary triangles to help you study these topics.

1. congruent figures
2. reflecting a figure
3. rotating a figure

After you complete this chapter, make summary triangles for the following topics.

4. similar figures
5. perimeters of similar figures
6. areas of similar figures
7. dilating a figure
8. transforming a figure

"I hope my owner sees my summary triangle. I just can't seem to learn 'roll over.' "

Sample Answers

1.

Congruent figures

Figures that have the same size and the same shape

Matching angles are called corresponding angles.

Matching sides are called corresponding sides.

Example: The figures are congruent.

A B C D   E F G H

Corresponding Angles	Corresponding Sides
$\angle A$ and $\angle G$	Side AB and Side GF
$\angle B$ and $\angle F$	Side BC and Side FE
$\angle C$ and $\angle E$	Side CD and Side EH
$\angle D$ and $\angle H$	Side DA and Side HG

2.

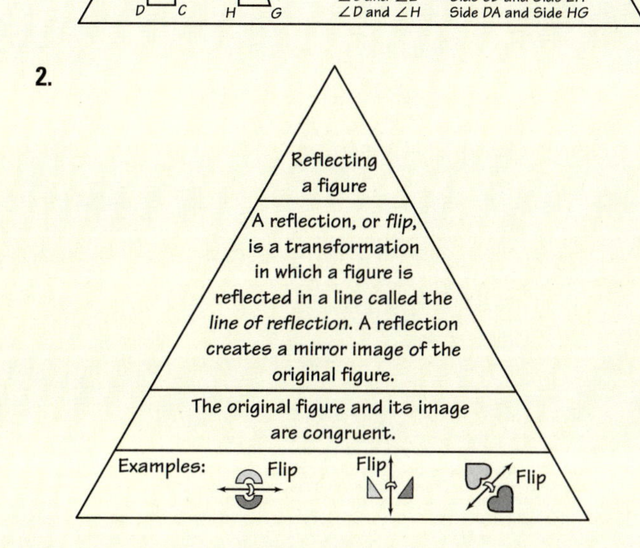

3.

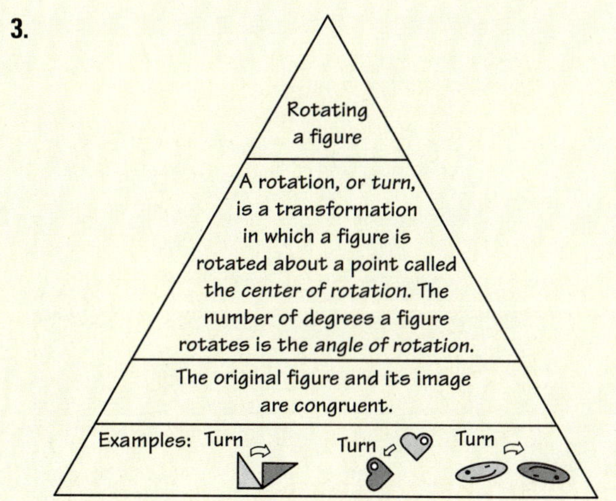

List of Organizers

Available at *BigIdeasMath.com*

Comparison Chart
Concept Circle
Example and Non-Example Chart
Formula Triangle
Four Square
Idea (Definition) and Examples Chart
Information Frame
Information Wheel
Notetaking Organizer
Process Diagram
Summary Triangle
Word Magnet
Y Chart

About this Organizer

A **Summary Triangle** can be used to explain a concept. Typically, the summary triangle is divided into 3 or 4 parts. In the top part, students write the concept being explained. In the middle part(s), students write any procedure, explanation, description, definition, theorem, and/or formula(s). In the bottom part, students write an example to illustrate the concept. A summary triangle can be used as an assessment tool, in which blanks are left for students to complete. Also, students can place their summary triangles on note cards to use as a quick study reference.

Technology for the Teacher

Editable Graphic Organizer

Answers

1. not congruent; Corresponding side lengths are not congruent.
2. congruent; Corresponding angles and side lengths are congruent.
3. no
4. yes
5. yes
6. no
7. *Sample answer:* rotate 90° clockwise about vertex $(-1, 1)$, translate 1 unit right and 1 unit down; rotate 270° counterclockwise about vertex $(-1, 1)$, translate 1 unit down and 1 unit right
8. *Sample answer:* rotate 180° clockwise about the origin, translate 1 unit right and 1 unit down; translate 1 unit left and 1 unit up, reflect in x-axis, reflect in y-axis
9. 6 units right and 4 units down
10. no; It will be 1 unit to the right of the hole.

Technology for the Teacher

Online Assessment
Assessment Book
ExamView® Assessment Suite

Alternative Quiz Ideas

100% Quiz
Error Notebook
Group Quiz
Homework Quiz
Math Log
Notebook Quiz
Partner Quiz
Pass the Paper

Math Log

Ask students to keep a math log for the chapter. Have them include diagrams, definitions, and examples. Everything should be clearly labeled. It might be helpful if they put the information in a chart. Students can add to the log as they are introduced to new topics.

Reteaching and Enrichment Strategies

If students need help...	If students got it...
Resources by Chapter • Practice A and Practice B • Puzzle Time Lesson Tutorials *BigIdeasMath.com*	Resources by Chapter • Enrichment and Extension • Technology Connection Game Closet at *BigIdeasMath.com* Start the next section

2.1–2.4 Quiz

Tell whether the two figures are congruent. Explain your reasoning. *(Section 2.1)*

1.

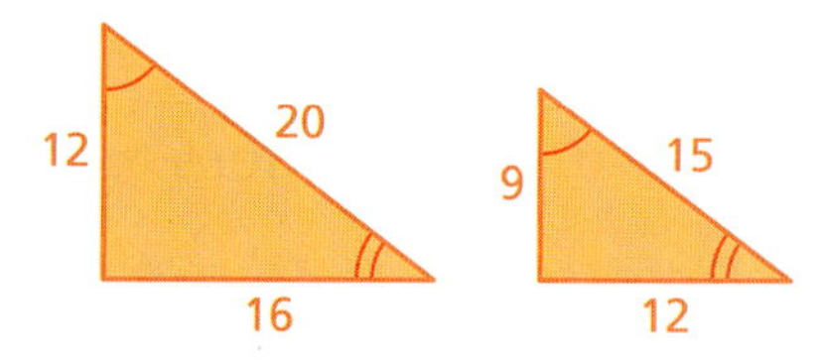

2.

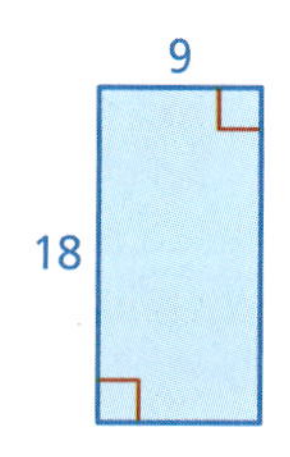

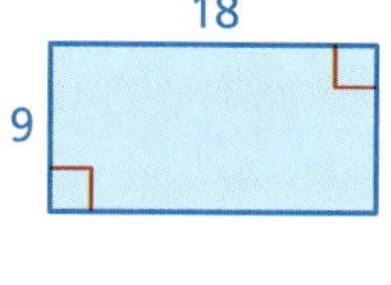

Tell whether the blue figure is a translation of the red figure. *(Section 2.2)*

3.

4.

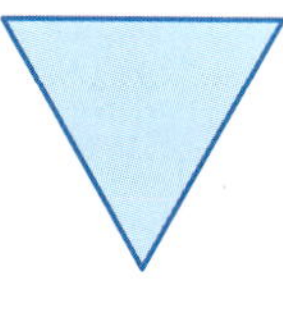

Tell whether the blue figure is a reflection of the red figure. *(Section 2.3)*

5.

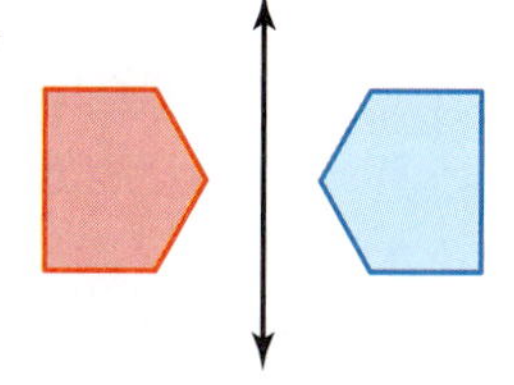

6.

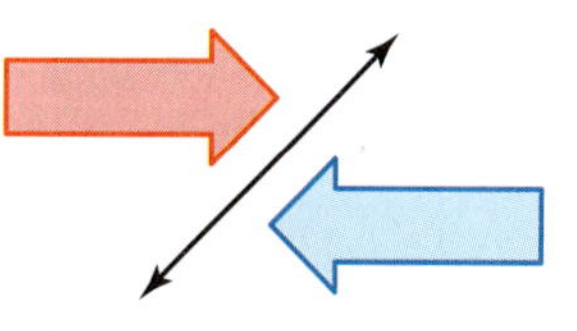

The red figure is congruent to the blue figure. Describe two different sequences of transformations in which the blue figure is the image of the red figure. *(Section 2.4)*

7.

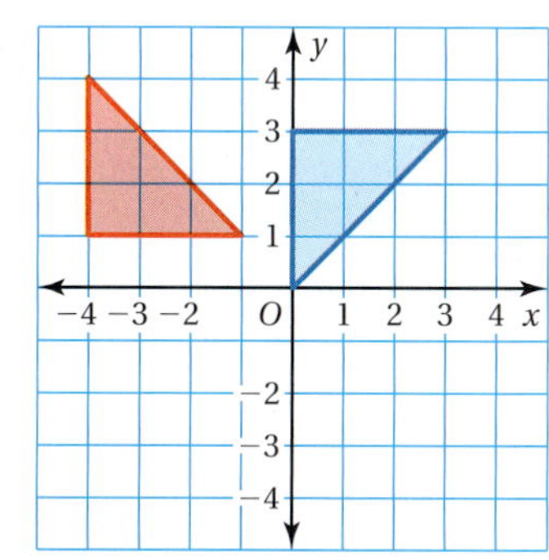

8.

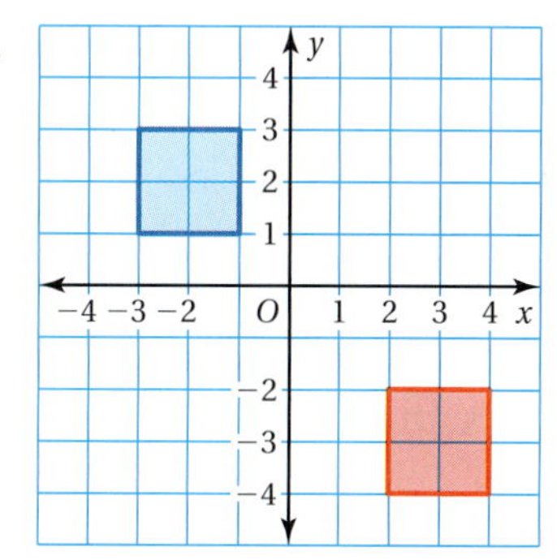

9. AIRPLANE Describe a translation of the airplane from point A to point B. *(Section 2.2)*

10. MINIGOLF You hit the golf ball along the red path so that its image will be a reflection in the y-axis. Does the golf ball land in the hole? Explain. *(Section 2.3)*

2.5 Similar Figures

Essential Question How can you use proportions to help make decisions in art, design, and magazine layouts?

Original photograph

In a computer art program, when you click and drag on a side of a photograph, you distort it.

But when you click and drag on a corner of the photograph, the dimensions remain proportional to the original.

Distorted

Distorted

Proportional

1 ACTIVITY: Reducing Photographs

Work with a partner. You are trying to reduce the photograph to the indicated size for a nature magazine. Can you reduce the photograph to the indicated size without distorting or cropping? Explain your reasoning.

a.

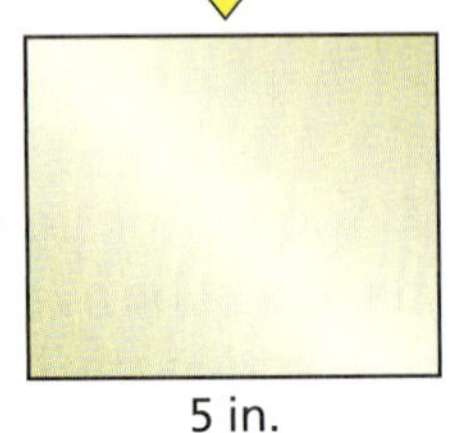

b.

Geometry

In this lesson, you will

- name corresponding angles and corresponding sides of similar figures.
- identify similar figures.
- find unknown measures of similar figures.

Preparing for Standard MACC.8.G.1.4

Laurie's Notes

Introduction

Standards for Mathematical Practice

- **MP6 Attend to Precision:** Students should use precise vocabulary to communicate their thinking to others. In working with similar figures, they should be able to state which ratios are in proportion and why.

Motivate

- Draw a simple stick figure or other image on a stretchable surface, such as a balloon, physical therapy elastic, or play putty.
- ? Ask students what they think will happen to the figure when you pull the picture to the right. Students should recognize that the image will be distorted. Pull one of the sides of the picture to confirm.
- Pull the top of the picture so that students see this result as the same.
- ? Ask students what they think will happen if the stretchable surface is pulled in both directions (right and up). Students should recognize that the image will enlarge proportionally.
- **Alternative:** If you can display a computer image to the class, you can drag a side to distort the image, or drag a corner to change the size of the image proportionally.

Activity Notes

Activity 1

- Photography is a good context to use to examine similarity. A common misconception is that standard photo sizes are proportional and, in fact, most are not. A 5" × 7" photo and a 4" × 6" photo are not proportional.
- **Common Misconception:** Students often believe that if you subtract the same amount from each dimension, the resulting ratio will be proportional to the first. For example, $\frac{5}{7} \neq \frac{5-1}{7-1} = \frac{4}{6}$.
- ? Ask questions about proportions and ratios:
  - "What is a proportion?" two equal ratios
  - "Are the two ratios 2:3 and 4:6 equal?" yes
  - "Are the ratios 2:3 and 8:9 equal? Explain." No, listen for students to get at the idea that $2 \times 4 = 8$, but 3×4 is 12, not 9.
- Remember, students have *not* learned a formal definition for similar figures. Remind students that the task is to decide if the photograph can be reduced to the new dimensions without distorting it. Therefore, students must use the information about keeping the side lengths proportional.

Words of Wisdom

- **MP6:** Listen to how students describe their proportions. There are many correct ways to set up a proportion and some students might hear one way and incorrectly think their way is wrong.
- ? Ask students, "Did anyone set up their proportions differently?" Here are two possibilities. The key is to make sure *like things* are being compared.

$$\frac{\text{length (original)}}{\text{width (original)}} = \frac{\text{length (new size)}}{\text{width (new size)}};\quad \frac{\text{length (original)}}{\text{length (new size)}} = \frac{\text{width (original)}}{\text{width (new size)}}$$

Florida Common Core Standards

MACC.8.G.1.4 Understand that a two-dimensional figure is similar to another if the second can be obtained from the first by a sequence of rotations, reflections, translations, and dilations; given two similar two-dimensional figures, describe a sequence that exhibits the similarity between them.

Previous Learning

Students should know how to write ratios and have a basic understanding of proportions.

2.5 Record and Practice Journal

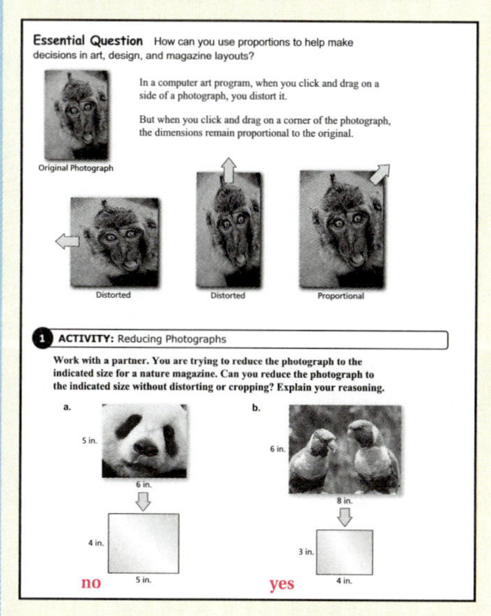

Essential Question How can you use proportions to help make decisions in art, design, and magazine layouts?

In a computer art program, when you click and drag on a side of a photograph, you distort it.

But when you click and drag on a corner of the photograph, the dimensions remain proportional to the original.

Original Photograph

Distorted

Distorted

Proportional

1 ACTIVITY: Reducing Photographs

Work with a partner. You are trying to reduce the photograph to the indicated size for a nature magazine. Can you reduce the photograph to the indicated size without distorting or cropping? Explain your reasoning.

a. 5 in. × 6 in. → 4 in. × 5 in. no

b. 6 in. × 8 in. → 3 in. × 4 in. yes

English Language Learners

Vocabulary

Ask students what *similar* means. Ask them if *similar* things are exactly alike. Explain that *similar figures* are not exactly alike. Similar figures have the same shape, but not necessarily the same size.

2.5 Record and Practice Journal

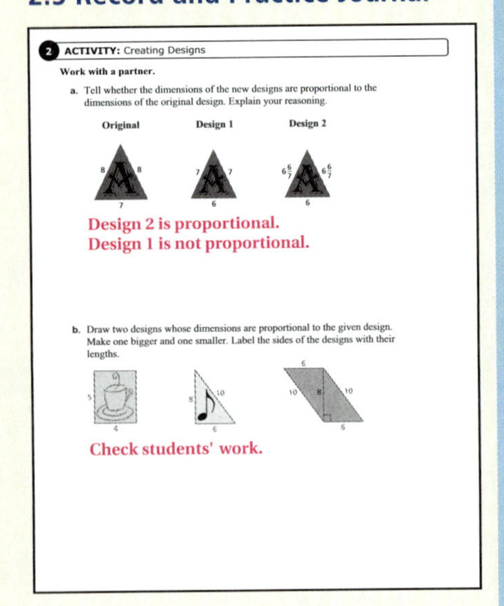

2 ACTIVITY: Creating Designs

Work with a partner.

a. Tell whether the dimensions of the new designs are proportional to the dimensions of the original design. Explain your reasoning.

Original | Design 1 | Design 2

Design 2 is proportional.
Design 1 is not proportional.

b. Draw two designs whose dimensions are proportional to the given design. Make one bigger and one smaller. Label the sides of the designs with their lengths.

Check students' work.

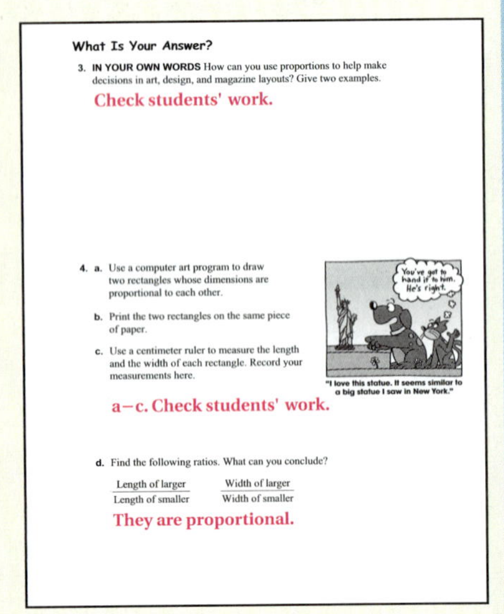

What Is Your Answer?

3. **IN YOUR OWN WORDS** How can you use proportions to help make decisions in art, design, and magazine layouts? Give two examples.

Check students' work.

4. a. Use a computer art program to draw two rectangles whose dimensions are proportional to each other.

b. Print the two rectangles on the same piece of paper.

c. Use a centimeter ruler to measure the length and the width of each rectangle. Record your measurements here.

"I love this statue. It seems similar to a big statue I saw in New York."

a–c. Check students' work.

d. Find the following ratios. What can you conclude?

$\frac{\text{Length of larger}}{\text{Length of smaller}}$ $\frac{\text{Width of larger}}{\text{Width of smaller}}$

They are proportional.

Laurie's Notes

Activity 2

- This activity is similar to the first activity. In Activity 1, students were asked whether the original figure would become distorted. In this activity, students are asked whether the dimensions of the designs are proportional.
- ? "What type of triangle is the original design?" Isosceles
- ? "Triangles don't have a *length* and *width* as rectangles do. What dimensions will you compare to decide whether the triangular designs are proportional?" Listen for language such as base and sides or base and legs.
- **Common Error:** Students think that subtracting 1 from each side length of the triangle produces a new triangle proportional to the original triangle.
- In completing part (b), students are not expected to make a scale drawing. They are drawing a figure that should look similar to the original. The rectangle should not look like a square. The right scalene triangle should not look equilateral.
- **MP3 Construct Viable Arguments and Critique the Reasoning of Others,** and **MP6:** Have students share the dimensions of their new figures. They should explain how they came up with the new dimensions. Listen for methods that use multiplication, not addition.

What Is Your Answer?

- **Technology:** Question 4 provides a great opportunity to have students work with a computer art program to enhance their understanding of similar figures.

Closure

- Are all three of these triangles proportional? Yes.

2 ACTIVITY: Creating Designs

Work with a partner.

Math Practice 4

Analyze Relationships

How can you use mathematics to determine whether the dimensions are proportional?

a. Tell whether the dimensions of the new designs are proportional to the dimensions of the original design. Explain your reasoning.

Original

Design 1

Design 2

b. Draw two designs whose dimensions are proportional to the given design. Make one bigger and one smaller. Label the sides of the designs with their lengths.

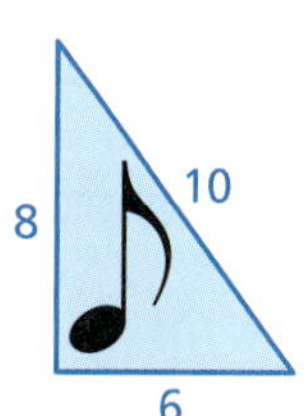

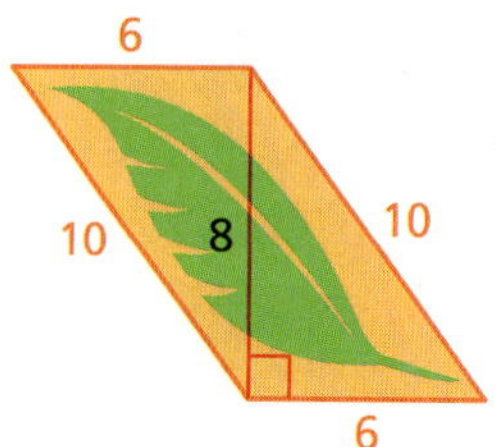

What Is Your Answer?

3. **IN YOUR OWN WORDS** How can you use proportions to help make decisions in art, design, and magazine layouts? Give two examples.

4. **a.** Use a computer art program to draw two rectangles whose dimensions are proportional to each other.

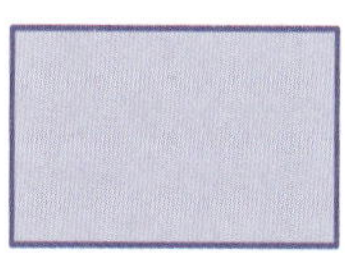

b. Print the two rectangles on the same piece of paper.

c. Use a centimeter ruler to measure the length and the width of each rectangle.

d. Find the following ratios. What can you conclude?

$$\frac{\text{Length of larger}}{\text{Length of smaller}} \qquad \frac{\text{Width of larger}}{\text{Width of smaller}}$$

You've got to hand it to him. He's right.

"I love this statue. It seems similar to a big statue I saw in New York."

Use what you learned about similar figures to complete Exercises 4 and 5 on page 74.

2.5 Lesson

Key Vocabulary
similar figures, *p. 72*

Key Idea

Similar Figures

Figures that have the same shape but not necessarily the same size are called **similar figures**.

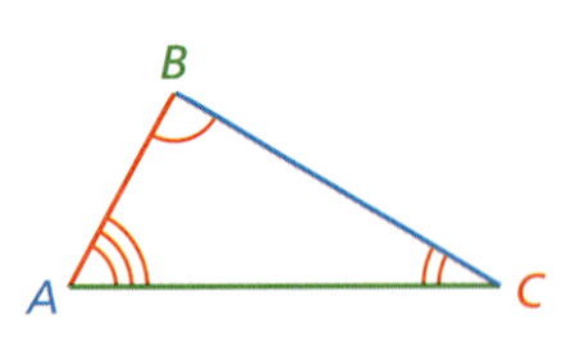

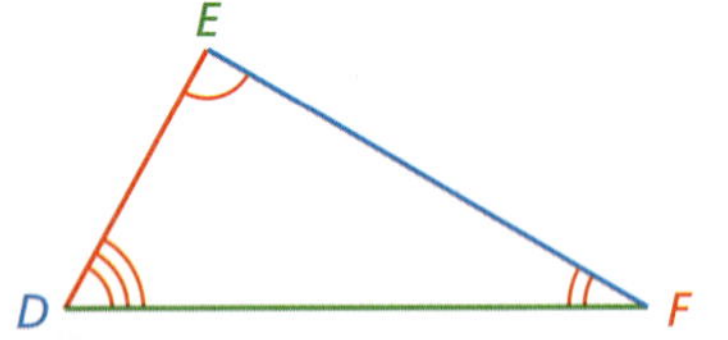

Triangle ABC is similar to Triangle DEF.

Words Two figures are similar when

- corresponding side lengths are proportional and
- corresponding angles are congruent.

Symbols

Side Lengths	*Angles*	*Figures*
$\frac{AB}{DE} = \frac{BC}{EF} = \frac{AC}{DF}$	$\angle A \cong \angle D$	$\triangle ABC \sim \triangle DEF$
	$\angle B \cong \angle E$	
	$\angle C \cong \angle F$	

Reading

The symbol $\sim$ means *is similar to*.

Common Error

When writing a similarity statement, make sure to list the vertices of the figures in the correct order.

EXAMPLE 1 Identifying Similar Figures

Which rectangle is similar to Rectangle A?

Rectangle A

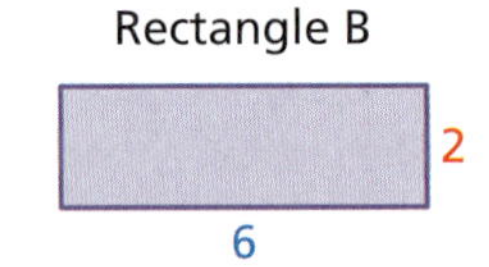

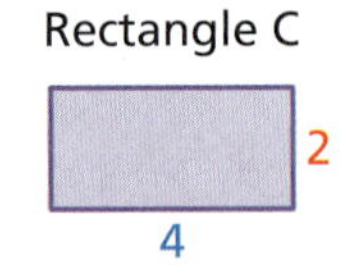

Rectangle A: 3, 6

Each figure is a rectangle. So, corresponding angles are congruent. Check to see if corresponding side lengths are proportional.

Rectangle A and Rectangle B

$\frac{\text{Length of A}}{\text{Length of B}} = \frac{6}{6} = 1$ $\frac{\text{Width of A}}{\text{Width of B}} = \frac{3}{2}$ Not proportional

Rectangle A and Rectangle C

$\frac{\text{Length of A}}{\text{Length of C}} = \frac{6}{4} = \frac{3}{2}$ $\frac{\text{Width of A}}{\text{Width of C}} = \frac{3}{2}$ Proportional

So, Rectangle C is similar to Rectangle A.

On Your Own

1. Rectangle D is 3 units long and 1 unit wide. Which rectangle is similar to Rectangle D?

Laurie's Notes

Introduction

Connect

- **Yesterday:** Students developed an intuitive understanding about proportional polygons. (MP3, MP6)
- **Today:** Students will use the formal definition of similar figures.

Motivate

- Place an item on an overhead projector, such as an index card, school ID, or other rectangular item. Ask questions about the actual item and its projected image.

? "How does the actual item compare to its projection?" Listen for: "They look alike," "They have the same shape," or "They're similar." It is unlikely that they know the mathematical definition of similar.

- Place a different-shaped item on the overhead.

? "There are two items and two projected images. Which projection goes with which item? How do you know?" Listen for students to say the items are the same shape but different sizes.

Lesson Notes

Key Idea

- Discuss the tilde symbol ~ that denotes similarity. Explain that the order in which the vertices of the triangle are written identifies how the sides and angles correspond.
- Remind students that *congruent* angles have the same measure.
- **Big Idea:** Discuss the need for two conditions to be met for two figures to be similar: corresponding side lengths are proportional *and* corresponding angles are congruent.
- **Representation:** Point out the color-coding, which should help students see the corresponding parts.
- Take your time in this section. There is a great deal of vocabulary, symbols, representations, *and* the fundamental concept of similarity. Give students time to ask questions and think about all that is being presented.

Example 1

? "What do you know about the angles of a rectangle?" 4 right angles

? "Are the corresponding angles congruent?" Yes.

? "What else must you check to know that the rectangles are similar?" corresponding side lengths are proportional

- Note that the problem has students focus on the dimensions of the rectangles, using the words *length* and *width,* without using the side names that can confuse students.

On Your Own

- Check that students correctly identify Rectangle B.

Goal Today's lesson is using proportions to determine if two figures are similar.

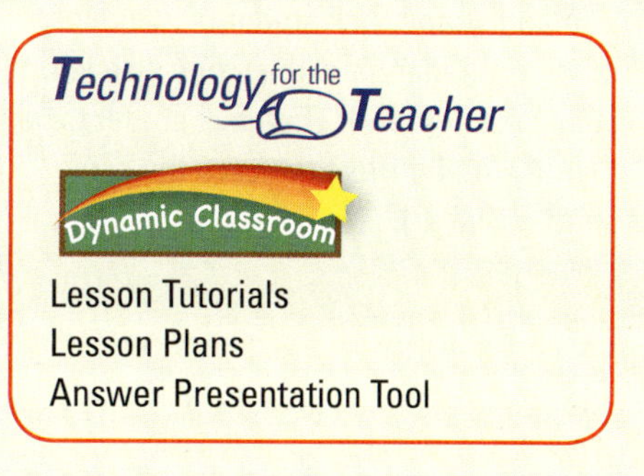

Extra Example 1

Which parallelogram is similar to Parallelogram A?

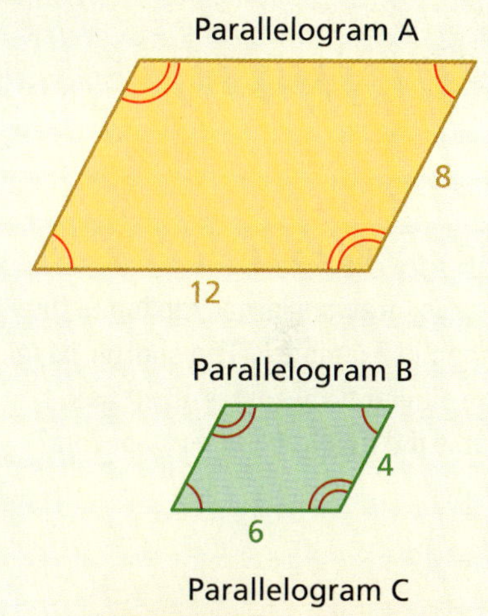

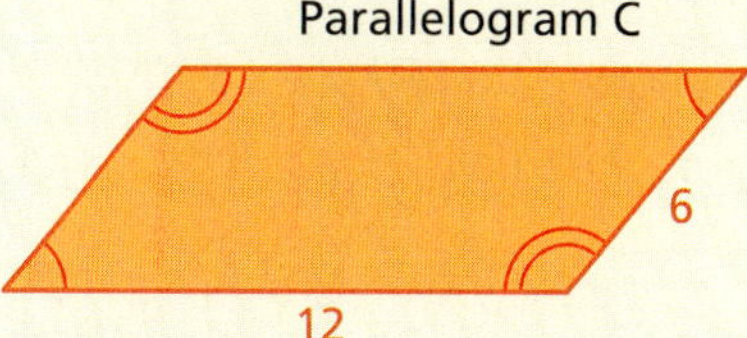

Parallelogram B

On Your Own

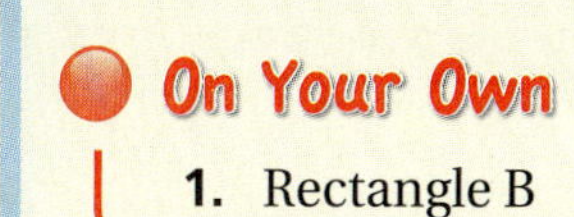

Differentiated Instruction

Visual

Bring in examples of figures that are similar and figures that have the same size and shape. Ask students to identify the figures that have the same size and shape (congruent). Then ask students to identify the figures that have the same shape (similar).

Extra Example 2

The triangles are similar. Find *x*.

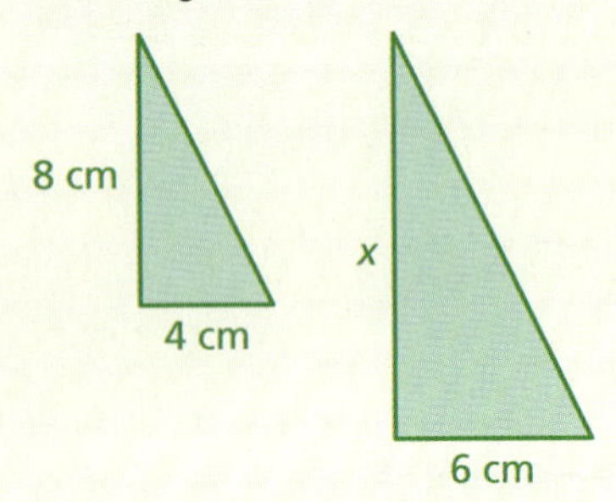

12 cm

On Your Own

2. $x = 4$ ft
3. $x = 24$ cm

Extra Example 3

The artist draws a larger replica of the painting in Example 3. The shorter base of the similar trapezoid is 10 inches. What is the height *h* of this trapezoid?
8 inches

On Your Own

4. 18 in.

Laurie's Notes

Example 2

- Draw the two triangles and state that they are similar.
- ? "Because the triangles are similar, what do you know?" Corresponding side lengths are proportional and corresponding angles are congruent.
- ? "The small triangle has a side that is 6 meters long. What is the length of the corresponding side in the large triangle?" 9 meters
- ? "The small triangle has a side that is 8 meters long. What is the labeled length of the corresponding side in the large triangle?" *x* meters
- Set up and solve the proportion.
- **MP6 Attend to Precision:** State the solution with the correct units.
- ? "Did anyone solve the proportion differently?" Some may say that because you add half of 6 to 6 to get 9, you can add half of 8 to 8 to get 12. Others may say that you can simplify the ratio $\frac{6}{8}$ as $\frac{3}{4}$ and then use mental math to solve for the missing side.
- **Connection:** Consider the larger triangle as a scale model (enlarged) of the smaller triangle. Ask students to find the scale factor. 1.5 : 1 or 3 : 2

On Your Own

- **Neighbor Check:** Have students work independently and then have their neighbors check their work. Have students discuss any discrepancies.

Example 3

- Share some information about the Berlin Wall.
  - It was built to keep East Germans from escaping to West Germany.
  - East German crews began tearing up streets and spreading barbed wire at midnight on August 13, 1961. By morning, the border with West Germany was closed. This is how the Berlin Wall began.
  - The wall was over a hundred miles long and went through four major changes. The final version of the Berlin Wall was 12 feet tall.
  - The Berlin Wall symbolized the boundary between communism and democracy until it was torn down in 1989.
  - Thierry Noir is one of the artists who turned the Berlin wall into the world's longest painting canvas. He wanted to "demystify" the wall with his bright-colored paintings.
- Draw the two trapezoids, one representing the painting and one representing the replica. Label the known dimensions.
- ? "The trapezoid in the actual painting has a side of length 15 inches. What is the length of the corresponding side in the replica?" 3.75 inches
- ? "You are to find the height *h* of the trapezoid in the replica. What is the length of the corresponding side in the actual painting?" 12 inches
- Set up and solve the proportion.
- **MP6:** State the solution with the correct units.

Closure

- Sketch two figures that look similar. Describe how you would determine whether your figures are actually similar.

EXAMPLE 2 Finding an Unknown Measure in Similar Figures

The triangles are similar. Find x.

Because the triangles are similar, corresponding side lengths are proportional. So, write and solve a proportion to find x.

6 m, 8 m; 9 m, x

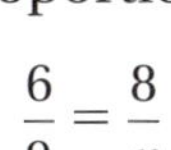

$$\frac{6}{9} = \frac{8}{x} \qquad \text{Write a proportion.}$$

$$6x = 72 \qquad \text{Cross Products Property}$$

$$x = 12 \qquad \text{Divide each side by 6.}$$

So, x is 12 meters.

On Your Own

The figures are similar. Find x.

2.

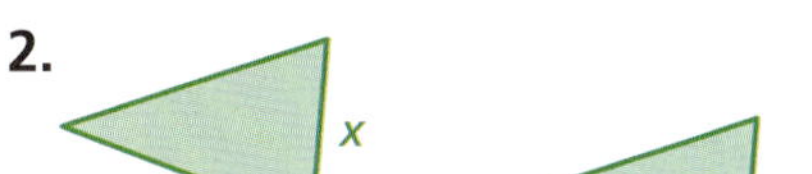
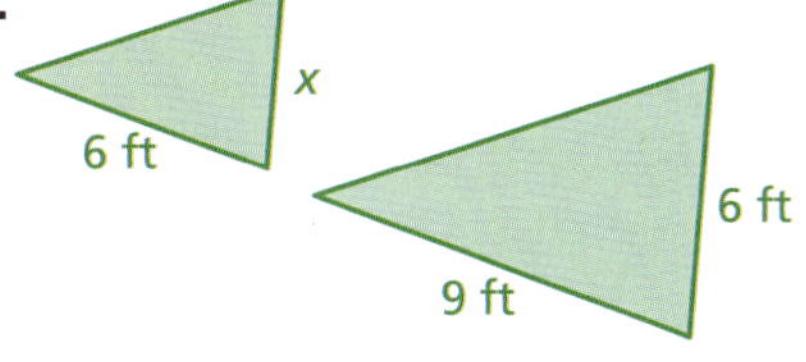

3.

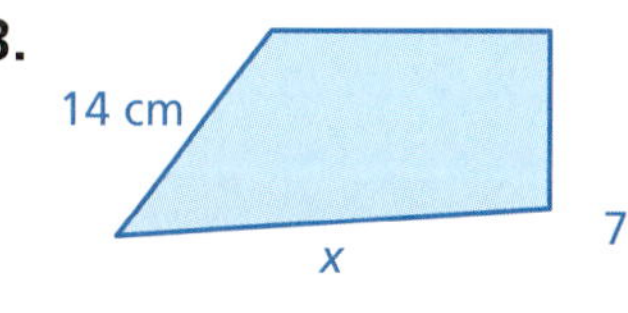

EXAMPLE 3 Real-Life Application

An artist draws a replica of a painting that is on the Berlin Wall. The painting includes a red trapezoid. The shorter base of the similar trapezoid in the replica is 3.75 inches. What is the height h of the trapezoid in the replica?

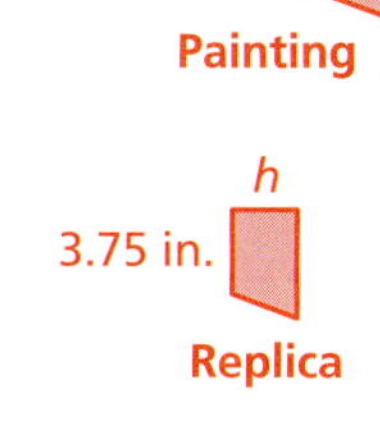

Because the trapezoids are similar, corresponding side lengths are proportional. So, write and solve a proportion to find h.

$$\frac{3.75}{15} = \frac{h}{12} \qquad \text{Write a proportion.}$$

$$12 \cdot \frac{3.75}{15} = 12 \cdot \frac{h}{12} \qquad \text{Multiplication Property of Equality}$$

$$3 = h \qquad \text{Simplify.}$$

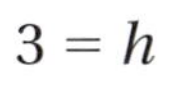

So, the height of the trapezoid in the replica is 3 inches.

On Your Own

4. **WHAT IF?** The longer base in the replica is 4.5 inches. What is the length of the longer base in the painting?

2.5 Exercises

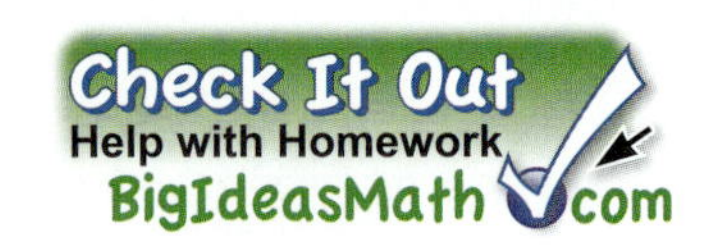

Vocabulary and Concept Check

1. **VOCABULARY** How are corresponding angles of two similar figures related?
2. **VOCABULARY** How are corresponding side lengths of two similar figures related?
3. **CRITICAL THINKING** Are two figures that have the same size and shape similar? Explain.

Practice and Problem Solving

Tell whether the two figures are similar. Explain your reasoning.

(1) 4.

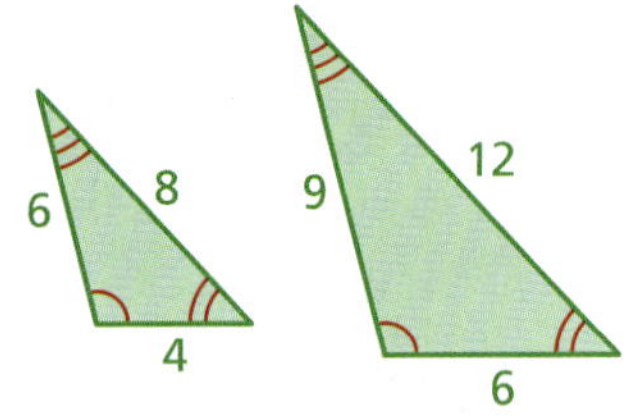

5.

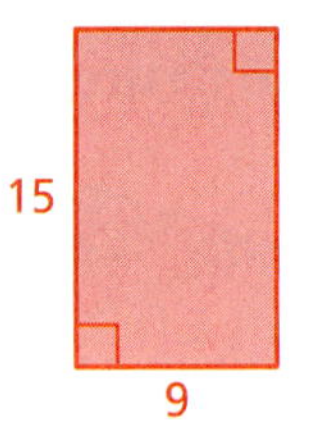

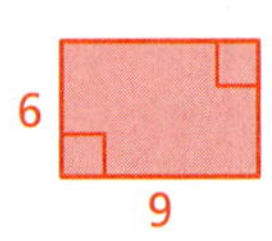

In a coordinate plane, draw the figures with the given vertices. Which figures are similar? Explain your reasoning.

6. Rectangle A: (0, 0), (4, 0), (4, 2), (0, 2)
   Rectangle B: (0, 0), (−6, 0), (−6, 3), (0, 3)
   Rectangle C: (0, 0), (4, 0), (4, 2), (0, 2)

7. Figure A: (−4, 2), (−2, 2), (−2, 0), (−4, 0)
   Figure B: (1, 4), (4, 4), (4, 1), (1, 1)
   Figure C: (2, −1), (5, −1), (5, −3), (2, −3)

The figures are similar. Find x.

(2) 8.

9.

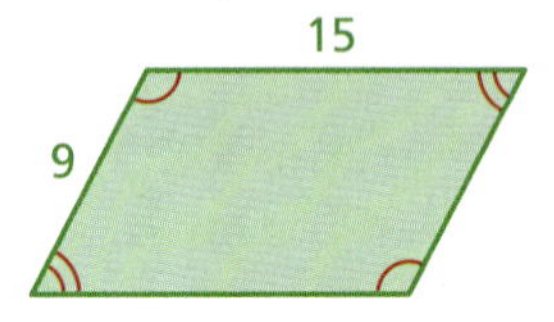

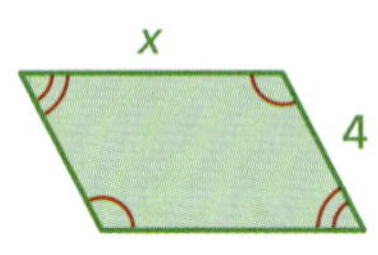

10.

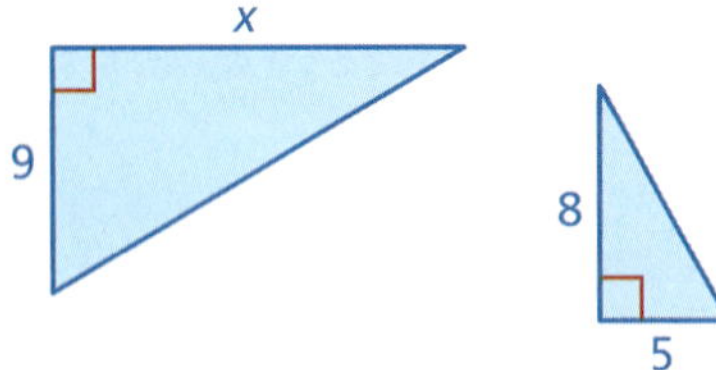

11.

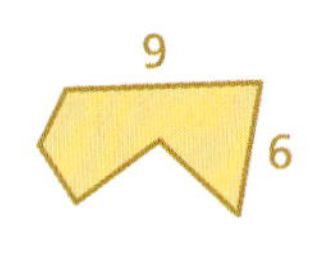

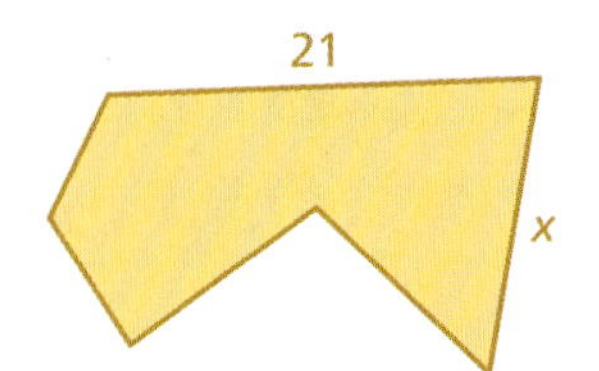

12. **MEXICO** A Mexican flag is 63 inches long and 36 inches wide. Is the drawing at the right similar to the Mexican flag?

13. **DESKS** A student's rectangular desk is 30 inches long and 18 inches wide. The teacher's desk is similar to the student's desk and has a length of 50 inches. What is the width of the teacher's desk?

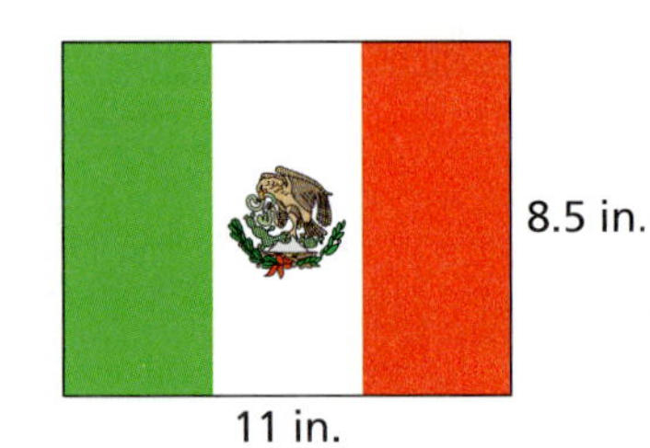

Assignment Guide and Homework Check

Level	Day 1 Activity Assignment	Day 2 Lesson Assignment	Homework Check
Basic	4, 5, 21–25	1–3, 7–19 odd	7, 9, 11, 13, 15
Average	4, 5, 21–25	1–3, 7–13 odd, 14–20 even	7, 9, 11, 13, 16
Advanced	1–5, 6–12 even, 13–25		6, 10, 16, 18, 20

Common Errors

- **Exercise 4** Students may think that the triangles are not similar because an optical illusion makes them appear to have different angles. Remind them to pay attention to the congruency markings on the figure.
- **Exercise 5** Students may be confused because the orientations of the rectangles are different. Tell them it is helpful to redraw the figures with the same orientations.
- **Exercises 8–11** Students may form the wrong proportions. Remind them to make sure like things are being compared.

2.5 Record and Practice Journal

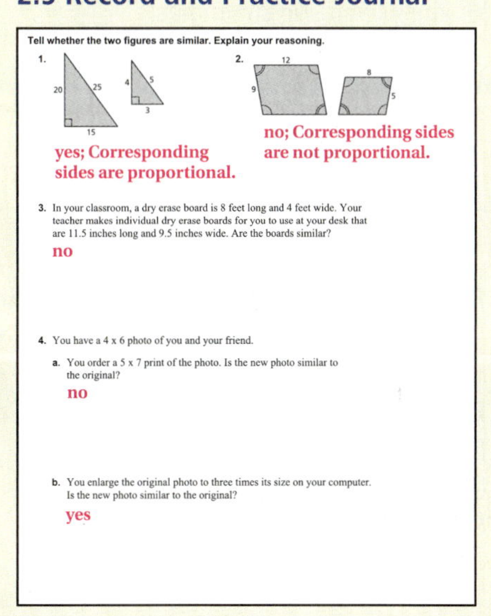
Tell whether the two figures are similar. Explain your reasoning.

1. yes; Corresponding sides are proportional.

2. no; Corresponding sides are not proportional.

3. In your classroom, a dry erase board is 8 feet long and 4 feet wide. Your teacher makes individual dry erase boards for you to use at your desk that are 11.5 inches long and 9.5 inches wide. Are the boards similar?
   no

4. You have a 4 x 6 photo of you and your friend.
   a. You order a 5 x 7 print of the photo. Is the new photo similar to the original?
   no
   b. You enlarge the original photo to three times its size on your computer. Is the new photo similar to the original?
   yes

Vocabulary and Concept Check

1. They are congruent.
2. They are proportional.
3. Yes, because the angles are congruent and the side lengths are proportional.

Practice and Problem Solving

4. similar; Corresponding angles are congruent. Because $\frac{4}{6} = \frac{6}{9} = \frac{8}{12}$, the corresponding side lengths are proportional.
5. not similar; Corresponding side lengths are not proportional.

6–7. See Additional Answers.

8. 15
9. $6\frac{2}{3}$
10. 14.4
11. 14
12. no
13. 30 in.
14. a. sometimes; They are similar only when corresponding side lengths are proportional and corresponding angles are congruent.
    b. always; All angles are congruent and all sides are proportional.
    c. sometimes; Corresponding angles are always congruent, but corresponding side lengths are not always proportional.
    d. never; They do not have the same shape.
15. See *Taking Math Deeper.*
16. a. yes b. no
17. 3 times

Practice and Problem Solving

18. yes; A scale drawing is a proportional drawing of an object, so corresponding angles are congruent and corresponding side lengths are proportional.

19–20. See Additional Answers.

Fair Game Review

21. $\frac{16}{81}$ **22.** $\frac{9}{64}$

23. $\frac{49}{16}$ **24.** $\frac{169}{16}$

25. C

Mini-Assessment

1.

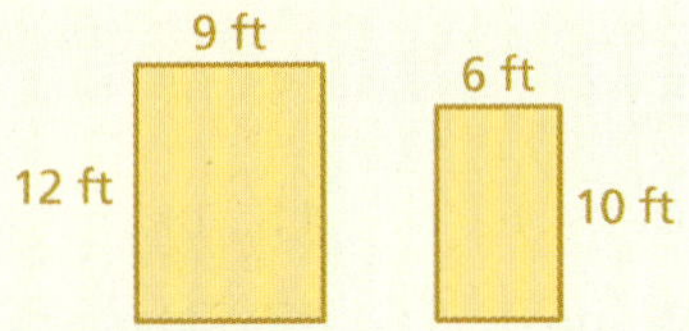

no; corresponding side lengths are not proportional

2. Are the two triangular stickers similar? Explain your reasoning.

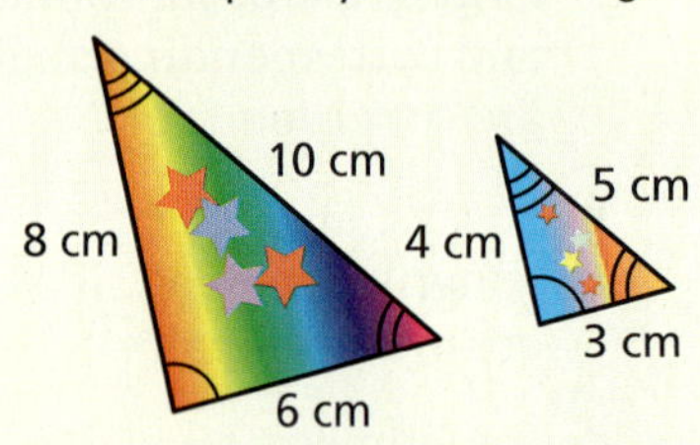

yes; corresponding side lengths are proportional and corresponding angles are congruent

3. The figures are similar. Find x.

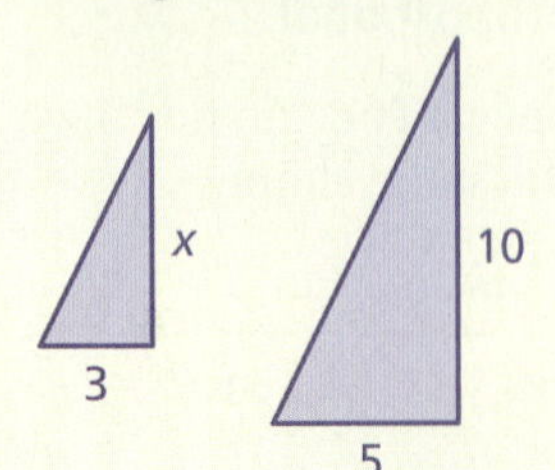

6

Taking Math Deeper

Exercise 15

Students may have difficulty finding an entry point when solving this problem. They could begin by reasoning about properties of different quadrilaterals.

Decide which type(s) of quadrilateral(s) to draw.

The quadrilaterals must each have two 130° angles and two 50° angles. So, you can eliminate rectangles because they can only have 90° angles.

One way to be sure that two quadrilaterals will not be similar is to use two quadrilaterals that have different shapes, such as trapezoids and parallelograms.

Construct a trapezoid and a parallelogram with the four given interior angle measures.

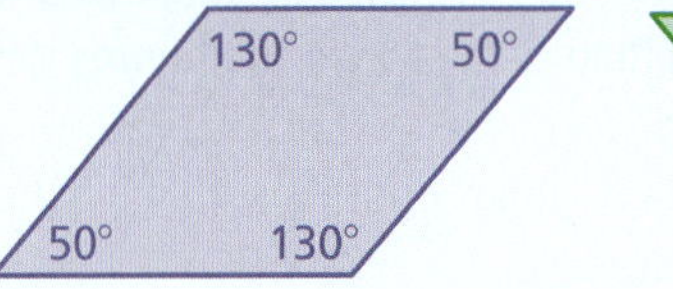

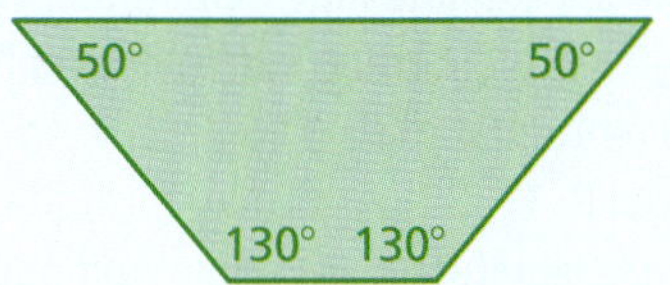

Students should pay close attention to orientation and size in problems like this. They may think that the two parallelograms below would justify their reasoning. However, the parallelograms are similar.

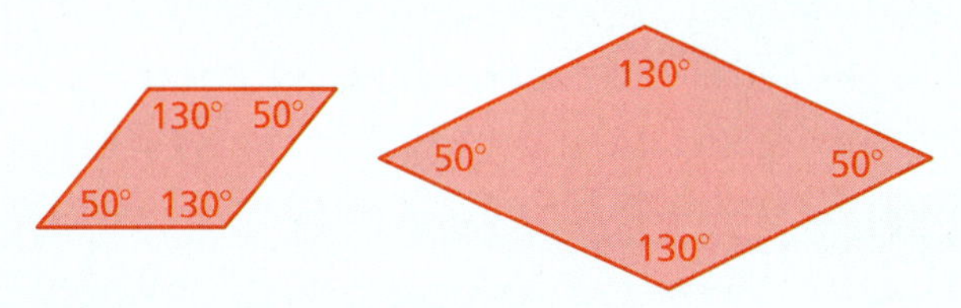

Project

Ask students if they can draw *three* different quadrilaterals each having two 100° angles and two 80° angles.

Reteaching and Enrichment Strategies

If students need help...	If students got it...
Resources by Chapter • Practice A and Practice B • Puzzle Time Record and Practice Journal Practice Differentiating the Lesson Lesson Tutorials Skills Review Handbook	Resources by Chapter • Enrichment and Extension • Technology Connection Start the next section

14. LOGIC Are the following figures *always*, *sometimes*, or *never* similar? Explain.

a. two triangles
b. two squares
c. two rectangles
d. a square and a triangle

15. CRITICAL THINKING Can you draw two quadrilaterals each having two 130° angles and two 50° angles that are *not* similar? Justify your answer.

16. SIGN All the angle measures in the sign are 90°.

a. You increase each side length by 20%. Is the new sign similar to the original?

b. You increase each side length by 6 inches. Is the new sign similar to the original?

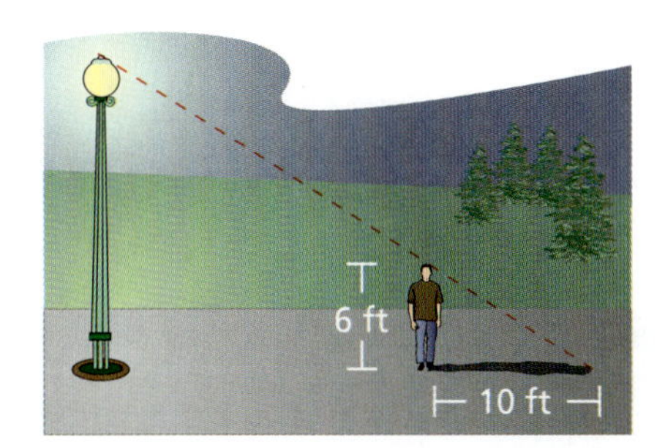

17. STREETLIGHT A person standing 20 feet from a streetlight casts a shadow as shown. How many times taller is the streetlight than the person? Assume the triangles are similar.

18. REASONING Is an object similar to a scale drawing of the object? Explain.

19. GEOMETRY Use a ruler to draw two different isosceles triangles similar to the one shown. Measure the heights of each triangle to the nearest centimeter.

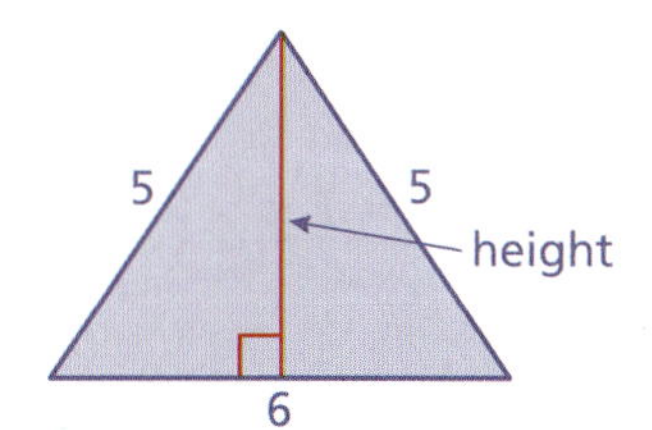

a. Is the ratio of the corresponding heights proportional to the ratio of the corresponding side lengths?

b. Do you think this is true for all similar triangles? Explain.

20. Critical Thinking Given $\triangle ABC \sim \triangle DEF$ and $\triangle DEF \sim \triangle JKL$, is $\triangle ABC \sim \triangle JKL$? Give an example or a non-example.

Fair Game Review What you learned in previous grades & lessons

Simplify. *(Skills Review Handbook)*

21. $\left(\frac{4}{9}\right)^2$
22. $\left(\frac{3}{8}\right)^2$
23. $\left(\frac{7}{4}\right)^2$
24. $\left(\frac{6.5}{2}\right)^2$

25. MULTIPLE CHOICE You solve the equation $S = \ell w + 2wh$ for w. Which equation is correct? *(Section 1.4)*

Ⓐ $w = \dfrac{S - \ell}{2h}$
Ⓑ $w = \dfrac{S - 2h}{\ell}$
Ⓒ $w = \dfrac{S}{\ell + 2h}$
Ⓓ $w = S - \ell - 2h$

2.6 Perimeters and Areas of Similar Figures

Essential Question How do changes in dimensions of similar geometric figures affect the perimeters and the areas of the figures?

1 ACTIVITY: Creating Similar Figures

Work with a partner. Use pattern blocks to make a figure whose dimensions are 2, 3, and 4 times greater than those of the original figure.

a. Square

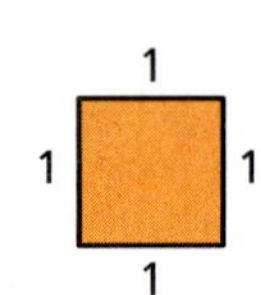

b. Rectangle

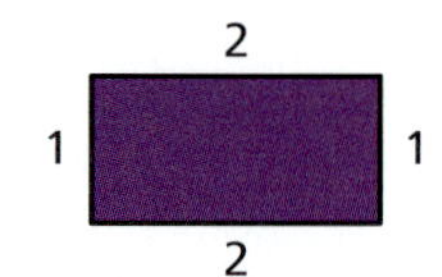

2 ACTIVITY: Finding Patterns for Perimeters

Work with a partner. Copy and complete the table for the perimeter *P* of each figure in Activity 1. Describe the pattern.

Figure	Original Side Lengths	Double Side Lengths	Triple Side Lengths	Quadruple Side Lengths
Square	$P =$			
Rectangle	$P =$			

3 ACTIVITY: Finding Patterns for Areas

Work with a partner. Copy and complete the table for the area *A* of each figure in Activity 1. Describe the pattern.

Figure	Original Side Lengths	Double Side Lengths	Triple Side Lengths	Quadruple Side Lengths
Square	$A =$			
Rectangle	$A =$			

COMMON CORE

Geometry

In this lesson, you will

- understand the relationship between perimeters of similar figures.
- understand the relationship between areas of similar figures.
- find ratios of perimeters and areas for similar figures.

Preparing for Standard MACC.8.G.1.4

Laurie's Notes

Introduction

Standards for Mathematical Practice

- **MP8 Look for and Express Regularity in Repeated Reasoning:** Students investigate how perimeters and areas of similar figures are related by finding a pattern.

Motivate

- Show various stages of the fractal known as The Sierpinski Triangle, which can be found in many places online. Ask students how many triangles (of various sizes) they see in each stage.

Activity Notes

Activity 1

- Use the pattern blocks provided in the Record and Practice Journal. Cut them out, store them in a reclosable bag, and reuse them each year.
- Circulate to make sure students are constructing the figures correctly.

Activities 2 and 3

? Ask questions to review perimeter and area.

- "What does perimeter mean?" the distance around a figure (you want an understanding of what perimeter *means*, not a formula)
- "How is it found?" Add the lengths of the sides of the figure.
- "What does area mean?" the amount of surface that a figure covers (you want an understanding of what area *means*, not a formula)
- "How is it found?" The type of figure determines which area formula you use.

? "When you make a scale drawing of a figure whose dimensions are twice the original dimensions, is the scale drawing similar to the figure? Explain." Yes, the corresponding sides are proportional and the corresponding angles are congruent.

? **MP8:** Ask questions to compare the perimeter of each new figure with the perimeter of the original figure.

- "When the dimensions of a figure are doubled, what happens to the perimeter?" The perimeter also doubles.
- "When the dimensions of a figure are tripled, what happens to the perimeter?" The perimeter also triples.
- "When the dimensions of a figure are quadrupled, what happens to the perimeter?" The perimeter also quadruples.

- Repeat the above questions for area. The correct answers are different!
- **Suggestion:** When students look at a table entry, they should consider statements such as the following:
  "I double the original side lengths and the perimeter is _____." double
  "I double the original side lengths and the area is _____." 4 times greater
- **MP8:** Students should recognize a pattern as they work through these activities.

Florida Common Core Standards

MACC.8.G.1.4 . . . Given two similar two-dimensional figures, describe a sequence that exhibits the similarity between them.

Previous Learning

Students should know how to plot ordered pairs. Students also need to remember how to solve a proportion.

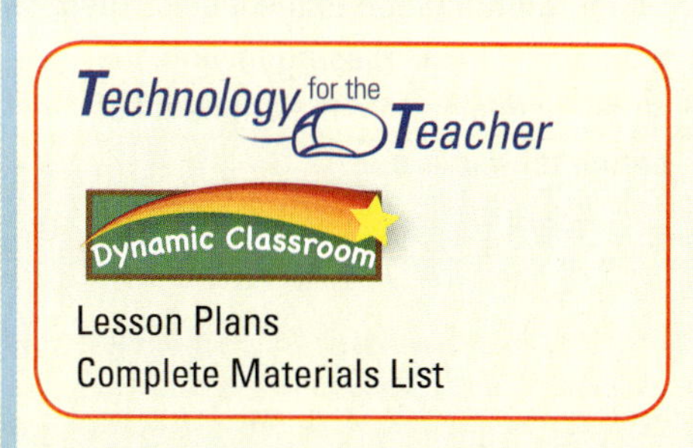

2.6 Record and Practice Journal

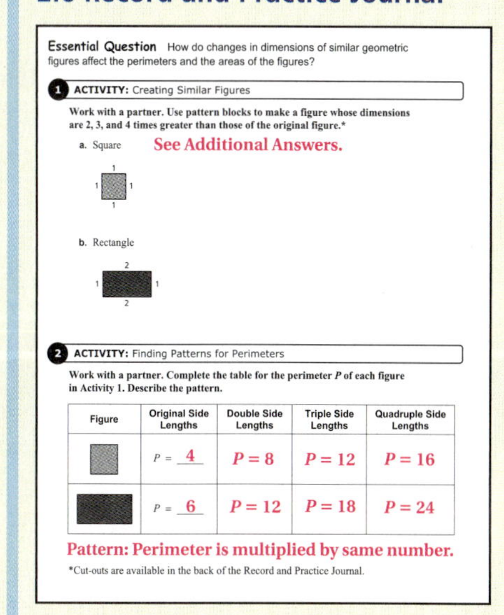

Essential Question How do changes in dimensions of similar geometric figures affect the perimeters and the areas of the figures?

1 ACTIVITY: Creating Similar Figures

Work with a partner. Use pattern blocks to make a figure whose dimensions are 2, 3, and 4 times greater than those of the original figure.*

a. Square See Additional Answers.

b. Rectangle

2 ACTIVITY: Finding Patterns for Perimeters

Work with a partner. Complete the table for the perimeter *P* of each figure in Activity 1. Describe the pattern.

Figure	Original Side Lengths	Double Side Lengths	Triple Side Lengths	Quadruple Side Lengths
(square)	$P = 4$	$P = 8$	$P = 12$	$P = 16$
(rectangle)	$P = 6$	$P = 12$	$P = 18$	$P = 24$

Pattern: Perimeter is multiplied by same number.

*Cut-outs are available in the back of the Record and Practice Journal.

English Language Learners

Vocabulary

Have students select objects around the classroom and use tape to mark the perimeters of the objects. Label the objects "perimeter." Have students select other objects in the classroom and cover them with square sheets of paper. Label these objects "area." Have students identify which units are best for measuring perimeter and area. Add this information to the labels. Keep the objects in the classroom until the students understand the concepts of perimeter and area.

2.6 Record and Practice Journal

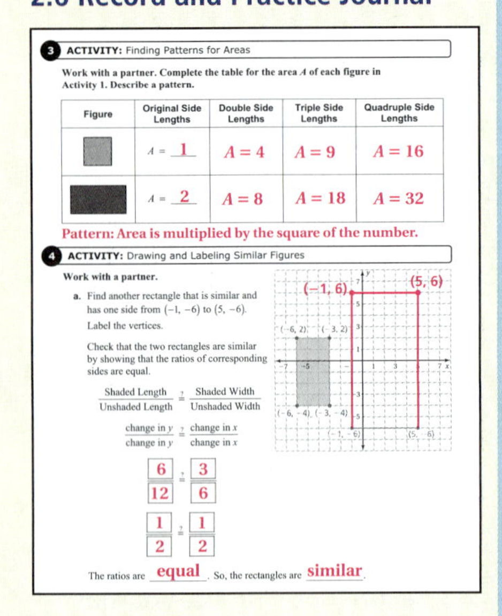

3 ACTIVITY: Finding Patterns for Areas

Work with a partner. Complete the table for the area A of each figure in Activity 1. Describe a pattern.

Figure	Original Side Lengths	Double Side Lengths	Triple Side Lengths	Quadruple Side Lengths
(square)	$A =$ 1	$A = 4$	$A = 9$	$A = 16$
(rectangle)	$A =$ 2	$A = 8$	$A = 18$	$A = 32$

Pattern: Area is multiplied by the square of the number.

4 ACTIVITY: Drawing and Labeling Similar Figures

Work with a partner.

a. Find another rectangle that is similar and has one side from $(-1, -6)$ to $(5, -6)$. Label the vertices.

Check that the two rectangles are similar by showing that the ratios of corresponding sides are equal.

$\dfrac{\text{Shaded Length}}{\text{Unshaded Length}} \stackrel{?}{=} \dfrac{\text{Shaded Width}}{\text{Unshaded Width}}$

$\dfrac{\text{change in } y}{\text{change in } y} \stackrel{?}{=} \dfrac{\text{change in } x}{\text{change in } x}$

$\dfrac{6}{12} \stackrel{?}{=} \dfrac{3}{6}$

$\dfrac{1}{2} \stackrel{?}{=} \dfrac{1}{2}$

The ratios are equal. So, the rectangles are similar.

b. Compare the perimeters and the areas of the figures. Are the results the same as your results from Activities 2 and 3? Explain.

See Additional Answers.

c. There are three other rectangles that are similar to the shaded rectangle and have the given side.

- Draw each one. Label the vertices of each.

See Additional Answers.

- Show that each is similar to the original shaded rectangle.

See Additional Answers.

What Is Your Answer?

5. IN YOUR OWN WORDS How do changes in dimensions of similar geometric figures affect the perimeters and the areas of the figures?

Perimeter is multiplied by the number. Area is multiplied by the square of the number.

6. What information do you need to know to find the dimensions of a figure that is similar to another figure? Give examples to support your explanation.

lengths of a pair of corresponding sides and the side length that corresponds to the unknown length

Laurie's Notes

Activity 4

? "What are the dimensions of the rectangle?" 3 units by 6 units

- It is okay for students to put their fingers on the sides and count units.
- **Common Error:** Students count the lattice points beginning with the vertex and end up with dimensions 4 by 7 instead of 3 by 6.
- **Connection:** Notice in part (a) the use of language that is also used for slope. To compute the "change in y" or the "change in x," students should just look at the diagram and count.
- Do not skip the last step of showing that the blue rectangles are similar to the original red rectangle. Students need the practice of writing proportions. Note that while the terms length and width are usually interchangeable, the length here is the longer of the two sides. The color reference is also easier for students to understand rather than saying "corresponding side in the left rectangle to the corresponding side in the right rectangle."
- Keep the language simple so students focus on the concept.
- Have students share their three solutions. Check to see that the coordinates are correctly labeled, for instance, $(5, -6)$ and not $(-6, 5)$.

? "What is the ratio of corresponding sides when the rectangles are the same size?" 1 : 1

What Is Your Answer?

- **Big Idea:** When the dimensions are doubled, tripled, or quadrupled, the resulting figure is similar to the original figure. Students should see a pattern in the perimeters and in the areas. The Key Ideas in the lesson will define these patterns further.

Closure

- **Exit Ticket:** Find the perimeter and area of a rectangle with dimensions 3 inches by 4 inches. If you double the dimensions, what will be the new perimeter and area? Original dimensions: perimeter is 14 in., area is 12 in.^{2} New dimensions: perimeter is $2(14) = 28$ in., area is $4(12) = 48$ in.^{2}

4 ACTIVITY: Drawing and Labeling Similar Figures

Work with a partner.

a. Find a blue rectangle that is similar to the red rectangle and has one side from $(-1, -6)$ to $(5, -6)$. Label the vertices.

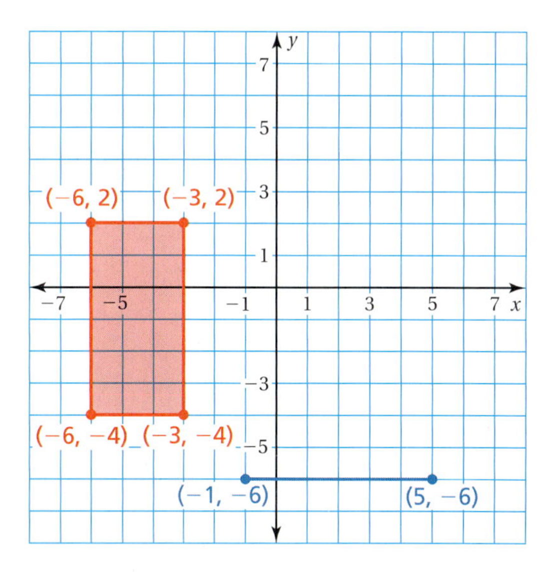

Check that the two rectangles are similar by showing that the ratios of corresponding sides are equal.

$$\frac{\text{Red Length}}{\text{Blue Length}} \stackrel{?}{=} \frac{\text{Red Width}}{\text{Blue Width}}$$

$$\frac{\text{change in } y}{\text{change in } y} \stackrel{?}{=} \frac{\text{change in } x}{\text{change in } x}$$

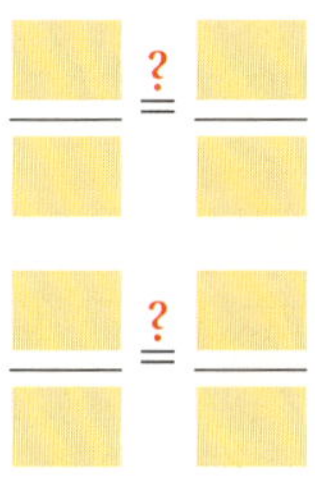

The ratios are equal. So, the rectangles are similar.

b. Compare the perimeters and the areas of the figures. Are the results the same as your results from Activities 2 and 3? Explain.

c. There are three other blue rectangles that are similar to the red rectangle and have the given side.

- Draw each one. Label the vertices of each.
- Show that each is similar to the original red rectangle.

Math Practice 1

Analyze Givens

What values should you use to fill in the proportion? Does it matter where each value goes? Explain.

What Is Your Answer?

5. IN YOUR OWN WORDS How do changes in dimensions of similar geometric figures affect the perimeters and the areas of the figures?

6. What information do you need to know to find the dimensions of a figure that is similar to another figure? Give examples to support your explanation.

Practice Use what you learned about perimeters and areas of similar figures to complete Exercises 8 and 9 on page 80.

2.6 Lesson

Perimeters of Similar Figures

When two figures are similar, the ratio of their perimeters is equal to the ratio of their corresponding side lengths.

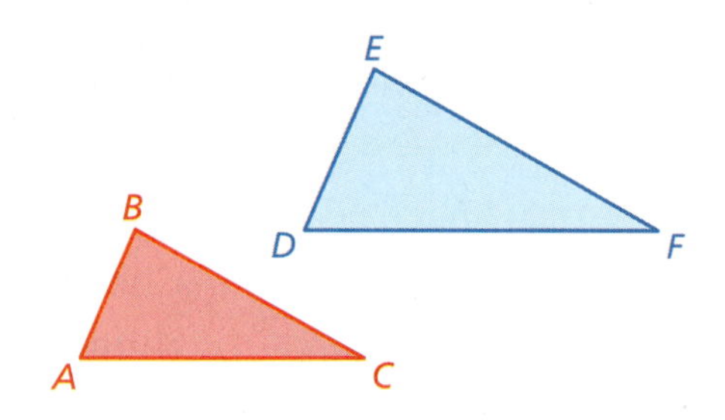

$$\frac{\text{Perimeter of } \triangle ABC}{\text{Perimeter of } \triangle DEF} = \frac{AB}{DE} = \frac{BC}{EF} = \frac{AC}{DF}$$

EXAMPLE 1 Finding Ratios of Perimeters

Find the ratio (red to blue) of the perimeters of the similar rectangles.

$$\frac{\text{Perimeter of red rectangle}}{\text{Perimeter of blue rectangle}} = \frac{4}{6} = \frac{2}{3}$$

The ratio of the perimeters is $\frac{2}{3}$.

On Your Own

1. The height of Figure A is 9 feet. The height of a similar Figure B is 15 feet. What is the ratio of the perimeter of A to the perimeter of B?

Areas of Similar Figures

When two figures are similar, the ratio of their areas is equal to the *square* of the ratio of their corresponding side lengths.

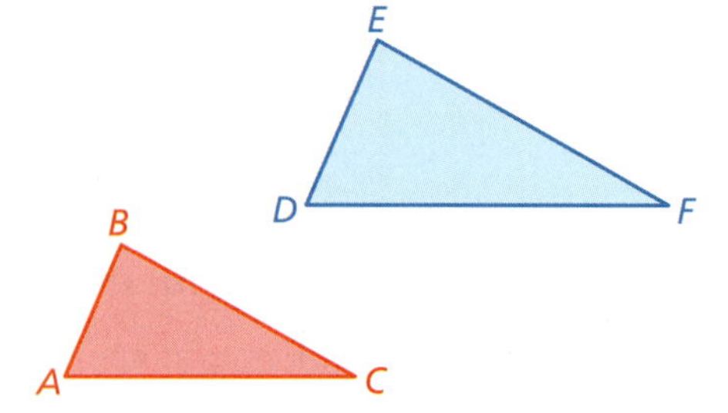

$$\frac{\text{Area of } \triangle ABC}{\text{Area of } \triangle DEF} = \left(\frac{AB}{DE}\right)^2 = \left(\frac{BC}{EF}\right)^2 = \left(\frac{AC}{DF}\right)^2$$

Laurie's Notes

Introduction

Connect

- **Yesterday:** Students used pattern blocks to investigate how changes in the dimensions of similar figures affect the perimeter and area of the figures. (MP8)
- **Today:** Students will use the stated relationships to solve problems.

Motivate

- **Story Time:** Tell students that your neighbor's lawn is twice the size of your lawn. In other words, it is twice as long and twice as wide. If it takes you one-half hour to mow your lawn, about how long does it take your neighbor to mow his lawn? You will answer this in the Closure activity.

Lesson Notes

Key Idea

? "How do you identify similar triangles?" Corresponding sides are proportional and corresponding angles are congruent.

- Write some side lengths on the two triangles, such as 3-4-5 and 6-8-10. Then find the two perimeters, 12 and 24, to show the same ratio of 1 : 2.

Example 1

? "What is the ratio of the corresponding sides?" 4 : 6 or 2 : 3 The ratio of the perimeters is also 2 : 3.

? **MP1 Make Sense of Problems and Persevere in Solving Them:** "Why do you not need to know both dimensions of one of the rectangles to find the ratio of the perimeters?" Listen for students to explain that because the rectangles are similar, only one pair of corresponding sides is necessary.

On Your Own

- Check that students have correctly written the ratio.

Key Idea

- **Representation:** Draw two triangles whose corresponding sides appear to have a ratio of 1 : 2.

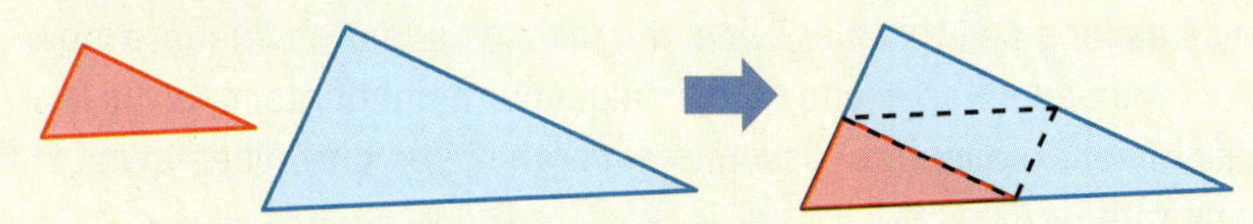

? "If the corresponding sides have a ratio of 1 : 2, then what will the ratio of the areas be?" 1 : 4; You can also use pattern blocks to show this relationship.

- **MP4 Model with Mathematics:** There are 4 copies of the smaller triangle inside the larger. Another way to state this relationship is that the larger triangle has an area 4 times greater than the area of the smaller triangle.

Goal Today's lesson is finding ratios of perimeters and areas of similar figures.

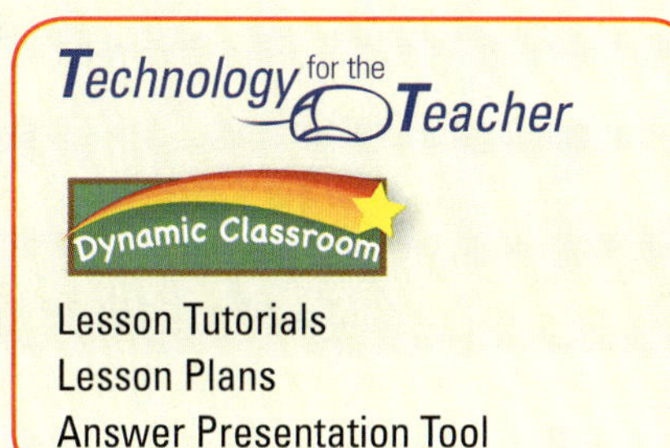

Lesson Tutorials
Lesson Plans
Answer Presentation Tool

Extra Example 1

Find the ratio (red to blue) of the perimeters of the similar trapezoids. $\frac{7}{5}$

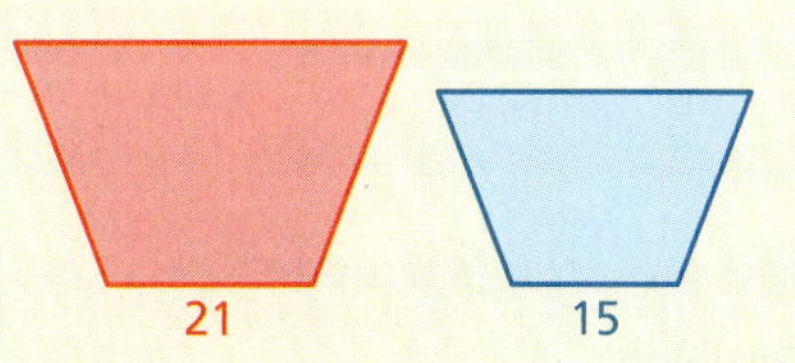

On Your Own

1. $\frac{3}{5}$

Extra Example 2

Find the ratio (red to blue) of the areas of the similar parallelograms. $\frac{81}{36} = \frac{9}{4}$

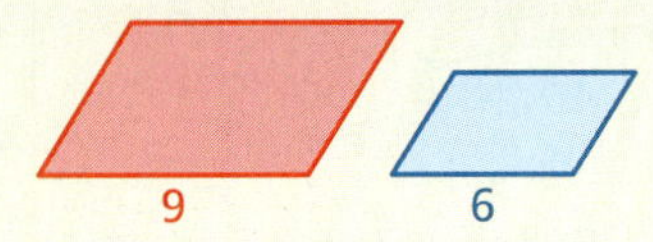

On Your Own

2. $\frac{64}{49}$

Extra Example 3

In Example 3, the width of the pool is 22 yards. Find the perimeter P and the area A of the pool. 132 yd; 968 yd^2

On Your Own

3. 96 yd; 512 yd^2

Differentiated Instruction

Kinesthetic

Provide students with large pieces of construction paper, rulers, and protractors. Have students work in pairs to draw a large right triangle on the paper. Record the lengths of the sides and the measures of the angles in a table. Connect the midpoints of each side of the large triangle and record the side lengths and angle measures of the second triangle in the table. Connect the midpoints of the sides of the second triangle to form a third triangle. Record the side lengths and angle measures of the third triangle in the table. Have students determine if the triangles are similar. If they are similar, find the ratios of the perimeters and areas of each pair of triangles.

Laurie's Notes

Example 2

- As with Example 1, only one pair of corresponding sides is given. The actual areas cannot be computed. Students must use the relationship stated in the Key Idea.
- Remind students that $\left(\frac{3}{5}\right)^2$ means $\left(\frac{3}{5}\right) \cdot \left(\frac{3}{5}\right) = \frac{9}{25}$.

On Your Own

- Have students work in pairs on Question 2.

Example 3

- Ask a volunteer to read the problem.
- ? "What information do you know?" one side length in each object, area and perimeter of the volleyball court, and the fact that the objects are similar
- ? "Are the sides whose lengths are given corresponding sides?" yes
- **MP1:** Give students time to think about solution strategies. There are many different strategies, and it is important for students to come up with an entry point that makes sense to them.
- One strategy, which makes use of the Key Ideas in the section, is given in the solution.
- Another strategy is to find the length ℓ of the volleyball court by solving $2(10) + 2\ell = 60$ for ℓ and then setting up a proportion to find the length of the pool. From there, you can find the area and perimeter of the pool.
- Discuss any other strategies that students might suggest. Ask students which strategy they prefer.
- **MP6 Attend to Precision:** Make sure students include the correct units in their answers.

On Your Own

- Have students work in pairs on Question 3.

Closure

- **Exit Ticket:** Return to the question used to motivate at the beginning of the lesson. Your neighbor's lawn is twice the size of your lawn, meaning twice as long and twice as wide. If it takes you one-half hour to mow your lawn, about how long does it take your neighbor to mow his lawn? Explain your reasoning. (Assume that both of you mow your lawns at the same rate.)
  2 hours; because the dimensions of your neighbor's lawn are double the dimensions of your lawn, you have the ratio neighbor : you $= 2 : 1$. The area of your neighbor's lawn is $\left(\frac{2}{1}\right)^2 = \frac{4}{1}$ times the area of your lawn, so it should take your neighbor 4 times longer to mow his lawn, $4\left(\frac{1}{2}\right) = 2$ hours.

EXAMPLE 2 Finding Ratios of Areas

Find the ratio (red to blue) of the areas of the similar triangles.

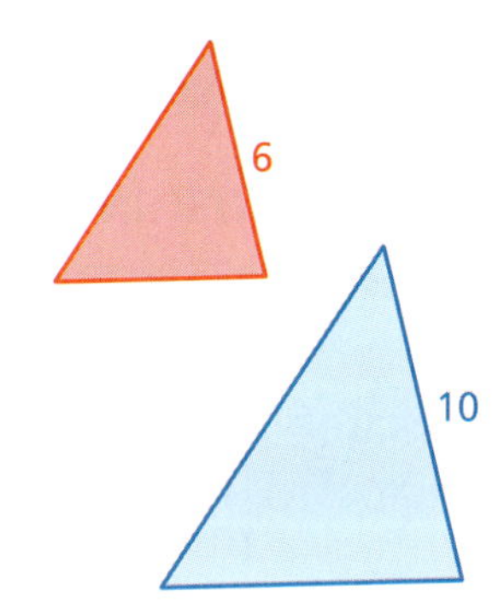

$$\frac{\text{Area of red triangle}}{\text{Area of blue triangle}} = \left(\frac{6}{10}\right)^2$$

$$= \left(\frac{3}{5}\right)^2 = \frac{9}{25}$$

The ratio of the areas is $\frac{9}{25}$.

On Your Own

2. The base of Triangle P is 8 meters. The base of a similar Triangle Q is 7 meters. What is the ratio of the area of P to the area of Q?

EXAMPLE 3 Using Proportions to Find Perimeters and Areas

A swimming pool is similar in shape to a volleyball court. Find the perimeter P and the area A of the pool.

18 yd

10 yd

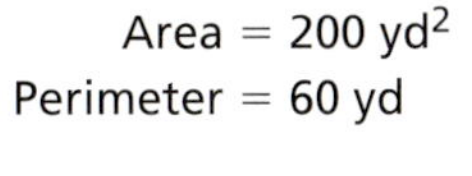

Area = 200 yd²
Perimeter = 60 yd

The rectangular pool and the court are similar. So, use the ratio of corresponding side lengths to write and solve proportions to find the perimeter and the area of the pool.

Perimeter

$$\frac{\text{Perimeter of court}}{\text{Perimeter of pool}} = \frac{\text{Width of court}}{\text{Width of pool}}$$

$$\frac{60}{P} = \frac{10}{18}$$

$$1080 = 10P$$

$$108 = P$$

Area

$$\frac{\text{Area of court}}{\text{Area of pool}} = \left(\frac{\text{Width of court}}{\text{Width of pool}}\right)^2$$

$$\frac{200}{A} = \left(\frac{10}{18}\right)^2$$

$$\frac{200}{A} = \frac{100}{324}$$

$$64{,}800 = 100A$$

$$648 = A$$

So, the perimeter of the pool is 108 yards, and the area is 648 square yards.

On Your Own

3. **WHAT IF?** The width of the pool is 16 yards. Find the perimeter P and the area A of the pool.

2.6 Exercises

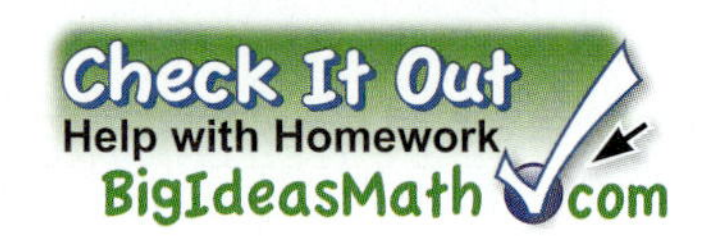

Vocabulary and Concept Check

1. **WRITING** How are the perimeters of two similar figures related?

2. **WRITING** How are the areas of two similar figures related?

3. **NUMBER SENSE** Rectangle *ABCD* is similar to Rectangle *WXYZ*. The area of *ABCD* is 30 square inches. Explain how to find the area of *WXYZ*.

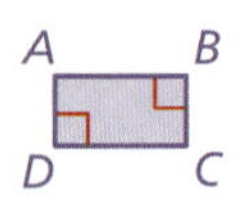

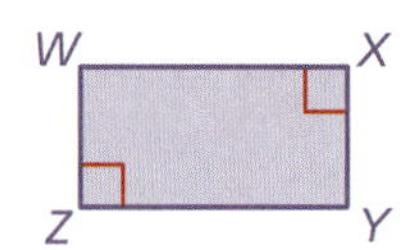

$\dfrac{AD}{WZ} = \dfrac{1}{2}$ $\qquad$ $\dfrac{AB}{WX} = \dfrac{1}{2}$

Practice and Problem Solving

The two figures are similar. Find the ratios (red to blue) of the perimeters and of the areas.

4.

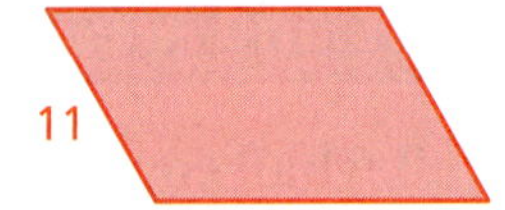

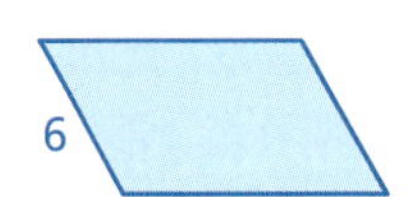

5.

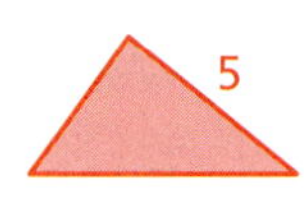

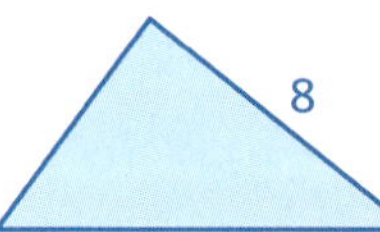

6.

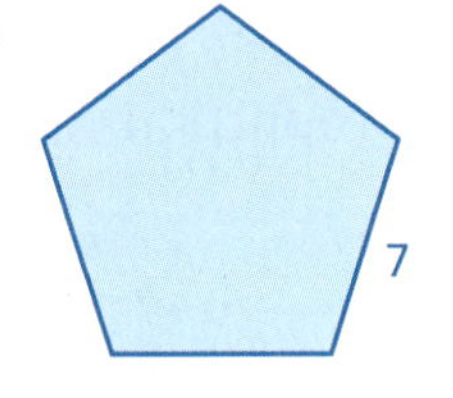

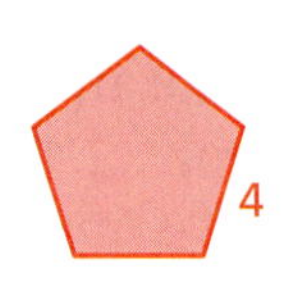

7.

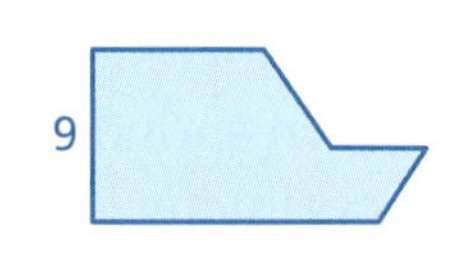

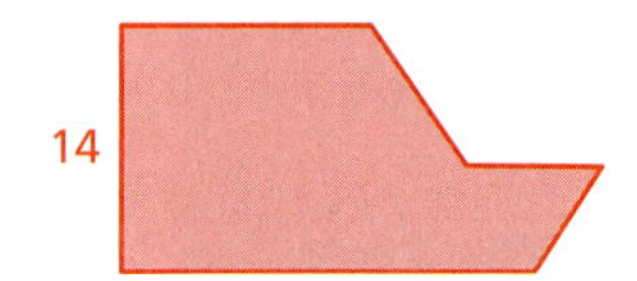

8. **PERIMETER** How does doubling the side lengths of a right triangle affect its perimeter?

9. **AREA** How does tripling the side lengths of a right triangle affect its area?

The figures are similar. Find *x*.

10. The ratio of the perimeters is 7 : 10.

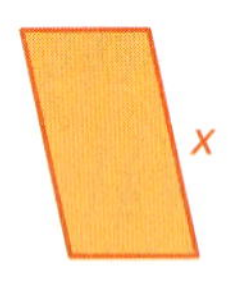

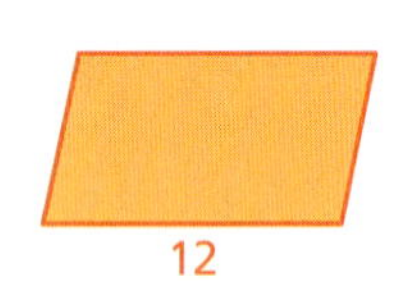

11. The ratio of the perimeters is 8 : 5.

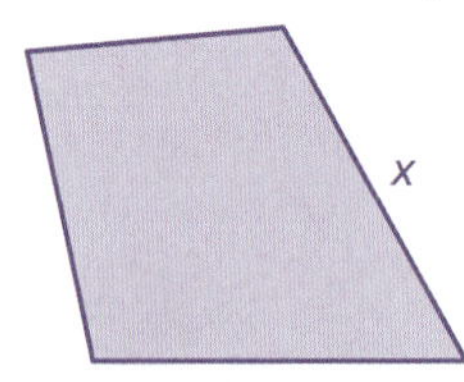

12. **FOOSBALL** The playing surfaces of two foosball tables are similar. The ratio of the corresponding side lengths is 10 : 7. What is the ratio of the areas?

13. **CHEERLEADING** A rectangular school banner has a length of 44 inches, a perimeter of 156 inches, and an area of 1496 square inches. The cheerleaders make signs similar to the banner. The length of a sign is 11 inches. What is its perimeter and its area?

Assignment Guide and Homework Check

Level	Day 1 Activity Assignment	Day 2 Lesson Assignment	Homework Check
Basic	8, 9, 21–24	1–3, 5, 7, 11–19 odd	5, 11, 13, 15
Average	8, 9, 21–24	1–3, 4, 6, 11–19 odd	6, 11, 13, 15, 17
Advanced	1–4, 6, 8, 9, 10–14 even, 15–24		6, 10, 16, 17, 19

Common Errors

- **Exercises 4–7** Students may find the reciprocal of the ratio. For example, they may find the ratio of blue to red instead of red to blue. Remind students to read the directions carefully.
- **Exercises 4–7** When finding the ratio of the areas, students may forget to square the ratio of the corresponding side lengths. Remind them of the Key Idea at the bottom of page 78.

2.6 Record and Practice Journal

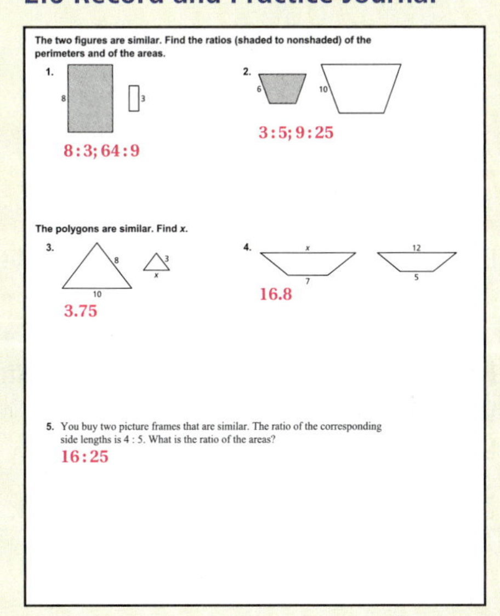

The two figures are similar. Find the ratios (shaded to nonshaded) of the perimeters and of the areas.

1. 8:3; 64:9
2. 3:5; 9:25

The polygons are similar. Find *x*.

3. 3.75
4. 16.8

5. You buy two picture frames that are similar. The ratio of the corresponding side lengths is 4 : 5. What is the ratio of the areas?
   16:25

Vocabulary and Concept Check

1. The ratio of the perimeters is equal to the ratio of the corresponding side lengths.
2. The ratio of the areas is equal to the square of the ratio of the corresponding side lengths.
3. Because the ratio of the corresponding side lengths is $\frac{1}{2}$, the ratio of the areas is equal to $\left(\frac{1}{2}\right)^2$. To find the area, solve the proportion $\frac{30}{x} = \frac{1}{4}$ to get $x = 120$ square inches.

Practice and Problem Solving

4. $\frac{11}{6}; \frac{121}{36}$
5. $\frac{5}{8}; \frac{25}{64}$
6. $\frac{4}{7}; \frac{16}{49}$
7. $\frac{14}{9}; \frac{196}{81}$
8. The perimeter doubles.
9. The area is 9 times larger.
10. 8.4
11. 25.6
12. 100 : 49
13. 39 in.; 93.5 in.^{2}

Practice and Problem Solving

14. *ABCD*: $P = 14, A = 12$; *WXYZ*: $P = 28, A = 48$; yes; Corresponding side lengths are proportional and corresponding angles are congruent.

15. 108 yd

16. See *Taking Math Deeper*.

17. **a.** 400 times greater; The ratio of the corresponding lengths is $\frac{120 \text{ in.}}{6 \text{ in.}} = \frac{20}{1}$. So, the ratio of the areas is $\left(\frac{20}{1}\right)^2 = \frac{400}{1}$.

b. 1250 ft^2

18. See Additional Answers.

19. 15 m **20.** 12.5 bottles

Fair Game Review

21. $x = -2$ **22.** $b = 1.6$

23. $n = -4$ **24.** B

Mini-Assessment

The two figures are similar. Find the ratio (red to blue) of the perimeters and of the areas.

1.

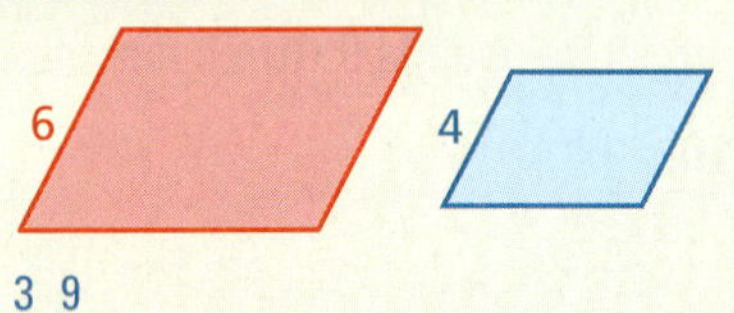

$\frac{3}{2}, \frac{9}{4}$

2.

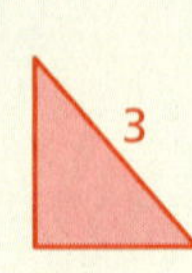

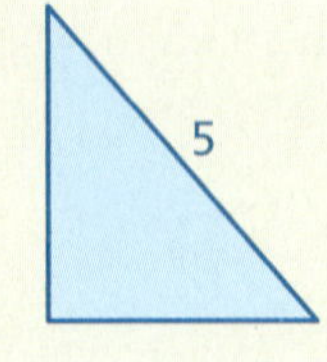

$\frac{3}{5}, \frac{9}{25}$

3. The ratio of the corresponding side lengths of two similar cellular phones is 3 : 4. The perimeter of the smaller phone is 9 inches. What is the perimeter of the larger phone? 12 in.

Taking Math Deeper

Exercise 16

There are several very different ways to solve this problem. This would be a good problem to encourage students to "think outside the box." You might have students work in pairs to see how many different ways they can solve the problem.

 Recognize that the smaller piece is one-fourth the size of the larger piece.

2 Use unit prices.

$$\frac{\text{Cost}}{\text{Area}} = \frac{1.31}{9 \times 21} \approx \$0.007 \text{ per in.}^2$$

Area of the 18×42 piece $= 18 \cdot 42 = 756 \text{ in.}^2$

Cost of the 18×42 piece $\approx 756 \cdot 0.007 \approx \$5.29$

 To find a more precise solution, use a proportion.

$$\frac{\text{Area}}{\text{Area}} = \frac{\text{Cost}}{\text{Cost}}$$

$$\frac{756}{189} = \frac{x}{1.31}$$

$$\$5.24 = x$$

Reteaching and Enrichment Strategies

If students need help. . .	If students got it. . .
Resources by Chapter • Practice A and Practice B • Puzzle Time Record and Practice Journal Practice Differentiating the Lesson Lesson Tutorials Skills Review Handbook	Resources by Chapter • Enrichment and Extension • Technology Connection Start the next section

14. REASONING The vertices of two rectangles are $A(-5, -1)$, $B(-1, -1)$, $C(-1, -4)$, $D(-5, -4)$ and $W(1, 6)$, $X(7, 6)$, $Y(7, -2)$, $Z(1, -2)$. Compare the perimeters and the areas of the rectangles. Are the rectangles similar? Explain.

15. SQUARE The ratio of the side length of Square A to the side length of Square B is 4 : 9. The side length of Square A is 12 yards. What is the perimeter of Square B?

16. FABRIC The cost of the fabric is $1.31. What would you expect to pay for a similar piece of fabric that is 18 inches by 42 inches?

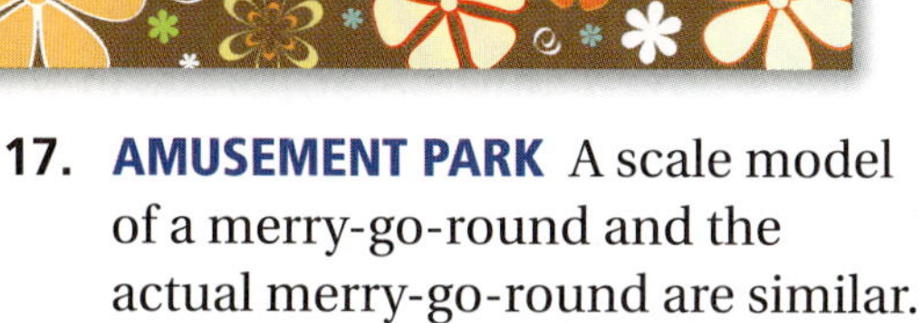

17. AMUSEMENT PARK A scale model of a merry-go-round and the actual merry-go-round are similar.

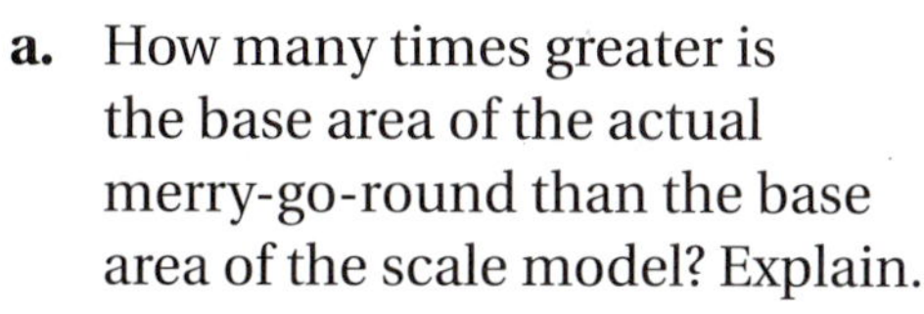

a. How many times greater is the base area of the actual merry-go-round than the base area of the scale model? Explain.

b. What is the base area of the actual merry-go-round in square feet?

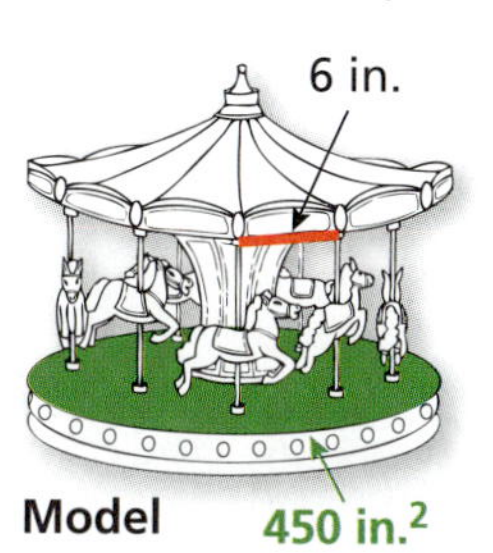

18. STRUCTURE The circumference of Circle K is π. The circumference of Circle L is 4π.

a. What is the ratio of their circumferences? of their radii? of their areas?

b. What do you notice?

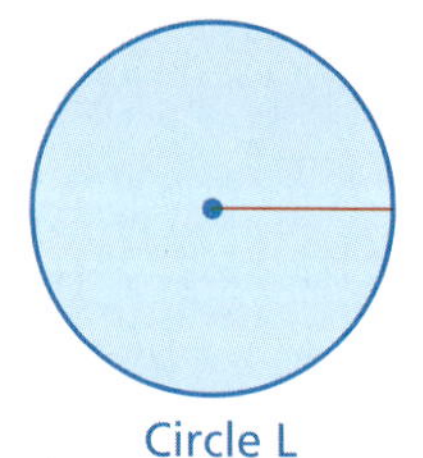

19. GEOMETRY A triangle with an area of 10 square meters has a base of 4 meters. A similar triangle has an area of 90 square meters. What is the *height* of the larger triangle?

20. Problem Solving You need two bottles of fertilizer to treat the flower garden shown. How many bottles do you need to treat a similar garden with a perimeter of 105 feet?

Fair Game Review What you learned in previous grades & lessons

Solve the equation. Check your solution. *(Section 1.3)*

21. $4x + 12 = -2x$ **22.** $2b + 6 = 7b - 2$ **23.** $8(4n + 13) = 6n$

24. MULTIPLE CHOICE Last week, you collected 20 pounds of cans for recycling. This week, you collect 25 pounds of cans for recycling. What is the percent of increase? *(Skills Review Handbook)*

Ⓐ 20% Ⓑ 25% Ⓒ 80% Ⓓ 125%

2.7 Dilations

Essential Question How can you enlarge or reduce a figure in the coordinate plane?

The Meaning of a Word ● Dilate

When you have your eyes checked, the optometrist sometimes **dilates** one or both of the pupils of your eyes.

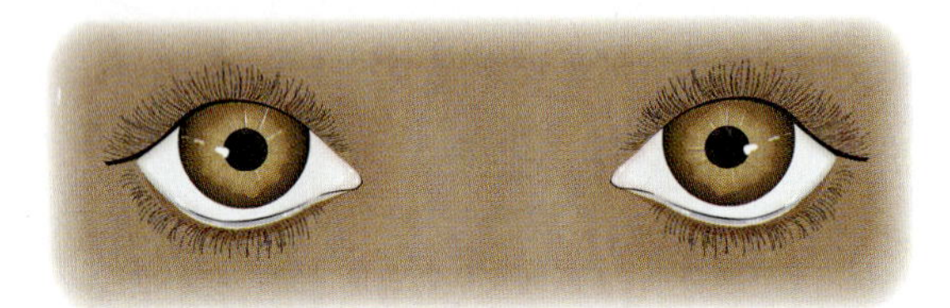

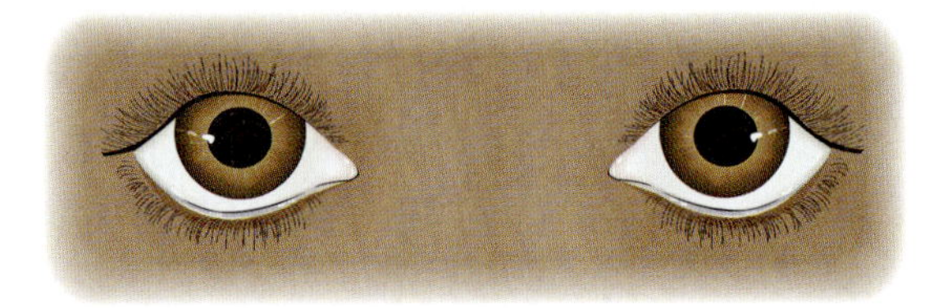

1 ACTIVITY: Comparing Triangles in a Coordinate Plane

Work with a partner. Write the coordinates of the vertices of the blue triangle. Then write the coordinates of the vertices of the red triangle.

a. How are the two sets of coordinates related?

b. How are the two triangles related? Explain your reasoning.

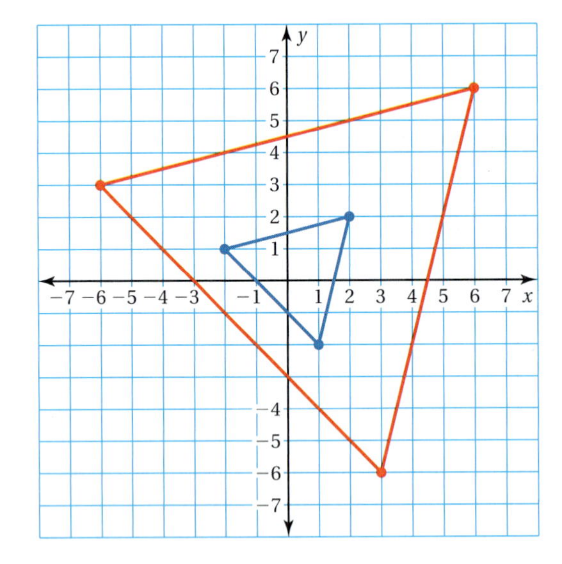

c. Draw a green triangle whose coordinates are twice the values of the corresponding coordinates of the blue triangle. How are the green and blue triangles related? Explain your reasoning.

d. How are the coordinates of the red and green triangles related? How are the two triangles related? Explain your reasoning.

Geometry

In this lesson, you will

- identify dilations.
- dilate figures in the coordinate plane.
- use more than one transformation to find images of figures.

Learning Standards
MACC.8.G.1.3
MACC.8.G.1.4

Laurie's Notes

Introduction

Standards for Mathematical Practice

- **MP4 Model with Mathematics:** Although students are not able to calculate the side lengths of the triangles in Activity 1, they analyze the relationships between the corresponding segments (slope) and between the corresponding vertices and conjecture that the triangles are similar.

Motivate

- Draw a figure on a transparency and display with the overhead projector. Trace over the projection on the board. Move the projector to enlarge the figure and also to reduce the figure. Each time trace over the projection.
- ? "What do you notice about the three figures?" Students should describe the figures as being the same shape, but different sizes.
- Alternatively, you can use a document camera.

Discuss

- ? "Did the size or shape of the figures change when you performed a translation, reflection, or rotation?" no
- "There is a fourth type of transformation. M.C. Escher used this type in creating his graphic art work."
- ? "At an eye appointment, the optometrist dilates your pupils. Who knows what it means to dilate your pupils?" Listen for enlarge your pupils.

Activity Notes

Activity 1

- Check that students record the coordinates correctly. It is common to interchange the *x*- and *y*-coordinates.
- When describing the relationship between the coordinates of the blue and red triangles, check that students state that both the *x*- and *y*-coordinates have been multiplied by a factor of 3.
- The red and blue triangles are similar— the same shape, but different size. Students are not able to determine side lengths or angle measures, but *may* say the red triangle is enlarged proportionally from the blue triangle. They may also say that the corresponding sides have the same slope.
- ? For part (c), ask, "How will you determine the coordinates of the green triangle?" Multiply the coordinates of the blue triangle by 2.
- **MP4:** Once the triangles are drawn, the students may observe that there is a line that can be drawn from the origin through 3 corresponding vertices. For example, the origin and the vertices (2, 2), (4, 4), and (6, 6) lie on a line.
- Students may also correctly observe that corresponding sides are parallel.
- Point out that there is an order implied when asked how the green and blue triangles are related. Green to blue is the ratio 2 : 1, while blue to green is the ratio 1 : 2.
- **MP6 Attend to Precision:** Students may incorrectly say that the red triangle is 3 times larger than the blue triangle. The ratios of the side lengths of the red triangle to the blue triangle are 3 : 1. The areas are not in the ratio of 3 : 1.

Florida Common Core Standards

MACC.8.G.1.3 Describe the effect of dilations, . . . on two-dimensional figures using coordinates.

MACC.8.G.1.4 Understand that a two-dimensional figure is similar to another if the second can be obtained from the first by a sequence of rotations, reflections, translations, and dilations; given two similar two-dimensional figures, describe a sequence that exhibits the similarity between them.

Previous Learning

Students should know how to multiply integers and plot points in the coordinate plane.

2.7 Record and Practice Journal

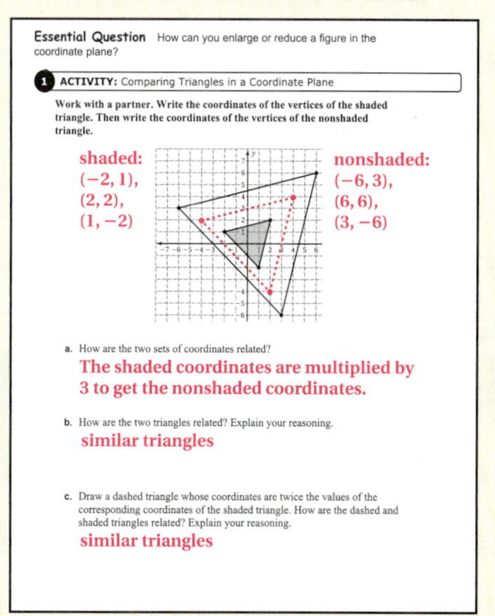

Essential Question How can you enlarge or reduce a figure in the coordinate plane?

1 ACTIVITY: Comparing Triangles in a Coordinate Plane

Work with a partner. Write the coordinates of the vertices of the shaded triangle. Then write the coordinates of the vertices of the nonshaded triangle.

shaded: (−2, 1), (2, 2), (1, −2)

nonshaded: (−6, 3), (6, 6), (3, −6)

a. How are the two sets of coordinates related?
The shaded coordinates are multiplied by 3 to get the nonshaded coordinates.

b. How are the two triangles related? Explain your reasoning.
similar triangles

c. Draw a dashed triangle whose coordinates are twice the values of the corresponding coordinates of the shaded triangle. How are the dashed and shaded triangles related? Explain your reasoning.
similar triangles

Differentiated Instruction

Visual

Have students use two different colored pencils when drawing a figure and its image after a dilation. This will make the figures easily identifiable.

2.7 Record and Practice Journal

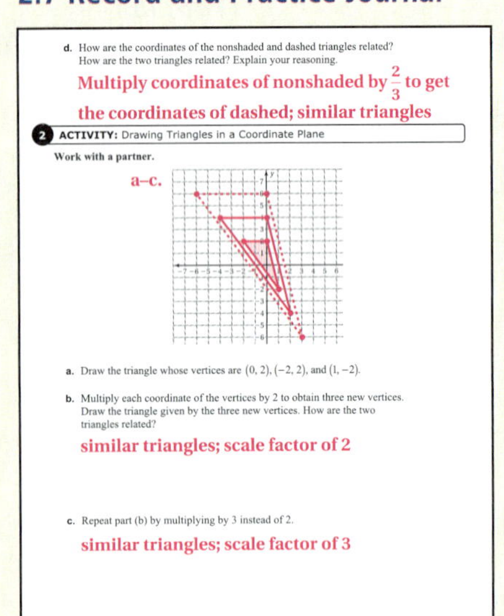

d. How are the coordinates of the nonshaded and dashed triangles related? How are the two triangles related? Explain your reasoning.

Multiply coordinates of nonshaded by $\frac{2}{3}$ to get the coordinates of dashed; similar triangles

2 ACTIVITY: Drawing Triangles in a Coordinate Plane

Work with a partner.

a–c.

a. Draw the triangle whose vertices are (0, 2), (−2, 2), and (1, −2).

b. Multiply each coordinate of the vertices by 2 to obtain three new vertices. Draw the triangle given by the three new vertices. How are the two triangles related?

similar triangles; scale factor of 2

c. Repeat part (b) by multiplying by 3 instead of 2.

similar triangles; scale factor of 3

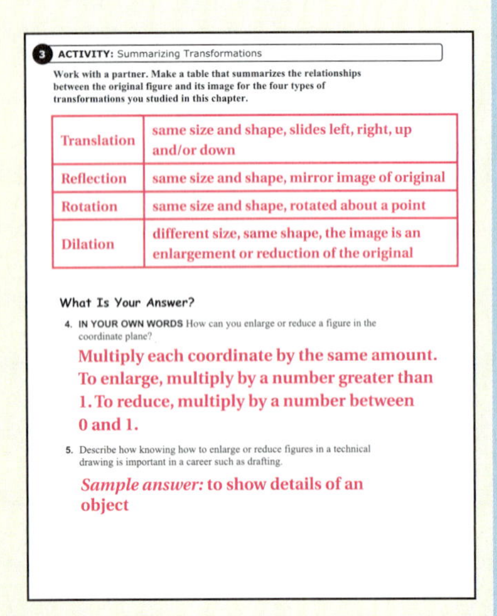

3 ACTIVITY: Summarizing Transformations

Work with a partner. Make a table that summarizes the relationships between the original figure and its image for the four types of transformations you studied in this chapter.

Translation	same size and shape, slides left, right, up and/or down
Reflection	same size and shape, mirror image of original
Rotation	same size and shape, rotated about a point
Dilation	different size, same shape, the image is an enlargement or reduction of the original

What Is Your Answer?

4. IN YOUR OWN WORDS How can you enlarge or reduce a figure in the coordinate plane?

Multiply each coordinate by the same amount. To enlarge, multiply by a number greater than 1. To reduce, multiply by a number between 0 and 1.

5. Describe how knowing how to enlarge or reduce figures in a technical drawing is important in a career such as drafting.

Sample answer: to show details of an object

Laurie's Notes

Activity 2

- The triangle formed is an obtuse scalene triangle. Given the distinctive shape of the triangle, the two new triangles formed will also be distinctive. They will be the same shape but not the same size.
- **MP3 Construct Viable Arguments and Critique the Reasoning of Others** and **MP6:** As you listen to students describing how the triangles are related, you should hear them mention previously learned concepts such as proportional dimensions, similar figures, and scale factor.

? **Extension:** "What do you think the triangle would look like if you had multiplied each coordinate by $\frac{1}{2}$?" Students should say the triangle would be smaller or reduced from the original triangle.

Activity 3

- You may need to give suggestions to students about what aspects or categories to write about. Students should be thinking about sides, angles, orientation, size, and shape.

What Is Your Answer?

- Students should work independently and then share their results with the class.

Closure

- **Writing Prompt:** To enlarge a trapezoid that is drawn in the coordinate plane you . . .

2 ACTIVITY: Drawing Triangles in a Coordinate Plane

Work with a partner.

a. Draw the triangle whose vertices are (0, 2), (−2, 2), and (1, −2).

b. Multiply each coordinate of the vertices by 2 to obtain three new vertices. Draw the triangle given by the three new vertices. How are the two triangles related?

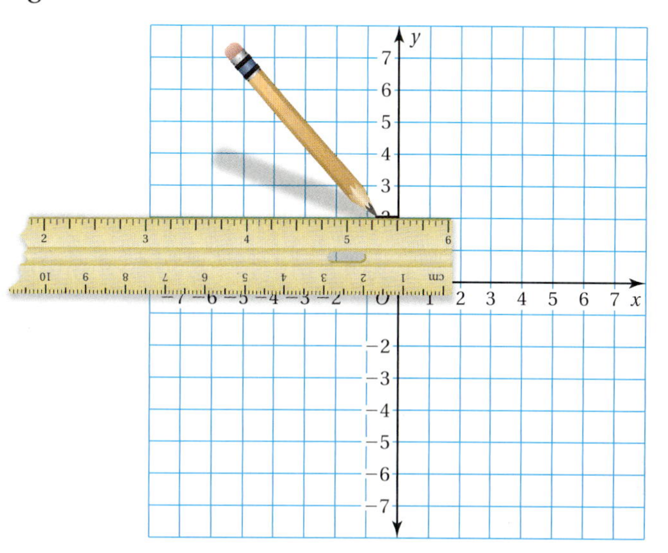

c. Repeat part (b) by multiplying by 3 instead of 2.

Math Practice 3

Use Prior Results

What are the four types of transformations you studied in this chapter? What information can you use to fill in your table?

3 ACTIVITY: Summarizing Transformations

Work with a partner. Make a table that summarizes the relationships between the original figure and its image for the four types of transformations you studied in this chapter.

What Is Your Answer?

4. **IN YOUR OWN WORDS** How can you enlarge or reduce a figure in the coordinate plane?

5. Describe how knowing how to enlarge or reduce figures in a technical drawing is important in a career such as drafting.

Practice Use what you learned about dilations to complete Exercises 4–6 on page 87.

2.7 Lesson

A **dilation** is a transformation in which a figure is made larger or smaller with respect to a point called the **center of dilation**.

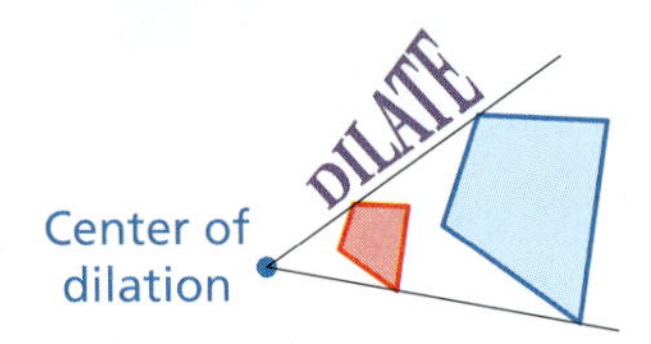

EXAMPLE 1 Identifying a Dilation

Key Vocabulary
dilation, *p. 84*
center of dilation, *p. 84*
scale factor, *p. 84*

Tell whether the blue figure is a dilation of the red figure.

a.

Lines connecting corresponding vertices meet at a point.

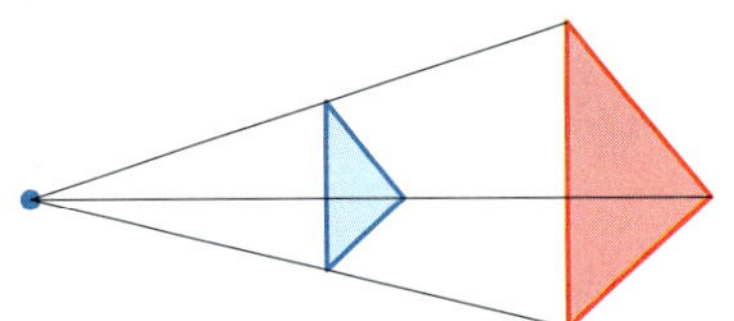

So, the blue figure is a dilation of the red figure.

b.

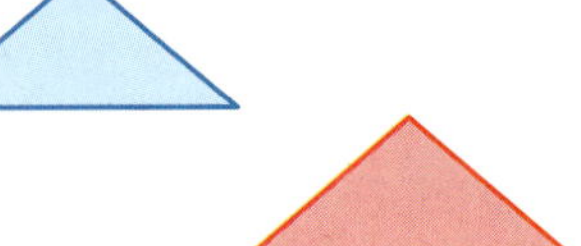

The figures have the same size and shape. The red figure *slides* to form the blue figure.

So, the blue figure is *not* a dilation of the red figure. It is a translation.

On Your Own

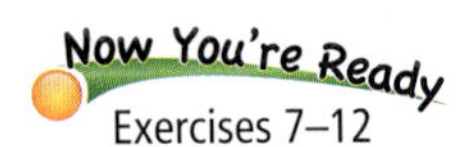

Exercises 7–12

Tell whether the blue figure is a dilation of the red figure. Explain.

1.

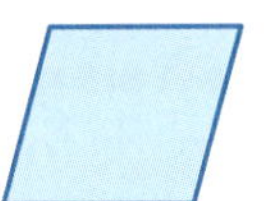

2.

In a dilation, the original figure and its image are similar. The ratio of the side lengths of the image to the corresponding side lengths of the original figure is the **scale factor** of the dilation.

Key Idea

Dilations in the Coordinate Plane

Words To dilate a figure with respect to the origin, multiply the coordinates of each vertex by the scale factor k.

Algebra $(x, y) \rightarrow (kx, ky)$

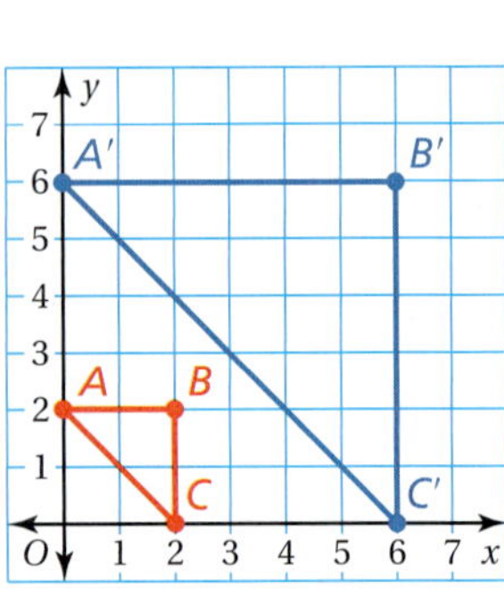

- When $k > 1$, the dilation is an enlargement.
- When $k > 0$ and $k < 1$, the dilation is a reduction.

Multi-Language Glossary at BigIdeasMath.com

Introduction

Connect
- **Yesterday:** Students explored enlarging and reducing triangles in the coordinate plane. (MP3, MP4, MP6)
- **Today:** Students will identify dilations and use scale factors to enlarge and reduce polygons in the coordinate plane.

Motivate
- Cut a rectangle out of heavier card stock. Use a flashlight to cast a shadow of the rectangle onto the wall.

? "What do you notice about the shadow?" similar to the rectangle
- Vary the distance between the bulb of the flashlight and the rectangle. Discuss how this changes the shadow.

? "Is the shadow always similar to the original figure?" It should be if the figure is held parallel to the wall and the flashlight is perpendicular to the wall.

Discuss
- Explain what a dilation and the center of a dilation are, based on the Motivate activity.

Lesson Notes

Example 1
- It is important to draw the line segments connecting the corresponding vertices. It enables you to locate the center of dilation.
- **Extension:** In part (a), if the blue triangle were rotated, you would not have a dilation. The triangles would still be similar, but there would not be a center of dilation.

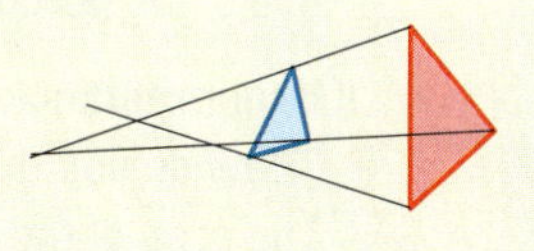

- In part (b), segments connecting corresponding vertices are parallel.

On Your Own
- Ask students to explain their answers.

Key Idea
- Write the Key Idea, including the example in the coordinate plane.

Discuss
- **Write:** Scale factor = side length of image : side length of original
- Put numbers on the vertical sides of the triangles in Example 1(a) to approximate the lengths, such as 4 (blue) and 8 (red). The blue triangle is reduced by an approximate scale factor of 4 : 8 or $\frac{1}{2}$.

? "Can the red triangle in Example 1(a) be a dilation of the blue triangle?" yes; It is an enlargement of the blue triangle.
- Discuss the difference between a scale factor for an enlargement, $k > 1$, and a scale factor for a reduction, $k > 0$ and $k < 1$.

? "What do you think the image would look like if $k = 1$?" The two figures would be congruent and there would be no center of dilation.

Goal Today's lesson is drawing **dilations** in the coordinate plane.

Lesson Tutorials
Lesson Plans
Answer Presentation Tool

English Language Learners
Vocabulary
Discussion about dilations brings up a lot of vocabulary words from previous lessons. Make sure that students are familiar with the definitions and uses of *ratio*, *similar figures*, and *scale factor*.

Extra Example 1
Tell whether the blue figure is a dilation of the red figure. Explain.

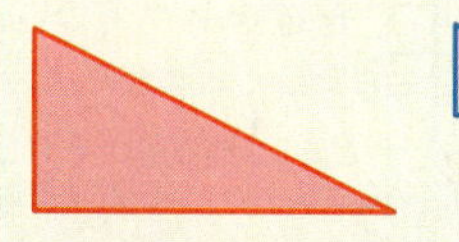
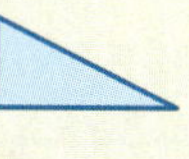

yes; Lines connecting corresponding vertices meet at a point.

On Your Own
1. no; It is a reflection.
2. yes; Lines connecting corresponding vertices meet at a point.

Laurie's Notes

Extra Example 2

The vertices of a triangle are $D(1, 4)$, $E(1, 1)$, and $F(3, 1)$. Draw the triangle and its image after a dilation with a scale factor of 2. Identify the type of dilation.

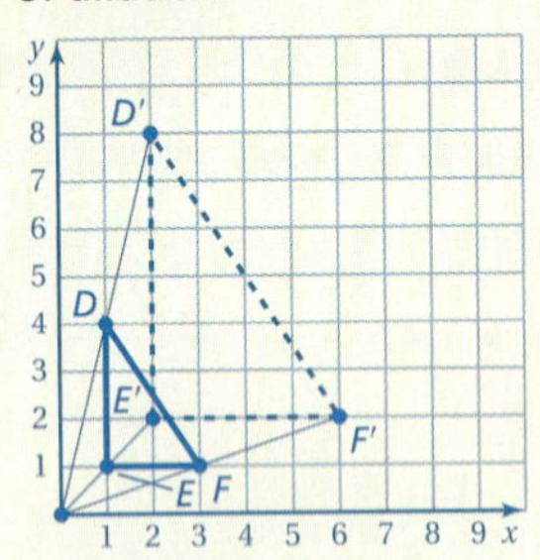

enlargement

Extra Example 3

The vertices of a rectangle are $J(-4, 2)$, $K(4, 2)$, $L(4, -2)$ and $M(-4, -2)$. Draw the rectangle and its image after a dilation with a scale factor of 0.5. Identify the type of dilation.

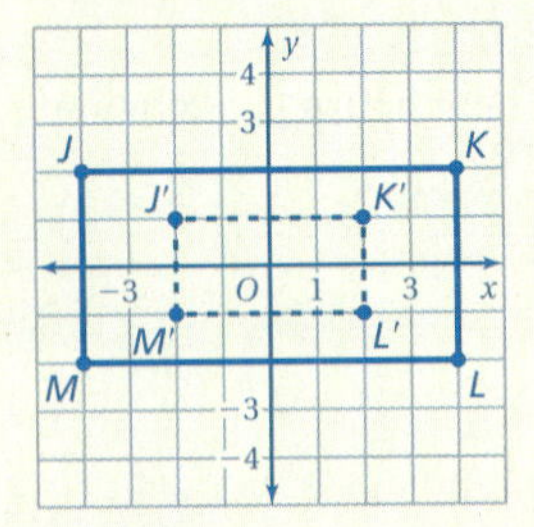

reduction

On Your Own

3. $A'(2, 6)$, $B'(4, 6)$, $C'(4, 2)$
4. $W'(-1, -1.5)$, $X'(-1, 2)$, $Y'(1, 2)$, $Z'(1, -1.5)$

Example 2

- Work through the example.
- Use color to identify the two different triangles and their vertices.
- Note that the two triangles are similar. Side AB is horizontal and its image side $A'B'$ is horizontal. Side BC is vertical and its image side $B'C'$ is vertical. Side AC and its image side $A'C'$ are parallel.
- Refer to the Study Tip.
- **MP3 Construct Viable Arguments and Critique the Reasoning of Others** and **MP6 Attend to Precision:** Ask the following questions.
  - ? "How do you think the perimeters of the two triangles compare? Explain." Because the two triangles are similar with a scale factor of 3, the perimeter of the larger triangle is 3 times the perimeter of the smaller triangle.
  - ? "How do you think the areas of the two triangles compare? Explain." The area of the larger triangle is the square of the scale factor, or $3^2 = 9$ times greater than the area of the smaller triangle.

Example 3

- Work through the example.
- Use color to identify the two different rectangles and their vertices.

? "What is the product of a positive integer and a negative integer?" negative integer

? "Where is the center of dilation?" the origin (0, 0)

- **MP3** and **MP6:** Ask the following questions.
  - ? "How do you think the perimeters of the two rectangles compare? Explain." Because the two rectangles are similar with a scale factor of 0.5, the perimeter of the smaller rectangle is $\frac{1}{2}$ the perimeter of the larger rectangle.
  - ? "How do you think the areas of the two rectangles compare? Explain." The area of the smaller rectangle is the square of the scale factor, or $\left(\frac{1}{2}\right)^2 = \frac{1}{4}$ times the area of the larger rectangle.

On Your Own

- Students should make a table to record the original coordinates and the coordinates of the dilation.
- Ask students to draw each dilation.

EXAMPLE 2 Dilating a Figure

Draw the image of Triangle *ABC* after a dilation with a scale factor of 3. Identify the type of dilation.

Multiply each x- and y-coordinate by the scale factor 3.

Vertices of *ABC*	$(3x, 3y)$	Vertices of *A′B′C′*
$A(1, 3)$	$(3 \cdot 1, 3 \cdot 3)$	$A'(3, 9)$
$B(2, 3)$	$(3 \cdot 2, 3 \cdot 3)$	$B'(6, 9)$
$C(2, 1)$	$(3 \cdot 2, 3 \cdot 1)$	$C'(6, 3)$

Study Tip

You can check your answer by drawing a line from the origin through each vertex of the original figure. The vertices of the image should lie on these lines.

∴ The image is shown at the right. The dilation is an *enlargement* because the scale factor is greater than 1.

EXAMPLE 3 Dilating a Figure

Draw the image of Rectangle *WXYZ* after a dilation with a scale factor of 0.5. Identify the type of dilation.

Multiply each x- and y-coordinate by the scale factor 0.5.

Vertices of *WXYZ*	$(0.5x, 0.5y)$	Vertices of *W′X′Y′Z′*
$W(-4, -6)$	$(0.5 \cdot (-4), 0.5 \cdot (-6))$	$W'(-2, -3)$
$X(-4, 8)$	$(0.5 \cdot (-4), 0.5 \cdot 8)$	$X'(-2, 4)$
$Y(4, 8)$	$(0.5 \cdot 4, 0.5 \cdot 8)$	$Y'(2, 4)$
$Z(4, -6)$	$(0.5 \cdot 4, 0.5 \cdot (-6))$	$Z'(2, -3)$

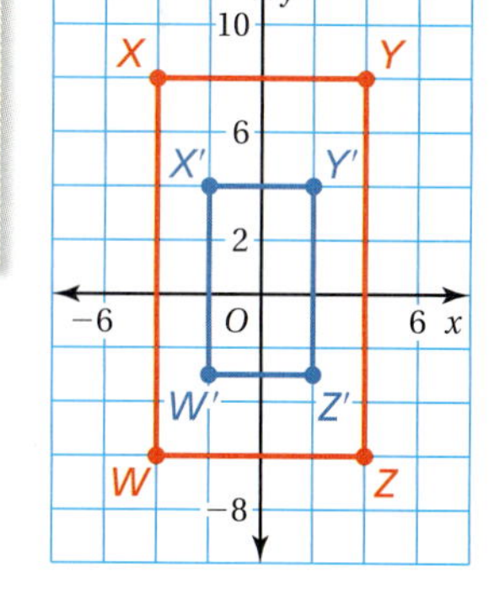

∴ The image is shown at the right. The dilation is a *reduction* because the scale factor is greater than 0 and less than 1.

On Your Own

3. **WHAT IF?** Triangle *ABC* in Example 2 is dilated by a scale factor of 2. What are the coordinates of the image?

4. **WHAT IF?** Rectangle *WXYZ* in Example 3 is dilated by a scale factor of $\frac{1}{4}$. What are the coordinates of the image?

EXAMPLE 4 Using More than One Transformation

The vertices of a trapezoid are $A(-2, -1)$, $B(-1, 1)$, $C(0, 1)$, and $D(0, -1)$. Dilate the trapezoid with respect to the origin using a scale factor of 2. Then translate it 6 units right and 2 units up. What are the coordinates of the image?

Draw *ABCD*. Then dilate it with respect to the origin using a scale factor of 2.

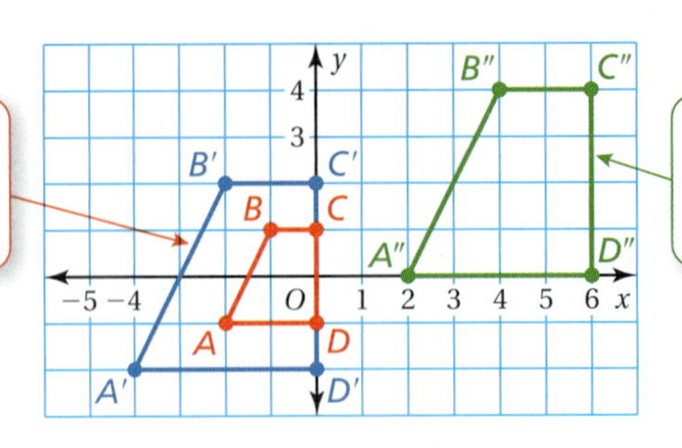

Translate the dilated figure 6 units right and 2 units up.

∴ The coordinates of the image are $A''(2, 0)$, $B''(4, 4)$, $C''(6, 4)$, and $D''(6, 0)$.

The image of a translation, reflection, or rotation is congruent to the original figure, and the image of a dilation is similar to the original figure. So, two figures are similar when one can be obtained from the other by a sequence of translations, reflections, rotations, and dilations.

EXAMPLE 5 Describing a Sequence of Transformations

The red figure is similar to the blue figure. Describe a sequence of transformations in which the blue figure is the image of the red figure.

From the graph, you can see that the blue figure is one-half the size of the red figure. So, begin with a dilation with respect to the origin using a scale factor of $\frac{1}{2}$.

After dilating, you need to flip the figure in the x-axis.

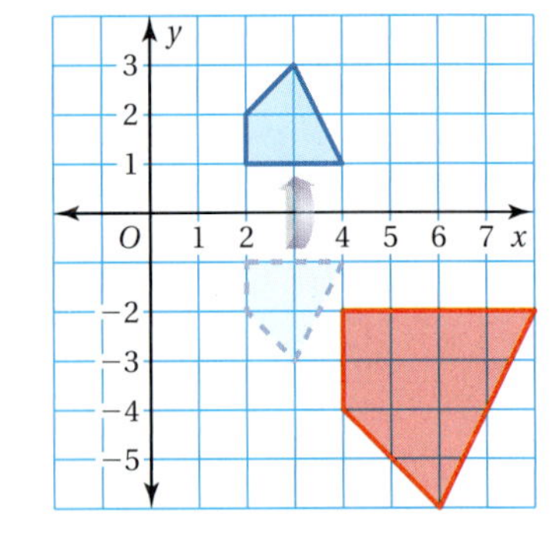

∴ So, one possible sequence of transformations is a dilation with respect to the origin using a scale factor of $\frac{1}{2}$ followed by a reflection in the x-axis.

On Your Own

5. In Example 4, use a scale factor of 3 in the dilation. Then rotate the figure 180° about the image of vertex C. What are the coordinates of the image?

6. In Example 5, can you reflect the red figure first, and then perform the dilation to obtain the blue figure? Explain.

Laurie's Notes

Example 4

- The legs of the original trapezoid are not congruent, and there is a right angle. Both of these attributes will help students with orientation questions.
- Plot the original trapezoid. Say that you will perform two transformations.
- ? "How will a dilation with a scale factor of 2 change the trapezoid?" It will double the size of the trapezoid.
- ? "What will a translation do to the dilated trapezoid?" It will shift the trapezoid to another location.
- Set up a table to organize the vertices, then draw each trapezoid.

Vertices of *ABCD*	Vertices of *A′B′C′D′*	Vertices of *A″B″C″D″*
$A(-2, -1)$	$A'(-4, -2)$	$A''(2, 0)$
$B(-1, 1)$	$B'(-2, 2)$	$B''(4, 4)$
$C(0, 1)$	$C'(0, 2)$	$C''(6, 4)$
$D(0, -1)$	$D'(0, -2)$	$D''(6, 0)$

- ? "How is each pair of trapezoids related?" *ABCD* and *A′B′C′D′* are similar; *A′B′C′D′* and *A″B″C″D″* are congruent; *ABCD* and *A″B″C″D″* are similar.
- ? **MP3** and **MP6:** "When is a figure and its image after a sequence of transformations congruent? Similar?" The figure and its image are congruent when there are translation(s), reflection(s), or rotation(s) with no dilations. They are similar when a dilation is involved.

Example 5

- Students should recognize that a dilation is involved.
- ? **MP3** and **MP6:** "What is the scale factor of the dilation and how do you know?" 0.5; The horizontal side of the blue trapezoid is 2 and the corresponding side of the red trapezoid is 4. So, the red figure must be reduced by a scale factor of $\frac{1}{2}$ to obtain the blue figure.
- Once the dilation is performed, students should recognize that the trapezoid must be reflected about the *x*-axis.
- **Discuss:** Make the connection between scale drawings, similarity, and dilation. Discuss scale factor in similar figures and how it relates to area and perimeter.
- **Extension:** Look back at Example 4 and ask students how the red figure can be transformed into the green figure using a single transformation. Use a dilation from a point other than the origin Then ask them what the center of dilation is and how to find it. $(-6, -2)$; Draw lines through vertices and find the point where the lines meet (intersect).

On Your Own

- Students can answer Question 6 without doing the transformations.

Closure

- **Exit Ticket:** The vertices of a triangle are $T(-2, 3)$, $R(3, 2)$, and $S(3, 1)$. Draw the triangle and its image after a dilation with a scale factor of 2. What are the coordinates of Triangle $T'R'S'$? $T'(-4, 6)$, $R'(6, 4)$, and $S'(6, 2)$

Extra Example 4

The vertices of a trapezoid are $A(-4, 0)$, $B(-2, 4)$, $C(2, 4)$, and $D(6, 0)$. Dilate the trapezoid with respect to the origin using a scale factor of 0.5. Then translate it 2 units right and 3 units down. What are the coordinates of the image?

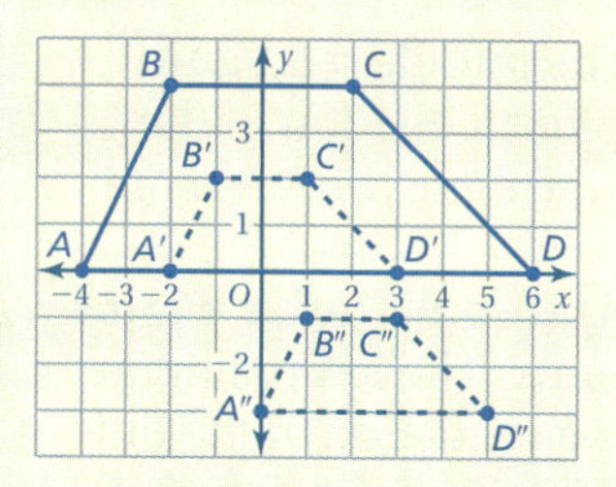

$A''(0, -3)$, $B''(1, -1)$, $C''(3, -1)$, $D''(5, -3)$

Extra Example 5

The red figure is similar to the blue figure. Describe a sequence of transformations in which the blue figure is the image of the red figure.

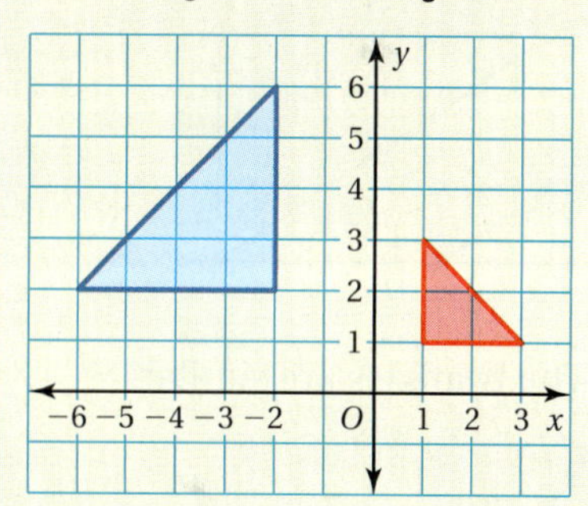

Sample answer: One possible sequence of transformations is a dilation with respect to the origin using a scale factor of 2 followed by a reflection in the *y*-axis.

On Your Own

5.

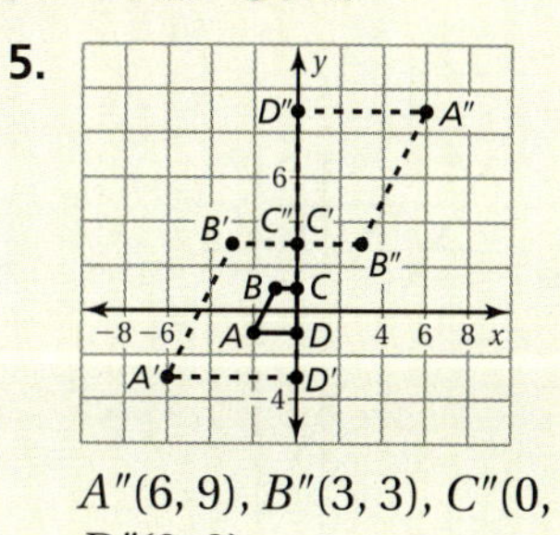

$A''(6, 9)$, $B''(3, 3)$, $C''(0, 3)$, $D''(0, 9)$

6. yes

Vocabulary and Concept Check

1. A dilation changes the size of a figure. The image is similar, not congruent, to the original figure.
2. $k > 1$; $k > 0$ and $k < 1$
3. The middle red figure is not a dilation of the blue figure because the height is half of the blue figure and the base is the same. The left red figure is a reduction of the blue figure and the right red figure is an enlargement of the blue figure.

Practice and Problem Solving

4.

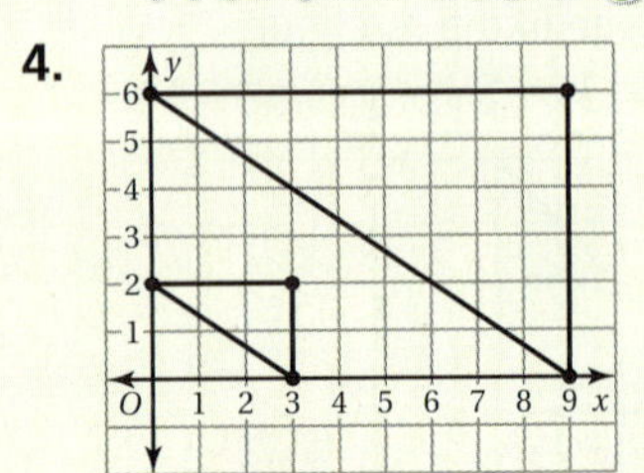

The triangles are similar.

5.

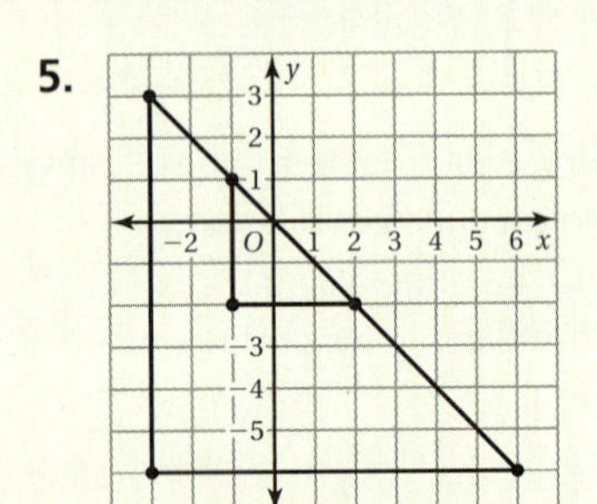

The triangles are similar.

6.

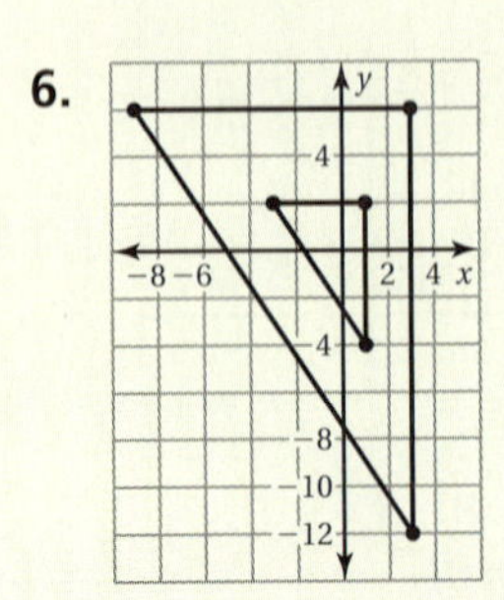

The triangles are similar.

7–18. See Additional Answers.

Assignment Guide and Homework Check

Level	Day 1 Activity Assignment	Day 2 Lesson Assignment	Homework Check
Basic	4–6, 38–41	1–3, 7–29 odd	9, 13, 17, 21, 25
Average	4–6, 38–41	1–3, 7–19 odd, 20–30 even	11, 15, 20, 24, 28
Advanced	1–6, 10–18 even, 19, 20–32 even, 38–41		10, 14, 18, 26, 30

Common Errors

- **Exercises 7–12** Students may think that a figure with the same shape as another is a dilation. Remind them that in order for a transformation to be a dilation, lines connecting corresponding vertices must meet at a point.
- **Exercises 13–18** Students may confuse the image of a dilation with the original polygon when both are drawn in the same coordinate plane. Remind them to pay attention to the prime notation when determining the type of dilation.

2.7 Record and Practice Journal

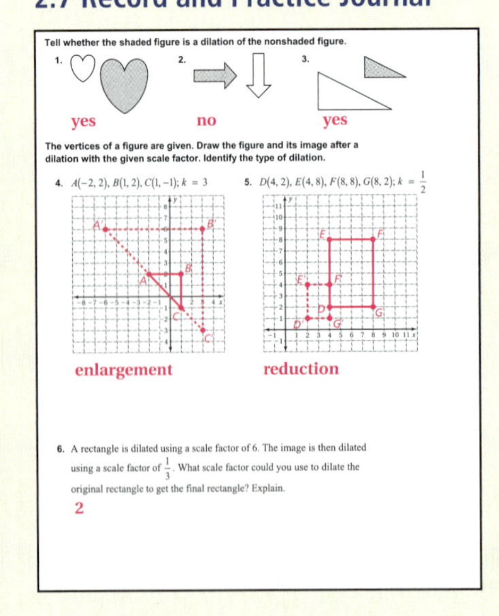

Tell whether the shaded figure is a dilation of the nonshaded figure.

1. yes
2. no
3. yes

The vertices of a figure are given. Draw the figure and its image after a dilation with the given scale factor. Identify the type of dilation.

4. $A(-2, 2)$, $B(1, 2)$, $C(1, -1)$; $k = 3$

enlargement

5. $D(4, 2)$, $E(4, 8)$, $F(8, 8)$, $G(8, 2)$; $k = \frac{1}{2}$

reduction

6. A rectangle is dilated using a scale factor of 6. The image is then dilated using a scale factor of $\frac{1}{3}$. What scale factor could you use to dilate the original rectangle to get the final rectangle? Explain.

2

2.7 Exercises

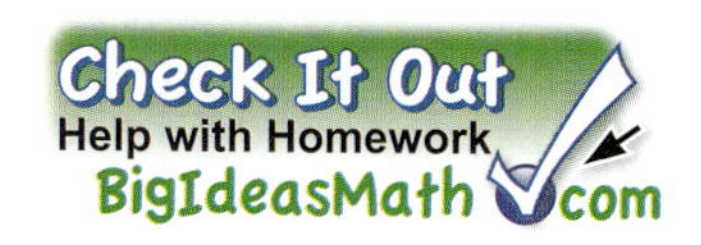

Vocabulary and Concept Check

1. **VOCABULARY** How is a dilation different from other transformations?

2. **VOCABULARY** For what values of scale factor k is a dilation called an *enlargement*? a *reduction*?

3. **REASONING** Which figure is *not* a dilation of the blue figure? Explain.

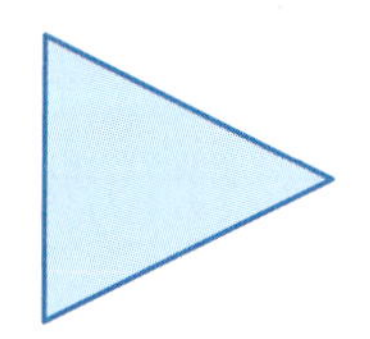 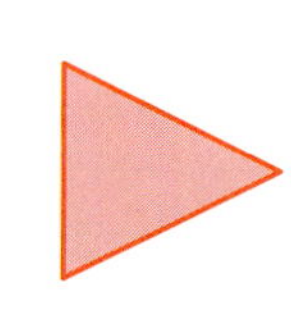  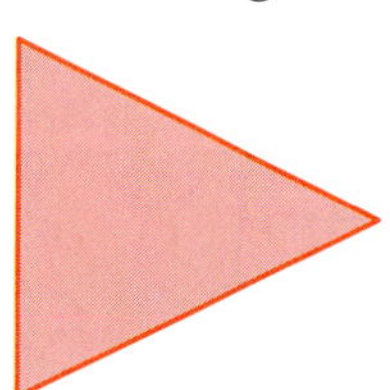

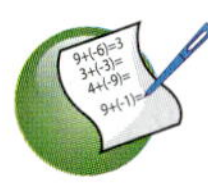

Practice and Problem Solving

Draw the triangle with the given vertices. Multiply each coordinate of the vertices by 3, and then draw the new triangle. How are the two triangles related?

4. (0, 2), (3, 2), (3, 0)

5. $(-1, 1)$, $(-1, -2)$, $(2, -2)$

6. $(-3, 2)$, $(1, 2)$, $(1, -4)$

Tell whether the blue figure is a dilation of the red figure.

7. 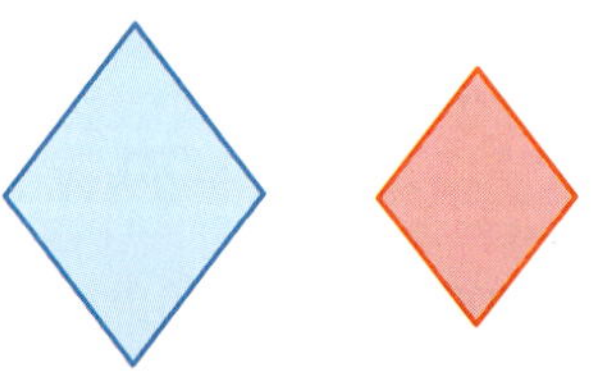

8. 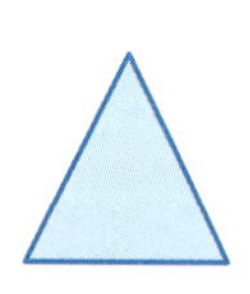 

9. 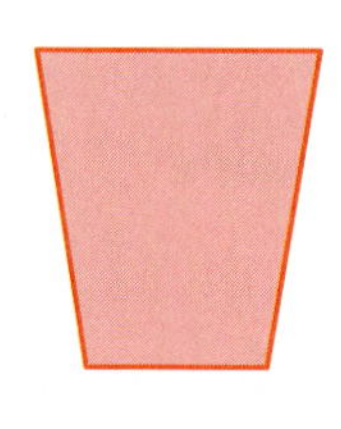 

10. 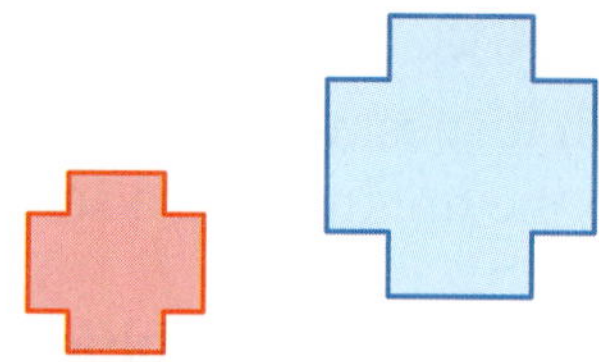

11. 

12. 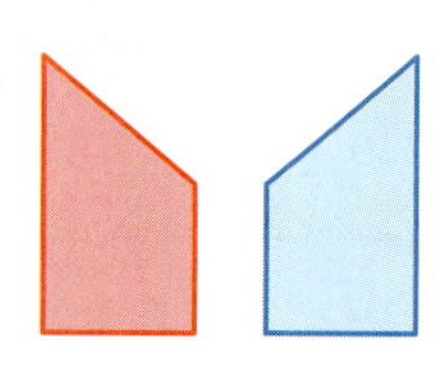

The vertices of a figure are given. Draw the figure and its image after a dilation with the given scale factor. Identify the type of dilation.

13. $A(1, 1)$, $B(1, 4)$, $C(3, 1)$; $k = 4$

14. $D(0, 2)$, $E(6, 2)$, $F(6, 4)$; $k = 0.5$

15. $G(-2, -2)$, $H(-2, 6)$, $J(2, 6)$; $k = 0.25$

16. $M(2, 3)$, $N(5, 3)$, $P(5, 1)$; $k = 3$

17. $Q(-3, 0)$, $R(-3, 6)$, $T(4, 6)$, $U(4, 0)$; $k = \frac{1}{3}$

18. $V(-2, -2)$, $W(-2, 3)$, $X(5, 3)$, $Y(5, -2)$; $k = 5$

19. ERROR ANALYSIS Describe and correct the error in listing the coordinates of the image after a dilation with a scale factor of $\frac{1}{2}$.

Vertices of ABC	$(2x, 2y)$	Vertices of $A'B'C'$
$A(2, 5)$	$(2 \cdot 2, 2 \cdot 5)$	$A'(4, 10)$
$B(2, 0)$	$(2 \cdot 2, 2 \cdot 0)$	$B'(4, 0)$
$C(4, 0)$	$(2 \cdot 4, 2 \cdot 0)$	$C'(8, 0)$

The blue figure is a dilation of the red figure. Identify the type of dilation and find the scale factor.

20.

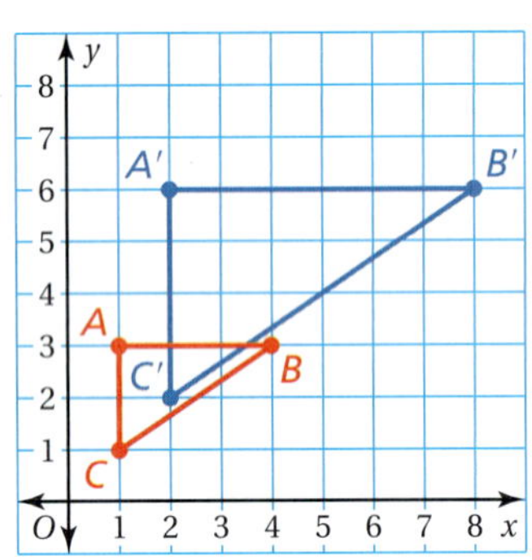

21.

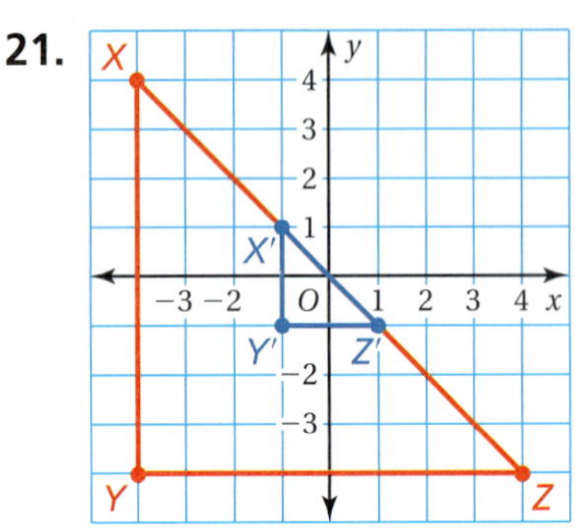

22.

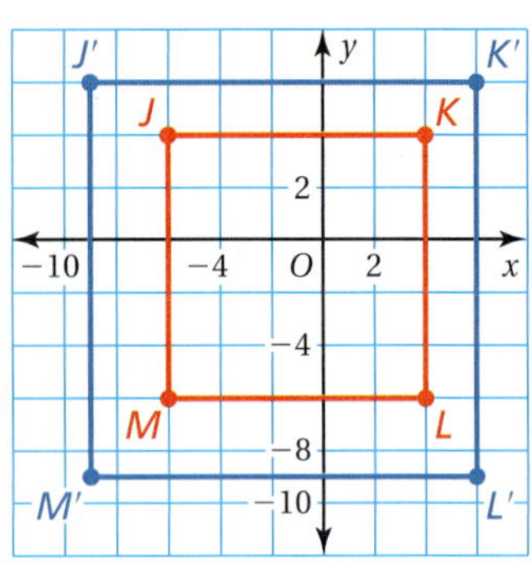

The vertices of a figure are given. Find the coordinates of the figure after the transformations given.

4 **23.** $A(-5, 3)$, $B(-2, 3)$, $C(-2, 1)$, $D(-5, 1)$

Reflect in the y-axis. Then dilate with respect to the origin using a scale factor of 2.

24. $F(-9, -9)$, $G(-3, -6)$, $H(-3, -9)$

Dilate with respect to the origin using a scale factor of $\frac{2}{3}$. Then translate 6 units up.

25. $J(1, 1)$, $K(3, 4)$, $L(5, 1)$

Rotate 90° clockwise about the origin. Then dilate with respect to the origin using a scale factor of 3.

26. $P(-2, 2)$, $Q(4, 2)$, $R(2, -6)$, $S(-4, -6)$

Dilate with respect to the origin using a scale factor of 5. Then dilate with respect to the origin using a scale factor of 0.5.

The red figure is similar to the blue figure. Describe a sequence of transformations in which the blue figure is the image of the red figure.

5 **27.**

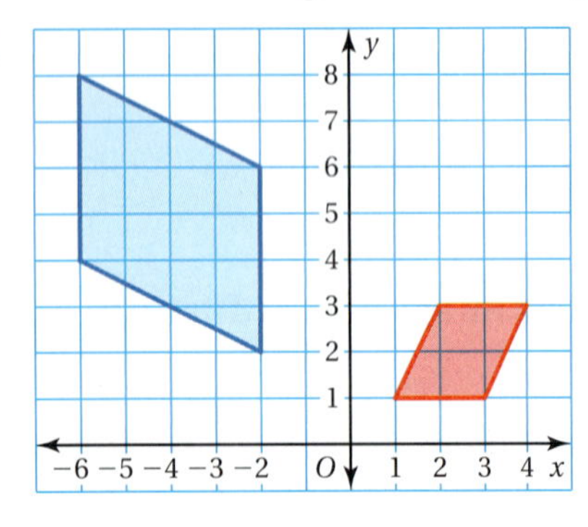

28.

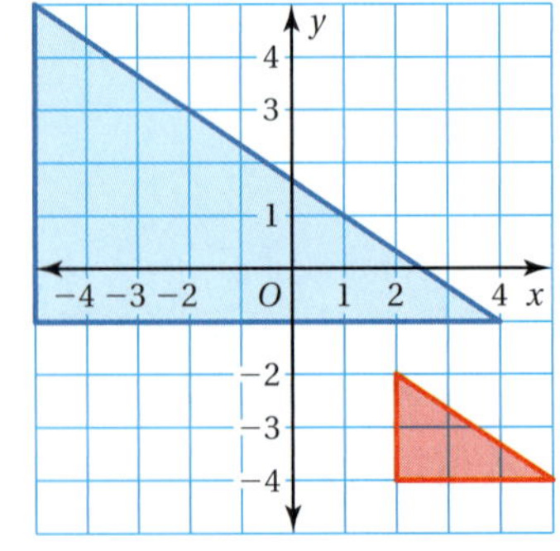

29. STRUCTURE In Exercises 27 and 28, is the blue figure still the image of the red figure when you perform the sequence in the opposite order? Explain.

Common Errors

- **Exercises 20–22** Students may indicate the wrong type of dilation and give the reciprocal of the scale factor. Remind students that you multiply each coordinate of the original figure by the scale factor to obtain the corresponding image coordinate.
- **Exercises 27 and 28** Students may give the sequence of transformations in reverse order. Encourage students to perform the sequence of transformations on the original figure to check their answers

Practice and Problem Solving

19. Each coordinate was multiplied by 2 instead of divided by 2. The coordinates should be $A'(1, 2.5)$, $B'(1, 0)$, and $C'(2, 0)$.
20. enlargement; 2
21. reduction; $\frac{1}{4}$
22. enlargement; $\frac{3}{2}$
23. $A''(10, 6)$, $B''(4, 6)$, $C''(4, 2)$, $D''(10, 2)$
24. $F''(-6, 0)$, $G''(-2, 2)$, $H''(-2, 0)$
25. $J''(3, -3)$, $K''(12, -9)$, $L''(3, -15)$
26. $P''(-5, 5)$, $Q''(10, 5)$, $R''(5, -15)$, $S''(-10, -15)$
27. *Sample answer:* Rotate 90° counterclockwise about the origin and then dilate with respect to the origin using a scale factor of 2
28. *Sample answer:* Dilate with respect to the origin using a scale factor of 3 and then translate 11 units left and 11 units up
29. Exercise 27: yes; Exercise 28: no; Explanations will vary based on sequences chosen in Exercises 27 and 28.

English Language Learners

Vocabulary

Help students remember the mathematical meaning of dilation. The everyday meaning is to enlarge or expand, such as the dilation of a pupil of an eye. In mathematics, a dilation makes a figure either larger or smaller, depending on the scale factor of the dilation.

Practice and Problem Solving

30. See *Taking Math Deeper*.

31–37. See Additional Answers.

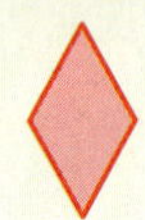

Fair Game Review

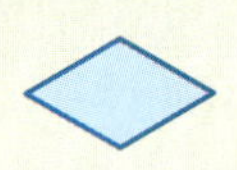

38. complementary; $x = 50$

39. supplementary; $x = 16$

40. complementary; $x = 9$

41. B

Mini-Assessment

1. Tell whether the blue figure is a dilation of the red figure.

no

2. The vertices of a triangle are $A(1, 2)$, $B(2, 4)$, $C(3, 1)$. Draw the triangle and its image after a dilation with a scale factor of 2. Identify the type of dilation.

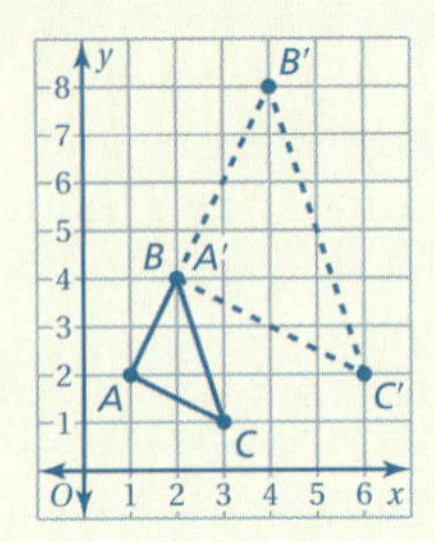

enlargement

3. The red figure is similar to the blue figure. Describe a sequence of transformations in which the blue figure is the image of the red figure.

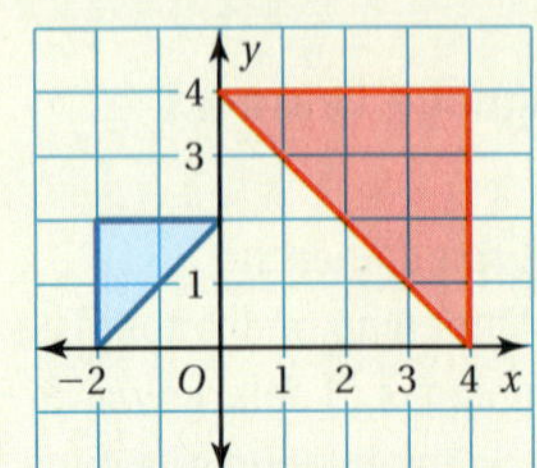

Sample Answer: A dilation with respect to the origin using a scale factor of 0.5 followed by a reflection in the y-axis.

Taking Math Deeper

Exercise 30

The concept seems to occur frequently on state tests. When you enlarge a figure by a scale factor of k:

1. the perimeter increases by a scale factor of k, and
2. the area increases by a scale factor of k^2.

1 Draw a diagram.

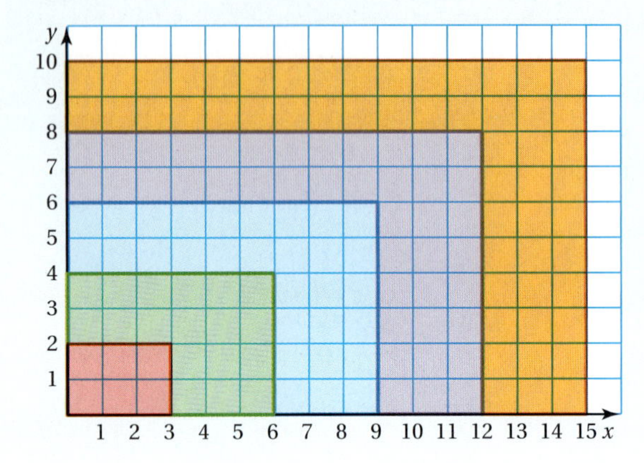

2 Compute the areas.

Area $= 3 \times 2 = 6$

$k = 2$: Area $= 6 \times 4 = 24$

$k = 3$: Area $= 9 \times 6 = 54$

$k = 4$: Area $= 12 \times 8 = 96$

$k = 5$: Area $= 15 \times 10 = 150$

3 How many times greater is the area?

$k = 2$: $\frac{24}{6} = 4$ times greater

$k = 3$: $\frac{54}{6} = 9$ times greater

$k = 4$: $\frac{96}{6} = 16$ times greater

$k = 5$: $\frac{150}{6} = 25$ times greater

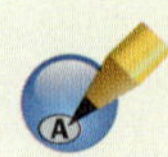

Reteaching and Enrichment Strategies

If students need help...	If students got it...
Resources by Chapter • Practice A and Practice B • Puzzle Time Record and Practice Journal Practice Differentiating the Lesson Lesson Tutorials Skills Review Handbook	Resources by Chapter • Enrichment and Extension • Technology Connection Start the next section

30. **OPEN-ENDED** Draw a rectangle on a coordinate plane. Choose a scale factor of 2, 3, 4, or 5, and then dilate the rectangle. How many times greater is the area of the image than the area of the original rectangle?

31. **SHADOW PUPPET** You can use a flashlight and a shadow puppet (your hands) to project shadows on the wall.

   a. Identify the type of dilation.

   b. What does the flashlight represent?

   c. The length of the ears on the shadow puppet is 3 inches. The length of the ears on the shadow is 4 inches. What is the scale factor?

   d. Describe what happens as the shadow puppet moves closer to the flashlight. How does this affect the scale factor?

32. **REASONING** A triangle is dilated using a scale factor of 3. The image is then dilated using a scale factor of $\frac{1}{2}$. What scale factor could you use to dilate the original triangle to get the final image? Explain.

CRITICAL THINKING The coordinate notation shows how the coordinates of a figure are related to the coordinates of its image after transformations. What are the transformations? Are the figure and its image similar or congruent? Explain.

33. $(x, y) \rightarrow (2x + 4, 2y - 3)$

34. $(x, y) \rightarrow (-x - 1, y - 2)$

35. $(x, y) \rightarrow \left(\frac{1}{3}x, -\frac{1}{3}y\right)$

36. **STRUCTURE** How are the transformations $(2x + 3, 2y - 1)$ and $(2(x + 3), 2(y + 1))$ different?

37. 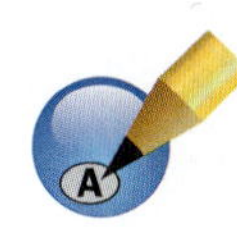

The vertices of a trapezoid are $A(-2, 3)$, $B(2, 3)$, $C(5, -2)$, and $D(-2, -2)$. Dilate the trapezoid with respect to vertex A using a scale factor of 2. What are the coordinates of the image? Explain the method you used.

Fair Game Review What you learned in previous grades & lessons

Tell whether the angles are *complementary* or *supplementary*. Then find the value of *x*. *(Skills Review Handbook)*

38. 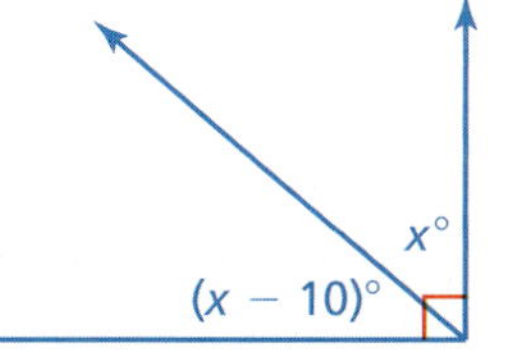

39. 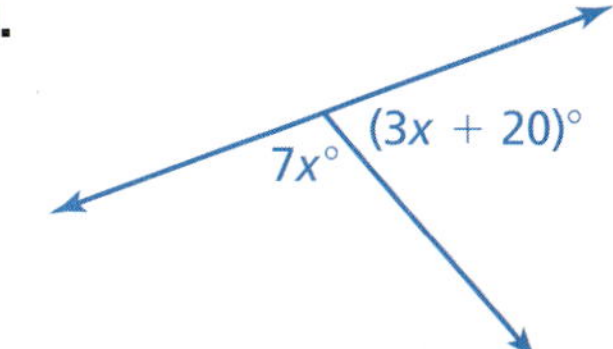

40. 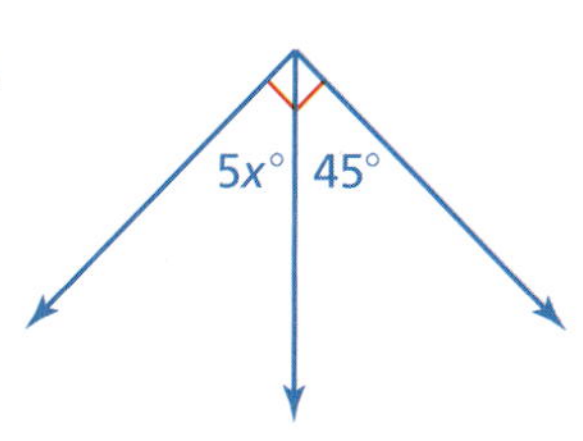

41. **MULTIPLE CHOICE** Which quadrilateral is *not* a parallelogram? *(Skills Review Handbook)*

   Ⓐ rhombus Ⓑ trapezoid Ⓒ square Ⓓ rectangle

2.5–2.7 Quiz

1. Tell whether the two rectangles are similar. Explain your reasoning. *(Section 2.5)*

The figures are similar. Find x. *(Section 2.5)*

2.

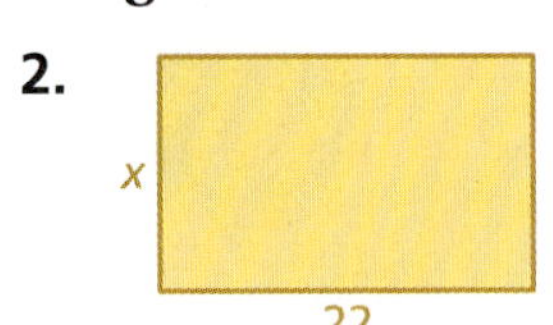

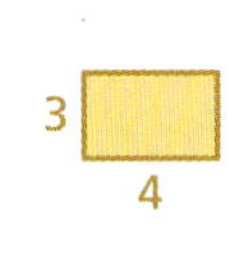

3.

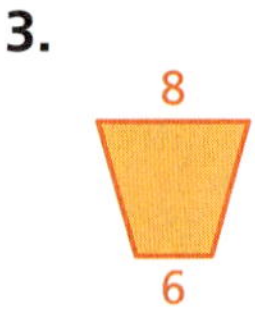

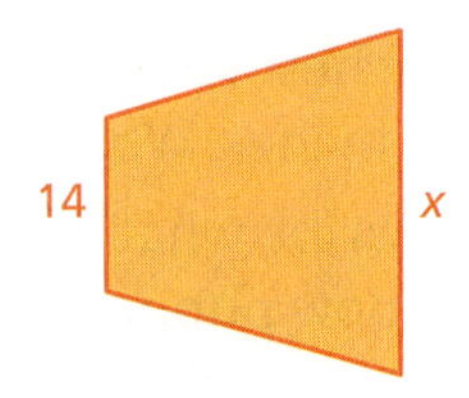

The two figures are similar. Find the ratios (red to blue) of the perimeters and of the areas. *(Section 2.6)*

4.

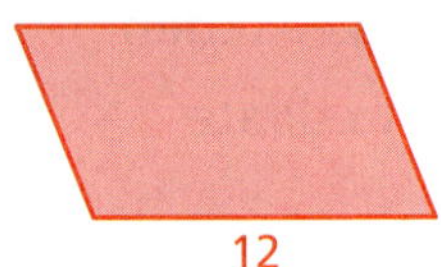

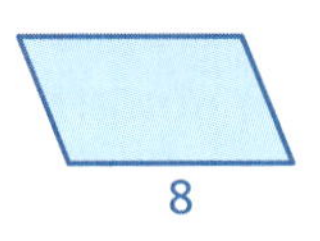

5.

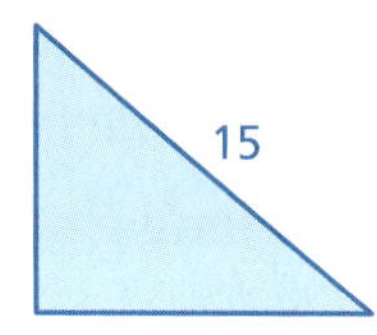

Tell whether the blue figure is a dilation of the red figure. *(Section 2.7)*

6.

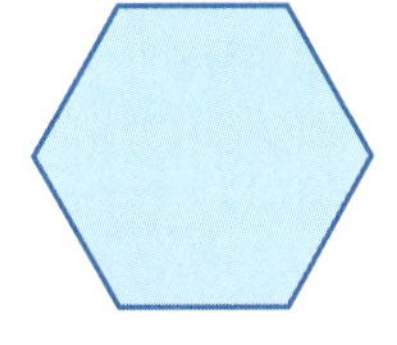

7.

8. **SCREENS** The TV screen is similar to the computer screen. What is the area of the TV screen? *(Section 2.6)*

9. **GEOMETRY** The vertices of a rectangle are $A(2, 4)$, $B(5, 4)$, $C(5, -1)$, and $D(2, -1)$. Dilate the rectangle with respect to the origin using a scale factor of $\frac{1}{2}$. Then translate it 4 units left and 3 units down. What are the coordinates of the image? *(Section 2.7)*

10. **TENNIS COURT** The tennis courts for singles and doubles matches are different sizes. Are the courts similar? Explain. *(Section 2.5)*

Singles

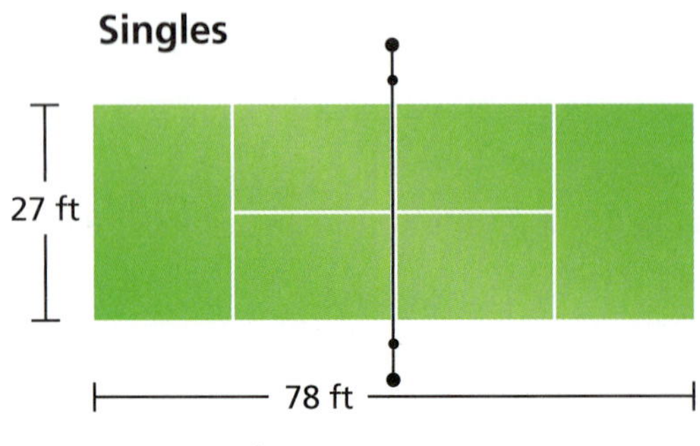

Doubles

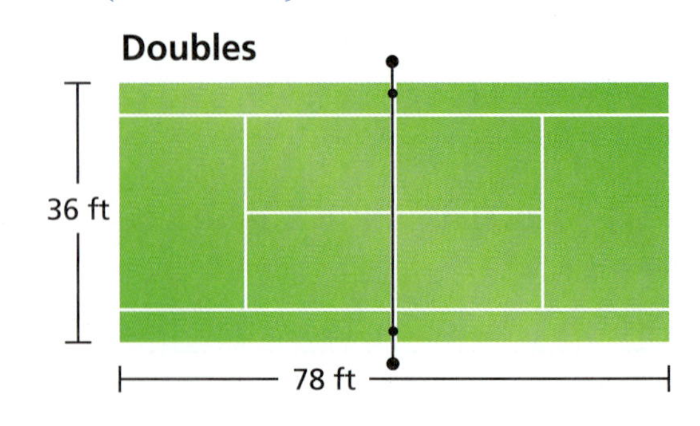

Alternative Assessment Options

Math Chat	Student Reflective Focus Question
Structured Interview	**Writing Prompt**

Writing Prompt

Ask students to write a few paragraphs about how transformations of figures are used in real life. They should include real-life examples of translations, reflections, rotations, and dilations. Then have students share what they have written with the class.

Study Help Sample Answers

Remind students to complete Graphic Organizers for the rest of the chapter.

4.

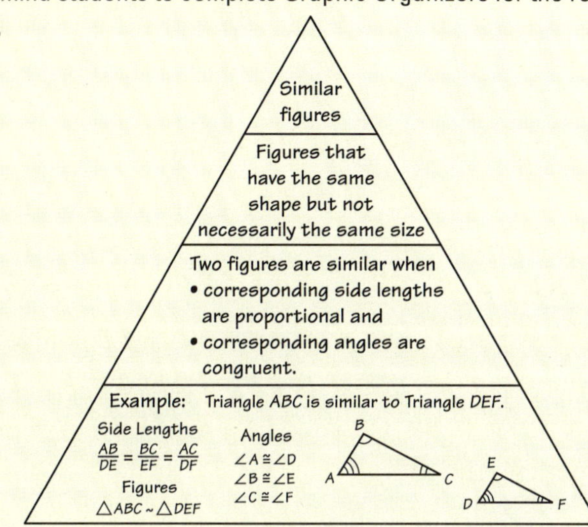

5–8. Available at *BigIdeasMath.com.*

Reteaching and Enrichment Strategies

If students need help. . .	If students got it. . .
Resources by Chapter • Practice A and Practice B • Puzzle Time Lesson Tutorials *BigIdeasMath.com*	Resources by Chapter • Enrichment and Extension • Technology Connection Game Closet at *BigIdeasMath.com* Start the Chapter Review

Answers

1. yes; Corresponding angles are congruent. Because $\frac{4}{10} = \frac{8}{20}$, the corresponding side lengths are proportional.
2. 16.5
3. $18\frac{2}{3}$
4. $\frac{3}{2}, \frac{9}{4}$
5. $\frac{4}{15}, \frac{16}{225}$
6. yes
7. no
8. 300 in.^{2}
9. $A''(-3, -1)$, $B''(-1.5, -1)$, $C''(-1.5, -3.5)$, $D''(-3, -3.5)$
10. no; Corresponding side lengths are not proportional.

Online Assessment
Assessment Book
ExamView® Assessment Suite

For the Teacher
Additional Review Options

- *BigIdeasMath.com*
- Online Assessment
- Game Closet at *BigIdeasMath.com*
- Vocabulary Help
- Resources by Chapter

Answers

1. 3 ft
2. 20 ft
3. $\angle A$ and $\angle K$, $\angle B$ and $\angle L$, $\angle C$ and $\angle M$; Side AB and Side KL, Side BC and Side LM, Side AC and Side KM
4. $\angle R$ and $\angle W$, $\angle Q$ and $\angle X$, $\angle T$ and $\angle Y$, $\angle S$ and $\angle Z$; Side RQ and Side WX, Side QT and Side XY, Side TS and Side YZ, Side SR and Side WZ

Review of Common Errors

Exercises 1 and 2

- Students may forget to include the units in their answers. Remind them that the units must be included to make their answers complete.

Exercises 3 and 4

- Students may forget to write the angle symbol with the name. Remind them that A is a point and $\angle A$ is the angle.

2 Chapter Review

Check It Out
Vocabulary Help
BigIdeasMath.com

Review Key Vocabulary

congruent figures, *p. 44*
corresponding angles, *p. 44*
corresponding sides, *p. 44*
transformation, *p. 50*
image, *p. 50*
translation, *p. 50*
reflection, *p. 56*
line of reflection, *p. 56*
rotation, *p. 62*
center of rotation, *p. 62*
angle of rotation, *p. 62*
similar figures, *p. 72*
dilation, *p. 84*
center of dilation, *p. 84*
scale factor, *p. 84*

Review Examples and Exercises

2.1 Congruent Figures *(pp. 42–47)*

Trapezoids *EFGH* and *QRST* are congruent.

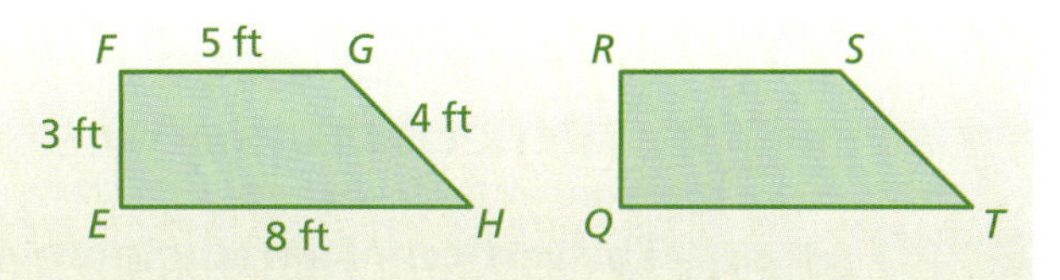

a. What is the length of side *QT*?

Side QT corresponds to side EH.

So, the length of side QT is 8 feet.

b. Which angle of *QRST* corresponds to $\angle H$?

$\angle T$ corresponds to $\angle H$.

Exercises

Use the figures above.

1. What is the length of side QR?
2. What is the perimeter of $QRST$?

The figures are congruent. Name the corresponding angles and the corresponding sides.

3. 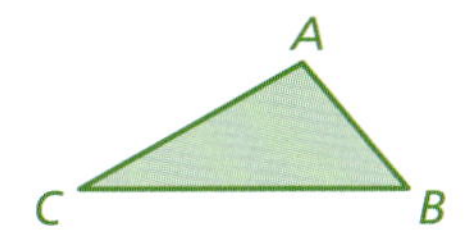

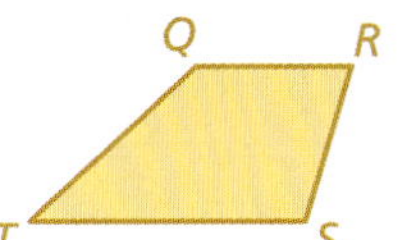

4. Q R T S W X Z Y

2.2 Translations *(pp. 48–53)*

Translate the red triangle 4 units left and 1 unit down. What are the coordinates of the image?

Move each vertex 4 units left and 1 unit down.

Connect the vertices. Label as A', B', and C'.

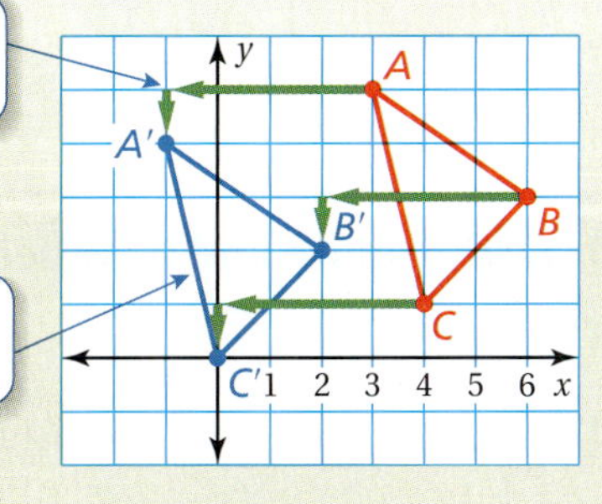

The coordinates of the image are $A'(-1, 4)$, $B'(2, 2)$, and $C'(0, 0)$.

Exercises

Tell whether the blue figure is a translation of the red figure.

5.

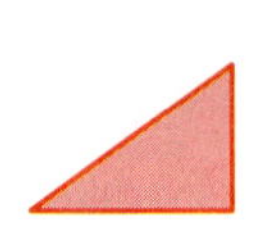

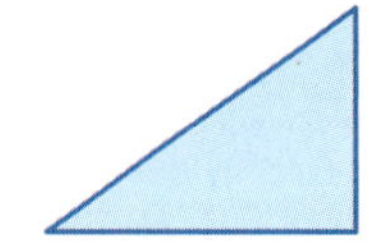

6.

7. The vertices of a quadrilateral are $W(1, 2)$, $X(1, 4)$, $Y(4, 4)$, and $Z(4, 2)$. Draw the figure and its image after a translation 3 units left and 2 units down.

8. The vertices of a triangle are $A(-1, -2)$, $B(-2, 2)$, and $C(-3, 0)$. Draw the figure and its image after a translation 5 units right and 1 unit up.

2.3 Reflections (pp. 54–59)

The vertices of a triangle are $A(-2, 1)$, $B(4, 1)$, and $C(4, 4)$. Draw the figure and its reflection in the x-axis. What are the coordinates of the image?

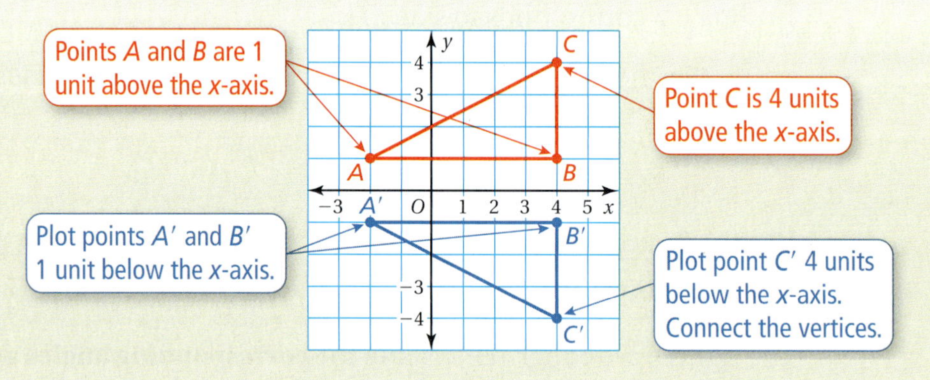

The coordinates of the image are $A'(-2, -1)$, $B'(4, -1)$, and $C'(4, -4)$.

Exercises

Tell whether the blue figure is a reflection of the red figure.

9.

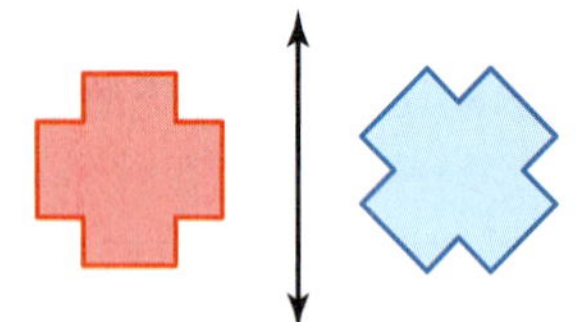

10.

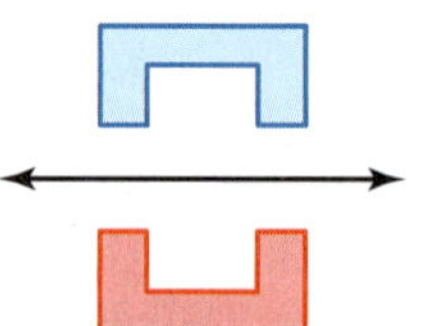

Draw the figure and its reflection in (a) the x-axis and (b) the y-axis.

11. $A(2, 0)$, $B(1, 5)$, $C(4, 3)$

12. $D(-5, -5)$, $E(-5, -1)$, $F(-2, -2)$, $G(-2, -5)$

13. The vertices of a rectangle are $E(-1, 1)$, $F(-1, 3)$, $G(-5, 3)$, and $H(-5, 1)$. Find the coordinates of the figure after reflecting in the x-axis, and then translating 3 units right.

Review of Common Errors (continued)

Exercises 5 and 6

- Students may forget that the objects must be the same size to be a translation. Remind them that the size stays the same. Tell students that when the size is different, it is a scale drawing.

Exercises 7 and 8

- Students may translate the shape the wrong direction or mix up the units for the translation. Tell them to write the direction of the translation on their drawings, using arrows to show the movement left, right, up, or down.

Exercises 9 and 10

- Some students may not be able to visualize whether the blue figure is a reflection of the red figure. Give them tracing paper and have them trace the objects. Then have them fold the paper to determine whether corresponding vertices of each figure align.

Exercises 11 and 12

- Some students may reflect the image obtained in part (a) to answer part (b). Point out that the directions are asking for a reflection of *the original figure* in each part of the exercise.

Exercises 11–13

- Students may confuse reflection in the x-axis with reflection in the y-axis (and vice versa). Have students label the axes in their drawings so it is clearer which way to reflect.

Answers

5. no **6.** yes

7.

8.

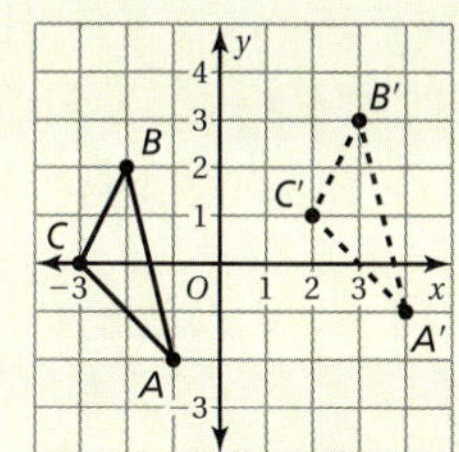

9. no **10.** yes

11. a.

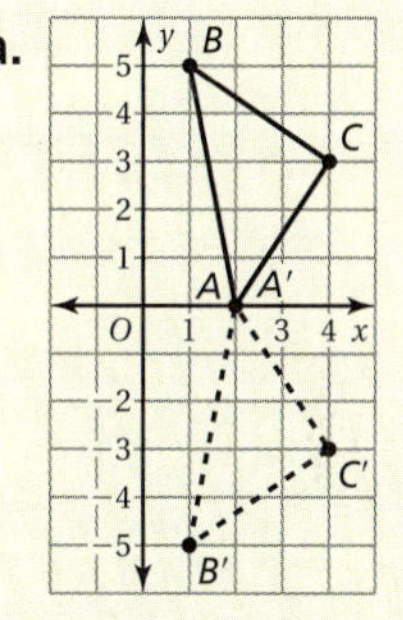

b.

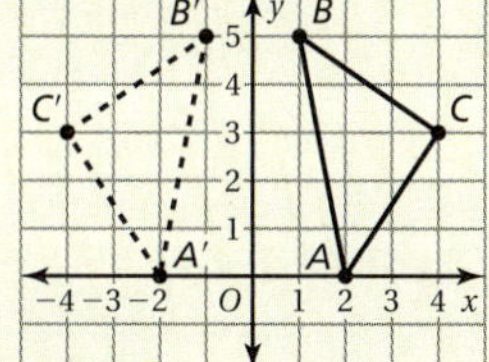

12. a.

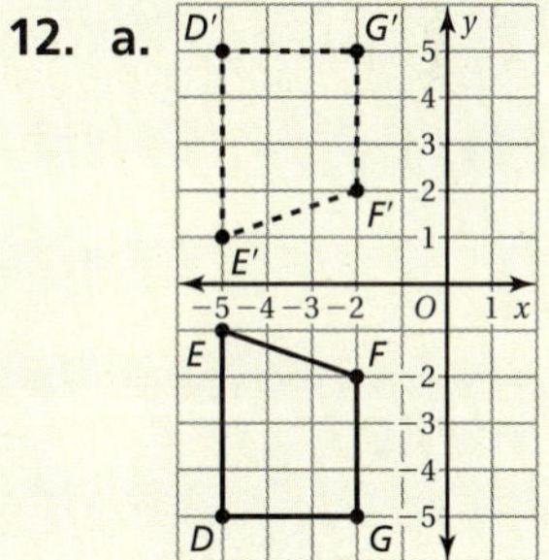

b.

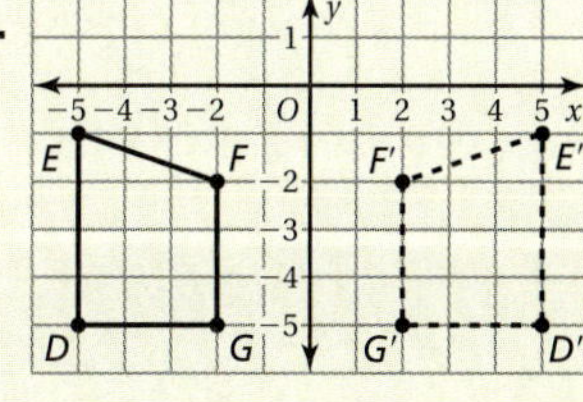

13. $E'(2, -1), F'(2, -3),$
$G'(-2, -3), H'(-2, -1)$

Answers

14. no

15. yes; 180° counterclockwise or clockwise

16. $A'(4, -2)$, $B'(2, -2)$, $C'(3, -4)$

17. $A'(-2, -4)$, $B'(-2, -2)$, $C'(-4, -3)$

Review of Common Errors (continued)

Exercises 14–17

- Students with minimal spatial skills may not be able to tell whether a figure is rotated or how to rotate a figure. Give them tracing paper and have them copy the figure and rotate it.

Exercises 18 and 19

- Students may think that because corresponding side lengths may have changed by the same amount, the figures are similar. Remind them that the *ratios* of corresponding side lengths must be the same for the two objects to be similar.

Exercises 20 and 21

- Students may write the proportion incorrectly. For example, for Exercise 20 they may write $\frac{14}{20} = \frac{x}{7}$ instead of $\frac{14}{20} = \frac{7}{x}$. Remind them that the lengths of corresponding sides should both be in the numerator or the denominator OR the side lengths of the larger shape or smaller shape should be in the numerators, and the corresponding side lengths of the other shape should be in the denominators.

Exercises 22–24

- Students may find the reciprocal of the ratio. For example, they may find the ratio of blue to red instead of red to blue. Remind students to read the directions carefully.
- When finding the ratio of the areas, students may forget to square the ratio of the corresponding side lengths. Remind them of the Key Idea at the bottom of page 78.

Exercises 25 and 26

- Students may think that a figure with the same shape as another is a dilation. Remind students that in order for a transformation to be a dilation, lines connecting corresponding vertices must meet at a point.

Exercises 27–29

- Students may confuse the image of a dilation with the original polygon when both are drawn in the same coordinate plane. Remind students to use and pay attention to the prime notation when drawing the image and determining the type of dilation.

2.4 Rotations (pp. 60–67)

The vertices of a triangle are $A(1, 1)$, $B(3, 2)$, and $C(2, 4)$. Rotate the triangle 90° counterclockwise about the origin. What are the coordinates of the image?

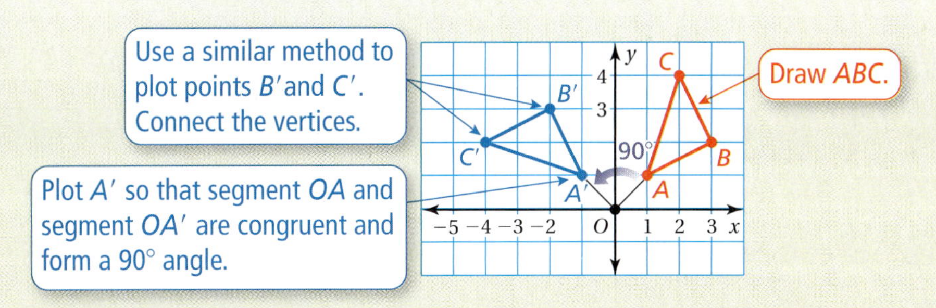

∴ The coordinates of the image are $A'(-1, 1)$, $B'(-2, 3)$, and $C'(-4, 2)$.

Exercises

Tell whether the blue figure is a rotation of the red figure about the origin. If so, give the angle and the direction of rotation.

14. 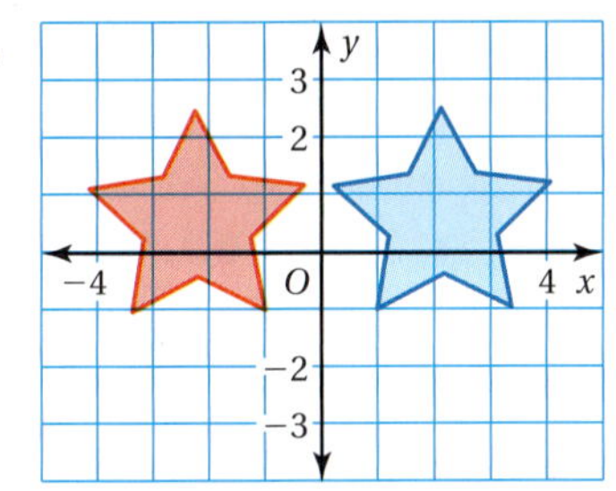

15. 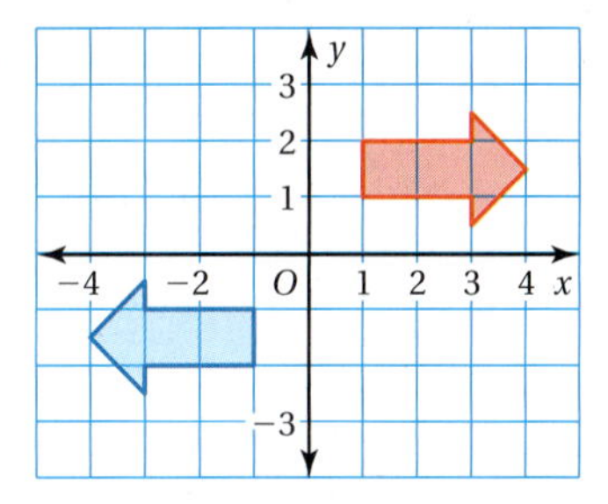

The vertices of a triangle are $A(-4, 2)$, $B(-2, 2)$, and $C(-3, 4)$. Rotate the triangle about the origin as described. Find the coordinates of the image.

16. 180°

17. 270° clockwise

2.5 Similar Figures (pp. 70–75)

a. Is Rectangle A similar to Rectangle B?

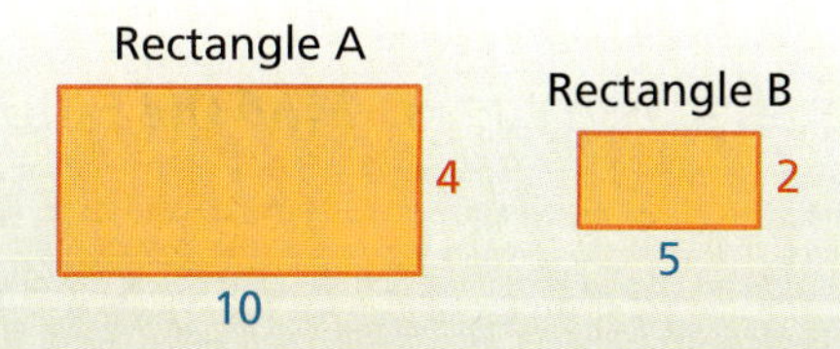

Each figure is a rectangle. So, corresponding angles are congruent. Check to see if corresponding side lengths are proportional.

$$\frac{\text{Length of A}}{\text{Length of B}} = \frac{10}{5} = 2 \qquad \frac{\text{Width of A}}{\text{Width of B}} = \frac{4}{2} = 2$$

Proportional

∴ So, Rectangle A is similar to Rectangle B.

b. The two rectangles are similar. Find x.

Because the rectangles are similar, corresponding side lengths are proportional. So, write and solve a proportion to find x.

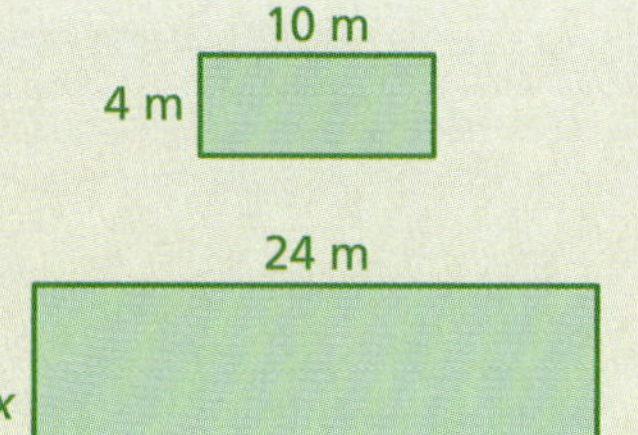

$\frac{10}{24} = \frac{4}{x}$	Write a proportion.
$10x = 96$	Cross Products Property
$x = 9.6$	Divide each side by 10.

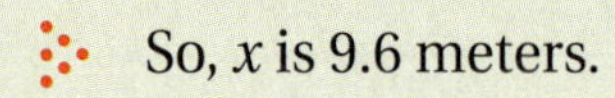

So, x is 9.6 meters.

Exercises

Tell whether the two figures are similar. Explain your reasoning.

18.

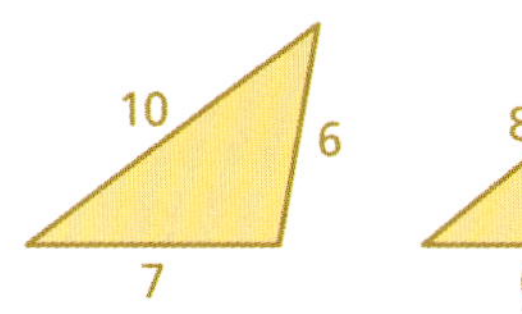

19.

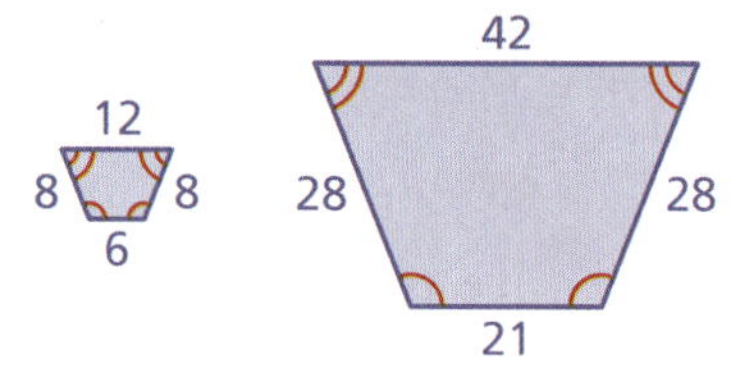

The figures are similar. Find x.

20.

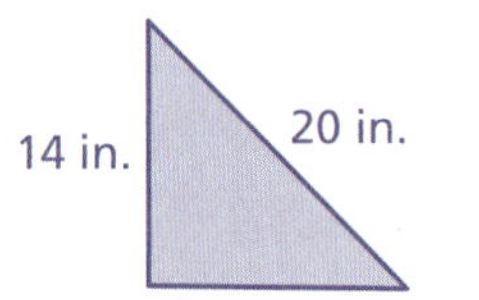

21.

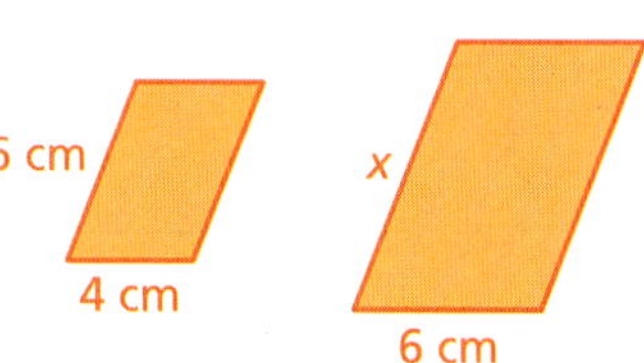

2.6 Perimeters and Areas of Similar Figures *(pp. 76–81)*

a. Find the ratio (red to blue) of the perimeters of the similar parallelograms.

$$\frac{\text{Perimeter of red parallelogram}}{\text{Perimeter of blue parallelogram}} = \frac{15}{9}$$

$$= \frac{5}{3}$$

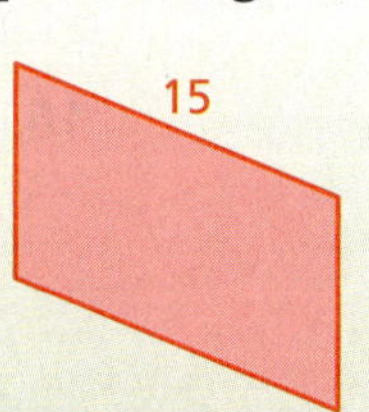

The ratio of the perimeters is $\frac{5}{3}$.

b. Find the ratio (red to blue) of the areas of the similar figures.

$$\frac{\text{Area of red figure}}{\text{Area of blue figure}} = \left(\frac{3}{4}\right)^2$$

$$= \frac{9}{16}$$

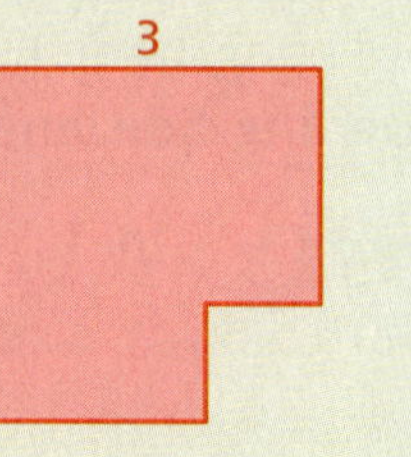

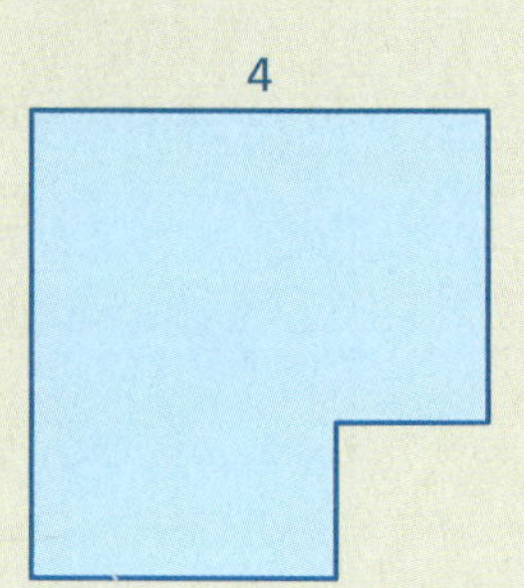

The ratio of the areas is $\frac{9}{16}$.

Review Game

Transformations

Materials
- masking tape

Players: 2

Directions
- Each player chooses an object in the classroom. The object must be something that can easily be moved by the student.
- Players use tape to create a coordinate plane beside the object.
- One player moves the object to a different position on the coordinate plane.
- The other player tries to guess whether the transformation is a translation, reflection, or rotation.
- The players switch spots and repeat the process.

Who Wins?
Each time a player guesses the correct transformation, they receive one point. The player with the most points at the end of the game wins.

For the Student Additional Practice
- Lesson Tutorials
- Multi-Language Glossary
- Self-Grading Progress Check
- *BigIdeasMath.com*
  Dynamic Student Edition
  Student Resources

Answers

18. no; The lengths of corresponding sides are not proportional.

19. yes; The lengths of corresponding sides are proportional and corresponding angles are congruent.

20. 10 in.

21. 9 cm

22. $\frac{3}{4}$; $\frac{9}{16}$

23. $\frac{7}{4}$; $\frac{49}{16}$

24. 9 : 16

25. no **26.** yes

27.
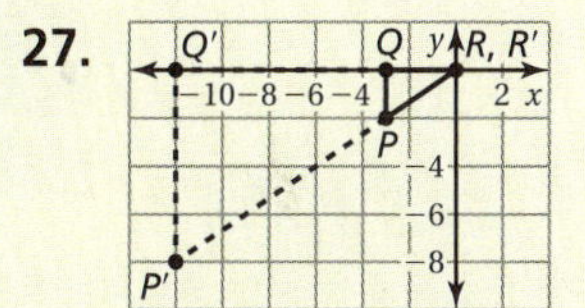

enlargement

28.
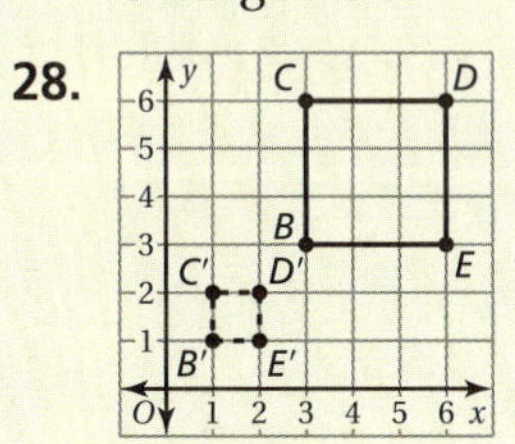

reduction

29. $Q''(-4, 2)$, $R''(14, 2)$, $S''(14, -7)$, $T''(-4, -7)$

My Thoughts on the Chapter

Teacher Tip
Not allowed to write in your teaching edition? Use sticky notes to record your thoughts.

What worked. . .

What did not work. . .

What I would do differently. . .

Exercises

The two figures are similar. Find the ratios (red to blue) of the perimeters and of the areas.

22.

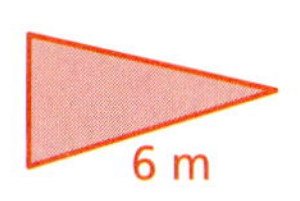

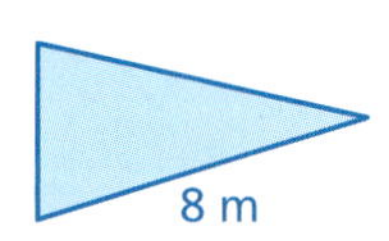

23.

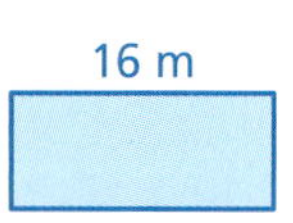

28 m

24. PHOTOS Two photos are similar. The ratio of the corresponding side lengths is 3 : 4. What is the ratio of the areas?

2.7 Dilations (pp. 82–89)

Draw the image of Triangle *ABC* after a dilation with a scale factor of 2. Identify the type of dilation.

Multiply each *x*- and *y*-coordinate by the scale factor 2.

Vertices of *ABC*	(2x, 2y)	Vertices of *A′B′C′*
$A(1, 1)$	$(2 \cdot 1, 2 \cdot 1)$	$A'(2, 2)$
$B(1, 2)$	$(2 \cdot 1, 2 \cdot 2)$	$B'(2, 4)$
$C(3, 2)$	$(2 \cdot 3, 2 \cdot 2)$	$C'(6, 4)$

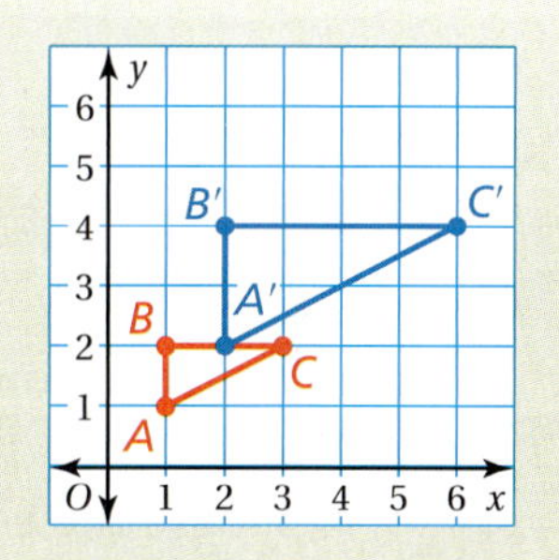

The image is shown at the above right. The dilation is an *enlargement* because the scale factor is greater than 1.

Exercises

Tell whether the blue figure is a dilation of the red figure.

25.

26.

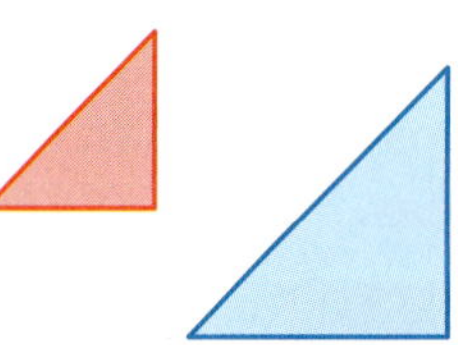

The vertices of a figure are given. Draw the figure and its image after a dilation with the given scale factor. Identify the type of dilation.

27. $P(-3, -2)$, $Q(-3, 0)$, $R(0, 0)$; $k = 4$

28. $B(3, 3)$, $C(3, 6)$, $D(6, 6)$, $E(6, 3)$; $k = \frac{1}{3}$

29. The vertices of a rectangle are $Q(-6, 2)$, $R(6, 2)$, $S(6, -4)$, and $T(-6, -4)$. Dilate the rectangle with respect to the origin using a scale factor of $\frac{3}{2}$. Then translate it 5 units right and 1 unit down. What are the coordinates of the image?

2 Chapter Test

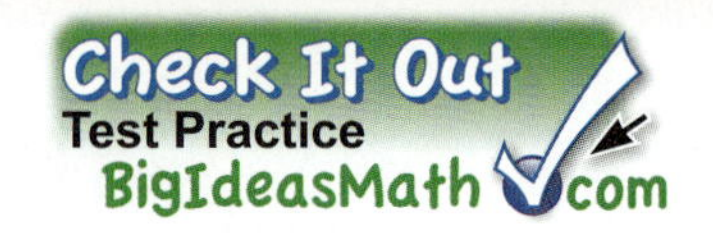

Triangles *ABC* and *DEF* are congruent.

1. Which angle of *DEF* corresponds to $\angle C$?

2. What is the perimeter of *DEF*?

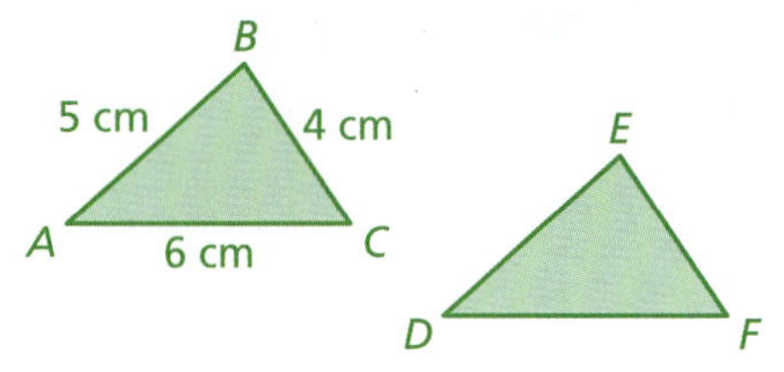

Tell whether the blue figure is a *translation, reflection, rotation,* or *dilation* of the red figure.

3. 

4. 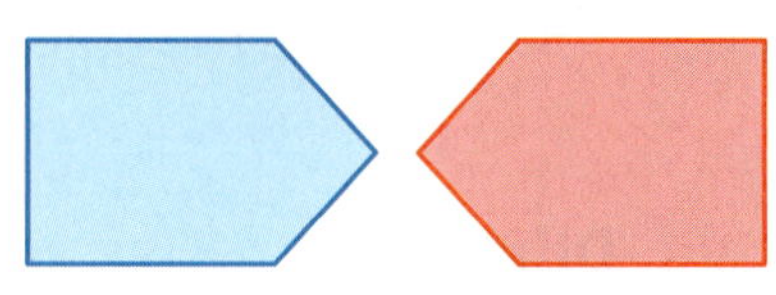

5. 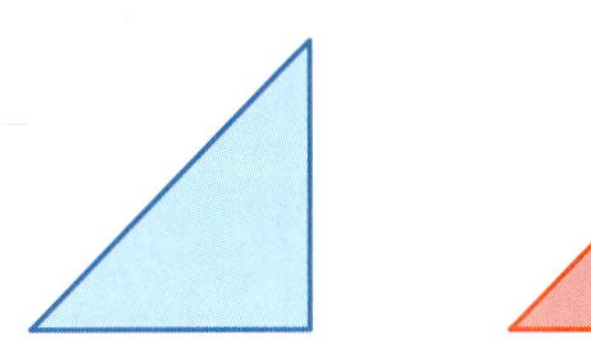

6. 

7. The vertices of a triangle are $A(2, 5)$, $B(1, 2)$, and $C(3, 1)$. Reflect the triangle in the x-axis, and then rotate the triangle 90° counterclockwise about the origin. What are the coordinates of the image?

8. The vertices of a triangle are $A(2, 4)$, $B(2, 1)$, and $C(5, 1)$. Dilate the triangle with respect to the origin using a scale factor of 2. Then translate the triangle 2 units left and 1 unit up. What are the coordinates of the image?

9. Tell whether the parallelograms are similar. Explain your reasoning.

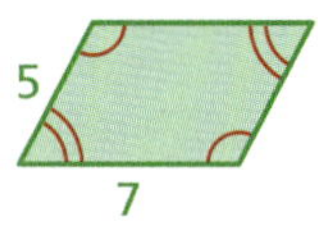

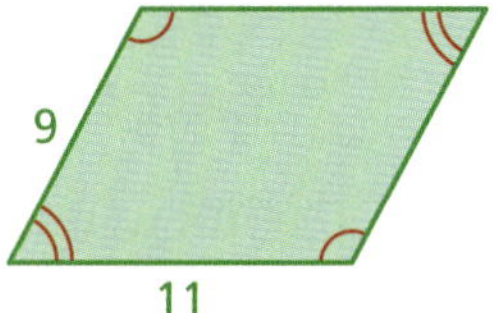

The two figures are similar. Find the ratios (red to blue) of the perimeters and of the areas.

10. 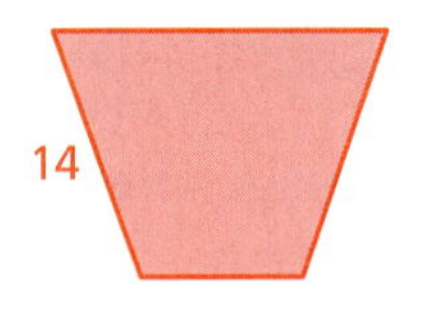

11. 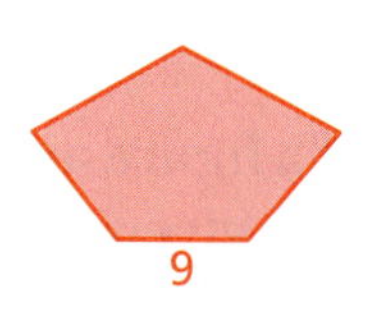

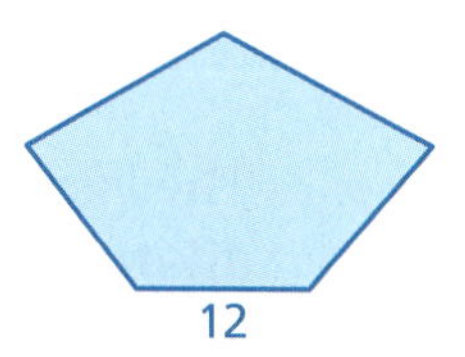

12. **SCREENS** A wide-screen television measures 36 inches by 54 inches. A movie theater screen measures 42 feet by 63 feet. Are the screens similar? Explain.

13. **CURTAINS** You want to use the rectangular piece of fabric shown to make a set of curtains for your window. Name the types of congruent shapes you can make with one straight cut. Draw an example of each type.

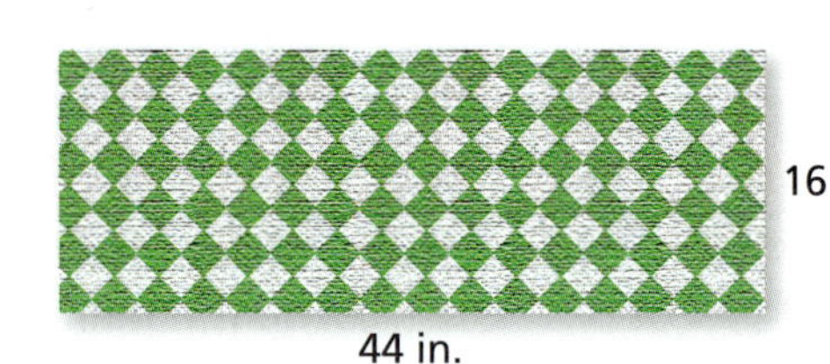

Test Item References

Chapter Test Questions	Section to Review	Florida Common Core Standards (MACC)
1, 2, 13	2.1	8.G.1.2
5	2.2	8.G.1.1, 8.G.1.2, 8.G.1.3
4	2.3	8.G.1.1, 8.G.1.2, 8.G.1.3
6, 7	2.4	8.G.1.1, 8.G.1.2, 8.G.1.3
9, 12	2.5	8.G.1.4
10, 11	2.6	8.G.1.4
3, 8	2.7	8.G.1.3, 8.G.1.4

Test-Taking Strategies

Remind students to quickly look over the entire test before they start so that they can budget their time. On this test it is very important for students to **Stop** and **Think**. When students hurry on a test dealing with transformations, they may end up with incorrect coordinates of transformed figures. Encourage students to work carefully and deliberately.

Common Errors

- **Exercise 1** Students may have difficulty determining the corresponding angle. Point out that the triangles are congruent figures and remind students that corresponding angles in congruent figures have matching positions.
- **Exercises 3–6** Students may confuse the terms *translation*, *reflection*, *rotation*, and *dilation*. Review these terms prior to the test.
- **Exercise 7** Students may reflect the triangle in the wrong axis or rotate it in the wrong direction. Refer them to the Key Idea on page 56 to review reflections. Also, remind them what *counterclockwise* means and suggest drawing an arrow on a graph of the triangle in the direction of rotation.
- **Exercise 9** Students may think that because the side lengths change by the same amount, the objects are similar. Remind them that the *ratios* of corresponding side lengths must be the same for the objects to be similar.
- **Exercises 10 and 11** When finding the ratio of the areas, students may forget to square the ratio of the corresponding side lengths. Refer them to the Key Idea at the bottom of page 78 to review ratios of areas of similar figures.

Reteaching and Enrichment Strategies

If students need help...	If students got it...
Resources by Chapter • Practice A and Practice B • Puzzle Time Record and Practice Journal Practice Differentiating the Lesson Lesson Tutorials *BigIdeasMath.com* Skills Review Handbook	Resources by Chapter • Enrichment and Extension • Technology Connection Game Closet at *BigIdeasMath.com* Start Standards Assessment

Answers

1. $\angle F$
2. 15 cm
3. dilation
4. reflection
5. translation
6. rotation
7. $A''(5, 2), B''(2, 1), C''(1, 3)$
8. $A''(2, 9), B''(2, 3), C''(8, 3)$
9. no; The lengths of corresponding sides are not proportional.
10. $\frac{7}{4}; \frac{49}{16}$
11. $\frac{3}{4}; \frac{9}{16}$
12. yes; Because both screens are rectangles, the corresponding angle measures are congruent. Corresponding side lengths are proportional.
13. 2 rectangles

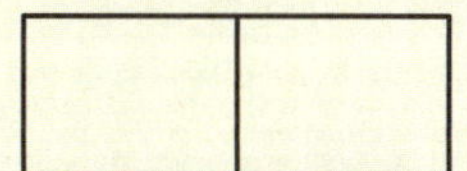

2 right triangles

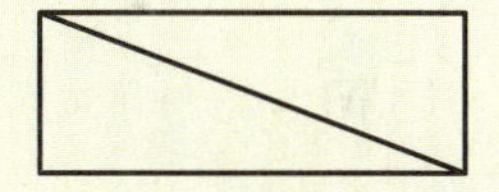

2 right trapezoids

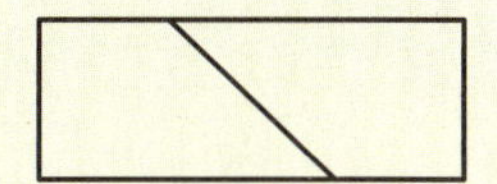

Technology for the Teacher

Online Assessment
Assessment Book
ExamView® Assessment Suite

Test-Taking Strategies

Available at *BigIdeasMath.com*

After Answering Easy Questions, Relax

Answer Easy Questions First
Estimate the Answer
Read All Choices before Answering
Read Question before Answering
Solve Directly or Eliminate Choices
Solve Problem before Looking at Choices
Use Intelligent Guessing
Work Backwards

About this Strategy

When taking a multiple choice test, be sure to read each question carefully and thoroughly. After skimming the test and answering the easy questions, stop for a few seconds, take a deep breath, and relax. Work through the remaining questions carefully, using your knowledge and test-taking strategies. Remember, you already completed many of the questions on the test!

Answers

1. 270°
2. D
3. I
4. C
5. G

Technology for the Teacher

Florida Common Core Standards Support Performance Tasks
Online Assessment
Assessment Book
ExamView® Assessment Suite

Item Analysis

1. **Gridded Response:** Correct answer: 270°

   Common Error: The student confuses a rotation of 90° with a rotation of 180° and thinks that a 90° clockwise rotation has the same result as a 90° counterclockwise rotation, getting an answer of 90°.

2. **A.** The student makes an operation error.
   B. The student makes an operation error.
   C. The student makes an operation error.
   D. Correct answer

3. **F.** The student does not apply inverses in the correct order.
   G. The student confuses inverse operations.
   H. The student confuses inverse operations.
   I. Correct answer

4. **A.** A translation *slides* a figure.
   B. A reflection *flips* a figure.
   C. Correct answer
   D. A dilation changes the size of a figure.

5. **F.** The student only translates the triangle 3 units right.
   G. Correct answer
   H. The student only translates the triangle 2 units down.
   I. The student translates the triangle 2 units right and 3 units down.

2 Standards Assessment

1. A clockwise rotation of 90° is equivalent to a counterclockwise rotation of how many degrees? *(MACC.8.G.1.2)*

2. The formula $K = C + 273.15$ converts temperatures from Celsius C to Kelvin K. Which of the following formulas is *not* correct? *(MACC.8.EE.3.7a)*

   A. $K - C = 273.15$

   B. $C = K - 273.15$

   C. $C - K = -273.15$

   D. $C = K + 273.15$

3. Joe wants to solve the equation $-3(x + 2) = 12x$. What should he do first? *(MACC.8.EE.3.7a)*

   F. Subtract 2 from each side.

   G. Add 3 to each side.

   H. Multiply each side by -3.

   I. Divide each side by -3.

4. Which transformation *turns* a figure? *(MACC.8.G.1.1)*

   A. translation

   B. reflection

   C. rotation

   D. dilation

5. A triangle is graphed in the coordinate plane below.

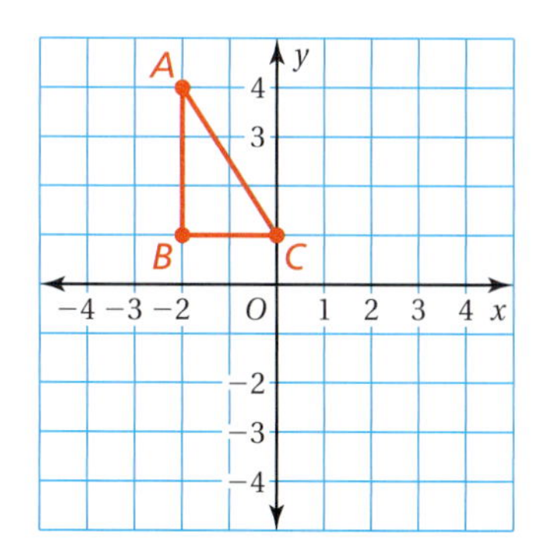

   Translate the triangle 3 units right and 2 units down. What are the coordinates of the image? *(MACC.8.G.1.3)*

   F. $A'(1, 4)$, $B'(1, 1)$, $C'(3, 1)$

   G. $A'(1, 2)$, $B'(1, -1)$, $C'(3, -1)$

   H. $A'(-2, 2)$, $B'(-2, -1)$, $C'(0, -1)$

   I. $A'(0, 1)$, $B'(0, -2)$, $C'(2, -2)$

6. Dale solved the equation in the box shown. What should Dale do to correct the error that he made? *(MACC.8.EE.3.7b)*

$$-\frac{x}{3} + \frac{2}{5} = -\frac{7}{15}$$
$$-\frac{x}{3} + \frac{2}{5} - \frac{2}{5} = -\frac{7}{15} - \frac{2}{5}$$
$$-\frac{x}{3} = -\frac{13}{15}$$
$$3 \cdot \left(-\frac{x}{3}\right) = 3 \cdot \left(-\frac{13}{15}\right)$$
$$x = -2\frac{3}{5}$$

A. Add $\frac{2}{5}$ to each side to get $-\frac{x}{3} = -\frac{1}{15}$.

B. Multiply each side by -3 to get $x + \frac{2}{5} = \frac{7}{5}$.

C. Multiply each side by -3 to get $x = 2\frac{3}{5}$.

D. Subtract $\frac{2}{5}$ from each side to get $-\frac{x}{3} = -\frac{5}{10}$.

7. Jenny dilates the rectangle below using a scale factor of $\frac{1}{2}$.

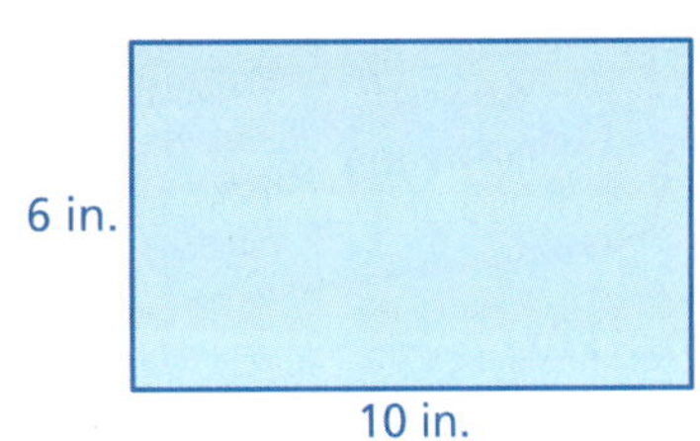

What is the area of the dilated rectangle in square inches? *(MACC.8.G.1.4)*

8. The vertices of a rectangle are $A(-4, 2)$, $B(3, 2)$, $C(3, -5)$, and $D(-4, -5)$. If the rectangle is dilated by a scale factor of 3, what will be the coordinates of vertex C'? *(MACC.8.G.1.3)*

F. $(9, -15)$

G. $(-12, 6)$

H. $(-12, -15)$

I. $(9, 6)$

9. In the figures, Triangle *EFG* is a dilation of Triangle *HIJ*.

Which proportion is *not* necessarily correct for Triangle *EFG* and Triangle *HIJ*? *(MACC.8.G.1.4)*

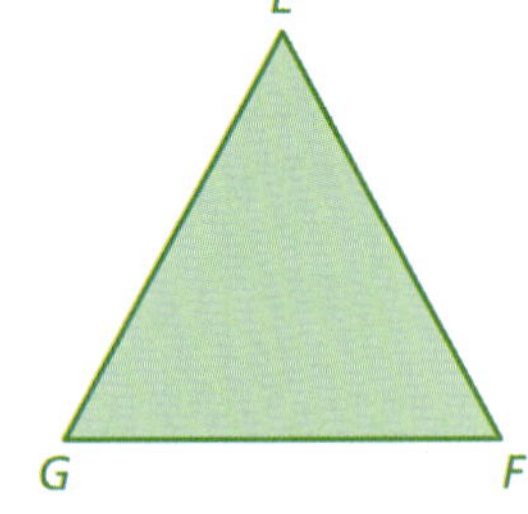

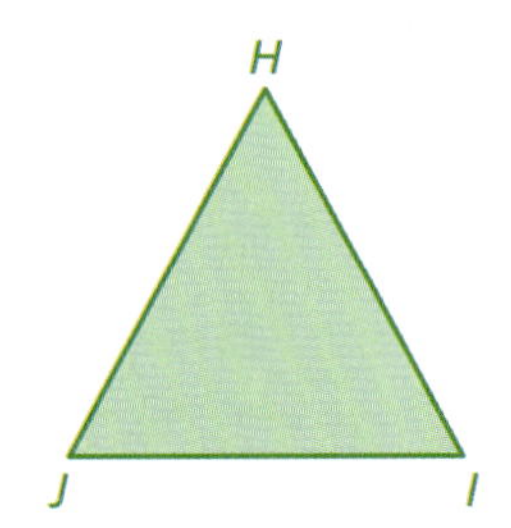

A. $\frac{EF}{FG} = \frac{HI}{IJ}$

B. $\frac{EG}{HI} = \frac{FG}{IJ}$

C. $\frac{GE}{EF} = \frac{JH}{HI}$

D. $\frac{EF}{HI} = \frac{GE}{JH}$

Item Analysis (continued)

6. **A.** The student *undoes* the addition of $\frac{2}{5}$ incorrectly by adding $\frac{2}{5}$ to each side, rather than subtracting.

   B. The student performs an order of operations error by not subtracting $\frac{2}{5}$ from each side first.

   C. Correct answer

   D. The student subtracts the numerators and denominators to get $-\frac{7-2}{15-5} = -\frac{5}{10}$.

7. **Gridded Response:** Correct answer: 15 square inches

   Common Error: The student finds half the area of the original rectangle, getting an answer of 30 square inches.

8. **F.** Correct answer

   G. The student used the coordinates of vertex A'.

   H. The student used the coordinates of vertex D'.

   I. The student used the coordinates of vertex B'.

9. **A.** The student chooses a proportion that correctly represents a relationship between pairs of corresponding sides of the triangles.

   B. Correct answer

   C. The student chooses a proportion that correctly represents a relationship between pairs of corresponding sides of the triangles.

   D. The student chooses a proportion that correctly represents a relationship between pairs of corresponding sides of the triangles.

Answers

6. C
7. 15
8. F
9. B

Answers

10. F

11. *Part A* translation up

Part B dilation (reduction)

Part C 2

12. A

Item Analysis (continued)

10. **F.** Correct answer

G. The student chooses an answer based only on visual approximation.

H. The student thinks that because 12 is 4 more than 8, x should be 4 more than 12.

I. The student thinks that because 8 is 4 less than 12, x should be 4 less than 21.

11. **2 points** The student demonstrates a thorough understanding of working with transformations. In Part A, the student correctly describes the transformation as a translation up. In Part B, the student correctly describes the transformation as a dilation (reduction). In Part C, the student correctly determines that the scale factor is 2. The student provides clear and complete work and explanations.

1 point The student demonstrates a partial understanding of working with transformations. The student provides some correct work and explanation.

0 points The student demonstrates insufficient understanding of working with transformations. The student is unable to make any meaningful progress toward a correct answer.

12. **A.** Correct answer

B. The student performs a reflection in the x-axis.

C. The student rotates the rectangle 90° clockwise.

D. The student thinks 180° causes the figure to end up where it started.

10. In the figures below, Rectangle *EFGH* is a dilation of Rectangle *IJKL*.

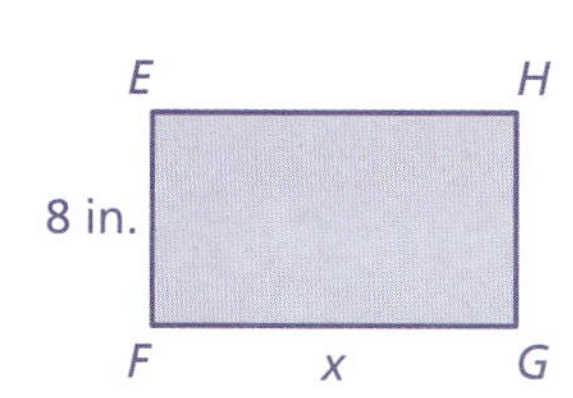

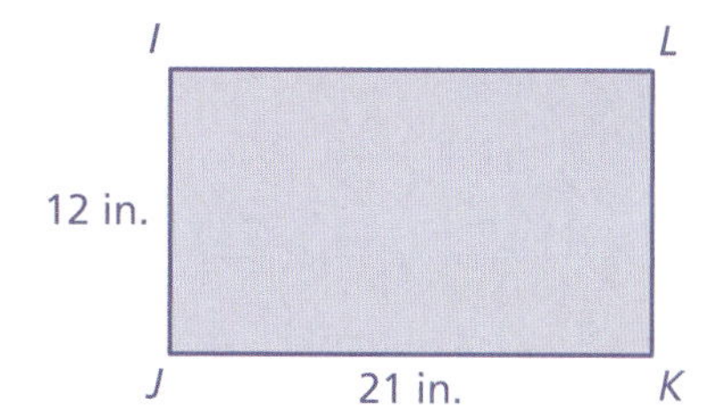

What is x? *(MACC.8.G.1.4)*

F. 14 in.

G. 15 in.

H. 16 in.

I. 17 in.

11. Several transformations are used to create the pattern. *(MACC.8.G.1.2, MACC.8.G.1.4)*

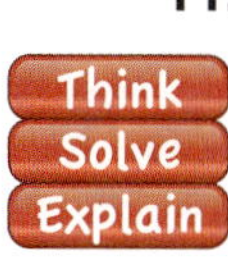

Part A Describe the transformation of Triangle *GLM* to Triangle *DGH*.

Part B Describe the transformation of Triangle *ALQ* to Triangle *GLM*.

Part C Triangle *DFN* is a dilation of Triangle *GHM*. Find the scale factor.

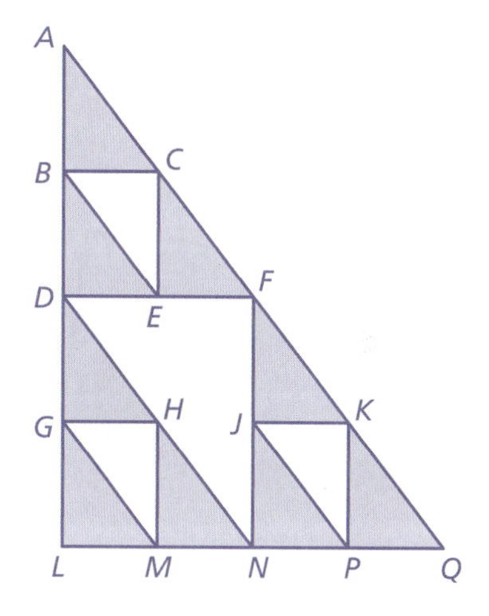

12. A rectangle is graphed in the coordinate plane below.

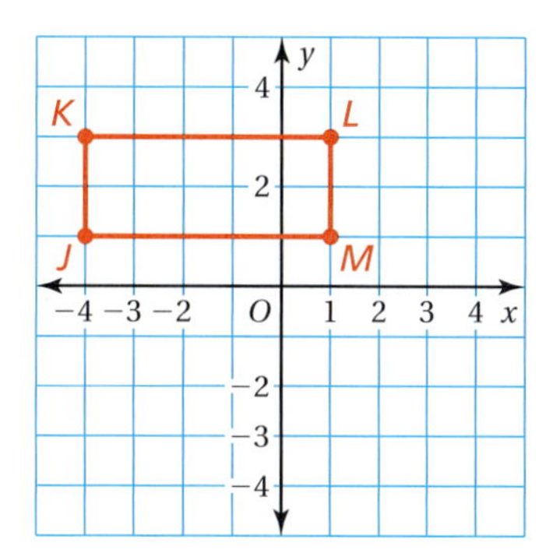

Rotate the triangle 180° about the origin. What are the coordinates of the image? *(MACC.8.G.1.3)*

A. $J'(4, -1)$, $K'(4, -3)$, $L'(-1, -3)$, $M'(-1, -1)$

B. $J'(-4, -1)$, $K'(-4, -3)$, $L'(1, -3)$, $M'(1, -1)$

C. $J'(1, 4)$, $K'(3, 4)$, $L'(3, -1)$, $M'(1, -1)$

D. $J'(-4, 1)$, $K'(-4, 3)$, $L'(1, 3)$, $M'(1, 1)$

3 Angles and Triangles

"Start with any triangle."

"Tear off the angles. You can always rearrange the angles so that they form a straight line."

"What does that prove?"

"Let's use shadows and similar triangles to indirectly measure the height of the giant hyena standing right behind you."

Florida Common Core Progression

6th Grade

- Use reasoning about multiplication and division to solve ratio and rate problems.

7th Grade

- Measure and describe relationships among vertical, adjacent, supplementary, and complementary angles.
- Use proportions to solve problems.

8th Grade

- Classify and determine the measures of angles created when parallel lines are cut by a transversal.
- Demonstrate that the sum of the interior angle measures of a triangle is 180° and apply this fact to find the unknown measures of angles and the sum of the angles of polygons.
- Use similar triangles to solve problems that include height and distance.

Pacing Guide for Chapter 3

Chapter Opener	1 Day
Section 1 Activity Lesson	 1 Day 1 Day
Section 2 Activity Lesson	 1 Day 1 Day
Study Help / Quiz	1 Day
Section 3 Activity Lesson	 1 Day 1 Day
Section 4 Activity Lesson	 1 Day 2 Days
Chapter Review/ Chapter Tests	2 Days
Total Chapter 3	13 Days
Year-to-Date	45 Days

Chapter Summary

Section	Florida Common Core Standard (MACC)	
3.1	Learning	8.G.1.5
3.2	Learning	8.G.1.5
3.3	Applying	8.G.1.5
3.4	Learning	8.G.1.5 ★

★ Teaching is complete. Standard can be assessed.

Technology for the Teacher

BigIdeasMath.com
Chapter at a Glance
Complete Materials List
Parent Letters: English, Spanish, and Haitian Creole

Florida Common Core Standards

MACC.7.G.2.5 Use facts about supplementary, complementary, vertical, and adjacent angles in a multi-step problem to write and solve simple equations for an unknown angle in a figure.

Additional Topics for Review

- Vocabulary of Angles and Triangles
- Similar Triangles
- Solving Multi-Step Equations

Try It Yourself

1. vertical; $x = 112$
2. adjacent; $x = 44$
3. complementary; $x = 78$
4. supplementary; $x = 50$

Record and Practice Journal Fair Game Review

1. vertical; 128
2. adjacent; 55
3. vertical; 37
4. adjacent; 15
5. 76°
6. supplementary; 63
7. complementary; 21
8. complementary; 49
9. supplementary; 14
10. 53

Math Background Notes

Vocabulary Review

- Adjacent Angles
- Vertical Angles
- Congruent Angles
- Complementary Angles
- Supplementary Angles

Adjacent and Vertical Angles

- Two intersecting lines will form four angles with a common vertex.
- *Adjacent angles* share a common side and have the same vertex.
- When two adjacent angles form a larger angle, the sum of the measures of the smaller angles is equal to the measure of the larger angle.
- The opposite angles formed by the intersection of two lines are *vertical angles*.
- Vertical angles are *congruent angles*, meaning they have the same measure.
- In Question 2, remind students of the corner mark used to designate a right angle.

Complementary and Supplementary Angles

- Two angles are complementary angles when the sum of their measures is 90°.
- Two angles are supplementary angles when the sum of their measures is 180°.
- In Question 3, remind students of the corner mark used to designate a right angle.

Reteaching and Enrichment Strategies

If students need help. . .	If students got it. . .
Record and Practice Journal • Fair Game Review Skills Review Handbook Lesson Tutorials	Game Closet at *BigIdeasMath.com* Start the next section

What You Learned Before

Complementary
Supplementary
30 degrees said to 60 degrees, "It's neat, you make me complete."

"I just remember that C comes before S and 90 comes before 180. That makes it easy."

Adjacent and Vertical Angles (MACC.7.G.2.5)

Example 1 **Tell whether the angles are *adjacent* or *vertical*. Then find the value of x.**

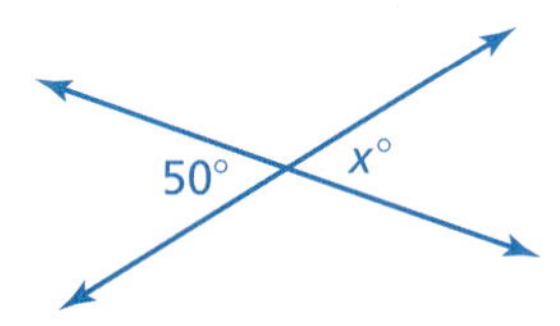

The angles are vertical angles. Because vertical angles are congruent, the angles have the same measure.

So, the value of x is 50.

Try It Yourself

Tell whether the angles are *adjacent* or *vertical*. Then find the value of x.

1.

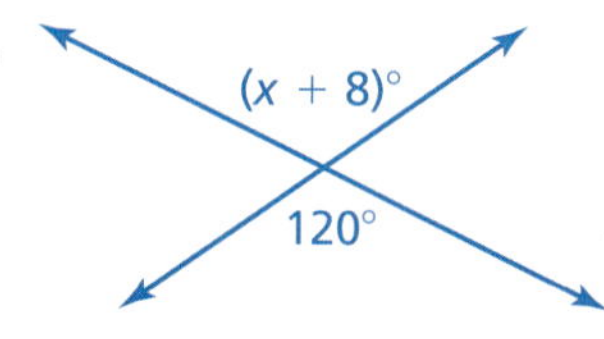

2.

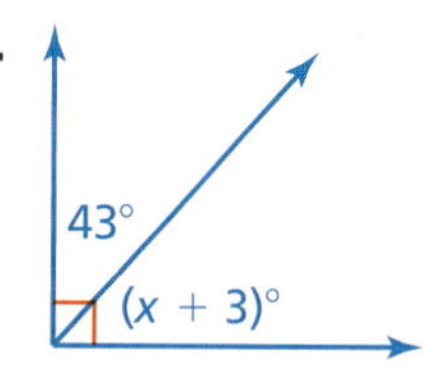

Complementary and Supplementary Angles (MACC.7.G.2.5)

Example 2 **Tell whether the angles are *complementary* or *supplementary*. Then find the value of x.**

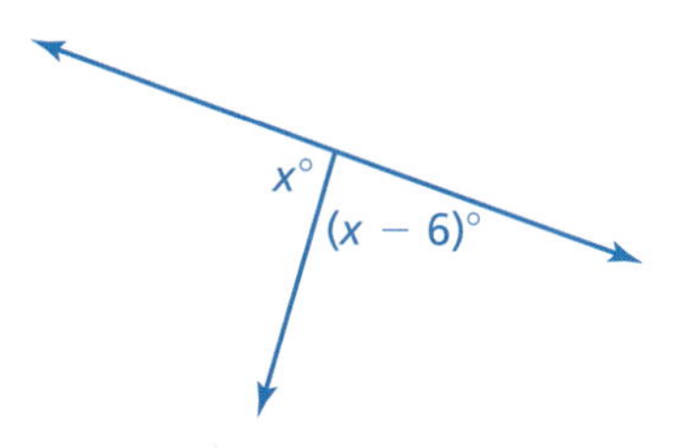

The two angles make up a straight angle. So, the angles are supplementary angles, and the sum of their measures is 180°.

$x + (x - 6) = 180$ — Write equation.

$2x - 6 = 180$ — Combine like terms.

$2x = 186$ — Add 6 to each side.

$x = 93$ — Divide each side by 2.

Try It Yourself

Tell whether the angles are *complementary* or *supplementary*. Then find the value of x.

3.

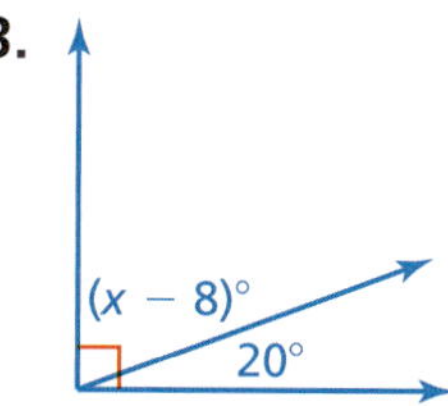

4.

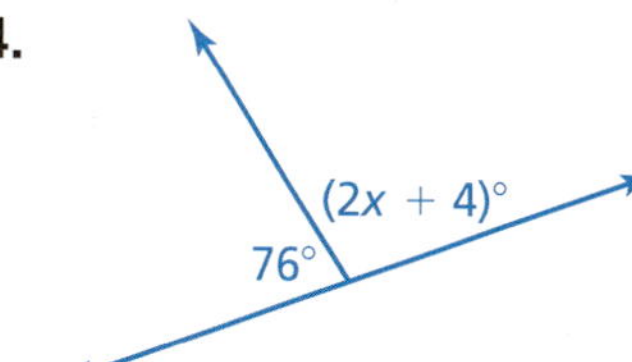

3.1 Parallel Lines and Transversals

Essential Question How can you describe angles formed by parallel lines and transversals?

The Meaning of a Word ● Transverse

When an object is **transverse**, it is lying or extending across something.

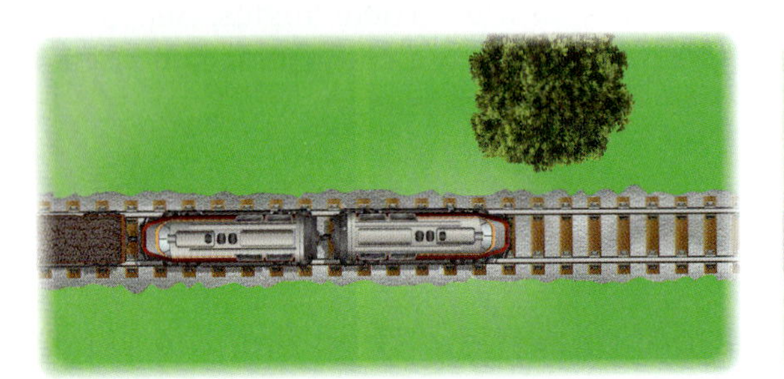

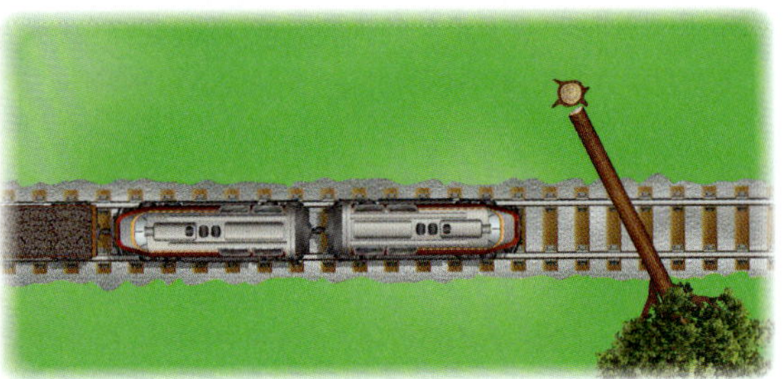

1 ACTIVITY: A Property of Parallel Lines

Work with a partner.

- **Discuss what it means for two lines to be parallel. Decide on a strategy for drawing two parallel lines. Then draw the two parallel lines.**
- **Draw a third line that intersects the two parallel lines. This line is called a transversal.**

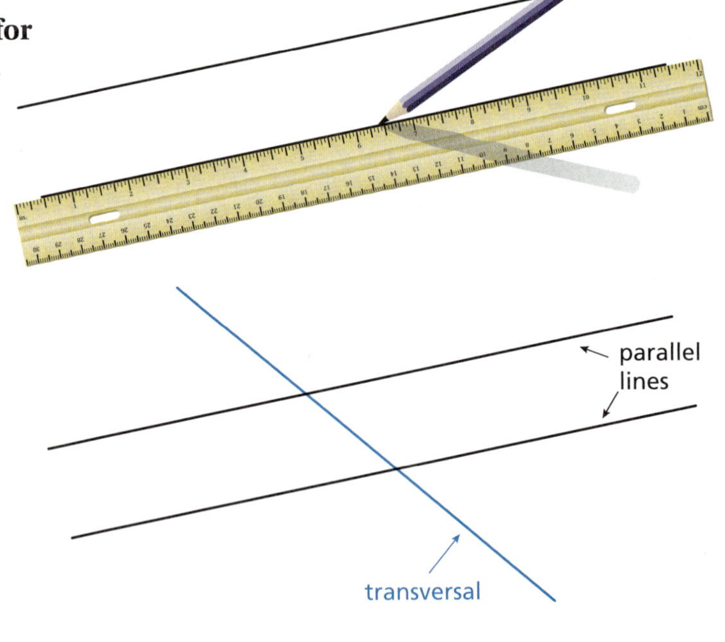

a. How many angles are formed by the parallel lines and the transversal? Label the angles.

b. Which of these angles have equal measures? Explain your reasoning.

Geometry

In this lesson, you will

- identify the angles formed when parallel lines are cut by a transversal.
- find the measures of angles formed when parallel lines are cut by a transversal.

Learning Standard
MACC.8.G.1.5

Laurie's Notes

Introduction

Standards for Mathematical Practice

- **MP5 Use Appropriate Tools Strategically:** There are many appropriate tools mathematically proficient students may use to gain an understanding of a new concept or to solve a problem. Several tools are useful in the investigations today.

Motivate

- **Preparation:** Use a transparency that has two parallel lines drawn on it on the overhead projector. Place a grid on top of the parallel lines so that the lines pass through obvious lattice points.
- ? "What appears to be true about the lines?" parallel
- Compute the slopes and conclude that the lines are parallel.
- Now remove the grid but keep the transparency with the (parallel) lines.
- ? "Are the lines still parallel? How do you know?" The point of these questions is not for students to give an answer, but for students to consider what it means for two lines to be parallel.

Discuss

- Ask students to give examples of parallel lines in real life, where the lines are
  - functional (as with railroad tracks or lanes of a highway).
  - aesthetic (on clothing).
  - coincidentally parallel (the edge of table and a pencil).
- Parallel lines are common. Often there is a reason for the parallelism.

Activity Notes

Activity 1

- Discuss the meaning of the word *transverse*. Give examples, such as a path that cuts through a field or a street that cuts through the downtown area of a city.
- **MP5:** Students should have tools available such as rulers, protractors, and tracing paper.
- Students will likely create parallel lines by tracing opposite edges of a rigid object they believe has parallel edges, such as a ruler or an index card. There are other methods. Ask students to share their methods, and give explanations as to why the lines are parallel.
- Students will use protractors to measure the angles formed. If tracing paper is available, some students may simply trace the acute angle and the obtuse angle and slide the tracing paper around.

Florida Common Core Standards

MACC.8.G.1.5 Use informal arguments to establish facts about . . . the angles created when parallel lines are cut by a transversal, and the angle-angle criterion for similarity of triangles.

Previous Learning

Students should know the definition of similar triangles.

3.1 Record and Practice Journal

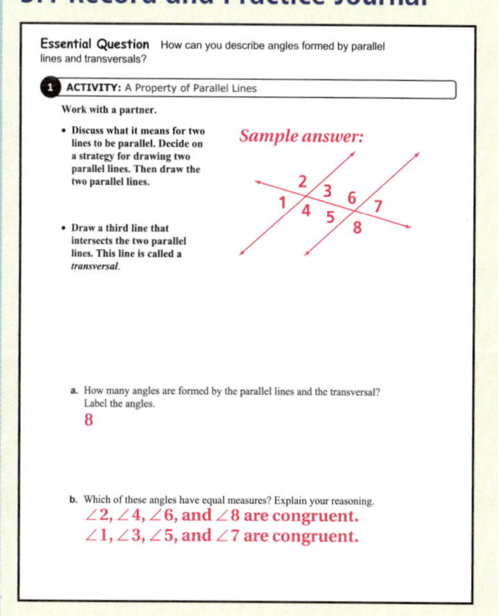

Essential Question How can you describe angles formed by parallel lines and transversals?

1 ACTIVITY: A Property of Parallel Lines

Work with a partner.

- Discuss what it means for two lines to be parallel. Decide on a strategy for drawing two parallel lines. Then draw the two parallel lines.
- Draw a third line that intersects the two parallel lines. This line is called a *transversal*.

Sample answer:

a. How many angles are formed by the parallel lines and the transversal? Label the angles.
8

b. Which of these angles have equal measures? Explain your reasoning.
$\angle 2, \angle 4, \angle 6$, and $\angle 8$ are congruent.
$\angle 1, \angle 3, \angle 5$, and $\angle 7$ are congruent.

Differentiated Instruction

Visual

In order to be successful in geometry, students need to be proficient with the vocabulary. Have them write vocabulary words in their math notebook glossaries. Next to the definitions students should illustrate the terms with color-coded diagrams.

3.1 Record and Practice Journal

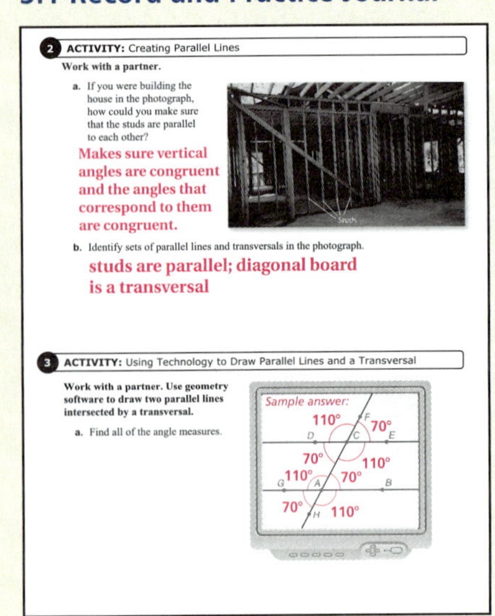

2 ACTIVITY: Creating Parallel Lines

Work with a partner.

a. If you were building the house in the photograph, how could you make sure that the studs are parallel to each other?

Makes sure vertical angles are congruent and the angles that correspond to them are congruent.

b. Identify sets of parallel lines and transversals in the photograph.

studs are parallel; diagonal board is a transversal

3 ACTIVITY: Using Technology to Draw Parallel Lines and a Transversal

Work with a partner. Use geometry software to draw two parallel lines intersected by a transversal.

a. Find all of the angle measures.

b. Adjust the figure by moving the parallel lines or the transversal to a different position. Describe how the angle measures and relationships change.

Check students' work.

What Is Your Answer?

4. **IN YOUR OWN WORDS** How can you describe angles formed by parallel lines and transversals? Give an example.

See Additional Answers.

5. Use geometry software to draw a transversal that is perpendicular to two parallel lines. What do you notice about the angles formed by the parallel lines and the transversal?

All eight angle measures = 90°.
See Additional Answers.

Laurie's Notes

Activity 2

- Students will have different thoughts about how to make sure the studs are parallel to each other. Be prepared for students who do not know much about construction.
- The transversals should be obvious, but remember that the horizontal boards perpendicular to the studs are also transversals.

Activity 3

- **MP5:** Seeing and being able to manipulate parallel lines and a transversal in a dynamic environment is very powerful. There is dynamic geometry software free for download. If a computer lab is not available, a class demonstration is an alternative.
- Be sure that students *construct* parallel lines versus drawing lines that look parallel. They should draw the first line, plot a point not on that line, and construct a line through the point parallel to the first line.
- Students should quickly see that when the lines are parallel the four acute angles are congruent and the four obtuse angles are congruent.
- When the lines are not parallel, only the four pairs of vertical angles will be congruent.

What Is Your Answer?

- In Question 5, make sure that students are constructing perpendicular lines versus using eyesight to draw two lines that appear perpendicular.

Closure

- Identify five pairs of parallel lines in your classroom. Note that the definition of *line* is modified to include things such as the metal molding on either side of the white board and the side casings on the door frame.

2 ACTIVITY: Creating Parallel Lines

Math Practice 6

Use Clear Definitions

What do the words *parallel* and *transversal* mean? How does this help you answer the question in part (a)?

Work with a partner.

a. If you were building the house in the photograph, how could you make sure that the studs are parallel to each other?

b. Identify sets of parallel lines and transversals in the photograph.

3 ACTIVITY: Using Technology

Work with a partner. Use geometry software to draw two parallel lines intersected by a transversal.

a. Find all the angle measures.

b. Adjust the figure by moving the parallel lines or the transversal to a different position. Describe how the angle measures and relationships change.

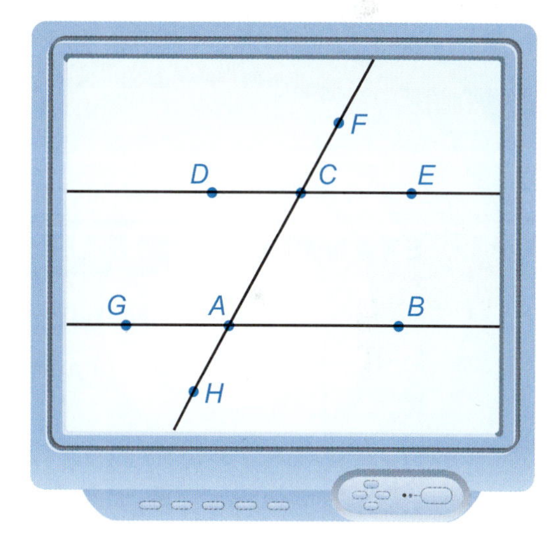

What Is Your Answer?

4. **IN YOUR OWN WORDS** How can you describe angles formed by parallel lines and transversals? Give an example.

5. Use geometry software to draw a transversal that is perpendicular to two parallel lines. What do you notice about the angles formed by the parallel lines and the transversal?

Use what you learned about parallel lines and transversals to complete Exercises 3–6 on page 107.

3.1 Lesson

Key Vocabulary

transversal, *p. 104*
interior angles, *p. 105*
exterior angles, *p. 105*

Lines in the same plane that do not intersect are called *parallel lines*. Lines that intersect at right angles are called *perpendicular lines*.

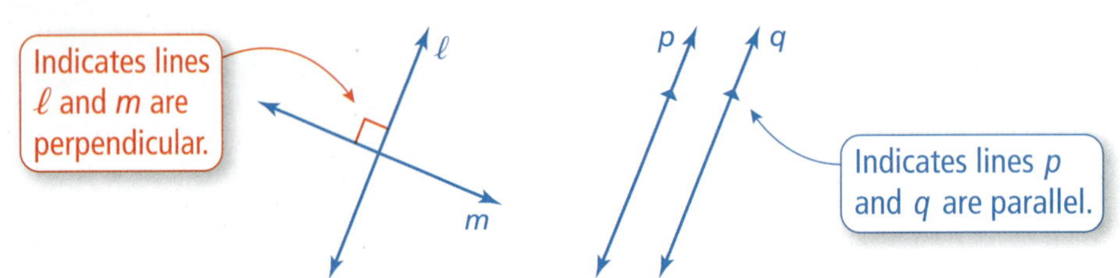

A line that intersects two or more lines is called a **transversal**. When parallel lines are cut by a transversal, several pairs of congruent angles are formed.

Key Idea

Corresponding Angles

When a transversal intersects parallel lines, corresponding angles are congruent.

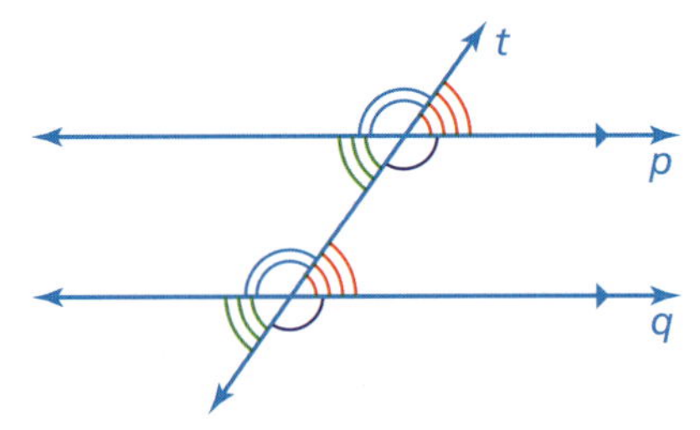

Corresponding angles

Study Tip

Corresponding angles lie on the same side of the transversal in corresponding positions.

EXAMPLE 1 Finding Angle Measures

Use the figure to find the measures of (a) $\angle 1$ and (b) $\angle 2$.

a. $\angle 1$ and the 110° angle are corresponding angles. They are congruent.

So, the measure of $\angle 1$ is 110°.

b. $\angle 1$ and $\angle 2$ are supplementary.

$\angle 1 + \angle 2 = 180^\circ$	Definition of supplementary angles
$110^\circ + \angle 2 = 180^\circ$	Substitute 110° for $\angle 1$.
$\angle 2 = 70^\circ$	Subtract 110° from each side.

So, the measure of $\angle 2$ is 70°.

On Your Own

Now You're Ready
Exercises 7–9

Use the figure to find the measure of the angle. Explain your reasoning.

1. $\angle 1$ **2.** $\angle 2$

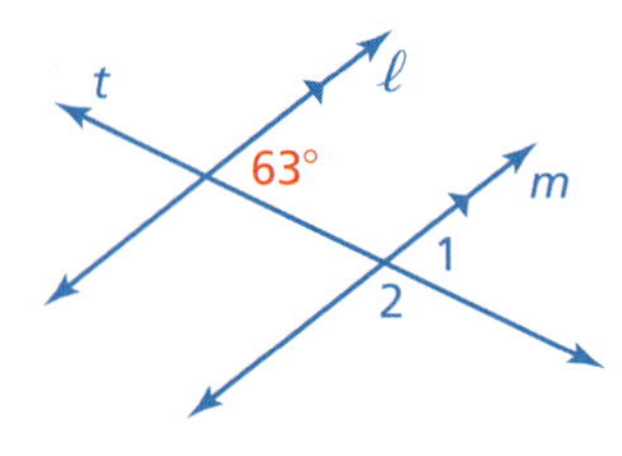

Laurie's Notes

Introduction

Connect

- **Yesterday:** Students explored angles formed when parallel lines are intersected by a transversal. (MP5)
- **Today:** Students will find the measures of many types of angles, all formed when parallel lines are cut by a transversal.

Motivate

- **Preparation:** Make a model to help discuss the big ideas of this lesson. Cut 3 strips of card stock; punch holes in the middle of two strips and punch two holes in the third strip. Attach the strips using brass fasteners.

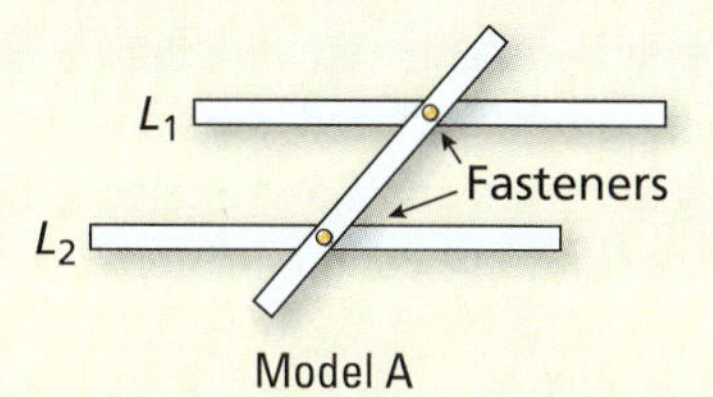

Model A

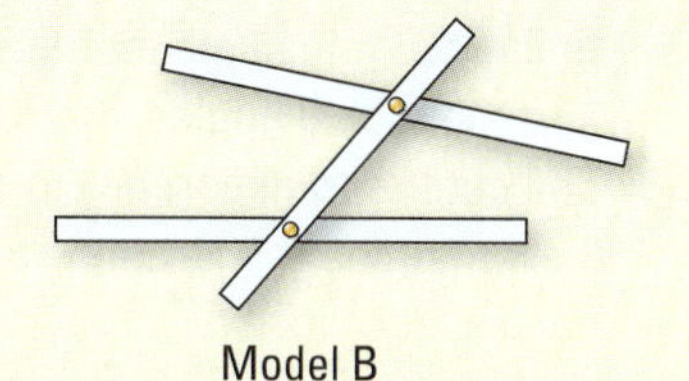
Model B

- Place the model on the overhead or document camera. Demonstrate to students that the pieces are moveable, by transforming from Model A to Model B.
- Focus students' attention on the connection between the 4 angles on L_1 and the 4 angles on L_2. Pairs of vertical angles will always be congruent whether or not L_1 and L_2 are parallel.
- Place the model on the overhead or document camera, and encourage students to point to the angles that they think are congruent. Use models A and B.

Lesson Notes

Key Idea

- Write the informal definitions of parallel lines and perpendicular lines. Draw examples of each and discuss the notation used in the diagram.
- Write the definition of transversal. Explain that a line that intersects two or more lines is called a **transversal** even if the lines are *not* parallel. Only when the lines are parallel are the pairs of angles in this lesson congruent.
- Write the Key Idea. Identify corresponding angles which are color-coded in the diagram. Mention the Study Tip.
- Students will ask what is meant by *corresponding position*. The corresponding angles are both above or below the parallel lines (when in horizontal position) and on the same side of the transversal (left or right).
- **MP6 Attend to Precision:** In a formal geometry course, the rules in the Key Ideas of this lesson would be stated in if-then (conditional) form. Discuss this with students.

Example 1

? "Are the lines parallel? How do you know?" yes; blue arrow marks

- Students will need to recall the definition of supplementary angles.

Goal Today's lesson is finding measures of angles formed by parallel lines and a **transversal**.

Lesson Tutorials
Lesson Plans
Answer Presentation Tool

Extra Example 1

Use the figure to find the measures of (a) $\angle 1$ and (b) $\angle 2$.

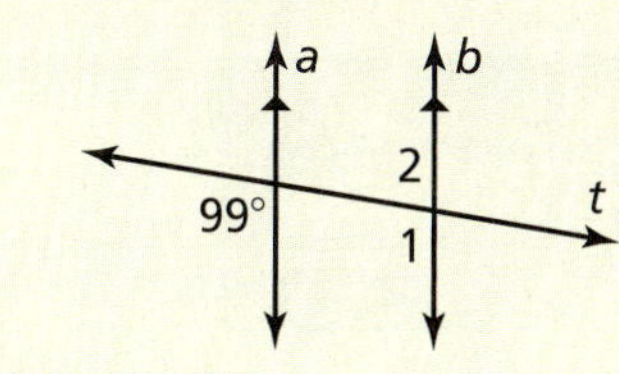

a. 99° **b.** 81°

On Your Own

1. 63°; Corresponding angles are congruent.
2. 117°; $\angle 1$ and $\angle 2$ are supplementary.

Laurie's Notes

Extra Example 2

Use the figure to find the measures of the numbered angles.

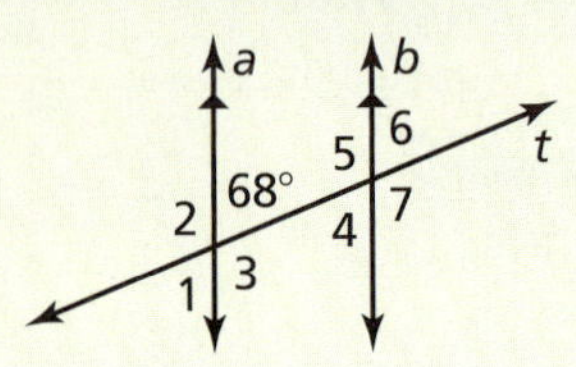

$\angle 1 = 68°$, $\angle 2 = 112°$, $\angle 3 = 112°$, $\angle 4 = 68°$, $\angle 5 = 112°$, $\angle 6 = 68°$, $\angle 7 = 112°$

Example 2

- Draw the figure on the board or overhead.

? "Can you find the measures of all the angles if you only know one angle?" Students may not know the answer at this point, but by the end of this example, they will see that they can.

- **MP8 Look for and Express Regularity in Repeated Reasoning** and **Big Idea:** If you know any angle when a transversal intersects two parallel lines, then you can use vertical, supplementary, and corresponding angles to find all 7 of the other measures. It is not necessary to learn other theorems about alternate exterior or interior angles. **Students should be able to do all of the homework after this example.**
- Once angles 1, 2, and 3 are found, you can use corresponding angles to find the remaining four angles. To help students visualize the corresponding angles, draw the figure on an overhead transparency and cut the transparency in half. Lay the given angle and angles 1, 2, and 3 over angles 4, 5, 6, and 7 to show that they are congruent corresponding angles.

On Your Own

3. $\angle 1 = 121°$, $\angle 2 = 59°$, $\angle 3 = 121°$, $\angle 4 = 121°$, $\angle 5 = 59°$, $\angle 6 = 121°$, $\angle 7 = 59°$

On Your Own

- **Question 3:** Students can say $\angle 3 = 121°$ because of vertical angles *or* because it is the supplement of $\angle 2$.

Discuss

- Use the model from the beginning of class to talk about other pairs of angles. Make the lines parallel even though the angles still have the same definition.
- Identify the four angles that are interior (between the two parallel lines) and the four angles that are exterior (outside the two parallel lines).

? "Are there pairs of interior angles that appear congruent?" yes; 3 & 6 and 4 & 5 (in the diagram)

? "Are there pairs of exterior angles that appear congruent?" yes; 1 & 8 and 2 & 7 (in the diagram)

Extra Example 3

The painting shows several parallel lines and transversals. What is the measure of $\angle 1$?

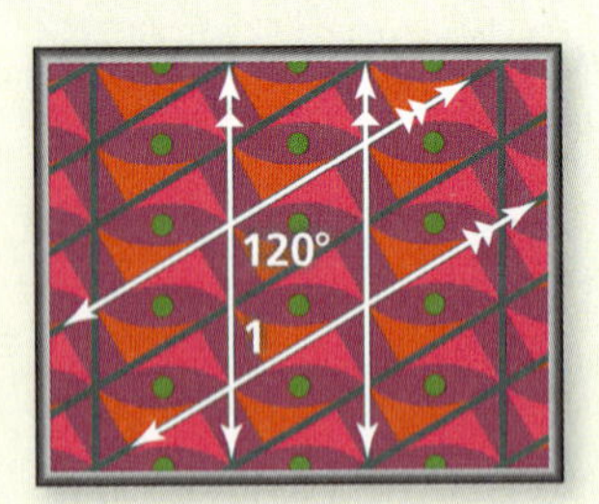

60°

Example 3

? "Are the two dashed lines parallel? How do you know?" Students may not be sure that they are parallel. Explain that because all the letters are slanted at an 80° angle, the lines are parallel.

- It is helpful to label other angles around $\angle 1$. For example:

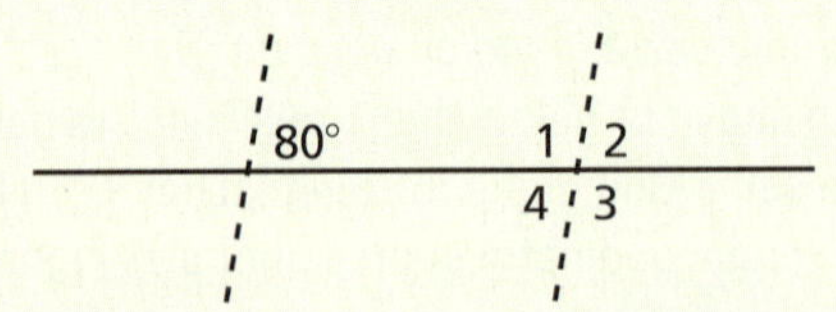

? "What angle is congruent to the 80° angle?" $\angle 2$

? "How can you find the measure of $\angle 1$?" Because $\angle 2$ and $\angle 1$ are supplementary and $\angle 2$ is congruent to the 80° angle, $\angle 1 = 180° - \angle 2 = 180° - 80° = 100°$.

English Language Learners

Vocabulary and Symbols

Make sure that students understand that the arrowhead marks *on* the lines indicate that the lines or line segments are parallel.

EXAMPLE 2 Using Corresponding Angles

Use the figure to find the measures of the numbered angles.

∠1: ∠1 and the 75° angle are vertical angles. They are congruent.

So, the measure of ∠1 is 75°.

∠2 and ∠3: The 75° angle is supplementary to both ∠2 and ∠3.

$75^\circ + \angle 2 = 180^\circ$	Definition of supplementary angles
$\angle 2 = 105^\circ$	Subtract 75° from each side.

So, the measures of ∠2 and ∠3 are 105°.

∠4, ∠5, ∠6, and ∠7: Using corresponding angles, the measures of ∠4 and ∠6 are 75°, and the measures of ∠5 and ∠7 are 105°.

On Your Own

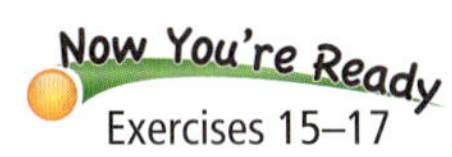

3. Use the figure to find the measures of the numbered angles.

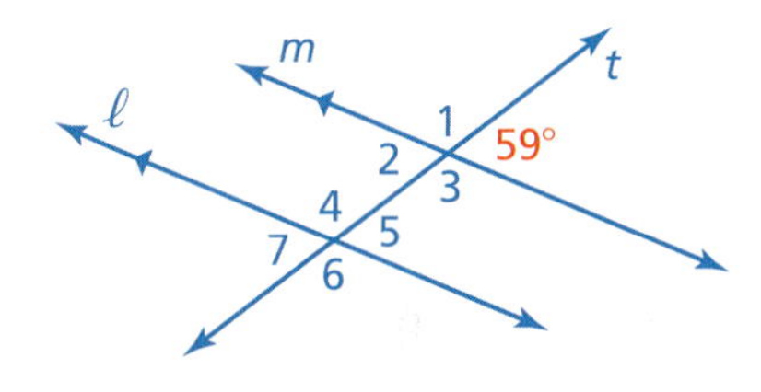

When two parallel lines are cut by a transversal, four **interior angles** are formed on the inside of the parallel lines and four **exterior angles** are formed on the outside of the parallel lines.

∠3, ∠4, ∠5, and ∠6 are interior angles.
∠1, ∠2, ∠7, and ∠8 are exterior angles.

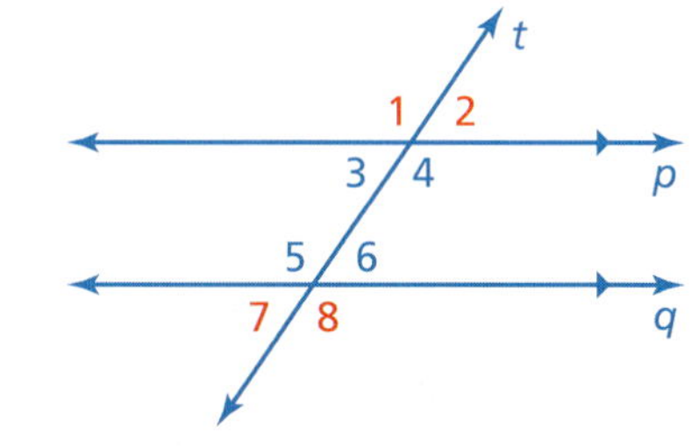

EXAMPLE 3 Using Corresponding Angles

A store owner uses pieces of tape to paint a window advertisement. The letters are slanted at an 80° angle. What is the measure of ∠1?

Ⓐ 80° Ⓑ 100° Ⓒ 110° Ⓓ 120°

Because all the letters are slanted at an 80° angle, the dashed lines are parallel. The piece of tape is the transversal.

Using corresponding angles, the 80° angle is congruent to the angle that is supplementary to ∠1, as shown.

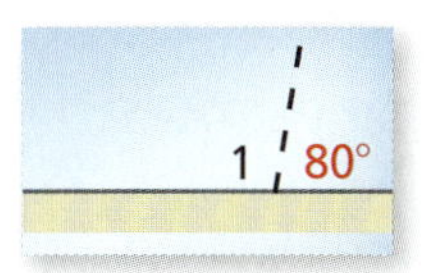

The measure of ∠1 is $180^\circ - 80^\circ = 100^\circ$. The correct answer is Ⓑ.

On Your Own

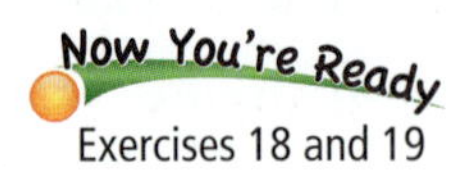

Exercises 18 and 19

4. **WHAT IF?** In Example 3, the letters are slanted at a 65° angle. What is the measure of ∠1?

Key Idea

Study Tip

Alternate interior angles and alternate exterior angles lie on opposite sides of the transversal.

Alternate Interior Angles and Alternate Exterior Angles

When a transversal intersects parallel lines, alternate interior angles are congruent and alternate exterior angles are congruent.

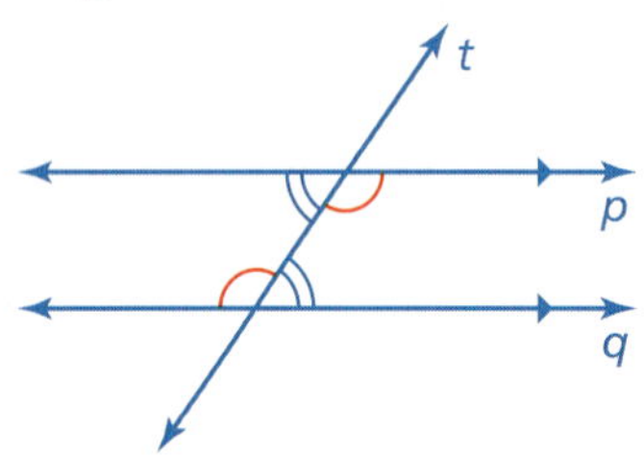

Alternate interior angles

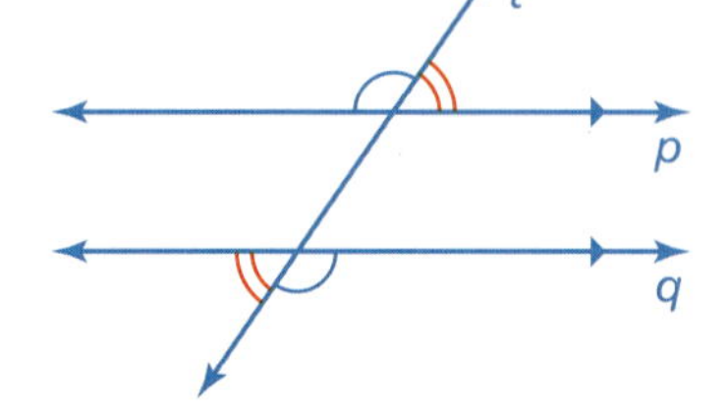

Alternate exterior angles

EXAMPLE 4 Identifying Alternate Interior and Alternate Exterior Angles

The photo shows a portion of an airport. Describe the relationship between each pair of angles.

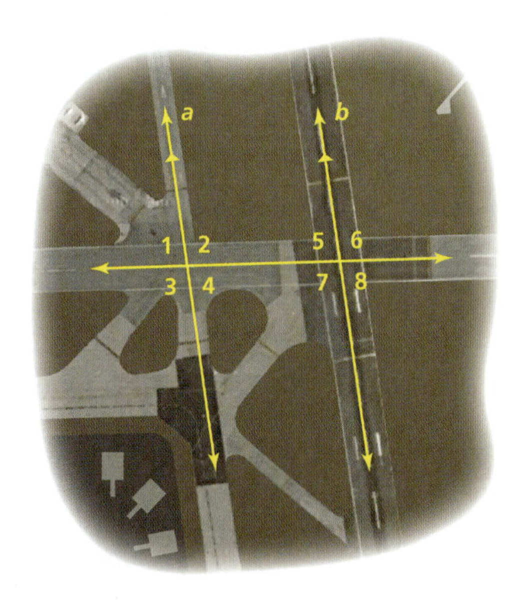

a. ∠3 and ∠6

∠3 and ∠6 are alternate exterior angles.

So, ∠3 is congruent to ∠6.

b. ∠2 and ∠7

∠2 and ∠7 are alternate interior angles.

So, ∠2 is congruent to ∠7.

On Your Own

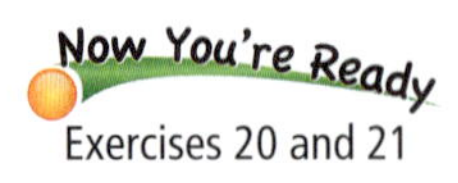

Exercises 20 and 21

In Example 4, the measure of ∠4 is 84°. Find the measure of the angle. Explain your reasoning.

5. ∠3

6. ∠5

7. ∠6

Laurie's Notes

On Your Own

- **Think-Pair-Share:** Students should read the question independently and then work in pairs to answer the question. When they have answered the question, the pair should compare their answer with another group and discuss any discrepancies.

Key Idea

- Write the *Key Idea*. Identify the angles which are marked congruent in the diagram. Mention the *Study Tip*.

Example 4

? "Are the lines parallel? How do you know?" yes; yellow arrow marks

- Work through the explanation as shown. This example helps students identify these new angle pairs.
- **Note:** There is a great deal of vocabulary in this section, so students may need extra practice. It is also important not to draw the parallel lines in the same orientation all of the time, particularly horizontal and vertical.

On Your Own

- Draw the diagram on the board. When students have finished, ask volunteers to come to the board to record their answers and explain their reasoning.
- **Question 7:** Students can say $\angle 6 = 96°$ because of alternate exterior angles *or* because it is the supplement of $\angle 8$.

Closure

- **Exit Ticket:** Find the measure of each angle. Explain your reasoning.

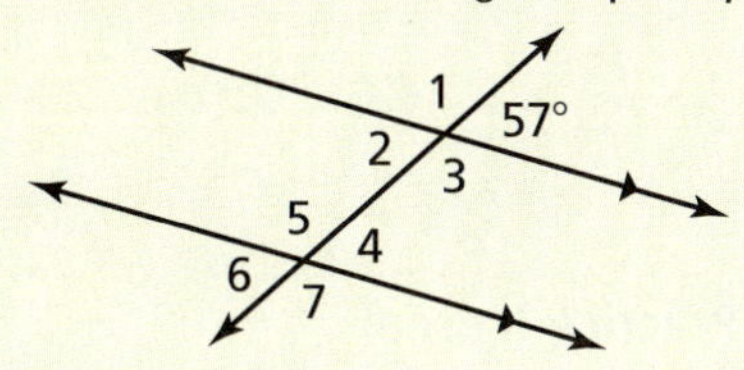

$\angle 1 = 123°$, $\angle 2 = 57°$, $\angle 3 = 123°$, $\angle 4 = 57°$, $\angle 5 = 123°$, $\angle 6 = 57°$, $\angle 7 = 123°$

On Your Own

4. 115°

Extra Example 4

Describe the relationship between each pair of angles.

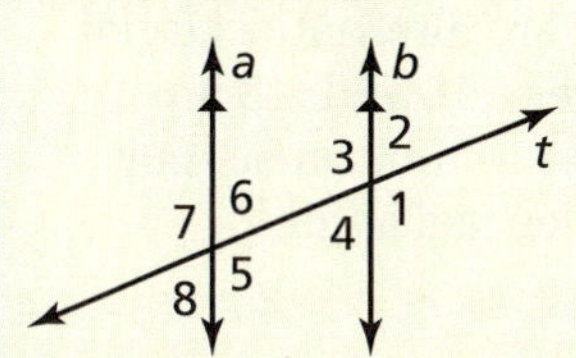

a. $\angle 1$ and $\angle 7$ $\angle 1$ and $\angle 7$ are alternate exterior angles. So, $\angle 1$ is congruent to $\angle 7$.

b. $\angle 3$ and $\angle 5$ $\angle 3$ and $\angle 5$ are alternate interior angles. So, $\angle 3$ is congruent to $\angle 5$.

On Your Own

5. 96°; $\angle 3$ and $\angle 4$ are supplementary.
6. 84°; Alternate interior angles are congruent.
7. 96°; $\angle 5$ and $\angle 6$ are supplementary.

Vocabulary and Concept Check

1. *Sample answer:*

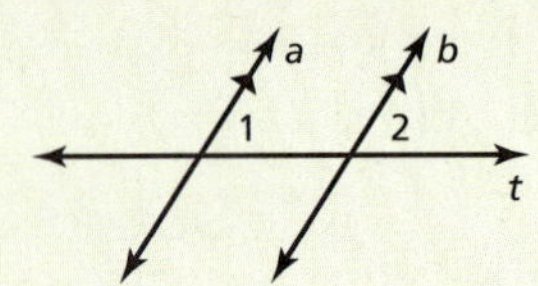

2. "The measure of $\angle 5$" doesn't belong because $\angle 2$, $\angle 6$, and $\angle 8$ are congruent and $\angle 5$ is not a corresponding, alternate interior, or alternate exterior angle with the other three angles. $\angle 2$ and $\angle 8$ are congruent because they are alternate exterior angles. $\angle 6$ and $\angle 8$ are congruent because they are vertical angles.

Practice and Problem Solving

3. *m* and *n*
4. *t*
5. 8
6. $\angle 5$, $\angle 7$, $\angle 1$, and $\angle 3$ are congruent. $\angle 8$, $\angle 6$, $\angle 4$, and $\angle 2$ are congruent.
7. $\angle 1 = 107°$, $\angle 2 = 73°$
8. $\angle 3 = 95°$, $\angle 4 = 85°$
9. $\angle 5 = 49°$, $\angle 6 = 131°$
10. The two lines are not parallel, so $\angle 5 \neq \angle 6$.
11. 60°; Corresponding angles are congruent.
12. *Sample answer:* Railroad tracks are parallel, and the out of bounds lines on a football field are parallel.

Assignment Guide and Homework Check

Level	Day 1 Activity Assignment	Day 2 Lesson Assignment	Homework Check
Basic	3–6, 31–35	1, 2, 7–11, 13–23 odd, 24	8, 15, 19, 24
Average	3–6, 31–35	1, 2, 7–27 odd, 10, 24	7, 15, 19, 24
Advanced	1–6, 10, 14–24 even, 25–35		14, 16, 22, 26, 28

Common Errors

- **Exercises 7–9** Students may mix up some of the definitions of congruent angles and find incorrect angle measures. Encourage them to look at the Key Ideas and color-code the figure they are given to determine what angles are congruent.
- **Exercise 11** Students may not realize that the line in front of the cars is the transversal. Remind them that lines are infinite and can be extended. Draw a diagram of the parallel parking spaces to help students visualize that $\angle 1$ and $\angle 2$ are corresponding angles.

3.1 Record and Practice Journal

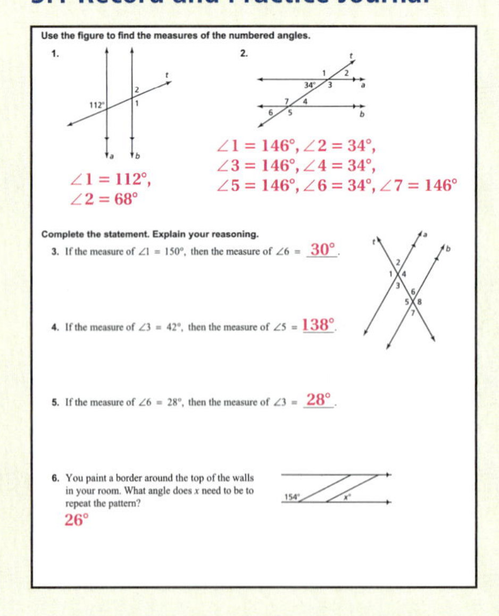

Use the figure to find the measures of the numbered angles.

1. $\angle 1 = 112°$, $\angle 2 = 68°$

2. $\angle 1 = 146°$, $\angle 2 = 34°$, $\angle 3 = 146°$, $\angle 4 = 34°$, $\angle 5 = 146°$, $\angle 6 = 34°$, $\angle 7 = 146°$

Complete the statement. Explain your reasoning.

3. If the measure of $\angle 1 = 150°$, then the measure of $\angle 6 =$ 30°.
4. If the measure of $\angle 3 = 42°$, then the measure of $\angle 5 =$ 138°.
5. If the measure of $\angle 6 = 28°$, then the measure of $\angle 3 =$ 28°.
6. You paint a border around the top of the walls in your room. What angle does *x* need to be to repeat the pattern?
26°

3.1 Exercises

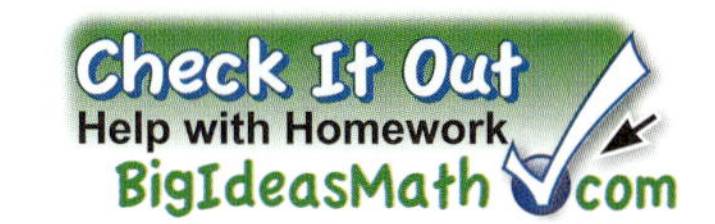

Vocabulary and Concept Check

1. **VOCABULARY** Draw two parallel lines and a transversal. Label a pair of corresponding angles.

2. **WHICH ONE DOESN'T BELONG?** Which statement does *not* belong with the other three? Explain your reasoning. Refer to the figure for Exercises 3–6.

The measure of $\angle 2$	The measure of $\angle 5$
The measure of $\angle 6$	The measure of $\angle 8$

Practice and Problem Solving

In Exercises 3–6, use the figure.

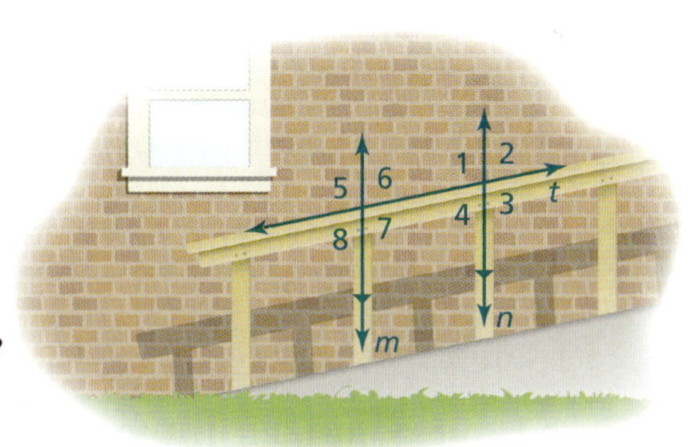

3. Identify the parallel lines.
4. Identify the transversal.
5. How many angles are formed by the transversal?
6. Which of the angles are congruent?

Use the figure to find the measures of the numbered angles.

7.

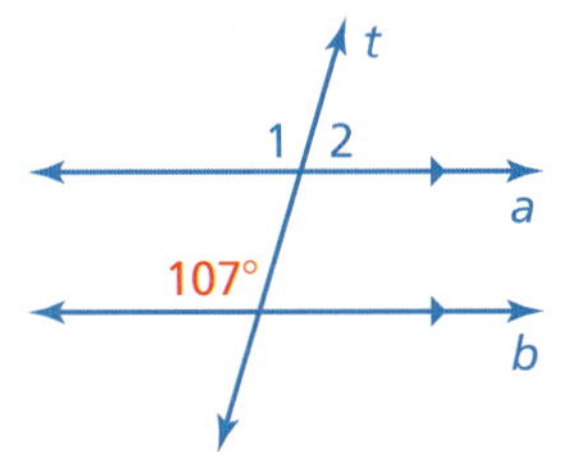

8.

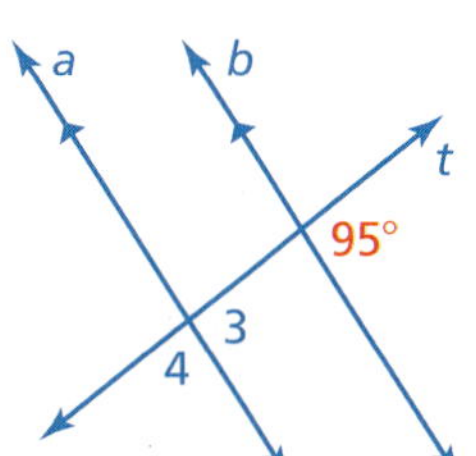

9.

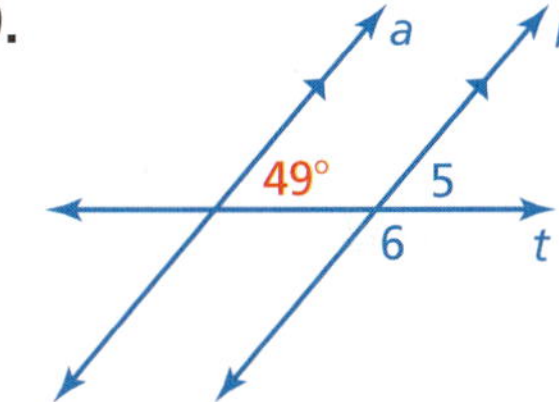

10. **ERROR ANALYSIS** Describe and correct the error in describing the relationship between the angles.

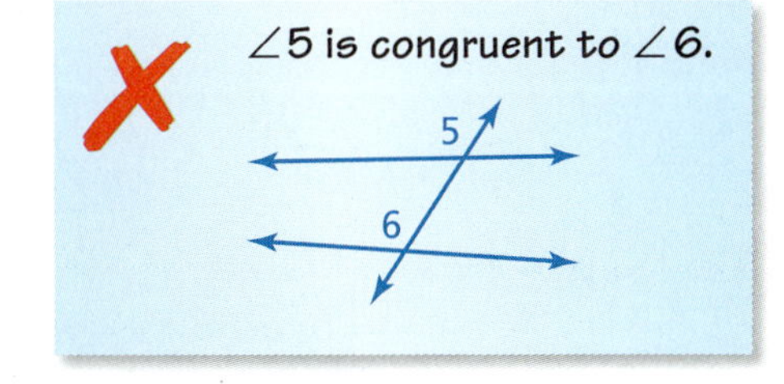

11. **PARKING** The painted lines that separate parking spaces are parallel. The measure of $\angle 1$ is 60°. What is the measure of $\angle 2$? Explain.

12. **OPEN-ENDED** Describe two real-life situations that use parallel lines.

13. **PROJECT** Trace line p and line t on a piece of paper. Label $\angle 1$. Move the paper so that $\angle 1$ aligns with $\angle 8$. Describe the transformations that you used to show that $\angle 1$ is congruent to $\angle 8$.

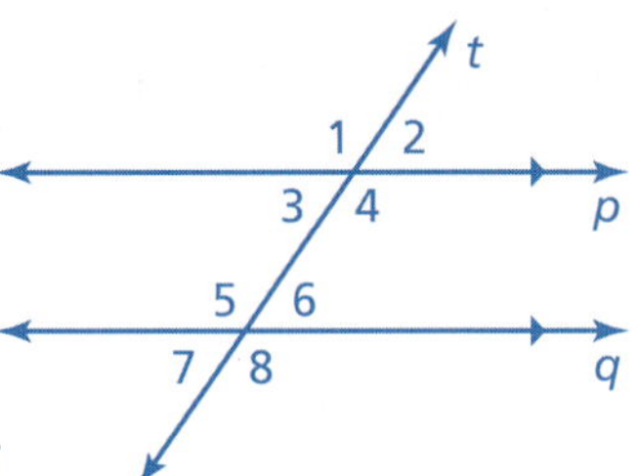

14. **REASONING** Two horizontal lines are cut by a transversal. What is the least number of angle measures you need to know in order to find the measure of every angle? Explain your reasoning.

Use the figure to find the measures of the numbered angles. Explain your reasoning.

2 15.

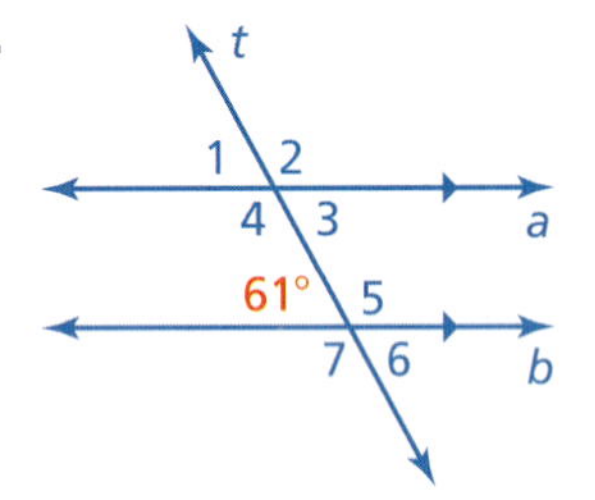

16.

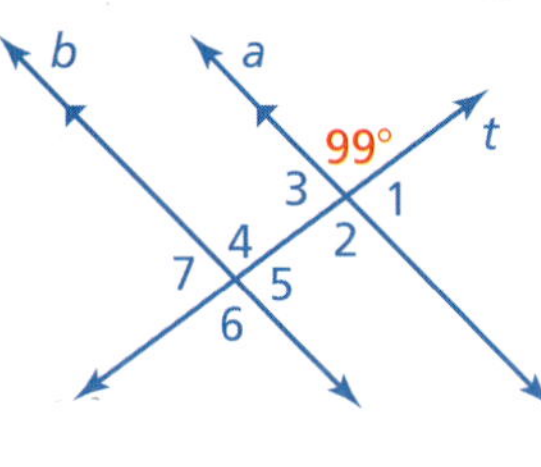

17.

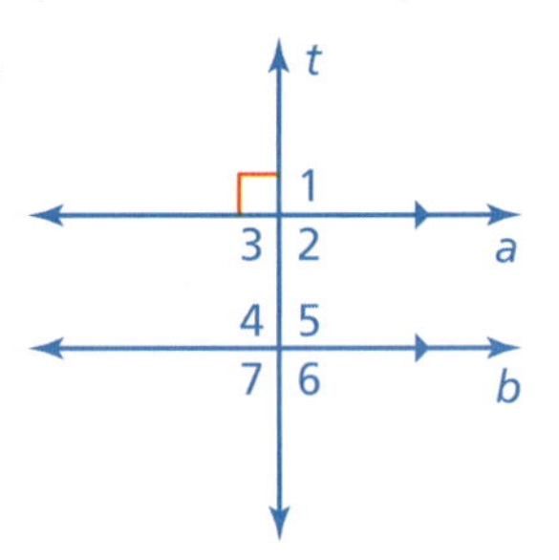

Complete the statement. Explain your reasoning.

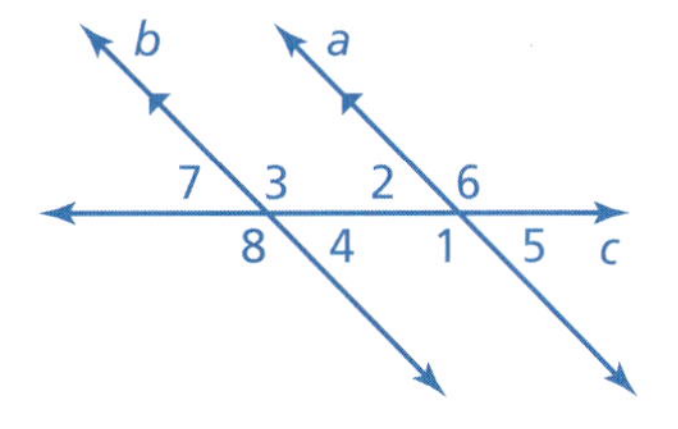

3 18. If the measure of $\angle 1 = 124°$, then the measure of $\angle 4 =$ ▢.

19. If the measure of $\angle 2 = 48°$, then the measure of $\angle 3 =$ ▢.

4 20. If the measure of $\angle 4 = 55°$, then the measure of $\angle 2 =$ ▢.

21. If the measure of $\angle 6 = 120°$, then the measure of $\angle 8 =$ ▢.

22. If the measure of $\angle 7 = 50.5°$, then the measure of $\angle 6 =$ ▢.

23. If the measure of $\angle 3 = 118.7°$, then the measure of $\angle 2 =$ ▢.

24. **RAINBOW** A rainbow forms when sunlight reflects off raindrops at different angles. For blue light, the measure of $\angle 2$ is 40°. What is the measure of $\angle 1$?

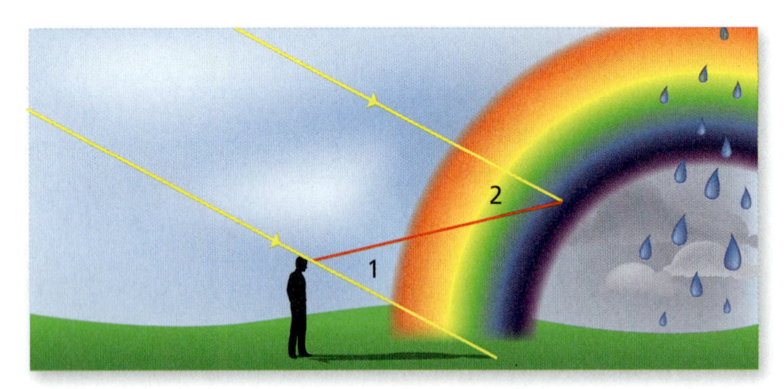

25. **REASONING** When a transversal is perpendicular to two parallel lines, all the angles formed measure 90°. Explain why.

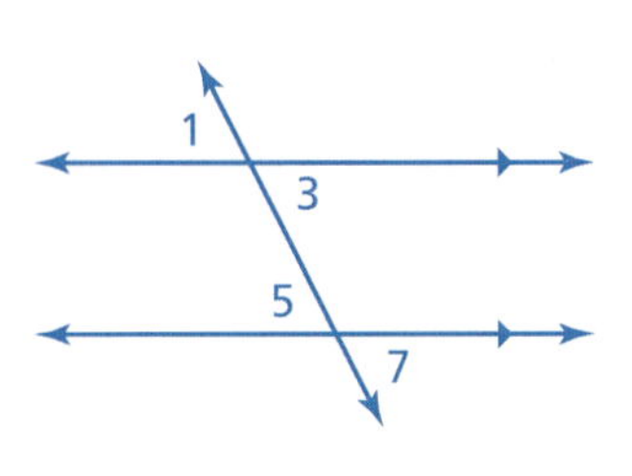

26. **LOGIC** Describe two ways you can show that $\angle 1$ is congruent to $\angle 7$.

Common Errors

- **Exercises 15–17** Students may not understand alternate interior and exterior angles and say that an exterior angle is congruent to the alternate interior angle. For example, in Exercise 15, a student may say the measure of $\angle 2$ is 61°. Use corresponding angles to show that this is not true.
- **Exercises 18–23** Students may use some of the definitions of congruent angles incorrectly in finding the angle measure of the unknown angle. Review the definitions and give an example with an adequate explanation of how to find the missing angle.
- **Exercises 27 and 28** Students may only see one set of parallel lines and think that they cannot find the measure of the missing angle. Point out the small arrows that denote that two lines are parallel. Encourage them to find the measure of an angle that is near the missing angle and then rotate the figure to help them visualize how to solve for the missing angle.

Differentiated Instruction

Auditory

Students may confuse the measures of *complementary angles* and *supplementary angles*. Show students that *complementary* comes before *supplementary* in the dictionary and that *90* comes before *180* numerically. So, the sum of complementary angles is 90° and the sum of supplementary angles is 180°.

13. *Sample answer:* rotate 180° and translate down.

14. You only need one angle because half of the angles are congruent to that angle and you can find the other angles using relationships.

15. $\angle 6 = 61°$; $\angle 6$ and the given angle are vertical angles.
$\angle 5 = 119°$ and $\angle 7 = 119°$; $\angle 5$ and $\angle 7$ are supplementary to the given angle.
$\angle 1 = 61°$; $\angle 1$ and the given angle are corresponding angles.
$\angle 3 = 61°$; $\angle 1$ and $\angle 3$ are vertical angles.
$\angle 2 = 119°$ and $\angle 4 = 119°$; $\angle 2$ and $\angle 4$ are supplementary to $\angle 1$.

16. $\angle 2 = 99°$; $\angle 2$ and the given angle are vertical angles.
$\angle 1 = 81°$ and $\angle 3 = 81°$; $\angle 1$ and $\angle 3$ are supplementary to the given angle.
$\angle 4 = 99°$; $\angle 2$ and $\angle 4$ are alternate interior angles.
$\angle 5 = 81°$ and $\angle 7 = 81°$; $\angle 5$ and $\angle 7$ are supplementary to $\angle 4$.
$\angle 6 = 99°$; $\angle 6$ and the given angle are alternate exterior angles.

17–26. See Additional Answers.

Practice and Problem Solving

27. 130

28. 115

29. **a.** no; They look like they are spreading apart.

b. Check students' work.

30. See *Taking Math Deeper*.

Fair Game Review

31. 13 **32.** 14

33. 51 **34.** 3

35. B

Mini-Assessment

Use the figure to find the measures of the numbered angles.

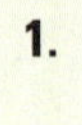

1.

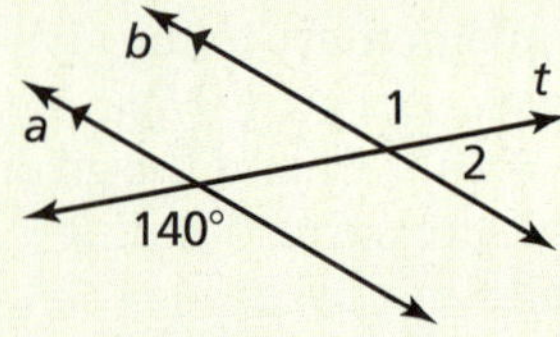

$\angle 1 = 140°$; $\angle 2 = 40°$

2.

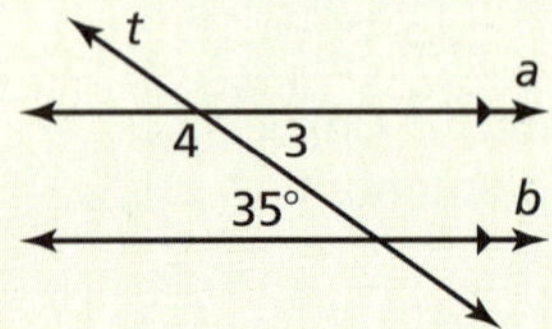

$\angle 3 = 35°$; $\angle 4 = 145°$

3.

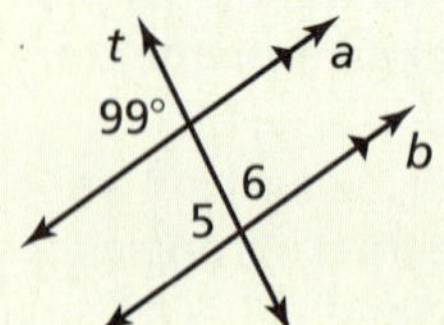

$\angle 5 = 99°$; $\angle 6 = 81°$

4.

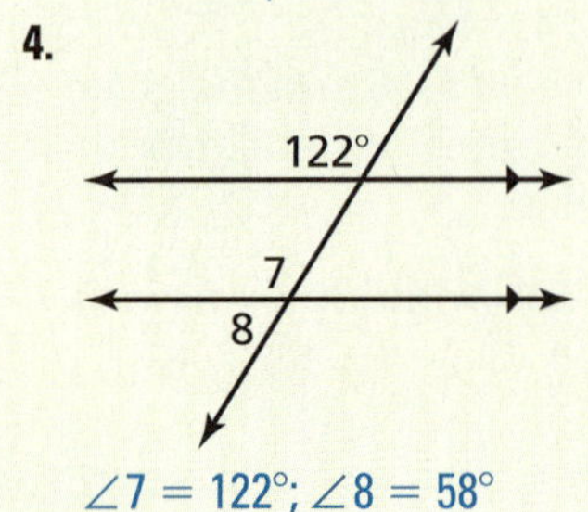

$\angle 7 = 122°$; $\angle 8 = 58°$

Taking Math Deeper

Exercise 30

This problem uses a well-known reflective property in physics. This property applies to mirrors, billiard tables, air hockey tables, and many other objects. The property states that when the hockey puck bounces off the side board, its out-going angle is equal to its in-coming angle.

Solve for m.

$$m + m + 64 = 180$$
$$2m = 116$$
$$m = 58$$

2 Solve for x.

a. Using the property of alternate interior angles, you can determine that $x = 64$.

3 Answer the question.

b. The goal is slightly wider than the hockey puck. So, there is some leeway allowed for the measure of x.

By studying the diagram, you can see that x cannot be much greater. However, x can be a little less and still have the hockey puck go into the goal.

Project

Write a report about at least two other games that use angles as part of their strategy.

Reteaching and Enrichment Strategies

If students need help. . .	If students got it. . .
Resources by Chapter • Practice A and Practice B • Puzzle Time Record and Practice Journal Practice Differentiating the Lesson Lesson Tutorials Skills Review Handbook	Resources by Chapter • Enrichment and Extension • Technology Connection Start the next section

CRITICAL THINKING Find the value of x.

27.

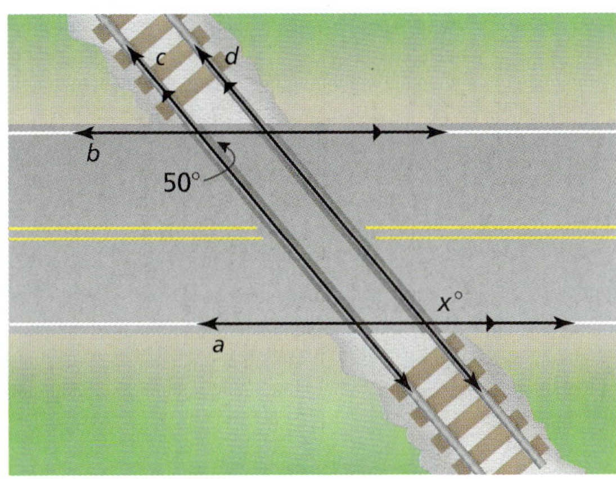

28.

29. OPTICAL ILLUSION Refer to the figure.

a. Do the horizontal lines appear to be parallel? Explain.

b. Draw your own optical illusion using parallel lines.

30. Geometry The figure shows the angles used to make a double bank shot in an air hockey game.

a. Find the value of x.

b. Can you still get the red puck in the goal when x is increased by a little? by a lot? Explain.

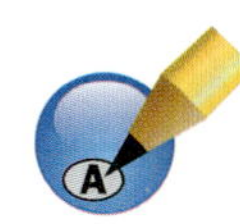

Fair Game Review What you learned in previous grades & lessons

Evaluate the expression. *(Skills Review Handbook)*

31. $4 + 3^2$ **32.** $5(2)^2 - 6$ **33.** $11 + (-7)^2 - 9$ **34.** $8 \div 2^2 + 1$

35. MULTIPLE CHOICE The triangles are similar. What length does x represent? *(Section 2.5)*

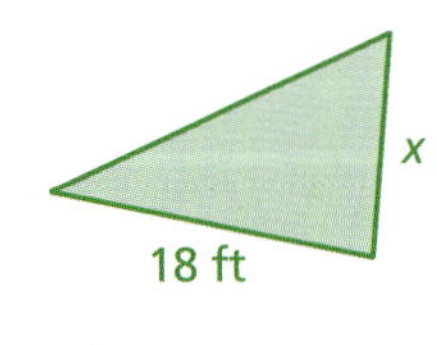

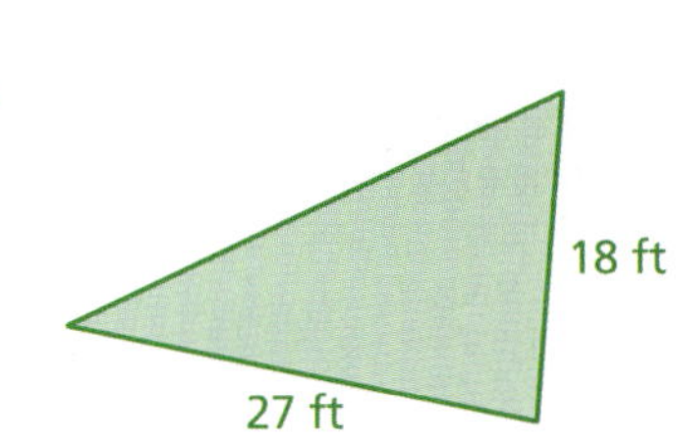

Ⓐ 2 ft Ⓑ 12 ft

Ⓒ 15 ft Ⓓ 27 ft

3.2 Angles of Triangles

Essential Question How can you describe the relationships among the angles of a triangle?

1 ACTIVITY: Exploring the Interior Angles of a Triangle

Work with a partner.

a. Draw a triangle. Label the interior angles A, B, and C.

b. Carefully cut out the triangle. Tear off the three corners of the triangle.

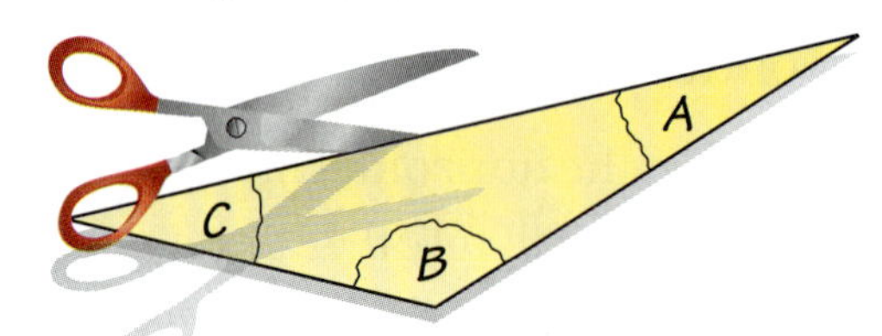

c. Arrange angles A and B so that they share a vertex and are adjacent.

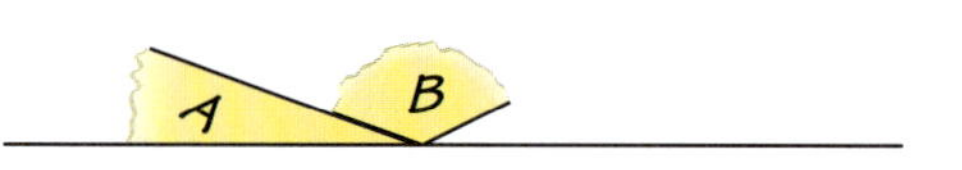

d. How can you place the third angle to determine the sum of the measures of the interior angles? What is the sum?

e. Compare your results with those of others in your class.

f. **STRUCTURE** How does your result in part (d) compare to the rule you wrote in Lesson 1.1, Activity 2?

2 ACTIVITY: Exploring the Interior Angles of a Triangle

Work with a partner.

a. Describe the figure.

b. **LOGIC** Use what you know about parallel lines and transversals to justify your result in part (d) of Activity 1.

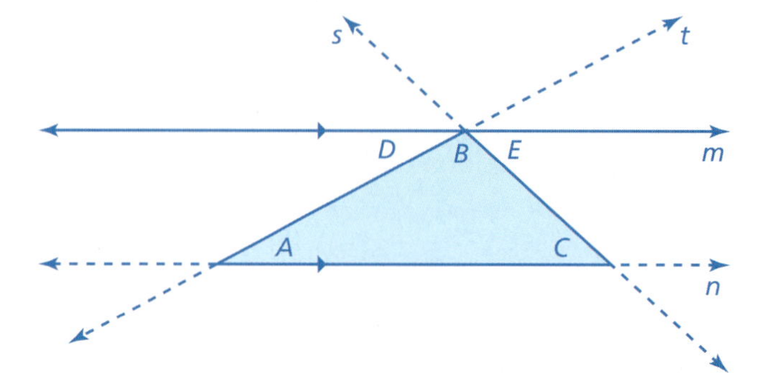

Geometry

In this lesson, you will

- understand that the sum of the interior angle measures of a triangle is 180°.
- find the measures of interior and exterior angles of triangles.

Learning Standard
MACC.8.G.1.5

Laurie's Notes

Introduction

Standards for Mathematical Practice

- **MP3 Construct Viable Arguments and Critique the Reasoning of Others:** Mathematically proficient students are able to make conjectures and construct arguments to explain their reasoning.

Motivate

- Make teams of three students. Give them 3 minutes to make a list of as many words as they can that begin with the prefix *tri-*.
- Some examples are: triangle, triathlon, tricycle, tri-fold, triangulate, triad, triaxial, trilogy, trimester, trinary, trinity, trio, trilingual, trillium.
- Provide dictionaries if necessary.
- The goal of this activity is to demonstrate that *tri-* is a common prefix.

Activity Notes

Activity 1

- Many students will have heard of the property they are investigating today. Having heard the property and internalizing it for all triangles are two different levels of knowledge.
- The sides of the triangle must be straight; otherwise the three angles will not lie adjacent to one another when placed about a point on the line.
- **Teaching Tip:** If you cannot gain access to enough pairs of scissors, you can cut out several triangles in advance using a paper cutter. It is okay to have multiple copies of the same triangle because different pairs of students will get one copy of the triangle.
- The conclusion, or rule, that students should discover is that the angle measures of any triangle will sum to 180°.
- **Management Tip:** There will be torn pieces of scrap paper resulting from this investigation. To help keep the room clean, cluster 4–6 desks together in a circle and tape a recycled paper or plastic bag to the front edge of one of the desks. Students are expected to put scraps of paper in the bags when they are finished with the investigation.

Activity 2

- **MP3** and **MP6 Attend to Precision:** In this activity, you want to hear students make statements based upon evidence they have. Instead of, "Angles *C* and *E* are the same measure because they look it," students should say, "Angles *C* and *E* are the same measure because lines *m* and *n* are parallel and angles *C* and *E* are alternate interior angles."
- If students are stuck, ask them to think about what they learned in the last lesson. They should mark the diagram with what they know.
- Students should make the connection that angles *B*, *D*, and *E* are the same as the three angles they placed about a point in Activity 1, where their conjecture was that the sum of the three angle measures is 180°. What is different in this activity is that students are asked to *justify* their answers.

Florida Common Core Standards

MACC.8.G.1.5 Use informal arguments to establish facts about the angle sum and exterior angle of triangles, ...

Previous Learning

Students should know basic vocabulary associated with angles and triangles.

3.2 Record and Practice Journal

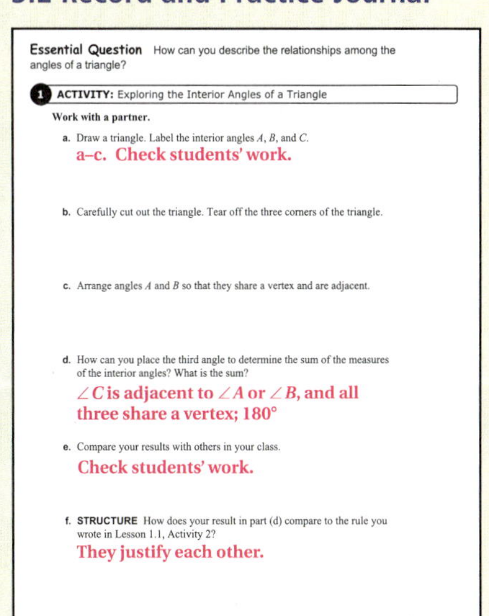

Essential Question How can you describe the relationships among the angles of a triangle?

1 ACTIVITY: Exploring the Interior Angles of a Triangle

Work with a partner.

a. Draw a triangle. Label the interior angles *A*, *B*, and *C*.
a–c. Check students' work.

b. Carefully cut out the triangle. Tear off the three corners of the triangle.

c. Arrange angles *A* and *B* so that they share a vertex and are adjacent.

d. How can you place the third angle to determine the sum of the measures of the interior angles? What is the sum?
$\angle C$ is adjacent to $\angle A$ or $\angle B$, and all three share a vertex; 180°

e. Compare your results with others in your class.
Check students' work.

f. STRUCTURE How does your result in part (d) compare to the rule you wrote in Lesson 1.1, Activity 2?
They justify each other.

Differentiated Instruction

Kinesthetic

When talking about right, acute, and obtuse angles of a triangle, ask students if it is possible to draw a triangle with 2 right angles. Students should see by drawing the two right angles with a common side that the remaining two sides of the right angles will never meet. So, no triangle can be formed with 2 right angles. Ask students if it is possible for a triangle to have 2 obtuse angles. Students should reach the same conclusion. No triangle can be formed with 2 obtuse angles.

3.2 Record and Practice Journal

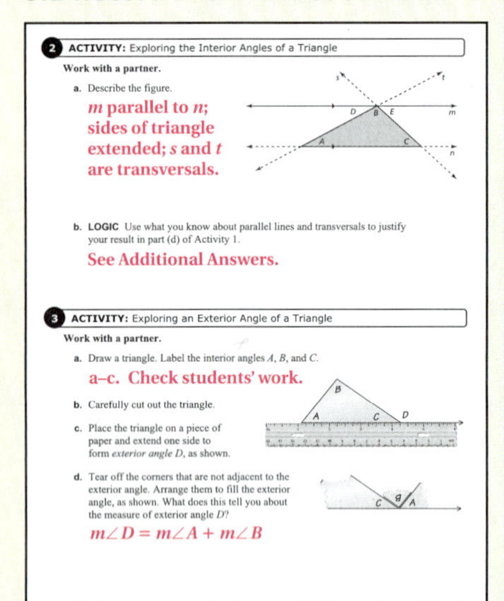

2 ACTIVITY: Exploring the Interior Angles of a Triangle

Work with a partner.

a. Describe the figure.

m parallel to *n*; sides of triangle extended; *s* and *t* are transversals.

b. **LOGIC** Use what you know about parallel lines and transversals to justify your result in part (d) of Activity 1.

See Additional Answers.

3 ACTIVITY: Exploring an Exterior Angle of a Triangle

Work with a partner.

a. Draw a triangle. Label the interior angles *A*, *B*, and *C*.

a–c. Check students' work.

b. Carefully cut out the triangle.

c. Place the triangle on a piece of paper and extend one side to form *exterior angle D*, as shown.

d. Tear off the corners that are not adjacent to the exterior angle. Arrange them to fill the exterior angle, as shown. What does this tell you about the measure of exterior angle *D*?

$m\angle D = m\angle A + m\angle B$

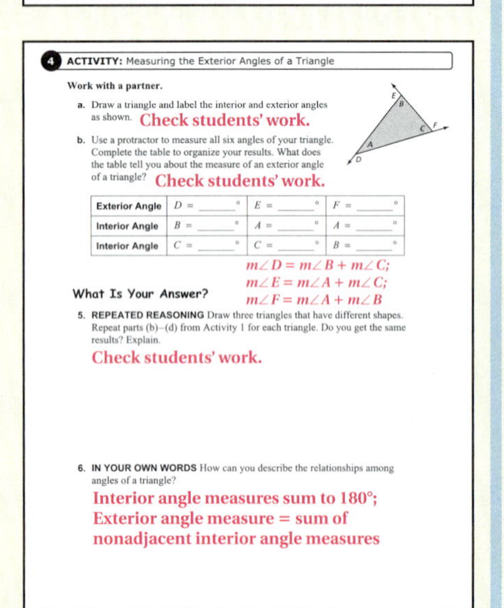

4 ACTIVITY: Measuring the Exterior Angles of a Triangle

Work with a partner.

a. Draw a triangle and label the interior and exterior angles as shown. Check students' work.

b. Use a protractor to measure all six angles of your triangle. Complete the table to organize your results. What does the table tell you about the measure of an exterior angle of a triangle? Check students' work.

Exterior Angle	$D =$ ____°	$E =$ ____°	$F =$ ____°
Interior Angle	$B =$ ____°	$A =$ ____°	$A =$ ____°
Interior Angle	$C =$ ____°	$C =$ ____°	$B =$ ____°

$m\angle D = m\angle B + m\angle C$;
$m\angle E = m\angle A + m\angle C$;
$m\angle F = m\angle A + m\angle B$

What Is Your Answer?

5. **REPEATED REASONING** Draw three triangles that have different shapes. Repeat parts (b)–(d) from Activity 1 for each triangle. Do you get the same results? Explain.

Check students' work.

6. **IN YOUR OWN WORDS** How can you describe the relationships among angles of a triangle?

Interior angle measures sum to 180°; Exterior angle measure = sum of nonadjacent interior angle measures

Laurie's Notes

Activity 3

- **Teaching Tip:** When asked to draw a triangle, suggest to students that they draw a scalene triangle. Draw the triangle large enough so that it is easy to see and measure.
- **Teaching Tip:** If gaining access to sufficient pairs of scissors is a problem, have pre-cut triangles available for students to use.
- If students are unclear about the direction "extend one side" in part (c), explain that they should draw a ray and place the triangle against the ray, as shown. The extended ray is one side of angle *D* and the second side of angle *D* is a side of the triangle *ABC*.
- **MP3:** Ask students to summarize their findings. Students are making a conjecture about an exterior angle of a triangle.

Activity 4

- Students should draw a large, scalene triangle.
- Students should extend the sides of the triangle in order to form one exterior angle at each vertex.
- **FYI:** In the diagram at the right, Angles 1 and 2 are exterior angles for interior angle *A*. There are two exterior angles at each vertex. These are congruent, vertical angles. It does not matter which of the two angles students draw and measure.

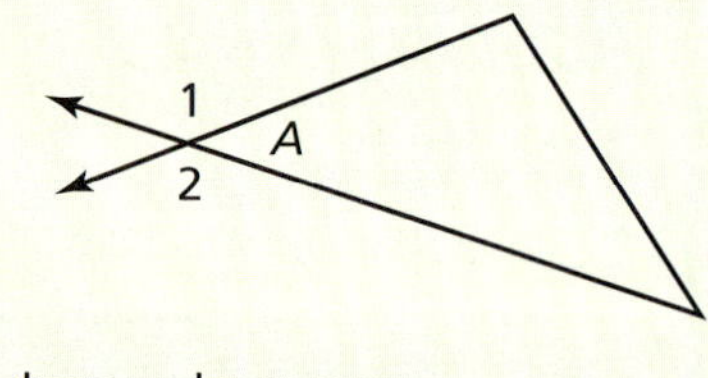

? "What are the interior angles of the triangle?" angles *A*, *B*, and *C*

? "What are the exterior angles of the triangle?" angles *D*, *E*, and *F*

? "What do you notice about the measure of an exterior angle of a triangle?" It is the same as the sum of the two non adjacent interior angles.

- **MP6 Attend to Precision:** Encourage students to be precise with their language. They may say that the exterior angle is the sum of two interior angles. Which two? How do they describe the two angles?
- **Extension:** Students may also observe that the sum of the exterior angle measures is 360°. Compare results with different groups of students. Is this true for the triangles drawn by all the students?

What Is Your Answer?

- **MP3:** Students should understand that conjectures need to be verified. Repeating an investigation on a different type of triangle gives additional evidence, or in some cases, it might become a counter-example.

Closure

- Draw the triangle shown and label the two angles. Find the measure of angles 1 and 2. 56°, 146°

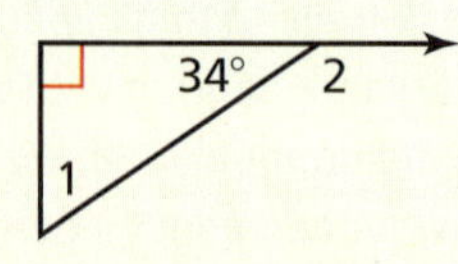

3 ACTIVITY: Exploring an Exterior Angle of a Triangle

Math Practice 8

Maintain Oversight

Do you think your conclusion will be true for the exterior angle of any triangle? Explain.

Work with a partner.

a. Draw a triangle. Label the interior angles *A*, *B*, and *C*.

b. Carefully cut out the triangle.

c. Place the triangle on a piece of paper and extend one side to form *exterior angle D*, as shown.

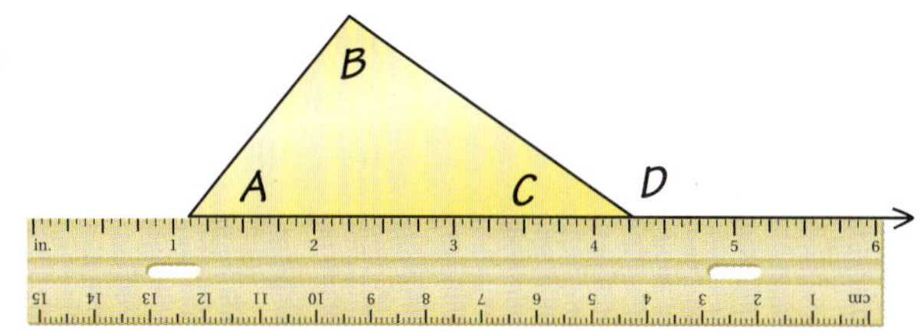

d. Tear off the corners that are not adjacent to the exterior angle. Arrange them to fill the exterior angle, as shown. What does this tell you about the measure of exterior angle *D*?

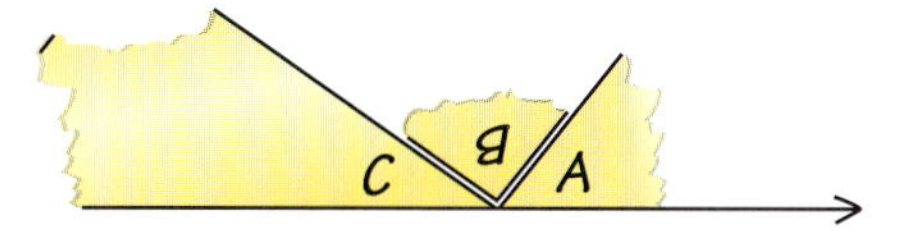

4 ACTIVITY: Measuring the Exterior Angles of a Triangle

Work with a partner.

a. Draw a triangle and label the interior and exterior angles, as shown.

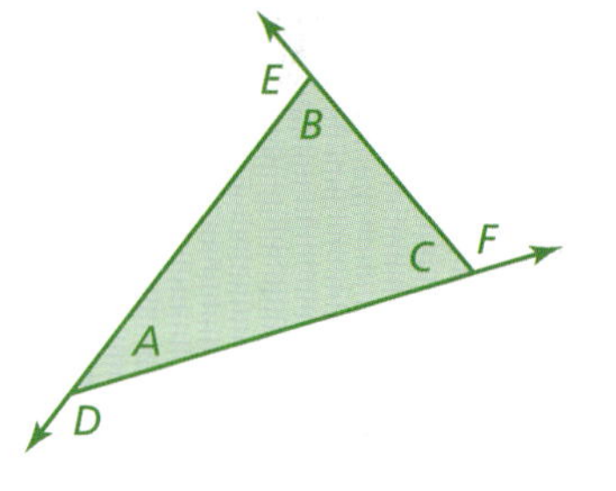

b. Use a protractor to measure all six angles. Copy and complete the table to organize your results. What does the table tell you about the measure of an exterior angle of a triangle?

Exterior Angle	$D = \square°$	$E = \square°$	$F = \square°$
Interior Angle	$B = \square°$	$A = \square°$	$A = \square°$
Interior Angle	$C = \square°$	$C = \square°$	$B = \square°$

What Is Your Answer?

5. **REPEATED REASONING** Draw three triangles that have different shapes. Repeat parts (b)–(d) from Activity 1 for each triangle. Do you get the same results? Explain.

6. **IN YOUR OWN WORDS** How can you describe the relationships among angles of a triangle?

Use what you learned about angles of a triangle to complete Exercises 4–6 on page 114.

3.2 Lesson

Key Vocabulary
interior angles of a polygon, *p. 112*
exterior angles of a polygon, *p. 112*

The angles inside a polygon are called **interior angles**. When the sides of a polygon are extended, other angles are formed. The angles outside the polygon that are adjacent to the interior angles are called **exterior angles**.

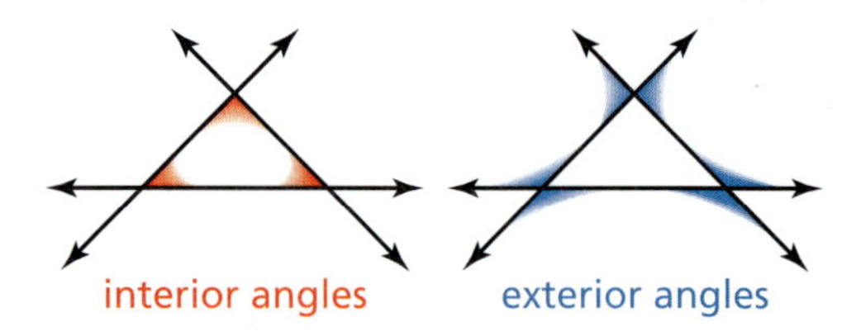

Key Idea

Interior Angle Measures of a Triangle

Words The sum of the interior angle measures of a triangle is 180°.

Algebra $x + y + z = 180$

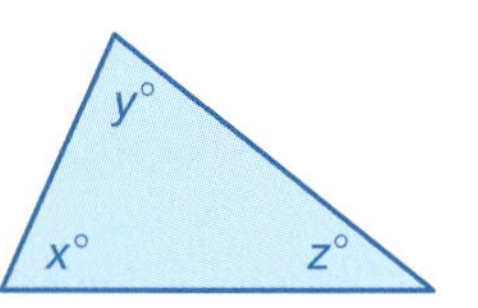

EXAMPLE 1 Using Interior Angle Measures

Find the value of x.

a.

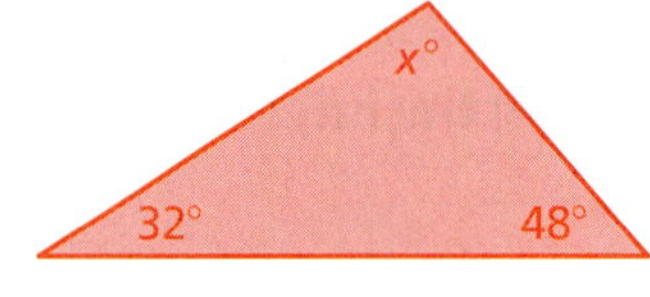

$$x + 32 + 48 = 180$$
$$x + 80 = 180$$
$$x = 100$$

b.

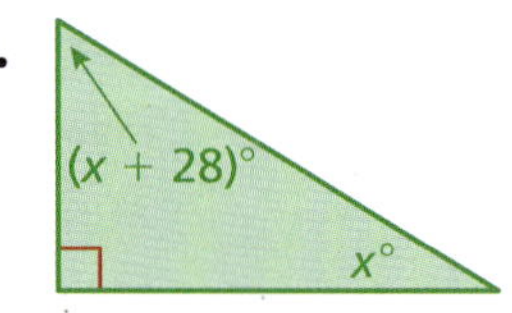

$$x + (x + 28) + 90 = 180$$
$$2x + 118 = 180$$
$$2x = 62$$
$$x = 31$$

On Your Own

Find the value of x.

1.

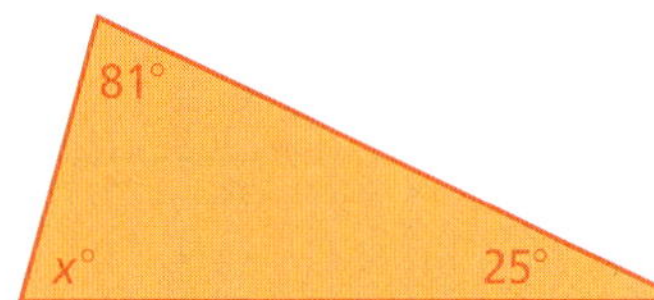

2.

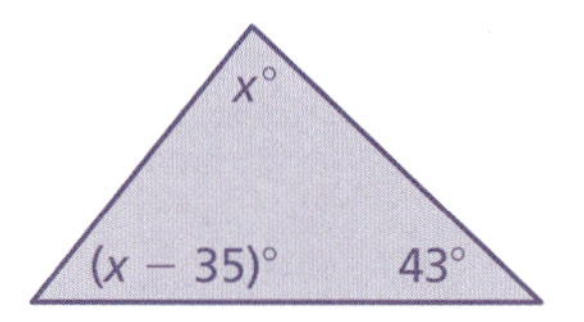

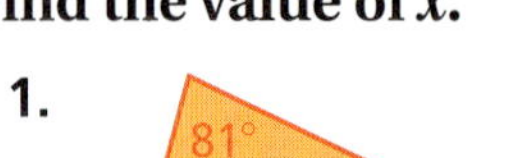

Key Idea

Exterior Angle Measures of a Triangle

Words The measure of an exterior angle of a triangle is equal to the sum of the measures of the two nonadjacent interior angles.

Algebra $z = x + y$

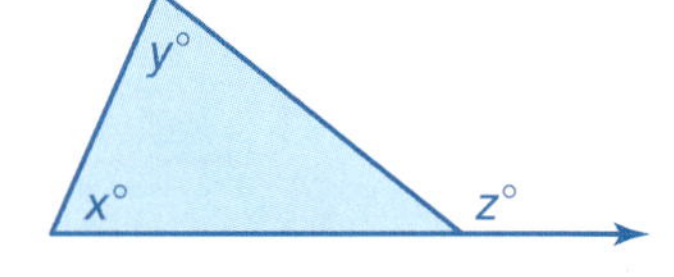

Laurie's Notes

Introduction

Connect

- **Yesterday:** Students explored the sum of the angle measures of a triangle and the vocabulary associated with triangles. (MP3, MP6)
- **Today:** Students will find the missing angle measure of a triangle and classify the triangle.

Motivate

- Discuss the Ohio State flag.
  The blue triangle represents hills and valleys. The red and white stripes represent roads and waterways. The 13 leftmost stars represent the 13 original colonies. The 4 stars on the right bring the total to 17, representing that Ohio was the 17th state admitted to the Union.

? "Do you know any other flags that contain a triangle?" Answers will vary.

Lesson Notes

Discuss

- Discuss interior and exterior angles of a triangle. Tell students that there are six exterior angles of a triangle, 2 at each vertex.

? "What is the relationship between the two exterior angles at each vertex?" They are congruent, because they are vertical angles.

Key Idea

- The property is written with variables to suggest that you can solve an equation to find the third angle when you know the other two angles. This is also called the *Triangle Sum Theorem*.

? "What type of angles are the remaining angles of a right triangle? a triangle with an obtuse angle?" Both are acute.

? "Do you think an obtuse triangle could have a right angle? Explain." no; The sum of the angle measures would be greater than 180°.

Example 1

- Some students may argue that all they need to do is add the angle measures and subtract from 180. Remind them that they are practicing a *process*, one that works when the three angle measures are given as algebraic expressions, such as $(x + 10)°$, $(x + 20)°$, and $(x + 30)°$.

Key Idea

- This is also called the *Exterior Angle Theorem*, and the two nonadjacent interior angles are also called the *remote interior angles*.

? "If the interior angle of a triangle is acute, what do you know about the exterior angle at that vertex?" obtuse and supplement of the acute angle

? "Could an exterior angle of a triangle be acute? Explain." yes; if the interior angle is obtuse

Goal Today's lesson is finding interior and exterior angle measures of triangles.

Technology for the Teacher

Dynamic Classroom

Lesson Tutorials
Lesson Plans
Answer Presentation Tool

Extra Example 1

Find the value of *x*.

a. Triangle with angles $x°$, $31°$, $42°$

107

b.

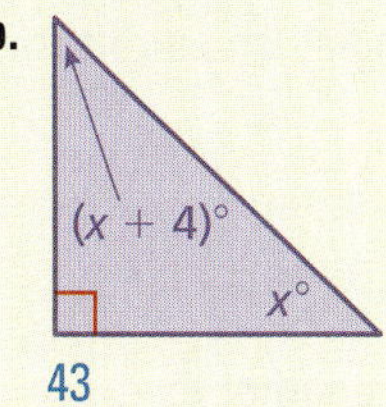

43

On Your Own

1. 74
2. 86

Extra Example 2

Find the measure of the exterior angle.

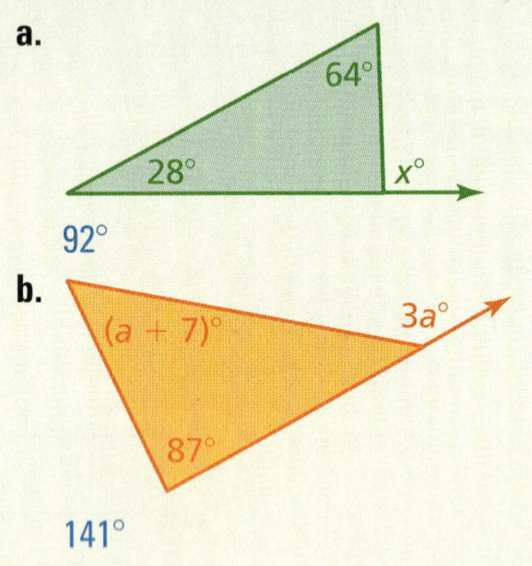

a. 92°

b. 141°

Extra Example 3

A car travels around the park shown below. What is the value of x?

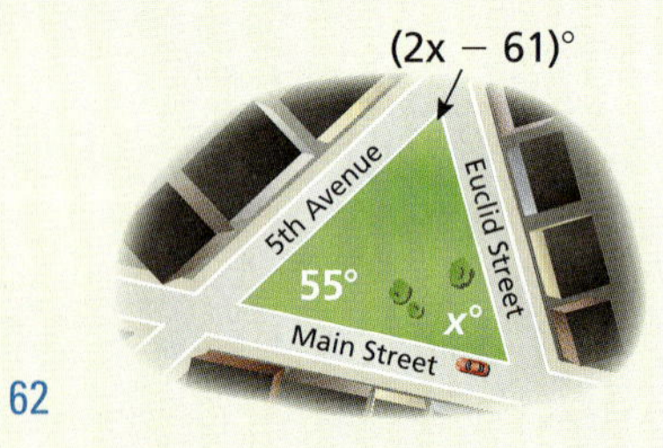

62

On Your Own

3. 70° **4.** 140°

5. 54.3

English Language Learners

Pair Activity

Create sets of index cards with the measures of three angles on each card, as shown below.

30°, 60°, 90°
40°, 50°, 60°
20°, 75°, 85°

Have students work together to determine whether the angles could be the angles of a triangle. If not, they should choose two of the angles and determine the measure of the third angle that would form a triangle.

Laurie's Notes

Example 2

- Write the problem in part (a).
- "How do you find the measure of the exterior angle?" The exterior angle will be the sum of the two remote interior angles, so 36° + 72°.
- **MP2 Reason Abstractly and Quantitatively:** "Explain two different ways to find the measure of the third angle of the triangle." Use the *Triangle Sum Theorem*, or find the supplement of the exterior angle.
- The problem in part (b) involves writing an equation versus doing mental math.
- Use the *Exterior Angle Theorem* to write the equation. Solve the equation as shown.
- **Common Error:** Students will solve for the variable a correctly and forget to answer the question asked, meaning they forget to substitute the value of a into the expression for the exterior angle.

Example 3

- Add a little interest by sharing information from the Department of the Navy website, *history.navy.mil*. The *Bermuda Triangle* is an imaginary area located off the southeastern Atlantic coast of the U.S. where a supposedly high incidence of unexplained disappearances of ships and aircraft occurs. The vertices of the triangle are Bermuda; Miami, Florida; and San Juan, Puerto Rico.
- Set up the equation and work through the problem as shown.
- This is a good review of equation solving and work with decimals.

On Your Own

- **Neighbor Check:** Have students work independently and then have their neighbors check their work. Have students discuss any discrepancies.

Closure

- **Exit Ticket:** Find the value of x. Then classify the triangle.

a. 75° x° 30°

75; acute triangle

b.

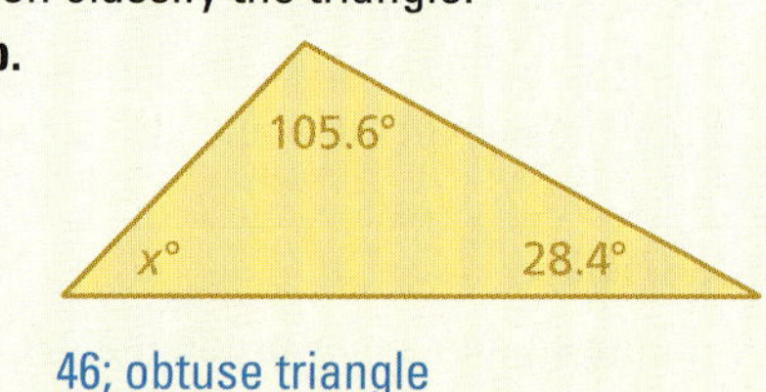

46; obtuse triangle

EXAMPLE 2 Finding Exterior Angle Measures

Study Tip

Each vertex has a pair of congruent exterior angles. However, it is common to show only one exterior angle at each vertex.

Find the measure of the exterior angle.

a.

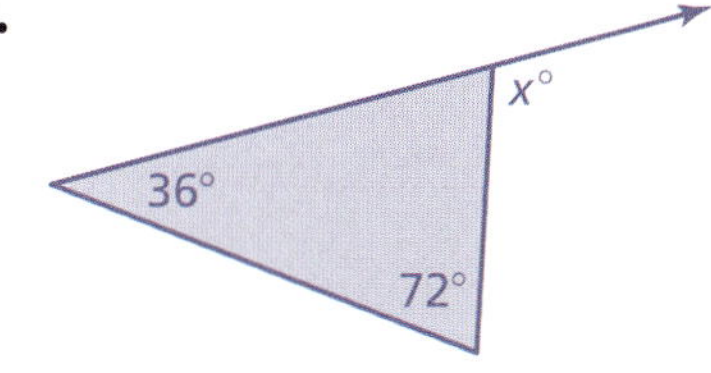

$x = 36 + 72$

$x = 108$

So, the measure of the exterior angle is 108°.

b.

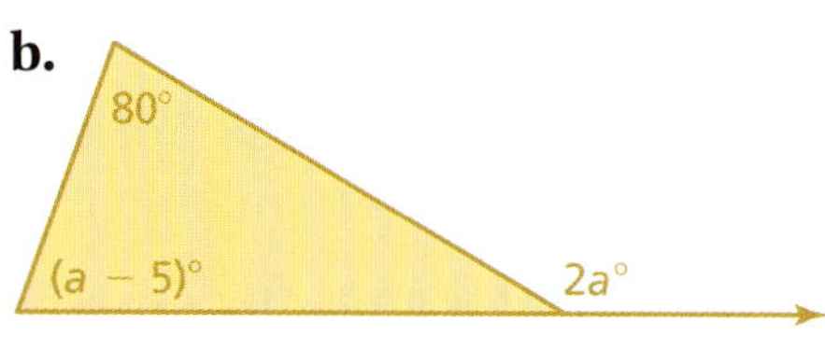

$2a = (a - 5) + 80$

$2a = a + 75$

$a = 75$

So, the measure of the exterior angle is $2(75)° = 150°$.

EXAMPLE 3 Real-Life Application

An airplane leaves from Miami and travels around the Bermuda Triangle. What is the value of x?

Ⓐ 26.8 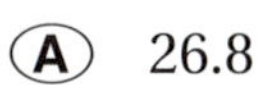 Ⓑ 27.2 Ⓒ 54 Ⓓ 64

Use what you know about the interior angle measures of a triangle to write an equation.

$x + (2x - 44.8) + 62.8 = 180$	Write equation.
$3x + 18 = 180$	Combine like terms.
$3x = 162$	Subtract 18 from each side.
$x = 54$	Divide each side by 3.

The value of x is 54. The correct answer is Ⓒ.

On Your Own

Now You're Ready
Exercises 12–14

Find the measure of the exterior angle.

3.

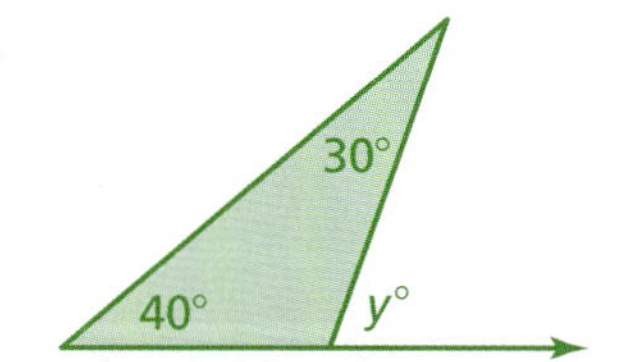

4.

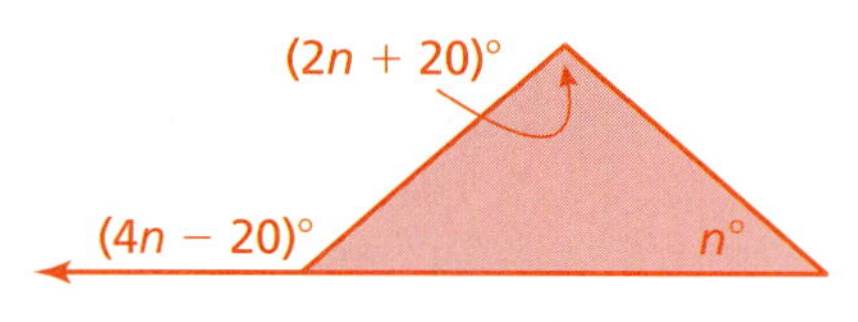

5. In Example 3, the airplane leaves from Fort Lauderdale. The interior angle measure at Bermuda is 63.9°. The interior angle measure at San Juan is $(x + 7.5)°$. Find the value of x.

3.2 Exercises

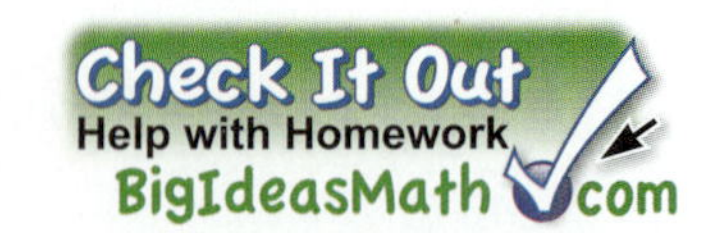

Vocabulary and Concept Check

1. **VOCABULARY** You know the measures of two interior angles of a triangle. How can you find the measure of the third interior angle?
2. **VOCABULARY** How many exterior angles does a triangle have at each vertex? Explain.
3. **NUMBER SENSE** List the measures of the exterior angles for the triangle shown at the right.

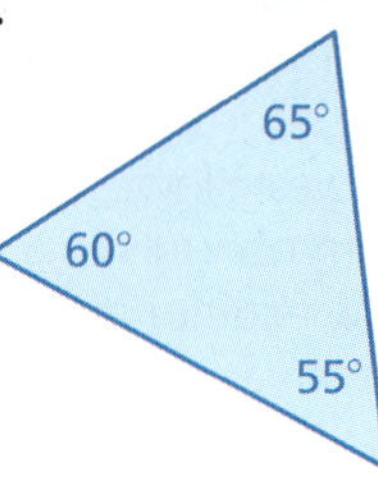

Practice and Problem Solving

Find the measures of the interior angles.

1 4.

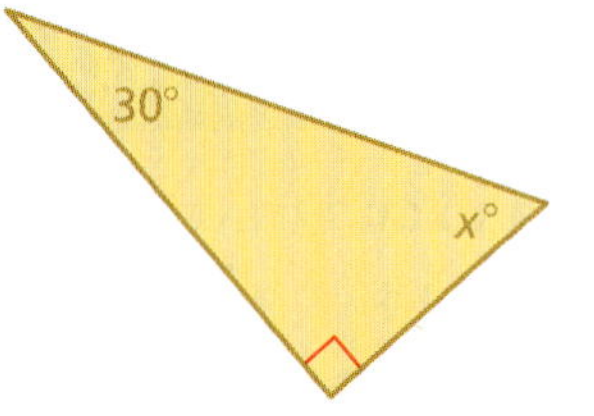

5.

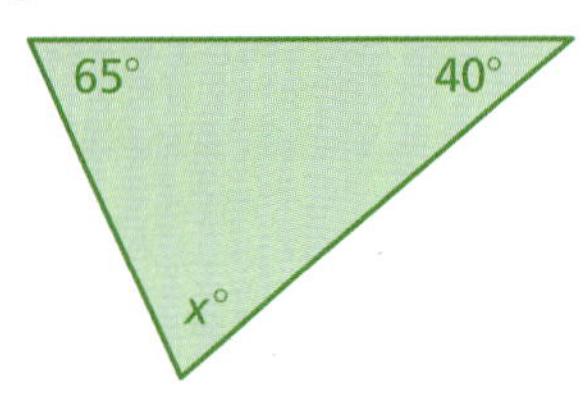

6.

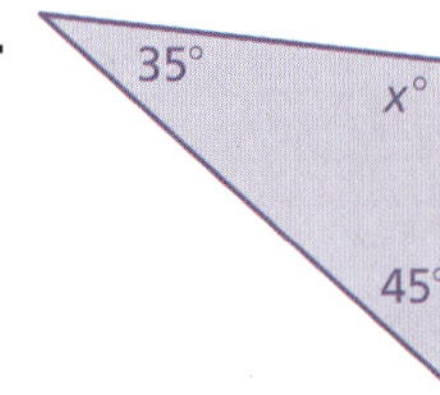

7.

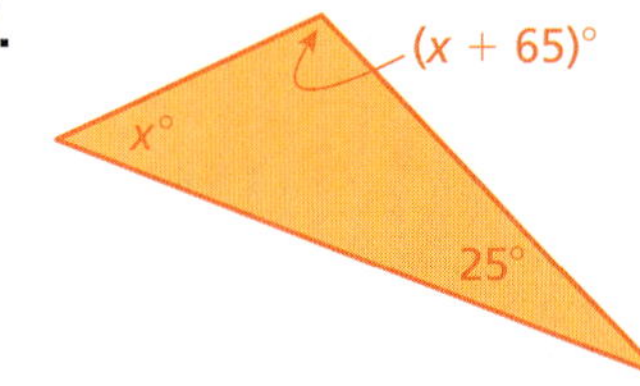

8.

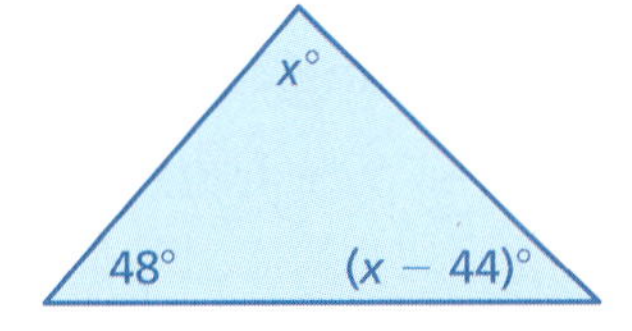

9.

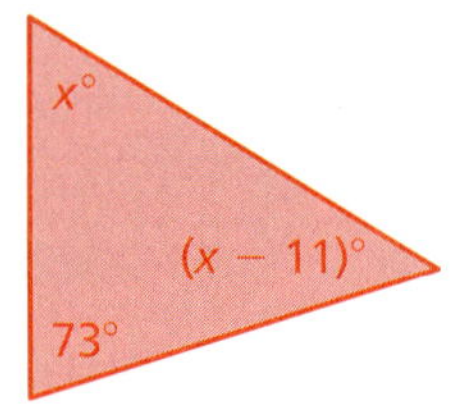

10. **BILLIARD RACK** Find the value of x in the billiard rack.

60°

$x°$ $x°$

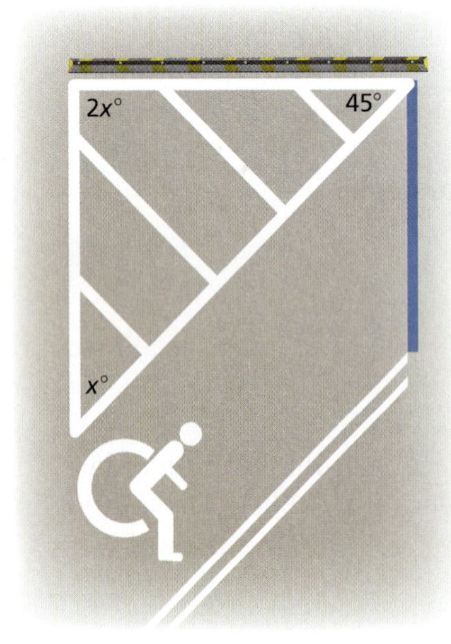

11. **NO PARKING** The triangle with lines through it designates a no parking zone. What is the value of x?

Assignment Guide and Homework Check

Level	Day 1 Activity Assignment	Day 2 Lesson Assignment	Homework Check
Basic	4–6, 23–26	1–3, 7–19 odd	7, 9, 11, 13, 17
Average	4–6, 23–26	1–3, 8–14 even, 15–21 odd	8, 10, 12, 14, 17
Advanced	1–6, 8–14 even, 15–26		8, 10, 14, 17, 18

Common Errors

- **Exercises 7–11** Students may forget to combine like terms when solving for x. Remind them that because there are two variables on the same side of the equal sign, they should start by combining like terms.
- **Exercises 7–9, 14, 17** Students may solve for the variable, but forget to find the measures of the angles. Remind them to read the directions carefully and to answer the question.

3.2 Record and Practice Journal

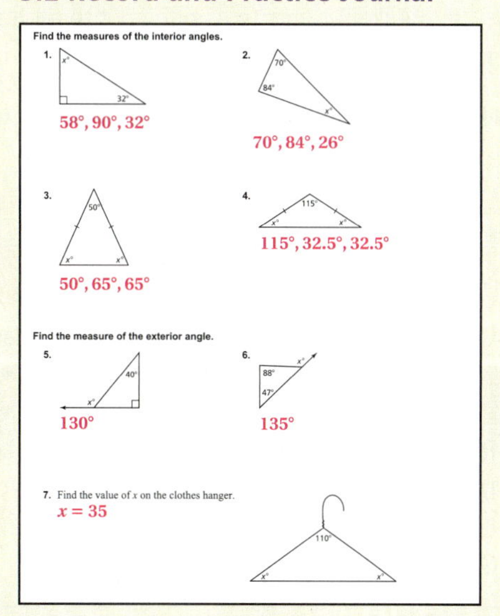
Find the measures of the interior angles.

1. $58°, 90°, 32°$
2. $70°, 84°, 26°$
3. $50°, 65°, 65°$
4. $115°, 32.5°, 32.5°$

Find the measure of the exterior angle.

5. $130°$
6. $135°$
7. Find the value of x on the clothes hanger.
$x = 35$

Vocabulary and Concept Check

1. Subtract the sum of the given measures from 180°.
2. 2; When the sides are extended, 2 angles are formed that are adjacent to the interior angle.
3. 115°, 120°, 125°

Practice and Problem Solving

4. 30°, 60°, 90°
5. 40°, 65°, 75°
6. 35°, 45°, 100°
7. 25°, 45°, 110°
8. 44°, 48°, 88°
9. 48°, 59°, 73°
10. 60
11. 45
12. 128°
13. 140°
14. 108°
15. The measure of the exterior angle is equal to the sum of the measures of the two nonadjacent interior angles. The sum of all three angles is not 180°;

$$(2x - 12) = x + 30$$
$$x = 42$$

The exterior angle is $(2(42) - 12)° = 72°$.

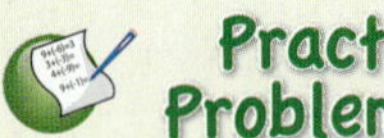

Practice and Problem Solving

16. See *Taking Math Deeper.*

17. 126°

18. no; The two nonadjacent interior angles could be any two angles that sum to 120°.

19–22. See Additional Answers.

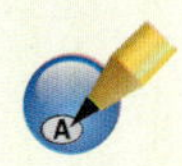

Fair Game Review

23. $x = -4$ **24.** $y = -1$

25. $n = -3$ **26.** A

Mini-Assessment

Find the value of *x*.

1.

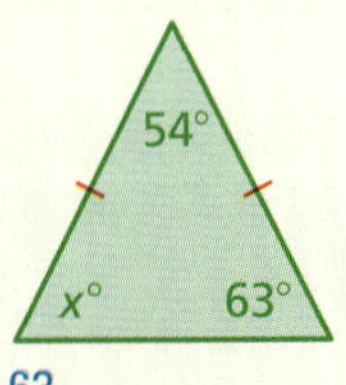

63

2.

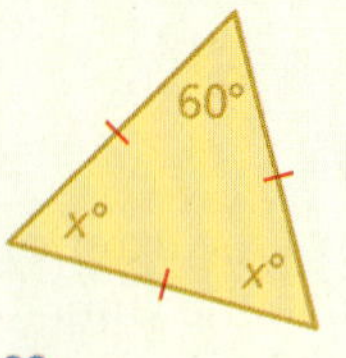

60

3.

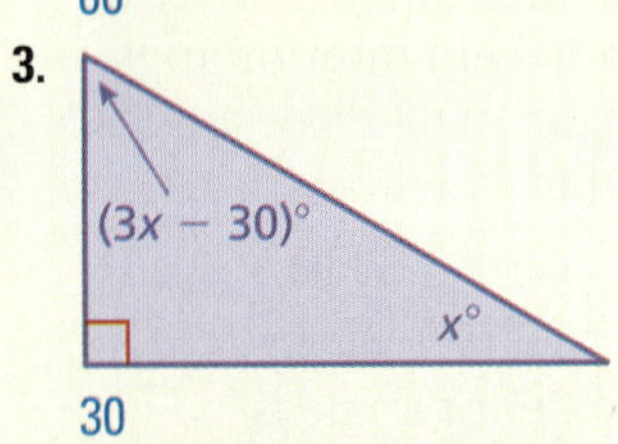

30

Find the measure of the exterior angle.

4. / **5.**

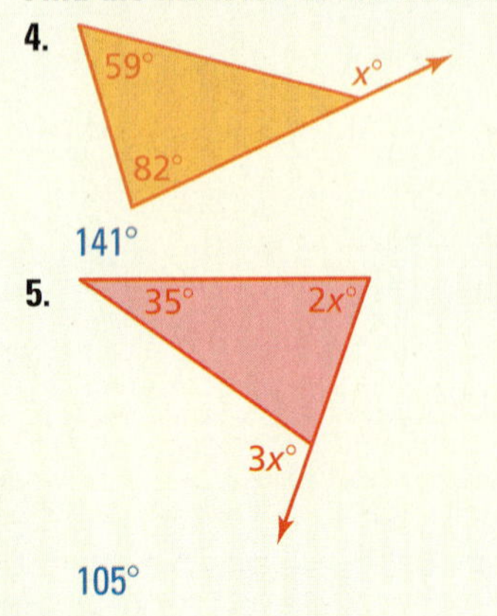

105°

Taking Math Deeper

Exercise 16

For this exercise, you can use a tape diagram to help visualize the relationship between the interior angle measures of the triangle.

Express the ratio 2 : 3 : 5 using a tape diagram.

Angle 1 — 2 parts

Angle 2 — 3 parts

Angle 3 — 5 parts

Interpret the tape diagram.

There are a total of $2 + 3 + 5 = 10$ parts, which represent the 180° in the triangle.

So, each part must represent $\frac{180}{10} = 18°$.

tape diagram

Answer the question.

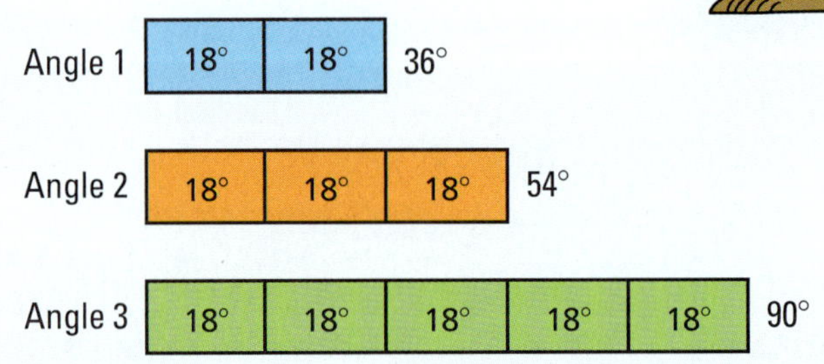

The interior angle measures of the triangle are 36°, 54°, and 90°.

Project

Tell students that the ratio of the interior angle measures of a triangle is $x : y : z$. Ask students to assign an integer between 1 and 10 to each variable, find the interior angle measures, and construct the triangle.

Reteaching and Enrichment Strategies

If students need help. . .	If students got it. . .
Resources by Chapter • Practice A and Practice B • Puzzle Time Record and Practice Journal Practice Differentiating the Lesson Lesson Tutorials Skills Review Handbook	Resources by Chapter • Enrichment and Extension • Technology Connection Start the next section

Find the measure of the exterior angle.

12\. 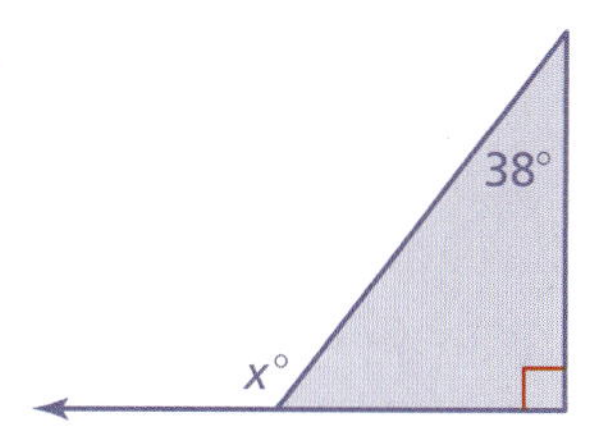

13\. 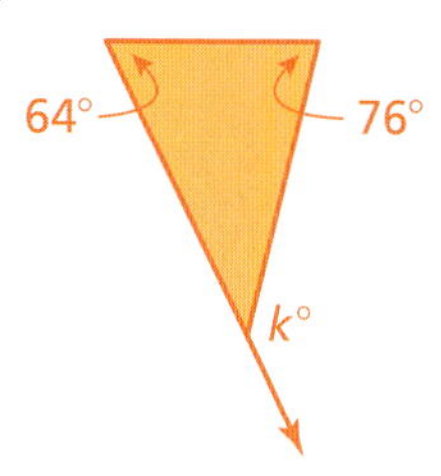

14\. 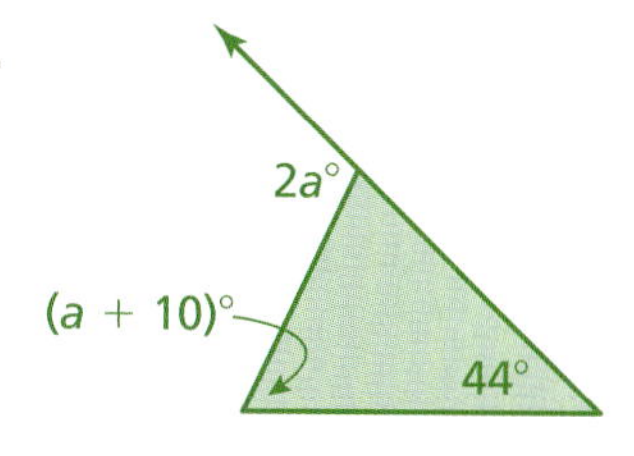

15\. **ERROR ANALYSIS** Describe and correct the error in finding the measure of the exterior angle.

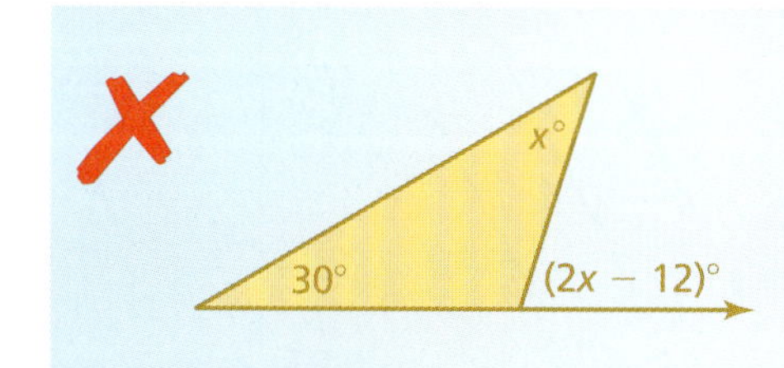

$(2x - 12) + x + 30 = 180$

$3x + 18 = 180$

$x = 54$

The exterior angle is $(2(54) - 12)° = 96°$.

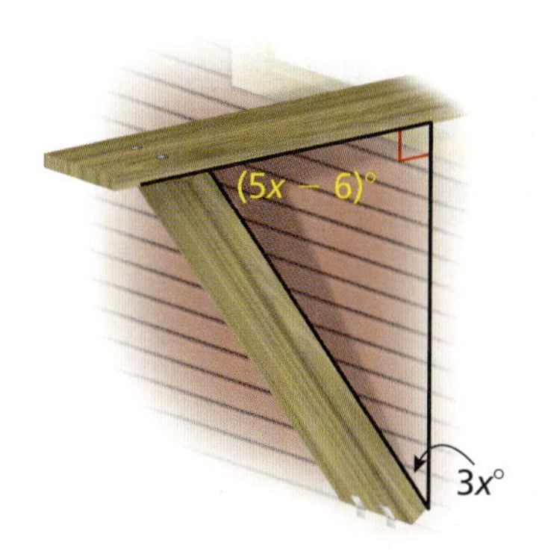

16\. **RATIO** The ratio of the interior angle measures of a triangle is 2 : 3 : 5. What are the angle measures?

17\. **CONSTRUCTION** The support for a window air-conditioning unit forms a triangle and an exterior angle. What is the measure of the exterior angle?

18\. **REASONING** A triangle has an exterior angle with a measure of 120°. Can you determine the measures of the interior angles? Explain.

Determine whether the statement is *always*, *sometimes*, or *never* true. Explain your reasoning.

19\. Given three angle measures, you can construct a triangle.

20\. The acute interior angles of a right triangle are complementary.

21\. A triangle has more than one vertex with an acute exterior angle.

22\. Precision Using the figure at the right, show that $z = x + y$. (*Hint:* Find two equations involving w.)

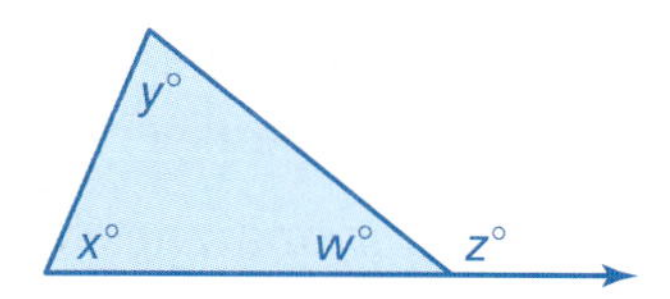

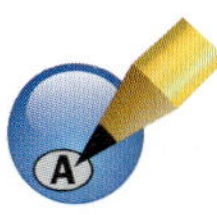

Fair Game Review What you learned in previous grades & lessons

Solve the equation. Check your solution. *(Section 1.2)*

23\. $-4x + 3 = 19$

24\. $2(y - 1) + 6y = -10$

25\. $5 + 0.5(6n + 14) = 3$

26\. **MULTIPLE CHOICE** Which transformation moves every point of a figure the same distance and in the same direction? *(Section 2.2)*

Ⓐ translation Ⓑ reflection Ⓒ rotation Ⓓ dilation

3 Study Help

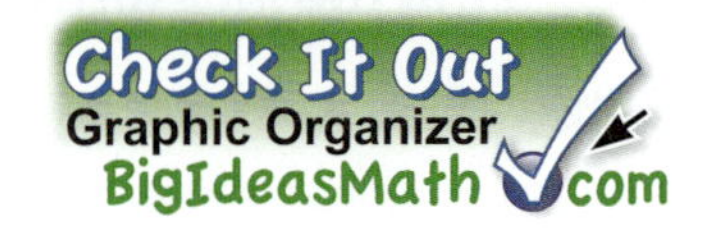

You can use an **example and non-example chart** to list examples and non-examples of a vocabulary word or item. Here is an example and non-example chart for transversals.

Transversals

Examples	Non-Examples
line p line q line r	line a line b line c
line a line b line c line d	line p line t

On Your Own

Make example and non-example charts to help you study these topics.

1. interior angles formed by parallel lines and a transversal
2. exterior angles formed by parallel lines and a transversal

After you complete this chapter, make example and non-example charts for the following topics.

3. interior angles of a polygon
4. exterior angles of a polygon
5. regular polygons
6. similar triangles

"What do you think of my example & non-example chart for popular cat toys?"

Sample Answers

1. Interior angles formed by parallel lines and a transversal

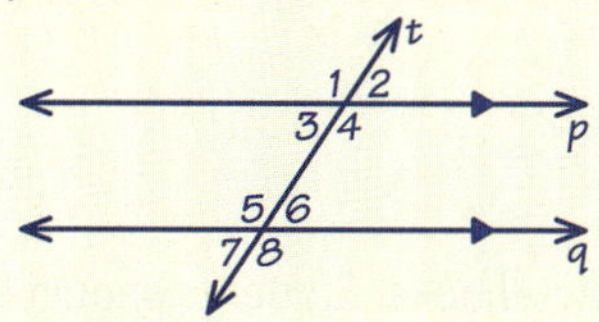

Examples	Non-Examples
∠3	∠1
∠4	∠2
∠5	∠7
∠6	∠8

2. Exterior angles formed by parallel lines and a transversal

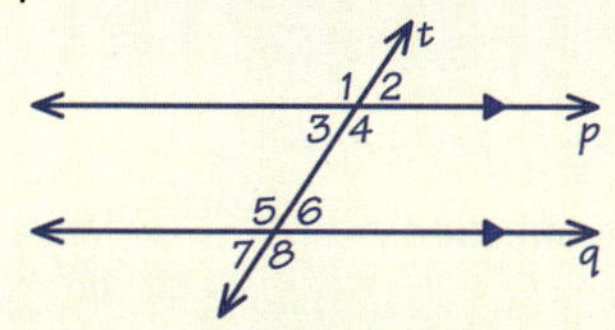

Examples	Non-Examples
∠1	∠3
∠2	∠4
∠7	∠5
∠8	∠6

List of Organizers

Available at *BigIdeasMath.com*

Comparison Chart
Concept Circle
Definition (Idea) and Example Chart
Example and Non-Example Chart
Formula Triangle
Four Square
Information Frame
Information Wheel
Notetaking Organizer
Process Diagram
Summary Triangle
Word Magnet
Y Chart

About this Organizer

An **Example and Non-Example Chart** can be used to list examples and non-examples of a vocabulary word or term. Students write examples of the word or term in the left column and non-examples in the right column. This type of organizer serves as a good tool for assessing students' knowledge of pairs of topics that have subtle but important differences, such as complementary and supplementary angles. Blank example and non-example charts can be included on tests or quizzes for this purpose.

Technology for the Teacher

Editable Graphic Organizer

Answers

1. $\angle 2 = 82°$; $\angle 2$ and the given angle are alternate exterior angles.
2. $\angle 6 = 82°$; $\angle 6$ and the given angle are vertical angles.
3. $\angle 4 = 82°$; $\angle 4$ and the given angle are corresponding angles.
4. $\angle 1 = 98°$; $\angle 4$ and $\angle 1$ are supplementary.
5. $123°$; $\angle 1$ and $\angle 7$ are alternate exterior angles.
6. $122°$; $\angle 2$ and $\angle 8$ are alternate interior angles and $\angle 8$ and $\angle 5$ are supplementary.
7. $119°$; $\angle 5$ and $\angle 3$ are alternate interior angles.
8. $60°$; $\angle 4$ and $\angle 6$ are alternate exterior angles.
9. $60°$; $60°$; $60°$
10. $115°$; $40°$; $25°$
11. $45°$; $45°$; $90°$
12. $105°$
13. $60°$
14. $\angle 1 = 108°$, $\angle 2 = 108°$; Because of alternate interior angles, the angle below $\angle 1$ is $72°$. This angle is supplementary to both $\angle 1$ and $\angle 2$.
15. Exterior angle with wall: $180 - 15 = 165°$; Exterior angle with ground: $180 - 5(15) = 105°$;

$$x + 5x + 90 = 180$$
$$x = 15$$

Online Assessment
Assessment Book
ExamView® Assessment Suite

Alternative Quiz Ideas

100% Quiz
Error Notebook
Group Quiz
Homework Quiz
Math Log
Notebook Quiz
Partner Quiz
Pass the Paper

Notebook Quiz

A notebook quiz is used to check students' notebooks. Students should be told at the beginning of the course what the expectations are for their notebooks: notes, class work, homework, date, problem number, goals, definitions, or anything else that you feel is important for your class. They also need to know that it is their responsibility to obtain the notes when they miss class.

1. On a certain day, what was the answer to the warm up question?
2. On a certain day, how was this vocabulary term defined?
3. For Section 3.1, what is the answer to On Your Own Question 1?
4. For Section 3.2, what is the answer to the Essential Question?
5. On a certain day, what was the homework assignment?

Give the students 5 minutes to answer these questions.

Reteaching and Enrichment Strategies

If students need help. . .	If students got it. . .
Resources by Chapter • Practice A and Practice B • Puzzle Time Lesson Tutorials *BigIdeasMath.com*	Resources by Chapter • Enrichment and Extension • Technology Connection Game Closet at *BigIdeasMath.com* Start the next section

3.1–3.2 Quiz

Check It Out
Progress Check
BigIdeasMath.com

Use the figure to find the measure of the angle. Explain your reasoning. *(Section 3.1)*

1. $\angle 2$

2. $\angle 6$

3. $\angle 4$

4. $\angle 1$

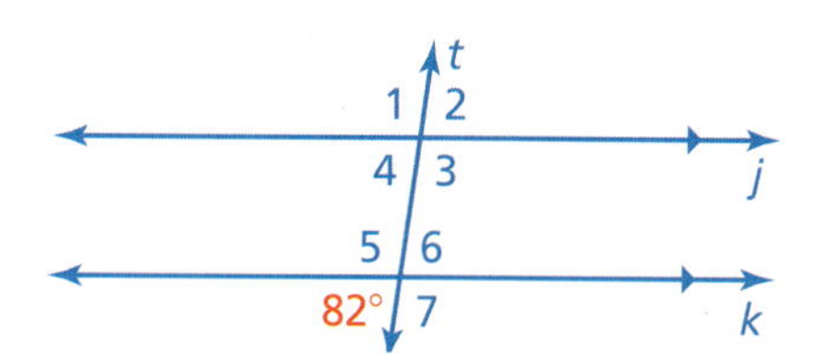

Complete the statement. Explain your reasoning. *(Section 3.1)*

5. If the measure of $\angle 1 = 123°$, then the measure of $\angle 7 =$ ____.

6. If the measure of $\angle 2 = 58°$, then the measure of $\angle 5 =$ ____.

7. If the measure of $\angle 5 = 119°$, then the measure of $\angle 3 =$ ____.

8. If the measure of $\angle 4 = 60°$, then the measure of $\angle 6 =$ ____.

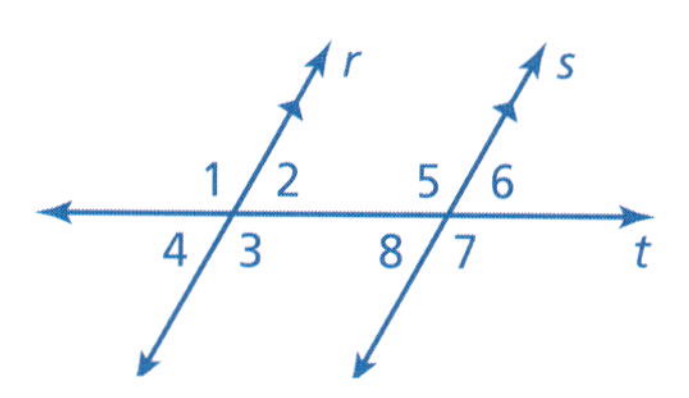

Find the measures of the interior angles. *(Section 3.2)*

9.

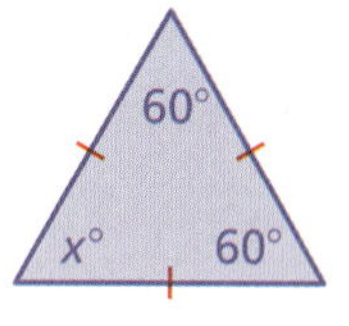

10.

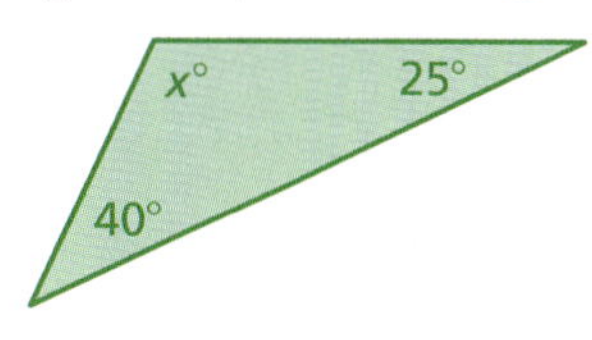

11.

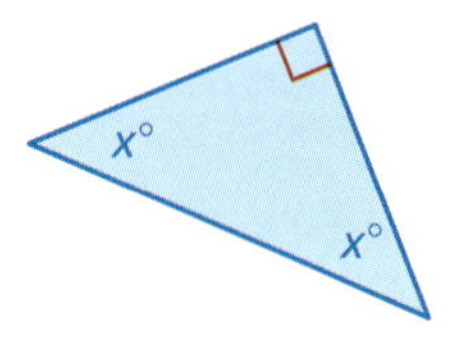

Find the measure of the exterior angle. *(Section 3.2)*

12.

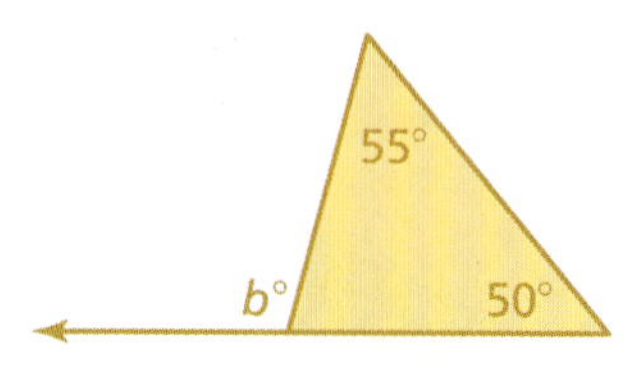

13.

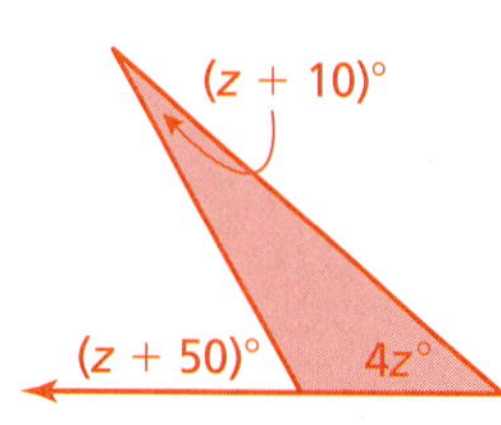

14. PARK In a park, a bike path and a horse riding path are parallel. In one part of the park, a hiking trail intersects the two paths. Find the measures of $\angle 1$ and $\angle 2$. Explain your reasoning. *(Section 3.1)*

15. LADDER A ladder leaning against a wall forms a triangle and exterior angles with the wall and the ground. What are the measures of the exterior angles? Justify your answer. *(Section 3.2)*

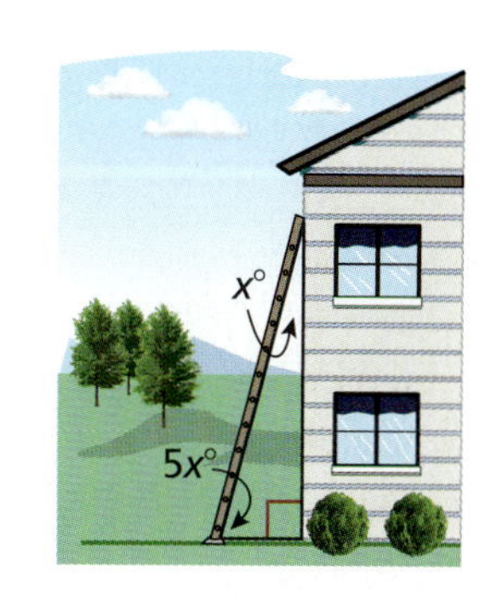

3.3 Angles of Polygons

Essential Question How can you find the sum of the interior angle measures and the sum of the exterior angle measures of a polygon?

1 ACTIVITY: Exploring the Interior Angles of a Polygon

Work with a partner. In parts (a) – (e), identify each polygon and the number of sides *n*. Then find the sum of the interior angle measures of the polygon.

a. Polygon: ▭ Number of sides: $n =$ ▭

Draw a line segment on the figure that divides it into two triangles. Is there more than one way to do this? Explain.

What is the sum of the interior angle measures of each triangle?

What is the sum of the interior angle measures of the figure?

b.

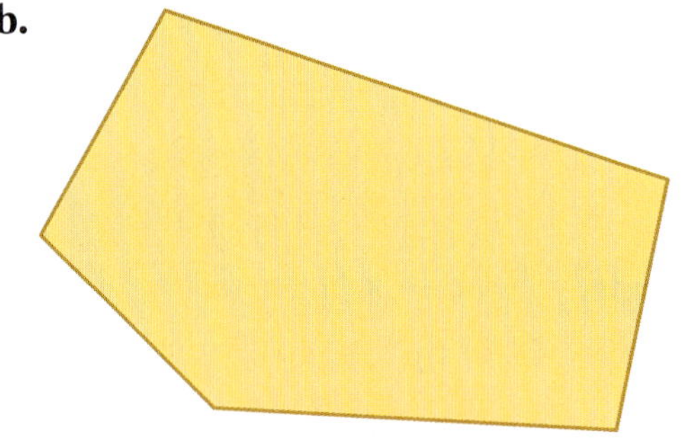

c.

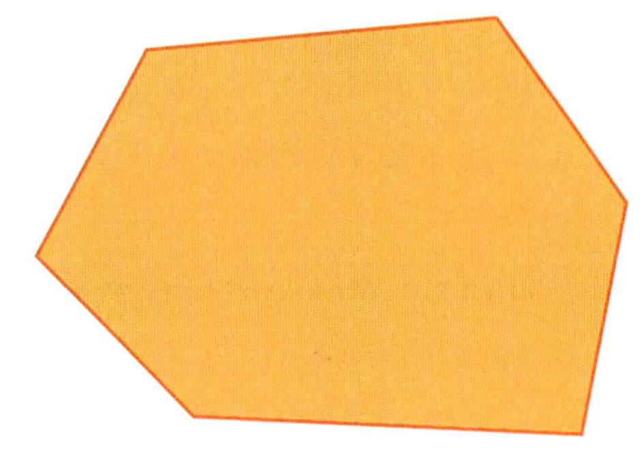

d.

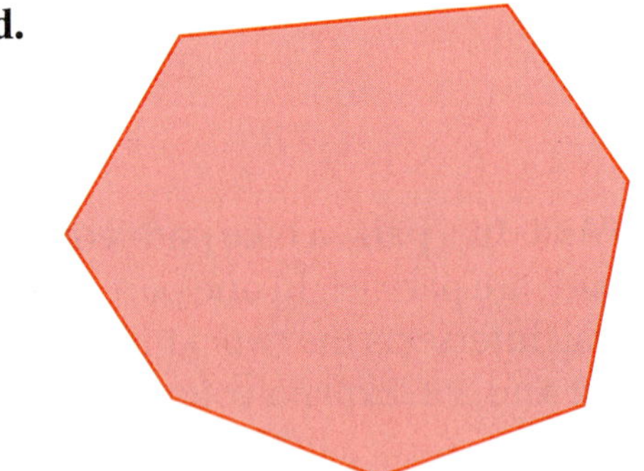

e.

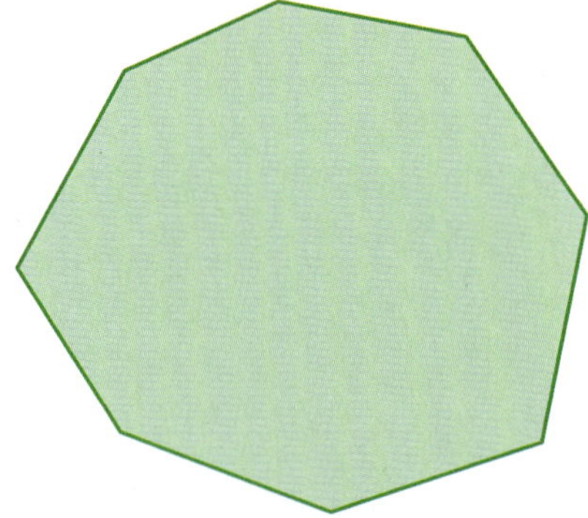

Geometry

In this lesson, you will

- find the sum of the interior angle measures of polygons.
- understand that the sum of the exterior angle measures of a polygon is 360°.
- find the measures of interior and exterior angles of polygons.

Applying Standard MACC.8.G.1.5

f. **REPEATED REASONING** Use your results to complete the table. Then find the sum of the interior angle measures of a polygon with 12 sides.

Number of Sides, *n*	3	4	5	6	7	8
Number of Triangles						
Angle Sum, *S*						

Laurie's Notes

Introduction

Standards for Mathematical Practice

- **MP8 Look for and Express Regularity in Repeated Reasoning:** In making sense of how to find the sum of the interior angles of a polygon, students will repeat a strategy, each time with a polygon having one more side than previously.

Motivate

? "How many of you are looking forward to getting your driver's license?"
- Tell them that they will likely be tested on road signs.
- Draw several shapes and ask students if they know the names of the shapes and what they are used for on highway signs.

Activity Notes

Activity 1

- You may need to guide students through the part (a) of the activity with the quadrilateral.
- Students recognize that the diagonal divides the quadrilateral into two triangles and that each triangle has interior angle measures that sum to 180°. The confusion is that the two triangles have a total of 6 angles and the quadrilateral has only 4 angles. Help students recognize that the diagonal divides two angles of the quadrilateral.
- The approach taken for the remaining polygons is slightly different. All of the diagonals are drawn from one vertex, forming triangles inside the polygon.

$n = 4$

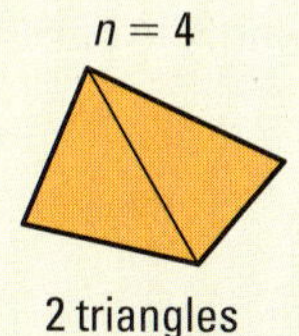

2 triangles

$n = 5$

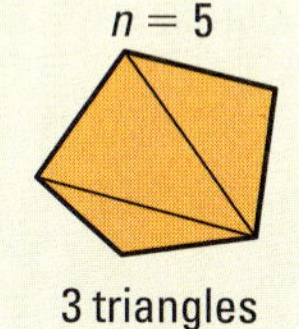

3 triangles

$n = 6$

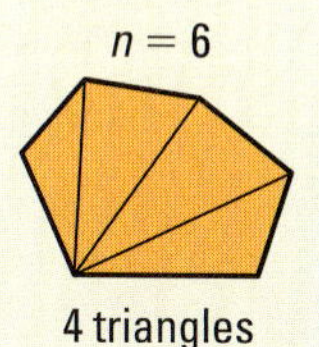

4 triangles

- **MP8:** Students will use repeated reasoning as they explore the remaining polygons: the pentagon has 3 interior triangles so 3×180, the hexagon has 4 interior triangles so 4×180, and so on.
- **MP3 Construct Viable Arguments and Critique the Reasoning of Others:** Some students may conjecture that the number of triangles is two less than the number of sides in the polygon. Other students may just notice that the sum of the interior angle measures is increasing by 180° each time.
- **Connection:** Students who snowboard or skateboard will recognize the number pattern quickly. They know the multiples of 180° well.
- **MP4 Model with Mathematics:** The table helps to organize the data. This also will help students make the connection between the number of sides and the number of triangles formed.
- Ask volunteers to share how they found the interior angle sum for a polygon with 12 sides.

Florida Common Core Standards

MACC.8.G.1.5 Use informal arguments to establish facts about the angle sum and exterior angle of triangles, about the angles created when parallel lines are cut by a transversal, and the angle-angle criterion for similarity of triangles.

Previous Learning

Students should know how to solve multi-step equations.

Lesson Plans
Complete Materials List

3.3 Record and Practice Journal

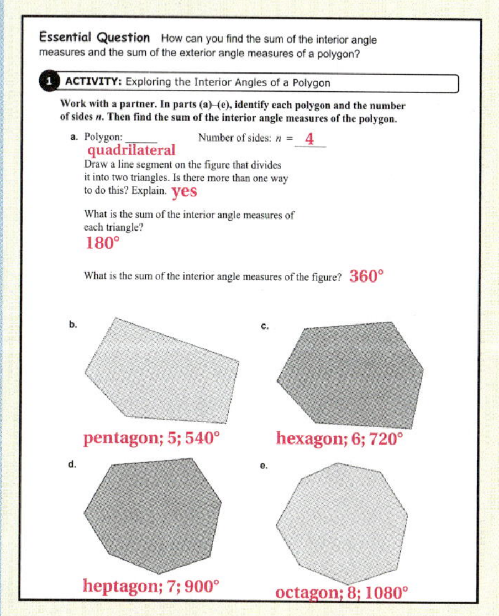

Essential Question How can you find the sum of the interior angle measures and the sum of the exterior angle measures of a polygon?

1 **ACTIVITY:** Exploring the Interior Angles of a Polygon

Work with a partner. In parts (a)–(e), identify each polygon and the number of sides n. Then find the sum of the interior angle measures of the polygon.

a. Polygon: quadrilateral Number of sides: $n =$ 4

Draw a line segment on the figure that divides it into two triangles. Is there more than one way to do this? Explain. yes

What is the sum of the interior angle measures of each triangle? 180°

What is the sum of the interior angle measures of the figure? 360°

b. pentagon; 5; 540°

c. hexagon; 6; 720°

d. heptagon; 7; 900°

e. octagon; 8; 1080°

Differentiated Instruction

Kinesthetic

Another way to discover the sum of the angle measures of a polygon is to have the students cut the polygon into triangles. This can be done for convex and concave polygons.

3.3 Record and Practice Journal

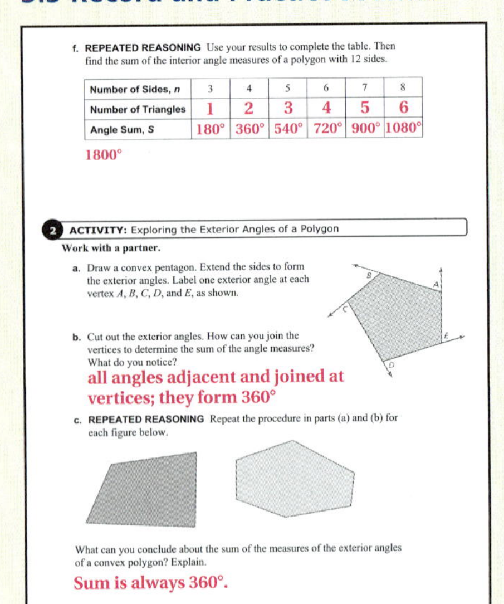

f. **REPEATED REASONING** Use your results to complete the table. Then find the sum of the interior angle measures of a polygon with 12 sides.

Number of Sides, n	3	4	5	6	7	8
Number of Triangles	1	2	3	4	5	6
Angle Sum, S	180°	360°	540°	720°	900°	1080°

1800°

2 ACTIVITY: Exploring the Exterior Angles of a Polygon

Work with a partner.

a. Draw a convex pentagon. Extend the sides to form the exterior angles. Label one exterior angle at each vertex A, B, C, D, and E, as shown.

b. Cut out the exterior angles. How can you join the vertices to determine the sum of the angle measures? What do you notice?

all angles adjacent and joined at vertices; they form 360°

c. **REPEATED REASONING** Repeat the procedure in parts (a) and (b) for each figure below.

What can you conclude about the sum of the measures of the exterior angles of a convex polygon? Explain.

Sum is always 360°.

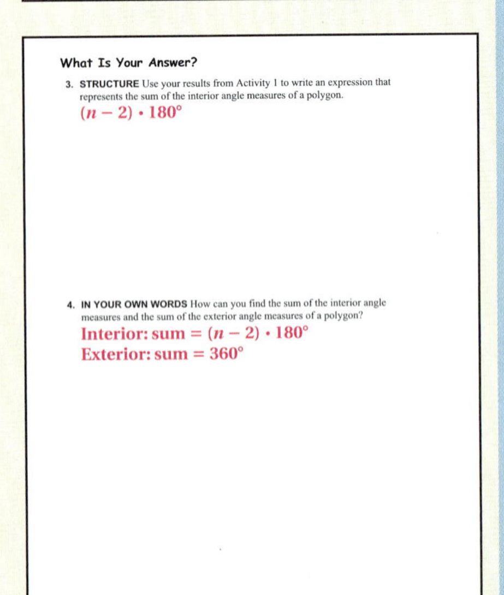

What Is Your Answer?

3. **STRUCTURE** Use your results from Activity 1 to write an expression that represents the sum of the interior angle measures of a polygon.

$(n - 2) \cdot 180°$

4. **IN YOUR OWN WORDS** How can you find the sum of the interior angle measures and the sum of the exterior angle measures of a polygon?

Interior: sum = $(n - 2) \cdot 180°$
Exterior: sum = 360°

Laurie's Notes

Discuss

- "Have you heard of the words convex and concave? If so, what do they mean?" Students may be familiar with the words from science class. They often say that a concave lens caves in.
- Define convex and concave polygons.
- Draw examples of polygons and have students identify them as being convex or concave.

Activity 2

- **Teaching Tip:** If sufficient pairs of scissors are not available, you could group students with 3 to 4 per group or have one student be "the cutter" at the front of the room.
- Remind students to draw their pentagon with a straightedge and large enough to be able to work with it easily. Students should also extend the sides far enough so that the lines aid the cutting process.
- **MP8:** Students will use repeated reasoning as they explore the two remaining polygons: the quadrilateral has 4 exterior angles to cut, the hexagon has 6 exterior angles to cut. Each time, the angles are placed about a single point, a technique students used in the previous lesson when they tore the angles off their paper triangle and placed them about a point.
- **MP3:** Some students may conjecture that the sum of the exterior angle measures of any polygon is 360°.

What Is Your Answer?

- In Question 3, students may have an expression that they use.
- In Question 4, they are using words to describe the formula they have discovered.

Closure

- Use the results of the investigations to find the sum of the interior angle measures and the sum of the exterior angle measures of a decagon, a polygon with 10 sides. 1440°; 360°

A polygon is **convex** when every line segment connecting any two vertices lies entirely inside the polygon. A polygon is **concave** when at least one line segment connecting any two vertices lies outside the polygon.

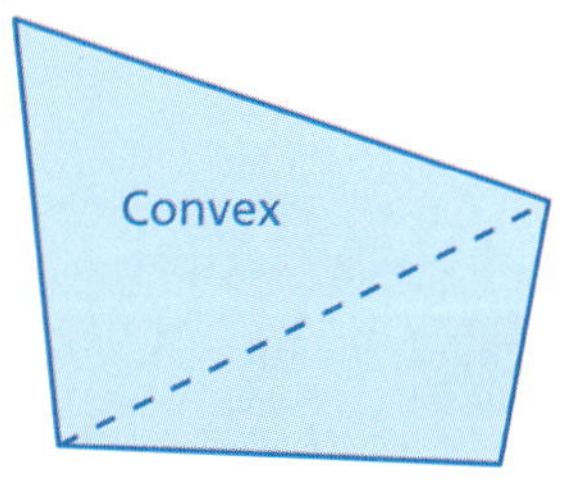

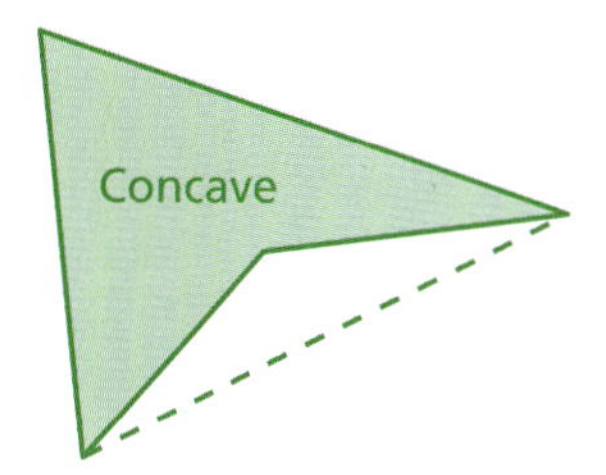

2 ACTIVITY: Exploring the Exterior Angles of a Polygon

Math Practice 3

Analyze Conjectures

Do your observations about the sum of the exterior angles make sense? Do you think they would hold true for any convex polygon? Explain.

Work with a partner.

a. Draw a convex pentagon. Extend the sides to form the exterior angles. Label one exterior angle at each vertex *A*, *B*, *C*, *D*, and *E*, as shown.

b. Cut out the exterior angles. How can you join the vertices to determine the sum of the angle measures? What do you notice?

c. **REPEATED REASONING** Repeat the procedure in parts (a) and (b) for each figure below.

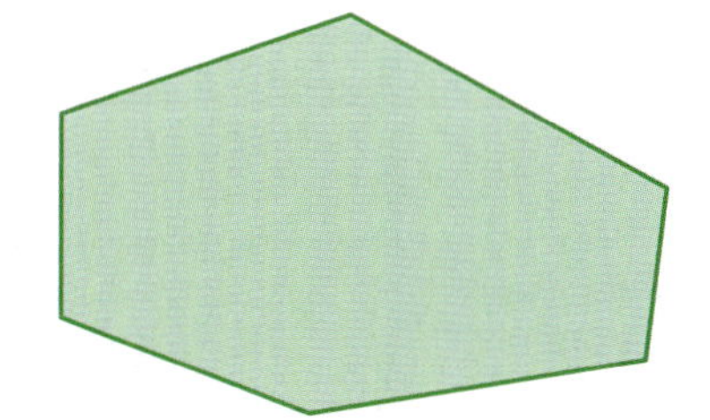

What can you conclude about the sum of the measures of the exterior angles of a convex polygon? Explain.

What Is Your Answer?

3. **STRUCTURE** Use your results from Activity 1 to write an expression that represents the sum of the interior angle measures of a polygon.

4. **IN YOUR OWN WORDS** How can you find the sum of the interior angle measures and the sum of the exterior angle measures of a polygon?

Use what you learned about angles of polygons to complete Exercises 4–6 on page 123.

3.3 Lesson

Key Vocabulary
convex polygon, *p. 119*
concave polygon, *p. 119*
regular polygon, *p. 121*

A *polygon* is a closed plane figure made up of three or more line segments that intersect only at their endpoints.

Polygons

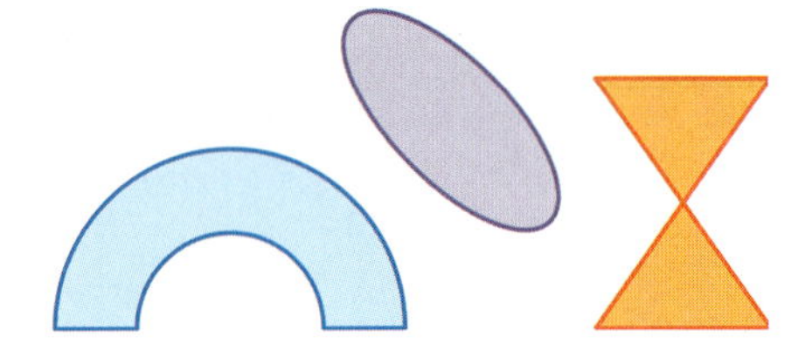

Not polygons

Key Idea

Interior Angle Measures of a Polygon

The sum S of the interior angle measures of a polygon with n sides is

$$S = (n - 2) \cdot 180^\circ.$$

EXAMPLE 1 Finding the Sum of Interior Angle Measures

Reading

For polygons whose names you have not learned, you can use the phrase "n-gon," where n is the number of sides. For example, a 15-gon is a polygon with 15 sides.

Find the sum of the interior angle measures of the school crossing sign.

The sign is in the shape of a pentagon. It has 5 sides.

$S = (n - 2) \cdot 180^\circ$	Write the formula.
$= (5 - 2) \cdot 180^\circ$	Substitute 5 for n.
$= 3 \cdot 180^\circ$	Subtract.
$= 540^\circ$	Multiply.

∴ The sum of the interior angle measures is 540°.

On Your Own

Now You're Ready
Exercises 7–9

Find the sum of the interior angle measures of the green polygon.

1.

2.

Laurie's Notes

Introduction

Connect

- **Yesterday:** Students explored finding the sums of interior and exterior angle measures of a polygon. (MP3, MP4, MP8)
- **Today:** Students will use a formula to find the sums of interior and exterior angle measures of a polygon.

Motivate

? "Did you ever wonder why bees use a hexagonal structure for their honeycomb? Why not squares? or circles? or octagons?"

- Draw a few cells of the honeycomb.

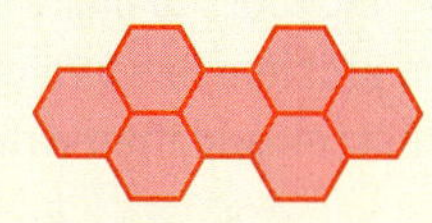

- Mathematicians have concluded that a hexagon is the most appropriate geometric form for the maximum use of a given area. This means that a hexagonal cell requires the minimum amount of wax for construction while it stores the maximum amount of honey.

Lesson Notes

Key Idea

- Write the definition of a polygon. Draw examples of shapes which are and are not polygons. Students should be able to explain why some are not polygons.
- In all of the samples shown, the interior is shaded. The polygon is the figure formed by the line segments. The polygonal region contains the interior of the polygon.
- Write the Key Idea. This is the same equation that students wrote yesterday, but in the more common form. This form highlights the fact that the sum is a multiple of 180°.

Example 1

- Review the names of common polygons: Triangle (3), Quadrilateral (4), Pentagon (5), Hexagon (6), Octagon (8), and Decagon (10). It is also common to say n-gon and replace n with 9 to talk about a 9-sided polygon.
- Read the problem. The polygon is a pentagon.
- Write the equation, substitute 5 for n, and solve.

On Your Own

- **Think-Pair-Share:** Students should read each question independently and then work in pairs to answer the questions. When they have answered the questions, the pair should compare their answers with another group and discuss any discrepancies.

? "What are the names of the polygons in Questions 1 and 2?"
7-gon or heptagon; hexagon

Goal Today's lesson is finding the interior and exterior angle measures of a polygon.

Lesson Tutorials
Lesson Plans
Answer Presentation Tool

Extra Example 1

Find the sum of the interior angle measures of the polygon.

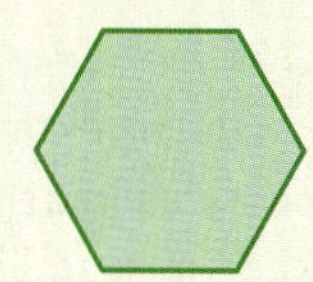

720°

1. 900°
2. 720°

Extra Example 2

Find the value of x.

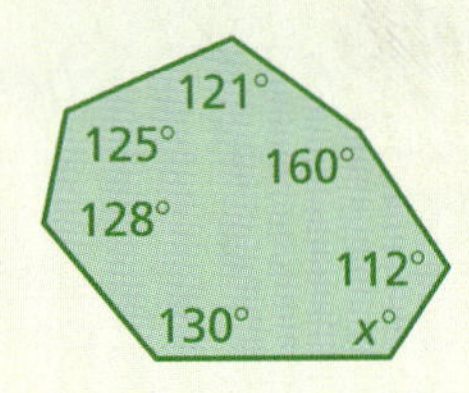

124

On Your Own

3. 105

4. 75

5. 35

Extra Example 3

What if the cloud system in Example 3 is in the approximate shape of a regular polygon with 12 sides? Find the measure of each interior angle of the polygon.
150°

English Language Learners

Vocabulary

Preview the *Key Vocabulary* in this chapter. Understanding geometry depends on understanding the terminology used. Have students write key vocabulary words in their notebooks. Include definitions and examples to help distinguish between words (e.g., convex polygon and concave polygon).

Laurie's Notes

Example 2

- **Connection:** This example integrates equation solving with finding a missing angle.
- ? "How many sides does the polygon have?" 7
- ? "How do you find the sum of the measures of all of the interior angles of a 7-gon?" Solve $(7 - 2)180 = 900$.
- Once the sum is known, write and solve the equation as shown. Caution students to be careful with their arithmetic.

On Your Own

- **MP1 Make Sense of Problems and Persevere in Solving Them:** Students should check with their neighbors to make sure they are setting up the equation correctly. Each problem has two parts: determining the sum of all of the interior angle measures and then writing the equation to solve for the missing angle.
- In Question 4, remind students that the symbol for a right angle means the angle measures 90°.
- In Question 5, two angles are missing, each with a measure of $2x°$. The sum of the interior angle measures of this pentagon is 540°, so $2x + 145 + 145 + 2x + 110 = 540$. The steps are to combine like terms, isolate the variable, and solve.
- **Common Error:** Students will solve for the variable correctly, but then forget to substitute this value back into the variable expression to solve for the angle measure. In Question 5, students were only asked to solve for x. If they had been asked to find the measure of the angle, there would be one last step. In this case, $x = 35$ and the two missing angle measures are each 70°.

Example 3

- Review the definition of a regular polygon. Point out to students that squares and equilateral triangles are examples of regular polygons.
- **MP1:** A regular hexagon has 6 congruent angles. If the angle measures of a hexagon sum to 720° and the 6 angles are congruent, it should make sense to students why they divide 720 by 6.
- Look back to the honeycomb you drew at the beginning of the lesson. There are three 120° angles about one point.
- You can show a video of the cloud system from the website *jpl.nasa.gov*.

EXAMPLE 2 Finding an Interior Angle Measure of a Polygon

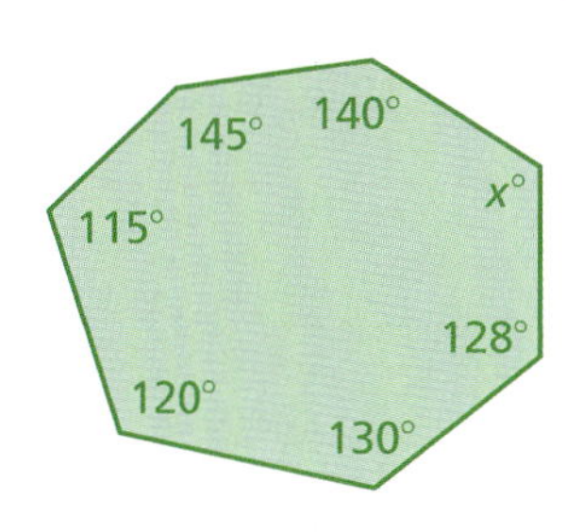

Find the value of x.

Step 1: The polygon has 7 sides. Find the sum of the interior angle measures.

$S = (n - 2) \cdot 180°$ Write the formula.

$= (7 - 2) \cdot 180°$ Substitute 7 for n.

$= 900°$ Simplify. The sum of the interior angle measures is 900°.

Step 2: Write and solve an equation.

$$140 + 145 + 115 + 120 + 130 + 128 + x = 900$$

$$778 + x = 900$$

$$x = 122$$

The value of x is 122.

On Your Own

Exercises 12–14

Find the value of x.

3. 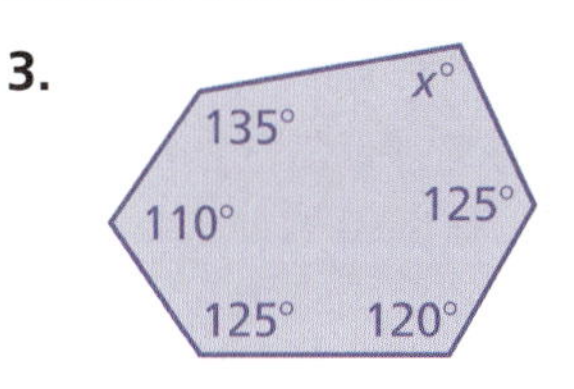

4. 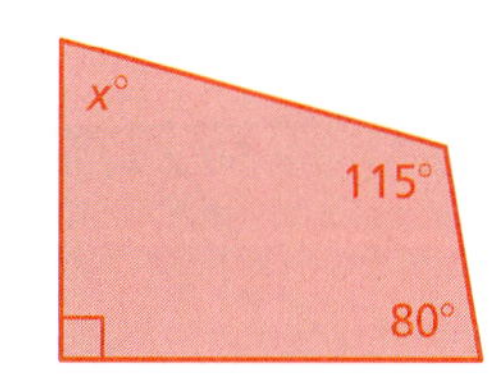

5. 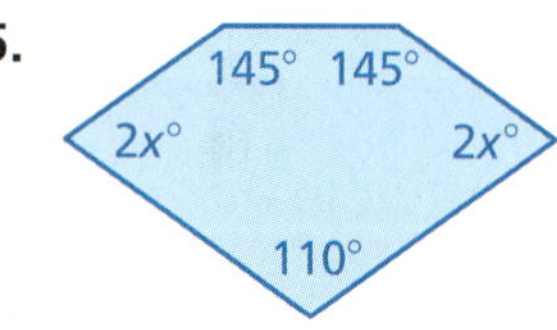

In a **regular polygon**, all the sides are congruent, and all the interior angles are congruent.

EXAMPLE 3 Real-Life Application

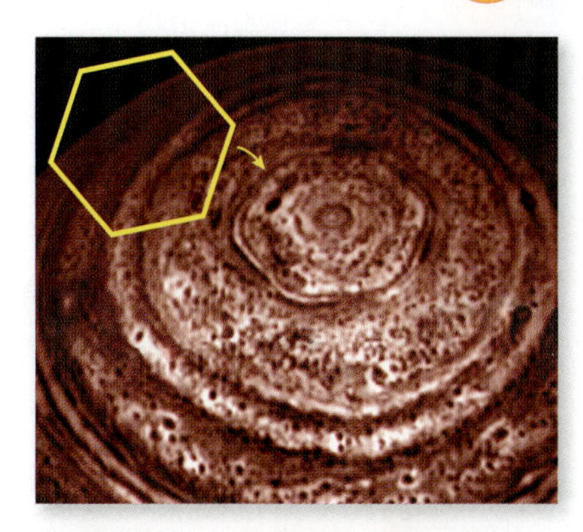

The hexagon is about 15,000 miles across. Approximately four Earths could fit inside it.

A cloud system discovered on Saturn is in the approximate shape of a regular hexagon. Find the measure of each interior angle of the hexagon.

Step 1: A hexagon has 6 sides. Find the sum of the interior angle measures.

$S = (n - 2) \cdot 180°$ Write the formula.

$= (6 - 2) \cdot 180°$ Substitute 6 for n.

$= 720°$ Simplify. The sum of the interior angle measures is 720°.

Step 2: Divide the sum by the number of interior angles, 6.

$$720° \div 6 = 120°$$

The measure of each interior angle is 120°.

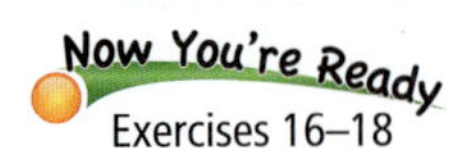

Find the measure of each interior angle of the regular polygon.

6. octagon **7.** decagon **8.** 18-gon

Exterior Angle Measures of a Polygon

Words The sum of the measures of the exterior angles of a convex polygon is 360°.

Algebra $w + x + y + z = 360$

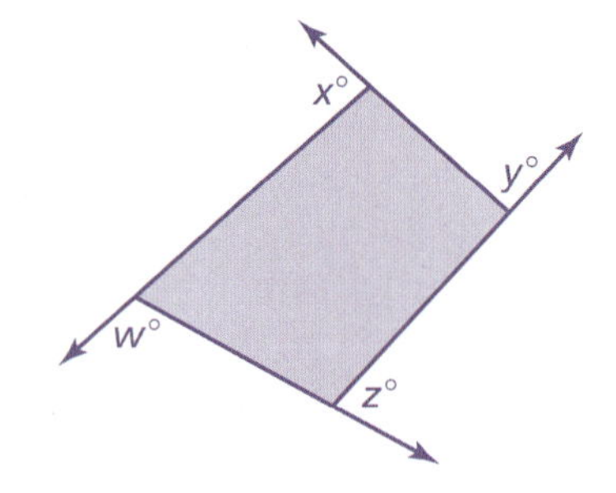

EXAMPLE 4 Finding Exterior Angle Measures

Find the measures of the exterior angles of each polygon.

a.

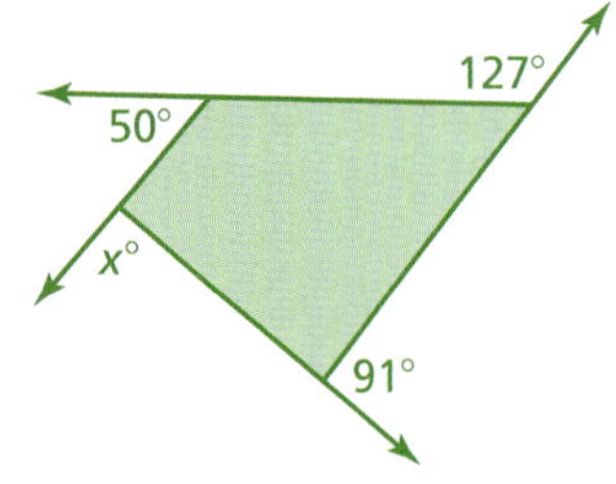

Write and solve an equation for x.

$$x + 50 + 127 + 91 = 360$$
$$x + 268 = 360$$
$$x = 92$$

So, the measures of the exterior angles are 92°, 50°, 127°, and 91°.

b.

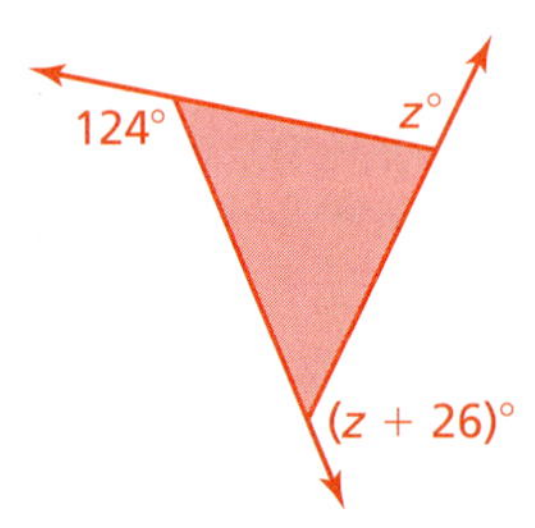

Write and solve an equation for z.

$$124 + z + (z + 26) = 360$$
$$2z + 150 = 360$$
$$z = 105$$

So, the measures of the exterior angles are 124°, 105°, and $(105 + 26)° = 131°$.

On Your Own

9. Find the measures of the exterior angles of the polygon.

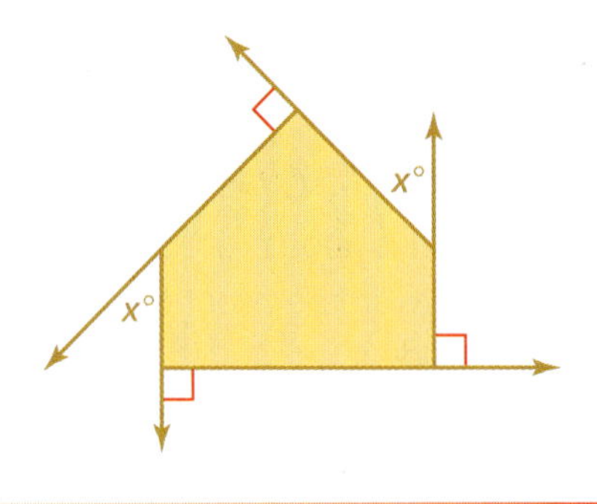

On Your Own

- **Think-Pair-Share:** Students should read each question independently and then work in pairs to answer the questions. When they have answered the questions, the pair should compare their answers with another group and discuss any discrepancies.

Key Idea

- Write the Key Idea. Draw a quadrilateral that does not have special attributes, and extend the sides so that an exterior angle is formed, one at each vertex.
- Remind students that although there are two exterior angles at each vertex, this formula considers only one exterior angle at each vertex. This is customary.
- Stress that the formula is true for *any* polygon, not just the quadrilateral.

Example 4

- Draw the example in part (a) on the board.
- ? "How can you find the measure of the missing exterior angle?" Set the sum of the four angle measures equal to 360°, and solve the equation for x.
- Draw the example in part (b) on the board.
- ? "How can we find the measures of the missing exterior angles?" Set the sum of the three angle measures equal to 360°, and solve the equation for z.
- The last step in this problem is to remember to substitute the value of z into the expression $z + 26$ to find the measure of last exterior angle.

On Your Own

- Students should recognize that three of the exterior angles are right angles and the other two angles are congruent.

Closure

- A pentagon has two interior right angles and the other three interior angles are all congruent. What is the measure of one of the missing angles? 120°

On Your Own

6. 135°
7. 144°
8. 160°

Extra Example 4

Find the measures of the exterior angles of each polygon.

a.

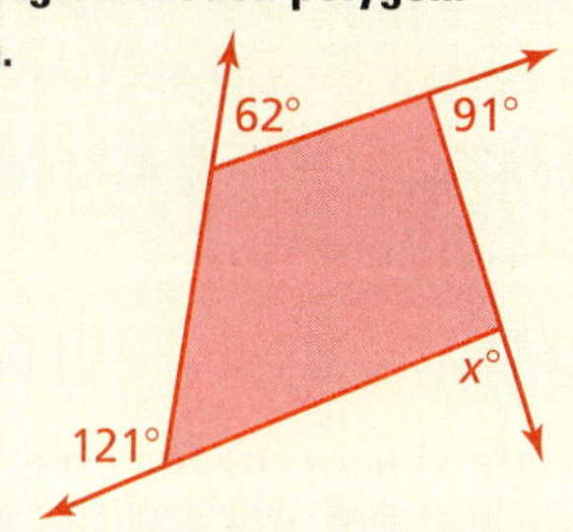

121°, 62°, 91°, and 86°

b.

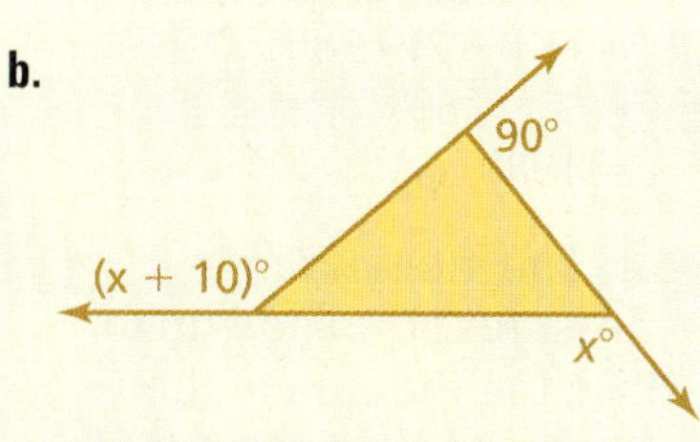

140°, 90°, and 130°

On Your Own

9. 90°, 45°, 90°, 45°, 90°

Vocabulary and Concept Check

1. *Sample answer:*

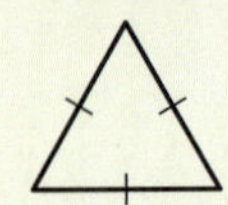

2. The second figure does not belong because it is not made up entirely of line segments.
3. What is the measure of an interior angle of a regular pentagon?; 108°; 540°

Practice and Problem Solving

4. 360°
5. 1260°
6. 900°
7. 360°
8. 1080°
9. 1260°
10. The right side of the formula is $(n - 2) \cdot 180°$, not $n \cdot 180°$.
$$\begin{aligned} S &= (n - 2) \cdot 180° \\ &= (13 - 2) \cdot 180° \\ &= 11 \cdot 180° \\ &= 1980° \end{aligned}$$
11. no; The interior angle measures given add up to 535°, but the sum of the interior angle measures of a pentagon is 540°.

Assignment Guide and Homework Check

Level	Day 1 Activity Assignment	Day 2 Lesson Assignment	Homework Check
Basic	4–6, 34–38	1–3, 7, 9, 10, 11–27 odd	7, 11, 13, 17, 21
Average	4–6, 34–38	1–3, 8–18 even, 19–31 odd	8, 14, 18, 21, 23
Advanced	1–6, 10–18 even, 19, 20–26 even, 28–38		14, 20, 24, 28, 30

Common Errors

- **Exercises 4–6** Students may struggle dividing the polygon into triangles. Encourage them to trace the polygon in pen in their notebooks and then to draw triangles with a pencil so that they can erase lines if necessary.
- **Exercises 7–9** Students may forget to subtract 2 from the number of sides when using the formula to find the sum of the interior angle measures. Remind them of the formula and encourage them to write the formula before substituting the number of sides.
- **Exercise 11** Students may say that because the sum of the interior angle measures is close to the value found when using the formula, a pentagon can have these angle measures. Remind them that the sum of the interior angle measures must be *exactly* the same as the sum found with the formula for the polygon to be drawn with the given angles.

3.3 Record and Practice Journal

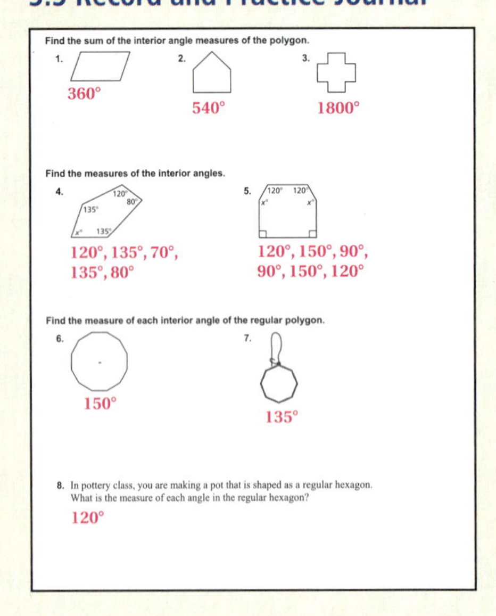
Find the sum of the interior angle measures of the polygon.

1. 360°
2. 540°
3. 1800°

Find the measures of the interior angles.

4. 120°, 135°, 70°, 135°, 80°
5. 120°, 150°, 90°, 90°, 150°, 120°

Find the measure of each interior angle of the regular polygon.

6. 150°
7. 135°

8. In pottery class, you are making a pot that is shaped as a regular hexagon. What is the measure of each angle in the regular hexagon?
120°

3.3 Exercises

Vocabulary and Concept Check

1. **VOCABULARY** Draw a regular polygon that has three sides.

2. **WHICH ONE DOESN'T BELONG?** Which figure does *not* belong with the other three? Explain your reasoning.

   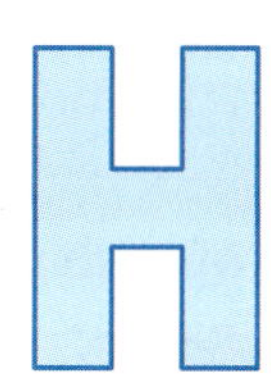

3. **DIFFERENT WORDS, SAME QUESTION** Which is different? Find "both" answers.

What is the measure of an interior angle of a regular pentagon?	What is the sum of the interior angle measures of a convex pentagon?
What is the sum of the interior angle measures of a regular pentagon?	What is the sum of the interior angle measures of a concave pentagon?

Practice and Problem Solving

Use triangles to find the sum of the interior angle measures of the polygon.

4. 

5. 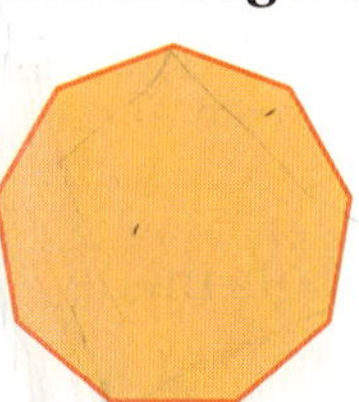

6. 

Find the sum of the interior angle measures of the polygon.

7. 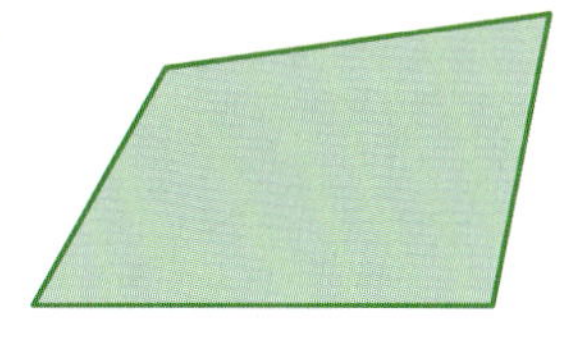

8. 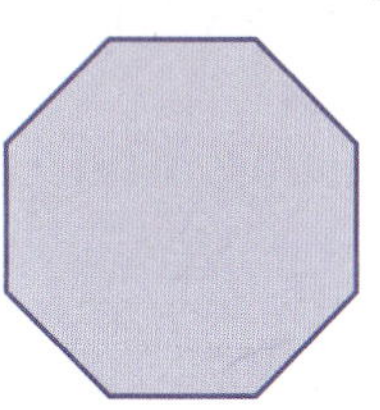

9. 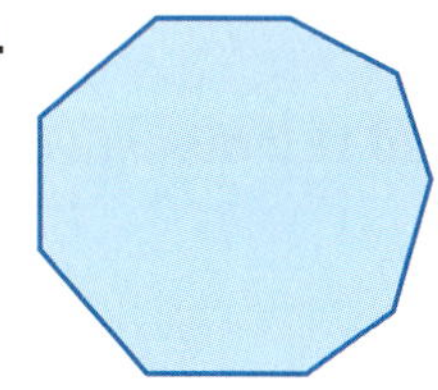

10. **ERROR ANALYSIS** Describe and correct the error in finding the sum of the interior angle measures of a 13-gon.

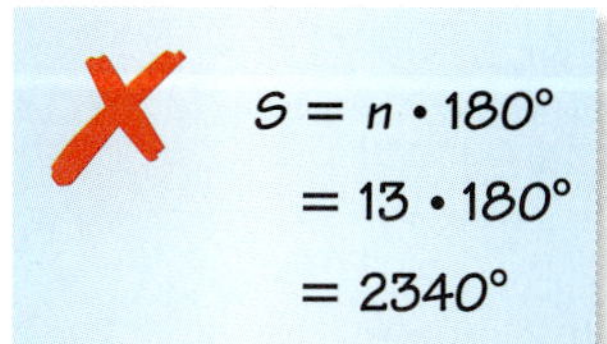

11. **NUMBER SENSE** Can a pentagon have interior angles that measure 120°, 105°, 65°, 150°, and 95°? Explain.

Find the measures of the interior angles.

12.

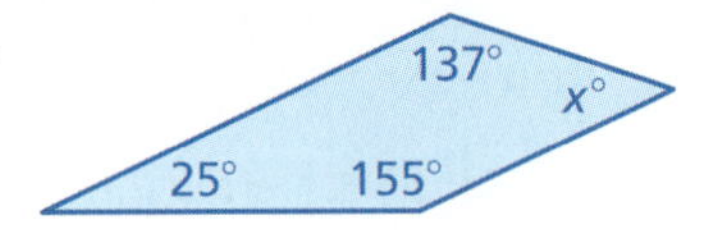

13.

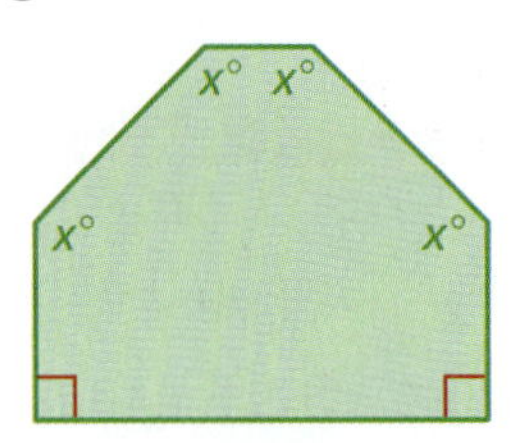

14.

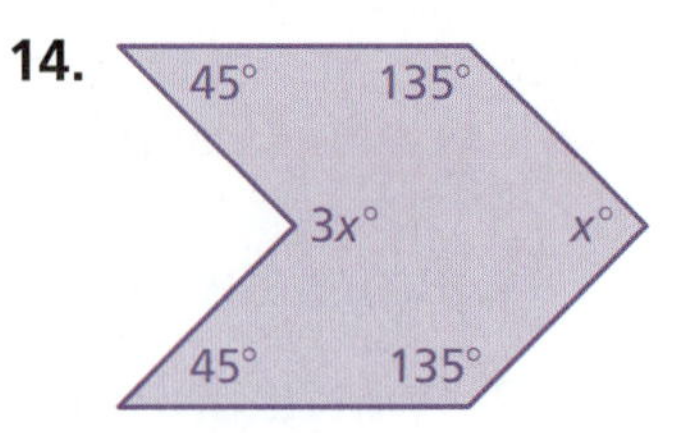

15. REASONING The sum of the interior angle measures in a regular polygon is 1260°. What is the measure of one of the interior angles of the polygon?

Find the measure of each interior angle of the regular polygon.

16.

17.

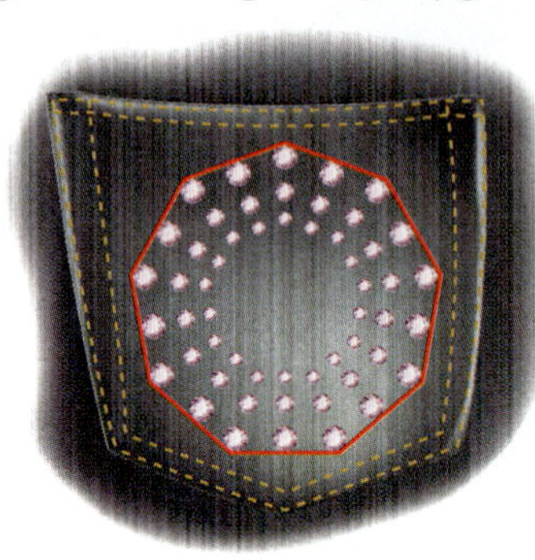

18.

19. ERROR ANALYSIS Describe and correct the error in finding the measure of each interior angle of a regular 20-gon.

✗

$S = (n - 2) \cdot 180°$

$= (20 - 2) \cdot 180°$

$= 18 \cdot 180°$

$= 3240°$

$3240° \div 18 = 180$

The measure of each interior angle is 180°.

20. FIRE HYDRANT A fire hydrant bolt is in the shape of a regular pentagon.

a. What is the measure of each interior angle?

b. Why are fire hydrants made this way?

21. PROBLEM SOLVING The interior angles of a regular polygon each measure 165°. How many sides does the polygon have?

Find the measures of the exterior angles of the polygon.

22.

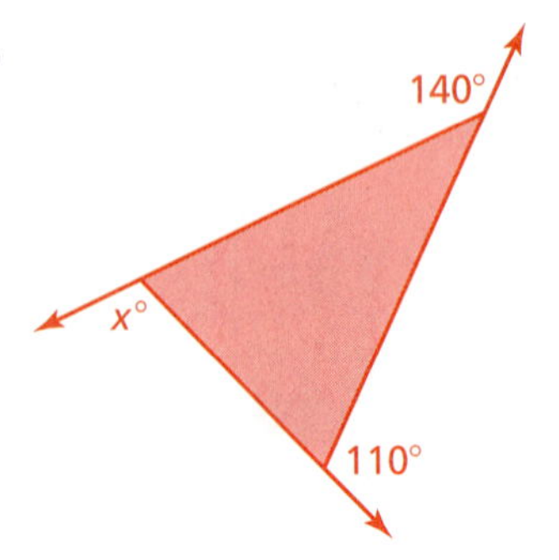

23.

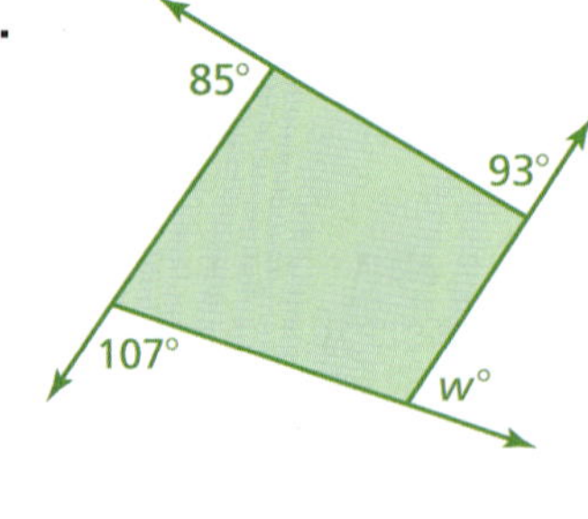

24.

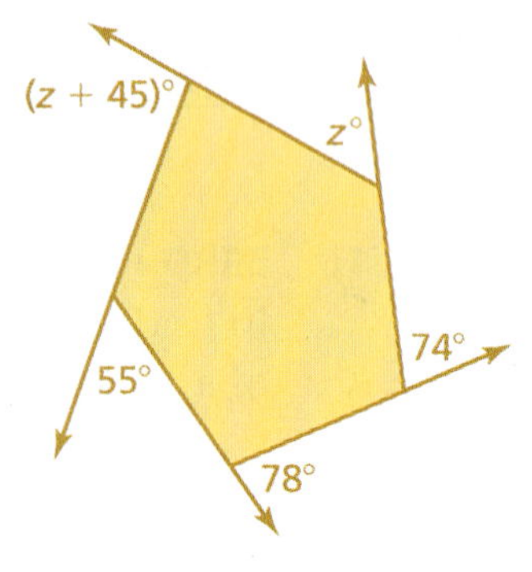

25. REASONING What is the measure of an exterior angle of a regular hexagon? Explain.

Common Errors

- **Exercises 12–14** Students may forget to include one or more of the given angles when writing an equation for the missing angles. For example, in Exercise 13, students may write $4x = 720$. Remind them to include all of the angles. Encourage them to write the equation and then count the number of terms to make sure that there are the same number of terms as angles before simplifying.
- **Exercises 16–18** Students may find the sum of the interior angle measures of the regular polygon, but forget to divide by the number of angles to answer the question. Remind them that they are finding the measure of *one* angle. Because all the angles are congruent (by the definition of a regular polygon), they can divide the sum of the interior angle measures by the number of angles.
- **Exercises 22–24, 26–28** Students may solve for the variable, but forget to find the measures of the angles. Remind them to read the directions carefully and to answer the question.

English Language Learners

Vocabulary

Many English mathematical names are influenced by Latin and Greek names for numbers.

English	Latin	Greek
one	unus	heis
two	duo	duo
three	tres	tria
four	quattuor	tettara
five	quinque	pente
six	sex	hex
seven	septe	hepta
eight	octo	okto
nine	nove	ennea
ten	dece	deka

Spanish speakers will recognize the similarity of numbers in Spanish to the numbers in Latin.

12. 25°, 137°, 43°, 155°

13. 90°, 135°, 135°, 135°, 135°, 90°

14. 45°, 135°, 90°, 135°, 45°, 270°

15. 140° **16.** 60°

17. 140° **18.** 150°

19. The sum of the interior angle measures should have been divided by the number of angles, 20. $3240° \div 20 = 162°$; The measure of each interior angle is 162°.

20. **a.** 108°

b. *Sample answer:* to deter people from tampering with fire hydrants, because most wrenches are hexagonal.

21. 24 sides

22. 110°, 110°, 140°

23. 75°, 93°, 85°, 107°

24. 54°, 74°, 78°, 55°, 99°

25. See Additional Answers.

Practice and Problem Solving

26. 90°, 45°, 45°, 90°, 45°, 45°

27. 120°, 120°, 120°

28. 125°, 125°, 55°, 55°

29. interior: 135°; exterior: 45°

30. See *Taking Math Deeper.*

31. 120°

32. **a.** 11 sides

b. 147°

33. See Additional Answers.

Fair Game Review

34. 9 **35.** 2

36. 3 **37.** 6

38. D

Mini-Assessment

Find the sum of the interior angle measures of the polygon.

1.

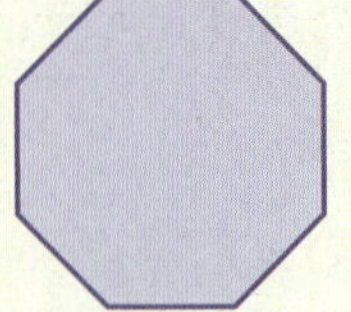

1080°

2.

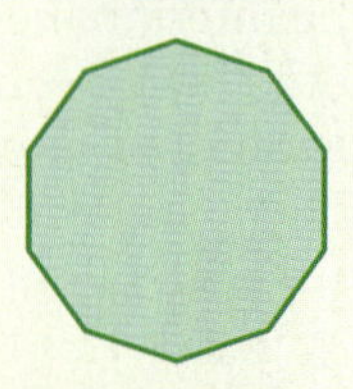

1440°

3. Find the measure of each interior angle of a 16-gon. 157.5°

4. Find the measures of the exterior angles of the polygon.

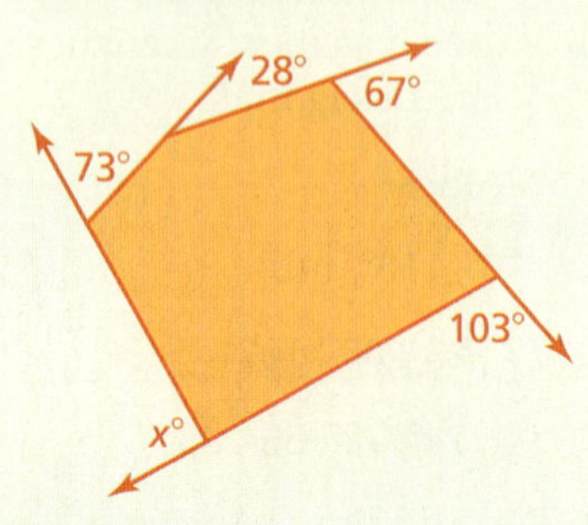

89°, 73°, 28°, 67°, and 103°

Taking Math Deeper

Exercise 30

Instead of trying to draw the angles one at a time, begin by drawing a different shape and manipulating it.

1. The desired pentagon has two right angles. So, you could start with a rectangle.

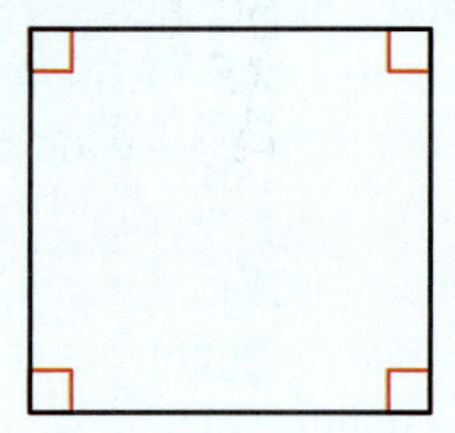

2. Erase one of the sides.

This leaves you with three sides and two right angles. You need two more sides and two 45° angles, so draw two sides that form 45° interior angles that are long enough to meet and create a fifth vertex.

3. Check the remaining angle.

Using a protractor, you can see that the remaining interior angle measure is 270°.

Notice that the result is a concave pentagon.

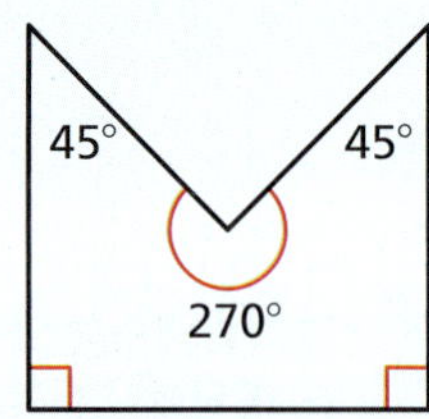

Project

Tell students to use the method above to draw a pentagon that has two right interior angles, two 15° interior angels, and one 330° angle. Ask them if they can start with *any* rectangle. If not, ask them to describe the dimensions of rectangles that would not work.

Reteaching and Enrichment Strategies

If students need help. . .	If students got it. . .
Resources by Chapter • Practice A and Practice B • Puzzle Time Record and Practice Journal Practice Differentiating the Lesson Lesson Tutorials Skills Review Handbook	Resources by Chapter • Enrichment and Extension • Technology Connection Start the next section

Find the measures of the exterior angles of the polygon.

26.

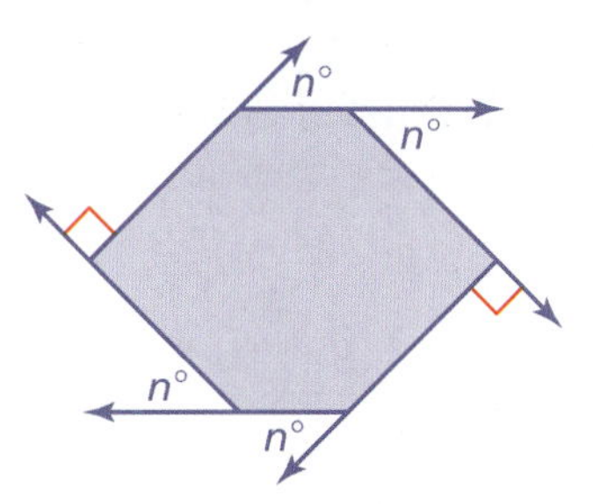

27.

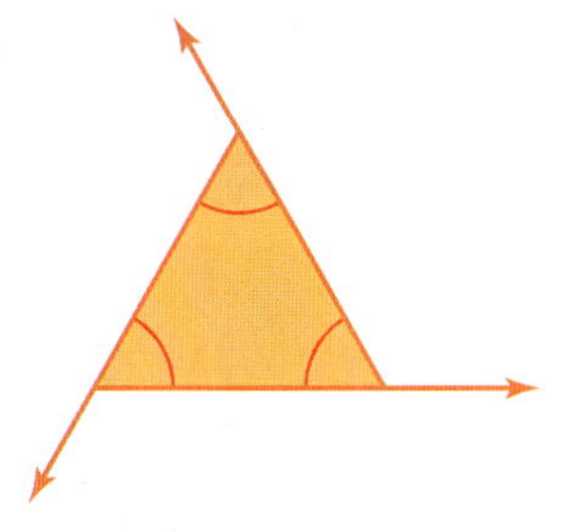

28.

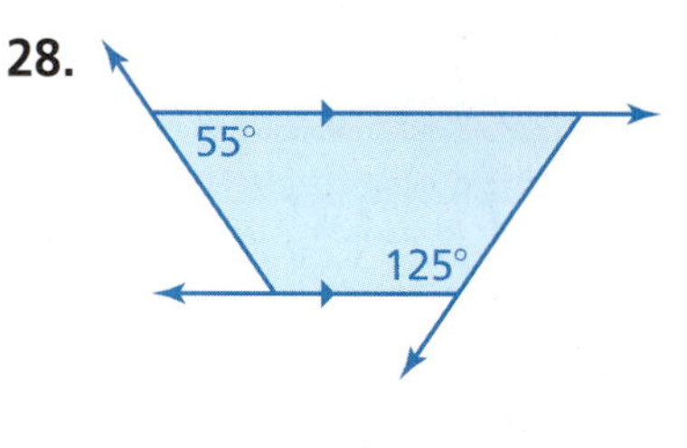

29. **STAINED GLASS** The center of the stained glass window is in the shape of a regular polygon. What is the measure of each interior angle of the polygon? What is the measure of each exterior angle?

30. **PENTAGON** Draw a pentagon that has two right interior angles, two 45° interior angles, and one 270° interior angle.

31. **GAZEBO** The floor of a gazebo is in the shape of a heptagon. Four of the interior angles measure 135°. The other interior angles have equal measures. Find their measures.

32. **MONEY** The border of a Susan B. Anthony dollar is in the shape of a regular polygon.

a. How many sides does the polygon have?

b. What is the measure of each interior angle of the border? Round your answer to the nearest degree.

33. Geometry When tiles can be used to cover a floor with no empty spaces, the collection of tiles is called a *tessellation*.

a. Create a tessellation using equilateral triangles.

b. Find two more regular polygons that form tessellations.

c. Create a tessellation that uses two different regular polygons.

d. Use what you know about interior and exterior angles to explain why the polygons in part (c) form a tessellation.

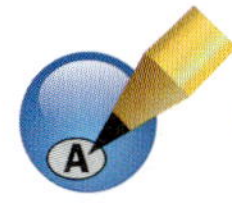

Fair Game Review What you learned in previous grades & lessons

Solve the proportion. *(Skills Review Handbook)*

34. $\frac{x}{12} = \frac{3}{4}$ **35.** $\frac{14}{21} = \frac{x}{3}$ **36.** $\frac{9}{x} = \frac{6}{2}$ **37.** $\frac{10}{4} = \frac{15}{x}$

38. **MULTIPLE CHOICE** The ratio of tulips to daisies is 3 : 5. Which of the following could be the total number of tulips and daisies? *(Skills Review Handbook)*

Ⓐ 6 Ⓑ 10 Ⓒ 15 Ⓓ 16

3.4 Using Similar Triangles

Essential Question How can you use angles to tell whether triangles are similar?

1 ACTIVITY: Constructing Similar Triangles

Work with a partner.

- **Use a straightedge to draw a line segment that is 4 centimeters long.**
- **Then use the line segment and a protractor to draw a triangle that has a 60° and a 40° angle, as shown. Label the triangle *ABC*.**

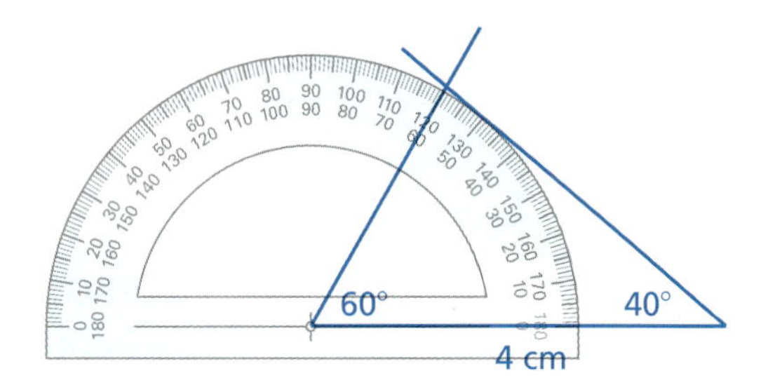

a. Explain how to draw a larger triangle that has the same two angle measures. Label the triangle *JKL*.

b. Explain how to draw a smaller triangle that has the same two angle measures. Label the triangle *PQR*.

c. Are all of the triangles similar? Explain.

2 ACTIVITY: Using Technology to Explore Triangles

Work with a partner. Use geometry software to draw the triangle below.

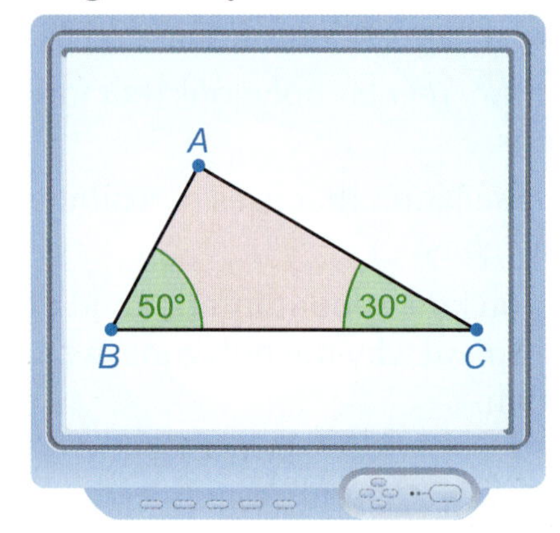

a. Dilate the triangle by the following scale factors.

$$2 \qquad \frac{1}{2} \qquad \frac{1}{4} \qquad 2.5$$

b. Measure the third angle in each triangle. What do you notice?

c. **REASONING** You have two triangles. Two angles in the first triangle are congruent to two angles in the second triangle. Can you conclude that the triangles are similar? Explain.

Geometry

In this lesson, you will

- understand the concept of similar triangles.
- identify similar triangles.
- use indirect measurement to find missing measures.

Learning Standard
MACC.8.G.1.5

Laurie's Notes

Introduction

Standards for Mathematical Practice

- **MP5 Use Appropriate Tools Strategically:** Similar triangles can be investigated using protractor and ruler or dynamic geometry software. Mathematically proficient students consider the available tools when solving a mathematics problem.

Motivate

- Each student will need a protractor and a ruler for today's activities.
- Ask students to work with their partners to construct a triangle with side lengths of 4 inches, 5 inches and 6 inches.
- Without a compass, they will need to work together, using both rulers to locate the third vertex.
- Have students measure the angles in the triangle they constructed.
- Have students hold up their constructions and look at the work of others. Discuss the results; namely, that all of the triangles are congruent.

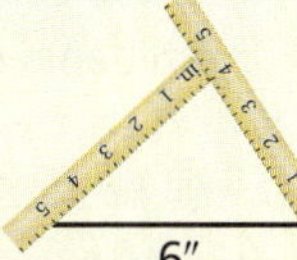

Activity Notes

Activity 1

- **MP5:** The goal of Activities 1 and 2 is the same. The approach is different depending upon the tools selected.
- After students have finished drawing the first triangle, stop to compare results. The orientations may be different, however all of the triangles should be congruent.

? "Do you know the measure of the third angle of your triangle without measuring? Explain." yes; 80° because the three angle measures need to sum to 180° and the first two angle measures already total 100°.

- Circulate as students work on parts (a) and (b). They should not have difficulty drawing the triangles.

? "Are all of the triangles similar? Explain." same shape, just a different size

? "How are all the triangles alike?" same angle measures, 40°, 60°, and 80°

- Discuss the definition of similar triangles. It is unlikely that any students checked to see that corresponding sides were in the same ratio. This is done easily using dynamic software. Otherwise, take the time for students to measure the sides, and allow for a bit of human error in measuring. Check to see that corresponding sides are proportional.

Activity 2

- The length of side BC is not specified.
- You may need to demonstrate how the dilate function of the software is used to complete part (a).
- Students can use the measure function to find the measure of angle A in each triangle. Students can also measure sides to see that corresponding sides are proportional.

Florida Common Core Standards

MACC.8.G.1.5 Use informal arguments to establish facts about the angle sum and exterior angle of triangles, about the angles created when parallel lines are cut by a transversal, and the angle-angle criterion for similarity of triangles.

Previous Learning

Students should know the definition of similar triangles.

Lesson Plans
Complete Materials List

3.4 Record and Practice Journal

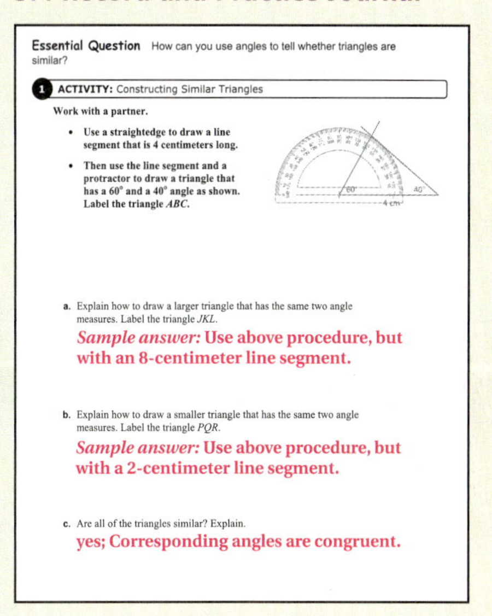

Essential Question How can you use angles to tell whether triangles are similar?

1 ACTIVITY: Constructing Similar Triangles

Work with a partner.

- **Use a straightedge to draw a line segment that is 4 centimeters long.**
- **Then use the line segment and a protractor to draw a triangle that has a 60° and a 40° angle as shown. Label the triangle *ABC*.**

60° 40° 4 cm

a. Explain how to draw a larger triangle that has the same two angle measures. Label the triangle *JKL*.

Sample answer: Use above procedure, but with an 8-centimeter line segment.

b. Explain how to draw a smaller triangle that has the same two angle measures. Label the triangle *PQR*.

Sample answer: Use above procedure, but with a 2-centimeter line segment.

c. Are all of the triangles similar? Explain.

yes; Corresponding angles are congruent.

Differentiated Instruction

Kinesthetic

When setting up a proportion, have students write each of the three known values and the one unknown value with their units on index cards. On a fifth index card, have the students write an equal sign. Students should then place the cards on their desks to set up the proportion. Discuss the different ways to set up a proportion.

3.4 Record and Practice Journal

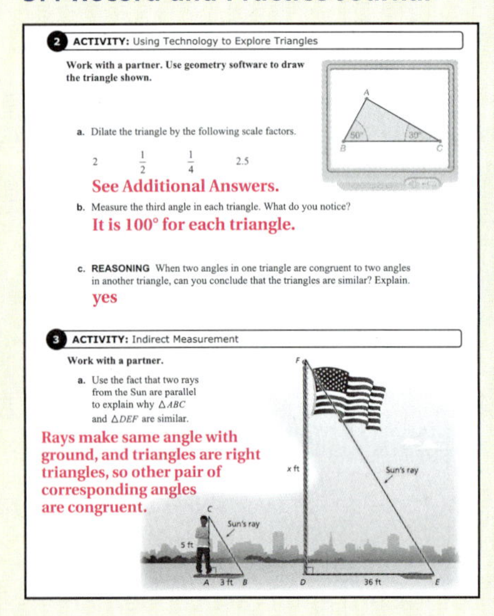

2 ACTIVITY: Using Technology to Explore Triangles

Work with a partner. Use geometry software to draw the triangle shown.

a. Dilate the triangle by the following scale factors.

$2 \quad \frac{1}{2} \quad \frac{1}{4} \quad 2.5$

See Additional Answers.

b. Measure the third angle in each triangle. What do you notice?

It is 100° for each triangle.

c. **REASONING** When two angles in one triangle are congruent to two angles in another triangle, can you conclude that the triangles are similar? Explain.

yes

3 ACTIVITY: Indirect Measurement

Work with a partner.

a. Use the fact that two rays from the Sun are parallel to explain why $\triangle ABC$ and $\triangle DEF$ are similar.

Rays make same angle with ground, and triangles are right triangles, so other pair of corresponding angles are congruent.

b. Explain how to use similar triangles to find the height of the flagpole.

Solve $\frac{x}{5} = \frac{36}{3}$.

What Is Your Answer?

4. **IN YOUR OWN WORDS** How can you use angles to tell whether triangles are similar?

When two angles of one triangle are congruent to two angles of another, the third angles are congruent and the triangles are similar.

5. **PROJECT** Work with a partner or in a small group.

a. Explain why the process in Activity 3 is called "indirect" measurement.

You are not measuring the flagpole directly.

b. **CHOOSE TOOLS** Use indirect measurement to measure the height of something outside your school (a tree, a building, a flagpole). Before going outside, decide what materials you need to take with you.

Check students' work.

c. **MODELING** Draw a diagram of the indirect measurement process you used. In the diagram, label the lengths that you actually measured and also the lengths that you calculated.

Check students' work.

6. **PRECISION** Look back at Exercise 17 in Section 2.5. Explain how you can show that the two triangles are similar.

See Additional Answers.

Laurie's Notes

Activity 3

- Students will need to use the results of the first two activities. When two triangles have two congruent angles, the triangles are similar.
- ? "Is your shadow shorter at noon or 5 P.M.? Explain." Noon; The sun is overhead, not at a lower position in the sky.
- ? "Do adjacent objects of different heights cast the same length shadow? Explain." no; Taller objects cast longer shadows.
- The triangles are similar because they both have a right angle and the parallel rays of the sun are at the same angle to the ground.

What Is Your Answer?

- Question 4 summarizes the results of today's activities.
- If you assign Question 5, it will take planning and discussions before students go outside.

Closure

- **Exit Ticket:** Are all 40°-60°-80° triangles congruent? Similar? Explain. no; For triangles to be congruent, corresponding sides must be congruent. yes; For similar triangles, only the corresponding angles need to be congruent.

3 ACTIVITY: Indirect Measurement

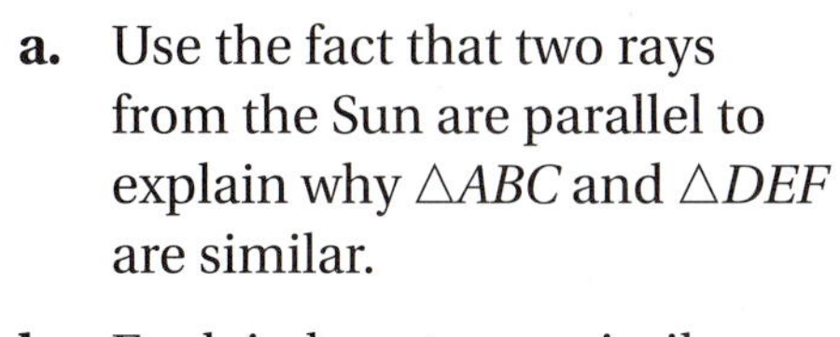

Work with a partner.

a. Use the fact that two rays from the Sun are parallel to explain why $\triangle ABC$ and $\triangle DEF$ are similar.

b. Explain how to use similar triangles to find the height of the flagpole.

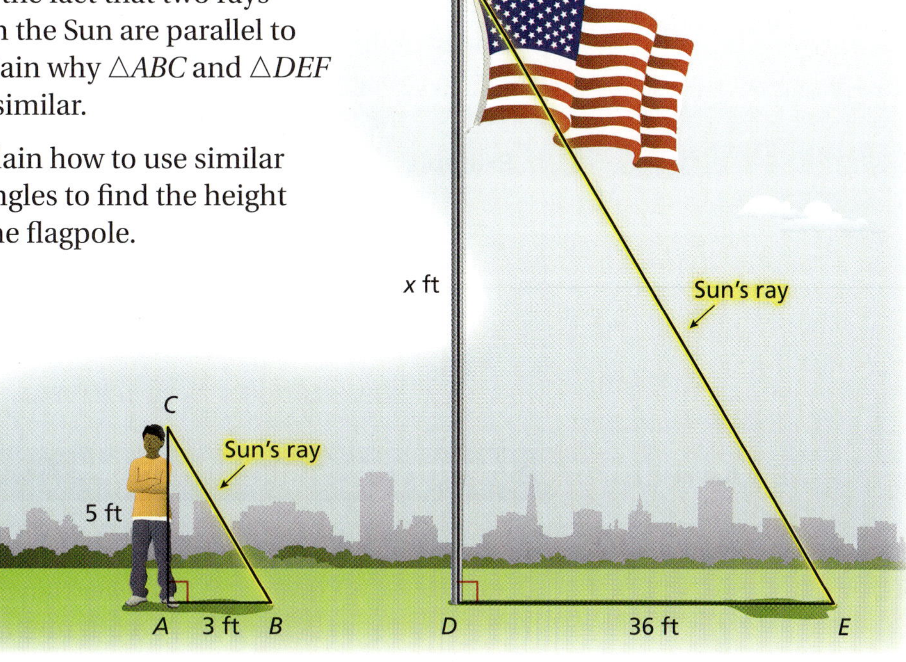

What Is Your Answer?

4. **IN YOUR OWN WORDS** How can you use angles to tell whether triangles are similar?

5. **PROJECT** Work with a partner or in a small group.

   a. Explain why the process in Activity 3 is called "indirect" measurement.

   b. **CHOOSE TOOLS** Use indirect measurement to measure the height of something outside your school (a tree, a building, a flagpole). Before going outside, decide what materials you need to take with you.

   c. **MODELING** Draw a diagram of the indirect measurement process you used. In the diagram, label the lengths that you actually measured and also the lengths that you calculated.

6. **PRECISION** Look back at Exercise 17 in Section 2.5. Explain how you can show that the two triangles are similar.

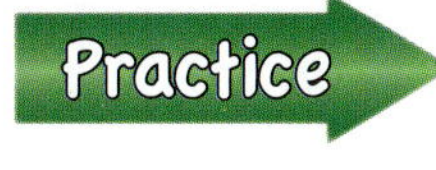

Use what you learned about similar triangles to complete Exercises 4 and 5 on page 130.

3.4 Lesson

Key Vocabulary

indirect measurement, *p. 129*

Angles of Similar Triangles

Words When two angles in one triangle are congruent to two angles in another triangle, the third angles are also congruent and the triangles are similar.

Example

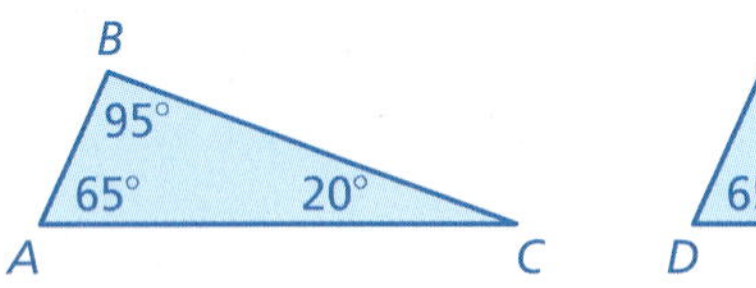

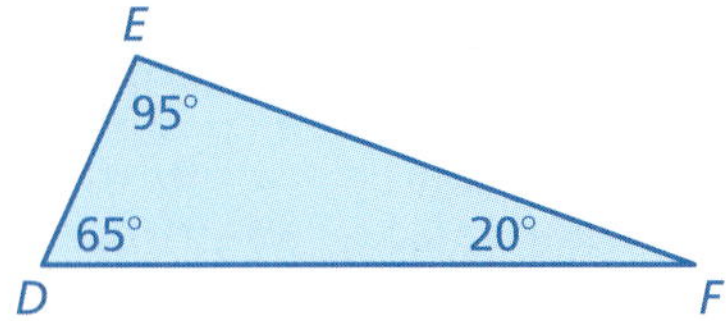

Triangle ABC is similar to Triangle DEF: $\triangle ABC \sim \triangle DEF$.

EXAMPLE 1 Identifying Similar Triangles

Tell whether the triangles are similar. Explain.

a.

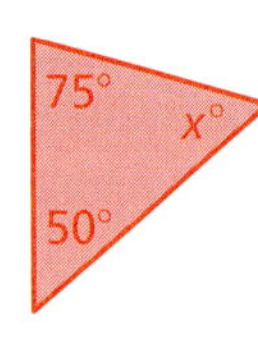

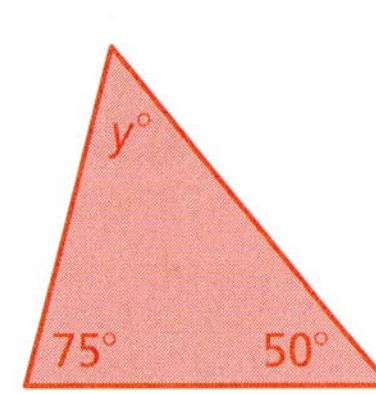

The triangles have two pairs of congruent angles.

So, the third angles are congruent, and the triangles are similar.

b.

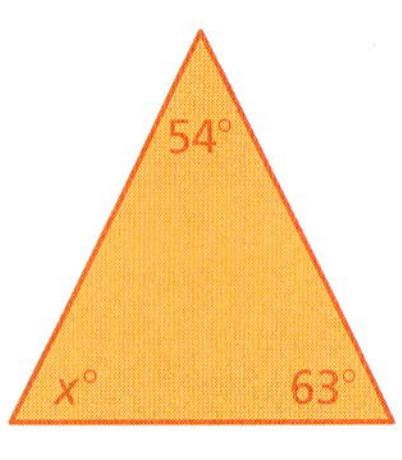

Write and solve an equation to find x.

$$x + 54 + 63 = 180$$

$$x + 117 = 180$$

$$x = 63$$

The triangles have two pairs of congruent angles.

So, the third angles are congruent, and the triangles are similar.

c.

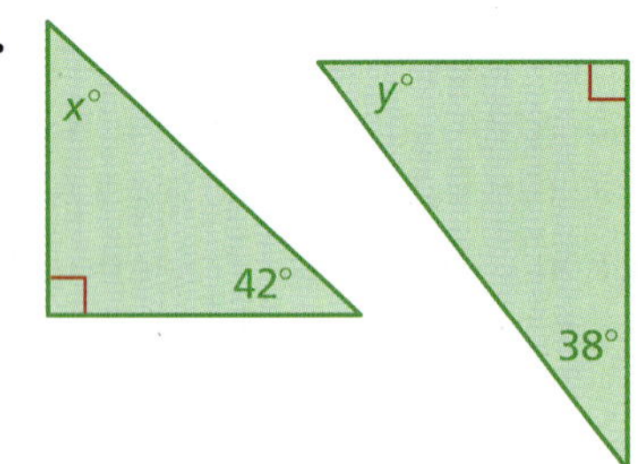

Write and solve an equation to find x.

$$x + 90 + 42 = 180$$

$$x + 132 = 180$$

$$x = 48$$

The triangles do not have two pairs of congruent angles.

So, the triangles are not similar.

Multi-Language Glossary at BigIdeasMath.com

Laurie's Notes

Introduction

Connect

- **Yesterday:** Students explored special properties of similar triangles. (MP5)
- **Today:** Students will use similar triangles to solve real-life problems.

Motivate

- Have pairs of students put a visual barrier (i.e., a notebook) between them. One student draws a triangle using a straightedge. This student now gives directions to the second student who will draw a triangle based on the information given. The only information the first student may give is angle measure! In fact, after the second angle measure is given, the second student should know the measure of the third angle.
- The triangles should be similar.

? "What do you notice about the triangles?" similar

? "How do you know they are similar?" Listen for same shape, different size; students may also mention yesterday's activities.

Lesson Notes

Key Idea

- Write the informal definition (same shape, not necessarily the same size), and draw examples of similar triangles.

? "What is the formal definition of similar triangles?" Similar triangles have corresponding sides that are proportional and corresponding angles that are congruent.

- Write the Key Idea.
- Make sure students understand that there are two parts stated in the Key Idea. When you have two angles in one triangle congruent to two angles in another triangle, then:
  1) the third angles are congruent.
  2) the two triangles are similar.

Example 1

- Draw the two triangles and label the given information. Ask students to solve for the missing angle measure of each triangle.

? "Are the triangles in part (a) similar? Explain." yes; The triangles have two pairs of congruent angles.

? "Are the triangles in part (b) similar? Explain." yes; First, solve for the missing angles. The triangles have two pairs of congruent angles, so they are similar.

? "Do you need to solve for y in part (b)?" no; Once you solve for x, you know that the triangles have two pairs of congruent angles.

? "What type of triangles are in part (c)?" Both are right triangles.

? "Are the triangles similar?" no; Solving for the missing angles, the triangles only have one pair of congruent angles. So the triangles are not similar.

Goal Today's lesson is using similar triangles to solve problems.

Extra Example 1

Tell whether the triangles are similar. Explain.

a. 58° x° 43° 43° y° 59°

no; The triangles do not have the same angle measures.

b.

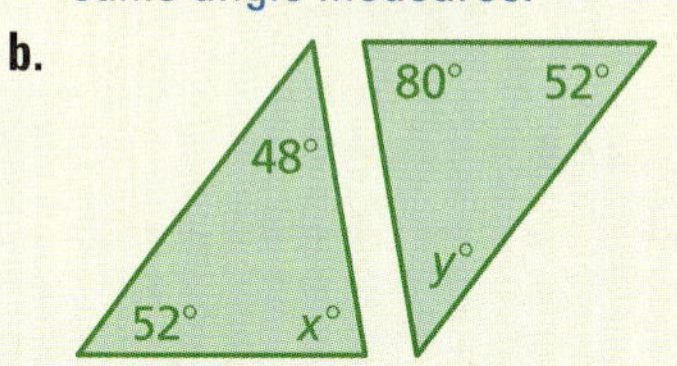

yes; The triangles have the same angle measures, 52°, 48°, and 80°.

c.

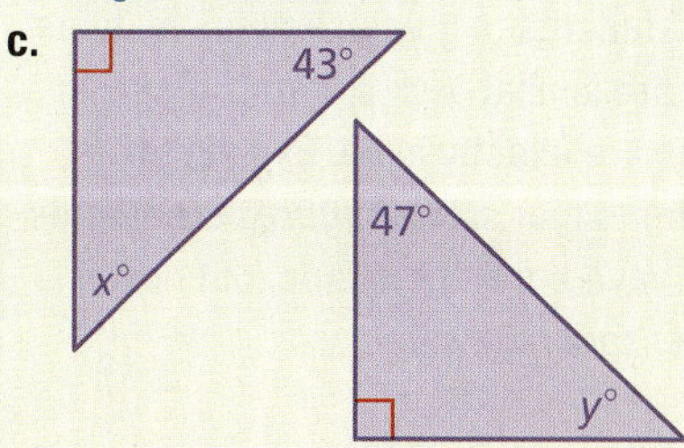

yes; The triangles have the same angle measures, 47°, 90°, and 43°.

On Your Own

1. no; The triangles do not have the same angle measures.
2. yes; The triangles have the same angle measures, 90°, 66°, and 24°.

Extra Example 2

You plan to cross a river and want to know how far it is to the other side. You take measurements on your side of the river and make the drawing shown.

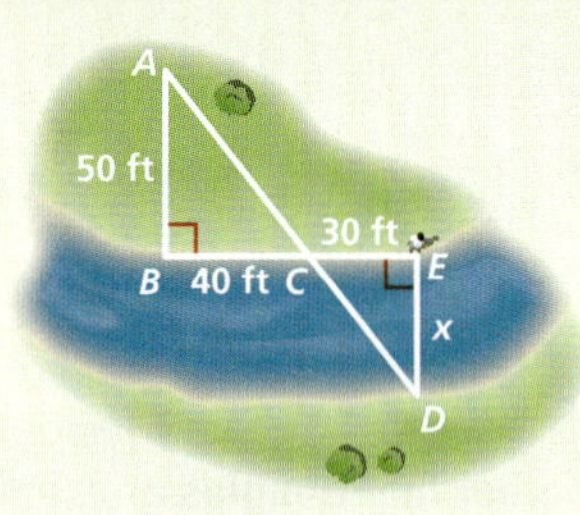

a. Explain why $\triangle ABC$ and $\triangle DEC$ are similar. Because two angles in $\triangle ABC$ are congruent to two angles in $\triangle DEC$, the third angles are also congruent and the triangles are similar.

b. What is the distance x across the river? 37.5 ft

On Your Own

3. 44 ft

English Language Learners

Build on Past Knowledge

Ask students to give examples of items that are similar. Ask students if similar items are exactly alike. Explain to students that *similar figures* are figures that have the same shape, but not necessarily the same size.

Laurie's Notes

On Your Own

- **Think-Pair-Share:** Students should read each question independently and then work in pairs to answer the questions. When they have answered the questions, the pair should compare their answers with another group and discuss any discrepancies.

Example 2

- Indirect measurement is used when you want to know the measurement of some length (or angle) and you cannot measure the object directly.
- Ask a volunteer to read the problem. Make a rough sketch of the diagram.

? "What do you know about the angles in either triangle?" $\angle B$ and $\angle E$ are right angles. The vertical angles are congruent (mark the diagram to show the congruent angles).

? "What do you know about the third angle in each triangle?" They are congruent.

- **MP6 Attend to Precision:** Because the triangles are similar, the corresponding sides will have the same ratio. Setting up the ratios is challenging for students. Talk about the sides in terms of being the shorter leg of the right triangle and the longer leg of the right triangle.
- Use the Multiplication Property of Equality or the Cross Products Property to solve. Check the reasonableness of the answer.

On Your Own

- **Think-Pair-Share:** Students should read the question independently and then work in pairs to answer the question. When they have answered the question, the pair should compare their answer with another group and discuss any discrepancies.

Closure

- **Exit Ticket:** Are the two triangles similar? Explain.

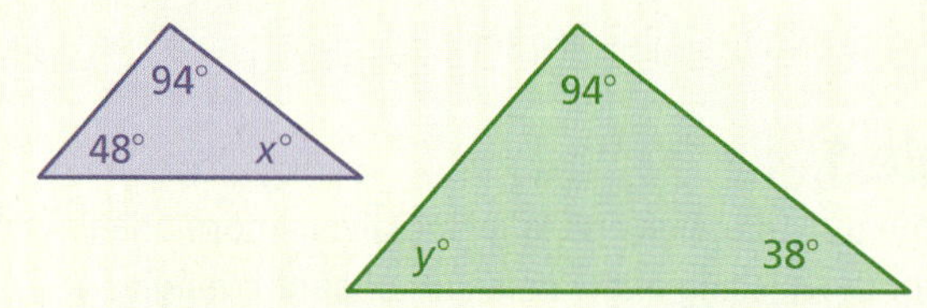

The triangles have the same angle measures, 94°, 48°, and 38°. So, the triangles are similar.

On Your Own

Tell whether the triangles are similar. Explain.

1. 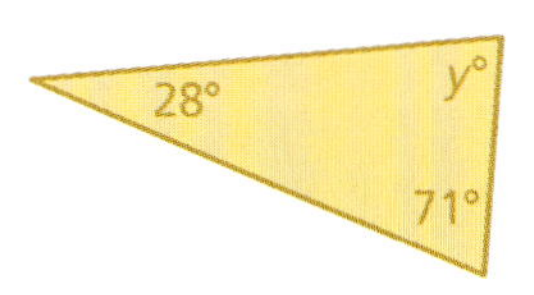

2. 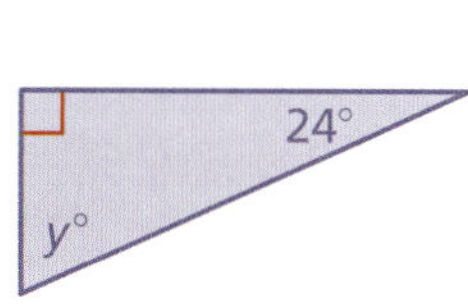

Indirect measurement uses similar figures to find a missing measure when it is difficult to find directly.

EXAMPLE 2 Using Indirect Measurement

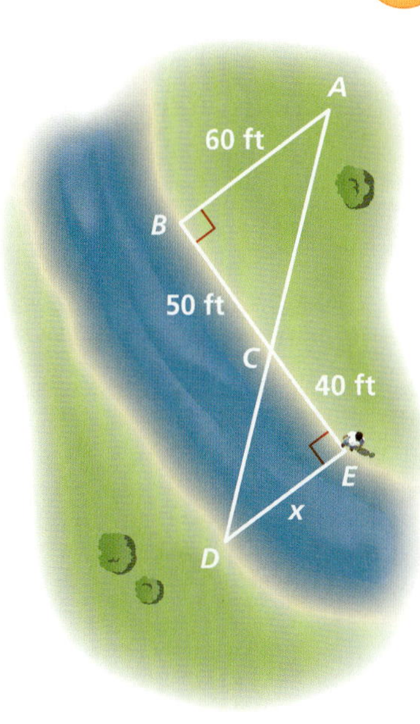

You plan to cross a river and want to know how far it is to the other side. You take measurements on your side of the river and make the drawing shown. (a) Explain why $\triangle ABC$ and $\triangle DEC$ are similar. (b) What is the distance x across the river?

a. $\angle B$ and $\angle E$ are right angles, so they are congruent. $\angle ACB$ and $\angle DCE$ are vertical angles, so they are congruent.

Because two angles in $\triangle ABC$ are congruent to two angles in $\triangle DEC$, the third angles are also congruent and the triangles are similar.

b. The ratios of the corresponding side lengths in similar triangles are equal. Write and solve a proportion to find x.

$$\frac{x}{60} = \frac{40}{50} \qquad \text{Write a proportion.}$$

$$60 \cdot \frac{x}{60} = 60 \cdot \frac{40}{50} \qquad \text{Multiplication Property of Equality}$$

$$x = 48 \qquad \text{Simplify.}$$

So, the distance across the river is 48 feet.

On Your Own

3. **WHAT IF?** The distance from vertex A to vertex B is 55 feet. What is the distance across the river?

3.4 Exercises

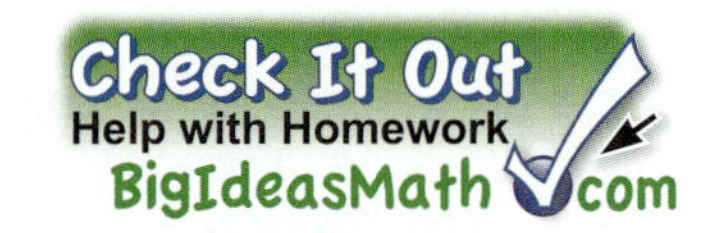

Vocabulary and Concept Check

1. **REASONING** How can you use similar triangles to find a missing measurement?
2. **WHICH ONE DOESN'T BELONG?** Which triangle does *not* belong with the other three? Explain your reasoning.

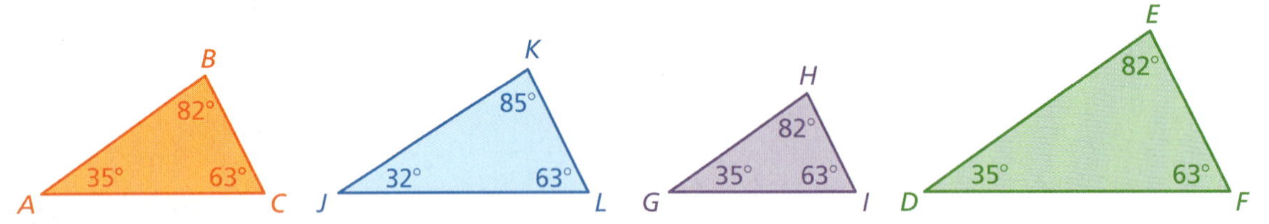

3. **WRITING** Two triangles have two pairs of congruent angles. In your own words, explain why you do not need to find the measures of the third pair of angles to determine that they are congruent.

Practice and Problem Solving

Make a triangle that is larger or smaller than the one given and has the same angle measures. Find the ratios of the corresponding side lengths.

4. 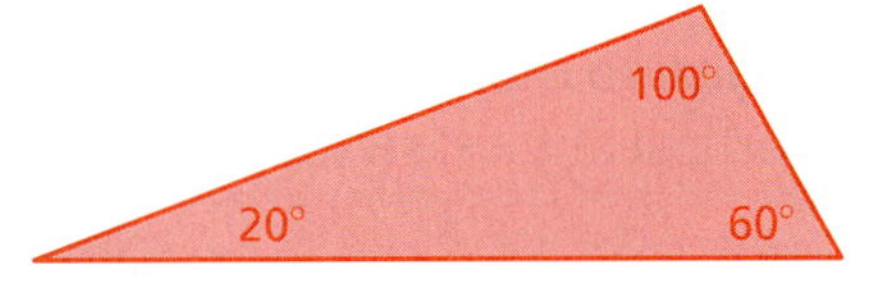

5. 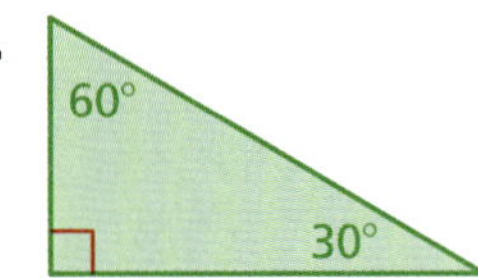

Tell whether the triangles are similar. Explain.

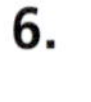

6. 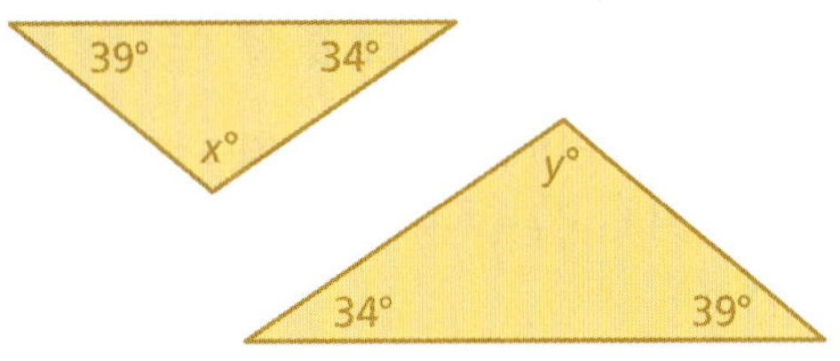

7. 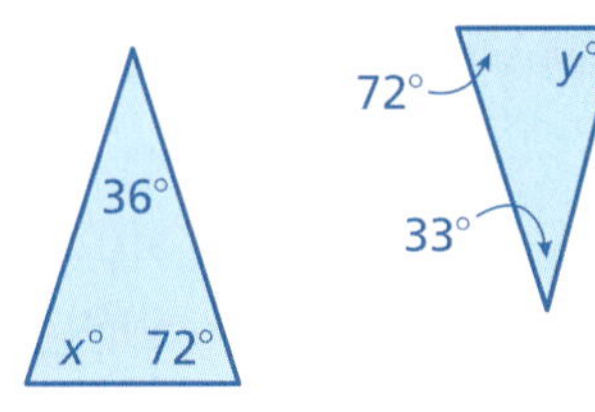

8. 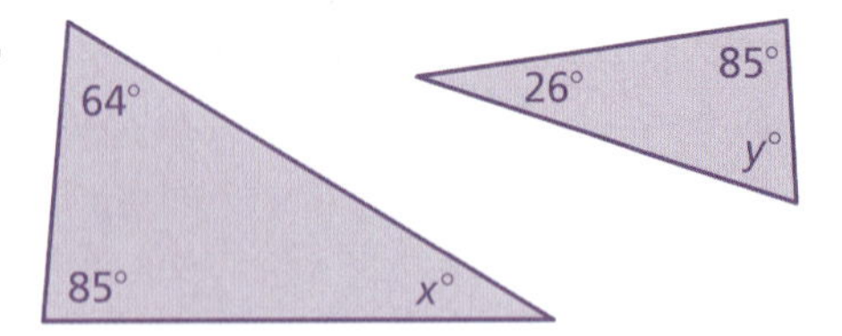

9. 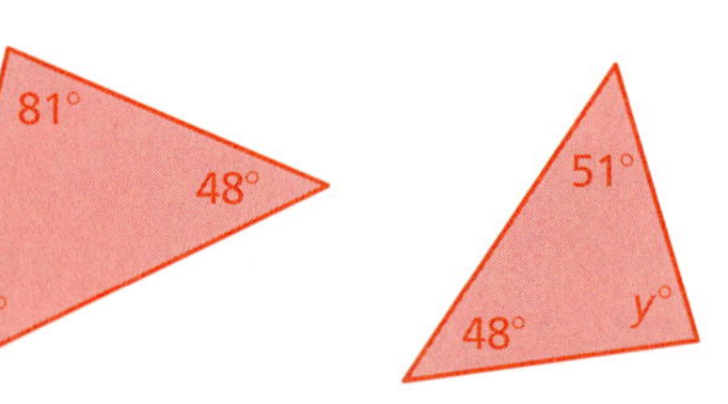

10. **RULERS** Which of the rulers are similar in shape? Explain.

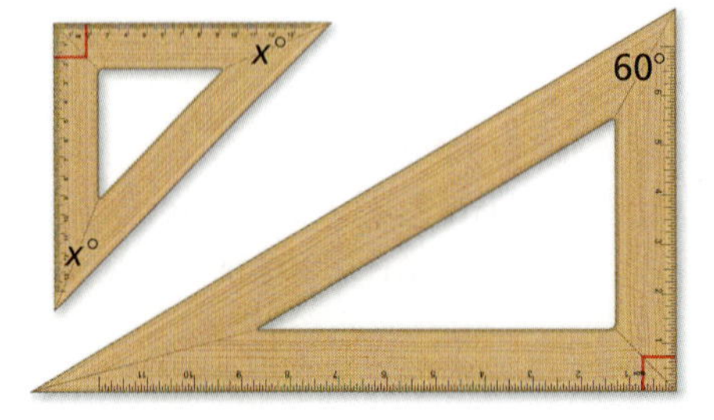

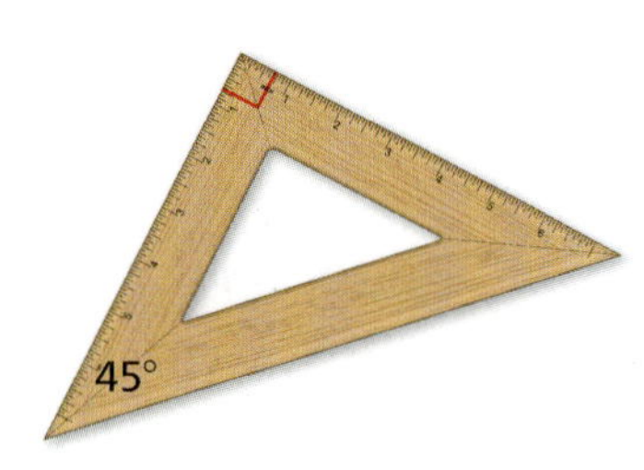

Assignment Guide and Homework Check

Level	Day 1 Activity Assignment	Day 2 Lesson Assignment	Homework Check
Basic	4, 5, 19–22	1–3, 6–14	7, 9, 10, 11, 13
Average	4, 5, 19–22	1–3, 6–16	8, 12, 13, 14, 16
Advanced	1–22		8, 12, 14, 15, 17

Common Errors

- **Exercises 6–9** Students may find the missing angle measure for one of the triangles and then make a decision about the similarity of the triangles. While it is possible to use this method, encourage them to find the missing angles of both triangles to verify that they are correct.

3.4 Record and Practice Journal

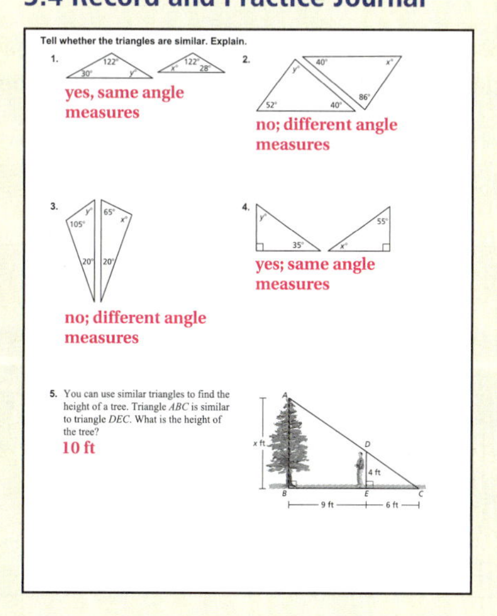
Tell whether the triangles are similar. Explain.

1. yes, same angle measures

2. no; different angle measures

3. no; different angle measures

4. yes; same angle measures

5. You can use similar triangles to find the height of a tree. Triangle *ABC* is similar to triangle *DEC*. What is the height of the tree?
10 ft

Vocabulary and Concept Check

1. Write a proportion that uses the missing measurement because the ratios of corresponding side lengths are equal.
2. $\triangle JKL$ because the other three triangles are similar.
3. *Sample answer:* Two of the angles are congruent, so they have the same sum. When you subtract this from 180°, you will get the same third angle.

Practice and Problem Solving

4–5. Student should draw a triangle with the same angle measures as the ones given in the textbook. If the student's triangle is larger than the one given, then the ratio of the corresponding side lengths, $\frac{\text{student's triangle length}}{\text{book's triangle length}}$, should be greater than 1. If the student's triangle is smaller than the one given, then the ratio of the corresponding side lengths, $\frac{\text{student's triangle length}}{\text{book's triangle length}}$, should be less than 1.

6. yes; The triangles have two pairs of congruent angles.

7. no; The triangles do not have two pairs of congruent angles.

8. no; The triangles do not have the same angle measures.

9. yes; The triangles have the same angle measures, 81°, 51°, and 48°.

10. the leftmost and rightmost; They both are right triangles with 45° angles.

Practice and Problem Solving

11. yes; The triangles have two pairs of congruent angles.

12. no; The triangles do not have two pairs of congruent angles.

13. See *Taking Math Deeper.*

14–18. See Additional Answers.

Fair Game Review

19. $y = 5x + 3$

20. $y = -\frac{2}{3}x + 2$

21. $y = 8x - 4$

22. B

Mini-Assessment

Tell whether the triangles are similar. Explain.

1.

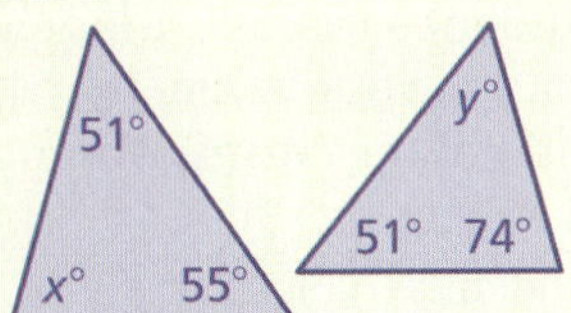

yes; The triangles have the same angle measures, 51°, 55°, and 74°.

2.

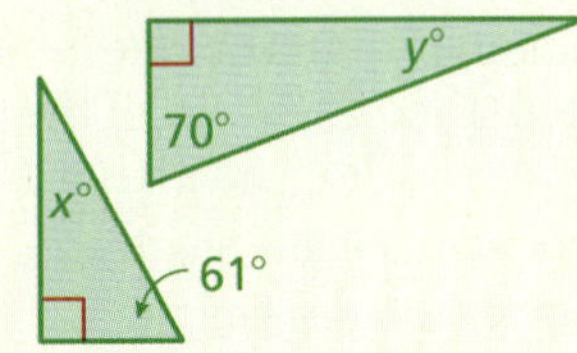

no; The triangles do not have the same angle measures.

3.

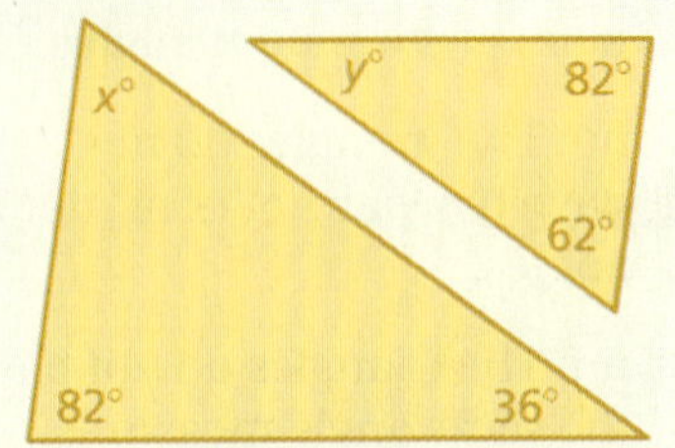

yes; The triangles have the same angle measures, 82°, 62°, and 36°.

4. A person that is 5 feet tall casts a 3-foot-long shadow. A nearby telephone pole casts a 12-foot-long shadow. What is the height h of the telephone pole? 20 ft

Taking Math Deeper

Exercise 13

In Chapter 2, students learned that two figures are similar when one can be obtained from the other by a sequence of translations, reflections, rotations, and dilations. So, one way to solve this problem is to draw the situation in a coordinate plane.

1 Explain why the triangles are similar.

Because the triangles have a pair of vertical angles and a pair of right angles, the triangles have the same interior angle measures. So, the triangles are similar.

2 Draw the larger triangle in a coordinate plane. Let the vertex of the vertical angles be located at (0, 0).

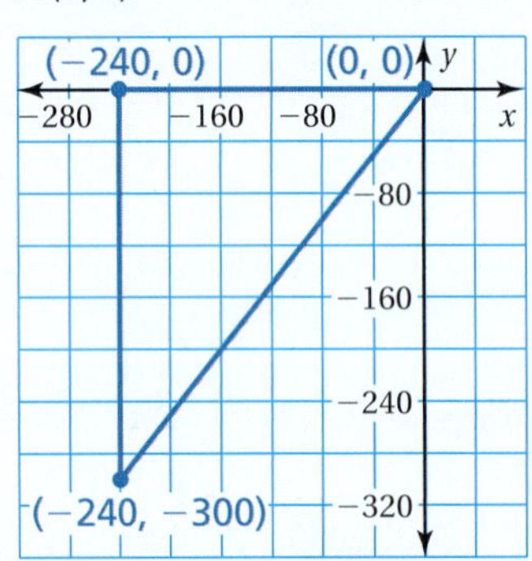

Rotate the triangle 180° about the origin. Then dilate the image using a scale factor of $\frac{1}{3}$ to obtain the smaller triangle.

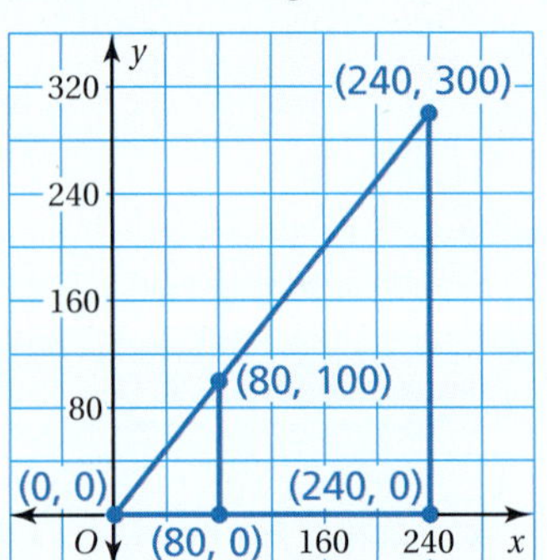

3 Answer the question.

By obtaining the smaller triangle through transformations, you can see that the treasure is located at (80, 100).

So, you take 100 steps from the pyramids to the treasure.

Project

Have students research the Pythagorean Theorem, and then use it to find the number of steps you would have by taking the "straight-line" approach.

Reteaching and Enrichment Strategies

If students need help. . .	If students got it. . .
Resources by Chapter • Practice A and Practice B • Puzzle Time Record and Practice Journal Practice Differentiating the Lesson Lesson Tutorials Skills Review Handbook	Resources by Chapter • Enrichment and Extension • Technology Connection Start the next section

Tell whether the triangles are similar. Explain.

11.

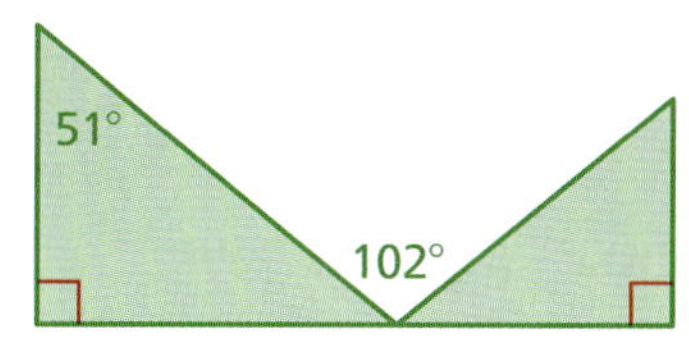

12.

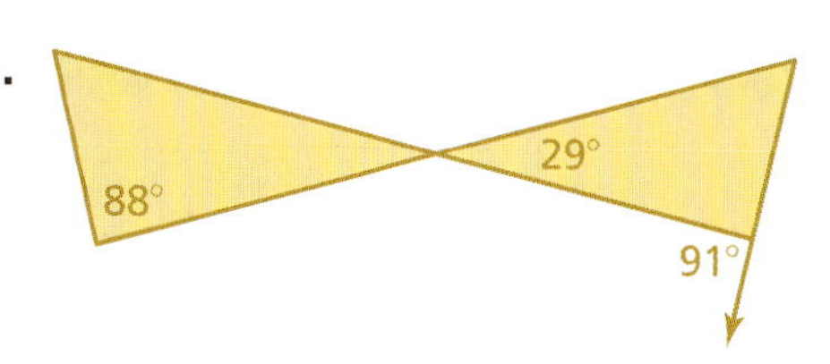

2 **13. TREASURE** The map shows the number of steps you must take to get to the treasure. However, the map is old, and the last dimension is unreadable. Explain why the triangles are similar. How many steps do you take from the pyramids to the treasure?

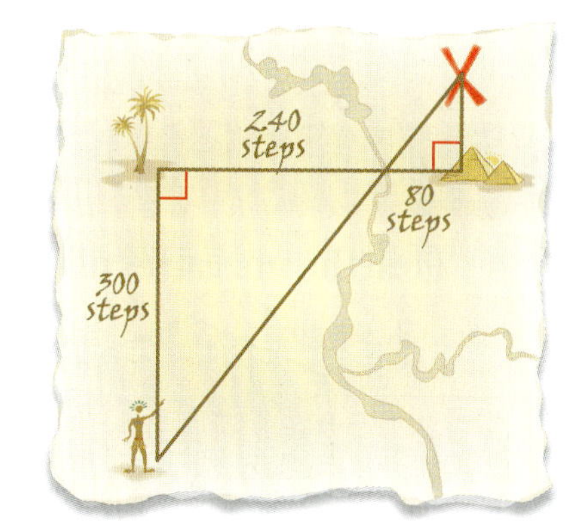

14. CRITICAL THINKING The side lengths of a triangle are increased by 50% to make a similar triangle. Does the area increase by 50% as well? Explain.

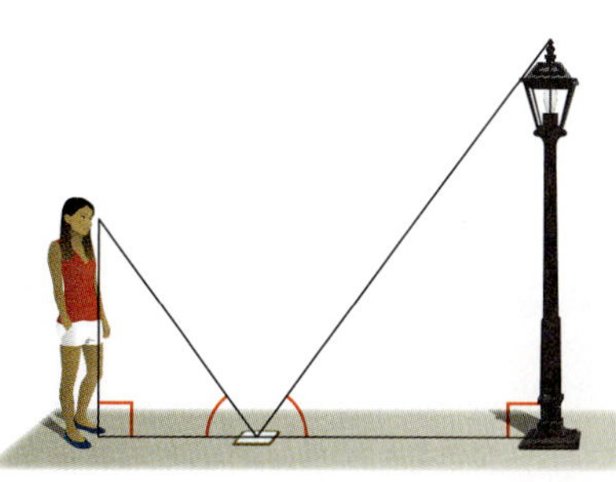

15. PINE TREE A person who is 6 feet tall casts a 3-foot-long shadow. A nearby pine tree casts a 15-foot-long shadow. What is the height h of the pine tree?

16. OPEN-ENDED You place a mirror on the ground 6 feet from the lamppost. You move back 3 feet and see the top of the lamppost in the mirror. What is the height of the lamppost?

17. REASONING In each of two right triangles, one angle measure is two times another angle measure. Are the triangles similar? Explain your reasoning.

18. Geometry In the diagram, segments BG, CF, and DE are parallel. The length of segment BD is 6.32 feet, and the length of segment DE is 6 feet. Name all pairs of similar triangles in the diagram. Then find the lengths of segments BG and CF.

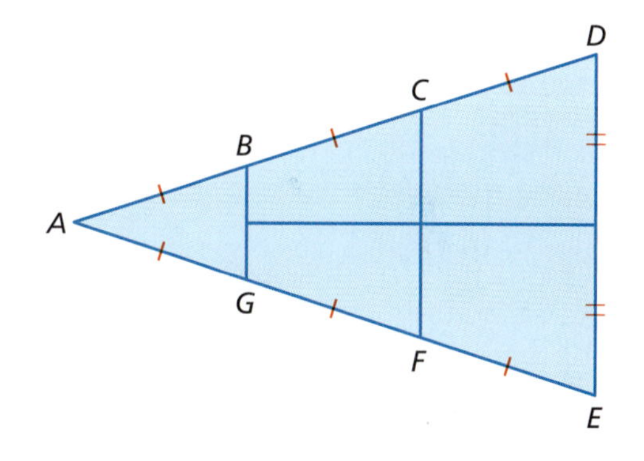

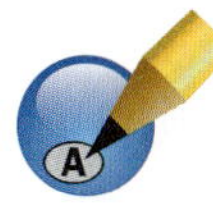

Fair Game Review What you learned in previous grades & lessons

Solve the equation for *y*. *(Section 1.4)*

19. $y - 5x = 3$

20. $4x + 6y = 12$

21. $2x - \frac{1}{4}y = 1$

22. MULTIPLE CHOICE What is the value of x? *(Section 3.2)*

Ⓐ 17 Ⓑ 62

Ⓒ 118 Ⓓ 152

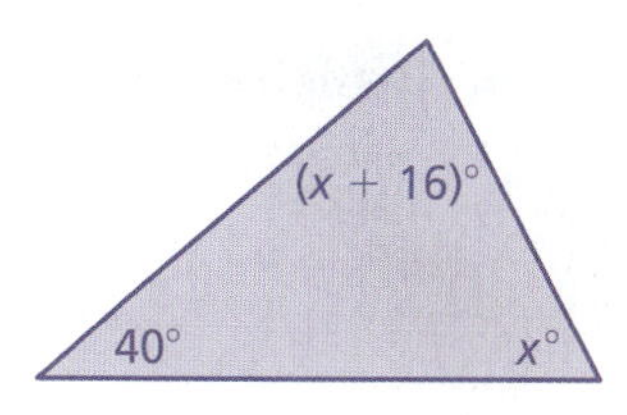

3.3–3.4 Quiz

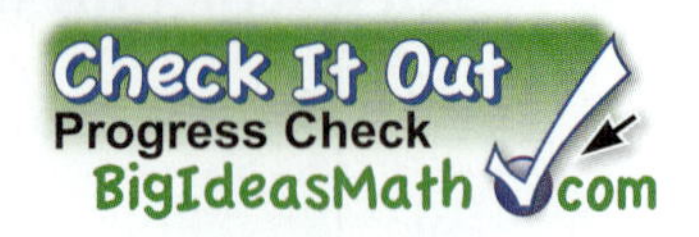

Find the sum of the interior angle measures of the polygon. *(Section 3.3)*

1.

2.

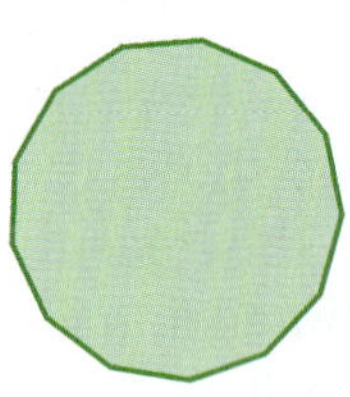

Find the measures of the interior angles of the polygon. *(Section 3.3)*

3.

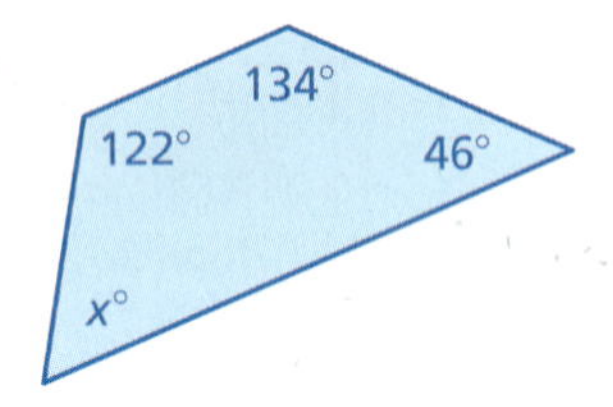

4.

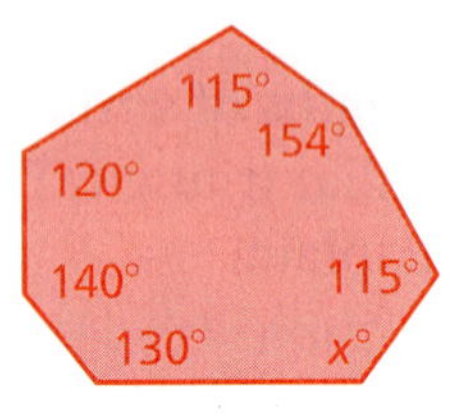

5.

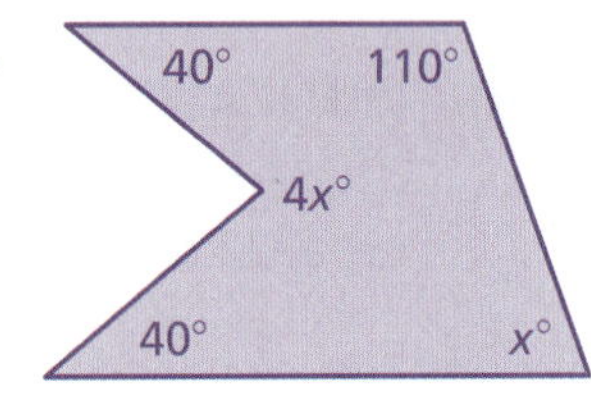

Find the measures of the exterior angles of the polygon. *(Section 3.3)*

6.

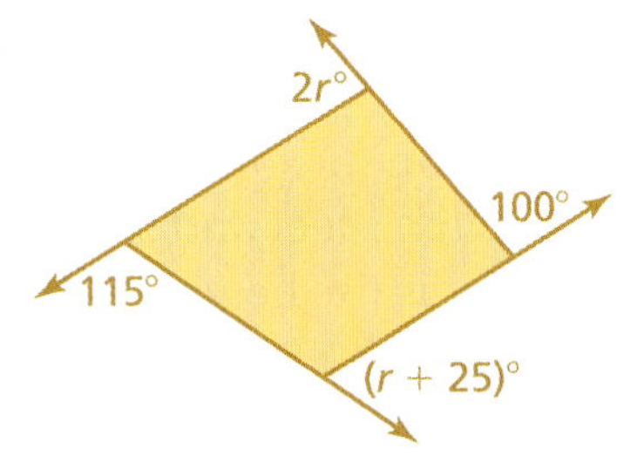

7.

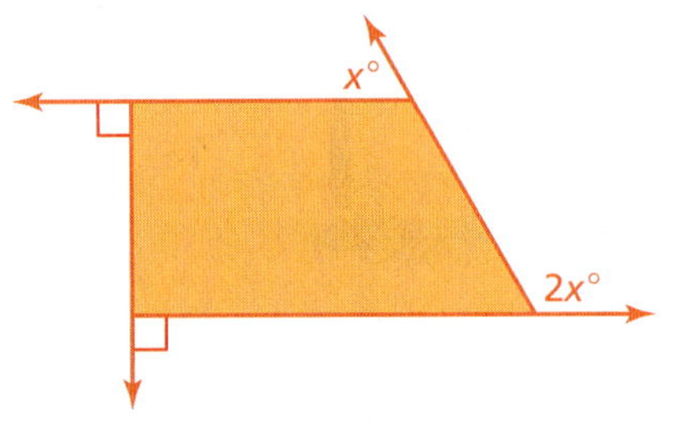

Tell whether the triangles are similar. Explain. *(Section 3.4)*

8.

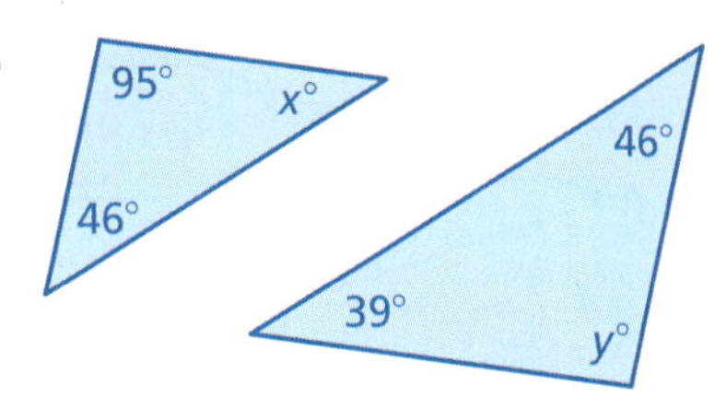

9.

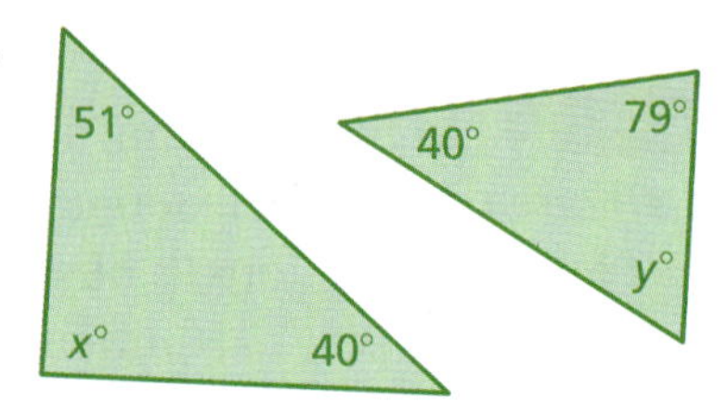

10. REASONING The sum of the interior angle measures of a polygon is 4140°. How many sides does the polygon have? *(Section 3.3)*

11. SWAMP You are trying to find the distance ℓ across a patch of swamp water. *(Section 3.4)*

a. Explain why $\triangle VWX$ and $\triangle YZX$ are similar.

b. What is the distance across the patch of swamp water?

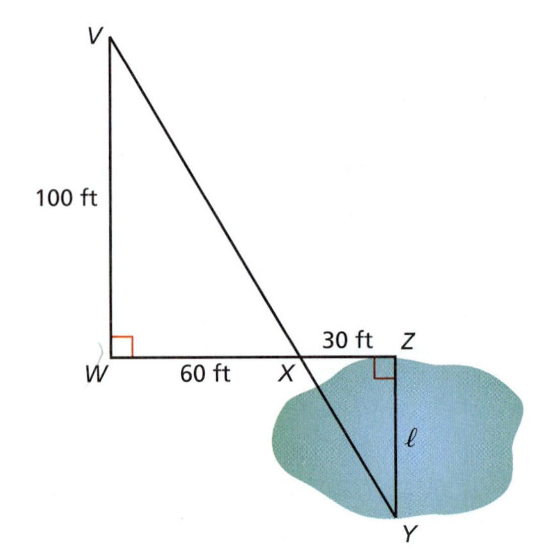

Alternative Assessment Options

Math Chat	Student Reflective Focus Question
Structured Interview	Writing Prompt

Math Chat

- Put students in pairs to complete and discuss the exercises from the quiz.
- The discussion should include terms such as sums and measures of interior angles, measures of exterior angles, and similar triangles.
- The teacher should walk around the classroom listening to the pairs and ask questions to ensure understanding.

Study Help Sample Answers

Remind students to complete Graphic Organizers for the rest of the chapter.

3. Interior angles of a polygon

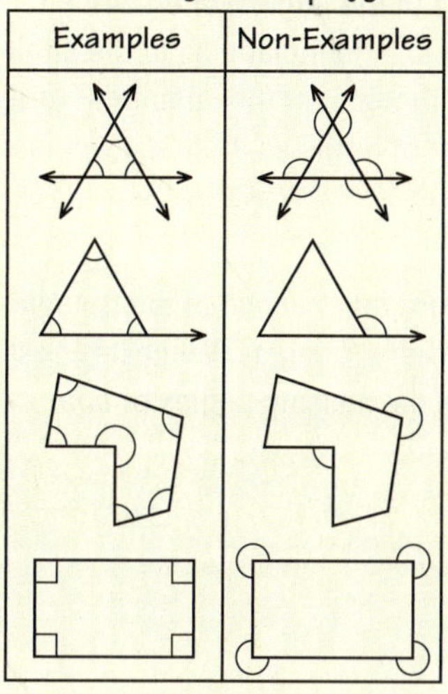

4–6. Available at *BigIdeasMath.com.*

Reteaching and Enrichment Strategies

If students need help...	If students got it...
Resources by Chapter • Practice A and Practice B • Puzzle Time Lesson Tutorials *BigIdeasMath.com*	Resources by Chapter • Enrichment and Extension • Technology Connection Game Closet at *BigIdeasMath.com* Start the Chapter Review

Answers

1. 1440°
2. 1980°
3. 58°; 122°; 134°; 46°
4. 126°; 130°; 140°; 120°; 115°; 154°; 115°
5. 280°; 40°; 110°; 70°; 40°
6. 80°; 100°; 65°; 115°
7. 60°; 120°; 90°; 90°
8. yes; The triangles have the same angle measures, 95°, 46°, and 39°.
9. no; The triangles do not have the same angle measures.
10. 25 sides
11. a. Angles W and Z are right angles, so they are congruent. Angles WXV and ZXY are vertical angles, so they are congruent. Because two angles in $\triangle VWX$ are congruent to two angles in $\triangle YZX$, the third angles are also congruent and the triangles are similar.

    b. 50 ft

Technology for the Teacher

Online Assessment
Assessment Book
ExamView® Assessment Suite

For the Teacher
Additional Review Options

- *BigIdeasMath.com*
- Online Assessment
- Game Closet at *BigIdeasMath.com*
- Vocabulary Help
- Resources by Chapter

Answers

1. 140°; ∠8 and the given angle are alternate exterior angles.
2. 140°; ∠8 and ∠5 are vertical angles.
3. 40°; ∠8 and ∠7 are supplementary.
4. 40°; ∠2 and the given angle are supplementary.

Review of Common Errors

Exercises 1–4

- Students may not understand alternate interior and exterior angles and think that an exterior angle is congruent to the alternate interior angle. Use corresponding angles to show that this is not necessarily true.

Exercise 8

- Students may forget to combine like terms when solving for t. Remind them that because there are two variables on the same side of the equal sign, they should start by combining like terms.
- Students may solve for the variable, but forget to find the measures of the angles. Remind them to read the directions carefully and to answer the question.

Exercises 9–11

- Students may forget to include one or more of the angles when writing an equation to find the value of x. Remind students to include all of the angles. Encourage them to write the equation and then count the number of terms to make sure that there is the same number of terms as there are angles before solving.

Exercises 14 and 15

- Students may find the missing angle measure for only one of the triangles and then make a decision about the similarity of the triangles. While it is possible to use this method, encourage them to find the missing angles of *both* triangles to verify that they are correct.

3 Chapter Review

Check It Out
Vocabulary Help
BigIdeasMath.com

Review Key Vocabulary

transversal, *p. 104*
interior angles, *p. 105*
exterior angles, *p. 105*
interior angles of a polygon, *p. 112*
exterior angles of a polygon, *p. 112*
convex polygon, *p. 119*
concave polygon, *p. 119*
regular polygon, *p. 121*
indirect measurement, *p. 129*

Review Examples and Exercises

3.1 Parallel Lines and Transversals *(pp. 102–109)*

Use the figure to find the measure of $\angle 6$.

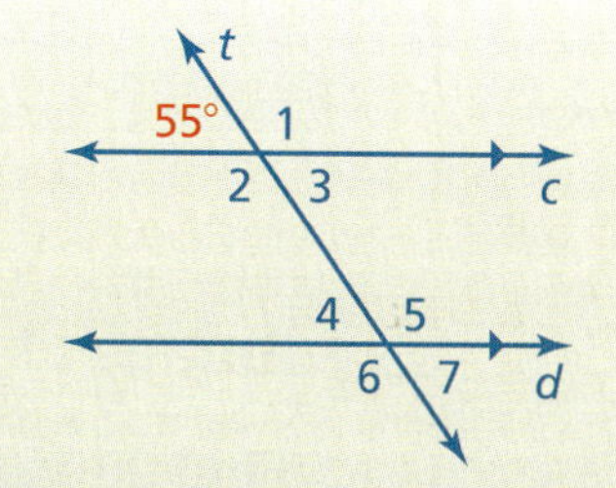

$\angle 2$ and the 55° angle are supplementary.
So, the measure of $\angle 2$ is $180° - 55° = 125°$.

$\angle 2$ and $\angle 6$ are corresponding angles.
They are congruent.

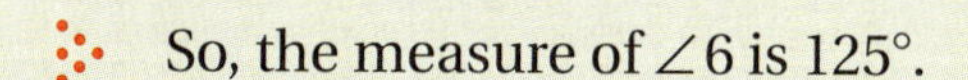

So, the measure of $\angle 6$ is 125°.

Exercises

Use the figure to find the measure of the angle. Explain your reasoning.

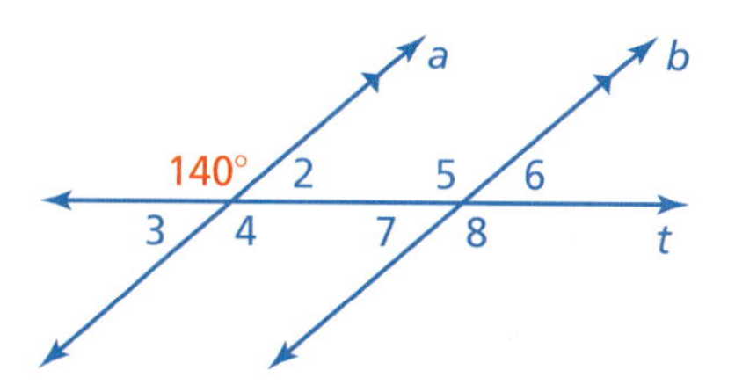

1. $\angle 8$
2. $\angle 5$
3. $\angle 7$
4. $\angle 2$

3.2 Angles of Triangles *(pp. 110–115)*

a. Find the value of x.

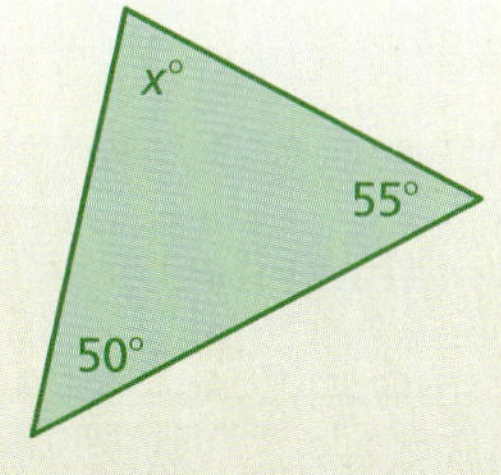

$$x + 50 + 55 = 180$$
$$x + 105 = 180$$
$$x = 75$$

The value of x is 75.

b. Find the measure of the exterior angle.

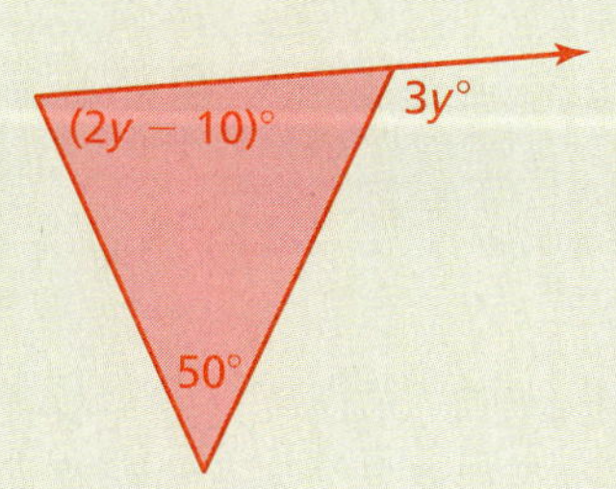

$$3y = (2y - 10) + 50$$
$$3y = 2y + 40$$
$$y = 40$$

So, the measure of the exterior angle is $3(40)° = 120°$.

Exercises

Find the measures of the interior angles.

5.

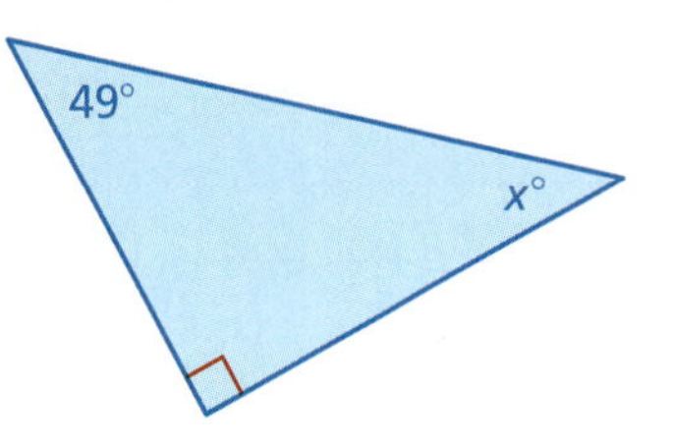

6.

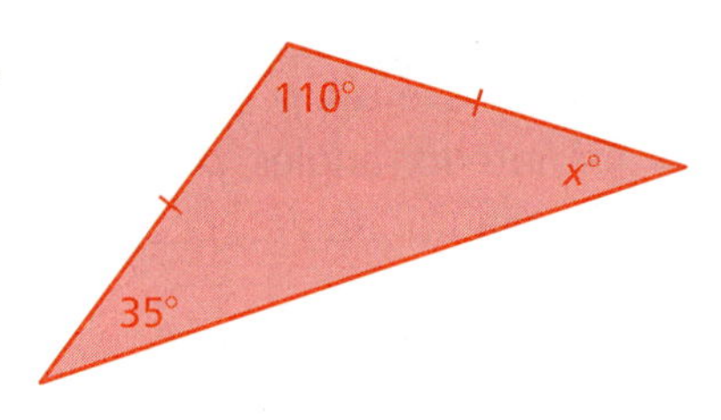

Find the measure of the exterior angle.

7.

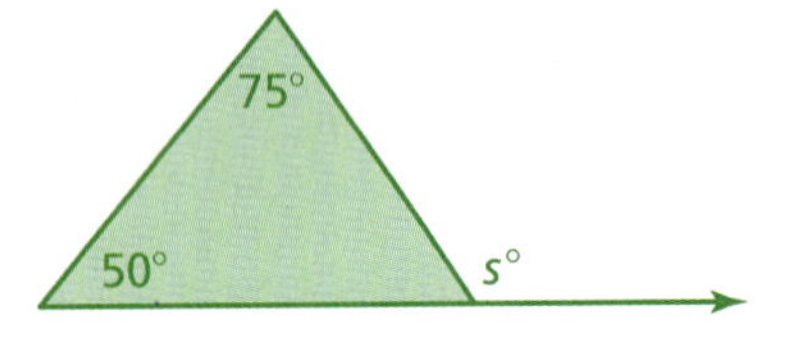

8.

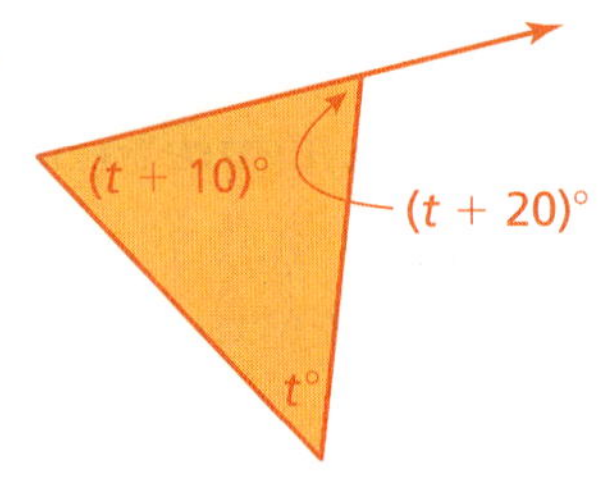

3.3 Angles of Polygons *(pp. 118–125)*

a. Find the value of x.

Step 1: The polygon has 6 sides. Find the sum of the interior angle measures.

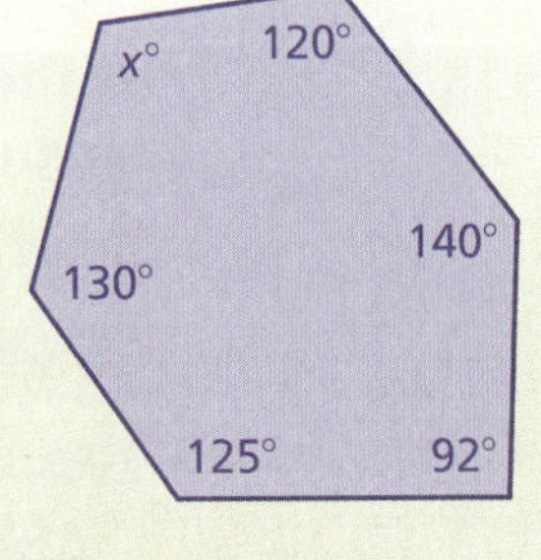

$S = (n - 2) \cdot 180°$ Write the formula.

$= (6 - 2) \cdot 180°$ Substitute 6 for n.

$= 720$ Simplify. The sum of the interior angle measures is 720°.

Step 2: Write and solve an equation.

$$130 + 125 + 92 + 140 + 120 + x = 720$$

$$607 + x = 720$$

$$x = 113$$

The value of x is 113.

b. Find the measures of the exterior angles of the polygon.

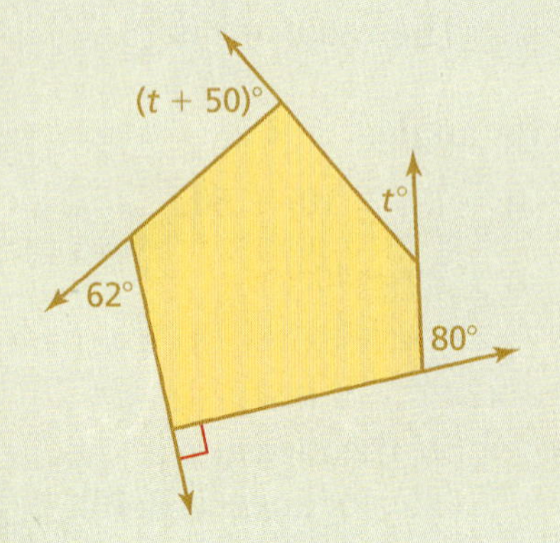

Write and solve an equation for t.

$$t + 80 + 90 + 62 + (t + 50) = 360$$

$$2t + 282 = 360$$

$$2t = 78$$

$$t = 39$$

So, the measures of the exterior angles are 39°, 80°, 90°, 62°, and $(39 + 50)° = 89°$.

Review Game

Finding Angle Measures

Materials per Group

- deck of playing cards
- paper
- pencil
- stopwatch

Directions

Divide the class into equally sized groups. A group member lays down two cards next to each other, and below this pair lays down another two cards next to each other. Then they multiply the values of the cards in each pair. (Count kings, queens, jacks, and aces as 10.) These are used to represent the measures of two angles of a triangle. The group member then finds the angle measure of the third angle. Other members time the one working and make sure the computed angle is correct. Each group member takes a turn going through the deck as fast as he or she can. If there is a combination that is impossible to use, they must identify this and move on.

Who wins?

The fastest member in a group after 2 rounds competes against the fastest members in the other groups. The winner is the fastest student.

For the Student Additional Practice

- Lesson Tutorials
- Multi-Language Glossary
- Self-Grading Progress Check
- *BigIdeasMath.com*
  Dynamic Student Edition
  Student Resources

Answers

5. 41°, 49°, 90°
6. 35°, 35°, 110°
7. 125°
8. 110°
9. 77°, 60°, 128°, 95°
10. 110°, 135°, 125°, 135°, 105°, 150°, 140°
11. 125°, 100°, 120°, 60°, 250°, 65°
12. 135°, 100°, 125°
13. 60°, 60°, 60°, 60°, 60°, 60°
14. yes; The triangles have the same angle measures, 90°, 68°, and 22°.
15. yes; The triangles have two pairs of congruent angles.

My Thoughts on the Chapter

Teacher Tip

Not allowed to write in your teaching edition? Use sticky notes to record your thoughts.

What worked. . .

What did not work. . .

What I would do differently. . .

Exercises

Find the measures of the interior angles of the polygon.

9.

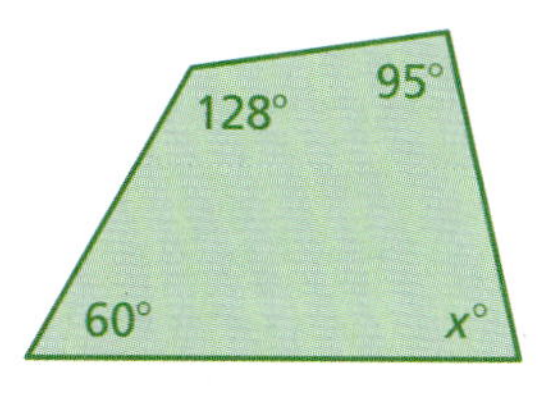

10.

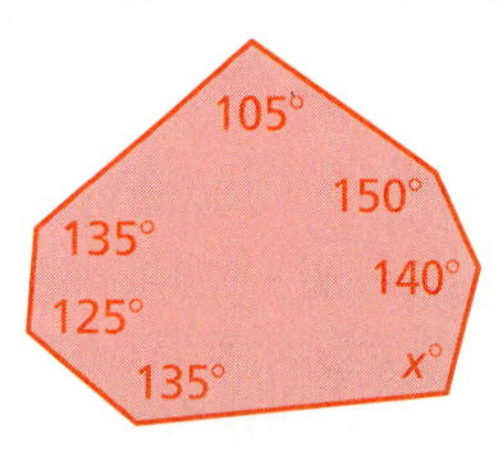

11.

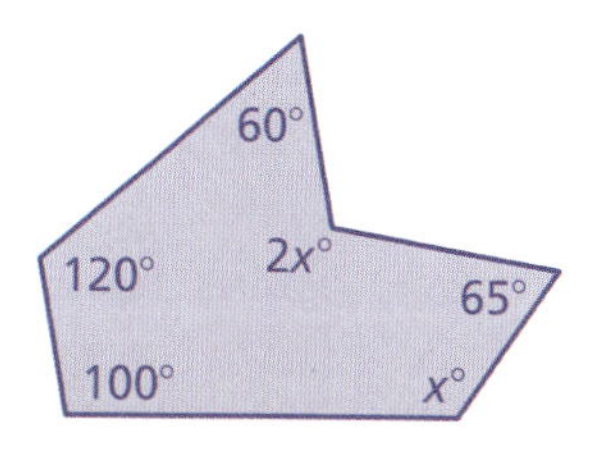

Find the measures of the exterior angles of the polygon.

12.

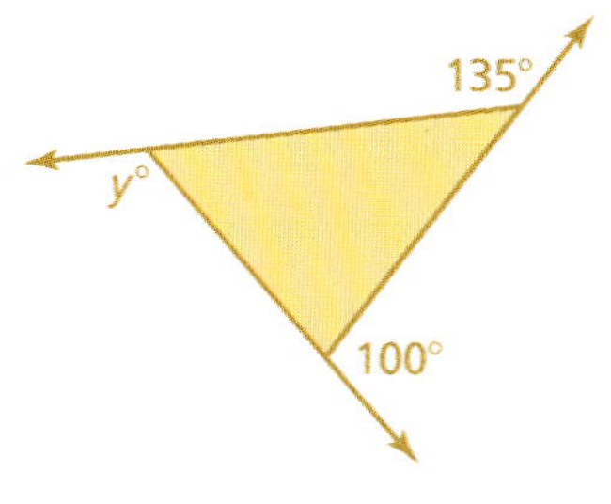

13.

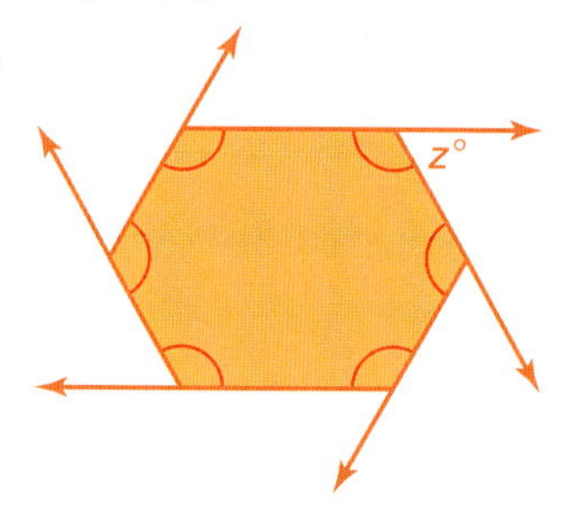

3.4 Using Similar Triangles *(pp. 126–131)*

Tell whether the triangles are similar. Explain.

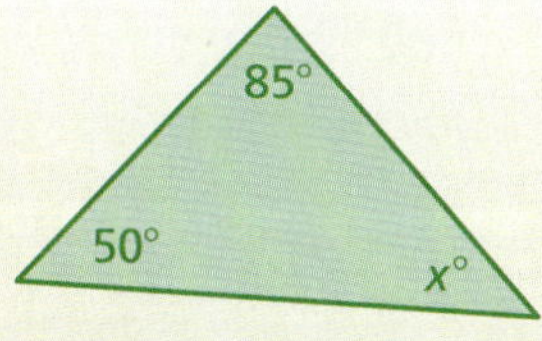

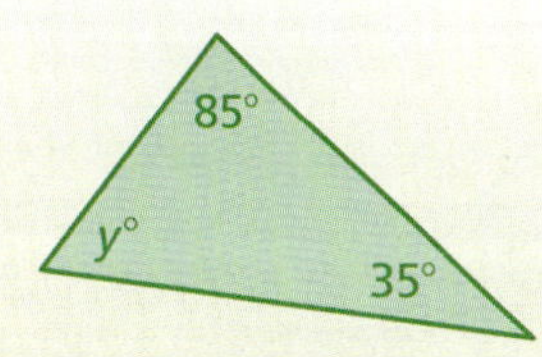

Write and solve an equation to find x.

$$50 + 85 + x = 180$$

$$135 + x = 180$$

$$x = 45$$

The triangles do not have two pairs of congruent angles. So, the triangles are not similar.

Exercises

Tell whether the triangles are similar. Explain.

14.

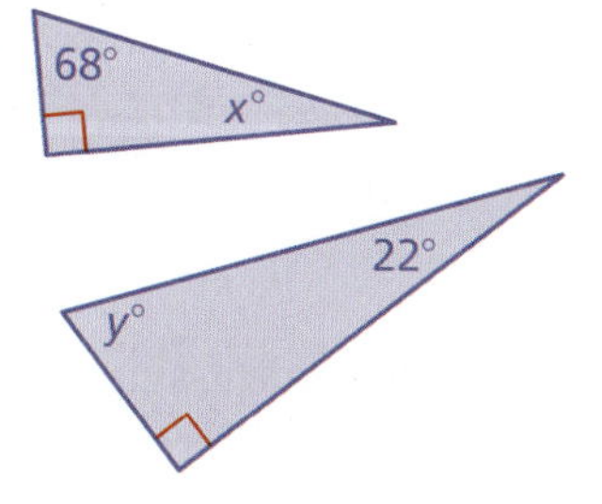

15.

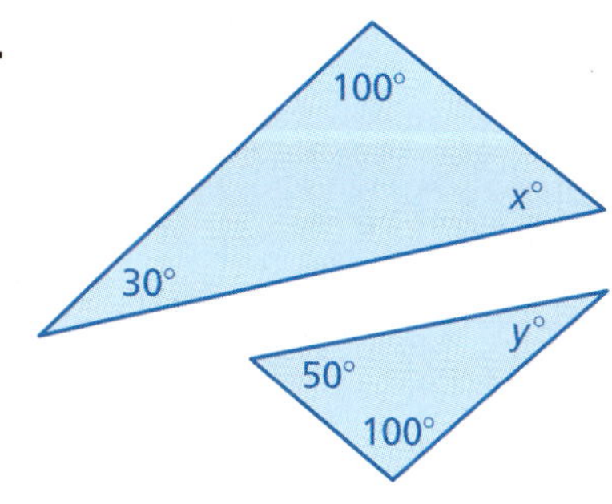

3 Chapter Test

Use the figure to find the measure of the angle. Explain your reasoning.

1. $\angle 1$ **2.** $\angle 8$

3. $\angle 4$ **4.** $\angle 5$

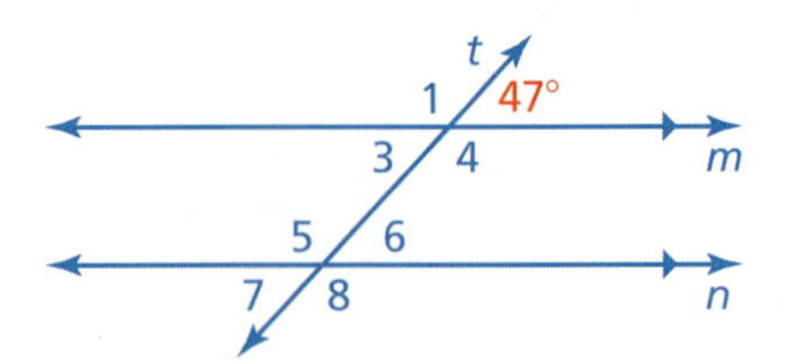

Find the measures of the interior angles.

5.

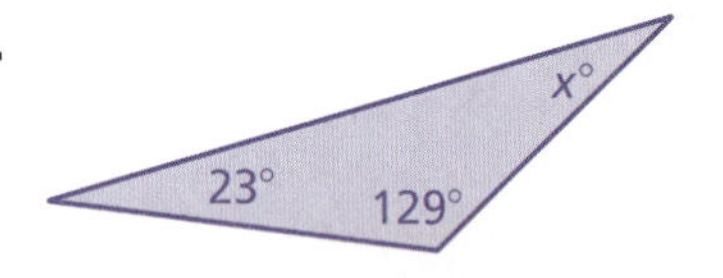

6.

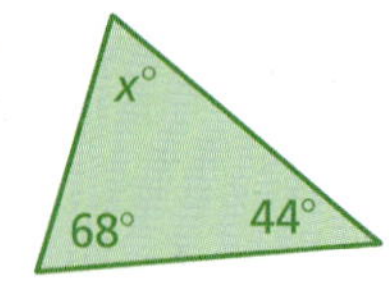

7.

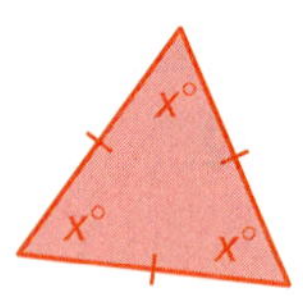

Find the measure of the exterior angle.

8.

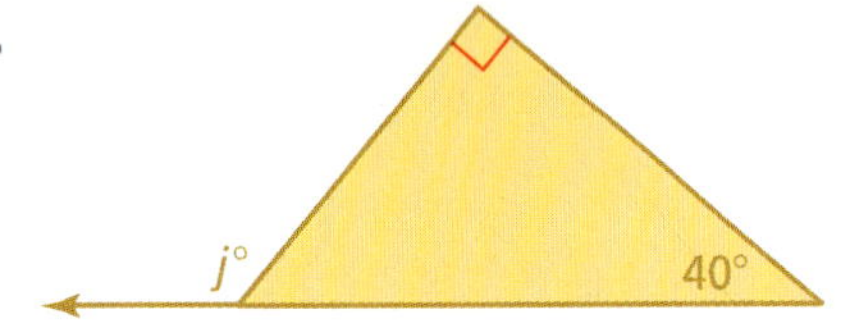

9.

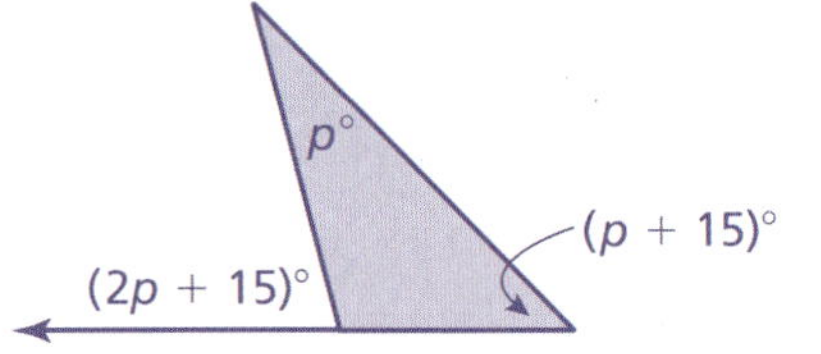

10. Find the measures of the interior angles of the polygon.

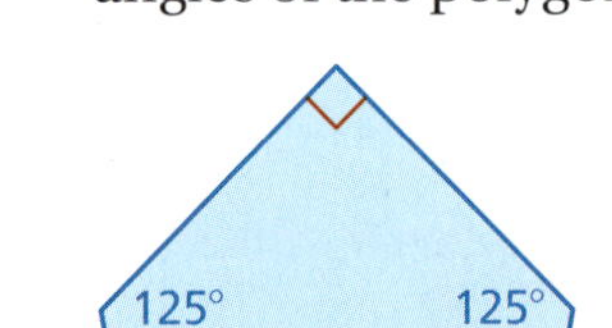

11. Find the measures of the exterior angles of the polygon.

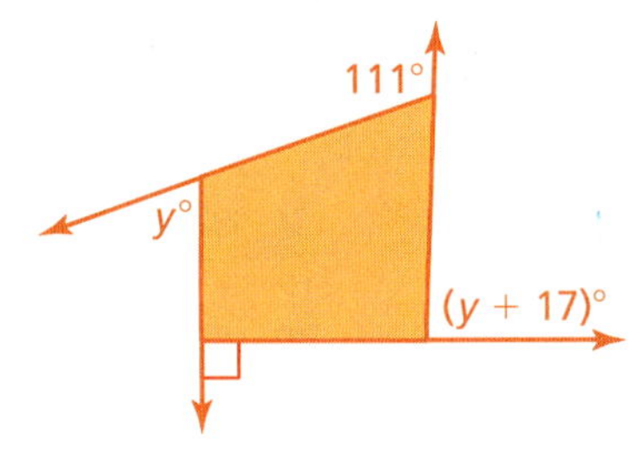

Tell whether the triangles are similar. Explain.

12.

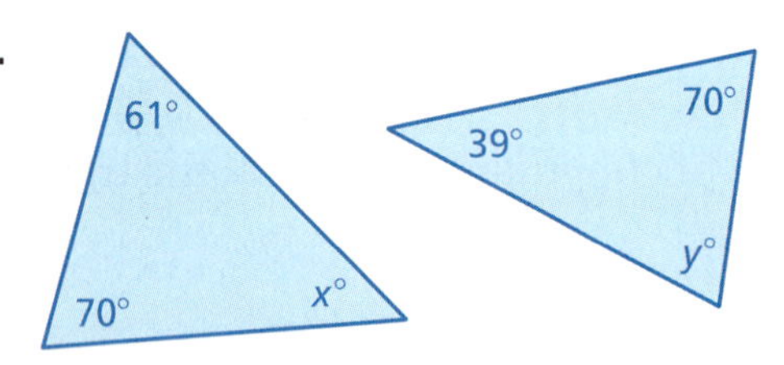

13.

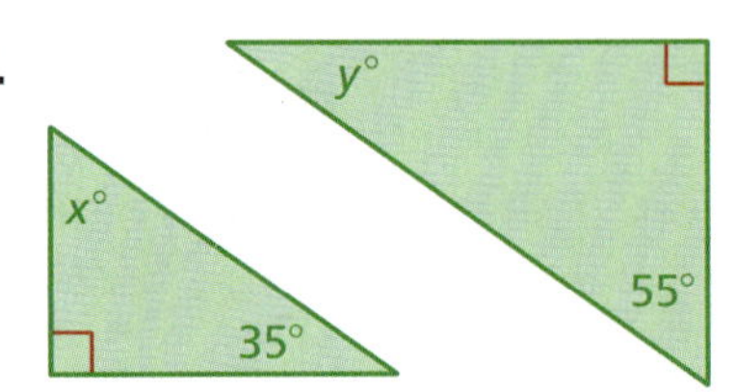

14. WRITING Describe two ways you can find the measure of $\angle 5$.

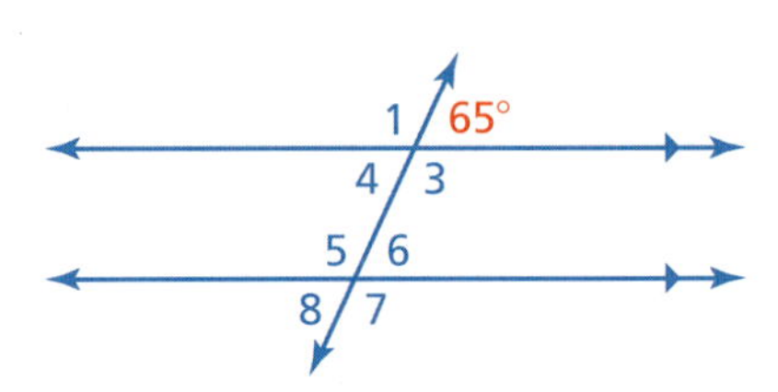

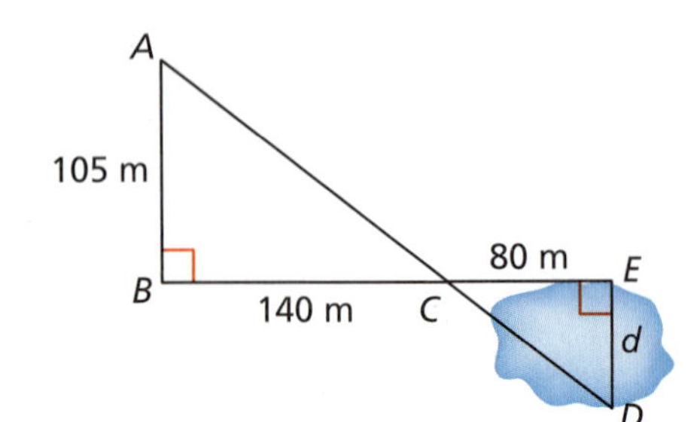

15. POND Use the given measurements to find the distance d across the pond.

Test Item References

Chapter Test Questions	Section to Review	Florida Common Core Standards (MACC)
1–4, 14	3.1	8.G.1.5
5–9	3.2	8.G.1.5
10, 11	3.3	8.G.1.5
12, 13, 15	3.4	8.G.1.5

Test-Taking Strategies

Remind students to quickly look over the entire test before they start so that they can budget their time. Students should jot down the formula for the sum of interior angles of a polygon on the back of the test before they begin. Students need to use the **Stop** and **Think** strategy before answering questions.

Common Errors

- **Exercises 9–11** Students may forget to combine like terms when solving for *x*. Remind them that because there are two variables on the same side of the equal sign, they should start by combining like terms.
- **Exercises 9–11** Students may solve for the variable, but forget to find the measures of the angles. Remind them to read the directions carefully and to answer the question.
- **Exercise 10** Students may forget to include one or more of the angles when writing an equation to find the value of *x*. Remind them to include all the angles. Encourage students to write the equation and then count the number of terms to make sure that there is the same number of terms as there are angles before solving.
- **Exercises 12 and 13** Students may find only one missing angle measure and then make a decision about the similarity of the triangles. While it is possible to use this method, encourage them to find *both* missing angle measures to verify their answers.

Reteaching and Enrichment Strategies

If students need help. . .	If students got it. . .
Resources by Chapter • Practice A and Practice B • Puzzle Time Record and Practice Journal Practice Differentiating the Lesson Lesson Tutorials *BigIdeasMath.com* Skills Review Handbook	Resources by Chapter • Enrichment and Extension • Technology Connection Game Closet at *BigIdeasMath.com* Start Standards Assessment

Answers

1. $133°$; $\angle 1$ and the given angle are supplementary.
2. $133°$; $\angle 8$ and $\angle 1$ are alternate exterior angles.
3. $133°$; $\angle 1$ and $\angle 4$ are vertical angles.
4. $133°$; $\angle 4$ and $\angle 5$ are alternate interior angles.
5. $28°, 129°, 23°$
6. $68°, 68°, 44°$
7. $60°, 60°, 60°$
8. $130°$
9. The exterior angle can have any measure greater than $15°$ and less than $180°$.
10. $90°, 125°, 100°, 100°, 125°$
11. $71°, 111°, 88°, 90°$
12. no; The triangles do not have the same angle measures.
13. yes; The two triangles have two pairs of congruent angles.
14. *Sample answer:*

    1) The given angle and $\angle 3$ are supplementary, so $\angle 3 = 115°$; $\angle 3$ and $\angle 5$ are alternate interior angles, so $\angle 3 = \angle 5 = 115°$.

    2) The given angle and $\angle 8$ are alternate exterior angles, so $\angle 8 = 65°$; $\angle 5$ and $\angle 8$ are supplementary, so $\angle 5 = 115°$.
15. 60 m

Online Assessment
Assessment Book
ExamView® Assessment Suite

Test-Taking Strategies

Available at *BigIdeasMath.com*

After Answering Easy Questions, Relax
Answer Easy Questions First
Estimate the Answer
Read All Choices before Answering
Read Question before Answering
Solve Directly or Eliminate Choices
Solve Problem before Looking at Choices
Use Intelligent Guessing
Work Backwards

About this Strategy

When taking a multiple choice test, be sure to read each question carefully and thoroughly. Sometimes it is easier to solve the problem and then look for the answer among the choices.

Answers

1. 147°
2. B
3. I
4. C

Technology for the Teacher

Florida Common Core Standards Support
Performance Tasks
Online Assessment
Assessment Book
ExamView® Assessment Suite

Item Analysis

1. **Gridded Response:** Correct answer: 147°

   Common Error: The student might divide 180 by 11.

2. **A.** The student adds 11 and 1.6 together before dividing.

   B. Correct answer

   C. The student divides first and then subtracts.

   D. The student subtracts 1.6 instead of dividing.

3. **F.** The student subtracts 20 from the right side instead of adding 20.

   G. The student did not distribute the 5 to both terms of $(x - 4)$.

   H. The student adds $3x$ to the left side instead of subtracting $3x$.

   I. Correct answer

4. **A.** The student does not correctly match corresponding side lengths, instead using the proportion $\frac{PQ}{QR} = \frac{US}{ST}$.

   B. The student does not correctly match corresponding side lengths, instead using the proportion $\frac{PQ}{QR} = \frac{TU}{ST}$.

   C. Correct answer

   D. The student does not correctly match corresponding side lengths, instead using the proportion $\frac{PQ}{QR} = \frac{ST}{US}$.

3 Standards Assessment

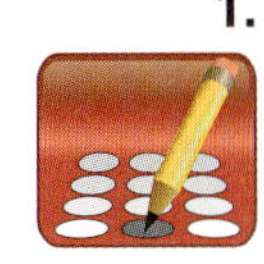

1. The border of a Canadian one-dollar coin is shaped like an 11-sided regular polygon. The shape was chosen to help visually impaired people identify the coin. How many degrees are in each angle along the border? Round your answer to the nearest degree. *(MACC.8.G.1.5)*

2. A public utility charges its residential customers for natural gas based on the number of therms used each month. The formula below shows how the monthly cost C in dollars is related to the number t of therms used.

$$C = 11 + 1.6t$$

Solve this formula for t. *(MACC.8.EE.3.7b)*

A. $t = \dfrac{C}{12.6}$

B. $t = \dfrac{C - 11}{1.6}$

C. $t = \dfrac{C}{1.6} - 11$

D. $t = C - 12.6$

3. What is the value of x? *(MACC.8.EE.3.7b)*

$$5(x - 4) = 3x$$

F. -10

G. 2

H. $2\frac{1}{2}$

I. 10

4. In the figures below, $\triangle PQR$ is a dilation of $\triangle STU$.

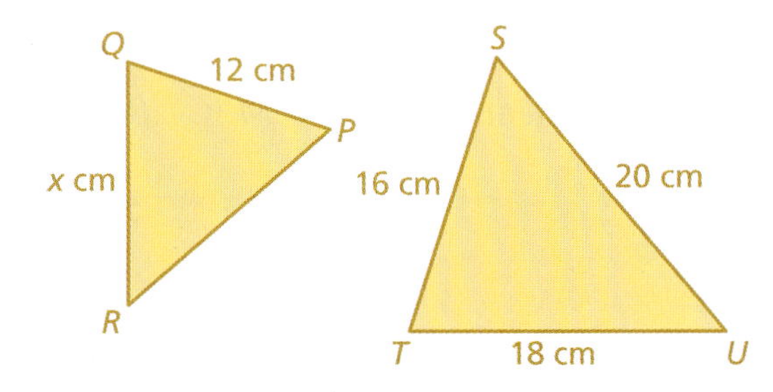

What is the value of x? *(MACC.8.G.1.4)*

A. 9.6

B. $10\frac{2}{3}$

C. 13.5

D. 15

5. What is the value of x? *(MACC.8.G.1.5)*

6. Olga was solving an equation in the box shown.

$$-\frac{2}{5}(10x - 15) = -30$$
$$10x - 15 = -30\left(-\frac{2}{5}\right)$$
$$10x - 15 = 12$$
$$10x - 15 + 15 = 12 + 15$$
$$10x = 27$$
$$\frac{10x}{10} = \frac{27}{10}$$
$$x = \frac{27}{10}$$

What should Olga do to correct the error that she made? *(MACC.8.EE.3.7b)*

F. Multiply both sides by $-\frac{5}{2}$ instead of $-\frac{2}{5}$.

G. Multiply both sides by $\frac{2}{5}$ instead of $-\frac{2}{5}$.

H. Distribute $-\frac{2}{5}$ to get $-4x - 6$.

I. Add 15 to -30.

Item Analysis (continued)

5. **Gridded Response:** Correct answer: 55

   Common Error: The student thinks the angles are congruent.

6. **F.** Correct answer

   G. The student thinks that multiplying by $\frac{2}{5}$ is the inverse operation of multiplying by $-\frac{2}{5}$.

   H. The student does not distribute the negative sign to the second term.

   I. The student makes an order of operations error by not first distributing the multiplication.

Answers

5. 55
6. F

Answers

7. B

8. *Part A* $S = (n - 2) \cdot 180$

Part B 80°

Part C
The sum of the angle measures of a triangle is 180 degrees. Because the pentagon can be divided into three triangles, the sum of the angle measures of a pentagon is

$180 + 180 + 180 = 540$ or

$(n - 2) \cdot 180 =$

$(5 - 2) \cdot 180 =$

$3 \cdot 180 = 540.$

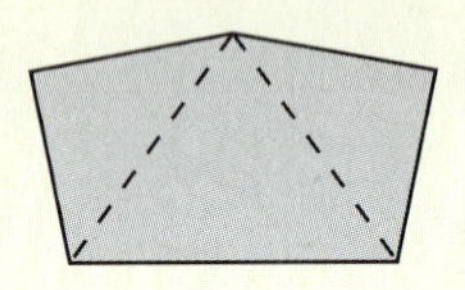

Item Analysis (continued)

7. **A.** The student reflects the figure in the *x*-axis.

B. Correct answer

C. The student translates the figure 7 units right.

D. The student translates the figure 6 units up.

8. **4 points** The student demonstrates a thorough understanding of writing and applying the angle sum formula for polygons, as well as how it relates to the fact that there are 180 degrees in a triangle. The student's work in Part B shows step-by-step how the fourth angle measures 80 degrees. The student's explanation in part C makes the algebraic-geometric connection clear.

3 points The student demonstrates an essential but less than thorough understanding. In particular, Parts A and B should be completed fully and clearly, but Part C may lack full explanation of the algebraic-geometric connection.

2 points The student's work and explanations demonstrate a lack of essential understanding. The formula in Part A should be properly stated, but Part B may show an error in application. Part C lacks any explanation.

1 point The student demonstrates limited understanding. The student's response is incomplete and exhibits many flaws, including, but not limited to, the inability to state the proper formula in Part A.

0 points The student provides no response, a completely incorrect or incomprehensible response, or a response that demonstrates insufficient understanding of writing, applying, and understanding the angle sum formula for polygons.

7. In the coordinate plane below, $\triangle XYZ$ is plotted and its vertices are labeled.

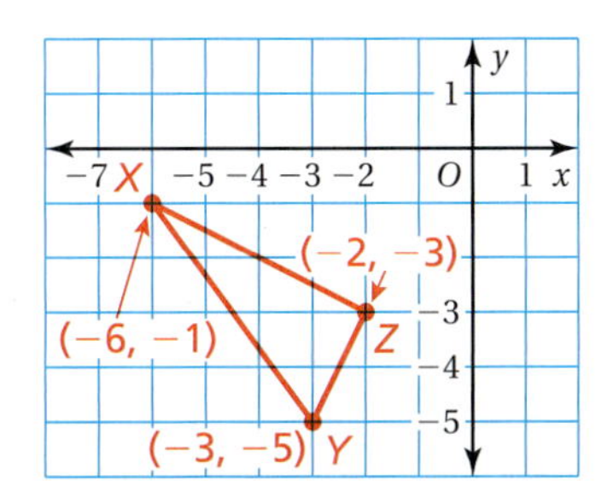

Which of the following shows $\triangle X'Y'Z'$, the image of $\triangle XYZ$ after it is reflected in the y-axis? *(MACC.8.G.1.3)*

A.

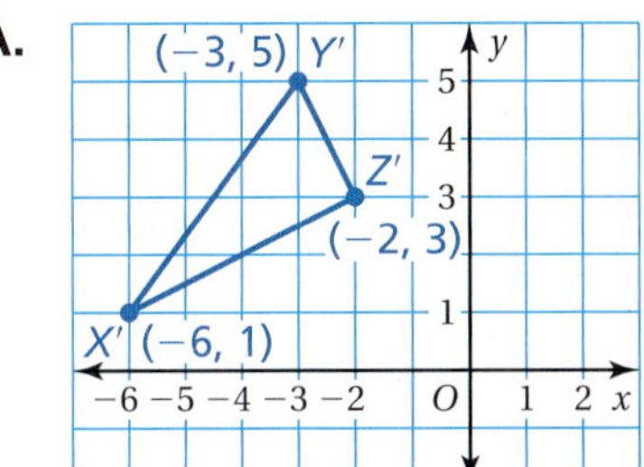

C.

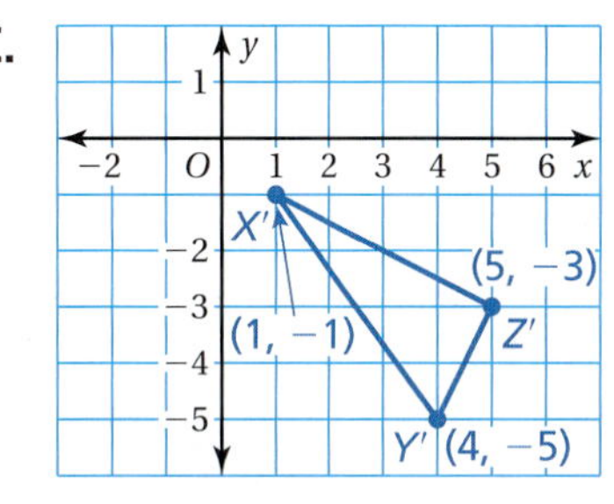

B.

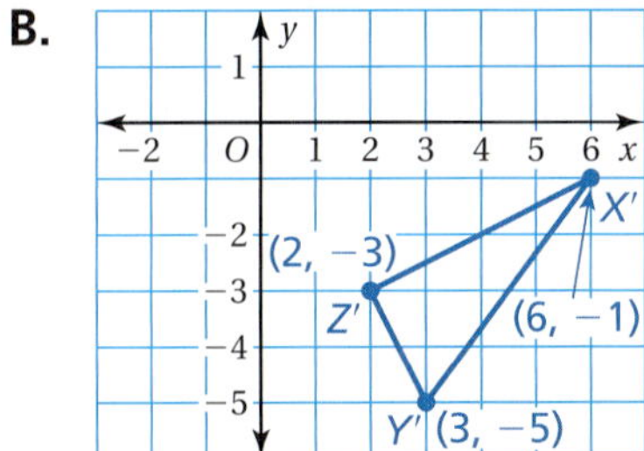

D.

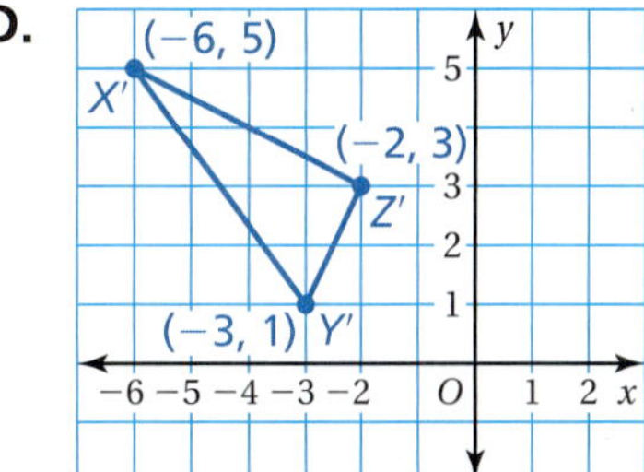

8. The sum S of the interior angle measures of a polygon with n sides can be found by using a formula. *(MACC.8.G.1.5)*

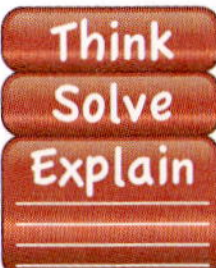

Part A Write the formula.

Part B A quadrilateral has angles measuring 100°, 90°, and 90°. Find the measure of its fourth angle. Show your work and explain your reasoning.

Part C The sum of the measures of the angles of the pentagon shown is 540°. Divide the pentagon into triangles to show why this must be true. Show your work and explain your reasoning.

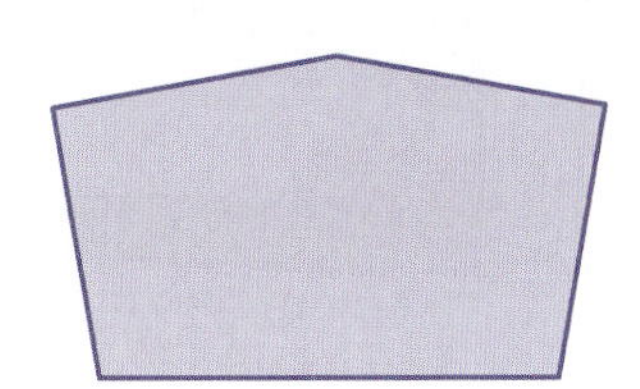

4 Graphing and Writing Linear Equations

"Okay Descartes, stand on the y-axis and try to intercept the pass when I throw."

"Here's an easy example of a line with a slope of 1."

"You eat one mouse treat the first day. Two treats the second day. And so on. Get it?"

Florida Common Core Progression

6th Grade

- Make tables of equivalent ratios and plot the pairs of values in a coordinate plane.
- Write an equation in two variables and analyze the relationship between the independent and dependent variables using graphs and tables.

7th Grade

- Identify the constant of proportionality (unit rate) in tables, graphs, equations, diagrams, and verbal descriptions.
- Represent proportional relationships with equations.

8th Grade

- Use similar triangles to explain why the slope is the same between any two points on a line.
- Graph proportional relationships, interpreting the unit rate as the slope.
- Compare proportional relationships represented in different ways.
- Derive $y = mx$ and $y = mx + b$.

Pacing Guide for Chapter 4

Chapter Opener	1 Day
Section 1 Activity Lesson	 1 Day 1 Day
Section 2 Activity Lesson Extension	 1 Day 1 Day 1 Day
Section 3 Activity Lesson	 1 Day 1 Day
Study Help / Quiz	1 Day
Section 4 Activity Lesson	 1 Day 1 Day
Section 5 Activity Lesson	 1 Day 1 Day
Section 6 Activity Lesson	 1 Day 1 Day
Section 7 Activity Lesson	 1 Day 1 Day
Chapter Review/ Chapter Tests	2 Days
Total Chapter 4	19 Days
Year-to-Date	64 Days

Chapter Summary

Section	Florida Common Core Standard (MACC)	
4.1	Preparing for	8.EE.2.5
4.2	Learning	8.EE.2.6
4.3	Learning	8.EE.2.5 ★, 8.EE.2.6
4.4	Learning	8.EE.2.6
4.5	Applying	8.EE.2.6 ★
4.6	Preparing for	8.F.2.4
4.7	Preparing for	8.F.2.4

★ Teaching is complete. Standard can be assessed.

Technology for the Teacher

BigIdeasMath.com
Chapter at a Glance
Complete Materials List
Parent Letters: English, Spanish, and Haitian Creole

Florida Common Core Standards

MACC.6.EE.1.2c Evaluate expressions at specific values of their variables. Include expressions that arise from formulas used in real-world problems. Perform arithmetic operations, including those involving whole-number exponents, in the conventional order when there are no parentheses to specify a particular order (Order of Operations).

MACC.6.NS.3.6c Find and position integers and other rational numbers on a horizontal or vertical number line diagram; find and position pairs of integers and other rational numbers on a coordinate plane.

Additional Topics for Review

- Order of Operations
- Exponents
- Similar Triangles
- Parallel and Perpendicular Lines
- Proportional Relationships
- Slope
- Unit Rate

Try It Yourself

1. -12 **2.** -23

3. 15 **4.** $4\frac{3}{4}$

5. (0, 4) **6.** (4, 2)

7. Point R **8.** Point N

Record and Practice Journal Fair Game Review

1. 5 **2.** 16

3. -5 **4.** $-38\frac{1}{2}$

5. 108 **6.** 65

7. $-3\frac{7}{19}$ **8.** 262

9. $50.00 **10.** $(-5, 0)$

11. $(3, -5)$ **12.** Point F

13. Point G

14. Point B, Point H

15. Point C, Point E

16–20. See Additional Answers.

Math Background Notes

Vocabulary Review

- Evaluate
- Expression
- Order of Operations
- Substitute
- Coordinates

Evaluating Expressions Using Order of Operations

- Students should know how to substitute values into algebraic expressions and evaluate the results using order of operations.
- **Teaching Tip:** Sometimes color coding substitutions can help students to evaluate expressions. Each time you want to substitute a number in place of a variable, you must substitute your lead pencil for a colored pencil.
- Remind students that after they substitute values in for the variables, they must use the correct order of operations to continue evaluating the expression.
- **Common Error:** Encourage students to use a set of parentheses whenever they do a substitution. This will help students distinguish between subtraction and multiplication.

Plotting Points

- Students should know how to plot points in all four quadrants.
- **Common Error:** Students may write the coordinates backwards. Remind them that coordinates are written in alphabetical order with the *x* move (horizontal) written before the *y* move (vertical).
- **Common Error:** Students may also have difficulty with the negative numbers associated with plotting outside Quadrant I. Remind them that the negatives are directional. A negative *x*-coordinate communicates a move to the left of the origin and a negative *y*-coordinate communicates a move downward from the origin.

Reteaching and Enrichment Strategies

If students need help. . .	If students got it. . .
Record and Practice Journal • Fair Game Review Skills Review Handbook Lesson Tutorials	Game Closet at *BigIdeasMath.com* Start the next section

What You Learned Before

"I estimate that we are on a slope of about −0.625. What do you think?"

Evaluating Expressions Using Order of Operations (MACC.6.EE.1.2c)

Example 1 **Evaluate $2xy + 3(x + y)$ when $x = 4$ and $y = 7$.**

$$2xy + 3(x + y) = 2(4)(7) + 3(4 + 7) \quad \text{Substitute 4 for } x \text{ and 7 for } y.$$
$$= 8(7) + 3(4 + 7) \quad \text{Use order of operations.}$$
$$= 56 + 3(11) \quad \text{Simplify.}$$
$$= 56 + 33 \quad \text{Multiply.}$$
$$= 89 \quad \text{Add.}$$

Try It Yourself

Evaluate the expression when $a = \frac{1}{4}$ and $b = 6$.

1. $-8ab$ **2.** $16a^2 - 4b$ **3.** $\frac{5b}{32a^2}$ **4.** $12a + (b - a - 4)$

Plotting Points (MACC.6.NS.3.6c)

Example 2 **Write the ordered pair that corresponds to point U.**

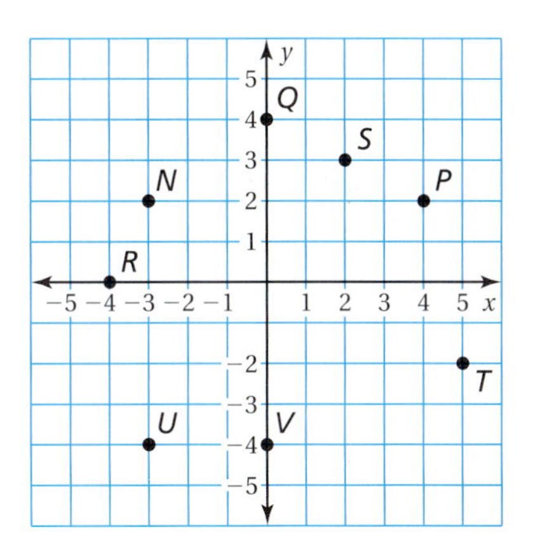

Point U is 3 units to the left of the origin and 4 units down. So, the x-coordinate is -3, and the y-coordinate is -4.

∴ The ordered pair $(-3, -4)$ corresponds to point U.

Example 3 **Which point is located at $(5, -2)$?**

Start at the origin. Move 5 units right and 2 units down.

∴ Point T is located at $(5, -2)$.

Try It Yourself

Use the graph to answer the question.

5. Write the ordered pair that corresponds to point Q.

6. Write the ordered pair that corresponds to point P.

7. Which point is located at $(-4, 0)$?

8. Which point is located in Quadrant II?

4.1 Graphing Linear Equations

Essential Question How can you recognize a linear equation? How can you draw its graph?

1 ACTIVITY: Graphing a Linear Equation

Work with a partner.

a. Use the equation $y = \frac{1}{2}x + 1$ to complete the table. (Choose any two x-values and find the y-values.)

	Solution Points	
x		
$y = \frac{1}{2}x + 1$		

b. Write the two ordered pairs given by the table. These are called *solution points* of the equation.

c. **PRECISION** Plot the two solution points. Draw a line *exactly* through the two points.

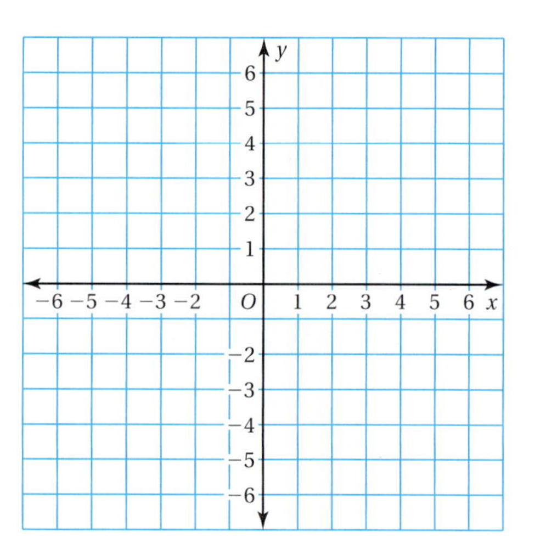

d. Find a different point on the line. Check that this point is a solution point of the equation $y = \frac{1}{2}x + 1$.

e. **LOGIC** Do you think it is true that *any* point on the line is a solution point of the equation $y = \frac{1}{2}x + 1$? Explain.

f. Choose five additional x-values for the table. (Choose positive and negative x-values.) Plot the five corresponding solution points. Does each point lie on the line?

	Solution Points				
x					
$y = \frac{1}{2}x + 1$					

g. **LOGIC** Do you think it is true that *any* solution point of the equation $y = \frac{1}{2}x + 1$ is a point on the line? Explain.

h. Why do you think $y = ax + b$ is called a *linear equation*?

COMMON CORE

Graphing Equations

In this lesson, you will

- understand that lines represent solutions of linear equations.
- graph linear equations.

Preparing for Standard MACC.8.EE.2.5

Laurie's Notes

Introduction

Standards for Mathematical Practice

- **MP4 Model with Mathematics:** The goal is for students to use a table of values and a graph to model a linear equation. Using multiple representations of linear equations deepens students' understanding and supports learning.
- Throughout this chapter, you may encounter applications that show a graph of discrete data with a line through the points. At this point in the text, we think it is acceptable for students to draw a line through discrete data points to help them solve an exercise. They will determine whether non-integer values or negative values are valid in various contexts. However, the terms discrete and continuous are not used at this time.

Motivate

- Play a game of coordinate BINGO.
- Distribute small coordinate grids to students. They should plot ten ordered pairs, where the *x*- and *y*-coordinates are integers between −4 and 4.
- Generate a random ordered pair in the grid. Write the integers from −4 to 4 on slips of paper and place them in a bag. Draw and replace two slips of paper to generate the ordered pair, then write it on the board.
- Each time you record a new ordered pair, the students check to see if it is one of their 10 ordered pairs. If it is, they put an X over the point on their grids. The goal is to be the first person with three Xs. A student who calls "BINGO" reads the three ordered pairs for you to check against the master list.

? "Are there ordered pairs that are not on lattice points, meaning the *x*- or *y*-coordinate is not an integer? Explain." yes; It's possible for the ordered pair to be $\left(3.5, \frac{1}{2}\right)$. Plot whatever examples students give.

- Remind students that the ordered pairs are always (x, y), where x is the horizontal direction and y is the vertical direction.

Activity Notes

Activity 1

- Some students will recognize right away that if they substitute an even number for *x*, the *y*-coordinate will not be a fraction. It is likely that students will only try positive *x*-values. Encourage them to try negative values for *x*.
- In part (d), suggest that students consider only those ordered pairs that appear to be lattice points.
- **MP3a Construct Viable Arguments:** Listen and discuss student responses to the generalizations in parts (e) and (g).
- **Big Idea:** The goal of this activity is for students to recognize and understand two related, but different, ideas. 1) *All* solution points of a linear equation lie on the same line. 2) *All* points on the line are solution points of the equation.

Florida Common Core Standards

MACC.8.EE.2.5 Graph proportional relationships, interpreting the unit rate as the slope of the graph. Compare two different proportional relationships represented in different ways.

Previous Learning

Students should know how to plot ordered pairs from a table.

Lesson Plans
Complete Materials List

4.1 Record and Practice Journal

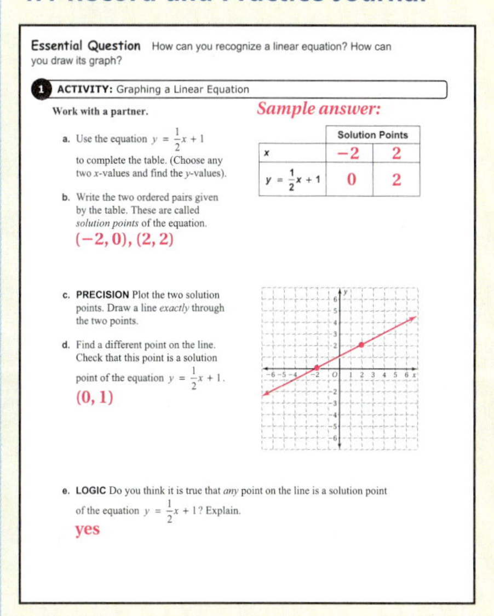

Essential Question How can you recognize a linear equation? How can you draw its graph?

1 ACTIVITY: Graphing a Linear Equation

Work with a partner.

Sample answer:

a. Use the equation $y = \frac{1}{2}x + 1$ to complete the table. (Choose any two *x*-values and find the *y*-values).

	Solution Points	
x	−2	2
$y = \frac{1}{2}x + 1$	0	2

b. Write the two ordered pairs given by the table. These are called *solution points* of the equation.
$(-2, 0), (2, 2)$

c. **PRECISION** Plot the two solution points. Draw a line *exactly* through the two points.

d. Find a different point on the line. Check that this point is a solution point of the equation $y = \frac{1}{2}x + 1$.
$(0, 1)$

e. **LOGIC** Do you think it is true that *any* point on the line is a solution point of the equation $y = \frac{1}{2}x + 1$? Explain.
yes

Differentiated Instruction

Kinesthetic

For students that are kinesthetic learners and have difficulty in plotting points in the coordinate plane, suggest they use a finger for tracing. Have each student place a finger at the origin and trace left or right along the x-axis to the first coordinate, then trace up or down to the second coordinate. Students should also practice writing the ordered pair of a plotted point. Guide students with questions such as, "Should you move left or right? How far? Should you move up or down? How far?"

4.1 Record and Practice Journal

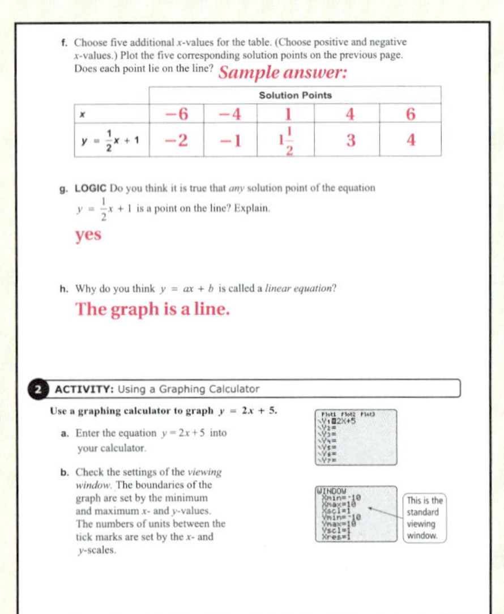

f. Choose five additional x-values for the table. (Choose positive and negative x-values.) Plot the five corresponding solution points on the previous page. Does each point lie on the line? *Sample answer:*

	Solution Points				
x	−6	−4	1	4	6
$y = \frac{1}{2}x + 1$	−2	−1	$1\frac{1}{2}$	3	4

g. **LOGIC** Do you think it is true that *any* solution point of the equation $y = \frac{1}{2}x + 1$ is a point on the line? Explain.

yes

h. Why do you think $y = ax + b$ is called a *linear equation*?

The graph is a line.

2 ACTIVITY: Using a Graphing Calculator

Use a graphing calculator to graph $y = 2x + 5$.

a. Enter the equation $y = 2x + 5$ into your calculator.

b. Check the settings of the *viewing window*. The boundaries of the graph are set by the minimum and maximum x- and y-values. The numbers of units between the tick marks are set by the x- and y-scales.

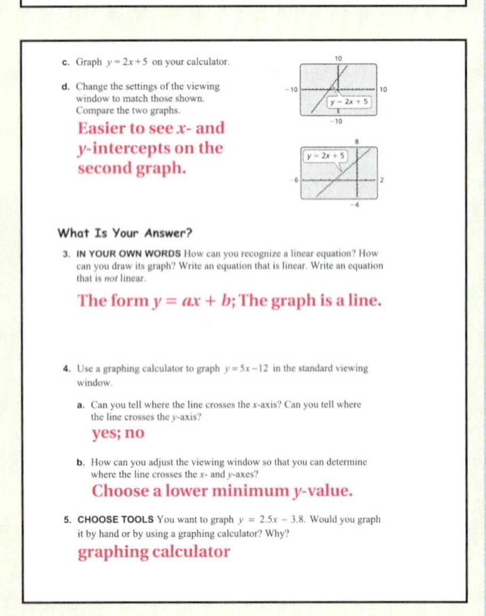

c. Graph $y = 2x + 5$ on your calculator.

d. Change the settings of the viewing window to match those shown. Compare the two graphs.

Easier to see x- and y-intercepts on the second graph.

What Is Your Answer?

3. **IN YOUR OWN WORDS** How can you recognize a linear equation? How can you draw its graph? Write an equation that is linear. Write an equation that is *not* linear.

The form $y = ax + b$; The graph is a line.

4. Use a graphing calculator to graph $y = 5x - 12$ in the standard viewing window.

a. Can you tell where the line crosses the x-axis? Can you tell where the line crosses the y-axis?

yes; no

b. How can you adjust the viewing window so that you can determine where the line crosses the x- and y-axes?

Choose a lower minimum y-value.

5. **CHOOSE TOOLS** You want to graph $y = 2.5x - 3.8$. Would you graph it by hand or by using a graphing calculator? Why?

graphing calculator

Laurie's Notes

Activity 2

- This may be a student's first experience with using a graphing calculator to graph a linear equation. Explain that the calculator can graph equations that are entered in the equation editor.
- Explain how to set the *standard viewing window*, or *standard viewing rectangle*.
- Because the viewing screen is a rectangle, one unit in the x-direction appears longer than one unit in the y-direction. When graphing by hand, you generally are using a square grid. It is important to point out this distinction to students because they will eventually graph $y = x$, which is a 45° line. It will not appear this way in the standard viewing rectangle.

? After the graph of $y = 2x + 5$ appears, ask, "Can you name a solution of this equation from looking at the graph?" Students may name solutions such as (0, 5) or (1, 7).

? After the viewing rectangle changes, ask, "Did the solutions of the equation change?" The graph appears less steep; however, it is the same graph in a new view. The solutions have not changed.

What Is Your Answer?

- Discuss students' responses to the first question.
- Have students share the viewing window they selected for Question 4b.
- **MP5 Use Appropriate Tools Strategically:** This year students will be asked to graph many linear equations. In certain contexts they will need an accurate graph, while in other contexts a rough sketch will be sufficient. Students should use appropriate tools (paper and pencil versus technology) strategically.

Closure

- Find three ordered pairs that are solutions of the equation $y = 2x - 3$. Draw the graph. *Sample answer:* (−1, −5), (0, −3), and (1, −1)

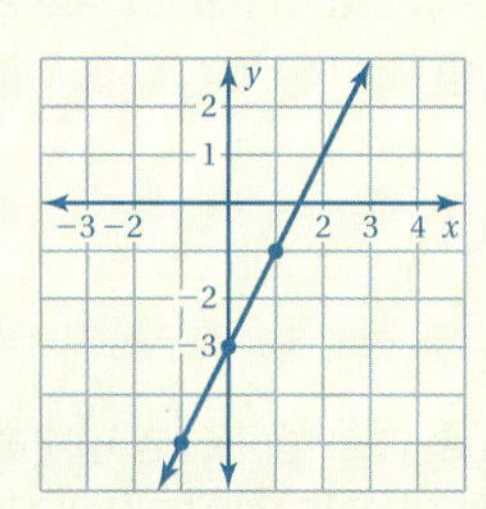

2 ACTIVITY: Using a Graphing Calculator

Use a graphing calculator to graph $y = 2x + 5$.

a. Enter the equation $y = 2x + 5$ into your calculator.

Plot1 Plot2 Plot3
\Y1 = 2X+5
\Y2=
\Y3=
\Y4=
\Y5=
\Y6=
\Y7=

Math Practice 5

Recognize Usefulness of Tools

What are some advantages and disadvantages of using a graphing calculator to graph a linear equation?

b. Check the settings of the *viewing window*. The boundaries of the graph are set by the minimum and the maximum x- and y-values. The numbers of units between the tick marks are set by the x- and y-scales.

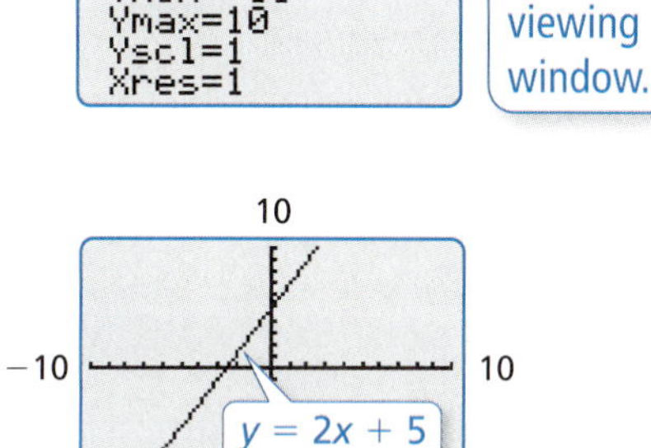

c. Graph $y = 2x + 5$ on your calculator.

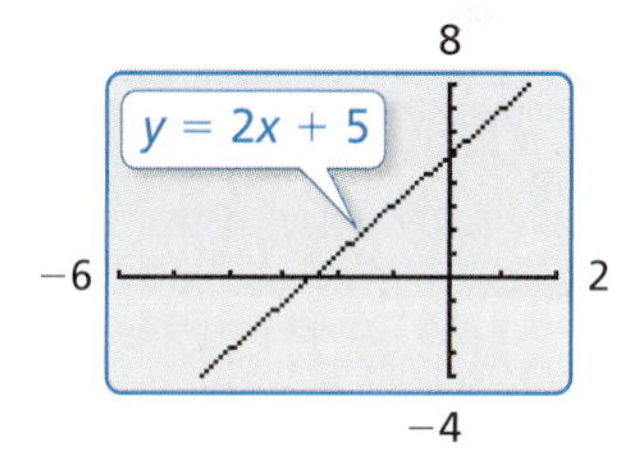

d. Change the settings of the viewing window to match those shown.

Compare the two graphs.

8
y = 2x + 5
−6
2
−4

What Is Your Answer?

3. **IN YOUR OWN WORDS** How can you recognize a linear equation? How can you draw its graph? Write an equation that is linear. Write an equation that is *not* linear.

4. Use a graphing calculator to graph $y = 5x - 12$ in the standard viewing window.

a. Can you tell where the line crosses the x-axis? Can you tell where the line crosses the y-axis?

b. How can you adjust the viewing window so that you can determine where the line crosses the x- and y-axes?

5. **CHOOSE TOOLS** You want to graph $y = 2.5x - 3.8$. Would you graph it by hand or by using a graphing calculator? Why?

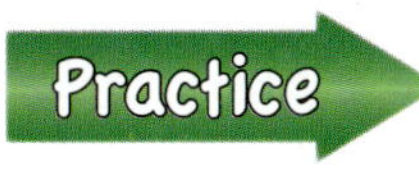

Use what you learned about graphing linear equations to complete Exercises 3 and 4 on page 146.

4.1 Lesson

Key Vocabulary

linear equation, p. 144

solution of a linear equation, p. 144

Key Idea

Linear Equations

A **linear equation** is an equation whose graph is a line. The points on the line are **solutions** of the equation.

You can use a graph to show the solutions of a linear equation. The graph below represents the equation $y = x + 1$.

x	y	(x, y)
-1	0	$(-1, 0)$
0	1	$(0, 1)$
2	3	$(2, 3)$

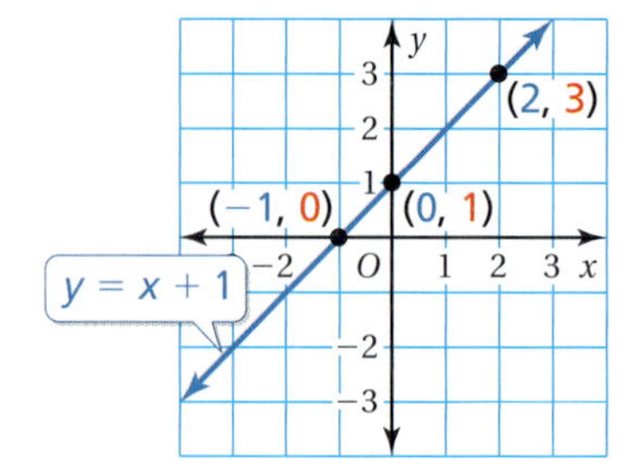

Remember

An ordered pair (x, y) is used to locate a point in a coordinate plane.

EXAMPLE 1 Graphing a Linear Equation

Graph $y = -2x + 1$.

Step 1: Make a table of values.

x	$y = -2x + 1$	y	(x, y)
-1	$y = -2(-1) + 1$	3	$(-1, 3)$
0	$y = -2(0) + 1$	1	$(0, 1)$
2	$y = -2(2) + 1$	-3	$(2, -3)$

Step 2: Plot the ordered pairs.

Step 3: Draw a line through the points.

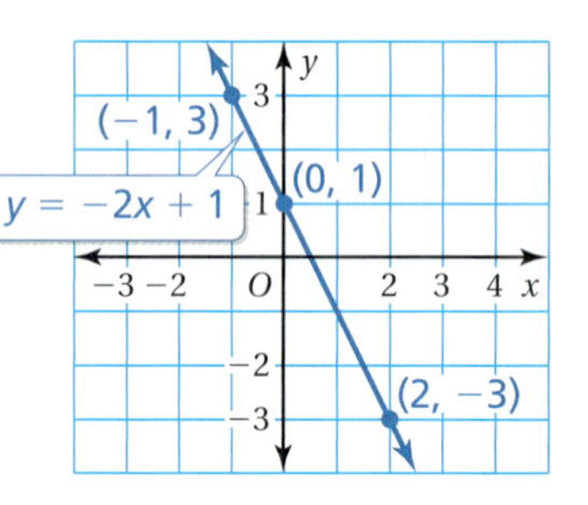

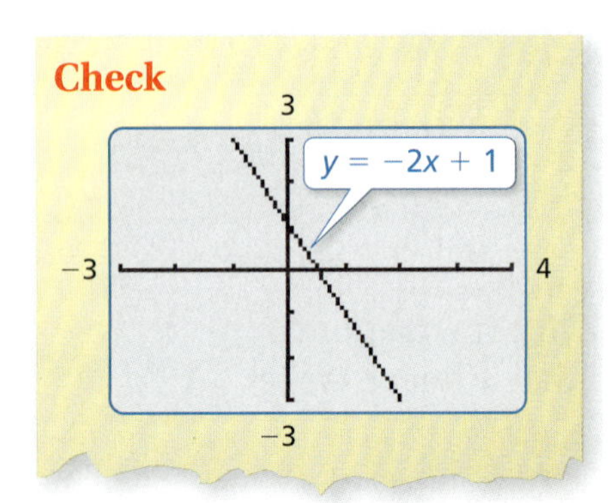

Key Idea

Graphing Horizontal and Vertical Lines

The graph of $y = b$ is a horizontal line passing through $(0, b)$.

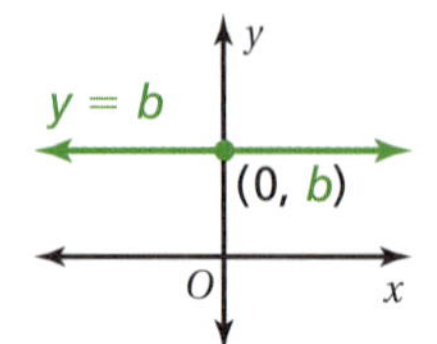

The graph of $x = a$ is a vertical line passing through $(a, 0)$.

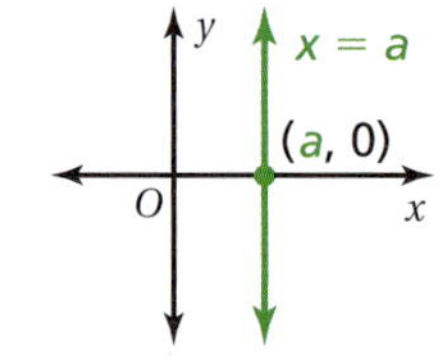

Introduction

Connect

- **Yesterday:** Students explored the graphs of linear equations. (MP3a, MP4, MP5)
- **Today:** Students will graph linear equations using a table of values.

Motivate

- Discuss a fact about wind speeds related to Example 3. During a wild April storm in 1934, a wind gust of 231 miles per hour (372 kilometers per hour) pushed across the summit of Mt. Washington in New Hampshire.

Lesson Notes

Key Idea

- Define *linear equation* and *solutions* of the equation.
- Note the use of color in the table. The equation used is a simple equation that helps students focus on the representation of the solutions as ordered pairs. The y-coordinate is always 1 greater than the x-coordinate, just as the equation states.

Example 1

? As a quick review, ask a volunteer to review the rules for integer multiplication. If the factors have the same sign, the product is positive. If the factors have different signs, the product is negative.

- Write the 4-column table. Take the time to show how the x-coordinate is being substituted in the second column. The number in blue is the only quantity that varies (variable); the other quantities are always the same (constant). Values from the first and third columns form the ordered pair.

? "From the graph, can you estimate the solution when $x = \frac{1}{2}$? Verify your answer by solving the equation when $x = \frac{1}{2}$." yes; $\left(\frac{1}{2}, 0\right)$

- **MP5 Use Appropriate Tools Strategically:** Students can check their graphs on a graphing calculator.

Key Idea

- Students are sometimes confused by the equations $x = a$ and $y = b$. Explain to students that a and b can equal any number.
- **Teaching Tip:** Another way to discuss the equation $y = b$ is to say that "y always equals a certain number, while x can equal anything." For example, if $y = -4$, the table of values may look like this:

x	−1	0	1	2
y	−4	−4	−4	−4

- **Teaching Tip:** Another way to discuss the equation $x = a$ is to say that "x always equals a certain number, while y can equal anything." For example, if $x = -2$, the table of values may look like this:

x	−2	−2	−2	−2
y	−1	0	1	2

Goal Today's lesson is graphing **linear equations**.

Lesson Tutorials
Lesson Plans
Answer Presentation Tool

Extra Example 1

Graph $y = \frac{1}{2}x - 3$.

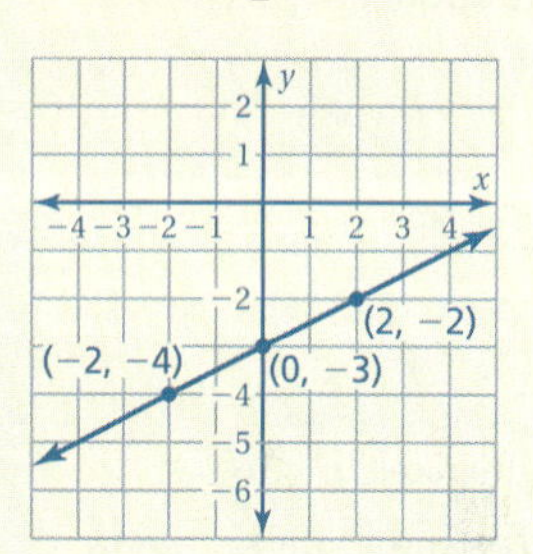

Extra Example 2

a. Graph $y = 4$.

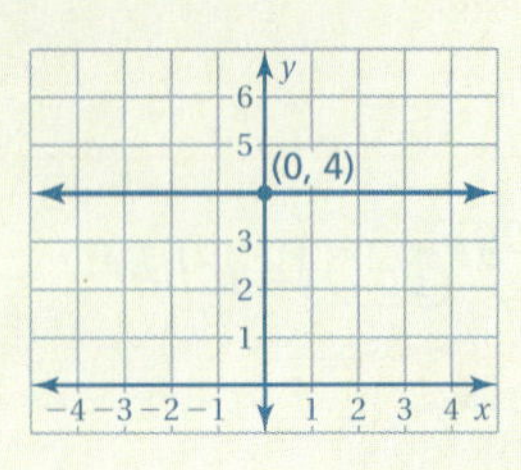

b. Graph $x = -1$.

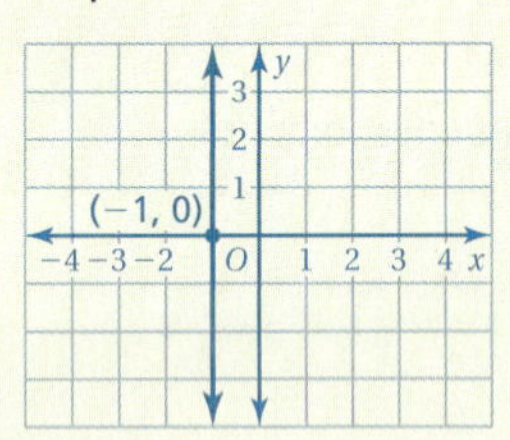

On Your Own

1–4. See Additional Answers.

Extra Example 3

The cost y (in dollars) for making friendship bracelets is $y = 0.5x + 2$, where x is the number of bracelets.

a. Graph the equation.

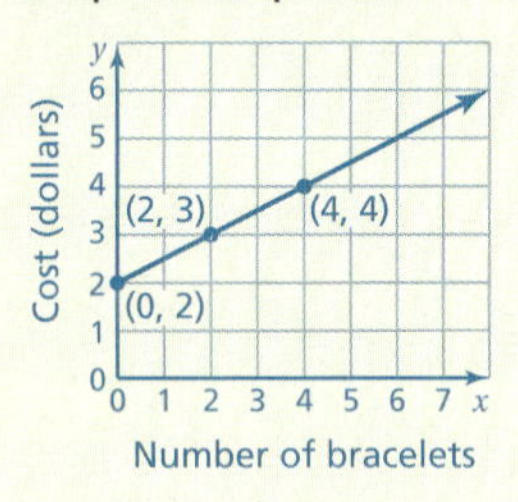

b. How many bracelets can be made for \$10? 16

On Your Own

5. 8 hours after it enters the Gulf of Mexico

English Language Learners

Vocabulary

Make sure students understand that the graph of a *linear* equation is a *line*. Only two points are needed to graph a line, but if one of the points is incorrect the wrong line will be graphed. Plotting three points for a line in the coordinate plane and making sure that the points form a line provide students with a check when graphing.

Laurie's Notes

Example 2

? "What are other points on the line $y = -3$?" *Sample answer:* (5, −3), or anything of the form $(x, -3)$

? "What are other points on the line $x = 2$?" *Sample answer:* (2, −3), or anything of the form $(2, y)$

On Your Own

- Ask volunteers to share their graphs at the board.
- Students may ask how to graph $x = -4$ on their calculators. To create a graph using a graphing calculator, the equation must begin with "$y =$." So, vertical lines cannot be graphed on a calculator.

Example 3

? "What does x represent in the problem? What does y represent?" x = number of hours after the storm enters the Gulf of Mexico; y = wind speed

- Work through the problem using the 4-column table to generate solutions of the equation.
- Note that the y-coordinate is much greater than the x-coordinate. For this reason, a broken vertical axis is used. Students should *not* scale the y-axis beginning at 0.

? "Why are only non-negative numbers substituted for x?" Because x equals the number of hours after the storm enters the Gulf of Mexico, you do not know if the equation makes sense for x-values before that.

- Note that the ordered pairs are all located in Quadrant I because x is non-negative. Even though this restriction was not stated explicitly, you know from reading the description of x that it needs to be non-negative.
- In part (b), help students read the graph. Starting with a y-value of 74 on the y-axis, trace horizontally until you reach the graph of the line, and then trace straight down (vertically) to the x-axis. The x-coordinate is 4.

On Your Own

- **Neighbor Check:** Have students work independently and then have their neighbors check their work. Have students discuss any discrepancies.

Closure

- Explain how you know if an equation is linear. *Sample answer:* The graph of the equation is a line.

EXAMPLE 2 Graphing a Horizontal Line and a Vertical Line

a. Graph $y = -3$.

The graph of $y = -3$ is a horizontal line passing through $(0, -3)$. Draw a horizontal line through this point.

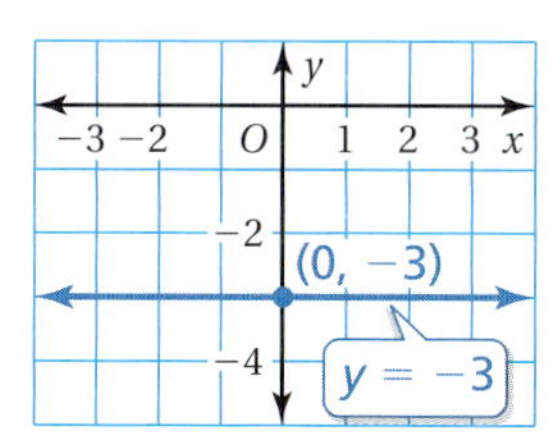

b. Graph $x = 2$.

The graph of $x = 2$ is a vertical line passing through $(2, 0)$. Draw a vertical line through this point.

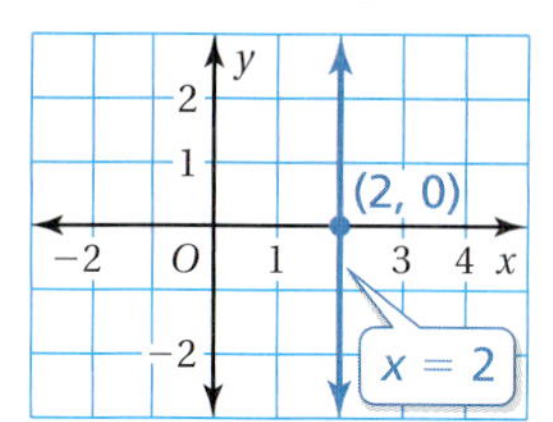

On Your Own

Graph the linear equation. Use a graphing calculator to check your graph, if possible.

1. $y = 3x$
2. $y = -\frac{1}{2}x + 2$
3. $x = -4$
4. $y = -1.5$

EXAMPLE 3 Real-Life Application

The wind speed y (in miles per hour) of a tropical storm is $y = 2x + 66$, where x is the number of hours after the storm enters the Gulf of Mexico.

a. Graph the equation.

b. When does the storm become a hurricane?

A tropical storm becomes a hurricane when wind speeds are at least 74 miles per hour.

a. Make a table of values.

x	$y = 2x + 66$	y	(x, y)
0	$y = 2(0) + 66$	66	(0, 66)
1	$y = 2(1) + 66$	68	(1, 68)
2	$y = 2(2) + 66$	70	(2, 70)
3	$y = 2(3) + 66$	72	(3, 72)

Plot the ordered pairs and draw a line through the points.

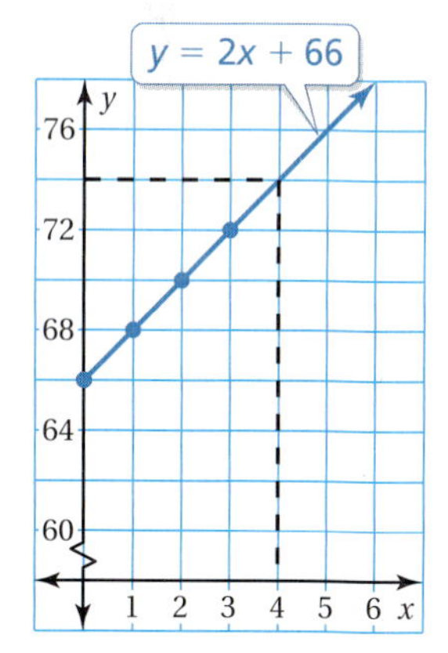

b. From the graph, you can see that $y = 74$ when $x = 4$. So, the storm becomes a hurricane 4 hours after it enters the Gulf of Mexico.

On Your Own

5. **WHAT IF?** The wind speed of the storm is $y = 1.5x + 62$. When does the storm become a hurricane?

4.1 Exercises

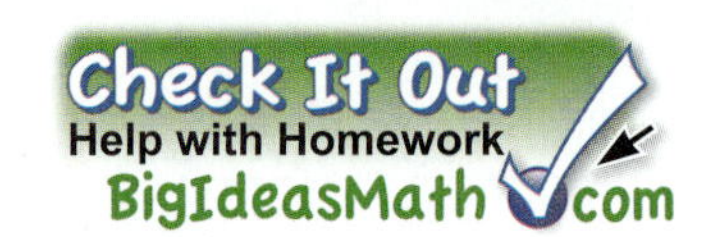

Vocabulary and Concept Check

1. **VOCABULARY** What type of graph represents the solutions of the equation $y = 2x + 4$?

2. **WHICH ONE DOESN'T BELONG?** Which equation does *not* belong with the other three? Explain your reasoning.

$y = 0.5x - 0.2$ $\quad$ $4x + 3 = y$ $\quad$ $y = x^2 + 6$ $\quad$ $\frac{3}{4}x + \frac{1}{3} = y$

Practice and Problem Solving

PRECISION Copy and complete the table. Plot the two solution points and draw a line *exactly* through the two points. Find a different solution point on the line.

3.

x		
$y = 3x - 1$		

4.

x		
$y = \frac{1}{3}x + 2$		

Graph the linear equation. Use a graphing calculator to check your graph, if possible.

5. $y = -5x$
6. $y = \frac{1}{4}x$
7. $y = 5$
8. $x = -6$
9. $y = x - 3$
10. $y = -7x - 1$
11. $y = -\frac{x}{3} + 4$
12. $y = \frac{3}{4}x - \frac{1}{2}$
13. $y = -\frac{2}{3}$
14. $y = 6.75$
15. $x = -0.5$
16. $x = \frac{1}{4}$

17. **ERROR ANALYSIS** Describe and correct the error in graphing the equation.

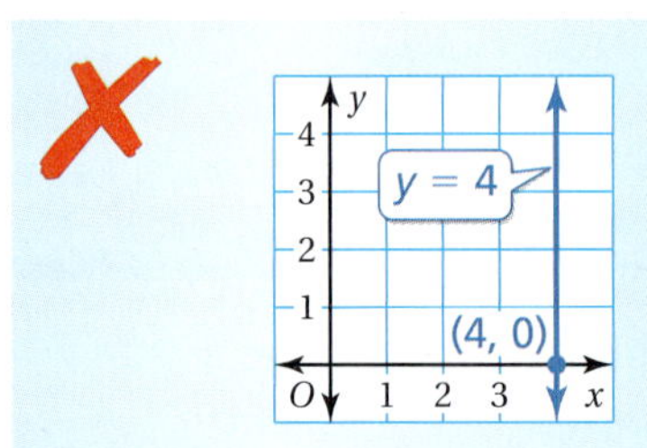

18. **MESSAGING** You sign up for an unlimited text-messaging plan for your cell phone. The equation $y = 20$ represents the cost y (in dollars) for sending x text messages. Graph the equation. What does the graph tell you?

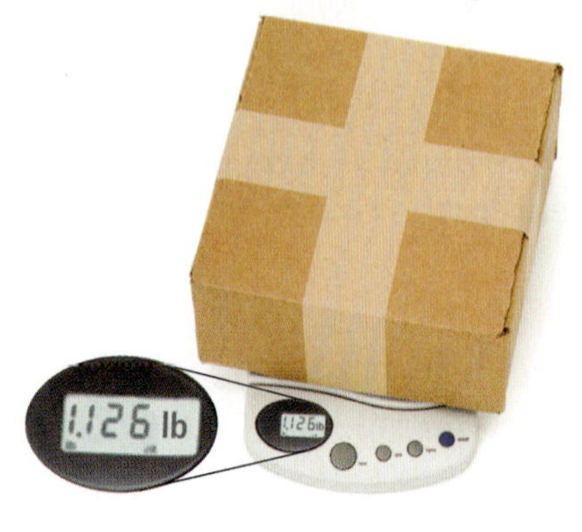

19. **MAIL** The equation $y = 2x + 3$ represents the cost y (in dollars) of mailing a package that weighs x pounds.

   a. Graph the equation.

   b. Use the graph to estimate how much it costs to mail the package.

   c. Use the equation to find exactly how much it costs to mail the package.

Assignment Guide and Homework Check

Level	Day 1 Activity Assignment	Day 2 Lesson Assignment	Homework Check
Basic	3, 4, 28–32	1, 2, 5–23 odd	7, 9, 11, 15, 19
Average	3, 4, 28–32	1, 2, 5–19 odd, 20–26 even	9, 11, 19, 20, 24
Advanced	1–4, 6–16 even, 17, 18–26 even, 27–32		12, 18, 22, 24, 26

Common Errors

- **Exercises 5–16** Students may make calculation errors when solving for ordered pairs. If they only find two ordered pairs for the graph, they may not recognize their mistakes. Encourage them to find at least three ordered pairs when drawing a graph.
- **Exercises 7, 13, and 14** Students may draw vertical lines through points on the x-axis. Remind them that the graph of the equation is a horizontal line. Ask them to identify the y-coordinate for several x-coordinates. For example, what is the y-coordinate for $x = 5$? $x = 6$? $x = -4$? Students should answer with the same y-coordinate each time.
- **Exercises 8, 15, and 16** Students may draw horizontal lines through points on the y-axis. Remind them that the graph of the equation is a vertical line. Ask them to identify the x-coordinate for several y-coordinates. For example, what is the x-coordinate for $y = 3$? $y = -1$? $y = 0$? Students should answer with the same x-coordinate each time.
- **Exercises 20–23** Students may make mistakes in solving for y, such as using the same operation instead of the opposite operation.

4.1 Record and Practice Journal

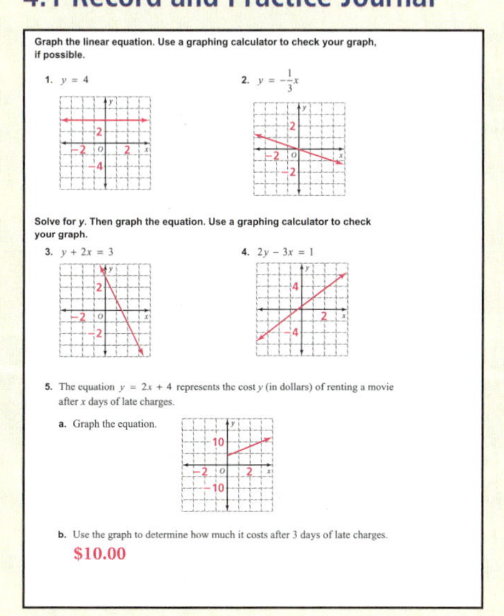

Graph the linear equation. Use a graphing calculator to check your graph, if possible.

1. $y = 4$
2. $y = -\frac{1}{3}x$

Solve for y. Then graph the equation. Use a graphing calculator to check your graph.

3. $y + 2x = 3$
4. $2y - 3x = 1$

5. The equation $y = 2x + 4$ represents the cost y (in dollars) of renting a movie after x days of late charges.

a. Graph the equation.

b. Use the graph to determine how much it costs after 3 days of late charges.
$10.00

Vocabulary and Concept Check

1. a line
2. $y = x^2 + 6$ does not belong because it is not a linear equation.

Practice and Problem Solving

3. *Sample answer:*

x	0	1
$y = 3x - 1$	−1	2

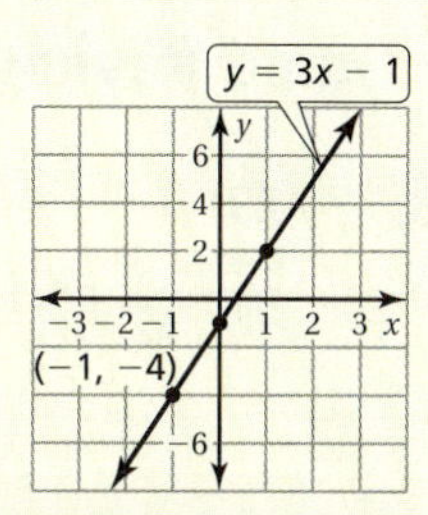

4. *Sample answer:*

x	0	3
$y = \frac{1}{3}x + 2$	2	3

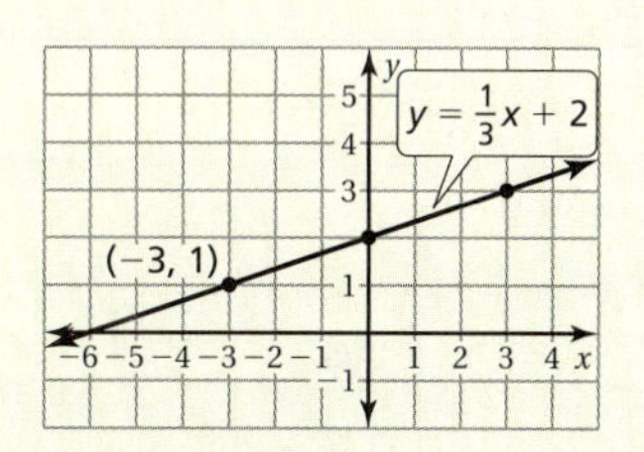

5.

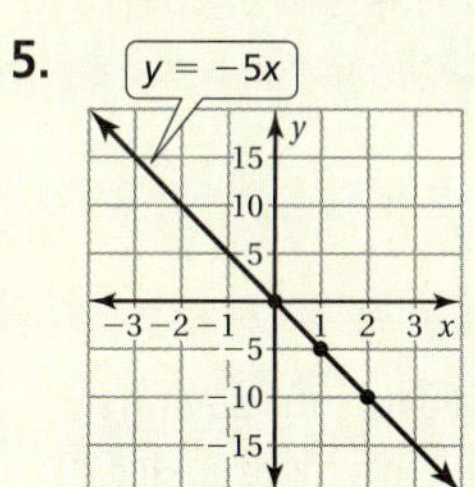

6.

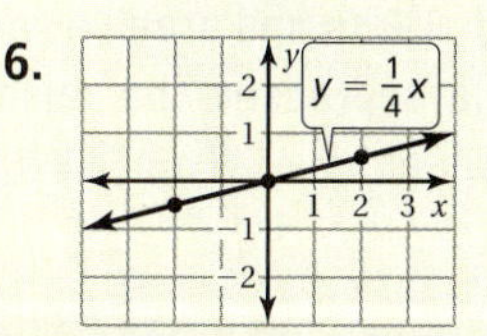

7–19. See Additional Answers.

Practice and Problem Solving

20. $y = 3x + 1$

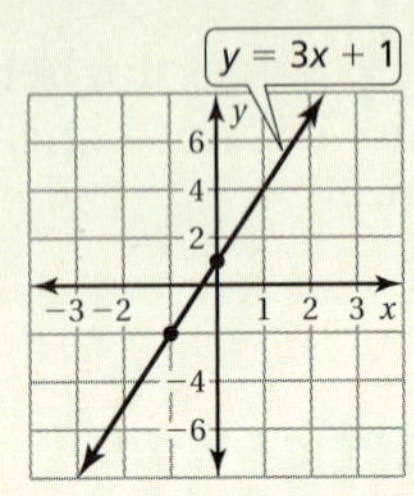

21–26. See Additional Answers.

27. See *Taking Math Deeper.*

Fair Game Review

28. (5, 3) **29.** (−6, 6)

30. (2, −2) **31.** (−4, −3)

32. B

Mini-Assessment

1. Graph $y = \frac{1}{2}x - 2$.

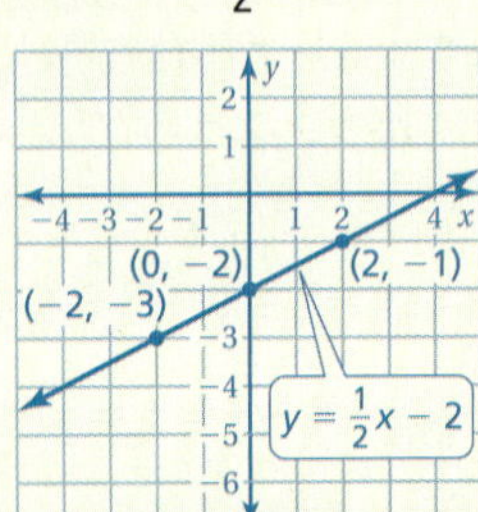

2. You have \$100 in your savings account and plan to deposit \$20 each month. Write and graph a linear equation that represents the balance in your account. $y = 20x + 100$

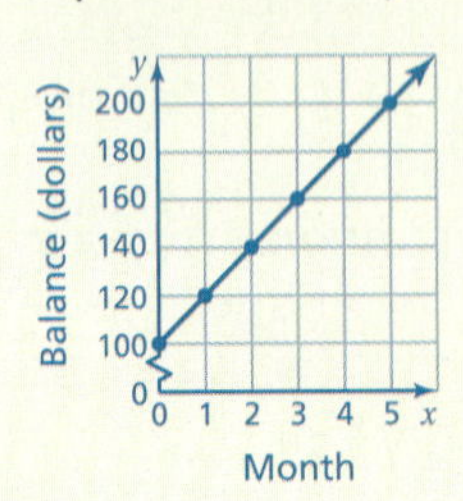

Taking Math Deeper

Exercise 27

Some of the information for this exercise is given in the photo and some is given in the text. It is a good idea to start by listing all of the given information.

List the given information.

- The camera can store 250 pictures.
- 1 second of video = 2 pictures.
- Video time used = 90 seconds.
- Let y = number of pictures.
- Let x = number of seconds of video.

a. Write and graph an equation for x and y.

$$y + 2x = 250$$
$$y = -2x + 250$$

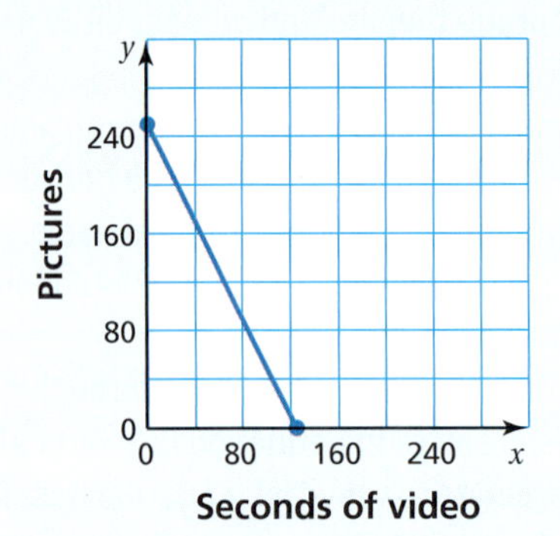

3 **b.** Answer the question.

When $x = 90$, the value of y is as follows.

$$y = -2x + 250$$
$$= -2(90) + 250$$
$$= 70$$

Your camera can store 70 pictures in addition to the 90-second video.

Project

Research digital cameras. Find the number of pictures that can be stored on five different cameras. Compare the prices of the cameras. Which do you consider to be the better buy? Why?

Reteaching and Enrichment Strategies

If students need help...	If students got it...
Resources by Chapter • Practice A and Practice B • Puzzle Time Record and Practice Journal Practice Differentiating the Lesson Lesson Tutorials Skills Review Handbook	Resources by Chapter • Enrichment and Extension • Technology Connection Start the next section

Solve for y. Then graph the equation. Use a graphing calculator to check your graph.

20. $y - 3x = 1$

21. $5x + 2y = 4$

22. $-\frac{1}{3}y + 4x = 3$

23. $x + 0.5y = 1.5$

24. SAVINGS You have \$100 in your savings account and plan to deposit \$12.50 each month.

a. Graph a linear equation that represents the balance in your account.

b. How many months will it take you to save enough money to buy 10 acres of land on Mars?

25. GEOMETRY The sum S of the interior angle measures of a polygon with n sides is $S = (n - 2) \cdot 180°$.

a. Plot four points (n, S) that satisfy the equation. Is the equation a linear equation? Explain your reasoning.

b. Does the value $n = 3.5$ make sense in the context of the problem? Explain your reasoning.

26. SEA LEVEL Along the U.S. Atlantic coast, the sea level is rising about 2 millimeters per year. How many millimeters has sea level risen since you were born? How do you know? Use a linear equation and a graph to justify your answer.

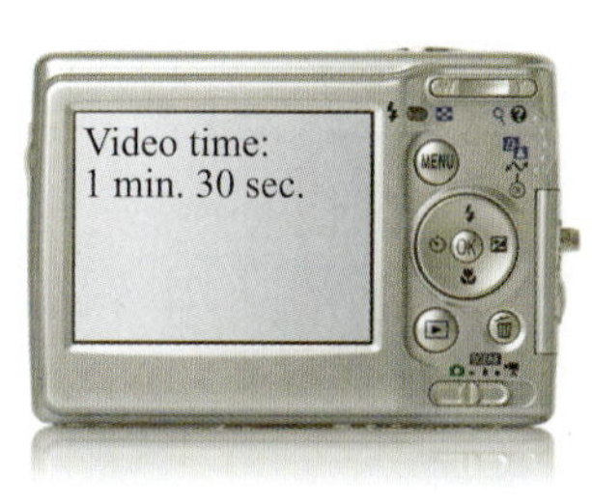

27. Problem Solving One second of video on your digital camera uses the same amount of memory as two pictures. Your camera can store 250 pictures.

a. Write and graph a linear equation that represents the number y of pictures your camera can store when you take x seconds of video.

b. How many pictures can your camera store in addition to the video shown?

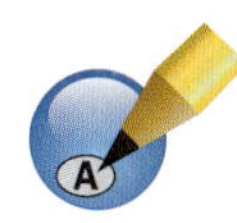

Fair Game Review What you learned in previous grades & lessons

Write the ordered pair corresponding to the point. *(Skills Review Handbook)*

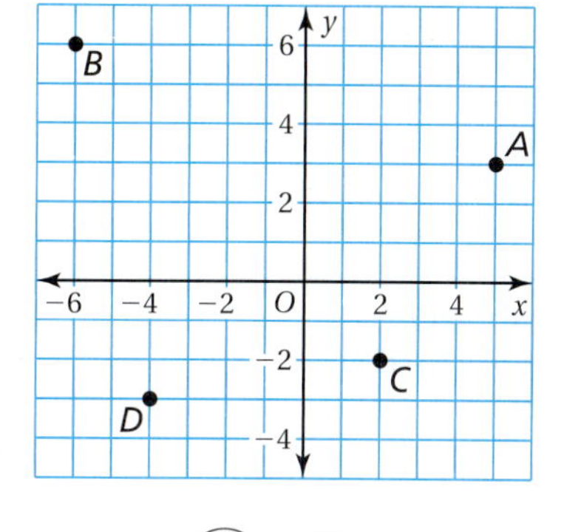

28. point A

29. point B

30. point C

31. point D

32. MULTIPLE CHOICE A debate team has 15 female members. The ratio of females to males is 3 : 2. How many males are on the debate team? *(Skills Review Handbook)*

 6  10  22 (D) 25

4.2 Slope of a Line

Essential Question How can you use the slope of a line to describe the line?

Slope is the rate of change between any two points on a line. It is the measure of the *steepness* of the line.

To find the slope of a line, find the ratio of the change in y (vertical change) to the change in x (horizontal change).

$$\text{slope} = \frac{\text{change in } y}{\text{change in } x}$$

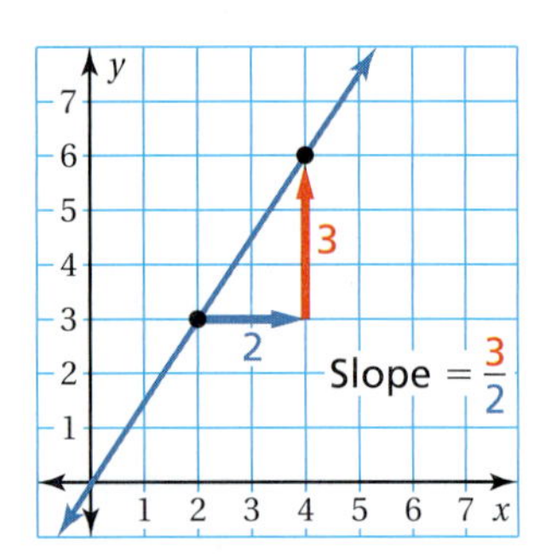

1 ACTIVITY: Finding the Slope of a Line

Work with a partner. Find the slope of each line using two methods.

Method 1: Use the two black points. ●

Method 2: Use the two pink points. ●

Do you get the same slope using each method? Why do you think this happens?

a.

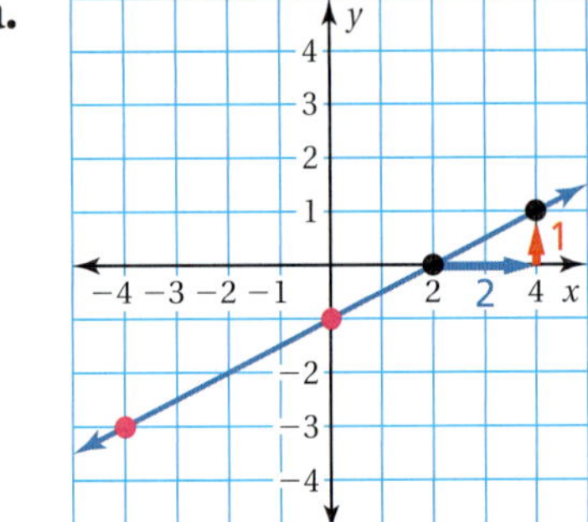

b.

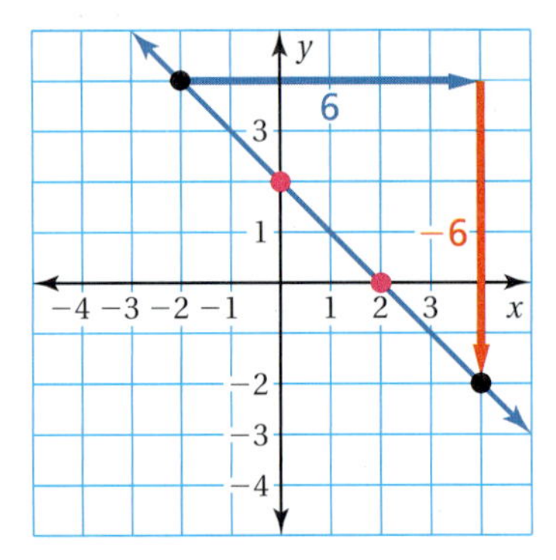

c.

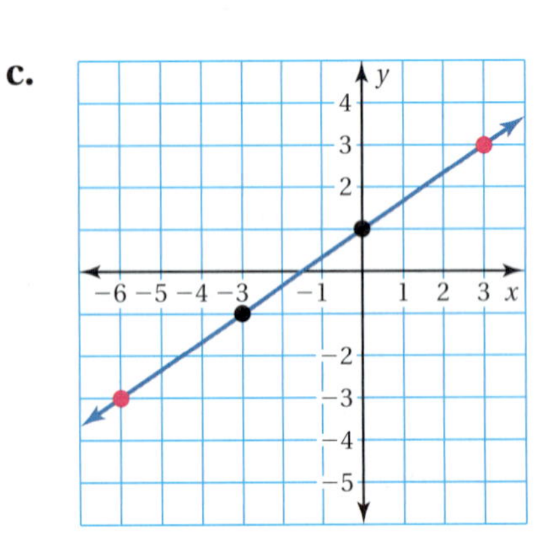

d.

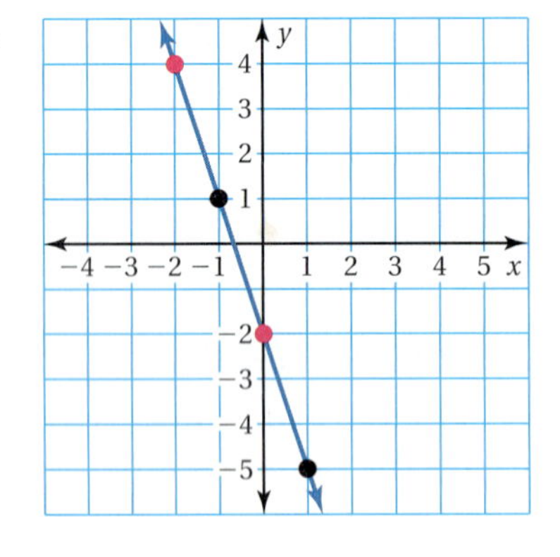

COMMON CORE

Graphing Equations

In this lesson, you will

- find slopes of lines by using two points.
- find slopes of lines from tables.

Learning Standard
MACC.8.EE.2.6

Laurie's Notes

Introduction

Standards for Mathematical Practice

- **MP1a Make Sense of Problems:** The goal is for students to use different pairs of points on a line to find the line's slope. Drawing the arrow diagrams will help students to visualize the *slope triangle*. Students will recognize the triangles are similar and proportions can be formed.

Motivate

? "How many of you have been on a roller coaster?"

- Discuss with students what makes one roller coaster more thrilling than another. Students will usually describe how quickly the coaster drops or the steepness of the hill. This is similar to the *change in y* of a line when finding the slope.

Activity Notes

Discuss

? "Does anyone remember what is meant by slope of a line?" At least one student should recall that it measures the steepness of a line.

- Write the definition for slope. Sketch the graph shown to demonstrate what is meant by change in y (red vertical arrow) and change in x (blue horizontal arrow).
- Remind students that slope is always the change in y in the numerator and the change in x in the denominator. This can be confusing for students because graphs are read from left to right, and we have a tendency to move in the x-direction first. For this reason, students want to write the change in x in the numerator.

? "Can the change in x be negative? Explain." yes; Moving to the left horizontally is negative.

? "Can the change in y be negative? Explain." yes; Moving down vertically is negative.

Activity 1

- Encourage students to draw the change arrows for each pair of points. Label the change in x (or y) next to the arrow.
- **Big Idea:** The slope of a line is always the same regardless of what two ordered pairs are selected.
- **Common Error:** Students may forget to make the change negative when moving downward in the y-direction.
- **MP2 Reason Abstractly and Quantitatively** and **MP3b Critique the Reasoning of Others:** Students are asked to reason why the slopes are the same. They will form their own conjectures and listen to reasons offered by other students. Students will use geometry to explain why this is true in Activity 2.

Florida Common Core Standards

MACC.8.EE.2.6 Use similar triangles to explain why the slope m is the same between any two distinct points on a non-vertical line in the coordinate plane; derive the equation $y = mx$ for a line through the origin and the equation $y = mx + b$ for a line intercepting the vertical axis at b.

Previous Learning

Students should know the relationship between corresponding sides of similar triangles.

Lesson Plans
Complete Materials List

4.2 Record and Practice Journal

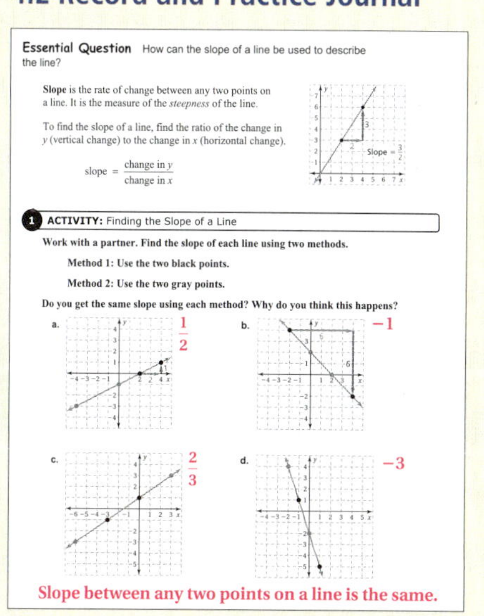

Essential Question How can the slope of a line be used to describe the line?

Slope is the rate of change between any two points on a line. It is the measure of the *steepness* of the line.

To find the slope of a line, find the ratio of the change in y (vertical change) to the change in x (horizontal change).

$$\text{slope} = \frac{\text{change in } y}{\text{change in } x}$$

Slope $= \frac{3}{2}$

1 ACTIVITY: Finding the Slope of a Line

Work with a partner. Find the slope of each line using two methods.

Method 1: Use the two black points.

Method 2: Use the two gray points.

Do you get the same slope using each method? Why do you think this happens?

a. $\frac{1}{2}$

b. -1

c. $\frac{2}{3}$

d. -3

Slope between any two points on a line is the same.

Differentiated Instruction

Kinesthetic

Help students develop number sense about slope. Have them draw lines in the coordinate plane through the following pairs of points.

(0, 0) and (3, 5) (0, 0) and (3, 4)
(0, 0) and (3, 3) (0, 0) and (3, 2)
(0, 0) and (3, 1)

Next have students find the slope of each line. Point out that the line passing through (3, 3) has a slope of 1. The lines with y-coordinates greater than 3 have a slope greater than 1. The lines with y-coordinates less than 3 have a slope less than 1. For positive slopes, the steeper lines will have a greater slope.

4.2 Record and Practice Journal

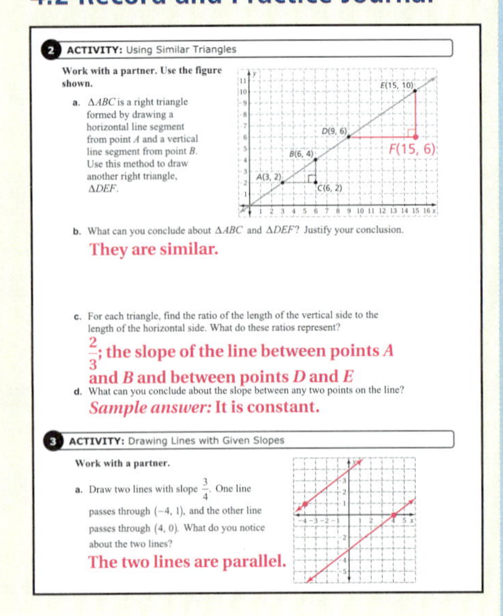

2 ACTIVITY: Using Similar Triangles

Work with a partner. Use the figure shown.

a. ΔABC is a right triangle formed by drawing a horizontal line segment from point A and a vertical line segment from point B. Use this method to draw another right triangle, ΔDEF.

b. What can you conclude about ΔABC and ΔDEF? Justify your conclusion.

They are similar.

c. For each triangle, find the ratio of the length of the vertical side to the length of the horizontal side. What do these ratios represent?

$\frac{2}{3}$; the slope of the line between points A and B and between points D and E

d. What can you conclude about the slope between any two points on the line?

Sample answer: It is constant.

3 ACTIVITY: Drawing Lines with Given Slopes

Work with a partner.

a. Draw two lines with slope $\frac{3}{4}$. One line passes through $(-4, 1)$, and the other line passes through $(4, 0)$. What do you notice about the two lines?

The two lines are parallel.

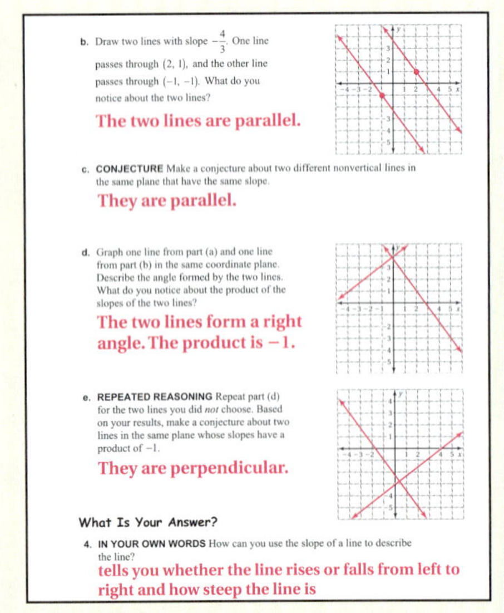

b. Draw two lines with slope $-\frac{4}{3}$. One line passes through (2, 1), and the other line passes through $(-1, -1)$. What do you notice about the two lines?

The two lines are parallel.

c. **CONJECTURE** Make a conjecture about two different nonvertical lines in the same plane that have the same slope.

They are parallel.

d. Graph one line from part (a) and one line from part (b) in the same coordinate plane. Describe the angle formed by the two lines. What do you notice about the product of the slopes of the two lines?

The two lines form a right angle. The product is -1.

e. **REPEATED REASONING** Repeat part (d) for the two lines you did *not* choose. Based on your results, make a conjecture about two lines in the same plane whose slopes have a product of -1.

They are perpendicular.

What Is Your Answer?

4. **IN YOUR OWN WORDS** How can you use the slope of a line to describe the line?

tells you whether the line rises or falls from left to right and how steep the line is

Laurie's Notes

Activity 2

- In part (a), check to see that students have located point F correctly.
- In part (b), students should use what they know about parallel lines and transversals to conclude that the triangles are similar.
- Explain that it is common to refer to $\triangle ABC$ and $\triangle DEF$ as *slope triangles.* They are formed by the line and the change in x and change in y arrows.
- Choose additional pairs of points on the line and repeat the example. Help students realize that they can pick any two points on the line and construct a right triangle that is similar to the ones shown. Then lead them to the conclusion that the ratio of the side lengths, which gives the slope, is the same regardless of which two points you choose.

Activity 3

- In addition to being able to determine the slope of a line, you want students to be able to draw a line that has a particular slope.

? "What does it mean for a line to have a slope of $\frac{3}{4}$?" For every 3 units of change in the y-direction, there is a change of 4 units in the x-direction.

? "What does it mean for a line to have a slope of $-\frac{4}{3}$?" For every -4 units of change in the y-direction, there is a change of 3 units in the x-direction. This is the same as 4 units in the y-direction and -3 units in the x-direction.

? "What do you notice about the lines drawn in part (a) and the lines drawn in part (b)?" Students should observe that the lines in part (a) are parallel and the lines in part (b) are parallel. They may also notice that positive slopes rise (from left to right) and negative slopes fall (from left to right).

- In part (c), students should realize that two different lines with the same slope are parallel.
- **MP6 Attend to Precision:** In parts (d) and (e), if students have graphed carefully they should observe that the lines form a right angle. They may describe the lines as being perpendicular. The product of the slopes is –1 and they may refer to the slopes as negative (or opposite) reciprocals.
- **Connection:** This activity provides the discovery for Extension 4.2.

What Is Your Answer?

- Discuss students' responses. Listen for the words steepness, rising, falling, and so on.

Closure

- Plot the point (0, 3). Draw the line through this point that has a slope of $\frac{1}{3}$.

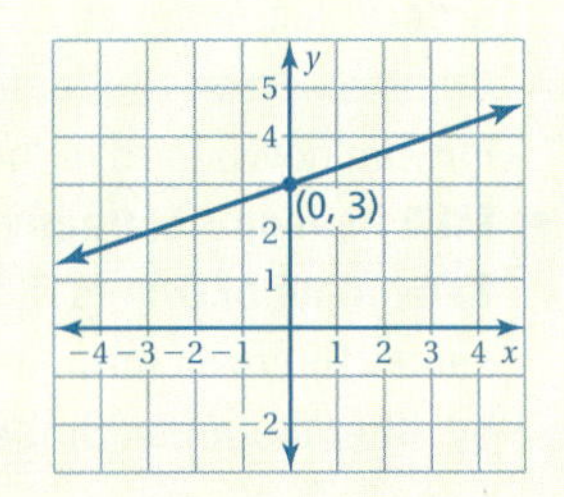

2 ACTIVITY: Using Similar Triangles

Work with a partner. Use the figure shown.

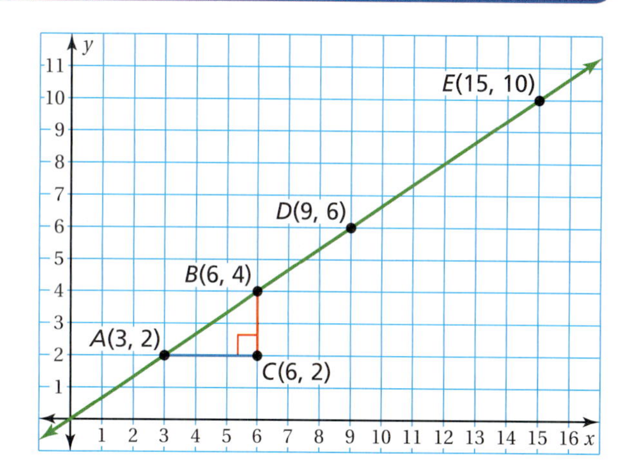

a. $\triangle ABC$ is a right triangle formed by drawing a horizontal line segment from point A and a vertical line segment from point B. Use this method to draw another right triangle, $\triangle DEF$.

b. What can you conclude about $\triangle ABC$ and $\triangle DEF$? Justify your conclusion.

c. For each triangle, find the ratio of the length of the vertical side to the length of the horizontal side. What do these ratios represent?

d. What can you conclude about the slope between any two points on the line?

3 ACTIVITY: Drawing Lines with Given Slopes

Work with a partner.

a. Draw two lines with slope $\frac{3}{4}$. One line passes through $(-4, 1)$, and the other line passes through $(4, 0)$. What do you notice about the two lines?

b. Draw two lines with slope $-\frac{4}{3}$. One line passes through $(2, 1)$, and the other line passes through $(-1, -1)$. What do you notice about the two lines?

c. **CONJECTURE** Make a conjecture about two different nonvertical lines in the same plane that have the same slope.

d. Graph one line from part (a) and one line from part (b) in the same coordinate plane. Describe the angle formed by the two lines. What do you notice about the product of the slopes of the two lines?

e. **REPEATED REASONING** Repeat part (d) for the two lines you did *not* choose. Based on your results, make a conjecture about two lines in the same plane whose slopes have a product of -1.

Math Practice 1

Interpret a Solution

What does the slope tell you about the graph of the line? Explain.

What Is Your Answer?

4. **IN YOUR OWN WORDS** How can you use the slope of a line to describe the line?

Practice

Use what you learned about the slope of a line to complete Exercises 4–6 on page 153.

Key Vocabulary
slope, *p. 150*
rise, *p. 150*
run, *p. 150*

Key Idea

Slope

The **slope** m of a line is a ratio of the change in y (the **rise**) to the change in x (the **run**) between any two points, (x_1, y_1) and (x_2, y_2), on the line.

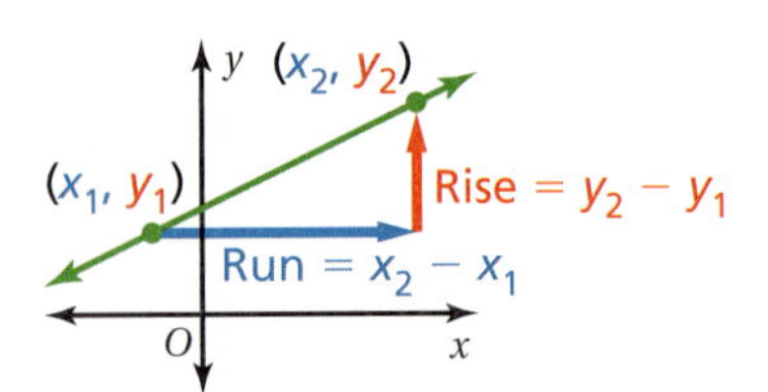

$$m = \frac{\text{rise}}{\text{run}} = \frac{\text{change in } y}{\text{change in } x} = \frac{y_2 - y_1}{x_2 - x_1}$$

Positive Slope

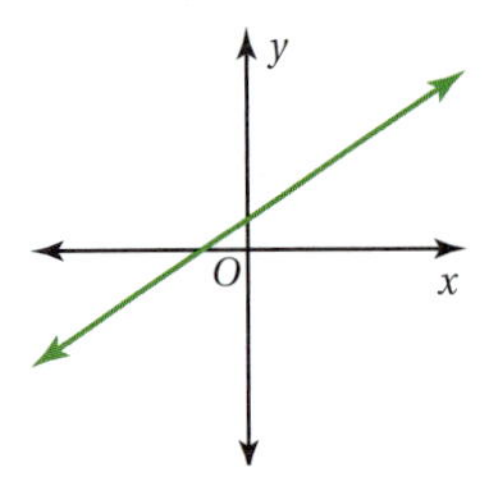

The line rises from left to right.

Negative Slope

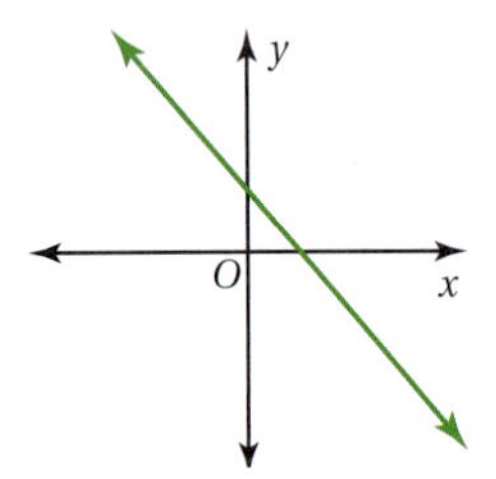

The line falls from left to right.

Reading

In the slope formula, x_1 is read as "x sub one," and y_2 is read as "y sub two." The numbers 1 and 2 in x_1 and y_2 are called *subscripts*.

EXAMPLE 1 Finding the Slope of a Line

Describe the slope of the line. Then find the slope.

a.

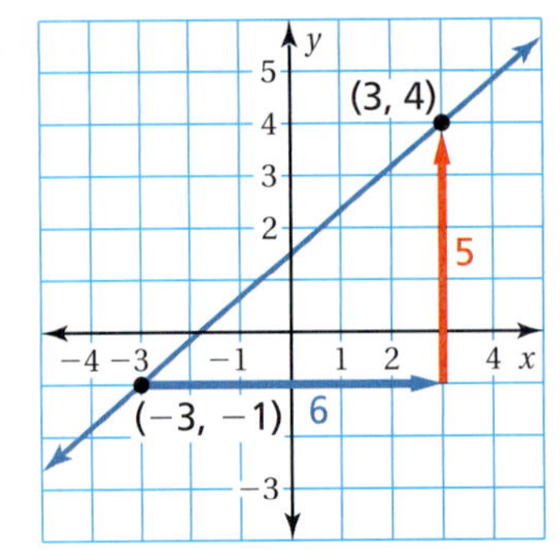

The line rises from left to right. So, the slope is positive. Let $(x_1, y_1) = (-3, -1)$ and $(x_2, y_2) = (3, 4)$.

$$m = \frac{y_2 - y_1}{x_2 - x_1}$$

$$= \frac{4 - (-1)}{3 - (-3)}$$

$$= \frac{5}{6}$$

b.

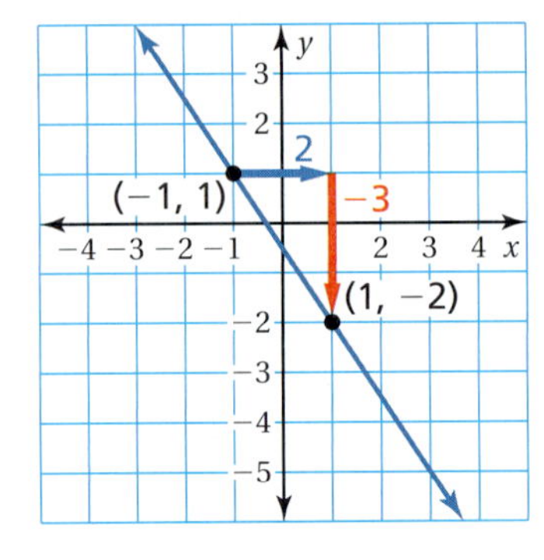

The line falls from left to right. So, the slope is negative. Let $(x_1, y_1) = (-1, 1)$ and $(x_2, y_2) = (1, -2)$.

$$m = \frac{y_2 - y_1}{x_2 - x_1}$$

$$= \frac{-2 - 1}{1 - (-1)}$$

$$= \frac{-3}{2}, \text{ or } -\frac{3}{2}$$

Study Tip

When finding slope, you can label either point as (x_1, y_1) and the other point as (x_2, y_2).

Laurie's Notes

Introduction

Connect

- **Yesterday:** Students explored slopes of lines. (MP1a, MP2, MP3b, MP6)
- **Today:** Students will find slopes of lines from graphs and tables.

Motivate

- Have students plot four points: $A(5, 0)$, $B(0, 5)$, $C(-5, 0)$, and $D(0, -5)$. Connect the points to form the quadrilateral $ABCD$.

? "What type of quadrilateral is $ABCD$?" Without proof, students should say square.

? "What is the slope of each side, meaning the slopes of the lines through AB, BC, CD, and DA?" Slopes of AB and CD are both -1. Slopes of BC and DA are both 1.

Lesson Notes

Key Idea

- Write the Key Idea. Define slope of a line.
- Tell students that it is traditional to use m to represent slope. They will also see this in future mathematics classes.
- Note the use of color in the definition and on the graph. The *change in y* and the *vertical change arrow* are both red. The *change in x* and the *horizontal change arrow* are both blue.
- Discuss the difference in positive and negative slopes, a concept students explored yesterday.
- Remind students that graphs are read from left to right.
- Explain to students that you can also subtract coordinates to find the rise and run in addition to finding rise and run graphically.

Example 1

- **MP1a Make Sense of Problems:** Drawing the arrow diagrams will help students visualize the *slope triangle*.
- Students often ask if they can move in the y-direction first, followed by the x-direction. The answer is yes. Demonstrate this on either graph.
  - In part (a), start at $(-3, -1)$ and move up 5 units in the y-direction and then to the right 6 units in the x-direction. You will end at $(3, 4)$.
  - In part (b), start at $(-1, 1)$ and move down 3 units in the y-direction and then to the right 2 units in the x-direction. You will end at $(1, -2)$.
- Discuss the Study Tip. You can move in either direction first, and the labeling of the ordered pairs is arbitrary. Either point can be (x_1, y_1).

Goal Today's lesson is finding the **slope** of a line.

Lesson Tutorials
Lesson Plans
Answer Presentation Tool

Extra Example 1

Describe the slope of the line. Then find the slope.

a.

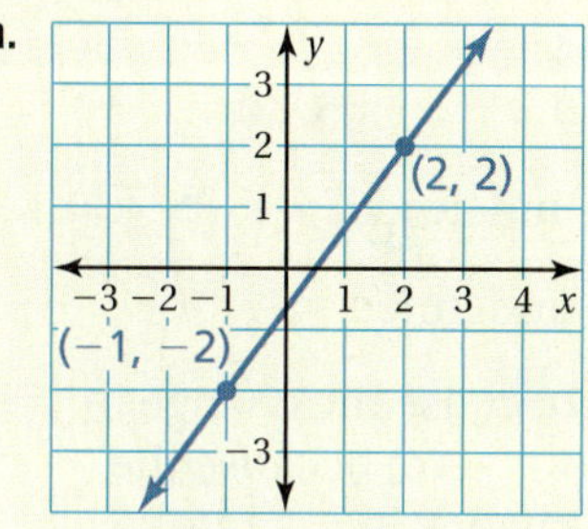

rises from left to right, so it is positive; $\frac{4}{3}$

b.

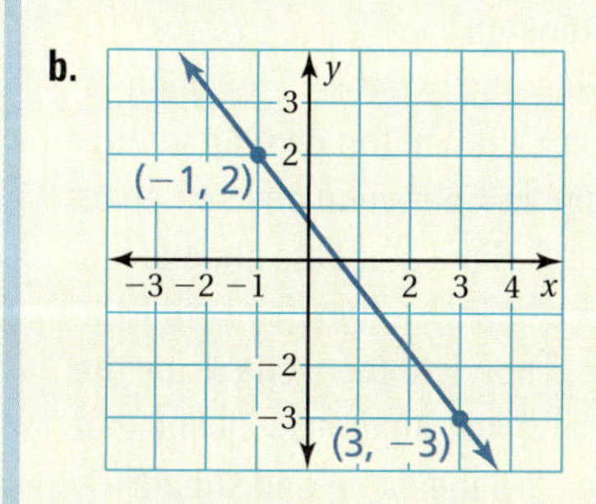

falls from left to right, so it is negative; $-\frac{5}{4}$

On Your Own

1. $-\frac{1}{5}$
2. $\frac{1}{3}$
3. $\frac{5}{2}$

Extra Example 2

Find the slope of the line.

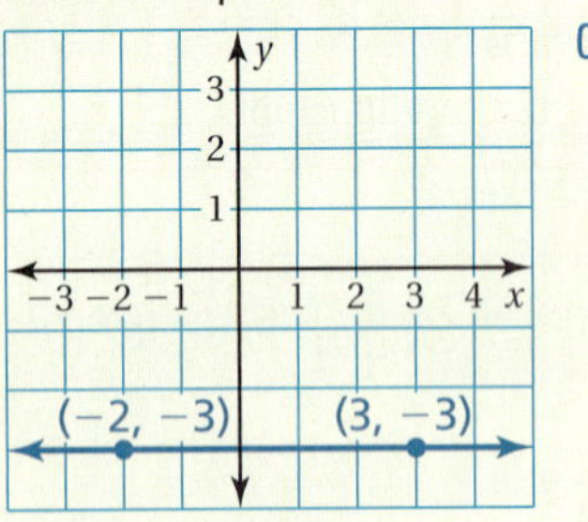

0

Extra Example 3

Find the slope of the line.

undefined

On Your Own

4. 0
5. 0
6. undefined
7. undefined
8. because the change in y is zero; because the change in x is zero

English Language Learners

Comprehension

The Key Idea box states, "The slope m of a line is a ratio of the change in y to the change in x between any two points on the line." Have students choose four points on a line. Use two points to find the slope. Then find the slope using the other two points. Students will find that the slopes are the same and should understand that the slope of the line is the same for the entire infinite length of the line.

Laurie's Notes

On Your Own

- **Neighbor Check:** Have students work independently and then have their neighbors check their work. Have students discuss any discrepancies.

Example 2

? "How does a slope of $\frac{1}{2}$ compare to a slope of $\frac{1}{5}$? Describe the lines." A slope of $\frac{1}{2}$ runs 2 units for every 1 unit it rises. A slope of $\frac{1}{5}$ runs 5 units for each 1 unit it rises. A slope of $\frac{1}{5}$ is not as steep.

? "What would a slope of $\frac{1}{10}$ look like?" A slope of $\frac{1}{10}$ is less steep than a slope of $\frac{1}{5}$, so it is almost flat.

? "How steep do you think a horizontal line is?" Listen for students to describe a horizontal line as having no rise. In this example, they will see it has a slope of 0.

- This progression of questions is to help students visualize that as slopes of lines get less steep, the lines become horizontal.
- Work through the example. There is no change in y. So, the change in y is 0.

Example 3

- Ask a series of questions similar to Example 2.

? "How does a slope of $\frac{9}{2}$ compare to a slope of $\frac{3}{2}$? Describe the lines." A slope of $\frac{9}{2}$ rises 9 units for every 2 units it runs. A slope of $\frac{3}{2}$ rises 3 units for every 2 units it runs. A slope of $\frac{9}{2}$ is steeper.

? "What would a slope of 10 look like?" A slope of 10 is steeper than a slope of $\frac{9}{2}$, so it is almost vertical.

? "How steep do you think a vertical line is?" Listen for students to describe a vertical line as having a slope of infinity.

- Work through the example. There is no change in x. So, the change in x is 0. Since division by zero is undefined, the slope of the line is undefined.

On Your Own

? Listen to students' explanations of Question 8. If necessary, you might ask,
  - "What is true about every y-coordinate for points on a horizontal line?" y-values are all the same
  - "When you compute the rise, what will you get?" 0
  - "What is true about every x-coordinate for points on a vertical line?" x-values are all the same
  - "When you compute the run, what will you get?" 0
- Students should recall that 0 can be divided by a non-zero number and the quotient is 0. Division by 0 is undefined.

On Your Own

Now You're Ready
Exercises 7–9

Find the slope of the line.

1.

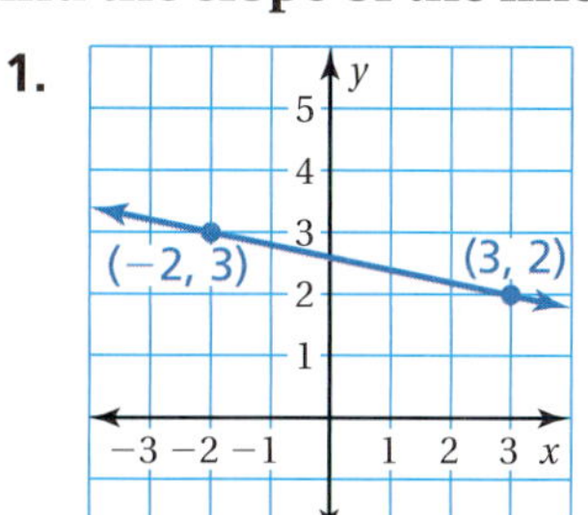

2.

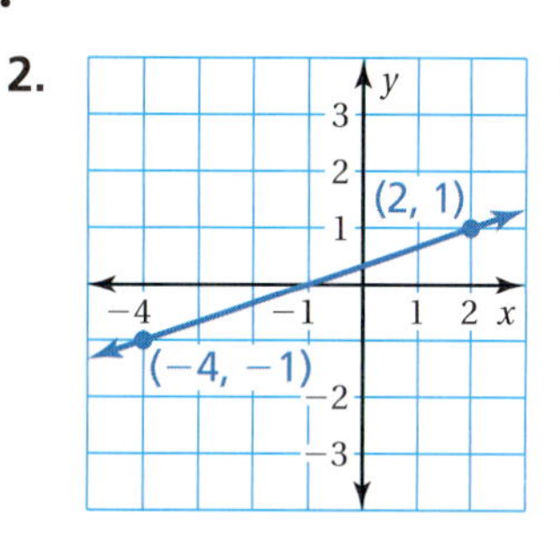

3.

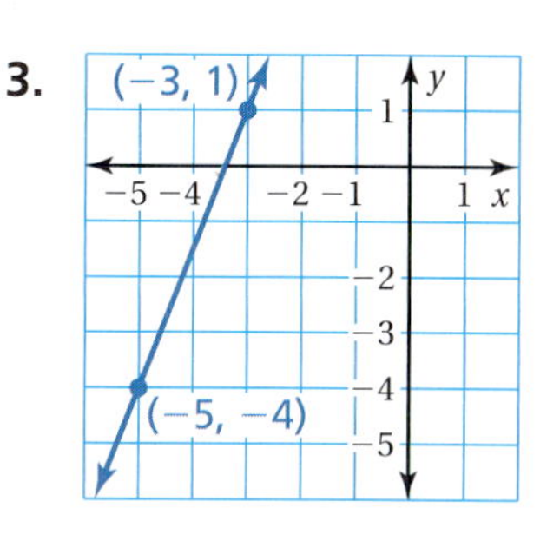

EXAMPLE 2 Finding the Slope of a Horizontal Line

Find the slope of the line.

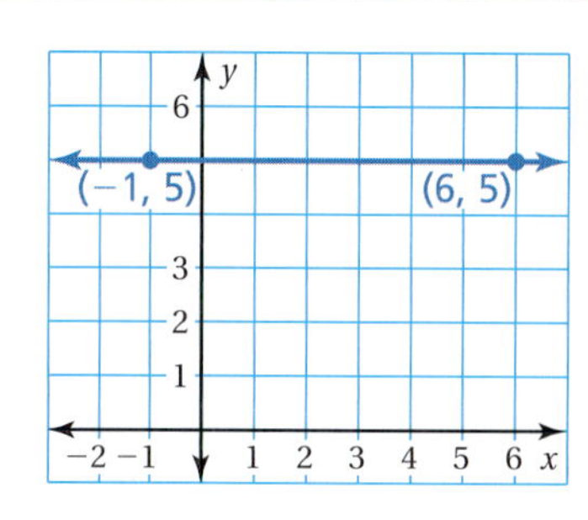

$$m = \frac{y_2 - y_1}{x_2 - x_1}$$

$$= \frac{5 - 5}{6 - (-1)}$$

$$= \frac{0}{7}, \text{ or } 0$$

The slope is 0.

EXAMPLE 3 Finding the Slope of a Vertical Line

Find the slope of the line.

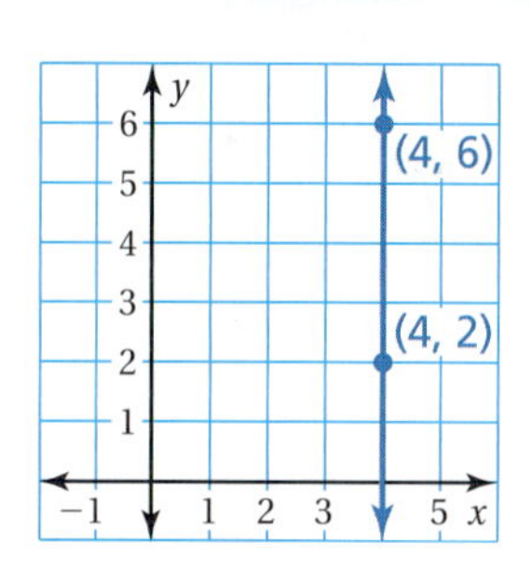

$$m = \frac{y_2 - y_1}{x_2 - x_1}$$

$$= \frac{6 - 2}{4 - 4}$$

$$= \frac{4}{0} \quad ✗$$

Because division by zero is undefined, the slope of the line is undefined.

Study Tip

The slope of every horizontal line is 0. The slope of every vertical line is undefined.

On Your Own

Now You're Ready
Exercises 13–15

Find the slope of the line through the given points.

4. (1, −2), (7, −2)

5. (−2, 4), (3, 4)

6. (−3, −3), (−3, −5)

7. (0, 8), (0, 0)

8. How do you know that the slope of every horizontal line is 0? How do you know that the slope of every vertical line is undefined?

EXAMPLE 4 Finding Slope from a Table

The points in the table lie on a line. How can you find the slope of the line from the table? What is the slope?

x	1	4	7	10
y	8	6	4	2

Choose any two points from the table and use the slope formula.

Use the points $(x_1, y_1) = (1, 8)$ and $(x_2, y_2) = (4, 6)$.

$$m = \frac{y_2 - y_1}{x_2 - x_1}$$

$$= \frac{6 - 8}{4 - 1}$$

$$= \frac{-2}{3}, \text{ or } -\frac{2}{3}$$

The slope is $-\frac{2}{3}$.

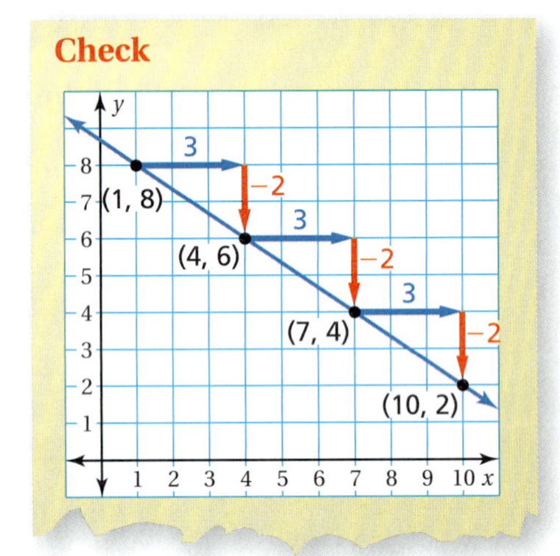

On Your Own

The points in the table lie on a line. Find the slope of the line.

9.

x	1	3	5	7
y	2	5	8	11

10.

x	−3	−2	−1	0
y	6	4	2	0

Summary

Slope

Positive Slope

The line rises from left to right.

Negative Slope

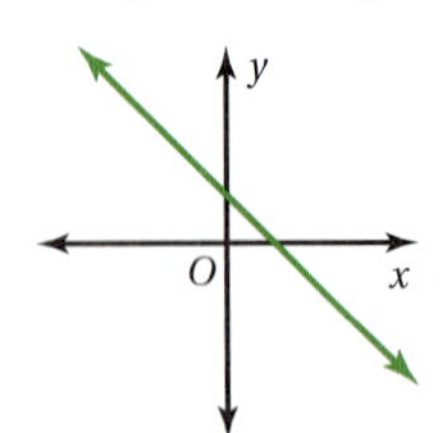

The line falls from left to right.

Slope of 0

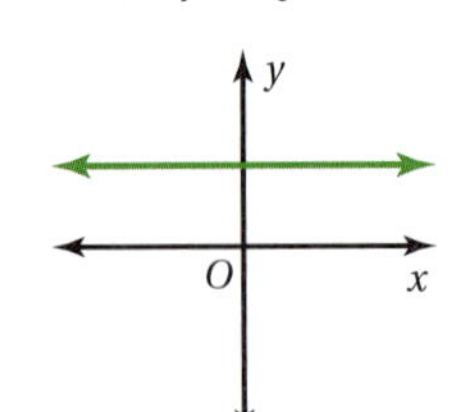

The line is horizontal.

Undefined Slope

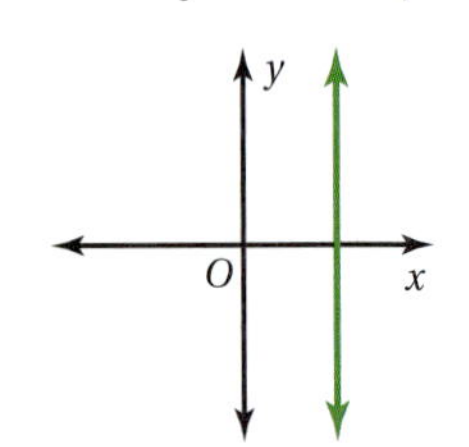

The line is vertical.

Laurie's Notes

Example 4

? "What do you notice about the x-values and the y-values?" The x-values are increasing by 3 and the y-values are decreasing by 2.

- **Connection:** Show the changes in x and y from each column to the next (i.e., $+3$, $+3$, $+3$ on the top and -2, -2, -2 on the bottom). Ask students to use the table and graph to determine if x and y are in a proportional relationship (no). Ask again in Question 10 (yes). This review of proportionality will help prepare for Section 4.3.
- Compute the slope between any two points in the table.
- Using (1, 8) and (4, 6): $m = \frac{y_2 - y_1}{x_2 - x_1} = \frac{6 - 8}{4 - 1} = -\frac{2}{3}$
- Using (1, 8) and (7, 4): $m = \frac{y_2 - y_1}{x_2 - x_1} = \frac{4 - 8}{7 - 1} = -\frac{4}{6} = -\frac{2}{3}$
- Remind students about the activity, where they found that the slope is the same regardless of which two points are selected. The slope triangles that are formed are similar.

? "The line has a negative slope. What do you notice about the line?" The line falls from left to right.

On Your Own

- **Neighbor Check:** Have students work independently and then have their neighbors check their work. Have students discuss any discrepancies.
- **Connection:** In Question 9, students may recognize from the table that both x and y are increasing and the slope is positive. In Question 10, as x increases, the y-values are decreasing and the slope is negative.
- Students should observe by inspection the change in x and the change in y. This is *reading* the table.

Summary

- Students have computed the slopes of many lines today. Discuss the Summary by referring to previous examples.

Closure

- The points in the table lie on a line. How can you find the slope of the line from the table? What is the slope?

x	-1	0	1	2
y	-3	-1	1	3

Choose any two points from the table and use the slope formula. $m = 2$

Extra Example 4

The points in the table lie on a line. How can you find the slope of the line from the table? What is the slope?

x	-2	-1	0	1
y	-8	-5	-2	1

Choose any two points from the table and use the slope formula; 3

9. $\frac{3}{2}$ 10. -2

Vocabulary and Concept Check

1. a. B and C

   b. A

   c. no; None of the lines are vertical.

2. *Sample answer:* When constructing a wheelchair ramp, you need to know the slope.

3. The line is horizontal.

Practice and Problem Solving

4. 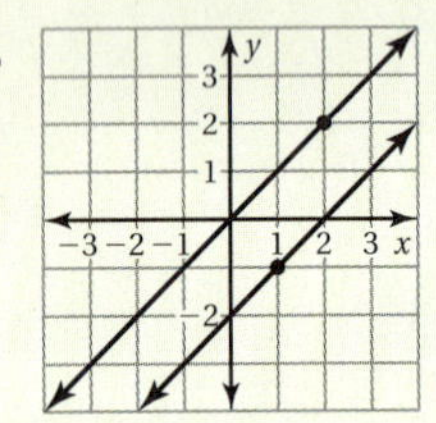

   The lines are parallel.

5. 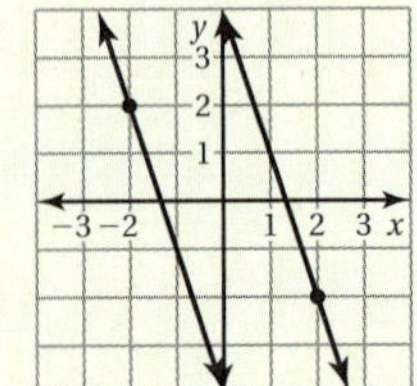

   The lines are parallel.

6. 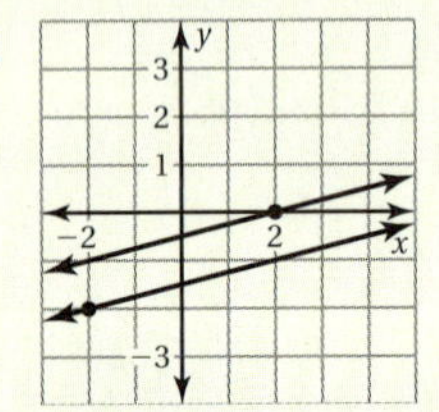

   The lines are parallel.

7. $\frac{3}{4}$ 8. $-\frac{5}{4}$

9. $-\frac{3}{5}$ 10. $\frac{1}{6}$

11. 0

12. undefined

Assignment Guide and Homework Check

Level	Day 1 Activity Assignment	Day 2 Lesson Assignment	Homework Check
Basic	4–6, 37–40	1–3, 7–29 odd	7, 13, 15, 21, 25
Average	4–6, 37–40	1–3, 11, 16–19, 22–30 even, 31–34	11, 17, 22, 26, 28
Advanced	1–6, 12, 18, 19, 20–30 even, 31–40		18, 24, 26, 34, 35

Common Errors

- **Exercises 7–12** Students may forget negatives or include them when they are not needed. Remind them that if the line rises from left to right the slope is positive, and if the line falls from left to right the slope is negative.
- **Exercises 7–18** Students may find the reciprocal of the slope because they mix up rise and run. Remind them that the change in y is the numerator and the change in x is the denominator.

4.2 Record and Practice Journal

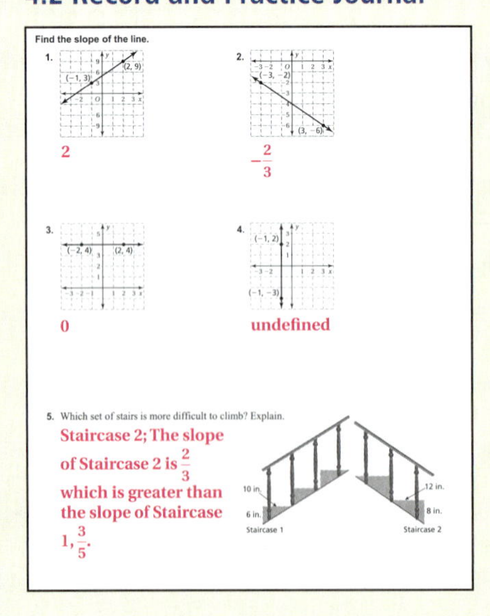

Find the slope of the line.

1. 2

2. $-\frac{2}{3}$

3. 0

4. undefined

5. Which set of stairs is more difficult to climb? Explain.

Staircase 2; The slope of Staircase 2 is $\frac{2}{3}$ which is greater than the slope of Staircase 1, $\frac{3}{5}$.

4.2 Exercises

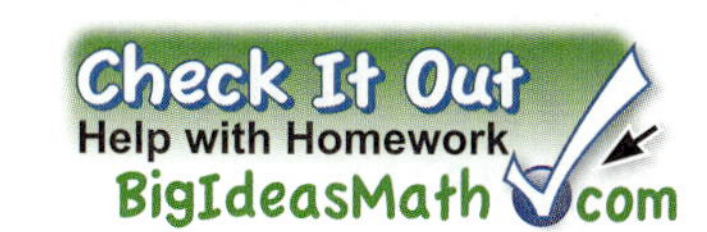

Vocabulary and Concept Check

1. **CRITICAL THINKING** Refer to the graph.

   a. Which lines have positive slopes?

   b. Which line has the steepest slope?

   c. Do any lines have an undefined slope? Explain.

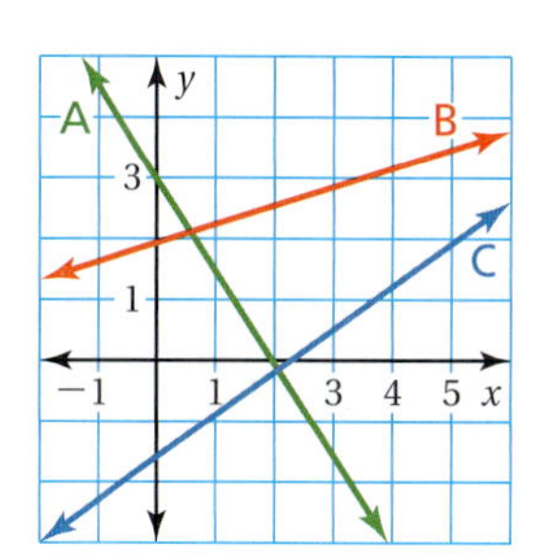

2. **OPEN-ENDED** Describe a real-life situation in which you need to know the slope.

3. **REASONING** The slope of a line is 0. What do you know about the line?

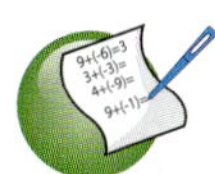

Practice and Problem Solving

Draw a line through each point using the given slope. What do you notice about the two lines?

4. slope $= 1$

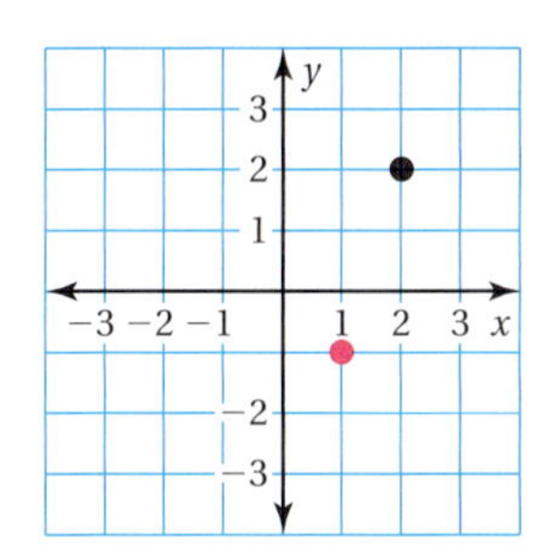

5. slope $= -3$

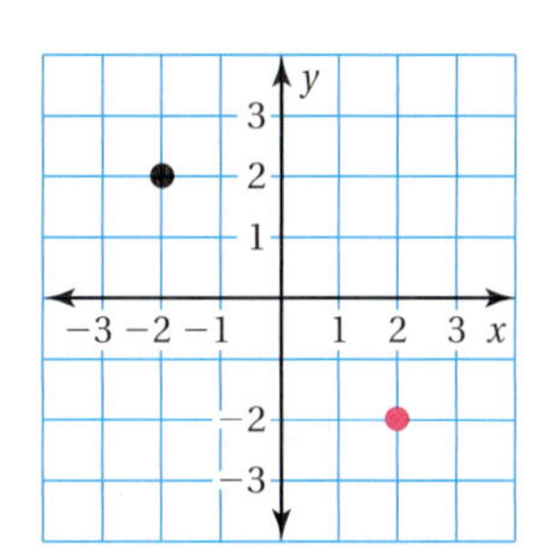

6. slope $= \frac{1}{4}$

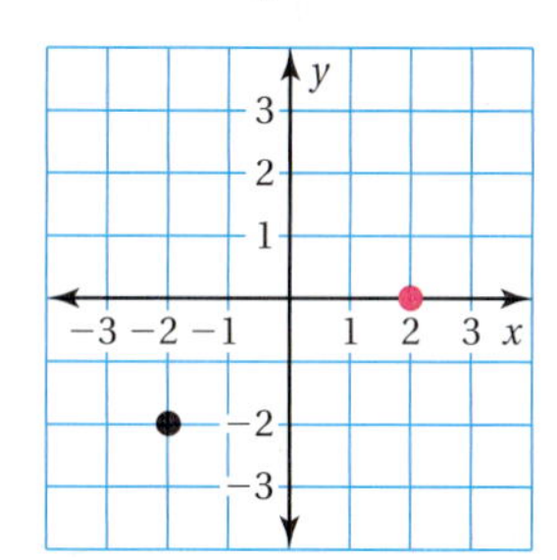

Find the slope of the line.

7.

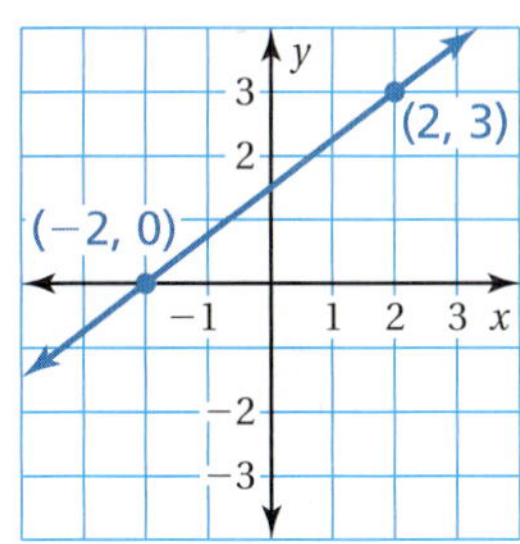

8.

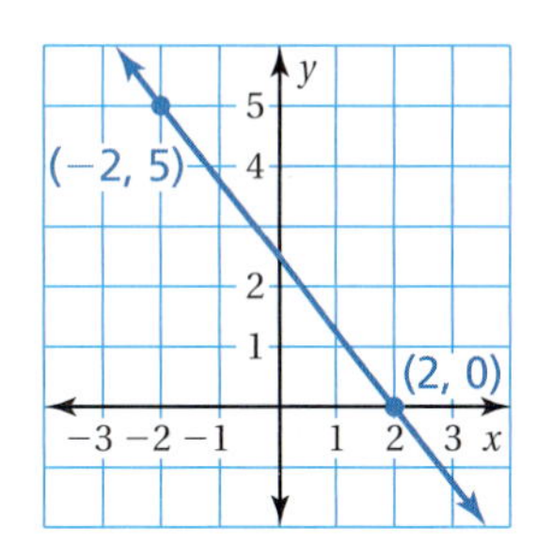

9.

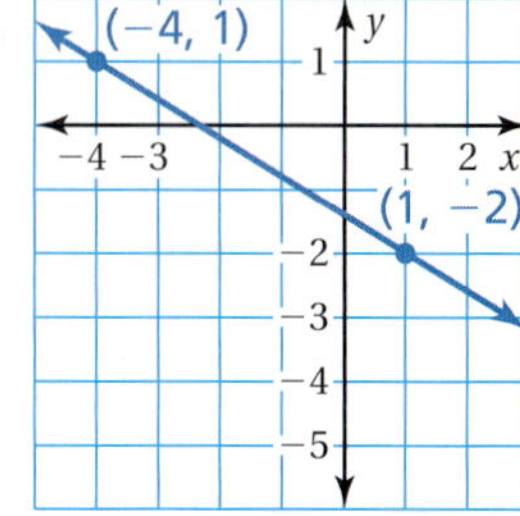

10.

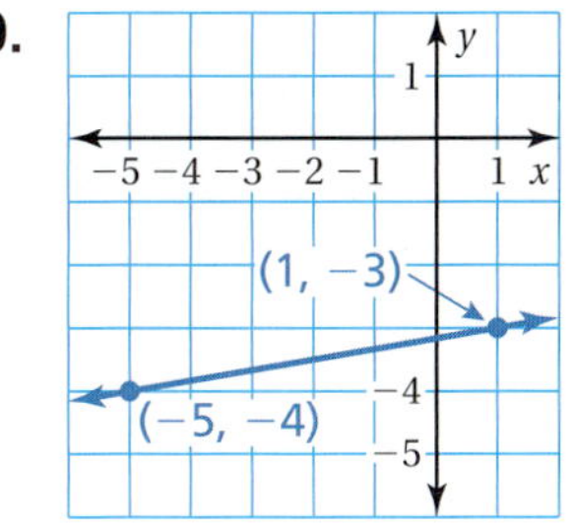

11.

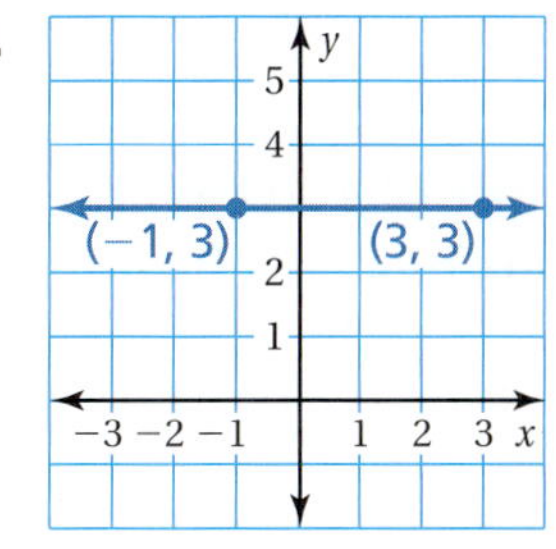

12.

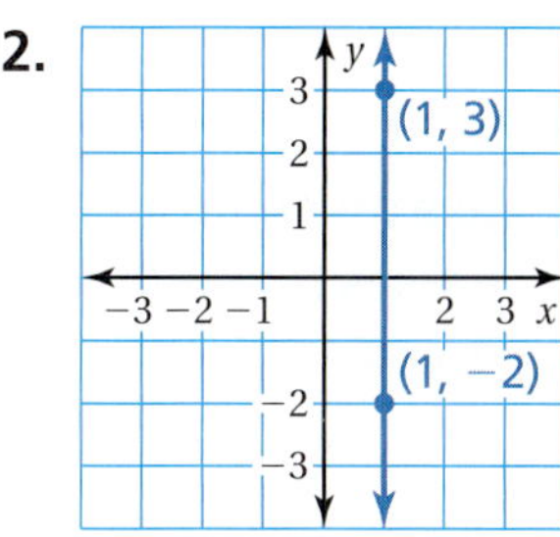

Find the slope of the line through the given points.

2 3 **13.** $(4, -1), (-2, -1)$ **14.** $(5, -3), (5, 8)$ **15.** $(-7, 0), (-7, -6)$

16. $(-3, 1), (-1, 5)$ **17.** $(10, 4), (4, 15)$ **18.** $(-3, 6), (2, 6)$

19. ERROR ANALYSIS Describe and correct the error in finding the slope of the line.

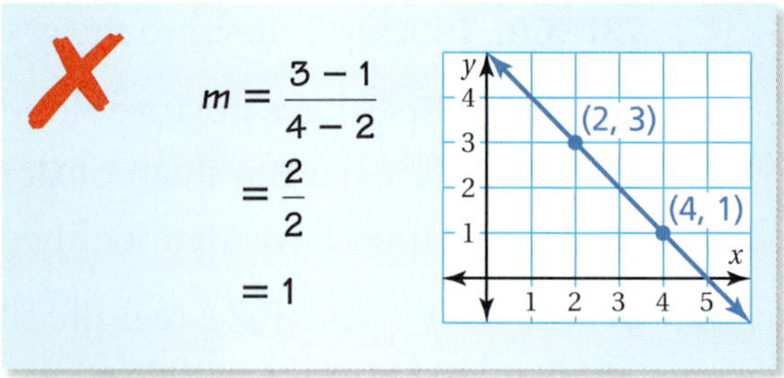

20. CRITICAL THINKING Is it more difficult to walk up the ramp or the hill? Explain.

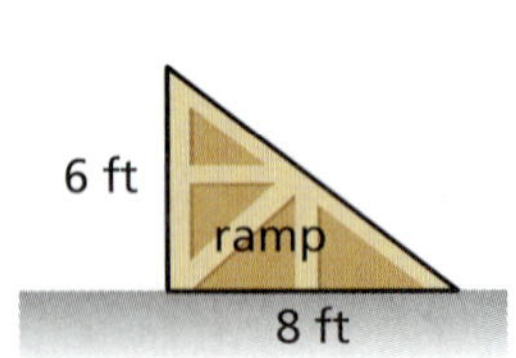

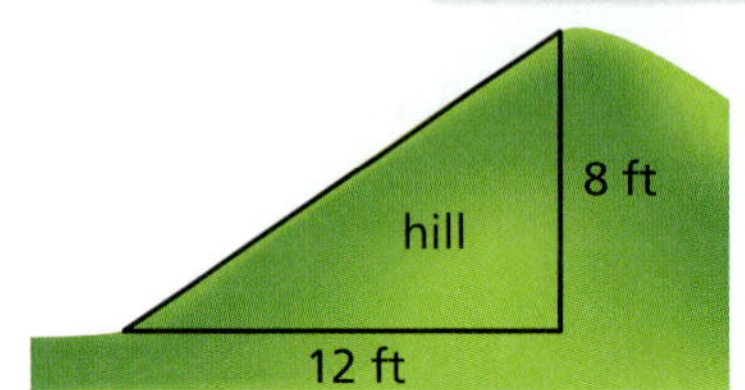

The points in the table lie on a line. Find the slope of the line.

4 **21.**

x	1	3	5	7
y	2	10	18	26

22.

x	-3	2	7	12
y	0	2	4	6

23.

x	-6	-2	2	6
y	8	5	2	-1

24.

x	-8	-2	4	10
y	8	1	-6	-13

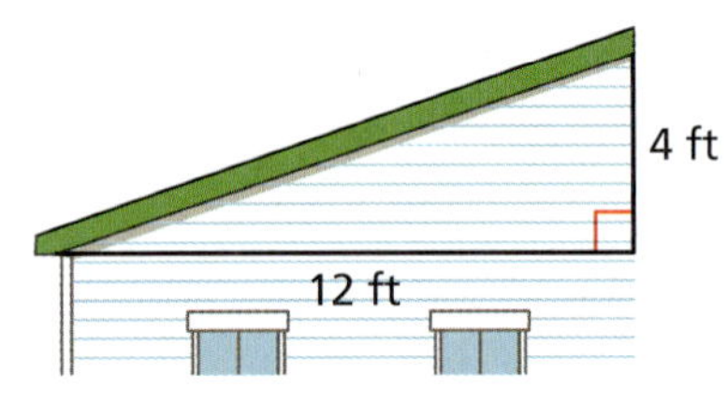

25. PITCH Carpenters refer to the slope of a roof as the *pitch* of the roof. Find the pitch of the roof.

26. PROJECT The guidelines for a wheelchair ramp suggest that the ratio of the rise to the run be no greater than 1 : 12.

a. CHOOSE TOOLS Find a wheelchair ramp in your school or neighborhood. Measure its slope. Does the ramp follow the guidelines?

b. Design a wheelchair ramp that provides access to a building with a front door that is 2.5 feet above the sidewalk. Illustrate your design.

Use an equation to find the value of k so that the line that passes through the given points has the given slope.

27. $(1, 3), (5, k); m = 2$ **28.** $(-2, k), (2, 0); m = -1$

29. $(-4, k), (6, -7); m = -\frac{1}{5}$ **30.** $(4, -4), (k, -1); m = \frac{3}{4}$

Common Errors

- **Exercise 20** Students may get confused because one of the slopes is negative and the other is positive. Tell them to think of the absolute values of the slopes when comparing. Encourage them to graph the slopes on a number line to check their answers.
- **Exercises 21–24** Students may find the change in *x* over the change in *y*. Remind them that slope is the change in *y* over the change in *x*.

Practice and Problem Solving

13. 0

14. undefined

15. undefined

16. 2

17. $-\frac{11}{6}$

18. 0

19. The denominator should be $2 - 4$.
$m = -1$

20. The ramp because its slope is steeper.

21. 4

22. $\frac{2}{5}$

23. $-\frac{3}{4}$

24. $-\frac{7}{6}$

25. $\frac{1}{3}$

26. See Additional Answers.

27. $k = 11$

28. $k = 4$

29. $k = -5$

30. $k = 8$

Differentiated Instruction

Auditory

Discuss how the rate of change in a rate problem is related to slope. For example, the cost to travel on a turnpike (cost per mile) can be expressed as $\frac{\text{cost (in dollars)}}{\text{miles driven}}$, where the cost is the change in *y*-values and the miles driven is the change in *x*-values.

Practice and Problem Solving

31. **a.** $\frac{3}{40}$

b. The cost increases by \$3 for every 40 miles you drive, or the cost increases by \$0.075 for every mile you drive.

32. The boat ramp, because it has a 16.67% grade.

33. yes; The slopes are the same between the points.

34. \$2750 per month

35. When you switch the coordinates, the differences in the numerator and denominator are the opposite of the numbers when using the slope formula. You still get the same slope.

36. See *Taking Math Deeper.*

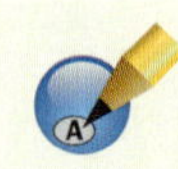

Fair Game Review

37. $b = 25$ **38.** $n = 56$

39. $x = 7.5$ **40.** B

Mini-Assessment

Find the slope of the line.

1.

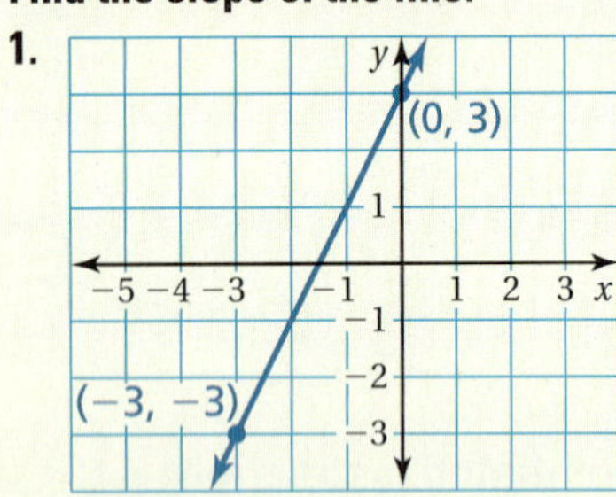

$m = 2$

2.

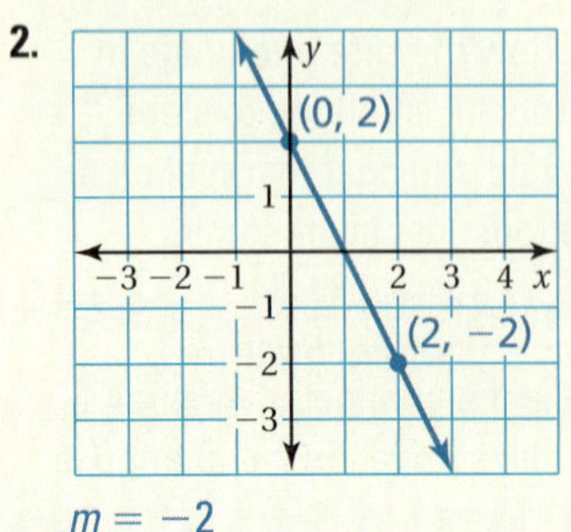

$m = -2$

Taking Math Deeper

Exercise 36

This exercise is a nice example of the power of a diagram. Instead of using the drawing of the slide, encourage students to draw the slide in a coordinate plane. Once that is done, the question is easier to answer.

Draw a diagram.

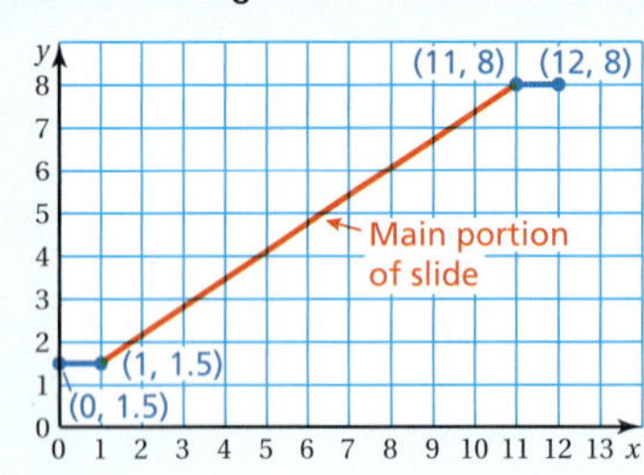

a. Find the slope of the slide.

$$m = \frac{y_2 - y_1}{x_2 - x_1} = \frac{8 - 1.5}{11 - 1} = \frac{6.5}{10} = 0.65$$

Compare the slopes.

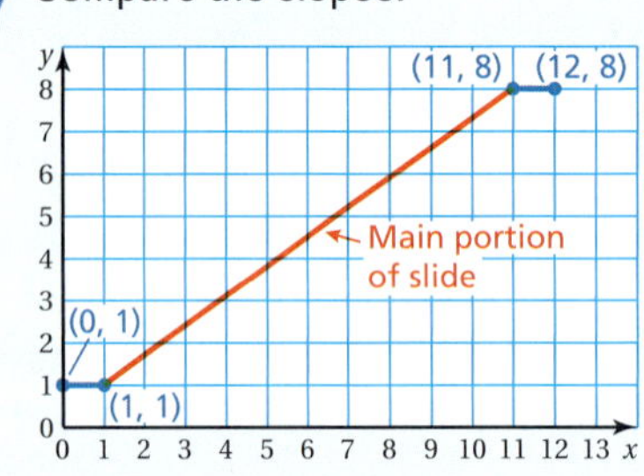

b. $$m = \frac{y_2 - y_1}{x_2 - x_1} = \frac{8 - 1}{11 - 1} = \frac{7}{10} = 0.7$$

Because $0.7 > 0.65$, the slope increased and the slide is steeper.

Project

Many water parks and amusement parks have water slides. Find the height of a slide and calculate the slope of the main part of the slide.

Reteaching and Enrichment Strategies

If students need help. . .	If students got it. . .
Resources by Chapter • Practice A and Practice B • Puzzle Time Record and Practice Journal Practice Differentiating the Lesson Lesson Tutorials Skills Review Handbook	Resources by Chapter • Enrichment and Extension • Technology Connection Start the next section

31. **TURNPIKE TRAVEL** The graph shows the cost of traveling by car on a turnpike.

  a. Find the slope of the line.

  b. Explain the meaning of the slope as a rate of change.

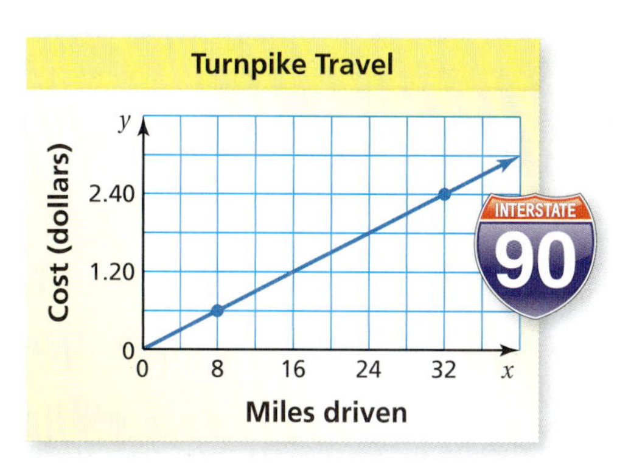

32. **BOAT RAMP** Which is steeper: the boat ramp or a road with a 12% grade? Explain. (*Note:* Road grade is the vertical increase divided by the horizontal distance.)

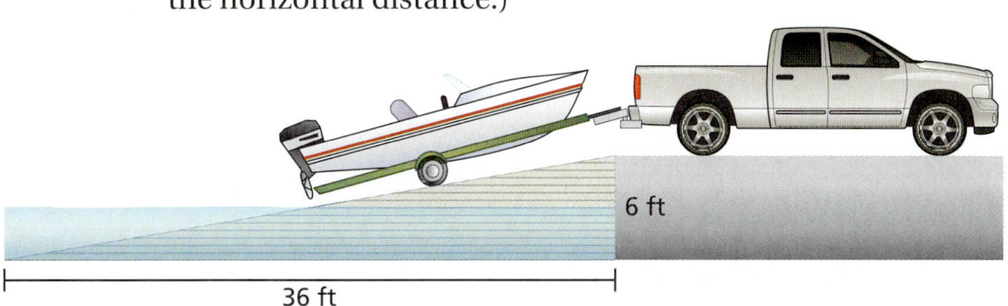

33. **REASONING** Do the points $A(-2, -1)$, $B(1, 5)$, and $C(4, 11)$ lie on the same line? Without using a graph, how do you know?

34. **BUSINESS** A small business earns a profit of $6500 in January and $17,500 in May. What is the rate of change in profit for this time period?

35. **STRUCTURE** Choose two points in the coordinate plane. Use the slope formula to find the slope of the line that passes through the two points. Then find the slope using the formula $\frac{y_1 - y_2}{x_1 - x_2}$. Explain why your results are the same.

36. 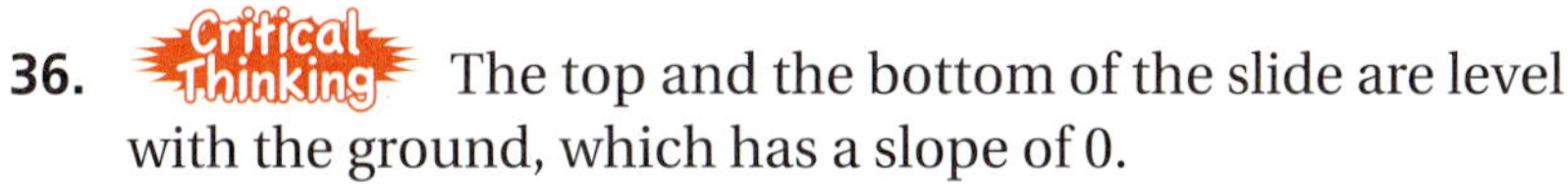 The top and the bottom of the slide are level with the ground, which has a slope of 0.

  a. What is the slope of the main portion of the slide?

  b. How does the slope change when the bottom of the slide is only 12 inches above the ground? Is the slide steeper? Explain.

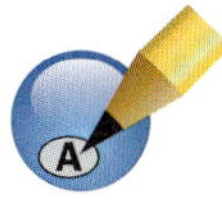

Fair Game Review What you learned in previous grades & lessons

Solve the proportion. *(Skills Review Handbook)*

37. $\frac{b}{30} = \frac{5}{6}$

38. $\frac{7}{4} = \frac{n}{32}$

39. $\frac{3}{8} = \frac{x}{20}$

40. **MULTIPLE CHOICE** What is the prime factorization of 84? *(Skills Review Handbook)*

Ⓐ $2 \times 3 \times 7$ Ⓑ $2^2 \times 3 \times 7$ Ⓒ $2 \times 3^2 \times 7$ Ⓓ $2^2 \times 21$

Slopes of Parallel and Perpendicular Lines

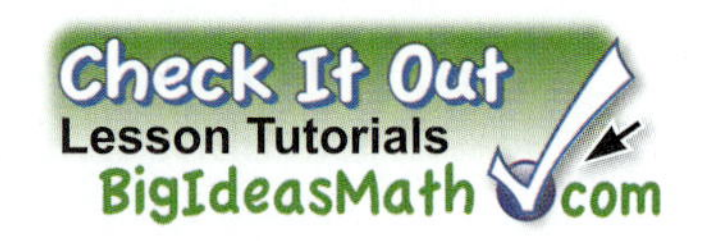

Graphing Equations

In this extension, you will

- identify parallel and perpendicular lines.

Applying Standard MACC.8.EE.2.6

Parallel Lines and Slopes

Lines in the same plane that do not intersect are parallel lines. Nonvertical parallel lines have the same slope.

All vertical lines are parallel.

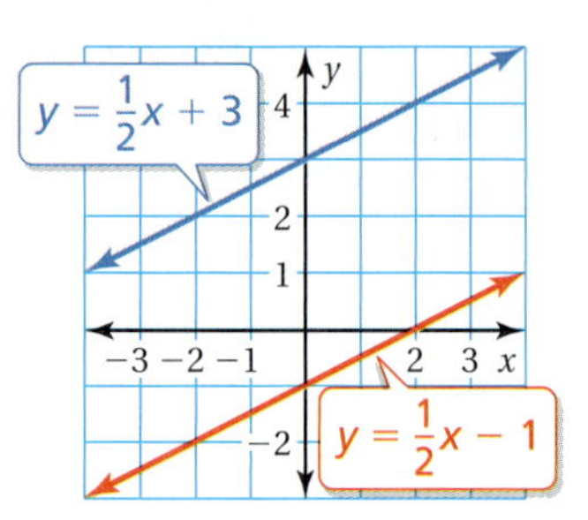

EXAMPLE 1 Identifying Parallel Lines

Which two lines are parallel? How do you know?

Find the slope of each line.

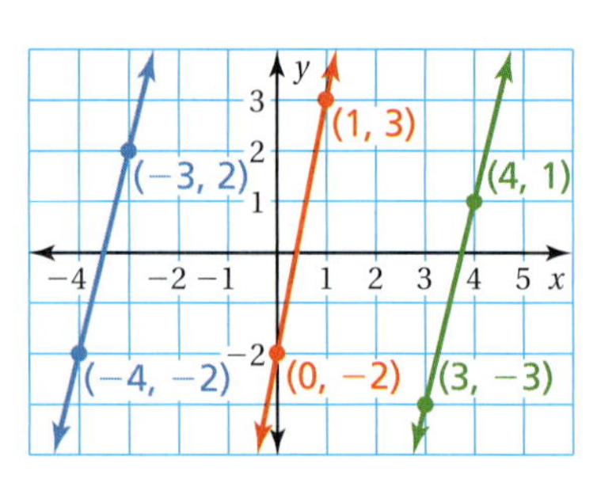

Blue Line

$$m = \frac{y_2 - y_1}{x_2 - x_1} = \frac{-2 - 2}{-4 - (-3)} = \frac{-4}{-1}, \text{ or } 4$$

Red Line

$$m = \frac{y_2 - y_1}{x_2 - x_1} = \frac{-2 - 3}{0 - 1} = \frac{-5}{-1}, \text{ or } 5$$

Green Line

$$m = \frac{y_2 - y_1}{x_2 - x_1} = \frac{-3 - 1}{3 - 4} = \frac{-4}{-1}, \text{ or } 4$$

The slopes of the blue and green lines are 4. The slope of the red line is 5.

The blue and green lines have the same slope, so they are parallel.

Practice

Which lines are parallel? How do you know?

1.

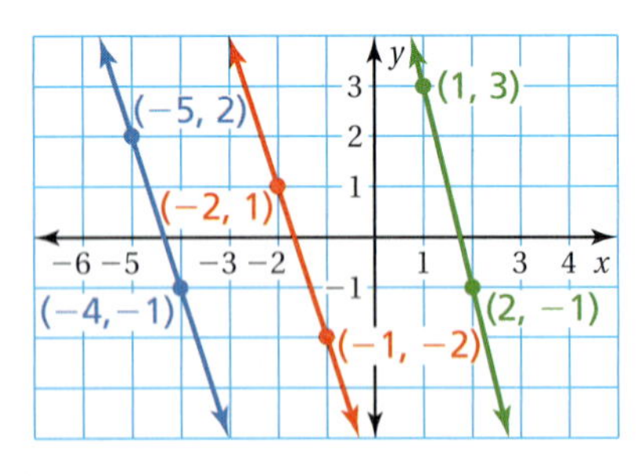

2.

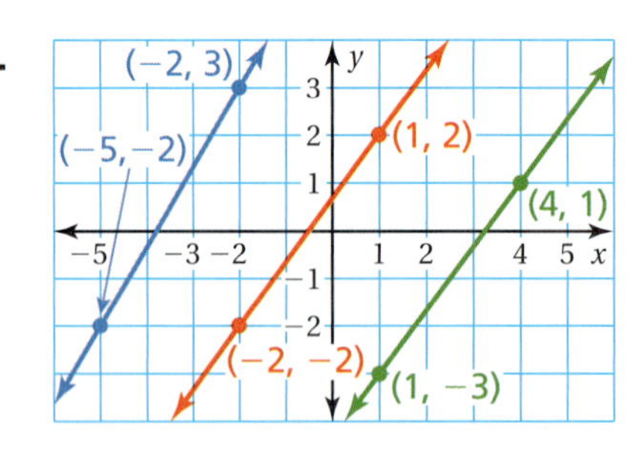

Are the given lines parallel? Explain your reasoning.

3. $y = -5, y = 3$

4. $y = 0, x = 0$

5. $x = -4, x = 1$

6. **GEOMETRY** The vertices of a quadrilateral are $A(-5, 3)$, $B(2, 2)$, $C(4, -3)$, and $D(-2, -2)$. How can you use slope to determine whether the quadrilateral is a parallelogram? Is it a parallelogram? Justify your answer.

Laurie's Notes

Introduction

Connect

- **Yesterday:** Students found slopes of lines. (MP1a)
- **Today:** Students will use slope to determine if lines are parallel or perpendicular.

Motivate

- The words *parallel* and *perpendicular* can be confusing for students and difficult to remember.
- The word parallel comes from para-, "beside," and allelois, "each other," so parallel lines are lines that are beside each other. For example, parallel bars in gymnastics have bars that are beside each other.
- The word perpendicular comes from perpendicularis, meaning "vertical, as a plumb line." A plumb line is a cord with a weight attached that is used to determine perpendicularity.
- You can refer back to the Motivate in Section 4.2 and ask about the angles of the quadrilateral and compare the slopes of the lines that form the quadrilateral.

Lesson Notes

Discuss

- The concepts presented in this extension were explored by students in Activity 3. Students graphed lines with a given slope through different ordered pairs.
- Refer to conjectures made regarding parallel and perpendicular lines.

Key Idea

- Write the Key Idea on the board.
- Model what a slope of $\frac{1}{2}$ means. Start at a point on the line and run 2 units for each unit you rise. Repeat for each line.

Example 1

- "How do you compute the slope for each line?" Students should use rise over run language, along with the formal definition, $\frac{y_2 - y_1}{x_2 - x_1}$.
- Work through the example. Students may look quickly and believe that all of the lines are parallel. They should compute the slope of each line to prove which lines are parallel.

Practice

- **MP3 Construct Viable Arguments and Critique the Reasoning of Others:** In Exercise 6, students are asked to make a conjecture and then justify their answers. Students need opportunities to construct viable arguments and share their thinking with other students.

Florida Common Core Standards

MACC.8.EE.2.6 Use similar triangles to explain why the slope m is the same between any two distinct points on a non-vertical line in the coordinate plane; derive the equation $y = mx$ for a line through the origin and the equation $y = mx + b$ for a line intercepting the vertical axis at b.

Goal Today's lesson is using slope to determine whether lines are parallel or perpendicular.

Lesson Tutorials
Lesson Plans
Answer Presentation Tool

Extra Example 1

Which lines are parallel? How do you know?

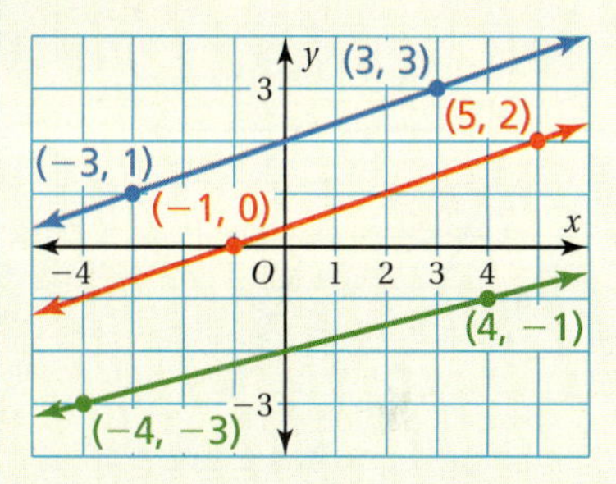

blue and red; They both have a slope of $\frac{1}{3}$.

Practice

1. blue and red; They both have a slope of -3.
2. red and green; They both have a slope of $\frac{4}{3}$.

3–6. See Additional Answers.

Record and Practice Journal Extension 4.2 Practice

1–10. See Additional Answers.

Extra Example 2

Which lines are perpendicular? How do you know?

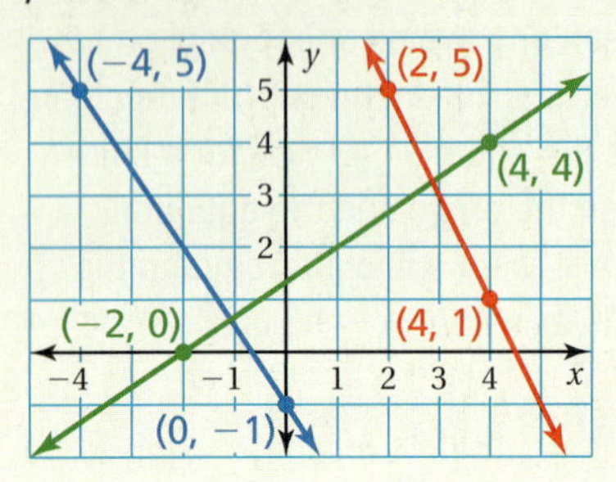

See Additional Answers.

Practice

7–12. See Additional Answers.

Mini-Assessment

1. Which two lines are parallel? How do you know?

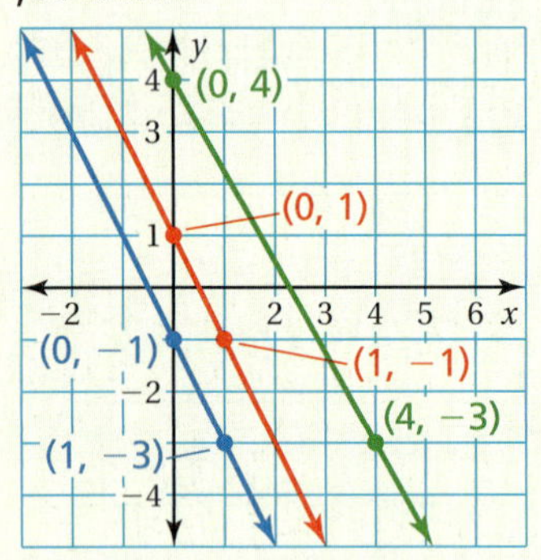

red and blue; Each has a slope of −2.

2. Which two lines are perpendicular? How do you know?

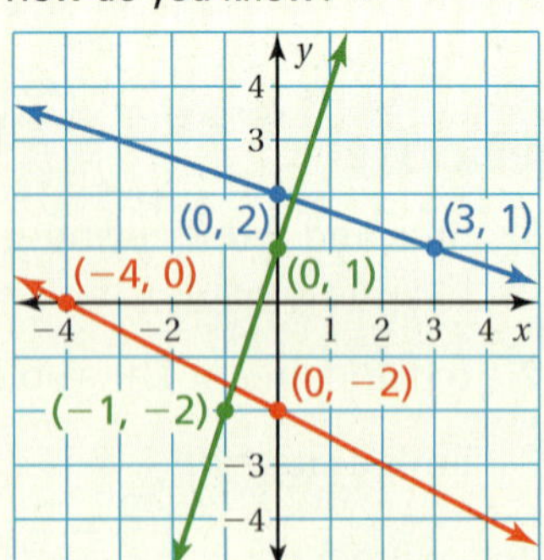

blue and green; The blue line has a slope of 3. The green line has a slope of $-\frac{1}{3}$. The product of their slopes is $3 \cdot -\frac{1}{3} = -1$.

Laurie's Notes

Key Idea

- Write the Key Idea on the board.
- **Teaching Tip:** Use the corner of a piece of paper, placed at the point of intersection, to provide a visual model of what perpendicular means.

Example 2

- Work through the example as shown.
- Students may also refer to the slopes of perpendicular lines as being opposite reciprocals.

Practice

- Because vertical lines have undefined slope, students cannot multiply slopes to determine if the lines are perpendicular. In Exercises 9 and 11, students should give reasoning that states horizontal and vertical lines are perpendicular. In Exercise 10, students should give reasoning that states that the lines are not perpendicular because they are both vertical lines.
- **MP3:** In Exercise 12, students are asked to make conjectures and then justify their answers. Students need opportunities to construct viable arguments and share their thinking with other students.

Closure

- **Exit Ticket:** Which lines are parallel? Which lines are perpendicular? How do you know?

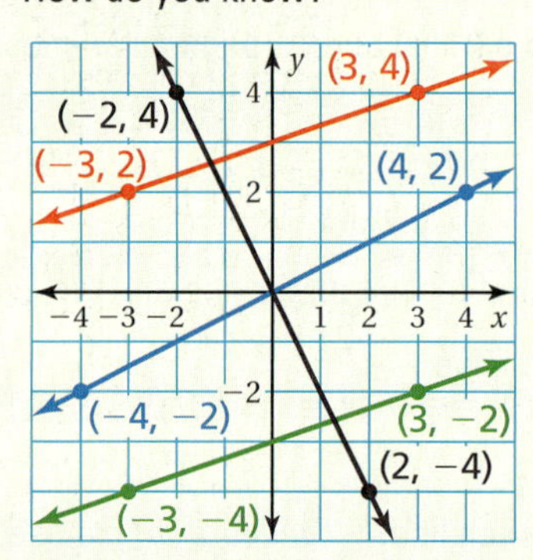

The red and green lines are parallel. Each has a slope of $\frac{1}{3}$. The black and blue lines are perpendicular. The product of their slopes is −1.

Perpendicular Lines and Slope

Lines in the same plane that intersect at right angles are perpendicular lines. Two nonvertical lines are perpendicular when the product of their slopes is -1.

Vertical lines are perpendicular to horizontal lines.

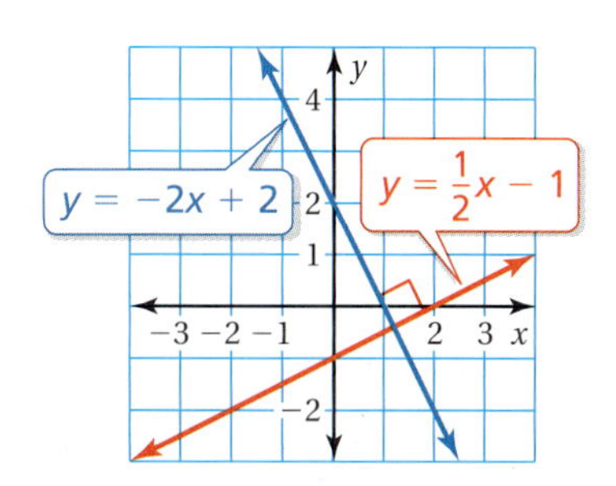

EXAMPLE 2 Identifying Perpendicular Lines

Which two lines are perpendicular? How do you know?

Find the slope of each line.

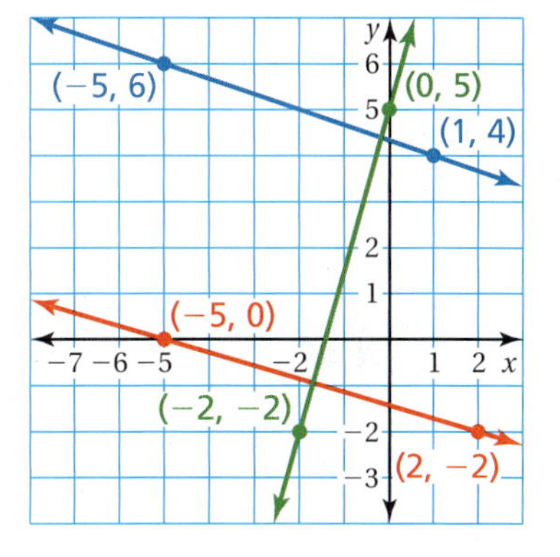

Blue Line	*Red Line*	*Green Line*
$m = \frac{y_2 - y_1}{x_2 - x_1}$	$m = \frac{y_2 - y_1}{x_2 - x_1}$	$m = \frac{y_2 - y_1}{x_2 - x_1}$
$= \frac{4 - 6}{1 - (-5)}$	$= \frac{-2 - 0}{2 - (-5)}$	$= \frac{5 - (-2)}{0 - (-2)}$
$= \frac{-2}{6}$, or $-\frac{1}{3}$	$= -\frac{2}{7}$	$= \frac{7}{2}$

The slope of the red line is $-\frac{2}{7}$. The slope of the green line is $\frac{7}{2}$.

∴ Because $-\frac{2}{7} \cdot \frac{7}{2} = -1$, the red and green lines are perpendicular.

Practice

Which lines are perpendicular? How do you know?

7.

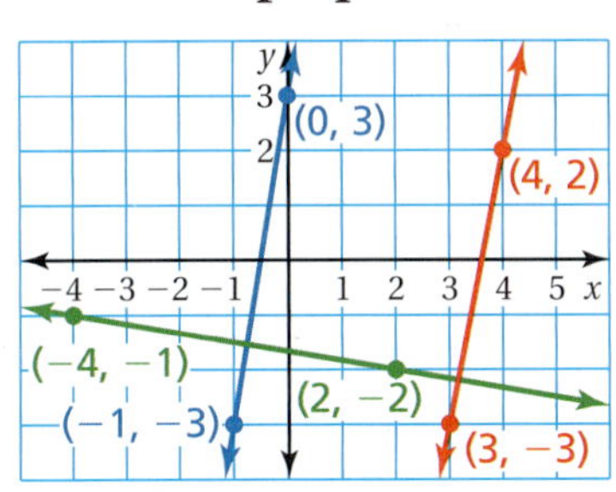

8.

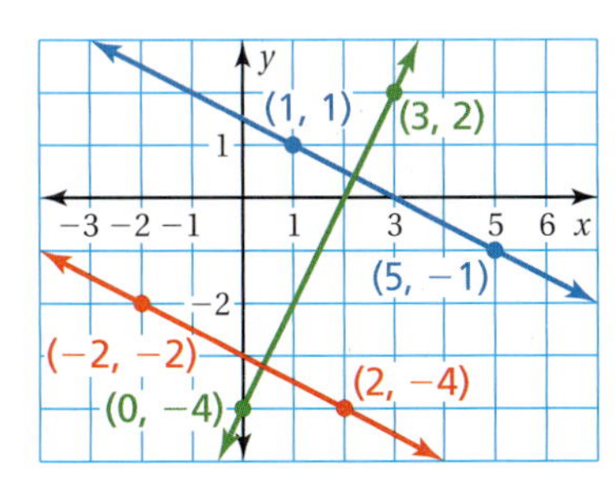

Are the given lines perpendicular? Explain your reasoning.

9. $x = -2, y = 8$

10. $x = -8, x = 7$

11. $y = 0, x = 0$

12. **GEOMETRY** The vertices of a parallelogram are $J(-5, 0)$, $K(1, 4)$, $L(3, 1)$, and $M(-3, -3)$. How can you use slope to determine whether the parallelogram is a rectangle? Is it a rectangle? Justify your answer.

4.3 Graphing Proportional Relationships

Essential Question How can you describe the graph of the equation $y = mx$?

1 ACTIVITY: Identifying Proportional Relationships

Work with a partner. Tell whether x and y are in a proportional relationship. Explain your reasoning.

a.

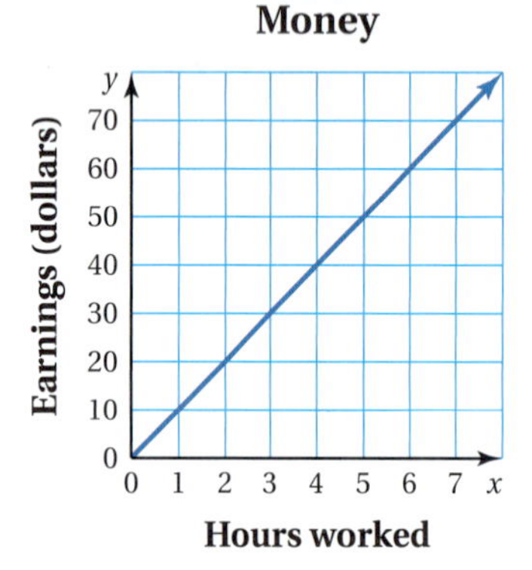

b.

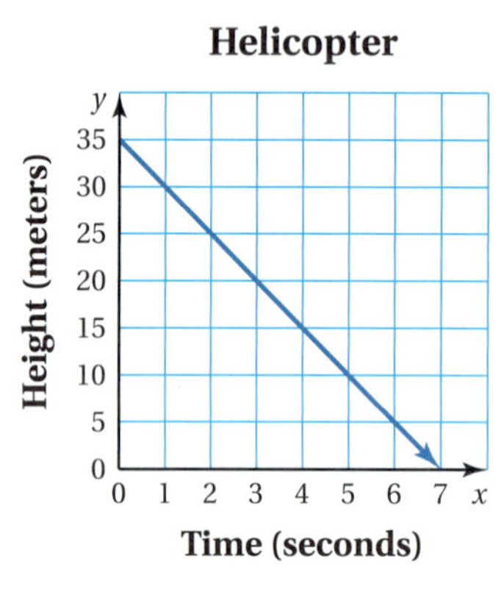

c.

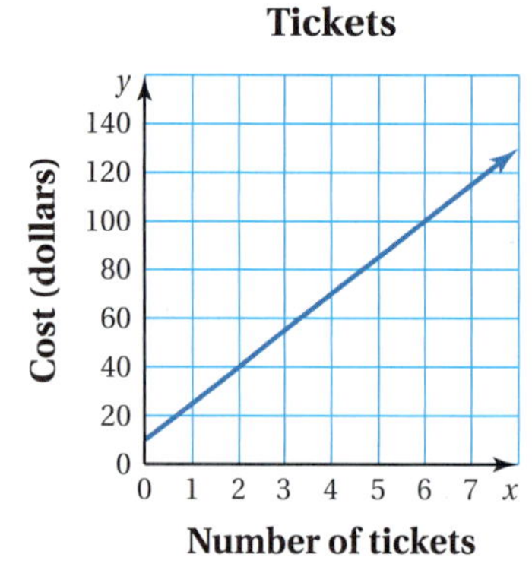

d.

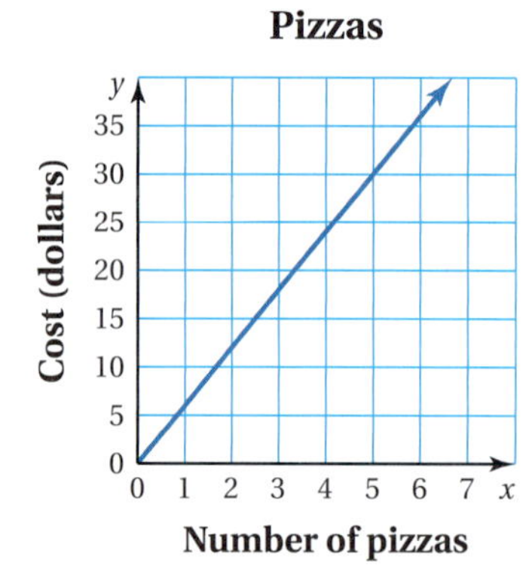

e.

Laps, x	1	2	3	4
Time (seconds), y	90	200	325	480

f.

Cups of Sugar, x	$\frac{1}{2}$	1	$1\frac{1}{2}$	2
Cups of Flour, y	1	2	3	4

COMMON CORE

Graphing Equations

In this lesson, you will

- write and graph proportional relationships.

Learning Standards
MACC.8.EE.2.5
MACC.8.EE.2.6

2 ACTIVITY: Analyzing Proportional Relationships

Work with a partner. Use only the proportional relationships in Activity 1 to do the following.

- **Find the slope of the line.**
- **Find the value of y for the ordered pair $(1, y)$.**

What do you notice? What does the value of y represent?

Laurie's Notes

Introduction

Standards for Mathematical Practice

- **MP2 Reason Abstractly and Quantitatively:** Mathematically proficient students make sense of the quantities and their relationships in problem situations. To develop this proficiency, students must be asked to interpret the meaning of a slope as a unit rate.

Motivate

- As a warm-up and to connect to prior content, tell students that x and y are in a proportional relationship. Have students fill in the ratio table.

Minutes, x	1	2	4	6
Gallons, y	8.5			

- **MP2** and **MP3a Construct Viable Arguments:** Ask volunteers to justify their procedures and explain why their procedures show a proportional relationship.

Activity Notes

Discuss

- You may wish to do a quick review before students start the activity.
- ? "What is a proportion?" an equation stating that two ratios are equivalent
- ? "How can you tell if two ratios form a proportion?" There are different methods. Students may mention the Cross Products Property.
- ? "Does anyone remember what a direct variation equation is?" an equation that can be written in the form $y = kx$; its graph passes through (0, 0)

Activity 1

- Expect students to solve parts (a)–(d) by observing whether the graph passes through the origin.
- **MP2:** Alternative reasoning in part (a) could be that the person earns \$10 in 1 hour, \$20 in 2 hours, \$30 in 3 hours, and so on. This is \$10 every hour, which describes a proportional relationship.
- Some students may want to graph the ordered pairs in parts (e) and (f). Consider asking them how they can solve without graphing (check for equivalent ratios).

Activity 2

- The goal of this activity is for students to connect the unit rate, the slope m, and the point $(1, y)$ on the graph of a proportional relationship.
- Students may correctly find the slope and say that the y-coordinate in $(1, y)$ is the slope, but they may not make the connection to the unit rate. One way to help them see the connection is to tell them to interpret the ordered pair. The ordered pair $(1, y)$ in part (a) means (1 hour, \$10 earned).

Florida Common Core Standards

MACC.8.EE.2.5 Graph proportional relationships, interpreting the unit rate as the slope of the graph. Compare two different proportional relationships represented in different ways.

MACC.8.EE.2.6 Use similar triangles to explain why the slope m is the same between any two distinct points on a non-vertical line in the coordinate plane; derive the equation $y = mx$ for a line through the origin and the equation $y = mx + b$ for a line intercepting the vertical axis at b.

Previous Learning

Students have determined whether two quantities are proportional. Students have also graphed linear equations and found slopes of lines.

Lesson Plans
Complete Materials List

4.3 Record and Practice Journal

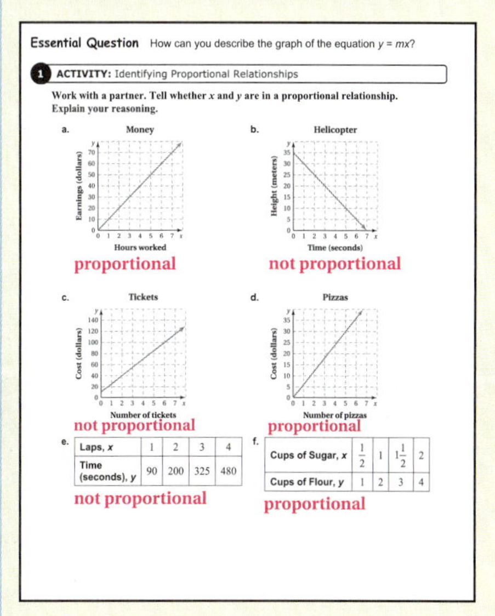

Essential Question How can you describe the graph of the equation $y = mx$?

1 ACTIVITY: Identifying Proportional Relationships

Work with a partner. Tell whether x and y are in a proportional relationship. Explain your reasoning.

a. Money — proportional

b. Helicopter — not proportional

c. Tickets — not proportional

d. Pizzas — proportional

e.

Laps, x	1	2	3	4
Time (seconds), y	90	200	325	480

not proportional

f.

Cups of Sugar, x	$\frac{1}{2}$	1	$1\frac{1}{2}$	2
Cups of Flour, y	1	2	3	4

proportional

English Language Learners

Word Problems

Most word problems follow a standard format that allows English learners to recognize key words that are integral to writing a mathematical statement of the problem. Most numbers given in a word problem are used. Analyzing the units in the mathematical statement and determining the units of the solution give students confidence that they are on the right path to solving the problem.

4.3 Record and Practice Journal

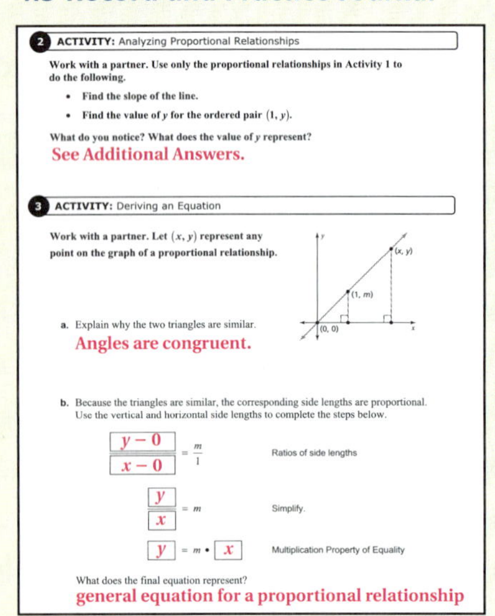

2 ACTIVITY: Analyzing Proportional Relationships

Work with a partner. Use only the proportional relationships in Activity 1 to do the following.

- **Find the slope of the line.**
- **Find the value of y for the ordered pair $(1, y)$.**

What do you notice? What does the value of y represent?

See Additional Answers.

3 ACTIVITY: Deriving an Equation

Work with a partner. Let (x, y) represent any point on the graph of a proportional relationship.

a. Explain why the two triangles are similar.

Angles are congruent.

b. Because the triangles are similar, the corresponding side lengths are proportional. Use the vertical and horizontal side lengths to complete the steps below.

$\dfrac{y - 0}{x - 0} = \dfrac{m}{1}$ Ratios of side lengths

$\dfrac{y}{x} = m$ Simplify.

$y = m \cdot x$ Multiplication Property of Equality

What does the final equation represent?

general equation for a proportional relationship

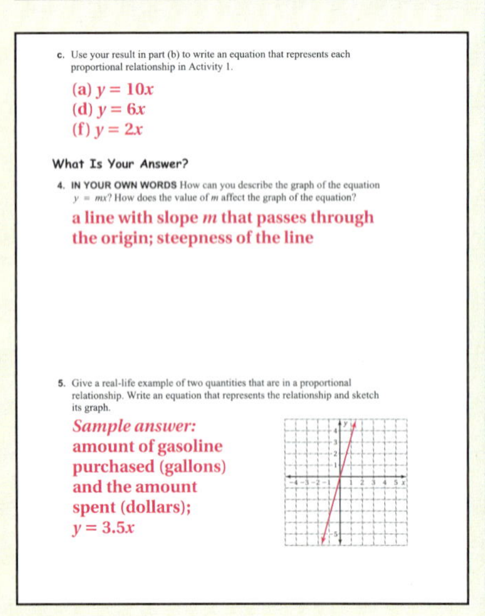

c. Use your result in part (b) to write an equation that represents each proportional relationship in Activity 1.

(a) $y = 10x$
(d) $y = 6x$
(f) $y = 2x$

What Is Your Answer?

4. IN YOUR OWN WORDS How can you describe the graph of the equation $y = mx$? How does the value of m affect the graph of the equation?

a line with slope m that passes through the origin; steepness of the line

5. Give a real-life example of two quantities that are in a proportional relationship. Write an equation that represents the relationship and sketch its graph.

Sample answer: amount of gasoline purchased (gallons) and the amount spent (dollars); $y = 3.5x$

Laurie's Notes

Activity 3

- This activity uses the result of Activity 2. If you know the direct variation equation representing a proportional relationship, then you can find the slope by solving the equation for y when $x = 1$. Therefore, you can label the ordered pair $(1, m)$.
- **MP2:** In part (a), students need to use what they learned in Section 3.4, that two triangles are similar when they have two pairs of congruent angles.
- The triangles in the diagram share an angle (at the origin) and both have a right angle. So, the triangles have two pairs of congruent angles and are similar.

? "When two triangles are similar, what do you know about their side lengths?" Corresponding side lengths are proportional.

- It may not be obvious to students that the vertical side of the larger triangle has a length of y units and the horizontal side of the larger triangle has a length of x units. It is less obvious to students because the ordered pair is represented by variables rather than numbers.
- Ask students to share their responses for part (c).

What Is Your Answer?

- **Think-Pair-Share:** Students should read each question independently and then work in pairs to answer the questions. When they have answered the questions, the pair should compare their answers with another group and discuss any discrepancies.

Closure

- **Exit Ticket:** Refer back to the ratio table in the Motivator with minutes and gallons.
  - **a.** Write an equation that represents this situation.
  - **b.** What is the value of x when y is 85 gallons? Explain.

3 ACTIVITY: Deriving an Equation

Work with a partner. Let (x, y) represent any point on the graph of a proportional relationship.

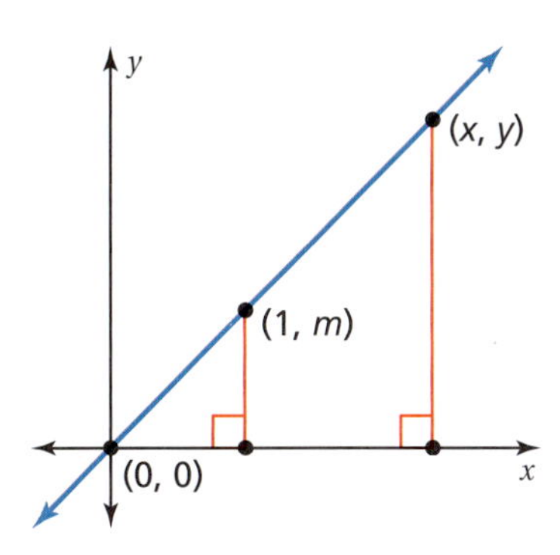

a. Explain why the two triangles are similar.

b. Because the triangles are similar, the corresponding side lengths are proportional. Use the vertical and horizontal side lengths to complete the steps below.

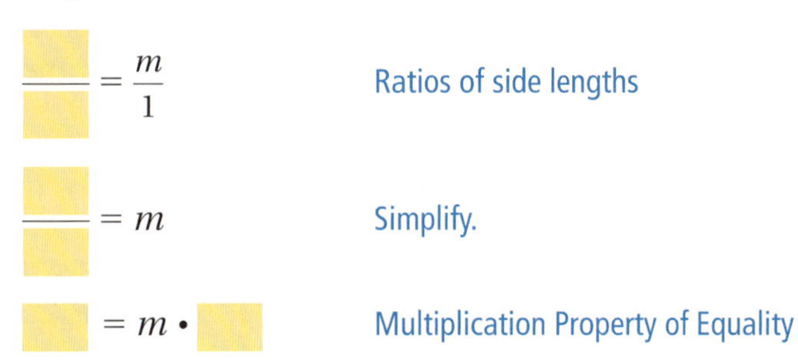

What does the final equation represent?

c. Use your result in part (b) to write an equation that represents each proportional relationship in Activity 1.

Math Practice 7

View as Components

What part of the graph can you use to find the side lengths?

What Is Your Answer?

4. IN YOUR OWN WORDS How can you describe the graph of the equation $y = mx$? How does the value of m affect the graph of the equation?

5. Give a real-life example of two quantities that are in a proportional relationship. Write an equation that represents the relationship and sketch its graph.

Use what you learned about proportional relationships to complete Exercises 3–6 on page 162.

Study Tip

In the direct variation equation $y = mx$, m represents the constant of proportionality, the slope, and the unit rate.

Direct Variation

Words When two quantities x and y are proportional, the relationship can be represented by the direct variation equation $y = mx$, where m is the constant of proportionality.

Graph The graph of $y = mx$ is a line with a slope of m that passes through the origin.

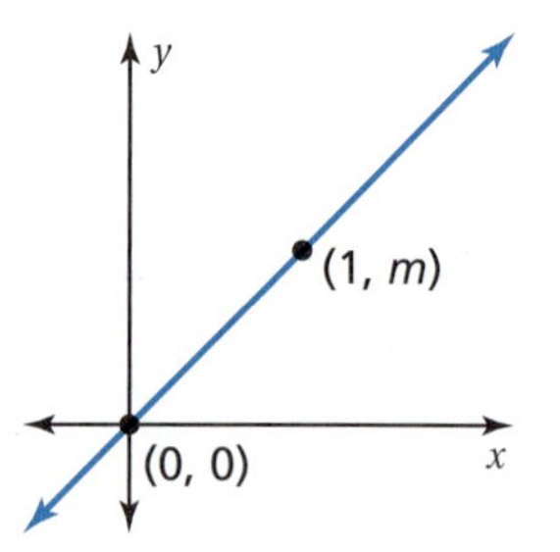

EXAMPLE 1 Graphing a Proportional Relationship

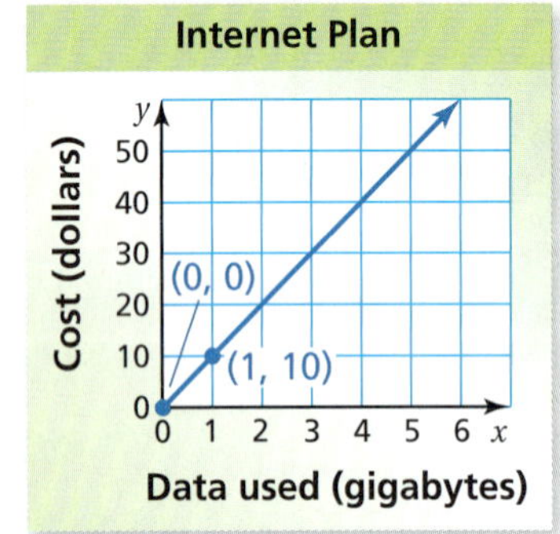

The cost y (in dollars) for x gigabytes of data on an Internet plan is represented by $y = 10x$. Graph the equation and interpret the slope.

The equation shows that the slope m is 10. So, the graph passes through (0, 0) and (1, 10).

Plot the points and draw a line through the points. Because negative values of x do not make sense in this context, graph in the first quadrant only.

The slope indicates that the unit cost is $10 per gigabyte.

EXAMPLE 2 Writing and Using a Direct Variation Equation

The weight y of an object on Titan, one of Saturn's moons, is proportional to the weight x of the object on Earth. An object that weighs 105 pounds on Earth would weigh 15 pounds on Titan.

a. Write an equation that represents the situation.

Use the point (105, 15) to find the slope of the line.

$y = mx$ — Direct variation equation

$15 = m(105)$ — Substitute 15 for y and 105 for x.

$\frac{1}{7} = m$ — Simplify.

So, an equation that represents the situation is $y = \frac{1}{7}x$.

b. How much would a chunk of ice that weighs 3.5 pounds on Titan weigh on Earth?

$3.5 = \frac{1}{7}x$ — Substitute 3.5 for y.

$24.5 = x$ — Multiply each side by 7.

So, the chunk of ice would weigh 24.5 pounds on Earth.

In Example 2, the slope indicates that the weight of an object on Titan is one-seventh its weight on Earth.

Laurie's Notes

Introduction

Connect
- **Yesterday:** Students explored slopes of proportional relationships. (MP2, MP3a)
- **Today:** Students will write and graph proportional relationships.

Motivate
- Ask if anyone has heard of Saturn's largest known moon, Titan. One of today's examples will be about Titan.
- Share some of the interesting information below. You can find more information about Titan at *www.nasa.gov*.
  - Titan is larger than our moon and the planet Mercury.
  - Titan has clouds and a thick atmosphere, like some planets.
  - It may rain "gasoline-like" liquids on Titan.

Lesson Notes

Key Idea
- Write the Key Idea. Draw the graph and discuss the two labeled points.
- Make sure students understand that a direct variation equation is a special kind of linear equation, one whose graph passes through the origin.
- Discuss the Study Tip to help students connect prior concepts. In the equation $y = mx$, you can think of the coefficient m as the unit rate (grades 6 and 7), constant of proportionality (grade 7), and slope (grades 7 and 8).

Example 1
? "Do x and y show direct variation?" Yes, the equation is of the form $y = mx$.

? "In the context of the problem, can x be negative?" no

- There are many points that can be graphed. Use (0, 0) and (1, m).
- **MP6 Attend to Precision:** Ask students to interpret the slope of the line. Students should precisely refer to the *unit* cost of $10 per gigabyte.
- **Extension:** Many of these plans are tiered such that once you exceed a whole number of gigabytes, the cost jumps $10 at once. Share this with students and ask if this is also a proportional relationship. Discuss the differences.

Example 2
- Read the problem. The first sentence explains how the ordered pairs will be written (weight on Earth, weight on Titan).

? "What two ordered pairs do we know?" (0, 0) and (105, 15)

- **MP2 Reason Abstractly and Quantitatively:** In part (a), students should recognize that they know an ordered pair that satisfies the equation $y = mx$. Substitute the x- and y-values and solve for m.
- In part (b), you need to find x when y is 3.5. Substitute and solve.
- **Connection:** Have students solve part (b) using a ratio table.

Goal Today's lesson is graphing proportional relationships.

Lesson Tutorials
Lesson Plans
Answer Presentation Tool

Differentiated Instruction
Visual
Encourage students to write down notes when solving word problems or to underline relevant information and cross out irrelevant information. Allow students to do this on handouts and tests.

Extra Example 1
The cost y (in dollars) to rent x video games is represented by $y = 4x$. Graph the equation and interpret the slope.

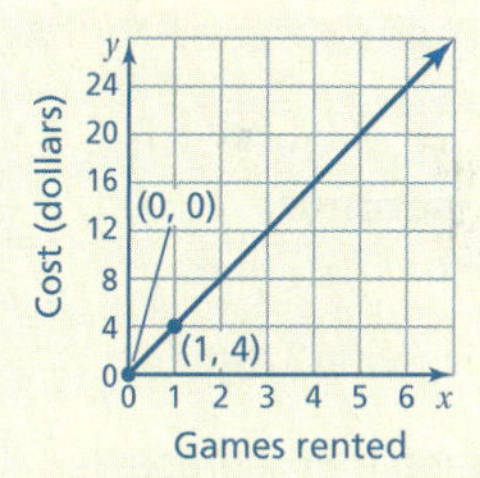

The slope indicates that the unit cost is $4 per video game.

Extra Example 2
The daily wage y (in dollars) of a factory worker is proportional to the number of parts x assembled in a day. A worker who assembles 250 parts in a day earns $75.

a. Write an equation that represents the situation. $y = \frac{3}{10}x$

b. How much does a worker earn who assembles 300 parts in a day? $90

On Your Own

1. 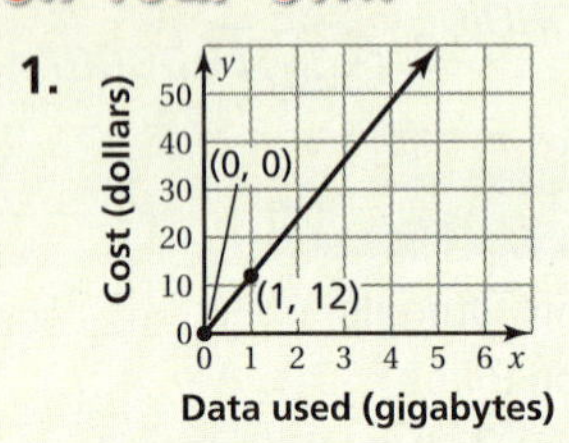

The slope indicates that unit cost is \$12 per gigabyte.

2. 500 kg

Extra Example 3

At a track event, the distance y (in meters) traveled by Student A in x seconds is represented by the equation $y = 7x$. The graph shows the distance traveled by Student B.

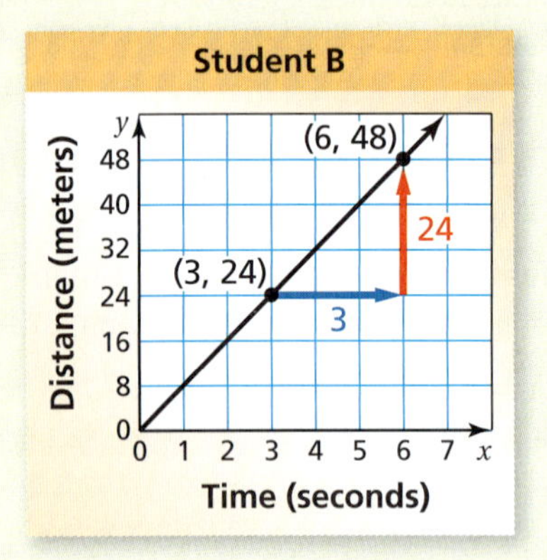

a. Which student is faster? Student B

b. Graph the equation that represents Student A in the same coordinate plane as Student B. Compare the steepness of the graphs. What does this mean in the context of the problem? See Additional Answers.

On Your Own

3. The T-bar ski lift speed of 2.25 meters per second is faster than the two-person lift, but slower than the four-person lift.

Laurie's Notes

On Your Own

- **Neighbor Check:** Have students work independently and then have their neighbors check their work. Have students discuss any discrepancies.

Example 3

- Discuss why these relationships are proportional. For instance, the two-person lift starts at 0, travels 2 meters in 1 second, 4 meters in 2 seconds, 6 meters in 3 seconds, and so on. $\frac{2 \text{ m}}{1 \text{ sec}} = \frac{4 \text{ m}}{2 \text{ sec}} = \frac{6 \text{ m}}{3 \text{ sec}}$.
- **MP6:** Make sure that students recognize the unit labels for each axis. The x-axis represents time (in seconds) and the y-axis represents distance (in meters).
- **Connection:** The question asks which ski lift is faster. Students are looking for the rate, or the speed of each lift. The rate is the slope of the line. For the two-person lift, the slope can be found using the ordered pairs in the graph. For the four-person lift, the slope is given in the equation.
- In part (b), point out to students that the graphs do not represent the steepness of the lifts, but rather the distance traveled (y-axis) over a period of time (x-axis).

? "If a vertical line is drawn through the graph in part (b) at $x = 4$, then it will intersect the two lines. What do these points of intersection mean in the context of the problem?" The y-value is the distance traveled by each lift in 4 seconds. The four-person lift travels farther in 4 seconds.

On Your Own

? "Is this a ratio table? Explain." yes; The ratios are all equivalent with a unit rate of 2.25 meters per second.

Closure

- **Writing Prompt:** "If a relationship is proportional, then . . ." the graph of the relationship goes through the origin and can be represented by an equation of the form $y = mx$, where m is the slope of the line.

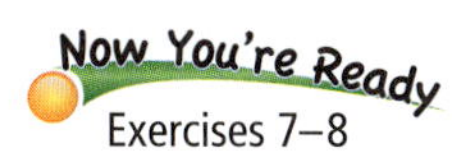

1. **WHAT IF?** In Example 1, the cost is represented by $y = 12x$. Graph the equation and interpret the slope.
2. In Example 2, how much would a spacecraft that weighs 3500 kilograms on Earth weigh on Titan?

EXAMPLE 3 Comparing Proportional Relationships

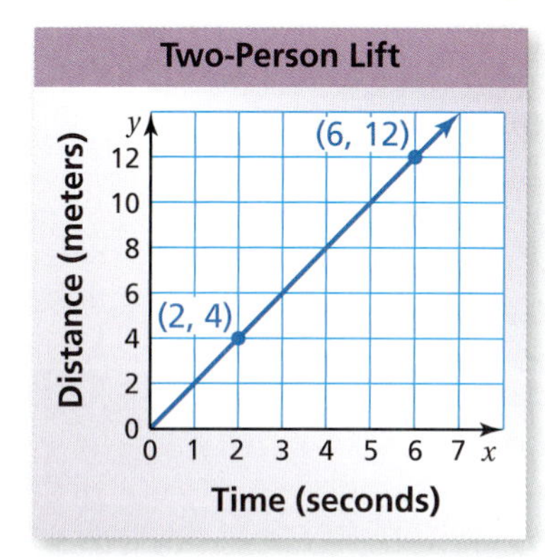

The distance y (in meters) that a four-person ski lift travels in x seconds is represented by the equation $y = 2.5x$. The graph shows the distance that a two-person ski lift travels.

a. Which ski lift is faster?

Interpret each slope as a unit rate.

Four-Person Lift

$y = 2.5x$

The slope is 2.5.

The four-person lift travels 2.5 meters per second.

Two-Person Lift

$$\text{slope} = \frac{\text{change in } y}{\text{change in } x}$$

$$= \frac{8}{4} = 2$$

The two-person lift travels 2 meters per second.

So, the four-person lift is faster than the two-person lift.

b. Graph the equation that represents the four-person lift in the same coordinate plane as the two-person lift. Compare the steepness of the graphs. What does this mean in the context of the problem?

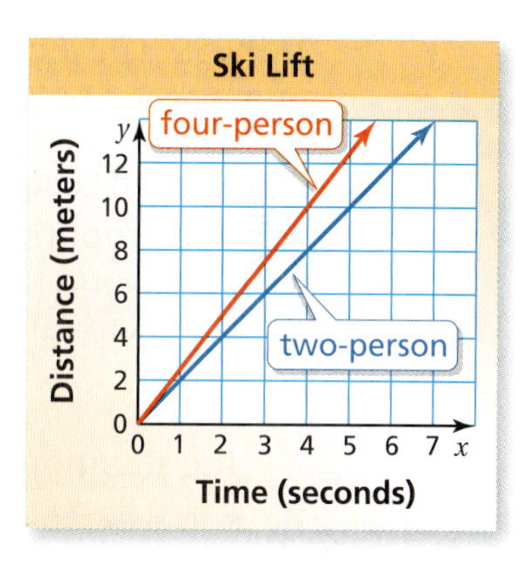

The graph that represents the four-person lift is steeper than the graph that represents the two-person lift. So, the four-person lift is faster.

On Your Own

3. The table shows the distance y (in meters) that a T-bar ski lift travels in x seconds. Compare its speed to the ski lifts in Example 3.

x (seconds)	1	2	3	4
y (meters)	$2\frac{1}{4}$	$4\frac{1}{2}$	$6\frac{3}{4}$	9

4.3 Exercises

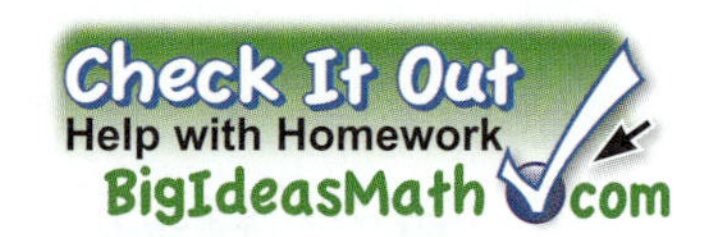

Vocabulary and Concept Check

1. **VOCABULARY** What point is on the graph of every direct variation equation?

2. **REASONING** Does the equation $y = 2x + 3$ represent a proportional relationship? Explain.

Practice and Problem Solving

Tell whether x and y are in a proportional relationship. Explain your reasoning. If so, write an equation that represents the relationship.

3.

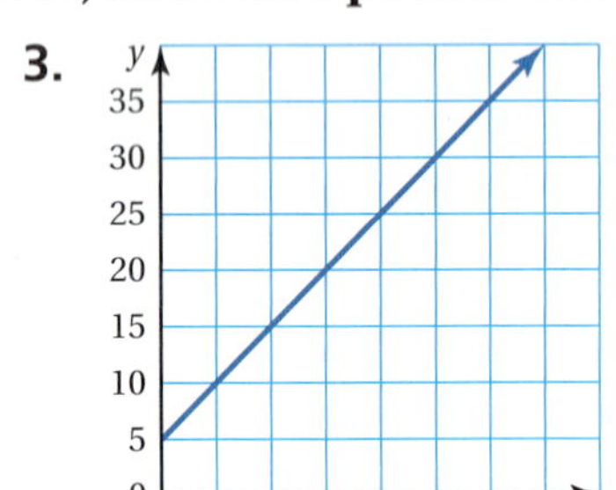

4.

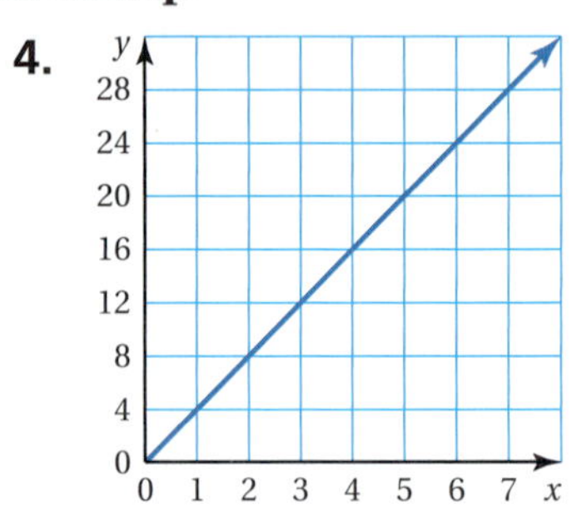

5.

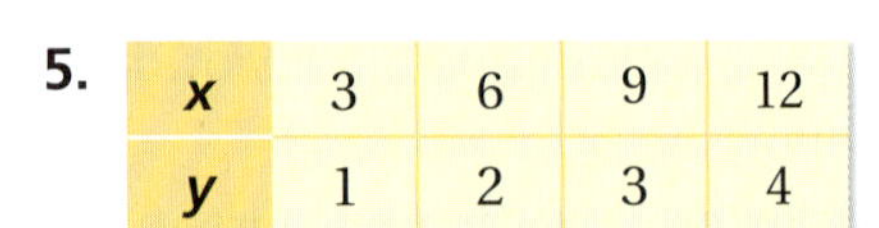

x	3	6	9	12
y	1	2	3	4

6.

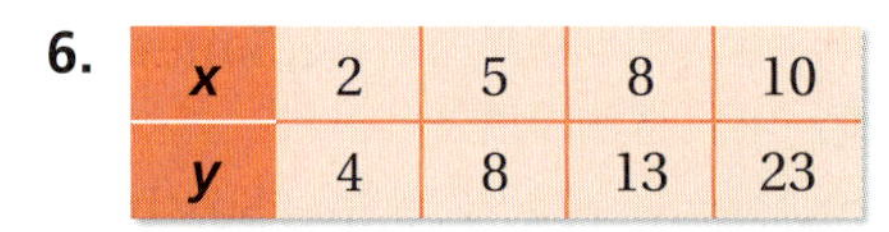

x	2	5	8	10
y	4	8	13	23

1 7. **TICKETS** The amount y (in dollars) that you raise by selling x fundraiser tickets is represented by the equation $y = 5x$. Graph the equation and interpret the slope.

2 8. **KAYAK** The cost y (in dollars) to rent a kayak is proportional to the number x of hours that you rent the kayak. It costs \$27 to rent the kayak for 3 hours.

a. Write an equation that represents the situation.

b. Interpret the slope.

c. How much does it cost to rent the kayak for 5 hours?

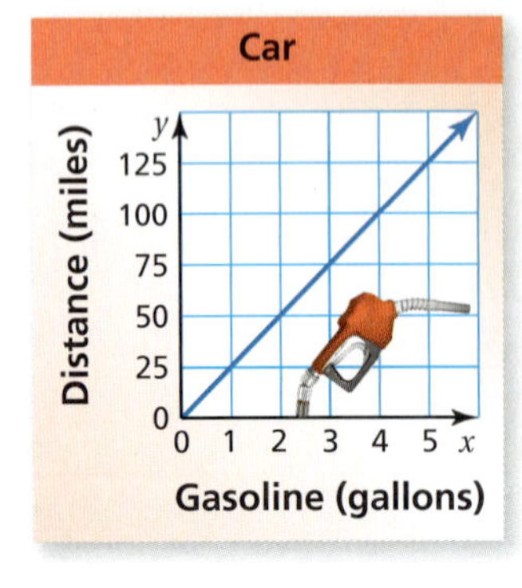

3 9. **MILEAGE** The distance y (in miles) that a truck travels on x gallons of gasoline is represented by the equation $y = 18x$. The graph shows the distance that a car travels.

a. Which vehicle gets better gas mileage? Explain how you found your answer.

b. How much farther can the vehicle you chose in part (a) travel than the other vehicle on 8 gallons of gasoline?

Assignment Guide and Homework Check

Level	Day 1 Activity Assignment	Day 2 Lesson Assignment	Homework Check
Basic	3–6, 14–17	1, 2, 7–10	7, 8, 9
Average	3–6, 14–17	1, 2, 7–12	8, 9, 10, 11
Advanced	1–17		8, 10, 11, 12

Common Errors

- **Exercises 3 and 4** Students may think that the slope of each line is 1. Remind them to pay attention to the scales on the axes.
- **Exercise 8** Students may switch the x- and y-values, substituting incorrectly in the equation $y = mx$ when finding the slope. Remind them to make sure they are substituting the correct value for each variable.

4.3 Record and Practice Journal

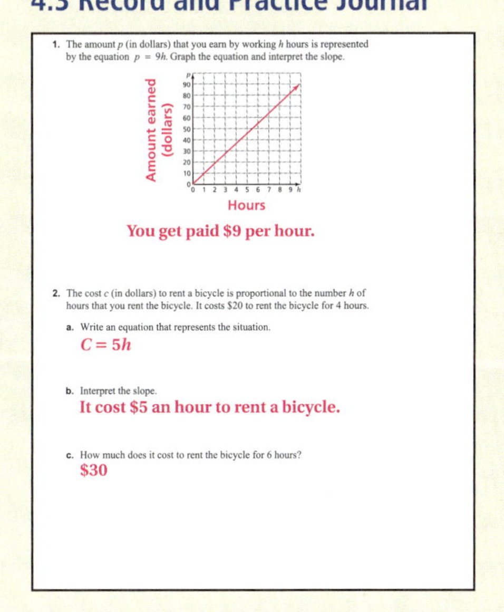

1. The amount p (in dollars) that you earn by working h hours is represented by the equation $p = 9h$. Graph the equation and interpret the slope.

You get paid $9 per hour.

2. The cost c (in dollars) to rent a bicycle is proportional to the number h of hours that you rent the bicycle. It costs $20 to rent the bicycle for 4 hours.

a. Write an equation that represents the situation.
$C = 5h$

b. Interpret the slope.
It cost $5 an hour to rent a bicycle.

c. How much does it cost to rent the bicycle for 6 hours?
$30

Vocabulary and Concept Check

1. (0, 0)
2. no; *Sample answer:* The graph of the equation does not pass through the origin.

Practice and Problem Solving

3. no; *Sample answer:* The graph of the equation does not pass through the origin.
4. yes; $y = 4x$; *Sample answer:* The graph is a line through the origin.
5. yes; $y = \frac{1}{3}x$; *Sample answer:* The rate of change in the table is constant.
6. no; *Sample answer:* The rate of change in the table is not constant.
7. 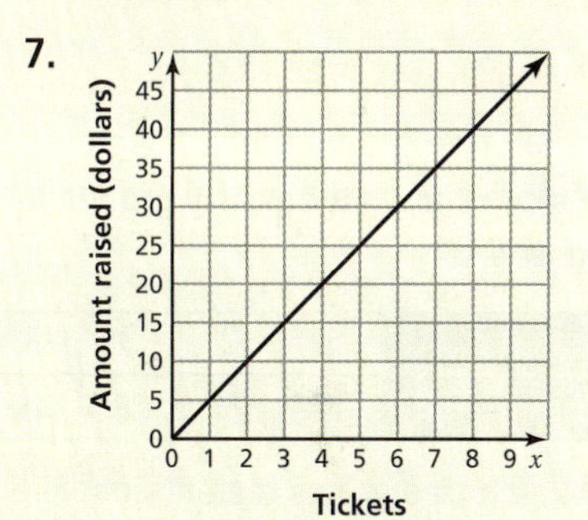

Each ticket costs $5.

8. **a.** $y = 9x$

   b. It costs $9 per hour to rent a kayak.

   c. $45

9. **a.** the car; *Sample answer:* The equation for the car is $y = 25x$. Because 25 is greater than 18, the car gets better gas mileage.

   b. 56 miles

Practice and Problem Solving

10. See Additional Answers.

11. See *Taking Math Deeper.*

12. **a.** yes; The line passes through the origin.

b. $y = -3.5x$; The temperature decreases by 3.5°F for each 1000-foot increase in altitude.

c. 54.75°F

13. See Additional Answers.

Fair Game Review

14–16. See Additional Answers.

17. B

Mini-Assessment

A maple tree grows 1.5 feet each year. The table shows the yearly growth for a pine tree.

Time (yr)	1	2	3	4
Growth (in.)	12	24	36	48

1. Which tree grows faster? maple
2. Write and graph equations that represent the growth rates of each tree. Compare the steepness of the graphs. What does this mean in the context of the problem?

   Maple tree: $y = 1.5x$, Pine tree: $y = x$

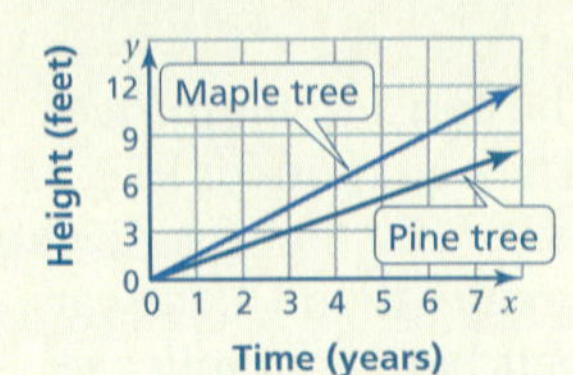

   The graph that represents the maple tree is steeper than the graph that represents the pine tree. So, the maple tree grows faster than the pine tree.

Taking Math Deeper

Exercise 11

This exercise is leading students to the realization that for any point (x, y) on the graph of a proportional relationship, the ratio y to x is constant. There are several ways that students could come to this conclusion.

Use the equation of a proportional relationship.

The quantities x and y are in a proportional relationship, so $y = mx$ for some constant m. Solve for m.

$y = mx$ — Write the equation.

$\frac{y}{x} = m$ — Divide each side by x.

This tells you that the ratio of y to x is equal to the slope of the line, m, which is constant.

Use the slope formula. Use the point (0, 0) because the graph of any proportional relationship passes through the origin.

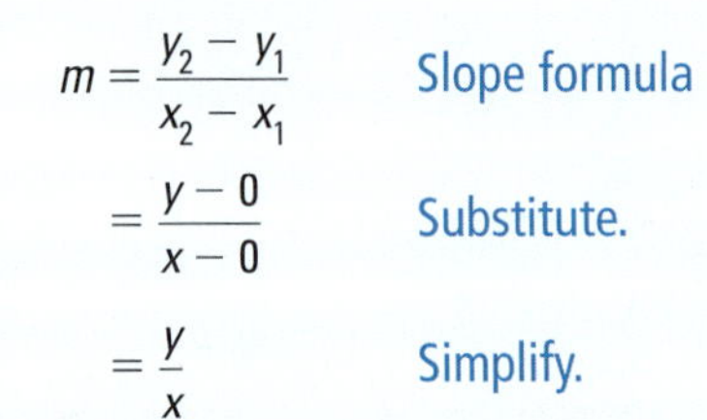

$m = \frac{y_2 - y_1}{x_2 - x_1}$ — Slope formula

$= \frac{y - 0}{x - 0}$ — Substitute.

$= \frac{y}{x}$ — Simplify.

The result is the same, showing the ratio of y to x is constant.

Make sure students understand that this means they can use any *single* point on the graph of a proportional relationship to find the slope, unit rate, or constant of proportionality.

Project

Have students research *inverse variation.* Does it represent a linear relationship? Ask students to compare and contrast direct variation and inverse variation.

Reteaching and Enrichment Strategies

If students need help. . .	If students got it. . .
Resources by Chapter • Practice A and Practice B • Puzzle Time Record and Practice Journal Practice Differentiating the Lesson Lesson Tutorials Skills Review Handbook	Resources by Chapter • Enrichment and Extension • Technology Connection Start the next section

10. **BIOLOGY** Toenails grow about 13 millimeters per year. The table shows fingernail growth.

Weeks	1	2	3	4
Fingernail Growth (millimeters)	0.7	1.4	2.1	2.8

 a. Do fingernails or toenails grow faster? Explain.

 b. In the same coordinate plane, graph equations that represent the growth rates of toenails and fingernails. Compare the steepness of the graphs. What does this mean in the context of the problem?

11. **REASONING** The quantities x and y are in a proportional relationship. What do you know about the ratio of y to x for any point (x, y) on the line?

12. **PROBLEM SOLVING** The graph relates the temperature change y (in degrees Fahrenheit) to the altitude change x (in thousands of feet).

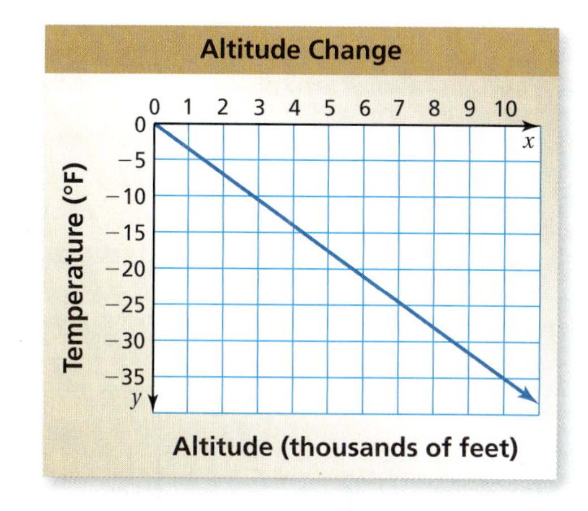

 a. Is the relationship proportional? Explain.

 b. Write an equation of the line. Interpret the slope.

 c. You are at the bottom of a mountain where the temperature is 74°F. The top of the mountain is 5500 feet above you. What is the temperature at the top of the mountain?

13. **Critical Thinking** Consider the distance equation $d = rt$, where d is the distance (in feet), r is the rate (in feet per second), and t is the time (in seconds).

 a. You run 6 feet per second. Are distance and time proportional? Explain. Graph the equation.

 b. You run for 50 seconds. Are distance and rate proportional? Explain. Graph the equation.

 c. You run 300 feet. Are rate and time proportional? Explain. Graph the equation.

 d. One of these situations represents *inverse variation*. Which one is it? Why do you think it is called inverse variation?

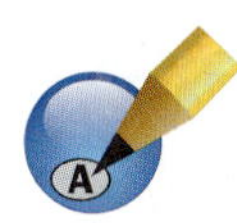

Fair Game Review What you learned in previous grades & lessons

Graph the linear equation. *(Section 4.1)*

14. $y = -\frac{1}{2}x$

15. $y = 3x - \frac{3}{4}$

16. $y = -\frac{x}{3} - \frac{3}{2}$

17. **MULTIPLE CHOICE** What is the value of x? *(Section 3.3)*

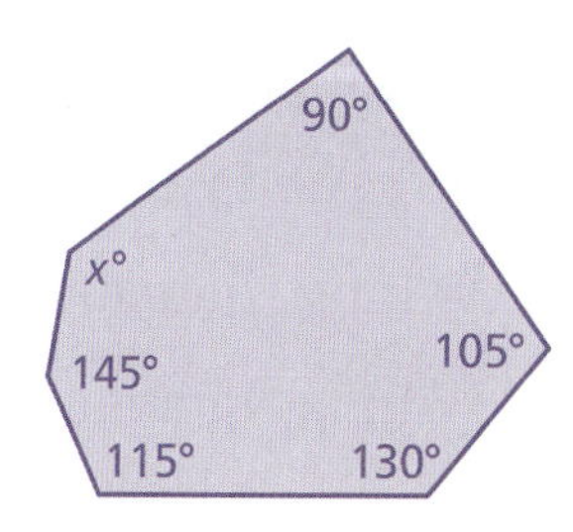

 (A) 110 (B) 135

 (C) 315 (D) 522

4 Study Help

You can use a **process diagram** to show the steps involved in a procedure. Here is an example of a process diagram for graphing a linear equation.

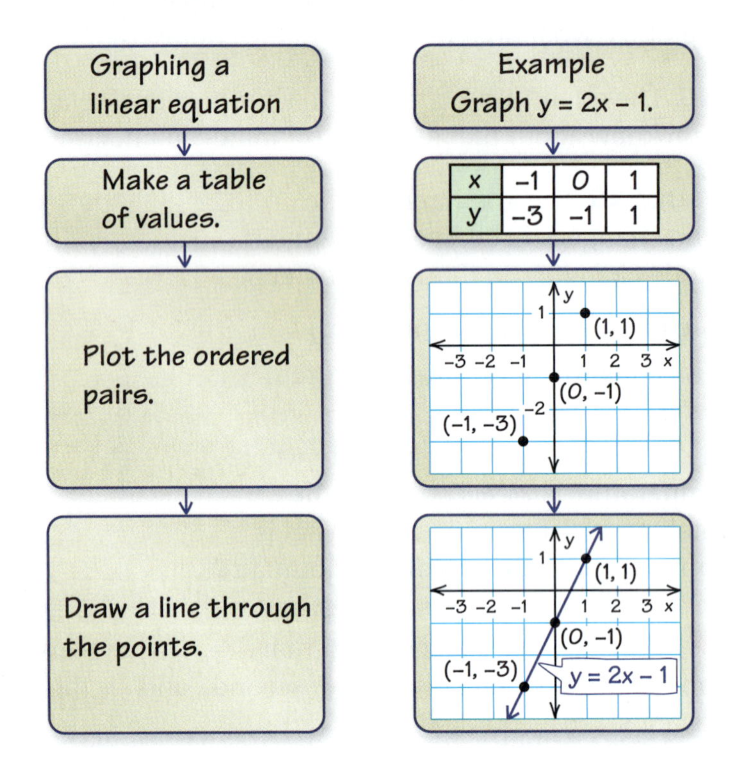

On Your Own

Make process diagrams with examples to help you study these topics.

1. finding the slope of a line
2. graphing a proportional relationship

After you complete this chapter, make process diagrams for the following topics.

3. graphing a linear equation using
   a. slope and y-intercept
   b. x- and y-intercepts
4. writing equations in slope-intercept form
5. writing equations in point-slope form

"Here is a process diagram with suggestions for what to do if a hyena knocks on your door."

Sample Answers

1.

Example

Finding the slope of a line → (3, 2), (−3, −2)

Determine whether the line rises or falls from left to right so you know whether the slope is positive or negative. → The line rises from left to right. So, the slope is positive.

Find the change in y or the rise. Find the change in x or the run. → (3, 2), (−3, −2), 4, 6

The slope is the ratio of the rise to the run. → $m = \frac{\text{rise}}{\text{run}} = \frac{4}{6} = \frac{2}{3}$

2.

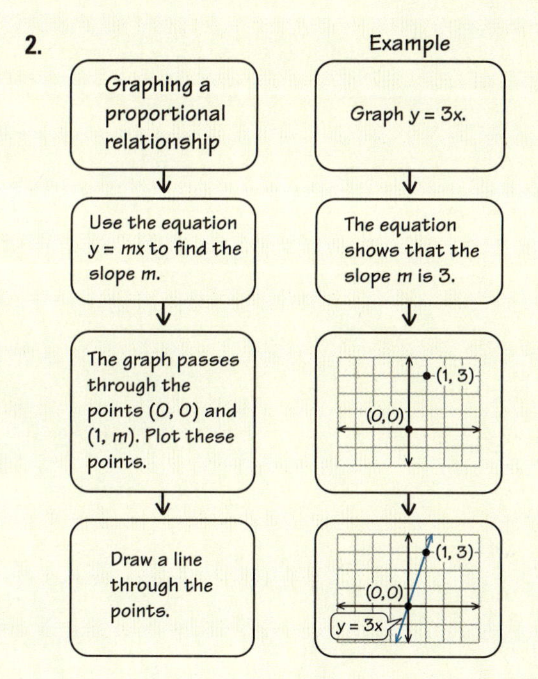

List of Organizers

Available at *BigIdeasMath.com*

Comparison Chart
Concept Circle
Definition (Idea) and Example Chart
Example and Non-Example Chart
Formula Triangle
Four Square
Information Frame
Information Wheel
Notetaking Organizer
Process Diagram
Summary Triangle
Word Magnet
Y Chart

About this Organizer

A **Process Diagram** can be used to show the steps involved in a procedure. Process diagrams are particularly useful for illustrating procedures with two or more steps, and they can have one or more branches. As shown, students' process diagrams can have two parallel parts, in which the procedure is stepped out in one part and an example illustrating each step is shown in the other part. Or, the diagram can be made up of just one part, with example(s) included in the last "bubble" to illustrate the steps that precede it.

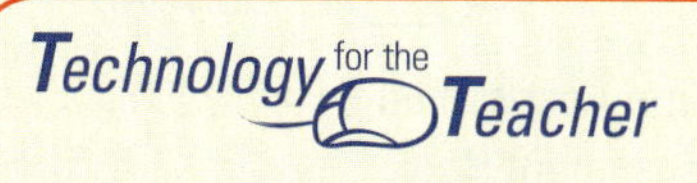

Editable Graphic Organizer

Answers

1–4. See Additional Answers.

5. $-\frac{1}{2}$

6. 2

7. undefined

8. 0

9. parallel slope: $-\frac{1}{2}$

perpendicular slope: 2

10. no; yes; The line $x = 1$ is vertical. The line $y = -1$ is horizontal. A vertical line is perpendicular to a horizontal line.

11.

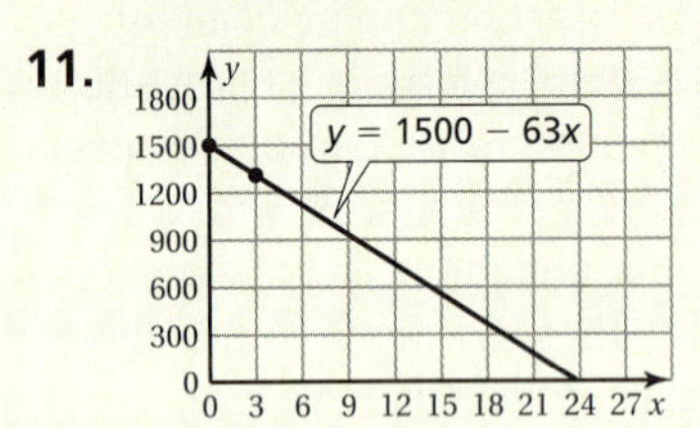

12.

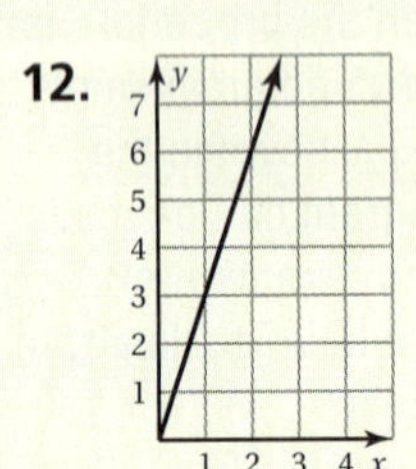

You take 3 hours of cello lessons per week.

13. a. $y = 7.50x$

b. The slope indicates that the unit cost is $7.50 per guest.

c. $75

Technology for the Teacher

Online Assessment
Assessment Book
ExamView® Assessment Suite

Alternative Quiz Ideas

100% Quiz
Error Notebook
Group Quiz
Homework Quiz
Math Log
Notebook Quiz
Partner Quiz
Pass the Paper

Pass the Paper

- Work in groups of four. The first student copies the problem and does a step, explaining his or her work.
- The paper is passed and the second student works through the next step, also explaining his or her work.
- This process continues until the problem is completed.
- The second member of the group starts the next problem. Students should be allowed to question and debate as they are working through the quiz.
- Student groups can be selected by the teacher, by students, through a random process, or any way that works for your class.
- The teacher walks around the classroom listening to the groups and asks questions to ensure understanding.

Reteaching and Enrichment Strategies

If students need help. . .	If students got it. . .
Resources by Chapter • Practice A and Practice B • Puzzle Time Lesson Tutorials *BigIdeasMath.com*	Resources by Chapter • Enrichment and Extension • Technology Connection Game Closet at *BigIdeasMath.com* Start the next section

4.1–4.3 Quiz

Check It Out
Progress Check
BigIdeasMath.com

Graph the linear equation. *(Section 4.1)*

1. $y = -x + 8$ **2.** $y = \frac{x}{3} - 4$ **3.** $x = -1$ **4.** $y = 3.5$

Find the slope of the line. *(Section 4.2)*

5.

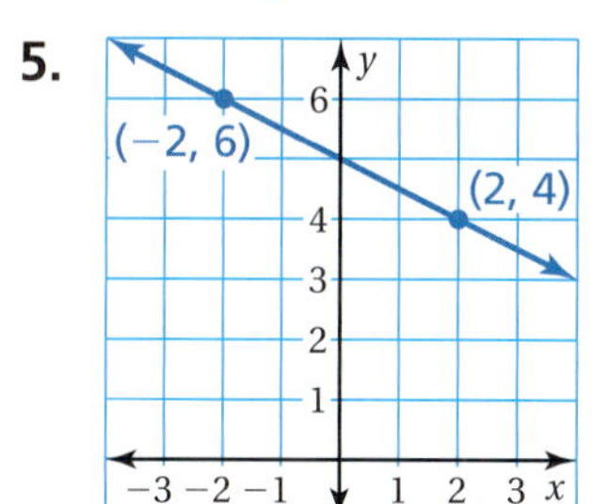

6.

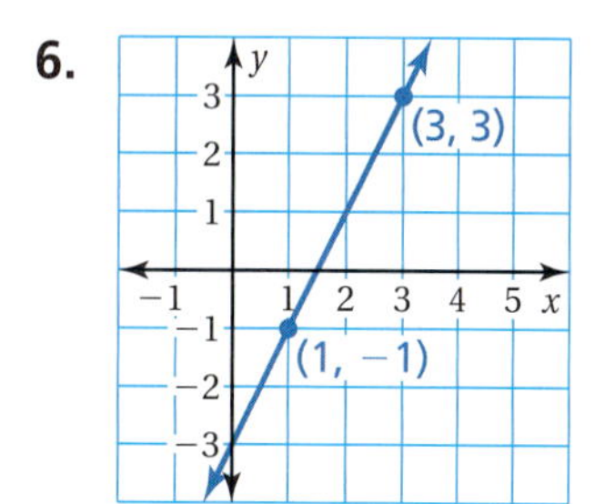

7.

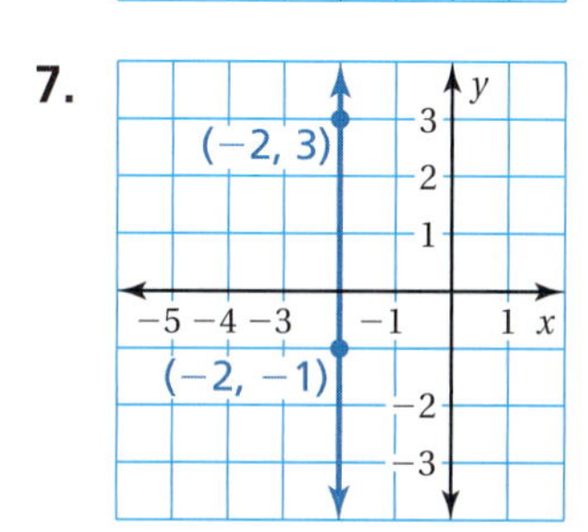

8.

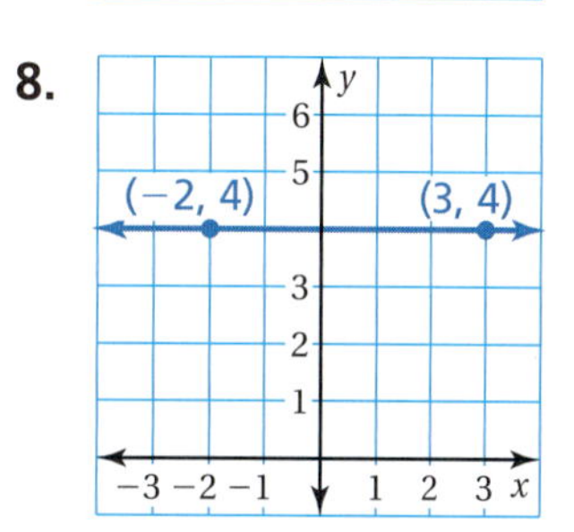

9. What is the slope of a line that is parallel to the line in Exercise 5? What is the slope of a line that is perpendicular to the line in Exercise 5? *(Section 4.2)*

10. Are the lines $y = -1$ and $x = 1$ parallel? Are they perpendicular? Justify your answer. *(Section 4.2)*

11. BANKING A bank charges $3 each time you use an out-of-network ATM. At the beginning of the month, you have $1500 in your bank account. You withdraw $60 from your bank account each time you use an out-of-network ATM. Graph a linear equation that represents the balance in your account after you use an out-of-network ATM x times. *(Section 4.1)*

12. MUSIC The number y of hours of cello lessons that you take after x weeks is represented by the equation $y = 3x$. Graph the equation and interpret the slope. *(Section 4.3)*

13. DINNER PARTY The cost y (in dollars) to provide food for guests at a dinner party is proportional to the number x of guests attending the party. It costs $30 to provide food for 4 guests. *(Section 4.3)*

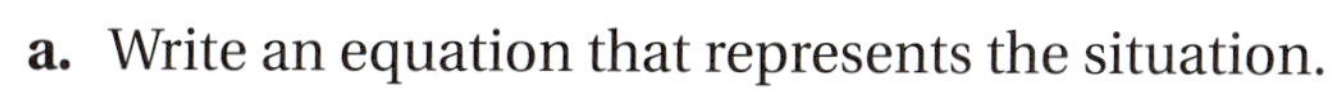

a. Write an equation that represents the situation.

b. Interpret the slope.

c. How much does it cost to provide food for 10 guests?

4.4 Graphing Linear Equations in Slope-Intercept Form

Essential Question How can you describe the graph of the equation $y = mx + b$?

1 ACTIVITY: Analyzing Graphs of Lines

Work with a partner.

- **Graph each equation.**
- **Find the slope of each line.**
- **Find the point where each line crosses the y-axis.**
- **Complete the table.**

Equation	Slope of Graph	Point of Intersection with y-axis
a. $y = -\frac{1}{2}x + 1$		
b. $y = -x + 2$		
c. $y = -x - 2$		
d. $y = \frac{1}{2}x + 1$		
e. $y = x + 2$		
f. $y = x - 2$		
g. $y = \frac{1}{2}x - 1$		
h. $y = -\frac{1}{2}x - 1$		
i. $y = 3x + 2$		
j. $y = 3x - 2$		

k. Do you notice any relationship between the slope of the graph and its equation? between the point of intersection with the y-axis and its equation? Compare the results with those of other students in your class.

Graphing Equations

In this lesson, you will

- find slopes and y-intercepts of graphs of linear equations.
- graph linear equations written in slope-intercept form.

Learning Standard
MACC.8.EE.2.6

Laurie's Notes

Introduction

Standards for Mathematical Practice

- **MP3a Construct Viable Arguments** and **MP8 Look for and Express Regularity in Repeated Reasoning:** The goal is for students to discover that when equations are written in slope-intercept form, the coefficient of the x-term is the slope and the constant is the y-intercept. Students graph the lines using a table of values, a good review of this skill.

Motivate

- **Preparation:** Make three demonstration cards on 8.5″x 11″ paper. The x-axis is labeled "time" and the y-axis is labeled "distance from home."
- Sample cards A, B, and C are shown.

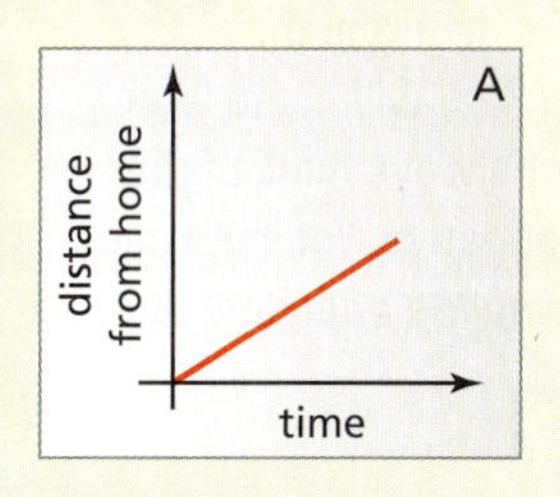

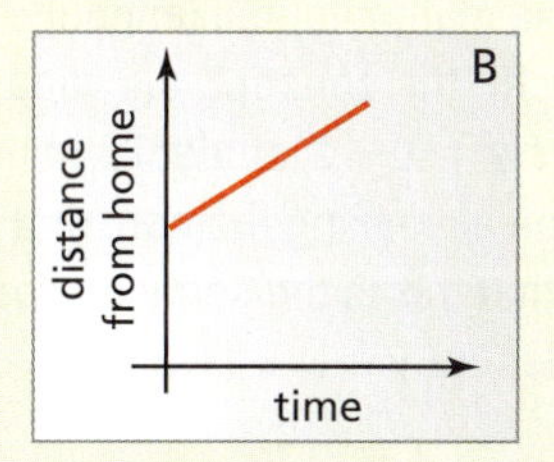

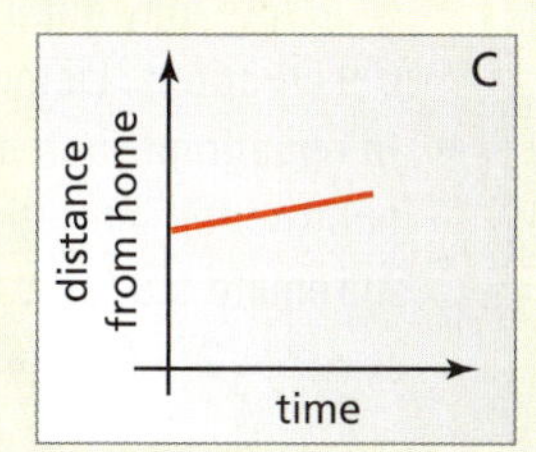

- Ask 3 students to hold the cards for the class to see.
- ? "Consider how the axes are labeled. What does the slope of the line represent?" $\frac{\text{distance}}{\text{time}} = \text{rate}$
- ? "What story does each card tell? How are the stories similar and different?" A: you begin at home; B: you travel at the same rate, but you start away from home; C: you start away from home, but you travel at a slower rate
- **Management Tip:** If you plan to use the demonstration cards again next year, laminate them.

Activity Notes

Activity 1

- When making a table of values, remind students to think about what values of x they should substitute. When the coefficient of x is a fraction, it is wise to select x-values that are multiples of the denominator. This may help eliminate fractional values.
- ? "How many points do you need in order to graph the equation?" Minimum is 2. Plot 3 to be safe.
- Remind students to solve the equation when $x = 0$. This will ensure that they find the point where the graph crosses the y-axis.
- Students should begin to observe patterns as they complete the table.
- Encourage students to draw the directed arrows in order to help them find the slope of the line.

Florida Common Core Standards

MACC.8.EE.2.6 Use similar triangles to explain why the slope m is the same between any two distinct points on a non-vertical line in the coordinate plane; derive the equation $y = mx$ for a line through the origin and the equation $y = mx + b$ for a line intercepting the vertical axis at b.

Previous Learning

Students should know how to graph linear equations and find slopes of lines.

4.4 Record and Practice Journal

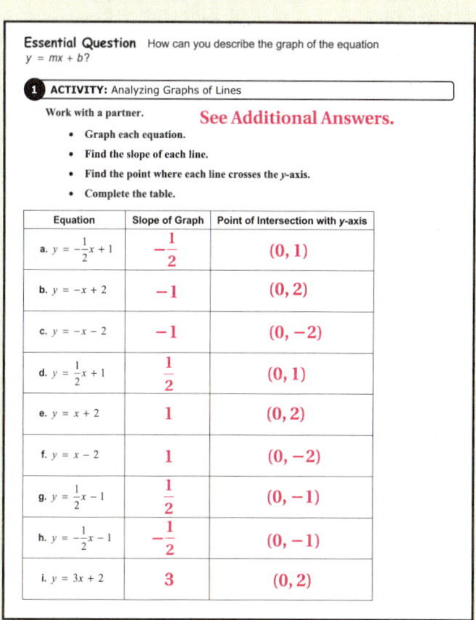

Essential Question How can you describe the graph of the equation $y = mx + b$?

1 ACTIVITY: Analyzing Graphs of Lines

Work with a partner. See Additional Answers.

- Graph each equation.
- Find the slope of each line.
- Find the point where each line crosses the y-axis.
- Complete the table.

Equation	Slope of Graph	Point of Intersection with y-axis
a. $y = -\frac{1}{2}x + 1$	$-\frac{1}{2}$	$(0, 1)$
b. $y = -x + 2$	-1	$(0, 2)$
c. $y = -x - 2$	-1	$(0, -2)$
d. $y = \frac{1}{2}x + 1$	$\frac{1}{2}$	$(0, 1)$
e. $y = x + 2$	1	$(0, 2)$
f. $y = x - 2$	1	$(0, -2)$
g. $y = \frac{1}{2}x - 1$	$\frac{1}{2}$	$(0, -1)$
h. $y = -\frac{1}{2}x - 1$	$-\frac{1}{2}$	$(0, -1)$
i. $y = 3x + 2$	3	$(0, 2)$

English Language Learners

Build on Past Knowledge

Remind students from their study of rational numbers that the slope -2 can be written as the fraction $\frac{-2}{1}$. By writing the integer as a fraction, students can see that the slope has a run of 1 and a rise of -2. This will help in graphing linear equations.

4.4 Record and Practice Journal

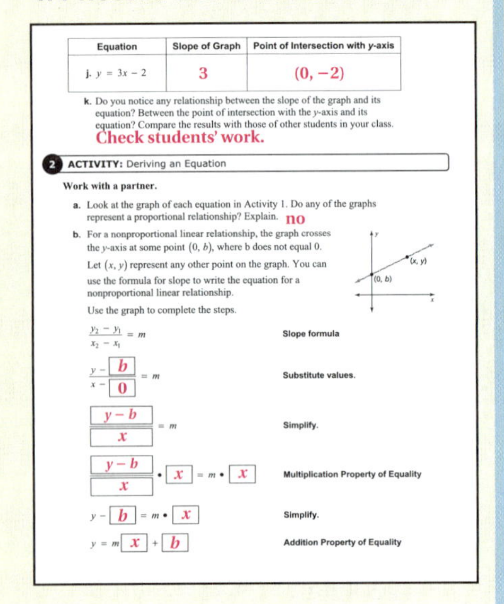

Equation	Slope of Graph	Point of Intersection with y-axis
j. $y = 3x - 2$	3	$(0, -2)$

k. Do you notice any relationship between the slope of the graph and its equation? Between the point of intersection with the y-axis and its equation? Compare the results with those of other students in your class.
Check students' work.

2 ACTIVITY: Deriving an Equation

Work with a partner.

a. Look at the graph of each equation in Activity 1. Do any of the graphs represent a proportional relationship? Explain. no

b. For a nonproportional linear relationship, the graph crosses the y-axis at some point $(0, b)$, where b does not equal 0.

Let (x, y) represent any other point on the graph. You can use the formula for slope to write the equation for a nonproportional linear relationship.

Use the graph to complete the steps.

$\frac{y_2 - y_1}{x_2 - x_1} = m$ Slope formula

$\frac{y - b}{x - 0} = m$ Substitute values.

$\frac{y - b}{x} = m$ Simplify.

$\frac{y - b}{x} \cdot x = m \cdot x$ Multiplication Property of Equality

$y - b = m \cdot x$ Simplify.

$y = mx + b$ Addition Property of Equality

c. What do m and b represent in the equation?
m represents the slope; b represents the y-coordinate of the point of intersection with the y-axis.

What Is Your Answer?

3. IN YOUR OWN WORDS How can you describe the graph of the equation $y = mx + b$?
a line with slope m that crosses the y-axis at $(0, b)$

a. How does the value of m affect the graph of the equation?
steepness of line and whether it rises or falls from left to right

b. How does the value of b affect the graph of the equation?
where the graph crosses the y-axis

c. Check your answers to parts (a) and (b) with three equations that are not in Activity 1.
Check students' work.

4. LOGIC Why do you think $y = mx + b$ is called the *slope-intercept form* of the equation of a line? Use drawings or diagrams to support your answer.
m is the slope and b is the y-intercept.

Laurie's Notes

Activity 1 (continued)

- Have students put a few graphs on the board to help facilitate discussion.

? In part (k), ask a series of summary questions.
  - "Compare certain pairs of graphs such as (e) and (f) or (i) and (j). What do you observe?" They have the same steepness (slope) and the number at the end of the equation is the y-coordinate of where the graph crosses the y-axis.
  - "Where does the equation $y = x + 7$ cross the y-axis?" at $(0, 7)$
  - "Compare certain groups of graphs such as (b), (e), and (i) or (c), (f), and (j). What do you observe?" They cross the y-axis at the same point, but the slopes are different; the coefficient of x is the slope of the line.
  - "What is the slope of the equation $y = 7x + 2$?" slope $= 7$

Words of Wisdom

- Students may not use mathematical language to describe their observations. Listen for the concept, the vocabulary will come later.
- In equations such as $y = x - 2$, students do not always think of the subtraction operation as making the constant negative. You may need to remind students that this is the same as *adding the opposite.* So, $y = x - 2$ is equivalent to $y = x + (-2)$.

Activity 2

- The goal of this activity is for students to derive the equation of a nonproportional linear relationship.
- **MP1a Make Sense of Problems:** Students are led through the steps necessary to derive the equation of the line. The ordered pairs $(0, b)$ and (x, y) are substituted into the slope formula.
- **MP2 Reason Abstractly and Quantitatively:** Students must reason abstractly to solve this equation for y. Working with the expression $(y - b)$ can feel very different to students than working with $(y - 4)$.

? "Assuming that there are no restrictions on x, does every nonvertical line pass through the y-axis?" yes

What Is Your Answer?

- These answers should follow immediately from discussing student observations.

Closure

- Refer back to the demonstration cards A, B, and C. Have students describe how the equations would be similar and how they would be different. *Sample answer:* A and B would have the same coefficient of x, but different constants. B and C would have different coefficients of x, but the same constant.

2 ACTIVITY: Deriving an Equation

Work with a partner.

a. Look at the graph of each equation in Activity 1. Do any of the graphs represent a proportional relationship? Explain.

b. For a nonproportional linear relationship, the graph crosses the y-axis at some point $(0, b)$, where b does not equal 0. Let (x, y) represent any other point on the graph. You can use the formula for slope to write the equation for a nonproportional linear relationship.

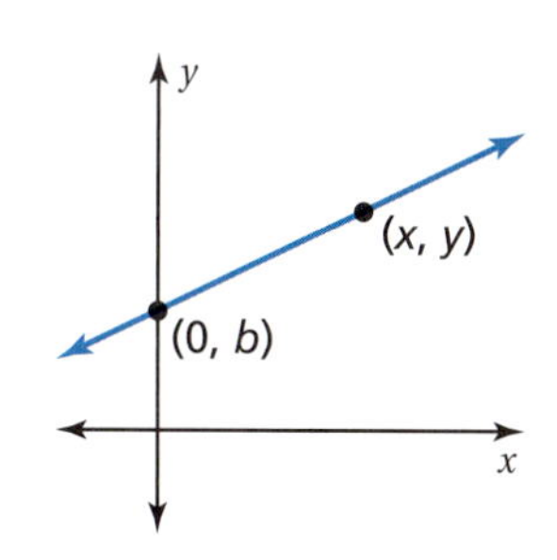

Use the graph to complete the steps.

$$\frac{y_2 - y_1}{x_2 - x_1} = m$$ Slope formula

$$\frac{y - \square}{x - \square} = m$$ Substitute values.

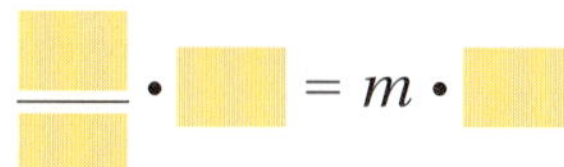

$$\frac{\square}{\square} = m$$ Simplify.

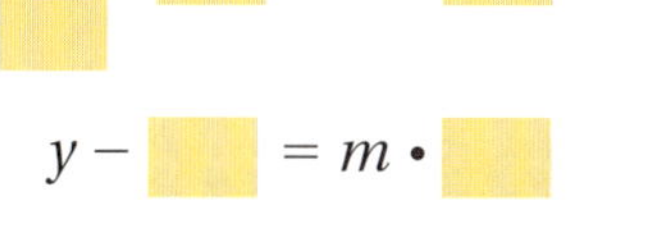

$$\frac{\square}{\square} \cdot \square = m \cdot \square$$ Multiplication Property of Equality

$$y - \square = m \cdot \square$$ Simplify.

$$y = m\square + \square$$ Addition Property of Equality

c. What do m and b represent in the equation?

Math Practice 3

Use Prior Results

How can you use the results of Activity 1 to help support your answer?

What Is Your Answer?

3. IN YOUR OWN WORDS How can you describe the graph of the equation $y = mx + b$?

a. How does the value of m affect the graph of the equation?

b. How does the value of b affect the graph of the equation?

c. Check your answers to parts (a) and (b) with three equations that are not in Activity 1.

4. LOGIC Why do you think $y = mx + b$ is called the *slope-intercept form* of the equation of a line? Use drawings or diagrams to support your answer.

Practice Use what you learned about graphing linear equations in slope-intercept form to complete Exercises 4–6 on page 170.

4.4 Lesson

Key Vocabulary

x-intercept, *p. 168*
y-intercept, *p. 168*
slope-intercept form, *p. 168*

Intercepts

The ***x*-intercept** of a line is the x-coordinate of the point where the line crosses the x-axis. It occurs when $y = 0$.

The ***y*-intercept** of a line is the y-coordinate of the point where the line crosses the y-axis. It occurs when $x = 0$.

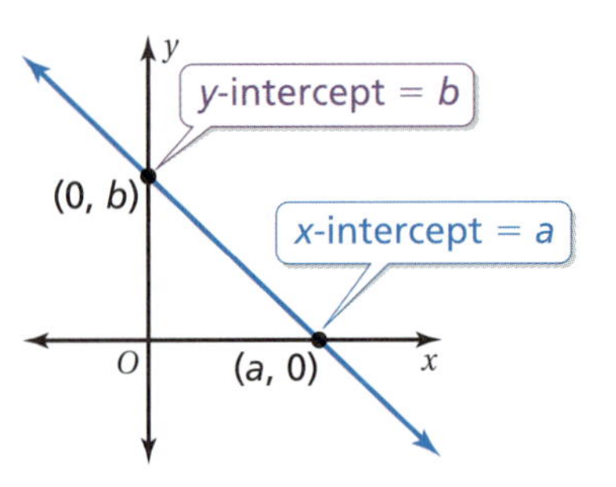

Study Tip

Linear equations can, but do not always, pass through the origin. So, proportional relationships are a special type of linear equation in which $b = 0$.

Slope-Intercept Form

Words A linear equation written in the form $y = mx + b$ is in **slope-intercept form**. The slope of the line is m, and the y-intercept of the line is b.

Algebra $y = mx + b$ (slope: m; y-intercept: b)

EXAMPLE 1 Identifying Slopes and y-Intercepts

Find the slope and the y-intercept of the graph of each linear equation.

a. $y = -4x - 2$

$y = -4x + (-2)$ Write in slope-intercept form.

∴ The slope is -4, and the y-intercept is -2.

b. $y - 5 = \frac{3}{2}x$

$y = \frac{3}{2}x + 5$ Add 5 to each side.

∴ The slope is $\frac{3}{2}$, and the y-intercept is 5.

On Your Own

Find the slope and the y-intercept of the graph of the linear equation.

1. $y = 3x - 7$

2. $y - 1 = -\frac{2}{3}x$

Laurie's Notes

Introduction

Connect

- **Yesterday:** Students explored the connection between the equation of a line and its graph. (MP1a, MP2, MP3a, MP8)
- **Today:** Students will graph equations in slope-intercept form.

Motivate

- Share the following taxi information. All trips start at a convention center.

Destination	Distance	Taxi Fare
Football stadium	18.7 mi	$39 approx.
Airport	12 mi	$32 flat fee
Shopping district	9.5 mi	$20 approx.

? "How do you think taxi fares are determined?" Answers will vary; listen for distance, number of passengers, tolls.

- Discuss why some locations, often involving airports, have flat fees associated with them.

Lesson Notes

Key Ideas

- Write the Key Ideas on the board. Draw the graph and discuss the vocabulary of this lesson: x-intercept, y-intercept, and slope-intercept form.
- Explain to students that the equation must be written with y in terms of x. This means that the equation must be solved for y.
- Discuss the Study Tip that connects to earlier lessons in this chapter.

Example 1

? "What is a linear equation?" an equation whose graph is a line

- Write part (a). This is written in the form $y = mx + b$, enabling students to quickly identify the slope and y-intercept.
- Write part (b).

? "Is $y - 5 = \frac{3}{2}x$ in slope-intercept form?" no "Can you rewrite it so that it is?" yes; Add 5 to each side of the equation.

- Make sure students understand that you can use the Commutative Property of Addition to write $y = b + mx$ as $y = mx + b$.

On Your Own

- **Think-Pair-Share:** Students should read each question independently and then work in pairs to answer the questions. When they have answered the questions, the pair should compare their answers with another group and discuss any discrepancies.

Goal Today's lesson is graphing the equation of a line written in **slope-intercept form**.

Technology for the Teacher

Lesson Tutorials
Lesson Plans
Answer Presentation Tool

Extra Example 1

Find the slope and y-intercept of the graph of each linear equation.

a. $y = \frac{3}{4}x - 5$

slope: $\frac{3}{4}$; y-intercept: -5

b. $y + \frac{1}{2} = -6x$

slope: -6; y-intercept: $-\frac{1}{2}$

On Your Own

1. slope: 3; y-intercept: -7
2. slope: $-\frac{2}{3}$; y-intercept: 1

Extra Example 2

Graph $y = -\frac{2}{3}x - 2$. Identify the x-intercept.

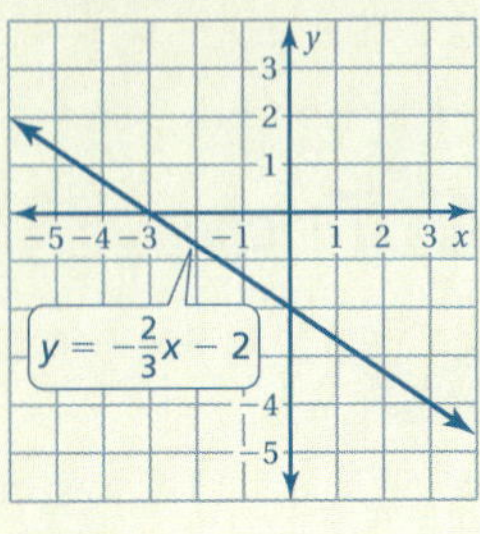

-3

Extra Example 3

The cost y (in dollars) for making friendship bracelets is $y = 0.5x + 2$, where x is the number of bracelets.

a. Graph the equation.

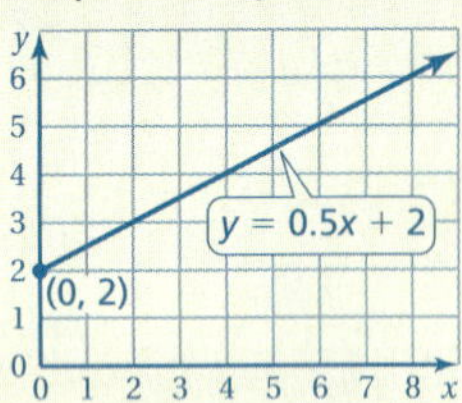

b. Interpret the slope and y-intercept.
The slope is 0.5. So, the cost per bracelet is \$0.50. The y-intercept is 2. So, there is an initial cost of \$2 to make the bracelets.

On Your Own

3–5. See Additional Answers.

Differentiated Instruction

Kinesthetic

When graphing a linear equation using the slope-intercept form, students must apply the slope correctly after plotting the point for the y-intercept. Have students plot (0, 3) in the coordinate plane. Then graph the lines $y = 4x + 3$, $y = \frac{1}{4}x + 3$, $y = -4x + 3$, and $y = -\frac{1}{4}x + 3$ in the same coordinate plane using (0, 3) as the starting point. Make sure students identify the correct rise and run for each line.

Laurie's Notes

Example 2

- ? "How can knowing the slope and the y-intercept help you graph a line?" Listen for student understanding of what slope and y-intercept mean.
- Remind students that a slope of -3 can be interpreted as $\frac{-3}{1} = \frac{3}{-1}$. Starting at the y-intercept, you can move to the right 1 unit and down 3 units, or to the left 1 unit and up 3 units. In both cases, you land on a point which satisfies the equation.
- ? **FYI:** "In this problem, you used the slope to plot a point that coincidentally landed on the x-axis. How would you find the x-intercept without using a graph?" Set $y = 0$ and solve for x.

Example 3

- Write the equation $y = 2.5x + 2$ on the board.
- ? "What is the slope and what does it mean in the context of this problem?" 2.5; It costs \$2.50 for each mile you travel in the taxi. "What is the y-intercept and what does it mean in the context of this problem?" 2; The initial fee is \$2 when you sit down in the taxi.
- **MP2 Reason Abstractly and Quantitatively:** Mathematically proficient students make sense of the quantities and their relationships in problem situations. To develop this proficiency, students must be asked to interpret the meaning of the symbols.
- **MP6 Attend to Precision:** Suggest to students that because the slope is 2.5, any ratio equivalent to 2.5 can also be used, such as $\frac{2.5}{1} = \frac{5}{2}$. Using whole numbers instead of decimals improves the accuracy of graphing.
- Explain that the graph of this equation will only be in Quadrant I because it does not make sense to have a negative number of miles or a negative cost.
- ? "What is the cost for a 2-mile taxi ride? a 10-mile taxi ride?" \$7; \$27

On Your Own

- Students are asked to check their answers using a graphing calculator. It is helpful to build proficiency with using the graphing calculator so that the calculator becomes a useful tool in problem solving.

Closure

- **Exit Ticket:** Graph $y - 4 = 2x$ and identify the slope and y-intercept.

  slope = 2; y-intercept = 4

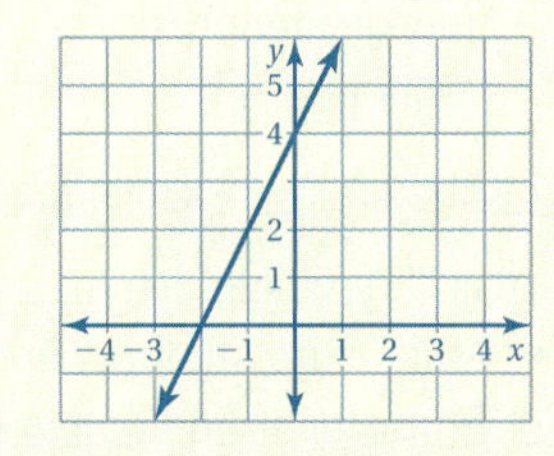

EXAMPLE 2 Graphing a Linear Equation in Slope-Intercept Form

Graph $y = -3x + 3$. Identify the x-intercept.

Step 1: Find the slope and the y-intercept.

$$y = -3x + 3$$

slope: -3 y-intercept: 3

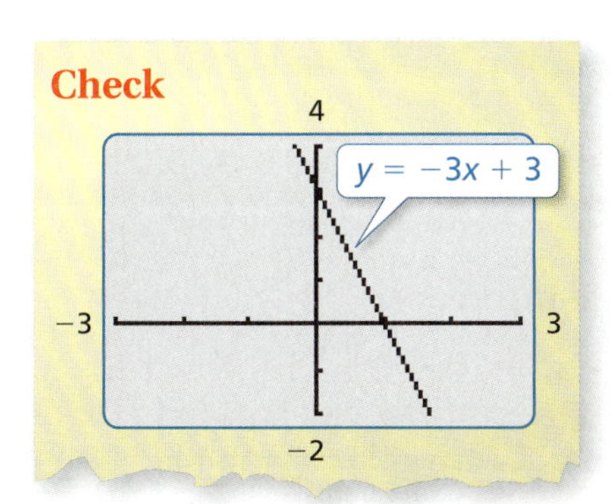

Step 2: The y-intercept is 3. So, plot (0, 3).

Step 3: Use the slope to find another point and draw the line.

$$m = \frac{\text{rise}}{\text{run}} = \frac{-3}{1}$$

Plot the point that is 1 unit right and 3 units down from (0, 3). Draw a line through the two points.

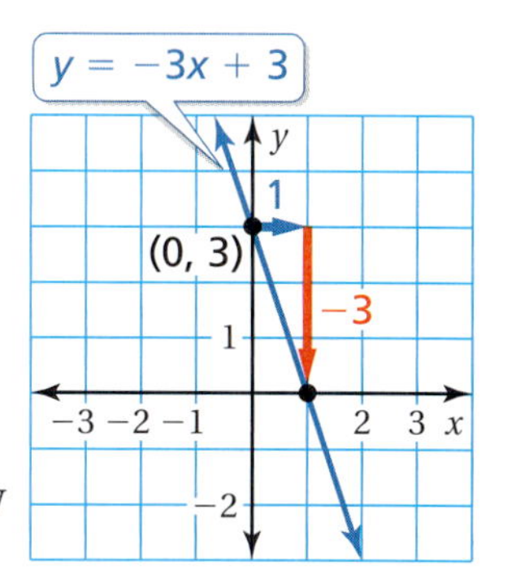

The line crosses the x-axis at (1, 0). So, the x-intercept is 1.

EXAMPLE 3 Real-Life Application

The cost y (in dollars) of taking a taxi x miles is $y = 2.5x + 2$. (a) Graph the equation. (b) Interpret the y-intercept and the slope.

a. The slope of the line is $2.5 = \frac{5}{2}$. Use the slope and the y-intercept to graph the equation.

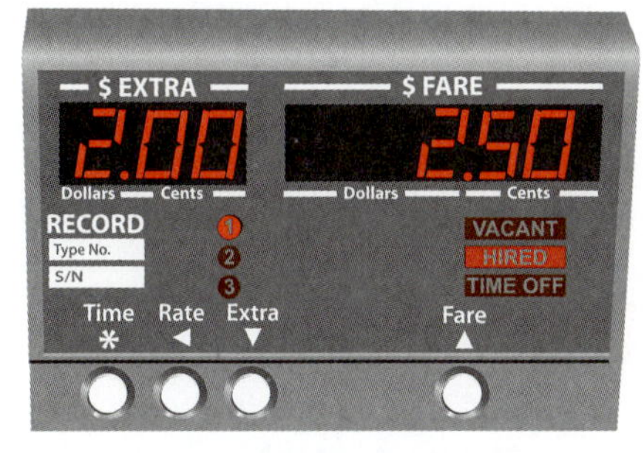

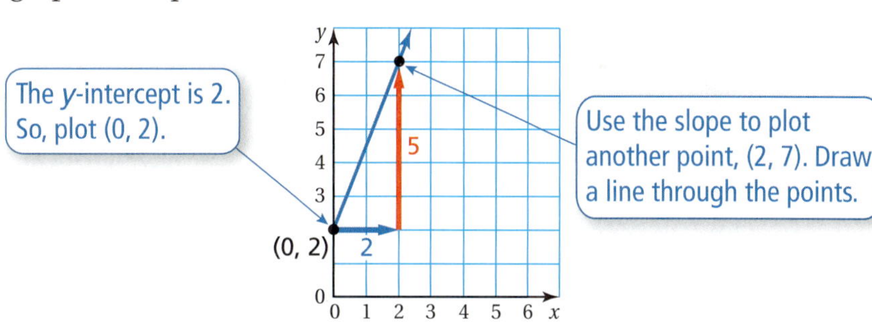

b. The slope is 2.5. So, the cost per mile is \$2.50. The y-intercept is 2. So, there is an initial fee of \$2 to take the taxi.

On Your Own

Now You're Ready
Exercises 18–23

Graph the linear equation. Identify the x-intercept. Use a graphing calculator to check your answer.

3. $y = x - 4$

4. $y = -\frac{1}{2}x + 1$

5. In Example 3, the cost y (in dollars) of taking a different taxi x miles is $y = 2x + 1.5$. Interpret the y-intercept and the slope.

4.4 Exercises

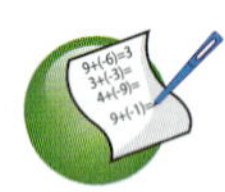

Vocabulary and Concept Check

1. **VOCABULARY** How can you find the x-intercept of the graph of $2x + 3y = 6$?
2. **CRITICAL THINKING** Is the equation $y = 3x$ in slope-intercept form? Explain.
3. **OPEN-ENDED** Describe a real-life situation that you can model with a linear equation. Write the equation. Interpret the y-intercept and the slope.

Practice and Problem Solving

Match the equation with its graph. Identify the slope and the y-intercept.

4. $y = 2x + 1$
5. $y = \frac{1}{3}x - 2$
6. $y = -\frac{2}{3}x + 1$

A.

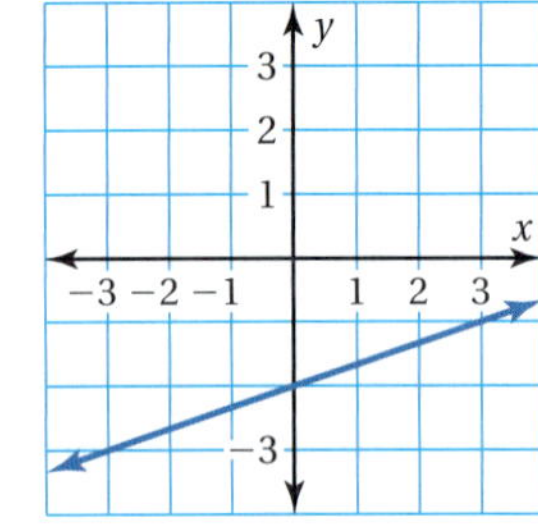

B.

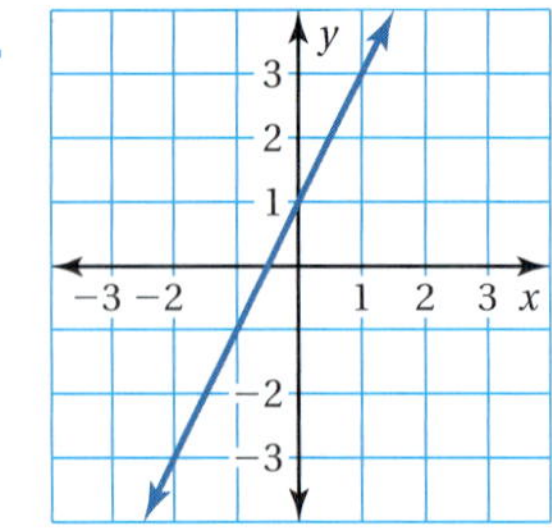

C.

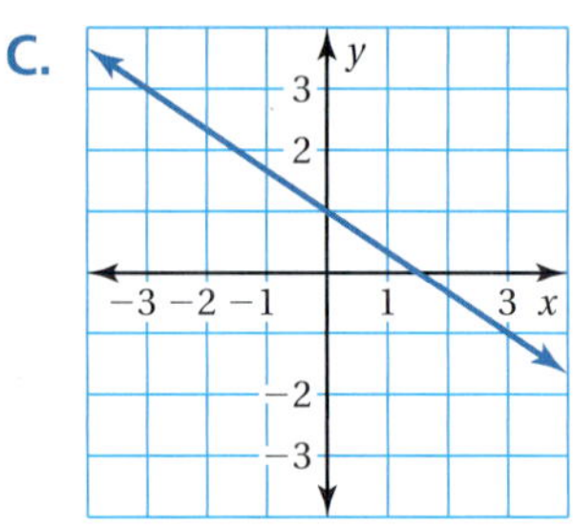

Find the slope and the y-intercept of the graph of the linear equation.

1 7. $y = 4x - 5$
8. $y = -7x + 12$
9. $y = -\frac{4}{5}x - 2$
10. $y = 2.25x + 3$
11. $y + 1 = \frac{4}{3}x$
12. $y - 6 = \frac{3}{8}x$
13. $y - 3.5 = -2x$
14. $y = -5 - \frac{1}{2}x$
15. $y = 11 + 1.5x$

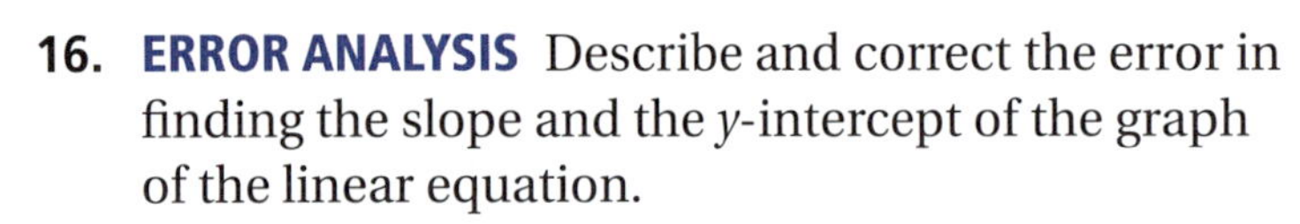

16. **ERROR ANALYSIS** Describe and correct the error in finding the slope and the y-intercept of the graph of the linear equation.

$y = 4x - 3$

The slope is 4, and the y-intercept is 3.

17. **SKYDIVING** A skydiver parachutes to the ground. The height y (in feet) of the skydiver after x seconds is $y = -10x + 3000$.
    a. Graph the equation.
    b. Interpret the x-intercept and the slope.

Assignment Guide and Homework Check

Level	Day 1 Activity Assignment	Day 2 Lesson Assignment	Homework Check
Basic	4–6, 27–31	1–3, 7–15 odd, 16, 17–25 odd	9, 11, 17, 23, 25
Average	4–6, 27–31	1–3, 8–16 even, 19–25 odd	10, 12, 23, 25
Advanced	1–6, 8–16 even, 18–31		12, 14, 22, 24, 25

Common Errors

- **Exercises 7–15** Students may forget to include negatives with the slope and/or y-intercept. Remind them to look at the sign in front of the slope and the y-intercept. Also remind students that slope-intercept form is $y = mx + b$. This means that if the linear equation has "minus b," then the y-intercept is negative.
- **Exercises 11–13** Students may identify the opposite y-intercept because they forget to solve for y. Remind them that slope-intercept form has y by itself, so they must solve for y before identifying the slope and y-intercept.
- **Exercises 18–23** Students may use the reciprocal of the slope when graphing and may find an incorrect x-intercept. Remind them that slope is *rise* over *run*, so the numerator represents vertical change, not horizontal.

4.4 Record and Practice Journal

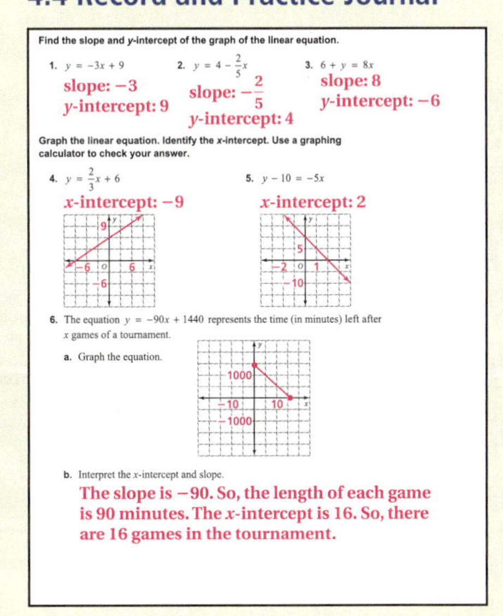

Find the slope and y-intercept of the graph of the linear equation.

1. $y = -3x + 9$
   slope: -3
   y-intercept: 9
2. $y = 4 - \frac{2}{5}x$
   slope: $-\frac{2}{5}$
   y-intercept: 4
3. $6 + y = 8x$
   slope: 8
   y-intercept: -6

Graph the linear equation. Identify the x-intercept. Use a graphing calculator to check your answer.

4. $y = \frac{2}{3}x + 6$
   x-intercept: -9
5. $y - 10 = -5x$
   x-intercept: 2

6. The equation $y = -90x + 1440$ represents the time (in minutes) left after x games of a tournament.
   a. Graph the equation.
   b. Interpret the x-intercept and slope.
   The slope is -90. So, the length of each game is 90 minutes. The x-intercept is 16. So, there are 16 games in the tournament.

Vocabulary and Concept Check

1. Find the x-coordinate of the point where the graph crosses the x-axis.
2. yes; The slope is 3 and the y-intercept is 0.
3. *Sample answer:* The amount of gasoline y (in gallons) left in your tank after you travel x miles is $y = -\frac{1}{20}x + 20$. The slope of $-\frac{1}{20}$ means the car uses 1 gallon of gas for every 20 miles driven. The y-intercept of 20 means there is originally 20 gallons of gas in the tank.

Practice and Problem Solving

4. B; slope: 2; y-intercept: 1
5. A; slope: $\frac{1}{3}$; y-intercept: -2
6. C; slope: $-\frac{2}{3}$; y-intercept: 1
7. slope: 4; y-intercept: -5
8. slope: -7; y-intercept: 12
9. slope: $-\frac{4}{5}$; y-intercept: -2
10. slope: 2.25; y-intercept: 3
11. slope: $\frac{4}{3}$; y-intercept: -1
12. slope: $\frac{3}{8}$; y-intercept: 6
13. slope: -2; y-intercept: 3.5
14. slope: $-\frac{1}{2}$; y-intercept: -5
15. slope: 1.5; y-intercept: 11
16. The y-intercept should be -3.
    $y = 4x - 3$
    The slope is 4 and the y-intercept is -3.
17. See Additional Answers.

Practice and Problem Solving

18.

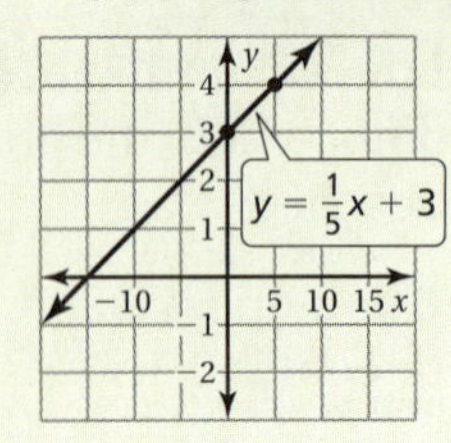

x-intercept: -15

19–25. See Additional Answers.

26. See *Taking Math Deeper*.

Fair Game Review

27. $y = 2x + 3$

28. $y = -\frac{4}{5}x + \frac{13}{5}$

29. $y = \frac{2}{3}x - 2$

30. $y = -\frac{7}{4}x + 2$

31. B

Mini-Assessment

Find the slope and y-intercept of the graph of the equation. Then graph the equation.

1. $y = -5x + 3$

slope $= -5$, y-intercept $= 3$

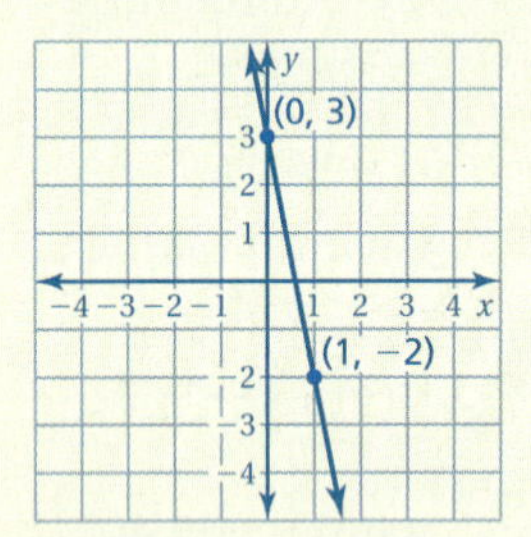

2. $y - 4 = \frac{1}{2}x$

slope $= \frac{1}{2}$, y-intercept $= 4$

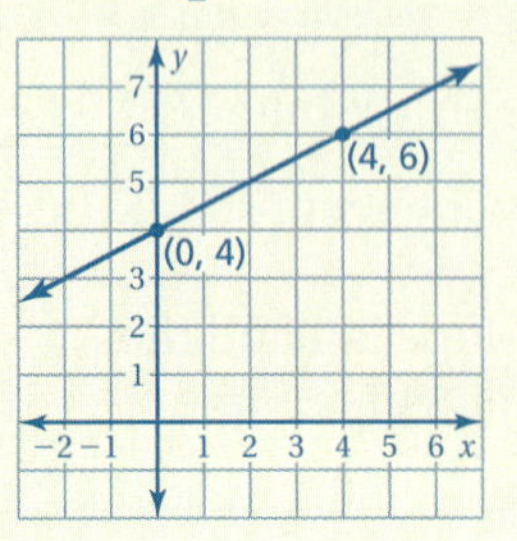

Taking Math Deeper

Exercise 26

This is a classic business problem. The business has monthly costs. The question is how many clicks are needed to cover the costs and start making a profit.

Organize the given information.

- The site has 5 banner ads.
- Monthly income is $0.005 per click.
- It costs $120 per month to run the site.
- Let y be the monthly income (in dollars).
- Let x be the number of clicks per month.

a. Write an equation for the income.

$y = 0.005x$

$y = 0.005x$

b. Graph the equation.

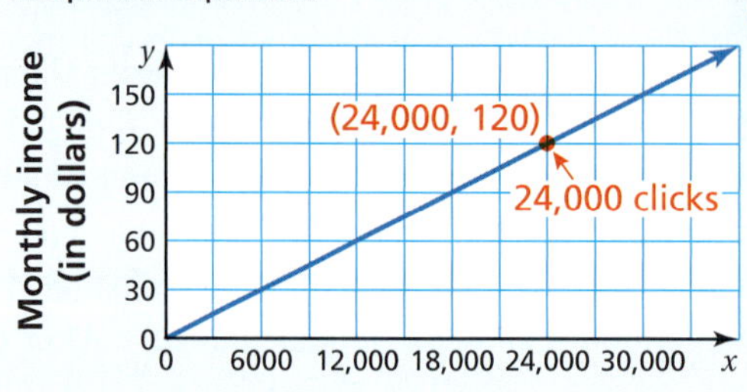

When the ads start to get 24,000 clicks a month, the income will be $120 per month. Each banner ad needs to average $\frac{24{,}000}{5} = 4800$ clicks. Any additional clicks per month will start earning a profit.

Project

Use the Internet or the school library to research methods for determining the number of clicks on a website.

Reteaching and Enrichment Strategies

If students need help. . .	If students got it. . .
Resources by Chapter • Practice A and Practice B • Puzzle Time Record and Practice Journal Practice Differentiating the Lesson Lesson Tutorials Skills Review Handbook	Resources by Chapter • Enrichment and Extension • Technology Connection Start the next section

Graph the linear equation. Identify the *x*-intercept. Use a graphing calculator to check your answer.

2 **18.** $y = \frac{1}{5}x + 3$ **19.** $y = 6x - 7$ **20.** $y = -\frac{8}{3}x + 9$

21. $y = -1.4x - 1$ **22.** $y + 9 = -3x$ **23.** $y = 4 - \frac{3}{5}x$

24. APPLES You go to a harvest festival and pick apples.

Admission: $5.00
Apples: $0.75 per lb

a. Which equation represents the cost (in dollars) of going to the festival and picking *x* pounds of apples? Explain.

$y = 5x + 0.75$ $y = 0.75x + 5$

b. Graph the equation you chose in part (a).

25. REASONING Without graphing, identify the equations of the lines that are (a) parallel and (b) perpendicular. Explain your reasoning.

$y = 2x + 4$	$y = -\frac{1}{3}x - 1$	$y = -3x - 2$	$y = \frac{1}{2}x + 1$
$y = 3x + 3$	$y = -\frac{1}{2}x + 2$	$y = -3x + 5$	$y = 2x - 3$

26. Critical Thinking Six friends create a website. The website earns money by selling banner ads. The site has 5 banner ads. It costs $120 a month to operate the website.

a. A banner ad earns $0.005 per click. Write a linear equation that represents the monthly income *y* (in dollars) for *x* clicks.

b. Graph the equation in part (a). On the graph, label the number of clicks needed for the friends to start making a profit.

Fair Game Review What you learned in previous grades & lessons

Solve the equation for *y*. *(Section 1.4)*

27. $y - 2x = 3$ **28.** $4x + 5y = 13$ **29.** $2x - 3y = 6$ **30.** $7x + 4y = 8$

31. MULTIPLE CHOICE Which point is a solution of the equation $3x - 8y = 11$? *(Section 4.1)*

Ⓐ $(1, 1)$ Ⓑ $(1, -1)$ Ⓒ $(-1, 1)$ Ⓓ $(-1, -1)$

4.5 Graphing Linear Equations in Standard Form

Essential Question How can you describe the graph of the equation $ax + by = c$?

1 ACTIVITY: Using a Table to Plot Points

Work with a partner. You sold a total of $16 worth of tickets to a school concert. You lost track of how many of each type of ticket you sold.

a. Let x represent the number of adult tickets.

Let y represent the number of student tickets.

Write an equation that relates x and y.

b. Copy and complete the table showing the different combinations of tickets you might have sold.

Number of Adult Tickets, x					
Number of Student Tickets, y					

c. Plot the points from the table. Describe the pattern formed by the points.

d. If you remember how many adult tickets you sold, can you determine how many student tickets you sold? Explain your reasoning.

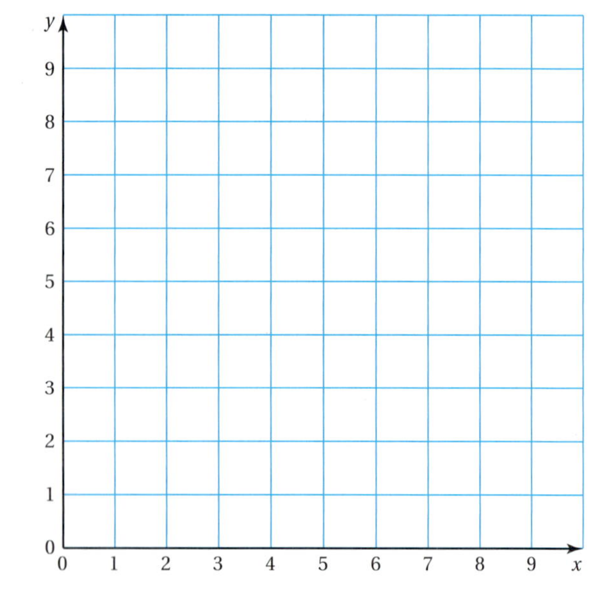

Graphing Equations

In this lesson, you will

- graph linear equations written in standard form.

Applying Standard MACC.8.EE.2.6

Laurie's Notes

Introduction

Standards for Mathematical Practice

- **MP7 Look for and Make Use of Structure:** In this lesson students will graph a linear equation in a new form. Mathematically proficient students discern a pattern or structure. Recognizing the equivalence of equations written in different forms requires that students be able to manipulate equations.

Motivate

- **Preparation:** Make a set of equation cards on strips of paper. The equations are all the same when simplified and need to be written large enough to be read by students sitting at the back of the classroom.
- Here is a sample set of equations: $y = 2x + 1$, $-2x + y = 1$, $2x - y = -1$, $4x - 2y = -2$
- Ask 4 students to stand at the front of the room and hold the cards so only they can see the equations.
- As you state an ordered pair, each student holding a card determines whether it is a solution to the equation on the card and if it is, raises his or her hand. If not, they do nothing. State several ordered pairs, four that are solutions and two that are not. Plot all of the points that you state. The four ordered pairs that are solutions will lie on a line.

? "How many lines can pass through any two points?" one "How many lines pass through the four solutions points?" Students will say 4; now is the time to discuss the idea of one line written in different forms.

- Have the students holding the cards reveal the equations to the class and read them aloud. Write each of the equations on the board.
- Explain to students that equations can be written in different forms. Today they will explore a new form of a linear equation.

Activity Notes

Activity 1

- Read the problem aloud. Note that a verbal model is shown for the equation $4x + 2y = 16$.
- Discuss what the variables x and y represent.

? "Could you have sold 5 adult tickets? Explain." No; 5 adult tickets would be \$20, which is too much.

- Students may say that they do not know how to figure out x and y. Students may not realize that there is more than one solution. Remind students that *Guess, Check, and Revise* would be an appropriate strategy to use.
- Discuss part (c). The points lie on a line.
- Discuss part (d). Students may not recognize that in knowing x, they can substitute and solve for y. This is not an obvious step for students.

? "Can $x = 1.5$? Explain." No, you cannot sell 1.5 tickets.

? "What are the different numbers of adult tickets that are possible to sell?" 0, 1, 2, 3, 4

- **Note:** This is an example of a discrete domain; there are only 5 possible values for the variable x. Discrete domains are not taught at this time.

Florida Common Core Standards

MACC.8.EE.2.6 Use similar triangles to explain why the slope m is the same between any two distinct points on a non-vertical line in the coordinate plane; derive the equation $y = mx$ for a line through the origin and the equation $y = mx + b$ for a line intercepting the vertical axis at b.

Previous Learning

Students should know how to graph linear equations in slope-intercept form.

4.5 Record and Practice Journal

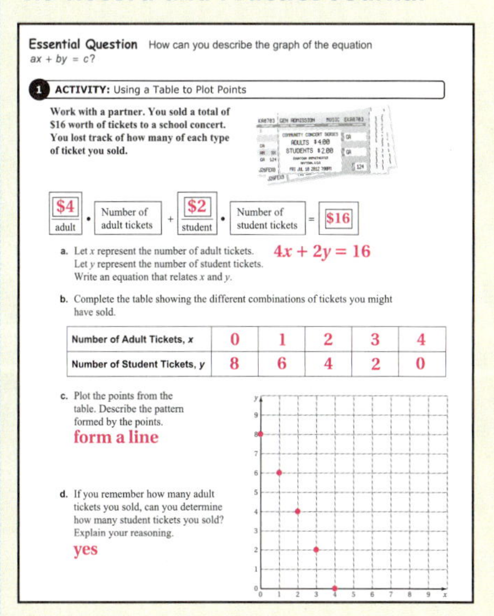

Essential Question How can you describe the graph of the equation $ax + by = c$?

1 ACTIVITY: Using a Table to Plot Points

Work with a partner. You sold a total of \$16 worth of tickets to a school concert. You lost track of how many of each type of ticket you sold.

$\frac{\$4}{\text{adult}}$ • Number of adult tickets + $\frac{\$2}{\text{student}}$ • Number of student tickets = \$16

a. Let x represent the number of adult tickets. Let y represent the number of student tickets. Write an equation that relates x and y. $4x + 2y = 16$

b. Complete the table showing the different combinations of tickets you might have sold.

Number of Adult Tickets, x	0	1	2	3	4
Number of Student Tickets, y	8	6	4	2	0

c. Plot the points from the table. Describe the pattern formed by the points. form a line

d. If you remember how many adult tickets you sold, can you determine how many student tickets you sold? Explain your reasoning. yes

Differentiated Instruction

Visual

Have students create charts in their notebooks of the equation forms and how to graph them.

Slope-intercept form $y = mx + b$	• Plot $(0, b)$. • Use the slope m to plot a second point. • Draw a line through the two points.
Horizontal line $y = c$	• Draw a horizontal line through $(0, c)$.
Vertical line $x = c$	• Draw a vertical line through $(c, 0)$.
Standard form $ax + by = c$	• Find the y-intercept. • Find the x-intercept. • Plot the associated points. Draw a line through the two points.

4.5 Record and Practice Journal

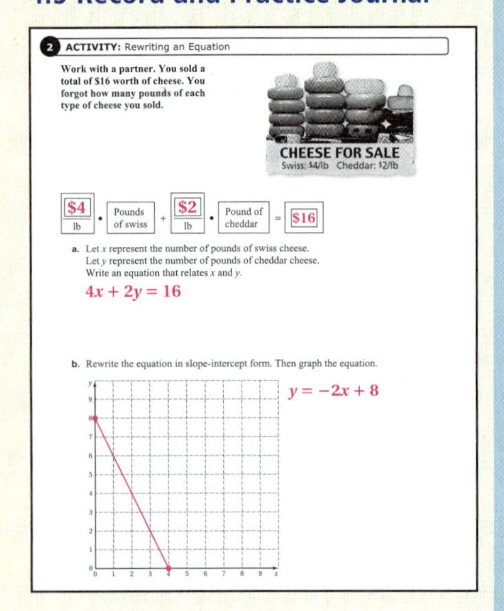

2 ACTIVITY: Rewriting an Equation

Work with a partner. You sold a total of \$16 worth of cheese. You forgot how many pounds of each type of cheese you sold.

\$4/lb • Pounds of swiss + \$2/lb • Pound of cheddar = \$16

a. Let x represent the number of pounds of swiss cheese. Let y represent the number of pounds of cheddar cheese. Write an equation that relates x and y.

$4x + 2y = 16$

b. Rewrite the equation in slope-intercept form. Then graph the equation.

$y = -2x + 8$

c. You sold 2 pounds of cheddar cheese. How many pounds of swiss cheese did you sell?

3 lb

d. Does the value $x = 2.5$ make sense in the context of the problem? Explain.

yes

What Is Your Answer?

3. **IN YOUR OWN WORDS** How can you describe the graph of the equation $ax + by = c$?

Sample answer: a line with a slope of $-\frac{a}{b}$ and y-intercept of $\frac{c}{b}$

4. Activities 1 and 2 show two different methods for graphing $ax + by = c$. Describe the two methods. Which method do you prefer? Explain.

Check students' work.

5. Write a real-life problem that is similar to those shown in Activities 1 and 2.

Check students' work.

6. Why do you think it might be easier to graph $x + y = 10$ without rewriting it in slope-intercept form and then graphing?

You can see that when $x = 0, y = 10$, and when $y = 0, x = 10$. You can graph the equation through its x- and y-intercepts.

Laurie's Notes

Activity 2

- Read the problem aloud. Note that a verbal model is shown for the equation $4x + 2y = 16$.
- Discuss what the variables x and y represent.

? "Could you have sold 5 pounds of swiss cheese? Explain." No; 5 pounds of swiss cheese would be \$20, which is too much.

- Give time for students to work with their partners. While this may be the same equation as Activity 1, the approach is different. Students are asked to write the equation in slope-intercept form. After the equation is in slope-intercept form, students can substitute a value for x and find y. This is generally not the case for equations written in standard form.
- Students may solve part (c) in several different ways. They may try to "read" the solution from their graphs or they may substitute $y = 2$ into either of the equations in parts (a) and (b).

? "Can $x = 2.5$? Explain." yes; You can buy a portion of a pound.

- **Note:** This is an example of a continuous domain; all numbers $0 \leq x \leq 4$ are possible. Continuous domains are not taught at this time.
- Students might observe that both examples have graphs in the first quadrant. This is common for real-life examples.

What Is Your Answer?

- **Question 3:** Students may guess that the graph is linear from Activity 1. However, some students may not be secure with this knowledge yet.
- Question 6 asks students to think about the process of graphing a line. Students consider the structure of the equation $x + y = 10$. When sharing their answers, listen for students to translate the equation as "the sum of two number is 10."

Closure

- Refer back to the equation cards. Rewrite the last three equations in slope-intercept form. $y = 2x + 1$

2 ACTIVITY: Rewriting an Equation

Work with a partner. You sold a total of $16 worth of cheese. You forgot how many pounds of each type of cheese you sold.

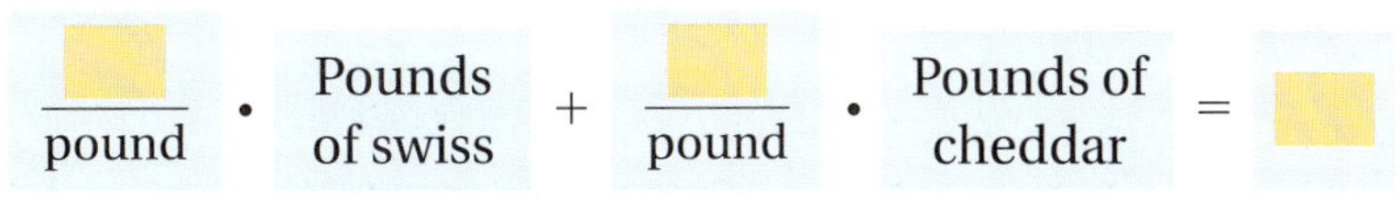

Math Practice 2

Understand Quantities

What do the equation and the graph represent? How can you use this information to solve the problem?

a. Let x represent the number of pounds of swiss cheese.

Let y represent the number of pounds of cheddar cheese.

Write an equation that relates x and y.

b. Rewrite the equation in slope-intercept form. Then graph the equation.

c. You sold 2 pounds of cheddar cheese. How many pounds of swiss cheese did you sell?

d. Does the value $x = 2.5$ make sense in the context of the problem? Explain.

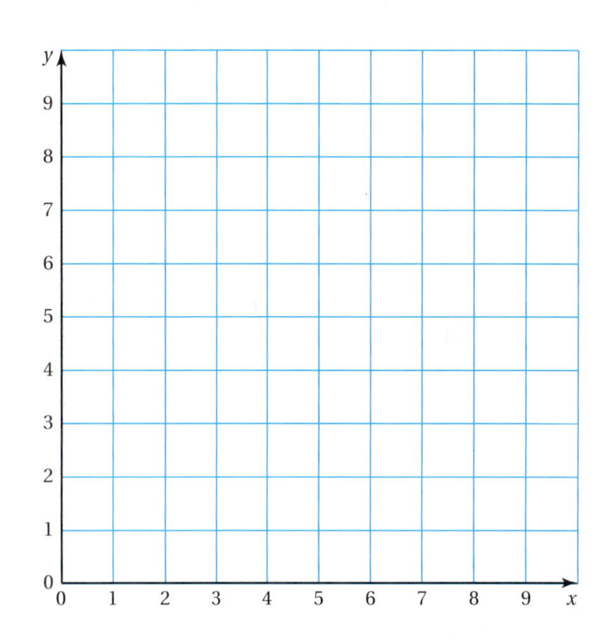

What Is Your Answer?

3. **IN YOUR OWN WORDS** How can you describe the graph of the equation $ax + by = c$?

4. Activities 1 and 2 show two different methods for graphing $ax + by = c$. Describe the two methods. Which method do you prefer? Explain.

5. Write a real-life problem that is similar to those shown in Activities 1 and 2.

6. Why do you think it might be easier to graph $x + y = 10$ without rewriting it in slope-intercept form and then graphing?

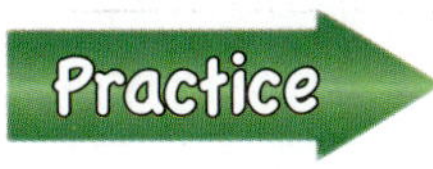

Use what you learned about graphing linear equations in standard form to complete Exercises 3 and 4 on page 176.

4.5 Lesson

Key Vocabulary
standard form, *p. 174*

Study Tip

Any linear equation can be written in standard form.

Standard Form of a Linear Equation

The **standard form** of a linear equation is

$$ax + by = c$$

where a and b are not both zero.

EXAMPLE 1 Graphing a Linear Equation in Standard Form

Graph $-2x + 3y = -6$.

Step 1: Write the equation in slope-intercept form.

$$-2x + 3y = -6 \qquad \text{Write the equation.}$$

$$3y = 2x - 6 \qquad \text{Add } 2x \text{ to each side.}$$

$$y = \frac{2}{3}x - 2 \qquad \text{Divide each side by 3.}$$

Step 2: Use the slope and the y-intercept to graph the equation.

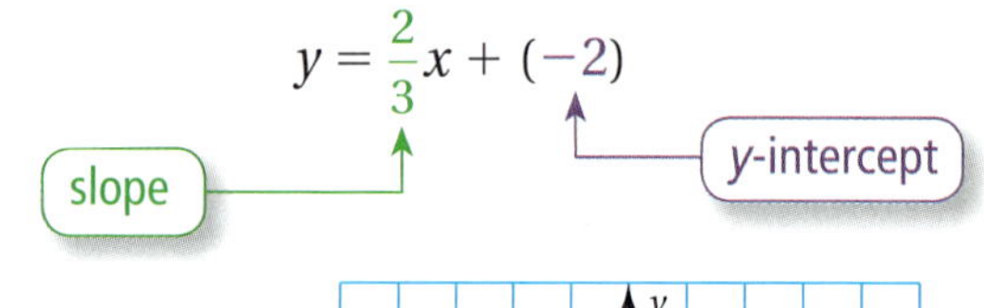

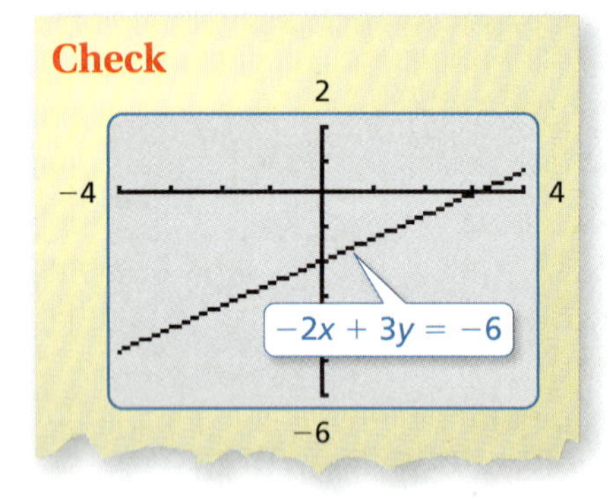

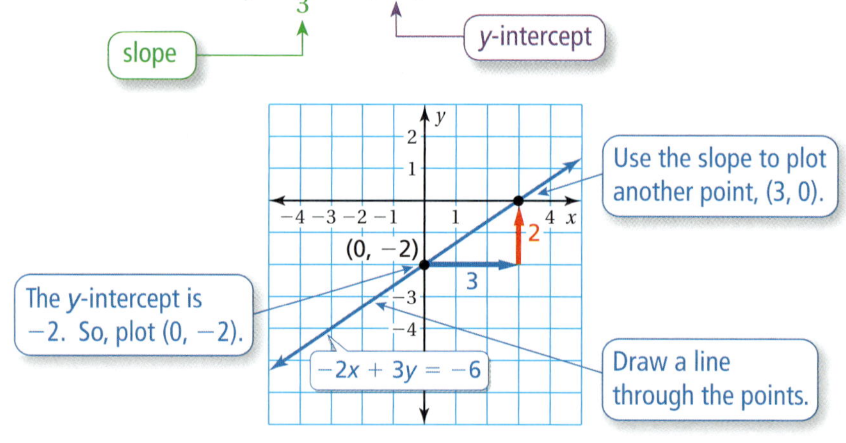

On Your Own

Now You're Ready
Exercises 5–10

Graph the linear equation. Use a graphing calculator to check your graph.

1. $x + y = -2$
2. $-\frac{1}{2}x + 2y = 6$
3. $-\frac{2}{3}x + y = 0$
4. $2x + y = 5$

Laurie's Notes

Introduction

Connect

- **Yesterday:** Students explored the graph of an equation written in standard form. (MP7)
- **Today:** Students will graph equations written in standard form.

Motivate

- ❓"How many pairs of numbers can you think of that have a sum of 5?" Encourage students to write their numbers on paper as ordered pairs. Example: (2, 3)
- ❓"Did any of you include numbers that are not whole numbers?" Check to see if anyone had negative numbers or rational numbers.
- Ask one student to name an x-coordinate and another student to provide the y-coordinate. Plot the ordered pairs in a coordinate plane.
- ❓"What do you think the equation of this line would be?" $x + y = 5$

Lesson Notes

Key Idea

- Define the standard form of a linear equation.
- Students may ask why both a and b cannot be zero. Explain that if $a = 0$ and $b = 0$, you would not have the equation of a line.
- **Teaching Tip:** Students are often confused when the standard form is written with parameters a, b, and c. Students see 5 variables. Show examples of equations written in standard form and identify a, b, and c.
- Ask students why they think $ax + by = c$ is called *standard* form. Students might suggest that the variables are on the left and a constant on the right.

Example 1

- Have students identify a, b, and c. $a = -2$, $b = 3$, and $c = -6$
- ❓"How do you solve for y?" Add $2x$ to each side, then divide both sides by 3.
- Explain that the reason for rewriting the equation in slope-intercept form is so that the slope and the y-intercept can be used to graph the equation.
- **Common Error:** Students only divide one of the two terms on the right side of the equation by 3. Relate this to fraction operations. You are separating the expression into two terms and then simplifying.
- ❓"Now that the equation is in slope-intercept form, explain how to graph the equation." Plot the ordered pair for the y-intercept. To plot another point, start at $(0, -2)$ and move to the right 3 units and up 2 units. Note that you can also move 3 units to the left and down 2 units. Connect these points with a line.
- Substitute the additional ordered pairs into the original equation to verify that they are solutions of the equation.

On Your Own

- In Questions 2 and 3, the fractional coefficients may present a problem.
- Remind students that equations must be solved for y in order to enter them in the equation editor of the graphing calculator.

Goal Today's lesson is graphing a linear equation written in **standard form**.

Lesson Tutorials
Lesson Plans
Answer Presentation Tool

Extra Example 1

Graph $3x - 2y = 2$.

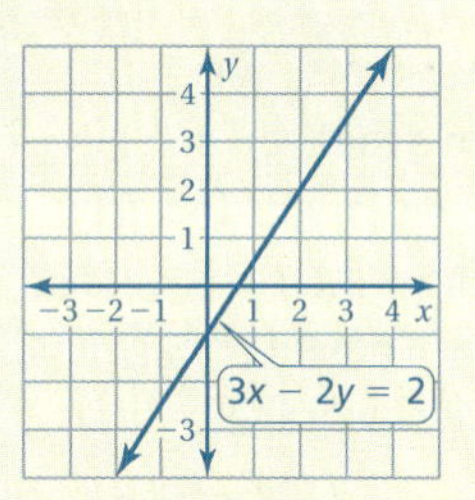

On Your Own

1. $x + y = -2$

2.

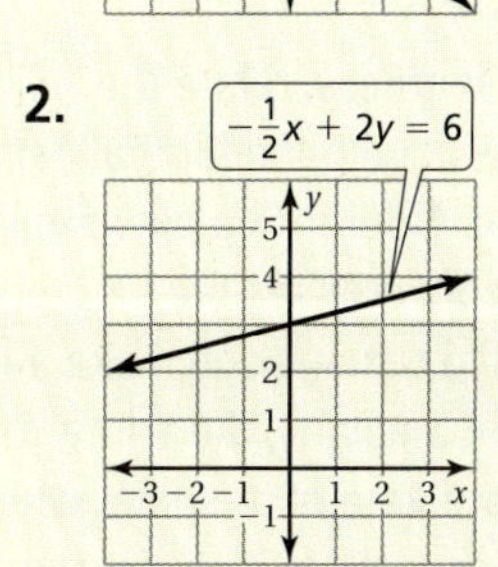

3–4. See Additional Answers.

Extra Example 2

Graph $5x - y = -5$ using intercepts.

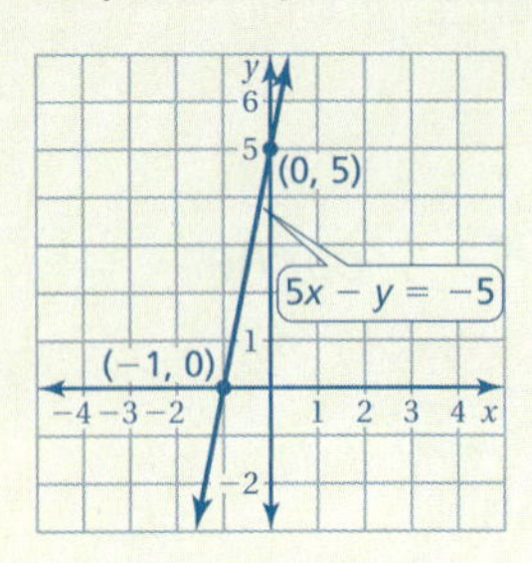

Extra Example 3

You have $2.40 to spend on grapes and bananas.

a. Graph the equation $1.2x + 0.6y = 2.4$, where x is the number of pounds of grapes and y is the number of pounds of bananas.

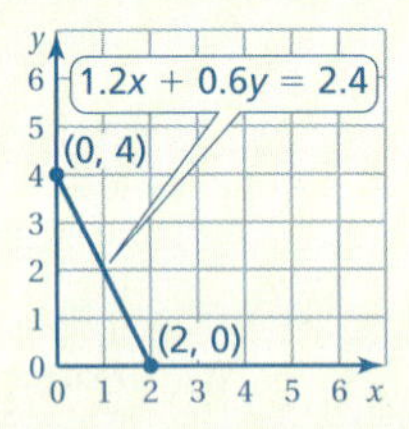

b. Interpret the intercepts. The x-intercept shows that you can buy 2 pounds of grapes, if you do not buy any bananas. The y-intercept shows that you can buy 4 pound of bananas, if you do not buy any grapes.

On Your Own

5–7. See Additional Answers.

English Language Learners

Vocabulary

For English learners, relate the word *intercept* with the football term *interception.* A defensive player on a football team crosses the path of the football to catch it and make an interception. Similarly, the y-intercept is the y-coordinate of the point where the line crosses the y-axis and the x-intercept is the x-coordinate of the point where the line crosses the x-axis.

Laurie's Notes

Example 2

- Start with a simple equation in standard form, such as $x + y = 4$. In this example, $a = 1$, $b = 1$, and $c = 4$. Explain to students that this could be solved for y by subtracting x from each side of the equation. Instead, you want to leave the equation as it was written.
- ? "Another way to think of this equation is *the sum of two numbers is 4.* Can you name some ordered pairs that would satisfy the equation?" Students should give many, including (0, 4) and (4, 0).
- Explain to students that sometimes an equation in standard form is graphed by using the two intercepts, instead of rewriting the equation in slope-intercept form.
- Write the equation shown: $x + 3y = -3$.
- ? "To find the x-intercept, what is the value of y? To find the y-intercept, what is the value of x?" 0; 0
- Finish the problem as shown.
- **Big Idea:** When the equation is in standard form, you can plot the points for the two intercepts and then draw the line through them.

Example 3

- Read the problem. Write the equation $1.5x + 0.6y = 6$ on the board.
- ? "What are the intercepts for this equation?" The x-intercept is 4 and the y-intercept is 10.
- Interpreting the intercepts in part (b) is an important step, particularly for real-life applications.
- Explain to students that negative values of x and y are not included in the graph because it does not make sense to have negative pounds of apples and bananas.
- ? "What is the cost of 2 pounds of apples and 5 pounds of bananas?" $6
- ? **MP1a Make Sense of Problems:** "What can you buy for $6?" *Sample answer:* 4 pounds of apples, or 10 pounds of bananas, or some other combination that is a solution. You are helping students make sense of the problem by asking them to interpret the symbolic representation.

On Your Own

- Students should work in pairs.

Closure

- **Writing Prompt:** To graph the equation $2x + y = 4$ … *Sample answer:* Find and plot the points for the x- and y-intercepts, then draw a line through these two points.

EXAMPLE 2 Graphing a Linear Equation in Standard Form

Graph $x + 3y = -3$ using intercepts.

Step 1: To find the x-intercept, substitute 0 for y.

$$x + 3y = -3$$
$$x + 3(0) = -3$$
$$x = -3$$

To find the y-intercept, substitute 0 for x.

$$x + 3y = -3$$
$$0 + 3y = -3$$
$$y = -1$$

Step 2: Graph the equation.

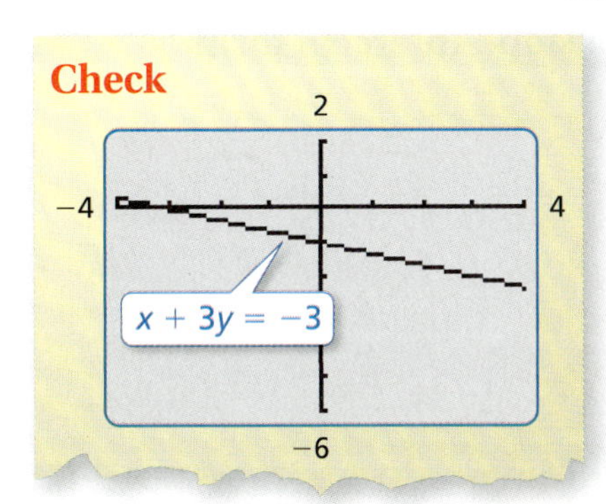

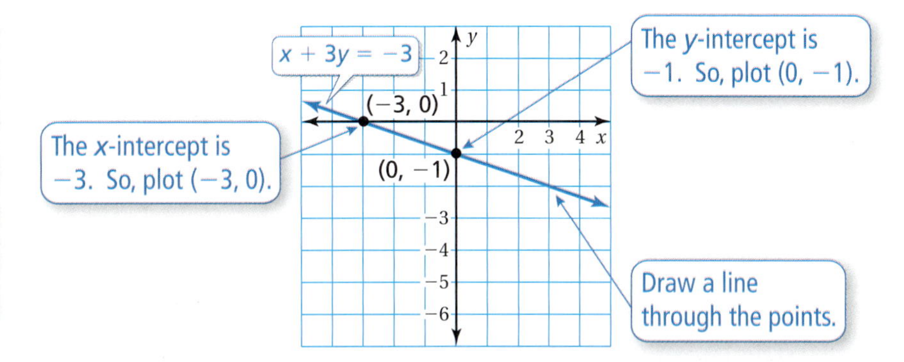

EXAMPLE 3 Real-Life Application

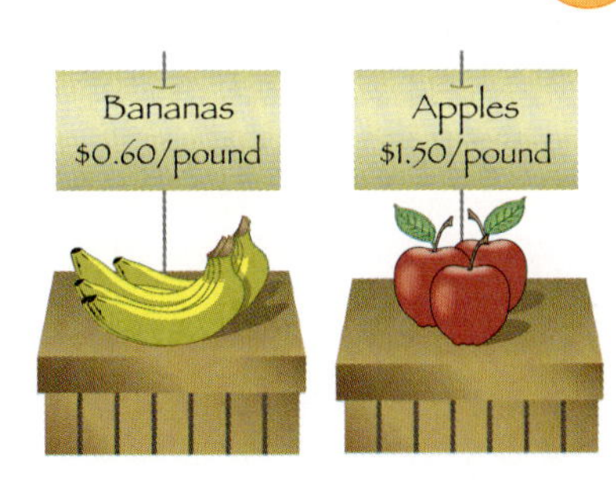

You have $6 to spend on apples and bananas. (a) Graph the equation $1.5x + 0.6y = 6$, where x is the number of pounds of apples and y is the number of pounds of bananas. (b) Interpret the intercepts.

a. Find the intercepts and graph the equation.

x-intercept	**y-intercept**
$1.5x + 0.6y = 6$	$1.5x + 0.6y = 6$
$1.5x + 0.6(0) = 6$	$1.5(0) + 0.6y = 6$
$x = 4$	$y = 10$

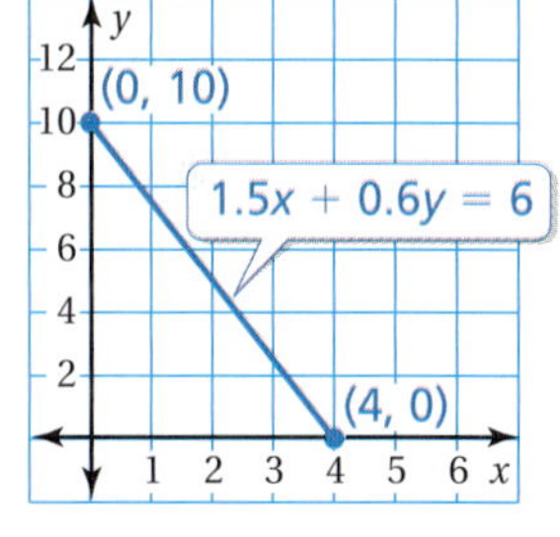

b. The x-intercept shows that you can buy 4 pounds of apples when you do not buy any bananas. The y-intercept shows that you can buy 10 pounds of bananas when you do not buy any apples.

On Your Own

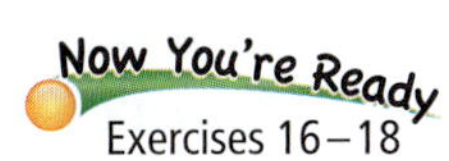

Graph the linear equation using intercepts. Use a graphing calculator to check your graph.

5. $2x - y = 8$
6. $x + 3y = 6$
7. **WHAT IF?** In Example 3, you buy y pounds of oranges instead of bananas. Oranges cost $1.20 per pound. Graph the equation $1.5x + 1.2y = 6$. Interpret the intercepts.

4.5 Exercises

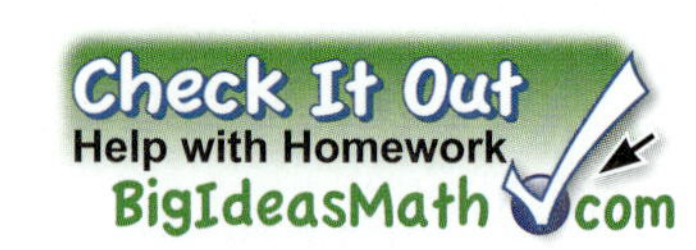

Vocabulary and Concept Check

1. **VOCABULARY** Is the equation $y = -2x + 5$ in standard form? Explain.
2. **WRITING** Describe two ways to graph the equation $4x + 2y = 6$.

Practice and Problem Solving

Define two variables for the verbal model. Write an equation in slope-intercept form that relates the variables. Graph the equation.

3. $\dfrac{\$2.00}{\text{pound}} \cdot$ Pounds of peaches $+ \dfrac{\$1.50}{\text{pound}} \cdot$ Pounds of apples $= \$15$

4. $\dfrac{16 \text{ miles}}{\text{hour}} \cdot$ Hours biked $+ \dfrac{2 \text{ miles}}{\text{hour}} \cdot$ Hours walked $=$ 32 miles

Write the linear equation in slope-intercept form.

5. $2x + y = 17$
6. $5x - y = \dfrac{1}{4}$
7. $-\dfrac{1}{2}x + y = 10$

Graph the linear equation. Use a graphing calculator to check your graph.

8. $-18x + 9y = 72$
9. $16x - 4y = 2$
10. $\dfrac{1}{4}x + \dfrac{3}{4}y = 1$

Match the equation with its graph.

11. $15x - 12y = 60$
12. $5x + 4y = 20$
13. $10x + 8y = -40$

A. 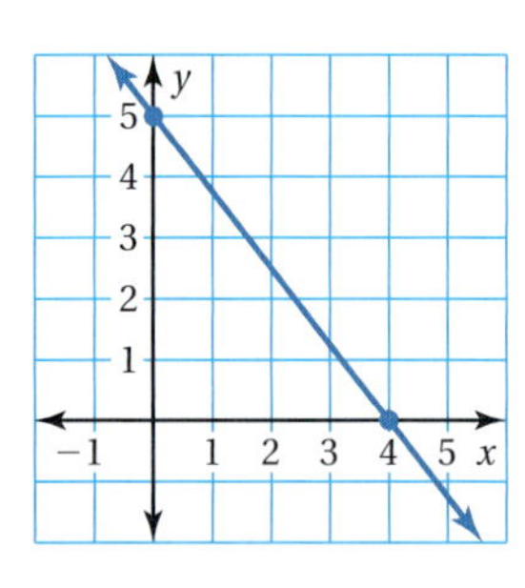

B. 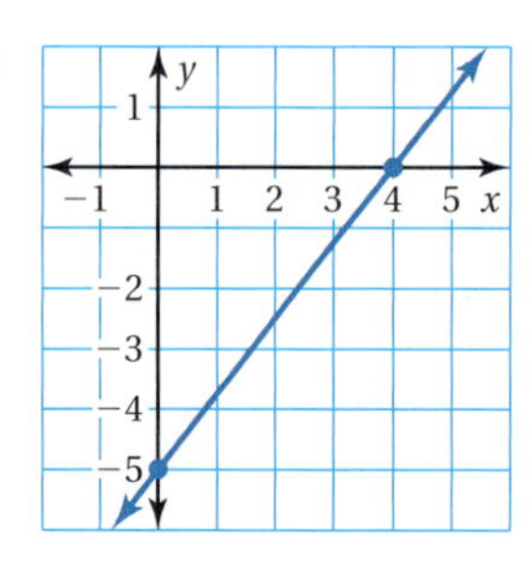

C. (graph: line through $(-4, 0)$ and $(0, -5)$)

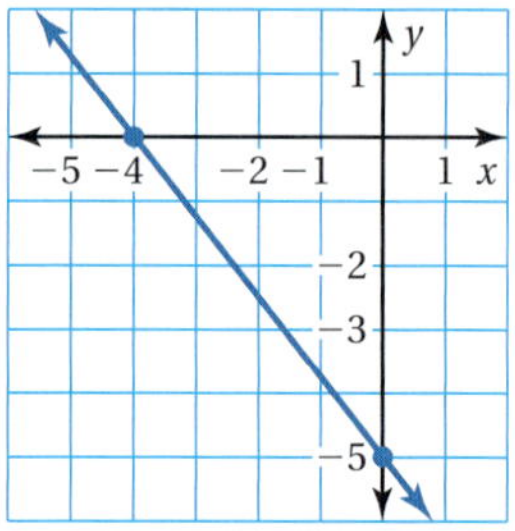

14. **ERROR ANALYSIS** Describe and correct the error in finding the x-intercept.

$$-2x + 3y = 12$$
$$-2(0) + 3y = 12$$
$$3y = 12$$
$$y = 4$$

15. **BRACELET** A charm bracelet costs \$65, plus \$25 for each charm. The equation $-25x + y = 65$ represents the cost y of the bracelet, where x is the number of charms.

    a. Graph the equation.

    b. How much does the bracelet shown cost?

Assignment Guide and Homework Check

Level	Day 1 Activity Assignment	Day 2 Lesson Assignment	Homework Check
Basic	3, 4, 24–26	1, 2, 5–13 odd, 14, 15–21 odd	7, 9, 15, 17, 19
Average	3, 4, 24–26	1, 2, 6–14 even, 15–23 odd	10, 15, 17, 19, 21
Advanced	1–4, 6–14 even, 16–26		10, 18, 19, 20, 22

Common Errors

- **Exercises 5–10** Students may use the same operation instead of the opposite operation when rewriting the equation in slope-intercept form.
- **Exercises 11–13, 16–18** Students may mix up the *x*- and *y*-intercepts. Remind them that the *x*-intercept is the *x*-coordinate of where the line crosses the *x*-axis and the *y*-intercept is the *y*-coordinate of where the line crosses the *y*-axis.

4.5 Record and Practice Journal

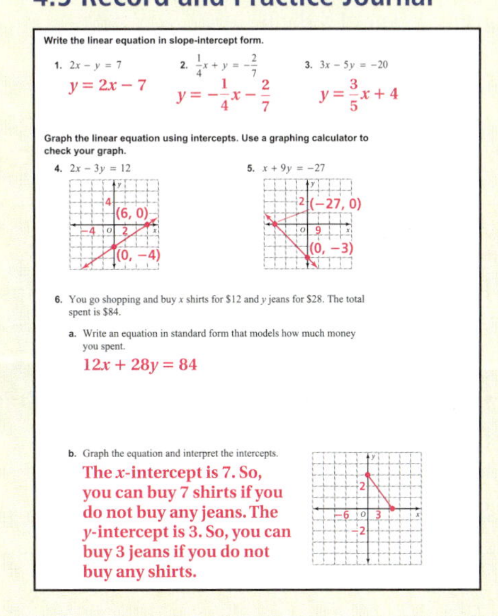

Write the linear equation in slope-intercept form.

1. $2x - y = 7$ $\quad y = 2x - 7$
2. $\frac{1}{4}x + y = -\frac{2}{7}$ $\quad y = -\frac{1}{4}x - \frac{2}{7}$
3. $3x - 5y = -20$ $\quad y = \frac{3}{5}x + 4$

Graph the linear equation using intercepts. Use a graphing calculator to check your graph.

4. $2x - 3y = 12$
5. $x + 9y = -27$

6. You go shopping and buy x shirts for \$12 and y jeans for \$28. The total spent is \$84.

   a. Write an equation in standard form that models how much money you spent.

   $12x + 28y = 84$

   b. Graph the equation and interpret the intercepts.

   The x-intercept is 7. So, you can buy 7 shirts if you do not buy any jeans. The y-intercept is 3. So, you can buy 3 jeans if you do not buy any shirts.

Vocabulary and Concept Check

1. no; The equation is in slope-intercept form.
2. *Sample answer:*
   1) Write the equation in slope-intercept form and use the slope and *y*-intercept to graph the equation.
   2) Find the *x*- and *y*-intercepts, plot the points representing the intercepts, and draw a line through the points.

Practice and Problem Solving

3. x = pounds of peaches
   y = pounds of apples
   $y = -\frac{4}{3}x + 10$

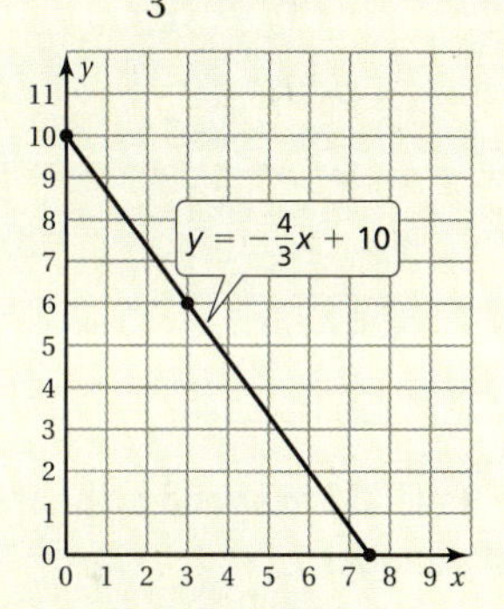

4. x = hours biked
   y = hours walked
   $y = -8x + 16$

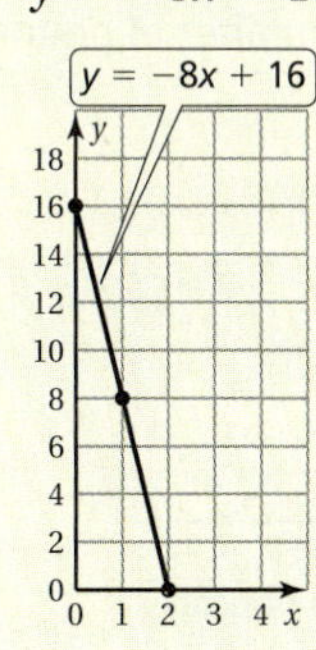

5. $y = -2x + 17$
6. $y = 5x - \frac{1}{4}$
7. $y = \frac{1}{2}x + 10$

8–15. See Additional Answers.

Practice and Problem Solving

16–19. See Additional Answers.

20. See *Taking Math Deeper.*

21–23. See Additional Answers.

Fair Game Review

24. 4 **25.** $\frac{1}{2}$

26. D

Mini-Assessment

1. Graph $-2x + 4y = 16$ using intercepts.

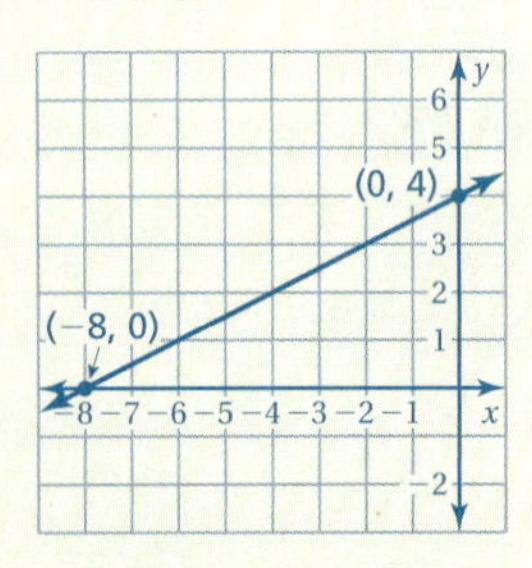

2. You have $12 to spend on pears and oranges.

a. Graph the equation $1.2x + 0.8y = 12$, where x is the number of pounds of pears and y is the number of pounds of oranges.

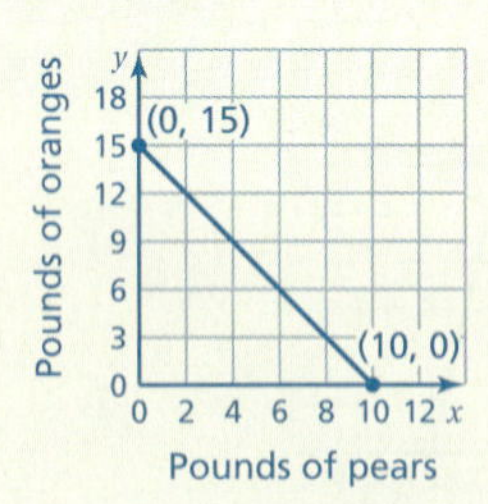

b. Interpret the intercepts.
The x-intercept shows that you can buy 10 pounds of pears if you do not buy any oranges. The y-intercept shows that you can buy 15 pounds of oranges if you do not buy any pears.

Taking Math Deeper

Exercise 20

As with many real-life problems, it helps to start by summarizing the given information.

 Summarize the given information.

- Let x = days for renting boat.
- Let y = days for renting scuba gear.
- Cost of boat = $250 per day.
- Cost of scuba gear = $50 per day.
- Total spent = $1000.

 **a.** Write an equation.

$$250x + 50y = 1000$$

 **b.** Graph the equation and interpret the intercepts.

$$y = -5x + 20$$

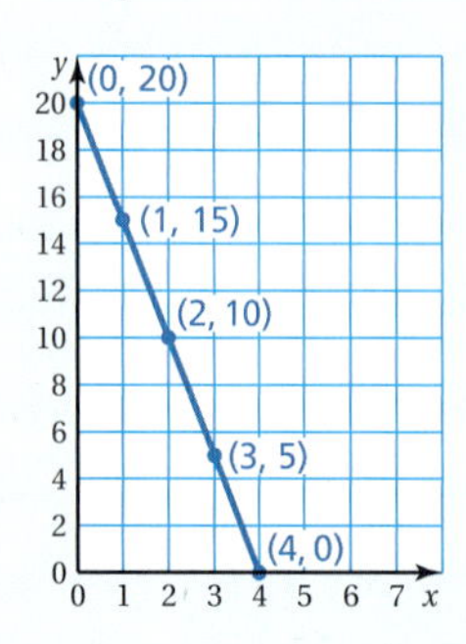

If $x = 0$, then the group rented only the scuba gear for 20 days.
If $y = 0$, then the group rented only the boat for 4 days.

Project

To go on a professional scuba diving tour, you need to be a certified diver. Use the school library or the Internet to research the requirements to become certified in scuba diving.

Reteaching and Enrichment Strategies

If students need help. . .	If students got it. . .
Resources by Chapter • Practice A and Practice B • Puzzle Time Record and Practice Journal Practice Differentiating the Lesson Lesson Tutorials Skills Review Handbook	Resources by Chapter • Enrichment and Extension • Technology Connection Start the next section

Graph the linear equation using intercepts. Use a graphing calculator to check your graph.

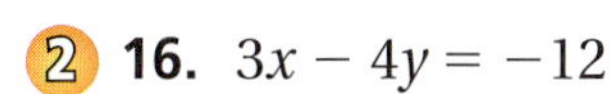

16. $3x - 4y = -12$ **17.** $2x + y = 8$ **18.** $\frac{1}{3}x - \frac{1}{6}y = -\frac{2}{3}$

19. SHOPPING The amount of money you spend on x CDs and y DVDs is given by the equation $14x + 18y = 126$. Find the intercepts and graph the equation.

20. SCUBA Five friends go scuba diving. They rent a boat for x days and scuba gear for y days. The total spent is $1000.

a. Write an equation in standard form that represents the situation.

b. Graph the equation and interpret the intercepts.

21. MODELING You work at a restaurant as a host and a server. You earn $9.45 for each hour you work as a host and $7.65 for each hour you work as a server.

a. Write an equation in standard form that models your earnings.

b. Graph the equation.

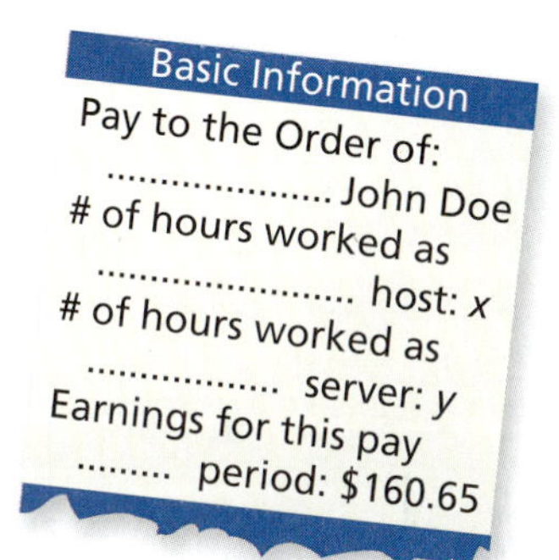

22. LOGIC Does the graph of every linear equation have an x-intercept? Explain your reasoning. Include an example.

23. Critical Thinking For a house call, a veterinarian charges $70, plus $40 an hour.

a. Write an equation that represents the total fee y (in dollars) the veterinarian charges for a visit lasting x hours.

b. Find the x-intercept. Does this value make sense in this context? Explain your reasoning.

c. Graph the equation.

Fair Game Review What you learned in previous grades & lessons

The points in the table lie on a line. Find the slope of the line. *(Section 4.2)*

24.

x	-2	-1	0	1
y	-10	-6	-2	2

25.

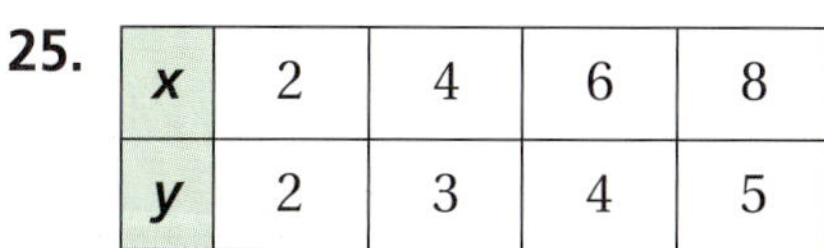

x	2	4	6	8
y	2	3	4	5

26. MULTIPLE CHOICE Which value of x makes the equation $4x - 12 = 3x - 9$ true? *(Section 1.3)*

Ⓐ -1 Ⓑ 0 Ⓒ 1 Ⓓ 3

4.6 Writing Equations in Slope-Intercept Form

Essential Question How can you write an equation of a line when you are given the slope and the y-intercept of the line?

1 ACTIVITY: Writing Equations of Lines

Work with a partner.

- **Find the slope of each line.**
- **Find the y-intercept of each line.**
- **Write an equation for each line.**
- **What do the three lines have in common?**

a.

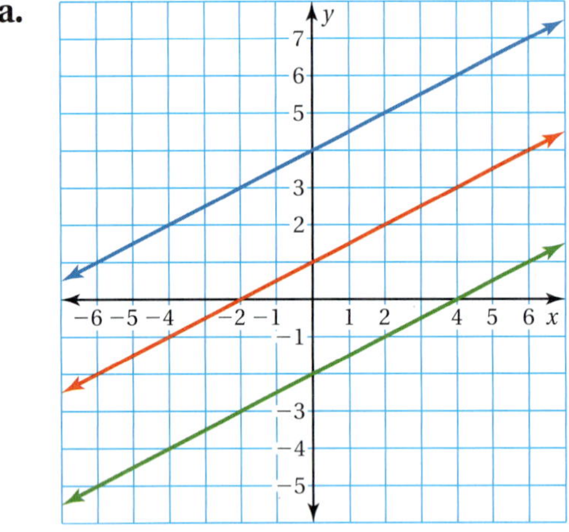

b.

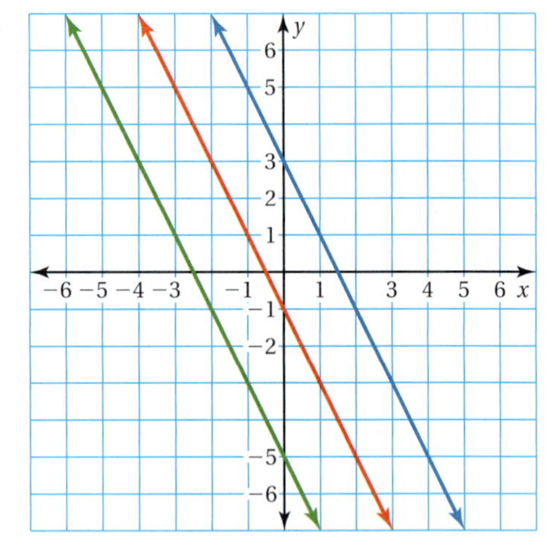

c.

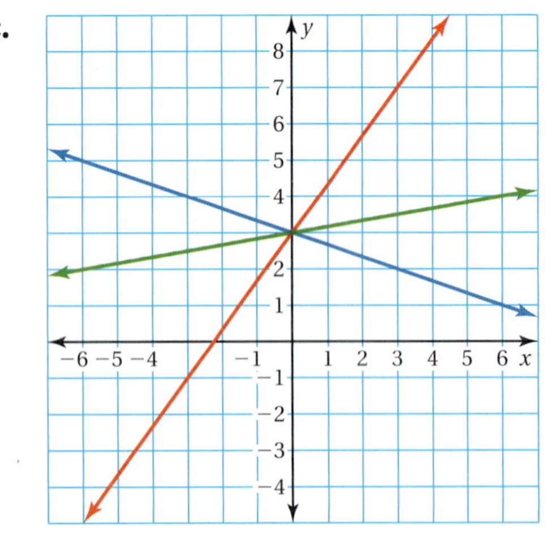

d.

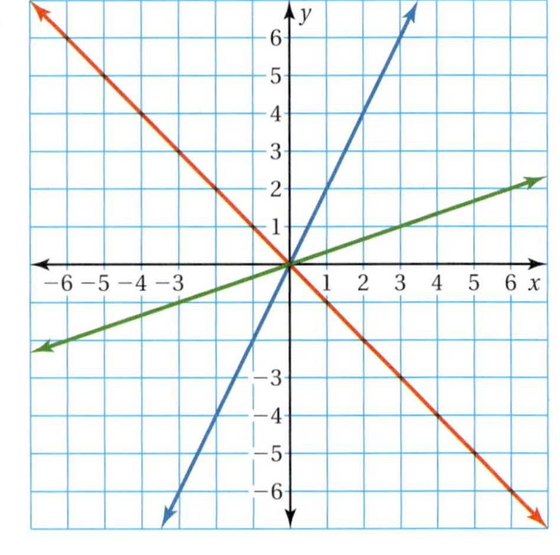

COMMON CORE

Writing Equations

In this lesson, you will

- write equations of lines in slope-intercept form.

Preparing for Standard MACC.8.F.2.4

Laurie's Notes

Introduction

Standards for Mathematical Practice

- **MP1 Make Sense of Problems and Persevere in Solving Them** and **MP4 Model with Mathematics:** The goal of this lesson is for students to write equations of lines by first determining the slope and *y*-intercept from a graph. The visual model helps students make sense of and solve the problem. The graphical representation (model) helps students identify the important features of the line.

Motivate

- If there is sufficient space in your classroom, hallway, or school foyer, make coordinate axes using masking tape. Use a marker to scale each axis with integers -5 through 5.
- Take turns having two students be the *rope anchors* who then will make a line on the coordinate axes while other students observe.
- Here are a series of directions you can give and some follow-up questions. Remind students that slope is rise over run and that the equation of a line in slope-intercept form is $y = mx + b$.
  - ? Make the line $y = x$. "What is the slope?" 1 "What is the *y*-intercept?" 0
  - ? Keep the same slope, but make the *y*-intercept 2. "What is the equation of this line?" $y = x + 2$
  - ? Use the *y*-intercept 2, but make the slope steeper. "What is the slope of this line?" Answers will vary.
  - ? Keep the same *y*-intercept, but make the slope $\frac{1}{2}$. "What is the equation?" $y = \frac{1}{2}x + 2$
- **Management Tip:** This activity can also be done by drawing the axes on the board and having the students hold the rope against the board.

Activity Notes

Activity 1

- ? "How do you determine the slope of a line drawn in a coordinate plane?" Use two points that you are sure are on the graph and find the rise and run between the points.
- ? "Does it matter whether you move left-to-right or right-to-left when you are finding the rise and run? Explain." no; Either way the slope will be the same.
- Students may have difficulty writing the equation in slope-intercept form. They think it should be harder to do!
- **FYI:** When the *y*-intercept is negative, students may leave their equation as $y = 3x + (-4)$ instead of $y = 3x - 4$. Remind students that it is more common to represent the equation as $y = 3x - 4$.
- **Teaching Tip:** If you have a student who is color blind, refer to the lines by a number or letter scheme (1, 2, 3 or A, B, C).
- Ask students to share what they found in common for each trio of lines.

Florida Common Core Standards

MACC.8.F.2.4 Construct a function to model a linear relationship between two quantities. Determine the rate of change and initial value of the function from a description of a relationship or from two (x, y) values, including reading these from a table or from a graph. Interpret the rate of change and initial value of a linear function in terms of the situation it models, and in terms of its graph or a table of values.

Previous Learning

Students should know how to find the slope of a line. Students should know about parallel lines.

4.6 Record and Practice Journal

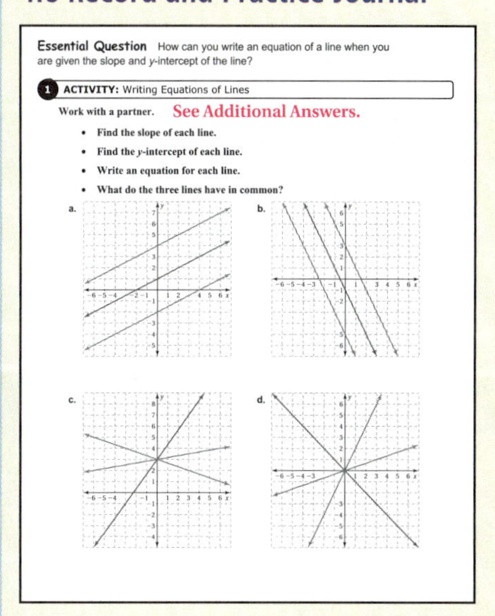

Essential Question How can you write an equation of a line when you are given the slope and *y*-intercept of the line?

1 ACTIVITY: Writing Equations of Lines

Work with a partner. See Additional Answers.

- **Find the slope of each line.**
- **Find the *y*-intercept of each line.**
- **Write an equation for each line.**
- **What do the three lines have in common?**

a. b. c. d.

Differentiated Instruction

Visual

To avoid mistakes when substituting the variables, have students color code the slope and y-intercept of an equation.

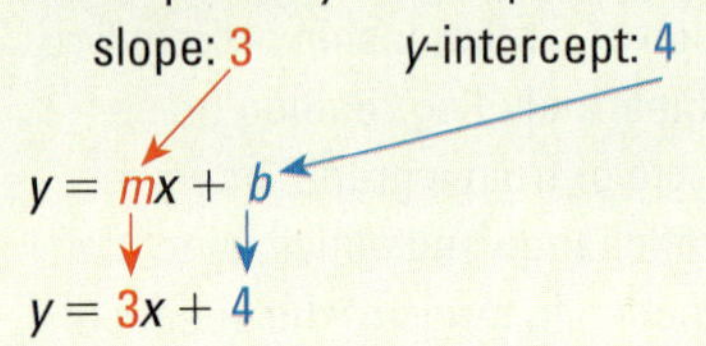

4.6 Record and Practice Journal

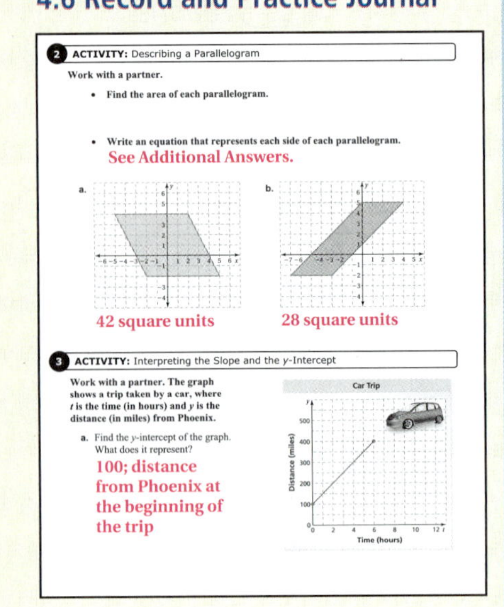

2 ACTIVITY: Describing a Parallelogram

Work with a partner.

- Find the area of each parallelogram.
- Write an equation that represents each side of each parallelogram.

See Additional Answers.

a. 42 square units

b. 28 square units

3 ACTIVITY: Interpreting the Slope and the y-Intercept

Work with a partner. The graph shows a trip taken by a car, where t is the time (in hours) and y is the distance (in miles) from Phoenix.

a. Find the y-intercept of the graph. What does it represent?

100; distance from Phoenix at the beginning of the trip

b. Find the slope of the graph. What does it represent?

50; the speed of the car in miles per hour

c. How long did the trip last?

6 hours

d. How far from Phoenix was the car at the end of the trip?

400 mi

e. Write an equation that represents the graph.

$y = 50x + 100$

What Is Your Answer?

4. IN YOUR OWN WORDS How can you write an equation of a line when you are given the slope and the y-intercept of the line? Give an example that is different from those in Activities 1, 2, and 3.

$y = mx + b$ with m as the slope and b as the y-intercept

5. Two sides of a parallelogram are represented by the equations $y = 2x + 1$ and $y = -x + 3$. Give two equations that can represent the other two sides.

Sample answer: $y = 2x + 5$ and $y = -x + 7$

Laurie's Notes

Activity 2

- ? "How do you find the area of a parallelogram?" area = base × height
- ? "Are the base and height the sides of the parallelogram?" They could be if it's a rectangle. Otherwise, height is the perpendicular distance between the two bases.
- **MP1:** This is a good example of where students have the necessary skills but they will need to make sense of the problem and then persevere in working through the problem.
- To find the base and height, students simply count the units on the diagram. Note that the height for the parallelogram in part (b) is outside the parallelogram.
- **Common Error:** The slope of the horizontal sides is zero. Students may say that you cannot find the slope *for a flat line*.
- ? "What is the equation of a horizontal line?" $y = b$
- The challenge in this activity is writing the equations for the diagonal sides of the figure in part (a). Suggest that by extending the sides using the slope, the students should be able to determine the y-intercept.
- This activity reviews positive, negative, and zero slope. Area of a parallelogram is also reviewed.

Activity 3

- The graph in this problem represents a real-life context. Mathematically proficient students are able to interpret the mathematical results in the context of the situation or problem.
- If students have difficulty getting started with this activity, remind them to read the labels on the axes and interpret the y-intercept. The car was 100 miles from Phoenix at the beginning of the trip.
- Discuss answers to each part of the problem as a class.
- ? **Extension:** Draw the segment from (6, 400) to (12, 0) and explain that this represents the return trip. Ask the following questions.
  - "What is the slope of this line segment? What does the slope mean in the context of the problem?" slope ≈ -67; returning at a rate of about 67 mi/h
  - "What does the point (12, 0) mean in the context of the problem?" You have arrived in Phoenix and drove 12 hours.
  - "What would the graph look like if the car had stopped for 1 hour?" horizontal segment of length 1 unit

What Is Your Answer?

- Discuss answers to Question 5 as a class.

Closure

- **Exit Ticket:** What is the slope and y-intercept of the equation $y = 2x + 4$? slope = 2, y-intercept = 4 Write an equation of a line with a slope of 3 and a y-intercept of 1. $y = 3x + 1$

2 ACTIVITY: Describing a Parallelogram

Math Practice 1
Analyze Givens
What do you need to know to write an equation?

Work with a partner.

- **Find the area of each parallelogram.**
- **Write an equation that represents each side of each parallelogram.**

a.

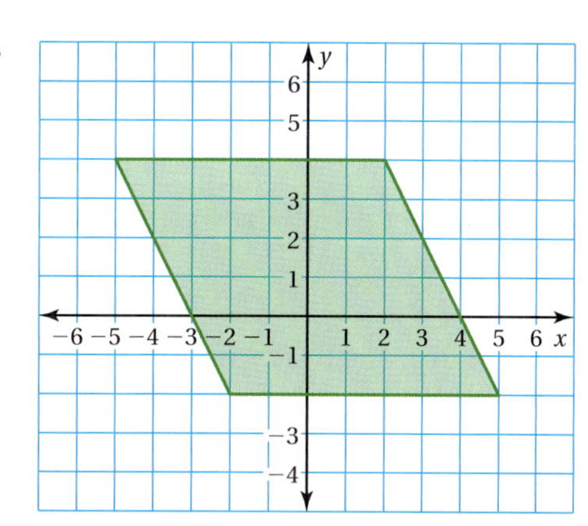

b.

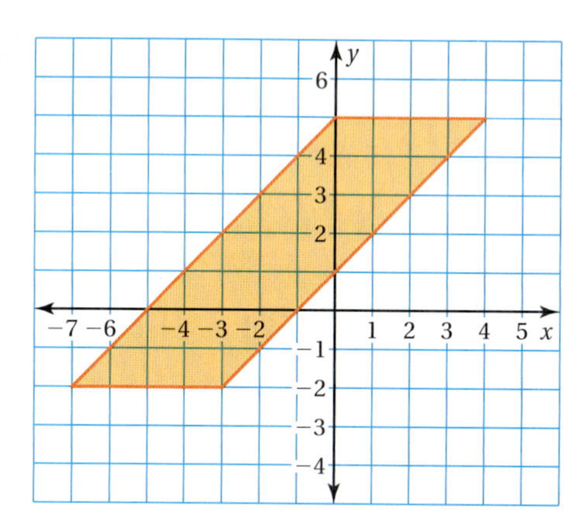

3 ACTIVITY: Interpreting the Slope and the *y*-Intercept

Work with a partner. The graph shows a trip taken by a car, where *t* is the time (in hours) and *y* is the distance (in miles) from Phoenix.

a. Find the y-intercept of the graph. What does it represent?

b. Find the slope of the graph. What does it represent?

c. How long did the trip last?

d. How far from Phoenix was the car at the end of the trip?

e. Write an equation that represents the graph.

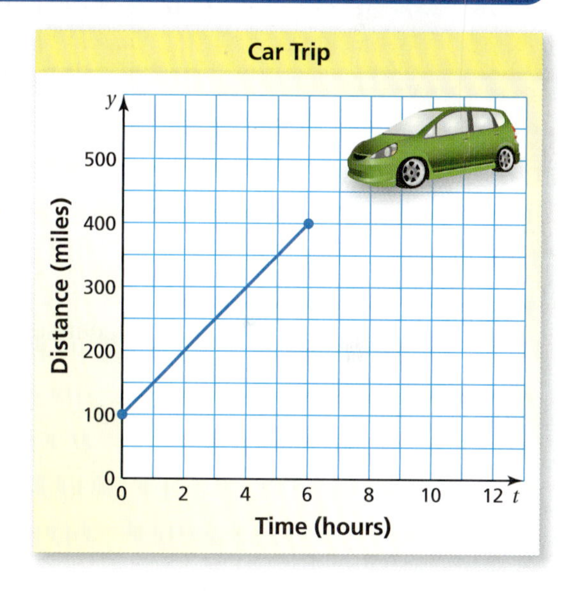

What Is Your Answer?

4. IN YOUR OWN WORDS How can you write an equation of a line when you are given the slope and the y-intercept of the line? Give an example that is different from those in Activities 1, 2, and 3.

5. Two sides of a parallelogram are represented by the equations $y = 2x + 1$ and $y = -x + 3$. Give two equations that can represent the other two sides.

Practice Use what you learned about writing equations in slope-intercept form to complete Exercises 3 and 4 on page 182.

4.6 Lesson

EXAMPLE 1 Writing Equations in Slope-Intercept Form

Write an equation of the line in slope-intercept form.

a.

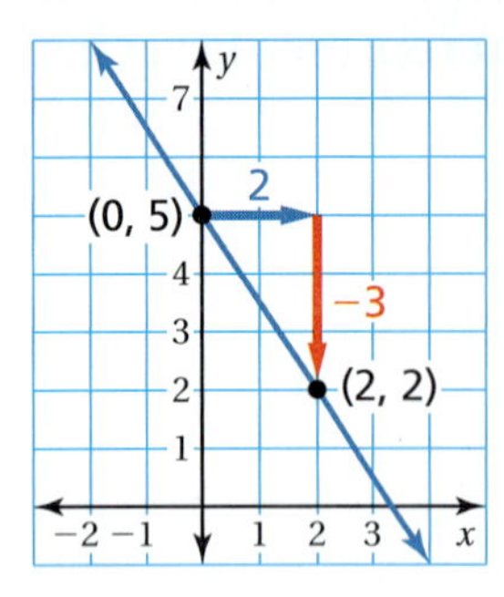

Find the slope and the y-intercept.

$$m = \frac{y_2 - y_1}{x_2 - x_1}$$

$$= \frac{2 - 5}{2 - 0}$$

$$= \frac{-3}{2}, \text{ or } -\frac{3}{2}$$

Because the line crosses the y-axis at (0, 5), the y-intercept is 5.

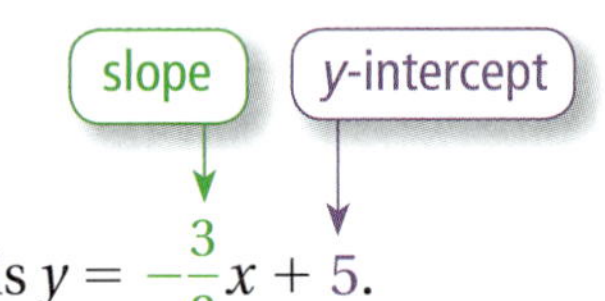

So, the equation is $y = -\frac{3}{2}x + 5$.

Study Tip

After writing an equation, check that the given points are solutions of the equation.

b.

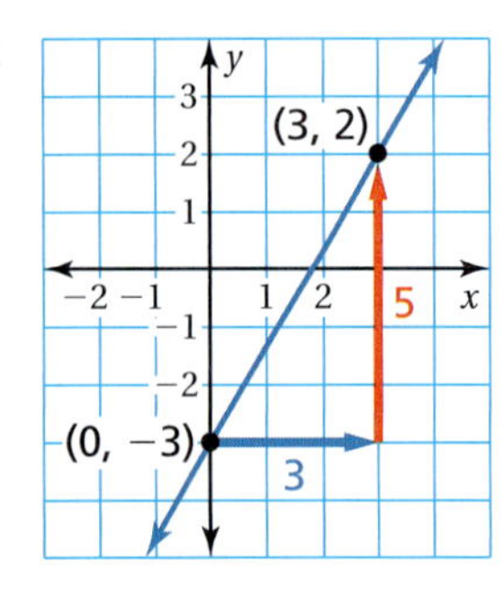

Find the slope and the y-intercept.

$$m = \frac{y_2 - y_1}{x_2 - x_1}$$

$$= \frac{-3 - 2}{0 - 3}$$

$$= \frac{-5}{-3}, \text{ or } \frac{5}{3}$$

Because the line crosses the y-axis at (0, −3), the y-intercept is −3.

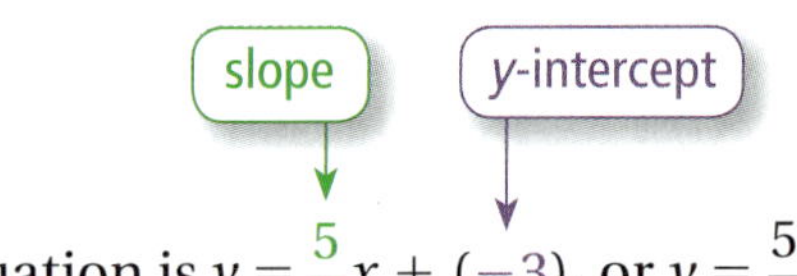

So, the equation is $y = \frac{5}{3}x + (-3)$, or $y = \frac{5}{3}x - 3$.

On Your Own

Now You're Ready
Exercises 5–10

Write an equation of the line in slope-intercept form.

1.

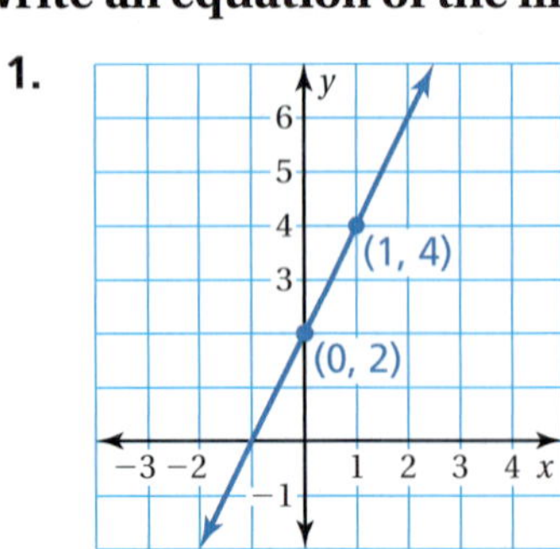

2.

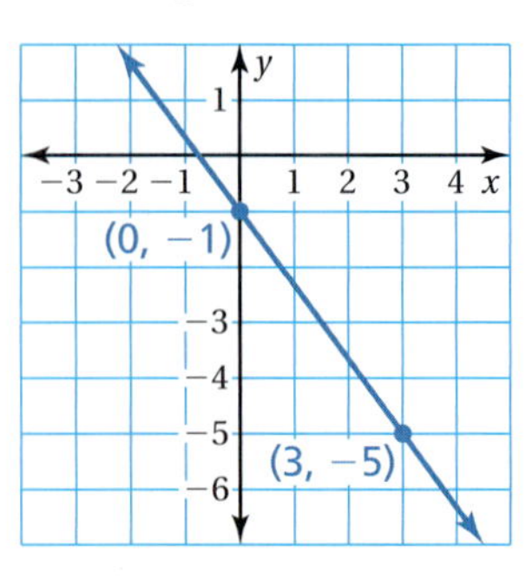

Laurie's Notes

Introduction

Connect

- **Yesterday:** Students developed an understanding of how to write an equation of a line using its slope and y-intercept. (MP1, MP4)
- **Today:** Students will write an equation of a line using the slope and y-intercept.

Motivate

- **Story Time:** Tell students that as a child you loved to dig tunnels in the sand. Ask if any of them like to dig tunnels or if they have traveled through tunnels. Hold a paper towel tube or other similar model to pique student interest. Share some facts about tunnels.
  - The world's longest overland tunnel is a 21-mile-long rail link under the Alps in Switzerland. The tunnel took eight years to build and cost $3.5 billion. It reduces the time trains need to cross between Germany and Italy from 3.5 hours to just under 2 hours.
  - The world's longest underwater tunnel is the Seikan Tunnel in Japan. It is about 33.5 miles long and runs under the Tsugaru Strait. It opened in 1988 and took 17 years to construct.
  - The Channel Tunnel (Chunnel) connects England and France. It is 31 miles long and travels under the English Channel.

Lesson Notes

Discuss

- **MP1a Make Sense of Problems** and **MP4 Model with Mathematics:** In this lesson students will make quick visual inspection of linear graphs to approximate the slope and the y-intercept. This approximation is a helpful check when the slope and y-intercept are computed.

Example 1

- Write the slope-intercept form of an equation, $y = mx + b$. Review with students that the coefficient of the x-term is the slope, and the constant b is the y-intercept. Also, review how to compute slope.
- ? "What do you know about the slope of the line in part (a) by inspection? Explain." Slope is negative because the graph falls from left to right.
- ? "In part (a), what are the coordinates of the point where the line crosses the y-axis?" (0, 5)
- Use the slope and the y-intercept to write the equation.
- Work through part (b). Remind students that you want the more simplified equation $y = \frac{5}{3}x - 3$ instead of $y = \frac{5}{3}x + (-3)$. Stress that while both forms are correct, the simplified version is preferred.

On Your Own

- Before students begin these two problems, they should do a visual inspection. They should make a note of the sign of the slope and y-intercept. It is very easy to have the wrong sign(s) when the equation is written.
- Encourage students to draw the slope triangle and label the horizontal and vertical lengths.

Goal Today's lesson is writing an equation of a line in slope-intercept form.

Lesson Tutorials
Lesson Plans
Answer Presentation Tool

Extra Example 1

Write an equation of the line in slope-intercept form.

a.

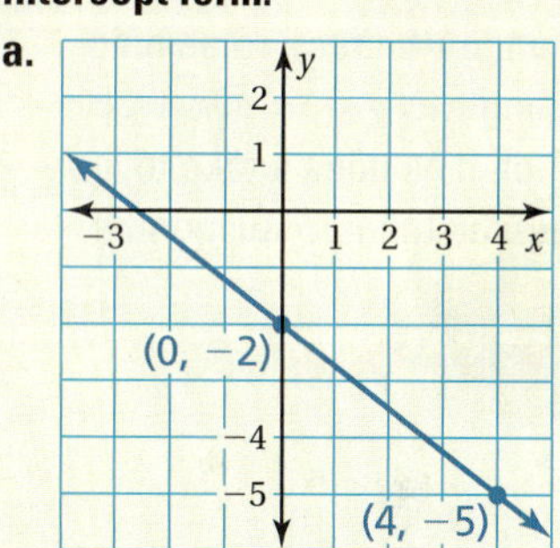

$y = -\frac{3}{4}x - 2$

b.

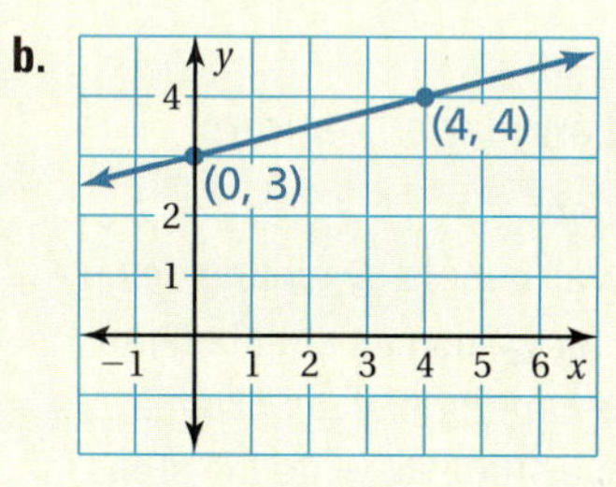

$y = \frac{1}{4}x + 3$

On Your Own

1. $y = 2x + 2$
2. $y = -\frac{4}{3}x - 1$

Laurie's Notes

Extra Example 2

Write an equation of the line that passes through the points (0, -1) and (4, -1). $y = -1$

Extra Example 3

In Example 3, the points are (0, 3500) and (5, 1750).

a. Write an equation that represents the distance y (in feet) remaining after x months. $y = -350x + 3500$

b. How much time does it take to complete the tunnel? 10 months

On Your Own

3. $y = 5$

4. $8\frac{3}{4}$ mo

English Language Learners

Organization

Students will benefit by writing down the steps for writing an equation in slope-intercept form when given a graph. Have students write the steps in their notebooks. A poster with the steps could be posted in the classroom.

Step 1: Write the slope-intercept form of an equation.

Step 2: Determine the slope of the line.

Step 3: Determine the y-intercept of the line.

Step 4: Write the equation in slope-intercept form.

Example 2

- Make a quick sketch of the graph to reference as you work the problem.
- When finding the slope, students are unsure of how to simplify $\frac{0}{3}$. This is a good time to review the difference between $\frac{0}{3}$ and $\frac{3}{0}$.
- **Teaching Tip:** To explain why $\frac{3}{0}$ is undefined, first write the problem $8 \div 4 = 2$ on the board. Then rewrite it as $4\overline{)8}$. To check, multiply the quotient (2) times the divisor (4) and you get the dividend (8). In other words, 2 multiplied by 4 is 8. Do the same thing with $\frac{3}{0}$. Rewrite it using long division, $0\overline{)3}$. What do you multiply 0 by to get 3? There is no quotient, so you say $\frac{3}{0}$ is undefined. You cannot divide by 0.
- **MP7 Look for and Make Use of Structure:** Students don't always recognize that $y = -4$ is a linear equation written in slope-intercept form. It helps to write the extra step of $y = (0)x + (-4)$ so students can see that the slope is 0. Students should recognize that $y = -4$ and $y = (0)x + (-4)$ are equivalent.

Example 3

- Ask a volunteer to read the problem. Discuss information that can be *read* from the graph.
- ? "By visual inspection, what do you know about the sign of the slope and the y-intercept in this problem?" The slope is negative. The y-intercept is positive.
- ? "What does a slope of -500 mean in the context of this problem?" A slope of -500 means that for each additional month of work, the distance left to complete is 500 feet less.
- The x-intercept for this graph is 7.
- Note that the graph is in Quadrant I. In the context of this problem, it doesn't make sense for time or distance to be negative.

On Your Own

- **MP6 Attend to Precision:** For Question 3, encourage students to sketch a graph of the line through the two points to give them a clue as to how to begin. The visual model is an approximation that can be used to check their final answer. This technique will help students start Question 4.

Closure

- **Writing Prompt:** For a line that has been graphed in a coordinate plane, you can write the equation by … finding the slope and y-intercept

EXAMPLE 2 Writing an Equation

Which equation is shown in the graph?

Ⓐ $y = -4$ Ⓑ $y = -3$

Ⓒ $y = 0$ Ⓓ $y = -3x$

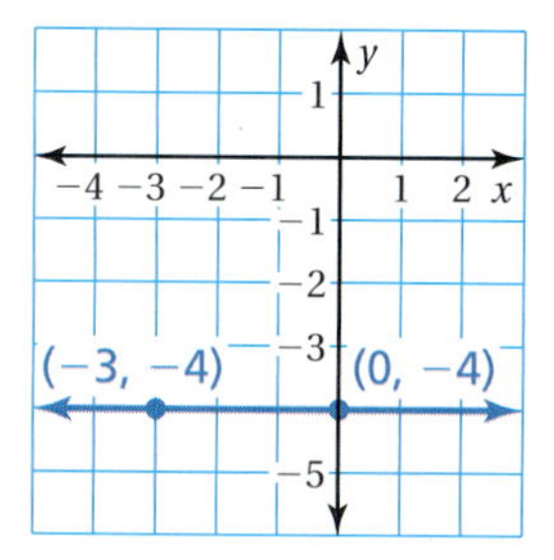

> **Remember**
>
> The graph of $y = a$ is a horizontal line that passes through $(0, a)$.

Find the slope and the y-intercept.

The line is horizontal, so the change in y is 0.

$$m = \frac{\text{change in } y}{\text{change in } x} = \frac{0}{3} = 0$$

Because the line crosses the y-axis at $(0, -4)$, the y-intercept is -4.

So, the equation is $y = 0x + (-4)$, or $y = -4$. The correct answer is Ⓐ.

EXAMPLE 3 Real-Life Application

The graph shows the distance remaining to complete a tunnel. (a) Write an equation that represents the distance y (in feet) remaining after x months. (b) How much time does it take to complete the tunnel?

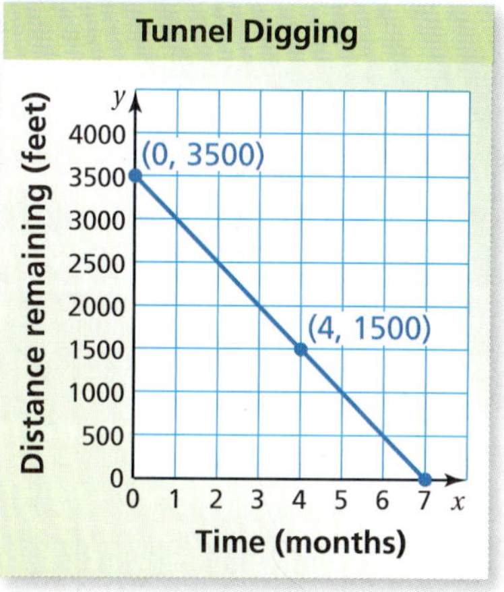

Engineers used tunnel boring machines like the ones shown above to dig an extension of the Metro Gold Line in Los Angeles. The new tunnels are 1.7 miles long and 21 feet wide.

Tunnel Digging

(Graph: Distance remaining (feet), y, from 0 to 4000, versus Time (months), x, from 0 to 7; line through (0, 3500) and (4, 1500).)

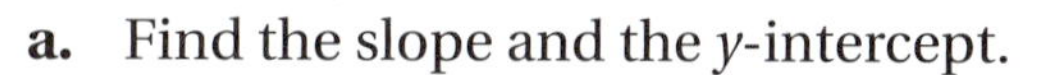

a. Find the slope and the y-intercept.

$$m = \frac{\text{change in } y}{\text{change in } x} = \frac{-2000}{4} = -500$$

Because the line crosses the y-axis at $(0, 3500)$, the y-intercept is 3500.

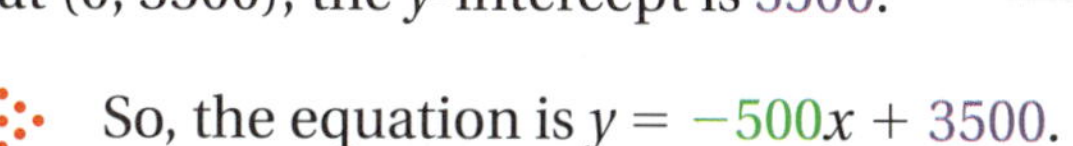

So, the equation is $y = -500x + 3500$.

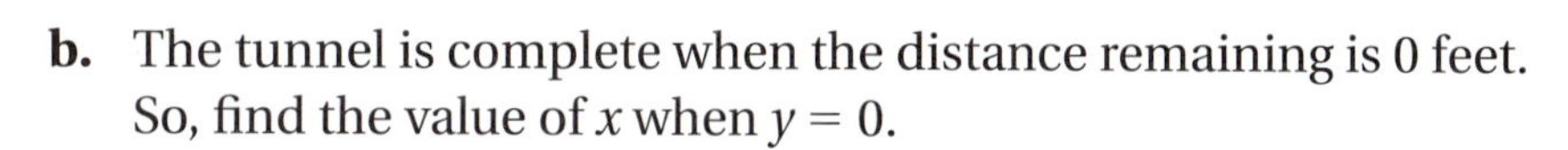

b. The tunnel is complete when the distance remaining is 0 feet. So, find the value of x when $y = 0$.

$y = -500x + 3500$	Write the equation.
$0 = -500x + 3500$	Substitute 0 for y.
$-3500 = -500x$	Subtract 3500 from each side.
$7 = x$	Divide each side by -500.

It takes 7 months to complete the tunnel.

On Your Own

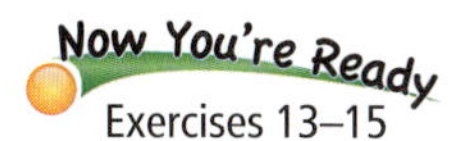

Now You're Ready
Exercises 13–15

3. Write an equation of the line that passes through $(0, 5)$ and $(4, 5)$.
4. **WHAT IF?** In Example 3, the points are $(0, 3500)$ and $(5, 1500)$. How long does it take to complete the tunnel?

4.6 Exercises

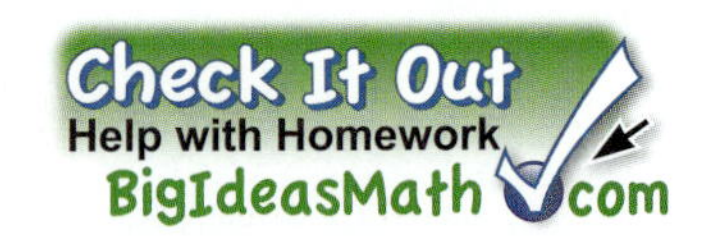

Vocabulary and Concept Check

1. **PRECISION** Explain how to find the slope of a line given the intercepts of the line.
2. **WRITING** Explain how to write an equation of a line using its graph.

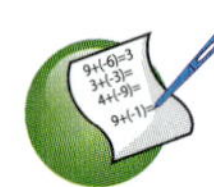

Practice and Problem Solving

Write an equation that represents each side of the figure.

3. 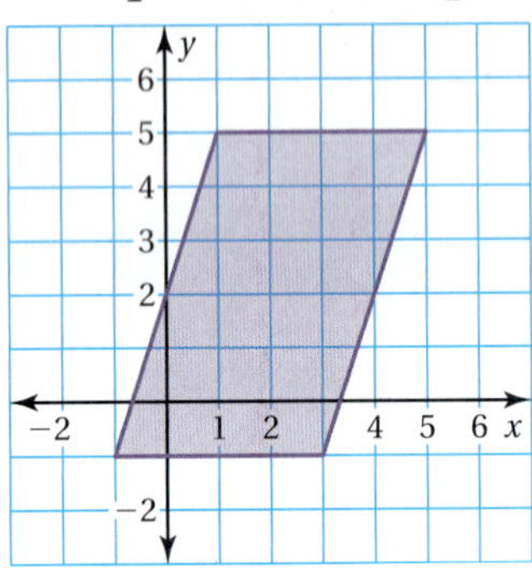

4. 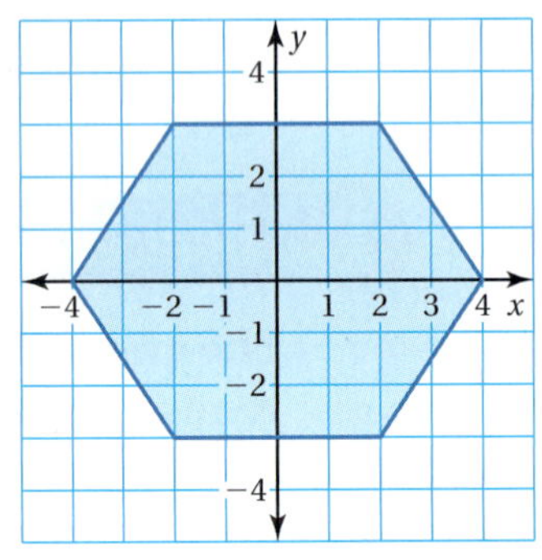

Write an equation of the line in slope-intercept form.

5. 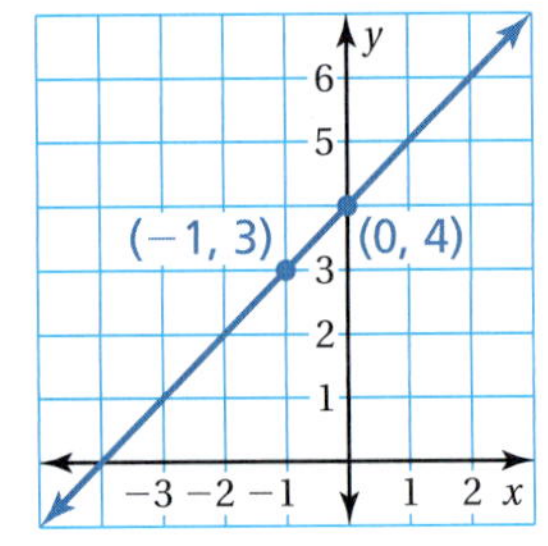

6. 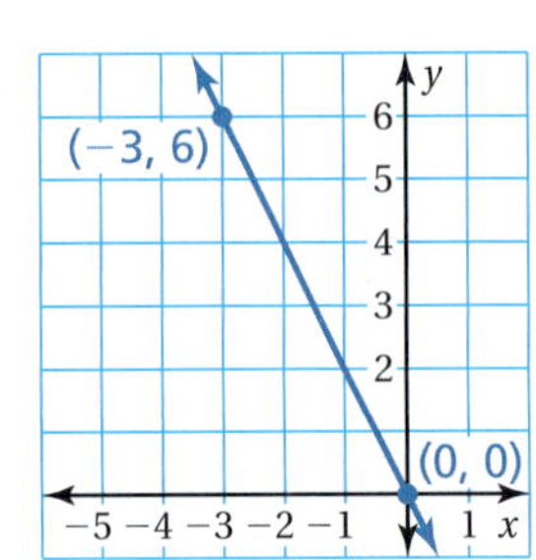

7. 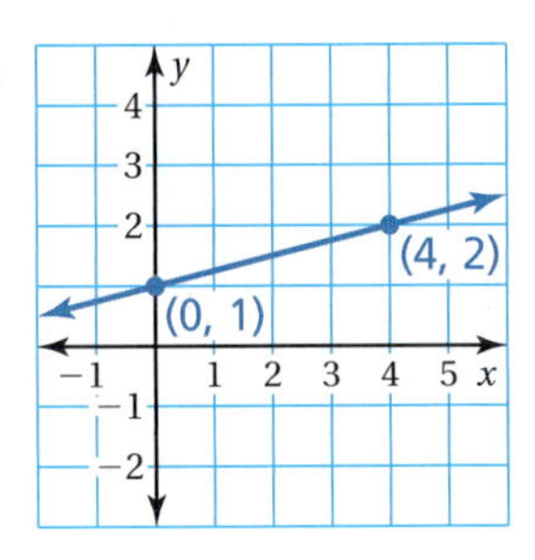

8. 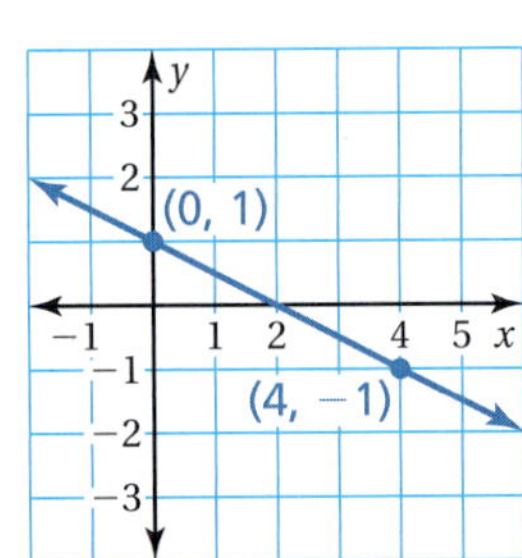

9. 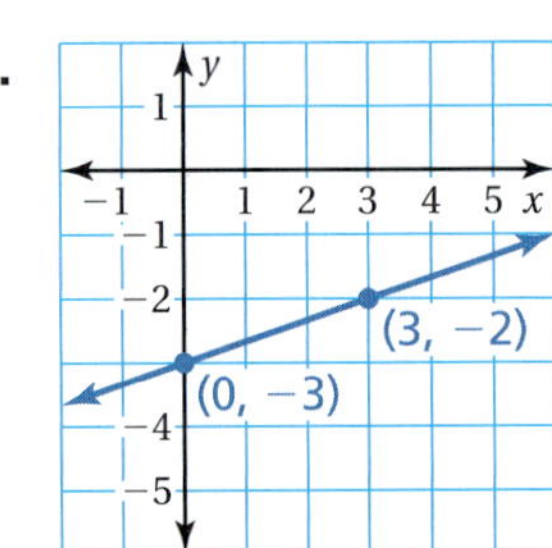

10. 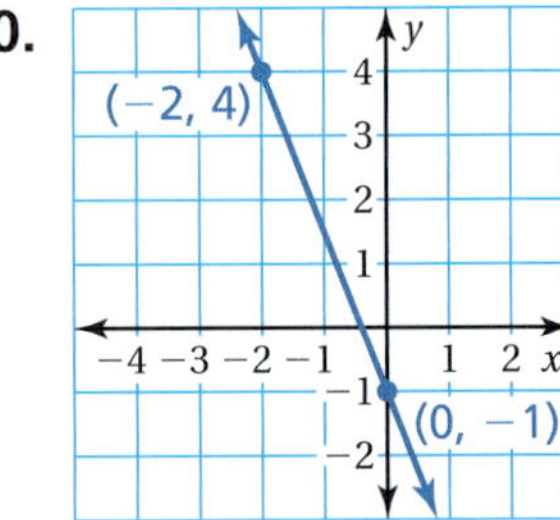

11. **ERROR ANALYSIS** Describe and correct the error in writing an equation of the line.

$y = \frac{1}{2}x + 4$

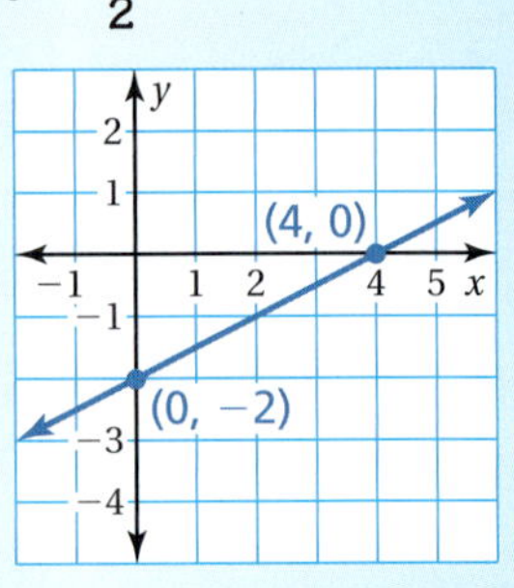

12. **BOA** A boa constrictor is 18 inches long at birth and grows 8 inches per year. Write an equation that represents the length y (in feet) of a boa constrictor that is x years old.

Assignment Guide and Homework Check

Level	Day 1 Activity Assignment	Day 2 Lesson Assignment	Homework Check
Basic	3, 4, 20–24	1, 2, 5–12	5, 8, 9, 10, 12
Average	3, 4, 20–24	1, 2, 5–11 odd, 12–18 even	7, 9, 12, 14, 16
Advanced	1–4, 6–10 even, 11–24		8, 12, 14, 17, 19

Common Errors

- **Exercises 5–10** Students may write the reciprocal of the slope or forget a negative sign. Remind them of the definition of slope. Ask students to predict the sign of the slope based on the rise or fall of the line.
- **Exercises 13–15** Students may write the wrong equation when the slope is zero. For example, instead of $y = 5$, students may write $x = 5$. Ask them what is the rise of the graph (zero) and write this in slope-intercept form with the y-intercept as well, such as $y = 0x + 5$. Then ask students what happens when a variable (or any number) is multiplied by zero. Rewrite the equation as $y = 5$.

4.6 Record and Practice Journal

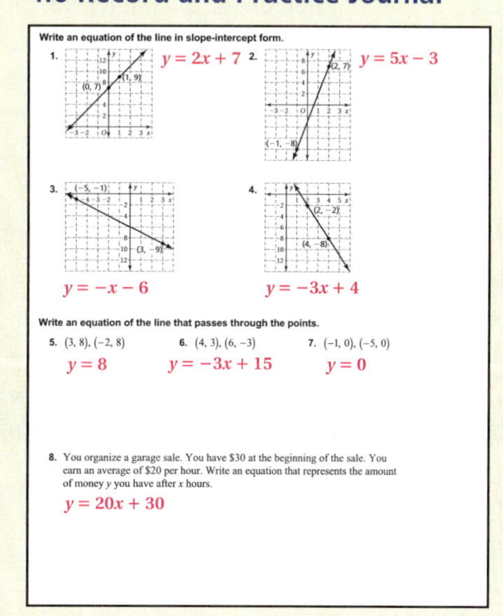

Write an equation of the line in slope-intercept form.

1. $y = 2x + 7$
2. $y = 5x - 3$
3. $y = -x - 6$
4. $y = -3x + 4$

Write an equation of the line that passes through the points.

5. $(3, 8), (-2, 8)$ $y = 8$
6. $(4, 3), (6, -3)$ $y = -3x + 15$
7. $(-1, 0), (-5, 0)$ $y = 0$

8. You organize a garage sale. You have $30 at the beginning of the sale. You earn an average of $20 per hour. Write an equation that represents the amount of money y you have after x hours.

$y = 20x + 30$

Vocabulary and Concept Check

1. *Sample answer:* Find the ratio of the rise to the run between the intercepts.
2. *Sample answer:* Find the slope of the line between any two points. Then find the y-intercept. The equation of the line is $y = mx + b$, where m is the slope and b is the y-intercept.

Practice and Problem Solving

3. $y = 3x + 2$;
   $y = 3x - 10$;
   $y = 5$;
   $y = -1$
4. $y = \frac{3}{2}x + 6$;
   $y = 3$;
   $y = -\frac{3}{2}x + 6$;
   $y = \frac{3}{2}x - 6$;
   $y = -3$;
   $y = -\frac{3}{2}x - 6$
5. $y = x + 4$
6. $y = -2x$
7. $y = \frac{1}{4}x + 1$
8. $y = -\frac{1}{2}x + 1$
9. $y = \frac{1}{3}x - 3$
10. $y = -\frac{5}{2}x - 1$
11. The x-intercept was used instead of the y-intercept.
    $y = \frac{1}{2}x - 2$
12. $y = \frac{2}{3}x + \frac{3}{2}$

Practice and Problem Solving

13–17. See Additional Answers.

18. $y = -140x + 500$

19. See *Taking Math Deeper*.

Fair Game Review

20–23.

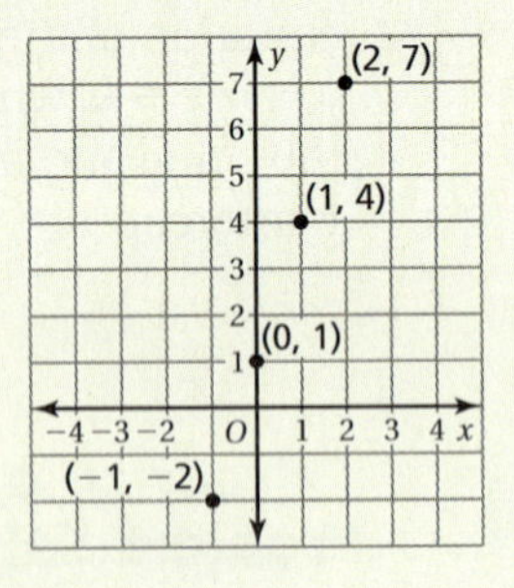

24. C

Mini-Assessment

Write an equation of the line in slope-intercept form.

1.

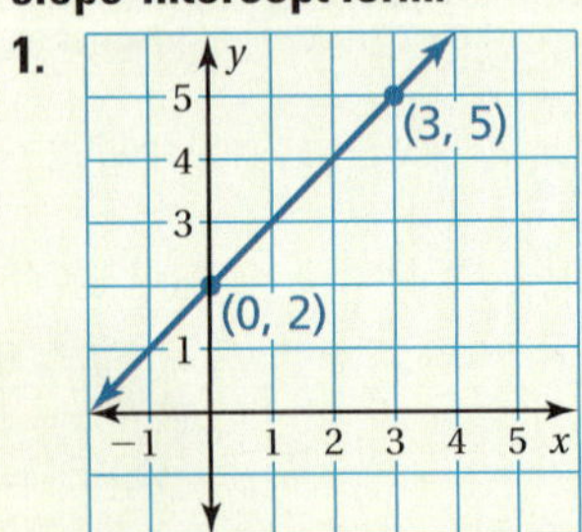

$y = x + 2$

2.

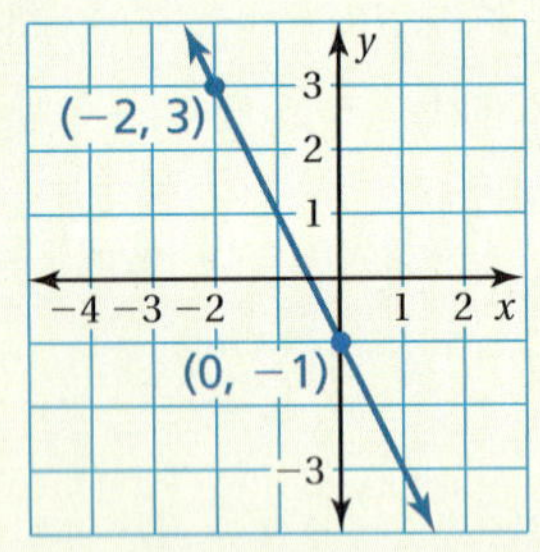

$y = -2x - 1$

3.

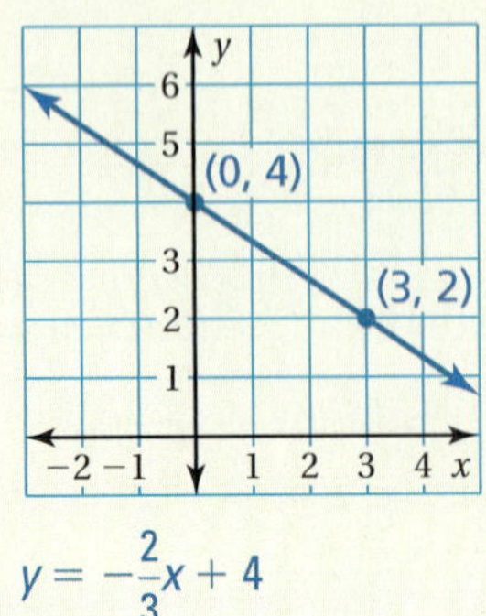

$y = -\frac{2}{3}x + 4$

Taking Math Deeper

Exercise 19

This is a nice real-life problem using estimation. For this problem, remember that you are not looking for exact solutions. You want to know *about* how much the trees grow each year so that you can predict their approximate heights.

Estimate the heights in the photograph.

a. Height of 10-year-old tree: about 18 ft
Height of 8-year-old tree: about 14 ft

b. Plot the points that represent the two trees.

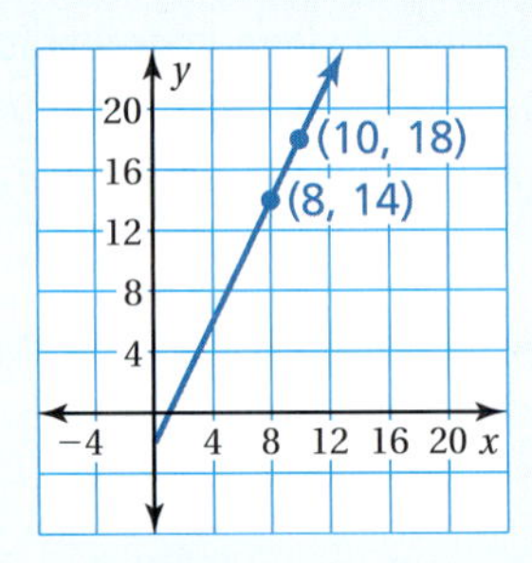

c. The trees are growing at a rate of about 2 feet per year. Because this would put the height of a 0-year-old tree at −2, it is better to adjust the rate of growth to be about 1.8 feet per year.

d. A possible equation for the growth rate is $y = 1.8x$.

Project

Research information about the palm tree. Pick any kind of palm tree that interests you. How old is the longest living palm tree?

Reteaching and Enrichment Strategies

If students need help. . .	If students got it. . .
Resources by Chapter • Practice A and Practice B • Puzzle Time Record and Practice Journal Practice Differentiating the Lesson Lesson Tutorials Skills Review Handbook	Resources by Chapter • Enrichment and Extension • Technology Connection Start the next section

Write an equation of the line that passes through the points.

2 **13.** (2, 5), (0, 5) **14.** (−3, 0), (0, 0) **15.** (0, −2), (4, −2)

16. **WALKATHON** One of your friends gives you $10 for a charity walkathon. Another friend gives you an amount per mile. After 5 miles, you have raised $13.50 total. Write an equation that represents the amount y of money you have raised after x miles.

17. **BRAKING TIME** During each second of braking, an automobile slows by about 10 miles per hour.

a. Plot the points (0, 60) and (6, 0). What do the points represent?

b. Draw a line through the points. What does the line represent?

c. Write an equation of the line.

18. **PAPER** You have 500 sheets of notebook paper. After 1 week, you have 72% of the sheets left. You use the same number of sheets each week. Write an equation that represents the number y of pages remaining after x weeks.

19. 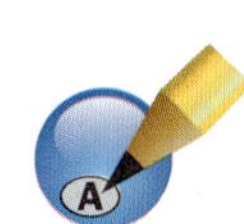

The palm tree on the left is 10 years old. The palm tree on the right is 8 years old. The trees grow at the same rate.

a. Estimate the height y (in feet) of each tree.

b. Plot the two points (x, y), where x is the age of each tree and y is the height of each tree.

c. What is the rate of growth of the trees?

d. Write an equation that represents the height of a palm tree in terms of its age.

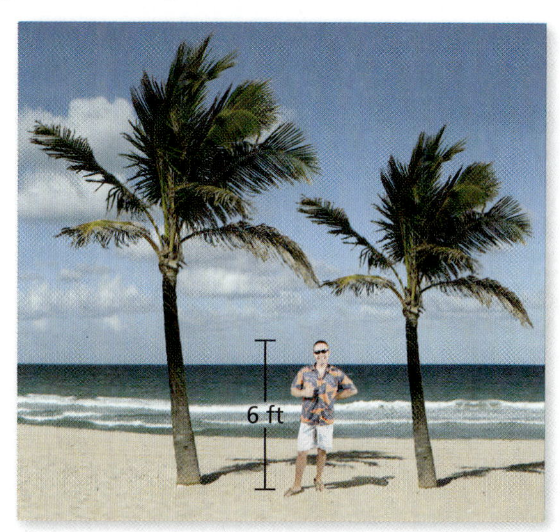

Fair Game Review What you learned in previous grades & lessons

Plot the ordered pair in a coordinate plane. *(Skills Review Handbook)*

20. (1, 4) **21.** (−1, −2) **22.** (0, 1) **23.** (2, 7)

24. **MULTIPLE CHOICE** Which of the following statements is true? *(Section 4.4)*

Ⓐ The x-intercept is 5.

Ⓑ The x-intercept is −2.

Ⓒ The y-intercept is 5.

Ⓓ The y-intercept is −2.

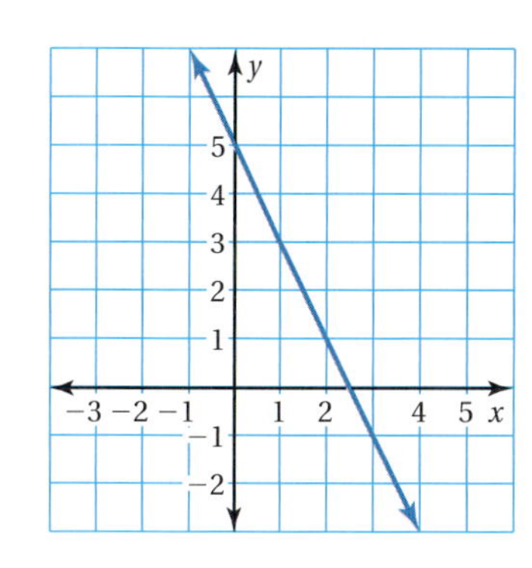

4.7 Writing Equations in Point-Slope Form

Essential Question How can you write an equation of a line when you are given the slope and a point on the line?

1 ACTIVITY: Writing Equations of Lines

Work with a partner.

- **Sketch the line that has the given slope and passes through the given point.**
- **Find the *y*-intercept of the line.**
- **Write an equation of the line.**

a. $m = -2$

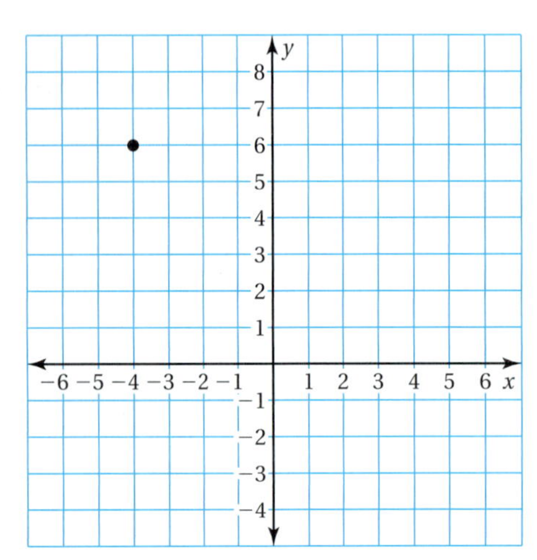

b. $m = \frac{1}{3}$

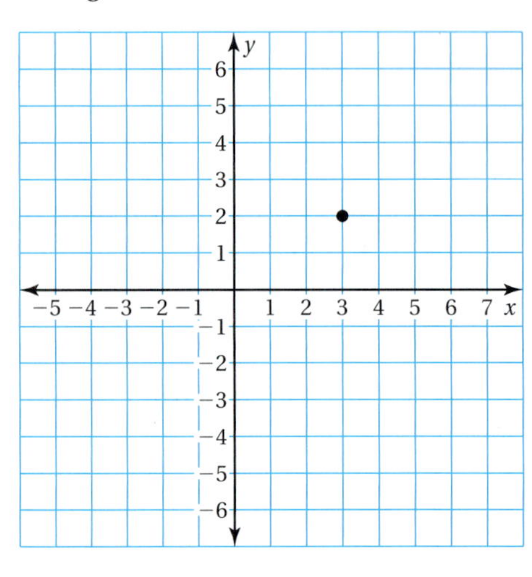

c. $m = -\frac{2}{3}$

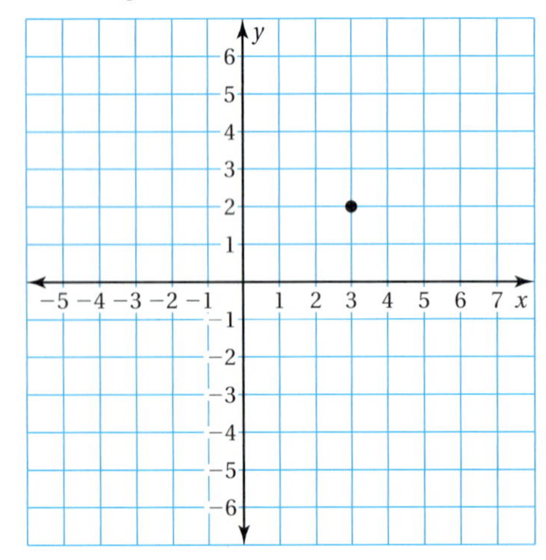

d. $m = \frac{5}{2}$

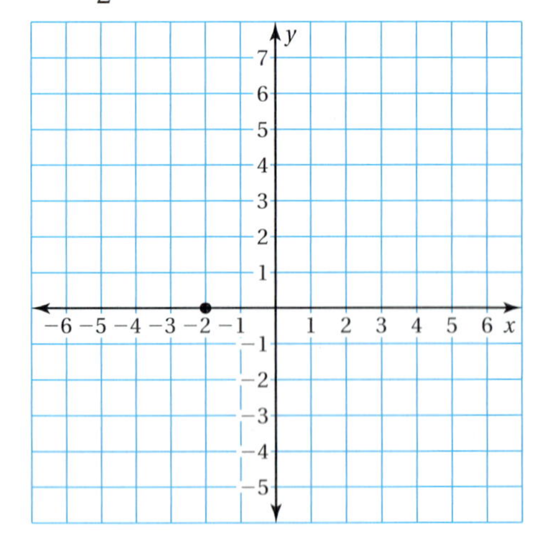

COMMON CORE

Writing Equations

In this lesson, you will

- write equations of lines using a slope and a point.
- write equations of lines using two points.

Preparing for Standard MACC.8.F.2.4

Laurie's Notes

Introduction

Standards for Mathematical Practice

- **MP1 Make Sense of Problems and Persevere in Solving Them:** The goal is for students to write equations of lines given a slope and a point. The slope may be stated explicitly or determined from a contextual setting. Students use the slope to graph the line and work backwards to find the *y*-intercept. There are different approaches students may use as they make sense of the problem.
- **MP3 Construct Viable Arguments and Critique the Reasoning of Others:** Take time for discussions and explanations so that students' reasoning is revealed.

Motivate

- Hold a piece of ribbon and a pair of scissors in your hands. Snip off a one-foot piece of ribbon. Repeat once or twice more.

? "Do you know how long my ribbon was when I first started?" no

- Your question should prompt students to ask two obvious questions: "How much are you cutting off each time?" and "How many times have you made a cut?" How much you cut off is the slope (-1). How many times you cut the ribbon helps students work backwards to find the length before any cuts were made, which is the *y*-intercept.

Activity Notes

Activity 1

? "What does it mean for a line to have a slope of -2? A slope of $\frac{1}{3}$?"

For every unit it runs, it falls 2 units. For every 3 units it runs, it rises 1.

- Students may also answer the last question by saying, "Over 1, down 2," and "Over 3, up 1." These geometric answers are fine. Students will need this level of understanding to locate additional points on a line in order to find the *y*-intercept.
- You cannot sketch the line immediately. You must first find additional points on the line. Students should start at the given point and use the slope to find additional points on the line. One of the points will give the *y*-intercept.
- For part (b), it might be helpful to think of the slope of $\frac{1}{3}$ as $\frac{-1}{-3}$. So, start at the point given and move left 3 units and then down 1 unit.
- **Common Error:** Students may interchange the rise and run. Have students look back at their graphs, to see if the slope looks correct to them.
- **Teaching Tip:** Encourage students to lightly trace the rise and run direction arrows with their pencils as they locate additional points.

? "What made it possible to write the equation of the line?" The slope was given and by using the slope, it was possible to find the *y*-intercept. Then substitute into the formula $y = mx + b$.

- **MP1:** Asking the last question about slope and having students state their understanding helps them make sense of the problem.

Florida Common Core Standards

MACC.8.F.2.4 Construct a function to model a linear relationship between two quantities. Determine the rate of change and initial value of the function from a description of a relationship or from two (x, y) values, including reading these from a table or from a graph. Interpret the rate of change and initial value of a linear function in terms of the situation it models, and in terms of its graph or a table of values.

Previous Learning

Students should know how to plot ordered pairs and apply the definition of slope.

4.7 Record and Practice Journal

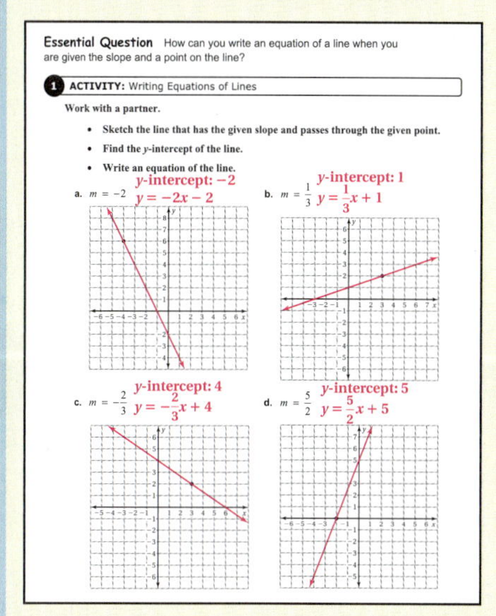

Essential Question How can you write an equation of a line when you are given the slope and a point on the line?

1 ACTIVITY: Writing Equations of Lines

Work with a partner.

- **Sketch the line that has the given slope and passes through the given point.**
- **Find the *y*-intercept of the line.**
- **Write an equation of the line.**

a. $m = -2$ *y*-intercept: -2; $y = -2x - 2$

b. $m = \frac{1}{3}$ *y*-intercept: 1; $y = \frac{1}{3}x + 1$

c. $m = -\frac{2}{3}$ *y*-intercept: 4; $y = -\frac{2}{3}x + 4$

d. $m = \frac{5}{2}$ *y*-intercept: 5; $y = \frac{5}{2}x + 5$

Differentiated Instruction

Kinesthetic

Write a list of linear equations on the board or overhead. Have students copy the equations onto index cards. On the back of each card students are to write the slope and y-intercept of the line. After the cards are completed, students can work in pairs to check each other's work. Finally, students can quiz each other with the flash cards they made.

4.7 Record and Practice Journal

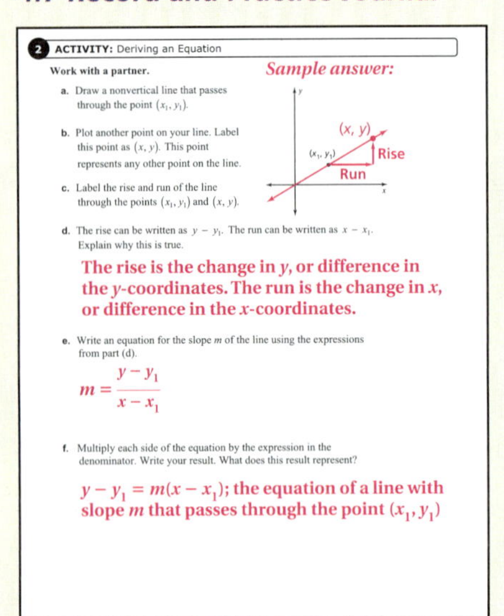

2 ACTIVITY: Deriving an Equation

Work with a partner. *Sample answer:*

a. Draw a nonvertical line that passes through the point (x_1, y_1).

b. Plot another point on your line. Label this point as (x, y). This point represents any other point on the line.

c. Label the rise and run of the line through the points (x_1, y_1) and (x, y).

d. The rise can be written as $y - y_1$. The run can be written as $x - x_1$. Explain why this is true.

The rise is the change in y, or difference in the y-coordinates. The run is the change in x, or difference in the x-coordinates.

e. Write an equation for the slope m of the line using the expressions from part (d).

$m = \dfrac{y - y_1}{x - x_1}$

f. Multiply each side of the equation by the expression in the denominator. Write your result. What does this result represent?

$y - y_1 = m(x - x_1)$; the equation of a line with slope m that passes through the point (x_1, y_1)

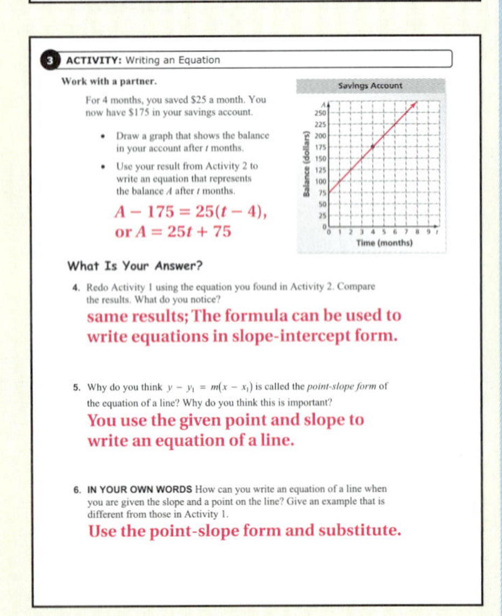

3 ACTIVITY: Writing an Equation

Work with a partner.

For 4 months, you saved \$25 a month. You now have \$175 in your savings account.

- Draw a graph that shows the balance in your account after t months.
- Use your result from Activity 2 to write an equation that represents the balance A after t months.

$A - 175 = 25(t - 4)$, or $A = 25t + 75$

What Is Your Answer?

4. Redo Activity 1 using the equation you found in Activity 2. Compare the results. What do you notice?
same results; The formula can be used to write equations in slope-intercept form.

5. Why do you think $y - y_1 = m(x - x_1)$ is called the *point-slope form* of the equation of a line? Why do you think this is important?
You use the given point and slope to write an equation of a line.

6. **IN YOUR OWN WORDS** How can you write an equation of a line when you are given the slope and a point on the line? Give an example that is different from those in Activity 1.
Use the point-slope form and substitute.

Laurie's Notes

Activity 2

- **Big Idea:** This activity derives the point-slope form of the equation of a line. Students should see the relationship between the slope formula and the point-slope form.
- **MP1:** The steps of the derivation are provided. Encourage students to read carefully, discuss with their partners, and think about the process. Do not jump in too quickly to rescue students!
- Allow sufficient time before having a class discussion of this activity. Ask volunteers to share their work on the board. Sharing their process aloud helps students become more confident in their reasoning.

Activity 3

- If students do not understand what the slope is, suggest they work backwards and make a table of values.

Month, t	0	1	2	3	4
Balance in Account, A	\$75	\$100	\$125	\$150	\$175

- Students can use the table to draw the graph.
- ? Ask a few questions to guide students' understanding:
  - "What is the slope for this problem? What point is given?" slope = 25; given point is (4, 175)
  - "Do you have enough information to write an equation?" yes; $A - 175 = 25(t - 4)$, or $A = 25t + 75$
  - "Explain why the slope is positive." You are putting money in the bank. Your account is growing.
- **MP1:** Take time for students to transform the equation into slope-intercept form. Students should interpret the slope and y-intercept in the context of the problem.

What Is Your Answer?

- **Neighbor Check:** Have students work independently and then have their neighbors check their work. Have students discuss any discrepancies.
- **MP7 Look for and Make Use of Structure:** In Question 4, the equations in point-slope form and slope-intercept form are equivalent, but structurally they look different. Take time for students to appreciate what information is known about the line from the form in which the equation is written.

Closure

- Refer back to the ribbon and scissors. If the ribbon is now 7 feet after making 4 equal cuts of 1-foot length, write an equation that gives the length of the ribbon R after n cuts. $R = 11 - n$ or $R = -n + 11$

2 ACTIVITY: Deriving an Equation

Work with a partner.

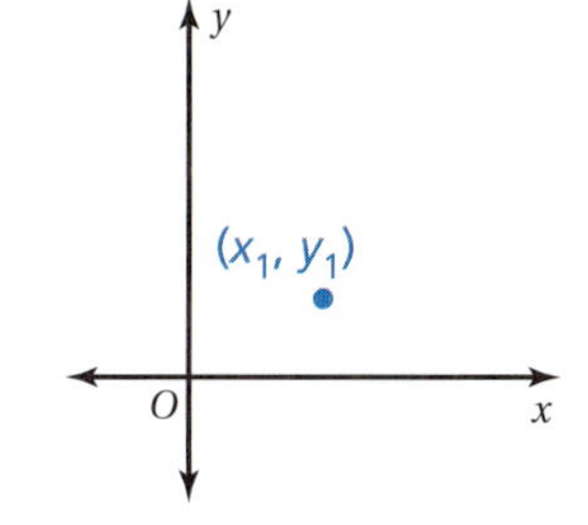

a. Draw a nonvertical line that passes through the point (x_1, y_1).

b. Plot another point on your line. Label this point as (x, y). This point represents any other point on the line.

c. Label the rise and the run of the line through the points (x_1, y_1) and (x, y).

d. The rise can be written as $y - y_1$. The run can be written as $x - x_1$. Explain why this is true.

e. Write an equation for the slope m of the line using the expressions from part (d).

f. Multiply each side of the equation by the expression in the denominator. Write your result. What does this result represent?

Math Practice 3

Construct Arguments

How does a graph help you derive an equation?

3 ACTIVITY: Writing an Equation

Work with a partner.

For 4 months, you saved \$25 a month. You now have \$175 in your savings account.

- Draw a graph that shows the balance in your account after t months.
- Use your result from Activity 2 to write an equation that represents the balance A after t months.

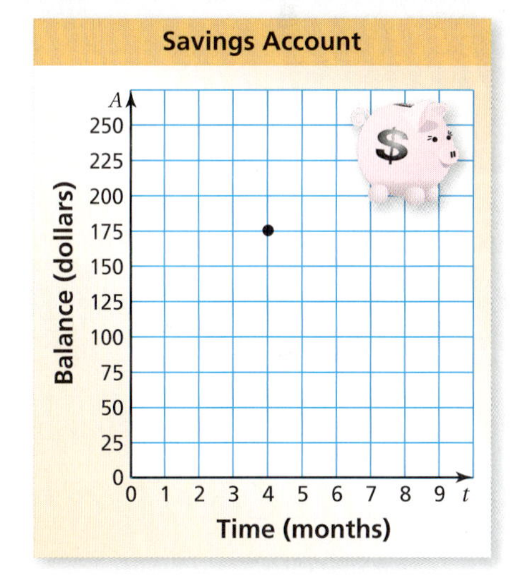

What Is Your Answer?

4. Redo Activity 1 using the equation you found in Activity 2. Compare the results. What do you notice?

5. Why do you think $y - y_1 = m(x - x_1)$ is called the *point-slope form* of the equation of a line? Why do you think it is important?

6. **IN YOUR OWN WORDS** How can you write an equation of a line when you are given the slope and a point on the line? Give an example that is different from those in Activity 1.

Practice Use what you learned about writing equations using a slope and a point to complete Exercises 3–5 on page 188.

Key Vocabulary
point-slope form, *p. 186*

Key Idea

Point-Slope Form

Words A linear equation written in the form $y - y_1 = m(x - x_1)$ is in **point-slope form**. The line passes through the point (x_1, y_1), and the slope of the line is m.

Algebra

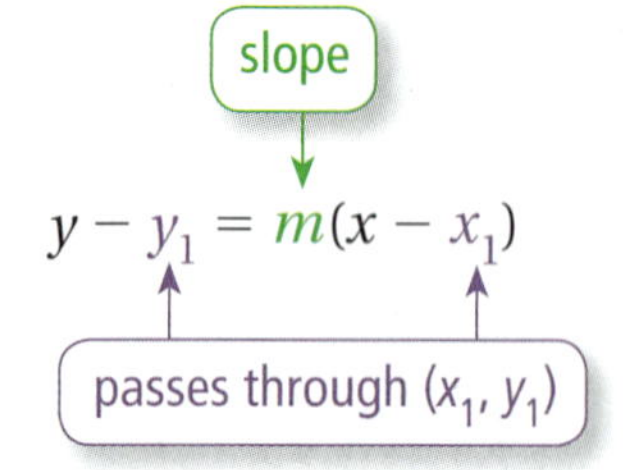

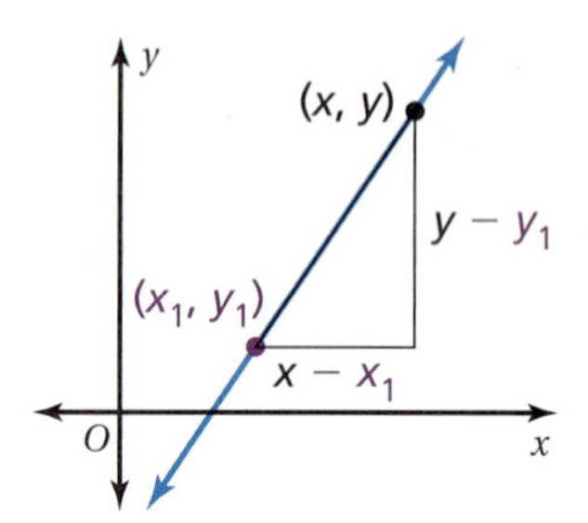

EXAMPLE 1 Writing an Equation Using a Slope and a Point

Write in point-slope form an equation of the line that passes through the point (−6, 1) with slope $\frac{2}{3}$.

$y - y_1 = m(x - x_1)$	Write the point-slope form.
$y - 1 = \frac{2}{3}[x - (-6)]$	Substitute $\frac{2}{3}$ for m, −6 for x_1, and 1 for y_1.
$y - 1 = \frac{2}{3}(x + 6)$	Simplify.

So, the equation is $y - 1 = \frac{2}{3}(x + 6)$.

Check Check that (−6, 1) is a solution of the equation.

$y - 1 = \frac{2}{3}(x + 6)$	Write the equation.
$1 - 1 \stackrel{?}{=} \frac{2}{3}(-6 + 6)$	Substitute.
$0 = 0$ ✓	Simplify.

On Your Own

Write in point-slope form an equation of the line that passes through the given point and has the given slope.

1. $(1, 2)$; $m = -4$ **2.** $(7, 0)$; $m = 1$ **3.** $(-8, -5)$; $m = -\frac{3}{4}$

Laurie's Notes

Introduction

Connect

- **Yesterday:** Students developed an intuitive understanding of how to write an equation of a line given the slope and a point. (MP1, MP3, MP7)
- **Today:** Students will write an equation of a line given the slope and a point.

Motivate

? "Have you seen an airplane come in for a landing either in real life, on television, or in movies?" Most will answer yes.

? "Can you describe in words or with a picture what it looks like?" Listen for a smooth approach, meaning a constant rate of descent.

? "If the plane descends 200 feet per second, what is its height 5 seconds before it lands?" 1000 ft

- Make a sketch of this scenario and ask if it's possible to write an equation that models the height h of the airplane, t seconds before it lands.

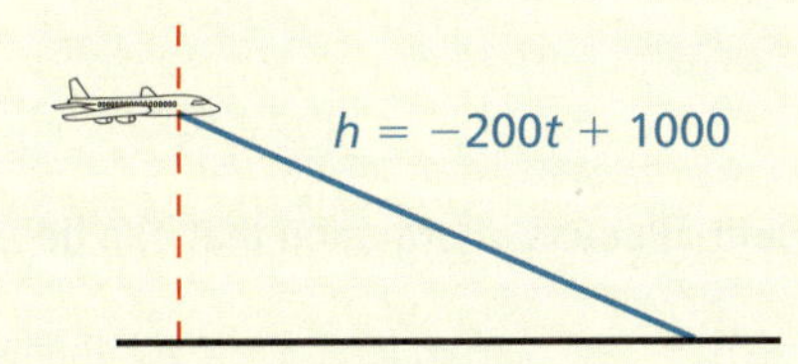

Lesson Notes

Key Idea

? Draw a coordinate plane and graph a point. "How many lines go through this point with a slope of $\frac{1}{2}$?" only one line

- Explain that the *point-slope form* of the equation of a line is equivalent to the slope-intercept form and is the equation of a unique line.
- Write the Key Idea on the board. Use of color is helpful.
- **Teaching Tip:** On a side board, write the formula for slope as $\frac{y - y_1}{x - x_1} = m$ so students are reminded of how point-slope form was derived.

Discuss

- **MP1a Make Sense of Problems:** Although students derived point-slope form in the activity, they will have lingering questions about the use of subscripts. They might ask why the first point was not labeled (x, y) and the second point (x_1, y_1). The labels are arbitrary, the line could be sloping downward, and the points could be located in any quadrant.

Example 1

- Write the point-slope form of a linear equation.

? "What is the slope of the line?" $\frac{2}{3}$ "What point do we know the line passes through?" $(-6, 1)$

- Substitute the known information. Remind students that they are subtracting a negative, so they have $x + 6$ inside the parentheses.

? "How can we check if our equation is reasonable?" Students might suggest a quick sketch or rewriting the equation in slope-intercept form to see if the y-intercept makes sense.

Goal Today's lesson is writing an equation of a line given a slope and a point or two points.

Technology for the Teacher

Lesson Tutorials
Lesson Plans
Answer Presentation Tool

Extra Example 1

Write in point-slope form an equation of the line that passes through the given point and has the given slope.

a. $(2, 2)$; $m = \frac{5}{2}$ $y - 2 = \frac{5}{2}(x - 2)$

b. $(3, -6)$; $m = -\frac{4}{3}$ $y + 6 = -\frac{4}{3}(x - 3)$

On Your Own

1. $y - 2 = -4(x - 1)$
2. $y - 0 = 1(x - 7)$
3. $y + 5 = -\frac{3}{4}(x + 8)$

Extra Example 2

Write in slope-intercept form an equation of the line that passes through the points $(-3, 0)$ and $(6, 3)$. $y = \frac{1}{3}x + 1$

Extra Example 3

You are pulling down your kite at a rate of 2 feet per second. After 3 seconds, your kite is 54 feet above you.

a. Write and graph an equation that represents the height y (in feet) of the kite above you after x seconds. $y = -2x + 60$

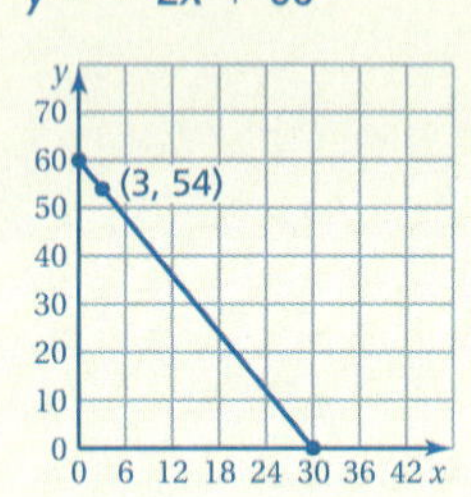

b. At what height was the kite flying before you began pulling it down? 60 ft

On Your Own

4. $y = -x - 1$
5. $y = 4x + 15$
6. $y = \frac{1}{2}x + 10$
7. $y = -10x + 55$

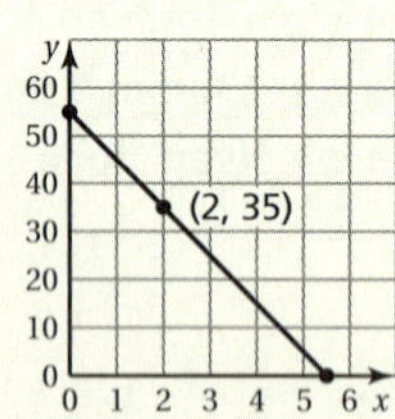

English Language Learners

Visual

Encourage English learners to plot the given point in the coordinate plane and then use the slope. The graph will give them a visual reference they can use when writing the equation.

Laurie's Notes

On Your Own

- **Neighbor Check:** Have students work independently and then have their neighbors check their work. Have students discuss any discrepancies.

Example 2

- Plot both points. Draw the line through the two points.
- ? "Is the slope positive or negative?" negative
- ? "How can you find the slope exactly?" Use the slope formula.
- ? "Can you estimate the y-intercept?" Listen for a positive number greater than 4.
- Continue to work through the problem as shown.
- ? "Do you think we would get the same equation if we had used $(5, -2)$ instead of $(2, 4)$? yes; Students may be unsure.
- **MP1a:** Work the problem again using $(5, -2)$ as shown in the Study Tip. Mathematically proficient students can make sense of why either point will result in the same equation.

Example 3

- Ask a volunteer to read the problem. Discuss information that can be *read* from the illustration.
- ? "Have any of you parasailed?" Wait for students to respond. Explain that when parasailing, you want a smooth descent, like an airplane.
- ? "What is the slope for this problem? How did you know?" Slope is -10. The arrow pointing down means the slope is negative.
- ? "Do we know a point that satisfies the equation?" yes, (2, 25)
- Write the point-slope form of a linear equation. Substitute the known information.
- **Extension:** Have students determine when you reach the boat, meaning $y = 0$.

On Your Own

- Discuss student solutions. Check that signs of numbers are correct for Questions 4–6.

Closure

- **Exit Ticket:** Write an equation of the line with a slope of 2 that passes through the point $(-1, 4)$ in point-slope form and slope-intercept form. $y - 4 = 2(x + 1)$; $y = 2x + 6$

EXAMPLE 2 Writing an Equation Using Two Points

You can use either of the given points to write the equation of the line.

Use $m = -2$ and $(5, -2)$.

$y - (-2) = -2(x - 5)$

$y + 2 = -2x + 10$

$y = -2x + 8$ ✓

Write in slope-intercept form an equation of the line that passes through the points (2, 4) and (5, −2).

Find the slope: $m = \dfrac{y_2 - y_1}{x_2 - x_1} = \dfrac{-2 - 4}{5 - 2} = \dfrac{-6}{3} = -2$

Then use the slope $m = -2$ and the point (2, 4) to write an equation of the line.

$y - y_1 = m(x - x_1)$	Write the point-slope form.
$y - 4 = -2(x - 2)$	Substitute −2 for m, 2 for x_1, and 4 for y_1.
$y - 4 = -2x + 4$	Distributive Property
$y = -2x + 8$	Write in slope-intercept form.

EXAMPLE 3 Real-Life Application

You finish parasailing and are being pulled back to the boat. After 2 seconds, you are 25 feet above the boat. (a) Write and graph an equation that represents your height *y* (in feet) above the boat after *x* seconds. (b) At what height were you parasailing?

a. You are being pulled down at the rate of 10 feet per second. So, the slope is −10. You are 25 feet above the boat after 2 seconds. So, the line passes through (2, 25). Use the point-slope form.

$y - 25 = -10(x - 2)$	Substitute for m, x_1, and y_1.
$y - 25 = -10x + 20$	Distributive Property
$y = -10x + 45$	Write in slope-intercept form.

So, the equation is $y = -10x + 45$.

b. You start descending when $x = 0$. The y-intercept is 45. So, you were parasailing at a height of 45 feet.

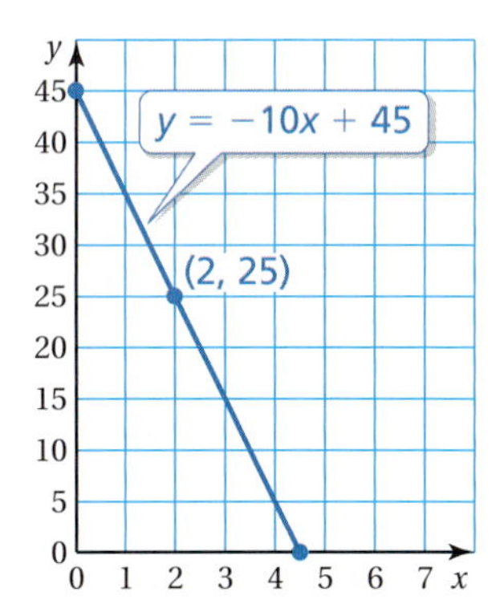

On Your Own

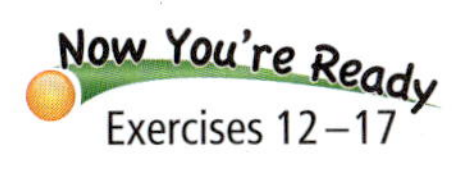

Write in slope-intercept form an equation of the line that passes through the given points.

4. (−2, 1), (3, −4) **5.** (−5, −5), (−3, 3) **6.** (−8, 6), (−2, 9)

7. WHAT IF? In Example 3, you are 35 feet above the boat after 2 seconds. Write and graph an equation that represents your height *y* (in feet) above the boat after *x* seconds.

4.7 Exercises

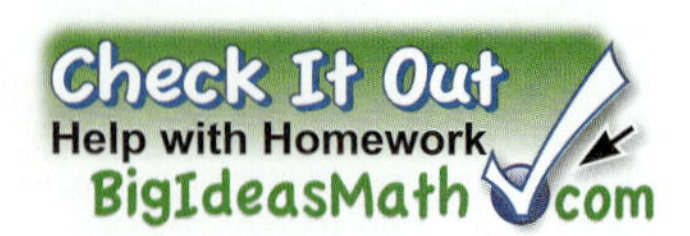

Vocabulary and Concept Check

1. **VOCABULARY** From the equation $y - 3 = -2(x + 1)$, identify the slope and a point on the line.
2. **WRITING** Describe how to write an equation of a line using (a) its slope and a point on the line and (b) two points on the line.

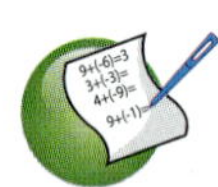

Practice and Problem Solving

Use the point-slope form to write an equation of the line with the given slope that passes through the given point.

3. $m = \frac{1}{2}$

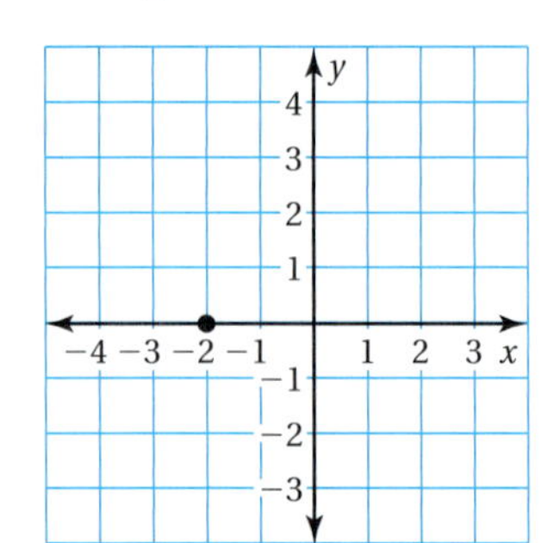

4. $m = -\frac{3}{4}$

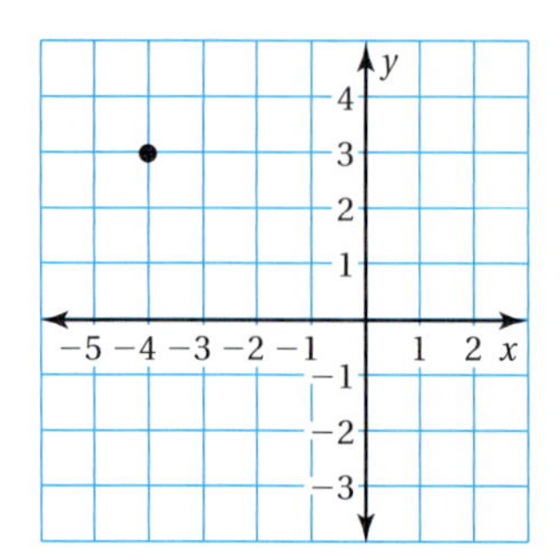

5. $m = -3$

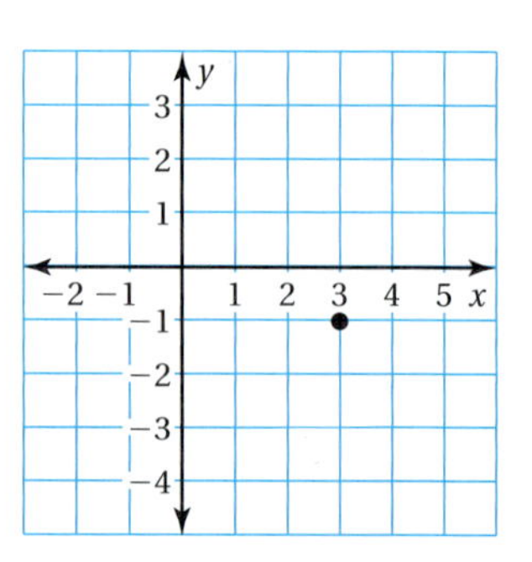

Write in point-slope form an equation of the line that passes through the given point and has the given slope.

① 6. $(3, 0)$; $m = -\frac{2}{3}$
7. $(4, 8)$; $m = \frac{3}{4}$
8. $(1, -3)$; $m = 4$
9. $(7, -5)$; $m = -\frac{1}{7}$
10. $(3, 3)$; $m = \frac{5}{3}$
11. $(-1, -4)$; $m = -2$

Write in slope-intercept form an equation of the line that passes through the given points.

② 12. $(-1, -1), (1, 5)$
13. $(2, 4), (3, 6)$
14. $(-2, 3), (2, 7)$
15. $(4, 1), (8, 2)$
16. $(-9, 5), (-3, 3)$
17. $(1, 2), (-2, -1)$

18. **CHEMISTRY** At 0 °C, the volume of a gas is 22 liters. For each degree the temperature T (in degrees Celsius) increases, the volume V (in liters) of the gas increases by $\frac{2}{25}$. Write an equation that represents the volume of the gas in terms of the temperature.

Assignment Guide and Homework Check

Level	Day 1 Activity Assignment	Day 2 Lesson Assignment	Homework Check
Basic	3–5, 24–27	1, 2, 7–21 odd	7, 11, 13, 17, 19
Average	3–5, 24–27	1, 2, 6–18 even, 19–23 odd	10, 14, 16, 18, 21
Advanced	1–5, 6–18 even, 19–27		16, 18, 19, 21, 22

Common Errors

- **Exercises 6–17** Students may forget to include negatives with the slope and coordinates, or they may apply them incorrectly. Remind them that when the coordinates are negative, they will be subtracting a negative after substituting in point-slope form, which results in adding a positive.
- **Exercises 12–17** Students may use the reciprocal of the slope when writing the equation. Remind them that slope is the change in *y* over the change in *x*.
- **Exercise 18** Students might have trouble knowing which variable can be compared with *x* and *y* and may write the given point backwards. Review what the words "in terms of" mean when writing an equation. In this problem, *V* could be replaced by *y* and *T* could be replaced by *x*. Remind students to check their equations by substituting the given point and checking that it is a solution of the equation.

4.7 Record and Practice Journal

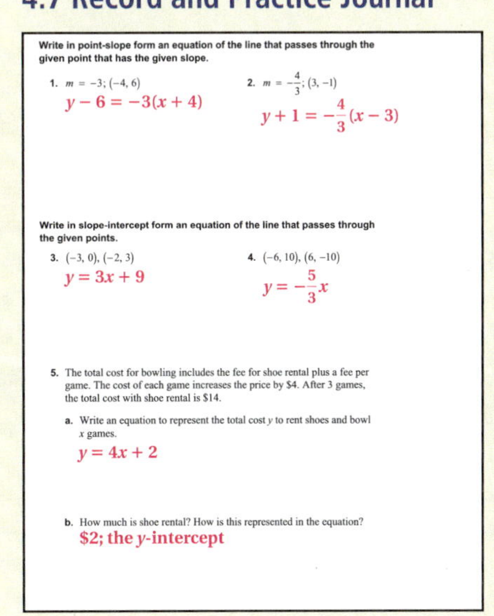

Write in point-slope form an equation of the line that passes through the given point that has the given slope.

1. $m = -3; (-4, 6)$
   $y - 6 = -3(x + 4)$
2. $m = -\frac{4}{3}; (3, -1)$
   $y + 1 = -\frac{4}{3}(x - 3)$

Write in slope-intercept form an equation of the line that passes through the given points.

3. $(-3, 0), (-2, 3)$
   $y = 3x + 9$
4. $(-6, 10), (6, -10)$
   $y = -\frac{5}{3}x$

5. The total cost for bowling includes the fee for shoe rental plus a fee per game. The cost of each game increases the price by \$4. After 3 games, the total cost with shoe rental is \$14.
   a. Write an equation to represent the total cost *y* to rent shoes and bowl *x* games.
   $y = 4x + 2$
   b. How much is shoe rental? How is this represented in the equation?
   \$2; the *y*-intercept

Vocabulary and Concept Check

1. $m = -2; (-1, 3)$
2. a. Write the point-slope form. Substitute the slope for m and the point for (x_1, y_1). Simplify and check your work.
   b. First use the two points to find the slope. Then write the point-slope form. Substitute the slope for m and one of the points for (x_1, y_1). Simplify and check your work.

Practice and Problem Solving

3. $y - 0 = \frac{1}{2}(x + 2)$
4. $y - 3 = -\frac{3}{4}(x + 4)$
5. $y + 1 = -3(x - 3)$
6. $y - 0 = -\frac{2}{3}(x - 3)$
7. $y - 8 = \frac{3}{4}(x - 4)$
8. $y + 3 = 4(x - 1)$
9. $y + 5 = -\frac{1}{7}(x - 7)$
10. $y - 3 = \frac{5}{3}(x - 3)$
11. $y + 4 = -2(x + 1)$
12. $y = 3x + 2$
13. $y = 2x$
14. $y = x + 5$
15. $y = \frac{1}{4}x$
16. $y = -\frac{1}{3}x + 2$
17. $y = x + 1$
18. $V = \frac{2}{25}T + 22$

Practice and Problem Solving

19. **a.** $V = -4000x + 30{,}000$

b. \$30,000

20. **a.** $y = 4x - 30$

b. $y = -\frac{1}{4}x + 4$

21. See *Taking Math Deeper*.

22. **a.** $y = -2x + 68$

b. 68 ounces

c. after 34 seconds

23. **a.** $y = 14x - 108.5$

b. 4 meters

Fair Game Review

24–26. See Additional Answers.

27. D

Mini-Assessment

Write in point-slope form an equation of the line that passes through the given point and has the given slope.

1. $(1, 4)$; $m = 3$ $y - 4 = 3(x - 1)$
2. $(-2, 1)$; $m = -2$ $y - 1 = -2(x + 2)$
3. $(3, 5)$; $m = 1$ $y - 5 = 1(x - 3)$
4. $(2, -1)$; $m = \frac{1}{2}$ $y + 1 = \frac{1}{2}(x - 2)$
5. **You rent a floor sander for \$24 per day. You pay \$82 for 3 days.**

   a. Write an equation that represents your total cost y (in dollars) after x days. $y = 24x + 10$

   b. Interpret the y-intercept. The y-intercept is 10. This means you paid an initial fee of \$10 to rent the sander.

Taking Math Deeper

Exercise 21

The challenge in this biology problem is to interpret the given information as a rate of change (or slope) and as an ordered pair.

Translate the given information into math.

T = temperature (°F)

x = chirps per minute

Rate of change = 0.25 degree per chirp

Write an equation.

Given point: $(x, T) = (40, 50)$

With a slope of 0.25, you can determine that the T-intercept of the line is 40. So, the equation is

a. $T = 0.25x + 40$.

Use the equation.

If $x = 100$ chirps per minute, then

$T = 0.25(100) + 40$

b. $= 65°F$.

If $T = 96$, then you can find the number of chirps per minute as follows.

$96 = 0.25x + 40$

$56 = 0.25x$

$224 = x$

c. So, you would expect the cricket to make 224 chirps in one minute.

This relationship between temperature and cricket chirps was first published by Amos Dolbear in 1897 in an article called *The Cricket as a Thermometer*.

Research other plants or animals that predict the temperature or weather.

Reteaching and Enrichment Strategies

If students need help. . .	If students got it. . .
Resources by Chapter • Practice A and Practice B • Puzzle Time Record and Practice Journal Practice Differentiating the Lesson Lesson Tutorials Skills Review Handbook	Resources by Chapter • Enrichment and Extension • Technology Connection Start the next section

19. **CARS** After it is purchased, the value of a new car decreases $4000 each year. After 3 years, the car is worth $18,000.

  a. Write an equation that represents the value V (in dollars) of the car x years after it is purchased.

  b. What was the original value of the car?

20. **REASONING** Write an equation of a line that passes through the point (8, 2) that is (a) parallel and (b) perpendicular to the graph of the equation $y = 4x - 3$.

21. **CRICKETS** According to Dolbear's law, you can predict the temperature T (in degrees Fahrenheit) by counting the number x of chirps made by a snowy tree cricket in 1 minute. For each rise in temperature of 0.25°F, the cricket makes an additional chirp each minute.

  a. A cricket chirps 40 times in 1 minute when the temperature is 50°F. Write an equation that represents the temperature in terms of the number of chirps in 1 minute.

  b. You count 100 chirps in 1 minute. What is the temperature?

  c. The temperature is 96°F. How many chirps would you expect the cricket to make?

Leaning Tower of Pisa

22. **WATERING CAN** You water the plants in your classroom at a constant rate. After 5 seconds, your watering can contains 58 ounces of water. Fifteen seconds later, the can contains 28 ounces of water.

  a. Write an equation that represents the amount y (in ounces) of water in the can after x seconds.

  b. How much water was in the can when you started watering the plants?

  c. When is the watering can empty?

23. **Problem Solving** The Leaning Tower of Pisa in Italy was built between 1173 and 1350.

  a. Write an equation for the yellow line.

  b. The tower is 56 meters tall. How far off center is the top of the tower?

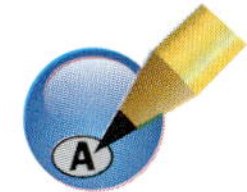

Fair Game Review What you learned in previous grades & lessons

Graph the linear equation. *(Section 4.4)*

24. $y = 4x$ 25. $y = -2x + 1$ 26. $y = 3x - 5$

27. **MULTIPLE CHOICE** What is the x-intercept of the equation $3x + 5y = 30$? *(Section 4.5)*

  (A) -10 (B) -6 (C) 6 (D) 10

4.4–4.7 Quiz

Check It Out
Progress Check
BigIdeasMath.com

Find the slope and the y-intercept of the graph of the linear equation. *(Section 4.4)*

1. $y = \frac{1}{4}x - 8$

2. $y = -x + 3$

Find the x- and y-intercepts of the graph of the equation. *(Section 4.5)*

3. $3x - 2y = 12$

4. $x + 5y = 15$

Write an equation of the line in slope-intercept form. *(Section 4.6)*

5.

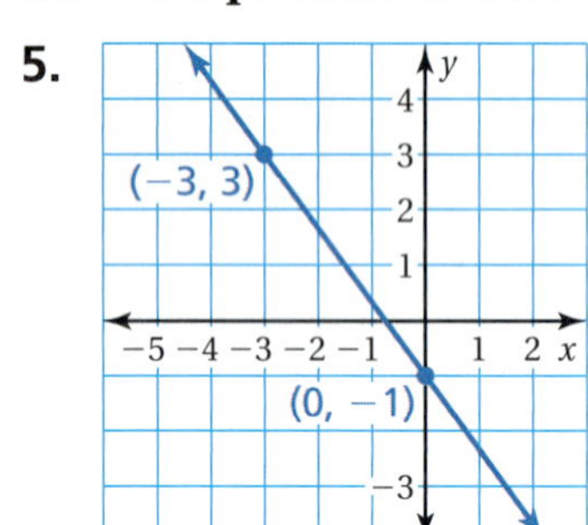

6.

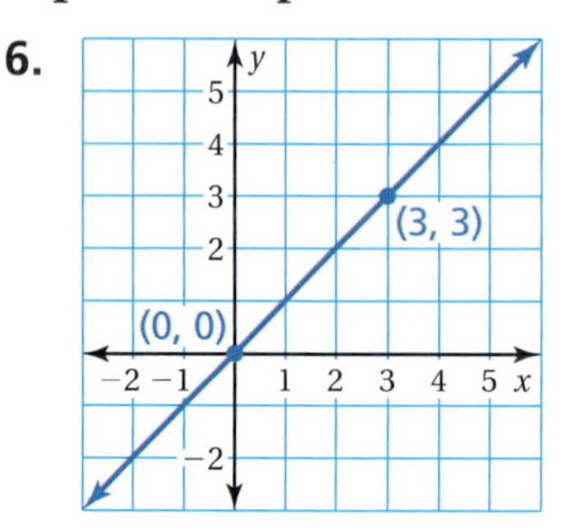

7.

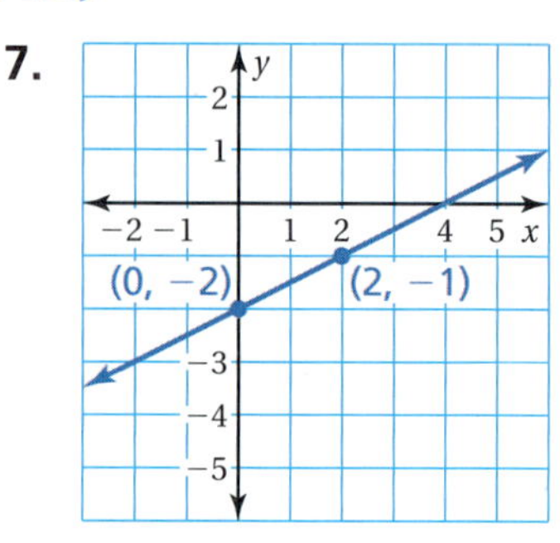

Write in point-slope form an equation of the line that passes through the given point and has the given slope. *(Section 4.7)*

8. $(1, 3);\ m = 2$

9. $(-3, -2);\ m = \frac{1}{3}$

10. $(-1, 4);\ m = -1$

11. $(8, -5);\ m = -\frac{1}{8}$

Write in slope-intercept form an equation of the line that passes through the given points. *(Section 4.7)*

12. $\left(0, -\frac{2}{3}\right), \left(-3, -\frac{2}{3}\right)$

13. $(4, 0), (0, 4)$

14. STATE FAIR The cost y (in dollars) of one person buying admission to a fair and going on x rides is $y = x + 12$. *(Section 4.4)*

a. Graph the equation.

b. Interpret the y-intercept and the slope.

15. PAINTING You used $90 worth of paint for a school float. *(Section 4.5)*

a. Graph the equation $18x + 15y = 90$, where x is the number of gallons of blue paint and y is the number of gallons of white paint.

b. Interpret the intercepts.

16. CONSTRUCTION A construction crew is extending a highway sound barrier that is 13 miles long. The crew builds $\frac{1}{2}$ of a mile per week. Write an equation that represents the length y (in miles) of the barrier after x weeks. *(Section 4.6)*

Alternative Assessment Options

Math Chat	Student Reflective Focus Question
Structured Interview	Writing Prompt

Student Reflective Focus Question

Ask students to summarize the similarities and differences of proportional relationships and nonproportional linear relationships. Be sure that they include examples. Select students at random to present their summaries to the class.

Study Help Sample Answers

Remind students to complete Graphic Organizers for the rest of the chapter.

3a.

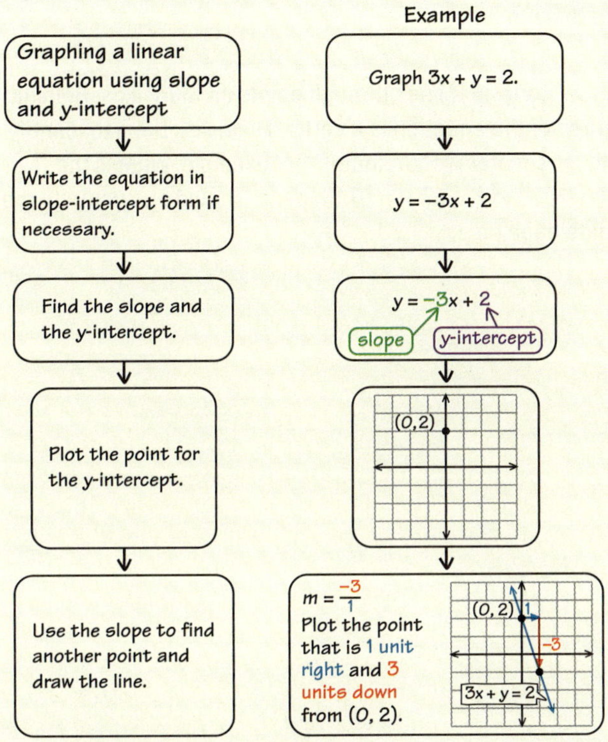

Example

Graphing a linear equation using slope and y-intercept → Graph 3x + y = 2.

Write the equation in slope-intercept form if necessary. → y = −3x + 2

Find the slope and the y-intercept. → y = −3x + 2 (slope: −3; y-intercept: 2)

Plot the point for the y-intercept. → (0, 2)

Use the slope to find another point and draw the line. → $m = \frac{-3}{1}$ Plot the point that is 1 unit right and 3 units down from (0, 2). (0, 2), 1, −3, 3x + y = 2

3b, 4–5. Available at *BigIdeasMath.com.*

Reteaching and Enrichment Strategies

If students need help. . .	If students got it. . .
Resources by Chapter • Practice A and Practice B • Puzzle Time Lesson Tutorials *BigIdeasMath.com*	Resources by Chapter • Enrichment and Extension • Technology Connection Game Closet at *BigIdeasMath.com* Start the Chapter Review

Answers

1. slope: $\frac{1}{4}$
   y-intercept: -8
2. slope: -1
   y-intercept: 3
3. x-intercept: 4
   y-intercept: -6
4. x-intercept: 15
   y-intercept: 3
5. $y = -\frac{4}{3}x - 1$
6. $y = x$
7. $y = \frac{1}{2}x - 2$
8. $y - 3 = 2(x - 1)$
9. $y + 2 = \frac{1}{3}(x + 3)$
10. $y - 4 = -1(x + 1)$
11. $y + 5 = -\frac{1}{8}(x - 8)$
12. $y = -\frac{2}{3}$
13. $y = -x + 4$
14. a. (graph: y from 12 to 20, x from 1 to 9)

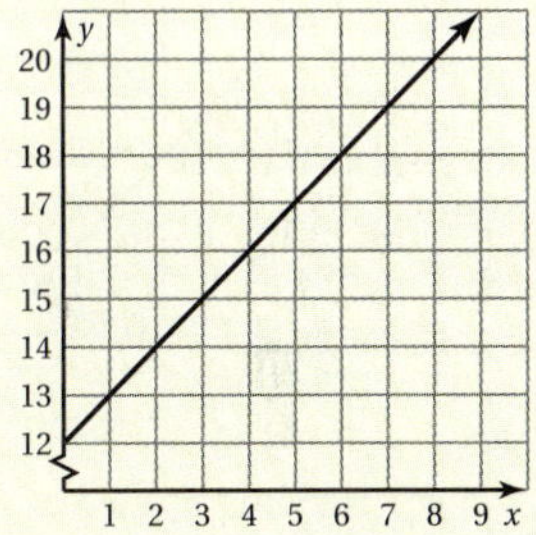

    b. The y-intercept represents the admission price of \$12 and the slope represents the unit cost of \$1 per ride.
15. See Additional Answers.
16. $y = \frac{1}{2}x + 13$

Technology for the Teacher

Online Assessment
Assessment Book
ExamView® Assessment Suite

For the Teacher
Additional Review Options

- *BigIdeasMath.com*
- Online Assessment
- Game Closet at *BigIdeasMath.com*
- Vocabulary Help
- Resources by Chapter

Answers

1.

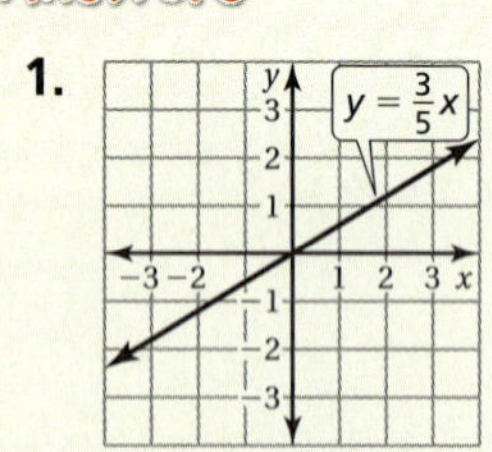

2.

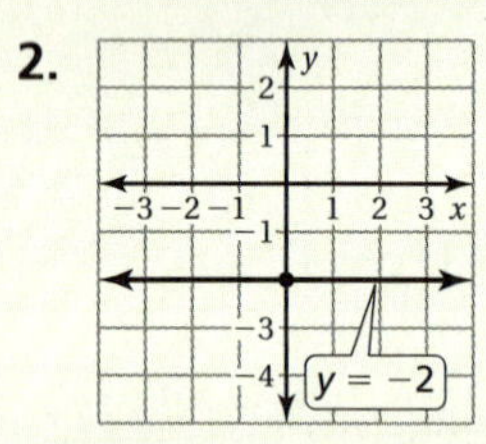

3.

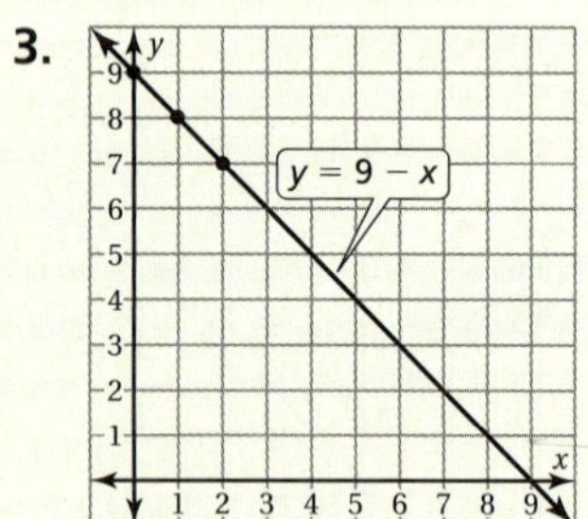

4.

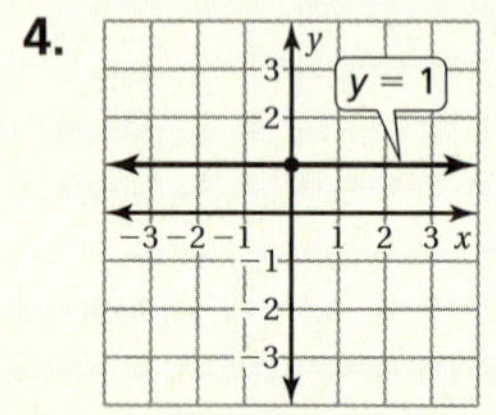

5.

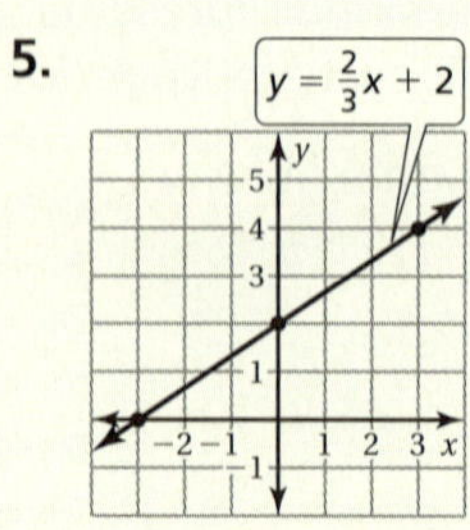

6.

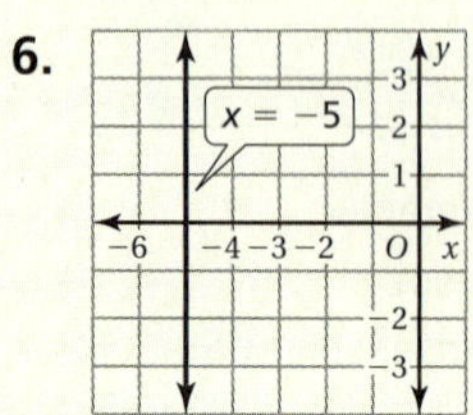

Review of Common Errors

Exercises 1–6

- Students may make calculation errors when solving for ordered pairs. If they only find two ordered pairs for the graph, they may not recognize their mistakes. Encourage them to find at least three ordered pairs when drawing a graph.

Exercises 2 and 4

- Students may draw vertical lines through points on the x-axis. Remind them that the graph of the equation is a horizontal line. Ask them to identify the y-coordinate for several x-coordinates. For example, what is the y-coordinate for $x = 3$? $x = 8$? $x = -5$? Students should answer with the same y-coordinate each time.

Exercise 6

- Students may draw horizontal lines through points on the y-axis. Remind them that the graph of the equation is a vertical line. Ask them to identify the x-coordinate for several y-coordinates. For example, what is the x-coordinate for $y = 2$? $y = -6$? $y = 0$? Students should answer with the same x-coordinate each time.

4 Chapter Review

Check It Out
Vocabulary Help
BigIdeasMath.com

Review Key Vocabulary

linear equation *p. 144*
solution of a linear equation, *p. 144*
slope, *p. 150*
rise, *p. 150*
run, *p. 150*
x-intercept, *p. 168*
y-intercept, *p. 168*
slope-intercept form, *p. 168*
standard form, *p. 174*
point-slope form, *p. 186*

Review Examples and Exercises

4.1 Graphing Linear Equations *(pp. 142–147)*

Graph $y = 3x - 1$.

Step 1: Make a table of values.

x	$y = 3x - 1$	y	(x, y)
-2	$y = 3(-2) - 1$	-7	$(-2, -7)$
-1	$y = 3(-1) - 1$	-4	$(-1, -4)$
0	$y = 3(0) - 1$	-1	$(0, -1)$
1	$y = 3(1) - 1$	2	$(1, 2)$

Step 2: Plot the ordered pairs.

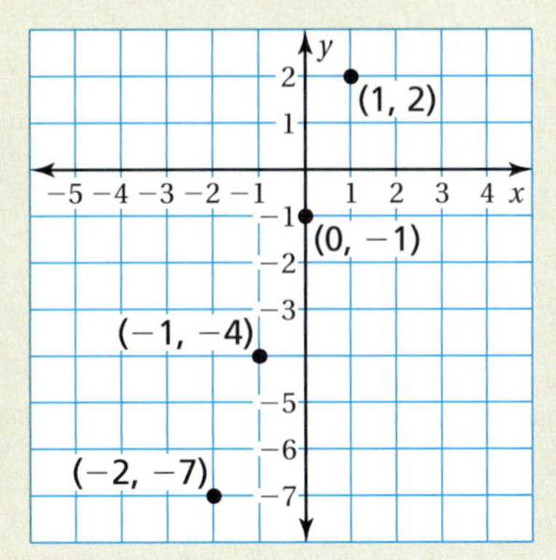

Step 3: Draw a line through the points.

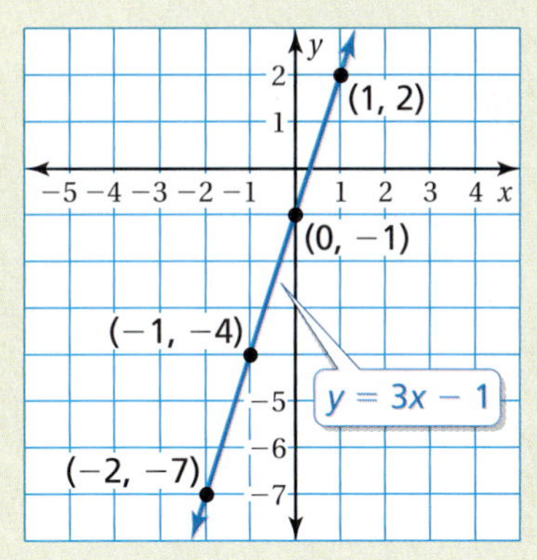

Exercises

Graph the linear equation.

1. $y = \frac{3}{5}x$
2. $y = -2$
3. $y = 9 - x$
4. $y = 1$
5. $y = \frac{2}{3}x + 2$
6. $x = -5$

4.2 Slope of a Line (pp. 148–157)

Find the slope of each line in the graph.

Red Line: $m = \dfrac{y_2 - y_1}{x_2 - x_1} = \dfrac{5 - (-3)}{2 - 2} = \dfrac{8}{0}$

The slope of the red line is undefined.

Blue Line: $m = \dfrac{y_2 - y_1}{x_2 - x_1} = \dfrac{-1 - 2}{4 - (-3)} = \dfrac{-3}{7}$, or $-\dfrac{3}{7}$

Green Line: $m = \dfrac{y_2 - y_1}{x_2 - x_1} = \dfrac{4 - 4}{5 - 0} = \dfrac{0}{5}$, or 0

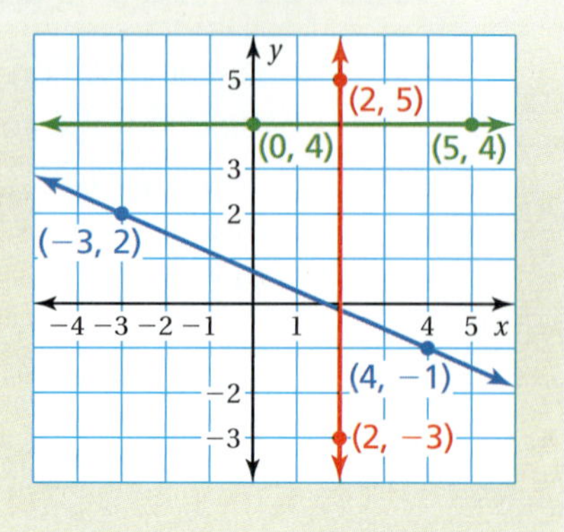

Exercises

The points in the table lie on a line. Find the slope of the line.

7.

x	0	1	2	3
y	−1	0	1	2

8.

x	−2	0	2	4
y	3	4	5	6

9. Are the lines $x = 2$ and $y = 4$ parallel? Are they perpendicular? Explain.

4.3 Graphing Proportional Relationships (pp. 158–163)

The cost y (in dollars) for x tickets to a movie is represented by the equation $y = 7x$. Graph the equation and interpret the slope.

The equation shows that the slope m is 7. So, the graph passes through (0, 0) and (1, 7).

Plot the points and draw a line through the points. Because negative values of x do not make sense in this context, graph in the first quadrant only.

The slope indicates that the unit cost is $7 per ticket.

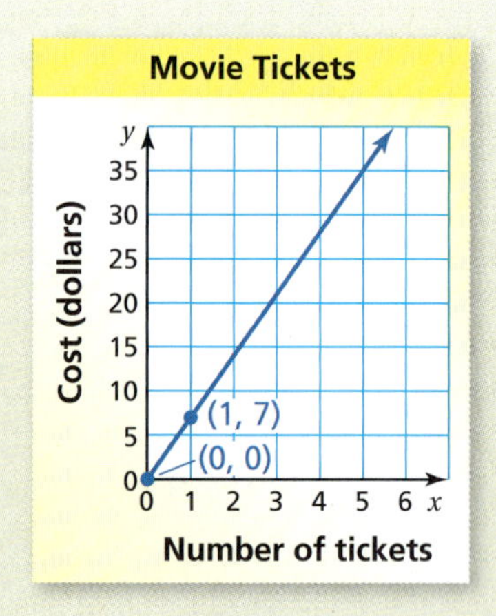

Exercises

10. RUNNING The number y of miles you run after x weeks is represented by the equation $y = 8x$. Graph the equation and interpret the slope.

11. STUDYING The number y of hours that you study after x days is represented by the equation $y = 1.5x$. Graph the equation and interpret the slope.

Review of Common Errors (continued)

Exercises 7 and 8

- Students may find the reciprocal of the slope instead of the slope. Remind them that slope is change in *y* over change in *x*.

Answers

7. 1

8. $\frac{1}{2}$

9. no; yes; The line $x = 2$ is vertical. The line $y = 4$ is horizontal. A vertical line is perpendicular to a horizontal line.

10.

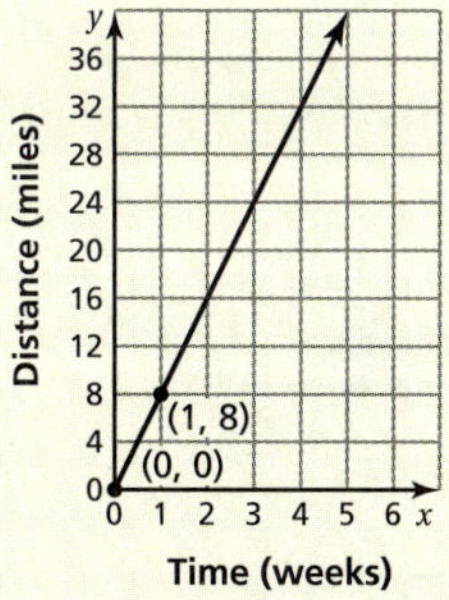

You run 8 miles per week.

11.

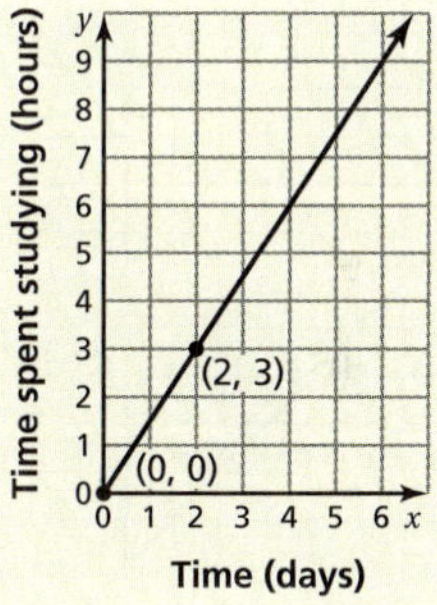

You study 1.5 hours per day.

Answers

12. 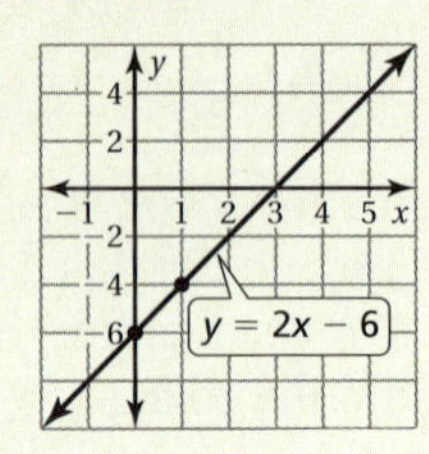

x-intercept: 3

13. 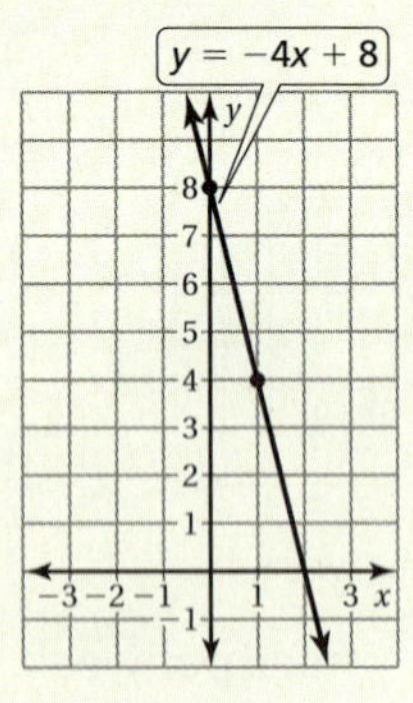

x-intercept: 2

14. 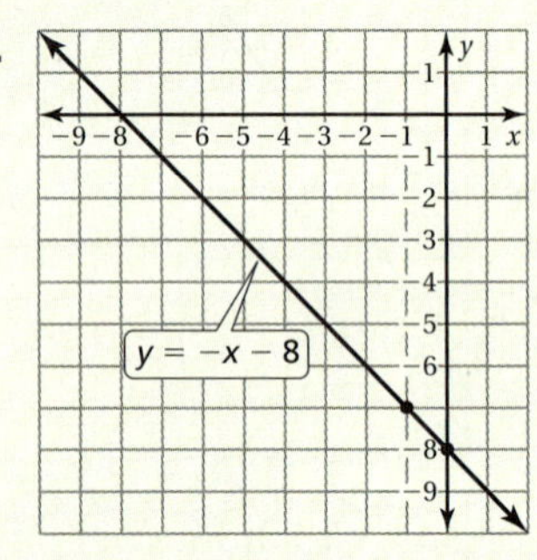

x-intercept: -8

15. 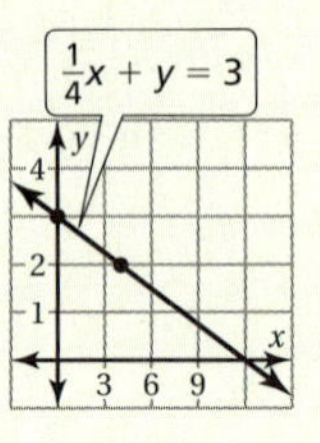

16. 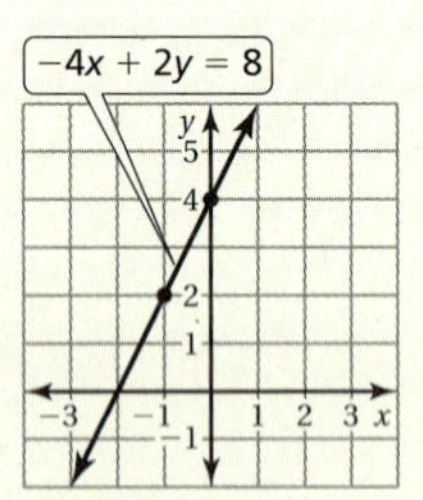

17. 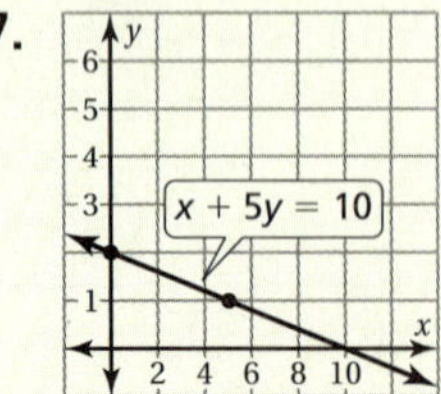

Review of Common Errors (continued)

Exercises 12–14

- Students may forget to include negatives with the slope and/or *y*-intercept. Remind them to look at the sign in front of the slope and the *y*-intercept. Also remind students that slope-intercept form is $y = mx + b$. This means that if the linear equation has "minus *b*," then the *y*-intercept is negative.
- Students may use the reciprocal of the slope when graphing and may find an incorrect *x*-intercept. Remind them that slope is *rise* over *run*, so the numerator represents vertical change, not horizontal.

Exercises 15–19

- Students may use the same operation instead of the inverse operation when rewriting the equation in slope-intercept form. Remind them of the steps to rewrite an equation.
- Students may mix up the *x*- and *y*-intercepts. Remind them that the *x*-intercept is the *x*-coordinate of where the line crosses the *x*-axis and the *y*-intercept is the *y*-coordinate of where the line crosses the *y*-axis.

Exercises 20–23

- Students may write the reciprocal of the slope or forget a negative sign. Remind them of the definition of slope. Ask them to predict the sign of the slope based on the rise or fall of the line.

Exercises 24 and 25

- Students may write the wrong equation when the slope is zero. For example, instead of $y = 5$, students may write $x = 5$. Ask them what is the rise of the graph (zero) and write this in slope-intercept form with the *y*-intercept as well, such as $y = 0x + 5$. Then ask students what happens when a variable (or any number) is multiplied by zero. Rewrite the equation as $y = 5$.

Exercises 26 and 27

- Students may use the reciprocal of the slope when writing the equation. Remind them that slope is the change in *y* over the change in *x*.

4.4 Graphing Linear Equations in Slope-Intercept Form (pp. 166–171)

Graph $y = 0.5x - 3$. Identify the x-intercept.

Step 1: Find the slope and the y-intercept.

$$y = 0.5x + (-3)$$

slope: 0.5; y-intercept: -3

Step 2: The y-intercept is -3. So, plot $(0, -3)$.

Step 3: Use the slope to find another point and draw the line.

$$m = \frac{\text{rise}}{\text{run}} = \frac{1}{2}$$

Plot the point that is 2 units right and 1 unit up from (0, –3). Draw a line through the two points.

∴ The line crosses the x-axis at (6, 0). So, the x-intercept is 6.

Exercises

Graph the linear equation. Identify the x-intercept. Use a graphing calculator to check your answer.

12. $y = 2x - 6$ **13.** $y = -4x + 8$ **14.** $y = -x - 8$

4.5 Graphing Linear Equations in Standard Form (pp. 172–177)

Graph $8x + 4y = 16$.

Step 1: Write the equation in slope-intercept form.

$8x + 4y = 16$ Write the equation.

$4y = -8x + 16$ Subtract $8x$ from each side.

$y = -2x + 4$ Divide each side by 4.

Step 2: Use the slope and the y-intercept to graph the equation.

$$y = -2x + 4$$

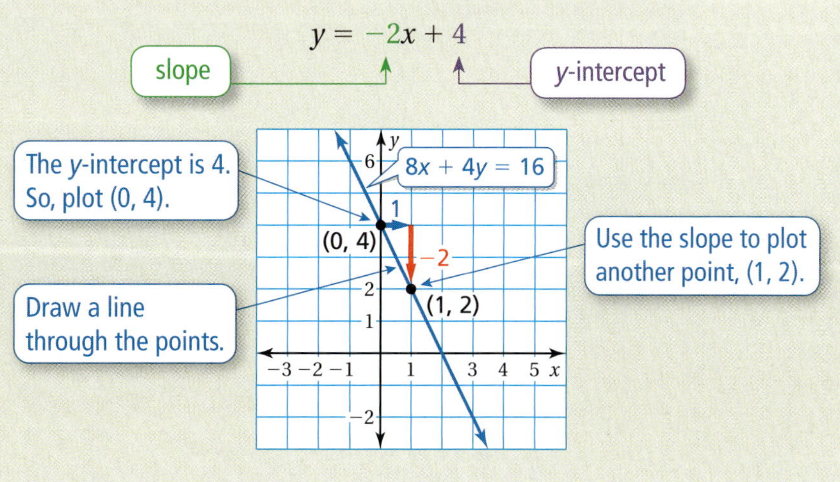

Exercises

Graph the linear equation.

15. $\frac{1}{4}x + y = 3$

16. $-4x + 2y = 8$

17. $x + 5y = 10$

18. $-\frac{1}{2}x + \frac{1}{8}y = \frac{3}{4}$

19. A dog kennel charges \$30 per night to board your dog and \$6 for each hour of playtime. The amount of money you spend is given by $30x + 6y = 180$, where x is the number of nights and y is the number of hours of playtime. Graph the equation and interpret the intercepts.

4.6 Writing Equations in Slope-Intercept Form *(pp. 178–183)*

Write an equation of the line in slope-intercept form.

a. Find the slope and the y-intercept.

$$m = \frac{y_2 - y_1}{x_2 - x_1} = \frac{4-2}{2-0} = \frac{2}{2}, \text{ or } 1$$

Because the line crosses the y-axis at (0, 2), the y-intercept is 2.

So, the equation is $y = 1x + 2$, or $y = x + 2$.

b. Find the slope and the y-intercept.

$$m = \frac{y_2 - y_1}{x_2 - x_1} = \frac{-4-(-2)}{3-0} = \frac{-2}{3}, \text{ or } -\frac{2}{3}$$

Because the line crosses the y-axis at (0, −2), the y-intercept is −2.

slope y-intercept

So, the equation is $y = -\frac{2}{3}x + (-2)$, or $y = -\frac{2}{3}x - 2$.

Review Game

Graphing Linear Equations

Materials per Group:

- map of the United States
- pencil
- straightedge

Directions:

On a map of the United States, students will place a coordinate plane with the origin located at Wichita, Kansas. The x-axis will go from -1700 miles to 1700 miles and the y-axis will go from -625 miles to 625 miles. These are roughly the dimensions of the United States.

The teacher will write equations and cities, in jumbled order, on the board. Students will work in groups and graph the equations to determine which line goes through which city.

Examples:

Dallas $y = \frac{156}{50}x - 156$

Denver $y = \frac{312}{625}x + 312$

Orlando $y = \frac{100}{200}x + 100$

Chicago $y = \frac{280}{600}x$

Las Vegas $y = \frac{625}{1275}x - 625$

Who Wins?

The first group to correctly graph the lines and match the cities wins.

For the Student
Additional Practice

- Lesson Tutorials
- Multi-Language Glossary
- Self-Grading Progress Check
- *BigIdeasMath.com*
  Dynamic Student Edition
  Student Resources

Answers

18.

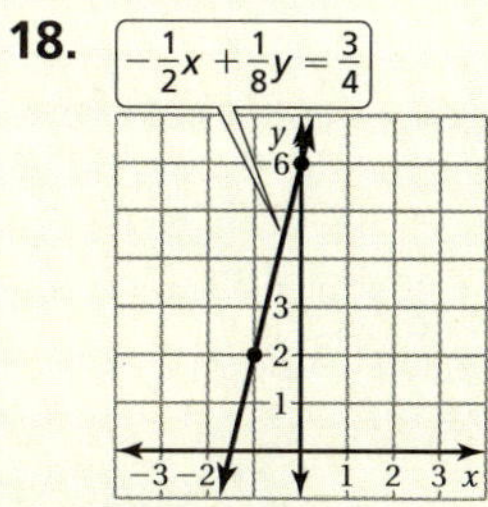

19.

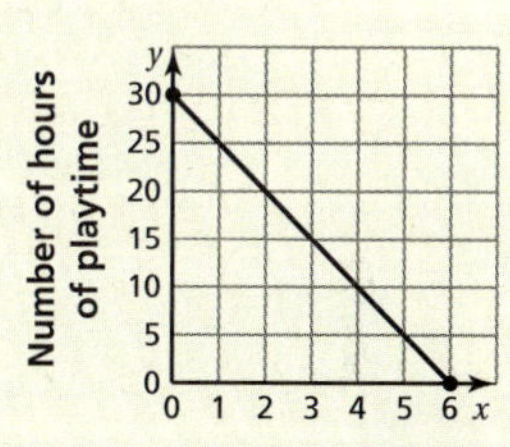

The x-intercept shows that you can board your dog for 6 nights when there are no hours of playtime. The y-intercept shows that you can have 30 hours of playtime for your dog when you do not leave your dog at the kennel for any nights.

20. $y = x - 2$

21. $y = -\frac{1}{2}x + 4$

22. $y = -2x + 1$

23. $y = 2x - 3$

24. $y = 8$

25. $y = -5$

26. $y - 4 = 3(x - 4)$

27. $y = -\frac{1}{2}x$

My Thoughts on the Chapter

Teacher Tip

Not allowed to write in your teaching edition? Use sticky notes to record your thoughts.

What worked. . .

What did not work. . .

What I would do differently. . .

Exercises

Write an equation of the line in slope-intercept form.

20.

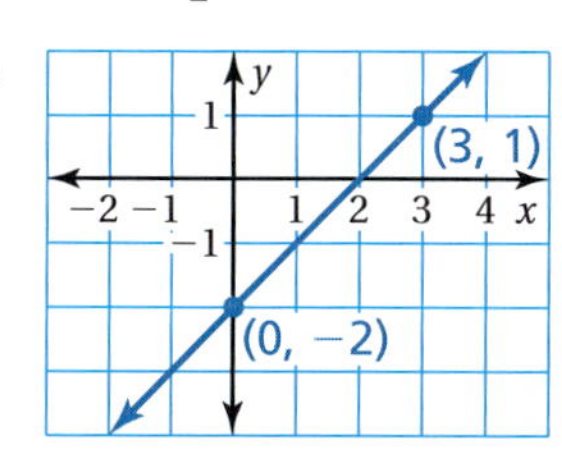

21.

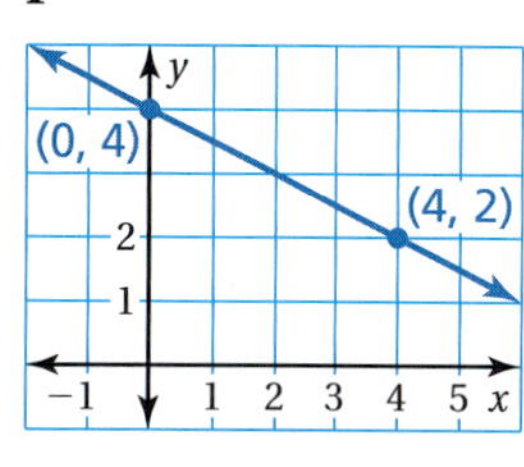

22.

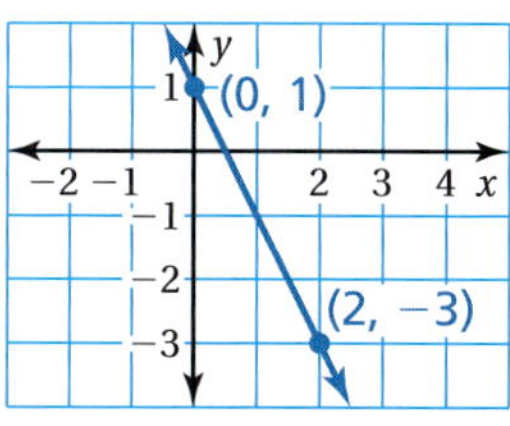

23.

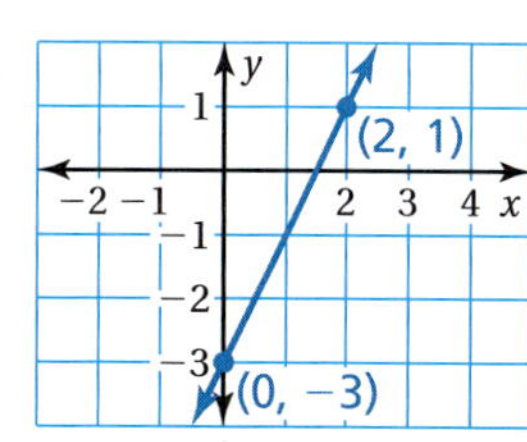

24. Write an equation of the line that passes through (0, 8) and (6, 8).

25. Write an equation of the line that passes through (0, −5) and (−5, −5).

4.7 Writing Equations in Point-Slope Form *(pp. 184–189)*

Write in slope-intercept form an equation of the line that passes through the points (2, 1) and (3, 5).

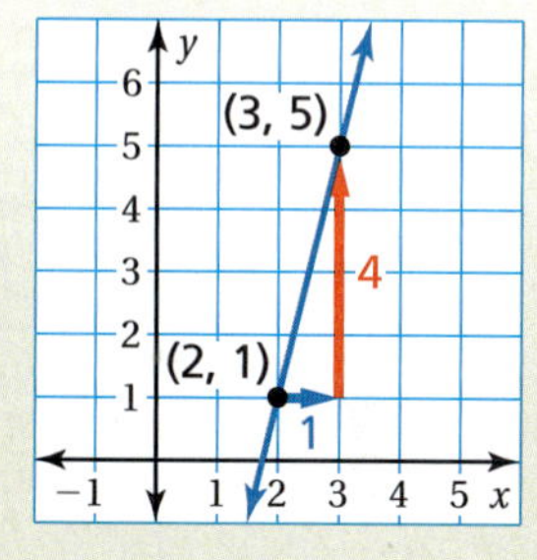

Find the slope.

$$m = \frac{y_2 - y_1}{x_2 - x_1} = \frac{5 - 1}{3 - 2} = \frac{4}{1}, \text{ or } 4$$

Then use the slope and one of the given points to write an equation of the line.

Use $m = 4$ and (2, 1).

$y - y_1 = m(x - x_1)$	Write the point-slope form.
$y - 1 = 4(x - 2)$	Substitute 4 for m, 2 for x_1, and 1 for y_1.
$y - 1 = 4x - 8$	Distributive Property
$y = 4x - 7$	Write in slope-intercept form.

So, the equation is $y = 4x - 7$.

Exercises

26. Write in point-slope form an equation of the line that passes through the point (4, 4) with slope 3.

27. Write in slope-intercept form an equation of the line that passes through the points (−4, 2) and (6, −3).

4 Chapter Test

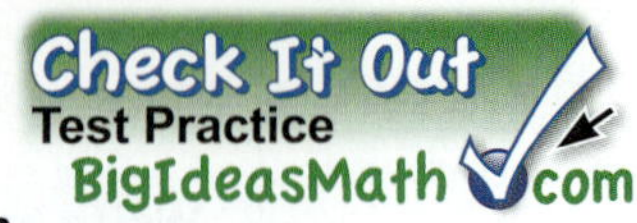

Find the slope and the y-intercept of the graph of the linear equation.

1. $y = 6x - 5$
2. $y = 20x + 15$
3. $y = -5x - 16$
4. $y - 1 = 3x + 8.4$
5. $y + 4.3 = 0.1x$
6. $-\frac{1}{2}x + 2y = 7$

Graph the linear equation.

7. $y = 2x + 4$
8. $y = -\frac{1}{2}x - 5$
9. $-3x + 6y = 12$

10. Which lines are parallel? Which lines are perpendicular? Explain.

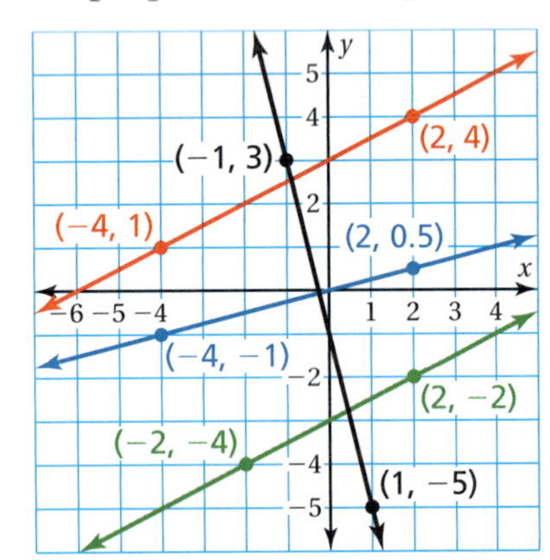

11. The points in the table lie on a line. Find the slope of the line.

x	y
−1	−4
0	−1
1	2
2	5

Write an equation of the line in slope-intercept form.

12.

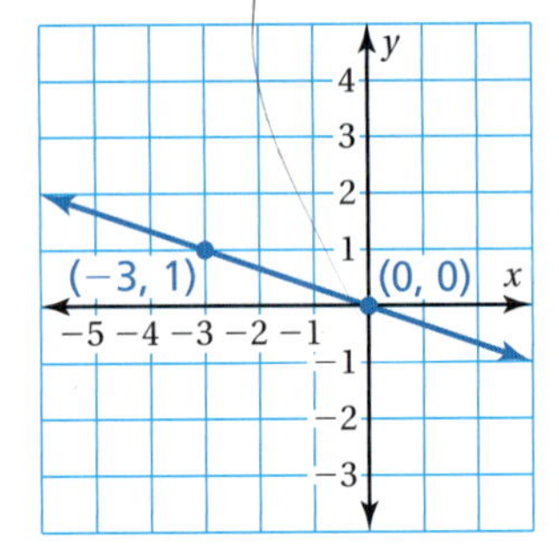

13.

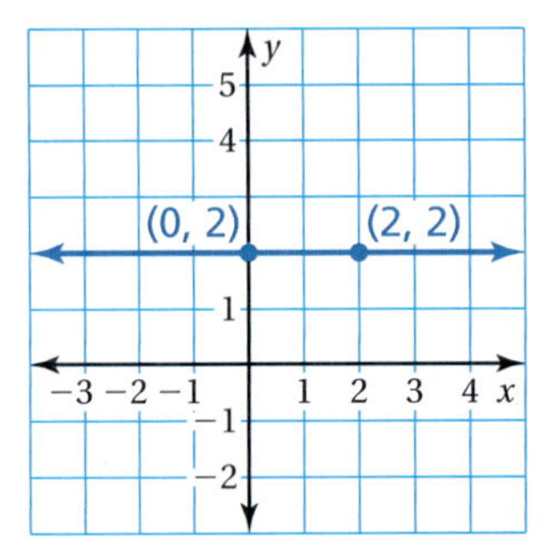

Write in slope-intercept form an equation of the line that passes through the given points.

14. $(-1, 5), (3, -3)$
15. $(-4, 1), (4, 3)$
16. $(-2, 5), (-1, 1)$

17. **VOCABULARY** The number y of new vocabulary words that you learn after x weeks is represented by the equation $y = 15x$.
    a. Graph the equation and interpret the slope.
    b. How many new vocabulary words do you learn after 5 weeks?
    c. How many more vocabulary words do you learn after 6 weeks than after 4 weeks?

Test Item References

Chapter Test Questions	Section to Review	Florida Common Core Standards (MACC)
7, 8	4.1	8.EE.2.5
10, 11	4.2	8.EE.2.6
17	4.3	8.EE.2.5, 8.EE.2.6
1–6	4.4	8.EE.2.6
9	4.5	8.EE.2.6
12, 13	4.6	8.F.2.4
14–16	4.7	8.F.2.4

Test-Taking Strategies

Remind students to quickly look over the entire test before they start so that they can budget their time. Students should jot down the formulas for slope-intercept form and point-slope form on the back of their test before they begin. Teach students to use the Stop and Think strategy before answering. **Stop** and carefully read the question, and **Think** about what the answer should look like.

Common Errors

- **Exercises 1–6** Students may use the reciprocal of the slope when graphing and may find an incorrect x-intercept. Remind them that slope is *rise* over *run*, so the numerator represents vertical change, not horizontal.
- **Exercises 7–9** Students may make calculation errors when solving for ordered pairs. If they only find two ordered pairs for the graph, they may not recognize their mistakes. Encourage them to find at least three ordered pairs when drawing a graph.
- **Exercise 12** Students may write the reciprocal of the slope or forget a negative sign. Ask them to predict the sign of the slope based on the rise or fall of the line.
- **Exercise 14–16** Students may use the reciprocal of the slope when writing the equation. Remind them that slope is the change in y over the change in x.

Reteaching and Enrichment Strategies

If students need help. . .	If students got it. . .
Resources by Chapter • Practice A and Practice B • Puzzle Time Record and Practice Journal Practice Differentiating the Lesson Lesson Tutorials *BigIdeasMath.com* Skills Review Handbook	Resources by Chapter • Enrichment and Extension • Technology Connection Game Closet at *BigIdeasMath.com* Start Standards Assessment

Answers

1. slope: 6; y-intercept: -5
2. slope: 20; y-intercept: 15
3. slope: -5; y-intercept: -16
4. slope: 3; y-intercept: 9.4
5. slope: 0.1; y-intercept: -4.3
6. slope: $\frac{1}{4}$; y-intercept: $\frac{7}{2}$

7–9. See Additional Answers.

10. The red and green lines are parallel. They both have a slope of $\frac{1}{2}$. The black and blue lines are perpendicular. The product of their slopes is -1.
11. 3
12. $y = -\frac{1}{3}x$
13. $y = 2$
14. $y = -2x + 3$
15. $y = \frac{1}{4}x + 2$
16. $y = -4x - 3$
17. a.

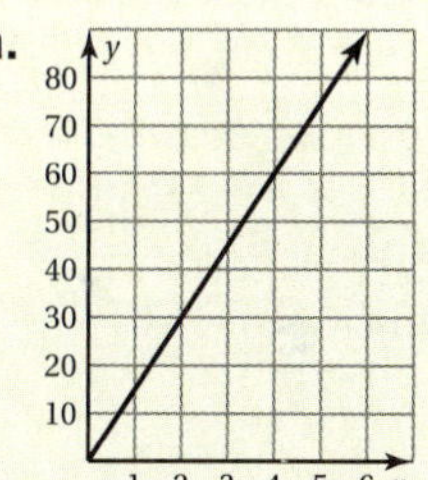

You learn 15 new vocabulary words per week.

b. 75 new vocabulary words

c. 30 more words

Technology for the Teacher

Online Assessment
Assessment Book
ExamView® Assessment Suite

Test-Taking Strategies

Available at *BigIdeasMath.com*

After Answering Easy Questions, Relax
Answer Easy Questions First
Estimate the Answer
Read All Choices before Answering
Read Question before Answering
Solve Directly or Eliminate Choices
Solve Problem before Looking at Choices
Use Intelligent Guessing
Work Backwards

About this Strategy

When taking a multiple choice test, be sure to read each question carefully and thoroughly. After reading the question, estimate the answer before trying to solve.

Answers

1. A
2. H
3. D

Technology for the Teacher

Florida Common Core Standards Support Performance Tasks
Online Assessment
Assessment Book
ExamView® Assessment Suite

Item Analysis

1. **A.** Correct answer

B. The student reads the slope correctly, but uses the wrong point to identify the y-intercept.

C. The student reads the y-intercept correctly, but miscalculates the slope.

D. The student finds the slope and y-intercept incorrectly.

2. **F.** The student interchanges correct values for x and y.

G. The student makes two errors: interchanging x and y, and assigning a negative sign incorrectly.

H. Correct answer

I. The student makes a mistake with a correct solution (4, 2) and assigns a negative value to 2, forgetting that there is already a minus sign in the equation.

3. **A.** The student mistakes slope for meaning that a line passes through (0, 0).

B. The student mistakes a vertical line for zero slope.

C. The student mistakes slope for meaning that a line passes through (0, 0).

D. Correct answer

4 Standards Assessment

1. Which equation matches the line shown in the graph? *(MACC.8.EE.2.6)*

   A. $y = 2x - 2$

   B. $y = 2x + 1$

   C. $y = x - 2$

   D. $y = x + 1$

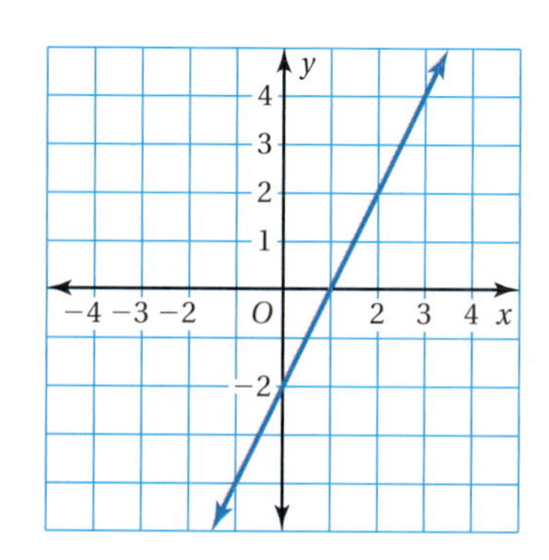

2. The equation $6x - 5y = 14$ is written in standard form. Which point lies on the graph of this equation? *(MACC.8.EE.2.6)*

   F. $(-4, -1)$

   G. $(-2, 4)$

   H. $(-1, -4)$

   I. $(4, -2)$

3. Which line has a slope of 0? *(MACC.8.EE.2.6)*

A.

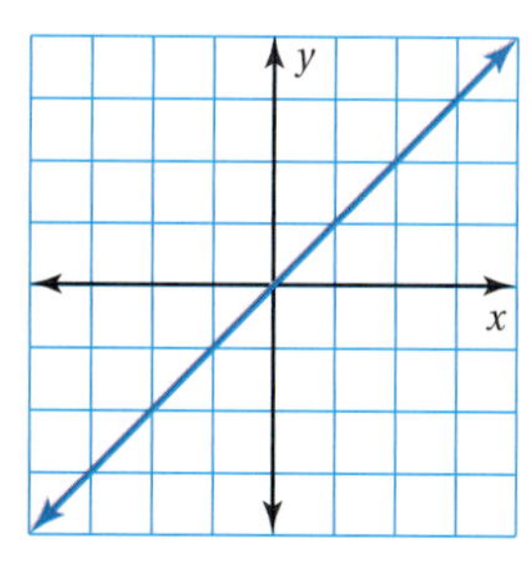

C.

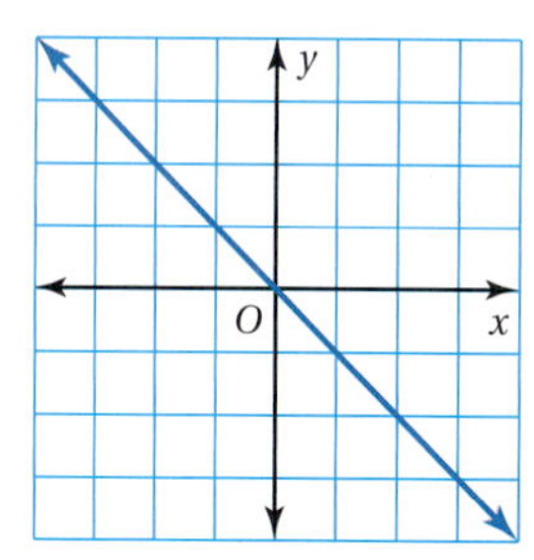

B.

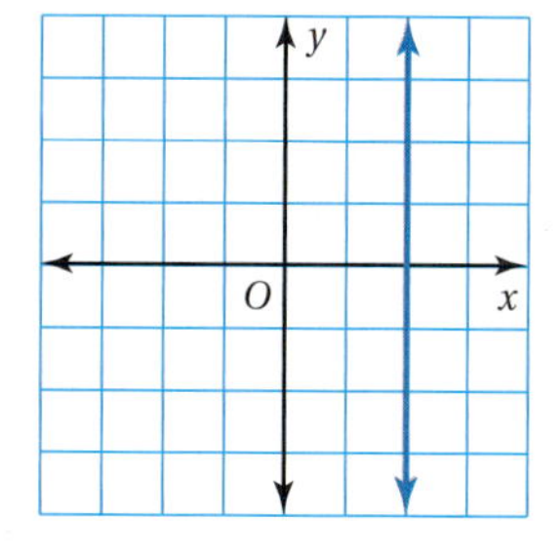

D.

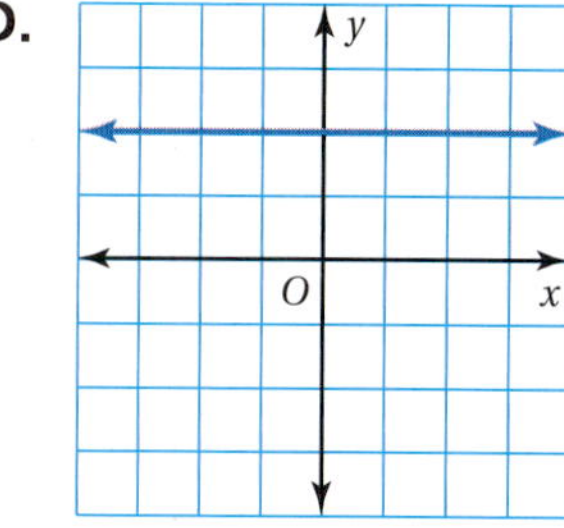

4. Which of the following is the equation of a line perpendicular to the line shown in the graph? *(MACC.8.EE.2.6)*

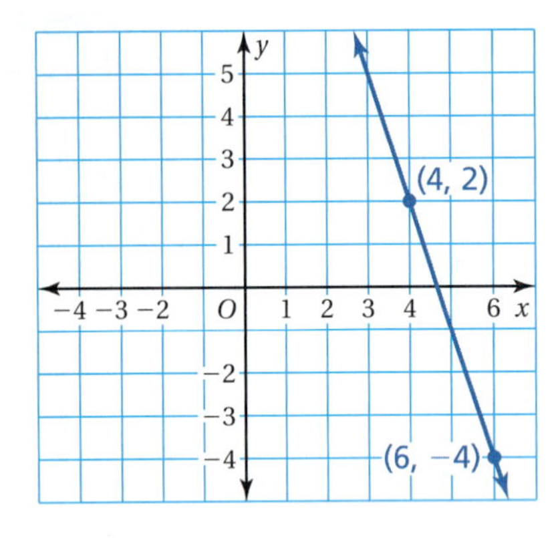

F. $y = 3x - 10$

G. $y = \frac{1}{3}x + 12$

H. $y = -3x + 5$

I. $y = -\frac{1}{3}x - 18$

5. What is the slope of the line that passes through the points $(2, -2)$ and $(8, 1)$? *(MACC.8.EE.2.6)*

6. A cell phone plan costs \$10 per month plus \$0.10 for each minute used. Last month, you spent \$18.50 using this plan. This can be modeled by the equation below, where m represents the number of minutes used.

$$0.1m + 10 = 18.5$$

How many minutes did you use last month? *(MACC.8.EE.3.7b)*

A. 8.4 min

B. 85 min

C. 185 min

D. 285 min

Think Solve Explain

7. It costs \$40 to rent a car for one day. In addition, the rental agency charges you for each mile driven, as shown in the graph. *(MACC.8.EE.2.6)*

Part A Determine the slope of the line joining the points on the graph.

Part B Explain what the slope represents.

Cost of Renting a Car

Rental cost (dollars)

y
100
80
60
40
20
0
(0, 40)
(100, 50)
(200, 60)
(300, 70)
(400, 80)
(500, 90)
0 100 200 300 400 500 x

Miles driven

Item Analysis (continued)

4. **F.** The student thinks perpendicular lines have slopes that are opposites.

   G. Correct answer

   H. The student thinks perpendicular lines have the same slope.

   I. The student thinks perpendicular lines have slopes that are reciprocals of each other.

5. **Gridded Response:** Correct answer: 0.5, or $\frac{1}{2}$

   Common Error: The student performs subtraction incorrectly for the *y*-terms, yielding an answer of $\frac{1}{6}$ or $-\frac{1}{6}$.

6. **A.** The student correctly subtracts 10 from both sides, but then subtracts 0.1 instead of dividing.

   B. Correct answer

   C. The student ignores 10 and simply divides by 0.1.

   D. The student adds 10 to both sides, and then divides by 0.1.

7. **2 points** The student demonstrates a thorough understanding of the slope of a line and what it represents, explains the work fully, and calculates the slope accurately. The slope of the line is $\frac{50-40}{100-0} = \frac{10}{100} = \frac{1}{10} = 0.10$. The slope represents the rental cost per mile driven, $0.10 per mile.

   1 point The student's work and explanations demonstrate a lack of essential understanding. The formula for the slope of a line is misstated, or the student incorrectly states what the slope of the line represents.

   0 points The student provides no response, a completely incorrect or incomprehensible response, or a response that demonstrates insufficient understanding of the slope of a line and what it represents.

Answers

4. G
5. $\frac{1}{2}$
6. B
7. *Part A* 0.10

   Part B $0.10 per mile

Answers

8. 6

9. F

10. B

11. H

Item Analysis (continued)

8. Gridded Response: Correct answer: 6

Common Error: The student correctly subtracts 7 from each side, but incorrectly adds $4x$ to $2x$ instead of subtracting.

9. **F.** Correct answer

G. The student reflects the figure in the y-axis.

H. The student rotates the figure 180°.

I. The student rotates the figure 90° counterclockwise.

10. **A.** The student divides M by 3, but fails to divide $(K + 7)$ by 3.

B. Correct answer

C. The student divides K by 3, but fails to divide 7 by 3.

D. The student subtracts 7 instead of adding it to both sides.

11. **F.** The student incorrectly sets up the proportion as $\frac{30}{100} = \frac{d}{12}$ or $\frac{100}{30} = \frac{12}{d}$.

G. The student thinks the corresponding side lengths are the same length.

H. Correct answer

I. The student incorrectly sets up the proportion as $\frac{d}{100} = \frac{30}{12}$ or $\frac{100}{d} = \frac{12}{30}$.

8. What value of x makes the equation below true? *(MACC.8.EE.3.7a)*

$$7 + 2x = 4x - 5$$

9. Trapezoid $KLMN$ is graphed in the coordinate plane shown.

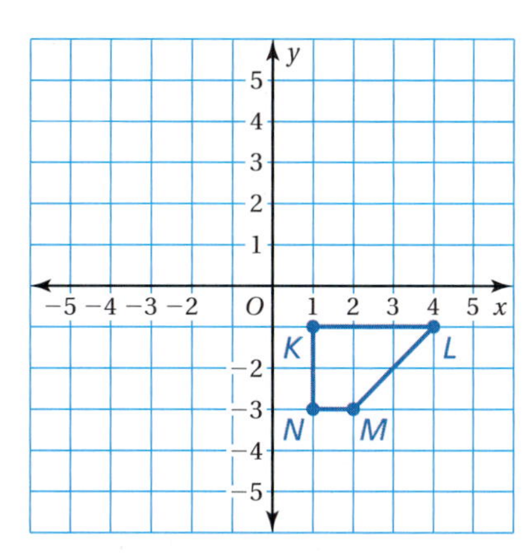

Rotate Trapezoid $KLMN$ 90° clockwise about the origin. What are the coordinates of point M', the image of point M after the rotation? *(MACC.8.G.1.3)*

F. $(-3, -2)$

G. $(-2, -3)$

H. $(-2, 3)$

I. $(3, 2)$

10. Solve the formula $K = 3M - 7$ for M. *(MACC.8.EE.3.7b)*

A. $M = K + 7$

B. $M = \frac{K+7}{3}$

C. $M = \frac{K}{3} + 7$

D. $M = \frac{K-7}{3}$

11. What is the distance d across the canyon? *(MACC.8.G.1.5)*

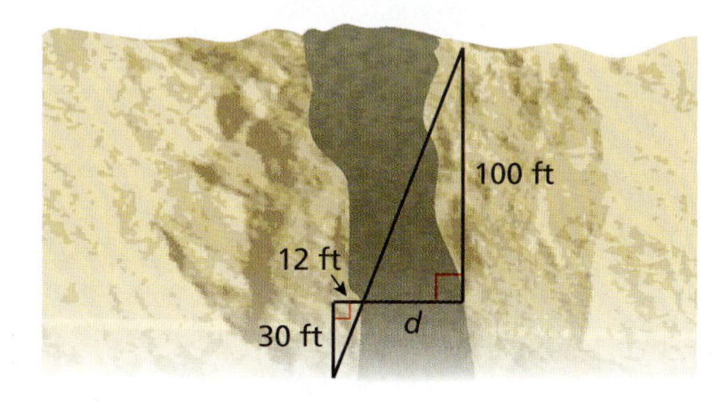

F. 3.6 ft

G. 12 ft

H. 40 ft

I. 250 ft

5 Systems of Linear Equations

Florida Common Core Progression

6th Grade

- Determine whether a value is a solution of an equation.
- Solve one-step equations.
- Write an equation to represent one quantity in terms of another quantity.
- Analyze the relationship between dependent and independent variables using graphs and tables.

7th Grade

- Write, graph, and solve one-step equations.
- Solve two-step equations.
- Construct simple equations.

8th Grade

- Show that a linear equation in one variable has one solution, infinitely many solutions, or no solution by transforming the equation into simpler forms.
- Solve multi-step equations.
- Understand that the solution of a system of two linear equations in two variables corresponds to the point of intersection of their graphs.
- Solve systems of two linear equations in two variables graphically and algebraically.
- Solve real-world mathematical problems leading to systems of two linear equations in two variables.

Pacing Guide for Chapter 5

Chapter Opener	1 Day
Section 1	
Activity	1 Day
Lesson	1 Day
Section 2	
Activity	1 Day
Lesson	1 Day
Study Help / Quiz	1 Day
Section 3	
Activity	1 Day
Lesson	1 Day
Section 4	
Activity	1 Day
Lesson	1 Day
Extension	1 Day
Chapter Review/ Chapter Tests	2 Days
Total Chapter 5	13 Days
Year-to-Date	77 Days

Chapter Summary

Section	Florida Common Core Standard (MACC)	
5.1	Learning	8.EE.3.8a, 8.EE.3.8b, 8.EE.3.8c
5.2	Learning	8.EE.3.8b, 8.EE.3.8c
5.3	Learning	8.EE.3.8b, 8.EE.3.8c
5.4	Learning	8.EE.3.7 ★, 8.EE.3.8a, 8.EE.3.8b, 8.EE.3.8c ★

★ Teaching is complete. Standard can be assessed.

Technology for the Teacher

BigIdeasMath.com
Chapter at a Glance
Complete Materials List
Parent Letters: English, Spanish, and Haitian Creole

Florida Common Core Standards

MACC.6.EE.1.3 Apply the properties of operations to generate equivalent expressions.

MACC.8.EE.3.7b Solve linear equations with rational number coefficients, including equations whose solutions require expanding expressions using the distributive property and collecting like terms.

Additional Topics for Review

- Writing Equations
- Solving Simple Equations
- Coordinate Plane
- Ordered Pairs
- Graphing Linear Equations
- Slope-Intercept Form
- Substitution
- Evaluating Algebraic Expressions
- Opposites

Try It Yourself

1. $2z + 5$ **2.** $6c + 43$

3. $x = 3$ **4.** $w = 4$

5. $z = 13$ **6.** $c = -17$

Record and Practice Journal Fair Game Review

1. $x + 5$ **2.** $-2d + 4$

3. $13y - 11$ **4.** $7z + 2$

5. $5s + 7$ **6.** $8x - 13$

7. $(12x - 2)$ feet

8. $y = 2$ **9.** $a = -3$

10. $k = 5$ **11.** $m = 6$

12. $t = -4$ **13.** $h = 9$

14. 45 calculators

Math Background Notes

Vocabulary Review

- Like Terms
- Expression
- Simplifying an Expression
- Properties of Addition and Multiplication
- Coefficient
- Equation
- Multi-Step Equation
- Checking a Solution

Combining Like Terms

- Students have been combining like terms to simplify expressions since Grade 6, so this topic should be very familiar to them.
- Students should also be familiar with the properties used in Example 1, including the Commutative Property of Addition, the Distributive Property, and the Multiplication Property of One. For those students who need a review, going through the solution steps of both parts of Example 1 should suffice.
- Some students may have learned the Multiplication Property of One by a different name: the Multiplicative Identity Property.
- **Common Error:** When combining like terms that have variables, you may still see students who want to change the exponent of the variable. For instance, they may write the answer to Example 1(a) as $9x^2 + 5$. Remind them that the exponents do not change when combining like terms. One analogy that seems to work for some students is to liken the variable to a tangible object, such as an apple. If you add 4 apples and 5 apples, do you get 9 apples or 9 *square* apples?

Solving Multi-Step Equations

- Students should know how to solve multi-step equations.
- To solve multi-step equations, use inverse operations to isolate the variable on one side of the equation.
- Remind students to isolate the variable term on one side of the equation before dividing each side by the coefficient of the variable term.
- **Common Error:** Students may use the Distributive Property incorrectly. Emphasize the second step in Example 2.

Reteaching and Enrichment Strategies

If students need help. . .	If students got it. . .
Record and Practice Journal • Fair Game Review Skills Review Handbook Lesson Tutorials	Game Closet at *BigIdeasMath.com* Start the next section

What You Learned Before

This hurts infinitely.

"Hold your tail a bit lower."

Combining Like Terms (MACC.6.EE.1.3)

Example 1 **Simplify each expression.**

a. $4x + 7 + 5x - 2$

$4x + 7 + 5x - 2 = 4x + 5x + 7 - 2$	Commutative Property of Addition
$= (4 + 5)x + 7 - 2$	Distributive Property
$= 9x + 5$	Simplify.

b. $z + z + z + z$

$z + z + z + z = 1z + 1z + 1z + 1z$	Multiplication Property of One
$= (1 + 1 + 1 + 1)z$	Distributive Property
$= 4z$	Add coefficients.

Try It Yourself

Simplify the expression.

1. $5 + 4z - 2z$

2. $5(c + 8) + c + 3$

Solving Multi-Step Equations (MACC.8.EE.3.7b)

Example 2 **Solve $4x - 2(3x + 1) = 16$.**

$4x - 2(3x + 1) = 16$	Write the equation.
$4x - 6x - 2 = 16$	Distributive Property
$-2x - 2 = 16$	Combine like terms.
$-2x = 18$	Add 2 to each side.
$x = -9$	Divide each side by -2.

The solution is $x = -9$.

Try It Yourself

Solve the equation. Check your solution.

3. $-5x + 8 = -7$

4. $7w + w - 15 = 17$

5. $-3(z - 8) + 10 = -5$

6. $2 = 10c - 4(2c - 9)$

5.1 Solving Systems of Linear Equations by Graphing

Essential Question How can you solve a system of linear equations?

1 ACTIVITY: Writing a System of Linear Equations

Work with a partner.

Your family starts a bed-and-breakfast. It spends $500 fixing up a bedroom to rent. The cost for food and utilities is $10 per night. Your family charges $60 per night to rent the bedroom.

a. Write an equation that represents the costs.

Cost, C (in dollars)	=	$10 per night	•	Number of nights, x	+	$500

b. Write an equation that represents the revenue (income).

Revenue, R (in dollars)	=	$60 per night	•	Number of nights, x

c. A set of two (or more) linear equations is called a **system of linear equations**. Write the system of linear equations for this problem.

2 ACTIVITY: Using a Table to Solve a System

Work with a partner. Use the cost and revenue equations from Activity 1 to find how many nights your family needs to rent the bedroom before recovering the cost of fixing up the bedroom. This is the *break-even point*.

a. Copy and complete the table.

x	0	1	2	3	4	5	6	7	8	9	10	11
C												
R												

b. How many nights does your family need to rent the bedroom before breaking even?

COMMON CORE

Systems of Equations

In this lesson, you will

- write and solve systems of linear equations by graphing.
- solve real-life problems.

Learning Standards
MACC.8.EE.3.8a
MACC.8.EE.3.8b
MACC.8.EE.3.8c

Laurie's Notes

Introduction

Standards for Mathematical Practice

- **MP1 Make Sense of Problems and Persevere in Solving Them:** Today, students will investigate a situation that can be represented by a system of equations. They will write a system of equations that represents the situation. They will have the opportunity to use different approaches to solve the system and discover the form and meaning of the solution. Encourage them to persevere in the different solution approaches.

Motivate

- Write the following "geometric equations" on the board. Explain that each square represents the same quantity, as does each triangle. Have students work with a partner to figure out what the square and triangle represent.

  $\square + \square + \square + \triangle = 47$

  $\square - \triangle = 1$

- Ask a pair of students to share their solution and explain how they figured out the answer. $\triangle = 11, \square = 12$; Explanations will vary.
- Share with students that this is the type of problem they will be working on in Chapter 5.

Activity Notes

Activity 1

- Discuss what is known about your costs and your income.
- **Financial Literacy:** Do not assume that students are knowledgeable about concepts such as *costs* (fixed and variable) and *revenue* (income). Explain these words as you use them.
- Point out to students that the units in the verbal model agree. This means that "dollars per night × nights" is equal to dollars. So, the units in the equation are dollars = dollars + dollars.
- Ask a pair of students to share the equations they wrote.
- Discuss the definition of a system of linear equations.

Activity 2

- Read the problem aloud. Define and discuss the break-even point.
- ? "Why would a business want to know the break-even point?" You want to know how many nights it will take before you start to make money.
- Give time for students to work with their partners to fill in the table. Make sure that all students are using correct equations.
- ? **Extension:** "What patterns do you observe in the table?" The two rows continue to get closer together until they are finally equal at $x = 10$. Then the revenue is greater than the cost.
- Make sure to interpret the answer to part (b). When you rent the room for 10 nights, the costs and the revenue both equal \$600. A solution of each equation is (10, 600).

Florida Common Core Standards

MACC.8.EE.3.8a Understand that solutions to a system of two linear equations in two variables correspond to points of intersection of their graphs, because points of intersection satisfy both equations simultaneously.

MACC.8.EE.3.8b Solve systems of two linear equations in two variables algebraically, and estimate solutions by graphing the equations. Solve simple cases by inspection.

MACC.8.EE.3.8c Solve real-world and mathematical problems leading to two linear equations in two variables.

Previous Learning

Students should know how to graph linear equations in two variables.

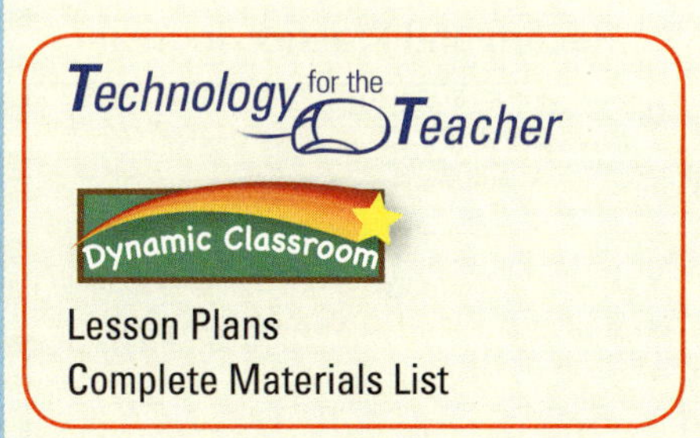

5.1 Record and Practice Journal

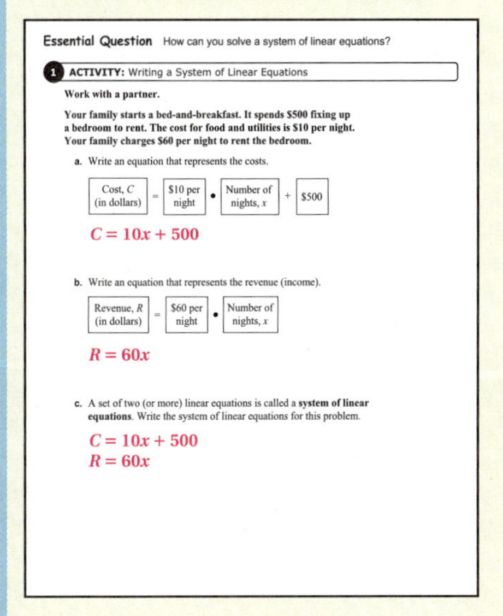

Essential Question How can you solve a system of linear equations?

1 ACTIVITY: Writing a System of Linear Equations

Work with a partner.

Your family starts a bed-and-breakfast. It spends \$500 fixing up a bedroom to rent. The cost for food and utilities is \$10 per night. Your family charges \$60 per night to rent the bedroom.

a. Write an equation that represents the costs.

Cost, C (in dollars) = \$10 per night • Number of nights, x + \$500

$C = 10x + 500$

b. Write an equation that represents the revenue (income).

Revenue, R (in dollars) = \$60 per night • Number of nights, x

$R = 60x$

c. A set of two (or more) linear equations is called a **system of linear equations.** Write the system of linear equations for this problem.

$C = 10x + 500$
$R = 60x$

English Language Learners

Pair Activity

Pair each English learner with an English speaker. Ask both students to solve the system graphically, but let one use a graphing utility while the other student makes the graphs by hand. Students then compare their answers. Partners should alternate solution methods as they continue to solve problems.

5.1 Record and Practice Journal

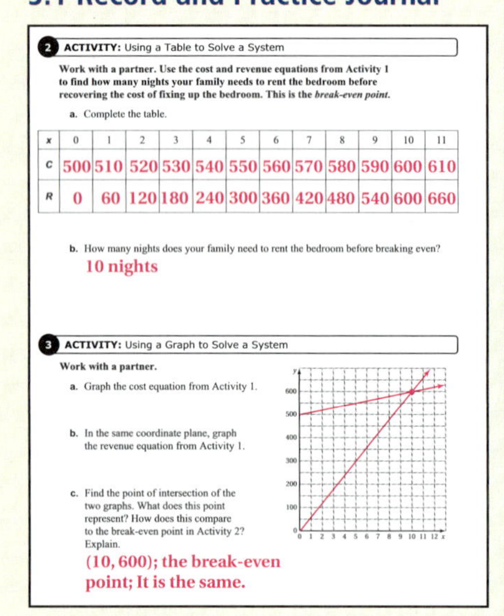

2 ACTIVITY: Using a Table to Solve a System

Work with a partner. Use the cost and revenue equations from Activity 1 to find how many nights your family needs to rent the bedroom before recovering the cost of fixing up the bedroom. This is the *break-even point*.

a. Complete the table.

x	0	1	2	3	4	5	6	7	8	9	10	11
C	500	510	520	530	540	550	560	570	580	590	600	610
R	0	60	120	180	240	300	360	420	480	540	600	660

b. How many nights does your family need to rent the bedroom before breaking even?
10 nights

3 ACTIVITY: Using a Graph to Solve a System

Work with a partner.

a. Graph the cost equation from Activity 1.

b. In the same coordinate plane, graph the revenue equation from Activity 1.

c. Find the point of intersection of the two graphs. What does this point represent? How does this compare to the break-even point in Activity 2? Explain.
(10, 600); the break-even point; It is the same.

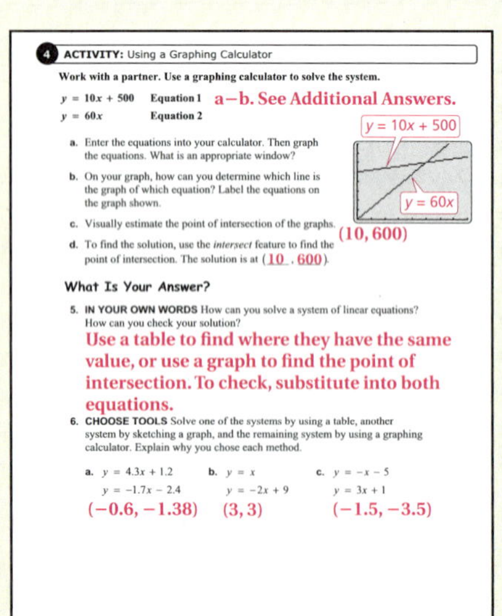

4 ACTIVITY: Using a Graphing Calculator

Work with a partner. Use a graphing calculator to solve the system.

$y = 10x + 500$ **Equation 1** a–b. See Additional Answers.
$y = 60x$ **Equation 2**

a. Enter the equations into your calculator. Then graph the equations. What is an appropriate window?

b. On your graph, how can you determine which line is the graph of which equation? Label the equations on the graph shown.

c. Visually estimate the point of intersection of the graphs. (10, 600)

d. To find the solution, use the *intersect* feature to find the point of intersection. The solution is at (10, 600).

What Is Your Answer?

5. **IN YOUR OWN WORDS** How can you solve a system of linear equations? How can you check your solution?
Use a table to find where they have the same value, or use a graph to find the point of intersection. To check, substitute into both equations.

6. **CHOOSE TOOLS** Solve one of the systems by using a table, another system by sketching a graph, and the remaining system by using a graphing calculator. Explain why you chose each method.

a. $y = 4.3x + 1.2$, $y = -1.7x - 2.4$ (−0.6, −1.38)

b. $y = x$, $y = -2x + 9$ (3, 3)

c. $y = -x - 5$, $y = 3x + 1$ (−1.5, −3.5)

Laurie's Notes

Activity 3

? "Look at the scaling of the axes for this activity. What are the units that will be used to graph each of these equations?" number of nights, dollars

- The cost equation is $C = 10x + 500$. The revenue equation is $R = 60x$.

? "In what form is the cost equation? What strategy can be used to graph the equation?" slope-intercept form; Students might say just plot the points from the table in Activity 2. Others might say plot the point for the *y*-intercept and then use a slope of 10 (which is equivalent to right 10 units, up 100 units).

? "In what form is the revenue equation? What strategy can be used to graph the equation?" similar response to previous question

- Provided that students have graphed the equations carefully, the lines should intersect at (10, 600).
- Make sure to interpret the answer to part (c). When you rent the room for 10 nights, the costs and the revenue both equal $600. A point on each line is (10, 600). It is the point of intersection for the graphs.

Activity 4

- Students have used a graphing calculator before to graph equations. They will enter both equations in the equation editor.
- Students should have insight from Activity 3 about how to set an appropriate viewing window.
- **MP2 Reason Abstractly and Quantitatively:** In part (b), students should be able to reason that the steeper graph has the greater slope.
- Students familiar with the *trace* feature of their graphing calculators can use it to answer part (c).
- If students have never used the *intersect* feature of their graphing calculators, then they may need help with part (d).
- **Extension:** As time allows, ask questions such as the following.

  ? "Name a solution of the first equation ($y = 10x + 500$) that is *not* a solution of the second equation ($y = 60x$)." *Sample answer:* (0, 500)

  ? "Name a solution of the second equation ($y = 60x$) that is *not* a solution of the first equation ($y = 10x + 500$)." *Sample answer:* (0, 0)

  ? "Name a solution of *both* equations." (10, 600)

  ? "Name an ordered pair that is *not* a solution of either equation." *Sample answer:* (10, 10)

What Is Your Answer?

- **MP6 Attend to Precision:** In Question 5, students should realize the importance of checking solutions of systems of equations.
- **MP5 Use Appropriate Tools Strategically:** In Question 6, students are choosing from the three methods shown in Activities 2, 3, and 4.

Closure

- **Phone call:** Write a brief script for a phone conversation with a friend who was not in class today. Explain what a system of linear equations is and how you solve a system of linear equations.

3 ACTIVITY: Using a Graph to Solve a System

Work with a partner.

a. Graph the cost equation from Activity 1.

b. In the same coordinate plane, graph the revenue equation from Activity 1.

c. Find the point of intersection of the two graphs. What does this point represent? How does this compare to the break-even point in Activity 2? Explain.

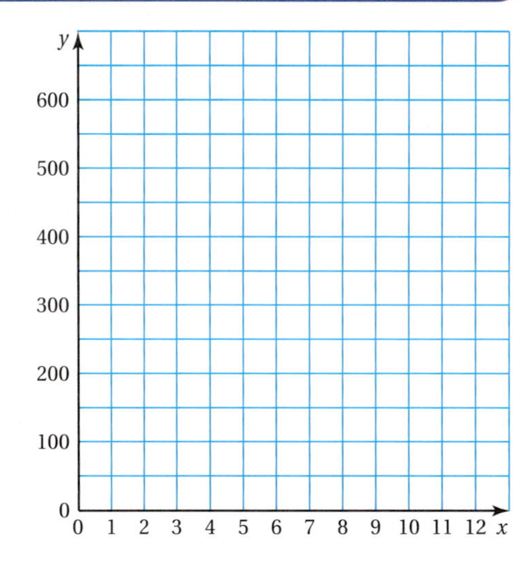

4 ACTIVITY: Using a Graphing Calculator

Work with a partner. Use a graphing calculator to solve the system.

$y = 10x + 500$ Equation 1

$y = 60x$ Equation 2

a. Enter the equations into your calculator. Then graph the equations. What is an appropriate window?

b. On your graph, how can you determine which line is the graph of which equation? Label the equations on the graph shown.

c. Visually estimate the point of intersection of the graphs.

d. To find the solution, use the *intersect* feature to find the point of intersection. The solution is (___ , ___).

Math Practice 5

Use Technology to Explore

How do you decide the values for the viewing window of your calculator? What other viewing windows could you use?

What Is Your Answer?

5. **IN YOUR OWN WORDS** How can you solve a system of linear equations? How can you check your solution?

6. **CHOOSE TOOLS** Solve one of the systems by using a table, another system by sketching a graph, and the remaining system by using a graphing calculator. Explain why you chose each method.

a. $y = 4.3x + 1.2$
$y = -1.7x - 2.4$

b. $y = x$
$y = -2x + 9$

c. $y = -x - 5$
$y = 3x + 1$

Practice Use what you learned about systems of linear equations to complete Exercises 4–6 on page 206.

5.1 Lesson

Key Vocabulary

system of linear equations, *p. 204*

solution of a system of linear equations, *p. 204*

A **system of linear equations** is a set of two or more linear equations in the same variables. An example is shown below.

$$y = x + 1 \qquad \text{Equation 1}$$
$$y = 2x - 7 \qquad \text{Equation 2}$$

A **solution of a system of linear equations** in two variables is an ordered pair that is a solution of each equation in the system. The solution of a system of linear equations is the point of intersection of the graphs of the equations.

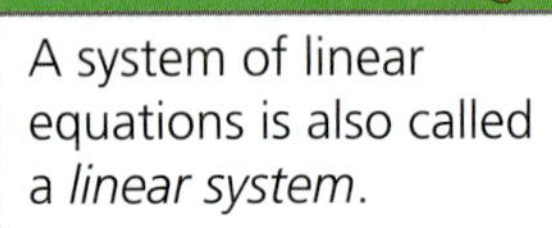

A system of linear equations is also called a *linear system*.

Solving a System of Linear Equations by Graphing

Step 1: Graph each equation in the same coordinate plane.

Step 2: Estimate the point of intersection.

Step 3: Check the point from Step 2 by substituting for x and y in each equation of the original system.

EXAMPLE  1 Solving a System of Linear Equations by Graphing

Solve the system by graphing. $\quad y = 2x + 5 \quad$ Equation 1

$\qquad\qquad\qquad\qquad\qquad\quad y = -4x - 1 \quad$ Equation 2

Step 1: Graph each equation.

Step 2: Estimate the point of intersection. The graphs appear to intersect at $(-1, 3)$.

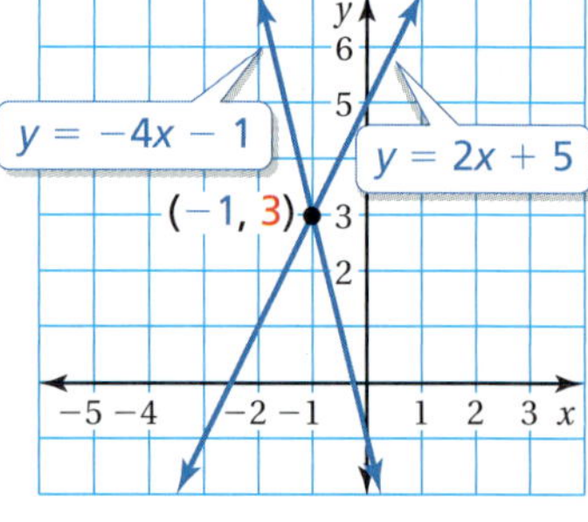

Step 3: Check the point from Step 2.

Equation 1	Equation 2
$y = 2x + 5$	$y = -4x - 1$
$3 \stackrel{?}{=} 2(-1) + 5$	$3 \stackrel{?}{=} -4(-1) - 1$
$3 = 3$ ✓	$3 = 3$ ✓

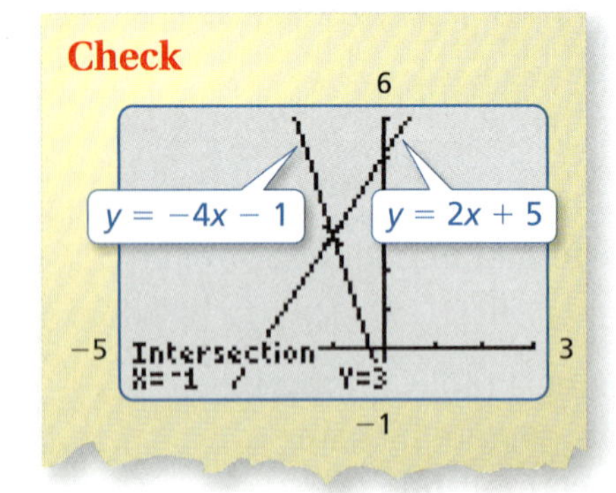

∴ The solution is $(-1, 3)$.

On Your Own

Now You're Ready
Exercises 10–12

Solve the system of linear equations by graphing.

1. $y = x - 1$
   $y = -x + 3$
2. $y = -5x + 14$
   $y = x - 10$
3. $y = x$
   $y = 2x + 1$

Multi-Language Glossary at BigIdeasMath.com

Laurie's Notes

Introduction

Connect

- **Yesterday:** Students investigated different approaches to solving a system of equations. (MP1, MP2, MP5, MP6)
- **Today:** Students will solve systems of linear equations by graphing.

Motivate

- Share a story about a trip to Indianapolis, where Market Street intersects Meridian Street at Monument Circle. Draw a sketch.

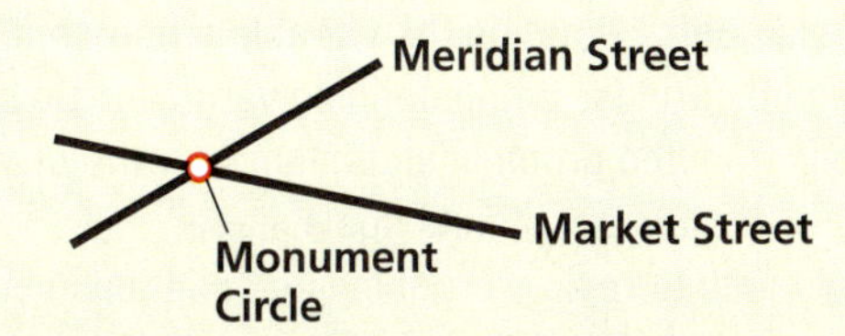

? Ask if students have ever visited a town or city where a monument was located in the middle of two streets.

- **Connection:** The monument is located on both streets. In other words, you will find the monument where the streets intersect.

Lesson Notes

Discuss

- Define a system of linear equations.

? "What is a solution of a linear equation in two variables?" an ordered pair that satisfies the equation

- Define a solution of a system of linear equations.

? "How many solutions do you think a system of linear equations can have and why?" Students are likely to suggest that there can be only one solution where the lines intersect. Do not correct this response—they will discover the rest of the possibilities in time.

Key Idea

- Discuss the steps for solving a system of linear equations by graphing.
- Checking solutions is important, especially if the graphing is done by hand.

? "Why do you have to check your answer in both equations?" It is possible for an ordered pair to satisfy only one of the equations.

Example 1

? "Do the lines intersect? Explain." yes; The slopes are different.

- This is a great time to review graphing an equation in slope-intercept form. One method is to plot the point for the y-intercept then find a second point on the graph using the slope. Another method is to make a table of values and plot ordered pairs.
- Work through the problem as shown.

? "Should you trust your eyes? What if the solution is actually $(-1.1, 3.2)$?" Checking the solution in the original equations will confirm your estimate.

Goal Today's lesson is solving a **system of linear equations** by graphing.

Technology for the Teacher

Lesson Tutorials
Lesson Plans
Answer Presentation Tool

Extra Example 1

Solve the system by graphing.

$y = -2x + 2$

$y = 3x - 3$

$(1, 0)$

On Your Own

1. $(2, 1)$
2. $(4, -6)$
3. $(-1, -1)$

Laurie's Notes

Extra Example 2

In Example 2, the kicker makes a total of 6 extra points and field goals and scores 12 points. Write and solve a system of equations to find the number of extra points and field goals. 3 extra points, 3 field goals

On Your Own

- **Neighbor Check:** Have students work independently and then have their neighbors check their work. Have students discuss any discrepancies.

Example 2

- **MP1 Make Sense of Problems and Persevere in Solving Them:** Ask a student to read the problem. Discuss vocabulary as needed. Ask questions to clarify the problem such as, "How many times did the kicker score points? How many of each type of kick was made? How many points were scored on field goals? What do the variables represent?"
- Students can be careless when defining variables. Be sure students use the definitions from the problem statement: $x =$ the *number* of extra points made and $y =$ the *number* of field goals made.
- This is a great time to review graphing an equation in standard form. One method is to graph the line using the intercepts $(0, y)$ and $(x, 0)$. A second method is to rewrite the equation in slope-intercept form.

 "What is your estimate for the point of intersection?" *Sample answer:* (6, 2)

- Remind students to check the solution in *both* equations.
- If time permits, use a graphing calculator to graph the equations.

On Your Own

4. $(-3, 5)$
5. $(-2, -7)$
6. $(4, -8)$
7. $x + y = 7$
   $x + 3y = 17$
   (2, 5); 2 extra points, 5 field goals

On Your Own

- Check students' work.
- **Question 6:** Ask a volunteer to share the solution. Ask students whether they worked with the fractional coefficient of the x-term in the first equation or they multiplied through by 2 to eliminate the fraction. In the second equation, did they divide through by 2 to simplify the equation before graphing? Students should be comfortable with these different approaches.

Closure

- **Exit Ticket:** Solve by graphing.
  $y = 2x + 3$
  $y = -x + 6$
  (1, 5)

Differentiated Instruction

Visual

Remind students that the graph of a linear equation in two variables is a line in which ordered pairs satisfy the equation. In a system of linear equations, the point of intersection of the two lines represents the ordered pair that satisfies both equations. Have students identify the x-value and the y-value of the solution and what the values represent in the problem.

EXAMPLE 2 Real-Life Application

A kicker on a football team scores 1 point for making an extra point and 3 points for making a field goal. The kicker makes a total of 8 extra points and field goals in a game and scores 12 points. Write and solve a system of linear equations to find the number x of extra points and the number y of field goals.

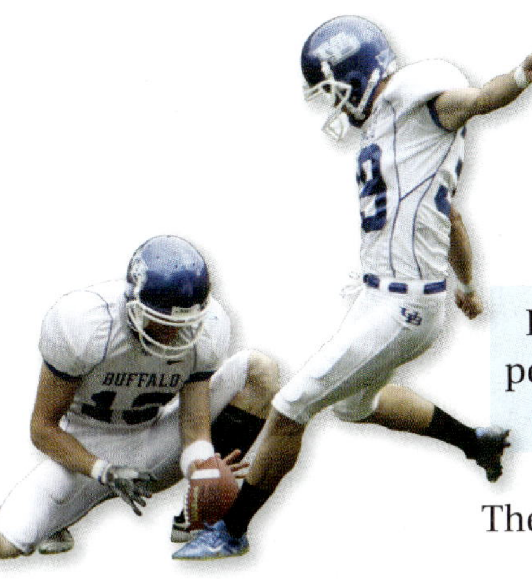

Use a verbal model to write a system of linear equations.

Number of extra points, x	+	Number of field goals, y	=	Total number of kicks

Points per extra point	•	Number of extra points, x	+	Points per field goal	•	Number of field goals, y	=	Total number of points

The system is: $x + y = 8$ Equation 1

$x + 3y = 12$ Equation 2

Step 1: Graph each equation.

Step 2: Estimate the point of intersection. The graphs appear to intersect at (6, 2).

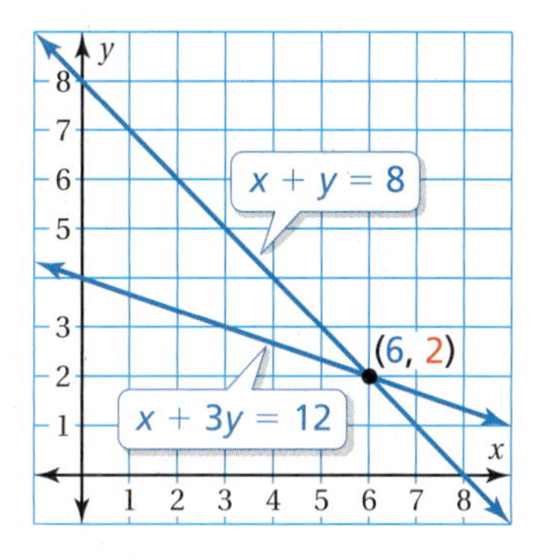

Step 3: Check your point from Step 2.

Equation 1	Equation 2
$x + y = 8$	$x + 3y = 12$
$6 + 2 \stackrel{?}{=} 8$	$6 + 3(2) \stackrel{?}{=} 12$
$8 = 8$ ✓	$12 = 12$ ✓

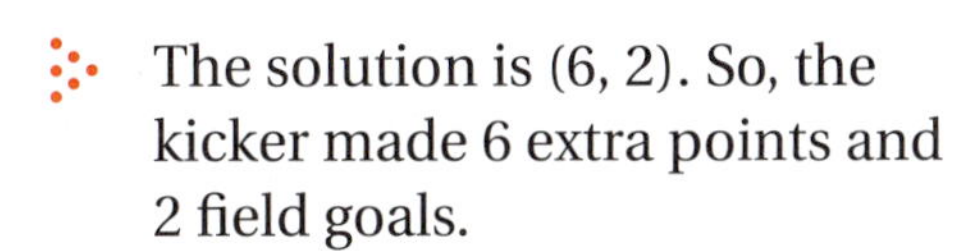

The solution is (6, 2). So, the kicker made 6 extra points and 2 field goals.

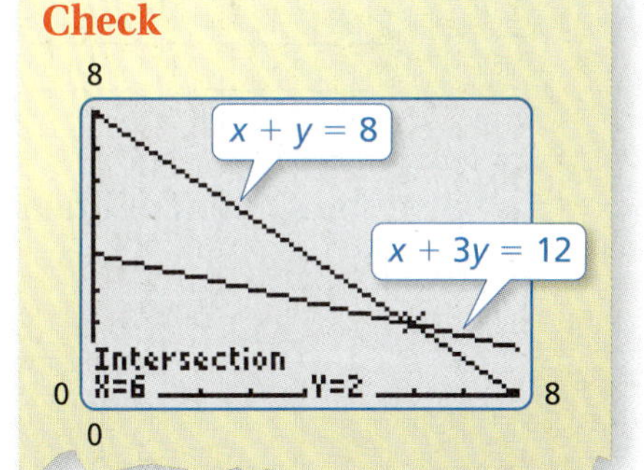

Study Tip

It may be easier to graph the equations in a system by rewriting the equations in slope-intercept form.

On Your Own

Now You're Ready
Exercises 13–15

Solve the system of linear equations by graphing.

4. $y = -4x - 7$
$x + y = 2$

5. $x - y = 5$
$-3x + y = -1$

6. $\frac{1}{2}x + y = -6$
$6x + 2y = 8$

7. WHAT IF? The kicker makes a total of 7 extra points and field goals and scores 17 points. Write and solve a system of linear equations to find the numbers of extra points and field goals.

5.1 Exercises

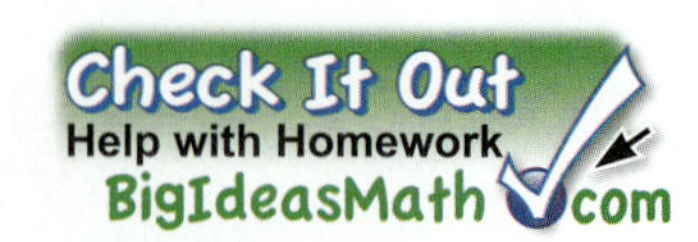

Vocabulary and Concept Check

1. **VOCABULARY** Do the equations $4x - 3y = 5$ and $7y + 2x = -8$ form a system of linear equations? Explain.
2. **WRITING** What does it mean to solve a system of equations?
3. **WRITING** You graph a system of linear equations, and the solution appears to be (3, 4). How can you verify that the solution is (3, 4)?

Practice and Problem Solving

Use a table to find the break-even point. Check your solution.

4. $C = 15x + 150$
   $R = 45x$
5. $C = 24x + 80$
   $R = 44x$
6. $C = 36x + 200$
   $R = 76x$

Match the system of linear equations with the corresponding graph. Use the graph to estimate the solution. Check your solution.

7. $y = 1.5x - 2$
   $y = -x + 13$
8. $y = x + 4$
   $y = 3x - 1$
9. $y = \frac{2}{3}x - 3$
   $y = -2x + 5$

A.

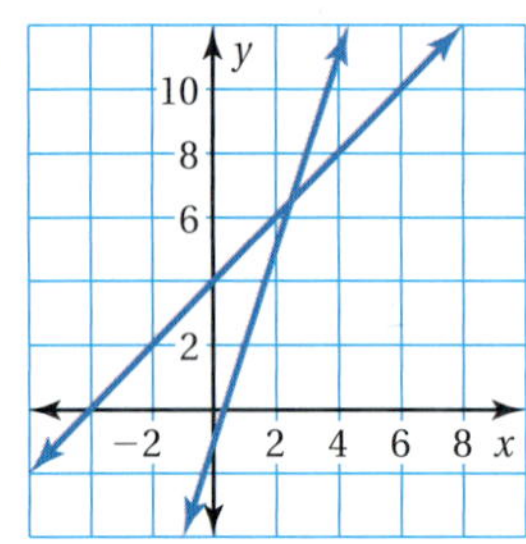

B.

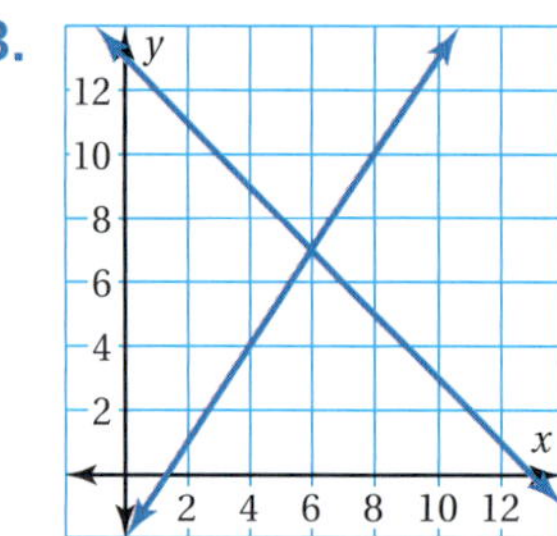

C.

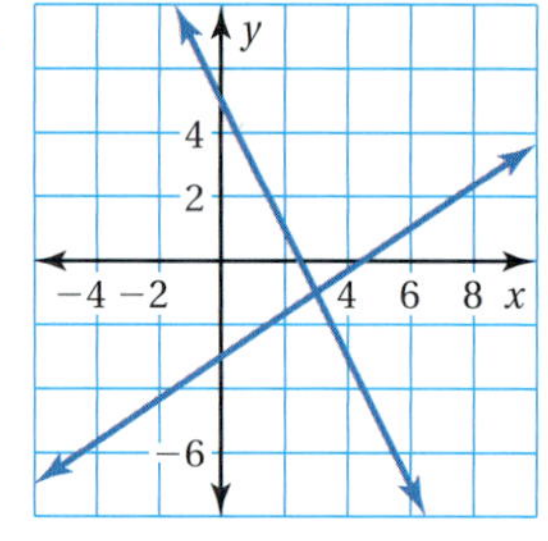

Solve the system of linear equations by graphing.

10. $y = 2x + 9$
    $y = 6 - x$
11. $y = -x - 4$
    $y = \frac{3}{5}x + 4$
12. $y = 2x + 5$
    $y = \frac{1}{2}x - 1$
13. $x + y = 27$
    $y = x + 3$
14. $y - x = 17$
    $y = 4x + 2$
15. $x - y = 7$
    $0.5x + y = 5$

16. **CARRIAGE RIDES** The cost C (in dollars) for the care and maintenance of a horse and carriage is $C = 15x + 2000$, where x is the number of rides.
    a. Write an equation for the revenue R in terms of the number of rides.
    b. How many rides are needed to break even?

Assignment Guide and Homework Check

Level	Day 1 Activity Assignment	Day 2 Lesson Assignment	Homework Check
Basic	4–6, 25–28	1–3, 7–19 odd, 20, 21, 23	11, 13, 15, 17, 23
Average	4–6, 25–28	1–3, 7–15 odd, 16–24 even	13, 15, 16, 18, 22
Advanced	1–6, 10–18 even, 20–28		14, 16, 18, 23, 24

Common Errors

- **Exercises 7–9** Students may use the graph directly below each system to estimate the solution. Remind them to first use the *y*-intercepts and slopes of the equations in each system to find the correct graph.
- **Exercises 10–15** Students may not show enough of the graph, so the lines will not intersect. Encourage them to extend their lines until they intersect.
- **Exercises 17–19** Students may try to visually estimate the point of intersection of the graphs on their graphing utility screens. Remind them to use the *intersect* feature to estimate the point of intersection.

5.1 Record and Practice Journal

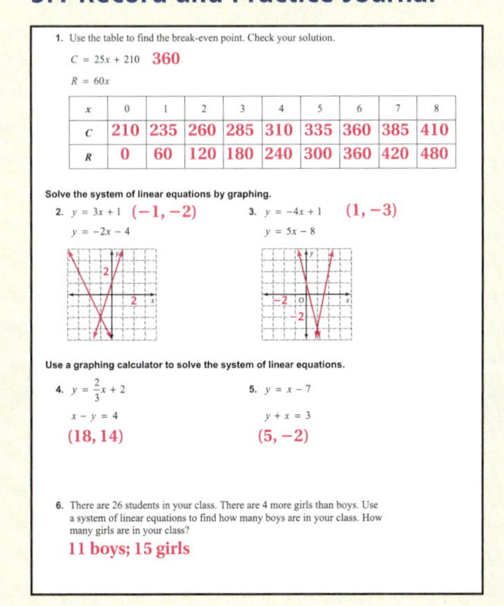

1. Use the table to find the break-even point. Check your solution.

$C = 25x + 210$ 360

$R = 60x$

x	0	1	2	3	4	5	6	7	8
C	210	235	260	285	310	335	360	385	410
R	0	60	120	180	240	300	360	420	480

Solve the system of linear equations by graphing.

2. $y = 3x + 1$ (−1, −2)
   $y = -2x - 4$

3. $y = -4x + 1$ (1, −3)
   $y = 5x - 8$

Use a graphing calculator to solve the system of linear equations.

4. $y = \frac{2}{3}x + 2$
   $x - y = 4$
   (18, 14)

5. $y = x - 7$
   $y + x = 3$
   (5, −2)

6. There are 26 students in your class. There are 4 more girls than boys. Use a system of linear equations to find how many boys are in your class. How many girls are in your class?
   11 boys; 15 girls

Vocabulary and Concept Check

1. yes; The equations are linear and in the same variables.
2. Find the ordered pair (x, y) that represents the point of intersection of the graphs of the equations in the system.
3. Check whether (3, 4) is a solution of each equation.

Practice and Problem Solving

4. (5, 225)
5. (4, 176)
6. (5, 380)
7. B; (6, 7)
8. A; (2.5, 6.5)
9. C; (3, −1)
10. (−1, 7)
11. (−5, 1)
12. (−4, −3)
13. (12, 15)
14. (5, 22)
15. (8, 1)
16. **a.** $R = 35x$
    b. 100 rides

Practice and Problem Solving

17. (5, 1.5)

18. (1, 3)

19. (−6, 2)

20. Only the x-value is given; The solution is (4, 3).

21. no; Two lines cannot intersect in exactly two points.

22. 26 math problems, 16 science problems

23. See *Taking Math Deeper.*

24. **a.** $y = 0.5x + 2.5$
$y = 0.4x + 5.8$

b. yes; month 33

Fair Game Review

25. $c = 8$ **26.** $y = 4$

27. $x = 11$ **28.** C

Mini-Assessment

Solve the system of linear equations by graphing.

1. $y = 2x + 6$
   $y = -2x - 2$ (−2, 2)
2. $y = 3x + 9$
   $y = -\frac{1}{4}x - 4$ (−4, −3)
3. $2x + y = 4$
   $y = x - 5$ (3, −2)
4. A wallet contains 23 bills. All the bills are $1 bills and $5 bills. There are 7 more $1 bills than $5 bills. How much money does the wallet contain? $55

Taking Math Deeper

Exercise 23

One way to look at this problem is to make a table to compare your cumulative distances at half-hour intervals.

Time	Your Distance	Friend's Distance
0 hour	0 mile	0.5 mile
0.5 hour	1.7 miles	2 miles
1 hour	3.4 miles	3.5 miles
1.5 hours	5.1 miles	5 miles
2 hours	6.8 miles	6.5 miles
2.5 hours	8.5 miles	8 miles

Because you are a half mile behind your friend, your distance is 0 mile and your friend's distance is 0.5 mile.

You paddle at 3.4 miles per hour, so your distance after 0.5 hour is half of that, or 1.7 miles. Use the rate of 1.7 miles per half-hour to complete your distances in the table.

Your friend paddles at 3 miles per hour, so your friend's distance after 0.5 hour is 1.5 miles. Use the rate of 1.5 miles per half-hour to complete your friend's distances in the table.

a. The table shows that you are 0.1 mile behind your friend after one hour and 0.1 mile ahead of your friend after 1.5 hours. You can deduce that you pass your friend halfway between those two times. So, you will catch up to your friend after paddling for 1.25 hours.

How far have you traveled?

By similar reasoning, the distance you have traveled when you catch up to your friend is the mean of 3.4 miles and 5.1 miles. So you catch up to your friend after traveling $(3.4 + 5.1)/2 = 4.25$ miles.

b. The table shows that your friend is 0.5 mile behind you when you finish the race.

Reteaching and Enrichment Strategies

If students need help. . .	If students got it. . .
Resources by Chapter • Practice A and Practice B • Puzzle Time Record and Practice Journal Practice Differentiating the Lesson Lesson Tutorials Skills Review Handbook	Resources by Chapter • Enrichment and Extension • Technology Connection Start the next section

Use a graphing calculator to solve the system of linear equations.

17. $2.2x + y = 12.5$
$1.4x - 4y = 1$

18. $2.1x + 4.2y = 14.7$
$-5.7x - 1.9y = -11.4$

19. $-1.1x - 5.5y = -4.4$
$0.8x - 3.2y = -11.2$

20. ERROR ANALYSIS Describe and correct the error in solving the system of linear equations.

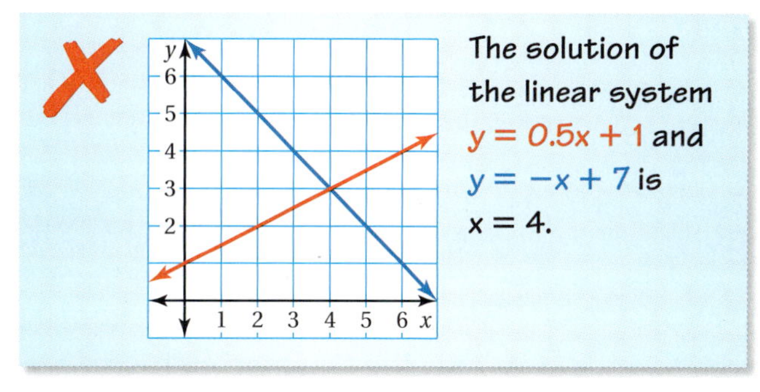

21. REASONING Is it possible for a system of two linear equations to have exactly two solutions? Explain your reasoning.

22. MODELING You have a total of 42 math and science problems for homework. You have 10 more math problems than science problems. How many problems do you have in each subject? Use a system of linear equations to justify your answer.

23. CANOE RACE You and your friend are in a canoe race. Your friend is a half mile in front of you and paddling 3 miles per hour. You are paddling 3.4 miles per hour.

a. You are 8.5 miles from the finish line. How long will it take you to catch up to your friend?

b. You both maintain your paddling rates for the remainder of the race. How far ahead of your friend will you be when you cross the finish line?

24. 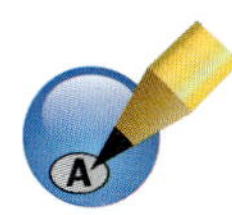

Your friend is trying to grow her hair as long as her cousin's hair. The table shows their hair lengths (in inches) in different months.

Month	Friend's Hair (in.)	Cousin's Hair (in.)
3	4	7
8	6.5	9

a. Write a system of linear equations that represents this situation.

b. Will your friend's hair ever be as long as her cousin's hair? If so, in what month?

Fair Game Review What you learned in previous grades & lessons

Solve the equation. Check your solution. *(Section 1.2)*

25. $\frac{3}{4}c - \frac{1}{4}c + 3 = 7$

26. $5(2 - y) + y = -6$

27. $6x - 3(x + 8) = 9$

28. MULTIPLE CHOICE What is the slope of the line that passes through $(-2, -2)$ and $(3, -1)$? *(Section 4.2)*

Ⓐ -5 Ⓑ $-\frac{1}{5}$ Ⓒ $\frac{1}{5}$ Ⓓ 5

5.2 Solving Systems of Linear Equations by Substitution

Essential Question How can you use substitution to solve a system of linear equations?

1 ACTIVITY: Using Substitution to Solve a System

Work with a partner. Solve each system of linear equations by using two methods.

$y = 6x - 11$

Method 1: Solve for x first.

Solve for x in one of the equations. Use the expression for x to find the solution of the system. Explain how you did it.

Method 2: Solve for y first.

Solve for y in one of the equations. Use the expression for y to find the solution of the system. Explain how you did it.

Is the solution the same using both methods?

a. $6x - y = 11$
$2x + 3y = 7$

b. $2x - 3y = -1$
$x - y = 1$

c. $3x + y = 5$
$5x - 4y = -3$

d. $5x - y = 2$
$3x - 6y = 12$

e. $x + y = -1$
$5x + y = -13$

f. $2x - 6y = -6$
$7x - 8y = 5$

2 ACTIVITY: Writing and Solving a System of Equations

Work with a partner.

a. Roll a pair of number cubes that have different colors. Then write the ordered pair shown by the number cubes. The ordered pair at the right is (3, 4).

x-value

y-value

b. Write a system of linear equations that has this ordered pair as its solution.

c. Exchange systems with your partner. Use one of the methods from Activity 1 to solve the system.

COMMON CORE

Systems of Equations

In this lesson, you will

- write and solve systems of linear equations by substitution.
- solve real-life problems.

Learning Standards
MACC.8.EE.3.8b
MACC.8.EE.3.8c

Laurie's Notes

Introduction

Standards for Mathematical Practice

- **MP1 Make Sense of Problems and Persevere in Solving Them:** Today, students will investigate another technique for solving a system of equations. Students have knowledge of systems, solutions of systems, and symbolic manipulation skills. Students will have the opportunity to make different attempts at solving a system. Encourage them to persevere and try different approaches.

Motivate

- Play a game of "Zip, Zap, Zoop," which is a combination of several games.
- **Directions:** Stand in a circle. Count around the circle. When your number is a *multiple* of 4, say, "**Zip**" instead of the number. When your number *contains* a 4, say, "**Zap.**" When your number is *both* a multiple of 4 *and* contains a 4, say, "**Zoop.**"
- The counting will go: 1, 2, 3, zoop, 5, 6, 7, zip, 9, 10, 11, zip, 13, zap, . . .
- They are *substituting* an expression for a number.
- The faster they count the funnier it becomes!

Activity Notes

Activity 1

- Given the equation $y = 6x - 11$, students are not familiar with the phrase, "use the expression for y." It may be helpful to highlight $6x - 11$ and identify it as "an expression for y."
- As a hint, tell students to think about the game they just played. They replaced certain numbers with words.
- ? After one of the equations has been solved for a variable ask, "What does it mean to solve a system of linear equations?" Find an ordered pair that satisfies both equations.
- ? "If there were only one linear equation with one variable, would you be able to solve it?" yes, provided that it has a solution
- ? "Can you see a way to use this system to get an equation with one variable?" Answers will vary.
- To save time have different groups solve different parts of the activity.
- Discuss whether the results for each method are the same.

Activity 2

- Students can struggle with trying to write any equation that passes through a given point. For the example (3, 4), ask students to think about the relationship between the two numbers: x is one less than y, so adding 1 to x is equal to y. This suggests the equation $y = x + 1$. A second equation can be generated by thinking, "You can double the x-value and subtract 2 to get the y-value." This suggests $y = 2x - 2$.
- Take time for students to share their strategies for generating their systems and to solve each other's system.

Florida Common Core Standards

MACC.8.EE.3.8b Solve systems of two linear equations in two variables algebraically, and estimate solutions by graphing the equations. Solve simple cases by inspection.

MACC.8.EE.3.8c Solve real-world and mathematical problems leading to two linear equations in two variables.

Previous Learning

Students should know how to solve linear equations and evaluate expressions at a specified value of the variable.

Lesson Plans
Complete Materials List

5.2 Record and Practice Journal

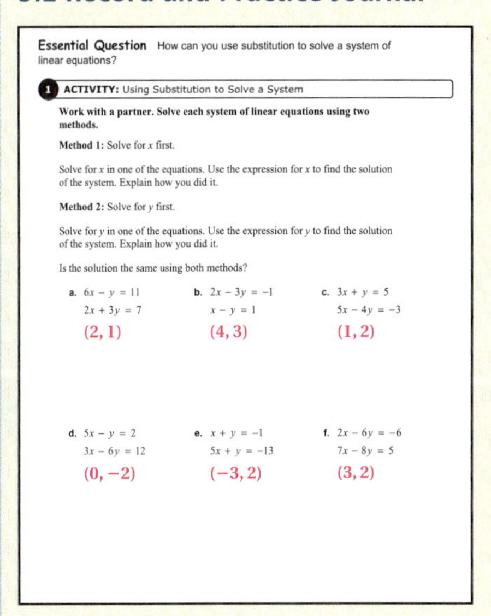

Essential Question How can you use substitution to solve a system of linear equations?

1 ACTIVITY: Using Substitution to Solve a System

Work with a partner. Solve each system of linear equations using two methods.

Method 1: Solve for x first.

Solve for x in one of the equations. Use the expression for x to find the solution of the system. Explain how you did it.

Method 2: Solve for y first.

Solve for y in one of the equations. Use the expression for y to find the solution of the system. Explain how you did it.

Is the solution the same using both methods?

a. $6x - y = 11$
$2x + 3y = 7$
(2, 1)

b. $2x - 3y = -1$
$x - y = 1$
(4, 3)

c. $3x + y = 5$
$5x - 4y = -3$
(1, 2)

d. $5x - y = 2$
$3x - 6y = 12$
(0, −2)

e. $x + y = -1$
$5x + y = -13$
(−3, 2)

f. $2x - 6y = -6$
$7x - 8y = 5$
(3, 2)

Differentiated Instruction

Pair Activity

Pair students. Each pair gets a set of two clue cards, one card to each student. Partners may not show each other their cards, but must communicate their clues verbally. Each pair must work together to answer the questions on their clue cards.

Prepare clue cards ahead of time.

Sample set:

Card 1: Three times Toni's age added to twice Sam's age is 88 years. How old is Toni?

Card 2: Sam's age reduced by half of Toni's age is 16 years. How old is Sam?

Answers: Toni: 14, Sam: 23

5.2 Record and Practice Journal

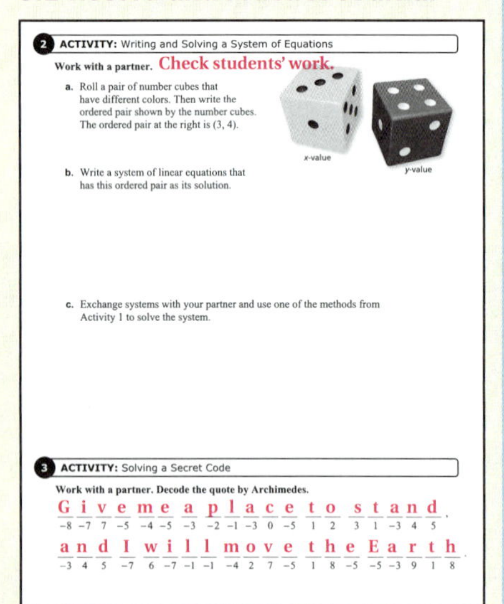

2 ACTIVITY: Writing and Solving a System of Equations

Work with a partner. Check students' work.

a. Roll a pair of number cubes that have different colors. Then write the ordered pair shown by the number cubes. The ordered pair at the right is (3, 4).

b. Write a system of linear equations that has this ordered pair as its solution.

c. Exchange systems with your partner and use one of the methods from Activity 1 to solve the system.

3 ACTIVITY: Solving a Secret Code

Work with a partner. Decode the quote by Archimedes.

G	i	v	e	m	e	a	p	l	a	c	e	t	o	s	t	a	n	d
−8	−7	7	−5	−4	−5	−3	−2	−1	−3	0	−5	1	2	3	1	−3	4	5

a	n	d	I	w	i	l	l	m	o	v	e	t	h	e	E	a	r	t	h
−3	4	5	−7	6	−7	−1	−1	−4	2	7	−5	1	8	−5	−5	−3	9	1	8

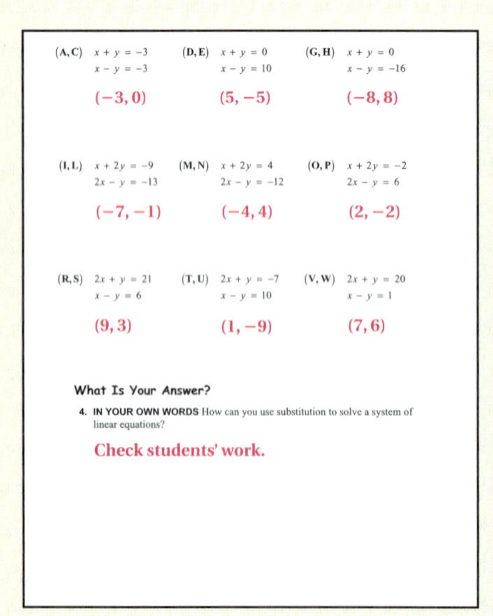

(A, C) $x + y = -3$, $x - y = -3$ — $(-3, 0)$

(D, E) $x + y = 0$, $x - y = 10$ — $(5, -5)$

(G, H) $x + y = 0$, $x - y = -16$ — $(-8, 8)$

(I, L) $x + 2y = -9$, $2x - y = -13$ — $(-7, -1)$

(M, N) $x + 2y = 4$, $2x - y = -12$ — $(-4, 4)$

(O, P) $x + 2y = -2$, $2x - y = 6$ — $(2, -2)$

(R, S) $2x + y = 21$, $x - y = 6$ — $(9, 3)$

(T, U) $2x + y = -7$, $x - y = 10$ — $(1, -9)$

(V, W) $2x + y = 20$, $x - y = 1$ — $(7, 6)$

What Is Your Answer?

4. **IN YOUR OWN WORDS** How can you use substitution to solve a system of linear equations?

Check students' work.

Laurie's Notes

Activity 3

- **FYI:** Archimedes is considered one of the greatest mathematicians in history. He was born in Syracuse, Greece in 287 B.C. He was killed during the siege of Syracuse in 212 B.C. by a Roman soldier who did not realize he was Archimedes. Some of his greatest contributions to mathematics were in the area of geometry. He was also an accomplished engineer and an inventor. He invented the screw pump for raising water up an inclined plane and explained how levers work.
- In this activity students may actually use trial and error to guess at the solution versus using the substitution method. They may also solve a limited number of problems, start to decode the message, and guess at the remaining letters.
- Encourage students to check each solution by substituting the ordered pair into both equations.

What Is Your Answer?

- **MP7 Look for and Make Use of Structure:** Students discovered in Activity 1 that given a system of two equations in two variables, they can solve one equation for one of the variables and then substitute the expression for that variable in the other equation to find the value of the other variable.

Closure

- Write a system of linear equations that could be solved easily by substitution. Write a system of linear equations that could *not* be solved easily by substitution. Explain your reasoning.

3 ACTIVITY: Solving a Secret Code

Math Practice 1

Check Progress

As you complete each system of equations, how do you know your answer is correct?

Work with a partner. Decode the quote by Archimedes.

$\underline{\quad}\,\underline{\quad}\,\underline{\quad}\,\underline{\quad}$ $\underline{\quad}\,\underline{\quad}$ $\underline{\quad}$ $\underline{\quad}\,\underline{\quad}\,\underline{\quad}\,\underline{\quad}\,\underline{\quad}$ $\underline{\quad}\,\underline{\quad}$ $\underline{\quad}\,\underline{\quad}\,\underline{\quad}\,\underline{\quad}\,\underline{\quad}$,

−8 −7 7 −5   −4 −5   −3   −2 −1 −3 0 −5   1 2   3 1 −3 4 5

$\underline{\quad}\,\underline{\quad}\,\underline{\quad}$ $\underline{\quad}$ $\underline{\quad}\,\underline{\quad}\,\underline{\quad}\,\underline{\quad}$ $\underline{\quad}\,\underline{\quad}\,\underline{\quad}\,\underline{\quad}$ $\underline{\quad}\,\underline{\quad}\,\underline{\quad}$ $\underline{\quad}\,\underline{\quad}\,\underline{\quad}\,\underline{\quad}\,\underline{\quad}$.

−3 4 5   −7   6 −7 −1 −1   −4 2 7 −5   1 8 −5   −5 −3 9 1 8

(A, C) $x + y = -3$, $x - y = -3$

(D, E) $x + y = 0$, $x - y = 10$

(G, H) $x + y = 0$, $x - y = -16$

(I, L) $x + 2y = -9$, $2x - y = -13$

(M, N) $x + 2y = 4$, $2x - y = -12$

(O, P) $x + 2y = -2$, $2x - y = 6$

(R, S) $2x + y = 21$, $x - y = 6$

(T, U) $2x + y = -7$, $x - y = 10$

(V, W) $2x + y = 20$, $x - y = 1$

What Is Your Answer?

4. **IN YOUR OWN WORDS** How can you use substitution to solve a system of linear equations?

Practice Use what you learned about systems of linear equations to complete Exercises 4–6 on page 212.

5.2 Lesson

Another way to solve systems of linear equations is to use substitution.

Key Idea

Solving a System of Linear Equations by Substitution

Step 1: Solve one of the equations for one of the variables.

Step 2: Substitute the expression from Step 1 into the other equation and solve for the other variable.

Step 3: Substitute the value from Step 2 into one of the original equations and solve.

EXAMPLE 1 Solving a System of Linear Equations by Substitution

Solve the system by substitution.

$$\mathbf{y = 2x - 4} \quad \text{Equation 1}$$

$$\mathbf{7x - 2y = 5} \quad \text{Equation 2}$$

Step 1: Equation 1 is already solved for y.

Step 2: Substitute $2x - 4$ for y in Equation 2.

$7x - 2y = 5$	Equation 2
$7x - 2(2x - 4) = 5$	Substitute $2x - 4$ for y.
$7x - 4x + 8 = 5$	Distributive Property
$3x + 8 = 5$	Combine like terms.
$3x = -3$	Subtract 8 from each side.
$x = -1$	Divide each side by 3.

Step 3: Substitute -1 for x in Equation 1 and solve for y.

$y = 2x - 4$	Equation 1
$= 2(-1) - 4$	Substitute -1 for x.
$= -2 - 4$	Multiply.
$= -6$	Subtract.

The solution is $(-1, -6)$.

Check

Equation 1

$$y = 2x - 4$$
$$-6 \stackrel{?}{=} 2(-1) - 4$$
$$-6 = -6 \checkmark$$

Equation 2

$$7x - 2y = 5$$
$$7(-1) - 2(-6) \stackrel{?}{=} 5$$
$$5 = 5 \checkmark$$

On Your Own

Now You're Ready
Exercises 10–15

Solve the system of linear equations by substitution. Check your solution.

1. $y = 2x + 3$
   $y = 5x$

2. $4x + 2y = 0$
   $y = \frac{1}{2}x - 5$

3. $x = 5y + 3$
   $2x + 4y = -1$

Laurie's Notes

Introduction

Connect

- **Yesterday:** Students discovered how to use substitution to solve a system of linear equations. (MP1, MP7)
- **Today:** Students will solve systems of linear equations by substitution.

Motivate

- Share a cooking story about salsa.

	Cilantro	Tomatoes	Onion
Summer Salsa	$\frac{1}{2}$ cup	3 cups	$\frac{3}{4}$ cup
Romero's Salsa	$1\frac{1}{4}$ cups	8 cups	2 cups

? "Do you think algebra can help a cook figure out how much salsa of each type can be made if you have 5 cups of cilantro and 40 cups of tomatoes?" Comments will vary.

- Explain to students that the techniques they will study today are used to solve problems of this type.

Lesson Notes

Discuss

- Students have solved a system of linear equations by graphing. Substitution is a second way to solve a system of linear equations.

Key Idea

- Discuss the steps in solving a system of linear equations by substitution.
- At the end of Step 2, you can add, "You now know either the x- or y-coordinate of the ordered pair that satisfies both equations. Next, you need to find the other coordinate."
- Remind students to check the solution in both equations.

Example 1

- Write Equation 1 and Equation 2.

? "In what form is Equation 1 written?" slope-intercept form

? "In what form is Equation 2 written?" standard form

- Say, "Equation 1 is already solved for y, so you can go right to Step 2. Substitute the expression $2x - 4$ for y in the second equation."
- **MP6 Attend to Precision:** Students may get sloppy and say they are "plugging in for y." "Plugging in" is not a mathematical operation or process. It is better to say that they are "substituting for y," so they become familiar with the math terminology they are expected to know.

? After the second step ask, "What does $x = -1$ mean in the context of this problem?" The x-coordinate of the solution is -1.

? "How do you determine the y-coordinate of the solution?" Substitute -1 for x in one of the original equations and solve for y.

- Check the solution.

Goal Today's lesson is solving a system of linear equations by substitution.

Technology for the Teacher

Lesson Tutorials
Lesson Plans
Answer Presentation Tool

Extra Example 1

Solve the system by substitution.

$y = 3x - 4$

$5x - 2y = 10$

$(-2, -10)$

On Your Own

1. $(1, 5)$
2. $(2, -4)$
3. $\left(\frac{1}{2}, -\frac{1}{2}\right)$

Laurie's Notes

On Your Own

- Note that in all of these systems, one of the variables has already been solved for explicitly.

Example 2

- ? **MP1 Make Sense of Problems and Persevere in Solving Them:** Ask a student to read the problem. Ask, "Is it possible that you bought only turkey burgers or only veggie burgers? Explain." no; You spent \$90, but 50 turkey burgers would cost \$100 and 50 veggie burgers would cost \$75.
- Students can be careless when defining variables. Be sure they use:
  $x =$ number of turkey burgers
  $y =$ number of veggie burgers
- ? "There are two equations, each with two variables. Can you solve for one of the variables in either equation?" yes
- ? "Is there a choice that might be easier than another? Explain." yes; It may be easiest to solve for x or y in the first equation because both coefficients are 1, and there are no decimals.
- Discuss the Study Tip.
- Work through the problem as shown.
- If time permits, solve the system again, solving for y in Step 1. Compare the answers.
- ? "Could this system be solved by graphing?" yes

On Your Own

- Observe student work. It is likely that some students will solve for x and some for y. When students have finished ask two volunteers to share their work at the board, demonstrating each method.

Closure

- **Exit Ticket:** Solve the system by substitution and by graphing.
  $y = 3x + 1$
  $y = x + 3$
  (1, 4)

Extra Example 2

A weightlifter uses a total of 12 plates to add 260 pounds to a bar. He uses 45-pound plates and 10-pound plates. Write and solve a system of equations to find the number x of 45-pound plates and the number y of 10-pound plates he uses.

$x + y = 12$
$45x + 10y = 260$
four 45-pound plates, eight 10-pound plates

On Your Own

4. $x + y = 100$
$2x + 3y = 240$
60 cups of lemonade, 40 cups of orange juice

English Language Learners

Simplifying the Language

To help your students understand and remember the steps for solving a system of linear equations by substitution, present a simplified version of the steps.

Step 1: Solve.

Step 2: Substitute and solve.

Step 3: Substitute and solve.

Ask students to explain what to solve and what to substitute in each step.

EXAMPLE 2 Real-Life Application

You buy a total of 50 turkey burgers and veggie burgers for $90. You pay $2 per turkey burger and $1.50 per veggie burger. Write and solve a system of linear equations to find the number x of turkey burgers and the number y of veggie burgers you buy.

Use a verbal model to write a system of linear equations.

Number of turkey burgers, x	+	Number of veggie burgers, y	=	Total number of burgers

Cost per turkey burger	•	Number of turkey burgers, x	+	Cost per veggie burger	•	Number of veggie burgers, y	=	Total cost

The system is: $x + y = 50$ — Equation 1

$2x + 1.5y = 90$ — Equation 2

Study Tip

It is easiest to solve for a variable that has a coefficient of 1 or -1.

Step 1: Solve Equation 1 for x.

$x + y = 50$ — Equation 1

$x = 50 - y$ — Subtract y from each side.

Step 2: Substitute $50 - y$ for x in Equation 2.

$2x + 1.5y = 90$ — Equation 2

$2(50 - y) + 1.5y = 90$ — Substitute $50 - y$ for x.

$100 - 2y + 1.5y = 90$ — Distributive Property

$-0.5y = -10$ — Simplify.

$y = 20$ — Divide each side by -0.5.

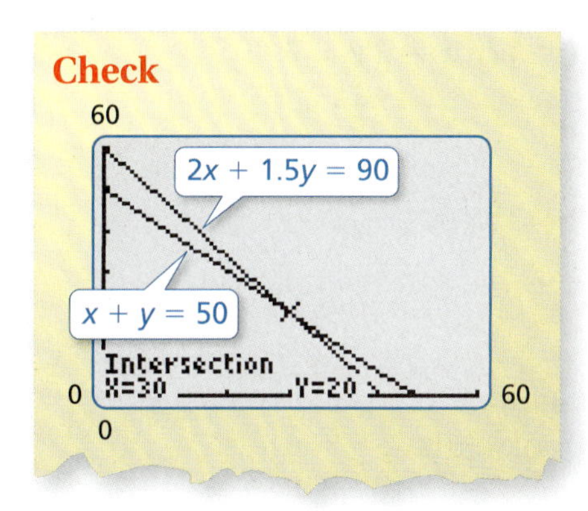

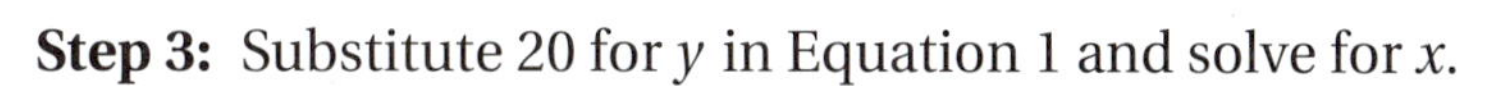

Step 3: Substitute 20 for y in Equation 1 and solve for x.

$x + y = 50$ — Equation 1

$x + 20 = 50$ — Substitute 20 for y.

$x = 30$ — Subtract 20 from each side.

∴ You buy 30 turkey burgers and 20 veggie burgers.

On Your Own

4. You sell lemonade for $2 per cup and orange juice for $3 per cup. You sell a total of 100 cups for $240. Write and solve a system of linear equations to find the number of cups of lemonade and the number of cups of orange juice you sold.

5.2 Exercises

Vocabulary and Concept Check

1. **WRITING** Describe how to solve a system of linear equations by substitution.
2. **NUMBER SENSE** When solving a system of linear equations by substitution, how do you decide which variable to solve for in Step 1?
3. **REASONING** Does solving a system of linear equations by graphing give the same solution as solving by substitution? Explain your reasoning.

Practice and Problem Solving

Write a system of linear equations that has the ordered pair as its solution. Use a method from Activity 1 to solve the system.

4. 

5. 

6.  

Tell which equation you would choose to solve for one of the variables when solving the system by substitution. Explain your reasoning.

7. $2x + 3y = 5$
   $4x - y = 3$

8. $\frac{2}{3}x + 5y = -1$
   $x + 6y = 0$

9. $2x + 10y = 14$
   $5x - 9y = 1$

Solve the system of linear equations by substitution. Check your solution.

① 10. $y = x - 4$
   $y = 4x - 10$

11. $y = 2x + 5$
   $y = 3x - 1$

12. $x = 2y + 7$
   $3x - 2y = 3$

13. $4x - 2y = 14$
   $y = \frac{1}{2}x - 1$

14. $2x = y - 10$
   $x + 7 = y$

15. $8x - \frac{1}{3}y = 0$
   $12x + 3 = y$

16. **SCHOOL CLUBS** There are a total of 64 students in a drama club and a yearbook club. The drama club has 10 more students than the yearbook club.

   a. Write a system of linear equations that represents this situation.

   b. How many students are in the drama club? the yearbook club?

17. **THEATER** A drama club earns \$1040 from a production. It sells a total of 64 adult tickets and 132 student tickets. An adult ticket costs twice as much as a student ticket.

   a. Write a system of linear equations that represents this situation.

   b. What is the cost of each ticket?

Assignment Guide and Homework Check

Level	Day 1 Activity Assignment	Day 2 Lesson Assignment	Homework Check
Basic	4–6, 26–29	1–3, 7–23 odd	11, 13, 17, 19, 23
Average	4–6, 26–29	1–3, 10–20 even, 21–25 odd	12, 14, 16, 18, 23
Advanced	1–6, 10–20 even, 21–29		14, 16, 20, 24, 25

Common Errors

- **Exercises 4–6** Students may write equations that do not give the correct solutions. For instance, in Exercise 4, a student might write one equation as $2x + 3y = 5$. Tell students to check each equation to make sure that the ordered pair represented by the dice is a solution.
- **Exercises 10–15, 18–20** Students may find one coordinate of the solution and stop there. Remind them that the answer is a coordinate pair representing both an x-value and a y-value.

5.2 Record and Practice Journal

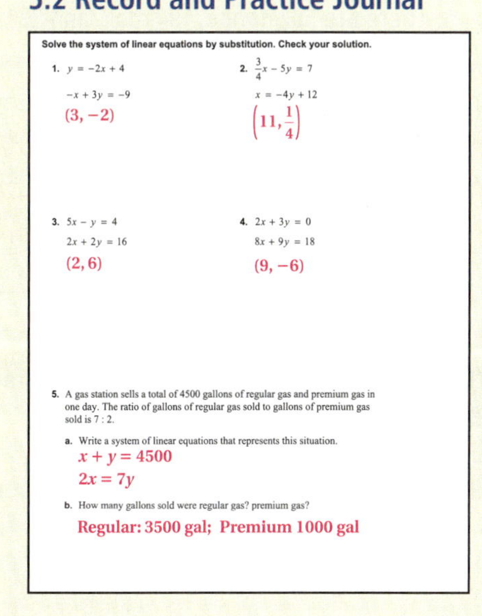

Solve the system of linear equations by substitution. Check your solution.

1. $y = -2x + 4$
   $-x + 3y = -9$
   $(3, -2)$

2. $\frac{3}{4}x - 5y = 7$
   $x = -4y + 12$
   $\left(11, \frac{1}{4}\right)$

3. $5x - y = 4$
   $2x + 2y = 16$
   $(2, 6)$

4. $2x + 3y = 0$
   $8x + 9y = 18$
   $(9, -6)$

5. A gas station sells a total of 4500 gallons of regular gas and premium gas in one day. The ratio of gallons of regular gas sold to gallons of premium gas sold is 7 : 2.

   a. Write a system of linear equations that represents this situation.
   $x + y = 4500$
   $2x = 7y$

   b. How many gallons sold were regular gas? premium gas?
   Regular: 3500 gal; Premium 1000 gal

Vocabulary and Concept Check

1. **Step 1:** Solve one of the equations for one of the variables.
   Step 2: Substitute the expression from Step 1 into the other equation and solve.
   Step 3: Substitute the value from Step 2 into one of the original equations and solve.
2. If possible, solve for a variable that has a coefficient of 1 or -1, or that is easy to solve.
3. sometimes; A solution obtained by graphing may not be exact.

Practice and Problem Solving

4. *Sample answer:* $x + y = 5$
   $x - y = -1$
5. *Sample answer:* $x + 2y = 6$
   $x - y = 3$
6. *Sample answer:*
   $x - y = 1$
   $2x - 3y = -3$
7. $4x - y = 3$; The coefficient of y is -1.
8. $x + 6y = 0$; The coefficient of x is 1, and there is no constant.
9. $2x + 10y = 14$; Dividing by 2 to solve for x yields integers.
10. $(2, -2)$
11. $(6, 17)$
12. $\left(-2, -\frac{9}{2}\right)$
13. $(4, 1)$
14. $(-3, 4)$
15. $\left(\frac{1}{4}, 6\right)$
16. **a.** $x + y = 64$
    $x = y + 10$
    b. 37 students; 27 students

Practice and Problem Solving

17. **a.** $x = 2y$

$64x + 132y = 1040$

b. adult tickets: \$8; student tickets: \$4

18. $(-3, -3)$

19. $(-2, 4)$ **20.** $(6, -3)$

21. The expression for y was substituted back into the same equation; solution: $(2, 1)$

22. $y = 2.5x$

$2x + y = 180$

base angles: 40°, third angle: 100°

23. 30 cats, 35 dogs

24. 26

25. See *Taking Math Deeper*.

Fair Game Review

26. $3x - 7y = 9$

27. $2x - 5y = -8$

28. $6x - y = 3$

29. B

Mini-Assessment

Solve the system of linear equations by substitution. Check your solution.

1. $y = 3x - 2$

   $y = -x + 6$ $(2, 4)$

2. $2y + 8 = x$

   $8x + y = -21$ $(-2, -5)$

3. $4x + 3y = 26$

   $2x - 3y = -14$ $(2, 6)$

4. You spent \$56 on food and clothes. You spent \$18 more on clothes than on food. Write and solve a system of equations to find how much you spent on each.

   $x + y = 56$ clothes: \$37

   $y - x = 18$ food: \$19

Taking Math Deeper

Exercise 25

One way to visualize the problem is to make a diagram.

Make a diagram.

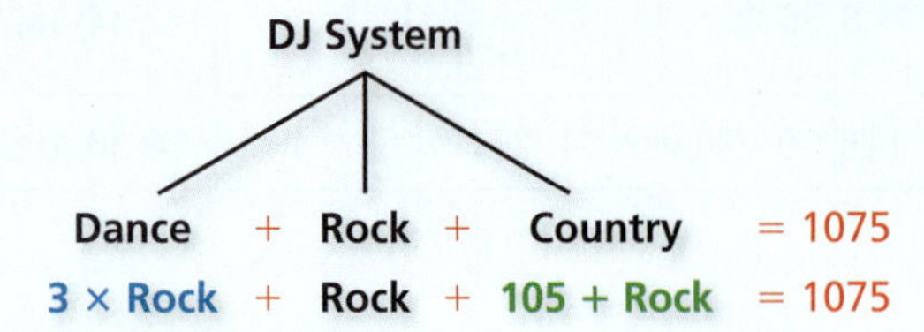

The DJ has 1075 songs on her system, so the sum of the numbers of dance, rock, and country songs is 1075.

The dance selection is 3 times the size of the rock selection, so you can substitute 3 × Rock for the number of dance songs.

There are 105 more country songs than rock songs, so you can substitute 105 + Rock for the number of country songs.

Now you can find the number r of rock songs.

Write and solve an equation.

$$3r + r + (105 + r) = 1075$$
$$5r + 105 = 1075$$
$$5r = 970$$
$$r = 194$$

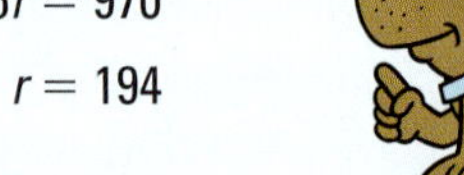

There are 194 rock songs.

Substitute and solve.

Dance $= 3 \times$ Rock $= 3 \times 194 = 582$

Country $= 105 +$ Rock $= 105 + 194 = 299$

There are 582 dance, 194 rock, and 299 country songs on the system.

Reteaching and Enrichment Strategies

If students need help. . .	If students got it. . .
Resources by Chapter • Practice A and Practice B • Puzzle Time Record and Practice Journal Practice Differentiating the Lesson Lesson Tutorials Skills Review Handbook	Resources by Chapter • Enrichment and Extension • Technology Connection Start the next section

Solve the system of linear equations by substitution. Check your solution.

2 **18.** $y - x = 0$
$2x - 5y = 9$

19. $x + 4y = 14$
$3x + 7y = 22$

20. $-2x - 5y = 3$
$3x + 8y = -6$

21. ERROR ANALYSIS Describe and correct the error in solving the system of linear equations.

✗		Step 1:	Step 2:
$2x + y = 5$	Equation 1	$2x + y = 5$	$2x + (-2x + 5) = 5$
$3x - 2y = 4$	Equation 2	$y = -2x + 5$	$2x - 2x + 5 = 5$
			$5 = 5$

22. STRUCTURE The measure of the obtuse angle in the isosceles triangle is two and a half times the measure of one base angle. Write and solve a system of linear equations to find the measures of all the angles.

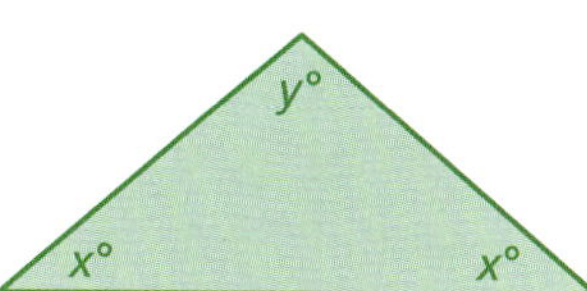

23. ANIMAL SHELTER An animal shelter has a total of 65 abandoned cats and dogs. The ratio of cats to dogs is 6 : 7. How many cats are in the shelter? How many dogs are in the shelter? Justify your answers.

24. NUMBER SENSE The sum of the digits of a two-digit number is 8. When the digits are reversed, the number increases by 36. Find the original number.

25. Repeated Reasoning A DJ has a total of 1075 dance, rock, and country songs on her system. The dance selection is three times the size of the rock selection. The country selection has 105 more songs than the rock selection. How many songs on the system are dance? rock? country?

Fair Game Review What you learned in previous grades & lessons

Write the equation in standard form. *(Section 4.5)*

26. $3x - 9 = 7y$

27. $8 - 5y = -2x$

28. $6x = y + 3$

29. MULTIPLE CHOICE Use the figure to find the measure of $\angle 2$. *(Section 3.1)*

Ⓐ 17° Ⓑ 73°

Ⓒ 83° Ⓓ 107°

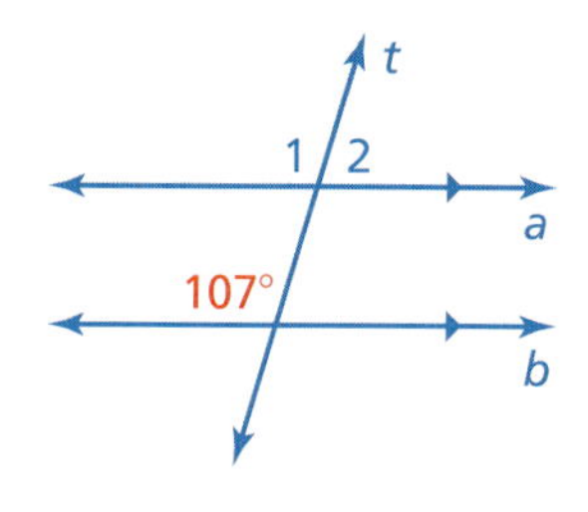

5 Study Help

You can use a **notetaking organizer** to write notes, vocabulary, and questions about a topic. Here is an example of a notetaking organizer for solving systems of linear equations by graphing.

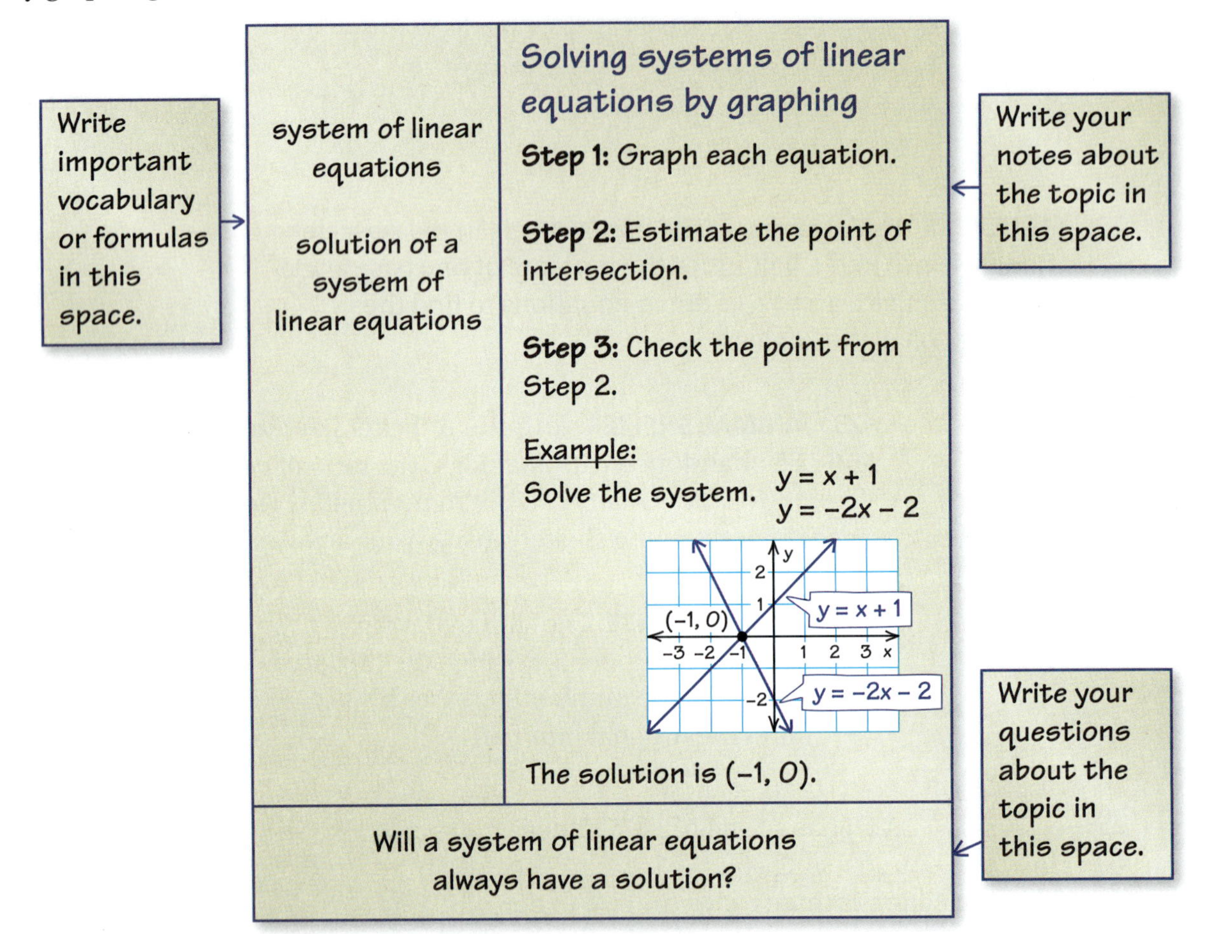

On Your Own

Make a notetaking organizer to help you study this topic.

1. solving systems of linear equations by substitution

After you complete this chapter, make notetaking organizers for the following topics.

2. solving systems of linear equations by elimination
3. solving systems of linear equations with no solution or infinitely many solutions
4. solving linear equations by graphing

"My notetaking organizer has me thinking about retirement when I won't have to fetch sticks anymore."

Sample Answer

1.

	Solving systems of linear equations by substitution
system of linear equations solution of a system of linear equations	Step 1: Solve one of the equations for one of the variables. Step 2: Substitute the expression from Step 1 into the other equation and solve. Step 3: Substitute the value from Step 2 into one of the original equations and solve. Example: Solve the system. $x - y = 3$ Equation 1 $4x + y = 2$ Equation 2 Step 1: Solve Equation 1 for x. $x - y = 3$ $x = y + 3$ Step 2: Substitute $y + 3$ for x in Equation 2, and solve for y. $4x + y = 2$ $4(y + 3) + y = 2$ $4y + 12 + y = 2$ $5y = -10$ $y = -2$ Step 3: Substitute -2 for y in Equation 1 and solve for x. $x - y = 3$ $x - (-2) = 3$ $x = 1$ The solution is $(1, -2)$.
Are there other ways to solve systems algebraically?	

List of Organizers

Available at *BigIdeasMath.com*

Comparison Chart
Concept Circle
Definition (Idea) and Example Chart
Example and Non-Example Chart
Formula Triangle
Four Square
Information Frame
Information Wheel
Notetaking Organizer
Process Diagram
Summary Triangle
Word Magnet
Y Chart

About this Organizer

A **Notetaking Organizer** can be used to write notes, vocabulary, and questions about a topic. In the space on the left, students write important vocabulary or formulas. In the space on the right, students write their notes about the topic. In the space at the bottom, students write their questions about the topic. A notetaking organizer can also be used as an assessment tool, in which blanks are left for students to complete.

Technology for the Teacher

Editable Graphic Organizer

Answers

1. B; $(1, -1)$
2. C; $(3, 0)$
3. A; $(-2, -3)$
4. $(4, 5)$
5. $(-1, 4)$
6. $(3, -5)$
7. $(6, -2)$
8. $(8, 3)$
9. $(-4, -1)$
10. a. $y = 2x + 15$ Members
    $y = 3x$ Nonmembers
    b. It is beneficial when you rent more than 15 new release movies per year.
11. 23 and 15; $(23, 15)$ is the solution of the system
    $x + y = 38$
    $x - y = 8$.
12. 60 ft by 30 ft
13. 63 nurses; 14 doctors

Technology for the Teacher

Online Assessment
Assessment Book
ExamView® Assessment Suite

Alternative Quiz Ideas

100% Quiz	Math Log
Error Notebook	Notebook Quiz
Group Quiz	Partner Quiz
Homework Quiz	Pass the Paper

Error Notebook

An error notebook provides an opportunity for students to analyze and learn from their errors. Have students make error notebooks for this chapter. They should work in their notebooks a little each day. Give students the following directions.

- Use a notebook and divide the page into three columns.
- Label the first column *problem*, second column *error*, and third column *correction*.
- In the first column, write down the problem on which the errors were made. Record the source of the problem (homework, quiz, in-class assignment).
- The second column should show the exact error that was made. Include a statement of why you think the error was made. This is where the learning takes place, so it is helpful to use a different color ink for the work in this column.
- The last column contains the corrected problems and comments that will help with future work.
- Separate each problem with horizontal lines.

Reteaching and Enrichment Strategies

If students need help. . .	If students got it. . .
Resources by Chapter • Practice A and Practice B • Puzzle Time Lesson Tutorials *BigIdeasMath.com*	Resources by Chapter • Enrichment and Extension • Technology Connection Game Closet at *BigIdeasMath.com* Start the next section

5.1–5.2 Quiz

Check It Out
Progress Check
BigIdeasMath.com

Match the system of linear equations with the corresponding graph. Use the graph to estimate the solution. Check your solution. *(Section 5.1)*

1. $y = x - 2$

$y = -2x + 1$

2. $y = x - 3$

$y = -\frac{1}{3}x + 1$

3. $y = \frac{1}{2}x - 2$

$y = 4x + 5$

A.

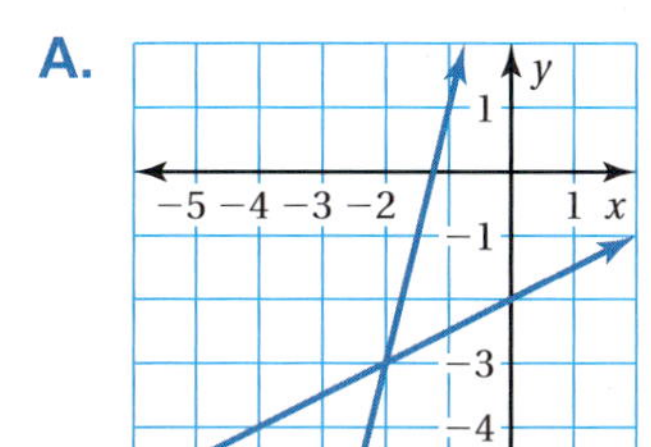

B.

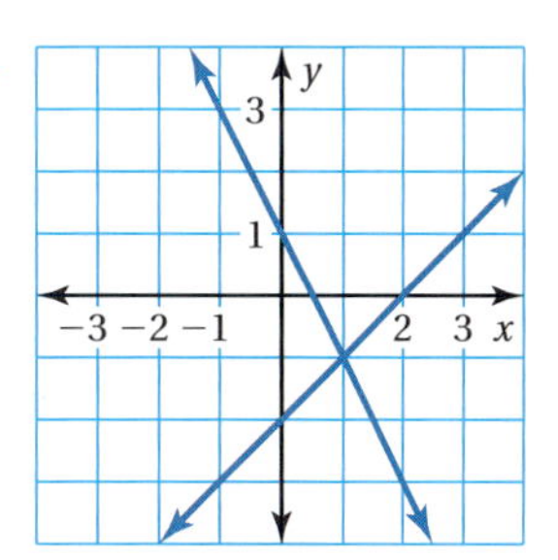

C.

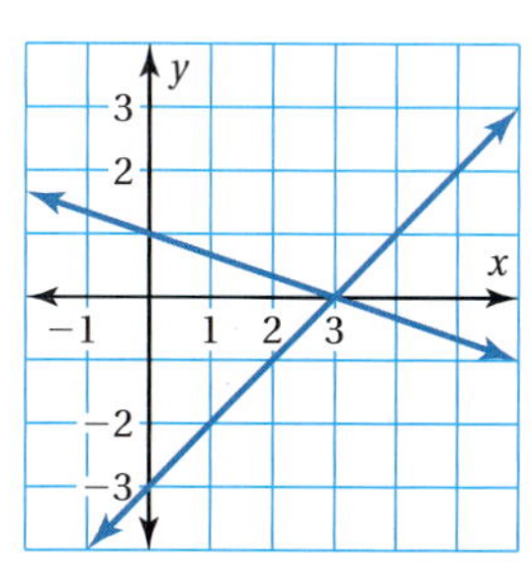

Solve the system of linear equations by graphing. *(Section 5.1)*

4. $y = 2x - 3$

$y = -x + 9$

5. $6x + y = -2$

$y = -3x + 1$

6. $4x + 2y = 2$

$3x = 4 - y$

Solve the system of linear equations by substitution. Check your solution. *(Section 5.2)*

7. $y = x - 8$

$y = 2x - 14$

8. $x = 2y + 2$

$2x - 5y = 1$

9. $x - 5y = 1$

$-2x + 9y = -1$

10. MOVIE CLUB Members of a movie rental club pay a \$15 annual membership fee and \$2 for new release movies. Nonmembers pay \$3 for new release movies. *(Section 5.1)*

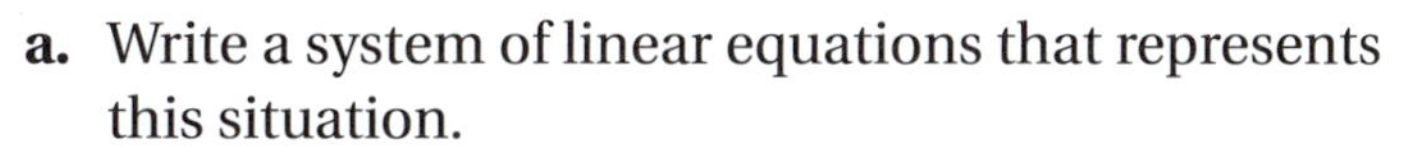

a. Write a system of linear equations that represents this situation.

b. When is it beneficial to have a membership?

11. NUMBER SENSE The sum of two numbers is 38. The greater number is 8 more than the other number. Find each number. Use a system of linear equations to justify your answer. *(Section 5.1)*

12. VOLLEYBALL The length of a sand volleyball court is twice its width. The perimeter is 180 feet. Find the length and width of the sand volleyball court. *(Section 5.2)*

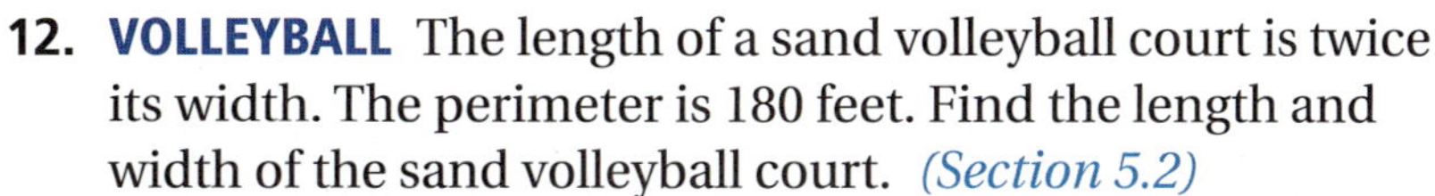

13. MEDICAL STAFF A hospital employs a total of 77 nurses and doctors. The ratio of nurses to doctors is 9 : 2. How many nurses are employed at the hospital? How many doctors are employed at the hospital? *(Section 5.2)*

5.3 Solving Systems of Linear Equations by Elimination

Essential Question How can you use elimination to solve a system of linear equations?

1 ACTIVITY: Using Elimination to Solve a System

Work with a partner. Solve each system of linear equations by using two methods.

Method 1: Subtract.

Subtract Equation 2 from Equation 1. What is the result? Explain how you can use the result to solve the system of equations.

Method 2: Add.

Add the two equations. What is the result? Explain how you can use the result to solve the system of equations.

Is the solution the same using both methods?

a. $2x + y = 4$
$2x - y = 0$

b. $3x - y = 4$
$3x + y = 2$

c. $x + 2y = 7$
$x - 2y = -5$

2 ACTIVITY: Using Elimination to Solve a System

Work with a partner.

$2x + y = 2$ Equation 1

$x + 5y = 1$ Equation 2

a. Can you add or subtract the equations to solve the system of linear equations? Explain.

b. Explain what property you can apply to Equation 1 in the system so that the y-coefficients are the same.

c. Explain what property you can apply to Equation 2 in the system so that the x-coefficients are the same.

d. You solve the system in part (b). Your partner solves the system in part (c). Compare your solutions.

e. Use a graphing calculator to check your solution.

COMMON CORE

Systems of Equations

In this lesson, you will

- write and solve systems of linear equations by elimination.
- solve real-life problems.

Learning Standards
MACC.8.EE.3.8b
MACC.8.EE.3.8c

Laurie's Notes

Introduction

Standards for Mathematical Practice

- **MP1 Make Sense of Problems and Persevere in Solving Them:** Today, students will investigate a third technique for solving a system of linear equations. Students have the prerequisite knowledge. Students will use more than one approach to solve a system. Encourage them to persevere and try to understand why these approaches work.
- There are multiple avenues for solving a system using the elimination technique. Students may solve the system using several different approaches, all of which can lead to the same answer.

Motivate

- Draw the following sketch on the board. Make it clear to students that the right sides represent 13 pounds of weight and 5 pounds of weight.

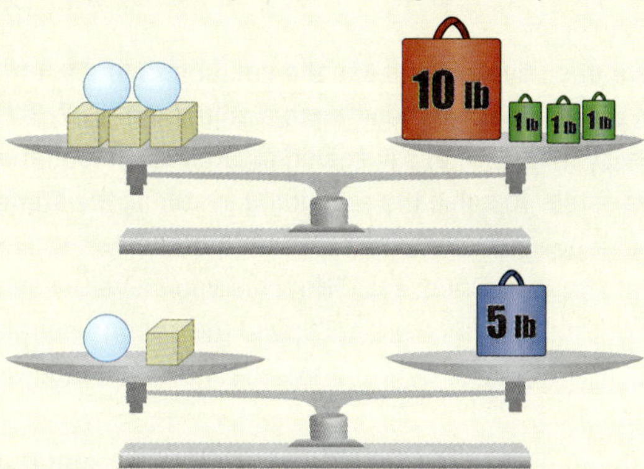

- Ask students to draw a sketch of a balance scale with 4 cubes and 3 balls at the left and 18 pounds at the right. Relate this to adding equations.
- Ask students to draw a sketch of a balance scale with 2 cubes and 1 ball at the left and 8 pounds at the right. Relate this to subtracting equations.
- Explain to students that today they will perform operations with equations similar to those they represented with the balance scales.

Activity Notes

Activity 1

- Students may not understand what it means to add or subtract equations. Tell them to think about the balance scales—they added the left sides of the scales and the right sides of the scales.
- As you walk around, you may need to remind students what it means to solve a system.
- ? After students find the value of one variable, ask, "How can you find the value of the other variable?" Substitute the known value into one of the original equations to solve for the other variable.
- Discuss the solutions found using each method (addition and subtraction).
- ? "Can you eliminate one variable by adding the equations for any system of two linear equations? Explain." no; One of the variables must have coefficients that are opposites in the two equations for this to work.

Florida Common Core Standards

MACC.8.EE.3.8b Solve systems of two linear equations in two variables algebraically, and estimate solutions by graphing the equations. Solve simple cases by inspection.

MACC.8.EE.3.8c Solve real-world and mathematical problems leading to two linear equations in two variables.

Previous Learning

Students should know how to solve linear equations and evaluate expressions at a specified value of the variable.

5.3 Record and Practice Journal

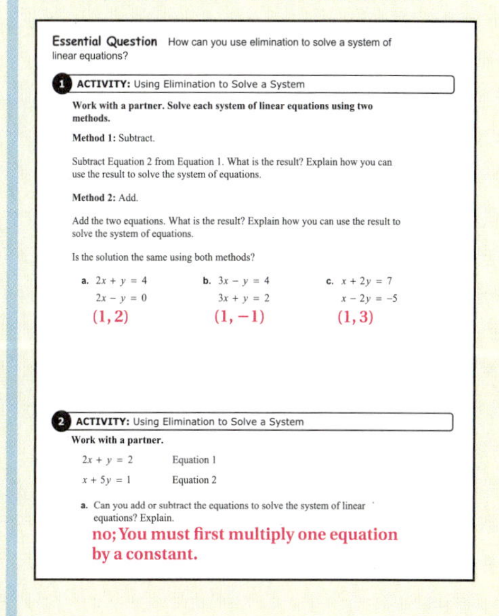

Essential Question How can you use elimination to solve a system of linear equations?

1 ACTIVITY: Using Elimination to Solve a System

Work with a partner. Solve each system of linear equations using two methods.

Method 1: Subtract.

Subtract Equation 2 from Equation 1. What is the result? Explain how you can use the result to solve the system of equations.

Method 2: Add.

Add the two equations. What is the result? Explain how you can use the result to solve the system of equations.

Is the solution the same using both methods?

a. $2x + y = 4$
$2x - y = 0$
$(1, 2)$

b. $3x - y = 4$
$3x + y = 2$
$(1, -1)$

c. $x + 2y = 7$
$x - 2y = -5$
$(1, 3)$

2 ACTIVITY: Using Elimination to Solve a System

Work with a partner.

$2x + y = 2$ Equation 1

$x + 5y = 1$ Equation 2

a. Can you add or subtract the equations to solve the system of linear equations? Explain.
no; You must first multiply one equation by a constant.

Differentiated Instruction

Kinesthetic

It may help students to write each term of a system on a small slip of paper. When a variable is eliminated, the student removes the terms related to that variable. This helps them to grasp the concept.

5.3 Record and Practice Journal

b. Explain what property you can apply to Equation 1 in the system so that the y coefficients are the same.

Multiply the equation by 5.

c. Explain what property you can apply to Equation 2 in the system so that the x coefficients are the same.

Multiply the equation by 2.

d. You solve the system in part (b). Your partner solves the system in part (c). Compare your solutions.

(1, 0); They are the same.

e. Use a graphing calculator to check your solution.

Check students' work.

3 ACTIVITY: Solving a Secret Code

Work with a partner. Solve the puzzle to find the name of a famous mathematician who lived in Egypt around 350 A.D.

4	B	W	R	M	F	Y	K	N
3	O	J	A	S	I	D	X	Z
2	Q	P	C	E	G	B	T	J
1	M	R	C	Z	N	O	U	W
0	K	X	U	H	L	Y	S	Q
−1	F	E	A	S	W	K	R	M
−2	G	J	Z	N	H	V	D	G
−3	E	L	X	L	F	Q	O	B

H Y P A T I A

H: $2x + y = 0$, $x - y = 3$

Y: $x + y = 2$, $2x - 2y = 4$

P: $3x + 3y = 0$, $2x - 2y = -8$

A: $2x + y = -3$, $x - y = 0$

T: $x + y = 5$, $x - y = 1$

I: $x + y = 4$, $x - y = -2$

A: $x + 2y = 5$, $2x - y = -5$

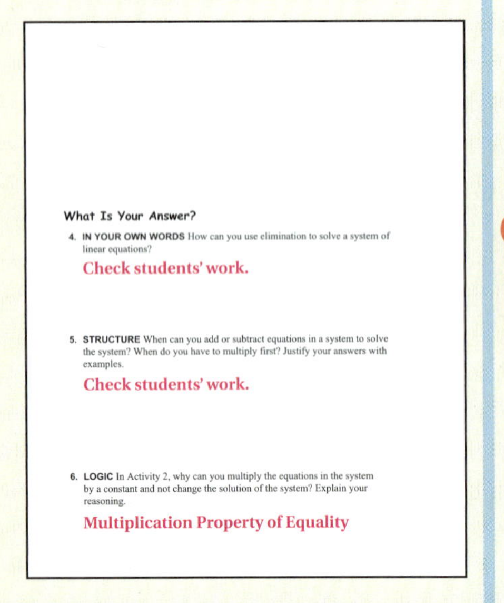

What Is Your Answer?

4. IN YOUR OWN WORDS How can you use elimination to solve a system of linear equations?

Check students' work.

5. STRUCTURE When can you add or subtract equations in a system to solve the system? When do you have to multiply first? Justify your answers with examples.

Check students' work.

6. LOGIC In Activity 2, why can you multiply the equations in the system by a constant and not change the solution of the system? Explain your reasoning.

Multiplication Property of Equality

Laurie's Notes

Activity 1 (continued)

- Discuss how elimination compares to the substitution method. With elimination, you add or subtract equations to eliminate a variable. With substitution, you write an equation in one variable by substituting an expression for the other variable. In both methods, you solve the resulting equation for one variable and use its value to find the other variable.

Activity 2

- ? **MP7 Look for and Make Use of Structure:** In parts (b) and (c), students may first think of properties such as the Commutative and Associative Properties. Refer to the first balance scale and ask, "If you double the number of cubes and balls on one side, and the weight on the other, will the scale balance?" yes "What will each side weigh?" 26 pounds
- ? "What property did you use?" Multiplication Property of Equality
- When students finish, discuss the process and the solutions.

Activity 3

- Students solve each system and use the solution to find a letter in the grid. For instance, the letter corresponding to a solution of (−2, 3) is "J."
- Several of the systems are easily solved by adding the equations to eliminate a variable. To solve the remaining systems, the students can multiply through by a constant.
- ? When students have finished, ask, "How did you solve the system for the third letter?" Students may have multiplied the second equation by 3/2, or multiplied the first equation by 2 and the second equation by 3. Compare the two approaches.
- Well-known quote associated with Hypatia: "Reserve your right to think, for even to think wrongly is better than not to think at all."
- Hypatia (370–415) was the daughter of a mathematician and is the first known woman mathematician. Hypatia wrote many mathematical treatises, and was noted for her ability to explain complex mathematical ideas clearly.

What Is Your Answer?

- **MP7:** Students need to realize that one pair of like variable terms in a system must have the same or opposite coefficients before you can eliminate a variable by adding or subtracting the equations.

Closure

- Write a system of linear equations that can be solved easily by elimination. Write a system of linear equations that *cannot* be solved easily by elimination. Explain your reasoning.

3 ACTIVITY: Solving a Secret Code

Math Practice 1

Find Entry Points

What is the first thing you do to solve a system of linear equations? Why?

Work with a partner. Solve the puzzle to find the name of a famous mathematician who lived in Egypt around 350 A.D.

4	B	W	R	M	F	Y	K	N
3	O	J	A	S	I	D	X	Z
2	Q	P	C	E	G	B	T	J
1	M	R	C	Z	N	O	U	W
0	K	X	U	H	L	Y	S	Q
−1	F	E	A	S	W	K	R	M
−2	G	J	Z	N	H	V	D	G
−3	E	L	X	L	F	Q	O	B
	−3	−2	−1	0	1	2	3	4

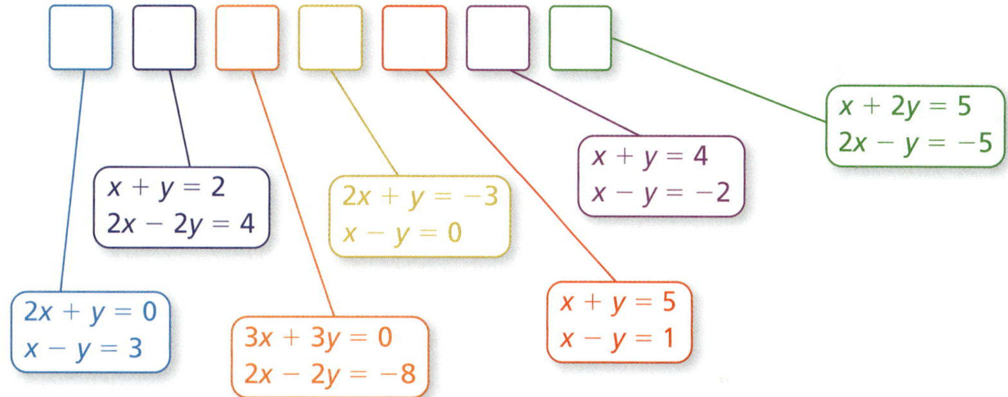

What Is Your Answer?

4. **IN YOUR OWN WORDS** How can you use elimination to solve a system of linear equations?

5. **STRUCTURE** When can you add or subtract equations in a system to solve the system? When do you have to multiply first? Justify your answers with examples.

6. **LOGIC** In Activity 2, why can you multiply equations in the system by a constant and not change the solution of the system? Explain your reasoning.

Practice Use what you learned about systems of linear equations to complete Exercises 4–6 on page 221.

5.3 Lesson

Solving a System of Linear Equations by Elimination

Step 1: Multiply, if necessary, one or both equations by a constant so at least 1 pair of like terms has the same or opposite coefficients.

Step 2: Add or subtract the equations to eliminate one of the variables.

Step 3: Solve the resulting equation for the remaining variable.

Step 4: Substitute the value from Step 3 into one of the original equations and solve.

EXAMPLE 1 Solving a System of Linear Equations by Elimination

Solve the system by elimination. $\mathbf{x + 3y = -2}$ Equation 1

$\mathbf{x - 3y = 16}$ Equation 2

Step 1: The coefficients of the y-terms are already opposites.

Step 2: Add the equations.

$x + 3y = -2$ Equation 1

$x - 3y = 16$ Equation 2

$2x = 14$ Add the equations.

Step 3: Solve for x.

$2x = 14$ Equation from Step 2

$x = 7$ Divide each side by 2.

Step 4: Substitute 7 for x in one of the original equations and solve for y.

$x + 3y = -2$ Equation 1

$7 + 3y = -2$ Substitute 7 for x.

$3y = -9$ Subtract 7 from each side.

$y = -3$ Divide each side by 3.

The solution is $(7, -3)$.

Study Tip

Because the coefficients of x are the same, you can also solve the system by subtracting in Step 2.

$$\begin{aligned} x + 3y &= -2 \\ x - 3y &= 16 \\ \hline 6y &= -18 \end{aligned}$$

So, $y = -3$.

Check

Equation 1

$x + 3y = -2$

$7 + 3(-3) \stackrel{?}{=} -2$

$-2 = -2$ ✓

Equation 2

$x - 3y = 16$

$7 - 3(-3) \stackrel{?}{=} 16$

$16 = 16$ ✓

On Your Own

Now You're Ready Exercises 7–12

Solve the system of linear equations by elimination. Check your solution.

1. $2x - y = 9$
   $4x + y = 21$
2. $-5x + 2y = 13$
   $5x + y = -1$
3. $3x + 4y = -6$
   $7x + 4y = -14$

Laurie's Notes

Introduction

Connect

- **Yesterday:** Students discovered how and when they can add or subtract the equations in a system to eliminate a variable. (MP1, MP7)
- **Today:** Students will solve systems of linear equations by elimination.

Motivate

- **Play the Opposites Game:** Pair students. Each student has a piece of paper and a pencil. Tell students that they will be given one minute to make a list of words that are opposites. One partner writes a word while the other partner writes the opposite of that word (example: hot and cold).
- Time the game for one minute.
- At the end of the game determine which pair has the longest list. Have them read their list of words, alternating between partners. Ask one or two others to share any new words from their lists.
- ? "What is the opposite of 4? -3.8? $7x$?" -4; 3.8; $-7x$
- Today students will use like terms whose coefficients are the same or opposite.

Lesson Notes

Discuss

- **MP3a Construct Viable Arguments:** Students have solved systems of linear equations by graphing and substitution. Elimination is a third method. Students should continue to focus on what method makes sense to use when solving a given system. They should choose a method based on the coefficients of the system and the form in which the equations are written. Students should be able to explain why they select a particular method.

Key Idea

- Discuss the steps for solving a system of linear equations by elimination. Students often find the steps very wordy, but the process is fairly simple and straightforward.
- As you work through each example, refer back to the steps listed.

Example 1

- Write Equation 1 and Equation 2.
- ? "What are the coefficients of the x-terms?" Both are 1.
- ? "What are the coefficients of the y-terms?" 3 and -3
- **Teaching Tip:** Line up like terms vertically so that you can add the terms in each column.
- ? "When you add the equations, what is the sum?" $2x = 14$
- ? "How do you determine the y-coordinate of the solution?" Substitute 7 for x in one of the original equations.
- Check the solution.

Goal Today's lesson is solving a system of linear equations by elimination.

Lesson Tutorials
Lesson Plans
Answer Presentation Tool

Extra Example 1

Solve the system by elimination.

$3x - y = 14$

$-3x + 4y = 16$

(8, 10)

On Your Own

1. (5, 1)
2. (−1, 4)
3. (−2, 0)

Laurie's Notes

Example 1 (continued)

- Discuss with students other methods they could use to solve the system and whether they would select that method. For instance, as shown in the Study Tip, another way to use elimination is to subtract one equation from the other to eliminate the variable x. Graphing might not be as efficient because the equations are in standard form. Substitution could be used.

On Your Own

- **Common Error:** Question 3 does not have coefficients that are opposites. Students may subtract the equations, but neglect to handle the subtraction properly on the right side. Make sure they subtract $-6 - (-14)$.
- Ask volunteers to share their work at the board.

Extra Example 2

Solve the system by elimination.

$3x - 4y = 33$

$4x + 3y = 19$

$(7, -3)$

Example 2

- Write the system of equations.
- ? "What do you notice about this system?" Listen for this answer: None of the coefficients of like variable terms are the same or opposite.
- ? "Can you think of a way to rewrite the equations so that either the x-terms or the y-terms have coefficients that are the same or opposite?" Multiply through one equation by a constant.
- ? "If you wanted to eliminate the x-term, what could you do?" Multiply the second equation by -3 and add, or multiply the second equation by 3 and subtract.
- Discuss the two options: Multiplying the second equation by -3 and adding or multiplying by 3 and subtracting.
- First work through the problem using one approach. Then show how the other approach gives the same result in Step 2.
- Although both approaches give you the same solution, sometimes fewer computation errors occur when equations are added rather than subtracted. Some students prefer adding equations.
- **Common Error:** When multiplying through by a constant, be sure that students multiply *every* term by the constant. It helps to use color:

  $(3)(-2x - 4y) = 14(3)$

  Remind students that they are using the Distributive Property.
- ? Before students try the problems on their own, ask, "What if this system had been written this way?"

  $5y - 6x = 25$

  $-2x - 4y = 14$

  Students should recognize that the like terms are not lined up in the same columns, so one of the equations should be rewritten.

On Your Own

- Have students check with their neighbors as they work through the problems. Remind students to use care when setting up the addition or subtraction. It is very frustrating to check your answer and find out that you made an error at the beginning of your solution process.

On Your Own

4. $(3, 2)$

5. $(-6, -1)$

6. $(0, 3)$

EXAMPLE 2 Solving a System of Linear Equations by Elimination

Solve the system by elimination.

$-6x + 5y = 25$ Equation 1

$-2x - 4y = 14$ Equation 2

Step 1: Multiply Equation 2 by 3.

$-6x + 5y = 25$ → $-6x + 5y = 25$ Equation 1

$-2x - 4y = 14$ **Multiply by 3.** → $-6x - 12y = 42$ Revised Equation 2

Study Tip

In Example 2, notice that you can also multiply Equation 2 by -3 and then add the equations.

Step 2: Subtract the equations.

$-6x + 5y = 25$ Equation 1

$-6x - 12y = 42$ Revised Equation 2

$17y = -17$ Subtract the equations.

Step 3: Solve for y.

$17y = -17$ Equation from Step 2

$y = -1$ Divide each side by 17.

Step 4: Substitute -1 for y in one of the original equations and solve for x.

$-2x - 4y = 14$ Equation 2

$-2x - 4(-1) = 14$ Substitute -1 for y.

$-2x + 4 = 14$ Multiply.

$-2x = 10$ Subtract 4 from each side.

$x = -5$ Divide each side by -2.

The solution is $(-5, -1)$.

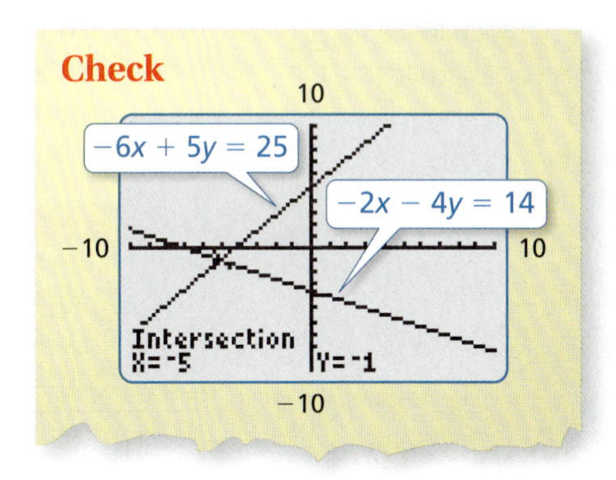

On Your Own

Solve the system of linear equations by elimination. Check your solution.

4. $3x + y = 11$
$6x + 3y = 24$

5. $4x - 5y = -19$
$-x - 2y = 8$

6. $5y = 15 - 5x$
$y = -2x + 3$

EXAMPLE 3 Real-Life Application

You buy 8 hostas and 15 daylilies for \$193. Your friend buys 3 hostas and 12 daylilies for \$117. Write and solve a system of linear equations to find the cost of each daylily.

Use a verbal model to write a system of linear equations.

Number of hostas	$\cdot$	Cost of each hosta, x	$+$	Number of daylilies	$\cdot$	Cost of each daylily, y	$=$	Total cost

The system is: $8x + 15y = 193$ — Equation 1 (You)

$3x + 12y = 117$ — Equation 2 (Your friend)

Step 1: To find the cost y of each daylily, eliminate the x-terms. Multiply Equation 1 by 3. Multiply Equation 2 by 8.

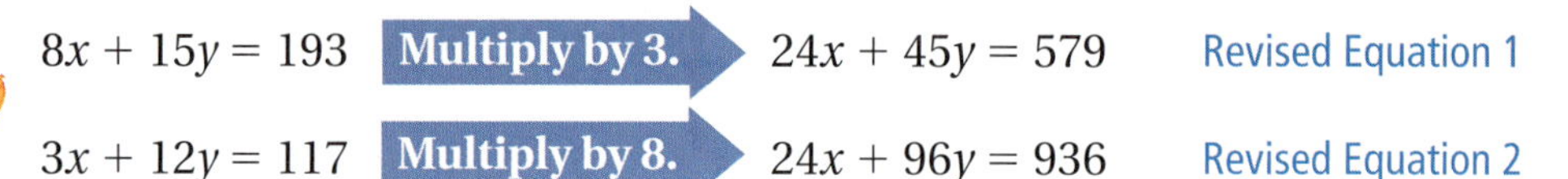

$8x + 15y = 193$ **Multiply by 3.** $24x + 45y = 579$ — Revised Equation 1

$3x + 12y = 117$ **Multiply by 8.** $24x + 96y = 936$ — Revised Equation 2

Step 2: Subtract the revised equations.

$$\begin{aligned} 24x + 45y &= 579 && \text{Revised Equation 1} \\ 24x + 96y &= 936 && \text{Revised Equation 2} \\ \hline -51y &= -357 && \text{Subtract the equations.} \end{aligned}$$

Step 3: Solving the equation $-51y = -357$ gives $y = 7$.

So, each daylily costs \$7.

On Your Own

7. A landscaper buys 4 peonies and 9 geraniums for \$190. Another landscaper buys 5 peonies and 6 geraniums for \$185. Write and solve a system of linear equations to find the cost of each peony.

Summary

Methods for Solving Systems of Linear Equations

Method	When to Use
Graphing *(Lesson 5.1)*	To estimate solutions
Substitution *(Lesson 5.2)*	When one of the variables in one of the equations has a coefficient of 1 or -1
Elimination *(Lesson 5.3)*	When at least 1 pair of like terms has the same or opposite coefficients
Elimination (Multiply First) *(Lesson 5.3)*	When one of the variables cannot be eliminated by adding or subtracting the equations

Laurie's Notes

Example 3

- Ask a volunteer to read and summarize the problem: Two people bought different numbers of two types of flowers and paid different amounts.
- Although the problem is to find the cost per daylily, you need a second variable for the cost per hosta.
- Discuss the verbal model used to generate each equation.
- ? "What does Equation 1 represent?" the amount of money you spend for your flowers
- ? "What does Equation 2 represent?" the amount of money your friend spends for the flowers
- ? "What do you notice about this system of equations that is different than the first two examples?" listen for this answer: Neither pair of like terms has a coefficient that is a multiple of the other.
- Students should recognize that 24 is the least common multiple of 8 and 3. Work through the problem as shown.
- Discuss alternative approaches to this problem.
  - Multiply Equation 2 by 8/3 and subtract to eliminate the x-terms.
  - Multiply Equation 1 by 4 and Equation 2 by 5 and subtract to eliminate the y-terms.
- **Extension:** Determine the cost of each hosta plant.

Summary

- **MP1a Make Sense of Problems** and **MP3a:** The summary box reviews the last three lessons. It is important for students to understand that the approach they use usually depends on the coefficients of the variables. Students should be able to make wise decisions about the approach to use and explain the reasoning behind their choices.

Closure

Write an example of a system for each condition relative to the solution by elimination.

- Adding or subtracting equations will eliminate one of the variables.
- You need to multiply one of the equations by an integer before you add or subtract.
- You need to multiply both of the equations by an integer (or one equation by a fraction) before you add or subtract.

Extra Example 3

There are 340 calories in 2 cups of cereal with 1 cup of milk. There are 570 calories in 3 cups of the cereal with 2 cups of milk. Write and solve a system of linear equations to find the number x of calories in 1 cup of the cereal without milk.

$2x + y = 340$

$3x + 2y = 570$;

110 calories

On Your Own

7. $4x + 9y = 190$

$5x + 6y = 185$

$x = 25$; $25 per peony

English Language Learners

Pair Activity

Pair English learners with English speakers. One partner uses an example covered in class to quiz the other about the process. The quizzer looks at the steps for solving a system by elimination in the Key Idea on page 218. Without revealing the step number, the quizzer reads one randomly chosen step at a time to the other student, and asks where the step was carried out in the example. This helps students get comfortable with the language used.

Vocabulary and Concept Check

1. **Step 1:** Multiply, if necessary, one or both equations by a constant so at least one pair of like terms has the same or opposite coefficients.

   Step 2: Add or subtract the equations to eliminate one of the variables.

   Step 3: Solve the resulting equation for the remaining variable.

   Step 4: Substitute the value from Step 3 into one of the original equations and solve.
2. Use multiplication when at least one pair of like terms are not the same or opposites.
3. $2x + 3y = 11$

   $3x - 2y = 10$;

   You have to use multiplication to solve the system by elimination.

Practice and Problem Solving

4. (2, 1)
5. (6, 2)
6. (−1, 3)
7. (2, 1)
8. (−1, 3)
9. (1, −3)
10. (4, −1)
11. (3, 2)
12. (−2, −5)
13. The student added y-terms, but subtracted x-terms and constants; solution: (1, 2)
14. **a.** $x + y = 58$

    $x - y = 14$

    b. You sell 36 tickets, your friend sells 22 tickets.
15. **a.** $2x + y = 10$

    $2x + 3y = 22$

    b. 6 minutes

Assignment Guide and Homework Check

Level	Day 1 Activity Assignment	Day 2 Lesson Assignment	Homework Check
Basic	4–6, 34–37	1–3, 7–21 odd, 22, 23–27 odd	9, 15, 17, 19, 25
Average	4–6, 34–37	1–3, 9–21 odd, 22–32 even	11, 15, 19, 21, 28
Advanced	1–6, 10, 12, 13, 14–26 even, 27–37		12, 14, 20, 30, 32

Common Errors

- **Exercises 4–12** Students may make careless errors. Remind students to line up like terms neatly in columns to avoid confusion about what terms to add or subtract. Also, take care with subtraction—it may help to change the sign of each term in the equation to be subtracted, so you can just add the terms.
- **Exercise 14** Students may struggle writing an equation to represent "You sell 14 more tickets than your friend sells." Explain that it helps to ask the question, "How does your number of tickets compare to your friend's number of tickets?"

5.3 Record and Practice Journal

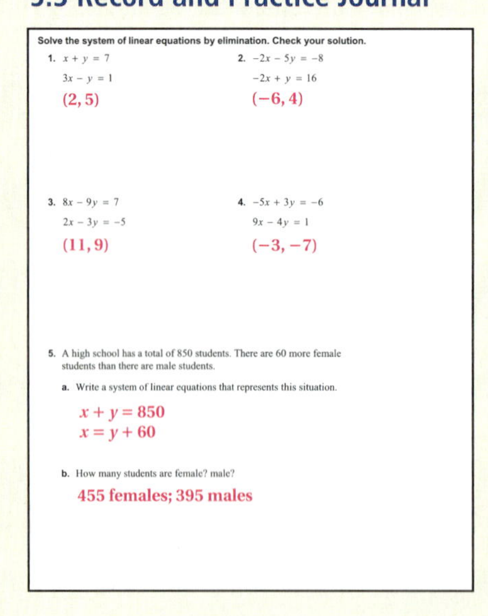

Solve the system of linear equations by elimination. Check your solution.

1. $x + y = 7$

   $3x - y = 1$

   (2, 5)
2. $-2x - 5y = -8$

   $-2x + y = 16$

   (−6, 4)
3. $8x - 9y = 7$

   $2x - 3y = -5$

   (11, 9)
4. $-5x + 3y = -6$

   $9x - 4y = 1$

   (−3, −7)
5. A high school has a total of 850 students. There are 60 more female students than there are male students.

   a. Write a system of linear equations that represents this situation.

   $x + y = 850$

   $x = y + 60$

   b. How many students are female? male?

   455 females; 395 males

5.3 Exercises

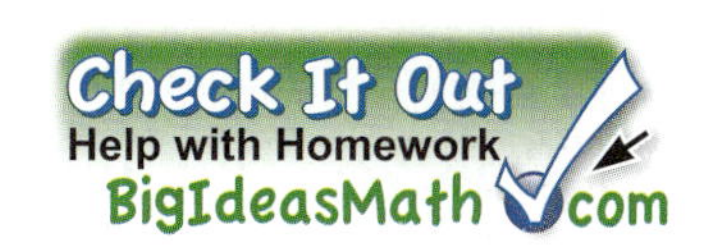

Vocabulary and Concept Check

1. **WRITING** Describe how to solve a system of linear equations by elimination.

2. **NUMBER SENSE** When should you use multiplication to solve a system of linear equations by elimination?

3. **WHICH ONE DOESN'T BELONG?** Which system of equations does *not* belong with the other three? Explain your reasoning.

$3x + 3y = 3$ $2x - 3y = 7$	$-2x + y = 6$ $2x - 3y = -10$	$2x + 3y = 11$ $3x - 2y = 10$	$x + y = 5$ $3x - y = 3$

Practice and Problem Solving

Use a method from Activity 1 to solve the system.

4. $x + y = 3$
   $x - y = 1$

5. $-x + 3y = 0$
   $x + 3y = 12$

6. $3x + 2y = 3$
   $3x - 2y = -9$

Solve the system of linear equations by elimination. Check your solution.

7. $x + 3y = 5$
   $-x - y = -3$

8. $x - 2y = -7$
   $3x + 2y = 3$

9. $4x + 3y = -5$
   $-x + 3y = -10$

10. $2x + 7y = 1$
    $2x - 4y = 12$

11. $2x + 5y = 16$
    $3x - 5y = -1$

12. $3x - 2y = 4$
    $6x - 2y = -2$

13. **ERROR ANALYSIS** Describe and correct the error in solving the system of linear equations.

$$\begin{aligned} 5x + 2y &= 9 \quad \text{Equation 1} \\ 3x - 2y &= -1 \quad \text{Equation 2} \\ \hline 2x \quad\ &= 10 \\ x &= 5 \end{aligned}$$

The solution is $(5, -8)$.

14. **RAFFLE TICKETS** You and your friend are selling raffle tickets for a new laptop. You sell 14 more tickets than your friend sells. Together, you and your friend sell 58 tickets.

    a. Write a system of linear equations that represents this situation.

    b. How many tickets does each of you sell?

15. **JOGGING** You can jog around your block twice and the park once in 10 minutes. You can jog around your block twice and the park 3 times in 22 minutes.

    a. Write a system of linear equations that represents this situation.

    b. How long does it take you to jog around the park?

Solve the system of linear equations by elimination. Check your solution.

2 3 **16.** $2x - y = 0$
$3x - 2y = -3$

17. $x + 4y = 1$
$3x + 5y = 10$

18. $-2x + 3y = 7$
$5x + 8y = -2$

19. $3x + 3 = 3y$
$2x - 6y = 2$

20. $2x - 6 = 4y$
$7y = -3x + 9$

21. $5x = 4y + 8$
$3y = 3x - 3$

22. **ERROR ANALYSIS** Describe and correct the error in solving the system of linear equations.

✗ $x + y = 1$	Equation 1	Multiply by −5.	$-5x + 5y = -5$
$5x + 3y = -3$	Equation 2		$5x + 3y = -3$
			$8y = -8$
			$y = -1$
			The solution is (2, −1).

23. **REASONING** For what values of a and b should you solve the system by elimination?

a. $4x - y = 3$
$ax + 10y = 6$

b. $x - 7y = 6$
$-6x + by = 9$

Determine whether the line through the first pair of points intersects the line through the second pair of points. Explain.

24. Line 1: (−2, 1), (2, 7)
Line 2: (−4, −1), (0, 5)

25. Line 1: (3, −2), (7, −1)
Line 2: (5, 2), (6, −2)

26. **AIRPLANES** Two airplanes are flying to the same airport. Their positions are shown in the graph. Write a system of linear equations that represents this situation. Solve the system by elimination to justify your answer.

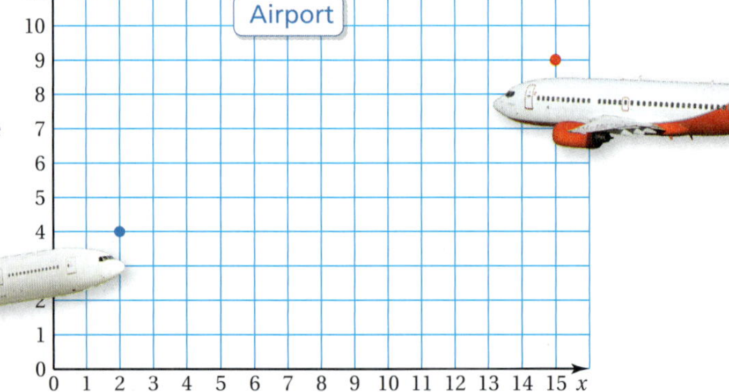

27. **TEST PRACTICE** The table shows the number of correct answers on a practice standardized test. You score 86 points on the test, and your friend scores 76 points.

	You	Your Friend
Multiple Choice	23	28
Short Response	10	5

a. Write a system of linear equations that represents this situation.

b. How many points is each type of question worth?

Common Errors

- **Exercises 16–21** Students may make errors in multiplying through the whole equation. Tell them to clearly show what they are multiplying each equation by so they do not forget during the process. They can check the sign and coefficient of each term after multiplying.
- **Exercise 26** Students may not write a correct system of equations for the situation. Ask them to consider what line each airplane has to follow to fly to the airport.
- **Exercise 27** Students may write a system of equations by reading across the table horizontally. Stress that your total number of points is equal to the sum of your points for multiple choice questions and for short response questions which are both listed in the column for "You."
- **Exercise 29** Students may use an incorrect problem-solving strategy. Make sure they realize that they first need to solve for the cost per hour for each activity, then use those rates to answer the question.
- **Exercise 31** Students may fail to write an equation for the situation using the gold percents. Explain that the total amount of gold contributed by each alloy is equal to the amount of gold in the final mixture. Each amount of gold is given by the percent times the number of grams.
- **Exercise 32** Students may not write a system of equations that represents the problem correctly. Tell them this is an extension of the formula $d = r \times t$ where the rate is increased or decreased by the speed of the current.
- **Exercise 33** Students may use an incorrect approach. Tell them to look for a way to eliminate two of the variables. This will allow them to solve for the third variable. They can then substitute the value of this variable back into two of the equations to write a new system of two equations in two variables, which they know how to solve.

Practice and Problem Solving

16. (3, 6) **17.** (5, −1)

18. (−2, 1) **19.** (−2, −1)

20. (3, 0) **21.** (4, 3)

22. The y-term was multiplied by 5 instead of −5; solution: (−3, 4)

23. **a.** ±4
b. ±7

24. no; The lines are parallel.

25. yes; The lines are perpendicular.

26. $y = 2x$
$y = -\frac{1}{3}x + 14$;
Solution: (6, 12)

27. **a.** $23x + 10y = 86$
$28x + 5y = 76$

b. Multiple choice: 2 points each; Short response: 4 points each

English Language Learners

Auditory

Ask students to explain how they used the table to set up the system of equations in Exercise 27. Then ask students to explain how they used the table to set up the system of equations in Exercise 29. Discuss the general factors you need to take into account in using a table to write the equations for a system of equations.

Practice and Problem Solving

28. no; You cannot sell −6 tickets.

29. \$95

30. *Sample answer:* Find the line perpendicular to $2x + y = 0$ through $(2, -4)$; $x - 2y = 10$; Solution: $(2, -4)$

31. 5 grams of 90% gold alloy, 3 grams of 50% gold alloy

32. See *Taking Math Deeper.*

33. $(-1, 2, 1)$

Fair Game Review

34. yes **35.** yes

36. no **37.** D

Mini-Assessment

Solve the system of linear equations by elimination. Check your solution.

1. $5x - 2y = 18$
$-5x + 3y = -22$ $(2, -4)$

2. $2x + 4y = 20$
$-3x + 4y = 30$ $(-2, 6)$

3. $4x - 2y = 2$
$7x - 3y = 6$ $(3, 5)$

4. You have 33 quarters and dimes in a jar. The jar contains a total of \$4.95. Write and solve a system of equations to find the number x of dimes and the number y of quarters.
$x + y = 33$
$0.1x + 0.25y = 4.95$;
$(22, 11)$ or 22 dimes and 11 quarters

Taking Math Deeper

Exercise 32

This is a classic riverboat problem. Students may find the problem complicated, so it helps to organize the information in a table.

 Make a table.

	Distance	Rate	Time
Downstream	10 mi	$(B + C)$ mi/h	1/2 h
Upstream	10 mi	$(B - C)$ mi/h	5/6 h

Distance: The boat travels 10 miles in each direction.

Time: The trip downstream takes 30 minutes and the trip upstream takes 50 minutes. Because boat speed is often measured in miles per hour, think of 30 minutes as 1/2 hour and 50 minutes as 5/6 hour.

Rate: The boat's speed B relative to the water is increased or decreased by the rate C of the current. So, the boat's actual rate of speed is $B + C$ downstream and $B - C$ upstream.

Now use the formula $d = r \times t$ to write and solve a set of equations.

2 Write a set of equations.

$10 = \frac{1}{2}(B + C)$ **Multiply by 2.** $20 = B + C$ Downstream

$10 = \frac{5}{6}(B - C)$ **Multiply by 6/5.** $12 = B - C$ Upstream

$32 = 2B$ Add equations.

The speed of the boat relative to the water is $B = 32/2 = 16$ miles per hour.

 Substitute.

Substitute 16 for B in $20 = B + C$ and solve for C.

$$20 = B + C$$
$$20 = 16 + C$$
$$4 = C$$

The speed of the current is 4 miles per hour.

Reteaching and Enrichment Strategies

If students need help. . .	If students got it. . .
Resources by Chapter • Practice A and Practice B • Puzzle Time Record and Practice Journal Practice Differentiating the Lesson Lesson Tutorials Skills Review Handbook	Resources by Chapter • Enrichment and Extension • Technology Connection Start the next section

28. **LOGIC** You solve a system of equations in which x represents the number of adult tickets sold and y represents the number of student tickets sold. Can $(-6, 24)$ be the solution of the system? Explain your reasoning.

29. **VACATION** The table shows the activities of two tourists at a vacation resort. You want to go parasailing for 1 hour and horseback riding for 2 hours. How much do you expect to pay?

	Parasailing	Horseback Riding	Total Cost
Tourist 1	2 hours	5 hours	\$205
Tourist 2	3 hours	3 hours	\$240

30. **REASONING** The solution of a system of linear equations is $(2, -4)$. One equation in the system is $2x + y = 0$. Explain how you could find a second equation for the system. Then find a second equation. Solve the system by elimination to justify your answer.

31. **JEWELER** A metal alloy is a mixture of two or more metals. A jeweler wants to make 8 grams of 18-carat gold, which is 75% gold. The jeweler has an alloy that is 90% gold and an alloy that is 50% gold. How much of each alloy should the jeweler use?

32. **PROBLEM SOLVING** A powerboat takes 30 minutes to travel 10 miles downstream. The return trip takes 50 minutes. What is the speed of the current?

33. **Critical Thinking** Solve the system of equations by elimination.

$$2x - y + 3z = -1$$
$$x + 2y - 4z = -1$$
$$y - 2z = 0$$

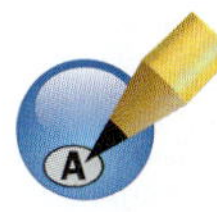

Fair Game Review What you learned in previous grades & lessons

Decide whether the two equations are equivalent. *(Section 1.2 and Section 1.3)*

34. $4n + 1 = n - 8$

    $3n = -9$

35. $2a + 6 = 12$

    $a + 3 = 6$

36. $7v - \frac{3}{2} = 5$

    $14v - 3 = 15$

37. **MULTIPLE CHOICE** Which line has the same slope as $y = \frac{1}{2}x - 3$? *(Section 4.4)*

    (A) $y = -2x + 4$ (B) $y = 2x + 3$ (C) $y - 2x = 5$ (D) $2y - x = 7$

5.4 Solving Special Systems of Linear Equations

Essential Question Can a system of linear equations have no solution? Can a system of linear equations have many solutions?

1 ACTIVITY: Writing a System of Linear Equations

Work with a partner. Your cousin is 3 years older than you. You can represent your ages by two linear equations.

$y = t$ Your age

$y = t + 3$ Your cousin's age

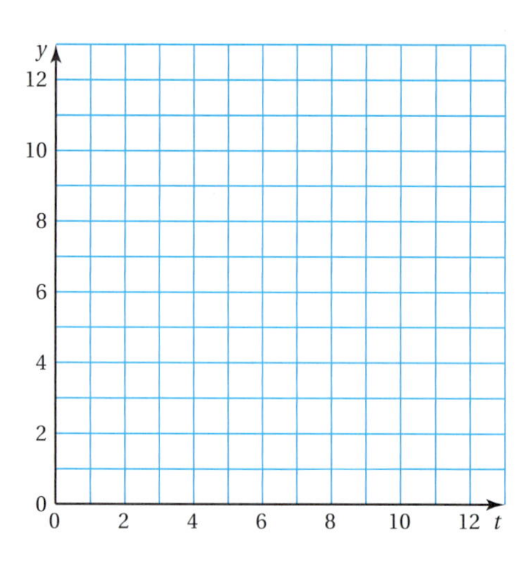

a. Graph both equations in the same coordinate plane.

b. What is the vertical distance between the two graphs? What does this distance represent?

c. Do the two graphs intersect? Explain what this means in terms of your age and your cousin's age.

2 ACTIVITY: Using a Table to Solve a System

Work with a partner. You invest $500 for equipment to make dog backpacks. Each backpack costs you $15 for materials. You sell each backpack for $15.

a. Copy and complete the table for your cost C and your revenue R.

x	0	1	2	3	4	5	6	7	8	9	10
C											
R											

b. When will you break even? What is wrong?

Systems of Equations

In this lesson, you will

- solve systems of linear equations with no solution or infinitely many solutions.

Learning Standards
MACC.8.EE.3.8a
MACC.8.EE.3.8b
MACC.8.EE.3.8c

Laurie's Notes

Introduction

Standards for Mathematical Practice

- **MP1 Make Sense of Problems and Persevere in Solving Them:** Students have learned three techniques for solving a system of linear equations. In this lesson, they use more than one technique to solve the same system, and in doing so recognize that not all systems have a single solution. Using a variety of strategies helps students make sense of the new possible outcomes for a solution of a system of equations.

Motivate

- If you have an overhead projector or document camera, place 2 pieces of uncooked spaghetti on display. Say, "These represent two lines. Right now they are intersecting."
- ? "Is there any other relationship they could have?" Listen for parallel (non-intersecting) and a reasonable description of coinciding lines.
- Now place a transparency of a coordinate grid on top of the spaghetti.
- Suggest that lines, when graphed, do not always have to intersect.
- **Management Tip:** I use uncooked spaghetti for many models throughout the year, so I keep a box in my desk.

Activity Notes

Activity 1

- ? "How do you measure the vertical distance between two graphs?" Use a vertical line on the graph to count the units from one graph to the other.
- **Part (c):** Students will likely say that the two are not the same age. The solution actually says more: you and your cousin will never be the same age at the same time. Your cousin will always be 3 years older.
- **Extension:** Students may notice that the lines have the same slope, so they are parallel.

Activity 2

- Have students write the cost equation and the revenue equation. Students should turn back to the activities in Section 5.1 as needed for help.
- Give time for students to work with their partners to fill in the table. Make sure that all students are using correct equations.
- ? **Extension:** "What patterns do you observe in the table?" Both rows of numbers are increasing by the same amount (15). The values are not getting closer together.
- Discuss student answers for part (b). Students should recognize that when you sell an item for the exact price that it costs, you will never make a profit, nor pay off the original investment of $500.
- ? **Extension:** "Does either of the equations (for cost or for revenue) represent a proportional relationship? Explain." yes; The revenue equation is a direct variation equation, $R = 15x$. Its graph is a line with a slope of 15 that passes through the origin.

Florida Common Core Standards

MACC.8.EE.3.8a Understand that solutions to a system of two linear equations in two variables correspond to points of intersection of their graphs, because points of intersection satisfy both equations simultaneously.

MACC.8.EE.3.8b Solve systems of two linear equations in two variables algebraically, and estimate solutions by graphing the equations. Solve simple cases by inspection.

MACC.8.EE.3.8c Solve real-world and mathematical problems leading to two linear equations in two variables.

Previous Learning

Students should know how to solve linear equations with variables on both sides.

5.4 Record and Practice Journal

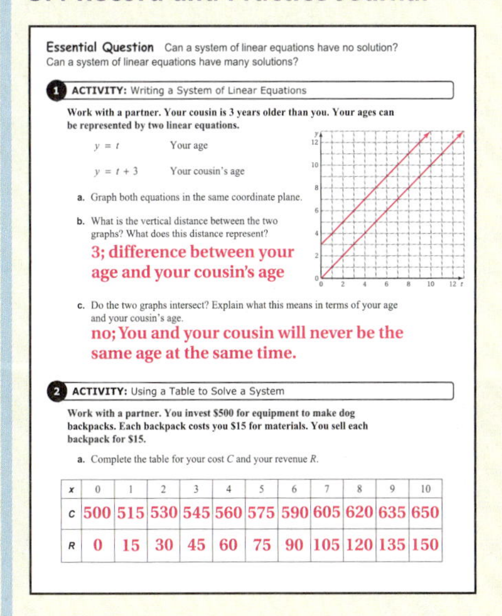

Essential Question Can a system of linear equations have no solution? Can a system of linear equations have many solutions?

1 ACTIVITY: Writing a System of Linear Equations

Work with a partner. Your cousin is 3 years older than you. Your ages can be represented by two linear equations.

$y = t$ Your age

$y = t + 3$ Your cousin's age

a. Graph both equations in the same coordinate plane.

b. What is the vertical distance between the two graphs? What does this distance represent?

3; difference between your age and your cousin's age

c. Do the two graphs intersect? Explain what this means in terms of your age and your cousin's age.

no; You and your cousin will never be the same age at the same time.

2 ACTIVITY: Using a Table to Solve a System

Work with a partner. You invest $500 for equipment to make dog backpacks. Each backpack costs you $15 for materials. You sell each backpack for $15.

a. Complete the table for your cost C and your revenue R.

x	0	1	2	3	4	5	6	7	8	9	10
C	500	515	530	545	560	575	590	605	620	635	650
R	0	15	30	45	60	75	90	105	120	135	150

Differentiated Instruction

Visual

To help students remember that lines with the same slope are parallel, write "If the slopes on the board are =, then the lines are parallel." Relate that the segments that make up the equal sign are parallel segments.

5.4 Record and Practice Journal

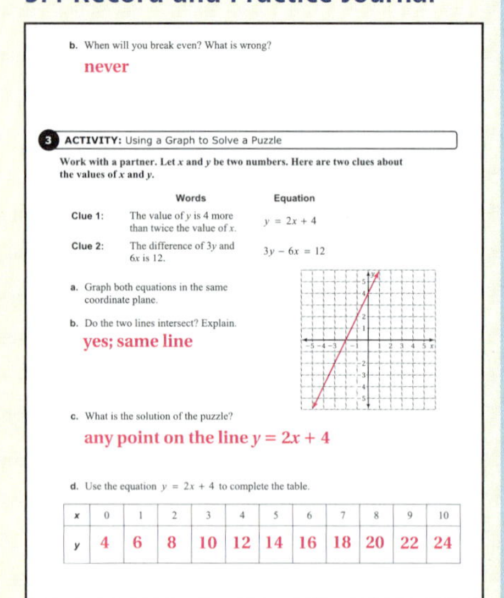

b. When will you break even? What is wrong?

never

3 ACTIVITY: Using a Graph to Solve a Puzzle

Work with a partner. Let x and y be two numbers. Here are two clues about the values of x and y.

	Words	Equation
Clue 1:	The value of y is 4 more than twice the value of x.	$y = 2x + 4$
Clue 2:	The difference of $3y$ and $6x$ is 12.	$3y - 6x = 12$

a. Graph both equations in the same coordinate plane.

b. Do the two lines intersect? Explain.

yes; same line

c. What is the solution of the puzzle?

any point on the line $y = 2x + 4$

d. Use the equation $y = 2x + 4$ to complete the table.

x	0	1	2	3	4	5	6	7	8	9	10
y	4	6	8	10	12	14	16	18	20	22	24

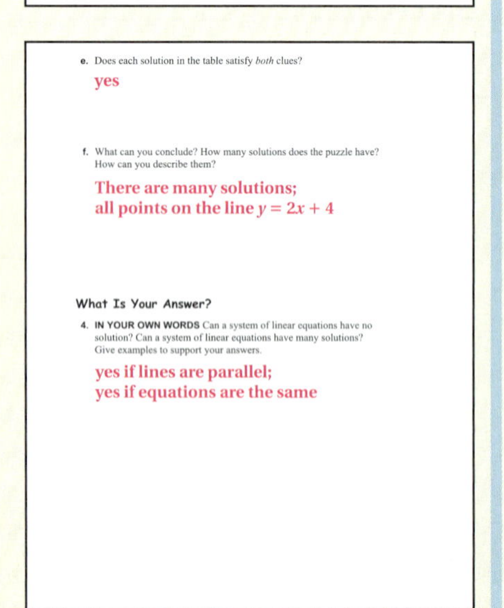

e. Does each solution in the table satisfy *both* clues?

yes

f. What can you conclude? How many solutions does the puzzle have? How can you describe them?

There are many solutions; all points on the line $y = 2x + 4$

What Is Your Answer?

4. **IN YOUR OWN WORDS** Can a system of linear equations have no solution? Can a system of linear equations have many solutions? Give examples to support your answers.

yes if lines are parallel; yes if equations are the same

Laurie's Notes

Activity 3

- Students enjoy puzzles. Present the next activity in this context.
- ? "Look at the words for each clue and then look at the equation. Does the translation make sense? Explain." Listen to student explanations.
- ? "In what form is each equation written? What strategy can be used to graph each equation?" Listen to student explanations.
- Provided students have graphed the equations carefully, the lines should coincide. There is only one graph.
- Discuss answers to part (f). The fact that the two graphs coincide means the equations are equivalent and have the same graph. There are many solutions to the puzzle! Help students recognize this by selecting two or three ordered pairs from the table. Talk through the substitution of the ordered pair into each equation. All of the ordered pairs will be solutions of both equations.
- **Extension:** Students may have heard of puzzles where you start with a number, do a few computations, and eventually end up with a number twice the original. Here's an example. (I have annotated the algebraic representation to the right.) Suggest to students that they start with a small number so that the computation is simple.

Step 1: Pick a number, perhaps your age.	x
Step 2: Add 10.	$x + 10$
Step 3: Multiply by 4.	$4(x + 10) = 4x + 40$
Step 4: Divide by 2.	$(4x + 40) \div 2 = 2x + 20$
Step 5: Subtract 20.	$2x + 20 - 20 = 2x$

Announce that you should now have a number that is twice the original number. In relation to this lesson, you would have:

$$y = \frac{4(x + 10)}{2} - 20 \text{ and } y = 2x.$$

The graph of both lines is $y = 2x$.

What Is Your Answer?

- This question tries to help students summarize the two additional cases for the solution of a system of linear equations.

Closure

- Sketch a graph of a system of equations that has no solution.
- Sketch a graph of a system of equations that has many solutions.

3 ACTIVITY: Using a Graph to Solve a Puzzle

Math Practice 4

Analyze Relationships

What do you know about the graphs of the two equations? How does this relate to the number of solutions?

Work with a partner. Let x and y be two numbers. Here are two clues about the values of x and y.

	Words	Equation
Clue 1:	y is 4 more than twice the value of x.	$y = 2x + 4$
Clue 2:	The difference of $3y$ and $6x$ is 12.	$3y - 6x = 12$

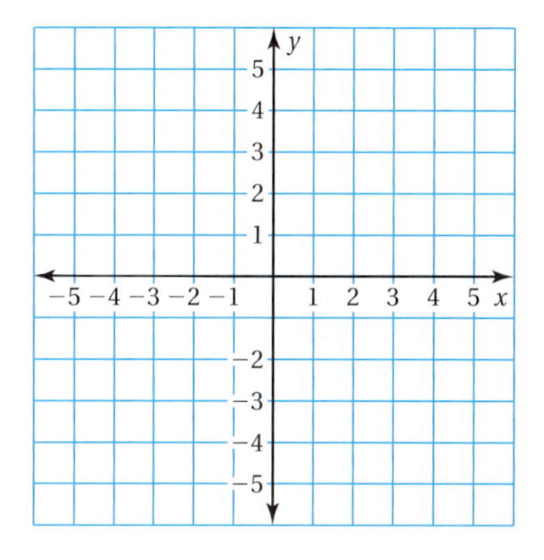

a. Graph both equations in the same coordinate plane.

b. Do the two lines intersect? Explain.

c. What is the solution of the puzzle?

d. Use the equation $y = 2x + 4$ to complete the table.

x	0	1	2	3	4	5	6	7	8	9	10
y											

e. Does each solution in the table satisfy *both* clues?

f. What can you conclude? How many solutions does the puzzle have? How can you describe them?

What Is Your Answer?

4. IN YOUR OWN WORDS Can a system of linear equations have no solution? Can a system of linear equations have many solutions? Give examples to support your answers.

Practice Use what you learned about special systems of linear equations to complete Exercises 3 and 4 on page 228.

Key Idea

Solutions of Systems of Linear Equations

A system of linear equations can have *one solution*, *no solution*, or *infinitely many solutions*.

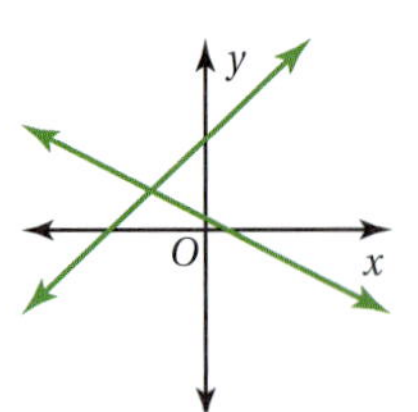

One solution

The lines intersect.

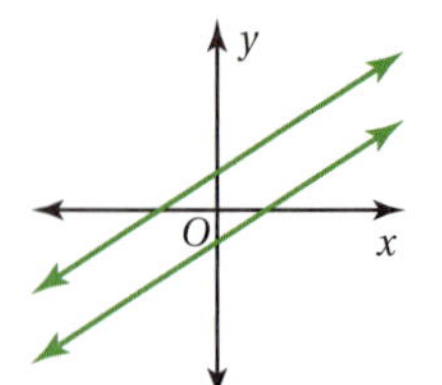

No solution

The lines are parallel.

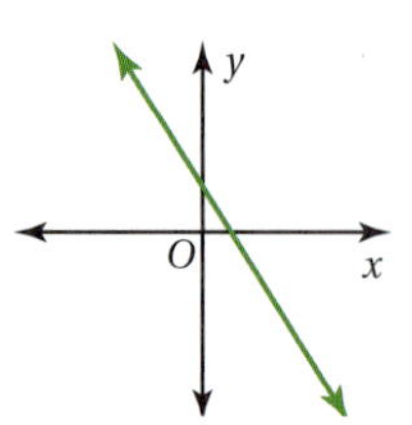

Infinitely many solutions

The lines are the same.

EXAMPLE 1 Solving a System: No Solution

Solve the system. $y = 3x + 1$ Equation 1

$y = 3x - 3$ Equation 2

Method 1: Solve by graphing.

Graph each equation. The lines have the same slope and different y-intercepts. So, the lines are parallel.

Because parallel lines do not intersect, there is no point that is a solution of both equations.

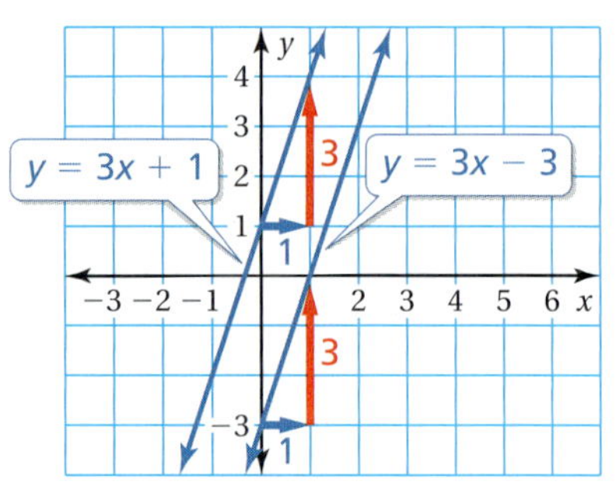

∴ So, the system of linear equations has no solution.

Method 2: Solve by substitution.

Substitute $3x - 3$ for y in Equation 1.

$y = 3x + 1$ Equation 1

$3x - 3 = 3x + 1$ Substitute $3x - 3$ for y.

$-3 = 1$ ✗ Subtract $3x$ from each side.

∴ The equation $-3 = 1$ is never true. So, the system of linear equations has no solution.

Study Tip

You can solve some linear systems by inspection. In Example 1, notice you can rewrite the system as

$-3x + y = 1$

$-3x + y = -3.$

This system has no solution because $-3x + y$ cannot be equal to 1 and -3 at the same time.

Solve the system of linear equations. Check your solution.

1. $y = -x + 3$
   $y = -x + 5$
2. $y = -5x - 2$
   $5x + y = 0$
3. $x = 2y + 10$
   $2x + 3y = -1$

Laurie's Notes

Introduction

Connect

- **Yesterday:** Students explored the graphs of two special systems of linear equations. (MP1)
- **Today:** Students will solve special systems of linear equations.

Motivate

- Make a set of cards in advance: equation card, slope card, and y-intercept card. Make enough cards so that everyone in the class has a card.

Examples:

$y = 3x - 4$	$m = 3$	$b = -4$
$3x + y = 4$	$m = -3$	$b = 4$
$y = 4x - 2$	$m = 4$	$b = -2$
$y = 2 - 4x$	$m = -4$	$b = 2$

? "What is the general form of an equation written in slope-intercept form?" $y = mx + b$

- Distribute all of the cards and have students form a matching set of 3.
- The goal of this quick activity is to review the slope-intercept form of an equation.

Lesson Notes

Key Idea

- Write the Key Idea. Connect this back to the spaghetti used yesterday to introduce special systems of linear equations.

Example 1

? "How do you graph equations in slope-intercept form?" Plot the point for the y-intercept, then use the slope to locate another point and draw the line.

- Because both equations are in slope-intercept form, students will observe the same slope for each equation and will not be surprised that the lines are parallel. Graphing is a visual check that the lines are parallel.

? "What if you did not notice that the lines had the same slope and you tried to solve the system algebraically? What do you think will happen?" Students are not likely to have a sense of what will happen.

- **MP7 Look for and Make Use of Structure:** Because both equations are solved for y, substitution in Method 2 is an approach that makes sense.
- Work through Method 2. Point out that solving the system algebraically leads to a statement $-5 = 1$ that is never true. You can interpret this false statement to mean that the system has no solution, because there are no ordered pairs that satisfy both equations. So, the lines must be parallel.
- **MP1 Make Sense of Problems and Persevere in Solving Them:** Discuss the Study Tip. Before starting any problem, students should consider what information is given and what approach makes sense.

Goal Today's lesson is solving a system of equations with no solution or infinitely many solutions.

Lesson Tutorials
Lesson Plans
Answer Presentation Tool

Extra Example 1

Solve the system.

$y = -2x + 5$

$y = -2x + 1$

no solution

On Your Own

1. no solution
2. no solution
3. $(4, -3)$

Extra Example 2

The perimeter of the trapezoid is 10 units. The perimeter of the triangle is 5 units. Write and solve a linear system to find the values of x and y.

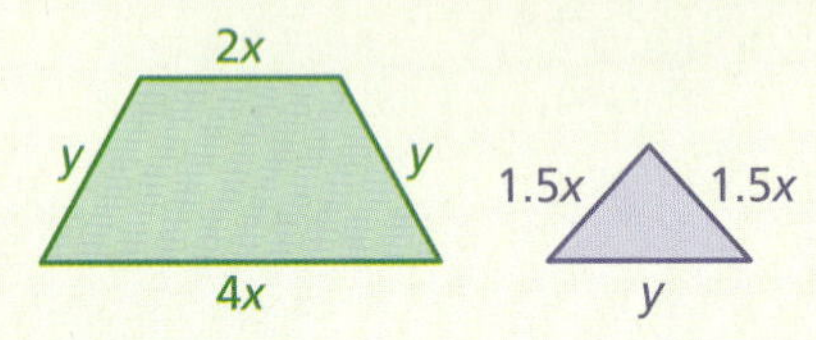

$6x + 2y = 10$
$3x + y = 5$
infinitely many solutions

On Your Own

4. (0, 3)
5. no solution
6. infinitely many solutions; all points on the line $y = \frac{1}{2}x - \frac{5}{2}$
7. There is no solution because the resulting equation $y = -\frac{1}{2}x + \frac{27}{4}$ is parallel to the equation of the perimeter of the triangle.

Differentiated Instruction

Visual

Make a poster of the Key Idea box on the number of solutions of systems of linear equations. Use color to help English learners make connections between concepts and language.

Laurie's Notes

On Your Own

- Students may quickly observe that the first two systems have no solution because the slopes are the same, so the lines are parallel. Make sure they base their answers on sound reasoning.

Example 2

- Ask a volunteer to read the problem. Check to see that students are comfortable with finding the perimeter of a triangle and a rectangle.
- Write the equations. Simplify each.
- Students can rewrite each equation in slope-intercept form to graph. They will notice that the equations are the same.

? "What do you think will happen if you try to solve the system algebraically?" Answers will vary.

- **MP8 Look for and Express Regularity in Repeated Reasoning:** Work through Method 2. Point out that solving the system algebraically leads to a statement that is always true, $0 = 0$. You can interpret this identity to mean that the system has infinitely many solutions, because there are infinitely many ordered pairs that satisfy both equations and the lines coincide.
- Note that although there are infinitely many solutions of this system, they are limited to Quadrant I because the side lengths cannot be negative or zero.

On Your Own

- **Think-Pair-Share:** Students should read each question independently and then work in pairs to answer the questions. When they have answered the questions, the pair should compare their answers with another group and discuss any discrepancies.

Closure

- **Exit Ticket:** Write a system of equations that has no solution.
  Sample answer: $y = 2x + 4$ and $y = 2x - 6$

EXAMPLE 2 Solving a System: Infinitely Many Solutions

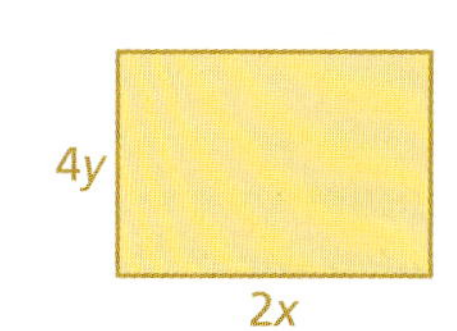

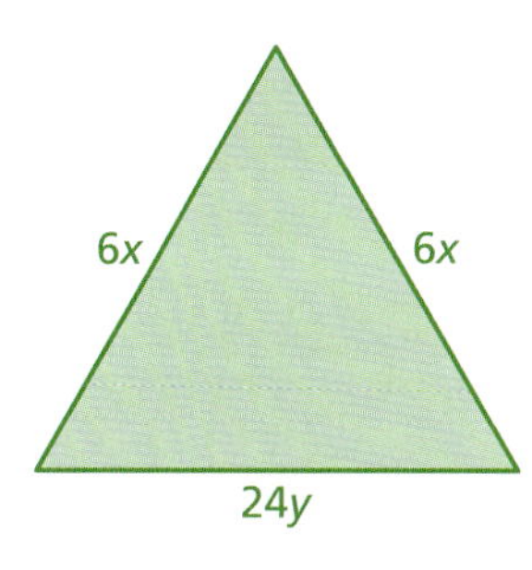

The perimeter of the rectangle is 36 units. The perimeter of the triangle is 108 units. Write and solve a system of linear equations to find the values of *x* and *y*.

Perimeter of Rectangle

$2(2x) + 2(4y) = 36$

$4x + 8y = 36$ Equation 1

Perimeter of Triangle

$6x + 6x + 24y = 108$

$12x + 24y = 108$ Equation 2

The system is: $4x + 8y = 36$ Equation 1

$12x + 24y = 108$ Equation 2

Method 1: Solve by graphing.

Graph each equation.

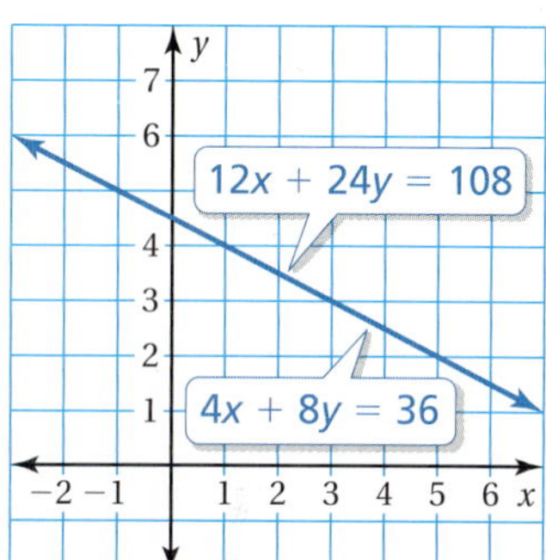

The lines have the same slope and the same y-intercept. So, the lines are the same.

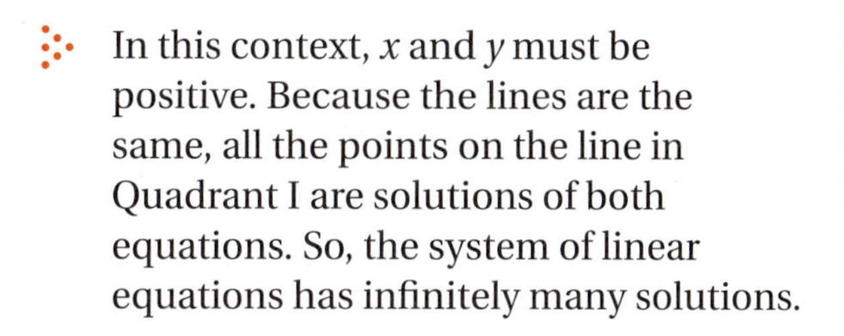

In this context, x and y must be positive. Because the lines are the same, all the points on the line in Quadrant I are solutions of both equations. So, the system of linear equations has infinitely many solutions.

Method 2: Solve by elimination.

Multiply Equation 1 by 3 and subtract the equations.

$4x + 8y = 36$ **Multiply by 3.** → $12x + 24y = 108$ Revised Equation 1

$12x + 24y = 108$ → $12x + 24y = 108$ Equation 2

$0 = 0$ Subtract.

The equation $0 = 0$ is always true. In this context, x and y must be positive. So, the solutions are all the points on the line $4x + 8y = 36$ in Quadrant I. The system of linear equations has infinitely many solutions.

On Your Own

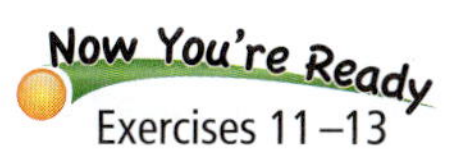

Solve the system of linear equations. Check your solution.

4. $x + y = 3$
$x - y = -3$

5. $2x + y = 5$
$4x + 2y = 0$

6. $2x - 4y = 10$
$-12x + 24y = -60$

7. **WHAT IF?** What happens to the solution in Example 2 if the perimeter of the rectangle is 54 units? Explain.

5.4 Exercises

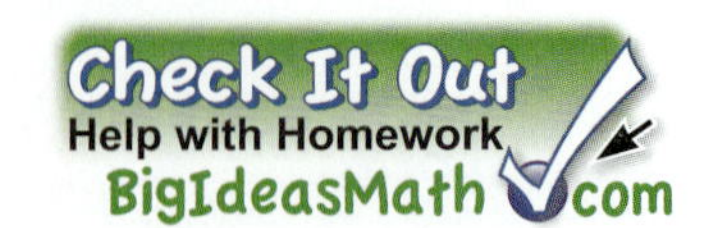

Vocabulary and Concept Check

1. **WRITING** Describe the difference between the graph of a system of linear equations that has *no solution* and the graph of a system of linear equations that has *infinitely many solutions*.

2. **REASONING** When solving a system of linear equations algebraically, how do you know when the system has *no solution*? *infinitely many solutions*?

Practice and Problem Solving

Let x and y be two numbers. Find the solution of the puzzle.

3. y is $\frac{1}{3}$ more than 4 times the value of x.

   The difference of $3y$ and $12x$ is 1.

4. $\frac{1}{2}$ of x plus 3 is equal to y.

   x is 6 more than twice the value of y.

Without graphing, determine whether the system of linear equations has *one solution, infinitely many solutions,* or *no solution.* Explain your reasoning.

5. $y = 5x - 9$
   $y = 5x + 9$

6. $y = 6x + 2$
   $y = 3x + 1$

7. $y = 8x - 2$
   $y - 8x = -2$

Solve the system of linear equations. Check your solution.

(1) 8. $y = 2x - 2$
   $y = 2x + 9$

9. $y = 3x + 1$
   $-x + 2y = -3$

10. $y = \frac{\pi}{3}x + \pi$
    $-\pi x + 3y = -6\pi$

(2) 11. $y = -\frac{1}{6}x + 5$
    $x + 6y = 30$

12. $\frac{1}{3}x + y = 1$
    $2x + 6y = 6$

13. $-2x + y = 1.3$
    $2(0.5x - y) = 4.6$

14. **ERROR ANALYSIS** Describe and correct the error in solving the system of linear equations.

$y = -2x + 4$
$y = -2x + 6$

The lines have the same slope, so, there are infinitely many solutions.

15. **PIG RACE** In a pig race, your pig gets a head start of 3 feet and is running at a rate of 2 feet per second. Your friend's pig is also running at a rate of 2 feet per second. A system of linear equations that represents this situation is $y = 2x + 3$ and $y = 2x$. Will your friend's pig catch up to your pig? Explain.

Assignment Guide and Homework Check

Level	Day 1 Activity Assignment	Day 2 Lesson Assignment	Homework Check
Basic	3, 4, 23–26	1, 2, 5–13 odd, 14, 15–21 odd	9, 11, 13, 15, 19
Average	3, 4, 23–26	1, 2, 6–14 even, 15–21 odd	10, 12, 17, 19, 21
Advanced	1–4, 6–16 even, 17–26		10, 12, 16, 19, 22

Common Errors

- **Exercises 5–13** Students may see that the slope is the same for both equations and immediately say that the system of linear equations has no solution. Remind them that they need to compare the slope *and* *y*-intercepts when determining the number of solutions.
- **Exercises 9–13** Students may make calculation errors when solving the equations for *y*. Encourage them to be careful when solving for *y*.

5.4 Record and Practice Journal

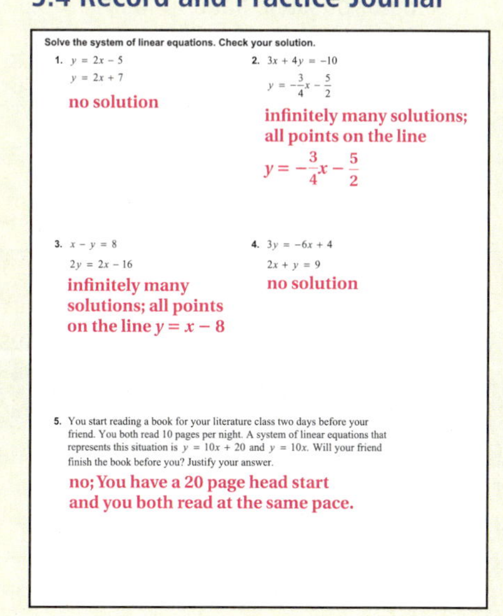
Solve the system of linear equations. Check your solution.

1. $y = 2x - 5$
   $y = 2x + 7$
   no solution

2. $3x + 4y = -10$
   $y = -\frac{3}{4}x - \frac{5}{2}$
   infinitely many solutions; all points on the line $y = -\frac{3}{4}x - \frac{5}{2}$

3. $x - y = 8$
   $2y = 2x - 16$
   infinitely many solutions; all points on the line $y = x - 8$

4. $3y = -6x + 4$
   $2x + y = 9$
   no solution

5. You start reading a book for your literature class two days before your friend. You both read 10 pages per night. A system of linear equations that represents this situation is $y = 10x + 20$ and $y = 10x$. Will your friend finish the book before you? Justify your answer.
   no; You have a 20 page head start and you both read at the same pace.

Vocabulary and Concept Check

1. The graph of a system with no solution is two parallel lines, and the graph of a system with infinitely many solutions is one line.
2. When solving a system of linear equations algebraically, you know the system has no solution when you reach an invalid statement such as $-7 = 2$; infinitely many solutions: a valid statement such as $2 = 2$.

Practice and Problem Solving

3. infinitely many solutions; all points on the line $y = 4x + \frac{1}{3}$
4. no solution
5. no solution; The lines have the same slope and different *y*-intercepts.
6. one solution; The lines have different slopes.
7. infinitely many solutions; The lines are identical.
8. no solution
9. $(-1, -2)$
10. no solution
11. infinitely many solutions; all points on the line $y = -\frac{1}{6}x + 5$
12. infinitely many solutions; all points on the line $y = -\frac{1}{3}x + 1$
13. $(-2.4, -3.5)$
14. There are different *y*-intercepts, so the lines are parallel; no solution
15. no; because they are running at the same speed and your pig had a head start

Practice and Problem Solving

16–17. See Additional Answers.

18. a. 6 h

b. You both work the same number of hours.

19. $y = 0.99x + 10$

$y = 0.99x$

no; Because you paid \$10 before buying the same number of songs at the same price, you spend \$10 more.

20. $a = b$: always; The lines are parallel; $a \geq b$: sometimes; The lines are parallel for $a = b$; $a < b$: never; The lines have different slopes and different y-intercepts.

21. See *Taking Math Deeper*.

22. $a = 2, b = 2$; yes; Both equations are the same.

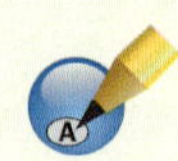

Fair Game Review

23. $y = 3x$

24. $y = 2x - 3$

25. $y = -\frac{1}{2}x + 2$

26. B

Mini-Assessment

Solve the system of linear equations. Check your solution, if possible.

1. $2x + 3y = 5$

$2x + 3y = 7$

no solution

2. $x + 2y = 12$

$y = -\frac{1}{2}x + 6$

infinitely many solutions; all points on the line $y = -\frac{1}{2}x + 6$

3. $-3x + 2y = 2$

$4x + 3y = 20$

one solution; (2, 4)

Taking Math Deeper

Exercise 21

You can approach this problem using the Guess, Test, and Revise method, by finding a set of possible costs for Group 1 and testing them for Group 2.

Guess.

Let L = the price of a lift ticket and S = the price of a ski rental. Make a guess using $L = \$10$. Find the corresponding value of S for Group 1.

$36L + 18S = 684$ — Equation for Group 1

$36(10) + 18S = 684$ — Substitute 10 for L.

$18S = 324$ — Subtract 360 from each side.

$S = 18$ — Divide each side by 18.

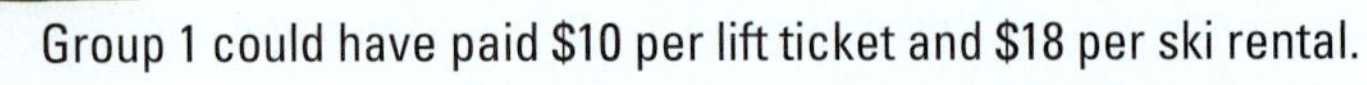

Group 1 could have paid \$10 per lift ticket and \$18 per ski rental.

Test.

Could Group 2 have paid \$10 per lift ticket and \$18 per ski rental?

$24L + 12S = 456$ — Equation for Group 2

$24(10) + 12(18) \stackrel{?}{=} 456$ — Substitute.

$456 = 456$ ✓ — Simplify.

So, Group 2 could have paid these prices. Are other prices possible?

Guess and test another set of prices.

Group 1: Guess $L = \$15$. Find S.

$36L + 18S = 684$

$36(15) + 18S = 684$

$18S = 144$

$S = 8$

Group 2: Test $L = \$15$ and $S = \$8$.

$24(15) + 12(8) \stackrel{?}{=} 456$

$456 = 456$

Another possible set of prices is \$15 per lift ticket and \$8 per ski rental.

Because the given information leads to more than one set of possible prices, it is not possible to determine how much each lift ticket costs.

Reteaching and Enrichment Strategies

If students need help. . .	If students got it. . .
Resources by Chapter • Practice A and Practice B • Puzzle Time Record and Practice Journal Practice Differentiating the Lesson Lesson Tutorials Skills Review Handbook	Resources by Chapter • Enrichment and Extension • Technology Connection Start the next section

16. **REASONING** One equation in a system of linear equations has a slope of -3. The other equation has a slope of 4. How many solutions does the system have? Explain.

17. **LOGIC** How can you use the slopes and the y-intercepts of equations in a system of linear equations to determine whether the system has *one solution, infinitely many solutions,* or *no solution*? Explain your reasoning.

$$4x + 8y = 64$$
$$8x + 16y = 128$$

18. **MONEY** You and a friend both work two different jobs. The system of linear equations represents the total earnings for x hours worked at the first job and y hours worked at the second job. Your friend earns twice as much as you.

    a. One week, both of you work 4 hours at the first job. How many hours do you and your friend work at the second job?

    b. Both of you work the same number of hours at the second job. Compare the number of hours each of you works at the first job.

19. **DOWNLOADS** You download a digital album for \$10. Then you and your friend download the same number of individual songs for \$0.99 each. Write a system of linear equations that represents this situation. Will you and your friend spend the same amount of money? Explain.

20. **REASONING** Does the system shown *always, sometimes,* or *never* have no solution when $a = b$? $a \geq b$? $a < b$? Explain your reasoning.

$$y = ax + 1$$
$$y = bx + 4$$

21. **SKIING** The table shows the number of lift tickets and ski rentals sold to two different groups. Is it possible to determine how much each lift ticket costs? Justify your answer.

Group	1	2
Number of Lift Tickets	36	24
Number of Ski Rentals	18	12
Total Cost (dollars)	684	456

22.  Find the values of a and b so the system shown has the solution (2, 3). Does the system have any other solutions? Explain.

$$12x - 2by = 12$$
$$3ax - by = 6$$

Fair Game Review What you learned in previous grades & lessons

Write an equation of the line that passes through the given points. *(Section 4.7)*

23. (0, 0), (2, 6)

24. (0, -3), (3, 3)

25. (-6, 5), (0, 2)

26. **MULTIPLE CHOICE** What is the solution of $-2(y + 5) \leq 16$? *(Skills Review Handbook)*

Ⓐ $y \leq -13$ Ⓑ $y \geq -13$ Ⓒ $y \leq -3$ Ⓓ $y \geq -3$

Extension 5.4 Solving Linear Equations by Graphing

Systems of Equations

In this extension, you will

- solve linear equations by graphing a system of linear equations.
- solve real-life problems.

Applying Standards
MACC.8.EE.3.7
MACC.8.EE.3.8

Solving Equations Using Graphs

Step 1: To solve the equation $ax + b = cx + d$, write two linear equations.

$$ax + b = cx + d$$

$y = ax + b$ and $y = cx + d$

Step 2: Graph the system of linear equations. The x-value of the solution of the system of linear equations is the solution of the equation $ax + b = cx + d$.

EXAMPLE 1 Solving an Equation Using a Graph

Solve $x - 2 = -\frac{1}{2}x + 1$ using a graph. Check your solution.

Step 1: Write a system of linear equations using each side of the equation.

$$x - 2 = -\frac{1}{2}x + 1$$

$y = x - 2$ $\qquad$ $y = -\frac{1}{2}x + 1$

Step 2: Graph the system.

$y = x - 2$

$y = -\frac{1}{2}x + 1$

The graphs intersect at (2, 0).

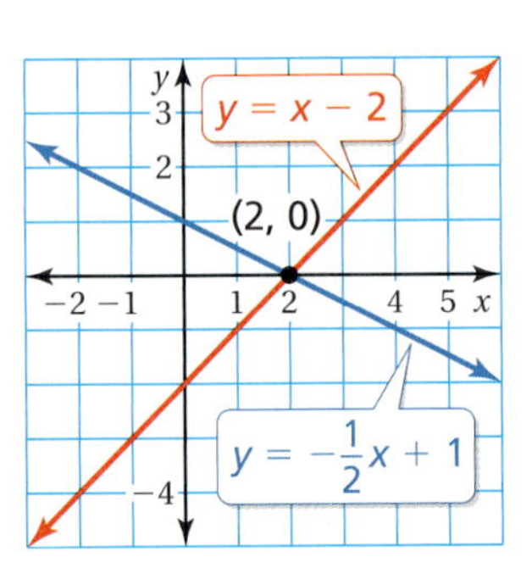

So, the solution of the equation is $x = 2$.

Check

$x - 2 = -\frac{1}{2}x + 1$

$2 - 2 \stackrel{?}{=} -\frac{1}{2}(2) + 1$

$0 = 0$ ✓

Practice

Use a graph to solve the equation. Check your solution.

1. $2x + 3 = 4$
2. $2x = x - 3$
3. $3x + 1 = 3x + 2$
4. $\frac{1}{3}x = x + 8$
5. $1.5x + 2 = 11 - 3x$
6. $3 - 2x = -2x + 3$
7. **STRUCTURE** Write an equation with variables on both sides that has no solution. How can you change the equation so that it has infinitely many solutions?

Laurie's Notes

Introduction

Connect

- **Yesterday:** Students solved special systems of linear equations both algebraically and by graphing. (MP1, MP7, MP8)
- **Today:** Students will solve equations with variables on both sides by graphing systems of equations.

Motivate

- ? "Have any of you seen the play *Little Shop of Horrors*? There is a flytrap-like alien plant that lives on human blood and eventually grows large enough to swallow people whole."
- In reality, the Venus flytrap does not grow rapidly. However, the Thuja Giant, a fast growing evergreen tree, grows 3–5 feet per year.
- Explain that one of the examples in today's lesson will explore the growth of plants.

Lesson Notes

Words of Wisdom

- Students often do not see the point of learning a new technique when the old technique worked well. The graphical approach helps to show the connection between the algebraic and geometric approaches. Some equations are easier to graph than they are to manipulate.

Example 1

- **MP5 Use Appropriate Tools Strategically** and **MP7 Look for and Make Use of Structure:** Students have solved systems by graphing previously. Today they will solve a linear equation with variables on both sides. The big idea is to think of each side of the equation as a linear equation by writing $y =$ left side of equation, and $y =$ right side of equation.
- Set y equal to each side of the equation to create the system of equations.
- The graph of each linear equation shows the value of the expression for different values of x. So, the intersection shows the value of x for which the expressions are equal.
- The check is algebraic. The correct value of x makes the original equation true.

Practice

- ? For Exercise 5 ask, "How do you graph a slope of 1.5?" Think of $\frac{rise}{run} = \frac{3}{2}$.
- Take time to have students share their thinking and work for Exercise 7. Students who found this problem difficult will benefit from hearing a variety of approaches.

Florida Common Core Standards

MACC.8.EE.3.7 Solve linear equations in one variable.

MACC.8.EE.3.8 Analyze and solve pairs of simultaneous linear equations.

Goal Today's lesson is solving equations with variables on both sides by graphing systems of equations.

Technology for the Teacher

Lesson Tutorials
Lesson Plans
Answer Presentation Tool

Extra Example 1

Solve $x + 3 = \frac{1}{2}x + 1$ using a graph. Check your solution. $x = -4$

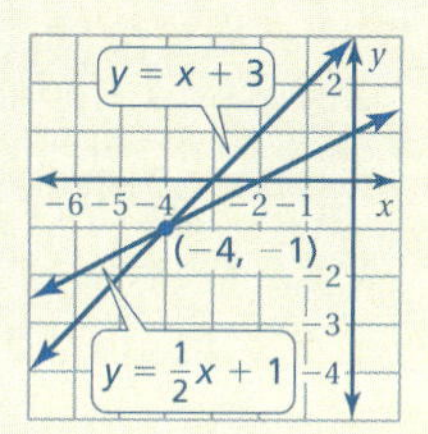

Practice

1. $x = \frac{1}{2}$
2. $x = -3$
3. no solution
4. $x = -12$
5. $x = 2$
6. infinitely many solutions
7. *Sample answer:* $6x - 3 = 6x$; Subtract 3 from the right side.

Record and Practice Journal Extension 5.4 Practice

1. $x = 1$
2. $x = 3$
3. $x = -10$
4. $x = 2$

5–7. See Additional Answers.

Laurie's Notes

Example 2

- Ask a volunteer to read the problem. Students will also need to read information from the diagram.
- The verbal model helps to focus attention on the goal, which is to determine *when* the two *plants will be the same height*.
- The growth rate is stated in terms of inches per month and is multiplied by months. So, you are adding inches to inches on each side of the equation.

$$\frac{\text{inches}}{\cancel{\text{months}}} \times \cancel{\text{months}} + \text{inches}$$

- Finish working through the problem as shown.

? "How tall are the plants after 5 months?" 15 inches

? "How can you check your solution?" Substitute $x = 5$ into the original equation; both sides equal 15.

- Discuss the Study Tip. This is another way to solve the equation.

Practice

- You may wish to have students use a graphing calculator to complete the problems.

Closure

- **Exit Ticket:** Solve the equation $3x - 4 = \frac{1}{2}x + 1$. $x = 2$

Extra Example 2

In Example 2, Plant A grows 0.4 inch per month. Plant B grows three times as fast.

a. Use the model to write an equation. $0.4x + 12 = 1.2x + 9$

b. After how many months x are the plants the same height? 3.75 mo

Practice

8. $x = 2.6$
9. $x = \frac{21}{2}$
10. $x = -20.5$
11. 6 mo

Mini-Assessment

Use a graph to solve the equation. Check your solution.

1. $-\frac{1}{3}x + 5 = 4x - 8$ $x = 3$

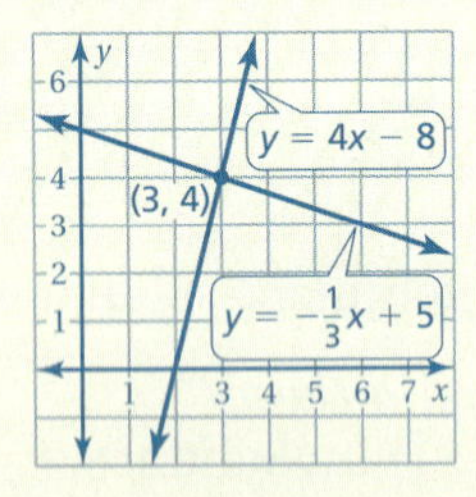

2. $-\frac{3}{2}x - 2 = -2x - 3$ $x = -2$

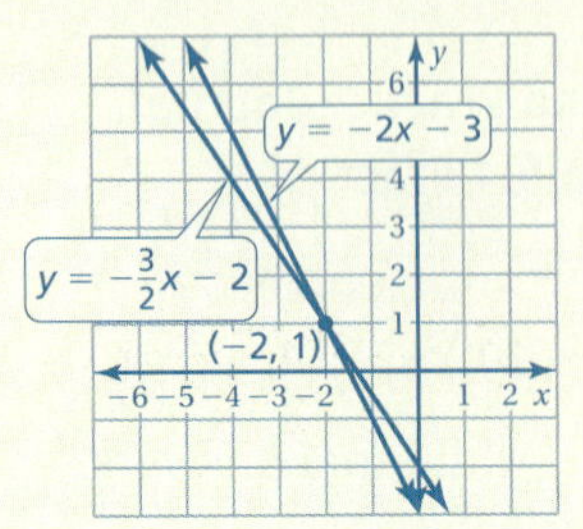

EXAMPLE 2 Real-Life Application

Plant A 12 in. Plant B 9 in.

Plant A grows 0.6 inch per month. Plant B grows twice as fast.

a. **Use the model to write an equation.**

b. **After how many months x are the plants the same height?**

Growth rate	$\cdot$	Months, x	+	Original height	=	Growth rate	$\cdot$	Months, x	+	Original height

a. The equation is $0.6x + 12 = 1.2x + 9$.

b. Write a system of linear equations using each side of the equation. Then use a graphing calculator to graph the system.

$$0.6x + 12 = 1.2x + 9$$

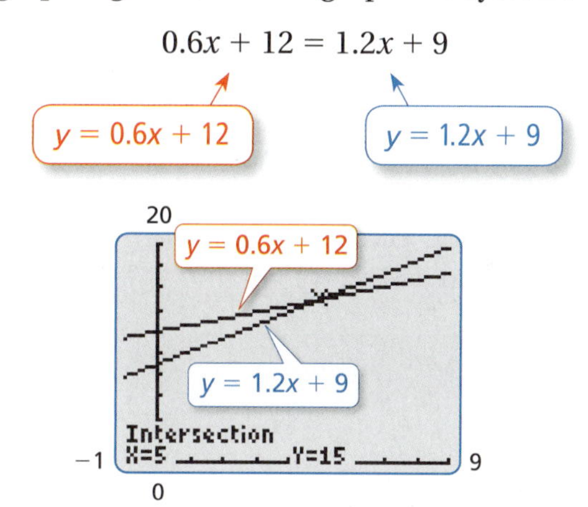

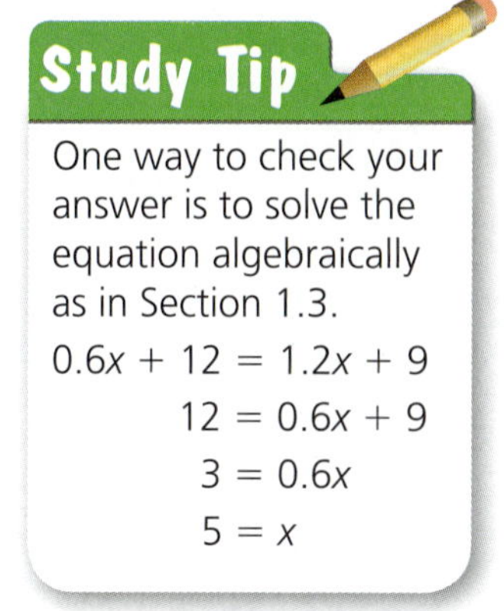

Study Tip

One way to check your answer is to solve the equation algebraically as in Section 1.3.

$$0.6x + 12 = 1.2x + 9$$
$$12 = 0.6x + 9$$
$$3 = 0.6x$$
$$5 = x$$

The solution of the system is (5, 15).

So, the plants are both 15 inches tall after 5 months.

Practice

Use a graph to solve the equation. Check your solution.

8. $6x - 2 = x + 11$

9. $\frac{4}{3}x - 1 = \frac{2}{3}x + 6$

10. $1.75x = 2.25x + 10.25$

11. **WHAT IF?** In Example 2, the growth rate of Plant A is 0.5 inch per month. After how many months x are the plants the same height?

5.3–5.4 Quiz

Solve the system of linear equations by elimination. Check your solution. *(Section 5.3)*

1. $x + 2y = 4$
$-x - y = 2$

2. $2x - y = 1$
$x + 3y - 4 = 0$

3. $3x = -4y + 10$
$4x + 3y = 11$

4. $2x + 5y = 60$
$2x - 5y = -20$

Solve the system of linear equations. Check your solution. *(Section 5.4)*

5. $3x - 2y = 16$
$6x - 4y = 32$

6. $4y = x - 8$
$-\frac{1}{4}x + y = -1$

7. $-2x + y = -2$
$3x + y = 3$

8. $3x = \frac{1}{3}y + 2$
$9x - y = -6$

Use a graph to solve the equation. Check your solution. *(Section 5.4)*

9. $4x - 1 = 2x$

10. $-\frac{1}{2}x + 1 = -x + 1$

11. $1 - 3x = -3x + 2$

12. $1 - 5x = 3 - 7x$

13. TOURS The table shows the activities of two visitors at a park. You want to take the boat tour for 2 hours and the walking tour for 3 hours. Can you determine how much you will pay? Explain. *(Section 5.4)*

	Boat Tour	Walking Tour	Total Cost
Visitor 1	1 hour	2 hours	\$19
Visitor 2	1.5 hours	3 hours	\$28.50

14. RENTALS A business rents bicycles and in-line skates. Bicycle rentals cost \$25 per day, and in-line skate rentals cost \$20 per day. The business has 20 rentals today and makes \$455. *(Section 5.3)*

a. Write a system of linear equations that represents this situation.

b. How many bicycle rentals and in-line skate rentals did the business have today?

Alternative Assessment Options

Math Chat
Structured Interview
Student Reflective Focus Question
Writing Prompt

Structured Interview

Interviews can occur formally or informally. Ask a student to perform a task and explain it, describing his or her thought process throughout the task. Probe the student for more information. Do not ask leading questions. Keep a rubric or notes.

Teacher Prompts	Student Answers	Teacher Notes
Tell me a story about earning money. Include: Job 1: \$9/h, Job 2: \$12/h Hours: at most 25 Earn: \$220/wk or more	Last week, I worked 12 hours at \$9 per hour and 10 hours at \$12 per hour for a total of \$228	Student understands the constraints of the system concerning hours and total earnings.

Study Help Sample Answers

Remind students to complete Graphic Organizers for the rest of the chapter.

2–4. Available at *BigIdeasMath.com*.

Reteaching and Enrichment Strategies

If students need help. . .	If students got it. . .
Resources by Chapter • Practice A and Practice B • Puzzle Time Lesson Tutorials *BigIdeasMath.com*	Resources by Chapter • Enrichment and Extension • Technology Connection Game Closet at *BigIdeasMath.com* Start the Chapter Review

Answers

1. $(-8, 6)$
2. $(1, 1)$
3. $(2, 1)$
4. $(10, 8)$
5. infinitely many solutions
6. no solution
7. $(1, 0)$
8. no solution
9. $x = 0.5$
10. $x = 0$
11. no solution
12. $x = 1$
13. The system has infinitely many solutions, so it is not possible to determine how much each tour costs.
14. **a.** $x + y = 20;\ 25x + 20y = 455$

    b. 11 bicycles, 9 in-line skates

Technology for the Teacher

Online Assessment
Assessment Book
ExamView® Assessment Suite

For the Teacher
Additional Review Options

- *BigIdeasMath.com*
- Online Assessment
- Game Closet at *BigIdeasMath.com*
- Vocabulary Help
- Resources by Chapter

Answers

1. $(5, 7)$

2. $(6, -2)$

3. $(-4, -2)$

Review of Common Errors

Exercises 1–3

- Students may not show enough of the graph, so the lines will not intersect. Encourage them to extend their lines until they intersect.

Exercises 4–6

- Students may find one coordinate of the solution and stop there. Remind them that the answer is a coordinate pair representing both an x-value and a y-value.

Exercise 7

- Students may make careless errors. Remind students to line up like terms neatly in columns to avoid confusion about what terms to add or subtract. Also, take care with subtraction—it may help to change the sign of each term in the equation to be subtracted, so you can just add the terms.

Exercises 8–10

- Students may see that the slope is the same for both equations and immediately say that the system of linear equations has no solution. Remind them that they need to compare the slope *and* y-intercepts when determining the number of solutions. Encourage them to check algebraically and graphically that the system of linear equations has the number of solutions they found.

Exercise 11

- Students may forget the process. Remind them to write a system of two equations by setting y equal to each side of the equation.

5 Chapter Review

Check It Out
Vocabulary Help
BigIdeasMath.com

Review Key Vocabulary

system of linear equations, *p. 204*

solution of a system of linear equations, *p. 204*

Review Examples and Exercises

5.1 Solving Systems of Linear Equations by Graphing *(pp. 202–207)*

Solve the system by graphing. $y = -2x$ Equation 1

$y = 3x + 5$ Equation 2

Step 1: Graph each equation.

Step 2: Estimate the point of intersection. The graphs appear to intersect at $(-1, 2)$.

Step 3: Check the point from Step 2.

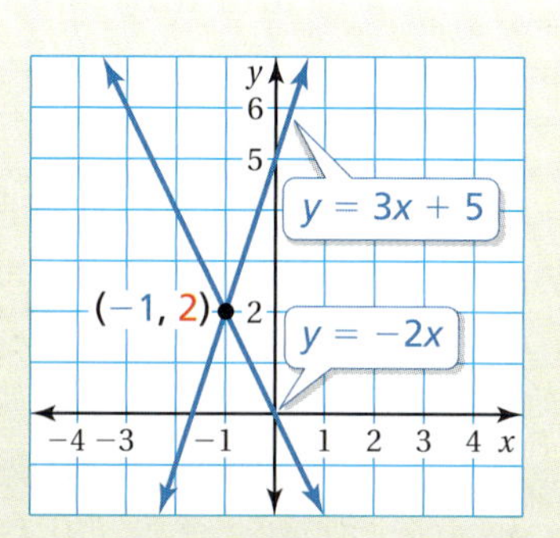

Equation 1	Equation 2
$y = -2x$	$y = 3x + 5$
$2 \stackrel{?}{=} -2(-1)$	$2 \stackrel{?}{=} 3(-1) + 5$
$2 = 2$ ✓	$2 = 2$ ✓

∴ The solution is $(-1, 2)$.

Exercises

Solve the system of linear equations by graphing.

1. $y = 2x - 3$
$y = x + 2$

2. $y = -x + 4$
$x + 3y = 0$

3. $x - y = -2$
$2x - 3y = -2$

5.2 Solving Systems of Linear Equations by Substitution *(pp. 208–213)*

Solve the system by substitution. $x = 1 + y$ Equation 1

$x + 3y = 13$ Equation 2

Step 1: Equation 1 is already solved for x.

Step 2: Substitute $1 + y$ for x in Equation 2.

$1 + y + 3y = 13$ Substitute $1 + y$ for x.

$y = 3$ Solve for y.

Step 3: Substituting 3 for y in Equation 1 gives $x = 4$.

∴ The solution is $(4, 3)$.

Exercises

Solve the system of linear equations by substitution. Check your solution.

4. $y = -3x - 7$
$y = x + 9$

5. $\frac{1}{2}x + y = -4$
$y = 2x + 16$

6. $-x + 5y = 28$
$x + 3y = 20$

5.3 Solving Systems of Linear Equations by Elimination *(pp. 216–223)*

You have a total of 5 quarters and dimes in your pocket. The value of the coins is \$0.80. Write and solve a system of linear equations to find the number x of dimes and the number y of quarters in your pocket.

Use a verbal model to write a system of linear equations.

Number of dimes, x + Number of quarters, y = Number of coins

Value of a dime • Number of dimes, x + Value of a quarter • Number of quarters, y = Total value

The system is $x + y = 5$ and $0.1x + 0.25y = 0.8$.

Step 1: Multiply Equation 2 by 10.

$x + y = 5$		$x + y = 5$	Equation 1
$0.1x + 0.25y = 0.8$	Multiply by 10.	$x + 2.5y = 8$	Revised Equation 2

Step 2: Subtract the equations.

$x + y = 5$	Equation 1
$x + 2.5y = 8$	Revised Equation 2
$-1.5y = -3$	Subtract the equations.

Step 3: Solving the equation $-1.5y = -3$ gives $y = 2$.

Step 4: Substitute 2 for y in one of the original equations and solve for x.

$x + y = 5$	Equation 1
$x + 2 = 5$	Substitute 2 for y.
$x = 3$	Subtract 2 from each side.

So, you have 3 dimes and 2 quarters in your pocket.

Exercises

7. GIFT BASKET A gift basket that contains jars of jam and packages of bread mix costs \$45. There are 8 items in the basket. Jars of jam cost \$6 each, and packages of bread mix cost \$5 each. Write and solve a system of linear equations to find the number of jars of jam and the number of packages of bread mix in the gift basket.

Review Game

Systems of Linear Equations

Materials per Group:

- 2 sheets of graph paper
- 2 pencils
- 2 books

Directions:

- Pair students. The game is for 2 players.
- Give each pair of students the same system of linear equations to solve. As students are working, they do not show each other their solutions. They each stand a book up to hide their work.
- The object of the game is to be the first to determine the correct solution of the system.

Round 1:

- Students are to solve a system by graphing.
- The first student to announce that he or she has determined the correct solution earns one point. If, after checking his or her work, it is determined that the student did not actually get the correct solution, then he or she loses the point. The second student then has the opportunity to "steal" the point by coming up with the correct solution. If neither student gets the correct solution, then neither student gets a point.
- Repeat for two more systems (or as many as time allows).

Round 2:

- Same as Round 1, except students are to solve a system by substitution.

Round 3:

- Same as Rounds 1 and 2, except students are to solve a system by elimination.

Who wins?

The student with the greatest total points after the three rounds wins. In the case of a tie, give the students one more system to solve, and specify the method they are to use to find the solution. Repeat as necessary until there is a clear winner.

For the Student Additional Practice

- Lesson Tutorials
- Multi-Language Glossary
- Self-Grading Progress Check
- *BigIdeasMath.com*
  Dynamic Student Edition
  Student Resources

Answers

4. $(-4, 5)$

5. $(-8, 0)$

6. $(2, 6)$

7. $x + y = 8$
$6x + 5y = 45$; 5 jars of jam, 3 packages of bread mix

8. $(-5, 0)$

9. infinitely many solutions; all points on the line $y = \frac{3}{2}x - \frac{1}{2}$

10. no solution

11. $x = -4$

My Thoughts on the Chapter

Teacher Tip

Not allowed to write in your teaching edition? Use sticky notes to record your thoughts.

What worked. . .

What did not work. . .

What I would do differently. . .

5.4 Solving Special Systems of Linear Equations *(pp. 224–231)*

a. Solve the system. $y = -5x - 8$ Equation 1

$y = -5x + 4$ Equation 2

Solve by substitution. Substitute $-5x + 4$ for y in Equation 1.

$y = -5x - 8$ Equation 1

$-5x + 4 = -5x - 8$ Substitute $-5x + 4$ for y.

$4 = -8$ ✗ Add $5x$ to each side.

∴ The equation $4 = -8$ is never true. So, the system of linear equations has no solution.

b. Solve $x + 1 = \frac{1}{3}x + 3$ using a graph. Check your solution.

Step 1: Write a system of linear equations using each side of the equation.

$$x + 1 = \frac{1}{3}x + 3$$

$y = x + 1$ $y = \frac{1}{3}x + 3$

Step 2: Graph the system.

$y = x + 1$

$y = \frac{1}{3}x + 3$

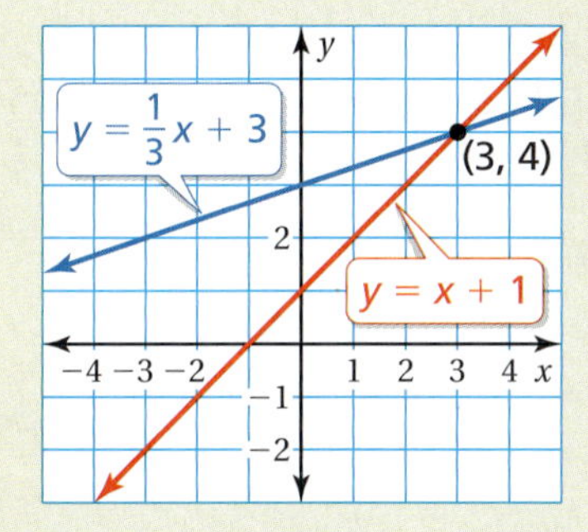

Check

$x + 1 = \frac{1}{3}x + 3$

$3 + 1 \stackrel{?}{=} \frac{1}{3}(3) + 3$

$4 = 4$ ✓

The graphs intersect at (3, 4).

∴ So, the solution is $x = 3$.

Exercises

Solve the system of linear equations. Check your solution.

8. $x + 2y = -5$
$x - 2y = -5$

9. $3x - 2y = 1$
$9x - 6y = 3$

10. $8x - 2y = 16$
$-4x + y = 8$

11. Use a graph to solve $2x - 9 = 7x + 11$. Check your solution.

5 Chapter Test

Solve the system of linear equations by graphing.

1. $y = 4 - x$
   $y = x - 4$

2. $y = \frac{1}{2}x + 10$
   $y = 4x - 4$

3. $y + x = 0$
   $3y + 6x = -9$

Solve the system of linear equations by substitution. Check your solution.

4. $-3x + y = 2$
   $-x + y - 4 = 0$

5. $x + y = 20$
   $y = 2x - 1$

6. $x - y = 3$
   $x + 2y = -6$

Solve the system of linear equations by elimination. Check your solution.

7. $2x + y = 3$
   $x - y = 3$

8. $x + y = 12$
   $3x = 2y + 6$

9. $-2x + y + 3 = 0$
   $3x + 4y = -1$

Without graphing, determine whether the system of linear equations has *one solution*, *infinitely many solutions*, or *no solution*. Explain your reasoning.

10. $y = 4x + 8$
    $y = 5x + 1$

11. $2y = 16x - 2$
    $y = 8x - 1$

12. $y = -3x + 2$
    $6x + 2y = 10$

Use a graph to solve the equation. Check your solution.

13. $\frac{1}{4}x - 4 = \frac{3}{4}x + 2$

14. $8x - 14 = -2x - 4$

15. **FRUIT** The price of 2 pears and 6 apples is \$14. The price of 3 pears and 9 apples is \$21. Can you determine the unit prices for pears and apples? Explain.

16. **BOUQUET** A bouquet of lilies and tulips has 12 flowers. Lilies cost \$3 each, and tulips cost \$2 each. The bouquet costs \$32. Write and solve a system of linear equations to find the number of lilies and tulips in the bouquet.

17. **DINNER** How much does it cost for 2 specials and 2 glasses of milk?

Test Item References

Chapter Test Questions	Section to Review	Florida Common Core Standards (MACC)
1–3	5.1	8.EE.3.8a, 8.EE.3.8b, 8.EE.3.8c
4–6, 16	5.2	8.EE.3.8b, 8.EE.3.8c
7–9, 17	5.3	8.EE.3.8b, 8.EE.3.8c
10–14, 15	5.4	8.EE.3.7, 8.EE.3.8a, 8.EE.3.8b, 8.EE.3.8c

Test-Taking Strategies

Remind students to quickly look over the entire test before they start so that they can budget their time. This test involves solving systems of equations, and the answers take on several different forms. So, it is important that students use the **Stop** and **Think** strategy before they answer questions.

Common Errors

- **Exercises 1–3** Students may not show enough of the graph, so the lines will not intersect. Encourage them to extend their lines until they intersect. All three exercises have a solution.
- **Exercises 4–6** Students may find one coordinate of the solution and stop there. Remind them that the answer is a coordinate pair representing both an x-value and a y-value.
- **Exercises 7–9** Students may make careless errors. Remind students to line up like terms neatly in columns to avoid confusion about what terms to add or subtract. Also, take care with subtraction—it may help to change the sign of each term in the equation to be subtracted, so you can just add the terms.
- **Exercises 10–12** Students may compare only the slopes. Remind them that they must also compare the y-intercepts to determine the number of solutions.
- **Exercises 13 and 14** Students may forget the process. Remind them to write a system of two equations by setting y equal to each side of the equation.

Reteaching and Enrichment Strategies

If students need help. . .	If students got it. . .
Resources by Chapter • Practice A and Practice B • Puzzle Time Record and Practice Journal Practice Differentiating the Lesson Lesson Tutorials *BigIdeasMath.com* Skills Review Handbook	Resources by Chapter • Enrichment and Extension • Technology Connection Game Closet at *BigIdeasMath.com* Start Standards Assessment

Answers

1. (4, 0)
2. (4, 12)
3. $(-3, 3)$
4. (1, 5)
5. (7, 13)
6. $(0, -3)$
7. $(2, -1)$
8. (6, 6)
9. $(1, -1)$
10. one solution; The lines have different slopes.
11. infinitely many solutions; The equations represent the same line.
12. no solution; The lines have the same slope and different y-intercepts.
13. $x = -12$
14. $x = 1$
15. no; The system has infinitely many solutions, so it is not possible to determine the unit prices.
16. $x + y = 12$;
$3x + 2y = 32$;
(8, 4); 8 lilies, 4 tulips
17. $16.10

Technology for the Teacher

Online Assessment
Assessment Book
ExamView® Assessment Suite

Test-Taking Strategies

Available at *BigIdeasMath.com*

After Answering Easy Questions, Relax
Answer Easy Questions First
Estimate the Answer
Read All Choices before Answering
Read Question before Answering
Solve Directly or Eliminate Choices
Solve Problem before Looking at Choices
Use Intelligent Guessing
Work Backwards

About this Strategy

When taking a multiple choice test, be sure to read each question carefully and thoroughly. Look closely for words that change the meaning of the question like not, never, all, every, and always.

Answers

1. D
2. 4
3. G

Technology for the Teacher

Florida Common Core Standards Support
Performance Tasks
Online Assessment
Assessment Book
ExamView® Assessment Suite

Item Analysis

1. **A.** The student finds a solution point for the first equation.

   B. The student finds a solution point for the first equation.

   C. The student makes an error rewriting the second equation in slope-intercept form.

   D. Correct answer

2. **Gridded Response:** Correct answer: 4

   Common error: The student uses the reciprocal of the slope instead of the opposite of the reciprocal of the slope.

3. **F.** The student translates the rectangle 4 units down.

   G. Correct answer

   H. The student reflects the rectangle in the y-axis.

   I. The student rotates the rectangle 180° about the origin.

5 Standards Assessment

1. What is the solution of the system of equations shown below? *(MACC.8.EE.3.8b)*

$$y = -\frac{2}{3}x - 1$$
$$4x + 6y = -6$$

A. $\left(-\frac{3}{2}, 0\right)$ **C.** no solution

B. $(0, -1)$ **D.** infinitely many solutions

2. What is the slope of a line that is perpendicular to the line $y = -0.25x + 3$? *(MACC.8.EE.2.6)*

3. On the grid below, Rectangle *EFGH* is plotted and its vertices are labeled.

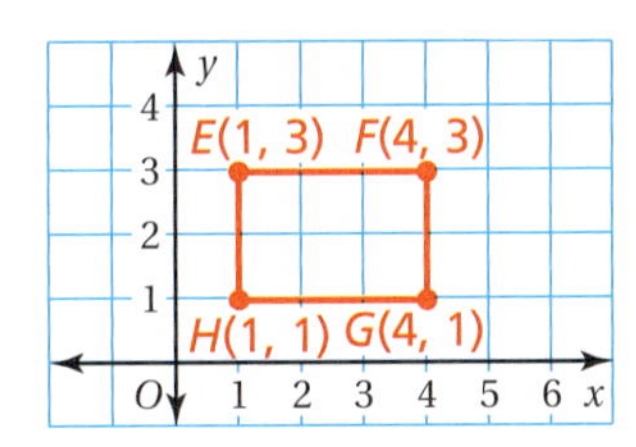

Which of the following shows Rectangle $E'F'G'H'$, the image of Rectangle *EFGH* after it is reflected in the *x*-axis? *(MACC.8.G.1.3)*

F.

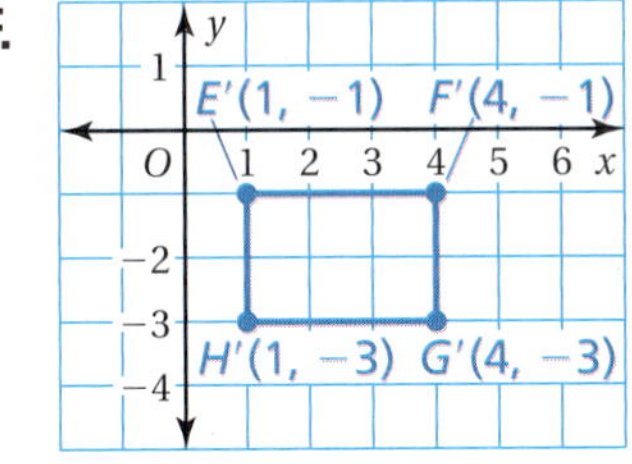

H.

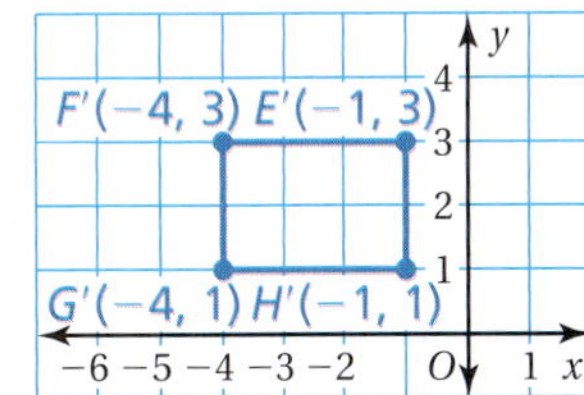

G.

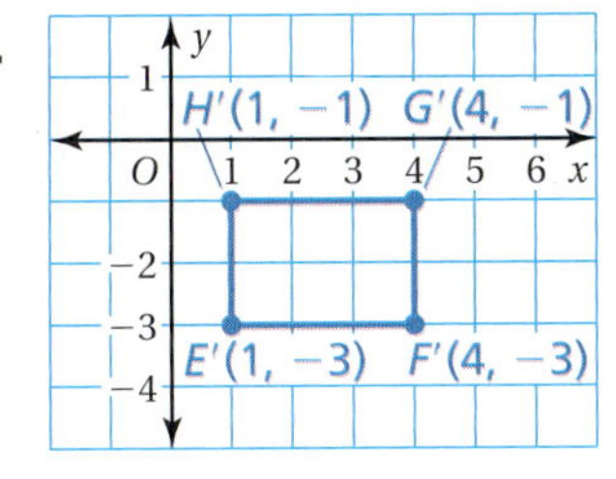

I.

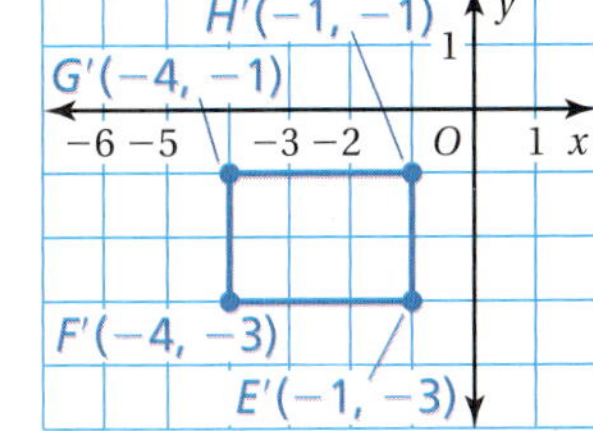

4. Which point is a solution of the system of equations shown below? *(MACC.8.EE.3.8b)*

$$x + 3y = 10$$
$$x = 2y - 5$$

A. $(1, 3)$

B. $(3, 1)$

C. $(55, -15)$

D. $(-35, -15)$

5. The graph of a system of two linear equations is shown. How many solutions does the system have? *(MACC.8.EE.3.8a)*

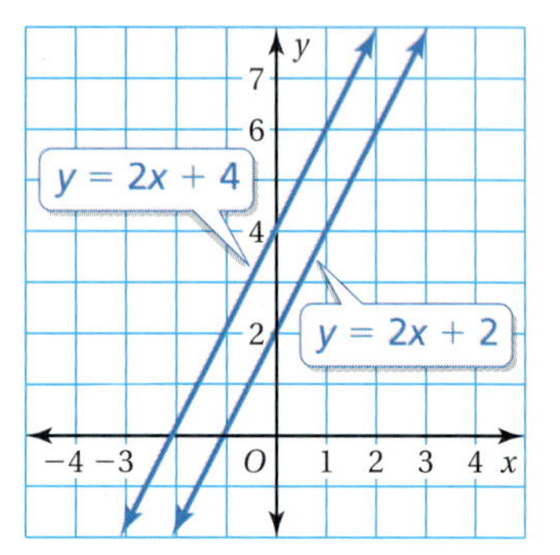

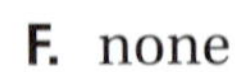

F. none

G. exactly one

H. exactly two

I. infinitely many

6. A scenic train ride has one price for adults and one price for children. One family of two adults and two children pays \$62 for the train ride. Another family of one adult and four children pays \$70. Which system of linear equations can you use to find the price x for an adult and the price y for a child? *(MACC.8.EE.3.8c)*

A. $2x + 2y = 70$
$x + 4y = 62$

B. $x + y = 62$
$x + y = 70$

C. $2x + 2y = 62$
$4x + y = 70$

D. $2x + 2y = 62$
$x + 4y = 70$

7. Which of the following is true about the graph of the linear equation $y = -7x + 5$? *(MACC.8.EE.2.6)*

F. The slope is 5, and the y-intercept is -7.

G. The slope is -5, and the y-intercept is -7.

H. The slope is -7, and the y-intercept is -5.

I. The slope is -7, and the y-intercept is 5.

8. What value of w makes the equation below true? *(MACC.8.EE.3.7b)*

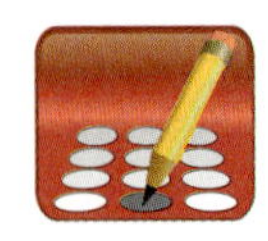

$$7w - 3w = 2(3w + 11)$$

Item Analysis (continued)

4. **A.** Correct answer
   B. The student checks solutions by substituting for x and y in reverse order.
   C. The student finds a solution for the first equation.
   D. The student finds a solution for the second equation.

5. **F.** Correct answer
   G. The student thinks parallel lines mean exactly one solution.
   H. The student thinks two parallel lines mean exactly two solutions.
   I. The student thinks parallel lines mean infinitely many solutions.

6. **A.** The student uses the wrong cost for each situation.
   B. The student defines x as the amount spent on adults and y as the amount spent on children.
   C. The student defines x as the amount per child and y as the amount per adult.
   D. Correct answer

7. **F.** The student switches the slope and the y-intercept in interpreting the slope-intercept form.
   G. The student confuses the characteristics of slope-intercept form.
   H. The student misinterprets the y-intercept as $-b$.
   I. Correct answer

8. **Gridded Response:** Correct answer: -11

   Common error: $-\frac{11}{2}$; The student fails to use the distributive property correctly by multiplying 11 by 2.

Answers

4. A
5. F
6. D
7. I
8. -11

Answers

9. C

10. $7.50

11. H

12. A

13. G

14. B

Item Analysis (continued)

9. **A.** The student forgets to reverse the sign of -1 when subtracting it from 4.

B. The student finds a perpendicular line instead of a parallel line.

C. Correct answer

D. The student determines the slope incorrectly as the change in x divided by the change in y.

10. **2 points** The student demonstrates a thorough understanding of how to solve a system of linear equations, explains the work fully, and applies the solution correctly to the context of the problem. The solution is (7.5, 10). The x-value represents the cost of each T-shirt, $7.50.

1 point The student's work and explanations demonstrate a lack of essential understanding. The system of linear equations was set up or solved incorrectly, or, if the solution is correct, was applied incorrectly to the problem.

0 points The student provides no response, a completely incorrect or incomprehensible response, or a response that demonstrates insufficient understanding of how to solve a system of linear equations.

11. **F.** The student fails to distribute 12 correctly.

G. The student incorrectly isolates y by dividing each side by $12y$ instead of subtracting $12y$.

H. Correct answer

I. The student first divides each side by 32, then fails to distribute $\frac{3}{8}$ correctly.

12. **A.** Correct answer

B. The student confuses a system with infinitely many solutions and a system with no solution.

C. The student thinks the lines must have different slopes to have infinitely many solutions.

D. The student misinterprets "infinitely many solutions."

13. **F.** The student adds $\frac{1}{3}$ instead of multiplying by $\frac{1}{3}$.

G. Correct answer

H. The student represents the sum of one-third of a number and ten as $\frac{1}{3}(n + 10)$.

I. The student adds 10 to the wrong side of the equation.

14. **A.** The student adds $7y$ to the right side instead of subtracting.

B. Correct answer

C. The student adds $7y$ to the right side instead of subtracting, and inverts the fraction.

D. The student fails to divide the right side by 4.

9. The graph of which equation is parallel to the line that passes through the points $(-1, 5)$ and $(4, 7)$? *(MACC.8.EE.2.6)*

A. $y = \frac{2}{3}x + 6$

B. $y = -\frac{5}{2}x + 4$

C. $y = \frac{2}{5}x + 1$

D. $y = \frac{5}{2}x - 1$

10. You buy 3 T-shirts and 2 pairs of shorts for \$42.50. Your friend buys 5 T-shirts and 3 pairs of shorts for \$67.50. Use a system of linear equations to find the cost of each T-shirt. Show your work and explain your reasoning. *(MACC.8.EE.3.8c)*

Think
Solve
Explain

11. The two figures have the same area. What is the value of y? *(MACC.8.EE.3.7b)*

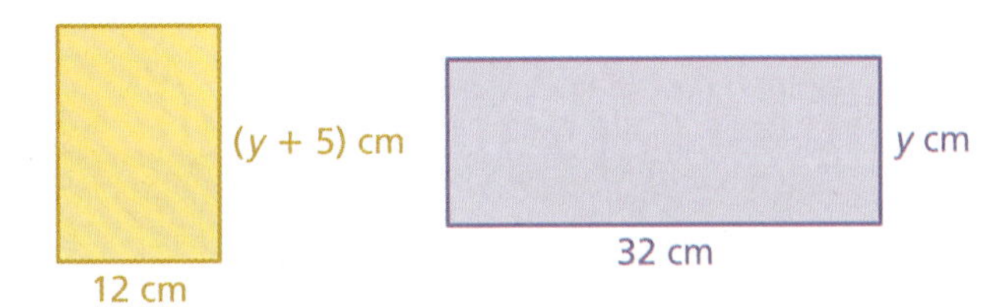

F. $\frac{1}{4}$

G. $\frac{15}{8}$

H. 3

I. 8

12. A system of two linear equations has infinitely many solutions. What can you conclude about the graphs of the two equations? *(MACC.8.EE.3.8a)*

A. The lines have the same slope and the same y-intercept.

B. The lines have the same slope and different y-intercepts.

C. The lines have different slopes and the same y-intercept.

D. The lines have different slopes and different y-intercepts.

13. The sum of one-third of a number and 10 is equal to 13. What is the number? *(MACC.8.EE.3.7b)*

F. $\frac{8}{3}$ **G.** 9 **H.** 29 **I.** 69

14. Solve the equation $4x + 7y = 16$ for x. *(MACC.8.EE.3.7b)*

A. $x = 4 + \frac{7}{4}y$

B. $x = 4 - \frac{7}{4}y$

C. $x = 4 + \frac{4}{7}y$

D. $x = 16 - 7y$

6 Functions

"Here's a math anagram."

"Here's another one."

"It is my treat-converter function machine. However many cat treats I input, the machine outputs TWICE that many dog biscuits. Isn't that cool?"

Florida Common Core Progression

6th Grade

- Write an equation in two variables, identifying the independent and dependent variables.
- Analyze the relationship between two variables using graphs and tables.

7th Grade

- Identify the constant of proportionality (unit rate) in tables, graphs, equations, diagrams, and verbal descriptions.
- Represent proportional relationships with equations.

8th Grade

- Understand the definition of a function.
- Compare and write functions represented in different ways (words, tables, graphs).
- Understand that $y = mx + b$ is a linear function and recognize nonlinear functions.
- Interpret the rate of change and initial value of a function.

Pacing Guide for Chapter 6

Chapter Opener	1 Day
Section 1 Activity Lesson	 1 Day 1 Day
Section 2 Activity Lesson	 1 Day 1 Day
Section 3 Activity Lesson	 1 Day 1 Day
Study Help / Quiz	1 Day
Section 4 Activity Lesson	 1 Day 1 Day
Section 5 Activity Lesson	 1 Day 1 Day
Chapter Review/ Chapter Tests	2 Days
Total Chapter 6	14 Days
Year-to-Date	91 Days

Chapter Summary

Section	Florida Common Core Standard (MACC)	
6.1	Learning	8.F.1.1
6.2	Learning	8.F.1.1 ★
6.3	Learning	8.F.1.2 ★, 8.F.1.3, 8.F.2.4 ★
6.4	Learning	8.F.1.3 ★
6.5	Learning	8.F.2.5 ★
★ Teaching is complete. Standard can be assessed.		

Technology for the Teacher

BigIdeasMath.com
Chapter at a Glance
Complete Materials List
Parent Letters: English, Spanish, and Haitian Creole

Florida Common Core Standards

MACC.5.OA.2.3 Generate two numerical patterns using two given rules. Identify apparent relationships between corresponding terms. Form ordered pairs consisting of corresponding terms from the two patterns, and graph the ordered pairs on a coordinate plane.

MACC.7.NS.1.3 Solve real-world and mathematical problems involving the four operations with rational numbers.

Additional Topics for Review

- Finding Area, Perimeter, and Circumference
- Slope of a Line
- Plotting Points in a Coordinate Plane
- Graphing and Writing Linear Equations in Slope-Intercept Form
- Operations with Decimals and Fractions

Try It Yourself

1. 80
2. 10
3. 20
4. −14
5. −6
6. −34

Record and Practice Journal Fair Game Review

1.	13	**2.**	36
3.	46	**4.**	19
5.	19.5	**6.**	31.2
7.	4	**8.**	−12
9.	32	**10.**	21
11.	3	**12.**	−3

Math Background Notes

Vocabulary Review

- Evaluate
- Algebraic Expression
- Substitute
- Order of Operations
- Opposite

Identifying Patterns

- Students should be able to identify relationships between corresponding elements of paired data.
- A pattern for each problem can be found in terms of the relationship between the corresponding *x* and *y* values.
- **Teaching Tip:** Some students may have difficulty identifying the pattern. Encourage these students to search for context clues. For example, are the *y*-values greater than the corresponding *x*-values? If so, then *y* is probably related to *x* by addition or multiplication.
- **Alternate Pattern:** Discuss another way to describe a pattern in Example 1: "As *x* increases by 10, *y* increases by 10."

Evaluating Algebraic Expressions

- Students should know how to substitute values into algebraic expressions and evaluate the results using order of operations.
- **Teaching Tip:** Sometimes color coding substitutions can help students to evaluate expressions. Each time you want to substitute a number in place of a variable, you must substitute a colored pencil for your lead pencil.
- Remind students that after they substitute for *x* or *y*, they must use the correct order of operations to continue simplifying the expression.
- **Common Error:** Encourage students to use a set of parentheses whenever they do a substitution. This will help students distinguish between subtracting 7 and multiplying by −7.

Reteaching and Enrichment Strategies

If students need help. . .	If students got it. . .
Record and Practice Journal • Fair Game Review Skills Review Handbook Lesson Tutorials	Game Closet at *BigIdeasMath.com* Start the next section

What You Learned Before

"Do you think the stripes in this shirt make me look too linear?"

Identifying Patterns (MACC.5.OA.2.3)

Example 1 Find the missing value in the table.

x	y
30	0
40	10
50	20
60	

Each y-value is 30 less than the x-value.

So, the missing value is $60 - 30 = 30$.

Try It Yourself

Find the missing value in the table.

1.

x	y
5	10
7	14
10	20
40	

2.

x	y
0.5	1
1.5	2
3	3.5
9.5	

3.

x	y
15	5
30	10
45	15
60	

Evaluating Algebraic Expressions (MACC.7.NS.1.3)

Example 2 Evaluate $2x - 12$ when $x = 5$.

$$2x - 12 = 2(5) - 12$$ Substitute 5 for x.

$$= 10 - 12$$ Using order of operations, multiply 2 and 5.

$$= 10 + (-12)$$ Add the opposite of 12.

$$= -2$$ Add.

Try It Yourself

Evaluate the expression when $y = 4$.

4. $-4y + 2$

5. $\frac{y}{2} - 8$

6. $-10 - 6y$

6.1 Relations and Functions

Essential Question How can you use a mapping diagram to show the relationship between two data sets?

1 ACTIVITY: Constructing Mapping Diagrams

Work with a partner. Copy and complete the mapping diagram.

a. Area A

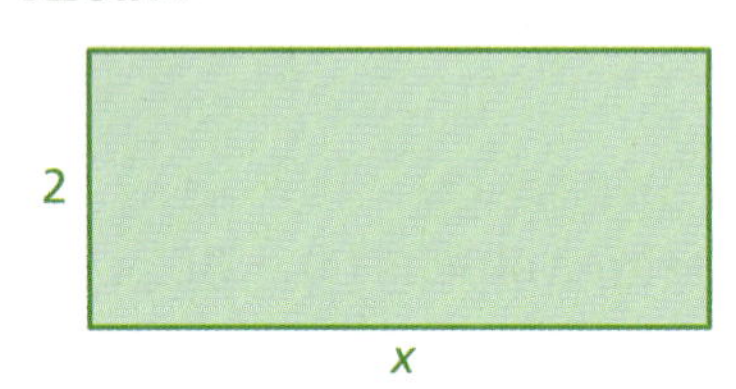

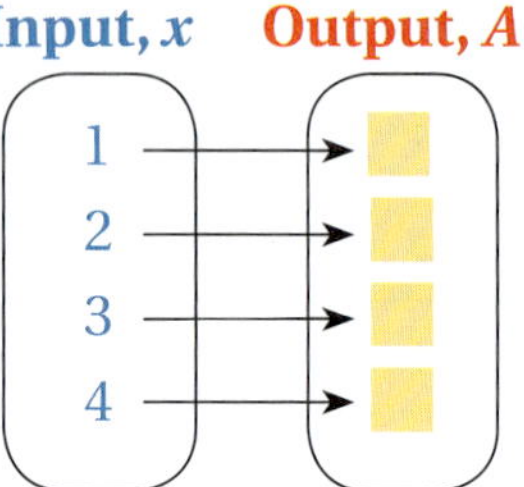

b. Perimeter P

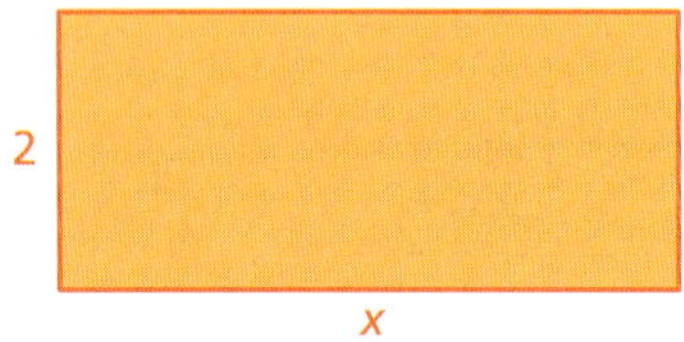

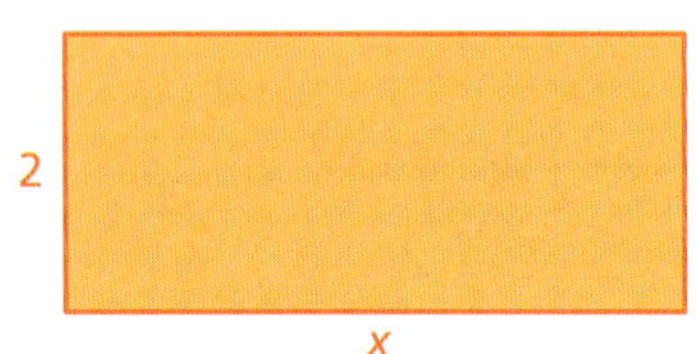

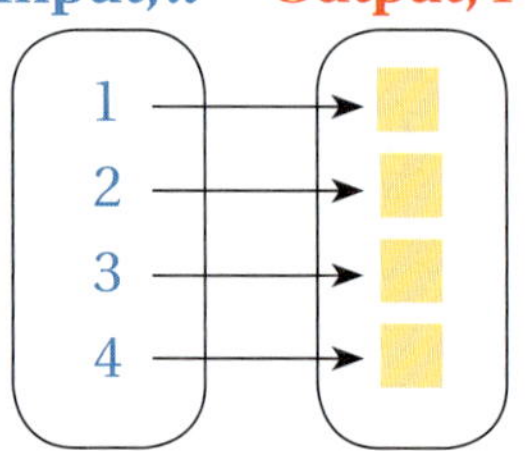

c. Circumference C

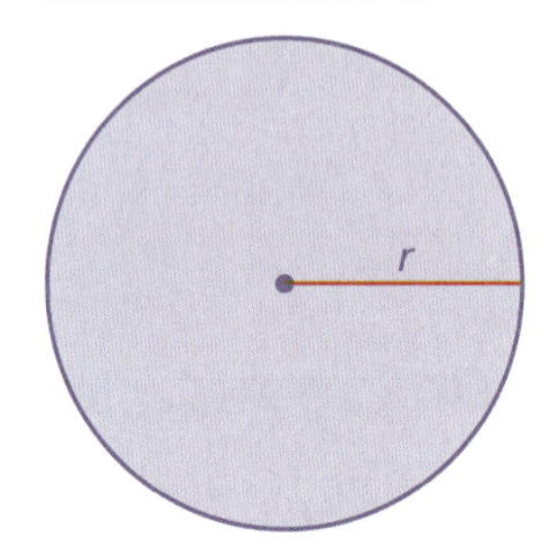

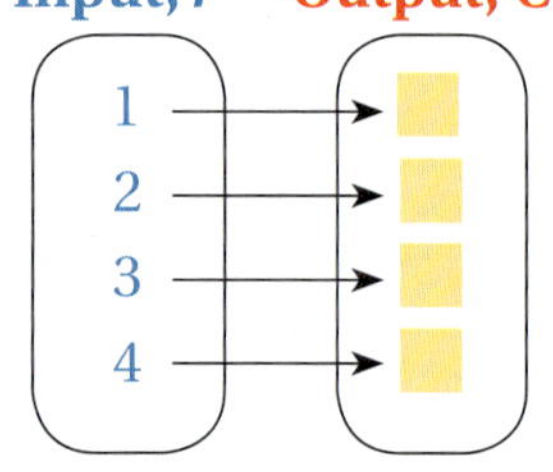

Functions

In this lesson, you will

- define relations and functions.
- determine whether relations are functions.
- describe patterns in mapping diagrams.

Learning Standard
MACC.8.F.1.1

d. Volume V

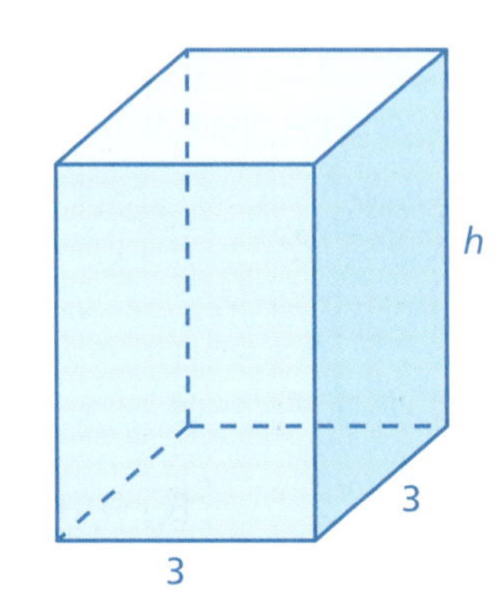

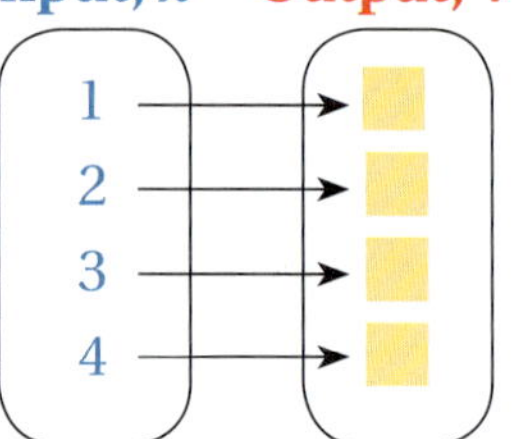

Laurie's Notes

Introduction

Standards for Mathematical Practice

- **MP2 Reason Abstractly and Quantitatively:** In these activities, students will use familiar formulas to complete a mapping diagram. Mathematically proficient students are able to make sense of quantities and their relationships in problem situations.

Motivate

? "What is a vending machine and what does it do?" Give students time to explain how vending machines operate. Have them discuss the idea of inserting money (input) and getting the desired item (output).

- Share some history of vending machines.
  - A Greek mathematician invented a machine in 215 B.C. to vend holy water in Egyptian temples.
  - During the early 1880s, the first commercial coin-operated vending machines were introduced in London, England, and dispensed postcards.
  - Vending machines soon offered other things, including stamps. In Philadelphia, a completely coin-operated restaurant called Horn & Hardart was opened in 1902 and served customers until 1962.

Activity Notes

Activity 1

? "What is meant by the area and perimeter of a rectangle? How do you compute each?" Area is the amount of surface covered. Perimeter is the distance around the rectangle.

- **MP2:** For each part, students will use a familiar formula. When they are given a particular dimension, they will solve for a certain measurement, such as area or perimeter.
- Students do not need to label the output with units as none are given for the input.
- **Big Idea:** A mapping diagram is one way of showing the result of evaluating an expression or formula for a set of numbers. The results are recorded in two ovals with an arrow connecting each input with its output(s). For the diagrams shown, students will know the formulas to use. A mapping diagram is similar to making a table of values.
- For each problem, have students state the formula used.

  a. $A = 2x$ **b.** $P = 2x + 4$ **c.** $C = 2\pi r$ **d.** $V = 9h$

? "Can the input of the mapping diagram be extended to other numbers, say 5, 6, and 7? Explain." yes; The variable dimension in each figure could be 5, 6, or 7.

? "Can the input of the mapping diagram be extended to other numbers, say 0, –1, and –2? Explain." no; For each figure, the variable dimension only makes sense for positive numbers.

Florida Common Core Standards

MACC.8.F.1.1 Understand that a function is a rule that assigns to each input exactly one output. The graph of a function is the set of ordered pairs consisting of an input and the corresponding output.

Previous Learning

Students should have an understanding of fractions, decimals, and percent operations.

6.1 Record and Practice Journal

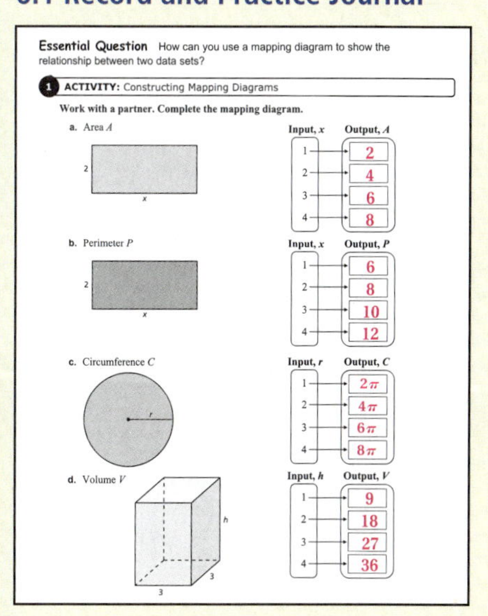
Essential Question How can you use a mapping diagram to show the relationship between two data sets?

1 ACTIVITY: Constructing Mapping Diagrams

Work with a partner. Complete the mapping diagram.

a. Area A

Input, x	Output, A
1	2
2	4
3	6
4	8

b. Perimeter P

Input, x	Output, P
1	6
2	8
3	10
4	12

c. Circumference C

Input, r	Output, C
1	2π
2	4π
3	6π
4	8π

d. Volume V

Input, h	Output, V
1	9
2	18
3	27
4	36

English Language Learners

Vocabulary

Students will find it helpful to relate the words *input* and *output* with the prepositions *in* and *out*.

6.1 Record and Practice Journal

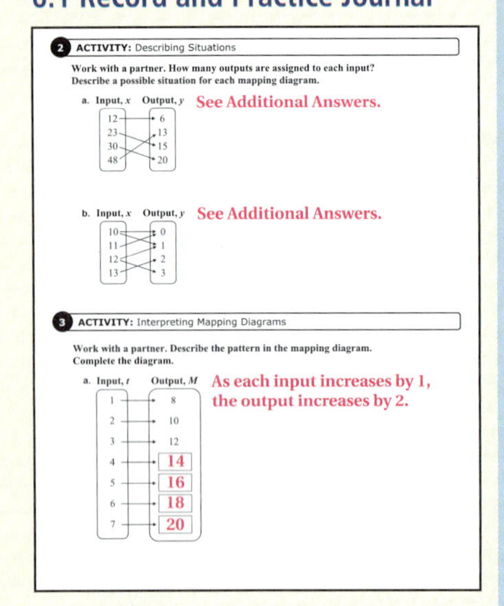

2 ACTIVITY: Describing Situations

Work with a partner. How many outputs are assigned to each input? Describe a possible situation for each mapping diagram.

a. Input, x Output, y See Additional Answers.

b. Input, x Output, y See Additional Answers.

3 ACTIVITY: Interpreting Mapping Diagrams

Work with a partner. Describe the pattern in the mapping diagram. Complete the diagram.

a. Input, t Output, M As each input increases by 1, the output increases by 2.

Input, t	Output, M
1	8
2	10
3	12
4	14
5	16
6	18
7	20

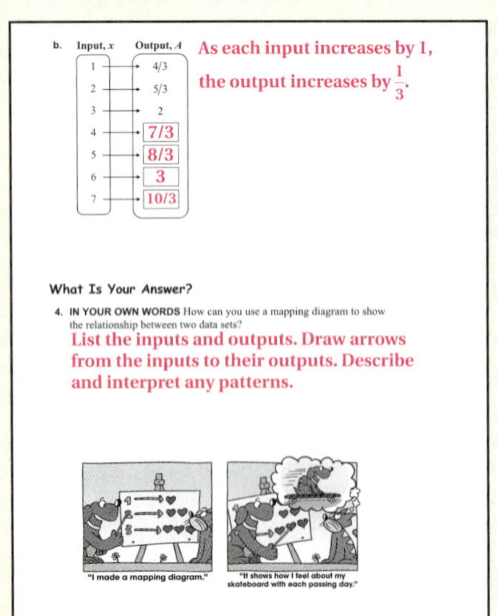

b. Input, x Output, A As each input increases by 1, the output increases by $\frac{1}{3}$.

Input, x	Output, A
1	4/3
2	5/3
3	2
4	7/3
5	8/3
6	3
7	10/3

What Is Your Answer?

4. IN YOUR OWN WORDS How can you use a mapping diagram to show the relationship between two data sets?
List the inputs and outputs. Draw arrows from the inputs to their outputs. Describe and interpret any patterns.

Laurie's Notes

Activity 2

- The first part of the activity is very literal. Students state how many outputs that there are for each input. In part (a), each input has exactly one output. In part (b), the inputs 10 and 12 each have two different outputs.
- Describing a situation for each diagram helps students realize how a mapping diagram can be used to show a relationship between two real-life quantities.
- **MP1b Persevere in Solving Problems:** Give students sufficient think time. Resist the urge to jump in and suggest a scenario. Clarify the direction line with additional words, but do not do the thinking for your students.
- Ask volunteers to share their situations for each mapping diagram.

Activity 3

- Students should be able to read a mapping diagram from the previous activity. In this activity, there are two things to do. First, determine what the pattern is for the given data. Then, use the pattern to finish the output column.
- **Note:** There is more than one correct pattern that can describe the data.

? "In either example, did any input have more than one output?" no

- **Extension:** Challenge students to think of two or more contexts for each problem.
- Possible contexts for part (a):
  - the even numbers, beginning with 8
  - the perimeter of a rectangle with side lengths of 3 and x
  - the total cost when admission is $6 and each ride you take costs $2
- Possible contexts for part (b):
  - counting by $\frac{1}{3}$, beginning with $\frac{4}{3}$
  - the perimeter of a triangle with side lengths of $\frac{1}{2}$, $\frac{1}{2}$, and $\frac{1}{3}x$

What Is Your Answer?

- Students should use the words *inputs* and *outputs* in their explanations.

Closure

- Describe the pattern in the mapping diagram. Complete the mapping diagram. As each input increases by 1, the output increases by 2. The last 2 numbers are 8 and 10.

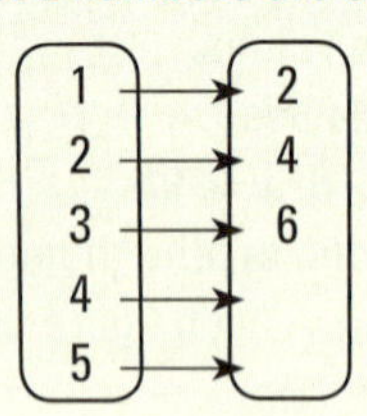

2 ACTIVITY: Describing Situations

Math Practice 7

View as Components

What are the input values? Do any of the input values point to more than one output value? How does this help you describe a possible situation?

Work with a partner. How many outputs are assigned to each input? Describe a possible situation for each mapping diagram.

a. Input, x Output, y

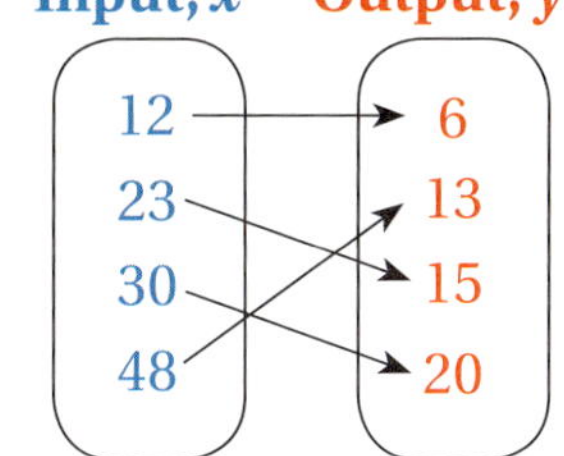

b. Input, x Output, y

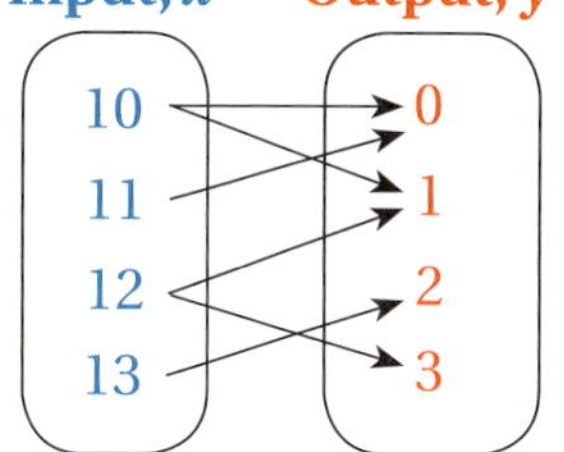

3 ACTIVITY: Interpreting Mapping Diagrams

Work with a partner. Describe the pattern in the mapping diagram. Copy and complete the diagram.

a. Input, t Output, M

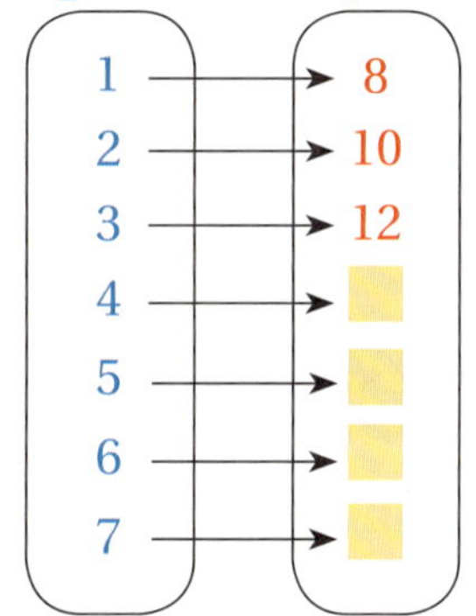

b. Input, x Output, A

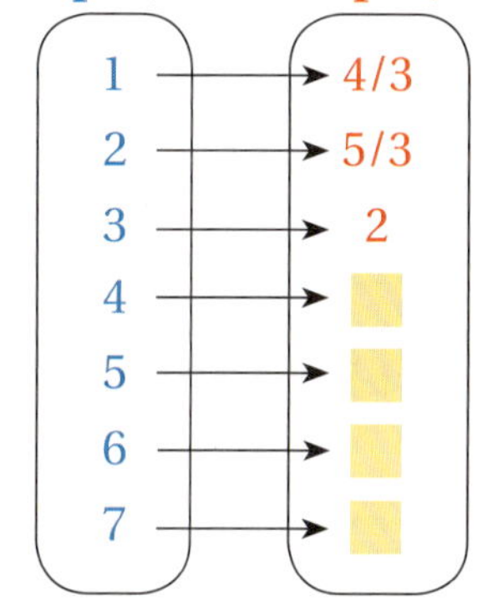

What Is Your Answer?

4. **IN YOUR OWN WORDS** How can you use a mapping diagram to show the relationship between two data sets?

"I made a mapping diagram."

"It shows how I feel about my skateboard with each passing day."

Practice Use what you learned about mapping diagrams to complete Exercises 3–5 on page 246.

6.1 Lesson

Key Vocabulary

input, *p. 244*
output, *p. 244*
relation, *p. 244*
mapping diagram, *p. 244*
function, *p. 245*

Ordered pairs can be used to show **inputs** and **outputs**.

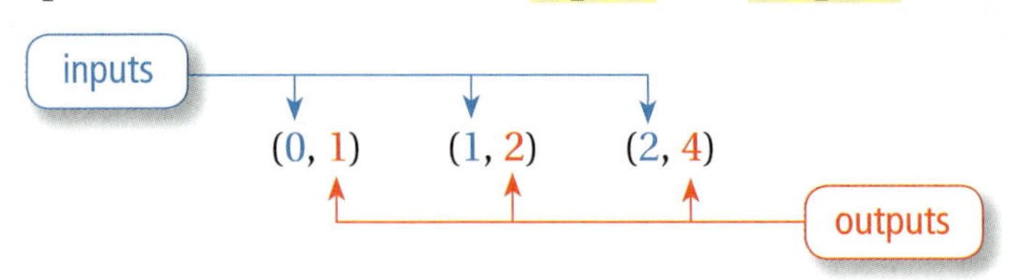

Key Idea

Relations and Mapping Diagrams

A **relation** pairs inputs with outputs. A relation can be represented by ordered pairs or a **mapping diagram**.

Ordered Pairs

(0, 1)

(1, 2)

(2, 4)

Mapping Diagram

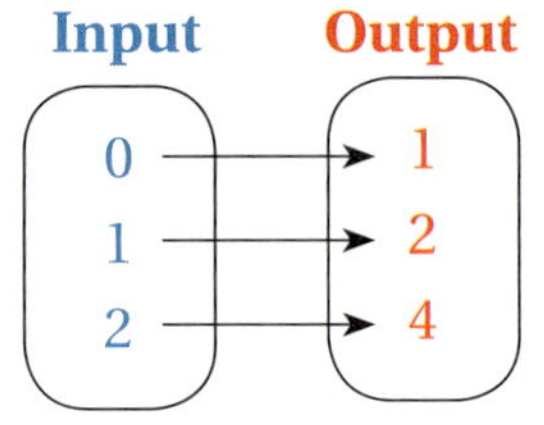

EXAMPLE 1 Listing Ordered Pairs of a Relation

List the ordered pairs shown in the mapping diagram.

a.

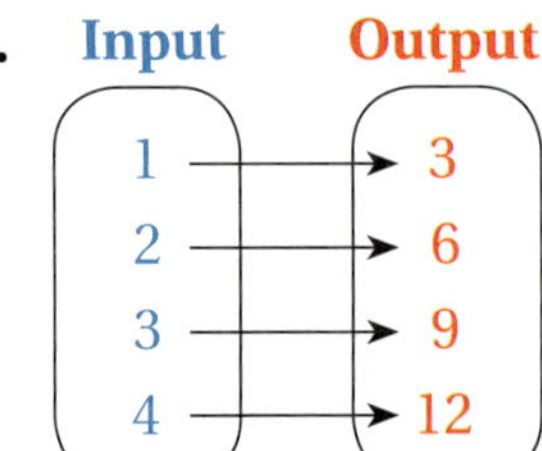

The ordered pairs are (1, 3), (2, 6), (3, 9), and (4, 12).

b.

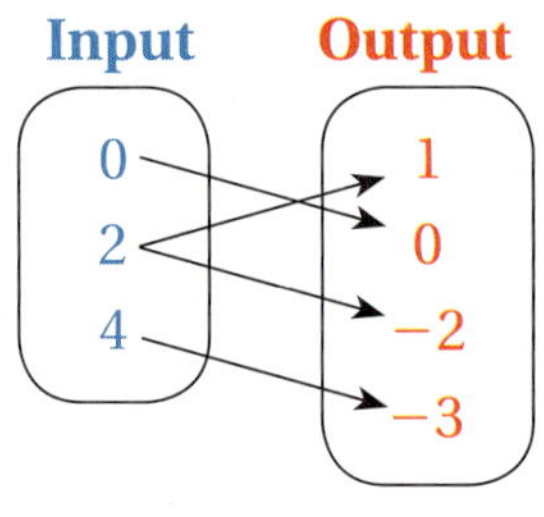

The ordered pairs are (0, 0), (2, 1), (2, −2), and (4, −3).

On Your Own

List the ordered pairs shown in the mapping diagram.

1. Input: 0, 2, 4, 6 → Output: 12, 10, 8, 6

2.

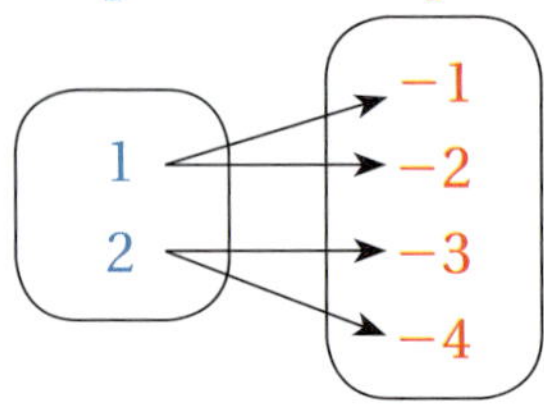

Laurie's Notes

Introduction

Connect

- **Yesterday:** Students explored mapping diagrams. (MP1b, MP2)
- **Today:** Students will use ordered pairs and mapping diagrams to display functions.

Motivate

- Time to play *Guess My Rule.*

Problem 1	Problem 2
(Santa Fe, New Mexico)	(Mexico, North America)
(Boise, Idaho)	(Japan, Asia)
(Frankfort, ?)	(Brazil, ?)
(Columbus, ?)	(Germany, ?)

? "If I gave another input for either problem, could you give the output?" depends upon how well students know their state capitals or continents

? "Could you make a mapping diagram from either problem?" yes

Lesson Notes

Key Idea

- Use either of the problems in the Motivate or the example in the text to explain the vocabulary.
- Stress that an ordered pair has an *order* associated with it, and the order matters! In the first problem, the ordered pairs are (state capital, state) and in the second problem, the ordered pairs are (country, continent).
- Spend time discussing the idea of a relation. A **relation** pairs each input with its output(s). A mapping diagram is a relation. A set of ordered pairs is also a relation.

Example 1

- **MP4 Model with Mathematics:** A mapping diagram is a helpful model to show the set of all the inputs and the set of all the outputs, while also showing the relationship between each input and its output(s).
- Note the use of color to differentiate between the input and output.
- **Common Question:** Why are the inputs (or outputs) not listed more than once when they are used more than once as in Example 1(b)? The inputs (or outputs) are a set of data. It is not customary to list elements in a data set more than once. The arrows in a mapping diagram show when an input (or output) is used more than once.

On Your Own

- Ask volunteers to put their ordered pairs on the board.

Today's lesson is representing a **relation** as ordered pairs or a **mapping diagram**, and determining whether a relation is a **function**.

Lesson Tutorials
Lesson Plans
Answer Presentation Tool

Differentiated Instruction

Auditory

Help students understand the concept of a function by discussing how functions are used in real life. For example, the number of plates to set on the dinner table is a function of the number of people expected to eat. A person earning an hourly wage has an income that is a function of the number of hours worked. Discuss with students other instances of functions in real life.

Extra Example 1

List the ordered pairs shown in the mapping diagram.

a.

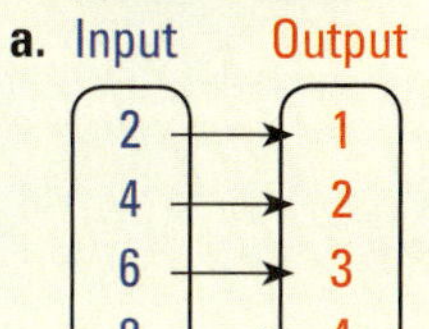

(2, 1), (4, 2), (6, 3), (8, 4)

b. Input: 1, 2, 3; Output: 0, 4, −5, −6

(1, 4), (2, 0), (2, −5), (3, −6)

On Your Own

1. (0, 12), (2, 10), (4, 8), (6, 6)
2. (1, −1), (1, −2), (2, −3), (2, −4)

Extra Example 2

Determine whether each relation is a function.

a. Input Output

Input	Output
0	2
1	4
2	6
	7

not a function

b. Input Output

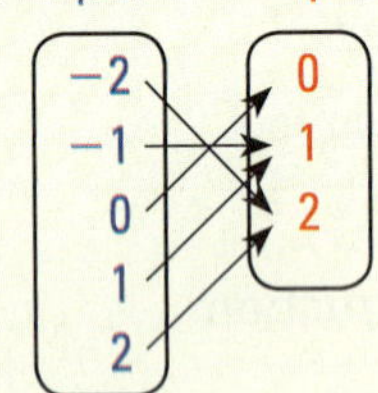

function

Extra Example 3

Consider the mapping diagram below.

Input Output

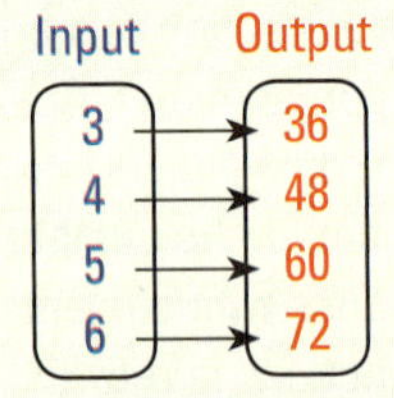

a. Determine whether the relation is a function. function

b. Describe the pattern of inputs and outputs in the mapping diagram. As each input increases by 1, the output increases by 12.

On Your Own

3. not a function
4. function
5. As each input increases by 2, the output decreases by 3.

Laurie's Notes

Discuss

- Write the definition of a *function*.
- Spend time discussing the idea of a function. Each input has exactly one output. For example, each state capital is in one state. Each country is in one continent.
- Show examples of functions represented as ordered pairs and as a mapping diagram. Discuss how the mapping diagram can be made from the ordered pairs and vice versa.

Example 2

- Make sure students can identify the ordered pairs for each mapping diagram.
- Part (a) shows that a relation is a function when output values are repeated, as long as no input values are repeated.

Example 3

? Write the mapping diagram and ask, "Is this relation a function? Explain." yes; Each input has exactly one output.

- Yesterday in Activity 3, students were asked to describe the pattern in the mapping diagram. Students may not be able to write an algebraic rule describing how to generate the output given an input. They should be able to describe in words what the pattern is.
- **Common Error:** Students often describe only what is happening to the output. In fact, the input is also changing. Encourage students to describe the change in output in terms of what is happening to the input. In the example, the description indicates that as the input increases by 1, the output increases by 15.
- The curved arrows are a convenient way to record what is changing as you move from one input to the next and from one output to the next.

On Your Own

- **MP2 Reason Abstractly and Quantitatively:** Students should be able to describe the pattern of inputs and outputs in Question 4.

Closure

- Describe the pattern. Make a mapping diagram. List the inputs. List the outputs.

  (40, Ronald Reagan) (41, George H.W. Bush) (42, William Clinton)
  (43, George W. Bush) (44, Barack Obama)

A relation that pairs each input with *exactly one* output is a **function**.

EXAMPLE 2 Determining Whether Relations Are Functions

Determine whether each relation is a function.

a.

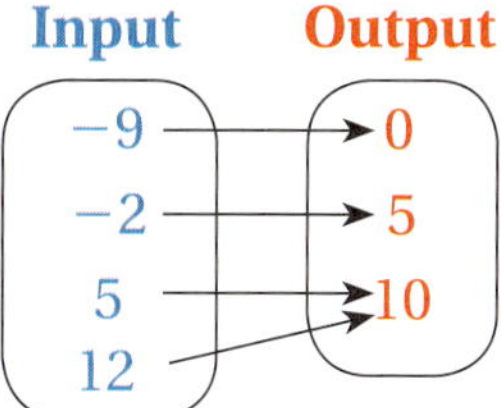

Each input has exactly one output. So, the relation is a function.

b.

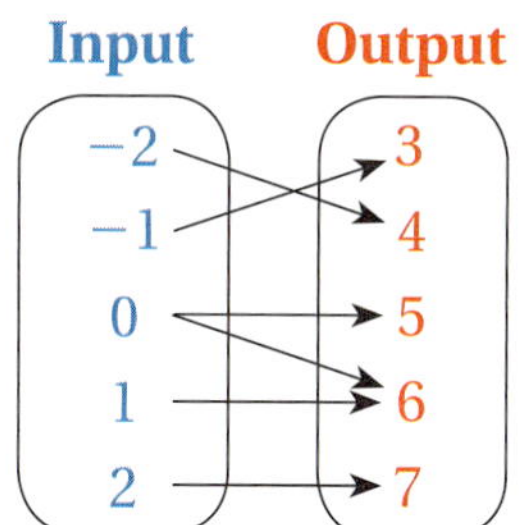

The input 0 has two outputs, 5 and 6. So, the relation is *not* a function.

EXAMPLE 3 Describing a Mapping Diagram

Input Output
1 → 15
2 → 30
3 → 45
4 → 60

Consider the mapping diagram at the left.

a. Determine whether the relation is a function.

Each input has exactly one output.

So, the relation is a function.

b. Describe the pattern of inputs and outputs in the mapping diagram.

Look at the relationship between the inputs and the outputs.

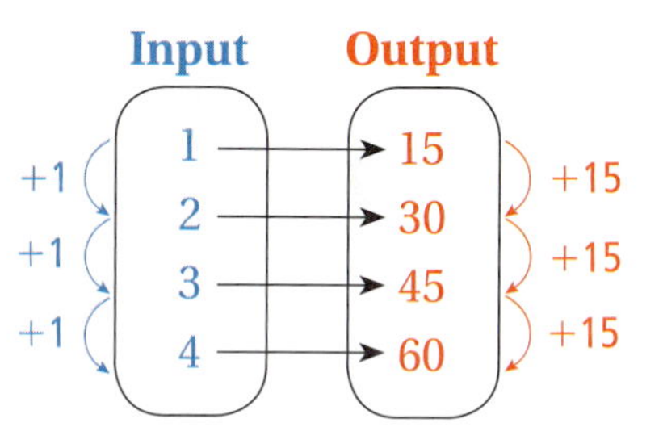

As each input increases by 1, the output increases by 15.

On Your Own

Now You're Ready
Exercises 9–11 and 13–15

Determine whether the relation is a function.

3.

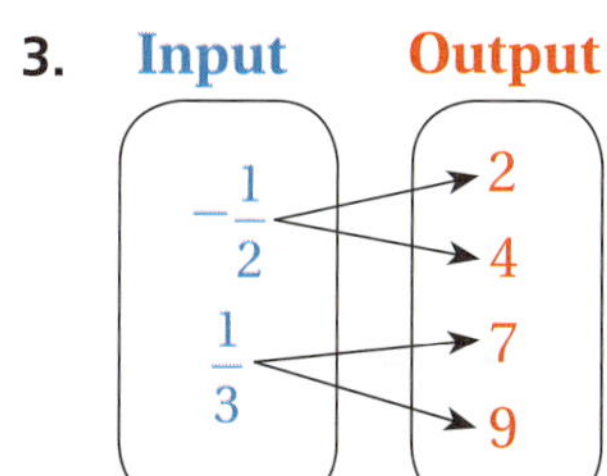

4.

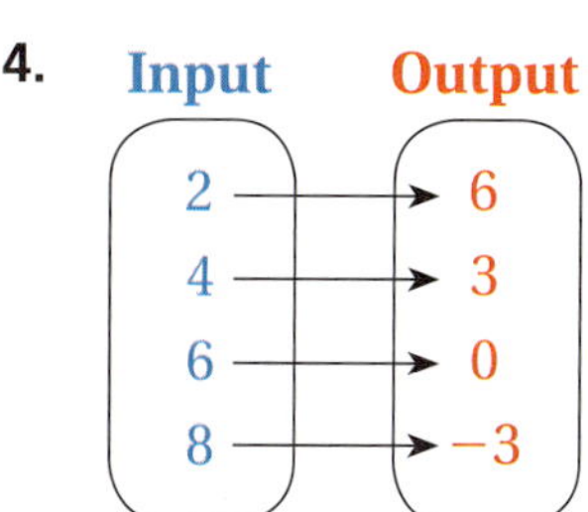

5. Describe the pattern of inputs and outputs in the mapping diagram in On Your Own 4.

6.1 Exercises

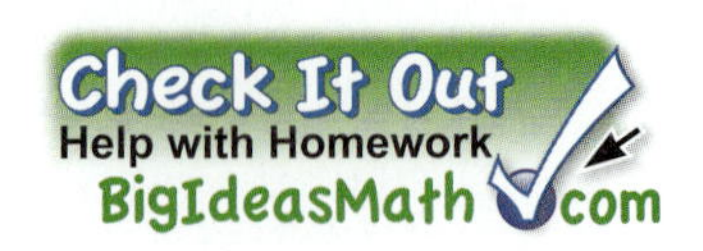

Vocabulary and Concept Check

1. **VOCABULARY** In an ordered pair, which number represents the input? the output?

2. **PRECISION** Describe how relations and functions are different.

Practice and Problem Solving

Describe the pattern in the mapping diagram. Copy and complete the diagram.

3. 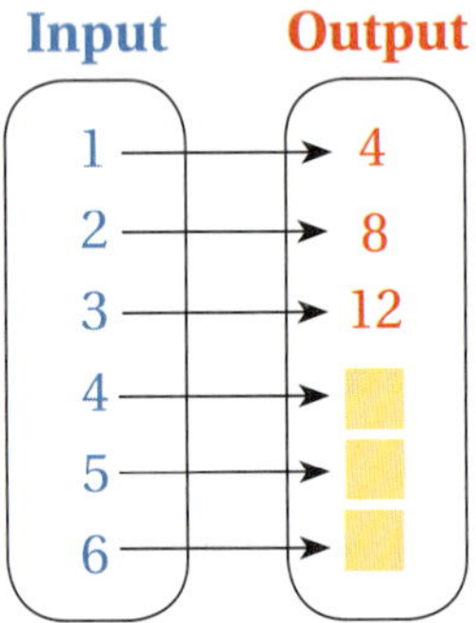

4. 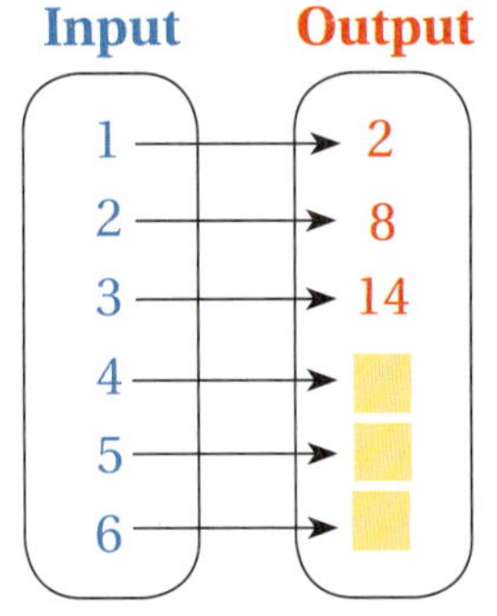

5. 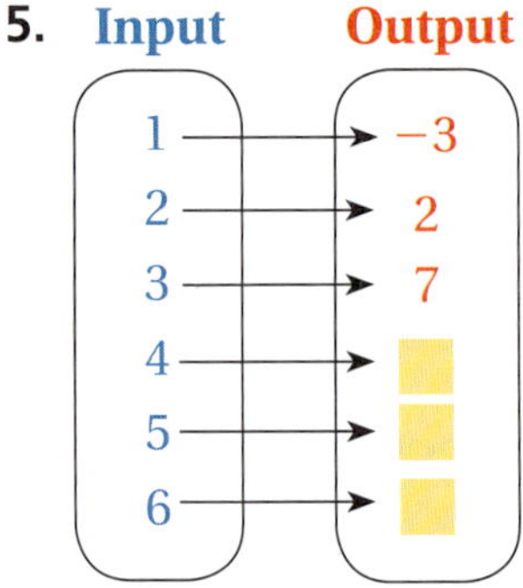

List the ordered pairs shown in the mapping diagram.

① 6. 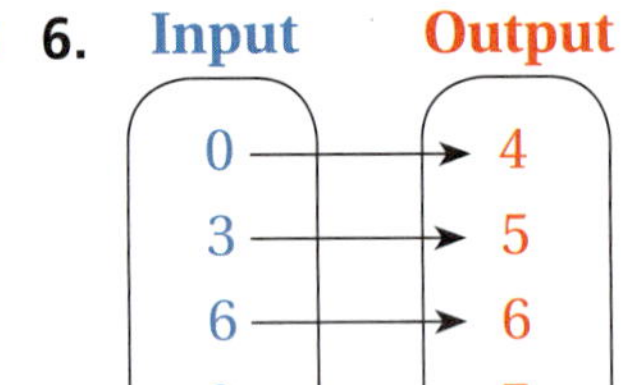

7. 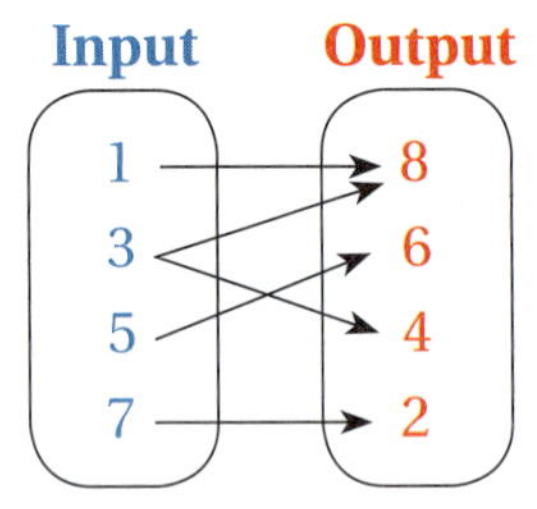

8. 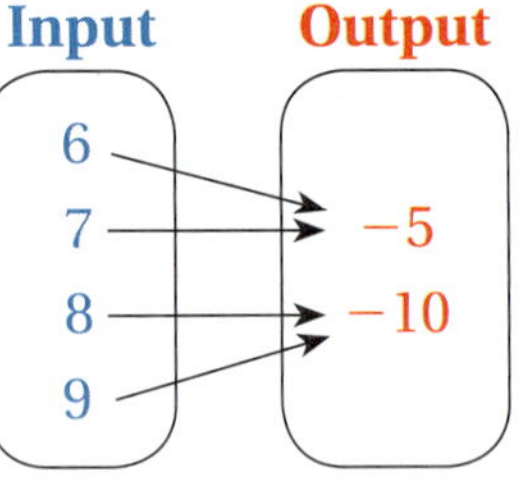

Determine whether the relation is a function.

② 9. 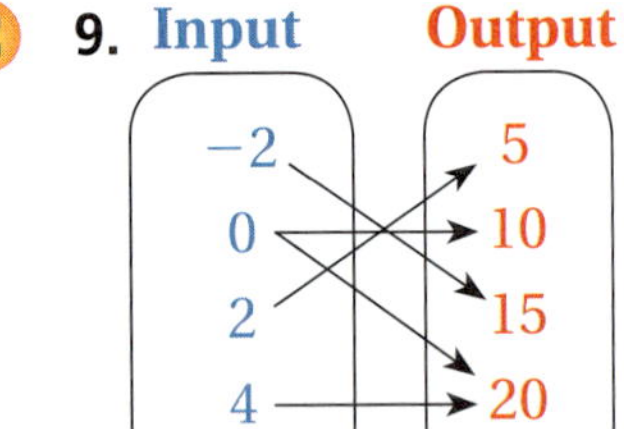

10. 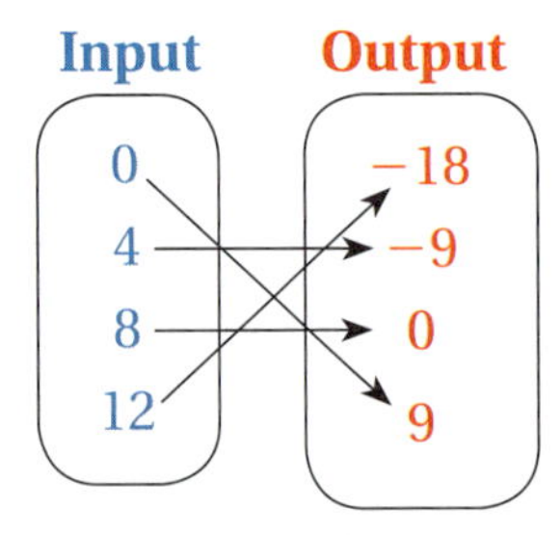

11. 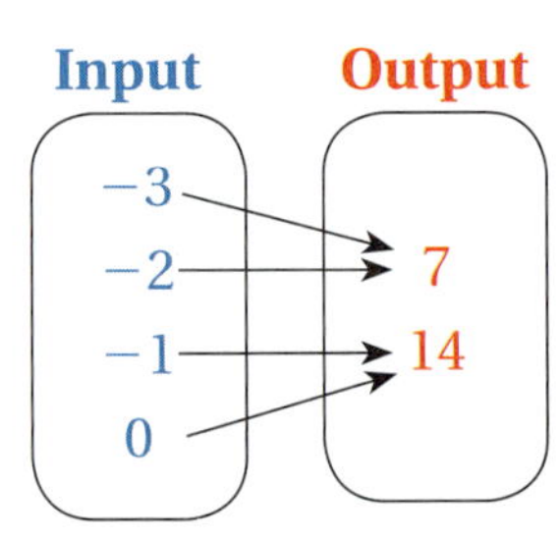

12. **ERROR ANALYSIS** Describe and correct the error in determining whether the relation is a function.

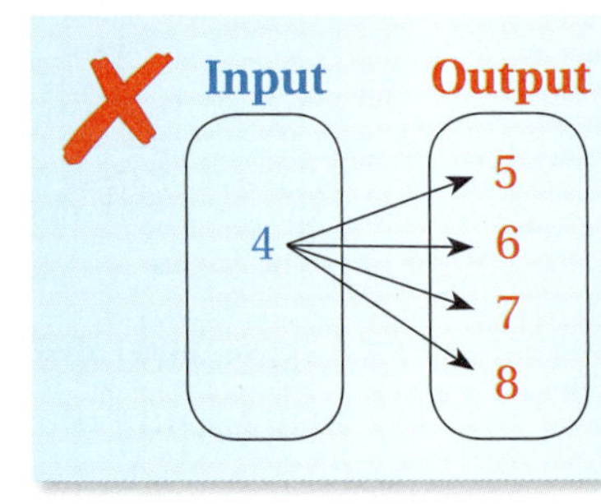

Assignment Guide and Homework Check

Level	Day 1 Activity Assignment	Day 2 Lesson Assignment	Homework Check
Basic	3–5, 19–22	1, 2, 7–11 odd, 12, 13–17 odd	7, 9, 13, 17
Average	3–5, 19–22	1, 2, 7–11 odd, 12, 13, 15–18	7, 11, 15, 16, 17
Advanced	1–5, 6–14 even, 16–22		8, 10, 14, 16, 18

Common Errors

- **Exercises 3–5, 13–15** Students may describe the pattern of the outputs only. Remind them that the inputs are changing as well, and it is important to describe the change in inputs that results in the specified change in outputs.
- **Exercises 6–8** Students may mix up the ordered pairs and write the output first and then the input. Encourage them to use the arrow as a guide. The arrow points from the first number in the ordered pair (the input) to the second number in the ordered pair (the output).

6.1 Record and Practice Journal

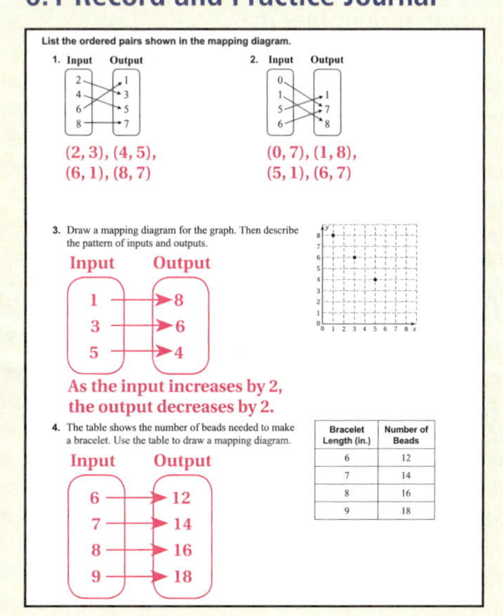
List the ordered pairs shown in the mapping diagram.

1. Input Output
(2, 3), (4, 5), (6, 1), (8, 7)

2. Input Output
(0, 7), (1, 8), (5, 1), (6, 7)

3. Draw a mapping diagram for the graph. Then describe the pattern of inputs and outputs.

Input Output
1 → 8
3 → 6
5 → 4

As the input increases by 2, the output decreases by 2.

4. The table shows the number of beads needed to make a bracelet. Use the table to draw a mapping diagram.

Bracelet Length (in.)	Number of Beads
6	12
7	14
8	16
9	18

Input Output
6 → 12
7 → 14
8 → 16
9 → 18

Vocabulary and Concept Check

1. the first number; the second number
2. A relation pairs inputs with outputs. A function is a relation that pairs each input with exactly one output.

Practice and Problem Solving

3. As each input increases by 1, the output increases by 4. 16; 20; 24
4. As each input increases by 1, the output increases by 6. 20; 26; 32
5. As each input increases by 1, the output increases by 5. 12; 17; 22
6. (0, 4), (3, 5), (6, 6), (9, 7)
7. (1, 8), (3, 8), (3, 4), (5, 6), (7, 2)
8. (6, −5), (7, −5), (8, −10), (9, −10)
9. no
10. yes
11. yes
12. In order for a relation to be a function, each input must be paired with exactly one output. So, the relation is not a function.
13. Input Output

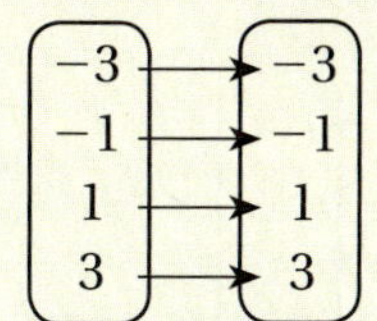

As each input increases by 2, the output increases by 2.

Practice and Problem Solving

14–15. See Additional Answers.

16. See *Taking Math Deeper*.

17. See Additional Answers.

18. 1025; *Sample answer:* As each input increases by 1, the output increases by 5. From 0 to 200 is an input increase of 200. So, the output can be found by $25 + 200(5)$.

Fair Game Review

19. y-axis **20.** x-axis

21. x-axis **22.** A

Mini-Assessment

1. List the ordered pairs shown in the mapping diagram, and determine whether the relation is a function.

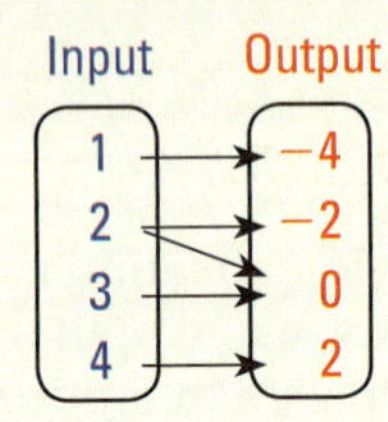

(1, −4), (2, −2), (2, 0), (3, 0), (4, 2); not a function

2. Consider the mapping diagram.

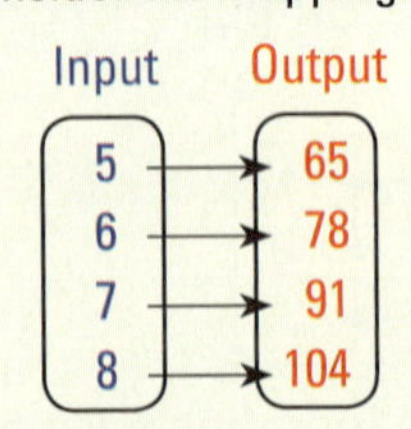

a. Determine whether the relation is a function. function

b. Describe the pattern of inputs and outputs in the mapping diagram. As each input increases by 1, the output increases by 13.

Taking Math Deeper

Exercise 16

This problem looks forward to the concept of a linear function. Students can see this in two ways.

(1) As the input increases by a constant amount, the output also increases by a constant amount.

(2) The graph of the inputs and outputs lie on a line.

a. Complete the mapping diagram.

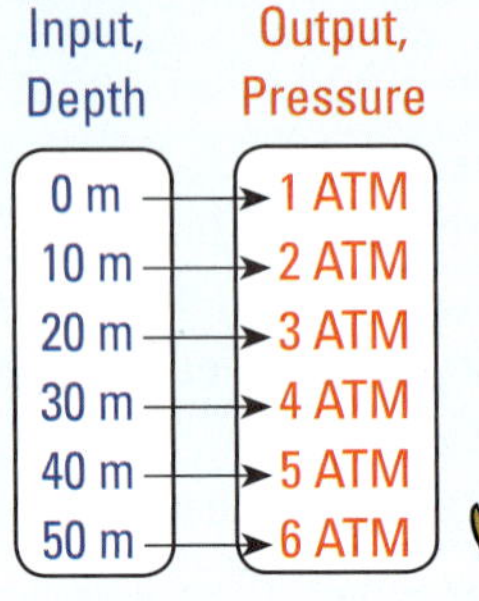

b. Each input has exactly one output, so the relation is a function.

c. List the ordered pairs. Then plot the ordered pairs in a coordinate plane.

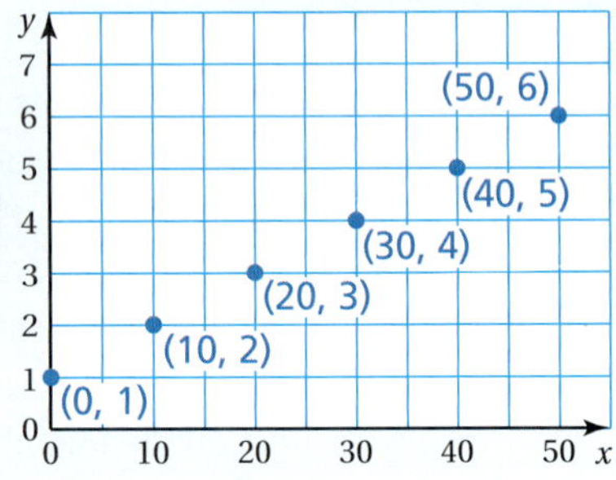

d. The graph is a visual display of the ordered pairs shown in the mapping diagram. Preferences will vary.

e. An amateur recreational diver should not go below 30 meters. The safe limit for a trained expert diver should be around 60 meters. Scuba diving can be dangerous. No one should attempt it without proper training.

Project

Use your school's library or the Internet to research scuba diving. Write a report that shows areas that are good for scuba diving. Include the cost of the sport in your report.

Reteaching and Enrichment Strategies

If students need help. . .	If students got it. . .
Resources by Chapter • Practice A and Practice B • Puzzle Time Record and Practice Journal Practice Differentiating the Lesson Lesson Tutorials Skills Review Handbook	Resources by Chapter • Enrichment and Extension • Technology Connection Start the next section

Draw a mapping diagram for the graph. Then describe the pattern of inputs and outputs.

3 **13.**

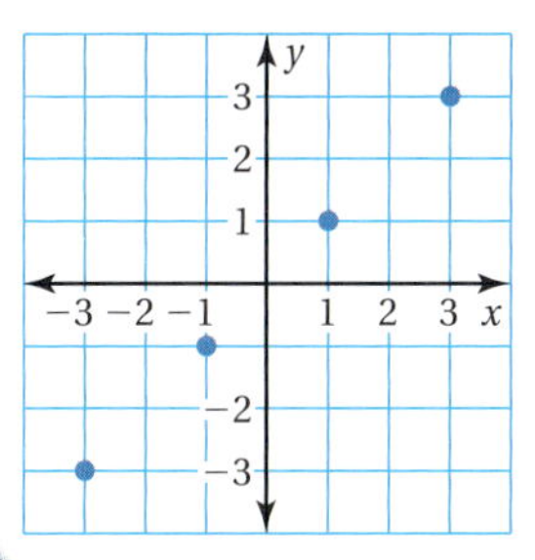

14.

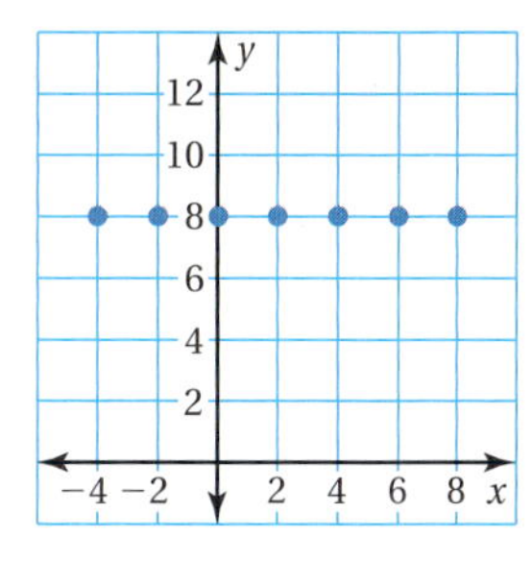

15.

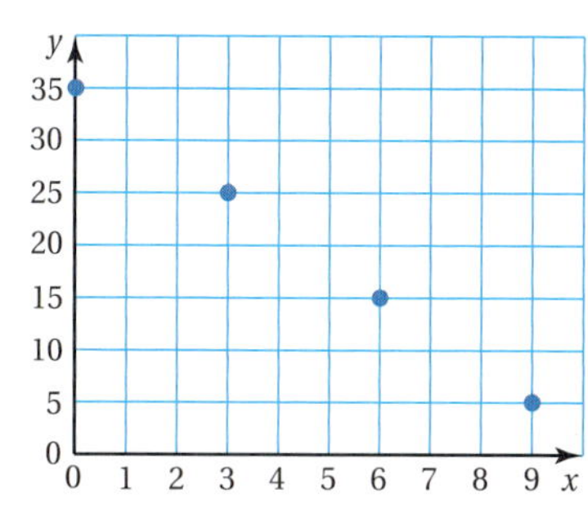

16. SCUBA DIVING The normal pressure at sea level is one atmosphere of pressure (1 ATM). As you dive below sea level, the pressure increases by 1 ATM for each 10 meters of depth.

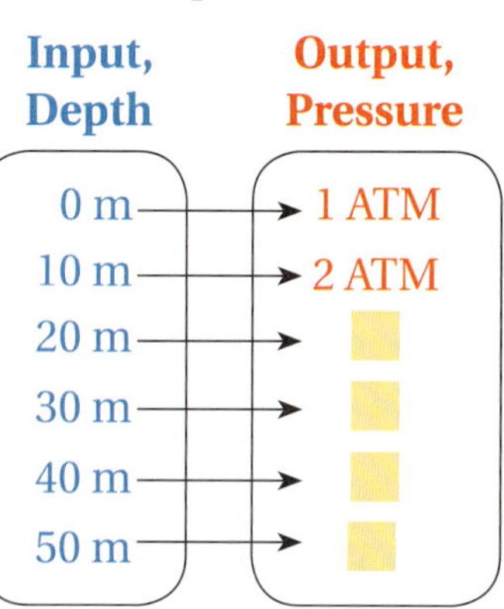

a. Complete the mapping diagram.

b. Is the relation a function? Explain.

c. List the ordered pairs. Then plot the ordered pairs in a coordinate plane.

d. Compare the mapping diagram and graph. Which do you prefer? Why?

e. RESEARCH What are common depths for people who are just learning to scuba dive? What are common depths for experienced scuba divers?

17. MOVIES A store sells previously viewed movies. The table shows the cost of buying 1, 2, 3, or 4 movies.

Movies	Cost
1	\$10
2	\$18
3	\$24
4	\$28

a. Use the table to draw a mapping diagram.

b. Is the relation a function? Explain.

c. Describe the pattern. How does the cost per movie change as you buy more movies?

18. Repeated Reasoning The table shows the outputs for several inputs. Use two methods to find the output for an input of 200.

Input, x	0	1	2	3	4
Output, y	25	30	35	40	45

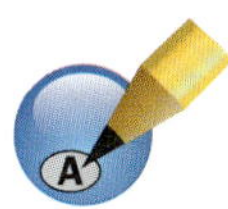

Fair Game Review What you learned in previous grades & lessons

The coordinates of a point and its image are given. Is the reflection in the *x*-axis or *y*-axis? *(Section 2.3)*

19. $(3, -3) \rightarrow (-3, -3)$ **20.** $(-5, 1) \rightarrow (-5, -1)$ **21.** $(-2, -4) \rightarrow (-2, 4)$

22. MULTIPLE CHOICE Which word best describes two figures that have the same size and the same shape? *(Section 2.1)*

Ⓐ congruent Ⓑ dilation Ⓒ parallel Ⓓ similar

6.2 Representations of Functions

Essential Question How can you represent a function in different ways?

1 ACTIVITY: Describing a Function

Work with a partner. Copy and complete the mapping diagram for the area of the figure. Then write an equation that describes the function.

a.

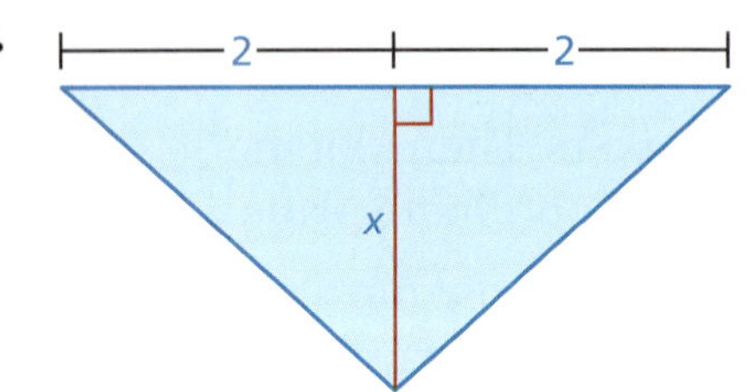

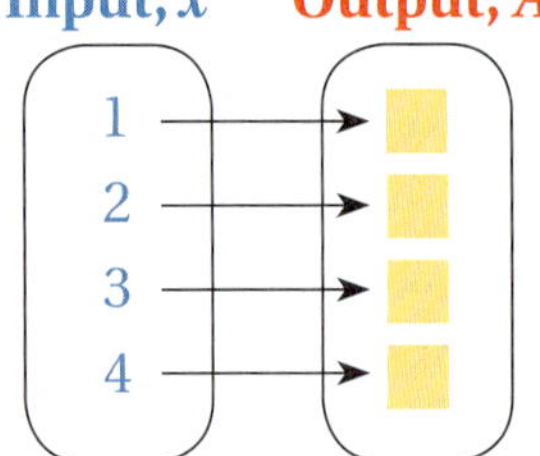

b.

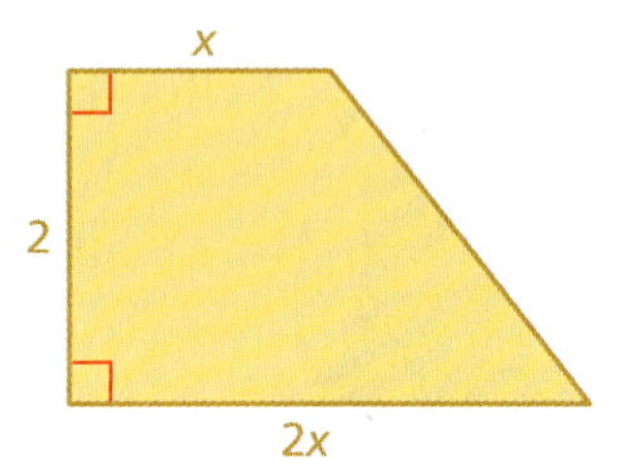

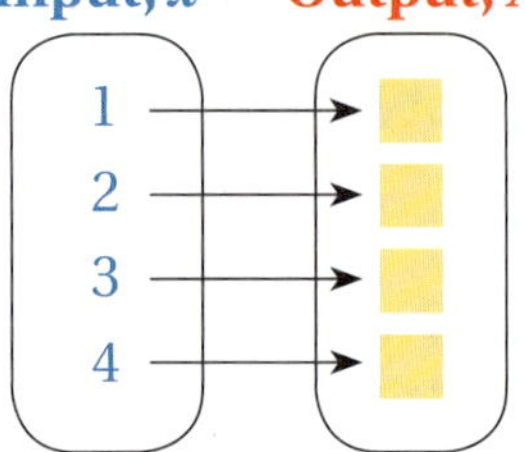

2 ACTIVITY: Using a Table

Work with a partner. Make a table that shows the pattern for the area, where the input is the figure number x and the output is the area A. Write an equation that describes the function. Then use your equation to find which figure has an area of 81 when the pattern continues.

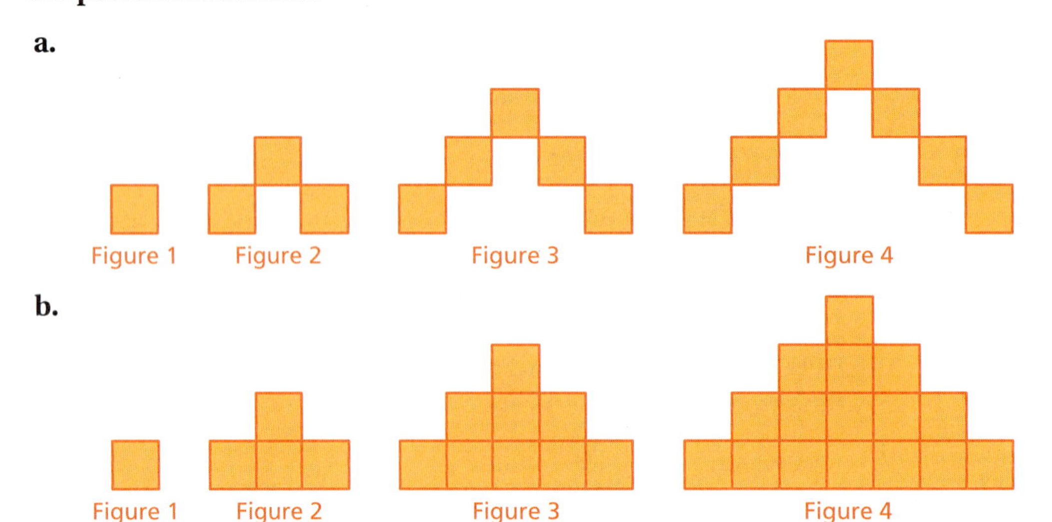

Functions

In this lesson, you will

- write function rules.
- use input-output tables to represent functions.
- use graphs to represent functions.

Learning Standard MACC.8.F.1.1

Laurie's Notes

Introduction

Standards for Mathematical Practice

- **MP3 Construct Viable Arguments and Critique the Reasoning of Others** and **MP4 Model with Mathematics:** In these activities, students write equations that represent functions presented in various ways, and they construct arguments about the data.

Motivate

- Make a separate card for each shape and each formula in the table.

(circle, radius r)	Area $= \pi r^2$	Circumference $= 2\pi r$
(square, side s)	Area $= s^2$	Perimeter $= 4s$

- Expand the table to include rectangle, triangle, parallelogram, and trapezoid.
- Give a card to each student. Students should walk around and match up with the two other cards in their set.
- Have each group share their set (aloud). Tape the sets on the board for students to view while they work on Activity 1.

Activity Notes

Activity 1

? "What information is needed in order to find the area of a triangle?" the length of the base and the height of the triangle "of a trapezoid?" the length of both bases and the height of the trapezoid

- I suggest to students, "Put x in the left input box. What expression goes in the right output box? Set that expression equal to A to write an equation."
- In writing the equation for the area of each figure, remind students to simplify expressions, if possible.

Activity 2

- **MP5 Use Appropriate Tools Strategically:** If available, provide square tiles to students. As students construct each figure with the tiles, they sometimes see the pattern that was not obvious to them from the figure.
- The inputs for each table are the figure numbers.
- It is easier for students to determine the pattern than the equation.
- After students enter the outputs ask, "for the xth figure, what is the area?"
- **FYI:** The outputs in part (b) are known as square numbers. A square number of tiles can be arranged into a square: 16 tiles can form a 4×4 square. Other numbers of tiles cannot be arranged into a square: 6 tiles can form a 1×6 rectangle or a 2×3 rectangle, but not a square.
- **Extension:** To help students understand why the numbers 1, 4, 9, 16, . . . are called square numbers, have them rearrange the tiles in each figure of part (b) to form a square.

Florida Common Core Standards

MACC.8.F.1.1 Understand that a function is a rule that assigns to each input exactly one output. The graph of a function is the set of ordered pairs consisting of an input and the corresponding output.

Previous Learning

Students should know how to find area and perimeter, write equations, and plot points in a coordinate plane.

6.2 Record and Practice Journal

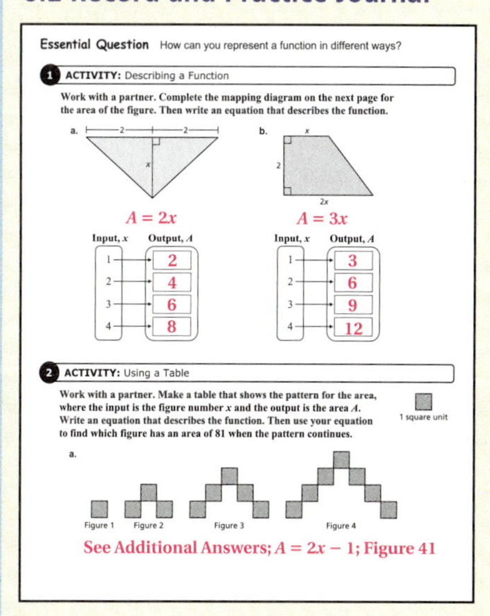

Essential Question How can you represent a function in different ways?

1 ACTIVITY: Describing a Function

Work with a partner. Complete the mapping diagram on the next page for the area of the figure. Then write an equation that describes the function.

a. $A = 2x$

Input, x	Output, A
1	2
2	4
3	6
4	8

b. $A = 3x$

Input, x	Output, A
1	3
2	6
3	9
4	12

2 ACTIVITY: Using a Table

Work with a partner. Make a table that shows the pattern for the area, where the input is the figure number x and the output is the area A. Write an equation that describes the function. Then use your equation to find which figure has an area of 81 when the pattern continues.

1 square unit

a. Figure 1 Figure 2 Figure 3 Figure 4

See Additional Answers; $A = 2x - 1$; Figure 41

English Language Learners

Auditory

Put the alphabet to use for your English learners. Because X comes before Y in the alphabet, students can alphabetize pairs of words and associate them with the *x*-coordinate and the *y*-coordinate. So, *input* comes before *output* and *horizontal* comes before *vertical*.

6.2 Record and Practice Journal

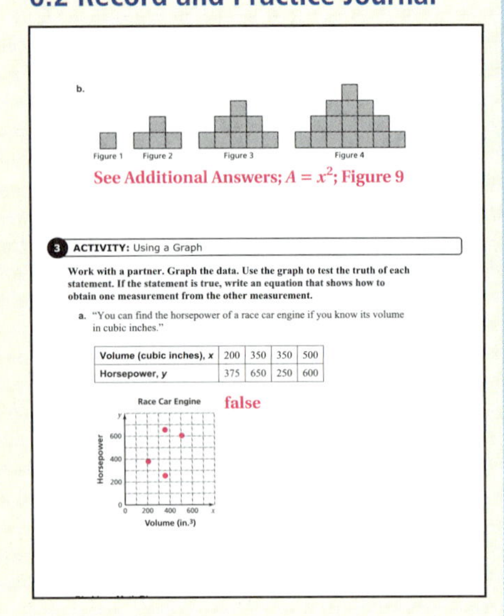

b.

Figure 1 Figure 2 Figure 3 Figure 4

See Additional Answers; $A = x^2$; Figure 9

3 ACTIVITY: Using a Graph

Work with a partner. Graph the data. Use the graph to test the truth of each statement. If the statement is true, write an equation that shows how to obtain one measurement from the other measurement.

a. "You can find the horsepower of a race car engine if you know its volume in cubic inches."

Volume (cubic inches), *x*	200	350	350	500
Horsepower, *y*	375	650	250	600

false

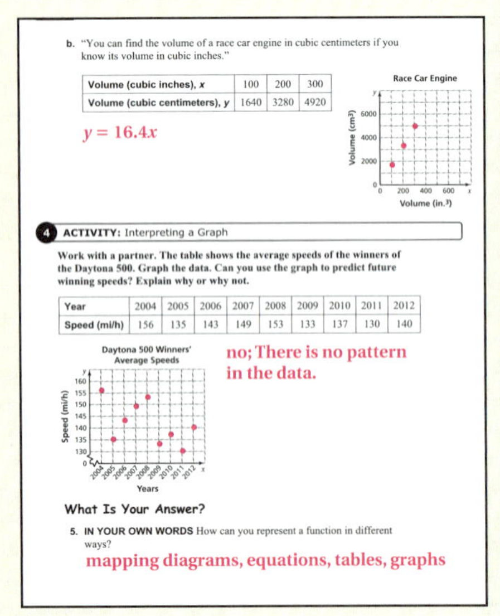

b. "You can find the volume of a race car engine in cubic centimeters if you know its volume in cubic inches."

Volume (cubic inches), *x*	100	200	300
Volume (cubic centimeters), *y*	1640	3280	4920

$y = 16.4x$

4 ACTIVITY: Interpreting a Graph

Work with a partner. The table shows the average speeds of the winners of the Daytona 500. Graph the data. Can you use the graph to predict future winning speeds? Explain why or why not.

Year	2004	2005	2006	2007	2008	2009	2010	2011	2012
Speed (mi/h)	156	135	143	149	153	133	137	130	140

no; There is no pattern in the data.

What Is Your Answer?

5. **IN YOUR OWN WORDS** How can you represent a function in different ways?

mapping diagrams, equations, tables, graphs

Laurie's Notes

Activity 3

- **MP1 Make Sense of Problems and Persevere in Solving Them:** The first step to understand each problem is to graph the data. Each set of data is in Quadrant I. Students should scale the axes with convenient values.
- The two graphs are different. The first has scattered points with no apparent pattern. The second appears to have a pattern.
- The challenge will be in writing the equation for part (b). Students may recognize and say, "Every 100 in.^{3} adds another 1640 cm^3," but not know how to write an equation for the pattern.

? "What if only 1 cubic inch was added; how many cubic centimeters would be added?" one-hundredth of 1640 or 16.4

- This may be enough to help students recognize the equation: $y = 16.4x$.
- Discuss student answers. Focus on how they determined whether the statement was true. Discuss the equation written for part (b).

Activity 4

- The challenging part of this activity is deciding how to scale each axis. You might want to guide students through this process. The years will be the input on the horizontal axis. The speed will be the output on the vertical axis. A common way to graph years on the *x*-axis is to let $x = 0$ represent 2000, so that $x = 4$ represents 2004.

? "What is the range of the speeds you need to graph?" 130 to 156

- **Big Idea:** This graph does not have a predictable pattern. Unlike part (a) in the previous activity, each input (year) has a unique output (average speed of winners). Even though the speeds repeat (143 miles per hour was the average speed in two different years), the years do not. This is a function, but not one that can be written as an equation.
- **Extension:** Ask how the average speed is determined. 500 miles divided by racing time in hours

What Is Your Answer?

- Students should describe a variety of ways in which a function can be represented: mapping diagram, ordered pairs, graph, area model, equation, table of values, and words.

Closure

- Describe the graph. Use the graph to predict the cost of 8 tickets. $48

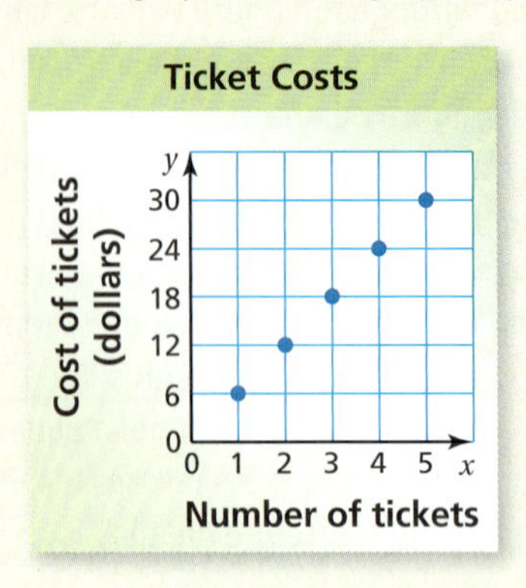

3 ACTIVITY: Using a Graph

Math Practice 3

Construct Arguments

How does the graph help you determine whether the statement is true?

Work with a partner. Graph the data. Use the graph to test the truth of each statement. If the statement is true, write an equation that shows how to obtain one measurement from the other measurement.

a. "You can find the horsepower of a race car engine if you know its volume in cubic inches."

Volume (cubic inches), x	200	350	350	500
Horsepower, y	375	650	250	600

b. "You can find the volume of a race car engine in cubic centimeters if you know its volume in cubic inches."

Volume (cubic inches), x	100	200	300
Volume (cubic centimeters), y	1640	3280	4920

4 ACTIVITY: Interpreting a Graph

Work with a partner. The table shows the average speeds of the winners of the Daytona 500. Graph the data. Can you use the graph to predict future winning speeds? Explain why or why not.

Year, x	2004	2005	2006	2007	2008	2009	2010	2011	2012
Speed (mi/h), y	156	135	143	149	153	133	137	130	140

What Is Your Answer?

5. **IN YOUR OWN WORDS** How can you represent a function in different ways?

"I graphed our profits."

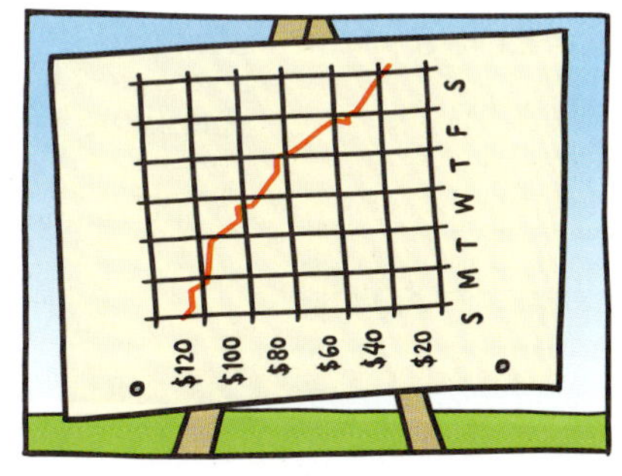

Think again.

"And I am happy to say that they are going up every day!"

Practice Use what you learned about representing functions to complete Exercises 4–6 on page 253.

6.2 Lesson

Key Vocabulary
function rule, *p. 250*

Remember

An independent variable represents a quantity that can change freely. A dependent variable *depends* on the independent variable.

Key Idea

Functions as Equations

A **function rule** is an equation that describes the relationship between inputs (independent variable) and outputs (dependent variable).

EXAMPLE 1 Writing Function Rules

a. Write a function rule for "The output is five less than the input."

Words The output is five less than the input.

Equation $y \quad = \quad x - 5$

A function rule is $y = x - 5$.

b. Write a function rule for "The output is the square of the input."

Words The output is the square of the input.

Equation $y \quad = \quad x^2$

A function rule is $y = x^2$.

EXAMPLE 2 Evaluating a Function

What is the value of $y = 2x + 5$ when $x = 3$?

$y = 2x + 5$ Write the equation.

$= 2(3) + 5$ Substitute 3 for x.

$= 11$ Simplify.

When $x = 3$, $y = 11$.

On Your Own

1. Write a function rule for "The output is one-fourth of the input."

Find the value of y when $x = 5$.

2. $y = 4x - 1$
3. $y = 10x$
4. $y = 7 - 3x$

Laurie's Notes

Introduction

Connect

- **Yesterday:** Students explored different ways to represent a function. (MP1, MP3, MP4, MP5)
- **Today:** Students will represent functions by writing equations in two variables, using input-output tables, and using graphs.

Motivate

- The daily admission at a theme park is \$73 for ages 10 and up.
- ? "What would it cost for a family of 4, ages 10 and up, to visit for one day?" $4 \times \$73 = \$292$
- ? "What would an equation be that calculates the cost for n people (all older than 10 years) to visit the theme park?" cost $= \$73n$

Lesson Notes

Key Idea

- The term **function rule** is used to introduce the idea of writing an equation to represent the relationship between the input (independent variable) and output (dependent variable), the two variables in the equation.
- It is common to write equations using the variables x and y, where x is the input and y is the output.
- **Discuss:** In the motivate section, the equation was written as cost $= 73n$. The equation could have been written as $c = 73n$. Notice the use of two variables. While any variable names could have been chosen, n and c were chosen to represent *n*umber of people and *c*ost of admission.
- Make a point of reading the Remember box.

Example 1

- If it helps, read and write the function rule as: The output, y, is five less than the input, x.
- In part (b), read and write the function rule as: The output, y, is the square of the input, x.
- Note the use of color to help students connect the words and symbols.

Example 2

- ? "For the equation $y = 2x + 5$, describe in words what operations are performed on the input x." The value of x is multiplied by 2 and then 5 is added.
- **Connection:** The rule in words for this function is "double the input and add 5," which students can make a mapping diagram for.

On Your Own

- **Connection:** Students already know how to evaluate algebraic expressions. Students would have been asked to evaluate $4x - 1$ when $x = 5$. Now the same problem is written as an equation with the answer called the output.

Goal Today's lesson is representing functions as equations, input-output tables, and graphs.

Lesson Tutorials
Lesson Plans
Answer Presentation Tool

Extra Example 1

a. Write a function rule for "The output is eight more than the input." $y = x + 8$

b. Write a function rule for "The output is four times the input." $y = 4x$

Extra Example 2

What is the value of $y = -2x + 7$ when $x = 2$? $y = 3$

On Your Own

1. $y = \frac{1}{4}x$
2. 19
3. 50
4. -8

Differentiated Instruction

Kinesthetic

When plotting an ordered pair in the coordinate plane, have students start with their fingers on (0, 0). Ask them how many units they are moving to the right. Then ask how many units they are moving up. Have students plot the point and write the ordered pair next to the point.

Extra Example 3

Graph the function $y = -x + 3$ using inputs of -1, 0, 1, and 2.

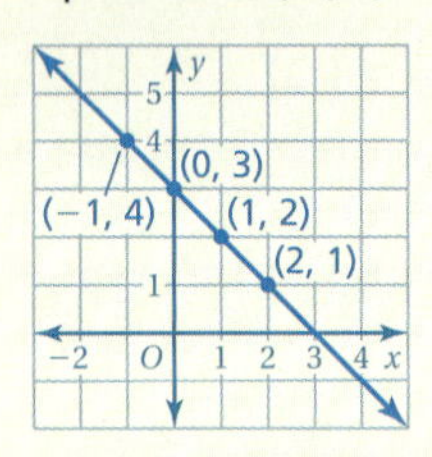

On Your Own

5.

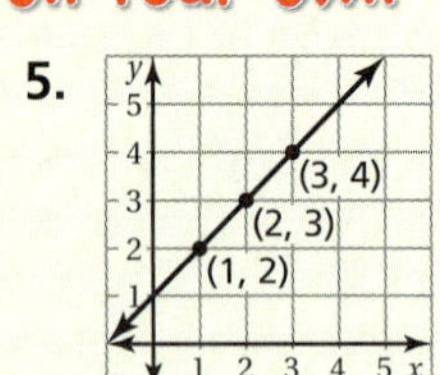

6.

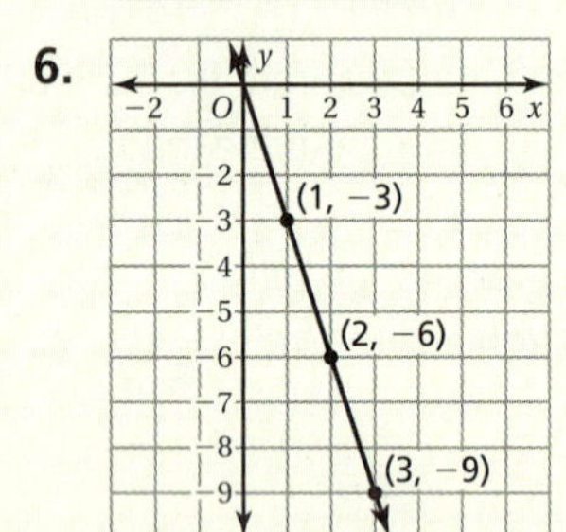

7.

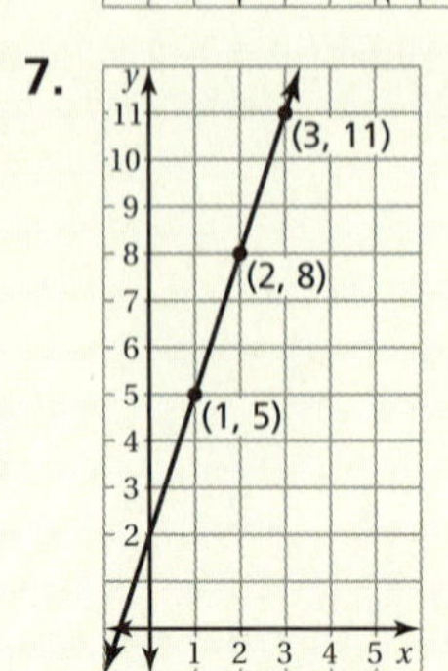

Laurie's Notes

Key Idea

- Write the Key Idea.
- The ordered pairs from the table are plotted. The input is x and is found along the horizontal axis. The output is y and is found along the vertical axis. The graph shows the relationship between inputs and outputs.
- Discuss with students why connecting the points with a line is the graph of all of the solutions of the equation $y = x + 2$. For instance, if $x = 1.5$, $y = 3.5$. Find this point on the graph.

? "Is the ordered pair (3, 1) a solution of $y = x + 2$?" no "Is (3, 1) on the graph?" no

Example 3

- The table is a good reminder of how equations are evaluated and how solutions can be recorded as ordered pairs.

? "Is the set of ordered pairs a function? Explain." yes; Each input is paired with exactly one output.

- Students may ask about calling the equation $y = -2x + 1$ a function rather than a linear equation in slope intercept form. Tell them that it is actually both! It is a linear function.
- Knowing the equation is a function assures you that each input has exactly one output. Linear equations are just one type of function that students will study.
- Students should also note that not all functions can be written as equations. Recall the data about the Daytona 500 winning speeds.

On Your Own

- Students should not have any difficulty graphing these functions.

Functions as Tables and Graphs

A function can be represented by an input-output table and by a graph. The table and graph below represent the function $y = x + 2$.

Input, x	Output, y	Ordered Pair, (x, y)
1	3	(1, 3)
2	4	(2, 4)
3	5	(3, 5)

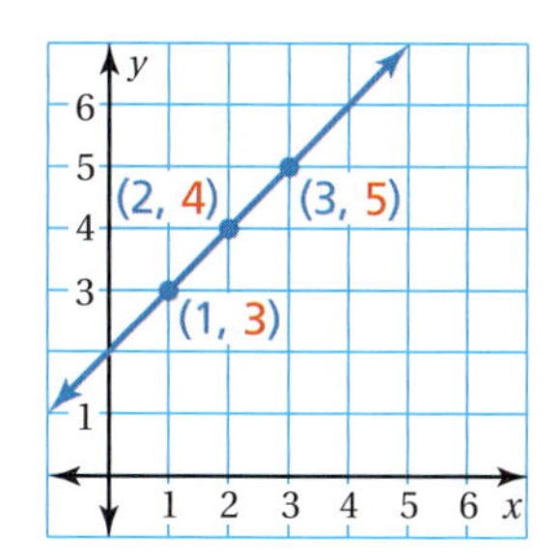

By drawing a line through the points, you graph *all* of the solutions of the function $y = x + 2$.

EXAMPLE 3 Graphing a Function

Graph the function $y = -2x + 1$ using inputs of -1, 0, 1, and 2.

Make an input-output table.

Input, x	$-2x + 1$	Output, y	Ordered Pair, (x, y)
-1	$-2(-1) + 1$	3	$(-1, 3)$
0	$-2(0) + 1$	1	(0, 1)
1	$-2(1) + 1$	-1	$(1, -1)$
2	$-2(2) + 1$	-3	$(2, -3)$

Plot the ordered pairs and draw a line through the points.

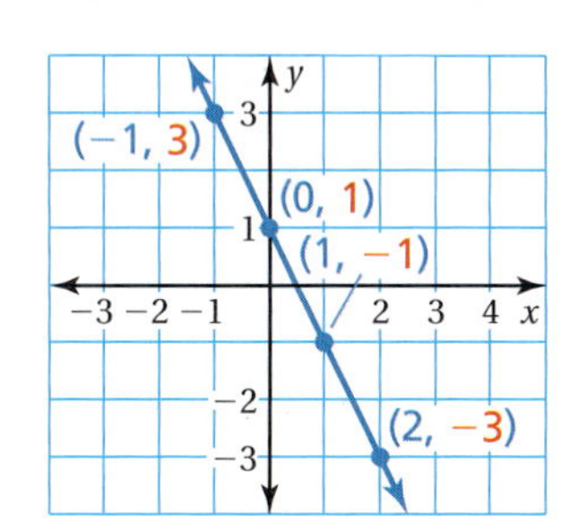

On Your Own

Now You're Ready
Exercises 19–24

Graph the function.

5. $y = x + 1$
6. $y = -3x$
7. $y = 3x + 2$

EXAMPLE 4 Real-Life Application

The number of pounds p of carbon dioxide produced by a car is 20 times the number of gallons g of gasoline used by the car. Write and graph a function that describes the relationship between g and p.

Write a function rule using the variables g and p.

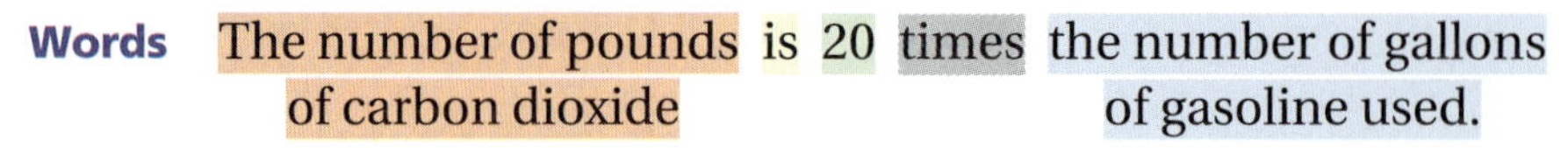

Words The number of pounds of carbon dioxide is 20 times the number of gallons of gasoline used.

Equation $p \quad = 20 \quad \cdot \quad g$

Make an input-output table that represents the function $p = 20g$.

Input, g	$20g$	Output, p	Ordered Pair, (g, p)
1	20(1)	20	(1, 20)
2	20(2)	40	(2, 40)
3	20(3)	60	(3, 60)

Plot the ordered pairs and draw a line through the points.

Because you cannot have a negative number of gallons, use only positive values of g.

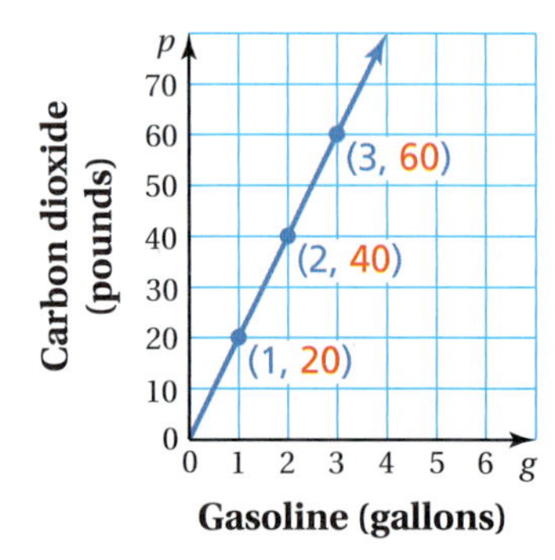

On Your Own

Exercise 26

8. **WHAT IF?** For a truck, p is 25 times g. Write and graph a function that describes the relationship between g and p.

Representations of Functions

Words An output is 2 more than the input.

Equation $y = x + 2$

Input-Output Table

Input, x	Output, y
−1	1
0	2
1	3
2	4

Mapping Diagram

Input, x → Output, y

−1 → 1

0 → 2

1 → 3

2 → 4

Graph

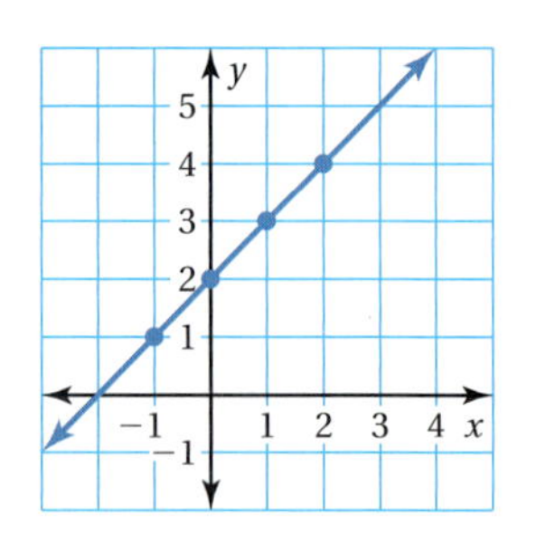

Laurie's Notes

Example 4

- **MP1 Make Sense of Problems and Persevere in Solving Them** and **MP4 Model with Mathematics:** Writing the verbal model helps students to make sense of the problem and to write an equation that represents the function.
- ? "What is the independent variable?" g "What is the dependent variable?" p
- The equation can be written as $y = 20x$. It is common to use variable names to remind us of what the variable represents. In this problem, the ordered pairs are: (gallons of gasoline, pounds of carbon dioxide), or (g, p). So, another way to write the equation is $p = 20g$.
- **Common Misconception:** Students sometimes believe that the axes have to be scaled in the same units.
- ? "Is it possible for $g = 1.5$?" yes "What is the output for $g = 1.5$?" 30 "Is the ordered pair (1.5, 30) on the graph?" yes
- **Extension:** Is it possible to estimate g when $p = 50$? yes; $g = 2.5$ by reading the graph

On Your Own

- Have one group graph their solution on a transparency to share with the class.

Summary

- Discuss the *Summary* box with students. They have represented functions using words, equations, input-output tables, mapping diagrams, and graphs.
- **Connection:** In future mathematics courses, students will study connections between the various forms of functions.
- **FYI:** It is not always possible to represent the function in a particular form. For instance, the data for the Daytona 500 winning speeds were represented in a table and graph, but could not be represented as an equation.

Closure

- Divide the class into groups of four. Say, "An output is 1 less than twice the input." Each group should represent the function in four different ways.

Sample answer:

Words: An output is 1 less than twice the input.

Equation: $y = 2x - 1$

Mapping Diagram:

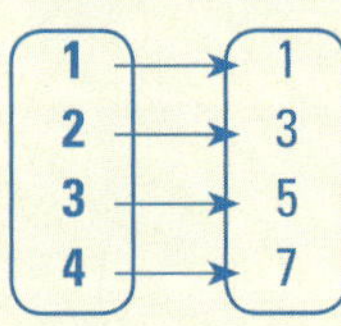

Graph:

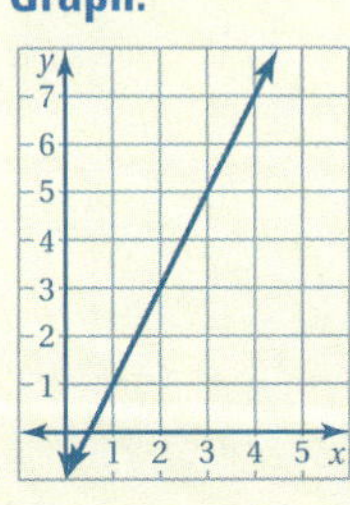

Extra Example 4

The distance m in miles traveled by a car is 35 times the number of gallons g of gasoline used by the car. Write and graph a function that describes the relationship between g and m. $m = 35g$

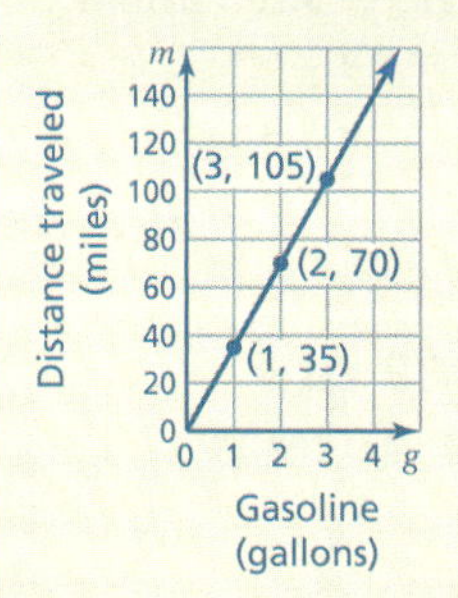

On Your Own

8. $p = 25g$

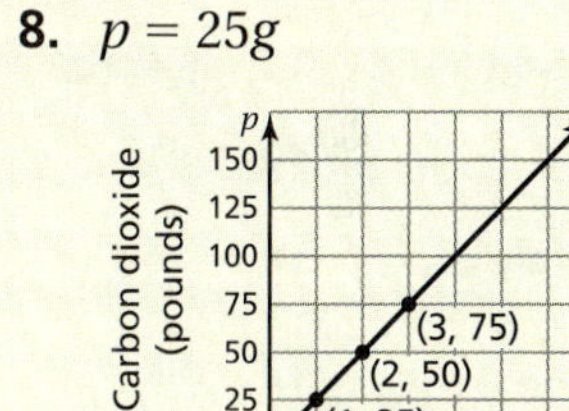

Vocabulary and Concept Check

1. input variable: x; output variable: y
2. words, equations, input-output tables, mapping diagrams, graphs
3. What output is twice the sum of the input 3 and 4?; $2(3 + 4) = 14$; $2(3) + 4 = 10$

Practice and Problem Solving

4. $y = 4x$
5. $y = x + 7$
6. $y = -x + 1$
7. $y = \frac{1}{2}x$
8. $y = x + 11$
9. $y = x - 3$
10. $y = x^3$
11. $y = 6x$
12. $y = 2x + 1$
13. 8
14. -35
15. -17
16. 3.5
17. 54
18. 3
19.

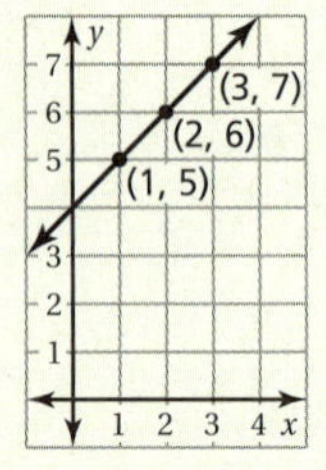

20.

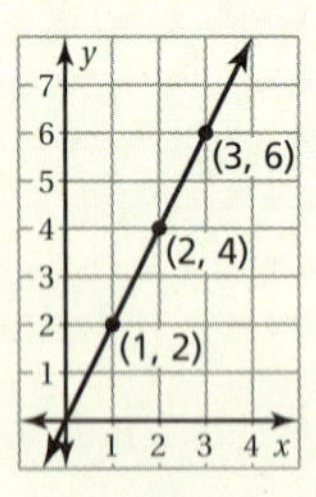

21.

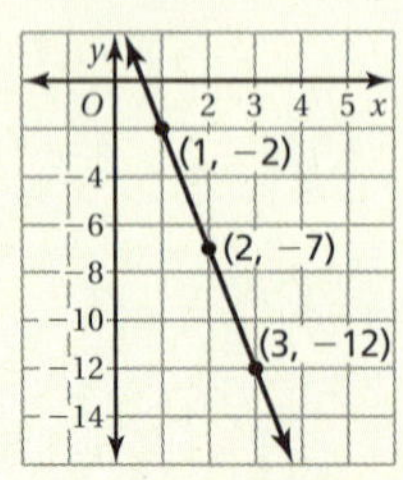

22–24. See Additional Answers.

Assignment Guide and Homework Check

Level	Day 1 Activity Assignment	Day 2 Lesson Assignment	Homework Check
Basic	4–6, 39–42	1–3, 7–33 odd	7, 15, 21, 31, 33
Average	4–6, 39–42	1–3, 9–25 odd, 26–34 even	9, 15, 23, 30, 34
Advanced	1–6, 12–24 even, 25, 26, 33–42		12, 18, 24, 34, 35

Common Errors

- **Exercises 7–12** Students may mix up the x and y variables when writing the equation. Remind them that x represents the input and y represents the output. Students may also need to be reminded how to write expressions from word phrases.
- **Exercises 13–18** Students may forget the order of operations for some of the exercises and try to add or subtract before multiplying or dividing after substituting the value of x. Remind them of the order of operations and how to evaluate an expression.
- **Exercises 19–24** Students may mix up their axes or label one or both axes inconsistently. Remind them that the input (x) values are horizontal and the output (y) values are vertical. Encourage students to label each axis in specific increments that easily show the points plotted and will give a line through the points.

6.2 Record and Practice Journal

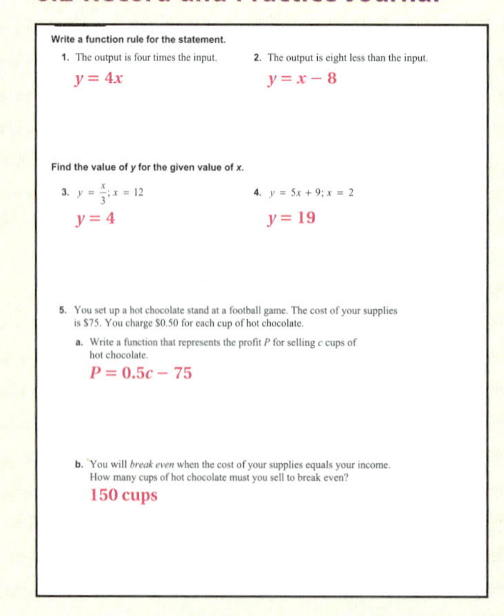

Write a function rule for the statement.

1. The output is four times the input. $y = 4x$
2. The output is eight less than the input. $y = x - 8$

Find the value of y for the given value of x.

3. $y = \frac{x}{3}; x = 12$ $y = 4$
4. $y = 5x + 9; x = 2$ $y = 19$

5. You set up a hot chocolate stand at a football game. The cost of your supplies is \$75. You charge \$0.50 for each cup of hot chocolate.
   a. Write a function that represents the profit P for selling c cups of hot chocolate. $P = 0.5c - 75$
   b. You will *break even* when the cost of your supplies equals your income. How many cups of hot chocolate must you sell to break even? 150 cups

6.2 Exercises

Vocabulary and Concept Check

1. **VOCABULARY** Identify the input variable and the output variable for the function rule $y = 2x + 5$.

2. **WRITING** Describe five ways to represent a function.

3. **DIFFERENT WORDS, SAME QUESTION** Which is different? Find "both" answers.

What output is 4 more than twice the input 3?	What output is twice the sum of the input 3 and 4?
What output is the sum of 2 times the input 3 and 4?	What output is 4 increased by twice the input 3?

Practice and Problem Solving

Write an equation that describes the function.

4. Input, x Output, y

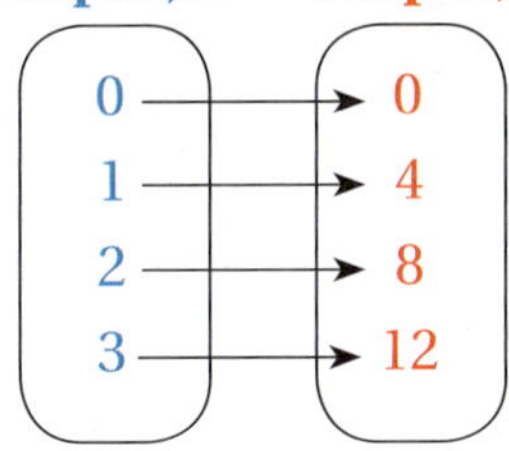

5.

Input, x	Output, y
1	8
2	9
3	10
4	11

6.

Input, x	Output, y
1	0
3	-2
5	-4
7	-6

Write a function rule for the statement.

(1) 7. The output is half of the input.

8. The output is eleven more than the input.

9. The output is three less than the input.

10. The output is the cube of the input.

11. The output is six times the input.

12. The output is one more than twice the input.

Find the value of y for the given value of x.

(2) 13. $y = x + 5;\ x = 3$

14. $y = 7x;\ x = -5$

15. $y = 1 - 2x;\ x = 9$

16. $y = 3x + 2;\ x = 0.5$

17. $y = 2x^3;\ x = 3$

18. $y = \frac{x}{2} + 9;\ x = -12$

Graph the function.

(3) 19. $y = x + 4$

20. $y = 2x$

21. $y = -5x + 3$

22. $y = \frac{x}{4}$

23. $y = \frac{3}{2}x + 1$

24. $y = 1 + 0.5x$

25. ERROR ANALYSIS Describe and correct the error in graphing the function represented by the input-output table.

Input, x	-4	-2	0	2
Output, y	-1	1	3	5

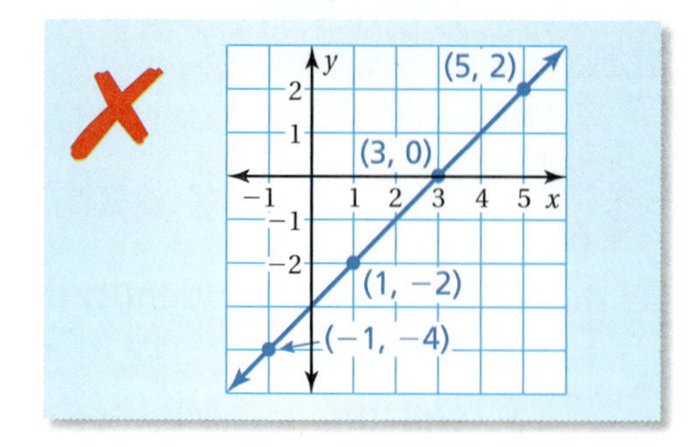

26. DOLPHIN A dolphin eats 30 pounds of fish per day.

a. Write and graph a function that relates the number of pounds p of fish that a dolphin eats in d days.

b. How many pounds of fish does a dolphin eat in 30 days?

Match the graph with the function it represents.

27. 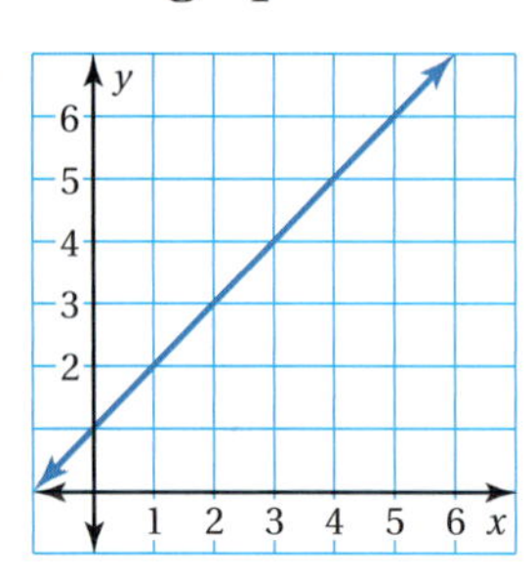

28. 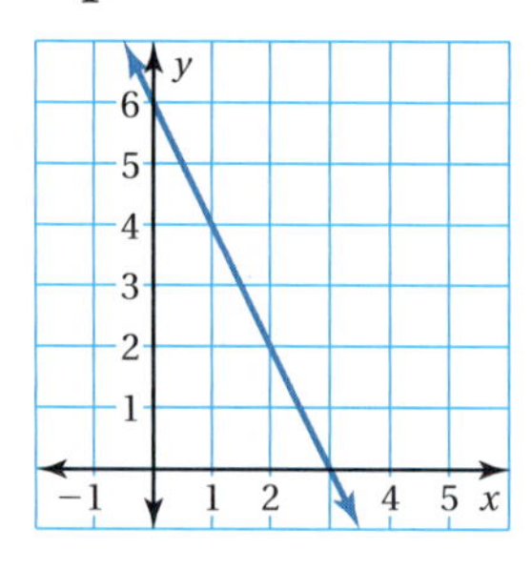

29. 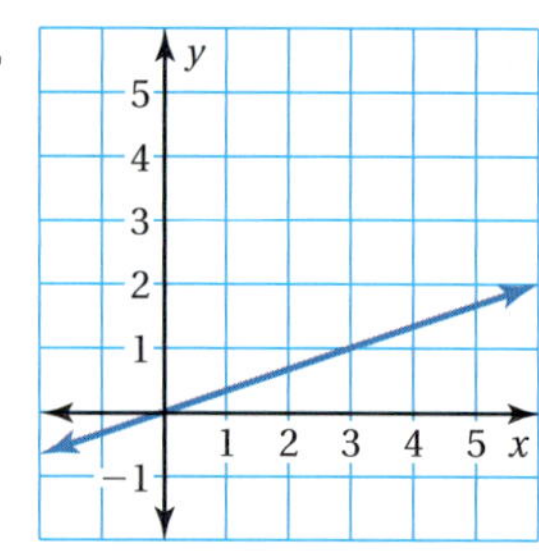

A. $y = \frac{x}{3}$ **B.** $y = x + 1$ **C.** $y = -2x + 6$

Find the value of x for the given value of y.

30. $y = 5x - 7;\ y = -22$ **31.** $y = 9 - 7x;\ y = 37$ **32.** $y = \frac{x}{4} - 7;\ y = 2$

33. BRACELETS You decide to make and sell bracelets. The cost of your materials is \$84. You charge \$3.50 for each bracelet.

a. Write a function that represents the profit P for selling b bracelets.

b. Which variable is independent? dependent? Explain.

c. You will *break even* when the cost of your materials equals your income. How many bracelets must you sell to break even?

34. SALE A furniture store is having a sale where everything is 40% off.

a. Write a function that represents the amount of discount d on an item with a regular price p.

b. Graph the function using the inputs 100, 200, 300, 400, and 500 for p.

c. You buy a bookshelf that has a regular price of \$85. What is the sale price of the bookshelf?

Common Errors (continued)

- **Exercises 27–29** Students may just guess which equation goes with each graph without testing any values. Encourage them to find a point or two on the graph and substitute the values into the equations to find which one matches.

Practice and Problem Solving

25. The order of the x- and y-coordinates is reversed in each coordinate pair.

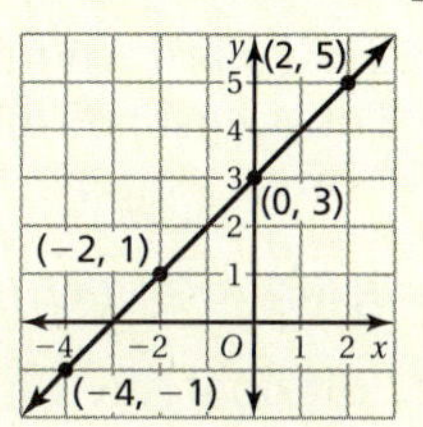

26. **a.** $p = 30d$

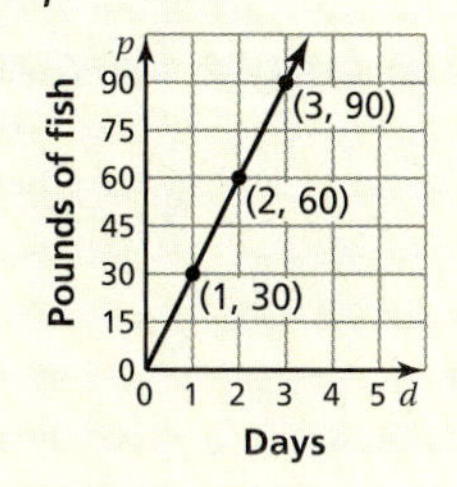

b. 900 pounds

27. B **28.** C

29. A **30.** -3

31. -4 **32.** 36

33. **a.** $P = 3.50b - 84$

b. independent variable: b; dependent variable: P; The profit depends on the number of bracelets sold.

c. 24 bracelets

34. **a.** $d = 0.40p$

b.

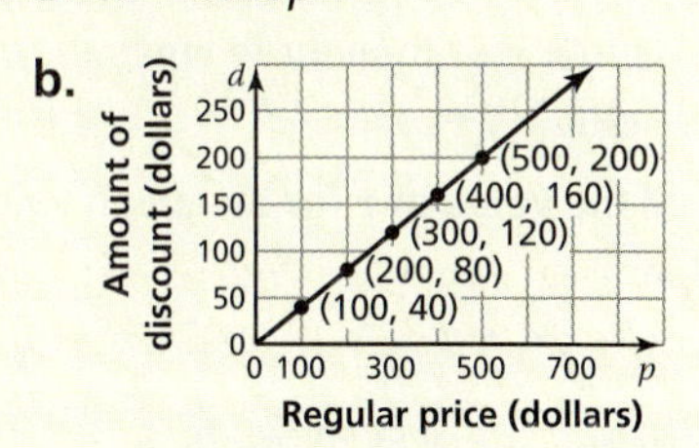

c. \$51

English Language Learners

Vocabulary

Have students add the key vocabulary words *input, output, relation, mapping diagram, function*, and *function rule* to their notebooks. Definitions, examples, and pictures should accompany the words.

Practice and Problem Solving

35. **a.** $G = 35 + 10h$

b. $S = 25h$

c. Snake Tours; For 2 hours, Gator Tours cost $55 and Snake Tours cost $50.

36. See *Taking Math Deeper*.

37–38. See Additional Answers.

Fair Game Review

39. 1 **40.** $-\frac{5}{2}$

41. $\frac{1}{3}$ **42.** C

Mini-Assessment

1. Write a function rule for "The output is one-third of the input." $y = \frac{1}{3}x$
2. Write a function rule for "The output is four less than three times the input." $y = 3x - 4$

Find the value of *y* for the given value of *x*.

3. $y = 6x$; $x = -4$ $y = -24$
4. $y = \frac{x}{5} - 8$; $x = 50$ $y = 2$
5. **You are selling magazines to raise money for your school. Each subscription you sell earns $8 for your school.**
   a. Write an equation you can use to find the total amount *d* in dollars that you can raise for your school after selling *s* subscriptions. $d = 8s$
   b. How much money will you raise if you sell 20 subscriptions? $160

Taking Math Deeper

Exercise 36

Because the slope of a line is the same between any two points on the line, find the slope and use it to find the value of *y*.

Find the slope of the line using (3, 2) and (5, 8).

$$m = \frac{y_2 - y_1}{x_2 - x_1} = \frac{8 - 2}{5 - 3} = \frac{6}{2} = 3$$

The slope is 3.

Write and solve a proportion to find the value of *y*.

$$m = \frac{y_2 - y_1}{x_2 - x_1}$$

$$3 = \frac{y - 2}{8 - 3}$$

$$15 = y - 2$$

$$17 = y$$

So, the value of *y* is 17.

Use a graph to check your solution.

Starting at (5, 8), repeatedly move 1 unit right and 3 units up until you arrive at a point whose *x* coordinate is 8. The *y*-coordinate is 17.

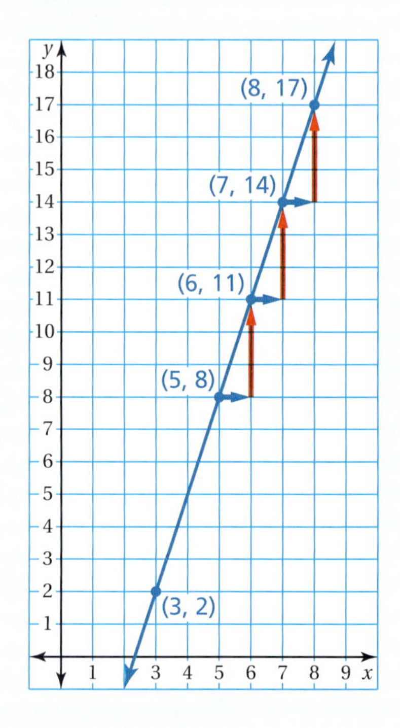

Reteaching and Enrichment Strategies

If students need help. . .	If students got it. . .
Resources by Chapter • Practice A and Practice B • Puzzle Time Record and Practice Journal Practice Differentiating the Lesson Lesson Tutorials Skills Review Handbook	Resources by Chapter • Enrichment and Extension • Technology Connection Start the next section

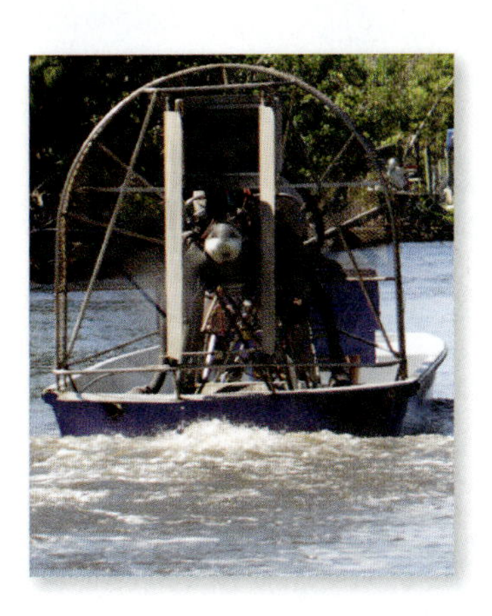

35. AIRBOAT TOURS You want to take a two-hour airboat tour.

a. Write a function that represents the cost G of a tour at Gator Tours.

b. Write a function that represents the cost S of a tour at Snake Tours.

c. Which is a better deal? Explain.

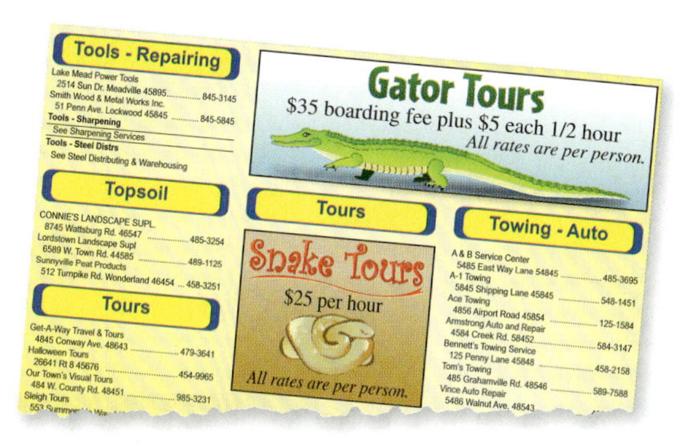

36. REASONING The graph of a function is a line that goes through the points (3, 2), (5, 8), and (8, y). What is the value of y?

37. CRITICAL THINKING Make a table where the independent variable is the side length of a square and the dependent variable is the *perimeter*. Make a second table where the independent variable is the side length of a square and the dependent variable is the *area*. Graph both functions in the same coordinate plane. Compare the functions and graphs.

38. Puzzle The blocks that form the diagonals of each square are shaded. Each block is one square unit. Find the "green area" of Square 20. Find the "green area" of Square 21. Explain your reasoning.

Square 1

Square 2

Square 3

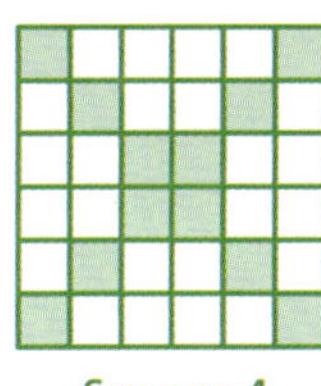
Square 4

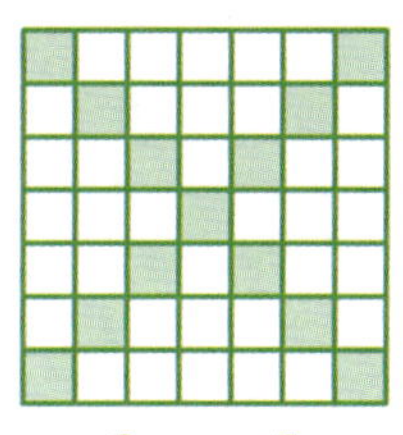

Square 5

Fair Game Review What you learned in previous grades & lessons

Find the slope of the line. *(Section 4.2)*

39.

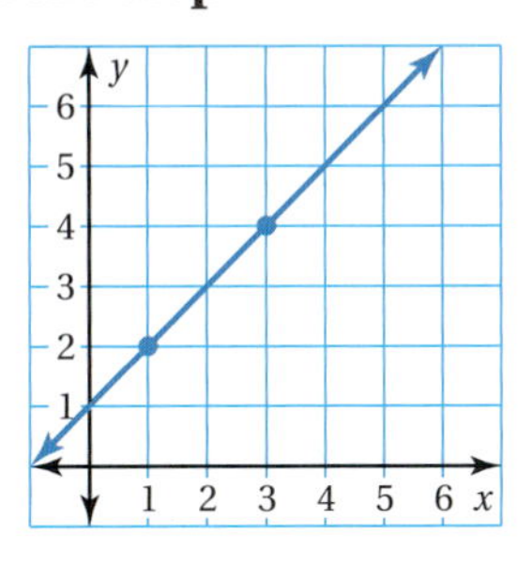

40.

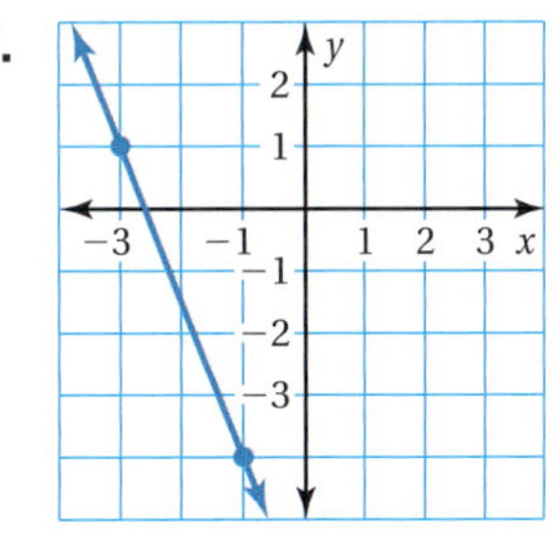

41.

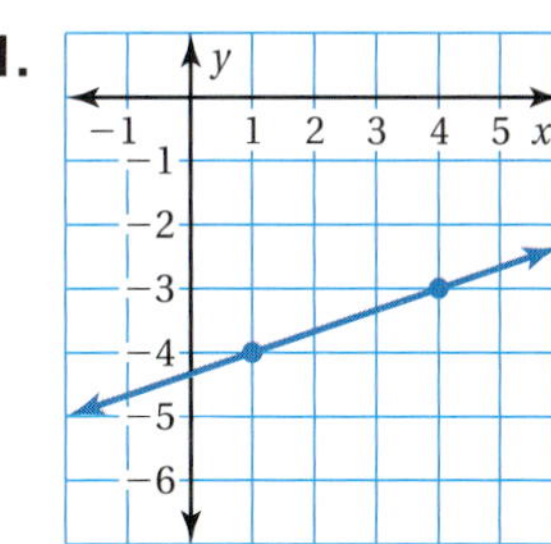

42. MULTIPLE CHOICE You want to volunteer for at most 20 hours each month. So far, you have volunteered for 7 hours this month. Which inequality represents the number of hours h you can volunteer for the rest of this month? *(Skills Review Handbook)*

Ⓐ $h \geq 13$ Ⓑ $h \geq 27$ Ⓒ $h \leq 13$ Ⓓ $h < 27$

6.3 Linear Functions

Essential Question How can you use a function to describe a linear pattern?

1 ACTIVITY: Finding Linear Patterns

Work with a partner.

- **Plot the points from the table in a coordinate plane.**
- **Write a linear equation for the function represented by the graph.**

a.

x	0	2	4	6	8
y	150	125	100	75	50

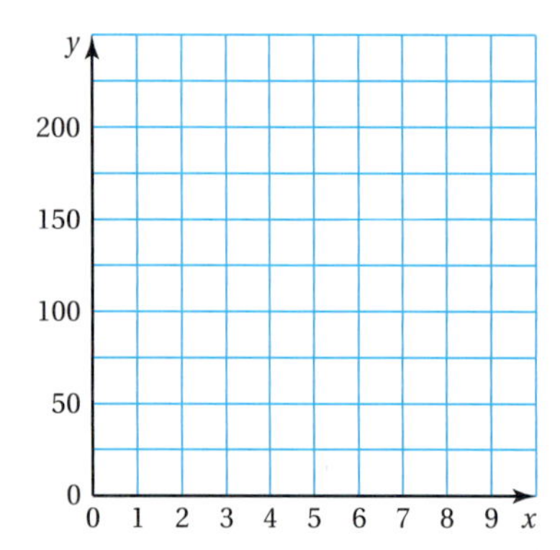

b.

x	4	6	8	10	12
y	15	20	25	30	35

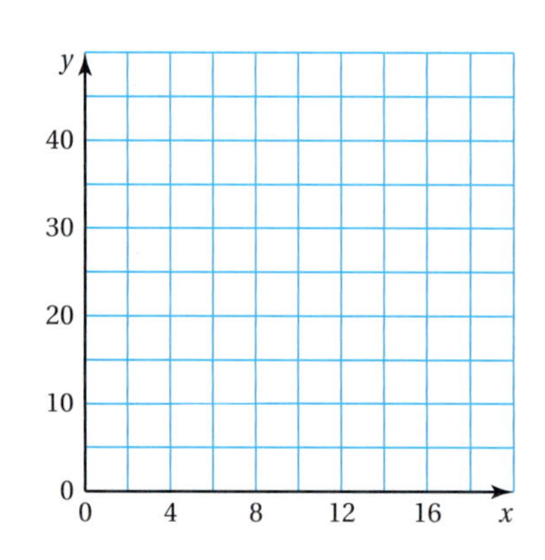

c.

x	−4	−2	0	2	4
y	4	6	8	10	12

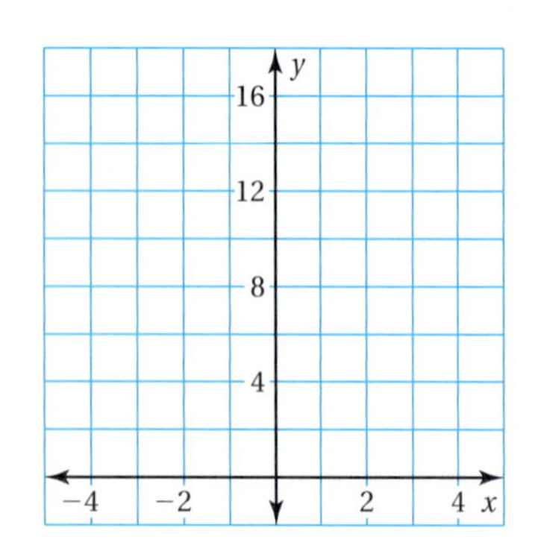

d.

x	−4	−2	0	2	4
y	1	0	−1	−2	−3

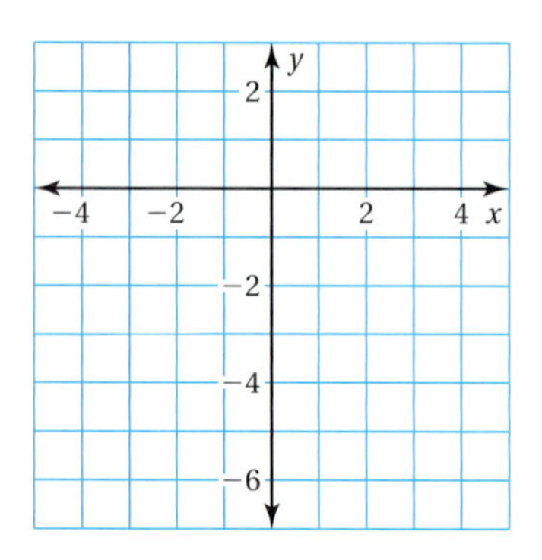

Functions

In this lesson, you will

- understand that the equation $y = mx + b$ defines a linear function.
- write linear functions using graphs or tables.
- compare linear functions.

Learning Standards
MACC.8.F.1.2
MACC.8.F.1.3
MACC.8.F.2.4

Laurie's Notes

Introduction

Standards for Mathematical Practice

- **MP8 Look for and Express Regularity in Repeated Reasoning:** The goal is for students to recognize that a linear pattern occurs when there is a constant rate of change in a table of values or in a graph. In Chapter 4, students wrote linear equations in slope-intercept form. What is new is the language—the linear equation is referred to as a function. Students will recognize a pattern (constant rate of change) in the data and write the function.

Motivate

- Do a quick matching game with students. Have 4–5 graphs on the board with slopes and y-intercepts that are different enough so that students can distinguish between them. Write the equations in a list. Have students work with partners to match the correct equation with each graph.
- Make sure that students are still focusing on key information from the graph. Is it increasing or decreasing from left to right? Is the slope steeper than 1 or close to 0? Is the y-intercept positive or negative?

Activity Notes

Activity 1

- ? "What do you notice about the scaling on the axes for each problem?" Answers will vary. Students should recognize that the x- and y-axes are scaled differently in parts (a)–(c).
- ? "For each problem, you are asked to write a linear equation for the function. How will you do this?" Find the slope and y-intercept.
- Give sufficient time for students to work through the four problems.
- From the graphs, students should be able to determine the slope. It is important that students pay attention to how the axes are scaled when they record values for rise and run.
- **MP1 Make Sense of Problems and Persevere in Solving Them** and **MP8 Look for and Express Regularity in Repeated Reasoning:** From the table of values, the y-intercept is given for 3 of the 4 problems. In part (a), the ordered pair (0, 150) gives the y-intercept, $b = 150$. To find the slope from the table, notice that every time x increases by 2, y decreases by 25. You can recognize this as a constant rate of change in which the run is 2 and the rise is -25. So, $m = \frac{\text{rise}}{\text{run}} = \frac{-25}{2} = -12.5$. Now write the equation in slope-intercept form, $y = -12.5x + 150$.
- When students have finished, check their equations.
- ? "What numeric patterns do you see in the table?" Listen for how the x- and y-values are changing.
- Make sure students recognize the connection between the numeric patterns in the table and the slope of the line.
- **FYI:** For students, recognizing a pattern in the table is the easy part. Helping students translate the pattern into a slope, and then into an equation, is the challenging part. This takes practice.

Florida Common Core Standards

MACC.8.F.1.2 Compare properties of two functions each represented in a different way (algebraically, graphically, numerically in tables, or by verbal descriptions).

MACC.8.F.1.3 Interpret the equation $y = mx + b$ as defining a linear function, whose graph is a straight line; . . ..

MACC.8.F.2.4 Construct a function to model a linear relationship between two quantities. Determine the rate of change and initial value of the function from a description of a relationship or from two (x, y) values, including reading these from a table or from a graph. Interpret the rate of change and initial value of a linear function in terms of the situation it models, and in terms of its graph or a table of values.

Previous Learning

Students should know how to write a linear equation in slope-intercept form. Students should know common geometric formulas, such as area and perimeter.

Technology for the Teacher

Lesson Plans
Complete Materials List

6.3 Record and Practice Journal

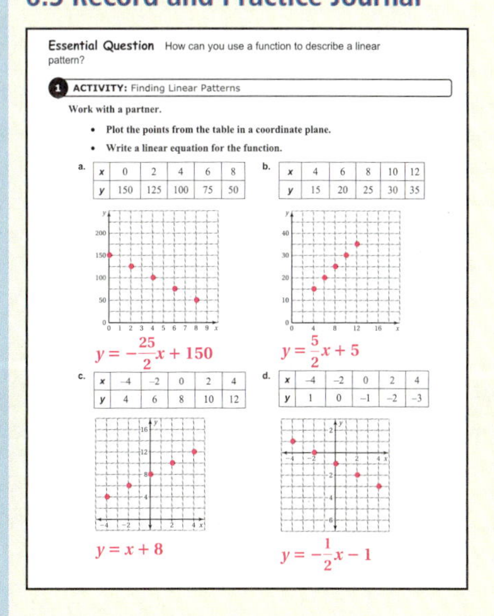

Essential Question How can you use a function to describe a linear pattern?

1 ACTIVITY: Finding Linear Patterns

Work with a partner.

- Plot the points from the table in a coordinate plane.
- Write a linear equation for the function.

a.

x	0	2	4	6	8
y	150	125	100	75	50

$y = -\frac{25}{2}x + 150$

b.

x	4	6	8	10	12
y	15	20	25	30	35

$y = \frac{5}{2}x + 5$

c.

x	−4	−2	0	2	4
y	4	6	8	10	12

$y = x + 8$

d.

x	−4	−2	0	2	4
y	1	0	−1	−2	−3

$y = -\frac{1}{2}x - 1$

Differentiated Instruction

Visual

Explain to students that representing a function table as a list of ordered pairs is for convenience. Once the function is represented by ordered pairs, it can be graphed in a coordinate plane. This is a visual representation of the function and is an excellent way to show students the connection between algebra and geometry.

6.3 Record and Practice Journal

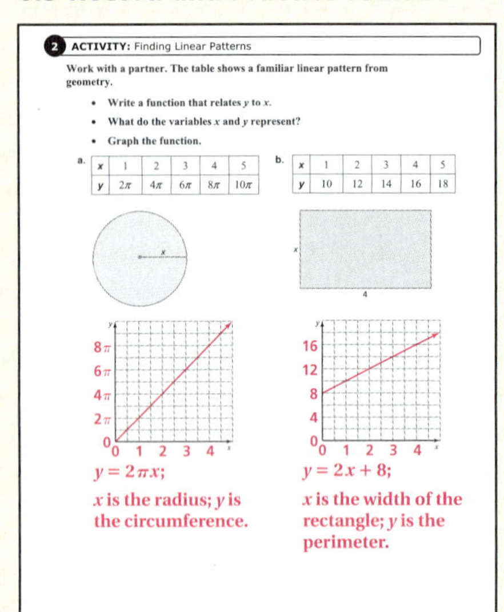

2 ACTIVITY: Finding Linear Patterns

Work with a partner. The table shows a familiar linear pattern from geometry.

- Write a function that relates y to x.
- What do the variables x and y represent?
- Graph the function.

a.

x	1	2	3	4	5
y	2π	4π	6π	8π	10π

b.

x	1	2	3	4	5
y	10	12	14	16	18

$y = 2\pi x$; x is the radius; y is the circumference.

$y = 2x + 8$; x is the width of the rectangle; y is the perimeter.

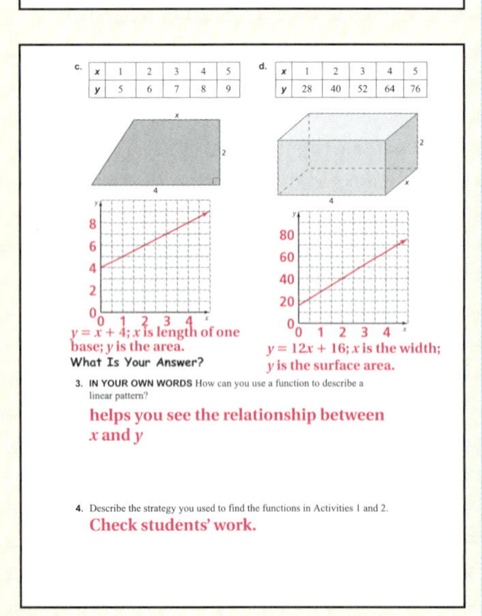

c.

x	1	2	3	4	5
y	5	6	7	8	9

d.

x	1	2	3	4	5
y	28	40	52	64	76

$y = x + 4$; x is length of one base; y is the area.

$y = 12x + 16$; x is the width; y is the surface area.

What Is Your Answer?

3. IN YOUR OWN WORDS How can you use a function to describe a linear pattern?
helps you see the relationship between x and y

4. Describe the strategy you used to find the functions in Activities 1 and 2.
Check students' work.

Laurie's Notes

Activity 2

- The challenge in these problems is that the equation relates to a geometric formula. The figure shown for each problem should provide a hint as to what the variables x and y represent in the problem.
- **Part (a):** Two formulas involving π and circles are circumference ($C = 2\pi r$) and area ($A = \pi r^2$). Substitute the value of x for the radius in each formula. The value of the circumference will match the y-values in the table.
- **Part (b):** Two formulas involving rectangles are perimeter ($P = 2\ell + 2w$) and area ($A = \ell w$). Substitute 4 for the length and the value of x for the width in each formula. The value of the perimeter will match the y-values in the table.

? "Could y represent the perimeter for part (c)? Explain." no; You only know 3 of the 4 side lengths, and the sum of the three sides you know is greater than y.

- **Part (c):** The formula for the area of a trapezoid is $A = (b + B)h \div 2$. Substitute 4 for B, 2 for h, and the value of x for the length of the shorter base. The value of the area will match the y-values in the table.
- **Part (d):** Two formulas involving a rectangular prism are surface area ($S = 2\ell w + 2wh + 2\ell h$) and volume ($V = \ell wh$). Substitute 4 for the length, 2 for the height, and the value of x for the width in each formula. The value of the surface area will match the y-values in the table.

? Extension: "In part (c), how does the diagram of the trapezoid change as the value of x increases?" When $x = 4$, the trapezoid becomes a rectangle. When $x > 4$, the upper base becomes the longer of the two bases.

- Note that in each of these problems, there is a numeric pattern in the table. Have students describe the numeric pattern. Encourage them to use language such as "as x increases by 1, y increases by 2π."

What Is Your Answer?

- **Think-Pair-Share:** Students should read each question independently and then work in pairs to answer the questions. When they have answered the questions, the pair should compare their answers with another group and discuss any discrepancies.

Closure

- **Exit Ticket:** Plot the points given in the table and write a linear equation for the function.

x	−2	0	2	4	6
y	2	3	4	5	6

$y = \frac{1}{2}x + 3$

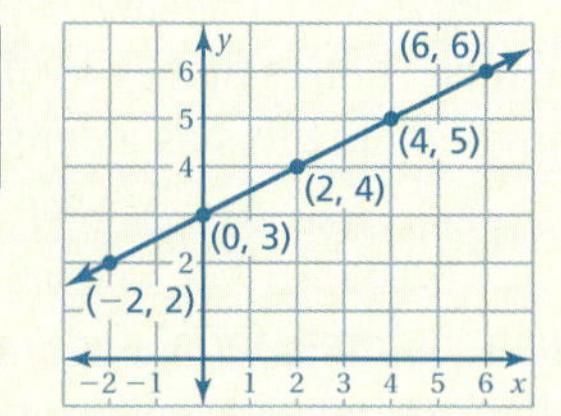

2 ACTIVITY: Finding Linear Patterns

Math Practice 6

Label Axes

How do you know what to label the axes? How does this help you accurately graph the data?

Work with a partner. The table shows a familiar linear pattern from geometry.

- **Write a function that relates *y* to *x*.**
- **What do the variables *x* and *y* represent?**
- **Graph the function.**

a.

x	1	2	3	4	5
y	2π	4π	6π	8π	10π

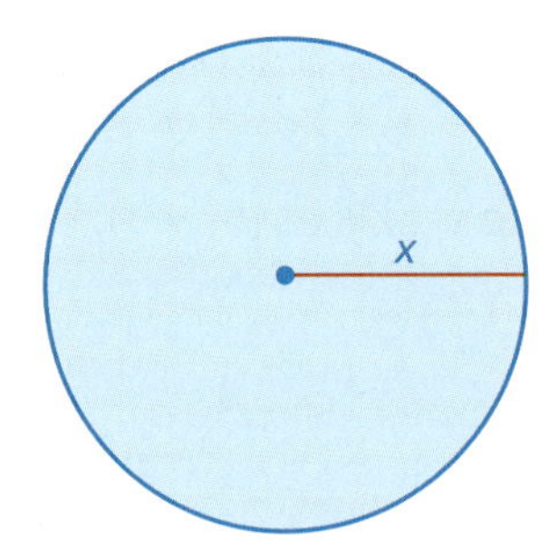

b.

x	1	2	3	4	5
y	10	12	14	16	18

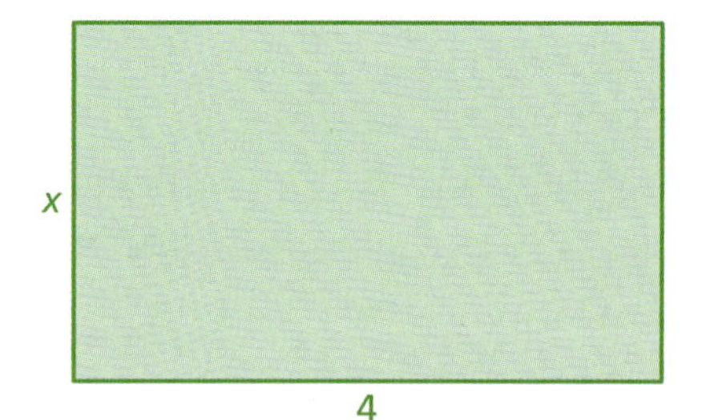

c.

x	1	2	3	4	5
y	5	6	7	8	9

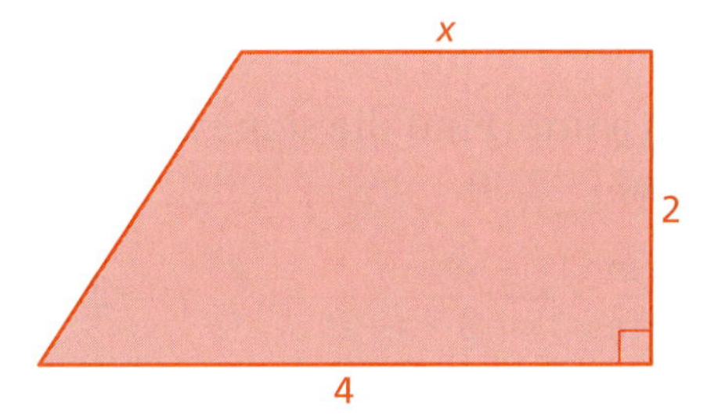

d.

x	1	2	3	4	5
y	28	40	52	64	76

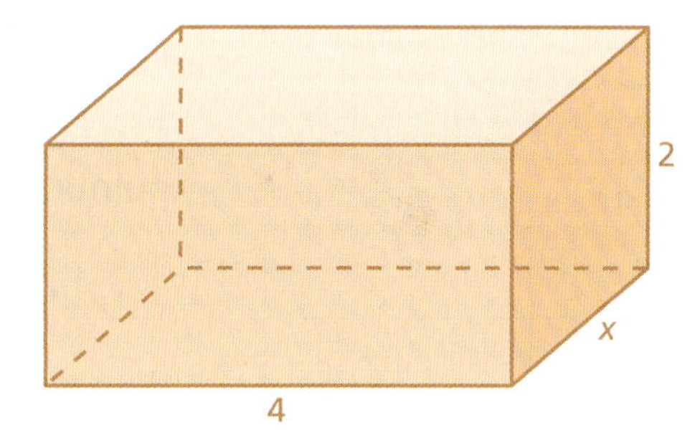

What Is Your Answer?

3. **IN YOUR OWN WORDS** How can you use a function to describe a linear pattern?

4. Describe the strategy you used to find the functions in Activities 1 and 2.

Practice Use what you learned about linear patterns to complete Exercises 3 and 4 on page 261.

6.3 Lesson

Key Vocabulary
linear function, p. 258

A **linear function** is a function whose graph is a nonvertical line. A linear function can be written in the form $y = mx + b$, where m is the slope and b is the y-intercept.

EXAMPLE 1 Writing a Linear Function Using a Graph

Use the graph to write a linear function that relates y to x.

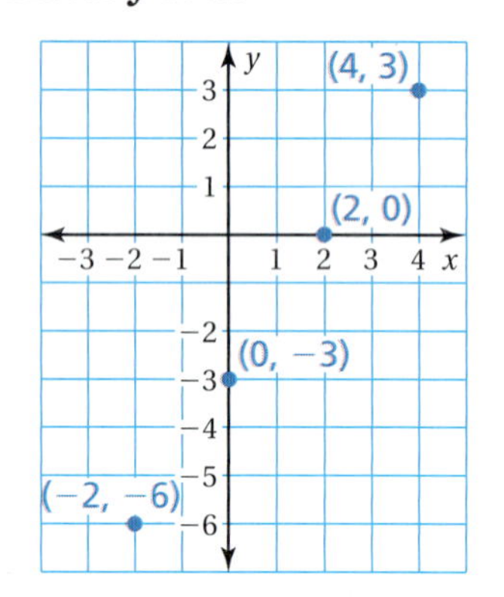

The points lie on a line. Find the slope by using the points (2, 0) and (4, 3).

$$m = \frac{\text{change in } y}{\text{change in } x} = \frac{3-0}{4-2} = \frac{3}{2}$$

Because the line crosses the y-axis at (0, −3), the y-intercept is −3.

So, the linear function is $y = \frac{3}{2}x - 3$.

EXAMPLE 2 Writing a Linear Function Using a Table

Use the table to write a linear function that relates y to x.

x	−3	−2	−1	0
y	9	7	5	3

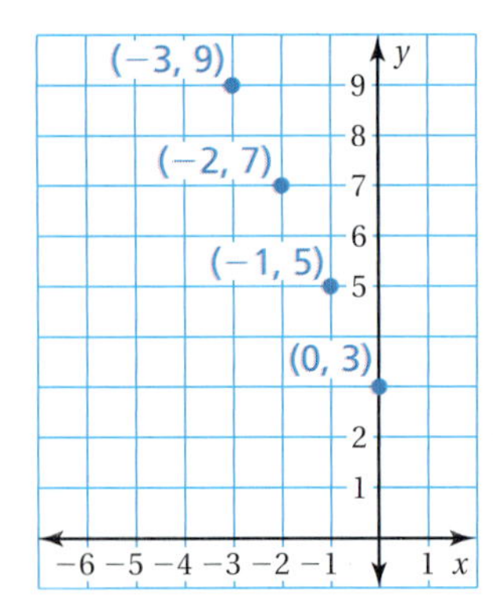

Plot the points in the table.

The points lie on a line. Find the slope by using the points (−2, 7) and (−3, 9).

$$m = \frac{\text{change in } y}{\text{change in } x} = \frac{9-7}{-3-(-2)} = \frac{2}{-1} = -2$$

Because the line crosses the y-axis at (0, 3), the y-intercept is 3.

So, the linear function is $y = -2x + 3$.

On Your Own

Now You're Ready
Exercises 5–10

Use the graph or table to write a linear function that relates y to x.

1.

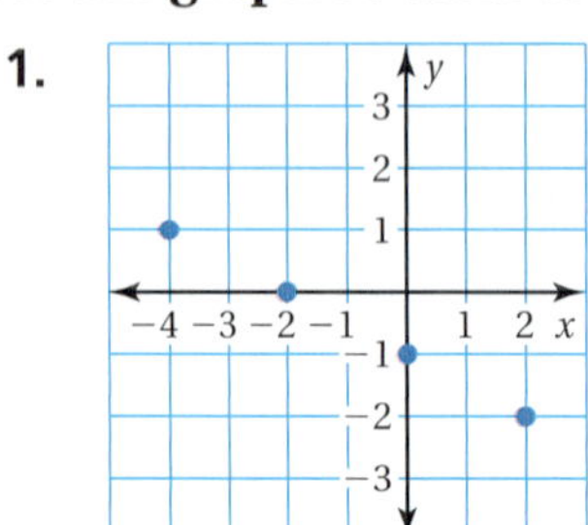

2.

x	−2	−1	0	1
y	2	2	2	2

Multi-Language Glossary at BigIdeasMath.com

Laurie's Notes

Introduction

Connect

- **Yesterday:** Students gained additional practice in writing linear equations. (MP1, MP8)
- **Today:** Students will write linear functions by recognizing patterns in graphical and tabular information.

Motivate

- Tell the story of Amos Dolbear, who in 1898 noticed that warmer crickets chirp faster. Dolbear made a detailed study of cricket chirp rates based on temperature and came up with the cricket chirping temperature formula known as Dolbear's Law. Remember that the formula is actually a linear function with a slope and y-intercept!

Lesson Notes

Example 1

- Students worked on similar problems in Chapter 4. The difference here is the function terminology.
- Review the definition of a function: a relation that pairs each input with exactly one output.
- Write the definition of a linear function.
- **Extension:** From the definition, students might guess that there are other types of functions besides linear functions. Draw a parabola or sine wave to make this connection.
- **Teaching Tip:** To find the slope, draw a right triangle with the hypotenuse between two of the points. Label the legs of the triangle to represent the rise and run. Then compute the slope.

? "Does it matter what two points you select to find the slope? Explain." no; The ratio of rise to run will be the same because the slope triangles are actually similar. It is unlikely students will say this; however, it is the case.

- **MP8 Look for and Express Regularity in Repeated Reasoning:** Demonstrate that it does not matter which two points are selected to compute the slope. The slope between (0, −3) and (2, 0) is $\frac{3}{2}$. The slope between (0, −3) and (4, 3) is $\frac{6}{4} = \frac{3}{2}$.

Example 2

? "Can you tell anything about the function without plotting the points? Explain." yes; As x increases by 1 (run), y decreases by 2 (rise). So, the function is linear and the graph has a slope of −2.

- Plot the ordered pairs and repeat the steps from Example 1.

On Your Own

- **Common Error:** Students may say the slope for Question 1 is −2 instead of $-\frac{1}{2}$. Remind students that slope is change in y over change in x, not the reverse.
- **Question 2:** Students may need to graph this function. Once graphed, they will recognize this as a horizontal line whose equation is $y = 2$.

Goal Today's lesson is writing a **linear function** from a graph or a table of values.

Technology for the Teacher

Lesson Tutorials
Lesson Plans
Answer Presentation Tool

Extra Example 1

Use the graph to write a linear function that relates y to x.

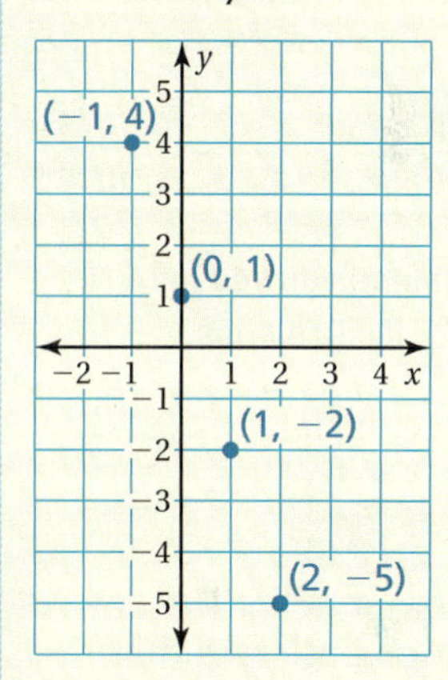

$y = -3x + 1$

Extra Example 2

Use the table to write a linear function that relates y to x.

x	−2	0	2	4
y	−2	−1	0	1

$y = \frac{1}{2}x - 1$

On Your Own

1. $y = -\frac{1}{2}x - 1$
2. $y = 2$

Extra Example 3

The table shows the number y of calories you burn in x hours of jogging.

Hours Jogging, x	Calories Burned, y
2	800
4	1600
6	2400
8	3200

a. Write a linear function that relates y to x. Interpret the slope and the y-intercept.
$y = 400x$; The slope indicates that you burn 400 calories per hour jogging. The y-intercept indicates that at the moment you begin to jog, you have not burned any calories yet.

b. Graph the linear function.

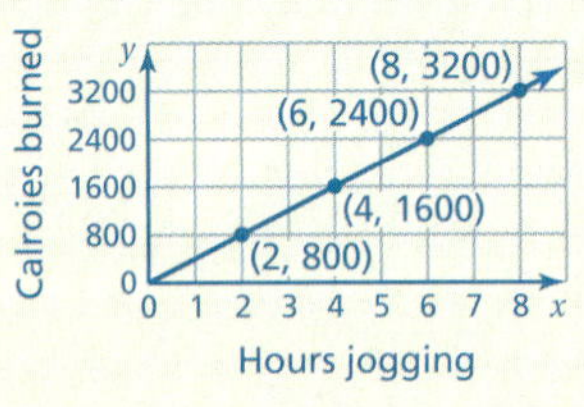

c. How many calories do you burn in 150 minutes? 1000 calories

On Your Own

3. a. $y = -x + 65$; The slope indicates that the height decreases 1000 feet per minute. The y-intercept indicates that the descent begins at a cruising altitude of 65,000 feet.

b.

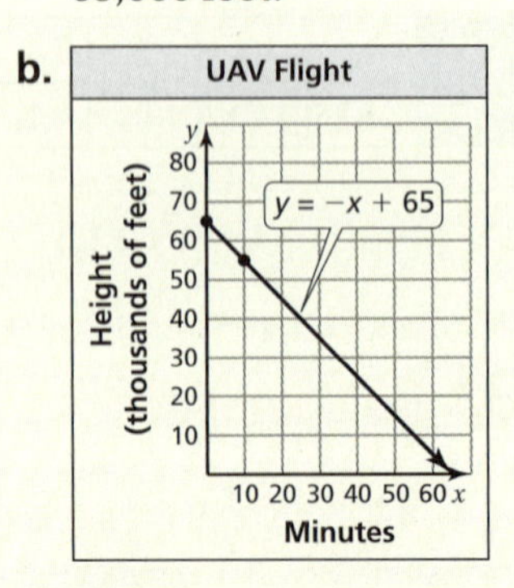

c. 5000 ft

Laurie's Notes

Example 3

- You could begin with a discussion of how altitude and height are related in aviation.
- Read the problem and write the table of values.
- ? Ask questions to check for understanding.
  - "Does the table of values represent a function? Explain." yes; Each input is associated with exactly one output.
  - "What is the independent variable (input)?" x (the time in minutes since the UAV started to descend)
  - "What is the dependent variable (output)? Explain." y (the height in thousands of feet of the UAV), is the dependent variable, because it depends on how many minutes x have passed.
  - "Is the function linear? How do you know?" yes; The table shows that as x increases by 10, y decreases by 5.
  - "Is the slope positive or negative? How do you know?" negative; As x increases, y decreases.
- Compute the slope using two ordered pairs.
- Because (0, 65) is in the data set, you know the y-intercept. You could use the slope and y-intercept to graph the function rather than simply plotting points from the table.
- ? "Why is the graph confined to Quadrant I?" In the context of the problem, x (time) cannot be negative, and y (height) cannot be negative.
- Note that when a real-life problem makes sense only for nonnegative values of x like this, the initial value is the y-intercept.
- Students should understand that once a function has been written, you can evaluate the function for values of x that are not in the table of values.
- ? "Does the function make sense for fractional values of x and y? Describe all the possible values of y for the function." (Assume you stop the descent after 1 hour.) yes; all real numbers between 0 and 35.

On Your Own

- **MP2 Reason Abstractly and Quantitatively:** Students should reason that by doubling the rate of descent, the amount of time it takes the UAV to reach the ground is cut in half.

EXAMPLE 3 Real-Life Application

Minutes, x	Height (thousands of feet), y
0	65
10	60
20	55
30	50

You are controlling an unmanned aerial vehicle (UAV) for surveillance. The table shows the height y (in thousands of feet) of the UAV x minutes after you start its descent from cruising altitude.

a. Write a linear function that relates y to x. Interpret the slope and the y-intercept.

You can write a linear function that relates the dependent variable y to the independent variable x because the table shows a constant rate of change. Find the slope by using the points (0, 65) and (10, 60).

$$m = \frac{\text{change in } y}{\text{change in } x} = \frac{60 - 65}{10 - 0} = \frac{-5}{10} = -0.5$$

Because the line crosses the y-axis at (0, 65), the y-intercept is 65.

Common Error

Make sure you consider the units when interpreting the slope and the y-intercept.

So, the linear function is $y = -0.5x + 65$. The slope indicates that the height decreases 500 feet per minute. The y-intercept indicates that the descent begins at a cruising altitude of 65,000 feet.

b. Graph the linear function.

Plot the points in the table and draw a line through the points.

Because time cannot be negative in this context, use only positive values of x.

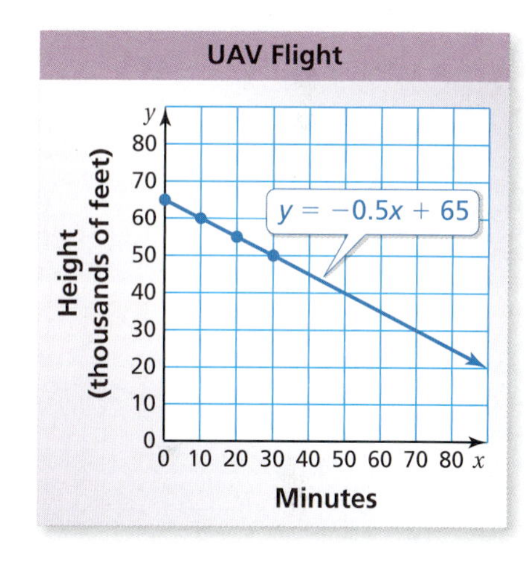

c. Find the height of the UAV when you stop the descent after 1 hour.

Because 1 hour = 60 minutes, find the value of y when $x = 60$.

$y = -0.5x + 65$ — Write the equation.

$= -0.5(60) + 65$ — Substitute 60 for x.

$= 35$ — Simplify.

So, the descent of the UAV stops at a height of 35,000 feet.

On Your Own

3. **WHAT IF?** You double the rate of descent. Repeat parts (a)–(c).

EXAMPLE 4 Comparing Linear Functions

The earnings y (in dollars) of a nighttime employee working x hours are represented by the linear function $y = 7.5x + 30$. The table shows the earnings of a daytime employee.

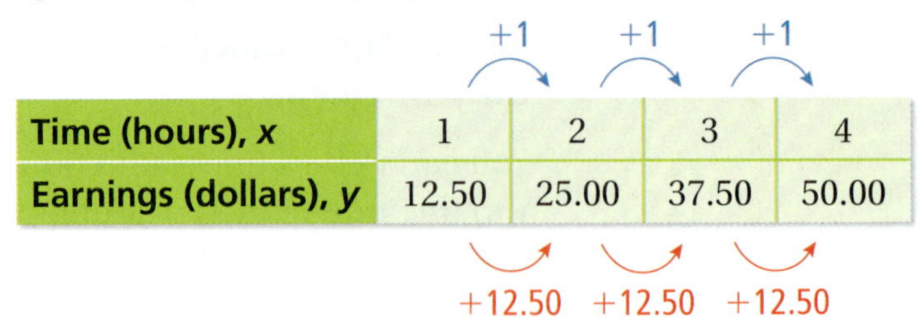

Time (hours), x	1	2	3	4
Earnings (dollars), y	12.50	25.00	37.50	50.00

a. Which employee has a higher hourly wage?

Nighttime Employee

$y = 7.5x + 30$

The nighttime employee earns \$7.50 per hour.

Daytime Employee

$\frac{\text{change in earnings}}{\text{change in time}} = \frac{\$12.50}{1 \text{ hour}}$

The daytime employee earns \$12.50 per hour.

So, the daytime employee has a higher hourly wage.

b. Write a linear function that relates the daytime employee's earnings to the number of hours worked. In the same coordinate plane, graph the linear functions that represent the earnings of the two employees. Interpret the graphs.

Use a verbal model to write a linear function that represents the earnings of the daytime employee.

$$\text{Earnings} = \text{Hourly wage} \cdot \text{Hours worked}$$

$$y = 12.5x$$

The graph shows that the daytime employee has a higher hourly wage but does not earn more money than the nighttime employee until each person has worked more than 6 hours.

On Your Own

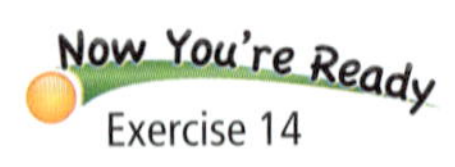

4. Manager A earns \$15 per hour and receives a \$50 bonus. The graph shows the earnings of Manager B.

   a. Which manager has a higher hourly wage?

   b. After how many hours does Manager B earn more money than Manager A?

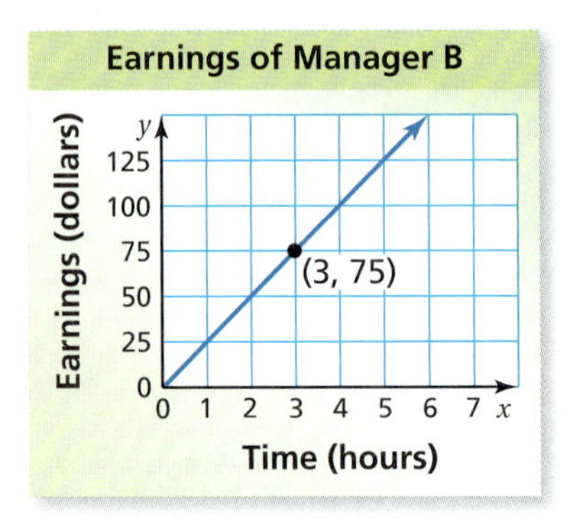

Laurie's Notes

Example 4

- Notice that the earnings information for a nighttime employee is given as an equation and the earnings information for a daytime employee is given in a table of values.
- ? "Are the earnings and number of hours worked proportional for either employee? Explain." yes; The daytime employee's earnings and number of hours worked are proportional because they can be related by a direct variation equation.
- The nighttime employee's earnings and hours worked are not proportional because they cannot be related by a direct variation equation.
- Discuss why both employees' earnings represent functions.
- **FYI:** Students may point out that the nighttime employee earns $30 for working 0 hours. This may seem strange, but employees working third shift sometimes earn a small bonus for working that particular shift.
- Part (b) reviews the important skill of writing a verbal model before trying to write the equation. Work through part (b) as shown.
- Point to different ordered pairs on each line and ask students to interpret the meaning of the ordered pair in the context of the problem. For example, the ordered pair (2, 45) on the blue line means that a nighttime employee earns $45 after 2 hours of work.

On Your Own

- **Common Error:** Students may incorrectly think that Manager A's bonus is actually the hourly wage.

Closure

- **Exit Ticket:** Write the table of values on the board and ask students to write the equation that relates the temperature to the number of cricket chirps. Acknowledge that this is an approximation and not every cricket will chirp exactly the same.

Chirps per minute	0	16	32	48	64
Temperature (°F)	40	44	48	52	56

$T = 0.25x + 40$ (Have students check this equation with the one they wrote for Exercise 21 in Section 4.7. It is the same equation.)

Extra Example 4

Your earnings y (in dollars) for working x hours are represented by the function $y = 6x + 12$. The table shows the earnings of your friend.

Time (hours)	1	2	3	4
Earnings ($)	9	18	27	36

a. Who has a higher hourly wage? your friend

b. Write a function that relates your friend's earnings to the number of hours worked. Graph both functions. Interpret the graphs. $y = 9x$

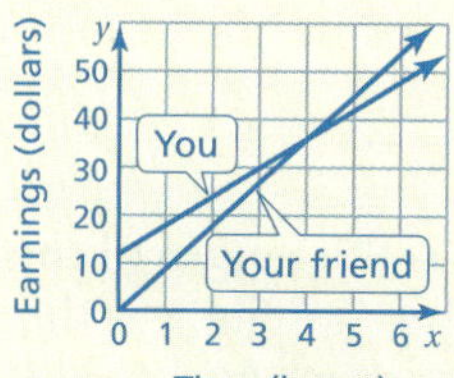

Your friend has a higher hourly wage, but does not earn more money than you until you both work for more than 4 hours.

On Your Own

4. **a.** Manager B
   b. after 5 hours

English Language Learners

Classroom

This chapter gives English learners a chance to share with the rest of the class and the opportunity to build their confidence. Many examples and exercises use tables and graphs giving English learners a rest from interpreting sentences.

Vocabulary and Concept Check

1. yes; The graph of $y = mx$ is a nonvertical line, so it is a linear function.
2. The vertical line has more than one output for the input $x = 3$.

Practice and Problem Solving

3. $y = \pi x$; x is the diameter; y is the circumference.

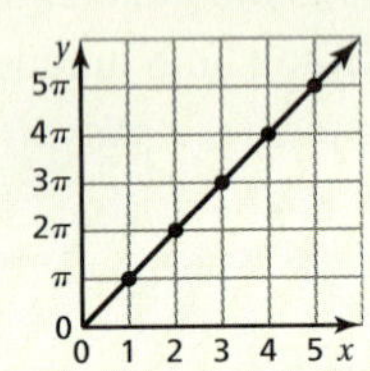

4. $y = 2x$; x is the length of the base of the triangle; y is the area of the triangle.

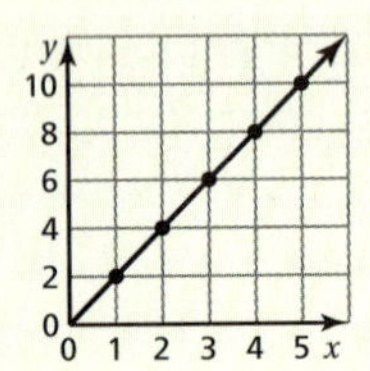

5. $y = \frac{4}{3}x + 2$
6. $y = -4x - 2$
7. $y = 3$
8. $y = 2x$
9. $y = -\frac{1}{4}x$
10. $y = \frac{2}{3}x + 5$
11. **a.** independent variable: x; dependent variable: y

    b. $y = 3x$; It costs \$3 to rent one movie.

    c.

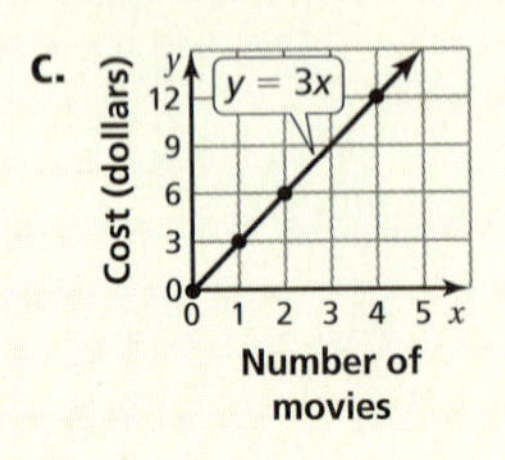

    d. \$9

Assignment Guide and Homework Check

Level	Day 1 Activity Assignment	Day 2 Lesson Assignment	Homework Check
Basic	3, 4, 20–23	1, 2, 5–17 odd	5, 9, 11, 15
Average	3, 4, 20–23	1, 2, 5–11 odd, 12–18 even	7, 9, 12, 14, 16
Advanced	1–4, 6–12 even, 13–23		8, 10, 14, 15, 18

Common Errors

- **Exercises 5 and 6** Students may find the wrong slope because they may misread the scale on an axis. Encourage them to label the points and to use the points they know to write the slope.
- **Exercise 7** Students may not remember how to write the equation for a horizontal line. They may write $x = 3$ instead of $y = 3$. Encourage them to think about the slope-intercept form of an equation.
- **Exercises 8–10** Students may write the reciprocal of the slope when writing the equation from the table. Encourage them to substitute a point into the equation and check to make sure that the equation is true for that point.

6.3 Record and Practice Journal

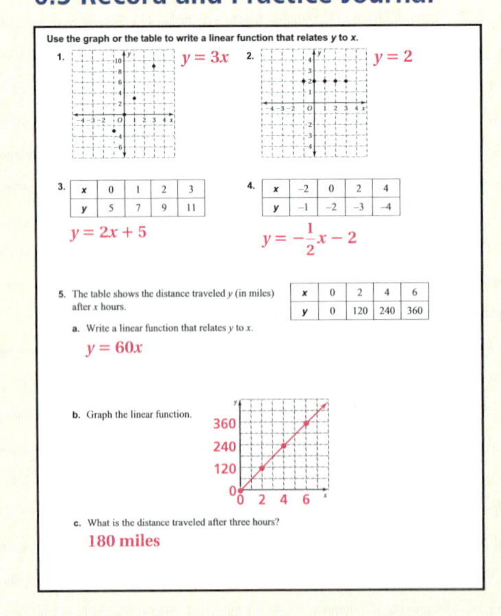

Use the graph or the table to write a linear function that relates y to x.

1. $y = 3x$
2. $y = 2$
3.

x	0	1	2	3
y	5	7	9	11

$y = 2x + 5$

4.

x	−2	0	2	4
y	−1	−2	−3	−4

$y = -\frac{1}{2}x - 2$

5. The table shows the distance traveled y (in miles) after x hours.

x	0	2	4	6
y	0	120	240	360

a. Write a linear function that relates y to x.

$y = 60x$

b. Graph the linear function.

c. What is the distance traveled after three hours?

180 miles

6.3 Exercises

Vocabulary and Concept Check

1. **STRUCTURE** Is $y = mx + b$ a linear function when $b = 0$? Explain.

2. **WRITING** Explain why the vertical line does not represent a linear function.

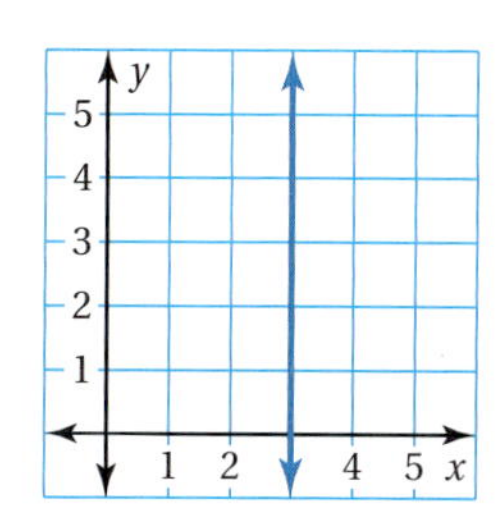

Practice and Problem Solving

The table shows a familiar linear pattern from geometry. Write a function that relates *y* to *x*. What do the variables *x* and *y* represent? Graph the function.

3.

x	1	2	3	4	5
y	π	2π	3π	4π	5π

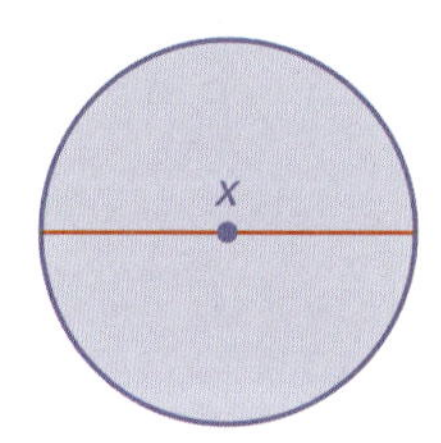

4.

x	1	2	3	4	5
y	2	4	6	8	10

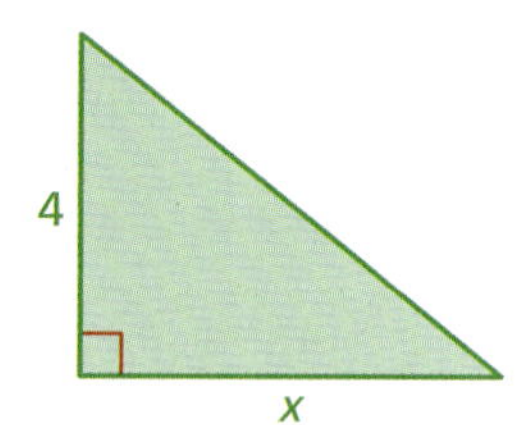

Use the graph or table to write a linear function that relates *y* to *x*.

5.

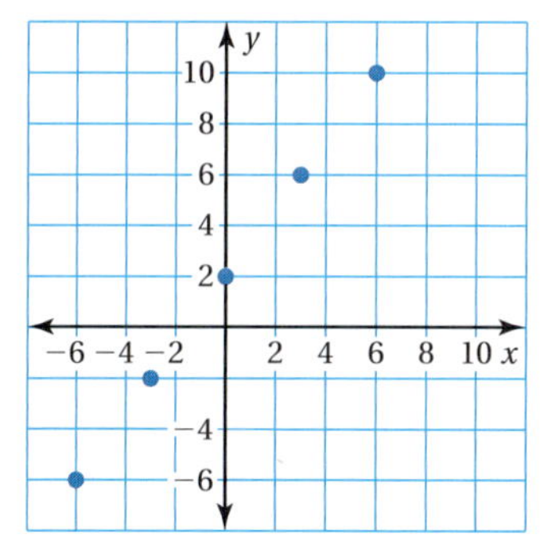

6.

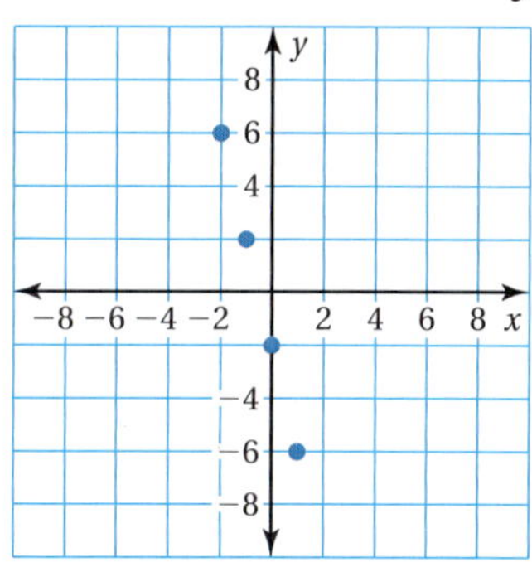

7.

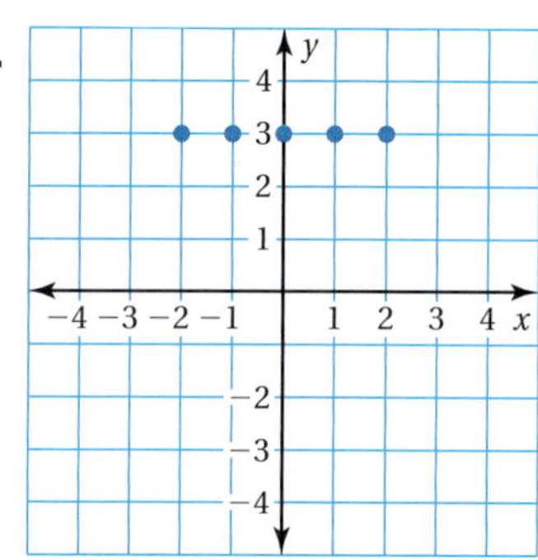

8.

x	−2	−1	0	1
y	−4	−2	0	2

9.

x	−8	−4	0	4
y	2	1	0	−1

10.

x	−3	0	3	6
y	3	5	7	9

3 11. **MOVIES** The table shows the cost *y* (in dollars) of renting *x* movies.

a. Which variable is independent? dependent?

b. Write a linear function that relates *y* to *x*. Interpret the slope.

c. Graph the linear function.

d. How much does it cost to rent three movies?

Number of Movies, *x*	0	1	2	4
Cost, *y*	0	3	6	12

12. **BIKE JUMPS** A *bunny hop* is a bike trick in which the rider brings both tires off the ground without using a ramp. The table shows the height y (in inches) of a bunny hop on a bike that weighs x pounds.

Weight (pounds), x	19	21	23
Height (inches), y	10.2	9.8	9.4

a. Write a linear function that relates the height of a bunny hop to the weight of the bike.

b. Graph the linear function.

c. What is the height of a bunny hop on a bike that weighs 21.5 pounds?

13. **BATTERY** The graph shows the percent y (in decimal form) of battery power remaining x hours after you turn on a laptop computer.

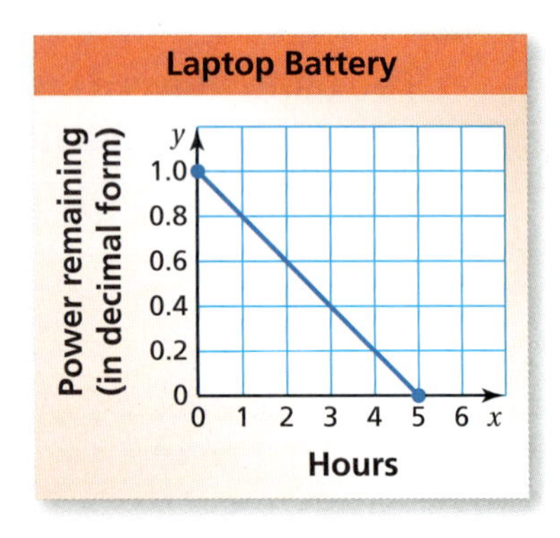

a. Write a linear function that relates y to x.

b. Interpret the slope, the x-intercept, and the y-intercept.

c. After how many hours is the battery power at 75%?

4 14. **RACE** You and a friend race each other. You give your friend a 50-foot head start. The distance y (in feet) your friend runs after x seconds is represented by the linear function $y = 14x + 50$. The table shows the distances you run.

Time (seconds), x	2	4	6	8
Distance (feet), y	38	76	114	152

a. Who runs at a faster rate? What is that rate?

b. Write a linear function that relates your distance to the number of seconds. In the same coordinate plane, graph the linear functions that represent the distances of you and your friend.

c. For what distances will you win the race? Explain.

15. **CALORIES** The number of calories burned y after x minutes of kayaking is represented by the linear function $y = 4.5x$. The graph shows the calories burned by hiking.

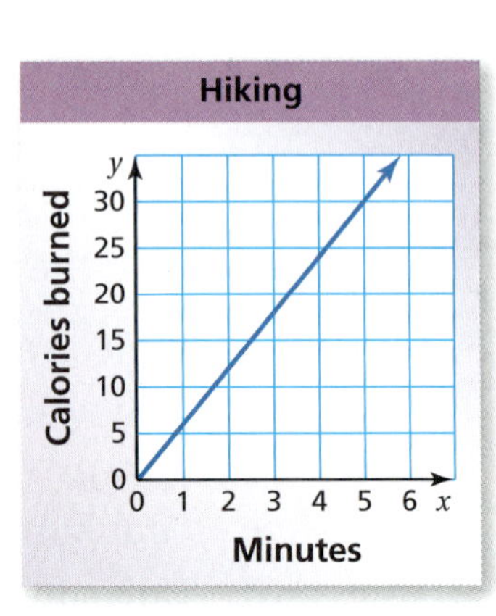

a. Which activity burns more calories per minute?

b. How many more calories are burned by doing the activity in part (a) than the other activity for 45 minutes?

Common Errors (continued)

- **Exercises 12 and 13** Students may incorrectly write an equation of the form $y = mx + b$ with a positive value of m. Remind students that in a situation where y decreases as x increases, the slope is negative.
- **Exercise 14** Students may incorrectly add 50 to the rate and conclude that the friend runs at a faster rate. Remind them that in $y = mx + b$, the slope m represents the rate of change.

12. **a.** $y = -0.2x + 14$

b.

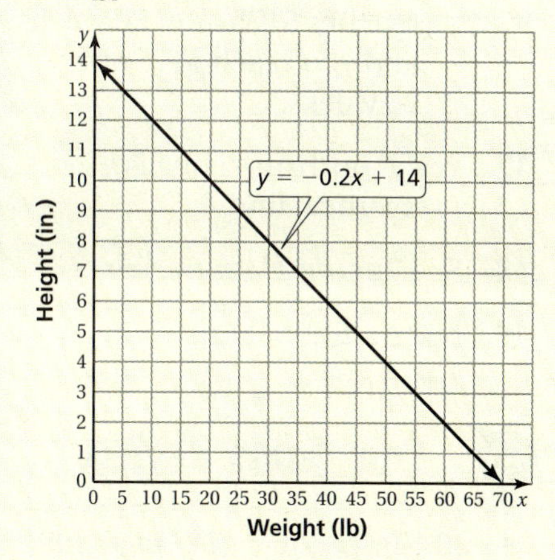

c. 9.7 in.

13. **a.** $y = -0.2x + 1$

b. The slope indicates that the power decreases by 20% per hour. The x-intercept indicates that the battery lasts 5 hours. The y-intercept indicates that the battery power is at 100% when you turn on the laptop.

c. 1.25 hours

14. See Additional Answers.

15. **a.** hiking

b. 67.5 calories

English Language Learners

Vocabulary

Make sure students understand that the graph of a *linear* equation is a *line*. Only two points are needed to graph a line, but if one of the points is incorrect the wrong line will be graphed. Plotting three points for a line in the coordinate plane and making sure that the points form a line provides students with a check when graphing.

Practice and Problem Solving

16. **a.** you; your friend

b. your friend; Your friend reaches $175 after 20 weeks and you reach $175 after 26 weeks.

17. yes; A horizontal line is a nonvertical line.

18. See *Taking Math Deeper*.

19. See Additional Answers.

Fair Game Review

20. $b = -2.6$ **21.** $w = 1.5$

22. $y = 2\frac{7}{20}$ **23.** C

Mini-Assessment

Use the graph or table to write a linear function that relates *y* to *x*.

1.

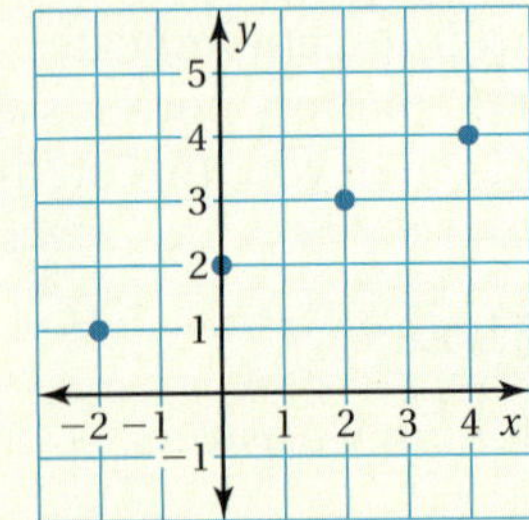

$y = \frac{1}{2}x + 2$

2.

x	−2	−1	0	1
y	9	4	−1	−6

$y = -5x - 1$

A maple tree grows 1.5 feet each year. The table shows the yearly growth for a pine tree.

Time (yr)	1	2	3	4
Growth (in.)	12	24	36	48

3. Which tree grows faster? maple

4. Write and graph equations that represent the growth rates of each tree. Compare the steepness of the graphs. What does this mean in the context of the problem? See Additional Answers.

Taking Math Deeper

Exercise 18

Students might find it interesting to discover that there is a correlation between years of education and salary. Of course, the correlation only relates annual salaries. There are many examples of people with no years of education beyond high school who have big salaries.

Graph the data. Describe the pattern.

a.

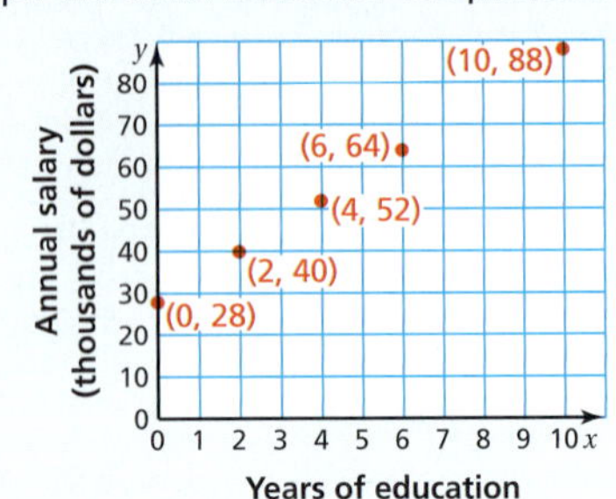

The pattern is that for every 2 years of additional education, the annual salary increases by $12,000.

Write a function.

Let x = years of education beyond high school.

Let y = annual salary.

y-intercept = 28

slope = 6

$y = 6x + 28$

Use the function.

For 8 years of education beyond high school, the annual salary is

b. $y = 6(8) + 28 = 76$, or $76,000.

For 30 years of education beyond high school, the annual salary is

c. $y = 6(30) + 28 = 208$, or $208,000.

No, the situation does not make sense. It is not realistic for a person to have 30 years of education beyond high school.

Reteaching and Enrichment Strategies

If students need help. . .	If students got it. . .
Resources by Chapter • Practice A and Practice B • Puzzle Time Record and Practice Journal Practice Differentiating the Lesson Lesson Tutorials Skills Review Handbook	Resources by Chapter • Enrichment and Extension • Technology Connection Start the next section

16. **SAVINGS** You and your friend are saving money to buy bicycles that cost \$175 each. The amount y (in dollars) you save after x weeks is represented by the equation $y = 5x + 45$. The graph shows your friend's savings.

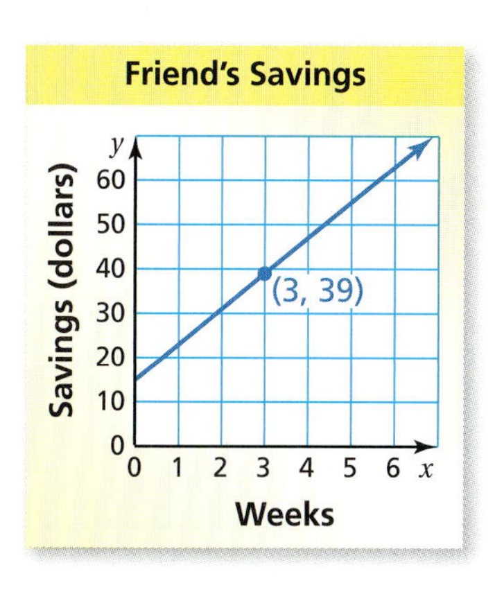

a. Who has more money to start? Who saves more per week?

b. Who can buy a bicycle first? Explain.

17. **REASONING** Can the graph of a linear function be a horizontal line? Explain your reasoning.

Years of Education, x	Annual Salary, y
0	28
2	40
4	52
6	64
10	88

18. **SALARY** The table shows a person's annual salary y (in thousands of dollars) after x years of education beyond high school.

a. Graph the data. Then describe the pattern.

b. What is the annual salary of the person after 8 years of education beyond high school?

c. Find the annual salary of a person with 30 years of education. Do you think this situation makes sense? Explain.

19. Problem Solving The Heat Index is calculated using the relative humidity and the temperature. For every 1 degree increase in the temperature from 94°F to 98°F at 75% relative humidity, the Heat Index rises 4°F.

a. On a summer day, the relative humidity is 75%, the temperature is 94°F, and the Heat Index is 122°F. Construct a table that relates the temperature t to the Heat Index H. Start the table at 94°F and end it at 98°F.

b. Identify the independent and dependent variables.

c. Write a linear function that represents this situation.

d. Estimate the Heat Index when the temperature is 100°F.

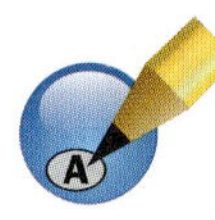

Fair Game Review What you learned in previous grades & lessons

Solve the equation. *(Section 1.1)*

20. $b - 1.6 \div 4 = -3$

21. $w + |-2.8| = 4.3$

22. $\frac{3}{4} = y - \frac{1}{5}(8)$

23. **MULTIPLE CHOICE** Which of the following describes the translation from the red figure to the blue figure? *(Section 2.2)*

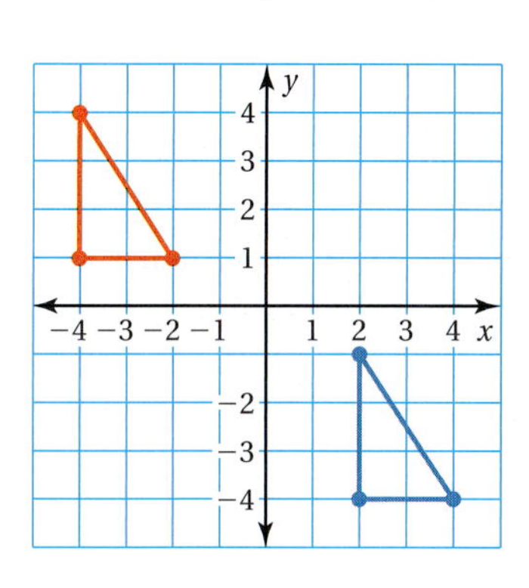

Ⓐ $(x - 6, y + 5)$ Ⓑ $(x - 5, y + 6)$

Ⓒ $(x + 6, y - 5)$ Ⓓ $(x + 5, y - 6)$

6 Study Help

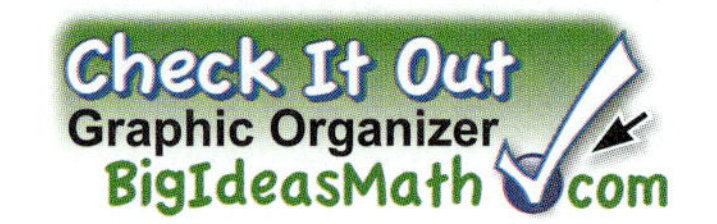

You can use a **comparison chart** to compare two topics. Here is an example of a comparison chart for relations and functions.

	Relations	Functions
Definition	A relation pairs inputs with outputs.	A relation that pairs each input with *exactly one* output is a function.
Ordered pairs	(1, 0) (3, −1) (3, 6) (7, 14)	(1, 0) (2, −1) (5, 7) (6, 20)
Mapping diagram	Input: 1, 3, 7 Output: −1, 0, 6, 14	Input: 1, 2, 5, 6 Output: −1, 0, 7, 20

On Your Own

Make comparison charts to help you study and compare these topics.

1. functions as tables and functions as graphs
2. linear functions with positive slopes and linear functions with negative slopes

After you complete this chapter, make comparison charts for the following topics.

3. linear functions and nonlinear functions
4. graphs with numerical values on the axes and graphs without numerical values on the axes

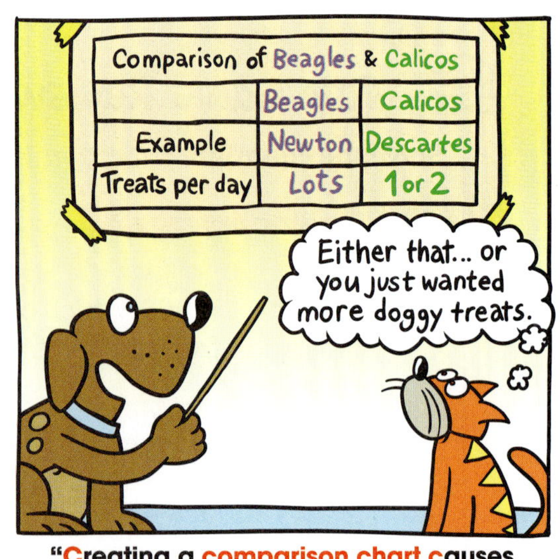

"Creating a comparison chart causes canines to crystalize concepts."

Sample Answers

1.

	Functions as tables	Functions as graphs
Description	A function can be represented by an input-output table.	A function can be represented by a graph. Make an input-output table for the function and plot the ordered pairs. Draw a line through the points to graph all of the solutions of the function.
Example: $y = x + 1$	Input, x | Output, y –1 | 0 0 | 1 1 | 2 2 | 3	(–1, 0), (0, 1), (1, 2), (2, 3)
Example: $y = \frac{1}{2}x$	Input, x | Output, y 0 | 0 1 | $\frac{1}{2}$ 2 | 1 3 | $\frac{3}{2}$	(0, 0), $(1, \frac{1}{2})$, (2, 1), $(3, \frac{3}{2})$

2.

	Linear functions with positive slopes	Linear functions with negative slopes
Algebra slope-intercept form: $y = mx + b$	*m* is positive	*m* is negative
Description	Graph is a line that rises from left to right (as x increases, y increases).	Graph is a line that falls from left to right (as x increases, y decreases).
Equations In slope-intercept form Not in slope-intercept form	$y = x - 1$ $y = 3x$ $x - 3y = 0$ $x + 3 = y - 3$	$y = -x + 1$ $y = -\frac{1}{2}x - 3$ $3x + 2y = 10$ $2x = 8 - \frac{2}{3}y$
Table	x | 0 | 1 | 2 | 3 y | –1 | 0 | 1 | 2	x | 0 | 1 | 2 | 3 y | 1 | 0 | –1 | –2
Graph		

List of Organizers

Available at *BigIdeasMath.com*

Comparison Chart
Concept Circle
Definition (Idea) and Example Chart
Example and Non-Example Chart
Formula Triangle
Four Square
Information Frame
Information Wheel
Notetaking Organizer
Process Diagram
Summary Triangle
Word Magnet
Y Chart

About this Organizer

A **Comparison Chart** can be used to compare two topics. Students list different aspects of the two topics in the left column. These can include *algebra, definition, description, equation(s), graph(s), table(s),* and *words*. Students write about or give examples illustrating these aspects in the other two columns for the topics being compared. Comparison charts are particularly useful with topics that are related but that have distinct differences. Students can place their comparison charts on note cards to use as a quick study reference.

Editable Graphic Organizer

Answers

1. (10, 0), (20, 0), (30, 10), (40, 5); yes

2. (0, −5), (0, −4), (1, −4), (2, −3), (3, −2); no

3. −30

4. −16

5. 7

6. 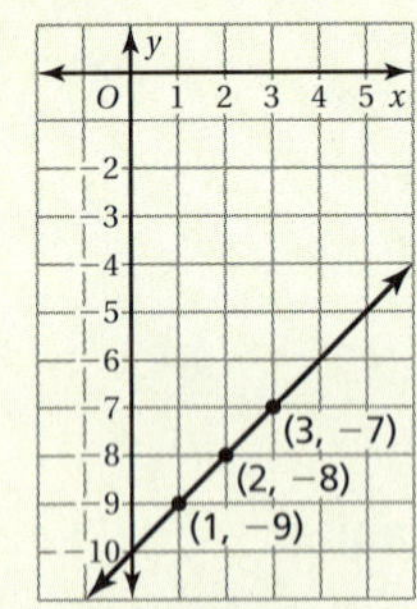

7. 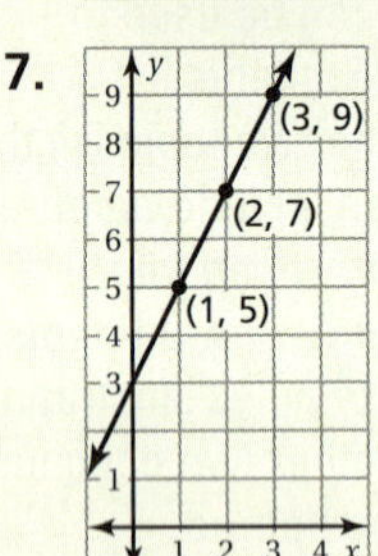

8. (2, 1), (4, 2), (6, 3)

9. $y = 2x - 4$

10. $y = \frac{2}{3}x - 1$

11. Input Output

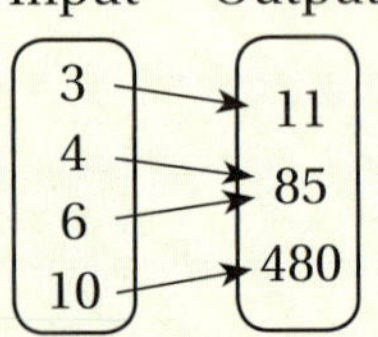

12. **a.** $C = 0.90s$

    b. $4.50

13. See Additional Answers.

Technology for the Teacher

Online Assessment
Assessment Book
ExamView® Assessment Suite

Alternative Quiz Ideas

100% Quiz	Math Log
Error Notebook	Notebook Quiz
Group Quiz	**Partner Quiz**
Homework Quiz	Pass the Paper

Partner Quiz

- Students should work in pairs. Each pair should have a small white board.
- The teacher selects certain problems from the quiz and writes one on the board.
- The pairs work together to solve the problem and write their answer on the white board.
- Students show their answers and, as a class, discuss any differences.
- Repeat for as many problems as the teacher chooses.
- For the word problems, teachers may choose to have students read them out of the book.

Reteaching and Enrichment Strategies

If students need help. . .	If students got it. . .
Resources by Chapter • Practice A and Practice B • Puzzle Time Lesson Tutorials *BigIdeasMath.com*	Resources by Chapter • Enrichment and Extension • Technology Connection Game Closet at *BigIdeasMath.com* Start the next section

6.1–6.3 Quiz

Check It Out
Progress Check
BigIdeasMath.com

List the ordered pairs shown in the mapping diagram. Then determine whether the relation is a function. *(Section 6.1)*

1. 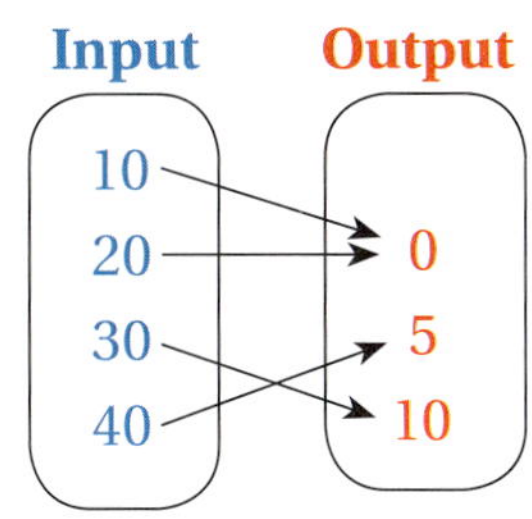

2. 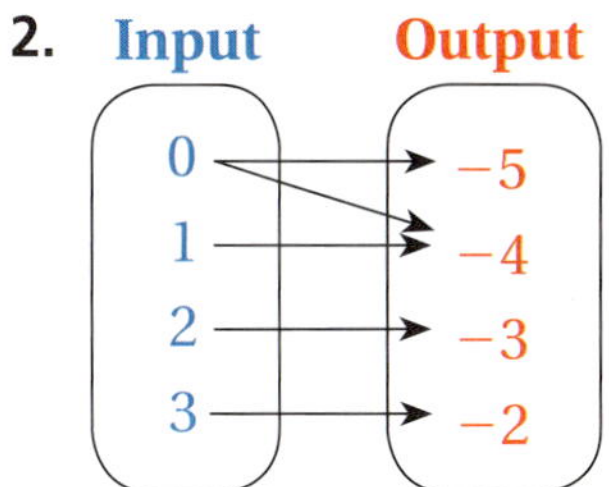

Find the value of *y* for the given value of *x*. *(Section 6.2)*

3. $y = 10x;\ x = -3$

4. $y = 6 - 2x;\ x = 11$

5. $y = 4x + 5;\ x = \frac{1}{2}$

Graph the function. *(Section 6.2)*

6. $y = x - 10$

7. $y = 2x + 3$

8. $y = \frac{x}{2}$

Use the graph or table to write a linear function that relates *y* to *x*. *(Section 6.3)*

9. 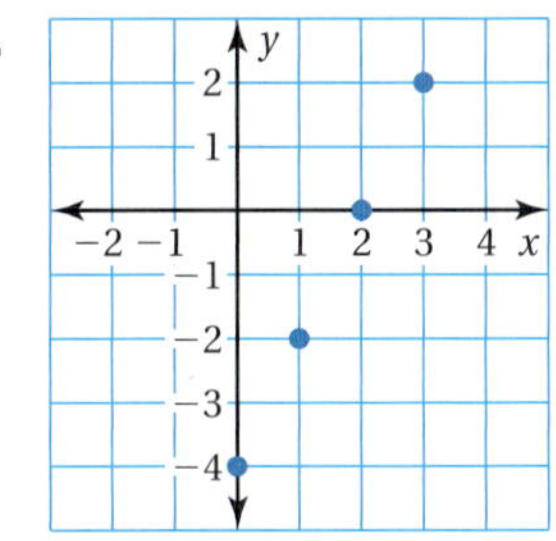

10.

x	y
−3	−3
0	−1
3	1
6	3

11. PUPPIES The table shows the ages of four puppies and their weights. Use the table to draw a mapping diagram. *(Section 6.1)*

Age (weeks)	Weight (oz)
3	11
4	85
6	85
10	480

12. MUSIC An online music store sells songs for \$0.90 each. *(Section 6.2)*

a. Write a function that you can use to find the cost C of buying s songs.

b. What is the cost of buying 5 songs?

13. ADVERTISING The table shows the revenue R (in millions of dollars) of a company when it spends A (in millions of dollars) on advertising. *(Section 6.3)*

a. Write and graph a linear function that relates the revenue to the advertising cost.

b. What is the revenue of the company when it spends \$15 million on advertising?

Advertising, *A*	Revenue, *R*
0	2
2	6
4	10
6	14
8	18

6.4 Comparing Linear and Nonlinear Functions

Essential Question How can you recognize when a pattern in real life is linear or nonlinear?

1 ACTIVITY: Finding Patterns for Similar Figures

Work with a partner. Copy and complete each table for the sequence of similar rectangles. Graph the data in each table. Decide whether each pattern is linear or nonlinear.

a. Perimeters of similar rectangles

x	1	2	3	4	5
P					

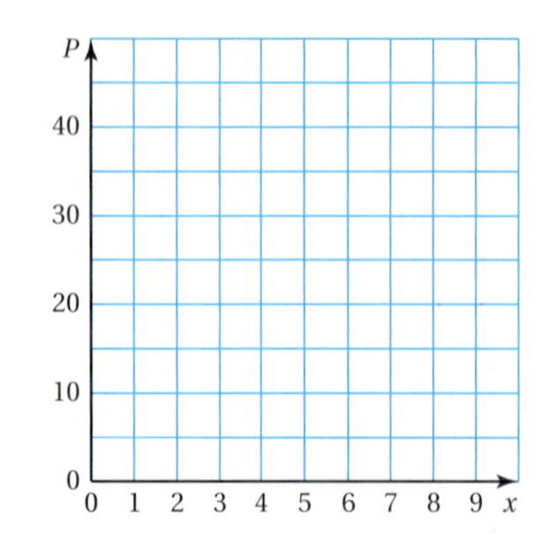

b. Areas of similar rectangles

x	1	2	3	4	5
A					

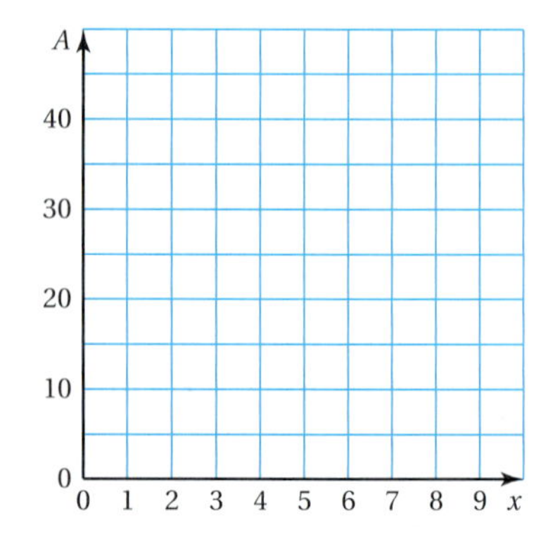

COMMON CORE

Functions

In this lesson, you will
- identify linear and nonlinear functions from tables or graphs.
- compare linear and nonlinear functions.

Learning Standard
MACC.8.F.1.3

Laurie's Notes

Introduction

Standards for Mathematical Practice

- **MP4 Model with Mathematics** and **MP8 Look for and Express Regularity in Repeated Reasoning:** The goal is for students to recognize when a pattern in real life is linear or nonlinear. Using familiar contexts—similar figures and falling objects, students will look for numeric patterns. The presence or absence of a *constant rate of change* will help students determine whether the data is linear or nonlinear.

Motivate

- ? "How many of you would like to try skydiving? Why?"
- Share with students that the first successful parachute jump made from a moving airplane was made by Captain Albert Berry in St. Louis, in 1912.
- The first parachute jump from a balloon was completed by André-Jacques Garnerin in 1797 over Monceau Park in Paris.
- Tell students that today they will explore whether the function that describes the height of a parachutist is linear or nonlinear.
- Students will study many types of nonlinear functions, such as quadratic functions, radical functions, and rational functions, in more detail later in the text.

Activity Notes

Activity 1

- ? "What does it mean for two rectangles to be similar?" Corresponding sides are proportional and corresponding angles have the same measure.
- ? "What is the relationship between the length and the width of the green rectangle?" The length is twice the width.
- ? "What is the relationship between the length and the width of the yellow rectangle? How do you know?" The length is twice the width. Because the rectangles are similar, the lengths of all the rectangles will be twice the widths.
- Explain to students that they will find the perimeter and area of each rectangle for the side lengths given, and then plot the results.
- **Teaching Tip:** It may be helpful to set up a table that includes a row for the second dimension as shown. The numeric pattern is more obvious when viewed in a table.

Width	x	1	2	3	4	5
Length	$2x$	2	4	6	8	10
Perimeter	P					

- **MP6 Attend to Precision:** Encourage students to be accurate with their graphing. Because only 5 points are being plotted for each graph, it is possible that students will not see the curvature of the area graph. Students should recognize, however, that the numeric data for area does not have a constant difference between A-values.

Florida Common Core Standards

MACC.8.F.1.3 Interpret the equation $y = mx + b$ as defining a linear function, whose graph is a straight line; give examples of functions that are not linear.

Previous Learning

Students should know common geometric formulas, such as area and perimeter.

Lesson Plans
Complete Materials List

6.4 Record and Practice Journal

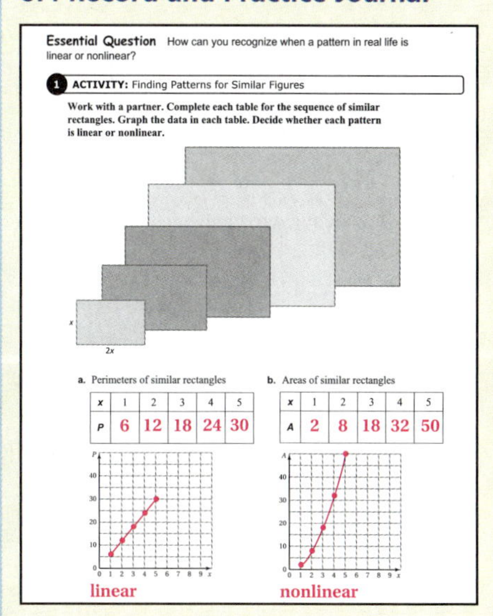

Essential Question How can you recognize when a pattern in real life is linear or nonlinear?

1 ACTIVITY: Finding Patterns for Similar Figures

Work with a partner. Complete each table for the sequence of similar rectangles. Graph the data in each table. Decide whether each pattern is linear or nonlinear.

a. Perimeters of similar rectangles

x	1	2	3	4	5
P	6	12	18	24	30

linear

b. Areas of similar rectangles

x	1	2	3	4	5
A	2	8	18	32	50

nonlinear

Differentiated Instruction

Visual

Students may be able to describe how the sequence of output numbers is changing, for example, *start with 2 and add 3*, but they may find it difficult to write a function rule for changing an input value to an output value. If students determine that the output increases or decreases by a constant value as the input increases, then the function is a linear equation that can be written in the form $y = mx + b$. Have students create function tables for equations such as $y = x + 3$, $y = 4x - 1$, and $y = 0.5x$ to see this pattern.

6.4 Record and Practice Journal

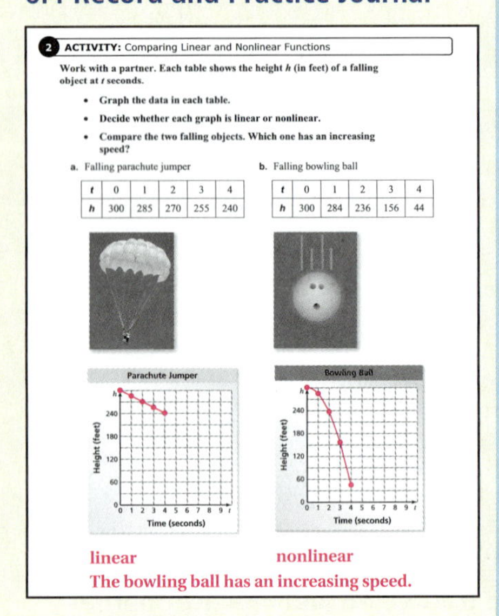

2 ACTIVITY: Comparing Linear and Nonlinear Functions

Work with a partner. Each table shows the height *h* (in feet) of a falling object at *t* seconds.

- **Graph the data in each table.**
- **Decide whether each graph is linear or nonlinear.**
- **Compare the two falling objects. Which one has an increasing speed?**

a. Falling parachute jumper

t	0	1	2	3	4
h	300	285	270	255	240

b. Falling bowling ball

t	0	1	2	3	4
h	300	284	236	156	44

linear

nonlinear

The bowling ball has an increasing speed.

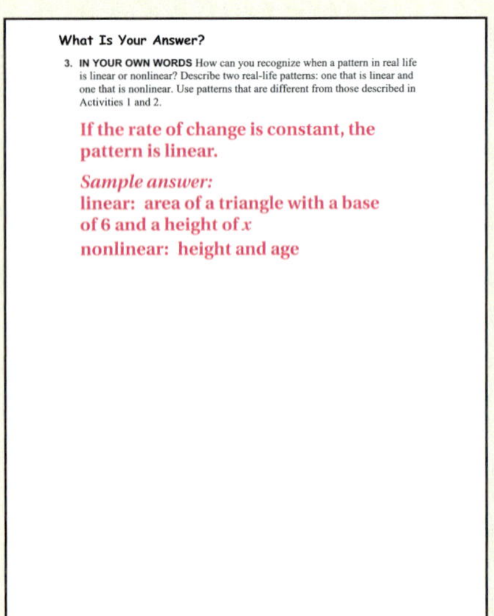

What Is Your Answer?

3. **IN YOUR OWN WORDS** How can you recognize when a pattern in real life is linear or nonlinear? Describe two real-life patterns: one that is linear and one that is nonlinear. Use patterns that are different from those described in Activities 1 and 2.

If the rate of change is constant, the pattern is linear.

Sample answer:
linear: area of a triangle with a base of 6 and a height of x
nonlinear: height and age

Laurie's Notes

Activity 2

- This activity is similar to Activity 1, except the ordered pairs are already given. Discuss the two falling objects—one with a parachute and one that is free falling.
- ? "Is there a difference in the rate at which two objects fall when one is attached to a parachute and the other is left to free fall? Explain." Listen for discussion of rate. It is unlikely students will bring up acceleration.
- You could make a small parachute using a handkerchief, tape, floss, and a small figurine to model a parachute-controlled fall and a free fall.
- **MP6:** Again, it is necessary for students to be accurate when plotting the ordered pairs given the scale on the y-axis.
- ? After students have plotted the points, ask about the two graphs. First note that the two graphs begin at the same height (y-intercept), 300 feet.
  - "How far has the jumper fallen after 4 seconds?" 60 ft "How far has the bowling ball fallen after 4 seconds?" 256 ft
  - "Describe the flight of the jumper." falling at a constant rate of 15 ft/sec
  - "Describe the flight of the bowling ball." Listen for students to describe that the bowling ball is picking up speed as it falls.
- **Extension:** Students could write a linear equation for the jumper, but not the bowling ball.

What Is Your Answer?

- Students may need help thinking of real-life patterns that are nonlinear. You might suggest area or volume relationships, or even simple story graphs about time and distance.

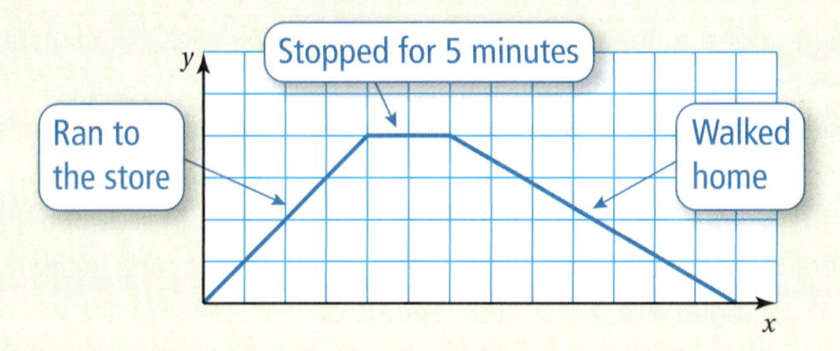

Closure

- Draw two functions with a domain of $x \geq 0$. Have one that is linear and one that is nonlinear. Describe how the graphs are alike and how they are different. Answers will vary.

2 ACTIVITY: Comparing Linear and Nonlinear Functions

Work with a partner. Each table shows the height *h* (in feet) of a falling object at *t* seconds.

- **Graph the data in each table.**
- **Decide whether each graph is linear or nonlinear.**
- **Compare the two falling objects. Which one has an increasing speed?**

a. Falling parachute jumper

t	0	1	2	3	4
h	300	285	270	255	240

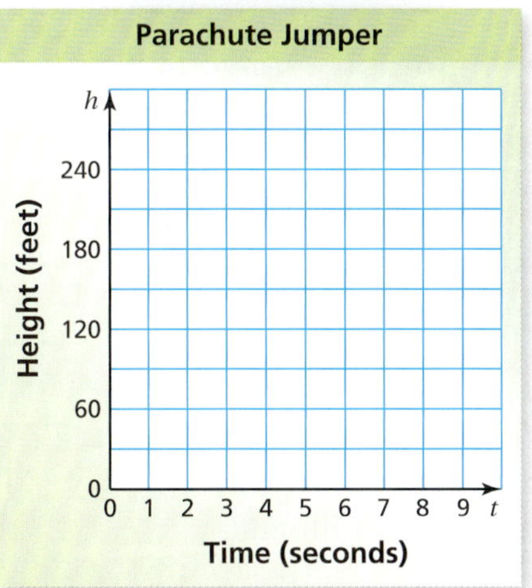

b. Falling bowling ball

t	0	1	2	3	4
h	300	284	236	156	44

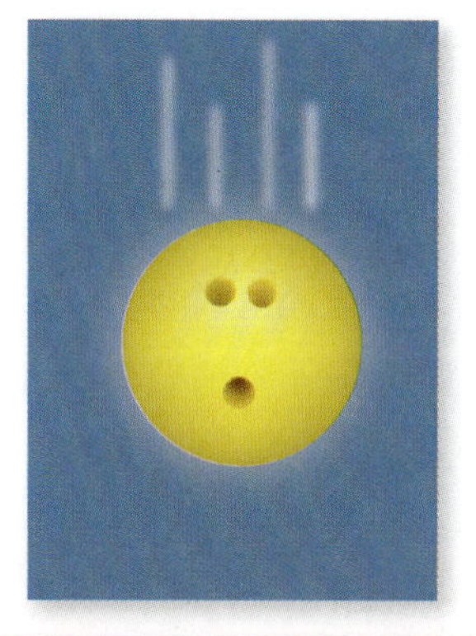

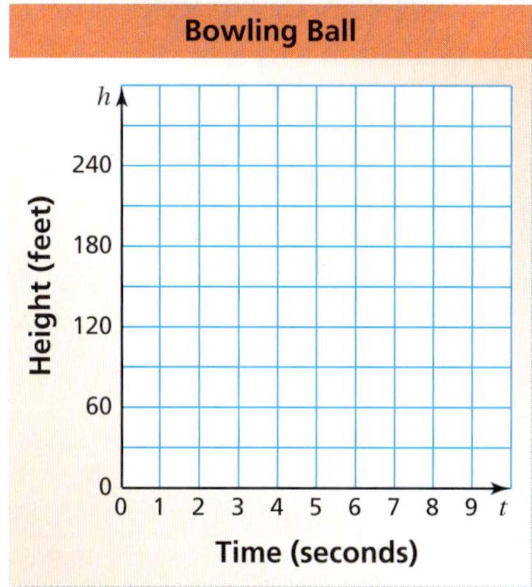

Apply Mathematics

What will the graph look like for an object that has a constant speed? an increasing speed? Explain.

What Is Your Answer?

3. **IN YOUR OWN WORDS** How can you recognize when a pattern in real life is linear or nonlinear? Describe two real-life patterns: one that is linear and one that is nonlinear. Use patterns that are different from those described in Activities 1 and 2.

Use what you learned about comparing linear and nonlinear functions to complete Exercises 3–6 on page 270.

6.4 Lesson

Key Vocabulary
nonlinear function, *p. 268*

The graph of a linear function shows a constant rate of change. A **nonlinear function** does not have a constant rate of change. So, its graph is *not* a line.

EXAMPLE 1 Identifying Functions from Tables

Does the table represent a *linear* or *nonlinear* function? Explain.

Study Tip

A constant rate of change describes a quantity that changes by equal amounts over equal intervals.

a.

+3 +3 +3

x	3	6	9	12
y	40	32	24	16

−8 −8 −8

As x increases by 3, y decreases by 8. The rate of change is constant. So, the function is linear.

b.

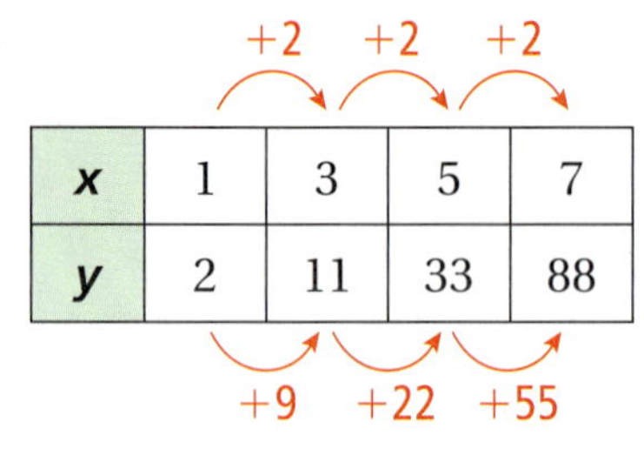

+2 +2 +2

x	1	3	5	7
y	2	11	33	88

+9 +22 +55

As x increases by 2, y increases by different amounts. The rate of change is *not* constant. So, the function is nonlinear.

EXAMPLE 2 Identifying Functions from Graphs

Does the graph represent a *linear* or *nonlinear* function? Explain.

a.

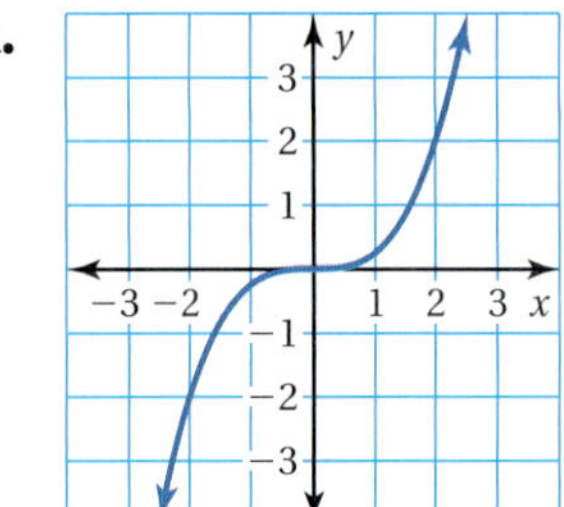

The graph is *not* a line. So, the function is nonlinear.

b.

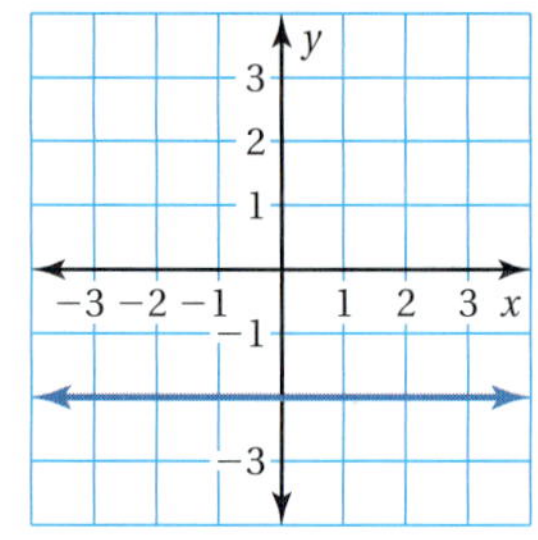

The graph is a line. So, the function is linear.

On Your Own

Now You're Ready
Exercises 7–10

Does the table or graph represent a *linear* or *nonlinear* function? Explain.

1.

x	*y*
0	25
7	20
14	15
21	10

2.

x	*y*
2	8
4	4
6	0
8	−4

3.

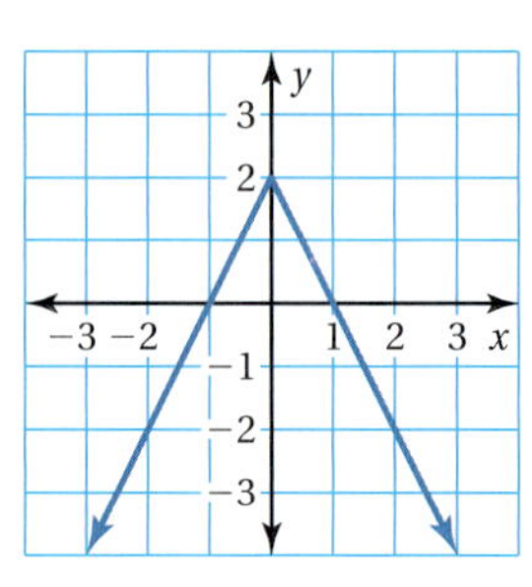

Laurie's Notes

Introduction

Connect

- **Yesterday:** Students explored the graphs of functions that were linear and nonlinear. (MP4, MP6, MP8)
- **Today:** Students will compare linear and nonlinear functions.

Motivate

- Ask 5 students to complete a table of values, where the domain is the same for 5 functions.

	−3	−2	−1	0	1	2	3
$y = x + 2$	−1	0	1	2	3	4	5
$y = x - 2$	−5	−4	−3	−2	−1	0	1
$y = 2x$	−6	−4	−2	0	2	4	6
$y = \frac{x}{2}$	$-\frac{3}{2}$	−1	$-\frac{1}{2}$	0	$\frac{1}{2}$	1	$\frac{3}{2}$
$y = x^2$	9	4	1	0	1	4	9

- Spend time discussing the many patterns in the table. Discuss one function at a time. Ask students for their observations about patterns, changes in *y*-values, slope, and *y*-intercept.
- For $y = x^2$, students want it to have a constant slope. Draw a quick plot of the points and show it is not a linear function.

Lesson Notes

Example 1

- Copy the first table of values. Draw attention to the change in *x* (increasing by 3 each time) and the change in *y* (decreasing by 8 each time). Because the rate of change is constant, the function is linear.
- Copy the second table of values. Draw attention to the change in *x* (increasing by 2 each time) and the change in *y* (increasing by different amounts each time). This is a nonlinear function.

Example 2

- Part (b) may seem obvious, but the horizontal line seems like a special case to students. They may not be sure it is a linear function.

? "What is the slope of this line?" 0 "What is the constant rate of change?" Each time *x* increases by 1, *y* stays the same.

On Your Own

? "What are the constant rates of change for Questions 1 and 2?" $-\frac{5}{7}$; −2

? "Why is Question 3 not a linear function?" There are two parts of this function. The rate of change is positive, then negative.

Goal Today's lesson is comparing linear and **nonlinear functions**.

Lesson Tutorials
Lesson Plans
Answer Presentation Tool

Extra Example 1

Does the table represent a *linear* or *nonlinear* function? Explain.

a.

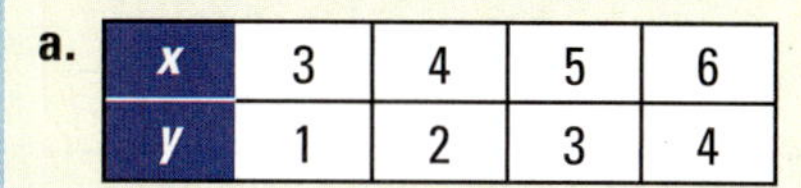

x	3	4	5	6
y	1	2	3	4

linear; As *x* increases by 1, *y* increases by 1.

b.

x	2	5	8	11
y	4	7	12	19

nonlinear; As *x* increases by 3, *y* increases by different amounts.

Extra Example 2

Does the graph represent a *linear* or *nonlinear* function? Explain.

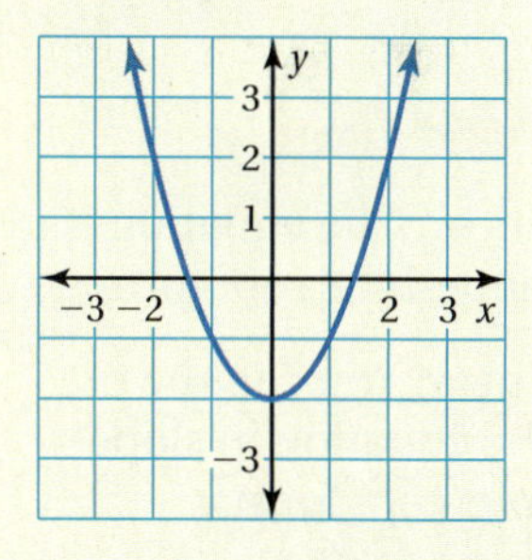

nonlinear; The graph is *not* a line.

On Your Own

1. linear; As *x* increases by 7, *y* decreases by 5.
2. linear; As *x* increases by 2, *y* decreases by 4.
3. nonlinear; The graph is not a line.

Extra Example 3

Does $y = 6x - 3$ represent a *linear* function? Yes, the equation is written in slope-intercept form.

Extra Example 4

Account A earns simple interest. Account B earns compound interest. The table shows the balances for 5 years. Graph the data and compare the graphs.

Year, *t*	Account A Balance	Account B Balance
0	\$50	\$50
1	\$55	\$55
2	\$60	\$60.50
3	\$65	\$66.55
4	\$70	\$73.21
5	\$75	\$80.53

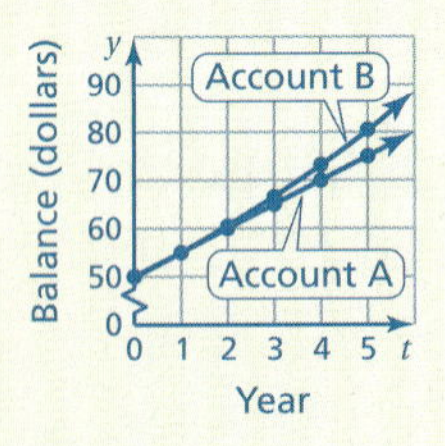

The function representing the balance of Account A is linear. The function representing the balance of Account B is nonlinear.

On Your Own

4. linear; The equation is in slope-intercept form.
5. linear; You can rewrite the equation in slope-intercept form.
6. nonlinear; You cannot rewrite the equation in slope-intercept form.

English Language Learners

Vocabulary

Begin the lesson by reviewing the terms *function* and *linear function.* Define *nonlinear function* and compare it to linear function.

Laurie's Notes

Example 3

- Discuss each equation. Remind students that all linear functions can be written in slope-intercept form.
- Example 3 can also be solved by using a table or graph.
- **MP7 Look for and Make Use of Structure:** Students often see $y = \frac{4}{x}$ and $y = \frac{x}{4}$ as *the same kind of function.* So, many students think this will be a linear function. Remind students of how fractions are multiplied, and use the examples $\frac{4}{x}$ and $\frac{x}{4}$.

  $\frac{4}{x} = \frac{4}{1} \cdot \frac{1}{x} = 4 \cdot \frac{1}{x}$ $\qquad$ $\frac{x}{4} = \frac{x}{1} \cdot \frac{1}{4} = x \cdot \frac{1}{4} = \frac{1}{4} \cdot x$

  So, $y = \frac{x}{4}$ is linear with a slope of $\frac{1}{4}$. The equation $y = \frac{4}{x}$ cannot be written as a linear equation.
- **Note:** The equation $y = \frac{4}{x}$ shows inverse variation.

Example 4

- **Financial Literacy:** Ask a volunteer to read the problem. In addition to looking at linear and nonlinear functions, you also want to integrate financial literacy skills when appropriate. It is a good idea for students to become familiar with the simple interest formula $I = Prt$.
- Point out that both functions have the same initial value. Explain to students that they should interpret the initial value as the starting balance, or the principal, in the context of the problem.

? "Each time the year increases by 1, what happens to the balance of Account A?" It increases by \$10.

? "Each time the year increases by 1, what happens to the balance of Account B?" It increases by a greater amount each year.

- **Extension:** Show students how to calculate the values in the table. Account A's balance is found using $I = Prt$, where $P = 100$, $r = 0.1$, and $t =$ year. Account B's balance can also be found using $I = Prt$, but the principal is changing each year.

On Your Own

- Students should see the exponent of 2 in Question 6 and quickly decide that the function is nonlinear.

Closure

- **Exit Ticket:** Describe how to determine if a function is linear or nonlinear from (a) the equation, (b) a table of values, and (c) a graph.

EXAMPLE 3 Identifying a Nonlinear Function

Which equation represents a *nonlinear* function?

Ⓐ $y = 4.7$　　Ⓑ $y = \pi x$

Ⓒ $y = \frac{4}{x}$　　Ⓓ $y = 4(x - 1)$

You can rewrite the equations $y = 4.7$, $y = \pi x$, and $y = 4(x - 1)$ in slope-intercept form. So, they are linear functions.

You cannot rewrite the equation $y = \frac{4}{x}$ in slope-intercept form. So, it is a nonlinear function.

∴ The correct answer is Ⓒ.

EXAMPLE 4 Real-Life Application

Account A earns simple interest. Account B earns compound interest. The table shows the balances for 5 years. Graph the data and compare the graphs.

Year, *t*	Account A Balance	Account B Balance
0	$100	$100
1	$110	$110
2	$120	$121
3	$130	$133.10
4	$140	$146.41
5	$150	$161.05

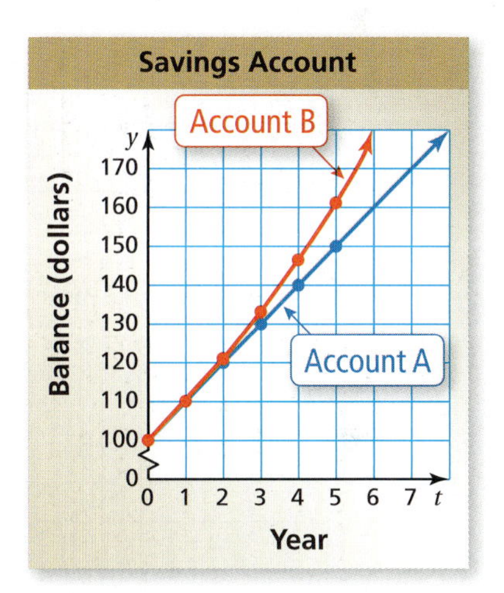

Both graphs show that the balances are positive and increasing.

The balance of Account A has a constant rate of change of $10. So, the function representing the balance of Account A is linear.

The balance of Account B increases by different amounts each year. Because the rate of change is not constant, the function representing the balance of Account B is nonlinear.

On Your Own

Now You're Ready
Exercises 12–14

Does the equation represent a *linear* or *nonlinear* function? Explain.

4. $y = x + 5$　　**5.** $y = \frac{4x}{3}$　　**6.** $y = 1 - x^2$

6.4 Exercises

Vocabulary and Concept Check

1. **VOCABULARY** Describe how linear functions and nonlinear functions are different.

2. **WHICH ONE DOESN'T BELONG?** Which equation does *not* belong with the other three? Explain your reasoning.

$5y = 2x$ $\quad y = \frac{2}{5}x$ $\quad 10y = 4x$ $\quad 5xy = 2$

Practice and Problem Solving

Graph the data in the table. Decide whether the graph is *linear* or *nonlinear*.

3.

x	0	1	2	3
y	4	8	12	16

4.

x	1	2	3	4
y	1	2	6	24

5.

x	6	5	4	3
y	21	15	10	6

6.

x	−1	0	1	2
y	−7	−3	1	5

Does the table or graph represent a *linear* or *nonlinear* function? Explain.

7.

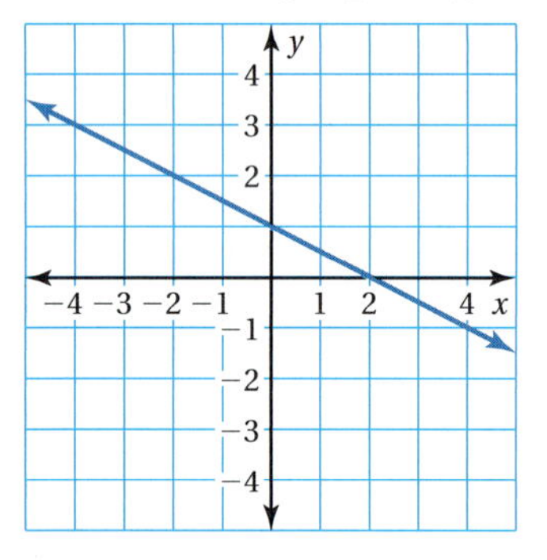

8.

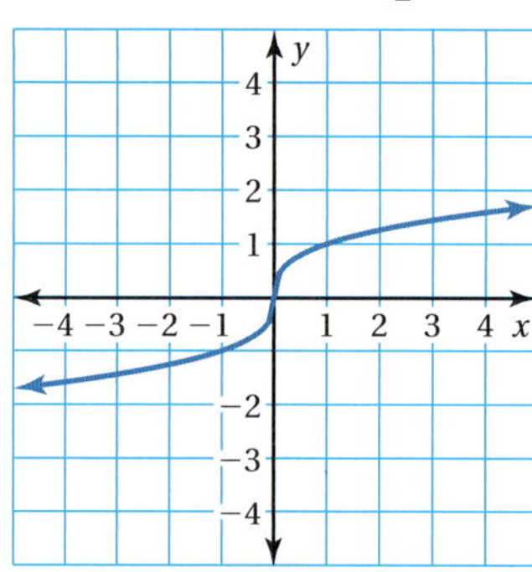

9.

x	5	11	17	23
y	7	11	15	19

10.

x	−3	−1	1	3
y	9	1	1	9

11. **VOLUME** The table shows the volume *V* (in cubic feet) of a cube with an edge length of *x* feet. Does the table represent a linear or nonlinear function? Explain.

Edge Length, *x*	1	2	3	4	5	6	7	8
Volume, *V*	1	8	27	64	125	216	343	512

Assignment Guide and Homework Check

Level	Day 1 Activity Assignment	Day 2 Lesson Assignment	Homework Check
Basic	3–6, 19–21	1, 2, 7–17 odd	7, 9, 13, 15, 17
Average	3–6, 19–21	1, 2, 7–11 odd, 12–18 even	7, 9, 12, 14, 16
Advanced	1–6, 8–14 even, 15–21		10, 14, 15, 17, 18

Common Errors

- **Exercises 3–6, 9, and 10** Students may say that the function is linear because the *x*-values are increasing or decreasing by the same amount each time. Encourage them to examine the *y*-values to see if the graph represents a line.
- **Exercises 12–14** Students may not rewrite the equation in slope-intercept form and will guess whether the equation is linear. Remind them to attempt to write the equation in slope-intercept form as a check.
- **Exercise 15** Students may try to graph the coordinate pairs to determine if the function is linear and make an incorrect assumption depending upon how they scale their axes. Encourage them to examine the change in *y* for each *x*-value.

6.4 Record and Practice Journal

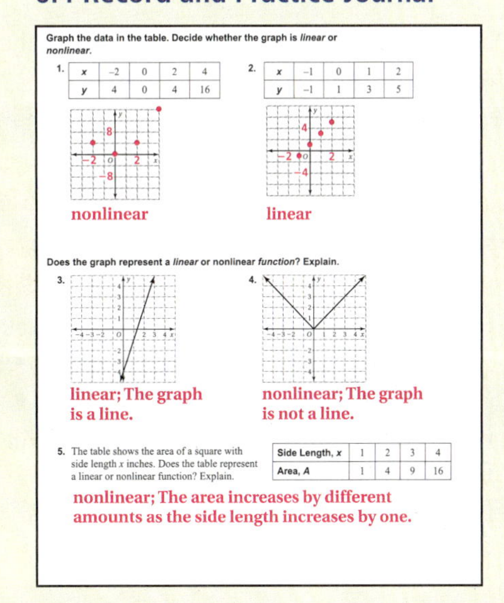
Graph the data in the table. Decide whether the graph is *linear* or *nonlinear*.

1.

x	−2	0	2	4
y	4	0	4	16

nonlinear

2.

x	−1	0	1	2
y	−1	1	3	5

linear

Does the graph represent a *linear* or nonlinear *function*? Explain.

3. linear; The graph is a line.

4. nonlinear; The graph is not a line.

5. The table shows the area of a square with side length x inches. Does the table represent a linear or nonlinear function? Explain.

Side Length, x	1	2	3	4
Area, A	1	4	9	16

nonlinear; The area increases by different amounts as the side length increases by one.

Vocabulary and Concept Check

1. A linear function has a constant rate of change. A nonlinear function does not have a constant rate of change.
2. $5xy = 2$; You cannot rewrite the equation in slope-intercept form.

Practice and Problem Solving

3. 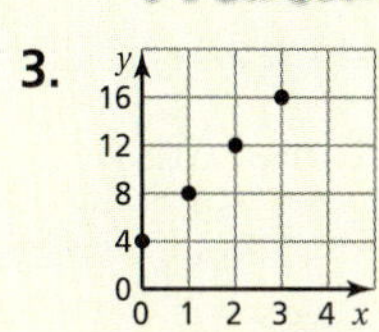

linear

4. 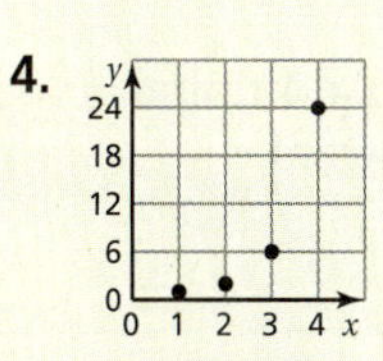

nonlinear

5. 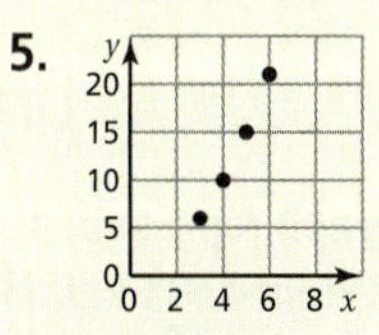

nonlinear

6. 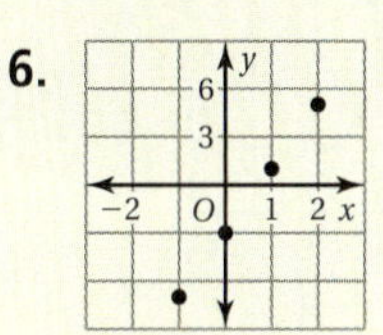

linear

7. linear; The graph is a line.
8. nonlinear; The graph is not a line.
9. linear; As x increases by 6, y increases by 4.
10. nonlinear; As x increases by 2, y changes by different amounts.
11. nonlinear; As x increases by 1, V increases by different amounts.

Practice and Problem Solving

12. linear; You can rewrite the equation in slope-intercept form.

13. linear; You can rewrite the equation in slope-intercept form.

14. nonlinear; You cannot rewrite the equation in slope-intercept form.

15. nonlinear; As x decreases by 65, y increases by different amounts.

16. See *Taking Math Deeper.*

17–18. See Additional Answers.

Fair Game Review

19–20. See Additional Answers.

21. C

Mini-Assessment

Does the table or graph represent a *linear* or *nonlinear* function? Explain.

1.

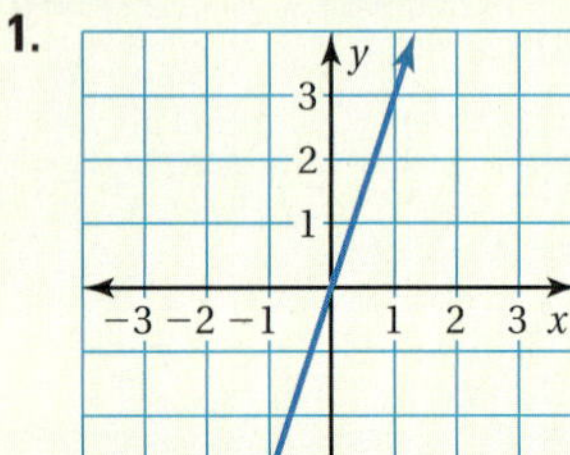

linear; The graph is a line.

2.

x	−2	0	2	4
y	8	0	8	64

nonlinear; As x increases by 2, y increases by different amounts.

3.

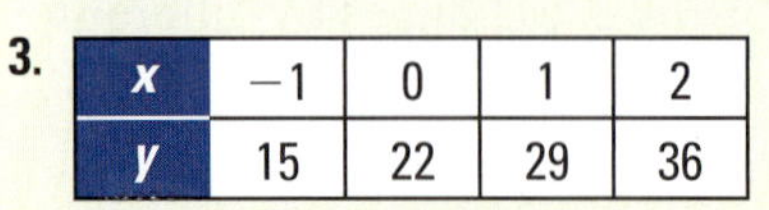

x	−1	0	1	2
y	15	22	29	36

linear; As x increases by 1, y increases by 7.

Taking Math Deeper

Exercise 16

Students can learn a valuable lesson about mathematics from this problem. Even though the problem does not specifically ask them to draw a graph, it is still a good idea. *Seeing* the relationship between pounds and cost is easier than simply finding the relationship using algebra.

Plot the two given points.

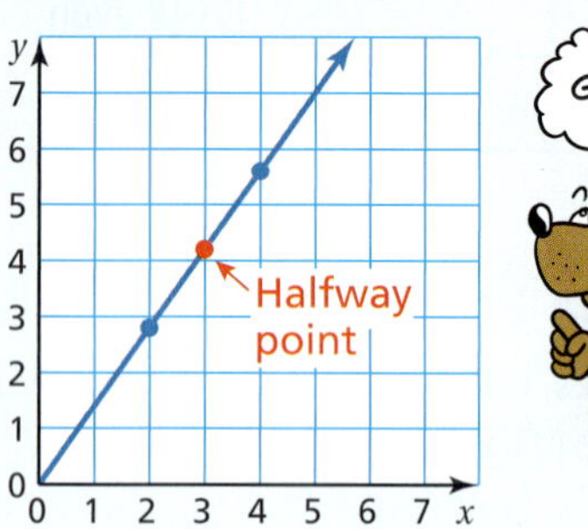

2 Find the halfway point.

Because you want the table to represent a linear function and 3 is halfway between 2 and 4, you need to find the number that is halfway between \$2.80 and \$5.60. This number is the mean of \$2.80 and \$5.60.

a. Mean $= \dfrac{2.80 + 5.60}{2} = 4.20$

Write a function.

Let $x =$ pounds of seeds.

Let $y =$ cost.

y-intercept $= 0$

slope $= \dfrac{5.60 - 2.80}{4 - 2} = 1.4$

b. $y = 1.4x$; The slope represents the cost per pound of sunflower seeds.

c. The function does not have a maximum value because you can always increase the cost by increasing the amount of sunflower seeds purchased.

Project

Plant some sunflower seeds. Keep track of the progress of the plants until they bloom.

Reteaching and Enrichment Strategies

If students need help. . .	If students got it. . .
Resources by Chapter • Practice A and Practice B • Puzzle Time Record and Practice Journal Practice Differentiating the Lesson Lesson Tutorials Skills Review Handbook	Resources by Chapter • Enrichment and Extension • Technology Connection Start the next section

Does the equation represent a *linear* or *nonlinear* function? Explain.

3 **12.** $2x + 3y = 7$ **13.** $y + x = 4x + 5$ **14.** $y = \frac{8}{x^2}$

15. **LIGHT** The frequency y (in terahertz) of a light wave is a function of its wavelength x (in nanometers). Does the table represent a linear or nonlinear function? Explain.

Color	Red	Yellow	Green	Blue	Violet
Wavelength, x	660	595	530	465	400
Frequency, y	454	504	566	645	749

16. **MODELING** The table shows the cost y (in dollars) of x pounds of sunflower seeds.

Pounds, x	Cost, y
2	2.80
3	?
4	5.60

a. What is the missing y-value that makes the table represent a linear function?

b. Write a linear function that represents the cost y of x pounds of seeds. Interpret the slope.

c. Does the function have a maximum value? Explain your reasoning.

17. **TREES** Tree A is 5 feet tall and grows at a rate of 1.5 feet per year. The table shows the height h (in feet) of Tree B after x years.

Years, x	Height, h
0	5
1	11
4	17
9	23

a. Does the table represent a linear or nonlinear function? Explain.

b. Which tree is taller after 10 years? Explain.

18. 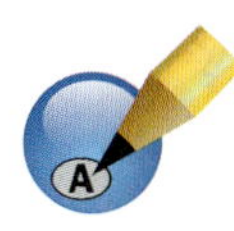

The ordered pairs represent a function.

(0, −1), (1, 0), (2, 3), (3, 8), and (4, 15)

a. Graph the ordered pairs and describe the pattern. Is the function linear or nonlinear?

b. Write an equation that represents the function.

Fair Game Review What you learned in previous grades & lessons

The vertices of a figure are given. Draw the figure and its image after a dilation with the given scale factor *k*. Identify the type of dilation. *(Section 2.7)*

19. $A(-3, 1), B(-1, 3), C(-1, 1); k = 3$

20. $J(-8, -4), K(2, -4), L(6, -10), M(-8, -10); k = \frac{1}{4}$

21. **MULTIPLE CHOICE** What is the value of x? *(Section 3.3)*

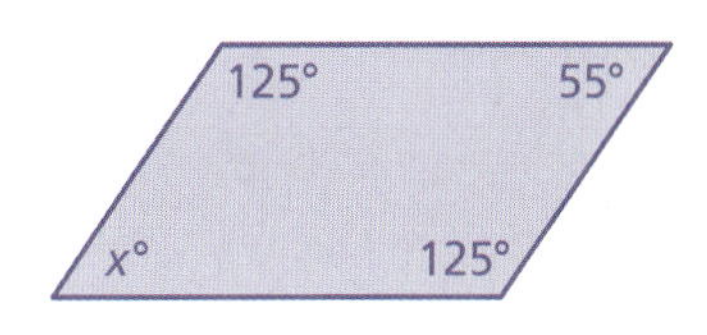

Ⓐ 25 Ⓑ 35

Ⓒ 55 Ⓓ 125

6.5 Analyzing and Sketching Graphs

Essential Question How can you use a graph to represent relationships between quantities without using numbers?

1 ACTIVITY: Interpreting a Graph

Work with a partner. Use the graph shown.

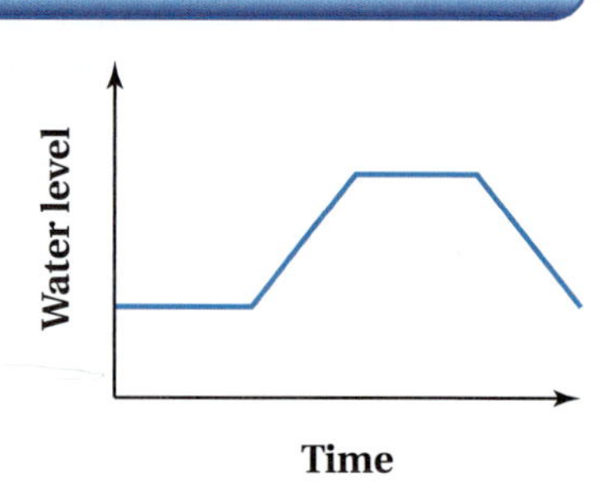

a. How is this graph different from the other graphs you have studied?

b. Write a short paragraph that describes how the water level changes over time.

c. What situation can this graph represent?

2 ACTIVITY: Matching Situations to Graphs

Work with a partner. You are riding your bike. Match each situation with the appropriate graph. Explain your reasoning.

A.

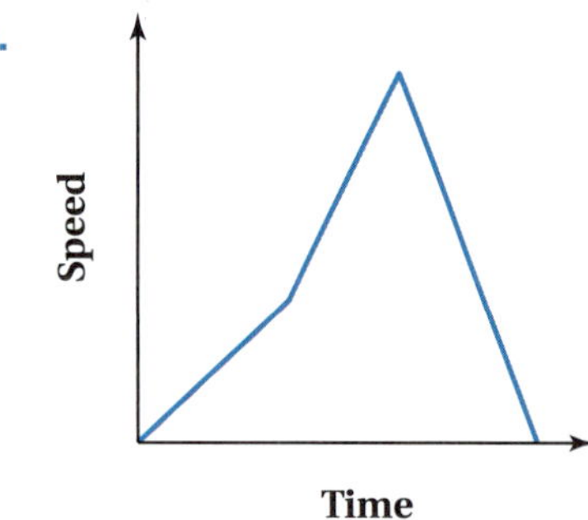

B.

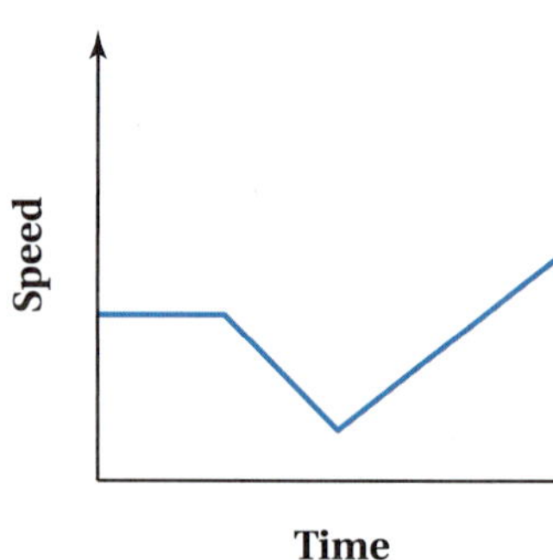

C.

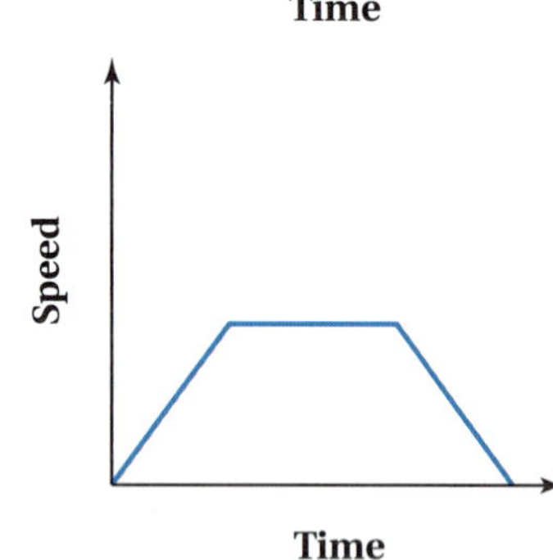

D.

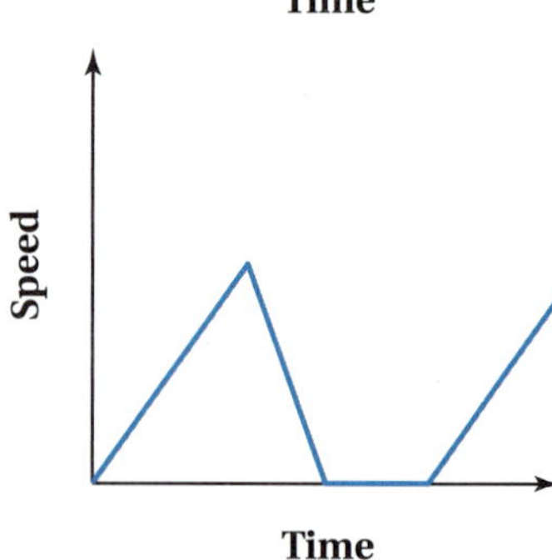

a. You gradually increase your speed, then ride at a constant speed along a bike path. You then slow down until you reach your friend's house.

b. You gradually increase your speed, then go down a hill. You then quickly come to a stop at an intersection.

c. You gradually increase your speed, then stop at a store for a couple of minutes. You then continue to ride, gradually increasing your speed.

d. You ride at a constant speed, then go up a hill. Once on top of the hill, you gradually increase your speed.

COMMON CORE

Functions

In this lesson, you will

- analyze the relationship between two quantities using graphs.
- sketch graphs to represent the relationship between two quantities.

Learning Standard
MACC.8.F.2.5

Laurie's Notes

Introduction

Standards for Mathematical Practice

- **MP2 Reason Abstractly and Quantitatively** and **MP3 Construct Viable Arguments and Critique the Reasoning of Others:** In the activities today, students will interpret and communicate their thinking about quantitative relationships represented as graphs without numbers. Expect students to critique the reasoning of their classmates as they use the slopes and y-intercepts to interpret the graphs.

Motivate

- **Alternate Approach:** Ask 3 students to volunteer to be *walkers*. Each will stand 5 feet from the front wall and do the following when you say, "Go."
  - Walker A: Walk slowly to the back wall at a constant rate
  - Walker B: Walk quickly to the back wall at a constant rate
  - Walker C: Stay where you are. Do not walk.
- Have the rest of the class sketch a graph representing the distance of each walker from the front wall over time. Ask what information they can show in the graphs without including a scale.

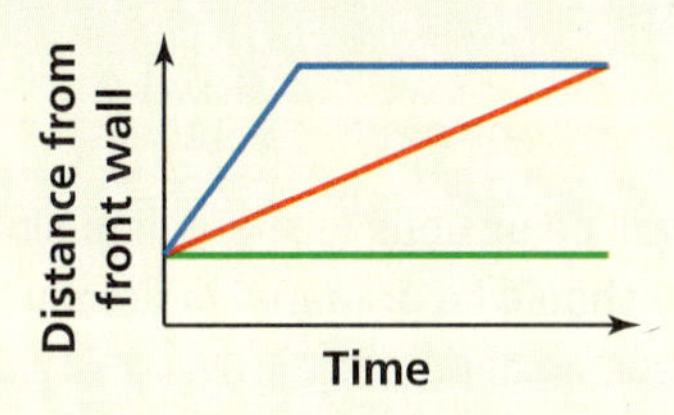

- **MP3:** Ask students to share and discuss their graphs.

Activity Notes

Activity 1

- Students will quickly comment that there are no numbers on the graph.
- **MP3:** Give students a few minutes to discuss the graph with their partners. Then have students share the different situations they come up with and why the graph can represent their situations.
- Features to listen for: The water level was constant for a time. Then it increased at a constant rate for a time. It stopped increasing and remained constant for a time, and then the level decreased at a constant rate.

Activity 2

- Give students sufficient time to discuss the graphs with their partners and then to do the matching.
- **MP6 Attend to Precision:** Students should use precise language to explain their reasoning. Listen for how the pattern of increasing, decreasing, and constant speeds represents the situation.
- Graph B is the only graph with a y-intercept not equal to 0. Do students point out this feature when they describe their graphs?
- **Extension:** Have students make graphs with numbers. Then swap graphs with classmates and describe reasonable scenarios for the graphs.

Florida Common Core Standards

MACC.8.F.2.5 Describe qualitatively the functional relationship between two quantities by analyzing a graph (e.g., where the function is increasing or decreasing, linear or nonlinear). Sketch a graph that exhibits the qualitative features of a function that has been described verbally.

Previous Learning

Students need to be familiar with graphing linear and nonlinear functions.

Lesson Plans
Complete Materials List

6.5 Record and Practice Journal

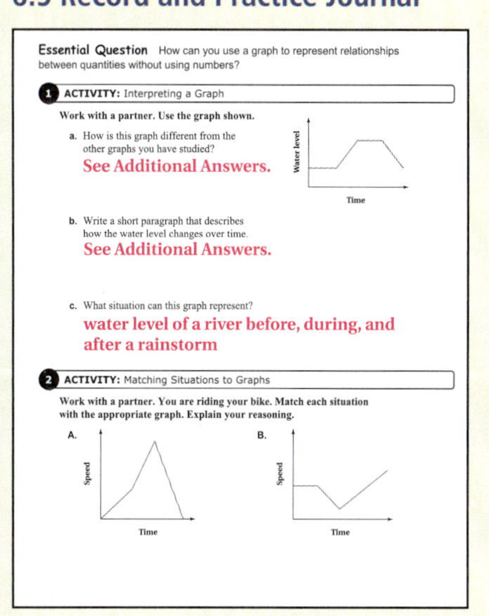

Essential Question How can you use a graph to represent relationships between quantities without using numbers?

1 ACTIVITY: Interpreting a Graph

Work with a partner. Use the graph shown.

a. How is this graph different from the other graphs you have studied?
See Additional Answers.

b. Write a short paragraph that describes how the water level changes over time.
See Additional Answers.

c. What situation can this graph represent?
water level of a river before, during, and after a rainstorm

Water level
Time

2 ACTIVITY: Matching Situations to Graphs

Work with a partner. You are riding your bike. Match each situation with the appropriate graph. Explain your reasoning.

A. Speed / Time

B. Speed / Time

Differentiated Instruction

Kinesthetic

Have students gather real-life data using a meter stick and a table tennis ball. Drop the ball 10 times from different initial heights and measure the initial bounce with the meter stick. Tell students to record the data in a table with the initial height as the input and the initial bounce as the output, and then have them graph the data.

6.5 Record and Practice Journal

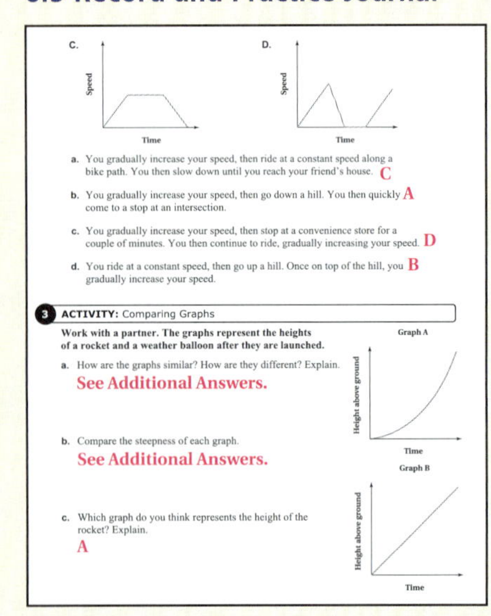

C. (graph: Speed vs. Time)

D. (graph: Speed vs. Time)

a. You gradually increase your speed, then ride at a constant speed along a bike path. You then slow down until you reach your friend's house. C

b. You gradually increase your speed, then go down a hill. You then quickly come to a stop at an intersection. A

c. You gradually increase your speed, then stop at a convenience store for a couple of minutes. You then continue to ride, gradually increasing your speed. D

d. You ride at a constant speed, then go up a hill. Once on top of the hill, you gradually increase your speed. B

3 ACTIVITY: Comparing Graphs

Work with a partner. The graphs represent the heights of a rocket and a weather balloon after they are launched.

a. How are the graphs similar? How are they different? Explain.
See Additional Answers.

b. Compare the steepness of each graph.
See Additional Answers.

c. Which graph do you think represents the height of the rocket? Explain.
A

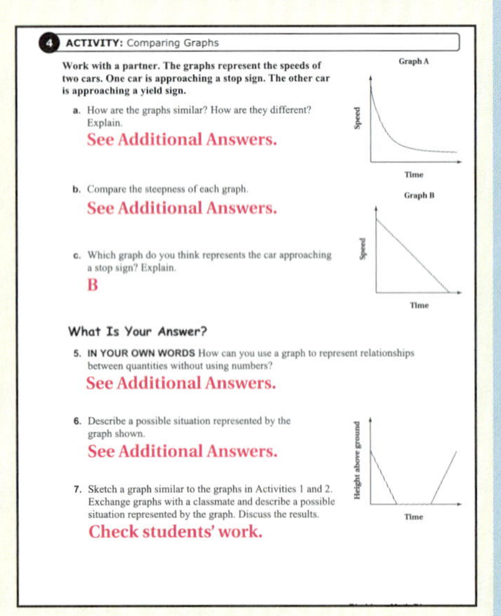

4 ACTIVITY: Comparing Graphs

Work with a partner. The graphs represent the speeds of two cars. One car is approaching a stop sign. The other car is approaching a yield sign.

a. How are the graphs similar? How are they different? Explain.
See Additional Answers.

b. Compare the steepness of each graph.
See Additional Answers.

c. Which graph do you think represents the car approaching a stop sign? Explain.
B

What Is Your Answer?

5. **IN YOUR OWN WORDS** How can you use a graph to represent relationships between quantities without using numbers?
See Additional Answers.

6. Describe a possible situation represented by the graph shown.
See Additional Answers.

7. Sketch a graph similar to the graphs in Activities 1 and 2. Exchange graphs with a classmate and describe a possible situation represented by the graph. Discuss the results.
Check students' work.

Laurie's Notes

Activity 3

- Expect a fair amount of discussion about how these two graphs are alike and how they differ. In addition, students should share what they know about rockets and weather balloons.
- The sixth graders at our school build and launch rockets in science class, so they have some knowledge about the behavior of small rockets!
- In talking about the steepness of Graph A, students often say that the initial portion of the graph is not too steep, then the graph curves, and finally it becomes quite steep.

? "Can you think of other contexts besides rockets and balloons that would match these graphs?" Answers vary.

- **Extension:** Have students consider reasonable numbers to use to scale the axes.

Activity 4

- This activity is similar to Activity 3 except the graphs are decreasing rather than increasing.
- The two contexts should be familiar to students, but just to be sure, you can ask what a yield sign is.
- **Extension:** Have students consider reasonable numbers to use to scale the axes.

What Is Your Answer?

- **MP3:** Students will be anxious to share their situations for the graph in Question 6. They should be prepared to defend their reasoning and to critique the reasoning of others.

Closure

- What could this graph represent?

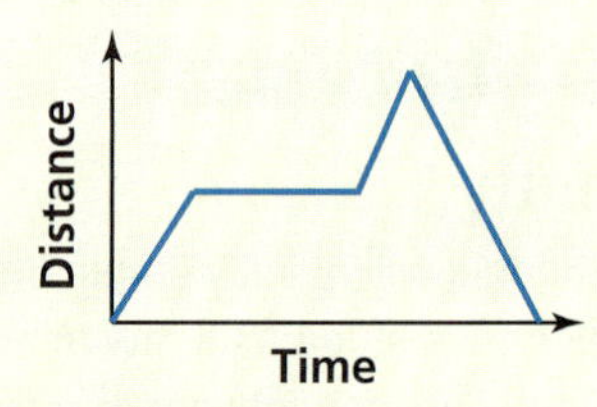

3 ACTIVITY: Comparing Graphs

Work with a partner. The graphs represent the heights of a rocket and a weather balloon after they are launched.

a. How are the graphs similar? How are they different? Explain.

b. Compare the steepness of each graph.

c. Which graph do you think represents the height of the rocket? Explain.

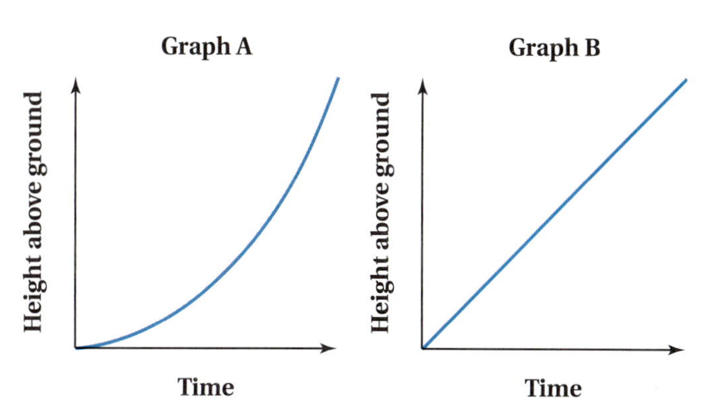

4 ACTIVITY: Comparing Graphs

Math Practice 1

Consider Similar Problems

How is this activity similar to the previous activity?

Work with a partner. The graphs represent the speeds of two cars. One car is approaching a stop sign. The other car is approaching a yield sign.

a. How are the graphs similar? How are they different? Explain.

b. Compare the steepness of each graph.

c. Which graph do you think represents the car approaching a stop sign? Explain.

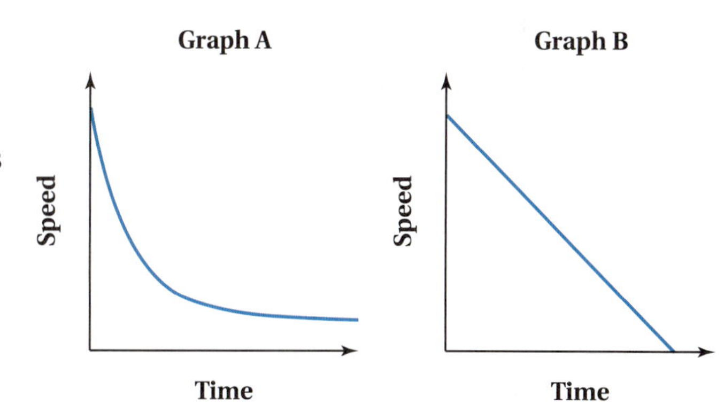

What Is Your Answer?

5. **IN YOUR OWN WORDS** How can you use a graph to represent relationships between quantities without using numbers?

6. Describe a possible situation represented by the graph shown.

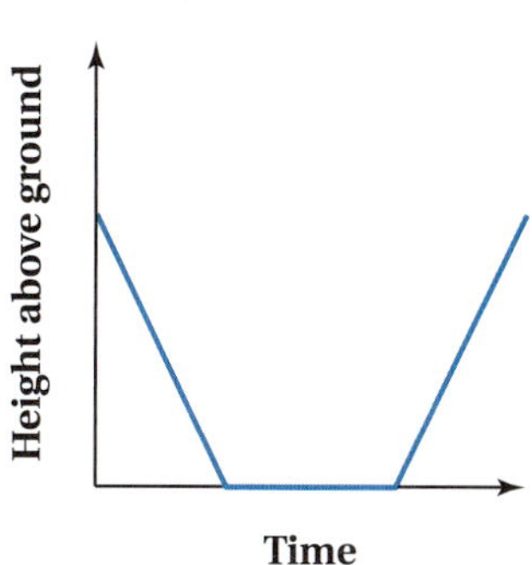

7. Sketch a graph similar to the graphs in Activities 1 and 2. Exchange graphs with a classmate and describe a possible situation represented by the graph. Discuss the results.

Practice Use what you learned about analyzing and sketching graphs to complete Exercises 7–9 on page 276.

6.5 Lesson

Graphs can show the relationship between quantities without using specific numbers on the axes.

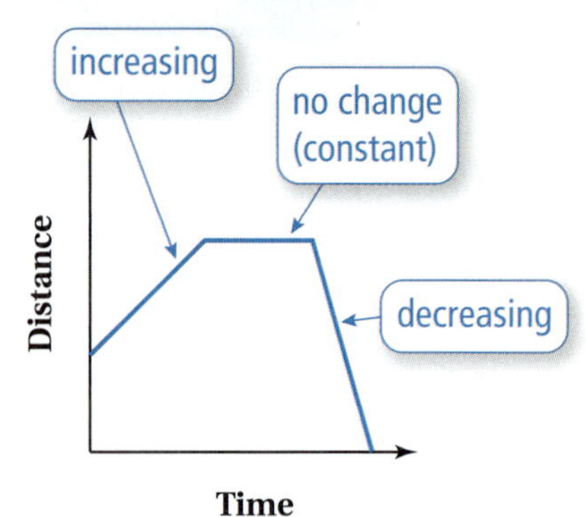

EXAMPLE 1 Analyzing Graphs

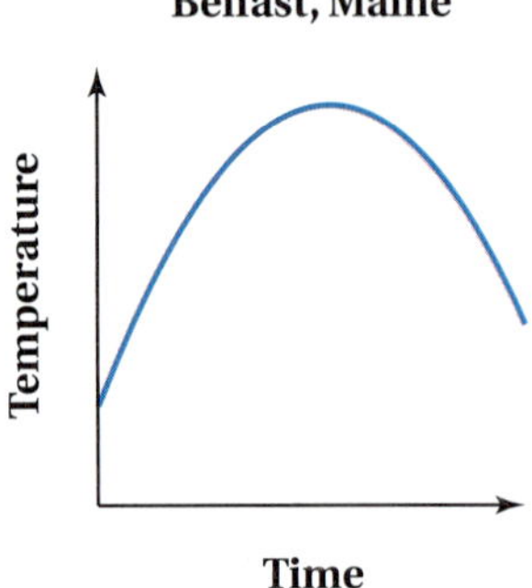

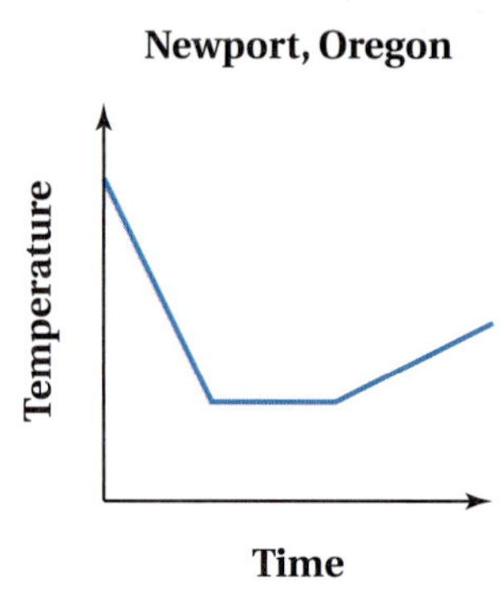

The graphs show the temperatures throughout the day in two cities.

a. Describe the change in temperature in each city.

Belfast: The temperature increases at the beginning of the day. Then the temperature begins to decrease at a faster and faster rate for the rest of the day.

Newport: The temperature decreases at a constant rate at the beginning of the day. Then the temperature stays the same for a while before increasing at a constant rate for the rest of the day.

b. Make three comparisons from the graphs.

Three possible comparisons follow:

- Both graphs show increasing and decreasing temperatures.
- Both graphs are nonlinear, but the graph of the temperatures in Newport consists of three linear sections.
- In Belfast, it was warmer at the end of the day than at the beginning. In Newport, it was colder at the end of the day than at the beginning.

Study Tip

The comparisons given in Example 1(b) are sample answers. You can make many other correct comparisons.

On Your Own

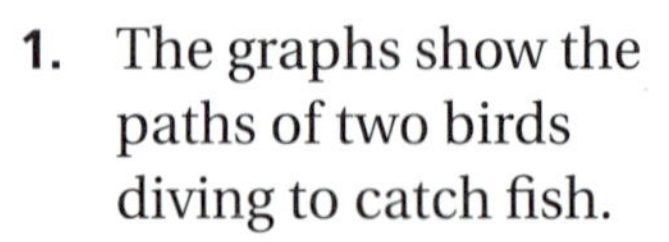

1. The graphs show the paths of two birds diving to catch fish.

   a. Describe the path of each bird.

   b. Make three comparisons from the graphs.

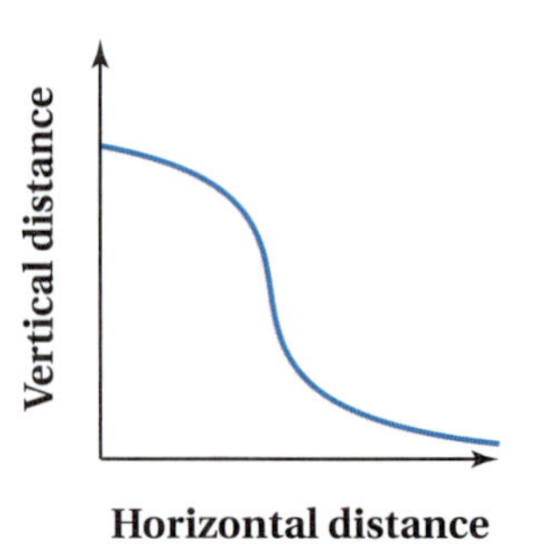

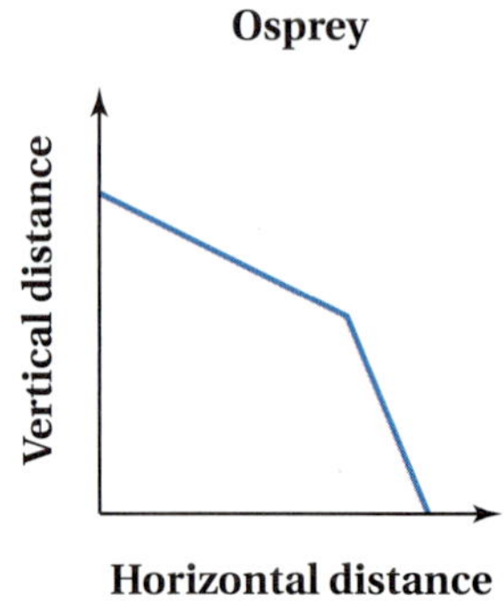

Laurie's Notes

Introduction

Connect

- **Yesterday:** Students explored graphs showing the relationship between quantities without using specific numbers. (MP2, MP3, MP6)
- **Today:** Students will use their knowledge of slope to analyze and sketch graphs without using specific numbers.

Motivate

? "How many of your families have a GPS unit?" Answers vary.

- Discuss how a GPS unit works. It determines its distances from four or more orbiting satellites and uses these distances to calculate its position on Earth. The position is given in coordinates of latitude (a measure of how far north or south) and longitude (a measure of how far east or west).

? "Will two locations with the same latitude have similar climates?" yes, although differences are likely because of geographical features

? "Will two locations with the same longitude have similar climates?" not necessarily; The greatest factor in climate is latitude, or the distance from the equator.

- Share the latitude and longitude of your location along with other locations that have the same latitude (but not the same longitude) and vice versa.

Lesson Notes

Discuss

- Draw the time-distance graph and discuss the three sections.

? "What does the increasing slope mean?" The distance is increasing.

? "What does the constant slope mean?" The distance is not changing.

? "What does the decreasing slope mean?" The distance is decreasing.

Example 1

- Draw the time-temperature graphs for the two cities.
- Both cities are at 44°N latitude, but are on opposite sides of the country.

? **MP6 Attend to Precision:** "How does the temperature change during the day in Belfast, Maine?" The temperature increases, but at a slowing rate, until it starts to decrease. It decreases more and more rapidly.

? **MP6:** "How does the temperature change during the day in Newport, Oregon?" The temperature decreases at a constant rate, then it stays constant for a time, and then it increases at a constant rate.

- Ask students to discuss the graphs with a neighbor, making at least three comparisons. After a few minutes, students should be ready to share their thinking about the graphs.
- If students do not mention linear versus non-linear features, bring it up.

On Your Own

- Give students time alone at first to write their thoughts, then have them share their thinking with their neighbors. Finally, discuss as a whole class.

Goal Today's lesson is using graphs without numbers to show the relationship between quantities.

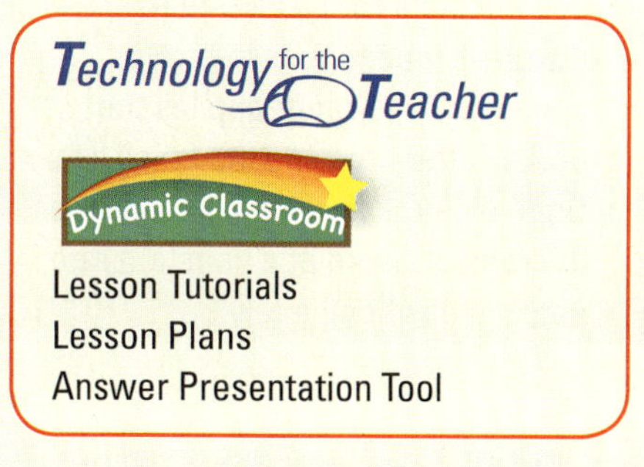

Lesson Tutorials
Lesson Plans
Answer Presentation Tool

Extra Example 1

The graphs show the temperatures throughout a day in two cities.

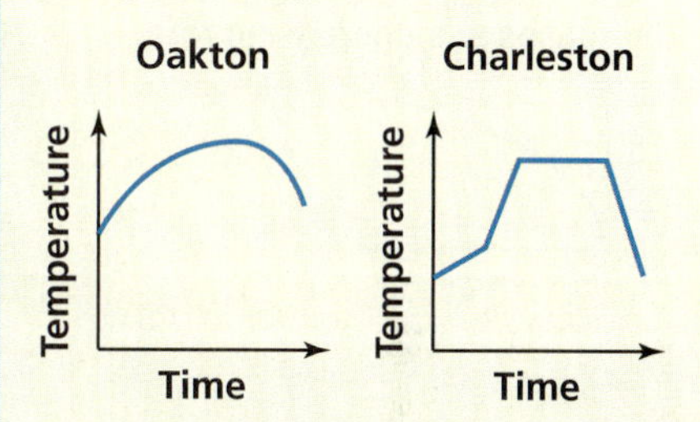

a. Describe the change in temperature in each city. **Oakton:** The temperature increases, then starts to decrease, and keeps decreasing more and more rapidly. **Charleston:** The temperature increases at a constant rate, then increases at a faster, constant rate. It stops changing for a time, then it decreases at a fast, constant rate.

b. Make three comparisons from the graphs. Both graphs show increasing and decreasing temperatures. Both are nonlinear, but Charleston's has four linear sections. Oakton's day is warmer at the end than at the beginning, whereas Charleston has about the same temperature at the beginning and at the end of the day.

On Your Own

1. See Additional Answers.

Extra Example 2

Sketch a graph that represents each situation.

a. A soapbox derby car picks up speed at a constant rate as it travels downhill. It continues to pick up speed until the racer applies the brakes after crossing the finish line. Once the brakes are applied, the car decreases speed at a constant rate until it stops completely.

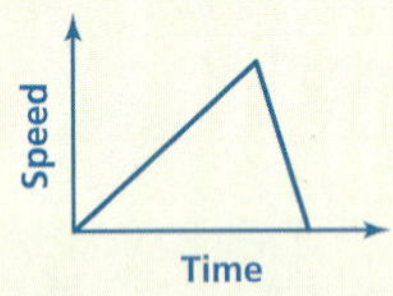

b. As the price of a product increases, the amount that producers supply increases at a decreasing rate.

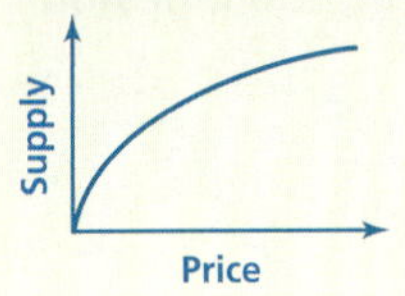

On Your Own

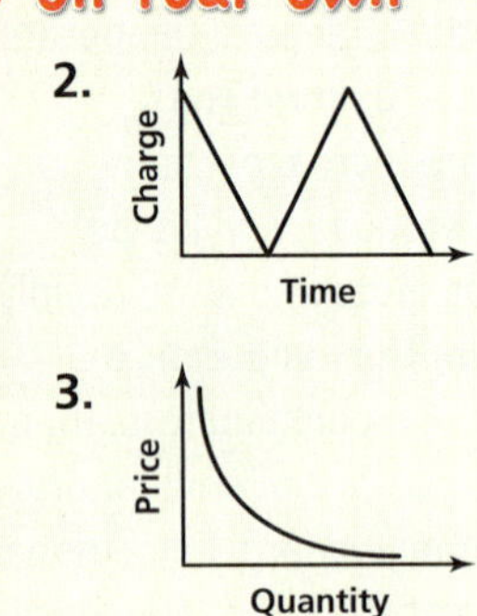

English Language Learners

Vocabulary

Give extra credit to students who can give acceptable definitions of words in this chapter. For instance, a definition of linear *function* could be "a linear function is a line when you graph it."

Laurie's Notes

Example 2

- Explain that a situation will now be described in words, and the problem is to sketch a graph that represents the situation.
- Write the part (a) problem statement, or display it on the overhead projector or document camera.
- ? "What does the first sentence tell you about the graph?" The train starts from a stop, so the graph begins at the origin. The train gains speed at a constant rate, which can be graphed as a line segment with a positive slope.
- ? "What does the second sentence tell you about the graph?" The train travels at its maximum speed for awhile, which can be graphed as a horizontal segment. Then it slows down at a constant rate until it comes to a stop. This can be graphed as a line segment with a negative slope and an endpoint on the x-axis.
- Write and display the second problem. Give students time to try this on their own before discussing it as a class.
- Because the price increases at an increasing rate, the graph is nonlinear, and students may say that it curves upward.

On Your Own

- **Think-Pair-Share:** Students should read each question independently and then work in pairs to answer the questions. When they have answered the questions, the pair should compare their answers with another group and discuss any discrepancies.

Closure

- Draw a graph without numbers that represents the temperature in your town or city over the last 24 hours (or for the daytime hours yesterday).

You can sketch graphs showing relationships between quantities that are described verbally.

EXAMPLE 2 Sketching Graphs

Sketch a graph that represents each situation.

a. A stopped subway train gains speed at a constant rate until it reaches its maximum speed. It travels at this speed for a while, and then slows down at a constant rate until coming to a stop at the next station.

Step 1: Draw the axes. Label the vertical axis "Speed" and the horizontal axis "Time."

Step 2: Sketch the graph.

Speed

Time

Words	Graph
A stopped subway train gains speed at a constant rate . . .	increasing line segment starting at the origin
until it reaches its maximum speed. It travels at this speed for a while, . . .	horizontal line segment
and then slows down at a constant rate until coming to a stop at the next station.	decreasing line segment ending at the horizontal axis

b. As television size increases, the price increases at an increasing rate.

Step 1: Draw the axes. Label the vertical axis "Price" and the horizontal axis "TV size."

Step 2: Sketch the graph.

The price *increases at an increasing rate*. So, the graph is nonlinear and becomes steeper and steeper as the TV size increases.

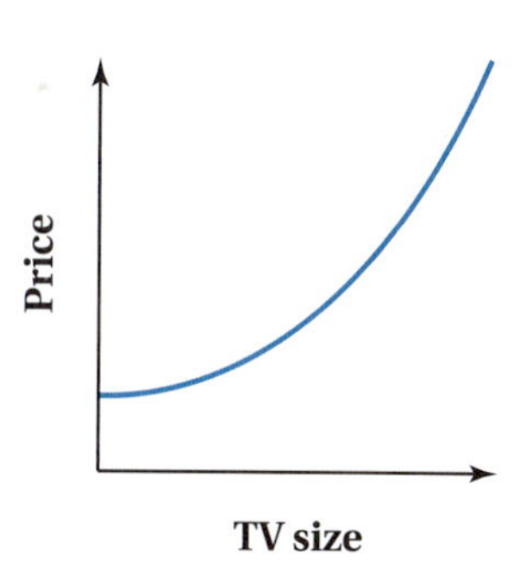

On Your Own

Sketch a graph that represents the situation.

2. A fully charged battery loses its charge at a constant rate until it has no charge left. You plug it in and recharge it fully. Then it loses its charge at a constant rate until it has no charge left.

3. As the available quantity of a product increases, the price decreases at a decreasing rate.

6.5 Exercises

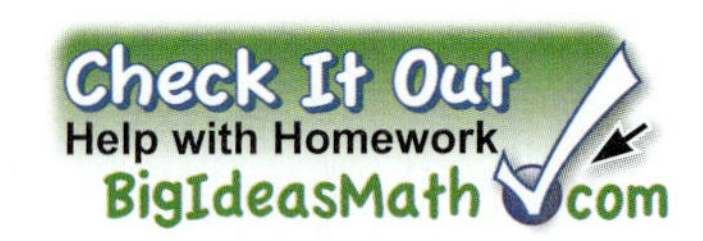

Vocabulary and Concept Check

MATCHING **Match the verbal description with the part of the graph it describes.**

1. stays the same
2. slowly decreases at a constant rate
3. slowly increases at a constant rate
4. increases at an increasing rate
5. quickly decreases at a constant rate
6. quickly increases at a constant rate

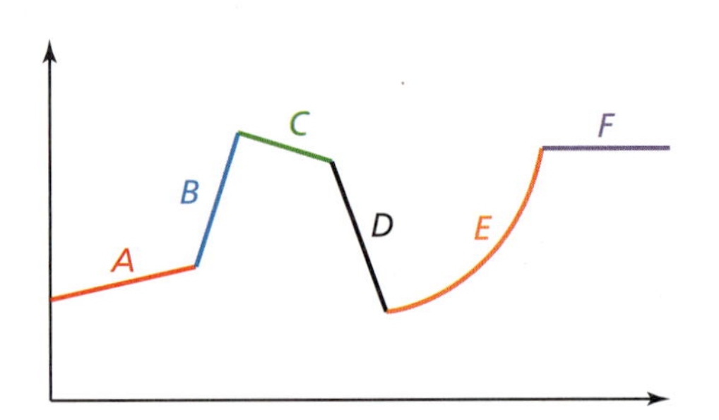

Practice and Problem Solving

Describe the relationship between the two quantities.

7.

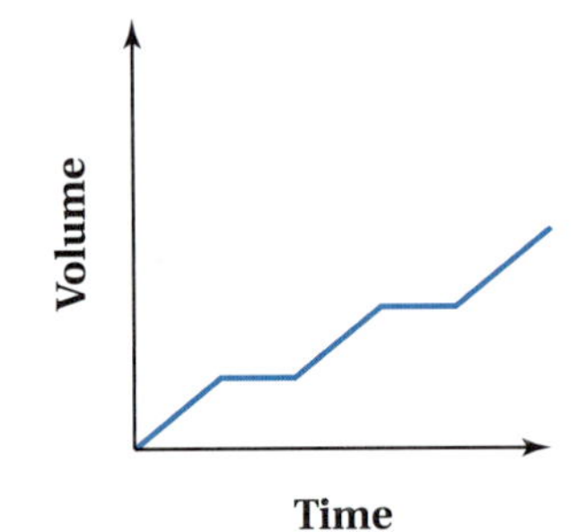

8.

9.

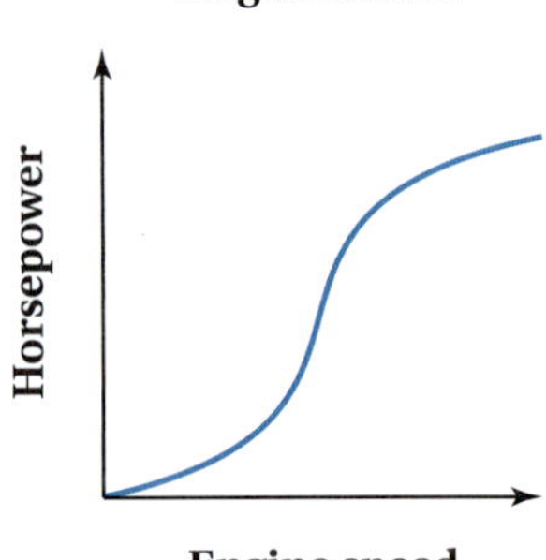

10.

11.

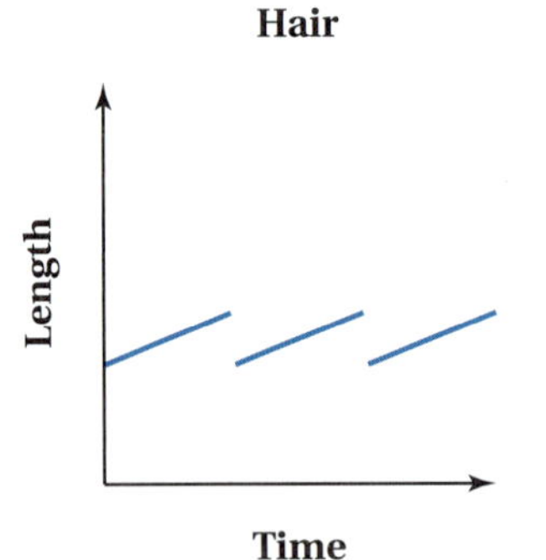

12.

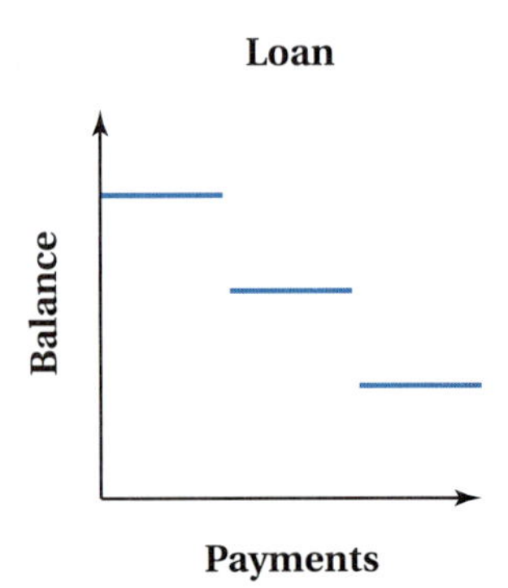

Usage

Jan Mar May July

13. **NATURAL GAS** The graph shows the natural gas usage for a house.

a. Describe the change in usage from January to March.

b. Describe the change in usage from March to May.

Assignment Guide and Homework Check

Level	Day 1 Activity Assignment	Day 2 Lesson Assignment	Homework Check
Basic	7–9, 20–23	1–6, 11–19 odd	11, 13, 15, 17
Average	7–9, 20–23	1–6, 10–14, 15–19 odd	10, 12, 13, 14, 19
Advanced	1–10, 12, 13–23		12, 13, 14, 18, 19

Common Errors

- **Exercise 11** Students may get confused because the graph has breaks in it. Remind them that because of growth, your hair length gradually increases until you get a haircut. Then the length is suddenly shorter.
- **Exercise 12** Students may get confused because the graph has breaks in it. If students are confused, remind them that the horizontal lines represent the balance owed. A break occurs when a payment is made.

6.5 Record and Practice Journal

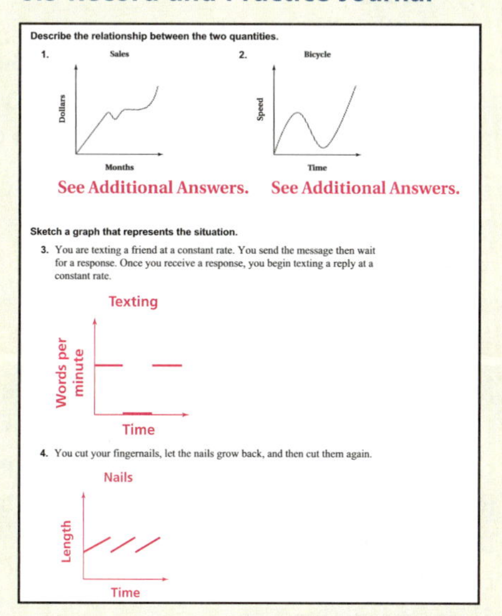
Describe the relationship between the two quantities.

1. Sales — Dollars / Months
See Additional Answers.

2. Bicycle — Speed / Time
See Additional Answers.

Sketch a graph that represents the situation.

3. You are texting a friend at a constant rate. You send the message then wait for a response. Once you receive a response, you begin texting a reply at a constant rate.

4. You cut your fingernails, let the nails grow back, and then cut them again.

Vocabulary and Concept Check

1. F
2. C
3. A
4. E
5. D
6. B

Practice and Problem Solving

7. The volume of the balloon increases at a constant rate, then stays constant, then increases at a constant rate, then stays constant, and then increases at a constant rate.
8. Sales decrease quickly at a constant rate and then increase slowly at a constant rate.
9. Horsepower increases at an increasing rate and then increases at a decreasing rate.
10. Grams decrease at a decreasing rate.
11. The hair length increases at a constant rate, then decreases instantly, then increases at a constant rate, then decreases instantly, and then increases at a constant rate.
12. The loan balance remains constant, then decreases instantly, then remains constant, then decreases instantly, and then remains constant.
13. **a.** The usage decreases at an increasing rate.

    b. The usage decreases at a decreasing rate.

Practice and Problem Solving

14. **a.** They both improved (increased scores) throughout the season.

b. Mark; Mike

15.

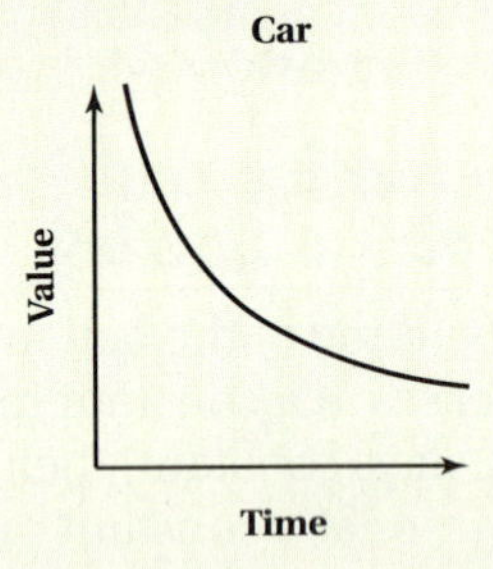

16–18. See Additional Answers.

19. See *Taking Math Deeper.*

Fair Game Review

20. (−3, −1) **21.** (2, −1)

22. (−2, 4) **23.** C

Mini-Assessment

1. Describe the change in the bacteria population over time.

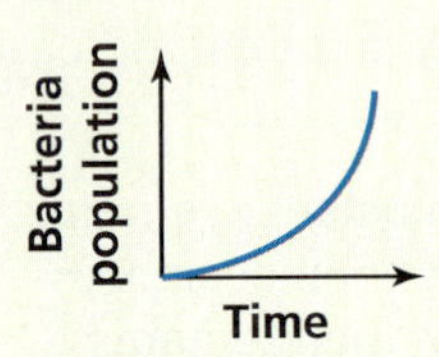

As time increases, the bacteria population increases at an increasing rate.

2. Sketch a graph that represents the situation: After takeoff, the altitude of an airplane increases at a constant rate, then remains constant for a time, then decreases at a constant rate until the airplane lands.

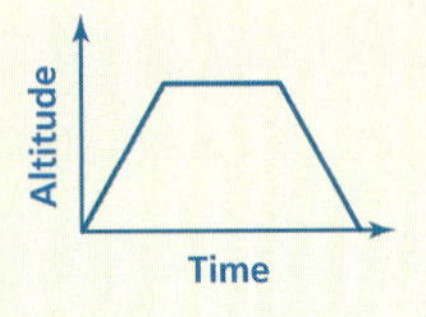

Taking Math Deeper

Exercise 19

This exercise highlights a key model in economics. Students should be able to use logic to make sense of the relationship between supply and demand.

a. Describe and interpret each curve.

The supply curve increases at an increasing rate, so suppliers produce more and more as price increases. The demand curve decreases at a decreasing rate, so consumers buy less and less as price increases.

b. Identify surplus and shortage (in quantity supplied) on the graph.

"Surplus" means *more than is needed.* So, identify the region of the graph in which quantity supplied is greater than quantity demanded for a given price.

"Shortage" means *less than is needed.* So, identify the region of the graph in which quantity supplied is less than quantity demanded for a given price.

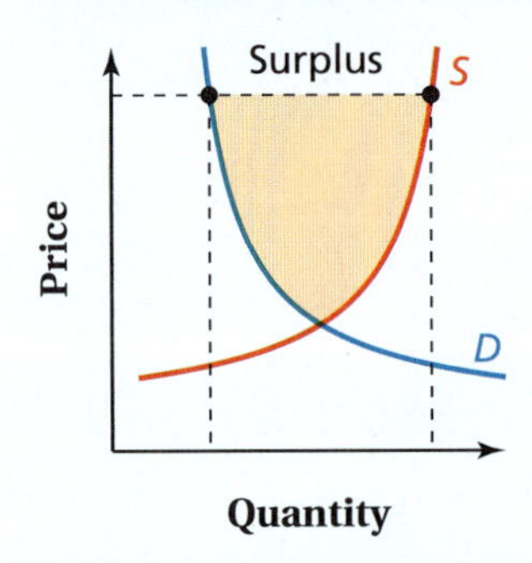

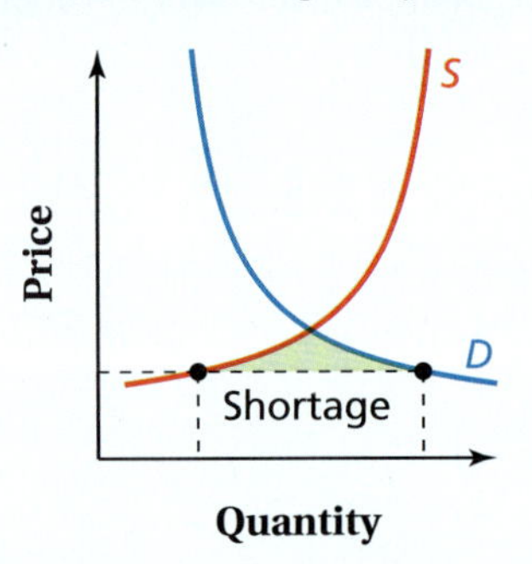

Shift the demand curve to the right.

3 **c.** The equilibrium point moves up and to the right, meaning that producers will supply more and the price will be higher.

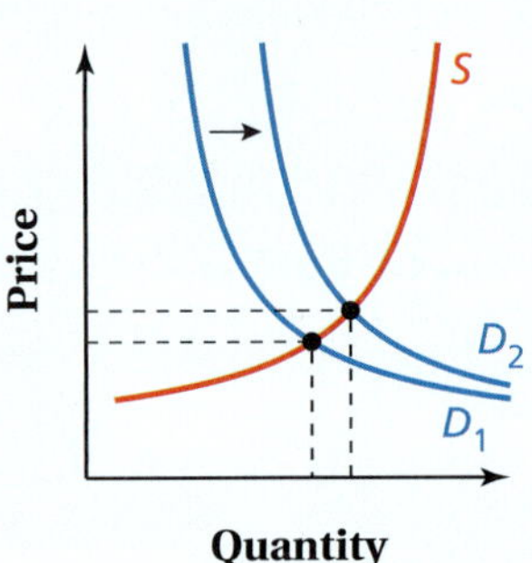

Project

Have students research *microeconomics* and *macroeconomics*, then write a few sentences describing the difference between the two. Tell them that if they attend college, they will likely have to take a course for each of these subjects.

Reteaching and Enrichment Strategies

If students need help. . .	If students got it. . .
Resources by Chapter • Practice A and Practice B • Puzzle Time Record and Practice Journal Practice Differentiating the Lesson Lesson Tutorials Skills Review Handbook	Resources by Chapter • Enrichment and Extension • Technology Connection Start the next section

14. **REASONING** The graph shows two bowlers' averages during a bowling season.

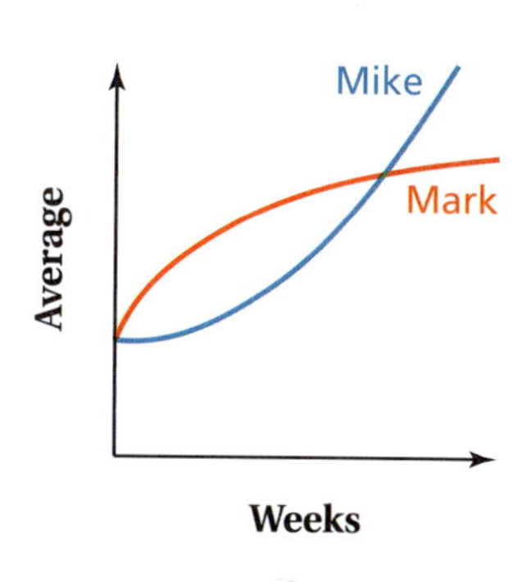

a. Describe each bowler's performance.

b. Who had a greater average most of the season? Who had a greater average at the end of the season?

Sketch a graph that represents the situation.

2 15. The value of a car depreciates. The value decreases quickly at first and then more slowly.

16. The distance from the ground changes as your friend swings on a swing.

17. The value of a rare coin increases at an increasing rate.

18. You are typing at a constant rate. You pause to think about your next paragraph, and then you resume typing at the same constant rate.

19. Economics You can use a *supply and demand model* to understand how the price of a product changes in a market. The *supply curve* of a particular product represents the quantity suppliers will produce at various prices. The *demand curve* for the product represents the quantity consumers are willing to buy at various prices.

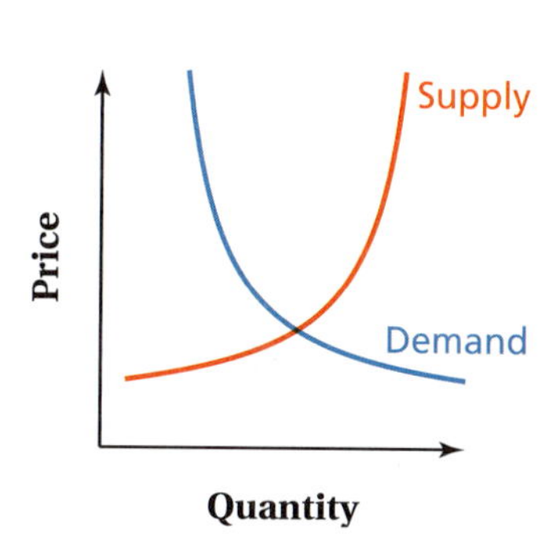

a. Describe and interpret each curve.

b. Which part of the graph represents a surplus? a shortage? Explain your reasoning.

c. The curves intersect at the *equilibrium point*, which is where the quantity produced equals the quantity demanded. Suppose that demand for a product suddenly increases, causing the entire demand curve to shift to the right. What happens to the equilibrium point?

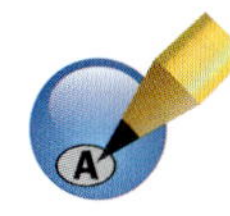

Fair Game Review What you learned in previous grades & lessons

Solve the system of linear equations by graphing. *(Section 5.1)*

20. $y = x + 2$
$y = -x - 4$

21. $x - y = 3$
$-2x + y = -5$

22. $3x + 2y = 2$
$5x - 3y = -22$

23. **MULTIPLE CHOICE** Which triangle is a rotation of Triangle D? *(Section 2.4)*

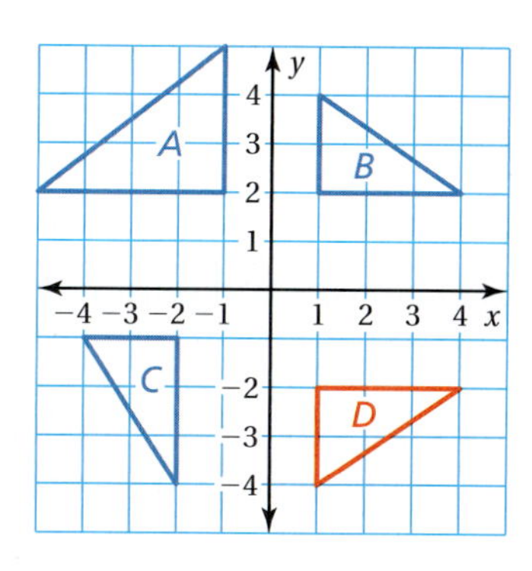

Ⓐ Triangle A

Ⓑ Triangle B

Ⓒ Triangle C

Ⓓ none

6.4–6.5 Quiz

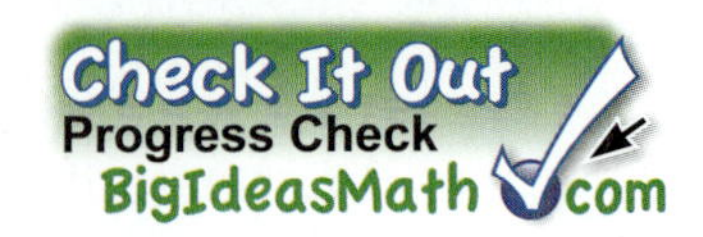

Does the table or graph represent a *linear* or *nonlinear* function? Explain. *(Section 6.4)*

1. 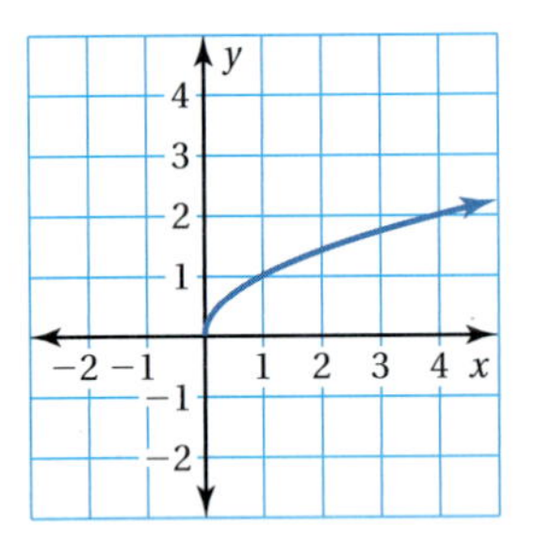

2. 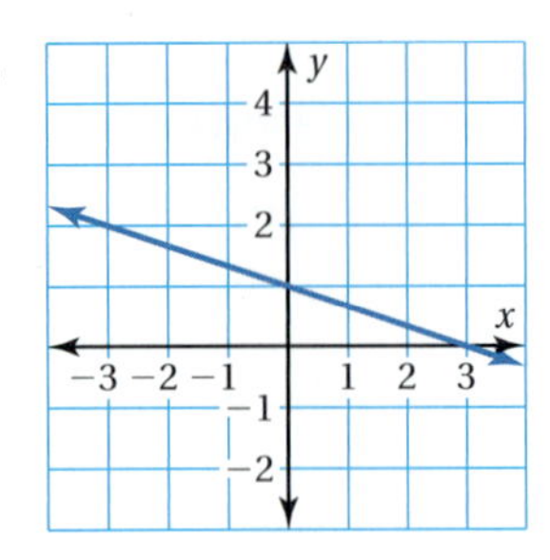

3.

x	y
0	3
3	0
6	3
9	6

4.

x	y
−1	3
1	7
3	11
5	15

5. **CHICKEN SALAD** The equation $y = 7.9x$ represents the cost y (in dollars) of buying x pounds of chicken salad. Does this equation represent a linear or nonlinear function? Explain. *(Section 6.4)*

6. **HEIGHTS** The graphs show the heights of two people over time. *(Section 6.5)*

a. Describe the change in height of each person.

b. Make three comparisons from the graphs.

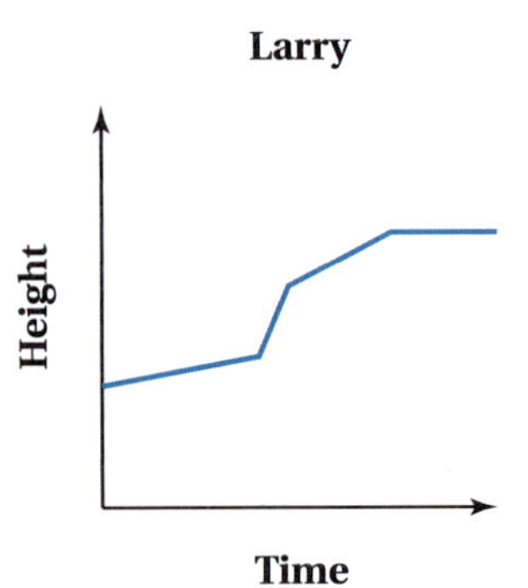

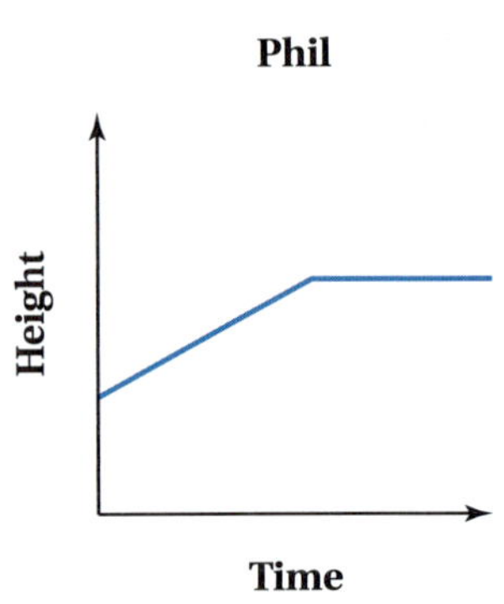

You are snowboarding down a hill. Sketch a graph that represents the situation. *(Section 6.5)*

7. You gradually increase your speed at a constant rate over time but fall about halfway down the hill. You take a short break, then get up, and gradually increase your speed again.

8. You gradually increase your speed at a constant rate over time. You come to a steep section of the hill and rapidly increase your speed at a constant rate. You then decrease your speed at a constant rate until you come to a stop.

Alternative Assessment Options

Math Chat	Student Reflective Focus Question
Structured Interview	Writing Prompt

Math Chat

- Have individual students work problems from the quiz on the board. The student explains the process used and justifies each step. Students in the class ask questions of the student presenting.
- The teacher probes the thought process of the student presenting, but does not teach or ask leading questions.

Study Help Sample Answers

Remind students to complete Graphic Organizers for the rest of the chapter.

3.

	Linear functions	Nonlinear functions
Definition/Description	A *linear function* is a function whose graph is a nonvertical line. The graph of a linear function shows a constant rate of change.	A *nonlinear function* is a function whose graph is *not* a line. A nonlinear function does not have a constant rate of change.
Equations	$y = \frac{1}{2}x + 1$ $y = 1$ $3x + 2y = 10$	$y = x^2$ $y = -\lvert x \rvert$ $y = \frac{1}{x}$
Tables	x: −2, −1, 0, 1, 2; y: 0, $\frac{1}{2}$, 1, $\frac{3}{2}$, 2 x: −2, −1, 0, 1, 2; y: 1, 1, 1, 1, 1	x: −2, −1, 0, 1, 2; y: 4, 1, 0, 1, 4 x: −2, −1, 0, 1, 2; y: −2, −1, 0, −1, −2
Graphs		

4. Available at *BigIdeasMath.com*.

Reteaching and Enrichment Strategies

If students need help. . .	If students got it. . .
Resources by Chapter • Practice A and Practice B • Puzzle Time Lesson Tutorials *BigIdeasMath.com*	Resources by Chapter • Enrichment and Extension • Technology Connection Game Closet at *BigIdeasMath.com* Start the Chapter Review

Answers

1. nonlinear; The graph is not a line.
2. linear; The graph is a line.
3. nonlinear; As x increases by 3, y changes by different amounts.
4. linear; as x increases by 2, y increases by 4
5. linear; The equation is in slope-intercept form.
6. a. Larry's height starts out steadily increasing at a constant rate. He has a growth spurt where his height rapidly increases at a constant rate. His height then steadily increases at a constant rate before flattening out and remaining unchanged.

   Phil's height starts out steadily increasing at a constant rate. His height then flattens out and remains unchanged.

   b. Both Larry and Phil are taller at the end of the time period.

   Both Larry and Phil stopped growing at some point during the time period.

   Both graphs are nonlinear. The graph that represents Larry's height has four linear sections. The graph that represents Phil's height has only two linear sections.

7–8. See Additional Answers.

Technology for the Teacher

Online Assessment
Assessment Book
ExamView® Assessment Suite

For the Teacher
Additional Review Options

- *BigIdeasMath.com*
- Online Assessment
- Game Closet at *BigIdeasMath.com*
- Vocabulary Help
- Resources by Chapter

Answers

1. no
2. yes
3. -11
4. -4
5. 7
6. 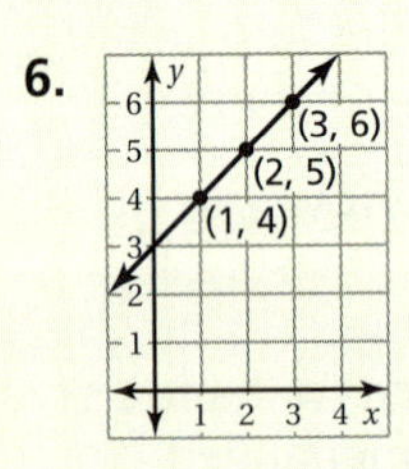

7. 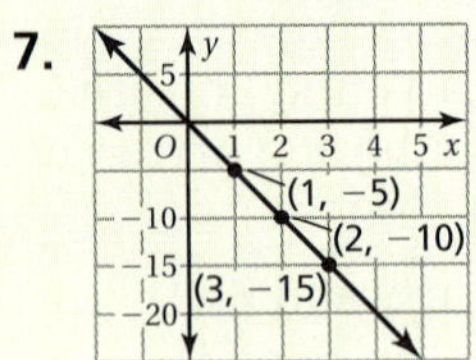

8. 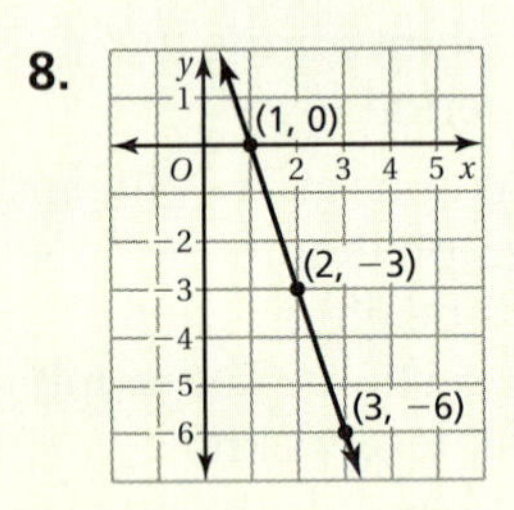

9. $y = \frac{1}{3}x + 3$
10. $y = -7$

Review of Common Errors

Exercises 3–5

- Students may substitute for the wrong variable.

Exercises 6–8

- Students may mix up their axes or label one or both axes with inconsistent increments.

Exercise 9

- Students may try to write the linear function without first finding the slope and *y*-intercept, or they may use the reciprocal of the slope in the function. Encourage them to check their work by making sure that all of the given points are solutions.

Exercise 10

- Students may notice that all of the values of *y* are the same and not be able to write the linear function. Encourage them to plot the points, and if necessary, remind them how to write the equation of a horizontal line.

Exercises 11 and 12

- Students may guess at the answer, or they may think that because the *x*-values are increasing by the same amount, the function is linear. Encourage them to examine the *y*-values or to plot the given points so that they can tell whether the table represents a linear or nonlinear function.

6 Chapter Review

Check It Out
Vocabulary Help
BigIdeasMath.com

Review Key Vocabulary

input, *p. 244*
output, *p. 244*
relation, *p. 244*
mapping diagram, *p. 244*
function, *p. 245*
function rule, *p. 250*
linear function, *p. 258*
nonlinear function, *p. 268*

Review Examples and Exercises

6.1 Relations and Functions *(pp. 242–247)*

Determine whether the relation is a function.

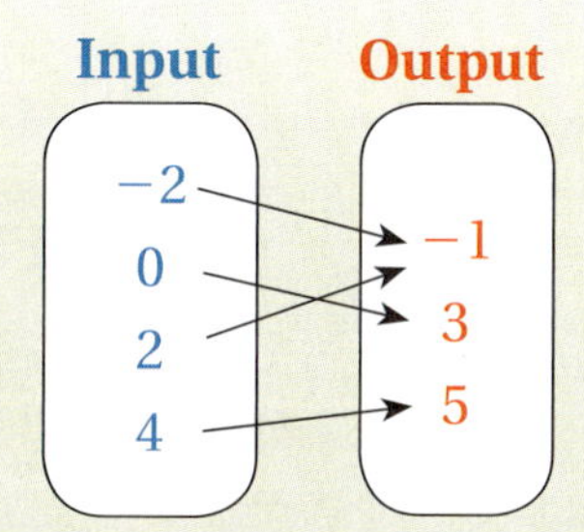

Each input has exactly one output.

∴ So, the relation is a function.

Exercises

Determine whether the relation is a function.

1. Input: 1, 3, 5, 7 Output: −4, 0, 6, 8

2. 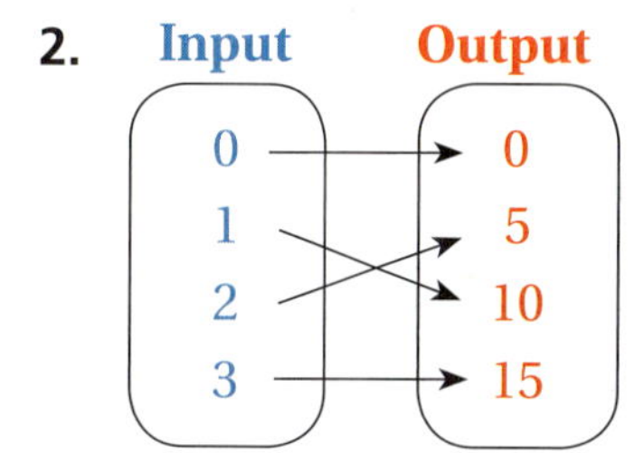

6.2 Representations of Functions *(pp. 248–255)*

Graph the function $y = x - 1$ using inputs of −1, 0, 1, and 2.

Make an input-output table.

Input, x	$x - 1$	Output, y	Ordered Pair, (x, y)
−1	−1 − 1	−2	(−1, −2)
0	0 − 1	−1	(0, −1)
1	1 − 1	0	(1, 0)
2	2 − 1	1	(2, 1)

Plot the ordered pairs and draw a line through the points.

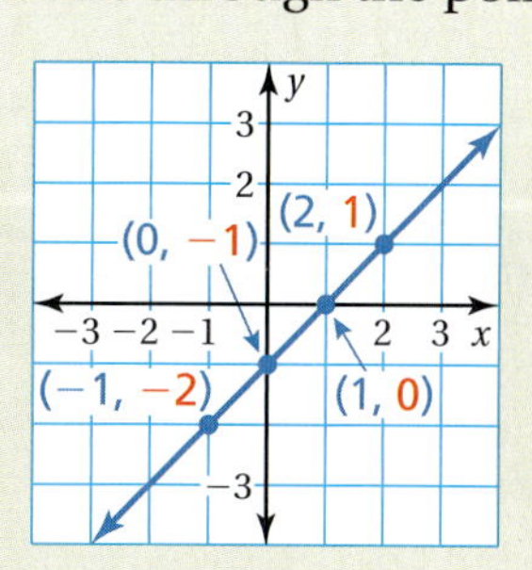

Exercises

Find the value of *y* for the given value of *x*.

3. $y = 2x - 3;\ x = -4$ **4.** $y = 2 - 9x;\ x = \frac{2}{3}$ **5.** $y = \frac{x}{3} + 5;\ x = 6$

Graph the function.

6. $y = x + 3$ **7.** $y = -5x$ **8.** $y = 3 - 3x$

6.3 Linear Functions (pp. 256–263)

Use the graph to write a linear function that relates *y* to *x*.

The points lie on a line. Find the slope by using the points (1, 1) and (2, 3).

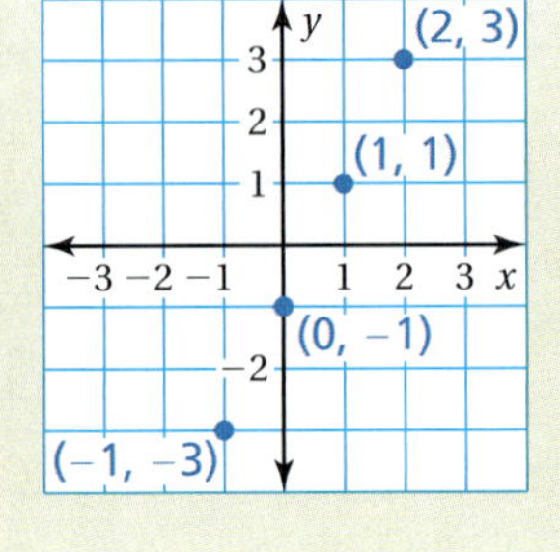

$$m = \frac{\text{change in } y}{\text{change in } x} = \frac{3 - 1}{2 - 1} = \frac{2}{1} = 2$$

Because the line crosses the *y*-axis at (0, −1), the *y*-intercept is −1.

So, the linear function is $y = 2x - 1$.

Exercises

Use the graph or table to write a linear function that relates *y* to *x*.

9.

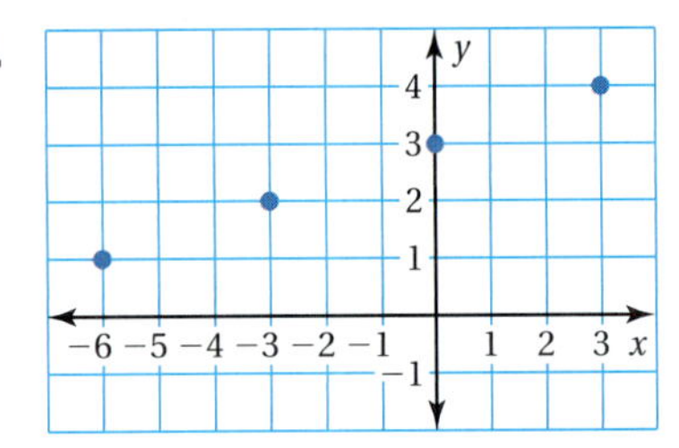

10.

x	−2	0	2	4
y	−7	−7	−7	−7

6.4 Comparing Linear and Nonlinear Functions (pp. 266–271)

Does the table represent a *linear* or *nonlinear* function? Explain.

a.

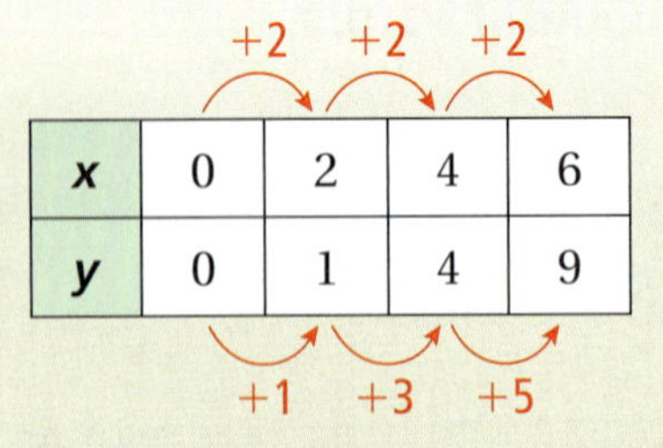

x	0	2	4	6
y	0	1	4	9

As *x* increases by 2, *y* increases by different amounts. The rate of change is *not* constant. So, the function is nonlinear.

b.

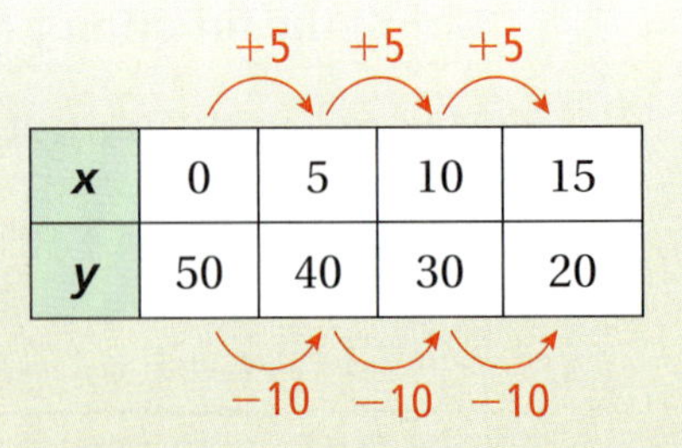

x	0	5	10	15
y	50	40	30	20

As *x* increases by 5, *y* decreases by 10. The rate of change is constant. So, the function is linear.

Review Game

Writing Linear Functions

Materials per Group:

- paper
- two yard sticks
- pencils

Directions:

- Divide the class into an even number of groups.
- Groups pair up to compete against each other.
- Each pair of groups makes a paper football.
- Students in each pair of groups take turns flicking the football with their fingers as high and as far as they can. Students from the other group measure and record the length and height that the football travels. Both groups have to agree on the measurements. Length can be measured after the football has come to rest, but height must be measured while it is moving.
- Each student writes the ordered pair representing the point where the football reached its maximum height during his or her turn. (See figure below.) Students write one linear function to approximate the ascent and another linear function to approximate the descent.

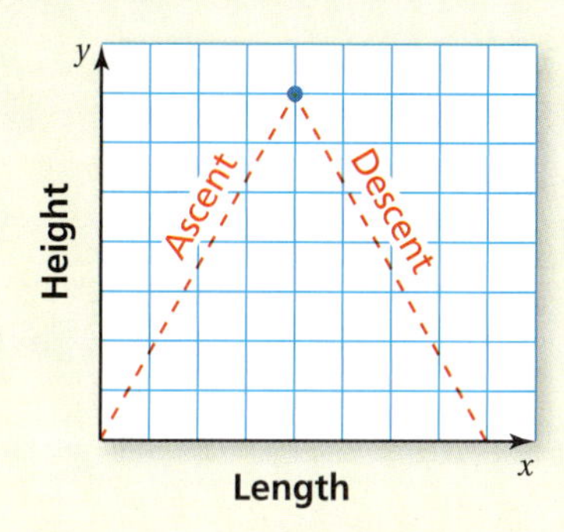

Who Wins?

Students earn their group a point for each inch the football travels horizontally, vertically up, and vertically down. Points only count if the linear functions correctly model the approximated lines of ascent and descent. The group with the most points wins.

For the Student Additional Practice

- Lesson Tutorials
- Multi-Language Glossary
- Self-Grading Progress Check
- *BigIdeasMath.com*
  Dynamic Student Edition
  Student Resources

Answers

11. linear; As x increases by 3, y increases by 9.

12. nonlinear; As x increases by 2, y changes by different amounts.

13. See Additional Answers.

14.

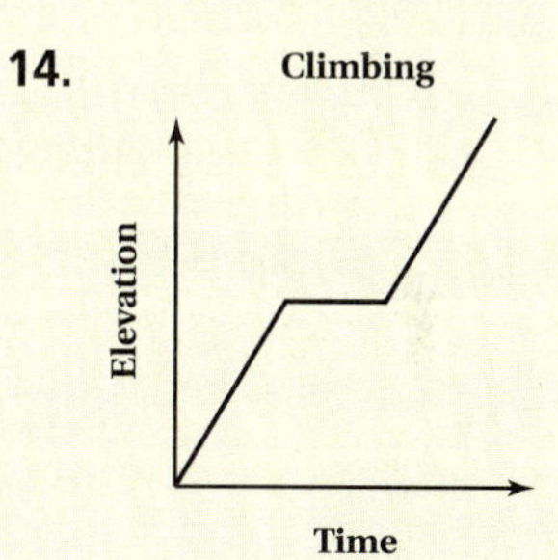

15.

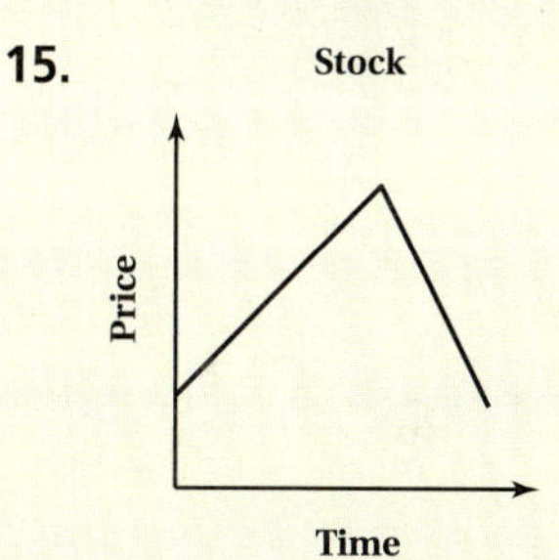

My Thoughts on the Chapter

Teacher Tip

Not allowed to write in your teaching edition? Use sticky notes to record your thoughts.

What worked. . .

What did not work. . .

What I would do differently. . .

Exercises

Does the table represent a *linear* or *nonlinear* function? Explain.

11.

x	3	6	9	12
y	1	10	19	28

12.

x	1	3	5	7
y	3	1	1	3

6.5 Analyzing and Sketching Graphs *(pp. 272–277)*

The graphs show the populations of two cities over several years.

a. Describe the change in population in each city.

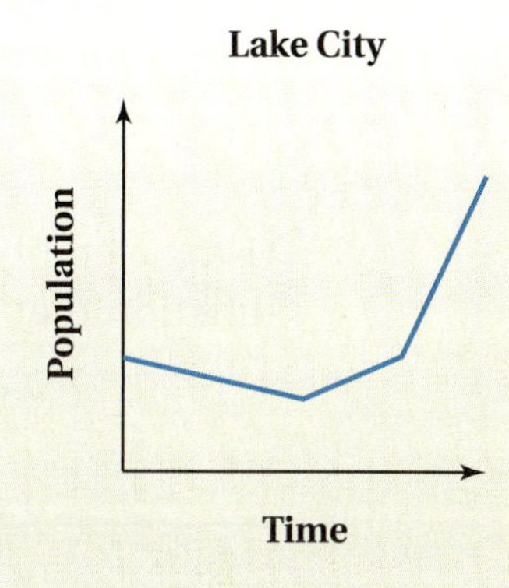

Lake City: The population gradually decreases at a constant rate, then gradually increases at a constant rate. Then the population rapidly increases at a constant rate.

Gold Point: The population rapidly increases at a constant rate. Then the population stays the same for a short period of time before gradually decreasing at a constant rate.

b. Make three comparisons from the graphs.

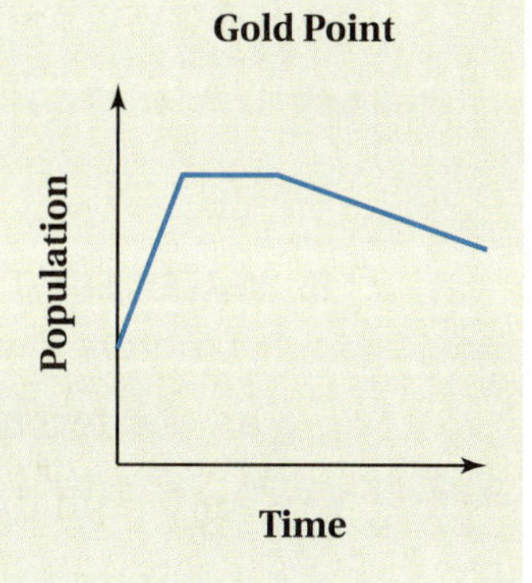

- Both graphs show increasing and decreasing populations.
- Both graphs are nonlinear, but both graphs consist of three linear sections.
- Both populations at the end of the time period are greater than the populations at the beginning of the time period.

Exercises

13. SALES The graphs show the sales of two companies.

a. Describe the sales of each company.

b. Make three comparisons from the graphs.

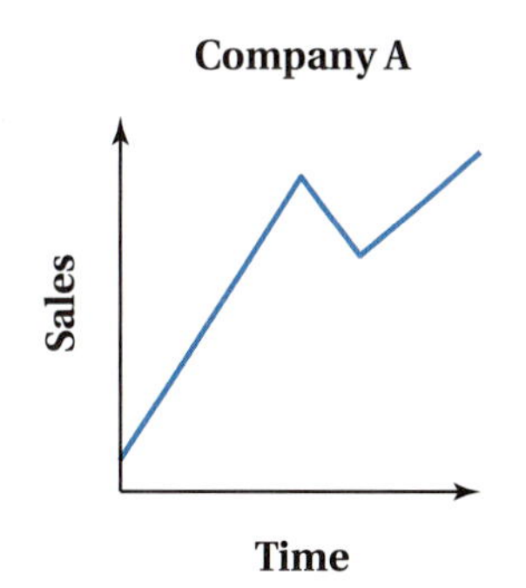

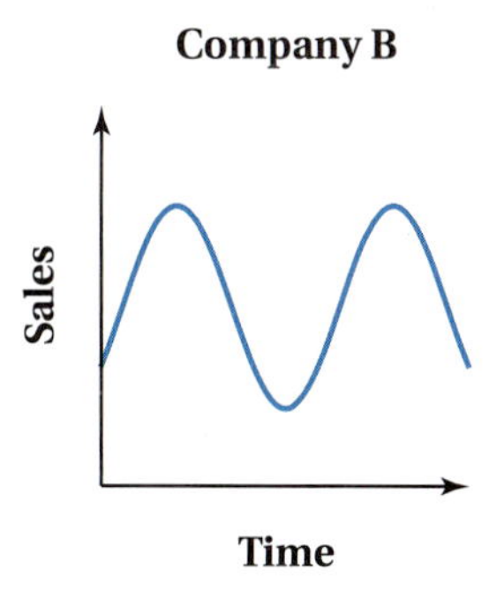

Sketch a graph that represents the situation.

14. You climb up a climbing wall. You gradually climb halfway up the wall at a constant rate, then stop and take a break. You then climb to the top of the wall at a constant rate.

15. The price of a stock steadily increases at a constant rate for several months before the stock market crashes. The price then quickly decreases at a constant rate.

6 Chapter Test

Determine whether the relation is a function.

1.

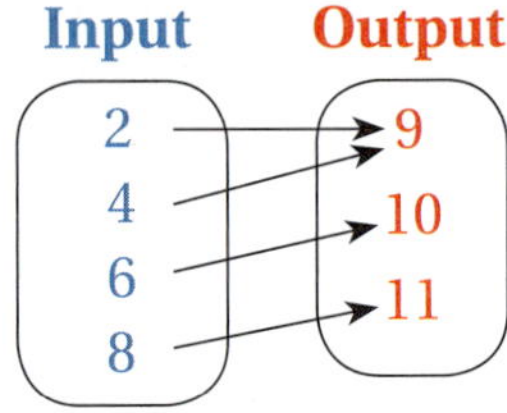

2.

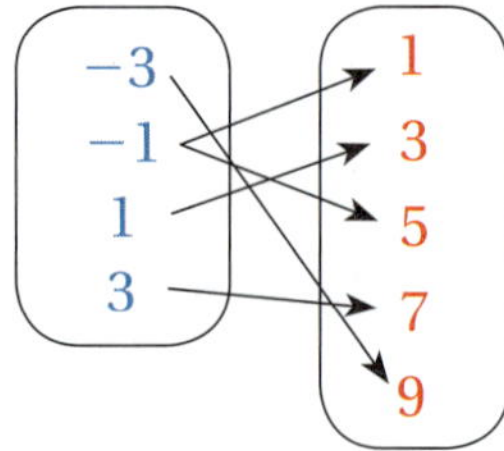

Graph the function.

3. $y = x + 8$

4. $y = 1 - 3x$

5. $y = x - 4$

6. Use the graph to write a linear function that relates y to x.

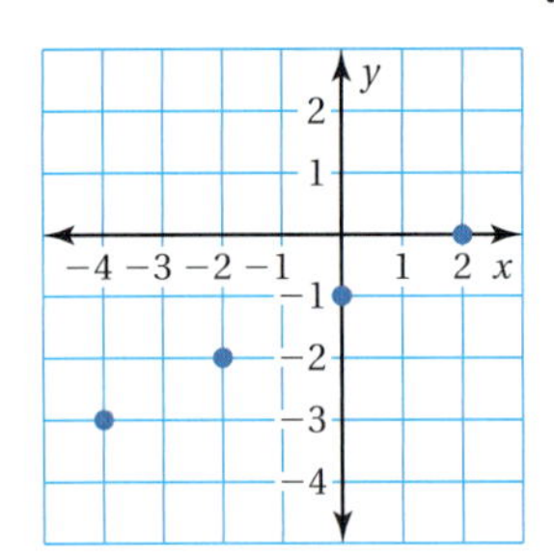

7. Does the table represent a *linear* or *nonlinear* function? Explain.

x	0	2	4	6
y	8	0	−8	−16

8. WATER SKI The table shows the number of meters a water skier travels in x minutes.

Minutes, x	1	2	3	4	5
Meters, y	600	1200	1800	2400	3000

a. Write a function that relates x to y.

b. Graph the linear function.

c. At this rate, how many *kilometers* would the water skier travel in 12 minutes?

9. STOCKS The graphs show the prices of two stocks during one day.

a. Describe the prices of each stock.

b. Make three comparisons from the graphs.

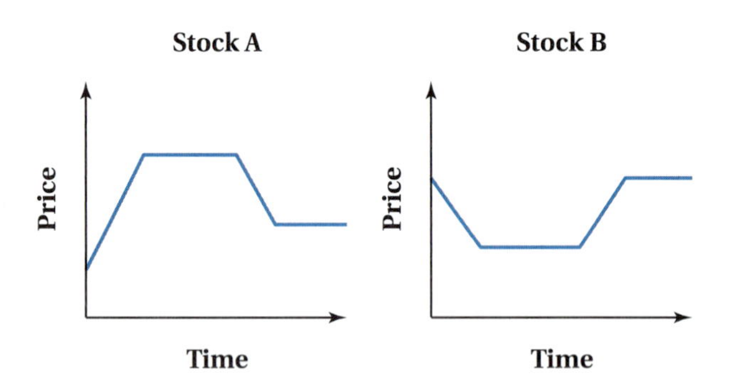

10. RACE You are competing in a race. You begin the race by increasing your speed at a constant rate. You then run at a constant speed until you get a cramp and have to stop. You wait until your cramp goes away before you start gradually increasing your speed again at a constant rate. Sketch a graph that represents the situation.

Test Item References

Chapter Test Questions	Section to Review	Florida Common Core Standards (MACC)
1, 2	6.1	8.F.1.1
3–5	6.2	8.F.1.1
6, 8	6.3	8.F.1.2, 8.F.1.3, 8.F.2.4
7	6.4	8.F.1.3
9, 10	6.5	8.F.2.5

Test-Taking Strategies

Remind students to quickly look over the entire test before they start so that they can budget their time. Have them use the **Stop** and **Think** strategy before they answer each question.

Common Errors

- **Exercises 3–5** When graphing a function, students may mix up their axes, or scale one or both axes with inconsistent increment widths. Remind them that the *x*-axis is horizontal and the *y*-axis is vertical. Encourage them to scale their axes consistently so that a line can be drawn through the ordered pairs that they plot.
- **Exercise 6** Students may try to write the linear function without first finding the slope and *y*-intercept, or they may use the reciprocal of the slope in the function. Encourage them to check their work by making sure that all of the given points are solutions.
- **Exercise 7** Students may guess that a table represents a linear function. Encourage them to plot the given points so they can tell for sure.

Reteaching and Enrichment Strategies

If students need help...	If students got it...
Resources by Chapter • Practice A and Practice B • Puzzle Time Record and Practice Journal Practice Differentiating the Lesson Lesson Tutorials *BigIdeasMath.com* Skills Review Handbook	Resources by Chapter • Enrichment and Extension • Technology Connection Game Closet at *BigIdeasMath.com* Start Standards Assessment

Answers

1. Each input has exactly one output. So, the relation is a function.

2. The input -1 has two outputs, 1 and 5. So, the relation is not a function.

3. 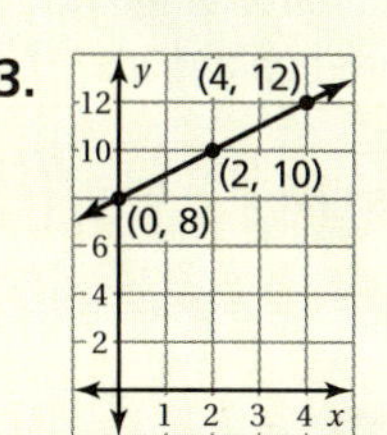

4. 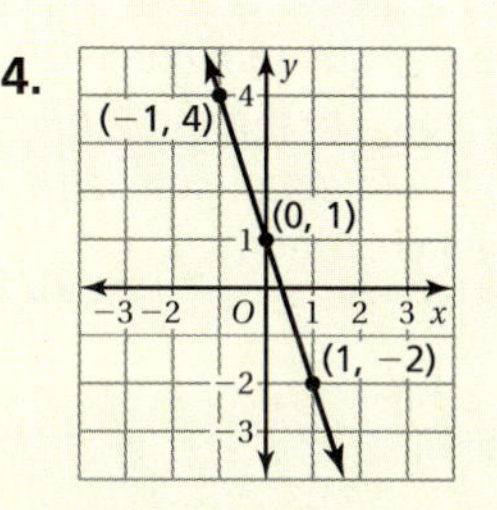

5. 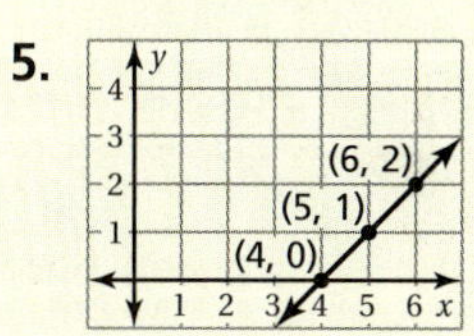

6. $y = \frac{1}{2}x - 1$

7. linear; As x increases by 2, y decreases by 8.

8. a. $y = 600x$

   b. 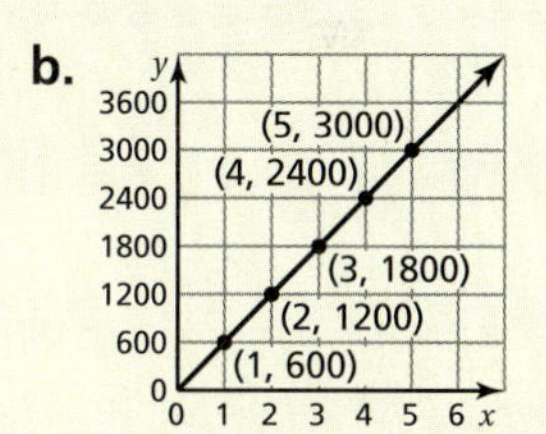

   c. 7.2 km

9. See Additional Answers.

10. 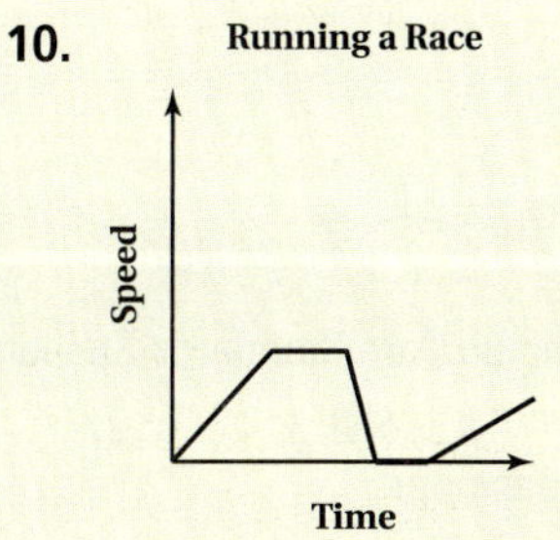

Item Analysis

1. **A.** The student subtracts −4 and 1 incorrectly.

   B. Correct answer

   C. The student subtracts 5 and −3 incorrectly, and subtracts −4 and 1 incorrectly.

   D. The student subtracts 5 and −3 incorrectly.

2. **F.** The student subtracts 9 from both sides instead of adding. The student then divides both sides by 3 instead of multiplying.

   G. The student divides both sides by 3 instead of multiplying.

   H. The student subtracts 9 from both sides instead of adding.

   I. Correct answer

3. **Gridded Response:** Correct answer: 7

   Common Error: The student does not divide the differences between the outputs by 3, getting an answer of 21.

4. **A.** The student reverses the coordinates.

   B. Correct answer

   C. The student reverses the coordinates and omits a negative sign.

   D. The student omits a negative sign.

Test-Taking Strategies

Available at *BigIdeasMath.com*

After Answering Easy Questions, Relax
Answer Easy Questions First
Estimate the Answer
Read All Choices before Answering
Read Question before Answering
Solve Directly or Eliminate Choices
Solve Problem before Looking at Choices
Use Intelligent Guessing
Work Backwards

About this Strategy

One way to answer the question is to work backwards. Try putting the responses into the question, one at a time, and see if you get a correct solution.

Answers

1. B
2. I
3. 7
4. B

Technology for the Teacher

Florida Common Core Standards Support Performance Tasks
Online Assessment
Assessment Book
ExamView® Assessment Suite

6 Standards Assessment

1. What is the slope of the line shown in the graph below? *(MACC.8.EE.2.6)*

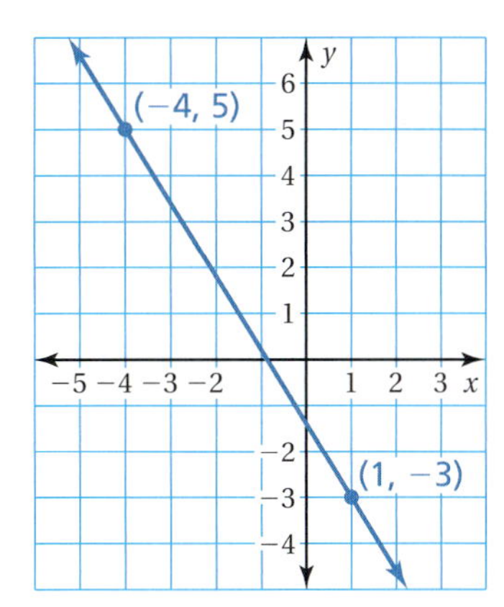

A. $-\frac{8}{3}$ **C.** $-\frac{2}{3}$

B. $-\frac{8}{5}$ **D.** $-\frac{2}{5}$

2. Which value of a makes the equation below true? *(MACC.8.EE.3.7b)*

$$24 = \frac{a}{3} - 9$$

F. 5 **H.** 45

G. 11 **I.** 99

3. A mapping diagram is shown.

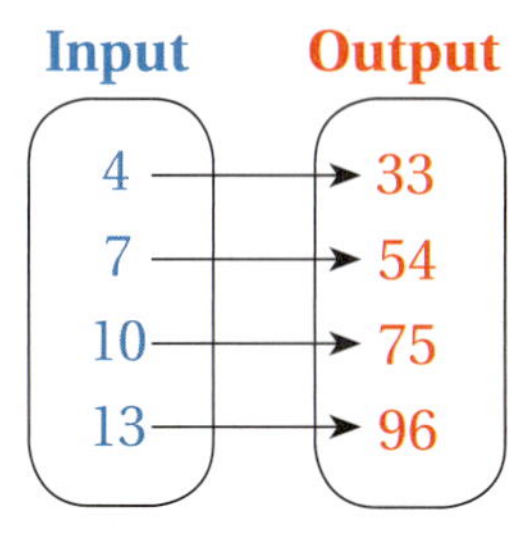

What number belongs in the box below so that the equation will correctly describe the function represented by the mapping diagram? *(MACC.8.F.1.1)*

$$y = \square x + 5$$

4. What is the solution of the system of linear equations shown below? *(MACC.8.EE.3.8b)*

$$y = 2x - 1$$

$$y = 3x + 5$$

A. $(-13, -6)$ **C.** $(-13, 6)$

B. $(-6, -13)$ **D.** $(-6, 13)$

5. A system of two linear equations has no solution. What can you conclude about the graphs of the two equations? *(MACC.8.EE.3.8a)*

F. The lines have the same slope and the same y-intercept.

G. The lines have the same slope and different y-intercepts.

H. The lines have different slopes and the same y-intercept.

I. The lines have different slopes and different y-intercepts.

6. Which graph shows a nonlinear function? *(MACC.8.F.1.3)*

A.

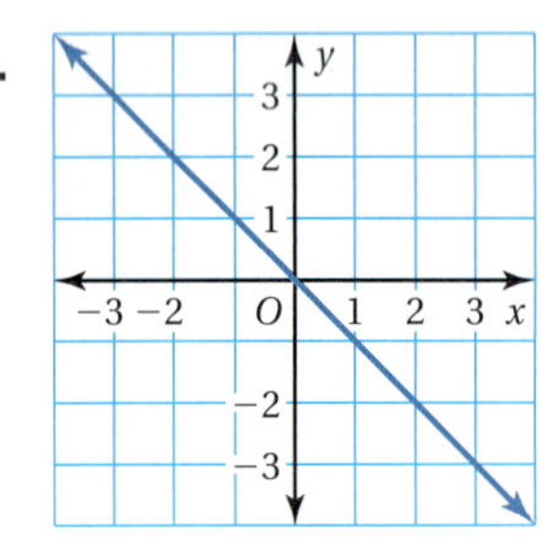

C.

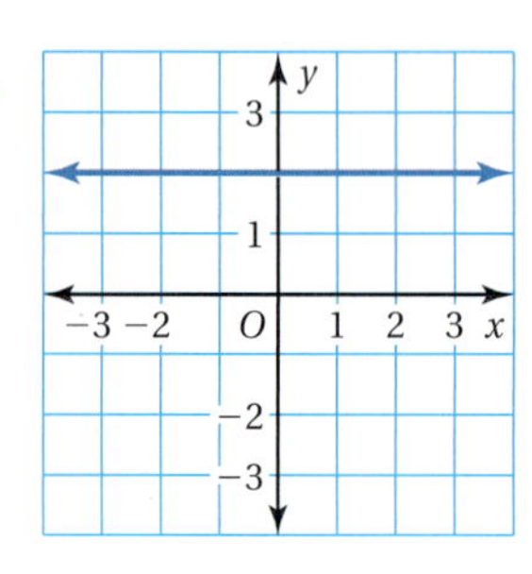

B.

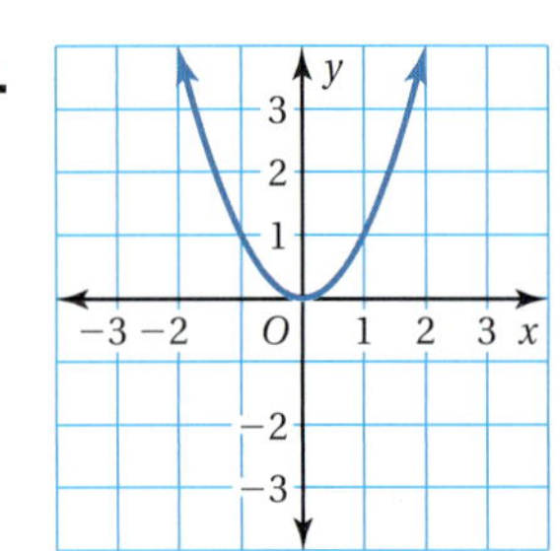

D.

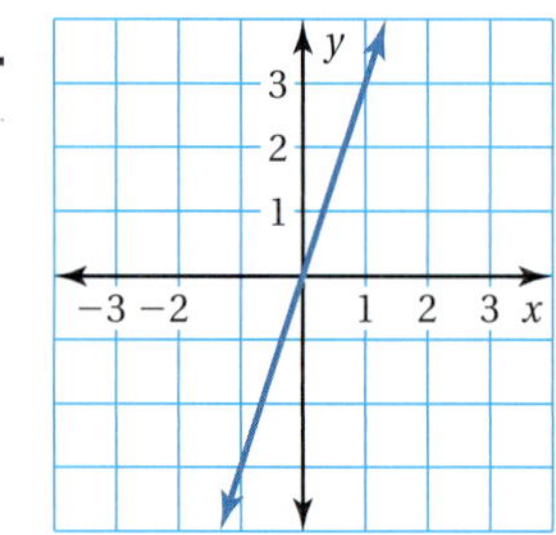

7. What is the value of x? *(MACC.8.G.1.5)*

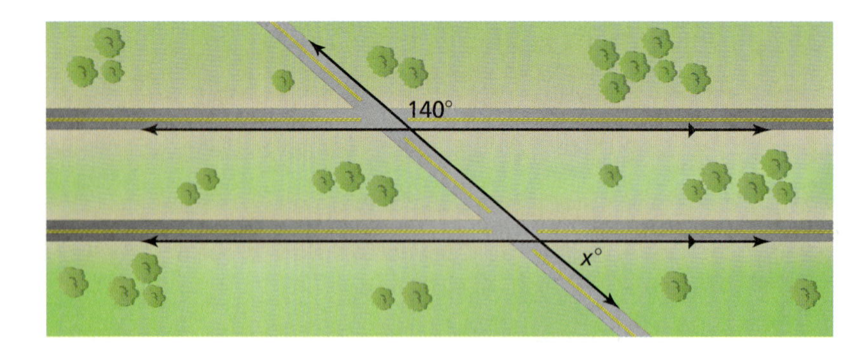

F. 40

G. 50

H. 140

I. 220

Item Analysis (continued)

5. **F.** The student thinks the lines must coincide.
   G. Correct answer
   H. The student reverses the idea that the slopes are the same and the *y*-intercepts are different.
   I. The student incorrectly reasons that both the slopes and *y*-intercepts must be different for there to be no solution.

6. **A.** The student thinks that a line with a negative slope is nonlinear.
   B. Correct answer
   C. The student thinks that a horizontal line is nonlinear.
   D. The student thinks that a steep line with a positive slope is nonlinear.

7. **F.** Correct answer
   G. The student calculated 140 − 90.
   H. The student used the measure of the given angle.
   I. The student calculated 360 − 140.

Answers

5. G
6. B
7. F

Answers

8. *Part A* yes; The sales change by the same amount each year.

Part B no; The sales do not change by the same amount each year.

9. C

10. 3

Item Analysis (continued)

8. 2 points The student demonstrates a thorough understanding of how to determine whether data show a linear function or a nonlinear function, explains the work fully, and relates perimeter to a linear function and area to a nonlinear function. The first table shows a linear function. The second table shows a nonlinear function.

1 point The student's work and explanations demonstrate a lack of essential understanding. The slope formula is used incorrectly or a graph of the data is incomplete.

0 points The student provides no response, a completely incorrect or incomprehensible response, or a response that demonstates insufficient understanding of linear functions and nonlinear functions.

9. A. The student associates 0 with a solution.

B. The student selects this choice because the first equation equals 0 when $x = 4$.

C. Correct answer

D. The student has no idea how to interpret the table or fails to see that the functions are equal at $x = 8$.

10. Gridded Response: Correct answer: 3

Common Error: Because $\overline{AC}$ is 4 cm shorter than $\overline{BC}$, the student subtracts 4 cm from 5 cm (the length of $\overline{EF}$) for an answer of 1 cm.

8. The tables show the sales (in millions of dollars) for two companies over a 5-year period. Examine the data in the tables. *(MACC.8.F.1.3)*

Think
Solve
Explain

Year	1	2	3	4	5
Sales	2	4	6	8	10

Year	1	2	3	4	5
Sales	1	1	2	3	5

Part A Does the first table show a linear function? Explain your reasoning.

Part B Does the second table show a linear function? Explain your reasoning.

9. The equations $y = -x + 4$ and $y = \frac{1}{2}x - 8$ form a system of linear equations. The table below shows the y-value for each equation at six different values of x. *(MACC.8.EE.3.8a)*

x	0	2	4	6	8	10
$y = -x + 4$	4	2	0	−2	−4	−6
$y = \frac{1}{2}x - 8$	−8	−7	−6	−5	−4	−3

What can you conclude from the table?

A. The system has one solution, when $x = 0$.

B. The system has one solution, when $x = 4$.

C. The system has one solution, when $x = 8$.

D. The system has no solution.

10. In the diagram below, Triangle ABC is a dilation of Triangle DEF. What is the value of x? *(MACC.8.G.1.4)*

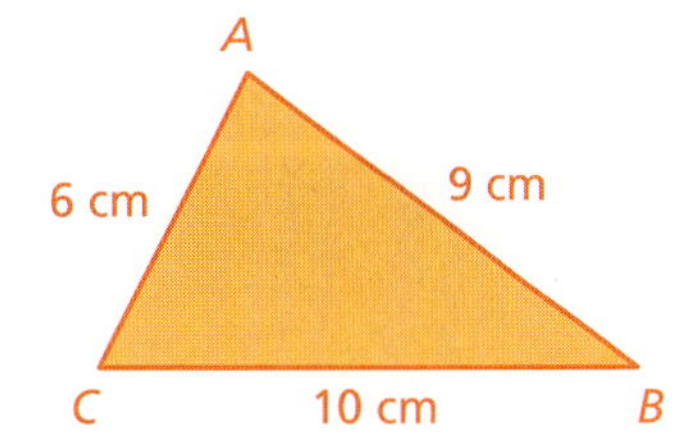

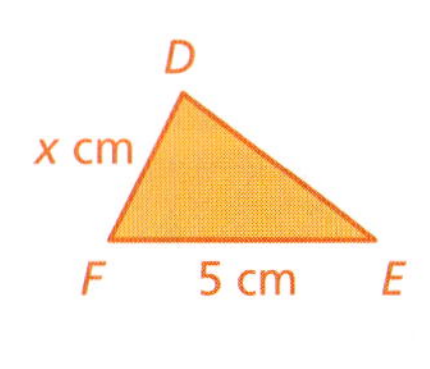

7 Real Numbers and the Pythagorean Theorem

7.1 Finding Square Roots
7.2 Finding Cube Roots
7.3 The Pythagorean Theorem
7.4 Approximating Square Roots
7.5 Using the Pythagorean Theorem

"I'm pretty sure that Pythagoras was a Greek."

"I said 'Greek,' not 'Geek.' "

"Here's how I remember the square root of 2."

"February is the 2nd month. It has 28 days. Split 28 into 14 and 14. Move the decimal to get 1.414."

Florida Common Core Progression

6th Grade

- Fluently divide whole numbers.
- Evaluate expressions with whole-number exponents.
- Divide fractions.
- Fluently add, subtract, multiply, and divide decimals.
- Find distances between points with the same *x*- or *y*-coordinate.

7th Grade

- Convert rational numbers to decimals using long division.
- Add, subtract, multiply, and divide rational numbers.
- Understand that every quotient of integers (non-zero divisor) is a rational number.

8th Grade

- Understand that every rational number has a decimal expansion that terminates or repeats.
- Understand that numbers that are not rational are irrational.
- Compare irrational numbers using rational approximations.
- Evaluate square roots and cube roots, including those resulting from solving equations.
- Explain a proof of the Pythagorean Theorem and its converse.
- Use the Pythagorean Theorem to find missing measures of right triangles and distances between points in the coordinate plane.

Chapter Summary

Section	Florida Common Core Standard (MACC)	
7.1	Learning	8.EE.1.2
7.2	Learning	8.EE.1.2
7.3	Learning	8.EE.1.2, 8.G.2.6, 8.G.2.7, 8.G.2.8
7.4	Learning	8.NS.1.1 ★, 8.NS.1.2 ★, 8.EE.1.2
7.5	Learning	8.EE.1.2 ★, 8.G.2.6 ★, 8.G.2.7 ★, 8.G.2.8 ★
★ Teaching is complete. Standard can be assessed.		

Pacing Guide for Chapter 7

Chapter Opener	1 Day
Section 1	
Activity	1 Day
Lesson	1 Day
Section 2	
Activity	1 Day
Lesson	1 Day
Section 3	
Activity	1 Day
Lesson	1 Day
Study Help / Quiz	1 Day
Section 4	
Activity	1 Day
Lesson	1 Day
Extension	1 Day
Section 5	
Activity	1 Day
Lesson	1 Day
Chapter Review/ Chapter Tests	2 Days
Total Chapter 7	15 Days
Year-to-Date	106 Days

Technology for the Teacher

BigIdeasMath.com
Chapter at a Glance
Complete Materials List
Parent Letters: English, Spanish, and Haitian Creole

Florida Common Core Standards

MACC.5.NBT.1.3b Compare two decimals . . . using >, =, and < symbols to record the results of comparisons.

MACC.7.NS.1.1 . . . add and subtract rational numbers . . .

MACC.7.NS.1.2 . . . multiply and divide rational numbers . . .

Additional Topics for Review

- Number line
- Converting decimals to fractions
- Order of operations
- Exponents
- Compare and order decimals and fractions

Try It Yourself

1. $=$ **2.** $<$

3. $<$

4. *Sample answer:* $-0.009, -0.001, 0.01$

5. *Sample answer:* $-1.75, -1.74, 1.74$

6. *Sample answer:* $-0.75, 0.74, 0.75$

7. -3 **8.** 181

9. 99

Record and Practice Journal Fair Game Review

1. $<$ **2.** $>$

3. $=$ **4.** $>$

5–8. Sample answers are given.

5. $-5.2, -5.3, -6.5$

6. $2.56, 2.3, -3.2$

7. $-3.18, -3.1, -2.05$

8. $0.05, 0.3, 1.55$

9. 12.49; 12.495; 12.55; 12.60; 12.63

10. 167 **11.** 3

12–16. See Additional Answers.

Math Background Notes

Vocabulary Review

- Greater Than
- Less Than
- Order of Operations

Comparing Decimals

- Students should know how to compare decimals.
- **Teaching Tip:** Some students will have difficulty determining which decimal is greater simply by looking. Encourage these students to convert the decimals to fractions with a common denominator and compare the numerators.
- **Common Error:** Some students will have difficulty with Example 3 because there is not one "right" answer. Remind them that any number that makes the comparison true is a correct answer. Encourage creativity and remind students their answers will not always match the teacher's answers!

Using Order of Operations

- Students should know how to use the order of operations.
- You may want to review the correct order of operations with students. Many students probably learned the pneumonic device *Please Excuse My Dear Aunt Sally.* Ask a volunteer to explain why this phrase is helpful.
- You may want to review exponents with students. Remind students that the exponent tells you how many times the base is a factor. Exponents express repeated multiplication.

Reteaching and Enrichment Strategies

If students need help. . .	If students got it. . .
Record and Practice Journal • Fair Game Review Skills Review Handbook Lesson Tutorials	Game Closet at *BigIdeasMath.com* Start the next section

What You Learned Before

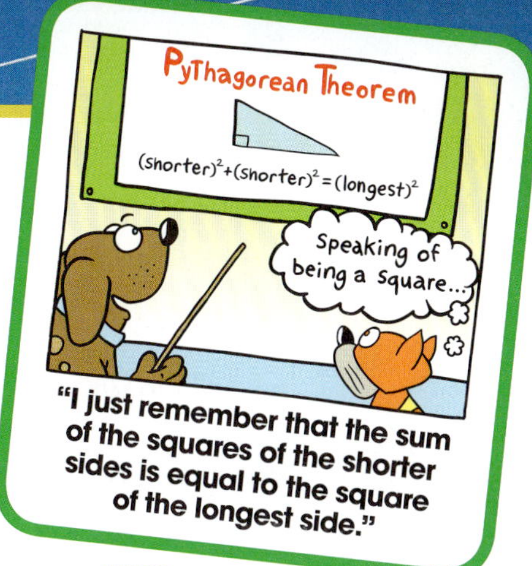

"I just remember that the sum of the squares of the shorter sides is equal to the square of the longest side."

Comparing Decimals (MACC.5.NBT.1.3b)

Complete the number sentence with <, >, or =.

Example 1 **1.1 $\square$ 1.01**

Because $\frac{110}{100}$ is greater than $\frac{101}{100}$, 1.1 is greater than 1.01.

So, $1.1 > 1.01$.

Example 2 **-0.3 $\square$ -0.003**

Because $-\frac{300}{1000}$ is less than $-\frac{3}{1000}$, -0.3 is less than -0.003.

So, $-0.3 < -0.003$.

Example 3 **Find three decimals that make the number sentence $-5.12 > \square$ true.**

Any decimal less than -5.12 will make the sentence true.

Sample answer: -10.1, -9.05, -8.25

Try It Yourself

Complete the number sentence with <, >, or =.

1. $2.10 \ \square \ 2.1$

2. $-4.5 \ \square \ -4.25$

3. $\pi \ \square \ 3.2$

Find three decimals that make the number sentence true.

4. $-0.01 \leq \square$

5. $1.75 > \square$

6. $0.75 \geq \square$

Using Order of Operations (MACC.7.NS.1.1, MACC.7.NS.1.2)

Example 4 **Evaluate $8^2 \div (32 \div 2) - 2(3 - 5)$.**

First:	Parentheses	$8^2 \div (32 \div 2) - 2(3 - 5) = 8^2 \div 16 - 2(-2)$
Second:	Exponents	$= 64 \div 16 - 2(-2)$
Third:	Multiplication and Division (from left to right)	$= 4 + 4$
Fourth:	Addition and Subtraction (from left to right)	$= 8$

Try It Yourself

Evaluate the expression.

7. $15\left(\frac{12}{3}\right) - 7^2 - 2 \cdot 7$

8. $3^2 \cdot 4 \div 18 + 30 \cdot 6 - 1$

9. $-1 + \left(\frac{4}{2}(6 - 1)\right)^2$

7.1 Finding Square Roots

Essential Question How can you find the dimensions of a square or a circle when you are given its area?

When you multiply a number by itself, you square the number.

Symbol for squaring is the exponent 2. → $4^2 = 4 \cdot 4$

$= 16$ 4 squared is 16.

To "undo" this, take the *square root* of the number.

Symbol for square root is a *radical sign*, $\sqrt{}$. → $\sqrt{16} = \sqrt{4^2} = 4$ The square root of 16 is 4.

1 ACTIVITY: Finding Square Roots

Work with a partner. Use a square root symbol to write the side length of the square. Then find the square root. Check your answer by multiplying.

a. Sample: $s = \sqrt{121} = 11$ ft

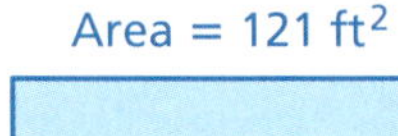

Area = 121 ft²

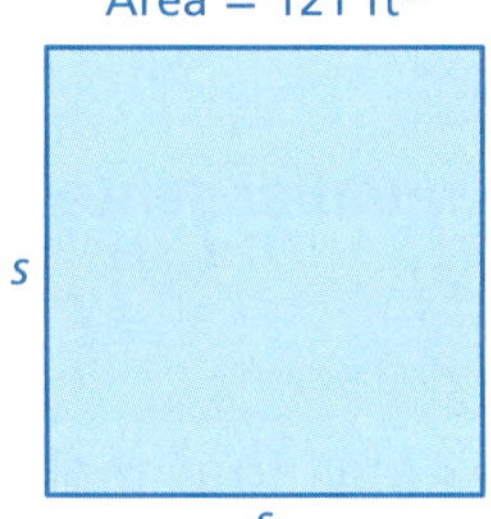

s, s

Check

$$\begin{array}{r} 11 \\ \times\, 11 \\ \hline 11 \\ 110 \\ \hline 121 \end{array} \checkmark$$

∴ The side length of the square is 11 feet.

b. Area = 81 yd²

s, s

c. Area = 324 cm²

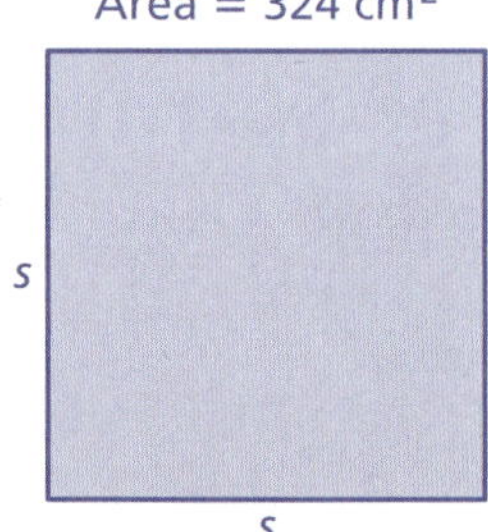

s, s

d. Area = 361 mi²

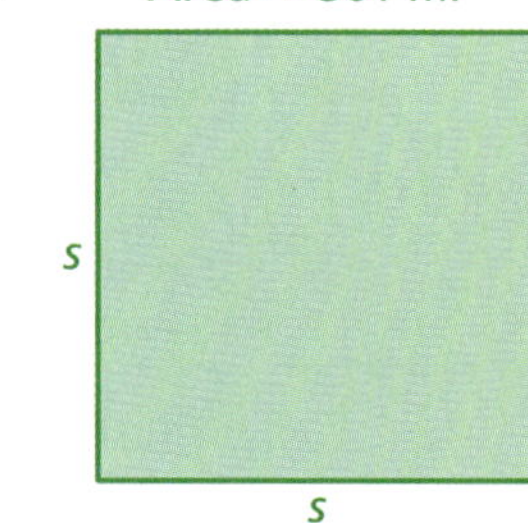

s, s

e. Area = 225 mi²

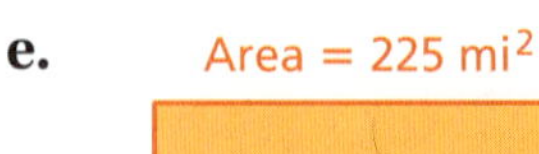

s, s

f. Area = 2.89 in.²

s, s

g. Area = $\frac{4}{9}$ ft²

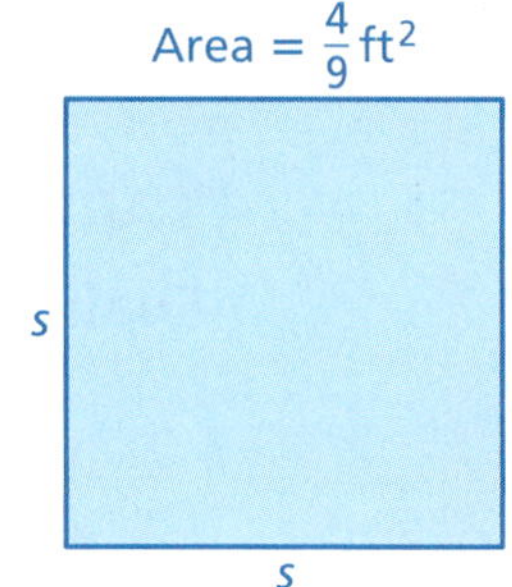

s, s

Square Roots

In this lesson, you will

- find square roots of perfect squares.
- evaluate expressions involving square roots.
- use square roots to solve equations.

Learning Standard
MACC.8.EE.1.2

Laurie's Notes

Introduction

Standards for Mathematical Practice

- **MP6 Attend to Precision:** In these activities, students will need to multiply fractions and decimals and recall perfect squares. When graphing, accuracy will be necessary in order to recognize that the plotted points are not linear.

Motivate

- **Preparation:** Make two (or more) pendulums of different lengths.
- Swing the two pendulums back and forth a few times while telling a story.
- ? "Does it take the same amount of time for the pendulum to go back and forth for each length?" Answers will vary.
- **Extension:** Have a student time you as you swing the pendulum through 10 periods. Divide the total time by 10 to find the time of one period. Repeat for a different length pendulum.

Activity Notes

Activity 1

- **Preparation:** Cut a number of squares from paper. Calculate the areas of several and write the areas on the figures. Make one of the squares 6 cm by 6 cm.
- ? "Can you find the area of any of these squares? Explain." Yes, multiply length by width.
- ? "There are no dimensions marked on the square, so how do you find the area?" Measure the side lengths, then multiply to find the area.
- Hold up the 6-cm-by-6-cm square and show students that it has Area = 36 cm^2 recorded on it.
- ? "If you know the area, how can you find the dimensions without measuring the side lengths?" Ideas will vary.
- **Common Error:** Students will say to divide by 4 (perimeter) or divide by 2. Using the 6-cm-by-6-cm square, remind students that the side lengths are the same and they are multiplied together to get 36. So, the side lengths must be 6 because $6(6) = 36$.
- Students should now be ready to begin the activity.
- When students have finished, discuss the answers and the strategies they used.

Florida Common Core Standards

MACC.8.EE.1.2 Use square root and cube root symbols to represent solutions to equations of the form $x^2 = p$ and $x^3 = p$, where p is a positive rational number. Evaluate square roots of small perfect squares and cube roots of small perfect cubes. Know that $\sqrt{2}$ is irrational.

Previous Learning

Students have found squares of numbers. They have also found areas of squares and circles.

7.1 Record and Practice Journal

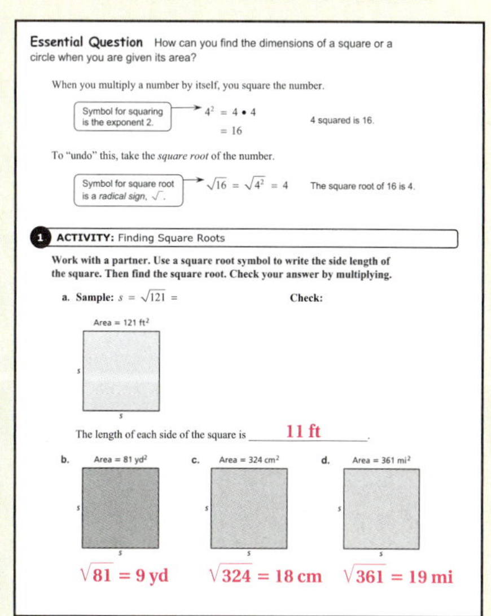

Essential Question How can you find the dimensions of a square or a circle when you are given its area?

When you multiply a number by itself, you square the number.

Symbol for squaring is the exponent 2. → $4^2 = 4 \bullet 4 = 16$ 4 squared is 16.

To "undo" this, take the *square root* of the number.

Symbol for square root is a *radical sign*, $\sqrt{\ }$. → $\sqrt{16} = \sqrt{4^2} = 4$ The square root of 16 is 4.

1 ACTIVITY: Finding Square Roots

Work with a partner. Use a square root symbol to write the side length of the square. Then find the square root. Check your answer by multiplying.

a. **Sample:** $s = \sqrt{121} =$ **Check:**

Area = 121 ft^2

The length of each side of the square is 11 ft.

b. Area = 81 yd^2 $\sqrt{81} = 9$ yd

c. Area = 324 cm^2 $\sqrt{324} = 18$ cm

d. Area = 361 mi^2 $\sqrt{361} = 19$ mi

English Language Learners

Build on Past Knowledge

Remind students of inverse operations. Addition and subtraction are inverse operations, as are multiplication and division. Taking the square root of a number is the inverse of squaring a number and squaring a number is the inverse of taking the square root of a number.

7.1 Record and Practice Journal

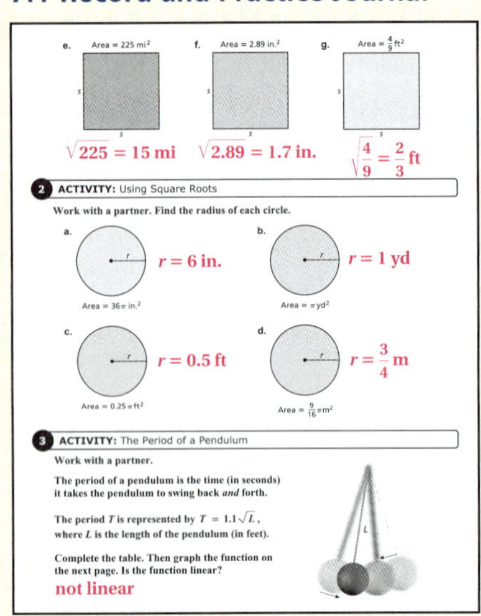

e. Area = 225 mi² f. Area = 2.89 in.² g. Area = $\frac{4}{9}$ ft²

$\sqrt{225} = 15$ mi $\sqrt{2.89} = 1.7$ in. $\sqrt{\frac{4}{9}} = \frac{2}{3}$ ft

2 ACTIVITY: Using Square Roots

Work with a partner. Find the radius of each circle.

a. Area = 36π in.² $r = 6$ in.

b. Area = π yd² $r = 1$ yd

c. Area = 0.25π ft² $r = 0.5$ ft

d. Area = $\frac{9}{16}\pi$ m² $r = \frac{3}{4}$ m

3 ACTIVITY: The Period of a Pendulum

Work with a partner.

The period of a pendulum is the time (in seconds) it takes the pendulum to swing back *and* forth.

The period T is represented by $T = 1.1\sqrt{L}$, where L is the length of the pendulum (in feet).

Complete the table. Then graph the function on the next page. Is the function linear?

not linear

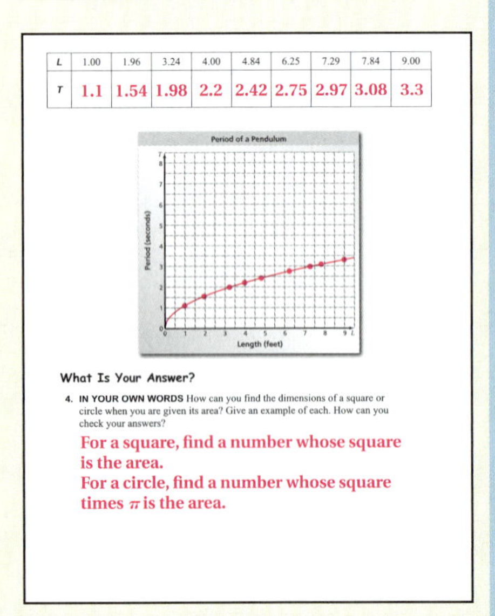

L	1.00	1.96	3.24	4.00	4.84	6.25	7.29	7.84	9.00
T	1.1	1.54	1.98	2.2	2.42	2.75	2.97	3.08	3.3

What Is Your Answer?

4. IN YOUR OWN WORDS How can you find the dimensions of a square or circle when you are given its area? Give an example of each. How can you check your answers?

For a square, find a number whose square is the area.
For a circle, find a number whose square times π is the area.

Laurie's Notes

Activity 2

- **MP1a Make Sense of Problems:** "How do you find the area of a circle?" $A = \pi r^2$
- "If you know the area of a circle, can you solve for the radius?" yes
- Do not give away too much at this point. Let students think through the problems. The first problem involves an obvious perfect square.
- **Common Error:** For part (c), students will often answer 0.05, forgetting how decimals are multiplied.

Activity 3

- Define and model the period of a pendulum, if you did not do the opening Motivate.
- Write the formula, $T = 1.1\sqrt{L}$. Explain that if the length is 9 feet, you can find out the time of one period by evaluating the equation for $L = 9$. So, $T = 1.1\sqrt{9} = 1.1(3) = 3.3$ seconds.
- For each of the values in the table, remind students to think about the whole numbers 100, 196, 324, 400, and so on, to help find $\sqrt{1.00}$, $\sqrt{1.96}$, $\sqrt{3.24}$, $\sqrt{4.00}$, and so on.
- **MP6:** When plotting the ordered pairs, students will need to be precise to see the curvature in the graph.
- **Connection:** The equation they are graphing is the function $y = 1.1\sqrt{x}$. This is not a linear function, so the graph is not a line. Find the slope between two different pairs of points on the graph. For (1, 1.1) and (4, 2.2), the slope is $\frac{1.1}{3} = 0.3\overline{6}$. For (4, 2.2) and (9, 3.3), the slope is $\frac{1.1}{5} = 0.22$. The slopes are not the same, so the graph cannot be a line.

What Is Your Answer?

- **Neighbor Check:** Have students work independently and then have their neighbors check their work. Have students discuss any discrepancies.

Closure

- **Matching Activity:** Match each square root with the correct answer.

1. $\sqrt{1600}$ D	**A.** 12
2. $\sqrt{400}$ B	**B.** 20
3. $\sqrt{144}$ A	**C.** 6
4. $\sqrt{36}$ C	**D.** 40

2 ACTIVITY: Using Square Roots

Work with a partner. Find the radius of each circle.

a.

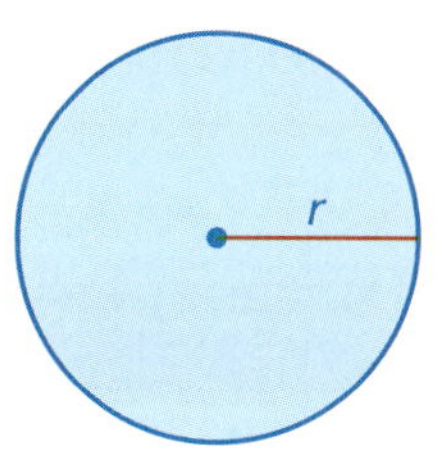

Area = 36π in.^{2}

b.

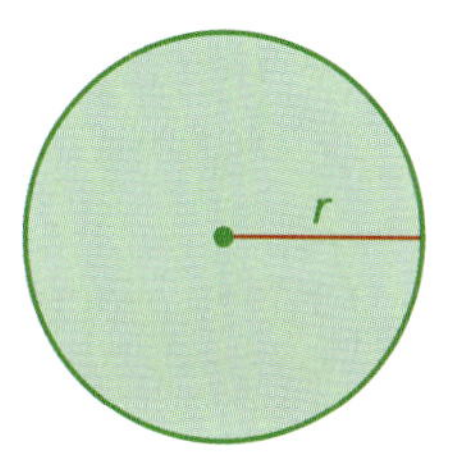

Area = π yd^2

c.

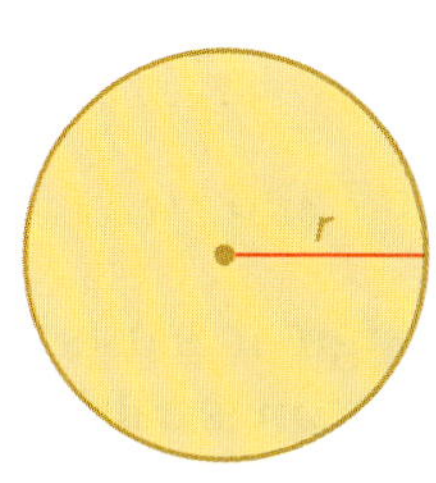

Area = 0.25π ft^2

d.

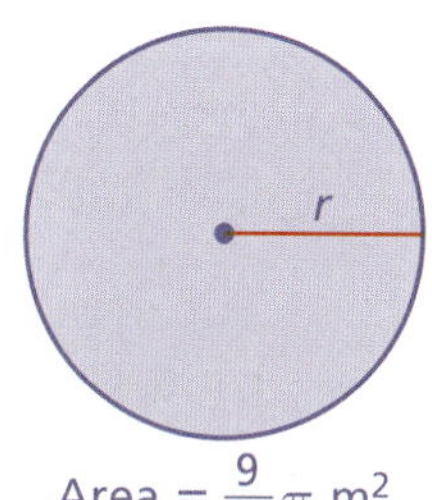

Area = $\frac{9}{16}\pi$ m^2

3 ACTIVITY: The Period of a Pendulum

Math Practice 6

Calculate Accurately

How can you use the graph to help you determine whether you calculated the values of T correctly?

Work with a partner.

The period of a pendulum is the time (in seconds) it takes the pendulum to swing back *and* forth.

The period T is represented by $T = 1.1\sqrt{L}$, where L is the length of the pendulum (in feet).

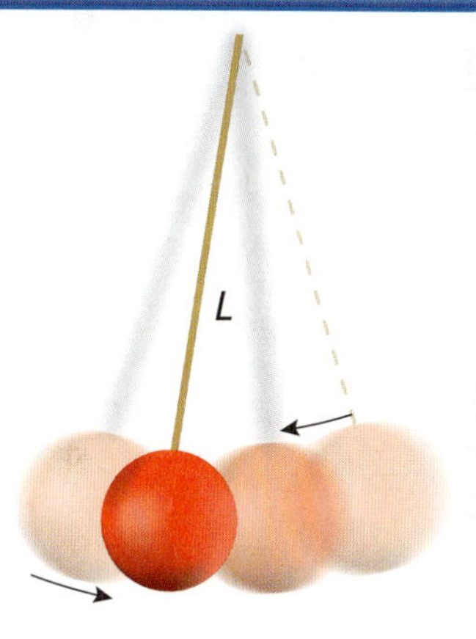

Copy and complete the table. Then graph the function. Is the function linear?

L	1.00	1.96	3.24	4.00	4.84	6.25	7.29	7.84	9.00
T									

What Is Your Answer?

4. **IN YOUR OWN WORDS** How can you find the dimensions of a square or a circle when you are given its area? Give an example of each. How can you check your answers?

Use what you learned about finding square roots to complete Exercises 4–6 on page 292.

7.1 Lesson

Key Vocabulary
square root, *p. 290*
perfect square, *p. 290*
radical sign, *p. 290*
radicand, *p. 290*

A **square root** of a number is a number that, when multiplied by itself, equals the given number. Every positive number has a positive *and* a negative square root. A **perfect square** is a number with integers as its square roots.

EXAMPLE 1 Finding Square Roots of a Perfect Square

Study Tip
Zero has one square root, which is 0.

Find the two square roots of 49.

$$7 \cdot 7 = 49 \text{ and } (-7) \cdot (-7) = 49$$

So, the square roots of 49 are 7 and −7.

The symbol $\sqrt{}$ is called a **radical sign**. It is used to represent a square root. The number under the radical sign is called the **radicand**.

Positive Square Root, $\sqrt{}$	Negative Square Root, $-\sqrt{}$	Both Square Roots, $\pm\sqrt{}$
$\sqrt{16} = 4$	$-\sqrt{16} = -4$	$\pm\sqrt{16} = \pm 4$

EXAMPLE 2 Finding Square Roots

Find the square root(s).

a. $\sqrt{25}$

$\sqrt{25}$ represents the *positive* square root.

Because $5^2 = 25$, $\sqrt{25} = \sqrt{5^2} = 5$.

b. $-\sqrt{\frac{9}{16}}$

$-\sqrt{\frac{9}{16}}$ represents the *negative* square root.

Because $\left(\frac{3}{4}\right)^2 = \frac{9}{16}$, $-\sqrt{\frac{9}{16}} = -\sqrt{\left(\frac{3}{4}\right)^2} = -\frac{3}{4}$.

c. $\pm\sqrt{2.25}$

$\pm\sqrt{2.25}$ represents both the *positive* and the *negative* square roots.

Because $1.5^2 = 2.25$, $\pm\sqrt{2.25} = \pm\sqrt{1.5^2} = 1.5$ and -1.5.

On Your Own

Find the two square roots of the number.

1. 36 **2.** 100 **3.** 121

Find the square root(s).

4. $-\sqrt{1}$ **5.** $\pm\sqrt{\frac{4}{25}}$ **6.** $\sqrt{12.25}$

Multi-Language Glossary at BigIdeasMath.com

Laurie's Notes

Introduction

Connect

- **Yesterday:** Students explored square roots. (MP1a, MP6)
- **Today:** Students will find square roots and evaluate expressions involving square roots.

Motivate

- Play the game *Keep it Going!*
- Give the students the first 3 to 4 numbers in a sequence and have them *Keep it Going*. If students are sitting in a row, each person in the row says the next number in the pattern. Keep the pattern going until it becomes too difficult to continue. For example, use the sequence 4, 400, 40,000, … (4,000,000, 400,000,000…)

Lesson Notes

Discuss

- Write and discuss the definitions of square root of a number and perfect square. Mention the *Study Tip*.
- Students are often confused when you say "every positive number has a positive and a negative square root." Use Example 1 to explain.

Example 1

- Note that the direction line is written in words without the square root symbol. The notation is introduced after this example.

? "What is the product of two positives? two negatives?" Both are positive.

Discuss

- **MP6 Attend to Precision:** It is important for students to use correct language and be familiar with the symbols $\sqrt{\ }$, $-\sqrt{\ }$, and $\pm\sqrt{\ }$.
- The square root symbol is called a **radical sign** and the number under the radical sign is the **radicand**.
- Write and discuss the three examples in the table. Explain to students that the symbol $\pm$ is read as *plus or minus*.

Example 2

- Remind students to pay attention to the signs that may precede the radical sign.

? "How do you multiply fractions?" Write the product of the numerators over the product of the denominators.

? "What fraction is multiplied by itself to get $\frac{9}{16}$?" $\frac{3}{4}$

On Your Own

- **Think-Pair-Share:** Students should read each question independently and then work in pairs to answer the questions. When they have answered the questions, the pair should compare their answers with another group and discuss any discrepancies.

Goal Today's lesson is finding **square roots**.

Technology for the Teacher

Lesson Tutorials
Lesson Plans
Answer Presentation Tool

Extra Example 1

Find the two square roots of 64.
8 and −8

Extra Example 2

Find the square root(s).

a. $-\sqrt{81}$ −9

b. $\pm\sqrt{\frac{9}{64}}$ $\frac{3}{8}$ and $-\frac{3}{8}$

c. $\sqrt{4.84}$ 2.2

On Your Own

1. 6 and −6
2. 10 and −10
3. 11 and −11
4. −1
5. $\pm\frac{2}{5}$
6. 3.5

Extra Example 3

Evaluate each expression.

a. $2\sqrt{144} - 30$ -6

b. $\sqrt{\frac{36}{4}} + \frac{1}{6}$ $3\frac{1}{6}$

c. $49 - (\sqrt{49})^2$ 0

Extra Example 4

What is the radius of the circle? Use 3.14 for π. about 4 in.

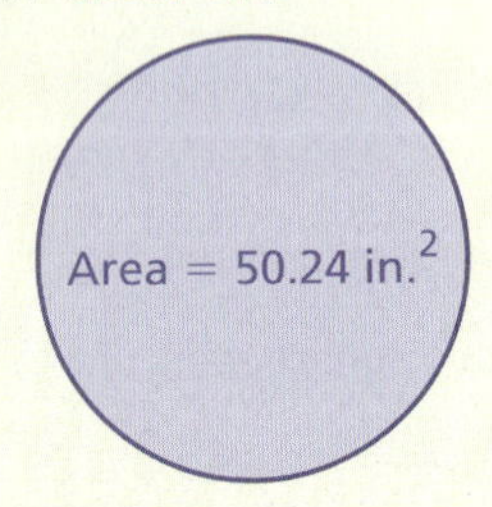

On Your Own

7. -3

8. 4.4

9. 11

10. $3.14r^2 = 2826$; about 30 ft

Differentiated Instruction

Auditory

Ask students to use mental math to answer the following verbal questions.

- "What is the sum of the square root of 9 and 3?" 6
- "What is the difference of 12 and the square root of 144?" 0
- "What is twice the square root of 16?" 8
- "What is one-fourth of the square root of 64?" 2

Laurie's Notes

Example 3

- **Teaching Tip:** In these examples, remind students that square roots are numbers, so you can evaluate numerical expressions that include square roots. Students see a *symbol*, think *variable*, and suddenly they forget things like the order of operations. In part (a), some students think of this as $5x + 7$ and will not know what to do.
- Write the expression in part (a).

? "Which operations are involved in this problem?" taking a square root, multiplication, and addition

- Work through parts (a) and (b) as shown.
- Write the expression in part (c). Discuss that squaring and taking the square root are inverse operations, as explained prior to the example.

? "What is $\sqrt{81}$?" 9 "What is 9^2?" 81

- Say, "So, taking the square root of 81 and then squaring the answer results in 81." Finish evaluating the expression as shown.

? "Can you name inverse operations?" addition and subtraction, multiplication and division, squaring and taking the square root

Example 4

? "How do you find the area of a circle?" $A = \pi r^2$

- The numbers involved may be overwhelming to students. Reassure them that this is an equation with one variable and that they know how to solve equations!
- Use a calculator or long division to divide 45,216 by 3.14.
- When you get to the step $14{,}400 = r^2$, remind students that whatever you do to one side of an equation, you must do to the other side. So, to get r by itself, you need to undo the squaring. So, take the square root of each side.
- Discuss with students why a negative square root does not make sense in this context. You cannot have a negative radius, so you can "ignore" the negative square root.

On Your Own

- Ask volunteers to share their work at the board for each of the problems.
- Question 8 looks more difficult because of the fraction. You may want to point out that $\frac{28}{7} = 4$.

Closure

- Write 3 numbers of which you know how to take the square root.
- Write 3 numbers of which you do not know how to take the square root.

Squaring a positive number and finding a square root are inverse operations. You can use this relationship to evaluate expressions and solve equations involving squares.

EXAMPLE 3 Evaluating Expressions Involving Square Roots

Evaluate each expression.

a. $5\sqrt{36} + 7 = 5(6) + 7$ Evaluate the square root.

$= 30 + 7$ Multiply.

$= 37$ Add.

b. $\frac{1}{4} + \sqrt{\frac{18}{2}} = \frac{1}{4} + \sqrt{9}$ Simplify.

$= \frac{1}{4} + 3$ Evaluate the square root.

$= 3\frac{1}{4}$ Add.

c. $\left(\sqrt{81}\right)^2 - 5 = 81 - 5$ Evaluate the power using inverse operations.

$= 76$ Subtract.

EXAMPLE 4 Real-Life Application

The area of a crop circle is 45,216 square feet. What is the radius of the crop circle? Use 3.14 for π.

$A = \pi r^2$ Write the formula for the area of a circle.

$45{,}216 \approx 3.14r^2$ Substitute 45,216 for A and 3.14 for π.

$14{,}400 = r^2$ Divide each side by 3.14.

$\sqrt{14{,}400} = \sqrt{r^2}$ Take positive square root of each side.

$120 = r$ Simplify.

The radius of the crop circle is about 120 feet.

On Your Own

Now You're Ready

Exercises 20–27

Evaluate the expression.

7. $12 - 3\sqrt{25}$

8. $\sqrt{\frac{28}{7}} + 2.4$

9. $15 - \left(\sqrt{4}\right)^2$

10. The area of a circle is 2826 square feet. Write and solve an equation to find the radius of the circle. Use 3.14 for π.

7.1 Exercises

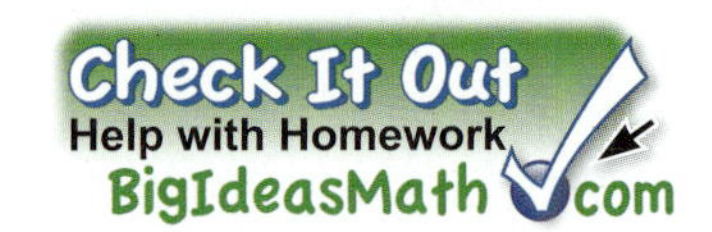

Vocabulary and Concept Check

1. **VOCABULARY** Is 26 a perfect square? Explain.
2. **REASONING** Can the square of an integer be a negative number? Explain.
3. **NUMBER SENSE** Does $\sqrt{256}$ represent the positive square root of 256, the negative square root of 256, or both? Explain.

Practice and Problem Solving

Find the dimensions of the square or circle. Check your answer.

4. Area = 441 cm^2

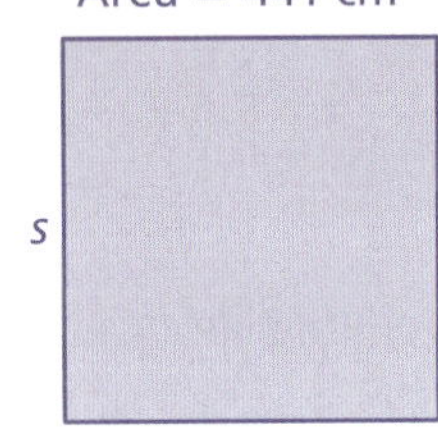

5. Area = 1.69 km^2

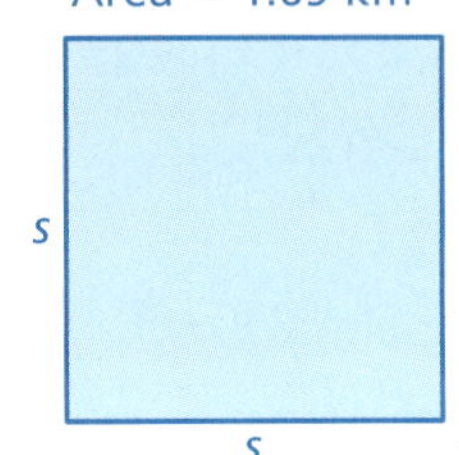

6. Area = 64π in.^{2}

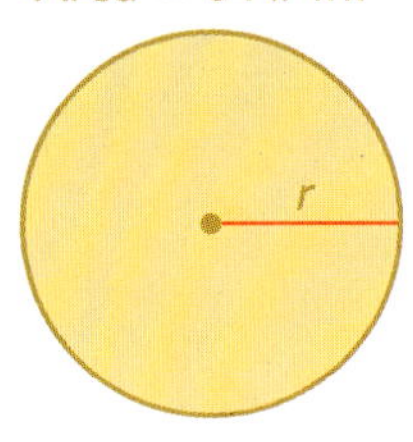

Find the two square roots of the number.

1 7. 9 8. 64 9. 4 10. 144

Find the square root(s).

2 11. $\sqrt{625}$ 12. $\pm\sqrt{196}$ 13. $\pm\sqrt{\frac{1}{961}}$ 14. $-\sqrt{\frac{9}{100}}$

15. $\pm\sqrt{4.84}$ 16. $\sqrt{7.29}$ 17. $-\sqrt{361}$ 18. $-\sqrt{2.25}$

19. **ERROR ANALYSIS** Describe and correct the error in finding the square roots.

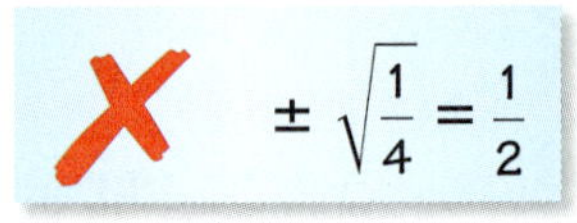

Evaluate the expression.

3 20. $(\sqrt{9})^2 + 5$ 21. $28 - (\sqrt{144})^2$ 22. $3\sqrt{16} - 5$ 23. $10 - 4\sqrt{\frac{1}{16}}$

24. $\sqrt{6.76} + 5.4$ 25. $8\sqrt{8.41} + 1.8$ 26. $2\left(\sqrt{\frac{80}{5}} - 5\right)$ 27. $4\left(\sqrt{\frac{147}{3}} + 3\right)$

28. **NOTEPAD** The area of the base of a square notepad is 2.25 square inches. What is the length of one side of the base of the notepad?

29. **CRITICAL THINKING** There are two square roots of 25. Why is there only one answer for the radius of the button?

$A = 25\pi$ mm^2

Assignment Guide and Homework Check

Level	Day 1 Activity Assignment	Day 2 Lesson Assignment	Homework Check
Basic	4–6, 39–42	1–3, 7–29 odd	9, 13, 15, 25, 29
Average	4–6, 39–42	1–3, 9, 13–29 odd, 30–36 even	9, 15, 23, 30, 36
Advanced	1–6, 10, 18, 19, 20–28 even, 30–42		18, 24, 32, 34, 37

Common Errors

- **Exercises 7–10** Students may only find the positive square root of the number given. Remind them that a square root can be positive or negative, and the question is asking for both answers.
- **Exercises 11–18** Students may divide the number by two instead of finding a number that, when multiplied by itself, gives the radicand. Remind them that taking the square root of a number is the inverse of squaring a number.
- **Exercises 20–27** Students may not follow the order of operations when evaluating the expression. Remind them of the order of operations. Because taking a square root is the inverse of squaring, it is evaluated before multiplication and division.

7.1 Record and Practice Journal

Find the two square roots of the number.

1. 16 — **4 and −4**
2. 100 — **10 and −10**
3. 196 — **14 and −14**

Find the square root(s).

4. $\sqrt{169}$ — **13**
5. $\sqrt{\frac{4}{225}}$ — $\frac{2}{15}$
6. $-\sqrt{12.25}$ — **−3.5**

Evaluate the expression.

7. $2\sqrt{36} + 9$ — **21**
8. $8 - 11\sqrt{\frac{25}{121}}$ — **3**
9. $3\left(\sqrt{\frac{125}{5}} - 8\right)$ — **−9**

10. A trampoline has an area of 49π square feet. What is the diameter of the trampoline?
14 ft

Vocabulary and Concept Check

1. no; There is no integer whose square is 26.
2. no; A positive number times a positive number is a positive number, and a negative number times a negative number is a positive number.
3. $\sqrt{256}$ represents the positive square root because there is not a − or a ± in front.

Practice and Problem Solving

4. $s = 21$ cm
5. $s = 1.3$ km
6. $r = 8$ in.
7. 3 and −3
8. 8 and −8
9. 2 and −2
10. 12 and −12
11. 25
12. ±14
13. $\frac{1}{31}$ and $-\frac{1}{31}$
14. $-\frac{3}{10}$
15. 2.2 and −2.2
16. 2.7
17. −19
18. −1.5
19. The positive and negative square roots should have been given.
$\pm\sqrt{\frac{1}{4}} = \frac{1}{2}$ and $-\frac{1}{2}$
20. 14
21. −116
22. 7
23. 9
24. 8
25. 25
26. −2
27. 40
28. 1.5 in.
29. because a negative radius does not make sense

Practice and Problem Solving

30. >

31. =

32. <

33. 9 ft

34. yes; *Sample answer:* Consider the perfect squares, a^2 and b^2. Their product can be written as $a^2b^2 = a \cdot a \cdot b \cdot b = (a \cdot b) \cdot (a \cdot b) = (a \cdot b)^2$.

35. 8 m/sec

36. See *Taking Math Deeper.*

37. 2.5 ft

38. 8 cm

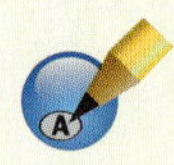

Fair Game Review

39. $y = 3x - 2$

40. $y = -2x + 5$

41. $y = \frac{3}{5}x + 1$

42. B

Mini-Assessment

Find the square root(s).

1. $\sqrt{169}$ 13
2. $\sqrt{225}$ 15
3. $\pm\sqrt{4.41}$ 2.1 and −2.1
4. $-\sqrt{\frac{16}{25}}$ $-\frac{4}{5}$
5. $\sqrt{\frac{512}{2}}$ 16

Taking Math Deeper

Exercise 36

In this problem, students are given the area of the smaller watch face and are asked to find the radius of the larger watch face.

Summarize the given information and find the radius of the smaller watch face.

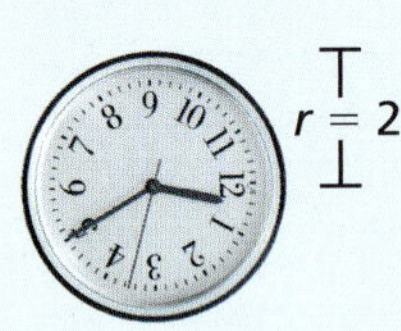

Area = 4π cm^2

Ratio of areas is 16 to 25.

Answer the question. The two watch faces are similar, so the ratio of their areas is equal to the square of the ratio of their radii.

$$\frac{\text{Area of small}}{\text{Area of large}} = \left(\frac{\text{Radius of small}}{\text{Radius of large}}\right)^2$$

$$\frac{16}{25} = \left(\frac{\text{Radius of small}}{\text{Radius of large}}\right)^2$$

$$\sqrt{\frac{16}{25}} = \frac{\text{Radius of small}}{\text{Radius of large}}$$

$$\frac{4}{5} = \frac{\text{Radius of small}}{\text{Radius of large}}$$

a. The ratio of the radius of the smaller watch face to the radius of the larger watch face is $\frac{4}{5}$.

3 **b.** Write and solve a proportion to find R.

$$\frac{4}{5} = \frac{r}{R}$$

$$\frac{4}{5} = \frac{2}{R}$$

$$R = \frac{10}{4}, \text{ or } \frac{5}{2}$$

The radius of the larger watch face is $\frac{5}{2}$, or 2.5 centimeters.

Reteaching and Enrichment Strategies

If students need help...	If students got it...
Resources by Chapter • Practice A and Practice B • Puzzle Time Record and Practice Journal Practice Differentiating the Lesson Lesson Tutorials Skills Review Handbook	Resources by Chapter • Enrichment and Extension • Technology Connection Start the next section

Copy and complete the statement with <, >, or =.

30. $\sqrt{81}$ ▭ 8 **31.** 0.5 ▭ $\sqrt{0.25}$ **32.** $\frac{3}{2}$ ▭ $\sqrt{\frac{25}{4}}$

33. SAILBOAT The area of a sail is $40\frac{1}{2}$ square feet. The base and the height of the sail are equal. What is the height of the sail (in feet)?

34. REASONING Is the product of two perfect squares always a perfect square? Explain your reasoning.

35. ENERGY The kinetic energy K (in joules) of a falling apple is represented by $K = \frac{v^2}{2}$, where v is the speed of the apple (in meters per second). How fast is the apple traveling when the kinetic energy is 32 joules?

36. PRECISION The areas of the two watch faces have a ratio of 16 : 25.

a. What is the ratio of the radius of the smaller watch face to the radius of the larger watch face?

b. What is the radius of the larger watch face?

37. WINDOW The cost C (in dollars) of making a square window with a side length of n inches is represented by $C = \frac{n^2}{5} + 175$. A window costs \$355. What is the length (in feet) of the window?

38. 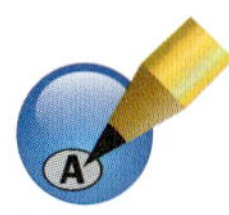

The area of the triangle is represented by the formula $A = \sqrt{s(s-21)(s-17)(s-10)}$, where s is equal to half the perimeter. What is the height of the triangle?

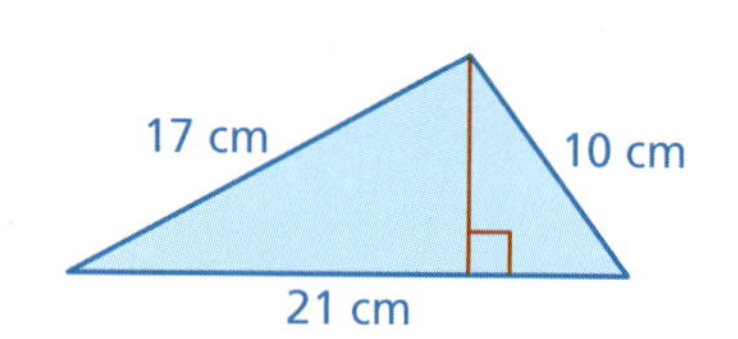

Fair Game Review What you learned in previous grades & lessons

Write in slope-intercept form an equation of the line that passes through the given points. *(Section 4.7)*

39. $(2, 4), (5, 13)$ **40.** $(-1, 7), (3, -1)$ **41.** $(-5, -2), (5, 4)$

42. MULTIPLE CHOICE What is the value of x? *(Section 3.2)*

Ⓐ 41 Ⓑ 44

Ⓒ 88 Ⓓ 134

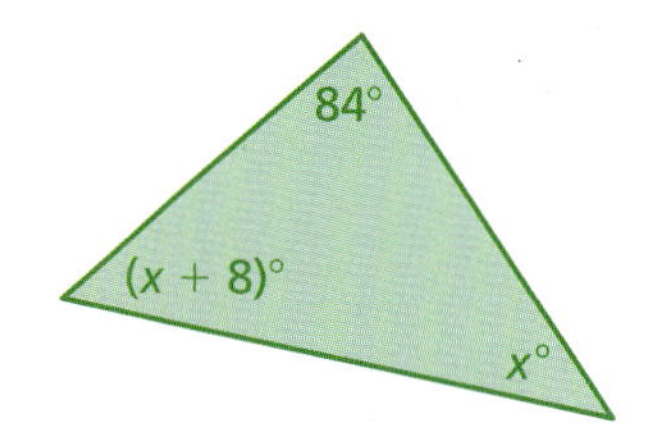

7.2 Finding Cube Roots

Essential Question How is the cube root of a number different from the square root of a number?

When you multiply a number by itself twice, you cube the number.

Symbol for cubing is the exponent 3. → $4^3 = 4 \cdot 4 \cdot 4$

$= 64$ 4 cubed is 64.

To "undo" this, take the *cube root* of the number.

Symbol for cube root is $\sqrt[3]{}$. → $\sqrt[3]{64} = \sqrt[3]{4^3} = 4$ The cube root of 64 is 4.

1 ACTIVITY: Finding Cube Roots

Work with a partner. Use a cube root symbol to write the edge length of the cube. Then find the cube root. Check your answer by multiplying.

a. Sample:

$s = \sqrt[3]{343} = \sqrt[3]{7^3} = 7$ inches

Volume = 343 in.^{3}

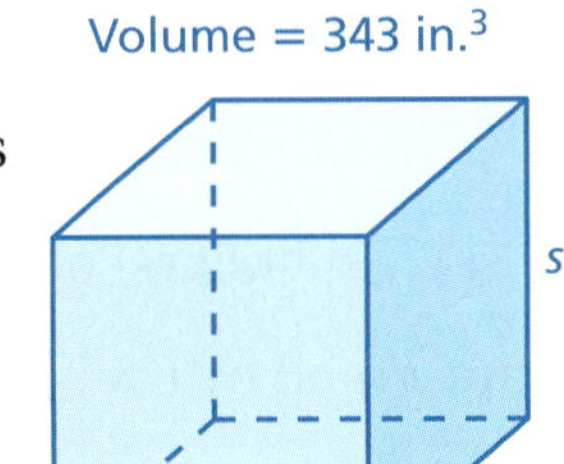

Check
$7 \cdot 7 \cdot 7 = 49 \cdot 7$
$= 343$ ✓

The edge length of the cube is 7 inches.

b. Volume = 27 ft^3

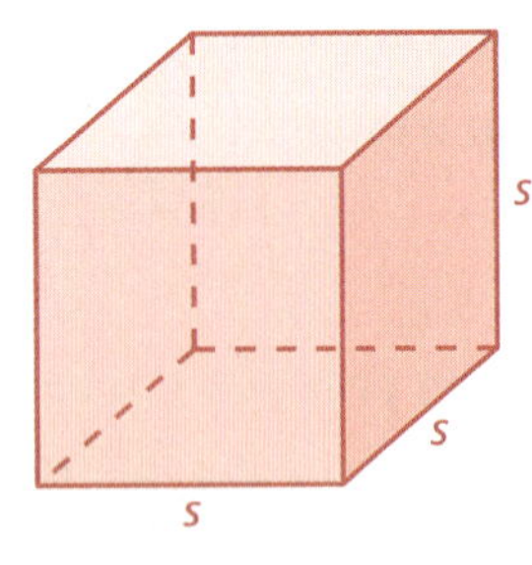

c. Volume = 125 m^3

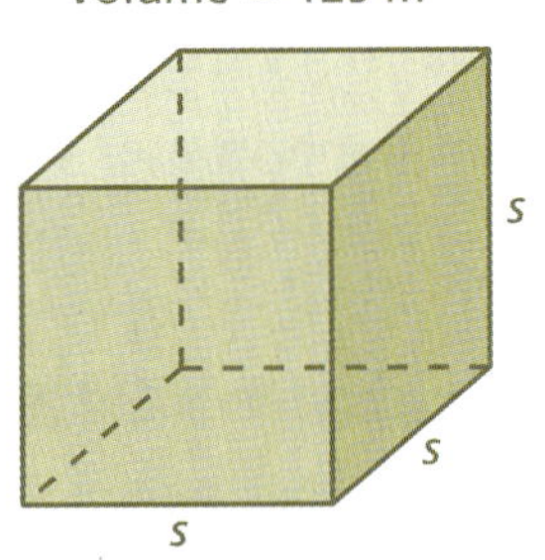

d. Volume = 0.001 cm^3

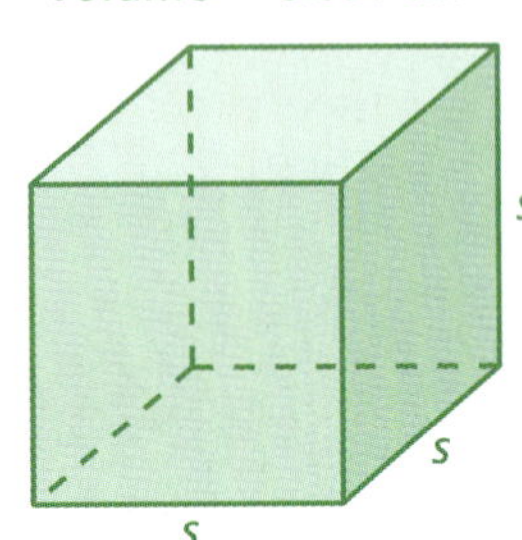

e. Volume = $\frac{1}{8}$ yd^3

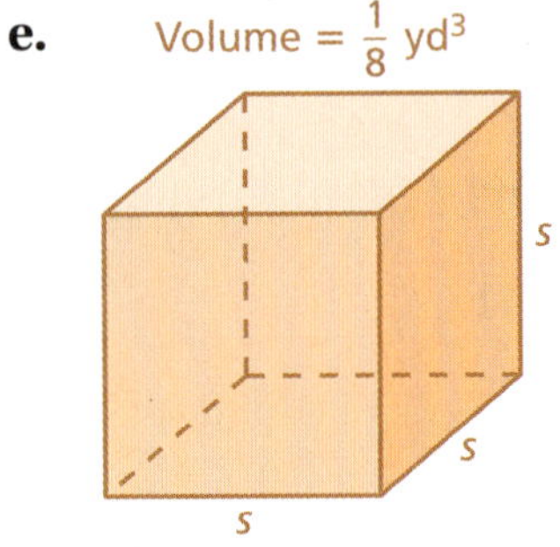

Cube Roots

In this lesson, you will

- find cube roots of perfect cubes.
- evaluate expressions involving cube roots.
- use cube roots to solve equations.

Learning Standard
MACC.8.EE.1.2

Laurie's Notes

Introduction

Standards for Mathematical Practice

- **MP2 Reason Abstractly and Quantitatively:** Students have found square roots of a number. When finding a cube root, they must think about multiplying a number by itself twice.

Motivate

- Use 8 cubes to build a $2 \times 2 \times 2$ cube. Inch-cubes or larger are best for visibility in the class. Display the large cube by holding it up on a rigid surface so that all students are able to see it.
- ? "Could someone describe this shape?" It is a cube.
- ? "Can you describe any of the numerical attributes of the cube?" The dimensions are $2 \times 2 \times 2$ and the volume is 8.
- ? "If the edge length of each smaller cube is 1 inch, what are the units for the volume of the large cube?" cubic inches
- Summarize by saying, "So, $8 = 2 \times 2 \times 2 = 2^3$. Cubing the edge length gives the volume."
- Hold another cube that is not made of smaller cubes, such as a cube of sticky notes.
- ? "If you knew the volume of this cube, do you think you could find the dimensions—without measuring?" Give students time to think about this question and then introduce the activities.

Activity Notes

Discuss

- Review the meaning of the notation in the expression 4^3. Introduce the cube root symbol, $\sqrt[3]{\ }$, and how it is read.

Activity 1

- **MP1 Make Sense of Problems and Persevere in Solving Them:** Following the sample, students should be able to do the rest of the activity. Before you jump in with answers, remind students to think about the Motivate problem and the relationship between edge length and volume.
- Students may think that a calculator is necessary to solve these problems, but they may not find the cube root key (if the calculator has one). If a calculator helps them explore the problem, let them use it.
- **MP6 Attend to Precision:** Remind students that answers need correct labels (units).
- **MP2:** When students have finished, ask volunteers to describe their thinking behind the solutions. They should describe thinking about what number could be multiplied by itself twice to get the volume.
- Students will often say that they need a number that, when multiplied by itself 3 times, will be the volume. Clarify that the number (edge length) is used as a factor 3 times, but it is multiplied by itself only 2 times.

Florida Common Core Standards

MACC.8.EE.1.2 Use square root and cube root symbols to represent solutions to equations of the form $x^2 = p$ and $x^3 = p$, where p is a positive rational number. Evaluate square roots of small perfect squares and cube roots of small perfect cubes. Know that $\sqrt{2}$ is irrational.

Previous Learning

Students have found cubes of numbers. They have also found volumes of cubes.

Lesson Plans
Complete Materials List

7.2 Record and Practice Journal

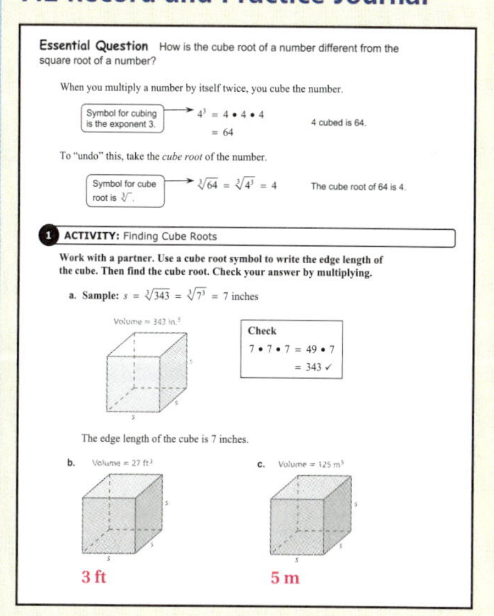

Essential Question How is the cube root of a number different from the square root of a number?

When you multiply a number by itself twice, you cube the number.

Symbol for cubing is the exponent 3. → $4^3 = 4 \bullet 4 \bullet 4 = 64$ 4 cubed is 64.

To "undo" this, take the *cube root* of the number.

Symbol for cube root is $\sqrt[3]{\ }$. → $\sqrt[3]{64} = \sqrt[3]{4^3} = 4$ The cube root of 64 is 4.

1 ACTIVITY: Finding Cube Roots

Work with a partner. Use a cube root symbol to write the edge length of the cube. Then find the cube root. Check your answer by multiplying.

a. **Sample:** $s = \sqrt[3]{343} = \sqrt[3]{7^3} = 7$ inches

Volume = 343 in.^{3}

Check
$7 \bullet 7 \bullet 7 = 49 \bullet 7 = 343$ ✓

The edge length of the cube is 7 inches.

b. Volume = 27 ft^3 — 3 ft

c. Volume = 125 m^3 — 5 m

English Language Learners

Vocabulary

For English learners, note the different meanings of the word *radical.* In everyday language, it can be used to describe a considerable departure from the usual or traditional view. In slang, *radical* is something that is excellent or cool. In mathematics, it is a symbol for roots—square roots, cube roots, and so on.

7.2 Record and Practice Journal

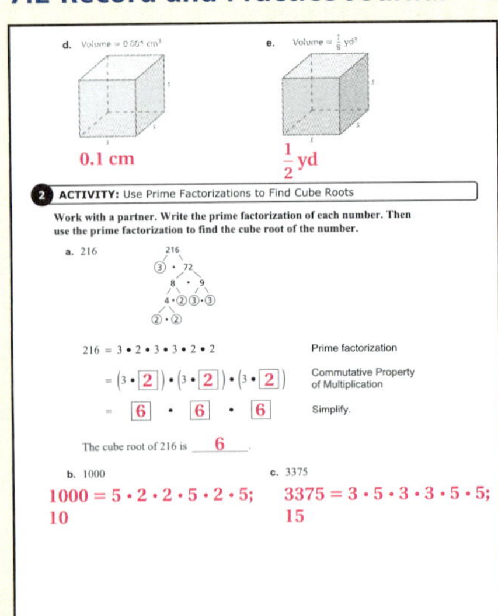

d. Volume = 0.001 cm³ — **0.1 cm**

e. Volume = $\frac{1}{8}$ yd³ — **$\frac{1}{2}$ yd**

2 ACTIVITY: Use Prime Factorizations to Find Cube Roots

Work with a partner. Write the prime factorization of each number. Then use the prime factorization to find the cube root of the number.

a. 216

$216 = 3 \cdot 2 \cdot 3 \cdot 3 \cdot 2 \cdot 2$ — Prime factorization

$= (3 \cdot 2) \cdot (3 \cdot 2) \cdot (3 \cdot 2)$ — Commutative Property of Multiplication

$= 6 \cdot 6 \cdot 6$ — Simplify.

The cube root of 216 is 6.

b. 1000 — $1000 = 5 \cdot 2 \cdot 2 \cdot 5 \cdot 2 \cdot 5$; 10

c. 3375 — $3375 = 3 \cdot 5 \cdot 3 \cdot 3 \cdot 5 \cdot 5$; 15

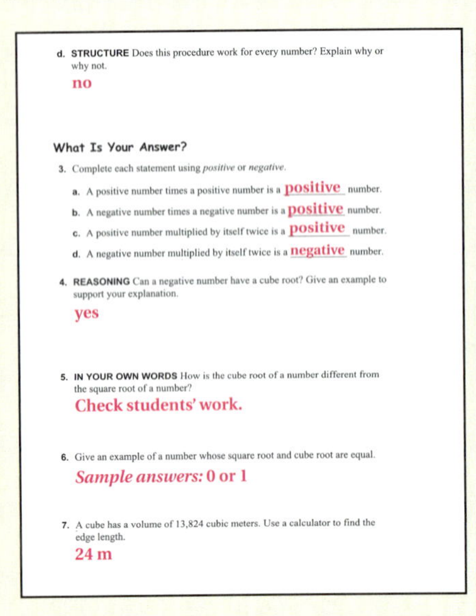

d. **STRUCTURE** Does this procedure work for every number? Explain why or why not.
no

What Is Your Answer?

3. Complete each statement using *positive* or *negative*.

a. A positive number times a positive number is a **positive** number.

b. A negative number times a negative number is a **positive** number.

c. A positive number multiplied by itself twice is a **positive** number.

d. A negative number multiplied by itself twice is a **negative** number.

4. **REASONING** Can a negative number have a cube root? Give an example to support your explanation.
yes

5. **IN YOUR OWN WORDS** How is the cube root of a number different from the square root of a number?
Check students' work.

6. Give an example of a number whose square root and cube root are equal.
Sample answers: 0 or 1

7. A cube has a volume of 13,824 cubic meters. Use a calculator to find the edge length.
24 m

Laurie's Notes

Activity 2

- **Connection:** This activity shows an interesting application of prime factorizations.

? "What is a prime number?" a number with exactly two factors, 1 and itself

? "What does it mean to write the prime factorization of a number?" Write the number as a product of only prime factors.

- Work through the factor tree in part (a), which will remind students of one method for finding the prime factorization of a number. The factor tree results in three 3s and three 2s.
- Make sure students understand why they rearrange the factors using the Commutative Property of Multiplication. You want to be able to write 216 as a number times itself twice. So, you want to group the prime factors into three identical groups.
- Work through the rest of part (a), showing three groups of $3 \cdot 2$, or $6 \cdot 6 \cdot 6$, meaning that 216 is a *perfect cube* whose cube root is 6. Perfect cube is not defined until the lesson, but students should understand what it means from their work with perfect squares.
- Students may quickly recognize 1000 as 10^3 without making a factor tree, and that is fine. It is unlikely that they will recognize 3375 as 15^3 without a bit of factoring!
- Part (d) may not be as obvious as you might imagine. Not all students will immediately understand that the procedure doesn't work for all numbers. Even if they do, they may not be able to explain correctly that not all numbers are *perfect cubes.*
- **Extension:** Ask students to describe how they could use a similar procedure to find square roots of large perfect squares.

What Is Your Answer?

- Questions 3–5 are all related. Take time to discuss students' responses and understanding of these problems.
- **MP2** and **MP5 Use Appropriate Tools Strategically:** Question 7 may be approached as a *Guess, Check, and Revise* problem or a prime factorization problem, versus students using the cube root key on their calculators.

Closure

- Find the edge length of the cube. $\frac{1}{3}$ in.

Volume = $\frac{1}{27}$ in.³

s s s

2 ACTIVITY: Using Prime Factorizations to Find Cube Roots

Math Practice 7

View as Components

When writing the prime factorizations in Activity 2, how many times do you expect to see each factor? Why?

Work with a partner. Write the prime factorization of each number. Then use the prime factorization to find the cube root of the number.

a. 216

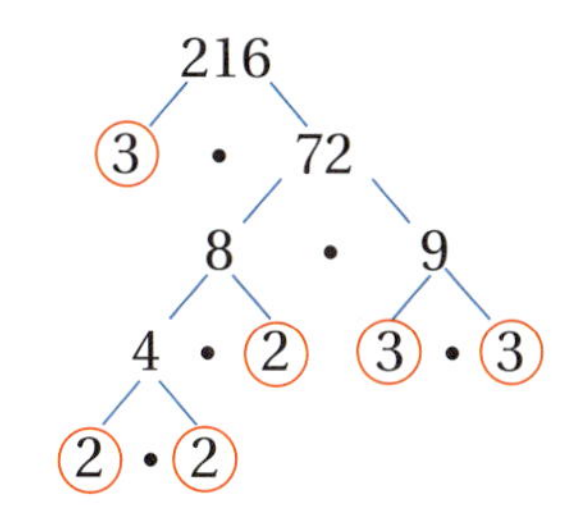

$216 = 3 \cdot 2 \cdot 3 \cdot 3 \cdot 2 \cdot 2$ Prime factorization

$= (3 \cdot \square) \cdot (3 \cdot \square) \cdot (3 \cdot \square)$ Commutative Property of Multiplication

$= \square \cdot \square \cdot \square$ Simplify.

The cube root of 216 is $\square$.

b. 1000

c. 3375

d. **STRUCTURE** Does this procedure work for every number? Explain why or why not.

What Is Your Answer?

3. Complete each statement using *positive* or *negative*.

a. A positive number times a positive number is a ________ number.

b. A negative number times a negative number is a ________ number.

c. A positive number multiplied by itself twice is a ________ number.

d. A negative number multiplied by itself twice is a ________ number.

4. **REASONING** Can a negative number have a cube root? Give an example to support your explanation.

5. **IN YOUR OWN WORDS** How is the cube root of a number different from the square root of a number?

6. Give an example of a number whose square root and cube root are equal.

7. A cube has a volume of 13,824 cubic meters. Use a calculator to find the edge length.

Practice Use what you learned about cube roots to complete Exercises 3–5 on page 298.

7.2 Lesson

Key Vocabulary
cube root, *p. 296*
perfect cube, *p. 296*

A **cube root** of a number is a number that, when multiplied by itself, and then multiplied by itself again, equals the given number. A **perfect cube** is a number that can be written as the cube of an integer. The symbol $\sqrt[3]{\ }$ is used to represent a cube root.

EXAMPLE 1 Finding Cube Roots

Find each cube root.

a. $\sqrt[3]{8}$

Because $2^3 = 8$, $\sqrt[3]{8} = \sqrt[3]{2^3} = 2$.

b. $\sqrt[3]{-27}$

Because $(-3)^3 = -27$, $\sqrt[3]{-27} = \sqrt[3]{(-3)^3} = -3$.

c. $\sqrt[3]{\dfrac{1}{64}}$

Because $\left(\dfrac{1}{4}\right)^3 = \dfrac{1}{64}$, $\sqrt[3]{\dfrac{1}{64}} = \sqrt[3]{\left(\dfrac{1}{4}\right)^3} = \dfrac{1}{4}$.

Cubing a number and finding a cube root are inverse operations. You can use this relationship to evaluate expressions and solve equations involving cubes.

EXAMPLE 2 Evaluating Expressions Involving Cube Roots

Evaluate each expression.

a. $2\sqrt[3]{-216} - 3 = 2(-6) - 3$ Evaluate the cube root.

$= -12 - 3$ Multiply.

$= -15$ Subtract.

b. $\left(\sqrt[3]{125}\right)^3 + 21 = 125 + 21$ Evaluate the power using inverse operations.

$= 146$ Add.

On Your Own

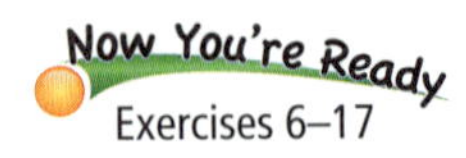

Find the cube root.

1. $\sqrt[3]{1}$
2. $\sqrt[3]{-343}$
3. $\sqrt[3]{-\dfrac{27}{1000}}$

Evaluate the expression.

4. $18 - 4\sqrt[3]{8}$
5. $\left(\sqrt[3]{-64}\right)^3 + 43$
6. $5\sqrt[3]{512} - 19$

Laurie's Notes

Introduction

Connect

- **Yesterday:** Students explored cube roots. (MP1, MP2, MP5, MP6)
- **Today:** Students will find cube roots and evaluate expressions involving cube roots.

Motivate

- Write the following sequences on the board:
  a. 1, 4, 9, 16, 25, . . . **b.** 1, 8, 27, 64, 125, . . .
- ? "Describe each sequence in words. What number comes next in each sequence?" Part (a) is a sequence of perfect squares and 36 is next; Part (b) is a sequence of perfect cubes and 216 is next.
- Explain to students that today they are going to be looking at perfect cubes and their cube roots.

Lesson Notes

Example 1

- Define the vocabulary: cube root and perfect cube.
- **FYI:** Every real number has three cube roots, one real and two imaginary. Students only need to be concerned with the real cube root in this lesson.
- ? Write part (a) and ask, "What number times itself twice equals 8?" 2
- ? Write part (b) and ask, "What number times itself twice equals -27?" -3
- ? Write part (c) and ask, "What number times itself twice equals $\frac{1}{64}$?" $\frac{1}{4}$

Example 2

- The cube root of a perfect cube is an integer and can be evaluated as part of an expression.
- ? Write part (a) and ask, "What is the cube root of -216?" -6
- Replace $\sqrt[3]{-216}$ with -6 and continue to evaluate the expression.
- Emphasize that cubing a number and finding a cube root are inverse operations, as stated before the example.
- Work through part (b), which helps students make the connection between cubing and taking the cube root.
- ? **MP2 Reason Abstractly and Quantitatively** and **MP3a Construct Viable Arguments:** "In general, do you think $\left(\sqrt[3]{n}\right)^3 = n$? Explain." Listen for an understanding of how taking the cube root of a number and then cubing the result gives the original number.

On Your Own

- Students are sometimes intimidated by the notation. Assure them that a calculator is not needed. The radicands are from the list of perfect cubes: 1, 8, 27, 64 125, ...
- Ask volunteers to share their work for Questions 4–6 at the board.

Goal Today's lesson is finding **cube roots**.

Technology for the Teacher

Lesson Tutorials
Lesson Plans
Answer Presentation Tool

Extra Example 1

Find each cube root.

a. $\sqrt[3]{512}$ 8

b. $\sqrt[3]{-729}$ -9

c. $\sqrt[3]{-\frac{125}{343}}$ $-\frac{5}{7}$

Extra Example 2

Evaluate each expression.

a. $3\sqrt[3]{125} - 8$ 7

b. $\left(\sqrt[3]{27}\right)^3 - 4$ 23

On Your Own

1. 1 **2.** -7

3. $-\frac{3}{10}$ **4.** 10

5. -21 **6.** 21

Extra Example 3

Evaluate $\frac{w}{30} - \sqrt[3]{\frac{w}{5}}$ when $w = 1080$. 30

On Your Own

7. 72 **8.** -3

Extra Example 4

Find the surface area of a cube with a volume of 27 cubic feet. 54 ft^2

On Your Own

9. 384 cm^2

Differentiated Instruction

Visual

Students may have a difficult time understanding that the square root or a cube root of a number between 0 and 1 is greater than the number. They may think that $\sqrt{\frac{1}{4}} = \frac{1}{16}$, when actually $\sqrt{\frac{1}{16}} = \frac{1}{4}$. Help students to understand this concept by using a 10-by-10 grid to represent the number 1. A square of $\frac{49}{100}$ square units has a side length of $\frac{7}{10}$. Because $\sqrt{\frac{49}{100}} = \frac{7}{10}$, $\frac{7}{10} > \frac{49}{100}$ $(0.7 > 0.49)$.

Laurie's Notes

Example 3

- Write the problem.
- "Is 192 a perfect cube?" no "Is $\frac{192}{3}$ a perfect cube?" Yes because it equals 64, which is a perfect cube.
- Substitute 192 for x and evaluate the expression as shown.
- "In general, do you think any value of x could be used in this problem? Explain." The correct answer is yes, but students may answer that you can only use values of x in which $\frac{1}{3}$ of x is a perfect cube. They are unaware that a cube root can be an *irrational number*, which is okay at this stage. They will learn about irrational numbers in Section 7.4.

On Your Own

- **Think-Pair-Share:** Students should read each question independently and then work in pairs to answer the questions. When they have answered the questions, the pair should compare their answers with another group and discuss any discrepancies.

Example 4

- **Teaching Tip:** Some tissue boxes are close to the shape of a cube. Use a tissue box or an actual cube as a prop. Pose the question: If you know the volume of the cube, can you find the surface area?
- "How do you find the volume of a cube?" Cube the edge length; $V = s^3$
- "How do you find the surface area of a cube?" Find the area of one face and multiply by 6; $S = 6s^2$
- The first step is to find the edge length of the cube. Students should be thinking, "What number multiplied by itself twice equals 125?"
- Continue to work through the problem as shown.

On Your Own

- **Neighbor Check:** Have students work independently and then have their neighbors check their work. Have students discuss any discrepancies.

Closure

- Explain the difference between $\sqrt{64}$ and $\sqrt[3]{64}$. $\sqrt{64}$ is a number that when multiplied by itself is equal to 64, and $\sqrt[3]{64}$ is a number that when multiplied by itself twice is equal to 64.

EXAMPLE 3 Evaluating an Algebraic Expression

Evaluate $\frac{x}{4} + \sqrt[3]{\frac{x}{3}}$ when $x = 192$.

$$\frac{x}{4} + \sqrt[3]{\frac{x}{3}} = \frac{192}{4} + \sqrt[3]{\frac{192}{3}}$$ Substitute 192 for *x*.

$$= 48 + \sqrt[3]{64}$$ Simplify.

$$= 48 + 4$$ Evaluate the cube root.

$$= 52$$ Add.

On Your Own

Now You're Ready
Exercises 18–20

Evaluate the expression for the given value of the variable.

7. $\sqrt[3]{8y} + y, y = 64$

8. $2b - \sqrt[3]{9b}, b = -3$

EXAMPLE 4 Real-Life Application

Find the surface area of the baseball display case.

The baseball display case is in the shape of a cube. Use the formula for the volume of a cube to find the edge length *s*.

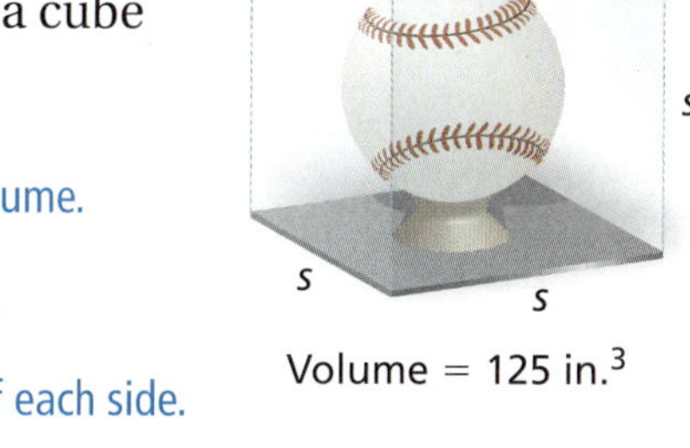

Volume = 125 in.^{3}

Remember

The volume V of a cube with edge length s is given by $V = s^3$. The surface area S is given by $S = 6s^2$.

$$V = s^3$$ Write formula for volume.

$$125 = s^3$$ Substitute 125 for *V*.

$$\sqrt[3]{125} = \sqrt[3]{s^3}$$ Take the cube root of each side.

$$5 = s$$ Simplify.

The edge length is 5 inches. Use a formula to find the surface area of the cube.

$$S = 6s^2$$ Write formula for surface area.

$$= 6(5)^2$$ Substitute 5 for *s*.

$$= 150$$ Simplify.

So, the surface area of the baseball display case is 150 square inches.

On Your Own

9. The volume of a music box that is shaped like a cube is 512 cubic centimeters. Find the surface area of the music box.

7.2 Exercises

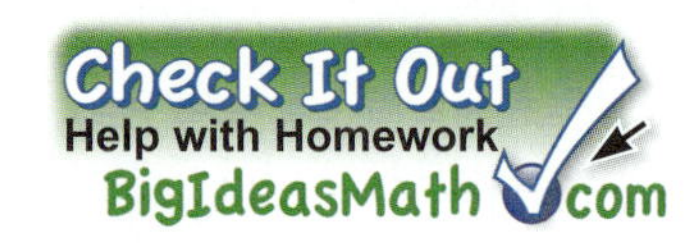

Vocabulary and Concept Check

1. **VOCABULARY** Is 25 a perfect cube? Explain.
2. **REASONING** Can the cube of an integer be a negative number? Explain.

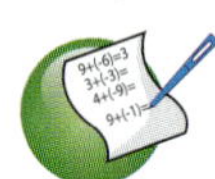

Practice and Problem Solving

Find the edge length of the cube.

3. Volume = 125,000 in.^{3}

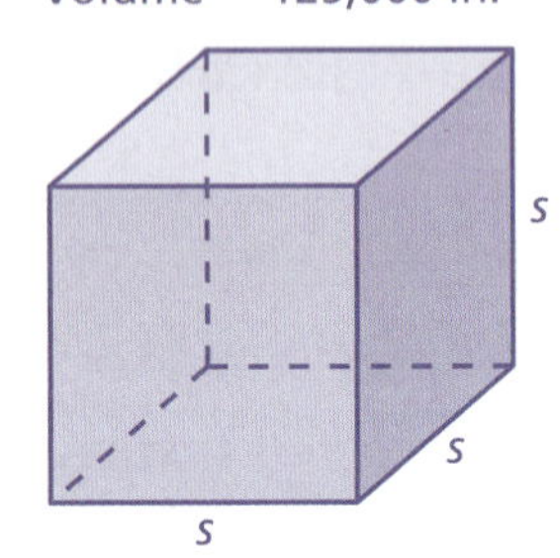

4. Volume = $\frac{1}{27}$ ft^3

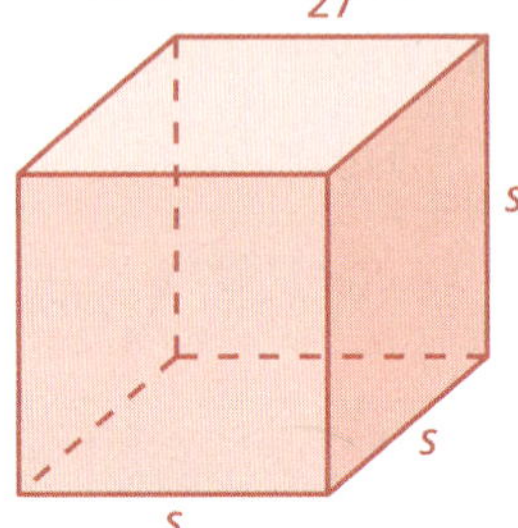

5. Volume = 0.064 m^3

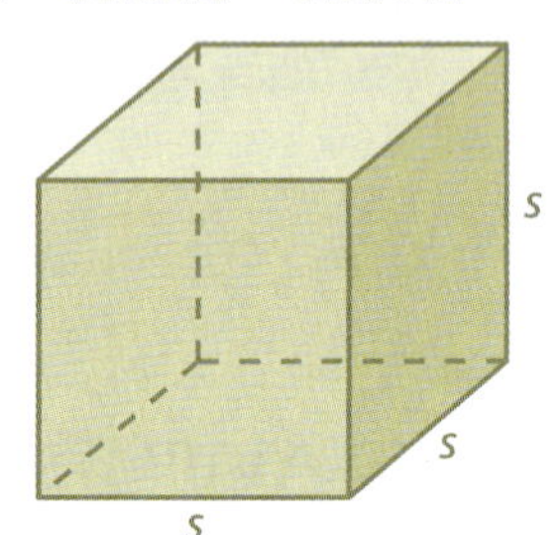

Find the cube root.

① 6. $\sqrt[3]{729}$
7. $\sqrt[3]{-125}$
8. $\sqrt[3]{-1000}$
9. $\sqrt[3]{1728}$
10. $\sqrt[3]{-\frac{1}{512}}$
11. $\sqrt[3]{\frac{343}{64}}$

Evaluate the expression.

② 12. $18 - \left(\sqrt[3]{27}\right)^3$
13. $\left(\sqrt[3]{-\frac{1}{8}}\right)^3 + 3\frac{3}{4}$
14. $5\sqrt[3]{729} - 24$
15. $\frac{1}{4} - 2\sqrt[3]{-\frac{1}{216}}$
16. $54 + \sqrt[3]{-4096}$
17. $4\sqrt[3]{8000} - 6$

Evaluate the expression for the given value of the variable.

③ 18. $\sqrt[3]{\frac{n}{4}} + \frac{n}{10}$, $n = 500$
19. $\sqrt[3]{6w} - w$, $w = 288$
20. $2d + \sqrt[3]{-45d}$, $d = 75$

21. **STORAGE CUBE** The volume of a plastic storage cube is 27,000 cubic centimeters. What is the edge length of the storage cube?

22. **ICE SCULPTURE** The volume of a cube of ice for an ice sculpture is 64,000 cubic inches.
    a. What is the edge length of the cube of ice?
    b. What is the surface area of the cube of ice?

Assignment Guide and Homework Check

Level	Day 1 Activity Assignment	Day 2 Lesson Assignment	Homework Check
Basic	3–5, 34–38	1, 2, 7–27 odd	9, 11, 15, 19, 21
Average	3–5, 34–38	1, 2, 6–22 even, 23–33 odd	10, 16, 20, 25, 29
Advanced	1–5, 10–24 even, 26–38		10, 16, 24, 26, 28

Common Errors

- **Exercises 6–11** Students may disregard negatives in the radicand. Remind them to pay attention to the sign of the radicand and to check the sign of their answers.
- **Exercises 12–17** Students may not follow the order of operations when evaluating the expression. Remind them of the order of operations. Because taking a cube root is the inverse of cubing, it is evaluated before multiplication and division.

7.2 Record and Practice Journal

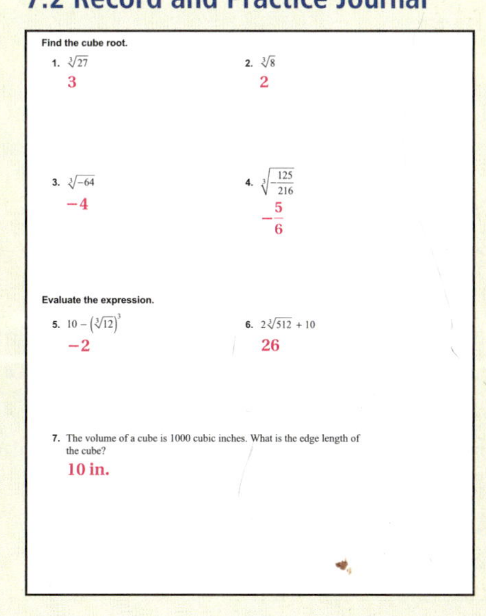
Find the cube root.

1. $\sqrt[3]{27}$
   3
2. $\sqrt[3]{8}$
   2
3. $\sqrt[3]{-64}$
   −4
4. $\sqrt[3]{-\frac{125}{216}}$
   $-\frac{5}{6}$

Evaluate the expression.

5. $10 - \left(\sqrt[3]{12}\right)^3$
   −2
6. $2\sqrt[3]{512} + 10$
   26
7. The volume of a cube is 1000 cubic inches. What is the edge length of the cube?
   10 in.

Vocabulary and Concept Check

1. no; There is no integer that equals 25 when cubed.
2. yes; When the integer is negative, the cube of the negative integer is negative. For example, $(-4)^3 = (-4) \cdot (-4) \cdot (-4) = -64$.

Practice and Problem Solving

3. 50 in.
4. $\frac{1}{3}$ ft
5. 0.4 m
6. 9
7. −5
8. −10
9. 12
10. $-\frac{1}{8}$
11. $\frac{7}{4}$
12. −9
13. $3\frac{5}{8}$
14. 21
15. $\frac{7}{12}$
16. 38
17. 74
18. 55
19. −276
20. 135
21. 30 cm
22. **a.** 40 in.

    b. 9600 in.^{2}

Practice and Problem Solving

23. >

24. >

25. <

26. 183 miles per hour

27. $-1, 0, 1$

28. **a.** not true; *Sample answer:* $\sqrt[3]{-8} = -2$

b. not true; *Sample answer:* 64 has only a positive cube root.

29. The side length of the square base is 18 inches and the height of the pyramid is 9 inches.

30. See *Taking Math Deeper*.

31. $x = 3$

32. $x = \frac{3}{2}$

33. $x = 4$

Fair Game Review

34. 25 **35.** 289

36. 144 **37.** 49

38. C

Mini-Assessment

1. Find $\sqrt[3]{-512}$. -8

Evaluate the expression.

2. $3 + 4\sqrt[3]{27}$ 15

3. $\sqrt[3]{-216} - 16$ -22

4. $\left(\sqrt[3]{-8}\right)^3 + 15$ 7

5. Evaluate $\left(\sqrt[3]{5b}\right) - \frac{3b}{5}$ when $b = 25$. -10

Taking Math Deeper

Exercise 30

One way to solve this problem is to write and solve a proportion. You can use what you know about factors to avoid calculations with large numbers.

Write a proportion.

$$\frac{125}{x} = \frac{x^2}{125}$$

Using the Cross Products Property, you obtain $x^3 = 15{,}625$. Rather than work with large numbers, rewrite the quantities in the original proportion as products of factors before solving.

$$\frac{125}{x} = \frac{x^2}{125} \quad \Rightarrow \quad \frac{5 \cdot 5 \cdot 5}{x} = \frac{x \cdot x}{5 \cdot 5 \cdot 5}$$

Solve the proportion. Multiply each side by the least common denominator, $5 \cdot 5 \cdot 5 \cdot x$.

$$(5 \cdot 5 \cdot 5 \cdot \cancel{x}) \cdot \frac{5 \cdot 5 \cdot 5}{\cancel{x}} = (\cancel{5} \cdot \cancel{5} \cdot \cancel{5} \cdot x) \cdot \frac{x \cdot x}{\cancel{5} \cdot \cancel{5} \cdot \cancel{5}}$$

$$5 \cdot 5 \cdot 5 \cdot 5 \cdot 5 \cdot 5 = x \cdot x \cdot x$$

$$(5 \cdot 5) \cdot (5 \cdot 5) \cdot (5 \cdot 5) = x \cdot x \cdot x$$

$$25 \cdot 25 \cdot 25 = x \cdot x \cdot x$$

$$25^3 = x^3$$

$$\sqrt[3]{25^3} = \sqrt[3]{x^3}$$

$$25 = x$$

So, x is 25.

Check the solution.

$$\frac{125}{x} = \frac{x^2}{125}$$

$$\frac{125}{25} \stackrel{?}{=} \frac{25^2}{125}$$

$$5 \stackrel{?}{=} \frac{625}{125}$$

$$5 = 5 \checkmark$$

Reteaching and Enrichment Strategies

If students need help...	If students got it...
Resources by Chapter • Practice A and Practice B • Puzzle Time Record and Practice Journal Practice Differentiating the Lesson Lesson Tutorials Skills Review Handbook	Resources by Chapter • Enrichment and Extension • Technology Connection Start the next section

Copy and complete the statement with <, >, or =.

23. $-\frac{1}{4}$ ☐ $\sqrt[3]{-\frac{8}{125}}$

24. $\sqrt[3]{0.001}$ ☐ 0.01

25. $\sqrt[3]{64}$ ☐ $\sqrt{64}$

26. DRAG RACE The estimated velocity v (in miles per hour) of a car at the end of a drag race is $v = 234\sqrt[3]{\frac{p}{w}}$, where p is the horsepower of the car and w is the weight (in pounds) of the car. A car has a horsepower of 1311 and weighs 2744 pounds. Find the velocity of the car at the end of a drag race. Round your answer to the nearest whole number.

27. NUMBER SENSE There are three numbers that are their own cube roots. What are the numbers?

28. LOGIC Each statement below is true for square roots. Determine whether the statement is also true for cube roots. Explain your reasoning and give an example to support your explanation.

a. You cannot find the square root of a negative number.

b. Every positive number has a positive square root and a negative square root.

29. GEOMETRY The pyramid has a volume of 972 cubic inches. What are the dimensions of the pyramid?

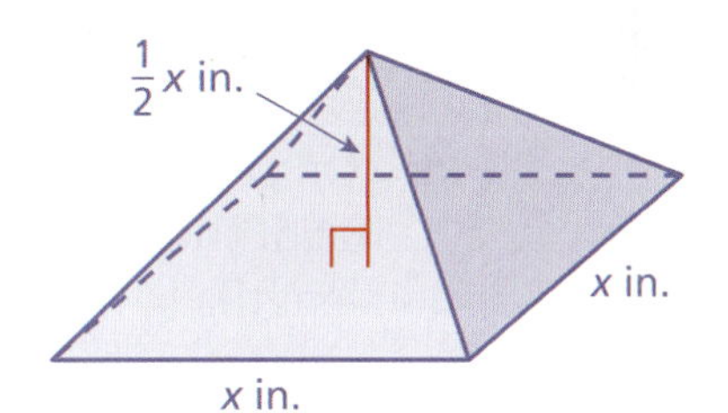

30. RATIOS The ratio $125 : x$ is equivalent to the ratio $x^2 : 125$. What is the value of x?

Solve the equation.

31. $(3x + 4)^3 = 2197$

32. $(8x^3 - 9)^3 = 5832$

33. $\left((5x - 16)^3 - 4\right)^3 = 216{,}000$

Fair Game Review What you learned in previous grades & lessons

Evaluate the expression. *(Skills Review Handbook)*

34. $3^2 + 4^2$

35. $8^2 + 15^2$

36. $13^2 - 5^2$

37. $25^2 - 24^2$

38. MULTIPLE CHOICE Which linear function is shown by the table? *(Section 6.3)*

x	1	2	3	4
y	4	7	10	13

Ⓐ $y = \frac{1}{3}x + 1$ Ⓑ $y = 4x$ Ⓒ $y = 3x + 1$ Ⓓ $y = \frac{1}{4}x$

7.3 The Pythagorean Theorem

Essential Question How are the lengths of the sides of a right triangle related?

Pythagoras was a Greek mathematician and philosopher who discovered one of the most famous rules in mathematics. In mathematics, a rule is called a **theorem**. So, the rule that Pythagoras discovered is called the Pythagorean Theorem.

Pythagoras
(c. 570–c. 490 B.C.)

1 ACTIVITY: Discovering the Pythagorean Theorem

Work with a partner.

a. On grid paper, draw any right triangle. Label the lengths of the two shorter sides a and b.

b. Label the length of the longest side c.

c. Draw squares along each of the three sides. Label the areas of the three squares a^2, b^2, and c^2.

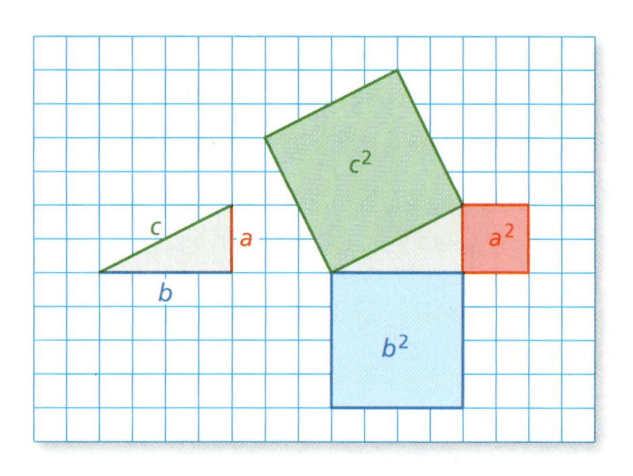

d. Cut out the three squares. Make eight copies of the right triangle and cut them out. Arrange the figures to form two identical larger squares.

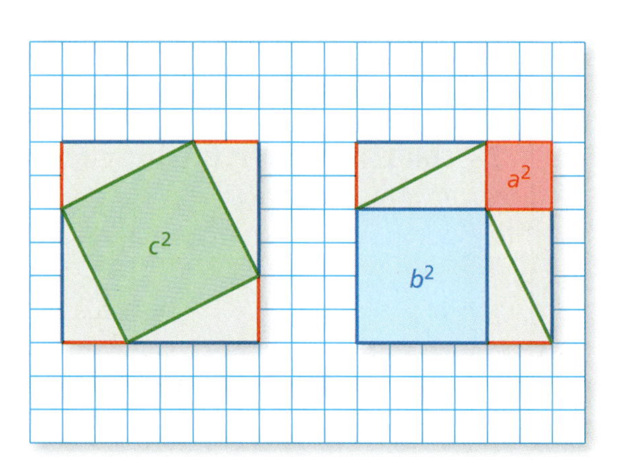

e. MODELING The Pythagorean Theorem describes the relationship among a^2, b^2, and c^2. Use your result from part (d) to write an equation that describes this relationship.

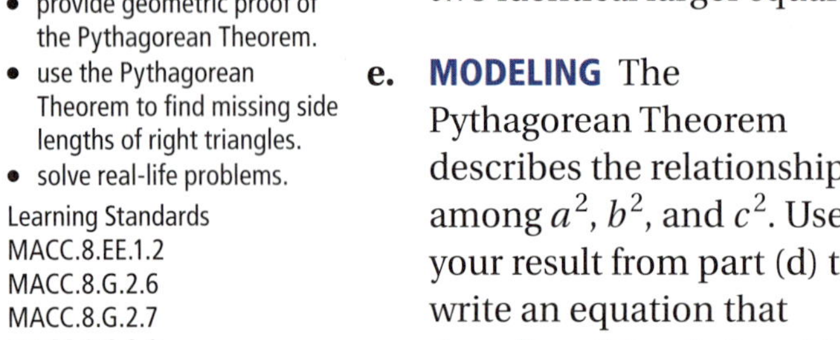

COMMON CORE

Pythagorean Theorem

In this lesson, you will

- provide geometric proof of the Pythagorean Theorem.
- use the Pythagorean Theorem to find missing side lengths of right triangles.
- solve real-life problems.

Learning Standards
MACC.8.EE.1.2
MACC.8.G.2.6
MACC.8.G.2.7
MACC.8.G.2.8

Laurie's Notes

Introduction

Standards for Mathematical Practice

- **MP4 Model with Mathematics:** In Activity 1, students will analyze the relationship between two models they create. This represents an "informal proof" of the Pythagorean Theorem.

Motivate

- Share information about Pythagoras, who was born in Greece in 569 B.C.
  - He is known as the *Father of Numbers*.
  - He traveled extensively in Egypt, learning math, astronomy, and music.
  - Pythagoras urged the citizens of Cretona to follow his religious, political, and philosophical goals.
  - His followers were known as Pythagoreans. They observed a rule of silence called *echemythia*. One had to remain silent for *five years* before he could contribute to the group. Breaking this silence was punishable by death!

Activity Notes

Activity 1

- **Suggestions:** Use centimeter grid paper for ease of manipulating the cut pieces. Suggest to students that they draw the original triangle in the upper left of the grid paper, and then make a working copy of the triangle towards the middle of the paper. This gives enough room for the squares to be drawn on each side of the triangle.
- Vertices of the triangle need to be on lattice points. You do not want every student in the room to use the same triangle. Suggest other lengths for the shorter sides (3 and 4, 3 and 6, 2 and 4, 2 and 3, and so on).
- **Model:** Drawing the square on the longest side of the triangle is the challenging step. Model one technique for accomplishing the task using a right triangle with shorter side lengths of 2 units and 5 units.
  - Notice that the longest side has a slope of "right 5 units, up 2 units."
  - Place your pencil on the upper right endpoint and rotate the paper 90° clockwise. Move your pencil right 5 units and up 2 units. Mark a point.
  - Repeat rotating and moving "right 5 units, up 2 units" until you get back to the longest side of the triangle.
  - Use a straightedge to connect the four points (two that you marked and two on the endpoints of the longest side) to form the square.
- Before students cut anything, check that they have 3 squares of the correct size.
- **Big Idea:** The two large squares in part (d) have equal area. Referring to areas, if $c^2 +$ (4 triangles) $= a^2 + b^2 +$ (4 triangles), then $c^2 = a^2 + b^2$ by subtracting the 4 triangles from each side of the equation.
- The work in this activity constitutes an "informal proof" of the Pythagorean Theorem. There are many proofs of this theorem, and this version is generally understood by middle school students.

Florida Common Core Standards

MACC.8.EE.1.2 Use square root and cube root symbols to represent solutions to equations of the form $x^2 = p$ and $x^3 = p$, where p is a positive rational number. Evaluate square roots of small perfect squares and cube roots of small perfect cubes. Know that $\sqrt{2}$ is irrational.

MACC.8.G.2.6 Explain a proof of the Pythagorean Theorem and its converse.

MACC.8.G.2.7 Apply the Pythagorean Theorem to determine unknown side lengths in right triangles in real-world and mathematical problems in two and three dimensions.

MACC.8.G.2.8 Apply the Pythagorean Theorem to find the distance between two points in a coordinate system.

Previous Learning

Students should know how to evaluate algebraic expressions for given values of the variables.

Lesson Plans
Complete Materials List

7.3 Record and Practice Journal

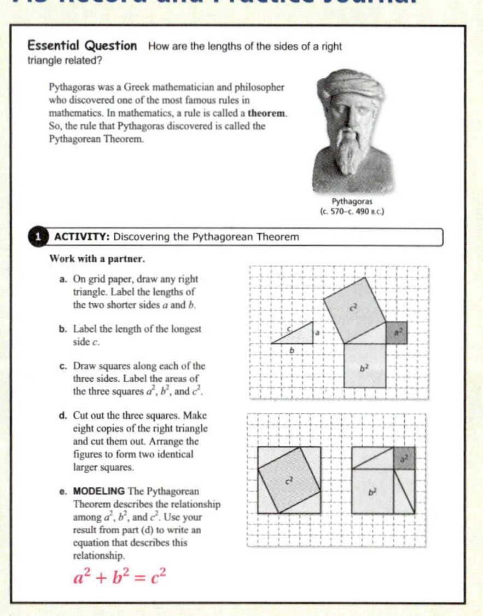

Essential Question How are the lengths of the sides of a right triangle related?

Pythagoras was a Greek mathematician and philosopher who discovered one of the most famous rules in mathematics. In mathematics, a rule is called a **theorem**. So, the rule that Pythagoras discovered is called the Pythagorean Theorem.

Pythagoras (c. 570–c. 490 B.C.)

1 ACTIVITY: Discovering the Pythagorean Theorem

Work with a partner.

a. On grid paper, draw any right triangle. Label the lengths of the two shorter sides a and b.

b. Label the length of the longest side c.

c. Draw squares along each of the three sides. Label the areas of the three squares a^2, b^2, and c^2.

d. Cut out the three squares. Make eight copies of the right triangle and cut them out. Arrange the figures to form two identical larger squares.

e. **MODELING** The Pythagorean Theorem describes the relationship among a^2, b^2, and c^2. Use your result from part (d) to write an equation that describes this relationship.

$a^2 + b^2 = c^2$

English Language Learners

Vocabulary

Help English learners understand the meanings of the words that make up a definition. Provide students with statements containing blanks and a list of the words used to fill in the blanks.

- In any right __, the __ is the side __ the right __.
  Word list: angle, hypotenuse, opposite, triangle
  triangle, hypotenuse, opposite, angle
- In any right __, the __ are the __ sides and the __ is always the __ side.
  Word list: hypotenuse, legs, longest, shorter, triangle
  triangle, legs, shorter, hypotenuse, longest

7.3 Record and Practice Journal

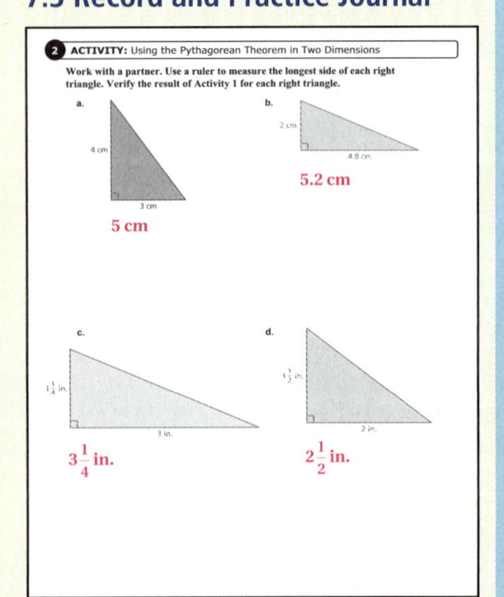

2 ACTIVITY: Using the Pythagorean Theorem in Two Dimensions

Work with a partner. Use a ruler to measure the longest side of each right triangle. Verify the result of Activity 1 for each right triangle.

a. 5 cm

b. 5.2 cm

c. $3\frac{1}{4}$ in.

d. $2\frac{1}{2}$ in.

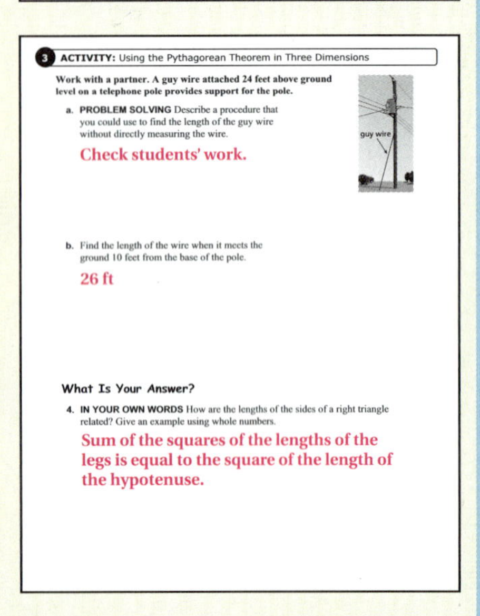

3 ACTIVITY: Using the Pythagorean Theorem in Three Dimensions

Work with a partner. A guy wire attached 24 feet above ground level on a telephone pole provides support for the pole.

a. **PROBLEM SOLVING** Describe a procedure that you could use to find the length of the guy wire without directly measuring the wire.

Check students' work.

b. Find the length of the wire when it meets the ground 10 feet from the base of the pole.

26 ft

What Is Your Answer?

4. **IN YOUR OWN WORDS** How are the lengths of the sides of a right triangle related? Give an example using whole numbers.

Sum of the squares of the lengths of the legs is equal to the square of the length of the hypotenuse.

Laurie's Notes

Activity 2

- Begin by asking students to summarize what they learned from Activity 1.
- Distribute rulers with both metric and customary measures.
- **MP6 Attend to Precision:** Caution students to pay attention to the units and to measure accurately. You may choose to have them begin by checking the accuracy of the side lengths already given.
- You may want students to use calculators to square some of the side lengths.
- When students have finished, ask them to share their measures for each of the longest sides *and* their calculations. Students should not simply measure. You want to ensure that they have performed the calculations to verify the results of Activity 1.

Activity 3

- Explain what a guy wire is if students are not familiar with the term.
- **MP3 Construct Viable Arguments and Critique the Reasoning of Others:** Students are asked to explain a method for finding the length of a guy wire without measuring directly. They may incorrectly assume that they can't measure anything. The length of one of the shorter sides of the right triangle is known (24 feet), and the other shorter side can be measured directly (distance from bottom of guy wire to the base of the pole).
- You may want to discuss the results of part (a), and then have students work through part (b) with a partner.

What Is Your Answer?

- **Neighbor Check:** Have students work independently and then have their neighbors check their work. Have students discuss any discrepancies.

Closure

- **Exit Ticket:** If you drew a right triangle with shorter side lengths of 4 and 6 on grid paper, what would be the area of the square drawn on the longest side of the triangle? 52 square units

2 ACTIVITY: Using the Pythagorean Theorem in Two Dimensions

Work with a partner. Use a ruler to measure the longest side of each right triangle. Verify the result of Activity 1 for each right triangle.

a.

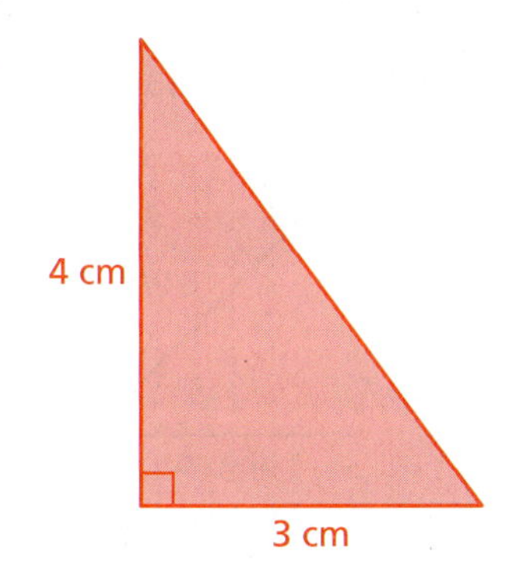

b.

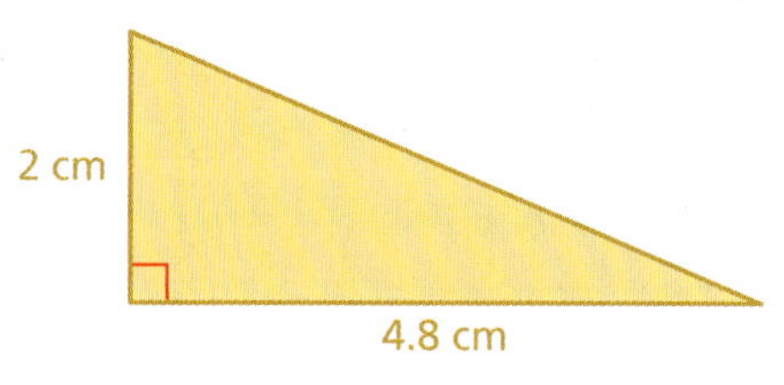

c.

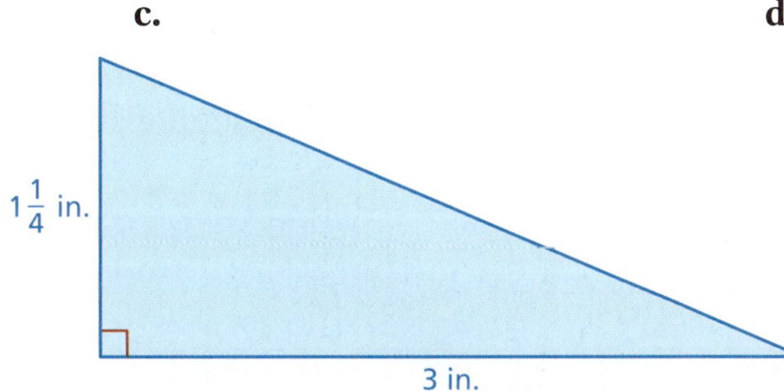

d.

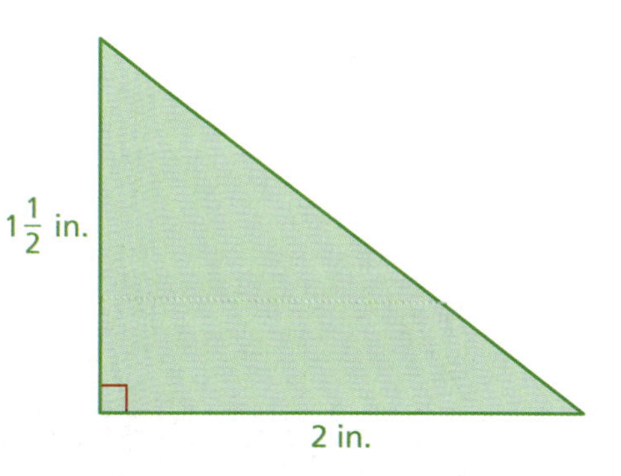

3 ACTIVITY: Using the Pythagorean Theorem in Three Dimensions

Math Practice 3

Use Definitions

How can you use what you know about the Pythagorean Theorem to describe the procedure for finding the length of the guy wire?

Work with a partner. A guy wire attached 24 feet above ground level on a telephone pole provides support for the pole.

a. **PROBLEM SOLVING** Describe a procedure that you could use to find the length of the guy wire without directly measuring the wire.

b. Find the length of the wire when it meets the ground 10 feet from the base of the pole.

What Is Your Answer?

4. **IN YOUR OWN WORDS** How are the lengths of the sides of a right triangle related? Give an example using whole numbers.

Practice Use what you learned about the Pythagorean Theorem to complete Exercises 3 and 4 on page 304.

7.3 Lesson

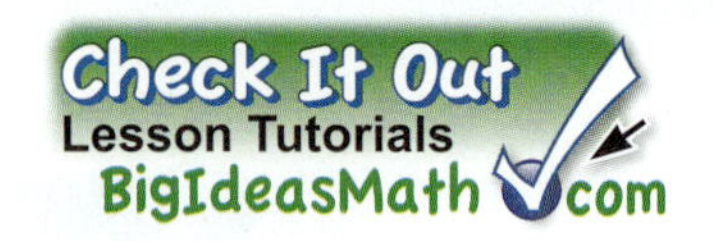

Key Vocabulary

theorem, *p. 300*
legs, *p. 302*
hypotenuse, *p. 302*
Pythagorean Theorem, *p. 302*

Key Ideas

Sides of a Right Triangle

The sides of a right triangle have special names.

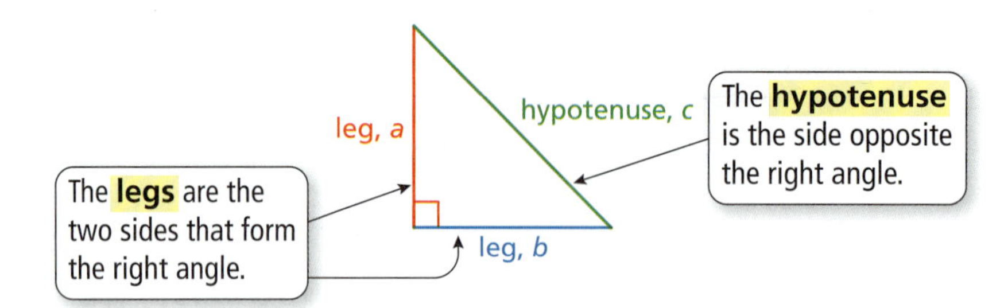

Study Tip

In a right triangle, the legs are the shorter sides and the hypotenuse is always the longest side.

The Pythagorean Theorem

Words In any right triangle, the sum of the squares of the lengths of the legs is equal to the square of the length of the hypotenuse.

Algebra $a^2 + b^2 = c^2$

EXAMPLE 1 Finding the Length of a Hypotenuse

Find the length of the hypotenuse of the triangle.

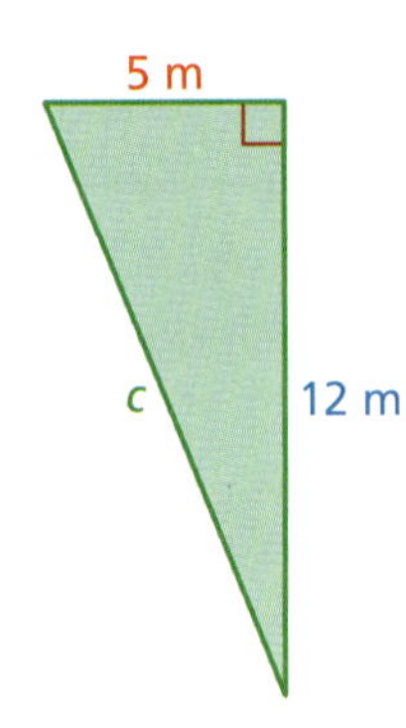

$a^2 + b^2 = c^2$	Write the Pythagorean Theorem.
$5^2 + 12^2 = c^2$	Substitute 5 for *a* and 12 for *b*.
$25 + 144 = c^2$	Evaluate powers.
$169 = c^2$	Add.
$\sqrt{169} = \sqrt{c^2}$	Take positive square root of each side.
$13 = c$	Simplify.

The length of the hypotenuse is 13 meters.

On Your Own

Now You're Ready
Exercises 3 and 4

Find the length of the hypotenuse of the triangle.

1. **2.**

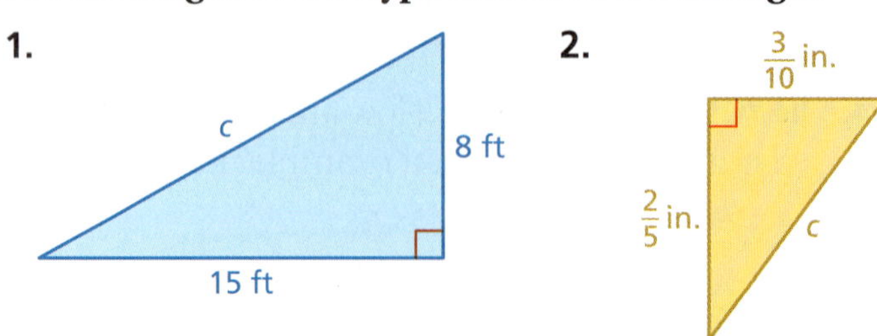

Laurie's Notes

Introduction

Connect

- **Yesterday:** Students investigated a visual proof of the Pythagorean Theorem. (MP3, MP4, MP6)
- **Today:** Students will use the Pythagorean Theorem to find missing side lengths of right triangles.

Motivate

- **Preparation:** Cut coffee stirrers (or carefully break spaghetti) so that triangles with the following side lengths can be made: 2-3-4; 3-4-5; 4-5-6.

? "What are consecutive numbers?" numbers in sequential order

- With student aid, use the coffee stirrers to make three triangles: 2-3-4; 3-4-5; and 4-5-6 on a document camera or overhead projector. If arranged carefully, all 3 will fit on the screen.
- Ask students to make observations about the 3 triangles. Students may mention that all triangles are scalene; one triangle appears to be acute, one right, one obtuse.
- They should observe that the change in the side lengths seems to have made a big change in the angle measures.

Lesson Notes

Key Ideas

- Draw a right triangle and label the *legs* and the *hypotenuse*. The **hypotenuse** is always opposite the right angle and is the longest side of a right triangle.
- Try not to have all right triangles in the same orientation.
- Write the Pythagorean Theorem.
- **Common Error:** Students often forget that the Pythagorean Theorem is a relationship that is *only* true for right triangles.

Example 1

- Draw and label the triangle. Review the symbol used to show that an angle is a right angle.

? "What information is known for this triangle?" The legs are 5 m and 12 m.

- Substitute and solve as shown. Explain that you disregard the negative square root because the length is positive.

On Your Own

- Give time for students to work the problems. Knowing their perfect squares is helpful.
- **MP2 Reason Abstractly and Quantitatively:** In Question 2, if students recognize that the decimal equivalents of the given fractions are 0.3 and 0.4, finding the hypotenuse may be quick for them.

Goal Today's lesson is using the **Pythagorean Theorem** to solve for side lengths of right triangles.

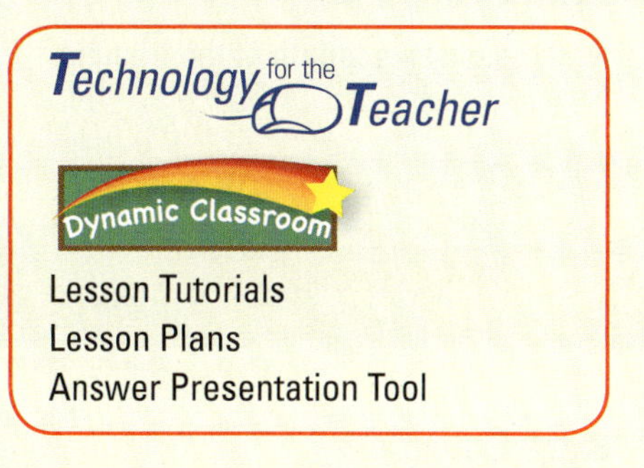

Lesson Tutorials
Lesson Plans
Answer Presentation Tool

Extra Example 1

Find the length of the hypotenuse of the triangle. 5 in.

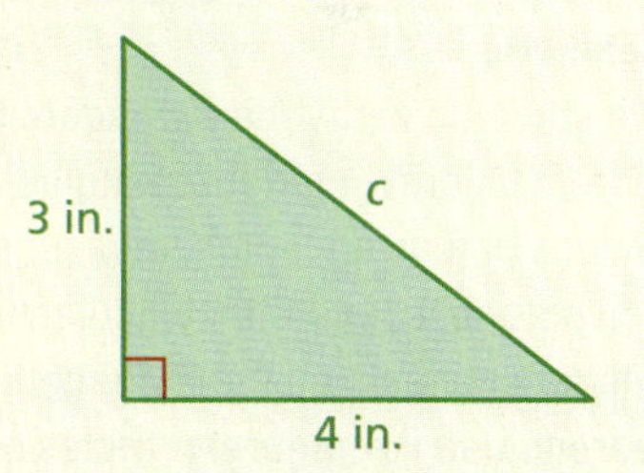

On Your Own

1. 17 ft
2. $\frac{1}{2}$ in.

Laurie's Notes

Extra Example 2

Find the missing length of the triangle. 24 ft

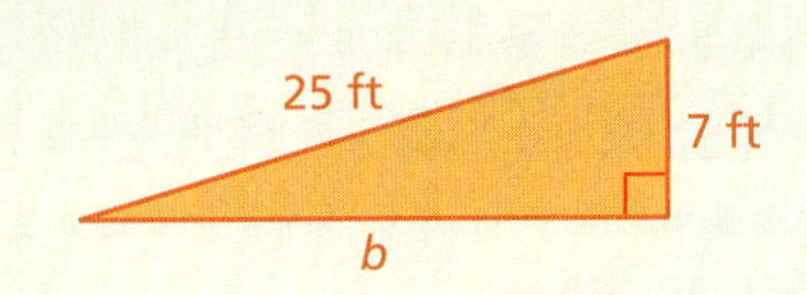

Extra Example 3

You and your cousin are planning to go to an amusement park. You live 36 miles south of the amusement park and 15 miles west of your cousin. How far away from the amusement park does your cousin live? 39 miles

On Your Own

3. 30 yd
4. 4 m
5. 100 yd

Differentiated Instruction

Kinesthetic

Have students verify the Pythagorean Theorem by drawing right triangles with legs of a given length, measuring the hypotenuse, and then calculating the hypotenuse using the Pythagorean Theorem. Use Pythagorean triples so that students work only with whole numbers.

Leg Lengths	*Hypotenuse Length*
3, 4	5
6, 8	10
5, 12	13
8, 15	17

Example 2

- ? "What information is known for this triangle?" One leg is 2.1 centimeters and the hypotenuse is 2.9 centimeters.
- Substitute and solve as shown.
- **Common Error:** Students need to be careful with decimal multiplication. It is very common for students to multiply the decimal by 2 instead of multiplying the decimal by itself.
- **FYI:** The triangle is similar to a 20-21-29 right triangle.

Example 3

- Ask a student to read the example.
- ? "Given the compass directions stated, what is a reasonable way to represent this information?" coordinate plane
- **MP4 Model with Mathematics:** Explain that east is the positive x-direction and north is the positive y-direction. Draw the situation in a coordinate plane.
- ? "Is there enough information to use the Pythagorean Theorem? Explain." yes; The legs of the triangle can be found and then used to solve for the hypotenuse.
- **FYI:** This example previews the *distance formula,* which will be presented in Section 7.5.

On Your Own

- **Think-Pair-Share:** Students should read each question independently and then work in pairs to answer the questions. When they have answered the questions, the pair should compare their answers with another group and discuss any discrepancies.

Closure

- **Exit Ticket:** Solve for the missing side length. $x = 15$ cm, $y = 0.5$ m

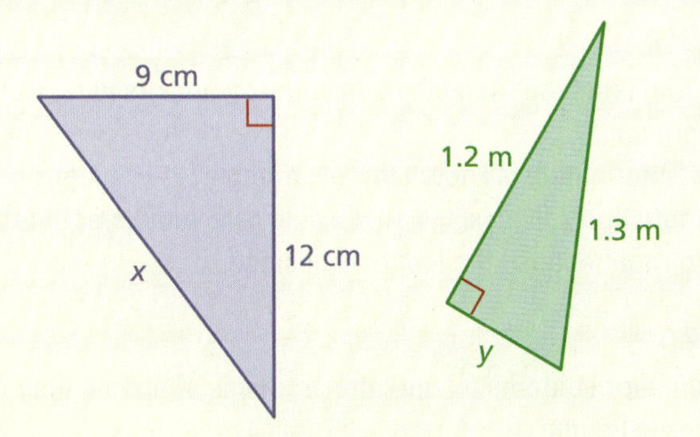

EXAMPLE 2 Finding the Length of a Leg

Find the missing length of the triangle.

a

2.1 cm

2.9 cm

$a^2 + b^2 = c^2$	Write the Pythagorean Theorem.
$a^2 + 2.1^2 = 2.9^2$	Substitute 2.1 for b and 2.9 for c.
$a^2 + 4.41 = 8.41$	Evaluate powers.
$a^2 = 4$	Subtract 4.41 from each side.
$a = 2$	Take positive square root of each side.

∴ The missing length is 2 centimeters.

EXAMPLE 3 Real-Life Application

You are playing capture the flag. You are 50 yards north and 20 yards east of your team's base. The other team's base is 80 yards north and 60 yards east of your base. How far are you from the other team's base?

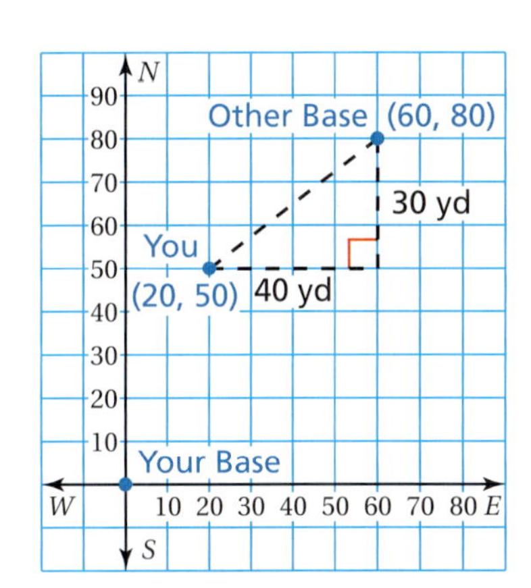

Step 1: Draw the situation in a coordinate plane. Let the origin represent your team's base. From the descriptions, you are at (20, 50) and the other team's base is at (60, 80).

Step 2: Draw a right triangle with a hypotenuse that represents the distance between you and the other team's base. The lengths of the legs are 30 yards and 40 yards.

Step 3: Use the Pythagorean Theorem to find the length of the hypotenuse.

$a^2 + b^2 = c^2$	Write the Pythagorean Theorem.
$30^2 + 40^2 = c^2$	Substitute 30 for a and 40 for b.
$900 + 1600 = c^2$	Evaluate powers.
$2500 = c^2$	Add.
$50 = c$	Take positive square root of each side.

∴ So, you are 50 yards from the other team's base.

On Your Own

Now You're Ready
Exercises 5–8

Find the missing length of the triangle.

3.

4.

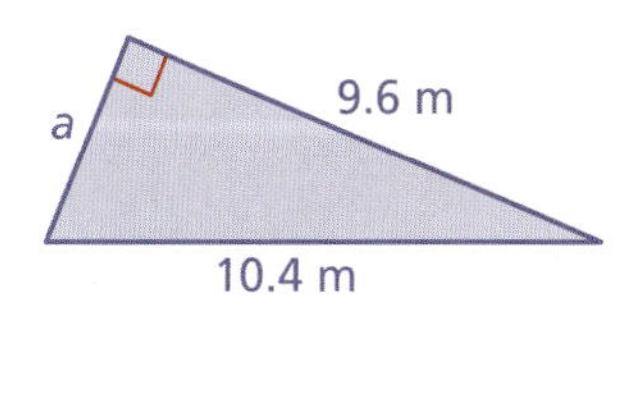

5. In Example 3, what is the distance between the bases?

7.3 Exercises

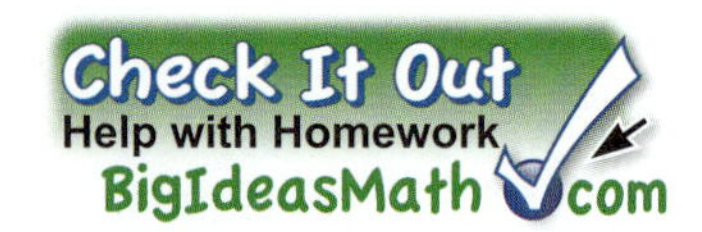

Vocabulary and Concept Check

1. **VOCABULARY** In a right triangle, how can you tell which sides are the legs and which side is the hypotenuse?

2. **DIFFERENT WORDS, SAME QUESTION** Which is different? Find "both" answers.

Which side is the hypotenuse?
Which side is the longest?
Which side is a leg?
Which side is opposite the right angle?

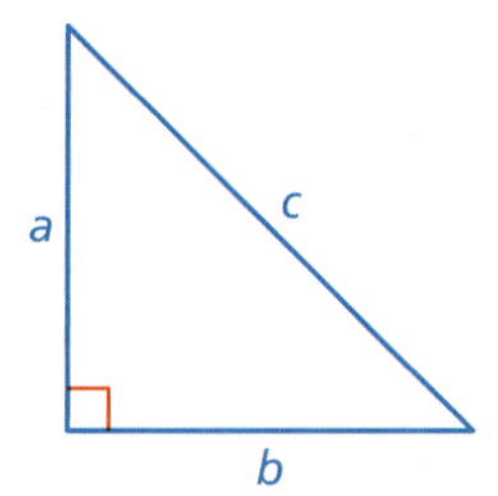

Practice and Problem Solving

Find the missing length of the triangle.

3. 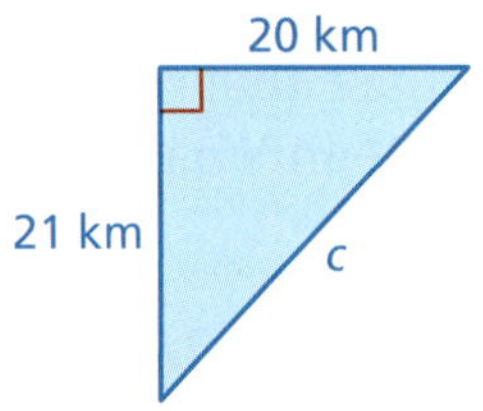

4. 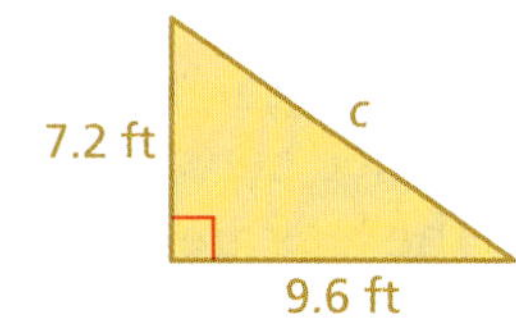

5. 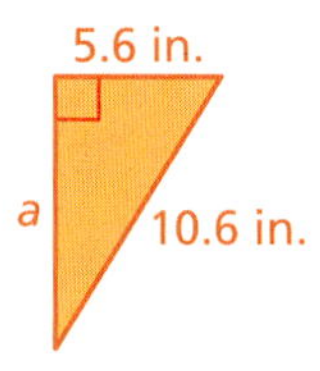

6. 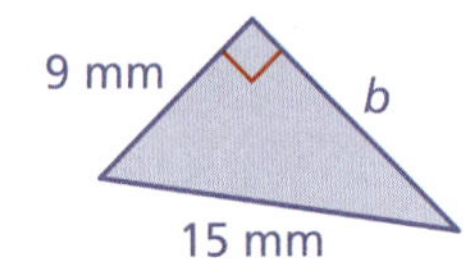

7. 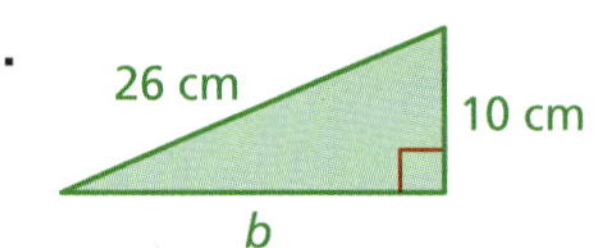

8. 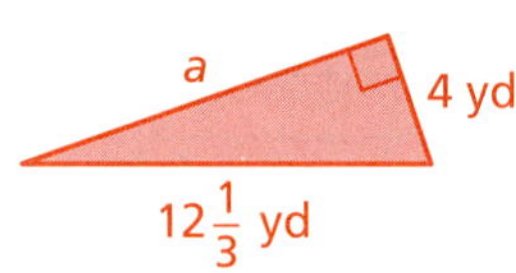

9. **ERROR ANALYSIS** Describe and correct the error in finding the missing length of the triangle.

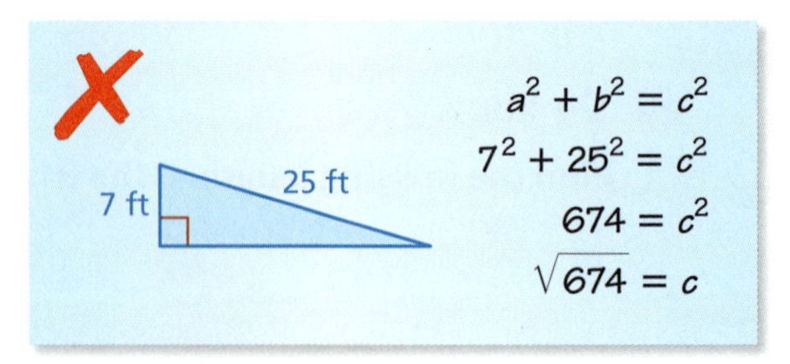

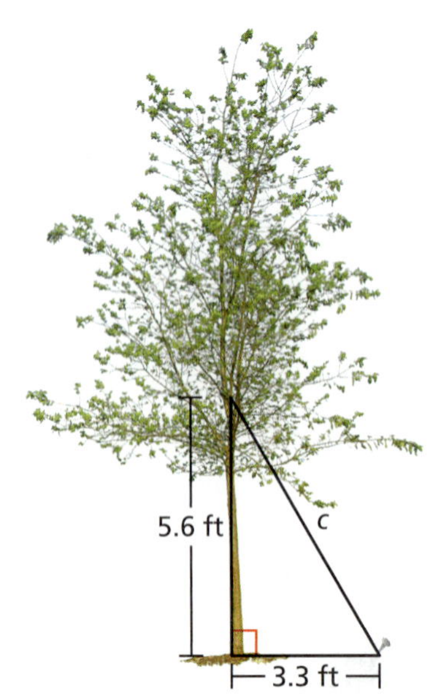

10. **TREE SUPPORT** How long is the wire that supports the tree?

Assignment Guide and Homework Check

Level	Day 1 Activity Assignment	Day 2 Lesson Assignment	Homework Check
Basic	3, 4, 19–23	1, 2, 5–17 odd	5, 7, 11, 13, 17
Average	3, 4, 19–23	1, 2, 5–9 odd, 10–18 even	5, 7, 10, 12, 14
Advanced	1–4, 6, 8–23		8, 10, 12, 14, 18

For Your Information

- **Exercise 17** There is more than one correct drawing for this exercise. Encourage students to start at the origin and move along an axis to begin.

Common Errors

- **Exercises 3–8** Students may substitute the given lengths in the wrong part of the formula. For example, if they are finding one of the legs, they may write $5^2 + 13^2 = c^2$ instead of $5^2 + b^2 = 13^2$. Remind them that the side opposite the right angle is the hypotenuse c.
- **Exercises 3–8** Students may multiply each side length by two instead of squaring the side length. Remind them of the definition of exponents.
- **Exercises 11 and 12** Students may think that there is not enough information to find the value of x. Tell them that it is possible to find x; however, they may have to make an extra calculation before writing an equation for x.

7.3 Record and Practice Journal

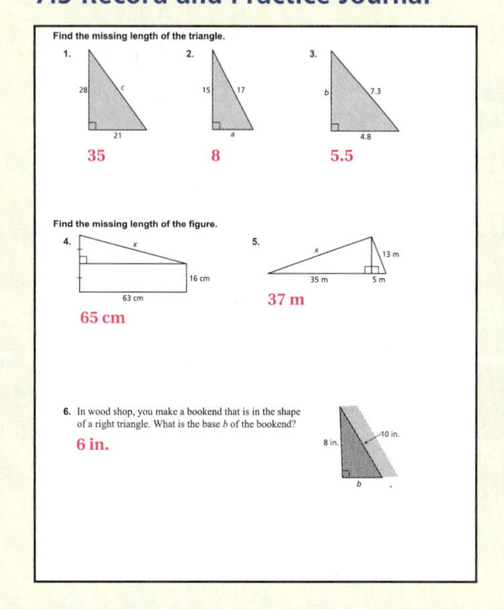
Find the missing length of the triangle.

1. 35
2. 8
3. 5.5

Find the missing length of the figure.

4. 65 cm
5. 37 m

6. In wood shop, you make a bookend that is in the shape of a right triangle. What is the base b of the bookend?

6 in.

Vocabulary and Concept Check

1. The hypotenuse is the longest side and the legs are the other two sides.
2. Which side is a leg?; a or b; c

Practice and Problem Solving

3. 29 km
4. 12 ft
5. 9 in.
6. 12 mm
7. 24 cm
8. $11\frac{2}{3}$ yd
9. The length of the hypotenuse was substituted for the wrong variable.

$$\begin{aligned} a^2 + b^2 &= c^2 \\ 7^2 + b^2 &= 25^2 \\ 49 + b^2 &= 625 \\ b^2 &= 576 \\ b &= 24 \end{aligned}$$

10. 6.5 ft

Practice and Problem Solving

11. 16 cm

12. 37 mm

13. See *Taking Math Deeper*.

14. yes; The distance from the player's mouth to the referee's ear is 25 feet.

15. *Sample answer:* length = 20 ft, width = 48 ft, height = 10 ft; $BC = 52$ ft, $AB = \sqrt{2804}$ ft

16. 7

17. See Additional Answers.

18. yes; The rod is 25 inches long and the diagonal from a top corner to the opposite bottom corner is 26 inches long.

Fair Game Review

19. 6 and −6 **20.** −11

21. 13 **22.** −15

23. C

Mini-Assessment

Find the missing length of the triangle.

1. 50 ft

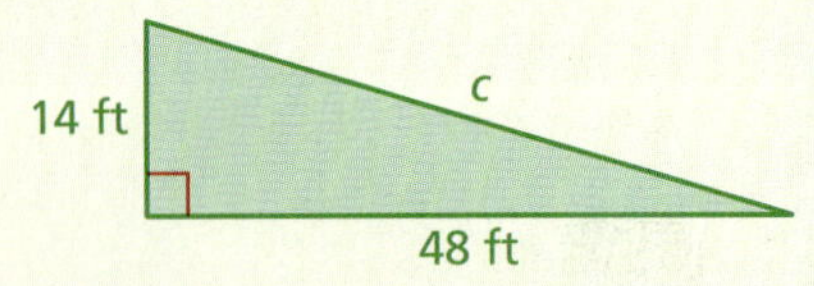

2. 24 mm

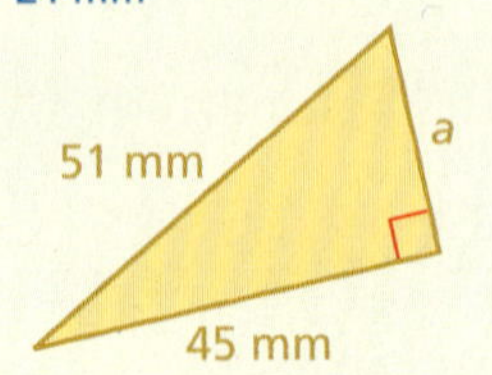

3. 12 in.

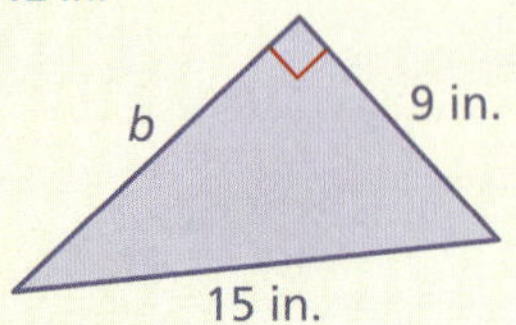

Taking Math Deeper

Exercise 13

The challenging part of this problem is realizing that the hypotenuse of the right triangle is given as 181 yards below the diagram.

 Draw and label a diagram with the given information.

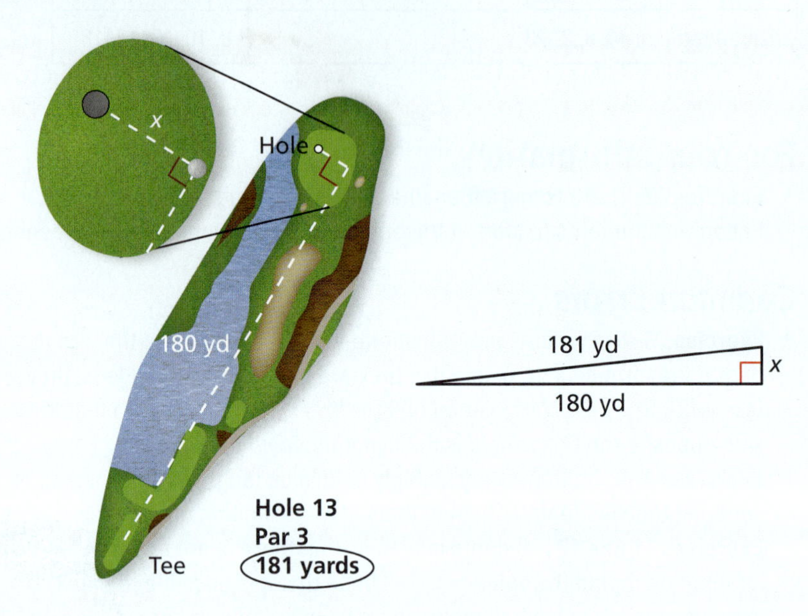

2 Use the Pythagorean Theorem.

$$180^2 + x^2 = 181^2$$
$$32{,}400 + x^2 = 32{,}761$$
$$x^2 = 361$$
$$x = 19$$

3 Answer the question.

The ball is 19 yards from the hole. Using the relationship of 3 feet = 1 yard, $19 \text{ yd} \times \frac{3 \text{ ft}}{1 \text{ yd}} = 57$ ft. So, the ball is 57 feet from the hole.

Reteaching and Enrichment Strategies

If students need help...	If students got it...
Resources by Chapter • Practice A and Practice B • Puzzle Time Record and Practice Journal Practice Differentiating the Lesson Lesson Tutorials Skills Review Handbook	Resources by Chapter • Enrichment and Extension • Technology Connection Start the next section

Find the missing length of the figure.

11.

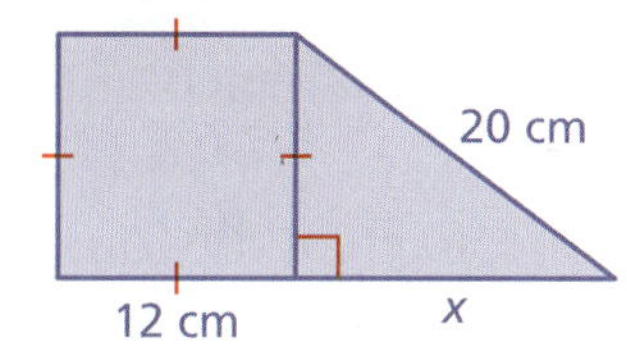

12.

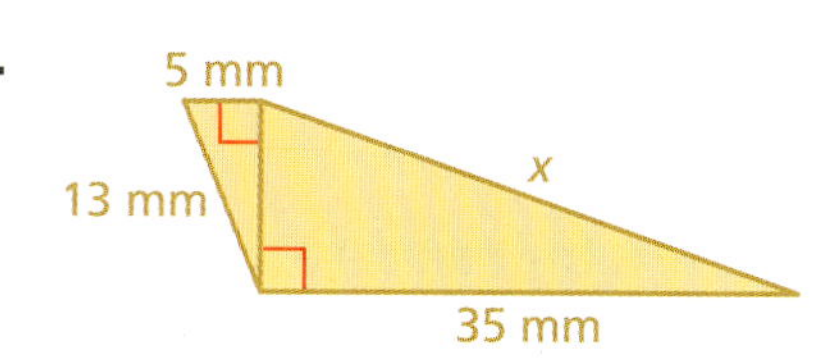

13. GOLF The figure shows the location of a golf ball after a tee shot. How many feet from the hole is the ball?

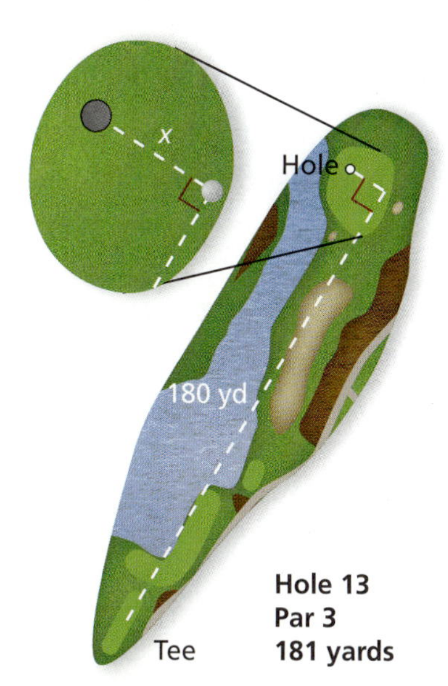

14. TENNIS A tennis player asks the referee a question. The sound of the player's voice travels only 30 feet. Can the referee hear the question? Explain.

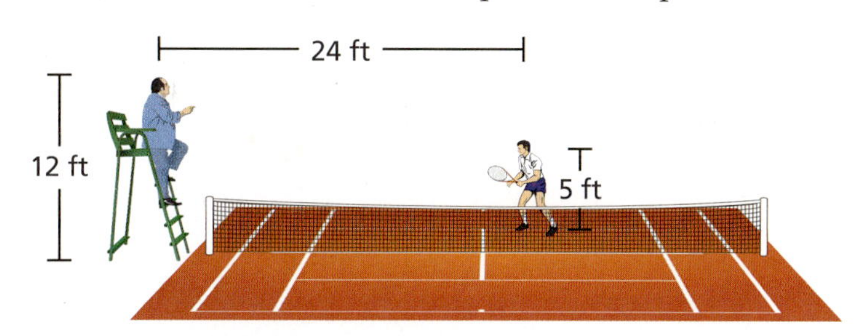

15. PROJECT Measure the length, width, and height of a rectangular room. Use the Pythagorean Theorem to find length *BC* and length *AB*.

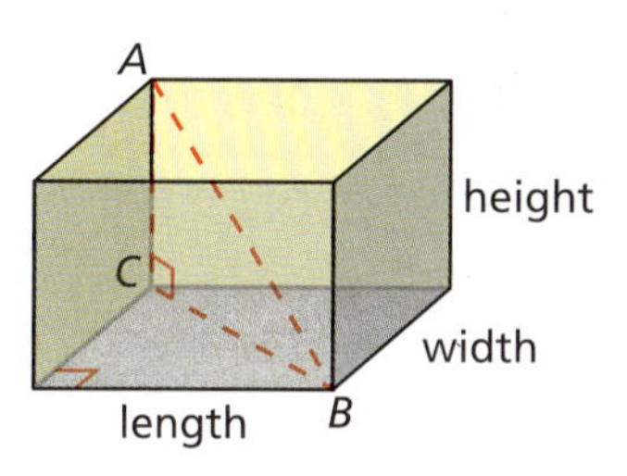

16. ALGEBRA The legs of a right triangle have lengths of 28 meters and 21 meters. The hypotenuse has a length of $5x$ meters. What is the value of x?

17. SNOWBALLS You and a friend stand back-to-back. You run 20 feet forward, then 15 feet to your right. At the same time, your friend runs 16 feet forward, then 12 feet to her right. She stops and hits you with a snowball.

a. Draw the situation in a coordinate plane.

b. How far does your friend throw the snowball?

18. Precision A box has a length of 6 inches, a width of 8 inches, and a height of 24 inches. Can a cylindrical rod with a length of 63.5 centimeters fit in the box? Explain your reasoning.

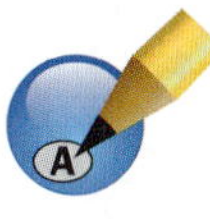

Fair Game Review What you learned in previous grades & lessons

Find the square root(s). *(Section 7.1)*

19. $\pm\sqrt{36}$ **20.** $-\sqrt{121}$ **21.** $\sqrt{169}$ **22.** $-\sqrt{225}$

23. MULTIPLE CHOICE What is the solution of the system of linear equations $y = 4x + 1$ and $2x + y = 13$? *(Section 5.2)*

Ⓐ $x = 1, y = 5$ Ⓑ $x = 5, y = 3$ Ⓒ $x = 2, y = 9$ Ⓓ $x = 9, y = 2$

7 Study Help

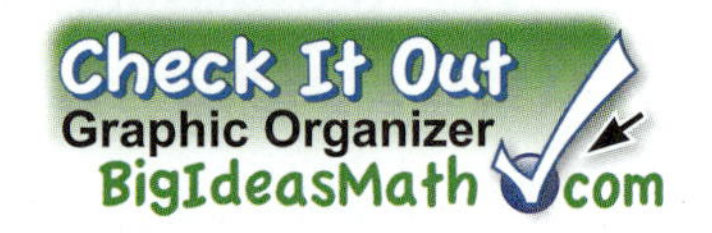

You can use a **four square** to organize information about a topic. Each of the four squares can be a category, such as *definition, vocabulary, example, non-example, words, algebra, table, numbers, visual, graph,* or *equation*. Here is an example of a four square for the Pythagorean Theorem.

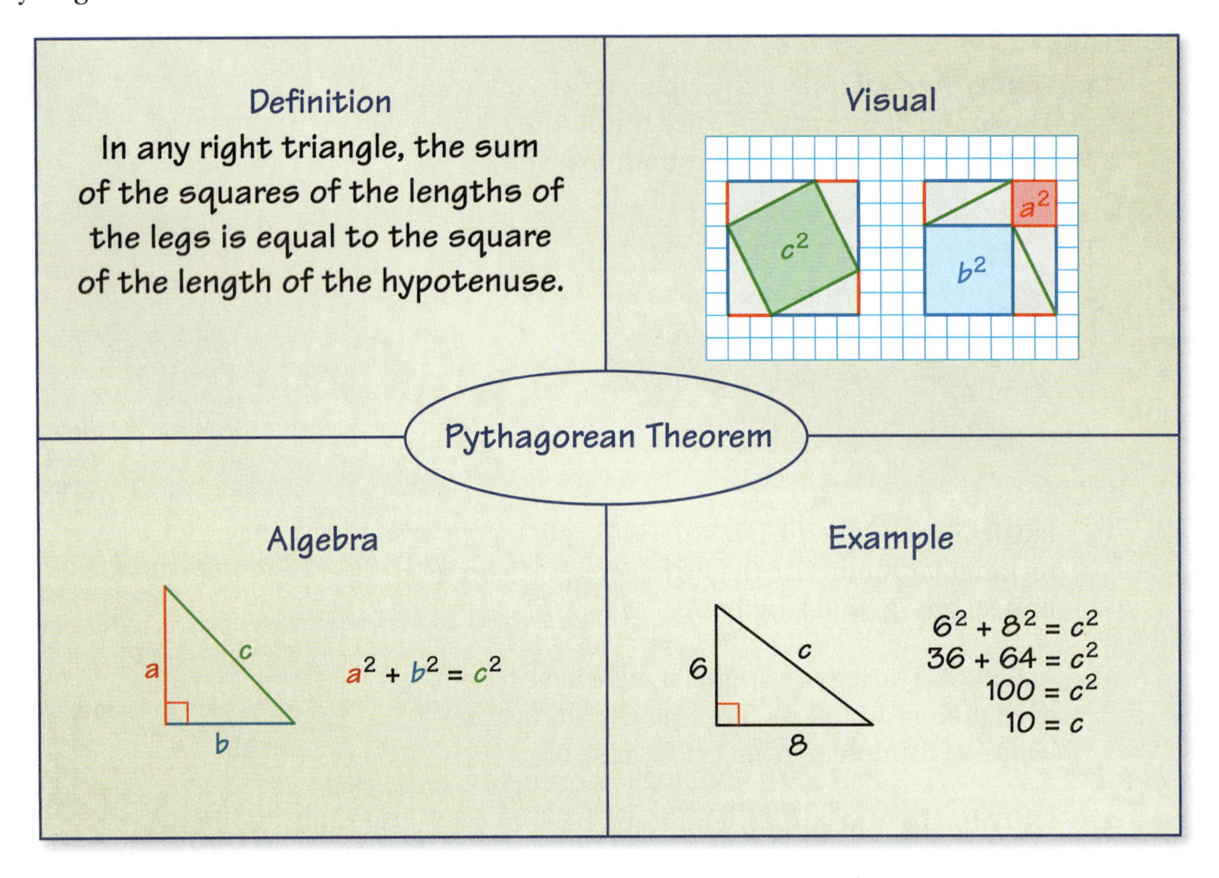

On Your Own

Make four squares to help you study these topics.

1. square roots
2. cube roots

After you complete this chapter, make four squares for the following topics.

3. irrational numbers
4. real numbers
5. converse of the Pythagorean Theorem
6. distance formula

"I'm taking a survey for my four square. How many fleas do you have?"

Sample Answers

1.

Definition	Symbol
A *square root* of a number is a number that, when multiplied by itself, equals the given number. Every positive number has a positive and a negative square root.	The symbol $\sqrt{}$ is called a *radical sign*. It is used to represent a square root. The number under the radical sign is called the *radicand*.
Square roots	
Example	**Example**
The two square roots of 100 are 10 and −10 because $10^2 = 100$ and $(-10)^2 = 100$. Positive square root: $\sqrt{100} = 10$ Negative square root: $-\sqrt{100} = -10$ Both square roots: $\pm\sqrt{100} = \pm 10$	Find $\sqrt{\frac{4}{9}}$. Because $\left(\frac{2}{3}\right)^2 = \frac{4}{9}$, $\sqrt{\frac{4}{9}} = \sqrt{\left(\frac{2}{3}\right)^2} = \frac{2}{3}$.

2.

Definition	Symbol
A *cube root* of a number is a number that, when multiplied by itself, and then multiplied by itself again, equals the given number.	The symbol $\sqrt[3]{}$ is used to represent a cube root.
Cube roots	
Algebra	**Examples**
If $a^3 = b$, then $a = \sqrt[3]{b}$, and a is a cube root of b.	a. Because $4^3 = 64$, $\sqrt[3]{64} = \sqrt[3]{4^3} = 4$. b. Because $(-5)^3 = -125$, $\sqrt[3]{-125} = \sqrt[3]{(-5)^3} = -5$. c. Because $\left(\frac{2}{3}\right)^3 = \frac{8}{27}$, $\sqrt[3]{\frac{8}{27}} = \sqrt[3]{\left(\frac{2}{3}\right)^3} = \frac{2}{3}$.

List of Organizers

Available at *BigIdeasMath.com*

Comparison Chart
Concept Circle
Definition (Idea) and Example Chart
Example and Non-Example Chart
Formula Triangle
Four Square
Information Frame
Information Wheel
Notetaking Organizer
Process Diagram
Summary Triangle
Word Magnet
Y Chart

About this Organizer

A **Four Square** can be used to organize information about a topic. Students write the topic in the "bubble" in the middle of the four square. Then students write concepts related to the topic in the four squares surrounding the bubble. Any concept related to the topic can be used. Encourage students to include concepts that will help them learn the topic. Students can place their four squares on note cards to use as a quick study reference.

Technology for the Teacher

Editable Graphic Organizer

Answers

1. -2
2. $\frac{4}{5}$
3. 2.5 and -2.5
4. 4
5. -6
6. $-\frac{7}{10}$
7. 26
8. -6
9. $5\frac{1}{4}$
10. 34
11. 30
12. -23
13. 41 ft
14. 28 in.
15. 6.3 cm
16. $\frac{1}{2}$ yd
17. $3.14r^2 = 314$; about 20 feet
18. 18 in.
19. 53 in.

Technology for the Teacher

Online Assessment
Assessment Book
ExamView® Assessment Suite

Alternative Quiz Ideas

100% Quiz	**Math Log**
Error Notebook	Notebook Quiz
Group Quiz	Partner Quiz
Homework Quiz	Pass the Paper

Math Log

Ask students to keep a math log for the chapter. Have them include diagrams, definitions, and examples. Everything should be clearly labeled. It might be helpful if they put the information in a chart. Students can add to the log as they are introduced to new topics.

Reteaching and Enrichment Strategies

If students need help. . .	If students got it. . .
Resources by Chapter • Practice A and Practice B • Puzzle Time Lesson Tutorials *BigIdeasMath.com*	Resources by Chapter • Enrichment and Extension • Technology Connection Game Closet at *BigIdeasMath.com* Start the next section

7.1–7.3 Quiz

Find the square root(s). *(Section 7.1)*

1. $-\sqrt{4}$

2. $\sqrt{\frac{16}{25}}$

3. $\pm\sqrt{6.25}$

Find the cube root. *(Section 7.2)*

4. $\sqrt[3]{64}$

5. $\sqrt[3]{-216}$

6. $\sqrt[3]{-\frac{343}{1000}}$

Evaluate the expression. *(Section 7.1 and Section 7.2)*

7. $3\sqrt{49} + 5$

8. $10 - 4\sqrt{16}$

9. $\frac{1}{4} + \sqrt{\frac{100}{4}}$

10. $\left(\sqrt[3]{-27}\right)^3 + 61$

11. $15 + 3\sqrt[3]{125}$

12. $2\sqrt[3]{-729} - 5$

Find the missing length of the triangle. *(Section 7.3)*

13.

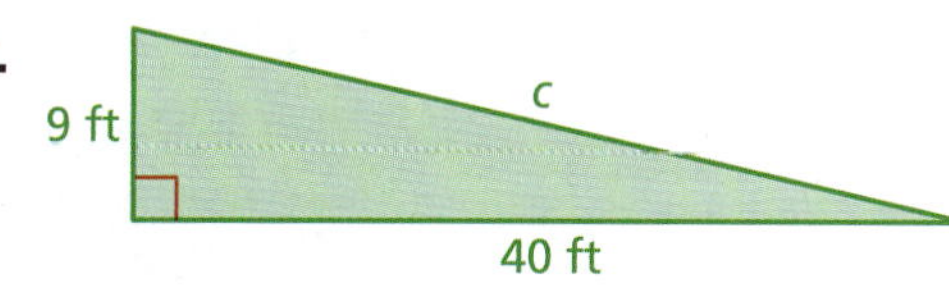

14.

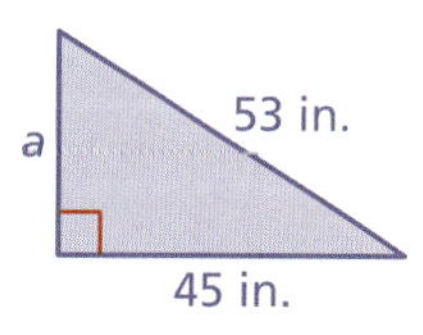

15.

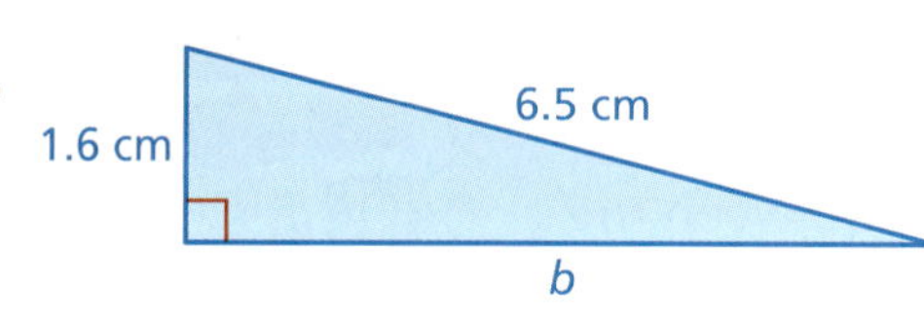

16.

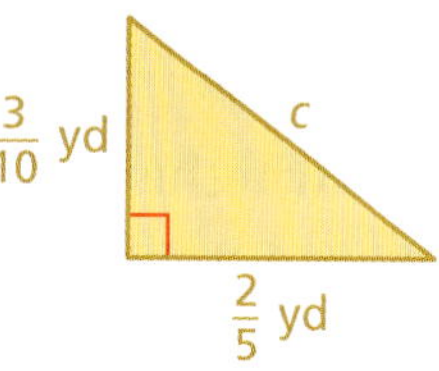

17. POOL The area of a circular pool cover is 314 square feet. Write and solve an equation to find the diameter of the pool cover. Use 3.14 for π. *(Section 7.1)*

18. PACKAGE A cube-shaped package has a volume of 5832 cubic inches. What is the edge length of the package? *(Section 7.2)*

19. FABRIC You are cutting a rectangular piece of fabric in half along the diagonal. The fabric measures 28 inches wide and $1\frac{1}{4}$ yards long. What is the length (in inches) of the diagonal? *(Section 7.3)*

7.4 Approximating Square Roots

Essential Question How can you find decimal approximations of square roots that are not rational?

1 ACTIVITY: Approximating Square Roots

Work with a partner. Archimedes was a Greek mathematician, physicist, engineer, inventor, and astronomer. He tried to find a rational number whose square is 3. Two that he tried were $\frac{265}{153}$ and $\frac{1351}{780}$.

a. Are either of these numbers equal to $\sqrt{3}$? Explain.

b. Use a calculator to approximate $\sqrt{3}$. Write the number on a piece of paper. Enter it into the calculator and square it. Then subtract 3. Do you get 0? What does this mean?

c. The value of $\sqrt{3}$ is between which two integers?

d. Tell whether the value of $\sqrt{3}$ is between the given numbers. Explain your reasoning.

1.7 and 1.8 | 1.72 and 1.73 | 1.731 and 1.732

2 ACTIVITY: Approximating Square Roots Geometrically

Work with a partner. Refer to the square on the number line below.

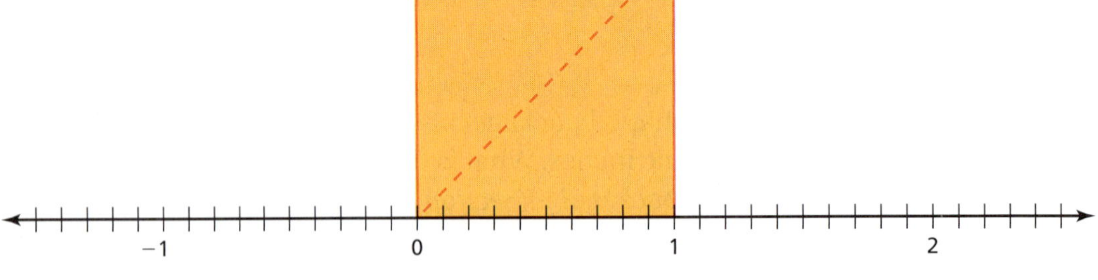

a. What is the length of the diagonal of the square?

b. Copy the square and its diagonal onto a piece of transparent paper. Rotate it about zero on the number line so that the diagonal aligns with the number line. Use the number line to estimate the length of the diagonal.

c. **STRUCTURE** How do you think your answers in parts (a) and (b) are related?

Square Roots

In this lesson, you will
- define irrational numbers.
- approximate square roots.
- approximate values of expressions involving irrational numbers.

Learning Standards
MACC.8.NS.1.1
MACC.8.NS.1.2
MACC.8.EE.1.2

Laurie's Notes

Introduction

Standards for Mathematical Practice

- **MP1a Make Sense of Problems:** To help students make sense of square roots they will use calculators and construction tools to estimate square roots.

Motivate

- Make a large Venn diagram based on student characteristics.
- The diagram can be made on the floor with yarn. Have students write their names on index cards.
- Use the diagram shown for students to place themselves. Sample labels for the groups: A = girls in our class, B = boys in our class, C = wears glasses/contacts, D = brown hair, E = taller than 5′ 4″, F = wearing a short sleeved T-shirt

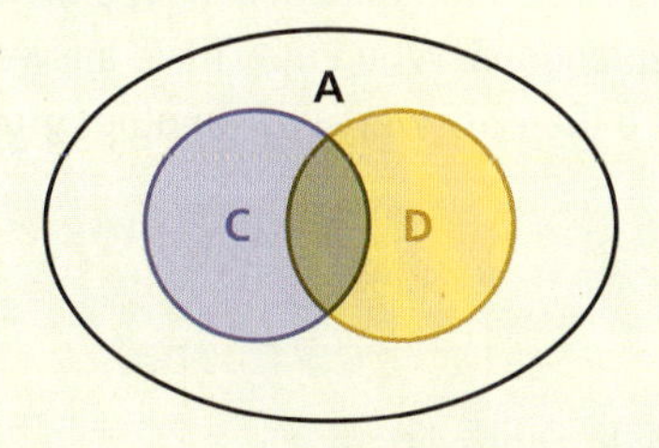

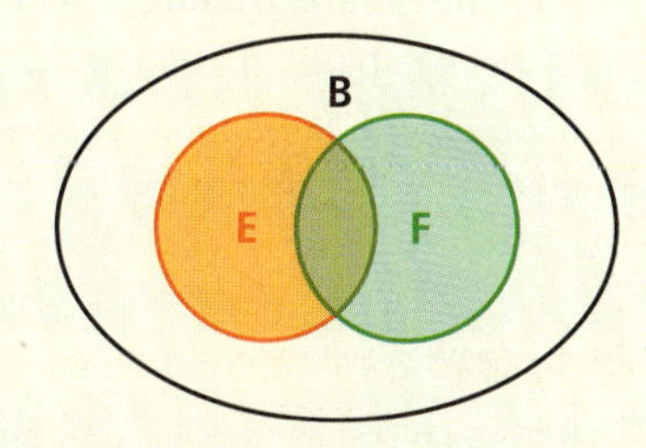

- Discuss what it means to be in certain sets and not in other sets.

Activity Notes

Activity 1

- **FYI:** Archimedes wanted a number x, such that $x^2 = 3$. If $x^2 = 3$, then $x = \pm\sqrt{3}$. Archimedes wanted to find a rational number that equals $\sqrt{3}$.
- **Part (a):** To square a fraction using a calculator, you can (1) write the fraction as a decimal by dividing, then square the decimal, or (2) write the fraction inside parentheses and use the exponent key.
- **Note:** In part (b), if you get anything other than 0, it means you did not have the exact value for $\sqrt{3}$, because $\sqrt{3}$ is irrational.
- **MP1 Make Sense of Problems and Persevere in Solving Them:** Part (d) checks understanding of place value as well as how much students persevered in approximating $\sqrt{3}$. For a given pair of numbers that $\sqrt{3}$ is not between, have students adjust the numbers so that it is between them.

Activity 2

- **Teaching Tip:** Students can use transparencies or tracing paper.
- The goal is for students to recognize that *irrational numbers* (defined in this lesson) can have a length associated with them. They are numbers that you can approximate. When students approximate the length of the diagonal, it should be close to 1.4 (approximation of $\sqrt{2}$).

Florida Common Core Standards

MACC.8.NS.1.1 Know that numbers that are not rational are called irrational. Understand informally that every number has a decimal expansion; for rational numbers show that the decimal expansion repeats eventually, and convert a decimal expansion which repeats eventually into a rational number.

MACC.8.NS.1.2 Use rational approximations of irrational numbers to compare the size of irrational numbers, locate them approximately on a number line diagram, and estimate the value of expressions (e.g., π^2).

MACC.8.EE.1.2 Use square root and cube root symbols to represent solutions to equations of the form $x^2 = p$ and $x^3 = p$, where p is a positive rational number. Evaluate square roots of small perfect squares and cube roots of small perfect cubes. Know that $\sqrt{2}$ is irrational.

Previous Learning

Students should know how to find square roots of perfect squares.

Lesson Plans
Complete Materials List

7.4 Record and Practice Journal

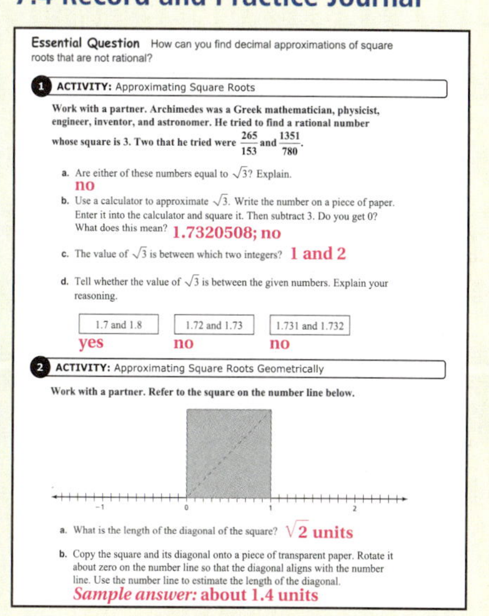

Essential Question How can you find decimal approximations of square roots that are not rational?

1 ACTIVITY: Approximating Square Roots

Work with a partner. Archimedes was a Greek mathematician, physicist, engineer, inventor, and astronomer. He tried to find a rational number whose square is 3. Two that he tried were $\frac{265}{153}$ and $\frac{1351}{780}$.

a. Are either of these numbers equal to $\sqrt{3}$? Explain.
no

b. Use a calculator to approximate $\sqrt{3}$. Write the number on a piece of paper. Enter it into the calculator and square it. Then subtract 3. Do you get 0? What does this mean? 1.7320508; no

c. The value of $\sqrt{3}$ is between which two integers? 1 and 2

d. Tell whether the value of $\sqrt{3}$ is between the given numbers. Explain your reasoning.

1.7 and 1.8	1.72 and 1.73	1.731 and 1.732
yes	no	no

2 ACTIVITY: Approximating Square Roots Geometrically

Work with a partner. Refer to the square on the number line below.

−1 0 1 2

a. What is the length of the diagonal of the square? $\sqrt{2}$ units

b. Copy the square and its diagonal onto a piece of transparent paper. Rotate it about zero on the number line so that the diagonal aligns with the number line. Use the number line to estimate the length of the diagonal.
Sample answer: about 1.4 units

Differentiated Instruction

Visual

On the board, create a large Venn diagram. Have students place numbers they are familiar with, such as natural numbers, whole numbers, integers, and rational numbers in the correct spaces on the Venn diagram. Reinforce that a number such as 5 is a natural number, as well as a whole number, an integer, and a rational number.

7.4 Record and Practice Journal

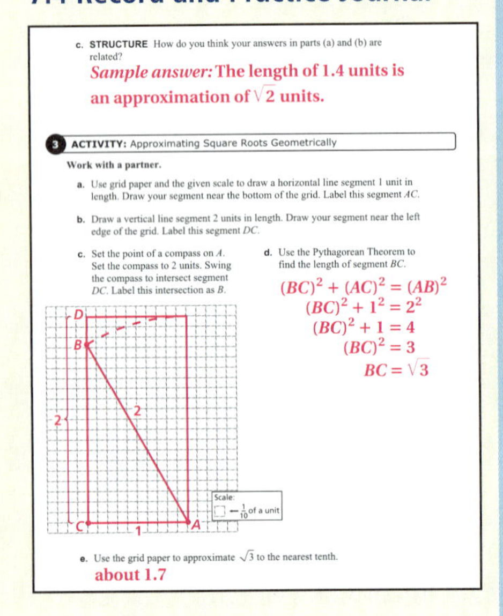

c. **STRUCTURE** How do you think your answers in parts (a) and (b) are related?

Sample answer: The length of 1.4 units is an approximation of $\sqrt{2}$ units.

3 ACTIVITY: Approximating Square Roots Geometrically

Work with a partner.

a. Use grid paper and the given scale to draw a horizontal line segment 1 unit in length. Draw your segment near the bottom of the grid. Label this segment *AC*.

b. Draw a vertical line segment 2 units in length. Draw your segment near the left edge of the grid. Label this segment *DC*.

c. Set the point of a compass on *A*. Set the compass to 2 units. Swing the compass to intersect segment *DC*. Label this intersection as *B*.

d. Use the Pythagorean Theorem to find the length of segment *BC*.

$(BC)^2 + (AC)^2 = (AB)^2$

$(BC)^2 + 1^2 = 2^2$

$(BC)^2 + 1 = 4$

$(BC)^2 = 3$

$BC = \sqrt{3}$

e. Use the grid paper to approximate $\sqrt{3}$ to the nearest tenth.

about 1.7

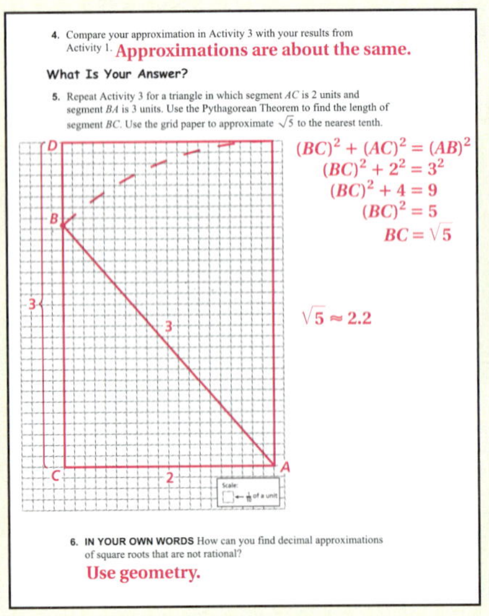

4. Compare your approximation in Activity 3 with your results from Activity 1. Approximations are about the same.

What Is Your Answer?

5. Repeat Activity 3 for a triangle in which segment *AC* is 2 units and segment *BA* is 3 units. Use the Pythagorean Theorem to find the length of segment *BC*. Use the grid paper to approximate $\sqrt{5}$ to the nearest tenth.

$(BC)^2 + (AC)^2 = (AB)^2$

$(BC)^2 + 2^2 = 3^2$

$(BC)^2 + 4 = 9$

$(BC)^2 = 5$

$BC = \sqrt{5}$

$\sqrt{5} \approx 2.2$

6. **IN YOUR OWN WORDS** How can you find decimal approximations of square roots that are not rational?

Use geometry.

Laurie's Notes

Activity 3

- **MP5 Use Appropriate Tools Strategically:** In Activity 1, students looked at $\sqrt{3}$ as a number. In this activity, students will look at $\sqrt{3}$ geometrically, as a length of a line segment.
- Students should be able to follow the written directions to construct segment *BC*.

? "Why does swinging the compass make *AB* equal 2 units?" *AB* is a radius of a circle that you know equals 2.

? "What type of triangle is *ABC*?" right

? "What information do you know about the triangle?" The hypotenuse equals 2 and one leg equals 1.

- Ask a pair of students to show how they used the Pythagorean Theorem to find the length of segment *BC*.

? "Is $\sqrt{3}$ greater than or less than 1.5?" greater than

What Is Your Answer?

- **Think-Pair-Share:** Students should read each question independently and then work in pairs to answer the questions. When they have answered the questions, the pair should compare their answers with another group and discuss any discrepancies.

Closure

- **Exit Ticket:** Describe how you would approximate $\sqrt{5}$.

  Listen for students to describe a procedure similar to that used in Activity 3, except with segments 2 units and 3 units in length.

3 ACTIVITY: Approximating Square Roots Geometrically

Math Practice 5

Recognize Usefulness of Tools

Why is the Pythagorean Theorem a useful tool when approximating a square root?

Work with a partner.

a. Use grid paper and the given scale to draw a horizontal line segment 1 unit in length. Label this segment *AC*.

b. Draw a vertical line segment 2 units in length. Label this segment *DC*.

c. Set the point of a compass on *A*. Set the compass to 2 units. Swing the compass to intersect segment *DC*. Label this intersection as *B*.

d. Use the Pythagorean Theorem to find the length of segment *BC*.

e. Use the grid paper to approximate $\sqrt{3}$ to the nearest tenth.

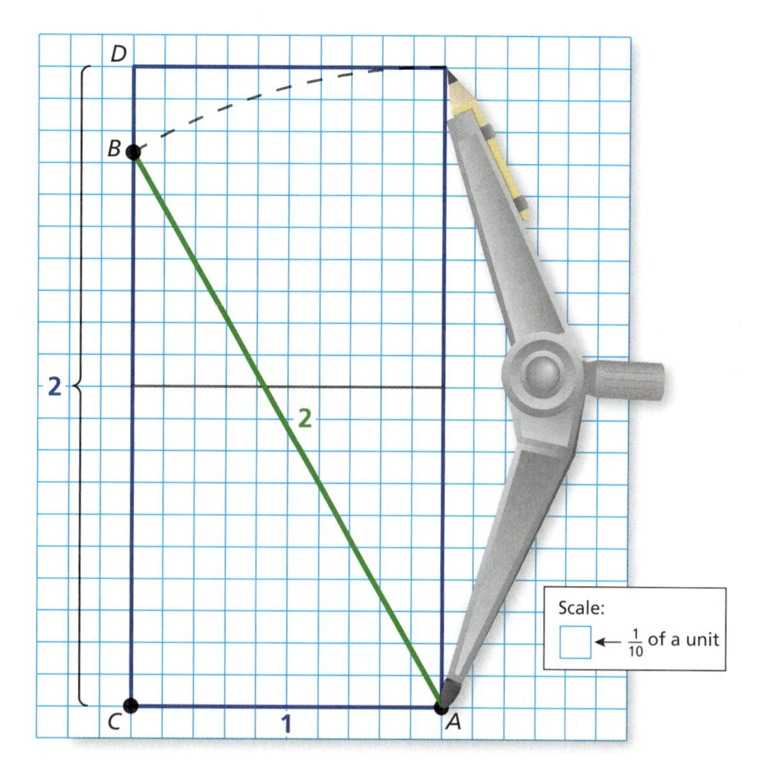

What Is Your Answer?

4. Compare your approximation in Activity 3 with your results from Activity 1.

5. Repeat Activity 3 for a triangle in which segment *AC* is 2 units and segment *BA* is 3 units. Use the Pythagorean Theorem to find the length of segment *BC*. Use the grid paper to approximate $\sqrt{5}$ to the nearest tenth.

6. **IN YOUR OWN WORDS** How can you find decimal approximations of square roots that are not rational?

Practice Use what you learned about approximating square roots to complete Exercises 5–8 on page 313.

7.4 Lesson

Key Vocabulary

irrational number, *p. 310*

real numbers, *p. 310*

A rational number is a number that can be written as the ratio of two integers. An **irrational number** cannot be written as the ratio of two integers.

- The square root of any whole number that is not a perfect square is irrational. The cube root of any integer that is not a perfect cube is irrational.
- The decimal form of an irrational number neither terminates nor repeats.

Key Idea

Real Numbers

Rational numbers and irrational numbers together form the set of **real numbers**.

Real Numbers

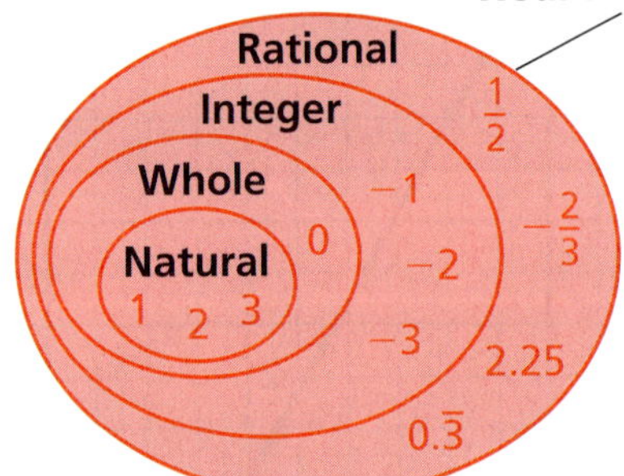

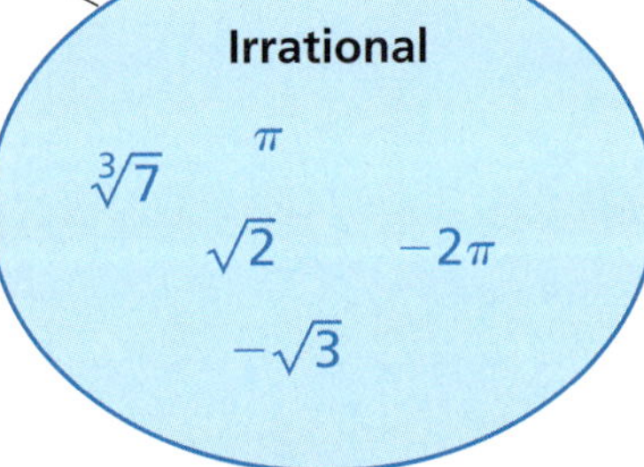

Remember

The decimal form of a rational number either terminates or repeats.

EXAMPLE 1 Classifying Real Numbers

Classify each real number.

	Number	Subset(s)	Reasoning
a.	$\sqrt{12}$	Irrational	12 is not a perfect square.
b.	$-0.\overline{25}$	Rational	$-0.\overline{25}$ is a repeating decimal.
c.	$-\sqrt{9}$	Integer, Rational	$-\sqrt{9}$ is equal to -3.
d.	$\frac{72}{4}$	Natural, Whole, Integer, Rational	$\frac{72}{4}$ is equal to 18.
e.	π	Irrational	The decimal form of π neither terminates nor repeats.

Study Tip

When classifying a real number, list all the subsets in which the number belongs.

On Your Own

Exercises 9–16

Classify the real number.

1. 0.121221222 . . . **2.** $-\sqrt{196}$ **3.** $\sqrt[3]{2}$

Laurie's Notes

Introduction

Connect

- **Yesterday:** Students investigated $\sqrt{3}$ numerically and geometrically. (MP1, MP5)
- **Today:** Students will approximate square roots.

Motivate

- Discuss applications of a periscope and share the following information.
  - A periscope is an optical device for conducting observations from a concealed or protected position.
  - Simple periscopes consist of reflecting mirrors and/or prisms at opposite ends of a tube container. The reflecting surfaces are parallel to each other and at a 45° angle to the axis of the tube.
  - The Navy attributes the invention of the periscope (1902) to Simon Lake and the perfection of the periscope to Sir Howard Grubb.

Lesson Notes

Key Idea

- Explain the definitions of *rational* and *irrational* numbers. Use the Venn diagram to give several examples of each.
- Write the Key Idea. It is important for students to understand that any real number is either rational or irrational. The two sets do not intersect.
- Explain that a *subset* is a set in which every element is contained within a larger set. The sets of rational numbers and irrational numbers are subsets of the set of real numbers. The sets of natural numbers, whole numbers, and integers are subsets of the set of rational numbers.
- Natural numbers are also called counting numbers because they are used to count objects. Whole numbers include the natural numbers and 0. Integers include the natural numbers, 0, and the opposites of the natural numbers.
- Mention that all square roots of perfect squares are rational.

? "Can you think of a repeating decimal and its fractional equivalent?"

Sample answer: $0.\overline{3} = \frac{1}{3}$

? "Can you think of a terminating decimal and its fractional equivalent?"

Sample answer: $0.5 = \frac{1}{2}$

Example 1

- Students will gain a better understanding of how to classify real numbers in this example.
- Discuss the Study Tip with students. Point out in part (c), for instance, that because $-\sqrt{9} = -3 = \frac{-3}{1}$, $-\sqrt{9}$ is an integer *and* a rational number.

On Your Own

- **Think-Pair-Share:** Students should read each question independently and then work in pairs to answer the questions. When they have answered the questions, the pair should compare their answers with another group and discuss any discrepancies.

Goal Today's lesson is approximating square roots.

Lesson Tutorials
Lesson Plans
Answer Presentation Tool

Extra Example 1

Classify each real number.

a. $\sqrt{15}$ irrational

b. 0.35 rational

On Your Own

1. irrational
2. integer, rational
3. irrational

Extra Example 2

Estimate $\sqrt{23}$ to the nearest (a) integer and (b) tenth.

a. 5

b. 4.8

On Your Own

4.	**a.** 3	**b.** 2.8
5.	**a.** -4	**b.** -3.6
6.	**a.** -5	**b.** -4.9
7.	**a.** 10	**b.** 10.5

Extra Example 3

Which is greater, $\sqrt{0.49}$ or 0.71? 0.71

English Language Learners

Vocabulary

Point out to students that the prefix *ir-* means *not*. An *irrational* number is a number that is not rational. Here are other common prefixes that also mean *not*.

dis-	disadvantage, disagree
il-	illiterate, illogical
im-	impolite, improper
in-	independent, indirect
ir-	irrational, irregular
un-	unfair, unfriendly

Laurie's Notes

Example 2

- It is important for students to make an estimate before using a calculator. Use reasoning first!
- ? "What are the first 10 perfect squares?" 1, 4, 9, 16, 25, 36, 49, 64, 81, 100
- ? "What type of number do you get when you take the square root of any of these perfect squares?" integer
- ? **MP3 Construct Viable Arguments and Critique the Reasoning of Others:** "Between what two whole numbers is $\sqrt{71}$, and how do you know?" 8 and 9 because $\sqrt{64} = 8$ and $\sqrt{81} = 9$, so $\sqrt{71}$ has to be a number between 8 and 9.
- ? "Is $\sqrt{71}$ closer to 8 or 9? Why?" It is closer to 8 because 71 is closer to 64 than to 81.
- You may wish to allow students to calculate squares of decimals using a calculator.
- You could explore more about square roots using a calculator approximation. For example, $\sqrt{71} \approx 8.4261498$. So, you can rationalize that $\sqrt{71}$ is between 8 and 9, between 8.4 and 8.5, between 8.42 and 8.43, etc., by truncating the decimal.

On Your Own

- Ask volunteers to share their thinking about each problem.

Example 3

- **MP4 Model with Mathematics:** A number line is used as a visual model.
- Students will ask where to place $\sqrt{5}$. Knowing that $\sqrt{5}$ is between $\sqrt{4}$ and $\sqrt{9}$ does not tell you where to graph it on the number line. Explain that you know it has to be closer to $\sqrt{4}$ than $\sqrt{9}$ because 5 is closer to 4 than to 9.
- The fraction $2\frac{2}{3}$ is greater than $2\frac{1}{2}$, so it is closer to $\sqrt{9}$.

EXAMPLE 2 Approximating a Square Root

Estimate $\sqrt{71}$ to the nearest (a) integer and (b) tenth.

a. Make a table of numbers whose squares are close to 71.

Number	7	8	9	10
Square of Number	49	64	81	100

The table shows that 71 is between the perfect squares 64 and 81. Because 71 is closer to 64 than to 81, $\sqrt{71}$ is closer to 8 than to 9.

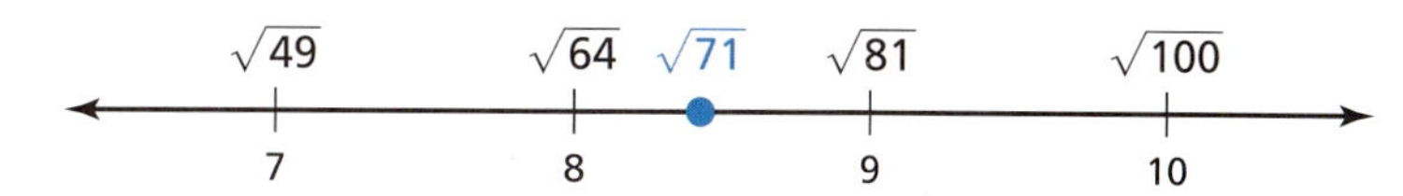

So, $\sqrt{71} \approx 8$.

b. Make a table of numbers between 8 and 9 whose squares are close to 71.

Number	8.3	8.4	8.5	8.6
Square of Number	68.89	70.56	72.25	73.96

Because 71 is closer to 70.56 than to 72.25, $\sqrt{71}$ is closer to 8.4 than to 8.5.

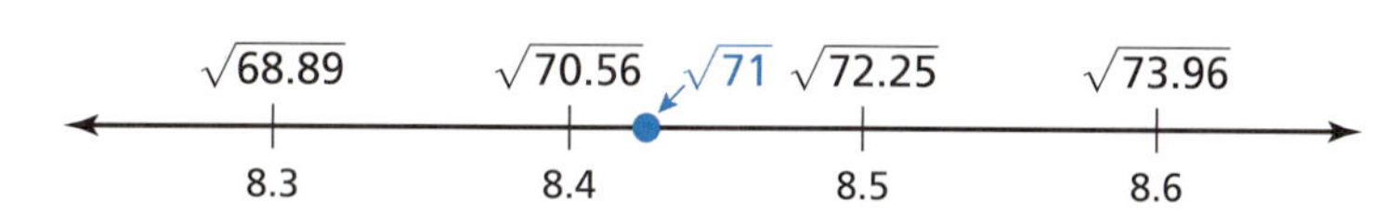

So, $\sqrt{71} \approx 8.4$.

Study Tip

You can continue the process shown in Example 2 to approximate square roots using more decimal places.

On Your Own

Now You're Ready
Exercises 20–25

Estimate the square root to the nearest (a) integer and (b) tenth.

4. $\sqrt{8}$ **5.** $-\sqrt{13}$ **6.** $-\sqrt{24}$ **7.** $\sqrt{110}$

EXAMPLE 3 Comparing Real Numbers

Which is greater, $\sqrt{5}$ or $2\frac{2}{3}$?

Estimate $\sqrt{5}$ to the nearest integer. Then graph the numbers on a number line.

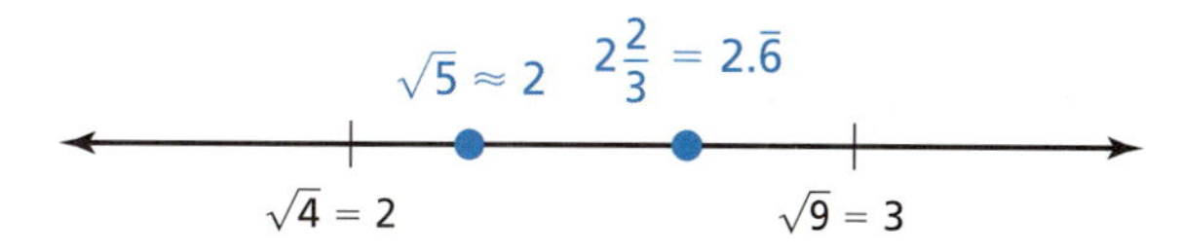

$2\frac{2}{3}$ is to the right of $\sqrt{5}$. So, $2\frac{2}{3}$ is greater.

EXAMPLE 4 Approximating the Value of an Expression

The radius of a circle with area A is approximately $\sqrt{\dfrac{A}{3}}$. The area of a circular mouse pad is 51 square inches. Estimate its radius to the nearest integer.

$$\sqrt{\frac{A}{3}} = \sqrt{\frac{51}{3}} \qquad \text{Substitute 51 for } A.$$

$$= \sqrt{17} \qquad \text{Divide.}$$

The nearest perfect square less than 17 is 16. The nearest perfect square greater than 17 is 25.

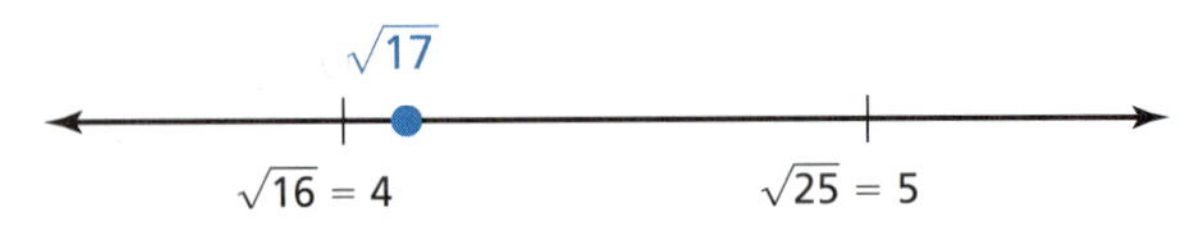

Because 17 is closer to 16 than to 25, $\sqrt{17}$ is closer to 4 than to 5.

So, the radius is about 4 inches.

EXAMPLE 5 Real-Life Application

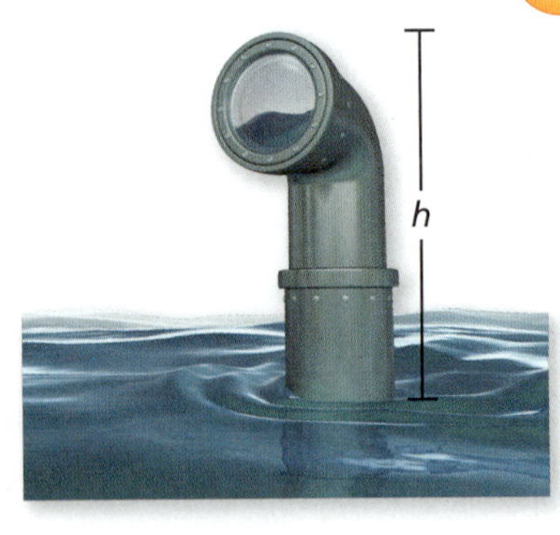

The distance (in nautical miles) you can see with a periscope is $1.17\sqrt{h}$, where h is the height of the periscope above the water. Can you see twice as far with a periscope that is 6 feet above the water than with a periscope that is 3 feet above the water? Explain.

```
1.17√(3)
        2.026499445
1.17√(6)
        2.865902999
```

Use a calculator to find the distances.

3 Feet Above Water		*6 Feet Above Water*
$1.17\sqrt{h} = 1.17\sqrt{3}$	Substitute for h.	$1.17\sqrt{h} = 1.17\sqrt{6}$
≈ 2.03	Use a calculator.	≈ 2.87

You can see $\dfrac{2.87}{2.03} \approx 1.41$ times farther with the periscope that is 6 feet above the water than with the periscope that is 3 feet above the water.

No, you cannot see twice as far with the periscope that is 6 feet above the water.

On Your Own

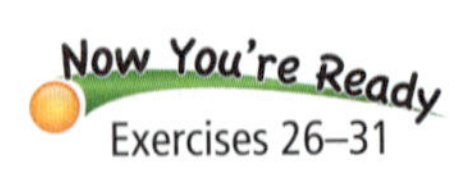

Which number is greater? Explain.

8. $4\frac{1}{5}, \sqrt{23}$ **9.** $\sqrt{10}, -\sqrt{5}$ **10.** $-\sqrt{2}, -2$

11. The area of a circular mouse pad is 64 square inches. Estimate its radius to the nearest integer.

12. In Example 5, you use a periscope that is 10 feet above the water. Can you see farther than 4 nautical miles? Explain.

Laurie's Notes

Example 4

- A simple diagram of a circle helps students focus on what is being asked.
- ? "What is the formula for the area of a circle?" $A = \pi r^2$
- Review with students how to solve this formula for r.

$A = \pi r^2$	Write the area formula.
$\frac{A}{\pi} = r^2$	Divide each side by π.
$\sqrt{\frac{A}{\pi}} = r$	Take positive square root of each side.

  Because π is close to 3, when approximating the radius, you can replace π with 3. This is the formula presented in this example.
- ? "What information is known in this problem?" You have a circle with an area of 51 square inches.
- ? "What are you trying to find?" an estimate for the radius
- Radicands that are fractions can be intimidating to students.
- Draw the number line and work the problem as shown.

Example 5

- Ask a student to read through the problem.
- You want to compare the distances for a periscope at two different heights above water, so the equation is used twice.
- ? "Do you think you can see twice as far at 6 feet than at 3 feet?" Most will say yes.
- Write the expression $1.17\sqrt{h}$ on the board and evaluate it for each height, as shown.

On Your Own

- Ask volunteers to share their work at the board.
- **Question 9:** A positive number is always greater than a negative number.
- **Question 10:** A number line is helpful for this question.
- **Question 11:** The quotient $\frac{64}{3}$ is not a whole number, but the question is still completed in the same fashion.

Closure

- Order the numbers from least to greatest: $\sqrt{38}$, $\sqrt{\frac{100}{3}}$, $6.\overline{5}$ $\sqrt{\frac{100}{3}}$, $\sqrt{38}$, $6.\overline{5}$

Extra Example 4

In Example 4, estimate the radius of a circular mouse pad with an area of 45 square inches. Round your answer to the nearest integer. about 4 in.

Extra Example 5

In Example 5, a periscope is 8 feet above the water. Can you see farther than 3 nautical miles? Explain. yes; You can see about 3.3 nautical miles.

On Your Own

8. $\sqrt{23}$; $\sqrt{23}$ is to the right of $4\frac{1}{5}$.
9. $\sqrt{10}$; $\sqrt{10}$ is positive and $-\sqrt{5}$ is negative.
10. $-\sqrt{2}$; $-\sqrt{2}$ is to the right of -2.
11. about 5 in.
12. no; You can only see about 3.7 nautical miles.

Vocabulary and Concept Check

1. A rational number can be written as the ratio of two integers. An irrational number cannot be written as the ratio of two integers.
2. 32 is between the perfect squares 25 and 36, but is closer to 36, so $\sqrt{32} \approx 6$.
3. all rational and irrational numbers; *Sample answer:* $-2, \frac{1}{8}, \sqrt{7}$
4. $\sqrt{8}$; $\sqrt{8}$ is irrational and the other three numbers are rational.

Practice and Problem Solving

5. yes
6. no
7. no
8. yes
9. whole, integer, rational
10. natural, whole, integer, rational
11. irrational
12. integer, rational
13. rational
14. natural, whole, integer, rational
15. irrational
16. irrational
17. 144 is a perfect square. So, $\sqrt{144}$ is rational.
18. no; 52 is not a perfect square.
19. **a.** If the last digit is 0, it is a whole number. Otherwise, it is a natural number.

    b. irrational number

    c. irrational number

20–25. See Additional Answers.

Assignment Guide and Homework Check

Level	Day 1 Activity Assignment	Day 2 Lesson Assignment	Homework Check
Basic	5–8, 51–54	1–4, 9–35 odd	11, 19, 23, 29, 35
Average	5–8, 51–54	1–4, 15, 17, 19, 20–42 even	15, 24, 28, 34, 40
Advanced	1–8, 17, 24–50 even, 51–54		24, 30, 34, 38, 50

Common Errors

- **Exercises 9–16** Students may not classify the real number in as many ways as possible. For instance, they may classify $\frac{52}{13}$ as rational only, because it is written as a fraction. Remind students that real numbers can have more than one classification. Point out that they should simplify the number, if possible, before classifying it.
- **Exercises 12 and 13** Students may think that all negative numbers are irrational. Remind them of the integers and that negative numbers can be rational or irrational.
- **Exercises 12, 15, and 16** Students may think that all square roots and cube roots are irrational. Remind them that square roots of perfect squares are rational and that cube roots of perfect cubes are also rational.
- **Exercises 20–25** Students may struggle with knowing what integer is closest to the given number. To help make comparisons, encourage them to write the first 10 perfect squares. If the number under the radical is greater than 100, then students should use *Guess, Check, and Revise* to find two integers on either side of the number. When determining which integer is closer to the rational number, encourage students to use a number line.

7.4 Record and Practice Journal

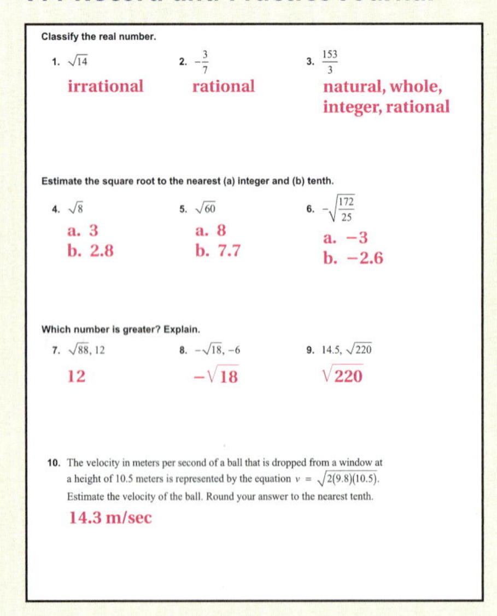

Classify the real number.

1. $\sqrt{14}$ — **irrational**
2. $-\frac{3}{7}$ — **rational**
3. $\frac{153}{3}$ — **natural, whole, integer, rational**

Estimate the square root to the nearest (a) integer and (b) tenth.

4. $\sqrt{8}$ — **a. 3 b. 2.8**
5. $\sqrt{60}$ — **a. 8 b. 7.7**
6. $-\sqrt{\frac{172}{25}}$ — **a. −3 b. −2.6**

Which number is greater? Explain.

7. $\sqrt{88}$, 12 — **12**
8. $-\sqrt{18}$, −6 — **$-\sqrt{18}$**
9. 14.5, $\sqrt{220}$ — **$\sqrt{220}$**

10. The velocity in meters per second of a ball that is dropped from a window at a height of 10.5 meters is represented by the equation $v = \sqrt{2(9.8)(10.5)}$. Estimate the velocity of the ball. Round your answer to the nearest tenth.

    14.3 m/sec

7.4 Exercises

Vocabulary and Concept Check

1. **VOCABULARY** How are rational numbers and irrational numbers different?
2. **WRITING** Describe a method of approximating $\sqrt{32}$.
3. **VOCABULARY** What are real numbers? Give three examples.
4. **WHICH ONE DOESN'T BELONG?** Which number does *not* belong with the other three? Explain your reasoning.

$-\frac{11}{12}$ 25.075   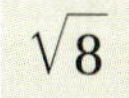 $\sqrt{8}$ $-3.\overline{3}$

Practice and Problem Solving

Tell whether the rational number is a reasonable approximation of the square root.

5. $\frac{559}{250}, \sqrt{5}$
6. $\frac{3021}{250}, \sqrt{11}$
7. $\frac{678}{250}, \sqrt{28}$
8. $\frac{1677}{250}, \sqrt{45}$

Classify the real number.

9. 0
10. $\sqrt[3]{343}$
11. $\frac{\pi}{6}$
12. $-\sqrt{81}$
13. -1.125
14. $\frac{52}{13}$
15. $\sqrt[3]{-49}$
16. $\sqrt{15}$
17. **ERROR ANALYSIS** Describe and correct the error in classifying the number.

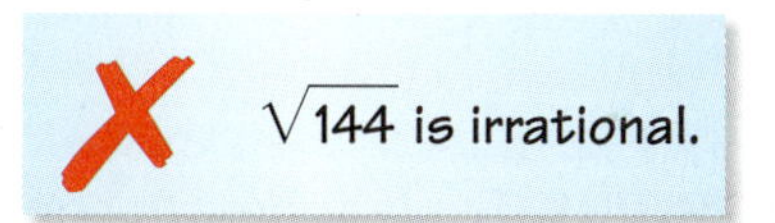

18. **SCRAPBOOKING** You cut a picture into a right triangle for your scrapbook. The lengths of the legs of the triangle are 4 inches and 6 inches. Is the length of the hypotenuse a rational number? Explain.

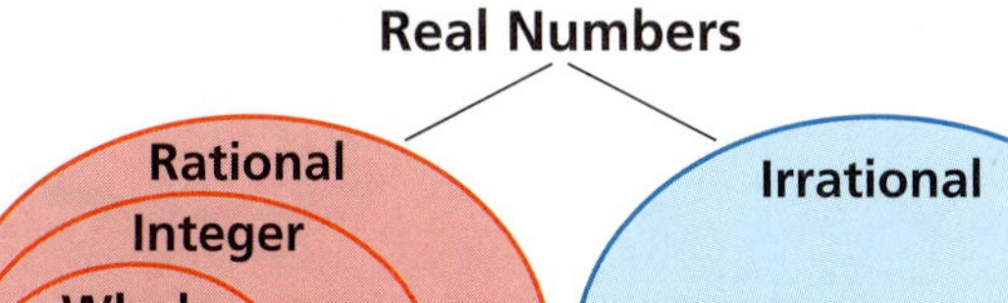
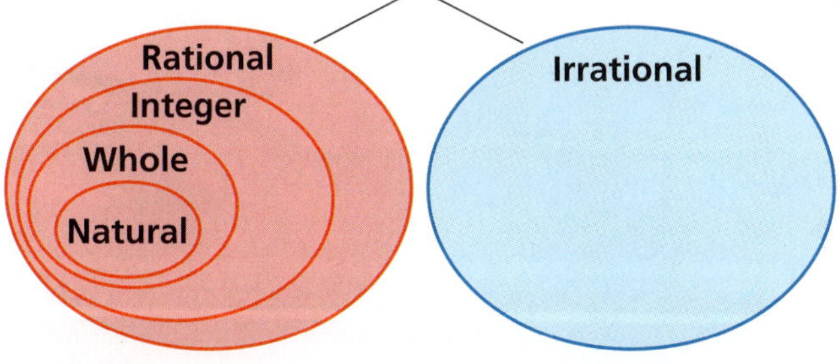

19. **VENN DIAGRAM** Place each number in the correct area of the Venn Diagram.
    a. the last digit of your phone number
    b. the square root of any prime number
    c. the ratio of the circumference of a circle to its diameter

Estimate the square root to the nearest (a) integer and (b) tenth.

2 20. $\sqrt{46}$
21. $\sqrt{685}$
22. $-\sqrt{61}$
23. $-\sqrt{105}$
24. $\sqrt{\frac{27}{4}}$
25. $-\sqrt{\frac{335}{2}}$

Which number is greater? Explain.

26. $\sqrt{20}$, 10

27. $\sqrt{15}$, -3.5

28. $\sqrt{133}$, $10\frac{3}{4}$

29. $\frac{2}{3}$, $\sqrt{\frac{16}{81}}$

30. $-\sqrt{0.25}$, -0.25

31. $-\sqrt{182}$, $-\sqrt{192}$

Use the graphing calculator screen to determine whether the statement is *true* or *false*.

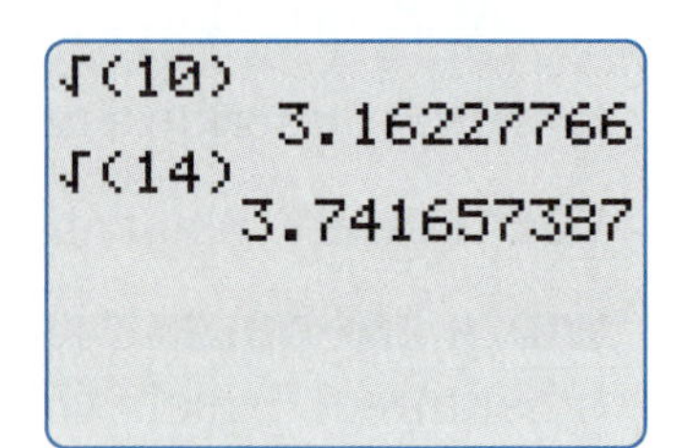

32. To the nearest tenth, $\sqrt{10} = 3.1$.

33. The value of $\sqrt{14}$ is between 3.74 and 3.75.

34. $\sqrt{10}$ lies between 3.1 and 3.16 on a number line.

35. FOUR SQUARE The area of a four square court is 66 square feet. Estimate the side length s to the nearest tenth of a foot.

36. CHECKERS A checkers board is 8 squares long and 8 squares wide. The area of each square is 14 square centimeters. Estimate the perimeter of the checkers board to the nearest tenth of a centimeter.

Approximate the length of the diagonal of the square or rectangle to the nearest tenth.

37.

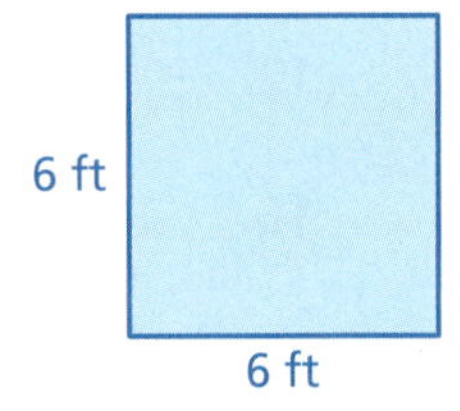

38.

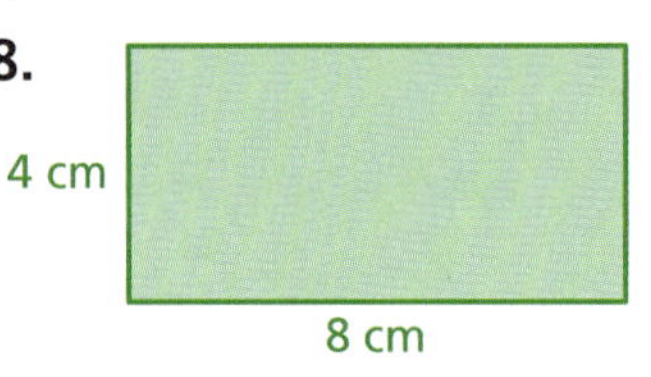

39.

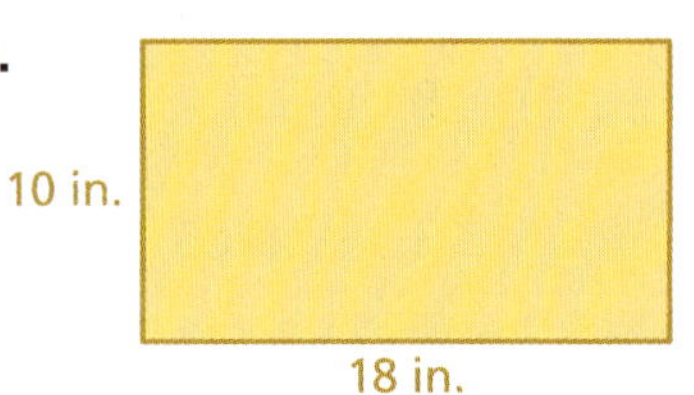

40. WRITING Explain how to continue the method in Example 2 to estimate $\sqrt{71}$ to the nearest hundredth.

41. REPEATED REASONING Describe a method that you can use to estimate a cube root to the nearest tenth. Use your method to estimate $\sqrt[3]{14}$ to the nearest tenth.

42. RADIO SIGNAL The maximum distance (in nautical miles) that a radio transmitter signal can be sent is represented by the expression $1.23\sqrt{h}$, where h is the height (in feet) above the transmitter.

Estimate the maximum distance x (in nautical miles) between the plane that is receiving the signal and the transmitter. Round your answer to the nearest tenth.

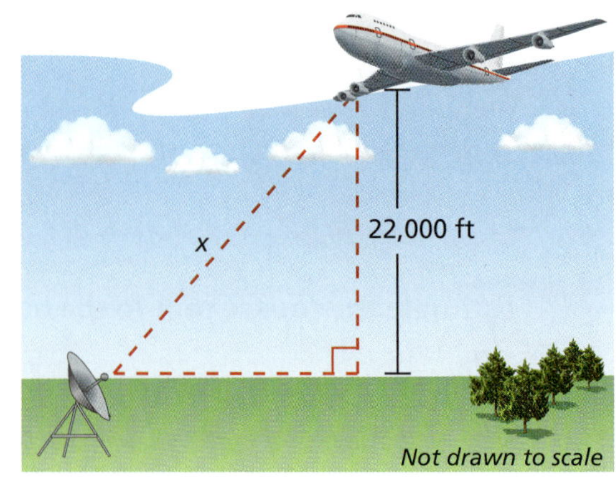

Common Errors

- **Exercises 26–31** Students may guess which is greater just by looking at the numbers. Encourage them to use a number line to compare the numbers. Also remind them to simplify and/or estimate the numbers so that they are easier to compare.
- **Exercises 44–46** Students may struggle estimating the square roots because of the decimals. Remind them of the method they used in Exercises 20–25. Tell them to use the list they wrote, but with the decimal points moved two places to the left. These new perfect squares will help to estimate the square roots.

English Language Learners

Illustrate

Draw a number line on the board. Have students locate $\sqrt{1}$, $\sqrt{4}$, and $\sqrt{9}$ on the number line. Now have them approximate the location of $\sqrt{5}$. Visualizing the $\sqrt{5}$ between 2 and 3 helps students form an approximation. Continue doing this with other square roots.

26. 10; 10 is to the right of $\sqrt{20}$.

27. $\sqrt{15}$; $\sqrt{15}$ is positive and -3.5 is negative.

28. $\sqrt{133}$; $\sqrt{133}$ is to the right of $10\frac{3}{4}$.

29. $\frac{2}{3}$; $\frac{2}{3}$ is to the right of $\sqrt{\frac{16}{81}}$.

30. -0.25; -0.25 is to the right of $-\sqrt{0.25}$.

31. $-\sqrt{182}$; $-\sqrt{182}$ is to the right of $-\sqrt{192}$.

32. false **33.** true

34. false **35.** 8.1 ft

36. 118.4 cm **37.** 8.5 ft

38. 8.9 cm **39.** 20.6 in.

40. Create a table of numbers between 8.4 and 8.5 whose squares are close to 71, and then determine which square is closest to 71.

41. Create a table of integers whose cubes are close to the radicand. Determine which two integers the cube root is between. Then create another table of numbers between those two integers whose cubes are close to the radicand. Determine which cube is closest to the radicand; 2.4

42. 182.4 nautical miles

Practice and Problem Solving

43. *Sample answer:* $a = 82$, $b = 97$

44. 0.6 **45.** 1.1

46. 1.2 **47.** 30.1 m/sec

48. yes; $\left(\frac{1}{2}\right)^2 = \frac{1}{4}$, so $\sqrt{\frac{1}{4}} = \frac{1}{2}$.
no; $\left(\frac{\sqrt{3}}{4}\right)^2 = \frac{3}{16}$, and $\sqrt{3}$ is irrational.

49. See *Taking Math Deeper.*

50. Sample answers are given.

- **a.** always; The product of two fractions is a fraction. $\frac{2}{3} \cdot \frac{3}{4} = \frac{1}{2}$
- **b.** sometimes; $\pi \cdot 0 = 0$ is rational, but $2 \cdot \sqrt{3}$ is irrational.
- **c.** sometimes; $\sqrt{2} \cdot \pi$ is irrational, but $\pi \cdot \frac{1}{\pi}$ is rational.

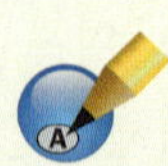

Fair Game Review

51. 40 m **52.** 24 in.

53. 9 cm **54.** D

Mini-Assessment

Estimate to the nearest integer and the nearest tenth.

1. $\sqrt{65}$ 8; 8.1
2. $\sqrt{99}$ 10; 9.9
3. $\sqrt{\frac{15}{2}}$ 3; 2.7

Which number is greater?

4. $2\frac{11}{12}$, $\sqrt{8}$ $2\frac{11}{12}$
5. $\frac{4}{5}$, $\sqrt{\frac{49}{64}}$ $\sqrt{\frac{49}{64}}$

Taking Math Deeper

Exercise 49

This is a nice science problem. Students learn from this problem that objects do not fall at a linear rate. Their speed increases with each second they are falling.

Understand the problem.

A water balloon is dropped from a height of 14 meters. How long does it take the balloon to fall to the ground?

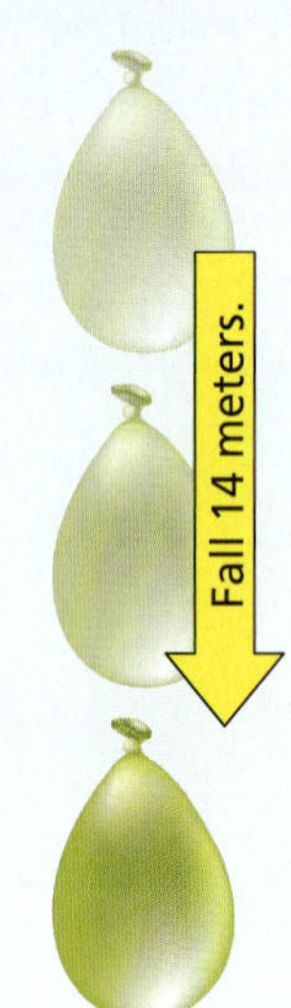

Use the given formula.

$$t = \sqrt{\frac{d}{4.9}} = \sqrt{\frac{14}{4.9}} \approx \sqrt{2.86} \approx 1.7$$

3 Answer the question.

The water balloon will hit the ground in about 1.7 seconds.

Project

Use a stop watch, a metric tape measure, and several basketballs. Measure the distance from the top of the bleachers at your school. Drop a ball and record the time it takes to fall to the ground. Use the formula in the problem to calculate the time it should take. Compare.

Reteaching and Enrichment Strategies

If students need help. . .	If students got it. . .
Resources by Chapter • Practice A and Practice B • Puzzle Time Record and Practice Journal Practice Differentiating the Lesson Lesson Tutorials Skills Review Handbook	Resources by Chapter • Enrichment and Extension • Technology Connection Start the next section

43. OPEN-ENDED Find two numbers a and b that satisfy the diagram.

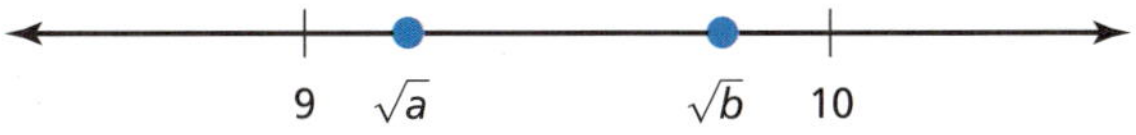

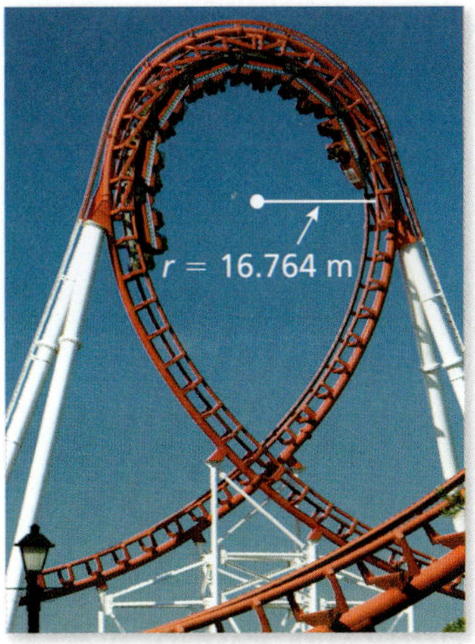

Estimate the square root to the nearest tenth.

44. $\sqrt{0.39}$ **45.** $\sqrt{1.19}$ **46.** $\sqrt{1.52}$

47. ROLLER COASTER The speed s (in meters per second) of a roller-coaster car is approximated by the equation $s = 3\sqrt{6r}$, where r is the radius of the loop. Estimate the speed of a car going around the loop. Round your answer to the nearest tenth.

48. STRUCTURE Is $\sqrt{\frac{1}{4}}$ a rational number? Is $\sqrt{\frac{3}{16}}$ a rational number? Explain.

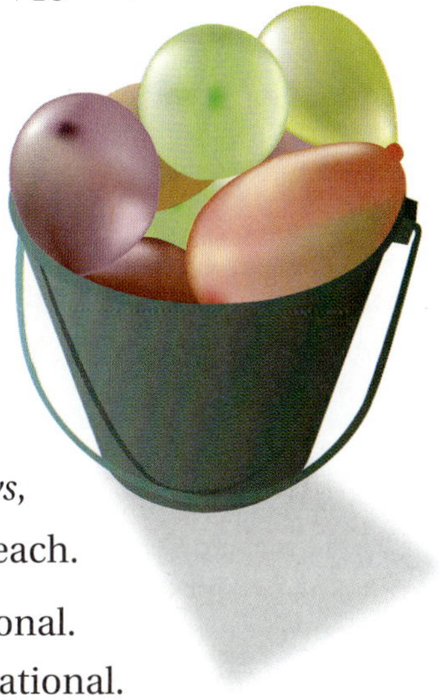

49. WATER BALLOON The time t (in seconds) it takes a water balloon to fall d meters is represented by the equation $t = \sqrt{\frac{d}{4.9}}$. Estimate the time it takes the balloon to fall to the ground from a window that is 14 meters above the ground. Round your answer to the nearest tenth.

50. **Number Sense** Determine if the statement is *sometimes, always,* or *never* true. Explain your reasoning and give an example of each.

a. A rational number multiplied by a rational number is rational.

b. A rational number multiplied by an irrational number is rational.

c. An irrational number multiplied by an irrational number is rational.

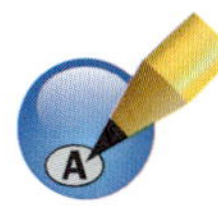

Fair Game Review What you learned in previous grades & lessons

Find the missing length of the triangle. *(Section 7.3)*

51.

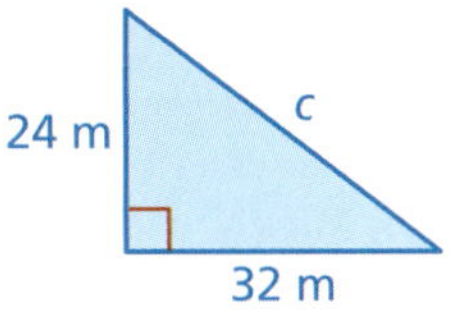

52.

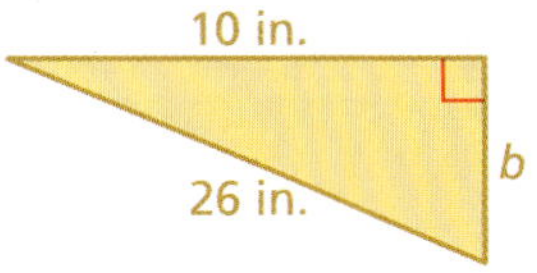

53.

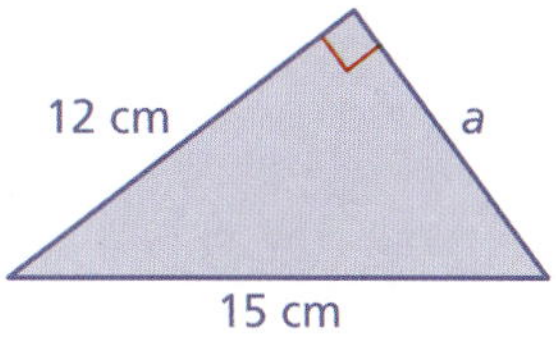

54. MULTIPLE CHOICE What is the ratio (red to blue) of the corresponding side lengths of the similar triangles? *(Section 2.5)*

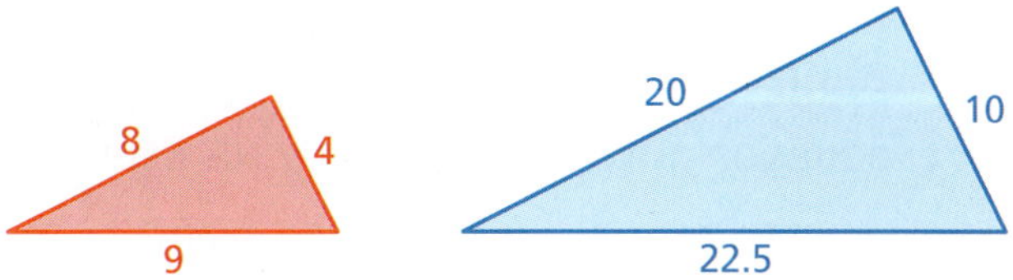

Ⓐ 1 : 3 Ⓑ 5 : 2

Ⓒ 3 : 4 Ⓓ 2 : 5

Extension 7.4 Repeating Decimals

You have written terminating decimals as fractions. Because repeating decimals are rational numbers, you can also write repeating decimals as fractions.

Key Idea

Writing a Repeating Decimal as a Fraction

Let a variable x equal the repeating decimal d.

Step 1: Write the equation $x = d$.

Step 2: Multiply each side of the equation by 10^n to form a new equation, where n is the number of repeating digits.

Step 3: Subtract the original equation from the new equation.

Step 4: Solve for x.

Rational Numbers

In this extension, you will

- write a repeating decimal as a fraction.

Learning Standard
MACC.8.NS.1.1

EXAMPLE 1 Writing a Repeating Decimal as a Fraction (1 Digit Repeats)

Write $0.\overline{4}$ as a fraction in simplest form.

Let $x = 0.\overline{4}$.

$x = 0.\overline{4}$ — Step 1: Write the equation.

$10 \cdot x = 10 \cdot 0.\overline{4}$ — Step 2: There is 1 repeating digit, so multiply each side by $10^1 = 10$.

$10x = 4.\overline{4}$ — Simplify.

$-(x = 0.\overline{4})$ — Step 3: Subtract the original equation.

$9x = 4$ — Simplify.

$x = \frac{4}{9}$ — Step 4: Solve for x.

So, $0.\overline{4} = \frac{4}{9}$.

Check

$$\begin{array}{r} 0.44\ldots \\ 9\overline{)4.00} \\ \underline{3\,6} \\ 40 \\ \underline{36} \\ 40 \end{array} \checkmark$$

Practice

Write the decimal as a fraction or a mixed number.

1. $0.\overline{1}$
2. $-0.\overline{5}$
3. $-1.\overline{2}$
4. $5.\overline{8}$

5. **STRUCTURE** In Example 1, why can you subtract the original equation from the new equation after multiplying by 10? Explain why these two steps are performed.

6. **REPEATED REASONING** Compare the repeating decimals and their equivalent fractions in Exercises 1–4. Describe the pattern. Use the pattern to explain how to write a repeating decimal as a fraction when only the tenths digit repeats.

Laurie's Notes

Introduction

Connect

- **Yesterday:** Students approximated irrational numbers. (MP3, MP4)
- **Today:** Students will write repeating decimals as fractions.

Motivate

- Ask students to use a calculator to help write the fractions $\frac{1}{11}, \frac{2}{11}, \frac{3}{11}, \ldots, \frac{10}{11}$ as decimals. To save time, have students work in groups on different fractions instead of trying to write decimal equivalents for all of them.
- Record the results on the board.
- ? "What patterns do you observe?" All of the decimals have two repeating digits, and the sum of those two digits is equal to 9. Also, the first of the two repeating digits is one less than the numerator of the fraction.
- Explain that today they are going to learn how to do the reverse of this process, begin with a repeating decimal, and write it as a fraction.

Lesson Notes

Key Idea

- It may be difficult for students to understand all of these steps until they actually try a problem. Once a problem has been completed, the steps may make more sense. Refer to these steps as you work through the examples.

Example 1

- Write the example. Some students may immediately tell you the correct fraction. When asked how they know, they may say to divide 4 by 9 to get $0.\overline{4}$. But knowing the answer is not the same as showing that $0.\overline{4} = \frac{4}{9}$.
- ? "If you let $x = 0.\overline{4}$, then what is $10x$?" $4.\overline{4}$; Because multiplying a decimal by 10 moves the decimal point one place value to the right.
- Say, "We now have two equations: $x = 0.\overline{4}$ and $10x = 4.\overline{4}$. We are going to subtract one equation from the other."
- **MP7 Look for and Make Use of Structure:** When subtracting the original equation, align the repeating decimal portion of each equation as shown.
- Explain that subtracting $0.\overline{4}$ from $4.\overline{4}$ gives a difference of 4.

Practice

- For Exercises 2 and 3, students can ignore the negative sign. The negative sign can be "put back in" after converting the decimal.
- For Exercises 3 and 4, students can "separate" the number into two parts: the integer and the repeating decimal. Then convert the repeating decimal to a fraction, and combine it back with the integer. Alternatively, they can leave the whole number and still begin by multiplying by 10.

Florida Common Core Standards

MACC.8.NS.1.1 Know that numbers that are not rational are called irrational. Understand informally that every number has a decimal expansion; for rational numbers show that the decimal expansion repeats eventually, and convert a decimal expansion which repeats eventually into a rational number.

Goal Today's lesson is writing repeating decimals as fractions.

Lesson Tutorials
Lesson Plans
Answer Presentation Tool

Extra Example 1

Write $1.\overline{7}$ as a fraction in simplest form.
$1\frac{7}{9}$

Practice

1. $\frac{1}{9}$
2. $-\frac{5}{9}$
3. $-1\frac{2}{9}$
4. $5\frac{8}{9}$
5. Because the solution does not change when adding/subtracting two equivalent equations; Multiply by 10 so that when you subtract the original equation, the repeating part is removed.
6. Write the digit that repeats in the numerator and use 9 in the denominator

Record and Practice Journal Extension 7.4 Practice

1–11. See Additional Answers.

Extra Example 2

Write $0.4\overline{6}$ as a fraction in simplest form. $\frac{7}{15}$

Extra Example 3

Write $5.\overline{12}$ as a mixed number. $5\frac{4}{33}$

Practice

7. $-\frac{13}{30}$
8. $2\frac{1}{15}$
9. $\frac{3}{11}$
10. $-4\frac{50}{99}$
11. Pattern: Digits that repeat are in the numerator and 99 is in the denominator; Use 9 as the integer part, 4 as the numerator, and 99 as the denominator of the fractional part.

Mini-Assessment

1. Write $0.\overline{2}$ as a fraction in simplest form. $\frac{2}{9}$
2. Write $0.7\overline{5}$ as a fraction in simplest form. $\frac{34}{45}$
3. Write $-3.\overline{81}$ as a mixed number. $-3\frac{9}{11}$

Laurie's Notes

Example 2

- In many cases, students may be unsure what to do when the repeating digit is not in the tenths place or what to do when there is more than one repeating digit. Examples 2 and 3 address these concerns.
- Write the example.
- ? "What portion of the decimal repeats?" only the digit 3
- ? "If you let $x = -0.2\overline{3}$, then what is $10x$?" $-2.\overline{3}$
- Write the two equations. Some students find it helpful when I write $-2.\overline{3}$ as $-2.3\overline{3}$ so that the bar is not over the tenths place.
- To help better understand Step 3, encourage students to write more digits for the repeating decimals to convince themselves that they will be left with a terminating decimal after subtracting.
- Subtract the equations and solve for x.
- Use a calculator to check the answer or check by dividing 7 by 30.

Example 3

- ? "What is 1.25 written as a fraction in simplest form?" $1\frac{1}{4}$
- ? "What does this fraction tell you?" The answer is approximately $1\frac{1}{4}$
- Let $x = 1.\overline{25}$. Because there are two repeating digits, multiply by 100.
- ? "If $x = 1.\overline{25}$ then what is $100x$?" $125.\overline{25}$
- Subtract the equations and solve for x.
- Use a calculator to check the answer.
- **Extension:** If time permits, you can explore with students what happens when you use the given steps for a repeating decimal that is not written in its "most abbreviated" form. For example, a repeating decimal written as $0.\overline{44}$ instead of $0.\overline{4}$. These steps still work, but the student must simplify a more difficult fraction.
- To challenge students, ask them why multiplying by 10^n works in Step 2. Tell them to focus on the portion that repeats and how the decimal point moves. Encourage them to try some examples with more repeating digits. They may see that it allows for the repeating portion to be "aligned" when subtracting in Step 3. So, it gets eliminated and produces a terminating decimal.

Practice

- Give students sufficient time to work through the exercises. Ask for volunteers to explain their work at the board.

Closure

- Explain the steps to write $2.\overline{35}$ as a mixed number.

EXAMPLE 2 Writing a Repeating Decimal as a Fraction (1 Digit Repeats)

Write $-0.2\overline{3}$ as a fraction in simplest form.

Let $x = -0.2\overline{3}$.

$x = -0.2\overline{3}$	Step 1: Write the equation.
$10 \cdot x = 10 \cdot (-0.2\overline{3})$	Step 2: There is 1 repeating digit, so multiply each side by $10^1 = 10$.
$10x = -2.\overline{3}$	Simplify.
$-(x = -0.2\overline{3})$	Step 3: Subtract the original equation.
$9x = -2.1$	Simplify.
$x = \frac{-2.1}{9}$	Step 4: Solve for x.

So, $-0.2\overline{3} = \frac{-2.1}{9} = -\frac{21}{90} = -\frac{7}{30}$.

Check

```
-7/30
   -.2333333333
```

EXAMPLE 3 Writing a Repeating Decimal as a Fraction (2 Digits Repeat)

Write $1.\overline{25}$ as a mixed number.

Let $x = 1.\overline{25}$.

$x = 1.\overline{25}$	Step 1: Write the equation.
$100 \cdot x = 100 \cdot 1.\overline{25}$	Step 2: There are 2 repeating digits, so multiply each side by $10^2 = 100$.
$100x = 125.\overline{25}$	Simplify.
$-(x = 1.\overline{25})$	Step 3: Subtract the original equation.
$99x = 124$	Simplify.
$x = \frac{124}{99}$	Step 4: Solve for x.

So, $1.\overline{25} = \frac{124}{99} = 1\frac{25}{99}$.

Check

```
124/99
    1.252525253
```

Practice

Write the decimal as a fraction or a mixed number.

7. $-0.4\overline{3}$ **8.** $2.0\overline{6}$ **9.** $0.\overline{27}$ **10.** $-4.\overline{50}$

11. **REPEATED REASONING** Find a pattern in the fractional representations of repeating decimals in which only the tenths and hundredths digits repeat. Use the pattern to explain how to write $9.\overline{04}$ as a mixed number.

7.5 Using the Pythagorean Theorem

Essential Question In what other ways can you use the Pythagorean Theorem?

The *converse* of a statement switches the hypothesis and the conclusion.

Statement: If p, then q.	Converse of the statement: If q, then p.

1 ACTIVITY: Analyzing Converses of Statements

Work with a partner. Write the converse of the true statement. Determine whether the converse is *true* or *false*. If it is true, justify your reasoning. If it is false, give a counterexample.

a. If $a = b$, then $a^2 = b^2$.

b. If $a = b$, then $a^3 = b^3$.

c. If one figure is a translation of another figure, then the figures are congruent.

d. If two triangles are similar, then the triangles have the same angle measures.

Is the converse of a true statement always true? always false? Explain.

2 ACTIVITY: The Converse of the Pythagorean Theorem

Work with a partner. The converse of the Pythagorean Theorem states: "If the equation $a^2 + b^2 = c^2$ is true for the side lengths of a triangle, then the triangle is a right triangle."

a. Do you think the converse of the Pythagorean Theorem is *true* or *false*? How could you use deductive reasoning to support your answer?

b. Consider $\triangle DEF$ with side lengths a, b, and c, such that $a^2 + b^2 = c^2$. Also consider $\triangle JKL$ with leg lengths a and b, where $\angle K = 90°$.

- What does the Pythagorean Theorem tell you about $\triangle JKL$?
- What does this tell you about c and x?
- What does this tell you about $\triangle DEF$ and $\triangle JKL$?
- What does this tell you about $\angle E$?
- What can you conclude?

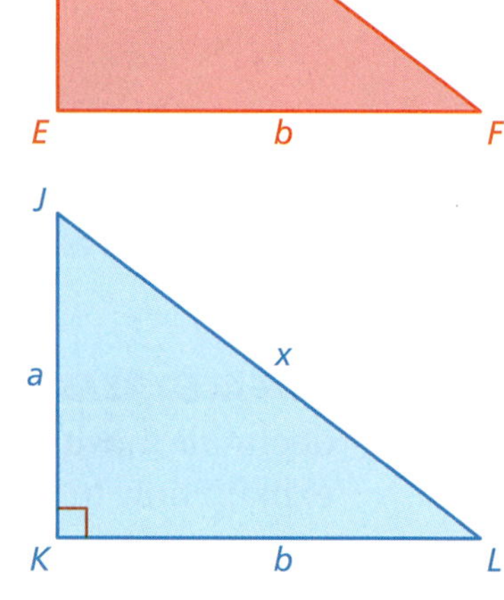

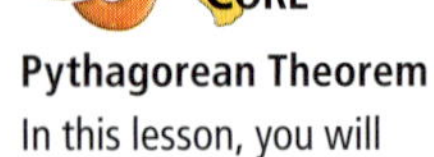

Pythagorean Theorem

In this lesson, you will

- use the converse of the Pythagorean Theorem to identify right triangles.
- use the Pythagorean Theorem to find distances in a coordinate plane.
- solve real-life problems.

Learning Standards
MACC.8.EE.1.2
MACC.8.G.2.6
MACC.8.G.2.7
MACC.8.G.2.8

Laurie's Notes

Introduction

Standards for Mathematical Practice

- **MP3 Construct Viable Arguments and Critique the Reasoning of Others:** Students will develop a "proof" of the converse of the Pythagorean Theorem, and they will derive the distance formula. It is important that students be able to explain the steps in their work and compare it to the reasoning of their classmates.

Motivate

- Explain what the converse of a statement is.
- **Example:** If I live in Moab, then I live in Utah.
  The converse is: If I live in Utah, then I live in Moab.
  The original statement is true, but the converse is false.
- Ask students to write four true if-then statements, two with true converses and two with false converses. Have students share some examples.

Activity Notes

Activity 1

- **MP3:** For each true statement, students construct a valid explanation versus simply saying the statement is true. Classmates are expected to listen and critique the explanation offered.
- Review the meaning of the word *counterexample*.
- It is important for students to understand that a statement might be false even if they cannot think of a counterexample.
- **Big Idea:** Even when a conditional statement is true, its converse does not have to be true. Students should keep this in mind for the next activity.

Activity 2

- Students often say that the Pythagorean Theorem is simply $a^2 + b^2 = c^2$. Tell them that it is actually a conditional statement. If a and b are the lengths of the legs and c is the length of the hypotenuse of a right triangle, then $a^2 + b^2 = c^2$.

? "What is the converse of the Pythagorean Theorem?" If $a^2 + b^2 = c^2$, then a triangle with side lengths a, b, and c is a right triangle.

? "Do you think the converse of the Pythagorean Theorem is true?" Students may simply guess at this point.

- **MP3:** In this activity, students use deductive reasoning to show that the converse is true. They may need guidance in linking their reasoning.
- **FYI:** Strategies for proofs are taught in later courses, so the framework of a proof is provided.
- In the second bullet, listen for $c^2 = x^2$ and then $c = x$. Students found this was not always true in Activity 1 part (a), but point out that it is true here because c and x must be positive.
- Students learned in Grade 7 that you can only construct one triangle given three side lengths. Review this concept for the third bullet.

Florida Common Core Standards

MACC.8.EE.1.2 Use square root and cube root symbols to represent solutions to equations of the form $x^2 = p$ and $x^3 = p$, where p is a positive rational number. Evaluate square roots of small perfect squares and cube roots of small perfect cubes. Know that $\sqrt{2}$ is irrational.

MACC.8.G.2.6 Explain a proof of the Pythagorean Theorem and its converse.

MACC.8.G.2.7 Apply the Pythagorean Theorem to determine unknown side lengths in right triangles in real-world and mathematical problems in two and three dimensions.

MACC.8.G.2.8 Apply the Pythagorean Theorem to find the distance between two points in a coordinate system.

Previous Learning

Students should know how to use the Pythagorean Theorem.

7.5 Record and Practice Journal

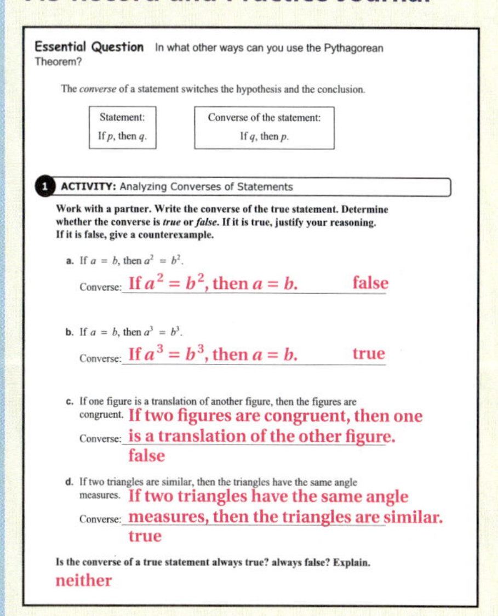

Essential Question In what other ways can you use the Pythagorean Theorem?

The *converse* of a statement switches the hypothesis and the conclusion.

Statement: If p, then q.	Converse of the statement: If q, then p.

1 ACTIVITY: Analyzing Converses of Statements

Work with a partner. Write the converse of the true statement. Determine whether the converse is *true* or *false*. If it is true, justify your reasoning. If it is false, give a counterexample.

a. If $a = b$, then $a^2 = b^2$.
Converse: If $a^2 = b^2$, then $a = b$. false

b. If $a = b$, then $a^3 = b^3$.
Converse: If $a^3 = b^3$, then $a = b$. true

c. If one figure is a translation of another figure, then the figures are congruent.
Converse: If two figures are congruent, then one is a translation of the other figure. false

d. If two triangles are similar, then the triangles have the same angle measures.
Converse: If two triangles have the same angle measures, then the triangles are similar. true

Is the converse of a true statement always true? always false? Explain.
neither

English Language Learners

Comprehension

English language learners may struggle with the concept of the converse of a statement. Some may think that the converse is always true. Give an example where the converse of a statement is not true.

Statement: *If a figure is a square, then it has four right angles.*

Converse of the Statement: *If a figure has four right angles, then the figure is a square.*

The converse of the statement is not always true. The figure could be a rectangle.

7.5 Record and Practice Journal

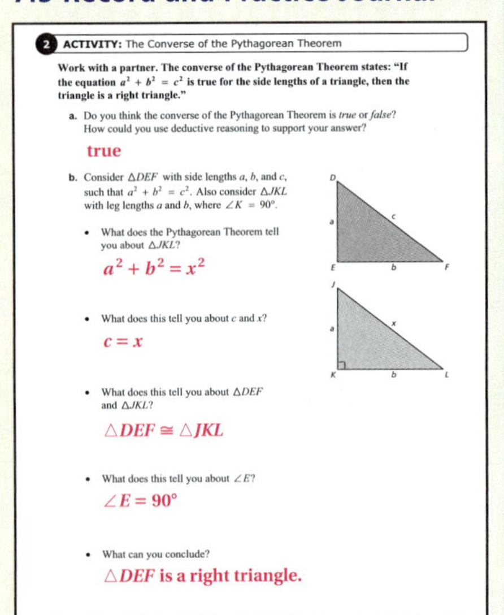

2 ACTIVITY: The Converse of the Pythagorean Theorem

Work with a partner. The converse of the Pythagorean Theorem states: "If the equation $a^2 + b^2 = c^2$ is true for the side lengths of a triangle, then the triangle is a right triangle."

a. Do you think the converse of the Pythagorean Theorem is *true* or *false*? How could you use deductive reasoning to support your answer?

true

b. Consider $\triangle DEF$ with side lengths a, b, and c, such that $a^2 + b^2 = c^2$. Also consider $\triangle JKL$ with leg lengths a and b, where $\angle K = 90°$.

- What does the Pythagorean Theorem tell you about $\triangle JKL$?

  $a^2 + b^2 = x^2$

- What does this tell you about c and x?

  $c = x$

- What does this tell you about $\triangle DEF$ and $\triangle JKL$?

  $\triangle DEF \cong \triangle JKL$

- What does this tell you about $\angle E$?

  $\angle E = 90°$

- What can you conclude?

  $\triangle DEF$ is a right triangle.

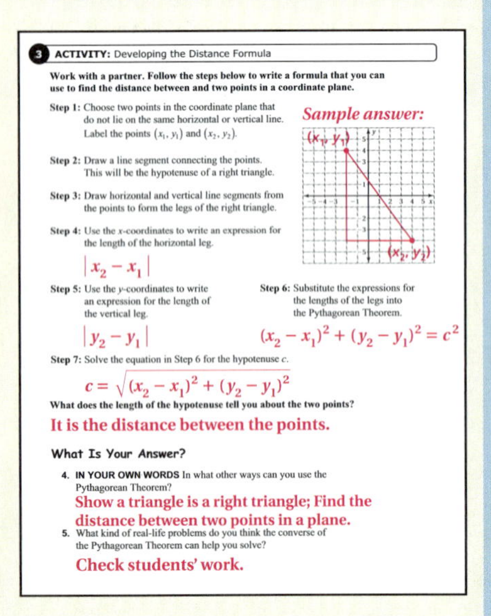

3 ACTIVITY: Developing the Distance Formula

Work with a partner. Follow the steps below to write a formula that you can use to find the distance between and two points in a coordinate plane.

Step 1: Choose two points in the coordinate plane that do not lie on the same horizontal or vertical line. Label the points (x_1, y_1) and (x_2, y_2).

Step 2: Draw a line segment connecting the points. This will be the hypotenuse of a right triangle.

Step 3: Draw horizontal and vertical line segments from the points to form the legs of the right triangle.

Step 4: Use the x-coordinates to write an expression for the length of the horizontal leg.

$|x_2 - x_1|$

Step 5: Use the y-coordinates to write an expression for the length of the vertical leg.

$|y_2 - y_1|$

Step 6: Substitute the expressions for the lengths of the legs into the Pythagorean Theorem.

$(x_2 - x_1)^2 + (y_2 - y_1)^2 = c^2$

Step 7: Solve the equation in Step 6 for the hypotenuse c.

$c = \sqrt{(x_2 - x_1)^2 + (y_2 - y_1)^2}$

What does the length of the hypotenuse tell you about the two points?

It is the distance between the points.

What Is Your Answer?

4. **IN YOUR OWN WORDS** In what other ways can you use the Pythagorean Theorem?

   Show a triangle is a right triangle; Find the distance between two points in a plane.

5. What kind of real-life problems do you think the converse of the Pythagorean Theorem can help you solve?

   Check students' work.

Laurie's Notes

Activity 3

- Visually, it will be helpful for students to select lattice points.
- In Step 3, point out that there are two distinct ways they can draw the legs. They can draw them either way because the two possible triangles are congruent.
- **MP7 Look for and Make Use of Structure:** In Steps 4 and 5, students may ask about the order in which the subtraction is performed. Tell them that in Step 6 these expressions are squared. So, the order in which the subtraction is performed does not matter.
- Give students adequate time to read carefully and work through the steps on their own. Resist the temptation to jump in and solve it for them.

What Is Your Answer?

- **Think-Pair-Share:** Students should read each question independently and then work in pairs to answer the questions. When they have answered the questions, the pair should compare their answers with another group and discuss any discrepancies.

Closure

- **Writing Prompt:** The Pythagorean Theorem can be used to . . . find missing side lengths of right triangles, find the distances between points in a coordinate plane, determine whether given side lengths form a right triangle, etc.

3 ACTIVITY: Developing the Distance Formula

Work with a partner. Follow the steps below to write a formula that you can use to find the distance between any two points in a coordinate plane.

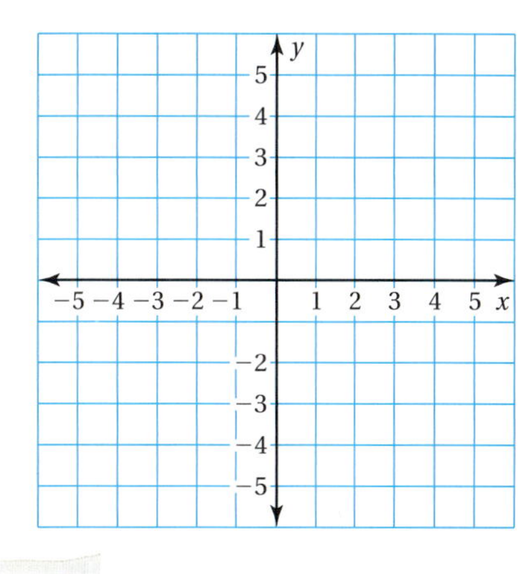

Math Practice 6

Communicate Precisely

What steps can you take to make sure that you have written the distance formula accurately?

Step 1: Choose two points in the coordinate plane that do not lie on the same horizontal or vertical line. Label the points (x_1, y_1) and (x_2, y_2).

Step 2: Draw a line segment connecting the points. This will be the hypotenuse of a right triangle.

Step 3: Draw horizontal and vertical line segments from the points to form the legs of the right triangle.

Step 4: Use the x-coordinates to write an expression for the length of the horizontal leg.

Step 5: Use the y-coordinates to write an expression for the length of the vertical leg.

Step 6: Substitute the expressions for the lengths of the legs into the Pythagorean Theorem.

Step 7: Solve the equation in Step 6 for the hypotenuse c.

What does the length of the hypotenuse tell you about the two points?

What Is Your Answer?

4. **IN YOUR OWN WORDS** In what other ways can you use the Pythagorean Theorem?

5. What kind of real-life problems do you think the converse of the Pythagorean Theorem can help you solve?

Practice Use what you learned about the converse of a true statement to complete Exercises 3 and 4 on page 322.

7.5 Lesson

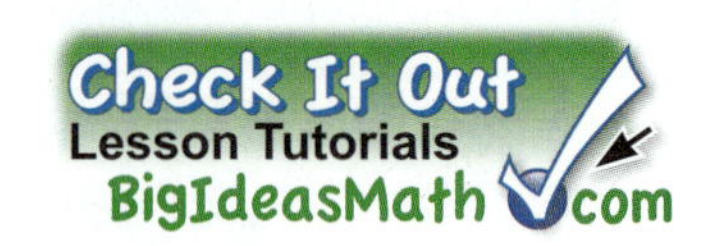

Key Vocabulary
distance formula, *p. 320*

Key Ideas

Converse of the Pythagorean Theorem

If the equation $a^2 + b^2 = c^2$ is true for the side lengths of a triangle, then the triangle is a right triangle.

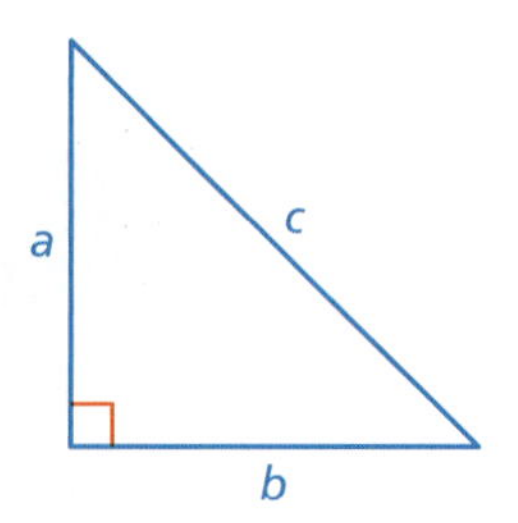

EXAMPLE 1 Identifying a Right Triangle

Study Tip
A *Pythagorean triple* is a set of three positive integers *a*, *b*, and *c*, where $a^2 + b^2 = c^2$.

Common Error
When using the converse of the Pythagorean Theorem, always substitute the length of the longest side for *c*.

Tell whether each triangle is a right triangle.

a.

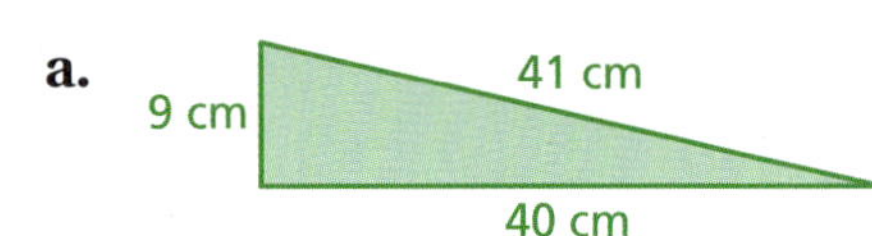

$$a^2 + b^2 = c^2$$
$$9^2 + 40^2 \stackrel{?}{=} 41^2$$
$$81 + 1600 \stackrel{?}{=} 1681$$
$$1681 = 1681 \checkmark$$

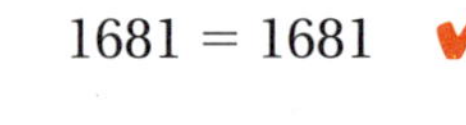

∴ It *is* a right triangle.

b.

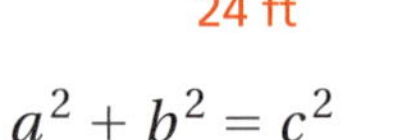

$$a^2 + b^2 = c^2$$
$$12^2 + 18^2 \stackrel{?}{=} 24^2$$
$$144 + 324 \stackrel{?}{=} 576$$
$$468 \neq 576 \quad ✗$$

∴ It is *not* a right triangle.

On Your Own

Tell whether the triangle with the given side lengths is a right triangle.

1. 28 in., 21 in., 20 in.
2. 1.25 mm, 1 mm, 0.75 mm

On page 319, you used the Pythagorean Theorem to develop the *distance formula*. You can use the **distance formula** to find the distance between any two points in a coordinate plane.

Key Idea

Distance Formula

The distance d between any two points (x_1, y_1) and (x_2, y_2) is given by the formula

$$d = \sqrt{(x_2 - x_1)^2 + (y_2 - y_1)^2}.$$

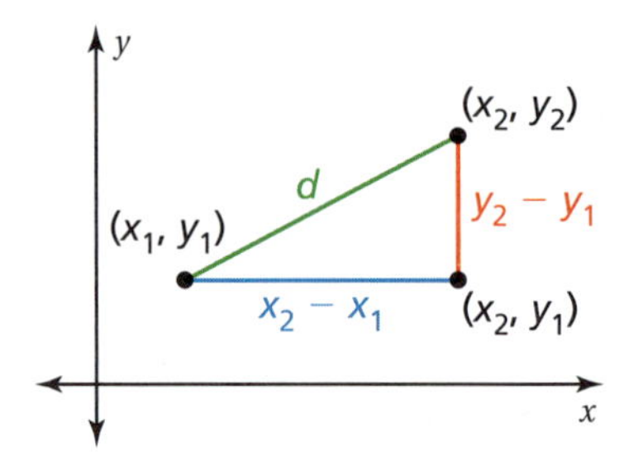

Laurie's Notes

Introduction

Connect

- **Yesterday:** Students proved the converse of the Pythagorean Theorem and developed the distance formula. (MP3, MP7)
- **Today:** Students will use the converse of the Pythagorean Theorem and the distance formula.

Motivate

- Write the following numbers on the board: 3-4-5, 5-12-13, and 8-15-17. Tell students that these are called *Pythagorean triples.* They are positive integers that satisfy the Pythagorean Theorem.
- Multiples of these examples also satisfy the Pythagorean Theorem. Have students try to name a few more Pythagorean triples.

Lesson Notes

Key Idea

- Write the Key Idea stating the converse of the Pythagorean Theorem. It is one way of determining whether given side lengths form a right triangle.

Example 1

- If the three numbers satisfy the theorem, then it is a right triangle. Explain that you are not using eyesight to decide if it is a right triangle.
- Substitute the side lengths for each triangle. Remind students that the longest side is substituted for c.
- Because 9, 40, and 41 satisfy the theorem, they form a Pythagorean triple and the triangle is a right triangle.
- Because 12, 18, and 24 do not satisfy the theorem, it is not a right triangle.
- **Extension:** The 12, 18, 24 triangle is similar to a triangle with side lengths 2, 3, and 4 (scale factor is 6). The 2, 3, 4 triangle is not a right triangle either.

On Your Own

- **Common Error:** Students may not substitute the longest side for c. This is particularly true because the measures are listed longest to shortest.

Key Idea

- Write the Key Idea that states the distance formula, derived from the Pythagorean Theorem.
- Note the use of colors in the diagram. The same colors can be used when writing the formula.
- **MP7 Look for and Make Use of Structure:** Discuss with students that the order in which the subtraction is performed is not important because the difference is squared.

Goal Today's lesson is using the converse of the Pythagorean Theorem and the **distance formula**.

Lesson Tutorials
Lesson Plans
Answer Presentation Tool

Extra Example 1

Tell whether each triangle is a right triangle.

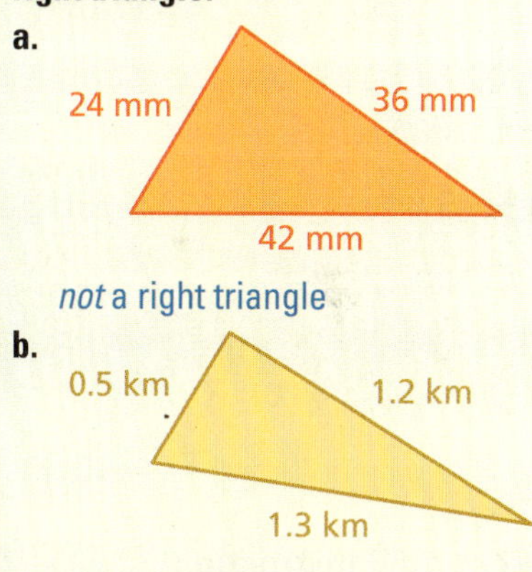

a. *not* a right triangle

b. right triangle

On Your Own

1. no
2. yes

Extra Example 2

Find the distance between (1, −5) and (7, 4). $\sqrt{117}$

Extra Example 3

In Example 3, your friend starts at (40, 40). Did your friend make a 90° turn? no

On Your Own

3. $\sqrt{41}$
4. $\sqrt{85}$
5. 5
6. no

Differentiated Instruction

Inclusion
Encourage students to learn the Pythagorean Theorem using the language of the triangle, $\text{leg}^2 + \text{leg}^2 = \text{hypotenuse}^2$. Have them label each side as a leg or hypotenuse before substituting the numbers into the equation.

Laurie's Notes

Example 2

- Although it is not necessary, you can plot the two points as a visual aid.
- The choice of which point is (x_1, y_1) and which point is (x_2, y_2) is arbitrary. If time permits, do the problem both ways to show that the result is the same.
- Caution students to be careful with the subtraction. It is easy to make a careless calculation mistake.
- There are no units associated with the answer. In a real-life example, students would need to label the units in their answers.

Example 3

- Ask a student to read the example.
- **MP4 Model with Mathematics:** Sketch the coordinate plane shown with the ordered pairs identified.

? "How do you determine if the receiver ran the play as designed?" Check whether the triangle is a right triangle.

? "How do you determine if the triangle is a right triangle?" Use the converse of the Pythagorean Theorem.

- Use the distance formula to find the length of each side of the triangle.
- Use the converse of the Pythagorean Theorem to show that it is a right triangle.
- Note that students could also use what they know about slopes of perpendicular lines to solve this problem.

On Your Own

- There are multiple steps required in each question. You may wish to divide the class into groups and assign a different question to each group.

Closure

- **Exit Ticket:** Find the distance between (−4, 6) and (3, −2). $\sqrt{113}$

EXAMPLE 2 Finding the Distance Between Two Points

Find the distance between (1, 5) and (−4, −2).

Let $(x_1, y_1) = (1, 5)$ and $(x_2, y_2) = (-4, -2)$.

$$d = \sqrt{(x_2 - x_1)^2 + (y_2 - y_1)^2}$$ Write the distance formula.

$$= \sqrt{(-4 - 1)^2 + (-2 - 5)^2}$$ Substitute.

$$= \sqrt{(-5)^2 + (-7)^2}$$ Simplify.

$$= \sqrt{25 + 49}$$ Evaluate powers.

$$= \sqrt{74}$$ Add.

EXAMPLE 3 Real-Life Application

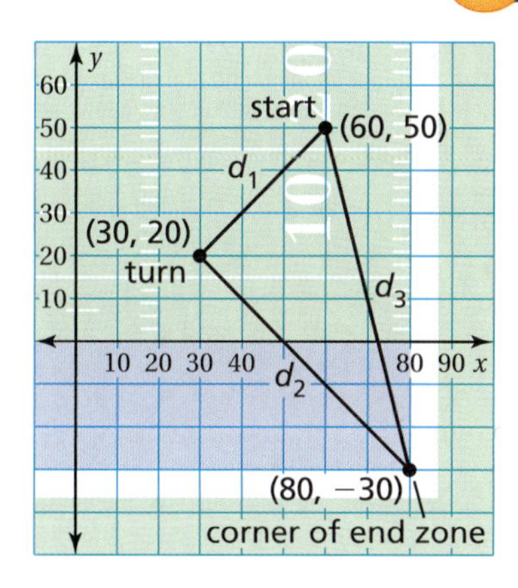

You design a football play in which a player runs down the field, makes a 90° turn, and runs to the corner of the end zone. Your friend runs the play as shown. Did your friend make a 90° turn? Each unit of the grid represents 10 feet.

Use the distance formula to find the lengths of the three sides.

$$d_1 = \sqrt{(60 - 30)^2 + (50 - 20)^2} = \sqrt{30^2 + 30^2} = \sqrt{1800} \text{ feet}$$

$$d_2 = \sqrt{(80 - 30)^2 + (-30 - 20)^2} = \sqrt{50^2 + (-50)^2} = \sqrt{5000} \text{ feet}$$

$$d_3 = \sqrt{(80 - 60)^2 + (-30 - 50)^2} = \sqrt{20^2 + (-80)^2} = \sqrt{6800} \text{ feet}$$

Use the converse of the Pythagorean Theorem to determine if the side lengths form a right triangle.

$$(\sqrt{1800})^2 + (\sqrt{5000})^2 \stackrel{?}{=} (\sqrt{6800})^2$$

$$1800 + 5000 \stackrel{?}{=} 6800$$

$$6800 = 6800 \checkmark$$

The sides form a right triangle.

∴ So, your friend made a 90° turn.

On Your Own

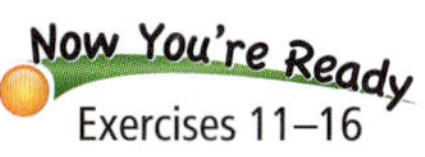

Find the distance between the two points.

3. (0, 0), (4, 5)
4. (7, −3), (9, 6)
5. (−2, −3), (−5, 1)
6. **WHAT IF?** In Example 3, your friend made the turn at (20, 10). Did your friend make a 90° turn?

7.5 Exercises

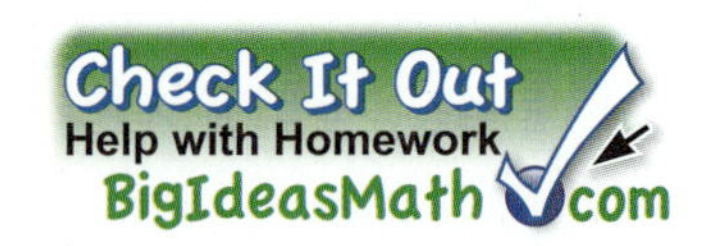

Vocabulary and Concept Check

1. **WRITING** Describe two ways to find the distance between two points in a coordinate plane.

2. **WHICH ONE DOESN'T BELONG?** Which set of numbers does *not* belong with the other three? Explain your reasoning.

3, 6, 8	6, 8, 10	5, 12, 13	7, 24, 25

Practice and Problem Solving

Write the converse of the true statement. Determine whether the converse is *true* or *false*. If it is true, justify your reasoning. If it is false, give a counterexample.

3. If a is an odd number, then a^2 is odd.

4. If $ABCD$ is a square, then $ABCD$ is a parallelogram.

Tell whether the triangle with the given side lengths is a right triangle.

1 5.

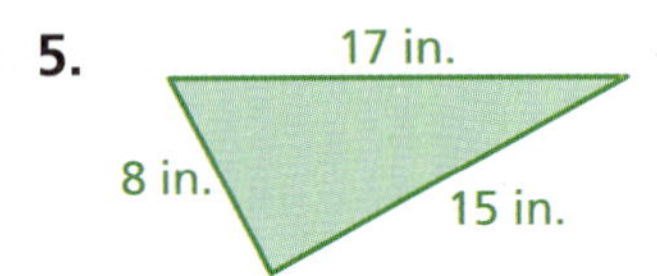

6.

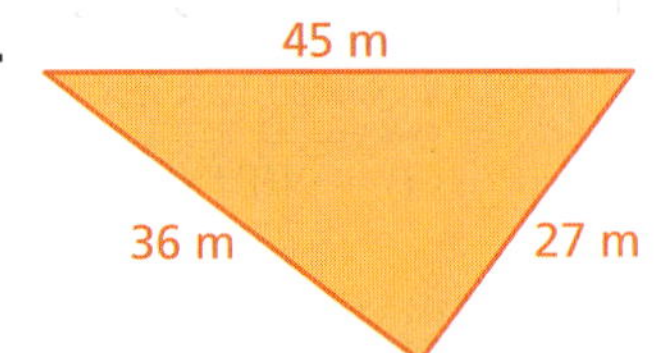

7.

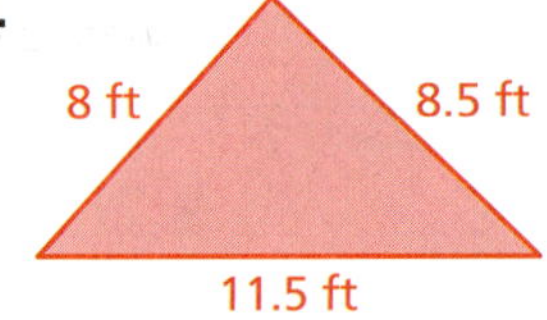

8. 14 mm, 19 mm, 23 mm

9. $\frac{9}{10}$ mi, $1\frac{1}{5}$ mi, $1\frac{1}{2}$ mi

10. 1.4 m, 4.8 m, 5 m

Find the distance between the two points.

2 11. (1, 2), (7, 6)

12. (4, −5), (−1, 7)

13. (2, 4), (7, 2)

14. (−1, −3), (1, 3)

15. (−6, −7), (0, 0)

16. (12, 5), (−12, −2)

17. **ERROR ANALYSIS** Describe and correct the error in finding the distance between the points (−3, −2) and (7, 4).

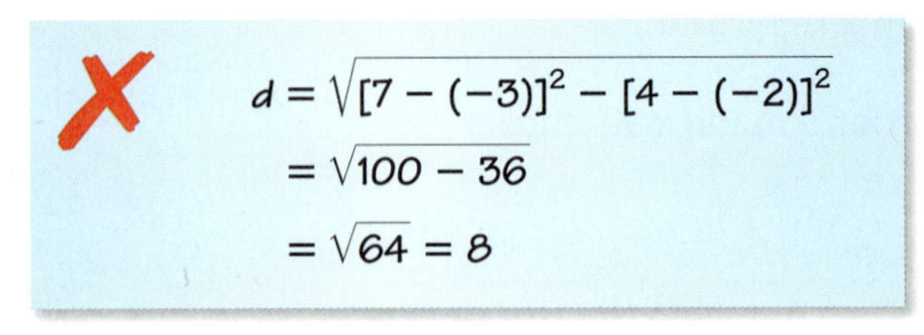

$$d = \sqrt{[7 - (-3)]^2 - [4 - (-2)]^2}$$
$$= \sqrt{100 - 36}$$
$$= \sqrt{64} = 8$$

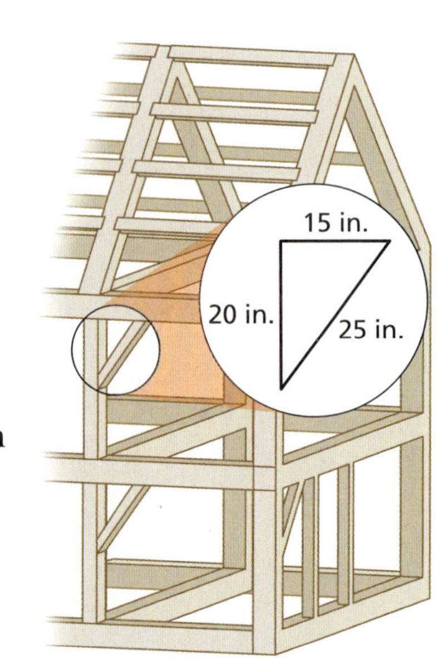

18. **CONSTRUCTION** A post and beam frame for a shed is shown in the diagram. Does the brace form a right triangle with the post and beam? Explain.

Assignment Guide and Homework Check

Level	Day 1 Activity Assignment	Day 2 Lesson Assignment	Homework Check
Basic	3, 4, 27–30	1, 2, 5–25 odd	7, 9, 13, 19, 23
Average	3, 4, 27–30	1, 2, 5–21 odd, 22–26 even	7, 9, 13, 21, 22
Advanced	1–4, 8–16 even, 17, 18, 20, 22–30		10, 16, 20, 25, 26

Common Errors

- **Exercises 5–10** Students may substitute the wrong value for *c* in the Pythagorean Theorem. Remind them that *c* will be the longest side, so they should substitute the greatest value for *c*.
- **Exercises 11–16** Students may mismatch the *x*-values and *y*-values when using the distance formula. This will result in students subtracting an *x* from a *y*, or vice versa. Encourage students to pair the numbers properly.
- **Exercises 11–16** Students may get careless when squaring negative numbers. Remind them that everything inside the parentheses is squared, including the minus sign, and that the square of a negative is a positive.
- **Exercise 25** Students may pick Plane A because it appears to be closer. Remind students that the drawing is not to scale. Tell them to calculate the distances before answering the question.

7.5 Record and Practice Journal

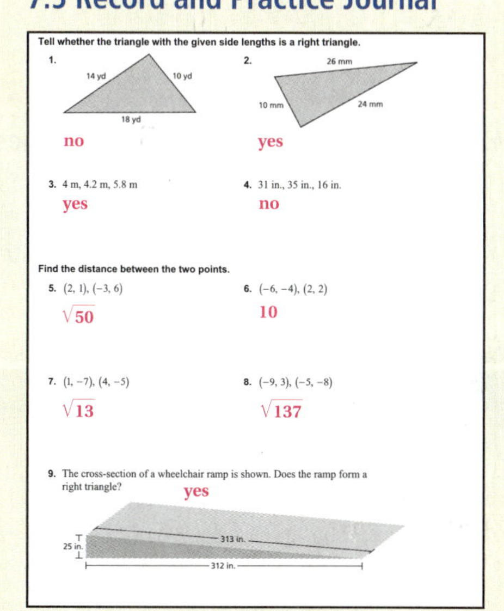

Tell whether the triangle with the given side lengths is a right triangle.

1. (14 yd, 10 yd, 18 yd) **no**
2. (26 mm, 10 mm, 24 mm) **yes**
3. 4 m, 4.2 m, 5.8 m **yes**
4. 31 in., 35 in., 16 in. **no**

Find the distance between the two points.

5. (2, 1), (−3, 6) $\sqrt{50}$
6. (−6, −4), (2, 2) **10**
7. (1, −7), (4, −5) $\sqrt{13}$
8. (−9, 3), (−5, −8) $\sqrt{137}$
9. The cross-section of a wheelchair ramp is shown. Does the ramp form a right triangle? **yes**

Vocabulary and Concept Check

1. the Pythagorean Theorem and the distance formula
2. 3, 6, 8; It is the only set that is not a Pythagorean triple.

Practice and Problem Solving

3. If a^2 is odd, then a is an odd number; true when a is an integer; A product of two integers is odd only when each integer is odd.
4. If *ABCD* is a parallelogram, then *ABCD* is a square; false; counterexample: any parallelogram that does not have right angles
5. yes
6. yes
7. no
8. no
9. yes
10. yes
11. $2\sqrt{13}$
12. 13
13. $\sqrt{29}$
14. $2\sqrt{10}$
15. $\sqrt{85}$
16. 25
17. The squared quantities under the radical should be added not subtracted; $2\sqrt{34}$
18. yes; The side lengths satisfy the converse of the Pythagorean Theorem.

Practice and Problem Solving

19. yes **20.** no

21. yes

22. yes; Use the distance formula to find the lengths of the three sides. Use the converse of the Pythagorean Theorem to show they form a right triangle.

23. no; The measures of the side lengths are $\sqrt{5000}$, $\sqrt{3700}$, and $\sqrt{8500}$ and $(\sqrt{5000})^2 + (\sqrt{3700})^2 \neq (\sqrt{8500})^2$.

24. yes; $\sqrt{58}$; Because you square the differences $(x_2 - x_1)$ and $(y_2 - y_1)$, it does not matter if the differences are positive or negative. The squares of opposite numbers are equivalent.

25. See *Taking Math Deeper.*

26. See Additional Answers.

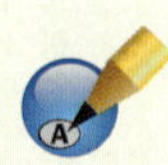

Fair Game Review

27. mean: 13; median: 12.5; mode: 12

28. mean: 21; median: 21; no mode

29. mean: 58; median: 59; mode: 59

30. B

Mini-Assessment

Tell whether the triangle with the given side lengths is a right triangle.

1. 32 m, 56 m, 64 m no

2. 1.8 mi, 8 mi, 8.2 mi yes

Find the distance between the two points.

3. $(-3, -1), (6, 2)$ $\sqrt{90}$

4. $(2, 10), (5, -4)$ $\sqrt{205}$

Taking Math Deeper

Exercise 25

At first this seems like a simple question. Students may think that Plane A is 5 kilometers from the base of the tower, and Plane B is 7 kilometers from the base of the tower. So, Plane A seems closer. However, on second glance, you see that Plane A is much higher than Plane B. So, to see which is closer, you need to compute the diagonal distance of each.

Find the distance for Plane A.

$$x^2 \approx 5^2 + 6.1^2$$
$$x^2 = 62.21$$
$$x \approx 7.89 \text{ km}$$

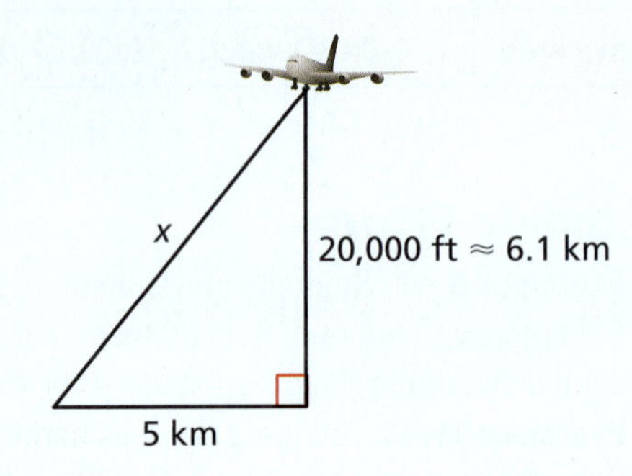

Find the distance for Plane B.

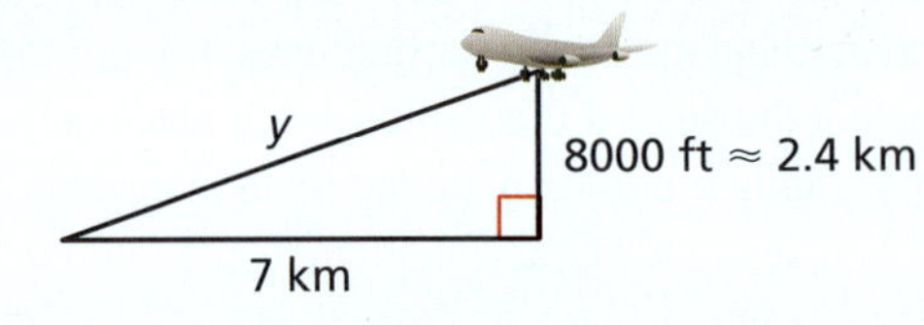

$$y^2 \approx 7^2 + 2.4^2$$
$$y^2 = 54.76$$
$$y = 7.4 \text{ km}$$

Answer the question.

Plane B is slightly closer to the base of the tower.

Reteaching and Enrichment Strategies

If students need help. . .	If students got it. . .
Resources by Chapter • Practice A and Practice B • Puzzle Time Record and Practice Journal Practice Differentiating the Lesson Lesson Tutorials Skills Review Handbook	Resources by Chapter • Enrichment and Extension • Technology Connection Start the next section

Tell whether a triangle with the given side lengths is a right triangle.

19. $\sqrt{63}$, 9, 12

20. 4, $\sqrt{15}$, 6

21. $\sqrt{18}$, $\sqrt{24}$, $\sqrt{42}$

22. REASONING Plot the points (−1, 3), (4, −2), and (1, −5) in a coordinate plane. Are the points the vertices of a right triangle? Explain.

23. GEOCACHING You spend the day looking for hidden containers in a wooded area using a Global Positioning System (GPS). You park your car on the side of the road, and then locate Container 1 and Container 2 before going back to the car. Does your path form a right triangle? Explain. Each unit of the grid represents 10 yards.

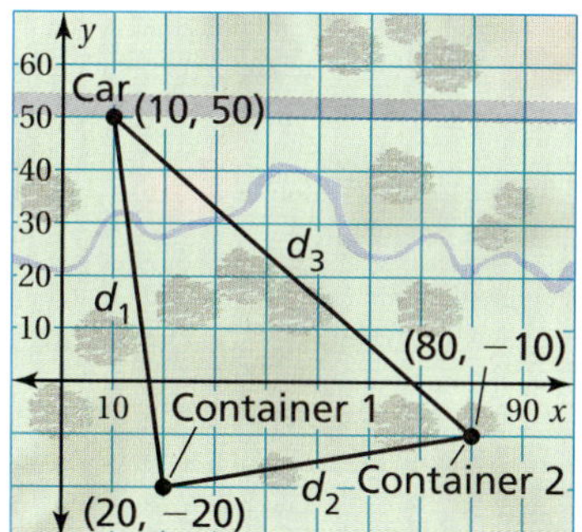

24. REASONING Your teacher wants the class to find the distance between the two points (2, 4) and (9, 7). You use (2, 4) for (x_1, y_1), and your friend uses (9, 7) for (x_1, y_1). Do you and your friend obtain the same result? Justify your answer.

25. AIRPORT Which plane is closer to the base of the airport tower? Explain.

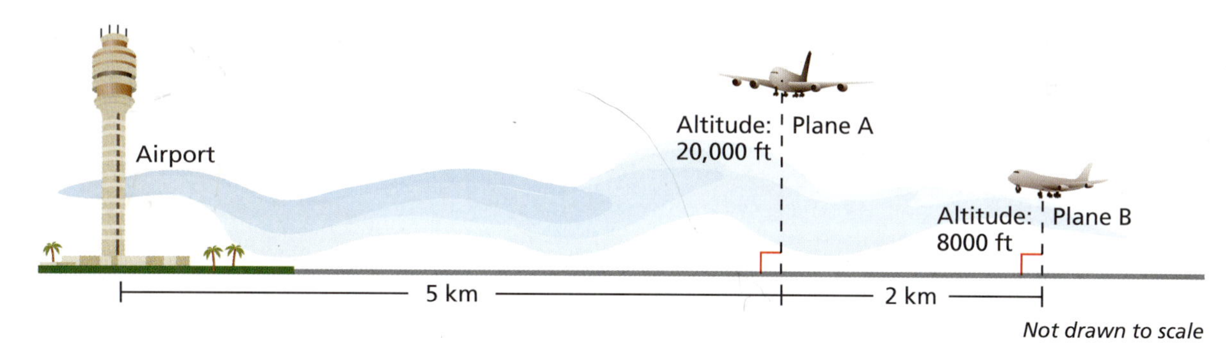

26. Structure Consider the two points (x_1, y_1) and (x_2, y_2) in the coordinate plane. How can you find the point (x_m, y_m) located in the middle of the two given points? Justify your answer using the distance formula.

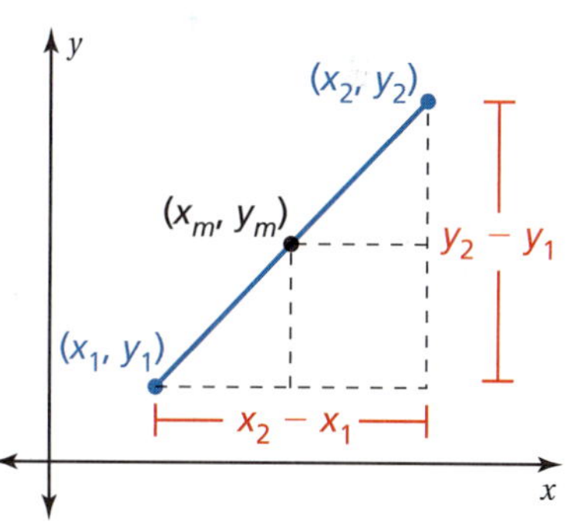

Fair Game Review What you learned in previous grades & lessons

Find the mean, median, and mode of the data. *(Skills Review Handbook)*

27. 12, 9, 17, 15, 12, 13

28. 21, 32, 16, 27, 22, 19, 10

29. 67, 59, 34, 71, 59

30. MULTIPLE CHOICE What is the sum of the interior angle measures of an octagon? *(Section 3.3)*

Ⓐ 720° Ⓑ 1080° Ⓒ 1440° Ⓓ 1800°

7.4–7.5 Quiz

Classify the real number. *(Section 7.4)*

1. $-\sqrt{225}$
2. $-1\frac{1}{9}$
3. $\sqrt{41}$
4. $\sqrt{17}$

Estimate the square root to the nearest (a) integer and (b) tenth. *(Section 7.4)*

5. $\sqrt{38}$
6. $-\sqrt{99}$
7. $\sqrt{172}$
8. $\sqrt{115}$

Which number is greater? Explain. *(Section 7.4)*

9. $\sqrt{11}, 3\frac{3}{5}$
10. $\sqrt{1.44}, 1.1\overline{8}$

Write the decimal as a fraction or a mixed number. *(Section 7.4)*

11. $0.\overline{7}$
12. $-1.\overline{63}$

Tell whether the triangle with the given side lengths is a right triangle. *(Section 7.5)*

13.

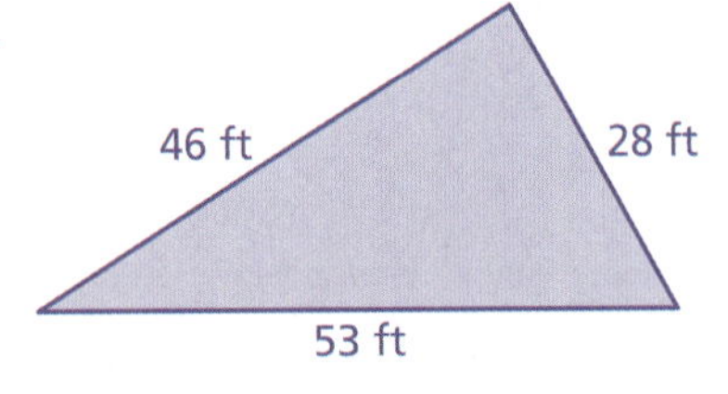

14.

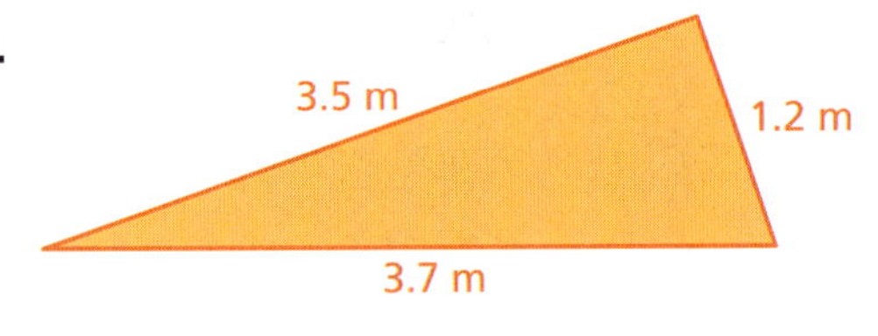

Find the distance between the two points. *(Section 7.5)*

15. $(-3, -1), (-1, -5)$
16. $(-4, 2), (5, 1)$
17. $(1, -2), (4, -5)$
18. $(-1, 1), (7, 4)$
19. $(-6, 5), (-4, -6)$
20. $(-1, 4), (1, 3)$

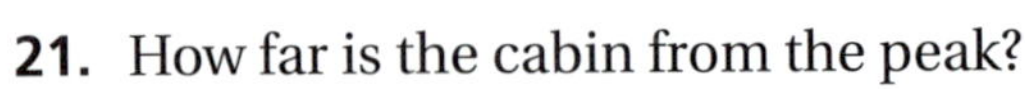

Use the figure to answer Exercises 21–24. Round your answer to the nearest tenth. *(Section 7.5)*

21. How far is the cabin from the peak?
22. How far is the fire tower from the lake?
23. How far is the lake from the peak?
24. You are standing at $(-5, -6)$. How far are you from the lake?

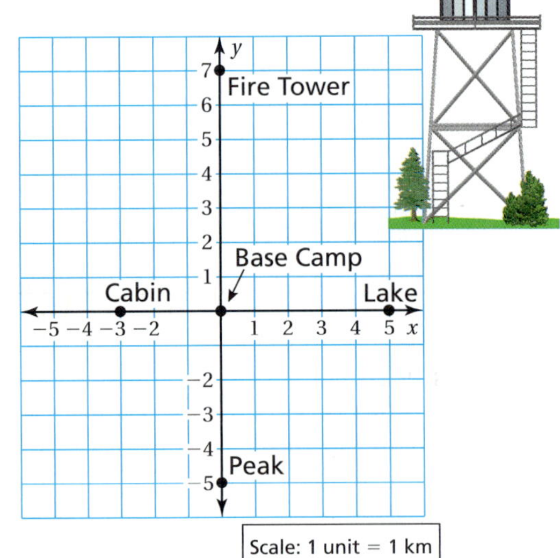

Alternative Assessment Options

Math Chat
Structured Interview
Student Reflective Focus Question
Writing Prompt

Student Reflective Focus Question

Ask students to summarize approximating square roots and using the Pythagorean Theorem. Be sure that they include examples. Select students at random to present their summaries to the class.

Study Help Sample Answers

Remind students to complete Graphic Organizers for the rest of the chapter.

3.

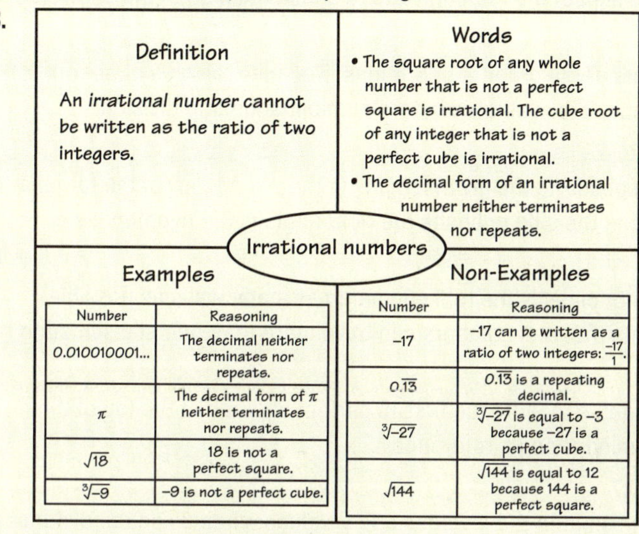

4–6. Available at *BigIdeasMath.com.*

Reteaching and Enrichment Strategies

If students need help. . .	If students got it. . .
Resources by Chapter • Practice A and Practice B • Puzzle Time Lesson Tutorials *BigIdeasMath.com*	Resources by Chapter • Enrichment and Extension • Technology Connection Game Closet at *BigIdeasMath.com* Start the Chapter Review

Answers

1. integer, rational
2. rational
3. irrational
4. irrational
5. a. 6
   b. 6.2
6. a. -10
   b. -9.9
7. a. 13
   b. 13.1
8. a. 11
   b. 10.7
9. $3\frac{3}{5}$; $3\frac{3}{5}$ is to the right of $\sqrt{11}$.
10. $\sqrt{1.44}$; $\sqrt{1.44}$ is to the right of $1.1\overline{8}$.
11. $\frac{7}{9}$
12. $-1\frac{7}{11}$
13. no
14. yes
15. $2\sqrt{5}$
16. $\sqrt{82}$
17. $3\sqrt{2}$
18. $\sqrt{73}$
19. $5\sqrt{5}$
20. $\sqrt{5}$
21. 5.8 km
22. 8.6 km
23. 7.1 km
24. 11.7 km

Technology for the Teacher

Online Assessment
Assessment Book
ExamView® Assessment Suite

For the Teacher
Additional Review Options

- *BigIdeasMath.com*
- Online Assessment
- Game Closet at *BigIdeasMath.com*
- Vocabulary Help
- Resources by Chapter

Answers

1. 1
2. $-\frac{3}{5}$
3. 1.3 and -1.3
4. -9
5. $3\frac{2}{3}$
6. -30
7. 9
8. $\frac{4}{7}$
9. $-\frac{2}{3}$
10. -13
11. 17
12. -42

Review of Common Errors

Exercises 1–3

- Students may divide the number by two instead of finding a number that, when multiplied by itself, gives the radicand. Remind them that taking the square root of a number is the inverse of squaring a number.

Exercises 4–6

- Remind students of the order of operations. Because taking a square root is the inverse of squaring, it is evaluated before multiplication and division.

Exercise 9

- Students may disregard the negative in the radicand. Remind them to pay attention to the sign of the radicand and to check the sign of the answer.

Exercises 10–12

- Remind students of the order of operations. Because taking a cube root is the inverse of cubing, it is evaluated before multiplication and division.

Exercises 13, 14, 24, 25

- Students may substitute the given lengths in the wrong part of the formula. Remind them that the side opposite the right angle is the hypotenuse c.

Exercises 15–17

- Students may not classify the real number in as many ways as possible. Remind students that real numbers can have more than one classification.

Exercises 18–20

- Encourage students to write the first 10 perfect squares at the top of their papers as a reminder and a reference.

Exercises 26 and 27

- Students may mismatch the x-values and y-values when using the distance formula. This will result in students subtracting an x from a y, or vice versa. Encourage students to pair the numbers properly.
- Students may get careless when squaring negative numbers. Remind them that everything inside the parentheses is squared, including the minus sign, and that the square of a negative is a positive.

7 Chapter Review

Check It Out
Vocabulary Help
BigIdeasMath.com

Review Key Vocabulary

square root, *p. 290*
perfect square, *p. 290*
radical sign, *p. 290*
radicand, *p. 290*
cube root, *p. 296*
perfect cube, *p. 296*
theorem, *p. 300*
legs, *p. 302*
hypotenuse, *p. 302*
Pythagorean Theorem, *p. 302*
irrational number, *p. 310*
real numbers, *p. 310*
distance formula, *p. 320*

Review Examples and Exercises

7.1 Finding Square Roots (*pp. 288–293*)

Find $-\sqrt{36}$.

$-\sqrt{36}$ represents the *negative* square root.

Because $6^2 = 36$, $-\sqrt{36} = -\sqrt{6^2} = -6$.

Exercises

Find the square root(s).

1. $\sqrt{1}$

2. $-\sqrt{\frac{9}{25}}$

3. $\pm\sqrt{1.69}$

Evaluate the expression.

4. $15 - 4\sqrt{36}$

5. $\sqrt{\frac{54}{6}} + \frac{2}{3}$

6. $10(\sqrt{81} - 12)$

7.2 Finding Cube Roots (*pp. 294–299*)

Find $\sqrt[3]{\frac{125}{216}}$.

Because $\left(\frac{5}{6}\right)^3 = \frac{125}{216}$, $\sqrt[3]{\frac{125}{216}} = \sqrt[3]{\left(\frac{5}{6}\right)^3} = \frac{5}{6}$.

Exercises

Find the cube root.

7. $\sqrt[3]{729}$

8. $\sqrt[3]{\frac{64}{343}}$

9. $\sqrt[3]{-\frac{8}{27}}$

Evaluate the expression.

10. $\sqrt[3]{27} - 16$

11. $25 + 2\sqrt[3]{-64}$

12. $3\sqrt[3]{-125} - 27$

7.3 The Pythagorean Theorem (pp. 300–305)

Find the length of the hypotenuse of the triangle.

$a^2 + b^2 = c^2$ Write the Pythagorean Theorem.

$7^2 + 24^2 = c^2$ Substitute.

$49 + 576 = c^2$ Evaluate powers.

$625 = c^2$ Add.

$\sqrt{625} = \sqrt{c^2}$ Take positive square root of each side.

$25 = c$ Simplify.

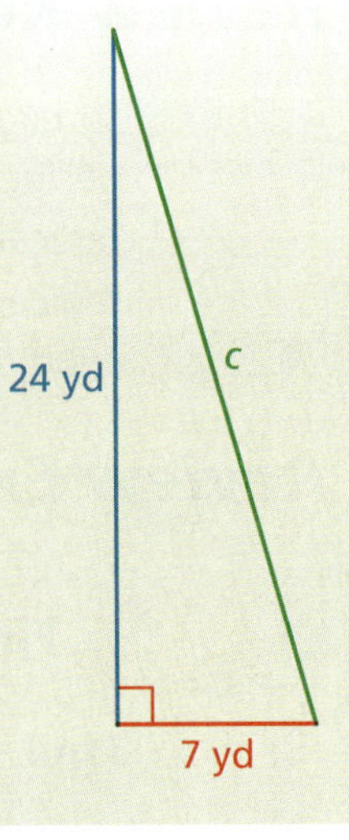

∴ The length of the hypotenuse is 25 yards.

Exercises

Find the missing length of the triangle.

13.

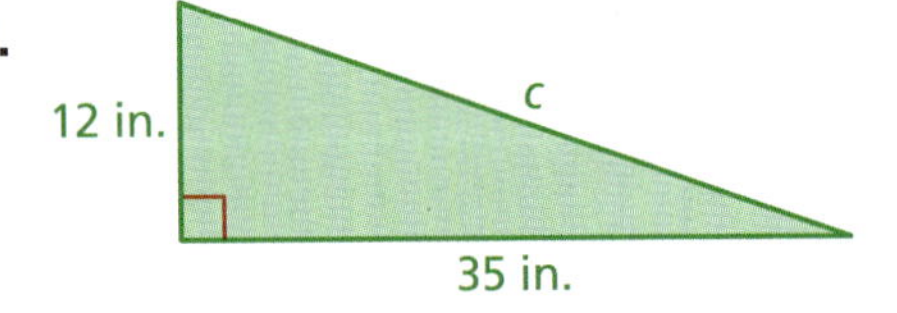

14.

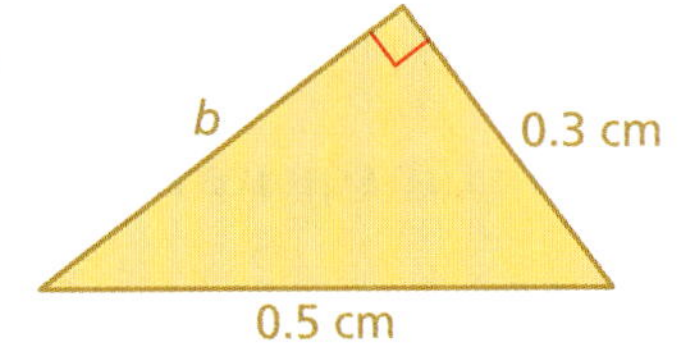

7.4 Approximating Square Roots (pp. 308–317)

a. Classify $\sqrt{19}$.

∴ The number $\sqrt{19}$ is irrational because 19 is not a perfect square.

b. Estimate $\sqrt{34}$ to the nearest integer.

Make a table of numbers whose squares are close to the radicand, 34.

Number	4	5	6	7
Square of Number	16	25	36	49

The table shows that 34 is between the perfect squares 25 and 36. Because 34 is closer to 36 than to 25, $\sqrt{34}$ is closer to 6 than to 5.

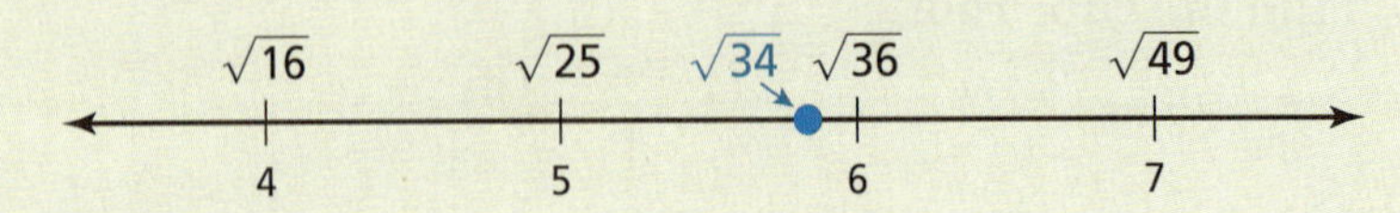

∴ So, $\sqrt{34} \approx 6$.

Review Game

Significant Square Roots

Materials per Group

- piece of paper
- pencil

Directions

Divide the class into groups of 3 or 4.

Each group is to come up with 5 significant numbers and compute the exact square root of each number.

Examples of significant numbers:

School address: 1764 Knowledge Road; $\sqrt{1764} = 42$

Year of presidential election: 1936—Franklin D. Roosevelt elected to his second term; $\sqrt{1936} = 44$

Age for driver's license: 16 years old; $\sqrt{16} = 4$

Who wins?

The first group to come up with five significant numbers and correct square roots wins.

For the Student Additional Practice

- Lesson Tutorials
- Multi-Language Glossary
- Self-Grading Progress Check
- *BigIdeasMath.com*
  Dynamic Student Edition
  Student Resources

Answers

13. 37 in.

14. 0.4 cm

15. rational

16. irrational

17. natural, whole, integer, rational

18. **a.** 4
b. 3.7

19. **a.** 9
b. 9.5

20. **a.** 13
b. 13.2

21. $\frac{8}{9}$

22. $\frac{4}{11}$

23. $-1\frac{2}{3}$

24. yes

25. no

26. $5\sqrt{5}$

27. $\sqrt{113}$

My Thoughts on the Chapter

Teacher Tip

Not allowed to write in your teaching edition? Use sticky notes to record your thoughts.

What worked. . .

What did not work. . .

What I would do differently. . .

Exercises

Classify the real number.

15. $0.81\overline{5}$ **16.** $\sqrt{101}$ **17.** $\sqrt{4}$

Estimate the square root to the nearest (a) integer and (b) tenth.

18. $\sqrt{14}$ **19.** $\sqrt{90}$ **20.** $\sqrt{175}$

Write the decimal as a fraction.

21. $0.\overline{8}$ **22.** $0.\overline{36}$ **23.** $-1.\overline{6}$

7.5 Using the Pythagorean Theorem *(pp. 318–323)*

a. Is the triangle formed by the rope and the tent a right triangle?

$$a^2 + b^2 = c^2$$

$$64^2 + 48^2 \stackrel{?}{=} 80^2$$

$$4096 + 2304 \stackrel{?}{=} 6400$$

$$6400 = 6400 \checkmark$$

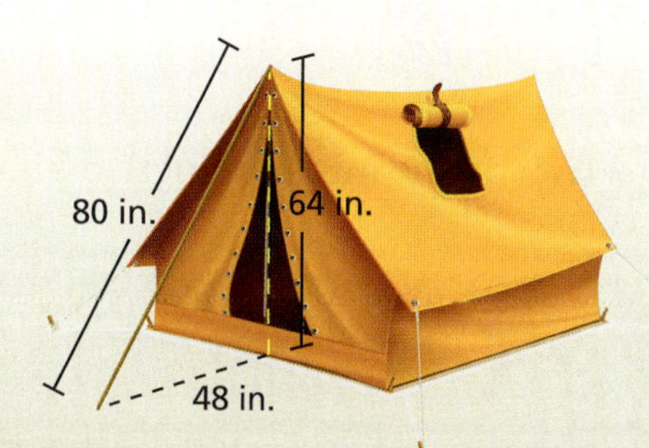

It *is* a right triangle.

b. Find the distance between (−3, 1) and (4, 7).

Let $(x_1, y_1) = (-3, 1)$ and $(x_2, y_2) = (4, 7)$.

$d = \sqrt{(x_2 - x_1)^2 + (y_2 - y_1)^2}$ Write the distance formula.

$= \sqrt{[4 - (-3)]^2 + (7 - 1)^2}$ Substitute.

$= \sqrt{7^2 + 6^2}$ Simplify.

$= \sqrt{49 + 36}$ Evaluate powers.

$= \sqrt{85}$ Add.

Exercises

Tell whether the triangle is a right triangle.

24.

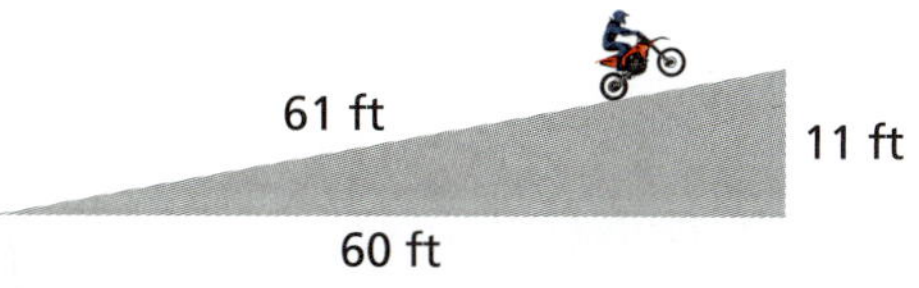

25.

Find the distance between the two points.

26. (−2, −5), (3, 5) **27.** (−4, 7), (4, 0)

7 Chapter Test

Find the square root(s).

1. $-\sqrt{1600}$
2. $\sqrt{\frac{25}{49}}$
3. $\pm\sqrt{\frac{100}{9}}$

Find the cube root.

4. $\sqrt[3]{-27}$
5. $\sqrt[3]{\frac{8}{125}}$
6. $\sqrt[3]{-\frac{729}{64}}$

Evaluate the expression.

7. $12 + 8\sqrt{16}$
8. $\frac{1}{2} + \sqrt{\frac{72}{2}}$
9. $\left(\sqrt[3]{-125}\right)^3 + 75$
10. $50\sqrt[3]{\frac{512}{1000}} + 14$
11. Find the missing length of the triangle.

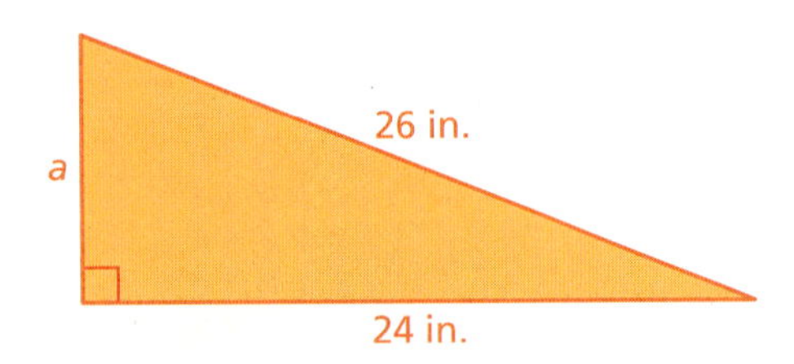

Classify the real number.

12. 16π
13. $-\sqrt{49}$

Estimate the square root to the nearest (a) integer and (b) tenth.

14. $\sqrt{58}$
15. $\sqrt{83}$

Write the decimal as a fraction or a mixed number.

16. $-0.\overline{3}$
17. $1.\overline{24}$
18. Tell whether the triangle is a right triangle.

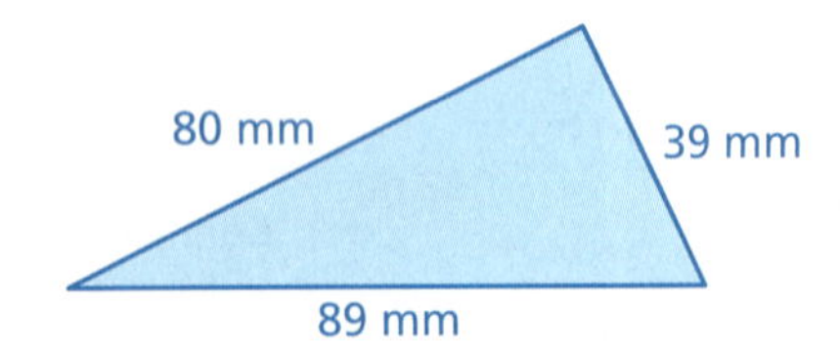

Find the distance between the two points.

19. $(-2, 3)$, $(6, 9)$
20. $(0, -5)$, $(4, 1)$
21. **SUPERHERO** Find the altitude of the superhero balloon.

Test Item References

Chapter Test Questions	Section to Review	Florida Common Core Standards (MACC)
1–3, 7, 8	7.1	8.EE.1.2
4–6, 9, 10	7.2	8.EE.1.2
11, 21	7.3	8.EE.1.2, 8.G.2.6, 8.G.2.7, 8.G.2.8
12–17	7.4	8.NS.1.1, 8.NS.1.2, 8.EE.1.2
18–20	7.5	8.EE.1.2, 8.G.2.6, 8.G.2.7, 8.G.2.8

Test-Taking Strategies

Remind students to quickly look over the entire test before they start so that they can budget their time. Students should estimate and check their answers for reasonableness as they work through the test. Teach the students to use the Stop and Think strategy before answering. **Stop** and carefully read the question, and **Think** about what the answer should look like.

Common Errors

- **Exercises 1–3** Remind students that a square root can be positive or negative.
- **Exercises 7 and 8** Remind students of the order of operations. Because taking a square root is the inverse of squaring, it is evaluated before multiplication and division.
- **Exercises 11, 18, 21** Students may substitute the given lengths in the wrong part of the formula. Remind them that the side opposite the right angle is the hypotenuse c.
- **Exercises 12 and 13** Students may not classify the real number in as many ways as possible. Remind students that real numbers can have more than one classification.

Reteaching and Enrichment Strategies

If students need help. . .	If students got it. . .
Resources by Chapter • Practice A and Practice B • Puzzle Time Record and Practice Journal Practice Differentiating the Lesson Lesson Tutorials *BigIdeasMath.com* Skills Review Handbook	Resources by Chapter • Enrichment and Extension • Technology Connection Game Closet at *BigIdeasMath.com* Start Standards Assessment

Answers

1. -40
2. $\frac{5}{7}$
3. $\frac{10}{3}$ and $-\frac{10}{3}$
4. -3
5. $\frac{2}{5}$
6. $-2\frac{1}{4}$
7. 44
8. $6\frac{1}{2}$
9. -50
10. 54
11. 10 in.
12. irrational
13. integer, rational
14. **a.** 8
    b. 7.6
15. **a.** 9
    b. 9.1
16. $-\frac{1}{3}$
17. $1\frac{8}{33}$
18. yes
19. 10
20. $2\sqrt{13}$
21. 66 ft

Technology for the Teacher

Online Assessment
Assessment Book
ExamView® Assessment Suite

Test-Taking Strategies

Available at *BigIdeasMath.com*

After Answering Easy Questions, Relax
Answer Easy Questions First
Estimate the Answer
Read All Choices before Answering
Read Question before Answering
Solve Directly or Eliminate Choices
Solve Problem before Looking at Choices
Use Intelligent Guessing
Work Backwards

About this Strategy

When taking a multiple choice test, be sure to read each question carefully and thoroughly. When taking a timed test, it is often best to skim the test and answer the easy questions first. Be careful that you record your answer in the correct position on the answer sheet.

Answers

1. D
2. H
3. D
4. F

Technology for the Teacher

Florida Common Core Standards Support Performance Tasks
Online Assessment
Assessment Book
ExamView® Assessment Suite

Item Analysis

1. **A.** The student adds 1.1 and 4.
   B. The student multiplies 1.1 by 4.
   C. The student adds 1.1 and 2.
   D. Correct answer

2. **F.** The student thinks that the parallelograms are similar because the corresponding angles are congruent. But, the corresponding sides are not proportional.
   G. The student thinks that the parallelograms are similar because the corresponding sides are proportional. But, the corresponding angles are not congruent.
   H. Correct answer
   I. The student thinks that the parallelograms are similar because the corresponding sides are proportional. But, the corresponding angles are not congruent.

3. **A.** The student overlooks the exponent, not realizing that it makes the function nonlinear.
   B. The student overlooks the fact that x is in the denominator, not realizing that it makes the function nonlinear.
   C. The student overlooks the fact that x and y are being multiplied, not realizing that it makes the function nonlinear.
   D. Correct answer

4. **F.** Correct answer
   G. The student has the correct slope but the wrong y-intercept, perhaps confusing the x- and y-intercepts.
   H. The student has the correct y-intercept but the wrong slope.
   I. The student has the wrong slope and the wrong y-intercept, perhaps confusing the x- and y-intercepts.

7 Standards Assessment

Test-Taking Strategy
Answer Easy Questions First

Was Fluffy in our kitchen?

There are $\sqrt{4}$ different tongue prints on the butter. How many cats licked the butter?

Ⓐ 1 Ⓑ 2 Ⓒ -2 Ⓓ 4

"Scan the test and answer the easy questions first. You know the square root of 4 is 2."

1. The period T of a pendulum is the time, in seconds, it takes the pendulum to swing back and forth. The period can be found using the formula $T = 1.1\sqrt{L}$, where L is the length, in feet, of the pendulum. A pendulum has a length of 4 feet. Find its period. *(MACC.8.EE.1.2)*

A. 5.1 sec **C.** 3.1 sec

B. 4.4 sec **D.** 2.2 sec

2. Which parallelogram is a dilation of parallelogram $JKLM$? (Figures not drawn to scale.) *(MACC.8.G.1.4)*

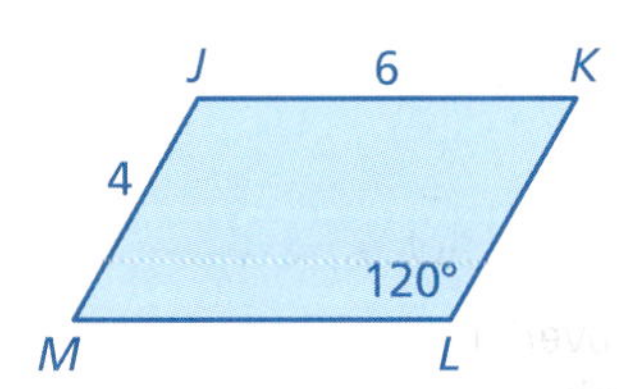

F.

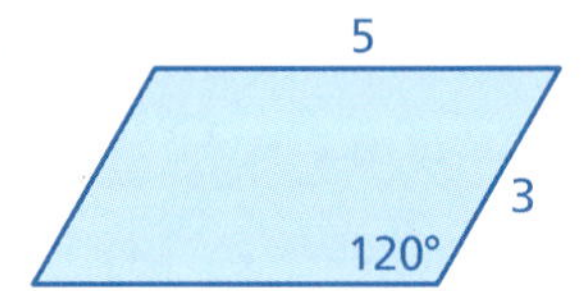

H.

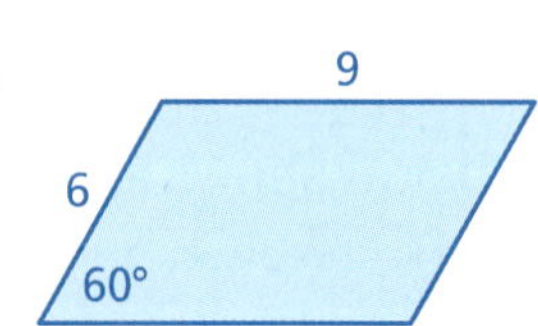

G.

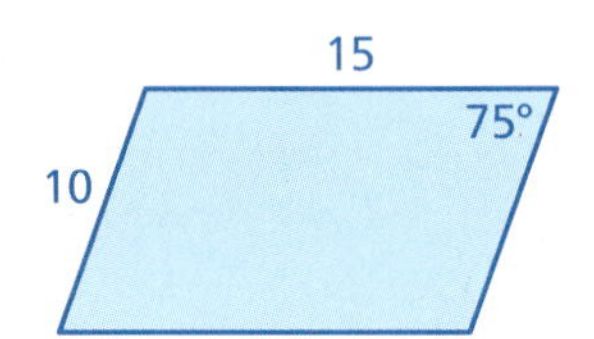

I.

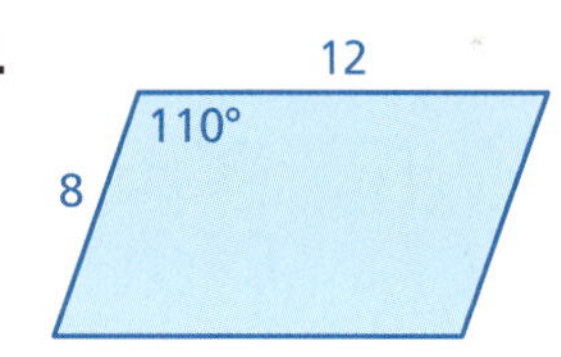

3. Which equation represents a linear function? *(MACC.8.F.1.3)*

A. $y = x^2$ **C.** $xy = 1$

B. $y = \frac{2}{x}$ **D.** $x + y = 1$

4. Which linear function matches the line shown in the graph? *(MACC.8.F.2.4)*

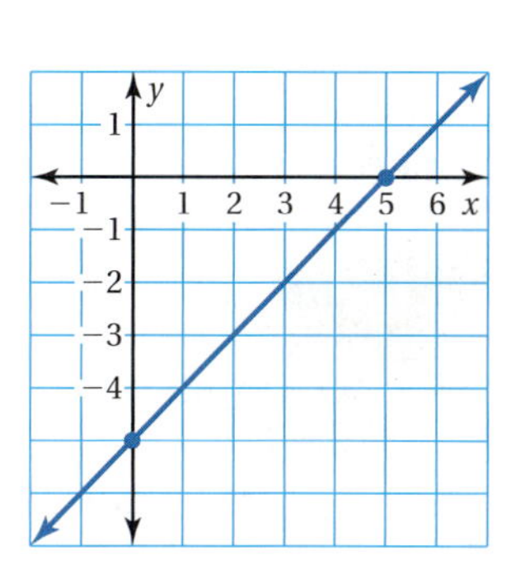

F. $y = x - 5$ **H.** $y = -x - 5$

G. $y = x + 5$ **I.** $y = -x + 5$

5. A football field is 40 yards wide and 120 yards long. Find the distance between opposite corners of the football field. Show your work and explain your reasoning. *(MACC.8.G.2.7)*

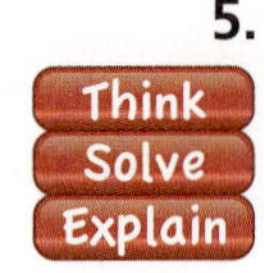

6. A computer consultant charges $50 plus $40 for each hour she works. The consultant charged $650 for one job. This can be represented by the equation below, where h represents the number of hours worked.

$$40h + 50 = 650$$

How many hours did the consultant work? *(MACC.8.EE.3.7b)*

7. You can use the formula below to find the sum S of the interior angle measures of a polygon with n sides. Solve the formula for n. *(MACC.8.EE.3.7b)*

$$S = 180(n - 2)$$

A. $n = 180(S - 2)$

B. $n = \frac{S}{180} + 2$

C. $n = \frac{S}{180} - 2$

D. $n = \frac{S}{180} + \frac{1}{90}$

8. The table below shows a linear pattern. Which linear function relates y to x? *(MACC.8.F.1.1)*

x	1	2	3	4	5
y	4	2	0	−2	−4

F. $y = 2x + 2$

G. $y = 4x$

H. $y = -2x + 2$

I. $y = -2x + 6$

9. An airplane flies from City 1 at (0, 0) to City 2 at (33, 56) and then to City 3 at (23, 32). What is the total number of miles it flies? Each unit of the coordinate grid represents 1 mile. *(MACC.8.G.2.8)*

10. What is the missing length of the right triangle shown? *(MACC.8.G.2.7)*

A. 16 cm

B. 18 cm

C. 24 cm

D. $\sqrt{674}$ cm

Item Analysis (continued)

5. **2 points** The student demonstrates a thorough understanding of how to apply the Pythagorean Theorem to the problem, explains the work fully, and calculates the distance accurately. The distance between opposite corners is $\sqrt{16{,}000} \approx 126.5$ yards.

   1 point The student's work and explanations demonstrate a lack of essential understanding. The Pythagorean Theorem is misstated or, if stated correctly, is applied incorrectly to the problem.

   0 points The student provides no response, a completely incorrect or incomprehensible response, or a response that demonstrates insufficient understanding of the Pythagorean Theorem.

6. **Gridded Response:** Correct answer: 15 hours

   Common Error: The student adds 50 to the right hand side of the equation instead of subtracting, yielding an answer of 17.5.

7. **A.** The student interchanges the roles of n and S.

   B. Correct answer

   C. The student subtracts 2 from S instead of adding it.

   D. The student adds 2 first, then divides through by 180.

8. **F.** The student has the wrong sign for the slope and makes a mistake finding the y-intercept.

   G. The student bases the equation on the first column only.

   H. The student finds the correct slope, but makes a mistake finding the y-intercept.

   I. Correct answer

9. **Gridded Response:** Correct answer: 91

   Common Error: The student only finds the distance between City 1 and City 2, yielding an answer of 65.

10. **A.** The student takes the average of the two sides.

    B. The student subtracts the shorter from the longer side.

    C. Correct answer

    D. The student treats the missing leg as the hypotenuse.

Answers

5. about 126.5 yards
6. 15 hours
7. B
8. I
9. 91 miles
10. C

Answers

11. H

12. D

13. F

Item Analysis (continued)

11. **F.** The student selects the point where one of the lines crosses the y-axis.

G. The student selects the point where one of the lines crosses the x-axis.

H. Correct answer

I. The student selects the point where one of the lines crosses the x-axis.

12. **A.** The student fails to see that these are supplementary, not congruent angles.

B. The student thinks these are alternate interior (or alternate exterior) angles, and believes they are congruent.

C. The student thinks these are corresponding angles, and believes they are congruent.

D. Correct answer

13. **F.** Correct answer

G. The student picks slope of 2 instead of -2.

H. The student picks y-intercept of 2 instead of -2.

I. The student picks y-intercept of 2 instead of -2 and slope of 2 instead of -2.

11. A system of linear equations is shown in the coordinate plane below. What is the solution for this system? *(MACC.8.EE.3.8a)*

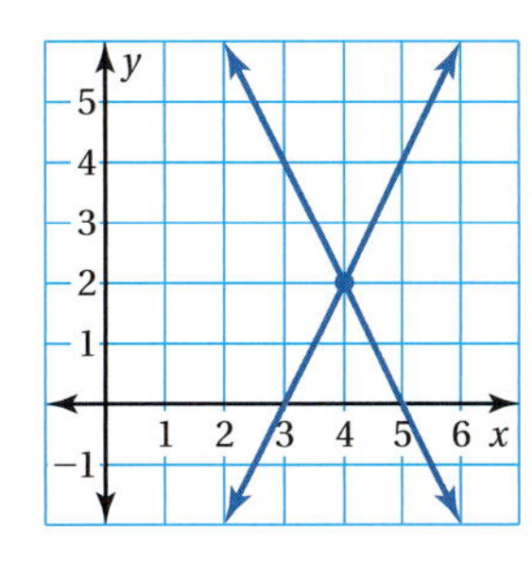

F. (0, 10)

G. (3, 0)

H. (4, 2)

I. (5, 0)

12. In the diagram, lines ℓ and m are parallel. Which angle has the same measure as $\angle 1$? *(MACC.8.G.1.5)*

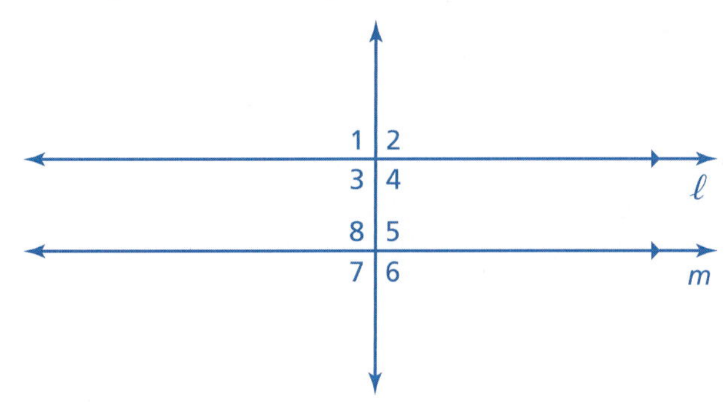

A. $\angle 2$

B. $\angle 5$

C. $\angle 7$

D. $\angle 8$

13. Which graph represents the linear equation $y = -2x - 2$? *(MACC.8.EE.2.6)*

F.

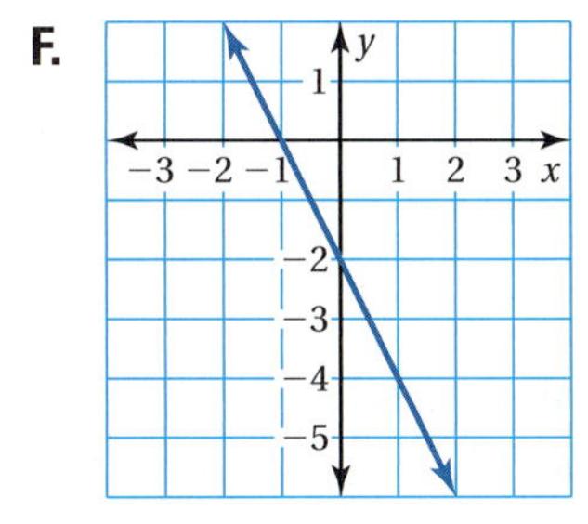

H.

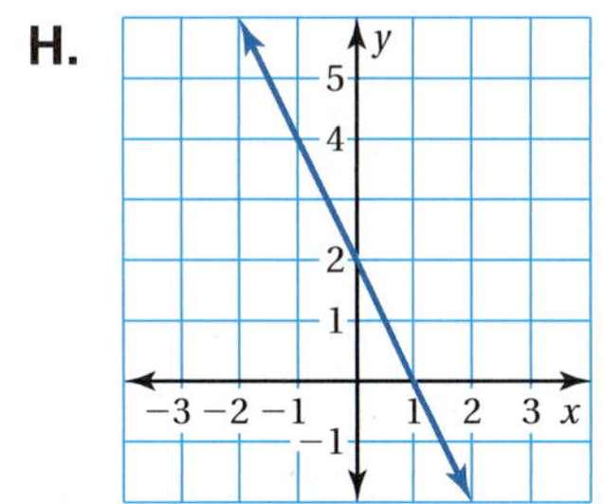

G.

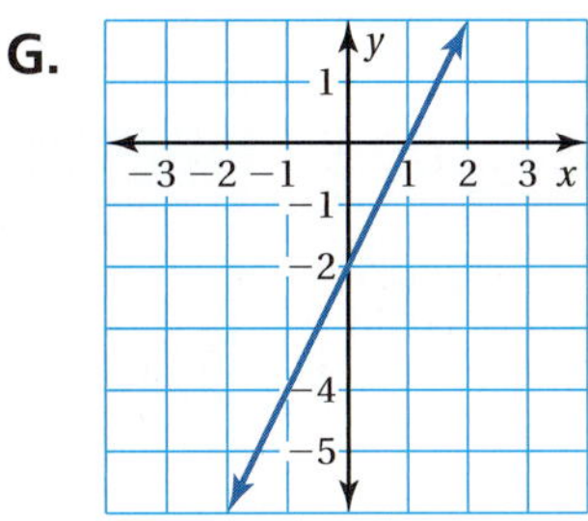

I.

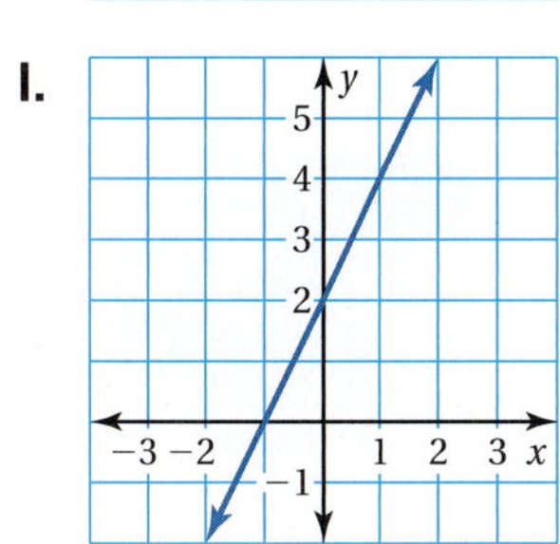

8 Volume and Similar Solids

"Dear Sir: Why do you sell dog food in tall cans and sell cat food in short cans?"

"Neither of these shapes is the optimal use of surface area when compared to volume."

"Do you know why the volume of a cone is one-third the volume of a cylinder with the same height and base?"

Florida Common Core Progression

6th Grade

- Find volumes of right rectangular prisms using formulas.
- Find areas of triangles, special quadrilaterals, and polygons.
- Use nets made up of rectangles and triangles to find surface areas.

7th Grade

- Find areas and circumferences of circles.
- Solve problems involving area, volume, and surface area of objects composed of triangles, quadrilaterals, cubes, and right prisms.
- Represent proportional relationships with equations.
- Decide whether two quantities are proportional.
- Use scale drawings to compute actual lengths and areas.

8th Grade

- Know and apply the formulas for the volumes of cones, cylinders, and spheres.
- Describe a sequence that exhibits similarity between two figures.

Chapter Summary

Section	Florida Common Core Standard (MACC)	
8.1	Learning	8.G.3.9
8.2	Learning	8.G.3.9
8.3	Learning	8.G.3.9
8.4	Applying	8.G.3.9 ★
★ Teaching is complete. Standard can be assessed.		

Pacing Guide for Chapter 8

Chapter Opener	1 Day
Section 1 Activity Lesson	 1 Day 1 Day
Section 2 Activity Lesson	 1 Day 1 Day
Study Help / Quiz	1 Day
Section 3 Activity Lesson	 1 Day 1 Day
Section 4 Activity Lesson	 1 Day 2 Days
Chapter Review/ Chapter Tests	2 Days
Total Chapter 8	13 Days
Year-to-Date	119 Days

Technology for the Teacher

BigIdeasMath.com
Chapter at a Glance
Complete Materials List
Parent Letters: English, Spanish, and Haitian Creole

Florida Common Core Standards

MACC.7.G.2.4 Know the formula for the area . . . of a circle and use it to solve problems . . ..

MACC.7.G.2.6 Solve real-world and mathematical problems involving area . . . of two- . . . dimensional objects composed of triangles, quadrilaterals, polygons . . ..

Additional Topics for Review

- Square Roots and Cube Roots
- Volumes of Prisms and Pyramids
- Cross Sections of Three-Dimensional Figures
- Writing and Solving Proportions
- Perimeters and Areas of Similar Figures
- Surface Areas of Prisms, Pyramids, and Cylinders

Try It Yourself

1. about 145.12 m^2
2. 86 cm^2
3. about 78.5 ft^2
4. about 530.66 $in.^2$
5. about 38.465 cm^2

Record and Practice Journal Fair Game Review

1. 51 m^2
2. about 146.93 m^2
3. 74 $in.^2$
4. 171 $in.^2$
5. 81 ft^2
6. 88 $in.^2$
7. $444
8. about 314 $in.^2$
9. about 113.04 m^2
10. about 452.16 cm^2
11. about 153.86 ft^2
12. about 490.625 yd^2
13. about 706.5 mm^2
14. about 502.4 cm^2

Math Background Notes

Vocabulary Review

- Area
- Composite figures
- Pi
- Radius
- Diameter

Finding the Area of a Composite Figure

- Students should be able to compute areas of composite figures.
- Remind students to identify the basic figures contained in the composite figure before they consider the area.
- Remind students that to find the area of a composite figure, all they need do is sum the areas of the basic figures.
- **Teaching Tip:** Sometimes students find it helpful to "break up" the composite figure. For instance, rather than working with the composite figure in Example 1, have students draw the triangle separately from the square and mark the dimensions on each figure. Ask students to find the area of each figure and then sum these quantities to determine the area of the composite figure.
- **Common Error:** Students will often think that the problem does not provide enough information to be solved. In Example 1, some students may think they have not been given the base of the triangle. Try to help students see that the basic shapes contained in the figure are just as important as how the shapes fit together. Because the base of the triangle stretches the same length as the top of the square, the base must measure 10 inches.

Finding the Areas of Circles

- Students should be able to compute areas of circles.
- You may wish to review the concept of pi with students. Pi is the ratio of a circle's circumference (perimeter) to its diameter. This ratio is constant regardless of the size of the circle. As a result of its frequent appearance in mathematics, the symbol π is used to represent the ratio. Students should be familiar with using $\frac{22}{7}$ or 3.14 as approximate values of pi.
- **Common Error:** You may want to review the relationship between a circle's diameter and radius before completing Example 3. Students will often substitute a circle's diameter rather than its radius into the formula.

Reteaching and Enrichment Strategies

If students need help. . .	If students got it. . .
Record and Practice Journal • Fair Game Review Skills Review Handbook Lesson Tutorials	Game Closet at *BigIdeasMath.com* Start the next section

What You Learned Before

"I just figured out how to find your volume. We'll immerse you in a barrel of water and measure the water that overflows."

Finding the Area of a Composite Figure (MACC.7.G.2.6)

Example 1 Find the area of the figure.

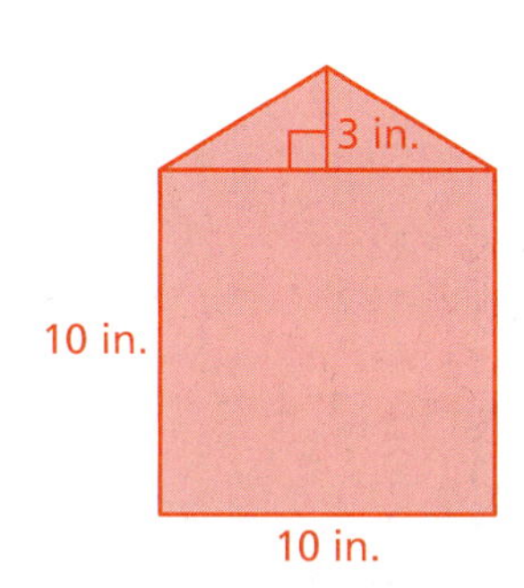

Area = Area of square + Area of triangle

$$A = s^2 + \frac{1}{2}bh$$

$$= 10^2 + \left(\frac{1}{2} \cdot 10 \cdot 3\right)$$

$$= 100 + 15$$

$$= 115 \text{ in.}^2$$

Try It Yourself

Find the area of the figure.

1.

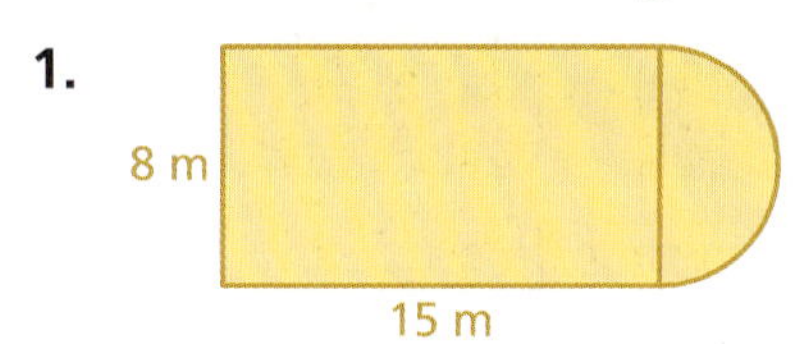

2.

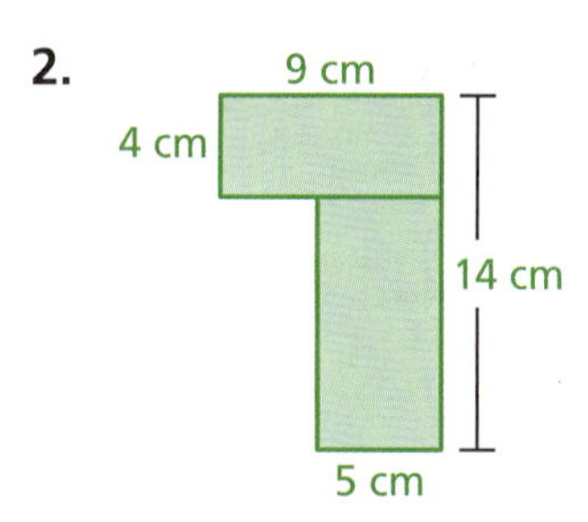

Finding the Areas of Circles (MACC.7.G.2.4)

Example 2 Find the area of the circle.

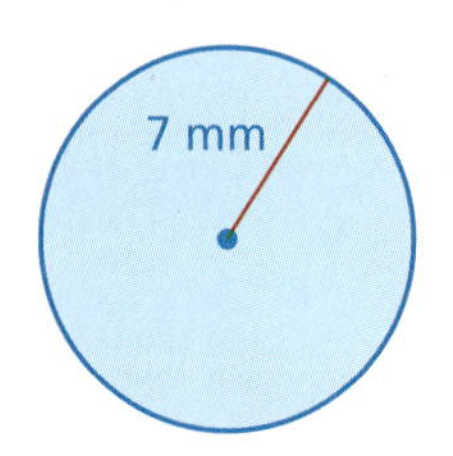

$$A = \pi r^2$$

$$\approx \frac{22}{7} \cdot 7^2$$

$$= \frac{22}{7} \cdot 49$$

$$= 154 \text{ mm}^2$$

Example 3 Find the area of the circle.

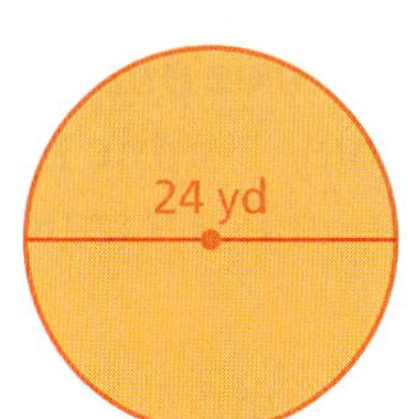

$$A = \pi r^2$$

$$\approx 3.14 \cdot 12^2$$

$$= 3.14 \cdot 144$$

$$= 452.16 \text{ yd}^2$$

Try It Yourself

Find the area of the circle.

3.

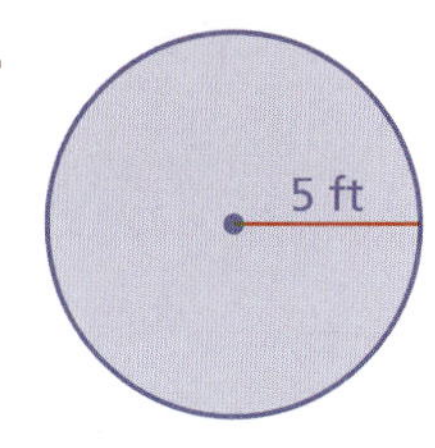

4.

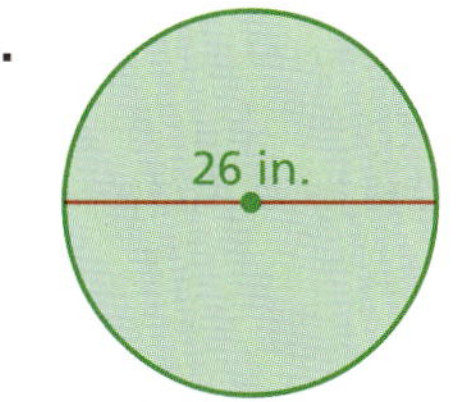

5.

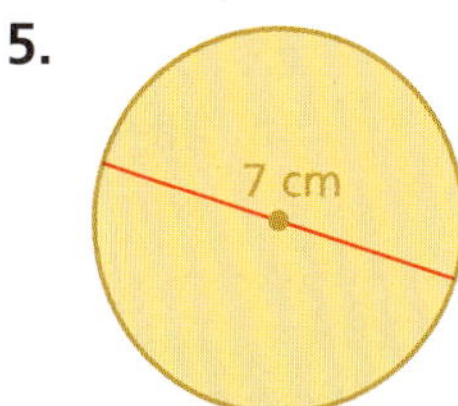

8.1 Volumes of Cylinders

Essential Question How can you find the volume of a cylinder?

1 ACTIVITY: Finding a Formula Experimentally

Work with a partner.

a. Find the area of the face of a coin.

b. Find the volume of a stack of a dozen coins.

c. Write a formula for the volume of a cylinder.

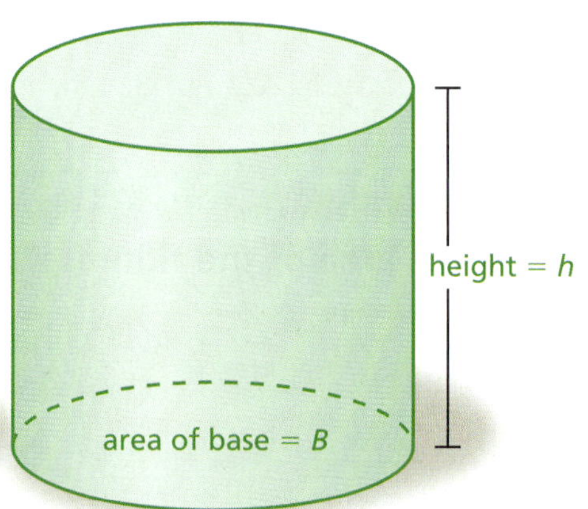

2 ACTIVITY: Making a Business Plan

Work with a partner. You are planning to make and sell three different sizes of cylindrical candles. You buy 1 cubic foot of candle wax for \$20 to make 8 candles of each size.

a. Design the candles. What are the dimensions of each size of candle?

b. You want to make a profit of \$100. Decide on a price for each size of candle.

c. Did you set the prices so that they are proportional to the volume of each size of candle? Why or why not?

Geometry

In this lesson, you will

- find the volumes of cylinders.
- find the heights of cylinders given the volumes.
- solve real-life problems.

Learning Standard
MACC.8.G.3.9

Laurie's Notes

Introduction

Standards for Mathematical Practice

- **MP4 Model with Mathematics:** In these activities, students will model and work with volumes of cylinders. In Activity 1, they will use a layering approach to write a formula for the volume of a cylinder. You want students to make a connection to the formula they learned for volume of a prism, where they found the area of the base and multiplied by the height.

Motivate

- Use round crackers, wafer candies, coins, or circular metal washers to model layers of very thin cylinders being stacked to make thicker cylinders.
- Display the models for students to see.

? "What do these items have in common?" *Sample answer:* They are cylinders made up of thin layers.

- Explain that today they will explore the volume of a cylinder using a similar approach.

Activity Notes

Activity 1

? "How will you find the area of the base?" Area of a circle $= \pi r^2$; Students will likely measure the diameter to the nearest tenth of a centimeter.

- Discuss with students how they found the volume of the stack of 12 coins.
- **Big Idea:** Area is measured in square units. Volume is measured in cubic units.

Activity 2

- **MP1 Make Sense of Problems and Persevere in Solving Them:** This is an open-ended activity that you can adjust to the skill level of your students.
- You can make the following assumptions: all of the candles have the same base with different heights: 2 inches, 3 inches, and 5 inches. Because there are 8 of each, the total height is 80 inches.

? There are 12^3 or 1728 cubic inches of wax available. "What does the radius of the candle need to be to use up the wax?"

- **MP2 Reason Abstractly and Quantitatively:** At this stage, students need to think through how they can find the volume of a cylinder. They need to find the area of the base (πr^2) and multiply by 80, the total height of the 24 candles. With a calculator, they will find that a radius of about 2.6 inches uses 1728 cubic inches of wax.
- Pricing the candles involves proportions. The 3-inch tall candle should cost 50% more than the 2-inch tall candle. The 5-inch tall candle should cost 150% more than the 2-inch tall candle.

Florida Common Core Standards

MACC.8.G.3.9 Know the formulas for the volumes of cones, cylinders, and spheres and use them to solve real-world and mathematical problems.

Previous Learning

Students should know that cylinders are composed of 2 circular bases and a rectangle.

Lesson Plans
Complete Materials List

8.1 Record and Practice Journal

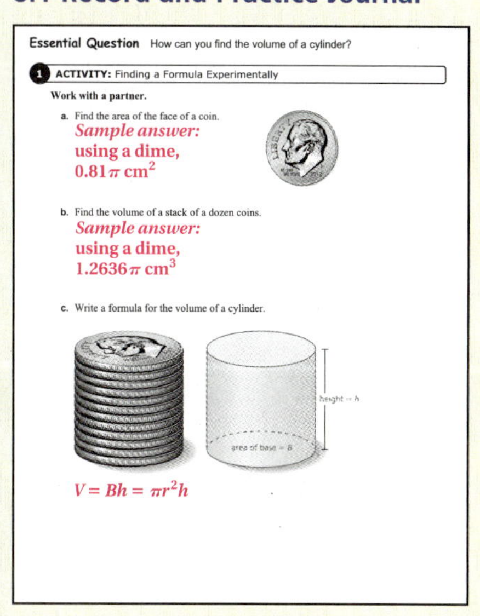
Essential Question How can you find the volume of a cylinder?

1 ACTIVITY: Finding a Formula Experimentally

Work with a partner.

a. Find the area of the face of a coin.
Sample answer: using a dime, 0.81π cm^2

b. Find the volume of a stack of a dozen coins.
Sample answer: using a dime, 1.2636π cm^3

c. Write a formula for the volume of a cylinder.

$V = Bh = \pi r^2 h$

Differentiated Instruction

Money

Have students bring in coins of different denominations from their country of origin. Repeat Activity 1 using the measurements of the coins.

8.1 Record and Practice Journal

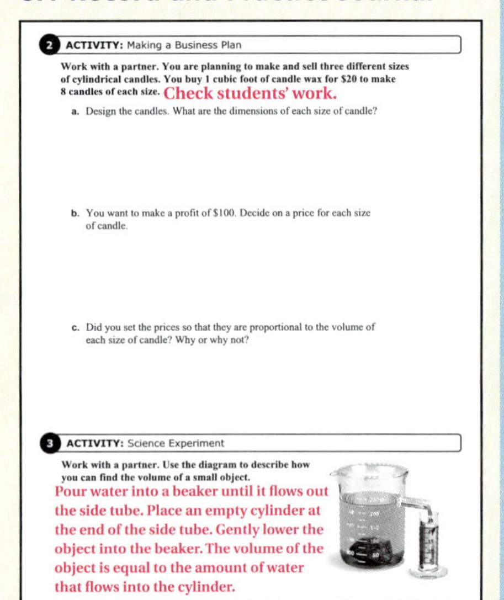

2 ACTIVITY: Making a Business Plan

Work with a partner. You are planning to make and sell three different sizes of cylindrical candles. You buy 1 cubic foot of candle wax for $20 to make 8 candles of each size. Check students' work.

a. Design the candles. What are the dimensions of each size of candle?

b. You want to make a profit of $100. Decide on a price for each size of candle.

c. Did you set the prices so that they are proportional to the volume of each size of candle? Why or why not?

3 ACTIVITY: Science Experiment

Work with a partner. Use the diagram to describe how you can find the volume of a small object.

Pour water into a beaker until it flows out the side tube. Place an empty cylinder at the end of the side tube. Gently lower the object into the beaker. The volume of the object is equal to the amount of water that flows into the cylinder.

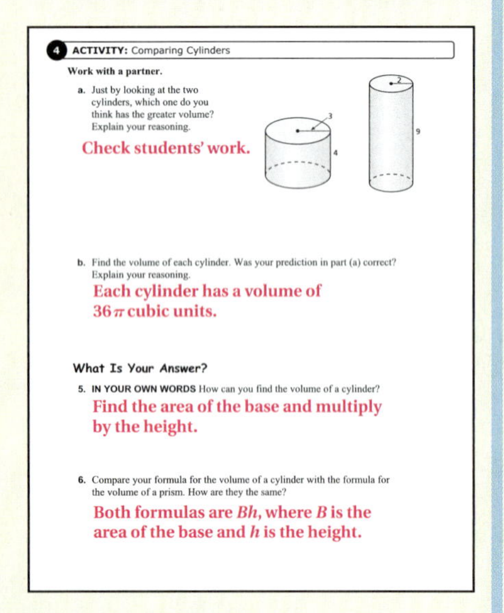

4 ACTIVITY: Comparing Cylinders

Work with a partner.

a. Just by looking at the two cylinders, which one do you think has the greater volume? Explain your reasoning.

Check students' work.

b. Find the volume of each cylinder. Was your prediction in part (a) correct? Explain your reasoning.

Each cylinder has a volume of 36π cubic units.

What Is Your Answer?

5. IN YOUR OWN WORDS How can you find the volume of a cylinder?

Find the area of the base and multiply by the height.

6. Compare your formula for the volume of a cylinder with the formula for the volume of a prism. How are they the same?

Both formulas are Bh, where B is the area of the base and h is the height.

Laurie's Notes

Activity 2 (continued)

- **MP3 Construct Viable Arguments and Critique the Reasoning of Others:** It is important for students to present their results to classmates. How did they make decisions about size and pricing? What assumptions did they make? What calculations did they perform and did they perform them accurately?

Activity 3

- Perhaps students have found volume by displacement in a science class.
- **Note:** This way of measuring volume is similar to the story of Archimedes who yelled "Eureka" when getting into a bathtub of water. He realized that he had displaced a volume of water equal to the volume of his body.
- If possible, have graduated cylinders and a large stone available to model this problem.

Activity 4

- Encourage students to take a guess even if their reasoning is no more than "it looks like it would hold a lot more."

? "What is the area of each base (leave answers in terms of π)?" 9π and 4π

? "So, which is greater, 4 layers of 9π or 9 layers of 4π?" They are the same.

What Is Your Answer?

- **Think-Pair-Share:** Students should read each question independently and then work in pairs to answer the questions. When they have answered the questions, the pair should compare their answers with another group and discuss any discrepancies.

Closure

- Refer to one of the cylinders used to motivate the activity and ask students to describe how they would find the volume.

3 ACTIVITY: Science Experiment

Work with a partner. Use the diagram to describe how you can find the volume of a small object.

4 ACTIVITY: Comparing Cylinders

Math Practice 1

Consider Similar Problems

How can you use the results of Activity 1 to find the volumes of the cylinders?

Work with a partner.

a. Just by looking at the two cylinders, which one do you think has the greater volume? Explain your reasoning.

b. Find the volume of each cylinder. Was your prediction in part (a) correct? Explain your reasoning.

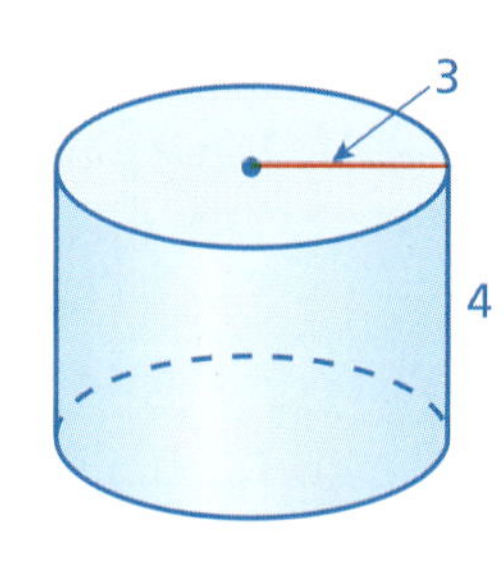

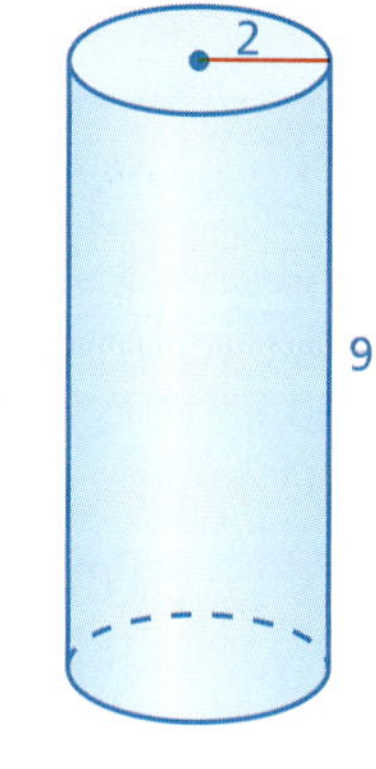

What Is Your Answer?

5. **IN YOUR OWN WORDS** How can you find the volume of a cylinder?

6. Compare your formula for the volume of a cylinder with the formula for the volume of a prism. How are they the same?

"Here's how I remember how to find the volume of any prism or cylinder."

"Base times tall, will fill 'em all."

Use what you learned about the volumes of cylinders to complete Exercises 3–5 on page 338.

Key Idea

Volume of a Cylinder

Words The volume V of a cylinder is the product of the area of the base and the height of the cylinder.

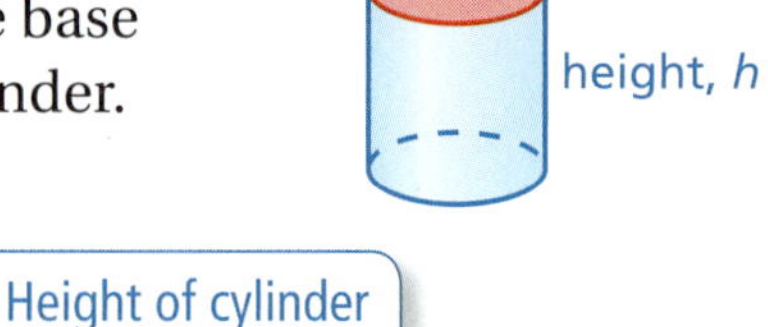

Algebra $V = Bh$

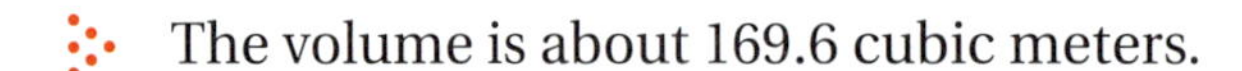

EXAMPLE 1 Finding the Volume of a Cylinder

Find the volume of the cylinder. Round your answer to the nearest tenth.

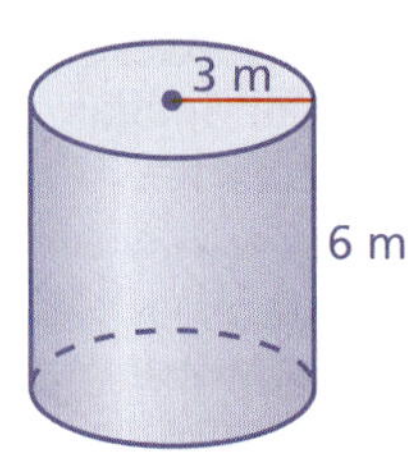

$V = Bh$ Write formula for volume.

$= \pi(3)^2(6)$ Substitute.

$= 54\pi \approx 169.6$ Use a calculator.

The volume is about 169.6 cubic meters.

Study Tip

Because $B = \pi r^2$, you can use $V = \pi r^2 h$ to find the volume of a cylinder.

EXAMPLE 2 Finding the Height of a Cylinder

Find the height of the cylinder. Round your answer to the nearest whole number.

The diameter is 10 inches. So, the radius is 5 inches.

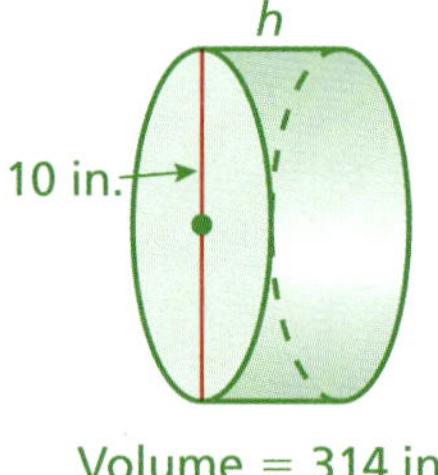

Volume = 314 in.^{3}

$V = Bh$ Write formula for volume.

$314 = \pi(5)^2(h)$ Substitute.

$314 = 25\pi h$ Simplify.

$4 \approx h$ Divide each side by 25π.

The height is about 4 inches.

On Your Own

Now You're Ready
Exercises 3–11 and 13–15

Find the volume V or height h of the cylinder. Round your answer to the nearest tenth.

1.

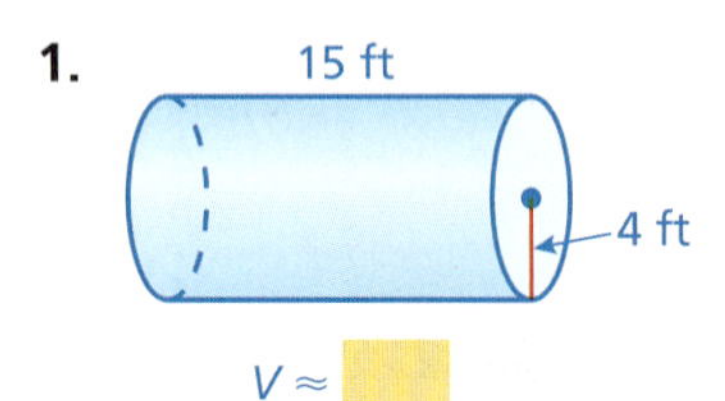

2.

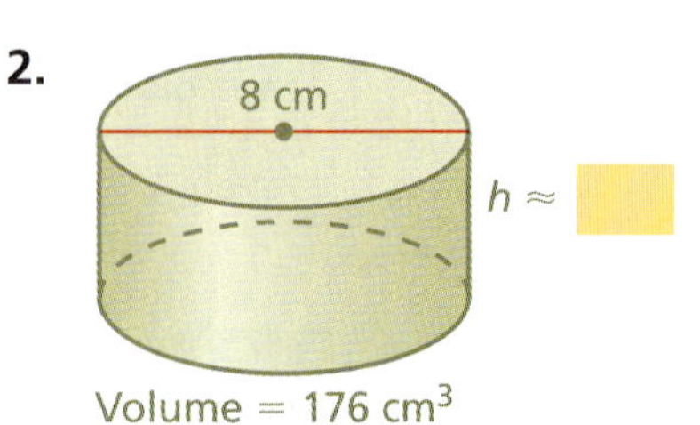

Volume = 176 cm^3

Laurie's Notes

Introduction

Connect

- **Yesterday:** Students discovered how to find the volume of a cylinder by considering the layers that make up a cylinder. (MP1, MP2, MP3, MP4)
- **Today:** Students will work with a formula for the volume of a cylinder.

Motivate

- Hold two different-sized cans for the class to see. I often use cans of whole tomatoes—not only are the dimensions what I want but the contents are the same, so you can do a little cost analysis at the end of class.
- ? "How do the volumes of these two cans compare?" The purpose here is to get students thinking about dimensions, not to do computations.

Lesson Notes

Key Idea

- ? "How are cylinders and rectangular prisms alike?" Each has two congruent bases and a lateral portion.
- ? "How are cylinders and rectangular prisms different?" Cylinders have circular bases, but rectangular prisms have rectangular bases.
- Write the formula in words.
- Before writing the formula in symbols, ask how to find the area of the base.
- Note the use of color to identify the base in the formula and the diagram.

Example 1

- Model good problem solving by writing the formula first.
- Notice that the values of the variables are substituted, simplified, and left in terms of π. The last step is to use the π key on a calculator. If the calculator doesn't have a π key, use 3.14. Note that if 3.14 is used, the answers may not agree exactly with the answers shown in this text.
- ? Write "$\approx$" on the board. "What does this symbol mean?" approximately equal to "Why do you use this symbol here?" π is an irrational number.
- **Extension:** Discuss how big this cylinder is. Its diameter and height (6 meters) are wider than my classroom and more than twice its height.
- **Big Idea:** Volume $= Bh$ is the general formula for both prisms and cylinders. The base of a cylinder is a circle, so the general formula can be rewritten as the specific formula $V = \pi r^2 h$.

Example 2

- **MP7 Look for and Make Use of Structure:** Students may be unsure of how to divide 314 by 25π. One way is to find the product 25π, and then divide 314 by the product. Another way is to divide 314 by π, and then divide that answer by 25.

On Your Own

- **Neighbor Check:** Have students work independently and then have their neighbors check their work. Have students discuss any discrepancies.

Goal Today's lesson is finding the volumes of cylinders.

Lesson Tutorials
Lesson Plans
Answer Presentation Tool

Extra Example 1

Find the volume of a cylinder with a radius of 6 feet and a height of 3 feet. Round your answer to the nearest tenth. $108\pi \approx 339.3 \text{ ft}^3$

Extra Example 2

Find the height of a cylinder with a diameter of 4 yards and a volume of 88 cubic yards. Round your answer to the nearest whole number. $\frac{22}{\pi} \approx 7$ yd

On Your Own

1. $240\pi \approx 754.0 \text{ ft}^3$
2. $\frac{11}{\pi} \approx 3.5$ cm

Extra Example 3

A jelly jar has a radius of 3 centimeters and a height of 8 centimeters. The jelly remaining in the jar has a height of 3 centimeters. How much jelly is missing from the jar? $45\pi \approx 141.4 \text{ cm}^3$

Extra Example 4

About how many gallons of water does the watercooler bottle in Example 4 contain if the bottle is 1.25 feet tall? about 7.4 gal

On Your Own

3. $125\pi \approx 392.7 \text{ cm}^3$
4. about 233,263 gal

English Language Learners

Vocabulary
Discuss the meanings of the words *volume* and *cubic units*. Have students add these words to their notebooks.

Laurie's Notes

Example 3

? "What percent of the salsa is missing and what percent remains?" 60%; 40%

- Work through the problem.
- **Extension:** Find the original volume without using the volume formula. *Hint:* You could use the percent equation or set up a proportion.

Example 4

- It is helpful to have 3 rulers to model what a cubic foot looks like. Hold the 3 rulers so they form 3 edges of a cube that meet at a vertex.

? "About how many gallons do you think would fill a cubic foot?" There will be a range of answers.

- Work through the problem, finding the volume of the watercooler in cubic feet.
- The second part of the problem involves dimensional analysis, a technique used earlier in the text.
- **MP6 Attend to Precision:** Students should be comfortable with the term "conversion factor."
- Estimate first. If $1 \text{ ft}^3 \approx 7.5$ gal, how many gallons would 1.3352 cubic feet be? An estimate of 10 gallons is reasonable and that is choice B.

On Your Own

- Give students sufficient time to do their work before asking volunteers to share their work at the board.

Closure

- Hold the two cans used to motivate the lesson and ask students to find the volume of each. If the contents are the same (or pretend that they are the same), how should the prices compare?

EXAMPLE 3 Real-Life Application

How much salsa is missing from the jar?

The empty space in the jar is a cylinder with a height of $10 - 4 = 6$ centimeters and a radius of 5 centimeters.

$V = Bh$	Write formula for volume.
$= \pi(5)^2(6)$	Substitute.
$= 150\pi \approx 471$	Use a calculator.

So, about 471 cubic centimeters of salsa are missing from the jar.

EXAMPLE 4 Real-Life Application

About how many gallons of water does the watercooler bottle contain? ($1\text{ ft}^3 \approx 7.5$ gal)

Ⓐ 5.3 gallons Ⓑ 10 gallons Ⓒ 17 gallons Ⓓ 40 gallons

Find the volume of the cylinder. The diameter is 1 foot. So, the radius is 0.5 foot.

$V = Bh$	Write formula for volume.
$= \pi(0.5)^2(1.7)$	Substitute.
$= 0.425\pi \approx 1.3352$	Use a calculator.

So, the bottle contains about 1.3352 cubic feet of water. To find the number of gallons it contains, multiply by the conversion factor $\frac{7.5\text{ gal}}{1\text{ ft}^3}$.

$$1.3352\ \cancel{\text{ft}^3} \times \frac{7.5\text{ gal}}{1\ \cancel{\text{ft}^3}} \approx 10\text{ gal}$$

The watercooler bottle contains about 10 gallons of water. So, the correct answer is Ⓑ.

On Your Own

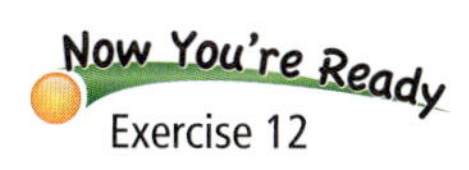

3. **WHAT IF?** In Example 3, the height of the salsa in the jar is 5 centimeters. How much salsa is missing from the jar?

4. A cylindrical water tower has a diameter of 15 meters and a height of 5 meters. About how many gallons of water can the tower contain? ($1\text{ m}^3 \approx 264$ gal)

8.1 Exercises

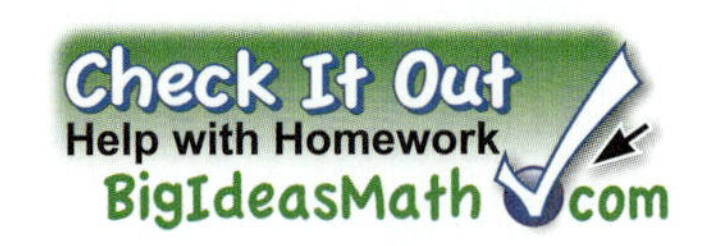

Vocabulary and Concept Check

1. **DIFFERENT WORDS, SAME QUESTION** Which is different? Find "both" answers.

   How much does it take to fill the cylinder?

   What is the capacity of the cylinder?

   How much does it take to cover the cylinder?

   How much does the cylinder contain?

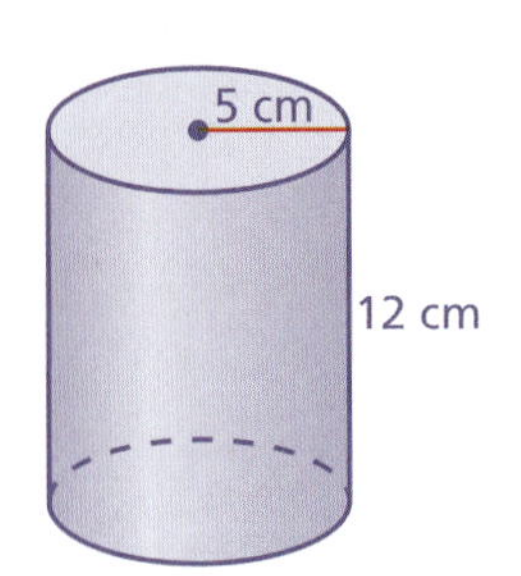

2. **REASONING** Without calculating, which of the solids has the greater volume? Explain.

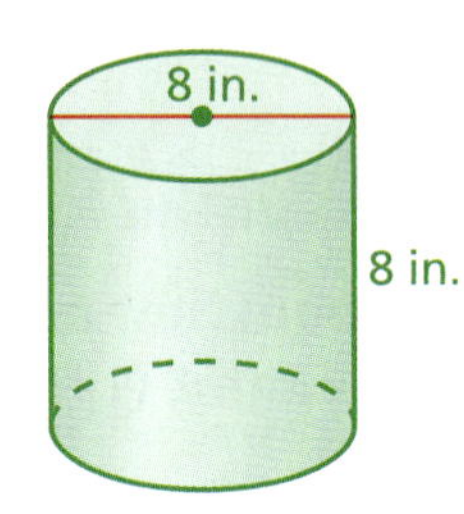

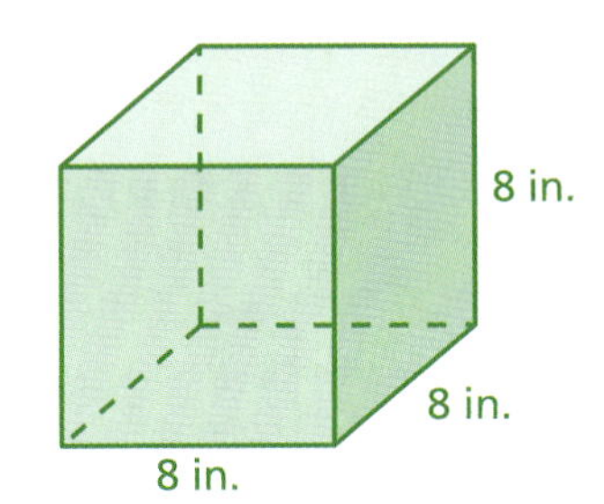

Practice and Problem Solving

Find the volume of the cylinder. Round your answer to the nearest tenth.

1 3.

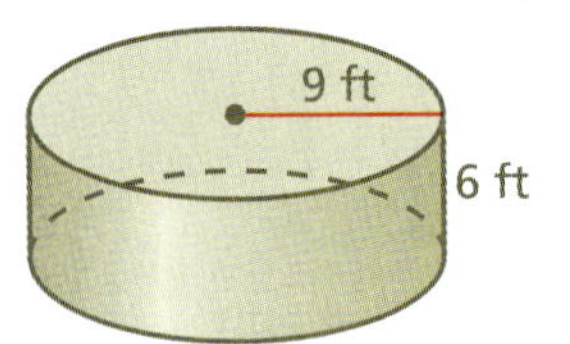

4.

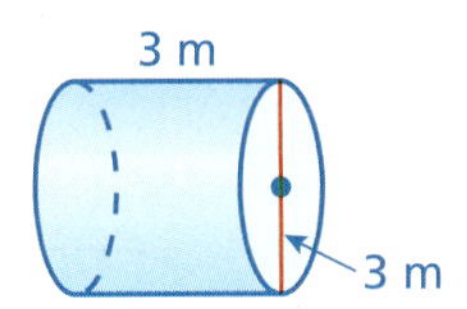

5.

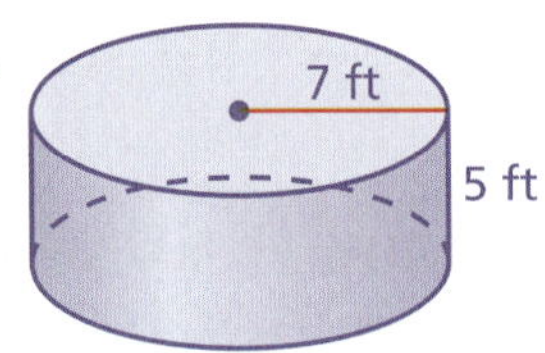

6.

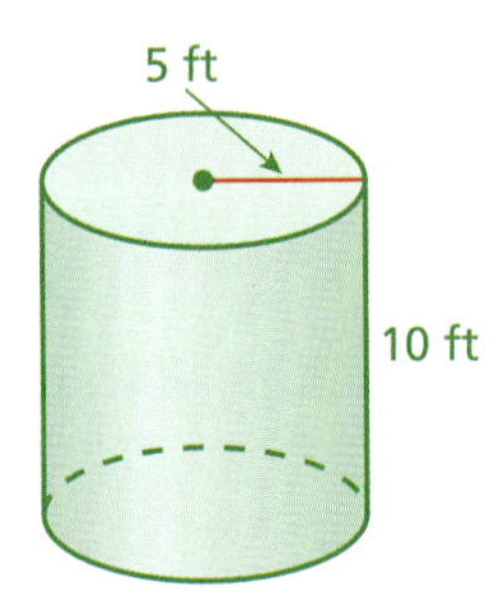

7.

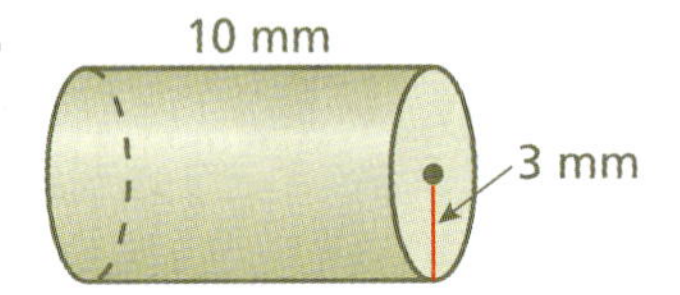

8.

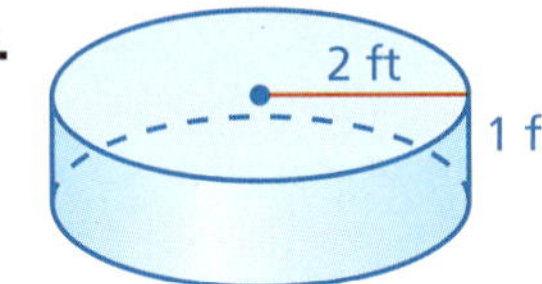

9.

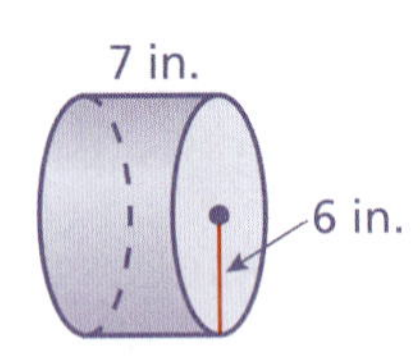

10.

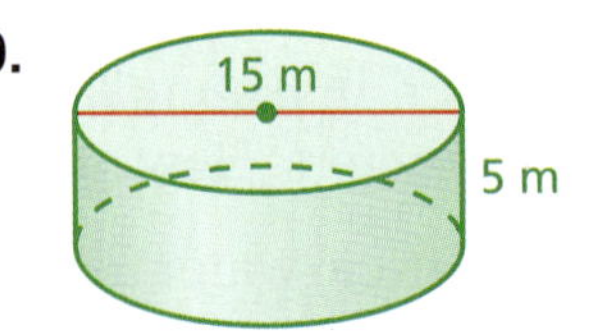

11.

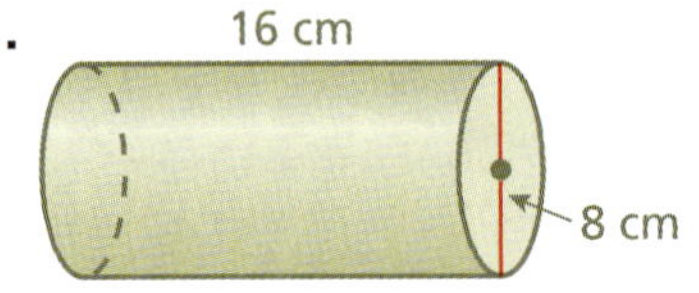

4 12. **SWIMMING POOL** A cylindrical swimming pool has a diameter of 16 feet and a height of 4 feet. About how many gallons of water can the pool contain? Round your answer to the nearest whole number. ($1 \text{ ft}^3 \approx 7.5 \text{ gal}$)

Assignment Guide and Homework Check

Level	Day 1 Activity Assignment	Day 2 Lesson Assignment	Homework Check
Basic	3–5, 21–24	1, 2, 6–12, 13, 15, 17	2, 6, 12, 13
Average	3–5, 21–24	1, 2, 6–11, 13, 15, 17, 18	2, 6, 13, 18
Advanced	1–5, 9–11, 14–24		2, 10, 14, 16

Common Errors

- **Exercises 3–11** Students may forget to square the radius when finding the area of the base. Remind them of the formula for the area of a circle.
- **Exercises 4, 10–14** Students may use the diameter in the formula for the area of a circle that calls for the radius. Encourage them to write the dimensions that they are given before attempting to find the volume. For example, in Exercise 4 a student would write: diameter = 3 m, height = 3 m.
- **Exercise 12** Students may find the volume of the pool, but forget to find how many gallons of water that the pool contains. Encourage them to write the information that they know about the problem and also what they are trying to find. This should help them answer each part of the question.

8.1 Record and Practice Journal

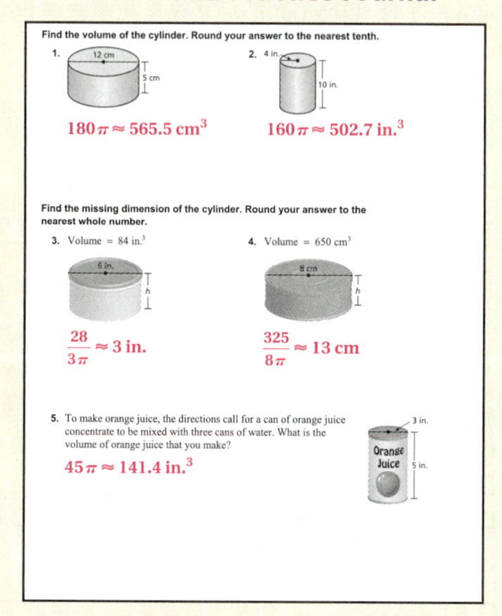

Find the volume of the cylinder. Round your answer to the nearest tenth.

1. (12 cm, 5 cm) $180\pi \approx 565.5 \text{ cm}^3$

2. (4 in., 10 in.) $160\pi \approx 502.7 \text{ in.}^3$

Find the missing dimension of the cylinder. Round your answer to the nearest whole number.

3. Volume = 84 in.³ (6 in., h) $\frac{28}{3\pi} \approx 3$ in.

4. Volume = 650 cm³ (8 cm, h) $\frac{325}{8\pi} \approx 13$ cm

5. To make orange juice, the directions call for a can of orange juice concentrate to be mixed with three cans of water. What is the volume of orange juice that you make? (3 in., 5 in., Orange Juice) $45\pi \approx 141.4 \text{ in.}^3$

Vocabulary and Concept Check

1. How much does it take to cover the cylinder?; $170\pi \approx 534.1 \text{ cm}^2$; $300\pi \approx 942.5 \text{ cm}^3$
2. The cube has a greater volume because the cylinder could fit inside the cube and there is still room in the corners of the cube that are not in the cylinder.

Practice and Problem Solving

3. $486\pi \approx 1526.8 \text{ ft}^3$
4. $\frac{27}{4}\pi \approx 21.2 \text{ m}^3$
5. $245\pi \approx 769.7 \text{ ft}^3$
6. $250\pi \approx 785.4 \text{ ft}^3$
7. $90\pi \approx 282.7 \text{ mm}^3$
8. $4\pi \approx 12.6 \text{ ft}^3$
9. $252\pi \approx 791.7 \text{ in.}^3$
10. $\frac{1125}{4}\pi \approx 883.6 \text{ m}^3$
11. $256\pi \approx 804.2 \text{ cm}^3$
12. about 6032 gal

Practice and Problem Solving

13. $\frac{125}{8\pi} \approx 5$ ft

14. $\frac{625}{16} \approx 39$ in.

15. $\sqrt{\frac{150{,}000}{19\pi}} \approx 50$ cm

16. The volume is $\frac{1}{4}$ of the original volume. Because the diameter is halved, the radius is also halved.

    So, $V = \pi\left(\frac{r}{2}\right)^2 h = \frac{1}{4}\pi r^2 h$.

17. See *Taking Math Deeper.*

18. about 4712 lb

19. $8325 - 729\pi \approx 6035$ m^3

20. **a.** $384\pi \approx 1206.37$ in.^{3}

    b. about 14.22 in.

    c. about 19 min

Fair Game Review

21. yes	**22.** yes
23. no	**24.** C

Mini-Assessment

Find the volume of the cylinder. Round your answer to the nearest tenth.

1.

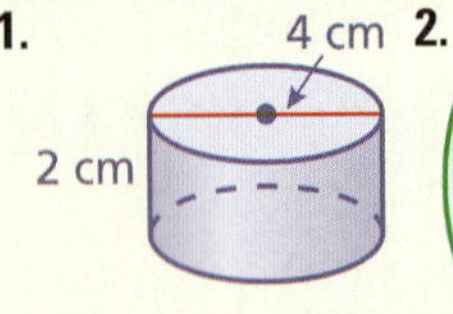

$8\pi \approx 25.1$ cm^3

2.

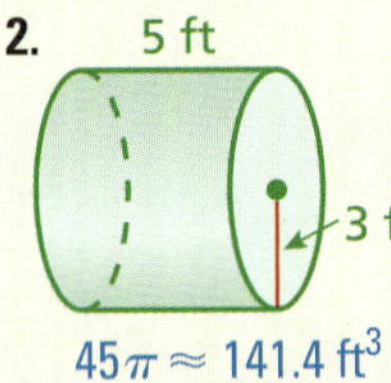

$45\pi \approx 141.4$ ft^3

3. Find the volume of the can of beans. Round your answer to the nearest whole number.

$\frac{81}{8}\pi \approx 32$ in.^{3}

Taking Math Deeper

Exercise 17

When driving through farm country, you can sometimes see hay and straw fields with large round bales. You seldom see the smaller *square bales*, but they are still used. The reason you don't see them is that they are usually moved to a storage shed as soon as they are baled.

1. Find the volume of the round bale.

$V = \pi r^2 h$
$= \pi \cdot 2^2 \cdot 5$
≈ 62.8 ft^3

2. Find the volume of the square bale.

$V = \ell wh$
$= 2 \cdot 2 \cdot 4$
$= 16$ ft^3

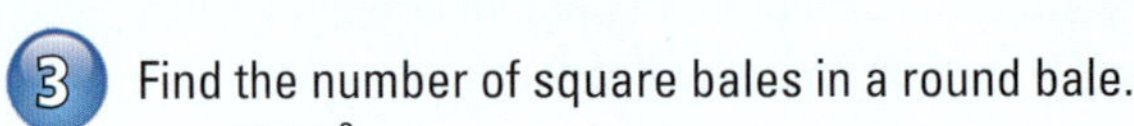

3. Find the number of square bales in a round bale.

$\frac{62.8 \text{ ft}^3}{16 \text{ ft}^3} = 3.925$

There are about 4 square bales in a round bale.

Both square and round bales vary in size. Suppose the square bale is only 18 in. by 18 in. by 3 ft. Its volume would be 6.75 cubic feet and there would be about 9.3 square bales in a round bale.

Project

In Germany, some farmers use large rolls of hay to make Mr. and Mrs. Hay people—like snow people. Draw a poster of two hay people. Assuming the rolls are the size of those in Exercise 17, how much hay would be needed to make your *people*?

Reteaching and Enrichment Strategies

If students need help. . .	If students got it. . .
Resources by Chapter • Practice A and Practice B • Puzzle Time Record and Practice Journal Practice Differentiating the Lesson Lesson Tutorials Skills Review Handbook	Resources by Chapter • Enrichment and Extension • Technology Connection Start the next section

Find the missing dimension of the cylinder. Round your answer to the nearest whole number.

2 **13.** Volume = 250 ft^3

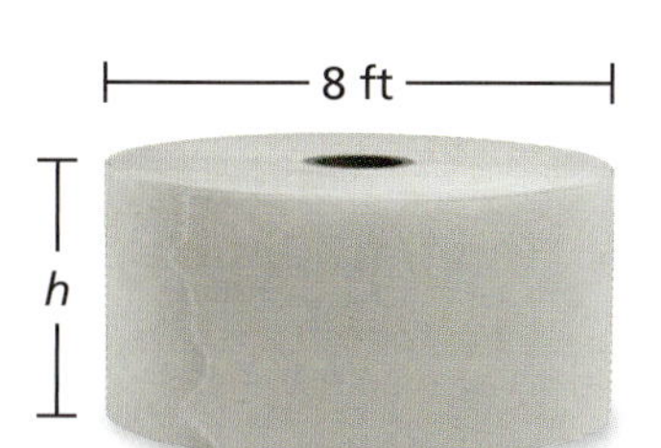

14. Volume = $10{,}000\pi$ $in.^3$

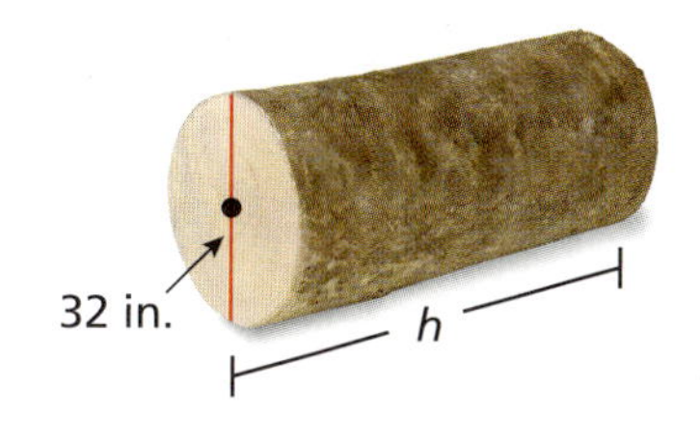

15. Volume = 600,000 cm^3

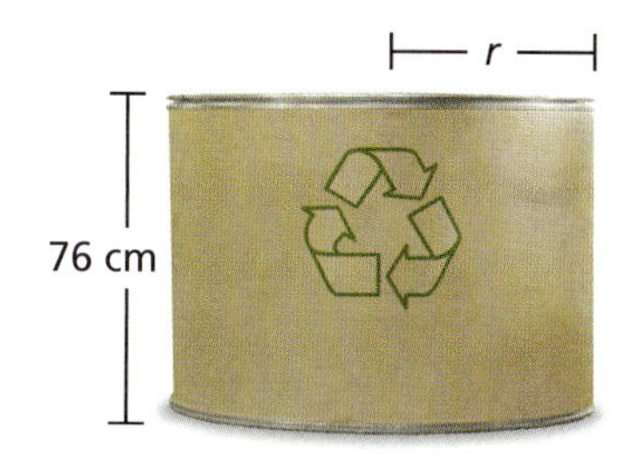

16. CRITICAL THINKING How does the volume of a cylinder change when its diameter is halved? Explain.

Round hay bale

17. MODELING A traditional "square" bale of hay is actually in the shape of a rectangular prism. Its dimensions are 2 feet by 2 feet by 4 feet. How many square bales contain the same amount of hay as one large "round" bale?

18. ROAD ROLLER A tank on a road roller is filled with water to make the roller heavy. The tank is a cylinder that has a height of 6 feet and a radius of 2 feet. One cubic foot of water weighs 62.5 pounds. Find the weight of the water in the tank.

19. VOLUME A cylinder has a surface area of 1850 square meters and a radius of 9 meters. Estimate the volume of the cylinder to the nearest whole number.

20. Problem Solving Water flows at 2 feet per second through a pipe with a diameter of 8 inches. A cylindrical tank with a diameter of 15 feet and a height of 6 feet collects the water.

a. What is the volume, in cubic inches, of water flowing out of the pipe every second?

b. What is the height, in inches, of the water in the tank after 5 minutes?

c. How many minutes will it take to fill 75% of the tank?

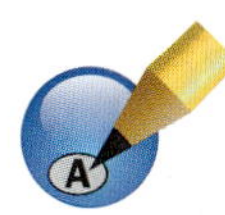

Fair Game Review What you learned in previous grades & lessons

Tell whether the triangle with the given side lengths is a right triangle. *(Section 7.5)*

21. 20 m, 21 m, 29 m

22. 1 in., 2.4 in., 2.6 in.

23. 5.6 ft, 8 ft, 10.6 ft

24. MULTIPLE CHOICE Which ordered pair is the solution of the linear system $3x + 4y = -10$ and $2x - 4y = 0$? *(Section 5.3)*

Ⓐ $(-6, 2)$ Ⓑ $(2, -6)$ Ⓒ $(-2, -1)$ Ⓓ $(-1, -2)$

8.2 Volumes of Cones

Essential Question How can you find the volume of a cone?

You already know how the volume of a pyramid relates to the volume of a prism. In this activity, you will discover how the volume of a cone relates to the volume of a cylinder.

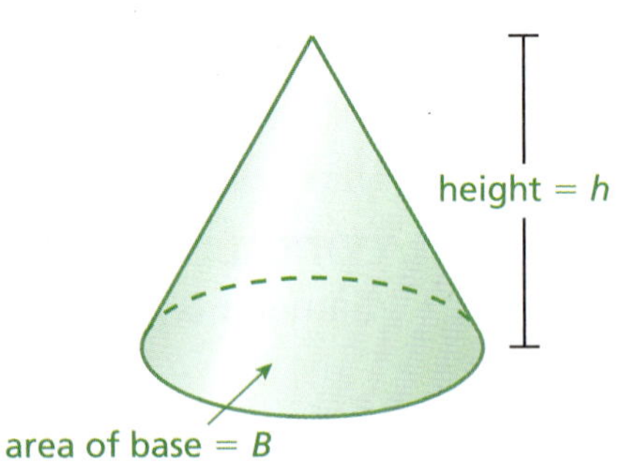

ACTIVITY: Finding a Formula Experimentally

Work with a partner. Use a paper cup that is shaped like a cone.

- Estimate the height of the cup.
- Trace the top of the cup on a piece of paper. Find the diameter of the circle.
- Use these measurements to draw a net for a cylinder with the same base and height as the paper cup.
- Cut out the net. Then fold and tape it to form an open cylinder.
- Fill the paper cup with rice. Then pour the rice into the cylinder. Repeat this until the cylinder is full. How many cones does it take to fill the cylinder?
- Use your result to write a formula for the volume of a cone.

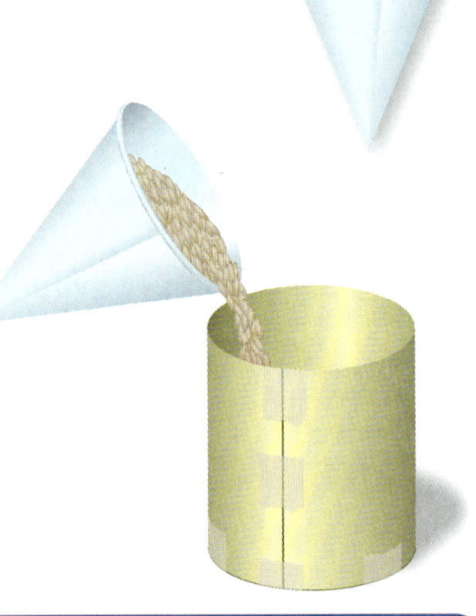

2 ACTIVITY: Summarizing Volume Formulas

Work with a partner. You can remember the volume formulas for prisms, cylinders, pyramids, and cones with just two concepts.

Volumes of Prisms and Cylinders

Volume = Area of base × 

Volumes of Pyramids and Cones

Volume = ☐ Volume of prism or cylinder with same base and height

Make a list of all the formulas you need to remember to find the area of a base. Talk about strategies for remembering these formulas.

Geometry

In this lesson, you will

- find the volumes of cones.
- find the heights of cones given the volumes.
- solve real-life problems.

Learning Standard
MACC.8.G.3.9

Laurie's Notes

Introduction

Standards for Mathematical Practice

- **MP4 Model with Mathematics:** In the first activity, students will explore the relationship between the volume of a cone and the volume of a cylinder. Modeling such relationships is a powerful teaching tool.
- **Big Idea:** It is not a big stretch for students to accept that the volume relationship between the cone and cylinder is the same as the volume relationship between the pyramid and prism.

  Volume of a Cylinder = (Area of Base)(Height)

  Volume of a Cone = $\frac{1}{3}$(Area of Base)(Height)

Motivate

- Give pairs of students two minutes to look around the classroom and make a list of all the geometric solids they see that are prisms, cylinders, pyramids, or cones.
- Make a column on the board for each type of solid. Ask one pair of students to list an item in each column. Continue to have pairs of students add to the lists, but only items that are not in the lists already.
- Was every group able to list 4 new items? If your classroom is like most, there are fewer pyramids and cones than prisms and cylinders.

Activity Notes

Activity 1

- **Management Tip:** Instead of having the whole class perform this activity, you may want to have a pair of students demonstrate it. If you have a cone and a cylinder with the same base and height from a commercially-available (fillable) geometric solids kit, then you could use these instead of having students make them. Or, you could show a video demonstration, such as the one at http://www.youtube.com/watch?v=QnVr_x7c79w.
- **MP4 Model with Mathematics:** Regardless, you want students to *see* the relationship and not just hear what it is.
- If necessary, have materials ready for students: heavy-weight paper, tape, scissors, ruler, and uncooked rice.
- **Common Error:** Students may measure the slant height of the cone. Point out that the height of a cone is shown in the diagram.
- The net students make for the cylinder will most likely have two bases. If so, then they can just tear off one of the bases to get the open cylinder.
- Have students guess the relationship before they see it. If they recall the relationship between the volume of a pyramid and the volume of a prism with the same base and height, then they may guess correctly.

? "How many times does it take to fill the cylinder with the contents of the cone?" three

? "How does the volume of the cone compare with the volume of the cylinder?" The volume of the cone is $\frac{1}{3}$ the volume of the cylinder.

Florida Common Core Standards

MACC.8.G.3.9 Know the formulas for the volumes of cones, cylinders, and spheres and use them to solve real-world and mathematical problems.

Previous Learning

Students should know how to find the surface area of a cone.

Lesson Plans
Complete Materials List

8.2 Record and Practice Journal

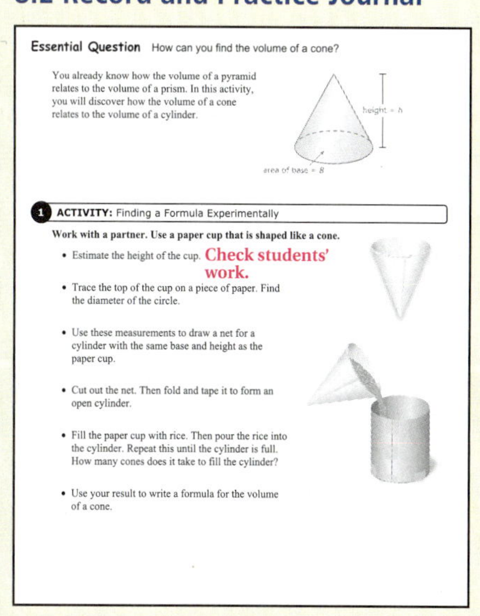

Essential Question How can you find the volume of a cone?

You already know how the volume of a pyramid relates to the volume of a prism. In this activity, you will discover how the volume of a cone relates to the volume of a cylinder.

1 **ACTIVITY:** Finding a Formula Experimentally

Work with a partner. Use a paper cup that is shaped like a cone.

- Estimate the height of the cup. Check students' work.
- Trace the top of the cup on a piece of paper. Find the diameter of the circle.
- Use these measurements to draw a net for a cylinder with the same base and height as the paper cup.
- Cut out the net. Then fold and tape it to form an open cylinder.
- Fill the paper cup with rice. Then pour the rice into the cylinder. Repeat this until the cylinder is full. How many cones does it take to fill the cylinder?
- Use your result to write a formula for the volume of a cone.

English Language Learners

Visual

Show students models of a pyramid and a prism. Then show them models of a cone and a cylinder. Have students note that the relationship between the volume of a cone and the volume of a cylinder with the same base and height is the same as the relationship between the volume of a pyramid and the volume of a prism with the same base and height.

8.2 Record and Practice Journal

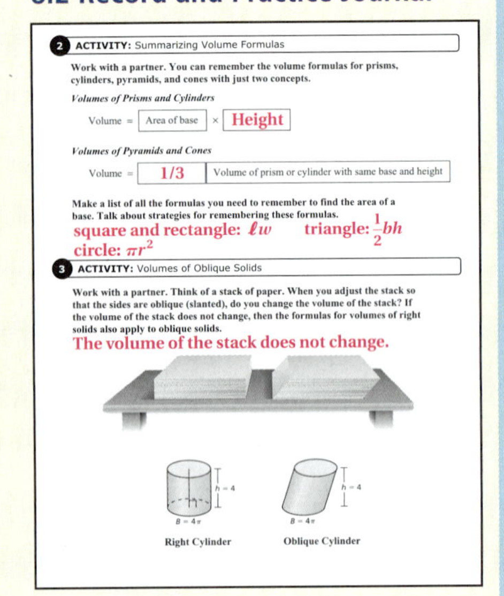

2 ACTIVITY: Summarizing Volume Formulas

Work with a partner. You can remember the volume formulas for prisms, cylinders, pyramids, and cones with just two concepts.

Volumes of Prisms and Cylinders

Volume = Area of base × **Height**

Volumes of Pyramids and Cones

Volume = **1/3** Volume of prism or cylinder with same base and height

Make a list of all the formulas you need to remember to find the area of a base. Talk about strategies for remembering these formulas.

square and rectangle: ℓw triangle: $\frac{1}{2}bh$
circle: πr^2

3 ACTIVITY: Volumes of Oblique Solids

Work with a partner. Think of a stack of paper. When you adjust the stack so that the sides are oblique (slanted), do you change the volume of the stack? If the volume of the stack does not change, then the formulas for volumes of right solids also apply to oblique solids.

The volume of the stack does not change.

$B = 4\pi$, $h = 4$ Right Cylinder

$B = 4\pi$, $h = 4$ Oblique Cylinder

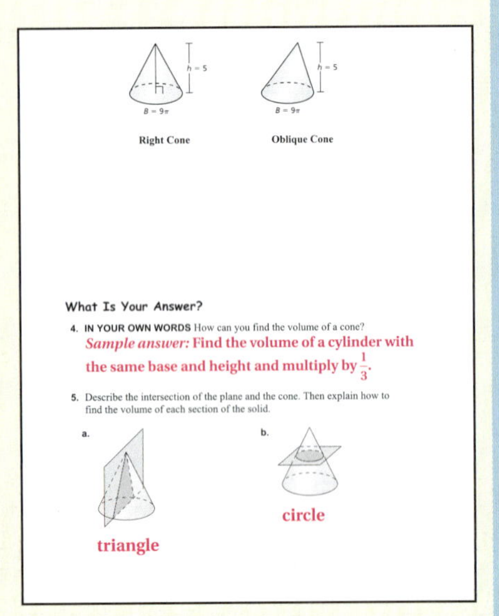

$B = 9\pi$, $h = 5$ Right Cone

$B = 9\pi$, $h = 5$ Oblique Cone

What Is Your Answer?

4. **IN YOUR OWN WORDS** How can you find the volume of a cone?
   Sample answer: **Find the volume of a cylinder with the same base and height and multiply by $\frac{1}{3}$.**

5. Describe the intersection of the plane and the cone. Then explain how to find the volume of each section of the solid.

   a. **triangle**

   b. **circle**

Laurie's Notes

Activity 2

- **MP2 Reason Abstractly and Quantitatively:** You do not want students to think that it is necessary to memorize a lot of formulas. Instead, students need to consider the structure of the shape. Prisms and cylinders have the same structure (two congruent bases and a lateral portion) and pyramids and cones have the same structure (one base and a lateral portion that contains a vertex). There is one general formula for each pair of solids. Moreover, the two general formulas have a 1 : 3 relationship.
- **Teaching Tip:** Make a poster of each of the two volume formulas for your classroom, or ask a student to make them.

? "How many general volume formulas are there?" two

? "In each formula, you need to find the area of a base. What types of bases have you studied?" Most were squares, rectangles, triangles, or circles.

- Have pairs of students share their lists and strategies. Collect information at the board.

Activity 3

- It is helpful to have a stack of paper or a deck of cards to model this activity.
- Give students time to discuss their thinking and then have them share their thoughts with the rest of the class.
- **Common Misconception:** Students may believe that the height decreases as the stack of paper is slanted to one side. The thickness doesn't change, so the height remains the same. The volume remains the same because no sheets are removed.

? "What is changing in this problem?" Some students may recognize that the surface area is changing because more area is being exposed.

What Is Your Answer?

- Have students work in pairs.

Closure

- Sketch a cube with edge length 2 centimeters, and a cylinder with height and diameter each 2 centimeters. Compare the volumes of the cube and cylinder.
- Sketch a cylinder and a cone, each with height and diameter of 3 centimeters. Compare the volumes of the cylinder and cone.

3 ACTIVITY: Volumes of Oblique Solids

Work with a partner. Think of a stack of paper. When you adjust the stack so that the sides are oblique (slanted), do you change the volume of the stack? If the volume of the stack does not change, then the formulas for volumes of right solids also apply to oblique solids.

Math Practice 2

Use Equations

What equation would you use to find the volume of the oblique solid? Explain.

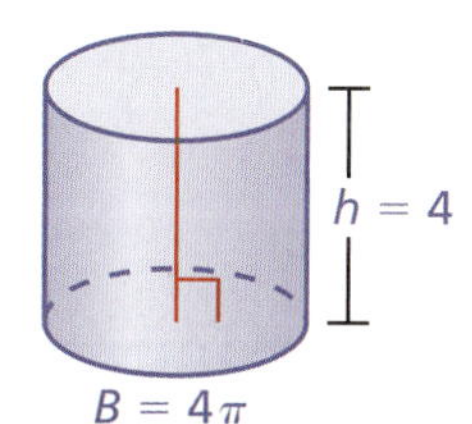

Right cylinder

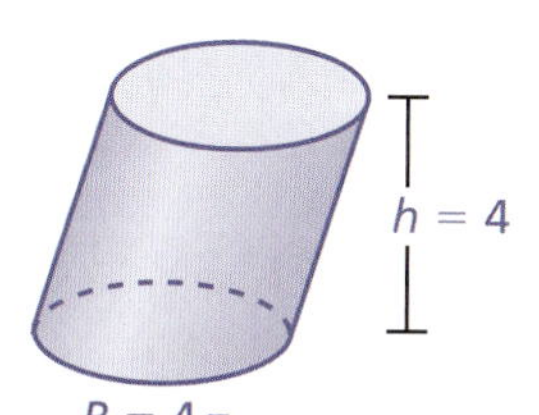

Oblique cylinder

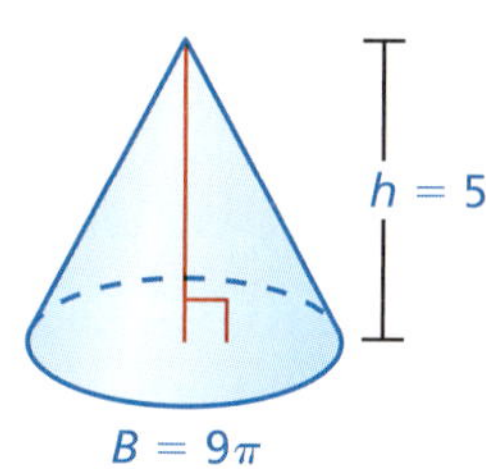

Right cone

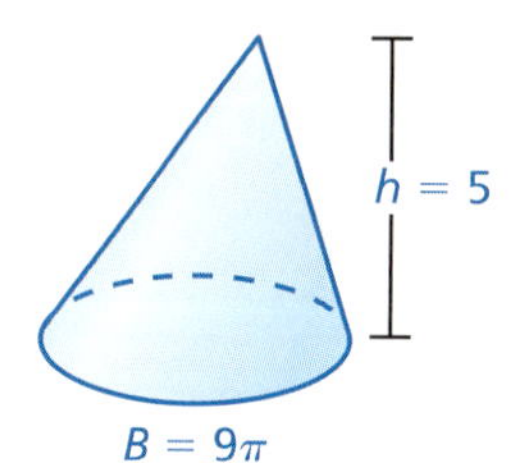

Oblique cone

What Is Your Answer?

4. **IN YOUR OWN WORDS** How can you find the volume of a cone?

5. Describe the intersection of the plane and the cone. Then explain how to find the volume of each section of the solid.

a.

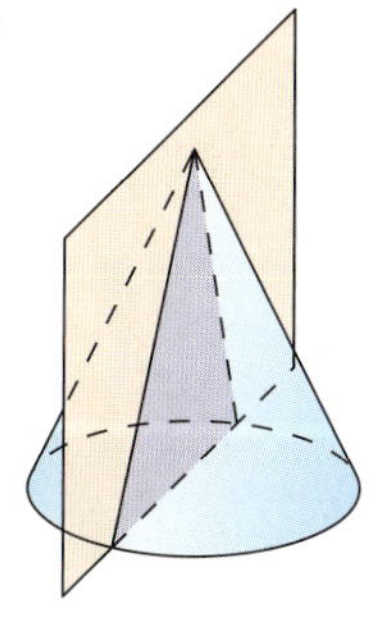

b.

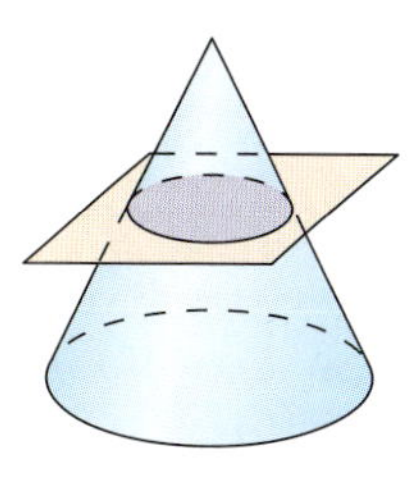

Practice Use what you learned about the volumes of cones to complete Exercises 4–6 on page 344.

8.2 Lesson

Study Tip

The *height* of a cone is the perpendicular distance from the base to the vertex.

Volume of a Cone

Words The volume V of a cone is one-third the product of the area of the base and the height of the cone.

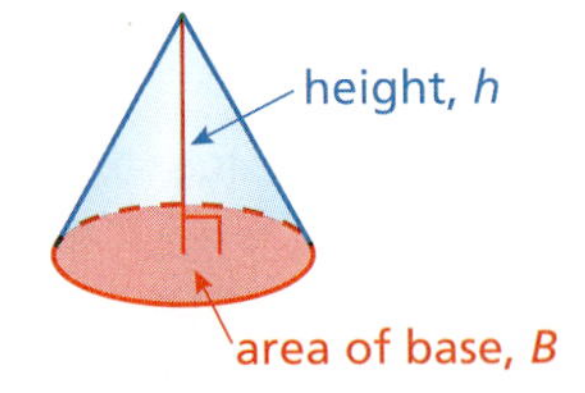

Algebra $V = \frac{1}{3}Bh$

Area of base

Height of cone

EXAMPLE 1 Finding the Volume of a Cone

Study Tip

Because $B = \pi r^2$, you can use $V = \frac{1}{3}\pi r^2 h$ to find the volume of a cone.

Find the volume of the cone. Round your answer to the nearest tenth.

The diameter is 4 meters. So, the radius is 2 meters.

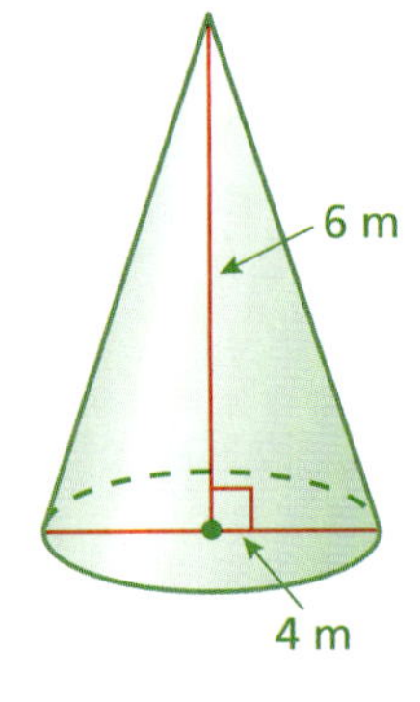

$V = \frac{1}{3}Bh$	Write formula for volume.
$= \frac{1}{3}\pi(2)^2(6)$	Substitute.
$= 8\pi \approx 25.1$	Use a calculator.

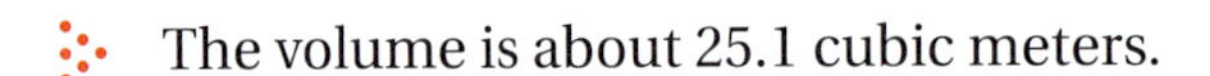

The volume is about 25.1 cubic meters.

EXAMPLE 2 Finding the Height of a Cone

Find the height of the cone. Round your answer to the nearest tenth.

$V = \frac{1}{3}Bh$	Write formula for volume.
$956 = \frac{1}{3}\pi(9)^2(h)$	Substitute.
$956 = 27\pi h$	Simplify.
$11.3 \approx h$	Divide each side by 27π.

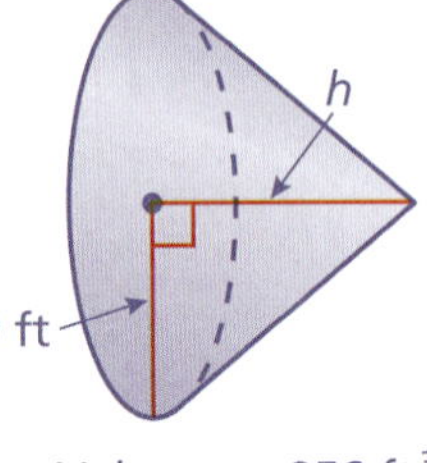

Volume = 956 ft^3

The height is about 11.3 feet.

Laurie's Notes

Introduction

Connect

- **Yesterday:** Students developed a strategy to summarize volume and surface area formulas. (MP2, MP4)
- **Today:** Students will work with the formula for the volume of a cone.

Motivate

- Bring an ice cream cone and ice cream scoop to class.
- **FYI:** An ice cream scoop with a radius of 1 inch will make a round (spherical) scoop of ice cream that is a little more than 4 cubic inches. Share this information with students. You may also want to mention that students will study volumes of spheres in the next section.

? "If I place a scoop of ice cream (with a 1 inch radius) on this cone, and the ice cream melts (because I received a phone call), will the ice cream overflow the cone?" This question is posed only to get students thinking about the volume of a cone. The volume of the cone will be found at the end of the lesson.

Lesson Notes

Key Idea

- Write the Key Idea.
- Write the formula in words. Draw the cone with the dimensions labeled.
- Write the symbolic formula.

? "What shape is the base?" circle "How do you find its area A?" $A = \pi r^2$

Example 1

- Notice that the work is done in terms of π. It is not until the last step that you use the π key on a calculator.
- **Representation:** Encourage students to use the parentheses to represent multiplication. Using the $\times$ symbol would make the expression confusing.
- **Common Misconception:** Remind students that π is a number and because multiplication is commutative and associative, this expression could be rewritten as $\frac{1}{3}(6)(2)^2\pi$, making the computation less confusing.

? "What is being squared in this expression?" only the 2

Example 2

- This example requires students to solve an equation for a variable.
- Work through the problem, annotating the steps as shown in the book.

? "How does $\frac{1}{3}\pi(9)^2h$ equal $27\pi h$?" Only the 9 is being squared, which is 81. One-third of 81 is 27. The order of the factors doesn't matter.

- **MP7 Look for and Make Use of Structure:** Students may have difficulty with the last step, dividing by 27π. It can be done in two steps—divide by 27 then divide by π. Or, divide 956 by the product 27π, which is about 84.82.

Goal Today's lesson is finding the volume of a cone.

Lesson Tutorials
Lesson Plans
Answer Presentation Tool

Extra Example 1

Find the volume of a cone with a diameter of 6 feet and a height of 3 feet. Round your answer to the nearest tenth. $9\pi \approx 28.3$ ft^3

Extra Example 2

Find the height of a cone with a radius of 6 yards and a volume of 75 cubic yards. Round your answer to the nearest whole number. $\frac{25}{4\pi} \approx 2$ yd

On Your Own

1. $180\pi \approx 565.5 \text{ cm}^3$

2. $\frac{96}{\pi} \approx 30.6 \text{ yd}$

Extra Example 3

In Example 3, the height of the sand is 36 millimeters and the radius is 15 millimeters. The sand falls at a rate of 150 cubic millimeters per second. How much time do you have to answer the question? about 57 sec

On Your Own

3. about 42 sec

4. about 6 sec

Differentiated Instruction

Organization

Some students might benefit from first finding the area of the base B of the cone. Then they can substitute this value into the formula, $V = \frac{1}{3}Bh$.

Laurie's Notes

On Your Own

- Ask volunteers to share their work at the board.

Example 3

- If you have a timer of this type, use it as a model.
- Ask a volunteer to read the problem. Ask for ideas as to how the problem can be solved.

? "How long is 30 millimeters?" 30 millimeters is equal to 3 centimeters, which is a little more than 1 inch. This helps students form a visual image of the actual size of the sand timer.

- **Teaching Tip**: Again, explain that $\frac{1}{3}(24)$ is a whole number. Then multiply $8(10)^2 = 800$.
- Be sure to use units in labeling answers. Dimensional analysis shows that the answer will have units of seconds.
- **Extension:** I have a sand timer in my classroom. Students calculate the volume, measure the amount of time it takes to fall to the bottom, and use this information to calculate the rate at which the sand is falling.

On Your Own

- **Extension:** Question 4 is a preview of an upcoming lesson. The height and radius have each been decreased by a factor of 2. (They are $\frac{1}{2}$ the original dimensions). What happens to the volume? It is decreased by a factor of 8, or 2^3.

Closure

- **Exit Ticket:** Have students find the volume of the ice cream cone used to motivate the lesson.

On Your Own

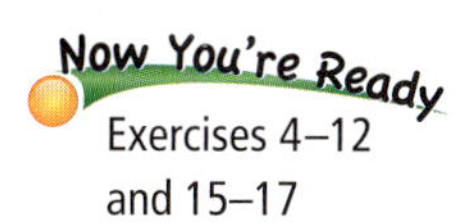

Exercises 4–12 and 15–17

Find the volume V or height h of the cone. Round your answer to the nearest tenth.

1. 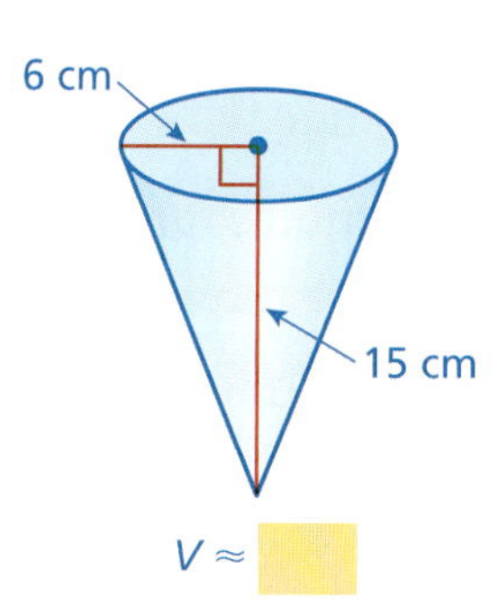

2. 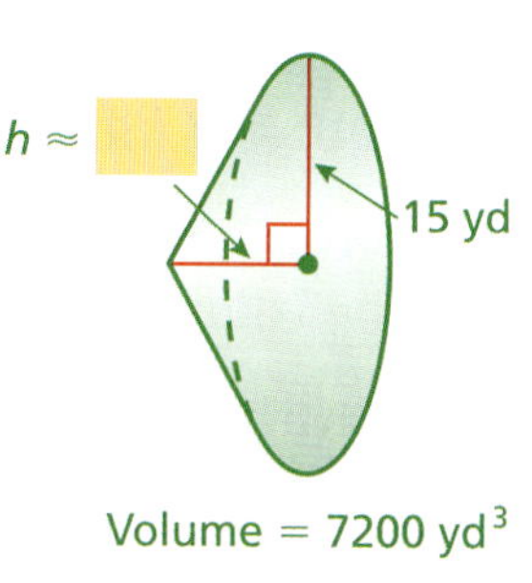

EXAMPLE 3 Real-Life Application

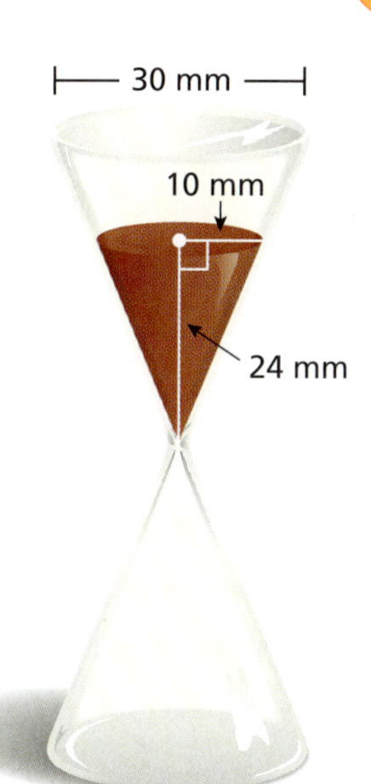

You must answer a trivia question before the sand in the timer falls to the bottom. The sand falls at a rate of 50 cubic millimeters per second. How much time do you have to answer the question?

Use the formula for the volume of a cone to find the volume of the sand in the timer.

$$V = \frac{1}{3}Bh \qquad \text{Write formula for volume.}$$

$$= \frac{1}{3}\pi(10)^2(24) \qquad \text{Substitute.}$$

$$= 800\pi \approx 2513 \qquad \text{Use a calculator.}$$

The volume of the sand is about 2513 cubic millimeters. To find the amount of time you have to answer the question, multiply the volume by the rate at which the sand falls.

$$2513 \cancel{\text{mm}^3} \times \frac{1 \text{ sec}}{50 \cancel{\text{mm}^3}} = 50.26 \text{ sec}$$

So, you have about 50 seconds to answer the question.

On Your Own

3. **WHAT IF?** The sand falls at a rate of 60 cubic millimeters per second. How much time do you have to answer the question?

4. **WHAT IF?** The height of the sand in the timer is 12 millimeters, and the radius is 5 millimeters. How much time do you have to answer the question?

8.2 Exercises

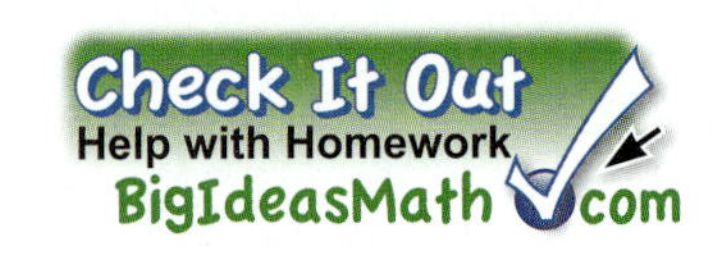

Vocabulary and Concept Check

1. **VOCABULARY** Describe the height of a cone.
2. **WRITING** Compare and contrast the formulas for the volume of a pyramid and the volume of a cone.
3. **REASONING** You know the volume of a cylinder. How can you find the volume of a cone with the same base and height?

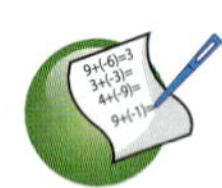

Practice and Problem Solving

Find the volume of the cone. Round your answer to the nearest tenth.

1 4. 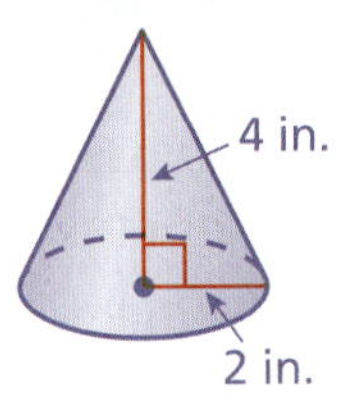

5. 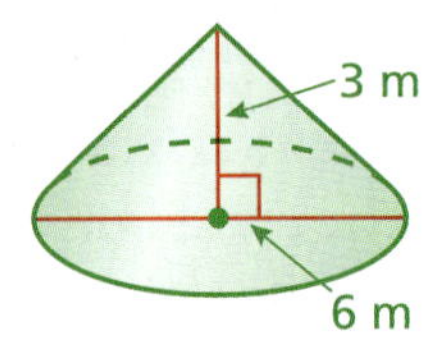

6. 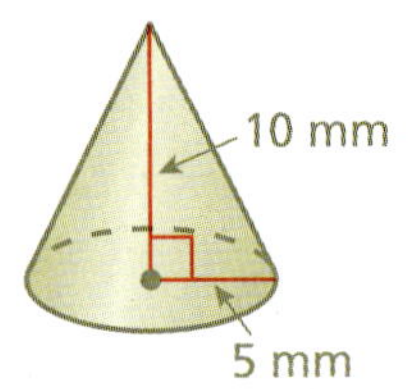

7. 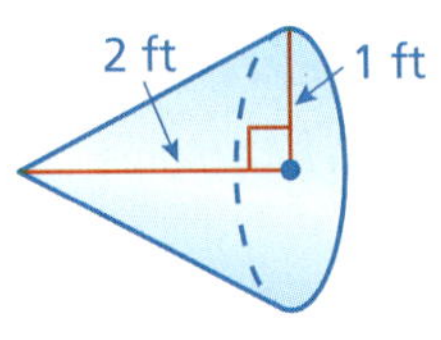

8. 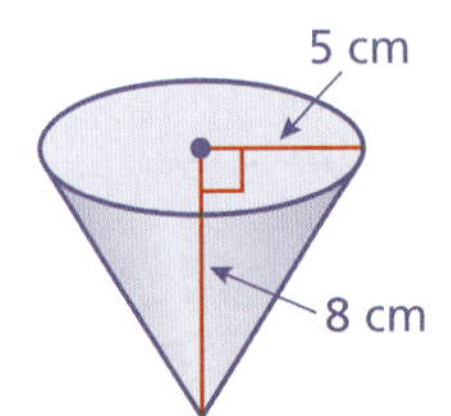

9. 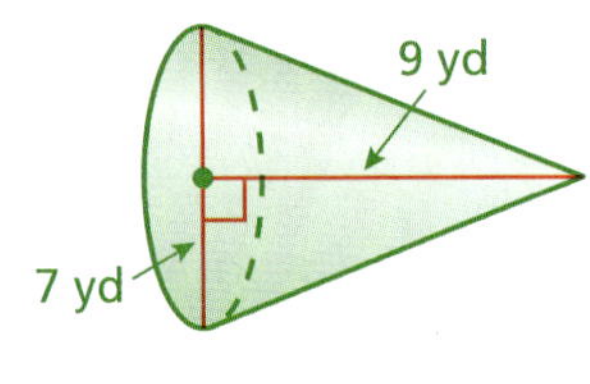

10. 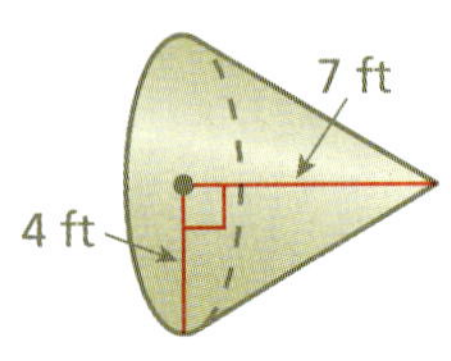

11. 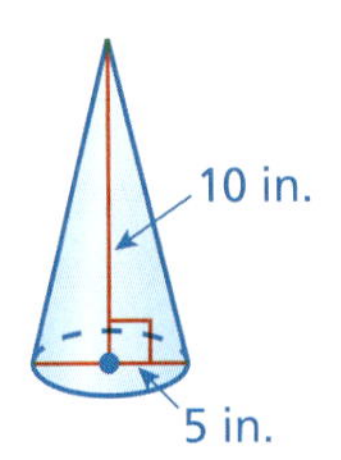

12. 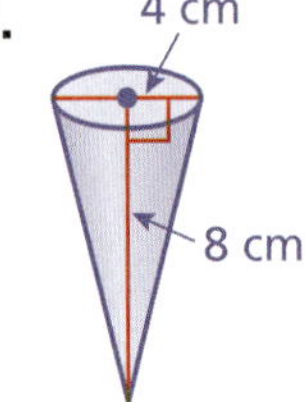

13. **ERROR ANALYSIS** Describe and correct the error in finding the volume of the cone.

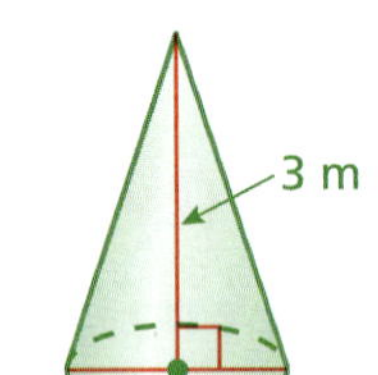

✗
$$V = \frac{1}{3}Bh$$
$$= \frac{1}{3}(\pi)(2)^2(3)$$
$$= 4\pi \text{ m}^3$$

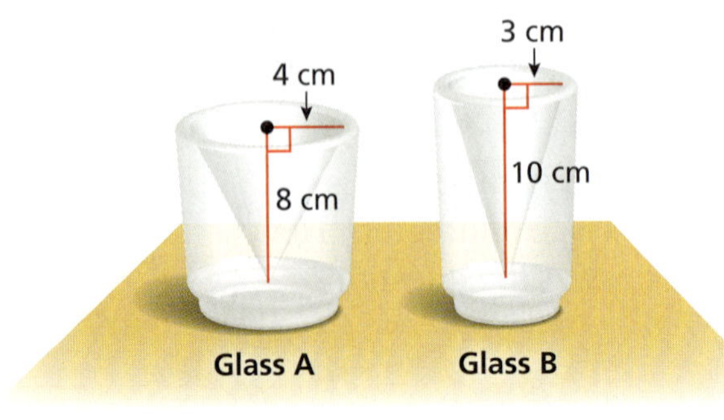

14. **GLASS** The inside of each glass is shaped like a cone. Which glass can hold more liquid? How much more?

Assignment Guide and Homework Check

Level	Day 1 Activity Assignment	Day 2 Lesson Assignment	Homework Check
Basic	4–6, 23–25	1–3, 7–17 odd, 14	2, 7, 14, 15
Average	4–6, 23–25	1–3, 7–19 odd, 20	2, 7, 15, 19
Advanced	1–6, 13, 16–25		2, 16, 18, 21

For Your Information

- **Exercise 15** Because the volume is given in terms of π, students should not substitute for π.

Common Errors

- **Exercises 4–12** Students may write linear or square units for volume rather than cubic units. Remind them that part of writing a correct answer is including the correct units.
- **Exercises 4–12** When finding the area of the base, students may not square the radius, or they may use the diameter when the formula calls for the radius. Remind them of the formula that they learned for the area of a circle.
- **Exercises 15 and 16** Students may try to use the Distributive Property before solving for h. For example, in Exercise 16, a student may incorrectly write $225 = \frac{1}{3}\pi(5^2) \cdot \frac{1}{3}h$. Remind them that factors are multiplied.
- **Exercise 17** The solution of this exercise has many parts that can be confusing to students. It may be helpful to go over it together in class.

8.2 Record and Practice Journal

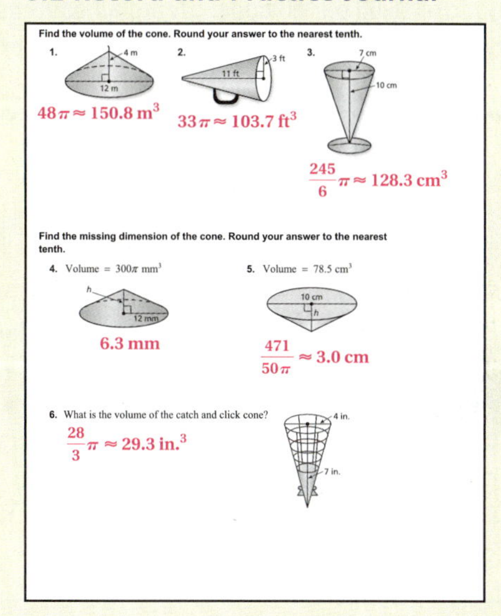
Find the volume of the cone. Round your answer to the nearest tenth.

1. $48\pi \approx 150.8 \text{ m}^3$

2. $33\pi \approx 103.7 \text{ ft}^3$

3. $\frac{245}{6}\pi \approx 128.3 \text{ cm}^3$

Find the missing dimension of the cone. Round your answer to the nearest tenth.

4. Volume = $300\pi \text{ mm}^3$

   6.3 mm

5. Volume = 78.5 cm^3

   $\frac{471}{50\pi} \approx 3.0 \text{ cm}$

6. What is the volume of the catch and click cone?

   $\frac{28}{3}\pi \approx 29.3 \text{ in.}^3$

Vocabulary and Concept Check

1. The height of a cone is the perpendicular distance from the base to the vertex.
2. Both formulas are $\frac{1}{3}Bh$, but the base of a cone is always a circle.
3. Divide by 3.

Practice and Problem Solving

4. $\frac{16\pi}{3} \approx 16.8 \text{ in.}^3$
5. $9\pi \approx 28.3 \text{ m}^3$
6. $\frac{250\pi}{3} \approx 261.8 \text{ mm}^3$
7. $\frac{2\pi}{3} \approx 2.1 \text{ ft}^3$
8. $\frac{200\pi}{3} \approx 209.4 \text{ cm}^3$
9. $\frac{147\pi}{4} \approx 115.5 \text{ yd}^3$
10. $\frac{112\pi}{3} \approx 117.3 \text{ ft}^3$
11. $\frac{125\pi}{6} \approx 65.4 \text{ in.}^3$
12. $\frac{32\pi}{3} \approx 33.5 \text{ cm}^3$
13. The diameter was used instead of the radius; $V = \frac{1}{3}(\pi)(1)^2(3) = \pi \text{ m}^3$
14. Glass A; $\frac{38\pi}{3} \approx 39.8 \text{ cm}^3$

Practice and Problem Solving

15. 1.5 ft

16. $\frac{27}{\pi} \approx 8.6$ cm

17. $2\sqrt{\frac{10.8}{4.2\pi}} \approx 1.8$ in.

18. 60π m^3

19. 24.1 min

20. See *Taking Math Deeper.*

21. $3y$

22. about 98 seconds

Fair Game Review

23. $A'(-1, 1), B'(-3, 4), C'(-1, 4)$

24. $E'(4, -1), F'(3, -3), G'(2, -3), H'(1, -1)$

25. D

Mini-Assessment

Find the volume of the cone. Round your answer to the nearest tenth.

1.

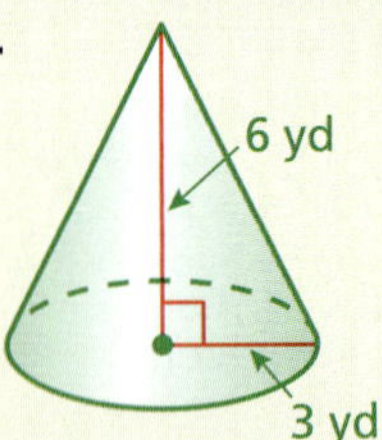

$18\pi \approx 56.5$ yd^3

2.

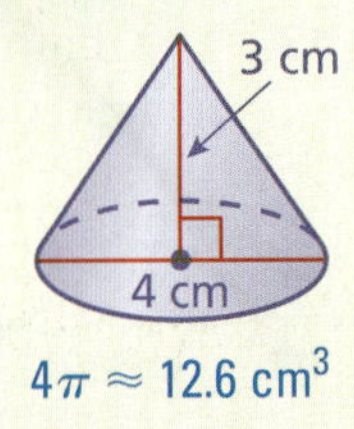

$4\pi \approx 12.6$ cm^3

3. The volume of the ice cream cone is 4.71 cubic inches. Find the height of the cone.

2 in.

$\frac{14.13}{\pi} \approx 4.5$ in.

Taking Math Deeper

Exercise 20

This is a great type of problem to help students understand the importance of *planning ahead.* Also, in planning, remind students that you can't plan *exactly* how many cups will be used, nor can you plan how full each cup will be. So, the answers to the questions are just "ball park" figures.

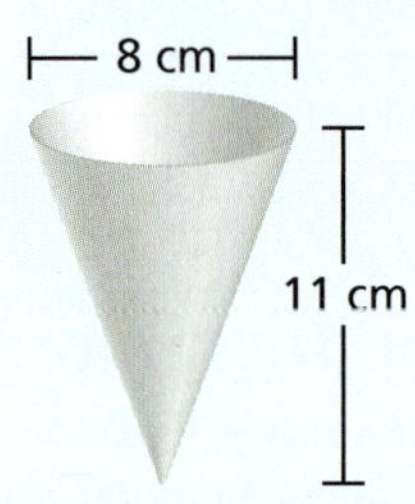

How many paper cups will you need?

$$\text{Volume of Cup} = \frac{1}{3}\pi(4)^2(11) \approx 184.3 \text{ cm}^3$$

$$\text{Amount of Lemonade} = (10 \text{ gal})\left(3785 \frac{\text{cm}^3}{\text{gal}}\right) = 37{,}850 \text{ cm}^3$$

$$\text{Number of Cups} \approx \frac{37{,}850}{184.3} \approx 205.4$$

a. You need about 206 cups.

Think Outside the Box

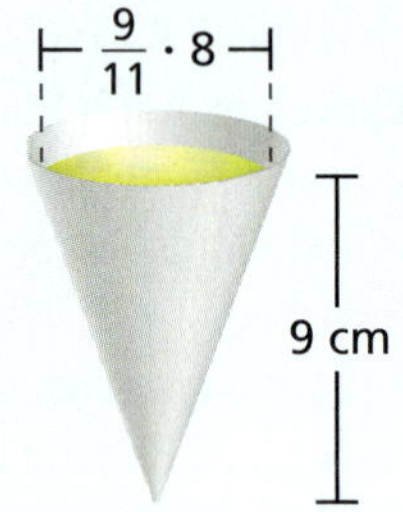

How many packs of 50 cups?

b. You should order 5 packs of 50 cups. This will give you 250 cups.

Suppose each cup is not filled to the brim, but only to a height of 9 centimeters. This could mean each cup has a volume of 101 cubic centimeters, which would imply that you would use about 375 cups. . . so 5 packs would *not* be enough.

How many cups are left over if you sell only 80% of the lemonade? 80% of 37,850 = 30,280 cm^3

$$\text{Number of Cups} \approx \frac{30{,}280}{184.3} \approx 164.3$$

c. You would have about 250 − 165 = 85 cups left over.

Sell 80%

Project

You open a lemonade stand. The lemonade costs you $5.00 per gallon and cups are $6.00 per 50 cups. Create an advertisement including the price of your lemonade. How did you determine the price to charge customers?

Reteaching and Enrichment Strategies

If students need help. . .	If students got it. . .
Resources by Chapter • Practice A and Practice B • Puzzle Time Record and Practice Journal Practice Differentiating the Lesson Lesson Tutorials Skills Review Handbook	Resources by Chapter • Enrichment and Extension • Technology Connection Start the next section

Find the missing dimension of the cone. Round your answer to the nearest tenth.

2 **15.** Volume $= \frac{1}{18}\pi$ ft^3

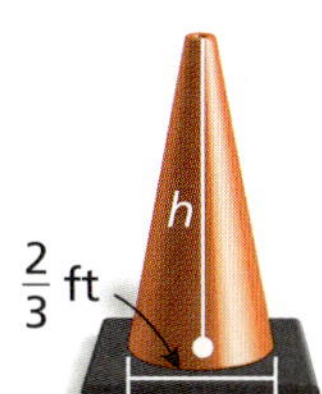

16. Volume $= 225$ cm^3

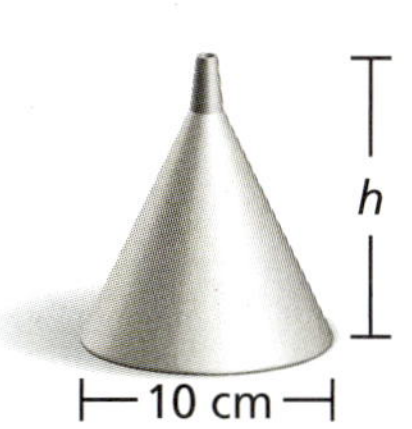

17. Volume $= 3.6$ in.^{3}

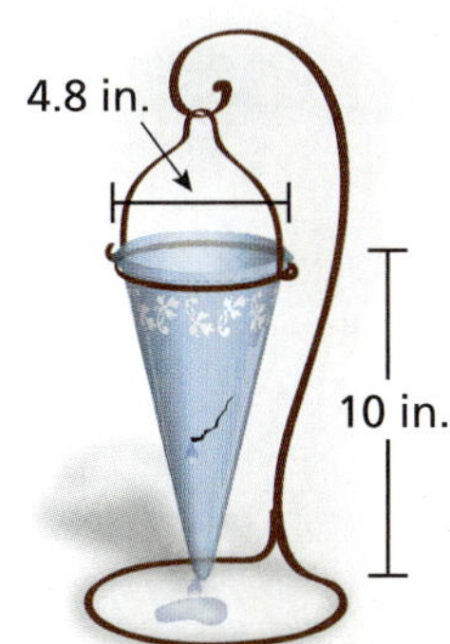

18. REASONING The volume of a cone is 20π cubic meters. What is the volume of a cylinder with the same base and height?

19. VASE Water leaks from a crack in a vase at a rate of 0.5 cubic inch per minute. How long does it take for 20% of the water to leak from a full vase?

20. LEMONADE STAND You have 10 gallons of lemonade to sell. (1 gal $\approx$ 3785 cm^3)

a. Each customer uses one paper cup. How many paper cups will you need?

b. The cups are sold in packages of 50. How many packages should you buy?

c. How many cups will be left over if you sell 80% of the lemonade?

21. STRUCTURE The cylinder and the cone have the same volume. What is the height of the cone?

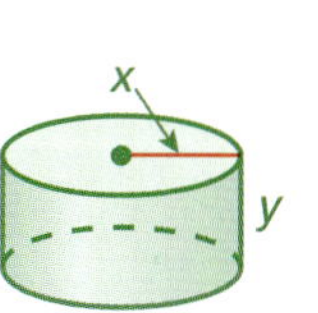

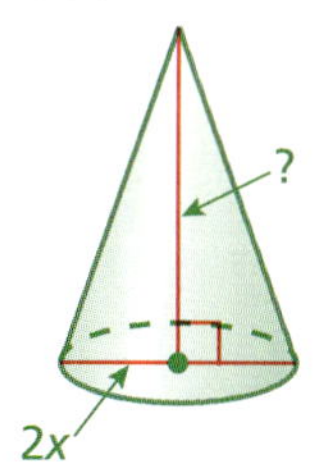

22. 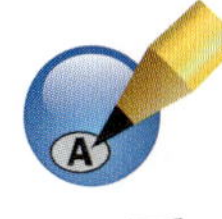 In Example 3, you use a different timer with the same dimensions. The sand in this timer has a height of 30 millimeters. How much time do you have to answer the question?

Fair Game Review What you learned in previous grades & lessons

The vertices of a figure are given. Rotate the figure as described. Find the coordinates of the image. *(Section 2.4)*

23. $A(-1, 1)$, $B(2, 3)$, $C(2, 1)$
90° counterclockwise about vertex A

24. $E(-4, 1)$, $F(-3, 3)$, $G(-2, 3)$, $H(-1, 1)$
180° about the origin

25. MULTIPLE CHOICE $\triangle ABC \sim \triangle XYZ$ by a scale factor of 3. How many times greater is the area of $\triangle XYZ$ than the area of $\triangle ABC$? *(Section 2.6)*

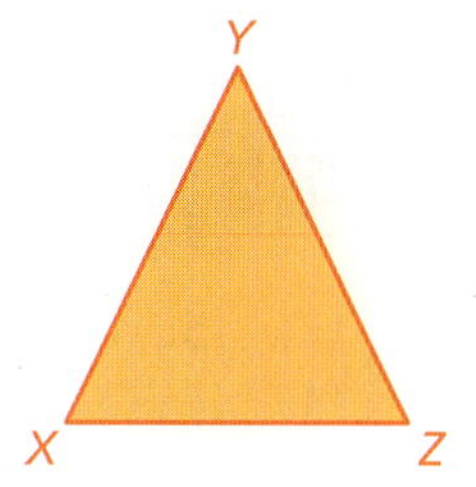

Ⓐ $\frac{1}{9}$ Ⓑ $\frac{1}{3}$

Ⓒ 3   Ⓓ 9

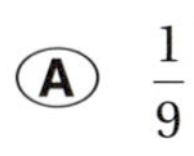

8 Study Help

Check It Out
Graphic Organizer
BigIdeasMath.com

You can use a **formula triangle** to arrange variables and operations of a formula. Here is an example of a formula triangle for the volume of a cylinder.

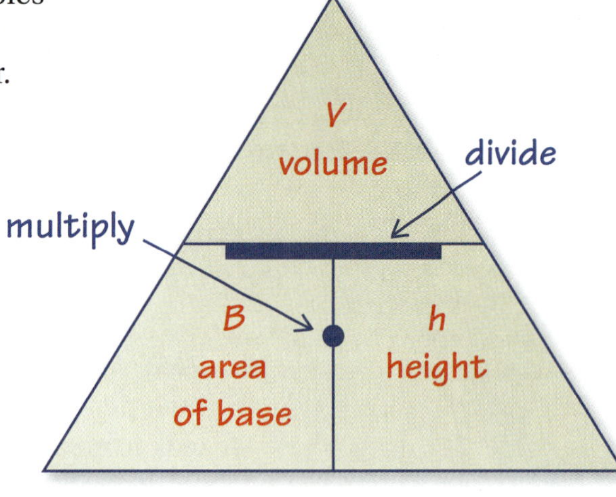

To find an unknown variable, use the other variables and the operation between them. For example, to find the area B of the base, cover up the B. Then you can see that you divide the volume V by the height h.

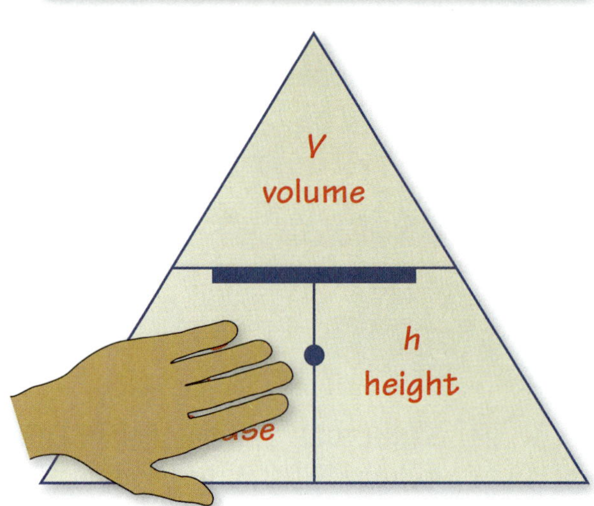

On Your Own

Make a formula triangle to help you study this topic. (*Hint:* Your formula triangle may have a different form than what is shown in the example.)

1. volume of a cone

After you complete this chapter, make formula triangles for the following topics.

2. volume of a sphere
3. volume of a composite solid
4. surface areas of similar solids
5. volumes of similar solids

"See how a formula triangle works? Cover any variable and you get its formula."

Sample Answer

1. Volume of a cone

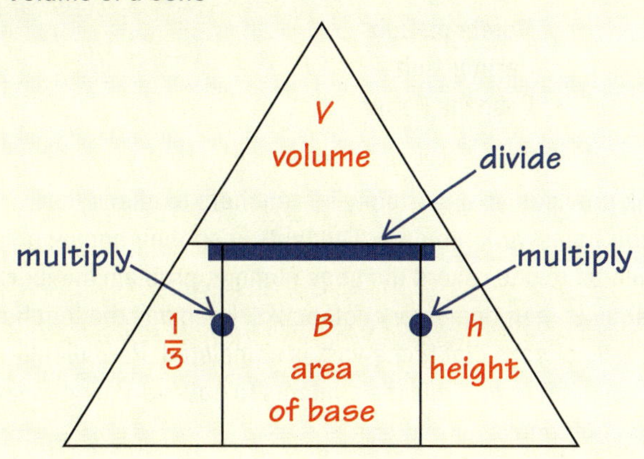

List of Organizers

Available at *BigIdeasMath.com*

Comparison Chart
Concept Circle
Example and Non-Example Chart
Formula Triangle
Four Square
Idea (Definition) and Examples Chart
Information Frame
Information Wheel
Notetaking Organizer
Process Diagram
Summary Triangle
Word Magnet
Y Chart

About this Organizer

A **Formula Triangle** can be used to arrange variables and operations of a formula. Students divide a triangle into the same number of parts as there are variables and factors in a formula. Then students write the variables and factors in the parts of the triangle and place either a multiplication or a division symbol, as appropriate, between the parts. This type of organizer can help students learn the formulas as well as see how the variables in the formulas are related. Students can place their formula triangles on note cards to use as a quick study reference.

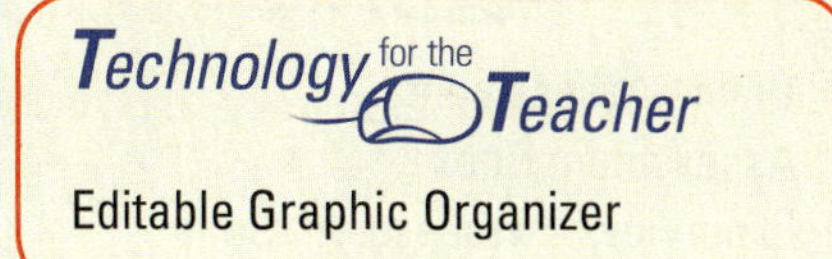

Editable Graphic Organizer

Answers

1. $14\pi \approx 44.0 \text{ yd}^3$
2. $36\pi \approx 113.1 \text{ ft}^3$
3. $50\pi \approx 157.1 \text{ cm}^3$
4. $132\pi \approx 414.7 \text{ in.}^3$
5. $\frac{340}{9\pi} \approx 12.0 \text{ ft}$
6. $\sqrt{\frac{2814}{4.7\pi}} \approx 13.8 \text{ cm}$
7. $\frac{28.26}{\pi} \approx 9 \text{ cm}$
8. The volume is 27 times greater.
9. about 42.45 in.^3
10. 13.5 in.

Technology for the Teacher

Online Assessment
Assessment Book
ExamView® Assessment Suite

Alternative Quiz Ideas

100% Quiz
Error Notebook
Group Quiz
Homework Quiz
Math Log
Notebook Quiz
Partner Quiz
Pass the Paper

Homework Quiz

A homework notebook provides an opportunity for teachers to check that students are doing their homework regularly. Students keep their homework in notebooks. They should be told to record the page number, problem number, and copy the problem exactly in their homework notebooks. Each day the teacher walks around and visually checks that homework is completed. Periodically, without advance notice, the teacher tells the students to put everything away except their homework notebooks.

Questions are from students' homework.

1. What are the answers to Exercises 13–15 on page 339?
2. What are the answers to Exercises 7–9 on page 344?
3. What are the answers to Exercises 3–5 on page 352?
4. What are the answers to Exercises 10–12 on page 360?

Reteaching and Enrichment Strategies

If students need help. . .	If students got it. . .
Resources by Chapter • Practice A and Practice B • Puzzle Time Lesson Tutorials *BigIdeasMath.com*	Resources by Chapter • Enrichment and Extension • Technology Connection Game Closet at *BigIdeasMath.com* Start the next section

8.1–8.2 Quiz

Find the volume of the solid. Round your answer to the nearest tenth. *(Section 8.1 and Section 8.2)*

1. 4 yd, 3.5 yd

2. 3 ft, 4 ft

3. 5 cm, 6 cm

4. 11 in., 12 in.

Find the missing dimension of the solid. Round your answer to the nearest tenth. *(Section 8.1 and Section 8.2)*

5. 3 ft, h, Volume = 340 ft^3

6. 4.7 cm, r, Volume = 938 cm^3

7. **PAPER CONE** The paper cone can hold 84.78 cubic centimeters of water. What is the height of the cone? *(Section 8.2)*

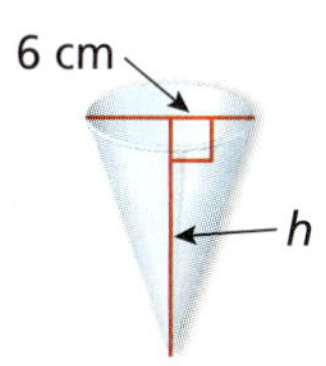

8. **GEOMETRY** Triple both dimensions of the cylinder. How many times greater is the volume of the new cylinder than the volume of the original cylinder? *(Section 8.1)*

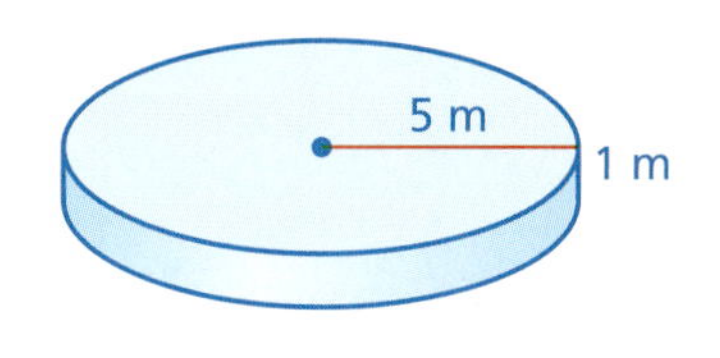

1.5 in.

16 in.

9. **SAND ART** There are 42.39 cubic inches of blue sand and 28.26 cubic inches of red sand in the cylindrical container. How many cubic inches of white sand are in the container? *(Section 8.1)*

10. **JUICE CAN** You are buying two cylindrical cans of juice. Each can holds the same amount of juice. What is the height of Can B? *(Section 8.1)*

8.3 Volumes of Spheres

Essential Question How can you find the volume of a sphere?

A **sphere** is the set of all points in space that are the same distance from a point called the *center*. The *radius r* is the distance from the center to any point on the sphere.

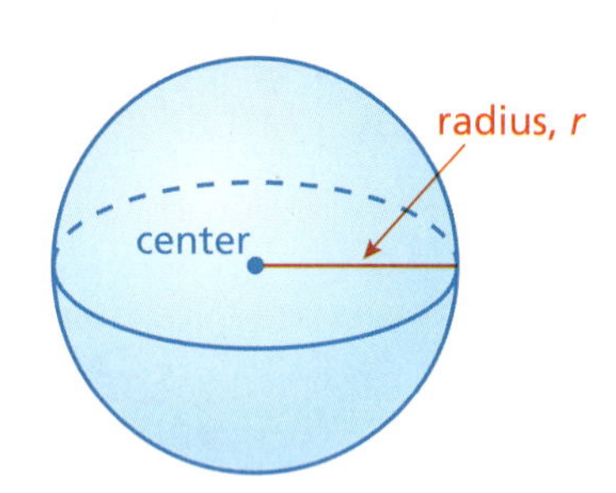

A sphere is different from the other solids you have studied so far because it does not have a base. To discover the volume of a sphere, you can use an activity similar to the one in the previous section.

1 ACTIVITY: Exploring the Volume of a Sphere

Work with a partner. Use a plastic ball similar to the one shown.

- Estimate the diameter and the radius of the ball.

- Use these measurements to draw a net for a cylinder with a diameter and a height equal to the diameter of the ball. How is the height h of the cylinder related to the radius r of the ball? Explain.

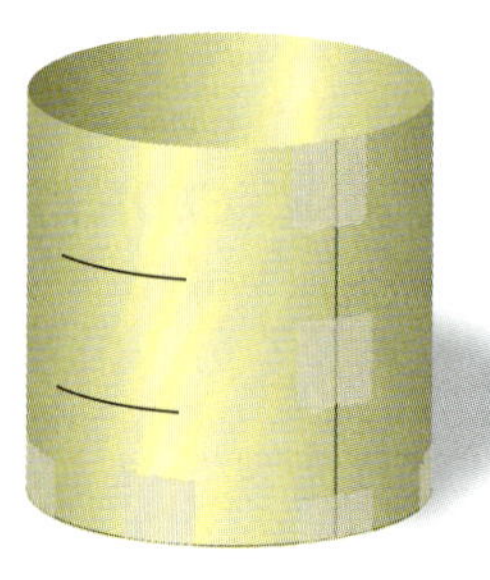

- Cut out the net. Then fold and tape it to form an open cylinder. Make two marks on the cylinder that divide it into thirds, as shown.

- Cover the ball with aluminum foil or tape. Leave one hole open. Fill the ball with rice. Then pour the rice into the cylinder. What fraction of the cylinder is filled with rice?

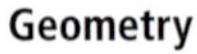

Geometry

In this lesson, you will

- find the volumes of spheres.
- find the radii of spheres given the volumes.
- solve real-life problems.

Learning Standard
MACC.8.G.3.9

Laurie's Notes

Introduction

Standards for Mathematical Practice

- **MP4 Model with Mathematics:** In the first activity, students will explore the relationship between the volume of a sphere and volume of a cylinder. Modeling such relationships is a powerful teaching tool.

Motivate

- Bring a collection of spherical objects, such as rubber balls, to class. The objects should be of different sizes.
- ? "What is the geometric name for these solids?" sphere
- ? "What linear dimension or dimensions does a sphere have?" radius
- Discuss with students the fact that spheres have only one linear dimension. Other solids they studied have two or three linear dimensions.
- Hold several of the objects and ask students which has the greatest volume and which has the least volume. Ask them to explain their reasoning. Because a sphere has only one linear dimension, the object with the greatest radius has the greatest volume and the object with the least radius has the least volume.

Activity Notes

Activity 1

- **Management Tip:** Instead of having the whole class perform this activity, you may want to have a pair of students demonstrate it. If you have a sphere and a cylinder from a commercially-available (fillable) geometric solids kit, then you could use these instead of having students make the cylinder. The sphere and the cylinder must have equal diameters and the height of the cylinder must also equal the diameter. Alternatively, you could show a video demonstration, such as the one at http://www.youtube.com/watch?v=aLyQddyY8ik.
- **MP4 Model with Mathematics:** Regardless, you want students to see the relationship and not just hear what it is.
- If necessary, have materials ready for students: heavy-weight paper, tape, scissors, ruler, and uncooked rice.
- The net students make for the cylinder will most likely have two bases. If so, then they can just tear off one of the bases to get the open cylinder.
- Be sure the cylinder(s) are marked by thirds so students can clearly see the relationship between the volume of the cylinder and the volume of the sphere.
- Students can use a piece of paper and shape it like a funnel to fill the ball with rice.
- Have students guess the relationship before they see it. Because the sphere fits inside the cylinder, students at least know that the cylinder has a greater volume.
- ? "How much of the cylinder is filled by the contents of the sphere?" $\frac{2}{3}$

Florida Common Core Standards

MACC.8.G.3.9 Know the formulas for the volumes of cones, cylinders, and spheres and use them to solve real-world and mathematical problems.

Previous Learning

Students should know how to find volumes of cylinders and cones.

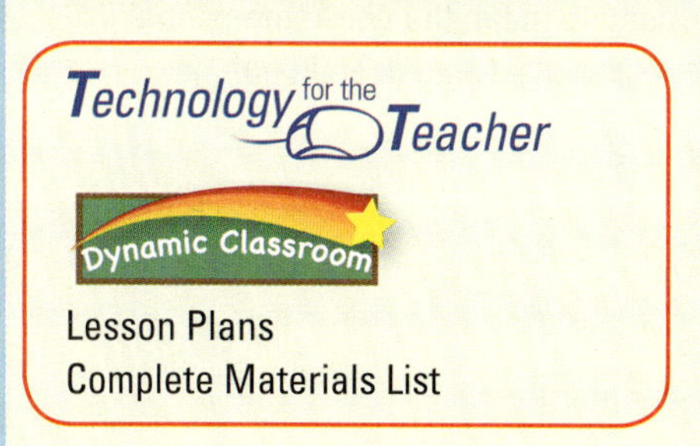

Lesson Plans
Complete Materials List

8.3 Record and Practice Journal

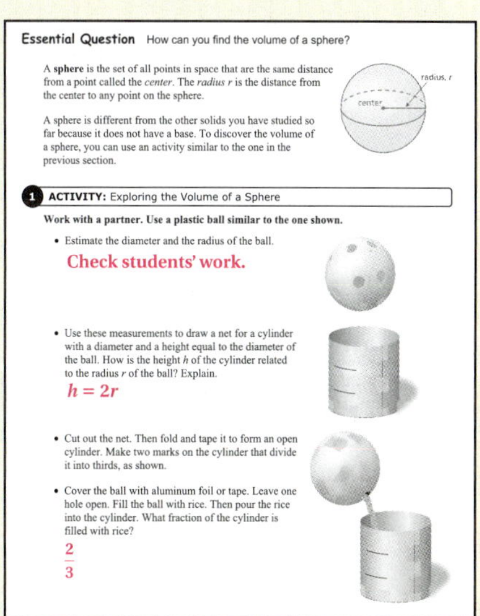

Essential Question How can you find the volume of a sphere?

A **sphere** is the set of all points in space that are the same distance from a point called the *center*. The *radius* r is the distance from the center to any point on the sphere.

A sphere is different from the other solids you have studied so far because it does not have a base. To discover the volume of a sphere, you can use an activity similar to the one in the previous section.

radius, r
center

1 ACTIVITY: Exploring the Volume of a Sphere

Work with a partner. Use a plastic ball similar to the one shown.

- Estimate the diameter and the radius of the ball.

  Check students' work.

- Use these measurements to draw a net for a cylinder with a diameter and a height equal to the diameter of the ball. How is the height h of the cylinder related to the radius r of the ball? Explain.

  $h = 2r$

- Cut out the net. Then fold and tape it to form an open cylinder. Make two marks on the cylinder that divide it into thirds, as shown.
- Cover the ball with aluminum foil or tape. Leave one hole open. Fill the ball with rice. Then pour the rice into the cylinder. What fraction of the cylinder is filled with rice?

  $\frac{2}{3}$

English Language Learners

Vocabulary

English learners may be familiar with the word *net* in everyday context. In finance, the net profit describes the bottom line of a financial transaction. In fishing, a net is a collection of knotted strings used to catch fish. In a mathematical context, such as in Activity 1, *net* or *geometric net* is used to mean the two-dimensional representation of a solid object.

8.3 Record and Practice Journal

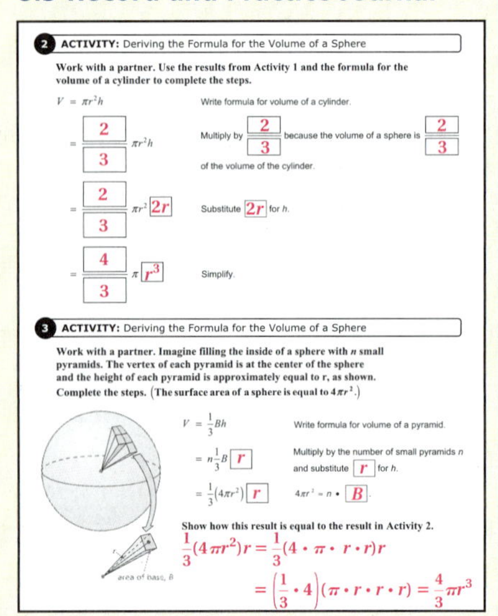

2 ACTIVITY: Deriving the Formula for the Volume of a Sphere

Work with a partner. Use the results from Activity 1 and the formula for the volume of a cylinder to complete the steps.

$V = \pi r^2 h$ — Write formula for volume of a cylinder.

$= \frac{2}{3}\pi r^2 h$ — Multiply by $\frac{2}{3}$ because the volume of a sphere is $\frac{2}{3}$ of the volume of the cylinder.

$= \frac{2}{3}\pi r^2 (2r)$ — Substitute $2r$ for h.

$= \frac{4}{3}\pi r^3$ — Simplify.

3 ACTIVITY: Deriving the Formula for the Volume of a Sphere

Work with a partner. Imagine filling the inside of a sphere with n small pyramids. The vertex of each pyramid is at the center of the sphere and the height of each pyramid is approximately equal to r, as shown. Complete the steps. (The surface area of a sphere is equal to $4\pi r^2$.)

$V = \frac{1}{3}Bh$ — Write formula for volume of a pyramid.

$= n\frac{1}{3}B\,r$ — Multiply by the number of small pyramids n and substitute r for h.

$= \frac{1}{3}(4\pi r^2)\,r$ — $4\pi r^2 = n \cdot B$.

Show how this result is equal to the result in Activity 2.

$\frac{1}{3}(4\pi r^2)r = \frac{1}{3}(4 \cdot \pi \cdot r \cdot r)r$

$= \left(\frac{1}{3} \cdot 4\right)(\pi \cdot r \cdot r \cdot r) = \frac{4}{3}\pi r^3$

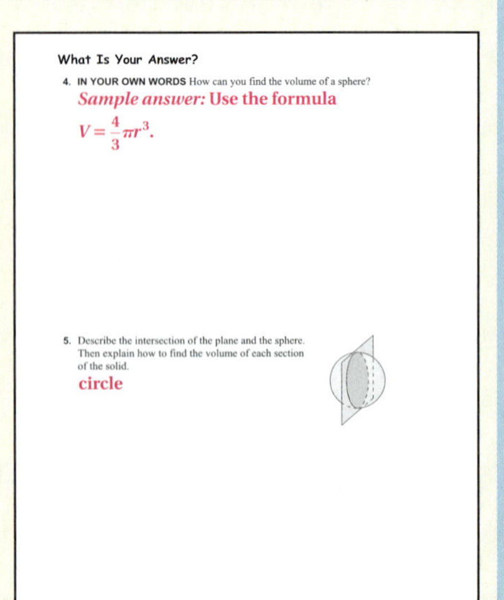

What Is Your Answer?

4. **IN YOUR OWN WORDS** How can you find the volume of a sphere?
Sample answer: Use the formula $V = \frac{4}{3}\pi r^3$.

5. Describe the intersection of the plane and the sphere. Then explain how to find the volume of each section of the solid.
circle

Laurie's Notes

Activity 2

- This activity uses the results from Activity 1, so it is important for students to complete that activity before moving on to this activity.
- ? "What is the formula for the volume V of a cylinder?" $V = Bh = 2\pi r^2 h$, where B is the area of the base, h is the height, and r is the radius of the cylinder.
- **MP4 Model with Mathematics:** Students will use the relationship between the volume of a sphere and the volume of a cylinder, and the formula for the volume of a cylinder, to derive the formula for the volume of a sphere.
- **MP1 Make Sense of Problems and Persevere in Solving Them:** Resist the urge to tell students how to fill in the blanks. Students should talk with their partners and use available resources to help them fill in the blanks.
- Students may struggle with substituting $2r$ for the height of the cylinder and then simplifying.
- After students perform the simplification in the last step, they will have the formula for the volume of a sphere.

Activity 3

- This activity, which is optional, takes students through an alternate derivation of the formula for the volume of a sphere.
- The symbolic manipulation should not be beyond the ability of your students. Students may, however, have difficulty visualizing a sphere composed of small pyramids.
- The curvature of the sphere is ignored because it is assumed that the base areas of the pyramids are small.
- Students are given the formula for the surface area of a sphere. Tell them that they will learn more about the surface area of a sphere in later courses.
- Animations of the approach shown in this activity can be viewed online, such as the one at http://www.youtube.com/watch?v=xuPl_8o_j7k. You may wish to view the animation yourself or show it to the class.

What Is Your Answer?

- Students should check their answers with their neighbors.
- Problem 5 suggests the idea of a sphere composed of two hemispheres. Point out that a hemisphere is half a sphere. Students will need to find the volume of a hemisphere in problems such as Example 3.

Closure

- **Exit Ticket:** The radius of a sphere is 2 centimeters. Find the volume of a cube that the sphere fits snugly within and find the volume of the sphere. 64 cm^3; $\frac{32}{3}\pi \approx 33.5\text{ cm}^3$

2 ACTIVITY: Deriving the Formula for the Volume of a Sphere

Math Practice 4

Analyze Relationships

What is the relationship between the volume of a sphere and the volume of a cylinder? How does this help you derive a formula for the volume of a sphere?

Work with a partner. Use the results from Activity 1 and the formula for the volume of a cylinder to complete the steps.

$V = \pi r^2 h$ — Write formula for volume of a cylinder.

$= \frac{\square}{\square} \pi r^2 h$ — Multiply by

because the volume of a sphere is

of the volume of the cylinder.

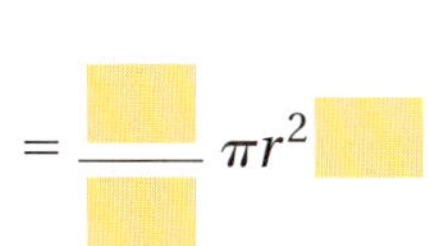

$= \frac{\square}{\square} \pi r^2 \square$ — Substitute $\square$ for h.

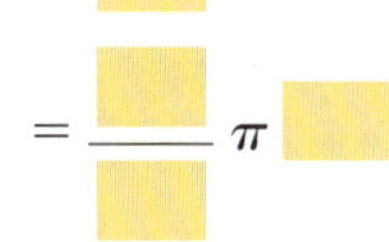

$= \frac{\square}{\square} \pi \square$ — Simplify.

3 ACTIVITY: Deriving the Formula for the Volume of a Sphere

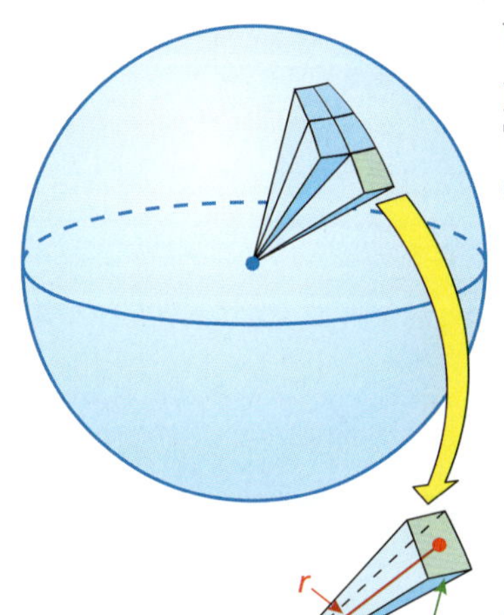

Work with a partner. Imagine filling the inside of a sphere with n small pyramids. The vertex of each pyramid is at the center of the sphere. The height of each pyramid is approximately equal to r, as shown. Complete the steps. (The surface area of a sphere is equal to $4\pi r^2$.)

$V = \frac{1}{3}Bh$ — Write formula for volume of a pyramid.

$= n\frac{1}{3}B\square$ — Multiply by the number of small pyramids n and substitute $\square$ for h.

$= \frac{1}{3}(4\pi r^2)\square$ — $4\pi r^2 \approx n \cdot \square$

Show how this result is equal to the result in Activity 2.

What Is Your Answer?

4. **IN YOUR OWN WORDS** How can you find the volume of a sphere?

5. Describe the intersection of the plane and the sphere. Then explain how to find the volume of each section of the solid.

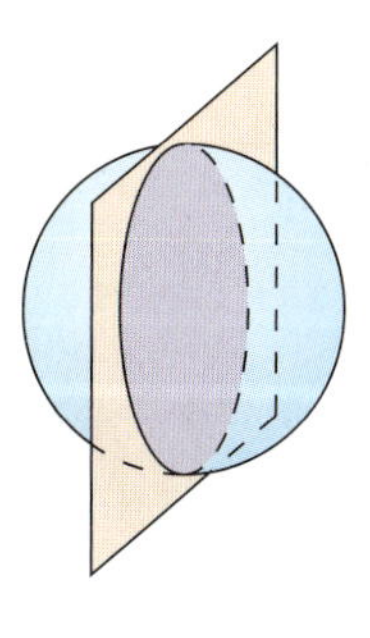

Practice Use what you learned about the volumes of spheres to complete Exercises 3–5 on page 352.

8.3 Lesson

Key Vocabulary
sphere, *p. 348*
hemisphere, *p. 351*

Volume of a Sphere

Words The volume V of a sphere is the product of $\frac{4}{3}\pi$ and the cube of the radius of the sphere.

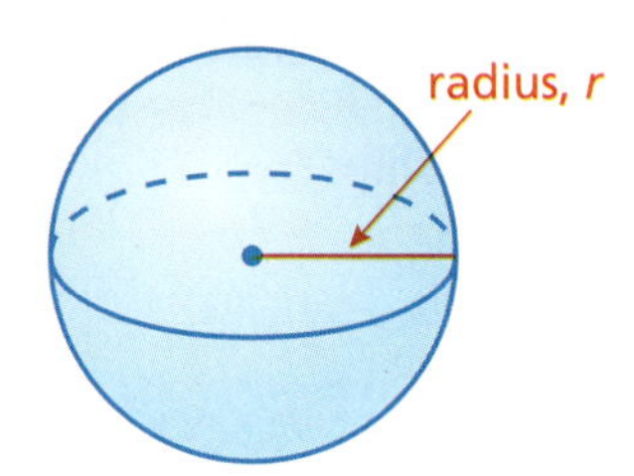

Algebra $V = \frac{4}{3}\pi r^3$ — Cube of radius of sphere

EXAMPLE 1 Finding the Volume of a Sphere

Find the volume of the sphere. Round your answer to the nearest tenth.

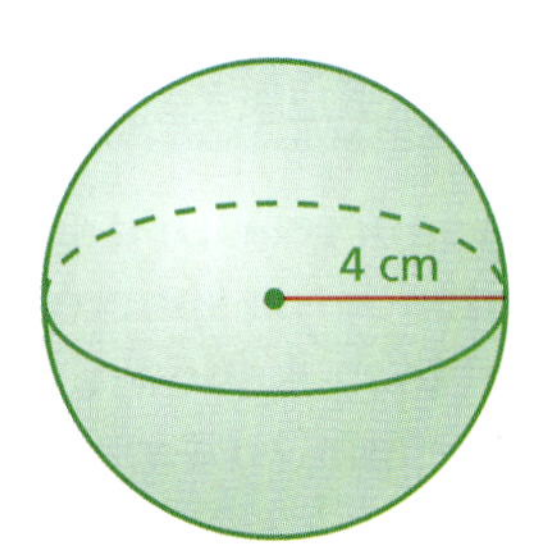

$V = \frac{4}{3}\pi r^3$ Write formula for volume.

$= \frac{4}{3}\pi(4)^3$ Substitute 4 for r.

$= \frac{256}{3}\pi$ Simplify.

≈ 268.1 Use a calculator.

The volume is about 268.1 cubic centimeters.

EXAMPLE 2 Finding the Radius of a Sphere

Find the radius of the sphere.

Volume = 288π in.^{3}

$V = \frac{4}{3}\pi r^3$ Write formula.

$288\pi = \frac{4}{3}\pi r^3$ Substitute.

$288\pi = \frac{4\pi}{3}r^3$ Multiply.

$\frac{3}{4\pi} \cdot 288\pi = \frac{3}{4\pi} \cdot \frac{4\pi}{3}r^3$ Multiplication Property of Equality

$216 = r^3$ Simplify.

$6 = r$ Take the cube root of each side.

The radius is 6 inches.

Laurie's Notes

Introduction

Connect

- **Yesterday:** Students derived the formula for the volume of a sphere. (MP1, MP4)
- **Today:** Students will find volumes of spheres.

Motivate

- In some places you can buy ground meat in a cylindrical casing. The casing with one pound of ground meat has a diameter of 3 inches and a height of 6 inches.
- Tell a story about making meatballs that have a 1-inch diameter using the ground meat from the cylindrical casing.
- ? "How can I figure out the number of meatballs I can make?" Find the volume of the ground meat in the cylindrical casing and divide it by the volume of a meatball.
- Explain that in today's lesson they will find the volumes of spheres.

Lesson Notes

Key Idea

- Have physical objects, such as a ball or a globe, available to reference.
- Draw the sphere and write the formula in words.
- Students may derive this formula again in a high school geometry course.

Example 1

- Notice that the work is done in terms of π. It is not until the last step that you use the π key on a calculator.
- ? "What information do you know?" The radius of the sphere is 4 centimeters.
- **Representation:** Encourage students to use the parentheses to represent multiplication. Using the $\times$ symbol would make the expression confusing.
- **Common Misconception:** Remind students that π is a number and because multiplication is commutative and associative, this expression could be rewritten as $\frac{4}{3}(4)^3\pi$, making the computation less confusing.
- ? "What is being cubed in this expression?" only the 4
- **Extension:** Ask students how big they think a cubic centimeter is. To help students visualize, tell them that a cubic centimeter is about the size of a sugar cube.

Example 2

- This example requires students to solve an equation for a variable.
- Work through the problem, annotating the steps as shown in the book.

Goal Today's lesson is finding volumes of **spheres**.

Technology for the Teacher

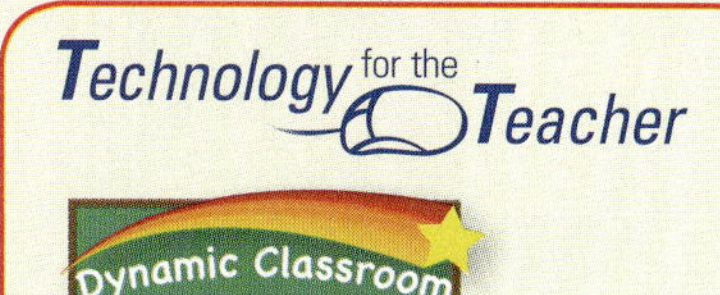

Lesson Tutorials
Lesson Plans
Answer Presentation Tool

Extra Example 1

Find the volume of a sphere with a radius of 11 meters. Round your answer to the nearest tenth.

$\frac{5324}{3}\pi \approx 5575.3 \text{ m}^3$

Extra Example 2

Find the radius of a sphere with a volume of 2304π cubic centimeters.

12 cm

On Your Own

1. $\frac{2048}{3}\pi \approx 2144.7 \text{ ft}^3$
2. 3 m

Extra Example 3

In Example 3, the radius of the silo is 9 feet and the overall height is 48 feet. What is the volume of the silo? Round your answer to the nearest thousand.
$3645\pi \approx 11{,}000 \text{ ft}^3$

On Your Own

3. $\frac{80}{3}\pi \approx 83.8 \text{ in.}^3$
4. $96\pi \approx 301.6 \text{ m}^3$

Differentiated Instruction

Visual

Students can check the volume of a sphere for reasonableness by finding the volume of a cube with side length $2r$, as suggested in the Closure on page T-349. An example is shown below.

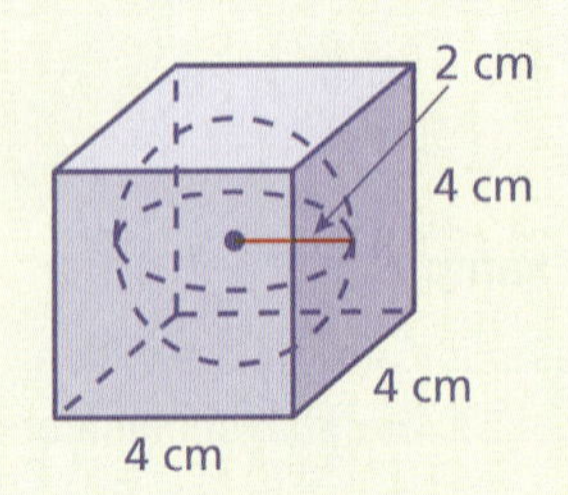

This will give an overestimate of the volume of the sphere. Also, it will help students remember to use the radius, not the diameter, in the given formula for volume of a sphere.

Laurie's Notes

Example 2 (continued)

- **MP7 Look for and Make Use of Structure:** Discuss why students can write $\frac{4}{3}\pi$ as $\frac{4\pi}{3}$. You want them to see the connection to fraction multiplication: $\frac{4}{3} \cdot 7 = \frac{4 \cdot 7}{3}$, so likewise $\frac{4}{3} \cdot \pi = \frac{4 \cdot \pi}{3}$.
- **MP7:** Students may have difficulty solving for r in this example. As shown, you can accomplish this by multiplying each side of the equation by the reciprocal of the variable term's coefficient and then taking the cube root of each side. Provide a quick review of cube roots if necessary.

On Your Own

- Ask volunteers to share their work at the board.

Example 3

- Students have previously found areas of composite figures. Students generally have little difficulty understanding how to find volumes of composite solids.
- Ask a student to read the problem. Sketch the solid as the student is reading.
- ? "What information do you know?" the height and radius of the silo
- Discuss the Study Tip.
- ? "How can you find the volume of the silo?" Find the volume of the cylinder and the volume of the hemisphere and add.
- Use the good problem solving technique of writing the formulas first and then substituting the values of the variables.
- Work through the problem as shown. Notice that the solution is kept in terms of π until the last step.

On Your Own

- **Think-Pair-Share:** Students should read each question independently and then work in pairs to answer the questions. When they have answered the questions, the pair should compare their answers with another group and discuss any discrepancies.

Closure

- Have students answer the following question:
  You have an ice cream scoop with a 2-inch diameter. You have an ice cream cone with a 2-inch diameter and a height of 5 inches. If you place one scoop of ice cream on the cone and let the ice cream melt, will it spill over the cone? Explain. no; The volume of the cone is greater than the volume of the ice cream.

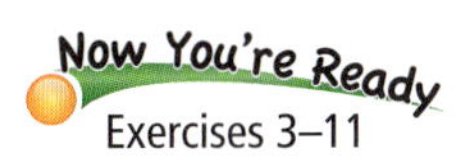

Find the volume V or radius r of the sphere. Round your answer to the nearest tenth, if necessary.

1.

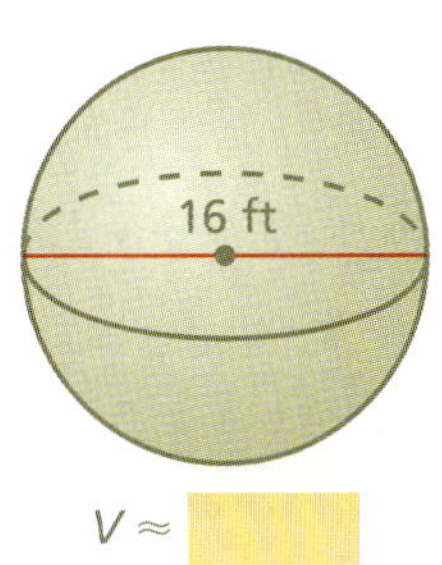

$V \approx$

2.

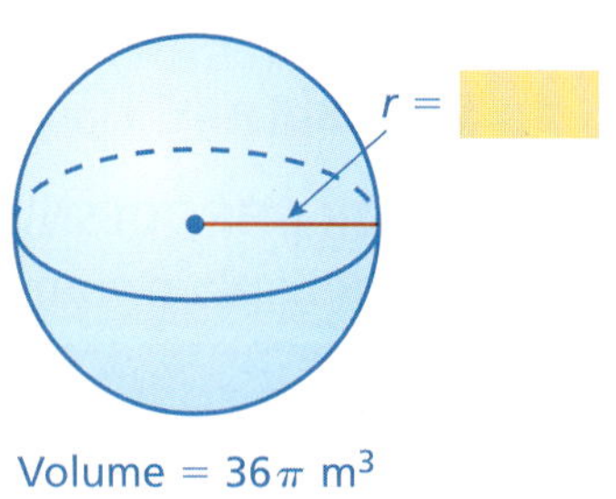

Volume $= 36\pi$ m^3

EXAMPLE 3 Finding the Volume of a Composite Solid

A hemisphere is one-half of a sphere. The top of the silo is a hemisphere with a radius of 12 feet. What is the volume of the silo? Round your answer to the nearest thousand.

The silo is made up of a cylinder and a hemisphere. Find the volume of each solid.

In Example 3, the height of the cylindrical part of the silo is the difference of the silo height and the radius of the hemisphere.

52 − 12 = 40 ft

Cylinder

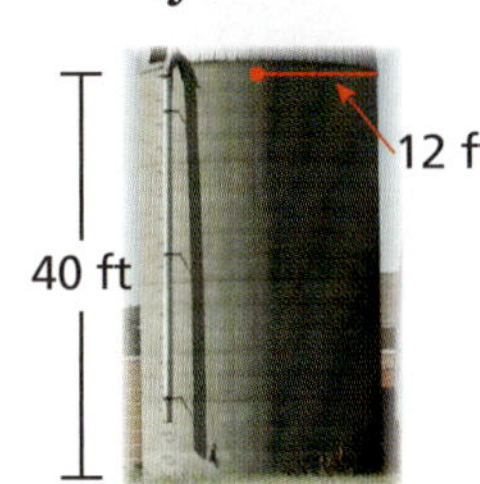

$V = Bh$

$= \pi(12)^2(40)$

$= 5760\pi$

Hemisphere

$V = \frac{1}{2} \cdot \frac{4}{3}\pi r^3$

$= \frac{1}{2} \cdot \frac{4}{3}\pi(12)^3$

$= 1152\pi$

So, the volume is $5760\pi + 1152\pi = 6912\pi \approx 22{,}000$ cubic feet.

On Your Own

Find the volume of the composite solid. Round your answer to the nearest tenth.

3.

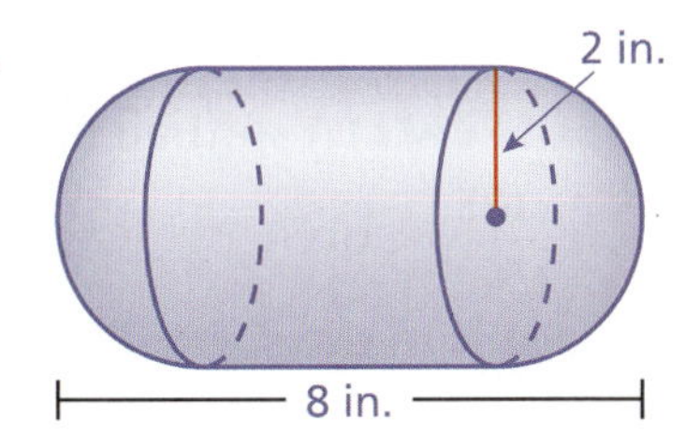

4.

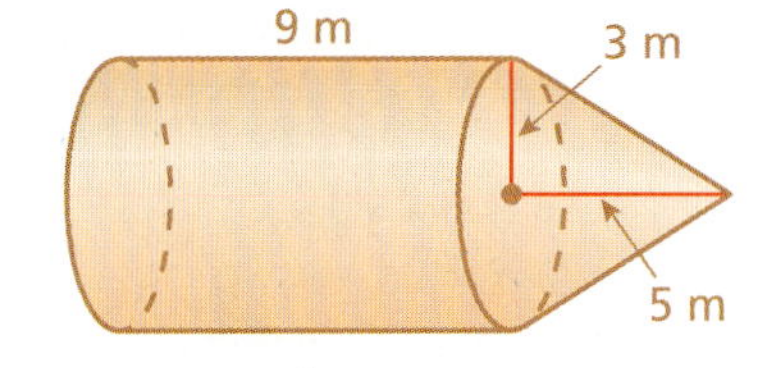

8.3 Exercises

Vocabulary and Concept Check

1. **VOCABULARY** How is a sphere different from a hemisphere?

2. **WHICH ONE DOESN'T BELONG?** Which figure does *not* belong with the other three? Explain your reasoning.

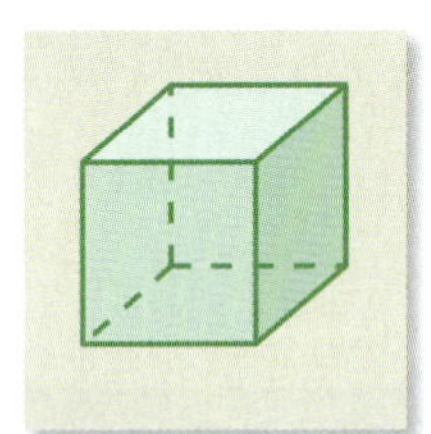

Practice and Problem Solving

Find the volume of the sphere. Round your answer to the nearest tenth.

1 3.

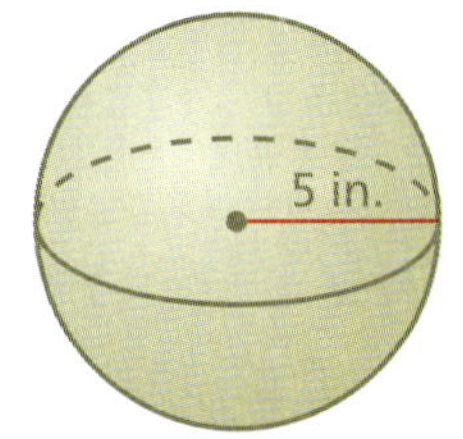

4.

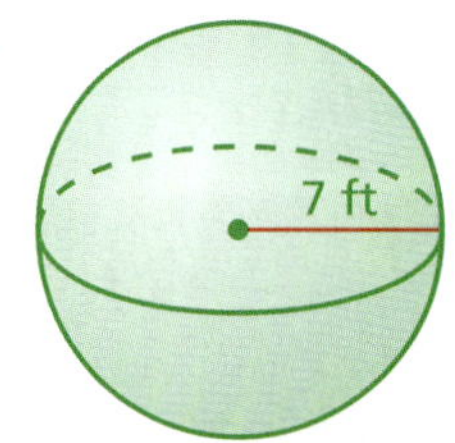

5.

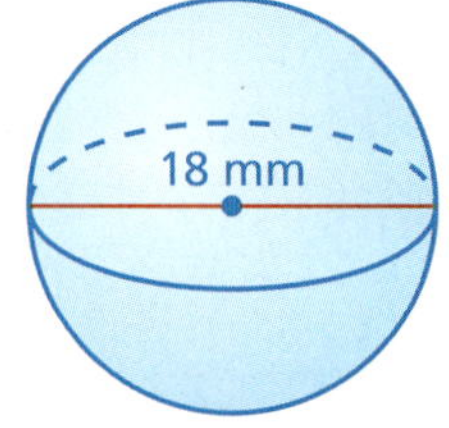

6.

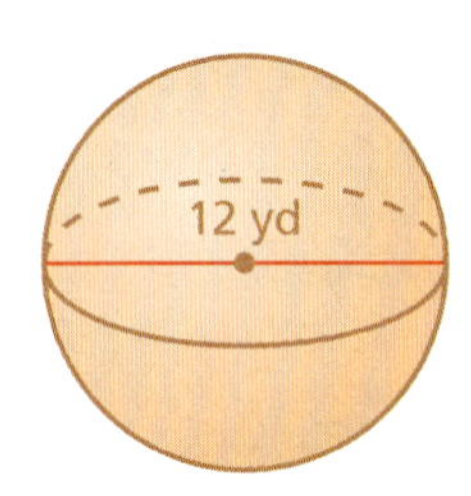

7.

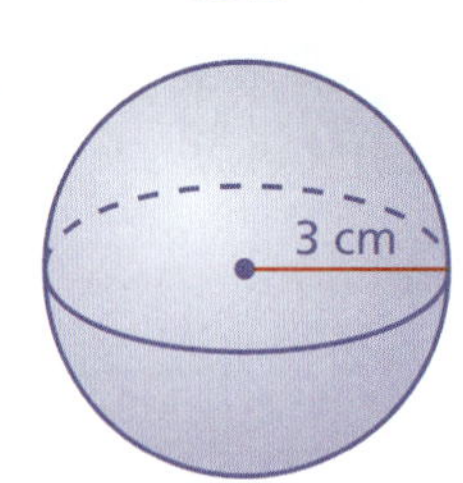

8.

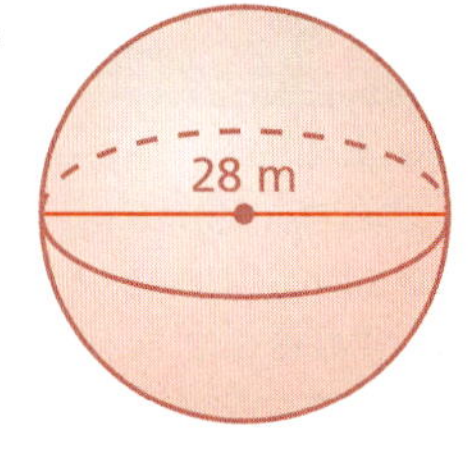

Find the radius of the sphere with the given volume.

9. Volume $= 972\pi\ \text{mm}^3$

10. Volume $= 4.5\pi\ \text{cm}^3$

11. Volume $= 121.5\pi\ \text{ft}^3$

12. **GLOBE** The globe of the Moon has a radius of 10 inches. Find the volume of the globe. Round your answer to the nearest whole number.

13. **SOFTBALL** A softball has a volume of $\frac{125}{6}\pi$ cubic inches. Find the radius of the softball.

Assignment Guide and Homework Check

Level	Day 1 Activity Assignment	Day 2 Lesson Assignment	Homework Check
Basic	3–5, 21–23	1, 2, 6–13, 15	7, 9, 13, 15
Average	3–5, 21–23	1, 2, 6–18	7, 11, 13, 15
Advanced	1–23		8, 10, 13, 15, 19

Common Errors

- **Exercises 3–8 and 12** When finding the volume of a sphere, students may forget to multiply by $\frac{4}{3}$ or cube the radius, they may use diameter when the formula calls for radius, or they may write the incorrect units. Remind them of the given formula for volume of a sphere and that part of writing a correct answer is including the correct units.
- **Exercises 9–11 and 13** Students may not complete the solution; they may solve for the *cube* of the radius instead of the radius. Point out that they need to find a cube root, which they learned in Section 7.2, and provide a quick review if necessary.
- **Exercises 14–16** Students may think that there is not enough information to solve the problem. It may help to have them "break up" the composite solid into two parts, whose volumes they know how to find, and mark the dimensions on each part.
- **Exercise 16** Students may add instead of subtract the volume of the hemisphere. Point out that in contrast to the solid in Example 3, this solid is made up of a cylinder with a hemisphere *removed*, not added.

8.3 Record and Practice Journal

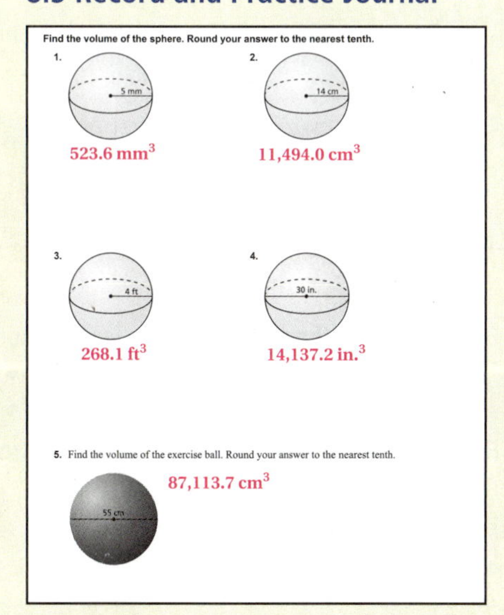
Find the volume of the sphere. Round your answer to the nearest tenth.

1. 5 mm — 523.6 mm^3
2. 14 cm — 11,494.0 cm^3
3. 4 ft — 268.1 ft^3
4. 30 in. — 14,137.2 in.^{3}
5. Find the volume of the exercise ball. Round your answer to the nearest tenth. 55 cm — 87,113.7 cm^3

Vocabulary and Concept Check

1. A hemisphere is one-half of a sphere.
2. sphere; It does not have a base.

Practice and Problem Solving

3. $\frac{500\pi}{3} \approx 523.6 \text{ in.}^3$
4. $\frac{1372\pi}{3} \approx 1436.8 \text{ ft}^3$
5. $972\pi \approx 3053.6 \text{ mm}^3$
6. $288\pi \approx 904.8 \text{ yd}^3$
7. $36\pi \approx 113.1 \text{ cm}^3$
8. $\frac{10{,}976\pi}{3} \approx 11{,}494.0 \text{ m}^3$
9. 9 mm
10. 1.5 cm
11. 4.5 ft
12. $\frac{4000\pi}{3} \approx 4189 \text{ in.}^3$
13. 2.5 in.
14. $512 + \frac{128\pi}{3} \approx 646.0 \text{ cm}^3$

Practice and Problem Solving

15. $256\pi + 128\pi = 384\pi \approx 1206.4 \text{ ft}^3$
16. $99\pi - 18\pi = 81\pi \approx 254.5 \text{ in.}^3$
17. $r = \frac{3}{4}h$
18. $162\pi - 108\pi = 54\pi \approx 170 \text{ cm}^3$
19. 5400 in.^2; $27{,}000 \text{ in.}^3$
20. See *Taking Math Deeper*.

Fair Game Review

21. enlargement; 2
22. reduction; $\frac{1}{3}$
23. A

Mini-Assessment

Find the volume of the sphere. Round your answer to the nearest tenth.

1. 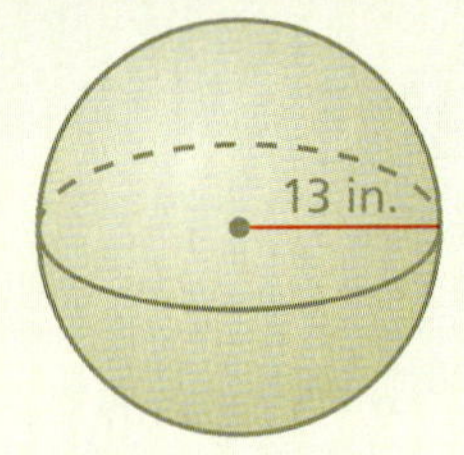

$\frac{8788\pi}{3} \approx 9202.8 \text{ in.}^3$

2. 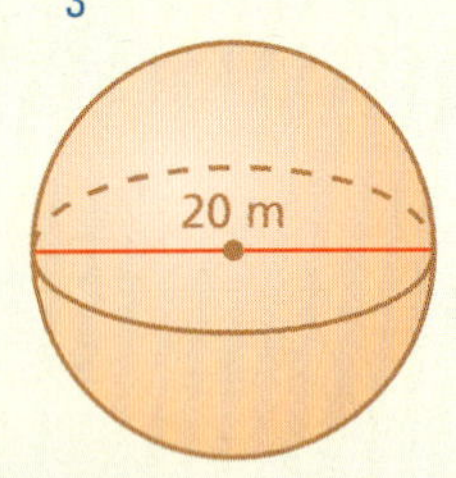

$\frac{4000\pi}{3} \approx 4188.8 \text{ m}^3$

3. Find the radius of a sphere with a volume of 972π cubic millimeters.
9 mm
4. In Example 3, the diameter of the silo is 18 meters and the overall height is 62 meters. What is the volume of the silo? Round your answer to the nearest thousand.
$4779\pi \approx 15{,}000 \text{ m}^3$

Taking Math Deeper

Exercise 20

This exercise is a good review of rewriting literal equations. Start by writing the volume formula for each solid.

Draw the solids and write the corresponding volume formulas.

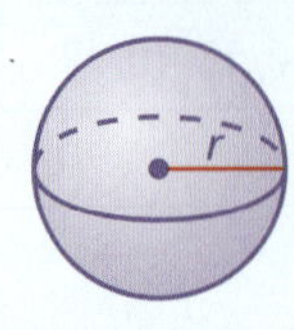

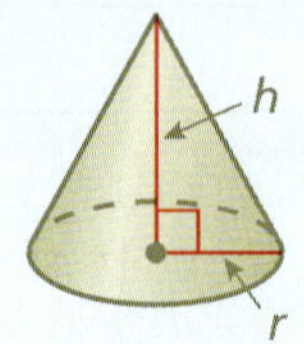

$V = \frac{4}{3}\pi r^3$ $\qquad$ $V = \frac{1}{3}\pi r^2 h$

Write and solve an equation.

Volume of sphere = 4 • Volume of cone

$$\frac{4}{3}\pi r^3 = 4 \cdot \frac{1}{3}\pi r^2 h$$
$$\frac{4}{3}\pi r^3 = \frac{4}{3}\pi r^2 h$$
$$r^3 = r^2 h$$
$$r \cdot r \cdot r = r \cdot r \cdot h$$
$$r = h$$

So, the volume of a sphere with radius r is four times the volume of a cone with radius r when the height of the cone is equal to the radius.

3. Check your solution by choosing a value for r. Let $r = 3$, so $h = 3$.

Sphere: $V = \frac{4}{3}\pi(3)^3 = 36\pi$ $\qquad$ **Cone:** $V = \frac{1}{3}\pi(3)^2(3) = 9\pi$

Because $36\pi = 4 \cdot 9\pi$, the solution checks.

Reteaching and Enrichment Strategies

If students need help. . .	If students got it. . .
Resources by Chapter • Practice A and Practice B • Puzzle Time Record and Practice Journal Practice Differentiating the Lesson Lesson Tutorials Skills Review Handbook	Resources by Chapter • Enrichment and Extension • Technology Connection Start the next section

Find the volume of the composite solid. Round your answer to the nearest tenth.

3 **14.**

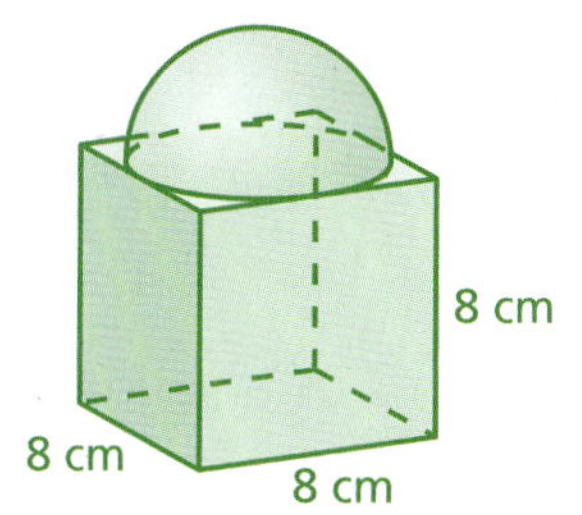

15.

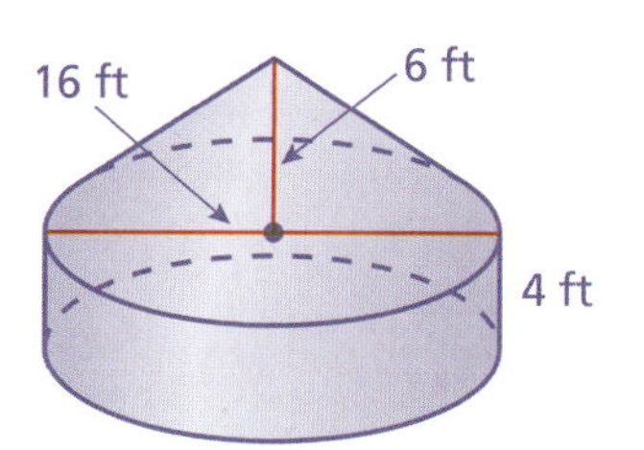

16.

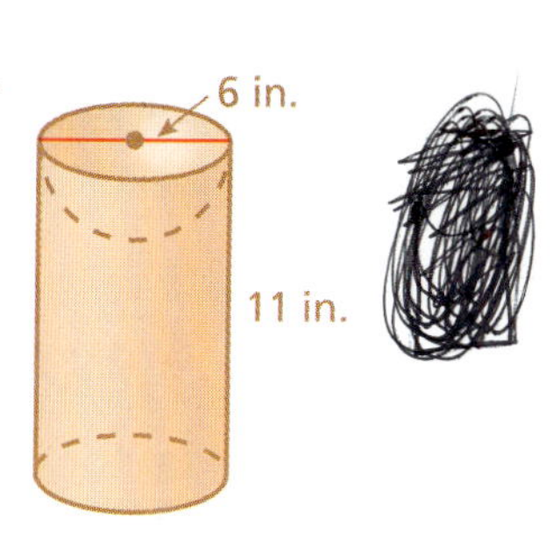

17. REASONING A sphere and a right cylinder have the same radius and volume. Find the radius r in terms of the height h of the cylinder.

18. PACKAGING A cylindrical container of three rubber balls has a height of 18 centimeters and a diameter of 6 centimeters. Each ball in the container has a radius of 3 centimeters. Find the amount of space in the container that is not occupied by rubber balls. Round your answer to the nearest whole number.

Volume = 4500π in.^{3}

19. BASKETBALL The basketball shown is packaged in a box that is in the shape of a cube. The edge length of the box is equal to the diameter of the basketball. What is the surface area and the volume of the box?

20. **Logic** Your friend says that the volume of a sphere with radius r is four times the volume of a cone with radius r. When is this true? Justify your answer.

Fair Game Review What you learned in previous grades & lessons

The blue figure is a dilation of the red figure. Identify the type of dilation and find the scale factor. *(Section 2.7)*

21.

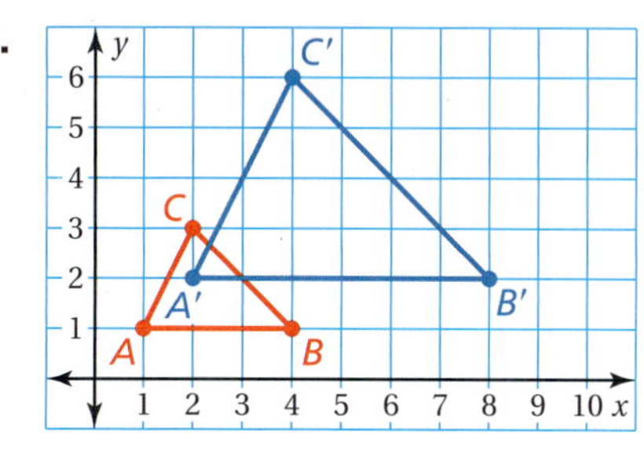

22.

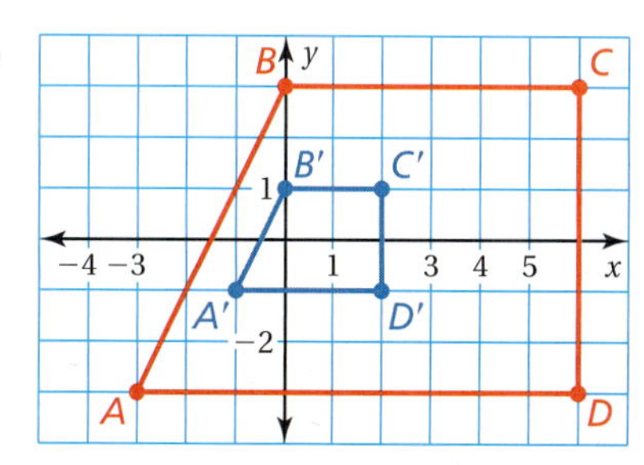

23. MULTIPLE CHOICE A person who is 5 feet tall casts a 6-foot-long shadow. A nearby flagpole casts a 30-foot-long shadow. What is the height of the flagpole? *(Section 3.4)*

Ⓐ 25 ft Ⓑ 29 ft Ⓒ 36 ft Ⓓ 40 ft

8.4 Surface Areas and Volumes of Similar Solids

Essential Question When the dimensions of a solid increase by a factor of k, how does the surface area change? How does the volume change?

1 ACTIVITY: Comparing Surface Areas and Volumes

Work with a partner. Copy and complete the table. Describe the pattern. Are the dimensions proportional? Explain your reasoning.

a.

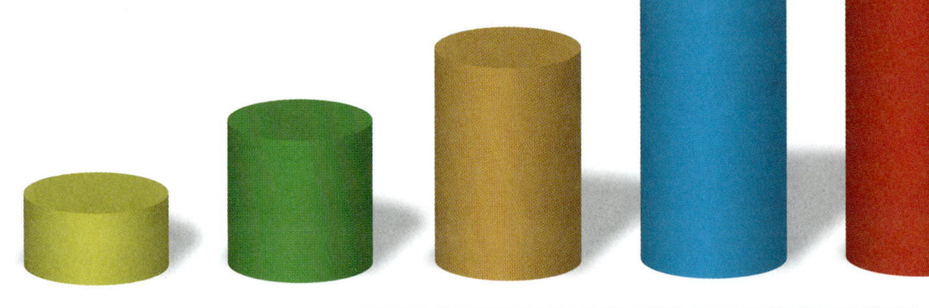

Radius	1	1	1	1	1
Height	1	2	3	4	5
Surface Area					
Volume					

b.

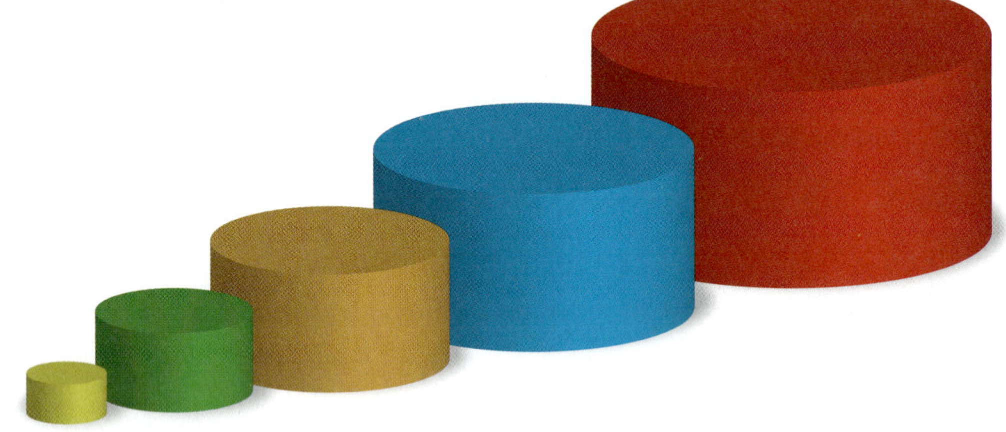

Radius	1	2	3	4	5
Height	1	2	3	4	5
Surface Area					
Volume					

COMMON CORE

Geometry

In this lesson, you will

- identify similar solids.
- use properties of similar solids to find missing measures.
- understand the relationship between surface areas of similar solids.
- understand the relationship between volumes of similar solids.
- solve real-life problems.

Applying Standard MACC.8.G.3.9

Laurie's Notes

Introduction

Standards for Mathematical Practice

- **MP8 Look for and Express Regularity in Repeated Reasoning:** Students may recognize and apply patterns in repeated computations they are performing, which will make these activities go quickly.

Motivate

- **Story Time:** Retell a portion of the story of *Goldilocks and the Three Bears.* Focus on the three sizes of porridge bowls, chairs, and beds.
- Share with students that Papa Bear's mattress was twice as long, twice as wide, and twice as high as Baby Bear's mattress. So, are there twice as many feathers in Papa Bear's feather bed mattress?
- This question will be answered at the end of the class.
- If you sense it is necessary, provide a review of how to find surface areas of cylinders, and surface areas and volumes of pyramids, before starting the activities.

Activity Notes

Activity 1

- Tell students to leave their answers in terms of π.
- ? To help students see the pattern in the first table, ask the following questions.
  - "Describe the changes in the dimensions." radius same, height increases by 1
  - "How does the surface area change?" increases by 2π
  - "How does the volume change?" increases by π
  - "Compare each figure's height to the original (first) figure's height. Do the same for surface areas and volumes. What do you notice?" The volumes are multiplied by the same number as the heights.
- ? To help students see the pattern in the second table, ask the following questions.
  - "Describe the changes in the dimensions." radius and height each increase by 1
  - "Compare each figure's height to the original figure's height. Do the same for radii, surface areas, and volumes. What do you notice?" The heights and radii are multiplied by the same number. The surface areas are multiplied by the square of this number, and the volumes are multiplied by the cube of this number.
- ? "Are the dimensions proportional in part (a)? Explain." No, only the height increases, not the radius.
- ? "Are the dimensions proportional in part (b)? Explain." Yes, both the radius and height increase by the same factor.

Florida Common Core Standards

MACC.8.G.3.9 Know the formulas for the volumes of cones, cylinders, and spheres and use them to solve real-world and mathematical problems.

Previous Learning

Students should be familiar with similar figures, surface area formulas, and volume formulas.

Lesson Plans
Complete Materials List

8.4 Record and Practice Journal

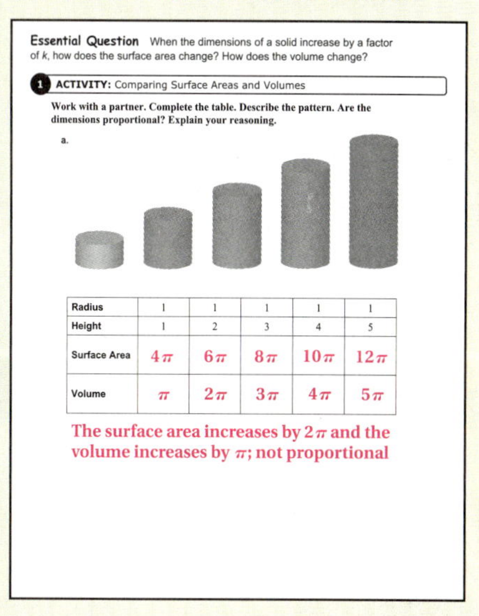

Essential Question When the dimensions of a solid increase by a factor of k, how does the surface area change? How does the volume change?

1 ACTIVITY: Comparing Surface Areas and Volumes

Work with a partner. Complete the table. Describe the pattern. Are the dimensions proportional? Explain your reasoning.

a.

Radius	1	1	1	1	1
Height	1	2	3	4	5
Surface Area	4π	6π	8π	10π	12π
Volume	π	2π	3π	4π	5π

The surface area increases by 2π and the volume increases by π; not proportional

Differentiated Instruction

In the examples, check to be sure that students are correctly identifying corresponding sides. Remind them that they have to identify corresponding linear measures to write proportions before solving them.

8.4 Record and Practice Journal

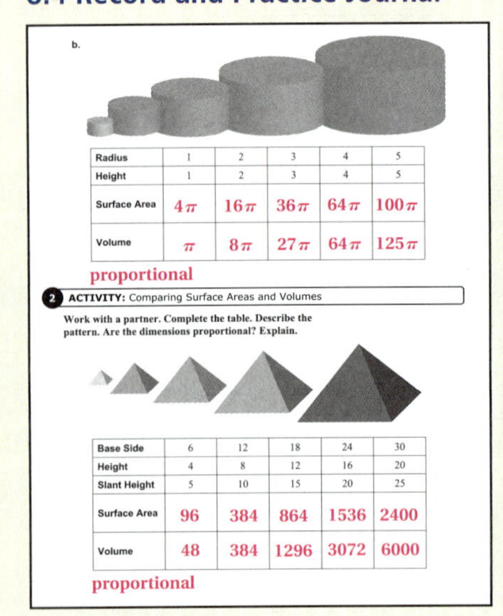

b.

Radius	1	2	3	4	5
Height	1	2	3	4	5
Surface Area	4π	16π	36π	64π	100π
Volume	π	8π	27π	64π	125π

proportional

2 ACTIVITY: Comparing Surface Areas and Volumes

Work with a partner. Complete the table. Describe the pattern. Are the dimensions proportional? Explain.

Base Side	6	12	18	24	30
Height	4	8	12	16	20
Slant Height	5	10	15	20	25
Surface Area	96	384	864	1536	2400
Volume	48	384	1296	3072	6000

proportional

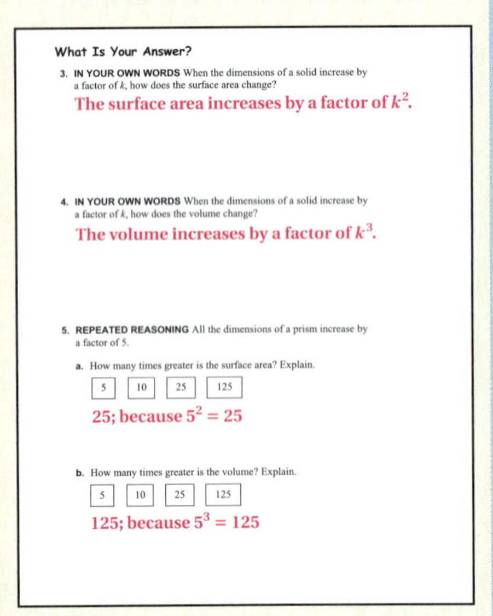

What Is Your Answer?

3. **IN YOUR OWN WORDS** When the dimensions of a solid increase by a factor of k, how does the surface area change?
The surface area increases by a factor of k^2.

4. **IN YOUR OWN WORDS** When the dimensions of a solid increase by a factor of k, how does the volume change?
The volume increases by a factor of k^3.

5. **REPEATED REASONING** All the dimensions of a prism increase by a factor of 5.

a. How many times greater is the surface area? Explain.
5 10 25 125
25; because $5^2 = 25$

b. How many times greater is the volume? Explain.
5 10 25 125
125; because $5^3 = 125$

Laurie's Notes

Activity 2

- You want students to see a pattern. It may be helpful for students to use a calculator.
- **MP8 Look for and Express Regularity in Repeated Reasoning:** Ask students to describe patterns they see. Remind them to think about factors (multiplication) versus addition. The first pyramid should be referred to as the original pyramid. Describe any patterns in terms of the original pyramid.

? "Are the dimensions proportional? Explain." yes; The three dimensions are all changing by factors of 2, 3, 4, and 5 times the original dimensions.

- To help students see the factor by which the surface areas and volumes are multiplied, they should divide the new surface area (or volume) by the original surface area (or volume).

Example for Blue Pyramid

Base Side	$24 \div 6 = 4$	Multiplied by a scale factor of 4
Height	$16 \div 4 = 4$	Multiplied by a scale factor of 4
Slant Height	$20 \div 5 = 4$	Multiplied by a scale factor of 4
Surface Area	$1536 \div 96 = 16$	Multiplied by a scale factor of 4^2
Volume	$3072 \div 48 = 64$	Multiplied by a scale factor of 4^3

- **Big Idea:** When the dimensions of a solid are all multiplied by a scale factor of k, the surface area is multiplied by a scale factor of k^2, and the volume is multiplied by a scale factor of k^3.

What Is Your Answer?

- **Think-Pair-Share:** Students should read each question independently and then work in pairs to answer the questions. When they have answered the questions, the pair should compare their answers with another group and discuss any discrepancies.

Closure

- Refer to Papa Bear's feather bed mattress. If the dimensions are all double Baby Bear's mattress, how many times more feathers are there? 8 times more feathers

2 ACTIVITY: Comparing Surface Areas and Volumes

Work with a partner. Copy and complete the table. Describe the pattern. Are the dimensions proportional? Explain.

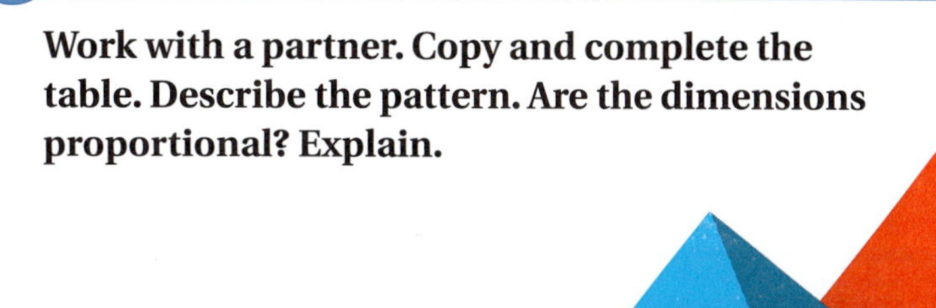

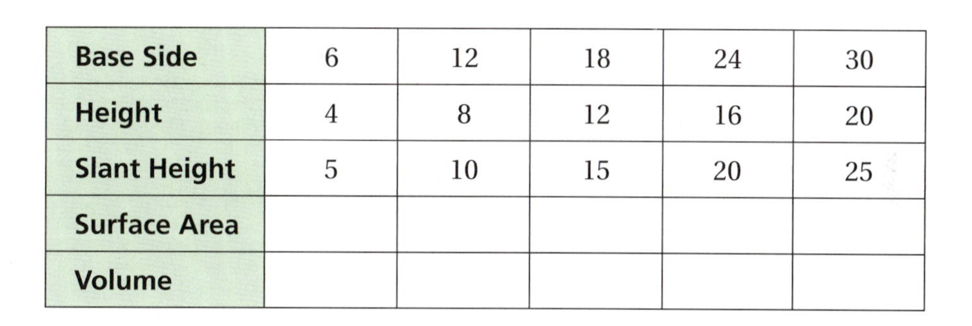

Math Practice 8

Repeat Calculations

Which calculations are repeated? How does this help you describe the pattern?

Base Side	6	12	18	24	30
Height	4	8	12	16	20
Slant Height	5	10	15	20	25
Surface Area					
Volume					

What Is Your Answer?

3. **IN YOUR OWN WORDS** When the dimensions of a solid increase by a factor of k, how does the surface area change?

4. **IN YOUR OWN WORDS** When the dimensions of a solid increase by a factor of k, how does the volume change?

5. **REPEATED REASONING** All the dimensions of a prism increase by a factor of 5.

   a. How many times greater is the surface area? Explain.

   5   10   25   125

   b. How many times greater is the volume? Explain.

   5   10   25   125

Practice Use what you learned about surface areas and volumes of similar solids to complete Exercise 3 on page 359.

8.4 Lesson

Key Vocabulary
similar solids, *p. 356*

Similar solids are solids that have the same shape and proportional corresponding dimensions.

EXAMPLE 1 Identifying Similar Solids

Which cylinder is similar to Cylinder A?

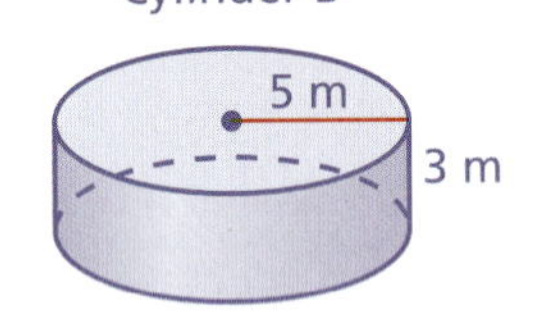

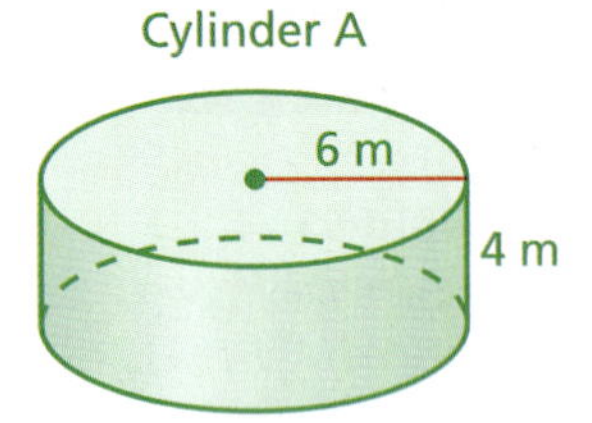

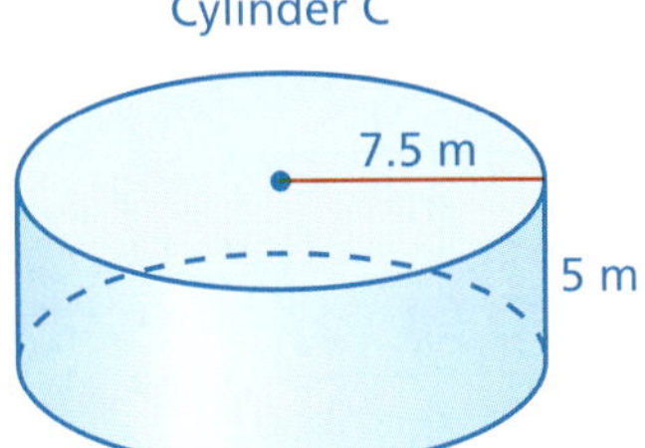

Check to see if corresponding dimensions are proportional.

Cylinder A and Cylinder B

$\dfrac{\text{Height of A}}{\text{Height of B}} = \dfrac{4}{3}$ $\qquad$ $\dfrac{\text{Radius of A}}{\text{Radius of B}} = \dfrac{6}{5}$ $\qquad$ Not proportional

Cylinder A and Cylinder C

$\dfrac{\text{Height of A}}{\text{Height of C}} = \dfrac{4}{5}$ $\qquad$ $\dfrac{\text{Radius of A}}{\text{Radius of C}} = \dfrac{6}{7.5} = \dfrac{4}{5}$ $\qquad$ Proportional

So, Cylinder C is similar to Cylinder A.

EXAMPLE 2 Finding Missing Measures in Similar Solids

The cones are similar. Find the missing slant height ℓ.

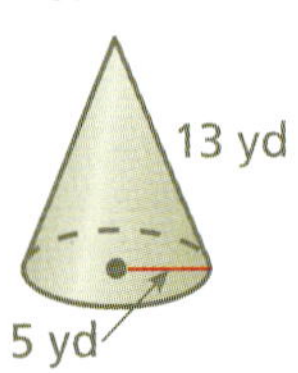

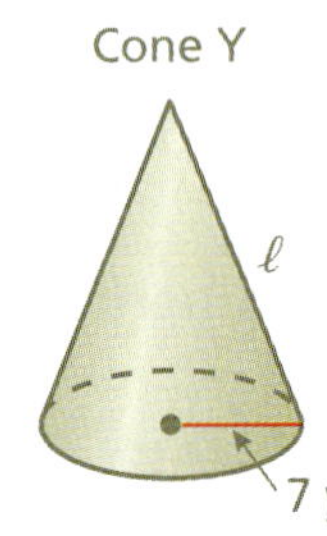

$$\frac{\text{Radius of X}}{\text{Radius of Y}} = \frac{\text{Slant height of X}}{\text{Slant height of Y}}$$

$\dfrac{5}{7} = \dfrac{13}{\ell}$ $\qquad$ Substitute.

$5\ell = 91$ $\qquad$ Cross Products Property

$\ell = 18.2$ $\qquad$ Divide each side by 5.

The slant height is 18.2 yards.

On Your Own

1. Cylinder D has a radius of 7.5 meters and a height of 4.5 meters. Which cylinder in Example 1 is similar to Cylinder D?
2. The prisms at the right are similar. Find the missing width and length.

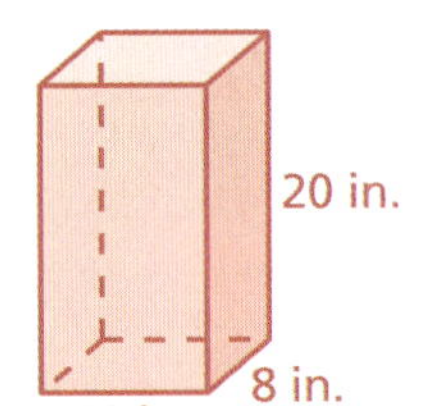

Multi-Language Glossary at BigIdeasMath.com

Laurie's Notes

Introduction

Connect

- **Yesterday:** Students explored what happens to the surface areas and volumes of solids when the dimensions are multiplied by a factor of k. (MP8)
- **Today:** Students will use properties of similar solids to solve problems.
- **FYI:** This is a long lesson and it may take more than one day to cover. Take your time and present the concepts well.

Motivate

- **Movie Time:** Hold an object that is miniature in size (model car, doll house item, statue, and so on). Tell students that this is a prop from a movie set. The movie plot is about giants. In order to make people look large, all of the props have been shrunk proportionally.
- Spend some time talking about movie making. Creating props larger than normal will make people appear smaller than normal, and vice versa.

Lesson Notes

Example 1

- Note that the definition simply states that the corresponding linear measures must be proportional. This means that similar solids are proportional in size.

? "What is a proportion?" an equation of two equal ratios

? "How do you know if two ratios are equal?" Students might say by eyesight; by simple arithmetic, like $\frac{1}{2} = \frac{2}{4}$; that the ratios simplify to the same ratio. Students should recall the Cross Products Property.

- Work through the example.
- Be sure to write the words and the numbers. Use language such as "the ratio of the height of A to the height of B is 4 to 3."

? "How do you know $\frac{6}{7.5} = \frac{4}{5}$?" Answers may vary depending upon students' number sense. By the Cross Products Property $6 \times 5 = 7.5 \times 4$.

Example 2

? "By the definition of similar solids, what can you determine about two similar cones?" Corresponding linear measures are proportional.

- Set up the proportion and solve for the missing slant height.

On Your Own

- **Think-Pair-Share:** Students should read each question independently and then work in pairs to answer the questions. When they have answered the questions, the pair should compare their answers with another group and discuss any discrepancies.
- Ask volunteers to put their work on the board.

Goal Today's lesson is finding the surface areas and volumes of **similar solids**.

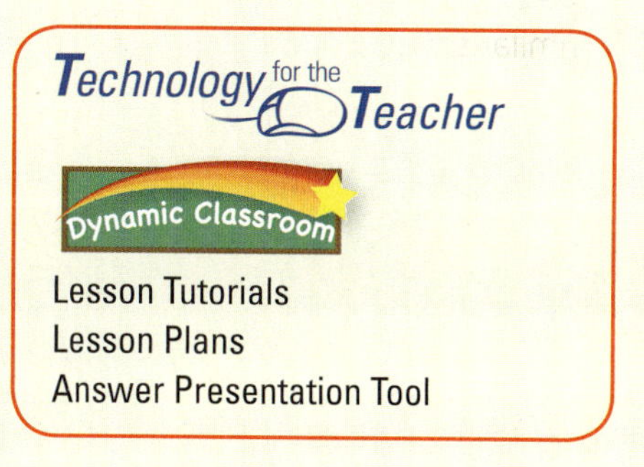

Extra Example 1

Which prism is similar to Prism A?

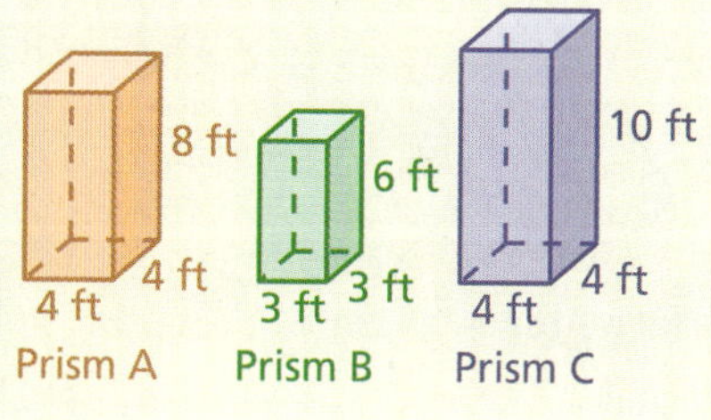

Prism B

Extra Example 2

The square pyramids are similar. Find the length of the base of Pyramid E.

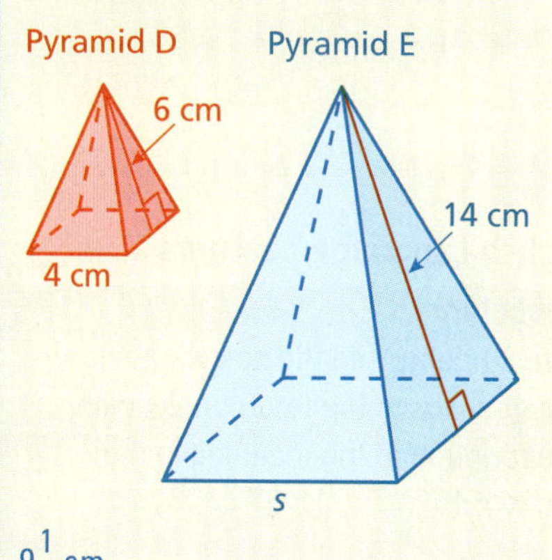

$9\frac{1}{3}$ cm

On Your Own

1. Cylinder B
2. $w = 3.2$ in.
   $\ell = 4.4$ in.

Laurie's Notes

Key Ideas

- Write the Key Ideas.
- **Example:** When the linear dimensions of B are double A, the dimensions are in the ratio of $\frac{1}{2}$, and the surface areas are in the ratio of $\left(\frac{1}{2}\right)^2$ or $\frac{1}{4}$.
- Refer to yesterday's activity to confirm that this relationship was found in Activity 1, part (b) and Activity 2.
- Students may not know how to find surface areas of cones or spheres. However, they do not need this skill to determine whether two cones or two spheres are similar. The relationship given in the Key Idea applies to all similar solids.

Extra Example 3

The cones are similar. What is the surface area of Cone G? Round your answer to the nearest tenth.

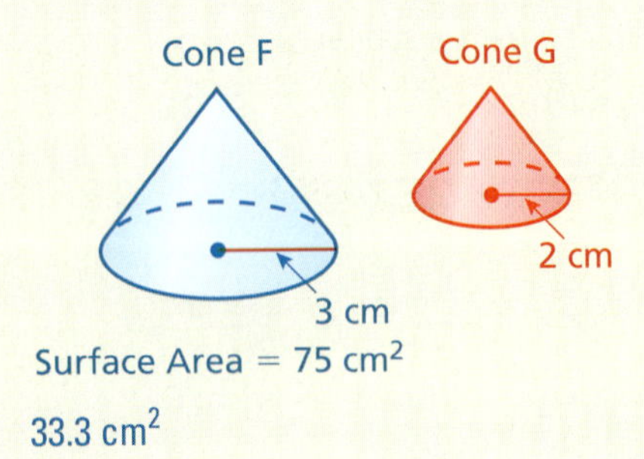

33.3 cm^2

Example 3

? "Do you have enough information to solve this problem? Explain." yes; The heights are in the ratio of $\frac{6}{10}$, so the surface areas are in the ratio of $\left(\frac{6}{10}\right)^2$.

- Set up the problem and solve.
- **FYI:** Notice that the problem is solved using the Multiplication Property of Equality. It could also be solved using the Cross Products Property.
- **Connection:** The ratio:

  $$\frac{\text{dimension of A}}{\text{dimension of B}}$$

  is the scale factor. The square of the scale factor is used to find the unknown surface area.

On Your Own

3. 237.5 m^2
4. 171.9 cm^2

On Your Own

- Students should first identify the ratio of the corresponding linear measurements. Question 3: $\frac{5}{8}$; Question 4: $\frac{5}{4}$
- Ask volunteers to share their work at the board.

English Language Learners

Vocabulary

Have students add the key vocabulary *similar solids* to their notebooks with a description of the meaning in their own words.

Key Ideas

Linear Measures

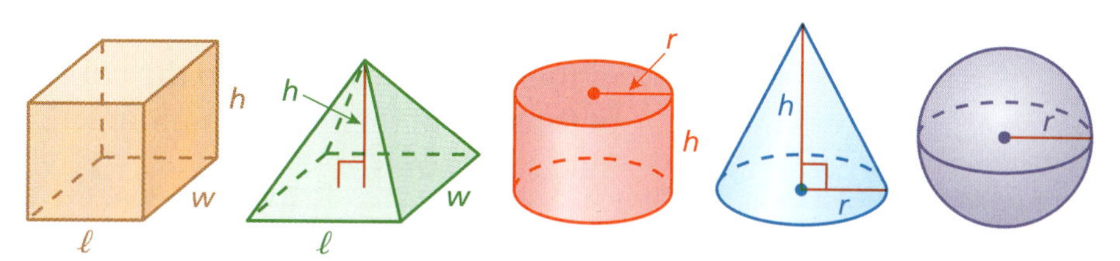

Surface Areas of Similar Solids

When two solids are similar, the ratio of their surface areas is equal to the square of the ratio of their corresponding linear measures.

$$\frac{\text{Surface Area of A}}{\text{Surface Area of B}} = \left(\frac{a}{b}\right)^2$$

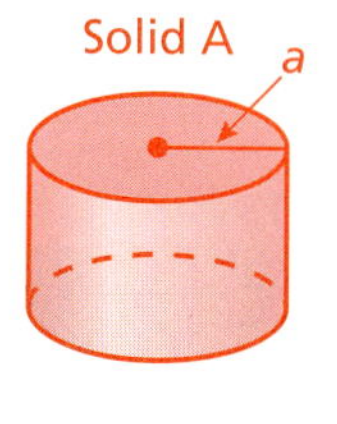

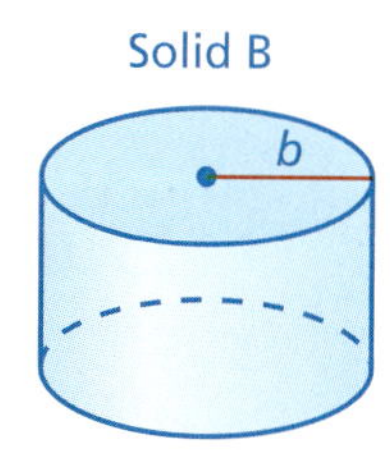

EXAMPLE 3 Finding Surface Area

Pyramid A

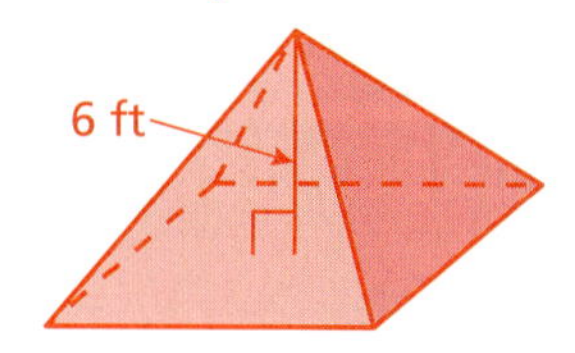

Pyramid B

10 ft

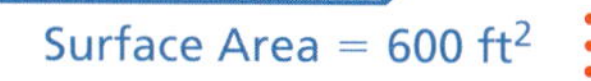

The pyramids are similar. What is the surface area of Pyramid A?

$$\frac{\text{Surface Area of A}}{\text{Surface Area of B}} = \left(\frac{\text{Height of A}}{\text{Height of B}}\right)^2$$

$$\frac{S}{600} = \left(\frac{6}{10}\right)^2 \qquad \text{Substitute.}$$

$$\frac{S}{600} = \frac{36}{100} \qquad \text{Evaluate.}$$

$$\frac{S}{600} \cdot 600 = \frac{36}{100} \cdot 600 \qquad \text{Multiplication Property of Equality}$$

$$S = 216 \qquad \text{Simplify.}$$

∴ The surface area of Pyramid A is 216 square feet.

On Your Own

The solids are similar. Find the surface area of the red solid. Round your answer to the nearest tenth.

3.

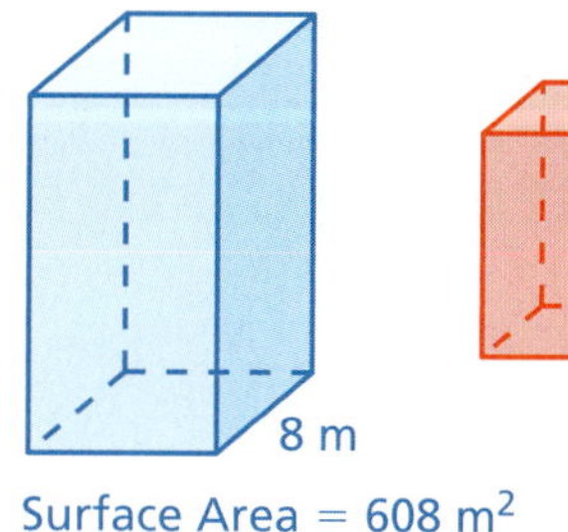

4.

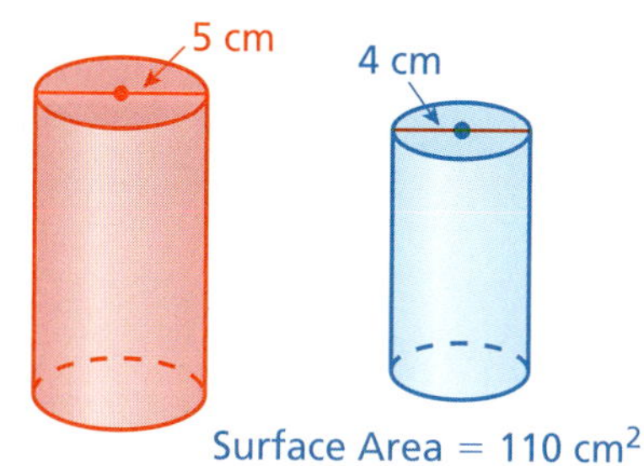

Volumes of Similar Solids

When two solids are similar, the ratio of their volumes is equal to the cube of the ratio of their corresponding linear measures.

$$\frac{\text{Volume of A}}{\text{Volume of B}} = \left(\frac{a}{b}\right)^3$$

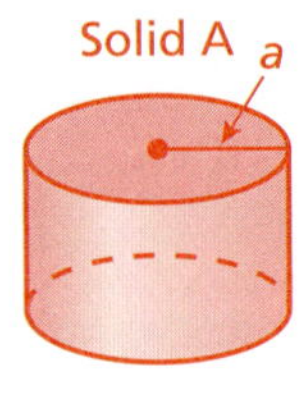

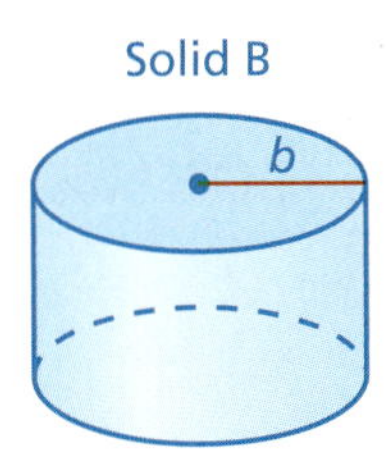

EXAMPLE 4 Finding Volume

Original Tank

Volume = 2000 ft^3

The dimensions of the touch tank at an aquarium are doubled. What is the volume of the new touch tank?

Ⓐ 150 ft^3 Ⓑ 4000 ft^3 Ⓒ 8000 ft^3 Ⓓ 16,000 ft^3

The dimensions are doubled, so the ratio of the dimensions of the original tank to the dimensions of the new tank is 1 : 2.

$$\frac{\text{Original volume}}{\text{New volume}} = \left(\frac{\text{Original dimension}}{\text{New dimension}}\right)^3$$

$$\frac{2000}{V} = \left(\frac{1}{2}\right)^3 \quad \text{Substitute.}$$

$$\frac{2000}{V} = \frac{1}{8} \quad \text{Evaluate.}$$

$$16{,}000 = V \quad \text{Cross Products Property}$$

∴ The volume of the new tank is 16,000 cubic feet. So, the correct answer is Ⓓ.

Study Tip

When the dimensions of a solid are multiplied by k, the surface area is multiplied by k^2 and the volume is multiplied by k^3.

On Your Own

Now You're Ready Exercises 10–13

The solids are similar. Find the volume of the red solid. Round your answer to the nearest tenth.

5.

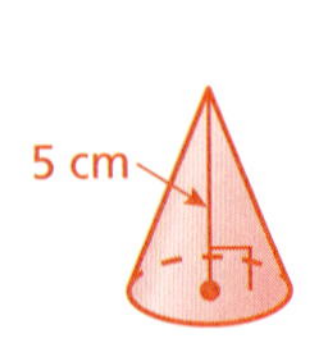

6.

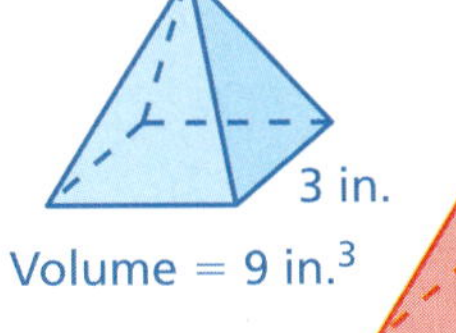

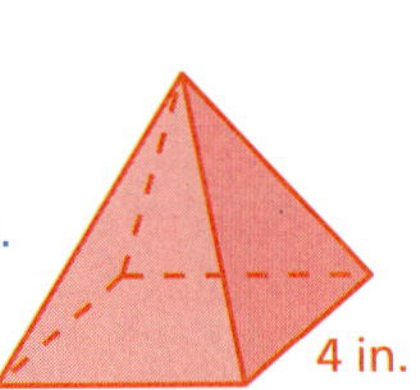

Laurie's Notes

Key Idea

- Write the Key Idea.
- **Example:** When the linear dimensions of B are double A, the dimensions are in the ratio of $\frac{1}{2}$, and the volumes are in the ratio of $\left(\frac{1}{2}\right)^3$ or $\frac{1}{8}$.
- Refer to the activity to confirm that this relationship was found in Activity 1, part (b) and Activity 2.

Example 4

- Write the problem in words to help students recognize how the numbers are being substituted.
- **Common Misconception:** Many students think that when you double the dimensions, the surface area and volume also double. This Big Idea takes time for students to fully understand.
- **Connection:** The ratio:

  $$\frac{\text{original dimension}}{\text{new dimension}}$$

  is the scale factor. The cube of the scale factor is used to find the new volume.

On Your Own

- Students should first identify the ratio of the corresponding linear measurements. Question 5: $\frac{5}{12}$; Question 6: $\frac{4}{3}$
- Ask volunteers to share their work at the board.

Closure

- Use one of the miniature items used to motivate the lesson and ask a question related to surface area or volume. Some miniature items have a scale printed on the item.

Extra Example 4

The cylinders are similar. Find the volume of Cylinder J. Round your answer to the nearest tenth.

Cylinder H

4 in.

Volume = 314 in.^{3}

Cylinder J

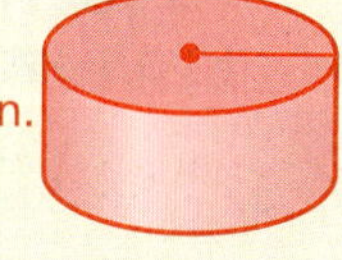

1059.8 in.^{3}

On Your Own

5. 20.8 cm^3
6. 21.3 in.^{3}

Vocabulary and Concept Check

1. Similar solids are solids of the same type that have proportional corresponding linear measures.

2. *Sample answer:*

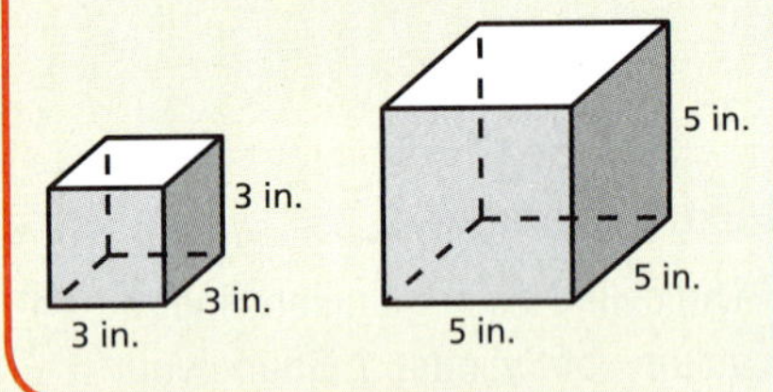

Practice and Problem Solving

3. **a.** $\frac{9}{4}$; because $\left(\frac{3}{2}\right)^2 = \frac{9}{4}$

   b. $\frac{27}{8}$; because $\left(\frac{3}{2}\right)^3 = \frac{27}{8}$

4. yes

5. no

6. yes

7. no

8. 25 in.

9. $b = 18$ m
   $c = 19.5$ m
   $h = 9$ m

Assignment Guide and Homework Check

Level	Day 1 Activity Assignment	Day 2 Lesson Assignment	Homework Check
Basic	3, 20–22	1, 2, 4–15	5, 9, 12, 15
Average	3, 20–22	1, 2, 6–17	7, 9, 12, 15, 17
Advanced	1–3, 7–22		7, 9, 13, 15, 19

Common Errors

- **Exercises 4–6** Students may compare only two pairs of corresponding dimensions instead of all three. The bases may be similar, but the heights may not be proportional. Remind them to check all of the corresponding dimensions when determining whether two solids are similar. Ask them how many ratios they need to write for each type of solid.
- **Exercises 8 and 9** Students may write the proportion incorrectly. For example, in Exercise 8, they may write $\frac{10}{4} = \frac{10}{d}$. Remind them to write the proportion correctly and to check their work to make sure it makes sense.

8.4 Record and Practice Journal

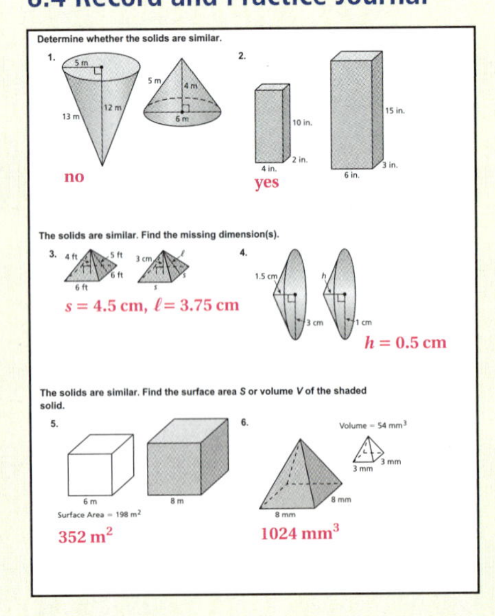

Determine whether the solids are similar.

1. no

2. yes

The solids are similar. Find the missing dimension(s).

3. $s = 4.5$ cm, $\ell = 3.75$ cm

4. $h = 0.5$ cm

The solids are similar. Find the surface area S or volume V of the shaded solid.

5. Surface Area = 198 m^2
   352 m^2

6. Volume = 54 mm^3
   1024 mm^3

8.4 Exercises

Vocabulary and Concept Check

1. **VOCABULARY** What are similar solids?
2. **OPEN-ENDED** Draw two similar solids and label their corresponding linear measures.

Practice and Problem Solving

3. **NUMBER SENSE** All the dimensions of a cube increase by a factor of $\frac{3}{2}$.
   - **a.** How many times greater is the surface area? Explain.
   - **b.** How many times greater is the volume? Explain.

Determine whether the solids are similar.

4.

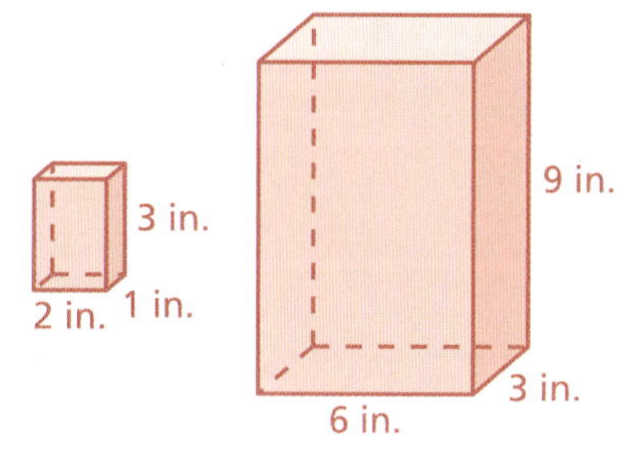

5.

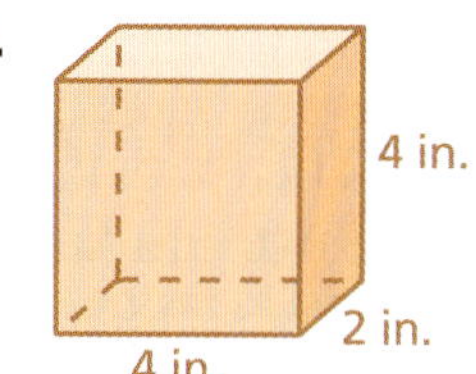

4 in.
2 in.
1 in.

6.

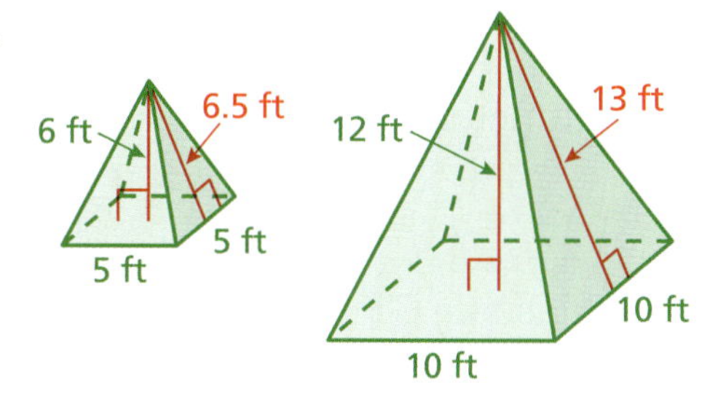

7.

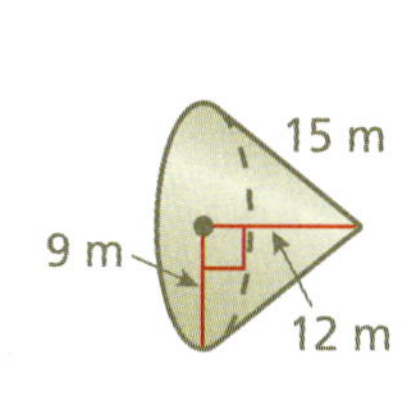

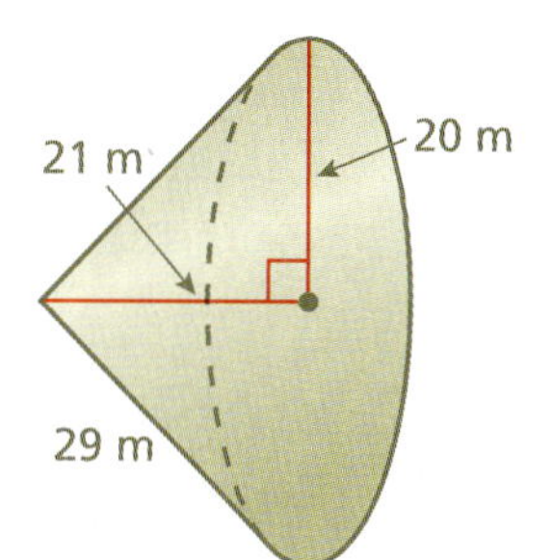

The solids are similar. Find the missing dimension(s).

8.

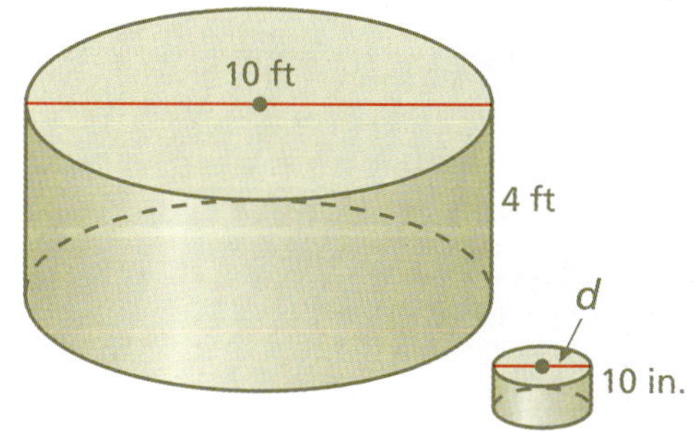

9.

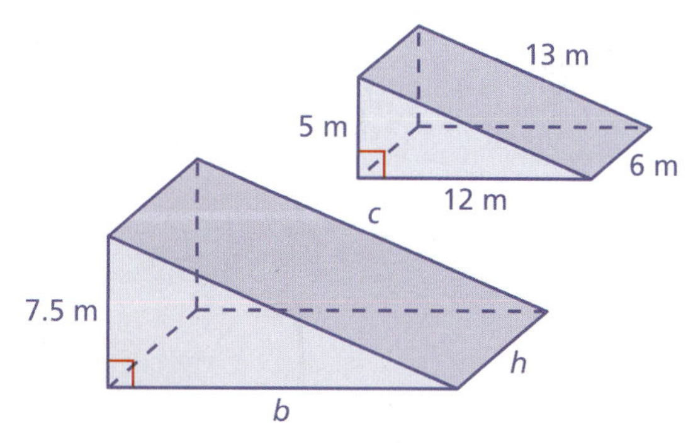

The solids are similar. Find the surface area S or volume V of the red solid. Round your answer to the nearest tenth.

10.

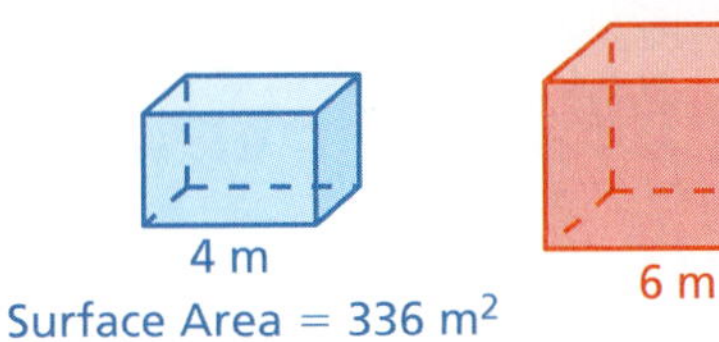

11.

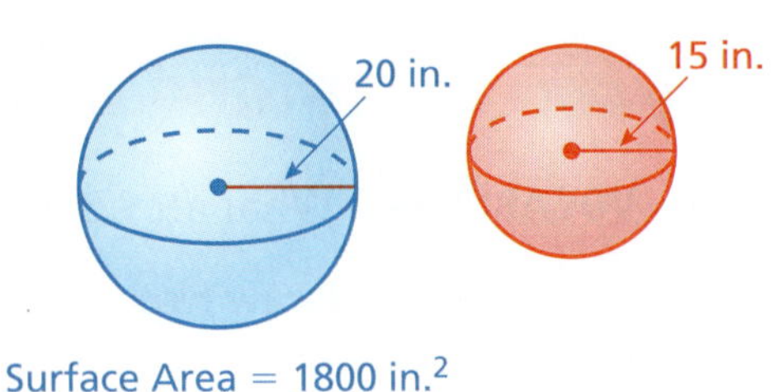

12.

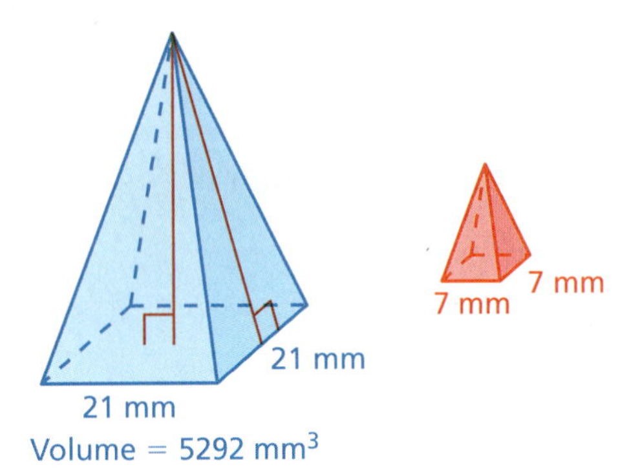

13.

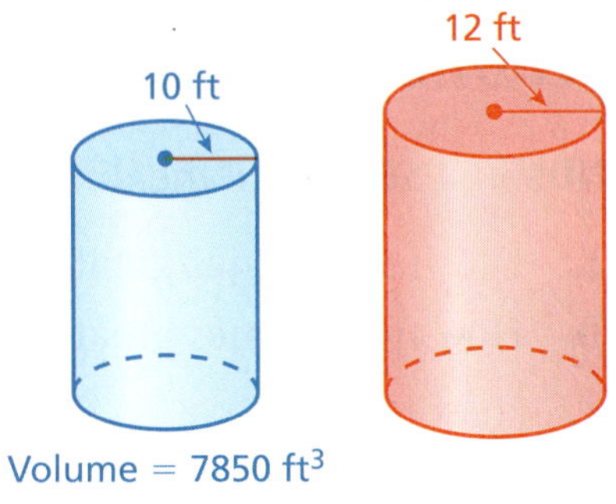

14. ERROR ANALYSIS The ratio of the corresponding linear measures of two similar solids is 3 : 5. The volume of the smaller solid is 108 cubic inches. Describe and correct the error in finding the volume of the larger solid.

$$\frac{108}{V} = \left(\frac{3}{5}\right)^2$$

$$\frac{108}{V} = \frac{9}{25}$$

$$300 = V$$

The volume of the larger solid is 300 cubic inches.

15. MIXED FRUIT The ratio of the corresponding linear measures of two similar cans of fruit is 4 to 7. The smaller can has a surface area of 220 square centimeters. Find the surface area of the larger can.

16. CLASSIC MUSTANG The volume of a 1968 Ford Mustang GT engine is 390 cubic inches. Which scale model of the Mustang has the greater engine volume, a 1 : 18 scale model or a 1 : 24 scale model? How much greater is it?

Common Errors

- **Exercises 10–13** Students may forget to square or cube the ratio of the corresponding linear measures when finding the surface area or volume of the red solid. Remind them of the Key Ideas in this section.
- **Exercises 10–13** Students may cube the ratio of corresponding linear measures when finding surface area or square the ratio of corresponding linear measures when finding volume. Remind them that the ratio of corresponding linear measures is squared for surface area and cubed for volume.
- **Exercise 15** Students may write the ratio of the surface areas incorrectly in the proportion. When they look at the ratio of corresponding linear measures as a fraction, ask whether the numerator or denominator corresponds to the smaller figure. This should help them write the ratio of the surface areas correctly.

10. 756 m^2

11. 1012.5 in.^2

12. 196 mm^3

13. $13{,}564.8 \text{ ft}^3$

14. The ratio of the volumes of two similar solids is equal to the cube of the ratio of their corresponding linear measures.

$$\frac{108}{V} = \left(\frac{3}{5}\right)^3$$

$$\frac{108}{V} = \frac{27}{125}$$

$$V = 500 \text{ in.}^3$$

15. 673.75 cm^2

16. 1 : 18 scale model; about 0.04 in.^3

English Language Learners

Pair Activity

As a confidence booster for English learners, have students write their own math problems. Place students in pairs, one English learner and one English speaker, and ask each pair to write a problem involving surface areas or volumes of similar solids. On a separate piece of paper, students solve their own problem. Next, students exchange their problem with another pair of students. Each pair then solves the new problem. Finally, the four students discuss the problems and the solution methods.

Practice and Problem Solving

17. See Additional Answers.

18. See *Taking Math Deeper.*

19. a. yes; Because all circles are similar, the slant height and the circumference of the base of the cones are proportional.

b. no; because the ratio of the volumes of similar solids is equal to the cube of the ratio of their corresponding linear measures

Fair Game Review

20–22. See Additional Answers.

Mini-Assessment

The solids are similar. Find the surface area *S* of the red solid. Round your answer to the nearest tenth.

1. 288 m^2

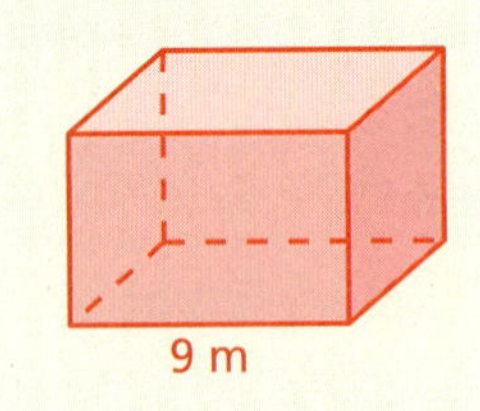

Surface Area = 32 m^2

2. 37.8 in.^{2}

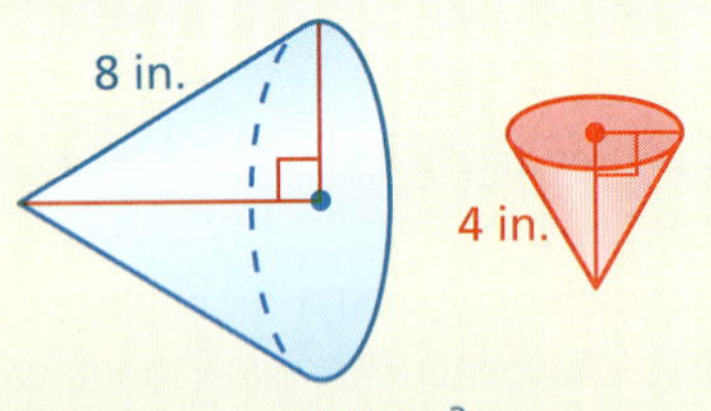

Surface Area = 151 in.^{2}

3. The cylinders are similar. Find the volume *V* of the red cylinder. 600 in.^{3}

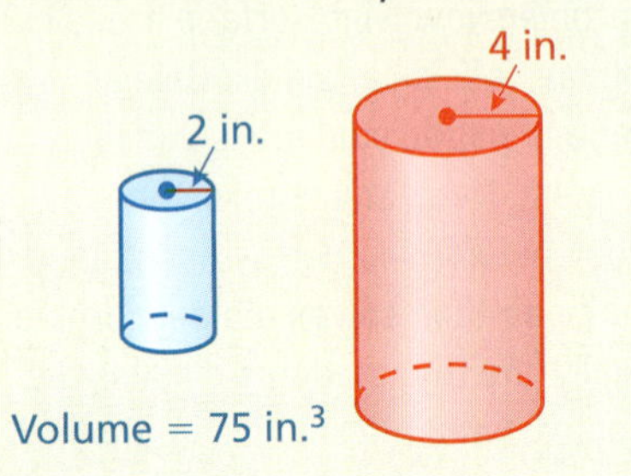

Volume = 75 in.^{3}

Taking Math Deeper

Exercise 18

This problem is a straightforward application of the two main concepts of the lesson. That is, with similar solids, the ratio of their surface areas is equal to the square of the ratio of their corresponding linear measures. Also, the ratio of their volumes is equal to the cube of the ratio of their corresponding linear measures. Even so, students have trouble with this problem because they do not see that they can let the surface area and volume of the shortest doll be *S* and *V*.

 Make a table. Include the height of each doll.

Height	1	2	3	4	5	6	7
Surface Area	S	$4S$	$9S$	$16S$	$25S$	$36S$	$49S$
Volume	V	$8V$	$27V$	$64V$	$125V$	$216V$	$343V$

2 Compare the surface areas of the dolls.

3 Compare the volumes of the dolls.

Matryoshka dolls, or Russian nested dolls, are also called stacking dolls. A set of matryoshkas consists of a wooden figure which can be pulled apart to reveal another figure of the same sort inside. It has, in turn, another figure inside, and so on. The number of nested figures is usually five or more.

Reteaching and Enrichment Strategies

If students need help. . .	If students got it. . .
Resources by Chapter • Practice A and Practice B • Puzzle Time Record and Practice Journal Practice Differentiating the Lesson Lesson Tutorials Skills Review Handbook	Resources by Chapter • Enrichment and Extension • Technology Connection Start the next section

17. **MARBLE STATUE** You have a small marble statue of Wolfgang Mozart. It is 10 inches tall and weighs 16 pounds. The original statue is 7 feet tall.

a. Estimate the weight of the original statue. Explain your reasoning.

b. If the original statue were 20 feet tall, how much would it weigh?

Wolfgang Mozart

18. **REPEATED REASONING** The largest doll is 7 inches tall. Each of the other dolls is 1 inch shorter than the next larger doll. Make a table that compares the surface areas and the volumes of the seven dolls.

19. **Precision** You and a friend make paper cones to collect beach glass. You cut out the largest possible three-fourths circle from each piece of paper.

a. Are the cones similar? Explain your reasoning.

b. Your friend says that because your sheet of paper is twice as large, your cone will hold exactly twice the volume of beach glass. Is this true? Explain your reasoning.

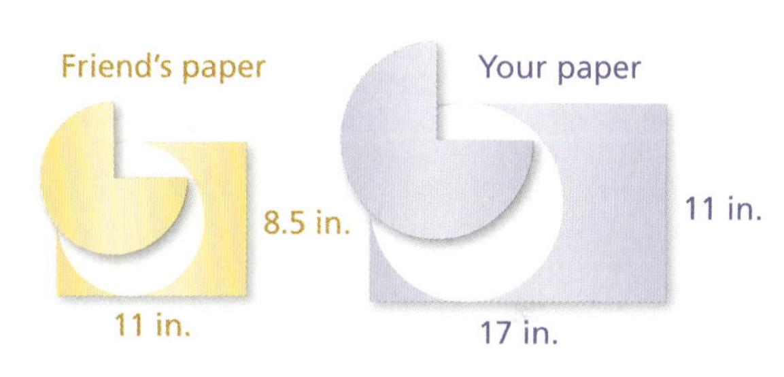

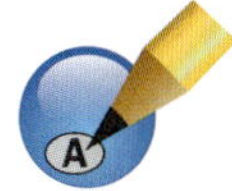

Fair Game Review What you learned in previous grades & lessons

Draw the figure and its reflection in the x-axis. Identify the coordinates of the image. *(Section 2.3)*

20. $A(1, 1)$, $B(3, 4)$, $C(4, 2)$

21. $J(-3, 0)$, $K(-4, 3)$, $L(-1, 4)$

22. **MULTIPLE CHOICE** Which system of linear equations has no solution? *(Section 5.4)*

Ⓐ $y = 4x + 1$
$y = -4x + 1$

Ⓑ $y = 2x - 7$
$y = 2x + 7$

Ⓒ $3x + y = 1$
$6x + 2y = 2$

Ⓓ $5x + y = 3$
$x + 5y = 15$

8.3–8.4 Quiz

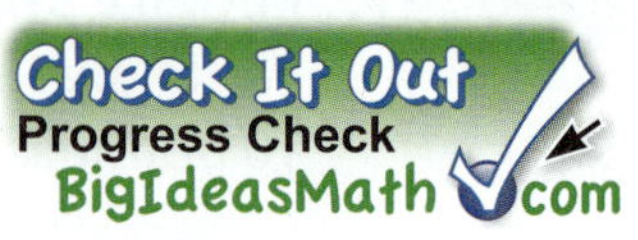

Find the volume of the sphere. Round your answer to the nearest tenth. *(Section 8.3)*

1.

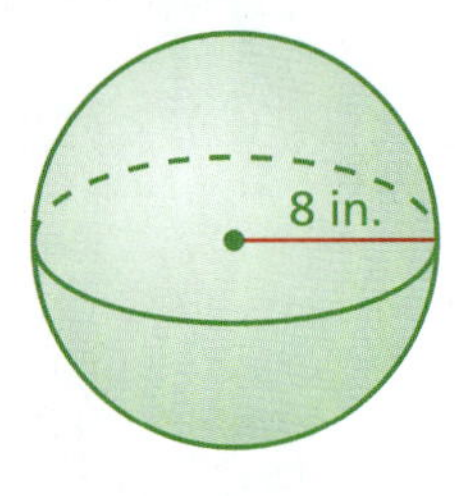

2.

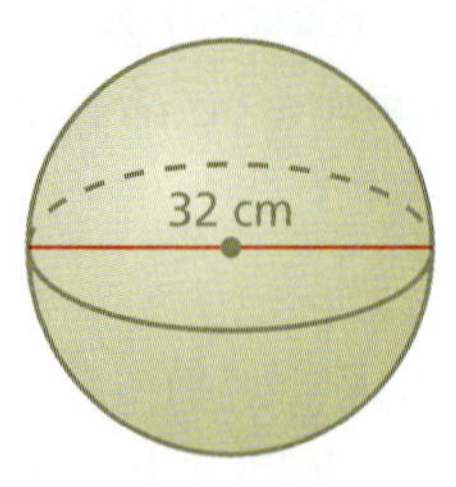

Find the radius of the sphere with the given volume. *(Section 8.3)*

3. Volume $= 4500\pi$ yd^3

4. Volume $= \frac{32}{3}\pi$ ft^3

5. Find the volume of the composite solid. Round your answer to the nearest tenth. *(Section 8.3)*

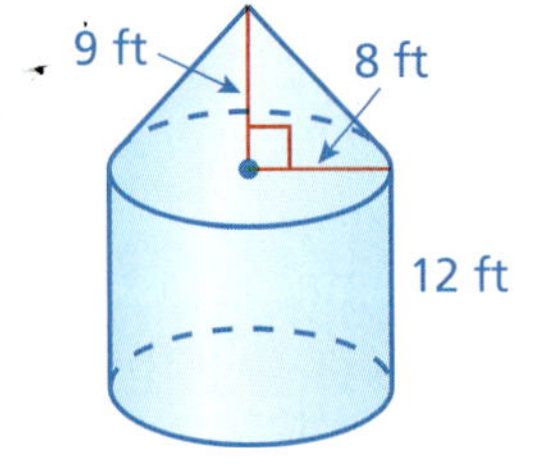

6. Determine whether the solids are similar. *(Section 8.4)*

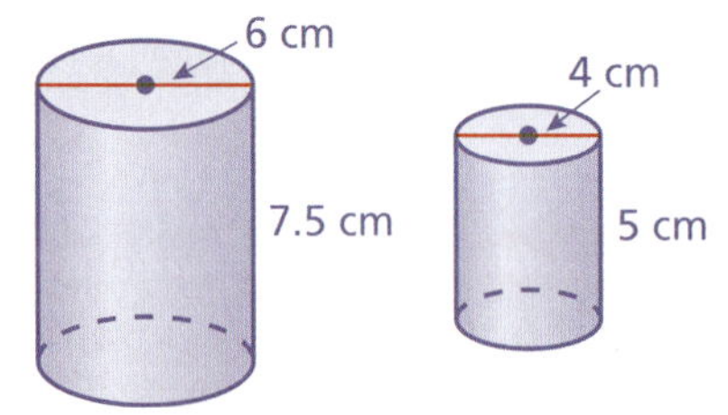

7. The prisms are similar. Find the missing width and height. *(Section 8.4)*

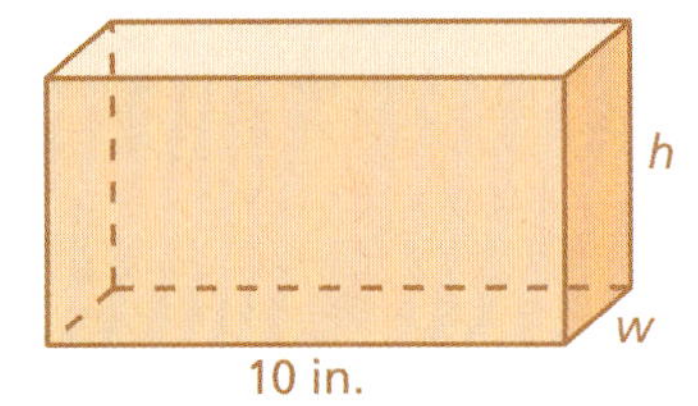

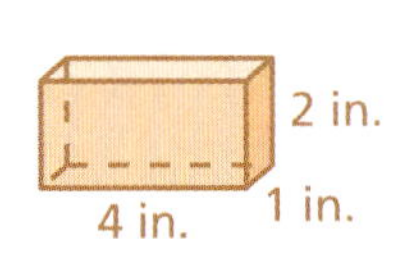

8. The solids are similar. Find the surface area of the red solid. *(Section 8.4)*

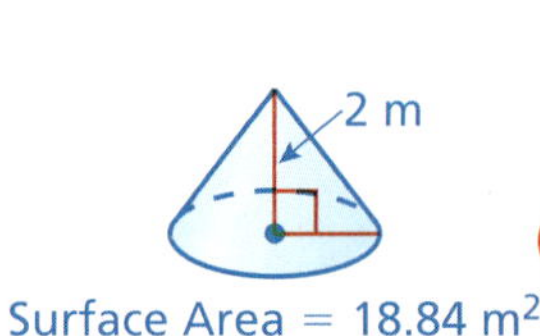

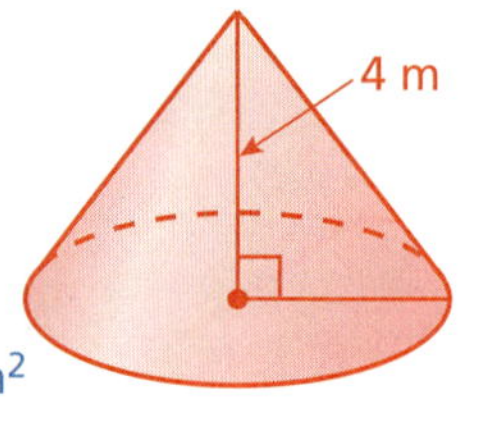

9. HAMSTER A hamster toy is in the shape of a sphere. What is the volume of the toy? Round your answer to the nearest whole number. *(Section 8.3)*

10. JEWELRY BOXES The ratio of the corresponding linear measures of two similar jewelry boxes is 2 to 3. The larger box has a volume of 162 cubic inches. Find the volume of the smaller jewelry box. *(Section 8.4)*

11. ARCADE You win a token after playing an arcade game. What is the volume of the gold ring? Round your answer to the nearest tenth. *(Section 8.3)*

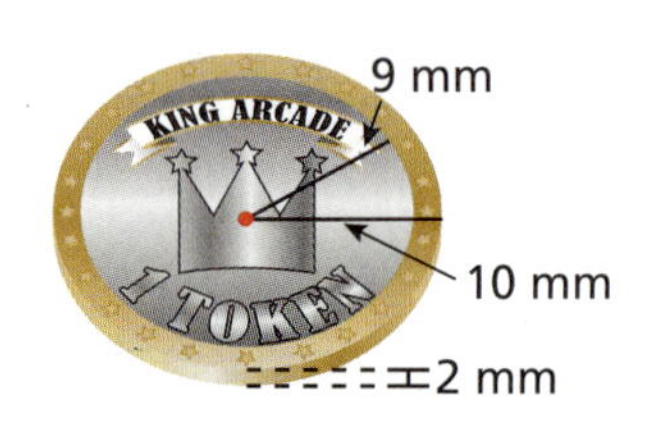

Alternative Assessment Options

Math Chat	Student Reflective Focus Question
Structured Interview	Writing Prompt

Math Chat

Ask students to use their own words to summarize what they know about finding volumes of spheres and finding surface areas and volumes of similar solids. Be sure that they include examples. Select students at random to present their summaries to the class.

Study Help Sample Answers

Remind students to complete Graphic Organizers for the rest of the chapter.

2. Volume of a sphere

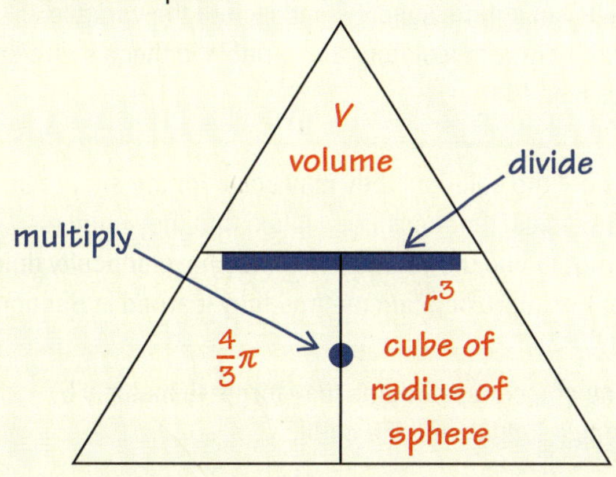

3. Volume of a composite solid

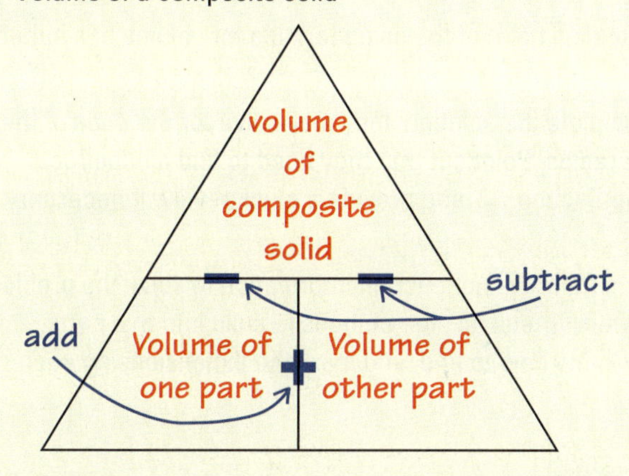

4–5. Available at *BigIdeasMath.com.*

Reteaching and Enrichment Strategies

If students need help. . .	If students got it. . .
Resources by Chapter • Practice A and Practice B • Puzzle Time Lesson Tutorials *BigIdeasMath.com*	Resources by Chapter • Enrichment and Extension • Technology Connection Game Closet at *BigIdeasMath.com* Start the Chapter Review

Answers

1. $\frac{2048\pi}{3} \approx 2144.7 \text{ in.}^3$
2. $\frac{16{,}384\pi}{3} \approx 17{,}157.3 \text{ cm}^3$
3. 15 yd
4. 2 ft
5. $768\pi + 192\pi = 960\pi$
   $\approx 3015.9 \text{ ft}^3$
6. yes
7. $w = 2.5$ in.
   $h = 5$ in.
8. 75.36 m^2
9. $\frac{32\pi}{3} \approx 34 \text{ cm}^3$
10. 48 in.^3
11. $38\pi \approx 119.4 \text{ mm}^3$

Technology for the Teacher

Online Assessment
Assessment Book
ExamView® Assessment Suite

For the Teacher
Additional Review Options

- *BigIdeasMath.com*
- Online Assessment
- Game Closet at *BigIdeasMath.com*
- Vocabulary Help
- Resources by Chapter

Answers

1. $\frac{1575\pi}{4} \approx 1237.0 \text{ ft}^3$
2. $40\pi \approx 125.7 \text{ in.}^3$
3. $108\pi \approx 339.3 \text{ yd}^3$
4. $1458\pi \approx 4580.4 \text{ in.}^3$
5. $\frac{25}{2.25\pi} \approx 4 \text{ in.}$
6. $\sqrt{\frac{7599}{20\pi}} \approx 11 \text{ m}$

Review of Common Errors

Exercises 1–4, 7, 8, 10, and 12–15

- Students may write linear or square units for volume rather than cubic units. Remind them that part of writing a correct answer is including the correct units.

Exercises 1–9, 12, and 14

- When finding the area of the circular base of a cylinder or a cone, students may not square the radius, or they may use the diameter when the formula calls for the radius. Remind them of the formula that they learned for the area of a circle.

Exercises 5, 6, 9, and 11

- Students may be confused about how to find the indicated dimension. Remind them to write the formula, substitute, simplify, and isolate the variable. Some students may need help with isolating the variable in these exercises, particularly in Exercises 6 and 11.

Exercise 6

- Students may not complete the solution; they may solve for the *square* of the radius instead of the radius. If necessary, explain that they need to approximate a square root to find the radius and if they have difficulty doing so, then remind them of the approximation method they learned in Section 7.4.

Exercises 7, 8, and 12

- When finding the volume of a cone, students may forget to multiply by $\frac{1}{3}$. Remind them of the formula for volume of a cone.

Exercise 10

- When finding the volume of the sphere, students may forget to multiply by $\frac{4}{3}$ or cube the radius. Remind them of the given formula for volume of a sphere.

Exercise 11

- Students may not complete the solution; they may solve for the *cube* of the radius instead of the radius. Point out that they need to find a cube root, which they learned in Section 7.2, and provide a quick review if necessary.

Exercises 12–14

- Students may think that there is not enough information to solve the problem. It may help to have them "break up" the composite solid into two parts, whose volumes they know how to find, and mark the dimensions on each part.

Exercise 14

- Students may not realize that the top part of the solid is a hemisphere. Point this out to students.

Exercises 15 and 16

- Students may raise the ratio of the linear measures to the wrong exponent, or forget to square or cube the ratio altogether. Discuss why squaring or cubing the ratio makes sense.

8 Chapter Review

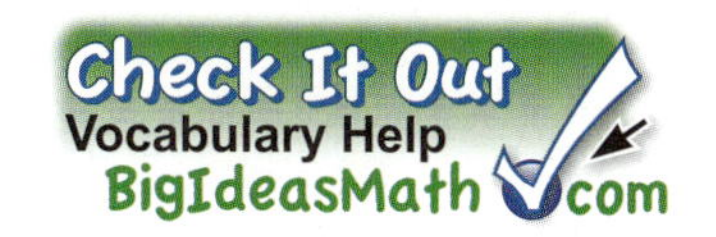

Review Key Vocabulary

sphere, *p. 348* hemisphere, *p. 351* similar solids, *p. 356*

Review Examples and Exercises

8.1 Volumes of Cylinders *(pp. 334–339)*

Find the volume of the cylinder. Round your answer to the nearest tenth.

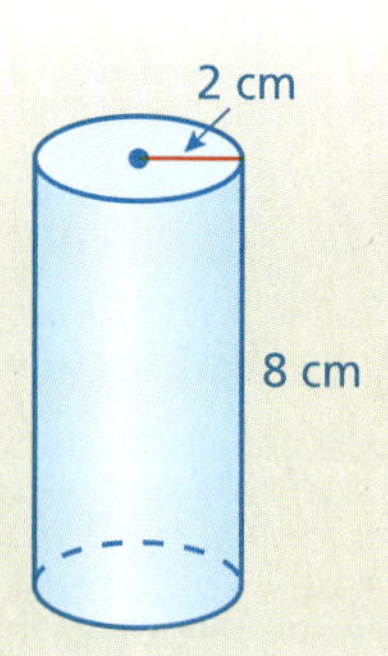

$$V = Bh \quad \text{Write formula for volume.}$$

$$= \pi(2)^2(8) \quad \text{Substitute.}$$

$$= 32\pi \approx 100.5 \quad \text{Use a calculator.}$$

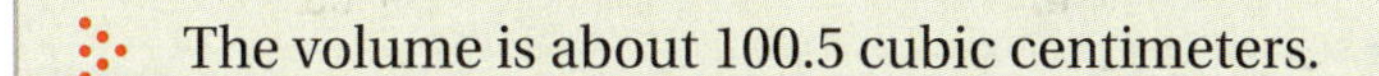

∴ The volume is about 100.5 cubic centimeters.

Exercises

Find the volume of the cylinder. Round your answer to the nearest tenth.

1. 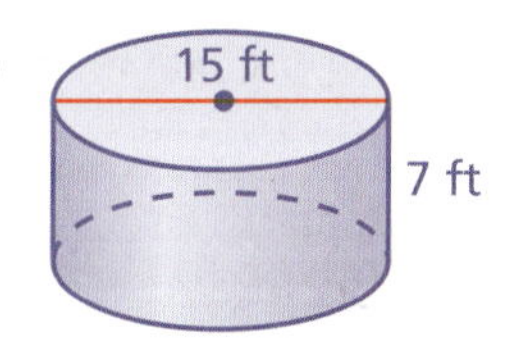

2. 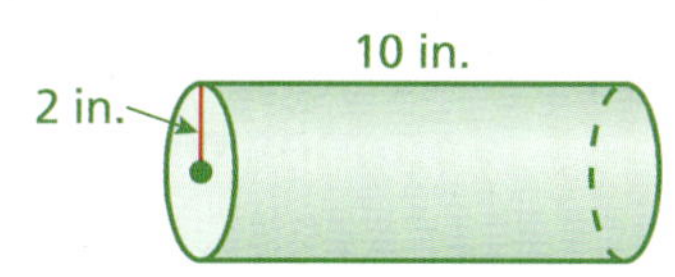

3. 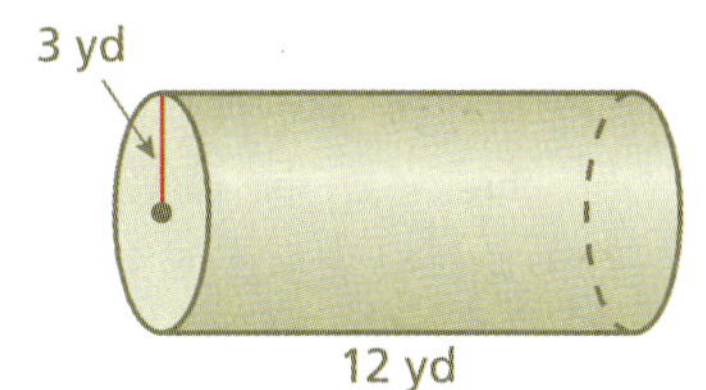

4. 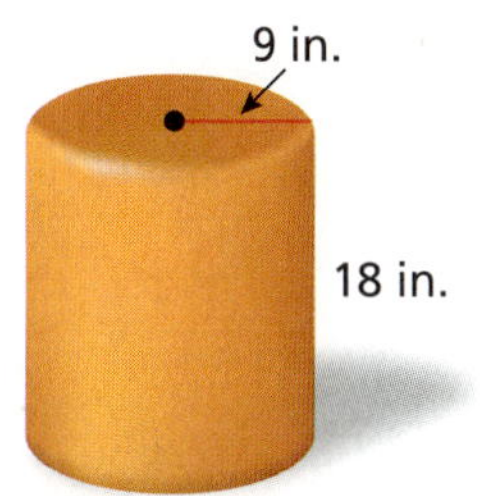

Find the missing dimension of the cylinder. Round your answer to the nearest whole number.

5. Volume = 25 in.^{3}

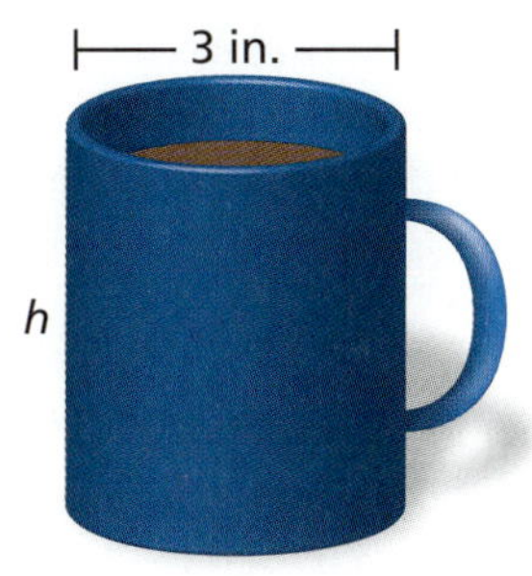

6. Volume = 7599 m^3

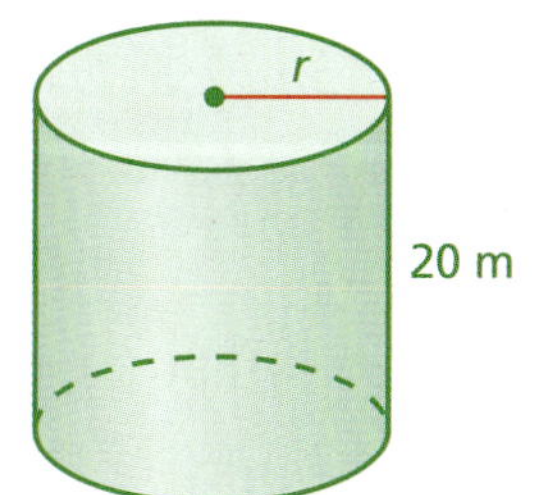

8.2 Volumes of Cones (pp. 340–345)

Find the height of the cone. Round your answer to the nearest tenth.

$V = \frac{1}{3}Bh$ — Write formula for volume.

$900 = \frac{1}{3}\pi(6)^2(h)$ — Substitute.

$900 = 12\pi h$ — Simplify.

$23.9 \approx h$ — Divide each side by 12π.

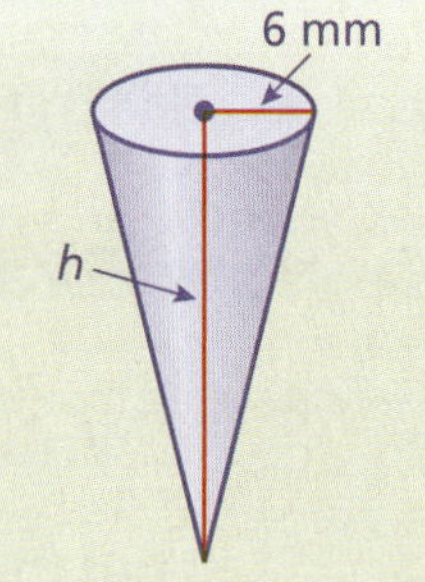

The height is about 23.9 millimeters.

Exercises

Find the volume V or height h of the cone. Round your answer to the nearest tenth.

7. 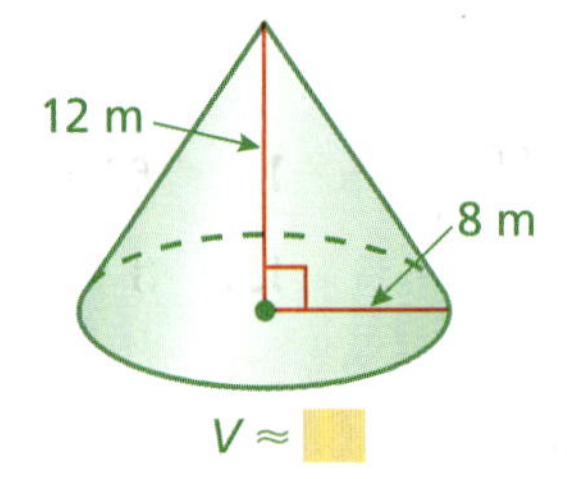

8. 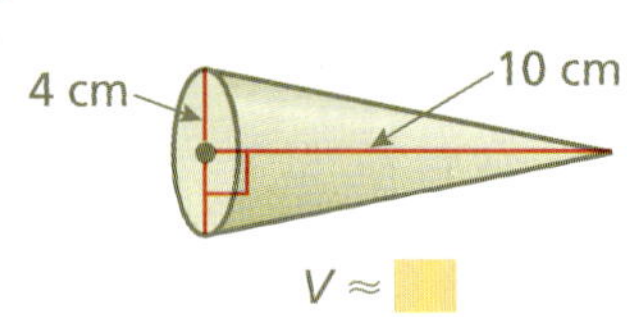

9. 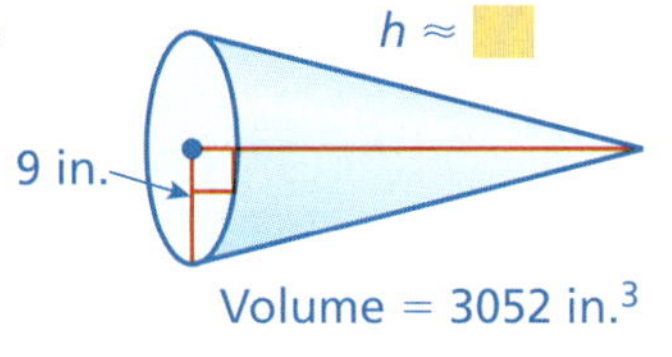

8.3 Volumes of Spheres (pp. 348–353)

a. Find the volume of the sphere. Round your answer to the nearest tenth.

$V = \frac{4}{3}\pi r^3$ — Write formula for volume.

$= \frac{4}{3}\pi(11)^3$ — Substitute 11 for r.

$= \frac{5324}{3}\pi$ — Simplify.

≈ 5575.3 — Use a calculator.

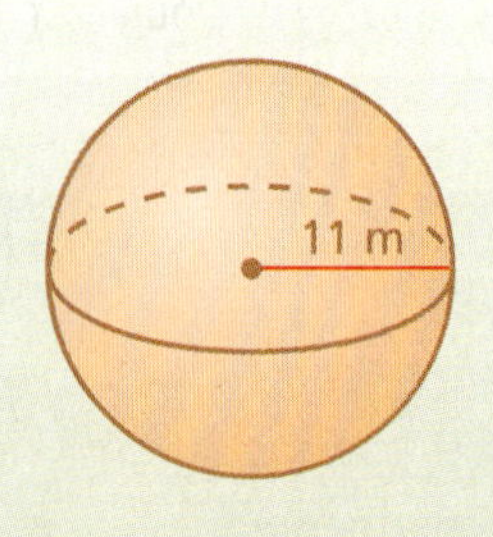

The volume is about 5575.3 cubic meters.

b. Find the volume of the composite solid. Round your answer to the nearest tenth.

Square Prism

$V = Bh$

$= (12)(12)(9)$

$= 1296$

Cylinder

$V = Bh$

$= \pi(5)^2(9)$

$= 225\pi \approx 706.9$

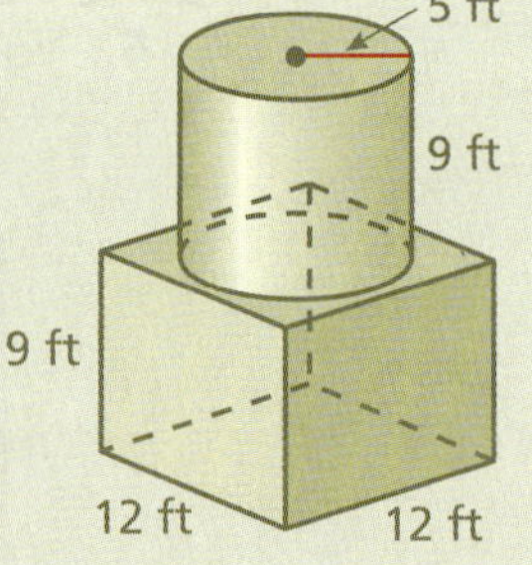

So, the volume is about 1296 + 706.9 = 2002.9 cubic feet.

Review Game

Volume

Materials

- a variety of containers of different shapes and sizes
- liquid measuring devices
- water

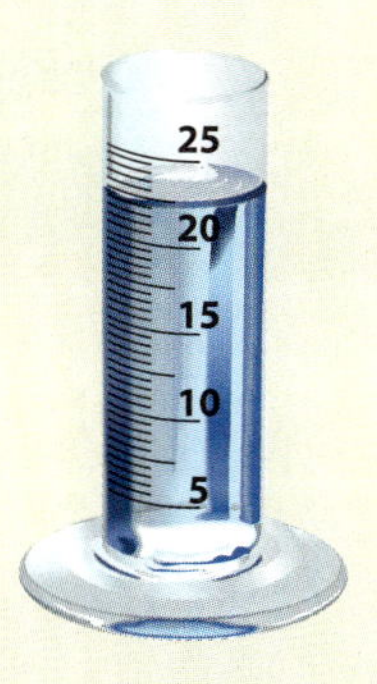

Directions

At the start of the chapter, have students bring in containers of different shapes and sizes. The containers should be shaped the same as the solids being studied in the chapter, they should be able to hold water, and they should be small enough so that they do not require a lot of water to fill. Collect containers until you have a sufficient variety of shapes and enough for the number of groups you want to have.

Work in groups. Give each group a container. Each group calculates the volume of the container and passes it to another group. Continue until all groups have calculated the volumes of the containers. When calculations are completed, each group measures the volume of a container using water. Measured volumes are shared with the class and compared to the calculated volumes.

Who Wins?

The group whose calculated volume is closest to the correct measured volume receives 1 point. The group with the most points wins.

For the Student Additional Practice

- Lesson Tutorials
- Multi-Language Glossary
- Self-Grading Progress Check
- *BigIdeasMath.com*
  Dynamic Student Edition
  Student Resources

Answers

7. $256\pi \approx 804.2 \text{ m}^3$
8. $\frac{40}{3}\pi \approx 41.9 \text{ cm}^3$
9. $\frac{3052}{27\pi} \approx 36.0 \text{ in.}$
10. $2304\pi \approx 7238.2 \text{ ft}^3$
11. 21 in.
12. $360\pi \approx 1131.0 \text{ m}^3$
13. 132 ft^3
14. $16\pi + \frac{16\pi}{3} = \frac{64\pi}{3} \approx 67.0 \text{ cm}^3$
15. 576 m^3
16. 86.6 yd^2

My Thoughts on the Chapter

Teacher Tip

Not allowed to write in your teaching edition? Use sticky notes to record your thoughts.

What worked. . .

What did not work. . .

What I would do differently. . .

Exercises

Find the volume V or radius r of the sphere. Round your answer to the nearest tenth, if necessary.

10.

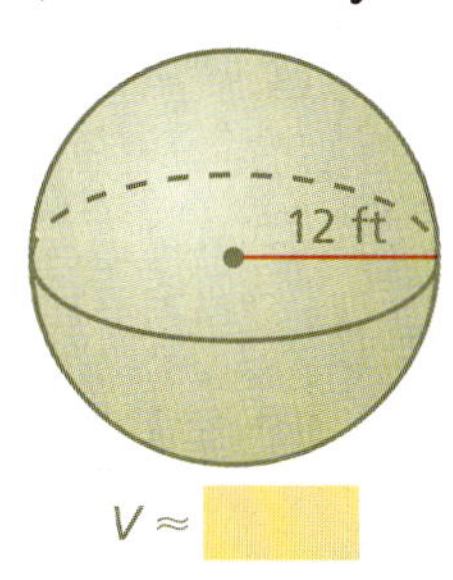

11.

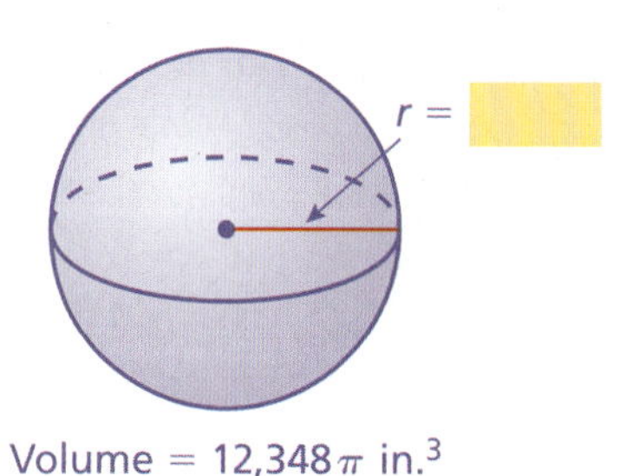

Find the volume of the composite solid. Round your answer to the nearest tenth.

12.

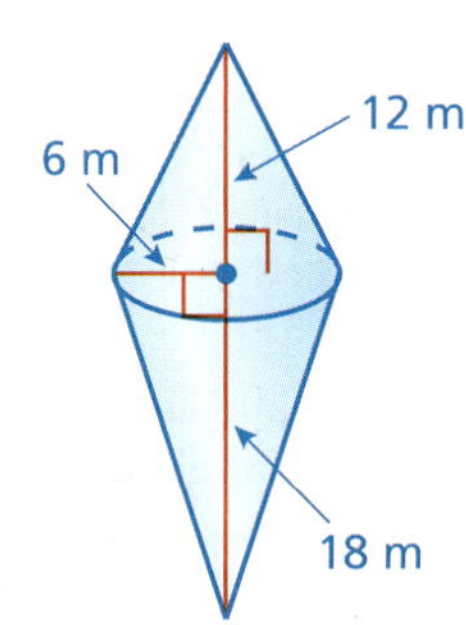

13.

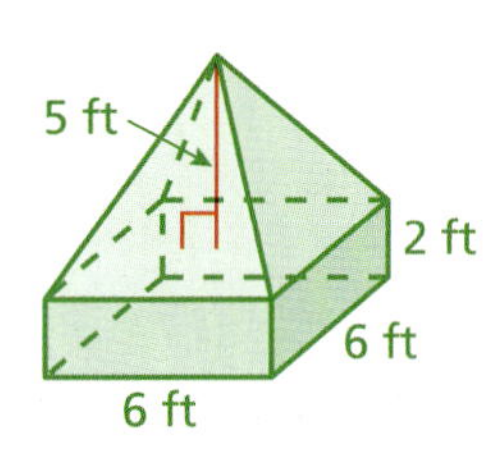

14.

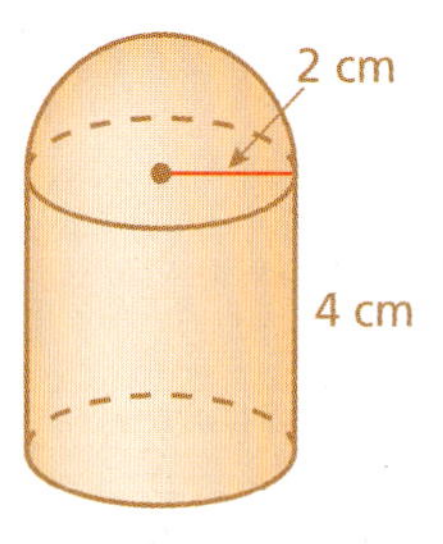

8.4 Surface Areas and Volumes of Similar Solids (pp. 354–361)

The cones are similar. What is the volume of the red cone? Round your answer to the nearest tenth.

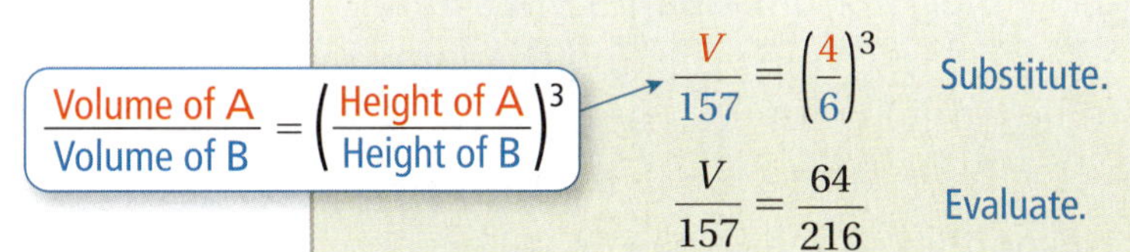

$$\frac{V}{157} = \left(\frac{4}{6}\right)^3 \quad \text{Substitute.}$$

$$\frac{V}{157} = \frac{64}{216} \quad \text{Evaluate.}$$

$$V \approx 46.5 \quad \text{Solve for } V.$$

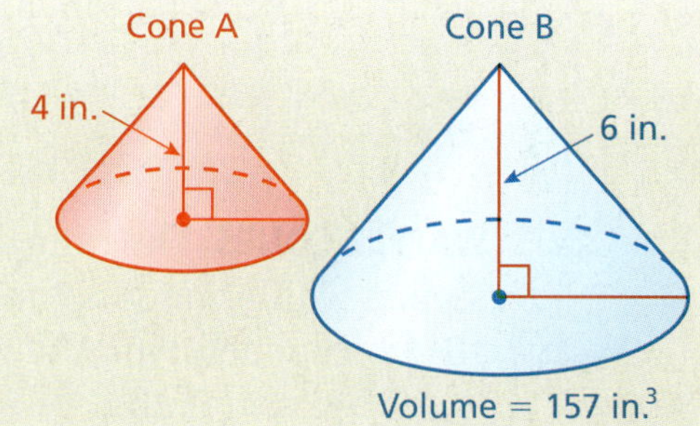

The volume is about 46.5 cubic inches.

Exercises

The solids are similar. Find the surface area S or volume V of the red solid. Round your answer to the nearest tenth.

15.

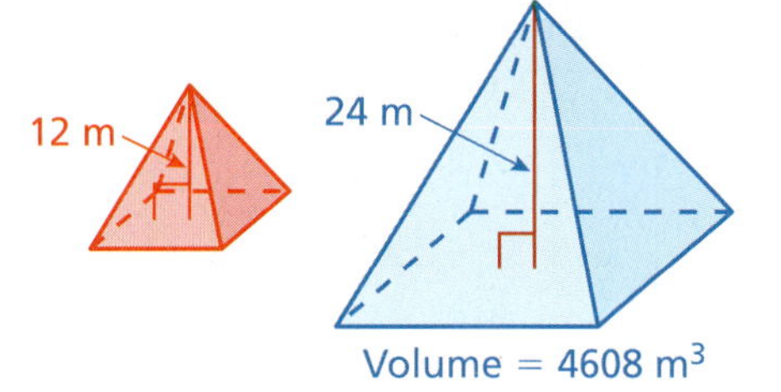

16.

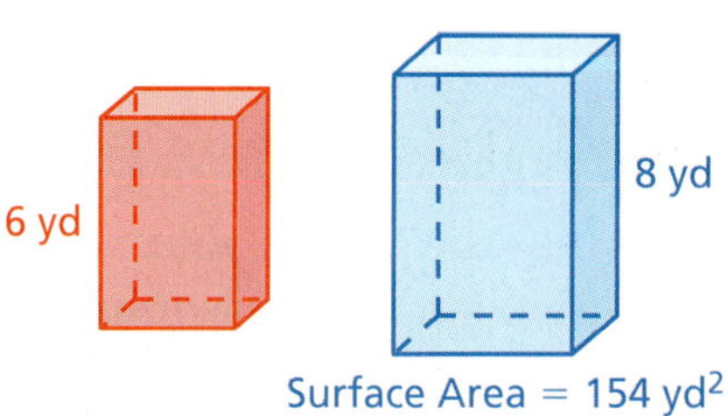

8 Chapter Test

Find the volume of the solid. Round your answer to the nearest tenth.

1.

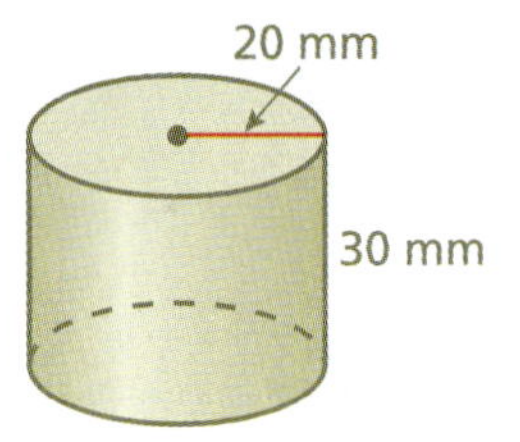

2.

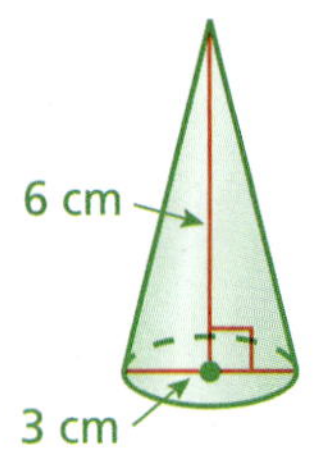

3.

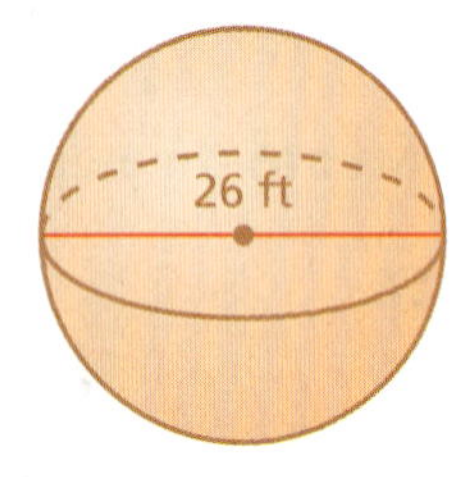

4.

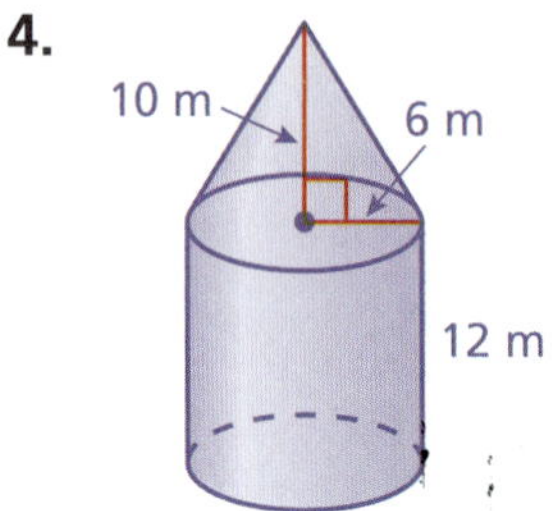

5. The pyramids are similar.

a. Find the missing dimension.

b. Find the surface area of the red pyramid.

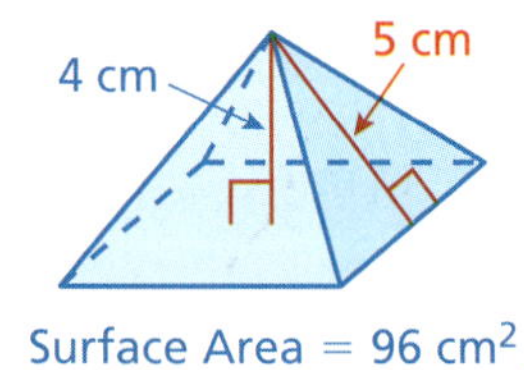

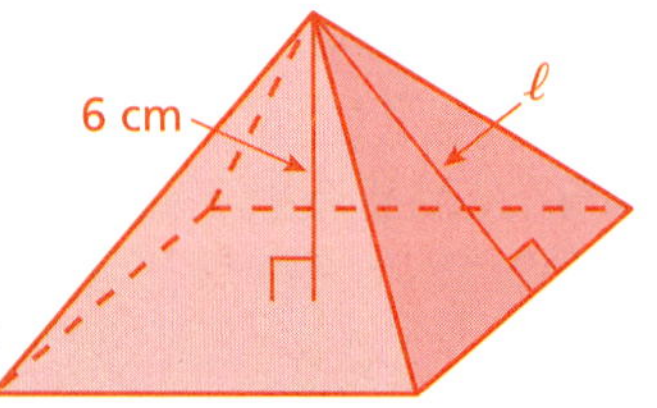

5 in.

5 in.

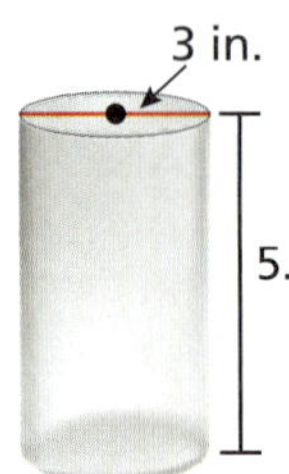

6. SMOOTHIES You are making smoothies. You will use either the cone-shaped glass or the cylindrical glass. Which glass holds more? About how much more?

7. WAFFLE CONES The ratio of the corresponding linear measures of two similar waffle cones is 3 to 4. The smaller cone has a volume of about 18 cubic inches. Find the volume of the larger cone. Round your answer to the nearest tenth.

8. OPEN-ENDED Draw two different composite solids that have the same volume but different surface areas. Explain your reasoning.

9. MILK Glass A has a diameter of 3.5 inches and a height of 4 inches. Glass B has a radius of 1.5 inches and a height of 5 inches. Which glass can hold more milk?

10. REASONING Which solid has the greater volume? Explain your reasoning.

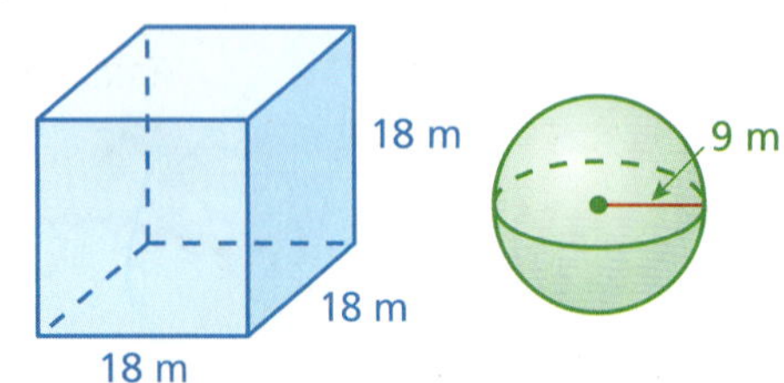

Test Item References

Chapter Test Questions	Section to Review	Florida Common Core Standards (MACC)
1, 6, 9	8.1	8.G.3.9
2, 6	8.2	8.G.3.9
3, 4, 8, 10	8.3	8.G.3.9
5, 7	8.4	8.G.3.9

Test-Taking Strategies

Remind students to quickly look over the entire test before they start so that they can budget their time. This test is very visual and requires that students remember many terms. It might be helpful for them to jot down some of the terms on the back of the test before they start. Students should make sketches and diagrams to help them.

Common Errors

- **Exercises 1–4** Students may write linear or square units for volume rather than cubic units. Remind them that part of writing a correct answer is including the correct units.
- **Exercises 1, 2, 4, 6, and 9** When finding the area of the circular base of a cylinder or a cone, students may not square the radius, or they may use the diameter when the formula calls for the radius. Remind them of the formula that they learned for the area of a circle.
- **Exercises 2, 4, and 6** When finding the volume of a cone, students may forget to multiply by $\frac{1}{3}$. Remind them of the formula for volume of a cone.
- **Exercises 3 and 10** When finding the volume of a sphere, students may forget to multiply by $\frac{4}{3}$ or cube the radius, they may use diameter when the formula calls for radius, or they may write the incorrect units. Remind them of the given formula for volume of a sphere and that part of writing a correct answer is including the correct units.
- **Exercises 5 and 7** Students may raise the ratio of the linear measures to the wrong exponent or forget to square or cube the ratio altogether. Discuss why squaring or cubing the ratio makes sense.

Reteaching and Enrichment Strategies

If students need help. . .	If students got it. . .
Resources by Chapter • Practice A and Practice B • Puzzle Time Record and Practice Journal Practice Differentiating the Lesson Lesson Tutorials *BigIdeasMath.com* Skills Review Handbook	Resources by Chapter • Enrichment and Extension • Technology Connection Game Closet at *BigIdeasMath.com* Start Standards Assessment

Answers

1. $12{,}000\pi \approx 37{,}699.1 \text{ mm}^3$
2. $4.5\pi \approx 14.1 \text{ cm}^3$
3. $\frac{8788\pi}{3} \approx 9202.8 \text{ ft}^3$
4. $552\pi \approx 1734.2 \text{ m}^3$
5. **a.** $\ell = 7.5$ cm
   b. 216 cm^2
6. cylindrical glass; about 6.2 in.^3
7. 42.7 in.^3
8. *Sample answer:*

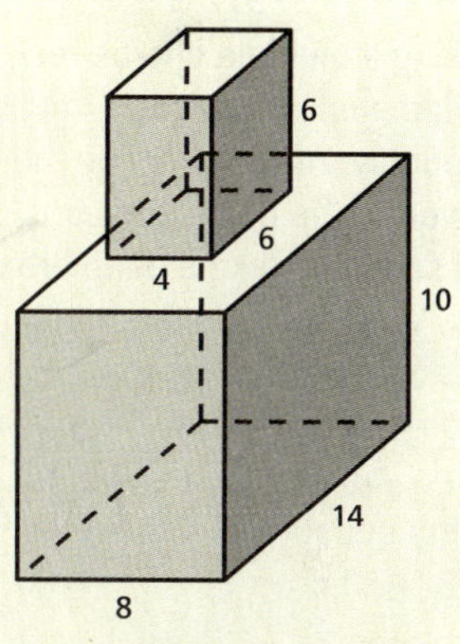

$V = 1264$ cubic units
$S = 784$ square units

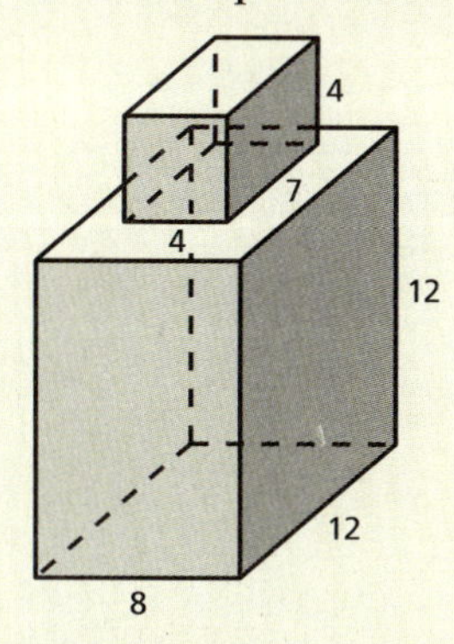

$V = 1264$ cubic units
$S = 760$ square units

9. Glass A
10. cube; The sphere could fit inside the cube and there would still be extra space outside the sphere but inside the cube.

Technology for the Teacher

Online Assessment
Assessment Book
ExamView® Assessment Suite

Test-Taking Strategies

Available at *BigIdeasMath.com*

After Answering Easy Questions, Relax

Answer Easy Questions First
Estimate the Answer
Read All Choices before Answering
Read Question before Answering
Solve Directly or Eliminate Choices
Solve Problem before Looking at Choices
Use Intelligent Guessing
Work Backwards

About this Strategy

When taking a multiple choice test, be sure to read each question carefully and thoroughly. After skimming the test and answering the easy questions, stop for a few seconds, take a deep breath, and relax. Work through the remaining questions carefully, using your knowledge and test-taking strategies. Remember, you already completed many of the questions on the test!

Answers

1. D
2. H
3. C

Technology for the Teacher

Florida Common Core Standards Support Performance Tasks
Online Assessment
Assessment Book
ExamView® Assessment Suite

Item Analysis

1. **A.** The student multiplies both sides by 3, distributes the 3 correctly, but does not distribute the 9 correctly. Alternatively, the student starts by distributing the 3 on the right side correctly. But when multiplying both sides by 3, the student does not distribute it across the expression $3w - 4$ correctly.

   B. The student starts by distributing the 3 on the right side incorrectly.

   C. The student multiplies both sides by 3, but fails to distribute the 3 correctly on the right side.

   D. Correct answer

2. **F.** The student incorrectly uses half the radius.

   G. The student incorrectly uses half the radius and also uses the formula for a right circular cylinder, neglecting to multiply by $\frac{1}{3}$.

   H. Correct answer

   I. The student uses the formula for a right circular cylinder, neglecting to multiply by $\frac{1}{3}$.

3. **A.** The student makes an order of operations error by not first distributing the multiplication.

   B. The student does not distribute the negative sign to the second term.

   C. Correct answer

   D. The student thinks that multiplying by $\frac{3}{2}$ is the inverse operation of multiplying by $-\frac{3}{2}$.

8 Standards Assessment

1. What value of w makes the equation below true? *(MACC.8.EE.3.7b)*

$$\frac{w}{3} = 3(w - 1) - 1$$

A. $\frac{1}{2}$ **C.** $\frac{5}{4}$

B. $\frac{3}{4}$ **D.** $\frac{3}{2}$

2. A right circular cone and its dimensions are shown below.

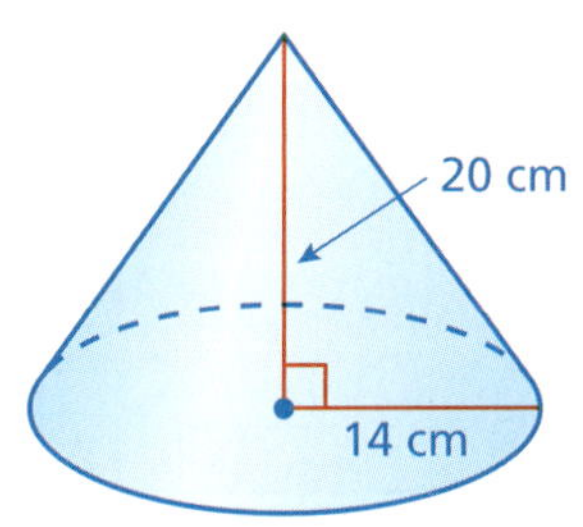

What is the volume of the right circular cone? $\left(\text{Use } \frac{22}{7} \text{ for } \pi.\right)$ *(MACC.8.G.3.9)*

F. $1{,}026\frac{2}{3}$ cm^3 **H.** $4{,}106\frac{2}{3}$ cm^3

G. 3,080 cm^3 **I.** 12,320 cm^3

3. Patricia solved the equation in the box shown.

$$-\frac{3}{2}(8x - 10) = -20$$
$$8x - 10 = -20\left(-\frac{3}{2}\right)$$
$$8x - 10 = 30$$
$$8x - 10 + 10 = 30 + 10$$
$$8x = 40$$
$$\frac{8x}{8} = \frac{40}{8}$$
$$x = 5$$

What should Patricia do to correct the error that she made? *(MACC.8.EE.3.7b)*

A. Add 10 to -20.

B. Distribute $-\frac{3}{2}$ to get $-12x - 15$.

C. Multiply both sides by $-\frac{2}{3}$ instead of $-\frac{3}{2}$.

D. Multiply both sides by $\frac{3}{2}$ instead of $-\frac{3}{2}$.

4. On the grid below, Rectangle *EFGH* is plotted and its vertices are labeled.

E (−1, 2)
F (3, 2)
H (−1, −3)
G (3, −3)

Which of the following shows Rectangle *E′F′G′H′*, the image of Rectangle *EFGH* after it is reflected in the *x*-axis? *(MACC.8.G.1.3)*

F. 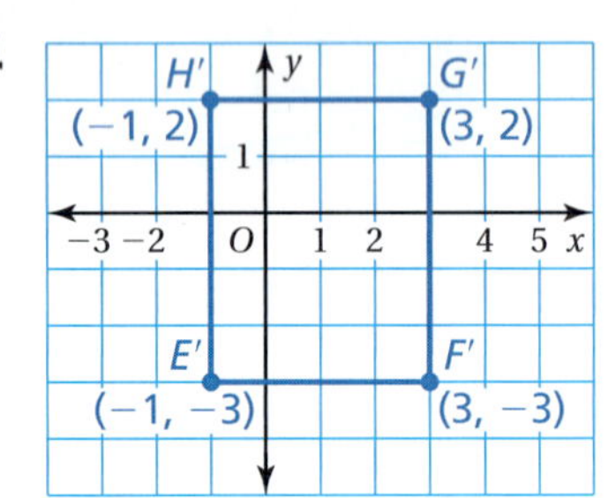

H. 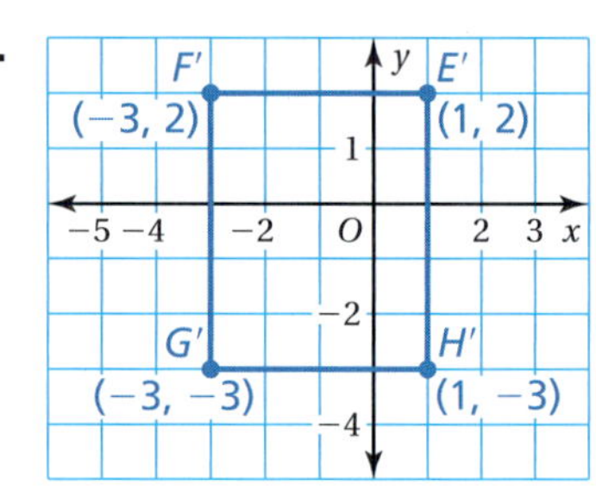

G. 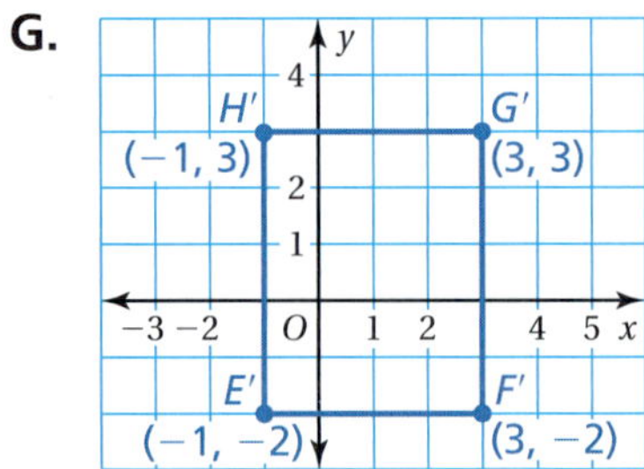

I. 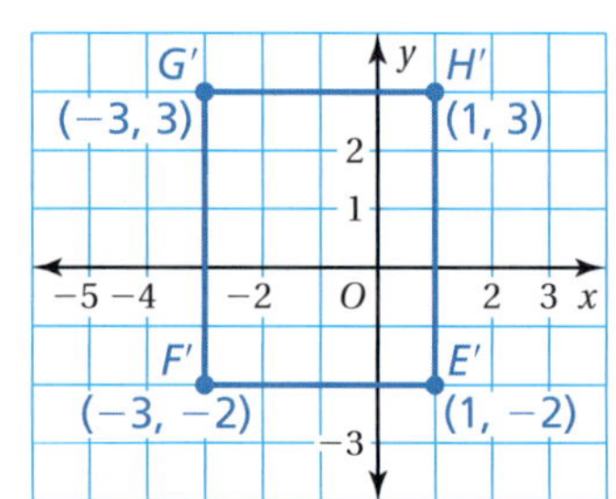

5. List the ordered pairs shown in the mapping diagram below. *(MACC.8.F.1.1)*

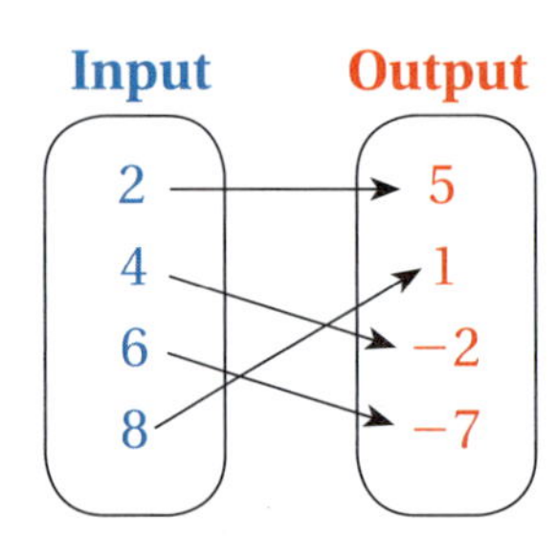

A. (2, 5), (4, −2), (6, −7), (8, 1)

B. (2, −7), (4, −2), (6, 1), (8, 5)

C. (2, 5), (4, 1), (6, −2), (8, −7)

D. (5, 2), (−2, 4), (−7, 6), (1, 8)

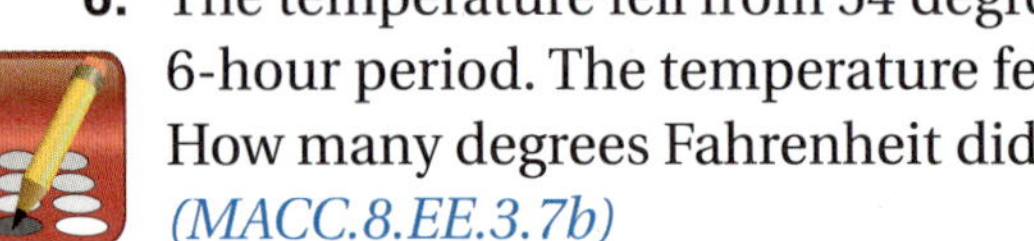

6. The temperature fell from 54 degrees Fahrenheit to 36 degrees Fahrenheit over a 6-hour period. The temperature fell by the same number of degrees each hour. How many degrees Fahrenheit did the temperature fall each hour? *(MACC.8.EE.3.7b)*

Item Analysis (continued)

4. **F.** The student chooses the same rectangle with the vertices labeled with different points.

   G. Correct answer

   H. The student reflects the rectangle in the *y*-axis.

   I. The student rotates the rectangle 180°.

5. **A.** Correct answer

   B. The student pairs the inputs and outputs by order from least to greatest.

   C. The student pairs each input with the output in the corresponding position.

   D. The student correctly pairs the inputs and outputs but switches the order so that the outputs are listed first.

6. **Gridded Response:** Correct answer: 3°F

   Common Error: The student subtracts 36 from 54, but fails to divide the result by 6.

Answers

4. G
5. A
6. 3°F

Answers

7. I
8. 9 in.^{3}
9. B
10. 113.04 in.^{3}

Item Analysis (continued)

7. **F.** The student performs the first step correctly, but then subtracts P rather than dividing.

   G. The student performs the first step correctly, but then divides A by P instead of dividing the entire expression by P.

   H. The student performs the first step correctly, but then divides P by P instead of dividing the entire expression by P.

   I. Correct answer

8. **Gridded Response:** Correct answer: 9 in.^{3}

   Common Error: The student divides the volume by r instead of r^2, getting an answer of 108 cubic inches.

9. **A.** The student thinks that a graph composed of line segments and a ray represents a linear function.

   B. Correct answer

   C. The student thinks that a graph with a continuous curve that approaches the x- and/or y-axis represents a linear function.

   D. The student thinks that a graph with a continuous curve represents a linear function.

10. **2 points** The student demonstrates a thorough understanding of finding the volume of a composite solid made up of a right circular cylinder and a right circular cone. The student correctly finds a volume of $36\pi \approx 113.04$ cubic inches. The student provides clear and complete work and explanations.

    1 point The student demonstrates a partial understanding of finding the volume of a composite solid made up of a right circular cylinder and a right circular cone. The student provides some correct work and explanation toward finding the volume.

    0 points The student demonstrates insufficient understanding of finding the volume of a composite solid made up of a right circular cylinder and a right circular cone. The student is unable to make any meaningful progress toward finding the volume.

7. Solve the formula below for I. *(MACC.8.EE.3.7b)*

$$A = P + PI$$

F. $I = A - 2P$

G. $I = \frac{A}{P} - P$

H. $I = A - \frac{P}{P}$

I. $I = \frac{A - P}{P}$

8. A right circular cylinder has a volume of 1296 cubic inches. If you divide the radius of the cylinder by 12, what would be the volume, in cubic inches, of the smaller cylinder? *(MACC.8.G.3.9)*

9. Which graph represents a linear function? *(MACC.8.F.1.3)*

A.

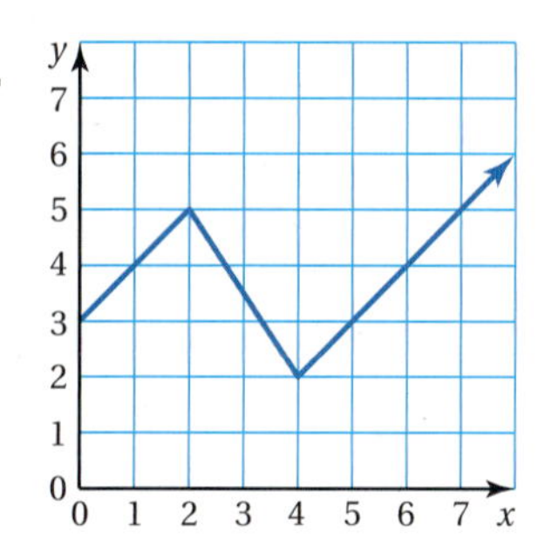

B.

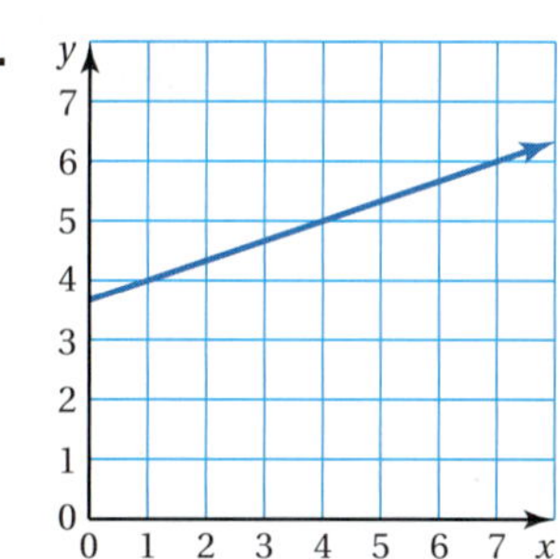

C.

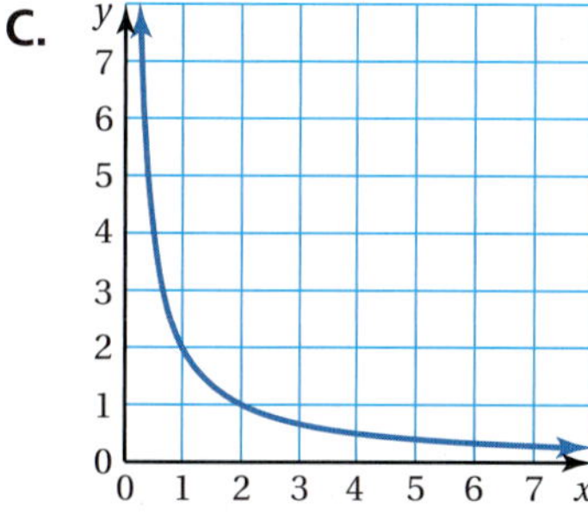

D.

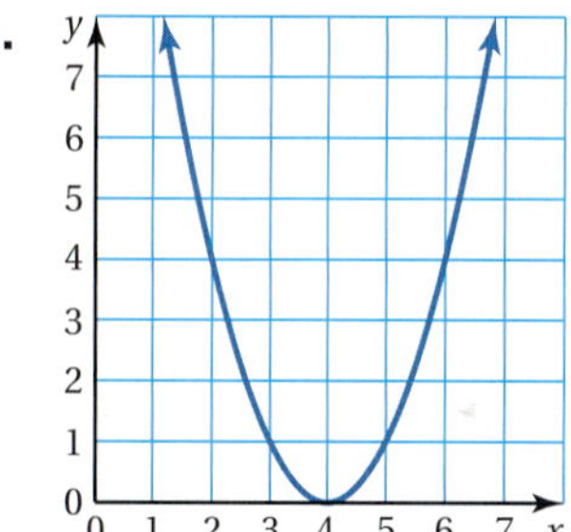

10. The figure below is a diagram for making a tin lantern.

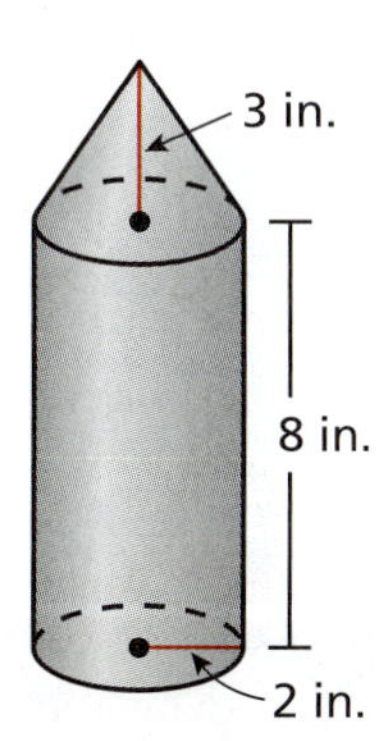

The figure consists of a right circular cylinder without its top base and a right circular cone without its base. What is the volume, in cubic inches, of the entire lantern? Show your work and explain your reasoning. (Use 3.14 for π.) *(MACC.8.G.3.9)*

9 Data Analysis and Displays

"Wow. The number of minutes I can dog paddle is growing like crazy!"

"The price of dog biscuits is up again this month."

"But I have a really good feeling about November."

Florida Common Core Progression

6th Grade

- Understand measures of center and variation for distributions of data.
- Use dot plots, histograms, box plots, and stem-and-leaf plots to display data.

7th Grade

- Use samples to draw inferences about populations.
- Use measures of center and variability from random samples to compare two populations.

8th Grade

- Construct and interpret scatter plots.
- Find and assess lines of fit for scatter plots.
- Use equations of lines to solve problems and interpret the slope and the *y*-intercept.
- Use two-way tables.
- Choose appropriate data displays.

Chapter Summary

Section	Florida Common Core Standard (MACC)	
9.1	Learning	8.SP.1.1
9.2	Learning	8.SP.1.1, 8.SP.1.2 ★, 8.SP.1.3 ★
9.3	Learning	8.SP.1.4 ★
9.4	Applying	8.SP.1.1 ★
★ Teaching is complete. Standard can be assessed.		

Pacing Guide for Chapter 9

Chapter Opener	1 Day
Section 1	
Activity	1 Day
Lesson	1 Day
Section 2	
Activity	1 Day
Lesson	1 Day
Study Help / Quiz	1 Day
Section 3	
Activity	1 Day
Lesson	1 Day
Section 4	
Activity	1 Day
Lesson	1 Day
Chapter Review/ Chapter Tests	2 Days
Total Chapter 9	12 Days
Year-to-Date	131 Days

Technology for the Teacher

BigIdeasMath.com
Chapter at a Glance
Complete Materials List
Parent Letters: English, Spanish, and Haitian Creole

Florida Common Core Standards

MACC.6.NS.3.6c Find and position integers and other rational numbers on a horizontal or vertical number line diagram; find and position pairs of integers and other rational numbers on a coordinate plane.

MACC.8.F.2.4 Construct a function to model a linear relationship between two quantities. Determine the rate of change and initial value of the function from . . . two (x, y) values, ...

Additional Topics for Review

- Plotting Points on a Number Line
- Evaluating Expressions
- Making Data Displays

Try It Yourself

1–4. See Additional Answers.

5. $y = x + 5$

6. $y = \frac{1}{8}x - 1$

7. $y = \frac{17}{3}x - 26$

Record and Practice Journal Fair Game Review

1.

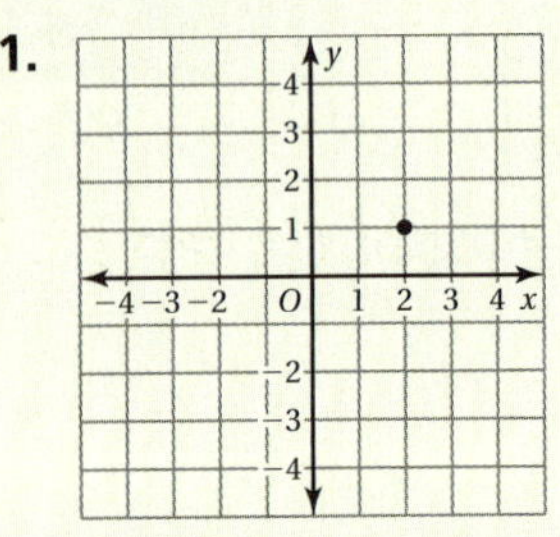

Quadrant I

2–4. See Additional Answers.

5. Point *A*: Quadrant II; Point *B*: Quadrant I; Point *C*: Quadrant IV

6. $y = 3x + 4$

7. $y = -\frac{6}{5}x + 5$

8. $y = \frac{4}{5}x + 8$

9. $y = -\frac{4}{3}x - 3$

10. $y = 3x$ **11.** $y = -2x$

Math Background Notes

Vocabulary Review

- Coordinate Plane
- Quadrant
- Ordered Pair
- Slope-Intercept Form

Plotting Points

- Students should know how to plot points in all four quadrants.
- **Common Error:** Students may plot the points backwards. Remind them that coordinates are written in alphabetical order with the *x* move (horizontal) written before the *y* move (vertical).
- **Common Error:** Students may also have difficulty with the negative numbers associated with plotting outside Quadrant I. Remind them that the negatives are directional. A negative *x*-value communicates a move to the left of the origin and a negative *y*-coordinate communicates a move downward from the origin.
- If students have difficulty remembering the quadrants, remind them that Quadrants II, III, and IV are arranged in a counterclockwise pattern about the origin from Quadrant I.

Writing an Equation Using Two Points

- Students should know how to write an equation of the line that passes through two given points.
- Review the formula for slope. Remind students that the formula represents the change in *y* over the change in *x*, or rise over run.
- **Common Error:** Students may substitute coordinates incorrectly when applying the slope formula. Remind them to label the points as (x_1, y_1) and (x_2, y_2), and then carefully substitute the appropriate coordinates into the slope formula.
- Remind students how to use one of the points and the slope to write the point-slope form of the equation of the line. Then solve for *y* to write the equation in slope-intercept form.

Reteaching and Enrichment Strategies

If students need help. . .	If students got it. . .
Record and Practice Journal • Fair Game Review Skills Review Handbook Lesson Tutorials	Game Closet at *BigIdeasMath.com* Start the next section

What You Learned Before

"Here's an interesting survey about favorite dog toys."

Plotting Points (MACC.6.NS.3.6c)

Example 1 **Plot (a) $(-3, 2)$ and (b) $(4, -2.5)$ in a coordinate plane. Describe the location of each point.**

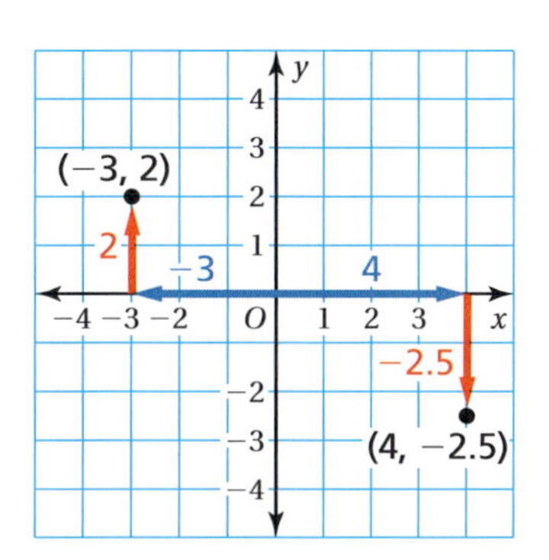

a. Start at the origin. Move 3 units left and 2 units up. Then plot the point.

The point is in Quadrant II.

b. Start at the origin. Move 4 units right and −2.5 units down. Then plot the point.

The point is in Quadrant IV.

Try It Yourself

Plot the ordered pair in a coordinate plane. Describe the location of the point.

1. $(1, 3)$ **2.** $(-2, 4)$ **3.** $(1, -3.5)$ **4.** $\left(-1\frac{3}{4}, -2\frac{1}{4}\right)$

Writing an Equation Using Two Points (MACC.8.F.2.4)

Example 2 **Write in slope-intercept form an equation of the line that passes through the points $(4, 2)$ and $(-1, -8)$.**

Find the slope:

$$m = \frac{y_2 - y_1}{x_2 - x_1} = \frac{-8 - 2}{-1 - 4} = \frac{-10}{-5} = 2$$

Then use the slope $m = 2$ and the point $(4, 2)$ to write an equation of the line.

$y - y_1 = m(x - x_1)$	Write the point-slope form.
$y - 2 = 2(x - 4)$	Substitute 2 for m, 4 for x_1, and 2 for y_1.
$y - 2 = 2x - 8$	Distributive Property
$y = 2x - 6$	Write in slope-intercept form.

Try It Yourself

Write in slope-intercept form an equation of the line that passes through the given points.

5. $(-1, 4), (3, 8)$ **6.** $(0, -1), (-8, -2)$ **7.** $(6, 8), (3, -9)$

9.1 Scatter Plots

Essential Question How can you construct and interpret a scatter plot?

1 ACTIVITY: Constructing a Scatter Plot

Work with a partner. The weights x (in ounces) and circumferences C (in inches) of several sports balls are shown.

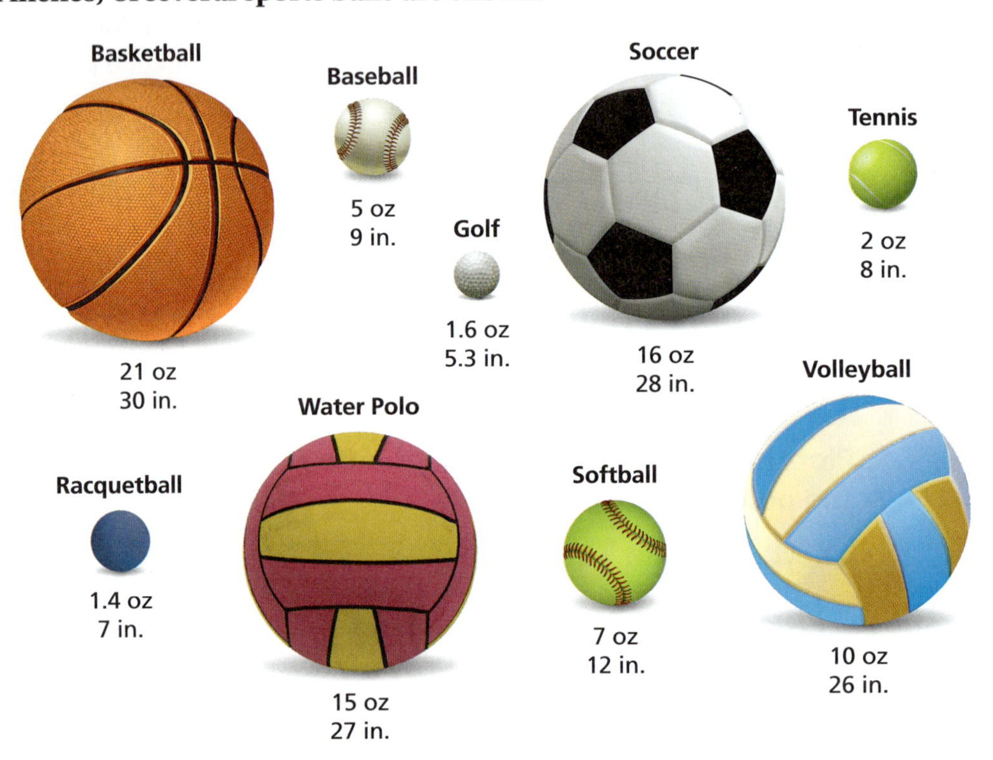

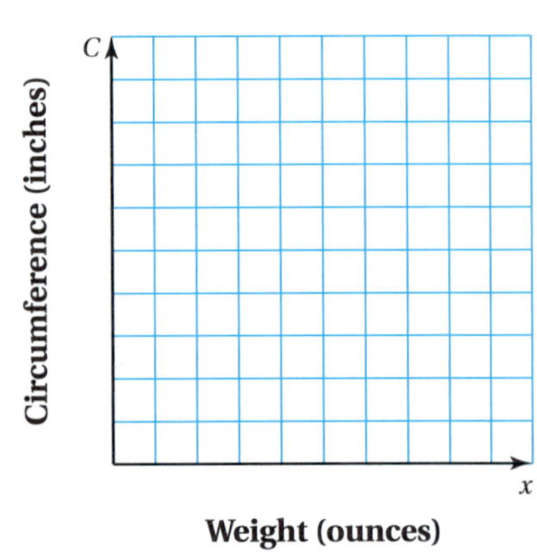

a. Choose a scale for the horizontal axis and the vertical axis of the coordinate plane shown.

b. Write the weight x and circumference C of each ball as an ordered pair. Then plot the ordered pairs in the coordinate plane.

c. Describe the relationship between weight and circumference. Are any of the points close together?

d. In general, do you think you can describe this relationship as *positive* or *negative*? *linear* or *nonlinear*? Explain.

e. A bowling ball has a weight of 225 ounces and a circumference of 27 inches. Describe the location of the ordered pair that represents this data point in the coordinate plane. How does this point compare to the others? Explain your reasoning.

COMMON CORE

Data Analysis

In this lesson, you will

- construct and interpret scatter plots.
- describe patterns in scatter plots.

Learning Standard
MACC.8.SP.1.1

Laurie's Notes

Introduction

Standards for Mathematical Practice

- **MP1a Make Sense of Problems:** A scatter plot shows the relationship between two data sets. Students will find that there may be a positive linear relationship, a negative linear relationship, a nonlinear relationship, or no relationship.

Motivate

- Discuss what students know about the construction, size, and weight of bowling balls.
- Ten-pin bowling balls range in weight from about 6 to 16 pounds and have a maximum circumference of 27 inches.
- ? "How are bowling balls like other sports balls?" *Sample answer:* A bowling ball is round and has a diameter close to that of a volleyball or soccer ball.
- ? "How are bowling balls different from other sports balls?" *Sample answer:* heavier than most, not filled with air, not intended to bounce
- Explain that you'll refer back to bowling balls at the end of the Activity 1.

Activity Notes

Activity 1

- ? "How do you find the circumference of a sphere?" Use the formula $C = \pi d$, where d is the diameter of the sphere.
- ? "Which sports ball shown has the least circumference? the greatest circumference?" golf ball; basketball
- ? "Which sports ball is the lightest? the heaviest?" racquetball; basketball
- Students should be able to plot the ordered pairs for the 9 sports balls.
- To write the ordered pairs, students may make a table.
- ? "Must the scale be the same for each axis?" no
- ? "What scale did you use for each axis?" Answers will vary.
- ? "Describe the scatter plot." *Sample answer:* There are two clusters of points.
- ? "What relationship between weight and circumference does the scatter plot suggest?" As weight increases, circumference tends to increase.
- To describe the relationship as positive or negative, think of lines with positive and negative slopes.
- ? "What do you think a scatter plot of data with a positive relationship looks like?" The points tend to rise from left to right.
- ? "Would you say that the point described in part (e) is an outlier? Explain." yes; The x-value (the weight of the bowling ball) is much greater than that of the other points.
- ? **MP1a** and **MP4 Model with Mathematics:** "Were the relationships between the data sets obvious before making the scatter plot? Explain." Answers will vary.
- ? **Extension:** "How would switching the axis labels affect the relationship?" The relationship is still positive and there are still two clusters.

Florida Common Core Standards

MACC.8.SP.1.1 Construct and interpret scatter plots for bivariate measurement data to investigate patterns of association between two quantities. Describe patterns such as clustering, outliers, positive or negative association, linear association, and nonlinear association.

Previous Learning

Students should be familiar with the concept of the slope of a line and know how to plot points in a coordinate plane.

Technology for the Teacher

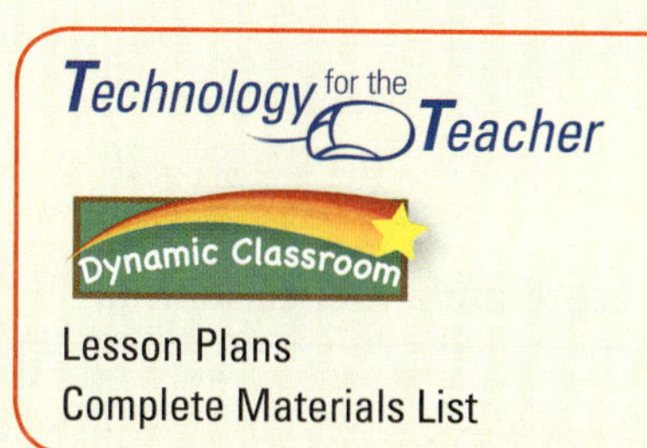

Lesson Plans
Complete Materials List

9.1 Record and Practice Journal

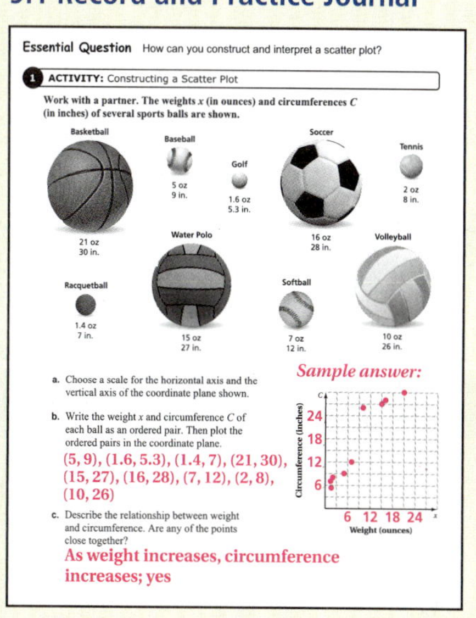

Essential Question How can you construct and interpret a scatter plot?

1 ACTIVITY: Constructing a Scatter Plot

Work with a partner. The weights x (in ounces) and circumferences C (in inches) of several sports balls are shown.

a. Choose a scale for the horizontal axis and the vertical axis of the coordinate plane shown.

b. Write the weight x and circumference C of each ball as an ordered pair. Then plot the ordered pairs in the coordinate plane.
(5, 9), (1.6, 5.3), (1.4, 7), (21, 30), (15, 27), (16, 28), (7, 12), (2, 8), (10, 26)

c. Describe the relationship between weight and circumference. Are any of the points close together?
As weight increases, circumference increases; yes

Sample answer:

English Language Learners

Class Activity

Provide English learners with an opportunity to interact while learning the concept. Draw a coordinate plane on poster board. Label the horizontal axis *shoe size* and the vertical axis *height*. Have each student place a sticker on the ordered pair that represents his or her shoe size and height. Then have the class discuss any trends in the data.

9.1 Record and Practice Journal

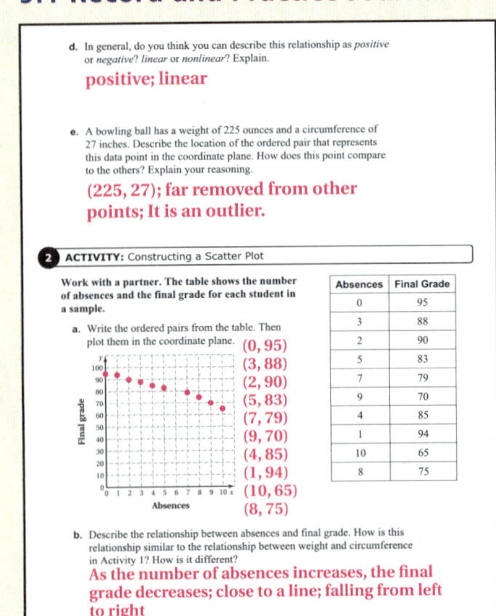

d. In general, do you think you can describe this relationship as *positive* or *negative*? *linear* or *nonlinear*? Explain.

positive; linear

e. A bowling ball has a weight of 225 ounces and a circumference of 27 inches. Describe the location of the ordered pair that represents this data point in the coordinate plane. How does this point compare to the others? Explain your reasoning.

(225, 27); far removed from other points; It is an outlier.

2 ACTIVITY: Constructing a Scatter Plot

Work with a partner. The table shows the number of absences and the final grade for each student in a sample.

Absences	Final Grade
0	95
3	88
2	90
5	83
7	79
9	70
4	85
1	94
10	65
8	75

a. Write the ordered pairs from the table. Then plot them in the coordinate plane.

(0, 95) (3, 88) (2, 90) (5, 83) (7, 79) (9, 70) (4, 85) (1, 94) (10, 65) (8, 75)

b. Describe the relationship between absences and final grade. How is this relationship similar to the relationship between weight and circumference in Activity 1? How is it different?

As the number of absences increases, the final grade decreases; close to a line; falling from left to right

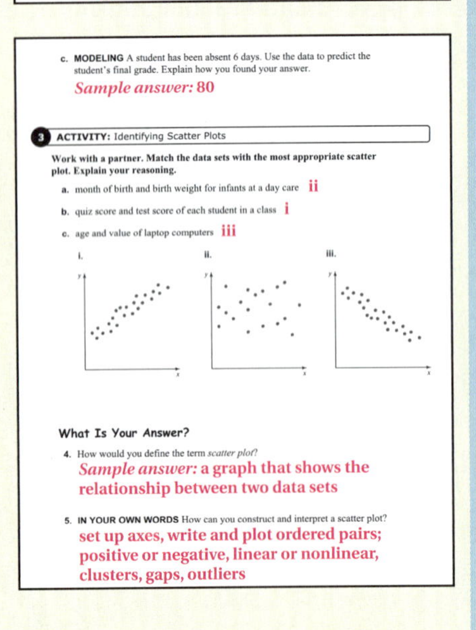

c. MODELING A student has been absent 6 days. Use the data to predict the student's final grade. Explain how you found your answer.

Sample answer: 80

3 ACTIVITY: Identifying Scatter Plots

Work with a partner. Match the data sets with the most appropriate scatter plot. Explain your reasoning.

a. month of birth and birth weight for infants at a day care ii

b. quiz score and test score of each student in a class i

c. age and value of laptop computers iii

What Is Your Answer?

4. How would you define the term *scatter plot*?

Sample answer: a graph that shows the relationship between two data sets

5. IN YOUR OWN WORDS How can you construct and interpret a scatter plot?

set up axes, write and plot ordered pairs; positive or negative, linear or nonlinear, clusters, gaps, outliers

Laurie's Notes

Activity 2

- The data in this activity are already organized in a table.
- Students may need help with scaling the axes. Students may want to use a broken vertical axis. You might let some students use a broken vertical axis and let others start their vertical axis at 0.
- When students have finished, ask volunteers to share their responses to part (b).
- If students used different scales for the vertical axis, have them display their scatter plots. Each plot will have a decreasing trend, but the steepness of the trend may look different.

Activity 3

- **MP1a:** For each scatter plot, students should interpret the trend of the plot in the context of the problem.
- Give students time to discuss the three descriptions and which scatter plots they match.
- **MP3 Construct Viable Arguments and Critique the Reasoning of Others:** Ask volunteers to explain how they matched the descriptions and scatter plots. If students disagree, they should give supporting arguments for their own answers and explain what they dislike about the other answers.

What Is Your Answer?

- Students should relate their understanding of linear functions to the positive and negative trends seen in today's activities.

Closure

- **Exit Ticket:** You empty a change purse and find the following: 9 pennies, 5 nickels, 4 dimes, 7 quarters, 1 half-dollar.
- Construct a scatter plot of (value of coin, number of coins).
- Interpret the scatter plot.

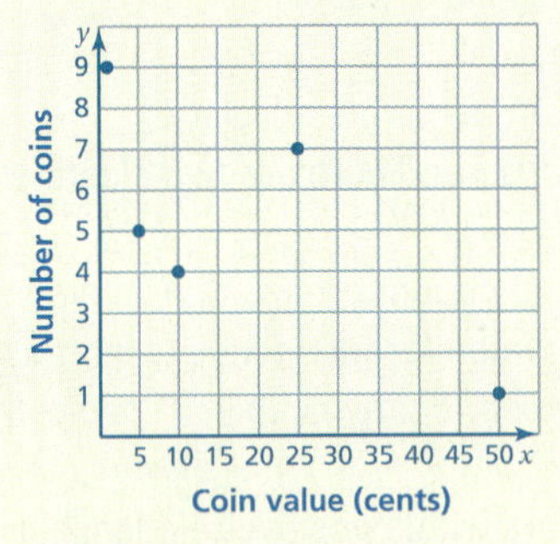

Answers will vary.

2 ACTIVITY: Constructing a Scatter Plot

Math Practice 5

Recognize Usefulness of Tools

How do you know when a scatter plot is a useful tool for making a prediction?

Work with a partner. The table shows the number of absences and the final grade for each student in a sample.

Absences	Final Grade
0	95
3	88
2	90
5	83
7	79
9	70
4	85
1	94
10	65
8	75

a. Write the ordered pairs from the table. Then plot them in a coordinate plane.

b. Describe the relationship between absences and final grade. How is this relationship similar to the relationship between weight and circumference in Activity 1? How is it different?

c. **MODELING** A student has been absent 6 days. Use the data to predict the student's final grade. Explain how you found your answer.

3 ACTIVITY: Identifying Scatter Plots

Work with a partner. Match the data sets with the most appropriate scatter plot. Explain your reasoning.

a. month of birth and birth weight for infants at a day care

b. quiz score and test score of each student in a class

c. age and value of laptop computers

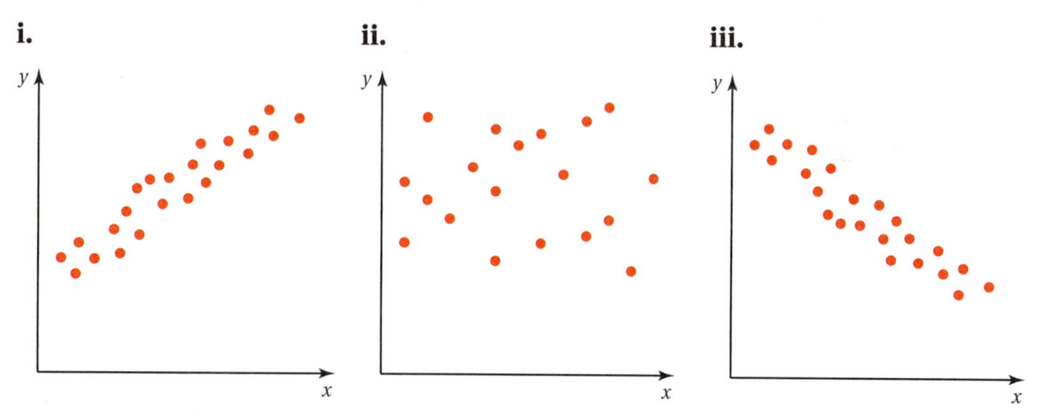

What Is Your Answer?

4. How would you define the term *scatter plot*?

5. **IN YOUR OWN WORDS** How can you construct and interpret a scatter plot?

Practice Use what you learned about scatter plots to complete Exercise 7 on page 376.

Key Vocabulary
scatter plot, *p. 374*

Scatter Plot

A **scatter plot** is a graph that shows the relationship between two data sets. The two sets of data are graphed as ordered pairs in a coordinate plane.

EXAMPLE 1 Interpreting a Scatter Plot

Restaurant Sandwiches

Calories (y): 0, 300, 350, 400, 450, 500, 550, 600, 650, 700, 750, 800
Fat (grams) (x): 0, 5, 10, 15, 20, 25, 30, 35, 40, 45

The scatter plot at the left shows the amounts of fat (in grams) and the numbers of calories in 12 restaurant sandwiches.

a. How many calories are in the sandwich that contains 17 grams of fat?

Draw a horizontal line from the point that has an x-value of 17. It crosses the y-axis at 400.

So, the sandwich has 400 calories.

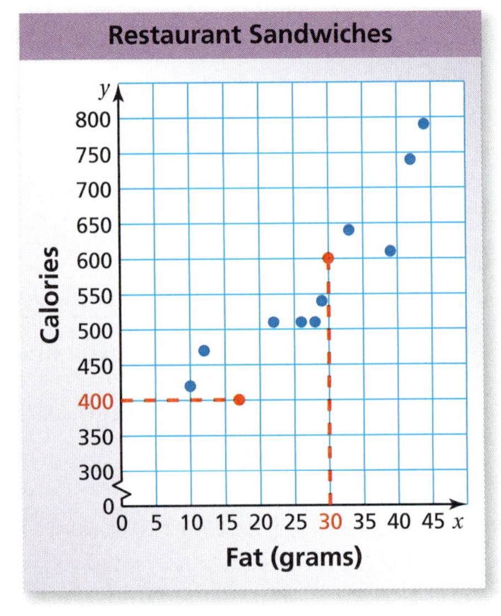

b. How many grams of fat are in the sandwich that contains 600 calories?

Draw a vertical line from the point that has a y-value of 600. It crosses the x-axis at 30.

So, the sandwich has 30 grams of fat.

c. What tends to happen to the number of calories as the number of grams of fat increases?

Looking at the graph, the plotted points go up from left to right.

So, as the number of grams of fat increases, the number of calories increases.

On Your Own

1. **WHAT IF?** A sandwich has 650 calories. Based on the scatter plot in Example 1, how many grams of fat would you expect the sandwich to have? Explain your reasoning.

Multi-Language Glossary at BigIdeasMath.com

Laurie's Notes

Introduction

Connect

- **Yesterday:** Students gained an intuitive understanding of how to construct and interpret scatter plots. (MP1a, MP3, MP4)
- **Today:** Students will construct scatter plots and identify relationships between the two data sets.

Motivate

- **Preparation:** Stop by any fast food restaurant to pick up a pamphlet, or go online to find nutritional information about the menu items.

? "Do you think there is a relationship between the grams of fat and number of calories in the sandwich?" yes

- Share the information about a few of the sandwiches from your pamphlet or printout to confirm students' opinions.

Lesson Notes

Key Idea

- Explain that the plot they are going to make today displays the relationship, if any, between two variables, such as grams of fat and calories.
- Define scatter plot.
- Discuss the two scatter plots made in the activities yesterday. In Activity 1, the two sets of data were the weights and circumferences of the balls. In Activity 2, the two sets of data were numbers of absences and final grades.

Example 1

- **MP1a Make Sense of Problems:** This example helps students understand how a scatter plot is read and interpreted. Discuss the labels on the axes and what an ordered pair represents: (grams of fat, number of calories). There are 12 different sandwiches that are represented.
- To read information from the scatter plot, find a point. Move vertically to the x-axis to find the x-value of the point. Then return to the point, and move horizontally to the y-axis to find the y-value of the point.
- A scatter plot allows you to see trends in the data. You read a scatter plot from left to right. As the x-coordinate increases, is the y-coordinate increasing, decreasing, staying the same, *or* is there no pattern?

On Your Own

- This question implies that you can use the trend of the data in the scatter plot to estimate information about other sandwiches. Although there are anywhere from 0 to 72 grams of fat in a 650-calorie food item, it is reasonable to use the trend of the sandwich data shown.

Goal Today's lesson is making a **scatter plot** and describing the relationship between the data.

Lesson Tutorials
Lesson Plans
Answer Presentation Tool

Extra Example 1

Use the scatter plot in Example 1.

a. How many grams of fat are in the sandwich that contains 740 calories? about 42 g

b. How many calories are in the sandwich that contains 33 grams of fat? about 640 calories

On Your Own

1. about 35 g; The point just below $y = 650$ has an x-value just below $x = 35$.

Differentiated Instruction

Kinesthetic

Form groups of 8 to 10 students who will create life-size models of a positive relationship, a negative relationship, and no relationship. Give two pairs of students 10-foot lengths of string and have them form the x- and y-axes of a coordinate plane. The remaining students will be the data points. Have these students represent a *positive relationship* in the coordinate plane. After students have had a few minutes, check their positions. Continue by having students represent a *negative relationship* and *no relationship*. Extend the activity by having two students hold a third string to show a line of fit.

Extra Example 2

Describe the relationship between the data. Identify any outliers, gaps, or clusters.

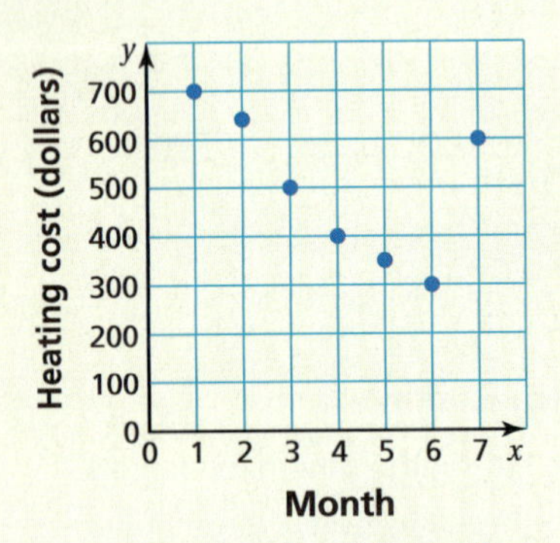

negative linear relationship; outlier: (7, 600), no obvious gaps or clusters

On Your Own

2. 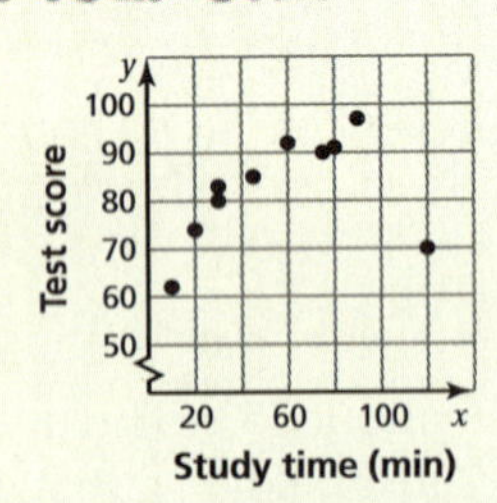

positive linear relationship; There is a gap in the data from $x = 90$ to $x = 120$; outlier: (120, 70); no obvious clusters

Laurie's Notes

Discuss

- There are four general cases that describe the relationship between two data sets. Draw a quick example of each case.
- Discuss the two scatter plots made in the activity. The sports balls data were an example of a positive relationship, and the absences/grades data were an example of a negative relationship.
- The focus at this grade level is recognizing positive linear relationships, negative linear relationships, and nonlinear relationships.

Example 2

- Have students review the two scatter plots shown.
- Ask students to complete this sentence. As the size of the television increases, the price increases. This is an example of a positive relationship.
- **Connection:** By this point, some students have made the connection between the slope of a line and the relationship shown in a scatter plot. A positive relationship is related to a positive slope.

? "Should there be a relationship between a person's age and the number of pets he or she owns?" no

- Part (b) makes sense to students. There should be no trend in the data.

On Your Own

- Give time for students to complete the scatter plot.
- **MP6 Attend to Precision:** A common difficulty for students is deciding how to scale the axes. Students should look at the range of numbers that need to be displayed, and then decide whether it is necessary to start their axes at 0 or if another starting point (broken axes) makes sense.
- Have transparency grids available so that results can be shared quickly as a class.

Closure

- **Exit Ticket:** Make a scatter plot that shows the relationship between the number of chocolate chip cookies eaten and the number of chocolate chips eaten.
  Answers will vary.

A scatter plot can show that a relationship exists between two data sets.

Positive Linear Relationship

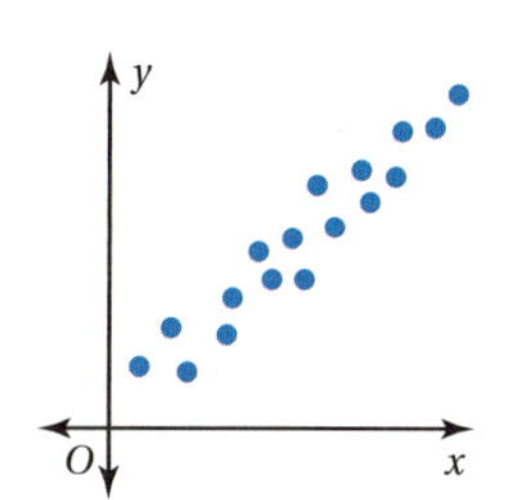

The points lie close to a line. As x increases, y increases.

Negative Linear Relationship

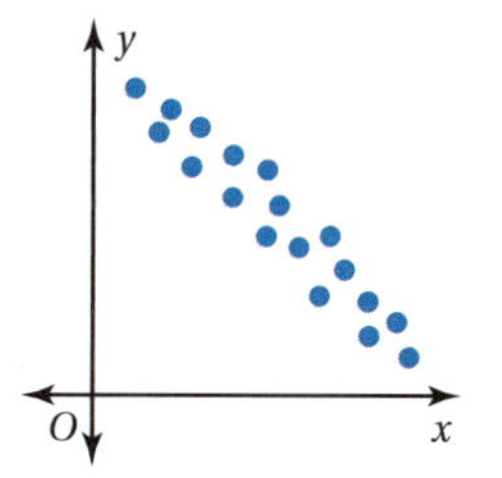

The points lie close to a line. As x increases, y decreases.

Nonlinear Relationship

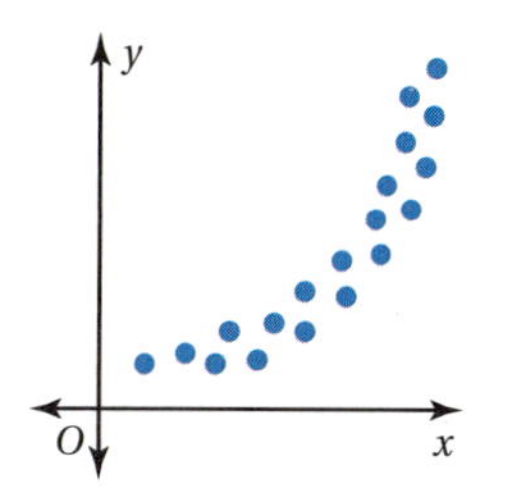

The points lie in the shape of a curve.

No Relationship

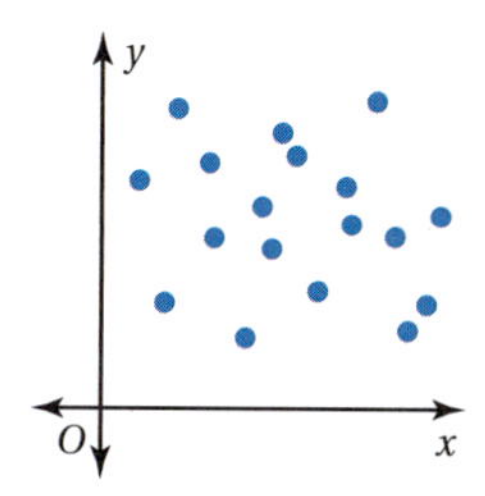

The points show no pattern.

EXAMPLE 2 Identifying Relationships

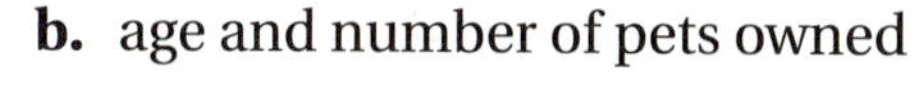

Describe the relationship between the data. Identify any outliers, gaps, or clusters.

a. television size and price

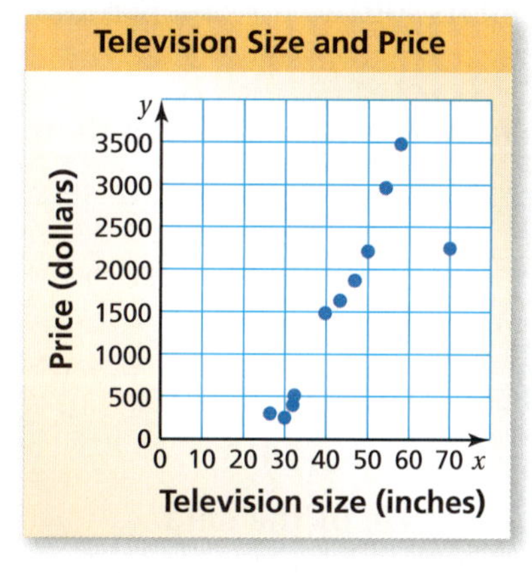

The points appear to lie close to a line. As x increases, y increases.

So, the scatter plot shows a positive linear relationship. There is an outlier at (70, 2250), a cluster of data under \$500, and a gap in the data from \$500 to \$1500.

b. age and number of pets owned

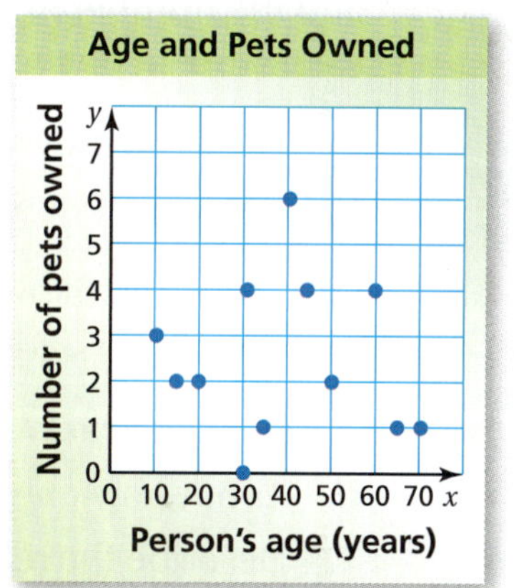

The points show no pattern.

So, the scatter plot shows no relationship. There are no obvious outliers, gaps, or clusters in the data.

On Your Own

2. Make a scatter plot of the data and describe the relationship between the data. Identify any outliers, gaps, or clusters.

Study Time (min), x	30	20	60	90	45	10	30	75	120	80
Test Score, y	80	74	92	97	85	62	83	90	70	91

9.1 Exercises

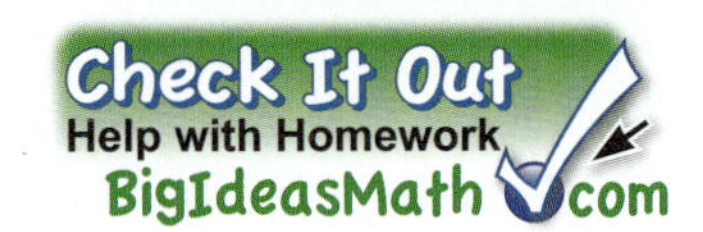

Vocabulary and Concept Check

1. **VOCABULARY** What type of data do you need to make a scatter plot? Explain.
2. **REASONING** How can you identify an outlier in a scatter plot?

LOGIC Describe the relationship you would expect between the data. Explain.

3. shoe size of a student and the student's IQ
4. time since a train's departure and the distance to its destination
5. height of a bouncing ball and the time since it was dropped
6. number of toppings on a pizza and the price of the pizza

Practice and Problem Solving

7. **JEANS** The table shows the average price (in dollars) of jeans sold at different stores and the number of pairs of jeans sold at each store in one month.

Average Price	22	40	28	35	46
Number Sold	152	94	134	110	81

   a. Write the ordered pairs from the table and plot them in a coordinate plane.
   b. Describe the relationship between the two data sets.

1 8. **SUVS** The scatter plot shows the numbers of sport utility vehicles sold in a city from 2009 to 2014.

   a. In what year were 1000 SUVs sold?
   b. About how many SUVs were sold in 2013?
   c. Describe the relationship shown by the data.

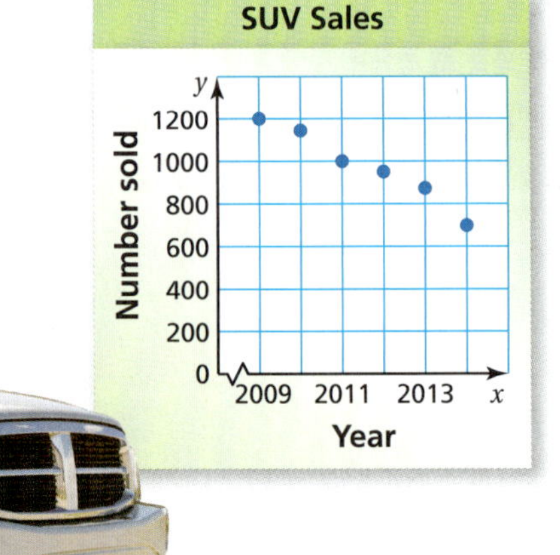

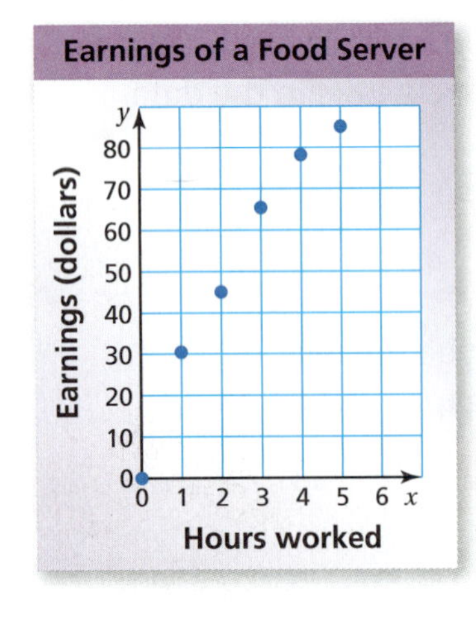

9. **EARNINGS** The scatter plot shows the total earnings (wages and tips) of a food server during one day.

   a. About how many hours must the server work to earn $70?
   b. About how much did the server earn for 5 hours of work?
   c. Describe the relationship shown by the data.

Assignment Guide and Homework Check

Level	Day 1 Activity Assignment	Day 2 Lesson Assignment	Homework Check
Basic	7, 18–21	1–6, 9–15 odd	9, 11, 13, 15
Average	7, 18–21	1–6, 8, 9, 10–16 even	9, 10, 14, 16
Advanced	1–7, 8–12 even, 13–21		8, 12, 14, 16, 17

Common Errors

- **Exercises 8 and 9** When finding values from the graph, students may accidentally shift over or up too far and get an answer that is off by an increment. Encourage them to start at the given value and trace the graph to where the point is, and then trace down or left to the other axis for the answer.
- **Exercises 10–12** Students may mix up positive and negative relationships. Remind them about slope. The slope is positive when the line rises from left to right and negative when it falls from left to right. The same is true for relationships in a scatter plot. When the data rise from left to right, it is a positive relationship. When the data fall from left to right, it is a negative relationship.

9.1 Record and Practice Journal

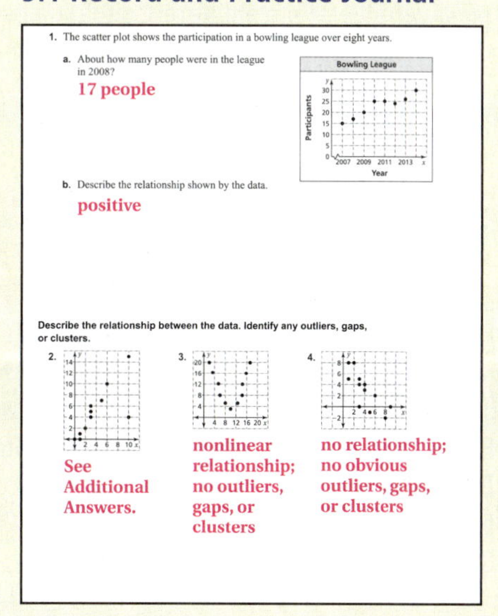
1. The scatter plot shows the participation in a bowling league over eight years.

   a. About how many people were in the league in 2008?
   17 people

   b. Describe the relationship shown by the data.
   positive

Describe the relationship between the data. Identify any outliers, gaps, or clusters.

2. **See Additional Answers.**
3. **nonlinear relationship; no outliers, gaps, or clusters**
4. **no relationship; no obvious outliers, gaps, or clusters**

Vocabulary and Concept Check

1. They must be ordered pairs so there are equal amounts of x- and y-values.
2. It is a data point that is far removed from the other points in a data set.
3. no relationship; A student's shoe size is not related to his or her IQ.
4. negative linear relationship; As time passes, the distance to the destination decreases.
5. nonlinear relationship; On each successive bounce, the ball rebounds to a height less than its previous bounce.
6. positive linear relationship; As the number of toppings increase, the price of the pizza will increase.

Practice and Problem Solving

7. a. (22, 152), (40, 94), (28, 134), (35, 110), (46, 81);

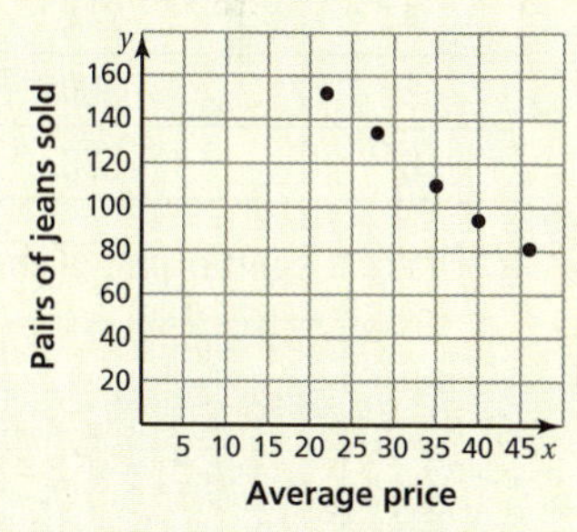

   b. As the average price of jeans increases, the number of pairs of jeans sold decreases.

8. a. 2011

   b. about 875 SUVs

   c. negative linear relationship

9. See Additional Answers.

Practice and Problem Solving

10. negative linear relationship; outlier at (15, 10), gap between x-values of 15 to 25 and y-values of 23 to 33

11. nonlinear relationship; no outliers, gaps, or clusters

12. no relationship; no obvious outliers, gaps, or clusters

13. positive linear relationship

14–16. See Additional Answers.

17. See *Taking Math Deeper*.

Fair Game Review

18. 2 **19.** 8

20. −4 **21.** B

Mini-Assessment

1. The table shows the distance you travel over a 6-hour period.

Hours, x	Distance (miles), y
1	60
2	130
3	186
4	244
5	300
6	378

a. Make a scatter plot of the data.

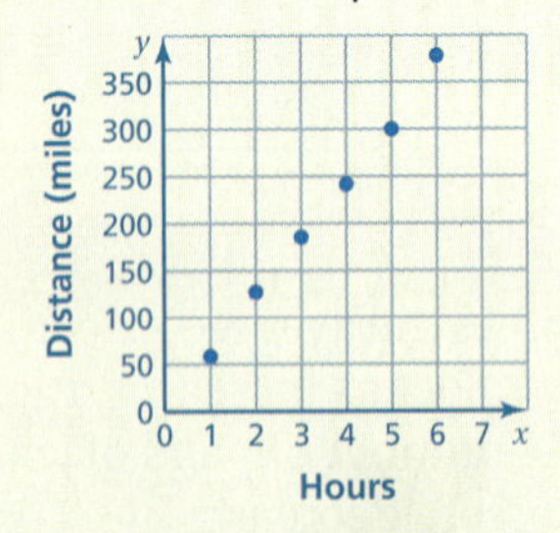

b. Describe the relationship between the data. positive linear relationship

c. Identify any outliers, gaps, or clusters in the data. There are no obvious outliers, gaps, or clusters in the data.

Taking Math Deeper

Exercise 17

Students may assume that when two events are correlated, one event *causes* the other event. However, other factors may be influencing the events. Students need to use reasoning to decide if there is *causation*.

The problem does not ask you to create a scatter plot, but it is one way to start thinking about the quantities involved. Because there is a positive linear relationship, the scatter plot could look like the one below.

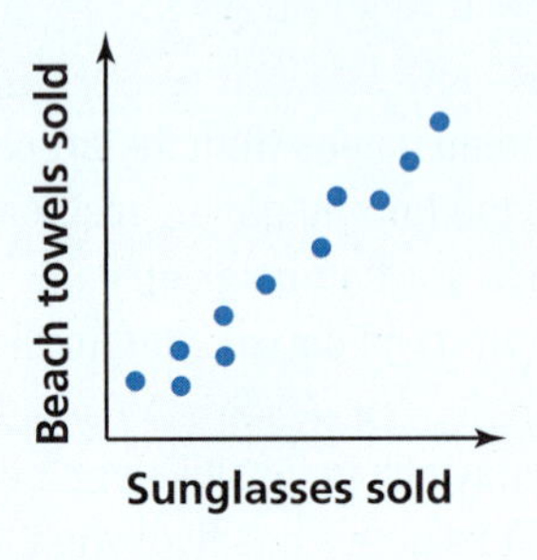

Imagine yourself in this context.

- Would buying sunglasses *cause* you to buy a beach towel?
- Would buying a beach towel *cause* you to buy sunglasses?

The answer to each question is no. It is more likely that summer weather causes the rise in sales of each item.

Note that this is not always the case. There can be correlation and causation, as in Example 14. It is likely that studying longer *causes* you to have a better test score.

Project

In this exercise, the fact that it was summer is an example of a *hidden* or *lurking variable*. Write a different situation in which there is correlation, but neither event causes the other due to a *lurking variable*.

Reteaching and Enrichment Strategies

If students need help...	If students got it...
Resources by Chapter • Practice A and Practice B • Puzzle Time Record and Practice Journal Practice Differentiating the Lesson Lesson Tutorials Skills Review Handbook	Resources by Chapter • Enrichment and Extension • Technology Connection Start the next section

Describe the relationship between the data. Identify any outliers, gaps, or clusters.

2 **10.**

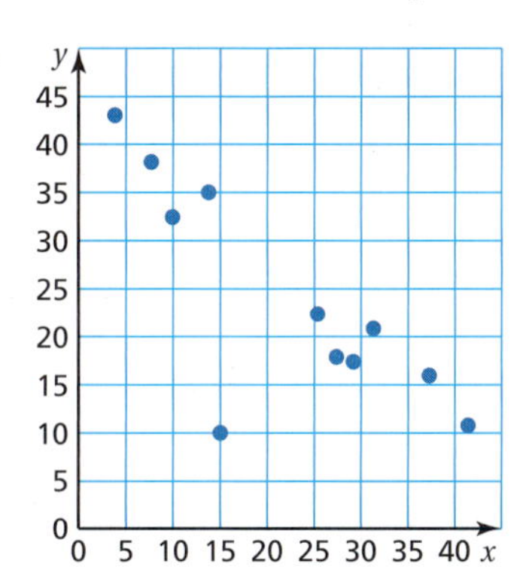

11.

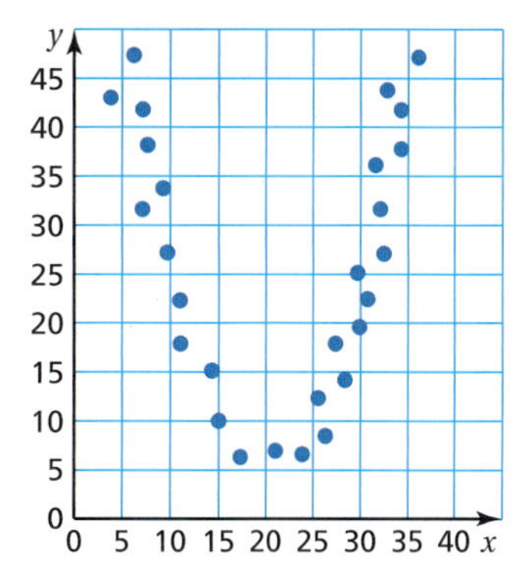

12.

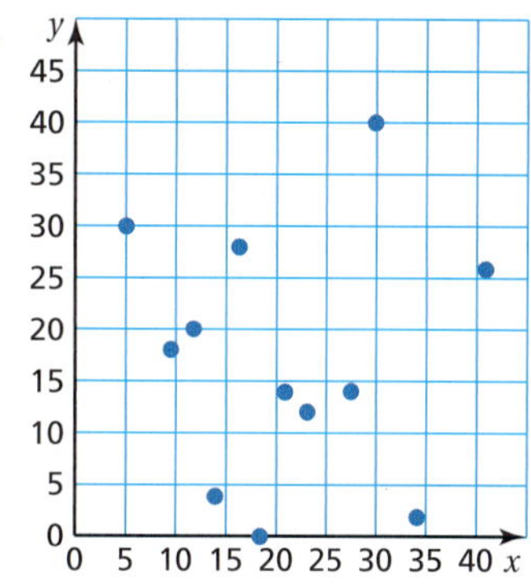

13. HONEY The table shows the average price per pound for honey in the United States from 2009 to 2012. What type of relationship do the data show?

Year, x	2009	2010	2011	2012
Average Price per Pound, y	\$4.65	\$4.85	\$5.15	\$5.53

14. TEST SCORES The scatter plot shows the numbers of minutes spent studying and the test scores for a science class. (a) What type of relationship do the data show? (b) Interpret the relationship.

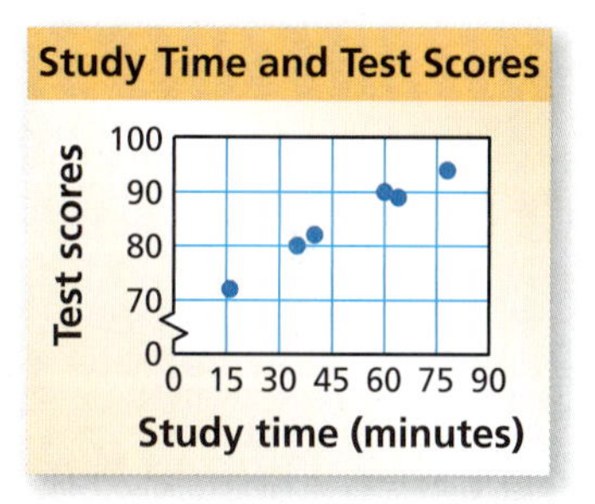

15. OPEN-ENDED Describe a set of real-life data that has a negative linear relationship.

16. PROBLEM SOLVING The table shows the memory capacities (in gigabytes) and prices (in dollars) of 7-inch tablet computers at a store. (a) Make a scatter plot of the data. Then describe the relationship between the data. (b) Identify any outliers, gaps, or clusters. Explain why you think they exist.

Memory (GB), x	8	16	4	32	4	16	4	8	16	8	16	8
Price (dollars), y	200	230	120	250	100	200	90	160	150	180	220	150

17. Reasoning Sales of sunglasses and beach towels at a store show a positive linear relationship in the summer. Does this mean that the sales of one item *cause* the sales of the other item to increase? Explain.

Fair Game Review What you learned in previous grades & lessons

Use a graph to solve the equation. Check your solution. *(Section 5.4)*

18. $5x = 2x + 6$ **19.** $7x + 3 = 9x - 13$ **20.** $\frac{2}{3}x = -\frac{1}{3}x - 4$

21. MULTIPLE CHOICE When graphing a proportional relationship represented by $y = mx$, which point is not on the graph? *(Section 4.3)*

(A) $(0, 0)$ (B) $(0, m)$ (C) $(1, m)$ (D) $(2, 2m)$

9.2 Lines of Fit

Essential Question How can you use data to predict an event?

1 ACTIVITY: Representing Data by a Linear Equation

Work with a partner. You have been working on a science project for 8 months. Each month, you measured the length of a baby alligator.

The table shows your measurements.

September (Month 0) — April (Month 7)

Month, x	0	1	2	3	4	5	6	7
Length (in.), y	22.0	22.5	23.5	25.0	26.0	27.5	28.5	29.5

Use the following steps to predict the baby alligator's length next September.

a. Graph the data in the table.

b. Draw a line that you think best approximates the points.

c. Write an equation for your line.

d. **MODELING** Use the equation to predict the baby alligator's length next September.

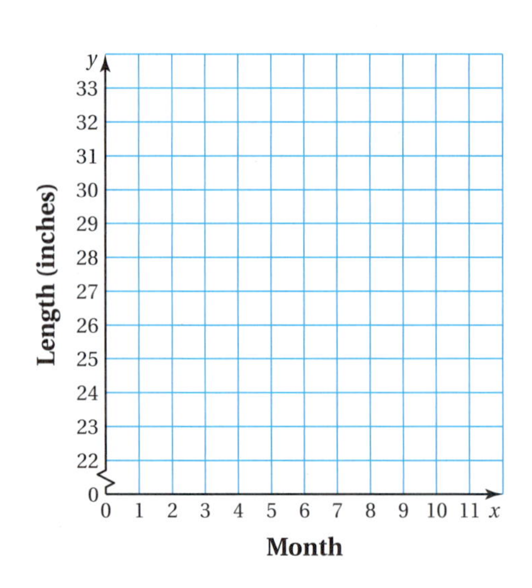

COMMON CORE

Data Analysis

In this lesson, you will

- find lines of fit.
- use lines of fit to solve problems.

Learning Standards
MACC.8.SP.1.1
MACC.8.SP.1.2
MACC.8.SP.1.3

Laurie's Notes

Introduction

Standards for Mathematical Practice

- **MP1a Make Sense of Problems:** Students will find a line of fit for data represented by a scatter plot and interpret what the equation means in the context of the problem.

Motivate

- Solicit information that students know about alligators. Share alligator facts with students as a warm-up. (See the next page.)
- That should be enough information to set the context for this first activity.

Activity Notes

Activity 1

? "Look at the table of values. What do the ordered pairs represent?" (month, length of alligator)

? "Do the data represent the first 7 months of growth of a baby alligator? Explain." No, alligators are only about 10–12 inches long when they hatch.

? "Are there any observations about the data in the table?" Months are increasing by 1. Lengths are increasing from 0.5 inch to 1.5 inches each month.

- **MP6 Attend to Precision:** Students will ask what drawing a line "that best approximates the points" means. You should explain that it is a line that passes as closely as possible to all the points. Use a straightedge to lightly draw the line.

? "What does the jagged symbol at the bottom of the *y*-axis mean?" broken axis

? "Do you think everyone in class drew the exact same line? Explain." no; They should be close, but they do not have to be exactly the same.

? "How did you write the equation for the line?" Listen for an approximation of the slope (rise over run) and the *y*-intercept (close to 22). Write the equation in slope-intercept form.

? "Does everyone have the same slope?" no; They should be relatively close, however, and should match the observations made about the data when looking at the table.

- **MP1a:** Have students interpret the slope and *y*-intercept in the context of the problem.

? "How does the equation help you answer part (d)?" Substitute 12 for *x*, and find *y*.

? "Without the equation, can you predict the length of the alligator next September?" yes; You need to extend the graph and use eyesight to approximate the ordered pair.

Florida Common Core Standards

MACC.8.SP.1.1 Construct and interpret scatter plots for bivariate measurement data to investigate patterns of association between two quantities. Describe patterns such as clustering, outliers, positive or negative association, linear association, and nonlinear association.

MACC.8.SP.1.2 Know that straight lines are widely used to model relationships between two quantitative variables. For scatter plots that suggest a linear association, informally fit a straight line, and informally assess the model fit by judging the closeness of the data points to the line.

MACC.8.SP.1.3 Use the equation of a linear model to solve problems in the context of bivariate measurement data, interpreting the slope and intercept.

Previous Learning

Students should know how to make scatter plots and write equations in slope-intercept form.

9.2 Record and Practice Journal

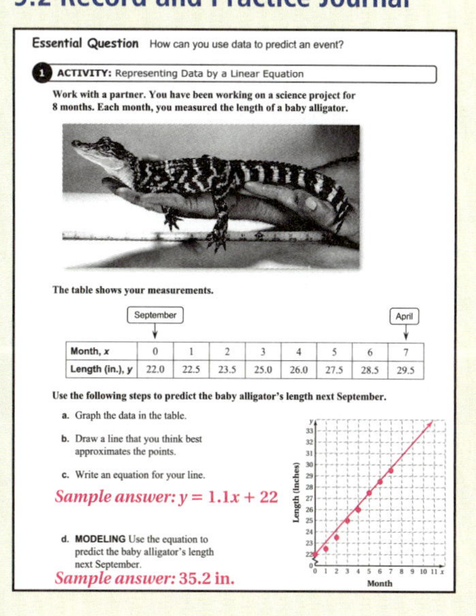

Essential Question How can you use data to predict an event?

1 ACTIVITY: Representing Data by a Linear Equation

Work with a partner. You have been working on a science project for 8 months. Each month, you measured the length of a baby alligator.

The table shows your measurements.

Month, x	0 (September)	1	2	3	4	5	6	7 (April)
Length (in.), y	22.0	22.5	23.5	25.0	26.0	27.5	28.5	29.5

Use the following steps to predict the baby alligator's length next September.

a. Graph the data in the table.

b. Draw a line that you think best approximates the points.

c. Write an equation for your line.

Sample answer: $y = 1.1x + 22$

d. **MODELING** Use the equation to predict the baby alligator's length next September.

Sample answer: 35.2 in.

Differentiated Instruction

Kinesthetic

Give students a piece of a transparency sheet. Have them use a ruler and permanent marker to draw a straight line about 4 inches long. Use a hole punch to make a hole at each end of the line. Students can move the transparency over a scatter plot until they find a line of fit, make pencil marks in the holes, and use the marks to draw the line on the scatter plot.

9.2 Record and Practice Journal

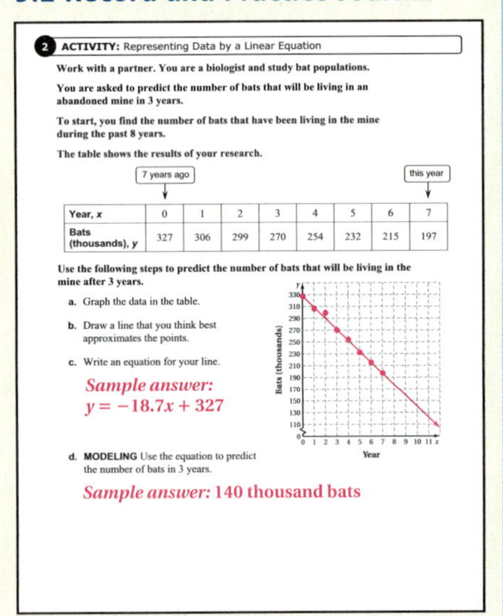

2 ACTIVITY: Representing Data by a Linear Equation

Work with a partner. You are a biologist and study bat populations.

You are asked to predict the number of bats that will be living in an abandoned mine in 3 years.

To start, you find the number of bats that have been living in the mine during the past 8 years.

The table shows the results of your research.

7 years ago — this year

Year, x	0	1	2	3	4	5	6	7
Bats (thousands), y	327	306	299	270	254	232	215	197

Use the following steps to predict the number of bats that will be living in the mine after 3 years.

a. Graph the data in the table.

b. Draw a line that you think best approximates the points.

c. Write an equation for your line.

Sample answer: $y = -18.7x + 327$

d. **MODELING** Use the equation to predict the number of bats in 3 years.

Sample answer: 140 thousand bats

What Is Your Answer?

3. **IN YOUR OWN WORDS** How can you use data to predict an event?

Plot data and line of fit.
Use equation of line to predict.

4. **MODELING** Use the Internet or some other reference to find data that appear to have a linear pattern. List the data in a table and graph the data. Use an equation that is based on the data to predict a future event.

Check students' work.

Laurie's Notes

Activity 2

- In this activity, the data have not been collected from an experiment. The data have been collected through research and recorded in the table.
- Read the introduction. The purpose of making a scatter plot is stated. You want to make a prediction about the future by examining known data.
- ? "Are there any observations about the data in the table?" Students may recognize that the number of bats is decreasing by about 15–20 (in thousands) per year.
- Discuss equations written by students. Record students' results on the board. There will likely be a bit more variation of results than in the first activity.
- **MP1a:** Have students interpret the slope and y-intercept in the context of the problem.
- **MP4 Model with Mathematics:** Discuss how the equation allows you to make predictions about the future.

What Is Your Answer?

- Question 4 can become a project due at the conclusion of the chapter.

Closure

- **Exit Ticket:** Describe the difference in the source of data for Activity 1 versus Activity 2. The data in Activity 1 are the result of gathering actual data from an experiment. The data in Activity 2 have been collected through research and recorded in a table.

More about Alligators

- The American alligator (*Alligator mississippiensis*) is the largest reptile in North America. The first reptiles appeared 300 million years ago. Ancestors of the American alligator appeared 200 million years ago.
- The name alligator comes from early Spanish explorers who called them "el lagarto" or "the lizard" when they first saw these giant reptiles.
- Louisiana and Florida have the most alligators. There are over one million wild alligators in each state with over a quarter million more on alligator farms.
- Alligators are about 10–12 inches in length when they are hatched from eggs. Growth rates vary from 2 inches per year to 12 inches per year, depending on the habitat, sex, size, and age of the alligator.
- Females can grow to about 9 feet in length and over 200 pounds. Males can grow to about 13 feet in length and over 500 pounds.
- The largest alligator was taken in Louisiana and measured 19 feet 2 inches.
- Alligators live about as long as humans, an average of 70 years.

2 ACTIVITY: Representing Data by a Linear Equation

Work with a partner. You are a biologist and study bat populations.

You are asked to predict the number of bats that will be living in an abandoned mine after 3 years.

To start, you find the number of bats that have been living in the mine during the past 8 years.

The table shows the results of your research.

Math Practice 4

Use a Graph

How can you draw a line that "fits" the collection of points? How should the points be positioned around the line?

7 years ago → Year 0; this year → Year 7

Year, x	0	1	2	3	4	5	6	7
Bats (thousands), y	327	306	299	270	254	232	215	197

Use the following steps to predict the number of bats that will be living in the mine after 3 years.

a. Graph the data in the table.

b. Draw a line that you think best approximates the points.

c. Write an equation for your line.

d. **MODELING** Use the equation to predict the number of bats in 3 years.

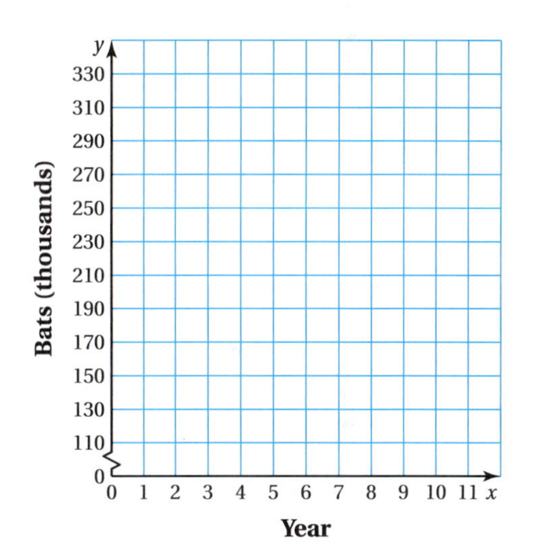

What Is Your Answer?

3. **IN YOUR OWN WORDS** How can you use data to predict an event?

4. **MODELING** Use the Internet or some other reference to find data that appear to have a linear pattern. List the data in a table, and then graph the data. Use an equation that is based on the data to predict a future event.

Practice Use what you learned about lines of fit to complete Exercise 4 on page 382.

9.2 Lesson

Key Vocabulary
line of fit, *p. 380*
line of best fit, *p. 381*

A **line of fit** is a line drawn on a scatter plot close to most of the data points. It can be used to estimate data on a graph.

EXAMPLE 1 Finding a Line of Fit

Month, x	Depth (feet), y
0	20
1	19
2	15
3	13
4	11
5	10
6	8
7	7
8	5

The table shows the depth of a river x months after a monsoon season ends. (a) Make a scatter plot of the data and draw a line of fit. (b) Write an equation of the line of fit. (c) Interpret the slope and the y-intercept of the line of fit. (d) Predict the depth in month 9.

a. Plot the points in a coordinate plane. The scatter plot shows a negative linear relationship. Draw a line that is close to the data points. Try to have as many points above the line as below it.

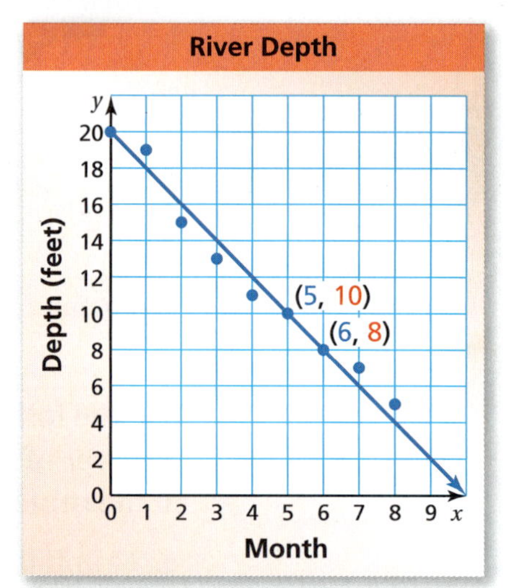

b. The line passes through (5, 10) and (6, 8).

$$\text{slope} = \frac{\text{rise}}{\text{run}} = \frac{-2}{1} = -2$$

Because the line crosses the y-axis at (0, 20), the y-intercept is 20.

So, an equation of the line of fit is $y = -2x + 20$.

c. The slope is -2, and the y-intercept is 20. So, the depth of the river is 20 feet at the end of the monsoon season and decreases by about 2 feet per month.

d. To predict the depth in month 9, substitute 9 for x in the equation of the line of fit.

$$y = -2x + 20 = -2(9) + 20 = 2$$

The depth in month 9 should be about 2 feet.

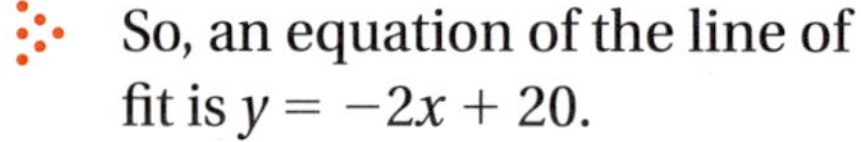

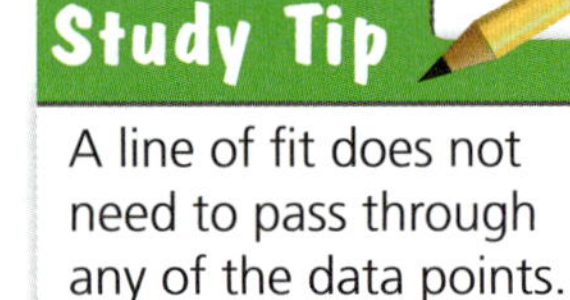

A line of fit does not need to pass through any of the data points.

On Your Own

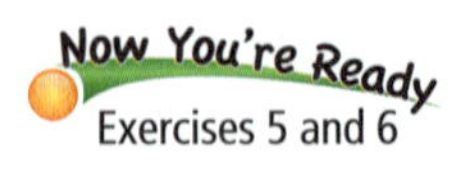

1. The table shows the numbers of people who have attended a festival over an 8-year period. (a) Make a scatter plot of the data and draw a line of fit. (b) Write an equation of the line of fit. (c) Interpret the slope and the y-intercept of the line of fit. (d) Predict the number of people who will attend the festival in year 10.

Year, x	1	2	3	4	5	6	7	8
Attendance, y	420	500	650	900	1100	1500	1750	2400

Laurie's Notes

Introduction

Connect

- **Yesterday:** Students gained an intuitive understanding of how to write an equation of a line of fit for a scatter plot. (MP1a, MP4, MP6)
- **Today:** Students will draw the line of fit for a scatter plot and analyze the equation of the line.

Motivate

- Display a scatter plot and ask students to draw a line of fit.
- Anticipate that it may not pass through any of the points.

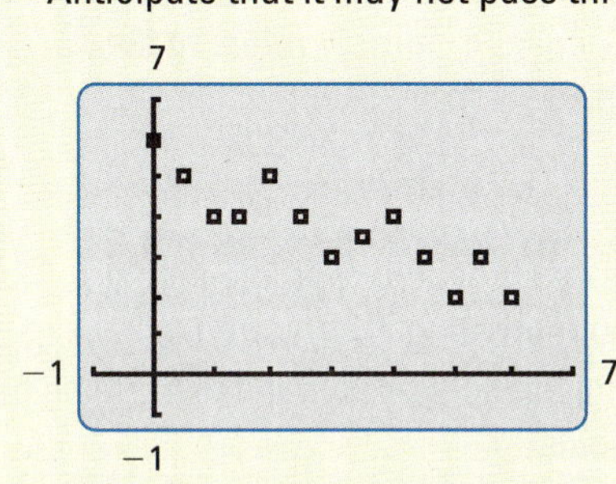

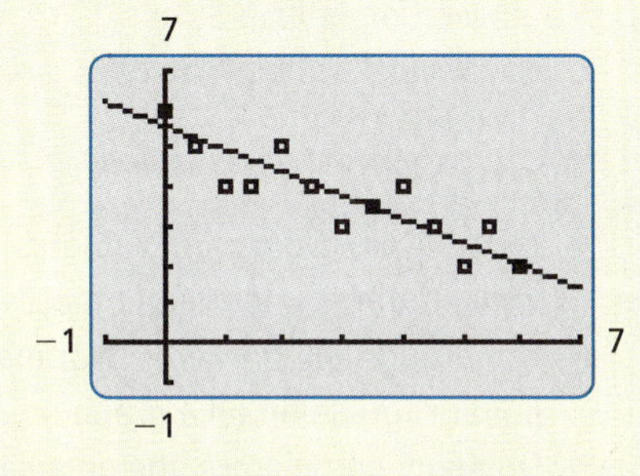

- Ask for observations about the line. Hopefully they will say that the line is a pretty good approximation for the data.
- You want students to focus on the line being the general trend of the data.

Lesson Notes

Discuss

- **MP5 Use Appropriate Tools Strategically:** Define and discuss a line of fit. It is helpful to model this with a piece of spaghetti. Use a scatter plot from the previous page that was completed on the transparency. Model how the spaghetti can approximate the trend of the data.
- Move the spaghetti so that it does *not* represent the data, and then move the spaghetti so that it does. You will use your eyesight when judging where to draw the line.

Example 1

? "How is the depth of the river changing?" The depth is decreasing.

- Note that a river can swell during monsoons and then gradually dry up.
- When drawing a line of fit, try to put as many points above the line as below it.
- Students may draw different lines of fit and still get reasonable answers.
- In this example, the line passes through two actual data points, (5, 10) and (6, 8). As noted in the *Study Tip*, a line of fit does not need to pass through any of the data points.
- **MP4 Model with Mathematics:** The purpose of writing the equation of the line of fit is to make predictions. The equation becomes a model for the data, describing its behavior.

On Your Own

- This is a nice summary problem. Students should quickly observe the positive relationship just from the table of values.

Goal Today's lesson is identifying a **line of fit** for a scatter plot.

Lesson Tutorials
Lesson Plans
Answer Presentation Tool

Extra Example 1

The table shows the number y of customers in line at a bank x minutes after noon.

Minutes, x	Customers, y
1	15
2	13
3	12
4	9
5	6
6	4
7	3

a. Make a scatter plot of the data and draw a line of fit.

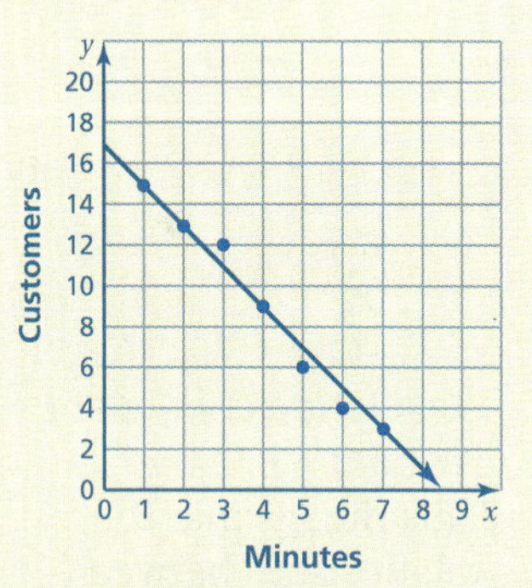

b. Write an equation of the line of fit. $y = -2x + 17$

c. Interpret the slope and the y-intercept of the line of fit. At noon, there are 17 customers in line. The number of people in line decreases by about 2 each minute.

d. Predict the number of customers in line 8 minutes after noon. 1

On Your Own

1. See Additional Answers.

English Language Learners

Vocabulary

Help English language learners with the meaning of *correlation coefficient*. The word *correlate* can be broken into two parts, *co-* and *relate*. The prefix *co-* means "with" or "together" as in *copilot*. The word *relate* means "to show or make a connection."

Extra Example 2

Use a graphing calculator to find an equation of the line of best fit for the data in Extra Example 1. Identify and interpret the correlation coefficient. $y = -2.1x + 17$; The correlation coefficient is about -0.991. This means that the relationship between minutes and customers in line is a strong negative correlation and the equation closely models the data.

On Your Own

2. $y = -1.9x + 20$; The correlation coefficient is about -0.988. This means that the relationship between months and depths is a strong negative correlation and that the equation closely models the data.

Laurie's Notes

Discuss

- Tell students that you can use a graphing calculator to determine the line of best fit. This is called linear regression.
- Explain that in addition to calculating the line of best fit, the calculator also gives a value called the correlation coefficient. This is a measure of how well the line fits the data.
- Describe positive, negative, and no correlation.
- Say, "The correlation coefficient is a number between -1 and 1." Draw and label the graphic shown.
- **Common Misconception:** Students may think that a strong negative correlation is *bad*. Stress that the "negative" merely refers to the downward trend of the data, or slope.

Example 2

- **MP1a Make Sense of Problems:** Ask a student to read the problem. Make sure students understand that $x = 0$ represents the year 2000.
- Go through the procedure for making the scatter plot and performing the linear regression on a graphing calculator.
- On many calculators, the correlation coefficient is a feature that can be turned on and off. Instruct students how to turn this feature on so that it is displayed when they perform the regression. An r^2-value may also be displayed, which students may learn about in a future course.

? "What is the line of best fit?" $y = 1.5x + 16$

? "What do the slope and y-intercept mean?" A slope of 1.5 means that ticket sales are increasing by about \$1.5 billion each year. A y-intercept of 16 means that in 2000, the ticket sales were about \$16 billion.

? "What is the correlation coefficient and what does it mean?" $r \approx 0.982$; It implies a strong positive correlation between years and ticket sales.

On Your Own

- Compare the line of best fit found in Question 2 to the line of fit found in Example 1.

Closure

- **Exit Ticket:** Give an example of a situation in which there is correlation between the variables. *Sample answer:* The number of teachers in a school and the number of students in the school

Study Tip

You know how to use two points to find an equation of a line of fit. When finding an equation of the line of best fit, every point in the data set is used.

Graphing calculators use a method called *linear regression* to find a precise line of fit called a **line of best fit**. This line best models a set of data. A calculator often gives a value r called the *correlation coefficient*. This value tells whether the correlation is positive or negative, and how closely the equation models the data. Values of r range from -1 to 1. When r is close to 1 or -1, there is a strong correlation between the variables. As r gets closer to 0, the correlation becomes weaker.

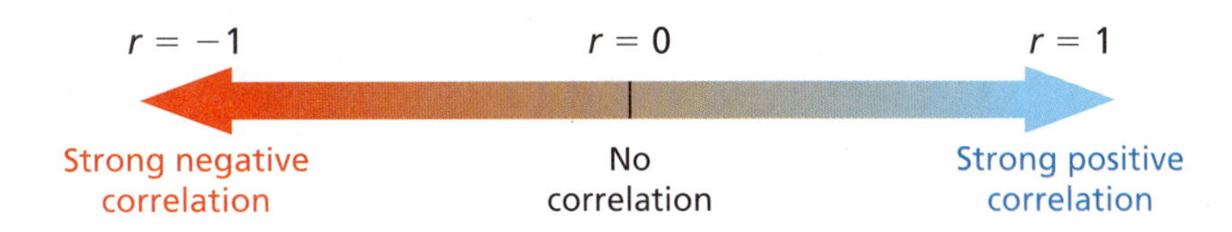

EXAMPLE 2 Finding a Line of Best Fit Using Technology

The table shows the worldwide movie ticket sales y (in billions of dollars) from 2000 to 2011, where $x = 0$ represents the year 2000. Use a graphing calculator to find an equation of the line of best fit. Identify and interpret the correlation coefficient.

Year, x	0	1	2	3	4	5	6	7	8	9	10	11
Ticket Sales, y	16	17	20	20	25	23	26	26	28	29	32	33

Step 1: Enter the data from the table into your calculator.

Step 2: Use the *linear regression* feature.

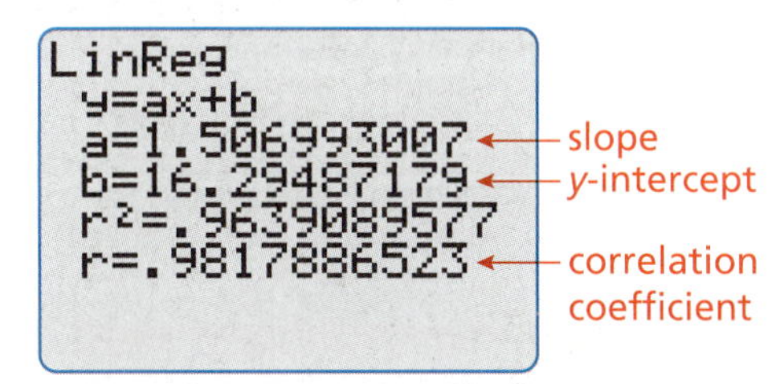

Study Tip

The slope of 1.5 indicates that sales are increasing by about \$1.5 billion each year. The y-intercept of 16 represents the ticket sales of \$16 billion for 2000.

An equation of the line of best fit is $y = 1.5x + 16$. The correlation coefficient is about 0.982. This means that the relationship between years and ticket sales is a strong positive correlation and that the equation closely models the data.

Check Use a graphing calculator to make a scatter plot and graph the line of best fit.

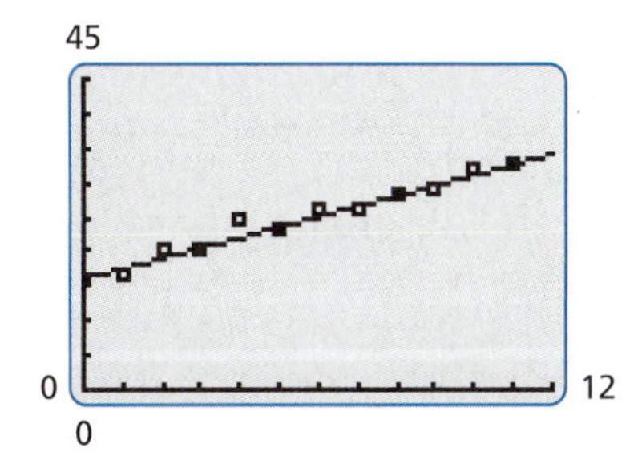

On Your Own

2. Use a graphing calculator to find an equation of the line of best fit for the data in Example 1. Identify and interpret the correlation coefficient.

9.2 Exercises

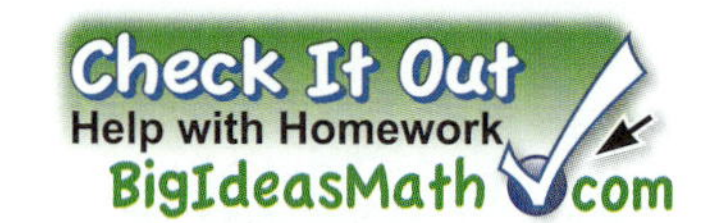

Vocabulary and Concept Check

1. **WRITING** Explain why a line of fit is helpful when analyzing data.
2. **REASONING** Tell whether the line drawn on the graph is a good fit for the data. Explain your reasoning.
3. **NUMBER SENSE** Which correlation coefficient indicates a stronger relationship: −0.98 or 0.91? Explain.

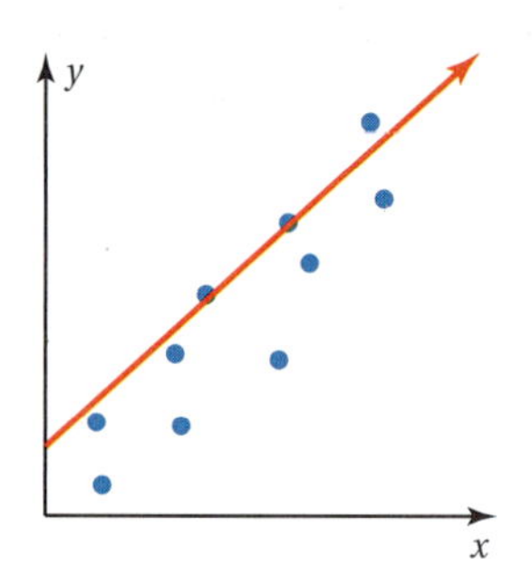

Practice and Problem Solving

4. **BLUEBERRIES** The table shows the weights y of x pints of blueberries.

Number of Pints, x	0	1	2	3	4	5
Weight (pounds), y	0	0.8	1.50	2.20	3.0	3.75

   a. Graph the data in the table.
   b. Draw a line that you think best approximates the points.
   c. Write an equation for your line.
   d. Use the equation to predict the weight of 10 pints of blueberries.
   e. Blueberries cost $2.25 per pound. How much do 10 pints of blueberries cost?

1 5. **HOT CHOCOLATE** The table shows the daily high temperature (°F) and the number of hot chocolates sold at a coffee shop for eight randomly selected days.

Temperature (°F), x	30	36	44	51	60	68	75	82
Hot Chocolates, y	45	43	36	35	30	27	23	17

   a. Make a scatter plot of the data and draw a line of fit.
   b. Write an equation of the line of fit.
   c. Interpret the slope and the y-intercept of the line of fit.
   d. Predict the number of hot chocolates sold when the high temperature is 20°F.

6. **VACATION** The table shows the distance you are away from home over a 6-hour period of your vacation.

   a. Make a scatter plot of the data and draw a line of fit.
   b. Write an equation of the line of fit.
   c. About how many miles per hour do you travel?
   d. About how far were you from home when you started?
   e. Predict the distance from home in 7 hours.

Hours, x	Distance (miles), y
1	62
2	123
3	188
4	228
5	280
6	344

Assignment Guide and Homework Check

Level	Day 1 Activity Assignment	Day 2 Lesson Assignment	Homework Check
Basic	4, 12–16	1–3, 5–9	2, 3, 5, 6
Average	4, 12–16	1–3, 5–10	3, 6, 7, 8
Advanced	1–16		6, 8, 9, 11

Common Errors

- **Exercise 4** Students may use inconsistent increments or forget to label their graphs. Students should use consistent increments to represent the data. Remind them to label the axes so that information can be read from the graph.
- **Exercise 5** Students may draw a line of fit that does not accurately reflect the data trend. Remind them that the line does not have to go through any of the data points. Also remind them that the line should go through the middle of the data so that about half of the data points are above the line and half are below. One strategy is to draw an oval around the data and then draw a line through the middle of the oval. For example:

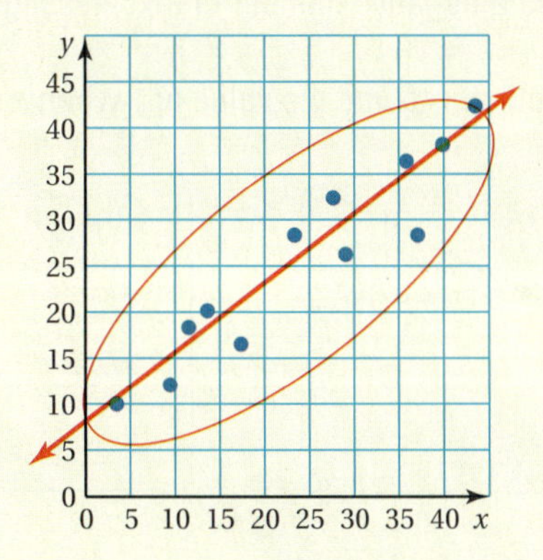

9.2 Record and Practice Journal

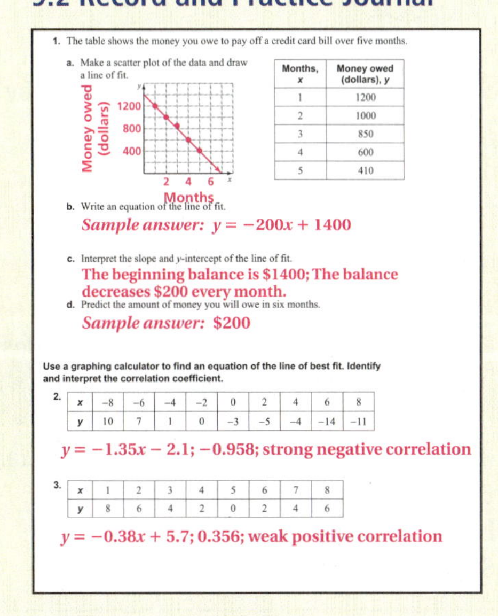

1. The table shows the money you owe to pay off a credit card bill over five months.

a. Make a scatter plot of the data and draw a line of fit.

Months, x	Money owed (dollars), y
1	1200
2	1000
3	850
4	600
5	410

b. Write an equation of the line of fit.
Sample answer: $y = -200x + 1400$

c. Interpret the slope and y-intercept of the line of fit.
The beginning balance is $1400; The balance decreases $200 every month.

d. Predict the amount of money you will owe in six months.
Sample answer: $200

Use a graphing calculator to find an equation of the line of best fit. Identify and interpret the correlation coefficient.

2.

x	−8	−6	−4	−2	0	2	4	6	8
y	10	7	1	0	−3	−5	−4	−14	−11

$y = -1.35x - 2.1$; -0.958; strong negative correlation

3.

x	1	2	3	4	5	6	7	8
y	8	6	4	2	0	2	4	6

$y = -0.38x + 5.7$; 0.356; weak positive correlation

Vocabulary and Concept Check

1. You can estimate and predict values.
2. *Sample answer:* not a good representation; Too many points in the data set lie below the line.
3. -0.98, because it is closer to -1 than 0.91 is to 1. $(|-0.98| > |0.91|)$

Practice and Problem Solving

4. a–b.

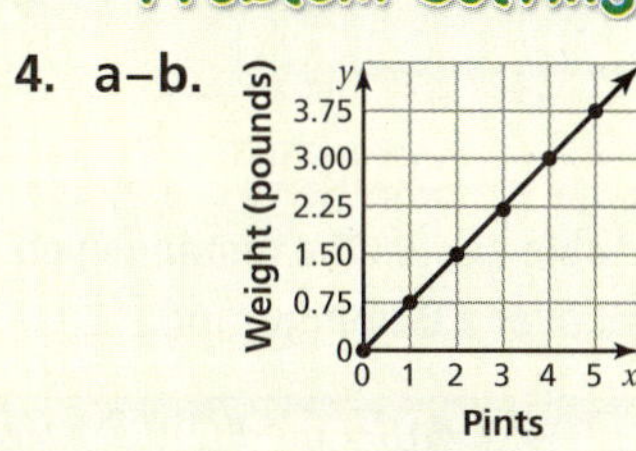

c. *Sample answer:* $y = 0.75x$

d. *Sample answer:* 7.5 lb

e. *Sample answer:* $16.88

5. a.

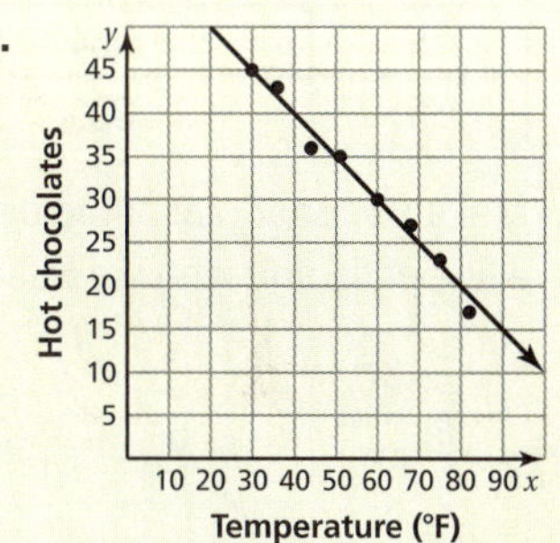

b. *Sample answer:* $y = -0.5x + 60$

c. *Sample answer:* The slope is -0.5 and the y-intercept is 60. So, you could predict that 60 hot chocolates are sold when the temperature is 0°F, and the sales decrease by about 1 hot chocolate for every 2°F increase in temperature.

d. 50 hot chocolates

6. See Additional Answers.

Practice and Problem Solving

7. no; There is no line that lies close to most of the points.

8–9. See Additional Answers.

10. See *Taking Math Deeper*.

11. See Additional Answers.

Fair Game Review

12. $\frac{2}{9}$

13. $-2\frac{7}{9}$

14. $-1\frac{7}{15}$

15. $\frac{9}{11}$

16. D

Mini-Assessment

1. The table shows the distance you travel over a 6-hour period.

Hours, x	Distance (miles), y
1	50
2	102
3	153
4	204
5	254
6	305

a. Make a scatter plot of the data, draw a line of fit, and write its equation.

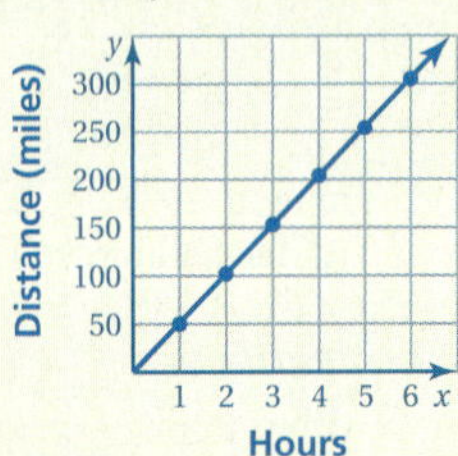

$y = 51x$

b. Interpret the slope of the line of fit.
You travel about 51 miles each hour.

c. Use a graphing calculator to find an equation of the line of best fit.
$y = 50.9x - 0.2$

d. Identify and interpret the correlation coefficient. The correlation coefficient is about 1.000. This means that the relationship between hours and distances traveled is a very strong positive correlation and the equation closely models the data.

Taking Math Deeper

Exercise 10

This exercise is a good use of technology. Students may realize that they can use a graphing calculator to evaluate a function in different ways.

Enter the data into a calculator and use the linear regression feature.

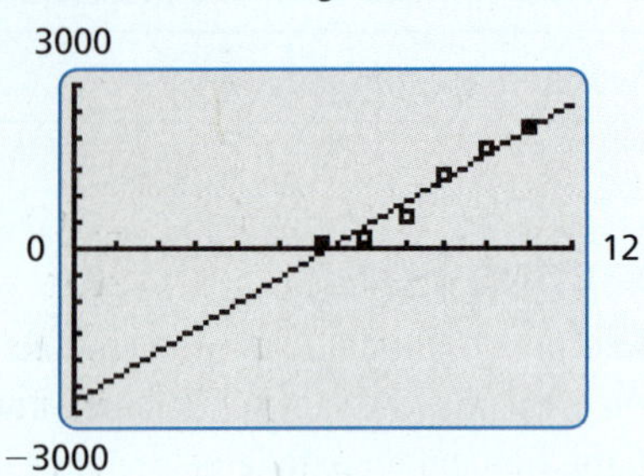

a. An equation of the line of best fit is $y = 454.8x - 2812$. The correlation coefficient of about 0.984 means that there is a strong positive correlation in the data and that the equation closely models the data.

2 b. The slope of 454.8 indicates that the number of text messages sent is increasing by about 454,800,000,000 messages each year. No, the y-intercept of -2812 does not make sense in this problem because there cannot be a negative number of text messages sent.

3 You can use a calculator to find the value of y when x is 15.

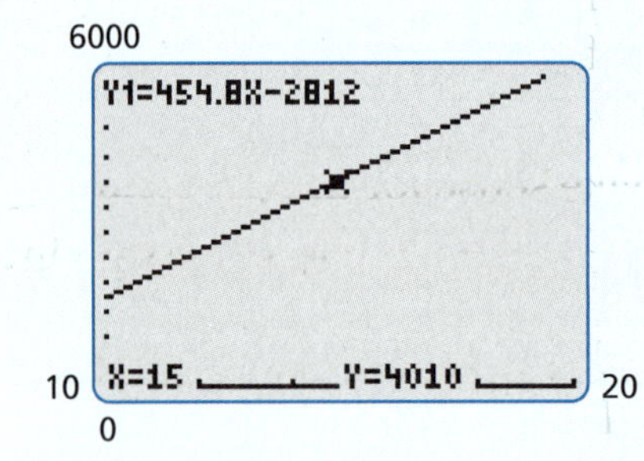

c. 4,010,000,000,000, or about 4 trillion text messages

Project

Have students find a set of real-life data that can be modeled by a linear equation. Have them justify their reasoning by using a graphing calculator.

Reteaching and Enrichment Strategies

If students need help. . .	If students got it. . .
Resources by Chapter • Practice A and Practice B • Puzzle Time Record and Practice Journal Practice Differentiating the Lesson Lesson Tutorials Skills Review Handbook	Resources by Chapter • Enrichment and Extension • Technology Connection Start the next section

7. **REASONING** A data set has no relationship. Is it possible to find a line of fit for the data? Explain.

② 8. **AMUSEMENT PARK** The table shows the attendance y (in thousands) at an amusement park from 2004 to 2013, where $x = 4$ represents the year 2004. Use a graphing calculator to find an equation of the line of best fit. Identify and interpret the correlation coefficient.

Year, x	4	5	6	7	8	9	10	11	12	13
Attendance (thousands), y	850	845	828	798	800	792	785	781	775	760

9. **SNOWSTORM** The table shows the total snow depth y (in inches) on the ground during a snowstorm x hours after it began. Use a graphing calculator to find an equation of the line of best fit. Identify and interpret the correlation coefficient. Use your equation to estimate how much snow was on the ground before the snowstorm began.

Hours, x	1	2	3	4	5	6	7	8
Snow Depth (inches), y	5	6	6.75	7.75	8.5	9.5	10.5	11.5

10. **TEXTING** The table shows the numbers y (in billions) of text messages sent from 2006 to 2011, where $x = 6$ represents the year 2006.

Year, x	Text Messages (billions), y
6	113
7	241
8	601
9	1360
10	1806
11	2206

a. Use a graphing calculator to find an equation of the line of best fit. Identify and interpret the correlation coefficient.

b. Interpret the slope of the line of best fit. Does the y-intercept make sense for this problem? Explain.

c. Predict the number of text messages sent in 2015.

11. **Modeling** The table shows the height y (in feet) of a baseball x seconds after it was hit.

Seconds, x	Height (feet), y
0	3
0.5	39
1	67
1.5	87
2	99

a. Use a graphing calculator to find an equation of the line of best fit. Identify and interpret the correlation coefficient.

b. Predict the height after 5 seconds.

c. The actual height after 5 seconds is about 3 feet. Why do you think this is different from your prediction?

Fair Game Review What you learned in previous grades & lessons

Write the decimal as a fraction or a mixed number. *(Section 7.4)*

12. $0.\overline{2}$ 13. $-2.\overline{7}$ 14. $-1.4\overline{6}$ 15. $0.\overline{81}$

16. **MULTIPLE CHOICE** Which expression represents the volume of a sphere with radius r? *(Section 8.3)*

Ⓐ $\frac{1}{3}\pi r^2 h$ Ⓑ $\pi r^2 h$ Ⓒ $4\pi r^2$ Ⓓ $\frac{4}{3}\pi r^3$

9 Study Help

You can use an **information frame** to help you organize and remember concepts. Here is an example of an information frame for scatter plots.

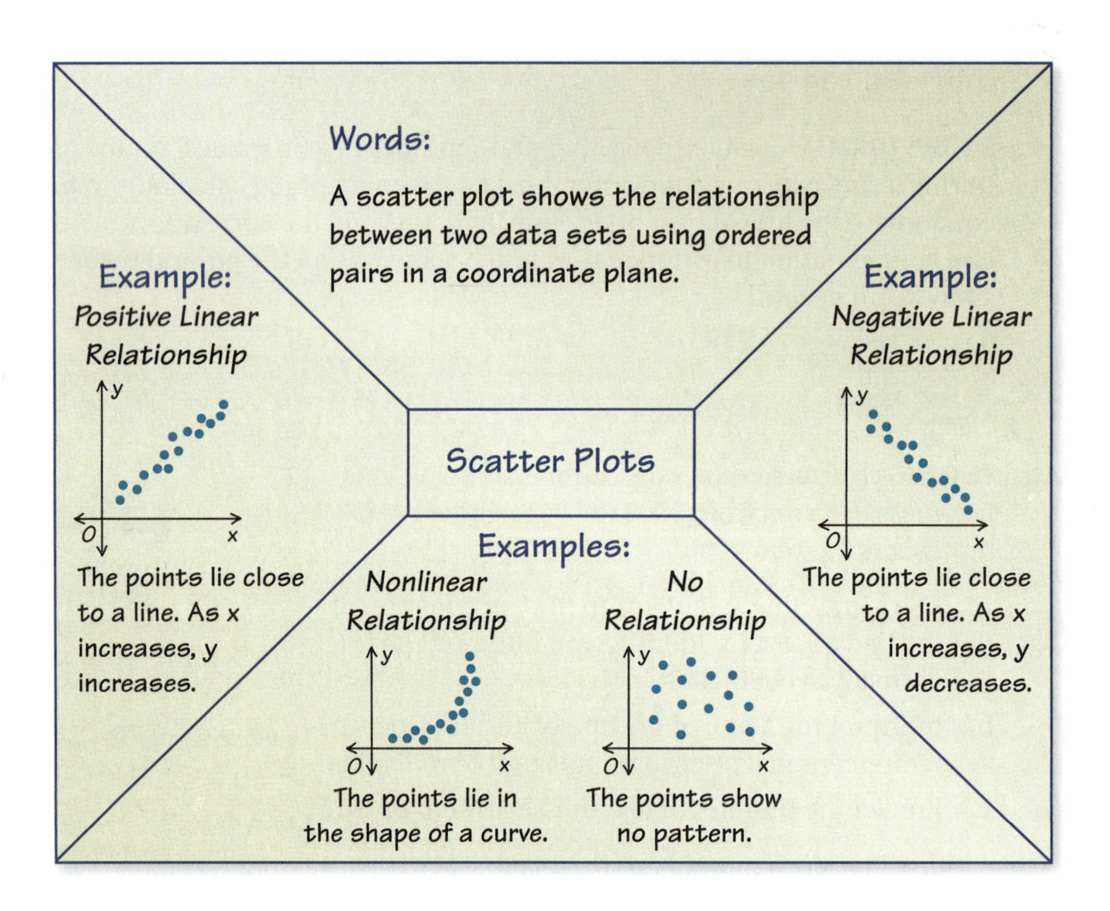

On Your Own

Make an information frame to help you study this topic.

1. lines of fit

After you complete this chapter, make information frames for the following topics.

2. two-way tables
3. data displays

"Dear Teacher, I am emailing my information frame showing the characteristics of circles."

Sample Answer

1.

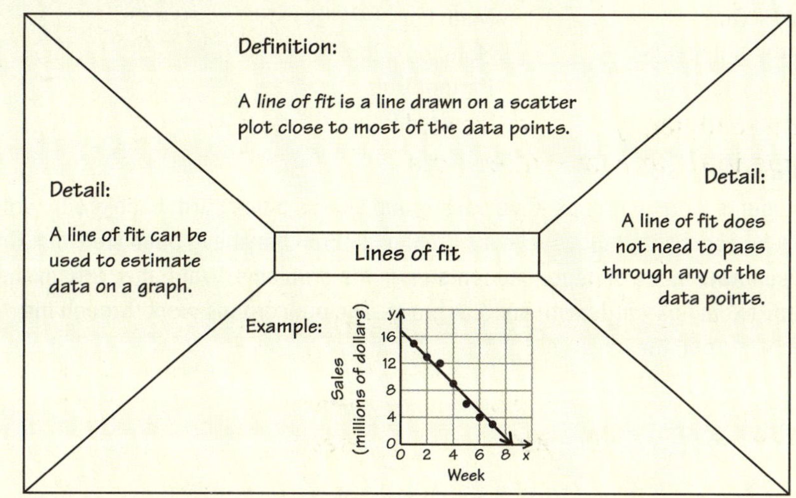

List of Organizers

Available at *BigIdeasMath.com*

Comparison Chart
Concept Circle
Definition (Idea) and Example Chart
Example and Non-Example Chart
Formula Triangle
Four Square
Information Frame
Information Wheel
Notetaking Organizer
Process Diagram
Summary Triangle
Word Magnet
Y Chart

About this Organizer

An **Information Frame** can be used to help students organize and remember concepts. Students write the topic in the middle rectangle. Then students write related concepts in the spaces around the rectangle. Related concepts can include *Words, Numbers, Algebra, Example, Definition, Non-Example, Visual, Procedure, Details,* and *Vocabulary.* Students can place their information frames on note cards to use as a quick study reference.

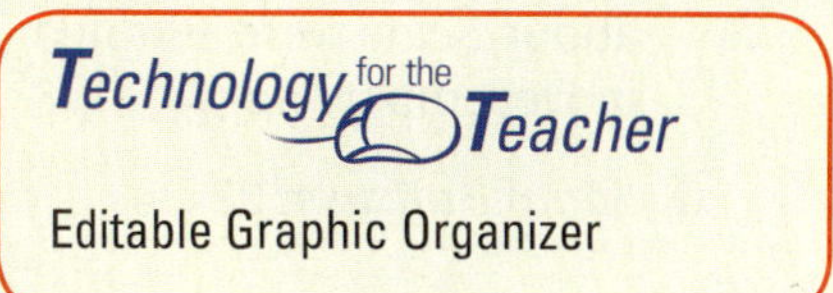

Editable Graphic Organizer

Answers

1. a. 2007 b. $120,000

   c. There is a negative relationship between year and amount donated.

2. negative linear relationship; outlier: (6, 10)

3. no relationship

4. positive linear relationship; cluster under $y = 16$

5. a.

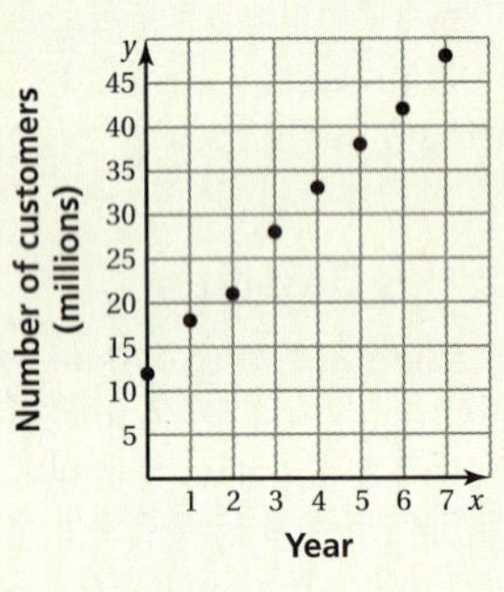

   positive linear relationship; There are no outliers, gaps, or clusters.

   b. $y = 5.1x + 12$; $r \approx 0.998$; This means that there is strong positive linear correlation between the year and the number of customers served.

6. a.

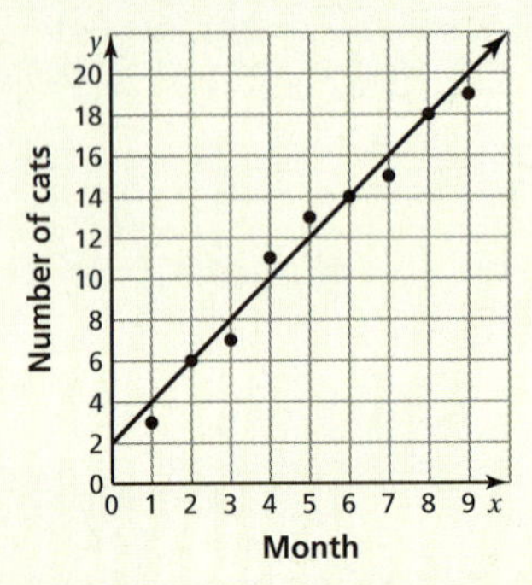

   b. *Sample answer:* $y = 2x + 2$

   c. The slope of the line of fit is 2. This means that the number of cats adopted increases by 2 cats per month. The y-intercept is about 2, which means about 2 cats were adopted in December.

   d. *Sample answer:* 22 cats

Alternative Quiz Ideas

100% Quiz
Error Notebook
Group Quiz
Homework Quiz
Math Log
Notebook Quiz
Partner Quiz
Pass the Paper

Group Quiz

Students work in groups. Give each group a large index card. Each group writes five questions that they feel evaluate the material they have been studying. On a separate piece of paper, students solve the problems. When they are finished, they exchange cards with another group. The new groups work through the questions on the card.

Reteaching and Enrichment Strategies

If students need help...	If students got it...
Resources by Chapter • Practice A and Practice B • Puzzle Time Lesson Tutorials *BigIdeasMath.com*	Resources by Chapter • Enrichment and Extension • Technology Connection Game Closet at *BigIdeasMath.com* Start the next section

9.1–9.2 Quiz

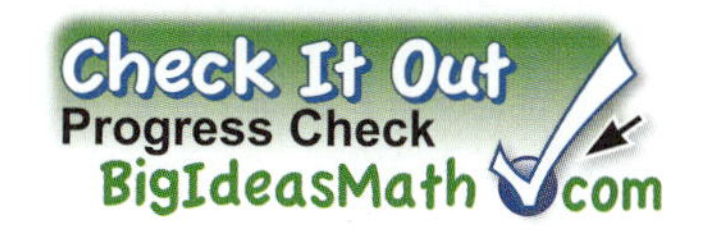

1. **CHARITY** The scatter plot shows the amount of money donated to a charity from 2007 to 2012. *(Section 9.1)*

   a. In what year did the charity receive $150,000?

   b. How much did the charity receive in 2010?

   c. Describe the relationship shown by the data.

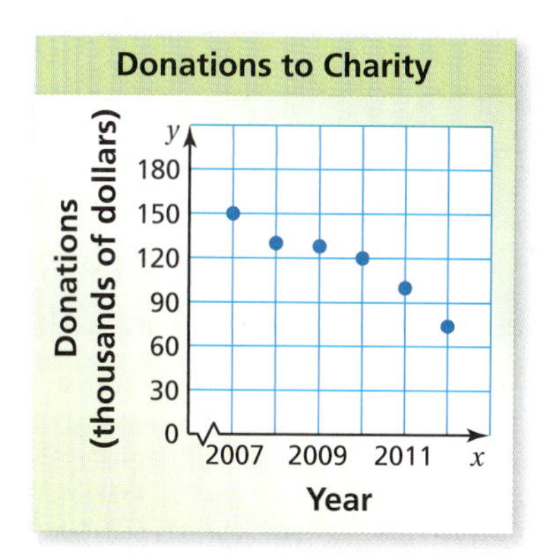

Describe the relationship between the data. Identify any outliers, gaps, or clusters. *(Section 9.1)*

2. 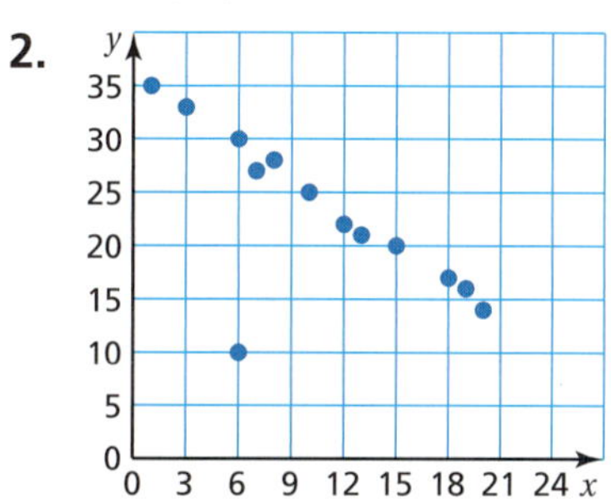

3. 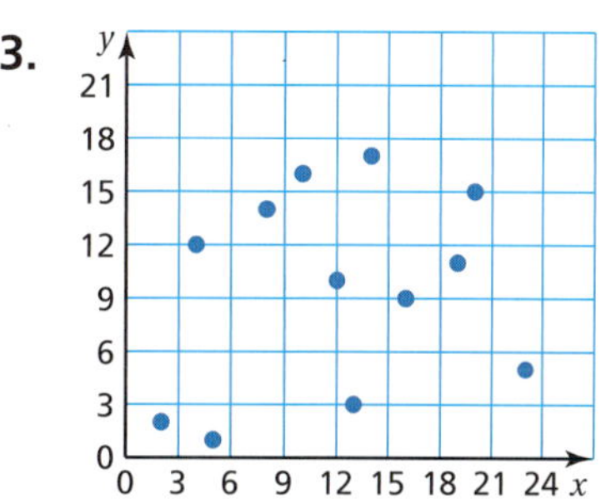

4. 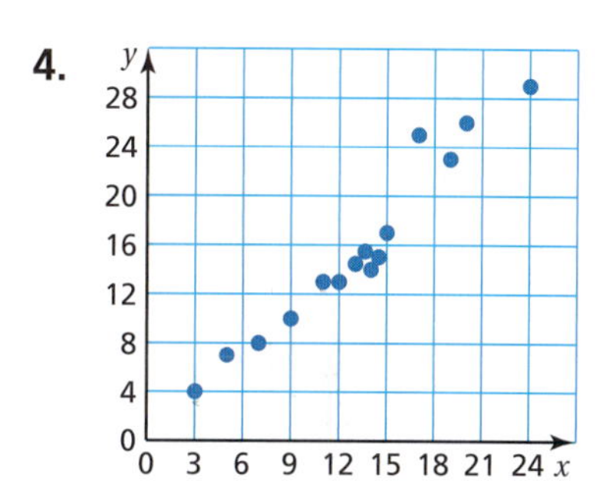

Year, x	Millions of Customers, y
0	12
1	18
2	21
3	28
4	33
5	38
6	42
7	48

5. **CUSTOMERS SERVED** The table shows the numbers of customers (in millions) served by a restaurant chain over an 8-year period. *(Section 9.1 and Section 9.2)*

   a. Make a scatter plot of the data. Then describe the relationship between the data. Identify any outliers, gaps, or clusters.

   b. Use a graphing calculator to find the equation of the line of best fit for the data. Identify and interpret the correlation coefficient.

6. **CATS** An animal shelter opens in December. The table shows the number of cats adopted from the shelter each month from January to September. *(Section 9.2)*

Month	1	2	3	4	5	6	7	8	9
Cats	3	6	7	11	13	14	15	18	19

   a. Make a scatter plot of the data and draw a line of fit.

   b. Write an equation of the line of fit.

   c. Interpret the slope and the y-intercept of the line of fit.

   d. Predict how many cats will be adopted in October.

9.3 Two-Way Tables

Essential Question How can you read and make a two-way table?

Two categories of data can be displayed in a *two-way table.*

1 ACTIVITY: Reading a Two-Way Table

Work with a partner. You are the manager of a sports shop. The two-way table shows the numbers of soccer T-shirts that your shop has left in stock at the end of the season.

		T-Shirt Size					
		S	M	L	XL	XXL	Total
Color	Blue/White	5	4	1	0	2	
	Blue/Gold	3	6	5	2	0	
	Red/White	4	2	4	1	3	
	Black/White	3	4	1	2	1	
	Black/Gold	5	2	3	0	2	
	Total						65

a. Complete the totals for the rows and columns.

b. Are there any black-and-gold XL T-shirts in stock? Justify your answer.

c. The numbers of T-shirts you ordered at the beginning of the season are shown below. Complete the two-way table.

		T-Shirt Size					
		S	M	L	XL	XXL	Total
Color	Blue/White	5	6	7	6	5	
	Blue/Gold	5	6	7	6	5	
	Red/White	5	6	7	6	5	
	Black/White	5	6	7	6	5	
	Black/Gold	5	6	7	6	5	
	Total						

d. **REASONING** How would you alter the numbers of T-shirts you order for next season? Explain your reasoning.

COMMON CORE

Data Analysis

In this lesson, you will

- read two-way tables.
- make and interpret two-way tables.

Learning Standard
MACC.8.SP.1.4

Laurie's Notes

Introduction

Standards for Mathematical Practice

- **MP2 Reason Abstractly and Quantitatively:** In this section, students are translating information into an organized table to make sense of the problem and to make observations and reason about the information. The goal is not to simply construct a table or read information from the table, but to reason about relationships that exist between categories in the table.

Motivate

- **Story Time:** Tell students about a few of the sessions and workshops you attended at a 3-day math conference. When you returned, you submitted your expenses.

	Day 1	Day 2	Day 3	Totals
Meals				A
Lodging				
Taxi				
Totals		B		C

- The school district does not want to share your expenses publicly so they are blacked out.
- ? "What do the numbers in A, B, and C represent?" A is the total amount spent on meals for 3 days; B is the total expenses for day 2; C is the total expenses for all 3 days.
- ? "How do you find C?" Find the sum of the last column or the sum of the last row.
- If students do not know the vocabulary—column and row—be sure to clarify.

Activity Notes

Activity 1

- Students should find that reading a two-way table is relatively easy. The term "two-way table" is new to students, yet they do not need a formal definition to make sense of the problem and what it is asking. In fact, some students will jump in and start adding the entries in the rows and columns without reading the introduction.
- When students have finished, discuss their responses to part (d).
- ? "What size(s) of shirts sold well?" XL and perhaps XXL "What size(s) of shirts did not sell well?" S and M
- ? "What color(s) of shirts sold well?" Black/White "What color(s) of shirts did not sell well?" Blue/Gold
- ? "Is there a way to quantify or rank the popular sizes and colors?" yes; You can compute the percent of each size or color that was sold.
- ? "Do you think merchants keep track of inventory in this manner?" Answers will vary. Successful merchants do track inventory to see what is selling.

Florida Common Core Standards

MACC.8.SP.1.4 Understand that patterns of association can also be seen in bivariate categorical data by displaying frequencies and relative frequencies in a two-way table. Construct and interpret a two-way table . . .. Use relative frequencies calculated for rows or columns to describe possible association between the two variables.

Previous Learning

Students should know how to display data using different types of displays, such as histograms, box-and-whisker plots, and scatter plots.

Lesson Plans
Complete Materials List

9.3 Record and Practice Journal

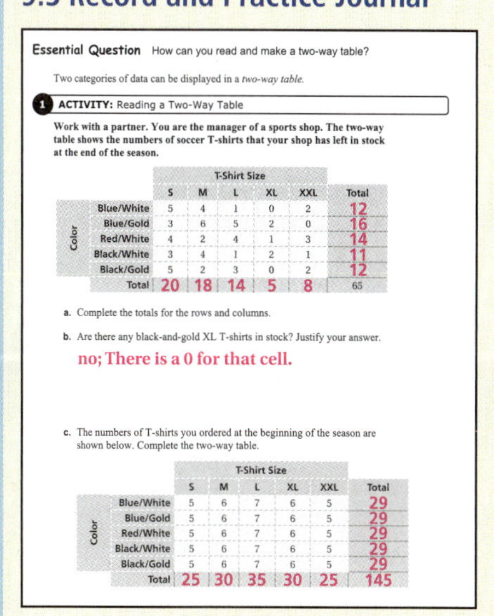

Essential Question How can you read and make a two-way table?

Two categories of data can be displayed in a *two-way table.*

1 ACTIVITY: Reading a Two-Way Table

Work with a partner. You are the manager of a sports shop. The two-way table shows the numbers of soccer T-shirts that your shop has left in stock at the end of the season.

Color	T-Shirt Size: S	M	L	XL	XXL	Total
Blue/White	5	4	1	0	2	12
Blue/Gold	3	6	5	2	0	16
Red/White	4	2	4	1	3	14
Black/White	3	4	1	2	1	11
Black/Gold	5	2	3	0	2	12
Total	20	18	14	5	8	65

a. Complete the totals for the rows and columns.

b. Are there any black-and-gold XL T-shirts in stock? Justify your answer.

no; There is a 0 for that cell.

c. The numbers of T-shirts you ordered at the beginning of the season are shown below. Complete the two-way table.

Color	T-Shirt Size: S	M	L	XL	XXL	Total
Blue/White	5	6	7	6	5	29
Blue/Gold	5	6	7	6	5	29
Red/White	5	6	7	6	5	29
Black/White	5	6	7	6	5	29
Black/Gold	5	6	7	6	5	29
Total	25	30	35	30	25	145

Differentiated Instruction

Advanced

Have students enter the data from a two-way table into a spreadsheet. Use the *chart* feature to create a 3-D column chart (as shown on page 387), a clustered column chart, a stacked column chart, and a 100% stacked column chart. Students should describe the relationship between the data values in each chart and give an example of what information could be gathered from reading the chart.

9.3 Record and Practice Journal

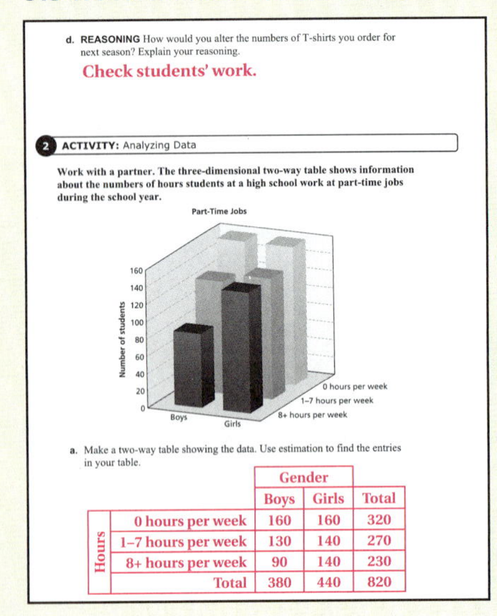

d. **REASONING** How would you alter the numbers of T-shirts you order for next season? Explain your reasoning.

Check students' work.

2 ACTIVITY: Analyzing Data

Work with a partner. The three-dimensional two-way table shows information about the numbers of hours students at a high school work at part-time jobs during the school year.

a. Make a two-way table showing the data. Use estimation to find the entries in your table.

		Gender		
		Boys	Girls	Total
Hours	0 hours per week	160	160	320
	1–7 hours per week	130	140	270
	8+ hours per week	90	140	230
	Total	380	440	820

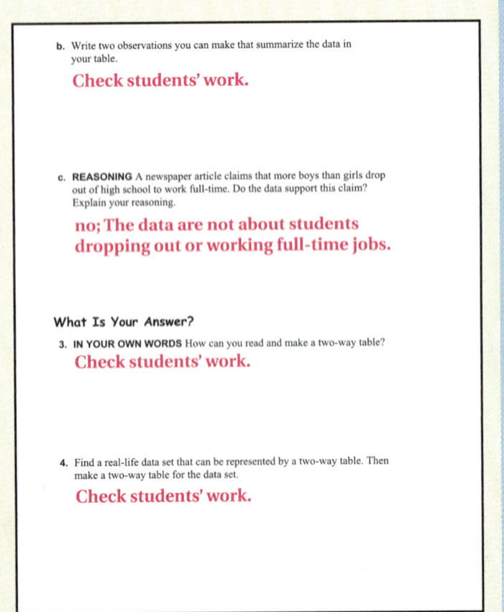

b. Write two observations you can make that summarize the data in your table.

Check students' work.

c. **REASONING** A newspaper article claims that more boys than girls drop out of high school to work full-time. Do the data support this claim? Explain your reasoning.

no; The data are not about students dropping out or working full-time jobs.

What Is Your Answer?

3. **IN YOUR OWN WORDS** How can you read and make a two-way table?

Check students' work.

4. Find a real-life data set that can be represented by a two-way table. Then make a two-way table for the data set.

Check students' work.

Laurie's Notes

Activity 2

- ? "Have you ever seen a three-dimensional two-way table?" Students who are familiar with spreadsheets may recognize this type of display. Students may have seen them on the Internet, or in newspapers or magazines.
- Ask students to explain what one of the prisms represents.
- ? "What information is represented by the taller red prism?" It represents the number of girls that work 8 or more hours per week at a part-time job.
- ? "How many girls are in that category?" about 140
- Check that students have the correct labels for the rows and columns of the two-way table: gender (boys, girls) and hours (3 intervals shown).
- Ask a volunteer to display the two-way table.
- ? "Can you determine how many students attend the school? Explain." yes; Find the sum of the last column or the last row in the table. They should be equal.
- **MP2:** In part (b), discuss student observations and listen for evidence of their statements. For instance, instead of saying that more girls have part-time jobs than boys, it is a stronger comparison to say that about 60 more girls have part-time jobs than boys. Encourage students to give quantitative evidence in their reasoning.
- Discuss part (c). Students may mention that there should be about the same number of boys and girls in the school district. Because there are fewer boys represented in the table, you might infer that more boys have dropped out. Students may offer a viable argument as to *why* more boys might have dropped out of high school than girls, and it may not be for the purpose of working full-time.
- **MP2** and **MP3a Construct Viable Arguments:** This type of discussion is important in developing reasoning habits and constructing a logical argument.

What Is Your Answer?

- Question 4 can become a project due at the conclusion of the chapter.

Closure

- Refer back to the information about your expenses at the math conference. What would an entry in the blacked-out area represent? type of expense and day of expense How do you know if your total amount of expenses is correct? The sum of the last column should be equal to the sum of the last row.

2 ACTIVITY: Analyzing Data

Math Practice 3

Construct Arguments

What are the advantages of using a table instead of a graph to analyze data?

Work with a partner. The three-dimensional two-way table shows information about the numbers of hours students at a high school work at part-time jobs during the school year.

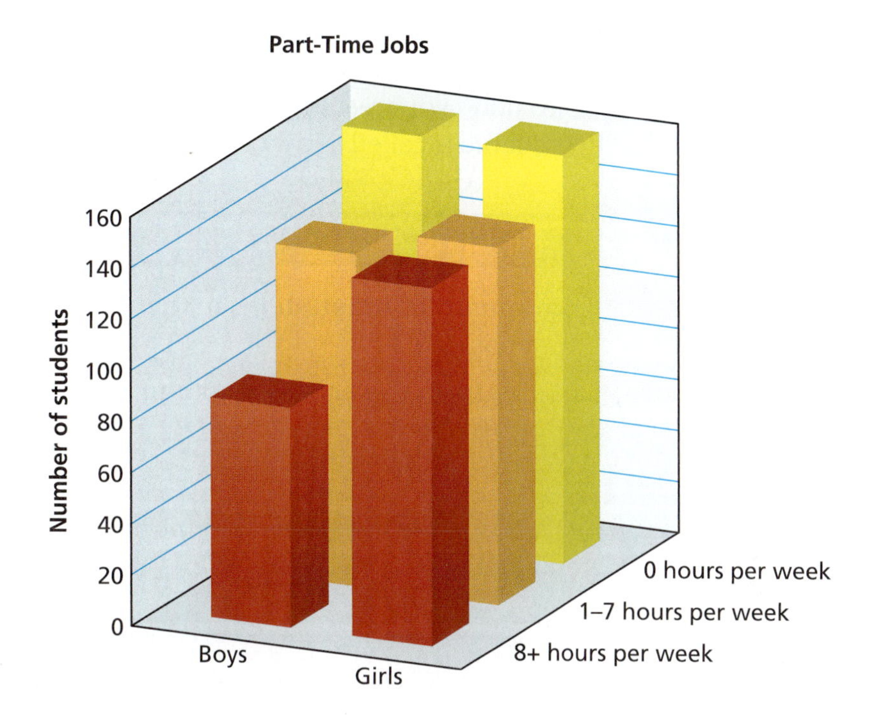

a. Make a two-way table showing the data. Use estimation to find the entries in your table.

b. Write two observations you can make that summarize the data in your table.

c. **REASONING** A newspaper article claims that more boys than girls drop out of high school to work full-time. Do the data support this claim? Explain your reasoning.

What Is Your Answer?

3. **IN YOUR OWN WORDS** How can you read and make a two-way table?

4. Find a real-life data set that you can represent by a two-way table. Then make a two-way table for the data set.

Practice Use what you learned about two-way tables to complete Exercises 3–6 on page 390.

9.3 Lesson

Key Vocabulary

two-way table, *p. 388*
joint frequency, *p. 388*
marginal frequency, *p. 388*

A **two-way table** displays two categories of data collected from the same source.

You randomly survey students in your school about their grades on the last test and whether they studied for the test. The two-way table shows your results. Each entry in the table is called a **joint frequency**.

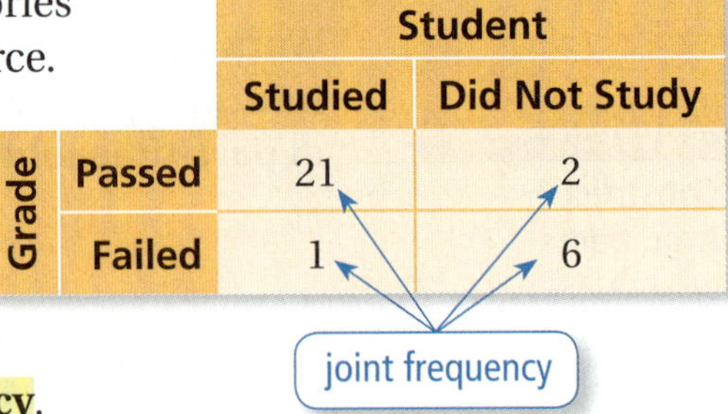

		Student	
		Studied	Did Not Study
Grade	Passed	21	2
	Failed	1	6

EXAMPLE 1 Reading a Two-Way Table

How many of the students in the survey above studied for the test and passed?

The entry in the "Studied" column and "Passed" row is 21.

∴ So, 21 of the students in the survey studied for the test and passed.

The sums of the rows and columns in a two-way table are called **marginal frequencies**.

EXAMPLE 2 Finding Marginal Frequencies

Find and interpret the marginal frequencies for the survey above.

Create a new column and a new row for the sums. Then add the entries.

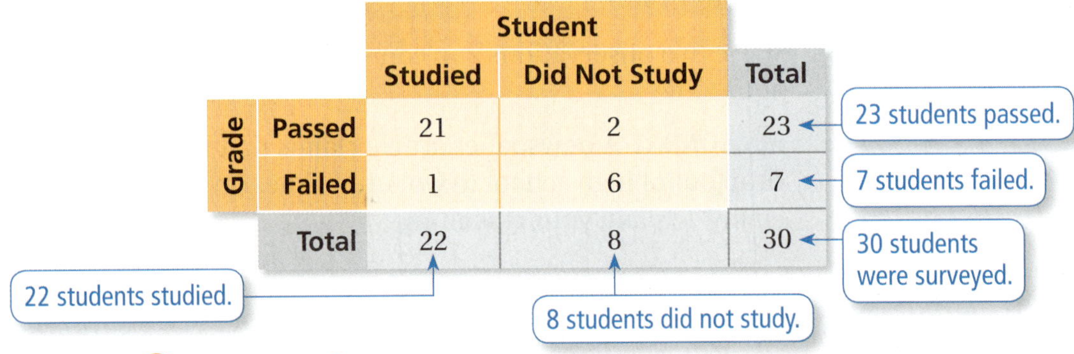

		Student		
		Studied	Did Not Study	Total
Grade	Passed	21	2	23
	Failed	1	6	7
	Total	22	8	30

On Your Own

1. You randomly survey students in a cafeteria about their plans for a football game and a school dance. The two-way table shows your results.
   a. How many students will attend the dance but not the football game?
   b. Find and interpret the marginal frequencies for the survey.

		Football Game	
		Attend	Not Attend
Dance	Attend	35	5
	Not Attend	16	20

Multi-Language Glossary at BigIdeasMath.com

Laurie's Notes

Introduction

Connect

- **Yesterday:** Students read two-way tables. (MP2, MP3a)
- **Today:** Students will construct two-way tables and identify relationships between categories of a two-way table.

Motivate

- Tell students about data you have collected from the faculty. You asked your coworkers who own both a computer and a cell phone if they use a Mac or a PC and if they use a smartphone or a basic cell phone.

? "How can you represent this data?" Answers will vary.

- Explain that today they will study a way in which this data can be displayed and analyzed.

Lesson Notes

Discuss

- Define two-way table. Emphasize that information is known about two categories from the same source. The focus in this lesson is drawing conclusions from the data in a two-way table.
- Refer to examples from yesterday in explaining "information about two categories from the same source," such as soccer shirts–size and color.
- Define joint frequency. Each entry in the two-way table is a frequency for two categories, hence the name joint frequency.

Example 1

? "What category do the rows represent?" test grade: passed or failed

? "What category do the columns represent?" preparation of the student: studied or did not study

Example 2

- Define marginal frequencies. The sums of the rows and columns appear on the *margins* of the two-way table.
- Expand the two-way table. Label the new row and new column "Total."
- Add the rows and columns. Identify the sums using the labels shown.
- Ask students general questions about the row and column totals.
- Make sure students understand that 30 students were surveyed, not 60. Because each student is tallied twice, once for each category, you do not add $22 + 8 + 23 + 7$ to find the number surveyed. The sum of the rows and the sum of the columns should be equal.

? "What can you conclude about the data?" *Sample answer:* Of the 30 students, all but one of those who studied for the test passed.

On Your Own

- **Extension:** Ask students percent questions such as, "What percent of the students in the survey are not planning to attend either event?"

Goal Today's lesson is constructing and analyzing **two-way tables**.

Technology for the Teacher

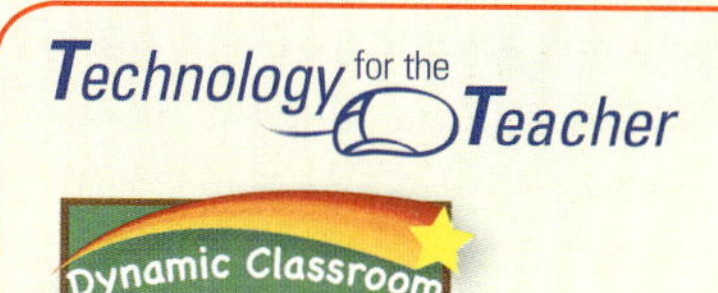

Lesson Tutorials
Lesson Plans
Answer Presentation Tool

Extra Example 1

You randomly survey students about whether they like orange juice. The two-way table shows your results. How many female students in the survey like orange juice? 29

		Gender	
		Male	Female
Orange juice	No	12	22
	Yes	37	29

Extra Example 2

Find and interpret the marginal frequencies for the survey above.

A total of 34 students do not like orange juice. A total of 66 students like orange juice. A total of 49 male students participated in the survey. A total of 51 female students participated in the survey.

On Your Own

1. a. 5 students

   b. A total of 51 students will attend the game. A total of 25 students will not attend the game. A total of 40 students will attend the dance. A total of 36 students will not attend the dance.

Extra Example 3

You randomly survey students in 6th, 7th, and 8th grade about whether they are going to try to join student council. The results are shown in the tally sheets. Make a two-way table that includes the marginal frequencies. See Additional Answers.

Join

Grade	Tally
6	𝍸 𝍸 |
7	𝍸 𝍸 |||
8	𝍸 ||

Not Join

Grade	Tally
6	𝍸 𝍸
7	𝍸 𝍸 ||||
8	𝍸 𝍸 𝍸

Extra Example 4

Use the two-way table in Extra Example 3.

a. For each grade, what percent of the students in the survey are going to try to join student council? not try to join student council? Organize the results in a two-way table. Explain what one of the entries represents. See Additional Answers.

b. Does the table in part (a) show a relationship between grade and whether students are going to try to join student council? Explain. See Additional Answers.

On Your Own

2. See Additional Answers.

English Language Learners

Class Activity

Form groups of 2 to 4 students with at least one English language learner and one English speaker. Have them work together to create and conduct a survey that includes two categories. The results of the survey should be presented to the class in a two-way table that includes the marginal frequencies.

Laurie's Notes

Example 3

- Guide students through the construction of the table and ask them to explain what several of the values represent.
- The amount of data in the two-way table may be overwhelming to some students. Make sure to talk through the problem, giving students time to stop and think about what each entry represents.
- **Teaching Tip:** Use two colors in the table, one for the joint frequencies and one for the marginal frequencies. This makes it easier to read.

? "What is an advantage of the two-way table over the tally marks on the sheets of paper?" Answers will vary. Students might say the table is more organized, easier to read, and more condensed.

Example 4

- Have a student read the problem. Explain that the problem is asking about percents within each *age group*—the data represented in the columns.

? "How many 12- to 13-year-olds ride the bus?" 24 "How many 12- to 13-year-olds are in the survey?" 40 "What percent of the 12- to 13-year-olds ride the bus?" $24/40 = 60\%$

- Guide students through the construction of the two-way table. Ask them to explain what each entry represents.
- **MP2 Reason Abstractly and Quantitatively:** Ask students to make an observation about the percents. As age increases, students are less likely to ride the bus to school. Ask students why this might be the case, encouraging them to reason about data.

? "In part (b), the sums of the columns are each 100%. Why don't the rows add up to 100%?" The base used to compute the percents referred to each of the age groups, not whether the student rides the bus.

? "Can percents in the table be found using the row totals?" yes "What would be the first entry in the table and what would it represent?" $24/50 = 48\%$; 48% of the students who ride the bus are 12–13 years old.

On Your Own

- **Common Error:** Students may find the percent of students who pack a lunch out of the total number of students who pack a lunch instead of the percent for each grade level.

Closure

- In the first example, is it likely that if you study for a test you will pass? Explain. yes; The table shows that the majority of students who studied for the test passed and the majority of students who did not study for the test failed.

EXAMPLE 3 Making a Two-Way Table

Rides Bus

Age	Tally
12-13	卌 卌 卌 卌 ||||
14-15	卌 卌 ||
16-17	卌 卌 ||||

Does Not Ride Bus

Age	Tally
12-13	卌 卌 卌 |
14-15	卌 卌 |||
16-17	卌 卌 卌 卌 |

You randomly survey students between the ages of 12 and 17 about whether they ride the bus to school. The results are shown in the tally sheets. Make a two-way table that includes the marginal frequencies.

The two categories for the table are the ages and whether or not they ride the bus. Use the tally sheets to calculate each joint frequency. Then add to find each marginal frequency.

		Age			
		12–13	14–15	16–17	Total
Student	Rides Bus	24	12	14	50
	Does Not Ride Bus	16	13	21	50
	Total	40	25	35	100

EXAMPLE 4 Finding a Relationship in a Two-Way Table

Use the two-way table in Example 3.

a. For each age group, what percent of the students in the survey ride the bus to school? do not ride the bus to school? Organize the results in a two-way table. Explain what one of the entries represents.

		Age		
		12–13	14–15	16–17
Student	Rides Bus	60%	48%	40%
	Does Not Ride Bus	40%	52%	60%

$\frac{14}{35} = 0.4$

So, 40% of the 16- and 17-year-old students in the survey ride the bus to school.

b. Does the table in part (a) show a relationship between age and whether students ride the bus to school? Explain.

Yes, the table shows that as age increases, students are less likely to ride the bus to school.

On Your Own

Exercise 10

Grade 6 Students
11 pack lunch, 9 buy school lunch

Grade 7 Students
23 pack lunch, 27 buy school lunch

Grade 8 Students
16 pack lunch, 14 buy school lunch

2. You randomly survey students in a school about whether they buy a school lunch or pack a lunch. Your results are shown.

   a. Make a two-way table that includes the marginal frequencies.

   b. For each grade level, what percent of the students in the survey pack a lunch? buy a school lunch? Organize the results in a two-way table. Explain what one of the entries represents.

   c. Does the table in part (b) show a relationship between grade level and lunch choice? Explain.

9.3 Exercises

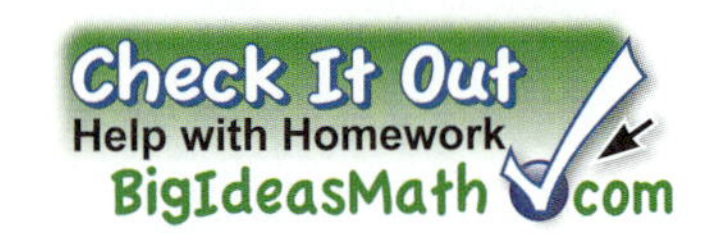

Vocabulary and Concept Check

1. **VOCABULARY** Explain the relationship between joint frequencies and marginal frequencies.
2. **OPEN-ENDED** Describe how you can use a two-way table to organize data you collect from a survey.

Practice and Problem Solving

You randomly survey students about participating in their class's yearly fundraiser. You display the two categories of data in the two-way table.

3. Find the total of each row.
4. Find the total of each column.

① 5. How many female students will be participating in the fundraiser?

6. How many male students will *not* be participating in the fundraiser?

		Fundraiser	
		No	Yes
Gender	Female	22	51
	Male	30	29

Find and interpret the marginal frequencies.

② 7.

		School Play	
		Attend	Not Attend
Class	Junior	41	30
	Senior	52	23

8.

		Cell Phone Minutes	
		Limited	Unlimited
Text Plan	Limited	78	0
	Unlimited	175	15

9. **GOALS** You randomly survey students in your school. You ask what is most important to them: grades, popularity, or sports. You display your results in the two-way table.

   a. How many 7th graders chose sports? How many 8th graders chose grades?

   b. Find and interpret the marginal frequencies for the survey.

   c. What percent of students in the survey are 6th graders who chose popularity?

		Goal		
		Grades	Popularity	Sports
Grade	6th	31	18	23
	7th	39	16	19
	8th	42	6	17

Assignment Guide and Homework Check

Level	Day 1 Activity Assignment	Day 2 Lesson Assignment	Homework Check
Basic	3–6, 14–17	1, 2, 7–11	2, 7, 9, 10
Average	3–6, 14–17	1, 2, 7–12	2, 7, 9, 10
Advanced	1–17		8, 9, 11, 12

Common Errors

- **Exercises 1, 7–11** Students may incorrectly identify joint frequencies as marginal frequencies or marginal frequencies as joint frequencies. Remind them of these definitions.
- **Exercise 11c** Students may find the percents based on all students surveyed, not just the students from each eye color group. Encourage them to read carefully.

9.3 Record and Practice Journal

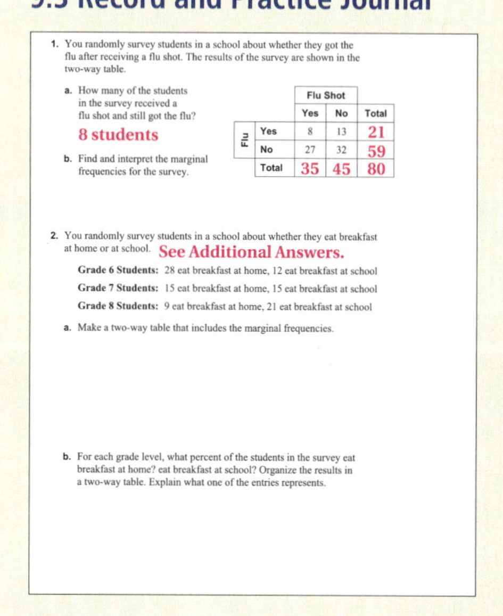

1. You randomly survey students in a school about whether they got the flu after receiving a flu shot. The results of the survey are shown in the two-way table.

   a. How many of the students in the survey received a flu shot and still got the flu?

   8 students

   b. Find and interpret the marginal frequencies for the survey.

		Flu Shot		
		Yes	No	Total
Flu	Yes	8	13	21
	No	27	32	59
	Total	35	45	80

2. You randomly survey students in a school about whether they eat breakfast at home or at school. **See Additional Answers.**

   Grade 6 Students: 28 eat breakfast at home, 12 eat breakfast at school

   Grade 7 Students: 15 eat breakfast at home, 15 eat breakfast at school

   Grade 8 Students: 9 eat breakfast at home, 21 eat breakfast at school

   a. Make a two-way table that includes the marginal frequencies.

   b. For each grade level, what percent of the students in the survey eat breakfast at home? eat breakfast at school? Organize the results in a two-way table. Explain what one of the entries represents.

Vocabulary and Concept Check

1. The joint frequencies are the entries in the two-way table that differentiate the two categories of data collected; The marginal frequencies are the sums of the rows and columns of the two-way table.
2. displays two categories of data collected from the same source

Practice and Problem Solving

3. total of females surveyed: 73; total of males surveyed: 59
4. total of "no" participants: 52; total of "yes" participants: 80
5. 51
6. 30
7. 71 students are juniors; 75 students are seniors; 93 students are attending the school play; 53 students are not attending the school play.
8. 78 people have limited cell phone texting plans; 190 people have unlimited cell phone texting plans; 253 people have limited cell phone minutes; 15 people have unlimited cell phone minutes.
9. a. 19; 42

   b. number of students surveyed: 72 6th-graders, 74 7th-graders, 65 8th-graders; 112 students chose grades, 40 students chose popularity, 59 students chose sports.

   c. about 8.5%

Practice and Problem Solving

10–12. See Additional Answers.

13. See *Taking Math Deeper*.

Fair Game Review

14. $y = 3x + 1$

15. $y = 5x - 2$

16. $y = \frac{1}{2}x + 3$

17. B

Mini-Assessment

1. You randomly survey students about whether they are involved in school sports.

Grade 5: 12 involved, 26 not involved
Grade 8: 23 involved, 19 not involved

a. Make a two-way table that includes the marginal frequencies. See Additional Answers.

b. For each grade level, what percent of the students in the survey are involved in school sports? are not involved in school sports? Organize the results in a two-way table. Explain what one of the entries represents. See Additional Answers.

c. Does the table in part (b) show a relationship between grade level and involvement in school sports? Explain. See Additional Answers.

Taking Math Deeper

Exercise 13

This problem can help students see the benefits of a two-way table and the many different questions that can be asked regarding the entries. Encourage students to pay close attention to the wording of a question.

When finding the percent of students that are either female or have green eyes, students may mistakenly count the number of females with green eyes twice.

		Eye Color			
		Green	Blue	Brown	Total
Gender	Male	5	16	27	48
	Female	3	19	18	40
	Total	8	35	45	88

$$\frac{5}{88} + \frac{3}{88} + \frac{19}{88} + \frac{18}{88} = \frac{45}{88} \approx 51.1\%$$

To find the percent of students that are males that do not have green eyes, divide the sum of the remaining two joint frequencies by the number of students in the survey.

$$\frac{16}{88} + \frac{27}{88} = \frac{43}{88} \approx 48.9\%$$

The sum of the percents is $51.1\% + 48.9\% = 100\%$.

These two percents account for everyone in the survey.

Note: You may want to challenge students by asking what type of data display could be used to display the information in a two-way table, such as a double bar graph.

Reteaching and Enrichment Strategies

If students need help. . .	If students got it. . .
Resources by Chapter • Practice A and Practice B • Puzzle Time Record and Practice Journal Practice Differentiating the Lesson Lesson Tutorials Skills Review Handbook	Resources by Chapter • Enrichment and Extension • Technology Connection Start the next section

③ ④ **10. SAVINGS** You randomly survey people in your neighborhood about whether they have at least \$1000 in savings. The results are shown in the tally sheets.

Have at Least \$1000 in Savings

Age	Tally
20-29	卌 卌 ||||
30-39	卌 卌 卌 卌 卌 ||
40-49	卌 卌 卌 卌 卌

Don't Have at Least \$1000 in Savings

Age	Tally
20-29	卌 卌 卌 卌 卌 卌 卌 |
30-39	卌 卌 卌 卌 卌 卌 |||
40-49	卌 卌 卌

a. Make a two-way table that includes the marginal frequencies.

b. For each age group, what percent of the people have at least \$1000 in savings? do not have at least \$1000 in savings? Organize the results in a two-way table.

c. Does the table in part (b) show a relationship between age and whether people have at least \$1000 in savings? Explain.

11. EYE COLOR You randomly survey students in your school about the color of their eyes. The results are shown in the tables.

Eye Color of Males Surveyed

Green	Blue	Brown
5	16	27

Eye Color of Females Surveyed

Green	Blue	Brown
3	19	18

a. Make a two-way table.

b. Find and interpret the marginal frequencies for the survey.

c. For each eye color, what percent of the students in the survey are male? female? Organize the results in a two-way table. Explain what two of the entries represent.

12. REASONING Use the information from Exercise 11. For each gender, what percent of the students in the survey have green eyes? blue eyes? brown eyes? Organize the results in a two-way table. Explain what two of the entries represent.

13. Precision What percent of students in the survey in Exercise 11 are either female or have green eyes? What percent of students in the survey are males who do not have green eyes? Find and explain the sum of these two percents.

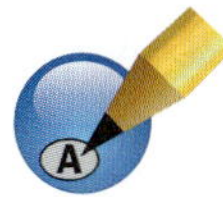

Fair Game Review What you learned in previous grades & lessons

Write an equation of the line that passes through the points. *(Section 4.6)*

14. (0, 1), (−2, −5) **15.** (0, −2), (3, 13) **16.** (−4, 1), (0, 3)

17. MULTIPLE CHOICE Which equation does not represent a linear function? *(Section 6.4)*

Ⓐ $y = 4x$ Ⓑ $xy = 8$ Ⓒ $y = -3$ Ⓓ $6x + 5y = -2$

9.4 Choosing a Data Display

Essential Question How can you display data in a way that helps you make decisions?

1 ACTIVITY: Displaying Data

Work with a partner. Analyze and display each data set in a way that best describes the data. Explain your choice of display.

a. ROADKILL A group of schools in New England participated in a 2-month study. They reported 3962 dead animals.

Birds: 307
Mammals: 2746
Amphibians: 145
Reptiles: 75
Unknown: 689

b. BLACK BEAR ROADKILL The data below show the numbers of black bears killed on a state's roads from 1993 to 2012.

1993:	30	2000:	47	2007:	99
1994:	37	2001:	49	2008:	129
1995:	46	2002:	61	2009:	111
1996:	33	2003:	74	2010:	127
1997:	43	2004:	88	2011:	141
1998:	35	2005:	82	2012:	135
1999:	43	2006:	109		

c. RACCOON ROADKILL A 1-week study along a 4-mile section of road found the following weights (in pounds) of raccoons that had been killed by vehicles.

13.4	14.8	17.0	12.9
21.3	21.5	16.8	14.8
15.2	18.7	18.6	17.2
18.5	9.4	19.4	15.7
14.5	9.5	25.4	21.5
17.3	19.1	11.0	12.4
20.4	13.6	17.5	18.5
21.5	14.0	13.9	19.0

d. What do you think can be done to minimize the number of animals killed by vehicles?

Data Analysis

In this lesson, you will
- choose appropriate data displays.
- identify and analyze misleading data displays.

Applying Standard
MACC.8.SP.1.1

Laurie's Notes

Introduction

Standards for Mathematical Practice

- **MP3 Construct Viable Arguments and Critique the Reasoning of Others:** In this section, students make decisions about how to display data. They will need to explain their reasoning for selecting a particular display. If two students select different data displays, it is important that they discuss the reasoning behind their choices.

Motivate

- The theme for the first activity is roadkill. While students may giggle at the thought, automobile accidents involving large animals can be serious. I had my first and only accident with a deer 5 years ago. I was 2 miles from home, and I was traveling 40 miles per hour. The deer was killed, my daughter and I were not injured, and repairs to my car were about $1400.
- Allow time for students to share personal stories.
- Use the Internet to research and share vehicular data with students, such as the number of miles of roads in the U.S., the number of registered vehicles, the number of accidents, and the number of animal-related accidents.

Activity Notes

Discuss

- Discuss the data displays with which students are familiar: pictograph, bar graph, line graph, circle graph, stem-and-leaf plot, histogram, dot plot, box-and-whisker plot, and scatter plot. Have students describe the feature(s) of each display.
- Discuss the different numerical tools they have for describing data: mean, median, mode, range, mean absolute deviation, quartile, and interquartile range.

Activity 1

- Students need to decide which display makes sense for the type of data that they have. There may be more than one appropriate answer.
- **MP3:** Discuss students' choices and their explanations.
- Possible data displays:
  - Part (a): a circle graph (what part of the whole set is each animal) or a bar graph (compare the different categories, although there is a large difference in bar heights: 75 to 2746)
  - Part (b): a scatter plot and line of best fit (pair data, show trend over time, and make predictions for the future) or a line graph
  - Part (c): a stem-and-leaf plot (spread of data), along with calculating the mean (about 16.7) and median (17.1)
  - Part (d): As a class, discuss students' ideas for minimizing the number of animals killed by vehicles.

Florida Common Core Standards

MACC.8.SP.1.1 Construct and interpret scatter plots for bivariate measurement data to investigate patterns of association between two quantities. Describe patterns such as clustering, outliers, positive or negative association, linear association, and nonlinear association.

Previous Learning

Students should know how to construct a variety of data displays from this year and past years.

9.4 Record and Practice Journal

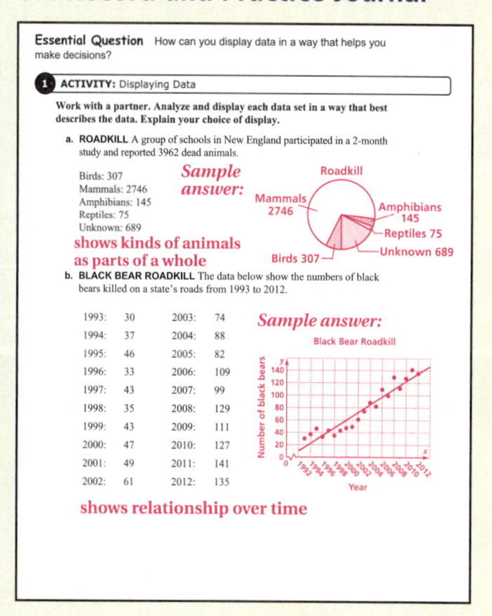

Essential Question How can you display data in a way that helps you make decisions?

1 ACTIVITY: Displaying Data

Work with a partner. Analyze and display each data set in a way that best describes the data. Explain your choice of display.

a. **ROADKILL** A group of schools in New England participated in a 2-month study and reported 3962 dead animals.

Birds: 307
Mammals: 2746
Amphibians: 145
Reptiles: 75
Unknown: 689

Sample answer: (Roadkill circle graph: Mammals 2746, Amphibians 145, Reptiles 75, Unknown 689, Birds 307)

shows kinds of animals as parts of a whole

b. **BLACK BEAR ROADKILL** The data below show the numbers of black bears killed on a state's roads from 1993 to 2012.

1993:	30	2003:	74
1994:	37	2004:	88
1995:	46	2005:	82
1996:	33	2006:	109
1997:	43	2007:	99
1998:	35	2008:	129
1999:	43	2009:	111
2000:	47	2010:	127
2001:	49	2011:	141
2002:	61	2012:	135

Sample answer: (Black Bear Roadkill scatter plot; Number of black bears vs. Year, 1992–2012)

shows relationship over time

English Language Learners

Vocabulary

English learners may need help understanding the word *scale*. There are several meanings in the English language. Some of the common meanings are:

a series of musical notes
the covering of a reptile
a device for weighing
a ratio
to climb

In bar graphs, the scale is a series of markings used for measuring. Most scales start at 0 and go to (at least) the greatest value of the data.

9.4 Record and Practice Journal

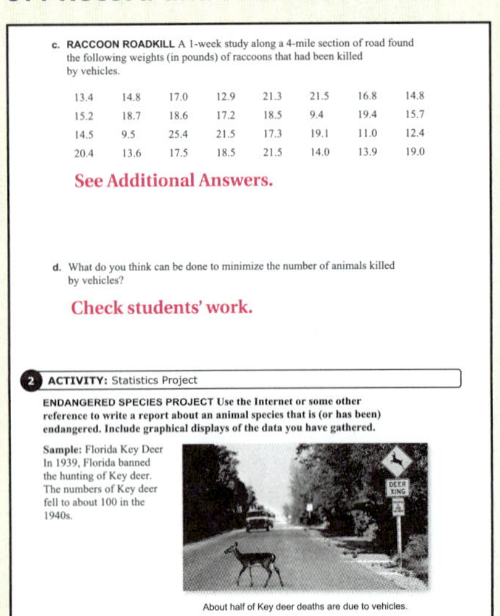

c. **RACCOON ROADKILL** A 1-week study along a 4-mile section of road found the following weights (in pounds) of raccoons that had been killed by vehicles.

13.4	14.8	17.0	12.9	21.3	21.5	16.8	14.8
15.2	18.7	18.6	17.2	18.5	9.4	19.4	15.7
14.5	9.5	25.4	21.5	17.3	19.1	11.0	12.4
20.4	13.6	17.5	18.5	21.5	14.0	13.9	19.0

See Additional Answers.

d. What do you think can be done to minimize the number of animals killed by vehicles?

Check students' work.

2 ACTIVITY: Statistics Project

ENDANGERED SPECIES PROJECT Use the Internet or some other reference to write a report about an animal species that is (or has been) endangered. Include graphical displays of the data you have gathered.

Sample: Florida Key Deer
In 1939, Florida banned the hunting of Key deer. The numbers of Key deer fell to about 100 in the 1940s.

About half of Key deer deaths are due to vehicles.

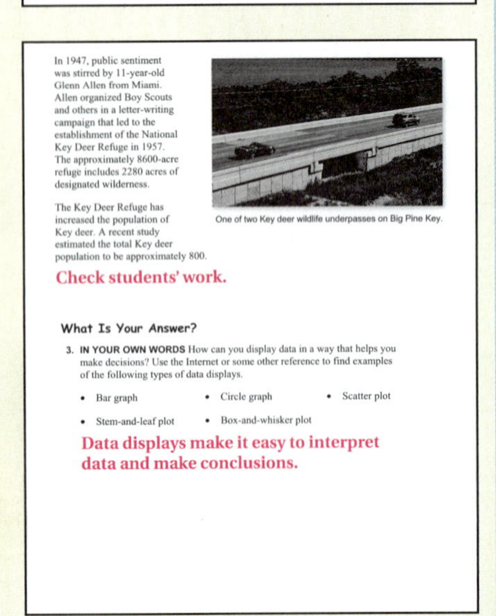

In 1947, public sentiment was stirred by 11-year-old Glenn Allen from Miami. Allen organized Boy Scouts and others in a letter-writing campaign that led to the establishment of the National Key Deer Refuge in 1957. The approximately 8600-acre refuge includes 2280 acres of designated wilderness.

The Key Deer Refuge has increased the population of Key deer. A recent study estimated the total Key deer population to be approximately 800.

One of two Key deer wildlife underpasses on Big Pine Key.

Check students' work.

What Is Your Answer?

3. **IN YOUR OWN WORDS** How can you display data in a way that helps you make decisions? Use the Internet or some other reference to find examples of the following types of data displays.

- Bar graph
- Circle graph
- Scatter plot
- Stem-and-leaf plot
- Box-and-whisker plot

Data displays make it easy to interpret data and make conclusions.

Laurie's Notes

Activity 2

- Ask a volunteer to read the information presented about Key deer. Discuss how the actions of one person can often make a big difference.
- It would be ideal if the library or computer room is available. If not, you or your students could bring in newspapers and magazines that contain graphical displays.
- If you assign this project, students will need several days.

What Is Your Answer?

- **MP4 Model with Mathematics:** Many students can make the displays, if they are told which display to use. It is equally important that students be able to select the display based upon the data and the question you hope to answer from making the display.
- The information gathered by students can be made into classroom posters.

Closure

- **Class Discussion:** Have students present their answers to Question 3. Then have students discuss features of each display, and what types of data lend itself to each data display.

2 ACTIVITY: Statistics Project

Math Practice 4

Use a Graph

How can you use a graph to represent the data you have gathered for your report? What does the graph tell you about the data?

ENDANGERED SPECIES PROJECT **Use the Internet or some other reference to write a report about an animal species that is (or has been) endangered. Include graphical displays of the data you have gathered.**

Sample: Florida Key Deer

In 1939, Florida banned the hunting of Key deer. The numbers of Key deer fell to about 100 in the 1940s.

In 1947, public sentiment was stirred by 11-year-old Glenn Allen from Miami. Allen organized Boy Scouts and others in a letter-writing campaign that led to the establishment of the National Key Deer Refuge in 1957. The approximately 8600-acre refuge includes 2280 acres of designated wilderness.

The Key Deer Refuge has increased the population of Key deer. A recent study estimated the total Key deer population to be approximately 800.

About half of Key deer deaths are due to vehicles.

One of two Key deer wildlife underpasses on Big Pine Key

What Is Your Answer?

3. **IN YOUR OWN WORDS** How can you display data in a way that helps you make decisions? Use the Internet or some other reference to find examples of the following types of data displays.

- Bar graph
- Circle graph
- Scatter plot
- Stem-and-leaf plot
- Box-and-whisker plot

Use what you learned about choosing data displays to complete Exercise 3 on page 397.

9.4 Lesson

Key Idea

Data Display	What does it do?
Pictograph	shows data using pictures
Bar Graph	shows data in specific categories
Circle Graph	shows data as parts of a whole
Line Graph	shows how data change over time
Histogram	shows frequencies of data values in intervals of the same size
Stem-and-Leaf Plot	orders numerical data and shows how they are distributed
Box-and-Whisker Plot	shows the variability of a data set by using quartiles
Dot Plot	shows the number of times each value occurs in a data set
Scatter Plot	shows the relationship between two data sets by using ordered pairs in a coordinate plane

EXAMPLE 1 Choosing an Appropriate Data Display

Choose an appropriate data display for the situation. Explain your reasoning.

a. the number of students in a marching band each year

A line graph shows change over time. So, a line graph is an appropriate data display.

b. a comparison of people's shoe sizes and their heights

You want to compare two different data sets. So, a scatter plot is an appropriate data display.

On Your Own

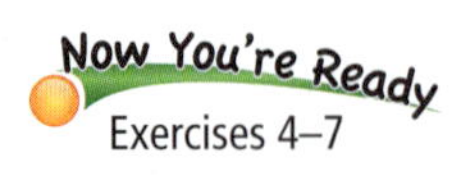

Choose an appropriate data display for the situation. Explain your reasoning.

1. the population of the United States divided into age groups
2. the percents of students in your school who play basketball, football, soccer, or lacrosse

Laurie's Notes

Introduction

Connect

- **Yesterday:** Students reviewed data displays. (MP3, MP4)
- **Today:** Students will choose and construct an appropriate data display.

Motivate

- Make a quick sketch of the two bar graphs shown and ask students to comment on each.

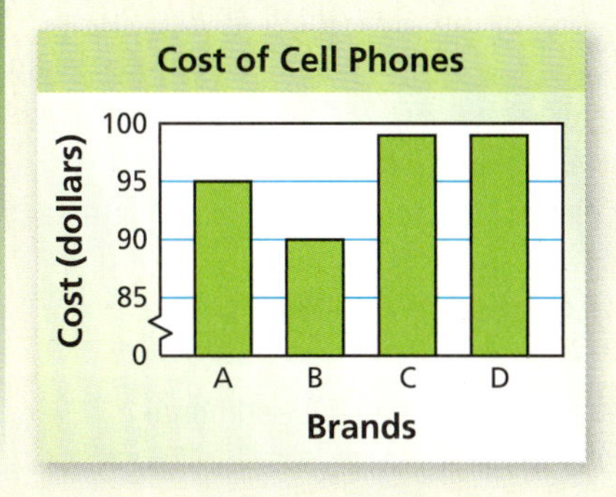

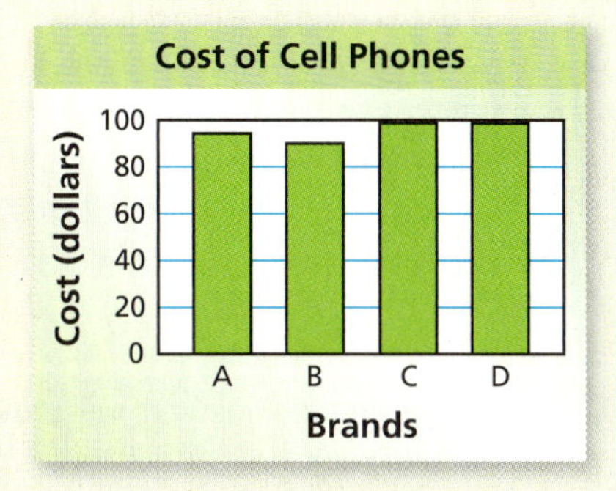

Lesson Notes

Key Idea

- Write the Key Idea. This is a terrific summary of data displays that students have learned to make.
- Emphasize that *choosing an appropriate display* is more art than science. But it is clearly possible to use any of the graphs in misleading ways.
- **MP3 Construct Viable Arguments and Critique the Reasoning of Others:** Students should be able to state their reasons for selecting a particular data display *and* why they did not select a different data display. If another student selected a different data display, students should compare their reasoning.
- There may be examples of each of these displays around your room.

Example 1

- Read each problem. Students should not have difficulty determining the appropriate data display for each problem.

On Your Own

- **Think-Pair-Share:** Students should read each question independently and then work in pairs to answer the questions. When they have answered the questions, the pair should compare their answers with another group and discuss any discrepancies.

Goal Today's lesson is choosing and constructing an appropriate data display.

Lesson Tutorials
Lesson Plans
Answer Presentation Tool

Extra Example 1

You conduct a survey at your school about insects that students fear the most. Choose an appropriate data display. Explain your reasoning. *Sample answers:* circle graph: shows data as parts of a whole; bar graph: shows data in specific categories; pictograph: shows data using pictures.

On Your Own

1. *Sample answer:* histogram; shows frequencies of ages (data values) in intervals of the same size
2. *Sample answer:* bar graph; shows data in specific categories

Extra Example 2

Tell whether the data display is appropriate for representing the data in Example 2. Explain your reasoning.

a. pictograph no; It would not be easy to show the number of hits for each month using pictures.

b. scatter plot yes; A scatter plot would show the number of hits for each month, with the months numbered from 1 to 5.

On Your Own

3. no; A dot plot would not show how the number of hits changed over time.
4. yes; A circle graph would show the fraction or percent of the total number of hits for each month.
5. no; A stem-and-leaf plot would not show how the number of hits changed over time.

English Language Learners

Vocabulary

Explain that a line graph and the graph of a line are *not* the same thing. A line graph is made up of line segments, each of which has endpoints. A line does not have endpoints.

Laurie's Notes

Example 2

- Write the data on the board and ask a volunteer to read the problem.
- ? "Looking at the data, how would you describe the change in website hits over the 5-month period?" increasing
- Remind students that they are looking for a data display that will show how the data changes during the 5 months.
- ? "How does the bar graph represent the change in the number of website hits?" by using different bar heights
- ? "Why are there only 3 bars in the histogram?" data have been grouped
- ? "What key information is lost in the histogram? Explain." time; You can no longer see the months.
- ? "How are the line graph and bar graph alike? Explain." They have the same shape. Connecting the midpoints of the tops of the bars creates a graph that looks like the line graph.

On Your Own

- Discussing the answers will provide a review of three additional data displays.

EXAMPLE 2 Identifying an Appropriate Data Display

You record the number of hits for your school's new website for 5 months. Tell whether the data display is appropriate for representing how the number of hits changed during the 5 months. Explain your reasoning.

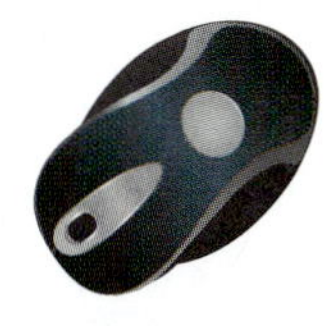

Month	Hits
August	250
September	320
October	485
November	650
December	925

a.

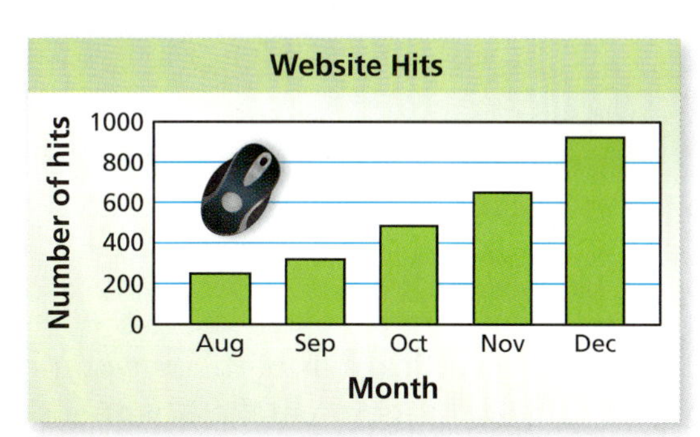

The bar graph shows the number of hits for each month. So, it is an appropriate data display.

b.

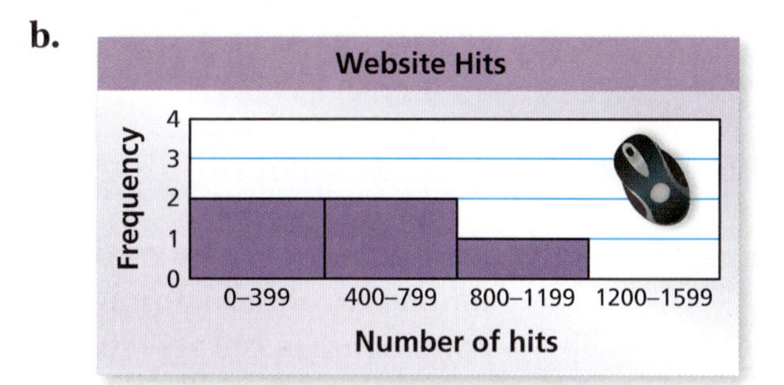

The histogram does not show the number of hits for each month or how the number of hits changes over time. So, it is *not* an appropriate data display.

c.

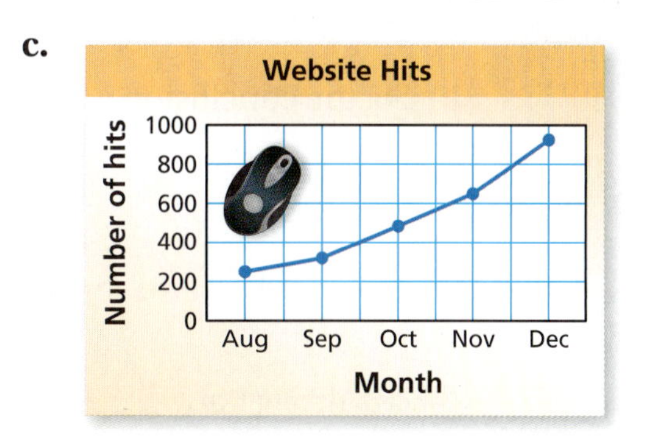

The line graph shows how the number of hits changes over time. So, it is an appropriate data display.

On Your Own

Tell whether the data display is appropriate for representing the data in Example 2. Explain your reasoning.

3. dot plot **4.** circle graph **5.** stem-and-leaf plot

EXAMPLE 3 Identifying a Misleading Data Display

Which line graph is misleading? Explain.

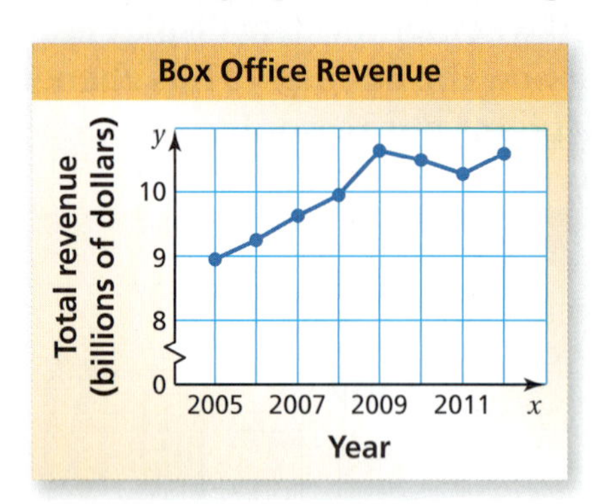

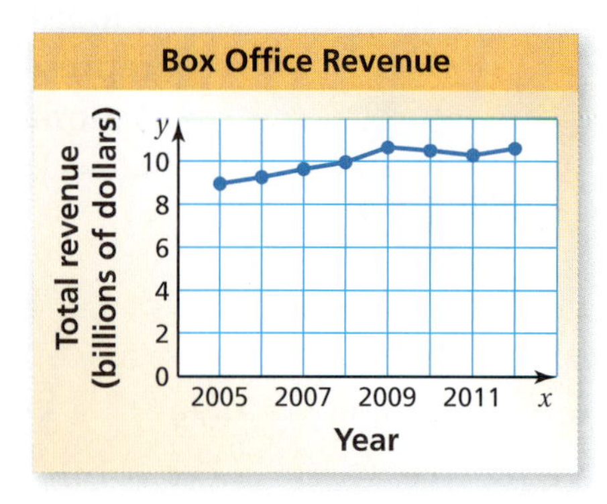

The vertical axis of the line graph on the left has a break (⚡) and begins at 8. This graph makes it appear that the total revenue increased rapidly from 2005 to 2009. The graph on the right has an unbroken axis. It is more honest and shows that the total revenue increased slowly.

∴ So, the graph on the left is misleading.

EXAMPLE 4 Analyzing a Misleading Data Display

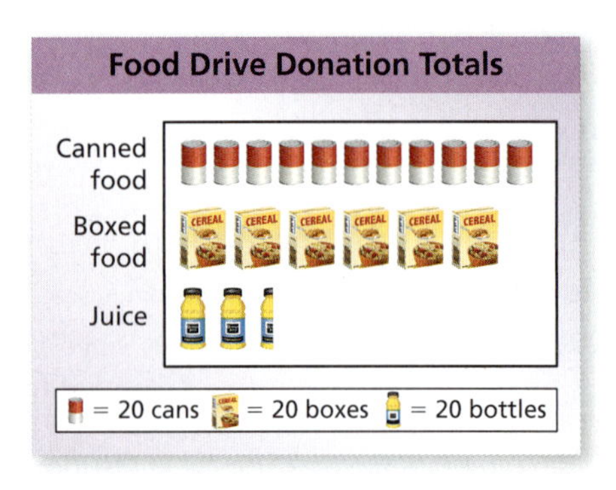

A volunteer concludes that the numbers of cans of food and boxes of food donated were about the same. Is this conclusion accurate? Explain.

Each icon represents the same number of items. Because the box icon is larger than the can icon, it looks like the number of boxes is about the same as the number of cans. But the number of boxes is actually about half of the number of cans.

∴ So, the conclusion is not accurate.

On Your Own

Now You're Ready
Exercises 11–14

Explain why the data display is misleading.

6.

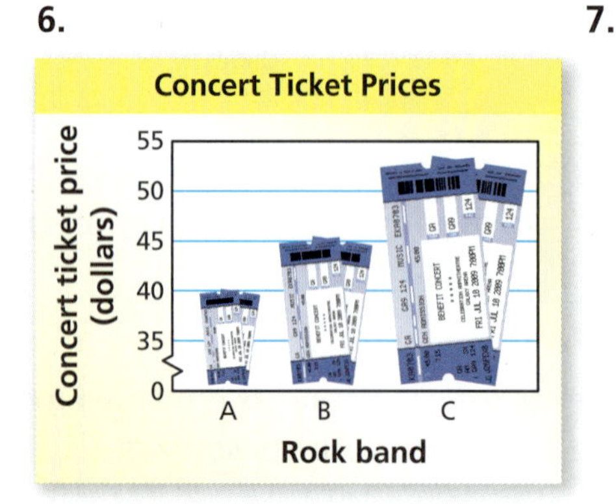

7.

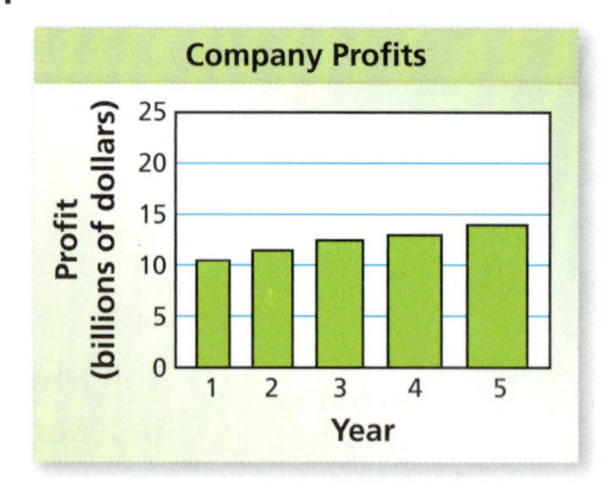

Laurie's Notes

Discuss

- I have a collection of misleading data displays. When you find a data display in the newspaper or magazine that is misleading, cut it out and save it for later use. Ask colleagues in your school to do the same.
- **MP6 Attend to Precision:** Often what makes a graph misleading is the scale selected for one, or both, of the axes. By spreading out the scale, or condensing it, the graph becomes misleading.
- As I always tell my students, the person who makes the data display influences how we will view it. They control the extent to which we can see, or not see, features of the data.

Example 3

? "The same data are displayed in each line graph. How do the graphs differ?" The vertical scale is different.

? "Which graph is misleading and why?" first graph; The differences in total revenue appear greater than they actually are.

- **Extension:** Have students pretend that both graphs appear in the newspaper with an article, and ask them what they would use for a headline for each article. What story does the author want readers to see when they look at each graph?

Example 4

- Have students "read" the pictograph and ask them to summarize what information it describes. Many students will conclude that the amount of cans and the amount of boxes is about the same due to the horizontal distance each set of icons takes up.

? "Approximately how many cans of food and boxes of food have been donated?" 11 cans $\times$ 20 = 220 cans; 6 boxes $\times$ 20 = 120 boxes

- Almost twice as many cans of food have been donated as boxes, so this is misleading. The box icon is too large. It should be the same width as the can.

On Your Own

- **Think-Pair-Share:** Students should read each question independently and then work in pairs to answer the questions. When they have answered the questions, the pair should compare their answers with another group and discuss any discrepancies.

Closure

- **Exit Ticket:** Make a pictograph for the data in Example 3 that is not misleading.

Extra Example 3

Which line graph is misleading? Explain.

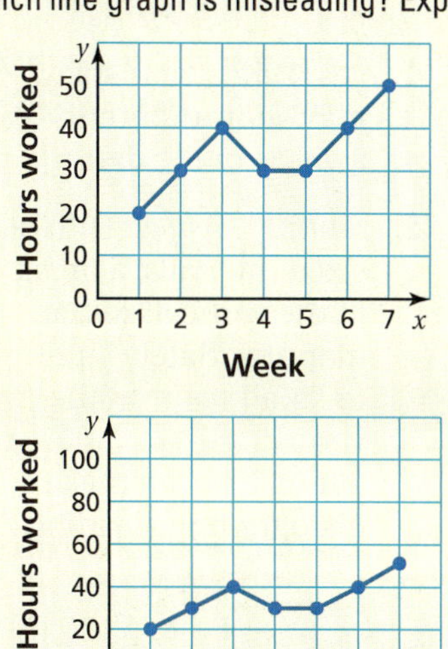

the second graph; The y-scale makes the change from week to week appear smaller.

Extra Example 4

Explain why the data display is misleading.

Favorite Pets

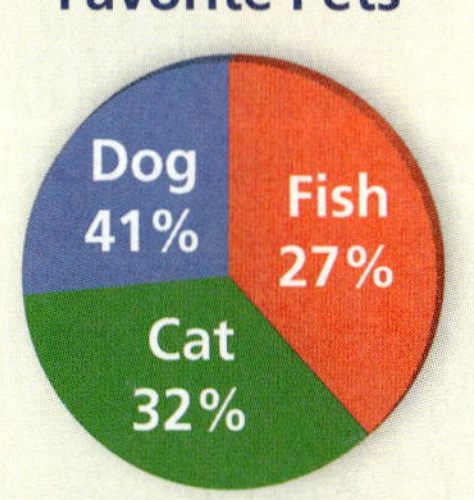

The size of each part of the circle is not proportional to the percent each choice represents.

On Your Own

6. The tickets vary in width and the break in the vertical axis makes the difference in ticket prices appear to be greater.
7. The bars become wider as the years progress, making the increase in profit appear greater.

Vocabulary and Concept Check

1. yes; Different displays may show different aspects of the data.
2. *Sample answer:* Different sized intervals, a break in the vertical axis, an inappropriately large or small scale on the vertical axis, or different bar widths can make a histogram misleading.

Practice and Problem Solving

3. *Sample answer:*

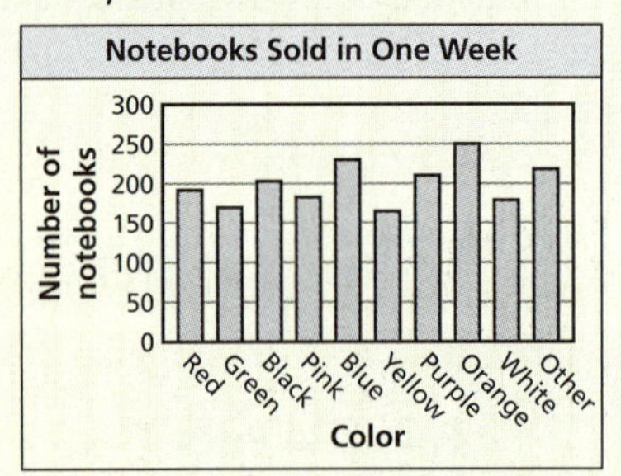

A bar graph shows the data in different color categories.

4. *Sample answer:* stem-and-leaf plot; shows how data is distributed
5. *Sample answer:* line graph; shows changes over time
6. *Sample answer:* dot plot; shows the number of times each outcome occurs
7. *Sample answer:* line graph; shows changes over time
8. a. yes; The pictograph shows the number of hours worked each month using pictures.

   b. yes; The bar graph shows the number of hours worked each month.

Assignment Guide and Homework Check

Level	Day 1 Activity Assignment	Day 2 Lesson Assignment	Homework Check
Basic	3, 20–23	1, 2, 4–7, 9–15	4, 9, 10, 12
Average	3, 20–23	1, 2, 4–8, 10–14	4, 8, 10, 12
Advanced	1–10, 12, 14–23		8, 10, 12, 16

Common Errors

- **Exercises 4–7** Students may confuse or forget the names of some of the data displays. For example, they may say that a dot plot should be used when they mean a scatter plot. Have them refer to the Key Idea for the names and descriptions of data displays.
- **Exercises 8 and 9** Students may guess whether a given data display is appropriate. Encourage them to carefully read the problem, and then carefully think about whether the display is appropriate.

9.4 Record and Practice Journal

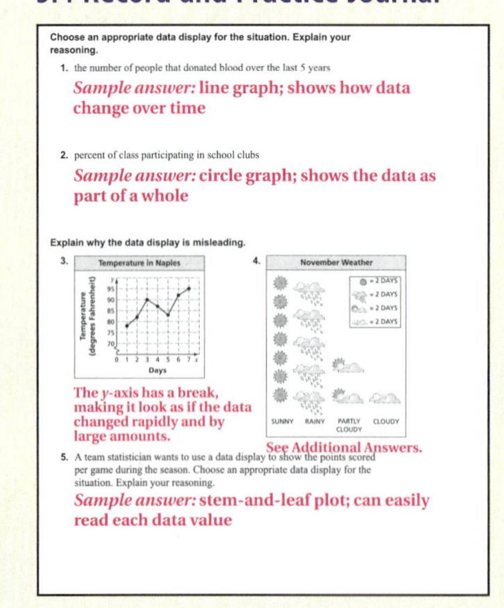

Choose an appropriate data display for the situation. Explain your reasoning.

1. the number of people that donated blood over the last 5 years

   Sample answer: **line graph; shows how data change over time**

2. percent of class participating in school clubs

   Sample answer: **circle graph; shows the data as part of a whole**

Explain why the data display is misleading.

3. Temperature in Naples

   The y-axis has a break, making it look as if the data changed rapidly and by large amounts.

4. November Weather

   See Additional Answers.

5. A team statistician wants to use a data display to show the points scored per game during the season. Choose an appropriate data display for the situation. Explain your reasoning.

   Sample answer: **stem-and-leaf plot; can easily read each data value**

9.4 Exercises

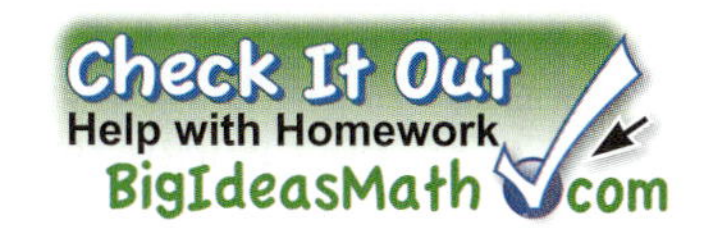

Vocabulary and Concept Check

1. **REASONING** Can more than one display be appropriate for a data set? Explain.
2. **OPEN-ENDED** Describe how a histogram can be misleading.

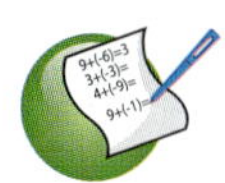

Practice and Problem Solving

3. Analyze and display the data in a way that best describes the data. Explain your choice of display.

Notebooks Sold in One Week				
192 red	170 green	203 black	183 pink	230 blue
165 yellow	210 purple	250 orange	179 white	218 other

Choose an appropriate data display for the situation. Explain your reasoning.

① 4. a student's test scores and how the scores are spread out

5. the distance a person drives each month
6. the outcome of rolling a number cube
7. homework problems assigned each day

② 8. **LIFEGUARD** The table shows how many hours you worked as a lifeguard from May to August. Tell whether the data display is appropriate for representing how the number of hours worked changed during the 4 months. Explain your reasoning.

Lifeguard Schedule	
Month	**Hours Worked**
May	40
June	80
July	160
August	120

a.

May
June
July
August

Key: = 20 hours

b.

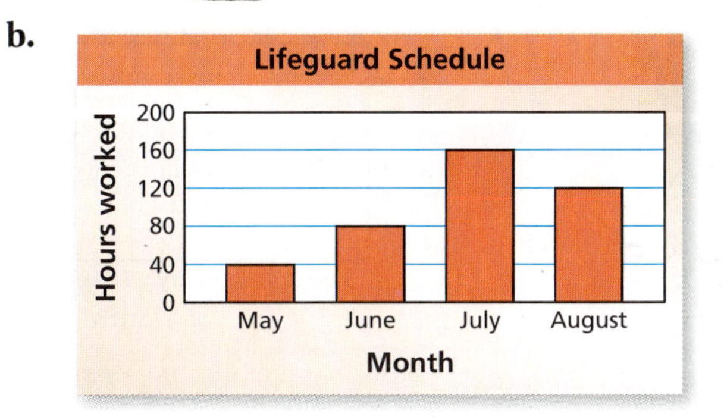

9. FAVORITE SUBJECT A survey asked 800 students to choose their favorite subject. The results are shown in the table. Tell whether the data display is appropriate for representing the portion of students who prefer math. Explain your reasoning.

Favorite School Subject	
Subject	**Number of Students**
Science	200
Math	160
Literature	240
Social Studies	120
Other	80

a.

Favorite School Subject

Science 25%
Math 20%
Social Studies 15%
Literature 30%
Other 10%

b.

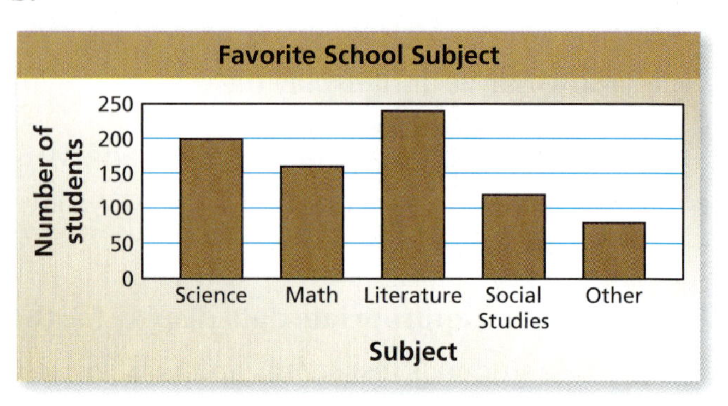

10. WRITING When should you use a histogram instead of a bar graph to display data? Use an example to support your answer.

Explain why the data display is misleading.

③ ④ **11.**

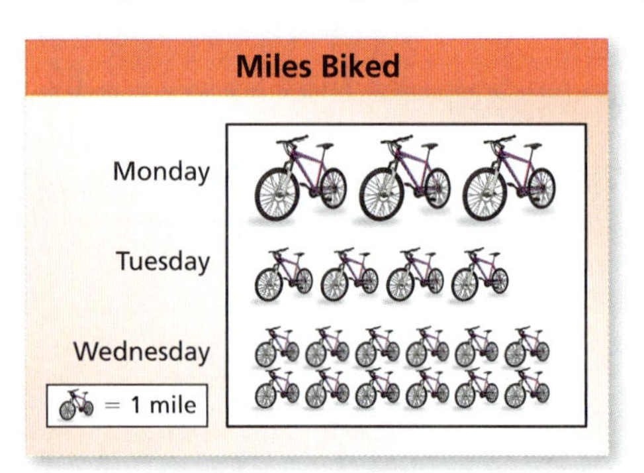

12.

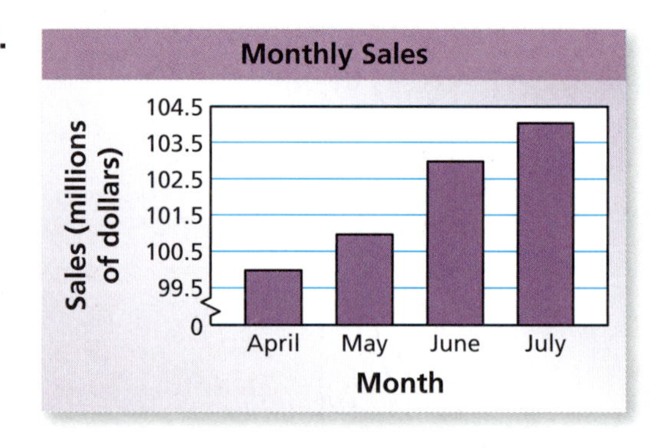

13.

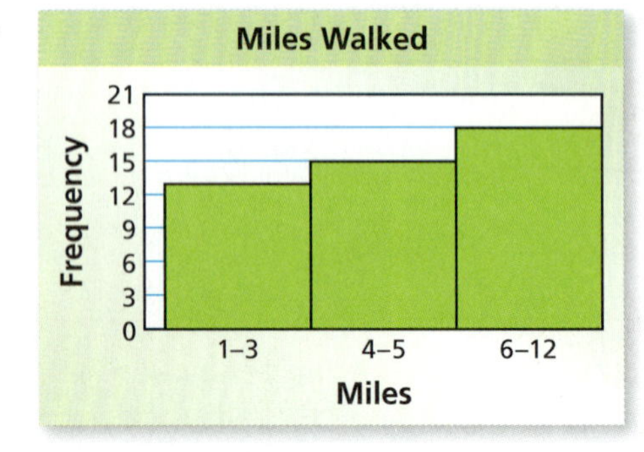

14.

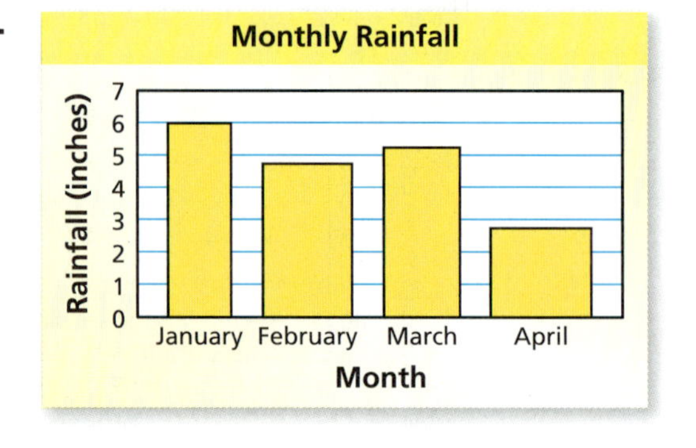

Common Errors

- **Exercises 11–14** Students may not be able to recognize why the data display is misleading. As a class, make a list of things to look for when analyzing a data display. For instance, students should check the increments or intervals for the axes, if applicable.
- **Exercise 17** Students may say that an appropriate data display for showing the mode is a stem-and-leaf plot, because the stem with the most leaves would be the mode. However, the stem with the most leaves may have data that is not part of the mode. Remind them that a dot plot isolates each data value and shows the frequency of each individual number, so this would be a more appropriate data display.

9. **a.** yes; The circle graph shows the data as parts of the whole.

   b. no; The bar graph shows the number of students, not the portion of students.

10. when the data are in terms of intervals of one category, as opposed to multiple categories; *Sample answer:* You can use a histogram to display the frequencies of voters in the last election by age group.

11. The pictures of the bikes are the largest on Monday and the smallest on Wednesday, which makes it seem like the distance is the same each day.

12. The break in the scale for the vertical axis makes it appear as though there is a greater difference in sales between months.

13. The intervals are not the same size.

14. The widths of the bars are different, so it looks like some months have more rainfall.

Differentiated Instruction

Auditory

Ask students which data display would best represent the given data.

- the number of baseball cards each boy in the class has box-and-whisker plot
- the number of hours studying for a test and the test scores of students in a class scatter plot

Practice and Problem Solving

15. *Sample answer:* bar graph; Each bar can represent a different vegetable.

16. yes; The vertical axis has a scale that increases by powers of 10, which makes the data appear to have a linear relationship.

17. *Sample answer:* dot plot

18. **a.** The percents do not sum to 100%.

b. *Sample answer:* bar graph; It would show the frequency of each sport.

19. See *Taking Math Deeper*.

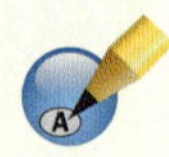

Fair Game Review

20. **a.** 4 **b.** 4.5

21. **a.** -9 **b.** -8.6

22. **a.** 12 **b.** 11.8

23. A

Mini-Assessment

Choose an appropriate data display for the situation. Explain your reasoning.

1. the outcome of flipping a coin
   Sample answers: pictograph, bar graph, or dot plot; all show the number of times you get heads or tails.
2. comparison of students' test scores and how long students studied
   Sample answer: scatter plot; You want to compare two data sets.
3. the number of students participating in after-school sports each year
   Sample answer: line graph; shows how data change over time

Taking Math Deeper

Exercise 19

This exercise introduces students to an amazing property of the number pi. Pi is an irrational number and therefore its decimal representation is not repeating. Even so, the ten digits from 0 to 9 each occur about 10 percent of the time, when one considers thousands of digits.

Display the data in a bar graph.

a.

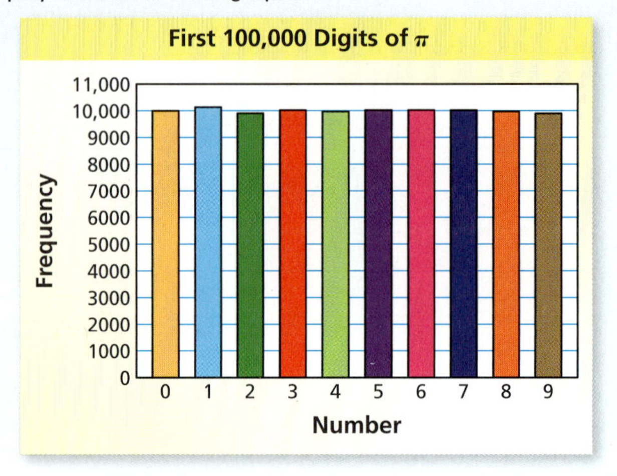

2 Display the data in a circle graph.

b.

3 **c. and d.** Compare the two displays.

Both graphs show that each digit occurs about 10% of the time, leading to a relatively *flat*, or *uniform* distribution. The bar graph has a slight advantage because it shows that some digits occur slightly more than others.

Reteaching and Enrichment Strategies

If students need help. . .	If students got it. . .
Resources by Chapter • Practice A and Practice B • Puzzle Time Record and Practice Journal Practice Differentiating the Lesson Lesson Tutorials Skills Review Handbook	Resources by Chapter • Enrichment and Extension • Technology Connection Start the next section

15. **VEGETABLES** A nutritionist wants to use a data display to show the favorite vegetables of the students at a school. Choose an appropriate data display for the situation. Explain your reasoning.

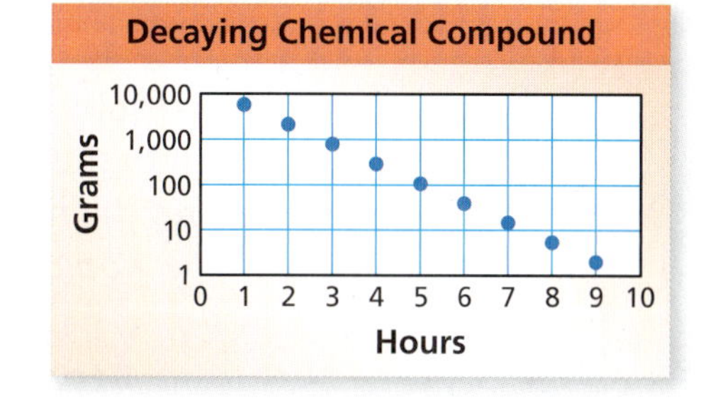

16. **CHEMICALS** A scientist gathers data about a decaying chemical compound. The results are shown in the scatter plot. Is the data display misleading? Explain.

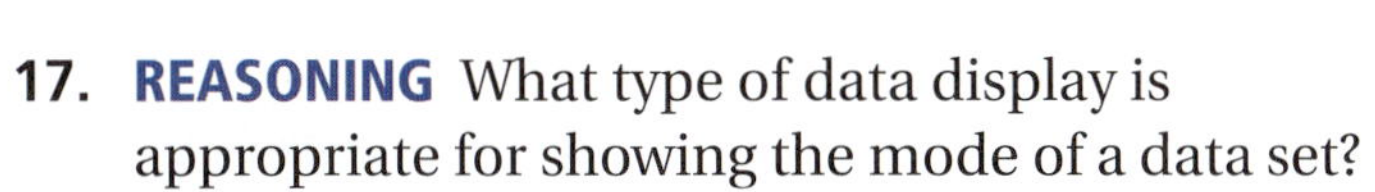

17. **REASONING** What type of data display is appropriate for showing the mode of a data set?

18. **SPORTS** A survey asked 100 students to choose their favorite sports. The results are shown in the circle graph.

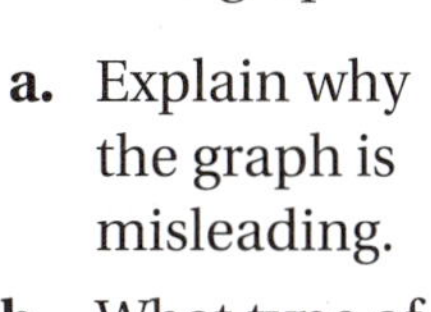

a. Explain why the graph is misleading.

b. What type of data display would be more appropriate for the data? Explain.

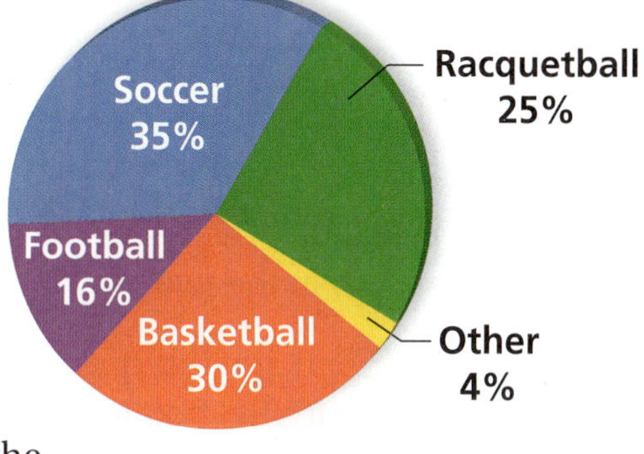

19. **Structure** With the help of computers, mathematicians have computed and analyzed billions of digits of the irrational number π. One of the things they analyze is the frequency of each of the numbers 0 through 9. The table shows the frequency of each number in the first 100,000 digits of π.

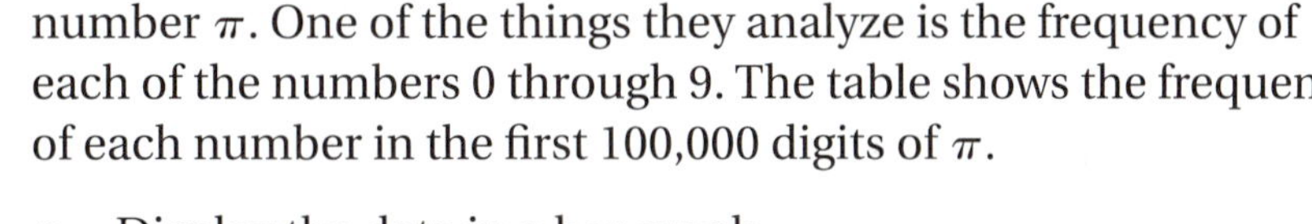

a. Display the data in a bar graph.

b. Display the data in a circle graph.

c. Which data display is more appropriate? Explain.

d. Describe the distribution.

Number	0	1	2	3	4	5	6	7	8	9
Frequency	9999	10,137	9908	10,025	9971	10,026	10,029	10,025	9978	9902

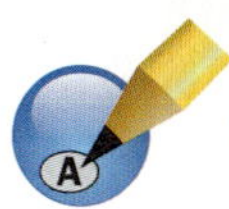

Fair Game Review What you learned in previous grades & lessons

Estimate the square root to the nearest (a) integer and (b) tenth. *(Section 7.4)*

20. $\sqrt{20}$

21. $-\sqrt{74}$

22. $\sqrt{140}$

23. **MULTIPLE CHOICE** What is 20% of 25% of 400? *(Skills Review Handbook)*

(A) 20 (B) 200 (C) 240 (D) 380

9.3–9.4 Quiz

1. **RECYCLING** The results of a recycling survey are shown in the two-way table. Find and interpret the marginal frequencies. *(Section 9.3)*

		Recycle	
		Yes	No
Gender	Female	28	9
	Male	24	14

2. **MUSIC** The results of a music survey are shown in the two-way table. Find and interpret the marginal frequencies. *(Section 9.3)*

		Jazz	
		Likes	Dislikes
Country	Likes	26	14
	Dislikes	17	8

3. **ELECTION** The results of a voting survey are shown in the two-way table. *(Section 9.3)*

		Voter's Age		
		18–34	35–64	65+
Candidate	Smith	36	25	6
	Jackson	12	32	24

   a. Find and interpret the marginal frequencies.

   b. For each age group, what percent of voters prefer Smith? prefer Jackson? Organize your results in a two-way table.

   c. Does your table in part (b) show a relationship between age and candidate preference? Explain.

Choose an appropriate data display for the situation. Explain your reasoning. *(Section 9.4)*

4. the percent of band students in each section of instruments

5. a company's profit for each week

6. **TURTLES** The tables show the weights (in pounds) of turtles caught in two ponds. Which type of data display would you use for this information? Explain. *(Section 9.4)*

Pond A			
12	13	15	6
7	8	12	7

Pond B			
9	12	5	8
12	15	16	19

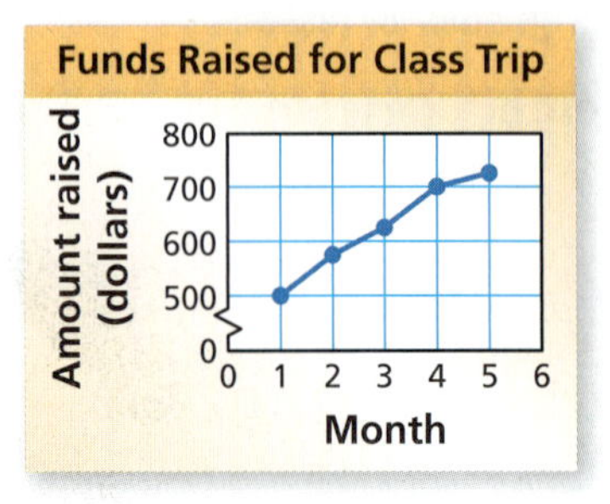

7. **FUNDRAISER** The line graph shows the amount of money that the eighth-grade students at a school raised each month to pay for a class trip. Is the graph misleading? Explain. *(Section 9.4)*

Alternative Assessment Options

Math Chat
Structured Interview
Student Reflective Focus Question
Writing Prompt

Student Reflective Focus Question

Ask students to summarize what line graphs, histograms, circle graphs, and two-way tables are used to show. Tell them to give an example of a real-life situation that can be represented by each type of display. Select students at random to present their summaries to the class.

Study Help Sample Answers

Remind students to complete Graphic Organizers for the rest of the chapter.

2.

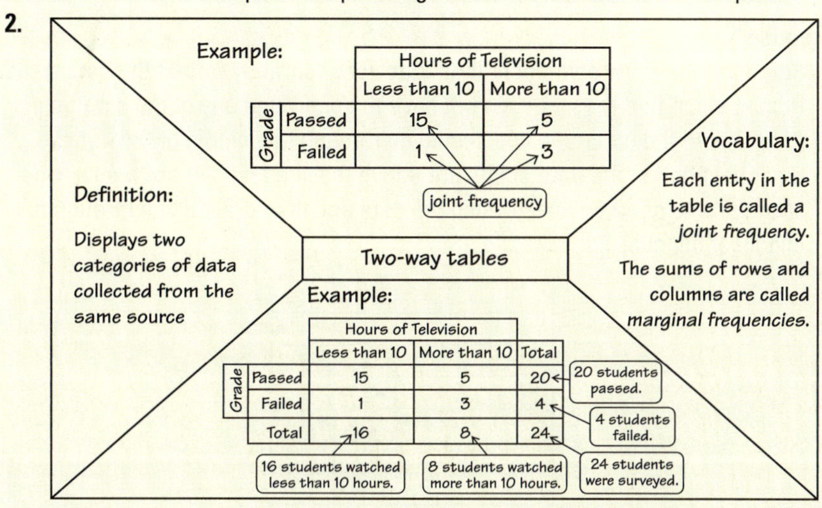

3.

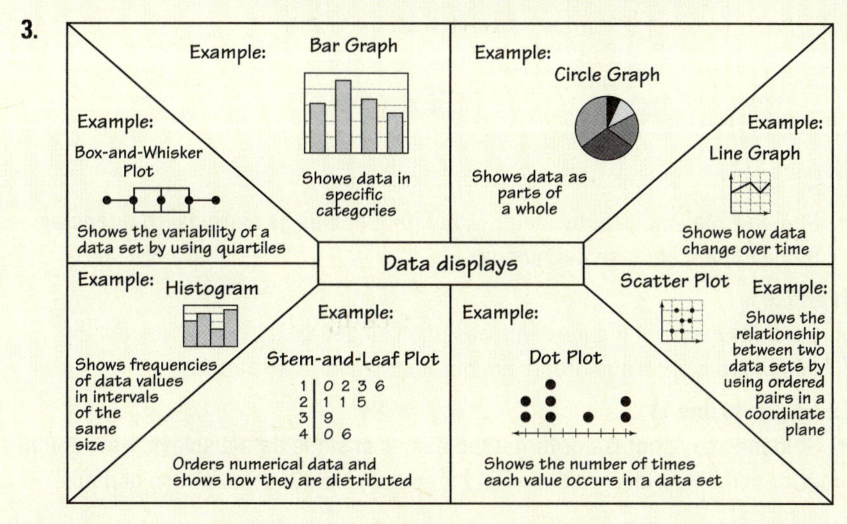

Reteaching and Enrichment Strategies

If students need help. . .	If students got it. . .
Resources by Chapter • Practice A and Practice B • Puzzle Time Lesson Tutorials *BigIdeasMath.com*	Resources by Chapter • Enrichment and Extension • Technology Connection Game Closet at *BigIdeasMath.com* Start the Chapter Review

Answers

1. 52 people recycle;
   23 people do not recycle;
   37 females were surveyed;
   38 males were surveyed.
2. 43 people like jazz;
   22 people dislike jazz;
   40 people like country;
   25 people dislike country.
3. See Additional Answers.
4. *Sample answer:* circle graph; shows data as parts of a whole
5. *Sample answer:* line graph; shows changes over time
6. *Sample answer:* box-and-whisker plot; shows the variability of the data
7. yes; The break in the vertical axis makes it appear that the amount of money raised increased very rapidly from month to month.

Online Assessment
Assessment Book
ExamView® Assessment Suite

For the Teacher

Additional Review Options

- *BigIdeasMath.com*
- Online Assessment
- Game Closet at *BigIdeasMath.com*
- Vocabulary Help
- Resources by Chapter

Answers

1. **a.** 2012

   b. 225 geese

   c. positive linear relationship

2. negative linear relationship; outlier: (21, 40)

3. no relationship; cluster around (12, 16)

4. positive linear relationship; gap from $x = 12$ to $x = 18$

5. **a.**

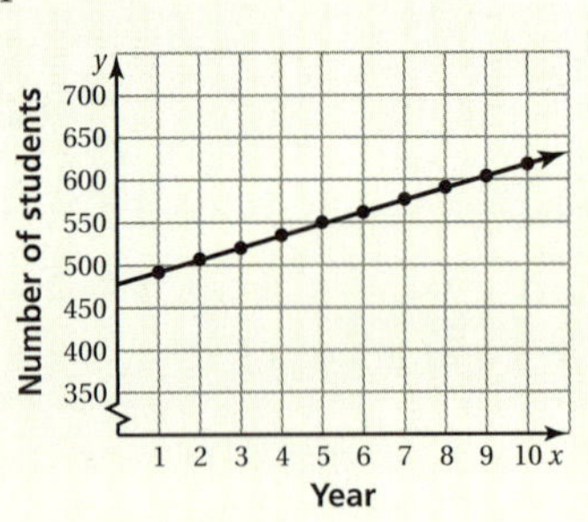

   b. *Sample answer:* $y = 14x + 478$

   c. *Sample answer:* The slope is 14 and the y-intercept is 478. So, the number of students in the year prior to the 10-year period was about 478 and the number of students is increasing by about 14 students per year.

   d. 632 students

6. $y = 14.0x + 478.7$; The correlation coefficient is about 1, which implies that there is a strong positive linear relationship between the year and the number of students.

Review of Common Errors

Exercise 1

- When finding values from the graph, students may accidentally shift over or up too far and get an answer that is off by an increment. Encourage them to start at the given value and trace the graph to where the point is, and then trace down or left to the other axis for the answer.

Exercises 2–4

- Students may mix up positive and negative relationships. Remind them about slope. The slope is positive when the line rises from left to right and negative when it falls from left to right. The same is true for relationships in a scatter plot. When the data rise from left to right, it is a positive relationship. When the data fall from left to right, it is a negative relationship.

Exercise 5

- Students may draw a line of fit that does not accurately reflect the data trend. Remind them that the line does not have to go through any of the data points. Also remind them that the line should go through the middle of the data so that about half of the data points are above the line and half are below. One strategy is to draw an oval around the data and then draw a line through the middle of the oval. For example:

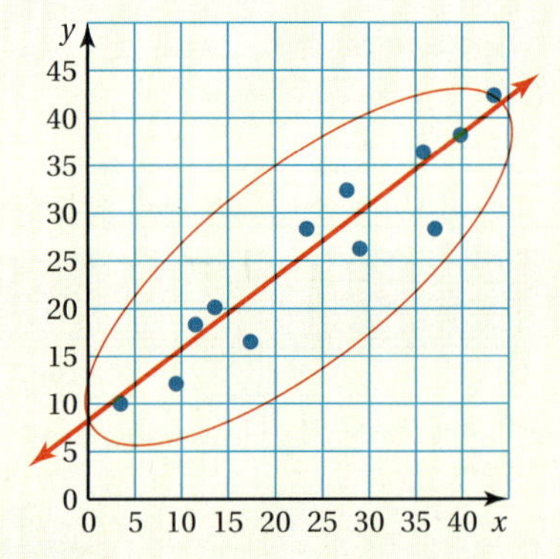

Exercise 7

- Students may incorrectly identify joint frequencies as marginal frequencies. Remind them of these definitions.

Exercise 8

- Students may find the percents based on all people surveyed, not just the people from each age group. Encourage them to read carefully.

Exercises 10 and 11

- Students may confuse or forget the names of some data displays. Have them refer to the Key Idea on page 394 for the names and descriptions of data displays.

9 Chapter Review

Review Key Vocabulary

scatter plot, *p. 374*
line of fit, *p. 380*
line of best fit, *p. 381*
two-way table, *p. 388*
joint frequency, *p. 388*
marginal frequency, *p. 388*

Review Examples and Exercises

9.1 Scatter Plots *(pp. 372–377)*

Your school is ordering custom T-shirts. The scatter plot shows the number of T-shirts ordered and the cost per shirt. What tends to happen to the cost per shirt as the number of T-shirts ordered increases?

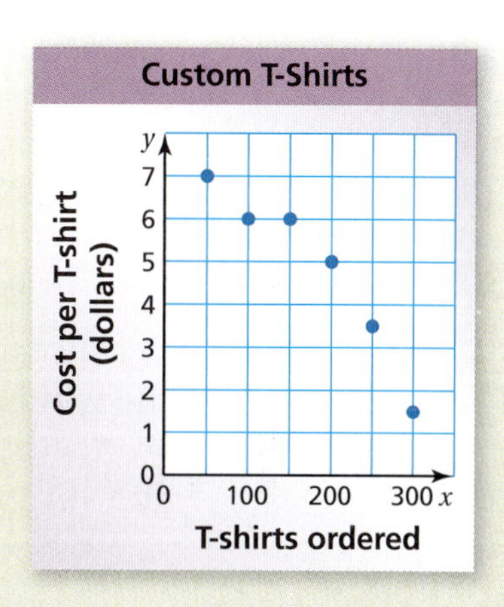

Looking at the graph, the plotted points go down from left to right.

∴ So, as the number of T-shirts ordered increases, the cost per shirt decreases.

Exercises

1. **MIGRATION** The scatter plot shows the number of geese that migrated to a park each season.

   a. In what year did 270 geese migrate?

   b. How many geese migrated in 2010?

   c. Describe the relationship shown by the data.

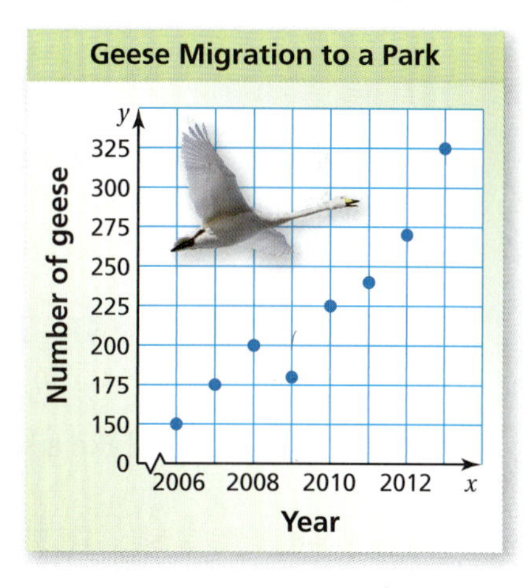

Describe the relationship between the data. Identify any outliers, gaps, or clusters.

2.

3.

4.

9.2 Lines of Fit (pp. 378–383)

The table shows the revenue (in millions of dollars) for a company over an 8-year period. (a) Make a scatter plot of the data and draw a line of fit. (b) Write an equation of the line of fit. (c) Interpret the slope and the y-intercept of the line of fit. (d) Predict what the revenue will be in year 9.

Year, x	1	2	3	4	5	6	7	8
Revenue (millions of dollars), y	20	35	46	56	68	82	92	108

a. Plot the points in a coordinate plane. The scatter plot shows a positive linear relationship. Draw a line that is close to the data points.

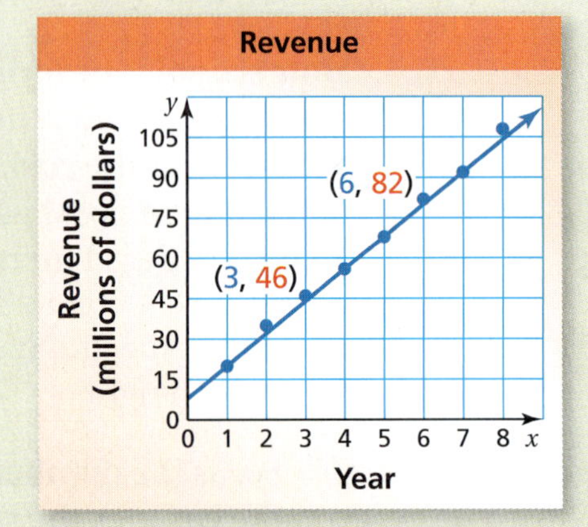

b. $\text{slope} = \frac{\text{rise}}{\text{run}} = \frac{36}{3} = 12$

Because the line crosses the y-axis at (0, 8), the y-intercept is 8.

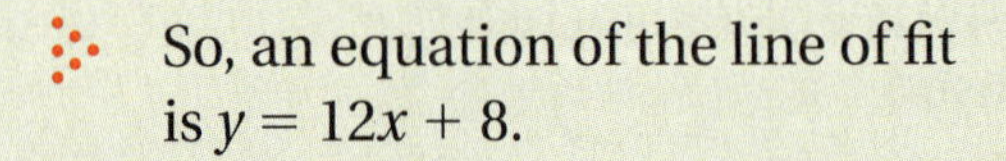
So, an equation of the line of fit is $y = 12x + 8$.

c. The slope is 12. So, the revenue increased by about $12 million each year. The y-intercept is 8. So, you can estimate that the revenue was $8 million in the year before this 8-year period.

d. $y = 12x + 8 = 12(9) + 8 = 116$

The revenue in year 9 will be about $116 million.

Exercises

5. **STUDENTS** The table shows the number of students at a middle school over a 10-year period.
   - **a.** Make a scatter plot of the data and draw a line of fit.
   - **b.** Write an equation of the line of fit.
   - **c.** Interpret the slope and the y-intercept of the line of fit.
   - **d.** Predict the number of students in year 11.

Year, x	Number of Students, y
1	492
2	507
3	520
4	535
5	550
6	562
7	577
8	591
9	604
10	618

6. **LINE OF BEST FIT** Use a graphing calculator to find an equation of the line of best fit for the data in Exercise 5. Identify and interpret the correlation coefficient.

Review Game

Rolling for Data

Materials per pair

- two number cubes
- paper
- pencil

Directions

Students should work in pairs. Students in each pair take turns rolling the two number cubes one at a time. The first number they roll represents the tens digit and the second number represents the ones digit of a whole number. For example, if a 1 is rolled and then a 6, the whole number is 16. Students record the whole numbers in a stem-and-leaf plot and keep rolling until they have 10 leaves for any one stem. Once a pair acquires the 10 leaves, they race to create a histogram and a circle graph for the data.

Who wins?

The first pair to finish all tasks wins 10 points, the second 9 points, the third 8 points, and so on. The game can be repeated as many times as desired. The pair with the most points after a predetermined number of rounds or amount of time wins.

For the Student Additional Practice

- Lesson Tutorials
- Multi-Language Glossary
- Self-Grading Progress Check
- *BigIdeasMath.com*
  Dynamic Student Edition
  Student Resources

Answers

7. See Additional Answers.

8.

		Food Court	
		Likes	Dislikes
Age Group	Teenagers	96%	4%
	Adults	21%	79%
	Senior Citizens	18%	82%

9. Yes, the table shows that teenages tend to like the food court, but adults and senior citizens tend to dislike the food court.

10. *Sample answer:* line graph; shows changes over time

11. *Sample answer:* scatter plot; shows the relationship between two data sets

My Thoughts on the Chapter

Teacher Tip

Not allowed to write in your teaching edition? Use sticky notes to record your thoughts.

What worked. . .

What did not work. . .

What I would do differently. . .

9.3 Two-Way Tables (pp. 386–391)

You randomly survey students in your school about whether they liked a recent school play. The results are shown. Make a two-way table that includes the marginal frequencies. What percent of the students surveyed liked the play?

Male Students
48 likes, 12 dislikes
Female Students
56 likes, 14 dislikes

Of the 130 students surveyed, 104 students liked the play.

		Student		
		Liked	Did Not Like	Total
Gender	Male	48	12	60
	Female	56	14	70
	Total	104	26	130

Because $\frac{104}{130} = 0.8$, 80% of the students in the survey liked the play.

Exercises

You randomly survey people at a mall about whether they like the new food court. The results are shown.

Teenagers
96 likes, 4 dislikes
Adults
21 likes, 79 dislikes
Senior Citizens
18 likes, 82 dislikes

7. Make a two-way table that includes the marginal frequencies.
8. For each group, what percent of the people surveyed like the food court? dislike the food court? Organize your results in a two-way table.
9. Does your table in Exercise 8 show a relationship between age and whether people like the food court?

9.4 Choosing a Data Display (pp. 392–399)

Choose an appropriate data display for the situation. Explain your reasoning.

a. the percent of votes that each candidate received in an election

A circle graph shows data as parts of a whole. So, a circle graph is an appropriate data display.

b. the distribution of the ages of U.S. presidents

A stem-and-leaf plot orders numerical data and shows how they are distributed. So, a stem-and-leaf plot is an appropriate data display.

Exercises

Choose an appropriate data display for the situation. Explain your reasoning.

10. the number of pairs of shoes sold by a store each week
11. a comparison of the heights of brothers and sisters

9 Chapter Test

1. **POPULATION** The graph shows the population (in millions) of the United States from 1960 to 2010.

   a. In what year was the population of the United States about 180 million?

   b. What was the approximate population of the United States in 1990?

   c. Describe the trend shown by the data.

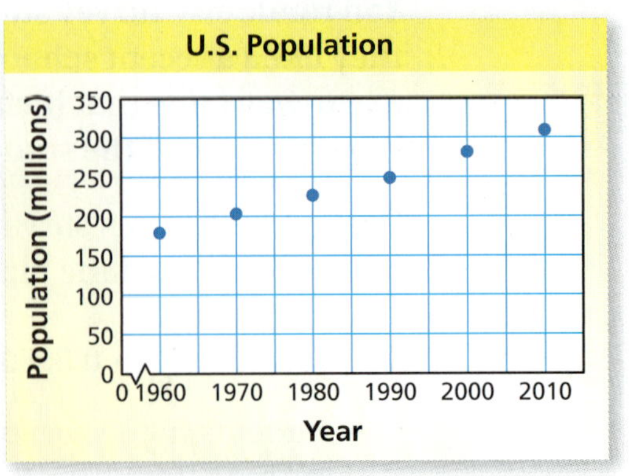

2. **WEIGHT** The table shows the weight of a baby over several months.

   a. Make a scatter plot of the data and draw a line of fit.

   b. Write an equation of the line of fit.

   c. Interpret the slope and the y-intercept of the line of fit.

   d. Predict how much the baby will weigh at 7 months.

Age (months)	Weight (pounds)
1	8
2	9.25
3	11.75
4	13
5	14.5
6	16

		Nonfiction	
		Likes	Dislikes
Fiction	Likes	26	20
	Dislikes	22	2

3. **READING** You randomly survey students at your school about what type of books they like to read. The two-way table shows your results. Find and interpret the marginal frequencies.

Choose an appropriate data display for the situation. Explain your reasoning.

4. magazine sales grouped by price

5. the distance a person hikes each week

6. **SAT** The table shows the numbers y of students (in thousands) who took the SAT from 2006 to 2010, where $x = 6$ represents the year 2006. Use a graphing calculator to find an equation of the line of best fit. Identify and interpret the correlation coefficient.

Year, x	6	7	8	9	10
Number of Students, y	1466	1495	1519	1530	1548

7. **RECYCLING** You randomly survey shoppers at a supermarket about whether they use reusable bags. Of 60 male shoppers, 15 use reusable bags. Of 110 female shoppers, 60 use reusable bags. Organize your results in a two-way table. Include the marginal frequencies.

Test Item References

Chapter Test Questions	Section to Review	Florida Common Core Standards (MACC)
1	9.1	8.SP.1.1
2, 6	9.2	8.SP.1.1, 8.SP.1.2, 8.SP.1.3
3, 7	9.3	8.SP.1.4
4, 5	9.4	8.SP.1.1

Test-Taking Strategies

Remind students to quickly look over the entire test before they start so that they can budget their time. Have them use the **Stop** and **Think** strategy before they answer each question.

Common Errors

- **Exercise 2** When finding values from the graph, students may accidentally shift over or up too far and get an answer that is off by an increment.
- **Exercise 3** Students may incorrectly identify joint frequencies as marginal frequencies. Remind them of these definitions.
- **Exercises 4 and 5** Students may confuse or forget the names of some data displays. Have them refer to the Key Idea on page 394 for the names and descriptions of data displays.

Reteaching and Enrichment Strategies

If students need help. . .	If students got it. . .
Resources by Chapter • Practice A and Practice B • Puzzle Time Record and Practice Journal Practice Differentiating the Lesson Lesson Tutorials *BigIdeasMath.com* Skills Review Handbook	Resources by Chapter • Enrichment and Extension • Technology Connection Game Closet at *BigIdeasMath.com* Start Standards Assessment

Answers

1. **a.** 1960

   b. about 250 million

   c. There is a positive linear relationship between year and population.

2. **a.**

   b. *Sample answer:* $y = 1.5x + 7$

   c. The slope is 1.5 and the y-intercept is 7. So, the baby is gaining 1.5 pounds per month and was born with a weight of 7 pounds.

   d. 17.5 pounds

3. 48 students like nonfiction; 22 students dislike nonfiction; 46 students like fiction; 24 students dislike fiction.

4. *Sample answer:* histogram; shows frequencies of data values in intervals of the same size

5. *Sample answer:* line graph; shows how data change over time

6. $y = 19.9x + 1352.4$; The correlation coefficient is about 0.986. This means that the relationship between years and the numbers of students taking the SAT is a strong positive correlation and the equation closely models the data.

7. See Additional Answers.

Online Assessment
Assessment Book
ExamView® Assessment Suite

Test-Taking Strategies

Available at *BigIdeasMath.com*

After Answering Easy Questions, Relax
Answer Easy Questions First
Estimate the Answer
Read All Choices before Answering
Read Question before Answering
Solve Directly or Eliminate Choices
Solve Problem before Looking at Choices
Use Intelligent Guessing
Work Backwards

About this Strategy

When taking a multiple choice test, be sure to read each question carefully and thoroughly. It is also very important to read each answer choice carefully. Do not pick the first answer that you think is correct! If two answer choices are the same, eliminate them both. There can only be one correct answer.

Answers

1. B
2. F
3. 12

Technology for the Teacher

Florida Common Core Standards Support Performance Tasks
Online Assessment
Assessment Book
ExamView® Assessment Suite

Item Analysis

1. **A.** The student subtracts the volume of the hemisphere from the volume of the cylinder.

   B. Correct answer

   C. The student finds the volume of a cylinder with a height of 18 inches.

   D. The student adds the volume of a whole sphere with a radius of 6 inches to the volume of the cylinder.

2. **F.** Correct answer

   G. The student confuses alternate interior angles with corresponding angles.

   H. The student picks the angle on the same side of the transversal as angle 6, but not the corresponding angle.

   I. The student confuses a linear pair with corresponding angles.

3. **Gridded Response:** Correct answer: 12

   Common Error: The student adds the entries in the "No" column for an answer of 29.

9 Standards Assessment

1. What is the volume of the trash bin? *(MACC.8.G.3.9)*

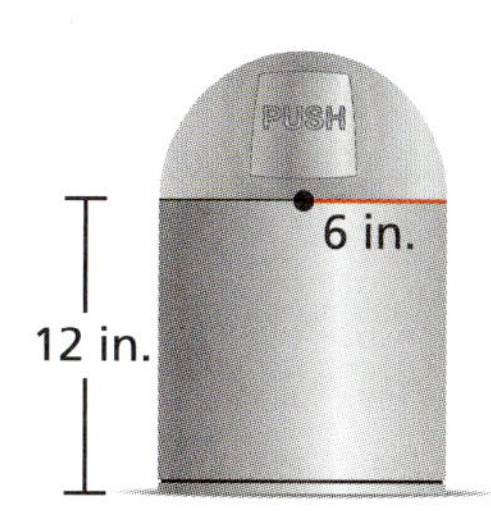

A. 288π in.^{3} **C.** 648π in.^{3}

B. 576π in.^{3} **D.** 720π in.^{3}

2. The diagram below shows parallel lines cut by a transversal. Which angle is the corresponding angle for $\angle 6$? *(MACC.8.G.1.5)*

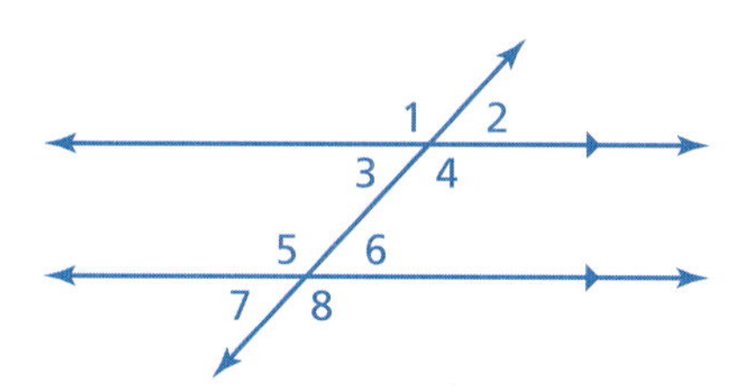

F. $\angle 2$ **H.** $\angle 4$

G. $\angle 3$ **I.** $\angle 8$

3. You randomly survey students in your school. You ask whether they have jobs. You display your results in the two-way table. How many male students do *not* have a job? *(MACC.8.SP.1.4)*

		Job	
		Yes	**No**
Gender	**Male**	27	12
	Female	31	17

4. Which scatter plot shows a negative relationship between x and y? *(MACC.8.SP.1.1)*

A. 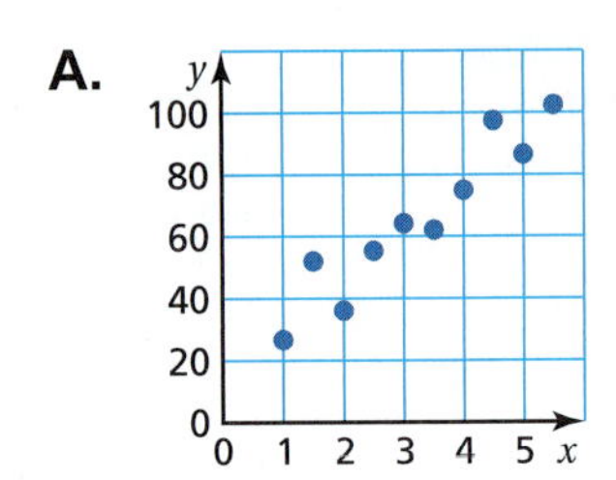

C. 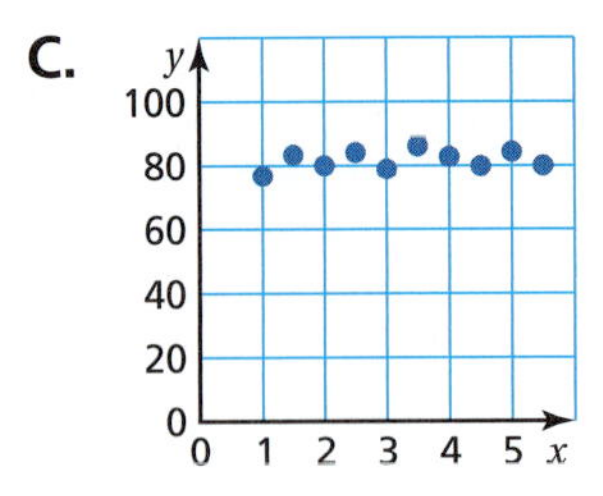

B. 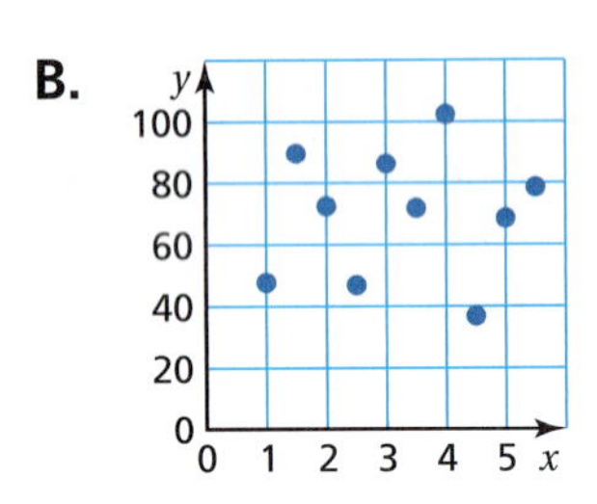

D. 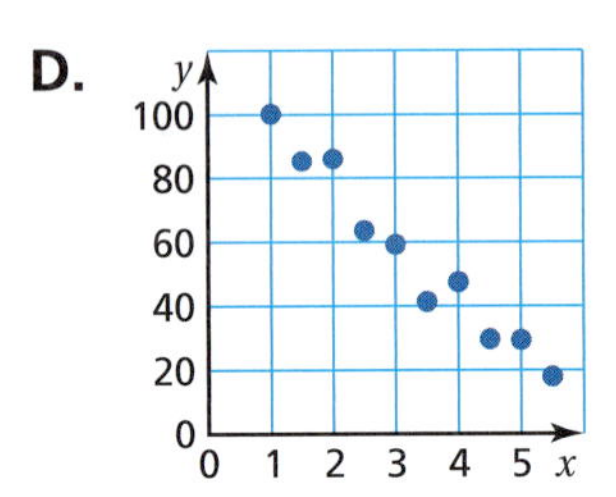

5. The legs of a right triangle have the lengths of 8 centimeters and 15 centimeters. What is the length of the hypotenuse, in centimeters? *(MACC.8.G.2.7)*

6. What is the solution of the equation? *(MACC.8.EE.3.7b)*

$$0.22(x + 6) = 0.2x + 1.8$$

F. $x = 2.4$

H. $x = 24$

G. $x = 15.6$

I. $x = 156$

7. Which triangle is *not* a right triangle? *(MACC.8.G.2.6)*

A. 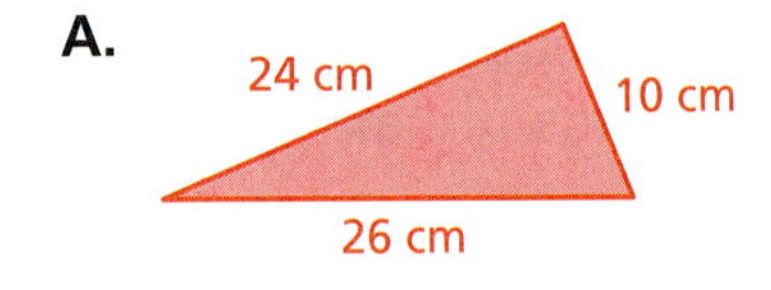

C. 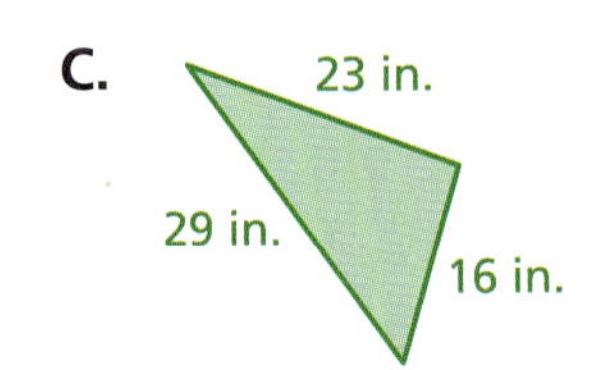

B. 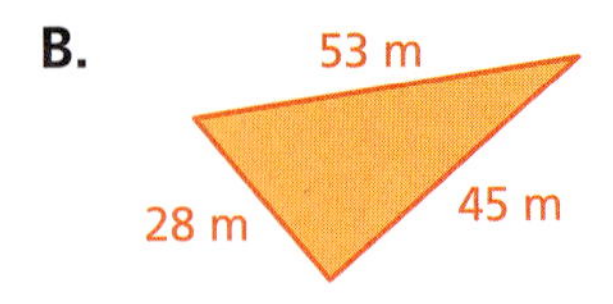

D. 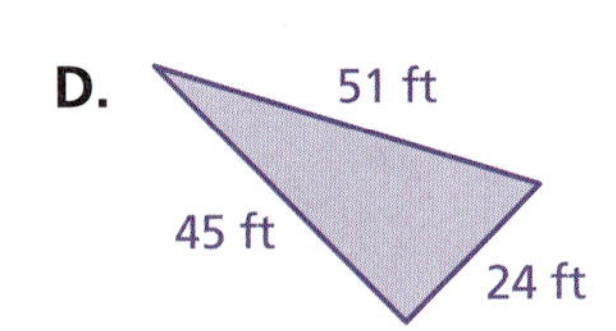

Item Analysis (continued)

4. **A.** The student confuses positive and negative relationships.
   B. The student confuses no relationship with a negative relationship.
   C. The student confuses a constant function with a negative relationship.
   D. Correct answer

5. **Gridded Response:** Correct answer: 17 cm

   Common Error: The student adds the legs for an answer of 23 centimeters.

6. **F.** The student makes an error placing a decimal point.
   G. The student makes an error applying the Addition Property of Equality and makes an error placing a decimal point.
   H. Correct answer
   I. The student makes an error applying the Addition Property of Equality.

7. **A.** The student makes a computation error.
   B. The student makes a computation error.
   C. Correct answer
   D. The student makes a computation error.

Answers

4. D
5. 17 cm
6. H
7. C

Answers

8. G

9. A

10. *Part A*

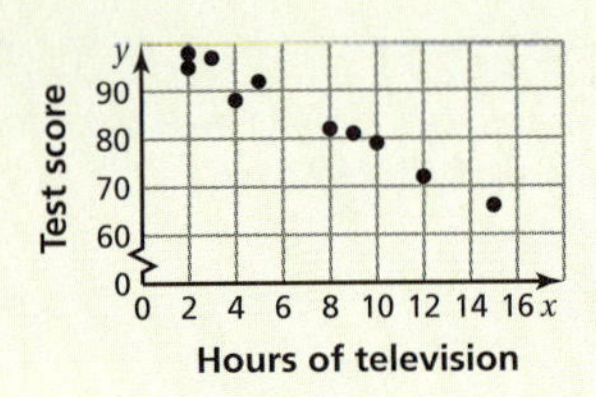

Part B There is a negative correlation between the number of hours of television watched and a student's test score.

Part C Enter the data in the calculator and find the correlation coefficient.

Item Analysis (continued)

8. F. The student does not realize that the underlying idea here is to show change over time, something of which a circle graph is not capable.

G. Correct answer

H. The student does not realize that the underlying idea here is to show change over time, something of which a histogram is not capable.

I. The student does not realize that the underlying idea here is to show change over time, something of which a stem-and-leaf plot is not capable.

9. A. Correct answer

B. The student reflects the figure in the y-axis.

C. The student rotates the figure 180°.

D. The student rotates the figure 90° counterclockwise.

10. 4 points: The student demonstrates a thorough understanding of using a scatter plot to determine the relationship between two data sets. In Part A, all the points are correctly plotted. In Part B, there is a negative relationship between hours of television watched and test score. In Part C, the line of best fit is $y = -2.4x + 102$ and the correlation coefficient is about -0.983, which is close to -1, confirming that there is a negative relationship.

3 points: The student demonstrates an essential but less than thorough understanding of using a scatter plot to determine the relationship between two data sets. The relationship between the data sets is stated correctly, but one or more points may be plotted incorrectly, or the correlation coefficient is not correctly interpreted.

2 points: The student demonstrates a partial understanding of using a scatter plot to determine the relationship between two data sets. There are errors in plotting the points and interpreting the scatter plot or the correlation coefficient.

1 point: The student demonstrates a limited understanding of using a scatter plot to determine the relationship between two data sets. Points are plotted incorrectly or not at all, and the interpretations are incorrect.

0 points: The student provides no response, a completely incorrect or incomprehensible response, or a response that demonstrates insufficient understanding of using a scatter plot to determine the relationship between two data sets.

8. A store has recorded total dollar sales each month for the past three years. Which type of graph would best show how sales have increased over this time period? *(MACC.8.SP.1.1)*

F. circle graph

G. line graph

H. histogram

I. stem-and-leaf plot

9. Trapezoid *KLMN* is graphed in the coordinate plane shown.

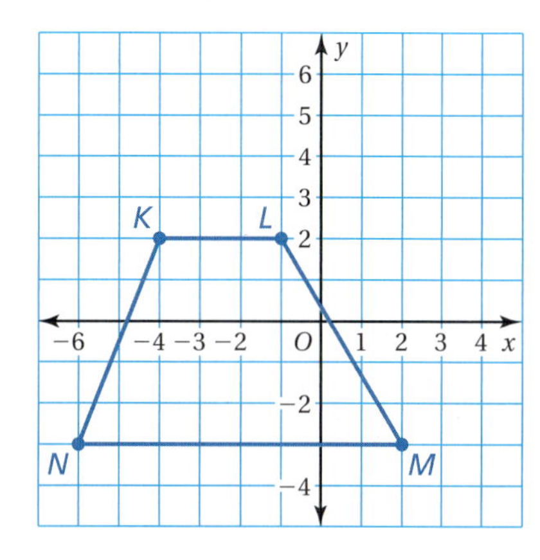

Rotate Trapezoid *KLMN* 90° clockwise about the origin. What are the coordinates of point M', the image of point M after the rotation? *(MACC.8.G.1.3)*

A. (−3, −2)

B. (−2, −3)

C. (−2, 3)

D. (3, 2)

10. The table shows the numbers of hours students spent watching television from Monday through Friday for one week and their scores on a test that Friday. *(MACC.8.SP.1.1, MACC.8.SP.1.2)*

Think Solve Explain

Hours of Television, x	5	2	10	15	3	4	8	2	12	9
Test Score, y	92	98	79	66	97	88	82	95	72	81

Part A Make a scatter plot of the data.

Part B Describe the relationship between hours of television watched and test score.

Part C Explain how to justify your answer in Part B using the linear regression feature of a graphing calculator.

10 Exponents and Scientific Notation

10.1 Exponents
10.2 Product of Powers Property
10.3 Quotient of Powers Property
10.4 Zero and Negative Exponents
10.5 Reading Scientific Notation
10.6 Writing Scientific Notation
10.7 Operations in Scientific Notation

"Here's how it goes, Descartes."

"The friends of my friends are my friends. The friends of my enemies are my enemies."

"The enemies of my friends are my enemies. The enemies of my enemies are my friends."

"If one flea had 100 babies, and each baby grew up and had 100 babies, ..."

"... and each of those babies grew up and had 100 babies, you would have 1,010,101 fleas."

Florida Common Core Progression

6th Grade

- Write and evaluate numerical expressions involving whole-number exponents.
- Perform arithmetic operations, including those involving whole-number exponents, using the order of operations.
- Apply the properties of operations to generate equivalent expressions.

7th Grade

- Solve problems involving operations with rational numbers.
- Understand that rewriting expressions in different forms can show how quantities are related.

8th Grade

- Use the properties of integer exponents to generate equivalent expressions.
- Use scientific notation to estimate very large or very small quantities.
- Perform operations with numbers expressed in scientific notation and other forms.
- Interpret scientific notation that has been generated by technology.

Chapter Summary

Section	Florida Common Core Standard (MACC)	
10.1	Learning	8.EE.1.1
10.2	Learning	8.EE.1.1
10.3	Learning	8.EE.1.1
10.4	Learning	8.EE.1.1 ★
10.5	Learning	8.EE.1.3, 8.EE.1.4
10.6	Learning	8.EE.1.3, 8.EE.1.4
10.7	Learning	8.EE.1.3 ★, 8.EE.1.4 ★

★ Teaching is complete. Standard can be assessed.

Pacing Guide for Chapter 10

Chapter Opener	1 Day
Section 1 Activity Lesson	 1 Day 1 Day
Section 2 Activity Lesson	 1 Day 1 Day
Section 3 Activity Lesson	 1 Day 1 Day
Section 4 Activity Lesson	 1 Day 1 Day
Study Help / Quiz	1 Day
Section 5 Activity Lesson	 1 Day 1 Day
Section 6 Activity Lesson	 1 Day 1 Day
Section 7 Activity Lesson	 1 Day 1 Day
Chapter Review/ Chapter Tests	2 Days
Total Chapter 10	18 Days
Year-to-Date	149 Days

Technology for the Teacher

BigIdeasMath.com
Chapter at a Glance
Complete Materials List
Parent Letters: English, Spanish, and Haitian Creole

Florida Common Core Standards

MACC.6.EE.1.1 Write and evaluate numerical expressions involving whole-number exponents.

MACC.6.NS.2.3 Fluently . . . multiply and divide multi-digit decimals using the standard algorithm for each operation.

Additional Topics for Review

- Factors
- Simplifying Expressions
- Multiplicative Inverse Property
- Place Value
- Reciprocals
- Converting Measures
- Distributive Property
- Commutative and Associative Properties of Multiplication

Try It Yourself

1. 13 **2.** 10

3. 48 **4.** 0.35

5. 0.84 **6.** 30.229

7. 0.1788 **8.** 60

9. 13.9 **10.** 24

11. 1800

Record and Practice Journal Fair Game Review

1. 14 **2.** 3

3. 394 **4.** 86

5. 76 **6.** 16

7. **a.** 386

b. $4(2) + 2(5^2) + 3^2(6^2) + 2^2 = 386$

8. 2.352 **9.** 0.1014

10. 6.0048 **11.** 9

12. 1.5 **13.** 2700

14. $6.93

Math Background Notes

Vocabulary Review

- Order of Operations
- Expression
- Evaluating an Expression
- Exponent
- Decimal
- Dividend
- Product
- Quotient

Using Order of Operations

- Students should know the order of operations, but you may want to review it.
- Many students probably learned the mnemonic device *Please Excuse My Dear Aunt Sally*. Ask a volunteer to explain why this phrase is helpful.
- **Common Error:** Students may misinterpret the exponent in Example 1. Watch for students who incorrectly compute $6^2 = 6 \cdot 2 = 12$.

Multiplying and Dividing Decimals

- Students should be able to multiply and divide decimals.
- Remind students to multiply as they would with whole numbers. After the multiplication is complete, remind them to count the total number of digits in both factors that appear to the right of the decimal point, and then put that many digits to the right of the decimal point in the answer.
- **Common Error:** In a horizontal division problem, some students have difficulty determining which number is the dividend and which number is the divisor. Encourage students to read the problem aloud as they rewrite it. Each time a student says the words "divided by," he or she should be trapping the first number inside the division box.
- **Common Error:** Some students may try to clear the decimal point from the dividend instead of from the divisor. Remind students that having a decimal inside the division box as part of the dividend is fine. Also remind students that the divisor and dividend must both be multiplied by the same power of 10, effectively multiplying by 1, so they do not change the problem.

Reteaching and Enrichment Strategies

If students need help. . .	If students got it. . .
Record and Practice Journal • Fair Game Review Skills Review Handbook Lesson Tutorials	Game Closet at *BigIdeasMath.com* Start the next section

What You Learned Before

"It's called the Power of Negative One, Descartes!"

Using Order of Operations (MACC.6.EE.1.1)

Example 1 Evaluate $6^2 \div 4 - 2(9 - 5)$.

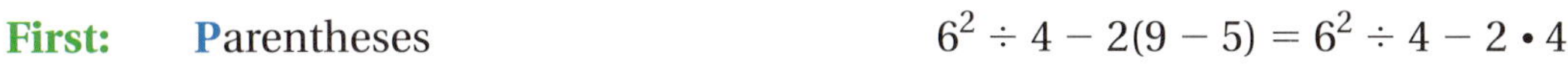

First: Parentheses $6^2 \div 4 - 2(9 - 5) = 6^2 \div 4 - 2 \cdot 4$

Second: Exponents $= 36 \div 4 - 2 \cdot 4$

Third: Multiplication and Division (from left to right) $= 9 - 8$

Fourth: Addition and Subtraction (from left to right) $= 1$

Try It Yourself

Evaluate the expression.

1. $15\left(\frac{8}{4}\right) + 2^2 - 3 \cdot 7$

2. $5^2 \cdot 2 \div 10 + 3 \cdot 2 - 1$

3. $3^2 - 1 + 2(4(3 + 2))$

Multiplying and Dividing Decimals (MACC.6.NS.2.3)

Example 2 Find $2.1 \cdot 0.35$.

$$\begin{array}{r} 2.1 \\ \times\ 0.35 \\ \hline 105 \\ 63 \\ \hline 0.735 \end{array}$$

- 2.1 ← 1 decimal place
- 0.35 ← + 2 decimal places
- 0.735 ← 3 decimal places

Example 3 Find $1.08 \div 0.9$.

$0.9\overline{)1.08}$ Multiply each number by 10.

$$\begin{array}{r} 1.2 \\ 9\overline{)10.8} \\ -\,9 \\ \hline 1\,8 \\ -\,1\,8 \\ \hline 0 \end{array}$$

Place the decimal point above the decimal point in the dividend 10.8.

Try It Yourself

Find the product or quotient.

4. $1.75 \cdot 0.2$

5. $1.4 \cdot 0.6$

6. $\begin{array}{r} 7.03 \\ \times\ 4.3 \\ \hline \end{array}$

7. $\begin{array}{r} 0.894 \\ \times\ 0.2 \\ \hline \end{array}$

8. $5.40 \div 0.09$

9. $4.17 \div 0.3$

10. $0.15\overline{)3.6}$

11. $0.004\overline{)7.2}$

10.1 Exponents

Essential Question How can you use exponents to write numbers?

The expression 3^5 is called a *power*. The *base* is 3. The *exponent* is 5.

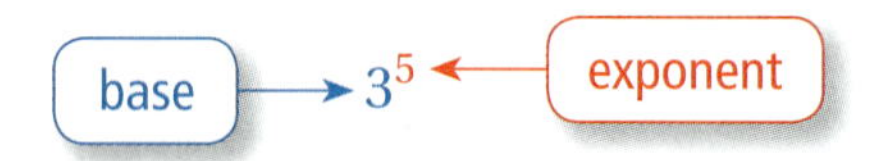

1 ACTIVITY: Using Exponent Notation

Work with a partner.

a. Copy and complete the table.

Power	Repeated Multiplication Form	Value
$(-3)^1$	-3	-3
$(-3)^2$	$(-3) \cdot (-3)$	9
$(-3)^3$		
$(-3)^4$		
$(-3)^5$		
$(-3)^6$		
$(-3)^7$		

b. **REPEATED REASONING** Describe what is meant by the expression $(-3)^n$. How can you find the value of $(-3)^n$?

2 ACTIVITY: Using Exponent Notation

Work with a partner.

a. The cube at the right has \$3 in each of its small cubes. Write a power that represents the total amount of money in the large cube.

b. Evaluate the power to find the total amount of money in the large cube.

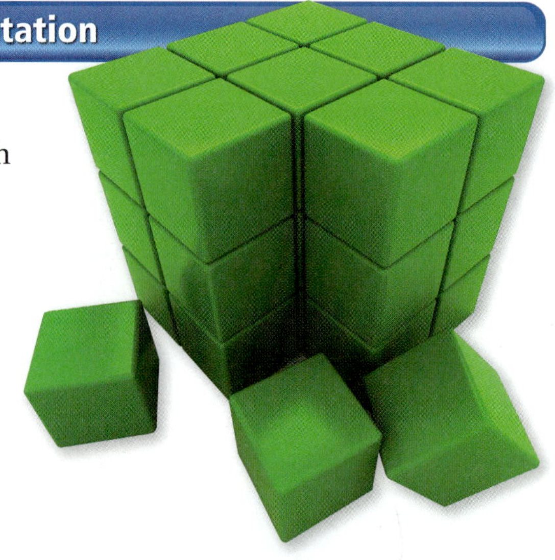

Exponents

In this lesson, you will

- write expressions using integer exponents.
- evaluate expressions involving integer exponents.

Learning Standard
MACC.8.EE.1.1

Laurie's Notes

Introduction

Standards for Mathematical Practice

- **MP8 Look for and Express Regularity in Repeated Reasoning:** In Activity 1, students will raise a negative number to whole number exponents and observe patterns that emerge.

Motivate

? "How big is a cubic millimeter?" A grain of salt is a reasonable estimate.
- Use a metric ruler and your fingers to show what a millimeter is. A cubic millimeter is 1 mm × 1 mm × 1 mm.

? "How big is a cubic meter?" about the size of a baby's play pen
- Use 3 meter sticks to demonstrate 1 m × 1 m × 1 m.

? "How many cubic millimeters are in a cubic meter?" Students may or may not have an answer.
- Give students time to think. Someone might ask how many millimeters there are in a meter. The prefix milli- means $\frac{1}{1000}$.
- The volume of a cubic meter in terms of cubic millimeters is $1000 \times 1000 \times 1000 = 1{,}000{,}000{,}000 = 1$ billion mm^3.

? "Can 1 billion be expressed using exponents?" 1000^3 or 10^9

Activity Notes

Activity 1

- **MP6 Attend to Precision:** Review the vocabulary associated with exponents.
- Have students work with their partners to complete the table. Students should recognize that a calculator is not necessary. They only need to multiply their previous product by −3.
- When students have finished, discuss the problem.

? "What do you notice about the values in the third column?" Values alternate—negative, positive, negative…
- **MP8:** Students might also mention: all values are odd and divisible by 3; last digits repeat in a cluster of 4: 3, 9, 7, 1, 3, 9, 7, …; there are two 1-digit numbers, two 2-digit numbers, two 3-digit numbers, and predict two 4-digit numbers
- **Part (b):** Listen for students to describe the exponent n in $(-3)^n$ as how many times −3 is multiplied by itself. Try to have students say that the exponent tells the number of times the base is used as a factor.

? "How can you find the value of $(-3)^n$?" Multiply −3 by itself n times.

Activity 2

- Although not all of the cubes are visible, students generally know that the cube contains 3^3 or 27 smaller cubes. So, at \$3 per small cube, $3 \times 3^3 = 3^4$.
- The expression is 3^4 and the answer is \$81.

Florida Common Core Standards

MACC.8.EE.1.1 Know and apply the properties of integer exponents to generate equivalent numerical expressions.

Previous Learning

Students should know how to raise a number to an exponent.

Lesson Plans
Complete Materials List

10.1 Record and Practice Journal

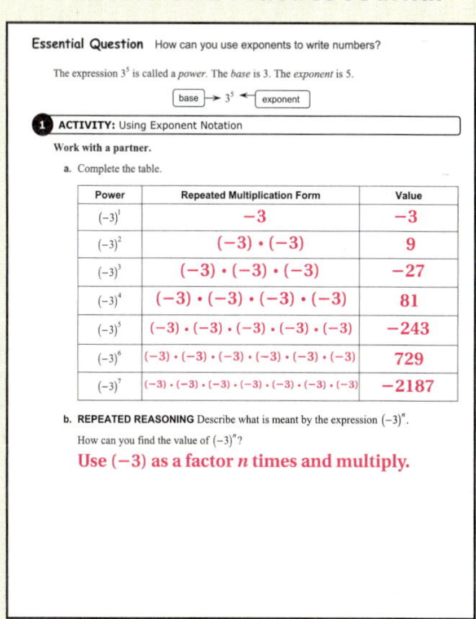

Essential Question How can you use exponents to write numbers?

The expression 3^5 is called a *power*. The *base* is 3. The *exponent* is 5.

base → 3^5 ← exponent

1 ACTIVITY: Using Exponent Notation

Work with a partner.

a. Complete the table.

Power	Repeated Multiplication Form	Value
$(-3)^1$	-3	-3
$(-3)^2$	$(-3) \cdot (-3)$	9
$(-3)^3$	$(-3) \cdot (-3) \cdot (-3)$	−27
$(-3)^4$	$(-3) \cdot (-3) \cdot (-3) \cdot (-3)$	81
$(-3)^5$	$(-3) \cdot (-3) \cdot (-3) \cdot (-3) \cdot (-3)$	−243
$(-3)^6$	$(-3) \cdot (-3) \cdot (-3) \cdot (-3) \cdot (-3) \cdot (-3)$	729
$(-3)^7$	$(-3) \cdot (-3) \cdot (-3) \cdot (-3) \cdot (-3) \cdot (-3) \cdot (-3)$	−2187

b. **REPEATED REASONING** Describe what is meant by the expression $(-3)^n$. How can you find the value of $(-3)^n$?

Use (−3) as a factor n times and multiply.

Differentiated Instruction

Visual

Have students create pyramids of factors −10, −5, −2, 2, 5, and 10. Ask them to write the exponential form and evaluate each row. Ask, “What is the product of 3 factors of −2?” and “What is the product of 4 factors of 5?” −8; 625

$$-2$$
$$(-2) \times (-2)$$
$$(-2) \times (-2) \times (-2)$$

$$5$$
$$5 \times 5$$
$$5 \times 5 \times 5$$
$$5 \times 5 \times 5 \times 5$$

10.1 Record and Practice Journal

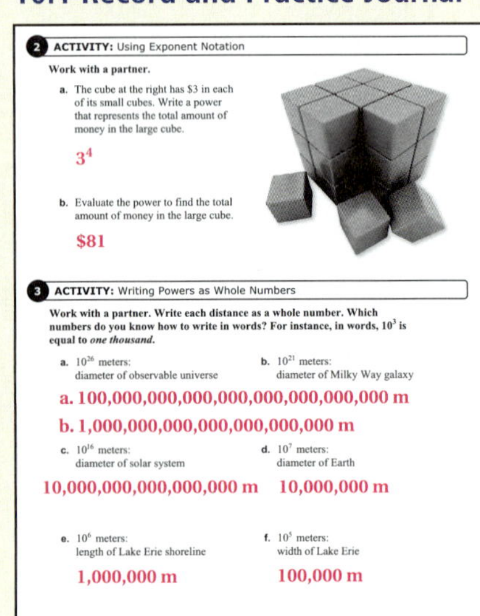

2 ACTIVITY: Using Exponent Notation

Work with a partner.

a. The cube at the right has \$3 in each of its small cubes. Write a power that represents the total amount of money in the large cube.

3^4

b. Evaluate the power to find the total amount of money in the large cube.

\$81

3 ACTIVITY: Writing Powers as Whole Numbers

Work with a partner. Write each distance as a whole number. Which numbers do you know how to write in words? For instance, in words, 10^3 is equal to *one thousand*.

a. 10^{26} meters: diameter of observable universe

b. 10^{21} meters: diameter of Milky Way galaxy

a. 100,000,000,000,000,000,000,000,000 m

b. 1,000,000,000,000,000,000,000 m

c. 10^{16} meters: diameter of solar system

d. 10^7 meters: diameter of Earth

10,000,000,000,000,000 m   10,000,000 m

e. 10^6 meters: length of Lake Erie shoreline

f. 10^5 meters: width of Lake Erie

1,000,000 m   100,000 m

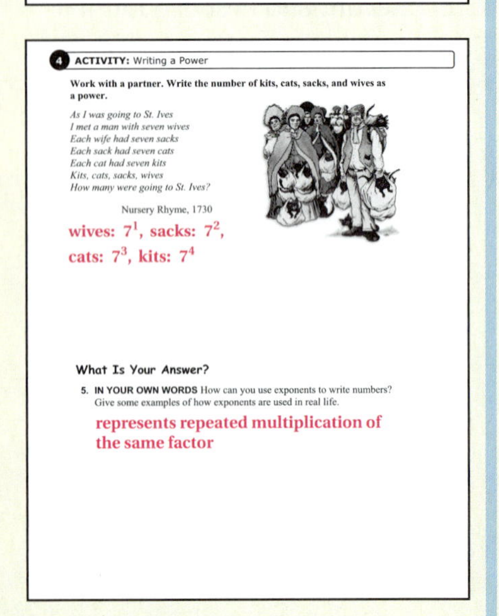

4 ACTIVITY: Writing a Power

Work with a partner. Write the number of kits, cats, sacks, and wives as a power.

As I was going to St. Ives
I met a man with seven wives
Each wife had seven sacks
Each sack had seven cats
Each cat had seven kits
Kits, cats, sacks, wives
How many were going to St. Ives?

Nursery Rhyme, 1730

wives: 7^1, sacks: 7^2, cats: 7^3, kits: 7^4

What Is Your Answer?

5. **IN YOUR OWN WORDS** How can you use exponents to write numbers? Give some examples of how exponents are used in real life.

represents repeated multiplication of the same factor

Laurie’s Notes

Activity 3

- The distances in this activity represent magnitudes, powers of 10, not exact distances.
- **FYI:** There is a classic video made by two designers in the late 1970s called *Powers of Ten,* which is easily available on the Internet.

? “Which numbers do you know the names for in parts (a)–(f)?” part (d): ten million; part (e): one million; part (f): one hundred thousand

- Share vocabulary that might be of interest:

million = 10^6	billion = 10^9	trillion = 10^{12}
quadrillion = 10^{15}	quintillion = 10^{18}	hexillion = 10^{21}
heptillion = 10^{24}	octillion = 10^{27}	nonillion = 10^{30}
decillion = 10^{33}	unodecillion = 10^{36}	duodecillion = 10^{39}

- Some of the above prefixes are the same ones used in naming polygons, so they should look familiar.

Activity 4

- This is a classic rhyme. In the original rhyme, the answer is one because the man and his wives were coming *from* St. Ives.
- For this problem, suggest to students that they draw a picture or diagram to help them solve the problem.
- **Summary:** wives = 7^1, sacks = 7^2, cats = 7^3, kits = 7^4

What Is Your Answer?

- **Neighbor Check:** Have students work independently and then have their neighbors check their work. Have students discuss any discrepancies.

Closure

- Compare the following powers using >, <, or =.

1. 2^{10} _____ 10^2 >   **2.** 10^3 _____ 3^{10} <

3 ACTIVITY: Writing Powers as Whole Numbers

Work with a partner. Write each distance as a whole number. Which numbers do you know how to write in words? For instance, in words, 10^3 is equal to *one thousand*.

a. 10^{26} meters: diameter of observable universe

b. 10^{21} meters: diameter of Milky Way galaxy

c. 10^{16} meters: diameter of solar system

d. 10^{7} meters: diameter of Earth

e. 10^{6} meters: length of Lake Erie shoreline

f. 10^{5} meters: width of Lake Erie

4 ACTIVITY: Writing a Power

Math Practice 1
Analyze Givens
What information is given in the poem? What are you trying to find?

Work with a partner. Write the number of kits, cats, sacks, and wives as a power.

As I was going to St. Ives
I met a man with seven wives
Each wife had seven sacks
Each sack had seven cats
Each cat had seven kits
Kits, cats, sacks, wives
How many were going to St. Ives?

Nursery Rhyme, 1730

What Is Your Answer?

5. **IN YOUR OWN WORDS** How can you use exponents to write numbers? Give some examples of how exponents are used in real life.

Use what you learned about exponents to complete Exercises 3–5 on page 414.

10.1 Lesson

Key Vocabulary
power, *p. 412*
base, *p. 412*
exponent, *p. 412*

A **power** is a product of repeated factors. The **base** of a power is the common factor. The **exponent** of a power indicates the number of times the base is used as a factor.

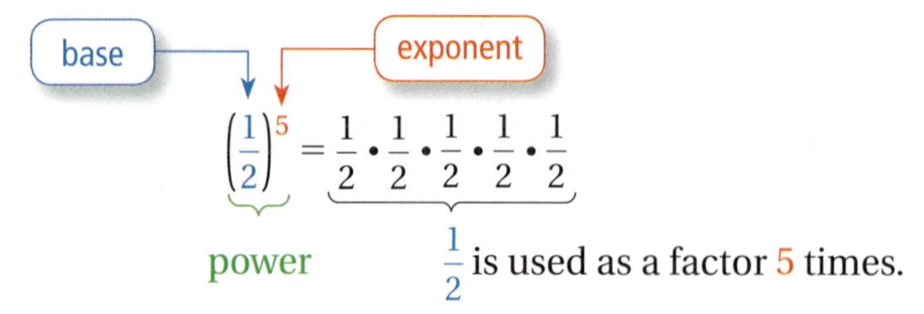

$$\left(\frac{1}{2}\right)^5 = \frac{1}{2} \cdot \frac{1}{2} \cdot \frac{1}{2} \cdot \frac{1}{2} \cdot \frac{1}{2}$$

$\frac{1}{2}$ is used as a factor 5 times.

EXAMPLE 1 Writing Expressions Using Exponents

Study Tip
Use parentheses to write powers with negative bases.

Write each product using exponents.

a. $(-7) \cdot (-7) \cdot (-7)$

Because -7 is used as a factor 3 times, its exponent is 3.

So, $(-7) \cdot (-7) \cdot (-7) = (-7)^3$.

b. $\pi \cdot \pi \cdot r \cdot r \cdot r$

Because π is used as a factor 2 times, its exponent is 2. Because r is used as a factor 3 times, its exponent is 3.

So, $\pi \cdot \pi \cdot r \cdot r \cdot r = \pi^2 r^3$.

On Your Own

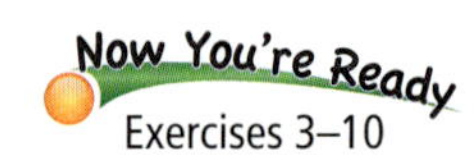

Write the product using exponents.

1. $\frac{1}{4} \cdot \frac{1}{4} \cdot \frac{1}{4} \cdot \frac{1}{4} \cdot \frac{1}{4}$

2. $0.3 \cdot 0.3 \cdot 0.3 \cdot 0.3 \cdot x \cdot x$

EXAMPLE 2 Evaluating Expressions

Evaluate each expression.

a. $(-2)^4$

$(-2)^4 = (-2) \cdot (-2) \cdot (-2) \cdot (-2)$ — Write as repeated multiplication.

$= 16$ — Simplify.

The base is -2.

b. -2^4

$-2^4 = -(2 \cdot 2 \cdot 2 \cdot 2)$ — Write as repeated multiplication.

$= -16$ — Simplify.

The base is 2.

Laurie's Notes

Introduction

Connect

- **Yesterday:** Students explored writing numbers with exponents. (MP6, MP8)
- **Today:** Students will write expressions involving exponents and evaluate powers.

Motivate

- Because U.S. currency has coins and bills which are powers of 10, find out what your students know about the people on the coins and bills. You can create a matching activity or have a few questions and answers ready for the class.

Lesson Notes

Example 1

- Write the definitions of power, base, and exponent. Note the use of *factor* instead of *multiplying the base by itself.*
- Write the example shown. When this power is evaluated, the answer is $\frac{1}{32}$. You say, "$\frac{1}{2}$ to the fifth power is $\frac{1}{32}$."
- Exponents are used to rewrite an expression involving repeated factors.

? "Is it necessary to write the multiplication dot between the factors in part (a)?" no

- **Common Error:** Parentheses *must* be used when you write a power with a negative base. This is a common error that students will make. For example: $(-2)^2 = (-2)(-2) = 4$
  $-2^2 = -(2)(2) = -4$
  Without the parentheses, the number being squared is 2, and then you multiply the product by -1. With the parentheses, the number being squared is -2, and the product is 4. The underlying property is the order of operations. Exponents are performed before multiplication.
- **Part (b):** Variables and constants are expressed using exponents in a similar fashion.

On Your Own

- **MP6 Attend to Precision:** In Question 1, $\frac{1}{4}$ needs to be written with parentheses so that both the numerator and the denominator are raised to an exponent. Without parentheses, it could be read as $\frac{1^5}{4}$.

Example 2

? "What is the base in each problem? What will be used as the factor in each problem?" Bases are -2 and 2. Factors are -2 and 2.

- **MP6:** This example addresses the need to write the base within parentheses when the base is negative.

Goal Today's lesson is writing and evaluating **powers**.

Technology for the Teacher

Lesson Tutorials
Lesson Plans
Answer Presentation Tool

Extra Example 1

Write each product using exponents.

a. $4 \cdot 4 \cdot 4 \cdot 4$ 4^4

b. $(-2) \cdot (-2) \cdot x \cdot x \cdot x$ $(-2)^2x^3$

On Your Own

1. $\left(\frac{1}{4}\right)^5$

2. $(0.3)^4x^2$

Extra Example 2

Evaluate each expression.

a. $(-3)^3$ -27

b. -3^3 -27

Laurie's Notes

Example 3

- You may wish to review the order of operations or wait to see how students evaluate the expressions.
- **Common Error:** Students may evaluate the problem left to right, performing the addition first. They need to be reminded of the order of operations.

? "How do you start to evaluate this expression?" powers first

- Continue to evaluate the problem as shown.
- In part (b), there are two powers to evaluate. After that is done, division is performed before subtraction.

On Your Own

- Encourage students to write out the steps in their solutions. Discourage them from performing multiple steps in their heads.

Example 4

- **FYI:** Ask if any of your students have heard of sphering. You can find information about this sport online at *zorb.com*.
- You want to find the volume of the inflated space.
- Write the formula for the volume V of a sphere with radius r: $V = \frac{4}{3}\pi r^3$.

  From the photo, the diameter of each sphere is known. The radius is half the diameter. Substitute the radius of each sphere and simplify.
- Let students use the context of the problem to decide how to round their answers.
- **Extension:** The inner sphere has a volume of a little more than 4 cubic meters. Find a region of your classroom that is approximately 4 cubic meters.

On Your Own

- Before students do any calculations, ask them whether the answer will be less than or greater than the answer to Example 4. greater than

Closure

- **Exit Ticket:** Evaluate.

  1. $4^2 - 8(2) + 3^3$ 27 **2.** $\left(-\frac{2}{3}\right)^3 + \left|5^2 - 2 \cdot 15\right|$ $\frac{127}{27} \approx 4.7$

Extra Example 3

Evaluate each expression.

a. $2 - 4 \cdot 5^2$ -98

b. $5 + 6^2 \div 4$ 14

On Your Own

3. -625 **4.** $-\frac{1}{216}$

5. 1 **6.** -7

Extra Example 4

In Example 4, the diameter of the inner sphere is 2.2 meters. What is the volume of the inflated space? about 8.56 m^3

On Your Own

7. about 11.08 m^3

English Language Learners

Labels

Have students label and practice reading statements of powers.

base
exponent

$2^3 = 2 \times 2 \times 2 = 8$

power
factor

"Two to the third equals two times two times two."

"The base is 2, the exponent is 3, and 2 is written as a factor 3 times."

EXAMPLE 3 Using Order of Operations

Evaluate each expression.

a. $3 + 2 \cdot 3^4 = 3 + 2 \cdot 81$ Evaluate the power.

$= 3 + 162$ Multiply.

$= 165$ Add.

b. $3^3 - 8^2 \div 2 = 27 - 64 \div 2$ Evaluate the powers.

$= 27 - 32$ Divide.

$= -5$ Subtract.

On Your Own

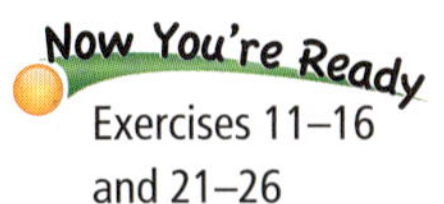

Exercises 11–16 and 21–26

Evaluate the expression.

3. -5^4 **4.** $\left(-\frac{1}{6}\right)^3$ **5.** $\left|-3^3 \div 27\right|$ **6.** $9 - 2^5 \cdot 0.5$

EXAMPLE 4 Real-Life Application

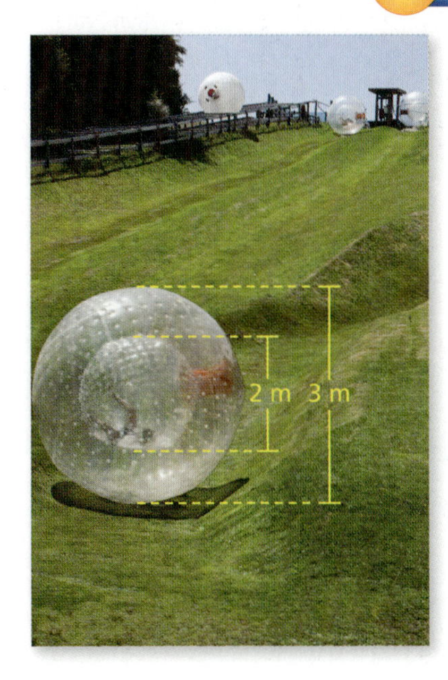

In sphering, a person is secured inside a small, hollow sphere that is surrounded by a larger sphere. The space between the spheres is inflated with air. What is the volume of the inflated space?

You can find the radius of each sphere by dividing each diameter given in the diagram by 2.

Outer Sphere		*Inner Sphere*
$V = \frac{4}{3}\pi r^3$	Write formula.	$V = \frac{4}{3}\pi r^3$
$= \frac{4}{3}\pi\left(\frac{3}{2}\right)^3$	Substitute.	$= \frac{4}{3}\pi(1)^3$
$= \frac{4}{3}\pi\left(\frac{27}{8}\right)$	Evaluate the power.	$= \frac{4}{3}\pi(1)$
$= \frac{9}{2}\pi$	Multiply.	$= \frac{4}{3}\pi$

So, the volume of the inflated space is $\frac{9}{2}\pi - \frac{4}{3}\pi = \frac{19}{6}\pi$, or about 10 cubic meters.

On Your Own

7. **WHAT IF?** The diameter of the inner sphere is 1.8 meters. What is the volume of the inflated space?

10.1 Exercises

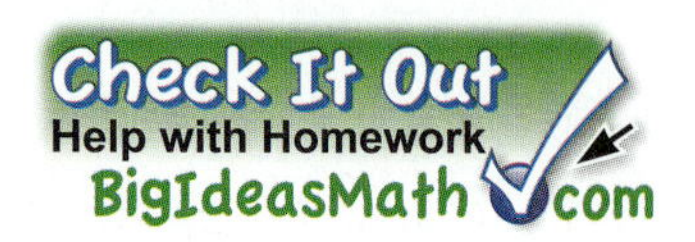

Vocabulary and Concept Check

1. **NUMBER SENSE** Describe the difference between -3^4 and $(-3)^4$.

2. **WHICH ONE DOESN'T BELONG?** Which one does *not* belong with the other three? Explain your reasoning.

5^3 The exponent is 3.	5^3 The power is 5.	5^3 The base is 5.	5^3 Five is used as a factor 3 times.

Practice and Problem Solving

Write the product using exponents.

3. $3 \cdot 3 \cdot 3 \cdot 3$

4. $(-6) \cdot (-6)$

5. $\left(-\frac{1}{2}\right) \cdot \left(-\frac{1}{2}\right) \cdot \left(-\frac{1}{2}\right)$

6. $\frac{1}{3} \cdot \frac{1}{3} \cdot \frac{1}{3}$

7. $\pi \cdot \pi \cdot \pi \cdot x \cdot x \cdot x \cdot x$

8. $(-4) \cdot (-4) \cdot (-4) \cdot y \cdot y$

9. $6.4 \cdot 6.4 \cdot 6.4 \cdot 6.4 \cdot b \cdot b \cdot b$

10. $(-t) \cdot (-t) \cdot (-t) \cdot (-t) \cdot (-t)$

Evaluate the expression.

2 11. 5^2

12. -11^3

13. $(-1)^6$

14. $\left(\frac{1}{2}\right)^6$

15. $\left(-\frac{1}{12}\right)^2$

16. $-\left(\frac{1}{9}\right)^3$

17. **ERROR ANALYSIS** Describe and correct the error in evaluating the expression.

✗ $-6^2 = (-6) \cdot (-6) = 36$

18. **PRIME FACTORIZATION** Write the prime factorization of 675 using exponents.

19. **STRUCTURE** Write $-\left(\frac{1}{4} \cdot \frac{1}{4} \cdot \frac{1}{4} \cdot \frac{1}{4}\right)$ using exponents.

20. **RUSSIAN DOLLS** The largest doll is 12 inches tall. The height of each of the other dolls is $\frac{7}{10}$ the height of the next larger doll. Write an expression involving a power for the height of the smallest doll. What is the height of the smallest doll?

Assignment Guide and Homework Check

Level	Day 1 Activity Assignment	Day 2 Lesson Assignment	Homework Check
Basic	3–5, 30–33	1, 2, 7–13, 17–27 odd	7, 8, 12, 21
Average	3–5, 30–33	1, 2, 8–16 even, 17–27 odd, 28	8, 12, 19, 21, 28
Advanced	1–5, 6–16 even, 17, 18–26 even, 27–33		8, 16, 18, 26, 28

Common Errors

- **Exercises 3–10** Students may count the wrong number of factors. Remind them to check their work.
- **Exercises 11–16** Students may have the wrong sign in their answers. Remind them that when the negative sign is inside the parentheses, it is part of the base. When the negative sign is outside the parentheses, it is not part of the base.
- **Exercises 21–26** Students may not remember the definition of absolute value or the correct order of operations. Review these topics with students.

10.1 Record and Practice Journal

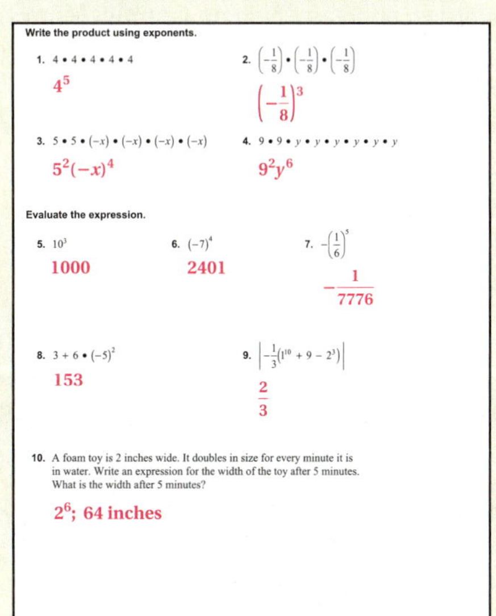
Write the product using exponents.

1. $4 \cdot 4 \cdot 4 \cdot 4 \cdot 4$
   4^5
2. $\left(-\frac{1}{8}\right) \cdot \left(-\frac{1}{8}\right) \cdot \left(-\frac{1}{8}\right)$
   $\left(-\frac{1}{8}\right)^3$
3. $5 \cdot 5 \cdot (-x) \cdot (-x) \cdot (-x) \cdot (-x)$
   $5^2(-x)^4$
4. $9 \cdot 9 \cdot y \cdot y \cdot y \cdot y \cdot y \cdot y$
   9^2y^6

Evaluate the expression.

5. 10^3
   1000
6. $(-7)^4$
   2401
7. $-\left(\frac{1}{6}\right)^5$
   $-\frac{1}{7776}$
8. $3 + 6 \cdot (-5)^2$
   153
9. $\left|-\frac{1}{3}(1^{10} + 9 - 2^3)\right|$
   $\frac{2}{3}$
10. A foam toy is 2 inches wide. It doubles in size for every minute it is in water. Write an expression for the width of the toy after 5 minutes. What is the width after 5 minutes?
   2^6; 64 inches

Vocabulary and Concept Check

1. -3^4 is the negative of 3^4, so the base is 3, the exponent is 4, and its value is -81. $(-3)^4$ has a base of -3, an exponent of 4, and a value of 81.
2. 5^3, The power is 5; The power is 5^3. Five is the base.

Practice and Problem Solving

3. 3^4
4. $(-6)^2$
5. $\left(-\frac{1}{2}\right)^3$
6. $\left(\frac{1}{3}\right)^3$
7. π^3x^4
8. $(-4)^3y^2$
9. $(6.4)^4b^3$
10. $(-t)^5$
11. 25
12. -1331
13. 1
14. $\frac{1}{64}$
15. $\frac{1}{144}$
16. $-\frac{1}{729}$
17. The negative sign is not part of the base; $-6^2 = -(6 \cdot 6) = -36$.
18. $3^3 \cdot 5^2$
19. $-\left(\frac{1}{4}\right)^4$
20. $12 \cdot \left(\frac{7}{10}\right)^3$; 4.116 in.

Practice and Problem Solving

21. 29

22. 65

23. 5

24. 5

25. 66

26. 2

27. See Additional Answers.

28. **a.** about 99.95 g

b. 99.95%

29. See *Taking Math Deeper.*

Fair Game Review

30. Commutative Property of Multiplication

31. Associative Property of Multiplication

32. Identity Property of Multiplication

33. B

Mini-Assessment

Evaluate the expression.

1. 7^2 49

2. -3^4 −81

3. $(-3)^4$ 81

4. $4 + 6 \cdot (-2)^3$ −44

5. $\frac{3}{4}\left(2^5 - 6 \div \left(\frac{1}{2}\right)^2\right)$ 6

Taking Math Deeper

Exercise 29

This exercise is based on the 12-note chromatic scale used in western music. Asian music uses a 9-note scale. Other cultures use 8-note and 10-note scales.

 Use a calculator or spreadsheet to calculate the frequency of each note.

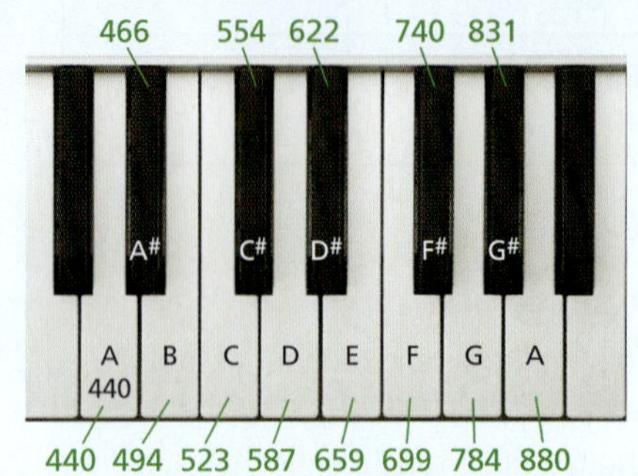

In the formula, the number 1.0595 is the 12th root of 2. A more accurate representation of this number is 1.0594630944.

 **a.** It takes 12 notes to travel from A-440 to A-880. That is why the western scale is called a 12-note scale. The term *octave,* which is based on the number 8, refers only to the white keys on the piano.

 **b.** The frequency of the A above A-440 is about 880 vibrations per second. Piano tuners use A-440 as the basic key to tune all the other notes. The A below A-440 has a frequency of 220 vibrations per second.

c. Because $(1.0594630944)^{12} \approx 2$, it follows that the A above A-440 has twice the frequency of A. For a 12-note increase, the frequency approximately doubles.

Note: The scale given by this formula is called the *equal temperament scale.* Notes whose frequencies have many common divisors "harmonize" more.

Project

Use the school library or the Internet to research instruments used in other countries, such as China and Japan. Many have different scales. Compare the scales of the instruments you find to that of the piano.

Reteaching and Enrichment Strategies

If students need help. . .	If students got it. . .
Resources by Chapter • Practice A and Practice B • Puzzle Time Record and Practice Journal Practice Differentiating the Lesson Lesson Tutorials Skills Review Handbook	Resources by Chapter • Enrichment and Extension • Technology Connection Start the next section

Evaluate the expression.

3 **21.** $5 + 3 \cdot 2^3$

22. $2 + 7 \cdot (-3)^2$

23. $(13^2 - 12^2) \div 5$

24. $\frac{1}{2}(4^3 - 6 \cdot 3^2)$

25. $\left| \frac{1}{2}(7 + 5^3) \right|$

26. $\left| \left(-\frac{1}{2}\right)^3 \div \left(\frac{1}{4}\right)^2 \right|$

27. MONEY You have a part-time job. One day your boss offers to pay you either $2^h - 1$ or 2^{h-1} dollars for each hour h you work that day. Copy and complete the table. Which option should you choose? Explain.

h	1	2	3	4	5
$2^h - 1$					
2^{h-1}					

28. CARBON-14 DATING Scientists use carbon-14 dating to determine the age of a sample of organic material.

a. The amount C (in grams) of a 100-gram sample of carbon-14 remaining after t years is represented by the equation $C = 100(0.99988)^t$. Use a calculator to find the amount of carbon-14 remaining after 4 years.

b. What percent of the carbon-14 remains after 4 years?

29. Critical Thinking The frequency (in vibrations per second) of a note on a piano is represented by the equation $F = 440(1.0595)^n$, where n is the number of notes above A-440. Each black or white key represents one note.

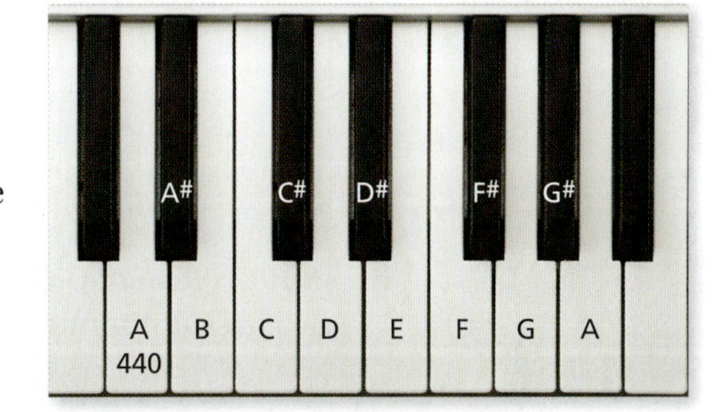

a. How many notes do you take to travel from A-440 to A?

b. What is the frequency of A?

c. Describe the relationship between the number of notes between A-440 and A and the increase in frequency.

Fair Game Review What you learned in previous grades & lessons

Tell which property is illustrated by the statement. *(Skills Review Handbook)*

30. $8 \cdot x = x \cdot 8$

31. $(2 \cdot 10)x = 2(10 \cdot x)$

32. $3(x \cdot 1) = 3x$

33. MULTIPLE CHOICE The polygons are similar. What is the value of x? *(Section 2.5)*

Ⓐ 15 Ⓑ 16

Ⓒ 17 Ⓓ 36

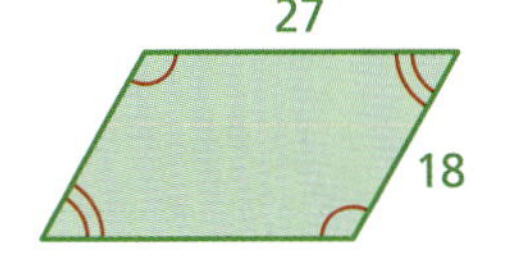

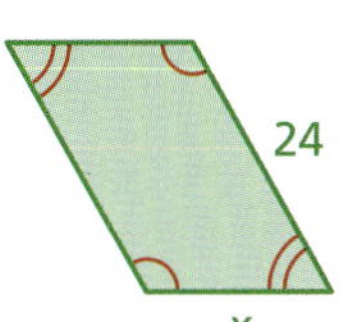

10.2 Product of Powers Property

Essential Question How can you use inductive reasoning to observe patterns and write general rules involving properties of exponents?

1 ACTIVITY: Finding Products of Powers

Work with a partner.

a. Copy and complete the table.

Product	Repeated Multiplication Form	Power
$2^2 \cdot 2^4$		
$(-3)^2 \cdot (-3)^4$		
$7^3 \cdot 7^2$		
$5.1^1 \cdot 5.1^6$		
$(-4)^2 \cdot (-4)^2$		
$10^3 \cdot 10^5$		
$\left(\frac{1}{2}\right)^5 \cdot \left(\frac{1}{2}\right)^5$		

b. **INDUCTIVE REASONING** Describe the pattern in the table. Then write a *general rule* for multiplying two powers that have the same base.

$$a^m \cdot a^n = a^{\square}$$

c. Use your rule to simplify the products in the first column of the table above. Does your rule give the results in the third column?

d. Most calculators have *exponent* keys that you can use to evaluate powers. Use a calculator with an exponent key to evaluate the products in part (a).

COMMON CORE

Exponents

In this lesson, you will

- multiply powers with the same base.
- find a power of a power.
- find a power of a product.

Learning Standard MACC.8.EE.1.1

2 ACTIVITY: Writing a Rule for Powers of Powers

Work with a partner. Write the expression as a single power. Then write a *general rule* for finding a power of a power.

a. $(3^2)^3 = (3 \cdot 3)(3 \cdot 3)(3 \cdot 3) = \square^{\square}$

b. $(2^2)^4 = $

c. $(7^3)^2 = $

d. $(y^3)^3 = $

e. $(x^4)^2 = $

Laurie's Notes

Introduction

Standards for Mathematical Practice

- **MP8 Look for and Express Regularity in Repeated Reasoning:** In the first three activities, students will find products of powers, powers of powers, and powers of products, and observe patterns that emerge.

Motivate

- **Story Time:** Tell students that the superintendent has agreed to put you on a special salary schedule for one month. On day 1 you will receive 1¢, on day 2 you will receive 2¢, day 3 is 4¢, and so on, with your salary doubling every school day for the month. There are 23 school days this month. Should you take the new salary?
- Give time for students to start the tabulation. Let them use a calculator for speed. The table below shows the daily pay.

1	$2 = 2^1$	$4 = 2^2$	$8 = 2^3$
$16 = 2^4$	$32 = 2^5$	$64 = 2^6$	$128 = 2^7$
$256 = 2^8$	$512 = 2^9$	$1024 = 2^{10}$	$2048 = 2^{11}$
$4096 = 2^{12}$	$8192 = 2^{13}$	$16{,}384 = 2^{14}$	$32{,}768 = 2^{15}$
$65{,}536 = 2^{16}$	$131{,}072 = 2^{17}$	$262{,}144 = 2^{18}$	$524{,}288 = 2^{19}$
$1{,}048{,}576 = 2^{20}$	$2{,}097{,}152 = 2^{21}$	$4{,}194{,}304 = 2^{22}$	

- If the superintendent is looking for additional math teachers, they will be lined up at the door.
- In this penny doubling problem, each day you are paid a power of 2. Your salary is actually the *sum* of all of these amounts.

Activity Notes

Activity 1

- Have students work with their partners to complete the table. It may help students if you point out that in the middle column, *repeated multiplication form* means the expanded form of each power in the product.

? "Why is the first column labeled **Product**?" Two powers are being multiplied.

? "What do you notice about the number of factors in the middle column and the exponent used to write the power?" same number

- **Part (b):** Students will recognize that the exponents are added together, but it may not be obvious to them how to write this fact using variables.
- **MP5 Use Appropriate Tools Strategically: Part (d):** Have students evaluate the products in the first column and the powers in the last column to confirm their answers and to confirm that they are using their calculators correctly.
- **Big Idea:** Write the summary statement: $a^m \cdot a^n = a^{m+n}$. Stress that the bases must be the same. That is why a is the base for both powers. This rule says nothing about how to simplify a product such as $3^3 \cdot 4^2$.

Florida Common Core Standards

MACC.8.EE.1.1 Know and apply the properties of integer exponents to generate equivalent numerical expressions.

Previous Learning

Students should know how to raise a number to an exponent.

Lesson Plans
Complete Materials List

10.2 Record and Practice Journal

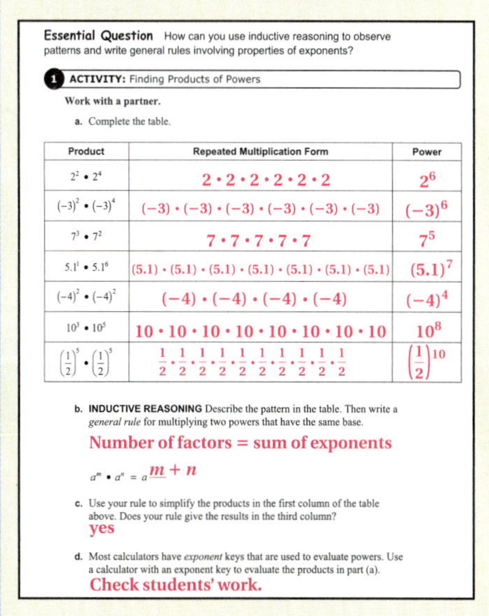

Essential Question How can you use inductive reasoning to observe patterns and write general rules involving properties of exponents?

1 ACTIVITY: Finding Products of Powers

Work with a partner.

a. Complete the table.

Product	Repeated Multiplication Form	Power
$2^2 \cdot 2^4$	$2 \cdot 2 \cdot 2 \cdot 2 \cdot 2 \cdot 2$	2^6
$(-3)^2 \cdot (-3)^4$	$(-3) \cdot (-3) \cdot (-3) \cdot (-3) \cdot (-3) \cdot (-3)$	$(-3)^6$
$7^3 \cdot 7^2$	$7 \cdot 7 \cdot 7 \cdot 7 \cdot 7$	7^5
$5.1^1 \cdot 5.1^6$	$(5.1) \cdot (5.1) \cdot (5.1) \cdot (5.1) \cdot (5.1) \cdot (5.1) \cdot (5.1)$	$(5.1)^7$
$(-4)^2 \cdot (-4)^2$	$(-4) \cdot (-4) \cdot (-4) \cdot (-4)$	$(-4)^4$
$10^3 \cdot 10^5$	$10 \cdot 10 \cdot 10 \cdot 10 \cdot 10 \cdot 10 \cdot 10 \cdot 10$	10^8
$\left(\frac{1}{2}\right)^5 \cdot \left(\frac{1}{2}\right)^5$	$\frac{1}{2} \cdot \frac{1}{2} \cdot \frac{1}{2} \cdot \frac{1}{2} \cdot \frac{1}{2} \cdot \frac{1}{2} \cdot \frac{1}{2} \cdot \frac{1}{2} \cdot \frac{1}{2} \cdot \frac{1}{2}$	$\left(\frac{1}{2}\right)^{10}$

b. **INDUCTIVE REASONING** Describe the pattern in the table. Then write a *general rule* for multiplying two powers that have the same base.

Number of factors = sum of exponents

$a^m \cdot a^n = a^{m+n}$

c. Use your rule to simplify the products in the first column of the table above. Does your rule give the results in the third column?

yes

d. Most calculators have *exponent* keys that are used to evaluate powers. Use a calculator with an exponent key to evaluate the products in part (a).

Check students' work.

English Language Learners

Pair Activity

Have students work in pairs to simplify exponential expressions. Each student simplifies a different expression. When both students are done, they take turns explaining the solution while the other person follows along.

10.2 Record and Practice Journal

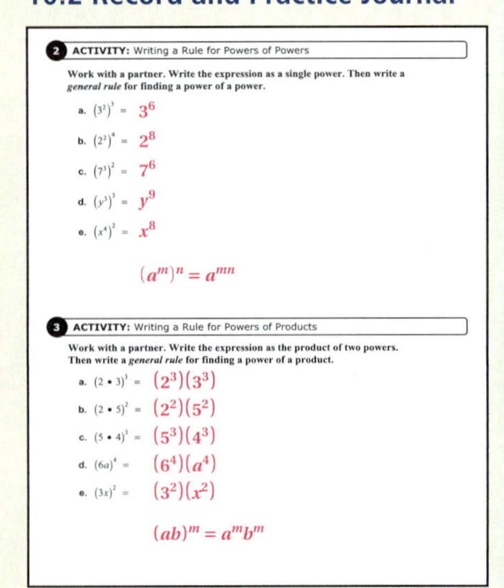

2 ACTIVITY: Writing a Rule for Powers of Powers

Work with a partner. Write the expression as a single power. Then write a *general rule* for finding a power of a power.

a. $(3^2)^3 = 3^6$

b. $(2^2)^4 = 2^8$

c. $(7^3)^2 = 7^6$

d. $(y^3)^3 = y^9$

e. $(x^4)^2 = x^8$

$(a^m)^n = a^{mn}$

3 ACTIVITY: Writing a Rule for Powers of Products

Work with a partner. Write the expression as the product of two powers. Then write a *general rule* for finding a power of a product.

a. $(2 \cdot 3)^3 = (2^3)(3^3)$

b. $(2 \cdot 5)^2 = (2^2)(5^2)$

c. $(5 \cdot 4)^3 = (5^3)(4^3)$

d. $(6a)^4 = (6^4)(a^4)$

e. $(3x)^2 = (3^2)(x^2)$

$(ab)^m = a^m b^m$

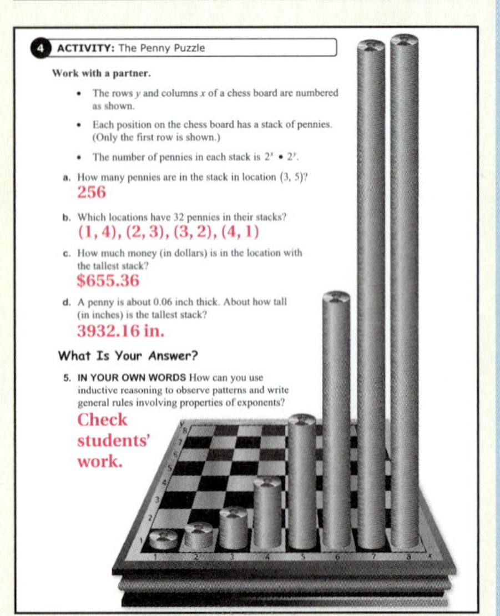

4 ACTIVITY: The Penny Puzzle

Work with a partner.

- The rows y and columns x of a chess board are numbered as shown.
- Each position on the chess board has a stack of pennies. (Only the first row is shown.)
- The number of pennies in each stack is $2^x \cdot 2^y$.

a. How many pennies are in the stack in location (3, 5)?
256

b. Which locations have 32 pennies in their stacks?
(1, 4), (2, 3), (3, 2), (4, 1)

c. How much money (in dollars) is in the location with the tallest stack?
$655.36

d. A penny is about 0.06 inch thick. About how tall (in inches) is the tallest stack?
3932.16 in.

What Is Your Answer?

5. **IN YOUR OWN WORDS** How can you use inductive reasoning to observe patterns and write general rules involving properties of exponents?
Check students' work.

Laurie's Notes

Activity 2

- Students often confuse this property with the Product of Powers Property (Activity 1). Make sure students can explain how $4^3 \cdot 4^2$ is different from $(4^3)^2$. Expanding the expressions helps to demonstrate this.
- When students have difficulty recognizing a pattern, have them expand the power in stages: $(2^2)^4 = 2^2 \cdot 2^2 \cdot 2^2 \cdot 2^2 = (2 \cdot 2)(2 \cdot 2)(2 \cdot 2)(2 \cdot 2) = 2^8$.
- **MP6 Attend to Precision:** Expect correct vocabulary, such as "When you raise a power to an exponent, multiply the exponents and keep the same base."

Activity 3

- "Products of powers" and "powers of products" sound very similar to students. Ask a volunteer to describe the difference.
- In an expression such as $(6a)^4$, students are often unsure how to simplify $6a \cdot 6a \cdot 6a \cdot 6a$. The Commutative and Associative Properties allow you to rewrite this as $6 \cdot 6 \cdot 6 \cdot 6 \cdot a \cdot a \cdot a \cdot a$.
- **MP6:** Expect correct vocabulary, such as "when you raise a product to an exponent, raise each base to the exponent and multiply the two powers."

Activity 4

- Take time to discuss the notation. In position (1, 1), the number of pennies is $2^1 \cdot 2^1 = 4$. In position (2, 1), the number of pennies is $2^2 \cdot 2^1 = 8$. Answer any questions about notation or how to find the number of pennies on any square.
- There are many patterns and interesting extensions to this problem that may surface as they explore the questions presented.
- **Part (a):** There are $2^3 \cdot 2^5 = 2^8 = 256$ pennies in location (3, 5).
- **Part (b):** Because $32 = 2^5$, the exponents need to sum to 5. The locations include (1, 4), (4, 1), (2, 3), and (3, 2).
- **Part (c):** The most money will be in the location where x and y have the greatest sum. This will occur at (8, 8), where the value is $2^8 \cdot 2^8 = 2^{16} = 65{,}536 = \$655.36$.
- **Part (d):** Multiply the number of pennies by the thickness, $65{,}536 \times 0.06 = 3932.16$ inches.

What Is Your Answer?

- **Neighbor Check:** Have students work independently and then have their neighbors check their work. Have students discuss any discrepancies.

Closure

- Refer back to the penny doubling problem from the beginning of the lesson. On what days was your salary more than $1000? days 18–23

3 ACTIVITY: Writing a Rule for Powers of Products

Work with a partner. Write the expression as the product of two powers. Then write a *general rule* for finding a power of a product.

a. $(2 \cdot 3)^3 = (2 \cdot 3)(2 \cdot 3)(2 \cdot 3) = \square^{\square} \cdot \square^{\square}$

b. $(2 \cdot 5)^2 = \square$

c. $(5 \cdot 4)^3 = \square$

d. $(6a)^4 = \square$

e. $(3x)^2 = \square$

4 ACTIVITY: The Penny Puzzle

Work with a partner.

- The rows y and columns x of a chessboard are numbered as shown.
- Each position on the chessboard has a stack of pennies. (Only the first row is shown.)
- The number of pennies in each stack is $2^x \cdot 2^y$.

Math Practice 7

Look for Patterns

What patterns do you notice? How does this help you determine which stack is the tallest?

a. How many pennies are in the stack in location (3, 5)?

b. Which locations have 32 pennies in their stacks?

c. How much money (in dollars) is in the location with the tallest stack?

d. A penny is about 0.06 inch thick. About how tall (in inches) is the tallest stack?

What Is Your Answer?

5. **IN YOUR OWN WORDS** How can you use inductive reasoning to observe patterns and write general rules involving properties of exponents?

Practice Use what you learned about properties of exponents to complete Exercises 3–5 on page 420.

10.2 Lesson

Key Ideas

Product of Powers Property

Words To multiply powers with the same base, add their exponents.

Numbers $4^2 \cdot 4^3 = 4^{2+3} = 4^5$ **Algebra** $a^m \cdot a^n = a^{m+n}$

Power of a Power Property

Words To find a power of a power, multiply the exponents.

Numbers $(4^6)^3 = 4^{6 \cdot 3} = 4^{18}$ **Algebra** $(a^m)^n = a^{mn}$

Power of a Product Property

Words To find a power of a product, find the power of each factor and multiply.

Numbers $(3 \cdot 2)^5 = 3^5 \cdot 2^5$ **Algebra** $(ab)^m = a^m b^m$

EXAMPLE 1 Multiplying Powers with the Same Base

a. $2^4 \cdot 2^5 = 2^{4+5}$ Product of Powers Property

$= 2^9$ Simplify.

b. $-5 \cdot (-5)^6 = (-5)^1 \cdot (-5)^6$ Rewrite -5 as $(-5)^1$.

$= (-5)^{1+6}$ Product of Powers Property

$= (-5)^7$ Simplify.

c. $x^3 \cdot x^7 = x^{3+7}$ Product of Powers Property

$= x^{10}$ Simplify.

Study Tip

When a number is written without an exponent, its exponent is 1.

EXAMPLE 2 Finding a Power of a Power

a. $(3^4)^3 = 3^{4 \cdot 3}$ Power of a Power Property

$= 3^{12}$ Simplify.

b. $(w^5)^4 = w^{5 \cdot 4}$ Power of a Power Property

$= w^{20}$ Simplify.

Laurie's Notes

Introduction

Connect

- **Yesterday:** Students explored exponents. (MP5, MP6, MP8)
- **Today:** Students will use the Product of Powers Property, the Power of a Power Property, and the Power of a Product Property to simplify expressions.

Motivate

- More money talk! The $10,000 bill, which is no longer in circulation, would be much easier to carry than the same amount in pennies.

? Ask a few questions about money.

  - "How many pennies equal $10,000?" $100 \times 10{,}000 = 1{,}000{,}000$ or 10^6
  - "How many dimes equal $10,000?" $10 \times 10{,}000 = 100{,}000$ or 10^5
  - "How many $10 bills equal $10,000?" $\frac{1}{10} \times 10{,}000 = 1000$ or 10^3
  - "How many $100 bills equal $10,000?" $\frac{1}{100} \times 10{,}000 = 100$ or 10^2

Lesson Notes

Key Ideas

- Write the Key Ideas. Explain the Words, Numbers, and Algebra.
- Explain that this and the following section mirror each other. However, this section contains more properties because the additional ones follow directly from the Product of Powers Property. Essentially, the properties in this section are the "multiplication properties" of exponents and the Quotient of Powers Property (in the next section) is the "division property" of exponents.

Example 1

- **Part (a):** Write and simplify the expression. The base is 2 for each power, so add the exponents.

? "In part (b), what is the base for each power? To what exponent is each base raised?" -5; 1 and 6

- **Common Error:** When the exponent is 1, it is not written. When it is not written, students will sometimes forget to add 1 in their answer.
- **Part (c):** The properties apply to variables as well as numbers.

Example 2

? "In the expression $(3^4)^3$, what does the exponent of 3 tell you to do?" Use 3^4 as a factor three times.

- Use the Power of a Power Property and multiply the exponents.

Goal Today's lesson is using the Product of Powers Property, the Power of a Power Property, and the Power of a Product Property to simplify expressions.

Lesson Tutorials
Lesson Plans
Answer Presentation Tool

Extra Example 1

Simplify each expression. Write your answer as a power.

a. $6^2 \cdot 6^7$ 6^9

b. $-2 \cdot (-2)^3$ $(-2)^4$

c. $x^2 \cdot x^5$ x^7

Extra Example 2

Simplify each expression. Write your answer as a power.

a. $(5^2)^3$ 5^6

b. $(y^4)^6$ y^{24}

Extra Example 3

Simplify each expression.

a. $(4x)^2$ $16x^2$

b. $(wz)^3$ w^3z^3

On Your Own

1. 6^6
2. $\left(-\frac{1}{2}\right)^9$
3. z^{13}
4. 4^{12}
5. y^8
6. $(-4)^6$
7. $625y^4$
8. a^5b^5
9. $0.25m^2n^2$

Extra Example 4

In Example 4, the total storage space of a computer is 32 gigabytes. How many bytes of total storage space does the computer have? 2^{35} bytes

On Your Own

10. 2^{34} bytes

Differentiated Instruction

Visual

Remind students that the Product of Powers Property can only be applied to powers having the same base. Have students highlight each unique base with a different color. Then add the exponents.

$$2^4 \cdot 2^5 - 3^2 \cdot 3^2 = 2^{4+5} - 3^{2+2}$$
$$= 2^9 - 3^4$$
$$= 512 - 81$$
$$= 431$$

Laurie's Notes

Example 3

- Be careful and deliberate with language when simplifying these expressions.

? "In part (a), what does the exponent of 3 tell you to do in the expression $(2x)^3$?" Use $2x$ as a factor three times.

- **MP7 Look for and Make Use of Structure** To verify the solution of part (a), you could write the factor $2x$ three times. Properties of Multiplication (Associative and Commutative) allow you to reorder the terms. You can identify six factors: three 2's and three x's. Use exponents to write the factors. Finally, 2^3 is rewritten as 8 and the final answer is $8x^3$.
- **Part (b):** Follow the same procedure to verify this solution, by writing $3xy$ as a factor twice. It is very common for students to write $x \cdot x = 2x$. Do not assume that students will see this error.

On Your Own

- Encourage students to write out the steps in their solutions.
- **Common Error:** In Question 9, $(0.5)^2 \neq 1$; $(0.5)^2 = 0.25$.
- Have volunteers write their solutions on the board.

Example 4

- Writing the verbal model is necessary in this problem because the terms gigabytes and bytes may not be familiar to all. The first sentence is a conversion fact: 1 GB $= 2^{30}$ bytes. There are 64 GB of total storage. Students may naturally think 64×2^{30} to solve the problem.
- Rewrite 64 as a power with a base of 2, $64 = 2^6$, and solve the problem.

On Your Own

- Students may respond with $\frac{1}{4}$ the total storage space. In fact, $\frac{1}{4}$ of $2^{36} = 2^{34}$. However, this is not an obvious step. Students should model the problem after Example 4.

Closure

- **Exit Ticket:** Simplify. $5^3 \cdot 5^4$ 5^7 $(-3x)^3$ $-27x^3$

EXAMPLE 3 Finding a Power of a Product

a. $(2x)^3 = 2^3 \cdot x^3$ — Power of a Product Property

$= 8x^3$ — Simplify.

b. $(3xy)^2 = 3^2 \cdot x^2 \cdot y^2$ — Power of a Product Property

$= 9x^2y^2$ — Simplify.

On Your Own

Now You're Ready
Exercises 3–14 and 17–22

Simplify the expression.

1. $6^2 \cdot 6^4$
2. $\left(-\frac{1}{2}\right)^3 \cdot \left(-\frac{1}{2}\right)^6$
3. $z \cdot z^{12}$
4. $(4^4)^3$
5. $(y^2)^4$
6. $((-4)^3)^2$
7. $(5y)^4$
8. $(ab)^5$
9. $(0.5mn)^2$

EXAMPLE 4 Simplifying an Expression

Details
Local Disk (C:)
Local Disk
Free Space: 16GB
Total Space: 64GB

A gigabyte (GB) of computer storage space is 2^{30} bytes. The details of a computer are shown. How many bytes of total storage space does the computer have?

(A) 2^{34} (B) 2^{36} (C) 2^{180} (D) 128^{30}

The computer has 64 gigabytes of total storage space. Notice that you can write 64 as a power, 2^6. Use a model to solve the problem.

Total number of bytes = Number of bytes in a gigabyte • Number of gigabytes

$= 2^{30} \cdot 2^6$ — Substitute.

$= 2^{30+6}$ — Product of Powers Property

$= 2^{36}$ — Simplify.

The computer has 2^{36} bytes of total storage space. The correct answer is (B).

On Your Own

10. How many bytes of free storage space does the computer have?

10.2 Exercises

Vocabulary and Concept Check

1. **REASONING** When should you use the Product of Powers Property?
2. **CRITICAL THINKING** Can you use the Product of Powers Property to multiply $5^2 \cdot 6^4$? Explain.

Practice and Problem Solving

Simplify the expression. Write your answer as a power.

① ② 3. $3^2 \cdot 3^2$ 4. $8^{10} \cdot 8^4$ 5. $(-4)^5 \cdot (-4)^7$

6. $a^3 \cdot a^3$ 7. $h^6 \cdot h$ 8. $\left(\frac{2}{3}\right)^2 \cdot \left(\frac{2}{3}\right)^6$

9. $\left(-\frac{5}{7}\right)^8 \cdot \left(-\frac{5}{7}\right)^9$ 10. $(-2.9) \cdot (-2.9)^7$ 11. $(5^4)^3$

12. $(b^{12})^3$ 13. $(3.8^3)^4$ 14. $\left(\left(-\frac{3}{4}\right)^5\right)^2$

ERROR ANALYSIS **Describe and correct the error in simplifying the expression.**

15.

$5^2 \cdot 5^9 = (5 \cdot 5)^{2+9}$
$= 25^{11}$

16.

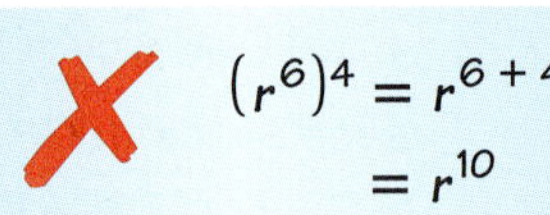

$(r^6)^4 = r^{6+4}$
$= r^{10}$

Simplify the expression.

③ 17. $(6g)^3$ 18. $(-3v)^5$ 19. $\left(\frac{1}{5}k\right)^2$

20. $(1.2m)^4$ 21. $(rt)^{12}$ 22. $\left(-\frac{3}{4}p\right)^3$

23. **PRECISION** Is $3^2 + 3^3$ equal to 3^5? Explain.

24. **ARTIFACT** A display case for the artifact is in the shape of a cube. Each side of the display case is three times longer than the width of the artifact.
    a. Write an expression for the volume of the case. Write your answer as a power.
    b. Simplify the expression.

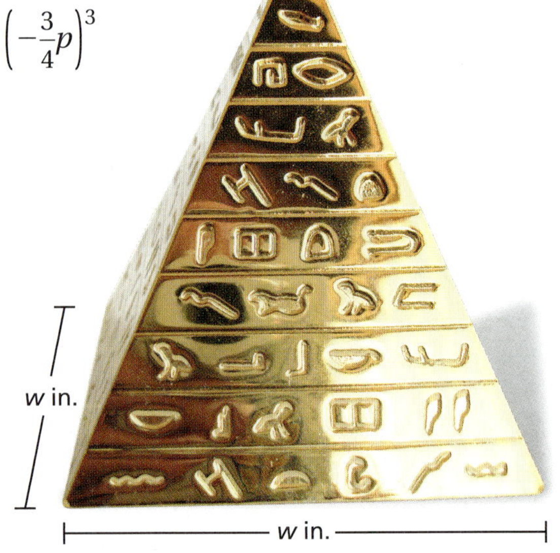

Assignment Guide and Homework Check

Level	Day 1 Activity Assignment	Day 2 Lesson Assignment	Homework Check
Basic	3–5, 33–37	1, 2, 6–13, 15–20, 23, 24	10, 12, 18, 23, 24
Average	3–5, 33–37	1, 2, 6–16 even, 17–23 odd, 24, 25, 27, 28, 29, 31	10, 12, 19, 25, 28
Advanced	1–5, 6–24 even, 25–37		14, 22, 26, 28, 30

Common Errors

- **Exercises 3–10** Students may multiply the bases. Remind them that the base stays the same and only the exponent changes when using the Product of Powers Property.
- **Exercises 3–14** Students may confuse the Product of Powers Property and the Power of a Power Property. Explain to them why it makes sense to add exponents when using the Product of Powers Property and multiply exponents when using the Power of a Power Property.
- **Exercises 17–22** Students may forget to find the power of the constant factor and write, for example, $(6g)^3 = 6g^3$. Remind them of the Power of a Product Property.

10.2 Record and Practice Journal

Simplify the expression. Write your answer as a power.

1. $(-6)^5 \cdot (-6)^4$ — $(-6)^9$
2. $x^1 \cdot x^9$ — x^{10}
3. $\left(\frac{4}{5}\right)^3 \cdot \left(\frac{4}{5}\right)^{12}$ — $\left(\frac{4}{5}\right)^{15}$
4. $(-1.5)^{11} \cdot (-1.5)^{11}$ — $(-1.5)^{22}$
5. $(y^{10})^{20}$ — y^{200}
6. $\left(\left(-\frac{2}{9}\right)^8\right)^7$ — $\left(-\frac{2}{9}\right)^{56}$

Simplify the expression.

7. $(2a)^6$ — $64a^6$
8. $(-4b)^4$ — $256b^4$
9. $\left(-\frac{9}{10}p\right)^2$ — $\frac{81}{100}p^2$
10. $(xy)^{15}$ — $x^{15}y^{15}$
11. $10^5 \cdot 10^3 - (10^1)^8$ — 0
12. $7^2(7^4 \cdot 7^4)$ — $282{,}475{,}249$
13. The surface area of the Sun is about $4 \times 3.141 \times (7 \times 10^5)^2$ square kilometers. Simplify the expression.
    6,156,360,000,000 square kilometers

Vocabulary and Concept Check

1. when multiplying powers with the same base
2. no; The bases are not the same.

Practice and Problem Solving

3. 3^4
4. 8^{14}
5. $(-4)^{12}$
6. a^6
7. h^7
8. $\left(\frac{2}{3}\right)^8$
9. $\left(-\frac{5}{7}\right)^{17}$
10. $(-2.9)^8$
11. 5^{12}
12. b^{36}
13. 3.8^{12}
14. $\left(-\frac{3}{4}\right)^{10}$
15. The bases should not be multiplied.
    $5^2 \cdot 5^9 = 5^{2+9}$
    $= 5^{11}$
16. The exponents should not be added. Write the expression as repeated multiplication.
    $(r^6)^4 = r^6 \cdot r^6 \cdot r^6 \cdot r^6$
    $= r^{6+6+6+6}$
    $= r^{24}$
17. $216g^3$
18. $-243v^5$
19. $\frac{1}{25}k^2$
20. $2.0736m^4$
21. $r^{12}t^{12}$
22. $-\frac{27}{64}p^3$
23. no; $3^2 + 3^3 = 9 + 27 = 36$ and $3^5 = 243$
24. a. $(3w)^3$
    b. $27w^3$

Practice and Problem Solving

25. 496

26. x^4

27. 78,125

28. 3^9 ft

29. **a.** $16\pi \approx 50.27$ in.^{3}

b. $192\pi \approx 603.19$ in.^{3} Squaring each of the dimensions causes the volume to be 12 times larger.

30. $V = \frac{3}{4}b^2h$

31. See *Taking Math Deeper.*

32. **a.** 3

b. 4

Fair Game Review

33. 4

34. 25

35. 3

36. 6

37. B

Mini-Assessment

Simplify the expression. Write your answer as a power.

1. $b^2 \cdot b^6$ b^8
2. $(-2)^3 \cdot (-2)^2$ $(-2)^5$
3. $(c^8)^3$ c^{24}

Simplify the expression.

4. $(-5w)^4$ $625w^4$
5. $(st)^{11}$ $s^{11}t^{11}$

Taking Math Deeper

Exercise 31

This exercise gives students some practice in representing large numbers as powers.

Summarize the given information.

Mail delivered each second: $2^8 \cdot 5^2 = 6400$
Seconds in 6 days: $2^8 \cdot 3^4 \cdot 5^2 = 518{,}400$
How many pieces of mail in 6 days?

Multiply to find the number of pieces of mail delivered in 6 days.

$$(2^8 \cdot 5^2)(2^8 \cdot 3^4 \cdot 5^2) = 2^8 \cdot 5^2 \cdot 2^8 \cdot 3^4 \cdot 5^2$$
$$= 2^{16} \cdot 3^4 \cdot 5^4$$

Write the number in normal decimal form.

If you expand this number, you find that the U.S. postal service delivers about 3 billion pieces of mail each week (6 days not counting Sunday). This is an average of 10 pieces of mail per week for each person in the United States!

Project

Research the price of a postage stamp. How many times has it changed? How often has it changed? What has been the range in the cost over the last one hundred years?

Reteaching and Enrichment Strategies

If students need help. . .	If students got it. . .
Resources by Chapter • Practice A and Practice B • Puzzle Time Record and Practice Journal Practice Differentiating the Lesson Lesson Tutorials Skills Review Handbook	Resources by Chapter • Enrichment and Extension • Technology Connection Start the next section

Simplify the expression.

25. $2^4 \cdot 2^5 - (2^2)^2$

26. $16\left(\frac{1}{2}x\right)^4$

27. $5^2(5^3 \cdot 5^2)$

28. CLOUDS The lowest altitude of an altocumulus cloud is about 3^8 feet. The highest altitude of an altocumulus cloud is about 3 times the lowest altitude. What is the highest altitude of an altocumulus cloud? Write your answer as a power.

29. PYTHON EGG The volume V of a python egg is given by the formula $V = \frac{4}{3}\pi abc$. For the python eggs shown, $a = 2$ inches, $b = 2$ inches, and $c = 3$ inches.

a. Find the volume of a python egg.

b. Square the dimensions of the python egg. Then evaluate the formula. How does this volume compare to your answer in part (a)?

30. PYRAMID A square pyramid has a height h and a base with side length b. The side lengths of the base increase by 50%. Write a formula for the volume of the new pyramid in terms of b and h.

31. MAIL The United States Postal Service delivers about $2^8 \cdot 5^2$ pieces of mail each second. There are $2^8 \cdot 3^4 \cdot 5^2$ seconds in 6 days. How many pieces of mail does the United States Postal Service deliver in 6 days? Write your answer as an expression involving powers.

32. Critical Thinking Find the value of x in the equation without evaluating the power.

a. $2^5 \cdot 2^x = 256$

b. $\left(\frac{1}{3}\right)^2 \cdot \left(\frac{1}{3}\right)^x = \frac{1}{729}$

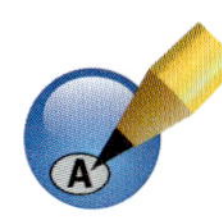

Fair Game Review What you learned in previous grades & lessons

Simplify. *(Skills Review Handbook)*

33. $\frac{4 \cdot 4}{4}$

34. $\frac{5 \cdot 5 \cdot 5}{5}$

35. $\frac{2 \cdot 3}{2}$

36. $\frac{8 \cdot 6 \cdot 6}{6 \cdot 8}$

37. MULTIPLE CHOICE What is the measure of each interior angle of the regular polygon? *(Section 3.3)*

Ⓐ 45° Ⓑ 135°

Ⓒ 1080° Ⓓ 1440°

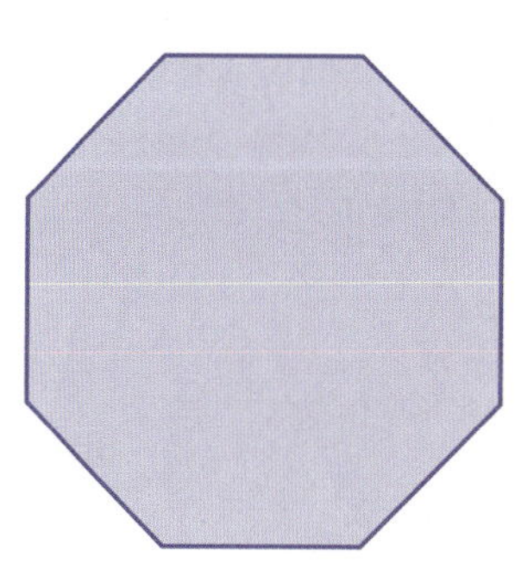

10.3 Quotient of Powers Property

Essential Question How can you divide two powers that have the same base?

1 ACTIVITY: Finding Quotients of Powers

Work with a partner.

a. Copy and complete the table.

Quotient	Repeated Multiplication Form	Power
$\frac{2^4}{2^2}$		
$\frac{(-4)^5}{(-4)^2}$		
$\frac{7^7}{7^3}$		
$\frac{8.5^9}{8.5^6}$		
$\frac{10^8}{10^5}$		
$\frac{3^{12}}{3^4}$		
$\frac{(-5)^7}{(-5)^5}$		
$\frac{11^4}{11^1}$		

Exponents

In this lesson, you will

- divide powers with the same base.
- simplify expressions involving the quotient of powers.

Learning Standard
MACC.8.EE.1.1

b. **INDUCTIVE REASONING** Describe the pattern in the table. Then write a rule for dividing two powers that have the same base.

$$\frac{a^m}{a^n} = a^{\square}$$

c. Use your rule to simplify the quotients in the first column of the table above. Does your rule give the results in the third column?

Laurie's Notes

Introduction

Standards for Mathematical Practice

- **MP8 Look for and Express Regularity in Repeated Reasoning:** In Activity 1, students will find quotients of powers and observe the pattern that emerges.
- Remember to use correct vocabulary in this lesson. The numbers are not *canceling*. The factors that are common in the numerator and the denominator are being divided out, similar to simplifying fractions. The fraction $\frac{2}{4} = \frac{1}{2}$ because there is a common factor of 2 in both the numerator and denominator that divide out. This same concept of dividing out common factors is why the Quotient of Powers Property works.

Motivate

- Tell students that you spent last evening working on a very long problem and you want them to give it a try. Write the problem on the board.

$$\frac{1}{2} \bullet \frac{2}{3} \bullet \frac{3}{4} \bullet \frac{4}{5} \bullet \frac{5}{6} \bullet \frac{6}{7} \bullet \frac{7}{8} \bullet \frac{8}{9} \bullet \frac{9}{10}$$

- It is likely that at least one of your students will recognize the answer immediately after you finish writing the problem. Act surprised and ask for their strategy...because you spent a long time on the problem.
- You want all students to recognize that the common factors in the numerator divide out with common factors in the denominator, leaving only $\frac{1}{10}$ as the final answer.

Activity Notes

Activity 1

- Have students work with their partners to complete the table. It may help students if you point out that in the middle column, *repeated multiplication form* means the expanded form of each power in the quotient.
- Notice that integers and decimals are used as bases.

? "Why is the first column labeled **Quotient**?" Two powers are being divided.

? "What do you notice about the number of factors in the numerator and denominator of the middle column, and the exponent used to write the power?" When you subtract the number of factors in the denominator from the number of factors in the numerator, it equals the exponent in the power.

- Students may need help in writing the summary statement: $\frac{a^m}{a^n} = a^{m-n}$. Stress that the bases must be the same in order to use this property.

Florida Common Core Standards

MACC.8.EE.1.1 Know and apply the properties of integer exponents to generate equivalent numerical expressions.

Previous Learning

Students should know how to simplify fractions by dividing out common factors.

Lesson Plans
Complete Materials List

10.3 Record and Practice Journal

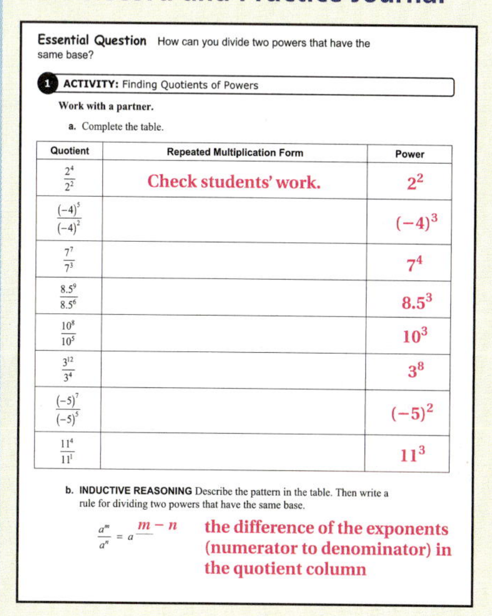

Essential Question How can you divide two powers that have the same base?

1 ACTIVITY: Finding Quotients of Powers

Work with a partner.

a. Complete the table.

Quotient	Repeated Multiplication Form	Power
$\frac{2^4}{2^2}$	Check students' work.	2^2
$\frac{(-4)^5}{(-4)^2}$		$(-4)^3$
$\frac{7^7}{7^3}$		7^4
$\frac{8.5^9}{8.5^6}$		8.5^3
$\frac{10^8}{10^5}$		10^3
$\frac{3^{12}}{3^4}$		3^8
$\frac{(-5)^7}{(-5)^5}$		$(-5)^2$
$\frac{11^4}{11^1}$		11^3

b. **INDUCTIVE REASONING** Describe the pattern in the table. Then write a rule for dividing two powers that have the same base.

$\frac{a^m}{a^n} = a^{m-n}$ the difference of the exponents (numerator to denominator) in the quotient column

Differentiated Instruction

Kinesthetic

Use algebra tiles or slips of paper to help students understand the Quotient of Powers Property. Have students model the quotient $\frac{x^4}{x^2}$.

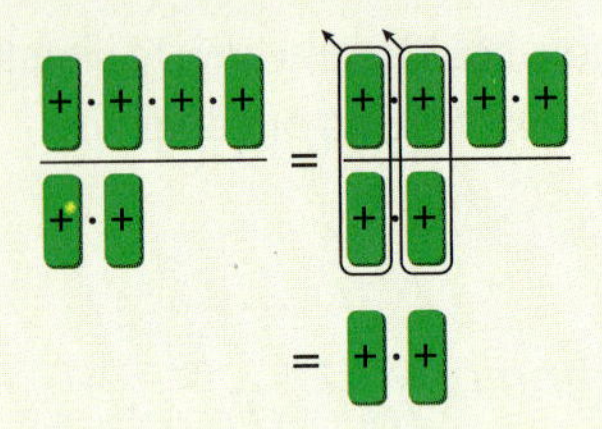

So, $\frac{x^4}{x^2} = x^2$.

10.3 Record and Practice Journal

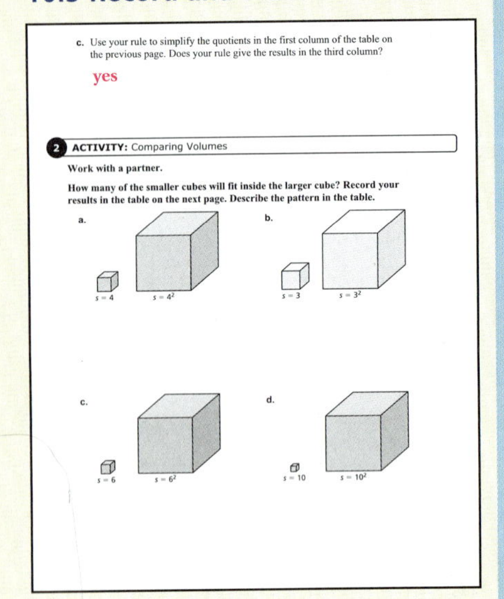

c. Use your rule to simplify the quotients in the first column of the table on the previous page. Does your rule give the results in the third column?

yes

2 ACTIVITY: Comparing Volumes

Work with a partner.

How many of the smaller cubes will fit inside the larger cube? Record your results in the table on the next page. Describe the pattern in the table.

a. $s = 4$, $s = 4^2$

b. $s = 3$, $s = 3^2$

c. $s = 6$, $s = 6^2$

d. $s = 10$, $s = 10^2$

	Volume of Smaller Cube	Volume of Larger Cube	$\frac{\text{Larger Volume}}{\text{Smaller Volume}}$	Answer
a.	4^3	$(4^2)^3 = 4^6$	$\frac{4^6}{4^3}$	4^3
b.	3^3	$(3^2)^3 = 3^6$	$\frac{3^6}{3^3}$	3^3
c.	6^3	$(6^2)^3 = 6^6$	$\frac{6^6}{6^3}$	6^3
d.	10^3	$(10^2)^3 = 10^6$	$\frac{10^6}{10^3}$	10^3

What Is Your Answer?

3. IN YOUR OWN WORDS How can you divide two powers that have the same base? Give two examples of your rule.

Subtract their exponents.

Laurie's Notes

Activity 2

- **MP4 Model with Mathematics:** If you have small wooden or plastic cubes available, model one of these problems or a similar problem to start.
- Point out to students that $s = 4$ means the edge (or side) length is 4. In part (a), the side length of the larger cube is 4^2 or 16, which is 4 times as long as the smaller cube.
- When completing the table, it is necessary for students to simplify a power raised to an exponent. Recall, $(4^2)^3 = 4^{2 \cdot 3} = 4^6$.
- Students will work with their partners to complete the table.

? "How do you find the volume of the smaller cube?" Cube the side length; s^3.

? "How do you find the volume of the larger cube?" Cube the side length; s^3; The side length for the larger cube, however, is expressed as a power.

- When finding the ratio of the volumes, students will need to divide out the common factors.

? "What do you notice about the volume of the smaller cube and the answer?" The answer is always the same as the volume of the smaller cube.

- **Extension:** Discuss how the relationship between volumes of similar solids is connected to this activity.

What Is Your Answer?

- **Think-Pair-Share:** Students should read the question independently and then work in pairs to answer the question. When they have answered the questions, the pair should compare their answers with another group and discuss any discrepancies.

Closure

- Simplify.

1. $\frac{2^2}{2} \cdot \frac{2^3}{2^2} \cdot \frac{2^4}{2^3}$ 2^3

2. $\frac{(-3)^7}{(-3)^4}$ $(-3)^3$

2 ACTIVITY: Comparing Volumes

Math Practice 8

Repeat Calculations

What calculations are repeated in the table?

Work with a partner.

How many of the smaller cubes will fit inside the larger cube? Record your results in the table. Describe the pattern in the table.

a.

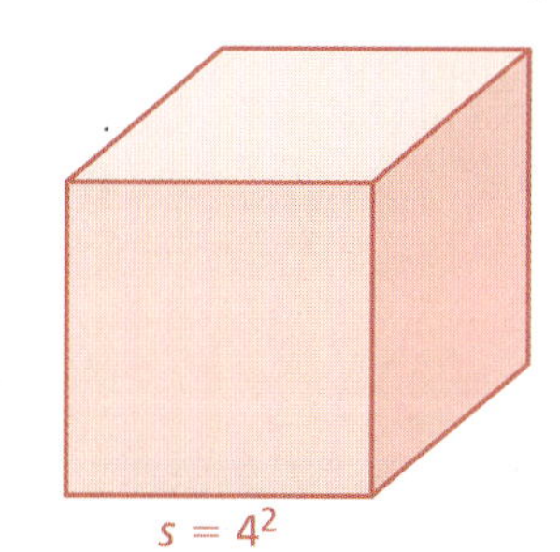

b.

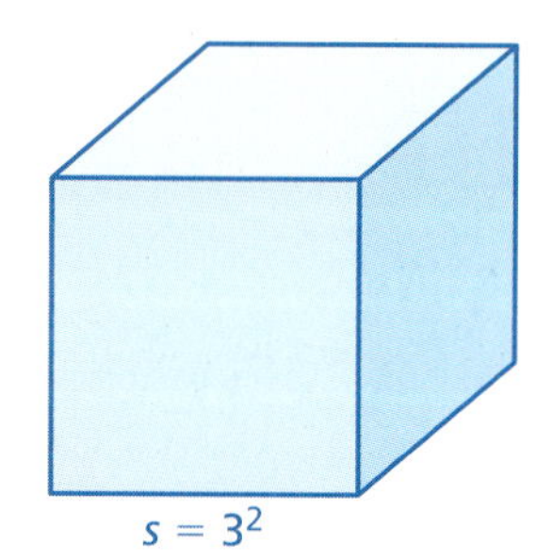

c.

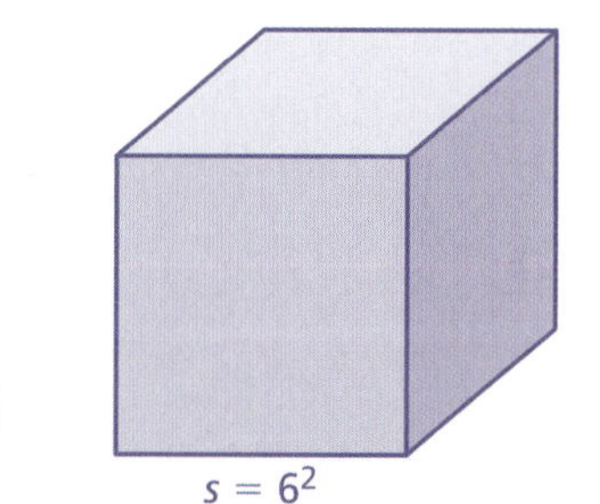

d.

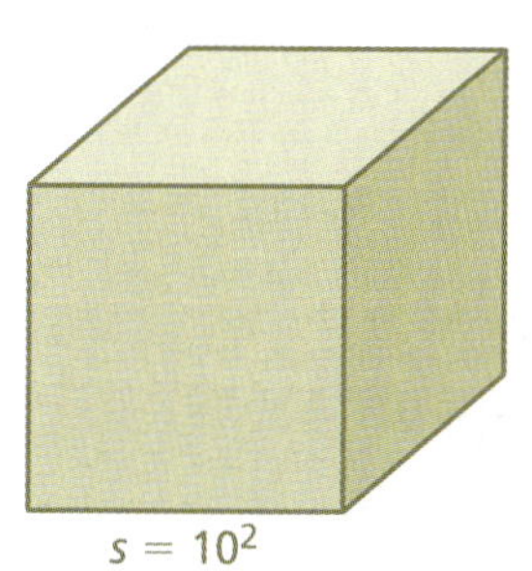

	Volume of Smaller Cube	Volume of Larger Cube	$\frac{\text{Larger Volume}}{\text{Smaller Volume}}$	Answer
a.				
b.				
c.				
d.				

What Is Your Answer?

3. **IN YOUR OWN WORDS** How can you divide two powers that have the same base? Give two examples of your rule.

Practice Use what you learned about dividing powers with the same base to complete Exercises 3–6 on page 426.

10.3 Lesson

Key Idea

Quotient of Powers Property

Words To divide powers with the same base, subtract their exponents.

Numbers $\dfrac{4^5}{4^2} = 4^{5-2} = 4^3$

Algebra $\dfrac{a^m}{a^n} = a^{m-n}$, where $a \neq 0$

EXAMPLE 1 Dividing Powers with the Same Base

a. $\dfrac{2^6}{2^4} = 2^{6-4}$ — Quotient of Powers Property

$= 2^2$ — Simplify.

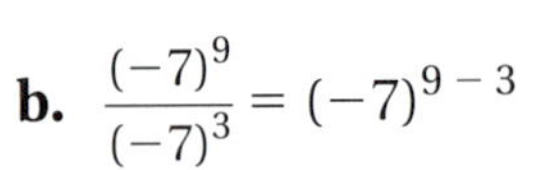

Common Error

When dividing powers, do not divide the bases.

$\dfrac{2^6}{2^4} = 2^2$, not 1^2.

b. $\dfrac{(-7)^9}{(-7)^3} = (-7)^{9-3}$ — Quotient of Powers Property

$= (-7)^6$ — Simplify.

c. $\dfrac{h^7}{h^6} = h^{7-6}$ — Quotient of Powers Property

$= h^1 = h$ — Simplify.

On Your Own

Now You're Ready
Exercises 7–14

Simplify the expression. Write your answer as a power.

1. $\dfrac{9^7}{9^4}$ **2.** $\dfrac{4.2^6}{4.2^5}$ **3.** $\dfrac{(-8)^8}{(-8)^4}$ **4.** $\dfrac{x^8}{x^3}$

EXAMPLE 2 Simplifying an Expression

Simplify $\dfrac{3^4 \cdot 3^2}{3^3}$. Write your answer as a power.

The numerator is a product of powers. Add the exponents in the numerator.

$\dfrac{3^4 \cdot 3^2}{3^3} = \dfrac{3^{4+2}}{3^3}$ — Product of Powers Property

$= \dfrac{3^6}{3^3}$ — Simplify.

$= 3^{6-3}$ — Quotient of Powers Property

$= 3^3$ — Simplify.

Laurie's Notes

Introduction

Connect

- **Yesterday:** Students explored exponents. (MP4, MP8)
- **Today:** Students will use the Quotient of Powers Property to simplify expressions.

Motivate

- **Preparation:** Find the area of your classroom in square feet. Select two smaller regions of your room that make logical sense given the shape of your room. My classroom is shown. I found the area of the entire room (A + B); the area of B; and the area of C.

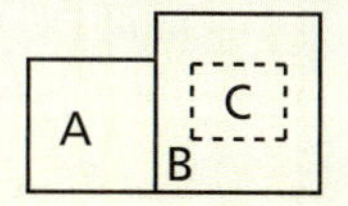

- First, have students stand around the room so that they are an arm's length away from everyone else.
- Then, ask them to stand only in region B (which includes C).
- Finally, ask them to move into region C. It should be very tight.
- Ask them to describe the three regions and how they felt about personal space. Then discuss population density and compute it for each region.
- Explain that one of the examples in today's lesson involves finding the population density of Tennessee.

Lesson Notes

Key Idea

- Write the Key Idea. Discuss the Words, Numbers, and Algebra.

Example 1

? **MP6 Attend to Precision: Part (a):** Write and simplify the expression. The base is 2 for each power. Ask the following questions to help develop correct vocabulary.
  - "How many factors of 2 are in the numerator?" 6
  - "How many factors of 2 are in the denominator?" 4
  - "How many factors of 2 are common in *both* the numerator and denominator?" 4
  - "How many factors of 2 remain after you divide out the common factors?" 2
- Repeat similar questions for parts (b) and (c).

On Your Own

- Ask students to explain their answers.

Example 2

? This example combines two properties. Ask the following questions.
  - "How many factors of 3 are in the numerator?" $4 + 2 = 6$
  - "How many factors of 3 are in the denominator?" 3
  - "How many factors of 3 are common in *both* the numerator and denominator?" 3
  - **MP7 Look for and Make Use of Structure:** "How many factors of 3 remain after you divide out the common factors?" 3

Goal Today's lesson is using the Quotient of Powers Property to simplify expressions.

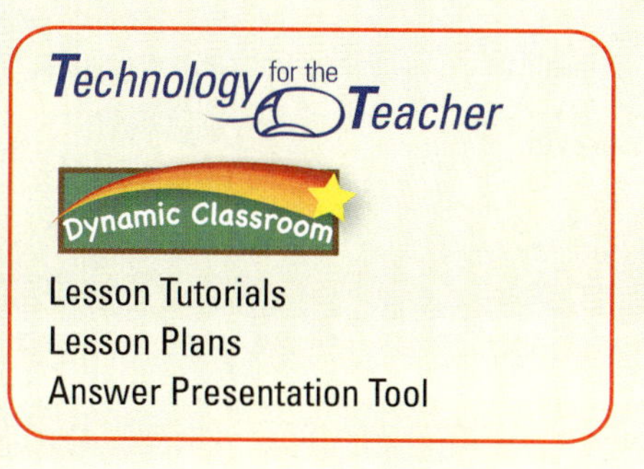

Lesson Tutorials
Lesson Plans
Answer Presentation Tool

Extra Example 1

Simplify each expression. Write your answer as a power.

a. $\frac{4^5}{4^2}$ 4^3

b. $\frac{(-2)^{10}}{(-2)^3}$ $(-2)^7$

c. $\frac{p^7}{p^6}$ p

On Your Own

1. 9^3 **2.** 4.2

3. $(-8)^4$ **4.** x^5

Extra Example 2

Simplify $\frac{5^6 \cdot 5^2}{5^4}$. Write your answer as a power. 5^4

Extra Example 3

Simplify $\frac{z^6}{z^2} \cdot \frac{z^8}{z^5}$. Write your answer as a power. z^7

On Your Own

5. 2^7
6. d^5
7. 5^8

Extra Example 4

The projected population of Hawaii in 2030 is about $5.59 \cdot 2^{18}$. The land area of Hawaii is about 2^{14} square kilometers. Predict the average number of people per square kilometer in 2030. about 89 people per km^2

On Your Own

8. 36 people per km^2

English Language Learners

Organization

Have students organize the *Key Ideas* of this chapter in their notebooks. This will provide them with easy access to the material and concepts of the chapter.

Laurie's Notes

Example 3

- This example also combines two properties.
- Work through the problem as shown.
- Discuss the approach with students. Each quotient was simplified first and then the product of the two expressions was found.

? "Will the answer be the same if the product of the two expressions is found and then the quotient is simplified? Explain." yes; It is similar to multiplying two fractions and then simplifying the answer.

- Simplify the expression using the alternate approach in the Study Tip.

$$\frac{a^{10}}{a^6} \cdot \frac{a^7}{a^4} = \frac{a^{10} \cdot a^7}{a^6 \cdot a^4} = \frac{a^{10+7}}{a^{6+4}} = \frac{a^{17}}{a^{10}} = a^{17-10} = a^7$$

On Your Own

- There is more than one way to simplify these expressions. Remind students to think about the number of factors as they work the problems.
- **Question 6:** Students may forget that $d = d^1$.
- Have volunteers write their solutions on the board.

Example 4

- This problem is about population density, the number of people per square unit. In this case, it is the projected number of people in Tennessee per square mile in 2030.
- When working through this problem, notice that the factor 5 in the numerator does not have the same base as the two powers.

? **MP7**: "Why can you move 5 out of the numerator and write it as a whole number times the quotient of $(5.9)^8$ and $(5.9)^6$?" definition of multiplying fractions

- Simplify the quotient and multiply by 5.
- Use local landmarks to help students visualize the size of a square mile.

On Your Own

- **Neighbor Check:** Have students work independently and then have their neighbors check their work. Have students discuss any discrepancies.

Closure

- Explain how the Quotient of Powers Property is related to simplifying fractions. You divide out the common factors.

EXAMPLE 3 Simplifying an Expression

Study Tip

You can also simplify the expression in Example 3 as follows.

$$\frac{a^{10}}{a^6} \cdot \frac{a^7}{a^4} = \frac{a^{10} \cdot a^7}{a^6 \cdot a^4}$$
$$= \frac{a^{17}}{a^{10}}$$
$$= a^{17-10}$$
$$= a^7$$

Simplify $\frac{a^{10}}{a^6} \cdot \frac{a^7}{a^4}$. Write your answer as a power.

$\frac{a^{10}}{a^6} \cdot \frac{a^7}{a^4} = a^{10-6} \cdot a^{7-4}$ — Quotient of Powers Property

$= a^4 \cdot a^3$ — Simplify.

$= a^{4+3}$ — Product of Powers Property

$= a^7$ — Simplify.

On Your Own

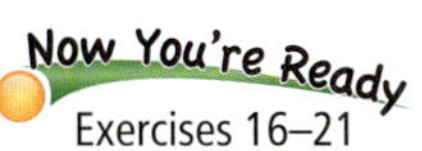

Simplify the expression. Write your answer as a power.

5. $\frac{2^{15}}{2^3 \cdot 2^5}$ **6.** $\frac{d^5}{d} \cdot \frac{d^9}{d^8}$ **7.** $\frac{5^9}{5^4} \cdot \frac{5^5}{5^2}$

EXAMPLE 4 Real-Life Application

The projected population of Tennessee in 2030 is about $5 \cdot 5.9^8$. Predict the average number of people per square mile in 2030.

Use a model to solve the problem.

$$\text{People per square mile} = \frac{\text{Population in 2030}}{\text{Land area}}$$

Land area: about 5.9^6 mi^2

$= \frac{5 \cdot 5.9^8}{5.9^6}$ — Substitute.

$= 5 \cdot \frac{5.9^8}{5.9^6}$ — Rewrite.

$= 5 \cdot 5.9^2$ — Quotient of Powers Property

$= 174.05$ — Evaluate.

So, there will be about 174 people per square mile in Tennessee in 2030.

On Your Own

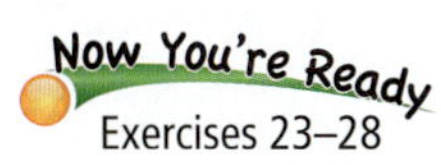

8. The projected population of Alabama in 2030 is about $2.25 \cdot 2^{21}$. The land area of Alabama is about 2^{17} square kilometers. Predict the average number of people per square kilometer in 2030.

10.3 Exercises

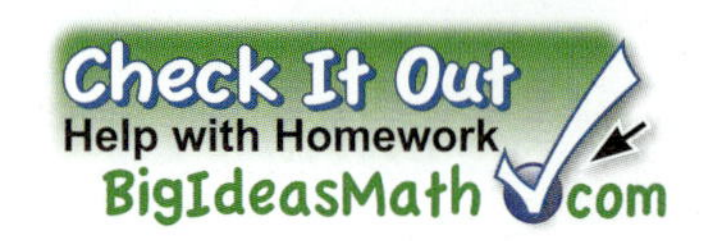

Vocabulary and Concept Check

1. **WRITING** Describe in your own words how to divide powers.

2. **WHICH ONE DOESN'T BELONG?** Which quotient does *not* belong with the other three? Explain your reasoning.

$\dfrac{(-10)^7}{(-10)^2}$ $\dfrac{6^3}{6^2}$ $\dfrac{(-4)^8}{(-3)^4}$ $\dfrac{5^6}{5^3}$

Practice and Problem Solving

Simplify the expression. Write your answer as a power.

3. $\dfrac{6^{10}}{6^4}$
4. $\dfrac{8^9}{8^7}$
5. $\dfrac{(-3)^4}{(-3)^1}$
6. $\dfrac{4.5^5}{4.5^3}$
7. $\dfrac{5^9}{5^3}$
8. $\dfrac{64^4}{64^3}$
9. $\dfrac{(-17)^5}{(-17)^2}$
10. $\dfrac{(-7.9)^{10}}{(-7.9)^4}$
11. $\dfrac{(-6.4)^8}{(-6.4)^6}$
12. $\dfrac{\pi^{11}}{\pi^7}$
13. $\dfrac{b^{24}}{b^{11}}$
14. $\dfrac{n^{18}}{n^7}$

15. **ERROR ANALYSIS** Describe and correct the error in simplifying the quotient.

Simplify the expression. Write your answer as a power.

16. $\dfrac{7^5 \cdot 7^3}{7^2}$
17. $\dfrac{2^{19} \cdot 2^5}{2^{12} \cdot 2^3}$
18. $\dfrac{(-8.3)^8}{(-8.3)^7} \cdot \dfrac{(-8.3)^4}{(-8.3)^3}$
19. $\dfrac{\pi^{30}}{\pi^{18} \cdot \pi^4}$
20. $\dfrac{c^{22}}{c^8 \cdot c^9}$
21. $\dfrac{k^{13}}{k^5} \cdot \dfrac{k^{17}}{k^{11}}$

22. **SOUND INTENSITY** The sound intensity of a normal conversation is 10^6 times greater than the quietest noise a person can hear. The sound intensity of a jet at takeoff is 10^{14} times greater than the quietest noise a person can hear. How many times more intense is the sound of a jet at takeoff than the sound of a normal conversation?

Assignment Guide and Homework Check

Level	Day 1 Activity Assignment	Day 2 Lesson Assignment	Homework Check
Basic	3–6, 33–37	1, 2, 7–21 odd, 22–25	9, 17, 21, 22, 25
Average	3–6, 33–37	1, 2, 7–25 odd, 29, 30	9, 17, 21, 25, 30
Advanced	1–6, 12, 14, 15, 16–28 even, 29–37		14, 20, 26, 30, 31

Common Errors

- **Exercises 3–14, 16–21** Students may divide the exponents when they should be subtracting them. Remind them that the Quotient of Powers Property states that the exponents are subtracted.
- **Exercises 3–14, 16–21** Students may multiply and/or divide the bases when simplifying the expression. Remind them that the base does not change when they use the Quotient of Powers or Product of Powers Property.
- **Exercises 23–28** Students may try to combine unlike bases when simplifying. Remind them that the Quotient of Powers and Product of Powers Properties only apply to powers with the same base.

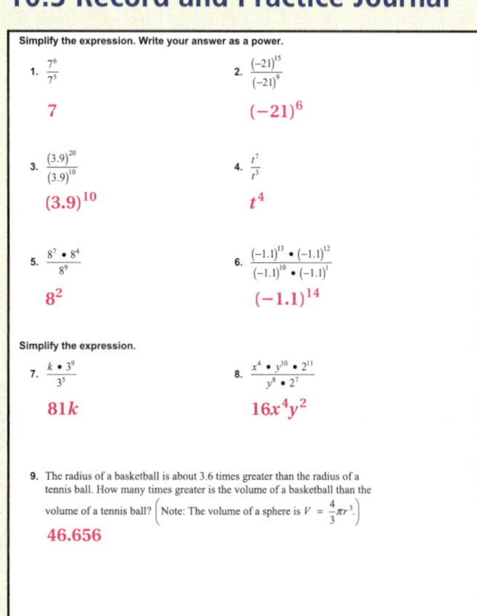

10.3 Record and Practice Journal

Simplify the expression. Write your answer as a power.

1. $\frac{7^6}{7^5}$ **7**
2. $\frac{(-21)^{15}}{(-21)^9}$ **$(-21)^6$**
3. $\frac{(3.9)^{20}}{(3.9)^{10}}$ **$(3.9)^{10}$**
4. $\frac{t^7}{t^3}$ **t^4**
5. $\frac{8^7 \cdot 8^4}{8^9}$ **8^2**
6. $\frac{(-1.1)^{13} \cdot (-1.1)^{12}}{(-1.1)^{10} \cdot (-1.1)^1}$ **$(-1.1)^{14}$**

Simplify the expression.

7. $\frac{k \cdot 3^9}{3^5}$ **$81k$**
8. $\frac{x^4 \cdot y^{10} \cdot 2^{11}}{y^8 \cdot 2^7}$ **$16x^4y^2$**
9. The radius of a basketball is about 3.6 times greater than the radius of a tennis ball. How many times greater is the volume of a basketball than the volume of a tennis ball? $\left(\text{Note: The volume of a sphere is } V = \frac{4}{3}\pi r^3.\right)$ **46.656**

Vocabulary and Concept Check

1. To divide powers means to divide out the common factors of the numerator and denominator. To divide powers with the same base, write the power with the common base and an exponent found by subtracting the exponent in the denominator from the exponent in the numerator.
2. $\frac{(-4)^8}{(-3)^4}$; The other quotients have powers with the same base.

Practice and Problem Solving

3. 6^6
4. 8^2
5. $(-3)^3$
6. 4.5^2
7. 5^6
8. 64
9. $(-17)^3$
10. $(-7.9)^6$
11. $(-6.4)^2$
12. π^4
13. b^{13}
14. n^{11}
15. You should subtract the exponents instead of dividing them.
$$\frac{6^{15}}{6^5} = 6^{15-5} = 6^{10}$$
16. 7^6
17. 2^9
18. $(-8.3)^2$
19. π^8
20. c^5
21. k^{14}
22. 10^8 times

Practice and Problem Solving

23. $64x$

24. $6w$

25. $125a^3b^2$

26. $125cd^2$

27. x^7y^6

28. m^9n

29. See *Taking Math Deeper.*

30. **a.** *Sample answer:* $m = 5$, $n = 3$

b. infinitely many solutions; Any two numbers that satisfy the equation $m - n = 2$ are solutions.

31. 10^{13} galaxies

32. 10; The difference in the exponents needs to be 9. To find x, solve the equation $3x - (2x + 1) = 9$.

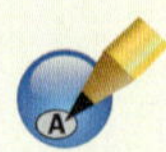

Fair Game Review

33. -9 **34.** -8

35. 61 **36.** -4

37. B

Mini-Assessment

Simplify the expression. Write your answer as a power.

1. $\dfrac{(-4)^3}{(-4)^1}$ $(-4)^2$
2. $\dfrac{9.7^7}{9.7^3}$ 9.7^4
3. $\dfrac{5^4 \cdot 5^2}{5^3}$ 5^3
4. $\dfrac{m^{10}}{m^5 \cdot m^2}$ m^3
5. $\dfrac{y^{17}}{y^{10}} \cdot \dfrac{y^6}{y^3}$ y^{10}

Taking Math Deeper

Exercise 29

This is an interesting problem. The memory in the different styles of MP3 players increases exponentially, but the price increases linearly.

Compare Player D with Player B.

a. $\dfrac{2^4}{2^2} = 2^2 = 4$ times more memory

2 Compare the memory with the price.

If you plot the five points representing the memory and the prices, you get the following graph.

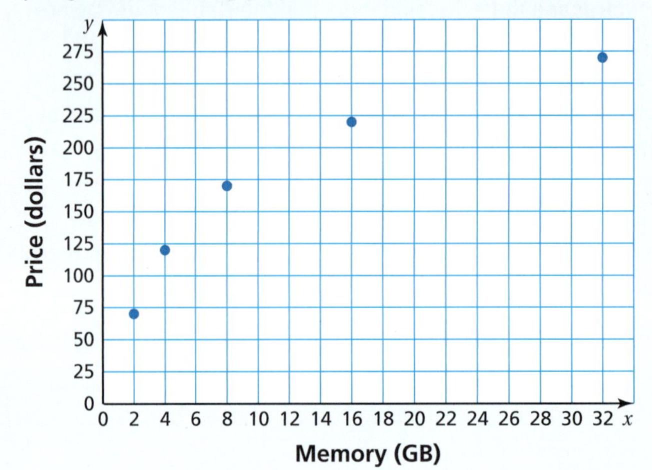

Answer the question.

b. This graph does not show a constant rate of change. So, memory and price do not show a linear relationship. However, the differences in price between consecutive sizes reflect a constant rate of change.

Project

What changes in technology have occurred over the past 50 years? What do you predict will change over the next 50 years?

Reteaching and Enrichment Strategies

If students need help. . .	If students got it. . .
Resources by Chapter • Practice A and Practice B • Puzzle Time Record and Practice Journal Practice Differentiating the Lesson Lesson Tutorials Skills Review Handbook	Resources by Chapter • Enrichment and Extension • Technology Connection Start the next section

Simplify the expression.

4 23. $\frac{x \cdot 4^8}{4^5}$

24. $\frac{6^3 \cdot w}{6^2}$

25. $\frac{a^3 \cdot b^4 \cdot 5^4}{b^2 \cdot 5}$

26. $\frac{5^{12} \cdot c^{10} \cdot d^2}{5^9 \cdot c^9}$

27. $\frac{x^{15}y^9}{x^8y^3}$

28. $\frac{m^{10}n^7}{m^1n^6}$

29. **MEMORY** The memory capacities and prices of five MP3 players are shown in the table.

MP3 Player	Memory (GB)	Price
A	2^1	\$70
B	2^2	\$120
C	2^3	\$170
D	2^4	\$220
E	2^5	\$270

a. How many times more memory does MP3 Player D have than MP3 Player B?

b. Do memory and price show a linear relationship? Explain.

30. **CRITICAL THINKING** Consider the equation $\frac{9^m}{9^n} = 9^2$.

a. Find two numbers m and n that satisfy the equation.

b. Describe the number of solutions that satisfy the equation. Explain your reasoning.

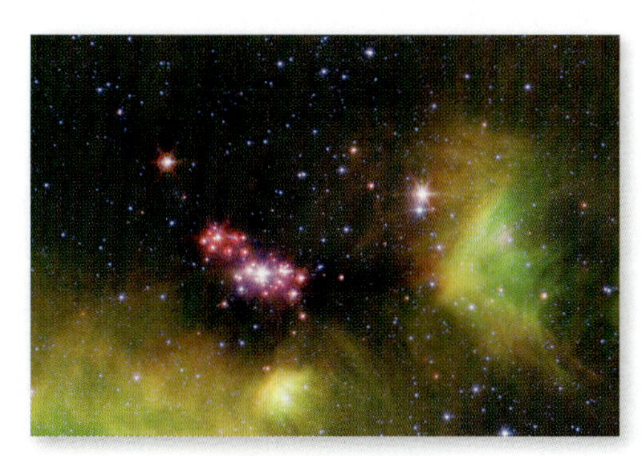

Milky Way galaxy
$10 \cdot 10^{10}$ stars

31. **STARS** There are about 10^{24} stars in the universe. Each galaxy has approximately the same number of stars as the Milky Way galaxy. About how many galaxies are in the universe?

32. Number Sense Find the value of x that makes $\frac{8^{3x}}{8^{2x+1}} = 8^9$ true. Explain how you found your answer.

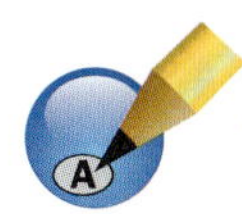

Fair Game Review What you learned in previous grades & lessons

Subtract. *(Skills Review Handbook)*

33. $-4 - 5$

34. $-23 - (-15)$

35. $33 - (-28)$

36. $18 - 22$

37. **MULTIPLE CHOICE** What is the value of x? *(Skills Review Handbook)*

(A) 20 (B) 30

(C) 45 (D) 60

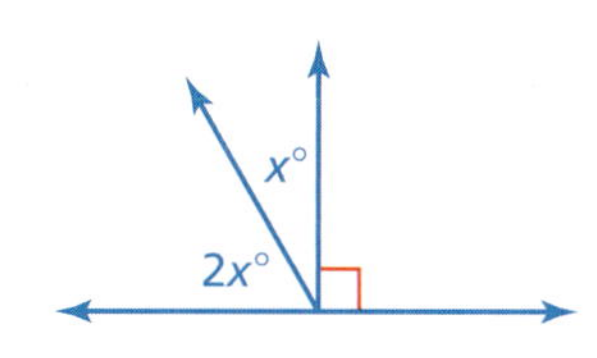

10.4 Zero and Negative Exponents

Essential Question How can you evaluate a nonzero number with an exponent of zero? How can you evaluate a nonzero number with a negative integer exponent?

1 ACTIVITY: Using the Quotient of Powers Property

Work with a partner.

a. Copy and complete the table.

Quotient	Quotient of Powers Property	Power
$\frac{5^3}{5^3}$		
$\frac{6^2}{6^2}$		
$\frac{(-3)^4}{(-3)^4}$		
$\frac{(-4)^5}{(-4)^5}$		

b. **REPEATED REASONING** Evaluate each expression in the first column of the table. What do you notice?

c. How can you use these results to define a^0 where $a \neq 0$?

2 ACTIVITY: Using the Product of Powers Property

Work with a partner.

a. Copy and complete the table.

Product	Product of Powers Property	Power
$3^0 \cdot 3^4$		
$8^2 \cdot 8^0$		
$(-2)^3 \cdot (-2)^0$		
$\left(-\frac{1}{3}\right)^0 \cdot \left(-\frac{1}{3}\right)^5$		

b. Do these results support your definition in Activity 1(c)?

COMMON CORE

Exponents

In this lesson, you will

- evaluate expressions involving numbers with zero as an exponent.
- evaluate expressions involving negative integer exponents.

Learning Standard MACC.8.EE.1.1

Laurie's Notes

Introduction

Standards for Mathematical Practice

- **MP8 Look for and Express Regularity in Repeated Reasoning:** In the first three activities, students will use previously learned properties to define zero and negative exponents.

Motivate

- Writing a number in expanded form should be familiar to students.
- ? "How do you write 234 in expanded form?" $200 + 30 + 4$
- ? "How do you write the expanded form using powers of 10?" $2 \times 10^2 + 3 \times 10 + 4 \times 1$
- ? "Do you think it is possible to write 234.56 in expanded form using powers of 10?" Answers will vary.
- Explain that in today's activities, students will explore zero and negative exponents, including zero and negative powers of 10, which can be used to write a decimal in expanded notation.

Activity Notes

Activity 1

- Have students work with their partners to complete the table. To fill in the middle column, students should use the Quotient of Powers Property to rewrite the quotients in the first column.
- As students work through the activity, they should discover that all of the powers in the last column of the table have an exponent of zero.
- **Part (b):** Students should notice that each quotient in the first column of the table is equivalent to 1.
- **Part (c):** Students should summarize their findings by writing $a^0 = 1$.
- Discuss the fact that some of the language used in earlier sections does not apply here. For instance, when referring to 5^0, you do not say, "5 used as a factor 0 times."

Activity 2

- This activity presents another way of showing that $a^0 = 1$ where $a \neq 0$.
- Have students work with their partners to complete the table. To fill in the middle column, students should use the Product of Powers Property to rewrite the products in the first column.
- **MP3 Construct Viable Arguments and Critique the Reasoning of Others: Part (b):** Students should explain their findings, particularly that the results support their definition in Activity 1(c). They should explain that each power in the last column of the table has the same exponent as the corresponding nonzero exponent in the first column.

Florida Common Core Standards

MACC.8.EE.1.1 Know and apply the properties of integer exponents to generate equivalent numerical expressions.

Previous Learning

Students should know how to apply the properties of exponents.

Technology for the Teacher

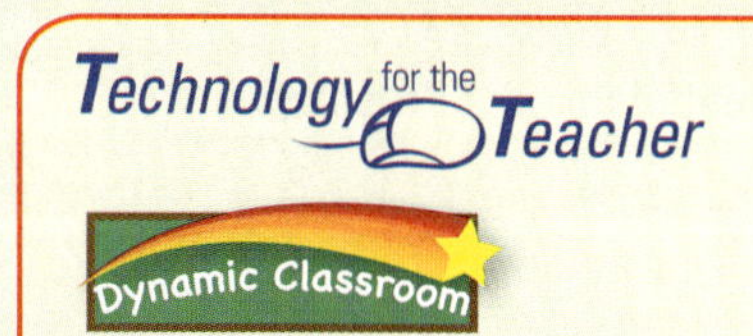

Lesson Plans
Complete Materials List

10.4 Record and Practice Journal

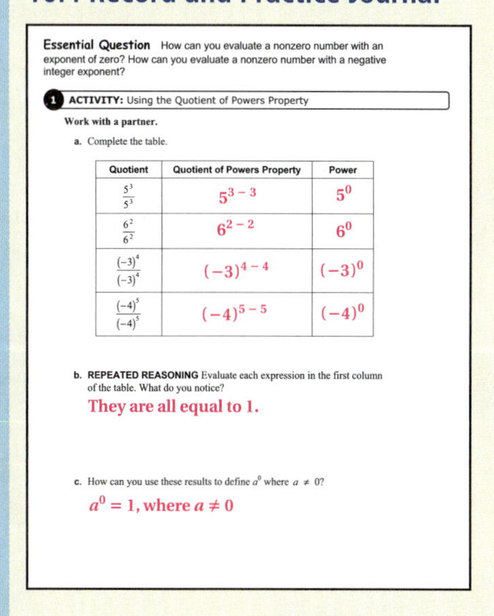

Essential Question How can you evaluate a nonzero number with an exponent of zero? How can you evaluate a nonzero number with a negative integer exponent?

1 **ACTIVITY:** Using the Quotient of Powers Property

Work with a partner.

a. Complete the table.

Quotient	Quotient of Powers Property	Power
$\frac{5^3}{5^3}$	5^{3-3}	5^0
$\frac{6^2}{6^2}$	6^{2-2}	6^0
$\frac{(-3)^4}{(-3)^4}$	$(-3)^{4-4}$	$(-3)^0$
$\frac{(-4)^5}{(-4)^5}$	$(-4)^{5-5}$	$(-4)^0$

b. **REPEATED REASONING** Evaluate each expression in the first column of the table. What do you notice?

They are all equal to 1.

c. How can you use these results to define a^0 where $a \neq 0$?

$a^0 = 1$, where $a \neq 0$

Differentiated Instruction

Visual

Help students to understand zero and negative exponents using methods already known to them.

Evaluating and then simplifying:

$$\frac{3^2}{3^3} = \frac{9}{27} = \frac{9 \div 9}{27 \div 9} = \frac{1}{3}$$

Dividing out common factors:

$$\frac{3^2}{3^3} = \frac{\cancel{3}^1 \cdot \cancel{3}^1}{\cancel{3}_1 \cdot \cancel{3}_1 \cdot 3} = \frac{1}{3}$$

Quotient of Powers Property:

$$\frac{3^2}{3^3} = 3^{2-3} = 3^{-1} = \frac{1}{3}$$

10.4 Record and Practice Journal

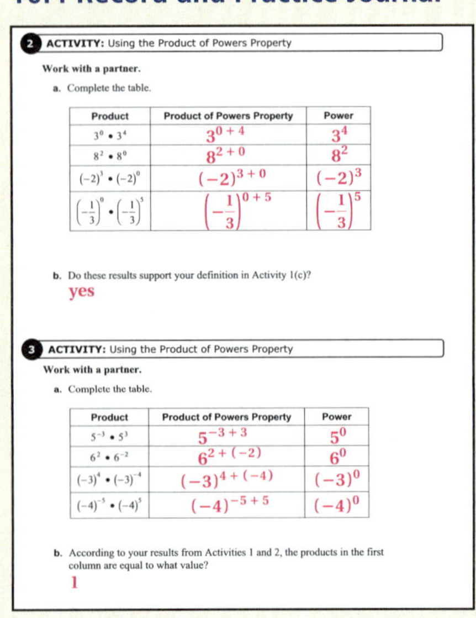

2 ACTIVITY: Using the Product of Powers Property

Work with a partner.

a. Complete the table.

Product	Product of Powers Property	Power
$3^0 \cdot 3^4$	3^{0+4}	3^4
$8^2 \cdot 8^0$	8^{2+0}	8^2
$(-2)^3 \cdot (-2)^0$	$(-2)^{3+0}$	$(-2)^3$
$\left(-\frac{1}{3}\right)^0 \cdot \left(-\frac{1}{3}\right)^5$	$\left(-\frac{1}{3}\right)^{0+5}$	$\left(-\frac{1}{3}\right)^5$

b. Do these results support your definition in Activity 1(c)?
yes

3 ACTIVITY: Using the Product of Powers Property

Work with a partner.

a. Complete the table.

Product	Product of Powers Property	Power
$5^{-3} \cdot 5^3$	5^{-3+3}	5^0
$6^2 \cdot 6^{-2}$	$6^{2+(-2)}$	6^0
$(-3)^4 \cdot (-3)^{-4}$	$(-3)^{4+(-4)}$	$(-3)^0$
$(-4)^{-5} \cdot (-4)^5$	$(-4)^{-5+5}$	$(-4)^0$

b. According to your results from Activities 1 and 2, the products in the first column are equal to what value?
1

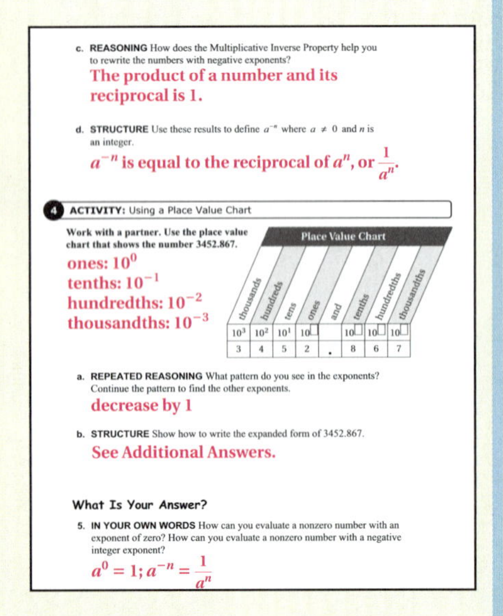

c. **REASONING** How does the Multiplicative Inverse Property help you to rewrite the numbers with negative exponents?
The product of a number and its reciprocal is 1.

d. **STRUCTURE** Use these results to define a^{-n} where $a \neq 0$ and n is an integer.
a^{-n} is equal to the reciprocal of a^n, or $\frac{1}{a^n}$.

4 ACTIVITY: Using a Place Value Chart

Work with a partner. Use the place value chart that shows the number 3452.867.

ones: 10^0
tenths: 10^{-1}
hundredths: 10^{-2}
thousandths: 10^{-3}

Place Value Chart

thousands	hundreds	tens	ones	and	tenths	hundredths	thousandths
10^3	10^2	10^1	$10^{\square}$		$10^{\square}$	$10^{\square}$	$10^{\square}$
3	4	5	2	.	8	6	7

a. **REPEATED REASONING** What pattern do you see in the exponents? Continue the pattern to find the other exponents.
decrease by 1

b. **STRUCTURE** Show how to write the expanded form of 3452.867.
See Additional Answers.

What Is Your Answer?

5. **IN YOUR OWN WORDS** How can you evaluate a nonzero number with an exponent of zero? How can you evaluate a nonzero number with a negative integer exponent?
$a^0 = 1; a^{-n} = \frac{1}{a^n}$

Laurie's Notes

Activity 3

- As in Activity 2, this activity uses the Product of Powers Property, but this time to help students come up with the definition of a negative exponent.
- Have students work with their partners to complete the table. To fill in the middle column, students should use the Product of Powers Property to rewrite the products in the first column.
- **Part (c):** The Multiplicative Inverse Property states that the product of a number and its reciprocal is 1. Make sure students see the connection between negative exponents and reciprocals.
- **MP3: Part (d):** Ask several students to explain their definitions.
- Discuss the fact that some of the language used in earlier sections does not apply here. For instance, when referring to 5^{-3}, you do not say, "5 used as a factor −3 times."

Activity 4

- Students should be familiar with place value charts.
- Students reviewed writing a whole number in expanded form at the beginning of class.
- By observing and continuing the pattern in the exponents in the given place value chart, students should be able to write the expanded form of a decimal.

What Is Your Answer?

- Have students share their understandings of zero and negative exponents.

Closure

- **Exit Ticket:** Write 234.56 in expanded notation using powers of 10.
$2 \times 10^2 + 3 \times 10^1 + 4 \times 10^0 + 5 \times 10^{-1} + 6 \times 10^{-2}$

3 ACTIVITY: Using the Product of Powers Property

Work with a partner.

a. Copy and complete the table.

Product	Product of Powers Property	Power
$5^{-3} \cdot 5^3$		
$6^2 \cdot 6^{-2}$		
$(-3)^4 \cdot (-3)^{-4}$		
$(-4)^{-5} \cdot (-4)^5$		

b. According to your results from Activities 1 and 2, the products in the first column are equal to what value?

c. **REASONING** How does the Multiplicative Inverse Property help you rewrite the numbers with negative exponents?

d. **STRUCTURE** Use these results to define a^{-n} where $a \neq 0$ and n is an integer.

4 ACTIVITY: Using a Place Value Chart

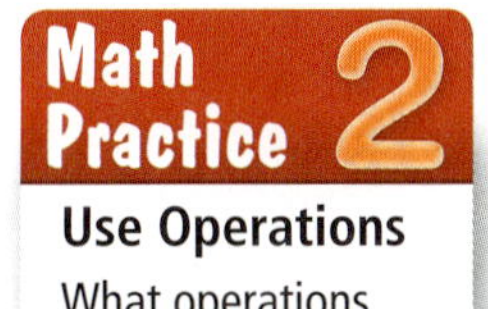

Use Operations

What operations are used when writing the expanded form?

Work with a partner. Use the place value chart that shows the number 3452.867.

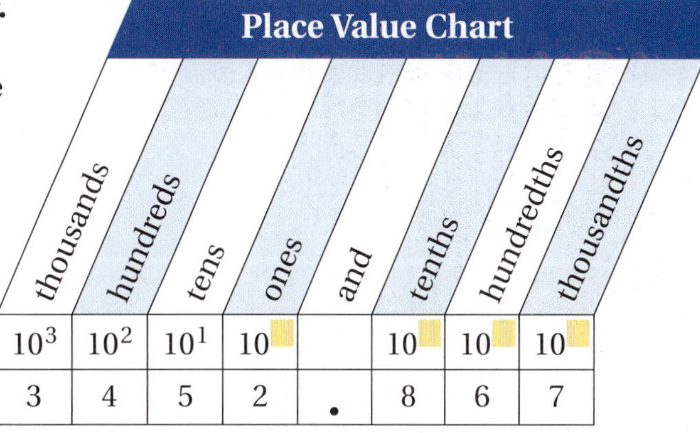

a. **REPEATED REASONING** What pattern do you see in the exponents? Continue the pattern to find the other exponents.

b. **STRUCTURE** Show how to write the expanded form of 3452.867.

What Is Your Answer?

5. **IN YOUR OWN WORDS** How can you evaluate a nonzero number with an exponent of zero? How can you evaluate a nonzero number with a negative integer exponent?

Practice Use what you learned about zero and negative exponents to complete Exercises 5–8 on page 432.

10.4 Lesson

Zero Exponents

Words For any nonzero number a, $a^0 = 1$. The power 0^0 is *undefined*.

Numbers $4^0 = 1$ **Algebra** $a^0 = 1$, where $a \neq 0$

Negative Exponents

Words For any integer n and any nonzero number a, a^{-n} is the reciprocal of a^n.

Numbers $4^{-2} = \frac{1}{4^2}$ **Algebra** $a^{-n} = \frac{1}{a^n}$, where $a \neq 0$

EXAMPLE 1 Evaluating Expressions

a. $3^{-4} = \frac{1}{3^4}$ Definition of negative exponent

$= \frac{1}{81}$ Evaluate power.

b. $(-8.5)^{-4} \cdot (-8.5)^4 = (-8.5)^{-4+4}$ Product of Powers Property

$= (-8.5)^0$ Simplify.

$= 1$ Definition of zero exponent

c. $\frac{2^6}{2^8} = 2^{6-8}$ Quotient of Powers Property

$= 2^{-2}$ Simplify.

$= \frac{1}{2^2}$ Definition of negative exponent

$= \frac{1}{4}$ Evaluate power.

On Your Own

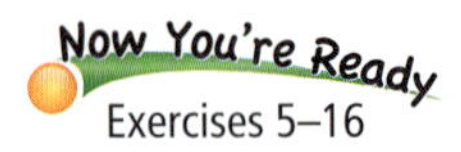

Exercises 5–16

Evaluate the expression.

1. 4^{-2} **2.** $(-2)^{-5}$ **3.** $6^{-8} \cdot 6^8$

4. $\frac{(-3)^5}{(-3)^6}$ **5.** $\frac{1}{5^7} \cdot \frac{1}{5^{-4}}$ **6.** $\frac{4^5 \cdot 4^{-3}}{4^2}$

Laurie's Notes

Introduction

Connect

- **Yesterday:** Students explored zero and negative exponents. (MP3, MP8)
- **Today:** Students will use the definitions of zero and negative exponents to evaluate and simplify expressions.

Motivate

? "Did you know that a faucet dripping 30 times per minute wastes 54 gallons of water per month?"

? "Can you visualize 54 gallons? Is that more than a standard kitchen sink? Is it more than a bathtub? To what does it compare?" Most students have difficulty estimating capacity. If there is a 50-gallon waste barrel in your school's cafeteria, you could compare it to that.

- Explain that one of the examples in today's lesson involves finding the amount of water that leaks from a faucet in a given period of time.

Lesson Notes

Key Ideas

- The definition of zero exponents is easily understood by most students. Writing a negative exponent as a unit fraction with a positive exponent in the denominator takes time for some students to understand. They need to see multiple examples of simplifying fractions using the Quotient of Powers Property and of dividing out common factors, where the exponent in the denominator is greater than the exponent in the numerator.
- Example: $\frac{5^2}{5^3} = 5^{2-3} = 5^{-1}$ $\qquad \frac{5^2}{5^3} = \frac{\cancel{5} \cdot \cancel{5}}{\cancel{5} \cdot \cancel{5} \cdot 5} = \frac{1}{5}$

  The Quotient of Powers Property is used on the left. Dividing out common factors is used on the right. Because both starting expressions are the same, the results must be equivalent. So, $5^{-1} = \frac{1}{5^1}$.
- Note that 0^0 is undefined.

Example 1

- **Part (a):** This is a direct application of the definition of negative exponents.
- **Part (b):** The bases are the same, so the exponents are added.
- **Part (c):** The Quotient of Powers Property is used first, resulting in a negative exponent. Use the definition of negative exponent and simplify.

On Your Own

- Question 5 can be done by thinking about simple fractions and how they are multiplied. The product of these two fractions is $\frac{1}{5^7 \cdot 5^{-4}} = \frac{1}{5^3}$.
- **MP2 Reason Abstractly and Quantitatively** and **MP7 Look for and Make Use of Structure:** Several of these expressions can be evaluated in more than one way. Ask students to share their methods.

Goal Today's lesson is evaluating and simplifying expressions with negative and zero exponents.

Technology for the Teacher

Lesson Tutorials
Lesson Plans
Answer Presentation Tool

Extra Example 1

Evaluate each expression.

a. 4^{-3} $\frac{1}{64}$

b. $(-3.7)^{-2} \cdot (-3.7)^2$ 1

c. $\frac{3^6}{3^9}$ $\frac{1}{27}$

On Your Own

1. $\frac{1}{16}$
2. $-\frac{1}{32}$
3. 1
4. $-\frac{1}{3}$
5. $\frac{1}{125}$
6. 1

Laurie's Notes

Extra Example 2

Simplify. If possible, write the expression using only positive exponents.

a. $-2x^0$ -2

b. $\frac{4b^{-4}}{b^7}$ $\frac{4}{b^{11}}$

On Your Own

7. $\frac{8}{x^2}$ **8.** $\frac{1}{b^{10}}$

9. $\frac{1}{15z^3}$

Extra Example 3

In Example 3, the faucet leaks water at a rate of 4^{-6} liter per second. How many liters of water leak from the faucet in 1 hour? about 0.88 L

On Your Own

10. 1.152 L

English Language Learners

Vocabulary

Remind English learners that when they see a negative exponent, they should think *reciprocal*. Review the meaning of the word reciprocal. Students often think that because the exponent is negative, the expression is negative. Remind them that a number of the form x^n cannot be negative unless the base is negative.

Example 2

- **MP6 Attend to Precision:** It is assumed that the variables are nonzero in expressions such as those given in this example, so that the rules for zero and negative exponents can be used. Discuss this with students.
- **Common Error:** In part (a), some students see the zero exponent and immediately think the answer is 1. Remind students that only the variable is being raised to the 0 exponent; -5 is not.
- **Common Error:** In part (b), the constant 9 is not being raised to an exponent, only the variables are. Students need to distinguish this. In the step where the expression has been simplified to $9y^{-8}$ ask, "What is being raised to the -8?" In other words, what is the base for the exponent?
- **MP1a Make Sense of Problems:** Work through the steps slowly. It takes time for students to make sense of all that is going on in each problem. Because there is often more than one approach to simplifying the expression, it can confuse students. Instead of seeing it as a way to show that the properties are all connected, students see it as a way of trying to confuse them.

On Your Own

- Have students share their work at the board, *and* explain aloud what they did. Students need to hear the words and see the work. It also helps students become better communicators when they have the opportunity to practice their skills.

Example 3

- Ask a student to read the problem. The information known in this problem is the rate of leaking (in drops per second) and the amount of liters in a drop (from the illustration).

 "What are you asked to find in this problem?" amount faucet leaks in 1 hour

- First, use unit analysis to convert 1 hour to 3600 seconds. Then, because the faucet leaks 50^{-2} liter every second and there are 3600 seconds in an hour, multiply to find how many liters leak in 1 hour.

On Your Own

- Before students begin, ask them which faucet leaks more, the faucet that leaks 50^{-2} liter per second or the faucet that leaks 5^{-5} liter per second.

Closure

- **Exit Ticket:** Simplify.

1. $4x^{-3}$ $\frac{4}{x^3}$ **2.** $\frac{6^3}{6^5}$ $\frac{1}{36}$ **3.** $\frac{4n^0}{n^2}$ $\frac{4}{n^2}$

EXAMPLE 2 Simplifying Expressions

a. $-5x^0 = -5(1)$ — Definition of zero exponent

$= -5$ — Multiply.

b. $\dfrac{9y^{-3}}{y^5} = 9y^{-3-5}$ — Quotient of Powers Property

$= 9y^{-8}$ — Simplify.

$= \dfrac{9}{y^8}$ — Definition of negative exponent

On Your Own

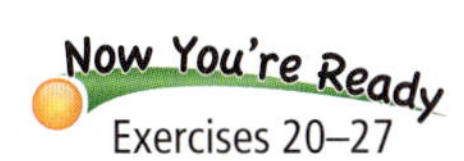

Exercises 20–27

Simplify. Write the expression using only positive exponents.

7. $8x^{-2}$ **8.** $b^0 \cdot b^{-10}$ **9.** $\dfrac{z^6}{15z^9}$

EXAMPLE 3 Real-Life Application

A drop of water leaks from a faucet every second. How many liters of water leak from the faucet in 1 hour?

Convert 1 hour to seconds.

$$1\text{ h} \times \frac{60\text{ min}}{1\text{ h}} \times \frac{60\text{ sec}}{1\text{ min}} = 3600\text{ sec}$$

Drop of water: 50^{-2} liter

Water leaks from the faucet at a rate of 50^{-2} liter per second. Multiply the time by the rate.

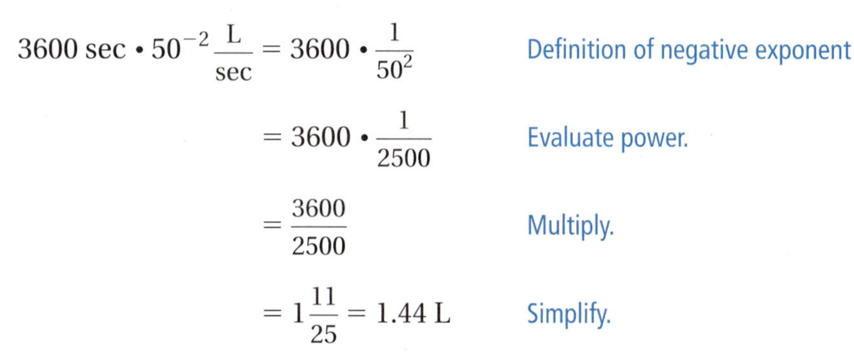

$3600\text{ sec} \cdot 50^{-2}\dfrac{\text{L}}{\text{sec}} = 3600 \cdot \dfrac{1}{50^2}$ — Definition of negative exponent

$= 3600 \cdot \dfrac{1}{2500}$ — Evaluate power.

$= \dfrac{3600}{2500}$ — Multiply.

$= 1\dfrac{11}{25} = 1.44\text{ L}$ — Simplify.

So, 1.44 liters of water leak from the faucet in 1 hour.

On Your Own

10. WHAT IF? The faucet leaks water at a rate of 5^{-5} liter per second. How many liters of water leak from the faucet in 1 hour?

10.4 Exercises

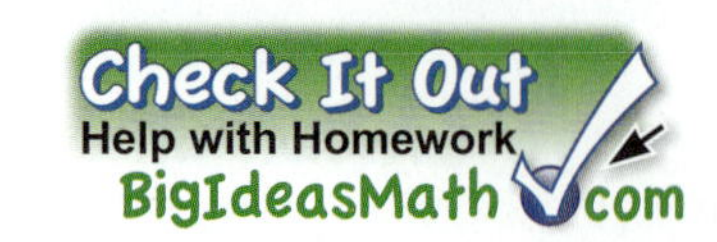

Vocabulary and Concept Check

1. **VOCABULARY** If a is a nonzero number, does the value of a^0 depend on the value of a? Explain.
2. **WRITING** Explain how to evaluate 10^{-3}.
3. **NUMBER SENSE** Without evaluating, order 5^0, 5^4, and 5^{-5} from least to greatest.
4. **DIFFERENT WORDS, SAME QUESTION** Which is different? Find "both" answers.

Rewrite $\frac{1}{3 \cdot 3 \cdot 3}$ using a negative exponent.	Write 3 to the negative third.
Write $\frac{1}{3}$ cubed as a power.	Write $(-3) \cdot (-3) \cdot (-3)$ as a power.

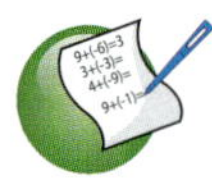

Practice and Problem Solving

Evaluate the expression.

5. $\frac{8^7}{8^7}$
6. $5^0 \cdot 5^3$
7. $(-2)^{-8} \cdot (-2)^8$
8. $9^4 \cdot 9^{-4}$
9. 6^{-2}
10. 158^0
11. $\frac{4^3}{4^5}$
12. $\frac{-3}{(-3)^2}$
13. $4 \cdot 2^{-4} + 5$
14. $3^{-3} \cdot 3^{-2}$
15. $\frac{1}{5^{-3}} \cdot \frac{1}{5^6}$
16. $\frac{(1.5)^2}{(1.5)^{-2} \cdot (1.5)^4}$

17. **ERROR ANALYSIS** Describe and correct the error in evaluating the expression.

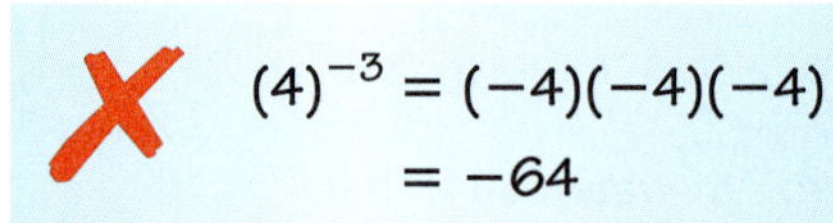

$$(4)^{-3} = (-4)(-4)(-4) = -64$$

18. **SAND** The mass of a grain of sand is about 10^{-3} gram. About how many grains of sand are in the bag of sand?

19. **CRITICAL THINKING** How can you write the number 1 as 2 to a power? 10 to a power?

Simplify. Write the expression using only positive exponents.

2 20. $6y^{-4}$
21. $8^{-2} \cdot a^7$
22. $\frac{9c^3}{c^{-4}}$
23. $\frac{5b^{-2}}{b^{-3}}$
24. $\frac{8x^3}{2x^9}$
25. $3d^{-4} \cdot 4d^4$
26. $m^{-2} \cdot n^3$
27. $\frac{3^{-2} \cdot k^0 \cdot w^0}{w^{-6}}$

Assignment Guide and Homework Check

Level	Day 1 Activity Assignment	Day 2 Lesson Assignment	Homework Check
Basic	5–8, 37–40	1–4, 9–31 odd	9, 15, 25, 27
Average	5–8, 37–40	1–4, 13–17, 19, 24–27, 31–33	14, 25, 27, 32
Advanced	1–8, 14, 16, 17, 18–28 even, 29–40		16, 27, 32, 34

Common Errors

- **Exercises 5–16** Students may think that a power with a zero exponent is equal to zero. Remind them of the definition of zero exponents.
- **Exercises 5–16** Students may think that a negative exponent makes the power negative. Remind them of the definition of negative exponents.
- **Exercises 5–16** Students may forget to complete the solution; they may simplify the expression, but leave the expression with exponents. Point out that they need to evaluate any powers to complete the solution.
- **Exercises 20–27** In an expression such as $6y^{-4}$, students may think that both the constant and the variable have the exponent -4. Make sure students understand that in such an expression, the base for the exponent is y and not $6y$.

10.4 Record and Practice Journal

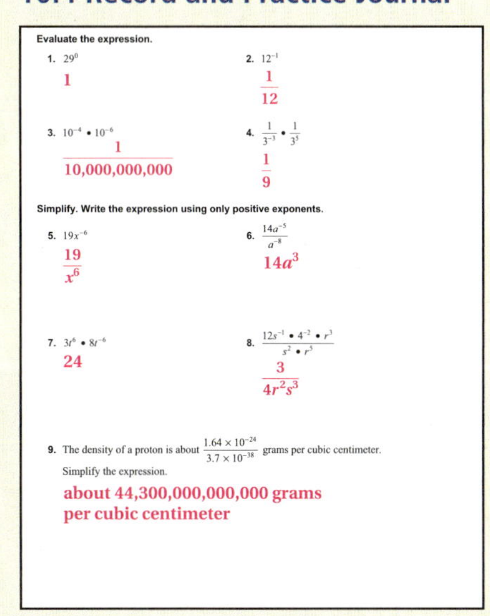

Evaluate the expression.

1. 29^0
   1
2. 12^{-1}
   $\frac{1}{12}$
3. $10^{-4} \cdot 10^{-6}$
   $\frac{1}{10{,}000{,}000{,}000}$
4. $\frac{1}{3^{-3}} \cdot \frac{1}{3^5}$
   $\frac{1}{9}$

Simplify. Write the expression using only positive exponents.

5. $19x^{-6}$
   $\frac{19}{x^6}$
6. $\frac{14a^{-5}}{a^{-8}}$
   $14a^3$
7. $3t^6 \cdot 8t^{-6}$
   24
8. $\frac{12s^{-1} \cdot 4^{-2} \cdot r^3}{s^2 \cdot r^5}$
   $\frac{3}{4r^2s^3}$
9. The density of a proton is about $\frac{1.64 \times 10^{-24}}{3.7 \times 10^{-38}}$ grams per cubic centimeter. Simplify the expression.
   about 44,300,000,000,000 grams per cubic centimeter

Vocabulary and Concept Check

1. no; Any nonzero base raised to the zero power is always 1.
2. Use the definition of negative exponents to rewrite it as $\frac{1}{10^3}$. Then evaluate the power to get $\frac{1}{1000}$.
3. $5^{-5}, 5^0, 5^4$
4. Write $(-3) \cdot (-3) \cdot (-3)$ as a power.; $(-3)^3$; 3^{-3}

Practice and Problem Solving

5. 1
6. 125
7. 1
8. 1
9. $\frac{1}{36}$
10. 1
11. $\frac{1}{16}$
12. $-\frac{1}{3}$
13. $5\frac{1}{4}$
14. $\frac{1}{243}$
15. $\frac{1}{125}$
16. 1
17. The negative sign goes with the exponent, not the base. $(4)^{-3} = \frac{1}{4^3} = \frac{1}{64}$
18. 10,000,000 grains of sand
19. 2^0; 10^0
20. $\frac{6}{y^4}$
21. $\frac{a^7}{64}$
22. $9c^7$
23. $5b$
24. $\frac{4}{x^6}$
25. 12
26. $\frac{n^3}{m^2}$
27. $\frac{w^6}{9}$

Practice and Problem Solving

28. *Sample answer:* 2^{-4}; 4^{-2}

29. 100 mm

30. 10,000 micrometers

31. 1,000,000 nanometers

32. 1,000,000 micrometers

33. **a.** 10^{-9} m

b. equal to

34. See *Taking Math Deeper.*

35. Write the power as 1 divided by the power and use a negative exponent. Justifications will vary.

36. If $a = 0$, then $0^n = 0$. Because you can not divide by 0, the expression $\frac{1}{0}$ is undefined.

Fair Game Review

37. 10^9 **38.** 10^3

39. 10^4 **40.** D

Mini-Assessment

Evaluate the expression.

1. 5^{-3} $\frac{1}{125}$
2. 9^0 1
3. $\frac{3^6}{3^{10}}$ $\frac{1}{81}$
4. $\frac{1}{2^{-2}} \cdot \frac{1}{2^5}$ $\frac{1}{8}$
5. $\frac{(2.3)^6}{(2.3)^{-3} \cdot (2.3)^8}$ 2.3

Taking Math Deeper

Exercise 34

To solve this problem, students need to notice that the blood sample shown has a volume of 500 milliliters. This is a good problem to help students with *unit analysis.*

a. How many white blood cells are in the donation?

$$500 \text{ mL} \cdot \frac{1 \text{ mm}^3}{10^{-3} \text{ mL}} \cdot \frac{10^4 \text{ white blood cells}}{1 \text{ mm}^3}$$
$$= 500 \cdot \frac{10^4 \text{ white blood cells}}{10^{-3}}$$
$$= 500 \cdot 10^7 \text{ white blood cells}$$
$$= 5{,}000{,}000{,}000 \text{ white blood cells}$$
$$= 5 \text{ billion white blood cells}$$

b. How many red blood cells are in the donation?

$$500 \text{ mL} \cdot \frac{1 \text{ mm}^3}{10^{-3} \text{ mL}} \cdot \frac{5 \cdot 10^6 \text{ red blood cells}}{1 \text{ mm}^3}$$
$$= 2500 \cdot \frac{10^6 \text{ red blood cells}}{10^{-3}}$$
$$= 2500 \cdot 10^9 \text{ red blood cells}$$
$$= 2{,}500{,}000{,}000{,}000 \text{ red blood cells}$$
$$= 2.5 \text{ trillion red blood cells}$$

c. The ratio of red blood cells to white blood cells is

$$\frac{2{,}500{,}000{,}000{,}000 \text{ red blood cells}}{5{,}000{,}000{,}000 \text{ white blood cells}} = \frac{500}{1}.$$

Red blood cells are responsible for picking up carbon dioxide from our blood and for transporting oxygen. White blood cells are responsible for fighting foreign organisms that enter the body.

Reteaching and Enrichment Strategies

If students need help. . .	If students got it. . .
Resources by Chapter • Practice A and Practice B • Puzzle Time Record and Practice Journal Practice Differentiating the Lesson Lesson Tutorials Skills Review Handbook	Resources by Chapter • Enrichment and Extension • Technology Connection Start the next section

28. OPEN-ENDED Write two different powers with negative exponents that have the same value.

METRIC UNITS In Exercises 29–32, use the table.

29. How many millimeters are in a decimeter?

30. How many micrometers are in a centimeter?

31. How many nanometers are in a millimeter?

32. How many micrometers are in a meter?

Unit of Length	Length (meter)
Decimeter	10^{-1}
Centimeter	10^{-2}
Millimeter	10^{-3}
Micrometer	10^{-6}
Nanometer	10^{-9}

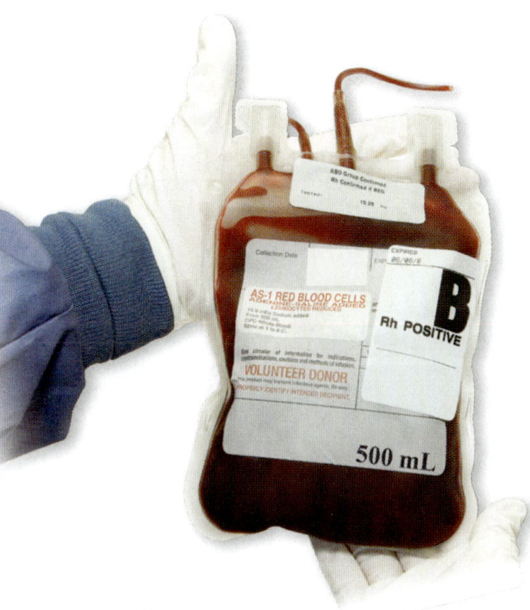

33. BACTERIA A species of bacteria is 10 micrometers long. A virus is 10,000 times smaller than the bacteria.

a. Using the table above, find the length of the virus in meters.

b. Is the answer to part (a) *less than, greater than,* or *equal to* one nanometer?

34. BLOOD DONATION Every 2 seconds, someone in the United States needs blood. A sample blood donation is shown. ($1 \text{ mm}^3 = 10^{-3}$ mL)

a. One cubic millimeter of blood contains about 10^4 white blood cells. How many white blood cells are in the donation? Write your answer in words.

b. One cubic millimeter of blood contains about 5×10^6 red blood cells. How many red blood cells are in the donation? Write your answer in words.

c. Compare your answers for parts (a) and (b).

35. PRECISION Describe how to rewrite a power with a positive exponent so that the exponent is in the denominator. Use the definition of negative exponents to justify your reasoning.

36. Reasoning The rule for negative exponents states that $a^{-n} = \frac{1}{a^n}$. Explain why this rule does not apply when $a = 0$.

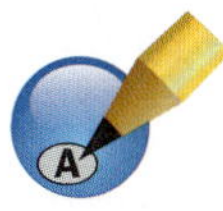

Fair Game Review What you learned in previous grades & lessons

Simplify the expression. Write your answer as a power. *(Section 10.2 and Section 10.3)*

37. $10^3 \cdot 10^6$

38. $10^2 \cdot 10$

39. $\frac{10^8}{10^4}$

40. MULTIPLE CHOICE Which data display best orders numerical data and shows how they are distributed? *(Section 9.4)*

Ⓐ bar graph

Ⓑ line graph

Ⓒ scatter plot

Ⓓ stem-and-leaf plot

10 Study Help

You can use an **information wheel** to organize information about a topic. Here is an example of an information wheel for exponents.

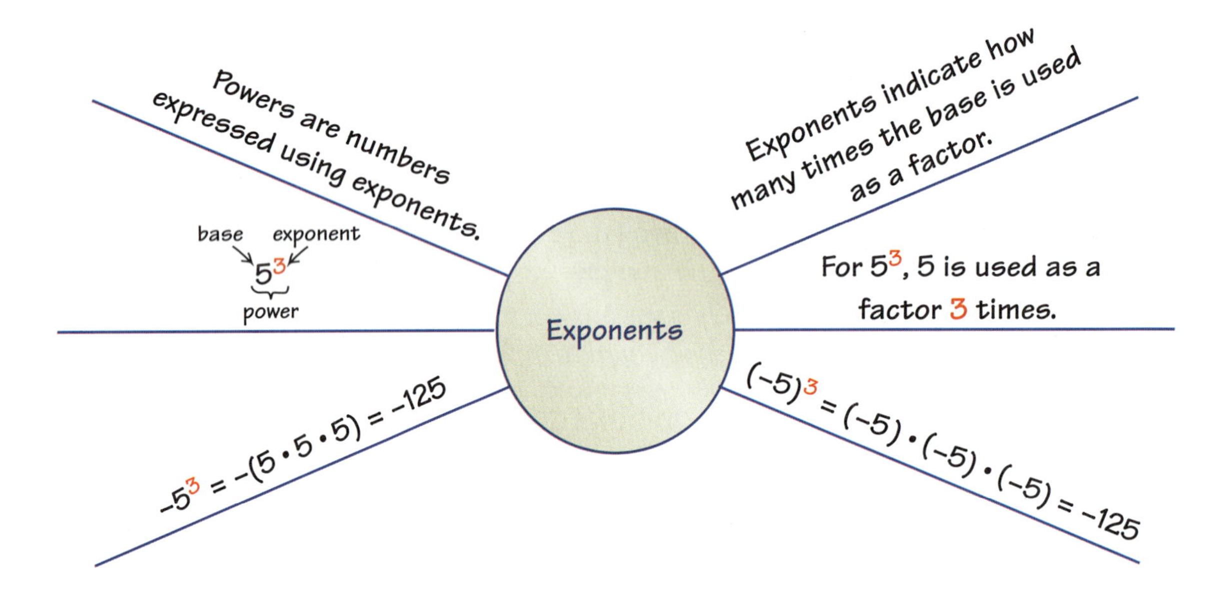

On Your Own

Make information wheels to help you study these topics.

1. Product of Powers Property
2. Quotient of Powers Property
3. zero and negative exponents

After you complete this chapter, make information wheels for the following topics.

4. writing numbers in scientific notation
5. writing numbers in standard form
6. adding and subtracting numbers in scientific notation
7. multiplying and dividing numbers in scientific notation
8. Choose three other topics you studied earlier in this course. Make an information wheel for each topic to summarize what you know about them.

"I decided to color code the different flavors in my information wheel."

Sample Answers

1.

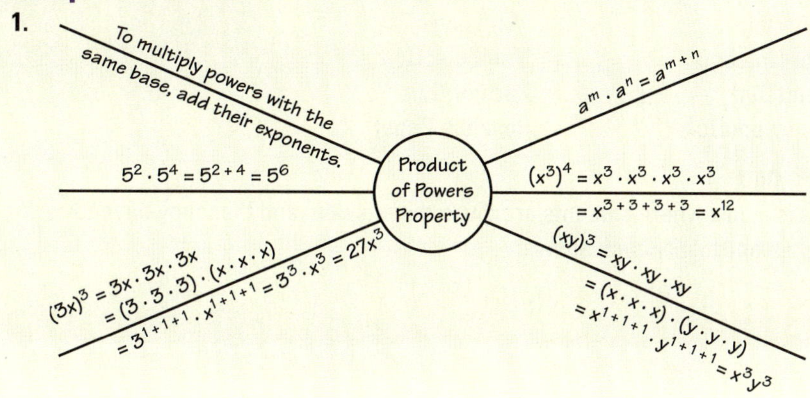

2.

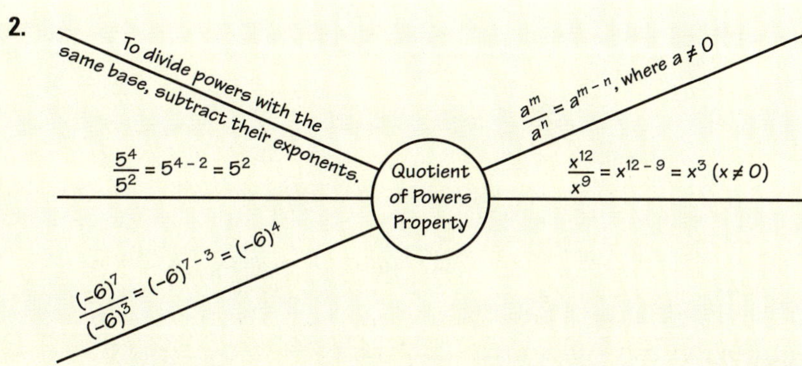

3.

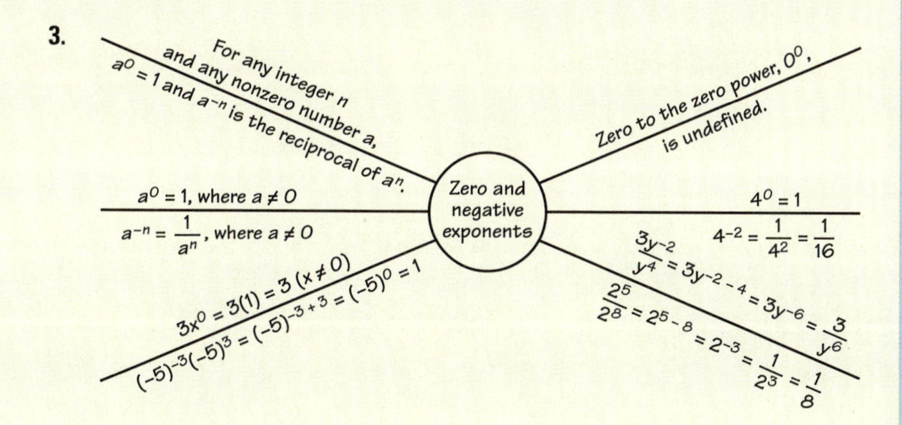

List of Organizers

Available at *BigIdeasMath.com*

Comparison Chart
Concept Circle
Definition (Idea) and Example Chart
Example and Non-Example Chart
Formula Triangle
Four Square
Information Frame
Information Wheel
Notetaking Organizer
Process Diagram
Summary Triangle
Word Magnet
Y Chart

About this Organizer

An **Information Wheel** can be used to organize information about a concept. Students write the concept in the middle of the "wheel." Then students write information related to the concept on the "spokes" of the wheel. Related information can include, but is not limited to: vocabulary words or terms, definitions, formulas, procedures, examples, and visuals. This type of organizer serves as a good summary tool because any information related to a concept can be included.

Editable Graphic Organizer

Answers

1. $(-5)^4$
2. 7^2m^3
3. 625
4. 64
5. 1
6. $\frac{1}{125}$
7. 3^9
8. a^{15}
9. $81c^4$
10. $\frac{4}{49}p^2$
11. 8^3
12. 6^8
13. π^3
14. t^{10}
15. $\frac{8}{d^6}$
16. $\frac{3}{x^2}$
17. **a** 10^{-3} m

    b. 1 millimeter; The length is less than 1 meter and a millimeter is smaller than a meter.
18. 10^5 times

Technology for the Teacher

Online Assessment
Assessment Book
ExamView® Assessment Suite

Alternative Quiz Ideas

100% Quiz
Error Notebook
Group Quiz
Homework Quiz
Math Log
Notebook Quiz
Partner Quiz
Pass the Paper

100% Quiz

This is a quiz where students are given the answers and then they have to explain and justify each answer.

Reteaching and Enrichment Strategies

If students need help. . .	If students got it. . .
Resources by Chapter • Practice A and Practice B • Puzzle Time Lesson Tutorials *BigIdeasMath.com*	Resources by Chapter • Enrichment and Extension • Technology Connection Game Closet at *BigIdeasMath.com* Start the next section

10.1–10.4 Quiz

Write the product using exponents. *(Section 10.1)*

1. $(-5) \cdot (-5) \cdot (-5) \cdot (-5)$

2. $7 \cdot 7 \cdot m \cdot m \cdot m$

Evaluate the expression. *(Section 10.1 and Section 10.4)*

3. 5^4

4. $(-2)^6$

5. $(-4.8)^{-9} \cdot (-4.8)^9$

6. $\dfrac{5^4}{5^7}$

Simplify the expression. Write your answer as a power. *(Section 10.2)*

7. $3^8 \cdot 3$

8. $(a^5)^3$

Simplify the expression. *(Section 10.2)*

9. $(3c)^4$

10. $\left(-\dfrac{2}{7}p\right)^2$

Simplify the expression. Write your answer as a power. *(Section 10.3)*

11. $\dfrac{8^7}{8^4}$

12. $\dfrac{6^3 \cdot 6^7}{6^2}$

13. $\dfrac{\pi^{15}}{\pi^3 \cdot \pi^9}$

14. $\dfrac{t^{13}}{t^5} \cdot \dfrac{t^8}{t^6}$

Simplify. Write the expression using only positive exponents. *(Section 10.4)*

15. $8d^{-6}$

16. $\dfrac{12x^5}{4x^7}$

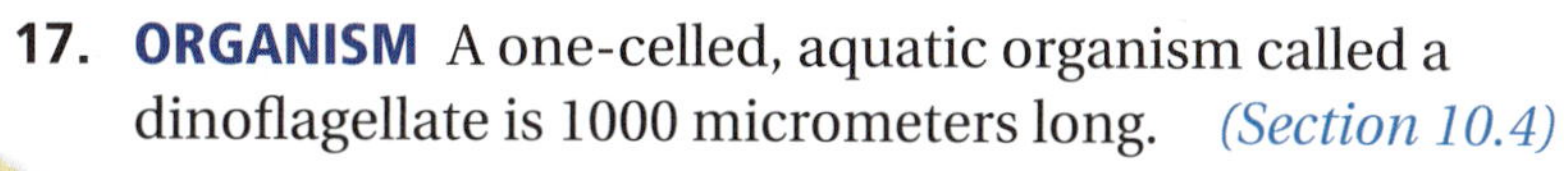

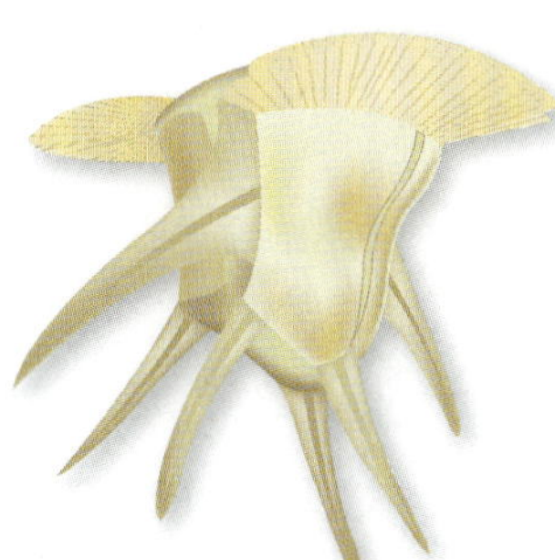

17. ORGANISM A one-celled, aquatic organism called a dinoflagellate is 1000 micrometers long. *(Section 10.4)*

a. One micrometer is 10^{-6} meter. What is the length of the dinoflagellate in meters?

b. Is the length of the dinoflagellate equal to 1 millimeter or 1 kilometer? Explain.

18. EARTHQUAKES An earthquake of magnitude 3.0 is 10^2 times stronger than an earthquake of magnitude 1.0. An earthquake of magnitude 8.0 is 10^7 times stronger than an earthquake of magnitude 1.0. How many times stronger is an earthquake of magnitude 8.0 than an earthquake of magnitude 3.0? *(Section 10.3)*

10.5 Reading Scientific Notation

Essential Question How can you read numbers that are written in scientific notation?

1 ACTIVITY: Very Large Numbers

Work with a partner.

- Use a calculator. Experiment with multiplying large numbers until your calculator displays an answer that is *not* in standard form.
- When the calculator at the right was used to multiply 2 billion by 3 billion, it listed the result as

  6.0E+18.

- Multiply 2 billion by 3 billion by hand. Use the result to explain what 6.0E+18 means.
- Check your explanation by calculating the products of other large numbers.
- Why didn't the calculator show the answer in standard form?
- Experiment to find the maximum number of digits your calculator displays. For instance, if you multiply 1000 by 1000 and your calculator shows 1,000,000, then it can display seven digits.

2 ACTIVITY: Very Small Numbers

Work with a partner.

- Use a calculator. Experiment with multiplying very small numbers until your calculator displays an answer that is *not* in standard form.
- When the calculator at the right was used to multiply 2 billionths by 3 billionths, it listed the result as

  6.0E–18.

- Multiply 2 billionths by 3 billionths by hand. Use the result to explain what 6.0E–18 means.
- Check your explanation by calculating the products of other very small numbers.

Scientific Notation

In this lesson, you will

- identify numbers written in scientific notation.
- write numbers in standard form.
- compare numbers in scientific notation.

Learning Standards
MACC.8.EE.1.3
MACC.8.EE.1.4

Laurie's Notes

Introduction

Standards for Mathematical Practice

- **MP5 Use Appropriate Tools Strategically:** In the first two activities, students will use calculators to multiply very large and very small numbers. Students will then explain the meanings of the resulting calculator displays.

Motivate

? "Have you had your millionth heartbeat?" Answers will vary.

- Assume that your heart beats once per second, and it has since you were born. Convert 1 million seconds to days.

$$10^6 \text{ sec} \cdot \frac{1 \text{ min}}{60 \text{ sec}} \cdot \frac{1 \text{ h}}{60 \text{ min}} \cdot \frac{1 \text{ day}}{24 \text{ h}} \approx 11.57 \text{ days}$$

- Clearly, all of your students have had their millionth heart beat. But have they had their billionth? Because there are 1000 million in 1 billion,

$$11.57 \text{ days} \times 1000 = 11{,}570 \text{ days and}$$

$$11{,}570 \text{ days} \cdot \frac{1 \text{ year}}{365 \text{ days}} \approx 31.7 \text{ years.}$$

Activity Notes

Activity 1

? "What does standard form mean in the context of this activity?" Numbers are written using digits; example: 123.

? "What does expanded form mean?" Numbers are written showing the value of each digit; example: $123 = 1 \times 100 + 2 \times 10 + 3 \times 1$.

- **MP5:** This activity gives students time to explore how scientific notation is displayed on their calculators.
- Students should be able to explain how they determined the number of digits the calculator displays.

? "When the display on a calculator reads 4.5 E+8, what does this mean?" The E+8 means that the decimal point should be moved 8 places to the right.

- Write on the board: 4.5 E+8 = 450,000,000.

Activity 2

- This activity is the same as Activity 1, except with very small numbers.
- Review place values less than 1.
- **Summary:** Check to see that everyone was successful in getting both large and small numbers to display. Has everyone figured out what notations such as E+4 and E−6 mean on the calculator? Do students know how many digits their calculators display?

? "When the display on a calculator reads 6.2 E−6, what does this mean?" The E−6 means that the decimal point should be moved 6 places to the left.

- Write on the board: 6.2 E−6 = 0.0000062.

Florida Common Core Standards

MACC.8.EE.1.3 Use numbers expressed in the form of a single digit times an integer power of 10 to estimate very large or very small quantities, and to express how many times as much one is than the other.

MACC.8.EE.1.4 Perform operations with numbers expressed in scientific notation, including problems where both decimal and scientific notation are used. Use scientific notation and choose units of appropriate size for measurements of very large or very small quantities. . . . Interpret scientific notation that has been generated by technology.

Previous Learning

Students should know the base 10 place value system.

Lesson Plans
Complete Materials List

10.5 Record and Practice Journal

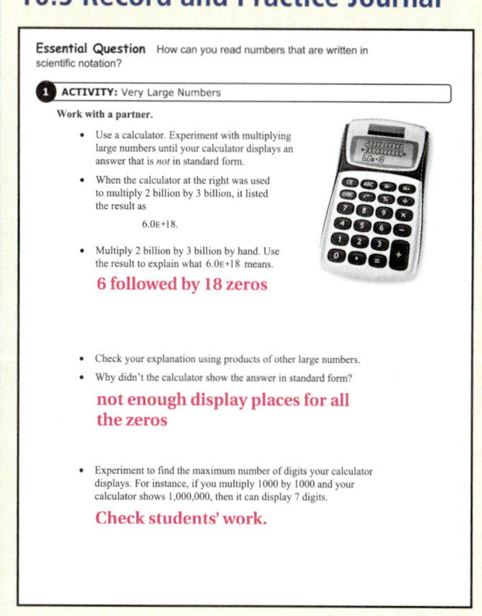

Essential Question How can you read numbers that are written in scientific notation?

1 ACTIVITY: Very Large Numbers

Work with a partner.

- Use a calculator. Experiment with multiplying large numbers until your calculator displays an answer that is *not* in standard form.
- When the calculator at the right was used to multiply 2 billion by 3 billion, it listed the result as

  6.0E+18.

- Multiply 2 billion by 3 billion by hand. Use the result to explain what 6.0E+18 means.

  6 followed by 18 zeros

- Check your explanation using products of other large numbers.
- Why didn't the calculator show the answer in standard form?

  not enough display places for all the zeros

- Experiment to find the maximum number of digits your calculator displays. For instance, if you multiply 1000 by 1000 and your calculator shows 1,000,000, then it can display 7 digits.

  Check students' work.

Differentiated Instruction

Kinesthetic

Have students use grid paper when converting numbers from scientific notation to standard form and from standard form to scientific notation. Write the number with one digit in each square. Place the decimal point on the line segment between the squares. Students may find it easier to count the number of squares than the number of digits.

10.5 Record and Practice Journal

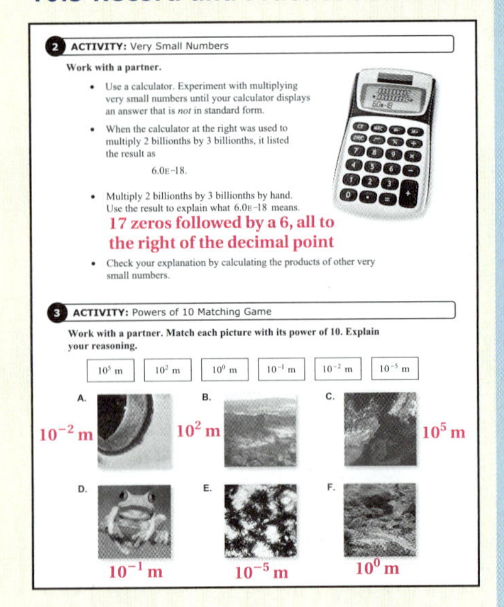

2 ACTIVITY: Very Small Numbers

Work with a partner.

- Use a calculator. Experiment with multiplying very small numbers until your calculator displays an answer that is *not* in standard form.
- When the calculator at the right was used to multiply 2 billionths by 3 billionths, it listed the result as

  6.0E−18.

- Multiply 2 billionths by 3 billionths by hand. Use the result to explain what 6.0E−18 means.

  17 zeros followed by a 6, all to the right of the decimal point

- Check your explanation by calculating the products of other very small numbers.

3 ACTIVITY: Powers of 10 Matching Game

Work with a partner. Match each picture with its power of 10. Explain your reasoning.

10^5 m | 10^2 m | 10^0 m | 10^{-1} m | 10^{-2} m | 10^{-5} m

A. 10^{-2} m
B. 10^2 m
C. 10^5 m
D. 10^{-1} m
E. 10^{-5} m
F. 10^0 m

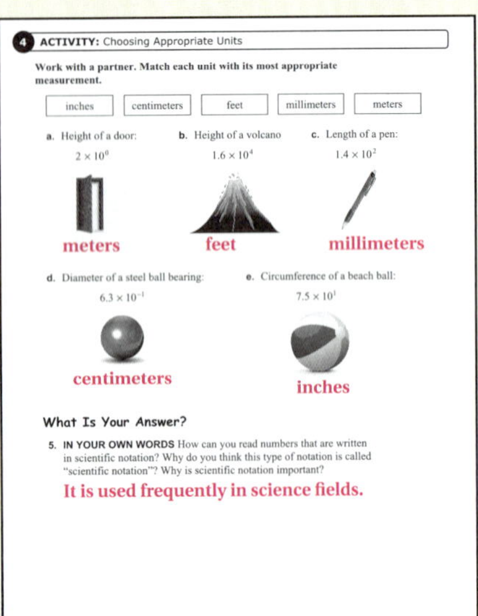

4 ACTIVITY: Choosing Appropriate Units

Work with a partner. Match each unit with its most appropriate measurement.

inches | centimeters | feet | millimeters | meters

a. Height of a door: 2×10^0 meters
b. Height of a volcano 1.6×10^4 feet
c. Length of a pen: 1.4×10^2 millimeters
d. Diameter of a steel ball bearing: 6.3×10^{-1} centimeters
e. Circumference of a beach ball: 7.5×10^1 inches

What Is Your Answer?

5. **IN YOUR OWN WORDS** How can you read numbers that are written in scientific notation? Why do you think this type of notation is called "scientific notation"? Why is scientific notation important?

   It is used frequently in science fields.

Laurie's Notes

Activity 3

- Most students will enjoy trying to figure out what is in each picture.
- Make a transparency of the six photos or display them under a document camera to facilitate discussion.

? "What clues did you use to order the photos?" open ended

Activity 4

- Most students will enjoy this activity.
- Students need to think about the standard form of each measurement *and* which unit makes sense.
- **MP2 Reason Abstractly and Quantitatively:** Students may successfully identify the standard form of each measurement but they may have difficulty reasoning about the appropriate unit. Ask questions such as, "Would a reasonable height of a door be 2 inches, 2 centimeters, 2 feet, 2 millimeters, or 2 meters?
- Ask a volunteer to share his or her answers.

What Is Your Answer?

- **Think-Pair-Share:** Students should read the question independently and then work in pairs to answer the questions. When they have answered the questions, the pair should compare their answer with another group and discuss any discrepancies.

Closure

- What examples have you read or heard about that involve very large or very small numbers?

3 ACTIVITY: Powers of 10 Matching Game

Math Practice 4

Analyze Relationships

How are the pictures related? How can you order the pictures to find the correct power of 10?

Work with a partner. Match each picture with its power of 10. Explain your reasoning.

10^5 m 10^2 m 10^0 m 10^{-1} m 10^{-2} m 10^{-5} m

A. 

B. 

C. 

D. 

E. 

F. 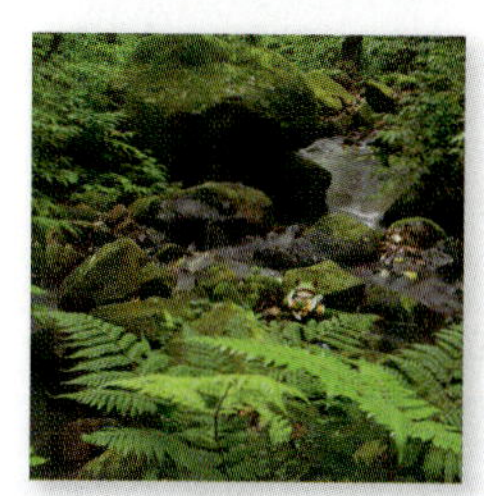

4 ACTIVITY: Choosing Appropriate Units

Work with a partner. Match each unit with its most appropriate measurement.

inches centimeters feet millimeters meters

A. Height of a door: 2×10^0

B. Height of a volcano: 1.6×10^4

C. Length of a pen: 1.4×10^2

D. Diameter of a steel ball bearing: 6.3×10^{-1}

E. Circumference of a beach ball: 7.5×10^1

What Is Your Answer?

5. **IN YOUR OWN WORDS** How can you read numbers that are written in scientific notation? Why do you think this type of notation is called *scientific notation*? Why is scientific notation important?

Practice Use what you learned about reading scientific notation to complete Exercises 3–5 on page 440.

10.5 Lesson

Key Vocabulary
scientific notation, *p. 438*

Study Tip
Scientific notation is used to write very small and very large numbers.

Key Idea

Scientific Notation

A number is written in **scientific notation** when it is represented as the product of a factor and a power of 10. The factor must be greater than or equal to 1 and less than 10.

The factor is greater than or equal to 1 and less than 10. → 8.3×10^{-7} ← The power of 10 has an integer exponent.

EXAMPLE 1 Identifying Numbers Written in Scientific Notation

Tell whether the number is written in scientific notation. Explain.

a. 5.9×10^{-6}

The factor is greater than or equal to 1 and less than 10. The power of 10 has an integer exponent. So, the number is written in scientific notation.

b. 0.9×10^{8}

The factor is less than 1. So, the number is not written in scientific notation.

Key Idea

Writing Numbers in Standard Form

The absolute value of the exponent indicates how many places to move the decimal point.

- If the exponent is negative, move the decimal point to the left.
- If the exponent is positive, move the decimal point to the right.

EXAMPLE 2 Writing Numbers in Standard Form

a. Write 3.22×10^{-4} in standard form.

$3.22 \times 10^{-4} = 0.000322$ Move decimal point $|-4| = 4$ places to the left.

4

b. Write 7.9×10^{5} in standard form.

$7.9 \times 10^{5} = 790{,}000$ Move decimal point $|5| = 5$ places to the right.

5

Laurie's Notes

Introduction

Connect

- **Yesterday:** Students explored very large and very small numbers written in scientific notation. (MP2, MP5)
- **Today:** Students will read numbers in scientific notation and write them in standard form.

Motivate

- Share some information about the Florida Keys.
- The Florida Keys are made up of approximately 1700 islands, or keys, that stretch about 126 miles from the mainland to the last key, Key West. Most of the islands are uninhabited, but the populated keys are connected by a highway that crosses 42 bridges.

? "Do you know how big a square foot is?" Answers will vary.

- The Florida Keys are approximately 3.83×10^9 square feet.

? "Is the area of the Keys more than or less than a billion square feet?" more than; Students will probably need to write 3.83×10^9 square feet in standard form to answer.

Lesson Notes

Key Idea

- Write the Key Idea. There are two parts to the definition; the factor is a number n, with $1 \leq n < 10$, and it is multiplied by a power of 10 with an integer exponent.

Example 1

- Work through each part of the example as shown.

Key Idea

- Write the Key Idea. From the activities yesterday, students should find it reasonable that the exponent of 10 is connected to place value. If the exponent is positive, the number will be larger, so the decimal point moves to the right. Conversely, if the exponent is negative, the number will be smaller, so the decimal point moves to the left.

? Have students fill in the blanks.

- "A power of 10 with a positive exponent is _____ 1." greater than or equal to
- "A power of 10 with a negative exponent is _____ 1." less than

Example 2

- In part (a), 3.22 is the factor and −4 is the exponent. The number in standard form will be less than 3.22, so the decimal point moves to the left 4 places.
- In part (b), 7.9 is the factor and 5 is the exponent. The number in standard form will be greater than 7.9, so the decimal point moves to the right 5 places.

Goal Today's lesson is reading numbers in **scientific notation** and writing them in standard form.

Technology for the Teacher

Lesson Tutorials
Lesson Plans
Answer Presentation Tool

Extra Example 1

Tell whether the number is written in scientific notation. Explain.

a. 2.5×10^{-9} yes; The factor is greater than or equal to 1 and less than 10. The power of 10 has an integer exponent.

b. 0.5×10^{6} no; The factor is less than 1.

Extra Example 2

a. Write 2.75×10^{-3} in standard form. 0.00275

b. Write 6.38×10^{7} in standard form. 63,800,000

On Your Own

1. no; The factor is greater than 10.
2. 60,000,000
3. 0.000099 **4.** 12,850

Extra Example 3

In Example 3, the density of an ear of corn is 7.21×10^2 kilograms per cubic meter. What happens when an ear of corn is placed in water? The ear of corn is less dense than water, so it will float.

Extra Example 4

In Example 4, a dog has 50 female fleas. How much blood do the fleas consume per day? 0.7 milliliter of blood per day

On Your Own

5. It will sink.
6. 1.05 mL

English Language Learners

Word Problems

Have students work in groups of 3 or 4, including both English learners and English speakers. Provide them with poster board and markers. Assign each group a problem-solving exercise with scientific notation. Each group is to solve their problem showing all of the steps and using scientific notation. English learners will benefit by having the opportunity to restate the problem and gain a deeper understanding of the concept.

Laurie's Notes

On Your Own

- ? **Extension:** "How could you write the number in Question 1 in scientific notation?" 1.2×10^5
- **Think-Pair-Share:** Students should read each question independently and then work in pairs to answer the questions. When they have answered the questions, the pair should compare their answers with another group and discuss any discrepancies.

Example 3

- Ask a student to read the problem. Discuss the density of a substance if they have encountered it in science class, otherwise give an explanation of what density means. Density equals mass divided by volume.
- Work through the example as shown.
- **MP7 Look for and Make Use of Structure:** Show students an alternate way to solve Example 3. You can compare numbers without writing them in standard form. Rewrite the numbers so that they all have the same exponent, then compare the factor of each number. For example, 1.0×10^3, 1.84×10^3, and 0.641×10^3.

Example 4

- Before students get too concerned about fleas, remind them that 1.4×10^{-5} liter is a small amount!
- Students should be comfortable multiplying by powers of 10.
- ? "Can you multiply 10^{-5} by 100 first, and then multiply by 1.4? Explain." yes; Multiplication is commutative.

On Your Own

- **Neighbor Check:** Have students work independently and then have their neighbors check their work. Have students discuss any discrepancies.

Closure

- **Exit Ticket:** Write the number in standard form.
  1. 1.56×10^7 15,600,000
  2. 6.3×10^{-5} 0.000063

1. Is 12×10^4 written in scientific notation? Explain.

Write the number in standard form.

2. 6×10^7
3. 9.9×10^{-5}
4. 1.285×10^4

EXAMPLE 3 Comparing Numbers in Scientific Notation

An object with a lesser density than water will float. An object with a greater density than water will sink. Use each given density (in kilograms per cubic meter) to explain what happens when you place a brick and an apple in water.

Water: 1.0×10^3 **Brick:** 1.84×10^3 **Apple:** 6.41×10^2

You can compare the densities by writing each in standard form.

Water	Brick	Apple
$1.0 \times 10^3 = 1000$	$1.84 \times 10^3 = 1840$	$6.41 \times 10^2 = 641$

The apple is less dense than water, so it will float. The brick is denser than water, so it will sink.

EXAMPLE 4 Real-Life Application

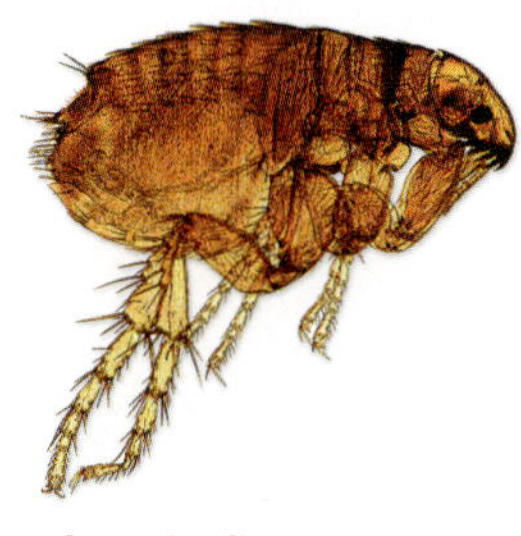
A female flea consumes about 1.4×10^{-5} liter of blood per day.

A dog has 100 female fleas. How much blood do the fleas consume per day?

$$1.4 \times 10^{-5} \cdot 100 = 0.000014 \cdot 100 \quad \text{Write in standard form.}$$
$$= 0.0014 \quad \text{Multiply.}$$

The fleas consume about 0.0014 liter, or 1.4 milliliters of blood per day.

On Your Own

5. **WHAT IF?** In Example 3, the density of lead is 1.14×10^4 kilograms per cubic meter. What happens when you place lead in water?
6. **WHAT IF?** In Example 4, a dog has 75 female fleas. How much blood do the fleas consume per day?

10.5 Exercises

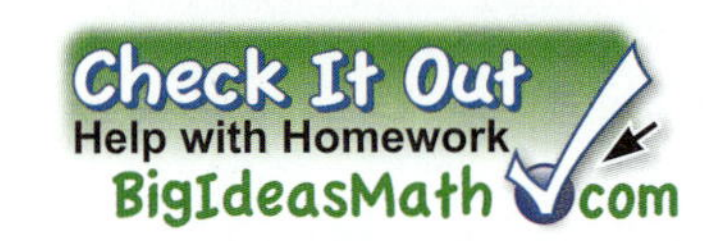

Vocabulary and Concept Check

1. **WRITING** Describe the difference between scientific notation and standard form.

2. **WHICH ONE DOESN'T BELONG?** Which number does *not* belong with the other three? Explain.

2.8×10^{15} | 4.3×10^{-30} | 1.05×10^{28} | 10×9.2^{-13}

Practice and Problem Solving

Write the number shown on the calculator display in standard form.

3. 5.6E12

4. 2.1E-10

5. 8.73E16

Tell whether the number is written in scientific notation. Explain.

① 6. 1.8×10^{9}

7. 3.45×10^{14}

8. 0.26×10^{-25}

9. 10.5×10^{12}

10. 46×10^{-17}

11. 5×10^{-19}

12. 7.814×10^{-36}

13. 0.999×10^{42}

14. 6.022×10^{23}

Write the number in standard form.

② 15. 7×10^{7}

16. 8×10^{-3}

17. 5×10^{2}

18. 2.7×10^{-4}

19. 4.4×10^{-5}

20. 2.1×10^{3}

21. 1.66×10^{9}

22. 3.85×10^{-8}

23. 9.725×10^{6}

24. **ERROR ANALYSIS** Describe and correct the error in writing the number in standard form.

$4.1 \times 10^{-6} = 4{,}100{,}000$

2.7×10^{8} platelets per milliliter

25. **PLATELETS** Platelets are cell-like particles in the blood that help form blood clots.

    a. How many platelets are in 3 milliliters of blood? Write your answer in standard form.

    b. An adult human body contains about 5 liters of blood. How many platelets are in an adult human body?

Assignment Guide and Homework Check

Level	Day 1 Activity Assignment	Day 2 Lesson Assignment	Homework Check
Basic	3–5, 33–36	1, 2, 7–23 odd, 24, 25, 27	9, 19, 25, 27
Average	3–5, 33–36	1, 2, 12–14, 21–29	12, 22, 25, 27
Advanced	1–5, 12–26 even, 27–36		12, 22, 27, 30, 31

Common Errors

- **Exercises 6–14** Students may think that all of the numbers are written in scientific notation because all of the exponents are integers. Remind them that the factor must be greater than or equal to 1 and less than 10 in order for the number to be in scientific notation.
- **Exercises 15–23** Students may move the decimal point in the wrong direction. Remind them that when the exponent is negative they move the decimal point to the left, and when it is positive they move the decimal point to the right.
- **Exercise 27** Students may order the surface temperatures by the factor without considering the power of 10. Encourage them to write the numbers in standard form, or rewrite the numbers so that they all have the same exponent, before comparing the numbers.

10.5 Record and Practice Journal

Tell whether the number is written in scientific notation. Explain.

1. 14×10^8 no; The factor is greater than 10.
2. 2.6×10^{12} yes; The factor is greater than 1 and less than 10.
3. 4.79×10^{-8} yes; The factor is greater than 1 and less than 10.
4. 3.99×10^{16} yes; The factor is greater than 1 and less than 10.
5. 0.15×10^{22} no; The factor is less than 1.
6. 6×10^3 yes; The factor is greater than 1 and less than 10.

Write the number in standard form.

7. 4×10^9 4,000,000,000
8. 2×10^{-5} 0.00002
9. 3.7×10^6 3,700,000
10. 4.12×10^{-3} 0.00412
11. 7.62×10^{10} 76,200,000,000
12. 9.908×10^{-12} 0.000000000009908
13. Light travels at 3×10^8 meters per second.
    a. Write the speed of light in standard form. 300,000,000
    b. How far has light traveled after 5 seconds? 1,500,000,000 meters

Vocabulary and Concept Check

1. Scientific notation uses a factor greater than or equal to 1 but less than 10 multiplied by a power of 10. A number in standard form is written out with all the zeros and place values included.
2. 10×9.2^{-13}; All of the other numbers are written in scientific notation.

Practice and Problem Solving

3. 5,600,000,000,000
4. 0.00000000021
5. 87,300,000,000,000,000
6. yes; The factor is greater than or equal to 1 and less than 10. The power of 10 has an integer exponent.
7. yes; The factor is greater than or equal to 1 and less than 10. The power of 10 has an integer exponent.
8. no; The factor is less than 1.
9. no; The factor is greater than 10.
10. no; The factor is greater than 10.
11. yes; The factor is greater than or equal to 1 and less than 10. The power of 10 has an integer exponent.
12. yes; The factor is greater than or equal to 1 and less than 10. The power of 10 has an integer exponent.
13. no; The factor is less than 1.
14. yes; The factor is greater than or equal to 1 and less than 10. The power of 10 has an integer exponent.

Practice and Problem Solving

15. 70,000,000
16. 0.008
17. 500
18. 0.00027
19. 0.000044
20. 2100
21. 1,660,000,000
22. 0.0000000385
23. 9,725,000
24. The negative exponent means the decimal point will move left, not right, when the number is written in standard form. $4.1 \times 10^{-6} = 0.0000041$
25. a. 810,000,000 platelets
    b. 1,350,000,000,000 platelets
26. 100 zeros
27. a. Bellatrix
    b. Betelgeuse
28. The value of the number is 10 times greater.
29. 1555.2 km^2
30. 5×10^{12} km^2
31. 35,000,000 km^3
32. See *Taking Math Deeper*.

Fair Game Review

33. 4^5
34. 3^3y^3
35. $(-2)^3$
36. B

Mini-Assessment

Write the number in standard form.

1. 5×10^{-4} 0.0005
2. 2.5×10^{-3} 0.0025
3. 1.66×10^3 1660
4. 3.89×10^{-5} 0.0000389
5. 4.576×10^8 457,600,000

Taking Math Deeper

Exercise 32

This is an interesting problem in physics. It is about the speed of light in different media. The problem gives students practice in unit analysis, and it also points out that to compare speeds, you need to compare apples to apples, not apples to oranges.

1 Make a table and convert measures.

Medium	Speed	Speed (m per sec)
Air	$\frac{6.7 \times 10^8 \text{ mi}}{\text{h}}$	$\frac{3.0 \times 10^8 \text{ m}}{\text{sec}}$
Glass	$\frac{6.6 \times 10^8 \text{ ft}}{\text{sec}}$	$\frac{2.0 \times 10^8 \text{ m}}{\text{sec}}$
Ice	$\frac{2.3 \times 10^5 \text{ km}}{\text{sec}}$	$\frac{2.3 \times 10^8 \text{ m}}{\text{sec}}$
Vacuum	$\frac{3.0 \times 10^8 \text{ m}}{\text{sec}}$	$\frac{3.0 \times 10^8 \text{ m}}{\text{sec}}$
Water	$\frac{2.3 \times 10^{10} \text{ cm}}{\text{sec}}$	$\frac{2.3 \times 10^8 \text{ m}}{\text{sec}}$

2 a. For the significant digits given, the speed of light is the same in air or in a vacuum. Light is fastest in these two media.

b. Of the five media listed, light travels the slowest in glass.

3 If students take a course in physics, they will learn that light is slowed down in transparent media such as air, water, ice, and glass. The ratio by which it is slowed is called the *refractive index* of the medium and is always greater than one. This was discovered by Jean Foucault in 1850. The refractive index of air is 1.0003, which means that light travels slightly slower in air than in a vacuum.

When people talk about "the speed of light" in a general context, they usually mean "the speed of light in a vacuum." This quantity is also refered to as c. It is famous from Einstein's equation $E = mc^2$.

Reteaching and Enrichment Strategies

If students need help. . .	If students got it. . .
Resources by Chapter • Practice A and Practice B • Puzzle Time Record and Practice Journal Practice Differentiating the Lesson Lesson Tutorials Skills Review Handbook	Resources by Chapter • Enrichment and Extension • Technology Connection Start the next section

26. **REASONING** A googol is 1.0×10^{100}. How many zeros are in a googol?

3 27. **STARS** The table shows the surface temperatures of five stars.

a. Which star has the highest surface temperature?

b. Which star has the lowest surface temperature?

Star	Betelgeuse	Bellatrix	Sun	Aldebaran	Rigel
Surface Temperature (°F)	6.2×10^3	3.8×10^4	1.1×10^4	7.2×10^3	2.2×10^4

28. **NUMBER SENSE** Describe how the value of a number written in scientific notation changes when you increase the exponent by 1.

29. **CORAL REEF** The area of the Florida Keys National Marine Sanctuary is about 9.6×10^3 square kilometers. The area of the Florida Reef Tract is about 16.2% of the area of the sanctuary. What is the area of the Florida Reef Tract in square kilometers?

30. **REASONING** A gigameter is 1.0×10^6 kilometers. How many square kilometers are in 5 square gigameters?

31. **WATER** There are about 1.4×10^9 cubic kilometers of water on Earth. About 2.5% of the water is fresh water. How much fresh water is on Earth?

32. 

The table shows the speed of light through five media.

a. In which medium does light travel the fastest?

b. In which medium does light travel the slowest?

Medium	Speed
Air	6.7×10^8 mi/h
Glass	6.6×10^8 ft/sec
Ice	2.3×10^5 km/sec
Vacuum	3.0×10^8 m/sec
Water	2.3×10^{10} cm/sec

Fair Game Review What you learned in previous grades & lessons

Write the product using exponents. *(Section 10.1)*

33. $4 \cdot 4 \cdot 4 \cdot 4 \cdot 4$

34. $3 \cdot 3 \cdot 3 \cdot y \cdot y \cdot y$

35. $(-2) \cdot (-2) \cdot (-2)$

36. **MULTIPLE CHOICE** What is the length of the hypotenuse of the right triangle? *(Section 7.3)*

Ⓐ $\sqrt{18}$ in. Ⓑ $\sqrt{41}$ in.

Ⓒ 18 in. Ⓓ 41 in.

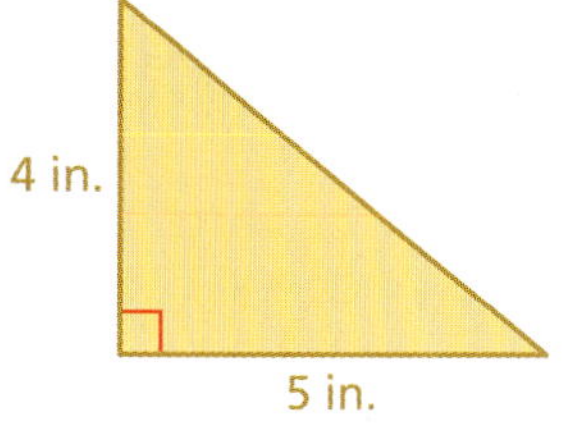

10.6 Writing Scientific Notation

Essential Question How can you write a number in scientific notation?

1 ACTIVITY: Finding pH Levels

Work with a partner. In chemistry, pH is a measure of the activity of dissolved hydrogen ions (H^+). Liquids with low pH values are called *acids*. Liquids with high pH values are called *bases*.

Find the pH of each liquid. Is the liquid a base, neutral, or an acid?

a. Lime juice:
$[H^+] = 0.01$

b. Egg:
$[H^+] = 0.00000001$

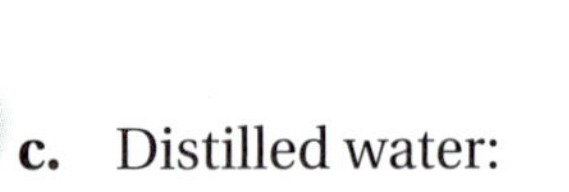

c. Distilled water:
$[H^+] = 0.0000001$

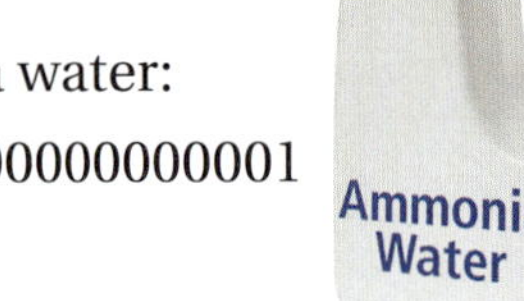

d. Ammonia water:
$[H^+] = 0.00000000001$

e. Tomato juice:
$[H^+] = 0.0001$

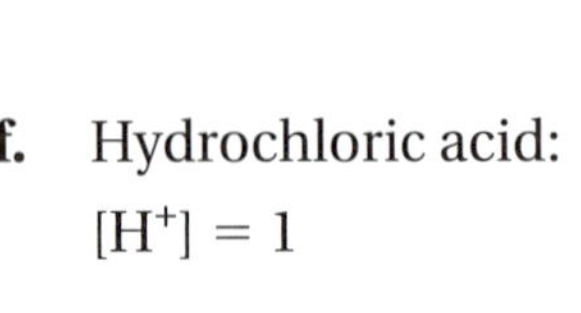

f. Hydrochloric acid:
$[H^+] = 1$

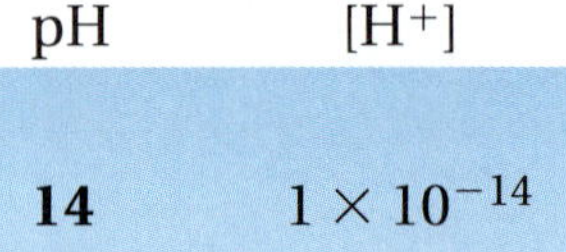

pH	$[H^+]$	
14	1×10^{-14}	Bases
13	1×10^{-13}	
12	1×10^{-12}	
11	1×10^{-11}	
10	1×10^{-10}	
9	1×10^{-9}	
8	1×10^{-8}	
7	1×10^{-7}	Neutral
6	1×10^{-6}	Acids
5	1×10^{-5}	
4	1×10^{-4}	
3	1×10^{-3}	
2	1×10^{-2}	
1	1×10^{-1}	
0	1×10^{0}	

Scientific Notation

In this lesson, you will

- write large and small numbers in scientific notation.
- perform operations with numbers written in scientific notation.

Learning Standards
MACC.8.EE.1.3
MACC.8.EE.1.4

Laurie's Notes

Introduction

Standards for Mathematical Practice

- **MP1a Make Sense of Problems** and **MP5 Use Appropriate Tools Strategically:** Visual models are used to help students make sense of numbers written in scientific notation.

Motivate

- **Preparation:** If possible, borrow a pH meter or a few strips of litmus paper from the science department. If these items are not available, move on to Activity 1.
- Without explanation, have 3 containers (coffee cups or paper cups) containing different liquids at the front of the room.
- Use the litmus paper or pH meter to test the liquids. Students should guess that you are doing a pH test, which leads into Activity 1.

Words of Wisdom

- **Safety:** For reasons of safety you should not consume, nor allow the students to consume, any of the liquids.

Activity Notes

Activity 1

? "How many of you have heard of pH or studied pH in science?"

? "What does pH level refer to? Can anyone explain?" Listen for the measure of concentration of dissolved hydrogen ions; liquids with low pH are called acids; liquids with high pH are called bases.

- **MP5:** Have students refer to the pH chart in the book. The pH level is a number from 0 to 14, the opposite of the exponent (0 to -14) measuring the concentration of the dissolved hydrogen ions. At the middle of the chart, a pH of 7 is called neutral.
- **FYI:** Pure water is neutral with a pH value of 7. Low on the scale are acids, which have a sour taste like lemons. High on the scale are bases, which have a bitter taste like soap.
- Notice that all of the pH values are given in standard form. Students will need to think about how these numbers would be written in scientific notation.

? "How did you compare the numbers in standard form with the scale in scientific notation?" Listen for idea of the number of place values away from 1.

Florida Common Core Standards

MACC.8.EE.1.3 Use numbers expressed in the form of a single digit times an integer power of 10 to estimate very large or very small quantities, and to express how many times as much one is than the other.

MACC.8.EE.1.4 Perform operations with numbers expressed in scientific notation, including problems where both decimal and scientific notation are used. Use scientific notation and choose units of appropriate size for measurements of very large or very small quantities. . . . Interpret scientific notation that has been generated by technology.

Previous Learning

Students should be familiar with the metric system and powers of ten.

Lesson Plans
Complete Materials List

10.6 Record and Practice Journal

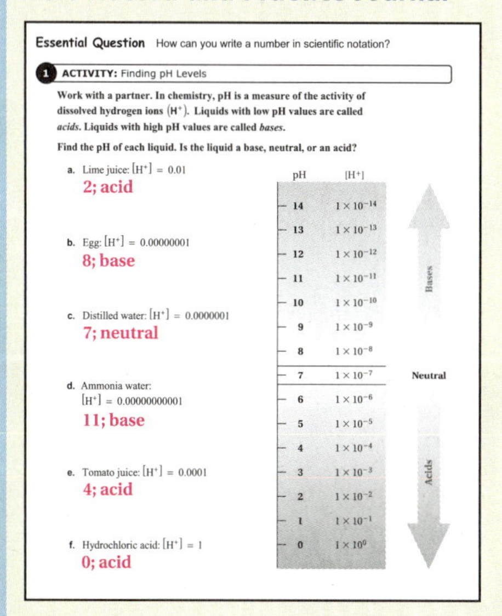

Essential Question How can you write a number in scientific notation?

1 ACTIVITY: Finding pH Levels

Work with a partner. In chemistry, pH is a measure of the activity of dissolved hydrogen ions (H^+). Liquids with low pH values are called *acids*. Liquids with high pH values are called *bases*.

Find the pH of each liquid. Is the liquid a base, neutral, or an acid?

a. Lime juice: $[H^+] = 0.01$
2; acid

b. Egg: $[H^+] = 0.00000001$
8; base

c. Distilled water: $[H^+] = 0.0000001$
7; neutral

d. Ammonia water: $[H^+] = 0.00000000001$
11; base

e. Tomato juice: $[H^+] = 0.0001$
4; acid

f. Hydrochloric acid: $[H^+] = 1$
0; acid

pH	$[H^+]$	
14	1×10^{-14}	Bases
13	1×10^{-13}	
12	1×10^{-12}	
11	1×10^{-11}	
10	1×10^{-10}	
9	1×10^{-9}	
8	1×10^{-8}	
7	1×10^{-7}	Neutral
6	1×10^{-6}	Acids
5	1×10^{-5}	
4	1×10^{-4}	
3	1×10^{-3}	
2	1×10^{-2}	
1	1×10^{-1}	
0	1×10^{0}	

English Language Learners

Culture

Have each student use the Internet or library resources to find the land area of his or her native country, the state where he or she currently lives, and two other states in the U.S. that have approximately the same land area as the native country. Each student can organize the information in a table giving the land area, the land area rounded to the nearest thousand square miles, and the land area in scientific notation. Create a classroom display of the tables.

10.6 Record and Practice Journal

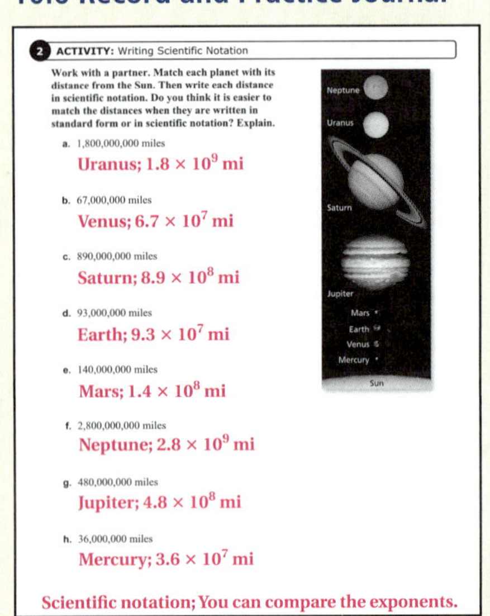

2 ACTIVITY: Writing Scientific Notation

Work with a partner. Match each planet with its distance from the Sun. Then write each distance in scientific notation. Do you think it is easier to match the distances when they are written in standard form or in scientific notation? Explain.

a. 1,800,000,000 miles
Uranus; 1.8×10^9 mi

b. 67,000,000 miles
Venus; 6.7×10^7 mi

c. 890,000,000 miles
Saturn; 8.9×10^8 mi

d. 93,000,000 miles
Earth; 9.3×10^7 mi

e. 140,000,000 miles
Mars; 1.4×10^8 mi

f. 2,800,000,000 miles
Neptune; 2.8×10^9 mi

g. 480,000,000 miles
Jupiter; 4.8×10^8 mi

h. 36,000,000 miles
Mercury; 3.6×10^7 mi

Scientific notation; You can compare the exponents.

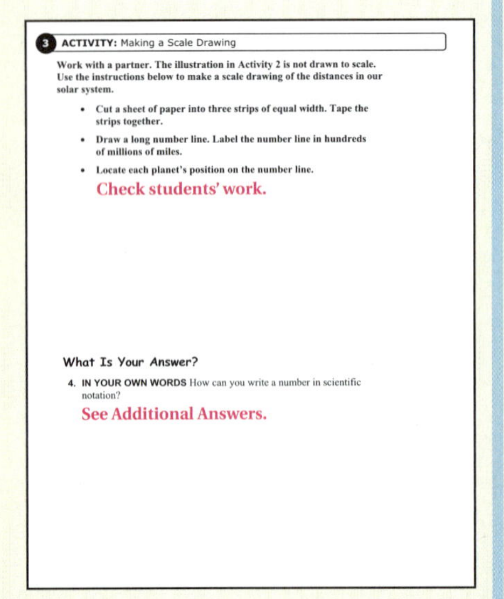

3 ACTIVITY: Making a Scale Drawing

Work with a partner. The illustration in Activity 2 is not drawn to scale. Use the instructions below to make a scale drawing of the distances in our solar system.

- **Cut a sheet of paper into three strips of equal width. Tape the strips together.**
- **Draw a long number line. Label the number line in hundreds of millions of miles.**
- **Locate each planet's position on the number line.**

Check students' work.

What Is Your Answer?

4. **IN YOUR OWN WORDS** How can you write a number in scientific notation?

See Additional Answers.

Laurie's Notes

Activity 2

- This activity demonstrates one way scientific notation is helpful. The distance of each planet from the sun is a very large number (in magnitude).
- Students should be able to get started right away without much explanation.

? "How do you match the planets with the distances?" Listen for ordering distances least to greatest and looking at the illustration.

? "How do you write the distances from the sun in scientific notation?" Move the decimal point to the left until you have a number that is at least 1 and less than 10. Count the number of places you moved the decimal point. This number becomes the exponent of the power of 10.

- **MP4 Model with Mathematics:** Because of the context, and the visual illustration, students should not have difficulty with this activity. They understand how to write the numbers in scientific notation.

Activity 3

- To help facilitate this activity, you may want to prepare strips of paper in advance.
- Even though the scale is given (hundreds of millions of miles), just writing multiples of this scale is a challenge for students. You may want to make a model on the front board.
- If students have studied the solar system, they should have some sense about how far away Neptune is.
- **MP1a** and **MP2 Reason Abstractly and Quantitatively:** To make a number line model of the distances to scale, you can write the distances in scientific notation. Each time the exponent increases by 1, the distance increases by a factor of 10.

What Is Your Answer?

- **Neighbor Check:** Have students work independently and then have their neighbors check their work. Have students discuss any discrepancies.

Closure

- **Writing Prompt:** Why is it useful to write very large or very small numbers in scientific notation?

2 ACTIVITY: Writing Scientific Notation

Work with a partner. Match each planet with its distance from the Sun. Then write each distance in scientific notation. Do you think it is easier to match the distances when they are written in standard form or in scientific notation? Explain.

a. 1,800,000,000 miles

b. 67,000,000 miles

c. 890,000,000 miles

d. 93,000,000 miles

e. 140,000,000 miles

f. 2,800,000,000 miles

g. 480,000,000 miles

h. 36,000,000 miles

3 ACTIVITY: Making a Scale Drawing

Math Practice 6

Calculate Accurately

How can you verify that you have accurately written each distance in scientific notation?

Work with a partner. The illustration in Activity 2 is not drawn to scale. Use the instructions below to make a scale drawing of the distances in our solar system.

- **Cut a sheet of paper into three strips of equal width. Tape the strips together to make one long piece.**
- **Draw a long number line. Label the number line in hundreds of millions of miles.**
- **Locate each planet's position on the number line.**

What Is Your Answer?

4. **IN YOUR OWN WORDS** How can you write a number in scientific notation?

Practice Use what you learned about writing scientific notation to complete Exercises 3–5 on page 446.

10.6 Lesson

Key Idea

Writing Numbers in Scientific Notation

Step 1: Move the decimal point so it is located to the right of the leading nonzero digit.

Step 2: Count the number of places you moved the decimal point. This indicates the exponent of the power of 10, as shown below.

Number Greater Than or Equal to 10

Use a positive exponent when you move the decimal point to the left.

$8600 = 8.6 \times 10^3$ (3)

Number Between 0 and 1

Use a negative exponent when you move the decimal point to the right.

$0.0024 = 2.4 \times 10^{-3}$ (3)

Study Tip

When you write a number greater than or equal to 1 and less than 10 in scientific notation, use zero as the exponent.

$6 = 6 \times 10^0$

EXAMPLE 1 Writing Large Numbers in Scientific Notation

Google purchased YouTube for $1,650,000,000. Write this number in scientific notation.

Move the decimal point 9 places to the left.

$1{,}650{,}000{,}000 = 1.65 \times 10^9$ (9)

The number is greater than 10. So, the exponent is positive.

EXAMPLE 2 Writing Small Numbers in Scientific Notation

The 2004 Indonesian earthquake slowed the rotation of Earth, making the length of a day 0.00000268 second shorter. Write this number in scientific notation.

Move the decimal point 6 places to the right.

$0.00000268 = 2.68 \times 10^{-6}$ (6)

The number is between 0 and 1. So, the exponent is negative.

On Your Own

Write the number in scientific notation.

1. 50,000
2. 25,000,000
3. 683
4. 0.005
5. 0.00000033
6. 0.000506

Laurie's Notes

Introduction

Connect

- **Yesterday:** Students explored very large and very small numbers written in standard form. (MP1a, MP2, MP4, MP5)
- **Today:** Students will convert numbers in standard form to scientific notation.

Motivate

- Share with students some information about the U.S. economy.
- Numbers related to the U.S. economy are often so large and so common, many become numb to their magnitude. The place values, millions, billions, and trillions, differ by only a letter or two, yet their magnitudes are significantly different.
- The reason to bring these numbers up is that after you move beyond a million, many calculators will not display the number in standard form.

Lesson Notes

Key Idea

- Review with students the definition of scientific notation.
- Write the Key Idea. Relate the steps to the activities from the investigation.
- ? "Why is the decimal point moved to the right of the first nonzero digit?" Need the factor n to be in the interval $1 \le n < 10$.

Example 1

- **Teaching Tip:** Have students underline the first nonzero digit and the digit to its right. In scientific notation, the decimal point is placed between these two digits.
- ? "How do you read the number?" one billion six hundred fifty million dollars
- **FYI:** Drawing the movement of the decimal point under the numbers helps students keep track of their counting.
- ? **MP3a Construct Viable Arguments:** "How do you know if the exponent for the power of 10 will be positive or negative?" If the standard form of the number is greater than or equal to 10, positive exponent; if the standard form of the number is between 0 and 1, negative exponent.

Example 2

- Discuss the context of this number.
- Note that reading the number takes time, and you have to count place values. The number is 268 hundred millionths.
- ? "Why is the exponent negative?" Original number is between 0 and 1.

On Your Own

- **Think-Pair-Share:** Students should read each question independently and then work in pairs to answer the questions. When they have answered the questions, the pair should compare their answers with another group and discuss any discrepancies.

Goal Today's lesson is writing numbers in scientific notation.

Lesson Tutorials
Lesson Plans
Answer Presentation Tool

Extra Example 1

Write 2,450,000 in scientific notation.
2.45×10^6

Extra Example 2

Write 0.0000045 in scientific notation.
4.45×10^{-6}

On Your Own

1. 5×10^4
2. 2.5×10^7
3. 6.83×10^2
4. 5×10^{-3}
5. 3.3×10^{-7}
6. 5.06×10^{-4}

Laurie's Notes

Extra Example 3

In Example 3, an album has sold 750,000 copies. How many more copies does it need to sell to receive the award? Write your answer in scientific notation. 9.25×10^6 copies

Extra Example 4

Order 6.8×10^4, 2.04×10^5, and 5.65×10^4 from least to greatest.
5.65×10^4, 6.8×10^4, 2.04×10^5

On Your Own

7. 9.045×10^6
8. Mesozoic era

Differentiated Instruction

Kinesthetic

Students may incorrectly count the number of zeros in a number and use that as the exponent in scientific notation. Have students write the number and place an arrow where the decimal point will be placed in the factor. Then have students count the number of places the decimal point moves.

$54{,}000 = 5.4 \times 10^4$
↑

$0.00000675 = 6.75 \times 10^{-6}$
↑

Example 3

- Encourage students to read the numbers when they are in standard form: 10 million; 8 million 780 thousand.
- The subtraction is performed on the numbers in standard form, with the result written in scientific notation.
- **MP7 Look for and Make Use of Structure:** Ask students if it is possible to subtract numbers in scientific notation. The answer is no, unless the powers of 10 are the same. Students will learn more about this in the next section.

Example 4

- Students have compared numbers written in scientific notation by writing each number in standard form. Another method is presented in this example. Students first compare the powers of ten, followed by comparing the decimal numbers.
- **Common Error:** Note that in order to use this method, *all* the numbers must be written in scientific notation.
- **Extension:** Create a proportionally correct timeline of the geologic periods. This can be done in a hallway or within the classroom. There are many resources online for this project.

On Your Own

- **Think-Pair-Share:** Students should read each question independently and then work in pairs to answer the questions. When they have answered the questions, the pair should compare their answers with another group and discuss any discrepancies.

Closure

- The land area of Virginia is about 39,500 square miles. The land area of Alaska is about 570,000 square miles. The United States land area is about 3,500,000 square miles. Write each of these in scientific notation.
  3.95×10^4, 5.7×10^5, 3.5×10^6

EXAMPLE 3 Using Scientific Notation

An album receives an award when it sells 10,000,000 copies.

An album has sold 8,780,000 copies. How many more copies does it need to sell to receive the award?

Ⓐ 1.22×10^{-7} Ⓑ 1.22×10^{-6}

Ⓒ 1.22×10^{6} Ⓓ 1.22×10^{7}

Use a model to solve the problem.

Remaining sales needed for award = Sales required for award − Current sales total

$= 10{,}000{,}000 - 8{,}780{,}000$

$= 1{,}220{,}000$

$= 1.22 \times 10^{6}$

∴ The album must sell 1.22×10^{6} more copies to receive the award. So, the correct answer is Ⓒ.

EXAMPLE 4 Real-Life Application

The table shows when the last three geologic eras began. Order the eras from earliest to most recent.

Era	Began
Paleozoic	5.42×10^{8} years ago
Cenozoic	6.55×10^{7} years ago
Mesozoic	2.51×10^{8} years ago

Step 1: Compare the powers of 10.

Because $10^{7} < 10^{8}$,
$6.55 \times 10^{7} < 5.42 \times 10^{8}$ and
$6.55 \times 10^{7} < 2.51 \times 10^{8}$.

Step 2: Compare the factors when the powers of 10 are the same.

Because $2.51 < 5.42$,
$2.51 \times 10^{8} < 5.42 \times 10^{8}$.

From greatest to least, the order is 5.42×10^{8}, 2.51×10^{8}, and 6.55×10^{7}.

∴ So, the eras in order from earliest to most recent are the Paleozoic era, Mesozoic era, and Cenozoic era.

To use the method in Example 4, the numbers must be written in scientific notation.

On Your Own

7. **WHAT IF?** In Example 3, an album has sold 955,000 copies. How many more copies does it need to sell to receive the award? Write your answer in scientific notation.

8. The *Tyrannosaurus rex* lived 7.0×10^{7} years ago. Consider the eras given in Example 4. During which era did the *Tyrannosaurus rex* live?

10.6 Exercises

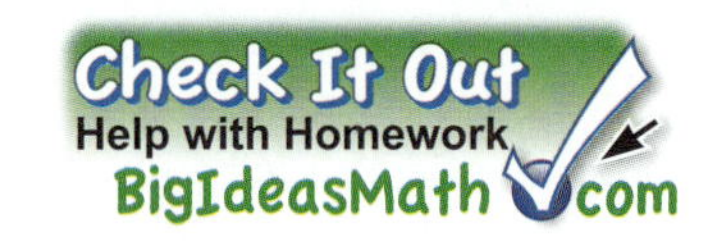

Vocabulary and Concept Check

1. **REASONING** How do you know whether a number written in standard form will have a positive or a negative exponent when written in scientific notation?
2. **WRITING** When is it appropriate to use scientific notation instead of standard form?

Practice and Problem Solving

Write the number in scientific notation.

(1) (2)

3. 0.0021
4. 5,430,000
5. 321,000,000
6. 0.00000625
7. 0.00004
8. 10,700,000
9. 45,600,000,000
10. 0.000000000009256
11. 840,000

ERROR ANALYSIS **Describe and correct the error in writing the number in scientific notation.**

12. 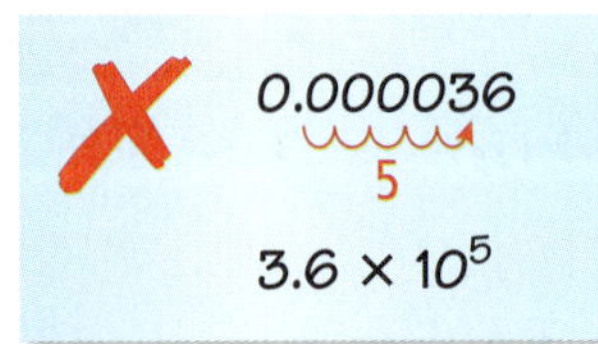

13. 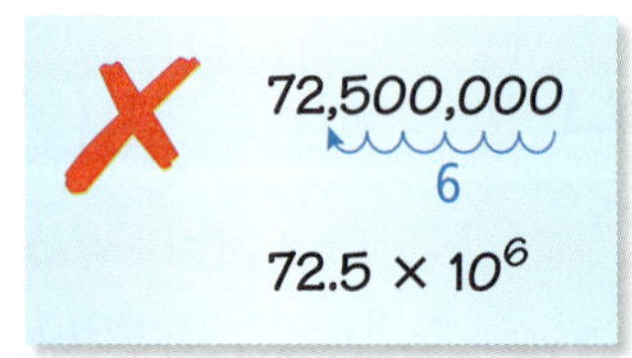

Order the numbers from least to greatest.

(4)

14. 1.2×10^8, 1.19×10^8, 1.12×10^8
15. 6.8×10^{-5}, 6.09×10^{-5}, 6.78×10^{-5}
16. 5.76×10^{12}, 9.66×10^{11}, 5.7×10^{10}
17. 4.8×10^{-6}, 4.8×10^{-5}, 4.8×10^{-8}
18. 9.9×10^{-15}, 1.01×10^{-14}, 7.6×10^{-15}
19. 5.78×10^{23}, 6.88×10^{-23}, 5.82×10^{23}

20. **HAIR** What is the diameter of a human hair written in scientific notation?

Diameter: 0.000099 meter

21. **EARTH** What is the circumference of Earth written in scientific notation?

Circumference at the equator: about 40,100,000 meters

22. **CHOOSING UNITS** In Exercise 21, name a unit of measurement that would be more appropriate for the circumference. Explain.

Assignment Guide and Homework Check

Level	Day 1 Activity Assignment	Day 2 Lesson Assignment	Homework Check
Basic	3–5, 31–34	1, 2, 7–25 odd, 12, 20, 22	7, 15, 20, 23
Average	3–5, 31–34	1, 2, 10–13, 18–21, 23–28	10, 18, 20, 24, 27
Advanced	1–5, 6–26 even, 27–34		10, 18, 24, 27, 29

Common Errors

- **Exercises 3–11** Students may write an exponent with the opposite sign of what is correct. Remind them that large numbers have a positive exponent in scientific notation and that small numbers have a negative exponent in scientific notation.
- **Exercises 14–19** Students may order the numbers without taking into account the power of 10. Encourage them to order the numbers by powers first and then by decimal factors.

10.6 Record and Practice Journal

Write the number in scientific notation.

1. 4,200,000 — 4.2×10^{6}
2. 0.038 — 3.8×10^{-2}
3. 600,000 — 6×10^{5}
4. 0.0000808 — 8.08×10^{-5}
5. 0.0007 — 7×10^{-4}
6. 29,010,000,000 — 2.901×10^{10}

Order the numbers from least to greatest.

7. $6.4 \times 10^{8}, 5.3 \times 10^{9}, 2.3 \times 10^{8}$ — 2.3×10^{8}, 6.4×10^{8}, 5.3×10^{9}
8. $9.1 \times 10^{-3}, 9.6 \times 10^{-3}, 9.02 \times 10^{-3}$ — 9.02×10^{-3}, 9.1×10^{-3}, 9.6×10^{-3}
9. $7.3 \times 10^{7}, 5.6 \times 10^{10}, 3.7 \times 10^{9}$ — 7.3×10^{7}, 3.7×10^{9}, 5.6×10^{10}
10. $1.4 \times 10^{-5}, 2.01 \times 10^{-15}, 6.3 \times 10^{-2}$ — 2.01×10^{-15}, 1.4×10^{-5}, 6.3×10^{-2}
11. A patient has 0.0000075 gram of iron in 1 liter of blood. The normal level is between 6×10^{-7} gram and 1.6×10^{-5} gram. Is the patient's iron level normal? Write the patient's amount of iron in scientific notation. — yes; 7.5×10^{-6}

Vocabulary and Concept Check

1. If the number is greater than or equal to 10, the exponent will be positive. If the number is less than 1 and greater than 0, the exponent will be negative.
2. It is appropriate to use scientific notation instead of standard form when a number is very large or very small.

Practice and Problem Solving

3. 2.1×10^{-3}
4. 5.43×10^{6}
5. 3.21×10^{8}
6. 6.25×10^{-6}
7. 4×10^{-5}
8. 1.07×10^{7}
9. 4.56×10^{10}
10. 9.256×10^{-12}
11. 8.4×10^{5}
12. The decimal point moved 5 places to the right, so the exponent should be negative. 3.6×10^{-5}
13. 72.5 is not less than 10. The decimal point needs to move one more place to the left. 7.25×10^{7}
14. $1.12 \times 10^{8}, 1.19 \times 10^{8}, 1.2 \times 10^{8}$
15. $6.09 \times 10^{-5}, 6.78 \times 10^{-5}, 6.8 \times 10^{-5}$
16. $5.7 \times 10^{10}, 9.66 \times 10^{11}, 5.76 \times 10^{12}$
17. $4.8 \times 10^{-8}, 4.8 \times 10^{-6}, 4.8 \times 10^{-5}$
18. $7.6 \times 10^{-15}, 9.9 \times 10^{-15}, 1.01 \times 10^{-14}$
19. $6.88 \times 10^{-23}, 5.78 \times 10^{23}, 5.82 \times 10^{23}$
20. 9.9×10^{-5} m
21. 4.01×10^{7} m

Practice and Problem Solving

22. *Sample answer:* kilometers or miles; They are both larger units of length, so the number would be smaller.

23. $680,\ 6.8 \times 10^3,\ \frac{68{,}500}{10}$

24. $0.02,\ \frac{5}{241},\ 2.1 \times 10^{-2}$

25. 6.25×10^{-3}, 6.3%, 0.625, $6\frac{1}{4}$

26. 305%, 3.3×10^2, 3033.4, $\frac{10{,}000}{3}$

27. 1.99×10^9 watts

28. *Sample answer:* Enter 1.174E10 − 9.75E9.

29. carat; Because 1 carat $= 1.2 \times 10^{23}$ atomic mass units and 1 milligram $= 6.02 \times 10^{20}$ atomic mass units, and $1.2 \times 10^{23} > 6.02 \times 10^{20}$.

30. See *Taking Math Deeper*.

Fair Game Review

31. natural, whole, integer, rational

32. integer, rational

33. irrational

34. D

Mini-Assessment

Write the number in scientific notation.

1. 0.00035 3.5×10^{-4}

2. 0.0000000000567 5.67×10^{-11}

3. 25,500,000 2.55×10^7

Order the numbers from least to greatest.

4. $3 \times 10^4,\ 6.1 \times 10^3,\ 1.6 \times 10^4$
$6.1 \times 10^3,\ 1.6 \times 10^4,\ 3 \times 10^4$

5. $5.8 \times 10^{-6},\ 2.8 \times 10^{-7},\ 5.9 \times 10^{-6}$
$2.8 \times 10^{-7},\ 5.8 \times 10^{-6},\ 5.9 \times 10^{-6}$

Taking Math Deeper

Exercise 30

This is an interesting question that deals with the different eras in Earth's geological history.

Order the eras from oldest to youngest. Write each beginning date in standard form.

Era	Began	Standard Form
Paleozoic era	5.42×10^8 years ago	542,000,000 years ago
Mesozoic era	2.51×10^8 years ago	251,000,000 years ago
Cenozoic era	6.55×10^7 years ago	65,500,000 years ago

Find the length of each era.

Paleozoic era = 542,000,000 − 251,000,000 = 291,000,000 yr
Mesozoic era = 251,000,000 − 65,500,000 = 185,500,000 yr
Cenozoic era = 65,500,000 − 0 = 65,500,000 yr

In scientific notation, these lengths from least to greatest are:

a. Cenozoic era $= 6.55 \times 10^7$ yr
Mesozoic era $= 1.855 \times 10^8$ yr
Paleozoic era $= 2.91 \times 10^8$ yr

b. You can make a time line that is either horizontal or vertical. Here is an example of a vertical time line.

c. For these three eras, the older the era, the longer it is. This is also true of the next three eras.
Neoproterozoic:
began 1.0×10^9 years ago
Mesoproterozoic:
began 1.6×10^9 years ago
Paleoproterozoic:
began 2.5×10^9 years ago

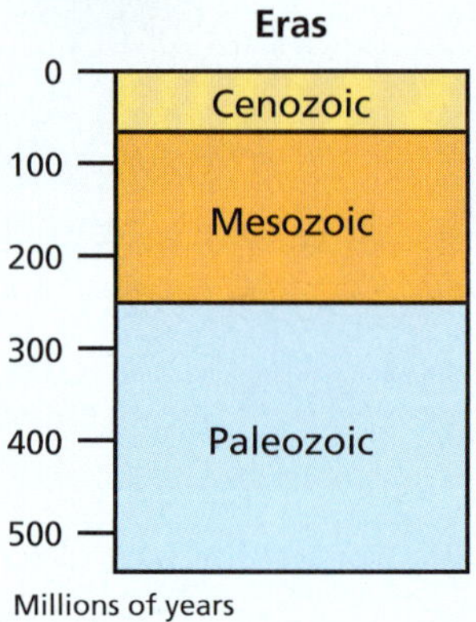

Reteaching and Enrichment Strategies

If students need help. . .	If students got it. . .
Resources by Chapter • Practice A and Practice B • Puzzle Time Record and Practice Journal Practice Differentiating the Lesson Lesson Tutorials Skills Review Handbook	Resources by Chapter • Enrichment and Extension • Technology Connection Start the next section

Order the numbers from least to greatest.

23. $\frac{68,500}{10}$, 680, 6.8×10^3

24. $\frac{5}{241}$, 0.02, 2.1×10^{-2}

25. 6.3%, 6.25×10^{-3}, $6\frac{1}{4}$, 0.625

26. 3033.4, 305%, $\frac{10,000}{3}$, 3.3×10^2

27. SPACE SHUTTLE The total power of a space shuttle during launch is the sum of the power from its solid rocket boosters and the power from its main engines. The power from the solid rocket boosters is 9,750,000,000 watts. What is the power from the main engines?

Total power = 1.174×10^{10} watts

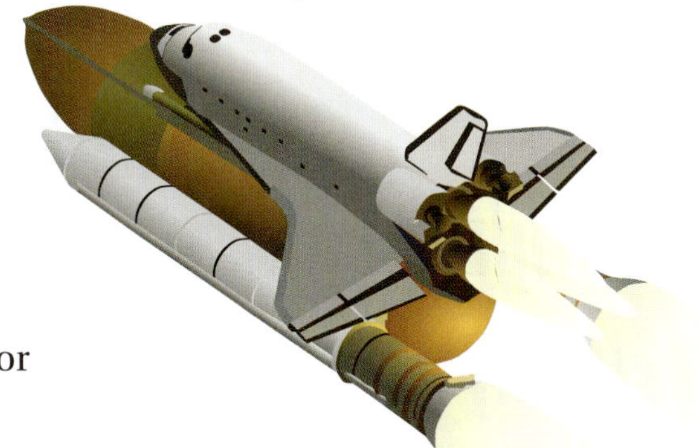

28. CHOOSE TOOLS Explain how to use a calculator to verify your answer to Exercise 27.

Equivalent to 1 Atomic Mass Unit
8.3×10^{-24} carat
1.66×10^{-21} milligram

29. ATOMIC MASS The mass of an atom or molecule is measured in atomic mass units. Which is greater, a *carat* or a *milligram*? Explain.

30. Reasoning In Example 4, the Paleozoic era ended when the Mesozoic era began. The Mesozoic era ended when the Cenozoic era began. The Cenozoic era is the current era.

a. Write the lengths of the three eras in scientific notation. Order the lengths from least to greatest.

b. Make a time line to show when the three eras occurred and how long each era lasted.

c. What do you notice about the lengths of the three eras? Use the Internet to determine whether your observation is true for *all* the geologic eras. Explain your results.

Fair Game Review What you learned in previous grades & lessons

Classify the real number. *(Section 7.4)*

31. 15

32. $\sqrt[3]{-8}$

33. $\sqrt{73}$

34. What is the surface area of the prism? *(Skills Review Handbook)*

Ⓐ 5 in.^{2}

Ⓑ 5.5 in.^{2}

Ⓒ 10 in.^{2}

Ⓓ 19 in.^{2}

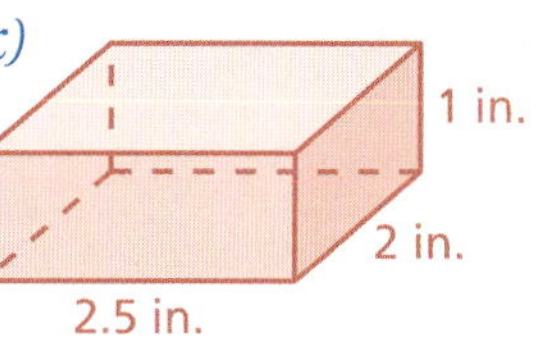

10.7 Operations in Scientific Notation

Essential Question How can you perform operations with numbers written in scientific notation?

1 ACTIVITY: Adding Numbers in Scientific Notation

Work with a partner. Consider the numbers 2.4×10^3 and 7.1×10^3.

a. Explain how to use order of operations to find the sum of these numbers. Then find the sum.

$$2.4 \times 10^3 + 7.1 \times 10^3$$

b. The factor ___ is common to both numbers. How can you use the Distributive Property to rewrite the sum $(2.4 \times 10^3) + (7.1 \times 10^3)$?

$(2.4 \times 10^3) + (7.1 \times 10^3) =$ ___ Distributive Property

c. Use order of operations to evaluate the expression you wrote in part (b). Compare the result with your answer in part (a).

d. **STRUCTURE** Write a rule you can use to add numbers written in scientific notation where the powers of 10 are the same. Then test your rule using the sums below.

- $(4.9 \times 10^5) + (1.8 \times 10^5) =$ ___
- $(3.85 \times 10^4) + (5.72 \times 10^4) =$ ___

2 ACTIVITY: Adding Numbers in Scientific Notation

Work with a partner. Consider the numbers 2.4×10^3 and 7.1×10^4.

a. Explain how to use order of operations to find the sum of these numbers. Then find the sum.

$$2.4 \times 10^3 + 7.1 \times 10^4$$

b. How is this pair of numbers different from the pairs of numbers in Activity 1?

c. Explain why you cannot immediately use the rule you wrote in Activity 1(d) to find this sum.

d. **STRUCTURE** How can you rewrite one of the numbers so that you can use the rule you wrote in Activity 1(d)? Rewrite one of the numbers. Then find the sum using your rule and compare the result with your answer in part (a).

e. **REASONING** Do these procedures work when subtracting numbers written in scientific notation? Justify your answer by evaluating the differences below.

- $(8.2 \times 10^5) - (4.6 \times 10^5) =$ ___
- $(5.88 \times 10^5) - (1.5 \times 10^4) =$ ___

Scientific Notation

In this lesson, you will

- add, subtract, multiply, and divide numbers written in scientific notation.

Learning Standards
MACC.8.EE.1.3
MACC.8.EE.1.4

Laurie's Notes

Introduction

Standards for Mathematical Practice

- **MP3a Construct Viable Arguments:** Mathematically proficient students are able to give explanations for how a computation is performed. They reference definitions and properties in establishing the validity of their argument.

Motivate

- Write the following problem on the board and ask student to evaluate it.
  $40 \times 10 + 8 \times 10 + 0.5 \times 10 + 0.07 \times 10$
- ? "How did you evaluate this expression?" It is likely that students found the sum of the four products.
- ? "Is there another method that could be used?" Students might mention factoring out the 10.
- **MP7 Look for and Make Use of Structure:** Underline the 10s and ask students how the Distributive Property could be used to evaluate the expression. Then write: $10(40 + 8 + 0.5 + 0.07)$.
- Tell students to use mental math to find the product: $10(48.57) = 485.7$.

Activity Notes

Activity 1

- After working through the Motivate, students should have a strategy for working through this activity. What students might not recognize is that 10^3 is a common factor, just like 10 was a common factor in the Motivate.
- ? "What is 10^3 in standard form?" 1000
- **MP3a:** In part (d), listen for students explaining the rule for adding numbers in scientific notation.

Activity 2

- The written problem looks very similar to the problem in Activity 1. What students will note when they examine it closely is that the powers of 10 are not the same.
- **MP1 Make Sense of Problems and Persevere in Solving Them:** Resist the urge to jump in and tell students how to proceed. Reassure them that they have the knowledge and skills to work through the problem.
- ? "Why can't you immediately use the rule you wrote in Activity 1?" There is no common factor because the powers of 10 are different.
- ? "What number or numbers did you rewrite in part (d)?" rewrite 2.4×10^3 as 0.24×10^4, or 7.1×10^4 as 71×10^3
- **Big Idea:** Students need to pay extra attention to the powers of 10 when adding and subtracting numbers in scientific notation.

Florida Common Core Standards

MACC.8.EE.1.3 Use numbers expressed in the form of a single digit times an integer power of 10 to estimate very large or very small quantities, and to express how many times as much one is than the other.

MACC.8.EE.1.4 Perform operations with numbers expressed in scientific notation, including problems where both decimal and scientific notation are used. Use scientific notation and choose units of appropriate size for measurements of very large or very small quantities. . . . Interpret scientific notation that has been generated by technology.

Previous Learning

Students have written numbers in scientific notation.

Lesson Plans
Complete Materials List

10.7 Record and Practice Journal

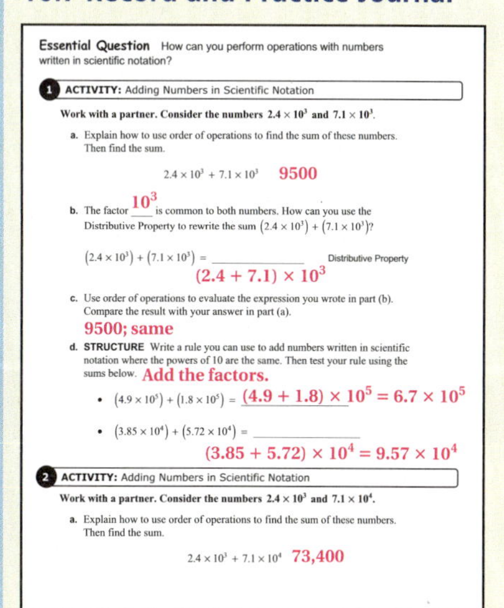

Essential Question How can you perform operations with numbers written in scientific notation?

1 ACTIVITY: Adding Numbers in Scientific Notation

Work with a partner. Consider the numbers 2.4×10^3 and 7.1×10^3.

a. Explain how to use order of operations to find the sum of these numbers. Then find the sum.

$2.4 \times 10^3 + 7.1 \times 10^3$ **9500**

b. The factor 10^3 is common to both numbers. How can you use the Distributive Property to rewrite the sum $(2.4 \times 10^3) + (7.1 \times 10^3)$?

$(2.4 \times 10^3) + (7.1 \times 10^3) =$ $(2.4 + 7.1) \times 10^3$ Distributive Property

c. Use order of operations to evaluate the expression you wrote in part (b). Compare the result with your answer in part (a).

9500; same

d. STRUCTURE Write a rule you can use to add numbers written in scientific notation where the powers of 10 are the same. Then test your rule using the sums below. **Add the factors.**

- $(4.9 \times 10^5) + (1.8 \times 10^5) = (4.9 + 1.8) \times 10^5 = 6.7 \times 10^5$
- $(3.85 \times 10^4) + (5.72 \times 10^4) = (3.85 + 5.72) \times 10^4 = 9.57 \times 10^4$

2 ACTIVITY: Adding Numbers in Scientific Notation

Work with a partner. Consider the numbers 2.4×10^3 and 7.1×10^4.

a. Explain how to use order of operations to find the sum of these numbers. Then find the sum.

$2.4 \times 10^3 + 7.1 \times 10^4$ **73,400**

English Language Learners

Group Activity

Have students work in groups that include both English learners and English speakers. Assign each group a scientific notation application problem. Have each group solve their problem showing all of the steps. English learners will benefit by having the opportunity to restate the problem and gain a deeper understanding of the concept.

10.7 Record and Practice Journal

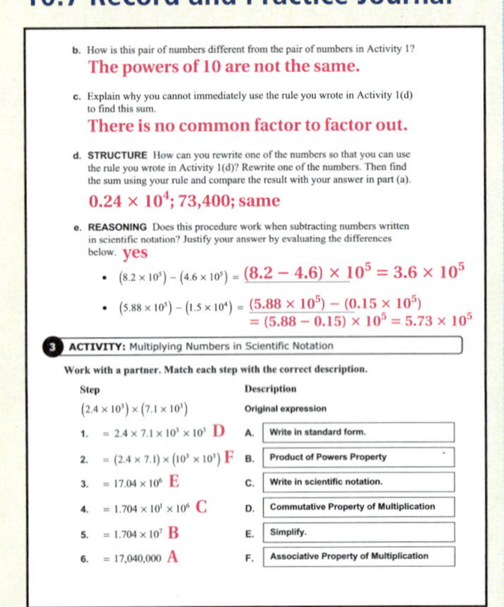

b. How is this pair of numbers different from the pair of numbers in Activity 1?
The powers of 10 are not the same.

c. Explain why you cannot immediately use the rule you wrote in Activity 1(d) to find this sum.
There is no common factor to factor out.

d. **STRUCTURE** How can you rewrite one of the numbers so that you can use the rule you wrote in Activity 1(d)? Rewrite one of the numbers. Then find the sum using your rule and compare the result with your answer in part (a).
0.24×10^4; 73,400; same

e. **REASONING** Does this procedure work when subtracting numbers written in scientific notation? Justify your answer by evaluating the differences below. yes

- $(8.2 \times 10^5) - (4.6 \times 10^5) = (8.2 - 4.6) \times 10^5 = 3.6 \times 10^5$
- $(5.88 \times 10^5) - (1.5 \times 10^4) = (5.88 \times 10^5) - (0.15 \times 10^5) = (5.88 - 0.15) \times 10^5 = 5.73 \times 10^5$

3 ACTIVITY: Multiplying Numbers in Scientific Notation

Work with a partner. Match each step with the correct description.

Step		Description
$(2.4 \times 10^3) \times (7.1 \times 10^3)$		Original expression
1. $= 2.4 \times 7.1 \times 10^3 \times 10^3$	D	A. Write in standard form.
2. $= (2.4 \times 7.1) \times (10^3 \times 10^3)$	F	B. Product of Powers Property
3. $= 17.04 \times 10^6$	E	C. Write in scientific notation.
4. $= 1.704 \times 10^1 \times 10^6$	C	D. Commutative Property of Multiplication
5. $= 1.704 \times 10^7$	B	E. Simplify.
6. $= 17{,}040{,}000$	A	F. Associative Property of Multiplication

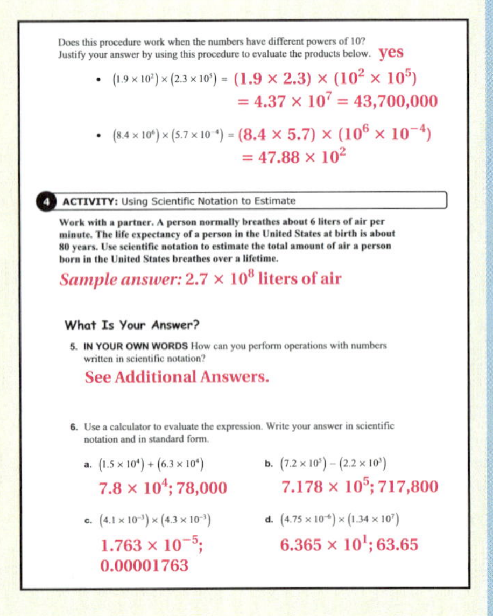

Does this procedure work when the numbers have different powers of 10? Justify your answer by using this procedure to evaluate the products below. yes

- $(1.9 \times 10^2) \times (2.3 \times 10^5) = (1.9 \times 2.3) \times (10^2 \times 10^5) = 4.37 \times 10^7 = 43{,}700{,}000$
- $(8.4 \times 10^6) \times (5.7 \times 10^{-4}) = (8.4 \times 5.7) \times (10^6 \times 10^{-4}) = 47.88 \times 10^2$

4 ACTIVITY: Using Scientific Notation to Estimate

Work with a partner. A person normally breathes about 6 liters of air per minute. The life expectancy of a person in the United States at birth is about 80 years. Use scientific notation to estimate the total amount of air a person born in the United States breathes over a lifetime.

Sample answer: 2.7×10^8 liters of air

What Is Your Answer?

5. **IN YOUR OWN WORDS** How can you perform operations with numbers written in scientific notation?
See Additional Answers.

6. Use a calculator to evaluate the expression. Write your answer in scientific notation and in standard form.

a. $(1.5 \times 10^4) + (6.3 \times 10^4)$
7.8×10^4; 78,000

b. $(7.2 \times 10^5) - (2.2 \times 10^3)$
7.178×10^5; 717,800

c. $(4.1 \times 10^{-3}) \times (4.3 \times 10^{-3})$
1.763×10^{-5}; 0.00001763

d. $(4.75 \times 10^{-6}) \times (1.34 \times 10^7)$
6.365×10^1; 63.65

Laurie's Notes

Activity 3

- Give students time to match the descriptions and steps with their partners. Remind students that the question they need to ask themselves is, "What was done in moving from one step to the next? What justifies, or explains each step?"
- This can be a challenging exercise for students. They can read through the steps and make sense of them, but they're unsure of what justifies each step.
- When you discuss this as a class, you could put the steps and descriptions on strips of paper. See if students can order the steps *and* the descriptions.

Activity 4

- This activity is similar to the heartbeat question students answered in the Activity Motivate in Section 10.5.
- **MP5 Use Appropriate Tools Strategically:** You may wish to give students access to calculators for this activity.
- **Teaching Tip:** Provide transparencies or blank paper so that students can display their answers on the overhead or at the document camera.
- Ask volunteers to share their work. Look for evidence of determining the total number of minutes in 80 years, or the number of liters of air breathed per year.

What Is Your Answer?

- **MP5:** Question 6 gives students extra practice with entering numbers written in scientific notation into a calculator.

Closure

- Evaluate:
  1. $6.4 \times 10^4 + 1.3 \times 10^4$ 7.7×10^4
  2. $(1.2 \times 10^3) \times (4.3 \times 10^3)$ 5.16×10^6

3 ACTIVITY: Multiplying Numbers in Scientific Notation

Math Practice 3

Justify Conclusions

Which step of the procedure would be affected if the powers of 10 were different? Explain.

Work with a partner. Match each step with the correct description.

Step	Description
$(2.4 \times 10^3) \times (7.1 \times 10^3)$	Original expression
1. $= 2.4 \times 7.1 \times 10^3 \times 10^3$	A. Write in standard form.
2. $= (2.4 \times 7.1) \times (10^3 \times 10^3)$	B. Product of Powers Property
3. $= 17.04 \times 10^6$	C. Write in scientific notation.
4. $= 1.704 \times 10^1 \times 10^6$	D. Commutative Property of Multiplication
5. $= 1.704 \times 10^7$	E. Simplify.
6. $= 17{,}040{,}000$	F. Associative Property of Multiplication

Does this procedure work when the numbers have different powers of 10? Justify your answer by using this procedure to evaluate the products below.

- $(1.9 \times 10^2) \times (2.3 \times 10^5) =$ ______
- $(8.4 \times 10^6) \times (5.7 \times 10^{-4}) =$ ______

4 ACTIVITY: Using Scientific Notation to Estimate

Work with a partner. A person normally breathes about 6 liters of air per minute. The life expectancy of a person in the United States at birth is about 80 years. Use scientific notation to estimate the total amount of air a person born in the United States breathes over a lifetime.

What Is Your Answer?

5. **IN YOUR OWN WORDS** How can you perform operations with numbers written in scientific notation?

6. Use a calculator to evaluate the expression. Write your answer in scientific notation and in standard form.

   a. $(1.5 \times 10^4) + (6.3 \times 10^4)$ **b.** $(7.2 \times 10^5) - (2.2 \times 10^3)$

   c. $(4.1 \times 10^{-3}) \times (4.3 \times 10^{-3})$ **d.** $(4.75 \times 10^{-6}) \times (1.34 \times 10^7)$

Practice Use what you learned about evaluating expressions involving scientific notation to complete Exercises 3–6 on page 452.

10.7 Lesson

To add or subtract numbers written in scientific notation with the same power of 10, add or subtract the factors. When the numbers have different powers of 10, first rewrite the numbers so they have the same power of 10.

EXAMPLE 1 Adding and Subtracting Numbers in Scientific Notation

Find the sum or difference. Write your answer in scientific notation.

a. $(4.6 \times 10^3) + (8.72 \times 10^3)$

$= (4.6 + 8.72) \times 10^3$ — Distributive Property

$= 13.32 \times 10^3$ — Add.

$= (1.332 \times 10^1) \times 10^3$ — Write 13.32 in scientific notation.

$= 1.332 \times 10^4$ — Product of Powers Property

Study Tip

In Example 1(b), you will get the same answer when you start by rewriting 3.5×10^{-2} as 35×10^{-3}.

b. $(3.5 \times 10^{-2}) - (6.6 \times 10^{-3})$

Rewrite 6.6×10^{-3} so that it has the same power of 10 as 3.5×10^{-2}.

$6.6 \times 10^{-3} = 6.6 \times 10^{-1} \times 10^{-2}$ — Rewrite 10^{-3} as $10^{-1} \times 10^{-2}$.

$= 0.66 \times 10^{-2}$ — Rewrite 6.6×10^{-1} as 0.66.

Subtract the factors.

$(3.5 \times 10^{-2}) - (0.66 \times 10^{-2})$

$= (3.5 - 0.66) \times 10^{-2}$ — Distributive Property

$= 2.84 \times 10^{-2}$ — Subtract.

On Your Own

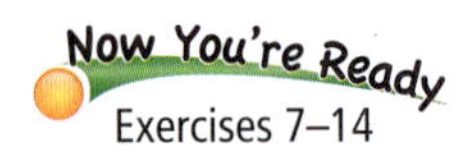

Exercises 7–14

Find the sum or difference. Write your answer in scientific notation.

1. $(8.2 \times 10^2) + (3.41 \times 10^{-1})$
2. $(7.8 \times 10^{-5}) - (4.5 \times 10^{-5})$

To multiply or divide numbers written in scientific notation, multiply or divide the factors and powers of 10 separately.

EXAMPLE 2 Multiplying Numbers in Scientific Notation

Find $(3 \times 10^{-5}) \times (5 \times 10^{-2})$. Write your answer in scientific notation.

$(3 \times 10^{-5}) \times (5 \times 10^{-2})$

$= 3 \times 5 \times 10^{-5} \times 10^{-2}$ — Commutative Property of Multiplication

$= (3 \times 5) \times (10^{-5} \times 10^{-2})$ — Associative Property of Multiplication

$= 15 \times 10^{-7}$ — Simplify.

$= 1.5 \times 10^1 \times 10^{-7}$ — Write 15 in scientific notation.

$= 1.5 \times 10^{-6}$ — Product of Powers Property

Study Tip

You can check your answer using standard form.

(3×10^{-5})
$\times (5 \times 10^{-2})$
$= 0.00003 \times 0.05$
$= 0.0000015$
$= 1.5 \times 10^{-6}$

Laurie's Notes

Introduction

Connect

- **Yesterday:** Students explored how to use properties to perform operations with numbers written in scientific notation. (MP1, MP3a, MP5, MP7)
- **Today:** Students will perform operations with numbers written in scientific notation.

Motivate

- Discuss the meaning of national debt. Compare the national debt of the United States x number of years ago with the national debt today.

Year	2000	2002	2004	2006	2008	2010	2012
Debt (in trillions)	5.7	6.2	7.4	8.5	10.0	13.6	16.1

You can also discuss each student's share of the U.S. national debt. Ask students how they might calculate their own share.

? "How do you write the 2000 and 2012 national debts in scientific notation?" 5.7×10^{12}; 1.61×10^{13}

- Pose questions about the national debt, such as the difference in the national debt between 2000 and 2012, and how many times greater the national debt is in 2012 compared to 2000. Explain that this type of arithmetic is the focus of this lesson.

Lesson Notes

Example 1

- **Note:** In part (a), students do not always recognize the Distributive Property when it is used to pull out a common factor, such as 10^3.

? "Why not leave the answer as 13.32×10^3?" It is not in scientific notation.

- In part (b), rewriting 6.6×10^{-3} can be confusing to students. Tell them this step is similar to rewriting 24×6 as $12 \times 2 \times 6$.
- Show students that this expression can also be simplified by rewriting 3.5×10^{-2} as 35×10^{-3}.

? "Why can't the numbers just be subtracted?" The factors are multiplied by different powers of 10, so you cannot use the Distributive Property.

- **Alternative Method:** In part (b) write the numbers in standard form, subtract, and then write the answer in scientific notation.

On Your Own

- **Neighbor Check:** Have students work independently and then have their neighbors check their work. Have students discuss any discrepancies.

Example 2

- Even though you cannot add or subtract numbers in scientific notation (unless they have the same power of 10), you can multiply them.

? "How would you multiply the numbers?" Most students will immediately suggest multiplying the factors, and then multiplying the powers of 10.

- **MP3a Construct Viable Arguments:** Make sure that students realize that the Commutative and Associative Properties allow this to happen. The Product of Powers Property is used to multiply the powers of 10.

Goal Today's lesson is adding, subtracting, multiplying, and dividing numbers in scientific notation.

Technology for the Teacher

Lesson Tutorials
Lesson Plans
Answer Presentation Tool

Extra Example 1

Find the sum or difference. Write your answer in scientific notation.

a. $(2.1 \times 10^{-4}) + (9.74 \times 10^{-4})$
1.184×10^{-3}

b. $(4.7 \times 10^{5}) - (7.2 \times 10^{3})$
4.628×10^{5}

On Your Own

1. 8.20341×10^2
2. 3.3×10^{-5}

Extra Example 2

Find $(2 \times 10^{-4}) \times (6 \times 10^{-3})$. Write your answer in scientific notation.
1.2×10^{-6}

Extra Example 3

Find $\frac{5.3 \times 10^8}{4 \times 10^{-3}}$. Write your answer in scientific notation. 1.325×10^{11}

On Your Own

3. 4.8×10^{-4}

4. 2.1×10^{8}

5. 2×10^{12}

6. 2×10^{-6}

Extra Example 4

The diameter of the Moon is about 3.48×10^3 kilometers. Using the information in Example 4, how many times greater is the diameter of the Sun than the diameter of the Moon? about 402 times greater

On Your Own

7. 693,600 km

Differentiated Instruction

Kinesthetic

When writing a number in scientific notation, have students underline the first nonzero digit and the digit to its right. The decimal point will be placed between these two digits to create the factor.

$\underline{27},000 = 2.7 \times 10^4$

$0.00000\underline{48} = 4.8 \times 10^{-6}$

Laurie's Notes

Example 3

- Before this example, work through a simple, but related problem such as: $\frac{2}{3} \cdot \frac{9}{10} = \frac{\cancel{2}^{1} \cdot \cancel{9}^{3}}{{}_{1}\cancel{3} \cdot \cancel{10}_{5}} = \frac{3}{5}$. Point out how the common factors divide out.
- Write the example and relate it to the problem above.

? "Why not leave the answer as 0.25×10^{-15}?" It is not in scientific notation.

On Your Own

- **Common Error:** Students may use the Quotient of Powers Property incorrectly when simplifying the fraction with the powers of 10.

Example 4

? "Does anyone know the approximate diameter of Earth? the Sun?" Answers will vary.

- While students may not know the diameters, they should know that the Sun's diameter is much greater than Earth's diameter. This example will determine how many times greater.
- Explain to students that the answer is written in standard form to make the comparison more meaningful. It is easier to understand that the Sun's diameter is about 109 times greater than Earth's diameter, instead of about 1.09×10^2 times greater.

On Your Own

- In Question 7, have students think about which form they should convert to first. Have them do it both ways. Ask them which they prefer and why.

Closure

- **Exit Ticket:** Add or divide. Write your answer in scientific notation.

  a. $(3.5 \times 10^4) + (7.6 \times 10^4)$ 1.11×10^5

  b. $\frac{8.4 \times 10^3}{4.2 \times 10^{-2}}$ 2×10^5

EXAMPLE 3 Dividing Numbers in Scientific Notation

Find $\dfrac{1.5 \times 10^{-8}}{6 \times 10^{7}}$. Write your answer in scientific notation.

$$\frac{1.5 \times 10^{-8}}{6 \times 10^{7}} = \frac{1.5}{6} \times \frac{10^{-8}}{10^{7}}$$ Rewrite as a product of fractions.

$$= 0.25 \times \frac{10^{-8}}{10^{7}}$$ Divide 1.5 by 6.

$$= 0.25 \times 10^{-15}$$ Quotient of Powers Property

$$= 2.5 \times 10^{-1} \times 10^{-15}$$ Write 0.25 in scientific notation.

$$= 2.5 \times 10^{-16}$$ Product of Powers Property

On Your Own

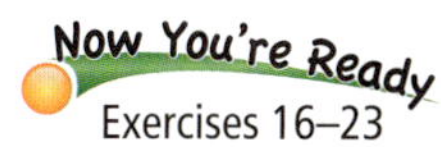

Find the product or quotient. Write your answer in scientific notation.

3. $6 \times (8 \times 10^{-5})$

4. $(7 \times 10^{2}) \times (3 \times 10^{5})$

5. $(9.2 \times 10^{12}) \div 4.6$

6. $(1.5 \times 10^{-3}) \div (7.5 \times 10^{2})$

EXAMPLE 4 Real-Life Application

How many times greater is the diameter of the Sun than the diameter of Earth?

Diameter = 1,400,000 km

Diameter = 1.28×10^4 km

Write the diameter of the Sun in scientific notation.

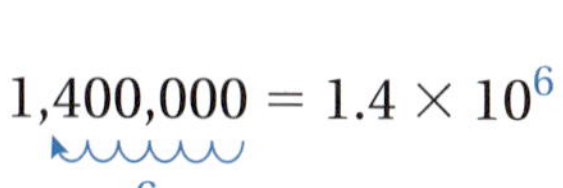

Divide the diameter of the Sun by the diameter of Earth.

$$\frac{1.4 \times 10^{6}}{1.28 \times 10^{4}} = \frac{1.4}{1.28} \times \frac{10^{6}}{10^{4}}$$ Rewrite as a product of fractions.

$$= 1.09375 \times \frac{10^{6}}{10^{4}}$$ Divide 1.4 by 1.28.

$$= 1.09375 \times 10^{2}$$ Quotient of Powers Property

$$= 109.375$$ Write in standard form.

The diameter of the Sun is about 109 times greater than the diameter of Earth.

On Your Own

7. How many more kilometers is the radius of the Sun than the radius of Earth? Write your answer in standard form.

10.7 Exercises

Vocabulary and Concept Check

1. **WRITING** Describe how to subtract two numbers written in scientific notation with the same power of 10.

2. **NUMBER SENSE** You are multiplying two numbers written in scientific notation with different powers of 10. Do you have to rewrite the numbers so they have the same power of 10 before multiplying? Explain.

Practice and Problem Solving

Evaluate the expression using two different methods. Write your answer in scientific notation.

3. $(2.74 \times 10^7) + (5.6 \times 10^7)$
4. $(8.3 \times 10^6) + (3.4 \times 10^5)$
5. $(5.1 \times 10^5) \times (9.7 \times 10^5)$
6. $(4.5 \times 10^4) \times (6.2 \times 10^3)$

Find the sum or difference. Write your answer in scientific notation.

1 7. $(2 \times 10^5) + (3.8 \times 10^5)$
8. $(6.33 \times 10^{-9}) - (4.5 \times 10^{-9})$
9. $(9.2 \times 10^8) - (4 \times 10^8)$
10. $(7.2 \times 10^{-6}) + (5.44 \times 10^{-6})$
11. $(7.8 \times 10^7) - (2.45 \times 10^6)$
12. $(5 \times 10^{-5}) + (2.46 \times 10^{-3})$
13. $(9.7 \times 10^6) + (6.7 \times 10^5)$
14. $(2.4 \times 10^{-1}) - (5.5 \times 10^{-2})$

15. **ERROR ANALYSIS** Describe and correct the error in finding the sum of the numbers.

$$(2.5 \times 10^9) + (5.3 \times 10^8) = (2.5 + 5.3) \times (10^9 \times 10^8)$$
$$= 7.8 \times 10^{17}$$

Find the product or quotient. Write your answer in scientific notation.

16. $5 \times (7 \times 10^7)$
17. $(5.8 \times 10^{-6}) \div (2 \times 10^{-3})$
18. $(1.2 \times 10^{-5}) \div 4$
19. $(5 \times 10^{-7}) \times (3 \times 10^6)$
20. $(3.6 \times 10^7) \div (7.2 \times 10^7)$
21. $(7.2 \times 10^{-1}) \times (4 \times 10^{-7})$
22. $(6.5 \times 10^8) \times (1.4 \times 10^{-5})$
23. $(2.8 \times 10^4) \div (2.5 \times 10^6)$

24. **MONEY** How many times greater is the thickness of a dime than the thickness of a dollar bill?

Thickness = 0.135 cm

Thickness = 1.0922×10^{-2} cm

Assignment Guide and Homework Check

Level	Day 1 Activity Assignment	Day 2 Lesson Assignment	Homework Check
Basic	3–6, 32–35	1, 2, 7–29 odd	11, 13, 17, 19, 27
Average	3–6, 32–35	1, 2, 7–23 odd, 24–30 even	11, 13, 17, 21, 28
Advanced	1–6, 8–14 even, 15, 16–26 even, 28–35		14, 20, 22, 26, 28

For Your Information

- **Exercise 30** Remind students that population density is the average number of people for some amount of land area. To find the population density, you divide the population by the land area.

Common Errors

- **Exercises 11–14** Students may incorrectly rewrite a number when adding or subtracting numbers with different powers of 10.
- **Exercises 16–23** Students may multiply the factors and leave the number greater than 10. Remind them that the factor in scientific notation must be at least one and less than 10. If the number is less than 10 and greater than or equal to 1, the exponent will be 0.
- **Exercises 16–23** Students may use the Quotient of Powers Property incorrectly when simplifying fractions containing powers of 10.

10.7 Record and Practice Journal

Find the sum or difference. Write your answer in scientific notation.

1. $(2 \times 10^4) + (7.2 \times 10^4)$
   9.2×10^4
2. $(3.2 \times 10^{-2}) + (9.4 \times 10^{-2})$
   1.26×10^{-1}
3. $(6.7 \times 10^5) - (4.3 \times 10^5)$
   2.4×10^5
4. $(8.9 \times 10^{-3}) - (1.9 \times 10^{-3})$
   7×10^{-3}

Find the product or quotient. Write your answer in scientific notation.

5. $(6 \times 10^8) \times (4 \times 10^6)$
   2.4×10^{15}
6. $(9 \times 10^{-3}) \times (9 \times 10^{-3})$
   8.1×10^{-5}
7. $(8 \times 10^3) \div (2 \times 10^2)$
   4.0×10^1
8. $(2.34 \times 10^5) \div (7.8 \times 10^5)$
   3×10^{-1}
9. How many times greater is the radius of a basketball than the radius of a marble?

Radius = 1.143×10^1 cm

Radius = 5×10^{-1} cm

about 23 times greater

Vocabulary and Concept Check

1. Use the Distributive Property to group the factors together. Then subtract the factors and write it with the power of 10. The number may need to be rewritten so that it is still in scientific notation.
2. no; You can use the Commutative and Associative Properties of Multiplication to group the factors and the powers of 10. Then, you multiply the factors and multiply the powers of 10.

Practice and Problem Solving

3. 8.34×10^7
4. 8.64×10^6
5. 4.947×10^{11}
6. 2.79×10^8
7. 5.8×10^5
8. 1.83×10^{-9}
9. 5.2×10^8
10. 1.264×10^{-5}
11. 7.555×10^7
12. 2.51×10^{-3}
13. 1.037×10^7
14. 1.85×10^{-1}
15. You have to rewrite the numbers so they have the same power of 10 before adding; 3.03×10^9
16. 3.5×10^8
17. 2.9×10^{-3}
18. 3×10^{-6}
19. 1.5×10^0
20. 5×10^{-1}
21. 2.88×10^{-7}
22. 9.1×10^3
23. 1.12×10^{-2}
24. about 12 times greater

Practice and Problem Solving

25. 4.006×10^9

26. 2.9×10^{-2}

27. 1.962×10^8 cm

28. 2.65×10^6 L;
1.85×10^8 L;
Justifications will vary.

29. See *Taking Math Deeper*.

30. *Answer should include, but is not limited to:* Make sure calculations using scientific notation are done correctly.

31. 3×10^8 m/sec

Fair Game Review

32. -9 **33.** $\frac{1}{8}$

34. $-\frac{5}{7}$ **35.** C

Mini-Assessment

Evaluate the expression. Write your answer in scientific notation.

1. $(3.4 \times 10^6) + (8.1 \times 10^6)$ 1.15×10^7
2. $(4.3 \times 10^{-3}) + (7.8 \times 10^{-4})$
   5.08×10^{-3}
3. $(5.6 \times 10^{-8}) - (1.9 \times 10^{-8})$
   3.7×10^{-8}
4. $(1.7 \times 10^2) \times (4.3 \times 10^4)$
   7.31×10^6
5. $(6.2 \times 10^5) \div (2 \times 10^{-4})$ 3.1×10^9
6. The mass of Earth is about 6.58×10^{21} tons. The mass of Mars is about 7.08×10^{20} tons. How much greater is the mass of Earth than the mass of Mars?
   about 5.872×10^{21} tons

Taking Math Deeper

Exercise 29

You are probably familiar with the look of a DVD, but do you know what the surface of a DVD looks like? This problem gives students some idea of what this digital storage device actually looks like.

1 Summarize the given information.

Width of each ridge = 0.000032 cm
Width of each valley = 0.000074 cm
Diameter of center portion = 4.26 cm

2 Find the diameter of the DVD.

$$\begin{aligned}\text{ridges} + \text{valleys} &= 73{,}000(0.000032) + 73{,}000(0.000074)\\ &= 73{,}000(0.000032 + 0.000074)\\ &= 73{,}000(0.000106)\\ &= 7.738 \text{ cm}\end{aligned}$$

OR using scientific notation

$$\begin{aligned}\text{ridges} + \text{valleys} &= (7.3 \times 10^4)(3.2 \times 10^{-5}) + (7.3 \times 10^4)(7.4 \times 10^{-5})\\ &= (7.3 \times 10^4)(3.2 \times 10^{-5} + 7.4 \times 10^{-5})\\ &= (7.3 \times 10^4)(10.6 \times 10^{-5})\\ &= 77.38 \times 10^{-1}\\ &= 7.738 \text{ cm}\end{aligned}$$

$$\begin{aligned}\text{diameter} &= \text{ridges} + \text{valleys} + \text{center portion}\\ &= 7.738 + 4.26\\ &= 11.998 \text{ cm}\\ &\approx 12 \text{ cm}\end{aligned}$$

3 Here's a fun fact.

The microscopic dimensions of the bumps make the spiral track on a DVD extremely long. If you could lift the data track off a single layer of a DVD, and stretch it out into a straight line, it would be almost 7.5 miles long!

Project

Write a report on the invention of the DVD and the DVD player.

Reteaching and Enrichment Strategies

If students need help...	If students got it...
Resources by Chapter • Practice A and Practice B • Puzzle Time Record and Practice Journal Practice Differentiating the Lesson Lesson Tutorials Skills Review Handbook	Resources by Chapter • Enrichment and Extension • Technology Connection Start the next section

Evaluate the expression. Write your answer in scientific notation.

25. $5{,}200{,}000 \times (8.3 \times 10^{2}) - (3.1 \times 10^{8})$

26. $(9 \times 10^{-3}) + (2.4 \times 10^{-5}) \div 0.0012$

27. GEOMETRY Find the perimeter of the rectangle.

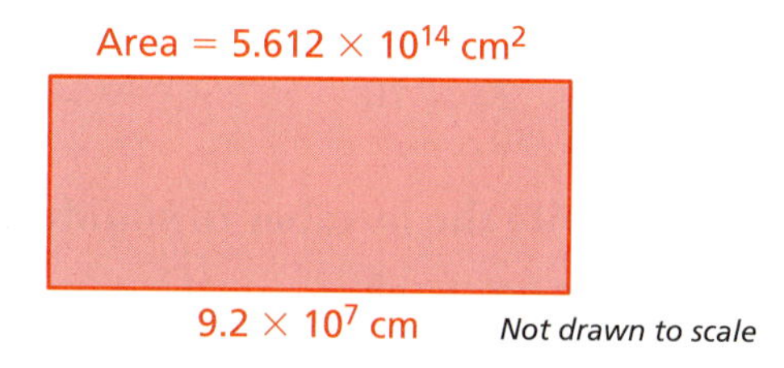

28. BLOOD SUPPLY A human heart pumps about 7×10^{-2} liter of blood per heartbeat. The average human heart beats about 72 times per minute. How many liters of blood does a heart pump in 1 year? in 70 years? Write your answers in scientific notation. Then use estimation to justify your answers.

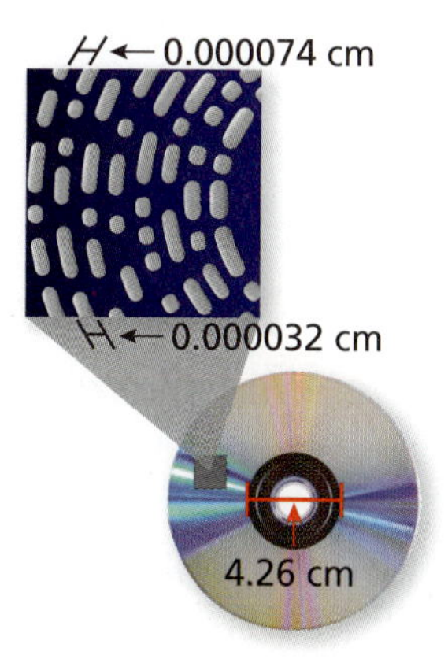

29. DVDS On a DVD, information is stored on bumps that spiral around the disk. There are 73,000 ridges (with bumps) and 73,000 valleys (without bumps) across the diameter of the DVD. What is the diameter of the DVD in centimeters?

30. PROJECT Use the Internet or some other reference to find the populations and areas (in square miles) of India, China, Argentina, the United States, and Egypt. Round each population to the nearest million and each area to the nearest thousand square miles.

a. Write each population and area in scientific notation.

b. Use your answers to part (a) to find and order the population densities (people per square mile) of each country from least to greatest.

31. 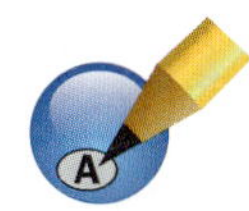

Albert Einstein's most famous equation is $E = mc^2$, where E is the energy of an object (in joules), m is the mass of an object (in kilograms), and c is the speed of light (in meters per second). A hydrogen atom has 15.066×10^{-11} joule of energy and a mass of 1.674×10^{-27} kilogram. What is the speed of light? Write your answer in scientific notation.

Fair Game Review What you learned in previous grades & lessons

Find the cube root. *(Section 7.2)*

32. $\sqrt[3]{-729}$

33. $\sqrt[3]{\frac{1}{512}}$

34. $\sqrt[3]{-\frac{125}{343}}$

35. MULTIPLE CHOICE What is the volume of the cone? *(Section 8.2)*

Ⓐ $16\pi \text{ cm}^3$

Ⓑ $108\pi \text{ cm}^3$

Ⓒ $48\pi \text{ cm}^3$

Ⓓ $144\pi \text{ cm}^3$

4 cm

9 cm

10.5–10.7 Quiz

Tell whether the number is written in scientific notation. Explain. *(Section 10.5)*

1. 23×10^9
2. 0.6×10^{-7}

Write the number in standard form. *(Section 10.5)*

3. 8×10^6
4. 1.6×10^{-2}

Write the number in scientific notation. *(Section 10.6)*

5. 0.00524
6. 892,000,000

Evaluate the expression. Write your answer in scientific notation. *(Section 10.7)*

7. $(7.26 \times 10^4) + (3.4 \times 10^4)$
8. $(2.8 \times 10^{-5}) - (1.6 \times 10^{-6})$
9. $(2.4 \times 10^4) \times (3.8 \times 10^{-6})$
10. $(5.2 \times 10^{-3}) \div (1.3 \times 10^{-12})$

11. **PLANETS** The table shows the equatorial radii of the eight planets in our solar system. *(Section 10.5)*

    a. Which planet has the second-smallest equatorial radius?

    b. Which planet has the second-largest equatorial radius?

Planet	Equatorial Radius (km)
Mercury	2.44×10^3
Venus	6.05×10^3
Earth	6.38×10^3
Mars	3.4×10^3
Jupiter	7.15×10^4
Saturn	6.03×10^4
Uranus	2.56×10^4
Neptune	2.48×10^4

12. **OORT CLOUD** The Oort cloud is a spherical cloud that surrounds our solar system. It is about 2×10^5 astronomical units from the Sun. An astronomical unit is about 1.5×10^8 kilometers. How far is the Oort cloud from the Sun in kilometers? *(Section 10.6)*

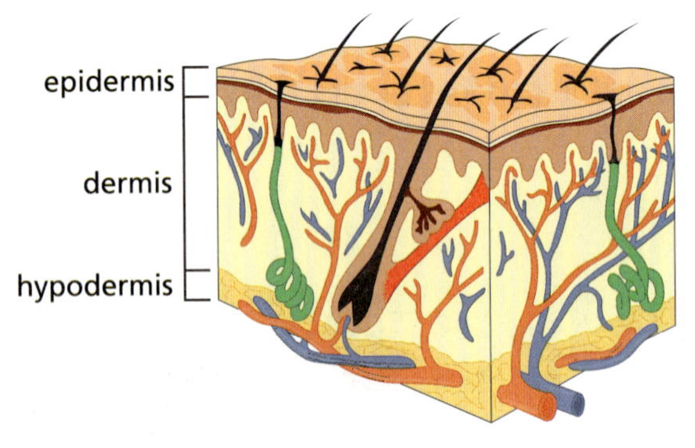

13. **EPIDERMIS** The outer layer of skin is called the *epidermis*. On the palm of your hand, the epidermis is 0.0015 meter thick. Write this number in scientific notation. *(Section 10.6)*

14. **ORBITS** It takes the Sun about 2.3×10^8 years to orbit the center of the Milky Way. It takes Pluto about 2.5×10^2 years to orbit the Sun. How many times does Pluto orbit the Sun while the Sun completes one orbit around the Milky Way? Write your answer in standard form. *(Section 10.7)*

Alternative Assessment Options

Math Chat	Student Reflective Focus Question
Structured Interview	Writing Prompt

Math Chat

- Have individual students work problems from the quiz on the board. The student explains the process used and justifies each step. Students in the class ask questions of the student presenting.
- The teacher probes the thought process of the student presenting, but does not teach or ask leading questions.

Study Help Sample Answers

Remind students to complete Graphic Organizers for the rest of the chapter.

4.

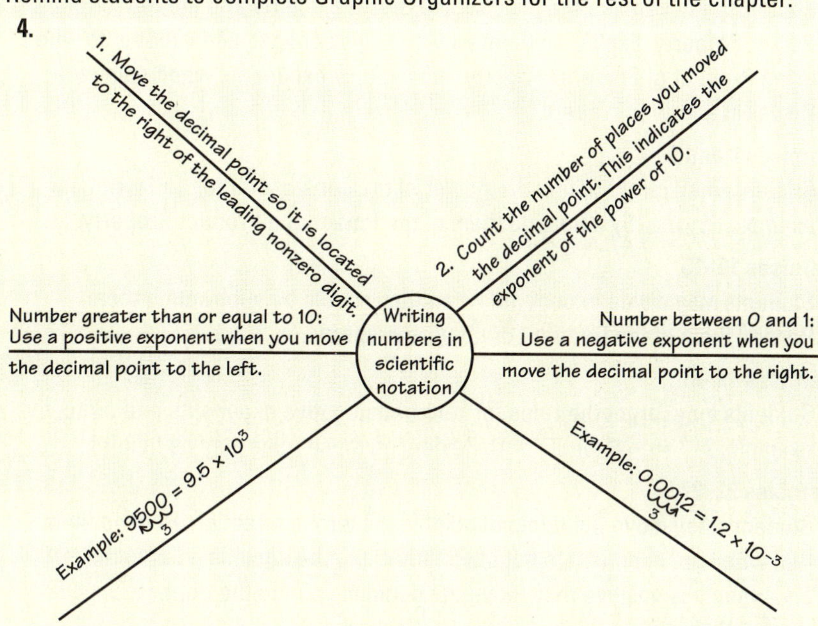

5–8. Available at *BigIdeasMath.com.*

Reteaching and Enrichment Strategies

If students need help...	If students got it...
Resources by Chapter • Practice A and Practice B • Puzzle Time Lesson Tutorials *BigIdeasMath.com*	Resources by Chapter • Enrichment and Extension • Technology Connection Game Closet at *BigIdeasMath.com* Start the Chapter Review

Answers

1. no; The factor is greater than 10.
2. no; The factor is less than 1.
3. 8,000,000
4. 0.016
5. 5.24×10^{-3}
6. 8.92×10^{8}
7. 1.066×10^{5}
8. 2.64×10^{-5}
9. 9.12×10^{-2}
10. 4×10^{9}
11. **a.** Mars
    b. Saturn
12. 3×10^{13} km
13. 1.5×10^{-3} m
14. 920,000

Online Assessment
Assessment Book
ExamView® Assessment Suite

For the Teacher
Additional Review Options

- *BigIdeasMath.com*
- Online Assessment
- Game Closet at *BigIdeasMath.com*
- Vocabulary Help
- Resources by Chapter

Answers

1. $(-9)^5$
2. 2^3n^2
3. 216
4. $-\frac{1}{16}$
5. 100
6. p^7
7. n^{22}
8. $125y^3$
9. $16k^4$

Review of Common Errors

Exercises 1 and 2

- Students may count the wrong number of factors. Remind them to check their work.

Exercises 3–5

- Students may treat the exponent as a factor. For example, some may think that 6^3 means 6×3. Remind them of the definition of an exponent.
- Students may have the wrong sign in their answers. Remind them that when a negative sign is outside the parentheses, it is not part of the base.
- Students may not use the correct order of operations when they evaluate the expression in Exercise 5. Review the order of operations with students.

Exercises 6 and 7

- Students may confuse the Product of Powers Property and the Power of a Power Property. Explain to them why it makes sense to add exponents when using Product of Powers Property and multiply exponents when using the Power of a Power Property.

Exercises 8 and 9

- Students may forget to find the power of the constant factor and write, for example, $(5y)^3 = 5y^3$. Remind them of the Power of a Product Property.

Exercises 10–15

- Students may divide exponents when they should be subtracting them. Remind them of the Quotient of Powers Property.

Exercises 16–21

- Students may forget the rules for zero and negative exponents and write, for example, $2^{-4} = -16$ or $95^0 = 0$. Reviewing these rules may be helpful.

Exercises 22–27

- Students may move the decimal point in the wrong direction. Remind them that when the exponent is negative they move the decimal point to the left, and when it is positive they move the decimal point to the right.

Exercises 28–30

- Students may write an exponent with the incorrect sign. Remind them that large numbers have a positive exponent in scientific notation and small numbers have a negative exponent in scientific notation.

Exercises 31–34

- Students may leave the decimal factor greater than 10 and/or combine exponents incorrectly, coming up with answers such as 10.1×10^{17} for Exercise 31, 4.1×10^0 for Exercise 32, 37.73×10^{-40} for Exercise 33, and $2 \times 10^{5/9}$ for Exercise 34. Remind them what scientific notation means and point out that the directions call for the answer in scientific notation. Also, if necessary, provide students with a quick review of the properties they learned in this chapter, as well as how to correctly perform operations in scientific notation. In particular, point out that in Exercise 31, students should rewrite one of the addends so that the powers of 10 are the same before adding.

10 Chapter Review

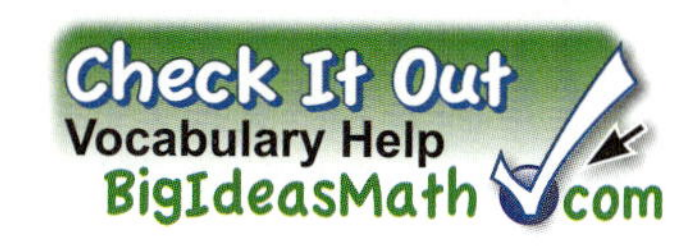

Review Key Vocabulary

power, *p. 412*
base, *p. 412*
exponent, *p. 412*
scientific notation, *p. 438*

Review Examples and Exercises

10.1 Exponents *(pp. 410–415)*

Write $(-4) \cdot (-4) \cdot (-4) \cdot y \cdot y$ using exponents.

Because -4 is used as a factor 3 times, its exponent is 3. Because y is used as a factor 2 times, its exponent is 2.

So, $(-4) \cdot (-4) \cdot (-4) \cdot y \cdot y = (-4)^3 y^2$.

Exercises

Write the product using exponents.

1. $(-9) \cdot (-9) \cdot (-9) \cdot (-9) \cdot (-9)$
2. $2 \cdot 2 \cdot 2 \cdot n \cdot n$

Evaluate the expression.

3. 6^3
4. $-\left(\frac{1}{2}\right)^4$
5. $\left|\frac{1}{2}(16 - 6^3)\right|$

10.2 Product of Powers Property *(pp. 416–421)*

a. $\left(-\frac{1}{8}\right)^7 \cdot \left(-\frac{1}{8}\right)^4 = \left(-\frac{1}{8}\right)^{7+4}$ — Product of Powers Property

$= \left(-\frac{1}{8}\right)^{11}$ — Simplify.

b. $(2.5^7)^2 = 2.5^{7 \cdot 2}$ — Power of a Power Property

$= 2.5^{14}$ — Simplify.

c. $(3m)^2 = 3^2 \cdot m^2$ — Power of a Product Property

$= 9m^2$ — Simplify.

Exercises

Simplify the expression.

6. $p^5 \cdot p^2$
7. $(n^{11})^2$
8. $(5y)^3$
9. $(-2k)^4$

10.3 Quotient of Powers Property (pp. 422–427)

a. $\frac{(-4)^9}{(-4)^6} = (-4)^{9-6}$ Quotient of Powers Property

$= (-4)^3$ Simplify.

b. $\frac{x^4}{x^3} = x^{4-3}$ Quotient of Powers Property

$= x^1$

$= x$ Simplify.

Exercises

Simplify the expression. Write your answer as a power.

10. $\frac{8^8}{8^3}$

11. $\frac{5^2 \cdot 5^9}{5}$

12. $\frac{w^8}{w^7} \cdot \frac{w^5}{w^2}$

Simplify the expression.

13. $\frac{2^2 \cdot 2^5}{2^3}$

14. $\frac{(6c)^3}{c}$

15. $\frac{m^8}{m^6} \cdot \frac{m^{10}}{m^9}$

10.4 Zero and Negative Exponents (pp. 428–433)

a. $10^{-3} = \frac{1}{10^3}$ Definition of negative exponent

$= \frac{1}{1000}$ Evaluate power.

b. $(-0.5)^{-5} \cdot (-0.5)^5 = (-0.5)^{-5+5}$ Product of Powers Property

$= (-0.5)^0$ Simplify.

$= 1$ Definition of zero exponent

Exercises

Evaluate the expression.

16. 2^{-4}

17. 95^0

18. $\frac{8^2}{8^4}$

19. $(-12)^{-7} \cdot (-12)^7$

20. $\frac{1}{7^9} \cdot \frac{1}{7^{-6}}$

21. $\frac{9^4 \cdot 9^{-2}}{9^2}$

Review Game

Comparing Values in Scientific Notation

Materials per Group

- pencil
- paper
- computer with Internet access

Directions

- The game can be completed in one to two class periods, but you may want to give the students one or two days to complete the necessary research.
- Each student comes up with three different values written in scientific notation. (Example: the length of an ant in meters, the weight of a person in ounces, the volume of a car's gas tank in cups)
- One value should be length, one value should be weight, and one value should be volume.
- Divide the class into an even number of groups.
- Randomly call on two groups and have them complete the following:
  - Each group writes the length (in scientific notation) of one of their items on the board.
  - The members of each group work together to write each number in standard form and determine which value is the least.
  - They write their answer on a piece of paper and submit it to the teacher.
  - One point is awarded to each group that answers correctly.
  - Note: Be sure students are aware that when they are comparing these values, the units of the values must be considered.
- Repeat this process for the remaining length, weight, and volume values.

Who wins?

After all groups have compared their values, the group(s) with the most points wins.

For the Student Additional Practice

- Lesson Tutorials
- Multi-Language Glossary
- Self-Grading Progress Check
- *BigIdeasMath.com*
  Dynamic Student Edition
  Student Resources

Answers

10. 8^5

11. 5^{10}

12. w^4

13. 16

14. $216c^2$

15. m^3

16. $\frac{1}{16}$

17. 1

18. $\frac{1}{64}$

19. 1

20. $\frac{1}{343}$

21. 1

22. 20,000,000

23. 0.034

24. 0.0000000015

25. 59,000,000,000

26. 0.0048

27. 625,000

28. 3.6×10^{-4}

29. 8×10^5

30. 7.92×10^7

31. 6.32×10^9

32. 4.1×10^{-4}

33. 3.773×10^4

34. 2×10^{-4}

My Thoughts on the Chapter

What worked. . .

Teacher Tip

Not allowed to write in your teaching edition? Use sticky notes to record your thoughts.

What did not work. . .

What I would do differently. . .

10.5 Reading Scientific Notation (pp. 436–441)

Write (a) 5.9×10^4 and (b) 7.31×10^{-6} in standard notation.

a. $5.9 \times 10^4 = 59{,}000$

4

Move decimal point $|4| = 4$ places to the right.

b. $7.31 \times 10^{-6} = 0.00000731$

6

Move decimal point $|-6| = 6$ places to the left.

Exercises

Write the number in standard form.

22. 2×10^7

23. 3.4×10^{-2}

24. 1.5×10^{-9}

25. 5.9×10^{10}

26. 4.8×10^{-3}

27. 6.25×10^5

10.6 Writing Scientific Notation (pp. 442–447)

Write (a) 309,000,000 and (b) 0.00056 in scientific notation.

a. $309{,}000{,}000 = 3.09 \times 10^8$

8

The number is greater than 10. So, the exponent is positive.

b. $0.00056 = 5.6 \times 10^{-4}$

4

The number is between 0 and 1. So, the exponent is negative.

Exercises

Write the number in scientific notation.

28. 0.00036

29. 800,000

30. 79,200,000

10.7 Operations in Scientific Notation (pp. 448–453)

Find $(2.6 \times 10^5) + (3.1 \times 10^5)$.

$(2.6 \times 10^5) + (3.1 \times 10^5) = (2.6 + 3.1) \times 10^5$ Distributive Property

$= 5.7 \times 10^5$ Add.

Exercises

Evaluate the expression. Write your answer in scientific notation.

31. $(4.2 \times 10^8) + (5.9 \times 10^9)$

32. $(5.9 \times 10^{-4}) - (1.8 \times 10^{-4})$

33. $(7.7 \times 10^8) \times (4.9 \times 10^{-5})$

34. $(3.6 \times 10^5) \div (1.8 \times 10^9)$

10 Chapter Test

Write the product using exponents.

1. $(-15) \cdot (-15) \cdot (-15)$

2. $\left(\frac{1}{12}\right) \cdot \left(\frac{1}{12}\right) \cdot \left(\frac{1}{12}\right) \cdot \left(\frac{1}{12}\right) \cdot \left(\frac{1}{12}\right)$

Evaluate the expression.

3. -2^3

4. $10 + 3^3 \div 9$

Simplify the expression. Write your answer as a power.

5. $9^{10} \cdot 9$

6. $(6^6)^5$

7. $(2 \cdot 10)^7$

8. $\dfrac{(-3.5)^{13}}{(-3.5)^9}$

Evaluate the expression.

9. $5^{-2} \cdot 5^2$

10. $\dfrac{-8}{(-8)^3}$

Write the number in standard form.

11. 3×10^7

12. 9.05×10^{-3}

Evaluate the expression. Write your answer in scientific notation.

13. $(7.8 \times 10^7) + (9.9 \times 10^7)$

14. $(6.4 \times 10^5) - (5.4 \times 10^4)$

15. $(3.1 \times 10^6) \times (2.7 \times 10^{-2})$

16. $(9.6 \times 10^7) \div (1.2 \times 10^{-4})$

17. **CRITICAL THINKING** Is $(xy^2)^3$ the same as $(xy^3)^2$? Explain.

18. **RICE** A grain of rice weighs about 3^3 milligrams. About how many grains of rice are in one scoop?

19. **TASTE BUDS** There are about 10,000 taste buds on a human tongue. Write this number in scientific notation.

One scoop of rice weighs about 3^9 milligrams.

20. **LEAD** From 1978 to 2008, the amount of lead allowed in the air in the United States was 1.5×10^{-6} gram per cubic meter. In 2008, the amount allowed was reduced by 90%. What is the new amount of lead allowed in the air?

Test Item References

Chapter Test Questions	Section to Review	Florida Common Core Standards (MACC)
1–4	10.1	8.EE.1.1
5–7, 17	10.2	8.EE.1.1
8, 18	10.3	8.EE.1.1
9, 10	10.4	8.EE.1.1
11, 12, 20	10.5	8.EE.1.3, 8.EE.1.4
19	10.6	8.EE.1.3, 8.EE.1.4
13–16	10.7	8.EE.1.3, 8.EE.1.4

Test-Taking Strategies

Remind students to quickly look over the entire test before they start so that they can budget their time. Have students use the **Stop** and **Think** strategy before they answer each question.

Common Errors

- **Exercises 3, 8, and 10** Students may have the wrong sign in their answers. Remind them that when the negative sign is inside parentheses, it is part of the base. Point out that in Exercise 3, the negative sign is not part of the base.
- **Exercise 4** Students may not use the correct order of operations when they evaluate the expression. Review the order of operations with students.
- **Exercises 5, 6, and 9** Students may confuse the Product of Powers Property and the Power of a Power Property. Explain to them why it makes sense to add exponents when using the Product of Powers Property and multiply exponents when using the Power of a Power Property.
- **Exercises 8 and 10** Students may divide the exponents instead of subtracting them. Remind them of the Quotient of Powers Property.
- **Exercises 11 and 12** Students may move the decimal point in the wrong direction. Remind them that when the exponent is negative they move the decimal point to the left, and when the exponent is positive they move the decimal point to the right.

Reteaching and Enrichment Strategies

If students need help. . .	If students got it. . .
Resources by Chapter • Practice A and Practice B • Puzzle Time Record and Practice Journal Practice Differentiating the Lesson Lesson Tutorials *BigIdeasMath.com* Skills Review Handbook	Resources by Chapter • Enrichment and Extension • Technology Connection Game Closet at *BigIdeasMath.com* Start Standards Assessment

Answers

1. $(-15)^3$
2. $\left(\frac{1}{12}\right)^5$
3. -8
4. 13
5. 9^{11}
6. 6^{30}
7. $2^7 \cdot 10^7$
8. $(-3.5)^4$
9. 1
10. $\frac{1}{64}$
11. 30,000,000
12. 0.00905
13. 1.77×10^8
14. 5.86×10^5
15. 8.37×10^4
16. 8×10^{11}
17. no; $(xy^2)^3 = (xy^2) \cdot (xy^2) \cdot (xy^2) = x \cdot x \cdot x \cdot y^2 \cdot y^2 \cdot y^2 = x^3y^6$
    $(xy^3)^2 = (xy^3) \cdot (xy^3) = x \cdot x \cdot y^3 \cdot y^3 = x^2y^6$
18. 3^6 or 729 grains
19. 1×10^4
20. 1.5×10^{-7} gram per cubic meter

Technology for the Teacher

Online Assessment
Assessment Book
ExamView® Assessment Suite

Test-Taking Strategies

Available at *BigIdeasMath.com*

After Answering Easy Questions, Relax
Answer Easy Questions First
Estimate the Answer
Read All Choices before Answering
Read Question before Answering
Solve Directly or Eliminate Choices
Solve Problem before Looking at Choices
Use Intelligent Guessing
Work Backwards

About this Strategy

When taking a multiple choice test, be sure to read each question carefully and thoroughly. Sometimes you don't know the answer. So . . . guess intelligently! Look at the choices and choose the ones that are possible answers.

Answers

1. C
2. I
3. D
4. 40

Technology for the Teacher

Florida Common Core Standards Support
Performance Tasks
Online Assessment
Assessment Book
ExamView® Assessment Suite

Item Analysis

1. **A.** The student assumes there should be 7 zeros.

   B. The student miscounts when adding zeros.

   C. Correct answer

   D. The student miscounts when adding zeros.

2. **F.** The student is finding the sum of the angle measures of a quadrilateral.

   G. The student is finding the sum of the measures of two acute angles of a right triangle.

   H. The student finds the measure of the wrong angle.

   I. Correct answer

3. **A.** The student multiplies exponents instead of adding them.

   B. The student multiplies the bases and adds the exponents.

   C. The student multiplies all of the numbers in the expression.

   D. Correct answer

4. **Gridded Response:** Correct answer: 40

   Common Error: The student applies the ratio of the corresponding sides to the corresponding angles, getting an answer of 80.

10 Standards Assessment

1. Mercury's distance from the Sun is approximately 5.79×10^7 kilometers. What is this distance in standard form? *(MACC.8.EE.1.4)*

A. 5,790,000,000 km **C.** 57,900,000 km

B. 579,000,000 km **D.** 5,790,000 km

2. The steps Jim took to answer the question are shown below. What should Jim change to correctly answer the question? *(MACC.8.G.1.5)*

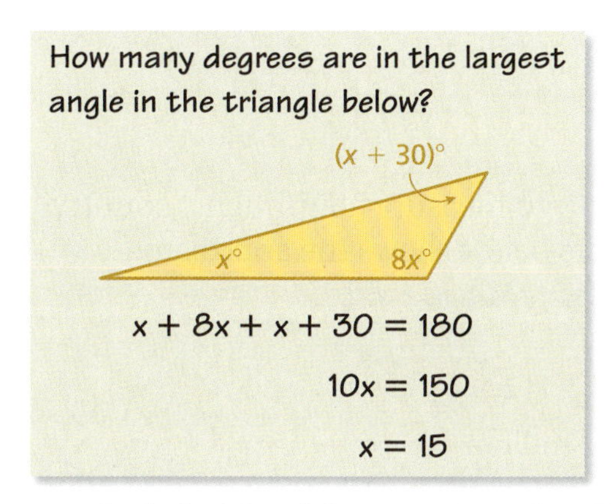

F. The left side of the equation should equal 360° instead of 180°.

G. The sum of the acute angles should equal 90°.

H. Evaluate the smallest angle when $x = 15$.

I. Evaluate the largest angle when $x = 15$.

3. Which expression is equivalent to the expression below? *(MACC.8.EE.1.1)*

$$2^4 2^3$$

A. 2^{12} **C.** 48

B. 4^7 **D.** 128

4. In the figure below, $\triangle ABC$ is a dilation of $\triangle DEF$.

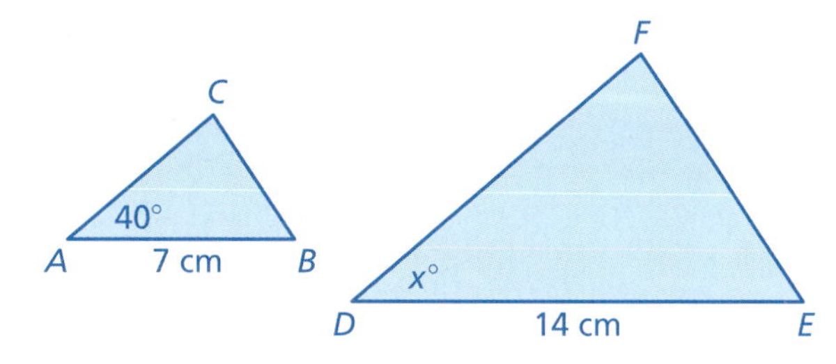

What is the value of x? *(MACC.8.G.1.4)*

5. A bank account pays interest so that the amount in the account doubles every 10 years. The account started with $5,000 in 1940. Which expression represents the amount (in dollars) in the account n decades later? *(MACC.8.EE.1.1)*

F. $2^n \cdot 5000$

G. $5000(n + 1)$

H. 5000^n

I. $2^n + 5000$

6. The formula for the volume V of a pyramid is $V = \frac{1}{3}Bh$. Solve the formula for the height h. *(MACC.8.EE.3.7b)*

A. $h = \frac{1}{3}VB$

B. $h = \frac{3V}{B}$

C. $h = \frac{V}{3B}$

D. $h = V - \frac{1}{3}B$

7. The gross domestic product (GDP) is a way to measure how much a country produces economically in a year. The table below shows the approximate population and GDP for the United States. *(MACC.8.EE.1.4)*

United States 2012	
Population	312 million (312,000,000)
GDP	15.1 trillion dollars ($15,100,000,000,000)

Part A Find the GDP per person for the United States. Show your work and explain your reasoning.

Part B Write the population and the GDP using scientific notation.

Part C Find the GDP per person for the United States using your answers from Part B. Write your answer in scientific notation. Show your work and explain your reasoning.

8. What is the equation of the line shown in the graph? *(MACC.8.EE.2.6)*

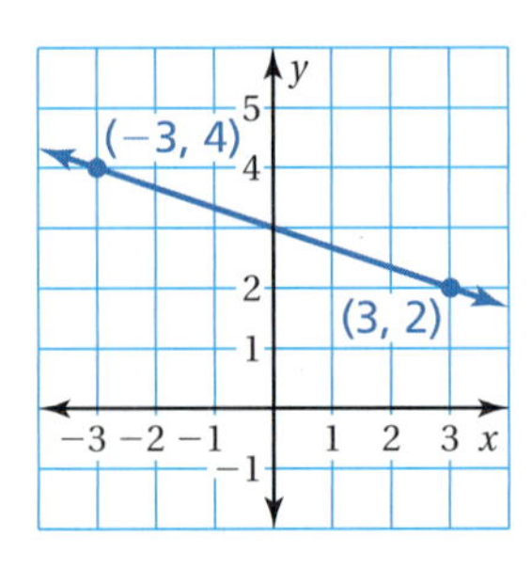

F. $y = -\frac{1}{3}x + 3$

G. $y = \frac{1}{3}x + 1$

H. $y = -3x + 3$

I. $y = 3x - \frac{1}{3}$

Item Analysis (continued)

5. F. Correct answer

   G. The student multiplies 5000 by one more than the number of decades.

   H. The student uses 5000 as the base.

   I. The student adds 2^n and 5000 instead of multiplying.

6. A. The student does not understand inverse operations.

   B. Correct answer

   C. The student does not understand inverse operations.

   D. The student subtracts $\frac{1}{3}B$ instead of dividing by $\frac{1}{3}B$.

7. **4 points** The student demonstrates a thorough understanding of how to work arithmetically with large numbers both in standard form and in scientific notation. In Part A, an answer of $48,397.44 is obtained. In Part B, the data are written as 3.12×10^8 and 1.51×10^{13}. In Part C, the student works out the quotient $\frac{1.51 \times 10^{13}}{3.12 \times 10^8}$ step-by-step.

   3 points The student demonstrates an essential but less than thorough understanding. In particular, Parts A and B should be completed correctly, but the steps taken in Part C may show gaps or be incomplete.

   2 points The student demonstrates an understanding of how to write the data using scientific notation, but is otherwise limited in understanding how to approach the problem arithmetically.

   1 point The student demonstrates limited understanding of working with large numbers arithmetically. The student's response is incomplete and exhibits many flaws.

   0 points The student provides no response, a completely incorrect or incomprehensible response, or a response that demonstrates insufficient understanding of how to work with large numbers.

8. F. Correct answer

   G. The student miscalculates slope as positive, and then chooses the equation that has (3, 2) as a solution.

   H The student reverses the roles of x and y in finding slope, and then uses the correct intercept, interpolated from the graph.

   I. The student reverses slope and y-intercept in the equation.

Answers

5. F

6. B

7. *Part A* about $48,397.44

   Part B 3.12×10^8; 1.51×10^{13} dollars

   Part C about 4.84×10^4 dollars

8. F

Answers

9. C

10. 0.16 or $\frac{4}{25}$

11. G

12. B

Item Analysis (continued)

9. A. The student does not square the radius.

B. The student multiplies by the diameter instead of the radius squared.

C. Correct answer

D. The student multiplies by the diameter squared instead of the radius squared.

10. Gridded Response: Correct answer: 0.16 or $\frac{4}{25}$

Common Error: The student writes the answer as a negative number.

11. F. The student mistakes the roles of slope and intercept, thinking same intercept means what same slope means.

G. Correct answer

H. The student chooses a conclusion for lines that have the same intercept and the same slope, overlooking that these lines have different slopes.

I. The student thinks that because the two lines have the same *y*-intercept, they must have the same slope.

12. A. The student fails to realize that a box-and-whisker plot is inappropriate. The display must allow a reader to compare parts of a whole.

B. Correct answer

C. The student fails to realize that a line graph is inappropriate. The display must allow a reader to compare parts of a whole.

D. The student fails to realize that a scatter plot is inappropriate. The display must allow a reader to compare parts of a whole.

9. A cylinder and its dimensions are shown below.

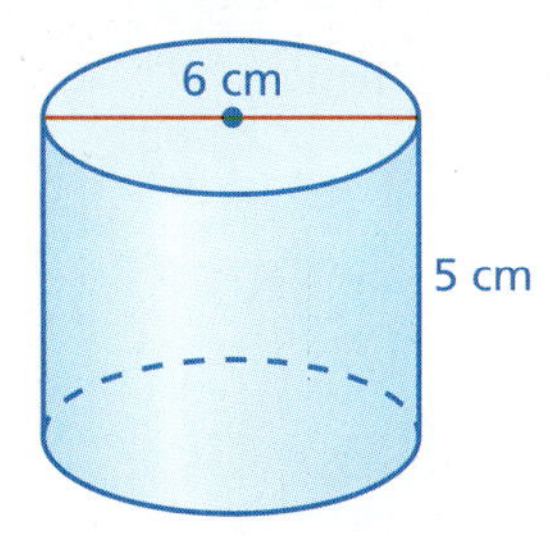

What is the volume of the cylinder? (Use 3.14 for π.) *(MACC.8.G.3.9)*

A. 47.1 cm^3

B. 94.2 cm^3

C. 141.3 cm^3

D. 565.2 cm^3

10. Find $(-2.5)^{-2}$. *(MACC.8.EE.1.1)*

11. Two lines have the same y-intercept. The slope of one line is 1, and the slope of the other line is -1. What can you conclude? *(MACC.8.EE.2.6)*

F. The lines are parallel.

G. The lines meet at exactly one point.

H. The lines meet at more than one point.

I. The situation described is impossible.

12. The director of a research lab wants to present data to donors. The data show how the lab uses a great deal of donated money for research and only a small amount of money for other expenses. Which type of display is best suited for showing these data? *(MACC.8.SP.1.1)*

A. box-and-whisker plot

B. circle graph

C. line graph

D. scatter plot

Appendix A
My Big Ideas Projects

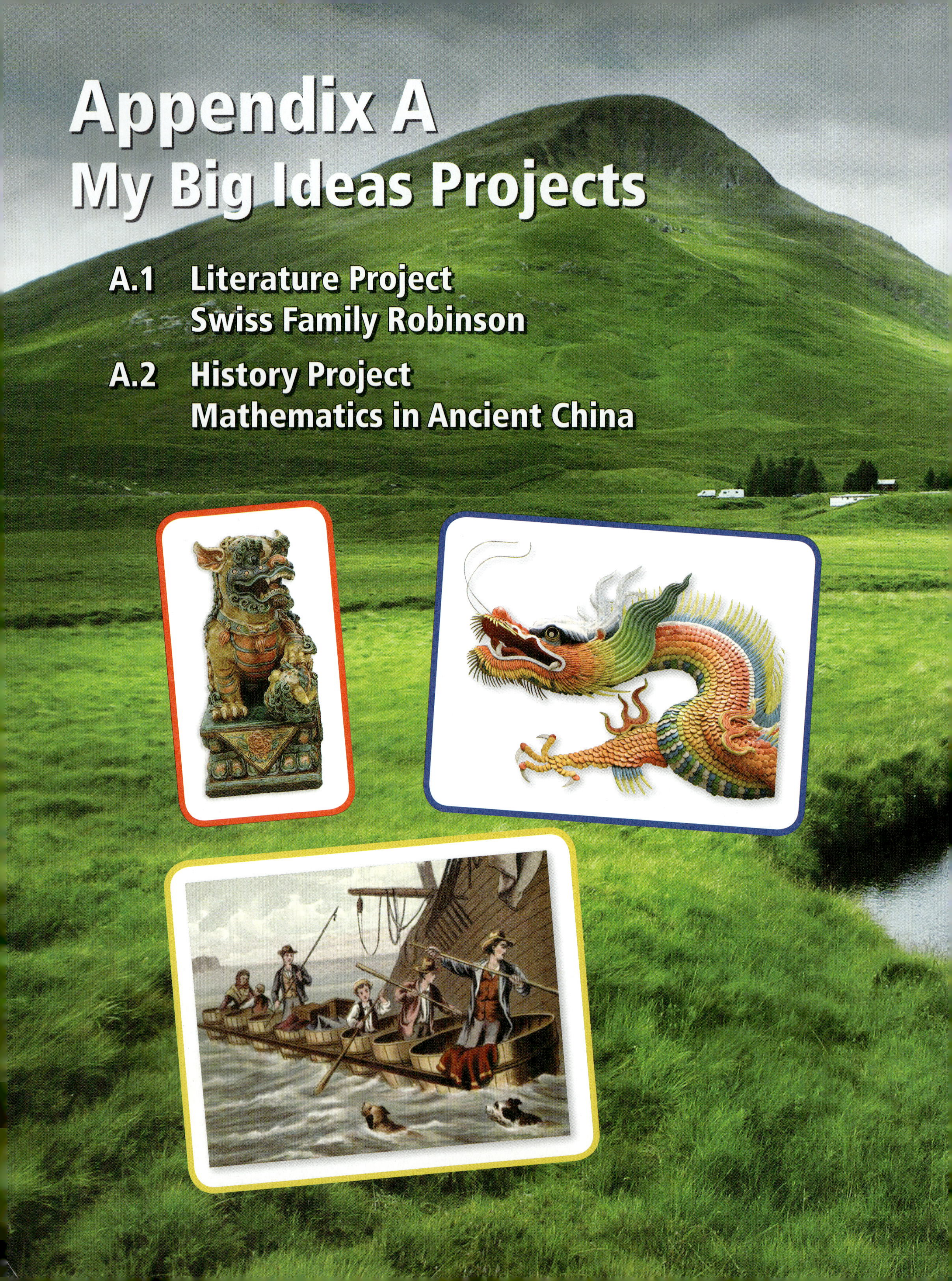

About the Appendix

- The interdisciplinary projects can be used anytime throughout the year.
- The projects offer students an opportunity to build on prior knowledge, to take mathematics to a deeper level, and to develop organizational skills.
- Students will use the Essential Questions to help them form "need to knows" to focus their research.

Essential Question

- **Literature Project**
  How does the knowledge of mathematics provide you and your family with survival tools?
- **History Project**
  How have tools and knowledge from the past influenced modern day mathematics?

Additional Resources

BigIdeasMath.com

Additional Resources

BigIdeasMath.com

Essential Question

- **Art Project**

  How does the knowledge of mathematics help you create a kaleidoscope?
- **Science Project**

  How do the characteristics of a planet influence whether or not it can sustain life?

My Big Ideas Projects

A.3 Art Project
Building a Kaleidoscope

A.4 Science Project
Our Solar System

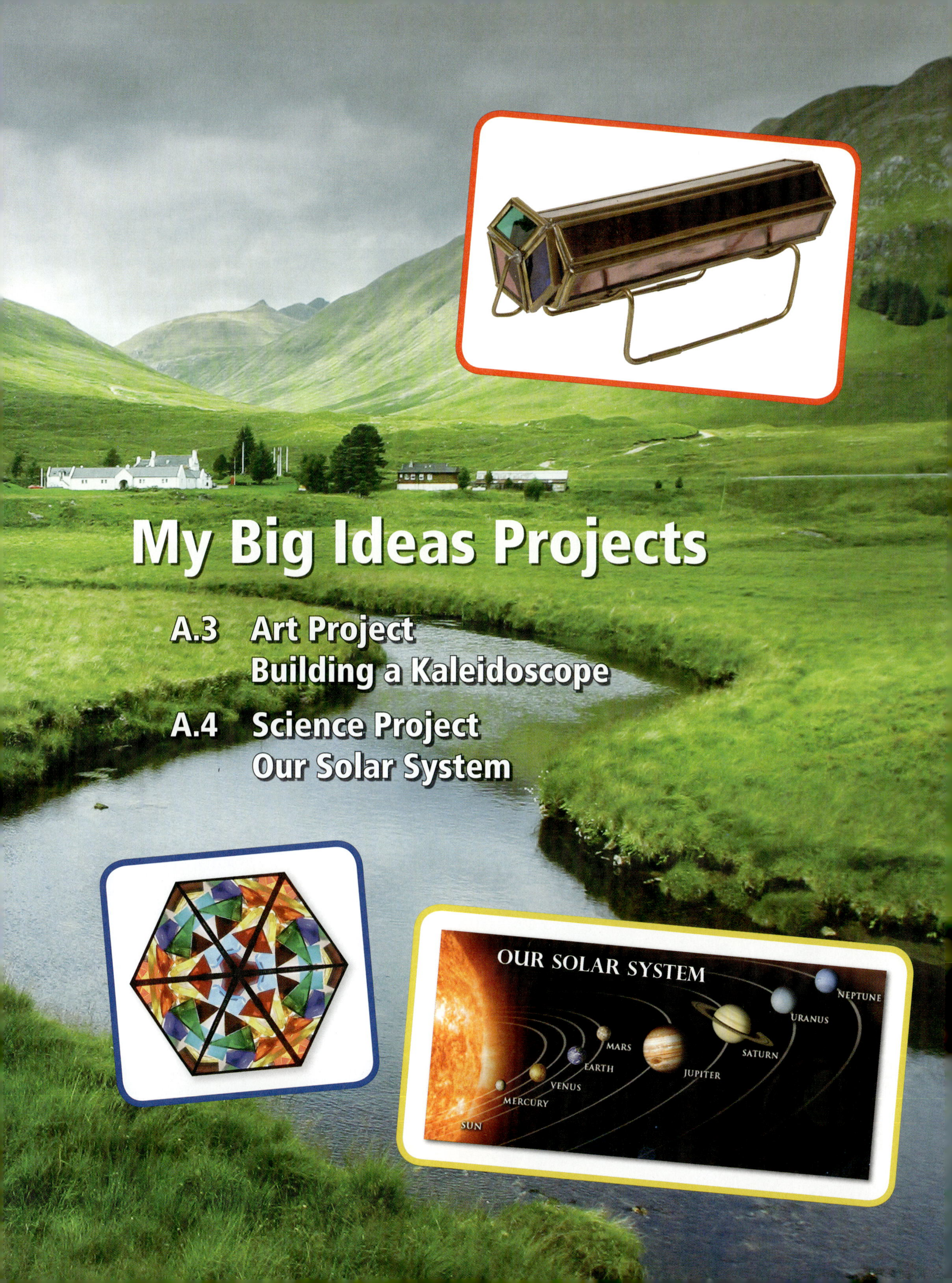

A.1 Literature Project

Swiss Family Robinson

1 Getting Started

Swiss Family Robinson is a novel about a Swiss family who was shipwrecked in the East Indies. The story was written by Johann David Wyss, and was first published in 1812.

Essential Question How does the knowledge of mathematics provide you and your family with survival tools?

Read *Swiss Family Robinson*. As you read the exciting adventures, think about the mathematics the family knew and used to survive.

Sample: The tree house built by the family was accessed by a long rope ladder. The ladder was about 30 feet long with a rung every 10 inches. To make the ladder, the family had to plan how many rungs were needed. They decided the number was $1 + 12(30) \div 10$. Why?

Project Notes

Introduction

For the Teacher

- **Goal:** Students will read *Swiss Family Robinson* by Johann David Wyss and write reports about the mathematics they find in the story. Samples of things that could be included in students' reports are discussed below.
- **Management Tip:** You may want to have students work together in groups.

Essential Question

- How does the knowledge of mathematics provide you and your family with survival tools?

Things to Think About

Summary of *Swiss Family Robinson*

- A husband, wife, and four young sons, were shipwrecked as a result of a turbulent storm. Abandoned by the crew, they were left to their own ingenuity. The ship was smashed on a rock, but within sight of land. They built a raft out of barrels to transport themselves and material from the ship to the shore. After repeated trips, they removed as much as possible from the ship, which was full of livestock, equipment, and food. There were even pieces of a sailboat onboard, which they assembled and sailed.
- The family established temporary quarters in a tent made with sailcloth. They hunted, planted, and explored. They built a treehouse to use as a safe place to live. They named the tree house *Falconhurst*. It was very comfortable until the rainy season.
- The family built several other structures to serve as shelters on their various expeditions. Each structure was well furnished using either the things from the ship or furniture that they had made.
- The father was extremely innovative, able to make anything they needed from the materials at hand. He seemed to know how to use every resource for either the appropriate or improvised purpose.
- The island was lush with both flowers and animals. Vegetables and fruits from potatoes to pineapples were abundant. The family planted additional crops using the seeds obtained from the ship. Animals, from apes to kangaroos, were captured and tamed or used for food. The family established farms with animals that they captured or rescued from the ship.
- The family considered themselves very fortunate, knowing that things could have been much worse. On the one-year anniversary of the shipwreck, they observed a day of thanksgiving.

References

Go to *BigIdeasMath.com* to access links related to this project.

Cross-Curricular Instruction

Meet with a reading or language arts teacher and review curriculum maps to identify whether students have or have yet to read *Swiss Family Robinson*. If the book has been read, you may want to discuss the work students have completed and review the book with them. If the book has not been read, perhaps you can both work simultaneously and share notes. Or, you may want to explore activities that the reading or language arts teacher has done in the past to support student learning in this particular area.

Project Notes

- Fritz, the eldest son, became restless after 10 years. So, he ventured out to explore, and discovered an albatross with a message tied to its leg. He found and rescued Jenny (Miss Montrose), an English woman who had also been shipwrecked. She lived alone on an island until Fritz noticed her signal fire. A rescue ship came shortly after Fritz found Jenny. Some of the occupants decided to stay at New Switzerland (as they liked to call their dominion). Others, including Jenny, Fritz, and Franz (the youngest son) decided to board the ship and go to England.

Mathematics Used in the Story

Some examples of mathematics used are illustrated in the following excerpts from the story.

- "This convinced me that we must not be far from the equator, for twilight results from the refraction of the sun's rays; the more obliquely these rays fall, the further does the partial light extend, while the more perpendicularly they strike the earth the longer do they continue their undiminished force, ..."
- "Jack showed me where he thought the bridge should be, and I certainly saw no better place, as the banks were at that point tolerably close to each other, steep, and of about equal height. 'How shall we find out if our planks are long enough to reach across? ... A surveyor's table would be useful now.' 'What do you say to a ball of string, father? ... Tie one end to a stone, throw it across, then draw it back, and measure the line!' ... we speedily ascertained the distance across to be eighteen feet. Then allowing three feet more at each side, I calculated twenty-four feet as the necessary length of the boards."
- In constructing the raft, they used empty water-casks, "arranging twelve of them side by side in rows of three" and placing planks on top for the floor.
- When making a canoe out of birch bark, they "cut the bark through in a circle" in two places, "took a narrow perpendicular slip of bark entirely out," and then loosened and separated the bark from the tree.
- While getting next to a whale they estimated, "the length being from sixty to sixty-five feet, and the girth between thirty and forty, while the weight could not have been less than 50,000 lbs. ...the enormous head about one-third the length of the entire hulk..."

Closure

- **Rubric** An editable rubric for this project is available at *BigIdeasMath.com.*
- Students may present their reports to the class or school as a television report or public information broadcast.

2 Things to Include

- Suppose you lived in the 18th century. Plan a trip from Switzerland to Australia. Describe your route. Estimate the length of the route and the number of miles you will travel each day. About how many days will the entire trip take?
- Suppose that your family is shipwrecked on an island that has no other people. What do you need to do to survive? What types of tools do you hope to salvage from the ship? Describe how mathematics could help you survive.
- Suppose that you are the oldest of four children in a shipwrecked family. Your parents have made you responsible for the education of your younger siblings. What type of mathematics would you teach them? Explain your reasoning.

3 Things to Remember

- You can download each part of the book at *BigIdeasMath.com*.
- Add your own illustrations to your project.
- Organize your math stories in a folder, and think of a title for your report.

A.2 History Project

Mathematics in Ancient China

1 Getting Started

Mathematics was developed in China independently of the mathematics that was developed in Europe and the Middle East. For example, the Pythagorean Theorem and the computation of pi were used in China prior to the time when China and Europe began communicating with each other.

Essential Question How have tools and knowledge from the past influenced modern day mathematics?

Sample: Here are the names and symbols that were used in ancient China to represent the digits from 1 through 10.

1	yi	一
2	er	二
3	san	三
4	si	四
5	wu	五
6	liu	六
7	qi	七
8	ba	八
9	jiu	九
10	shi	十

Life-size Terra-cotta Warriors

A Chinese Abacus

Project Notes

Introduction

For the Teacher

- **Goal:** Students will discover how mathematics was used in ancient China.
- **Management Tip:** Students can work in groups to research the required topics and generate reports.

Essential Question

- How have tools and knowledge from the past influenced modern day mathematics?

Things to Think About

What is the ancient Chinese book *The Nine Chapters on the Mathematical Art* (c. 100 B.C.)?

- The book is one of the earliest surviving mathematics texts of ancient China. It is a collection of scholarly math writings that were written over a span of more than a thousand years. It assisted the ancient Chinese in solving problems dealing with trade, taxation, surveying, and engineering. It consists of 246 problems, a solution to each problem, and an explanation of how to get each solution.

What types of mathematics are contained in *The Nine Chapters on the Mathematical Art*?

- **Chapter One:** Areas of triangles, rectangles, circles, and trapeziums; operations on fractions; finding greatest common divisor
- **Chapter Two:** Rates of exchange; proportions; percentages
- **Chapter Three:** Direct, inverse, and compound proportions
- **Chapter Four:** Keeping the area of a rectangle the same while increasing its width; square roots and cube roots
- **Chapter Five:** Volumes of prisms, pyramids, cylinders, and cones
- **Chapter Six:** Ratios and proportions
- **Chapter Seven:** Linear equations
- **Chapter Eight:** Systems of linear equations
- **Chapter Nine:** Right triangles; similar triangles; Pythagorean Theorem (called the Gougu rule)

What is an abacus?

- The Chinese abacus consists of a wooden frame with 13 vertical rods and a horizontal wooden divider. There are seven beads on each rod. Five of the seven beads on each rod are located below the wooden divider, which the Chinese called the earth. The other two beads on each rod are located above the wooden divider, called the heaven. Earth beads are worth one point each, while the heaven beads are worth five points each.

How is a number represented on the abacus?

- To start, all of the beads in both sections are pushed away from the divider. To represent numbers on the abacus, the beads are pushed toward the divider. The rod on the far right is the ones rod, then the one to its left is the tens rod, then the hundreds rod, then the thousands rod, etc.

References

Go to *BigIdeasMath.com* to access links related to this project.

Cross-Curricular Instruction

Meet with a history teacher and review curriculum maps to identify whether students have covered or have yet to discuss ancient China. If the topic has been covered, you may want to discuss the work students have completed and review prior knowledge with them. If the history teacher has not discussed these concepts, perhaps you can both work simultaneously on these concepts and share notes. Or, you may want to explore activities that the history teacher has done in the past to support student learning in this particular area.

Project Notes

- Suppose you want to represent the number 827 on the abacus. On the ones rod, move one of the heaven beads to the divider (5 points) and two of the earth beads to the divider (2 points). On the tens rod, move none of the heaven beads (0 points) and two of the earth beads (2 points). On the hundreds rod, move one of the heaven beads (5 points) and three of the earth beads (3 points).

? How is an abacus used to add or subtract numbers?

- Suppose you want to add 827 and 122. With 827 already on the abacus, add 122, digit by digit, starting with the ones rod. On the ones rod, move two more earth beads to the divider, making that rod represent a nine. On the tens rod, move two more earth beads to the divider, making that rod represent a four. On the hundreds rod, move one more earth bead to the divider, making that rod represent a nine. So, $827 + 122 = 949$.
- To subtract two numbers, follow the same format, except in reverse, moving beads away from the divider instead of toward it.

? How did the ancient Chinese write numbers that are greater than 10?

- The ancient Chinese had symbols to represent each of the numbers from zero through nine.

- They also had symbols to represent 10, 100, 1000, and 10,000.

丅 日 丅 禺

10 100 1000 10,000

- To write a number such as 467, they had to write the number as $4 \cdot 100 + 6 \cdot 10 + 7$.

467 = 四百六十七

Closure

- **Rubric** An editable rubric for this project is available at *BigIdeasMath.com*.
- You may hold a class debate where students can compare, defend, and discuss their findings with other students.

2 Things to Include

Ancient Chinese Teapot

- Describe the ancient Chinese book *The Nine Chapters on the Mathematical Art* (c. 100 B.C.). What types of mathematics are contained in this book?
- How did the ancient Chinese use the abacus to add and subtract numbers? How is the abacus related to base 10?
- How did the ancient Chinese use mathematics to build large structures, such as the Great Wall and the Forbidden City?
- How did the ancient Chinese write numbers that are greater than 10?
- Describe how the ancient Chinese used mathematics. How does this compare with the ways in which mathematics is used today?

The Great Wall of China

3 Things to Remember

- Add your own illustrations to your project.
- Organize your math stories in a folder, and think of a title for your report.

Chinese Guardian Fu Lions

A.3 Art Project

Building a Kaleidoscope

1 Getting Started

Mirrors set at 60°

A kaleidoscope is a tube of mirrors containing loose colored beads, pebbles, or other small colored objects. You look in one end and light enters the other end, reflecting off the mirrors.

Essential Question How does the knowledge of mathematics help you create a kaleidoscope?

If the angle between the mirrors is 45°, you see 8 duplicate images. If the angle is 60°, you see 6 duplicate images. If the angle is 90°, you see 4 duplicate images. As the tube is rotated, the colored objects tumble, creating various patterns.

Write a report about kaleidoscopes. Discuss the mathematics you need to know in order to build a kaleidoscope.

Sample: A kaleidoscope whose mirrors meet at 60° angles has reflective symmetry and rotational symmetry.

Reflect

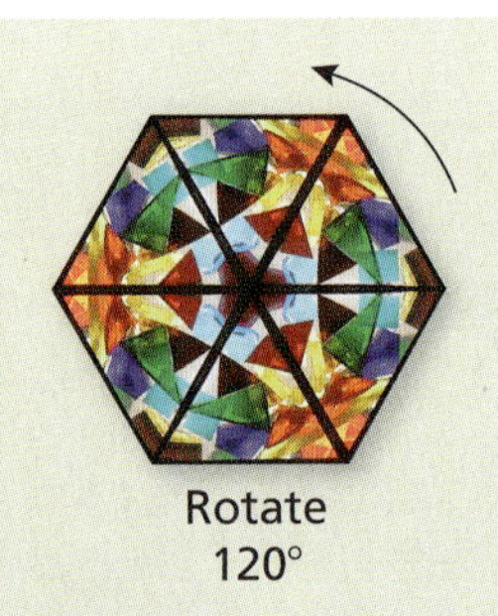
Rotate 120°

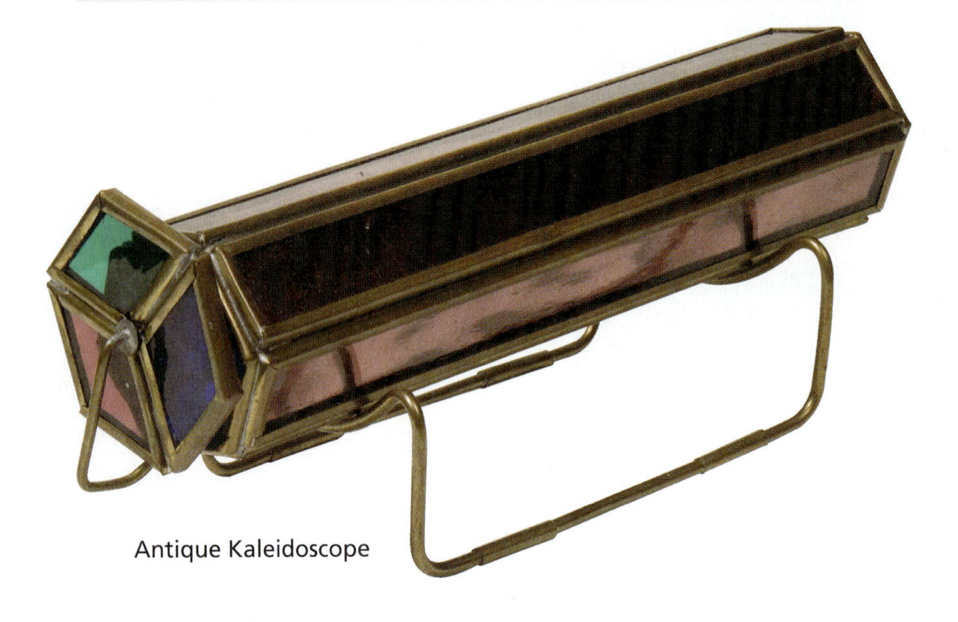
Antique Kaleidoscope

Project Notes

Introduction

For the Teacher

- **Goal:** Students will discover how a kaleidoscope functions.
- **Management Tip:** Students may wish to search online for a virtual kaleidoscope. If time permits, students can visit *BigIdeasMath.com* for links to websites containing instructions on how to make a kaleidoscope. Students can create a kaleidoscope on their own or in groups.

Essential Question

- How does the knowledge of mathematics help you create a kaleidoscope?

Things to Think About

? What is a kaleidoscope?

- A kaleidoscope has two, three, or four mirrors inside of a tube. It also has a collection of glass pieces, beads, pebbles, water, or other substances that reflect light.
- When the viewer looks through one end of the tube, light is reflected through the other end. The results are multiple reflections of the materials, making beautiful patterns. As the tube is rotated, the patterns change. Whatever the pattern, the reflections and symmetry are determined by the angle(s) of the mirrors. The intersection of the lines of symmetry becomes the center of the rotational symmetry.

? Who invented the kaleidoscope?

- The ancient Greeks contemplated the use of reflections from mirrors. The ancient Egyptians placed two highly polished pieces of limestone together at different angles to see different reflections.
- Sir David Brewster of Scotland (the inventor of the modern lighthouse) is credited with the invention of the kaleidoscope. In 1816, he was working with optical tools, including prisms, and noticed the beautiful reflections and symmetrical patterns that were formed. In 1817, he named his invention the kaleidoscope, which means 'beautiful form to see.' Brewster filed for a patent of the kaleidoscope, but an error in the paperwork cost Brewster the financial rewards he deserved. His initial design was a tube with pairs of mirrors at one end, pairs of translucent disks at the other end, and beads between the two ends. Initially intended as a science tool, the kaleidoscope was quickly copied as a toy.
- In America, Charles Bush popularized the kaleidoscope. Although the early kaleidoscopes sold for around $2.00, today they often sell for $2000.00.

References

Go to *BigIdeasMath.com* to access links related to this project.

Cross-Curricular Instruction

Meet with an art teacher and review curriculum maps to identify whether students have covered kaleidoscopes. If the topic has already been covered, you may want to discuss the work students have completed and then review prior knowledge with them. If the art teacher has not discussed these concepts, perhaps you can both work simultaneously on these concepts and share notes. Or, you may want to explore activities that the art teacher has done in the past to support student learning in this particular area.

Project Notes

? How does a kaleidoscope work?

- Kaleidoscopes can be made of two, three, or four mirrors. The angle between the mirrors determines the number of reflections. The measures of the angles between the mirrors must divide evenly into 360°. The number of reflections is determined by dividing the number of degrees into 360°.
- When two mirrors are used, they are placed in a V formation. The table shows the number of reflections in a kaleidoscope based upon the angle formed by the two mirrors.

Degrees	Number of reflections
90	$360 \div 90 = 4$
60	$360 \div 60 = 6$
45	$360 \div 45 = 8$
36	$360 \div 36 = 10$
30	$360 \div 30 = 12$
22.5	$360 \div 22.5 = 16$
15	$360 \div 15 = 24$
10	$360 \div 10 = 36$
1	$360 \div 1 = 360$

- When three mirrors are used, they are placed in a triangular formation. This system works similarly to the two-mirror system except the third mirror replaces the side of the V formation that does not have a mirror. This results in a continuation of reflections throughout the field of view. With the triangular formation, it is important that the measure of each angle divides into 360°, but the sum of the angles must be 180°. Only three combinations work: 60°-60°-60°, 45°-45°-90°, and 30°-60°-90°.

Closure

- **Rubric** An editable rubric for this project is available at *BigIdeasMath.com.*
- Students may present their reports to a parent panel or community members.

2 Things to Include

- How does the angle at which the mirrors meet affect the number of duplicate images that you see?
- What angles can you use other than 45°, 60°, and 90°? Explain your reasoning.
- Research the history of kaleidoscopes. Can you find examples of kaleidoscopes being used before they were patented by David Brewster in 1816?
- Make your own kaleidoscope.
- Describe the mathematics you used to create your kaleidoscope.

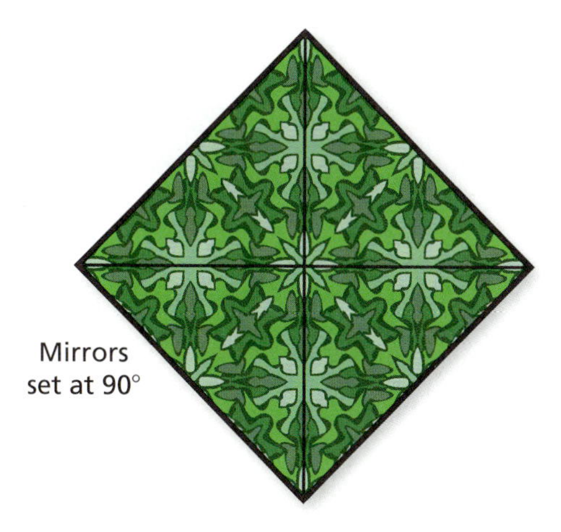

Mirrors set at 90°

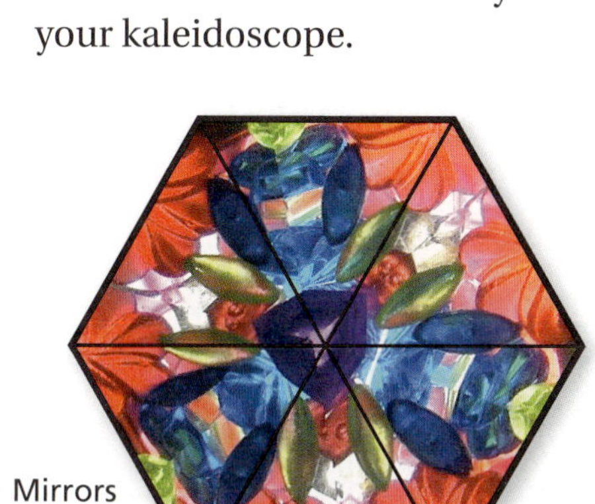

Mirrors set at 60°

3 Things to Think About

- Add your own drawings and pattern creations to your project.
- Organize your report in a folder, and think of a title for your report.

Mirrors set at 45°

Giant Kaleidoscope, San Diego harbor

A.4 Science Project

Our Solar System

1 Getting Started

Our solar system consists of four inner planets, four outer planets, dwarf planets such as Pluto, several moons, and many asteroids and comets.

Essential Question How do the characteristics of a planet influence whether or not it can sustain life?

Sample: The average temperatures of the eight planets in our solar system are shown in the graph.

The average temperature tends to drop as the distance between the Sun and the planet increases.

An exception to this rule is Venus. It has a higher average temperature than Mercury, even though Mercury is closer to the Sun.

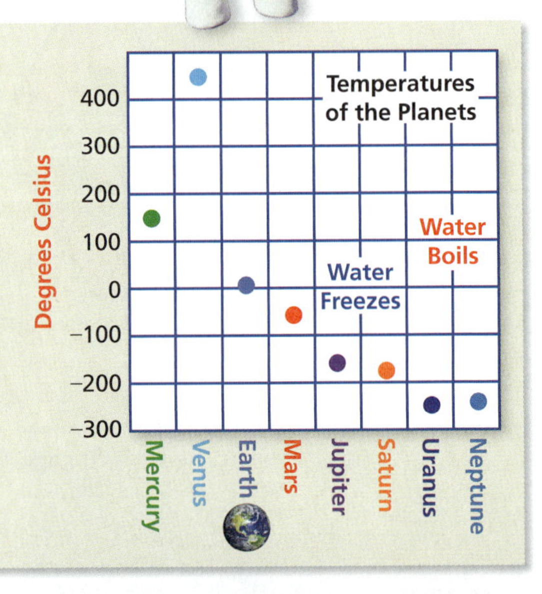

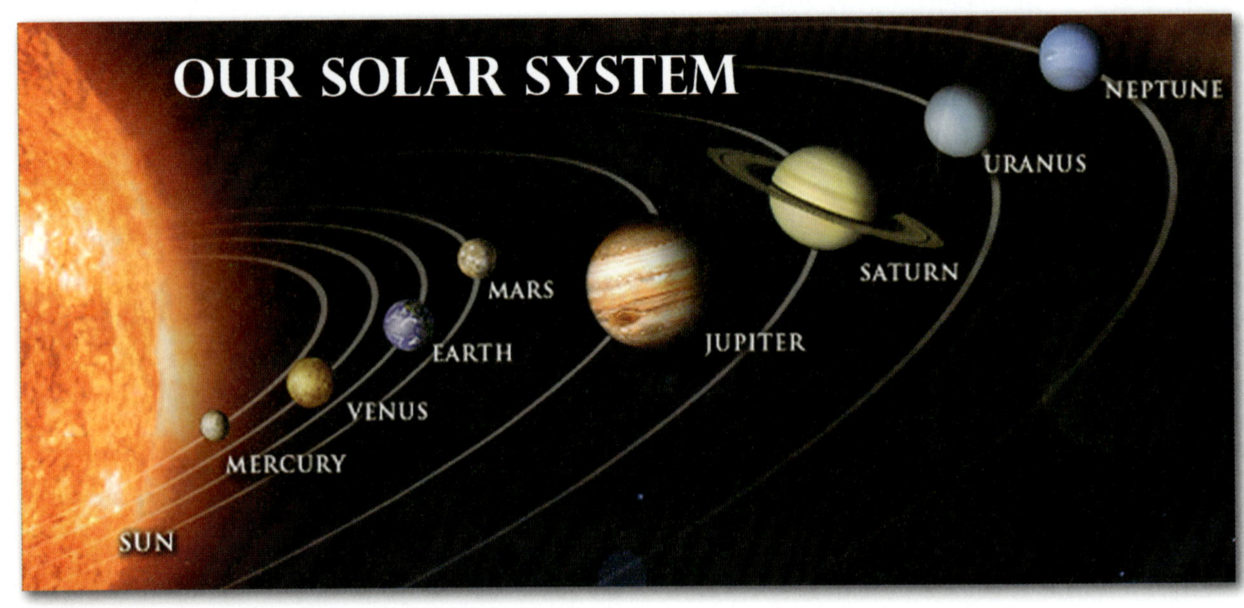

Project Notes

Introduction

For the Teacher

- **Goal:** Students will discover facts about objects in our solar system.
- **Management Tip:** Students can work in groups to create reports on our solar system.

Essential Question

- How do the characteristics of a planet influence whether or not it can sustain life?

Things to Think About

Inner System

	Mercury	Venus	Earth	Mars
Distance from the Sun	5.8×10^7 km	1.08×10^8 km	1.5×10^8 km	2.3×10^8 km
Diameter	4.8×10^3 km	1.2×10^4 km	1.27×10^4 km	6.7×10^3 km
Mass	3.3×10^{23} kg	4.9×10^{24} kg	6.0×10^{24} kg	6.4×10^{23} kg
Gravitational pull	38% of Earth	91% of Earth	100% of Earth	38% of Earth
Length of day	1407 hours, 30 minutes	5832 hours,	1 Earth Day	24 hours, 37 minutes
Length of year	88 days	225 days	365.25 days	687 days
Range or average temperature	−279°F to 801°F	864°F	45°F	−125°F to 23°F
Support human life?	No	No	Yes	No

? Have the planets been visited by humans?

- **Mercury** has been visited by two spacecrafts; *Mariner 10* in 1974 and 1975 and *Messenger* which was launched by NASA in 2004. *Messenger* did several "flybys" in 2008 and began to orbit Mercury in 2011.

References

Go to *BigIdeasMath.com* to access links related to this project.

Cross-Curricular Instruction

Meet with a science teacher and review curriculum maps to identify whether students have covered or have yet to discuss the planets in our solar system. If the topic has been covered, you may want to discuss the work students have completed and review prior knowledge with them. If the science teacher has not discussed these concepts, perhaps you can both work simultaneously on these concepts and share notes. Or, you may want to explore activities that the science teacher has done in the past to support student learning in this particular area.

Project Notes

- **Venus** has been visited by at least 20 spacecrafts. *Mariner 2* was the first in 1962. Others have included *Pioneer Venus*, the Soviet's *Venera 7* and *Venera 9*, and ESA's *Venus Express*.
- **Mars** has been visited by several spacecrafts, the first being *Mariner 4* in 1965. Spacecrafts that have landed on Mars include: *Mars 2*, two *Viking* landers, *Mars Pathfinder*, the twin Mars Expedition Rovers *Spirit* and *Opportunity*, and Mars rover *Curiosity*.
- **Jupiter** has been visited by the spacecrafts *Pioneer 10*, *Pioneer 11*, *Voyager 1*, *Voyage 2*, *Ulysses*, *Galileo*, and *JUNO*.
- **Saturn** has been visited by *Pioneer 11* in 1979, *Voyager 1*, *Voyager 2*, and *Cassini*.
- **Uranus** *Voyager 2* is the only spacecraft to have visited Uranus.
- **Neptune** *Voyager 2* is the only spacecraft to have visited Neptune.

Outer System

	Jupiter	Saturn	Uranus	Neptune
Distance from the Sun	7.7×10^8 km	1.4×10^9 km	2.8×10^9 km	4.5×10^9 km
Diameter	1.4×10^5 km	1.2×10^5 km	5.1×10^4 km	4.9×10^4 km
Mass	2.0×10^{27} kg	5.7×10^{26} kg	8.7×10^{25} kg	1.0×10^{26} kg
Gravitational pull	260% of Earth	117% of Earth	85% of Earth	120% of Earth
Length of day	9 hours, 56 minutes	10 hours, 39 minutes	17 hours, 5 minutes	16 hours, 7 minutes
Length of year	4331 days	10,759 days	30,687 days	60,190 days
Range or average temperature	$-234°$F	$-288°$F	$-357°$F	$-353°$F
Support human life?	No	No	No	No

Closure

- **Rubric** An editable rubric for this project is available at *BigIdeasMath.com*.
- Students may present their reports to the class or compare their reports with other students' reports.

2 Things to Include

- Compare the masses of the planets.
- Compare the gravitational forces of the planets.
- How long is a "day" on each planet? Why?
- How long is a "year" on each planet? Why?
- Which planets or moons have humans explored?
- Which planets or moons could support human life? Explain your reasoning.

Mars Rover

3 Things to Remember

- Add your own drawings or photographs to your report. You can download photographs of the solar system and space travel at *NASA.gov*.
- Organize your report in a folder, and think of a title for your report.

Hubble Image of Space

Hubble Spacecraft

Key Vocabulary Index

Mathematical terms are best understood when you see them used and defined *in context.* This index lists where you will find key vocabulary. A full glossary is available in your Record and Practice Journal and at *BigIdeasMath.com*.

Student Index

This student-friendly index will help you find vocabulary, key ideas, and concepts. It is easily accessible and designed to be a reference for you whether you are looking for a definition, real-life application, or help with avoiding common errors.

Student Index

Q

R

S

T

Additional Answers

Chapter 1

Section 1.2

Record and Practice Journal

2. indigo: 45°, 45°, 90° yellow: 25°, 60°, 95°
violet: 60°, 60°, 60° blue: 75°, 75°, 30°
orange: 75°, 65°, 40° green: 15°, 135°, 30°

Section 1.3

Record and Practice Journal

5. *Sample answer:*
$$\begin{aligned} 4(x+2) &= x-1 \\ 4x+8 &= x-1 \\ 4x-x+8 &= -1 \\ 3x+8 &= -1 \\ 3x &= -9 \\ x &= -3 \end{aligned}$$

Practice and Problem Solving

43. a. 40 ft

b. no;

$$\begin{aligned} 2(\text{white area}) &= \text{black area} \\ 2[5(6x)] &= 4[6(x+1)] \\ 60x &= 24x+24 \\ 36x &= 24 \\ x &= \frac{2}{3} \end{aligned}$$

$$5x + 4(x+1) \stackrel{?}{=} 40$$

$$\text{Length of hallway is } 5\left(\frac{2}{3}\right) + 4\left(\frac{2}{3}+1\right) \stackrel{?}{=} 40$$

$$10 \neq 40$$

Chapter 2

Try It Yourself

5. a. (0, −1) **b.** (0, 1)

6. a. (−5, 0) **b.** (5, 0)

7. a. (4, 6.5) **b.** (−4, −6.5)

8. a. $\left(-3\frac{1}{2}, 4\right)$ **b.** $\left(3\frac{1}{2}, -4\right)$

9.

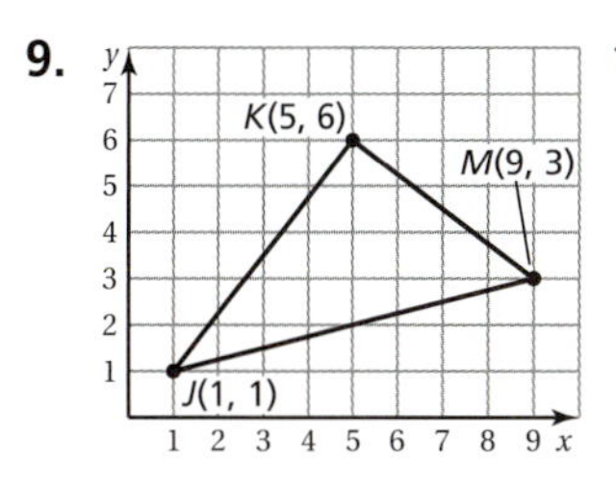

10.

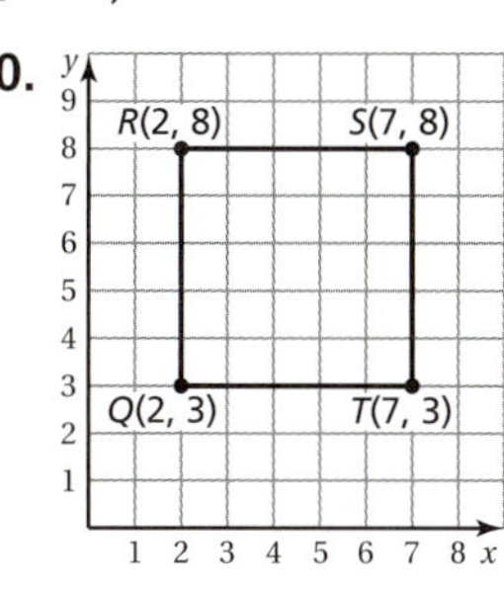

Record and Practice Journal Fair Game Review

5. (−1, −2); (1, 2) **6.** (3, −2); (−3, 2)

7.

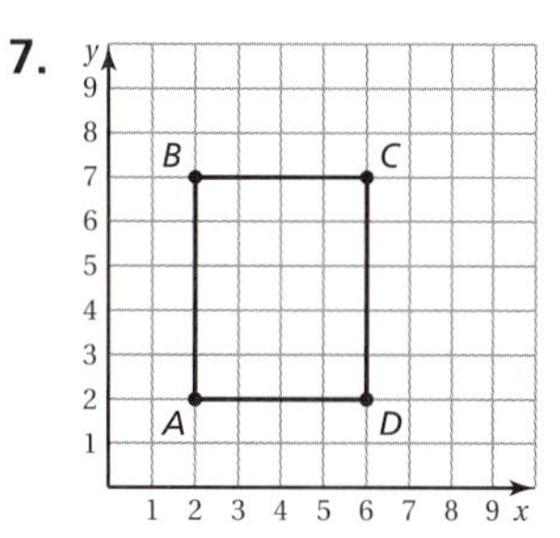

8.

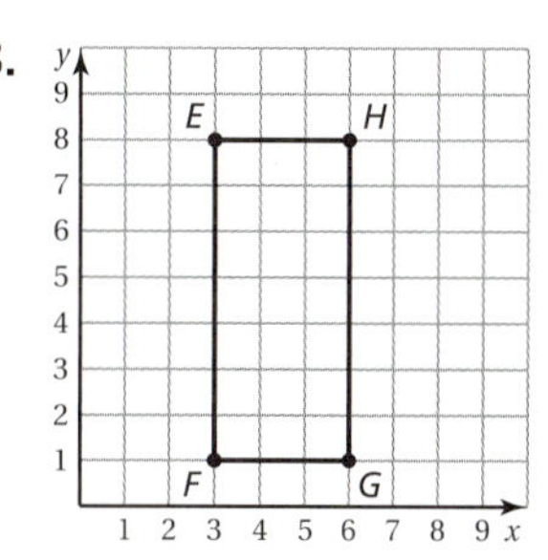

9.

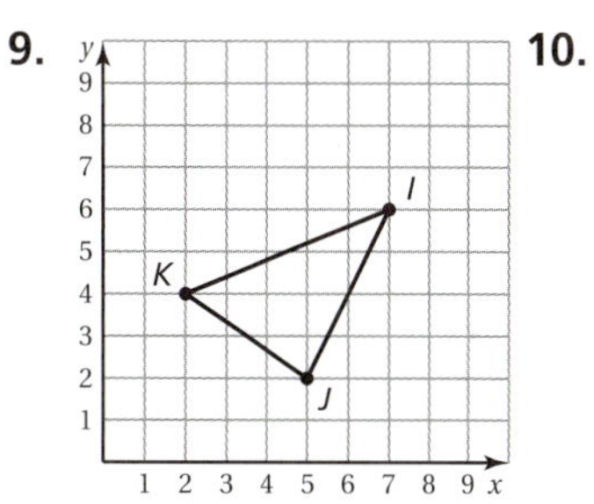

10.

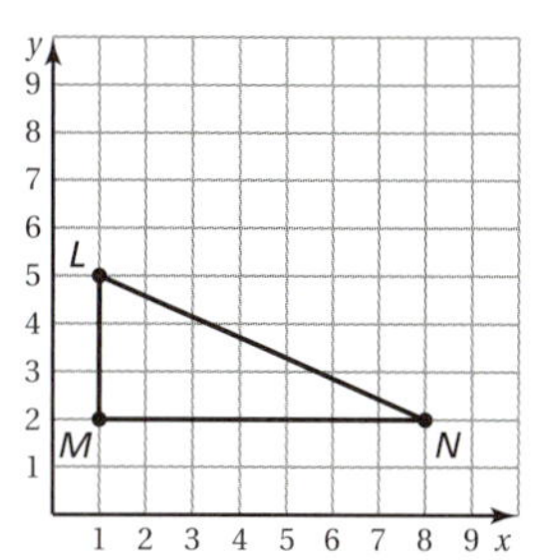

11.

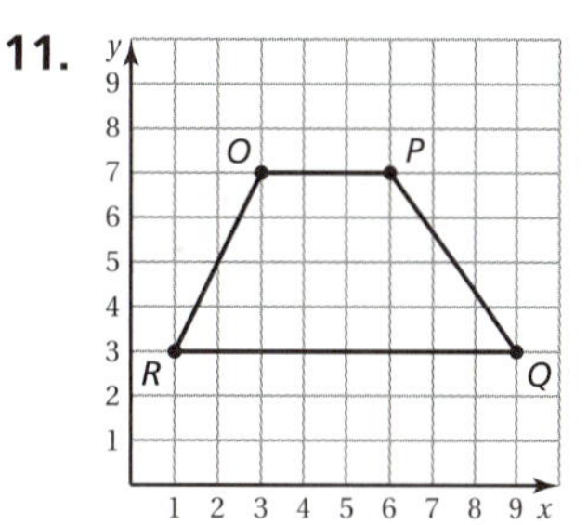

12.

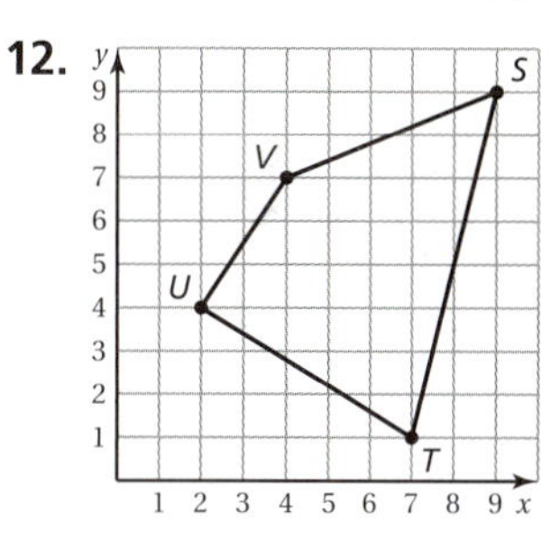

Section 2.1

Record and Practice Journal

2. a.

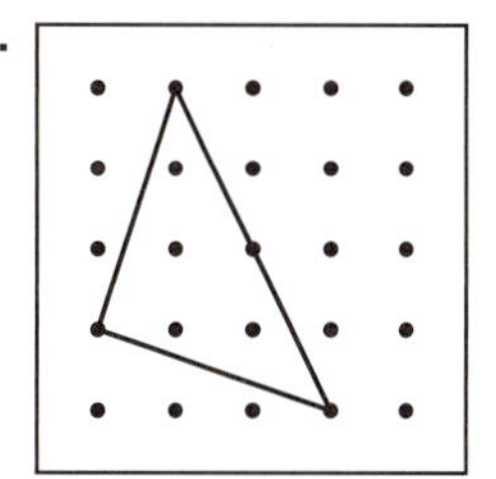

c.

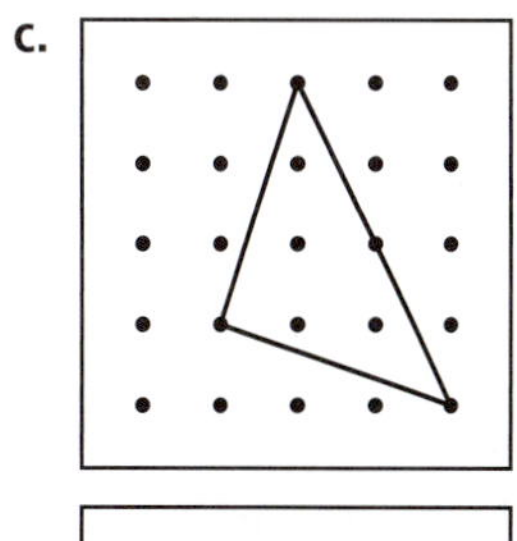

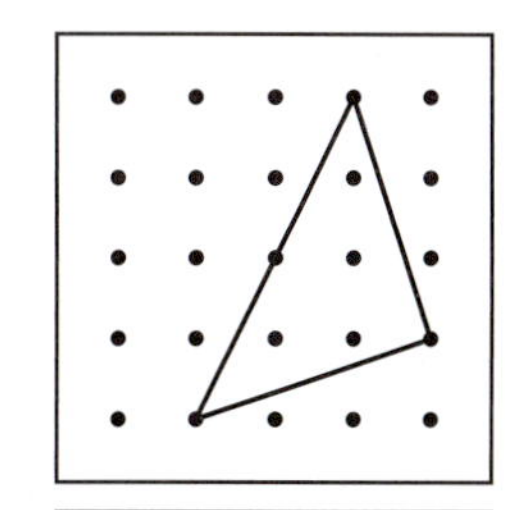

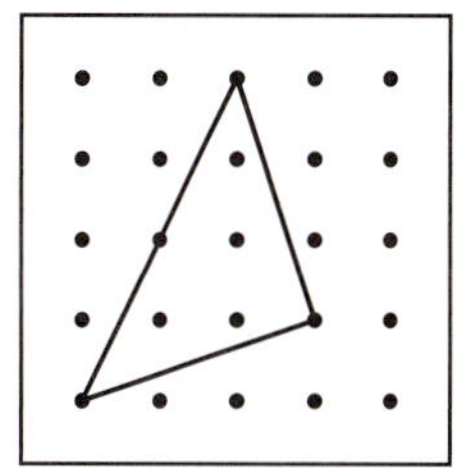

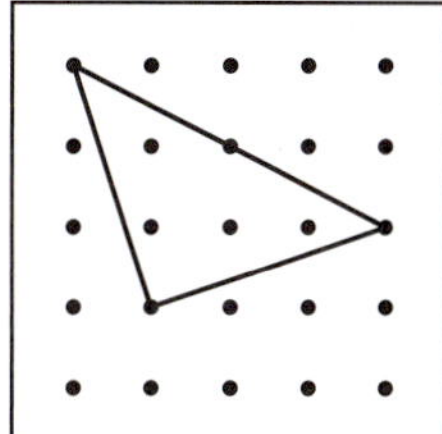

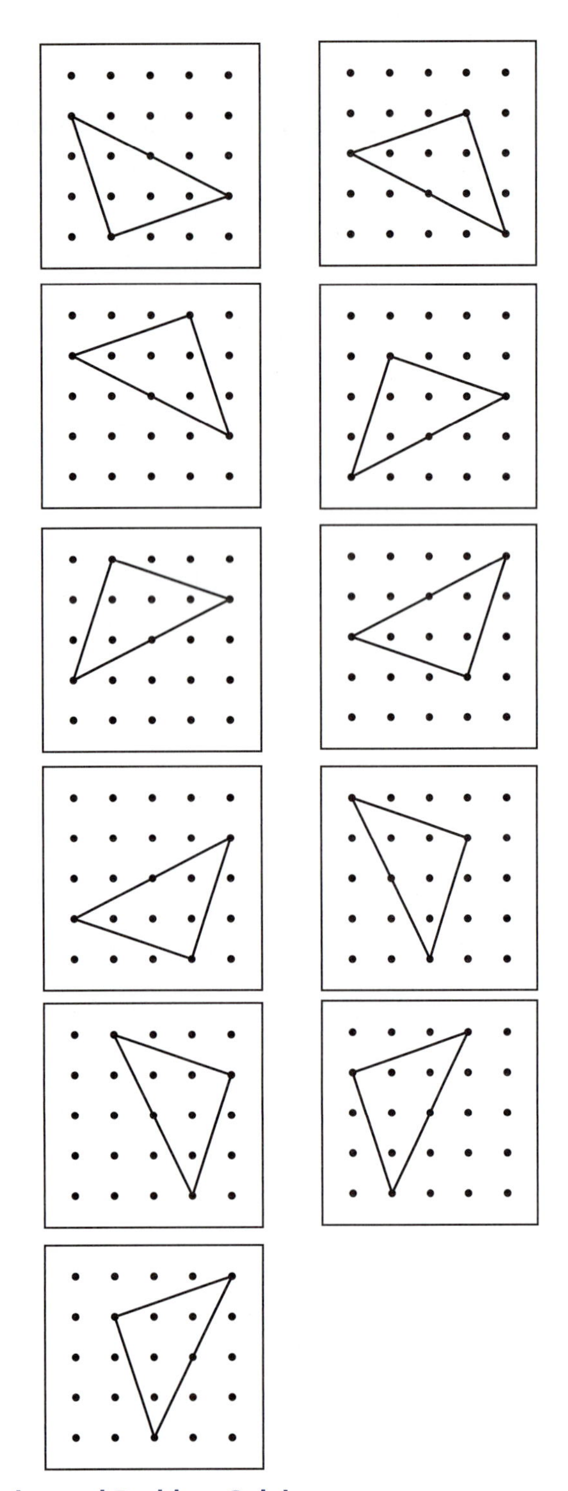

Practice and Problem Solving

12. **a.** 32 ft **b.** $\angle M$

c. 20 ft **d.** 96 ft

14. yes; The dimensions of congruent figures are equal, so the areas of the figures are equal.

15. **a.** true; Side AB corresponds to Side YZ.

b. true; $\angle A$ and $\angle X$ have the same measure.

c. false; $\angle A$ corresponds to $\angle Y$.

d. true; The measure of $\angle A$ is 90°, the measure of $\angle B$ is 140°, the measure of $\angle C$ is 40°, and the measure of $\angle D$ is 90°. So, the sum of the angle measures of $ABCD$ is $90° + 140° + 40° + 90° = 360°$.

Fair Game Review

16–19.

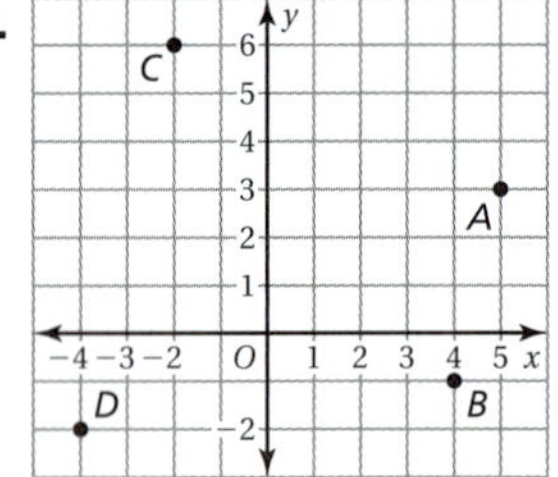

20. B

Record and Practice Journal Practice

1. Corresponding angles: $\angle A$ and $\angle J$, $\angle B$ and $\angle K$, $\angle C$ and $\angle L$, $\angle D$ and $\angle M$

Corresponding sides: Side AB and Side JK, Side BC and Side KL, Side CD and Side LM, Side DA and Side MJ

2. Corresponding angles: $\angle J$ and $\angle Q$, $\angle K$ and $\angle P$, $\angle L$ and $\angle T$, $\angle M$ and $\angle S$, $\angle N$ and $\angle R$

Corresponding sides: Side JK and Side QP, Side KL and Side PT, Side LM and Side TS, Side MN and Side SR, Side NJ and Side RQ

Section 2.2

On Your Own

5.

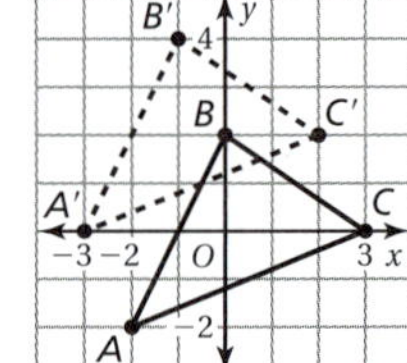

Practice and Problem Solving

15.

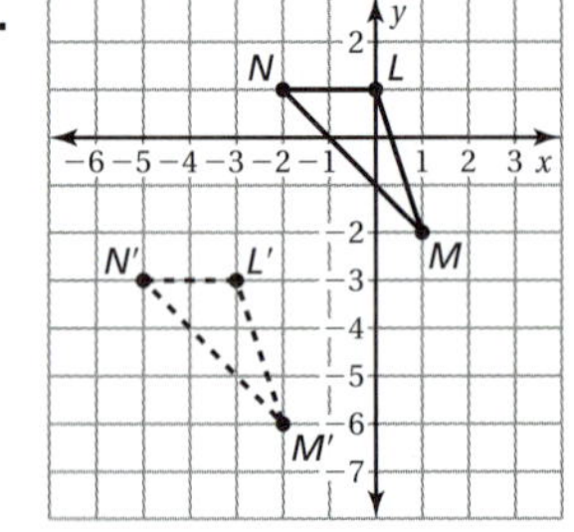

19. 6 units right and 3 units down

20. 5 units left and 2 units down

21. **a.** 5 units right and 1 unit up

b. no; It would hit the island.

c. 4 units up and 4 units right

22. yes; You can write one translation to get from the original triangle to the final triangle, which is $(x + 2, y - 10)$. So, the triangles are congruent. You can also measure the sides and angles to determine that the triangles are congruent.

Fair Game Review

24. yes
25. no
26. no
27. yes
28. B

Section 2.3

On Your Own

4. **a.**

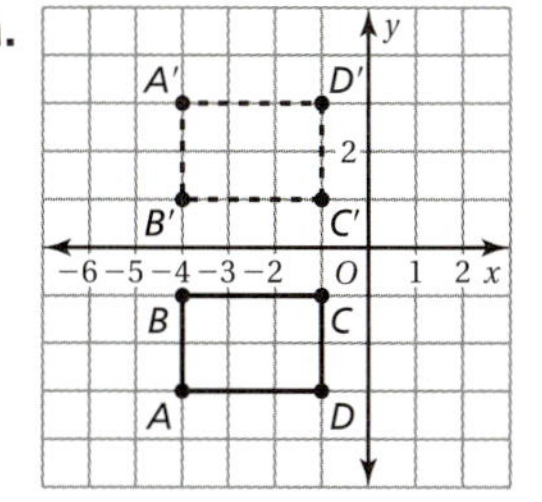

b.

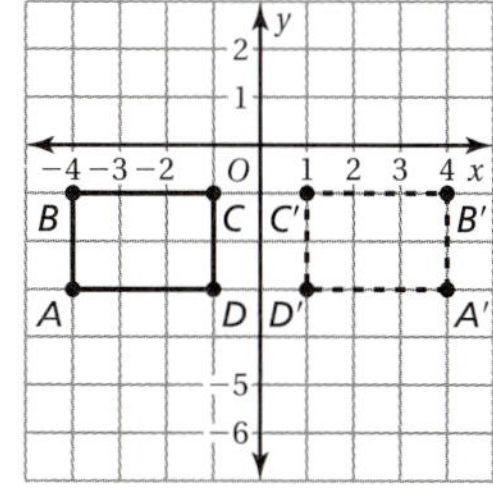

c. yes; They are all rectangles of the same size and shape.

Practice and Problem Solving

13.

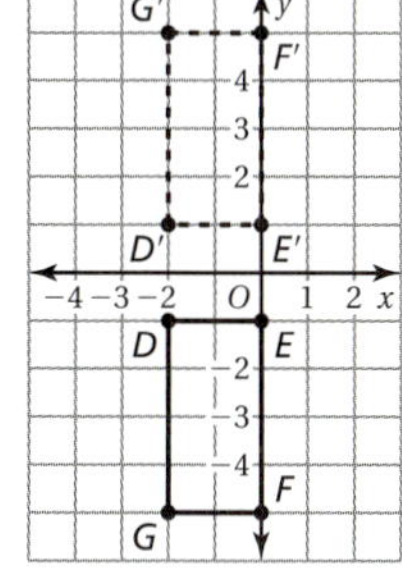

$D'(-2, 1)$, $E'(0, 1)$, $F'(0, 5)$, $G'(-2, 5)$

14.

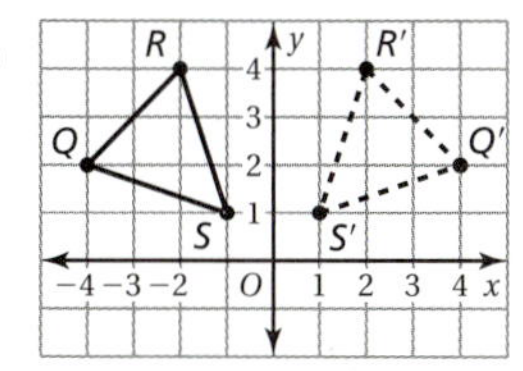

$Q'(4, 2)$, $R'(2, 4)$, $S'(1, 1)$

15.

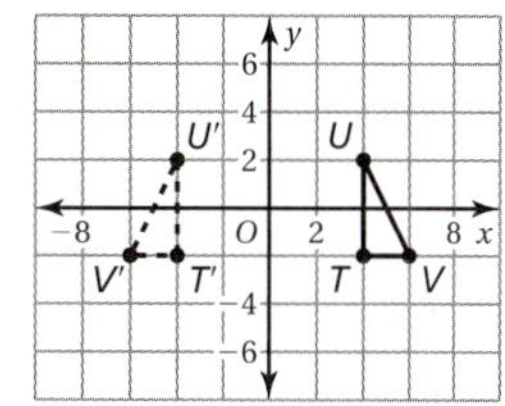

$T'(-4, -2)$, $U'(-4, 2)$, $V'(-6, -2)$

16.

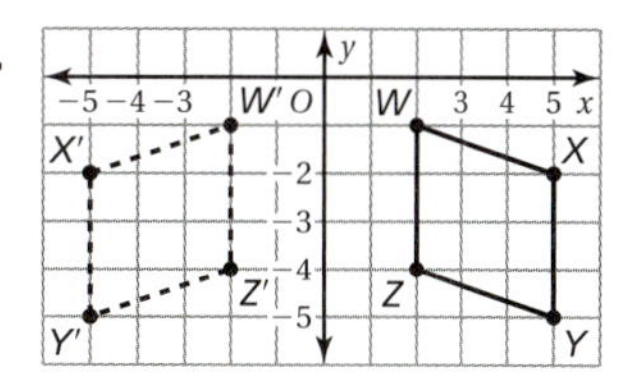

$W'(-2, -1)$, $X'(-5, -2)$, $Y'(-5, -5)$, $Z'(-2, -4)$

17.

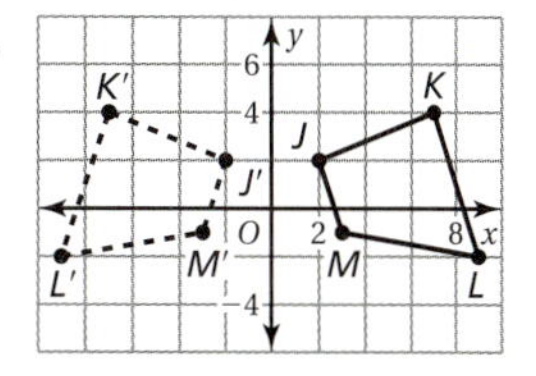

$J'(-2, 2)$, $K'(-7, 4)$, $L'(-9, -2)$, $M'(-3, -1)$

28.

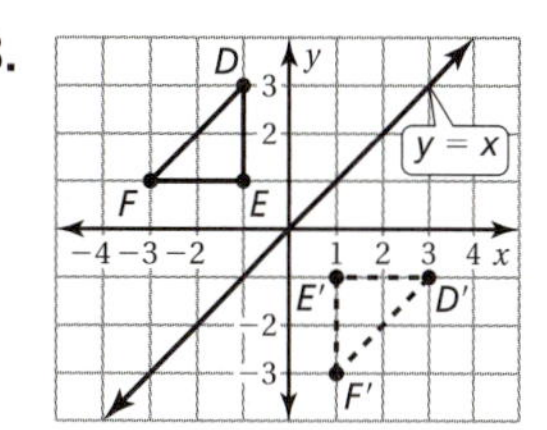

The x-coordinate and y-coordinate for each point are switched in the image.

Section 2.4

On Your Own

3. **a.**

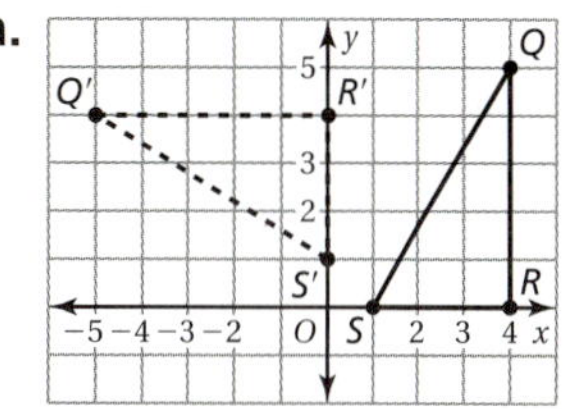

b.

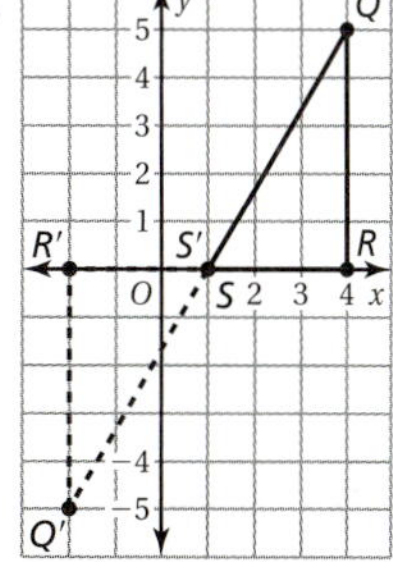

c. yes; In a rotation, the original figure and its image are congruent. Because both images are congruent to the same (original) figure, they are congruent to each other.

Practice and Problem Solving

28. **a.**

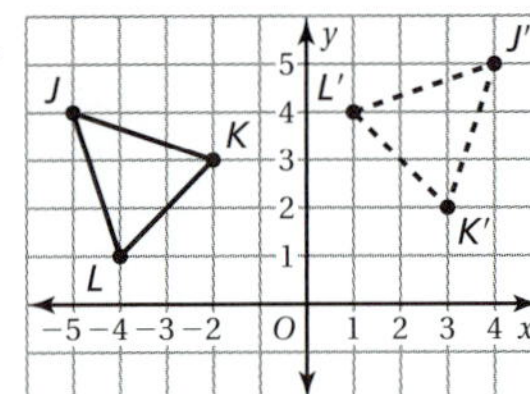

The x-coordinates of $\triangle J'K'L'$ are the same as the y-coordinates of $\triangle JKL$. The y-coordinates of $\triangle J'K'L'$ are the opposite of the x-coordinates of $\triangle JKL$.

b.

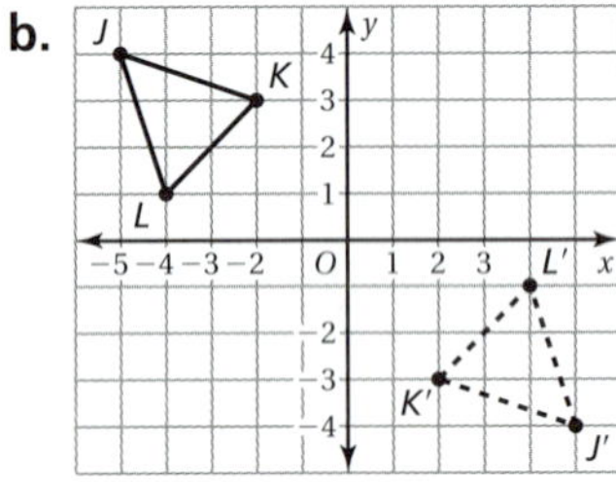

The x-coordinates of $\triangle J'K'L'$ are the opposite of the x-coordinates of $\triangle JKL$. The y-coordinates of $\triangle J'K'L'$ are the opposite of the y-coordinates of $\triangle JKL$.

c. yes; Explanations will vary.

Section 2.5

Practice and Problem Solving

6.

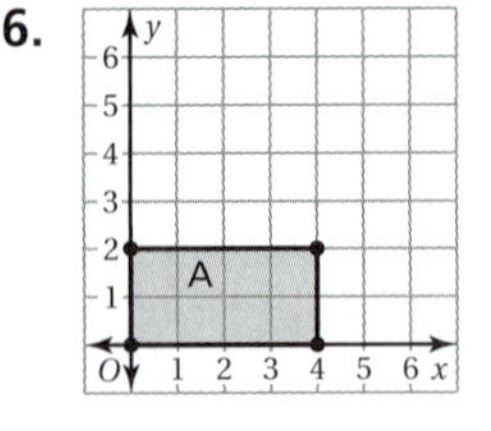

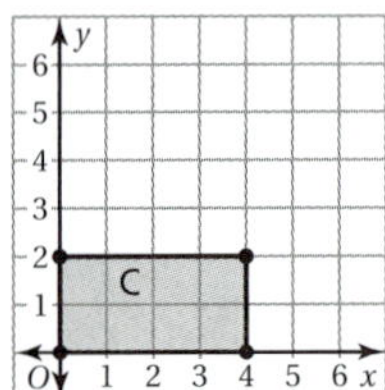

A, B and C; Corresponding side lengths are proportional and corresponding angles are congruent.

7.

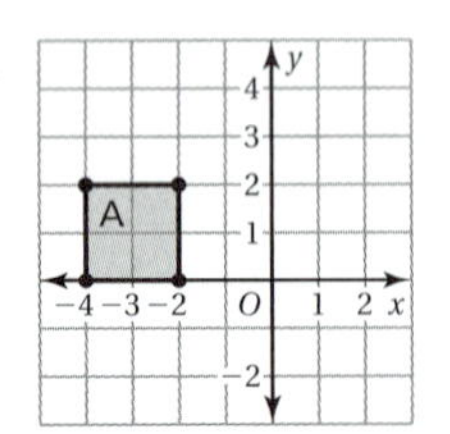

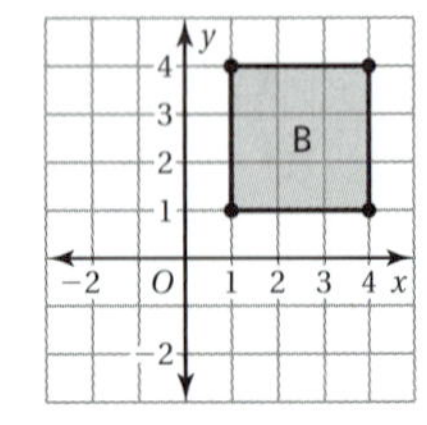

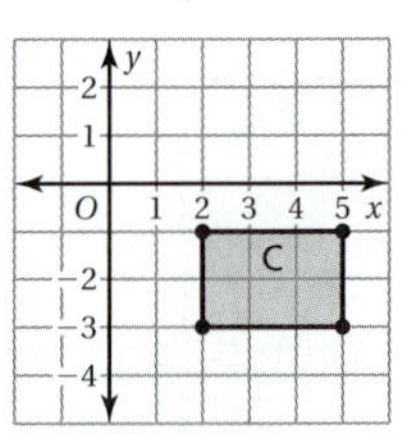

A and B; Corresponding side lengths are proportional and corresponding angles are congruent.

19. **a.** yes

b. yes; It represents the fact that the sides are proportional because you can split the isosceles triangles into smaller right triangles that will be similar.

20. yes; *Sample answer:*

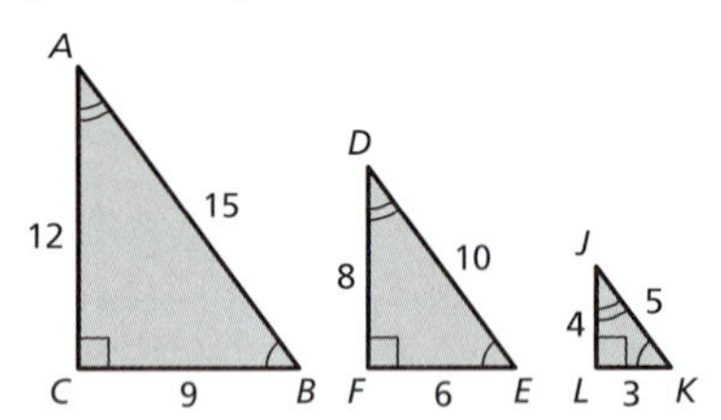

Because corresponding angles are congruent and corresponding side lengths are proportional, $\triangle ABC$ is similar to $\triangle JKL$.

Section 2.6

Record and Practice Journal

1. **a.**

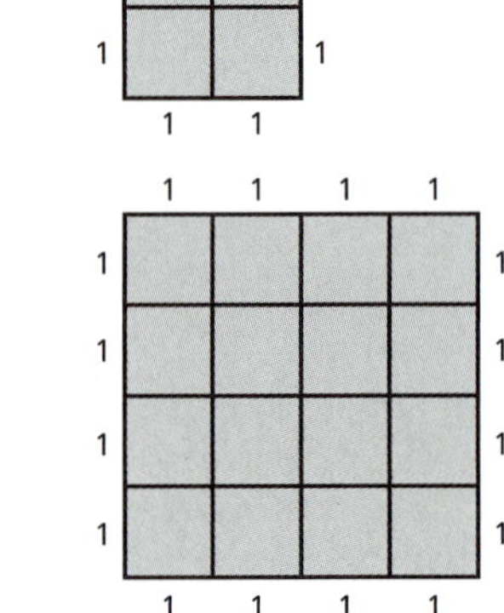

b.

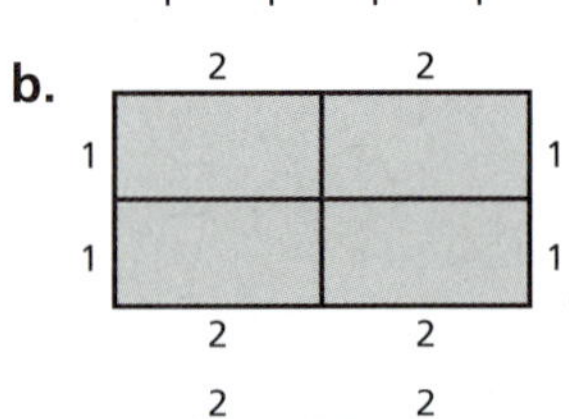

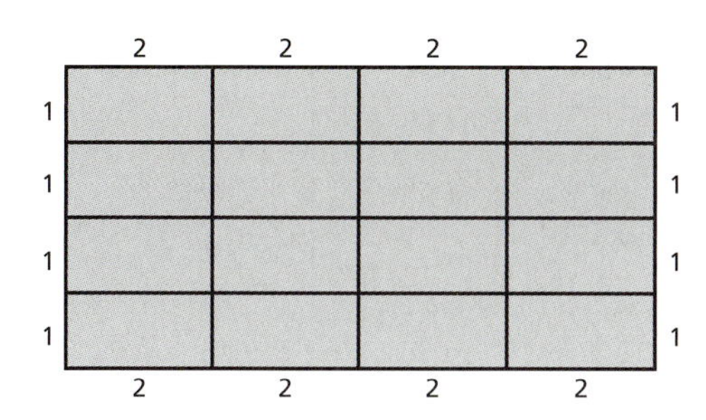

4. b. perimeter of shaded rectangle: 18 units; area of shaded rectangle: 18 units^2;

perimeter of unshaded rectangle: 36 units; area of unshaded rectangle: 72 units;

yes; The dimensions are doubled. So, the perimeter of the unshaded rectangle is twice the perimeter of the shaded rectangle and the area of the unshaded rectangle is 4 times the area of the shaded rectangle.

c.

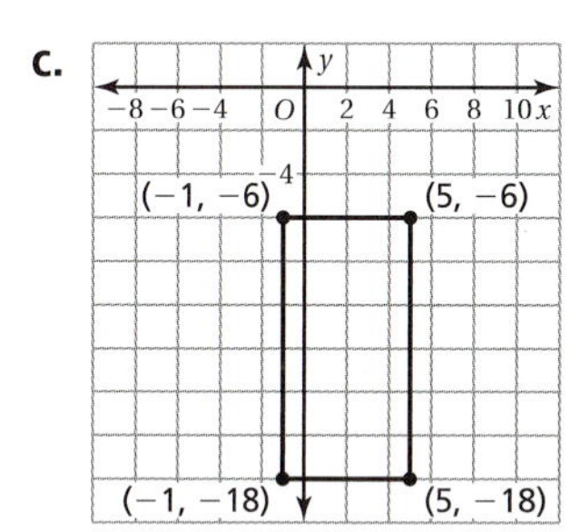

$$\frac{\text{Shaded Length}}{\text{Unshaded Length}} \stackrel{?}{=} \frac{\text{Shaded Width}}{\text{Unshaded Width}}$$

$$\frac{6}{12} \stackrel{?}{=} \frac{3}{6}$$

$$\frac{1}{2} = \frac{1}{2}$$

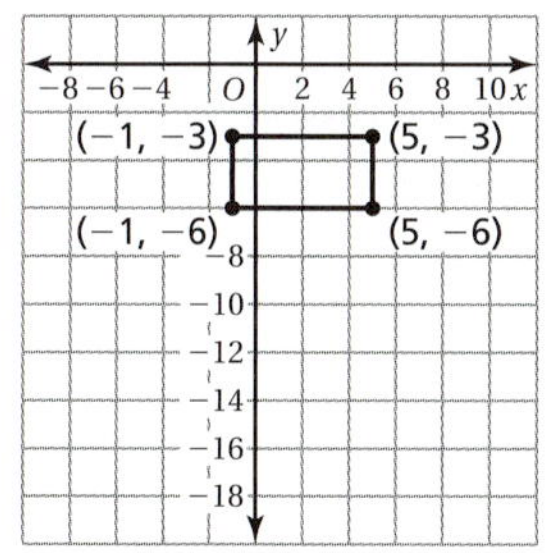

$$\frac{\text{Shaded Length}}{\text{Unshaded Length}} \stackrel{?}{=} \frac{\text{Shaded Width}}{\text{Unshaded Width}}$$

$$\frac{6}{6} \stackrel{?}{=} \frac{3}{3}$$

$$1 = 1$$

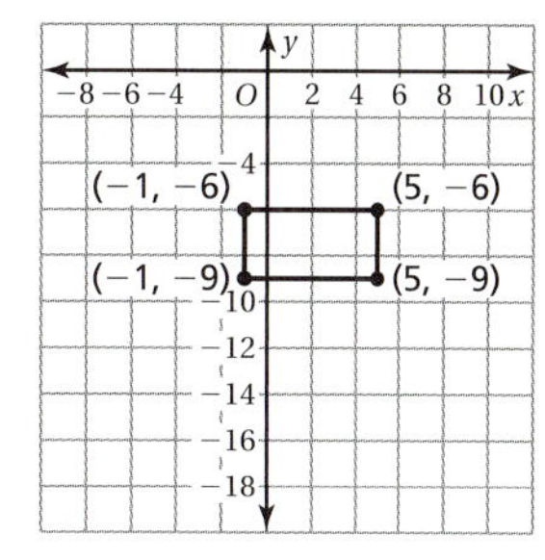

$$\frac{\text{Shaded Length}}{\text{Unshaded Length}} \stackrel{?}{=} \frac{\text{Shaded Width}}{\text{Unshaded Width}}$$

$$\frac{6}{6} \stackrel{?}{=} \frac{3}{3}$$

$$1 = 1$$

Practice and Problem Solving

18. a. $\frac{1}{4}; \frac{1}{4}; \frac{1}{16}$

b. The ratio of the circumferences is equal to the ratio of the radii. The ratio of the square of the radii is equal to the ratio of the areas. These are the same proportions that are used for similar figures.

Section 2.7

Practice and Problem Solving

7. yes

8. yes

9. no

10. yes

11. yes

12. no

13.

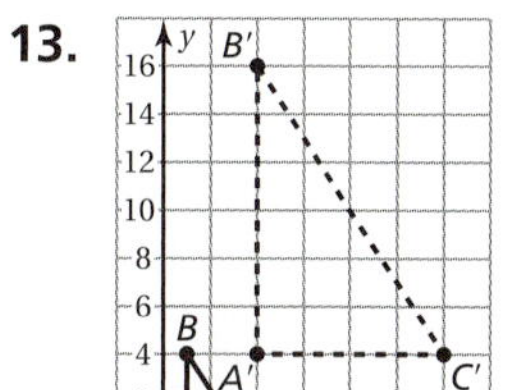

enlargement

14.

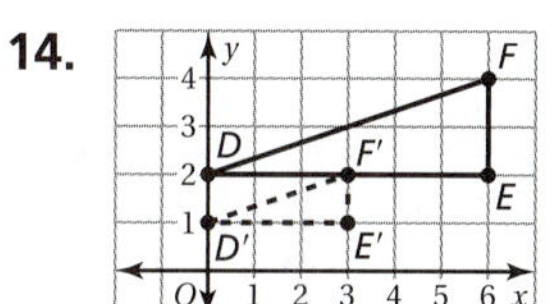

reduction

15.

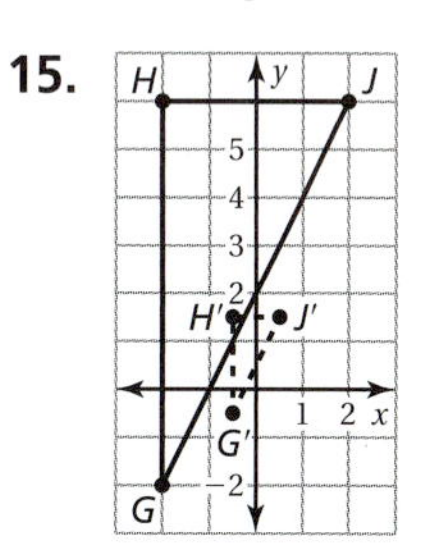

reduction

16. enlargement

17. 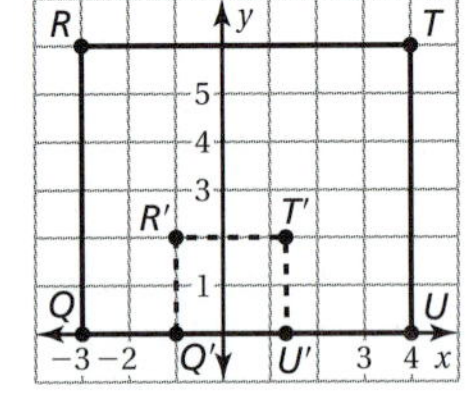

reduction

18. 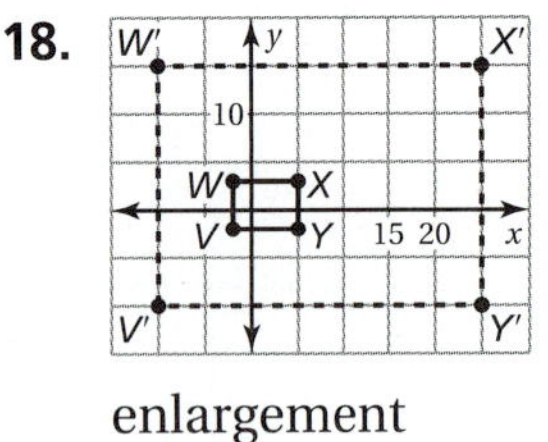

enlargement

31. **a.** enlargement **b.** center of dilation

c. $\frac{4}{3}$

d. The shadow on the wall becomes larger. The scale factor will become larger.

32. $\frac{3}{2}$; Multiply the two scale factors.

33. The transformations are a dilation using a scale factor of 2 and then a translation of 4 units right and 3 units down; similar; A dilation produces a similar figure and a translation produces a congruent figure, so the final image is similar.

34. The transformations are a reflection in the y-axis and then a translation of 1 unit left and two units down; congruent; A reflection produces a congruent figure and a translation produces a congruent figure, so the final image is congruent.

35. The transformations are a dilation using a scale factor of $\frac{1}{3}$ and then a reflection in the x-axis; similar; A dilation produces a similar figure and a reflection produces a congruent figure, so the final image is similar.

36. $(2x + 3, 2y - 1)$ is a dilation using a scale factor of 2 followed by a translation 3 units right and 1 unit down. $(2(x + 3), 2(y + 1))$ is a translation 3 units right and 1 unit down followed by a dilation using a scale factor of 2.

37. $A'(-2, 3)$, $B'(6, 3)$, $C'(12, -7)$, $D'(-2, -7)$; Methods will vary.

Chapter 3

Section 3.1

Record and Practice Journal

4. *Sample answer:* When two parallel lines are intersected by a transversal, eight angles are formed. In the figure $\angle 1$, $\angle 3$, $\angle 5$, and $\angle 7$ are congruent and $\angle 2$, $\angle 4$, $\angle 6$, and $\angle 8$ are congruent.

5. 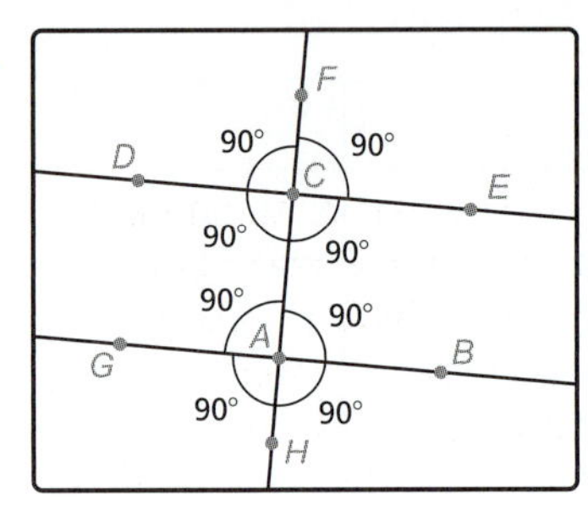

Practice and Problem Solving

17. $\angle 2 = 90°$; $\angle 2$ and the given angle are vertical angles.
$\angle 1 = 90°$ and $\angle 3 = 90°$; $\angle 1$ and $\angle 3$ are supplementary to the given angle.
$\angle 4 = 90°$; $\angle 4$ and the given angle are corresponding angles.
$\angle 6 = 90°$; $\angle 4$ and $\angle 6$ are vertical angles.
$\angle 5 = 90°$ and $\angle 7 = 90°$; $\angle 5$ and $\angle 7$ are supplementary to $\angle 4$.

18. 56°; *Sample answer:* $\angle 1$ and $\angle 8$ are corresponding angles and $\angle 8$ and $\angle 4$ are supplementary.

19. 132°; *Sample answer:* $\angle 2$ and $\angle 4$ are alternate interior angles and $\angle 4$ and $\angle 3$ are supplementary.

20. 55°; *Sample answer:* $\angle 4$ and $\angle 2$ are alternate interior angles.

21. 120°; *Sample answer:* $\angle 6$ and $\angle 8$ are alternate exterior angles.

22. 129.5°; *Sample answer:* $\angle 7$ and $\angle 5$ are alternate exterior angles and $\angle 5$ and $\angle 6$ are supplementary.

23. 61.3°; *Sample answer:* $\angle 3$ and $\angle 1$ are alternate interior angles and $\angle 1$ and $\angle 2$ are supplementary.

24. 40°

25. They are all right angles because perpendicular lines form 90° angles.

26. *Sample answer:* 1) $\angle 1$ and $\angle 7$ are congruent because they are alternate exterior angles.
2) $\angle 1$ and $\angle 5$ are corresponding angles and $\angle 5$ and $\angle 7$ are vertical angles. So, $\angle 1$ and $\angle 7$ are congruent.

Section 3.2

Record and Practice Journal

2. b. *Sample answer:* The sum of the measures of angle D, angle B, and angle E is 180° by definition of a straight line. Angle A is congruent to angle D and angle C is congruent to angle E because they are alternate interior angles of a transversal through parallel lines. Then by substitution, the sum of the measures of angle A, angle B, and angle C is 180°.

Practice and Problem Solving

19. sometimes; The sum of the angle measures must equal 180°.

20. always; Because the sum of the interior angle measures must equal 180° and one of the interior angles is 90°, the other two interior angles must sum to 90°.

21. never; If a triangle had more than one vertex with an acute exterior angle, then it would have to have more than one obtuse interior angle which is impossible.

22. You know that $x + y + w = 180$ and $w + z = 180$. Substitute $w + z$ for 180 in the first equation and you get $x + y + w = w + z$. Now subtract w from each side to get $x + y = z$.

Section 3.3

Practice and Problem Solving

25. 60°; The sum of the interior angle measures of a hexagon is 720°. Because it is regular, each angle has the same measure. So, each interior angle is $720° \div 6 = 120°$ and each exterior angle is 60°.

33. a. *Sample answer:*

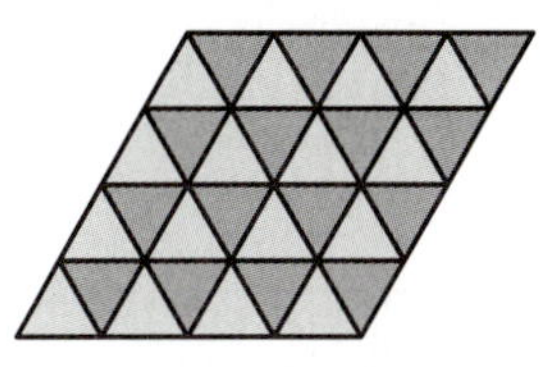

b. *Sample answer:* square, regular hexagon

c. *Sample answer:*

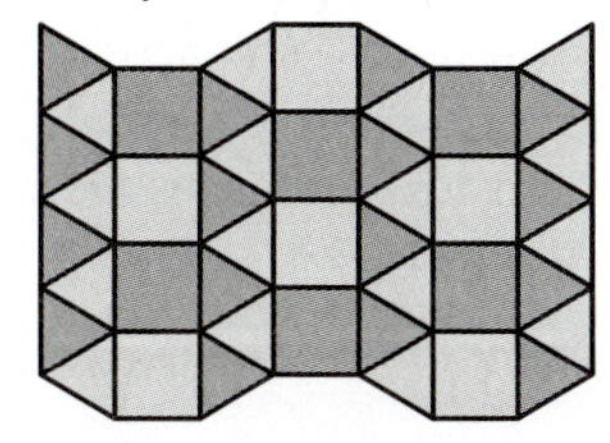

d. *Answer should include, but is not limited to:* a discussion of the interior and exterior angles of the polygons in the tessellation and how they add to 360° where the vertices meet.

Section 3.4

Record and Practice Journal

2. a. *Sample answer:*

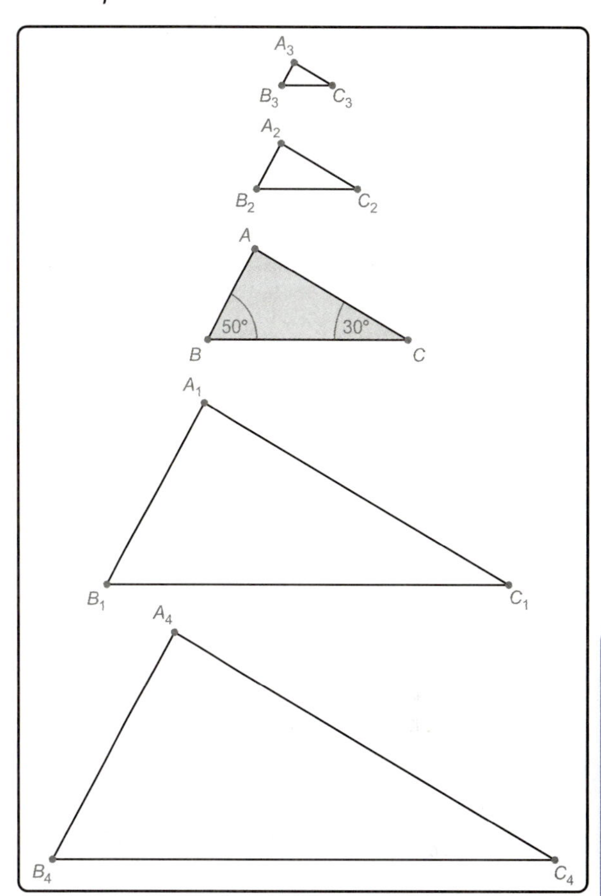

6. *Sample answer:* In the figure, you know that the streetlight forms a right angle with the ground and the person forms a right angle with the ground. Each triangle shares the same angle formed by the ground and the top of the streetlight. Because two angles in one triangle are congruent to two angles in another triangle, the third angles are also congruent. So, the triangles are similar.

Sample answer: Students can also use properties of parallel lines cut by a transversal to show that the two triangles are similar.

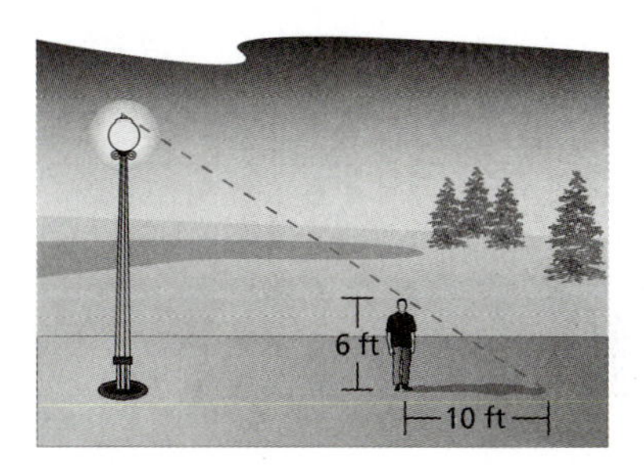

Additional Answers

Practice and Problem Solving

14. no; Each side increases by 50%, so each side is multiplied by a factor of $\frac{3}{2}$. The area is $\frac{3}{2}\left(\frac{3}{2}\right) = \frac{9}{4}$ or 225% of the original area, which is a 125% increase.

15. 30 ft

16. *Sample answer:* 10 ft

Assume that you are 5 feet tall.

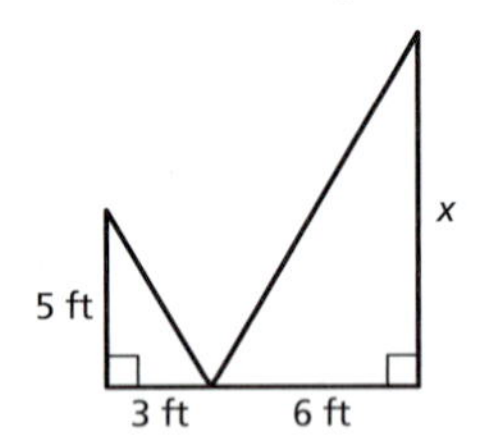

17. maybe; They are similar when both have measures of 30°, 60°, 90° or both have measures of 45°, 45°, 90°. They are not similar when one has measures of 30°, 60°, 90° and the other has measures of 45°, 45°, 90°.

18. $\triangle ABG \sim \triangle ACF$, $\triangle ABG \sim \triangle ADE$, $\triangle ACF \sim \triangle ADE$; 2 ft; 4 ft

Chapter 4

Record and Practice Journal Fair Game Review

16–20.

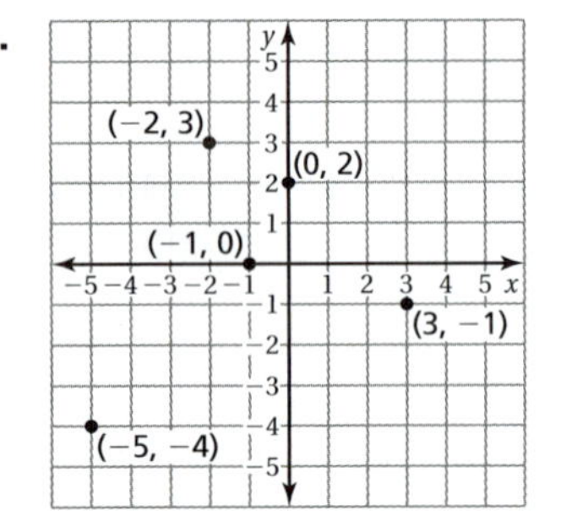

Section 4.1

On Your Own

1.

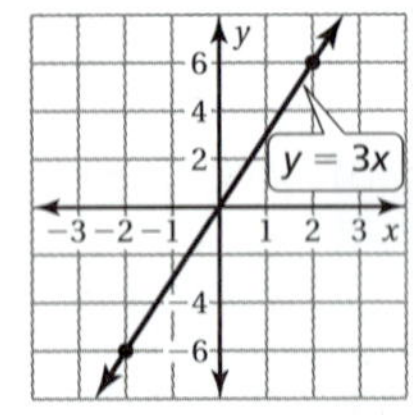

2.

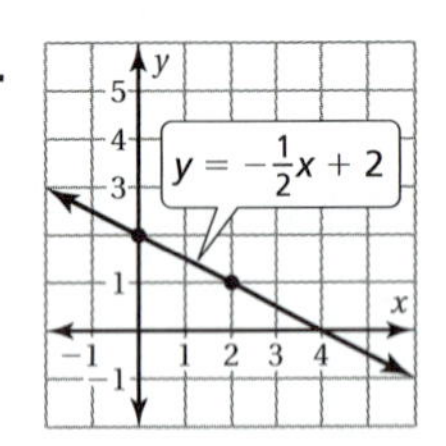

3.

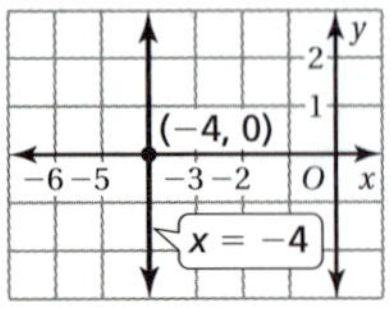

4.

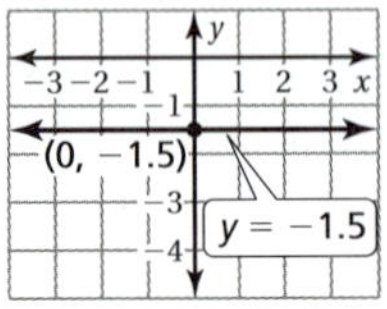

Practice and Problem Solving

7.

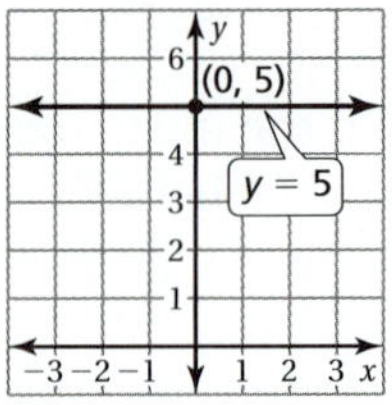

8.

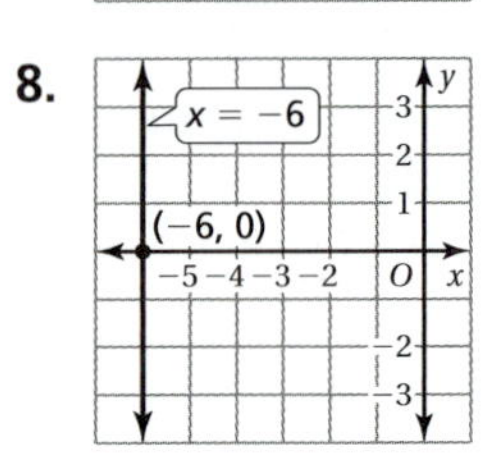

9.

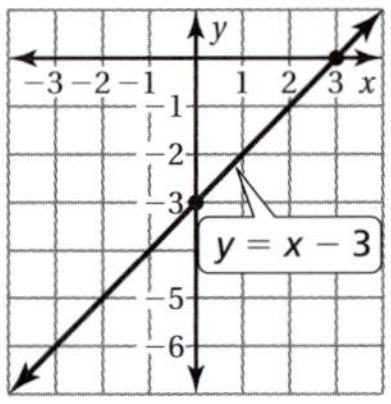

10.

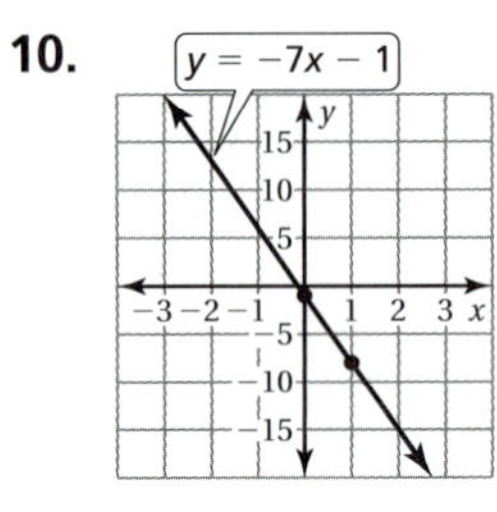

11.

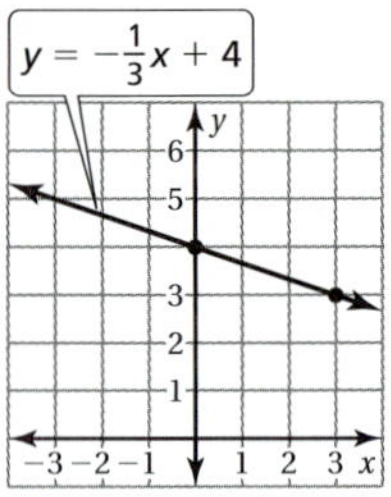

12.

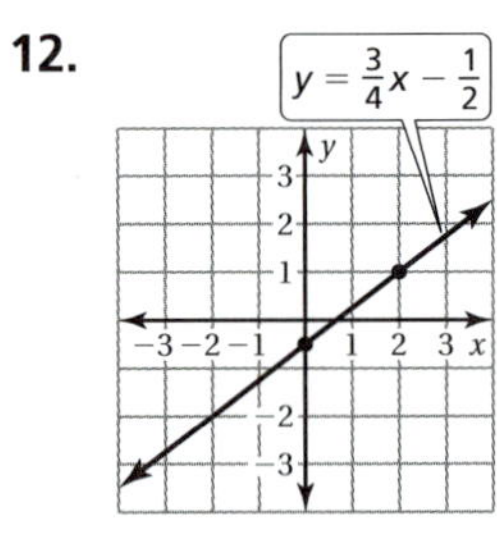

13.

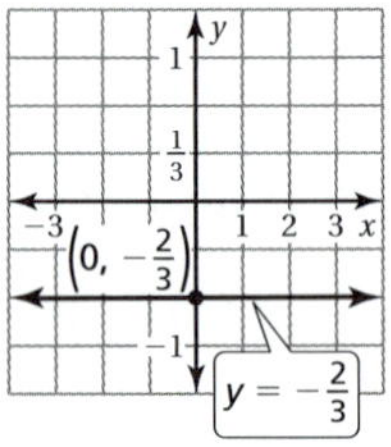

14.

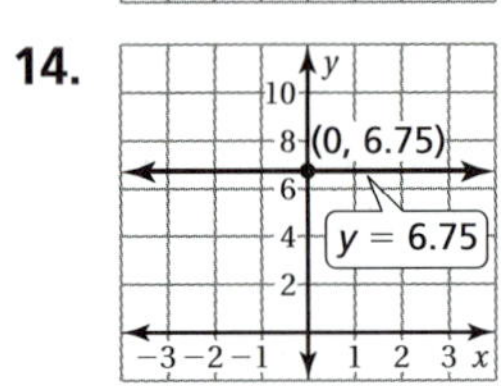

15.

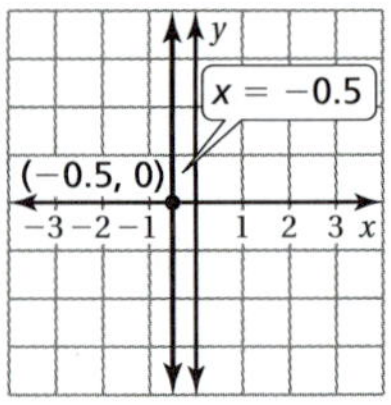

16.

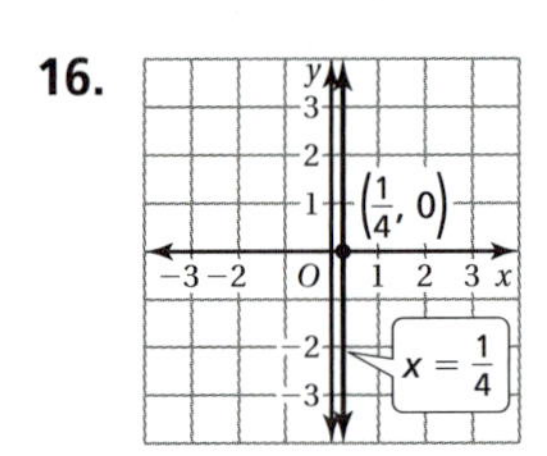

17. The equation $x = 4$ is graphed, not $y = 4$.

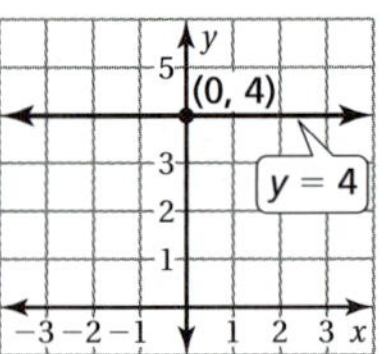

18.

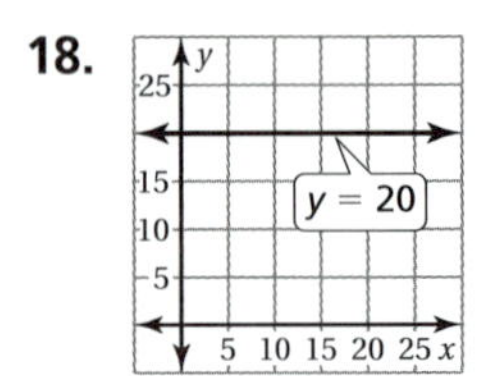

Sample answer: No matter how many text messages are sent, the cost is $20.

19. **a.** $y = 2x + 3$

b. about \$5

c. \$5.25

21. $y = -\frac{5}{2}x + 2$

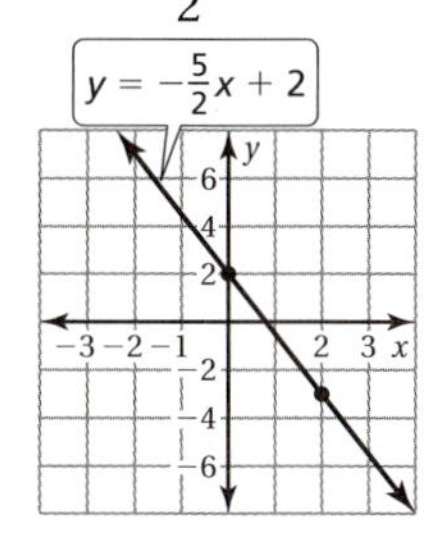

22. $y = 12x - 9$

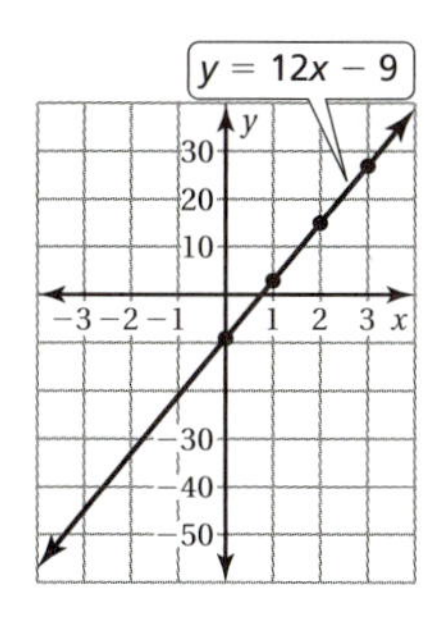

23. $y = -2x + 3$

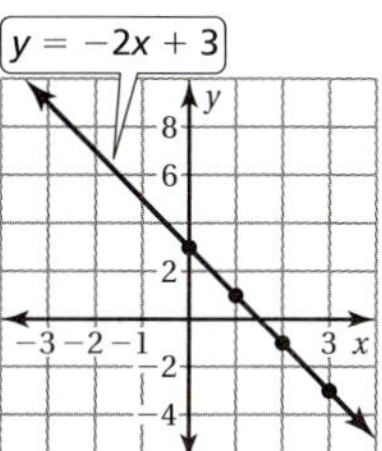

24. **a.**

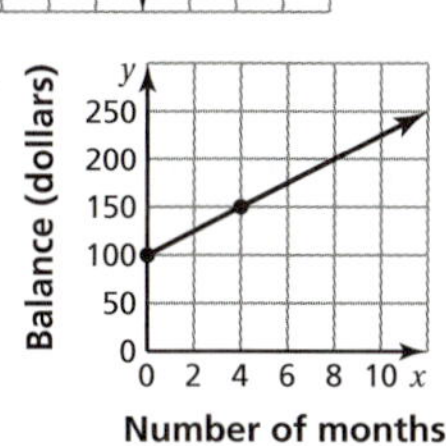

b. 6 mo

25. **a.** *Sample answer:*

yes; The graph of the equation is a line.

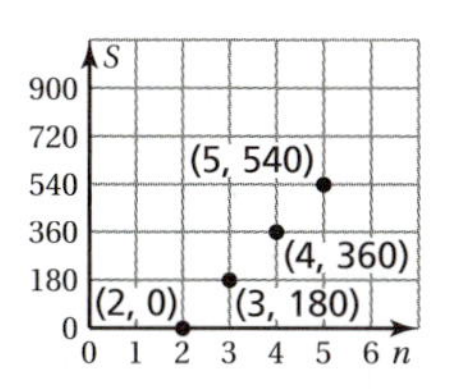

b. no; $n = 3.5$ does not make sense because a polygon cannot have half a side.

26. *Sample answer:* If you are 13 years old, the sea level has risen 26 millimeters since you were born.

$y = 2x$

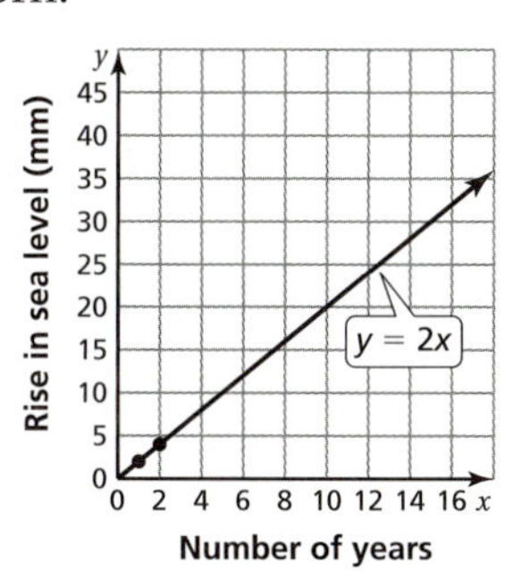

Section 4.2

Practice and Problem Solving

26. *Sample answer:*

a. Yes, it follows the guidelines.

b.

2.5 ft

30 ft

Extension 4.2

Extra Example 2

blue and green; The blue line has a slope of $-\frac{3}{2}$. The green line has a slope of $\frac{2}{3}$. The product of their slopes is $-\frac{3}{2} \cdot \frac{2}{3} = -1$.

Practice

3. yes; Both lines are horizontal and have a slope of 0.

4. no; $y = 0$ has a slope of 0 and $x = 0$ has an undefined slope.

5. yes; Both lines are vertical and have an undefined slope.

6. Use the vertices of the quadrilateral to find the slope of each side. When opposite sides are parallel (have the same slope), the quadrilateral is a parallelogram.

slope of $AB = -\frac{1}{7}$; slope of $BC = -\frac{5}{2}$;

slope of $CD = -\frac{1}{6}$; slope of $DA = -\frac{5}{3}$;

No, it is not a parallelogram. Because they have different slopes, opposite sides are not parallel.

7. blue and green; The blue line has a slope of 6. The green line has a slope of $-\frac{1}{6}$. The product of their slopes is $6 \cdot \left(-\frac{1}{6}\right) = -1$.

8. blue and green, red and green; The blue and red lines both have a slope of $-\frac{1}{2}$. The green line has a slope of 2. The product of their slopes is $2 \cdot \left(-\frac{1}{2}\right) = -1$.

9. yes; The line $x = -2$ is vertical. The line $y = 8$ is horizontal. A vertical line is perpendicular to a horizontal line.

10. no; Both lines are vertical and have undefined slopes.

11. yes; The line $x = 0$ is vertical. The line $y = 0$ is horizontal. A vertical line is perpendicular to a horizontal line.

12. Use the vertices of the quadrilateral to find the slope of each side. When adjacent sides are perpendicular, then the parallelogram is a rectangle. Note that the quadrilateral is a parallelogram, so you already know that opposite sides are parallel.

slope of $JK = \frac{2}{3}$; slope of $KL = -\frac{3}{2}$

slope of $LM = \frac{2}{3}$; slope of $MJ = -\frac{3}{2}$

Yes, it is a rectangle.

JK is perpendicular to KL because $\frac{2}{3} \cdot \left(-\frac{3}{2}\right) = -1$.

KL is perpendicular to LM because $-\frac{3}{2} \cdot \frac{2}{3} = -1$.

LM is perpendicular to MJ because

$\frac{2}{3} \cdot \left(-\frac{3}{2}\right) = -1$.

MJ is perpendicular to JK because $-\frac{3}{2} \cdot \frac{2}{3} = -1$.

Record and Practice Journal Practice

1. line B and line G; they both have a slope of $\frac{5}{3}$.
2. line B and line R; They both have a slope of 9.
3. yes; Both lines are vertical and have undefined slopes.
4. no; The line $x = 3$ has an undefined slope and the line $y = -3$ has a slope of 0.
5. yes; Because opposite sides have the same slope, they are parallel. Because opposite sides are parallel, the quadrilateral is a parallelogram.
6. line B and line R; Line B has a slope of 1. Line R has a slope of -1. the product of their slopes is $1 \cdot (-1) = -1$.
7. line R and line G; Line R has a slope of 4. Line G has a slope of $-\frac{1}{4}$. The product of their slopes is $4 \cdot \left(-\frac{1}{4}\right) = -1$.
8. yes; The line $x = 0$ is vertical. The line $y = 3$ is horizontal. A vertical line is perpendicular to a horizontal line.
9. no; Both lines are horizontal and have a slope of 0.
10. yes; Because the products of the slopes of intersecting sides are equal to -1, the parallelogram is a rectangle.

Section 4.3

Record and Practice Journal

2. The quantities in parts (a), (d), and (f) are in a proportional relationship.

For part (a): slope = 10; The value of y for $(1, y)$ is 10.

For part (d): slope = 6; the value of y for $(1, y)$ is 6.

For part (f): slope = 2; The value of y for $(1, y)$ is 2.

The value of y is equal to the slope of the line. The value of y represents the unit rate.

Extra Example 3

3. b.

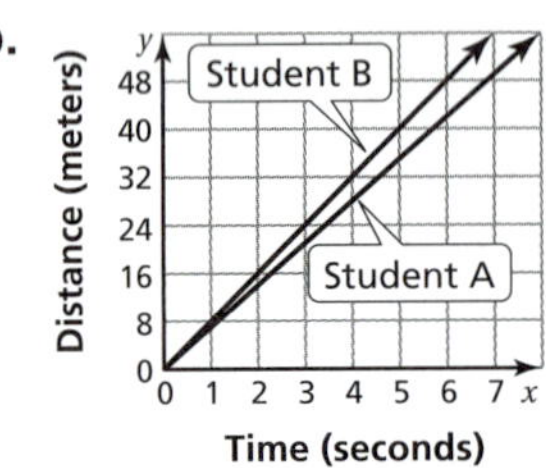

The graph that represents Student B is steeper than the graph that represents Student A. So, Student B is faster than Student A.

Practice and Problem Solving

10. a. fingernails; Fingernails grow about 0.7 millimeter per week and toenails grow about 0.25 millimeter per week.

b.

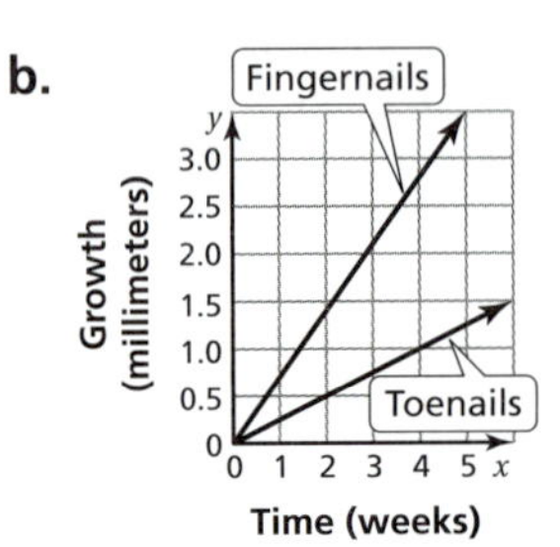

The graph that represents fingernails is steeper than the graph that represents toenails. So, fingernails grow faster than toenails.

13. **a.** yes; The equation is $d = 6t$, which represents a proportional relationship.

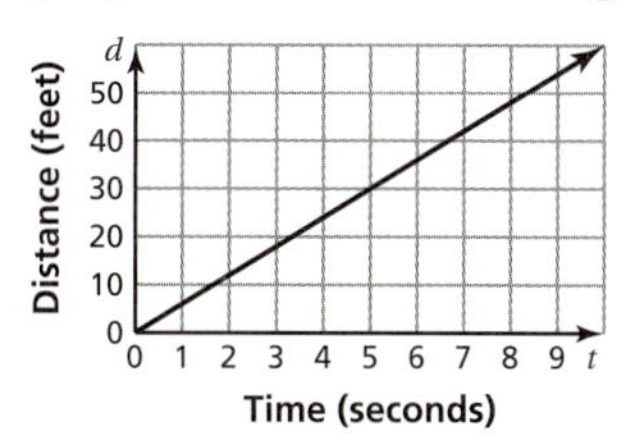

b. yes; The equation is $d = 50r$, which represents a proportional relationship.

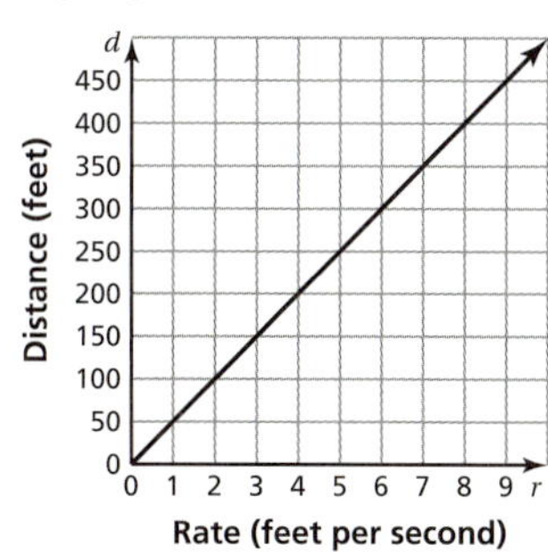

c. no; The equation is $t = \dfrac{300}{r}$, which does not represent a proportional relationship.

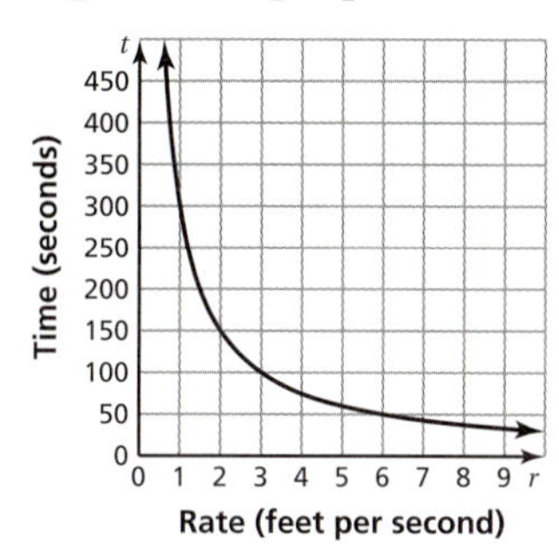

d. part c; It is called inverse variation because when the rate increases, the time decreases, and when the rate decreases, the time increases.

Fair Game Review

14.

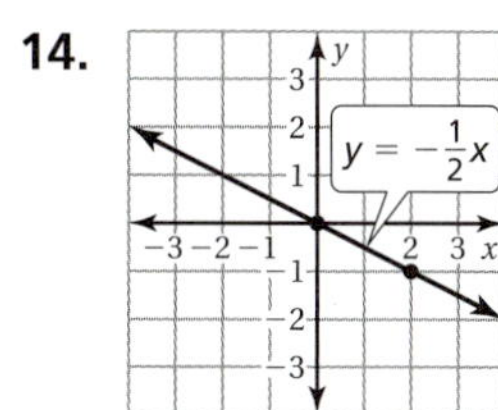

15.

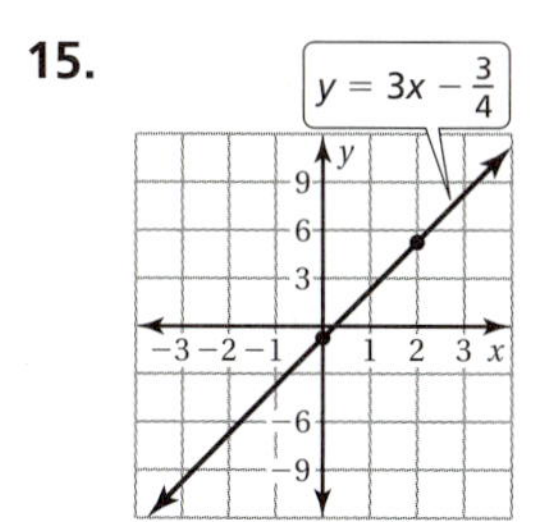

16.

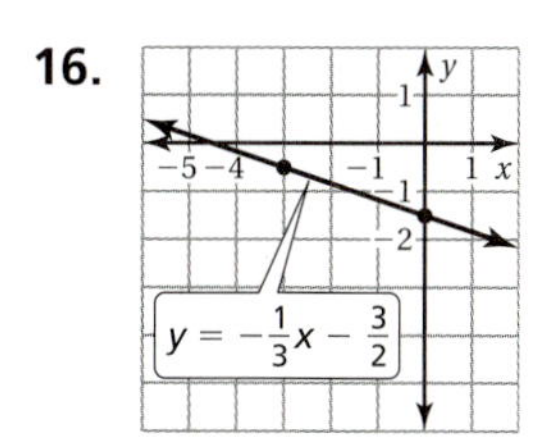

4.1–4.3 Quiz

1.

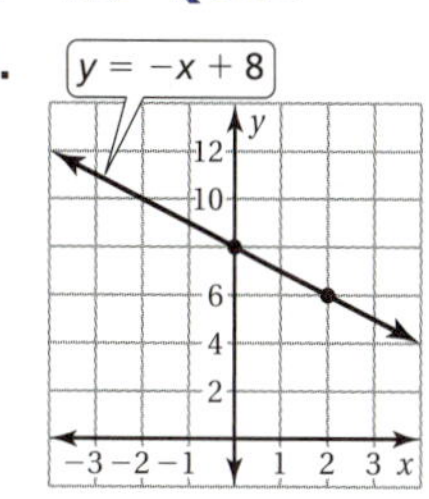

2.

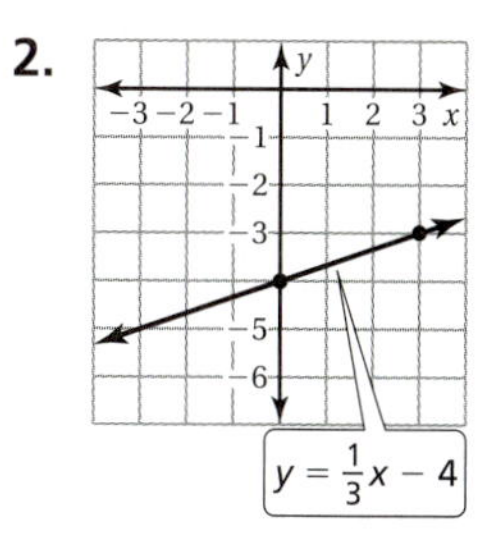

3.

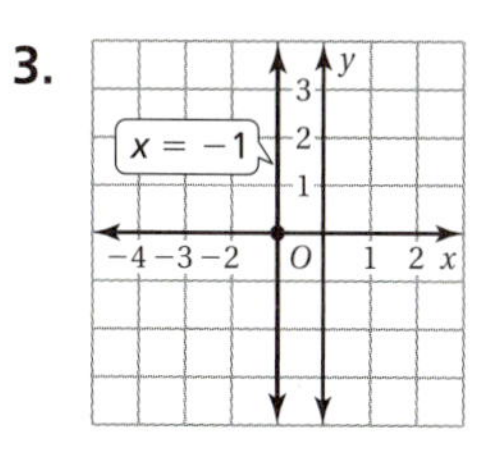

4.

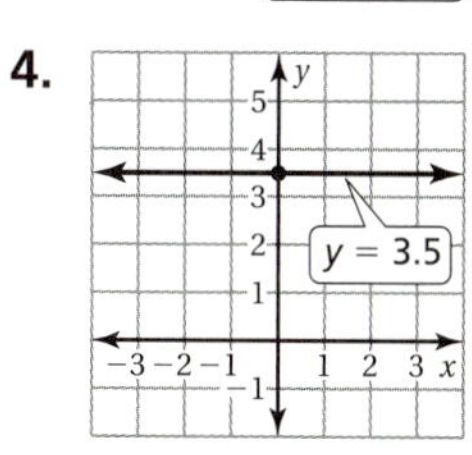

Section 4.4

Record and Practice Journal

1. **a.**

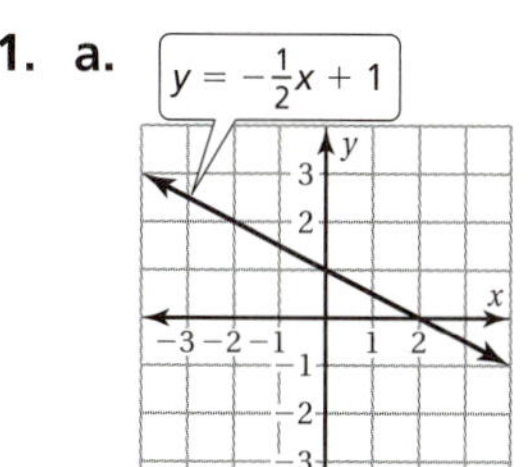

b.

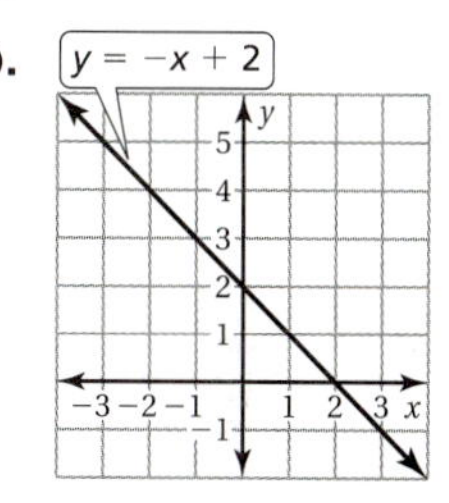

c.

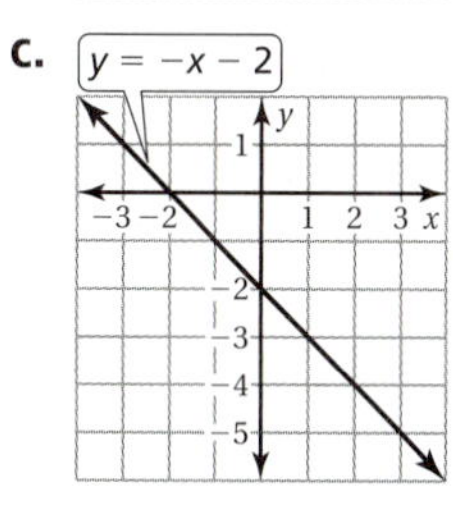

d.

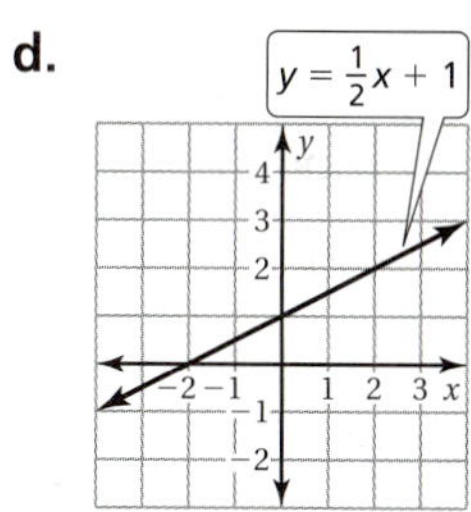

e.

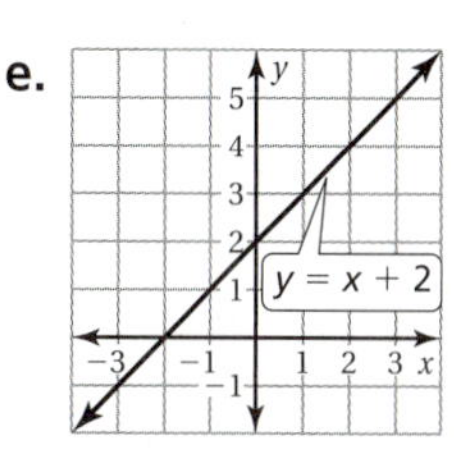

f.

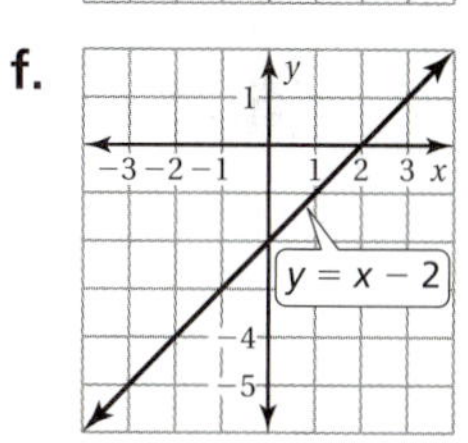

g.

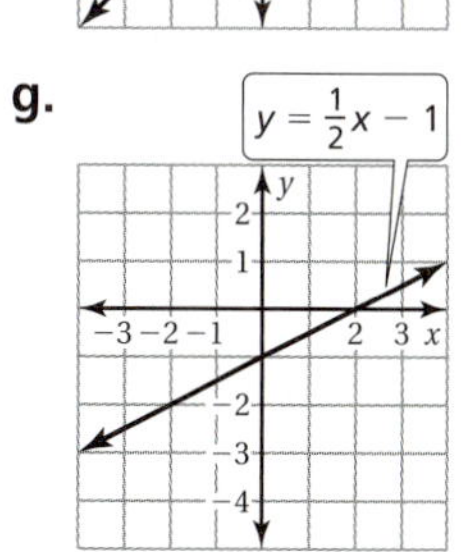

h.

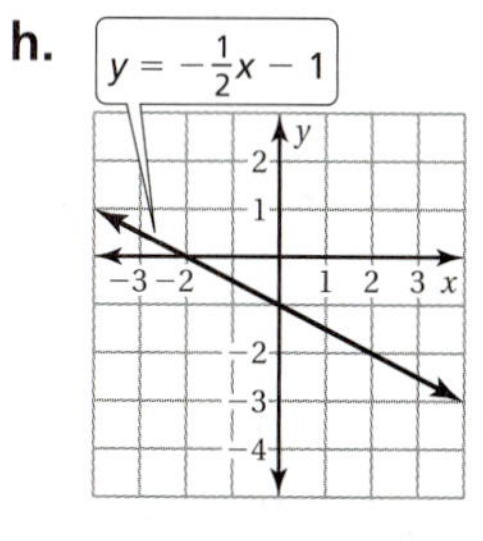

i.

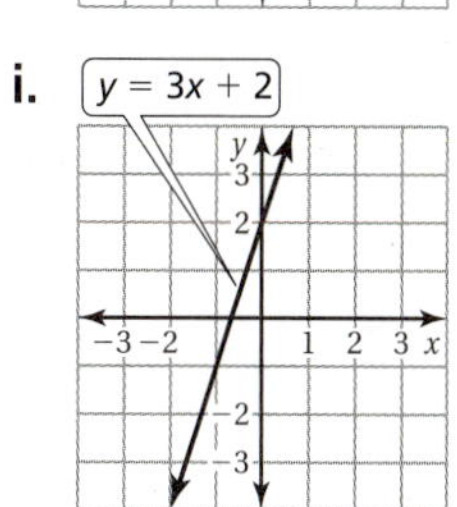

j.

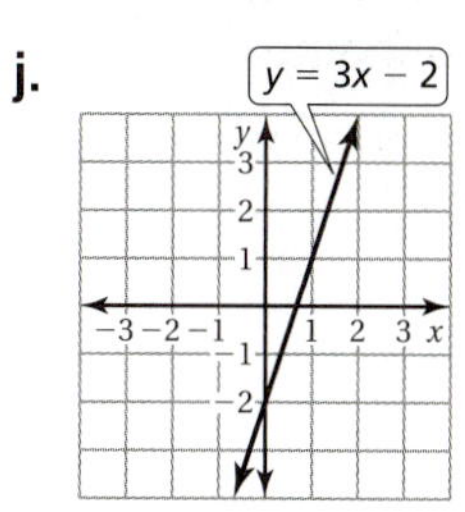

On Your Own

3.

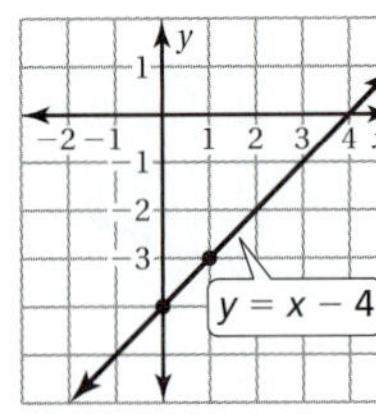

x-intercept: 4

4.

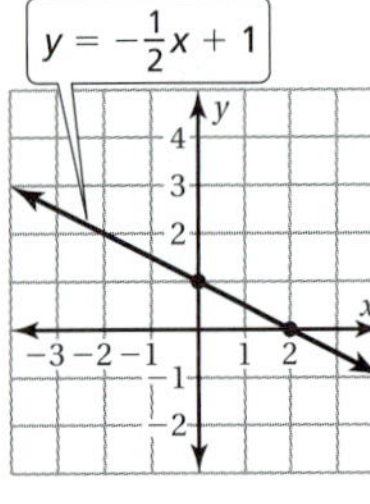

x-intercept: 2

5. The y-intercept means that the taxi has an initial fee of \$1.50. The slope means the taxi charges \$2 per mile.

Practice and Problem Solving

17. a.

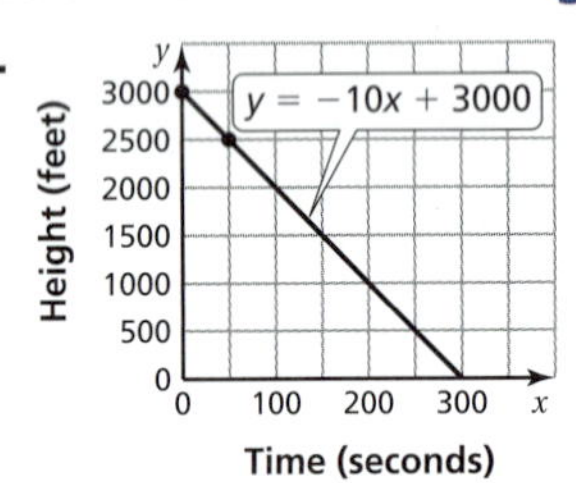

b. The x-intercept of 300 means the skydiver lands on the ground after 300 seconds. The slope of -10 means that the skydiver falls to the ground at a rate of 10 feet per second.

19.

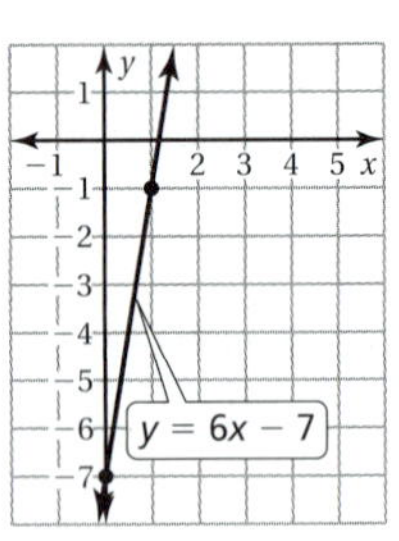

x-intercept: $\frac{7}{6}$

20.

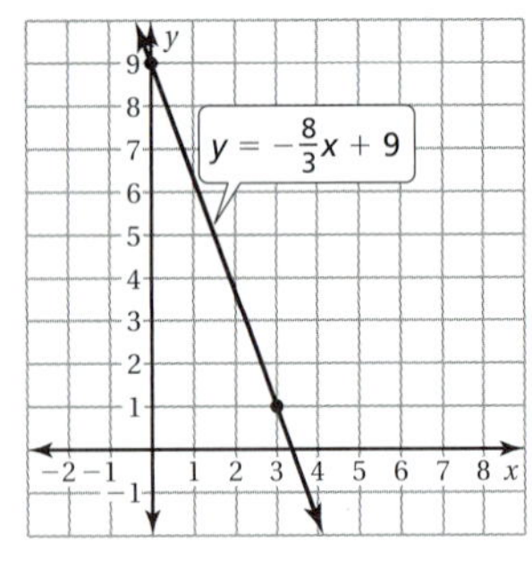

x-intercept: $\frac{27}{8}$

21.

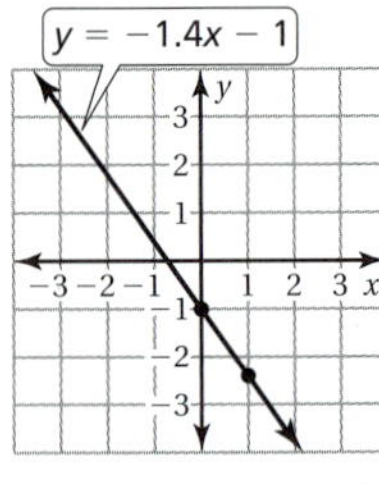

x-intercept: $-\frac{5}{7}$

22.

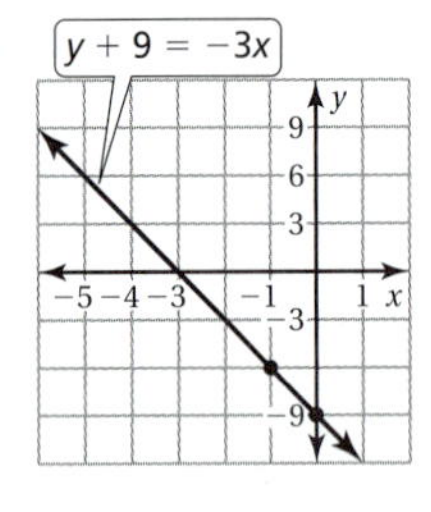

x-intercept: -3

23.

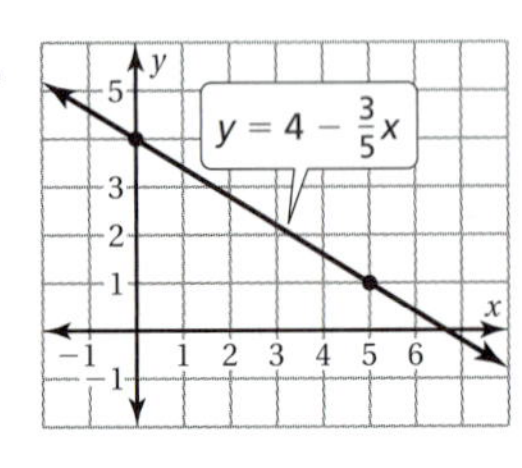

x-intercept: $\frac{20}{3}$

24. a. $y = 0.75x + 5$; The cost of going to the festival is the sum of the cost of picking x pounds of apples, $0.75x$, and the cost of admission, 5.

b.

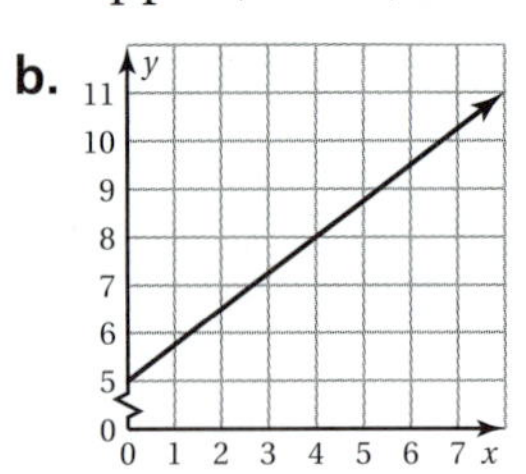

25. a. $y = 2x + 4$ and $y = 2x - 3$ are parallel because the slope of each line is 2; $y = -3x - 2$ and $y = -3x + 5$ are parallel because the slope of each line is -3.

b. $y = 2x + 4$ and $y = -\frac{1}{2}x + 2$ are perpendicular because the product of their slopes is -1; $y = 2x - 3$ and $y = -\frac{1}{2}x + 2$ are perpendicular because the product of their slopes is -1; $y = -\frac{1}{3}x - 1$ and $y = 3x + 3$ are perpendicular because the product of their slopes is -1.

Section 4.5

On Your Own

3.

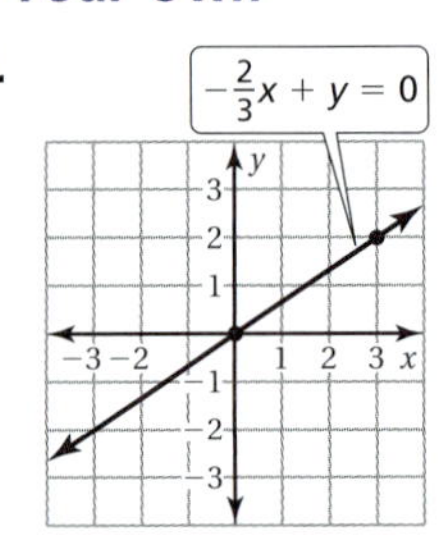

4.

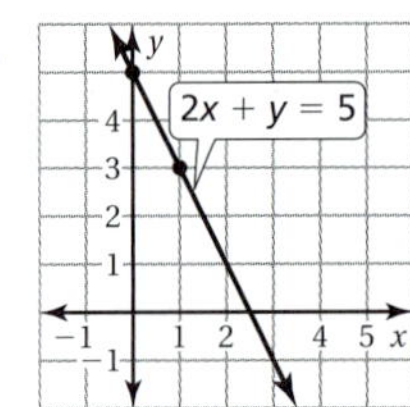

5.

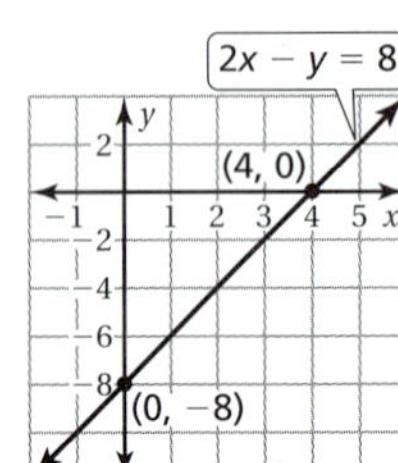

6.

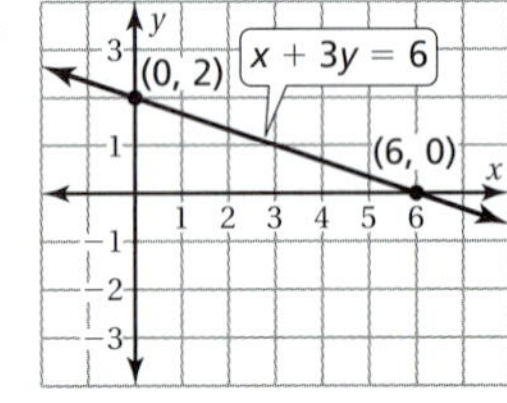

7. 1.5x + 1.2y = 6; (0, 5); (4, 0)

The x-intercept shows that you can buy 4 pounds of apples if you do not buy any oranges. The y-intercept shows that you can buy 5 pounds of oranges if you do not buy any apples.

Practice and Problem Solving

8. 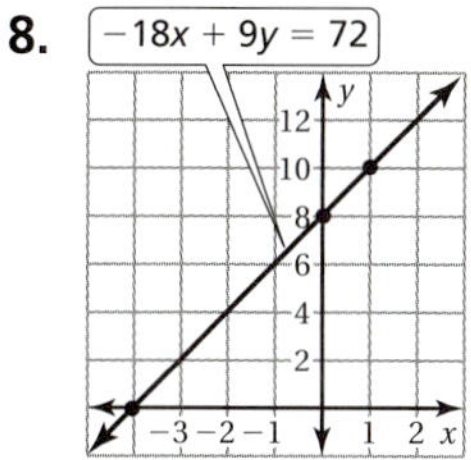

9. 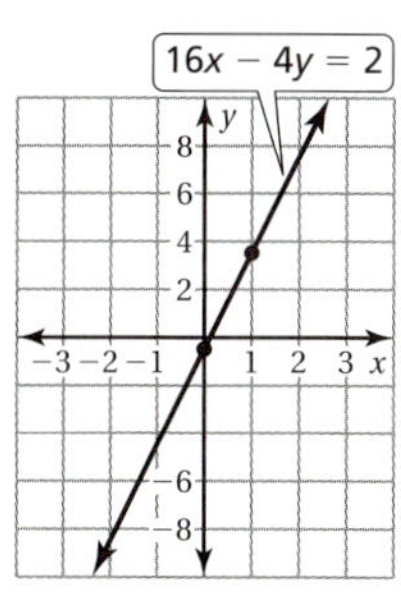

10. 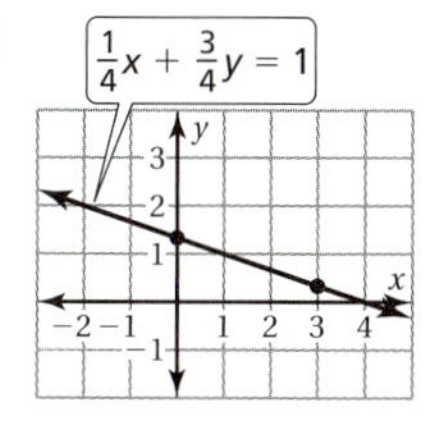

11. B

12. A

13. C

14. They should have let $y = 0$, not $x = 0$.

$$-2x + 3y = 12$$
$$-2x + 3(0) = 12$$
$$-2x = 12$$
$$x = -6$$

15. a. 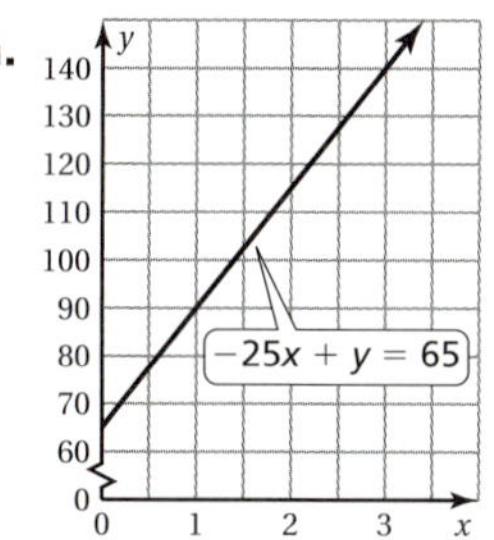

b. \$390

16. 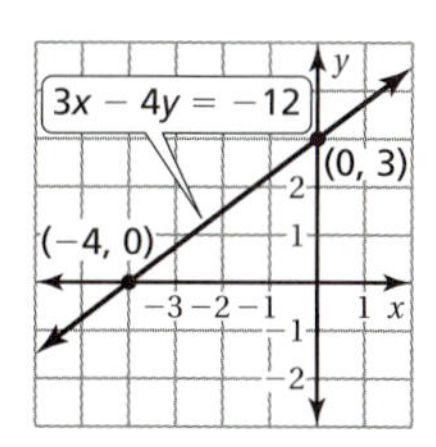

17. 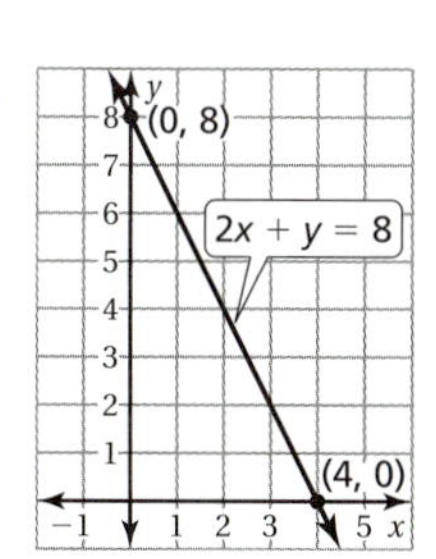

18. 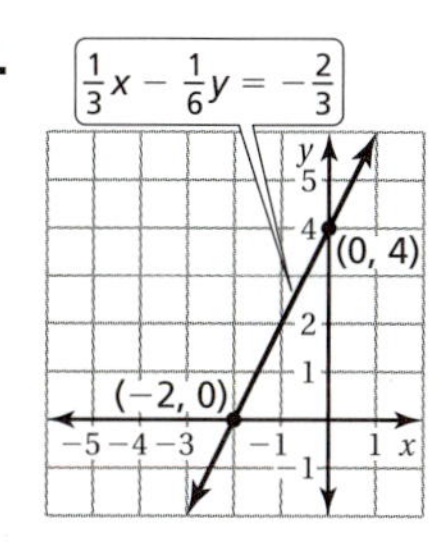

19. x-intercept: 9

y-intercept: 7

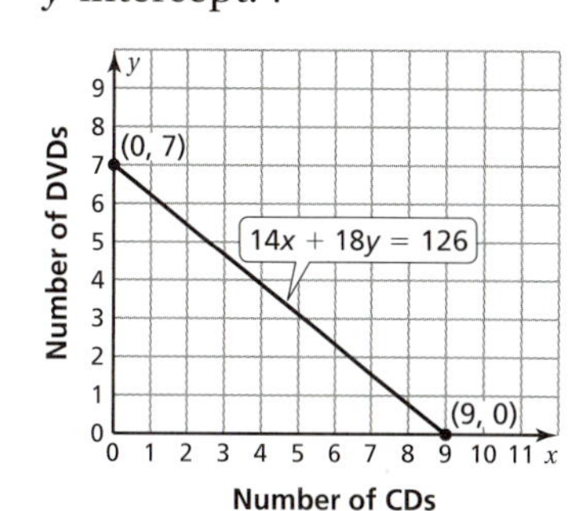

21. a. $9.45x + 7.65y = 160.65$

b. 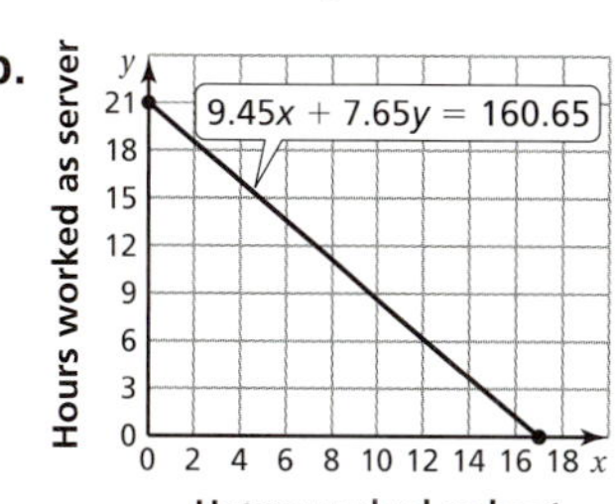

22. no; For example, $y = 5$ does not have an x-intercept, neither do any horizontal lines except $y = 0$.

23. a. $y = 40x + 70$

b. x-intercept: $-\frac{7}{4}$; no; You cannot have a negative time.

c. 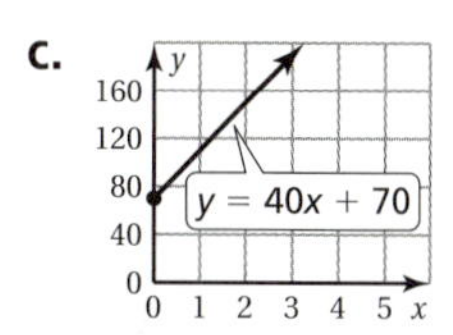

Section 4.6

Record and Practice Journal

1. a. top line: slope: $\frac{1}{2}$; y-intercept: 4; $y = \frac{1}{2}x + 4$

middle line: slope: $\frac{1}{2}$; y-intercept: 1; $y = \frac{1}{2}x + 1$

bottom line: slope: $\frac{1}{2}$; y-intercept: -2;

$$y = \frac{1}{2}x - 2$$

The lines are parallel.

b. right line: slope: -2; y-intercept: 3;
$y = -2x + 3$

middle line: slope: -2; y-intercept: -1;
$y = -2x - 1$

left line: slope: -2; y-intercept: -5;
$y = -2x - 5$

The lines are parallel.

c. line passing through (3, 2):

slope: $-\frac{1}{3}$; y-intercept: 3; $y = -\frac{1}{3}x + 3$

line passing through (3, 7):

slope: $\frac{4}{3}$; y-intercept: 3; $y = \frac{4}{3}x + 3$

line passing through (6, 4):

slope: $\frac{1}{6}$; y-intercept: 3; $y = \frac{1}{6}x + 3$

The lines have the same y-intercept.

d. line passing through (1, 2):

slope: 2; y-intercept: 0; $y = 2x$

line passing through (1, -1):

slope: -1; y-intercept: 0; $y = -x$

line passing through (3, 1):

slope: $\frac{1}{3}$; y-intercept: 0; $y = \frac{1}{3}x$

The lines have the same y-intercept.

2. a. $y = 4$; $y = -2$; $y = -2x + 8$; $y = -2x - 6$

b. $y = 5$; $y = -2$; $y = x + 5$; $y = x + 1$

Practice and Problem Solving

13. $y = 5$

14. $y = 0$

15. $y = -2$

16. $y = 0.7x + 10$

17. a–b.

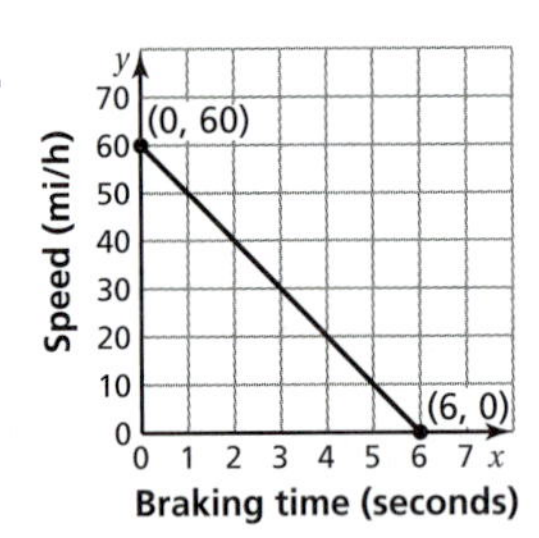

(0, 60) represents the speed of the automobile before braking. (6, 0) represents the amount of time it takes to stop. The line represents the speed y of the automobile after x seconds of braking.

c. $y = -10x + 60$

Section 4.7

Fair Game Review

24.

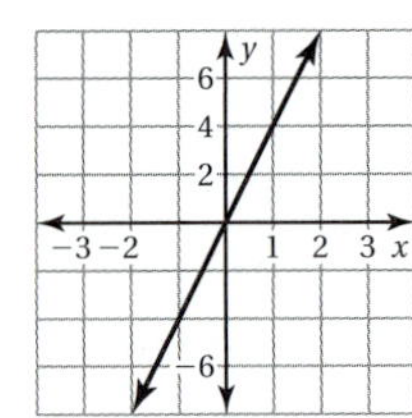

25.

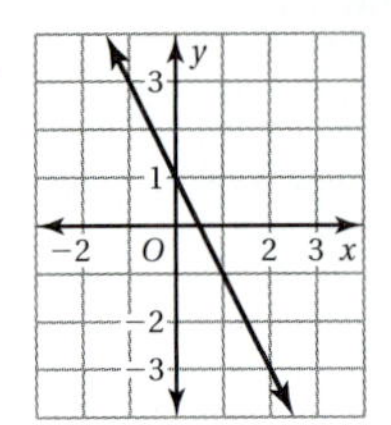

26.

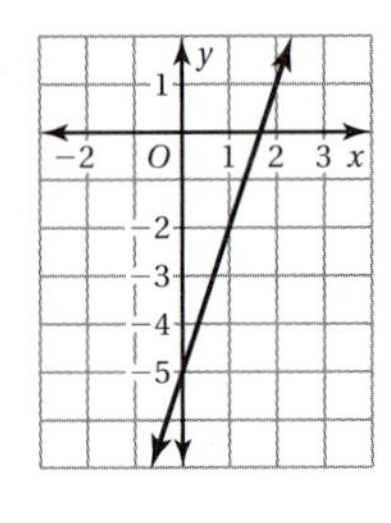

4.4–4.7 Quiz

15. a.

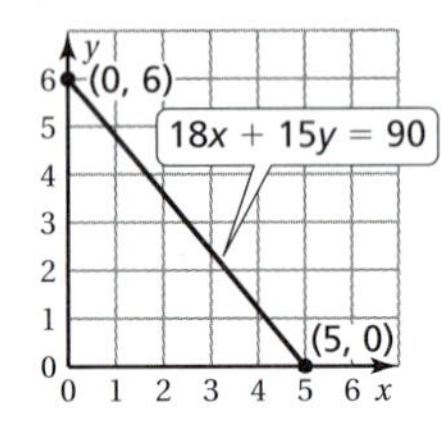

b. The x-intercept, 5, shows that you can buy 5 gallons of blue paint if you do not buy any white paint. The y-intercept, 6, shows that you can buy 6 gallons of white paint if you do not buy any blue paint.

Chapter 4 Test

7.

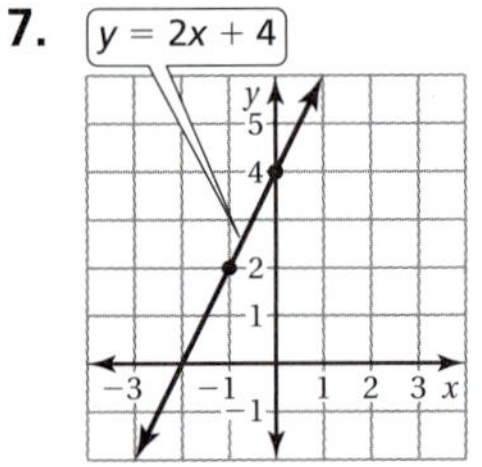

8.

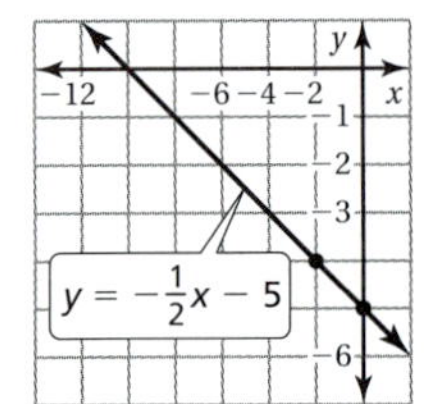

9.

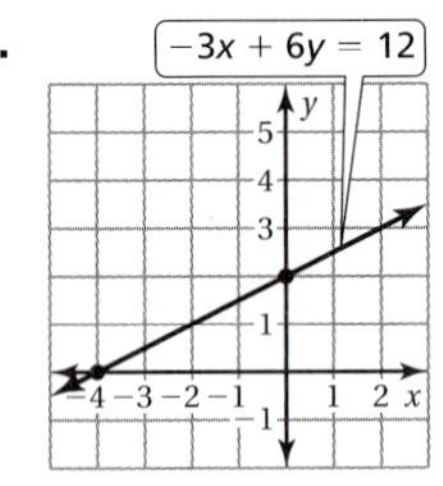

Chapter 5

Section 5.1

Record and Practice Journal

4. **a.** *Sample answer:* Xmin = 0, Xmax = 16, Ymin = 0, Ymax = 800

   b. Equation 2 has a greater slope and passes through the origin.

Section 5.4

Practice and Problem Solving

16. one solution; Because the lines have different slopes, they will intersect in one point.

17. When the slopes are different, there is one solution. When the slopes are the same, there is no solution if the y-intercepts are different and infinitely many solutions if the y-intercepts are the same.

Extension 5.4

Record and Practice Journal Practice

5. yes; You earn the same amount each day if $x = \frac{9}{10}$.

6. 4 min

7. **a.** $25x + 500 = 15x + 750$

   b. 25 years

Practice and Problem Solving

14. 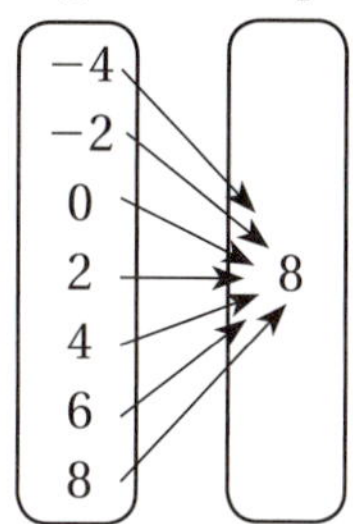

As each input increases by 2, the output is 8.

15. 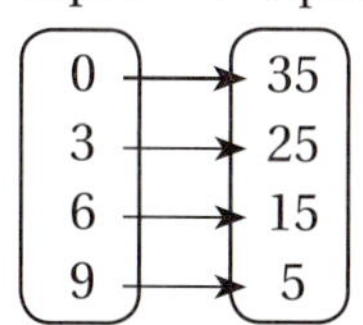

As each input increases by 3, the output decreases by 10.

17. **a.** 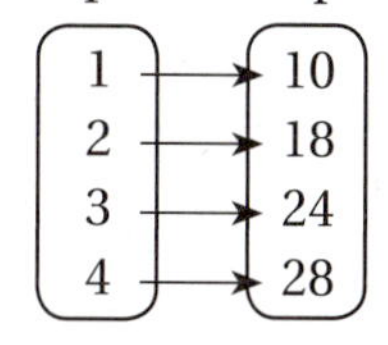

   b. yes; Each input has exactly one output.

   c. The pattern is that for each input increase of 1, the output increases by \$2 less than the previous increase. For each additional movie you buy, your cost per movie decreases by \$1.

Chapter 6

Section 6.1

Record and Practice Journal

2. **a.** Each input has one output. *Sample answer:* This relationship is possible if the input values represent the jersey numbers of basketball players and the output values represent the number of points each player scored in a basketball game.

   b. Input 10 has two outputs. Input 11 has one output. Input 12 has two outputs. Input 13 has one output. *Sample answer:* This relationship is possible if the input values represent the ages of players on a little league team and the output values represent the number of homeruns hit.

Section 6.2

Record and Practice Journal

2. **a.**

Input, x	1	2	3	4
Output, A	1	3	5	7

   b.

Input, x	1	2	3	4
Output, A	1	4	9	16

Practice and Problem Solving

22. 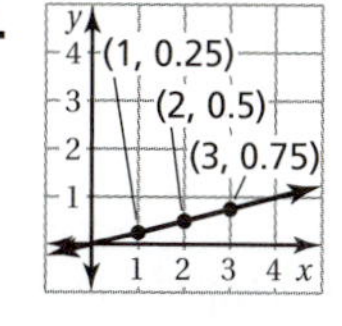

23. 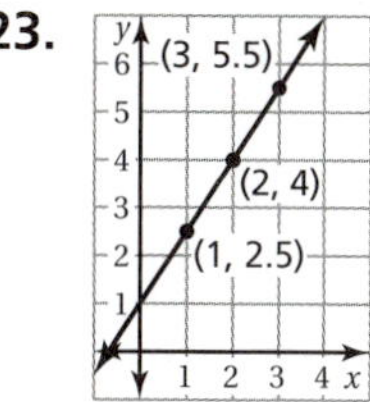

24. 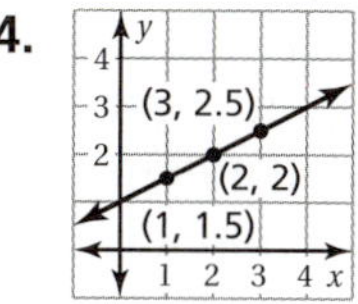

37. *Sample answer:*

Side Length	1	2	3	4	5
Perimeter	4	8	12	16	20

Side Length	1	2	3	4	5
Area	1	4	9	16	25

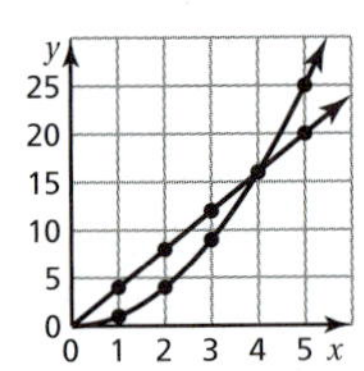

Sample answer: The perimeter function appears to form a line, and the area function appears to form a curve. When the side length is less than 4, the perimeter function is greater. When the side length is greater than 4, the area function is greater. When the side length is 4, the two functions are equal.

38. 44 square units; 45 square units; *Sample answer:* The "green area" of an even numbered square is equal to 4 more than the product of 2 and the square number. The "green area" of an odd numbered square is equal to 3 more than the product of 2 and the square number.

Section 6.3

Practice and Problem Solving

14. a. you; 19 feet per second

b. $y = 19x$;

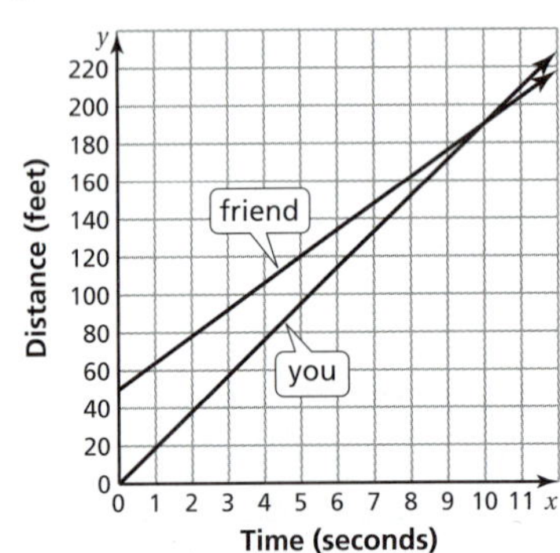

c. any distance greater than 190 feet; You catch up to your friend after you run 190 feet, so you will be ahead of him for any distance greater than 190 feet.

19. a.

Temperature (°F), *t*	94	95	96	97	98
Heat Index (°F), *H*	122	126	130	134	138

b. independent variable: t; dependent variable: H

c. $H = 4t - 254$

d. 146°F

Mini-Assessment

4. Maple tree: $y = 1.5x$; Pine tree: $y = x$

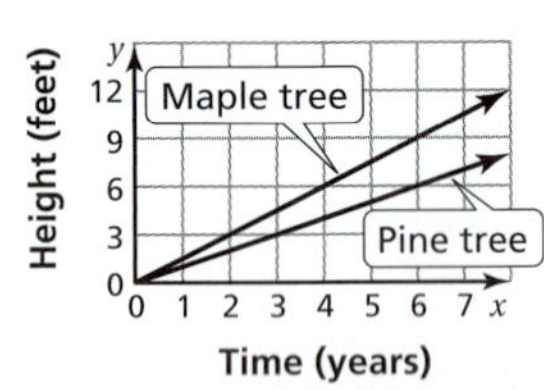

The graph that represents the maple tree is steeper than the graph that represents the pine tree. So, the maple tree grows faster than the pine tree.

6.1–6.3 Quiz

13. a. $R = 2A + 2$

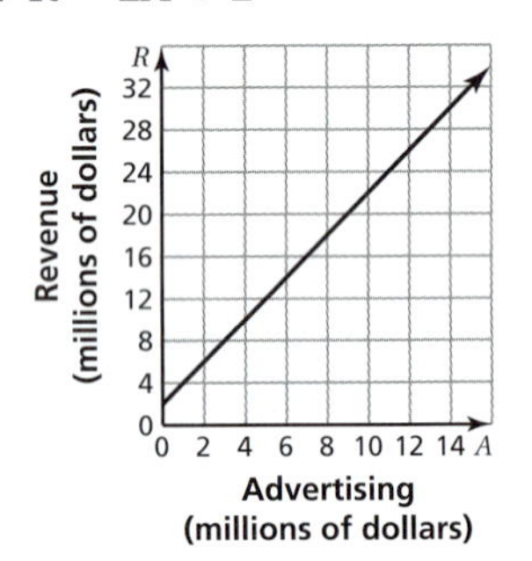

b. $32 million

Section 6.4

Practice and Problem Solving

17. a. nonlinear; When graphing the points, they do not lie on a line.

b. Tree B; After ten years, the height of Tree A is 20 feet and the height of Tree B is at least 23 feet.

18. a.

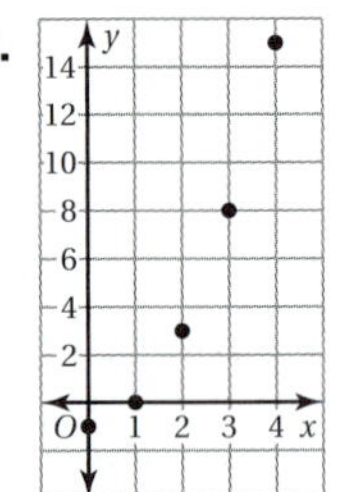

As x increases by 1, y increases by 2 more than the previous increase; nonlinear

b. $y = x^2 - 1$

Fair Game Review

19.

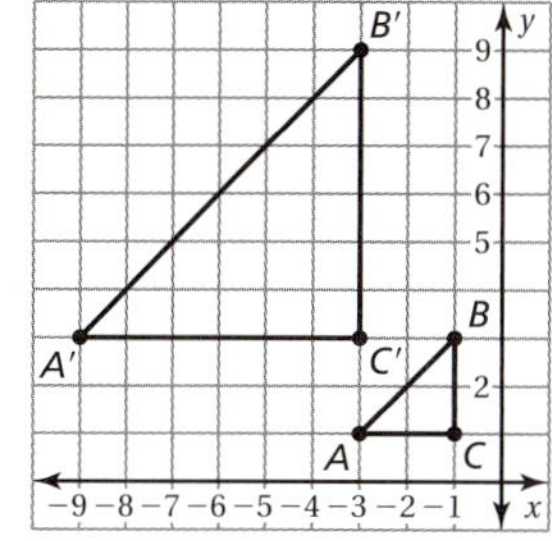

enlargement

20.

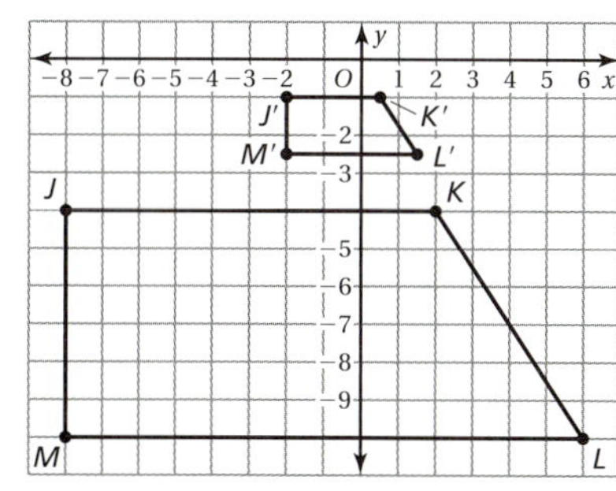

reduction

Section 6.5

Record and Practice Journal

1. a. The graph has no numerical values or tick marks on the axes. The graph itself has no points plotted or labeled. There are no equations shown. It is made up of several different linear graphs.

b. At first, the graph is constant (horizontal line segment), so the water level is not changing. It then increases at a constant rate, so the water level rises. The graph is then constant again, so the water level peaks and does not change. Finally the graph decreases at a constant rate, so the water level falls back down.

3. a. Both graphs are increasing over time. Graph B is linear and increasing at a constant rate of change. Graph A is nonlinear and increasing at a faster rate, compared to Graph B.

b. At first, Graph B is steeper and increasing at a faster rate than Graph A. But over time, Graph A gets much steeper than Graph B.

4. a. Both graphs are decreasing over time. Graph B is linear and decreasing at a constant rate of change. Graph A is nonlinear and decreasing at a faster rate, then it flattens out.

b. At first, Graph A is steeper and decreasing at a faster rate than Graph B. But over time, Graph A flattens out and is decreasing less and less, compared to Graph B.

5. Determine which variables you want the horizontal and vertical axes to represent. Identify the important elements of the situation and determine whether those elements in graphical form will be increasing, decreasing, or constant.

6. An airplane lands at an airport, has a short layover, or break, then takes off again.

On Your Own

1. a. Pelican: descends slowly at first, and then more and more quickly before slowing again and gradually approaching ground level
Osprey: descends at a constant rate and then descends at a faster constant rate

b. Both graphs are decreasing; Both are nonlinear, but the osprey's has two linear sections; The pelican's descent is steepest near the middle of its path, and the osprey's is steepest at the end.

Practice and Problem Solving

16.

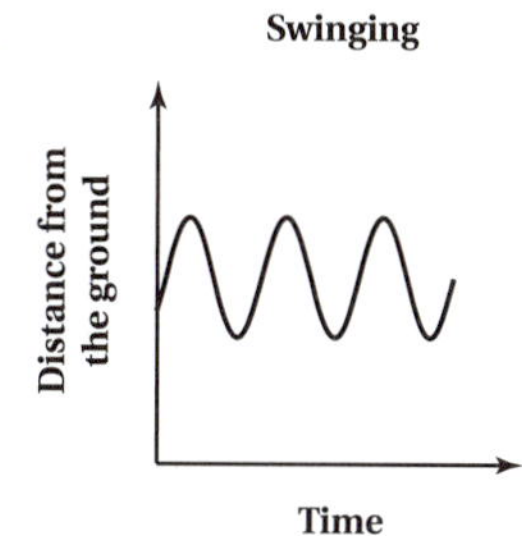

17.

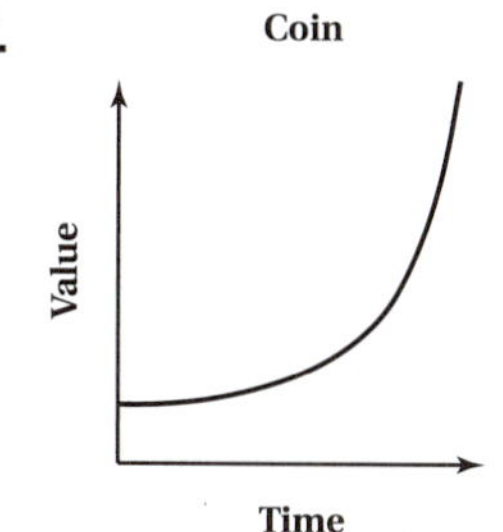

18.

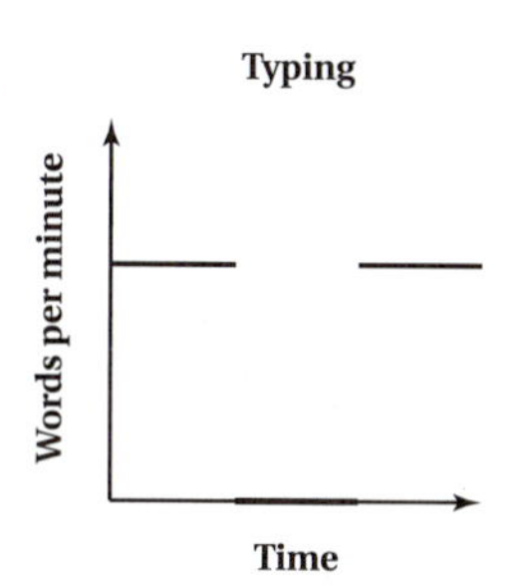

Record and Practice Journal Practice

1. Sales dollars increase quickly at a constant rate, decrease quickly at a constant rate, increase quickly at a constant rate, stay constant, and then increase quickly at a constant rate.

2. Speed increases quickly at a constant rate, decreases quickly at a constant rate, and then increases quickly at a constant rate.

6.4–6.5 Quiz

7. 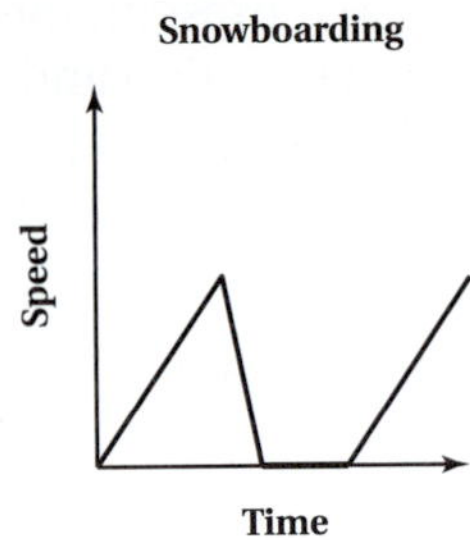

8. 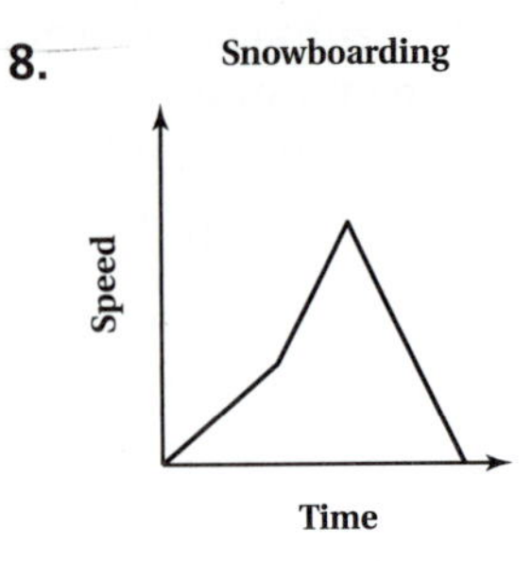

Chapter 6 Review

13. a. The sales of Company A increase at a constant rate, then decrease at a constant rate, then increase at a constant rate. The sales of Company B increase then decrease, then increase and decrease again. None of the rates of increase or decrease are constant.

    b. Overall, the sales of Company A increased over the time period. The sales of Company B appear to be the same at the beginning and end of the time period.

    The sales of Company A increased and decreased at a constant rate over the time period. The sales of Company B did not increase or decrease at a constant rate.

    Both graphs are nonlinear. The graph for Company A consists of three linear sections. The graph for Company B has no linear sections.

Chapter 6 Test

9. a. The price of Stock A increases at a constant rate, then stays the same, then decreases at a constant rate, then stays the same again.

    The price of Stock B decreases at a constant rate, then stays the same, then increases at a constant rate, then stays the same again.

    b. At the end of the day, the price of Stock A is more than it was at the beginning of the day. The price of Stock B is about the same at the beginning and end of the day.

    Both graphs are nonlinear. The graph for each stock consists of four linear sections.

    At the beginning of the day, the price of Stock A increased, and the price of Stock B decreased.

Chapter 7

Record and Practice Journal Fair Game Review

12. 63
13. 116
14. -51
15. 1
16. $\frac{24 + 32 + 30 + 28}{2}$; 57

Section 7.3

Practice and Problem Solving

17. a. *Sample answer:*

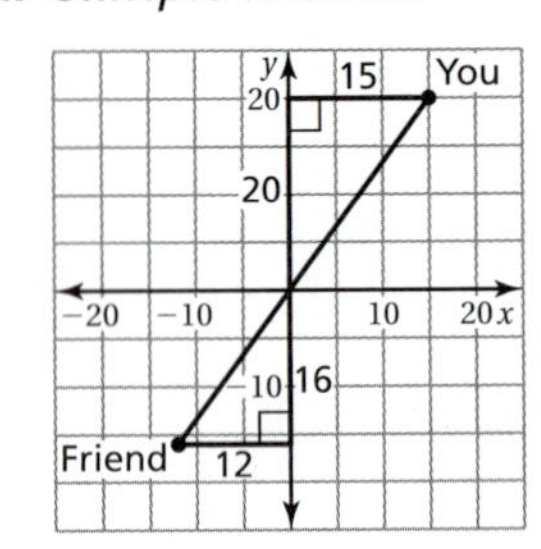

    b. 45 ft

Section 7.4

Practice and Problem Solving

20. a. 7
    b. 6.8
21. a. 26
    b. 26.2
22. a. -8
    b. -7.8
23. a. -10
    b. -10.2
24. a. 3
    b. 2.6
25. a. -13
    b. -12.9

Extension 7.4

Record and Practice Journal Practice

1. $\frac{1}{3}$
2. $-\frac{2}{9}$
3. $1\frac{7}{9}$
4. $-2\frac{2}{3}$
5. $\frac{7}{15}$
6. $-1\frac{15}{18}$
7. $-\frac{11}{15}$
8. $\frac{2}{11}$
9. $-3\frac{8}{33}$
10. $1\frac{1}{11}$
11. $1\frac{17}{30}$ in.

Section 7.5

Practice and Problem Solving

26.

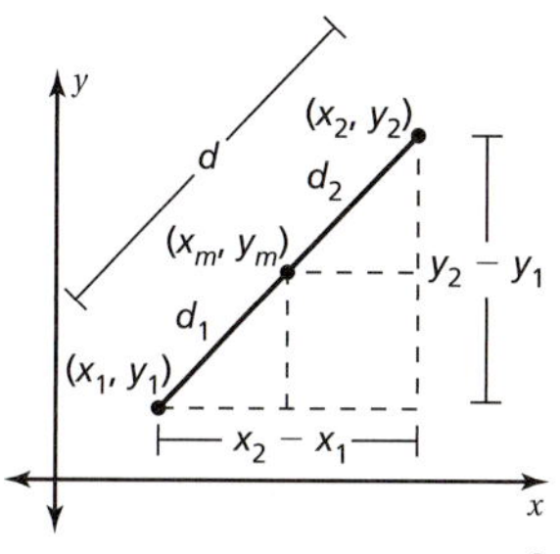

$$x_m = x_1 + \frac{1}{2}(x_2 - x_1) = \frac{2x_1 + x_2 - x_1}{2} = \frac{x_1 + x_2}{2}$$

Similarly, $y_m = \dfrac{y_1 + y_2}{2}$

$$d_1 = \sqrt{\left(x_1 - \frac{x_1 + x_2}{2}\right)^2 + \left(y_1 - \frac{y_1 + y_2}{2}\right)^2}$$

$$= \sqrt{\left(\frac{x_1 - x_2}{2}\right)^2 + \left(\frac{y_1 - y_2}{2}\right)^2}$$

$$= \frac{1}{2}\sqrt{(x_1 - x_2)^2 + (y_1 - y_2)^2}$$

$$d_2 = \sqrt{\left(\frac{x_1 + x_2}{2} - x_2\right)^2 + \left(\frac{y_1 + y_2}{2} - y_2\right)^2}$$

$$= \sqrt{\left(\frac{x_1 - x_2}{2}\right)^2 + \left(\frac{y_1 - y_2}{2}\right)^2}$$

$$= \frac{1}{2}\sqrt{(x_1 - x_2)^2 + (y_1 - y_2)^2}$$

So, $d_1 + d_2 = \frac{1}{2}\sqrt{(x_1 - x_2)^2 + (y_1 - y_2)^2} + \frac{1}{2}\sqrt{(x_1 - x_2)^2 + (y_1 - y_2)^2}$

$$= \sqrt{(x_1 - x_2)^2 + (y_1 - y_2)^2}$$

$$= d$$

Chapter 8

Section 8.4

Practice and Problem Solving

17. a. 9483 pounds; The ratio of the height of the original statue to the height of the small statue is 8.4 : 1. So, the ratio of the weights, or volumes is $\left(\frac{8.4}{1}\right)^3$.

b. 221,184 lb

Fair Game Review

20.

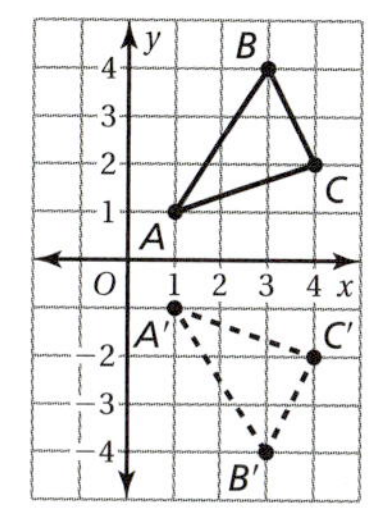

$A'(1, -1), B'(3, -4), C'(4, -2)$

21.

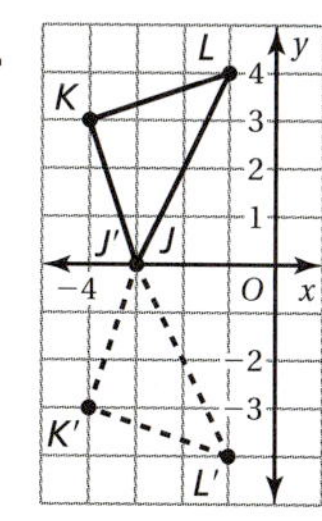

$J'(-3, 0), K'(-4, -3), L'(-1, -4)$

22. B

Chapter 9

Try It Yourself

1.

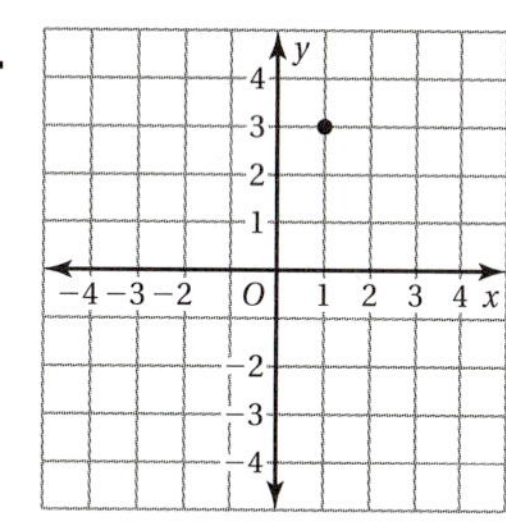

Quadrant I

2.

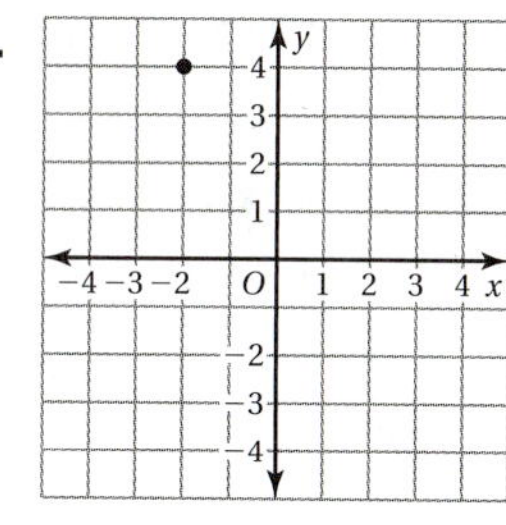

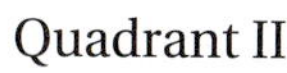

Quadrant II

3.

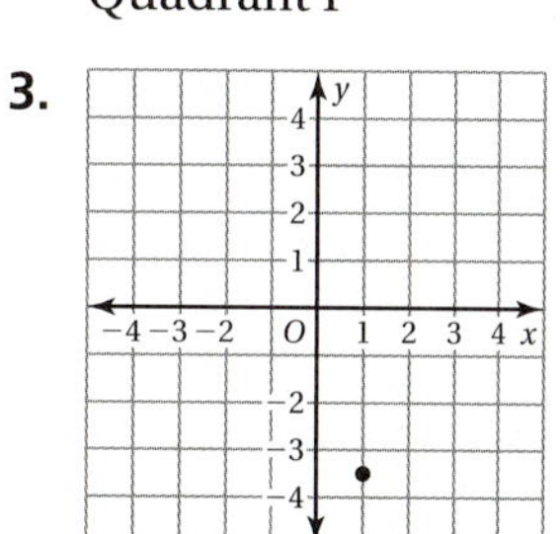

Quadrant IV

4.

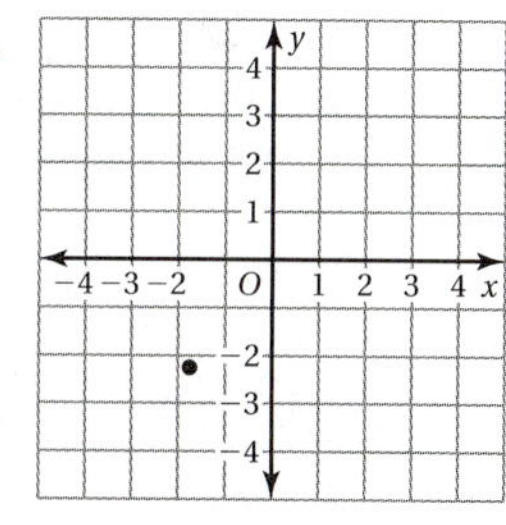

Quadrant III

Record and Practice Journal Fair Game Review

2.

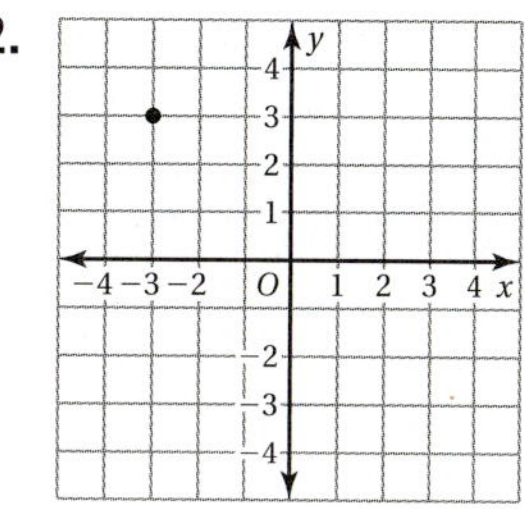

Quadrant II

3.

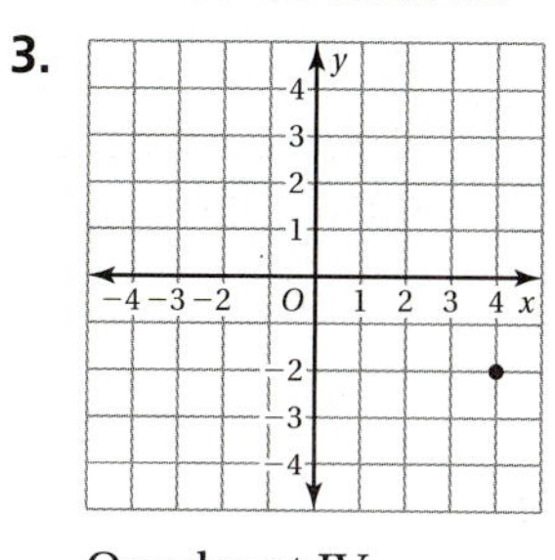

Quadrant IV

4. 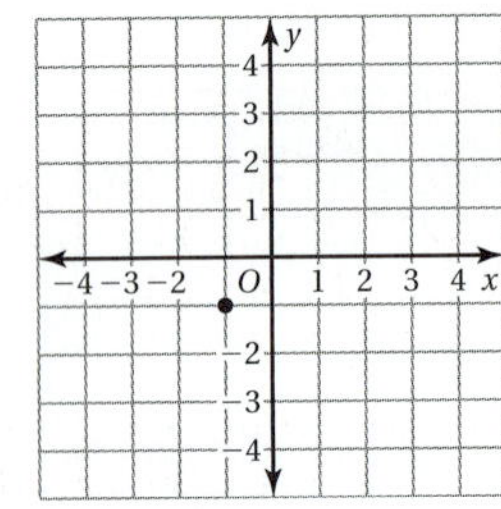

Quadrant III

Section 9.1

Practice and Problem Solving

9. a. 3.5 h

   b. \$85

   c. There is a positive linear relationship between hours worked and earnings.

14. a. positive linear relationship

    b. The more time spent studying, the better the test score.

15. *Sample answer:* bank account balance during a shopping spree

16. a. 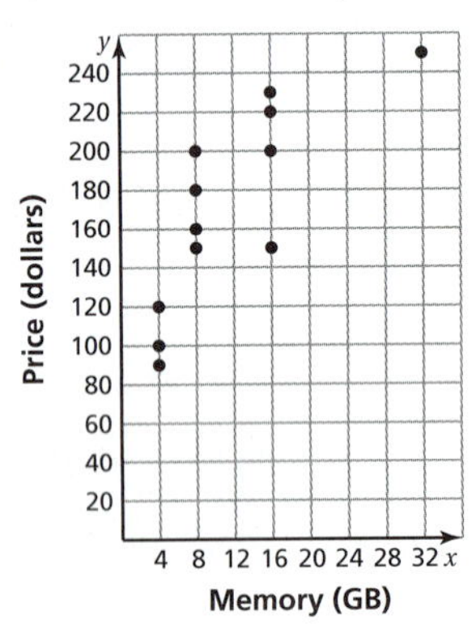

    The data show a weak positive linear relationship.

    b. *Sample answer:* The point (32, 250) is an outlier because the store only offers one 32 GB 7-inch tablet and 32 GB is significantly greater than the other options. There are gaps between $x = 4$, $x = 8$, and $x = 16$ because these are the only available options for memory. There are clusters along $x = 4$, $x = 8$, and $x = 16$ because these are the only available options for memory.

Record and Practice Journal Practice

2. positive linear relationship; outlier at (10, 4), gap between x-values of 6 to 10 and y-values of 10 to 15

Section 9.2

On Your Own

1. a. 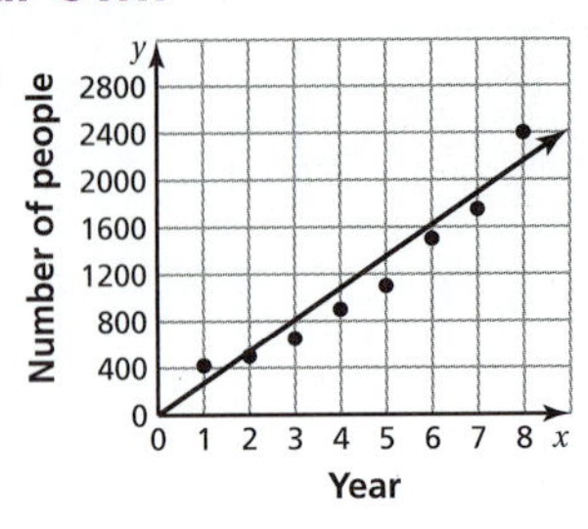

   b. *Sample answer:* $y = 270x$

   c. *Sample answer:* The slope shows that the attendance increased by about 270 people each year. The y-intercept shows that the attendance in the year prior to this 8-year period was 0.

   d. *Sample answer:* about 2700 people

Practice and Problem Solving

6. a. 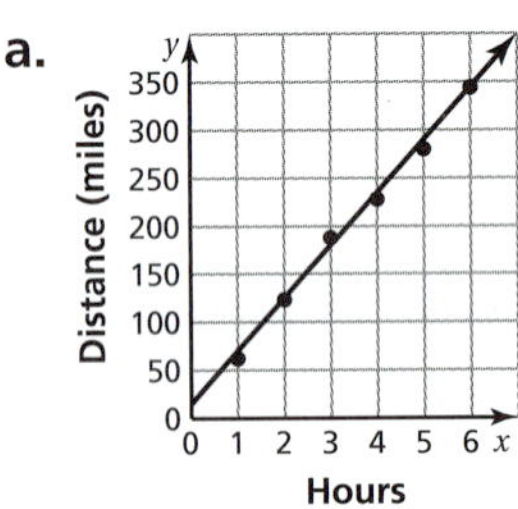

   b. *Sample answer:* $y = 55x + 15$

   c. 55 miles

   d. 15 miles

   e. *Sample answer:* 400 mi

8. $y = -9.6x + 883$; $r \approx -0.96$; The relationship between x and y is a strong negative correlation and the equation closely models the data.

9. $y = 0.9x + 4$; $r \approx 0.999$; The relationship between x and y is a strong positive correlation and the equation closely models the data; 4 in.

11. a. $y = 48x + 11$; $r \approx 0.98$; The relationship between x and y is a strong positive correlation and the equation closely models the data.

    b. 251 feet

    c. The height of a hit baseball is not linear. The best fit line from part (a) only models a small part of the data.

Section 9.3

Extra Example 3

		Grade			
		6	7	8	Total
Join Student Council	Yes	11	13	7	31
	No	10	14	15	39
	Total	21	27	22	70

Extra Example 4

a.

		Grade		
		6	7	8
Join Student Council	Yes	52%	48%	32%
	No	48%	52%	68%

Sample answer: About 32% of 8th grade students in the survey are going to try to join student council.

b. The table shows that as grade level increases, students are less likely to try to join student council.

On Your Own

2. a.

		Grade			
		6	7	8	Total
Lunch	Pack	11	23	16	50
	Buy	9	27	14	50
	Total	20	50	30	100

b.

		Grade		
		6	7	8
Lunch	Pack	55%	46%	53%
	Buy	45%	54%	47%

Sample answer: 45% of the 6th grade students in the survey buy a school lunch.

c. no; About half of the students in each grade buy a school lunch.

Practice and Problem Solving

10. a.

		Age			
		20–29	30–39	40–49	Total
Saved at Least $1000	Yes	14	27	25	66
	No	36	33	15	84
	Total	50	60	40	150

b.

		Age		
		20–29	30–39	40–49
Saved at Least $1000	Yes	28%	45%	62.5%
	No	72%	55%	37.5%

c. Yes, the table shows that as age increases, people are more likely to have at least $1000 in savings.

11. a.

		Eye Color			
		Green	Blue	Brown	Total
Gender	Male	5	16	27	48
	Female	3	19	18	40
	Total	8	35	45	88

b. 48 males were surveyed.
40 females were surveyed.
8 students have green eyes.
35 students have blue eyes.
45 students have brown eyes.

c.

		Eye Color		
		Green	Blue	Brown
Gender	Male	63%	46%	60%
	Female	38%	54%	40%

Sample answer: About 63% of the students with green eyes are male. 40% of the students with brown eyes are female.

12.

		Eye Color		
		Green	Blue	Brown
Gender	Male	10.4%	33.3%	56.3%
	Female	7.5%	47.5%	45%

Sample answers: About 10.4% of the males surveyed have green eyes. 7.5% of the females surveyed have green eyes.

Record and Practice Journal Practice

2. a.

		Grade 6	Grade 7	Grade 8	Total
Student	Eats Breakfast at Home	28	15	9	52
	Eats Breakfast at School	12	15	21	48
	Total	40	30	30	100

b.

		Grade 6	Grade 7	Grade 8
Student	Eats Breakfast at Home	70%	50%	30%
	Eats Breakfast at School	30%	50%	70%

$\frac{9}{30} = 0.3$

So, 30% of the grade 8 students in the survey eat breakfast at home.

Mini-Assessment

1. a.

		Grade 5	Grade 8	Total
School Sports	Involved	12	23	35
	Not Involved	26	19	45
	Total	38	42	80

b.

		Grade 5	Grade 8
School Sports	Involved	32%	55%
	Not Involved	68%	45%

Sample answer: About 55% of the 8th grade students in the survey are involved in school sports.

c. yes; Students in Grade 8 are more likely to be involved in school sports than students in Grade 5.

Section 9.4

Record and Practice Journal

1. c. *Sample answer:*

Raccoon Roadkill Weights

Stem	Leaf
9	4 5
10	
11	0
12	4 9
13	4 6 9
14	0 5 8 8
15	2 7
16	8
17	0 2 3 5
18	5 5 6 7
19	0 1 4
20	4
21	3 5 5 5
22	
23	
24	
25	4

Key: 9|4 = 9.4 pounds

The stem-and-leaf plot shows how the raccoon weights are distributed.

Record and Practice Journal Practice

4. Because the rain icon is larger than the sun icon, it makes it look as if there were equal amounts of sunny and rainy days when there were not.

9.3–9.4 Quiz

3. a. 48 voters are 18–34; 57 voters are 35–64; 30 voters are 65+; 67 voters prefer Smith; 68 voters prefer Jackson.

b.

		Voter's Age 18–34	35–64	65+
Candidate	Smith	75%	44%	20%
	Jackson	25%	56%	80%

c. Yes, the table shows that as age increases, preference for Jackson increases.

Chapter 9 Review

7.

		Food Court		
		Likes	Dislikes	Total
Age Group	Teenagers	96	4	100
	Adults	21	79	100
	Senior Citizens	18	82	100
	Total	135	165	300

Chapter 9 Test

7.

		Use reuseable bags?		
		Yes	No	Total
Gender	Male	15	45	60
	Female	60	50	110
	Total	75	95	170

Chapter 10

Section 10.1

Practice and Problem Solving

27.

h	1	2	3	4	5
$2^h - 1$	1	3	7	15	31
2^{h-1}	1	2	4	8	16

$2^h - 1$; The option $2^h - 1$ pays you more money when $h > 1$.

Section 10.4

Record and Practice Journal

4. b.
$$3452.867 = 3 \cdot 10^3 + 4 \cdot 10^2 + 5 \cdot 10^1 + 2 \cdot 10^0 + 8 \cdot 10^{-1} + 6 \cdot 10^{-2} + 7 \cdot 10^{-3}$$
$$= 3 \cdot 1000 + 4 \cdot 100 + 5 \cdot 10 + 2 \cdot 1 + 8 \cdot \frac{1}{10} + 6 \cdot \frac{1}{100} + 7 \cdot \frac{1}{1000}$$
$$= 3000 + 400 + 50 + 2 + 0.8 + 0.06 + 0.007$$

Section 10.6

Record and Practice Journal

4. *Sample answer:* Move the decimal point left or right so the number is at least 1 but less than 10. Then multiply by ten raised to the number of times you moved the decimal. If you moved the decimal point to the left, the exponent will be positive. If you moved the decimal point to the right, the exponent should be negative.

Section 10.7

Record and Practice Journal

5. *Sample answer:* To add or subtract numbers written in scientific notation, add or subtract the factors when the powers of 10 are the same. When the powers of 10 are different, first use properties of exponents to rewrite the numbers so the powers are the same.

   To multiply or divide numbers written in scientific notation, use the properties of multiplication to rewrite the expressions, then use the Product of Powers Property and the Quotient of Powers Property to simplify.

Photo Credits

Cover

Pavelk/Shutterstock.com, Pincasso/Shutterstock.com, valdis torms/Shutterstock.com

Front matter

i Pavelk/Shutterstock.com, Pincasso/Shutterstock.com, valdis torms/Shutterstock.com; **iv** Big Ideas Learning, LLC; **vi** *top* ©iStockphoto.com/Lisa Thornberg, ©iStockphoto.com/Ann Marie Kurtz; *bottom* ©iStockphoto.com/Jane norton; **vii** *top* Kasiap/Shutterstock.com, ©iStockphoto.com/Ann Marie Kurtz; *bottom* wavebreakmedia ltd/Shutterstock.com; **viii** *top* ©iStockphoto.com/sumnersgraphicsinc, ©iStockphoto.com/Ann Marie Kurtz; *bottom* Odua Images/Shutterstock.com; **ix** *top* ©iStockhphoto.com/Jonathan Larsen; *bottom* James Flint/Shutterstock.com; **x** *top* stephan kerkhofs/Shutterstock.com, Cigdem Sean Cooper/Shutterstock.com, ©iStockphoto.com/Andreas Gradin; *bottom* william casey/Shutterstock.com; **xi** *top* ©iStockphoto.com/ALEAIMAGE, ©iStockphoto.com/Ann Marie Kurtz; *bottom* Edyta Pawlowska/Shutterstock.com; **xii** *top* ©iStockphoto/Michael Flippo, ©iStockphoto.com/Ann Marie Kurtz; *bottom* PETER CLOSE/Shutterstock.com; **xiii** *top* ©iStockphoto.com/ALEAIMAGE, ©iStockphoto.com/Ann Marie Kurtz; *bottom* Kharidehal Abhirama Ashwin/Shutterstock.com; **xiv** *top* ©iStockphoto.com/Alistair Cotton; *bottom* ©iStockphoto.com/Noraznen Azit; **xv** *top* Varina and Jay Patel/Shutterstock.com, ©iStockphoto.com/Ann Marie Kurtz; *bottom* ©iStockphoto.com/Thomas Perkins; **xxviii** *top left* AVAVA/Shutterstock.com; *bottom right* Tomasz/Shutterstock.com, Nemida/Shutterstock.com; **xxix** *top left* Konstantin Chagin/Shutterstock.com; *top right* Roger Jegg - Fotodesign-Jegg.de/Shutterstock.com; **xxxvi** Ljupco Smokovski/Shutterstock.com

Chapter 1

1 ©iStockphoto.com/Lisa Thornberg, ©iStockphoto.com/Ann Marie Kurtz; **6** ©iStockphoto.com/David Freund; **7** ©iStockphoto.com/nicolas hansen; **8** amskad/Shutterstock.com; **9** ©iStockphoto.com/Ryan Lane; **12** ©iStockphoto.com/Harley McCabe; **13** ©iStockphoto.com/Jacom Stephens; **14** ©iStockphoto.com/Harry Hu; **15** ©iStockphoto.com/Ralf Hettler, Vibrant Image Studio/Shutterstock.com; **23** ©iStockphoto.com/Andrey Krasnov; **24** Shawn Hempel/Shutterstock.com; **31** *top right* ©iStockphoto.com/Alan Crawford; *center left* ©iStockphoto.com/Julio Yeste; *bottom right* ©iStockphoto.com/Mark Stay; **36** *center right* Ljupco Smokovski/Shutterstock.com; *bottom left* emel82/Shutterstock.com

Chapter 2

40 Kasiap/Shutterstock.com, ©iStockphoto.com/Ann Marie Kurtz; **48** Azat1976/Shutterstock.com; **52** ©iStockphoto.com/Er Ten Hong; **T-53** ©iStockphoto.com/Tryfonov levgenii; **53** *center left* ©iStockphoto.com/Sergey Galushko; *center right* ©iStockphoto.com/Tryfonov levgenii; **54** ©iStockphoto.com/ingmar wesemann; **T-59** ©iStockphoto.com/Hazlan Abdul Hakim; **59** ©iStockphoto.com/Hazlan Abdul Hakim; **67** ©iStockphoto.com/Maksim Shmeljov; **70** *top* ©iStockphoto.com/Viatcheslav Dusaleev; *bottom left* ©iStockphoto.com/Jason Mooy; *bottom right* ©iStockphoto.com/Felix Möckel; **73** gary718/Shutterstock.com; **83** Diego Cervo/Shutterstock.com; **90** *center left* Antonio Jorge Nunes/Shutterstock.com, Tom C Amon/Shutterstock.com; *center right* ©iStockphoto.com/Alex Slobodkin

Chapter 3

100 ©iStockphoto.com/sumnersgraphicsinc, ©iStockphoto.com/Ann Marie Kurtz; **102** PILart/Shutterstock.com, Wildstyle/Shutterstock.com; **103** Estate Craft Homes, Inc.; **114** Marc Dietrich/Shutterstock.com; **120** *bottom left* ©iStockphoto.com/Evgeny Terentev; *bottom right* ©iStockphoto.com/Vadym Volodin; **121** NASA; **124** iStockphoto.com/Evelyn Peyton; **125** *top right* ©iStockphoto.com/Terraxplorer; *top left* ©iStockphoto.com/Lora Clark; *center right* ©iStockphoto.com/Jennifer Morgan

Chapter 4

140 ©iStockphoto.com/Jonathan Larsen; **145** NASA; **146** ©iStockphoto.com/David Morgan; **147** *top right* NASA; *center left* ©iStockphoto.com/jsemeniuk; **154** ©iStockphoto.com/Amanda Rohde; **155** Julian Rovagnati/Shutterstock.com; **159** RyFlip/Shutterstock.com; **162** Luke Wein/Shutterstock.com; **165** AVAVA/Shutterstock.com; **170** ©iStockphoto.com/Dreamframer; **171** *top right* Jerry Horbert/Shutterstock.com; *center left* ©iStockphoto.com/Chris Schmidt; **173** ©iStockphoto.com/biffspandex; **176** ©iStockphoto.com/Stephen Pothier; **177** *top left* Gina Smith/Shutterstock.com; *center left* Dewayne Flowers/Shutterstock.com; **181** Herrenknecht AG; **182** ©iStockphoto.com/Adam Mattel; **T-183** ©iStockphoto.com/marcellus2070, ©iStockphoto.com/beetle8; **183** *top left* ©iStockphoto.com/Gene Chutka; *center right* ©iStockphoto.com/marcellus2070, ©iStockphoto.com/beetle8; **187** ©iStockphoto.com/Connie Maher; **188** ©iStockphoto.com/Jacom Stephens; **189** *top right* ©iStockphoto.com/Petr Podzemny; *bottom left* ©iStockphoto.com/adrian beesley; **190** Richard Goldberg/Shutterstock.com; **196** Thomas M Perkins/Shutterstock.com

Chapter 5

200 stephan kerkhofs/Shutterstock.com, Cigdem Sean Cooper/Shutterstock.com, ©iStockphoto.com/Andreas Gradin; **202** Howard Sandler/Shutterstock.com, ©iStockphoto.com/Dori OConnell; **205** Richard Paul Kane/Shutterstock.com; **206** ©iStockphoto.com/Kathy Hicks; **208** *top right* YuriyZhuravov/Shutterstock.com; *bottom right* Talvi/Shutterstock.com; **211** aguilarphoto/Shutterstock.com; **212** Kiselev Andrey Valerevich/Shutterstock.com; **213** *center left* Susan Schmitz/Shutterstock.com; *center right* akva/Shutterstock.com; **215** Andrey Yurlov/Shutterstock.com; **216** Steve Cukrov/Shutterstock.com; **220** Le Do/Shutterstock.com, Quang Ho/Shutterstock.com, SergeyIT/Shutterstock.com, jon Le-Bon/Shutterstock.com; **221** Ariwasabi/Shutterstock.com; **222** Ewa/Shutterstock.com; **223** *top left* Gordana Sermek/Shutterstock.com; *center right* Rashevskyi Viacheslav/Shutterstock.com; **224** ©iStockphoto.com/walik; **228** ©iStockphoto.com/Corina Estepa; **229** ©iStockphoto.com/Tomislav Forgo; **231** Kateryna Larina/Shutterstock.com; **232** Selena/Shutterstock.com; **236** kostudio/Shutterstock.com

Chapter 6

240 ©iStockphoto.com/ALEAIMAGE, ©iStockphoto.com/Ann Marie Kurtz; **247** ©iStockphoto.com/Kevin Panizza; **249** ©iStockphoto.com/Jacom Stephens; **252** ©iStockphoto.com/DivaNir4a; **254** *top left* ©iStockphoto.com/Manuel Angel Diaz Blanco; *bottom right* ©iStockphoto.com/Sergey Lemeshencko; **255** ©iStockphoto.com/Robert Rushton; **259** General Atomics Aeronautical Systems, Inc.; **262** ©iStockphoto.com/Mlenny Photography; **263** ©iStockphoto.com/medobear; **267** ©iStockphoto.com/PeskyMonkey; **271** ©iStockphoto.com/Tom Buttle; **278** gillmar/Shutterstock.com

Chapter 7

286 ©iStockphoto/Michael Flippo, ©iStockphoto.com/Ann Marie Kurtz; **291** Perfectblue97; **292** ©iStockphoto.com/Benjamin Lazare; **T-293** *top left* ©iStockphoto.com/Jill Chen; *top right* Oleksiy Mark/Shutterstock.com; **293** *top right* ©iStockphoto.com/iShootPhotos, LLC; *center left* ©iStockphoto.com/Jill Chen, Oleksiy Mark/Shutterstock.com; **298** Gary Whitton/Shutterstock.com; **299** Michael Stokes/Shutterstock.com; **300** ©Oxford Science Archive/Heritage Images/Imagestate; **304** ©iStockphoto.com/Melissa Carroll; **307** *center left* ©iStockphoto.com/Yvan Dubé; *bottom right* Snvv/Shutterstock.com; **308** ©iStockphoto.com/Kais Tolmats; **312** *top left* ©iStockphoto.com/Don Bayley; *center left* ©iStockphoto.com/iLexx; **315** ©iStockphoto.com/Marcio Silva; **319** Monkey Business Images/Shutterstock.com; **327** LoopAll/Shutterstock.com; **328** CD Lanzen/Shutterstock.com

Chapter 8

332 ©iStockphoto.com/ALEAIMAGE, ©iStockphoto.com/Ann Marie Kurtz; **334** ©iStockphoto.com/Jill Chen; **337** ©iStockphoto.com/camilla wisbauer; **T-339** ©iStockphoto.com/Matthew Dixon; **339** *Exercises 13 and 14* ©iStockphoto.com/Prill Mediendesigns & Fotografie; *Exercise 15* ©iStockphoto.com/subjug; *center left* ©iStockphoto.com/Matthew Dixon; *center right* ©iStockphoto.com/nilgun bostanci; **345** ©iStockphoto.com/Stefano Tiraboschi; **351** Donald Joski/Shutterstock.com; **352** ©iStockphoto.com/Yury Kosourov; **353** Carlos Caetano/Shutterstock.com; **360** Courtesy of Green Light Collectibles; **T-361**©iStockphoto.com/ivanastar; **361** *top right* ©iStockphoto.com/wrangel; *center left* ©iStockphoto.com/ivanastar; *bottom left* ©iStockphoto.com/Daniel Cardiff; **362** Eric Isselée/Shutterstock.com; **366** ©iStockphoto.com/Daniel Loiselle

Chapter 9

370 ©iStockphoto.com/Alistair Cotton; **372** *baseball* Kittisak/Shutterstock.com; *golf ball* tezzstock/Shutterstock.com; *basketball* vasosh/Shutterstock.com; *tennis ball* UKRID/Shutterstock.com; *water polo ball* John Kasawa/Shutterstock.com; *softball* Ra Studio/Shutterstock.com; *volleyball* vberla/Shutterstock.com; **376** ©iStockphoto.com/Jill Fromer; **377** ©iStockphoto.com/Janis Litavnieks; **378** Gina Brockett; **379** ©iStockphoto.com/Craig Dingle; **381** Sashkin/Shutterstock.com; **382** ©iStockphoto.com/Brian McEntire; **385** Dwight Smith/Shutterstock.com; **386** Aptyp_koK/Shutterstock.com; **391** Alberto Zornetta/Shutterstock.com; **392** *center left* ©iStockphoto.com/Tony Campbell; *bottom right* Eric Isselee/Shutterstock.com; **393** *top right* Larry Korhnak; *bottom right* Photo by Andy Newman; **399** *top left* ©iStockphoto.com/Jane norton; *bottom right* ©iStockphoto.com/Krzysztof Zmij; **400** IrinaK/Shutterstock.com; **404** Lim Yong Hian/Shutterstock.com

Chapter 10

408 Varina and Jay Patel/Shutterstock.com, ©iStockphoto.com/Ann Marie Kurtz; **410** ©iStockphoto.com/Franck Boston; **411** *Activity 3a* ©iStockphoto.com/Manfred Konrad; *Activity 3b* NASA/JPL-Caltech/R.Hurt (SSC); *Activity 3c and d* NASA; *bottom right* Stevyn Colgan; **413** ©iStockphoto.com/Philippa Banks; **414** ©iStockphoto.com/clotilde hulin; **T-415** ©iStockphoto.com/Boris Yankov; **415** ©iStockphoto.com/Boris Yankov; **420** ©iStockphoto.com/VIKTORIIA KULISH; **421** *top right* ©iStockphoto.com/Paul Tessier; *center left* ©iStockphoto.com/subjug, ©iStockphoto.com/Valerie Loiseleux, ©iStockphoto.com/Linda Steward; **426** ©iStockphoto.com/Petrovich9; **427** *top right* Dash/Shutterstock.com; *center left* NASA/JPL-Caltech/L.Cieza (UT Austin); **431** ©iStockphoto.com/Aliaksandr Autayeu; **432** EugeneF/Shutterstock.com; **433** ©iStockphoto.com/Nancy Louie; **435** ©iStockphoto.com/Dan Moore; **436** ©iStockphoto.com/Kais Tolmats; **437** *Activity 3a and d* Tom C Amon/Shutterstock.com; *Activity 3b* Olga Gabay/Shutterstock.com; *Activity 3c* NASA/MODIS Rapid Response/Jeff Schmaltz; *Activity 3f* HuHu/Shutterstock.com; *Activity 4a* PILart/Shutterstock.com; *Activity 4b* Matthew Cole/Shutterstock.com; *Activity 4c* Yanas/Shutterstock.com; *Activity 4e* unkreativ/Shutterstock.com; **439** *top left* ©iStockphoto.com/Mark Stay; *top center* ©iStockphoto.com/Frank Wright; *top right* ©iStockphoto.com/Evgeniy Ivanov; *bottom left* ©iStockphoto.com/Oliver Sun Kim; **440** ©iStockphoto.com/Christian Jasiuk; **441** Microgen/Shutterstock.com; **442** *Activity 1a* ©iStockphoto.com/Susan Trigg; *Activity 1b* ©iStockphoto.com/subjug; *Activity 1c* ©iStockphoto.com/camilla wisbauer; *Activity 1d* ©iStockphoto.com/Joe Belanger; *Activity 1e* ©iStockphoto.com/thumb; *Activity 1f* ©iStockphoto.com/David Freund; **443** NASA; **444** *center* Google and YouTube logos are registered trademarks of Google Inc., used with permission.; **445** *top left* Elaine Barker/Shutterstock.com; *center right* ©iStockphoto.com/breckeni; **446** *bottom left* ©iStockphoto.com/Max Delson Martins Santos; *bottom right* ©iStockphoto.com/Jan Rysavy; **447** *top right* BORTEL Pavel/Shutterstock.com; *center right* ©iStockphoto.com/breckeni; **451** *center left* Sebastian Kaulitzki/Shutterstock.com; *center right* ©iStockphoto.com/Jan Rysavy; **453** ©iStockphoto.com/Boris Yankov; **454** mmutlu/Shutterstock.com; **458** *bottom right* ©iStockphoto.com/Eric Holsinger; *bottom left* TranceDrumer/Shutterstock.com

Appendix A

A0 *background* ©iStockphoto.com/Björn Kindler; *top left* ©iStockphoto.com/mika makkonen; *top right* ©iStockphoto.com/Hsing-Wen Hsu; **A1** *top right* ©iStockphoto.com/toddmedia; *bottom left* ©iStockphoto.com/Loretta Hostettler; *bottom right* NASA; **A4** *top right* ©iStockphoto.com/Hsing-Wen Hsu; *bottom left* ©iStockphoto.com/Thomas Kuest; *bottom right* Lim ChewHow/Shutterstock.com; **A5** *top right* ©iStockphoto.com/Richard Cano; *bottom left* ©iStockphoto.com/best-photo; *bottom right* ©iStockphoto.com/mika makkonen; **A6** *top right* ©iStockphoto.com/Loretta Hostettler; *bottom* ©iStockphoto.com/toddmedia; **A7** *top right* LudmilaM/Shutterstock.com; *center left and bottom right* ©iStockphoto.com/Clayton Hansen; *bottom left* Billwhittaker at en.wikipedia; **A8 and A9** NASA

Cartoon illustrations Tyler Stout

Florida Common Core Standards

Kindergarten

Counting and Cardinality	– Count to 100 by Ones and Tens; Compare Numbers
Operations and Algebraic Thinking	– Understand and Model Addition and Subtraction
Number and Operations in Base Ten	– Work with Numbers 11–19 to Gain Foundations for Place Value
Measurement and Data	– Describe and Compare Measurable Attributes; Classify Objects into Categories
Geometry	– Identify and Describe Shapes

Grade 1

Operations and Algebraic Thinking	– Represent and Solve Addition and Subtraction Problems
Number and Operations in Base Ten	– Understand Place Value for Two-Digit Numbers; Use Place Value and Properties to Add and Subtract
Measurement and Data	– Measure Lengths Indirectly; Write and Tell Time; Represent and Interpret Data
Geometry	– Draw Shapes; Partition Circles and Rectangles into Two and Four Equal Shares

Grade 2

Operations and Algebraic Thinking	– Solve One- and Two-Step Problems Involving Addition and Subtraction; Build a Foundation for Multiplication
Number and Operations in Base Ten	– Understand Place Value for Three-Digit Numbers; Use Place Value and Properties to Add and Subtract
Measurement and Data	– Measure and Estimate Lengths in Standard Units; Work with Time and Money
Geometry	– Draw and Identify Shapes; Partition Circles and Rectangles into Two, Three, and Four Equal Shares

Grade 3

Operations and Algebraic Thinking	– Represent and Solve Problems Involving Multiplication and Division; Solve Two-Step Problems Involving Four Operations
Number and Operations in Base Ten	– Round Whole Numbers; Add, Subtract, and Multiply Multi-Digit Whole Numbers
Number and Operations—Fractions	– Understand Fractions as Numbers
Measurement and Data	– Solve Time, Liquid Volume, and Mass Problems; Understand Perimeter and Area
Geometry	– Reason with Shapes and Their Attributes

Grade 4

Operations and Algebraic Thinking	– Use the Four Operations with Whole Numbers to Solve Problems; Understand Factors and Multiples
Number and Operations in Base Ten	– Generalize Place Value Understanding; Perform Multi-Digit Arithmetic
Number and Operations—Fractions	– Build Fractions from Unit Fractions; Understand Decimal Notation for Fractions
Measurement and Data	– Convert Measurements; Understand and Measure Angles
Geometry	– Draw and Identify Lines and Angles; Classify Shapes

Grade 5

Operations and Algebraic Thinking	– Write and Interpret Numerical Expressions
Number and Operations in Base Ten	– Perform Operations with Multi-Digit Numbers and Decimals to Hundredths
Number and Operations—Fractions	– Add, Subtract, Multiply, and Divide Fractions
Measurement and Data	– Convert Measurements within a Measurement System; Understand Volume
Geometry	– Graph Points in the First Quadrant of the Coordinate Plane; Classify Two-Dimensional Figures

Mathematics Reference Sheet

Conversions

U.S. Customary

1 foot = 12 inches
1 yard = 3 feet
1 mile = 5280 feet
1 acre ≈ 43,560 square feet
1 cup = 8 fluid ounces
1 pint = 2 cups
1 quart = 2 pints
1 gallon = 4 quarts
1 gallon = 231 cubic inches
1 pound = 16 ounces
1 ton = 2000 pounds
1 cubic foot ≈ 7.5 gallons

U.S. Customary to Metric

1 inch = 2.54 centimeters
1 foot ≈ 0.3 meter
1 mile ≈ 1.61 kilometers
1 quart ≈ 0.95 liter
1 gallon ≈ 3.79 liters
1 cup ≈ 237 milliliters
1 pound ≈ 0.45 kilogram
1 ounce ≈ 28.3 grams
1 gallon ≈ 3785 cubic centimeters

Time

1 minute = 60 seconds
1 hour = 60 minutes
1 hour = 3600 seconds
1 year = 52 weeks

Temperature

$$C = \frac{5}{9}(F - 32)$$

$$F = \frac{9}{5}C + 32$$

Metric

1 centimeter = 10 millimeters
1 meter = 100 centimeters
1 kilometer = 1000 meters
1 liter = 1000 milliliters
1 kiloliter = 1000 liters
1 milliliter = 1 cubic centimeter
1 liter = 1000 cubic centimeters
1 cubic millimeter = 0.001 milliliter
1 gram = 1000 milligrams
1 kilogram = 1000 grams

Metric to U.S. Customary

1 centimeter ≈ 0.39 inch
1 meter ≈ 3.28 feet
1 kilometer ≈ 0.62 mile
1 liter ≈ 1.06 quarts
1 liter ≈ 0.26 gallon
1 kilogram ≈ 2.2 pounds
1 gram ≈ 0.035 ounce
1 cubic meter ≈ 264 gallons

Number Properties

Commutative Properties of Addition and Multiplication

$a + b = b + a$

$a \cdot b = b \cdot a$

Associative Properties of Addition and Multiplication

$(a + b) + c = a + (b + c)$

$(a \cdot b) \cdot c = a \cdot (b \cdot c)$

Addition Property of Zero

$a + 0 = a$

Multiplication Properties of Zero and One

$a \cdot 0 = 0$

$a \cdot 1 = a$

Distributive Property:

$a(b + c) = ab + ac$

$a(b - c) = ab - ac$

Properties of Equality

Addition Property of Equality

If $a = b$, then $a + c = b + c$.

Subtraction Property of Equality

If $a = b$, then $a - c = b - c$.

Multiplication Property of Equality

If $a = b$, then $a \cdot c = b \cdot c$.

Multiplicative Inverse Property

$n \cdot \frac{1}{n} = \frac{1}{n} \cdot n = 1, n \neq 0$

Division Property of Equality

If $a = b$, then $a \div c = b \div c, c \neq 0$.

Squaring both sides of an equation

If $a = b$, then $a^2 = b^2$.

Cubing both sides of an equation

If $a = b$, then $a^3 = b^3$.

Properties of Exponents

Product of Powers Property: $a^m \cdot a^n = a^{m+n}$

Quotient of Powers Property: $\frac{a^m}{a^n} = a^{m-n}, a \neq 0$

Power of a Power Property: $(a^m)^n = a^{mn}$

Power of a Product Property: $(ab)^m = a^m b^m$

Zero Exponents: $a^0 = 1, a \neq 0$

Negative Exponents: $a^{-n} = \frac{1}{a^n}, a \neq 0$

Slope

$$m = \frac{\text{rise}}{\text{run}}$$

$$= \frac{\text{change in } y}{\text{change in } x}$$

$$= \frac{y_2 - y_1}{x_2 - x_1}$$

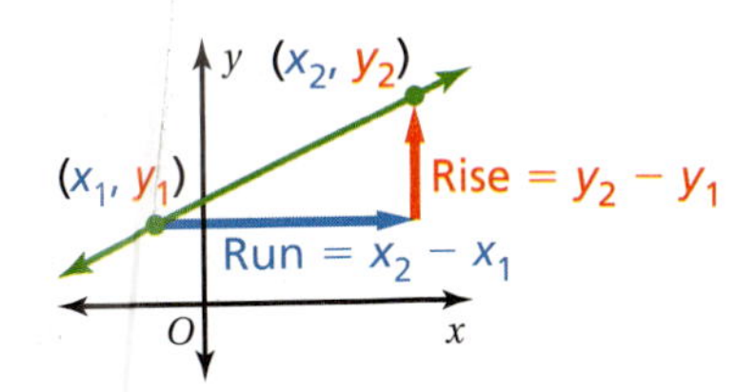

Equations of Lines

Slope-intercept form

$y = mx + b$

Standard form

$ax + by = c, a, b \neq 0$

Point-slope form

$y - y_1 = m(x - x_1)$

Volume

Cylinder

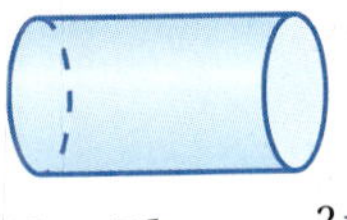

$V = Bh = \pi r^2 h$

Cone

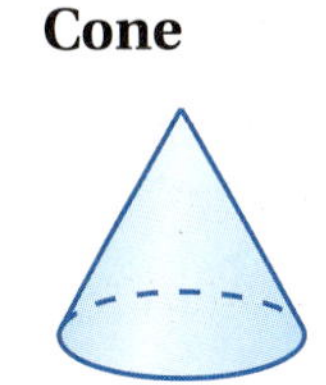

$V = \frac{1}{3}Bh = \frac{1}{3}\pi r^2 h$

Sphere

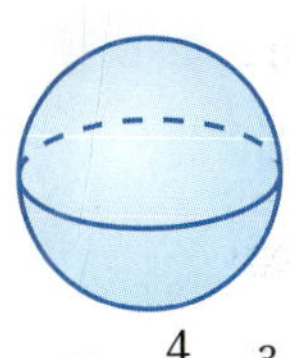

$V = \frac{4}{3}\pi r^3$

Pythagorean Theorem

$a^2 + b^2 = c^2$

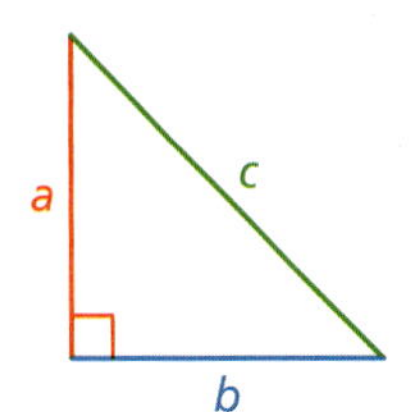

Converse of the Pythagorean Theorem

If the equation $a^2 + b^2 = c^2$ is true for the side lengths of a triangle, then the triangle is a right triangle.

Distance Formula

$$d = \sqrt{(x_2 - x_1)^2 + (y_2 - y_1)^2}$$

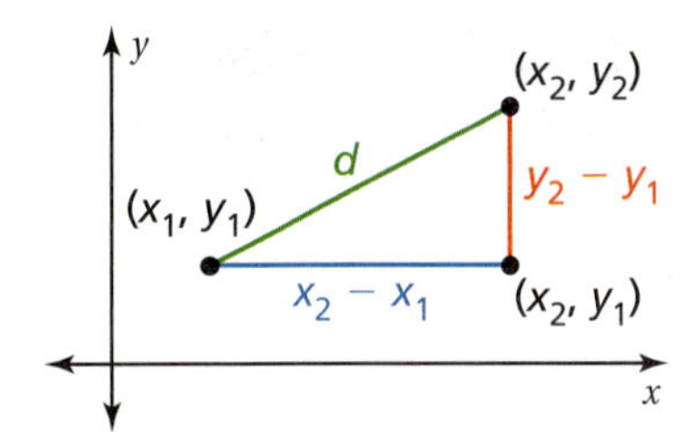

Angles of Polygons

Interior Angle Measures of a Triangle

$x + y + z = 180$

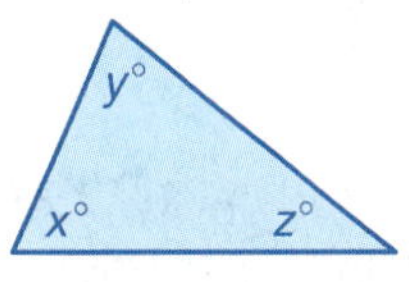

Interior Angle Measures of a Polygon

The sum S of the interior angle measures of a polygon with n sides is $S = (n - 2) \cdot 180°$.

Exterior Angle Measures of a Polygon

$w + x + y + z = 360$

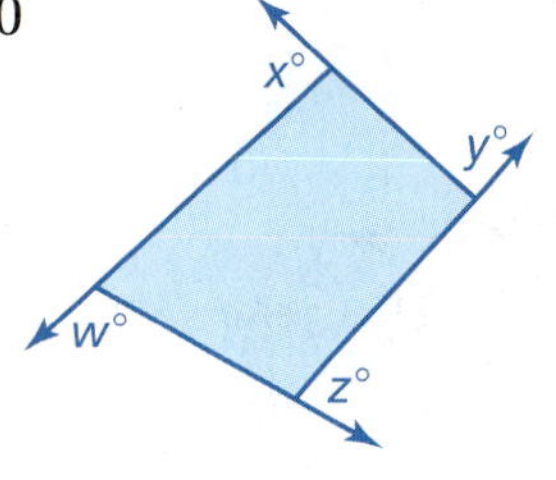